McGraw Hill Glencoe

New York, New York Columbus, Ohio Chicago, Illinois Peoria, Illinois Woodland Hills, California

Glencoe Science

Level Red

The cover shows the Colorado River in the Grand Canyon. It is the primary river of the American Southwest. This forked lightning bolt shows a large static discharge between the dark clouds and the ground. This red-eyed tree frog can be found in rain forests in South and Central America, and as far north as Mexico.

Glencoe

The McGraw-Hill Companies

Send all inquiries to:
Glencoe/McGraw-Hill
8787 Orion Place
Columbus, OH 43240-4027

ISBN 0-07-860049-9 (Student Edition)
ISBN 0-07-860048-0 (Teacher Wraparound Edition)

Printed in the United States of America.

1 2 3 4 5 6 7 8 9 10 071/055 09 08 07 06 05 04

Glencoe Science Level Red
Features and Benefits

		Pages
Dynamic Instructional Strategies	**. . . present clear and comprehensive coverage of middle school science.**	
	■ Each engaging chapter opener includes a *Launch Lab* and *Foldables*™.	5
	■ *National Geographic Visualizing* features illustrate important concepts in middle school science.	46
A Strong Reading Strand	**. . . encourages active reading and learning for students of all reading levels. In the Student Edition:**	
	■ as you read gives students a preview of learning objectives and vocabulary for each section;	70
	■ *Reading Checks* help students check their reading comprehension; and	131
	■ *Caption questions* ask students to interpret visuals.	164
	■ *Reading Essentials, An Interactive Student Textbook* is designed to help struggling readers comprehend core content. It is written at a reading level of about two to three grades below the Student Edition.	
Meeting the Needs of All Students	**. . . facilitates understanding of science concepts for students of all learning levels. In the Teacher Wraparound Edition:**	
	■ *Differentiated Instruction* strategies help meet the needs of students with learning disabilities and physical challenges, or create opportunities to enrich and extend students' learning;	229
	■ *Daily Intervention* provides intervention strategies for struggling students; and	200
	■ *Identifying Misconceptions* helps uncover and address common science misconceptions.	257
	■ The *English/Spanish Glossary*, also in the Student Edition, helps English-language learners comprehend science terms.	731–749
Extensive Standardized Test Practice	**. . . gives students the opportunity to practice for state and national exams.**	
	■ Each chapter ends with a variety of standardized test practice questions, including *Multiple Choice, Short Response/Grid In*, and *Open Ended.*	312–313
A Variety of Labs	**. . . gets students excited about being involved in science. The Student Edition provides:**	
	■ *MiniLABs,* traditional labs, and *Design Your Own, Model and Invent,* and *Use the Internet* labs; and	319, 322, 332
	■ *Extra Try at Home Labs* provide opportunities for students to practice their science skills at home with adult supervision using materials from the kitchen, junk drawer, or backyard.	690–700
	■ *Virtual Labs* CD-ROM contains an interactive virtual lab for each chapter.	340
	■ *Video Labs* (VHS) reinforce lab techniques and safety skills, offer troubleshooting tips, and give expected outcomes.	379
	■ The *Science Lab Manual, Probeware Lab Manual,* and *Science Inquiry Lab Manual* provide additional opportunities to practice laboratory techniques.	
Multi-Level Review	**. . . presents multiple opportunities for all students to review and master content.**	
	■ Each section ends with a review that contains a *Summary* of the section's major concepts and a *Self Check* that has questions to assess student learning and practice math or science skills.	413
	■ The *Study Guide* at the end of each chapter can preview, review, summarize, and visualize the chapter's main ideas.	467
	■ *Study Guide* and *Reinforcement* help students grasp core content.	
Teacher Resources	**. . . provide innovative strategies to help new and experienced teachers.**	
	■ *Chapter Resources Fast File*™ contains important reproducible masters.	474B
	■ Section Focus, Assessment, and Teaching transparencies accompany each chapter.	496C
	■ *Performance Assessment in the Science Classroom* has assessment guidelines, strategies, sample rubrics, and more.	
Online Resources	**. . . enrich the learning experience with the click of a mouse.**	
	■ For prescreened Web links, standardized test practice, self-check quizzes, chapter tests, Vocabulary PuzzleMaker extra math practice, science career information, current science news, and WebQuest interactive projects, visit **red.msscience.com**.	
	■ The complete interactive Student Edition is available at The McGraw-Hill Learning Network Web site, **mhln.com**.	
Technology	**. . . provides timesaving products to help teachers creatively engage their students.**	
	■ *MindJogger Videoquizzes* (VHS & DVD) provide a game-show style interactive quiz for each chapter.	
	■ Easy to edit *Interactive Chalkboard* Microsoft® PowerPoint® presentations include step-by-step lessons, an image bank, chapter and section review questions, standardized test practice, and transparencies.	
	■ *ExamView*® *Pro Testmaker* CD-ROM in English or Spanish allows you to customize assessments.	
	■ *TeacherWorks* CD-ROM is your all-in-one resource center that helps you plan and organize lessons.	
	■ *StudentWorks* CD-ROM solves the heavy backpack problem.	

SAFETY SYMBOLS

SAFETY SYMBOLS	HAZARD	EXAMPLES	PRECAUTION	REMEDY
DISPOSAL	Special disposal procedures need to be followed.	certain chemicals, living organisms	Do not dispose of these materials in the sink or trash can.	Dispose of wastes as directed by your teacher.
BIOLOGICAL	Organisms or other biological materials that might be harmful to humans	bacteria, fungi, blood, unpreserved tissues, plant materials	Avoid skin contact with these materials. Wear mask or gloves.	Notify your teacher if you suspect contact with material. Wash hands thoroughly.
EXTREME TEMPERATURE	Objects that can burn skin by being too cold or too hot	boiling liquids, hot plates, dry ice, liquid nitrogen	Use proper protection when handling.	Go to your teacher for first aid.
SHARP OBJECT	Use of tools or glassware that can easily puncture or slice skin	razor blades, pins, scalpels, pointed tools, dissecting probes, broken glass	Practice common-sense behavior and follow guidelines for use of the tool.	Go to your teacher for first aid.
FUME	Possible danger to respiratory tract from fumes	ammonia, acetone, nail polish remover, heated sulfur, moth balls	Make sure there is good ventilation. Never smell fumes directly. Wear a mask.	Leave foul area and notify your teacher immediately.
ELECTRICAL	Possible danger from electrical shock or burn	improper grounding, liquid spills, short circuits, exposed wires	Double-check setup with teacher. Check condition of wires and apparatus.	Do not attempt to fix electrical problems. Notify your teacher immediately.
IRRITANT	Substances that can irritate the skin or mucous membranes of the respiratory tract	pollen, moth balls, steel wool, fiberglass, potassium permanganate	Wear dust mask and gloves. Practice extra care when handling these materials.	Go to your teacher for first aid.
CHEMICAL	Chemicals can react with and destroy tissue and other materials	bleaches such as hydrogen peroxide; acids such as sulfuric acid, hydrochloric acid; bases such as ammonia, sodium hydroxide	Wear goggles, gloves, and an apron.	Immediately flush the affected area with water and notify your teacher.
TOXIC	Substance may be poisonous if touched, inhaled, or swallowed.	mercury, many metal compounds, iodine, poinsettia plant parts	Follow your teacher's instructions.	Always wash hands thoroughly after use. Go to your teacher for first aid.
FLAMMABLE	Flammable chemicals may be ignited by open flame, spark, or exposed heat.	alcohol, kerosene, potassium permanganate	Avoid open flames and heat when using flammable chemicals.	Notify your teacher immediately. Use fire safety equipment if applicable.
OPEN FLAME	Open flame in use, may cause fire.	hair, clothing, paper, synthetic materials	Tie back hair and loose clothing. Follow teacher's instruction on lighting and extinguishing flames.	Notify your teacher immediately. Use fire safety equipment if applicable.

Eye Safety
Proper eye protection should be worn at all times by anyone performing or observing science activities.

Clothing Protection
This symbol appears when substances could stain or burn clothing.

Animal Safety
This symbol appears when safety of animals and students must be ensured.

Handwashing
After the lab, wash hands with soap and water before removing goggles.

Authors

NATIONAL GEOGRAPHIC
Education Division
Washington, D.C.

Alton Biggs
Retired Biology Teacher
Allen High School
Allen, TX

Lucy Daniel, PhD
Teacher/Consultant
Rutherford County Schools
Rutherfordton, NC

Ralph M. Feather Jr., PhD
Assistant Professor
Geoscience Department
Indiana University of Pennsylvania
Indiana, PA

Edward Ortleb
Science Consultant
St. Louis, MO

Susan Leach Snyder
Retired Teacher, Consultant
Jones Middle School
Upper Arlington, OH

Dinah Zike
Educational Consultant
Dinah-Might Activities, Inc.
San Antonio, TX

Series Consultants

CONTENT

Alton J. Banks, PhD
Director of the Faculty Center
for Teaching and Learning
North Carolina State University
Raleigh, NC

Jack Cooper
Ennis High School
Ennis, TX

Sandra K. Enger, PhD
Associate Director,
Associate Professor
UAH Institute for Science Education
Huntsville, AL

David G. Haase, PhD
North Carolina State University
Raleigh, NC

Michael A. Hoggarth, PhD
Department of Life and
Earth Sciences
Otterbein College
Westerville, OH

Jerome A. Jackson, PhD
Whitaker Eminent Scholar in Science
Program Director
Center for Science, Mathematics,
and Technology Education
Florida Gulf Coast University
Fort Meyers, FL

William C. Keel, PhD
Department of Physics
and Astronomy
University of Alabama
Tuscaloosa, AL

Linda McGaw
Science Program Coordinator
Advanced Placement Strategies, Inc.
Dallas, TX

Madelaine Meek
Physics Consultant Editor
Lebanon, OH

Robert Nierste
Science Department Head
Hendrick Middle School, Plano ISD
Plano, TX

Connie Rizzo, MD, PhD
Depatment of Science/Math
Marymount Manhattan College
New York, NY

Dominic Salinas, PhD
Middle School Science Supervisor
Caddo Parish Schools
Shreveport, LA

Cheryl Wistrom, PhD
St. Joseph's College
Rensselaer, IN

Carl Zorn, PhD
Staff Scientist
Jefferson Laboratory
Newport News, VA

MATH

Michael Hopper, DEng
Manager of Aircraft Certification
L-3 Communications
Greenville, TX

Teri Willard, EdD
Mathematics Curriculum Writer
Belgrade, MT

READING

Elizabeth Babich
Special Education Teacher
Mashpee Public Schools
Mashpee, MA

Barry Barto
Special Education Teacher
John F. Kennedy Elementary
Manistee, MI

Carol A. Senf, PhD
School of Literature,
Communication, and Culture
Georgia Institute of Technology
Atlanta, GA

Rachel Swaters-Kissinger
Science Teacher
John Boise Middle School
Warsaw, MO

SAFETY

Aileen Duc, PhD
Science 8 Teacher
Hendrick Middle School, Plano ISD
Plano, TX

Sandra West, PhD
Department of Biology
Texas State University-San Marcos
San Marcos, TX

ACTIVITY TESTERS

Nerma Coats Henderson
Pickerington Lakeview Jr. High School
Pickerington, OH

Mary Helen Mariscal-Cholka
William D. Slider Middle School
El Paso, TX

Science Kit and Boreal Laboratories
Tonawanda, NY

Reviewers

Deidre Adams
West Vigo Middle School
West Terre Haute, IN

Sharla Adams
IPC Teacher
Allen High School
Allen, TX

Maureen Barrett
Thomas E. Harrington Middle School
Mt. Laurel, NJ

John Barry
Seeger Jr.-Sr. High School
West Lebanon, IN

Desiree Bishop
Environmental Studies Center
Mobile County Public Schools
Mobile, AL

William Blair
Retired Teacher
J. Marshall Middle School
Billerica, MA

Tom Bright
Concord High School
Charlotte, NC

Lois Burdette
Green Bank Elementary-Middle School
Green Bank, WV

Marcia Chackan
Pine Crest School
Boca Raton, FL

Obioma Chukwu
J.H. Rose High School
Greenville, NC

Karen Curry
East Wake Middle School
Raleigh, NC
Merrilville, IN

Joanne Davis
Murphy High School
Murphy, NC

Robin Dillon
Hanover Central High School
Cedar Lake, IN

Anthony J. DiSipio, Jr.
8th Grade Science
Octorana Middle School
Atglen, PA

Sandra Everhart
Dauphin/Enterprise Jr. High Schools
Enterprise, AL

Mary Ferneau
Westview Middle School
Goose Creek, SC

Cory Fish
Burkholder Middle School
Henderson, NV

Linda V. Forsyth
Retired Teacher
Merrill Middle School
Denver, CO

George Gabb
Great Bridge Middle School
Chesapeake Public Schools
Chesapeake, VA

Annette D'Urso Garcia
Kearney Middle School
Commerce City, CO

Nerma Coats Henderson
Pickerington Lakeview Jr. High School
Pickerington, OH

Lynne Huskey
Chase Middle School
Forest City, NC

Maria E. Kelly
Principal
Nativity School
Catholic Diocese of Arlington
Burke, VA

Michael Mansour
Board Member
National Middle Level Science Teacher's Association
John Page Middle School
Madison Heights, MI

Mary Helen Mariscal-Cholka
William D. Slider Middle School
El Paso, TX

Michelle Mazeika
Whiting Middle School
Whiting, IN

Joe McConnell
Speedway Jr. High School
Indianapolis, IN

Sharon Mitchell
William D. Slider Middle School
El Paso, TX

Amy Morgan
Berry Middle School
Hoover, AL

Norma Neely, EdD
Associate Director for Regional Projects
Texas Rural Systemic Initiative
Austin, TX

Annette Parrott
Lakeside High School
Atlanta, GA

Nora M. Prestinari Burchett
Saint Luke School
McLean, VA

Mark Sailer
Pioneer Jr.-Sr. High School
Royal Center, IN

Joanne Stickney
Monticello Middle School
Monticello, NY

Dee Stout
Penn State University
University Park, PA

Darcy Vetro-Ravndal
Hillsborough High School
Tampa, FL

Karen Watkins
Perry Meridian Middle School
Indianapolis, IN

Clabe Webb
Permian High School
Ector County ISD
Odessa, TX

Alison Welch
William D. Slider Middle School
El Paso, TX

Kim Wimpey
North Gwinnett High School
Suwanee, GA

Kate Ziegler
Durant Road Middle School
Raleigh, NC

Teacher Advisory Board

The Teacher Advisory Board gave the authors, editorial staff, and design team feedback on the content and design of the Student Edition. They provided valuable input in the development of the 2005 edition of ***Glencoe Science Level Red.***

John Gonzales
Challenger Middle School
Tucson, AZ

Rachel Shively
Aptakisic Jr. High School
Buffalo Grove, IL

Roger Pratt
Manistique High School
Manistique, MI

Kirtina Hile
Northmor Jr. High/High School
Galion, OH

Marie Renner
Diley Middle School
Pickerington, OH

Nelson Farrier
Hamlin Middle School
Springfield, OR

Jeff Remington
Palmyra Middle School
Palmyra, PA

Erin Peters
Williamsburg Middle School
Arlington, VA

Rubidel Peoples
Meacham Middle School
Fort Worth, TX

Kristi Ramsey
Navasota Jr. High School
Navasota, TX

Student Advisory Board

The Student Advisory Board gave the authors, editorial staff, and design team feedback on the design of the Student Edition. We thank these students for their hard work and creative suggestions in making the 2005 edition of ***Glencoe Science Level Red*** student friendly.

Jack Andrews
Reynoldsburg Jr. High School
Reynoldsburg, OH

Peter Arnold
Hastings Middle School
Upper Arlington, OH

Emily Barbe
Perry Middle School
Worthington, OH

Kirsty Bateman
Hilliard Heritage Middle School
Hilliard, OH

Andre Brown
Spanish Emersion Academy
Columbus, OH

Chris Dundon
Heritage Middle School
Westerville, OH

Ryan Manafee
Monroe Middle School
Columbus, OH

Addison Owen
Davis Middle School
Dublin, OH

Teriana Patrick
Eastmoor Middle School
Columbus, OH

Ashley Ruz
Karrar Middle School
Dublin, OH

Aaron Haupt

The Glencoe middle school science Student Advisory Board taking a timeout at COSI, a science museum in Columbus, Ohio.

Teacher Handbook

Table of Contents

Dynamic Instruction

The consistent instructional strategies in each chapter strengthen students' learning—from the beginning of each chapter where students see "Chapter Preview," to the end where they have a chance to test the knowledge they have acquired and prepare for the next lesson.

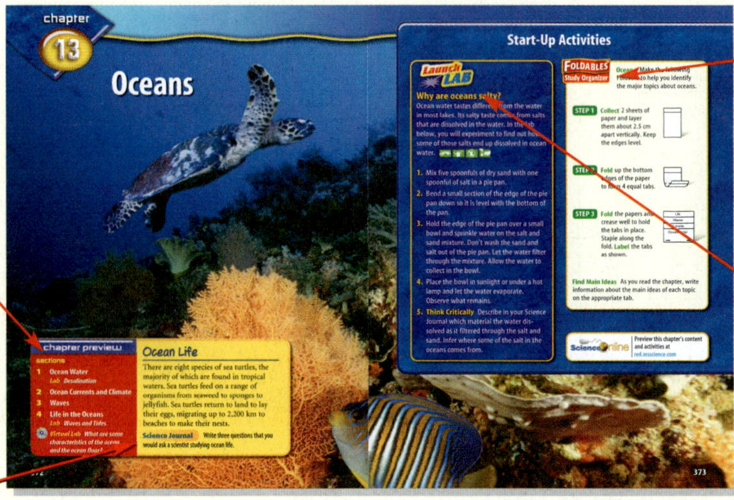

Chapter Preview at the beginning of each chapter introduces the main concepts.

Dinah Zike's Foldables™ let students create interactive study guides.

Launch Lab gives students an opportunity to explore new ideas at the beginning of the chapter.

Science Journal promotes writing and critical-thinking skills.

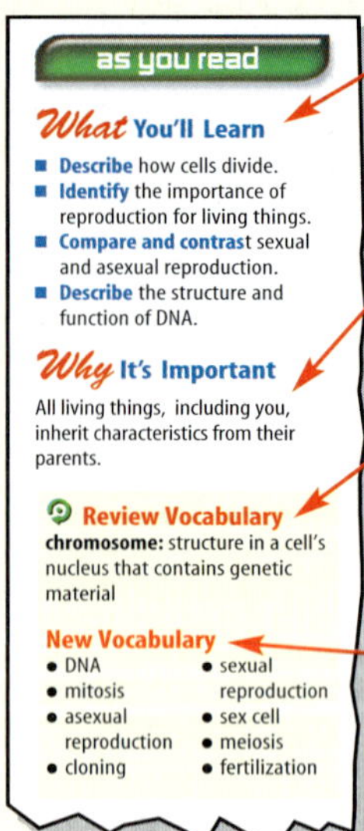
as you read

What You'll Learn
- **Describe** how cells divide.
- **Identify** the importance of reproduction for living things.
- **Compare and contrast** sexual and asexual reproduction.
- **Describe** the structure and function of DNA.

Why It's Important

All living things, including you, inherit characteristics from their parents.

Review Vocabulary

chromosome: structure in a cell's nucleus that contains genetic material

New Vocabulary
- DNA
- mitosis
- asexual reproduction
- cloning
- sexual reproduction
- sex cell
- meiosis
- fertilization

What You'll Learn at the beginning of each section introduces the main concepts.

Why It's Important provides an answer to "Why do we have to learn this?"

Review Vocabulary reviews a term that helps students better understand section content.

New Vocabulary highlights new terms students will learn in the section.

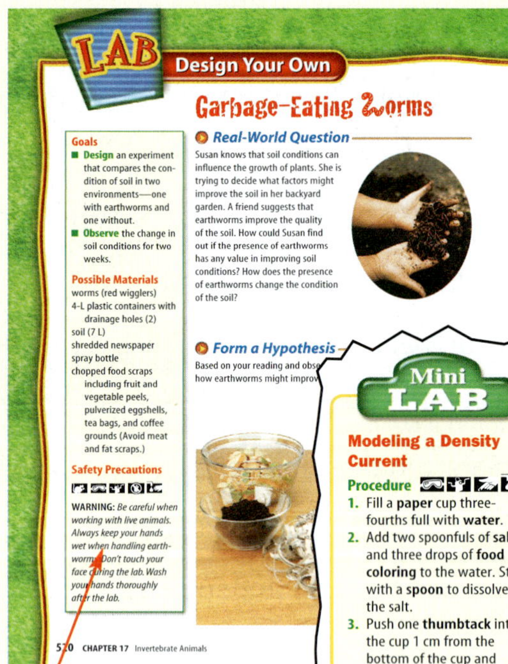

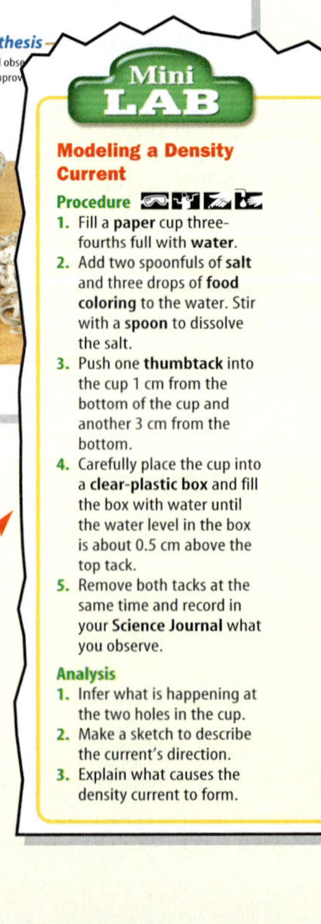
Mini LAB

Modeling a Density Current

Procedure
1. Fill a **paper** cup three-fourths full with **water.**
2. Add two spoonfuls of **salt** and three drops of **food coloring** to the water. Stir with a **spoon** to dissolve the salt.
3. Push one **thumbtack** into the cup 1 cm from the bottom of the cup and another 3 cm from the bottom.
4. Carefully place the cup into a **clear-plastic box** and fill the box with water until the water level in the box is about 0.5 cm above the top tack.
5. Remove both tacks at the same time and record in your **Science Journal** what you observe.

Analysis
1. Infer what is happening at the two holes in the cup.
2. Make a sketch to describe the current's direction.
3. Explain what causes the density current to form.

Labs Whether students are designing their own experiments or following well-tested procedures, they'll learn and practice science processes.

MiniLABs offer students quick and easy-to-do ways to clarify concepts and reinforce skills.

Multi-Level Review and Assessment

Each chapter provides five pages of review and testing to help you evaluate students' knowledge and ability to apply science concepts.

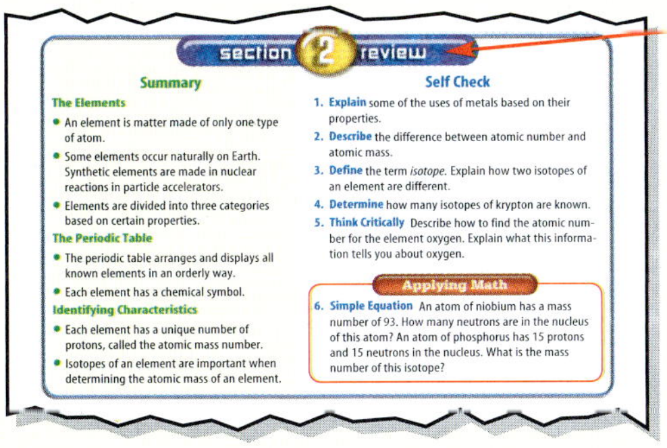

Section Review

- Summary pinpoints important concepts in the section.
- Skill-based questions promote critical thinking skills.

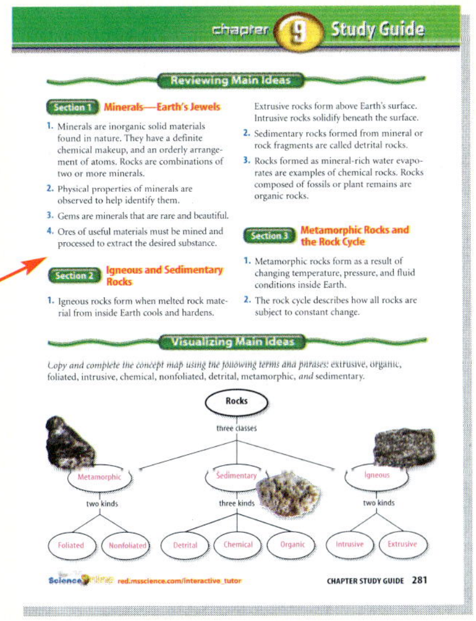

Study Guide

- Main idea summary of each section
- Concept mapping activity to help students visualize the main ideas

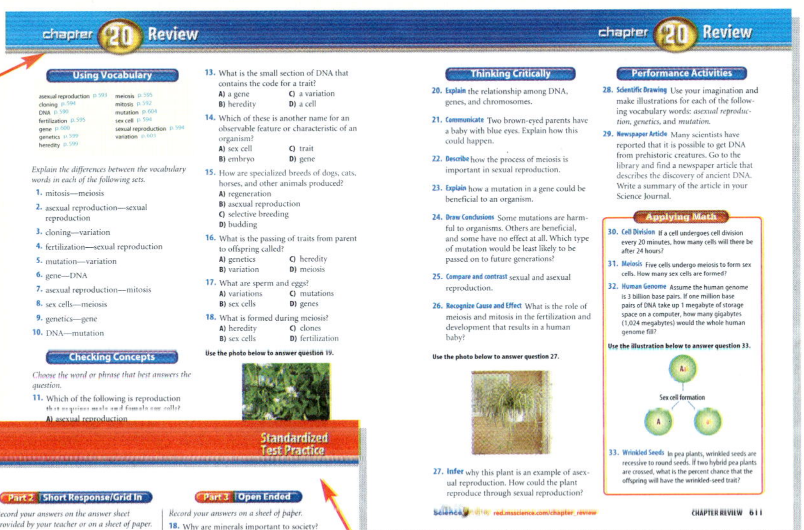

Chapter Review

- Using Vocabulary
- Checking Concepts
- Thinking Critically
- Performance Activities
- Applying Math

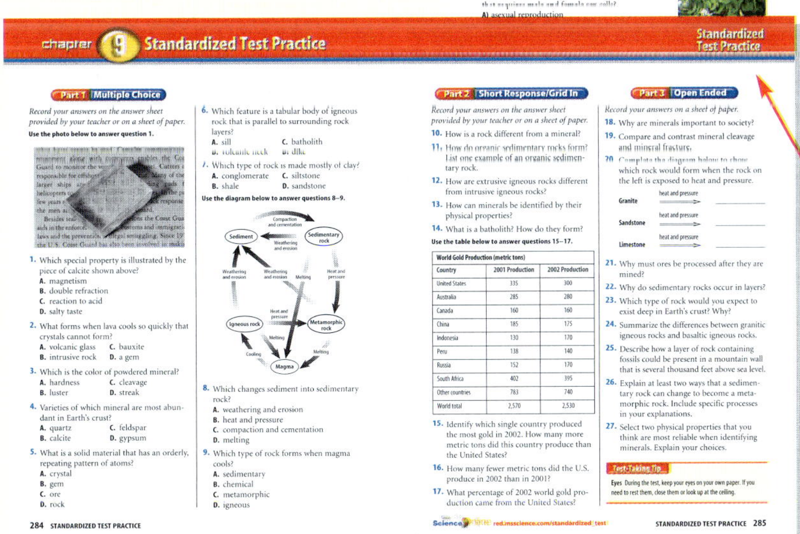

Standardized Test Practice

- Multiple Choice
- Short Response/Grid In
- Open-Ended Questions

Time-Saving Teacher Resources

Glencoe Science provides an extensive array of support materials and resources designed to help you create and customize your science course quickly and easily.

Time-Saving Teacher Resources

FAST FILE Chapter Resources

FAST FILE

For each chapter, Chapter Resources contain key reproducible masters along with additional teaching strategies, teacher support, and answer keys.

Reproducible Student Resources include:

- Worksheets for all MiniLABs and Labs
- *Directed Reading for Content Mastery* (in English and Spanish), *Reinforcement* and *Enrichment* worksheets for each chapter
- *Section Focus, Teaching,* and *Assessment Transparencies* masters
- Chapter Review and Chapter Test masters

Teacher Support and Planning includes:

- Additional student preparation materials for each chapter
- Spanish Resources provide translations of all objectives, vocabulary, and main ideas
- Answer keys and teaching strategies

Teacher Wraparound Edition

The *Teacher Wraparound Edition* is your key to the teaching resources available. In addition to teaching strategies and suggestions, the *Teacher Wraparound Edition* provides a guide for all print and software materials available for each lesson.

Transparencies

- Section Focus Transparency
- Assessment Transparency
- Teaching Transparency
- Transparency masters and worksheets

Exam*View*® Pro Testmaker CD-ROM

- Create, edit, and customize tests
- Create multiple versions of tests
- Translate tests from English to Spanish and vice versa
- Build tests aligned with state standards

Video Labs

- Step-by-step lab procedures for selected *Student Edition* labs
- Lab safety skills
- Teacher support
- Troubleshooting advice

Technology Support

TeacherWorks™

This CD-ROM is your all-in-one teacher resource center. Personalize a lesson plan, access resources from the *Teacher Wraparound Edition*, connect to the Internet, or make a to-do list. These are only a few of the many features that can assist you in the planning and organizing of your lessons.

Includes:

- A calendar feature
- Access to all program blackline masters
- Standards correlations and more

StudentWorks™

This CD-ROM is a valuable resource for students to access content online and use online resources to continue learning chapter concepts.

Includes:

- Complete *Student Edition* on CD-ROM
- Links to online activities and resources
- Access to all student worksheets

Interactive Chalkboard

PowerPoint® Presentations

This CD-ROM brings Microsoft® PowerPoint® presentations right to your door. With the large number of graphics provided, students can use a visual approach to learning chapter content.

Includes:

- A pre-made presentation for every chapter
- Image bank and interactive graphics
- Animations and audio clips
- All new section and chapter questions
- Standardized test practice
- Transparencies
- Pre-lab questions for all labs
- Foldables™ directions
- Links to red.msscience.com

Virtual Labs CD-ROM Program

The Virtual Labs CD-ROM contains a collection of labs that allow students to complete labs that are too expensive, take too long to complete, or might be too dangerous in a classroom laboratory.

Science Online

red.msscience.com

For students:

- Prescreened Internet sites (Web links) that correlate to text chapters
- Interactive activities that review chapter concepts
- Internet Labs where students can share data with each other
- Access to *Student Edition* online
- Interactive self-check quizzes for each chapter section as well as the entire chapter

For teachers:

- *WebQuest* activities
- Teacher Forum for teachers to share activities and ideas

McGraw Hill Learning Network mhln.com

mhln.com is an online teaching and learning space for teachers, students, and parents.

- For students, mhln.com provides an interactive online textbook. This means a lighter student backpack.
- For teachers, mhln.com provides a customizable online learning space for posting assignments, web-based instruction, student resources, and a class calendar.
- For parents, mhln.com provides tools to help their child succeed in school. Parents have instant access to their child's progress and daily assignments.

Help Your Students with Reading and Writing

Glencoe Science increases science literacy, improves reading comprehension, and deepens students' understanding of ideas and concepts. The reading strategies are active, constructive, and engaging.

Reading and Writing

In the *Student Edition*

Reading Checks stimulate quick recall and keep students focused on the main idea.

Reading Check

Caption Questions offer a visual approach to learning. Students are asked to recall what they have read by interpreting the visual.

> **Figure 18** Three allotropes of carbon are depicted here.
> **Identify** *the geometric shapes that make up each allotrope.*

Science Journals provide opportunities to write responses to questions that require critical thinking, to conduct research and write about it, or to practice creative writing skills.

Skill Words such as *compare and contrast, describe, explain,* and *state* are included in the Self Check at the end of each section, in the goals of each Lab, and in the chapter reviews.

Technology Support

Vocabulary PuzzleMaker allows you to create crossword puzzles, jumble puzzles, or word searches in minutes to review chapter vocabulary. The puzzles can be printed or played on the computer screen.

Guided Reading Audio Program (available on MP3-formatted CD-ROMs) provides a comprehensive overview in Spanish and English for auditory learners, students with reading difficulties, and English-language learners (ELL).

In the *Teacher Edition*

Project CRISS™ (Creating Independence through Student-Owned Strategies) strategies are highlighted in the Unit Openers where an activity is supplied that relates to unit content. For more information about Project CRISS™, visit www.projectcriss.com.

Use Science Words features encourage students to look up biology terms. Use these as a pre-reading activity, or have students keep a vocabulary journal.

> **Use Science Words**
> **Word Usage** The distinction between distance and displacement can be confusing. Have students use each of these words correctly in a sentence. Possible response: When I go to school and then back home, my displacement is zero, even though the distance from home to school is 2 km.

Additional Science Journal writing activities promote writing and critical thinking skills and extend the *Student Edition* journaling activity.

Active Reading Strategies utilize a variety of learning styles, and encourage cooperative learning and intrapersonal reflection on chapter content.

Additional Teacher Resources

Reading Essentials provides summaries of each section of the textbook, focusing on the most important scientific concepts.

Reading and Writing in the Science Classroom provides teachers with effective strategies for building students' skills in reading and writing about science.

Help Your Students with Reading and Writing

Making Concept Maps and Charts

Bubble Map Students brainstorm and organize words in clusters to describe concepts.

Flow Chart Students logically analyze and draw a sequence of events.

Cause-and-Effect Chart Students visually represent the causes and effects of an event or process.

Supporting-Idea Chart Students make a concept map to analyze relationships between a whole and its parts.

Network-Tree Concept Map Students show a hierarchy and use branching procedures.

Events-Chain Concept Map Students order steps or stages in a linear procedure.

Cycle Concept Map Students show how a series of events interact.

Spider Concept Map Students use for brainstorming and grouping nonrelated terms to a central idea.

Using the Science Journal

Double-Reference Journal Students read and record ideas.

Metacognition Students analyze what and how they have learned.

Learning Journal Students create notes and reflect on the content.

Problem-Solution Journal Students analyze problems and suggest workable solutions.

Speculation About Effects/Prediction Journal Students examine events and speculate about their possible long-term effects.

Synthesis Journal Students reflect on a project, a paper, or a performance task and plan how to apply what they have learned to their own lives.

Reflective Journal Students identify what they learned in an activity and record responses.

Quickwrites Students use spontaneous writing to discover what they already know.

Collaborative Learning Strategies

Pair of Pairs Partners respond to a question and compare their response to other pairs and to the class.

Write-Draw-Discuss Students write about and draw a picture of a concept, then share it with the class.

Active Reading

Learning Reading Have students divide a sheet of paper in half. Have them record research notes, lecture notes, and vocabulary terms related to the laws of motion in the left column. Tell them to use the right column to respond, interpret, question, and analyze left-column entries. L2 **Intrapersonal**

Four-Corner Discussion The class works in four groups to debate a complex issue.

Jigsaw Students work in groups to become experts on a portion of text and share what they've learned with the class.

Buddy Interviews Students interview one another to find out what strategies they use to learn the text.

Reciprocal Teaching Students take turns reading a section of text, retelling it in their own words, then asking questions about it.

News Summary Students are given several minutes to summarize, retell, or analyze an activity for a "TV" audience.

ReQuest The teacher reads aloud an article or story. Student pairs then construct discussion questions and review the content.

Foldables™

Foldables™ are easy-to-make, three-dimensional, interactive graphic organizers that students create out of simple sheets of paper. These unique hands-on tools for studying and reviewing were created exclusively for Glencoe by education specialist Dinah Zike.

Research Behind Foldables™

According to research (Bransford, 1979; Corno, 1994), study strategies help students understand, organize, remember and apply new information presented in science textbooks. Some study strategies include concept mapping, highlighting, outlining, note taking, summarizing, and underlining (Peverly, Brobst, Graham & Shaw, 2003). Glencoe Science offers Dinah Zike's Foldables™ Study Organizers as an organizational tool and study guide for students.

Build Prereading Skills

- Encourages students to prepare for what they will be learning
- Gives students an opportunity to recall what they already know about a subject

Encourage Active Reading and Writing

- Practices basic reading and writing skills
- Develops skills in finding and reporting main ideas
- Organizes information
- Reviews key vocabulary terms

Summarize Content for Review

- Creates a comprehensive, interactive snapshot of the chapter
- Provides preparation support for chapter, unit, and end-of-course exams, as well as standardized tests

FOLDABLES™ Study Organizer

Invertebrates Make the following Foldable to compare and contrast the characteristics of water and land invertebrates.

STEP 1 **Fold** one sheet of paper lengthwise.

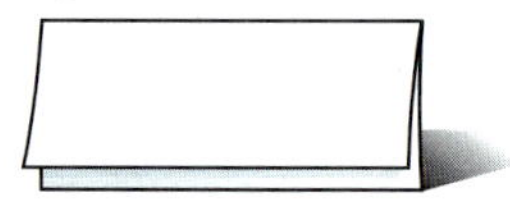

STEP 2 **Fold** into thirds.

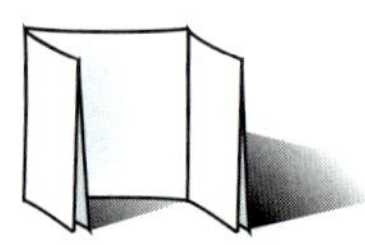

STEP 3 **Unfold and draw** overlapping ovals. Cut the top sheet along the folds.

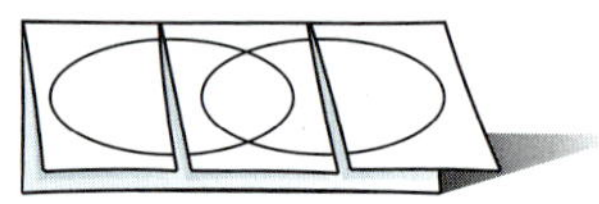

STEP 4 **Label** the ovals as shown.

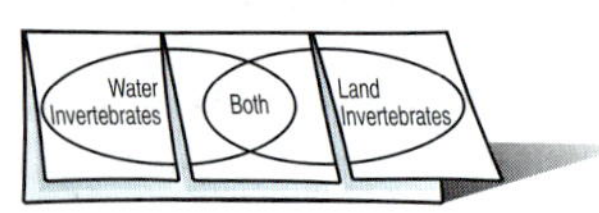

Construct a Venn Diagram As you read this chapter, list the characteristics unique to water invertebrates under the left tab, those unique to land invertebrates under the right tab, and those characteristics common to both under the middle tab.

Dinah Zike's *Teaching Science with Foldables™*

- Comprehensive presentations of Foldables™ instructions
- More ideas on how to incorporate Foldables™ into your lessons
- Easy-to-read folding instruction pages in color

Course Planning Guide

Glencoe Science programs provide a complete selection of core concepts that can be presented in a way that meets the needs of all your students. As the teacher, you are in the best position to design a science course that sets the pace at which the content is covered and determines what material should be given the most emphasis. To assist you in planning the course, the following Course Planning Guide is provided.

The chart shown below offers general suggestions for pacing your students through the book. Pacing for both standard class periods and block schedule class periods is given. The total number of days in each level of pacing is the typical 188-day school year and 95-day semester.

Chapter	Single-Class (188 days)	Block (95 days)
1	9	4.5
2	13	6.5
3	10	5
4	8	4
5	9	4.5
6	9	4.5
7	10	5
8	7	3.5
9	7	3.5
10	8	4
11	7	3.5
12	9	4.5
13	9	4.5
14	7	3.5
15	7	3.5
16	7	3.5
17	9	4.5
18	9	4.5
19	5	3.5
20	12	6
21	7	3.5
22	10	5

Assessment

Glencoe Science offers the Glencoe Assessment Advantage, a system designed to give you all the tools you need to prepare your students for success in any testing situation.

In the *Student Edition*

Section Review and **Applying Math** questions appear in every chapter.

Chapter Review questions help you evaluate students' knowledge and ability to apply science concepts.

Standardized Test Practice questions at the end of each chapter provide students with additional opportunities to practice their test-taking skills.

In the *Teacher Wraparound Edition*

Assessments located throughout the *Teacher Wraparound Edition* provide methods for assessing students' comprehension with Performance, Process, and Content exercises.

Teacher Classroom Resources

Performance Assessment in the Science Classroom

- Guidelines for assessing the performance of a task
- Reproducible activities for evaluating students
- Sample rubrics and checklists

***Fast File* Chapter Resources** provides six pages of assessement for every chapter including *Testing Concepts, Applying Concepts,* and *Writing Skills.*

Technology Support

MindJogger Videoquizzes are interactive video quizzes set in game show format. Each is designed for the full range of student learning styles.

Exam*View*® Pro Testmaker CD-ROM for Windows® and Macintosh® provides an easy way to create, edit, and customize your tests. Select your own test items by objective from two different levels of difficulty, or write and edit your own. Translate tests from English to Spanish and vice versa.

Rubrics

The following rubrics are sample scoring devices for short response and open-ended questions.

Short Response

Points	Description
2	The student demonstrates a thorough understanding of the science of the task. The response may contain minor flaws that do not detract from the demonstration of a thorough understanding.
1	The student has provided a response that is only partially correct.
0	The student has provided a completely incorrect solution or no response at all.

Open Ended

Points	Description
4	The student demonstrates a thorough understanding of the science of the task. The response may contain minor flaws that do not detract from the demonstration of a thorough understanding.
3	The student demonstrates an understanding of the science of the task. The response is essentially correct and demonstrates an essential but less than thorough understanding of the science.
2	The student demonstrates only a partial understanding of the science of the task. Although the student may have used the correct approach to a solution or may have provided a correct solution, the work lacks an essential understanding of the underlying science concepts.
1	The student demonstrates a very limited understanding of the science of the task. The response is incomplete and exhibits many flaws.
0	The student provides a completely incorrect solution or no response at all.

Educational Partnerships

NATIONAL GEOGRAPHIC

Some topics in the chapter either require or benefit from a larger, more detailed visual explanation. The National Geographic Society has created *Visualizing* features that call out an important concept from the chapter and illustrate it in a way that will inform, excite, and motivate your students.

TIME

TIME magazine brings science topics and history together to further explain the chapter's main ideas and show how science relates to real life.

Differentiated Instruction

Teaching Strategies

Following each suggested assessment and activity, ability levels are supplied to accommodate all students.

- LS Multiple Learning Styles logos are used throughout the text to indicate strategies that address different learning styles.
- L1 Level 1 activities should be appropriate for students with learning difficulties.
- L2 Level 2 activities should be within the ability range of all students.
- L3 Level 3 activities are designed for above-average students.
- ELL English-Language Learners activities should be within the ability range of English-Language Learners or students who speak English as a second language (ESL).
- COOP LEARN Cooperative Learning activities are designed for small group work.
- PBL Problem-Based Learning activities apply real-world situations to learning.
- P Portfolio strategies represent student products that can be placed into a best-work portfolio.

Identifying Misconceptions

These short, diagnostic, and perscriptive lessons target common science misconceptions.

IDENTIFYING Misconceptions

Necessary Metals Students may think it strange that metals are needed by the body. Explain that the metals calcium, potassium, and sodium, for example, help transmit electrical signals within the nervous system.

Multiple Learning Styles

Look for these italicized designations under various activities to help you target your lessons to each student's preferred learning style.

- ***Kinesthetic*** learners learn through touch, movement, and manipulating objects.
- ***Visual-Spatial*** learners think in terms of images, illustrations, and models.
- ***Interpersonal*** learners understand and work well with other people.
- ***Intrapersonal*** learners can analyze their own strengths and weaknesses and may prefer to work on their own.
- ***Linguistic*** learners write clearly and easily understand the written word.
- ***Logical-Mathematical*** learners understand numbers easily and have highly-developed reasoning skills.

Daily Interventions

Found at the end of each chapter section, this feature is designed to intercept students who are struggling and prescribe a system to help them get back on track. *Reteach* provides reinforcement of the section's concepts through visual activities.

DAILY INTERVENTION

Check for Understanding

Kinesthetic Have a student light a match and allow it to burn. Also, place an ice cube in a glass container and allow it to melt. Have students explain what type of changes took place with each item and describe the final properties of each.

Differentiated Instruction

These activities present various teaching strategies designed to help you meet the special needs of students with learning disabilities, physical challenges, visual impairment, and hearing impairment. Challenge activities provide opportunities for students who excel to engage in activities and research projects that extend the chapter's concepts. English-language learners in the classroom will also find exercises that bridge the gap between language barriers and the chapter content.

Differentiated Instruction

English-Language Learners Have students make flashcards with the name of a gland on one side and the hormone it produces on the other. Student pairs can quiz each other using the flashcards.

Cultural Diversity

These readings provide insights into the unique ways in which people of different ethnicities and cultural heritage have approached science. The intent of these features is to build awareness and appreciation for the global community in which we live.

Cultural Diversity

Tsunami! Tsunamis occur in many areas around the world and have been reported since ancient times. One of the earliest record tsunamis struck Syria around 2,000 B.C. Thera, one of the Cyclades Islands in the Mediterranean, may be the remnant of a volcano that erupted—causing tsunamis that ended the Minoan civilization on Crete. Tsunami is a Japanese word for "harbor wave." Many have struck the Japanese shore. Because Japan is an island nation, the threat of tsunamis is a national safety concer. Today, by using expected tsunami characteristics, the Japan Meteorological Agency can forecast tsunami heights for the Japanese coastline. This provides residents with the knowledge necessary to move a safe distance away from the shore.

Inquiry-Based Science

The call for more inquiry-based science by the *National Science Education Standards* has been met by Glencoe Science.

Glencoe Science recognizes the importance of conducting inquiry-based science activities in the classroom. The process of doing inquiry models actual science practice, encouraging problem-solving strategies and developing critical thinking skills. Inquiry gets students actively involved in the learning process by allowing them to determine materials, procedures, or the topics and questions they want to investigate.

Inquiry can range from a very structured activity for those students who need more guidance to a more open-ended approach where students lead the investigations. Glencoe Science recognizes that the inquiry activities suggested will not look the same in every classroom. We encourage teachers to modify the suggested activities in a manner that best supports your students.

Glencoe also provides teachers with Alternative Inquiry Labs, teaching strategies or suggestions for making existing labs more inquiry-based.

Alternative Inquiry Lab

Extend the Experience To make this Lab an Inquiry Lab, have students examine the magnetic field around the motor, using a compass or iron fillings. Test the armature and rotor before final assembly, as it will be difficult to fit the compass inside the motor. Encourage students to make various adaptations to the motor. Ask them to figure out how to make the motor spin the other way, spin faster, and spin slower. How could the motor be turned into a generator? How could they make the motor consume less power? Finally, students might research how these issues are handled by industry or other questions they find interesting.

Research-Based Learning Strategies

Glencoe Science incorporates the most current and applicable educational research on science learning and follows recommendations from the American Association for the Advancement of Science and the National Science Teachers Association. The following research-based strategies can be found throughout the text.

Research-Based Learning Strategies

Learning Strategies

The following research-based strategies can be found throughout the text:

- **Using Prior Knowledge** Glencoe Science encourages students to use their prior knowledge to learn information because this adds relevance to the material. Students are referred back to other parts of the text or to their own real-life experiences.
- **Practicing Important Tasks** By offering students an opportunity to practice important tasks using a variety of labs and activities in the *Student Edition, Teacher Wraparound Edition,* ancillaries and technology, Glencoe Science makes learning fun and relevant for students.
- **Using Visuals to Communicate, Organize, and Reinforce Learning** High-quality art and photos throughout the text communicate concepts more efficiently and reinforce learning, while allowing students to organize information.
- **Motivating Students to Achieve** Active strategies and real-world experiences motivate students to achieve. Throughout Glencoe's programs, students are encouraged to apply their knowledge in ways that will motivate them to learn.
- **Developing Decoding and Reading Comprehension Strategies** Throughout the text, students are supplied with caption questions, reading checks, and other strategies to aid in comprehension.
- **Using Study Strategies** Through the use of highlighting, outlining, note-taking, summarizing, and other such strategies, students can monitor their own progress and organize information more effectively, thereby increasing their scientific literacy. These strategies are found throughout the text and ancillaries.

The use of these strategies within Glencoe Science will help teachers to achieve the goals set forth by the *National Science Education Standards.*

White Paper

The Glencoe Science White Paper outlines the educational strategies on which this program was based. This document provides specific examples from the *Student Edition, Teacher Wraparound Edition,* ancillary program, and technology resources, highlighting extensive use of educationally sound strategies that help students learn science.

Spencer Grant/PhotoEdit, Inc.

Field Research and Testing

Feedback from students, teachers, curriculum supervisors, department chairpersons, parents, learning specialists, and science content experts was invaluable in the development of this program. The following pre-publication and post-publication research was conducted.

Prior to Publication

- Detailed classroom teacher and curriculum supervisor **surveys** were conducted by independently contracted researchers.
- A **nationwide panel** of science teachers, curriculum supervisors, and department chairpersons provided countless hours of feedback and assistance throughout program development.
- A wide range of **educator and content reviewers** provided in-depth reviews of and suggestions for manuscripts and pre-publication versions of the program.
- **Face-to-face interviews** with science teachers provided insight into teachers' day-to-day challenges.

After Publication

- Field tests were conducted in which students and teachers used a pre-publication manuscript in the classroom.
- Follow-up interviews, observations, and surveys of Glencoe Science users provide ongoing opportunities for program development and verification of program success.

Field-Test Results

- Field-test research indicates that test scores increased among students using Glencoe Science programs.
- Nine out of ten students earned higher scores after using Glencoe programs.
- Scores improved among both male and female students.
- Scores improved among both minority and non-minority students.
- Overall, the gap between the average pre-test score and a perfect score closed by 33 percent. Stated differently, on average, **scores increased 77 percent after students used the Glencoe program.**

Field Research and Testing

KS Studios

National Education Standards

Correlation of *Glencoe Science Level Red* to the National Science Education Standards.

Science Standard	Chapter and Section
(UCP) Unifying Concepts and Processes	
1. Systems, order, and organization	1-2, 1-3, 1-4, 2-1, 2-2, 2-3, 3-1, 3-2, 4-1, 4-2, 4-3, 5-1, 5-2, 5-3, 6-1, 6-2, 6-3, 8-1, 8-2, 8-3, 9-1, 9-2, 9-3, 10-1, 10-2, 12-1, 12-2, 12-3, 13-1, 13-2, 13-3, 13-4, 14-1, 14-2, 14-3, 15-1, 15-2, 15-3, 16-1, 16-2, 17-1, 17-2, 17-3, 17-4, 18-1, 18-2, 18-3, 18-4, 19-1, 19-2, 20-1, 20-2, 21-1, 21-2, 21-3
2. Evidence, models, and explanation	1-1, 1-2, 1-3, 1-4, 2-1, 2-2, 2-3, 3-1, 3-2, 4-1, 4-2, 4-3, 5-1, 5-2, 5-3, 6-1, 6-2, 6-3, 7-1, 7-2, 7-3, 8-1, 8-2, 8-3, 9-1, 9-2, 9-3, 10-1, 10-2, 12-1, 12-2, 12-3, 13-1, 13-2, 13-3, 13-4, 14-1, 14-2, 14-3, 15-1, 15-2, 15-3, 16-1, 16-2, 17-1, 17-2, 17-3, 17-4, 18-1, 18-2, 18-3, 18-4, 19-1, 19-2, 20-1, 20-2, 21-1, 21-2, 21-3, 22-1, 22-2, 22-3
3. Change, constancy, and measurement	1-3, 1-4, 2-1, 2-2, 3-1, 3-2, 4-1, 4-2, 4-3, 5-1, 5-2, 5-3, 6-1, 6-2, 6-3, 7-1, 7-2, 7-3, 8-1, 8-2, 8-3, 9-1, 9-2, 9-3, 10-1, 10-2, 11-1, 11-2, 12-1, 12-2, 12-3, 13-1, 13-2, 13-3, 13-4, 14-1, 14-2, 14-3, 15-1, 15-2, 15-3, 16-1, 16-2, 17-1, 17-2, 17-3, 17-4, 18-1, 18-2, 18-3, 18-4, 19-1, 19-2, 20-1, 20-2, 21-1, 21-2, 21-3, 22-1, 22-2, 22-3
4. Evolution and equilibrium	11-1, 20-1, 20-2
5. Form and function	3-1, 3-2, 4-1, 4-2, 4-3, 5-1, 5-2, 5-3, 6-2, 6-3, 7-1, 7-2, 7-3, 8-1, 8-2, 8-3, 9-1, 9-2, 9-3, 10-1, 10-2, 12-1, 12-2, 12-3, 13-1, 13-2, 13-3, 13-4, 14-1, 14-2, 14-3, 15-1, 15-2, 15-3, 16-1, 16-2, 17-1, 17-2, 17-3, 17-4, 18-1, 18-2, 18-3, 18-4, 19-1, 19-2, 20-1, 21-1, 21-2, 21-3
(A) Science as Inquiry	
1. Abilities necessary to do scientific inquiry	1-1, 1-2, 1-3, 1-4, 2-1, 2-2, 2-3, 3-1, 3-2, 4-1, 4-2, 4-3, 5-1, 5-2, 5-3, 6-1, 6-2, 6-3, 7-1, 7-2, 7-3, 8-1, 8-2, 8-3, 9-1, 9-2, 9-3, 10-1, 10-2, 11-1, 11-2, 12-1, 12-2, 12-3, 13-1, 13-2, 13-3, 13-4, 14-1, 14-2, 14-3, 15-1, 15-2, 15-3, 16-1, 16-2, 17-1, 17-2, 17-3, 17-4, 18-1, 18-2, 18-3, 18-4, 19-1, 19-2, 20-1, 20-2, 21-1, 21-2, 21-3, 22-1, 22-2, 22-3
2. Understandings about scientific theory	1-1, 1-2, 1-3, 1-4, 2-1, 2-2, 2-3, 3-1, 3-2, 4-1, 4-2, 4-3, 5-1, 5-2, 5-3, 6-1, 6-2, 6-3, 7-1, 7-2, 7-3, 8-1, 8-2, 8-3, 9-1, 9-2, 9-3, 10-1, 10-2, 11-2, 12-1, 12-2, 12-3, 13-1, 13-2, 13-3, 13-4, 14-1, 14-2, 14-3, 15-1, 15-2, 15-3, 16-1, 16-2, 17-1, 17-2, 17-3, 17-4, 18-1, 18-2, 18-3, 18-4, 19-1, 19-2, 20-1, 20-2, 21-1, 21-2, 21-3, 22-1, 22-2, 22-3
(B) Physical Science	
1. Properties and changes of properties in matter	1-1, 3-1, 3-2, 4-1, 4-2, 4-3, 6-1, 6-2, 6-3, 7-1, 7-2, 7-3, 11-1
2. Motions and forces	3-1, 5-1, 5-2, 5-3, 6-1, 6-2, 6-3, 7-1, 7-3, 8-1, 8-3, 9-2, 9-3, 10-1, 10-2, 11-2, 12-2, 12-3, 13-3, 15-1, 18-1
3. Transfer of energy	3-1, 3-2, 5-1, 5-2, 5-3, 6-1, 6-2, 6-3, 7-1, 8-1, 8-3, 9-2, 9-3, 10-1, 10-2, 12-2, 12-3, 13-4
(C) Life Science	
1. Structure and function in living systems	11-1, 13-1, 13-2, 13-3, 13-4, 16-1, 16-2, 17-1, 17-2, 17-3, 17-4, 18-1, 18-2, 18-3, 18-4, 19-1, 19-2, 20-1, 21-1, 21-2, 21-3
2. Reproduction and heredity	17-2, 17-3, 17-4, 18-1, 18-2, 18-3, 18-4, 19-1, 19-2, 20-1, 20-2
3. Regulation and behavior	13-4, 16-1, 16-2, 17-2, 17-3, 17-4, 18-1, 18-2, 18-3, 18-4, 19-1, 19-2, 20-1, 20-2, 21-2
4. Populations and ecosystems	13-4, 18-4, 21-1, 21-2, 21-3
5. Diversity and adaptations of organisms	11-1, 13-4, 16-1, 16-2, 17-2, 17-3, 17-4, 18-1, 18-2, 18-3, 18-4, 21-2
(D) Earth and Space Science	
1. Structure of the Earth system	7-3, 8-1, 8-2, 8-3, 9-1, 9-2, 9-3, 10-1, 10-2, 12-1, 12-2, 12-3, 13-1, 13-2, 13-3, 13-4, 15-1
2. Earth's history	9-1, 9-2, 9-3, 10-1, 10-2, 13-3, 13-4
3. Earth and the solar system	14-1, 14-2, 14-3, 15-1, 15-2
(E) Science and Technology	
1. Abilities of technological design	18-1, 22-3
2. Understandings about science and technology	11-2

Science Standard	Chapter and Section
(F) Science in Personal and Social Perspectives	
1. Personal health	7-1
2. Populations, resources, and environments	19-1, 22-2
3. Natural hazards	12-3
4. Risks and benefits	6-3
5. Science and technology in society	1-1, 11-2, 14-3, 16-2, 21-2, 22-1, 22-2, 22-3
(G) History and Nature of Science	
1. Science as a human endeavor	4-1, 4-3, 18-1
2. Nature of science	1-1, 1-2, 1-3, 1-4, 4-1, 4-3
3. History of science	4-1, 4-2, 4-3, 7-1, 9-1

How Glencoe Science Aligns with the National Science Education Standards

The correlations at the left show the close alignment between Glencoe Science and the grade-appropriate standards. Glencoe Science allows students to discover concepts within each of the content standards and gives students opportunities to make connections among the science disciplines. Hands-on activities and inquiry-based lessons reinforce the science processes emphasized in the standards.

How Glencoe Science Aligns with the NCTM Standards for Grades 6–8

Throughout Glencoe Science, each Applying Math activity provides students with the opportunity to practice and apply some of the mathematical concepts and applications described in the NCTM Standards. These activities serve to reinforce mathematical skills in real-life situations, thus preparing students to meet their needs in an ever-changing world.

Correlation of *Glencoe Science Level Red* to NCTM Standards

Math Standard	Page
1. Number and Operations	37, 48, 65, 93, 123, 132, 134, 140, 145, 157, 207, 249, 311, 337, 353, 369, 401, 435, 469, 485, 493, 525, 547, 555, 585, 603, 611, 641, 673
2. Algebra	93, 123, 131, 132, 134, 140, 145, 157, 207, 221, 249, 311, 337, 353, 369, 401, 469, 485, 585, 603, 611, 641, 673
3. Geometry	401, 416, 641
4. Measurement	37, 65, 93, 189, 283, 369, 401, 435, 485, 493, 641
5. Data Analysis and Probability	17, 65, 93, 189, 283, 311, 337, 401, 435, 469, 485, 493, 555, 585, 641
6. Problem Solving	17, 157, 547
7. Reasoning and Proof	601, 607
8. Communication	55
9. Connections	93, 131, 132, 134, 140, 145, 157, 189, 207, 221, 249, 283, 311, 337, 353, 369, 401, 435, 469, 485, 493, 525, 555, 585, 603, 611, 673
10. Representation	58, 59, 65

Learning in the Laboratory

In the *Student Edition*

Working in the lab is perhaps the most exciting part of science. Labs and MiniLABs give your students the hands-on opportunity to create, investigate, and explore science. Students will learn new material, while having fun.

Traditional Labs provide guided inquiry as students hypothesize, plan investigations, and collect and analyze data.

Design Your Own Labs challenge students to design their own experiments that will provide answers to lab problems.

Model and Invent Labs allow students to create a model or invent a product that can demonstrate scientific concepts.

Use the Internet Labs help students share and retrieve data from others around the country and world, by posting data they have collected on red.msscience.com. These activities show students the possible range of data, importance of collecting large amounts of data, and data analysis.

MiniLABs and **Applying Science** activities are a quick way for students to practice specific process skills as they learn science concepts. Many of them can be classroom demonstrations or homework assignments.

In the *Teacher Wraparound Edition*

Quick Demos allow students to visualize biology concepts.

Inquiry Labs in each chapter can be used to support inquiry-based science or as alternatives to traditional labs.

Technology Support

Virtual Labs CD-ROM includes a collection of interactive activities involving major chapter themes that allow students to understand science concepts in a virtual setting. These labs provide a way to conduct experiments without classroom time, cost, and safety constraints.

Video Labs show students step-by-step lab procedures while emphasizing important lab safety skills. The lab videos also include teacher support, troubleshooting advice, and expected outcomes.

Additional Resources

Laboratory Activities offer a variety of traditional laboratory experiences that reinforce the biology principles in the text. The *Teacher Wraparound Edition* provides full support.

Probeware Labs present activities for students to explore scientific concepts using a probeware data collection system. These hand-held systems provide a fast and simple way to collect, view, and analyze data in the classroom or during a field investigation. Integrating technology in the classroom is made simple with step-by-step instructions for setting up and using probeware.

Science Inquiry Lab Manual gives students the opportunity to become increasingly independent in stating hypotheses, designing and performing experiments, and collecting and analyzing data.

Safety in the Laboratory

All activities are designed to minimize dangers in the laboratory. Careful laboratory planning and management by both the instructor and the student are essential to a safe laboratory experience. **Local, state, and federal laboratory safety laws and regulations must be strictly followed.** The information provided here is one of the many resources to which you can refer for information about laboratory safety.

Classroom and Laboratory Preparation

1. Label and store chemicals properly and securely. See p. 20T.
2. Store equipment properly and securely and other thing.
   a. Clean and dry all equipment before storing.
   b. Protect electronic equipment and microscopes from dust, humidity, and extreme temperatures.
   c. Number, catalog, and organize equipment.
3. Ensure adequate work space for each student.
4. Ensure adequate classroom and storeroom ventilation.
5. Explain and post safety and evacuation guidelines along with expectations of conduct.
6. Ensure that all safety equipment is functioning properly and is clearly visible.
7. Provide hot plates as a heat source whenever possible. If gas burners are used, know where the central gas supply shutoff valve is located.
8. Ensure that each workstation has a GFCI-protected electrical source.
9. Provide safety goggles consistent with ANSI Standard Z87.1 for each student, including students who wear corrective lenses.

First Day of Class (with students)

1. Distribute and discuss safety rules, safety symbols, and first aid guidelines. Have students to review safety symbols and guidelines.
2. Review safe use of equipment and chemicals.
3. Review use and location of safety equipment.
4. Discuss safe disposal of materials and laboratory cleanup policy.
5. Discuss proper laboratory attitude and conduct.
6. Document students' understanding of the preceding points. Have students sign a safety contract and return it.

Before Each MiniLAB or Lab

1. Perform each investigation yourself before assigning it.
2. Arrange the lab in such a way that equipment and supplies are clearly labeled and easily accessible.
3. Have available only equipment and supplies needed to complete the assigned investigation.
4. Review the procedure with students, emphasizing any caution statements or safety symbols that appear.
5. Be sure all students know the proper procedures to follow if an accident should occur.
6. Provide containers for disposing of chemicals, waste products, and biological specimens. Disposal methods should meet local guidelines.

During the MiniLAB or Lab

1. Make sure the lab is clean and free of clutter.
2. Insist that students wear goggles and aprons.
3. Never allow a student to work alone in the lab.
4. Never allow students to use a cutting device with more than one edge.
5. Students should not point the open end of a heated test tube toward anyone.
6. Remove broken glassware or frayed cords from use. Also clean up any spills immediately. Dilute solutions with water before removing.
7. Be sure all glassware that is to be heated is of a heat-treated type that will not shatter.
8. Remind students that hot glassware looks cool.
9. Prohibit eating and drinking in the lab.

After the MiniLAB or Lab

1. Be sure that the lab is clean.
2. Be certain that students have returned all equipment and disposed of broken glassware and chemicals properly.
3. Be sure that all hot plates and electrical connections are off.
4. Insist that each student wash his or her hands when lab work is completed.

Chemical Storage and Disposal

Chemical Storage and Disposal

General Guidelines

Be sure to store all chemicals properly. The following are guidelines commonly used. Your school, city, county, or state may have additional requirements for handling chemicals. It is the responsibility of each teacher to become informed of the rules or guidelines in effect in his or her area.

1. Separate chemicals by reaction type. Strong acids should be stored together. Likewise, strong bases should be stored together and should be separated from acids. Oxidants should be stored away from easily oxidized materials, and so on.
2. Be sure all chemicals are stored in labeled containers indicating contents, concentration, source, date purchased (or prepared), any precautions for handling and storage, and expiration date.
3. Dispose of any outdated or waste chemicals properly according to accepted disposal procedures.
4. Do not store chemicals above eye level.
5. Wood shelving is preferable to metal. All shelving should be firmly attached to the wall and should have anti-roll edges.
6. Store only those chemicals that you plan to use.
7. Hazardous chemicals require special storage containers and conditions. Be sure to know which chemicals those are and the accepted practices for your area. Some substances must be stored outside the building.
8. When working with chemicals or preparing solutions, observe the same general safety precautions that you would expect from students. These include wearing an apron and goggles. Wear gloves and use the fume hood when necessary. Students will want to do as you do whether they admit it or not.
9. If you are a new teacher in a particular laboratory, it is your responsibility to survey the chemicals stored there to be sure they are stored properly. If not, they should be disposed of. Consult the rules and laws in your area concerning which chemicals can be kept in your classroom. For disposal, consult up-to-date disposal information from state and federal governments.

Disposal of Chemicals

Local, state, and federal laws regulate the proper disposal of chemicals. These laws should be consulted before chemical disposal is attempted. Although many substances encountered in the science classroom can be flushed down the drain with plenty of water, it is not safe to assume that this is always true. Teachers who use chemicals should consult the following book from the National Research Council:

Prudent Practices in the Laboratory. Washington, DC: National Academy Press, 1995. This book is useful and was revised in 1995. Current laws in your area would, of course, supersede the information in this book.

DISCLAIMER

Glencoe Publishing Company makes no claims to the completeness of this discussion of laboratory safety and chemical storage. The material presented is not all-inclusive, nor does it address all of the hazards associated with handling, storing, and disposing of chemicals, or with laboratory management.

Clement Mok/PictureQuest

Preparation of Solutions

It is important to use safe laboratory techniques when handling all chemicals. Always check the MSDS (Material Safety Data Sheet) for each chemical before using it in the classroom. Many substances might appear harmless, but might be toxic, corrosive, or very reactive. Chemicals should never be ingested. Use proper techniques to smell any chemical, wear safety goggles and an apron in the laboratory, and observe the following precautions.

1. **Dilution of Acids and Bases** When diluting acids with water, always add the acids to the water. Never add water to acids. When sulfuric acid and sodium hydroxide are added to water, a large amount of thermal energy is released. Use extra care when handling these substances.

2. **Poisonous and Corrosive Liquids or Vapors** Use a fume hood if possible. Examples include hydrochloric acid, acetic acid, nitric acid, and ammonium hydroxide.

3. **Poisonous and Corrosive to Eyes, Lungs, and Skin** Examples include acids, bases, silver nitrate, iodine, and potassium permanganate.

Bromthymol blue: Add 0.5 g bromthymol blue powder to 500 mL distilled water to make a BTB stock solution. Dilute 40 mL BTB stock solution to 2 L with distilled water. Solution should be bright blue. If not, add one drop of NaOH at a time, swirling to mix. Check color.

Hydrochloric acid (HCL) solution: To make a 5% solution, add 13.6 mL concentrated HCl to 73 mL water while stirring. To make a 0.1*M* solution, add 1 mL concentrated hydrochloric acid to 100 mL water while stirring.

Iodine solution/Iodine stain: Dilute 1 part Lugol's solution with 15 parts water.

Lugol's solution: Dissolve 10 g potassium iodide in 100 mL distilled water. Then add and dissolve 5 g iodine. Store in dark bottle. Keeps indefinitely.

Phenolphthalein indicator: From a drug store, buy a package of any laxative that contains phenolphthalein. To make 1% solution, mash 4 tablets and pour the powder into 10 mL of rubbing alcohol. Let mixture soak for 15 minutes. Pour liquid into and store in a dropper bottle.

Potassium permanganate: For a 0.01*M* solution of potassium permanganate, dissolve 0.15 g $KMnO_4$ in 100 mL water.

Red cabbage concentrate: Put 5 leaves of red cabbage in a pot. Add 1 L of water, bring to a boil, and simmer until water turns a deep purple. Pour liquid through a strainer or piece of cheesecloth into a storage bottle. Keep refrigerated.

Salt solution: For a 3.5% salt (NaCl) solution that simulates the concentration of ocean water, dissolve 35 g of salt (NaCl) in 965 mL of water. For a 1% solution (weak), dissolve 1 g of salt (NaCl) in 99 mL of water. For a 6% solution, dissolve 6 g of salt (NaCl) in 94 mL of water.

Silver nitrate solution: To make a 10% solution, put 5 g of silver nitrate in 50 mL of distilled water.

Sugar solution: Add 1 tablespoon of sugar to 1 cup of warm water in a deep jar or flask. Stir to dissolve.

Sodium hydroxide (dilute): To make a 1% solution, dissolve 1 g NaOH in 99 mL of water.

Equipment and Materials List

These easy-to-use tables of equipment and consumable materials can help you prepare for your science classes for the year. Refer to the Chapter Organizer in front of each chapter for a list of equipment and materials used for each laboratory activity in the chapter.

Consumables

Material	Launch Lab (Chapter)	MiniLAB (Chapter-Section)	Lab (Chapter-Section)
alcohol, rubbing		4-3	3-2, 20-1
aluminum foil			10-1, 21-3
ammonia, household			19-2
apples			3-2, 17-4
baking soda			4-3
balloon	7	7-1, 17-3	
bananas			3-2
battery, 1.5-V			8-3
battery, D-cell		7-3	8-3
bean seeds			21-3
bicarbonate of soda			19-2
bran or oatmeal			17-4
bulbs, light, 1.5-V			8-3
candle, birthday		12-3	
cardboard	5	6-1, 20-1	
cardboard cylinder		15-3	
cardboard tube		14-1	
cardboard, stiff			6-3
celery, with leaves, fresh			20-2
chalk	15	11-1	
cheesecloth			17-4
clay	6, 10	6-1, 10-2, 12-3, 20-1	
clay, modeling			9-3
cloth, cotton		18-3	
coffee filter			19-2
confetti			12-3
corn syrup			3-2
cornstarch			4-3
cup, paper		13-2	
diet soda		1-2	
dishwashing liquid	12		
dried bread or cookie crumbs			17-4
envelopes, large	17		
fish food			21-1
flour			17-4
food coloring		1-3, 6-2, 13-2	10-1
food coloring, red			20-2
food scraps			17-4
freezer bags, self-sealing			6-3
glue		9-2, 22-3	
glycerin			3-2
grass sod			11-2
gravel, aquarium			21-1
gravel, washed			18-2
hydrochloric acid		11-1	
hydrochloric acid, dilute		9-1	
index card(s)	16		4-2
iodine solution			4-3
labels			21-3
lemon juice			19-2
lemon/water solution			3-2
lettuce, boiled			18-2
limestone		11-1	

Equipment and Materials List

Equipment and Materials List

Material	Launch Lab (Chapter)	MiniLAB (Chapter-Section)	Lab (Chapter-Section)
macaroni		9-2, 20-1	
mailing tubes, cardboard			1-4
markers			15-3
newspaper		11-1	17-4
oil, corn			3-2
oil, salad		4-3	
oil, vegetable		1-3	
onion			20-1
onion skin	20		
orange	20	6-1	
orange juice			19-2
pancake syrup			6-3
paper	7	20-1	4-2, 10-1
paper towel	4		17-4
paper, black	15	15-3	
paper, drawing		15-1	15-3
paper, graph			2-2, 5-3, 22-3
pasta wheels	18		
pears			3-2
pencils, colored			20-1, 22-3
petroleum jelly		18-3	
plastic bag, self-sealing		18-4	
plastic wrap	14, 20		13-1
powdered sugar			4-3
red cabbage			19-2
rice	15		9-3
rock, large			18-2
rope			1-4
rubber band			10-1, 17-4
salt	13	13-2	13-1
sand	11, 13		13-4, 21-1
shoe box		4-1	
soda		1-2	
soda straw, white		7-3	
soil		11-1, 21-1	11-1, 11-2, 17-4, 21-3
string		14-2	
sugar		3-2, 4-3, 20-1	
taffy		10-1	
tape	1, 4, 19	15-3	7-3, 8-3, 13-1
toothpick		13-3	20-1
vinegar		11-1	4-3, 19-2
water, pond			18-2, 20-1
wire, 22-gauge		7-3	
yarn		20-1	7-2
yeast		3-2, 20-1	
Nonconsumables			
aluminum pie pan		9-2	
anemometer			12-3
aquarium, 4-L or jar			18-2
ball		10-2	
ball, rubber			3-2
bar magnet	7	7-3	
barometer			12-3
baseball	6	6-1	6-3
basin, large			13-4
bathroom scale		5-2	
beaker(s)	6, 12	2-1, 12-1	3-2, 4-3, 6-3, 10-1, 11-2, 13-1, 20-1
block, brick or wood	11		11-2
blocks, wooden			10-2
boards			13-4

Equipment and Materials List

Equipment and Materials List

Material	Launch Lab (Chapter)	MiniLAB (Chapter-Section)	Lab (Chapter-Section)
books	5	5-2	
books, planetary			15-3
bottle(s), plastic	12		13-4
bottle, 2-L			21-1
bowl(s), mixing, plastic or glass	4	13-3, 19-1	3-2
bowl, large			13-1
box, clear-plastic		13-2	
bread pan	11		
bricks			13-4
broomstick		5-3	
bulletin board	17		4-2
buttons		20-1	
calcite		9-1	
calculator			11-2, 22-1
camera		15-2	
chair		5-2	
coat, heavy		5-2	
coffee can			12-3
coiled spring toy, long			7-2, 7-3
coin		20-2	
comb		7-1	
compass			8-3, 12-3
computer probe			6-3
computer temperature probe		6-1	
container, flat		17-4	
containers, transparent		6-2	
cooking pot			19-2
cork		14-2	
cork or straw	7		
coverslip		20-1	20-1
cup, plastic	4	21-1	10-1
drinking glass, opaque		7-3	
drinking glasses		4-3, 12-1	19-2, 20-2
dropper	6-7	6-2, 9-1, 11-1, 20-1	20-1
dropper bottles			4-3
encyclopedia			4-2
fan (battery-operated)			13-4
fishnet, small			18-2
flashlight	6	15-3	15-1
fossil limestone sample		9-2	
globe		15-1	
gneiss and granite samples			9-3
golf ball		6-1	
graduated cylinder		3-1	3-2
hornblende		9-1	
hot lamp	13		
hot plate	6		4-3
jar or aquarium, 4-L			18-2
jar, large			20-1
jar, large-mouth, or fish bowl			17-4
jars, peanut butter and olive			12-3
jar, tall and narrow		12-3	
knife, plastic		10-2	
lamp, shaded		15-1	
light source			21-3
magnet		9-1	8-3
magnetite		9-1	
magnifying lens	9	3-2, 11-1	9-3, 14-1, 20-1
marble	5		5-3, 13-1
marker, permanent			10-2, 12-3
markers, colored			4-2
measuring cup, liquid, plastic or glass	2-2		

Equipment and Materials List

Material	Launch Lab (Chapter)	MiniLAB (Chapter-Section)	Lab (Chapter-Section)
Merck Index			4-2
meterstick			2-2, 2-3, 5-3, 6-3, 7-2, 7-3, 12-3
metric ruler			2-2, 12-3, 19-1, 21-1
microscope		20-1	20-1
microscope slide	16	20-1	20-1
minerals, set			9-3
mirrors, concave and flat			14-1
Mohs scale of hardness			9-3
pails			11-2
paint trays			11-2
paper clip	7		
paring knife			3-2
pennies	3	12-3, 13-3	3-2
pie pan	13	12-3	4-3
planting containers, small			21-3
plastic container, 4-L			17-4
plate, clear plastic	7		
pot, large		1-2	
putty knife			9-3
quartz		9-1	
refrigerator or heater			21-3
rock, sparkling	9		
rocks			13-4
rolling pin			6-3
rope		5-3	
ruler	15	7-1	10-2
scissors			1-4, 20-2, 21-1
scoops, small			4-3
sockets, mini-bulb		8-3	
softball			15-1
sponge		17-4	
spoon or stirring rod		13-2, 13-3	6-3, 19-2, 21-3
spoon(s)		13-2, 13-3	19-2, 21-3
stereomicroscope			11-1, 18-2
stopwatch			2-3, 5-3, 6-3, 7-2, 7-3
storage box, clear-plastic			10-2
streak plate			9-3
table		15-1	
table tennis ball			6-3
tennis ball			6-3
test tubes		1-3	4-3
test-tube holder			4-3
thermometers		2-1	12-3, 6-3
thumbtacks or pins		13-2	4-2
triple-beam balance			6-3, 11-2
wash tub, plastic	11		
watch			11-2
watch glass			18-2
wind vane			12-3
wood sphere			3-2
wool	7	7-1	
Biological			
algae			20-1
aquatic plants			18-2
Elodea plant			21-1
frog egg mass			18-2
guppy			21-1
human cells		20-2	
mealworms			17-4
sow bugs		17-4	
worms (red wigglers)			17-4

Suppliers

Equipment Suppliers

American Science & Surplus
P.O. Box 1030
Skokie, IL 60076
(847) 647-0011
www.sciplus.com

Bio-Rad Laboratories
2000 Alfred Nobel Dr.
Life Science Group
Hercules, CA 94547
(800) 424-6723
www.bio-rad.com

Carolina Biological Supply Co.
2700 York Road
Burlington, NC 27215
(800) 334-5551
www.carolina.com

Edmund Scientific Company
60 Pearce Ave.
Tonawanda, NY 14150
(800) 728-6999
scientificsonline.com

Fisher Science Education
Educational Materials Division
4500 Turnberry Dr.
Hanover Park, IL 60133
(800) 955-1177
www.fisheredu.com

Nasco Science
901 Janesville Avenue
P.O. Box 901
Fort Atkinson, WI 53538-0901
(800) 558-9595
www.enasco.com

Nebraska Scientific
3823 Leavenworth St.
Omaha, NE 68105-1180
(800) 228-7117
nebraskascientific.com

PASCO Scientific
10101 Foothills Blvd.
P.O. Box 619011
Roseville, CA 95747
(800) 772-8700
www.pasco.com

Sargent-Welch/VWR Scientific Products
P.O. Box 5229
Buffalo Grove, IL 60089-5229
(800) SAR-GENT 727-4368
www.SargentWelch.com

Science Kit and Boreal Laboratories
777 East Park Dr.
P.O. Box 5003
Tonawanda, NY 14150
(800) 828-7777
www.sciencekit.com

Ward's Natural Science Est.
5100 W. Henrietta Road
P.O. Box 92912
Rochester, NY 14692-9012
(800) 962-2660
www.wardsci.com

Audiovisual Distributors

Bullfrog Films
P.O. Box 149
Oley, PA 19547
(800) 543-FROG
www.bullfrogfilms.com

Coronet/MTI Film & Video
2349 Chaffee Dr.
St. Louis, MO 63146
(800) 221-1274

Discovery Channel School
1 Discovery Place
Silver Springs, MD 20910
(240) 662-2000
www.discoveryschool.com

Films for the Humanities and Sciences
P.O. Box 2053
Princeton, NJ 08543
(800) 257-5126

Flinn Scientific
P.O. Box 219
770 N. Raddant Rd.
Batavia, IL 60510
(800) 452-1261
www.flinnsci.com

Frey Scientific, Div. of Beckley Cardy
100 Paragon Parkway
Mansfield, OH 44903
(800) 225-FREY
www.freyscientific.com

Media Design Associates
1731 15th St.
Suite 220
Boulder, CO 80302
(866) 546-9151
www.indra.com

National Geographic Society Educational Services
1145 17th Street, N.W.
Washington, DC 20036
(800) 368-2728
www.nationalgeographic.com

Scholastic, Inc.
555 Broadway
New York, NY 10012-3999
(800) 325-6149
www.scholastic.com

Videodiscovery Inc.
920 N. 34th St.
Suite 300
Seattle, WA 98103
(800) 548-3472
www.videodiscovery.com

Software Distributors

Boreal Laboratories, Ltd.
399 Vansickle Rd.
St. Catharines, Ontario,
L2S 3T4
Canada
(800) 387-9393
boreal.com

Educational Activities, Inc.
1937 Grand Ave.
P.O. Box 87
Baldwin, NY 11510
(800) 645-3739
www.edact.com

IBM Education
www.solutions.ibm.com/k12

J. Weston Walch, Publisher
321 Valley St.
P.O. Box 658
Portland, ME 04104-0658
(800) 341-6094
www.walch.com

Scholastic, Inc.
555 Broadway
New York, NY 10012-3999
(800) 325-6149
www.scholastic.com

Sunburst Technology
1550 Executive Drive
Elgin, IL 60123
(800) 321-7511
www.SUNBURST.com

Contents In Brief

HOW TO...

Use Your Science Book

Why do I need my science book?

Have you ever been in class and not understood all of what was presented? Or, you understood everything in class, but at home, got stuck on how to answer a question? Maybe you just wondered when you were ever going to use this stuff?

These next few pages are designed to help you understand everything your science book can be used for . . . besides a paperweight!

Before You Read

- **Chapter Opener** Science is occurring all around you, and the opening photo of each chapter will preview the science you will be learning about. The **Chapter Preview** will give you an idea of what you will be learning about, and you can try the **Launch Lab** to help get your brain headed in the right direction. The **Foldables** exercise is a fun way to keep you organized.
- **Section Opener** Chapters are divided into two to four sections. The **As You Read** in the margin of the first page of each section will let you know what is most important in the section. It is divided into four parts. **What You'll Learn** will tell you the major topics you will be covering. **Why It's Important** will remind you why you are studying this in the first place! The **Review Vocabulary** word is a word you already know, either from your science studies or your prior knowledge. The **New Vocabulary** words are words that you need to learn to understand this section. These words will be in **boldfaced** print and highlighted in the section. Make a note to yourself to recognize these words as you are reading the section.

As You Read

- **Headings** Each section has a title in large red letters, and is further divided into blue titles and small red titles at the beginnings of some paragraphs. To help you study, make an outline of the headings and subheadings.
- **Margins** In the margins of your text, you will find many helpful resources. The **Science Online** exercises and **Integrate** activities help you explore the topics you are studying. **MiniLabs** reinforce the science concepts you have learned.
- **Building Skills** You also will find an **Applying Math** or **Applying Science** activity in each chapter. This gives you extra practice using your new knowledge, and helps prepare you for standardized tests.
- **Student Resources** At the end of the book you will find **Student Resources** to help you throughout your studies. These include **Science, Technology,** and **Math Skill Handbooks,** an **English/Spanish Glossary,** and an **Index.** Also, use your **Foldables** as a resource. It will help you organize information, and review before a test.
- **In Class** Remember, you can always ask your teacher to explain anything you don't understand.

FOLDABLES™ Study Organizer

Science Vocabulary Make the following Foldable to help you understand the vocabulary terms in this chapter.

STEP 1 **Fold** a vertical sheet of notebook paper from side to side.

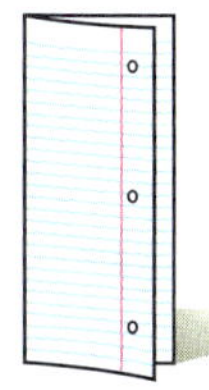

STEP 2 **Cut** along every third line of only the top layer to form tabs.

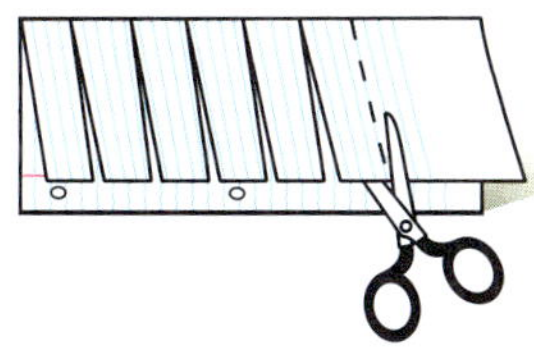

STEP 3 **Label** each tab with a vocabulary word from the chapter.

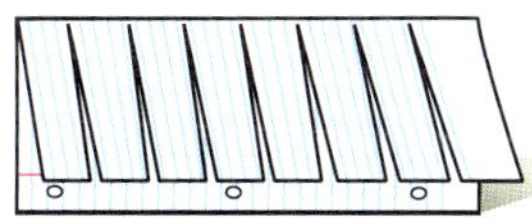

Build Vocabulary As you read the chapter, list the vocabulary words on the tabs. As you learn the definitions, write them under the tab for each vocabulary word.

In Lab

Working in the laboratory is one of the best ways to understand the concepts you are studying. Your book will be your guide through your laboratory experiences, and help you begin to think like a scientist. In it, you not only will find the steps necessary to follow the investigations, but you also will find helpful tips to make the most of your time.

- Each lab provides you with a **Real-World Question** to remind you that science is something you use every day, not just in class. This may lead to many more questions about how things happen in your world.
- Remember, experiments do not always produce the result you expect. Scientists have made many discoveries based on investigations with unexpected results. You can try the experiment again to make sure your results were accurate, or perhaps form a new hypothesis to test.
- Keeping a **Science Journal** is how scientists keep accurate records of observations and data. In your journal, you also can write any questions that may arise during your investigation. This is a great method of reminding yourself to find the answers later.

Before a Test

Admit it! You don't like to take tests! However, there *are* ways to review that make them less painful. Your book will help you be more successful taking tests if you use the resources provided to you.

- Review all of the **New Vocabulary** words and be sure you understand their definitions.
- Review the notes you've taken on your **Foldables,** in class, and in lab. Write down any question that you still need answered.
- Review the **Summaries** and **Self Check questions** at the end of each section.
- Study the concepts presented in the chapter by reading the **Study Guide** and answering the questions in the **Chapter Review.**

Look For...

- **Reading Checks** and **caption questions** throughout the text.
- the **Summaries** and **Self Check questions** at the end of each section.
- the **Study Guide** and **Review** at the end of each chapter.
- the **Standardized Test Practice** after each chapter.

Scavenger HUNT

Let's Get Started

To help you find the information you need quickly, use the Scavenger Hunt below to learn where things are located in Chapter 1.

1. What is the title of this chapter?
2. What will you learn in Section 1?
3. Sometimes you may ask, "Why am I learning this?" State a reason why the concepts from Section 2 are important.
4. What is the main topic presented in Section 2?
5. How many reading checks are in Section 1?
6. What is the Web address where you can find extra information?
7. What is the main heading above the sixth paragraph in Section 2?
8. There is an integration with another subject mentioned in one of the margins of the chapter. What subject is it?
9. List the new vocabulary words presented in Section 2.
10. List the safety symbols presented in the first Lab.
11. Where would you find a Self Check to be sure you understand the section?
12. Suppose you're doing the Self Check and you have a question about concept mapping. Where could you find help?
13. On what pages are the Chapter Study Guide and Chapter Review?
14. Look in the Table of Contents to find out on which page Section 2 of the chapter begins.
15. You complete the Chapter Review to study for your chapter test. Where could you find another quiz for more practice?

Contents

unit 1 The Nature of Matter—2

In each chapter, look for these opportunities for review and assessment:

- Reading Checks
- Caption Questions
- Section Review
- Chapter Study Guide
- Chapter Review
- Standardized Test Practice
- Online practice at red.msscience.com

unit 2 Interactions of Matter—126

In each chapter, look for these opportunities for review and assessment:
- Reading Checks
- Caption Questions
- Section Review
- Chapter Study Guide
- Chapter Review
- Standardized Test Practice
- Online practice at red.msscience.com

unit 3 Earth's Changing Surface—252

chapter 9 Rocks and Minerals—254

chapter 10 Forces Shaping Earth—286

chapter 11 Weathering and Erosion—314

chapter 12 The Atmosphere in Motion—340

Contents

Contents

chapter 13 Oceans—372

unit 4 Beyond Earth—404

chapter 14 Exploring Space—406

chapter 15 The Solar System and Beyond—438

In each chapter, look for these opportunities for review and assessment:

- Reading Checks
- Caption Questions
- Section Review
- Chapter Study Guide
- Chapter Review
- Standardized Test Practice
- Online practice at red.msscience.com

unit 5 Life's Diversity—472

chapter 16 Cells—The Units of Life—474

chapter 17 Invertebrate Animals—496

chapter 18 Vertebrate Animals—528

chapter 19 The Human Body—558

Contents

In each chapter, look for these opportunities for review and assessment:

- Reading Checks
- Caption Questions
- Section Review
- Chapter Study Guide
- Chapter Review
- Standardized Test Practice
- Online practice at red.msscience.com

Student Resources—676

Cross-Curricular Readings

NATIONAL GEOGRAPHIC Unit Openers

NATIONAL GEOGRAPHIC VISUALIZING

TIME SCIENCE AND Society

TIME SCIENCE AND HISTORY

Oops! Accidents in SCIENCE

Science and Language Arts

SCIENCE Stats

LABS

[video lab icon] available as a video lab

Launch LAB

Mini LAB

Mini LAB Try at Home

LABS

available as a video lab

One-Page Labs

LABS

Two-Page Labs

Design Your Own Labs

Model and Invent Labs

Use the Internet Labs

Activities

Applying Math

Applying Science

INTEGRATE

Astronomy: 51, 507
Career: 13, 43, 45, 104, 418, 442, 485, 538, 604
Chemistry: 43, 290, 318, 382, 479, 502, 564
Earth Science: 117, 132, 210, 232, 658
Health: 9, 81, 199, 234, 344, 409, 595, 660
History: 99, 174, 198, 629
Language Arts: 75, 304
Life Science: 77, 137, 145, 164, 257, 349, 420
Physics: 230, 267, 288, 325, 458, 462, 477, 517, 532, 577, 634
Social Studies: 266, 353, 391, 507, 564, 569

Science Online

18, 22, 47, 58, 71, 72, 102, 107, 116, 140, 146, 168, 181, 199, 202, 235, 241, 262, 274, 293, 302, 326, 330, 350, 357, 361, 377, 381, 387, 392, 420, 425, 427, 444, 449, 478, 487, 507, 513, 536, 542, 571, 576, 591, 605, 620, 629, 653, 656

Standardized Test Practice

38–39, 66–67, 94–95, 124–125, 158–159, 190–191, 222–223, 250–251, 284–285, 312–313, 338–339, 370–371, 402–403, 436–437, 470–471, 494–495, 526–527, 556–557, 586–587, 612–613, 642–643, 674–675

unit 1

Unit Contents

WebQuest ***The Nature of Science: Evaluating Bias in Advertisements*** is an exploration of specific internet sites. Students become informed about the techniques of advertising as well as evaluating bias in advertising. They use their new evaluation skills to take a critical look at a print advertisement.

unit 1 The Nature of Matter

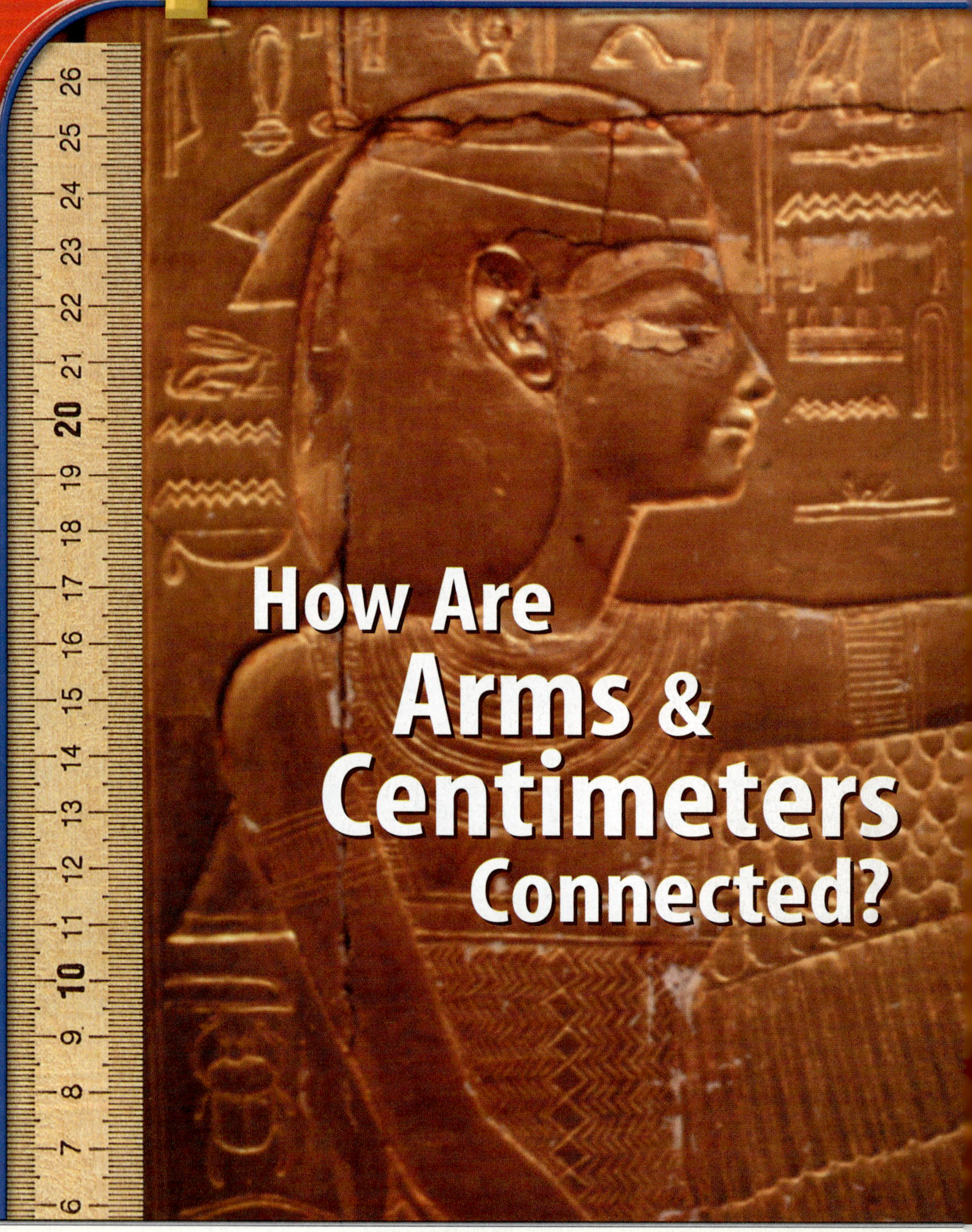

How Are Arms & Centimeters Connected?

PROJECT CRISS[SM]

Study Skills

Discussion The sticky-note strategy is used to stimulate discussion and to encourage students to ask questions. Provide students with sticky-notes. As they read this unit, have them use the sticky-notes to mark information they find interesting or sentences about which they have questions. Follow this activity with a class discussion.

About 5,000 years ago, the Egyptians developed one of the earliest recorded units of measurement—the cubit, which was based on the length of the arm from elbow to fingertip. The Egyptian measurement system probably influenced later systems, many of which also were based on body parts such as arms and feet. Such systems, however, could be problematic, since arms and feet vary in length from one person to another. Moreover, each country had its own system, which made it hard for people from different countries to share information. The need for a precise, universal measurement system eventually led to the adoption of the meter as the basic international unit of length. A meter is defined as the distance that light travels in a vacuum in a certain fraction of a second—a distance that never varies. Meters are divided into smaller units called centimeters, which are seen on the rulers here.

unit projects

Visit red.msscience.com/unit_project to find project ideas and resources. Projects include:

- **History** Brainstorm characteristics of science fields, and then design a collage with science ideas for a ceiling tile or book cover.
- **Technology** Convert a family recipe from English measurement to SI units of measurement. Enjoy a classroom bake-off!
- **Model** Create a character in an SI world. Develop a picture storybook or comic book to demonstrate your knowledge of SI measurement.

WebQuest *The Nature of Science: Evaluating Bias in Advertisements* helps students to become informed about the techniques of advertising and evaluate bias in print media.

3

NATIONAL GEOGRAPHIC **How Are Arms & Centimeters Connected?** As ancient people traveled and trade began, a need for accuracy in measuring became necessary. Early weight units may have been derived from containers or calculations of what a person or animal could haul.

Early linear measurements were often based on body parts. Because of its accuracy, the Egyptian cubit was the standard of linear measurement. All cubit sticks used in Egypt were measured at regular intervals against the royal master cubit made of black granite.

There are seven basic SI units, one of which is the meter. The centimeter is a subunit of the meter.

unit projects

History Have students brainstorm characteristics of different branches of science. Working in small groups, ask students to design a collage to represent one branch of science. Students may paint their design on bulletin board paper which may be placed over ceiling tiles in the classroom or science hallway. It may be preferable for individuals to use their design on personal science book covers.

Technology Ask students to supply a copy of a favorite family recipe. Have students rewrite their recipe using SI units: volume, weight, length, and temperature. Some students may want to bake their food for class presentation. Discuss what makes it difficult for people to switch from one system of measurement to another for baking, building, or car repair.

Model Have students create a unique character with a simple SI measurement topic. Themes might include an octopus in water (volume), an elephant in a strong-elephant contest (mass), an inchworm in a metric world (length), or a hotheaded fire ant (temperature). Have students design a problem situation for their character, and then solve it using SI measurements. Ask students to create a small story book to illustrate their science knowledge through literature.

Additional Resources For more information, resources, and assessment rubrics, visit red.msscience.com/unit_project

Section/Objectives	Standards		Labs/Features
	National	State/Local	
Chapter Opener	See p. 16T–17T for a Key to Standards.		**Launch Lab:** Observe How Gravity Accelerates Objects, p. 5 **Foldables,** p. 5
Section 1 What is science? 2 sessions 1 block 1. **Define** science and identify questions that science cannot answer. 2. **Compare and contrast** theories and laws. 3. **Identify** a system and its components. 4. **Identify** the three branches of science.	National Content Standards: UCP.1, UCP.2, A.1, A.2, B.1, F.5, G.2		**MiniLAB:** Classify Parts of a System, p. 8 **Integrate Health,** p. 9
Section 2 Science in Action 2 sessions 1 block 5. **Identify** some skills scientists use. 6. **Define** hypothesis. 7. **Recognize** the difference between observation and inference.	National Content Standards: UCP.1, UCP.2, A.1, A.2, G.2		**Integrate Career,** p. 13 **MiniLAB:** Forming a Hypothesis, p. 14 **Applying Math:** Seasonal Temperatures, p. 17 **Science Online,** p. 18
Section 3 Models in Science 2 sessions 1 block 8. **Describe** various types of models. 9. **Discuss** limitations of models.	National Content Standards: UCP.1, UCP.2, UCP.3, A.1, A.2, G.2		**Science Online,** p. 22 **MiniLAB:** Thinking Like a Scientist, p. 23 **Visualizing the Modeling of King Tut,** p. 24
Section 4 Evaluating Scientific Explanation 3 sessions 1.5 blocks 10. **Evaluate** scientific explanations. 11. **Evaluate** promotional claims.	National Content Standards: UCP.1, UCP.2, UCP.3, A.1, A.2, G.2		**Lab:** What is the right answer?, p. 31 **Lab:** Identifying Parts of an Investigation, pp. 32–33 **Science and History:** Women in Science, p. 34

Lab Materials	Reproducible Resources	Section Assessment	Technology
Launch Lab: 3 identical unsharpened pencils, tape	**Chapter *FAST FILE* Resources** Foldables Worksheet, p. 19 Note-taking Worksheets, pp. 37–39	Glencoe's Assessment Advantage	TeacherWorks includes: • Interactive Teacher Edition • Lesson Planner with calendar • Access to all program blacklines • Correlations to standards • Web links
MiniLAB: paper, pencil or pen *Need materials?* Contact Science Kit at 1-800-828-7777 or www.sciencekit.com on the Internet.	**Chapter *FAST FILE* Resources** Transparency Activity, p. 48 MiniLAB, p. 3 Enrichment, p. 33 Reinforcement, p. 29 Directed Reading, p. 22 **Life Science Critical Thinking/ Problem Solving,** p. 1	**Portfolio** Assessment, p. 8 **Performance** Assessment, p. 11 Applying Skills, p. 11 **Content** Section Review, p. 11	Section Focus Transparency Virtual Labs CD-ROM Guided Reading Audio Program Interactive Chalkboard CD-ROM
MiniLAB: large pot of water, unopened can of diet soda, unopened can of regular soda	**Chapter *FAST FILE* Resources** Transparency Activity, p. 49 MiniLAB, p. 4 Enrichment, p. 34 Reinforcement, p. 30 Directed Reading, p. 23 Lab Activity, pp. 11–14 Transparency Activity, pp. 53–54	**Portfolio** Active Reading, p. 16 **Performance** MiniLAB, p. 14 Applying Math, p. 17 **Content** Section Review, p. 20	Section Focus Transparency Teaching Transparency Virtual Labs CD-ROM Guided Reading Audio Program Interactive Chalkboard CD-ROM
MiniLAB: 15 mL water, test tube, 5 mL vegetable oil, 2 drops food coloring, graduated cylinder, watch or clock	**Chapter *FAST FILE* Resources** Transparency Activity, p. 50 MiniLAB, p. 5 Enrichment, p. 35 Reinforcement, p. 31 Directed Reading, p. 23 Lab Activity, pp. 15–18	**Portfolio** Curriculum Connection, p. 22 **Performance** MiniLAB, p. 23 Applying Math, p. 26 **Content** Section Review, p. 26	Section Focus Transparency Guided Reading Audio Program Interactive Chalkboard CD-ROM
Lab: cardboard mailing tubes, length of rope, scissors **Lab:** description of fertilizer experiment, data table from the fertilizer experiment	**Chapter *FAST FILE* Resources** Transparency Activity, p. 51 Enrichment, p. 36 Reinforcement, p. 32 Directed Reading, pp. 23, 24 Lab Worksheet, pp. 7–8, 9–10	**Portfolio** Assessment, p. 30 **Performance** Assessment, p. 30 Applying Skills, p. 30 **Content** Section Review, p. 30	Section Focus Transparency Virtual Labs CD-ROM Guided Reading Audio Program Interactive Chalkboard CD-ROM Video Lab

End of Chapter Assessment

Blackline Masters	Technology	Professional Series
Chapter *FAST FILE* Resources Chapter Review, pp. 41–42 Chapter Tests, pp. 43–46 **Standardized Test Practice,** pp. 7–10	MindJogger Videoquiz Virtual Labs CD-ROM Exam*View*® Pro Testmaker TeacherWorks CD-ROM Interactive Chalkboard CD-ROM	**Performance Assessment in the Science Classroom (PASC)**

Transparencies

Section Focus

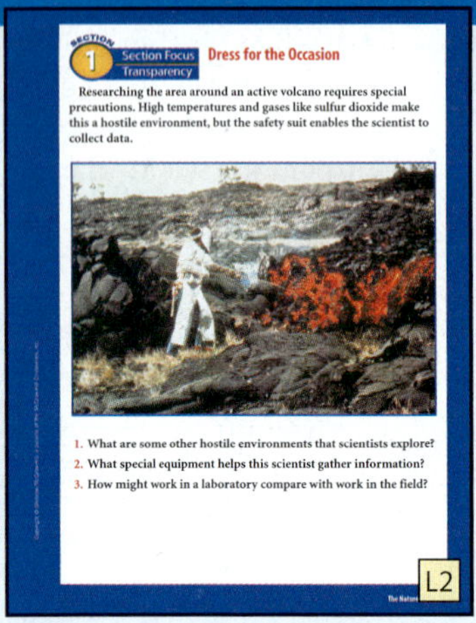
Section 1 Section Focus Transparency

Dress for the Occasion

Researching the area around an active volcano requires special precautions. High temperatures and gases like sulfur dioxide make this a hostile environment, but the safety suit enables the scientist to collect data.

1. What are some other hostile environments that scientists explore?
2. What special equipment helps this scientist gather information?
3. How might work in a laboratory compare with work in the field?

L2

Section 2 Section Focus Transparency

Did we skip a step?

These are the remains of the *Vasa*, the largest warship of its time. Due to improper testing, it sank on its maiden voyage in 1625. In 1961, the ship was salvaged and a museum built to house it.

1. Prior to the *Vasa's* maiden voyage an admiral ordered a test for stability. The ship failed the test, but the admiral kept the result a secret. What would a scientist have done with this result?
2. How could you test a new ship design without actually building the ship?

L2

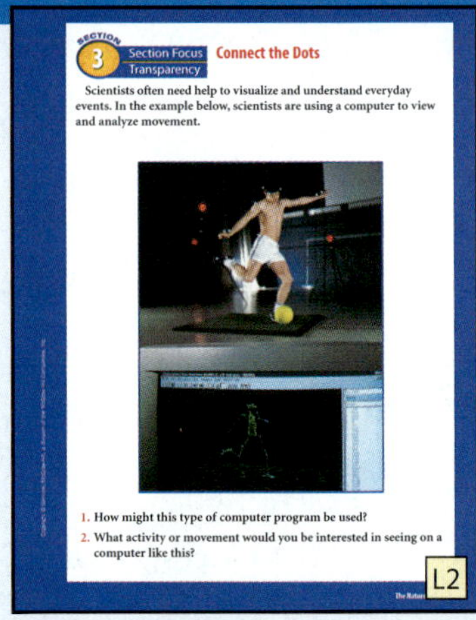
Section 3 Section Focus Transparency

Connect the Dots

Scientists often need help to visualize and understand everyday events. In the example below, scientists are using a computer to view and analyze movement.

1. How might this type of computer program be used?
2. What activity or movement would you be interested in seeing on a computer like this?

L2

This is a representation of key blackline masters available in the Teacher Classroom Resources. See Resource Manager boxes within the chapter for additional information.

Assessment

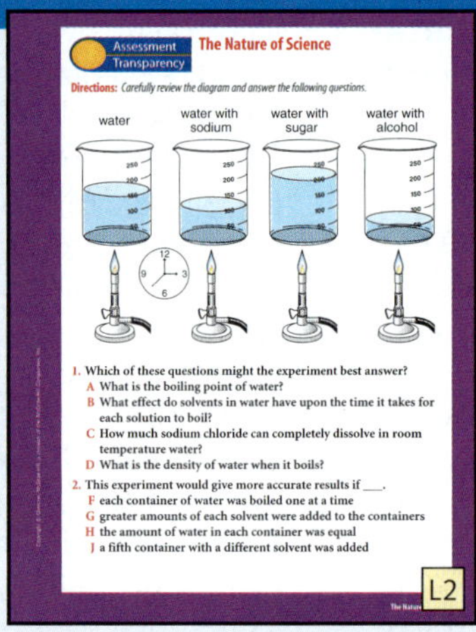
Assessment Transparency

The Nature of Science

Directions: *Carefully review the diagram and answer the following questions.*

1. Which of these questions might the experiment best answer?
   - A What is the boiling point of water?
   - B What effect do solvents in water have upon the time it takes for each solution to boil?
   - C How much sodium chloride can completely dissolve in room temperature water?
   - D What is the density of water when it boils?
2. This experiment would give more accurate results if ____.
   - F each container of water was boiled one at a time
   - G greater amounts of each solvent were added to the containers
   - H the amount of water in each container was equal
   - J a fifth container with a different solvent was added

L2

Teaching

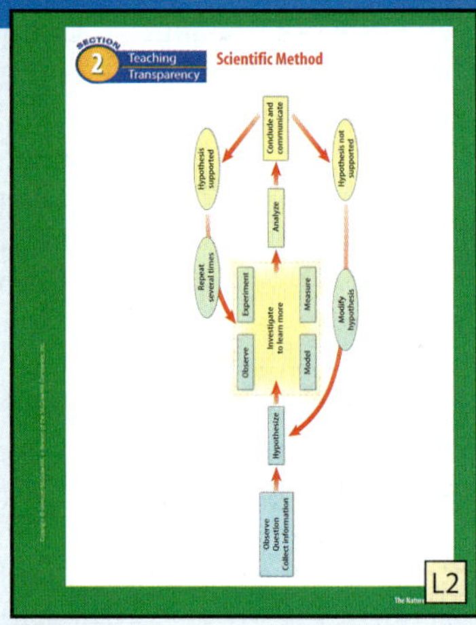
Section 2 Teaching Transparency

Scientific Method

L2

Key to Teaching Strategies

The following designations will help you decide which activities are appropriate for your students.

- L1 Level 1 activities should be appropriate for students with learning difficulties.
- L2 Level 2 activities should be within the ability range of all students.
- L3 Level 3 activities are designed for above-average students.
- ELL ELL activities should be within the ability range of English Language Learners.
- COOP LEARN Cooperative Learning activities are designed for small group work.
- LS Multiple Learning Styles logos, as described on page 12T, are used throughout to indicate strategies that address different learning styles.
- P These strategies represent student products that can be placed into a best-work portfolio.
- PBL Problem-Based Learning activities apply real-world situations to learning.

Hands-on Activities

Student Text Lab Worksheet

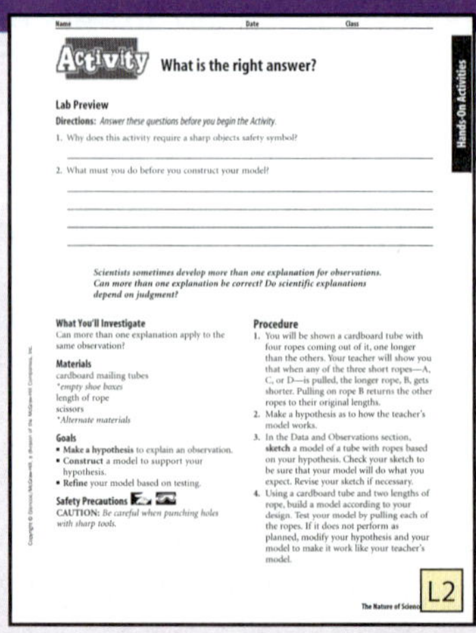
Name Date Class

Activity What is the right answer?

Hands-On Activities

Lab Preview

Directions: *Answer these questions before you begin the Activity.*

1. Why does this activity require a sharp objects safety symbol?
2. What must you do before you construct your model?

Scientists sometimes develop more than one explanation for observations. Can more than one explanation be correct? Do scientific explanations depend on judgment?

What You'll Investigate

Can more than one explanation apply to the same observation?

Materials

cardboard mailing tubes
**empty shoe boxes*
length of rope
scissors
**Alternate materials*

Goals

- **Make a hypothesis** to explain an observation.
- **Construct** a model to support your hypothesis.
- **Refine** your model based on testing.

Safety Precautions

CAUTION: *Be careful when punching holes with sharp tools.*

Procedure

1. You will be shown a cardboard tube with four ropes coming out of it, one longer than the others. Your teacher will show you that when any of the three short ropes—A, C, or D—is pulled, the longer rope, B, gets shorter. Pulling on rope B returns the other ropes to their original lengths.
2. Make a hypothesis as to how the teacher's model works.
3. In the Data and Observations section, **sketch** a model of a tube with ropes based on your hypothesis. Check your sketch to be sure that your model will do what you expect. Revise your sketch if necessary.
4. Using a cardboard tube and two lengths of rope, build a model according to your design. Test your model by pulling each of the ropes. If it does not perform as planned, modify your hypothesis and your model to make it work like your teacher's model.

The Nature of Science

L2

Laboratory Activities

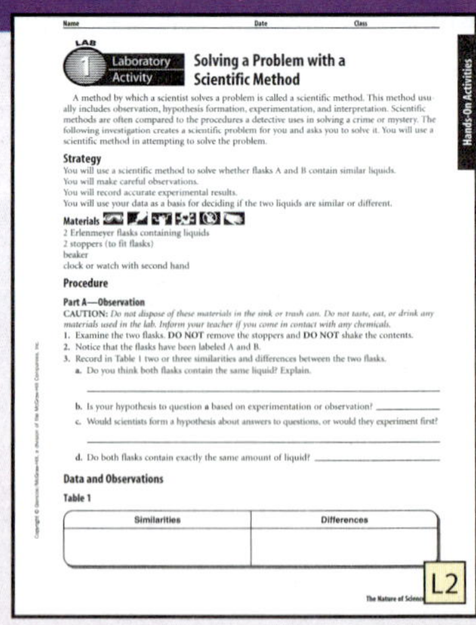
Name Date Class

LAB 1 Laboratory Activity

Solving a Problem with a Scientific Method

Hands-On Activities

A method by which a scientist solves a problem is called a scientific method. This method usually includes observation, hypothesis formation, experimentation, and interpretation. Scientific methods are often compared to the procedures a detective uses in solving a crime or mystery. The following investigation creates a scientific problem for you and asks you to solve it. You will use a scientific method in attempting to solve the problem.

Strategy

You will use a scientific method to solve whether flasks A and B contain similar liquids.
You will make careful observations.
You will record accurate experimental results.
You will use your data as a basis for deciding if the two liquids are similar or different.

Materials

2 Erlenmeyer flasks containing liquids
2 stoppers (to fit flasks)
beaker
clock or watch with second hand

Procedure

Part A—Observation

CAUTION: *Do not dispose of these materials in the sink or trash can. Do not taste, eat, or drink any materials used in the lab. Inform your teacher if you come in contact with any chemicals.*

1. Examine the two flasks. **DO NOT** remove the stoppers and **DO NOT** shake the contents.
2. Notice that the flasks have been labeled A and B.
3. Record in Table 1 two or three similarities and differences between the two flasks.
   - a. Do you think both flasks contain the same liquid? Explain.
   - b. Is your hypothesis to question **a** based on experimentation or observation?
   - c. Would scientists form a hypothesis about answers to questions, or would they experiment first?
   - d. Do both flasks contain exactly the same amount of liquid?

Data and Observations

Table 1

Similarities	Differences

The Nature of Science

L2

Meeting Different Ability Levels

Content Outline

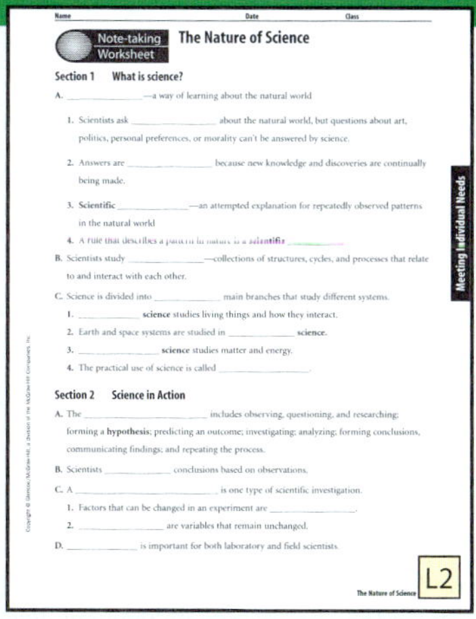

Reinforcement

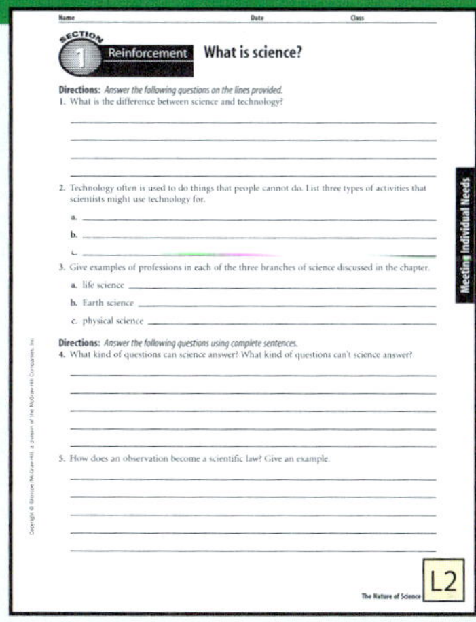

Enrichment

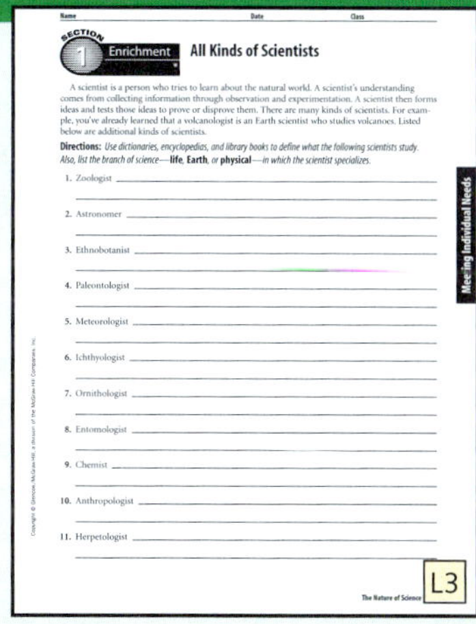

Directed Reading (English/Spanish)

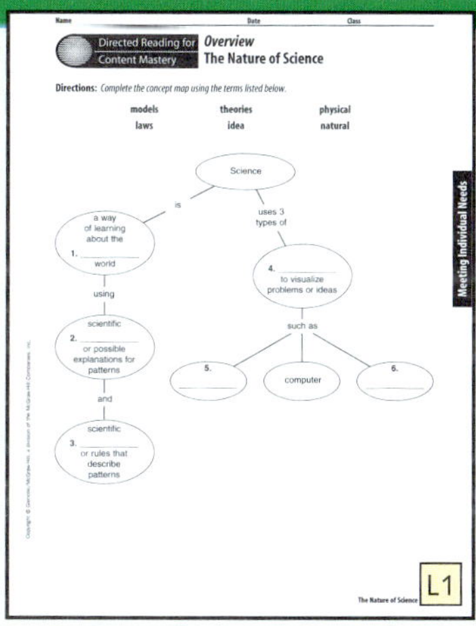

Study Guide

Reading Essentials

Assessment

Test Practice Workbook

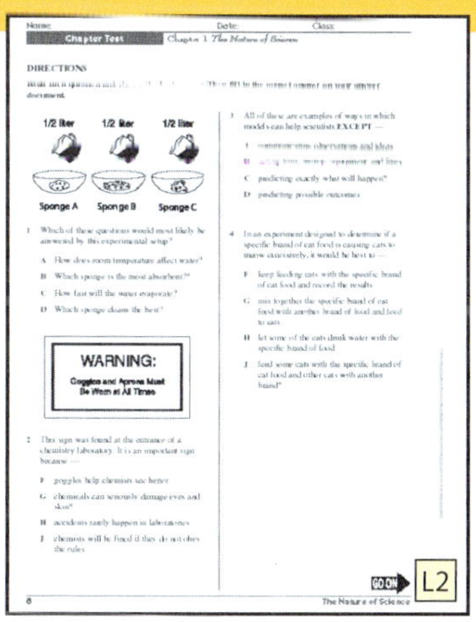

Chapter Review

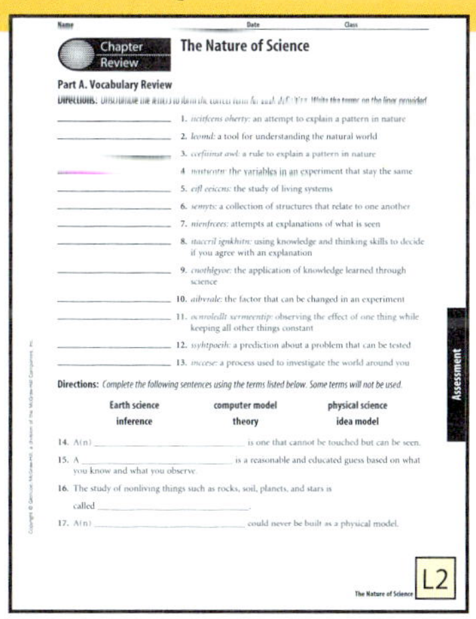

Chapter Tests

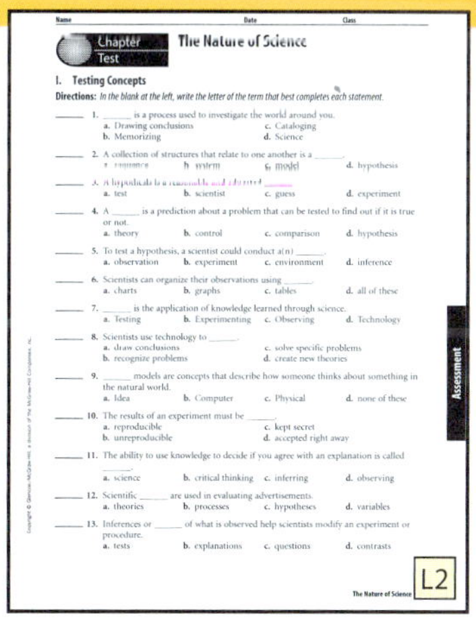

Science Content Background

section 1 What is science?

Learning About the World

Science has a dual nature. On one hand it can be thought of as an organized body of agreed upon descriptions and explanations about the natural world. On the other hand, it can be thought of as the methods that lead to the development of descriptions and explanations about the world.

When solving a problem, scientists often break it down into parts, or questions. Scientists spend a great deal of time researching and studying background information about the problem before deciding on a specific new question to investigate. They determine what research has been done already. What will need to be done first? Asking well-researched questions instead of random questions can save time and funds.

Scientists often study the interaction of components in a system. Systems in science include solar systems, ecosystems, and weather systems. Systems can be narrowly defined such as the lymphatic system or broadly defined such as in the Earth system.

The Branches of Science

Often the most fertile ground for investigation is at the periphery of two or more branches of science. Biochemistry originated at the borders of biology and chemistry and today is its own branch of science. In a similar way, biophysics is arising from biology and physics.

section 2 Science in Action

Science Skills

Science skills or science process skills are abilities used to investigate the natural world. Observations are the basis of scientific investigation, in that any hypothesis, theory, or law must be consistent with all observations. Other process skills include classifying, predicting, and communicating.

The term *The Scientific Method* is an inaccurate description of the work of scientists. The term *The* makes it seem as though all scientists use one method, which is certainly not the case. There are differences between the methods of an astronomer and a medical researcher as there are between a theoretical physicist and a conservation biologist. Furthermore, science does not proceed in a linear fashion. False starts, dead ends, backtracking, and detours can occur often in the path. Thus the term *Science Methods* is preferred, because it is more inclusive of the diverse methods of science. It is, however, still useful to talk about components of methods of science, such as problems and hypotheses, because they shed light on aspects of

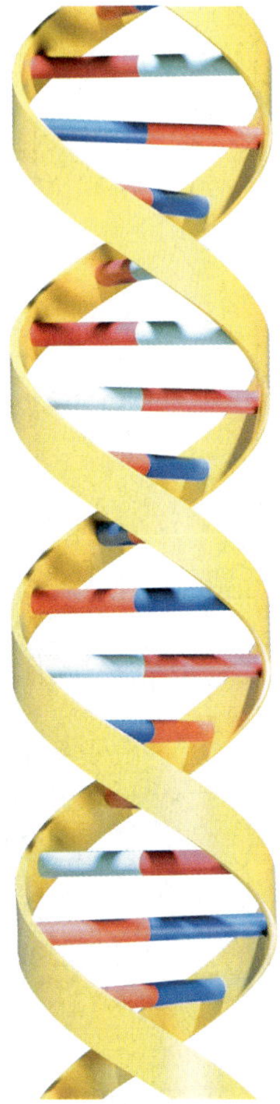

Micheal Simpson/FPG International

Teacher to Teacher
Diedre Adams
West Vigo Middle School
West Terre Haute, Indiana

"To help students calculate speed, I provide them with a variety of wind-up, pull-back toys. By placing these toys next to a meterstick, students can measure the time it takes the toy to reach the end of the stick. By using both fast and slow toys, students can also measure the distance that different toys stop. Advanced students can convert their measurements to m/s."

Diedre Adams

scientific endeavors and because scientific papers often use this format to report results.

The difference between observations and inferences is shown in the discovery of different solar systems. The observation of unusual movements in stars is used to infer the presence of a large planet near that star.

Experiments

Scientists usually apply the term *experiment* to situations in which they manipulate a variable to see the effect. Observing a star would not be an experiment because no variable was manipulated.

In past experiments, many sick people who believed they were receiving medicine, but in fact were given only sugar pills, actually got better. For this reason, in drug trials, the participants never know whether or not they are actually being given a drug. The treatment group is given the drug while the control group is given a placebo—something that resembles the drug but, like a sugar pill, has no effect.

section 3

Models in Science

Making Models

An important part of science is to construct models that approximate the real world. Models come in many forms such as physical, conceptual, or mathematical. In many cases, models are analogies: balls and springs, for example, can be presented as analogous to molecules and forces. Models may be similar to the real world, but they are never perfect matches. Perhaps for this reason it is common to have different models attempting to describe the same phenomenon.

section 4

Evaluating Scientific Explanation

Believe it or not?

Many of the most common fallacies of logic also afflict scientific inquiry, and even the most experienced scientist must be aware of the various errors that can invalidate an investigation. The design of an experiment can be flawed from the beginning—for instance, a variable that will affect the result is not rigorously controlled or is completely overlooked. Even in a well-designed experiment, the scientist's observations might be biased, however subtly. If the data collected from a perfectly designed and properly executed experiment are based on too small a sample, conclusions drawn from the data may not be reliable for wider application. And it is even possible that an incorrect conclusion will be drawn from completely reliable data. A successful scientific investigation is the result of careful thought and painstaking procedure.

chapter content resources

Internet Resources

For additional content background, visit **red.msscience.com** to:

- access your book online
- find references to related articles in popular science magazines
- access Web links with related content background
- access current events with science journal topics

Print Resources

Resources for Teaching Middle School Science, by National Science Resources Center of the National Academy of Sciences, National Academy of Engineering, Institute of Medicine, and the Smithsonian Institution, 1998

Classroom Critters and the Scientific Method, by Sally Kneidel, Fulcrum Pub., 1999

Oh, Yuck: The Encyclopedia of Everything Nasty, by Joy Masoff and Terry Sirrell, Workman Publishing Company, 2000

chapter 1

Chapter Vocabulary

science, p. 6
scientific theory, p. 7
scientific law, p. 7
system, p. 8
life science, p. 9
Earth science, p. 10
physical science, p. 10
technology, p. 11
hypothesis, p. 14
infer, p. 16
controlled experiment, p. 18
variable, p. 18
independent variable, p. 18
dependent variable, p. 18
constant, p. 18
model, p. 21
critical thinking, p. 27
data, p. 28

Science Journal Student responses may vary, but may include cooking food, heating your home, or food digestion.

INTERACTIVE CHALKBOARD with Image Bank

PowerPoint® Presentations

This CD-ROM is an editable Microsoft® PowerPoint® presentation that includes:

- a pre-made presentation for every chapter
- interactive graphics
- animations
- audio clips
- image bank
- all new section and chapter questions
- Standardized Test Practice
- transparencies
- pre-lab questions for all labs
- Foldables directions
- links to red.msscience.com

chapter 1

The Nature of Science

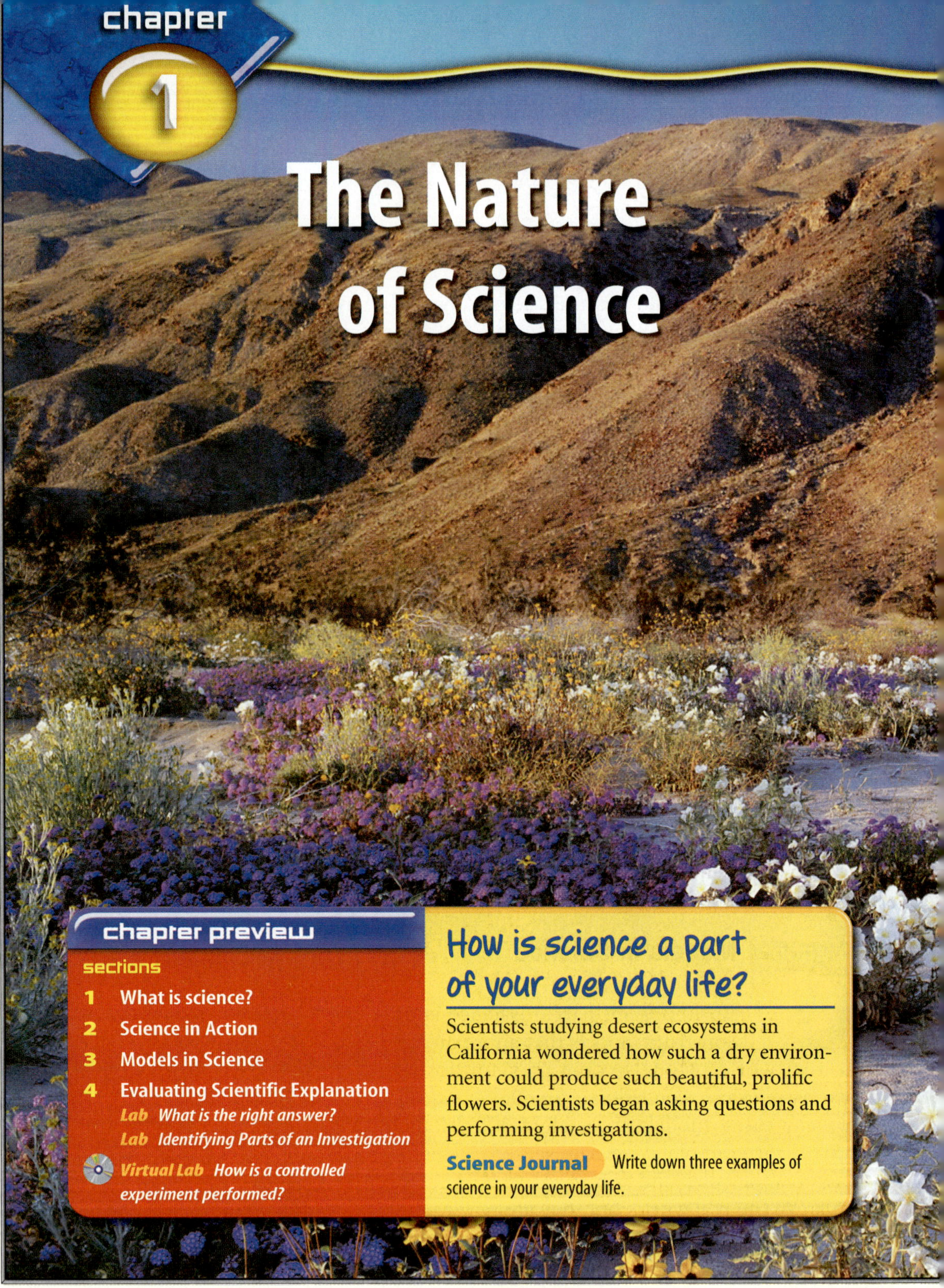

chapter preview

sections

1. What is science?
2. Science in Action
3. Models in Science
4. Evaluating Scientific Explanation
   - *Lab* *What is the right answer?*
   - *Lab* *Identifying Parts of an Investigation*
   - *Virtual Lab* *How is a controlled experiment performed?*

How is science a part of your everyday life?

Scientists studying desert ecosystems in California wondered how such a dry environment could produce such beautiful, prolific flowers. Scientists began asking questions and performing investigations.

Science Journal Write down three examples of science in your everyday life.

Theme Connection

Systems and Interactions Science is a systematic way of studying the world to find relationships and interactions among the processes, cycles, and structures that make up the systems of the world.

About the Photo

Wonders of Nature In this photo, wildflowers are growing in the desert while very little vegetation is growing in the mountain. Ask students why they think this occurs. Giving possible explanations as to why something occurs allow scientists to share information.

Start-Up Activities

Launch LAB

Observe How Gravity Accelerates Objects

Gravity is a familiar natural force that keeps you anchored on Earth, but how does it work? Scientists learn about gravity and other concepts by asking questions and making observations. By observing things in action scientists can study nature. Perform the lab below to see how gravity affects objects.

1. Collect three identical, unsharpened pencils.
2. Tape two of the pencils together.
3. Hold all the pencils at the same height as high as you can. Drop them together and observe what happens as they fall.
4. **Think Critically** Did the single pencil fall faster or slower than the pair? Predict in your Science Journal what would happen if you taped 30 pencils together and dropped them at the same time as you dropped a single pencil.

Science Online | Preview this chapter's content and activities at red.msscience.com

FOLDABLES™ Study Organizer

Science Make the following Foldable to help identify what you already know, what you want to know, and what you learned about science.

STEP 1 **Fold** a vertical sheet of paper from side to side. Make the front edge about 1/2 inch shorter than the back edge.

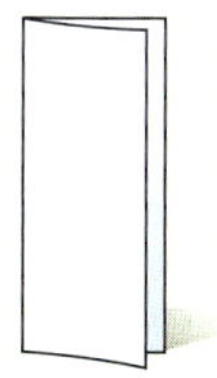

STEP 2 **Turn** lengthwise and fold into thirds.

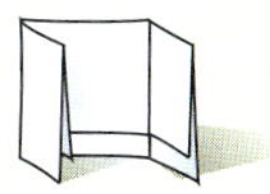

STEP 3 **Unfold and cut** only the top layer along both folds to make three tabs. **Label** each tab.

Identify Questions Before you read the chapter, write what you already know about science under the left tab of your Foldable, and write questions about what you'd like to know under the center tab. After you read the chapter, list what you learned under the right tab.

5

Launch LAB

Purpose Students explore how gravity affects objects of different masses. L2 LS **Kinesthetic**

Materials 3 identical, unsharpened pencils; tape

Teaching Strategy Make sure students drop the pencils from the greatest height possible.

Think Critically

The single pencil and the pair of pencils fall at the same rate. From this, students can infer that 30 pencils also should fall just as quickly as one pencil.

Assessment

Process Have students design other experiments to test how quickly objects of different sizes and weights fall. Encourage them to compare objects of different shapes and sizes. Use **Performance Assessment in the Science Classroom,** p. 95.

FOLDABLES™ Study Organizer **Dinah Zike Study Fold**

Student preparation materials for this Foldable are available in the Chapter *FAST FILE* Resources.

section 1 What is science?

1 Motivate

INTERACTIVE CHALKBOARD PowerPoint® Presentations

Bellringer

Section Focus Transparencies also are available on the Interactive Chalkboard CD-ROM. L2 ELL

SECTION 1 Section Focus Transparency **Dress for the Occasion**

Researching the area around an active volcano requires special precautions. High temperatures and gases like sulfur dioxide make this a hostile environment, but the safety suit enables the scientist to collect data.

1. What are some other hostile environments that scientists explore?
2. What special equipment helps this scientist gather information?
3. How might work in a laboratory compare with work in the field?

The Nature of Science

Tie to Prior Knowledge

Everyday Science Ask students to name events that occur every day, such as the Sun rising in the morning. Tell students that in this section they will learn how science uses regularities like these to make sense of the world. L2

as you read

What You'll Learn

- **Define** science and identify questions that science cannot answer.
- **Compare** and contrast theories and laws.
- **Identify** a system and its components.
- **Identify** the three main branches of science.

Why It's Important

Science can be used to learn more about the world you live in.

Review Vocabulary

theory: explanation of things or events that is based on knowledge gained from many observations and experiments

New Vocabulary

- science
- scientific theory
- scientific law
- system
- life science
- Earth science
- physical science
- technology

Learning About the World

When you think of a scientist, do you imagine a person in a laboratory surrounded by charts, graphs, glass bottles, and bubbling test tubes? It might surprise you to learn that anyone who tries to learn something about the natural world is a scientist. **Science** is a way of learning more about the natural world. Scientists want to know why, how, or when something occurred. This learning process usually begins by keeping your eyes open and asking questions about what you see.

Asking Questions Scientists ask many questions. How do things work? What do things look like? What are they made of? Why does something take place? Science can attempt to answer many questions about the natural world, but some questions cannot be answered by science. Look at the situations in **Figure 1.** Who should you vote for? What does this poem mean? Who is your best friend? Questions about art, politics, personal preference, or morality can't be answered by science. Science can't tell you what is right, wrong, good, or bad.

Figure 1 Questions about politics, literature, and art cannot be answered by science.

Section 1 Resource Manager

Chapter *FAST FILE* Resources
- Transparency Activity, p. 48
- Directed Reading for Content Mastery, p. 22
- Note-taking Worksheets, pp. 37–39
- MiniLAB, p. 3
- Enrichment, p. 33

Life Science Critical Thinking/Problem Solving, p. 1

Earth Science Critical Thinking/Problem Solving, p. 14

Physical Science Critical Thinking/Problem Solving, p. 1

Figure 2 As new information becomes available, explanations can be modified or discarded and new explanations can be made.

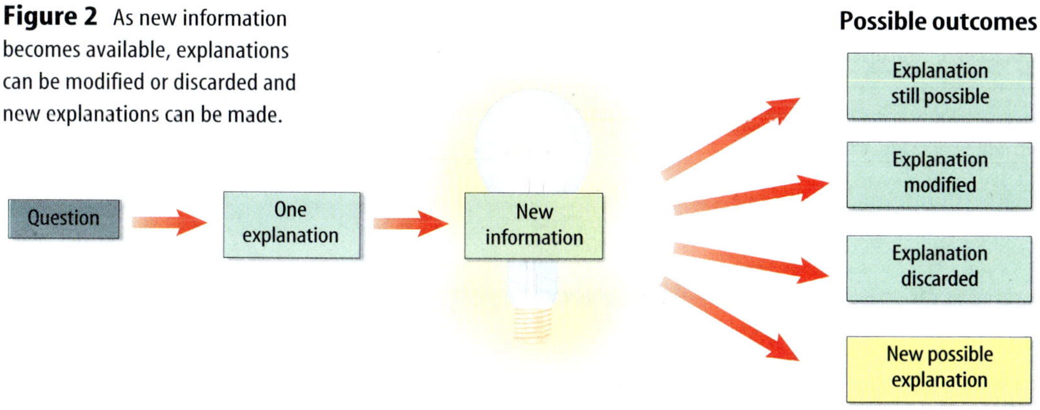

Possible Explanations If learning about your world begins with asking questions, can science provide answers to these questions? Science can answer a question only with the information available at the time. Any answer is uncertain because people will never know everything about the world around them. With new knowledge, they might realize that some of the old explanations no longer fit the new information. As shown in **Figure 2,** some observations might force scientists to look at old ideas and think of new explanations. Science can only provide possible explanations.

Reading Check *Why can't science answer questions with certainty?*

Scientific Theories An attempt to explain a pattern observed repeatedly in the natural world is called a **scientific theory.** Theories are not simply guesses or someone's opinions, nor are theories vague ideas. Theories in science must be supported by observations and results from many investigations. They are the best explanations that have been found so far. However, theories can change. As new data become available, scientists evaluate how the new data fit the theory. If enough new data do not support the theory, the theory can be changed to fit the new observations better.

Scientific Laws A rule that describes a pattern in nature is a **scientific law.** For an observation to become a scientific law, it must be observed repeatedly. The law then stands until someone makes observations that do not follow the law. A law helps you predict that an apple dropped from arm's length will always fall to Earth. The law, however, does not explain why gravity exists or how it works. A law, unlike a theory, does not attempt to explain why something happens. It simply describes a pattern.

2 Teach

Use Science Words

Word Origins The word *scientist* was first used by William Whewell who suggested this term be used for those who studied the natural world. He thought of it as a counterpart to the word *artist.* An artist is a person skilled in one of the arts. Have students look up the words *science* and *scientist* in the dictionary and suggest how the word *scientist* relates to the word *artist.* The word *science* comes from the Latin word *scire,* which means "to know." A scientist is a person learned in science like an artist is a person skilled in art. L2 LS **Linguistic**

Reading Check

Answer New information is being discovered all the time. People will never know everything about the world around them.

Discussion

Theories and Laws Have students state things they know about the world and determine whether they are theories or laws. Possible statements: "The Sun rises in the east every morning" is a law because it describes an observation. "The Sun rises every day because gravity keeps Earth in orbit around the Sun" is a theory because it explains something. L2

Visual Learning

Figure 2 This flow chart demonstrates how as new information becomes available, scientific explanations can be modified and discarded. How could you use this flow chart to answer the question proposed in the Explore Activity about how gravity affects objects? Students observed that a single pencil fell at the same rate that two pencils fell. They may infer then that 30 pencils would fall at the same rate; however, new information would point out that heavier objects accelerate more rapidly.

Caption Answer

Figure 3 the desks and their arrangements, the pictures on the wall, the students, the teacher, the chalkboard, the lights

Mini LAB

Purpose Students study a familiar system. L2 LS **Logical-Mathematical**

Materials paper, pencil or pen

Teaching Strategy Have a cafeteria worker come to class to answer questions about his or her job and how the food is prepared.

Analysis

Structures in the cafeteria include the tables, serving plates, warmers, ovens, spoons, and forks. Processes include preparing and cooking the food. Cycles include the different jobs a worker does during the course of a day, from cooking to serving to cleaning up afterward.

Assessment

Process Have each student make a timeline showing the sequence of events during a typical day in the cafeteria. Use **Performance Assessment in the Science Classroom,** p. 163. P

Try at Home

Reading Check

Answer a collection of structures, cycles, and processes that relate to and interact with one another

Figure 3 Systems are a collection of structures, cycles, and processes.
Infer *What systems can you identify in this classroom?*

Systems in Science

Scientists can study many different things in nature. Some might study how the human body works or how planets move around the Sun. Others might study the energy carried in a lightning bolt. What do all of these things have in common? All of them are systems. A **system** is a collection of structures, cycles, and processes that relate to and interact with each other. The structures, cycles, and processes are the parts of a system, just like your stomach is one of the structures of your digestive system.

Reading Check *What is a system?*

Systems are not found just in science. Your school is a system with structures such as the school building, the tables and chairs, you, your teacher, the school bell, your pencil, and many other things. **Figure 3** shows some of these structures. Your school day also has cycles. Your daily class schedule and the calendar of holidays are examples of cycles. Many processes are at work during the school day. When you take a test, your teacher has a process. You might be asked to put your books and papers away and get out a pencil before the test is distributed. When the time is over, you are told to put your pencil down and pass your test to the front of the room.

Parts of a System Interact In a system, structures, cycles, and processes interact. Your daily schedule influences where you go and what time you go. The clock shows the teacher when the test is complete, and you couldn't complete the test without a pencil.

Mini LAB

Classifying Parts of a System

Procedure

Think about how your school's cafeteria is run. Consider the physical structure of the cafeteria. How many people run it? Where does the food come from? How is it prepared? Where does it go? What other parts of the cafeteria system are necessary?

Analysis

Classify the parts of your school cafeteria's system as structures, cycles, or processes.

Try at Home

Science Journal

Systems in Your Day Have students write in their Science Journals about the different systems that are important to their day. They can choose to concentrate on structures, processes, or cycles, or they can describe the parts of each system in chronological order. L2 LS **Linguistic**

Quick Demo

Ecosystems

Estimated Time 10–15 minutes

Procedure Discuss different ecosystems in the schoolyard with students. Explain that puddles and ponds are recycled back into the water cycle. Explain that an area such as a playing field can be altered over time by planting seedlings. Have students describe how ecosystems interact with one another and how a slight change in one system can affect another.

Parts of a Whole All systems are made up of other systems. For example, you are part of your school. The human body is a system—within your body are other systems. Your school is part of a system—district, state, and national. You have your regional school district. Your district is part of a statewide school system. Scientists often break down problems by studying just one part of a system. A scientist might want to learn about how construction of buildings affects the ecosystem. Because an ecosystem has many parts, one scientist might study a particular animal, and another might study the effect of construction on plant life.

The Branches of Science

Science often is divided into three main categories, or branches—life science, Earth science, and physical science. Each branch asks questions about different kinds of systems.

Life Science The study of living systems and the ways in which they interact is called **life science.** Life scientists attempt to answer questions like "How do whales navigate the ocean?" and "How do vaccines prevent disease?" Life scientists can study living organisms, where they live, and how they interact. Dian Fossey, **Figure 4,** was a life scientist who studied gorillas, their habitat, and their behaviors.

People who work in the health field know a lot about the life sciences. Physicians, nurses, physical therapists, dietitians, medical researchers, and others focus on the systems of the human body. Some other examples of careers that use life science include biologists, zookeepers, botanists, farmers, and beekeepers.

INTEGRATE Health

Health Integration Systems The human body is composed of many different systems that all interact with one another to perform a function. The heart is like the control center. Even though not all systems report directly to the heart, they all interact with its function. If the heart is not working, the other systems fail as well. Research human body systems and explain how one system can affect another.

Figure 4 Over a span of 18 years, life scientist Dian Fossey spent much of her time observing mountain gorillas in Rwanda, Africa. She was able to interact with them as she learned about their behavior.

INTEGRATE Health

Health Integration Systems Body systems can fail at any time without warning. Doctors have developed different tests to try and predict health problems, but these tests are not guaranteed.

Research Have students research some of the tests that have been developed to test body systems. How can these tests lead to treatment and prevention of body system failure? What advancements have been made to adjust for the body system failure in order to keep the body working? L2

Use Science Words

Word Meaning Have students look up the names of various branches of science. Have them determine how the name came to be applied to that branch of science. Possible examples: Zoology comes from the Greek word *zoe,* which means "life." Zoology is the study of living animals. Medicine derives from the Latin word *medicina,* which means "remedy." L2 LS **Linguistic**

Discussion

Science Activities Ask students to name their favorite activities and discuss the types of science that are important to these activities. Possible responses: Reading—chemistry is involved in the processes for making the ink and the paper; physics was applied in the designing of the printing press. Sports—athletes regularly consult physicians and trainers who use their knowledge of life science to help athletes stay in good condition.

L2 LS **Logical-Mathematical**

Teacher FYI

Gorillas Dian Fossey was an occupational therapist. She went to Africa to study and observe gorillas without any formal training. Her research led to a new view of these animals, once believed to be savage. She found the gorillas to be gentle vegetarians, some even sacrificing themselves in defense of the group.

Fun Fact

Ecotourists to Rwanda can visit the mountain gorillas that Dian Fossey worked so hard to study and save. Small groups are led on a remote trek to a family of gorillas. The guides warn the visitors not to look the male silverback in the eyes and, if he attacks, not to run.

Reading Check

Answer nonliving things, such as rocks, soil, clouds, rivers, oceans, planets, stars, meteors, black holes, weather, and climate, on Earth and in space

IDENTIFYING Misconceptions

Branches of Science Students may think that each branch of science is isolated and that the branches never overlap. In fact, a scientist doing work in one of the branches of science often must rely on information obtained from the other branches. Have students think of investigations in which the different sciences would interact. Possible examples: To study how the Sun produces heat, an Earth scientist would need to know about physical science. To study an organism's interaction with its environment, a life scientist would have to know something about the rocks, soil, and water the organism comes into contact with. L3 LS **Logical-Mathematical**

Activity

Professional Scientists Ask professionals in Earth, life, and physical science to come to class and tell students about the things they do. L2 LS **Auditory-Musical**

Figure 5 These volcanologists are studying the temperature of the lava flowing from a volcano.

Earth Science The study of Earth systems and the systems in space is **Earth science.** It includes the study of nonliving things such as rocks, soil, clouds, rivers, oceans, planets, stars, meteors, and black holes. Earth science also covers the weather and climate systems that affect Earth. Earth scientists ask questions like "How can an earthquake be detected?" or "Is water found on other planets?" They make maps and investigate how geologic features formed on land and in the oceans. They also use their knowledge to search for fuels and minerals. Meteorologists study weather and climate. Geologists study rocks and geologic features. **Figure 5** shows a volcanologist—a person who studies volcanoes—measuring the temperature of lava.

Reading Check *What do Earth scientists study?*

Physical Science The study of matter and energy is **physical science.** Matter is anything that takes up space and has mass. The ability to cause change in matter is energy. Living and nonliving systems are made of matter. Examples include plants, animals, rocks, the atmosphere, and the water in oceans, lakes, and rivers. Physical science can be divided into two general fields—chemistry and physics. Chemistry is the study of matter and the interactions of matter. Physics is the study of energy and its ability to change matter. Figure 6 shows physical scientists at work.

Figure 6 Physical scientists study a wide range of subjects.

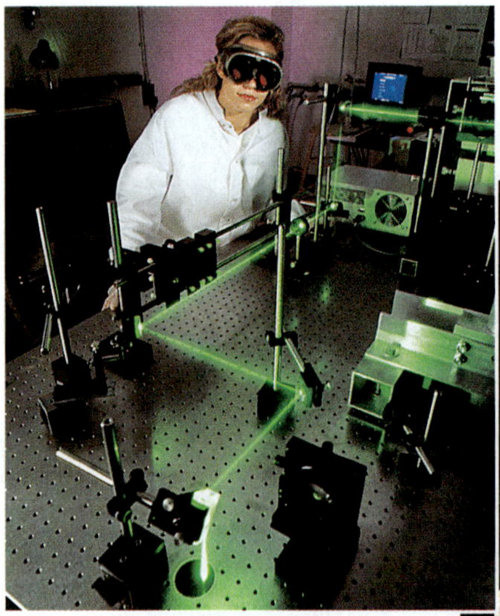

This chemist is studying the light emitted by certain compounds.

This physicist is studying light as it travels through optical fibers.

Differentiated Instruction

Challenge Have each student research careers that are of interest to him or her and then make a table showing how each career relies on the different branches of science. L3 LS **Linguistic**

Learning Disabled Have each student make three flash cards, one saying life science, one saying Earth science, and one saying physical science. Name different structures or processes, and have students raise the card naming the branch of science that studies those structures or processes. Ask students to explain their responses. L1

Careers Chemists ask questions such as "How can I make plastic stronger?" or "What can I do to make aspirin more effective?" Physicists might ask other types of questions, such as "How does light travel through glass fibers?" or "How can humans harness the energy of sunlight for their energy needs?"

Many careers are based on the physical sciences. Physicists and chemists are some obvious careers. Ultrasound and X-ray technicians working in the medical field study physical science because they study the energy in ultrasound or X rays and how it affects a living system.

Science and Technology Although learning the answers to scientific questions is important, these answers do not help people directly unless they can be applied in some way. **Technology** is the practical use of science, or applied science, as illustrated in **Figure 7.** Engineers apply science to develop technology. The study of how to use the energy of sunlight is science. Using this knowledge to create solar panels is technology. The study of the behavior of light as it travels through thin, glass, fiber-optic wires is science. The use of optical fibers to transmit information is technology. A scientist uses science to study how the skin of a shark repels water. The application of this knowledge to create a material that helps swimmers slip through the water faster is technology.

Figure 7 Solar-powered cars and the swimsuits worn in the Olympics are examples of technology—the application of science.

section 1 review

Summary

Learning About the World

- Scientists ask questions to learn how, why, or when something occurred.
- A theory is a possible explanation for observations that is supported by many investigations.
- A scientific law describes a pattern but does not explain why things happen.

Systems in Science

- A system is composed of structures, cycles, and processes that interact with each other.

The Branches of Science

- Science is divided into three branches—life science, Earth science, and physical science.
- Technology is the application of science in our everyday lives.

Self Check

1. **Compare and contrast** scientific theory and scientific law. Explain how a scientific theory can change.
2. **Explain** why science can answer some questions, but not others.
3. **Classify** the following statement as a theory or a law: Heating the air in a hot-air balloon causes the balloon to rise.
4. **Think Critically** Describe the importance of technology and how it relates to science.

Applying Skills

5. **Infer** Scientists ask questions and make observations. What types of questions and observations would you make if you were a scientist studying schools of fish in the ocean?

3 Assess

DAILY INTERVENTION

Check for Understanding

Logical-Mathematical Have students propose scientific laws. These do not have to be accepted laws but can be any general statement. Possible response: On the second Tuesday of each month, there is a science test. Next, have them provide theories to explain why the law holds. Possible response: There's a science test on the second Tuesday of each month, because the teacher plans it that way. Accept multiple explanations and have the class decide which makes the most sense. L3 LS

Reteach

Baseball Systems Ask students to name the different systems in a baseball game and classify them as structures, processes, or cycles. The bat, ball, bases, and field are structures; the rules and different strategies of play are processes; and the batting lineup, the nine innings, and each team's turn at bat are cycles. L2 LS **Logical-Mathematical**

Assessment

Performance Have students act out a play in which a group of scientists proposes a scientific law. Be sure the play includes the scientists making detailed observations to determine the exact nature of the regularity they are seeing. Use **Performance Assessment in the Science Classroom,** p. 147. L2

section 1 review

1. A scientific theory tries to explain why something happens. A scientific law describes a regularity in nature. As new information becomes available, scientific theories can change.
2. Science can answer questions about the natural world through observations and investigation. Opinions and questions about art, books, and music cannot be answered by science.
3. scientific law
4. Science answers scientific questions, and technology applies those answers to make products that help people.
5. Answers will vary. For example, "Why do fish swim in schools?" "Are fish swimming in schools more protected from enemies or are they more vulnerable?"

section 2 Science in Action

1 Motivate

Bellringer

INTERACTIVE CHALKBOARD PowerPoint® Presentations

Section Focus Transparencies also are available on the Interactive Chalkboard CD-ROM. L2 ELL

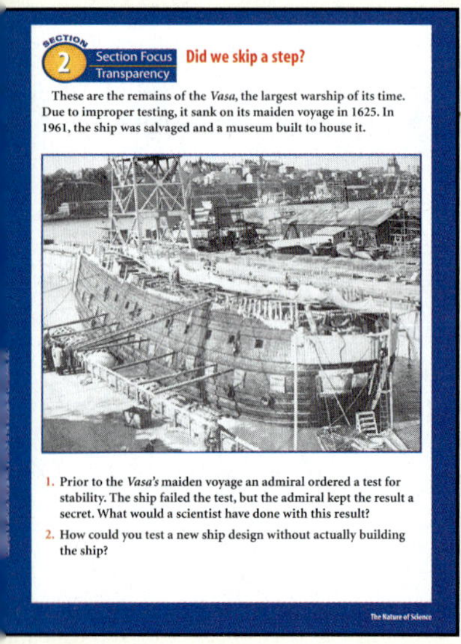
Section 2 Section Focus Transparency **Did we skip a step?**

These are the remains of the *Vasa*, the largest warship of its time. Due to improper testing, it sank on its maiden voyage in 1625. In 1961, the ship was salvaged and a museum built to house it.

1. Prior to the *Vasa's* maiden voyage an admiral ordered a test for stability. The ship failed the test, but the admiral kept the result a secret. What would a scientist have done with this result?
2. How could you test a new ship design without actually building the ship?

The Nature of Science

Tie to Prior Knowledge

Science Process Ask students whether they've ever tried to move a big piece of furniture through a doorway. Ask them to describe the process they used. They probably looked at the piece first, thought of a way to get it through the doorway, and then tried moving it. If this didn't work, they then might have put it down and rethought the problem. Explain that this is similar to the process scientists use when they try to solve a problem in science. L2

Text Question Answer

The scientific method can be applied to any investigation. Observations lead to hypotheses. Once you have developed a hypothesis, you must perform an investigation that provides you data to analyze.

---

as you read

What **You'll Learn**

- **Identify** some skills scientists use.
- **Define** hypothesis.
- **Recognize** the difference between observation and inference.

Why **It's Important**

Science can be used to learn more about the world you live in.

Review Vocabulary

observation: a record or description of an occurrence or pattern in nature

New Vocabulary

- hypothesis
- infer
- controlled experiment
- variable
- independent variable
- dependent variable
- constant

Science Skills

You know that science involves asking questions, but how does asking questions lead to learning? Because no single way to gain knowledge exists, a scientist doesn't start with step one, then go to step two, and so on. Instead, scientists have a huge collection of skills from which to choose. Some of these skills include thinking, observing, predicting, investigating, researching, modeling, measuring, analyzing, and inferring. Science also can advance with luck and creativity.

Science Methods Investigations often follow a general pattern. As illustrated in **Figure 8,** most investigations begin by seeing something and then asking a question about what was observed. Scientists often perform research by talking with other scientists. They read books and scientific magazines to learn as much as they can about what is already known about their question. Usually, scientists state a possible explanation for their observation. To collect more information, scientists almost always make more observations. They might build a model of what they study or they might perform investigations. Often, they do both. How might you combine some of these skills in an investigation?

Figure 8 Although there are different scientific methods for investigating a specific problem, most investigations follow a general pattern.

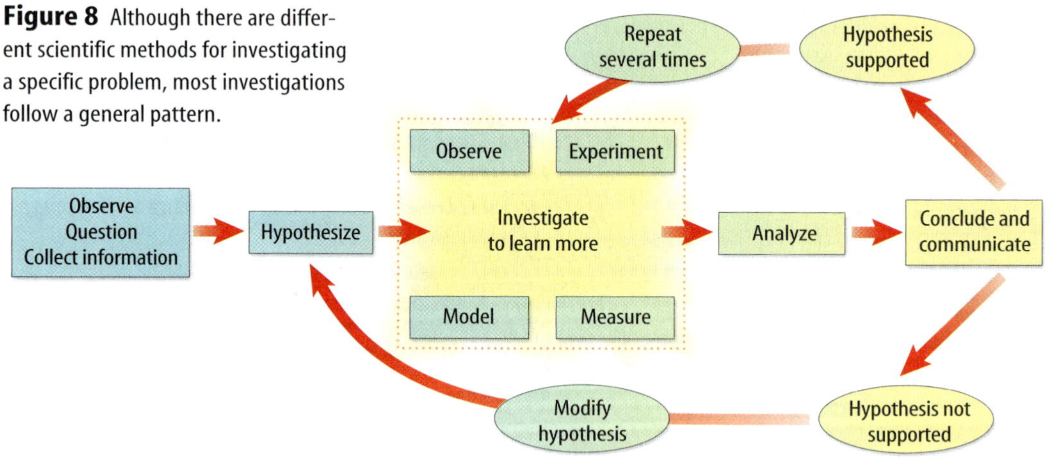

Section 2 Resource Manager

Chapter *Fast File* Resources

Transparency Activity, pp. 42, 49–50
Directed Reading for Content Mastery, p. 23
Enrichment, p. 34
MiniLAB, p. 4
Lab Activity, pp. 11–14
Reinforcement, p. 30

Physical Science Critical Thinking/Problem Solving, p. 2
Mathematics Skill Activities, p. 29
Reading and Writing Skill Activities, pp. 7, 41
Life Science Critical Thinking/Problem Solving, p. 7
Performance Assessment in the Science Classroom, p. 42

Figure 9 Investigations often begin by making observations and asking questions.

Questioning and Observing Ms. Clark placed a sealed shoe box on the table at the front of the laboratory. Everyone in the class noticed the box. Within seconds the questions flew. "What's in the box?" "Why is it there?"

Ms. Clark said she would like the class to see how they used some science skills without even realizing it.

"I think that she wants us to find out what's in it," Isabelle said to Marcus.

"Can we touch it?" asked Marcus.

"It's up to you," Ms. Clark said.

Marcus picked up the box and turned it over a few times.

"It's not heavy," Marcus observed. "Whatever is inside slides around." He handed the box to Isabelle.

Isabelle shook the box. The class heard the object strike the sides of the box. With every few shakes, the class heard a metallic sound. The box was passed around for each student to make observations and write them in his or her Science Journal. Some observations are shown in **Figure 9.**

Taking a Guess "I think it's a pair of scissors," said Marcus.

"Aren't scissors lighter than this?" asked Isabelle, while shaking the box. "I think it's a stapler."

"What makes you think so?" asked Ms. Clark.

"Well, staplers are small enough to fit inside a shoe box, and it seems to weigh about the same," said Isabelle.

"We can hear metal when we shake it," said Enrique.

"So, you are guessing that a stapler is in the box?"

"Yes," they agreed.

"You just stated a hypothesis," exclaimed Ms. Clark.

"A what?" asked Marcus.

INTEGRATE Career

Biologist Some naturalists study the living world, using mostly their observational skills. They observe animals and plants in their natural environment, taking care not to disturb the organisms they are studying. Make observations of organisms in a nearby park or backyard. Record your observations in your Science Journal.

2 Teach

Visual Learning

Figure 8 Although there is no one right way to carry out a scientific investigation, most follow the pattern outlined in this flow chart. Think of a scientific investigation as trying to make a tasty pot of soup. How would the steps outlined in the figure help you do this? Possible response: You select ingredients and decide which ones should go in the soup (hypothesize). You add them and sample the soup to see how it tastes (observe and analyze). You keep adding to and taking away from the recipe until it tastes the best (experiment). You write down the recipe so others can make it in exactly the same way (conclude and communicate). L2

INTEGRATE Career

Biologist Have a local biologist come to the classroom and discuss the organisms that they study. Ask the biologist to explain what type of observations that they make. The biologist can work for a zoo, local or federal wildlife protection agency, or botanical garden.

Differentiated Instruction

Visually Impaired Glue string to a sheet of construction paper to make a three-dimensional map of **Figure 7.** Place Braille labels on the different skills. Visually impaired students can use this for later reference.

Mini LAB

Purpose Students form a hypothesis based on observations. L2 LS **Kinesthetic**

Materials large pot, water, can of regular soda, can of diet soda

Teaching Strategy Not all regular soda cans will sink. If regular soda cans float, have students suggest possible reasons this occurs.

Safety Precautions Caution students not to shake the cans or handle them roughly.

Analysis

1. The can of regular soda did not float, but the can of diet soda did.
2. Possible answer: It led me to think that the cans behaved differently because they contained different ingredients.
3. Answers may vary. The regular soda uses a lot of sugar for sweetening. The diet soda uses only a little artificial sweetener.

Assessment

Process Have students measure the mass and volume of the soda in each can and determine the density of the soda. Ask them to suggest how the soda's density affected the behavior of the can in the water. Use **Performance Assessment in the Science Classroom,** p. 97.

Try at Home

Mini LAB

Forming a Hypothesis

Procedure

1. Fill a large **pot** with **water.** Place an **unopened can of diet soda** and an **unopened can of regular soda** into the pot of water and observe what each can does.
2. In your **Science Journal,** make a list of the possible explanations for your observation. Select the best explanation and write a hypothesis.
3. Read the nutritional facts on the back of each can and compare their ingredients.
4. Revise your hypothesis based on this new information.

Analysis

1. What did you observe when you placed the cans in the water?
2. How did the nutritional information on the cans change your hypothesis?
3. Infer why the two cans behaved differently in the water.

Try at Home

The Hypothesis "A **hypothesis** is a reasonable and educated possible answer based on what you know and what you observe."

"We know that a stapler is small, it can be heavy, and it is made of metal," said Isabelle.

"We observed that what is in the box is small, heavier than a pair of scissors, and made of metal," continued Marcus.

Analyzing Hypotheses "What other possible explanations fit with what you observed?" asked Ms. Clark.

"Well, it has to be a stapler," said Enrique.

"What if it isn't?" asked Ms. Clark. "Maybe you're overlooking explanations because your minds are made up. A good scientist keeps an open mind to every idea and explanation. What if you learn new information that doesn't fit with your original hypothesis? What new information could you gather to verify or disprove your hypothesis?"

"Do you mean a test or something?" asked Marcus.

"I know," said Enrique, "We could get an empty shoe box that is the same size as the mystery box and put a stapler in it. Then we could shake it and see whether it feels and sounds the same." Enrique's test is shown in **Figure 10.**

Making a Prediction "If your hypothesis is correct, what would you expect to happen?" asked Ms. Clark.

"Well, it would be about the same weight and it would slide around a little, just like the other box," said Enrique.

"It would have that same metallic sound when we shake it," said Marcus.

"So, you predict that the test box will feel and sound the same as your mystery box. Go ahead and try it," said Ms. Clark.

Figure 10 Comparing the known information with the unknown information can be valuable even though you cannot see what is inside the closed box.

LAB DEMONSTRATION

Purpose to make predictions based on sounds

Materials 5 identical glasses, spoon, water

Preparation Put a different amount of water in each glass.

Procedure Tap on the glass containing the most water, and tell students this glass contains the most water. Do the same with the glass containing the least water. Then tap on two of the other glasses and ask which has more water in it. Repeat this process until students have ranked the five glasses from the one containing the most water to the one with the least water.

Assessment

State the relationship between the amount of water in a glass and the pitch produced when the glass is hit. The more water in the glass, the higher the pitch. Is this a hypothesis, a theory, or a law? Why? It is a law—it is based on observation.

Testing the Hypothesis Ms. Clark gave the class an empty shoe box that appeared to be identical to the mystery box. Isabelle found a metal stapler. Enrique put the stapler in the box and taped the box closed. Marcus shook the box.

"The stapler does slide around but it feels just a little heavier than what's inside the mystery box," said Marcus. "What do you think?" he asked Isabelle as he handed her the box.

"It is heavier," said Isabelle "and as hard as I shake it, I can't get a metallic sound. What if we find the mass of both boxes? Then we'll know the exact mass difference between the two."

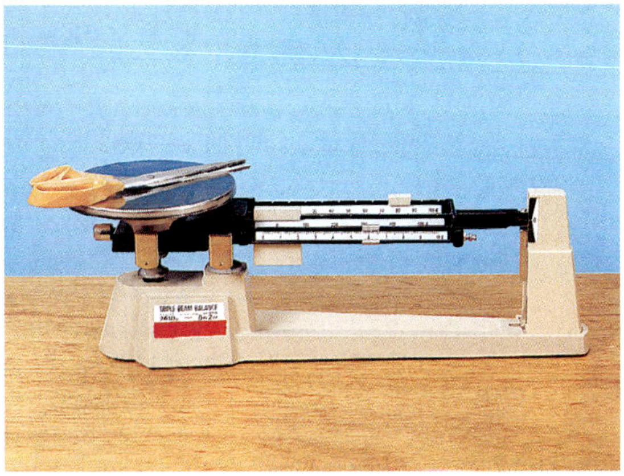

Figure 11 Laboratory balances are used to find the mass of objects.

Using a balance, as shown in **Figure 11,** the class found that the test box had a mass of 410 g, and the mystery box had a mass of 270 g.

Organizing Your Findings "Okay. Now you have some new information," said Ms. Clark. "But before you draw any conclusions, let's organize what we know. Then we'll have a summary of our observations and can refer back to them when we are drawing our conclusions."

"We could make a chart of our observations in our Science Journals," said Marcus.

"We could compare the observations of the mystery box with the observations of the test box," said Isabelle. The chart that the class made is shown in **Table 1.**

Table 1 Observation Chart

Questions	Mystery Box	Our Box
Does it roll or slide?	It slides and appears to be flat.	It slides and appears to be flat.
Does it make any sounds?	It makes a metallic sound when it strikes the sides of the box.	The stapler makes a thudding sound when it strikes the sides of the box.
Is the mass evenly distributed in the box?	No. The object doesn't completely fill the box.	No. The mass of the stapler is unevenly distributed.
What is the mass of the box?	270 g	410 g

Fun Fact

One of the first known stapling machines was built in the 1700s for King Louis XV of France. The staples it used were handmade and had the insignia of the Royal Court on them. Staplers only came into general use in the early 1900s.

Identifying Misconceptions

Changing Hypotheses Students may think that a hypothesis, once made, is never changed. Explain that many of the great scientists of the past made hypotheses that later were abandoned. A good hypothesis does not have to be one that is proven correct. It only has to be a reasonable explanation that is based upon observations and allows you to make predictions.

Discussion

Collecting Data Why is it important to collect data accurately? Possible response: Recording data inaccurately can cause you to draw inaccurate conclusions. It would be like studying for a test using the wrong book. L2 LS **Logical-Mathematical**

Curriculum Connection

History A Scientific Revolution occurred in western Europe from about the late fifteenth century to the mid-seventeenth century. Have students use an encyclopedia to find out more about the Scientific Revolution. This Scientific Revolution was important because it introduced a new way of thinking about how to carry out scientific investigations. L2 LS **Linguistic**

SECTION 2 Science in Action 15

Caption Answer

Figure 12 The pot was knocked to the floor. You might infer that the cat did it. However, you can't be sure given the limited information in the photo.

Teacher FYI

Inductive and Deductive Reasoning The students in the story are doing what is called inductive reasoning, in which a generalized conclusion is drawn from specific circumstances. They are making observations and eliminating possibilities until they arrive at a suitable answer to the question. Deductive reasoning, on the other hand, is the kind of reasoning in which a conclusion about particulars is drawn from general or universal premises. This is the type of reasoning that says that if you do the steps of a math problem correctly, the process must give you the correct answer.

Differentiated Instruction

Challenge Have students critique the investigation of the students in the story. Ask them to suggest ways in which the students in the story might have done things differently and to describe other things the students might do to get a better idea of what is inside the box. Possible responses: They could try to guess at the size of the object in the box by noticing how long it takes to slide. They also could find objects of the same weight as the object in the box and see how the objects sound when shaken inside a box. They could experiment with objects to find one that makes a metallic sound like the sound made by the object in the box. L3 LS **Logical-Mathematical**

Drawing Conclusions

"What have you learned from your investigation so far?" asked Ms. Clark.

"The first thing that we learned was that our hypothesis wasn't correct," answered Marcus.

"Would you say that your hypothesis was entirely wrong?" asked Ms. Clark.

"The boxes don't weigh the same, and the box with the stapler doesn't make the same sound as the mystery box. But there could be a difference in the kind of stapler in the box. It could be a different size or made of different materials."

"So you infer that the object in the mystery box is not exactly the same type of stapler, right?" asked Ms. Clark.

"What does *infer* mean?" asked Isabelle.

"To **infer** something means to draw a conclusion based on what you observe," answered Ms. Clark.

"So we inferred that the things in the boxes had to be different because our observations of the two boxes are different," said Marcus.

"I guess we're back to where we started," said Enrique. "We still don't know what's in the mystery box."

"Do you know more than you did before you started?" asked Ms. Clark.

"We eliminated one possibility," Isabelle added.

"Yes. We inferred that it's not a stapler, at least not like the one in the test box," said Marcus.

"So even if your observations don't support your hypothesis, you know more than you did when you started," said Ms. Clark.

Continuing to Learn "So when do we get to open the box and see what it is?" asked Marcus.

"Let me ask you this," said Ms. Clark. "Do you think scientists always get a chance to look inside to see if they are right?"

"If they are studying something too big or too small to see, I guess they can't," replied Isabelle. "What do they do in those cases?"

"As you learned, your first hypothesis might not be supported by your investigation. Instead of giving up, you continue to gather information by making more observations, making new hypotheses, and by investigating further. Some scientists have spent lifetimes researching their questions. Science takes patience and persistence," said Ms. Clark.

Figure 12 Observations can be used to draw inferences.
Infer *Looking at both of these photos, what do you infer has taken place?*

Active Reading

Learning Journal This strategy encourages students to interact with the reading, allowing personal responses. Students should draw a vertical line down the center of each page of the Learning Journal. The left column entries can be research notes, lecture notes, or vocabulary terms. The right column entries are the students' responses to, interpretations of, questions about, or analysis of the left column entries. Have students write a Learning Journal that incorporates all the scientific methods described in this section. L3 P

Communicating Your Findings It is not unusual for one scientist to continue the work of another or to try to duplicate the work of another scientist. It is important for scientists to communicate to others not only the results of the investigation, but also the methods by which the investigation was done. Scientists often publish reports in journals, books, and on the Internet to show other scientists the work that was completed. They also might attend meetings where they make speeches about their work.

Like the science-fair student in **Figure 13** demonstrates, an important part of doing science is the ability to communicate methods and results to others.

Reading Check *Why do scientists share information?*

Figure 13 Presentations are one way people in science communicate their findings.

Applying Math — Make a Data Table

SEASONAL TEMPERATURES Suppose you were given the average temperatures in a city for the four seasons over a three-year period: spring 1997 was 11°C; summer 1997 was 25°C; fall 1997 was 5°C; winter 1997 was −5°C; spring 1998 was 9°C; summer 1998 was 36°C; fall 1998 was 10°C; winter 1998 was −3°C; spring 1999 was 10°C; summer 1999 was 30°C; fall 1999 was 9°C; and winter 1999 was −2°C. How can you tell in which of the years each season had its coldest average?

Solution

1. *This is what you know:* Temperatures were: 1997: 11°C, 25°C, 5°C, −5°C
   1998: 9°C, 36°C, 10°C, −3°C
   1999: 10°C, 30°C, 9°C, −2°C
2. *This is what you need to find out:* Which of the years each season had its coldest temperature?
3. *This is the procedure you need to use:*
   - Create a table with rows for seasons and columns for the years.
   - Insert the values you were given.
4. *Check your answer:* The four coldest seasons were spring 1998, summer 1997, fall 1997, and winter 1997.

Practice Problems

Use your table to find out which season had the greatest difference in temperatures over the three years from 1997 through 1999.

Science Online For more practice, visit red.msscience.com/math_practice

Reading Check

Answer so other scientists can continue the work or try to duplicate it

Inquiry Lab

Understanding the Scientific Method

Purpose To understand the scientific method, have students work through the steps and design an experiment based on an observation made in the schoolyard.

Possible Materials paper, pen, magnifying glass, thermometer, rain gauge

Estimated Time 1 class session (Students may need to collect research for a few days before performing their investigation.)

Teaching Strategies

- Students can place thermometers around the schoolyard and record the temperature at each location several times throughout the day. They could hypothesize why the temperature varies from one location to another.
- Students can set rain gauges up around the schoolyard to collect rainwater. They could hypothesize which areas will receive more water than others.
- Students could compare the pH levels of the rainwater with that of a nearby lake or stream. L2

For additional inquiry activities, see *Science Inquiry Labs*.

Applying Math

National Math Standards

Correlation to Mathematics Objectives

5, 6

Answers to Practice Problems

1. Summer; it had a range of 11°C.
2. Possible answers: You can say that spring temperatures changed the least and that fall and spring temperatures are the closest to each other.

Differentiated Instruction

Learning Disabled To help students organize the data given in the Problem-Solving Activity, suggest that before they construct their tables, they write the data as a list. L1

Spring 1997: 11°C   Fall 1997: 5°C
Summer 1997: 25°C   Winter 1997: –5°C
Spring 1998: 9°C , and so on.

Quick Demo

Variables and Constants

Materials salt, sugar, 2 tall glasses of water, measuring spoon, stirring spoon, thermometer

Estimated Time 10 minutes

Procedure Pour 1 tsp of salt into one glass and 1 tsp of sugar into another glass. Ask the students to record each time you add salt and sugar on a piece of paper. Stir the contents until all the granules have dissolved. Now add another tsp. of salt to the glass and sugar to the other glass. Stir until it is all dissolved. Continue doing this until no more salt or sugar dissolves. Ask the students what the variables and constants are in this experiment. How would the results of the experiment change if you left one glass unstirred and stirred the other? Ask the students what other variables and constants could be added to this experiment. L2 **Kinesthetic**

Discussion

Expanding Knowledge How can the things you are learning in this section help you do things other than science experiments? Possible answer: Making observations and forming hypotheses about technique and practice could help students play baseball or a musical instrument better. L2 LS **Linguistic**

IDENTIFYING Misconceptions

Variables Students might think that the term *variable* refers only to the factors that *are* changed in an experiment when, in fact, all factors that *can be* changed in an experiment are called variables. In a controlled experiment, an attempt is made to keep all variables constant except the dependent and independent variables.

Science Online

Topic: Scientific Method
Visit red.msscience.com for Web links to information about the scientific method.

Activity Identify the three variables needed in every experiment and summarize the differences between them.

Experiments

Different types of questions call for different types of investigations. Ms. Clark's class made many observations about their mystery box and about their test box. They wanted to know what was inside. To answer their question, building a model—the test box—was an effective way to learn more about the mystery box. Some questions ask about the effects of one factor on another. One way to investigate these kinds of questions is by doing a controlled experiment. A **controlled experiment** involves changing one factor and observing its effect on another while keeping all other factors constant.

Variables and Constants Imagine a race in which the lengths of the lanes vary. Some lanes are 102 m long, some are 98 m long, and a few are 100 m long. When the first runner crosses the finish line, is he or she the fastest? Not necessarily. The lanes in the race have different lengths.

Variables are factors that can be changed in an experiment. Reliable experiments, like the race shown in **Figure 14,** attempt to change one variable and observe the effect of this change on another variable. The variable that is changed in an experiment is called the **independent variable.** The **dependent variable** changes as a result of a change in the independent variable. It usually is the dependent variable that is observed in an experiment. Scientists attempt to keep all other variables constant—or unchanged.

The variables that are not changed in an experiment are called **constants.** Examples of constants in the race include track material, wind speed, and distance. This way it is easier to determine exactly which variable is responsible for the runners' finish times. In this race, the runners' abilities were varied. The runners' finish times were observed.

Figure 14 The 400-m race is an example of a controlled experiment. The distance, track material, and wind speed are constants. The runners' abilities and their finish times are varied.

18

Cultural Diversity

Ancient Greek Science Ancient Greeks did not do experiments. They believed that everything that was important to know about the world could be learned through reasoning. One of the greatest contributions of the Scientific Revolution of the sixteenth and seventeenth centuries was the idea that experiments were an important part of science.

Figure 15 Safety is the most important aspect of any investigation.

Laboratory Safety

In your science class, you will perform many types of investigations. However, performing scientific investigations involves more than just following specific steps. You also must learn how to keep yourself and those around you safe by obeying the safety symbol warnings, shown in **Figure 16.**

In a Laboratory When scientists work in a laboratory, as shown in **Figure 15,** they take many safety precautions.

The most important safety advice in a science lab is to think before you act. Always check with your teacher several times in the planning stage of any investigation. Also make sure you know the location of safety equipment in the laboratory room and how to use this equipment, including the eyewashes, thermal mitts, and fire extinguisher.

Good safety habits include the following suggestions. Before conducting any investigation, find and follow all safety symbols listed in your investigation. You always should wear an apron and goggles to protect yourself from chemicals, flames, and pointed objects. Keep goggles on until activity, cleanup, and handwashing are complete. Always slant test tubes away from yourself and others when heating them. Never eat, drink, or apply makeup in the lab. Report all accidents and injuries to your teacher and always wash your hands after working with lab materials.

In the Field Investigations also take place outside the lab, in streams, farm fields, and other places. Scientists must follow safety regulations there, as well, such as wearing eye goggles and any other special safety equipment that is needed. Never reach into holes or under rocks. Always wash your hands after you've finished your field work.

Eye Safety
Clothing Protection
Disposal
Biological
Extreme Temperature
Sharp Object
Fume
Irritant
Toxic
Animal Safety
Flammable
Electrical
Chemical
Open Flame
Handwashing

Figure 16 Safety symbols are present on nearly every investigation you will do this year.
List *the safety symbols that should be on the lab the student is preparing to do in* ***Figure 15.***

Discussion

Safety Symbols What are some of the advantages of using standardized safety symbols in addition to writing separate safety precautions for each activity? Safety symbols are easy to recognize. You can see quickly what safety precautions the activity requires. Even those who can't read very well can see what precautions need to be taken. The symbols also take up less space. L2

Visual Learning

Figure 15 Ask students to name each safety precaution this boy is taking as he works in the lab and describe a possible problem that could occur without each precaution. Possible answers: If he were not wearing goggles, he could get harmful chemicals in his eyes. If he were not wearing gloves, or an apron he could get harmful chemicals on his hands or his clothing. L2 LS **Visual-Spatial**

Activity

Scientific Articles Have students look in a major library for copies of scientific journals such as *Journal of the American Chemical Society*, *Nature*, and *Physical Review*. They also can look at journal Web sites on the Internet. Ask students to note features about them that make them appear scientific. Possible answers: The design is simple and not visually catchy. Many articles have complex tables, diagrams, graphs, and charts. The articles are written very dryly and use a lot of big words. L2 LS **Visual-Spatial**

Caption Answer

Figure 16 symbols for eye safety, irritant, clothing protection

Curriculum Connection

Art Safety symbols are icons—images that convey information. Have students design icons to alert students to the laboratory safety hazards of long hair, clothing with loose sleeves, jewelry, or any other safety hazard they can think of. L2 LS **Visual-Spatial**

3 Assess

DAILY INTERVENTION

Check for Understanding

Logical-Mathematical Ask students what might be some of the variables and constants in a study of local wildlife. Variables might include the amount of daylight at the time of year and the time an animal can spend looking for food. A constant might be the physical features of the land the animal lives on. L2 LS

Reteach

Scientific Investigations Have students draw their own diagrams showing the steps in a scientific investigation. Encourage them to incorporate safety notes into their diagrams. L2 LS **Visual-Spatial**

Assessment

Content Have each student write an ending to the story in this section. Have groups of students act the story out as a play, incorporating the different endings written by the students in the group. Use **Performance Assessment in the Science Classroom,** p. 147. L2

Figure 17 Accidents are not planned. Safety precautions must be followed to prevent injury.

Why have safety rules? Doing science in the class laboratory or in the field can be much more interesting than reading about it. However, safety rules must be strictly followed, so that the possibility of an accident greatly decreases. However, you can't predict when something will go wrong.

Think of a person taking a trip in a car. Most of the time when someone drives somewhere in a vehicle, an accident, like the one shown in **Figure 17,** does not occur. But to be safe, drivers and passengers always should wear safety belts. Likewise, you always should wear and use appropriate safety gear in the lab—whether you are conducting an investigation or just observing. The most important aspect of any investigation is to conduct it safely.

section 2 review

Summary

Science Skills

- The scientific method was developed to help scientists investigate their questions.
- Hypotheses are possible explanations for why something occurs.

Drawing Conclusions

- Scientists communicate with one another to share important information.

Experiments

- Controlled experiments test the effect of one factor on another.

Laboratory Safety

- Safety precautions must be followed when conducting any investigation.

Self Check

1. **Explain** the difference between an inference and an observation.
2. **Explain** the differences between independent and dependent variables.
3. **Think Critically** A classroom investigation lists bleach as an ingredient. Bleach can irritate your skin, damage your eyes, and stain your clothes. What safety symbols should be listed with this investigation? Explain.

Applying Skills

4. **Describe** the different types of safety equipment found in a scientific laboratory. From your list, which equipment should you use when working with a flammable liquid in the lab?

Science Online red.msscience.com/self_check_quiz

section 2 review

1. An observation is something that comes from your senses. An inference is a conclusion your mind draws.
2. An independent variable is the variable that changes in an experiment. The dependent variable changes as a result of the changing independent variable.
3. Symbols listed: disposal alert, irritant alert, fume safety, clothing protection, eye safety, toxic alert, and chemical safety.
4. You should wear an apron, goggles, and gloves. You also should wear long, tight sleeves to protect your arms. You should make sure an eyewash and a fire extinguisher are nearby.

section 3

Models in Science

Why are models necessary?

Just as you can take many different paths in an investigation, you can test a hypothesis in many different ways. Ms. Clark's class tested their hypothesis by building a model of the mystery box. A model is one way to test a hypothesis. In science, a **model** is any representation of an object or an event used as a tool for understanding the natural world.

Models can help you visualize, or picture in your mind, something that is difficult to see or understand. Ms. Clark's class made a model because they couldn't see the item inside the box. Models can be of things that are too small or too big to see. They also can be of things that can't be seen because they don't exist anymore or they haven't been created yet. Models also can show events that occur too slowly or too quickly to see. **Figure 18** shows different kinds of models.

as you read

What You'll Learn

- **Describe** various types of models.
- **Discuss** limitations of models.

Why It's Important

Models can be used to help understand difficult concepts.

Review Vocabulary

scientific method: processes scientists use to collect information and answer questions

New Vocabulary

- model

Figure 18 Models help scientists visualize and study complex things and things that can't be seen.

Solar system model

Prototype model

Cell model

Dinosaur model

Section 3 Resource Manager

Chapter *FAST FILE* Resources

Transparency Activity, pp. 50, 51

MiniLAB, p. 5

Enrichment, p. 35

Directed Reading for Content Mastery, pp. 23, 24

Reinforcement, p. 31

Lab Activity, pp. 15–18

Reading and Writing Skill Activities, p. 43

1 Motivate

INTERACTIVE CHALKBOARD PowerPoint® Presentations

Bellringer

Section Focus Transparencies also are available on the Interactive Chalkboard CD-ROM. L2 ELL

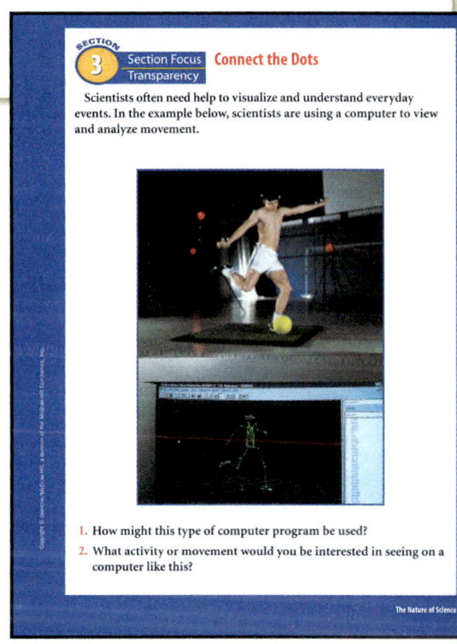

Section 3 Section Focus Transparency — Connect the Dots

Scientists often need help to visualize and understand everyday events. In the example below, scientists are using a computer to view and analyze movement.

1. How might this type of computer program be used?
2. What activity or movement would you be interested in seeing on a computer like this?

The Nature of Science

Tie to Prior Knowledge

Modeling Events Ask students whether they've ever described an event to someone and had a hard time doing it. To make things easier, they may try to act out the event. If it was the winning shot in a basketball game, for example, they may have begun "Okay, there's the basket and you're the player guarding me. . . " Tell them that if they did this, they made a model of the event. L1

2 Teach

Use an Analogy

Sports When you practice playing a sport, you are making a model. Good coaches know the kinds of plays another team uses and will simulate those plays in practice to prepare the players. This is a type of physical model.

Fun Fact

Pilots and astronauts receive some of their most critical flight training without ever leaving the ground. They use complicated, realistic flight simulators that allow them to practice emergency procedures without risking an accident in the air.

Reading Check

Answer show things that occur too slowly or too quickly to be seen and show motions and positions of things that would take hours or days to calculate by hand or using a calculator

Discussion

Topographic Maps Show students a topographic map and point out the contour lines that show changes in elevation. Explain that when the contour lines get closer together, elevation changes are much steeper. Point out the different symbols for mountains, rivers, and lakes. Have students find a topographic map of their home state and analyze the symbols and elevation gradients.

Types of Models

Most models fall into three basic types—physical models, computer models, and idea models. Depending on the reason that a model is needed, scientists can choose to use one or more than one type of model.

Physical Models Models that you can see and touch are called physical models. Examples include things such as a tabletop solar system, a globe of Earth, a replica of the inside of a cell, or a gumdrop-toothpick model of a chemical compound. Models show how parts relate to one another. They also can be used to show how things appear when they change position or how they react when an outside force acts on them.

Computer Models Computer models are built using computer software. You can't touch them, but you can view them on a computer screen. Some computer models can model events that take a long time or take place too quickly to see. For example, a computer can model the movement of large plates in the Earth and might help predict earthquakes.

Computers also can model motions and positions of things that would take hours or days to calculate by hand or even using a calculator. They can also predict the effect of different systems or forces. **Figure 19** shows how computer models are used by scientists to help predict the weather based on the motion of air currents in the atmosphere.

Reading Check *What do computer models do?*

Science Online
Topic: Topographic Maps
Visit red.msscience.com for Web links to information about topographic maps.

Activity List some of the different features found on topographic maps and explain their importance when reading and interpreting maps.

Figure 19 A weather map is a computer model showing weather patterns over large areas. Scientists can use this information to predict the weather and to alert people to potentially dangerous weather on the way.

Curriculum Connection

Social Studies You may have heard that a TV program is the number-one-rated show, but what exactly does that mean? Have students find out and report on how TV viewing statistics are obtained. TV viewing statistics are obtained by selecting a scientifically identified representative sample (a model) of the entire country's population. These households' viewing habits are monitored by special boxes on their sets and the results extended to the whole country. In the United States, about 25,000 homes rate the viewing habits of the entire country. L3 LS Linguistic P

22 CHAPTER 1 The Nature of Science

Figure 20 Models can be created using various types of tools.

Idea Models Some models are ideas or concepts that describe how someone thinks about something in the natural world. Albert Einstein is famous for his theory of relativity, which involves the relationship between matter and energy. One of the most famous models Einstein used for this theory is the mathematical equation $E = mc^2$. This explains that mass, m, can be changed into energy, E. Einstein's idea models never could be built as physical models, because they are basically ideas.

Making Models

The process of making a model is something like a sketch artist at work, as shown in **Figure 20.** The sketch artist attempts to draw a picture from the description given by someone. The more detailed the description is, the better the picture will be. Like a scientist who studies data from many sources, the sketch artist can make a sketch based on more than one person's observation. The final sketch isn't a photograph, but if the information is accurate, the sketch should look realistic. Scientific models are made much the same way. The more information a scientist gathers, the more accurate the model will be. The process of constructing a model of King Tutankhamun, who lived more than 3,000 years ago, is shown in **Figure 21.**

Reading Check *How are sketches like specific models?*

Using Models

When you think of a model, you might think of a model airplane or a model of a building. Not all models are for scientific purposes. You use models, and you might not realize it. Drawings, maps, recipes, and globes are all examples of models.

Mini LAB

Thinking Like a Scientist

Procedure

1. Pour 15 mL of **water** into a **test tube.**
2. Slowly pour 5 mL of **vegetable oil** into the test tube.
3. Add two drops of **food coloring** and observe the liquid for 5 min.

Analysis

1. Record your observations of the test tube's contents before and after the oil and the food coloring were added to it.
2. Infer a scientific explanation for your observations.

Quick Demo

Star Gazing

Materials star chart

Estimated Time 10 minutes

Procedure Obtain a star chart and demonstrate how it can be used to make predictions of the night sky on different days of the year.

Mini LAB

Purpose Students use scientific thinking as they make observations and draw inferences. L2

LS **Kinesthetic**

Materials water, vegetable oil, graduated cylinder, test tube, food coloring, watch or clock

Teaching Strategies

- Before they do the experiment, explain to students that some liquids do not mix with each other.
- Tell the students to remember the phrase "like dissolves like."
- Have students discuss what the wisest choice would be for disposal of the vegetable oil following the experiment.

Troubleshooting Have students keep the test tubes as still as possible.

Analysis

1. Students' records should show that the drops remained intact in the oil and sat on the oil and water border for several seconds until they pushed through the layer. Once in the water, the food dye exploded into swirls of color.
2. Food coloring does not mix with oil. Food coloring mixes with water because it is like water and "like dissolves like."

Assessment

Performance Extend the MiniLab and have students experiment what happens when you mix soap in with the oil and water. How does the use of soapy water help clean dishes?

Assessment

Process Have students suggest ways in which the observations made in this activity could be used to design an experiment that determines the identity of a substance. Use **Performance Assessment in the Science Classroom,** p. 95. L2

Reading Check

Answer The more detailed the information used to make the model or the sketch, the better it will be.

SECTION 3 Models in Science 23

NATIONAL GEOGRAPHIC

Visualizing the Modeling of King Tut

Have students examine the pictures and read the captions. Then ask the following questions.

Why might scientists want to make a model of the face of a person from a skull? Possible answer: to help discover the identity of victims of accidents, fires, plane crashes, war, or murder.

How is the model of King Tut like other models used by scientists? Like other models, it is used by scientists to describe things they cannot see directly. Models can help scientists explain data and predict outcomes.

Activity

Models Have small groups of students make models of objects or ideas related to science. After the models are complete, have each group present and explain its model to the class. Discuss the accuracy of each model and how well it serves its purpose. L2

NATIONAL GEOGRAPHIC **VISUALIZING THE MODELING OF KING TUT**

Figure 21

More than 3,000 years ago, King Tutankhamun ruled over Egypt. His reign was a short one, and he died when he was just 18. In 1922, his mummified body was discovered, and in 1983 scientists recreated the face of this most famous of Egyptian kings. Some of the steps in building the model are shown here.

This is the most familiar image of the face of King Tut—the gold funerary mask that was found covering his skeletal face.

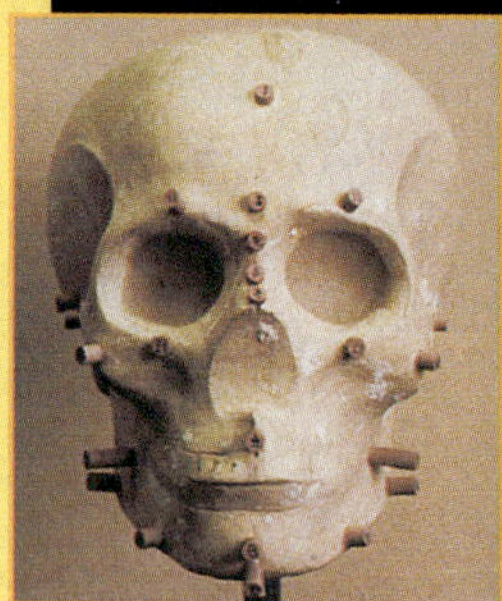

A First, a scientist used measurements and X rays to create a cast of the young king's skull. Depth markers (in red) were then glued onto the skull to indicate the likely thickness of muscle and other tissue.

B Clay was applied to fill in the area between the markers.

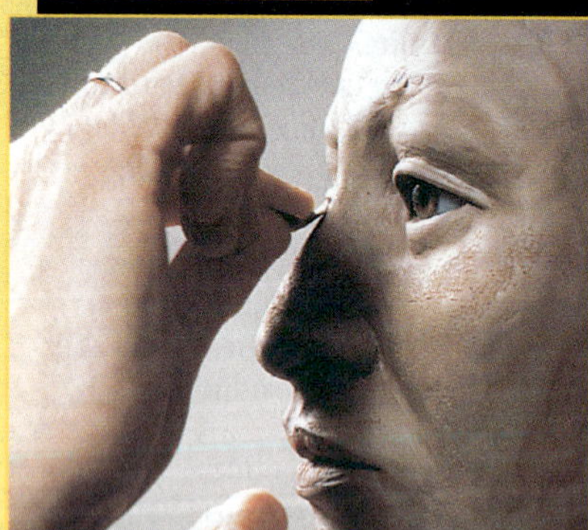

C Next, the features were sculpted. Here, eyelids are fashioned over inlaid prosthetic, or artificial, eyes.

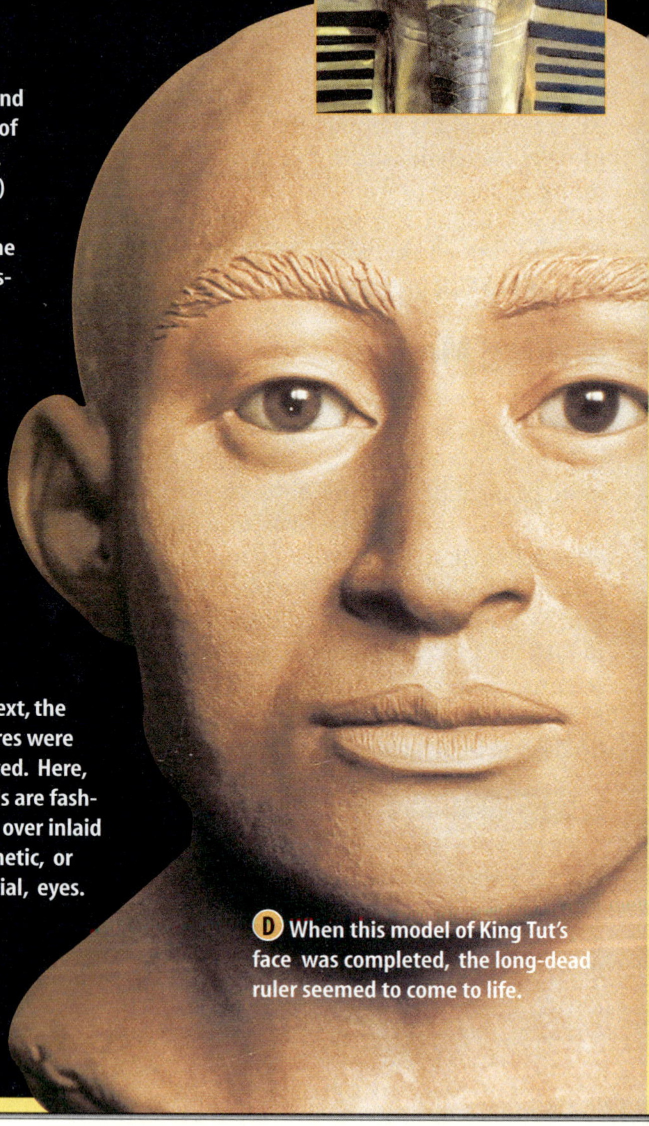

D When this model of King Tut's face was completed, the long-dead ruler seemed to come to life.

Differentiated Instruction

Challenge Have students research how forensic anthropologists use the technique described in this feature to identify skeletal remains. Students could present their information in written or oral reports. L3

Visually Impaired Have students use clay to construct a simple model of something familiar to them. Ask the students to share their models with one another. Rather than viewing the model with open eyes, ask the students to close their eyes and gently feel the model. Do they recognize what the model is?

Models Communicate Some models are used to communicate observations and ideas to other people. Often, it is easier to communicate ideas you have by making a model instead of writing your ideas in words. This way others can visualize them, too.

Models Test Predictions Some models are used to test predictions. Ms. Clark's class predicted that a box with a stapler in it would have characteristics similar to their mystery box. To test this prediction, the class made a model. Automobile and airplane engineers use wind tunnels to test predictions about how air will interact with their products.

Models Save Time, Money, and Lives Other models are used because working with and testing a model can be safer and less expensive than using the real thing. For example, the crash-test dummies shown in **Figure 22** are used in place of people when testing the effects of automobile crashes. To help train astronaunts in the conditions they will encounter in space, NASA has built a special airplane. This airplane flies in an arc that creates the condition of freefall for 20 to 25 seconds. Making several trips in the airplane is easier, safer, and less expensive than making a trip into space.

Figure 22 Models are a safe and relatively inexpensive way to test ideas.

Wind tunnels can be used to test new airplane designs or changes made to existing airplanes.

Crash-test dummies are used to test vehicles without putting people in danger.

Astronauts train in a special aircraft that models the conditions of space.

Make a Model

Different Models Have students make models of the classroom. Have some draw pictures or take photographs of the room. Have others make paper cut outs or use blocks to represent the different classroom structures, paying close attention to scale. Ask which model would better represent the classroom to someone who hadn't seen it and which would better assist in making predictions about how the classroom could be arranged differently. L2 LS **Kinesthetic**

Teacher FYI

Movie Models Before computerized animation, the models used to put special effects in movies had to be creative and diverse. To film a tornado scene in one early movie, for example, a string of muslin stockings was sewn together, stuffed, and swirled above a small-scale version of the movie set. For close-up views of the tornado, the filmmakers used a dust-filled burlap bag that released dark clouds when beaten.

Visual Learning

Figure 23 Have students consider the Earth-centered and Sun-centered models. Standing here on Earth, is there any reason not to believe Earth is at the center of things? Not really; we don't feel any motion and there is no way of proving just by sight that Earth is moving. L2 LS **Logical-Mathematical**

Teacher FYI

"Weightless Wonder" Plane The "weightless wonder" plane used to train NASA astronauts is a very busy machine. Most of the time, it is being used by scientists doing microgravity research. Also, when the Space Shuttle is being transported back to the Kennedy Space Center, it flies fifteen minutes ahead of the 747 plane carrying the shuttle to warn of the possibility of bad weather that might damage the shuttle's delicate tiles.

3 Assess

DAILY INTERVENTION

Check for Understanding

Visual-Spatial Have students use a video camera to model a special effect in a motion picture. Possible effects might include someone flying, a giant river, or a tornado. L3 LS

Reteach

Models Ask students which type of model—physical, computer, or idea—they'd use to represent the following: how a building will look when it's completed; the relationship between mass and gravity; how Earth's landscape has changed over time. physical or computer, idea, computer. L2 LS **Logical-Mathematical**

Assessment

Process Students probably had some knowledge of models before reading the section. Have them write in their Science Journals how their ideas about models have changed since reading the section. Use **Performance Assessment in the Science Classroom,** p. 157. L2

Figure 23 The model of Earth's solar system changed as new information was gathered.

An early model of the solar system had Earth in the center with everything revolving around it.

Later on, a new model had the Sun in the center with everything revolving around it.

Limitations of Models

The solar system is too large to be viewed all at once, so models are made to understand it. Many years ago, scientists thought that Earth was the center of the universe and the sky was a blanket that covered the planet.

Later, through observation, it was discovered that the objects you see in the sky are the Sun, the Moon, stars, and other planets. This new model explained the solar system differently. Earth was still the center, but everything else orbited it as shown in **Figure 23.**

Models Change Still later, through more observation, it was discovered that the Sun is the center of the solar system. Earth, along with the other planets, orbits the Sun. In addition, it was discovered that other planets also have moons that orbit them. A new model was developed to show this.

Earlier models of the solar system were not meant to be misleading. Scientists made the best models they could with the information they had. More importantly, their models gave future scientists information to build upon. Models are not necessarily perfect, but they provide a visual tool to learn from.

section 3 review

Summary

Why are models necessary?

- Scientists develop models to help them visualize complex concepts.

Types of Models

- There are three types of models—physical models, computer models, and idea models.

Making Models

- The more information you have when creating a model, the more accurate the model will be.

Using Models

- Models are used to convey important information such as maps and schedules.

Limitations of Models

- Models can be changed over time as new information becomes available.

Self Check

1. **Infer** what types of models can be used to model weather. How are they used to predict weather patterns?
2. **Explain** how models are used in science.
3. **Describe** how consumer product testing services use models to ensure the safety of the final products produced.
4. **Describe** the advantages and limitations of the three types of models.
5. **Think Critically** Explain why some models are better than others for certain situations. Give one example.

Applying Math

6. **Use Proportions** On a map of a state, the scale shows that 1 cm is approximately 5 km. If the distance between two cities is 1.7 cm on the map, how many kilometers separate them?

Science Online red.msscience.com/self_check_quiz

section 3 review

1. Computer models and paper drawings can be used to model weather. Each can show past events but computer models can predict future events.
2. Models are used in science to study things that are too big, too small, take too long, or happen too quickly.
3. A model product is tested using a wide range of conditions in which the product is expected to operate.
4. physical model—can take a while to make and be difficult to modify; computer model—can show slow or fast changes and can be run many times but can't show spatial relations very well; idea model—can show relationships between abstract concepts but can be hard to understand
5. Possible response: It would be difficult to make a physical model of weather changes because there are too many factors to consider.
6. $1.7 \text{ cm} \times \frac{5 \text{ km}}{1 \text{ cm}} = 8.5 \text{ km}$

section 4

Evaluating Scientific Explanation

Believe it or not?

Look at the photo in **Figure 24.** Do you believe what you see? Do you believe everything you read or hear? Think of something that someone told you that you didn't believe. Why didn't you believe it? Chances are you looked at the facts you were given and decided that there wasn't enough proof to make you believe it. What you did was evaluate, or judge the reliability of what you heard. When you hear a statement, you ask the question "How do you know?" If you decide that what you are told is reliable, then you believe it. If it seems unreliable, then you don't believe it.

Critical Thinking When you evaluate something, you use critical thinking. **Critical thinking** means combining what you already know with the new facts that you are given to decide if you should agree with something. You can evaluate an explanation by breaking it down into two parts. First you can look at and evaluate the observations. Based upon what you know, are the observations accurate? Then you can evaluate the inferences—or conclusions made about the observations. Do the conclusions made from the observations make sense?

as you read

What You'll Learn

- **Evaluate** scientific explanations.
- **Evaluate** promotional claims.

Why It's Important

Evaluating scientific claims can help you make better decisions.

Review Vocabulary

prediction: an educated guess as to what is going to happen based on observation

New Vocabulary

- critical thinking
- data

Figure 24 In science, observations and inferences are not always agreed upon by everyone.
Compare *Do you see the same things your classmates see in this photo?*

1 Motivate

INTERACTIVE CHALKBOARD PowerPoint® Presentations

Bellringer

Section Focus Transparencies also are available on the Interactive Chalkboard CD-ROM. L2 ELL

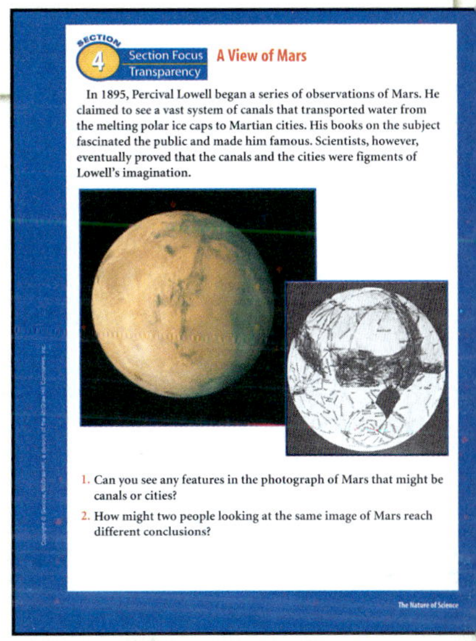
Section 4 Section Focus Transparency A View of Mars

In 1895, Percival Lowell began a series of observations of Mars. He claimed to see a vast system of canals that transported water from the melting polar ice caps to Martian cities. His books on the subject fascinated the public and made him famous. Scientists, however, eventually proved that the canals and the cities were figments of Lowell's imagination.

1. Can you see any features in the photograph of Mars that might be canals or cities?
2. How might two people looking at the same image of Mars reach different conclusions?

The Nature of Science

Tie to Prior Knowledge

Be a Smart Consumer Ask students whether they believe everything they hear. People usually don't, especially if they don't trust the source of the information. Explain that the fact that someone says something is supported by science also doesn't mean you should believe it without question. To be a smart consumer, you need to know how to evaluate the evidence. L1

Section 4 Resource Manager

Chapter *FAST FILE* Resources

Transparency Activity, p. 51
Directed Reading, pp. 23, 24
Enrichment, p. 36
Lab Worksheet, pp. 7–8, 9–10
Reinforcement, p. 32

Home and Community Involvement, p. 32
Reading and Writing Skill Activities, p. 47
Life Science Critical Thinking/Problem Solving, p. 4
Lab Management and Safety, p. 71

2 Teach

Caption Answer

Figure 24 Answers will vary.

Discussion

Evaluating Ads Have students discuss different advertisements they've heard on the radio or seen on TV. Do any of these ads use scientific evidence to back up their claims? Students may mention ads that say "Four out of five doctors prefer . . ." or "Hospitals recommend . . ." Do you trust this evidence? Answers will vary. L2

Activity

Percentages Ask students to express the data given in Table 2 as percentages. What percent of the people surveyed preferred pizza? $37 \div (37 + 28) \times 100\% = 57\%$ What percent preferred hamburgers? $28 \div (37 + 28) \times 100\% = 43\%$ How many people in this survey would make up 50% of the people surveyed? $x \div (37 + 28) \times 100\% = 50\%$; $x = 50\% \times (37 + 28) \div 100\% = 32.5$ Point out that the statement "More people surveyed preferred pizza than preferred hamburgers" would be true if anywhere from 33 to 65 people in the survey preferred pizza. L2 LS **Logical-Mathematical**

Caption Answer

Figure 25 It will increase the reliability of their data, because they will be less likely to forget something important.

Table 2 Favorite Foods

People's Preference	Tally	Frequency
Pepperoni pizza	𝍸 𝍸 𝍸 𝍸 𝍸 𝍸 𝍸 ||	37
Hamburgers with ketchup	𝍸 𝍸 𝍸 𝍸 𝍸 |||	28

Evaluating the Data

A scientific investigation always contains observations—often called **data.** Data are gathered during a scientific investigation and can be recorded in the form of descriptions, tables, graphs, or drawings. When evaluating a scientific claim, you might first look to see whether any data are given. You should be cautious about believing any claim that is not supported by data.

Are the data specific? The data given to back up a claim should be specific. That means they need to be exact. What if your friend tells you that many people like pizza more than they like hamburgers? What else do you need to know before you agree with your friend? You might want to hear about a specific number of people rather than unspecific words like *many* and *more.* You might want to know how many people like pizza more than hamburgers. How many people were asked about which kind of food they liked more? When you are given specific data, a statement is more reliable and you are more likely to believe it. An example of data in the form of a frequency table is shown in **Table 2.** A frequency table shows how many times types of data occur. Scientists must back up their scientific statements with specific data.

Take Good Notes Scientists must take thorough notes at the time of an investigation, as the scientists shown in **Figure 25** are doing. Important details can be forgotten if you wait several hours or days before you write down your observations. It is also important for you to write down every observation, including ones that you don't expect. Often, great discoveries are made when something unexpected happens in an investigation.

Figure 25 These scientists are writing down their observations during their investigation rather than waiting until they are back on land.

Draw Conclusions *Do you think this will increase or decrease the reliability of their data?*

Visual Learning

Figure 24 Many times it's no accident that pictures of supposed UFOs appear out of focus. Why might this make the claim more believable? Because the object is blurry, you can't say for sure that it's not a UFO. How is this similar to using words like many and most? The words *many* and *most* sometimes blur ideas like a camera's lens can blur an object. L2 LS **Visual-Spatial**

Differentiated Instruction

English-Language Learners Have students observe something outside for 5 min. A bird, students on the playground, traffic in front of the schoolyard. Ask them to write down their observations emphasizing that their notes should be specific. Pair students up and have them compare their notes and compose a paragraph describing what they observed.

Your Science Journal During this course, you will be keeping a science journal. You will write down what you do and see during your investigations. Your observations should be detailed enough that another person could read what you wrote and repeat the investigation exactly as you performed it. Instead of writing "the stuff changed color," you might say "the clear liquid turned to bright red when I added a drop of food coloring." Detailed observations written down during an investigation are more reliable than sketchy observations written from memory. Practice your observation skills by describing what you see in **Figure 26.**

Can the data be repeated? If your friend told you he could hit a baseball 100 m, but couldn't do it when you were around, you probably wouldn't believe him. Scientists also require repeatable evidence. When a scientist describes an investigation, as shown in **Figure 27,** other scientists should be able to do the investigation and get the same results. The results must be repeatable. When evaluating scientific data, look to see whether other scientists have repeated the data. If not, the data might not be reliable.

Evaluating the Conclusions

When you think about a conclusion that someone has made, you can ask yourself two questions. First, does the conclusion make sense? Second, are there any other possible explanations? Suppose you hear on the radio that your school will be running on a two-hour delay in the morning because of snow. You look outside. The roads are clear of snow. Does the conclusion that snow is the cause for the delay make sense? What else could cause the delay? Maybe it is too foggy or icy for the buses to run. Maybe there is a problem with the school building. The original conclusion is not reliable unless the other possible explanations are proven unlikely.

Figure 26 Detailed observations are important in order to get reliable data.

Observe *Use ten descriptive words to describe what you see happening in this photo.*

Figure 27 Working together is an important part of science. Several scientists must repeat an experiment and obtain the same results before data are considered reliable.

Caption Answer

Figure 26 Answers will vary. Examples include: A spiral-shaped, copper-colored metal is in a blue solution. Sparkly silver stuff is also in the blue solution. It looks like the sparkly stuff is coming from the metal.

Discussion

Multiple Trials Tell students that, when doing experiments, scientists often run multiple trials of the same test and average the results. How does this affect the way you evaluate the scientists' data? Because the results are based on the averages of repeated trials, they are more likely to be repeatable and reliable. L2 LS Logical-Mathematical

Quick Demo

Packaging Claims

Materials cereal boxes, food labels, product wrapping

Estimated Time 10 minutes

Procedure Share the products with the students and have them write down five claims made on the packaging. These can include ingredients, uses for the product, age requirements, allergic concerns, photos, etc. Which claims do they believe and which do they not believe? Why? L2

Differentiated Instruction

Challenge Have students research the consequences companies face when they advertise false claims. How do corporations try and cover their mistake when it becomes public knowledge? What sort of things would a corporation need to do to make the public believe their advertising after being identified as having made false claims? L3

Teacher FYI

False Advertising Most of the time, advertisers do not intend to defraud the consumer, but some advertisers go too far. If you feel you've been the victim of false advertising, you can contact the Federal Trade Commission in Washington, D.C. In one year, the FTC may log more than 300,000 complaints from consumers who didn't get what had been promised to them in ads. Make sure students understand that they also need to be cautious of advertising they see on the Internet.

Caption Answer

Figure 28 Answers will vary.

3 Assess

DAILY INTERVENTION

Check for Understanding

Linguistic Have students read product reviews for several specific products from a noncommercial consumer magazine. Then tell them to compare the reviews to advertisements for the products. Ask them to write short essays in their Science Journals discussing the accuracy of the ads. L2 LS

Reteach

Reliable Data Read to students a list of ten items such as the names of cities. Ask students to recite these back to you in the order you gave them. How does this demonstrate the importance of writing down data? Possible response: It's hard to remember many specific things, even when you've just heard them. L2 LS **Auditory-Musical**

Assessment

Performance Have each student write an advertisement for a household product. The ad should be upbeat and complimentary, but not make any false claims. Use **Performance Assessment in the Science Classroom**, p. 157. L2 P

Fun Fact

Puffery is exaggerated advertising or boasting. It includes subjective claims about products that cannot be proven true or false. Claims of "thorough research" or "countless hours of research" are merely puffery and are not subject to legal action.

Figure 28 All material should be read with an analytical mind. **Explain** *what this advertisement means.*

Evaluating Promotional Materials

Scientific processes are not used only in the laboratory. Suppose you saw an advertisement in the newspaper like the one in **Figure 28.** What would you think? First, you might ask, "Does this make sense?" It seems unbelievable. You would probably want to hear some of the scientific data supporting the claim before you would believe it. How was this claim tested? How is the amount of wrinkling in skin measured? You might also want to know if an independent laboratory repeated the results. An independent laboratory is one that is not related in any way to the company that is selling the product or service. It has nothing to gain from the sales of the product. Results from an independent laboratory usually are more reliable than results from a laboratory paid by the selling company. Advertising materials are designed to get you to buy a product or service. It is important that you carefully evaluate advertising claims and the data that support them before making a quick decision to spend your money.

section 4 review

Summary

Believe it or not?

- By combining what you already know with new information as it becomes available, you can decide whether something is fact or fiction.
- Explanations should be evaluated by looking at both the observations and the conclusions the explanation is based on.

Evaluating the Data

- It is important to take thorough notes during any investigation.

Evaluating the Conclusions

- In order for a conclusion to be reliable, it must make sense.

Evaluating Promotional Materials

- Independent laboratories test products in order to provide more reliable results.

Self Check

1. **Describe** why it is important that scientific experiments be repeated.
2. **List** what types of scientific claims should be verified.
3. **Explain** how vague claims in advertising can be misleading.
4. **Think Critically** An advertisement on a food package claims it contains Glistain, a safe taste enhancer. Make a list of ten questions you would ask when evaluating this claim.

Applying Skills

5. **Classify** Watch three television commercials and read three magazine advertisements. Record the claims that each advertisement made. Classify each claim as being vague, misleading, reliable, and/or scientific.

Science Online red.msscience.com/self_check_quiz

section 4 review

1. In order for scientific evidence to be reliable, experiments need to be repeatable by other scientists.
2. All scientific claims should be verified by some kind of specific data.
3. Vague claims are misleading because they contain terms that can have more than one meaning.
4. Possible questions include: What is Glistain? Has Glistain been tested? What were the results? Has it been tested on humans? Over how long a period of time was it tested? Was it tested by an independent lab?
5. Answers will vary. Make sure students select ads from each medium.

LAB

What is the right answer?

Scientists sometimes develop more than one explanation for observations. Can more than one explanation be correct? Do scientific explanations depend on judgment?

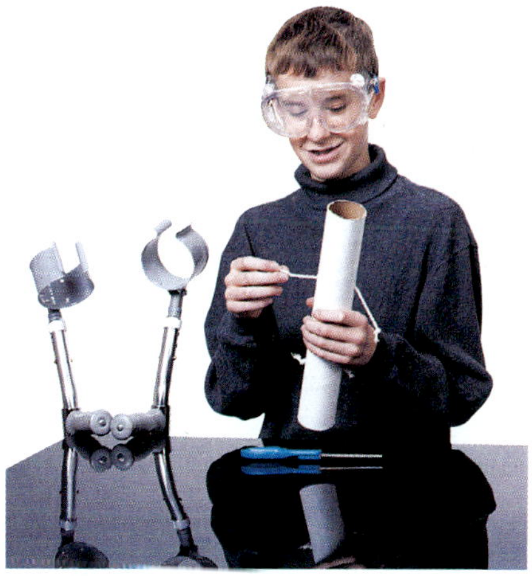

Real-World Question

Can more than one explanation apply to the same observation?

Goals

- **Make a hypothesis** to explain an observation.
- **Construct** a model to support your hypothesis.
- **Refine** your model based on testing.

Materials

cardboard mailing tubes
*empty shoe boxes
length of rope
scissors

*Alternate materials

Safety Precautions

WARNING: *Be careful when punching holes with sharp tools.*

Procedure

1. You will be shown a cardboard tube with four ropes coming out of it, one longer than the others. Your teacher will show you that when any of the three short ropes—A, C, or D—are pulled, the longer rope, B, gets shorter. Pulling on rope B returns the other ropes to their original lengths.
2. Make a hypothesis as to how the teacher's model works.
3. **Sketch** a model of a tube with ropes based on your hypothesis. Using a cardboard tube and two lengths of rope, build a model according to your design. Test your model by pulling each of the ropes. If it does not perform as planned, modify your hypothesis and your model to make it work like your teacher's model.

Conclude and Apply

1. **Compare** your model with those made by others in your class.
2. Can more than one design give the same result? Can more than one explanation apply to the same observation? Explain.
3. Without opening the tube, can you tell which model is exactly like your teacher's?

Communicating Your Data

Make a display of your working model. Include sketches of your designs. **For more help, refer to the** Science Skill Handbook.

BENCH TESTED

Real-World Question

Purpose Students show that multiple explanations sometimes can apply to the same observations. L2 LS **Kinesthetic**

Process Skills Form a hypothesis, design an experiment to test a hypothesis, control variables, observe, infer, predict, recognize cause and effect, communicate

Time Required 45 minutes

Procedure

Alternate Materials Empty shoe boxes with lids can be used in place of the tube allowing easy access to the ropes inside.

Teaching Strategy Demonstrate a way in which one rope may affect another rope.

Conclude and Apply

1. Have students note differences and similarities in their designs and demonstrate that all models work.
2. Yes; yes; the ropes may be configured differently but perform the same.
3. No

Communicating Your Data

Suggest that students draw different designs and show how they can act similarly. If students have access to a computer animation program, encourage them to make animated drawings of the inside of the tube. L3

Assessment

Content Students can show their rope trick to students in lower grades. Encourage students to present the trick, and then clearly explain how it works. Use **Performance Assessment in the Science Classroom,** p. 143. L2

BENCH TESTED

Real-World Question

Purpose Students identify the steps of a scientific investigation. L2

Process Skills recognize cause and effect, observe, infer, make and use graphs

Time Required 45 minutes

Procedure

Teaching Strategies Write the steps of the scientific method on the board for students to use as a reference as they complete this lab. Students should refer to the Science Skill Handbook.

Analyze Your Data

Expected Outcome Students should see that fertilizer B gave the best results. They should also see that, after starting out more slowly, fertilizer A did produce about the same results from week to week as did fertilizer B.

LAB

Identifying Parts of an Investigation

Goals

- **Identify** parts of an experiment.
- **Identify** constants, variables, and controls in the experiment.
- **Graph** the results of the experiment and draw appropriate conclusions.

Materials

description of fertilizer experiment

Real-World Question

Science investigations contain many parts. How can you identify the various parts of an investigation? In addition to variables and constants, many experiments contain a control. A control is one test, or trial, where everything is held constant. A scientist compares the control trial to the other trials. What are the various parts of an experiment to test which fertilizer helps a plant grow best?

Procedure

1. **Read** the description of the fertilizer experiment.
2. **List** factors that remained constant in the experiment.
3. **Identify** any variables in the experiment.
4. **Identify** the control in the experiment.
5. **Identify** one possible hypothesis that the gardener could have tested in her investigation.
6. **Describe** how the gardener went about testing her hypothesis using different types of fertilizers.
7. **Graph** the data that the gardener collected in a line graph.

A gardener was interested in helping her plants grow faster. When she went to the nursery, she found three fertilizers available for her plants. One of those fertilizers, fertilizer A, was recommended to her. However, she decided to conduct a test to determine which of the three fertilizers, if any, helped her plants grow fastest. The gardener planted four seeds, each in a separate pot. She used the same type of pot and the same type of soil in each pot. She fertilized one seed

Alternative Inquiry Lab

Real-World Connection To make this Lab an Inquiry Lab, tell students that they are landscape designers who have been hired by the community center to landscape the entrance. The community center wants to use plants that are native to the area and that require relatively little care and maintenance. The students will need to investigate the growing and watering requirements of the plants. What plant suggestions could the students provide? What type of experiment could they design to verify the information they provide to the community center? Encourage students to research how landscape designers determine what plants to use in landscapes and what the variables and constants are in landscape environments. L2 PBL

with fertilizer A, one with fertilizer B, and one with fertilizer C. She did not fertilize the fourth seed. She placed the four pots near one another in her garden. She made sure to give each plant the same amount of water each day. She measured the height of the plants each week and recorded her data. After eight weeks of careful observation and record keeping, she had the following table of data.

Plant Height (cm)

Week	Fertilizer A	Fertilizer B	Fertilizer C	No Fertilizer
1	0	0	0	0
2	2	4	1	1
3	5	8	5	4
4	9	13	8	7
5	14	18	12	10
6	20	24	15	13
7	27	31	19	16
8	35	39	22	20

Analyze Your Data

1. **Describe** the results indicated by your graph. What part of an investigation have you just done?
2. **Infer** Based on the results in the table and your graph, which fertilizer do you think the gardener should use if she wants her plants to grow the fastest? What part of an investigation have you just done?
3. **Define** Suppose the gardener told a friend who also grows these plants about her results. What is this an example of?

Conclude and Apply

1. **Interpret Data** Suppose fertilizer B is much more expensive than fertilizers A and C. Would this affect which fertilizer you think the gardener should buy? Why or why not?
2. **Explain** Does every researcher need the same hypothesis for an experiment? What is a second possible hypothesis for this experiment (different from the one you wrote in step 5 in the Procedure section)?
3. **Explain** if the gardener conducted an adequate test of her hypothesis.

Communicating Your Data

Compare your conclusions with those of other students in your class. **For more help, refer to the** Science Skill Handbook.

Answers to Questions

1. The graph shows that the plant grew the most with fertilizer B, that it grew only slightly better with fertilizer C than with no fertilizer, and that with fertilizer A the plant started slowly, but then grew as fast as the plant with fertilizer B. This step is analyzing data.
2. She'd use fertilizer B. This is hypothesizing, or making a guess based on data.
3. This is communicating results.

Error Analysis Other factors might have affected the rate of plant growth. There may have been some problem with the soil in fertilizer C's pot. Maybe the seed in fertilizer A just didn't grow as much.

Conclude and Apply

1. Probably; if the gardener cannot afford the extra money she should buy fertilizer A because it helps the plant almost as much as fertilizer B.
2. No, but the procedures must be the same or you cannot compare the results. One possible hypothesis is that fertilizers stop having an effect after eight weeks.
3. She worked hard to control variables and she did her experiment over a good period of time. To confirm her results, she could try the experiment more times and for longer periods of time.

Assessment

Performance Have students act out the gardener going to buy the fertilizer, the store clerk's comments to her, and her comments back to the store clerk after her investigation. Make sure each character argues using scientific evidence. Use **Performance Assessment in the Science Classroom**, p. 147. L2

Communicating Your Data

Students can use a computer graphing program to make their graphs look more professional. The programs also may allow them to predict seed growth in upcoming weeks. Using these predictions, students can argue for or against a specific fertilizer. L2

TIME

Content Background

All three women featured here made sacrifices to work in their chosen fields. Maria Goeppert Mayer often had to work "under" or "for" men in order to make her famous discovery. After earning her doctorate in 1930, Mayer worked as a lecturer and on secondary projects that helped back up the findings of men. In 1960, the University of Chicago offered her a full professorship and Mayer was finally considered for her talents as a physicist.

Politics and geography inhibited Montalcini's work. A 1936 Manifesto forbid non-Aryan Italian citizens from holding academic and professional jobs. In order to continue her work, Montalcini built a research unit in her bedroom.

Discussion

Overcoming Obstacles Each of these women had to overcome many obstacles in pursuing their careers. Discuss some obstacles that students and their families have had to overcome. Answers will vary. May include moving from a different country, new school, and special needs requirements. L2

Historical Significance

Focusing on women in science is one way of turning attention to important scientific discoveries made by women. It is also a way to inform students that women have not always been considered capable of such rigorous pursuits. Engage students in a discussion about the ways women have been excluded from many disciplines throughout history. As an example, tell them that the first woman senator, Hattie Wyatt Caraway, was not elected into office until 1932. In 2001, only 13 of the 100 U.S. Senators were women. L2

TIME SCIENCE AND HISTORY — SCIENCE CAN CHANGE THE COURSE OF HISTORY!

Women in Science

Is your family doctor a man or a woman? To your great-grandparents, such a question would likely have seemed odd. Why? Because 100 years ago, women weren't encouraged to study science as they are today. But that does not mean that there were no female scientists back in your great-grandparents' day. Many women managed to overcome great barriers and made discoveries that changed the world.

Maria Goeppert Mayer

"To my surprise, winning the prize wasn't half as exciting as doing the work itself. That was the fun—seeing it work out." Dr. Maria Goeppert Mayer won the Nobel Prize in Physics in 1963 for her work on the structure of an atom. Her model greatly increased human understanding of atoms, which make up all forms of matter.

Rita Levi-Montalcini

In 1986, Dr. Rita Levi-Montalcini was awarded the Nobel Prize in Medicine for her discovery of growth factors. Growth factors regulate the growth of cells and organs in the body. Because of her work, doctors are better able to understand why tumors form and wounds heal.

Rosalyn Sussman Yalow

"The world cannot afford the loss of the talents of half its people if we are to solve the many problems which beset us," Dr. Rosalyn Sussman Yalow said upon winning the Nobel Prize in Medicine in 1977 for discovering a way to measure tiny substances in the blood, such as hormones and drugs.

Her discovery made it possible for doctors to diagnose problems that they could not detect before.

Research Visit the link to the right to research some recent female Nobel prizewinners in physics, chemistry, and medicine. Write a short biography about their lives. How did their discoveries impact their scientific fields or people in general?

Science Online — For more information, visit red.msscience.com/time

Research Answers will vary. Marie Curie is best known for her discovery of the radioactive elements polonium and radium. Rosalind Franklin was the first to recognize the helix shape of DNA.

Resources for Teachers and Students

Nobel Prize Women in Science: Their Lives, Struggles, and Momentous Discoveries, by Sharon Bertsch McGrayne, Kensington Pub. Corp., 2001

Tech camps inspire girls but may not narrow growing gender gap, by the Associated Press, USA Today, July 2003

Nobel Prize Winners (Women in Profile Service), by Carlotta Hacker. Crabtree Publishing, 1998

Reviewing Main Ideas

Section 1 What is science?

1. Science is a way of learning more about the natural world. It can provide possible explanations for why and how things happen.
2. Systems are made up of structures, cycles, and processes that interact with one another.

Section 2 Science in Action

1. A hypothesis is a possible explanation based on what you know and what you observe.
2. It is important to always follow laboratory safety symbols and to wear and use appropriate gear during an experiment.

Section 3 Models in Science

1. Models are a graphic representation of an object or an event used to communicate ideas; test predictions; and save time, money, and lives.

Section 4 Evaluating Scientific Explanation

1. Reliable data are specific and repeatable by other scientists.
2. In order for a conclusion to be considered reliable, it must make sense and be the most likely explanation.

Visualizing Main Ideas

Copy and complete the following concept map.

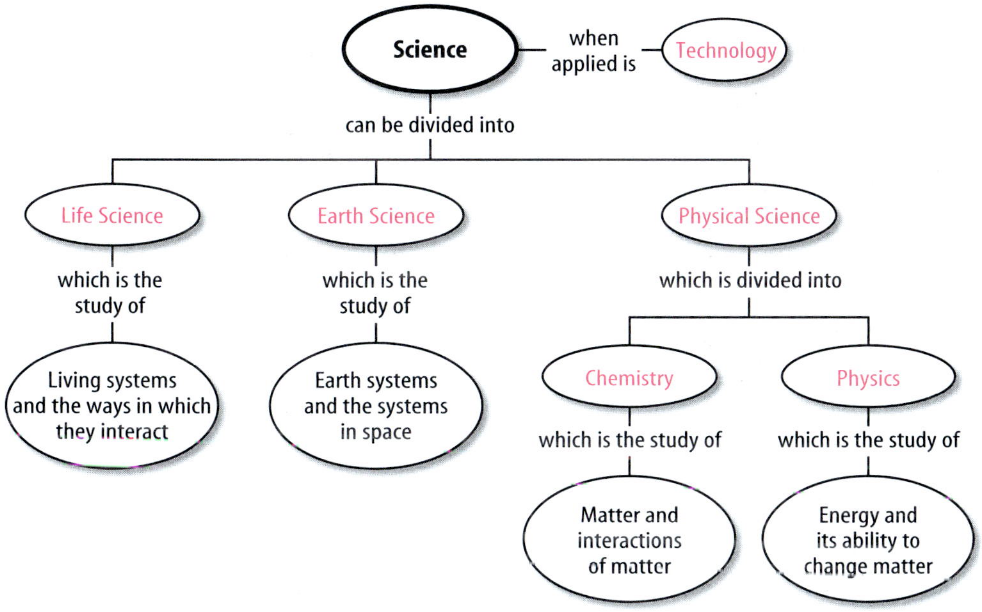

Reviewing Main Ideas

Summary statements can be used by students to review the major concepts of the chapter.

Visualizing Main Ideas

See student page.

Science Online

Visit red.msscience.com
/self_check_quiz
/interactive_tutor
/vocabulary_puzzlemaker
/chapter_review
/standardized_test

Assessment Transparency

For additional assessment questions, use the *Assessment Transparency* located in the transparency book.

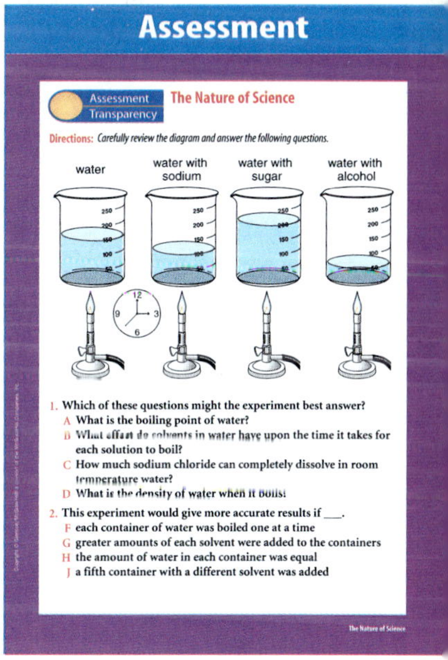
Assessment

Assessment Transparency **The Nature of Science**

Directions: *Carefully review the diagram and answer the following questions.*

1. Which of these questions might the experiment best answer?
   A What is the boiling point of water?
   B What effect do solvents in water have upon the time it takes for each solution to boil?
   C How much sodium chloride can completely dissolve in room temperature water?
   D What is the density of water when it boils?
2. This experiment would give more accurate results if ___.
   F each container of water was boiled one at a time
   G greater amounts of each solvent were added to the containers
   H the amount of water in each container was equal
   J a fifth container with a different solvent was added

The Nature of Science

Using Vocabulary

1. A hypothesis is a guess scientists make based upon what they observe. A scientific theory is an attempt to explain a pattern seen repeatedly in the natural world.
2. Variables are the parts of an experiment that change. A constant is the one variable in an experiment that does not change.
3. Science is a way of learning about the world. Technology is the use of this knowledge.
4. Science is a way of learning about the world. A system is a collection of structures, cycles, or processes in the world that relate to one another.
5. Earth science is the study of Earth systems and the systems in space. Physical science is the study of matter and energy and their relationships.
6. Critical thinking is used to evaluate evidence. To infer is to draw a conclusion based on that evidence.
7. A scientific law is a rule that describes some pattern seen in nature. An observation can be anything you notice about the natural world.
8. A model is any representation of an object or event. A system is a collection of processes, cycles, or structures in the world.
9. A controlled experiment is an experiment in which everything but one part stays the same. Variables are the things in a controlled experiment that can change.
10. A scientific theory is an explanation of why something happens in the world. A scientific law is a statement of some regularity about something seen in the world.

Checking Concepts

11. B	14. C	17. A
12. C	15. A	18. C
13. C	16. C	19. A

chapter 1 Review

Using Vocabulary

constant p. 18	life science p. 9
controlled experiment p. 18	model p. 21
critical thinking p. 27	physical science p. 10
data p. 28	science p. 6
dependent variable p. 18	scientific law p. 7
Earth science p. 10	scientific theory p. 7
hypothesis p. 14	system p. 8
independent variable p. 18	technology p. 11
infer p. 16	variable p. 18

Explain the relationship between the words in the following sets.

1. hypothesis—scientific theory
2. constant—variable
3. science—technology
4. science—system
5. Earth science—physical science
6. critical thinking—infer
7. scientific law—observation
8. model—system
9. controlled experiment—variable
10. scientific theory—scientific law

Checking Concepts

Choose the word or phrase that best answers the question.

11. What does it mean to make an inference?
    A) make observations
    B) draw a conclusion
    C) replace
    D) test
12. Which of the following CANNOT protect you from splashing acid?
    A) goggles **C)** fire extinguisher
    B) apron **D)** gloves
13. If the results from your investigation do not support your hypothesis, what should you do?
    A) Should not do anything.
    B) Repeat the investigation until it agrees with the hypothesis.
    C) Modify your hypothesis.
    D) Change your data to fit your hypothesis.
14. Which of the following is NOT an example of a scientific hypothesis?
    A) Earthquakes happen because of stresses along continental plates.
    B) Some animals can detect ultrasound frequencies caused by earthquakes.
    C) Paintings are prettier than sculptures.
    D) Lava takes different forms depending on how it cools.
15. Using a computer to make a three-dimensional picture of a building is a type of which of the following?
    A) model **C)** constant
    B) hypothesis **D)** variable
16. Which of the following increases the reliability of a scientific explanation?
    A) vague statements
    B) notes taken after an investigation
    C) repeatable data
    D) several likely explanations
17. Which is an example of technology?
    A) a squirt bottle **C)** a cat
    B) a poem **D)** physical science
18. What explains something that takes place in the natural world?
    A) scientific law **C)** scientific theory
    B) technology **D)** experiments
19. An airplane model is an example of what type of model?
    A) physical **C)** idea
    B) computer **D)** mental

Science Online red.msscience.com/vocabulary_puzzlemaker

Use the Exam*View*® Pro Testmaker CD-ROM to:

- create multiple versions of tests
- create modified tests with one mouse click for inclusion students
- edit existing questions and add your own questions
- build tests aligned with state standards using built-in State Curriculum Tags
- change English tests to Spanish with one mouse click and vice versa

Thinking Critically

20. **Draw Conclusions** When scientists study how well new medicines work, one group of patients receives the medicine while a second group does not. Why?

21. **Predict** How is using a rock hammer an example of technology?

22. **Compare and Contrast** How are scientific theories and scientific laws similar? How are they different?

Use the table below to answer question 23.

Hardness

Object	Mohs Scale
copper	3.5
diamond	10
fingernail	2.5
glass	5.5
quartz	7
steel file	6.5

23. **Use Tables** Mohs hardness scale measures how easily an object can be scratched. The higher the number, the harder the material is. Use the table above to identify which material is the hardest and which is the softest.

24. **Make Operational Definitions** How does a scientific law differ from a state law? Give some examples of both types of laws.

25. **Infer** Why it is important to record and measure data accurately during an experiment?

26. **Predict** the quickest way to get to school in the morning. List some ways you could test your prediction.

Performance Activities

27. **Hypothesize** Using a basketball and a tennis ball, make a hypothesis about the number of times each ball will bounce when it hits the ground. Drop each ball from shoulder height five times, recording the number of bounces in a table. Which ball bounced more? Make a hypothesis to explain why.

28. **Observe** Pour some water in a small dish and sprinkle some pepper on top. Notice how the pepper floats on the water. Now add a few drops of liquid soap to the water. Write down your observations as you watch what happens to the pepper.

Applying Math

Use the illustration below to answer question 29.

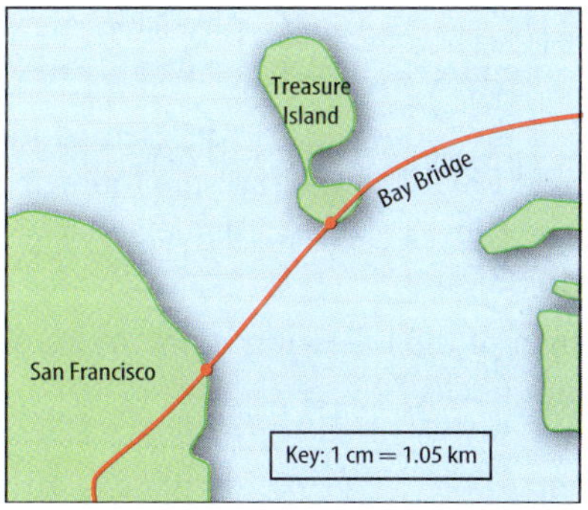

29. **Use Proportions** The map above shows the distance between two points. The scale shows that 1 cm is approximately 1.05 km. What is the approximate distance between Point A and Point B?

chapter Review 1

Thinking Critically

20. By not giving the medicine to one group of patients, scientists can see the effects of that medicine on the other group.
21. The study of levers and the study of materials are applied to create a rock hammer that uses a material that will break rocks and a shape that acts as a lever.
22. Both scientific theories and scientific laws are generalizations. A scientific theory explains why something happens, but a scientific law only says that something happens.
23. Diamond is the hardest and a fingernail is the softest.
24. A state law tells you what you are forbidden to do. A scientific law just tells you that things happen. A state law could forbid a driver to turn right at a red light. A scientific law would tell the driver that the force of gravity keeps the car on the road.
25. You need to record data accurately so you can be sure of the results of your experiment when you go back later to draw your conclusions.
26. You could test the hypothesis by timing your method and comparing it to the times it takes to get to school using other methods.

Performance Activities

27. The tennis ball bounced more times than the basketball. The surface area of the tennis ball is less than the basketball and therefore retains more energy while it bounces.
28. The pepper spreads to the outer edges of the dish when soap is added to the water.

Applying Math

National Math Standards

1, 4

29. The distance between the two points is 2.8 cm. The distance is $2.8\text{ cm} \cdot \frac{1.05\text{ km}}{1\text{ cm}} = 2.94\text{ km}$.

Assessment Resources

Reproducible Masters

Chapter *Fast File* Resources
Chapter Review, pp. 41–42
Chapter Tests, pp. 43–46
Assessment Transparency Activity, p. 55

Glencoe Science Web site
Chapter Review Test
Standardized Test Practice

Glencoe Technology

Assessment Transparency
Exam*View*® Pro Testmaker
MindJogger Videoquiz
Interactive Chalkboard

FAST FILE

Answer Sheet A practice answer sheet can be found at red.msscience.com/answer_sheet.

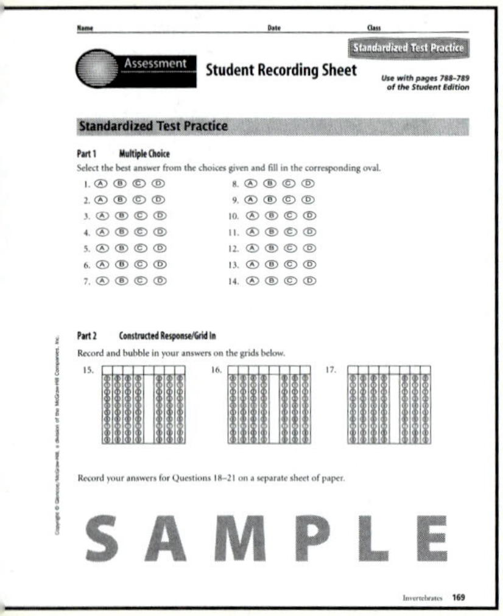
Name Date Class

Assessment Student Recording Sheet

Standardized Test Practice

Use with pages 788–789 of the Student Edition

Standardized Test Practice

Part 1 Multiple Choice

Select the best answer from the choices given and fill in the corresponding oval.

Part 2 Constructed Response/Grid In

Record and bubble in your answers on the grids below.

Record your answers for Questions 18–21 on a separate sheet of paper.

SAMPLE

Part 1 Multiple Choice

1. B	4. A	7. C
2. C	5. D	8. B
3. D	6. D	9. A

Part 2 Short Response

10. Answers will vary.
11. goggles, apron, thermal mitts, tie back hair, caution should be used around the sharp objects
12. Science answers questions with the information available at the time. As new information becomes available, explanations for why things happen can change and be modified.
13. Scientists share information with other scientists because experiments must be repeatable and the results must be the same in order for the data to be reliable.
14. An observation is when you watch something and make notes of its movements or contents. An inference means you draw a conclusion based on what you have observed.
15. Science studies why or how things happen in the natural world. Technology applies that information to everyday life.
16. Scientists follow the steps of the scientific method. In general, they observe, question, and collect information, then hypothesize, followed by investigations to learn more, then analyze their information, and finally conclude and communicate.
17. The three branches of science are life science, Earth science, and physical science. Life science asks questions like: "How do whales navigate the ocean?" and "How do vaccines prevent disease?" Earth science asks questions like "How can an earthquake be detected?" or "Is water found on other planets?" Physical science asks questions like: "How can plastic be made stronger?" or "How can humans harness the energy of sunlight for their energy needs?"
18. Models allow scientists to visualize something that is difficult to see or understand.

Part 1 Multiple Choice

Record your answers on the answer sheet provided by your teacher or on a sheet of paper.

1. What is a rule describing a pattern in nature called?
   A. possible explanation
   B. scientific law
   C. scientific theory
   D. technology

Use the illustration below to answer questions 2–3.

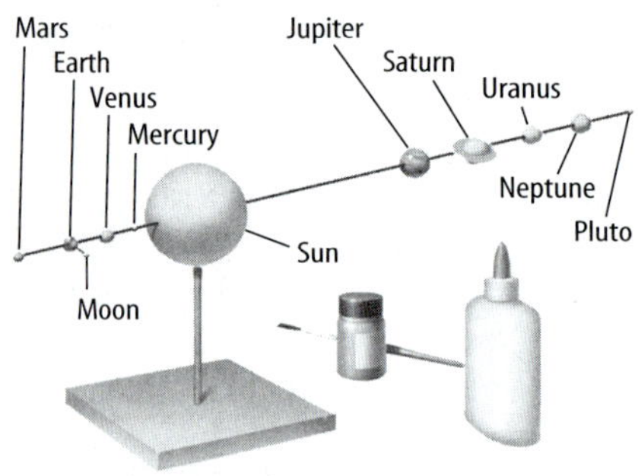

2. The model of the solar system best represents which kind of scientific model?
   A. idea **C.** physical
   B. computer **D.** realistic
3. All of the following are represented in the model EXCEPT which of the following?
   A. the sun **C.** planets
   B. the moon **D.** stars
4. Which of the following is not an example of a model?
   A. CD **C.** recipe
   B. map **D.** drawing

Test-Taking Tip

Practice Practice Remember that test taking skills can improve with practice. If possible, take at least one practice test and familiarize yourself with the test format and instructions.

5. Which of the following questions can science NOT answer?
   A. Why do the leaves on trees change colors in the fall?
   B. Why do bears hibernate in the winter?
   C. Where do waves in the ocean form?
   D. What is the most popular book?
6. What is it called when you combine what you already know with new facts?
   A. estimate **C.** inference
   B. hypothesis **D.** critical thinking
7. What are the variables that do not change in an experiment called?
   A. independent variables
   B. dependent variables
   C. constants
   D. inferences
8. An educated guess based on what you know and what you observe is called which of the following?
   A. prediction. **C.** conclusion.
   B. hypothesis. **D.** data.

Use the photo below to answer question 9.

9. What type of scientist could the person above be classified as?
   A. life scientist
   B. physical scientist
   C. Earth scientist
   D. medical doctor

Part 2 Short Response/Grid In

Record your answers on the answer sheet provided by your teacher or on a sheet of paper.

Use the photo below to answer questions 10 and 11.

10. Look at the photo above and write down your immediate observations.

11. What safety precautions might this student want to take?

12. Explain why science can only provide possible explanations for occurrences in nature.

13. In class it is sometimes not common for students to share their answers on a test. Why is it important for scientists to share information?

14. Compare and contrast observation and inference.

15. Explain the relationship between science and technology.

16. What steps do scientists follow when investigating a problem?

17. List the three branches of science and give examples of questions that they ask.

18. What is the importance of scientific models?

Part 3 Open Ended

Record your answers on a sheet of paper.

19. You want to know whether plants grow faster if there is music playing in their environment. How would you conduct this experiment? Be sure to identify the independent and dependent variables, and the constants.

20. Many outdoor clothing products are coated in a special waterproofing agent to protect the material from rain and snow. The manufacturers of the waterproofing agent hire independent field-testers to use their product in the field before marketing it to the public. Why would you want to know the results of the field-testers tests?

Use the illustrations below to answer questions 21–23.

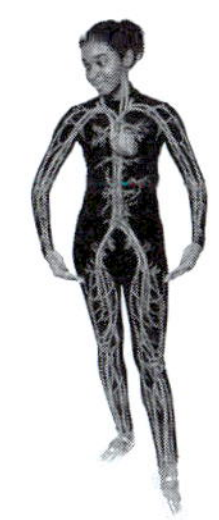

21. What are the above drawings outlining?

22. Body systems interact with one another in order to function. What would happen if one system failed?

23. What is the importance of systems in science?

24. Make a frequency table from the following data. Make two observations about the data. 15 students prefer cold pizza for lunch; 10 students enjoy peanut butter with jelly; 3 students bring ham and cheese; and 5 students eat hot dogs and chips.

Part 3 Open Ended

19. Experiment setup: materials: two plants, radio, watering can

The plants should be placed in the exact same environment. Light exposure needs to be the same for both plants. They need to be given the same amount of water at the same interval. The radio is placed by one of the plants but not anywhere near the other plant. Music should be played softly at all times.

Independent variable: radio

Dependent variable: growth of the plants

Constants: amount of water provided to each plant and timing of water, lighting exposure

20. The company is trying to sell a product and could enhance the results of their tests in order to make it sell. Independent testers are not paid by the company and would not get in trouble or lose their job for being honest when testing the product. If the creators of Gore-Tex say their product is waterproof and then sell it to the public and it gets wet, they will go out of business. However, if independent testers take the product out into the field and find that the material does not get wet, the company can stand behind its claim that the material is indeed waterproof.

21. body systems

22. The body systems all interact with one another and if one was to fail or stop working, the other systems would suffer. Some systems would work harder to compensate for the failed system while others would simply stop working.

23. Scientists are usually interested in problems within a system. They must break down the parts of the system to see how one part affects another.

24. The frequency table should be two columns listing food preference in one column and frequency in the other. Possible observations include: more students prefer cold pizza than other foods. Ham and cheese is the least preferred food. Twice as many students prefer peanut butter and jelly than hot dogs and chips.

Rubrics

For more help evaluating open-ended assessment questions, see the rubric on p. 10T.

chapter 2 Organizer

Section/Objectives	Standards		Labs/Features
	National	State/Local	
Chapter Opener	See pp. 16T–17T for a Key to Standards.		**Launch Lab:** Measuring Accurately, p. 41 **Foldables,** p. 41 A data-collection lab using Probeware technology can be found in the **Probeware Lab Manual,** pp. 41–43
Section 1 Description and Measurement 4 sessions 2 blocks 1. **Determine** how reasonable a measurement is by estimating. 2. **Identify** and use the rules for rounding a number. 3. **Distinguish** between precision and accuracy in measurements.	National Content Standards: UCP.1, UCP.2, UCP.3, A.1, A.2		**Integrate Career,** p. 43 **MiniLAB:** Measuring Accuracy, p. 44 **Visualizing Precision and Accuracy,** p. 46 **Science Online,** p. 47 **Applying Math:** Rounding, p. 48
Section 2 SI Units 4 sessions 2 blocks 4. **Identify** the purpose of SI. 5. **Identify** the SI units of length, volume, mass, temperature, time, and rate.	National Content Standards: UCP.1, UCP.2, UCP.3, A.1, A.2		**Integrate Astronomy,** p. 51 **MiniLAB:** Measuring Volume, p. 52 **Lab:** Scale Drawing, p. 55
Section 3 Drawings, Tables, and Graphs 5 sessions 2.5 blocks 6. **Describe** how to use pictures and tables to give information. 7. **Identify** and use three types of graphs. 8. **Distinguish** the correct use of each type of graph.	National Content Standards: UCP.1, UCP.2, UCP.3, A.1, A.2		**Science Online,** p. 58 **Lab:** Pace Yourself, pp. 60–61 **Science Stats:** Biggest, Tallest, Loudest, p. 62

Glencoe Exclusive!
TeacherWorks™
All-In-One Planner and Resource Center

Lab Materials	Reproducible Resources	Section Assessment	Technology
Launch Lab: none	**Chapter *Fast File* Resources** Foldables Worksheet, p. 13 Directed Reading Overview, p. 15 Note-taking Worksheets, pp. 29–31		TeacherWorks includes: • Interactive Teacher Edition • Lesson Planner with calendar • Access to all program blacklines • Correlations to standards • Web links
MiniLAB: 400-mL beaker, crushed ice, water, computer temperature probe, alcohol, thermometer, paper towel, Science Journal	**Chapter *Fast File* Resources** Transparency Activity, p. 40 MiniLAB, p. 3 Enrichment, p. 26 Reinforcement, p. 23 Directed Reading, p. 16 **Cultural Diversity,** p. 29 **Physical Science Critical Thinking/Problem Solving,** p. 11	**Portfolio** Science Journal, p. 45 Assessment, p. 49 **Performance** MiniLAB, p. 44 Applying Math, p. 48 Applying Math, p. 49 **Content** Section Review, p. 49	Section Focus Transparency Guided Reading Audio Program Interactive Chalkboard CD-ROM
MiniLAB: measuring cup, water, solid object that will fit in cup, pencil, Science Journal **Lab:** 1-cm graph paper, pencil, metric ruler, meterstick	**Chapter *Fast File* Resources** Transparency Activity, p. 41 MiniLAB, p. 4 Enrichment, p. 27 Reinforcement, p. 24 Directed Reading, pp. 16, 17 Lab Worksheet, pp. 5–6 Lab Activity, pp. 9–10, 11–12 Transparency Activity, pp. 43–44	**Portfolio** Assessment, p. 54 **Performance** MiniLAB, p. 52 Applying Math, p. 54 **Content** Section Review, p. 54	Section Focus Transparency Teaching Transparency Guided Reading Audio Program Interactive Chalkboard CD-ROM
Lab: meterstick, stopwatch or watch with a second hand *Need materials?* Contact Science Kit at 1-800-828-7777 or www.sciencekit.com on the Internet.	**Chapter *Fast File* Resources** Transparency Activity, p. 42 Enrichment, p. 28 Reinforcement, p. 25 Directed Reading, pp. 17, 18 Lab Worksheet, pp. 7–8 **Lab Management and Safety,** pp. 70, 71	**Portfolio** Activity, p. 57 **Performance** Applying Skills, p. 59 **Content** Section Review, p. 59	Section Focus Transparency Virtual Labs CD-ROM Guided Reading Audio Program Interactive Chalkboard CD-ROM Video Lab Probeware Lab

End of Chapter Assessment

Blackline Masters	Technology	Professional Series
Chapter *Fast File* Resources Chapter Review, pp. 33–34 Chapter Tests, pp. 35–38 **Standardized Test Practice,** pp. 11–14	MindJogger Videoquiz Virtual Labs CD-ROM Exam*View*® Pro Testmaker TeacherWorks CD-ROM Interactive Chalkboard CD-ROM	**Performance Assessment in the Science Classroom (PASC)**

Transparencies

Section Focus

Section 1 Section Focus Transparency **Maybe He'll Go Away**

Do you think that the teacher really wanted his students to measure the length of the battery? Look at the objects that are scattered on the lab table. What do you suppose is the focus of this lab activity?

FOXTROT © Bill Amend. Reprinted with permission of UNIVERSAL PRESS SYNDICATE. All rights reserved.

1. What are the students using to measure the battery? Does this give them useful information?
2. What are some measurements that people make in their everyday lives?
3. When is it important to measure accurately? When is an estimate acceptable?

L2

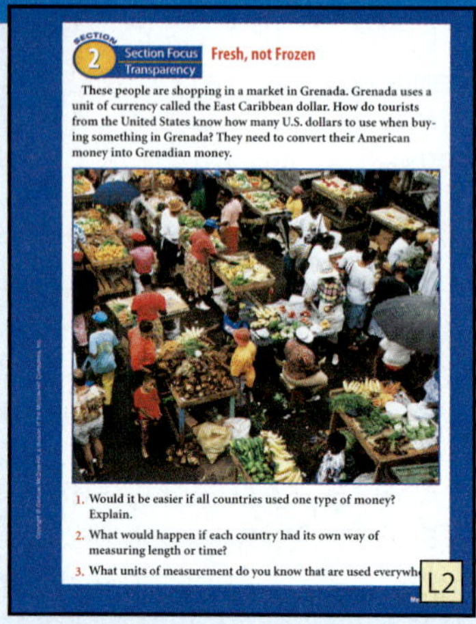
Section 2 Section Focus Transparency **Fresh, not Frozen**

These people are shopping in a market in Grenada. Grenada uses a unit of currency called the East Caribbean dollar. How do tourists from the United States know how many U.S. dollars to use when buying something in Grenada? They need to convert their American money into Grenadian money.

1. Would it be easier if all countries used one type of money? Explain.
2. What would happen if each country had its own way of measuring length or time?
3. What units of measurement do you know that are used everywh...

L2

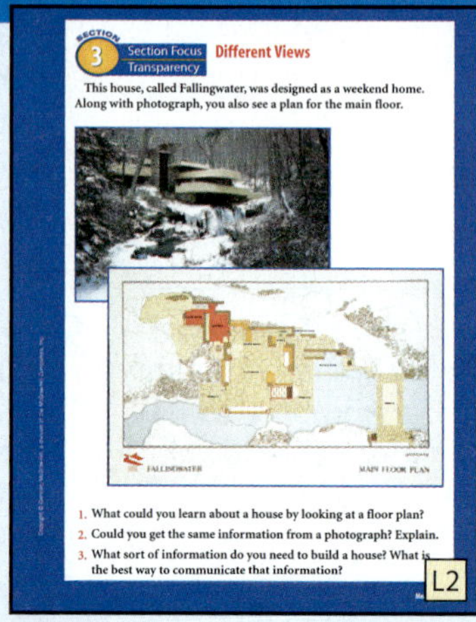
Section 3 Section Focus Transparency **Different Views**

This house, called Fallingwater, was designed as a weekend home. Along with photograph, you also see a plan for the main floor.

1. What could you learn about a house by looking at a floor plan?
2. Could you get the same information from a photograph? Explain.
3. What sort of information do you need to build a house? What is the best way to communicate that information?

L2

This is a representation of key blackline masters available in the Teacher Classroom Resources. See Resource Manager boxes within the chapter for additional information.

Assessment

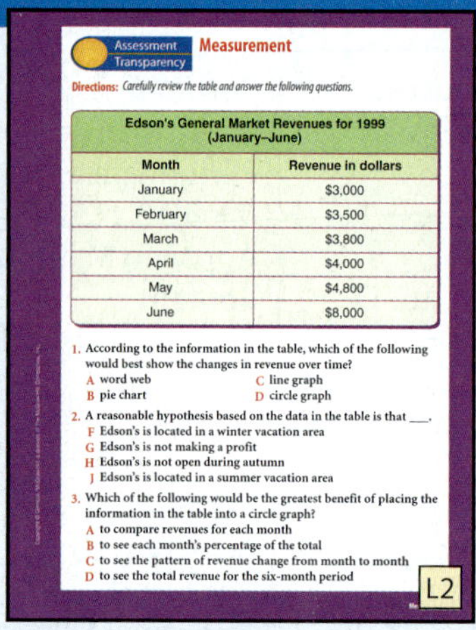
Assessment Transparency **Measurement**

Directions: *Carefully review the table and answer the following questions.*

Edson's General Market Revenues for 1999 (January–June)	
Month	**Revenue in dollars**
January	$3,000
February	$3,500
March	$3,800
April	$4,000
May	$4,800
June	$8,000

1. According to the information in the table, which of the following would best show the changes in revenue over time?
   A word web C line graph
   B pie chart D circle graph
2. A reasonable hypothesis based on the data in the table is that ___.
   F Edson's is located in a winter vacation area
   G Edson's is not making a profit
   H Edson's is not open during autumn
   J Edson's is located in a summer vacation area
3. Which of the following would be the greatest benefit of placing the information in the table into a circle graph?
   A to compare revenues for each month
   B to see each month's percentage of the total
   C to see the pattern of revenue change from month to month
   D to see the total revenue for the six-month period

L2

Teaching

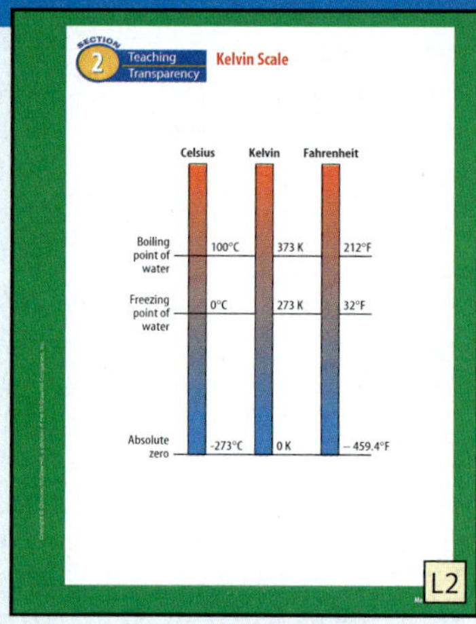
Section 2 Teaching Transparency **Kelvin Scale**

L2

Key to Teaching Strategies

The following designations will help you decide which activities are appropriate for your students.

L1 Level 1 activities should be appropriate for students with learning difficulties.

L2 Level 2 activities should be within the ability range of all students.

L3 Level 3 activities are designed for above-average students.

ELL ELL activities should be within the ability range of English Language Learners.

COOP LEARN Cooperative Learning activities are designed for small group work.

LS Multiple Learning Styles logos, as described on page 12T, are used throughout to indicate strategies that address different learning styles.

P These strategies represent student products that can be placed into a best-work portfolio.

PBL Problem-Based Learning activities apply real-world situations to learning.

Hands-on Activities

Student Text Lab Worksheet

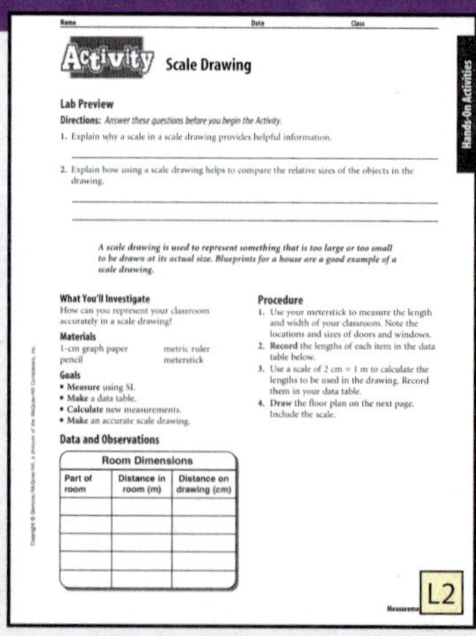
Name Date Class

Activity **Scale Drawing**

Hands-On Activities

Lab Preview

Directions: *Answer these questions before you begin the Activity.*

1. Explain why a scale in a scale drawing provides helpful information.
2. Explain how using a scale drawing helps to compare the relative sizes of the objects in the drawing.

A scale drawing is used to represent something that is too large or too small to be drawn at its actual size. Blueprints for a house are a good example of a scale drawing.

What You'll Investigate

How can you represent your classroom accurately in a scale drawing?

Materials

1-cm graph paper, metric ruler, pencil, meterstick

Goals

- **Measure** using SI.
- **Make** a data table.
- **Calculate** new measurements.
- **Make** an accurate scale drawing.

Procedure

1. Use your meterstick to measure the length and width of your classroom. Note the locations and sizes of doors and windows.
2. **Record** the lengths of each item in the data table below.
3. Use a scale of 2 cm = 1 m to calculate the lengths to be used in the drawing. Record them in your data table.
4. **Draw** the floor plan on the next page. Include the scale.

Data and Observations

Room Dimensions		
Part of room	Distance in room (m)	Distance on drawing (cm)

L2

Laboratory Activities

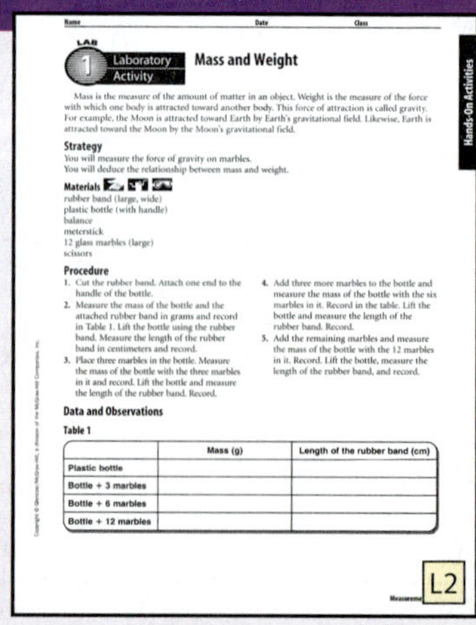
Name Date Class

LAB 1 Laboratory Activity **Mass and Weight**

Hands-On Activities

Mass is the measure of the amount of matter in an object. Weight is the measure of the force with which one body is attracted toward another body. This force of attraction is called gravity. For example, the Moon is attracted toward Earth by Earth's gravitational field. Likewise, Earth is attracted toward the Moon by the Moon's gravitational field.

Strategy

You will measure the force of gravity on marbles.
You will deduce the relationship between mass and weight.

Materials

rubber band (large, wide)
plastic bottle (with handle)
balance
meterstick
12 glass marbles (large)
scissors

Procedure

1. Cut the rubber band. Attach one end to the handle of the bottle.
2. Measure the mass of the bottle and the attached rubber band in grams and record in Table 1. Lift the bottle using the rubber band. Measure the length of the rubber band in centimeters and record.
3. Place three marbles in the bottle. Measure the mass of the bottle with the three marbles in it and record. Lift the bottle and measure the length of the rubber band. Record.
4. Add three more marbles to the bottle and measure the mass of the bottle with the six marbles in it. Record in the table. Lift the bottle and measure the length of the rubber band. Record.
5. Add the remaining marbles and measure the mass of the bottle with the 12 marbles in it. Record. Lift the bottle, measure the length of the rubber band, and record.

Data and Observations

Table 1

	Mass (g)	Length of the rubber band (cm)
Plastic bottle		
Bottle + 3 marbles		
Bottle + 6 marbles		
Bottle + 12 marbles		

L2

Meeting Different Ability Levels

Content Outline

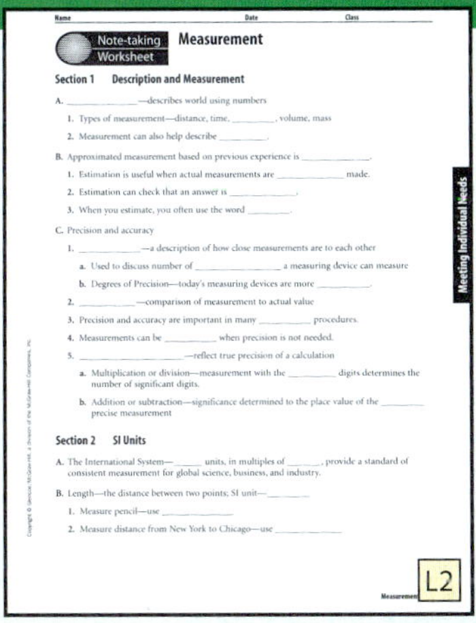

Reinforcement

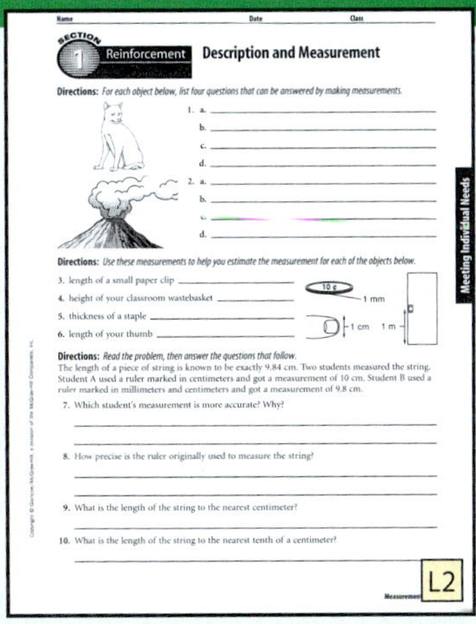

Enrichment

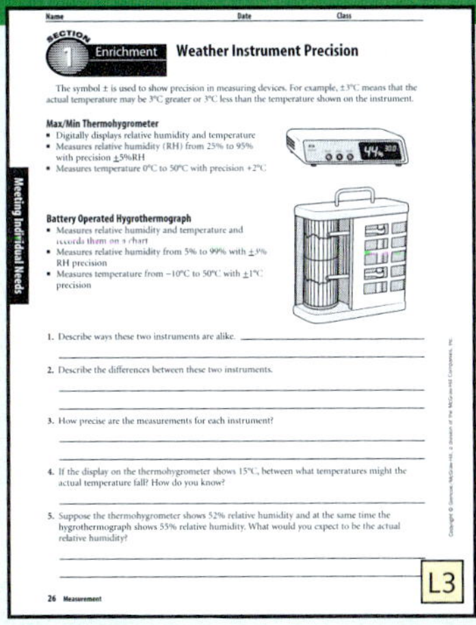

Directed Reading (English/Spanish)

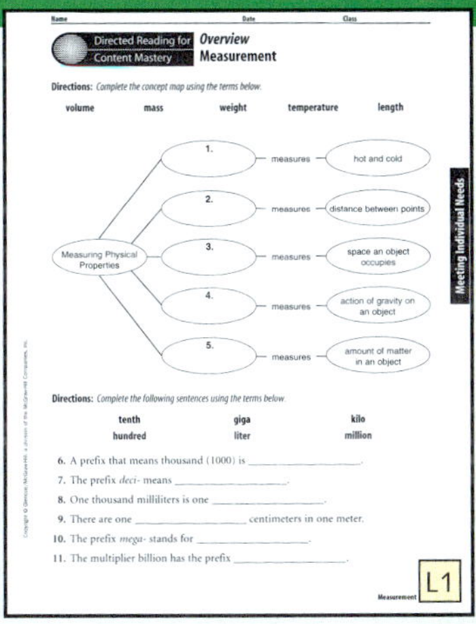

Study Guide

Reading Essentials

Assessment

Test Practice Workbook

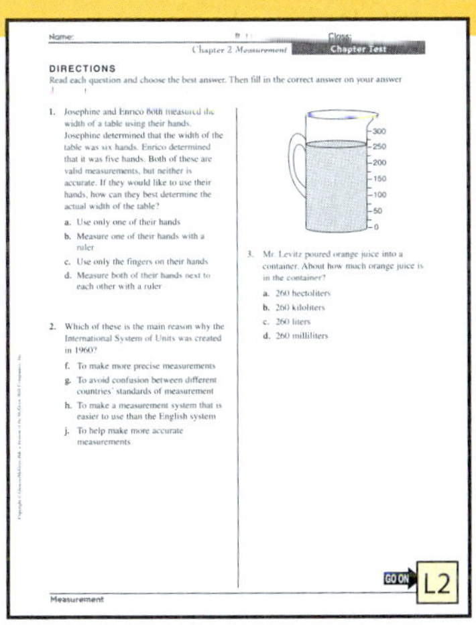

Chapter Review

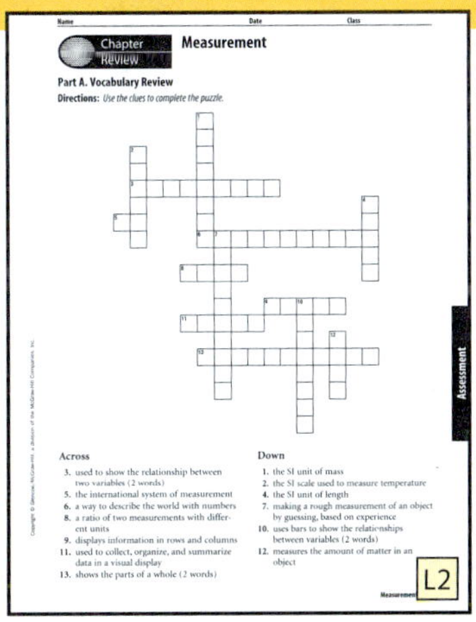

Chapter Tests

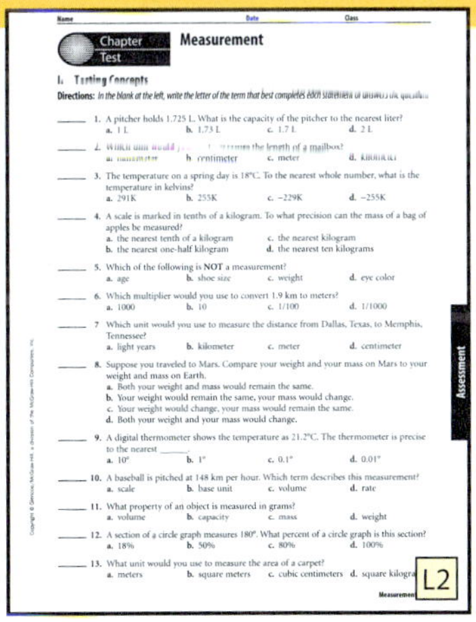

Science Content Background

section 1 Description and Measurement

Degrees of Precision

Scientists use physical and chemical properties to describe matter. Measurement describes properties with numbers. It describes features such as how long, how far, and how many.

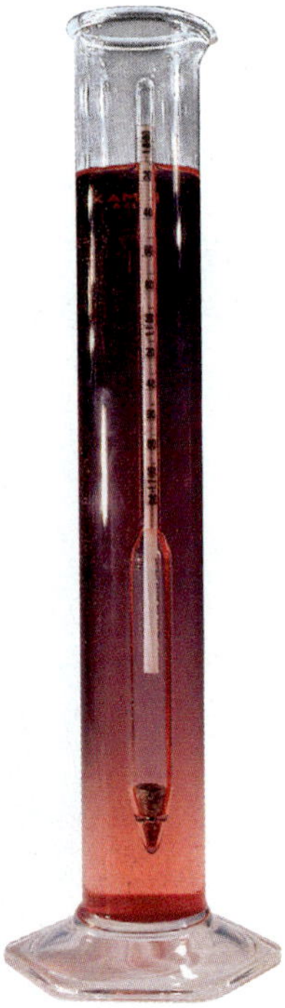

Runk/Schoenberger/Grant Heilman Photography, Inc.

Precision and Accuracy

Precision is related to the degree of exactness that is used to measure an item. Although measurements of 4.2 and 4.20 may seem to be the same, the 4.20 measurement shows that it is a more precise measurement. When numbers are rounded off in calculations, they are rounded to show the precision of the measuring instruments.

section 2 SI Units

The International System

The initials SI stand for the International System of Units for measurement. These units are used all over the world for science, commerce, and communication.

SI standards are defined by the prefixes, of which the most commonly used measurement units are: *kilo-*, *centi-*, and *milli-*. All units differ by factors of ten, so to convert the same unit with different prefixes simply requires the sliding of the decimal point to the left or right. To convert between units, use dimensional analysis, also called unit analysis. Cancel units the same way you cancel numbers.

- 3 km × (1,000 m/1 km) = 3,000 m
- 45 cm (1 m/100 cm) = 0.45 m

Length

Old units of length students might have heard of include the cubit and the roman mile. Accurate length measurement was important for ancient engineering projects such as the Nazca lines, pyramids in Central America and Africa, and the great public buildings found in many cultures.

Mass

Mass and weight are often confused. When moving an object to different locations, the mass is constant while the weight—a measure of the amount of gravity on an object—can change. Because the force of gravity is nearly constant on Earth's surface, mass and weight are often used interchangeably. The weight on Earth is found by multiplying the mass by the acceleration due to gravity, which averages 9.8 m/s^2.

chapter content resources

Internet Resources

For additional content background, visit **red.msscience.com** to:

- access your book online
- find references to related articles in popular science magazines
- access Web links with related content background
- access current events with science journal topics

Print Resources

Hershey's Milk Chocolate Weights and Measures, Jerry Pallota and Rob Bolster, Scholastic, Inc., 2003

Take a Guess: A Look at Estimation, Janine Scott and Mary Beth Fletcher, Compass Point Books, 2003

Weights and Measures, SparkCharts, Spark Publishing, 2002

section 3 Drawings, Tables, and Graphs

Scientific Illustrations

Each science develops a visual language, a standard way to illustrate and communicate important information. A geologist, for example, will learn to "read," or interpret, illustrations of geological strata as part of his or her training. Meteorologists learn to interpret images captured by satellites.

Tables and Graphs

A line graph is used to show the relationship between two numerical variables. As a rule, the independent variable is on the horizontal axis and the dependent variable is on the vertical axis. For example, distance traveled depends on time. While it is common to use the left and bottom sides of a grid as axes, the top or right side can also be used. Bar graphs and circle graphs often are used to present the results of opinion polls.

> **Teacher to Teacher**
> Eva Vanliere
> Waubay School
> Waubay, South Dakota
>
> "I have the students measure the basketball court in feet then convert it to inches. They then convert those to m, cm, and mm. Then I have students measure the width and do the same conversions. Then the students measure the distance to the three-point line, and to the free-throw line. Students learn conversion of English measure to metric and learn math skills."
>
> *Eva Vanliere*

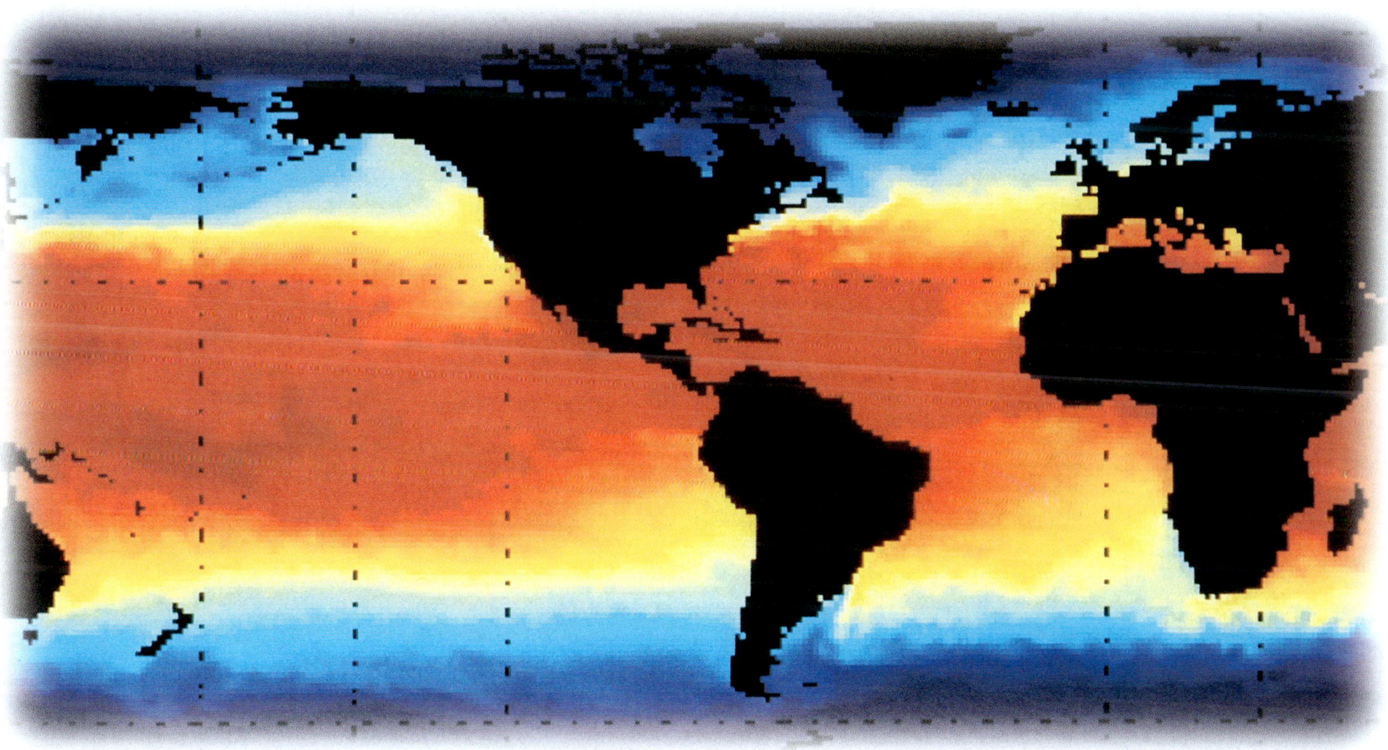

Warren Faidley/International Stock

chapter 2

Chapter Vocabulary

measurement, p. 42
estimation, p. 43
precision, p. 44
accuracy, p. 45
SI, p. 50
meter, p. 51
volume, p. 52
mass, p. 53
kilogram, p. 53
weight, p. 53
kelvin, p. 54
rate, p. 54
table, p. 57
graph, p. 57
line graph, p. 57
bar graph, p. 58
circle graph, p. 58

Science Journal Student responses will vary, but should include answers on how to calculate the distance the driver drove and calculate how much gas was used.

INTERACTIVE CHALKBOARD with Image Bank

PowerPoint® Presentations

This CD-ROM is an editable Microsoft® PowerPoint® presentation that includes:

- a pre-made presentation for every chapter
- interactive graphics
- animations
- audio clips
- image bank
- all new section and chapter questions
- Standardized Test Practice
- transparencies
- pre-lab questions for all labs
- Foldables directions
- links to red.msscience.com

chapter 2

Measurement

chapter preview

sections

1. Description and Measurement
2. SI Units
   Lab *Scale Drawing*
3. Drawings, Tables, and Graphs
   Lab *Pace Yourself*

Virtual Lab *How are graphs used to represent data?*

The Checkered Flag

Race car drivers win or lose by tenths of a second. The driver has to monitor his fuel usage, speed, and oil temperature in order to win the race. In this chapter, you will learn how scientists measure things like distance, time, volume, and temperature.

Science Journal As a member of the pit crew, how can you determine the miles per gallon the car uses? Write in your Science Journal how you would calculate this.

Theme Connection

Systems and Interactions The SI system is used throughout the world. Because it is based on powers of 10, it is easy to convert between units and do calculations.

About the Photo

Does the expression "winning by a nose" mean anything to you? If you have ever "won by a nose," that means the race was close. Sometimes car races, such as this one, are so close the winner has to be determined by a photograph. But there is more to measure than just how close the race was. How fast did the car drive? In this chapter, you will learn how scientists measure things like distance, time, volume, and temperature.

Start-Up Activities

Launch LAB

Measuring Accurately

You make measurements every day. If you want to communicate those measurements to others, how can you be sure that they will understand exactly what you mean? Using vague words without units won't work. Do the lab below to see how confusion can result from using measurements that aren't standard.

1. As a class, choose six objects to measure in your classroom.
2. Measure each object using the width of your hand and write your measurements in your Science Journal.
3. Compare your measurements to those of your classmates.
4. **Think Critically** Describe in your Science Journal why it is better to switch from using hands to using units of measurement that are the same all the time.

Science Online | Preview this chapter's content and activities at red.msscience.com

FOLDABLES™ Study Organizer

Measurement Make the following Foldable to help you organize information about measurements.

STEP 1 **Fold** a sheet of paper in half two times lengthwise. Unfold

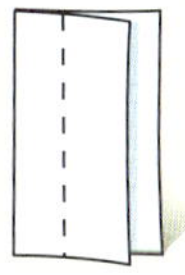

STEP 2 **Fold** the paper widthwise in equal thirds and then in half.

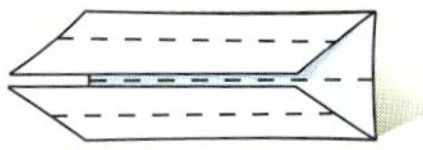

STEP 3 **Unfold,** lay the paper lengthwise, and draw lines along the folds. **Label** your table as shown.

	Estimate It	Measure It	Round It
Length of			
Volume of			
Mass of			
Temperature of			
Rate of			

Estimates Before you read the chapter, select objects to measure and estimate their measurements. As you read the chapter, complete the table.

41

Launch LAB

Purpose Students discover what happens when measurements aren't standardized. L1 ELL COOP LEARN LS **Kinesthetic**

Preparation Before doing this lab, practice using your hand as a unit of length and measure the length of several objects.

Materials something to measure, such as a tabletop, a bookshelf, and a door, in addition to smaller objects

Teaching Strategy Model for the students by measuring your arm or desk with your hand.

Think Critically

No; because hands vary in size, a more standard unit of measurement is needed.

Assessment

Process Have students work in pairs. Each pair should choose a limb of the body, then use it to measure the length of something. Have each student of the pair do the measurement using his or her own limb, then have students compare their measurements with one another. Use **Performance Assessment in the Science Classroom,** p. 169.

FOLDABLES™ Study Organizer — Dinah Zike Study Fold

Student preparation materials for this Foldable are available in the Chapter *FAST FILE* Resources.

Probeware Labs

This chapter's data-collection lab using Probeware technology can be found in the *Probeware Lab Manual* pages 41–43.

section 1

Description and Measurement

1 Motivate

INTERACTIVE CHALKBOARD
PowerPoint® Presentations

Bellringer

Section Focus Transparencies also are available on the Interactive Chalkboard CD-ROM. L2 ELL

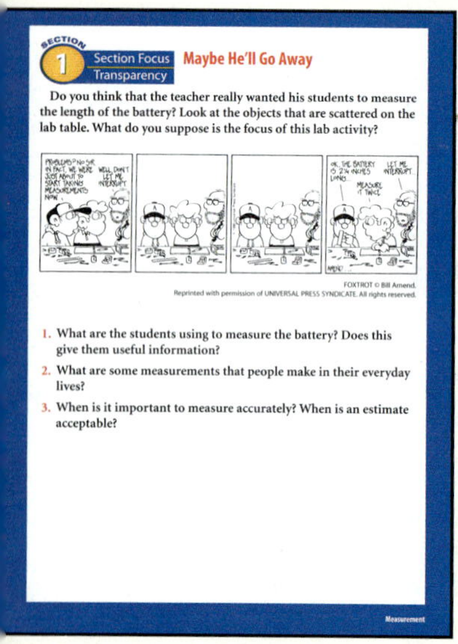
Section 1 Section Focus Transparency **Maybe He'll Go Away**

Do you think that the teacher really wanted his students to measure the length of the battery? Look at the objects that are scattered on the lab table. What do you suppose is the focus of this lab activity?

FOXTROT © Bill Amend. Reprinted with permission of UNIVERSAL PRESS SYNDICATE. All rights reserved.

1. What are the students using to measure the battery? Does this give them useful information?
2. What are some measurements that people make in their everyday lives?
3. When is it important to measure accurately? When is an estimate acceptable?

Measurement

Tie to Prior Knowledge

Measuring Tools Discuss with students times when they have used measuring tools such as rulers or scales.

as you read

What You'll Learn

- **Determine** how reasonable a measurement is by estimating.
- **Identify** and use the rules for rounding a number.
- **Distinguish** between precision and accuracy in measurements.

Why It's Important

Measurement helps you communicate information and ideas.

Review Vocabulary

description: an explanation of an observation

New Vocabulary

- measurement
- estimation
- precision
- accuracy

Measurement

How would you describe what you are wearing today? You might start with the colors of your outfit and perhaps you would even describe the style. Then you might mention sizes—size 7 shoes, size 14 shirt. Every day you are surrounded by numbers. **Measurement** is a way to describe the world with numbers. It answers questions such as how much, how long, or how far. Measurement can describe the amount of milk in a carton, the cost of a new compact disc, or the distance between your home and your school. It also can describe the volume of water in a swimming pool, the mass of an atom, or how fast a penguin's heart pumps blood.

The circular device in **Figure 1** is designed to measure the performance of an automobile in a crash test. Engineers use this information to design safer vehicles. In scientific endeavors, it is important that scientists rely on measurements instead of the opinions of individuals. You would not know how safe the automobile is if this researcher turned in a report that said, "Vehicle did fairly well in head-on collision when traveling at a moderate speed." What does "fairly well" mean? What is a "moderate speed"?

Figure 1 This device measures the range of motion of a seat-belted mannequin in a simulated accident.

Section 1 Resource Manager

Chapter *FAST FILE* Resources

- Transparency Activity, p. 40
- Directed Reading for Content Mastery, pp. 15, 16
- MiniLAB, p. 3
- Note-taking Worksheets, pp. 29–31
- Enrichment, p. 26
- Reinforcement, p. 23

Earth Science Critical Thinking/Problem Solving, p. 11

Cultural Diversity, p. 29

Physical Science Critical Thinking/Problem Solving, p. 11

Figure 2 Accurate measurement of distance and time is important for competitive sports like track and field.
Infer *Why wouldn't a clock that measured in minutes be precise enough for this race?*

Describing Events Measurement also can describe events such as the one shown in **Figure 2.** In the 1956 summer Olympics, sprinter Betty Cuthbert of Australia came in first in the women's 200-m dash. She ran the race in 23.4 s. In the 2000 summer Olympics, Marion Jones of the United States won the 100-m dash in a time of 10.75 s. In this example, measurements convey information about the year of the race, its length, the finishing order, and the time. Information about who competed and in what event are not measurements but help describe the event completely.

Estimation

What happens when you want to know the size of an object but you can't measure it? Perhaps it is too large to measure or you don't have a ruler handy. **Estimation** can help you make a rough measurement of an object. When you estimate, you can use your knowledge of the size of something familiar to estimate the size of a new object. Estimation is a skill based on previous experience and is useful when you are in a hurry and exact numbers are not required. Estimation is a valuable skill that improves with experience, practice, and understanding.

Reading Check *When should you not estimate a value?*

How practical is the skill of estimation? In many instances, estimation is used on a daily basis. A caterer prepares for each night's crowd based on an estimation of how many will order each entree. A chef makes her prize-winning chili. She doesn't measure the cumin; she adds "just that much." Firefighters estimate how much hose to pull off the truck when they arrive at a burning building.

INTEGRATE Career

Precision and Accuracy A pharmacist has a very important job: making sure that patients receive the right medication at the correct dosage. Any error in dosage or type of pill could harm the patient. Explain how precision and accuracy play a role in the pharmacist's job. If a patient receives the wrong medication or an extra pill, how could that affect their health? Research some other careers that rely on precision and accuracy? How could errors in a measurement affect the professional's finished product?

2 Teach

Activity

Ruler Measurement Have students measure a common item to the nearest 0.1 cm using a ruler. Ask them to share the measurements and to discuss possible reasons for variations. differences in the rulers, differences in the way students align the ruler with the object, differences in the angle at which they look at the ruler L1 ELL LS **Kinesthetic**

Caption Answer

Figure 2 Olympic runners often win by fractions of a second.

INTEGRATE Career

Precision and Accuracy A pharmacist has to be both accurate and precise when dispensing medicines. Accuracy is needed when selecting the right medicine that the doctor has ordered. Precision is needed when weighing and mixing the correct drugs. Examples of careers that rely on precision and accuracy are athletes and scientists. By not measuring correctly, the product would be different every time it was made.

Reading Check

Answer when you need to know the exact measurement

Curriculum Connection

History Length was once measured using units such as the width of the palm or the length of the foot. A cubit is the distance from the elbow to the tip of the middle finger. A pace is the length of a walking step from one heel to another. Have students measure a distance of ten paces, and compare their results. L2 ELL LS **Kinesthetic**

Caption Answer

Figure 3 About 4.5 meters (15 feet); accept all reasonable estimates.

Mini LAB

Purpose Students will practice measurement skills with computer temperature probes and thermometers.

Materials crushed ice, cold water, 400-mL beaker, computer temperature probe, alcohol thermometer

Teaching Strategies

- Demonstrate how to make accurate temperature measurements by not letting the thermometer touch the sides or bottom of the beaker.
- Show students how to read an alcohol thermometer.
- Make sure that the beaker is mostly ice for each measurement.

Analysis

1. Review the process for calculating an average.
2. Possible answers: The most precise instrument will give the set of measurements that are closest together. The computer probe tends to be more precise. To determine the accuracy of the measurements, the actual temperature needs to be known. If the students know that the water should be 0°C, then they can determine which device is more accurate by seeing which average is closest to 0°C.

Assessment

Oral Have students work in small groups to define measurement and show how it is done. Have students demonstrate and present their results to the class. Use **Performance Assessment in the Science Classroom**, p. 143. L1 ELL LS **Kinesthetic**

Figure 3 This student is about 1.5 m tall.
Estimate *the height of the tree in the photo.*

Mini LAB

Measuring Accurately

Procedure

1. Fill a **400-mL beaker** with **crushed ice.** Add enough **cold water** to fill the beaker.
2. Make three measurements of the temperature of the ice water using a **computer temperature probe.** Remove the computer probe and dry it with a **paper towel.** Record the measurement in your **Science Journal**. Allow the probe to warm to room temperature between each measurement.
3. Repeat step two using an **alcohol thermometer.**

Analysis

1. Average each set of measurements.
2. Which measuring device is more precise? Explain. Can you determine which is more accurate? How?

Using Estimation You can use comparisons to estimate measurements. For example, the tree in **Figure 3** is too tall to measure easily, but because you know the height of the student next to the tree, you can estimate the height of the tree. When you estimate, you often use the word *about.* For example, doorknobs are about 1 m above the floor, a sack of flour has a mass of about 2 kg, and you can walk about 5 km in an hour.

Estimation also is used to check that an answer is reasonable. Suppose you calculate your friend's running speed as 47 m/s. You are familiar with how long a second is and how long a meter is. Think about it. Can your friend really run a 50-m dash in 1 s? Estimation tells you that 47 m/s is unrealistically fast and you need to check your work.

Precision and Accuracy

One way to evaluate measurements is to determine whether they are precise. **Precision** is a description of how close measurements are to each other. Suppose you measure the distance between your home and your school five times with an odometer. Each time, you determine the distance to be 2.7 km. Suppose a friend repeated the measurements and measured 2.7 km on two days, 2.8 km on two days, and 2.6 km on the fifth day. Because your measurements were closer to each other than your friend's measurements, yours were more precise. The term *precision* also is used when discussing the number of decimal places a measuring device can measure. A clock with a second hand is considered more precise than one with only an hour hand.

Differentiated Instruction

Learning Disabled Make sure that learning disabled students handle measuring equipment and make measurements, rather than just reading and observing.

Fun Fact

The earliest known calculating device is probably the abacus. It dates back to at least 1100 B.C. and it is still in use today, particularly in Asia. It consists of a rectangular frame with thin parallel rods strung with beads. Merchants trading goods not only needed a way to count goods bought and sold, but also to calculate the cost of those goods.

Degrees of Precision The timing for Olympic events has become more precise over the years. Events that were measured in tenths of a second 100 years ago are measured to the hundredth of a second today. Today's measuring devices are more precise. **Figure 4** shows an example of measurements of time with varying degrees of precision.

Accuracy When you compare a measurement to the real, actual, or accepted value, you are describing **accuracy.** A watch with a second hand is more precise than one with only an hour hand, but if it is not properly set, the readings could be off by an hour or more. Therefore, the watch is not accurate. However, measurements of 1.03 m, 1.04 m, and 1.06 m compared to an actual value of 1.05 m are accurate, but not precise. **Figure 5** illustrates the difference between precision and accuracy.

Reading Check *What is the difference between precision and accuracy?*

Figure 4 Each of these clocks provides a different level of precision. **Infer** *which of the three you could use to be sure to make the 3:35 bus.*

For centuries, analog clocks—the kind with a face—were the standard.

Before the invention of clocks, as they are known today, a sundial was used. As the Sun passes through the sky, a shadow moves around the dial.

Digital clocks are now as common as analog ones.

Discussion

Timekeeping Tools Discuss with students timekeeping tools used at school and at home. These include a stopwatch, clock, kitchen timer, alarm clock, and timer on a microwave oven. What are some of the limitations of these tools for timing an event? Possible answers: Many timers read only to the second and sometimes it is necessary to measure tenths or hundredths of seconds. The time it takes a timekeeper to start and stop a stopwatch affects its precision. L2 LS **Linguistic**

Reading Check

Answer Precision describes the exactness of a measure. Accuracy compares a measurement to the actual or accepted value.

Caption Answer

Figure 4 the analog or digital clock

Visual Learning

Figure 4 Have small groups discuss ways each of the clocks could be used. Possible answers: sundial to see if it is before or after noon; analog clock to make it on time for school or for after-school activities; The digital clock could be used to determine if exactly three minutes has elapsed. L2 ELL COOP LEARN LS **Interpersonal**

Teacher FYI

Cesium Atomic Clock The cesium atomic clock uses the radiation generated by the transition between two states of cesium-133 atoms to operate a clock that is both precise and accurate. The clock is used for the basic unit of time in the International System of Units. It has an error of plus or minus one second in one million years.

Science Journal

Using Measurements In their Science Journals, have students describe a situation in which both accuracy and precision in measurement are necessary. Have students share examples in a small group. Sample answers: shooting a basketball into a hoop, landing on a balance beam after doing a flip, throwing a strike in bowling L2 LS **Linguistic**

NATIONAL GEOGRAPHIC

Visualizing Precision and Accuracy

Have students examine the pictures and read the captions. Then ask the following questions.

How would you describe the accuracy and the precision of a basketball player who makes 97 out of 100 free throws? good accuracy and good precision

How would you describe the accuracy and precision of a basketball player who has 99 out of 100 free throws hit the front rim of the basket and bounce off? good precision, poor accuracy

How would you describe the accuracy and precision of a basketball player who makes 33 out of 100 free throws, while the others miss the basket completely? poor precision, poor accuracy

Activity

Sport Display Have small groups of students select another sport and make a similar display that illustrates precision and accuracy. Have each group present its display to the class. L2 ELL COOP LEARN LS **Visual-Spatial and Interpersonal**

Make a Model

Estimate Measurement Collect commonly used containers such as one- and two-liter bottles. Pour given amounts of water into each bottle. Use a permanent ink marker to draw a line at the water level and to mark metric amounts on the outside of the bottle. For example, pour 100 mL of water into a two-liter bottle, draw a line around the container at the water level, and write *100 mL* above the line. Have students estimate where the 200-mL mark would be on each bottle. L2 ELL LS **Visual-Spatial**

NATIONAL GEOGRAPHIC **VISUALIZING PRECISION AND ACCURACY**

Figure 5

From golf to gymnastics, many sports require precision and accuracy. Archery—a sport that involves shooting arrows into a target—clearly shows the relationship between these two factors. An archer must be accurate enough to hit the bull's-eye and precise enough to do it repeatedly.

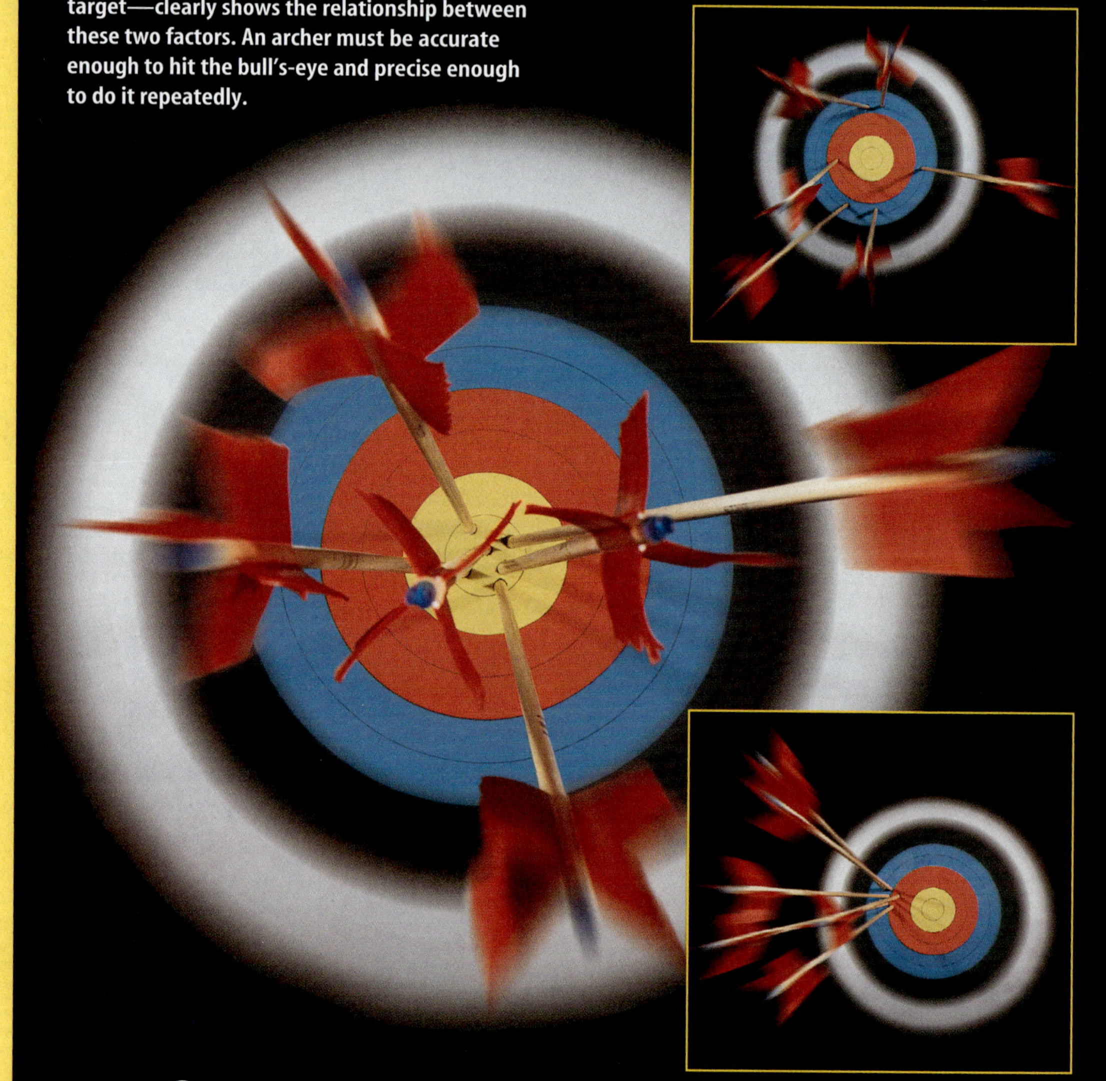

A The archer who shot these arrows is neither accurate nor precise—the arrows are scattered all around the target.

B This archer's attempt demonstrates precision but not accuracy—the arrows were shot consistently to the left of the target's center.

C Here we have a winner! All of the arrows have hit the bull's-eye, a result that is both precise and accurate.

Teacher FYI

GPS Satellites Twenty-four Global Positioning System (GPS) satellites orbit Earth. Each satellite measures time precisely and accurately using four atomic clocks. A coded signal telling the time, the satellite's location, and other data is continuously broadcasted toward Earth. GPS receivers pick up signals and can measure the distance from any GPS satellite based on the time it took the signal to arrive. The receiver compares data from four different satellites to calculate its own latitude, longitude, and elevation, as well as the correct time. If the receiver is moving, it also can determine its own velocity.

INTEGRATE Health

Precision and accuracy are important in many medical procedures. One of these procedures is the delivery of radiation in the treatment of cancerous tumors. Because radiation damages cells, it is important to limit the radiation to only the cancerous cells that are to be destroyed. A technique called Stereotactic Radiotherapy (SRT) allows doctors to be accurate and precise in delivering radiation to areas of the brain. The patient makes an impression of his or her teeth on a bite plate that is then attached to the radiation machine. This same bite plate is used for every treatment to position the patient precisely the same way each time. A CAT scan locates the tumor in relation to the bite plate, and the doctors can pinpoint with accuracy and precision where the radiation should go.

Science Online

Topic: Measurement

Visit red.msscience.com for Web links to information about the importance of accuracy and precision in the medical field.

Activity Research a topic of interest on the Internet and present the topic and numeric data to your class. How might your classmates' understanding of the topic be affected if you presented crucial information inaccurately?

Rounding a Measurement Not all measurements have to be made with instruments that measure with great precision like the scale in **Figure 6.** Suppose you need to measure the length of the sidewalk outside your school. You could measure it to the nearest millimeter. However, you probably would need to know the length only to the nearest meter or tenth of a meter. So, if you found that the length was 135.841 m, you could round off that number to the nearest tenth of a meter and still be considered accurate. How would you round this number? To round a given value, follow these steps:

1. Look at the digit to the right of the place being rounded to.
   - If the digit to the right is 0, 1, 2, 3, or 4, the digit being rounded to remains the same.
   - If the digit to the right is 5, 6, 7, 8, or 9, the digit being rounded to increases by one.
2. The digits to the right of the digit being rounded to are deleted if they are also to the right of a decimal. If they are to the left of a decimal, they are changed to zeros.

Look back at the sidewalk example. If you want to round the sidewalk length of 135.841 to the tenths place, you look at the digit to the right of the 8. Because that digit is a 4, you keep the 8 and round it off to 135.8 m. If you want to round to the ones place, you look at the digit to the right of the 5. In this case you have an 8, so you round up, changing the 5 to a 6, and your answer is 136 m.

Figure 6 This laboratory scale measures to the nearest hundredth of a gram.

Quick Demo

Rounding Measurements

Materials bag of candy, laboratory scale

Estimated Time 10–15 minutes

Procedure To demonstrate rounding measurements to students, you could use a bag of M&Ms (any type of candy or popcorn) and divide the contents amongst the students. First, the M&Ms would need to be counted and then divided by the number of students in the class. If the total number of M&Ms divides into the number of students, remove some from the bag. Record the numbers on the chalkboard and have students round to the number of M&Ms they will receive. As they eat the candy, you could measure the weight of an individual piece on a laboratory scale and round to the nearest hundredth of a gram.

INTEGRATE Health

Measurement Accuracy Ask students to identify which parts of the procedure make the measurement precise and which make it accurate. Precise: the instruments and the techniques used make measurements with a high degree of exactness; also, the bite plate positions the patient exactly the same way each time. Accurate: the CAT scan locates the tumor accurately.

Curriculum Connection

History The first chemists were the alchemists of ancient Egypt, China, Greece, and Rome, and of medieval Arabia and Europe. Have students find out what alchemists did and what kinds of systems they used for measurement. Alchemists tried to change metals into gold and helped develop perfumes, cosmetics, and the gilding of metals. They did not have precise tools or a standard measuring system.

Applying Math

National Math Standards

Correlation to Mathematics Objectives

1

Teaching Strategies

To round to a certain place, remember to ignore all of the numbers more than one place to the right.

Answers to Practice Problems

1. first object = 6.9 g; second object = 20.2 g
2. 26, 3.4, 5.98, 9.8

Inquiry Lab

Purpose To practice measurement, estimation, and prediction skills, have students develop an experiment and collect data.

Possible Materials tape measure, calculator, string, ruler, balance, trees, paper clip

Estimated Time 3–5 days to collect data (a few minutes at the beginning or end of class), one class to complete experiment

Teaching Strategies

- Students could perform research on tree rings and growth patterns. They could measure trees around the school and make predictions about the age of the tree based on their research and data.
- Students could make a paper airplane and throw it as far as possible, measuring the distance it flew. Then they could attach a paper clip and throw the airplane, measuring the distance it flew. These measurements could be compared. Ask students why the measurements vary.

For additional inquiry activities, see *Science Inquiry Labs*.

Precision and Number of Digits When might you need to round a number? Suppose you want to divide a 2-L bottle of soft drink equally among seven people. When you divide 2 by 7, your calculator display reads as shown in **Figure 7.** Will you measure exactly 0.285 714 285 L for each person? No. All you need to know is that each person gets about 0.3 L of soft drink.

Using Precision and Significant Digits The number of digits that truly reflect the precision of a number are called the significant digits or significant figures. They are figured as follows:

- Digits other than zero are always significant.
- Final zeros after a decimal point (6.545 600 g) are significant.
- Zeros between any other digits (507.0301 g) are significant.
- Initial zeros (0.000 2030 g) are NOT significant.
- Zeros in a whole number (1650) may or may not be significant.
- A number obtained by counting instead of measuring, such as the number of people in a room or the number of meters in a kilometer, has infinite significant figures.

Applying Math — Rounding

ROUNDED VALUES The mass of one object is 6.941 g. The mass of a second object is 20.180 g. You need to know these values only to the nearest whole number to solve a problem. What are the rounded values?

Solution

1 *This is what you know:*	• mass of first object = 6.941 g • mass of second object = 20.180 g
2 *This is what you need to find out:*	• the number to the right of the one's place • first object: 9; second object: 1
3 *This is the procedure you need to use:*	digits 0, 1, 2, 3, 4 remain the same for digits 5, 6, 7, 8, 9, round up
4 *Check your answer:*	• first object: 9 makes the 6 round up = 7 • second object: 1 makes the 0 remain the same = 20

Practice Problems

1. What are the rounded masses of the objects to the nearest tenth of a unit?
2. Round the following numbers: 25.643 to the ones place, 3.429 to the tenths place, 5.982 to the hundredths place, and 9.8210 to the tenths place.

Science Online — For more practice, visit red.msscience.com/math_practice

Cultural Diversity

Social Time While time's passage can be accurately measured, its perceived importance, or social time, varies from culture to culture. In the U.S., people call and apologize if they expect to be late. In Latin American and Arab countries, people may arrive an hour late with no apology expected or given. One study of social time rated Japan highest and the U.S. second in accuracy of bank clocks and pace of life.

Differentiated Instruction

Challenge Discuss this example with your students. The mass of a substance is determined to be 0.0045 kilograms. How many significant digits are in this measurement? There are two significant digits. The zeros in 0.0045 are used to show only the place value of the decimal and are not counted as significant digits. L3

Following the Rules In the soft drink example you have an exact number, seven, for the number of people. This number has infinite significant digits. You also have the number two, for how many liters of soft drink you have. This has only one significant digit.

There are also rules to follow when deciding the number of significant digits in the answer to a calculation. They depend on what kind of calculation you are doing.

- For multiplication and division, you determine the number of significant digits in each number in your problem. The significant digits of your answer are determined by the number with fewer digits.

$$6.14 \times 5.6 = \boxed{34}.384$$

3 digits 2 digits 2 digits

- For addition and subtraction, you determine the place value of each number in your problem. The significant digits of the answer are determined by the number that is least precise.

6.14	to the hundredths
+ 5.6	to the tenths
$\boxed{11.7}$4	to the tenths

In the soft drink example you are dividing and the number of significant digits is determined by the amount of soft drink, 2 L. There is one significant digit there, therefore, the amount of soft drink each person gets is rounded to 0.3 L.

Figure 7 Sometimes considering the size of each digit will help you realize they are unneeded. In this calculation, the seven ten-thousandths of a liter represents just a few drops of soft drink.

section 1 review

Summary

Measurement

- Measurement is used to answer questions such as how much, how long, or how far.

Estimation

- When making an estimate, rely on previous knowledge to make an educated guess about the size of an object.

Precision and Accuracy

- Precision is the ability to remain consistent. Accuracy compares a measurement to the real value of an object.
- Significant digits affect precision when calculating an answer and are determined by rules based on calculation.

Self Check

1. **Estimate** the distance between your desk and your teacher's desk. Explain the method you used.
2. **Infer** John's puppy has chewed on his ruler. Will John's measurements be accurate or precise? Why?
3. **Think Critically** Would the sum of 5.7 cm and 6.2 cm need to be rounded? Why or why not? Would the sum of 3.28 cm and 4.1 cm need to be rounded? Why or why not?

Applying Math

4. **Calculate** Perform the following calculations and express the answer using the correct number of significant digits: 42.35 + 214; 225/12. **For more help, refer to the Math Skill Handbook.**

Science Online red.msscience.com/self_check_quiz

3 Assess

DAILY INTERVENTION

Check for Understanding

Ask students to write paragraphs explaining whether they think it is possible for experimental data to have good accuracy but poor precision. Encourage them to explain their reasoning. L2 ELL COOP LEARN LS **Linguistic and Logical-Mathematical**

Reteach

Measuring Tools Obtain measuring tools and items to measure. Have students describe how they can use the tools to collect data that are accurate and precise. L2 LS **Visual-Spatial**

Assessment

Performance Have each student make a poster that shows the difference between estimation and precise measurement. Students should present their posters to a small group of students or to the class. Use **Performance Assessment in the Science Classroom,** p. 145. P

section 1 review

1. Possible answer: About 3 m; it looks like two students could almost lie down head to toe in this space.
2. John's measurements will not be precise in the area in which measuring lines have been destroyed by the dog. They may still be accurate, depending on what he's measuring and what the dog destroyed.
3. The sum of 5.7 and 6.2 would not need to be rounded because both are measured to the same place—the tenths place. The sum of 3.28 and 4.1 would be rounded to the place of the least precise measurement—the tenths place.
4. 256; 19

1 Motivate

Bellringer

INTERACTIVE CHALKBOARD PowerPoint® Presentations

Section Focus Transparencies also are available on the Interactive Chalkboard CD-ROM. L2 ELL

SECTION 2 Section Focus Transparency **Fresh, not Frozen**

These people are shopping in a market in Grenada. Grenada uses a unit of currency called the East Caribbean dollar. How do tourists from the United States know how many U.S. dollars to use when buying something in Grenada? They need to convert their American money into Grenadian money.

1. Would it be easier if all countries used one type of money? Explain.
2. What would happen if each country had its own way of measuring length or time?
3. What units of measurement do you know that are used everywhere?

Measurement

Tie to Prior Knowledge

Precise Tools Recall from the previous section the importance of using precise tools. In this section, students will learn about the tools and units used for scientific measurement.

Reading Check

Answer to provide a worldwide standard of physical measurement for science, industry, and commerce

Fun Fact

SI stands for *Système International d'Unités*, which is the French term for International System of Units.

section 2

SI Units

as you read

What You'll Learn

- **Identify** the purpose of SI.
- **Identify** the SI units of length, volume, mass, temperature, time, and rate.

Why It's Important

The SI system is used throughout the world, allowing you to measure quantities in the exact same way as other students around the world.

Review Vocabulary

variable: factors that can be changed in an experiment

New Vocabulary

- SI
- meter
- volume
- mass
- kilogram
- weight
- kelvin
- rate

The International System

Can you imagine how confusing it would be if people in every country used different measuring systems? Sharing data and ideas would be complicated. To avoid confusion, scientists established the International System of Units, or **SI,** in 1960 as the accepted system for measurement. It was designed to provide a worldwide standard of physical measurement for science, industry, and commerce. SI units are shown in **Table 1.**

Reading Check *Why was SI established?*

The SI units are related by multiples of ten. Any SI unit can be converted to a smaller or larger SI unit by multiplying by a power of 10. For example, to rewrite a kilogram measurement in grams, you multiply by 1,000.

$$\text{Ex. } 5.67 \text{ kg} \times 1000 = 5670 \text{ grams}$$

The new unit is renamed by changing the prefix, as shown in **Table 2.** For example, one millionth of a meter is one *micro*meter. One thousand grams is one *kilo*gram. **Table 3** shows some common objects and their measurements in SI units.

Table 1 SI Base Units

Quantity	Unit	Symbol
length	meter	m
mass	kilogram	kg
temperature	kelvin	K
time	second	s
electric current	ampere	A
amount of substance	mole	mol
intensity of light	candela	cd

Table 2 SI Prefixes

Prefix	Multiplier
giga-	1,000,000,000
mega-	1,000,000
kilo-	1,000
hecto-	100
deka-	10
[unit]	1
deci-	0.1
centi-	0.01
milli-	0.001
micro-	0.000 001
nano-	0.000 000 001

Section 2 Resource Manager

Chapter *FAST FILE* Resources

Transparency Activity, p. 42
Directed Reading for Content Mastery, pp. 16, 17
Lab Activity, pp. 9–10
MiniLAB, p. 4
Enrichment, p. 27
Reinforcement, p. 24
Lab Activity, pp. 11–12
Lab Worksheet, pp. 5–6

Length

Length is defined as the distance between two points. Lengths measured with different tools can describe a range of things from the distance from Earth to Mars to the thickness of a human hair. In your laboratory activities, you usually will measure length with a metric ruler or meterstick.

The **meter** (m) is the SI unit of length. One meter is about the length of a baseball bat. The size of a room or the dimensions of a building would be measured in meters. For example, the height of the Washington Monument in Washington, D.C. is 169 m.

Smaller objects can be measured in centimeters (cm) or millimeters (mm). The length of your textbook or pencil would be measured in centimeters. A twenty-dollar bill is 15.5 cm long. You would use millimeters to measure the width of the words on this page. To measure the length of small things such as blood cells, bacteria, or viruses, scientists use micrometers (millionths of a meter) and nanometers (billionths of a meter).

Figure 8 These runners have just completed a 10-kilometer race—known as a 10K.
Estimate *how many kilometers is the distance between your home and your school.*

A Long Way Sometimes people need to measure long distances, such as the distance a migrating bird travels or the distance from Earth to the Moon. To measure such lengths, you use kilometers. Kilometers might be most familiar to you as the distance traveled in a car or the measure of a long-distance race, as shown in **Figure 8.** The course of a marathon is measured carefully so that the competitors run 42.2 km. When you drive from New York to Los Angeles, you cover 4,501 km.

Table 3 Common Objects in SI Measurements

Object	Type of Measurement	Measurement
can of soft drink	volume	355 mL
bag of potatoes	mass	4.5 kg
fluorescent tube	length	1.2 m
refrigerator	temperature	276 K

INTEGRATE Astronomy

Measurement Accuracy How important are accurate measurements? In 1999, the *Mars Climate Orbiter* disappeared as it was to begin orbiting Mars. NASA later discovered that a unit system error caused the flight path to be incorrect and the orbiter to be lost. Research the error and determine what systems of units were involved. How can using two systems of units cause errors?

2 Teach

Caption Answer

Figure 8 Answers will vary. Have students explain how they arrived at their answers.

Use Science Words

Word Origin The prefix *centi-*, the word *percent*, and the American unit of money called *cent* all derive from the same French and Latin root, which means 100. Can you think of other ways we use the root word *cent?* centimeter, century, centigrade, centipede L2 LS **Linguistic**

INTEGRATE Astronomy

Measurement Accuracy The error occurred because the common unit *pound • seconds* was used instead of the SI unit *Newton • seconds* in the computer program that determined trajectories of the orbiter.

Research Have students research other errors that resulted from using two different systems of units. Ask students to explain how these errors can be minimized and how scientists can predict the possibility of error before the error is made.

Activity

Measurement Accuracy Have students use rulers to measure the length of a small item on their desks while sitting. Then have them remeasure the item while standing. Discuss which measurement is more accurate. The measurement made while looking directly at it from above is more accurate. L2 ELL LS **Kinesthetic**

Visual Learning

Tables 1 and 2 Have students use the information in the tables to determine the meaning of the following units: kilogram—1,000 grams; centimeter—1/100 meter. Have pairs take turns using the prefixes to make their own units and identifying their meanings. Sample answer: gigagram (1 billion grams) and megameter (1 million meters) L2 ELL LS **Logical-Mathematical**

Caption Answer

Figure 9 There are 1,000,000 cubic centimeters in a cubic meter.

Quick Demo

Familiar Measuring Units

Materials ruler, coins

Estimated Time 5 minutes

Procedure Units of metric measurement can be understood using common items that students are familiar with. Show students that a meter is about the height of a doorknob, a centimeter is about the width of their little fingertip, and a millimeter is about the thickness of a dime.

Text Question Answer

A compact disc case is about 1 cm × 12.5 cm × 14.2 cm. Its volume is 177.5 cm^3.

Mini LAB

Purpose to measure volume by immersion

Materials transparent measuring cup, water, small object

Teaching Strategies

- Make sure students do not spill water from the measuring cup when lowering the object into the water.
- Do not use an object that floats, since pushing will cause errors.

Analysis

1. the difference between the first volume of water and the second
2. Check students' work.

Assessment

Process Have students redo the MiniLAB using a small rectangular solid then measure the sides of the object and calculate its volume. Are the two volumes the same? Use **Performance Assessment in the Science Classroom,** p. 89.

Try at Home

Figure 9 A cubic meter equals the volume of a cube 1 m by 1 m by 1 m.
Infer *how many cubic centimeters are in a cubic meter.*

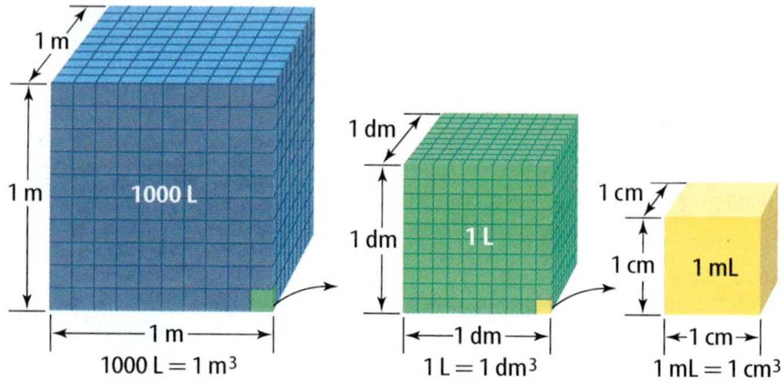

Mini LAB

Measuring Volume

Procedure

1. Fill a plastic or glass **liquid measuring cup** until half full with **water.** Measure the volume.
2. Find an **object,** such as a rock, that will fit in your measuring cup.
3. Carefully lower the object into the water. If it floats, push it just under the surface with a **pencil.**
4. Record in your **Science Journal** the new volume of the water.

Analysis

1. How much space does the object occupy?
2. If 1 mL of water occupies exactly 1 cm^3 of space, what is the volume of the object in cm^3?

Try at Home

Volume

The amount of space an object occupies is its **volume.** The cubic meter (m^3), shown in **Figure 9,** is the SI unit of volume. You can measure smaller volumes with the cubic centimeter (cm^3 or cc). To find the volume of a square or rectangular object, such as a brick or your textbook, measure its length, width, and height and multiply them together. What is the volume of a compact disc case?

You are probably familiar with a 2-L bottle. A liter is a measurement of liquid volume. A cube 10 cm by 10 cm by 10 cm holds 1 L (1,000 cm^3) of water. A cube 1 cm on each side holds 1 mL (1 cm^3) of water.

Volume by Immersion Not all objects have an even, regular shape. How can you find the volume of something irregular like a rock or a piece of metal?

Have you ever added ice cubes to a nearly full glass of water only to have the water overflow? Why did the water overflow? Did you suddenly have more water? The volume of water did not increase at all, but the water was displaced when the ice cubes were added. Each ice cube takes up space or has volume. The difference in the volume of water before and after the addition of the ice cubes equals the volume of the ice cubes that are under the surface of the water.

The ice cubes took up space and caused the total volume in the glass to increase. When you measure the volume of an irregular object, you do the same thing. You start with a known volume of water and drop in, or immerse, the object. The increase in the volume of water is equal to the volume of the object.

52 CHAPTER 2 Measurement

Differentiated Instruction

Learning Disabled While presenting information about measurement, use physical models as you discuss each concept. For example, a meterstick and centimeter ruler should be shown while learning about length, a pan balance or digital scale should be shown while discussing mass, and a stopwatch should be shown while discussing rate. L2 ELL LS **Visual-Spatial**

Teacher FYI

Metric System An international measurement system was originally established during the French Revolution by representatives from seventeen nations, including the United States. In 1902, a law requiring the United States to use the metric system was defeated by one vote.

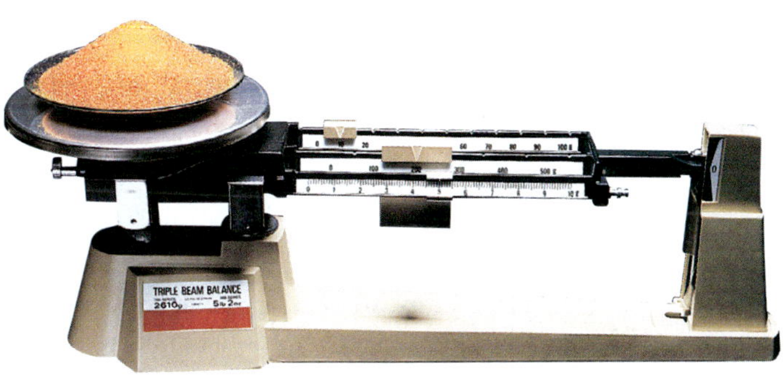

Figure 10 A triple beam balance compares an unknown mass to known masses.

Mass

The **mass** of an object measures the amount of matter in the object. The **kilogram** (kg) is the SI unit for mass. One liter of water has a mass of about 1 kg. Smaller masses are measured in grams (g). One gram is about the mass of a large paper clip.

You can determine mass with a triple-beam balance, shown in **Figure 10.** The balance compares an object to a known mass. It is balanced when the known standard mass of the slides on the balance is equal to the object on the pan.

Why use the word *mass* instead of *weight?* Weight and mass are not the same. Mass depends only on the amount of matter in an object. If you ride in an elevator in the morning and then ride in the space shuttle later that afternoon, your mass is the same. Mass does not change when only your location changes.

Weight **Weight** is a measurement of force. The SI unit for weight is the newton (N). Weight depends on gravity, which can change depending on where the object is located. A spring scale measures how a planet's gravitational force pulls on objects. Several spring scales are shown in **Figure 11.**

If you were to travel to other planets, your weight would change, even though you would still be the same size and have the same mass. This is because gravitational force is different on each planet. If you could take your bathroom scale, which uses a spring, to each of the planets in this solar system, you would find that you weigh much less on Mars and much more on Jupiter. A mass of 75 pounds, or 34 kg, on Earth is a weight of 332 N. On Mars, the same mass is 126 N, and on Jupiter it is 782 N.

Reading Check *What does weight measure?*

Figure 11 A spring scale measures an object's weight by how much it stretches a spring.

Discussion

Moon Changes Have students discuss what would happen to their mass and weight if they visited the moon. What might happen if they tried jumping while on the moon? They would have the same mass but less weight, so they would be able to jump higher and farther. L2 LS **Logical-Mathematical**

Use Science Words

Word Usage Have students use the words mass and weight in sentences describing how the concepts are different. Possible answer: Mass is the amount of matter in an object; it is not affected by gravity. Weight is a measure of the force of gravity; it changes with the force of gravity. L2 ELL LS **Linguistic**

Use an Analogy

Seesaw A pan balance is like a seesaw. When the weights on each side are equal, it is balanced.

Reading Check

Answer the force of gravity on an object

Activity

Weight Have students suspend two 2-packs of fig cookies from a spring scale calibrated in newtons and record the weight. Have them determine the weight of a single cookie by dividing by 4. Then have them weigh a single cookie on the spring scale to check their answers.

Differentiated Instruction

Challenge Have students research the measuring standards in other countries. For example, only the United States uses miles to describe distance. Most other countries rely on kilometers and yards to define distance. Students could construct a conversion chart of the different measurements to familiarize themselves with international standards. L3

Science Journal

Rates Have students use the information in the student text to write several rates. Possible answers: kilometers per hour, kilometers per liter of gas. Ask them to explain how these rates are used.

Teacher FYI

Absolute Zero Temperature is a measure of the kinetic energy of molecules and atoms. The temperature 0 K is called absolute zero. Scientists believe that absolute zero is the lowest temperature possible because at this temperature molecules and atoms have the lowest amount of energy possible.

3 Assess

DAILY INTERVENTION

Check for Understanding

Visual-Spatial Have each student make a chart to organize the information he or she learned in this section. The chart should include a title with International System in it. It should also include pictures and information about measuring length, volume, mass, temperature, and time. L2 ELL LS

Reteach

Tools Collect a set of measurement tools. Display each tool and have a student volunteer tell the class its name, what it measures, and the unit(s) it uses. L1 ELL LS **Visual-Spatial**

Assessment

Process Ask each student to make an illustration of a spring scale, showing how it works. Use **Performance Assessment in the Science Classroom,** p. 127. P

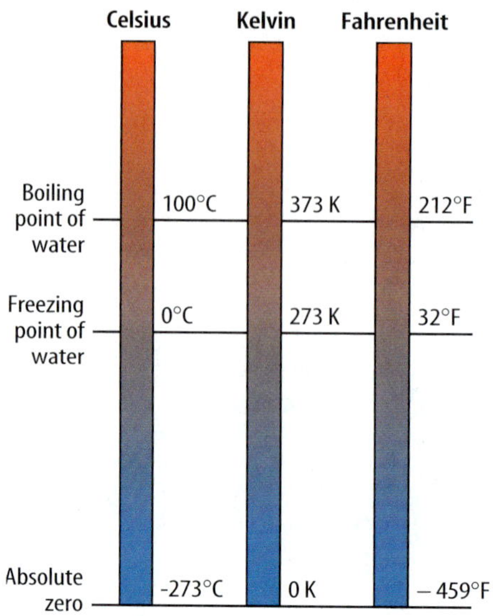

Figure 12 The kelvin scale starts at 0 K. In theory, 0 K is the coldest temperature possible in nature.

Temperature

The physical property of temperature is related to how hot or cold an object is. Temperature is a measure of the kinetic energy, or energy of motion, of the particles that make up matter.

Temperature is measured in SI with the **kelvin** (K) scale. The Fahrenheit and Celsius temperature scales are the two most common scales used on thermometers and in classroom laboratories. These two scales do not start at zero, as shown in **Figure 12.** A 1 K difference in temperature is the same as a 1°C difference in temperature.

Time and Rates

Time is the interval between two events. The SI unit of time is the second (s). Time also is measured in hours (h). Can you imagine hearing that a marathon was run in 7,620 s instead of 2 h and 7 min?

A **rate** is the amount of change of one measurement in a given amount of time. One rate you are familiar with is speed, which is the distance traveled in a given time. Speeds often are measured in kilometers per hour (km/h).

The unit that is changing does not necessarily have to be an SI unit. For example, you can measure the number of cars that pass through an intersection per hour in cars/h.

section 2 review

Summary

The International System

- The International System of Units, SI, was established to provide a standard of physical measurement and to reduce international confusion when comparing measurements.

Measurement

- Length is the distance between two points.
- Volume is the amount of space an object occupies.
- To calculate volume, multiply length by width by height.
- The amount of matter in an object is its mass.
- Weight is determined by gravitational pull.
- Celsius temperature scales are more common in laboratories than kelvin scales.

Self Check

1. **Describe** a situation in which different units of measure could cause confusion.
2. **Define** what type of quantity the cubic meter measures.
3. **Explain** how you would change a measurement in centimeters to kilometers.
4. **Identify** what SI unit replaces the pound. What does this measure?
5. **Think Critically** How would you find the mass of a metal cube?

Applying Math

6. **Measure** A block of wood is 0.2 m by 0.1 m by 0.5 m. Find its dimensions in centimeters. Then find its volume in cubic centimeters.

Science Online red.msscience.com/self_check_quiz

section 2 review

1. Possible answer: Someone thinks a temperature is given in Fahrenheit degrees, but it is actually in Celsius degrees.
2. volume
3. divide by 100,000
4. Newton; the force of gravity on an object
5. Use a pan balance or scale. The description should provide step-by-step instructions for using the instrument.
6. 20 cm × 10 cm × 50 cm = 10,000 cm^3

LAB

Scale Drawing

A scale drawing is used to represent something that is too large or too small to be drawn at its actual size. Blueprints for a house are a good example of a scale drawing.

Real-World Question

How can you represent your classroom accurately in a scale drawing?

Goals

- **Measure** using SI.
- **Make** a data table.
- **Calculate** new measurements.
- **Make** an accurate scale drawing.

Materials

1-cm graph paper
pencil
metric ruler
meterstick

Procedure

1. Use your meterstick to measure the length and width of your classroom. Note the locations and sizes of doors and windows.
2. **Record** the lengths of each item in a data table similar to the one below.
3. Use a scale of 2 cm = 1 m to calculate the lengths to be used in the drawing. Record them in your data table.
4. **Draw** the floor plan. Include the scale.

Room Dimensions

Part of Room	Distance in Room (m)	Distance on Drawing (cm)

Conclude and Apply

1. How did you calculate the lengths to be used on your drawing? Did you put a scale on your drawing?
2. **Infer** what your scale drawing would look like if you chose a different scale?
3. **Sketch** your room at home, estimating the distances. Compare this sketch to your scale drawing of the classroom. When would you use each type of illustration?
4. What measuring tool simplifies this task?

Communicating Your Data

Measure your room at home and compare it to the estimates on your sketch. Explain to someone at home what you did and how well you estimated the measurements. **For more help, refer to the Science Skill Handbook.**

BENCH TESTED

Real-World Question

Purpose to learn how to make and use a scale drawing L2 ELL COOP LEARN LS **Visual-Spatial**

Process Skills compare and contrast, interpret data, make and use tables, use numbers, make models

Time Required one 45-minute period to measure the classroom, one 30-minute period to draw the room to scale (could be homework)

Procedure

Teaching Strategy

- Demonstrate how to draw the length of a wall to scale.

Conclude and Apply

1. Length was measured with a meterstick. The scale of 2 cm/1 m should be noted on the drawing.
2. The shape of the drawing would remain the same, but it would be larger or smaller, depending on the new scale chosen.
3. Check students' work. A rough sketch might be used to give an idea of how to design a room. A scale drawing might be used to determine if the furniture will actually fit in the space available.
4. Possible tools to use: meterstick, tape measure, retractable metal tape measure, measuring wheel

Communicating Your Data

Have students use a drawing to discuss scale and features with a small group or the class, in addition to explaining these things to someone at home.

Assessment

Portfolio Have students add to their work a detailed explanation of how to make and use a scale drawing. Use **Performance Assessment in the Science Classroom,** p. 177.

section 3 Drawings, Tables, and Graphs

1 Motivate

INTERACTIVE CHALKBOARD PowerPoint® Presentations

Bellringer

Section Focus Transparencies also are available on the Interactive Chalkboard CD-ROM. L2 ELL

Section 3 Section Focus Transparency **Different Views**

This house, called Fallingwater, was designed as a weekend home. Along with photograph, you also see a plan for the main floor.

FALLINGWATER MAIN FLOOR PLAN

1. What could you learn about a house by looking at a floor plan?
2. Could you get the same information from a photograph? Explain.
3. What sort of information do you need to build a house? What is the best way to communicate that information?

Measurement

Tie to Prior Knowledge

Publication Charts Ask students to recall publications other than science books in which they have seen graphs and charts. Newspapers and news magazines are good sources of examples.

as you read

What You'll Learn

- **Describe** how to use pictures and tables to give information.
- **Identify** and use three types of graphs.
- **Distinguish** the correct use of each type of graph.

Why It's Important

Illustrations, tables, and graphs help you communicate data about the world around you in an organized and efficient way.

Review Vocabulary
model: a representation of an object or event used as a tool for understanding the natural world

New Vocabulary
- table
- graph
- line graph
- bar graph
- circle graph

Scientific Illustrations

Most science books include pictures. Photographs and drawings model and illustrate ideas and sometimes make new information clearer than written text can. For example, a drawing of an airplane engine shows how all the parts fit together much better than several pages of text could describe it.

Drawings A drawing is sometimes the best choice to show details. For example, a canyon cut through red rock reveals many rock layers. If the layers are all shades of red, a drawing can show exactly where the lines between the layers are. A drawing can emphasize only the things that are necessary to show.

A drawing also can show things you can't see. You can't see the entire solar system, but drawings show you what it looks like. Also, you can make quick sketches to help model problems. For example, you could draw the outline of two continents to show how they might have fit together at one time.

Drawings can show hidden things, as well. A drawing can show the details of the water cycle, as in **Figure 13.** Architects use drawings to show what the inside of a building will look like. Biologists use drawings to show where the nerves in your arm are found.

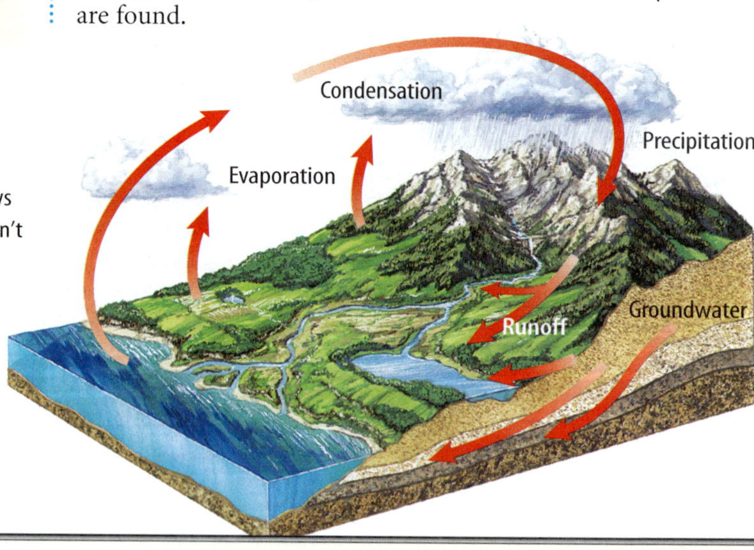

Figure 13 This drawing shows details of the water cycle that can't be seen in a photograph.

Section 3 Resource Manager

Chapter *FAST FILE* Resources
- Transparency Activity, p. 42
- Directed Reading for Content Mastery, pp. 17, 18
- Enrichment, p. 28
- Reinforcement, p. 25
- Lab Worksheet, pp. 7–8

Photographs A still photograph shows an object exactly as it is at a single moment in time. Movies show how an object moves and can be slowed down or sped up to show interesting features. In your schoolwork, you might use photographs in a report. For example, you could show the different types of trees in your neighborhood for a report on ecology.

Tables and Graphs

Everyone who deals with numbers and compares measurements needs an organized way to collect and display data. A **table** displays information in rows and columns so that it is easier to read and understand, as seen in **Table 4.** The data in the table could be presented in a paragraph, but it would be harder to pick out the facts or make comparisons.

A **graph** is used to collect, organize, and summarize data in a visual way. The relationships between the data often are seen more clearly when shown in a graph. Three common types of graphs are line, bar, and circle graphs.

Line Graph A **line graph** shows the relationship between two variables. A variable is something that can change, or vary, such as the temperature of a liquid or the number of people in a race. Both variables in a line graph must be numbers. An example of a line graph is shown in **Figure 14.** One variable is shown on the horizontal axis, or *x*-axis, of the graph. The other variable is placed along the vertical axis, or *y*-axis. A line on the graph shows the relationship between the two variables.

Table 4 Endangered Animal Species in the United States

Year	Number of Endangered Animal Species
1984	192
1986	213
1988	245
1990	263
1992	284
1994	321
1996	324
1998	357
2000	379
2002	389

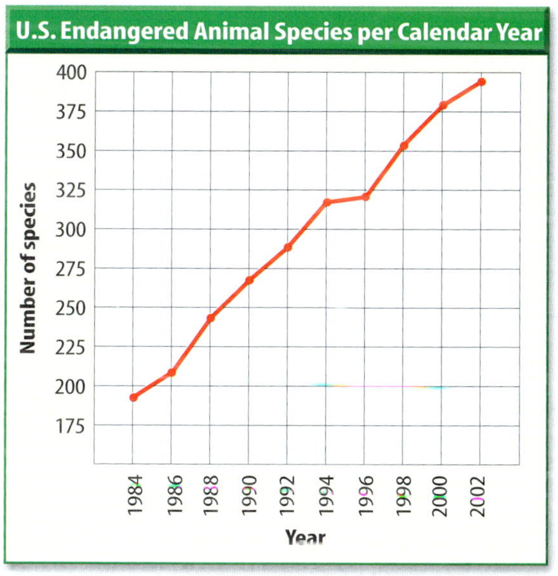

Figure 14 To find the number of endangered animal species in 1992, find that year on the *x*-axis and see what number corresponds to it on the *y*-axis.

Interpret Data *How many species were endangered in 1998?*

2 Teach

Activity

Cross Sections Ask each student to choose an area of science in which he or she is interested. The student should locate a drawing that shows a cross section of an item related to that science. Examples may include the bones inside the arm, the inner structure of a plant, or the gears inside a watch. Students can show and explain their drawings to the class. L3 LS **Visual-Spatial** P

Teacher FYI

Species Count Be sure to point out to students that the numbers for endangered animal species used in the graphs and table on these pages reflect only a small portion of the total number of species in the United States. For example, there are 1,043 species of birds native to the U.S. and only 77 are on the U.S. Endangered Species list; about 7.4%.

Differentiated Instruction

Challenge Have students calculate the percent increase in the number of endangered animal species from 1984 to 2002. L2

Caption Answer

Figure 14 350 species

LAB DEMONSTRATION

Purpose to show how to collect information and organize it as a table and a line graph

Materials graph paper, ten books (of varying weight), centimeter ruler, meterstick

Preparation Set up a table on the board for data collected in this demonstration.

Procedure Measure and record the height of one book placed on a table. Add books one at a time, and measure and record the height of the stack after each addition. Finally, demonstrate how to use the data to make a line graph.

Expected Outcome The graph shows that the height of the stack of books increases.

Assessment

What does the graph tell about the relationship between the number of books and the height of the stack? The height of the stack increases with each additional book, but the rate of increase is not even.

Caption Question Answer

Figure 15 birds

Quick Demo

Materials chalk, chalkboard, some data to graph

Estimated Time five minutes

Procedure Draw two line graphs on the chalkboard having the same data but two different scales for the *y*-axis. Point out to students that one of the lines looks much steeper than the other. Have students study the units on the axes of the graphs to see that the information in both graphs is the same.

Visual Learning

Figure 15 Ask students to calculate the percent values of each endangered species.

Arachnids = 5.9%
Amphibians = 9.8%
Reptiles = 13.7%
Crustaceans = 17.7%
Snails = 19.6%
Insects = 29.4%
Clams = 59.8%
Mammals = 61.8%
Fishes = 66.7%

Discussion

Everyday Graphs Discuss how graphs play a very important role in conveying information to the public. Have students look through a newspaper and find references to graphs. Ask students where else they may find graphs. Possible answers: magazines, doctor's offices, informational brochures, textbooks, reference materials, stores Ask students how these graphs help them understand the product or information. The graphs display the data as an easy-to-understand visual reference. Sometimes the translation of data into text becomes confusing. Graphs make it easier to understand complex information or view the results of an experiment.

Reading Check

pie graph

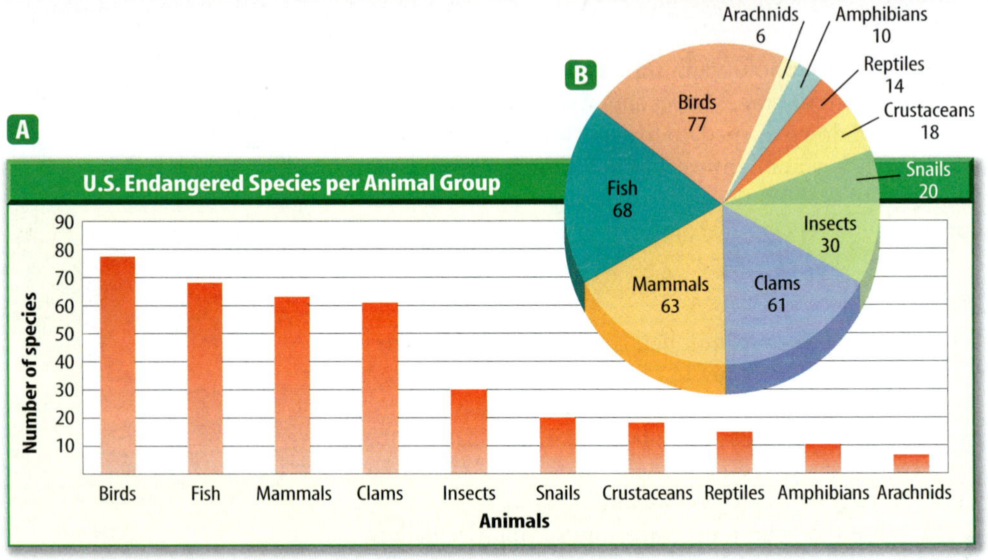

Figure 15 **A** Bar graphs allow you to picture the results easily. **B** On this circle graph, you can see what part of the whole each animal represents.

Infer *Which category of animals has the most endangered species?*

Bar Graph A **bar graph** uses rectangular blocks, or bars, of varying sizes to show the relationships among variables. One variable is divided into parts. It can be numbers, such as the time of day, or a category, such as an animal. The second variable must be a number. The bars show the size of the second variable. For example, if you made a bar graph of the endangered species data from **Figure 14,** the bar for 1990 would represent 263 species. An example of a bar graph is shown in **Figure 15A.**

Circle Graph Suppose you want to show the relationship among the types of endangered species. A **circle graph** shows the parts of a whole. Circle graphs are sometimes called pie graphs. Each piece of pie visually represents a fraction of the total. Looking at the circle graph in **Figure 15B,** you see quickly which animals have the highest number of endangered species by comparing the sizes of the pieces of pie.

A circle has a total of 360°. To make a circle graph, you need to determine what fraction of 360 each part should be. First, determine the total of the parts. In **Figure 15B,** the total of the parts, or endangered species, is 367. One fraction of the total, *Mammals,* is 63 of 367 species. What fraction of 360 is this? To determine this, set up a ratio and solve for *x:*

$$\frac{63}{367} = \frac{x}{360°} \qquad x = 61.8°$$

Mammals will have an angle of 61.8° in the graph. The other angles in the circle are determined the same way.

Reading Check *What is another name for a circle graph?*

Science Online

Topic: Scientific Data

Visit red.msscience.com for Web links to information about scientific illustrations, tables, and graphs and their importance in the scientific community.

Activity Create a table or graph using data collected from a classroom observation.

Active Reading

Metacognition Journal In this strategy, each student analyzes his or her own thought processes. Have students divide the paper in half. On the left, have them record what they have learned about a topic. On the right, have them record the reason they learned it. Have students write a Metacognition Journal about graphs.

Differentiated Instruction

English-Language Learners Have students pair up and compose quick surveys to gather information from the class. They can write their questions on a sheet of paper and either conduct them in front of the class or have students fill them out on paper. When the survey is complete, have the students choose a graph that will best display their data. ELL

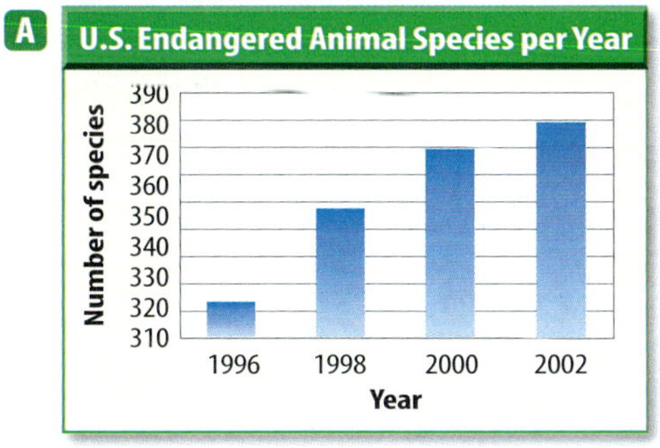

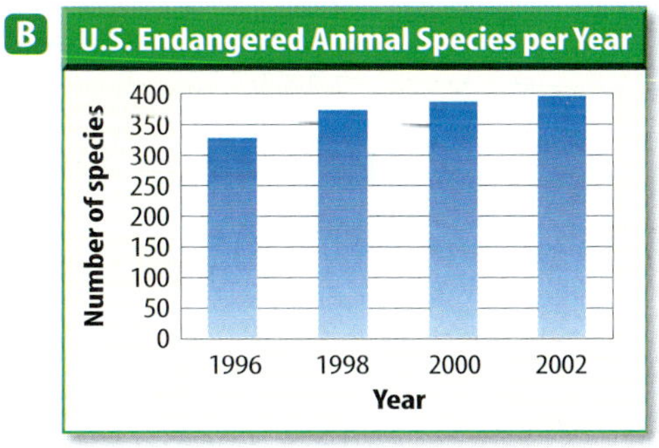

Reading Graphs When you are using or making graphs to display data, be careful—the scale of a graph can be misleading. The way the scale on a graph is marked can create the wrong impression, as seen in **Figure 16A.** Until you see that the *y*-axis doesn't start at zero, it appears that the number of endangered species has quadrupled in just six years.

This is called a broken scale and is used to highlight small but significant changes, just as an inset on a map draws attention to a small area of a larger map. **Figure 16B** shows the same data on a graph that does not have a broken scale. The number of species has only increased 20 percent from 1996 to 2002. Both graphs have correct data, but must be read carefully. Always analyze the measurements and graphs that you come across. If there is a surprising result, look closer at the scale.

Figure 16 Careful reading of graphs is important. **A** This graph does not start at zero, which makes it appear that the number of species has more than quadrupled from 1996–2002. **B** The actual increase is about 20 percent, as you can see from this full graph. The broken scale must be noted in order to interpret the results correctly.

section 3 review

Summary

Scientific Illustrations

- Drawings and illustrations can help people visualize complex concepts.
- A drawing can include details you see and those that are hidden.
- Photographs are an exact representation of an object at a single moment in time.

Tables and Graphs

- Tables display information while graphs are used to summarize data.
- A line graph shows the relationship between two variables, a bar graph shows the relationship among variables, and a circle graph shows the parts of a whole.
- It is important to pay close attention to the scale on graphs in order to analyze the information.

Self Check

1. **Explain** how to use **Figure 16** to find the number of endangered species in 1998.
2. **Infer** what type of graph you would use to display data gathered in a survey about students' after-school activities.
3. **Think Critically** Why is it important to be careful when making or using graphs?
4. **Describe** a time when an illustration would be helpful in everyday activities.
5. **Identify** when you would use a broken scale.

Applying Skills

6. **Use a Spreadsheet** Make a spreadsheet to display how the total mass of a 500-kg elevator changes as 50-kg passengers are added one at a time.

Science Online red.msscience.com/self_check_quiz

3 Assess

DAILY INTERVENTION

Check for Understanding

Visual-Spatial Have students collect tables and graphs from newspapers or magazines. Display their information on a bulletin board with the following headings: table, line graph, bar graph, circle graph, other graphs, misleading graphs. Ask students to place their graphs in the appropriate category. Discuss placement of each or some of the graphs. L2 COOP LEARN IS

Reteach

Types of Graphs Have each student make a chart of the different types of graphs defined in the section. Then have them write the key words that define each under the type of graph. line graph: shows relationship between two number variables; bar graph: shows relationship between a number variable and a number or category variable; circle graph: shows relationship among parts of a whole L2 IS **Visual-Spatial**

Assessment

Performance Have students collect weather data for a week and display it in a table and a graph. Use **Performance Assessment in the Science Classroom,** p. 111.

Virtual Labs

Graphs *How are graphs used to represent data?*

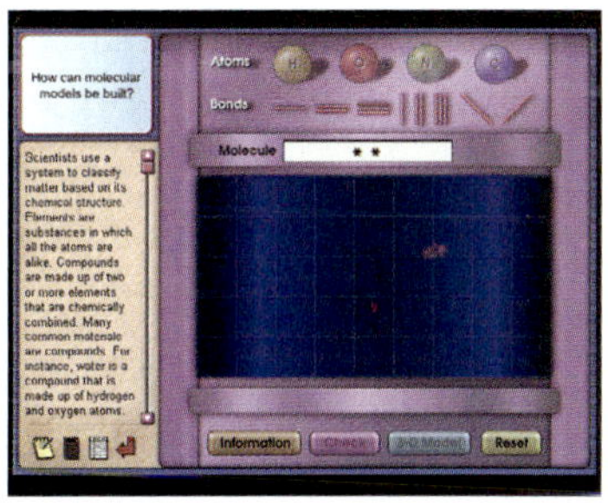

section 3 review

1. Possible answer: Find the bar for 1998. Use a ruler or visually align the top of the bar with the time scale. This tells you the number of endangered species for that year.
2. A bar graph or a circle graph would be appropriate. A circle graph would show the percent of students who prefer each activity. A bar graph would show the number of students who prefer each activity.
3. It is important to be careful when making graphs because the data is displayed visually. It must be as accurate as possible. There is often a key to help understand the data displayed in the graph, but the lines and numbers must be clearly identified and labels should be used to help others understand the graph.
4. Illustrations could be used to display daily schedules, maps and directions, and confusing instructions.
5. A broken scale is used to highlight small but important changes.
6. Accept all reasonable spreadsheets

BENCH TESTED

Real-World Question

Purpose to apply the concepts of precision, measurement, and graphing to a specific problem L2 ELL LS **Kinesthetic**

Process Skills design an experiment, form a hypothesis, communicate, observe and infer, make and use tables, interpret data

Time Required 45 minutes each to plan and do the experiment

Materials meterstick, stopwatch

Alternate Materials clock with a second hand

Safety Precautions Work in an area where it is safe to run. If this is not possible, measure walking speed only. Check that students do not have health problems that could prevent them from exercising. Allow students to decline running if it is physically uncomfortable or embarrassing for them.

Form a Hypothesis

Possible Hypotheses A course on a flat, straight surface will be the most accurately measurable with the given tools. Accuracy will be better if more than one group member times each walker or runner.

LAB Design Your Own

PACE YOURSELF

Goals

- **Design** an experiment that allows you to measure speed for each member of your group accurately.
- **Display** data in a table and a graph.

Possible Materials

meterstick
stopwatch
**watch with a second hand*

**Alternate materials*

Safety Precautions

Work in an area where it is safe to run. Participate only if you are physically able to exercise safely. As you design your plan, make a list of all the specific safety and health precautions you will take as you perform the investigation. Get your teacher's approval of the list before you begin.

Real-World Question

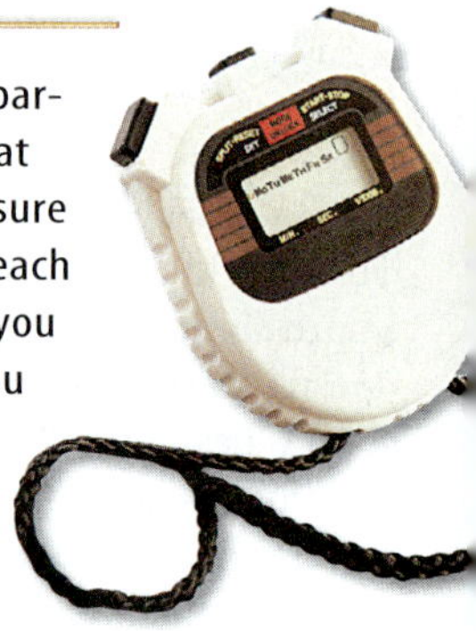

Track meets and other competitions require participants to walk, run, or wheel a distance that has been precisely measured. Officials make sure all participants begin at the same time, and each person's time is stopped at the finish line. If you are practicing for a local marathon or 10K, you need to know your speed or pace in order to compare it with those of other participants. How can your performance be measured accurately? How will you measure the speed of each person in your group? How will you display these data?

Form a Hypothesis

Think about the information you have learned about precision, measurement, and graphing. In your group, make a hypothesis about a technique that will provide you with the most precise measurement of each person's pace.

Test Your Hypothesis

Possible Procedures Measure the distance to be walked. Mark the starting and ending points for the walk. For each student, use a stopwatch to measure walking time to the nearest tenth of a second. Record each student's time on a data table. Graph the results. Repeat these steps, replacing the walk with a run.

Time to Walk and Run 18 m

Student	Walking Time	Running Time	Walking Speed	Running Speed
Carlos	21.3 s	10.6 s	0.845 m/s	1.70 m/s
Brianna	19.6 s	10.9 s	0.918 m/s	1.65 m/s

Using Scientific Methods

Test Your Hypothesis

Make a Plan

1. As a group, decide what materials you will need.
2. How far will you travel? How will you measure that distance? How precise can you be?
3. How will you measure time? How precise can you be?
4. List the steps and materials you will use to test your hypothesis. Be specific. Will you try any part of your test more than once?
5. Before you begin, create a data table. Your group must decide on its design. Be sure to leave enough room to record the results for each person's time. If more than one trial is to be run for each measurement, include room for the additional data.

Follow Your Plan

1. Make sure that your teacher approves your plan before you start.
2. Carry out the experiment as planned and approved.
3. Be sure to record your data in the data table as you proceed with the measurements.

Analyze Your Data

1. **Graph** your data. What type of graph would be best?
2. Are your data table and graph easy to understand? Explain.
3. How do you know that your measurements are precise?
4. Do any of your data appear to be out of line with the rest?

Conclude and Apply

1. **Explain** how it is possible for different members of a group to find different times while measuring the same event.
2. **Infer** what tools would help you collect more precise data.
3. What other data displays could you use? What are the advantages and disadvantages of each?

Communicating Your Data

Make a larger version of your graph to display in your classroom with the graphs of other groups. For more help, refer to the Science Skill Handbook.

Teaching Strategy Demonstrate to students how to use the meterstick to measure walking/running distance and how to stop, start, and reset the stopwatch.

Expected Outcome Most results will show students moving at different speeds from each other. Often, two students recording the same event will measure different times.

Analyze Your Data

Answers to Questions

1. bar graph
2. Check students' work. A person unfamiliar with the experiment should be able to quickly grasp the results.
3. Students should explain their efforts to measure precisely.
4. Students should use their bar graphs to answer this question.

Error Analysis Have students compare their results and their hypotheses and explain why differences occurred.

Conclude and Apply

1. People's reaction times with the stopwatch may be different. People watching the race from different angles may see the runner pass the finish line at slightly different times. Stopwatches may vary.
2. Sample response: global positioning data, a timer that used infrared or laser light to determine start and stop times
3. Sample response: Graph with a different scale, bar graph or data table comparing average time for male students versus average time for female students; each arranges the information in different ways, so more relationships can be seen.

Assessment

Performance Ask students to design a method to measure another speed, such as the speed of a person swimming two laps in a pool. Use **Performance Assessment in the Science Classroom,** p. 95.

Communicating Your Data

Have each group use its graph to explain its data to the class.

Content Background

The rafflesia is a parasitic plant that grows in the mountains of Malaysia. This plant has adapted to feed off of the roots of large vines that are in the grape family. The large, fleshy flower remains open for five to seven days. The flower emits a very strong, offensive odor that attracts carrion-feeding flies. It is thought that the flies are the source of pollination.

The eruption of Krakatau in 1883 was so loud that the explosion was heard 3,500 km (2,175 mi) away. A series of eruptions discharged so much ash and dust into the air that the area surrounding the volcano was plunged into darkness for days.

Discussion

Magnitude Students may not have a good grasp of the magnitude of these features. Make the conversion calculations to English units on the board as a class to give students a better idea of how large these items are. Useful conversion factors (numbers are rounded off): 1 kg = 2.2 lbs; 1 m = 3.281 ft. Converted Numbers: 11 kg = 24.2 lbs; 1 m = 39.37 in; 1,800 m = 5,900 ft; 110 m = 360 ft; 452 m = 1,480 ft; 442 m = 1,450 ft; 33.5 m = 110 ft (three significant figures); 1.65 m = 5.41 ft

Applying Math

Answer 33.5 or 34

SCIENCE Stats

Biggest, Tallest, Loudest

Did you know...

... The world's most massive flower belongs to a species called *Rafflesia* (ruh FLEE zhee uh) and has a mass of up to 11 kg. The diameter, or the distance across the flower's petals, can measure up to 1 m.

... The world's tallest building is the Petronus Towers in Kuala Lumpur, Malaysia. It is 452 m tall. The tallest building in the United States is Chicago's Sears Tower, shown here, which measures 442 m.

Applying Math How many of the largest rafflesia petals would you have to place side by side to equal the height of the Sears Tower?

...One of the loudest explosions on Earth was the 1883 eruption of Krakatau (krah kuh TAHEW), an Indonesian volcano. It was heard from more than 3,500 km away.

Write About It

Visit red.msscience.com/science_stats to find facts that describe some of the shortest, smallest, or fastest things on Earth. Create a class bulletin board with the facts you and your classmates find.

62 CHAPTER 2 Measurement

Write About It

Shortest, Smallest, Fastest Have students research some of the shortest, smallest, or fastest things on Earth and record this information on a bulletin board. A good source of information for this may be *The Guinness Book of World Records.*

Differentiated Instruction

Challenge Have students research the heights of different historical features worldwide. The features should include both human-made and natural landforms. Students can graph the measurements they find and then compare the differences in height, length, and width. L3

Reviewing Main Ideas

Section 1 Description and Measurement

1. Length, volume, mass, temperature, and rates are used to describe objects and events.
2. Estimation is used to make an educated guess at a measurement.
3. Accuracy describes how close a measurement is to the true value. Precision describes how close measurements are to each other.

Section 2 SI Units

1. The international system of measurement is called SI. It is used throughout the world for communicating data.
2. The SI unit of length is the meter. Volume—the amount of space an object occupies—can be measured in cubic meters. The mass of an object is measured in kilograms.

Section 3 Drawings, Tables, and Graphs

1. Tables, photographs, drawings, and graphs are tools used to collect, organize, summarize, and display data in a way that is easy to use and understand.
2. Line graphs show the relationship between two variables that are numbers on an x-axis and a y-axis. Bar graphs divide a variable into parts to show a relationship. Circle graphs show the parts of a whole like pieces of a pie.

Visualizing Main Ideas

Copy and complete the following concept map.

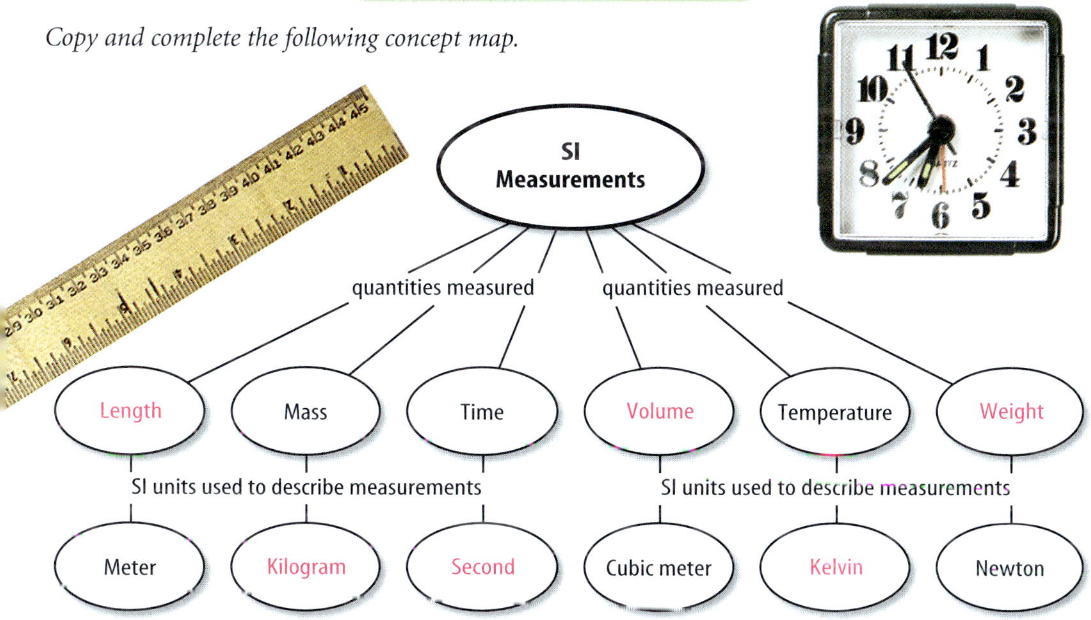

Reviewing Main Ideas

Summary statements can be used by students to review the major concepts of the chapter.

Visualizing Main Ideas

See student page.

Science Online

Visit red.msscience.com
/self_check_quiz
/interactive_tutor
/vocabulary_puzzlemaker
/chapter_review
/standardized_test

Assessment Transparency

For additional assessment questions, use the *Assessment Transparency* located in the transparency book.

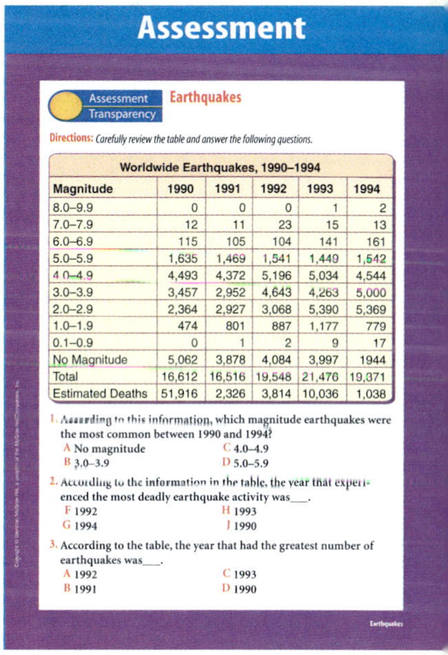

Assessment

Assessment Transparency **Earthquakes**

Directions: *Carefully review the table and answer the following questions.*

Worldwide Earthquakes, 1990–1994					
Magnitude	1990	1991	1992	1993	1994
8.0–9.9	0	0	0	1	2
7.0–7.9	12	11	23	15	13
6.0–6.9	115	105	104	141	161
5.0–5.9	1,635	1,469	1,541	1,449	1,542
4.0–4.9	4,493	4,372	5,196	5,034	4,544
3.0–3.9	3,457	2,952	4,643	4,263	5,000
2.0–2.9	2,364	2,927	3,068	5,390	5,369
1.0–1.9	474	801	887	1,177	779
0.1–0.9	0	1	2	9	17
No Magnitude	5,062	3,878	4,084	3,997	1944
Total	16,612	16,516	19,548	21,476	19,371
Estimated Deaths	51,916	2,326	3,814	10,036	1,038

1. According to this information, which magnitude earthquakes were the most common between 1990 and 1994?
   A No magnitude
   B 3.0–3.9
   C 4.0–4.9
   D 5.0–5.9
2. According to the information in the table, the year that experienced the most deadly earthquake activity was___.
   F 1992
   G 1994
   H 1993
   J 1990
3. According to the table, the year that had the greatest number of earthquakes was___.
   A 1992
   B 1991
   C 1993
   D 1990

Earthquakes

Using Vocabulary

1. meter
2. measurement
3. estimation
4. mass
5. circle graph
6. precision
7. Kelvin
8. SI
9. volume

Checking Concepts

10. D
11. D
12. C
13. A
14. D
15. C
16. A
17. C
18. B

chapter 2 Review

Using Vocabulary

accuracy p. 45	measurement p. 42
bar graph p. 58	meter p. 51
circle graph p. 58	precision p. 44
estimation p. 43	rate p. 54
graph p. 57	SI p. 50
kelvin p. 54	table p. 57
kilogram p. 53	volume p. 52
line graph p. 57	weight p. 53
mass p. 53	

Each phrase below describes a vocabulary word. Write the word that matches the phrase describing it.

1. the SI unit for length
2. a description with numbers
3. a method of making a rough measurement
4. the amount of matter in an object
5. a graph that shows parts of a whole
6. a description of how close measurements are to each other
7. the SI unit for temperature
8. an international system of units
9. the amount of space an object occupies

Checking Concepts

Choose the word or phrase that best answers the question.

10. The measurement 25.81 g is precise to the nearest
    A) gram.
    B) kilogram.
    C) tenth of a gram.
    D) hundredth of a gram.
11. What is the SI unit of mass?
    A) kilometer **C)** liter
    B) meter **D)** kilogram
12. What would you use to measure length?
    A) graduated cylinder
    B) balance
    C) meterstick
    D) spring scale
13. The cubic meter is the SI unit of what?
    A) volume **C)** mass
    B) weight **D)** distance
14. Which term describes how close measurements are to each other?
    A) significant digits
    B) estimation
    C) accuracy
    D) precision
15. Which of the following is a temperature scale?
    A) volume **C)** Celsius
    B) mass **D)** Mercury
16. Which of the following is used to organize data?
    A) table **C)** precision
    B) rate **D)** meterstick

17. To show the number of wins for each football team in your district, which of the following would you use?
    A) photograph **C)** bar graph
    B) line graph **D)** SI
18. To show 25 percent on a circle graph, the section must measure what angle?
    A) 25° **C)** 180°
    B) 90° **D)** 360°

ScienceOnline red.msscience.com/vocabulary_puzzlemaker

Use the ExamView® Pro Testmaker CD-ROM to:
- create multiple versions of tests
- create modified tests with one mouse click for inclusion students
- edit existing questions and add your own questions
- build tests aligned with state standards using built-in State Curriculum Tags
- change English tests to Spanish with one mouse click and vice versa

Thinking Critically

19. Infer How would you estimate the volume your backpack could hold?

20. Explain Why do scientists in the United States use SI rather than the English system (feet, pounds, pints, etc.) of measurement?

21. List the following in order from smallest to largest: 1 m, 1 mm, 10 km, 100 mm.

22. Describe Give an example of an instance when you would use a line graph. Could you use a bar graph for the same purpose?

23. Compare and contrast volume, length, and mass. How are they similar? Different? Give several examples of units that are used to measure each quantity. Which units are SI?

24. Infer Computer graphics artists can specify the color of a point on a monitor by using characters for the intensities of three colors of light. Why was this method of describing color invented?

Use the photo below to answer question 25.

25. Interpreting Scientific Illustrations What does the figure show? How has this drawing been simplified?

Performance Activities

26. Newspaper Search Look through a week's worth of newspapers and evaluate any graphs or tables that you find.

Applying Math

Use the table below to answer question 27.

Areas of Bodies of Water

Body of Water	Area (km^2)
Currituck Sound (North Carolina)	301
Pocomoke Sound (Maryland/Virginia)	286
Chincoteague Bay (Maryland/Virginia)	272
Core Sound (North Carolina)	229

27. Make and Use Graphs The table shows the area of several bodies of water. Make a bar graph of the data.

Use the illustration below to answer question 28.

28. Travel Distances The map above shows the driving distance from New York City to Denver, Colorado in kilometers. Convert the distance to meters and then find out how many meters are in a mile and convert the distance to miles.

29. Round Digits Round the following numbers to the correct number of significant digits.

42.86 kg × 38.703 kg

10 g × 25.05 g

5.8972 nm × 34.15731 nm

Science Online red.msscience.com/chapter_review

Assessment Resources

Reproducible Masters

Chapter *Fast File* Resources
Chapter Review, pp. 33–34
Chapter Tests, pp. 35–38
Assessment Transparency Activity, p. 45

Glencoe Science Web site
Chapter Review Test
Standardized Test Practice

Glencoe Technology

- Assessment Transparency
- Exam*View*® Pro Testmaker
- MindJogger Videoquiz
- Interactive Chalkboard

Thinking Critically

19. Multiply its length times its width times its height.

20. The SI system of measurement is understood internationally, while the English system is understood in only a limited number of countries.

21. 1 mm, 100 mm, 1 m, 10 km

22. Accept any reasonable answer. Sample answer: To show distance traveled versus time of travel; no, you cannot use a bar graph for this.

23. Volume, length, and mass are all properties that can be measured. Volume is the amount of space an object occupies; length is the distance between two points; mass is the amount of matter in an object. Sample units: cubic meter (volume); meter (length); kilogram (mass). These are all SI units.

24. Sample response: to allow for precision and reproducibility in design and printing; for example, ads for a given product all use exactly the same colors.

25. It shows Earth, the Moon, and the Sun. Size and distance are not to scale, and the inner planets are missing.

Performance Activities

26. Use **Performance Assessment in the Science Classroom,** p. 99.

Applying Math

National Math Standards

1, 4, 5

27. Check students' graphs.

28. 1000 m = 1 km; 0.62 miles = 1 km; 2,859.85 km × 1000 m/1 km = 2,859,850 m

2,859.85 km × 0.62 miles/1 km = 1773.11 miles

29. 1,658.81; 250; 201.4325; traveled versus time of travel; no, you cannot use a bar graph for this.

FAST FILE

Answer Sheet A practice answer sheet can be found at red.msscience.com/answer_sheet.

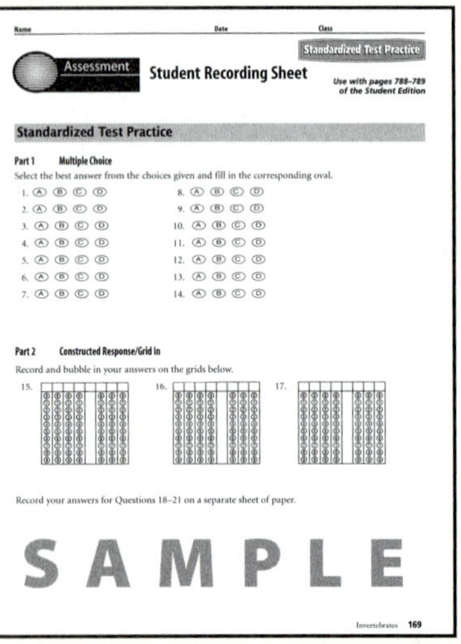
Name Date Class

Assessment Student Recording Sheet

Standardized Test Practice

Use with pages 788–789 of the Student Edition

Standardized Test Practice

Part 1 Multiple Choice

Select the best answer from the choices given and fill in the corresponding oval.

Part 2 Constructed Response/Grid In

Record and bubble in your answers on the grids below.

Record your answers for Questions 18–21 on a separate sheet of paper.

SAMPLE

Part 1 Multiple Choice

1. A	5. B
2. C	6. C
3. B	7. A
4. D	8. D

Part 2 Short Response

9. 95.3, with three significant digits, is less precise than 35.77, with four significant digits. The answer is only significant to the tenths place, so it must be rounded from 131.07 g to 131.1 g.
10. From least to most precise: sundial, wall clock, stopwatch, atomic clock.
11. Volume is the amount of space occupied by an object. The volume of the cube is 4 cm x 4 cm x 4 cm, or 64 cm^3. 64 cm^3 = 64 mL.
12. Measure and record a volume of water. Immerse the cube in the water, and record the new volume. Subtract the initial volume from the new volume to find the volume of the cube.

chapter 2 Standardized Test Practice

Part 1 Multiple Choice

Record your answers on the answer sheet provided by your teacher or on a sheet of paper.

1. Which best describes measurements that are accurate?
   A. They are very close to an accepted value.
   B. They are based on an estimate.
   C. They are very close to each other.
   D. They are not based on numbers.

2. The mass of a sample of calcium chloride is 33.755 grams. Round to the nearest hundredth of a gram.
   A. 33.8 g
   B. 34 g
   C. 33.76 g
   D. 33.75 g

Use the illustration below to answer questions 3 and 4.

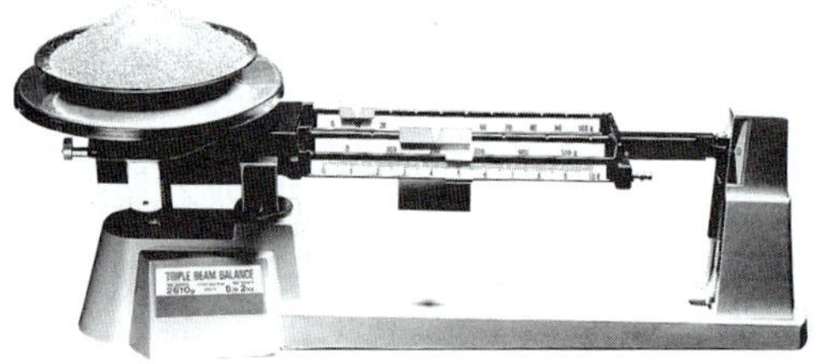

3. Which quantity is measured using this tool?
   A. weight **C.** volume
   B. mass **D.** length

4. Which measurement does this balance show?
   A. about 315 g **C.** about 325 g
   B. about 326 g **D.** about 215 g

Test-Taking Tip

Take Your Time Read carefully and make notes of the units used in any measurement.

Use the graph below to answer questions 5 and 6.

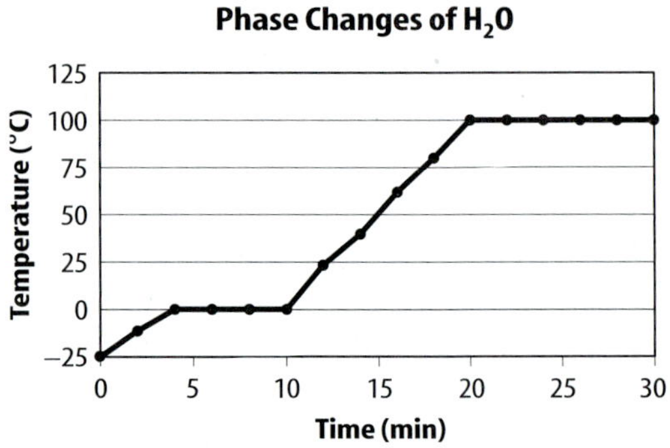

5. The graph shows data from an experiment in which ice was heated until it melted, then became steam. In what phase is the H_2O at 16 minutes?
   A. solid **C.** gas
   B. liquid **D.** plasma

6. Which statement describes the trend evident in the data?
   A. Temperature continually increased as time increased.
   B. Temperature did not change as time increased.
   C. Temperature increases were divided by plateaus as time increased.
   D. Temperature continually decreased as time increased.

7. Which of these represents 1/1000th of a meter?
   A. mm **C.** ms
   B. km **D.** dm

8. Which is NOT a unit of volume?
   A. milliliter
   B. cubic centimeter
   C. deciliter
   D. kelvin

13.

Mass	Weight
Measures the amount of matter in an object	A measurement of force
Does not change with location	Can change with location
Is not affected by gravitational changes	Depends on gravity
Measured using a balance	Measured using a scale

14. Relationships between data are often more clear when the data is shown on a graph.
15. Drawings can be used to communicate ideas and information, sometimes more clearly than written text. Many times, it is easier to visualize how all the parts of an item fit together than reading about it. Drawings can show things that are too large to see at once, as well as the details of objects, which are very small. Drawings can show things which are usually hidden.

Part 2 Short Response/Grid In

Record your answers on the answer sheet provided by your teacher or on a sheet of paper.

9. 35.77 g of Solid A are mixed with 95.3 g of Solid B. Write the mass of the mixture with the correct number of significant digits. Explain your answer.

10. Arrange these measuring tools in order from least to most precise: stopwatch measuring to 1/100ths of a second, atomic clock, sundial, wall clock with 2 hands.

Use the illustration below to answer questions 11 and 12.

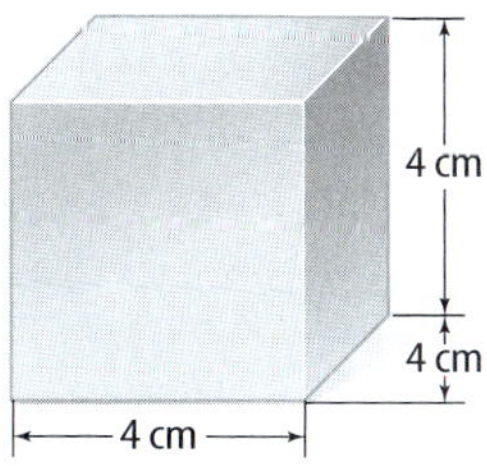

11. Define the term volume. Calculate the volume of the cube shown above. Give your answer in cm^3 and mL.

12. Describe how you would find the volume of the cube using the immersion method.

13. Create a table which shows the differences between mass and weight.

14. How do graphs make it easier to analyze data?

15. Why are drawings an effective way to communicate information?

16. Explain the difference between precision and accuracy.

Part 3 Open Ended

Record your answers on a sheet of paper.

17. A recipe calls for 1 cup sugar, 1 cup flour, and 1 cup milk. Define the term precision, and explain how to measure these ingredients precisely using kitchen tools.

18. While shopping, you find a rug for your room. Without a ruler, you must estimate the rug's size to determine if it will fit in your room. How will you proceed?

19. Identify the most appropriate SI length unit to measure the following: your height, the distance between two cities, the width of a computer screen, the radius of a coin, the length of a muscle cell. How are units converted in the SI system?

Use the figure below to answer questions 20 and 21.

Composition of Earth's Atmosphere

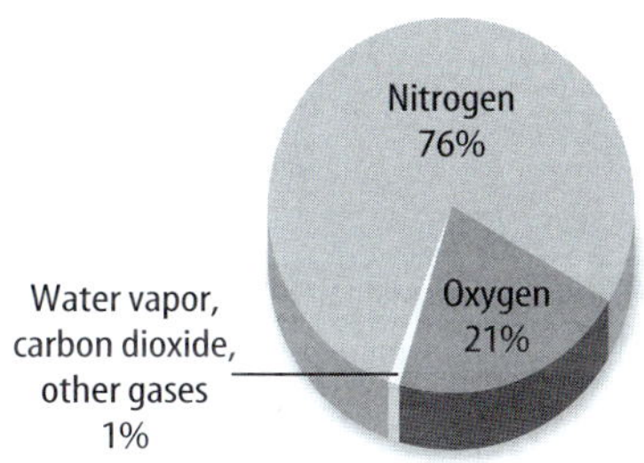

20. Calculate the angle of each section in this circle graph.

21. Create a bar graph using this data.

22. Measurement is a part of everyday life. Describe measurements someone might make as part of his or her normal activities.

23. You must decide what items to pack for a hiking trip. Space is limited, and you must carry all items during hikes. What measurements are important in your preparation?

Science Online red.msscience.com/standardized_test

16. Precision is how close measurements are to each other. Accuracy is a comparison of a measurement to the actual or accepted value.

Part 3 Open Ended

17. Precision is a description of how close measurements are to each other. To measure the ingredients precisely, use the same measuring cup each time, pouring each ingredient to the exact same height in the cup.

18. Possible techniques: estimate the length of a hand, arm, or foot, and use this body part to mark and add increments along the rug; estimate the length of an object like a book or piece of string, and use this object to mark and add increments along the rug.

19. Height: meter; distance between 2 cities: kilometer; width of computer screen: centimeter or decimeter; radius of a coin: millimeter or centimeter; length of a muscle cell: micrometer or nanometer. SI units are converted by multiplying by the appropriate power of 10.

20. Nitrogen: $\frac{78}{100} = \frac{x}{360°}$; $x = 280.8°$

Oxygen: $\frac{21}{100} = \frac{x}{360°}$; $x = 75.6°$

Other: $\frac{1}{100} = \frac{x}{360°}$; $x = 3.6°$

21. check student's work

22. Student responses will vary but may include examples such as cooking, buying paint, clothes, or carpet. Accept all reasonable answers.

23. The length, width, and height, or total amount of volume in the duffel bag should determine which items would fit. Mass is important, as massive objects, which must be carried on hikes will be difficult to manage.

Rubrics

For more help evaluating open-ended assessment questions, see the rubric on p. 10T.

chapter 3 Organizer

Section/Objectives	Standards		Labs/Features
	National	**State/Local**	
Chapter Opener	See pp. 16T–17T for a Key to Standards.		**Launch Lab:** Can you classify pennies by their properties?, p. 69 **Foldables,** p. 69
Section 1 Physical Properties and Changes 5 sessions 2 blocks 1. **Identify** physical properties of matter. 2. **Explain** why materials with different masses have different densities. 3. **Observe** water displacement to determine volume. 4. **Describe** the states of matter. 5. **Determine** how temperature changes affect substances. 6. **Classify** matter using physical properties.	National Content Standards: UCP.1, UCP.2, UCP.3, UCP.5, A.1, A.2, B.1, B.2, B.3		**Science Online,** p. 71 **Science Online,** p. 72 **MiniLAB:** Determining Volume, p. 73 **Integrate Language Arts,** p. 75 **Visualizing Dichotomous Keys,** p. 78
Section 2 Chemical Properties and Changes 5 sessions 3 blocks 7. **Recognize** chemical properties. 8. **Identify** chemical changes. 9. **Classify** matter according to chemical properties. 10. **Describe** the law of conservation of mass.	National Content Standards: UCP.1, UCP.2, UCP.3, UCP.5, A.1, A.2, B.1, B.3		**Integrate Health,** p. 81 **MiniLAB:** Observing Yeast, p. 84 **Applying Science:** Do light sticks conserve mass?, p. 85 **Lab:** Liquid Layers, p. 87 **Lab:** Fruit Salad Favorites, p. 88 **Science and History:** The Road to Understanding Matter, p. 90

Glencoe Exclusive! TeacherWorks™ All-In-One Planner and Resource Center

Lab Materials	Reproducible Resources	Section Assessment	Technology
Launch Lab: collection of pennies	**Chapter *Fast File* Resources** Foldables Worksheet, p. 17 Directed Reading Overview, p. 19 Note-taking Worksheets, pp. 31–32	Glencoe's Assessment Advantage	TeacherWorks includes: • Interactive Teacher Edition • Lesson Planner with calendar • Access to all program blacklines • Correlations to standards • Web links
MiniLAB: three objects of the same size such as a marble, rubber ball, and wood sphere; 100-mL graduated cylinder; water *Need materials?* Contact Science Kit at 1-800-828-7777 or www.sciencekit.com on the Internet.	**Chapter *Fast File* Resources** Transparency Activity, p. 42 MiniLAB, p. 3 Enrichment, p. 29 Reinforcement, p. 27 Directed Reading, p. 20 Lab Activity, pp. 9–11 Transparency Activity, pp. 45–46 **Science Inquiry Labs,** pp. 43–44 **Reading and Writing Skill Activities,** p. 17	**Portfolio** Curriculum Connection, p. 74 **Performance** MiniLAB, p. 73 Applying Math, p. 79 **Content** Section Review, p. 79	Section Focus Transparency Teaching Transparency Virtual Labs CD-ROM Guided Reading Audio Program Interactive Chalkboard CD-ROM
MiniLAB: dry yeast, magnifying lens, warm water, sugar **Lab:** 250-mL beaker, graduated cylinder, corn syrup, glycerin, water, corn oil, rubbing alcohol, penny, wood sphere, rubber ball **Lab:** bananas, apples, pears, 2 mixing bowls, lemon/water solution, paring knife	**Chapter *Fast File* Resources** Transparency Activity, p. 43 MiniLAB, p. 4 Enrichment, p. 30 Reinforcement, p. 28 Directed Reading, pp. 21, 22 Lab Worksheet, pp. 5–6, 7–8 **Mathematics Skill Activities,** p. 9 **Reading and Writing Skill Activities,** p. 1 **Life Science Critical Thinking/ Problem Solving,** p. 8 **Lab Management and Safety,** p. 38	**Portfolio** Science Journal, p. 83 Curriculum Connection, p. 84 **Performance** MiniLAB, p. 84 Applying Science, p. 85 Applying Math, p. 86 **Content** Section Review, p. 86	Section Focus Transparency Virtual Labs CD-ROM Guided Reading Audio Program Interactive Chalkboard CD-ROM Video Lab

End of Chapter Assessment

Glencoe's Assessment Advantage

Blackline Masters	Technology	Professional Series
Chapter *Fast File* Resources Chapter Review, pp. 35–36 Chapter Tests, pp. 37–40 **Standardized Test Practice,** pp. 15–18	MindJogger Videoquiz Virtual Labs CD-ROM Exam*View*® Pro Testmaker TeacherWorks CD-ROM Interactive Chalkboard CD-ROM	**Performance Assessment in the Science Classroom (PASC)**

Transparencies

Section Focus

Section 1 Section Focus Transparency **Water, Water Everywhere**

How much water can you find here? Actually, water is almost everywhere.

1. Where do you see water as a solid? A liquid?
2. What happens to snow on a sunny, warm day? What happens to water in a kettle if you light a burner underneath? How are these events similar?

L2

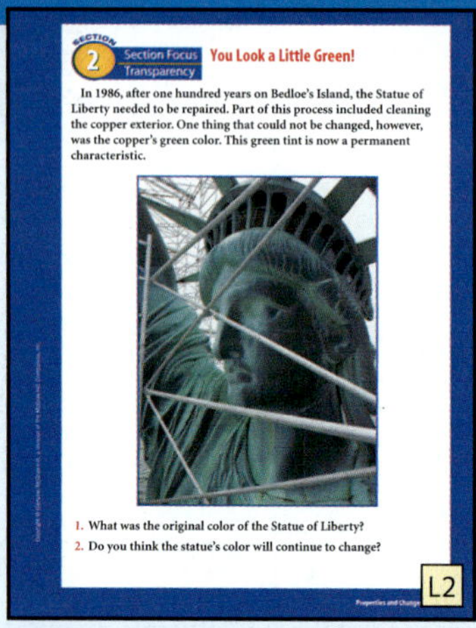
Section 2 Section Focus Transparency **You Look a Little Green!**

In 1986, after one hundred years on Bedloe's Island, the Statue of Liberty needed to be repaired. Part of this process included cleaning the copper exterior. One thing that could not be changed, however, was the copper's green color. This green tint is now a permanent characteristic.

1. What was the original color of the Statue of Liberty?
2. Do you think the statue's color will continue to change?

L2

This is a representation of key blackline masters available in the Teacher Classroom Resources. See Resource Manager boxes within the chapter for additional information.

Assessment

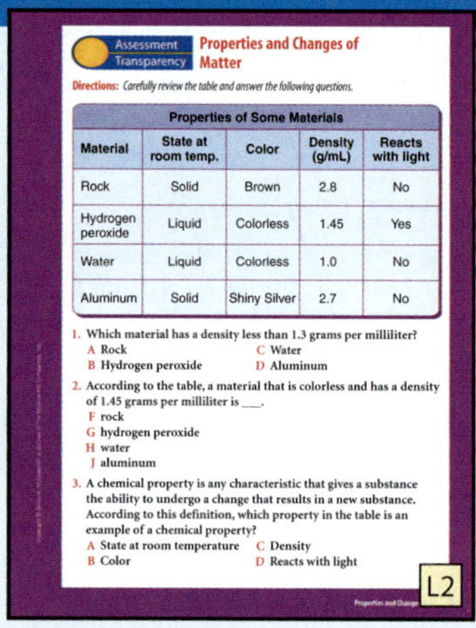
Assessment Transparency **Properties and Changes of Matter**

Directions: Carefully review the table and answer the following questions.

Properties of Some Materials

Material	State at room temp.	Color	Density (g/mL)	Reacts with light
Rock	Solid	Brown	2.8	No
Hydrogen peroxide	Liquid	Colorless	1.45	Yes
Water	Liquid	Colorless	1.0	No
Aluminum	Solid	Shiny Silver	2.7	No

1. Which material has a density less than 1.3 grams per milliliter?
   A Rock C Water
   B Hydrogen peroxide D Aluminum
2. According to the table, a material that is colorless and has a density of 1.45 grams per milliliter is ___.
   F rock
   G hydrogen peroxide
   H water
   J aluminum
3. A chemical property is any characteristic that gives a substance the ability to undergo a change that results in a new substance. According to this definition, which property in the table is an example of a chemical property?
   A State at room temperature C Density
   B Color D Reacts with light

L2

Teaching

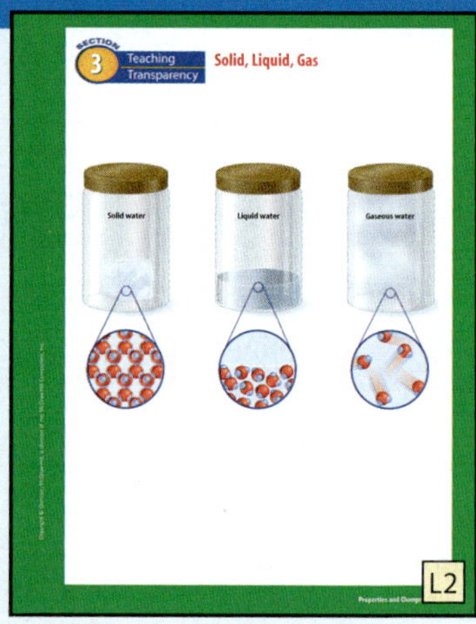
Section 3 Teaching Transparency **Solid, Liquid, Gas**

L2

Key to Teaching Strategies

The following designations will help you decide which activities are appropriate for your students.

- **L1** Level 1 activities should be appropriate for students with learning difficulties.
- **L2** Level 2 activities should be within the ability range of all students.
- **L3** Level 3 activities are designed for above-average students.
- **ELL** ELL activities should be within the ability range of English Language Learners.
- **COOP LEARN** Cooperative Learning activities are designed for small group work.
- **LS** Multiple Learning Styles logos, as described on page 12T, are used throughout to indicate strategies that address different learning styles.
- **P** These strategies represent student products that can be placed into a best-work portfolio.
- **PBL** Problem-Based Learning activities apply real-world situations to learning.

Hands-on Activities

Student Text Lab Worksheet

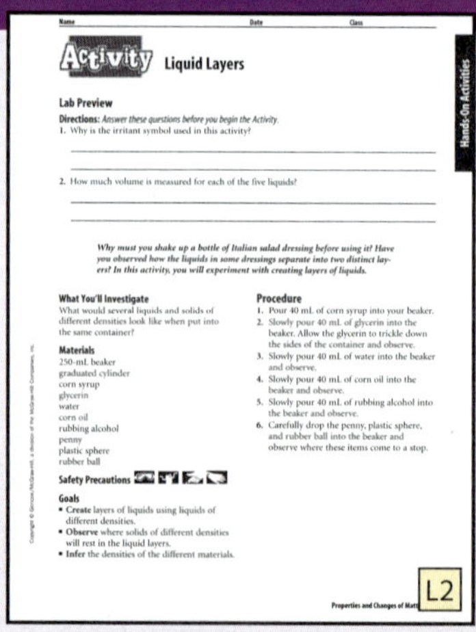
Name Date Class

Activity **Liquid Layers**

Lab Preview

Directions: *Answer these questions before you begin the Activity.*

1. Why is the irritant symbol used in this activity?
2. How much volume is measured for each of the five liquids?

Why must you shake up a bottle of Italian salad dressing before using it? Have you observed how the liquids in some dressings separate into two distinct layers? In this activity, you will experiment with creating layers of liquids.

What You'll Investigate

What would several liquids and solids of different densities look like when put into the same container?

Materials

250-mL beaker
graduated cylinder
corn syrup
glycerin
water
corn oil
rubbing alcohol
penny
plastic sphere
rubber ball

Safety Precautions

Goals

- **Create** layers of liquids using liquids of different densities.
- **Observe** where solids of different densities will rest in the liquid layers.
- **Infer** the densities of the different materials.

Procedure

1. Pour 40 mL of corn syrup into your beaker.
2. Slowly pour 40 mL of glycerin into the beaker. Allow the glycerin to trickle down the sides of the container and observe.
3. Slowly pour 40 mL of water into the beaker and observe.
4. Slowly pour 40 mL of corn oil into the beaker and observe.
5. Slowly pour 40 mL of rubbing alcohol into the beaker and observe.
6. Carefully drop the penny, plastic sphere, and rubber ball into the beaker and observe where these items come to a stop.

Hands-On Activities

L2

Laboratory Activities

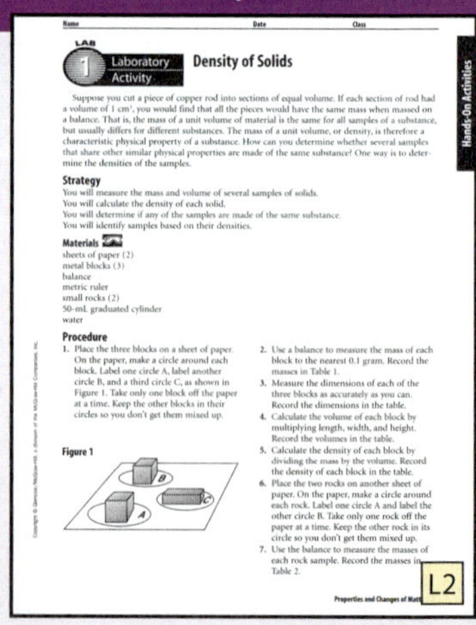
Name Date Class

Lab 1 Laboratory Activity **Density of Solids**

Suppose you cut a piece of copper rod into sections of equal volume. If each section of rod had a volume of 1 cm^3, you would find that all the pieces would have the same mass when massed on a balance. That is, the mass of a unit volume of material is the same for all samples of a substance, but usually differs for different substances. The mass of a unit volume, or density, is therefore a characteristic physical property of a substance. How can you determine whether several samples that share other similar physical properties are made of the same substance? One way is to determine the densities of the samples.

Strategy

You will measure the mass and volume of several samples of solids.
You will calculate the density of each solid.
You will determine if any of the samples are made of the same substance.
You will identify samples based on their densities.

Materials

sheets of paper (2)
metal blocks (3)
balance
metric ruler
small rocks (2)
50-mL graduated cylinder
water

Procedure

1. Place the three blocks on a sheet of paper. On the paper, make a circle around each block. Label one circle A, label another circle B, and a third circle C, as shown in Figure 1. Take only one block off the paper at a time. Keep the other blocks in their circles so you don't get them mixed up.
2. Use a balance to measure the mass of each block to the nearest 0.1 gram. Record the masses in Table 1.
3. Measure the dimensions of each of the three blocks as accurately as you can. Record the dimensions in the table.
4. Calculate the volume of each block by multiplying length, width, and height. Record the volumes in the table.
5. Calculate the density of each block by dividing the mass by the volume. Record the density of each block in the table.
6. Place the two rocks on another sheet of paper. On the paper, make a circle around each rock. Label one circle A and label the other circle B. Take only one rock off the paper at a time. Keep the other rock in its circle so you don't get them mixed up.
7. Use the balance to measure the masses of each rock sample. Record the masses in Table 2.

Figure 1

Hands-On Activities

L2

Meeting Different Ability Levels

Content Outline

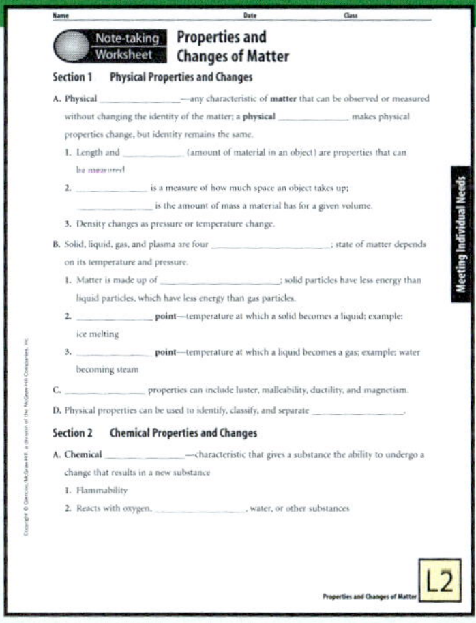

Reinforcement

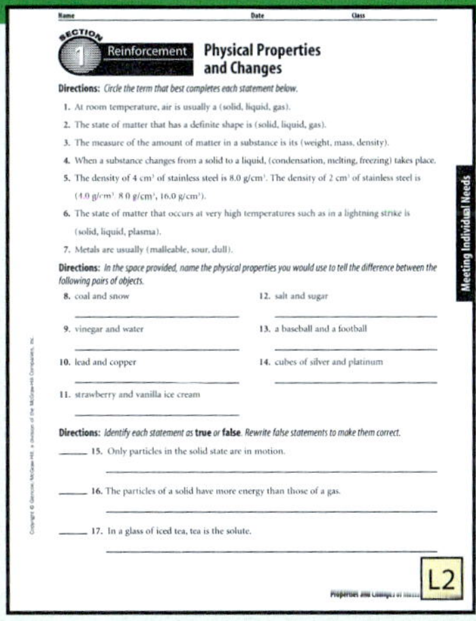

Enrichment

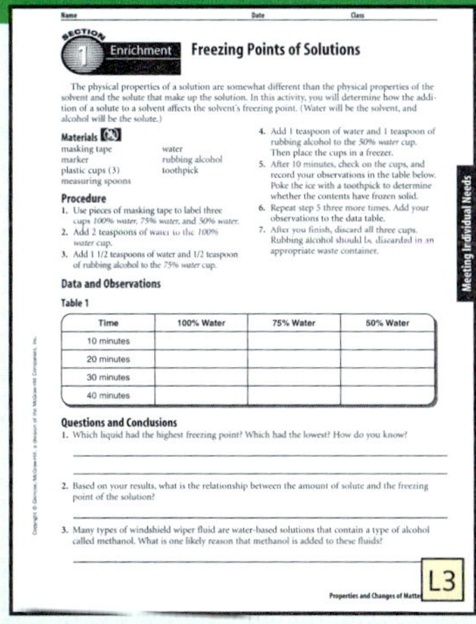

Directed Reading (English/Spanish)

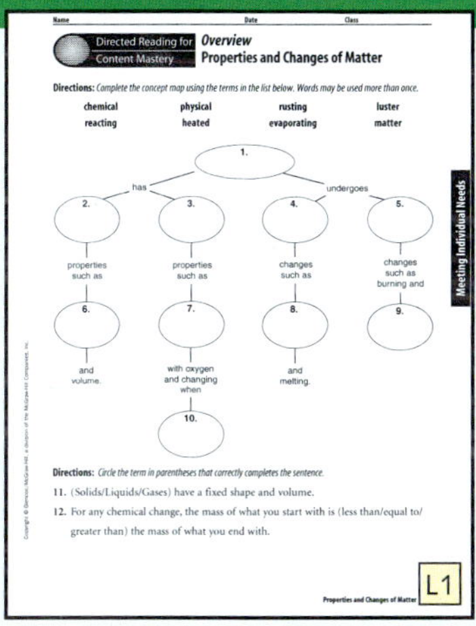

Study Guide

Reading Essentials

Assessment

Test Practice Workbook

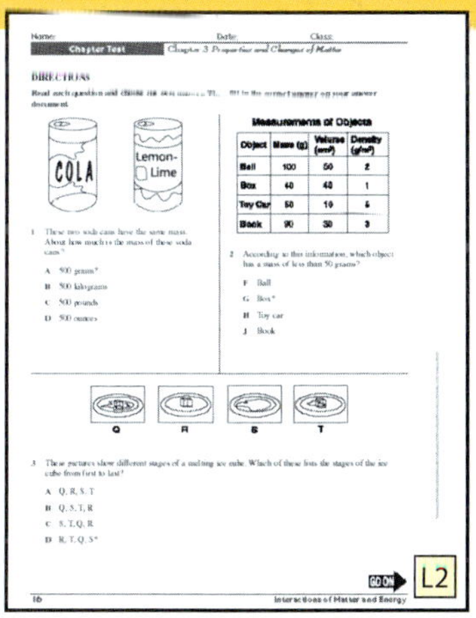

Chapter Review

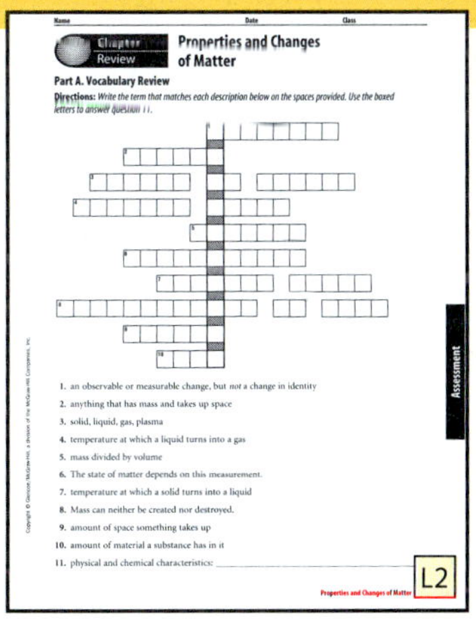

Chapter Tests

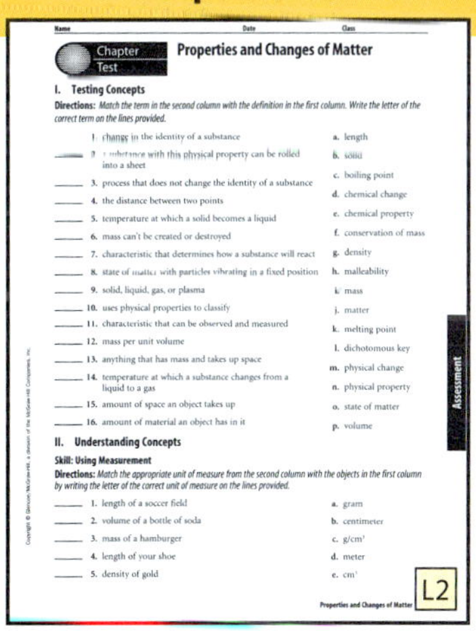

Science Content Background

section 1 Physical Properties and Changes

Physical Properties

Care must be taken in describing some physical properties, particularly properties such as solubility. Describing solubility can involve a reaction and, for that reason, solubility may not be a true physical property. For example, if you say that sodium is soluble in water, that implies that the sodium simply dissolves in water. Instead, sodium reacts with water to form hydrogen gas and an aqueous sodium hydroxide solution.

Physical Changes

Knowledge of physical changes and the circumstances under which they occur can be used to separate mixtures. Distillation is a change-of-state operation that is used to separate substances with different boiling points. Distillation is used to separate drinking water from seawater, and hydrocarbons such as gasoline and kerosene from petroleum.

Intensive and Extensive Properties

Properties can also be classified as either intensive or extensive, depending on whether the value changes with the size of the sample. Intensive properties, such as melting point and temperature, have values that do not depend on the amount of the sample. Extensive properties, such as mass, length, and volume, have values that do depend on the sample size. Density is an intensive physical property that relates the mass of an object to its volume. Density is temperature dependent because most substances change in volume when heated or cooled.

Anna E. Zuckerman/PhotoEdit

chapter content resources

Internet Resources

For additional content background, visit **red.msscience.com** to:

- access your book online
- find references to related articles in popular science magazines
- access Web links with related content background
- access current events with science journal topics

Print Resources

Chemistry; The Molecular Nature of Matter and Change, by Martin S. Silberberg, McGraw-Hill, 2003

Solid, Liquid, Gas: What is Matter?, by Erica Smith, The Rosen Publishing Group, Incorporated, 2002

Matter, by Ann Fullick, Heinemann Library, 1999

Teacher to Teacher
Joe Kiser,
Ervinton High School
Nora, Virginia

"Add about 100 cm^3 of tap water to a 250 cm^3 beaker. Determine the temperature. Slowly add concentrated HCl to the tap water with constant stirring. Observe the temperature effect. Add 100 cm^3 of ice water to a 250 cm^3 beaker. Record temperature. Slowly add concentrated HCl to the ice water with constant stirring. Observe the temperature effect. Students will be surprised at the opposite effect."

Joe Kiser

section 2 Chemical Properties and Changes

Chemical Properties

A substance's chemical properties describe its ability to react with other substances or to decompose. For example, one chemical property of limestone, also known as calcium carbonate, is its ability to react with hydrochloric acid, producing carbon dioxide, water, and calcium chloride. Another chemical property of calcium carbonate is the ability to decompose, forming calcium oxide and carbon dioxide when heated.

The ability to corrode many metals is a chemical property of some acids. For example, when sulfuric acid corrodes aluminum, the aluminum atoms oxidize (lose electrons) to form positively charged aluminum ions, while the hydrogen ions in the sulfuric acid reduce (gain electrons) to form hydrogen atoms.

Chemical Changes

Chemical reaction is another term for chemical change. In a chemical change, atoms are rearranged and may lose or gain electrons. A rule of thumb states that if a precipitate, gas, color change, or energy change occurs, a chemical change has taken place. Almost all chemical changes involve either the taking in (endothermic) or giving off (exothermic) of energy. Cold packs used by athletic trainers for sports injuries take advantage of endothermic changes.

Charles D. Winters/Photo Researchers, Inc.

chapter 3

Chapter Vocabulary

physical property, p. 70
matter, p. 71
physical change, p. 71
density, p. 72
states of matter, p. 73
melting point, p. 75
boiling point, p. 75
chemical property, p. 80
chemical change, p. 81
law of conservation of mass, p. 85

Science Journal Solids include coral, oxygen tank, and the torch. Liquids are the water. Gases include the bubbles from the torch and the bubbles exhaled from Wendy Craig Duncan.

INTERACTIVE CHALKBOARD with Image Bank

PowerPoint® Presentations

This CD-ROM is an editable Microsoft® PowerPoint® presentation that includes:

- a pre-made presentation for every chapter
- interactive graphics
- animations
- audio clips
- image bank
- all new section and chapter questions
- Standardized Test Practice
- transparencies
- pre-lab questions for all labs
- Foldables directions
- links to red.msscience.com

chapter 3

Matter and Its Changes

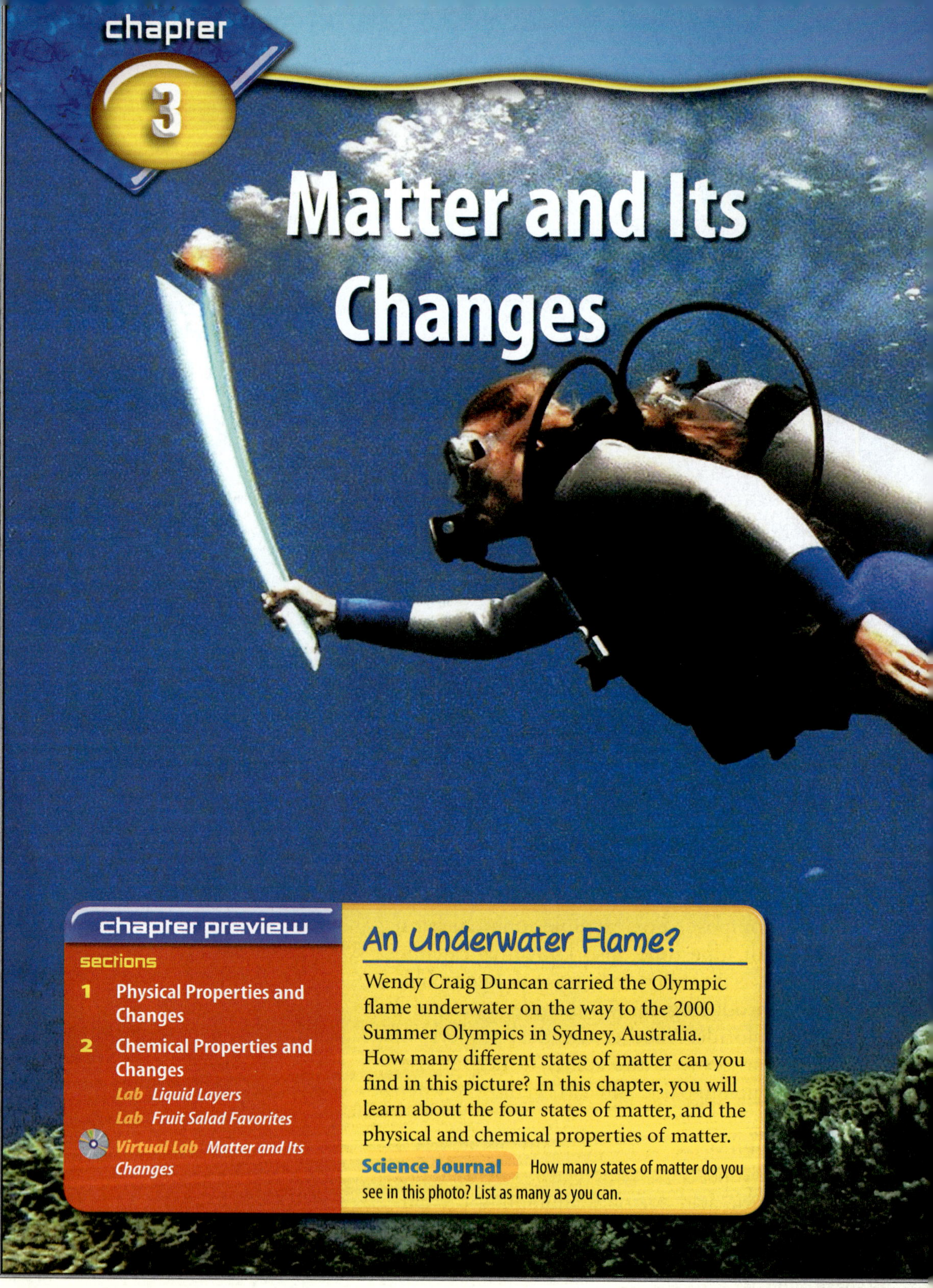

chapter preview

sections

An Underwater Flame?

Wendy Craig Duncan carried the Olympic flame underwater on the way to the 2000 Summer Olympics in Sydney, Australia. How many different states of matter can you find in this picture? In this chapter, you will learn about the four states of matter, and the physical and chemical properties of matter.

Science Journal How many states of matter do you see in this photo? List as many as you can.

Theme Connection

Stability and Change Matter can be described by its physical and chemical properties. Matter is capable of undergoing physical and chemical changes based upon these properties.

About the Photo

Olympic Torch A special Olympic torch was designed for the 3-minute underwater trip along the Great Barrier Reef. The torch burns a mixture of chemicals that produces oxygen and nitrogen gas. It also burns light producing magnesium. The pressure generated by the gases keeps water from entering the torch and extinguishing the flame.

Start-Up Activities

Launch Lab

Can you classify pennies by their properties?

Your teacher has given you a collection of pennies. It is your task to separate these pennies into groups. In this chapter, you will learn how to identify things based on their physical and chemical properties. With an understanding of these principles of matter, you will discover how things are classified or put into groups.

1. Observe the collection of pennies.
2. Choose a characteristic that will allow you to separate the pennies into groups.
3. Classify and sort each penny based on the chosen feature. Tally your data in a frequency table.
4. Explain how you classified the pennies. Compare your system of classification with those of others in the classroom.
5. **Think Critically** Write a paragraph in your Science Journal explaining how your group classified its pennies. What other requirements could have been used to classify the pennies?

Science Online | Preview this chapter's content and activities at red.msscience.com

FOLDABLES™ Study Organizer

Properties of Matter Make the following Foldable to help you organize your thoughts into clear categories about properties of matter.

STEP 1 **Draw** a mark at the midpoint of a sheet of paper along the side edge. Then **fold** the top and bottom edges in to touch the midpoint.

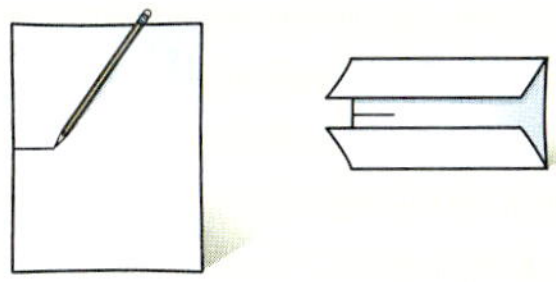

STEP 2 **Fold** in half from side to side.

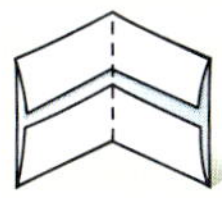

STEP 3 **Turn** the paper vertically. **Open and cut** along the inside fold lines to form four tabs.

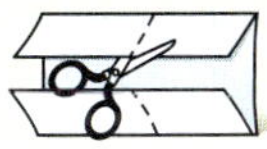

STEP 4 **Label** each tab as shown.

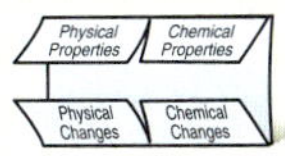

Classify Before you read the chapter, define each term on the front of the tabs. As you read the chapter, correct your definitions and write about each under the appropriate tab. Use the information in your Foldable to compare and contrast physical and chemical properties of matter. Write about each on the back of the tabs.

69

Launch Lab

Purpose Use the Launch Lab to help students understand that observable properties are used to identify and classify matter. L1 ELL COOP LEARN LS **Interpersonal**

Preparation Collect coins for each group of students.

Materials one set of approximately 20 pennies for each group

Teaching Strategy Ask students to share possible categories they could use to classify the pennies. Possible answers: date, amount of wear, shininess, place minted

Think Critically

Descriptions should be clear and should include any difficulties experienced in classifying specific pennies.

Assessment

Oral What can you learn about a penny from how you classify it? Categories such as amount of wear and shininess reflect amount of use and how that use has affected the substance from which pennies are made; date and place minted reflect the original manufacture. How could descriptions of groups of pennies assist a person in identifying a particular coin? The person could use the information to match it with a group, thus narrowing the possible choices for the penny's identity. Use **Performance Assessment in the Science Classroom**, p. 89.

FOLDABLES™ Study Organizer **Dinah Zike Study Fold**

Student preparation materials for this Foldable are available in the Chapter *Fast File* Resources.

1 Motivate

INTERACTIVE CHALKBOARD
PowerPoint® Presentations

Bellringer

Section Focus Transparencies also are available on the Interactive Chalkboard CD-ROM. L2 ELL

Section 1 Section Focus Transparency **Water, Water Everywhere**

How much water can you find here? Actually, water is almost everywhere.

1. Where do you see water as a solid? A liquid?
2. What happens to snow on a sunny, warm day? What happens to water in a kettle if you light a burner underneath? How are these events similar?

Properties and Changes of Matter

Tie to Prior Knowledge

Sorting Have students recall past experiences of being asked to sort things into groups. They may have sorted laundry into piles of white, dark, and other, or they may have gone through their own clothing to separate items that fit from items that were too small. In this section, students will learn how matter can be described according to properties and how changes in these properties occur.

section 1

Physical Properties and Changes

as you read

What You'll Learn

- **Identify** physical properties of matter.
- **Explain** why materials with different masses have different densities.
- **Observe** water displacement to determine volume.
- **Describe** the states of matter.
- **Determine** how temperature changes affect substances.
- **Classify** matter using physical properties.

Why It's Important

Observing physical properties will help you interpret the world around you.

Review Vocabulary
mass: amount of matter in an object

New Vocabulary
- physical property
- matter
- physical change
- density
- states of matter
- melting point
- boiling point

Using Your Senses

As you look in your empty wallet and realize that your allowance isn't coming anytime soon, you decide to get an after-school job. You've been hired at the new grocery store that will open next month. They are getting everything ready for the grand opening, and you will be helping make decisions about where things will go and how they will be arranged.

When you come into a new situation or have to make any kind of decision, what do you usually do first? Most people would make some observations. Observing involves seeing, hearing, tasting, touching, and smelling.

Whether in a new job or in the laboratory, you use your senses to observe materials. Any characteristic of a material that can be observed or measured without changing the identity of the material is a **physical property.** However, it is important to never taste, touch, or smell any of the materials being used in the lab without guidance, as noted in **Figure 1.** For safety reasons you will rely mostly on other observations.

Figure 1 For safety reasons, in the laboratory you usually use only two of your senses—sight and hearing. Many chemicals can be dangerous to touch, taste, and smell.

Section 1 Resource Manager

Chapter *FAST FILE* Resources

Transparency Activity, pp. 42, 45–46
Directed Reading for Content Mastery, pp. 19, 20
MiniLAB, p. 3
Note-taking Worksheets, pp. 31–32
Enrichment, p. 29
Lab Activity, pp. 9–11
Reinforcement, p. 27

Reading and Writing Skill Activities, p. 17
Science Inquiry Labs, pp. 43–44
Physical Science Critical Thinking/Problem Solving, p. 10

Figure 2 The identity of the material does not necessarily depend on its color. Each of these bottles is made of high-density polyethylene (HDPE).
Describe *a physical change that can be applied to the bottles.*

Physical Properties

On the first day of your new job, the boss gives you an inventory list and a drawing of the store layout. She explains that every employee is going to give his or her input as to how the merchandise should be arranged. Where will you begin?

You decide that the first thing you'll do is make some observations about the items on the list. One of the key senses used in observing physical properties is sight, so you go shopping to look at what you will be arranging.

Color and Shape Everything that you can see, touch, smell, or taste is matter. **Matter** is anything that has mass and takes up space. What things do you observe about the matter on your inventory list? The list already is organized by similarity of products, so you go to an aisle and look.

Color is the first thing you notice. The laundry detergent bottles you are looking at come in every color. Maybe you will organize them in the colors of the rainbow. You make a note and look more closely. Each bottle or box has a different shape. Some are square, some rectangular, and some are a free-form shape. You could arrange the packages by their shape.

When the plastic used to make the packaging is molded, it changes shape. However, the material is still plastic. This type of change is called a physical change. It is important to realize that in a **physical change,** the physical properties of a substance change, but the identity of the substance does not change. Notice **Figure 2.** The detergent bottles are made of high-density polyethylene regardless of the differences in the physical properties of color or shape.

Reading Check *What is matter?*

Science Online

Topic: Physical Properties
Visit red.msscience.com for Web links to information about classifying matter by its physical properties.

Activity Choose three objects in the room around you. Try to describe them using as many different physical properties as you can. Pass your description to another classmate and see if they are able to identify the object.

2 Teach

Caption Answer

Figure 2 The bottles can be crushed, cut up, or melted.

Quick Demo

Acid-Washed Jeans

Materials beaker of $6M$ hydrochloric acid, beaker of water, small piece of denim, gloves, goggles, apron

Estimated Time 15 minutes

Procedure Without identifying the contents, set the HCl and water side by side for students to look at. Explain that although the liquids in both beakers look like water, only one of them is water. Put a piece of denim fabric into each liquid. The water will not react with the denim, but the acid will. Explain to students that in a lab, dangerous substances can look like safe substances. Therefore, they should never taste, touch, or smell lab materials. LS **Visual-Spatial**

Quick Demo

Paper Changes

Materials sheet of notebook paper

Estimated Time 10 minutes

Procedure Show students a sheet of notebook paper, and have them describe its size and shape. Crumple the paper into a ball, and have students again describe its size and shape. Explain that while the description of the paper has changed, the paper remains paper. Crumpling paper is therefore an example of a physical change. ELL LS **Visual-Spatial**

Reading Check

Answer anything that has mass and takes up space

SECTION 1 Physical Properties and Changes 71

Use Science Words

Word Usage Have students use the term *physical property* as they describe an object in their backpacks or desks. Possible answer: My ruler has the physical properties of being shaped long, narrow, and thin. L1 ELL LS **Linguistic**

IDENTIFYING Misconceptions

Mass v. Weight Students may think that mass and weight are the same. Mass is a measure of the amount of matter in a substance, while weight is the measure of the force of gravity on it. Ask students what would happen to the mass and weight of a desk if it were transported to the moon. Its mass would stay the same, but its weight would decrease to one-sixth of its weight on Earth.

Fun Fact

Icebergs float because solid ice is less dense than liquid water. Hydrogen is used to inflate weather balloons because it is much less dense than air.

Science Online

Topic: Density

Visit red.msscience.com for Web links to information about density.

Activity Find three objects in the room that are about the same size. This might be a pencil eraser, a necklace pendant, and a small rock sample. Determine the density of each object. Is density dependent on size?

Length and Mass Some properties of matter can be identified by using your senses, and other properties can be measured. How much is there? How much space does it take up?

One useful and measurable physical property is length. Length is measured using a ruler, meterstick, or tape measure, as shown in **Figure 3.** Objects can be classified by their length. For example, you could choose to organize the French bread in the bakery section of your store by the length of the loaf. But, even though the dough has been shaped in different lengths, it is still French bread.

Back in the laundry aisle, you notice a child struggling to lift one of the boxes of detergent. That raises a question. How much detergent is in each box? Mass is a physical property that describes the amount of material in an object. Some of the boxes are heavy, but, the formula of the detergent hasn't changed from the small box to the large box. Organizing the boxes by mass is another option.

Volume and Density Mass isn't the only physical property that describes how much of something you have. Another measurement is volume. Volume measures the amount of space an object takes up. Liquids usually are measured by volume. The juice bottles on your list could be organized by volume.

Another measurable physical property related to mass and volume is **density**—the amount of mass a material has in a given volume. You notice this property when you try to lift two things of equal volume that have different masses. Density is found by dividing the mass of an object by its volume.

$$\text{density} = \text{mass}/\text{volume}, \text{ or } D = m/V$$

Figure 3 The length of any object can be measured with the appropriate tool.
Describe *how you would measure the length of your school building.*

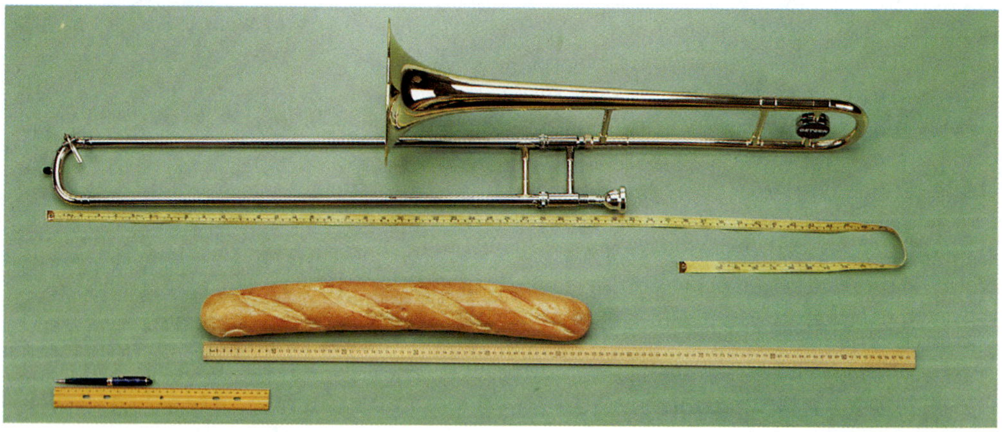

LAB DEMONSTRATION

Purpose to demonstrate volume and density as properties of matter

Materials water, 100-mL graduated cylinder, 10 g clay, string

Preparation Pour water into the graduated cylinder to the 50-mL mark. Shape the clay to fit into the graduated cylinder.

Procedure Cut a piece of string and attach it to the clay. Use the string to lower the clay into the graduated cylinder. Be careful not to splash out any water.

Expected Outcome The volume and density of the clay will vary.

Assessment

What is the volume of the clay? (the volume of the water + clay) − 50 mL What is the density of the clay? 10 g/volume of clay How would the density change if you used a larger piece of clay? It would not change.

Figure 4 These balls take up about the same space, but the bowling ball on the left has more mass than the kickball on the right. Therefore, the bowling ball is more dense.

Same Volume, Different Mass **Figure 4** shows two balls that are the same size but not the same mass. The bowling ball is more dense than the kickball. The customers of your grocery store will notice the density of their bags of groceries if the baggers load all of the canned goods in one bag and put all of the cereal and napkins in the other.

The density of a material stays the same as long as pressure and temperature stay the same. Water at room temperature has a density of 1.00 g/cm^3. However, when you do change the temperature or pressure, the density of a material can change. Water kept in the freezer at 0°C is in the form of ice. The density of that ice is 0.9168 g/cm^3. Has the identity of water changed? No, but something has changed.

Reading Check *What two properties are related in the measurement of density?*

States of Matter

How does water change when it goes from 20°C to 0°C? It changes from a liquid to a solid. The four **states of matter** are solid, liquid, gas, and plasma (PLAZ muh). The state of matter of a substance depends on its temperature and pressure. Three of these states of matter are things you talk about or experience every day, but the term *plasma* might be unfamiliar. The plasma state occurs at very high temperatures and is found in fluorescent (floo RE sunt) lightbulbs, the atmosphere, and in lightning strikes.

As you look at the products to shelve in your grocery store, you might make choices of classification based on the state of matter. The state of matter of a material is another physical property. The liquid juices all will be in one place, and the solid, frozen juice concentrates will be in another.

Mini LAB

Determining Volume

Procedure

1. Find **three objects of the same size.** For example: a marble, a rubber ball, and a wood sphere.
2. Fill a **100-mL graduated cylinder** with 50 mL of **water.**
3. Submerge one object into the graduated cylinder and record the new water level. Empty the graduated cylinder.
4. Repeat steps 2 and 3 for the remaining two objects.

Analysis

1. Which of the three items displaced the most water? Which displaced the least?
2. What does this tell you about the volume of the objects?
3. What other quantities would you measure to determine the density of each object?

Mini LAB

Purpose to determine the volume of an object by water displacement. L2

Materials three objects of similar size such as a marble, rubber ball, and a wood sphere, 100-mL graduated cylinder, water

Teaching Strategies

- Objects should be able to fit inside the graduated cylinder as well as be similar in size.
- Students should submerge the wood sphere with a pencil until it is just below the water's surface.

Analysis

1. Answer will vary based on the objects used.
2. The object that displaced the most water has the greatest volume. The object that displaced the least water has the least volume.
3. Mass is the other quantity that would have to be measured to determine the density of an object.

Assessment

Performance Have students design an experiment that would test their volume by displacement. Use **Performance Assessment in the Science Classroom,** p. 95.

Reading Check

Answer mass and volume

Caption Answer

Figure 3 Possible answer: Two students using a tape measure would begin at one end of the school. They would pull the tape measure to its greatest length and place a marker. Then repeat the procedure from the marker until they reach the opposite end of the school.

Visual Learning

Figure 4 What physical properties do the two balls share? shape, size, volume, state What physical properties are different for these two items? mass, density, color L2 LS **Visual-Spatial**

Teacher FYI

Plasma Adding a large amount of heat or energy to a neutral gas causes the atoms to break apart and release electrons. When enough gas molecules are broken, it is then called plasma.

SECTION 1 Physical Properties and Changes 73

Visual Learning

Figure 5 Point out to students the illustrations showing the motion of particles in solids, liquids, and gases. Ask a volunteer to explain how the motion of particles in a gas is different from the motion of particles in a liquid. The particles in a gas move faster than the particles in a liquid and move independently of one another. Also point out to students that the amount of gaseous water in the third jar is much less than the amount of solid or liquid water in the first two jars because gaseous water takes up much more volume than solid or liquid water. L2 LS **Visual-Spatial**

Make a Model

Composition of Matter Have students use gumdrops and toothpicks to make models that show how particles in solids, liquids, and gases are arranged and how they can move. For a solid, students should connect the gumdrops with the toothpicks to form a rigid structure. For a liquid, gumdrops should be connected to form layers. The gumdrops in each layer should be connected, but the layers should not be connected to each other. For a gas, the gumdrops should not be connected at all. L2 LS **Kinesthetic**

Moving Particles Matter is made up of moving particles. The state of matter is determined by how much energy the particles have. The particles of a solid vibrate in a fixed position. They remain close together and give the solid a definite shape and volume. The particles of a liquid are moving much faster and have enough energy to slide past one another. This allows a liquid to take the shape of its container. The particles of a gas are moving so quickly that they have enough energy to move freely away from other particles. The particles of a gas take up as much space as possible and will spread out to fill any container. **Figure 5** illustrates the differences in the states of water.

Particles of matter move faster as higher temperatures are applied. To demonstrate this, fill one beaker with cold water and another with very hot water. Add ten drops of food coloring. Observe in which beaker the color becomes uniform first.

Changes of State You witness a change of state when you place ice cubes in a cup and they melt. You still have water but in another form. The opposite physical change happens when you put liquid water in ice-cube trays and pop them in your freezer. The water doesn't change identity—only the state it is in.

For your job, you will need to make some decisions based on the ability of materials to change state. You don't want all the frozen items thawing out and becoming slushy liquid. You also don't want some of the liquids to get so cold that they freeze.

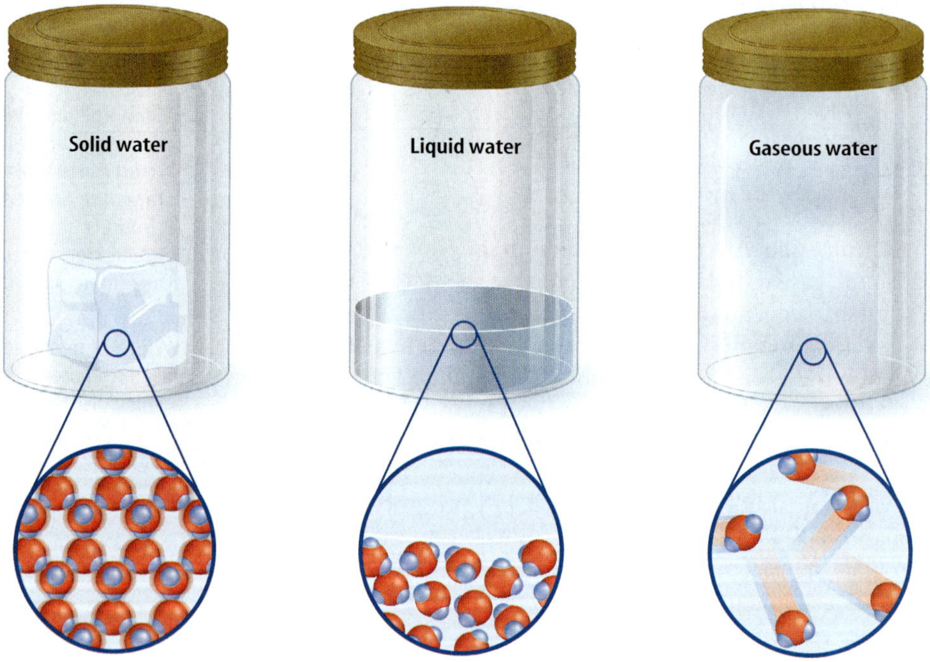

Figure 5 Water can be in three different states: solid, liquid, and gas. The molecules in ice are tightly packed and vibrate in place, but in liquid water they can slip past each other because they have more energy to move. In water vapor, they move freely all around the container with even more energy.

74 CHAPTER 3 Matter and Its Changes

Curriculum Connection

Art Glass is a solid that has been used to make artistic and practical objects since before 2000 B.C. To make these objects, glass must be heated until it changes state and becomes a liquid. In the liquid state, it can be blown or molded into almost any form, which it keeps when it cools down and again becomes a solid. Have students bring in pictures of works of art made from glass. L2 LS **Visual-Spatial** P

Melting and Boiling Points At what temperature will water in the form of ice change into a liquid? The temperature at which a solid becomes a liquid is its **melting point.** The melting point of a pure substance does not change with the amount of the substance. This means that a small sliver of ice and a block of ice the size of a house both will melt at 0°C. Lead always melts at 327.5°C. When a substance melts, it changes from a solid to a liquid. This is a physical change, and the melting point is a physical property.

At what temperature will liquid water change to a gas? The **boiling point** is the temperature at which a substance in the liquid state becomes a gas. Each pure substance has a unique boiling point at atmospheric pressure. The boiling point of water is 100°C at atmospheric pressure. The boiling point of nitrogen is −195.8°C, so it changes to a gas when it warms after being spilled into the open air, as shown in **Figure 6.** The boiling point, like the melting point, does not depend on the amount of the substance.

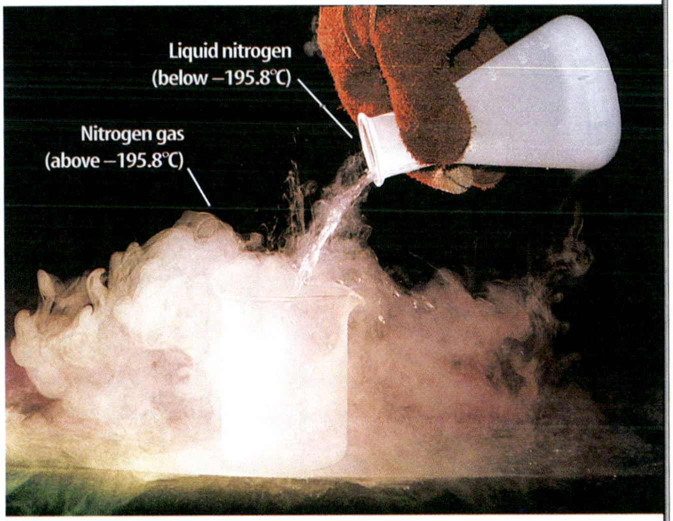

Figure 6 When liquid nitrogen is poured from a flask, you see an instant change to gas because nitrogen's boiling point is −195.8°C, which is much lower than room temperature.

Reading Check *What physical change takes place at the boiling point?*

However, the boiling point and melting point can help to identify a substance. If you observe a clear liquid that boils at 56.1°C at atmospheric pressure, it is not water. Water boils at 100°C. If you know the boiling points and melting points of substances, you can classify substances based on those properties.

Metallic Properties

Other physical properties allow you to classify substances as metals. You already have seen how you can classify things as solids, liquids, or gases or according to color, shape, length, mass, volume, or density. What properties do metals have?

How do metals look? Often the first thing you notice about something that is a metal is its shiny appearance. This is due to the way light is reflected from the surface of the metal. This shine is called luster. New handlebars on a bike have a metallic luster. Words to describe the appearance of nonmetallic objects are *pearly, milky,* or *dull.*

INTEGRATE Language Arts

Rock Descriptions When geologists describe rocks, they use specific terms that have meaning to all other scientists who read their descriptions. To describe the appearance of a rock or mineral, they use the following terms: *metallic, adamantine, vitreous, resinous, pearly, silky,* and *greasy.* Research these terms and write a definition and example of each in your Science Journal.

Quick Demo

Melting Ice

Materials 200 mL crushed ice, beaker, thermometer, Bunsen burner, tongs, watch or timepiece

Estimated Time 15 minutes

Procedure Place 200 mL of crushed ice into a beaker. Place a thermometer into the beaker and note the temperature. Heat the beaker using a Bunsen burner and tongs. Note the temperature every 30 seconds as the ice melts. Explain that as the ice melts, the temperature remains constant at the melting point of ice. If heat is added after all the ice has melted, the temperature of the water will start to rise again. LS **Visual-Spatial**

Reading Check

Answer change of state from liquid to gas

INTEGRATE Language Arts

Rock Descriptions metallic: shines like a metal—pyrite, galena; adamantine: shines like a diamond—cassiterite; vitreous: glassy—fluorite, calcite, quartz; resinous: less brilliant than greasy—sphalerite; pearly: shining, silvery—muscovite mica; silky: finely fibrous with many parallel needles—gypsum; greasy: a little less brilliant than diamond—halite

Discussion

Metallic Luster Have students identify items in the classroom that have a metallic luster. Possible answers: chrome legs on classroom chairs, frame around the board, parts of a wristwatch Have students think of other shiny metal objects not found in the classroom. Possible answers: car, faucet, refrigerator door handle, aluminum foil L2 ELL LS **Visual-Spatial**

Cultural Diversity

Bronze Age As early as 4000 B.C., people in what is now the Middle East discovered they could melt together copper and tin to make bronze. Bronze is strong, hard, resistant to corrosion, and has a low melting point. Metal workers shaped and pounded it into jewelry, weapons, coins, tools, and art works. Women of ancient Egypt, Greece, and Rome used mirrors of polished bronze.

Visual Learning

Figure 7 Name other objects that use the properties of ductility or malleability in their construction. Possible answers: bridges, playground equipment, power lines, appliance cords, brass musical instruments L1

Use Science Words

Word Usage *Malleable* means "capable of being shaped by hammering, rolling, or pressing." Use the word *malleable* in a sentence describing a metal object found in the classroom. Sample answer: The file cabinet is made from malleable metal sheets. L2 LS **Linguistic**

Discussion

Bent Metal Ask students to think of everyday instances when they either observed bent metal or personally bent a metal object. Possible answers: dent in car or filing cabinet, paper clip, staple, bent spoon, coat hanger, electric cord, steel wool L2

Figure 7 This artist has taken advantage of the ductility of metal by choosing wire as the medium for this sculpture.

Uses of Metals Metals can be used in unique ways because of some of the physical properties they have. For example, many metals can be hammered, pressed, or rolled into thin sheets. This property of metals is called malleability (mal lee uh BIH luh tee). The malleability of copper makes it an ideal choice for artwork such as the Statue of Liberty. Many metals can be drawn into wires as shown in **Figure 7.** This property is called ductility (duk TIH luh tee). The wires in buildings and most electrical equipment and household appliances are made from copper. Silver and platinum are also ductile.

You probably observe another physical property of some metals every day when you go to the refrigerator to get milk or juice for breakfast. Your refrigerator door is made of metal. Some metals respond to magnets. Most people make use of that property and put reminder notes, artwork, and photos on their refrigerators. Some metals have groups of atoms that can be affected by the force of a magnet, and they are attracted to the magnet because of that force. The magnet in **Figure 8** is being used to select metallic objects.

Figure 8 This junkyard magnet pulls scrap metal that can be salvaged from the rest of the debris. It is sorting by a physical property.

At the grocery store, your employer might think about these properties of metals as she looks at grocery carts and thinks about shelving. Malleable carts can be dented. How could the shelf's attraction of magnets be used to post advertisements or weekly specials? Perhaps the prices could be fixed to the shelves with magnetic numbers. After you observe the physical properties of an object, you can make use of those properties.

Teacher FYI

Nitinol The wire used in braces is nitinol, an alloy made from nickel and titanium. Nitinol wire is made in a preset shape that can be deformed to connect to brackets on teeth. The wire attempts to return to its original shape, putting pressure on the brackets and gradually moving the attached teeth.

Differentiated Instruction

Challenge Ask groups of students to make sculptures using wire or aluminum foil. They should identify the physical properties of their projects, including color, luster, shape, length, mass, volume, state, and whether they show malleability, ductility, and/or attraction to magnets. L3 COOP LEARN LS **Kinesthetic**

Using Physical Properties

In the previous pages, many physical properties were discussed. These physical properties—such as appearance, state, shape, length, mass, volume, ability to attract a magnet, density, melting point, boiling point, malleability, and ductility—can be used to help you identify, separate, and classify substances.

For example, salt can be described as a white solid. Each salt crystal, if you look at it under a microscope, could be described as having a three-dimensional cubic structure. You can measure the mass, volume, and density of a sample of salt or find out if it would attract a magnet. These are examples of how physical properties can be used to identify a substance.

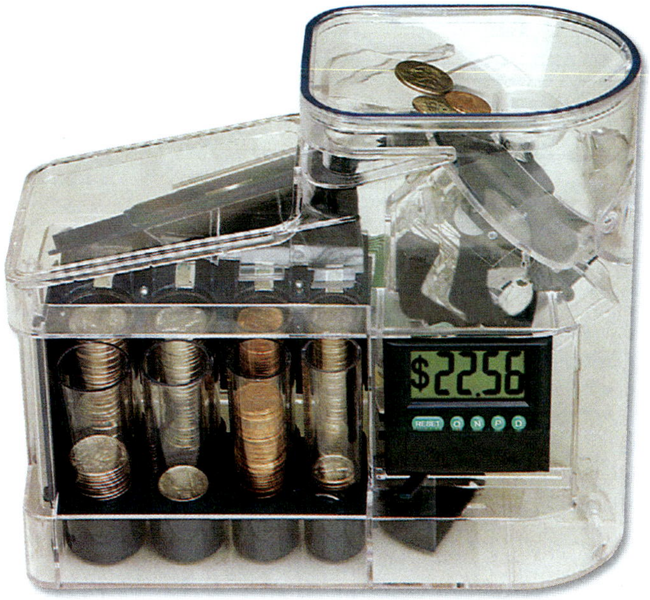

Figure 9 Coins can be sorted by their physical properties. Sorting by size is used here.
Identify *three other properties that can be used to sort coins.*

Sorting and Separating When you do laundry, you sort according to physical properties. Perhaps you sort by color. When you select a heat setting on an iron, you classify the clothes by the type of fabric. When miners during the Gold Rush panned for gold, they separated the dirt and rocks by the density of the particles. **Figure 9** shows a coin sorter that separates the coins based on their size. Iron filings can be separated from sand by using a magnet.

INTEGRATE Life Science Scientists who work with animals use physical properties or characteristics to determine the identity of a specimen. They do this by using a tool called a dichotomous (di KAH tuh mus) key. The term *dichotomous* refers to two parts or divisions. Part of a dichotomous key for identifying hard-shelled crabs is shown on the next page in **Figure 10.** To begin the identification of your unknown animal, you are given two choices. Your animal will match only one of the choices. In the key in **Figure 10,** you are to determine whether or not your crab lives in a borrowed shell. Based on your answer, you are either directed to another set of choices or given the name of the crab you are identifying.

Caption Answer

Figure 9 Coins can be sorted by weight, density, width, or the material that composes the coins.

Activity

Physical Properties Provide groups of students with a tile streak plate, a magnet, and the following unidentified numbered samples: graphite (1), magnetite (2), talc (3), and fluorite (4). Explain that they will identify the samples using physical properties. First, have students rub each sample on the streak plate and record the color of each mineral's streak. Then have students try to scratch each sample with a fingernail. Finally, have students determine which is attracted to the magnet. When students have completed the activity, reveal the identity of each sample Graphite: gray/black streak, can be scratched, is not attracted by magnet; magnetite: black streak, cannot be scratched, attracted by magnet; talc: white streak, scratched by a fingernail, not attracted by a magnet; fluorite: white streak, cannot be scratched, not attracted by magnet. L2 COOP LEARN LS **Kinesthetic**

Differentiated Instruction

Visually Impaired Ask these students to describe physical properties that can be identified by nonvisual means. Possible answers: solid—shape, size, texture, density, smell; liquid—thickness, slipperiness, smell; gas—smell L2 LS **Kinesthetic**

NATIONAL GEOGRAPHIC

Visualizing Dichotomous Keys

Have students examine the pictures and read the captions, and then ask the following questions.

Have students identify each crab pictured. What physical property did you use to identify each one? The crab in the lower right has a borrowed shell; the crab in the upper right has a shell that overlaps its walking legs; the crab on the left has walking legs that are exposed.

What are some things other than animals that could be identified using a dichotomous key? Possible answers: trees; flowers; rocks; minerals; or kitchen utensils

Activity

Dichotomous Keys Divide the class into small groups. Give each group five or six different objects, such as a pen, a pencil, a piece of chalk, a paint brush, a ruler, and an eraser. Ask each group to develop a dichotomous key that could be used to identify the objects, based on their physical properties. Have each group present its key to the class and have other groups critique the key. L2 COOP LEARN LS **Visual-Spatial and Interpersonal**

NATIONAL GEOGRAPHIC **VISUALIZING DICHOTOMOUS KEYS**

Figure 10

Whether in the laboratory or in the field, scientists often encounter substances or organisms that they cannot immediately identify. One approach to tracking down the identity of such "unknowns" is to use a dichotomous key, such as the one shown. The key is designed so a user can compare physical properties or characteristics of the unknown substance or organism—in this case, a crab—with characteristics of known organisms in a stepwise manner. With each step, a choice must be made. Each choice leads to subsequent steps that guide the user through the key until a positive identification is made.

Dichotomous Key		
1.	A. Lives in a "borrowed" shell (usually some type of snail shell)	Hermit Crab
	B. Does not live in a "borrowed" shell	go to #2
2.	A. Shell completely overlaps the walking legs	Box Crab
	B. Walking legs are exposed	Kelp Crab

Can you identify the three crabs shown here by following this dichotomous key?

Differentiated Instruction

Challenge Have students research the physical properties of the elements gold, silver, copper, mercury, iodine, and sulfur. Challenge them to make a dichotomous key that other students could use to identify these elements. L3 LS **Logical-Mathematical**

Everyday Examples Identification by physical properties is a subject in science that is easy to observe in the real world. Suppose you volunteer to help your friend choose a family pet. While visiting the local animal shelter, you spot a cute dog. The dog looks like the one in **Figure 11.** You look at the sign on the cage. It says that the dog is male, one to two years old, and its breed is unknown. You and your friend wonder what breed of dog he is. What kind of information do you and your friend need to figure out the dog's breed? First, you need a thorough description of the physical properties of the dog. What does the dog look like? Second, you need to know the descriptions of various breeds of dogs. Then you can match up the description of the dog with the correct breed. The dog you found is a white, medium-sized dog with large black spots on his back. He also has black ears and a black mask around his eyes. The manager of the shelter tells you that the dog is close to full-grown. What breed is the dog?

Narrowing the Options To find out, you may need to research the various breeds of dogs and their descriptions. Often, determining the identity of something that is unknown is easiest by using the process of elimination. You figure out all of the breeds the dog can't be. Then your list of possible breeds is smaller. Upon looking at the descriptions of various breeds, you eliminate small dog and large dog breeds. You also eliminate breeds that do not contain white dogs. With the remaining breeds, you might look at photos to see which ones most resemble your dog. Scientists use similar methods to determine the identities of living and nonliving things.

Figure 11 Physical descriptions are used to determine the identities of unknown things.
Observe *What physical properties can be used to describe this dog?*

section 1 review

Summary

Physical Properties

- Physical properties include color, shape, length, mass, volume, and density.

States of Matter

- There are four states of matter.
- Matter can change from one state of matter to another.
- State of matter is determined by how much energy the particles have.

Using Physical Properties

- Substances can be classified according to their physical properties.

Self Check

1. **Identify** the physical properties of this textbook.
2. **List** the four states of matter. Describe each and give an example.
3. **Explain** how water might have two different densities.
4. **Think Critically** Which evaporates more quickly—rubbing alcohol that has been refrigerated or unrefrigerated?

Applying Math

5. **Solve One-Step Equations** Nickel has a density of 9.8 g/cm^3. Lead has a density of 11.3 g/cm^3. If both samples have a volume of 4 cm^3, what are the masses of each?

Caption Answer

Figure 11 Possible Answers: floppy ears, large spots on back, small spots on legs, white stripe on face, and so on.

3 Assess

DAILY INTERVENTION

Check for Understanding

Visual-Kinesthetic In groups of 4 or 5, have students demonstrate how atoms move in a solid, a liquid, and a gas. L2

Reteach

Calculate Volume and Density Ask each student to select a rectangular solid and calculate its volume and density. They can do this by measuring the length, width, and height of the object and using the formula $V = lwh$ to calculate volume. Have them then determine the object's mass and calculate its density with the formula $D = m/v$. L2

Assessment

Oral Have students choose objects in the classroom and describe their physical properties. Ask students to explain physical changes the objects might undergo and how those changes would alter the objects' physical properties. Use **Performance Assessment in the Science Classroom**, p. 89.

section 1 review

1. Answers may include shape, dimensions, color, and state.
2. The four states of matter are solid, which has a definite shape and volume, such as ice; liquid, which has a definite volume but takes the shape of its container, such as water; gas, which fills the shape and volume of its container, such as air; and plasma, which occurs at very high temperatures and is found in fluorescent lightbulbs.
3. by changing the temperature or pressure
4. The unrefrigerated rubbing alcohol evaporates more quickly due to its higher temperature. The higher the temperature, the faster the particles move and thus, change state from liquid to gas.
5. mass of nickel = 9.8 g/cm^3 × 4 cm^3 = 39.2 g; mass of lead = 11.3 g/cm^3 × 4 cm^3 = 45.2 g

section 2

Chemical Properties and Changes

as you read

What You'll Learn

- **Recognize** chemical properties.
- **Identify** chemical changes.
- **Classify** matter according to chemical properties
- **Describe** the law of conservation of mass.

Why It's Important

Knowing the chemical properties will allow you to distinguish differences in matter.

Review Vocabulary

heat: a form of energy that flows from a warmer object to a cooler object

New Vocabulary

- chemical property
- chemical change
- law of conservation of mass

Ability to Change

It is time to celebrate. You and your coworkers have cooperated in classifying all of the products and setting up the shelves in the new grocery store. The store manager agrees to a celebration party and campfire at the nearby park. Several large pieces of firewood and some small pieces of kindling are needed to start the campfire. After the campfire, all that remains of the wood is a small pile of ash. Where did the wood go? What property of the wood is responsible for this change?

All of the properties that you observed and used for classification in the first section were physical properties that you could observe easily. In addition, even when those properties changed, the identity of the object remained the same. Something different seems to have happened in the bonfire example.

Some properties do indicate a change of identity for the substances involved. A **chemical property** is any characteristic that gives a substance the ability to undergo a change that results in a new substance. **Figure 12** shows some properties of substances that can be observed only as they undergo a chemical change.

Reading Check *What does a chemical property give a substance the ability to do?*

Figure 12 These are four examples of chemical properties.

Flammability | Reacts with oxygen | Reacts with light | Reacts with water

1 Motivate

INTERACTIVE CHALKBOARD PowerPoint® Presentations

Bellringer

Section Focus Transparencies also are available on the Interactive Chalkboard CD-ROM. L2 ELL

Section 2 Section Focus Transparency **You Look a Little Green!**

In 1986, after one hundred years on Bedloe's Island, the Statue of Liberty needed to be repaired. Part of this process included cleaning the copper exterior. One thing that could not be changed, however, was the copper's green color. This green tint is now a permanent characteristic.

1. What was the original color of the Statue of Liberty?
2. Do you think the statue's color will continue to change?

Properties and Changes of Matter

Tie to Prior Knowledge

Physical and Chemical Properties Review the concepts of physical properties and changes. Ask students to give examples of each. Explain that substances also have chemical properties and undergo chemical changes that result in new substances.

Reading Check

Answer undergo a change that results in a new substance

Section 2 Resource Manager

Chapter *FAST FILE* Resources

- Transparency Activity, p. 43
- Directed Reading for Content Mastery, pp. 21, 22
- Enrichment, p. 30
- MiniLAB, p. 4
- Lab Activity, pp. 13–15
- Reinforcement, p. 28
- Lab Worksheet, pp. 5–6, 7–8

Life Science Critical Thinking/Problem Solving, p. 8
Reading and Writing Skill Activities, p. 1
Home and Community Involvement, p. 47
Mathematical Skill Activities, p. 9
Science Inquiry Labs, pp. 43–44

An untreated iron gate will rust.

Silver dishes develop tarnish.

Figure 13 Many kinds of interactions with oxygen can occur. Copper sculptures develop a green patina, which is a mixture of copper compounds.

Common Chemical Properties

You don't have to be in a laboratory to see the changes that take place because of chemical properties. These are called chemical changes. A **chemical change** is a change in the identity of a substance due to the chemical properties of that substance. A new substance or substances are formed as a result of such a change.

The campfire you enjoyed to celebrate the opening of the grocery store resulted in chemical changes. The oxygen in the air reacted with the wood to form a new substance called ash. Wood can burn. This chemical property is called flammability. Some products have warnings on their labels about keeping them away from heat and flame because of the flammability of the materials. Sometimes after a campfire you see stones that didn't burn around the edge of the ashes. These stones have the chemical property of being incombustible.

Common Reactions An unpainted iron gate, such as the one shown in **Figure 13,** will rust in time. The rust is a result of oxygen in the air reacting with the iron and causing corrosion. The corrosion produces a new substance called iron oxide, also known as rust. Other chemical reactions occur when metals interact with other elements. The middle photo shows tarnish, the grayish-brown film that develops on silver when it reacts with sulfur in the air. The ability to react with oxygen or sulfur is a chemical property. The photo on the right shows another example of this chemical property.

Have you ever sliced an apple or banana and left it sitting on the table? The brownish coloring that you notice is a chemical change that occurs between the fruit and the oxygen in the air. Those who work in the produce department at the grocery store must be careful with any fruit they slice to use as samples. Although nothing is wrong with brown apples, they don't look appetizing.

INTEGRATE Health

Enzyme Research
Researchers have discovered an enzyme in fruit that is involved in the browning process. They are doing experiments to try to grow grapevines in which the level of this enzyme, polyphenol oxidase (PPO), is reduced. This could result in grapes that do not brown as quickly. Write a paragraph in your Science Journal about why this would be helpful to fruit growers, store owners, and customers.

2 Teach

IDENTIFYING Misconceptions

Physical Change When students see a substance such as water change state, they may think there has been a chemical change, resulting in a different substance. Point out that ice, water, and steam (water vapor) all have the same chemical makeup and formula, H_2O.

INTEGRATE Health

Enzyme Research Cells of fruit damaged by cutting or rubbing release polyphenol oxidase. This enzyme catalyzes the conversion of monophenols to quinones, which oxidize to form brown polyphenolic pigments. In the grapevines, the gene for polyphenol oxidase has been switched off by genetic modification. Students may suggest that these grapes may be more attractive to consumers, and that growers and store owners would suffer less loss from fruit damage.

Research Find as many ways to prevent fruit browning after the fruit is purchased as you can. vitamin C (lemon juice) after cutting, heat blanching, sulfur dioxide dip, sugar syrup dip L2

Teacher FYI

Galvanizing One way to prevent iron from rusting is to coat it with another material that will corrode first. *Galvanizing* with zinc works because the zinc coating reacts with the oxygen in the air to form a thin protective layer of zinc oxide.

Use an Analogy

Regrouping Materials During a chemical change, atoms are regrouped into new, different substances. As an analogy, have students imagine taking apart a car and re-shaping and regrouping its components to make a lawn mower, two sofas, and a shed. What had been a car is now several entirely different things.

Quick Demo

Disappearing Penny

Materials two pennies, sandpaper, glass beakers (2), 3*M* hydrochloric acid, bunsen burner, tongs, cold water

Estimated Time 15 minutes

Procedure Use the sandpaper to scratch the edge of one penny, exposing the zinc center. Place the penny in a small amount of the hydrochloric acid in a beaker. Students will observe a chemical reaction between the zinc and the acid, while the outer layer of the penny will remain intact. Now scratch the other penny with sandpaper to expose the zinc center. Using tongs, grip the penny and place it near the flame. The zinc will melt, which is a physical change that causes a separation between the zinc and the outside covering of the penny. Have a beaker with cold water nearby to catch the molten metal. **WARNING:** *The inside of the penny will melt and fall out. It will be extremely hot.* ELL

LS **Visual-Spatial**

Visual Learning

Figure 14 Many chemical reactions, such as this one, are potentially dangerous. A photograph can be used to observe the reaction safely. Teachers are not to demonstrate this experiment in a middle-school classroom, nor should students attempt this at school or at home. L2 LS **Visual-Spatial**

Heat and Light Vitamins often are dispensed in dark-brown bottles. Do you know why? Many vitamins will change when exposed to light. This is a chemical property. They are protected in those colored bottles from undergoing a chemical change with light.

Some substances are sensitive to heat and will undergo a chemical change only when heated or cooled. One example is limestone. Limestone is generally thought of as unreactive. Some limestone formations have been around for centuries without changing. However, if limestone is heated, it goes through a chemical change and produces carbon dioxide and lime, a chemical used in many industrial processes. The chemical property in this case is the ability to change when heated.

Another chemical property is the ability to change with electrical contact. Electricity can cause a change in some substances and decompose some compounds. Water is one of those compounds that can be broken down with electricity.

Something New

The important difference in a chemical change is that a new substance is formed. Because of chemical changes, you can enjoy many things in life that you would not have experienced without them. What about that perfect, browned marshmallow you roasted at the campfire? A chemical change occurred as a result of the fire to make the taste and the appearance different.

Sugar is normally a white, crystalline substance, but after you heat it over a flame, it turns to a dark-brown caramel. A new substance has been formed. Sugar also can undergo a chemical change when sulfuric acid is added to it. The new substance has obviously different properties from the original, as shown in **Figure 14.**

If eggs, sugar, flour, and other ingredients didn't change chemically through baking, you couldn't enjoy birthday cake. Cake begins as liquid and ends as solid. The baked cake clearly has different properties.

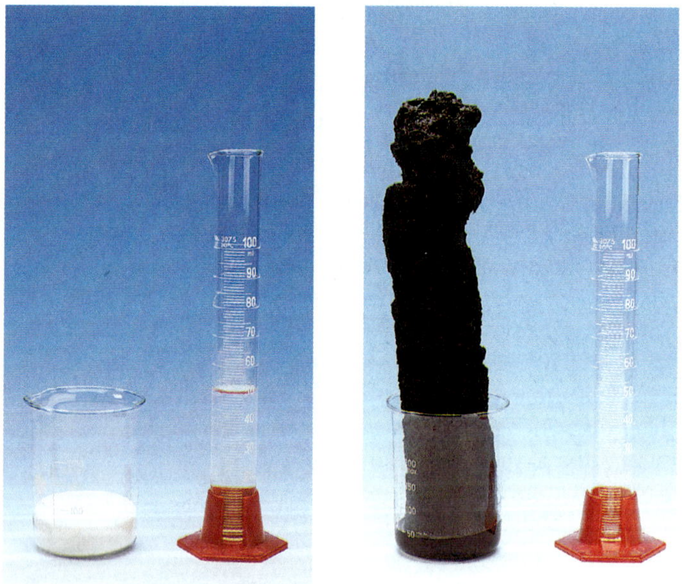

Figure 14 When sugar and sulfuric acid combine, a chemical change occurs and a new substance forms. During this reaction, the mixture foams and a toxic gas is released, leaving only water and air-filled carbon behind.

Active Reading

Double-Entry Journal In this strategy, the student takes notes and adds his or her own reflections while reading the student text. Students are encouraged to explore ideas, make responses, and take risks in giving opinions about the reading. Have them divide a sheet of paper in half. On the left, identify a particular passage or quotation of significance in the reading. The reader records anything luminous, enigmatic, stimulating, or disturbing. On the right, the reader responds, questions, elaborates, makes personal connections, evaluates, reflects, analyzes, or interprets. Have students make a Double-Entry Journal for the discussion of chemical properties and chemical changes in this section. L2

Signs of Change How do you know that you have a new substance? Is it just because it looks different? You could put a salad in a blender and it would look different, but a chemical change would not have occurred. You still would have lettuce, carrots, and any other vegetables that were there to begin with.

You can look for signs when evaluating whether you have a new substance as a result of a chemical change. Look at the piece of birthday cake in **Figure 15.** When a cake bakes, gas bubbles form and grow within the ingredients. Bubbles are a sign that a chemical change has taken place. When you look closely at a piece of cake, you can see the airholes left from the bubbles.

Other signs of change include the production of heat, light, smoke, change in color, and sound. Which of these signs of change would you have seen or heard during the campfire?

Figure 15 The evidence of a chemical change in the cake is the holes left by the air bubbles that were produced during baking. **Identify** *other examples of a chemical change.*

Is it reversible? One other way to determine whether a physical change or a chemical change has occurred is to decide whether or not you can reverse the change by simple physical means. Physical changes usually can be reversed easily. For example, melted butter can become solid again if it is placed in the refrigerator. A figure made of modeling clay, like the one in **Figure 16,** can be smashed to fit back into a container. However, chemical changes can't be reversed using physical means. For example, the ashes in a fireplace cannot be put back together to make the logs that you had to start with. Can you find the egg in a cake? Where is the white flour?

Reading Check *What kind of change can be reversed easily?*

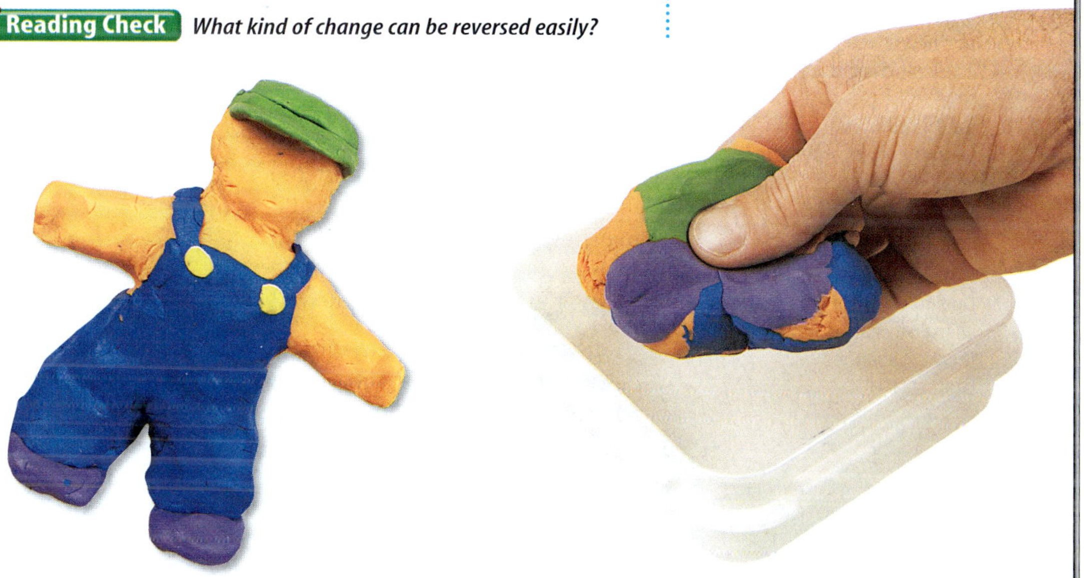

Figure 16 A change such as molding clay can be undone easily.

Caption Answer

Figure 15 production of heat, light, smoke, change in color, and sound

Fun Fact

Pizza is produced by a combination of chemical and physical changes. For example, the yeast in the dough reacts chemically with the sugar and other substances, causing the dough to rise while it bakes. The melting of the cheese is a physical change.

Text Question Answer

You'd have heard crackling sounds, seen light, felt heat, seen smoke, and seen a color change in the wood as it burned.

Activity

Chemical Change Have student groups make posters for the classroom about chemical change. Each poster should include a colorful illustration demonstrating one or more signs that a chemical change is occurring or has occurred. L1 ELL COOP LEARN LS **Visual-Spatial**

Reading Check

Answer a physical change

Differentiated Instruction

Visually Impaired Give students effervescent cold tablets. Explain that breaking the tablets is a physical change because the starting and ending materials are the same. Then have them drop one of the tablets into a beaker of water. Students will be able to hear the "fizz" of the gas being produced. That is the chemical change. L1

Science Journal

Chemistry Notes Ask students to record their observations of chemical changes they encounter at home. They should write down the initial substances, the resulting substances, and the signs that the change was chemical rather than physical. L2 LS **Intrapersonal** P

Mini LAB

Purpose Students observe gas formation during a chemical reaction. L2 ELL COOP LEARN

LS **Kinesthetic**

Materials paper plate or lab tray, dry yeast, magnifying lens, warm water, sugar, clear cup or beaker

Teaching Strategy Have students place a small amount of dry yeast on the paper plate or lab tray to make their initial observation.

Analysis

1. The bubbles that form show that a new substance, a gas, is formed when the sugar, water, and yeast are combined.
2. This is a chemical change because gas is produced.

Assessment

Oral Ask students to explain the results of their investigation. Have them use information given in the text under the heading *Signs of Change* to support their answer. Use **Performance Assessment in the Science Classroom,** p. 143.

Try at Home

Discussion

Rusting Have students identify items in the school grounds or at home that have rusted. Possible answers include nails, bike chains, cars, or the flag pole. Ask them to explain, in their own words, how the metal object became rusty. Students should discuss how the metal rusted over time as it was exposed to the oxygen in the air. L2

Table 1 Comparing Properties

Physical Properties	color, shape, length, mass, volume, density, state, ability to attract a magnet, melting point, boiling point, malleability, ductility
Chemical Properties	flammability; ability to react with: oxygen, electricity, light, water, heat, vinegar, etc.

Mini LAB

Observing Yeast

Procedure

1. Observe a **tablespoon** of **dry yeast** with a **magnifying lens.** Draw and describe what you observe.
2. Put the yeast in 50 mL of **warm,** not hot, **water.**
3. Compare your observations of the dry yeast with those of the wet yeast.
4. Put a pinch of **sugar** in the water and observe for 15 minutes.
5. Record your observations.

Analysis

1. Are new substances formed when sugar is added to the water and yeast? Explain
2. Do you think this is a chemical change or a physical change? Explain.

Try at Home

Classifying According to Chemical Properties

Classifying according to physical properties is often easier than classifying according to chemical properties. **Table 1** summarizes the two kinds of properties. The physical properties of a substance are easily observed, but the chemical properties can't be observed without changing the substance. However, once you know the chemical properties, you can classify and identify matter based on those properties. For example, if you try to burn what looks like a piece of wood but find that it won't burn, you can rule out the possibility that it is wood.

In a grocery store, the products sometimes are separated according to their flammability or sensitivity to light or heat. You don't often see the produce section in front of big windows where heat and light come in. The fruit and vegetables would undergo a chemical change and ripen too quickly. You also won't find the lighter fluid and rubbing alcohol near the bakery or other places where heat and flame could be present.

Architects and product designers have to take into account the chemical properties of materials when they design buildings and merchandise. For example, children's sleepwear and bedding can't be made of a flammable fabric. Also, some of the architects designing the most modern buildings are choosing materials like titanium because it does not react with oxygen like many other metals do.

The Law of Conservation of Mass

It was so convenient to turn the firewood into the small pile of ash left after the campfire. You began with many kilograms of flammable substances but ended up with just a few kilograms of ash. Could this be a solution to the problems with landfills and garbage dumps? Why not burn all the trash? If you could make such a reduction without creating undesirable materials, this would be a great solution.

Curriculum Connection

Language Arts Have students write a diamond poem about physical and chemical properties. Form: Line 1, first topic; Line 2, two adjectives describing Line 1; Line 3, three participles related to Line 1; Line 4, four nouns—the first two related to Line 1 and the next two to Line 7; Line 5, three participles related to Line 7; Line 6, two adjectives describing Line 7; Line 7, second topic. L2 P

Example:
physical
color, texture
smelling, feeling, seeing
rock, pillow, fire, bubble
burning, baking, rusting
flammable, reactive
chemical

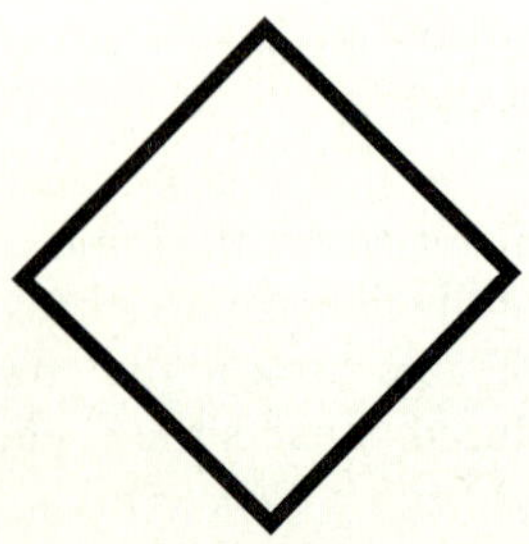

Mass Is Not Destroyed Before you celebrate your discovery, think this through. Did mass really disappear during the fire? It appears that way when you compare the mass of the pile of ashes to the mass of the firewood you started with. The **law of conservation of mass** states that the mass of what you end with is always the same as the mass of what you start with.

This law was first investigated about 200 years ago, and many investigations since then have proven it to be true. One experiment done by French scientist Antoine Lavoisier was a small version of a campfire. He determined that a fire does not make mass disappear or truly get rid of anything. The question, however, remains. Where did the mass go? The ashes aren't heavy enough to account for the mass of all of the pieces of firewood.

Where did the mass go? If you look at the campfire example more closely, you see that the law of conservation of mass is true. When flammable materials burn, they combine with oxygen. Ash, smoke, and gases are produced. The smoke and gases escape into the air. If you could measure the mass of the oxygen and all of the original firewood that was burned and compare it to the remaining mass of the ash, smoke, and gas, they would be equal.

Applying Science

Do light sticks conserve mass?

Light sticks often are used on Halloween to light the way for trick-or-treaters. They make children visible to drivers. They also are used as toys, for camping, marking trails, emergency traffic problems, by the military, and they work well underwater. A light stick contains two chemicals in separate tubes. When you break the inner tube, the two chemicals react producing a greenish light. The chemicals are not toxic, and they will not catch fire.

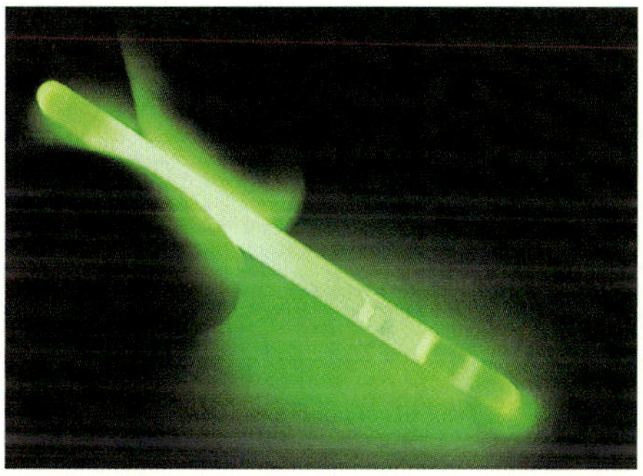

Identifying the Problem

In all reactions that occur in the world, mass is never lost or gained. This is the law of conservation of mass. An example of this phenomenon is the light stick. How can you prove this?

Solving the Problem

Describe how you could show that a light stick does not gain or lose mass when you allow the reaction to take place. Is this reaction a chemical or physical change? What is your evidence?

Applying Science

Teaching Strategy

Students might think that because light is produced, the tube should weigh less when the reaction is finished. Explain that light has no mass; it is energy.

Answer

You could weigh a light stick before you break the inner tube and after the reaction is finished. The mass should be the same both times. During the reaction, no chemicals left the tube and no chemicals were added. The reaction is a chemical change. The evidence is the light energy emitted.

Inquiry Lab

Missing Mass

Purpose To show students how much mass is changed into immeasurable substances during a chemical reaction. L2 PBL

Possible Materials non-galvanized nail, wire, sheet metal, water, clear beaker, balance

Estimated Time Initially 20 minutes, then 5 minutes a day for a week

Teaching Strategies

- Students should pick a metal object to test. Encourage them to develop a test to determine how much of the original mass is "lost" to the environment during a chemical reaction.
- They should leave their metal piece in the water for a week, measuring the mass of the metal each day and recording any observations they make.
- After the final measurement, have the students hypothesize where the "missing" mass went during the chemical reaction.

For additional inquiry activities, see *Science Inquiry Labs*.

Differentiated Instruction

Challenge Provide students with a small piece of marble or fossiliferous limestone. Ask them to weigh the sample and record the mass. Provide a small beaker and a small amount of 1*M* dilute hydrochloric acid (explain the dangers of the acid). Ask them to predict whether mass will be conserved as the acid reacts with the rock. Have them design an experiment to test their hypothesis. L3

English-Language Learners Define the word *conservation* for students who are learning the English language. Assign students to small groups and ask them to come up with different examples of conservation. Explain how conserving something, the environment, energy, or water, means to save it up. L2 ELL

3 Assess

DAILY INTERVENTION

Check for Understanding

Interpersonal Have students work in small groups to think of as many ways as they can to put out a fire. Ask each group to choose one and explain why it works. L2

Reteach

Chemical Changes Have students work in groups to name signs that indicate that a chemical change has occurred. Ask each group then to identify one situation in which one of these signs is present but in which no chemical change has occurred. Signs include the production of heat, light, smoke, color change, gas bubbles, sound, and the ease with which the change can be reversed. Gas bubbles are produced when water boils, but this is a physical change, not a chemical change. L2 LS

Assessment

Oral Have students think of chemical changes in which the product appears larger than the original substances and in which the product appears smaller. bread rising; wood burning Have them explain how mass is conserved in these chemical changes. The expansion of bread is caused by the formation of a gas, which pushes on the bread and makes it rise. Combustion produces a large amount of gas, which in this case is able to move freely away from the fire, leaving only a small amount of solid products behind. Use **Performance Assessment in the Science Classroom,** p. 89.

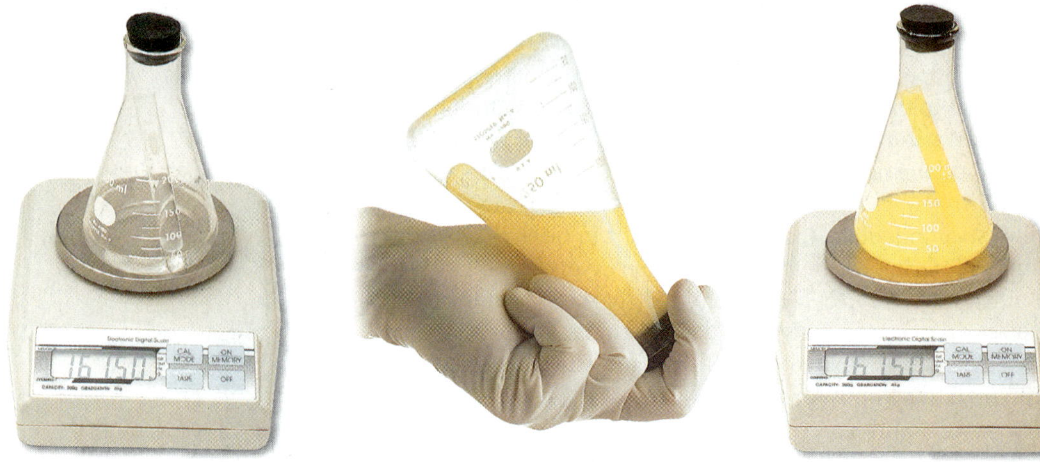

Figure 17 This reaction demonstrates the law of conservation of mass. Although a chemical change has occurred and new substances were made, the mass remained constant.

Before and After Mass is not destroyed or created during any chemical change. The law of conservation of mass is demonstrated in **Figure 17.** In the first photo, you see one substance in the flask and a different substance contained in a test tube inside the flask. The total mass is 16.150 g. In the second photo, the flask is turned upside down. This allows the two substances to mix and react. Because the flask is sealed, nothing is allowed to escape. In the third photo, the flask is placed on the balance again and the total mass is determined to be 16.150 g. If no mass is lost or gained, what happens in a reaction? Instead of disappearing or appearing, the particles in the substances rearrange into different combinations with different properties.

section 2 review

Summary

Common Chemical Properties

- A new substance, or substances, form(s) as a result of a chemical change.
- Exposure to oxygen, heat, and light can cause chemical reactions.

Something New

- Physical changes can be reversed. Chemical changes cannot be reversed.
- Substances can be classified according to their chemical properties.

The Law of Conservation of Mass

- Mass is not gained or lost during a chemical reaction.

Self Check

1. **Define** What is a chemical property? Give four examples.
2. **Identify** some of the signs that a chemical change has occurred.
3. **Think Critically** You see a bright flash and then flames during a class demonstration. Is this an example of a physical change or a chemical change? Explain.

Applying Math

4. **Solving One-Step Equations** A student heats 4.00 g of a blue compound to produce 2.56 g of a white compound and an unknown amount of colorless gas. What is the mass of this gas?

 Science Online red.msscience.com/self_check_quiz

section 2 review

1. Any characteristic that gives a substance the ability to undergo a change that results in a new substance. Possible answers: ability to burn, to react with O_2, to react to heat or light, to break down with electricity
2. Possible answers: production of heat, light, smoke, color change, gas bubbles, sound
3. Production of heat and light are signs of a chemical change.
4. The mass of the gas is 1.44 g.

LAB

Liquid Layers

Why must you shake up a bottle of Italian salad dressing before using it? Have you observed how the liquids in some dressings separate into two distinct layers? In this lab, you will experiment with creating layers of liquids.

Real-World Question

What would several liquids and solids of different densities look like when put into the same container?

Goals

- **Create** layers of liquids using liquids of different densities.
- **Observe** where solids of different densities will rest in the liquid layers.
- **Infer** the densities of the different materials.

Materials

250-mL beaker
graduated cylinder
corn syrup
glycerin
water
corn oil
rubbing alcohol
penny
wood sphere
rubber ball

Safety Precautions

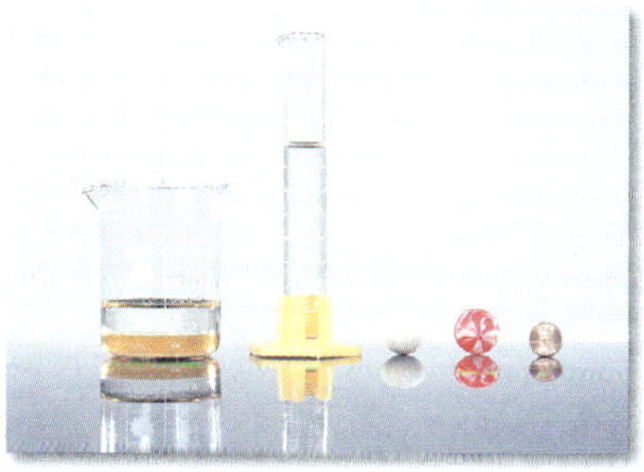

Procedure

1. Pour 40 mL of corn syrup into your beaker.
2. Slowly pour 40 mL of glycerin into the beaker. Allow the glycerin to trickle down the sides of the container and observe.
3. Slowly pour 40 mL of water into the beaker and observe.
4. Repeat step 3 with 40 mL of corn oil and then 40 mL of rubbing alcohol.
5. Carefully drop the penny, wood sphere, and rubber ball into the beaker and observe where these items come to a stop.

Conclude and Apply

1. **Draw and Label** In your Science Journal, draw a picture of the liquids and solids in your beaker. Label your diagram.
2. **Describe** what happened to the five liquids when you poured them into the beaker. Why did the liquids behave this way?
3. If water has a density of 1 g/cm^3, infer the relative densities of the rest of the materials.
4. **List** the liquids and solids in order from the highest density to the lowest density.

Communicating Your Data

Draw a labeled poster of the substances you placed in your beaker. Research the densities of each substance and include these densities on your poster. **For more help, refer to the** Science Skill Handbook.

BENCH TESTED

Real-World Question

Purpose Students observe the behavior of liquids and solids of different densities. L2 ELL LS **Kinesthetic**

Process Skills observe, sequence, infer, communicate, recognize cause and effect

Time Required 20 minutes

Safety Precautions Remind students not to taste any of the materials used for this lab.

Procedure

Teaching Strategy Pouring the liquids in such a way that the layers remain intact can be tricky. Before having students do this lab, practice pouring the liquids yourself so you can show them how to do it.

Troubleshooting Make sure the beaker is clean and dry at the beginning of this lab.

Conclude and Apply

1. Check students' work.
2. The liquids formed separate layers, one on top of the other. They behave this way because they have different densities.
3. The densities of the penny, corn syrup, and glycerin are greater than 1 g/cm^3, and the densities of the corn oil, rubbing alcohol, wood sphere, and rubber ball are less than 1 g/cm^3.
4. penny, corn syrup, glycerin, water, corn oil, rubbing alcohol, wood sphere, rubber ball

Assessment

Performance Ask them to clean the beaker and then pour 40 mL of corn oil followed by 40 mL of water. Use **Performance Assessment in the Science Classroom,** p. 97.

Communicating Your Data

Densities: corn syrup, 1.360 g/cm^3; glycerin, 1.260 g/cm^3; water, 1.0 g/cm^3; corn oil, 0.91 g/cm^3; rubbing alcohol, 0.791 g/cm^3; penny (copper and zinc), about 8.0 g/cm^3. The densities of the wood sphere and the rubber ball will vary.

BENCH TESTED

Real-World Question

Purpose Students observe and participate in controlling a chemical change. L2 ELL COOP LEARN
IS Interpersonal

Process Skills design an experiment, form a hypothesis, observe and infer, communicate, recognize cause and effect, separate and control variables, interpret data

Time Required two 45-minute periods

Possible Materials bananas, apples, pears, plastic or glass mixing bowls, 500 mL lemon/water solution, paring knife

Safety Precautions Be careful when handling the paring knife. If students will use the knife, show them the safe way to use the knife. Number the knives and collect them before the class ends.

Form a Hypothesis

Possible Hypotheses Most students' hypotheses will reflect that a chemical change can be controlled.

Test Your Hypothesis

Possible Procedures Slice fruit and place an even amount in each of two containers. In one container place 500 mL lemon/water solution. Make observations of each container every 5 minutes.

LAB Design Your Own

Fruit Salad Favorites

Goals

- **Design** an experiment that identifies physical changes and chemical changes in fruit.
- **Observe** whether chemical changes can be controlled.

Possible Materials

bananas
apples
pears
plastic or glass mixing bowls (2)
lemon/water solution (500 mL)
paring knife

Safety Precautions

WARNING: *Be careful when working with sharp objects. Always keep hands away from sharp blades. Never eat anything in the laboratory.*

Real-World Question

When you are looking forward to enjoying a tasty, sweet fruit salad at a picnic, the last thing you want to see is brown fruit in the bowl. What can you do about this problem? Your teacher has given you a few different kinds of fruit. It is your task to perform a test in which you will observe a physical change and a chemical change. Can a chemical change be controlled?

Form a Hypothesis

Based on your reading and observations, state a hypothesis about whether you can control a chemical change.

Test Your Hypothesis

Make a Plan

1. As a group, agree upon the hypothesis and decide how you will test it. Identify what results will confirm the hypothesis.
2. **List** each of the steps you will need in order to test your hypothesis. Be specific. Describe exactly what you will do in each step. List all of your materials.
3. Prepare a data table in your Science Journal or on a computer for your observations.
4. Read the entire investigation to make sure all steps are in logical order.
5. **Identify** all constants, variables, and controls of the investigation.

Alternative Inquiry Lab

Spoiled Fruit Make this Lab an Inquiry Lab by building on the experience. Ask students what sort of factors may lead to more rapid chemical or physical changes in their fruit salad. Instruct them to continue thinking of their fruit salad in the context of picnic food. Have students brainstorm in small groups the factors that may increase the rate of the changes they have observed. Students should choose one of the factors and design an experiment, with a control, to minimize the impact of that environmental factor. Conduct the experiments after all the necessary supplies have been gathered. L2 PBL COOP LEARN

Follow Your Plan

1. Ask your teacher to approve your plan and choice of constants, variables and controls before you start.
2. Perform the investigation as planned.
3. While doing the investigation, record your observations and complete the data table you prepared in your Science Journal.

Analyze Your Data

1. **Compare and contrast** the changes you observe in the control and the test fruit.
2. **Compare** your results with those of other groups.
3. What was your control in this investigation?
4. What are your variables?
5. Did you encounter any problems carrying out the investigation?
6. Do you have any suggestions for changes in a future investigation?

Conclude and Apply

1. Did the results support your hypothesis? Explain.
2. **Describe** what effect refrigerating the two salads would have on the fruit.
3. What will you do with the fruit from this experiment? Could it be eaten?

Communicating Your Data

Write a page for an illustrated cookbook explaining the benefits you found in this experiment. Include drawings and a step-by-step procedure. **For more help, refer to the Science Skill Handbook.**

Teaching Strategies

Tie to Prior Knowledge Most students know that fresh fruit, when cut, will begin to brown in a short period of time.

Troubleshooting Fruit may be cut by the teacher prior to the investigation. If so, cut fruit while students are creating and recording their plan, keeping in mind that it will soon begin to brown. Distribute fruit once a group's procedure is approved.

Expected Outcome Fresh fruit, when cut, will react quickly with the oxygen in the air, resulting in a brown tint on the freshly cut surface of the fruit. A lemon/water solution will retard the oxidation process and leave the fruit looking fresh and crisp.

Analyze Your Data

1. The fruit with the lemon/water solution did not brown; the fruit without the solution turned brown.
2. Results should be similar.
3. the freshly cut fruit that was not in the lemon/water solution
4. the presence of lemon/water, different types of fruit
5. Answers will vary.
6. Answers will vary.

Conclude and Apply

1. Answers will vary.
2. Refrigerating the two salads would cause the fruit without the lemon/water solution to brown more slowly.
3. Encourage students to think of practical and safe disposal methods, possibly including feeding it to classroom animals (if appropriate), composting, etc. It is most important that they articulate that no students should eat anything in a lab situation.

Assessment

Oral Have students explain what chemical reaction is inhibited in this experiment. Freshly cut fruit reacts with oxygen in the air to turn brown.

Communicating Your Data

Students may base their directions on the procedure used in their experiments. Have them review an illustrated cookbook for ideas on how to communicate what they have learned. LS **Linguistic**

TIME

Content Background

Researchers have identified hundreds of subatomic particles using particle accelerators. The building blocks of atoms are classified into three major categories, including fermions, hadrons, and bosons. Fermion particles include leptons, such as electrons and neutrinos, and quarks. Leptons do not play a role in holding the nucleus together, while quarks do participate in the strong nuclear force. Antimatter particles are also classified as fermions, including positrons, antiquarks, and antileptons. Hadrons, including protons and neutrons, are formed by combinations of quarks. Four boson particles have now been identified, including the gluon and the photon.

Discussion

Discoveries Through Time Create a timeline showing the major discoveries discussed in the article. What is the most important factor affecting the large gaps which appear on this timeline? Possible answer: Technological advances have increased the ability to study the properties of matter on a small scale. L2

Historical Significance

Prior to experimentation by scientists in the 1700's, matter as a gas was viewed as mysterious and impossible to collect, confine, and study. Joseph Black first showed in 1754 that air was not a simple substance, but a mixture of at least two gases. The importance of the work of 18th century chemists who identified and experimented with gases cannot be overstated. Does it make sense that Lavoisier's law of conservation of mass would result from experiments with gases? Why?

TIME SCIENCE AND HISTORY

SCIENCE CAN CHANGE THE COURSE OF HISTORY!

The Road to Understanding Matter

What a Ride!

Front wheel drive, a powerful motor, and a smooth ride are characteristics to look for in an automobile. This car, developed by Nippondenso, packs it all under a gold plated hood. At 4.78 mm long, this car is about the size of a grain of rice! Created to show the power and potential of technology applied to unimaginably small objects, this car demonstrates a fraction of the knowledge scientists have gained in exploring matter on a very small scale.

Matter Mileposts

Philosophers and scientists have speculated about the building blocks of matter for centuries. Around 425 B.C., the Greek Democritis used the term "atomos" to describe the indivisible particles making up matter of all types.

While early thinkers typically lacked the ability to test their theories, later technological advances applied to the study of matter moved science from the realm of the philosophical to the quantitative. In the 1700's, scientists experimented with gases, a type of matter difficult to confine and hard to study. Their findings eliminated the last of the old Greek notions and laid the foundation for modern chemistry.

It was the work of French scientist Antoine Lavoisier (1743–1794) which earned him the title "Father of Modern Chemistry." By focusing on measurable, quantifiable data, he forever changed the way science was conducted. Lavoisier's experiments with gases led to the development of the law of conservation of mass, a cornerstone of modern chemistry which helps explain what happens to matter during chemical change.

A Changing Road Map

In the 1930's, scientists used the first particle accelerators to reveal the composition of the atom. These machines accelerate subatomic particles, like electrons, to speeds close to the speed of light. Collisions at this speed cause these particles to shatter, and provide the opportunity to detect and analyze the smaller particles which comprise them.

Once thought to be the smallest building blocks, the proton, neutron, and electron are now joined by other subatomic particles groups, including quarks. Scientists currently believe the quark is the most fundamental particle. Studying particles created in particle accelerators is difficult because most exist for less than a billionth of a second.

As the technology behind these powerful machines advances, current hypotheses will undergo revision. The nature of scientific study is to build upon and extend, while sometimes uprooting, commonly held theories. Experimentation to discover the building blocks of matter is no exception.

Investigate Research the two types of particle accelerators. Compare how they work and their sizes. Describe what scientists learn about atomic structure using these machines. Use the link to the right or your school's media center to get started.

Science Online

For more information, visit red.msscience.com/time

Investigate Linear accelerators are huge, underground facilities, in which particles are accelerated down an extremely long, straight track. In circular accelerators, particles are propelled around a circular track, which occupies a fraction of the space required by a linear accelerator.

Resources for Teachers and Students

A Tour of the Subatomic Zoo: A Guide to Particle Physics, by Cindy Schwarz, Springer Verlag, 1997

A Brief History of Time, by Stephen Hawking, Bantam Books, 1988

Reviewing Main Ideas

Section 1 Physical Properties and Changes

1. Any characteristic of a material that can be observed or measured is a physical property.
2. The four states of matter are solid, liquid, gas, and plasma. The state of matter is determined by the energy the particles have.
3. Color, shape, length, mass, volume, density, melting point, boiling point, are common physical properties.
4. In a physical change the properties of a substance change but the identity of the substance always stays the same.
5. You can classify materials according to their physical properties.

Section 2 Chemical Properties and Changes

1. Chemical properties give a substance the ability to undergo a chemical change.
2. Common chemical properties include: ability to burn, reacts with oxygen, reacts with heat or light, and breaks down with electricity.
3. In a chemical change substances combine to form a new material.
4. The mass of the products of a chemical change is always the same as the mass of what you started with.
5. A chemical change results in a substance with a new identity, but matter is not created or destroyed.

Visualizing Main Ideas

Copy and complete the following table comparing properties of different objects.

Properties of Matter		
Type of Matter	Physical Properties	Chemical Properties
Log	brown, round, might have length, solid	flammability
Pillow	soft, might describe color, size, solid	flammability
Bowl of cookie dough	color, gooey, chunks in it, solid	reacts with heat
Book	might describe size, rectangle, color, solid	flammability
Glass of orange juice	might describe volume, liquid, color	reacts with bleach

chapter 3 Study Guide

Reviewing Main Ideas

Summary statements can be used by students to review the major concepts of the chapter.

Visualizing Main Ideas

See student page.

Science Online

Visit red.msscience.com
/self_check_quiz
/interactive_tutor
/vocabulary_puzzlemaker
/chapter_review
/standardized_test

Assessment Transparency

For additional assessment questions, use the *Assessment Transparency* located in the transparency book.

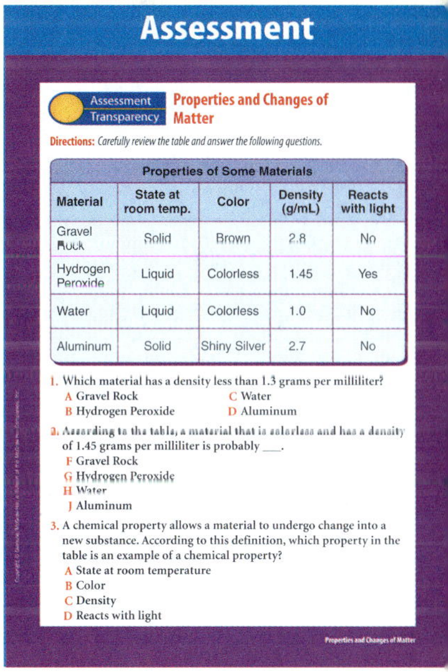

Assessment

Assessment Transparency **Properties and Changes of Matter**

Directions: *Carefully review the table and answer the following questions.*

Properties of Some Materials				
Material	State at room temp.	Color	Density (g/mL)	Reacts with light
Gravel Rock	Solid	Brown	2.8	No
Hydrogen Peroxide	Liquid	Colorless	1.45	Yes
Water	Liquid	Colorless	1.0	No
Aluminum	Solid	Shiny Silver	2.7	No

1. Which material has a density less than 1.3 grams per milliliter?
   A Gravel Rock
   B Hydrogen Peroxide
   C Water
   D Aluminum
2. According to the table, a material that is colorless and has a density of 1.45 grams per milliliter is probably ___.
   F Gravel Rock
   G Hydrogen Peroxide
   H Water
   J Aluminum
3. A chemical property allows a material to undergo change into a new substance. According to this definition, which property in the table is an example of a chemical property?
   A State at room temperature
   B Color
   C Density
   D Reacts with light

Properties and Changes of Matter

Using Vocabulary

1. The <u>melting point</u> is the temperature at which matter in a solid state changes to a liquid.
2. <u>Density</u> is a measure of the mass of an object in a given volume.
3. A <u>physical property</u> is easily observed or measured without changing the object.
4. <u>Chemical changes</u> result in a new substance and cannot be reversed by physical means.
5. Solid, liquid, and gas are all examples of <u>states of matter</u>.

Checking Concepts

6. D	11. D
7. A	12. C
8. B	13. B
9. C	14. A
10. C	15. D

Thinking Critically

16. All of the atoms of the reacting substances become part of the new substances formed.
17. A solid has a definite shape and volume, a liquid has definite volume but takes the shape of its container, a gas has the shape and volume of its container, and plasma occurs at very high temperatures.

chapter 3 Review

Using Vocabulary

boiling point p. 75	matter p. 71
chemical change p. 81	melting point p. 75
chemical property p. 80	physical change p. 71
density p. 72	physical property p. 70
law of conservation of mass p. 85	states of matter p. 73

Fill in the blanks with the correct vocabulary word or words.

1. The _________ is the temperature at which matter in a solid state changes to a liquid.
2. _________ is a measure of the mass of an object in a given volume.
3. A(n) _________ is easily observed or measured without changing the object.
4. _________ result in a new substance and cannot be reversed by physical means.
5. Solid, liquid, and gas are all examples of _________.

Checking Concepts

Choose the word or phrase that best answers the question.

6. Which of the following is an example of a physical change?
   A) tarnishing **C)** burning
   B) rusting **D)** melting

7. Which of the following is a sign that a chemical change has occurred?
   A) smoke **C)** change in shape
   B) broken pieces **D)** change in state

8. When iron reacts with oxygen, what substance is produced?
   A) tarnish **C)** patina
   B) rust **D)** ashes

9. What statement describes the physical property of density?
   A) the distance between two points
   B) how light is reflected from an object's surface
   C) the amount of mass for a given volume
   D) the amount of space an object takes up

10. Which of the choices below describes a boiling point?
    A) a chemical property
    B) a chemical change
    C) a physical property
    D) a color change

11. What property is described by the ability of metals to be hammered into sheets?
    A) mass
    B) density
    C) volume
    D) malleability

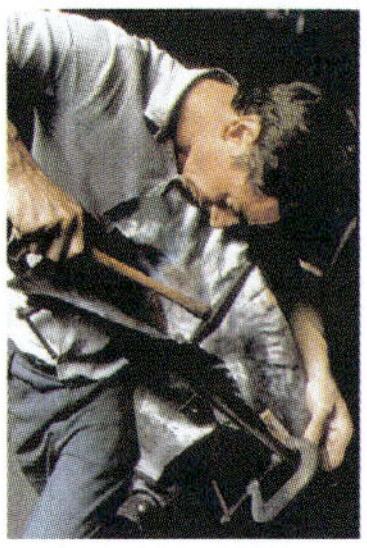

12. Which of these is a chemical property?
    A) size **C)** flammability
    B) density **D)** volume

13. Which describes what volume is?
    A) the area of a square
    B) the amount of space an object takes up
    C) the distance between two points
    D) the temperature at which boiling begins

14. What kind of change results in a new substance being produced?
    A) chemical **C)** physical
    B) mass **D)** change of state

15. What is conserved during any change?
    A) color **C)** identity
    B) volume **D)** mass

Science Online red.msscience.com/vocabulary_puzzlemaker

Use the Exam*View*® Pro Testmaker CD-ROM to:

- create multiple versions of tests
- create modified tests with one mouse click for inclusion students
- edit existing questions and add your own questions
- build tests aligned with state standards using built-in State Curriculum Tags
- change English tests to Spanish with one mouse click and vice versa

Thinking Critically

16. **Explain** Use the law of conservation of mass to explain what happens to atoms when they combine to form a new substance.
17. **Describe** the four states of matter. How are they different?
18. **Observe** A globe is placed on your desk and you are asked to identify its physical properties. How would you describe the globe?

19. **Evaluate** What information do you need to know about a material to find its density?
20. **Classify** the following as a chemical or physical change: an egg breaks, a newspaper burns in the fireplace, a dish of ice cream is left out and melts, and a loaf of bread is baked.
21. **Draw Conclusions** List the physical and chemical properties and changes that describe the process of scrambling eggs.
22. **Infer** Concrete is formed through a chemical reaction of sand, gravel, crushed stones, and water. Do the starting materials have the same properties as the end materials? Give two examples to support your response.
23. **Describe** In terms of particle movement explain how increasing temperature changes water in the solid state.
24. **Concept Map** Use a spider map to organize and define physical properties of matter. Include the concepts of color, shape, length, density, mass, states of matter, volume, density, melting point, and boiling point.

Performance Activities

25. **Comic Strip** Create a comic strip demonstrating a chemical change in a substance. Include captions and drawings that demonstrate your understanding of the law of conservation of mass.

Applying Math

26. **Measure in SI** Find the density of the piece of lead that has a mass of 49.01 g and a volume of 4.5 cm^3.

Use the table below to answer question 27.

States of Matter

Type	Definition	Examples
Solid	Particles vibrate in a fixed position.	books, desk, chair, ice cubes
Liquid	Particles do not stay in one position. They move past each other.	water, juice, tea, soft drink, etc.
Gas	Particles move freely with energy. They move all around.	oxygen, helium, vapor.

27. **States of Matter** Copy and complete the table by supplying the missing information.
28. **Density** Using the formula for density evaluate if two samples with the same volume, but different densities will have the same mass. Give two sample calculations to support your answer.

Science Online red.msscience.com/chapter_review

Assessment Resources

Reproducible Masters

Chapter *Fast File* Resources
- Chapter Review, pp. 35–36
- Chapter Tests, pp. 37–40
- Assessment Transparency Activity, p. 47

Glencoe Science Web site
- Chapter Review Test
- Standardized Test Practice

Glencoe Technology
- Assessment Transparency
- Exam*View*® Pro Testmaker
- MindJogger Videoquiz
- Interactive Chalkboard

18. The globe is spherical. Without changing its state, one could determine its mass, diameter, volume, density, color, and ability to attract a magnet.
19. volume and mass
20. chemical changes: a newspaper burns in the fireplace, a loaf of bread is baked; physical changes: an egg breaks, a dish of ice cream is left out and melts
21. The yolk and white of the eggs are mixed (physical change), the eggs are heated and cooked (chemical change to form bonds that harden the eggs).
22. No. Sand is pourable, but hardened concrete is not. Gravel and crushed stones have different volumes than hardened concrete. Water is a liquid but hardened concrete is a solid.
23. Increasing the temperature of water in the solid state will make the particles move faster. If enough heat is added, the particles will have enough energy to move past each other and become a liquid.
24. Check students' work.

Performance Activities

25. Comic strips will vary but should identify the substances both before and after the chemical change and should indicate that matter has been conserved. Use **Performance Assessment in the Science Classroom,** p. 133.

Applying Math

National Math Standards

1, 2, 4, 5, 9

26. $49.01\ g/4.5\ cm^3 = 10.89\ g/cm^3$
27. See student page.
28. Two samples with different densities and the same volume will have different masses.

FAST FILE

Answer Sheet A practice answer sheet can be found at red.msscience.com/answer_sheet.

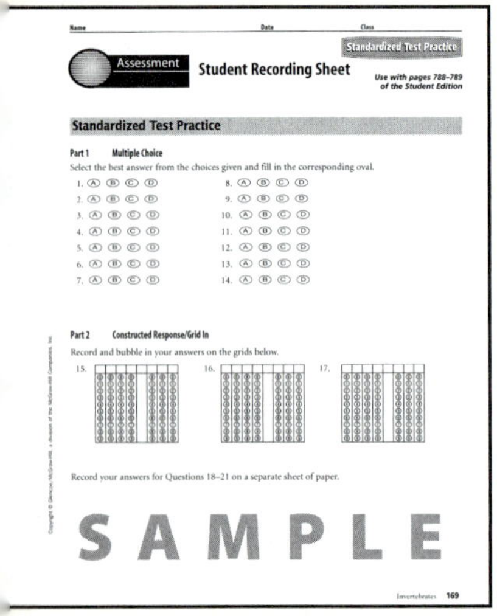
Name Date Class
Assessment
Student Recording Sheet
Standardized Test Practice
Use with pages 788–789 of the Student Edition

Standardized Test Practice

Part 1 Multiple Choice
Select the best answer from the choices given and fill in the corresponding oval.

1. Ⓐ Ⓑ Ⓒ Ⓓ
2. Ⓐ Ⓑ Ⓒ Ⓓ
3. Ⓐ Ⓑ Ⓒ Ⓓ
4. Ⓐ Ⓑ Ⓒ Ⓓ
5. Ⓐ Ⓑ Ⓒ Ⓓ
6. Ⓐ Ⓑ Ⓒ Ⓓ
7. Ⓐ Ⓑ Ⓒ Ⓓ
8. Ⓐ Ⓑ Ⓒ Ⓓ
9. Ⓐ Ⓑ Ⓒ Ⓓ
10. Ⓐ Ⓑ Ⓒ Ⓓ
11. Ⓐ Ⓑ Ⓒ Ⓓ
12. Ⓐ Ⓑ Ⓒ Ⓓ
13. Ⓐ Ⓑ Ⓒ Ⓓ
14. Ⓐ Ⓑ Ⓒ Ⓓ

Part 2 Constructed Response/Grid In
Record and bubble in your answers on the grids below.
15. 16. 17.

Record your answers for Questions 18–21 on a separate sheet of paper.

SAMPLE

Part 1 Multiple Choice

1. D
2. C
3. B
4. A
5. D
6. A
7. B
8. D

Part 2 Short Response

9. Answers will vary. Descriptions of the object may include color and shape, as well as approximate length, mass, and volume measurements.
10. The dry sponge (A) has a density of 0.3 g/cm^3. The moist sponge (B) has a greater density, at 0.5 g/cm^3.
11. Metals are shiny, or have luster. They are malleable, meaning they can be pressed into sheets, and they are ductile, meaning they can be drawn into wire. Possible examples include jewelry or a bicycle, which rely on metallic properties for beauty and function.
12. All three are chemical changes. Burning in picture A, rusting in picture B, and reacting with water in picture C.

Part 1 Multiple Choice

Record your answers on the answer sheet provided by your teacher or on a sheet of paper.

1. Which of these is NOT a physical property?
   A. volume **C.** density
   B. mass **D.** flammability

Use the illustration below to answer questions 2 and 3.

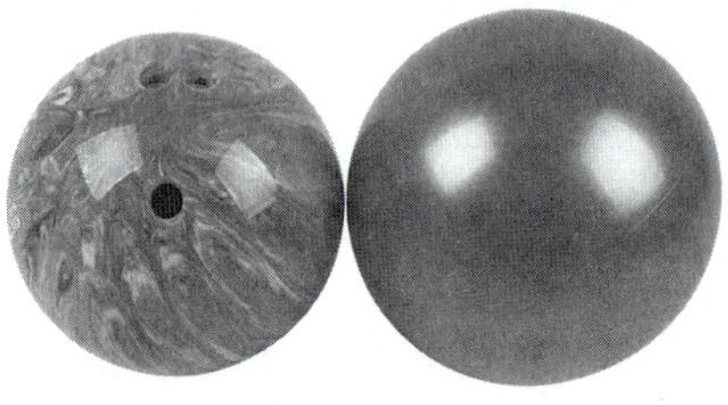

2. Which statement about the bowling ball and the kickball shown above is TRUE?
   A. These balls have nearly equal densities.
   B. These balls have nearly equal masses.
   C. These balls have nearly equal volumes.
   D. One of these balls exists as a different state of matter than the other.
3. A hole is punched in the kickball. Most of the air escapes and the ball collapses and shrinks. What happens to the ball?
   A. The mass of the ball increases.
   B. The volume of the ball decreases.
   C. The volume of the ball increases.
   D. The mass of the ball does not change.
4. Which term is a physical property that describes the amount of matter in an object?
   A. mass **C.** volume
   B. density **D.** state of matter

Test-Taking Tip

Be Relaxed and Focused Stay calm during the test. If you feel yourself getting nervous, close your eyes, and take five slow, deep breaths.

5. Which step in the process of making a cake results in a chemical change?
   A. breaking an egg and removing the contents from the shell
   B. melting butter
   C. mixing sugar and flour
   D. baking a cake in the oven

Use the illustration below to answer questions 6 and 7.

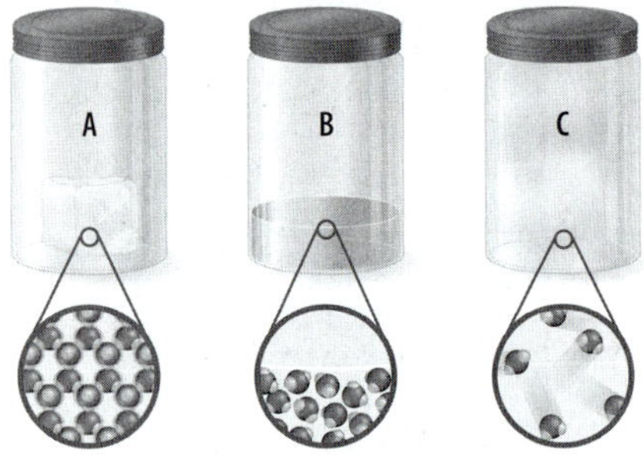

6. The particles in the jar labeled A represent a
   A. solid. **C.** gas.
   B. liquid. **D.** plasma.
7. If the material in the jars are all a form of H_2O, then jar C must be
   A. liquid water. **C.** ice.
   B. water vapor. **D.** pure oxygen.
8. When mercury (II) oxide, HgO, is heated, liquid mercury (Hg) and oxygen (O_2) are produced.

$2HgO \rightarrow 2Hg + O_2$	
Beginning mass of HgO	216 grams
Mass of Hg after heating	200 grams
Mass of O_2 after heating	? grams

According to the law of conservation of mass, what mass of O_2 is generated?
   A. 0 g **C.** 200 g
   B. 216 g **D.** 16 g

13. In a physical change, the physical properties of the substance change while the identity of the substance does not. In a chemical change, a new substance is formed. Signs of a chemical change include production of heat, light, smoke, color change, gas bubbles, and sound. Signs of a physical change include changing volume, length, or shape.

Part 3 Open Ended

14. Describe as many properties of the element as possible. Qualitative observations including color, luster, malleability, magnetic attraction, and shape may be helpful. Measure the mass and calculate the volume of the solid, then calculate its density. Attempt to determine the melting point of the material through heating. To determine the element's identity, compare data collected to known properties of elements.
15. In ice, the particles vibrate in a fixed position. Particles in liquid water move much more quickly, sliding past each other. As particle speed increases, water molecules

Part 2 Short Response/Grid In

Record your answers on the answer sheet provided by your teacher or on a sheet of paper.

9. Choose an object in the room and describe its physical properties as completely as possible.

10. Compare the density of two sponges: Dry Sponge A has a mass of 60 g. Moist Sponge B has a mass of 90 g. The volume of each sponge is 180 cm^3.

11. Describe the key properties of metals. Identify something you own which is made of metal. How has a metallic property made it possible to create or use this item?

Use the photos below to answer questions 12 and 13.

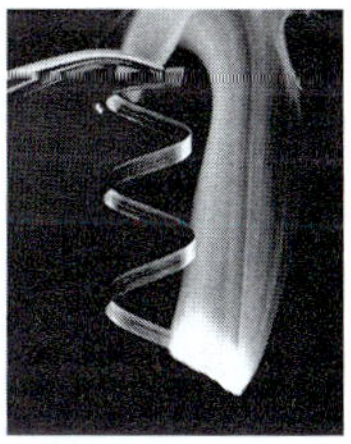

A

B

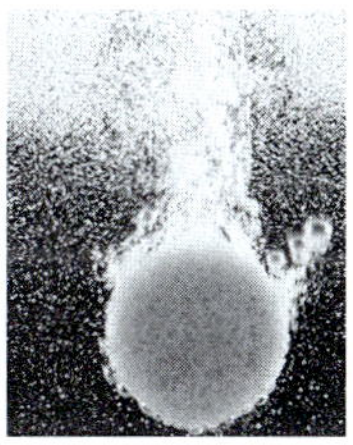

C

12. What type of change is occurring in each picture? Describe the signs of the change.

13. How do chemical and physical changes differ? What signs indicate a chemical change? What signs indicate a physical change?

Part 3 Open Ended

Record your answers on a sheet of paper.

14. You are asked to determine the identity of an unknown element. How will you use the properties of the element to discover its identity?

15. Based on the way particles move in solid, liquid, and gas phases of matter, describe what happens when solid water (ice) gradually changes to water as a gas (water vapor).

16. Making bread is an example of a chemical change. Describe the properties of the starting materials and the finished product.

Use the graph below to answer questions 17 and 18.

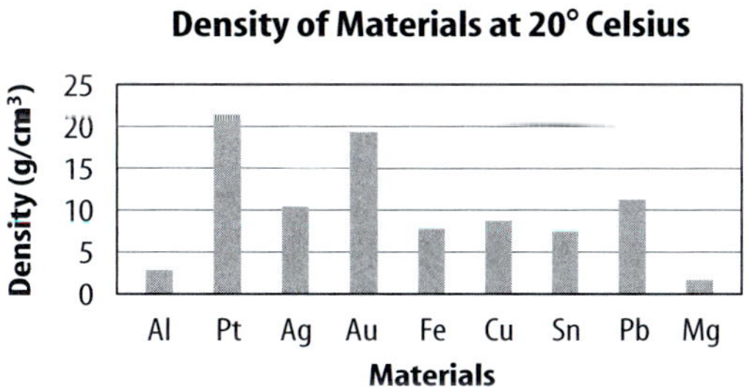

17. Rank the materials shown from most to least dense.

18. Imagine the graph had the title: *Density of Materials at 5° Celsius.* How and why would the graph look different?

19. Your teacher asks you to change a piece of paper both physically and chemically. Describe what you could do to show these two things.

20. An oil and vinegar salad dressing recipe calls for 20 mL of oil and 20 mL of vinegar. Would the oil and vinegar have equal densities? Predict if the oil and vinegar have the same mass.

have enough energy to separate, moving as far from each other as possible as molecules of gas.

16. Starting materials include solids and liquids such as flour, yeast, eggs, and milk. The finished product is a solid.

17. platinum, gold, lead, silver, copper, iron, tin, aluminum, magnesium

18. If the temperature were lowered to 5° Celsius, the particles would slow their vibration and pack more tightly together. The graph would show a slightly greater density for each material.

19. A physical change is crumbling the paper into a ball. A chemical change is burning the paper.

20. Vinegar forms a layer on top of oil showing that they have different densities. Different densities, but the same volume will have different masses.

Rubrics

The following rubrics are sample scoring devices for short response and open-ended questions.

Short Response

Points	Description
2	The student demonstrates a thorough understanding of the science of the task. The response may contain minor flaws that do not detract from the demonstration of a thorough understanding.
1	The student has provided a response that is only partially correct.
0	The student has provided a completely incorrect solution or no response at all.

Open Ended

Points	Description
4	The student demonstrates a thorough understanding of the science of the task. The response may contain minor flaws that do not detract from the demonstration of a thorough understanding.
3	The student demonstrates an understanding of the science of the task. The response is essentially correct and demonstrates an essential but less than thorough understanding of the science.
2	The student demonstrates only a partial understanding of the science of the task. Although the student may have used the correct approach to a solution or may have provided a correct solution, the work lacks an essential understanding of the underlying science concepts.
1	The student demonstrates a very limited understanding of the science of the task. The response is incomplete and exhibits many flaws.
0	The student provides a completely incorrect solution or no response at all.

chapter 4 Organizer

Section/Objectives	Standards		Labs/Features
	National	**State/Local**	
Chapter Opener	See p. 31T for a Key to Standards.		**Launch Lab:** Observe Matter, p. 97 **Foldables,** p. 97
Section 1 Structure of Matter 2 sessions 1 block 1. **Describe** the characteristics of matter. 2. **Identify** what makes up matter. 3. **Identify** the parts of an atom. 4. **Compare** the models that are used for atoms.	National Content Standards: UCP.1, UCP.2, UCP.3, UCP.5, A.1, A.2, B.1, G.1, G.2, G.3		**Integrate History,** p. 99 **MiniLAB:** Investigating the Unseen, p. 100 **Science Online,** p. 102 **Integrate Career,** p. 104
Section 2 The Simplest Matter 2 sessions 1 block 5. **Describe** the relationship between elements and the periodic table. 6. **Explain** the meaning of atomic mass and atomic number. 7. **Identify** what makes an isotope. 8. **Contrast** metals, metalloids, and nonmetals.	National Content Standards: UCP.1, UCP.2, UCP.3, UCP.5, A.1, A.2, B.1, G.3		**Science Online,** p. 107 **Visualizing the Periodic Table,** p. 108 **Lab:** Elements and the Periodic Table, p. 112
Section 3 Compounds and Mixtures 4 sessions 2 blocks 9. **Identify** the characteristics of a compound. 10. **Compare and contrast** different types of mixtures.	National Content Standards: UCP.1, UCP.2, UCP.3, UCP.5, A.1, A.2, B.1, G.1, G.2, G.3		**MiniLAB:** Comparing Compounds, p. 114 **Applying Science:** What's the best way to desalt ocean water?, p. 115 **Science Online,** p. 116 **Integrate Earth Science,** p. 117 **Lab:** Mystery Mixture, p. 118 **Science and History:** Ancient Views of Matter, p. 120

Glencoe Exclusive! TeacherWorks™ All-In-One Planner and Resource Center

Lab Materials	Reproducible Resources	Section Assessment	Technology
Launch Lab: dry paper towel or tissue, tape, plastic cup, bowl, or sink water	**Chapter *FAST FILE* Resources** Foldables Worksheet, p. 13 Note-taking Worksheets, pp. 29–31	Glencoe's Assessment Advantage	TeacherWorks includes: • Interactive Teacher Edition • Lesson Planner with calendar • Access to all program blacklines • Correlations to standards • Web links
MiniLAB: sealed shoe box, one or more items *Need materials?* Contact Science Kit at 1-800-828-7777 or www.sciencekit.com on the Internet.	**Chapter *FAST FILE* Resources** Transparency Activity, p. 40 MiniLAB, p. 3 Enrichment, p. 26 Reinforcement, p. 23 Directed Reading, p. 16 Transparency Activity, pp. 43–44 **Reading and Writing Skill Activities,** p. 41 **Mathematics Skill Activities,** p. 39	**Portfolio** Activity, p. 102 **Performance** MiniLAB, p. 100 Applying Skills, p. 105 **Content** Section Review, p. 105	Section Focus Transparency Teaching Transparency Virtual Labs CD-ROM Guided Reading Audio Program Interactive Chalkboard CD-ROM
Lab: colored markers, large index cards, Merck index, encyclopedia, large bulletin board, 8 1/2- × 14-inch paper, thumbtacks	**Chapter *FAST FILE* Resources** Transparency Activity, p. 41 Enrichment, p. 27 Reinforcement, p. 24 Directed Reading, p. 17 Activity Worksheet, pp. 5–6 **Science Inquiry Labs,** pp. 49–50	**Portfolio** Science Journal, p. 110 Assessment, p. 112 **Performance** Applying Math, p. 111 **Content** Section Review, p. 111	Section Focus Transparency Guided Reading Audio Program Interactive Chalkboard CD-ROM
MiniLAB: sugar, rubbing alcohol, salad oil, spoon, glasses, hot water **Lab:** test tubes and holder, cornstarch, sugar, baking soda, mystery mixture, small scoops, dropper bottles, iodine solution, white vinegar, hot plate, 250-mL beaker, water, small pie pan, matches	**Chapter *FAST FILE* Resources** Transparency Activity, p. 42 MiniLAB, p. 4 Enrichment, p. 28 Reinforcement, p. 25 Directed Reading, pp. 17, 18 Lab Worksheet, pp. 7–8 Lab Activity, pp. 9–10, 11–12 **Lab Management and Safety,** p. 44	**Portfolio** MiniLAB Assessment, p. 114 **Performance** MiniLAB, p. 114 Applying Science, p. 115 Applying Skills, p. 117 **Content** Section Review, p. 117	Section Focus Transparency Guided Reading Audio Program Interactive Chalkboard CD-ROM Video Lab

End of Chapter Assessment

Glencoe's Assessment Advantage

Blackline Masters	Technology	Professional Series
Chapter *FAST FILE* Resources Chapter Review, pp. 33–34 Chapter Tests, pp. 35–38 **Standardized Test Practice,** pp. 19–22	MindJogger Videoquiz Virtual Labs CD-ROM Exam*View*® Pro Testmaker TeacherWorks CD-ROM Interactive Chalkboard CD-ROM	**Performance Assessment in the Science Classroom (PASC)**

Transparencies

Section Focus

Section 1 Section Focus Transparency — **What's the matter in this picture?**

This unusual image is made with a scanning tunneling microscope (STM). This microscope actually images individual atoms.

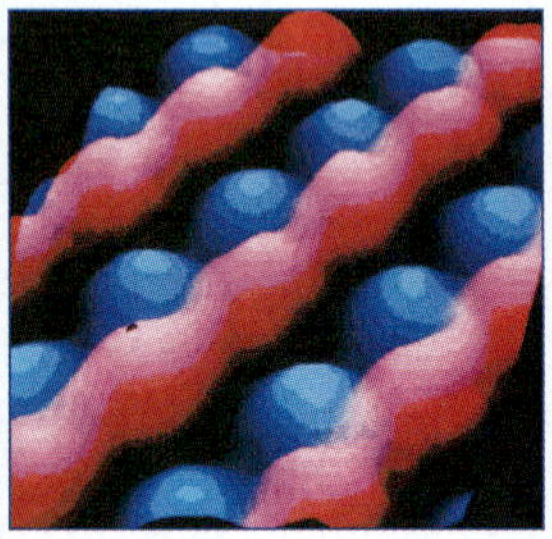

1. Why could this image not be from a traditional microscope?
2. How many different types of atoms are shown?

L2

Section 2 Section Focus Transparency — **Some Call It Quicksilver**

This element is a liquid at room temperature. It is probably best known as the liquid in traditional thermometers; however, it is so toxic that people often choose digital thermometers that do not contain this element.

1. Describe what you see in the picture. What characteristics does this element have?
2. Which element do you think this is? Do you know of any other uses for it?

L2

Section 3 Section Focus Transparency — **Here Today, Rust Tomorrow**

This was seaworthy metal once, but now it's mostly rust. Before it rusted, it was a sturdy ship for work and transportation. As rust, it is falling apart. What happened?

1. What properties of metals are useful for making ships?
2. Is rust the same material as the original metal? Why or why not?

L2

This is a representation of key blackline masters available in the Teacher Classroom Resources. See Resource Manager boxes within the chapter for additional information.

Assessment

Assessment Transparency — **Matter**

Directions: *Carefully review the table and answer the following questions.*

Characteristics of Some Elements

Element	Atomic number	Mass number	Number of protons	Number of electrons	Number of neutrons
Boron	5	11	5	5	6
Carbon	6	12	6	6	6
Nitrogen	7	14	7	7	7
Oxygen	8	16	8	8	8
Fluorine	9	19	9	9	10

1. According to the table, which element has an atomic number of 7?
   A Boron C Nitrogen
   B Carbon D Oxygen
2. According to the table, which element has a mass number less than 12?
   F Boron H Oxygen
   G Carbon J Fluorine
3. According to the table, an atom with 9 protons in its nucleus is ___.
   A carbon C oxygen
   B nitrogen D fluorine

L2

Teaching

Section 1 Teaching Transparency — **Fire and Ash**

L2

Key to Teaching Strategies

The following designations will help you decide which activities are appropriate for your students.

- L1 Level 1 activities should be appropriate for students with learning difficulties.
- L2 Level 2 activities should be within the ability range of all students.
- L3 Level 3 activities are designed for above-average students.
- ELL ELL activities should be within the ability range of English Language Learners.
- COOP LEARN Cooperative Learning activities are designed for small group work.
- LS Multiple Learning Styles logos, as described on page 12T, are used throughout to indicate strategies that address different learning styles.
- P These strategies represent student products that can be placed into a best-work portfolio.
- PBL Problem-Based Learning activities apply real-world situations to learning.

Hands-on Activities

Student Text Lab Worksheet

Name Date Class

Activity — **Elements and the Periodic Table**

Hands-On Activities

Lab Preview

Directions: *Answer these questions before you begin the Activity.*

1. What safety practice is indicated by the safety symbol with the hand that is included in this activity?
2. List at least two research sources that might be useful in this activity.

The periodic table organizes the elements, but what do they look like? What are they used for? In this activity, you'll examine some elements and share your findings with your classmates.

What You'll Investigate

What are some of the characteristics and purposes of the chemical elements?

Goals

- **Classify** the chemical elements.
- **Design** your own periodic table that shows the classification of the elements.

Materials

colored markers
large index cards
Merck Index
encyclopedia
**other reference materials*
large bulletin board
$8\frac{1}{2} \times 14$-inch paper
thumbtacks
**pushpins*
**Alternate materials*

Safety Precautions

Use care when handling sharp objects.

Procedure

1. Select the assigned number of elements from the list provided by your teacher.
2. **Design** an index card for each of your selected elements. On each card, mark the element's atomic number in the upper left-hand corner and write its symbol and name in the upper right-hand corner.
3. **Research** each of the elements and write several sentences on the card about its appearance, its other properties, and its uses.
4. **Classify** each of your elements as a metal, a metalloid, or a nonmetal based upon its properties.
5. **Write** the appropriate classification on each of your cards using the colored marker chosen by your teacher.
6. Work with your classmates to make a large periodic table. Use thumbtacks to attach your cards to a bulletin board in their proper positions on the periodic table.
7. **Draw** your own periodic table. Place the elements' symbols and atomic numbers in the proper locations on your table.

L2

Laboratory Activities

Name Date Class

LAB 1 Laboratory Activity — **Mixtures and Compounds**

Hands-On Activities

Matter is anything that has mass and occupies space. Matter exists in different forms. Three classifications of matter are well known to us: elements, mixtures, and compounds. Elements are the basic materials of our world. Elements in a mixture have recognizable boundaries and can be separated by mechanical means. Elements that form a chemical compound can be separated only by a chemical process. The element oxygen(O) combines with the element hydrogen (H) to form water (H_2O), which is a compound. Salt water is a mixture of two compounds, water and salt.

Strategy

You will separate a mixture into its parts.
You will compare the characteristics of a compound and a mixture.

Materials

magnifying glass
sand (coarse)
granite rock
granite (crushed)
water
disposable pie pans (2)
rock salt
heat source

Procedure

1. Use the magnifying glass to observe the sand and granite rock. Sketch the granite with the minerals in it, and the shapes of the sand grains under Sketch A.
2. Sort the crushed granite into separate piles according to color.
3. Sketch the general shape of a piece from each pile of the sorted granite, and label each as to color under Sketch B.
4. Mix a spoonful of sand in some water in a pie pan. Sketch what you observed under Sketch C.
5. Examine and sketch the salt crystals under Sketch D. **CAUTION:** *Do not ingest rock salt. It might contain harmful impurities.*
6. Mix a spoonful of salt in some water in the second pie pan. Record your observations.
7. Heat both pans until the water is evaporated. Sketch what is left in each pan under Sketch E. **CAUTION:** *Be careful not to get clothes or hair close to the heat source.*

Data and Observations

Sketch A

Sketch B

L2

Resource Manager

Meeting Different Ability Levels

Content Outline

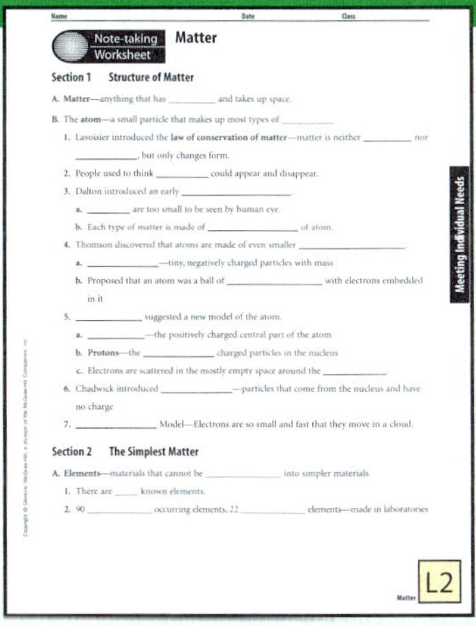

Reinforcement

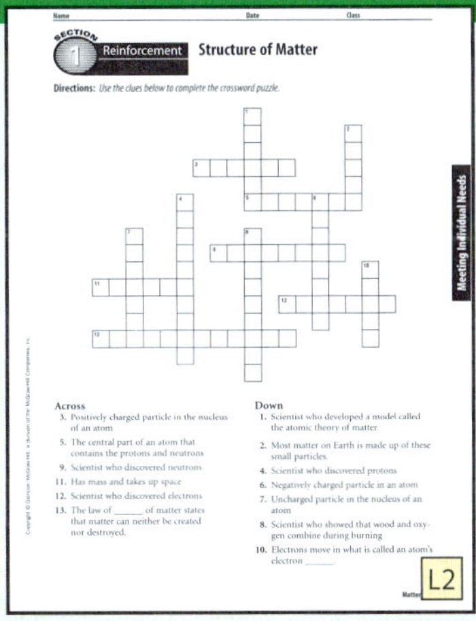

Enrichment

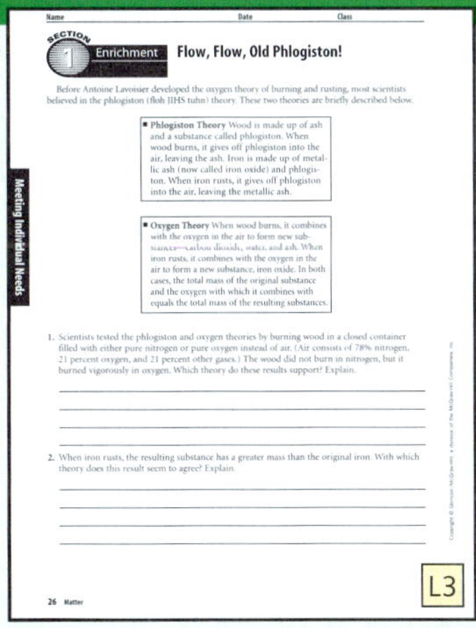

Directed Reading (English/Spanish)

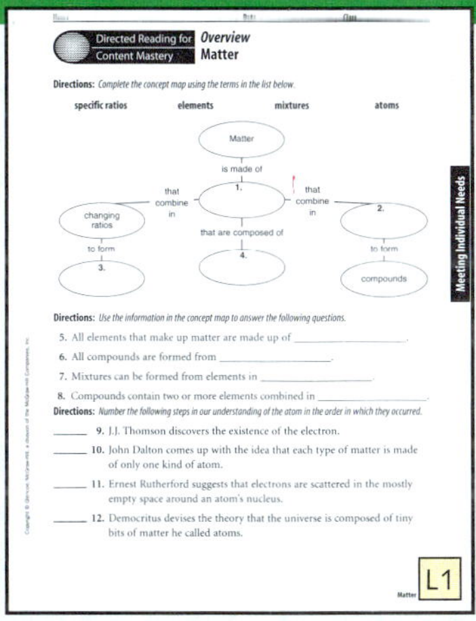

Study Guide

Reading Essentials

Assessment

Test Practice Workbook

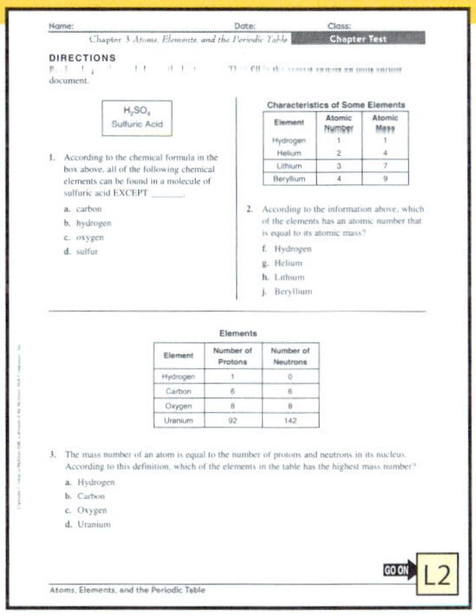

Chapter Review

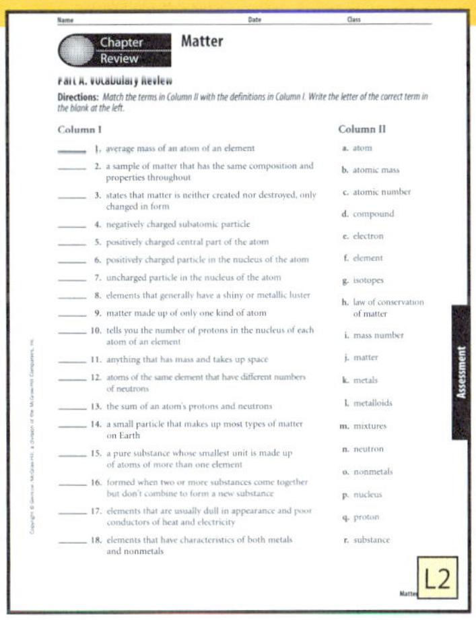

Chapter Tests

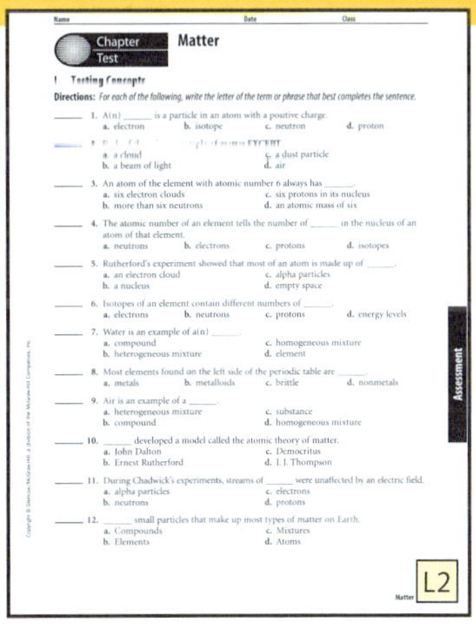

Science Content Background

section 1 Structure of Matter

Matter Basics

Solids, liquids, and gases are matter—they take up space and have mass. The amount of space an object occupies is its volume. The mass of an object refers to the amount of matter it contains. On Earth the amount of mass is directly proportional to the weight of the object. Dividing matter into smaller pieces reduces its volume and mass, and because the mass is reduced, so is the weight.

Ions

Many substances are ionic. That is, they are made up of charged particles called ions rather than neutral atoms. For example, sodium chloride is made up of sodium ions that have 1+ charges and chloride ions that have 1− charges. A sodium ion is formed when a sodium atom loses one of its 11 electrons. A chloride ion is formed when a chlorine atom gains an additional electron to go with the 17 it had as an atom.

Plasma

Plasma, another fluid state of matter, is a gaslike state of negatively charged electrons and positively charged ions. Most matter in the universe is plasma. Stars are mostly plasma, and the hydrogen ions that occur in interstellar space exist in the plasma state.

section 2 The Simplest Matter

Elements

The arrangement of electrons in the energy levels and sublevels in the atoms of an element determines the properties of that element. Elements whose atoms have identical arrangements of electrons in their highest energy levels have similar properties and make up a family of elements on the periodic table.

section 3 Compounds and Mixtures

Compounds

Electron sharing is called covalent bonding. Covalent bonding between atoms of different elements produces molecular compounds. The molecules of molecular compounds attract each other to varying degrees. In the solid state, these attractions sometimes lead to an orderly arrangement of molecules called a molecular crystal.

When atoms react by losing or gaining electrons, called ionic bonding, ions are formed. The formula for an ionic compound such as sodium chloride ($NaCl$) is called an empirical formula and gives the simplest ratio of ions in the compound, not the formula for a molecule.

Mixtures

Solid solutions, made up of uniform mixtures of metals and sometimes nonmetals, are called alloys. Brass and stainless steel are common alloys.

chapter content resources

Internet Resources

For additional content background, visit **red.msscience.com** to:

- access your book online
- find references to related articles in popular science magazines
- access Web links with related content background
- access current events with science journal topics

Print Resources

Chemistry, Steve S. Zumdahl, Susan A. Zumdahl, Houghton Mifflin Company, 2003

Chemistry, Raymond Chang, McGraw-Hill College, 2001

General Chemistry, Ralph Petrucci, William Harwood, Prentice Hall, 1997

IDENTIFYING Misconceptions

Find Out What Students Think

Students may think that . . .

Objects have mass only if you can feel them.

If you keep dividing matter, you will eventually get small pieces with no mass.

Students may have these misconceptions because they are relying on their senses. In this case they are relying on their sense of touch. They may think that if an object does not press down on their hands and feel heavy, it has no mass.

Demonstration

Hold up a plastic foam cup, and ask students if the cup has mass. Establish that it does. Break a chunk off the cup, and ask how the mass of this part of the cup compares with the mass of the original cup. Establish that it has less mass. Give pieces of the cup to four different students sitting in different parts of the classroom. Ask students to break their pieces and hold up the biggest piece of cup that has no mass. When they hold these up ask other students if they agree that the pieces have no mass.

Promote Understanding

Activity

Divide the class into groups. Give a foam cup to each group. Have them discuss whether the cup is matter. Establish that it is because it has mass and takes up space. Explain that on Earth, the mass of a cup is directly proportional to its weight. This means that if a cup has a large mass, it has a large weight, and if it has a small mass, it has a small weight. Perform the following:

- Using a balance that can measure mass to at least 0.1 g, measure the mass of the cup.
- Draw a picture of the cup, and write its mass.
- Use scissors to cut the cup exactly in half.
- Draw a picture of the half cup, calculate its mass (original mass divided by two), and write the new mass by the picture.
- Take one half of the cup and cut it into two equal parts.
- Draw a picture of one of the new pieces, and write its mass next to it.
- Repeat this procedure four more times, each time cutting the piece in half, drawing it, and calculating its mass.
- Make a graph of the data. The x-axis should be labeled *Mass (g)*. It should start with zero and have a range that includes the mass of the whole cup. The y-axis should be labeled *Fraction of cup*. It should begin with 1/64th of a cup and go up in the following increments: 1/32, 1/16, 1/8, 1/4, 1/2, and 1.

Instruct students to examine their data and discuss whether they will ever get a piece of cup that has absolutely no mass. Make sure students understand that as pieces get smaller and smaller, their masses and their weights get smaller and smaller, but the pieces always have some mass.

Assess

After completing the chapter, see *Identifying Misconceptions* in the Study Guide at the end of the chapter.

chapter 4

Chapter Vocabulary

matter, p. 98
atom, p. 99
law of conservation of matter, p. 100
electron, p. 102
nucleus, p. 103
proton, p. 103
neutron, p. 104
element, p. 106
atomic number, p. 109
isotope, p. 109
mass number, p. 109
atomic mass, p. 110
metal, p. 110
nonmetal, p. 111
metalloid, p. 111
substance, p. 113
compound, p. 113
mixture, p. 115

Science Journal Student responses will vary but may include questions about flight, balloons, and how matter is involved with flight.

INTERACTIVE CHALKBOARD with Image Bank

PowerPoint® Presentations

This CD-ROM is an editable Microsoft® PowerPoint® presentation that includes:

- a pre-made presentation for every chapter
- interactive graphics
- animations
- audio clips
- image bank
- all new section and chapter questions
- Standardized Test Practice
- transparencies
- pre-lab questions for all labs
- Foldables directions
- links to red.msscience.com

chapter 4

Atoms, Elements, and the Periodic Table

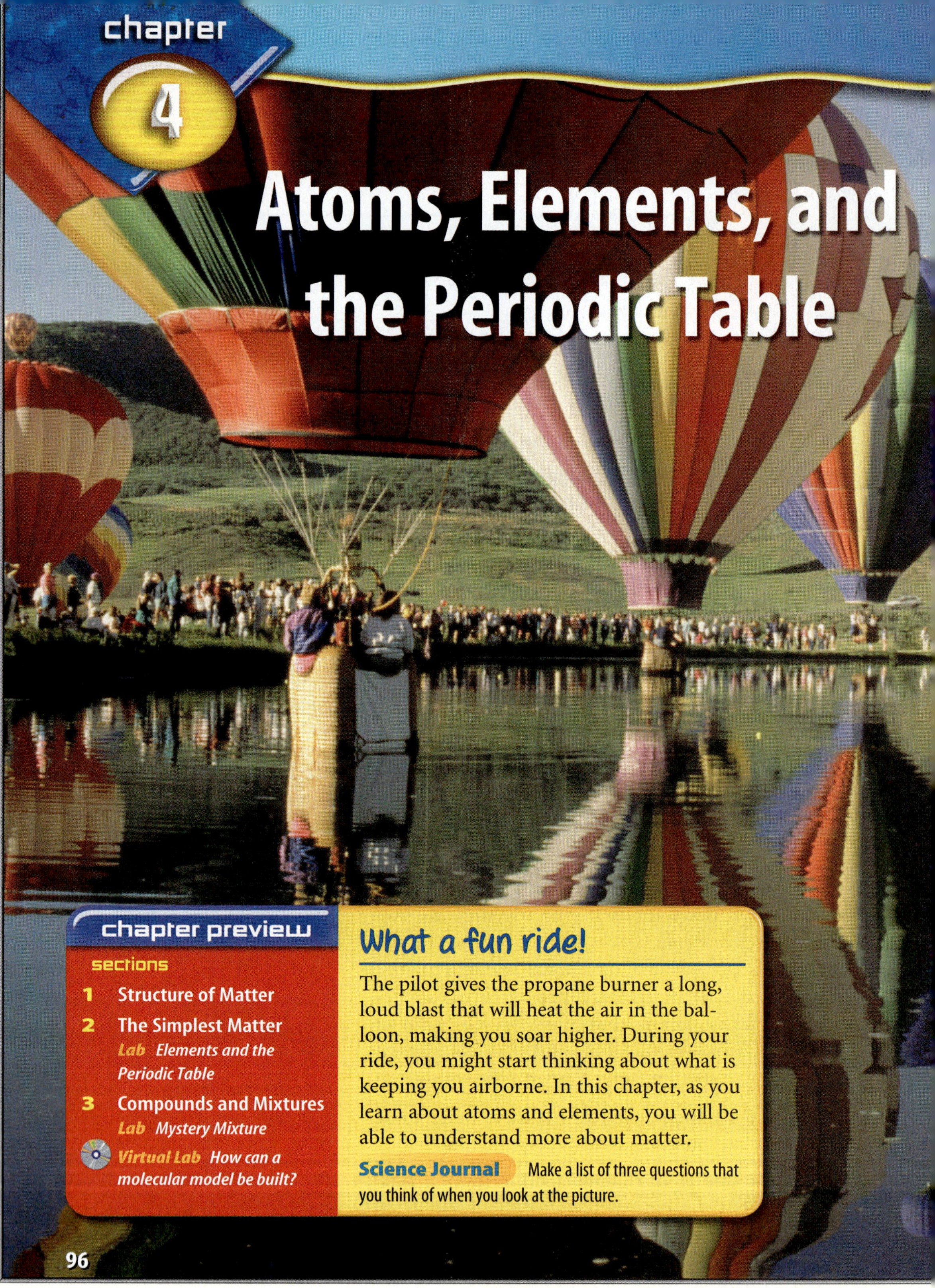

chapter preview

sections

1 Structure of Matter
2 The Simplest Matter
Lab Elements and the Periodic Table
3 Compounds and Mixtures
Lab Mystery Mixture
Virtual Lab How can a molecular model be built?

What a fun ride!

The pilot gives the propane burner a long, loud blast that will heat the air in the balloon, making you soar higher. During your ride, you might start thinking about what is keeping you airborne. In this chapter, as you learn about atoms and elements, you will be able to understand more about matter.

Science Journal Make a list of three questions that you think of when you look at the picture.

96

Theme Connection

Scale and Structure Tiny particles called atoms make up elements. Elements are the building blocks of all matter.

About the Photo

Rise and Sink What keeps a hot air balloon flying? It is the same principle that keeps food frozen in the open chest freezers at the grocery store. The principle is that hot air rises and cold air sinks. While the super-cooled air in a open chest freezer settles down around the food, the hot air in a hot air balloon pushes up, keeping the balloon afloat.

Start-Up Activities

Launch LAB

Observe Matter

You've just finished playing basketball. You're hot and thirsty. You reach for your bottle of water and take a drink. Releasing your grip, you notice that the bottle is nearly empty. Is the bottle really almost empty? According to the dictionary, *empty* means "containing nothing." When you have finished all the water in the bottle, will it be empty or full?

1. Wad up a dry paper towel or tissue and tape it to the inside of a plastic cup as shown.
2. Fill a bowl or sink with water. Turn the cup upside down and slowly push the cup straight down into the water as far as you can.
3. Slowly raise the cup straight up and out of the water. Remove the paper towel or tissue paper and examine it.
4. **Think Critically** In your Science Journal, describe the lab and its results. Explain what you think happened. Was anything in the cup besides the paper? If so, what was it?

FOLDABLES Study Organizer

Atoms, Elements, and the Periodic Table Make the following Foldable to help you identify the main ideas about atoms, elements, compounds, and mixtures.

STEP 1 **Draw** a mark at the midpoint of a sheet of paper along the side edge. Then **fold** the top and bottom edges in to touch the midpoint.

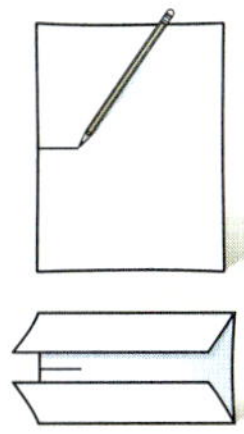

STEP 2 **Fold** in half from side to side.

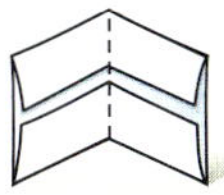

STEP 3 **Open and cut** along the inside fold lines to form four tabs.

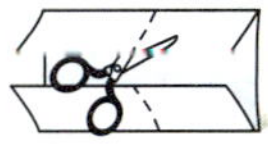

STEP 4 **Label** each tab as shown.

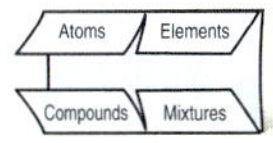

Read and Write As you read the chapter, list several everyday examples of atoms, elements, compounds, and mixtures on the back of the appropriate tab.

Science Online — Preview this chapter's content and activities at red.msscience.com

Launch LAB

Purpose Use the Launch Lab to help students discover that air is matter because it occupies space. L2 ELL LS **Kinesthetic**

Preparation Obtain water if sinks are not available in your classroom. The sink or bowl used in the lab must be deep enough for the cup to be submerged in water.

Materials paper towel or tissue, plastic cup, sink or bowl, water, tape

Teaching Strategy In Step 2, remind students to *slowly* push the cup into the water to avoid spills.

Think Critically

Students should observe that the paper towel was dry when removed from the cup. They should infer that air in the cup took up space and did not allow water to enter.

Assessment

Process Have students work in small groups to make a mini picture book that explains to third graders how air is matter. Use **Performance Assessment in the Science Classroom,** p. 129.

FOLDABLES Study Organizer — Dinah Zike Study Fold

Student preparation materials for this Foldable are available in the Chapter *FAST FILE* Resources.

1 Motivate

Bellringer

INTERACTIVE CHALKBOARD PowerPoint® Presentations

Section Focus Transparencies also are available on the Interactive Chalkboard CD-ROM. L2 ELL

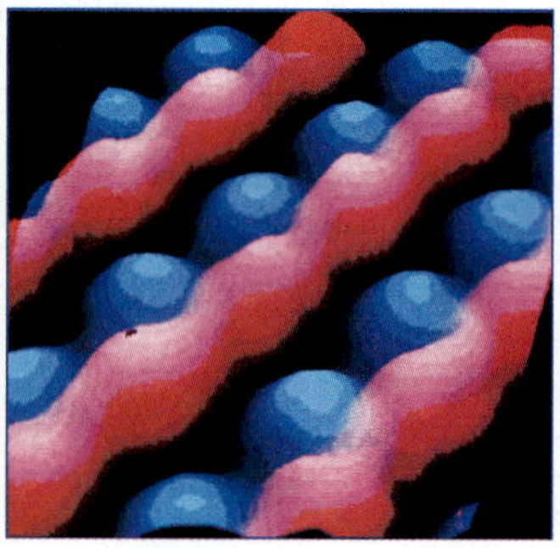
Section 1 Section Focus Transparency **What's the matter in this picture?**

This unusual image is made with a scanning tunneling microscope (STM). This microscope actually images individual atoms.

1. Why could this image not be from a traditional microscope?
2. How many different types of atoms are shown?

Matter

Tie to Prior Knowledge

Gases in Balloons Have students recall what they know about gases. Then demonstrate that air has mass. Tie an inflated balloon to each end of a meterstick. Balance the meterstick on top of a ring stand or on your fingertip, and pop one of the balloons.

Caption Answer

Figure 1 The sculpture and pumpkin are matter; the drops the light filters through to make the rainbow are matter, but the rainbow itself is light, and thus not matter.

Reading Check

Answer Air has mass and takes up space; light does not.

section 1

Structure of Matter

as you read

What You'll Learn

- **Describe** characteristics of matter.
- **Identify** what makes up matter.
- **Identify** the parts of an atom.
- **Compare** the models that are used for atoms.

Why It's Important

Matter makes up almost everything we see—and much of what we can't see.

Review Vocabulary

density: the mass of an object divided by its volume

New Vocabulary

- matter
- atom
- law of conservation of matter
- electron
- nucleus
- proton
- neutron

What is matter?

Is a glass with some water in it half empty or half full? Actually, neither is correct. The glass is completely full—half full of water and half full of air. What is air? Air is a mixture of several gases, including nitrogen and oxygen, which are kinds of matter. **Matter** is anything that has mass and takes up space. So, even though you can't see it or hold it in your hand, air is matter. What about all the things you can see, taste, smell, and touch? Most are made of matter, too. Look at the things pictured in **Figure 1** and determine which of them are matter.

What isn't matter?

You can see the words on this page because of the light from the Sun or from a fixture in the room. Does light have mass or take up space? What about the warmth from the Sun or the heat from the heater in your classroom? Light and heat do not take up space, and they have no mass. Therefore, they are not forms of matter. Emotions, thoughts, and ideas are not matter either. Does this information change your mind about the items in **Figure 1?**

Reading Check *Why is air matter, but light is not?*

Figure 1 A rainbow is formed when light filters through the raindrops, a plant grows from a seed in the ground, and a statue is sculpted from bronze.
Identify *which are matter.*

Section 1 Resource Manager

Chapter *FAST FILE* Resources
- Transparency Activity, pp. 40, 43–44
- Note-taking Worksheets, pp. 29–31
- Directed Reading for Content Mastery, p. 16
- MiniLAB, p. 3
- Enrichment, p. 26
- Reinforcement, p. 23

Mathematics Skill Activities, p. 39
Cultural Diversity, p. 59
Reading and Writing Skill Activities, p. 41

Figure 2 Early Beliefs About the Composition of Matter

Many Indian Philosophers (1,000 B.C.)	Kashyapa, an Indian Philosopher (1,000 B.C.)	Many Greek Philosophers (500–300 B.C.)	Democritus (380 B.C.)	Aristotle (330 B.C.)	Chinese Philosophers (300 B.C.)
• Ether—an invisible substance that filled the heavens • Earth • Water • Air • Fire	• Five elements broken down into smaller units called parmanu • Parmanu of earth elements are heavier than air elements	• Earth • Water • Air • Fire	• Tiny individual particles he called *atomos* • Empty space through which atoms move • Each substance composed of one type of *atomos*	• Empty space could not exist • Earth • Water • Air • Fire	• Metal • Earth • Water • Air • Fire

What makes up matter?

Suppose you cut a chunk of wood into smaller and smaller pieces. Do the pieces seem to be made of the same matter as the large chunk you started with? If you could cut a small enough piece, would it still have the same properties as the first chunk? Would you reach a point where the last cut resulted in a piece that no longer resembled the first chunk? Is there a limit to how small a piece can be? For centuries, people have asked questions like these and wondered what matter is made of.

An Early Idea Democritus, who lived from about 460 B.C. to 370 B.C., was a Greek philosopher who thought the universe was made of empty space and tiny bits of stuff. He believed that the bits of stuff were so small they could no longer be divided into smaller pieces. He called these tiny pieces atoms. The term *atom* comes from a Greek word that means "cannot be divided." Today an **atom** is defined as a small particle that makes up most types of matter. **Figure 2** shows the difference between Democritus's ideas and those of other early scientists and philosophers. Democritus thought that different types of atoms existed for every type of matter and that the atom's identity explained the characteristics of each type of matter. Democritus's ideas about atoms were a first step toward understanding matter. However, his ideas were not accepted for over 2,000 years. It wasn't until the early 1800s that scientists built upon the concept of atoms to form the current atomic theory of matter.

INTEGRATE History

Atomism Historians note that Leucippus developed the idea of the atom around 440 B.C. He and his student, Democritus, refined the idea of the atom years later. Their concept of the atom was based on five major points: (1) all matter is made of atoms, (2) there are empty spaces between atoms, (3) atoms are complete solids, (4) atoms do not have internal structure, and (5) atoms are different in size, shape, and weight.

2 Teach

IDENTIFYING Misconceptions

Weight/Mass Students may think that objects have weight only if you can feel them. Refer to the F page at the beginning of the chapter for teaching strategies that address this misconception.

Activity

Classifying Matter Take students outside to make a list of all the things they observe. Return to the classroom and have them classify the items as "Matter" and "Not Matter." What is matter? Possible answers: Matter is anything that has mass and takes up space and includes things that we see, taste, smell, and touch. What isn't matter? Possible answers: Emotions, thoughts, ideas, and light are not matter because they do not have mass and do not take up space. L2

INTEGRATE History

Atomism Democritus (460–370 B.C.) was born into a noble and wealthy family. The Magi instructed him in astronomy and theology. After the death of his father, Democritus traveled to search for wisdom. It was said that he visited many countries, such as Egypt and Ethiopia. Democritus was known as "The Laughing Philosopher" because he expressed disapproval for human foolishness while laughing. He lived for more than one hundred years.

Curriculum Connection

Physics Ancient astronomers had no understanding of what stars were. This is because they lacked a framework of physics in which to understand how stars worked and what they constituted. The study of modern physics includes a wide variety of fields that are used to understand the origin and evolution of stars, matter, energy and the forces of nature.

Caption Answer

Figure 3 the air

Mini LAB

Purpose Students infer the number and types of objects in a sealed box to model how scientists study atoms. L2 LS **Logical-Mathematical**

Materials sealed shoe box containing various items (eraser, coin, paper clip, ruler)

Teaching Strategy Emphasize that students should not open the box.

Analysis

1. Answers will vary depending on items in the box.
2. Scientists perform experiments on matter to make models of the atom, but they cannot actually see inside an atom to verify that their models are correct.

Assessment

Process Have students open the box and examine the items contained inside. Ask them to infer why some objects were easier to identify than others. Use **Performance Assessment in the Science Classroom,** p. 89.

Teacher FYI

Creating Matter According to the law of conservation of matter, matter can neither be created nor destroyed. But experimental data indicate that a high energy X ray can collide with the nucleus of an atom to create an electron and a positron in its place. Also, an electron and positron can collide with and annihilate each other to produce X rays. Although matter is created and destroyed in these instances, the amount of total energy stays the same, but the energy can change its form from electromagnetic radiation to matter.

Figure 3 When wood burns, matter is not lost. The total mass of the wood and the oxygen it combines with during a fire equals the total mass of the ash, water vapor, carbon dioxide, and other gases produced.
Infer *When you burn wood in a fireplace, what is the source of oxygen?*

Mini LAB

Investigating the Unseen

Procedure

1. Your teacher will give you a **sealed shoe box** that contains **one or more items.**
2. Try to find out how many and what kinds of items are inside the box. You cannot look inside the box. The only observations you can make are by handling the box.

Analysis

1. How many items do you infer are in the box? Sketch the apparent shapes of the items and identify them if you can.
2. Compare your procedure with how scientists perform experiments and make models to find out more about the atom.

Lavoisier's Contribution Lavoisier (la VWAH see ay), a French chemist who lived about 2,000 years after Democritus, also was curious about matter—especially when it changed form. Before Lavoisier, people thought matter could appear and disappear because of the changes they saw as matter burned or rusted. You might have thought that matter can disappear if you've ever watched wood burn in a fireplace or at a bonfire. Lavoisier showed that wood and the oxygen it combines with during burning have the same mass as the ash, water, carbon dioxide, and other gases that are produced, as shown in **Figure 3.** In a similar way, an iron bar, oxygen, and water have the same mass as the rust that forms when they interact. From Lavoisier's work came the **law of conservation of matter,** which states that matter is not created or destroyed—it only changes form.

Models of the Atom

Models are often used for things that are too small or too large to be observed or that are too difficult to be understood easily. One way to make a model is to make a smaller version of something large. If you wanted to design a new sailboat, would you build a full-sized boat and hope it would float? It would be more efficient, less expensive, and safer to build and test a smaller version first. Then, if it didn't float, you could change your design and build another model. You could keep trying until the model worked.

In the case of atoms, scientists use large models to explain something that is too small to be looked at. These models of the atom were used to explain data or facts that were gathered experimentally. As a result, these models are also theories.

Visual Learning

Figure 3 Why could being in a closed space with a fire be hazardous, even if you are in no danger of being burned? As the fuel burns, it consumes oxygen and produces smoke. Thus, being confined with a fire can lead to suffocation, not only from lack of oxygen, but also from smoke inhalation.

100 CHAPTER 4 Atoms, Elements, and the Periodic Table

Dalton's Atomic Model In the early 1800s, an English schoolteacher and chemist named John Dalton studied the experiments of Lavoisier and others. Dalton thought he could design an atomic model that explained the results of those experiments. Dalton's atomic model was a set of ideas—not a physical object. Dalton believed that matter was made of atoms that were too small to be seen by the human eye. He also thought that each type of matter was made of only one kind of atom. For example, gold atoms make up a gold nugget and give a gold ring its shiny appearance. Likewise, iron atoms make up an iron bar and give it unique properties, and so on. Because predictions using Dalton's model were supported by data, the model became known as the atomic theory of matter.

Sizes of Atoms Atoms are so small it would take about 1 million of them lined up in a row to equal the thickness of a human hair. For another example of how small atoms are, look at **Figure 4.** Imagine you are holding an orange in your hand. If you wanted to be able to see the individual atoms on the orange's surface, the size of the orange would have to be increased to the size of Earth. Then, imagine the Earth-sized orange covered with billions and billions of marbles. Each marble would represent one of the atoms on the skin of the orange. No matter what kind of model you use to picture it, the result is the same—an atom is an extremely small particle of matter.

Figure 4 If this orange were as large as Earth, each of its atoms would be marble-sized.

Discussion

Educational Models Show students several educational models. Examples include plant, solar system, or anatomy models. How can models be helpful? Models can help us understand the structure of things and how they behave. L1

Fun Fact

Our knowledge of Democritus' atomic theory comes from a poem written by the Roman Lucretius, who lived about 300 years after Democritus.

Virtual Labs

Atomic Structure *How can a molecular model be built?*

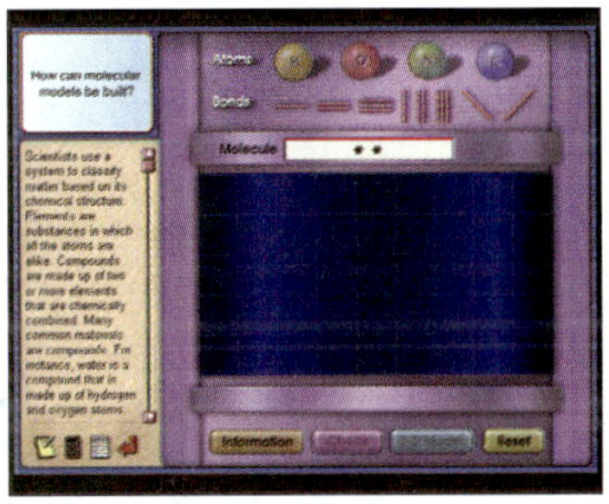

LAB DEMONSTRATION

Purpose to demonstrate that matter is neither created nor destroyed

Materials fine steel wool, heat-proof mitt, safety goggles, crucible tongs, burner, pan balance

Preparation Wear goggles and heat-proof mitt while performing this activity.

Procedure Obtain the mass of a piece of steel wool. Hold the steel wool in a burner flame using tongs. After a few fibers burn, ask question 1. After students have responded, obtain the mass of the burned steel wool and then ask question 2.

Expected Outcome Mass increases.

Assessment

1. Will what's left have a mass greater or less than the original? Most students will likely think the mass has decreased. 2. Why is there a gain in mass? The iron in the steel wool reacted with O_2 in the air to form iron oxide, so the mass of the O_2 has been added to the original mass.

Caption Answers

Figure 5 The rays would straighten.

Figure 6 Because matter is neutral, if it contains particles with negative charges, it must also contain particles with positive charges.

Figure 7 in the nucleus

Quick Demo

Magnetic Field

Materials iron filings, overhead projector, transparency sheet, bar magnet

Estimated Time five minutes

Procedure Demonstrate the presence and shape of a magnet field by placing the iron filings on a transparency sheet and placing the sheet over a bar magnet on an overhead projector.

Use Science Words

Word Usage Have students use the word *electron* in a sentence describing its characteristics. Sentences should include that an electron has a small mass, a negative charge, and is found outside the atomic nucleus. L2

Reading Check

Answer The particles in the rays were charged, and thus were affected by the magnet's magnetic field.

Activity

Parts of an Atom Have students fold a piece of paper into four sections. Instruct them to write one of the following terms in each box: electron, nucleus, proton, neutron. As you read and discuss this section, have students illustrate and write key facts for each part of the atom. Illustrations will vary. Key facts should include: electron–negatively charged particles scattered outside the nucleus; nucleus–positively charged central part of the atom; proton–positively charged particle located in the nucleus; neutron–a particle with no charge found in the nucleus of the atom. P

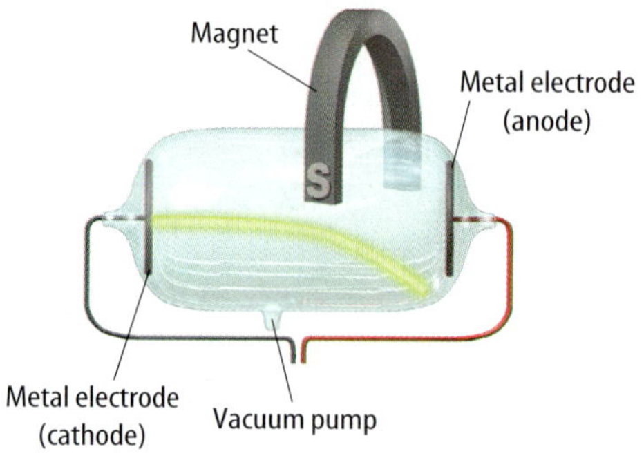

Figure 5 In Thomson's experiment, the magnet caused the cathode rays inside the tube to bend. **Describe** *what you think would happen to the cathode rays if the magnet were removed.*

Science Online

Topic: Subatomic Particles

Visit red.msscience.com for Web links to information about particles that make up atoms.

Activity Can any of the particles be divided further? Display your data in a table.

Discovering the Electron One of the many pioneers in the development of today's atomic model was J.J. Thomson, an English scientist. He conducted experiments using a cathode ray tube, which is a glass tube sealed at both ends out of which most of the air has been pumped. Thomson's tube had a metal plate at each end. The plates were connected to a high-voltage electrical source that gave one of the plates—the anode—a positive charge and the other plate—the cathode—a negative charge. During his experiments, Thomson observed rays that traveled from the cathode to the anode. These cathode rays were bent by a magnet, as seen in **Figure 5,** showing that they were made up of particles that had mass and charge. Thomson knew that like charges repel each other and opposite charges attract each other. When he saw that the rays traveled toward a positively charged plate, he concluded that the cathode rays were made up of negatively charged particles. These invisible, negatively charged particles are called **electrons.**

Reading Check *Why were the cathode rays in Thomson's cathode ray tube bent by a magnet?*

Try to imagine Thomson's excitement at this discovery. He had shown that atoms are not too tiny to divide after all. Rather, they are made up of even smaller subatomic particles. Other scientists soon built upon Thomson's results and found that the electron had a small mass. In fact, an electron is 1/1,837 the mass of the lightest atom, the hydrogen atom. In 1906, Thomson received the Nobel Prize in Physics for his work on the discovery of the electron.

Matter that has an equal amount of positive and negative charge is said to be neutral—it has no net charge. Because most matter is neutral, Thomson pictured the atom as a ball of positive charge with electrons embedded in it. It was later determined that neutral atoms contained an equal number of positive and negative charges.

Differentiated Instruction

Learning Disabled Have students make up new lyrics to a familiar tune using the vocabulary and definitions from the chapter. The song can focus on one concept, such as the parts of an atom, or cover the entire chapter's content. Some examples for the familiar tune are "Row, Row, Row Your Boat," "Old McDonald," and "Itsy Bitsy Spider." LS **Auditory-Musical**

Active Reading

Reflective Journal Have students write reflective journals about their experiences in this activity. Have students divide sheets of paper into several columns and record their thoughts under headings such as *What I Did, What I Learned, What Questions I Have, What Surprises I Experienced,* and *Overall Response.* Suggest that volunteers share their reflections with the class.

Thomson's Model Thomson's model, shown in **Figure 6,** can be compared to chocolate chips spread throughout a ball of cookie dough. However, the model did not provide all the answers to the questions that puzzled scientists about atoms.

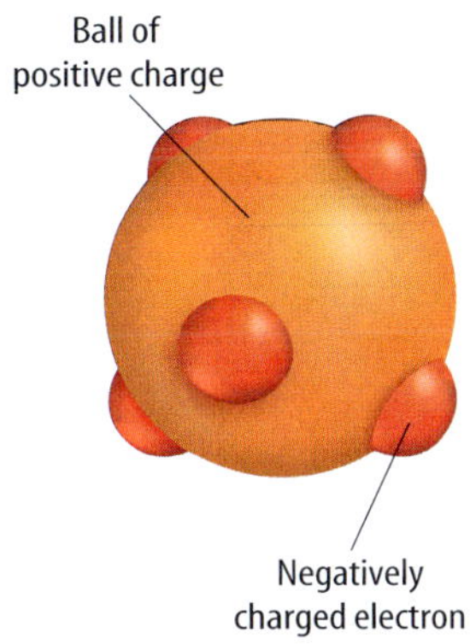

Figure 6 Thomson's model shows the atom as electrons embedded in a ball of positive charge.
Explain *how Thomson knew atoms contained positive and negative charges.*

Rutherford—The Nucleus Scientists still had questions about how the atom was arranged and about the presence of positively charged particles. In about 1910, a team of scientists led by Ernest Rutherford worked on these questions. In their experiment, they bombarded an extremely thin piece of gold foil with alpha particles. Alpha particles are tiny, high-energy, positively charged particles that he predicted would pass through the foil. Most of the particles passed straight through the foil as if it were not there at all. However, other particles changed direction, and some even bounced back. Rutherford thought the result was so remarkable that he later said, "It was almost as incredible as if you had fired a 15-inch shell at a piece of tissue paper, and it came back and hit you."

Positive Center Rutherford concluded that because so many of the alpha particles passed straight through the gold foil, the atoms must be made of mostly empty space. However, because some of the positively charged alpha particles bounced off something, the gold atoms must contain some positively charged object concentrated in the midst of this empty space. Rutherford called the positively charged, central part of the atom the **nucleus** (NEW klee us). He named the positively charged particles in the nucleus **protons**. He also suggested that electrons were scattered in the mostly empty space around the nucleus, as shown in **Figure 7.**

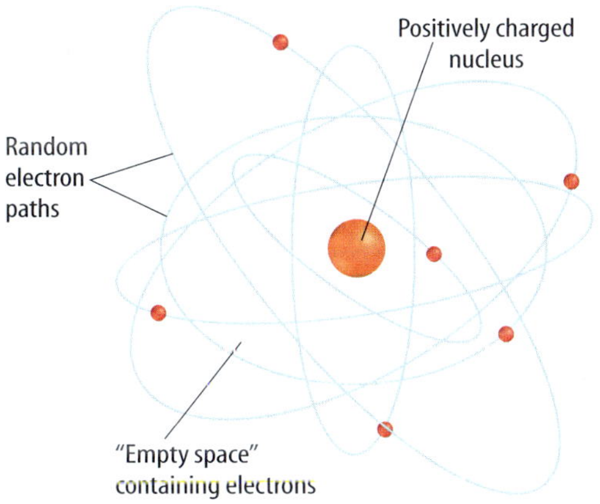

Figure 7 Rutherford concluded that the atom must be mostly empty space in which electrons travel in random paths around the nucleus. He also thought the nucleus of the atom must be small and positively charged.
Identify *where most of the mass of an atom is concentrated.*

Discussion

Protons and Neutrons How are protons and neutrons similar and how are they different? Protons and neutrons are both found in atomic nuclei. Both particles are small but massive, and their masses are nearly equal. The proton is positively charged, while the neutron has no charge.

Inquiry Lab

Touching Atoms

Purpose Students will work with models of atoms to gain a better understanding of their structure and properties.

Possible Materials nuts or bolts in two different sizes, long-distance measuring tape, ground stake

Estimated Time one or two class sessions

Teaching Strategies

- Size: Have students calculate where the first electron would be if the nucleus of an atom was the size of the head of a pin (1mm diameter). The average atom is of the order of 10^{-10} m, while the average nucleus is of the order of 10^{-15} m. It will be about 100 m away.
- Isotopes: Gather a handful of small and large bolts. Tell students it is the element 'bolt' and you want to find its average mass, just like atoms have average masses for all of their real isotopes. Have students weigh a large bolt and a small bolt and record the mass. Then, multiply the mass of the object by the number of objects, add it to the mass of the other size object times the number of objects, and divide by the total number of objects. This is the same method for calculating isotopic mass of elements. You could add a medium-sized bolt to the calculation, too.
- Have students brainstorm other ways to represent this information.

For additional inquiry activities, see *Science Inquiry Labs*.

Active Reading

Buddy Interviews This strategy helps students to understand and clarify the reading. Having students interview one another to find out what helps them to understand what they are reading, how they find answers, and how they assimilate new vocabulary. Have students use Buddy Interviews to help them master the concept of matter and the atomic theory of matter.

Visual Learning

Figures 6 and 7 Have students compare Thomson's model with Rutherford's model. How are they the same? Both are spherical and contain particles. How are they different? Thomson's model is like a solid ball, while Rutherford's is mostly empty space, like a cloud.

Make a Model

Atomic Model Students may find this abstract information difficult to understand unless models and visual examples are used during reading and discussion. Have students construct a teaching model by using a standard paper plate as the electron cloud. Have them use different colors for electrons (negative symbols) and protons (positive symbols).

IDENTIFYING Misconceptions

Three-Dimensional Students may think that atoms and their structures are two-dimensional or flat because the models and illustrations are flat. In fact, an atom is a three-dimensional object in which the electrons travel in a random pattern within a given distance around the nucleus.

INTEGRATE Career

Physicists and Chemists Science is a team effort. Even though physicists and chemists may study different subjects, often they are working on a joint project. One of the newest areas of study is nanotechnology. This is one field where physicists and chemists can work jointly to develop new products.

Research Have students investigate the various places where physicists and chemists can work.

INTEGRATE Career

Physicists and Chemists Physicists generally study the physical atom. The physical atom includes the inner components of an atom such as protons and neutrons, the forces that hold or change their positions in space and the bulk properties of elements such as melting point. Chemists, on the other hand, study the chemical atom. The chemical atom refers to the manner in which different elements relate to each other and the new substances formed by their union.

Discovering the Neutron Rutherford had been puzzled by one observation from his experiments with alpha particles. After the collisions, the nuclei seemed to be heavier. Where did this extra mass come from? James Chadwick, a student of Rutherford's, answered this question. The alpha particles themselves were not heavier, but the atoms that had been bombarded had given off new particles. Chadwick experimented with these new particles and found that, unlike electrons, the paths of these particles were not affected by an electric field. To explain his observations, he said that these particles came from the nucleus and had no charge. Chadwick called these uncharged particles **neutrons** (NEW trahn). His proton-neutron model of the atomic nucleus is still accepted today.

Improving the Atomic Model

Early in the twentieth century, a scientist named Niels Bohr found evidence that electrons in atoms are arranged according to energy levels. The lowest energy level is closest to the nucleus and can hold only two electrons. Higher energy levels are farther from the nucleus and can contain more electrons. To explain these energy levels, some scientists thought that the electrons might orbit an atom's nucleus in paths that are specific distances from the nucleus, as shown in **Figure 8.** This is similar to how the planets orbit the Sun.

The Modern Atomic Model As a result of continuing research, scientists now realize that because electrons have characteristics that are similar to waves and particles, their energy levels are not defined, planet-like orbits around the nucleus. Rather, it seems most likely that electrons move in what is called the atom's electron cloud, as shown in **Figure 9.**

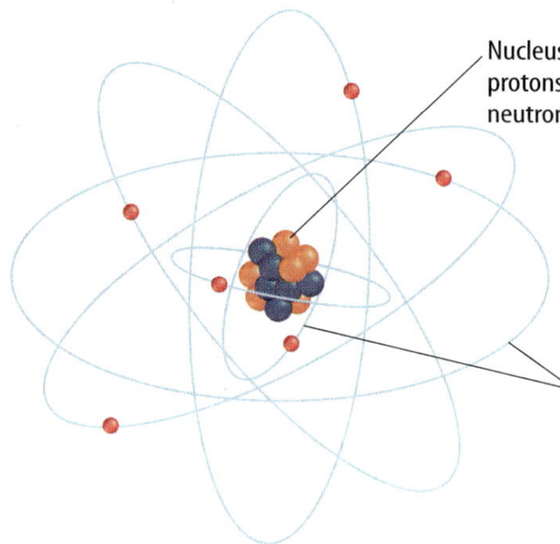

Figure 8 This simplified Bohr model shows a nucleus of protons and neutrons and electron paths based on energy levels.

Differentiated Instruction

Challenge Have students find out more about the people involved in the discovery of the parts of the atom—where they lived and worked, when they made their discoveries, how the discoveries were made, and other scientists they worked with. Have them combine their reports to make a class newspaper called "Great Moments in Atomic History." L3 COOP LEARN LS **Linguistic**

Teacher FYI

Alpha Particles An alpha particle consists of two protons and two neutrons and is identical to the nucleus of a helium atom. Alpha particles are emitted by many radioactive nuclei, such as the ones that Rutherford used in his experiment.

The Electron Cloud The electron cloud is a spherical cloud of varying density surrounding the nucleus. The varying density shows where an electron is more or less likely to be. Atoms with electrons in higher energy levels have electron clouds of different shapes that also show where those electrons are likely to be. Generally, the electron cloud has a radius 10,000 times that of the nucleus.

Further Research By the 1930s, it was recognized that matter was made up of atoms, which were, in turn, made up of protons, neutrons, and electrons. But scientists, called physicists, continued to study the basic parts of this atom. Today, they have succeeded in breaking down protons and neutrons into even smaller particles called quarks. These particles can combine to make other kinds of tiny particles, too. The six types of quarks are *up, down, strange, charmed, top,* and *bottom*. Quarks have fractional electric charges of +2/3 or −1/3, unlike the +1 charge of a proton or the −1 charge of an electron. Research will continue as new discoveries are made about the structure of matter.

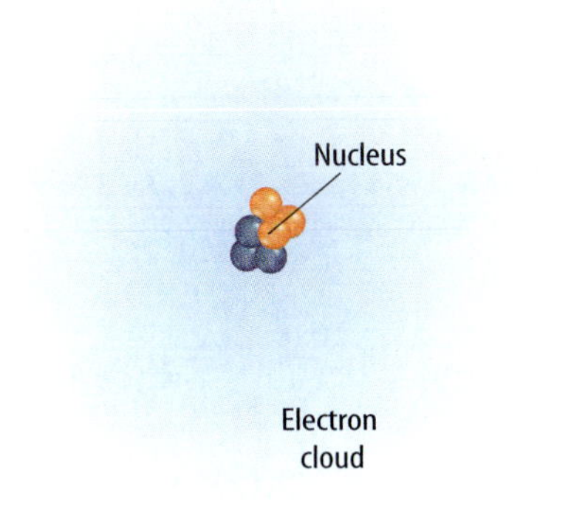

Figure 9 This model of the atom shows the electrons moving around the nucleus in a region called an electron cloud. The dark cloud of color represents the area where the electron is more likely to be found.
Infer *What does the intensity of color near the nucleus suggest?*

section 1 review

Summary

What is matter?

- Matter is anything that has mass and takes up space.
- Matter is composed of atoms.

Models of the Atom

- Democritus introduced the idea of an atom. Lavoisier showed matter is neither created nor destroyed, just changed.
- Dalton's ideas led to the atomic theory of matter.
- Thomson discovered the electron.
- Rutherford discovered protons exist in the nucleus.
- Chadwick discovered the neutron.

Improving the Atomic Model

- Niels Bohr suggested electrons move in energy levels.
- More recent physicists introduced the idea of the electron cloud and were able to break down protons and neutrons into smaller particles called quarks.

Self Check

1. **List** five examples of matter and five examples that are not matter. Explain your answers.
2. **Describe** and name the parts of the atom.
3. **Explain** why the word *atom* was an appropriate term for Democritus's idea.
4. **Think Critically** When neutrons were discovered, were these neutrons created in the experiment? How does Lavoisier's work help answer this question?
5. **Explain** the law of conservation of matter using your own examples.
6. **Think Critically** What made Niels Bohr predict that electrons were arranged in energy levels?

Applying Skills

7. **Classify** each scientist and his contribution according to the type of discovery each person made. Explain why you grouped certain scientists together.
8. **Evaluate Others' Data and Conclusions** Analyze, review, and critique the strengths and weaknesses of Thomson's "cookie dough" theory using the results of Rutherford's gold foil experiment.

Science Online red.msscience.com/self_check_quiz

3 Assess

DAILY INTERVENTION

Check for Understanding

Kinesthetic Have students use common materials to make models that correspond to today's model of the atom. Have a contest and award prizes to the models that are judged the most accurate or creative. L2 ELL COOP LEARN LS

Reteach

Diagram an Atom Have students make a labeled diagram of an atom. Diagrams should include the nucleus, protons, neutrons, and electrons. Labels should include brief definitions. L1 LS **Visual-Spatial**

Assessment

Performance Have students work in small groups to create presentations that illustrate the basic structure of matter. Use **Performance Assessment in the Science Classroom,** p. 143.

Caption Answer

Figure 9 The intensity of color near the nucleus suggests that the electrons are more likely to be found here.

section 1 review

1. Possible answers: matter: things that have mass and occupy space; not matter: ideas, emotions
2. A nucleus consists of positively charged protons and neutral neutrons; negatively charged electrons move outside the nucleus.
3. *Atom* means cannot be divided.
4. No; According to the law of the conservation of matter cannot be created or destroyed.
5. Students can discuss any examples that show evaporation, condensation, freezing, sublimation, melting or any other rational answer.
6. To develop a model of the atom so that it did not contradict the laws of electricity and magnetism.
7. Classification schemes will vary. Students could group scientists by theories or by particles.
8. Thomson: atom is a ball of positive charge with electrons embedded in it. Rutherford: atom is mostly empty space with electrons scattered around small nucleus. Strength: shows atom is divisible and composed of charged particles. Weakness: does not show atom is mostly space and has "structure".

1 Motivate

Bellringer

INTERACTIVE CHALKBOARD
PowerPoint® Presentations

Section Focus Transparencies also are available on the Interactive Chalkboard CD-ROM. L2 ELL

Tie to Prior Knowledge

Chemical Formulas Many students are familiar with the chemical formula for water, H_2O. Ask whether they know what the symbols *H* and *O* represent. *H represents the element hydrogen; O represents the element oxygen.* Have students list the names and symbols of any other elements they know.

section 2 The Simplest Matter

as you read

What You'll Learn

- **Describe** the relationship between elements and the periodic table.
- **Explain** the meaning of atomic mass and atomic number.
- **Identify** what makes an isotope.
- **Contrast** metals, metalloids, and nonmetals.

Why It's Important

Everything on Earth is made of the elements that are listed on the periodic table.

Review Vocabulary

mass: a measure of the amount of matter an object has

New Vocabulary

- element
- atomic number
- isotope
- mass number
- atomic mass
- metal
- nonmetal
- metalloid

The Elements

Have you watched television today? TV sets are common, yet each one is a complex system. The outer case is made mostly of plastic, and the screen is made of glass. Many of the parts that conduct electricity are metals or combinations of metals. Other parts in the interior of the set contain materials that barely conduct electricity. All of the different materials have one thing in common: they are made up of even simpler materials. In fact, if you had the proper equipment, you could separate the plastics, glass, and metals into these simpler materials.

One Kind of Atom Eventually, though, you would separate the materials into groups of atoms. At that point, you would have a collection of elements. An **element** is matter made of only one kind of atom. At least 115 elements are known and about 90 of them occur naturally on Earth. These elements make up gases in the air, minerals in rocks, and liquids such as water. Examples of naturally occurring elements include the oxygen and nitrogen in the air you breathe and the metals gold, silver, aluminum, and iron. The other elements are known as synthetic elements. These elements have been made in nuclear reactions by scientists with machines called particle accelerators, like the one shown in **Figure 10.** Some synthetic elements have important uses in medical testing and are found in smoke detectors and heart pacemaker batteries.

Figure 10 The Tevatron has a circumference of 6.3 km—a distance that allows particles to accelerate to high speeds. These high-speed collisions can create synthetic elements.

106

Section 2 Resource Manager

Chapter FAST FILE Resources
- Transparency Activity, p. 41
- Directed Reading for Content Mastery, p. 27
- Enrichment, p. 27
- Reinforcement, p. 24
- Lab Worksheet, pp. 5–6

Physical Science Critical Thinking/Problem Solving, p. 13
Science Inquiry Labs, pp. 49–50
Earth Science Critical Thinking/Problem Solving, p. 2

Figure 11 When you look for information in the library, a system of organization called the Dewey Decimal Classification System helps you find a book quickly and efficiently.

Dewey Decimal Classification System	
000	Computers, information and general reference
100	Philosophy and psychology
200	Religion
300	Social sciences
400	Languages
500	Science
600	Technology
700	Arts and recreation
800	Literature
900	History and geography

The Periodic Table

Suppose you go to a library, like the one shown in **Figure 11,** to look up information for a school assignment. How would you find the information? You could look randomly on shelves as you walk up and down rows of books, but the chances of finding your book would be slim. To avoid such haphazard searching, some libraries use the Dewey Decimal Classification System to categorize and organize their volumes and to help you find books quickly and efficiently.

Charting the Elements Chemists have created a chart called the periodic table of the elements to help them organize and display the elements. **Figure 12** shows how scientists changed their model of the periodic table over time.

On the inside back cover of this book, you will find a modern version of the periodic table. Each element is represented by a chemical symbol that contains one to three letters. The symbols are a form of chemical shorthand that chemists use to save time and space—on the periodic table as well as in written formulas. The symbols are an important part of an international system that is understood by scientists everywhere.

The elements are organized on the periodic table by their properties. There are rows and columns that represent relationships between the elements. The rows in the table are called periods. The elements in a row have the same number of energy levels. The columns are called groups. The elements in each group have similar properties related to their structure. They also tend to form similar bonds.

Science Online

Topic: New Elements

Visit red.msscience.com for Web links to information about new elements.

Activity Research physical properties of two synthetic elements.

2 Teach

IDENTIFYING Misconceptions

Unknown Aspects Students may think that all aspects of the structure and building blocks of matter are completely known. Point out that much remains to be learned about atoms and elements.

Activity

Familiar Elements Have students skim the periodic table for symbols and names of familiar elements. Have them share with the class the things that these elements with similar properties are found in. Example: Ca—calcium is found in milk and other food products.

L2 LS **Interpersonal**

Use an Analogy

Calendar Describe a month's calendar as being like the periodic table. The days are arranged from left to right, as periods of elements are arranged from left to right. Days with the same name are arranged in columns, as groups of elements are arranged in the table.

Teacher FYI

Ancient Elements Elements such as gold, silver, tin, copper, lead, and mercury have been known since ancient times. As more elements were discovered, people began to recognize patterns in their properties. Later, scientists used the patterns to classify the elements.

Fun Fact

The Fermilab, which is home to the Tevatron particle accelerator, is about 20,000 times longer than Thompson's cathode-ray tube.

Differentiated Instruction

English-Language Learners Ask the students to pick 8–10 elements and share with the class both the English and native pronunciation of each element.

NATIONAL GEOGRAPHIC

Visualizing the Periodic Table

Have students examine the pictures and read the captions. Then ask the following questions.

Why do you think the periodic table is so useful to scientists? It helps them organize and understand the chemical properties of the elements.

One difference between Mendeleev's table and the one used today is that his table lacks the column containing the elements helium through radon. Why do you think this is? None of the elements in that column had been discovered in Mendeleev's time. This is because they are not common and they don't readily undergo chemical reactions.

Activity

Periodic Groups Divide the class into eighteen groups. Assign each group a different group of elements from the periodic table. Have them research the similarities and differences among the elements in their respective groups and make a poster illustrating the properties of these elements. L2

COOP LEARN LS **Interpersonal**

Fun Fact

Mendeleev left gaps in his periodic table to locate the next element in its proper family and predicted properties of undiscovered elements that would fit these spaces. His predictions were confirmed when three predicted elements, gallium, germanium, and scandium were discovered.

NATIONAL GEOGRAPHIC **VISUALIZING THE PERIODIC TABLE**

Figure 12

The familiar periodic table that adorns many science classrooms is based on a number of earlier efforts to identify and classify the elements. In the 1790s, one of the first lists of elements and their compounds was compiled by French chemist Antoine-Laurent Lavoisier, who is shown in the background picture with his wife and assistant, Marie Anne. Three other tables are shown here.

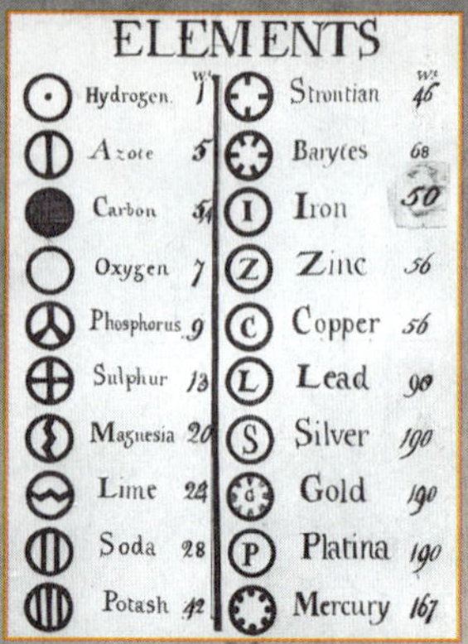

John Dalton (Britain, 1803) used symbols to represent elements. His table also assigned masses to each element.

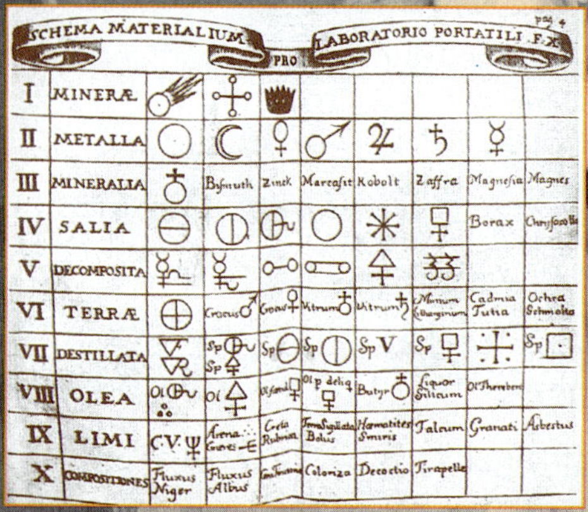

An early alchemist put together this table of elements and compounds. Some of the symbols have their origin in astrology.

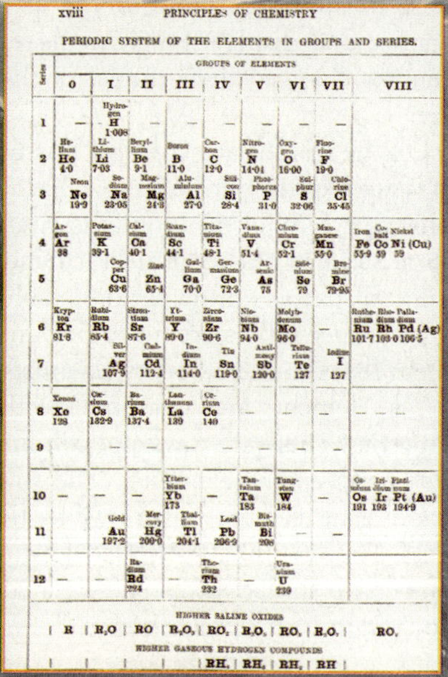

Dmitri Mendeleev (Russia, 1869) arranged the 63 elements known to exist at that time into groups based on their chemical properties and atomic weights. He left gaps for elements he predicted were yet to be discovered.

108 CHAPTER 4 Atoms, Elements, and the Periodic Table

Teacher FYI

Mendeleev's Periodic Table In 1869, Dmitri Mendeleev published a periodic table based on properties and atomic masses of 60 known elements. He made a card for each element. Each card contained the element's symbol, atomic mass, and its characteristic chemical and physical properties. By arranging the cards on a table, in order of increasing atomic mass, and by grouping elements of similar properties together, the resultant "periodic table" showed vertical, horizontal, and diagonal relationships. Mendeleev left gaps in his table for future discovery of some of the then unknown elements.

Identifying Characteristics

Each element is different and has unique properties. These differences can be described in part by looking at the relationships between the atomic particles in each element. The periodic table contains numbers that describe these relationships.

Number of Protons and Neutrons Look up the element chlorine on the periodic table found on the inside back cover of your book. Cl is the symbol for chlorine, as shown in **Figure 13,** but what are the two numbers? The top number is the element's **atomic number.** It tells you the number of protons in the nucleus of each atom of that element. Every atom of chlorine, for example, has 17 protons in its nucleus.

Reading Check *What are the atomic numbers for Cs, Ne, Pb, and U?*

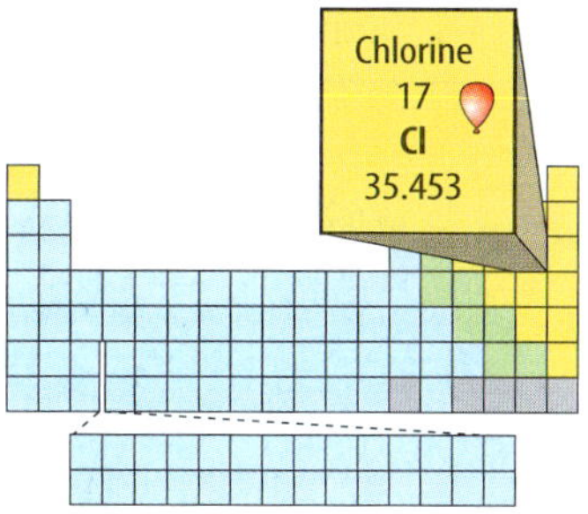

Figure 13 The periodic table block for chlorine shows its symbol, atomic number, and atomic mass.
Determine *if chlorine atoms are more or less massive than carbon atoms.*

Isotopes Although the number of protons changes from element to element, every atom of the same element has the same number of protons. However, the number of neutrons can vary even for one element. For example, some chlorine atoms have 18 neutrons in their nucleus while others have 20. These two types of chlorine atoms are chlorine-35 and chlorine-37. They are called **isotopes** (I suh tohps), which are atoms of the same element that have different numbers of neutrons.

You can tell someone exactly which isotope you are referring to by using its mass number. An atom's **mass number** is the number of protons plus the number of neutrons it contains. The numbers 35 and 37, which were used to refer to chlorine, are mass numbers. Hydrogen has three isotopes with mass numbers of 1, 2, and 3. They are shown in **Figure 14.** Each hydrogen atom always has one proton, but in each isotope the number of neutrons is different.

Figure 14 Three isotopes of hydrogen are known to exist. They have zero, one, and two neutrons in addition to their one proton. Protium, with only the one proton, is the most abundant isotope.

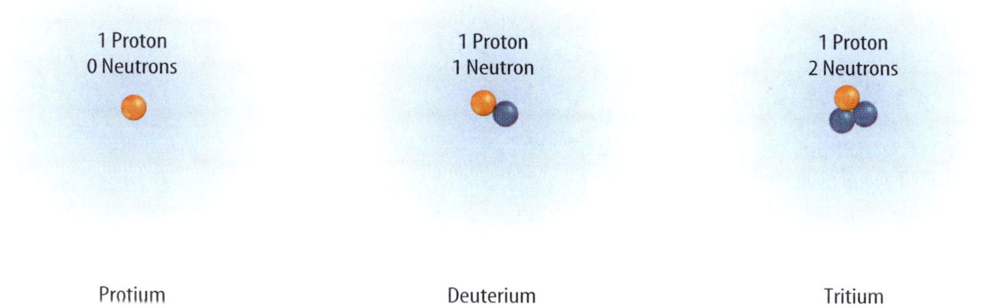

Caption Answer

Figure 13 more massive

Fun Fact

To get an idea of how small the mass of an atom is, consider that there are roughly 560 quintillion (560,000,000,000,000,000,000) atoms of hydrogen and oxygen combined in one drop of water.

Reading Check

Answer Cs–55; Ne–10; Pb–82; U–92

Visual Learning

Figure 14 Have students identify the differences among the three isotopes of hydrogen. Then explain that almost all elements have some isotopes. As an extension, have students research the isotopes of some elements and draw the different nuclei. Carbon is an example: C-12: 6 p, 6 n; C-13: 6 p, 7 n; C-14: 6 p, 8 n. Remind students that all of the isotopes of one element have the same number of protons.

Science Journal

Heavy Water Have students write a summary about what heavy water is. Have them explain how it differs from normal water. Normal water is made of two hydrogen atoms and an oxygen atom covalently bonded together. This gives the famous H_20 formula. Heavy water has exactly the same structure, except the hydrogen atoms are isotopes of hydrogen called deuterium. Standard hydrogen has one proton in its nucleus while deuterium has one proton and one neutron in it nucleus; hence it is 'heavier' than normal hydrogen. This leads to the water being heavier. L2

Discussion

Cobalt and Nickel Have students use the information on these pages to explain how cobalt and nickel got their current positions on the periodic table. Position on the current periodic table is based on the atomic number of an element. L2

Differentiated Instruction

Learning Disabled While presenting information about the atom and its structure, use colored markers to illustrate the parts of the atom and their location on an overhead projector. For example, protons are orange and neutrons are blue in **Figure 14.** Remind students that these parts are not really colored. L1

Activity

Classifying by Atomic Mass Mendeleev grouped elements based on their atomic masses. Have students find three pairs of elements on the periodic table at the back of the book that would have been switched in Mendeleev's table based on atomic masses. Cobalt is element 9 and has an atomic mass of 58.933; nickel is element 10 and has an atomic mass of 58.693. Other pairs of elements that would have been switched on Mendeleev's table are iodine and tellurium, and argon and potassium. L2 LS **Logical-Mathematical**

Quick Demo

Testing for Metals

Materials copper wire; 1.5-V mini lightbulb, D-size battery, electrical tape, wire cutter, different metallic and nonmetallic objects or materials to test

Estimated Time 15 minutes

Procedure Cut a piece of copper wire and tape one end to the positive end of the battery and wrap the other end around the bottom of the lightbulb. Cut another piece of copper wire and tape that end to the negative end of the battery. Cut another piece of copper wire and tape one end to the bottom of the lightbulb. Take one of the objects you are testing and place the wire that is connected to the negative side of the battery to one end of the object and place the wire that is connected to the bottom of the lightbulb to the other end of the object. Repeat the last step until you have tested all the objects. Observe whether the lightbulb lights up or not for each object that is tested. A property of metals is that they conduct electricity. This experiment is based on that property. Make students notice that metallic materials made the lightbulb light up while other nonmetals didn't.

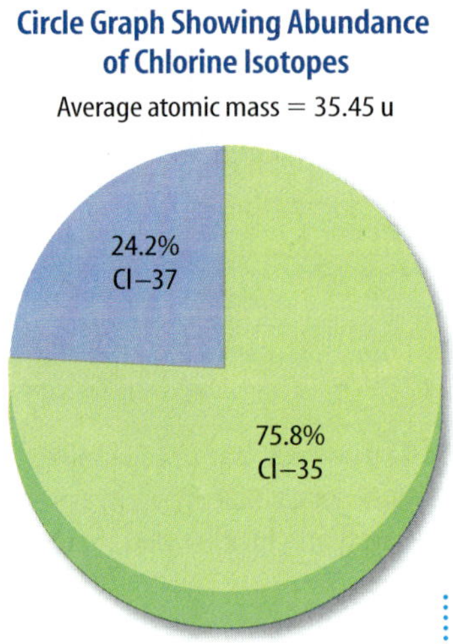

Figure 15 If you have 1,000 atoms of chlorine, about 758 will be chlorine-35 and have a mass of 34.97 u each. About 242 will be chlorine-37 and have a mass of 36.97 u each. The total mass of the 1,000 atoms is 35,454 u, so the average mass of one chlorine atom is about 35.45 u.

Atomic Mass The **atomic mass** is the weighted average mass of the isotopes of an element. The atomic mass is the number found below the element symbol in **Figure 13.** The unit that scientists use for atomic mass is called the atomic mass unit, which is given the symbol u. It is defined as 1/12 the mass of a carbon-12 atom.

The calculation of atomic mass takes into account the different isotopes of the element. Chlorine's atomic mass of 35.45 u could be confusing because there aren't any chlorine atoms that have that exact mass. About 76 percent of chlorine atoms are chlorine-35 and about 24 percent are chlorine-37, as shown in **Figure 15.** The weighted average mass of all chlorine atoms is 35.45 u.

Classification of Elements

Elements fall into three general categories—metals, metalloids (ME tuh loydz), and nonmetals. The elements in each category have similar properties.

Metals generally have a shiny or metallic luster and are good conductors of heat and electricity. All metals, except mercury, are solids at room temperature. Metals are malleable (MAL yuh bul), which means they can be bent and pounded into various shapes. The beautiful form of the shell-shaped basin in **Figure 16** is a result of this characteristic. Metals are also ductile, which means they can be drawn into wires without breaking. If you look at the periodic table, you can see that most of the elements are metals.

Figure 16 The artisan is chasing, or chiseling, the malleable metal into the desired form.

Science Journal

Medical Uses of Isotopes Have students research and write essays on isotopes that are used to diagnose and treat diseases. Some isotopes students might research include ^{67}Cu, ^{111}In, ^{186}Re, ^{82}Sr, ^{68}Ge, ^{153}Sm, and ^{103}Pd. L3 P

Differentiated Instruction

Challenge Have students research the following questions: How were alchemists' ideas about matter different from modern ideas? How is this reflected in their version of the periodic table? How did modern chemistry develop from the methods of alchemy? Ask students to share their findings with the class. L3 LS **Linguistic**

110 CHAPTER 4 Atoms, Elements, and the Periodic Table

Other Elements **Nonmetals** are elements that are usually dull in appearance. Most are poor conductors of heat and electricity. Many are gases at room temperature, and bromine is a liquid. The solid nonmetals are generally brittle, meaning they cannot change shape easily without breaking. The nonmetals are essential to the chemicals of life. More than 97 percent of your body is made up of various nonmetals, as shown in **Figure 17.** You can see that, except for hydrogen, the nonmetals are found on the right side of the periodic table.

Metalloids are elements that have characteristics of metals and nonmetals. On the periodic table, metalloids are found between the metals and nonmetals. All metalloids are solids at room temperature. Some metalloids are shiny and many are conductors, but they are not as good at conducting heat and electricity as metals are. Some metalloids, such as silicon, are used to make the electronic circuits in computers, televisions, and other electronic devices.

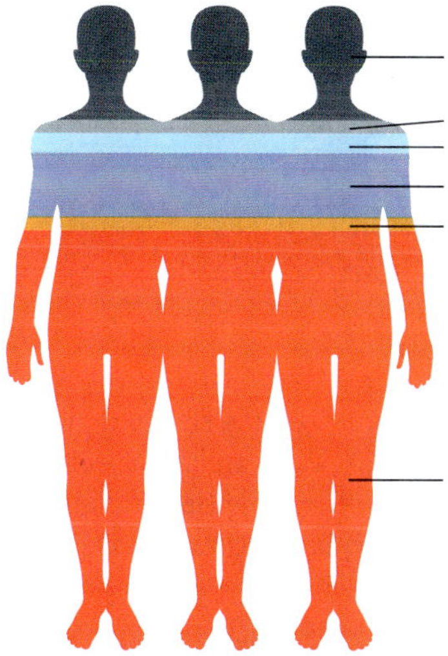

Figure 17 You are made up of mostly nonmetals.

Reading Check *What is a metalloid?*

section 2 review

Summary

The Elements

- An element is matter made of only one type of atom.
- Some elements occur naturally on Earth. Synthetic elements are made in nuclear reactions in particle accelerators.
- Elements are divided into three categories based on certain properties.

The Periodic Table

- The periodic table arranges and displays all known elements in an orderly way.
- Each element has a chemical symbol.

Identifying Characteristics

- Each element has a unique number of protons, called the atomic mass number.
- Isotopes of an element are important when determining the atomic mass of an element.

Self Check

1. **Explain** some of the uses of metals based on their properties.
2. **Describe** the difference between atomic number and atomic mass.
3. **Define** the term *isotope.* Explain how two isotopes of an element are different.
4. **Determine** how many isotopes of krypton are known.
5. **Think Critically** Describe how to find the atomic number for the element oxygen. Explain what this information tells you about oxygen.

Applying Math

6. **Simple Equation** An atom of niobium has a mass number of 93. How many neutrons are in the nucleus of this atom? An atom of phosphorus has 15 protons and 15 neutrons in the nucleus. What is the mass number of this isotope?

Science Online red.msscience.com/self_check_quiz

SECTION 2 The Simplest Matter 111

Reading Check

Answer an element with characteristics of both metals and nonmetals

3 Assess

DAILY INTERVENTION

Check for Understanding

Linguistic Have small groups of students classify classroom objects as being made mostly of metals or nonmetals and record their answers on a chart. Then have students write a statement that can be used to determine if their chart is correct.

Possible Answers:

Metal	Nonmetal
File cabinet	Carpet
Chair legs	Plastic seat

Metal items are shiny, solid at room temperature, conduct heat and electricity, and can be pounded into different shapes without breaking. Nonmetals do not have those properties.

Reteach

Periodic Table Write all the information found on the periodic table for one element on the board. Have students tell what each symbol or number represents. L2 LS **Visual-Spatial**

Assessment

Oral Name elements on the periodic table at random. Ask students to tell whether each selected element is a metal, a metalloid, or a nonmetal.

section 2 review

1. good conductors of heat and electricity (electrical circuits and cookware), malleable (formed into tools), ductile (drawn into wires)
2. atomic number: the number of protons in the nucleus of each atom of an element; mass number: the sum of an atom's protons and neutrons
3. Isotopes are atoms of the same element that contain different numbers of neutrons.
4. The isotopes of hydrogen are protium, deuterium, and tritium.
5. Locate oxygen on the periodic table. Oxygen's atomic number (8) is the large number in the element's box and means that every atom of oxygen contains eight protons in its nucleus.
6. $93 - 41 = 52$ n; 15 p + 15 n = 30

BENCH TESTED

Real-World Question

Purpose Students classify elements and build a periodic table that shows their classifications. L2 LS **Visual-Spatial**

Process Skills classify, compare and contrast, make and use tables, interpret data, predict

Time Required two to three class periods

Procedure

Teaching Strategy Because a large amount of information is available for most elements, advise students to focus their efforts and collect only key information.

Conclude and Apply

1. The table students make will look like the periodic table, but only certain elements will be present.
2. This element should be shiny in the solid state, have such metallic characteristics as high electrical and thermal conductivities, and be reactive with nonmetals to form compounds.

Assessment

Portfolio Take a photograph of the class periodic table and have a copy made for each student. Have students include the index cards for their elements and the photograph of the class periodic table in their portfolios. Use **Performance Assessment in the Science Classroom,** pp. 135; 169. P

LAB

Elements and the Periodic Table

The periodic table organizes the elements, but what do they look like? What are they used for? In this lab, you'll examine some elements and share your findings with your classmates.

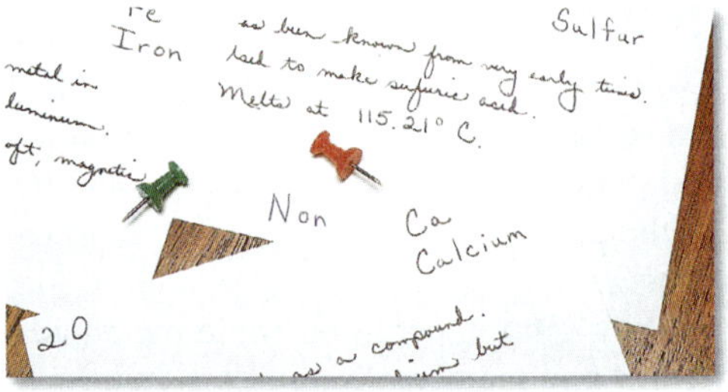

Real-World Question

What are some of the characteristics and purposes of the chemical elements?

Goals

- **Classify** the chemical elements.
- **Organize** the elements into the groups and periods of the periodic table.

Materials

colored markers	large bulletin board
large index cards	$8^1/_2$-in × 14-in paper
Merck Index	thumbtacks
encyclopedia	**pushpins*
**other reference materials*	

*Alternate materials

Safety Precaution

WARNING: *Use care when handling sharp objects.*

Procedure

1. Select the assigned number of elements from the list provided by your teacher.
2. **Design** an index card for each of your selected elements. On each card, mark the element's atomic number in the upper left-hand corner and write its symbol and name in the upper right-hand corner.
3. **Research** each of the elements and write several sentences on the card about its appearance, its other properties, and its uses.
4. **Classify** each of your elements as a metal, a metalloid, or a nonmetal based upon its properties.
5. **Write** the appropriate classification on each of your cards using the colored marker chosen by your teacher.
6. Work with your classmates to make a large periodic table. Use thumbtacks to attach your cards to a bulletin board in their proper positions on the periodic table.
7. **Draw** your own periodic table. Place the elements' symbols and atomic numbers in the proper locations on your table.

Conclude and Apply

1. **Interpret** the class data and classify the elements into the categories metal, metalloid, and nonmetal. Highlight each category in a different color on your periodic table.
2. **Predict** the properties of a yet-undiscovered element located directly under francium on the periodic table.

Communicating Your Data

Have students use a database program to display their results. The printed database can be shared with a small group or with the class.

Using Science Words

Word Meaning Have students use the dictionary to determine the meaning of the word *element.* Ask them why chemical particles are referred to as elements. In chemistry, the basic chemical particles are called elements because elements are fundamental components of matter. L2

112 CHAPTER 4 Atoms, Elements, and the Periodic Table

section 3

Compounds and Mixtures

Substances

Scientists classify matter in several ways that depend on what it is made of and how it behaves. For example, matter that has the same composition and properties throughout is called a **substance.** Elements, such as a bar of gold or a sheet of aluminum, are substances. When different elements combine, other substances are formed.

Compounds What do you call the colorless liquid that flows from the kitchen faucet? You probably call it water, but maybe you've seen it written H_2O. The elements hydrogen and oxygen exist as separate, colorless gases. However, these two elements can combine, as shown in **Figure 18,** to form the compound water, which is different from the elements that make it up. A **compound** is a substance whose smallest unit is made up of atoms of more than one element bonded together.

Compounds often have properties that are different from the elements that make them up. Water is distinctly different from the elements that make it up. It is also different from another compound made from the same elements. Have you ever used hydrogen peroxide (H_2O_2) to disinfect a cut? This compound is a different combination of hydrogen and oxygen and has different properties from those of water.

Water is a nonirritating liquid that is used for bathing, drinking, cooking, and much more. In contrast, hydrogen peroxide carries warnings on its labels such as *Keep Hydrogen Peroxide Out of the Eyes.* Although it is useful in solutions for cleaning contact lenses, it is not safe for your eyes as it comes from the bottle.

as you read

***What* You'll Learn**

- **Identify** the characteristics of a compound.
- **Compare and contrast** different types of mixtures.

***Why* It's Important**

The food you eat, the materials you use, and all matter can be classified by compounds or mixtures.

Review Vocabulary

formula: shows which elements and how many atoms of each make up a compound

New Vocabulary

- substance
- mixture
- compound

Figure 18 A space shuttle is powered by the reaction between liquid hydrogen and liquid oxygen. The reaction produces a large amount of energy and the compound water.
Explain *why a car that burns hydrogen rather than gasoline would be friendly to the environment.*

1 Motivate

INTERACTIVE CHALKBOARD
PowerPoint® Presentations

Bellringer

Section Focus Transparencies also are available on the Interactive Chalkboard CD-ROM. L2 ELL

Section 3 Section Focus Transparency — Here Today, Rust Tomorrow

This was seaworthy metal once, but now it's mostly rust. Before it rusted, it was a sturdy ship for work and transportation. As rust, it is falling apart. What happened?

1. What properties of metals are useful for making ships?
2. Is rust the same material as the original metal? Why or why not?

Matter

Tie to Prior Knowledge

Combining Elements Have students recall chemical formulas for substances such as water or salt. H_2O and NaCl Show a periodic table and point out that these substances represent a combination of elements from the periodic table.

Caption Answer

Figure 18 It would release water as a by-product rather than pollution.

Section 3 Resource Manager

Chapter *Fast File* Resources

Transparency Activity, p. 42
Directed Reading for Content Mastery, pp. 17, 18
MiniLAB, p. 4
Enrichment, p. 28
Lab Worksheet, pp. 7–8
Lab Activity, pp. 9–10, 11–12

2 Teach

Make a Model

H_2O_2 and H_2O While discussing the law of definite proportions, have students make models of the compounds in **Figure 19** using materials such as gumdrops and toothpicks. Have them share their models with the class. L1 ELL LS **Kinesthetic**

Mini LAB

Purpose Students observe and compare the properties of three compounds. L2 ELL LS **Visual-Spatial**

Materials sugar, rubbing alcohol, salad oil, glasses (3), hot water

Teaching Strategy Warn students to use hot tap water, not boiling water, in Step 3.

Analysis

1. rubbing alcohol—colorless liquid, strong chemical smell, dissolves in water; salad oil—golden liquid, more viscous than rubbing alcohol, little or no odor, floats on water; sugar—white granular solid, little or no odor, dissolves easily in water
2. The number of atoms of each type of element and their arrangement account for the different properties.

Assessment

Content Have students draw pictures showing how the substances mix with each other in the MiniLAB. Use **Performance Assessment in the Science Classroom**, p. 127. P

Try at Home

Reading Check

Answer C_3H_8

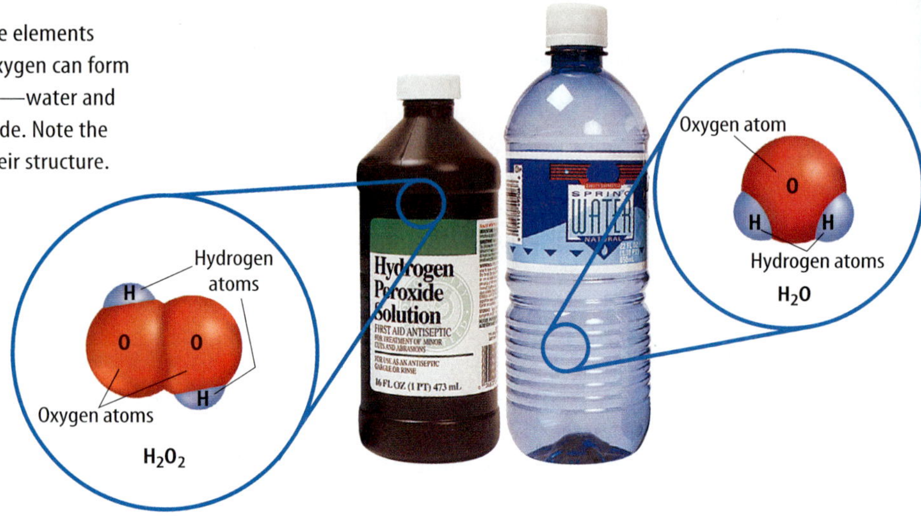

Figure 19 The elements hydrogen and oxygen can form two compounds—water and hydrogen peroxide. Note the differences in their structure.

Mini LAB

Comparing Compounds

Procedure

1. Collect the following substances—**granular sugar, rubbing alcohol,** and **salad oil.**
2. Observe the color, appearance, and state of each substance. Note the thickness or texture of each substance.
3. Stir a spoonful of each substance into separate **glasses** of **hot tap water** and observe.

Analysis

1. Compare the different properties of the substances.
2. The formulas of the three substances are made of only carbon, hydrogen, and oxygen. Infer how they can have different properties.

Try at Home

Compounds Have Formulas What's the difference between water and hydrogen peroxide? H_2O is the chemical formula for water, and H_2O_2 is the formula for hydrogen peroxide. The formula tells you which elements make up a compound as well as how many atoms of each element are present. Look at **Figure 19.** The subscript number written below and to the right of each element's symbol tells you how many atoms of that element exist in one unit of that compound. For example, hydrogen peroxide has two atoms of hydrogen and two atoms of oxygen. Water is made up of two atoms of hydrogen and one atom of oxygen.

Carbon dioxide, CO_2, is another common compound. Carbon dioxide is made up of one atom of carbon and two atoms of oxygen. Carbon and oxygen also can form the compound carbon monoxide, CO, which is a gas that is poisonous to all warm-blooded animals. As you can see, no subscript is used when only one atom of an element is present. A given compound always is made of the same elements in the same proportion. For example, water always has two hydrogen atoms for every oxygen atom, no matter what the source of the water is. No matter what quantity of the compound you have, the formula of the compound always remains the same. If you have 12 atoms of hydrogen and six atoms of oxygen, the compound is still written H_2O, but you have six molecules of H_2O (6 H_2O), not $H_{12}O_6$. The formula of a compound communicates its identity and makeup to any scientist in the world.

Reading Check *Propane has three carbon and eight hydrogen atoms. What is its chemical formula?*

Visual Learning

Figure 19 Ask students to write the formula for hydrogen peroxide. (H_2O_2) Point out how much easier it is to write the chemical formulas than it is to write out all of the words.

Teacher FYI

Chemical Formulas Structural formulas show the bonding relationship (e.g. H-O-O-H); molecular formulas show how many atoms of each element are present (e.g. H_2O_2); empirical formulas show kinds of atoms are present, and gives the simplest whole number ratio of atoms in the compound (e.g. HO).

114 CHAPTER 4 Atoms, Elements, and the Periodic Table

Mixtures

When two or more substances (elements or compounds) come together but don't combine to make a new substance, a **mixture** results. Unlike compounds, the proportions of the substances in a mixture can be changed without changing the identity of the mixture. For example, if you put some sand into a bucket of water, you have a mixture of sand and water. If you add more sand or more water, it's still a mixture of sand and water. Its identity has not changed. Air is another mixture. Air is a mixture of nitrogen, oxygen, and other gases, which can vary at different times and places. Whatever the proportion of gases, it is still air. Even your blood is a mixture that can be separated, as shown in **Figure 20,** by a machine called a centrifuge.

Reading Check *How do the proportions of a mixture relate to its identity?*

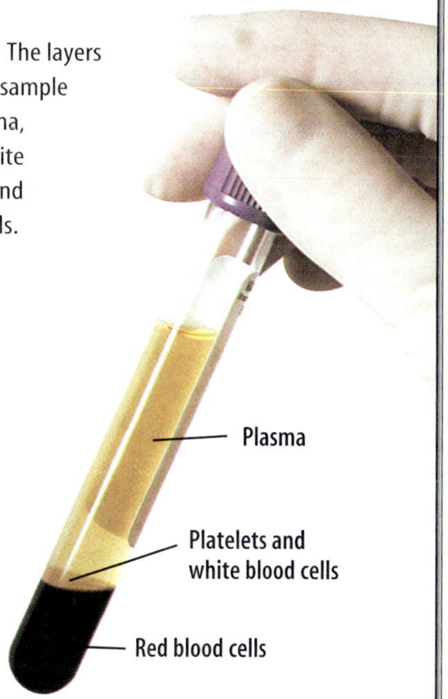

Figure 20 The layers in this blood sample include plasma, platelets, white blood cells, and red blood cells.

Applying Science

What's the best way to desalt ocean water?

You can't drink ocean water because it contains salt and other suspended materials. Or can you? In many areas of the world where drinking water is in short supply, methods for getting the salt out of salt water are being used to meet the demand for fresh water. Use your problem-solving skills to find the best method to use in a particular area.

Methods for Desalting Ocean Water

Process	Amount of Water a Unit Can Desalt in a Day (m^3)	Special Needs	Number of People Needed to Operate
Distillation	1,000 to 200,000	lots of energy to boil the water	many
Electrodialysis	10 to 4,000	stable source of electricity	1 to 2 persons

Identifying the Problem

The table above compares desalting methods. In distillation, the ocean water is heated. Pure water boils off and is collected, and the salt is left behind. Electrodialysis uses an electric current to pull salt particles out of water.

Solving the Problem

1. What method(s) might you use to desalt the water for a large population where energy is plentiful?
2. What method(s) would you choose to use in a single home?

Discussion

Compounds Remind students that a compound is a pure substance whose smallest unit is made up of atoms of more than one element. What compounds do you use in your daily activities? Examples include table salt, water, and sugar.

Reading Check

Answer They can change without changing the identity of a mixture.

Applying Science

Answers

1. Distillation would be best to use for a large population. Although the unit needs many operators and lots of energy, it produces the largest amount of fresh water.
2. Electrodialysis works in a home or small resort hotel. This method produces smaller amounts of fresh water and does not require many skilled operators. It only needs a stable source of electricity.

Differentiated Instruction

Challenge Have students research the following question and develop answers. If there were a mixture of sand and salt in a beaker, how would you separate the sand from the salt? Ask students to share their findings with the class. The salt can be dissolved using water and later evaporated. L3

Cultural Diversity

Ocher Ocher, a naturally occurring iron oxide (Fe_2O_3), is formed by the weathering of rocks. Historically, ocher has been used to produce rock paintings and continues to be used for artwork today. It comes in a variety of colors, depending on the percentage of iron oxide present. Hematite ore is one source of ocher that can be ground to produce a bright red color. Ocher also has been found in burial plots of several different cultures. Have students find out the other colors produced by iron oxide ores. Other colors produced include black (magnetite), brown to dark yellow (limonite), and brown (siderite and pyrite).

Activity

Food Labels Use labels from popular food items and work with students to determine whether each of the ingredients is an element, a compound, or a mixture. L3 ELL COOP LEARN

Use Science Words

Word Usage Ask students to classify these mixtures as heterogeneous or homogeneous: clear apple juice homogeneous; oil and vinegar heterogeneous. L2

Quick Demo

Separating Sawdust and Iron

Materials sawdust, iron filings, bar magnet, non-stick spatula, 500-mL beaker

Estimated Time 15 minutes

Procedure Mix the sawdust and the iron filings in the beaker using the non-stick spatula. Explain that this is a mixture. Use the bar magnet to attract the iron filings and separate the filings from the mixture. Let the students know that there are many ways of separating mixtures. Challenge the students to develop another method to separate the mixture of iron filing and the sawdust. Note that by adding water to the beaker containing the mixture you can separate the sawdust from the iron filing because the sawdust will float.

Figure 21 Mixtures are part of your everyday life.

Science Online

Topic: Mixtures

Visit red.msscience.com for Web links to information about separating mixtures.

Activity Describe the difference between mixtures and compounds.

INTEGRATE Life Science Your blood is a mixture made up of elements and compounds. It contains white blood cells, red blood cells, water, and a number of dissolved substances. The different parts of blood can be separated and used by doctors in different ways. The proportions of the substances in your blood change daily, but the mixture does not change its identity.

Separating Mixtures Sometimes you can use a liquid to separate a mixture of solids. For example, if you add water to a mixture of sugar and sand, only the sugar dissolves in the water. The sand then can be separated from the sugar and water by pouring the mixture through a filter. Heating the remaining solution will separate the water from the sugar.

At other times, separating a mixture of solids of different sizes might be as easy as pouring them through successively smaller sieves or filters. A mixture of marbles, pebbles, and sand could be separated in this way.

Curriculum Connection

Art Have students make collages that demonstrate the difference between a homogeneous mixture and a heterogeneous mixture. Have them search through magazines to collect illustrations for their collages. Images could include: homogeneous—any type of solution; heterogeneous: strawberry ice cream, salad, chocolate-chip cookie. L1 P

Differentiated Instruction

Challenge Have students prepare a table detailing the differences between mixtures and compounds. Note that mixtures may have components in any proportion and that components may be separated by physical means. Also, note that components in compounds are present in fixed proportions and that compounds may be separated by chemical means. L3

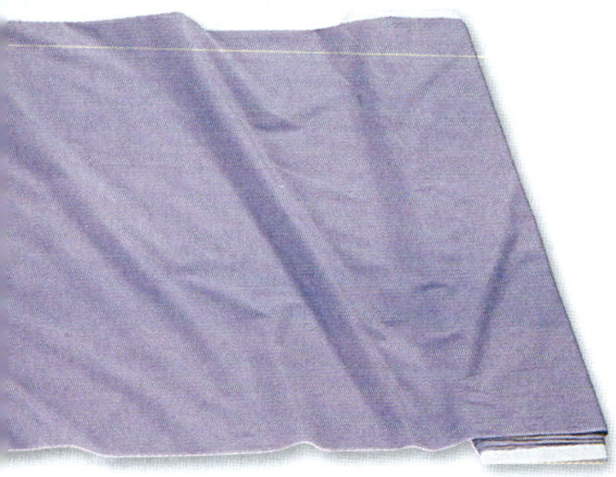

Homogeneous or Heterogeneous Mixtures, such as the ones shown in **Figure 21,** can be classified as homogeneous or heterogeneous. *Homogeneous* means "the same throughout." You can't see the different parts in this type of mixture. In fact, you might not always know that homogeneous mixtures are mixtures because you can't tell by looking. Which mixtures in **Figure 21** are homogeneous? No matter how closely you look, you can't see the individual parts that make up air or the parts of the mixture called brass in the lamp shown. Homogeneous mixtures can be solids, liquids, or gases.

A heterogeneous mixture has larger parts that are different from each other. You can see the different parts of a heterogeneous mixture, such as sand and water. How many heterogeneous mixtures are in **Figure 21?** A pepperoni and mushroom pizza is a tasty kind of heterogeneous mixture. Other examples of this kind of mixture include tacos, vegetable soup, a toy box full of toys, or a toolbox full of nuts and bolts.

INTEGRATE Earth Science

Rocks and Minerals Scientists called geologists study rocks and minerals. A mineral is composed of a pure substance. Rocks are mixtures and can be described as being homogeneous or heterogeneous. Research to learn more about rocks and minerals and note some examples of homogeneous and heterogeneous rocks in your Science Journal.

section 3 review

Summary

Substances

- A substance can be either an element or a compound.
- A compound contains more than one kind of element bonded together.
- A chemical formula shows which elements and how many atoms of each make up a compound.

Mixtures

- A mixture contains substances that are not chemically bonded together.
- There are many ways to separate mixtures, based on their physical properties.
- Homogeneous mixtures are those that are the same throughout. These types of mixtures can be solids, liquids, or gases.
- Heterogeneous mixtures have larger parts that are different from each other.

Self Check

1. **List** three examples of compounds and three examples of mixtures. Explain your choices.
2. **Determine** Ethanol is an alcohol made up of carbon, hydrogen, and oxygen atoms. How many atoms of each element does ethanol have?
3. **Think Critically** Explain whether your breakfast was a compound, a homogeneous mixture, or a heterogeneous mixture.

Applying Skills

4. **Compare and contrast** compounds and mixtures based on what you have learned from this section.
5. **Use a Database** Use a computerized card catalog or database to find information about one element from the periodic table. Include information about the properties and uses of the mixtures and/or compounds in which the element is frequently found.

Science Online red.msscience.com/self_check_quiz

INTEGRATE Earth Science

Rocks and Minerals Strictly speaking, the term *rocks* applies to the solid materials that make up Earth's crust. These materials are not homogeneous and have no definite chemical composition. *Minerals,* on the other hand, have more or less definite chemical compositions, crystal structures, and properties.

3 Assess

DAILY INTERVENTION

Check for Understanding

Visual-Spatial Have students develop a demonstration that differentiates between homogeneous and heterogeneous mixtures. Possible answer: Homogeneous mixtures can include substances such as apple juice or non-carbonated soft drinks. Heterogeneous mixtures can include substances such as a tossed salad, or fruit salad.

Reteach

Blood Ask a medical technologist to make a class presentation explaining the different parts of the mixture known as human blood and describing how these parts can be separated by physical means. L1

Assessment

Process Have students design an experimental procedure that they could use to separate a mixture of white sand and sugar. Use **Performance Assessment in the Science Classroom,** p. 95.

section 3 review

1. Possible answers: compounds— water, carbon dioxide, table salt; mixtures— air, ocean water, brass; mixtures can be separated by physical means; compounds can't.
2. C:2; H:6; O:1
3. Answers will vary, but most breakfasts, such as orange juice and cold cereal with milk, will be heterogeneous mixtures.
4. Possible answer: Compounds are pure substances whose properties vary from the original elements that they are made from. Mixtures are made from different parts and can be separated.
5. Sample answer: Carbon is found in the Sun, stars, comets, and atmospheres of most planets. It is present as carbon dioxide in the atmosphere and dissolved in water. It is a component of rocks as carbonates of calcium (limestone), magnesium, and iron.

BENCH TESTED

Real-World Question

Purpose Students test for certain compounds and decide which are present in a mystery mixture. L2 ELL LS Logical-Mathematical

Process Skills form a hypothesis, communicate, observe and infer, recognize cause and effect, separate and control variables, interpret data

Time Required 45 minutes

Procedure

Materials Compounds used are cornstarch, powdered sugar, and baking soda. Combine any two of these compounds to make the mystery mixture. Prepare iodine test solution by adding 7 g of iodine and 5 g of potassium iodide to 5 mL of water, then diluting to 100 mL with denatured alcohol.

Safety Precautions

- Students must wear goggles and aprons. Remind students that while heating the test tube, the opening of the tube should never be pointed at themselves or others.
- Iodine is poisonous and will stain clothing.
- Remind students that test tubes should not be touched while they are hot.

LAB

Mystery Mixture

Goals

- **Test** for the presence of certain compounds.
- **Decide** which of these compounds are present in an unknown mixture.

Materials

test tubes (4)
cornstarch
powdered sugar
baking soda
mystery mixture
small scoops (3)
dropper bottles (2)
iodine solution
white vinegar
hot plate
250-mL beaker
water (125 mL)
test-tube holder
small pie pan

Safety Precautions

WARNING: *Use caution when handling hot objects. Substances could stain or burn clothing. Be sure to point the test tube away from your face and your classmates while heating.*

Real-World Question

You will encounter many compounds that look alike. For example, a laboratory stockroom is filled with white powders. It is important to know what each is. In a kitchen, cornstarch, baking powder, and powdered sugar are compounds that look alike. To avoid mistaking one for another, you can learn how to identify them. Different compounds can be identified by using chemical tests. For example, some compounds react with certain liquids to produce gases. Other combinations produce distinctive colors. Some compounds have high melting points. Others have low melting points. How can the compounds in an unknown mixture be identified by experimentation?

Alternative Inquiry Lab

Separating Compounds To extend this Lab into an Inquiry Lab, have students brainstorm and/or research the difference between mixtures and compounds. Chemical reactions are required to separate compounds. You may have students try reducing copper oxide ore by combining it with carbon powder and heating it in a crucible—they should get pure shiny copper metal as a result (keep the lid on and heat it as hot as possible for at least fifteen minutes). Special instruments (such as magnetic resonance imaging and infrared spectrometers) are available to find out what elements are in compounds, but tell students that these machines involve complex processes. Some students may want to research the MRI and IR processes.

118 CHAPTER 4 Atoms, Elements, and the Periodic Table

Using Scientific Methods

Procedure

1. Copy the data table into your Science Journal. Record your results carefully for each of the following steps.
2. Place a small scoopful of cornstarch on the pie pan. Do the same for the sugar and baking soda making separate piles. Add a drop of vinegar to each. Wash and dry the pan after you record your observations.
3. Again, place a small scoopful of cornstarch, sugar, and baking soda on the pie pan. Add a drop of iodine solution to each one. Wash and dry the pan after you record your observations.
4. Again place a small scoopful of each compound in a separate test tube. Hold the test tube with the test-tube holder and with an oven mitt. Gently heat the test tube in a beaker of boiling water on a hot plate.
5. Follow steps 2 through 4 to test your mystery mixture for each compound.

Identifying Presence of Compounds

Substance to Be Tested	Fizzes with Vinegar	Turns Blue with Iodine	Melts When Heated
Cornstarch	no	yes	no
Sugar	no	no	yes
Baking soda	yes	no	no
Mystery mix	yes	yes	yes

Analyze Your Data

Identify from your data table which compound(s) you have.

Conclude and Apply

1. **Describe** how you decided which substances were in your unknown mixture.
2. **Explain** how you would be able to tell if all three compounds were not in your mystery substance.
3. **Draw a Conclusion** What would you conclude if you tested baking powder from your kitchen and found that it fizzed with vinegar, turned blue with iodine, and did not melt when heated?

Communicating Your Data

Make a different data table to display your results in a new way. **For more help, refer to the Science Skill Handbook.**

Teaching Strategies

- In Steps 2 and 3, have students use only sufficient amounts of the solids to produce satisfactory test results. Test your scoops to determine how much to use.
- In Step 4, have students use only enough solid to fill the bottom, rounded portion of the test tubes.

Analyze Your Data

Expected Outcome Cornstarch will react with iodine to make a blue color. Sugar will melt when heated. Baking soda will fizz in the presence of vinegar.

Error Analysis If students let their experiments run together or do not clean their equipment, they may get erroneous results. Have students compare their results to the results of other groups. If results differ, discuss errors that could have caused the differences.

Conclude and Apply

1. Answers will vary depending on the two compounds used for the mystery mixture. In describing their conclusions, students should include the data they gathered.
2. If all three compounds were absent from the mystery mixture, the mixture would not turn blue in the presence of iodine, it would not melt when heated, and it would not fizz with vinegar.
3. It contained baking soda and cornstarch.

Assessment

Performance After students have determined the two compounds that were present in the mystery mixture, have them determine other possible two-material combinations and explain how the combinations would react when tested. Use **Performance Assessment in the Science Classroom,** p. 89.

Communicating Your Data

Students could use illustrations, chemical symbols, or color to revise their data table.

TIME

Content Background

Kashyapa lived during the 6th century B.C., and was the first to use the word *parmanu;* derived from the two Sanskrit words *param*, meaning "beyond", and *anu*, meaning "atom." Among other things, he proposed that two different classes of parmanu could combine to form what he called *dwinuka*, or binary molecules, of new substances.

The Chinese term for the five material agents of matter, as they called the five elements, is *wu hsing*. The Chinese believed that these material agents took part in cyclical patterns of change that resulted in the opposites, yin and yang.

Discussion

Atomic Theory Do you think Kashyapa's ideas were known to Democritus when he developed his theory of the atom? Why or why not? Accept all well-supported answers. Point out to students on a globe the relative locations of Greece and India and remind them of the means of transportation and communication available at the time these two men lived.

Historical Significance

Both the Chinese and Indian philosophies have roots in ancient history. Encourage students to choose a current scientific theory and use a variety of research tools to trace the idea back to its earliest origins. Make sure students understand that one theory may have origins in ideas that arose independently in different cultures. This activity will help students become aware that our modern theories have had a long history of development.

TIME SCIENCE AND HISTORY

SCIENCE CAN CHANGE THE COURSE OF HISTORY!

Ancient Views of Matter

Two cultures observed the world around them differently

The world's earliest scientists were people who were curious about the world around them and who tried to develop explanations for the things they observed. This type of observation and inquiry flourished in ancient cultures such as those found in India and China. Read on to see how the ancient Indians and Chinese defined matter.

Indian Ideas

To Indians living about 3,000 years ago, the world was made up of five elements: fire, air, earth, water, and ether, which they thought of as an unseen substance that filled the heavens. Building upon this concept, the early Indian philosopher Kashyapa (kah SHI ah pah) proposed that the five elements could be broken down into smaller units called parmanu (par MAH new). Parmanu were similar to atoms in that they were too small to be seen but still retained the properties of the original element. Kashyapa also believed that each type of parmanu had unique physical and chemical properties.

Parmanu of earth elements, for instance, were heavier than parmanu of air elements. The different properties of the parmanu determined the characteristics of a substance. Kashyapa's ideas about matter are similar to those of the Greek philosopher Democritus, who lived centuries after Kashyapa.

Chinese Ideas

The ancient Chinese also broke matter down into five elements: fire, wood, metal, earth, and water. Unlike the early Indians, however, the Chinese believed that the elements constantly changed form. For example, wood can be burned and thus changes to fire. Fire eventually dies down and becomes ashes, or earth. Earth gives forth metals from the ground. Dew or water collects on these metals, and the water then nurtures plants that grow into trees, or wood.

This cycle of constant change was explained in the fourth century B.C. by the philosopher Tsou Yen. Yen, who is known as the founder of Chinese scientific thought, wrote that all changes that took place in nature were linked to changes in the five elements.

Research **Write a brief paragraph that compares and contrasts the ancient Indian and Chinese views of matter. How are they different? Similar? Which is closer to the modern view of matter? Explain.**

Science Online

For more information, visit red.msscience.com/time

Resources for Teachers and Students

Chemistry in Action, by Nina Morgan, Oxford University Press, 1995

The Periodic Kingdom, by P.W. Atkins, Basic Books, 1995

Research Students will need to research their information by using the school's media center, the library, or visiting the link on the student page.

Reviewing Main Ideas

Section 1 Structure of Matter

1. Matter is anything that occupies space and has mass.
2. Matter is made up of atoms.
3. Atoms are made of smaller parts called protons, neutrons, and electrons.
4. Many models of atoms have been created as scientists try to discover and define the atom's internal structure. Today's model has a central nucleus with the protons and neutrons, and an electron cloud surrounding it.

Section 2 The Simplest Matter

1. Elements are the building blocks of matter.
2. An element's atomic number tells how many protons its atoms contain, and its atomic mass tells the average mass of its atoms.
3. Isotopes are two or more atoms of the same element that have different numbers of neutrons.

Section 3 Compounds and Mixtures

1. Compounds are substances that are produced when elements combine. Compounds contain specific proportions of the elements that make them up.
2. Mixtures are combinations of compounds and elements that have not formed new substances. Their proportions can change.

Visualizing Main Ideas

Copy and complete the following concept map.

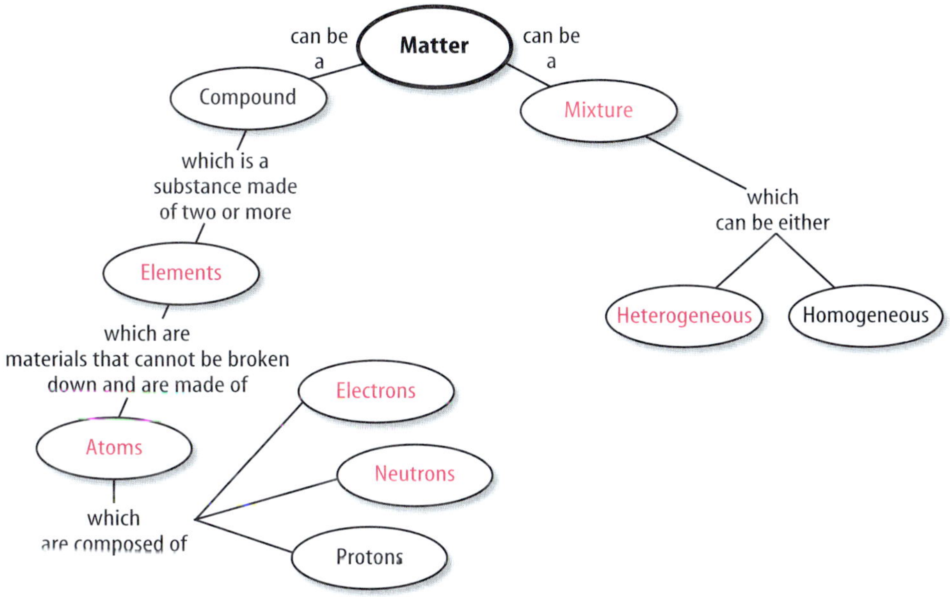

Identifying Misconceptions

Assess

Use the assessment as follow-up to page F at the beginning of the chapter after students have completed the chapter.

Materials 50-mL graduated cylinder, beaker, water, droppers, pan balance

Procedure Find the mass of the empty graduated cylinder, then pour 50 mL of water into it and find the mass again. Use the dropper to decrease the volume of water by one half and calculate the mass of the water. Repeat this process three more times to obtain volumes of 50 mL, 25 mL, 12.5 mL, 6.25 mL, and 3.125 mL. Will you ever get a drop of water that has no mass?

Expected Outcome No matter how small the volume, the water has mass and therefore weight.

chapter 4 Study Guide

Reviewing Main Ideas

Summary statements can be used by students to review the major concepts of the chapter.

Visualizing Main Ideas

See student page.

Science Online

Visit red.msscience.com
/self_check_quiz
/interactive_tutor
/vocabulary_puzzlemaker
/chapter_review
/standardized_test

Assessment Transparency

For additional assessment questions, use the *Assessment Transparency* located in the transparency book.

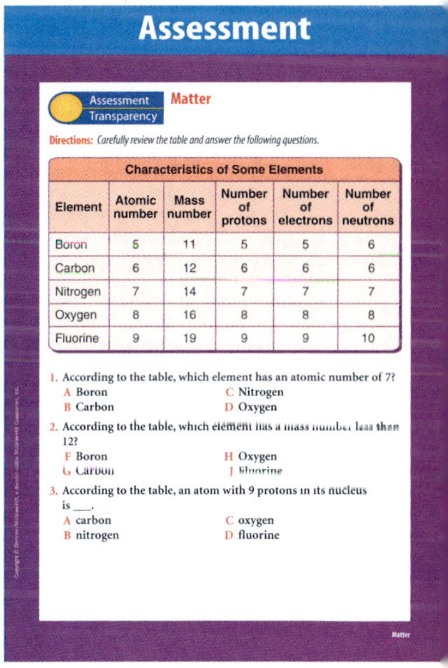

Assessment

Assessment Transparency **Matter**

Directions: *Carefully review the table and answer the following questions.*

Characteristics of Some Elements					
Element	Atomic number	Mass number	Number of protons	Number of electrons	Number of neutrons
Boron	5	11	5	5	6
Carbon	6	12	6	6	6
Nitrogen	7	14	7	7	7
Oxygen	8	16	8	8	8
Fluorine	9	19	9	9	10

1. According to the table, which element has an atomic number of 7?
   A Boron
   B Carbon
   C Nitrogen
   D Oxygen
2. According to the table, which element has a mass number less than 12?
   F Boron
   G Carbon
   H Oxygen
   J Fluorine
3. According to the table, an atom with 9 protons in its nucleus is ___.
   A carbon
   B nitrogen
   C oxygen
   D fluorine

Matter

Using Vocabulary

1. proton
2. compound
3. matter
4. neutrons
5. metals

Checking Concepts

6. D
7. C
8. B
9. C
10. D
11. B
12. C
13. C
14. B
15. D

Thinking Critically

16. one sulfur atom to two oxygen atoms
17. An atom with seven protons is nitrogen.
18. The element loses its characteristics and a new substance is formed.
19. Cobalt-60 and cobalt-59 are isotopes because they both have 27 protons in their nuclei.
20. Rutherford's gold foil experiment told scientists that the atom contains a small, dense, positively charged structure.
21. Because the nuclei in all atoms are composed of densely packed positive charge (protons), the positive alpha particle would have been deflected by the aluminum foil. Students also may add that the angle of deflection would not be as great because an aluminum nucleus is less dense than a gold nucleus.

chapter 4 Review

Using Vocabulary

atom p. 99	matter p. 98
atomic mass p. 110	metal p. 110
atomic number p. 109	metalloid p. 111
compound p. 113	mixture p. 115
electron p. 102	neutron p. 104
element p. 106	nonmetal p. 111
isotope p. 109	nucleus p. 103
law of conservation of matter p. 100	proton p. 103
mass number p. 109	substance p. 113

Fill in the blanks with the correct vocabulary word or words.

1. The ________ is the particle in the nucleus of the atom that carries a positive charge and is counted to identify the atomic number.
2. The new substance formed when elements combine chemically is a(n) ________.
3. Anything that has mass and takes up space is ________.
4. The particles in the atom that account for most of the mass of the atom are protons and ________.
5. Elements that are shiny, malleable, ductile, good conductors of heat and electricity, and make up most of the periodic table are ________.

Checking Concepts

Choose the word or phrase that best answers the question.

6. What is a solution an example of?
   - **A)** element
   - **B)** heterogeneous mixture
   - **C)** compound
   - **D)** homogeneous mixture
7. The nucleus of one atom contains 12 protons and 12 neutrons, while the nucleus of another atom contains 12 protons and 16 neutrons. What are the atoms?
   - **A)** chromium atoms
   - **B)** two different elements
   - **C)** two isotopes of an element
   - **D)** negatively charged
8. What is a compound?
   - **A)** a mixture of chemicals and elements
   - **B)** a combination of two or more elements
   - **C)** anything that has mass and occupies space
   - **D)** the building block of matter
9. What does the atom consist of?
   - **A)** electrons, protons, and alpha particles
   - **B)** neutrons and protons
   - **C)** electrons, protons, and neutrons
   - **D)** elements, protons, and electrons
10. In an atom, where is an electron located?
    - **A)** in the nucleus with the proton
    - **B)** on the periodic table of the elements
    - **C)** with the neutron
    - **D)** in a cloudlike formation surrounding the nucleus
11. How is matter defined?
    - **A)** the negative charge in an atom
    - **B)** anything that has mass and occupies space
    - **C)** the mass of the nucleus
    - **D)** sound, light, and energy
12. What are two atoms that have the same number of protons called?
    - **A)** metals
    - **B)** nonmetals
    - **C)** isotopes
    - **D)** metalloids
13. Which is a heterogeneous mixture?
    - **A)** air
    - **B)** brass
    - **C)** a salad
    - **D)** apple juice

Science Online red.msscience.com/vocabulary_puzzlemaker

Use the ExamView® Pro Testmaker CD-ROM to:

- create multiple versions of tests
- create modified tests with one mouse click for inclusion students
- edit existing questions and add your own questions
- build tests aligned with state standards using built-in State Curriculum Tags
- change English tests to Spanish with one mouse click and vice versa

122 CHAPTER REVIEW

Use the illustration below to answer questions 14 and 15.

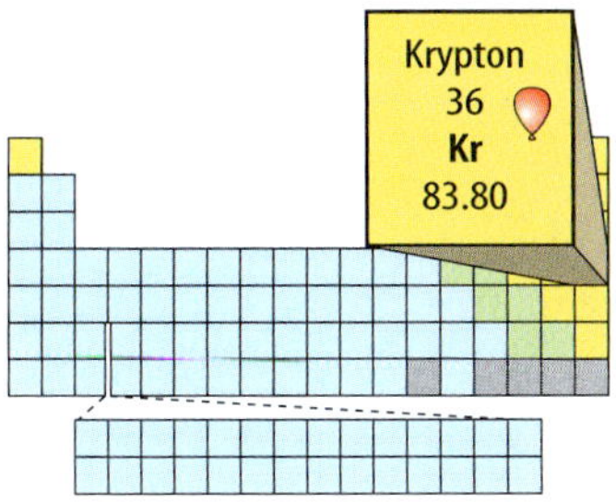

14. Using the figure above, krypton has

A) an atomic number of 84.
B) an atomic number of 36.
C) an atomic mass of 36.
D) an atomic mass of 72.

15. From the figure, the element krypton is

A) a solid. **C)** a mixture.
B) a liquid. **D)** a gas.

Thinking Critically

16. Analyze Information A chemical formula is written to indicate the makeup of a compound. What is the ratio of sulfur atoms to oxygen atoms in SO_2?

17. Determine which element contains seven electrons and seven protons.

18. Describe what happens to an element when it becomes part of a compound.

19. Explain how cobalt-60 and cobalt-59 can be the same element but have different mass numbers.

20. Analyze Information What did Rutherford's gold foil experiment tell scientists about atomic structure?

21. Predict Suppose Rutherford had bombarded aluminum foil with alpha particles instead of the gold foil he used in his experiment. What observations do you predict Rutherford would have made? Explain your prediction.

22. Draw Conclusions You are shown a liquid that looks the same throughout. You're told that it contains more than one type of element and that the proportion of each varies throughout the liquid. Is this an element, a compound, or a mixture?

Use the illustrations below to answer question 23.

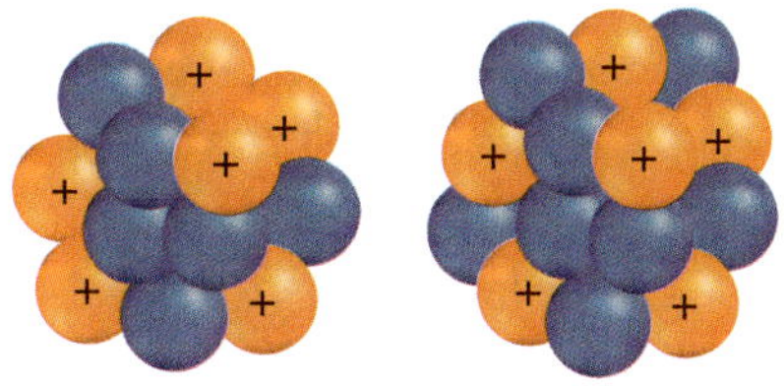

23. Interpret Scientific Illustrations Look at the two carbon atoms above. Explain whether or not the atoms are isotopes.

24. Explain how the atomic mass of krypton was determined.

Performance Activities

25. Newspaper Article Research the source, composition, and properties of asbestos. Why was it used in the past? Why is it a health hazard now? What is being done about it? Write a newspaper article to share your findings.

Applying Math

26. Atomic Mass Krypton has six naturally occurring isotopes with atomic masses of 78, 80, 82, 83, 84, and 86. Make a table of the number of protons, electrons, and neutrons in each isotope.

27. Atomic Ratio A researcher is analyzing two different compounds, sulfuric acid (H_2SO_4) and hydrogen peroxide (H_2O_2). What is the ratio of hydrogen to oxygen in hydrogen peroxide? What is the ratio of oxygen to hydrogen in sulfuric acid?

chapter Review 4

22. The liquid is a mixture. Students also might add that it is homogeneous.

23. The atoms are isotopes because they differ in their number of neutrons. The atom on the left has 6 p and 6 n whereas the one on the right has 6 p and 8 n.

24. The atomic mass of krypton was determined by using the weighted average of the isotopes of krypton.

Performance Activities

25. Asbestos is a hydrous magnesium silicate—$Mg_3Si_2O_5(OH)_4$. It occurs as fibers, which are spun into yarn or mixed with other materials to form cement board or other fireproof building materials. When asbestos fibers become airborne, they can be inhaled. The sharp fibers penetrate the lungs, and can cause a form of lung cancer. Use **Performance Assessment in the Science Classroom,** p. 141.

Applying Math

National Math Standards

1, 2

26.

Isotope/Atomic Mass	P	E	N
Kr_{78}	36	36	42
Kr_{80}	36	36	44
Kr_{82}	36	36	46
Kr_{83}	36	36	47
Kr_{84}	36	36	48
Kr_{86}	36	36	50

27. 1:1; 2:1.

Assessment Resources

Reproducible Masters

Chapter *Fast File* Resources
Chapter Review, pp. 33–34
Chapter Tests, pp. 35–38
Assessment Transparency Activity, p. 45

Glencoe Science Web site
Chapter Review Test
Standardized Test Practice

Glencoe Technology
Assessment Transparency
Exam*View*® Pro Testmaker
MindJogger Videoquiz
Interactive Chalkboard

FAST FILE

Answer Sheet A practice answer sheet can be found at red.msscience.com/answer_sheet.

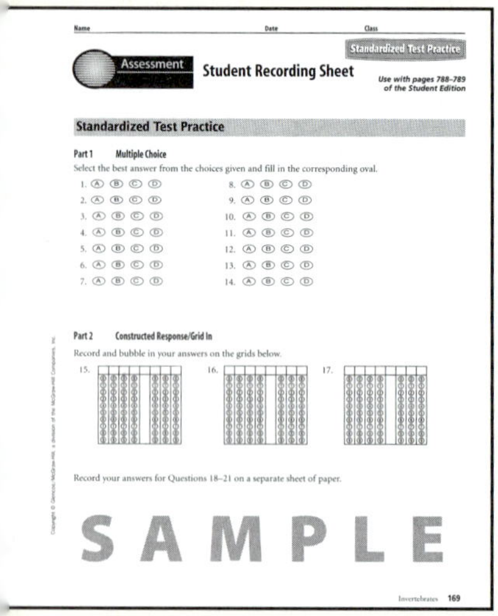

Part 1 Multiple Choice

1. D
2. D
3. A
4. C
5. D
6. D
7. C
8. B

Part 2 Short Response

9. They are more likely to be close to the nucleus because they are attracted to the positive charges of the protons.
10. 90
11. nonmetals
12. 6 H_2O_2
13. electrons
14. mixture
15. substance

chapter 4 Standardized Test Practice

Part 1 Multiple Choice

Record your answers on the answer sheet provided by your teacher or on a sheet of paper.

1. Which of the scientists below introduced the idea that matter is made up of tiny, individual bits called atoms?
   A. Arrhenius **C.** Chadwick
   B. Avogadro **D.** Democritus

Use the illustration below to answer questions 2 and 3.

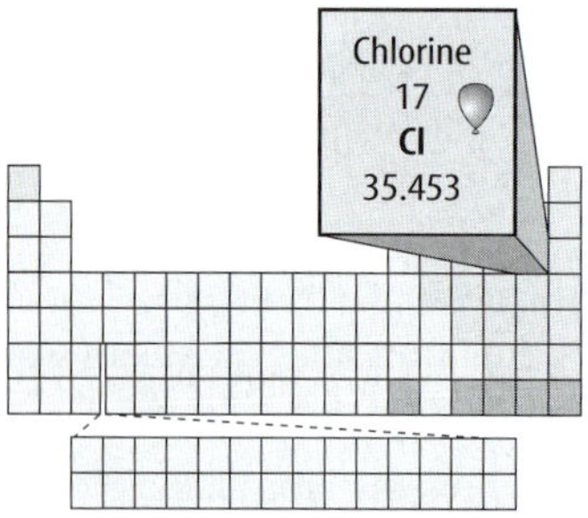

2. The periodic table block shown above lists properties of the element chlorine. What does the number 35.453 mean?
   A. the number of neutrons and in every chlorine atom
   B. the number of neutrons and protons in every chlorine atom
   C. the average number of neutrons in a chlorine atom
   D. the average number of neutrons and protons in a chlorine atom
3. According to the periodic table block, how many electrons does an uncharged atom of chlorine have?
   A. 17 **C.** 35
   B. 18 **D.** 36

Test-Taking Tip

Full Understanding Read each question carefully for full understanding.

4. Which of the following scientists envisioned the atom as a ball of positive charge with electrons embedded in it, much like chocolate chips spread through cookie dough?
   A. Crookes **C.** Thomson
   B. Dalton **D.** Rutherford

Use the illustration below to answer questions 5 and 6.

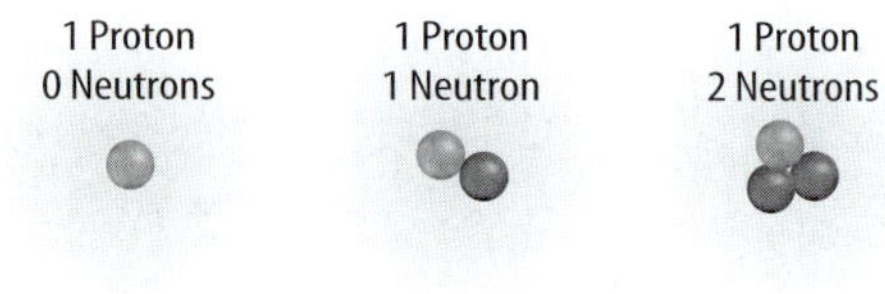

5. Which of the following correctly identifies the three atoms shown in the illustration above?
   A. hydrogen, lithium, sodium
   B. hydrogen, helium, lithium
   C. hydrogen, helium, helium
   D. hydrogen, hydrogen, hydrogen
6. What is the mass number for each of the atoms shown in the illustration?
   A. 0, 1, 2
   B. 1, 1, 1
   C. 1, 2, 2
   D. 1, 2, 3
7. Which of the following are found close to the right side of the periodic table?
   A. metals **C.** nonmetals
   B. lanthanides **D.** metalloids
8. Which of the following is a characteristic that is typical of a solid, nonmetal element?
   A. shiny
   B. brittle
   C. good heat conductor
   D. good electrical conductor

Part 3 Open Ended

16. Dalton thought that matter was made up of atoms, and the atoms could not be divided into smaller pieces. He proposed that all atoms of an element are exactly alike. Different elements, he said, are made of different kinds of atoms. Dalton pictured the atom as a hard sphere that was the same throughout.
17. A source of alpha particles was aimed at a thin sheet of gold foil. The foil was surrounded by a fluorescent screen that emitted a flash of light each time it was hit by a charged particle. Rutherford thought most of the alpha particles would pass straight through the foil. He didn't think the foil contained enough matter to stop or deflect the particles. He thought the positive charge in the gold might cause a few small changes in the paths of the alpha particles.
18. The particles that reflected at large angles showed that Thomson's model of the atom was incorrect. The positive charge in the gold was

Part 2 Short Response/Grid In

Record your answers on the answer sheet provided by your teacher or on a sheet of paper.

9. Are electrons more likely to be in an energy level close to the nucleus or far away from the nucleus? Why?

10. How many naturally-occurring elements are listed on the periodic table?

11. Is the human body made of mostly metal, nonmetals, or metalloids?

12. A molecule of hydrogen peroxide is composed of two atoms of hydrogen and two atoms of oxygen. What is the formula for six molecules of hydrogen peroxide?

13. What is the modern-day name for cathode rays?

Use the illustration below to answer questions 14 and 15.

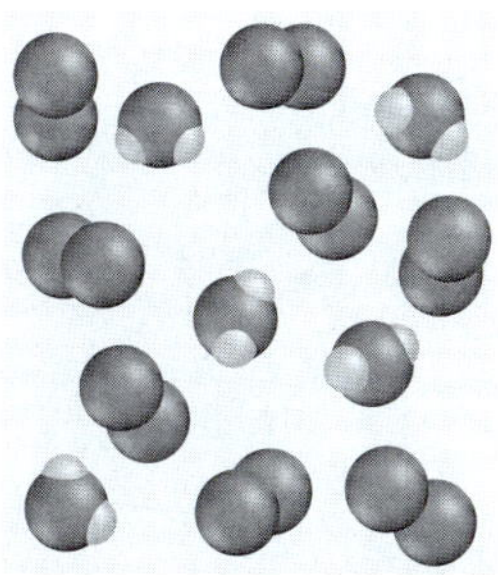

14. The illustration above shows atoms of an element and molecules of a compound that are combined without making a new compound. What term describes a combination such as this?

15. If the illustration showed only the element or only the compound, what term would describe it?

Part 3 Open Ended

Record your answers on a sheet of paper.

16. Describe Dalton's ideas about the composition of matter, including the relationship between atoms and elements.

Use the illustration below to answer questions 17 and 18.

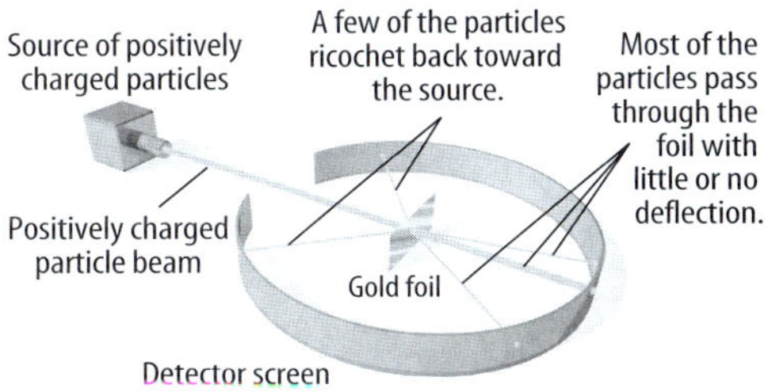

17. The illustration above shows Rutherford's gold foil experiment. Describe the setup shown. What result did Rutherford expect from his experiment?

18. What is the significance of the particles that reflected back from the gold foil? How did Rutherford explain his results?

19. Describe three possible methods for separating mixtures. Give an example for each method.

20. What are the rows and columns on the periodic table called? How are elements in the rows similar, and how are elements in the columns similar?

21. Describe how Thomson was able to show that cathode rays were streams of particles, not light.

22. Describe how the mass numbers, or atomic masses, listed on the periodic table for the elements are calculated.

reflecting the particles. Rutherford proposed that almost all the mass of the atom and all of its positive charge was contained in a small nucleus of the atom.

19. A mixture of sugar and sand can be separated by adding water. The sugar dissolves in the water and the sand can then be filtered out. A mixture of water and sugar can be separated by heating the water so that the water evaporates, leaving the sugar. Mixtures of different-sized solids, such as pebbles and sand, can be separated by filtering out the larger solid.

20. The rows are called periods. Elements in a row have the same number of energy levels. The columns are called groups. Elements in each group have similar properties related to their structure. They also tend to form similar bonds.

21. Thomson repeated Crookes' experiment so that cathode rays moved from the negative electrode to the positive electrode. However, when Thomson placed a magnet near the cathode ray tube, the cathode rays were deflected. Since light is not deflected by a magnet, Thomson deduced that the rays must be made of charged particles.

22. Each atom's mass number is the number of protons plus the number of neutrons. Each isotope has an integer mass number. The mass number for the element is calculated by calculating a weighted average. The weighted average is the sum of each isotope's mass number multiplied by the percentage of isotopes of that kind.

Rubrics

For more help evaluating open-ended assessment questions, see the rubric on p. 10T.

unit 2

Unit Contents

WebQuest ***MagLev Trains: Floating Locomotives*** is an investigation of magnets and how they work. Students will study different types of magnetic levitation, design and build a maglev train, then test and refine their models. Class presentations reflect student designs, procedures, refinements, and results.

unit 2

Interactions of Matter

How Are Train Schedules & Oil Pumps Connected?

PROJECT CRISS[SM]

Study Skills

Illustrate Key Concepts When students actively engage in meaningful conversation and activities, deeper understanding of topics takes place. As a post-reading exercise, divide the class into cooperative groups, and ask them to describe and draw pictures that illustrate the conversion of one kind of energy to another. Have them share these free-form maps with the class followed by a whole-class discussion.

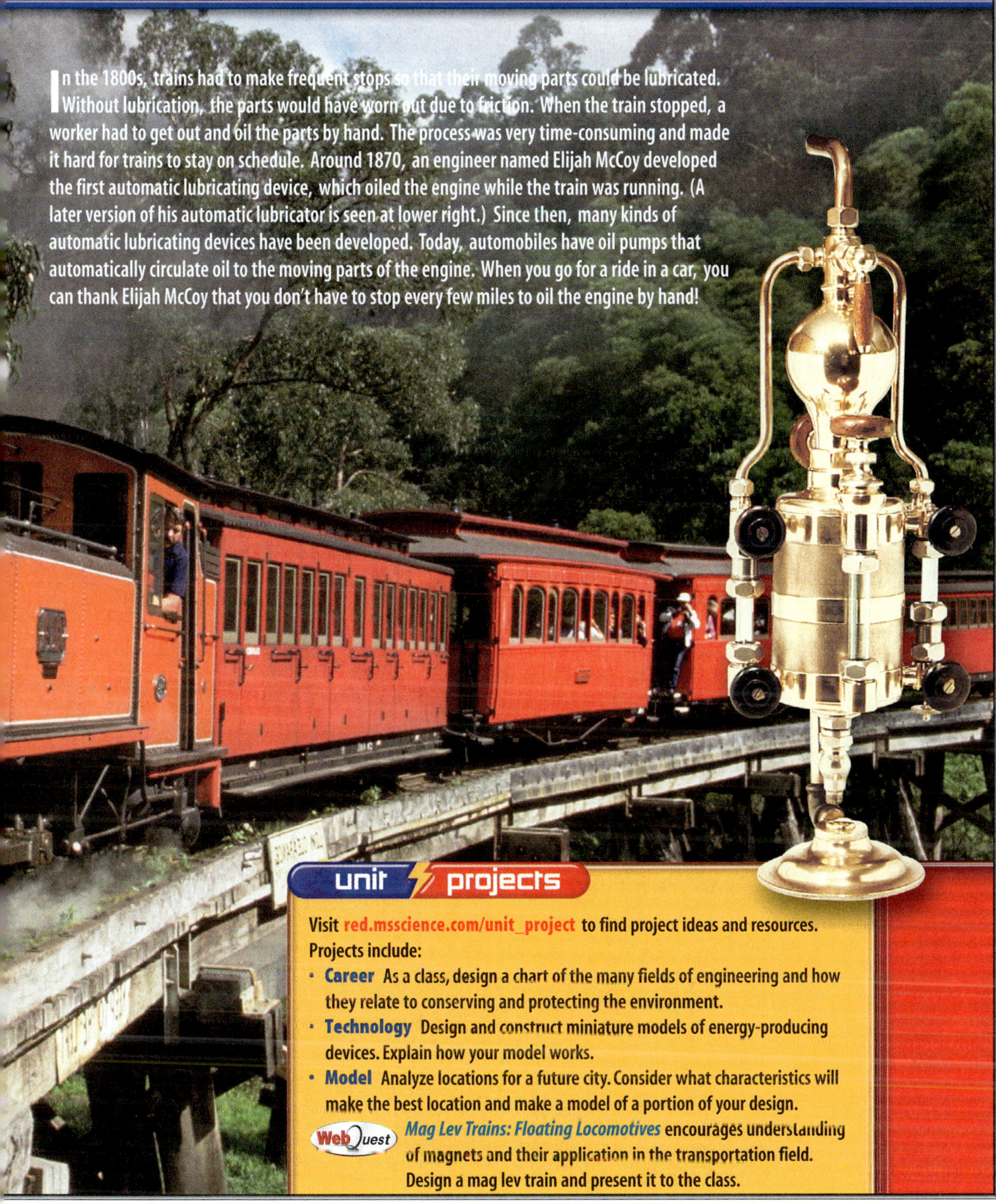

In the 1800s, trains had to make frequent stops so that their moving parts could be lubricated. Without lubrication, the parts would have worn out due to friction. When the train stopped, a worker had to get out and oil the parts by hand. The process was very time-consuming and made it hard for trains to stay on schedule. Around 1870, an engineer named Elijah McCoy developed the first automatic lubricating device, which oiled the engine while the train was running. (A later version of his automatic lubricator is seen at lower right.) Since then, many kinds of automatic lubricating devices have been developed. Today, automobiles have oil pumps that automatically circulate oil to the moving parts of the engine. When you go for a ride in a car, you can thank Elijah McCoy that you don't have to stop every few miles to oil the engine by hand!

unit projects

Visit red.msscience.com/unit_project to find project ideas and resources.
Projects include:

- **Career** As a class, design a chart of the many fields of engineering and how they relate to conserving and protecting the environment.
- **Technology** Design and construct miniature models of energy-producing devices. Explain how your model works.
- **Model** Analyze locations for a future city. Consider what characteristics will make the best location and make a model of a portion of your design.

WebQuest *Mag Lev Trains: Floating Locomotives* encourages understanding of magnets and their application in the transportation field. Design a mag lev train and present it to the class.

How Are Train Schedules & Oil Pumps Connected?

A rough surface has high spots and low spots that are large enough that you can feel or sometimes see. If you try to slide these surfaces past each other, the high spots hit each other, and impede the motion of the surfaces.

However, friction also exists between smooth surfaces. On a microscopic scale, the surface is rough. Lubricants can reduce friction by forming a thin film between the surfaces. This film separates the surfaces so that the high spots no longer make contact with each other. The result is longer-lasting parts and a more efficient use of fuels to produce motion.

unit projects

Career Have students study aspects of the field of engineering and how this career option influences resources, energy, and development. Careers include environmental, mechanical, electrical, civil, and ceramic engineers, as well as hydrologists, geologists, and metallurgists. As a class, design a chart that lists the field of engineering and how it relates to conserving and protecting the environment.

Technology Have students research, design, and construct energy resource saving models of windmills, wave and water mills, solar panels, dams, geothermal stations, and composting gas for energy. Students should be able to explain how their model works, saves energy resources, money, and the environment.

Model Working in small groups, have students analyze different locations to build a city. Student decisions should take into account natural resources, climate, food and water sources, economic development, and environmental concerns and how energy can best be used to conserve those resources. Using recycled materials, have students build a portion of their city as a model. Students should consider size limits and scale, and how to best present their energy efficient community.

Additional Resources For more information, resources, and assessment rubrics, visit red.msscience.com/unit_project

chapter 5 Organizer

Section/Objectives	Standards		Labs/Features
	National	**State/Local**	
Chapter Opener	See pp. 16T–17T for a Key to Standards.		**Launch Lab:** Model Halfpipe Motion, p. 129 **Foldables,** p. 129
Section 1 Motion 3 sessions 1.5 blocks 1. **Define** speed and acceleration. 2. **Relate** acceleration to change in speed. 3. **Calculate** distance, speed, and acceleration.	National Content Standards: UCP.1–3, UCP.5, A.1, A.2, B.2, B.3		**Applying Math:** Bicycle Speed, p. 131 **Integrate Earth Science,** p. 132 **Applying Math:** Family Trip Distance, p. 132 **Applying Math:** Acceleration Down a Hill, p. 134
Section 2 Newton's Laws of Motion 2 sessions 1 block 4. **Describe** how forces affect motion. 5. **Calculate** acceleration using Newton's second law of motion. 6. **Explain** Newton's third law of motion.	National Content Standards: UCP.1–3, UCP.5, A.1, A.2, B.2, B.3		**Integrate Life Science,** p. 137 **MiniLAB:** Determining Weights in Newtons, p. 138 **Science Online,** p. 140 **Applying Math:** Acceleration of a Basketball, p. 140 **Visualizing Newton's Laws and Space Travel,** p. 142
Section 3 Work and Simple Machines 4 sessions 2 blocks 7. **Define** work. 8. **Distinguish** the different types of simple machines. 9. **Explain** how machines make work easier.	National Content Standards: UCP.1–3, UCP.5, A.1, A.2, B.2, B.3		**Integrate Life Science,** p. 145 **Applying Math:** Weight Lifting, p. 145 **Science Online,** p. 146 **MiniLAB:** Observing Mechanical Advantage—Pulleys, p. 147 **Lab:** Motion, p. 151 **Lab:** Methods of Travel, p. 152 **Science Stats,** Fastest Facts, p. 154

Glencoe Exclusive! TeacherWorks™ All-In-One Planner and Resource Center

Lab Materials	Reproducible Resources	Section Assessment	Technology
Launch Lab: 2–4 books, heavy paper or cardboard, marble	**Chapter *FAST FILE* Resources** Foldables Worksheet, p. 15 Note-taking Worksheets, pp. 31–33 Directed Reading Overview, p. 17	Glencoe's Assessment Advantage	TeacherWorks includes: • Interactive Teacher Edition • Lesson Planner with calendar • Access to all program blacklines • Correlations to standards • Web links
Need materials? Contact Science Kit at 1-800-828-7777 or www.sciencekit.com on the Internet.	**Chapter *FAST FILE* Resources** Transparency Activity, p. 42 Enrichment, p. 28 Reinforcement, p. 25 Directed Reading, p. 18 Lab Activity, pp. 9–11 **Mathematics Skill Activities,** p. 9 **Reading and Writing Skill Activities,** p. 35	**Performance** Applying Math, p. 131 Applying Math, p. 132 Applying Math, p. 134 Applying Math, p. 135 **Content** Section Review, p. 135	Section Focus Transparency Guided Reading Audio Program Interactive Chalkboard CD-ROM
MiniLAB: bathroom scale, large book, chair, heavy coat, another heavy object, calculator	**Chapter *FAST FILE* Resources** Transparency Activity, p. 43 MiniLAB, p. 3 Enrichment, p. 29 Reinforcement, p. 26 Directed Reading, p. 18 Lab Activity, pp. 13–14 Transparency Activity, pp. 45–46 **Physical Science Critical Thinking/ Problem Solving,** p. 3 **Cultural Diversity,** p. 63	**Portfolio** Differentiated Instruction, p. 138 **Performance** MiniLAB, p. 138 Applying Math, p. 140 Applying Math, p. 143 **Content** Section Review, p. 143	Section Focus Transparency Teaching Transparency Virtual Labs CD-ROM Guided Reading Audio Program Interactive Chalkboard CD-ROM
MiniLAB: 2 broomsticks or 2 dowels, rope (3-m long) **Lab:** small ball or marble, meterstick or tape measure, stopwatch or clock with second hand, graph paper **Lab:** no materials needed	**Chapter *FAST FILE* Resources** Transparency Activity, p. 44 MiniLAB, p. 4 Enrichment, p. 30 Reinforcement, p. 27 Directed Reading, pp. 19, 20 Lab Worksheet, pp. 5–6, 7–8 **Lab Management and Safety,** p. 71	**Portfolio** Cultural Diversity, p. 145 MiniLAB Assessment, p. 147 **Performance** Applying Math, p. 145 MiniLAB, p. 147 Applying Skills, p. 150 **Content** Section Review, p. 150	Section Focus Transparency Guided Reading Audio Program Interactive Chalkboard CD-ROM Video Lab

End of Chapter Assessment

Glencoe's Assessment Advantage

Blackline Masters	Technology	Professional Series
Chapter *FAST FILE* Resources Chapter Review, pp. 35–36 Chapter Tests, pp. 37–40 **Standardized Test Practice,** pp. 23–26	MindJogger Videoquiz Virtual Labs CD-ROM Exam*View*® Pro Testmaker TeacherWorks CD-ROM Interactive Chalkboard CD-ROM	**Performance Assessment in the Science Classroom (PASC)**

chapter 5 Motion, Forces, and Simple Machines

Transparencies

Section Focus

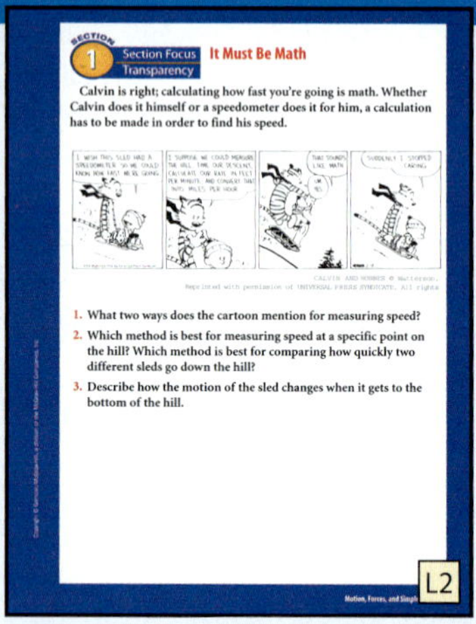

1 Section Focus Transparency **It Must Be Math**

Calvin is right; calculating how fast you're going is math. Whether Calvin does it himself or a speedometer does it for him, a calculation has to be made in order to find his speed.

CALVIN AND HOBBES © Watterson. Reprinted with permission of UNIVERSAL PRESS SYNDICATE. All rights

1. What two ways does the cartoon mention for measuring speed?
2. Which method is best for measuring speed at a specific point on the hill? Which method is best for comparing how quickly two different sleds go down the hill?
3. Describe how the motion of the sled changes when it gets to the bottom of the hill.

L2

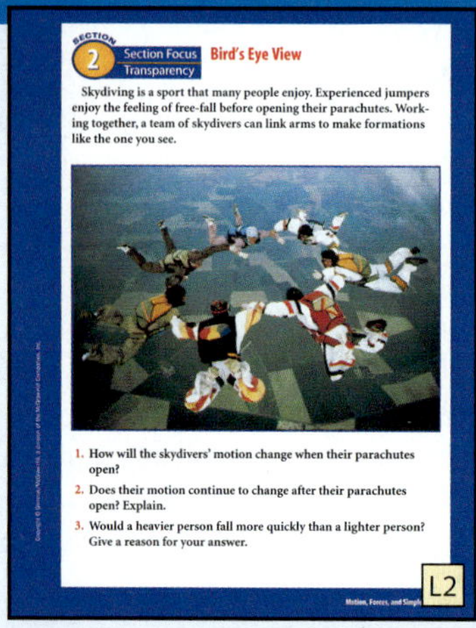

2 Section Focus Transparency **Bird's Eye View**

Skydiving is a sport that many people enjoy. Experienced jumpers enjoy the feeling of free-fall before opening their parachutes. Working together, a team of skydivers can link arms to make formations like the one you see.

1. How will the skydivers' motion change when their parachutes open?
2. Does their motion continue to change after their parachutes open? Explain.
3. Would a heavier person fall more quickly than a lighter person? Give a reason for your answer.

L2

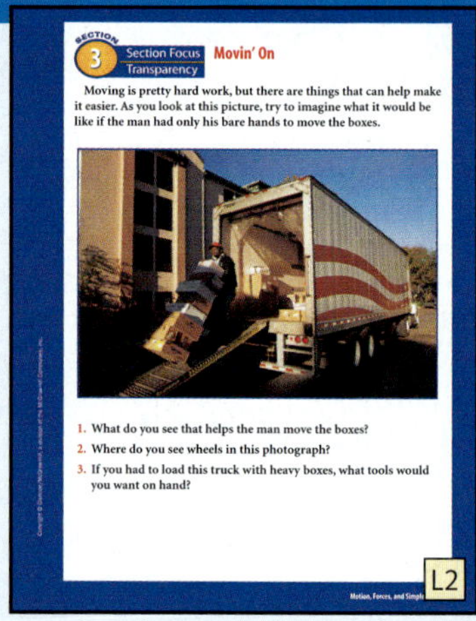

3 Section Focus Transparency **Movin' On**

Moving is pretty hard work, but there are things that can help make it easier. As you look at this picture, try to imagine what it would be like if the man had only his bare hands to move the boxes.

1. What do you see that helps the man move the boxes?
2. Where do you see wheels in this photograph?
3. If you had to load this truck with heavy boxes, what tools would you want on hand?

L2

This is a representation of key blackline masters available in the Teacher Classroom Resources. See Resource Manager boxes within the chapter for additional information.

Assessment

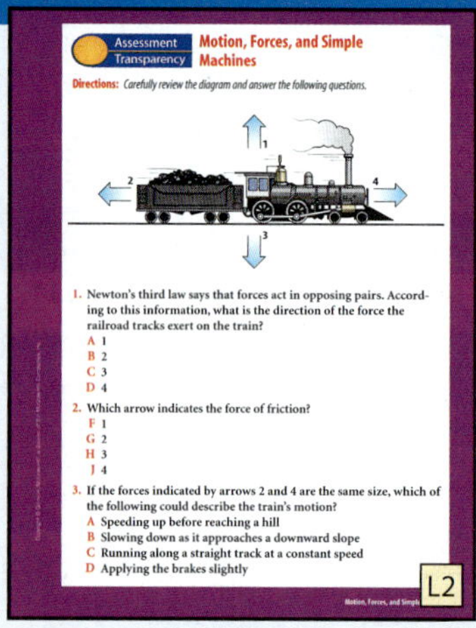

Assessment Transparency **Motion, Forces, and Simple Machines**

Directions: *Carefully review the diagram and answer the following questions.*

1. Newton's third law says that forces act in opposing pairs. According to this information, what is the direction of the force the railroad tracks exert on the train?
   A 1
   B 2
   C 3
   D 4
2. Which arrow indicates the force of friction?
   F 1
   G 2
   H 3
   J 4
3. If the forces indicated by arrows 2 and 4 are the same size, which of the following could describe the train's motion?
   A Speeding up before reaching a hill
   B Slowing down as it approaches a downward slope
   C Running along a straight track at a constant speed
   D Applying the brakes slightly

L2

Teaching

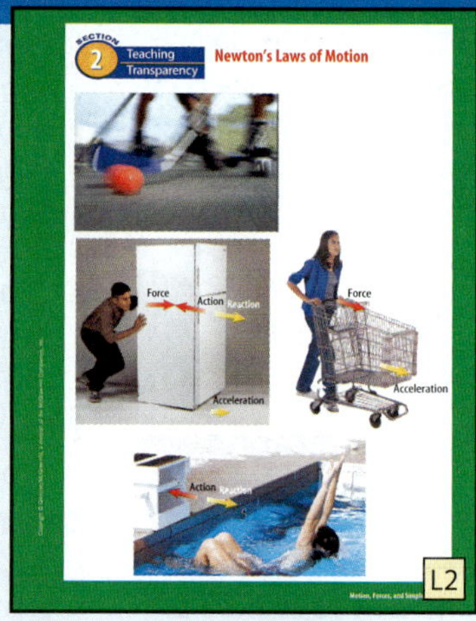

2 Teaching Transparency **Newton's Laws of Motion**

L2

Key to Teaching Strategies

The following designations will help you decide which activities are appropriate for your students.

L1 Level 1 activities should be appropriate for students with learning difficulties.

L2 Level 2 activities should be within the ability range of all students.

L3 Level 3 activities are designed for above-average students.

ELL ELL activities should be within the ability range of English Language Learners.

COOP LEARN Cooperative Learning activities are designed for small group work.

LS Multiple Learning Styles logos, as described on page 12T, are used throughout to indicate strategies that address different learning styles.

P These strategies represent student products that can be placed into a best-work portfolio.

PBL Problem-Based Learning activities apply real-world situations to learning.

Hands-on Activities

Student Text Lab Worksheet

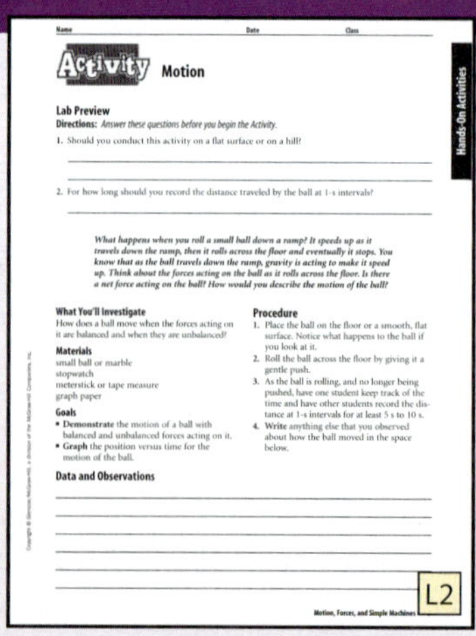

Name Date Class

Activity **Motion**

Hands-On Activities

Lab Preview

Directions: *Answer these questions before you begin the Activity.*

1. Should you conduct this activity on a flat surface or on a hill?
2. For how long should you record the distance traveled by the ball at 1-s intervals?

What happens when you roll a small ball down a ramp? It speeds up as it travels down the ramp, then it rolls across the floor and eventually it stops. You know that as the ball travels down the ramp, gravity is acting to make it speed up. Think about the forces acting on the ball as it rolls across the floor. Is there a net force acting on the ball? How would you describe the motion of the ball?

What You'll Investigate

How does a ball move when the forces acting on it are balanced and when they are unbalanced?

Materials

small ball or marble
stopwatch
meterstick or tape measure
graph paper

Goals

- **Demonstrate** the motion of a ball with balanced and unbalanced forces acting on it.
- **Graph** the position versus time for the motion of the ball.

Procedure

1. Place the ball on the floor or a smooth, flat surface. Notice what happens to the ball if you look at it.
2. Roll the ball across the floor by giving it a gentle push.
3. As the ball is rolling, and no longer being pushed, have one student keep track of the time and have other students record the distance at 1-s intervals for at least 5 s to 10 s.
4. **Write** anything else that you observed about how the ball moved in the space below.

Data and Observations

Motion, Forces, and Simple Machines L2

Laboratory Activities

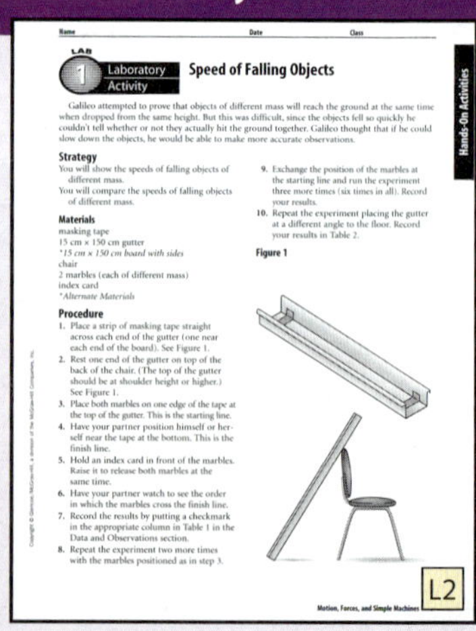

Name Date Class

LAB 1 Laboratory Activity **Speed of Falling Objects**

Hands-On Activities

Galileo attempted to prove that objects of different mass will reach the ground at the same time when dropped from the same height. But this was difficult, since the objects fell so quickly he couldn't tell whether or not they actually hit the ground together. Galileo thought that if he could slow down the objects, he would be able to make more accurate observations.

Strategy

You will show the speeds of falling objects of different mass.
You will compare the speeds of falling objects of different mass.

Materials

masking tape
15 cm × 150 cm gutter
**15 cm × 150 cm board with sides*
chair
2 marbles (each of different mass)
index card
**Alternate Materials*

Procedure

1. Place a strip of masking tape straight across each end of the gutter (one near each end of the board). See Figure 1.
2. Rest one end of the gutter on top of the back of the chair. (The top of the gutter should be at shoulder height or higher.) See Figure 1.
3. Place both marbles on one edge of the tape at the top of the gutter. This is the starting line.
4. Have your partner position himself or herself near the tape at the bottom. This is the finish line.
5. Hold an index card in front of the marbles. Raise it to release both marbles at the same time.
6. Have your partner watch to see the order in which the marbles cross the finish line.
7. Record the results by putting a checkmark in the appropriate column in Table 1 in the Data and Observations section.
8. Repeat the experiment two more times with the marbles positioned as in step 3.
9. Exchange the position of the marbles at the starting line and run the experiment three more times (six times in all). Record your results.
10. Repeat the experiment placing the gutter at a different angle to the floor. Record your results in Table 2.

Figure 1

Motion, Forces, and Simple Machines L2

Meeting Different Ability Levels

Content Outline

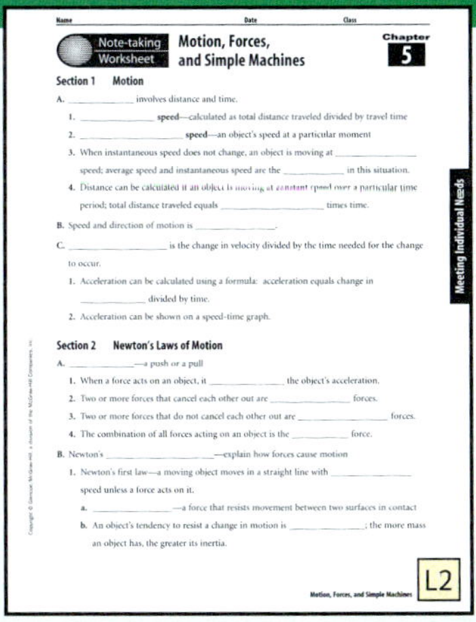

Reinforcement

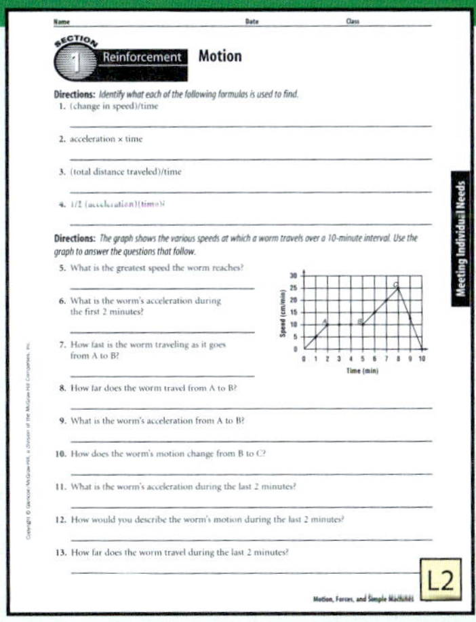

Enrichment

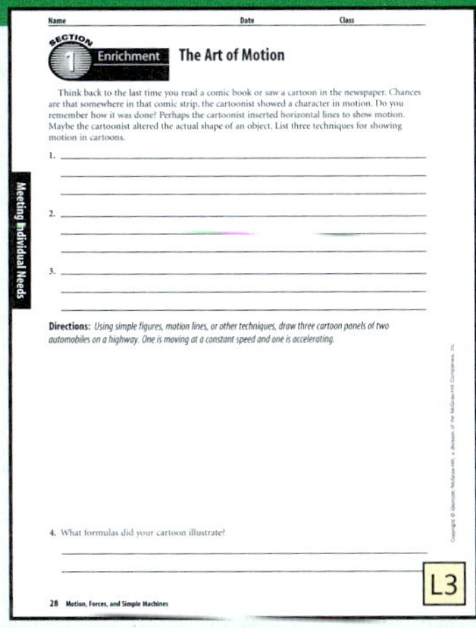

Directed Reading (English/Spanish)

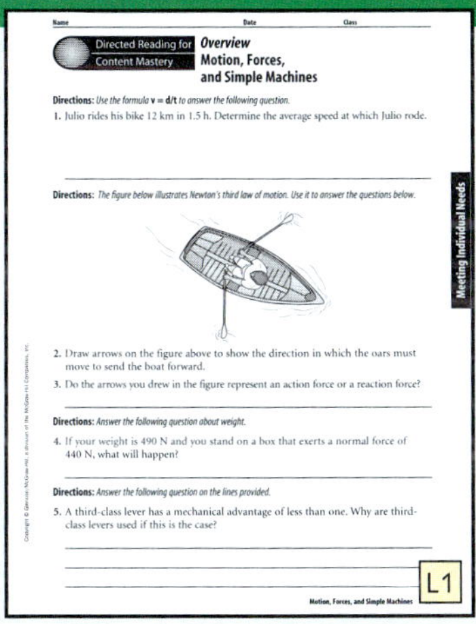

Study Guide

Reading Essentials

Assessment

Test Practice Workbook

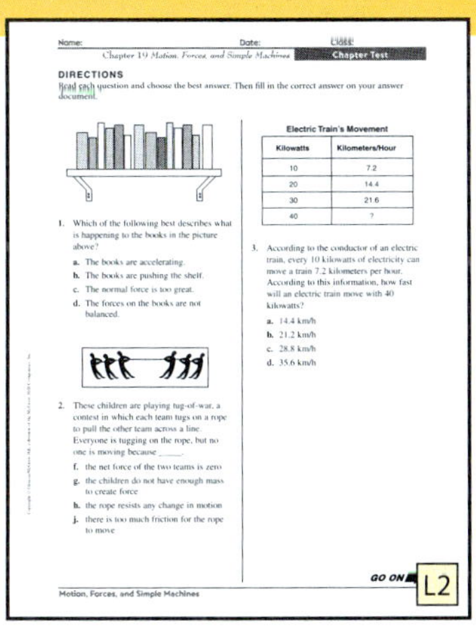

Chapter Review

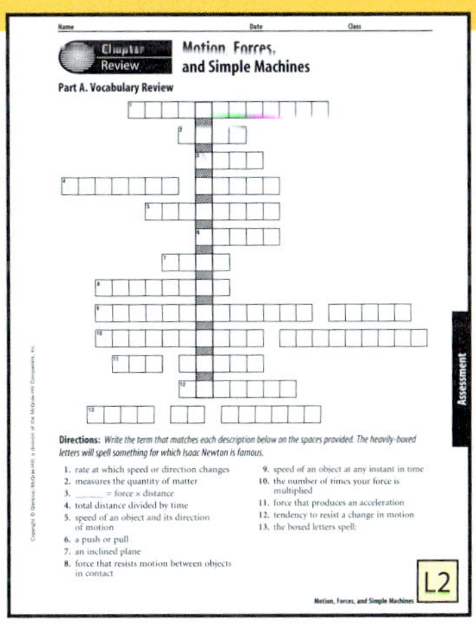

Chapter Tests

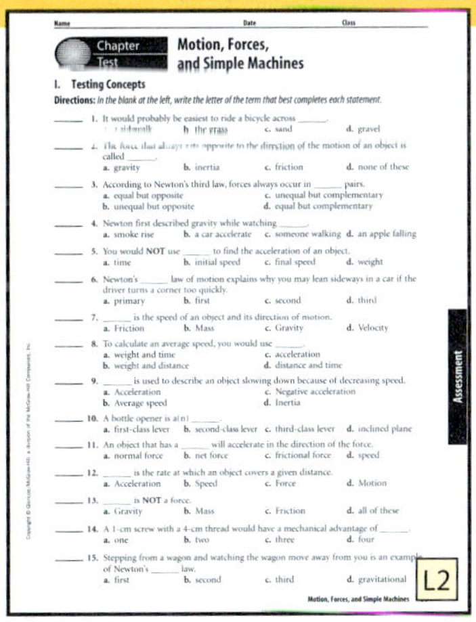

Science Content Background

section 1 Motion

Acceleration

When an object's motion changes, the object is accelerating. Acceleration is the rate of change of an object's velocity. Acceleration always has units of speed divided by time, such as meters per second divided by seconds. Because the velocity of an object has a direction, an object is accelerating when it changes its direction of motion, as well as when it speeds up or slows down.

section 2 Newton's Laws of Motion

Newton's First Law

Straight-line motion with constant speed is just as natural as rest. This motion can be approximated in a variety of ways, but it is impossible to produce a completely friction-free environment where no kinetic energy is lost. An air-hockey puck skimming across a cushion of air is an example of nearly uniform, straight-line motion.

When a ball is thrown upward, an upward force is exerted on the ball that is greater than the downward force of gravity. This upward force is exerted on the ball only while the ball is in contact with the thrower's hand.

Newton's Second Law

When an unbalanced force acts on an object, the object will accelerate in the direction of this force according to $F_{\text{net}} = ma$. If air resistance is negligible, the net force on a falling object is just the weight of the object. On the surface of the Earth, weight can be calculated using $w = mg$, where g is the gravitational constant, 9.8 m/s^2. When this expression is substituted as F_{net} in Newton's second law, the masses on the two sides of the equation cancel. This is a very profound result that bothered Newton. Are inertia and gravitational mass the same quantity? Einstein later resolved the issue when he stated his principle for equivalence, which says that an object falling in a gravitational field is identical to an object being accelerated by any other means. Therefore, the inertial and gravitational masses are the same.

section 3 Work and Simple Machines

Work

Simple machines often enable less effort to be used to do a job, such as moving a heavy object. However, machines do not reduce the amount of work needed to do a job. Even though less effort may be needed, this effort must be exerted over a longer distance.

What is a machine?

In all real machines, friction converts some of the input work into heat. As a result, the output work is always less than the input work, and no machine is 100 percent efficient.

chapter content resources

Internet Resources

For additional content background, visit **red.msscience.com** to:

- access your book online
- find references to related articles in popular science magazines
- access Web links with related content background
- access current events with science journal topics

Print Resources

Methods of Motion: An Introduction to Mechanics, by Jack E. Gartrell, Jr., NSTA, 1992

Hands-On Physics Activities with Real-Life Applications, by James Cunningham and Norman Herr, Jossey-Bass, 2002

Basic Physics: A Self-Teaching Guide, 2nd edition, by Karl F. Kuhn, John Wiley & Sons, 1996

IDENTIFYING Misconceptions

Find Out What Students Think

Students may think that . . .

If an object is moving, there is a force acting on it in the direction of the motion.

Friction makes it seem that an object will move only while a force is being applied. If a box on the floor is given a push, it slides in the direction of the push and then stops. If the box were placed on an icy surface and pushed, it would slide much farther. Without friction, the box would continue moving with constant velocity, even though no force was being exerted on it.

Activity

Give each group of students a marble. Ask students to practice rolling the marble on a flat surface until they can get it to roll about 1 m and then stop. When they are able to do this, ask why the marble stops after they have started it rolling. Have students in each group discuss this question, and then ask groups to explain their answers. Some students will explain that the marble stopped because it ran out of force.

Promote Understanding

Activity

Give each student a piece of clear plastic tubing and a marble that can roll easily inside the tubing.

- Demonstrate for students how to curl the tubing so that it forms part of a circle. Lift up one end of the tubing a little so gravity helps the marble roll down the circle.
- Ask students to each draw a diagram showing how the ball will roll after it leaves the tubing.
- Have students hold the tubing so that it forms part of a circle and then roll the marble through it. Tell students to repeat the process until they are confident of their results.
- Students should see that the marble continues in a straight line after it leaves the tubing. Ask students why the marble doesn't curve.
- Make sure they realize that the marble moved in a curve inside the tubing because the walls of the tubing exerted forces on the marble that pushed in the curve.
- Once the marble was free of the tubing, those forces were no longer acting on it, so it continued in straight-line motion. The tubing could exert forces on the marble only while the marble and tubing were in contact.

Mark M. Lawrence/The Stock Market

Assess

After completing the chapter, see *Identifying Misconceptions* in the Study Guide at the end of the chapter.

chapter 5

Chapter Vocabulary

average speed, p. 130
instantaneous speed, p. 131
velocity, p. 133
acceleration, p. 133
force, p. 136
Newton's laws of motion, p. 138
friction, p. 139
inertia, p. 139
work, p. 144
simple machine, p. 146
compound machine, p. 146
mechanical advantage, p. 146
pulley, p. 147
lever, p. 148
inclined plane, p. 149

Science Journal Accept any reasonable response, but encourage students to be detailed in their descriptions.

INTERACTIVE CHALKBOARD with Image Bank

PowerPoint® Presentations

This CD-ROM is an editable Microsoft® PowerPoint® presentation that includes:

- a pre-made presentation for every chapter
- interactive graphics
- animations
- audio clips
- image bank
- all new section and chapter questions
- Standardized Test Practice
- transparencies
- pre-lab questions for all labs
- Foldables directions
- links to red.msscience.com

chapter 5

Motion, Forces, and Simple Machines

chapter preview

sections

1 Motion
2 Newton's Laws of Motion
3 Work and Simple Machines
Lab *Motion*
Lab *Methods of Travel*
Virtual Lab *How is momentum conserved in a vehicle collision?*

Catching Some Air

This skateboarder pauses briefly in the air as he changes direction and begins his descent. How does his motion change as he reaches the bottom of the halfpipe and starts up the other side? In this chapter, you'll learn how forces affect motion.

Science Journal Write a paragraph comparing the motion of a ball and a paper airplane being thrown high in the air and returning to the ground.

128

Theme Connection

Systems and Interactions The study of the motion of macroscopic objects involves applying Newton's laws to a system of objects. The interactions between objects in this system can be described by Newton's second and third laws. If the objects are not interacting, Newton's first law applies.

About the Photo

Halfpipes Halfpipes are very popular in skateboard parks. Their sloping sides allow boarders to reach great speeds that can then carry them up the other side of the pipe, even beyond its rim. After "catching air" like this, the boarders can perform quick tricks to demonstrate their creativity and skill. How surprised they would be to know that their spectacular feats are completely governed by Newton's laws of motion!

Start-Up Activities

Launch LAB

Model Halfpipe Motion

Skateboarders who can ride halfpipes make it look easy. They race down one side and up the other. They rise above the ledge and appear to float as they spin and return. They practice these tricks many times until they get them right. In this chapter, you'll learn how this complicated motion can be explained by forces such as gravity.

1. Use heavy paper or cardboard between two stacks of books to make a U-shaped ramp to model a halfpipe like the one in the picture. A marble will model the skateboard.
2. Release the marble from a point near the bottom of the curve. Observe the motion. How high does it go? When is its speed greatest?
3. Release the marble from a point near the top of the curve. Observe the motion. Compare this to the marble's motion in step 2.
4. **Think Critically** How did the different starting points affect how high the marble rolled up the other side?

Science Online | Preview this chapter's content and activities at red.msscience.com

FOLDABLES™ Study Organizer

Describing and Explaining Motion Make the following Foldable to help you understand motion, forces, and simple machines.

STEP 1 **Fold** a vertical sheet of paper from side to side. Make the front edge about 1 cm shorter than the back edge.

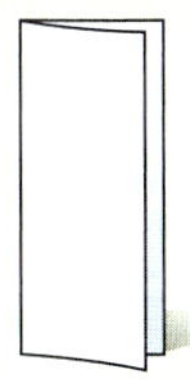

STEP 2 **Turn** lengthwise and **fold** into thirds.

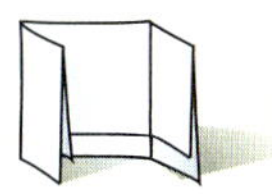

STEP 3 **Unfold and cut** only the top layer along both folds to make three tabs.

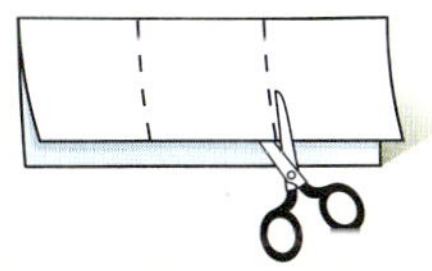

STEP 4 **Label** each tab.

Identify Questions Before you read the chapter, write what you already know about motion, forces, and simple machines under the left tab of your Foldable, and write questions about what you'd like to know under the center tab. After you read the chapter, list what you learned under the right tab.

Launch LAB

Purpose Students model the motion of an object rolling down and up a ramp. They will observe and manipulate the variables affected by gravity. L1 ELL LS **Kinesthetic**

Preparation Have different sizes of heavy paper on hand.

Materials various sizes of heavy paper or thin cardboard, marbles, tape, marking pens, books

Teaching Strategies

Suggest that students mark a scale on the paper to help measure the height the marble rises up the side of the halfpipe.

Think Critically

The farther up the side of the ramp it is released, the higher it rises on the other side. The height to which the marble rises is nearly the same as the height from which it was released.

Assessment

Performance Have students use their results to predict where they must release the marble to reach a height that you choose on the other side. Use **Performance Assessment in the Science Classroom**, p. 89.

FOLDABLES™ Study Organizer Dinah Zike Study Fold

Student preparation materials for this Foldable are available in the Chapter *FAST FILE* Resources.

section 1

Motion

as you read

What You'll Learn

- **Define** speed and acceleration.
- **Relate** acceleration to change in speed.
- **Calculate** distance, speed, and acceleration.

Why It's Important

Motion can be described using distance, time, speed, and acceleration.

Review Vocabulary

meter: SI unit of distance, abbreviated "m," equal to approximately 39.37 in

New Vocabulary

- average speed
- instantaneous speed
- velocity
- acceleration

Speed

Imagine you're a snowboarder speeding down the side of a halfpipe. Your heart pounds as you move faster. As you reach the bottom, you are going fast, and you feel excitement and maybe even fear. You flow through the change in direction as you start up the other side. Your speed decreases as you move higher up the slope. When you reach the top, you are at a near standstill. If you think fast, you can grab hold of the ledge and take a break. Otherwise, you change direction and back down you go, speeding up again as you make your way along the U-shaped ramp.

To understand how to describe even complicated motion like this, think about the simpler movement of the bicycle in **Figure 1.** To describe how fast the bicycle is traveling, you have to know two things about its motion. One is the distance it has traveled, or how far it has gone. The other is how much time it took to travel that distance.

Average Speed A bike rider can speed up and slow down several times in a certain time period. One way to describe the bike rider's motion over this time period is to give the average speed. To calculate **average speed,** divide the distance traveled by the time it takes to travel that distance.

Speed Equation

$$\textbf{average speed} \text{ (in m/s)} = \frac{\textbf{distance traveled} \text{ (in m)}}{\textbf{travel time} \text{ (in s)}}$$

$$s = \frac{d}{t}$$

Because average speed is calculated by dividing distance by time, its units always will be a distance unit divided by a time unit. For example, the average speed of a bicycle is usually given in meters per second. The speed of a car usually has units of kilometers per hour instead.

Figure 1 To find the biker's average speed, divide the distance traveled down the hill by the time taken to cover that distance.
Infer *what would happen to the average speed if the hill were steeper.*

1 Motivate

INTERACTIVE CHALKBOARD PowerPoint® Presentations

Bellringer

Section Focus Transparencies also are available on the Interactive Chalkboard CD-ROM. L2 ELL

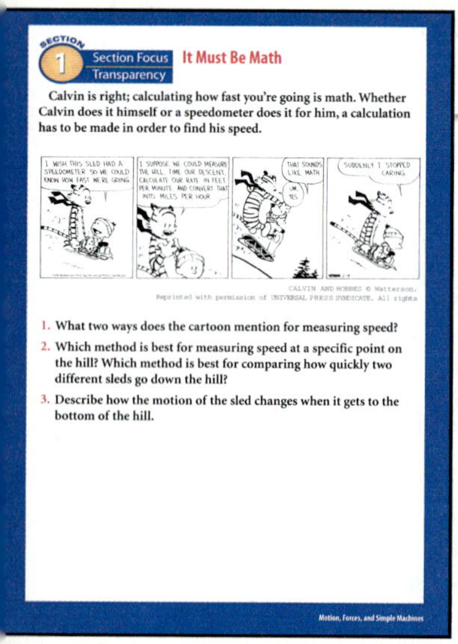

Section 1 Section Focus Transparency **It Must Be Math**

Calvin is right; calculating how fast you're going is math. Whether Calvin does it himself or a speedometer does it for him, a calculation has to be made in order to find his speed.

1. **What two ways does the cartoon mention for measuring speed?**
2. **Which method is best for measuring speed at a specific point on the hill? Which method is best for comparing how quickly two different sleds go down the hill?**
3. **Describe how the motion of the sled changes when it gets to the bottom of the hill.**

Motion, Forces, and Simple Machines

Tie to Prior Knowledge

Fastest Speed Ask students what they think their fastest speed is on their bike or skateboard. A typical bike speed is about 4 m/s, and a typical skateboard speed is about 3 m/s. Tell them that in this section they will learn different ways of describing speed.

2 Teach

Caption Answer

Figure 1 It would increase.

Section 1 Resource Manager

Chapter *FAST FILE* Resources

Transparency Activity, p. 42
Note-taking Worksheets, pp. 31–33
Enrichment, p. 28
Lab Activity, pp. 9–11
Reinforcement, p. 25
Directed Reading for Content Mastery, pp. 17, 18

Mathematics Skill Activities, p. 9
Reading and Writing Skill Activities, p. 35
Performance Assessment in the Science Classroom, p. 35

Applying Math — Solve a Simple Equation

BICYCLE SPEED Riding your bike, it takes you 30 min to get to your friend's house, which is 9 km away. What is your average speed?

Solution

1. *This is what you know:*
   - distance: $d = 9$ km
   - time: $t = 30$ min $= 0.5$ h
2. *This is what you need to find:* speed: $s = ?$ m/s
3. *This is the procedure you need to use:* Substitute the known values for distance and time into the speed equation and calculate the speed:
   $s = \frac{d}{t}, \frac{9\text{ km}}{0.5\text{ h}} = 18$ km/h
4. *Check your answer:* Multiply your answer by the time. You should calculate the distance that was given.

Practice Problems

1. If an airplane travels 1,350 km in 3 h, what is its average speed?
2. Determine the average speed, in km/h, of a runner who finishes a 5-km race in 18 min.

Science Online — For more practice, visit red.msscience.com/math_practice

Instantaneous Speed Average speed is useful if you don't care about the details of the motion. For example, suppose you went on a long road trip and traveled 640 km in 8 h. Your average speed was 80 km/h, even though you might have been stuck in a traffic jam for some of the time.

When your motion is speeding up and slowing down, it might be useful to know how fast you are going at a certain time. For example, suppose the speed limit on a part of the above trip was 50 km/h. Does your average speed of 80 km/h mean you were speeding during that part of the trip?

To keep from exceeding the speed limit, the driver would need to know the **instantaneous speed**—the speed of an object at any instant of time. When you ride in a car, the instantaneous speed is given by the speedometer, as shown in **Figure 2.** How does your instantaneous speed change as you coast on a bicycle down one hill and then up another one?

Reading Check *How is instantaneous speed different from average speed?*

Figure 2 The odometer in a car measures the distance traveled. The speedometer measures instantaneous speed.
Describe *how you could use this speedometer to measure average speed.*

Applying Math

National Math Standards

Correlation to Mathematics Objectives
1, 2, 9

Answers to Practice Problems

1. speed $= \frac{1350\text{ km}}{3\text{ h}} = 450$ km/h
2. speed $= \frac{5\text{ km}}{0.3\text{ h}} = 1.7$ km/h

Caption Answer

Figure 2 Measure the distance traveled using the odometer and divide that distance by the time needed to travel the distance.

Reading Check

Answer Instantaneous speed is speed at any instant; average speed is speed over a time interval.

Fun Fact

The world's fastest train is in Japan. It has reached a speed of 552 km/h (342 miles per hour).

Differentiated Instruction

Challenge Position and speed are always measured relative to a reference point. Have students find out the average speed of a student sitting in a chair when the reference point is the Sun. Earth travels through the solar system around the Sun, a distance of about 940,100,000 km, in one year, which is equal to about 365 days, or about 8,766 hours. Earth's average speed, and therefore the average speed of the student sitting at the desk, is about 107,200 km/h. L3 LS **Logical-Mathematical**

INTEGRATE Earth Science

Movement of Earth's Crust The rate of plate motion varies widely. The slowest movement is along the Arctic Ridge, which moves less than 2.5 cm/yr. The fastest movement is at the East Pacific Rise, west of South America. Here the Pacific Plate and the Nazca Plate move about 15 cm/yr relative to one another. Occasionally a plate will have a sudden movement much faster than its normal rate. For example, the Pacific Plate normally moves northwest relative to the North American Plate at a rate of about 5 cm per year. In 1906, however, the Pacific Plate suddenly moved about 6 m, resulting in the famous San Francisco earthquake.

Research Have students research the natural phenomena that occur because of the movements of the plates. Have students find the plate outlines on a world map and relate this to natural phenomena that occur in these areas.

Applying Math

National Math Standards

Correlation to Mathematics Objectives

1, 2, 9

Answers to Practice Problems

1. distance = 5 km/h × 6 h = 30 km
2. distance = 830 km/h × 6 h = 5,000 km

INTEGRATE Earth Science

Movement of Earth's Crust The outer part of Earth is the crust. Earth's crust is broken into huge pieces called plates that move slowly. Research how fast plates can move. In your Science Journal, make a table showing the speeds of some plates.

Constant Speed Sometimes an object is moving such that its instantaneous speed doesn't change. When the instantaneous speed doesn't change, an object is moving with constant speed. Then the average speed and the instantaneous speed are the same.

Calculating Distance If an object is moving with constant speed, then the distance it travels over any period of time can be calculated using the equation for average speed. When both sides of this equation are multiplied by the time, you have the following new equation.

Distance Equation

distance traveled (in m) = **average speed** (in m/s) × **time** (in s)

$$d = st$$

Notice that the units of time in the speed, s, and in the time, t, have to be the same. Otherwise, these units of time won't cancel.

Applying Math — Solve a Simple Equation

FAMILY TRIP DISTANCE It takes your family 2 h to drive to an amusement park at an average speed of 73 km/h. How far away is the amusement park?

Solution

1. *This is what you know:*
   - speed: s = 73 km/h
   - time: t = 2 h
2. *This is what you need to know:* distance: d = ? m
3. *This is the procedure you need to use:* Substitute the known values for speed and time into the distance equation and calculate the distance: $d = st = (73\text{ km/h})(2\text{ h}) = 146\text{ km}$
4. *Check your answer:* Divide your answer by the time. You should get the speed that was given.

Practice Problems

1. You and your friends walk at an average speed of 5 km/h on a nature hike. After 6 h, you reach the ranger station. How far did you hike?
2. An airplane flying from Boston to San Francisco traveled at an average speed of 830 km/h for 6 h. What distance did it fly?

Science Online — For more practice, visit red.msscience.com/math_practice

Science Journal

Speeds Have students measure the speed of their movements several times in a day. Ask them to write in their Science Journals at least one example each of a constant speed, an instantaneous speed, and an average speed at which they moved that day. L2 LS **Naturalist**

Active Reading

Reflective Journal Have students divide sheets of paper into several columns and record their thoughts under headings such as *What I did*, *What I learned*, *What questions I have*, *What surprises I experienced*, and *Overall response*. Have students write a Reflective Journal for the Applying Math activities in this section.

Velocity

Suppose you are walking at a constant speed on a street, headed north. You turn when you reach an intersection and start walking at the same speed, but you now are headed east, as shown in **Figure 3.** Your motion has changed, even though your speed has remained constant. To completely describe your movement, you would have to tell not only how fast you were moving, but also your direction. The **velocity** of an object is the speed of the object and its direction of motion.

Velocity changes when the speed changes, the direction of motion changes, or both change. When you turned the corner at the intersection, your direction of motion changed, even though your speed remained constant. Therefore, your velocity changed.

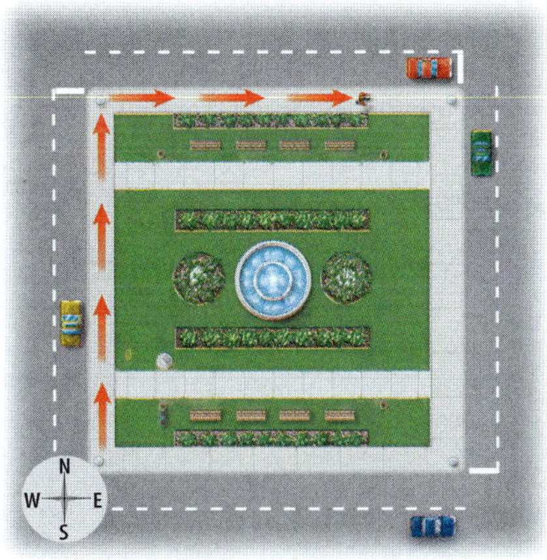

Figure 3 If you are walking north at a constant speed and then turn east, continuing at the same speed, you have changed your velocity.
Identify *another way to change your velocity.*

Acceleration

At the top of a skateboard halfpipe, you are at rest. Your speed is zero. When you start down, you smoothly speed up, going faster and faster. If the angle of the halfpipe were steeper, you would speed up at an even greater rate.

How could you describe how your speed is changing? If you changed direction, how could you describe how your velocity was changing? Just as speed describes how the distance traveled changes with time, acceleration describes how the velocity changes with time. **Acceleration** is the change in velocity divided by the time needed for the change to occur. **Figure 4** shows some examples of acceleration when the speed changes but the direction of motion stays the same.

Reading Check *The motion of an object can change in what two ways when it accelerates?*

Figure 4 If the speed but not direction of an object is changing, the object is accelerating. The direction of the acceleration depends on whether the object is speeding up or slowing down.

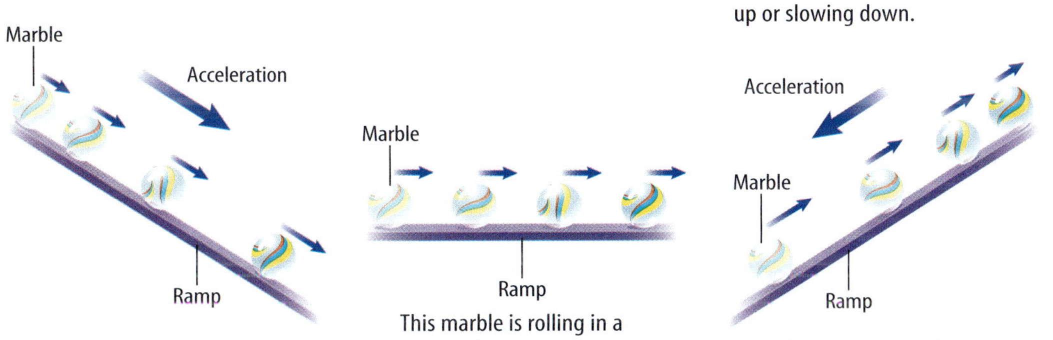

A marble rolling in a straight line down a hill speeds up. Its motion and acceleration are in the same direction.

This marble is rolling in a straight line on a level surface with constant velocity. Its acceleration is zero.

A marble rolling in a straight line up a hill slows down. Its motion and acceleration are in opposite directions.

Activity

Truck Math Give students the following scenario. At 5:00 A.M., a newspaper truck leaves the printer's office with two loads of newspapers to deliver. It travels on a highway at 80 km/h for 30 min. Then it travels on side roads for 20 min at 15 km/h, reaching its first delivery at 5:50 A.M. It takes the driver 10 min to deliver the first load of papers. Then the truck travels 15 min at 20 km/h to reach its second delivery. There it takes the driver 15 min to unload the papers. At 6:30 A.M., the empty truck heads back to the printer. What was its instantaneous speed at 5:10 A.M.? 80 km/h How far did the truck travel to deliver its two loads? Use the equation $d = s \times t$ for each leg of the trip. Total distance = 80 km/h × 1/2 h + 15 km/h × 1/3 h + 0 km/h × 1/6 h + 20 km/h × 1/4 h = 40 + 5 + 0 + 5 km = 50 km What was the truck's average speed? 50 km/1.5 h = 33.33 km/h (about 33 km/h)

Caption Answer

Figure 3 increase or decrease your speed

Reading Check

Answer changing speed or changing direction

Differentiated Instruction

English-Language Learners Help these students by assigning each one a partner to provide support in understanding new concepts. Encourage partners to demonstrate or illustrate ideas to add meaning to verbal explanations.

Visual Learning

Figure 4 The marble's position is shown at fixed time intervals. A larger separation indicates a higher speed. Because of gravity, the marble going downhill has a steady increase in velocity, so the acceleration is constant. The marble going uphill has a steady decrease in velocity, which indicates a constant deceleration. Have students sketch simple velocity-time and acceleration-time graphs for these marbles. L2 LS **Visual-Spatial**

SECTION 1 Motion 133

Applying Math

National Math Standards

Correlation to Mathematics Objectives

1, 2, 9

Answers to Practice Problems

1. $a = \frac{25\text{ m/s} - 10\text{ m/s}}{5\text{s}} = \frac{15\text{ m/s}}{5\text{ s}} = 3\text{ m/s}^2$
2. $a = \frac{0\text{ m/s} - 16\text{ m/s}}{9\text{ s}} = \frac{-16\text{ m/s}}{9\text{s}} = -1.67\text{ m/s}^2$ (about -1.7 m/s^2)

Discussion

Gravity Tell students that gravity gives the same acceleration to all objects falling near Earth's surface. Have students discuss why this fact might have been difficult to discover and why some people find it hard to believe.

Quick Demo

Instantaneous Speed

Materials two toy cars, several lengths of track, blocks or books

Estimated Time 15 minutes

Procedure To demonstrate instantaneous speed, set up one track flat along the floor. Using blocks or books, set up the other track with two small hills in the middle. Push the two cars along the tracks with the same force. Point out that the car on the flat track moves at constant speed, whereas the other car slows down going up the hills and speeds up going down the hills.

Calculating Acceleration If an object changes speed but not direction then its acceleration can be calculated from the following formula.

Acceleration Equation

$$\textbf{acceleration}\text{ (in m/s}^2\text{)} = \frac{\textbf{final speed}\text{ (in m/s)} - \textbf{initial speed}\text{ (in m/s)}}{\textbf{time}\text{ (in s)}}$$

$$a = \frac{(s_f - s_i)}{t}$$

The SI units for acceleration are m/s^2, which means meters/(seconds × seconds). The units m/s^2 are the result when the units m/s are divided by the unit s.

Applying Math — Calculate Acceleration

ACCELERATION DOWN A HILL You are sliding on a snow-covered hill at a speed of 8 m/s. There is a drop that increases your speed to 18 m/s in 5 s. Find your acceleration.

Solution

1. *This is what you know:*
   - initial speed: $s_i = 8$ m/s
   - final speed: $s_f = 18$ m/s
   - time: $t = 5$ s
2. *This is what you need to know:* acceleration: $a = ?$ m/s^2
3. *This is the procedure you need to use:* Substitute the known values for initial speed, final speed, and time into the acceleration equation

   $$a = \frac{(s_f - s_i)}{t} = \frac{18\text{ m/s} - 8\text{ m/s}}{5\text{ s}} = \frac{10\text{ m/s}}{5\text{ s}} = 2\text{ m/s}^2$$
4. *Check your answer:* Multiply your answer by the time. Add the initial speed. You should get the final speed that was given.

Practice Problems

1. The roller coaster you are on is moving at 10 m/s. 5 s later it does a loop-the-loop and is moving at 25 m/s. What is the roller coaster's acceleration over this time?
2. A car you're riding in is slowing down for a stoplight. It was initially traveling at 16 m/s and comes to a stop in 9 s. What is the car's acceleration?

Science Online — For more practice, visit red.msscience.com/math_practice

Science Journal

Sports Science Have students write descriptions in their Science Journals of how speed, velocity, and acceleration are important in several sports. Descriptions might include the velocity (speed and direction) with which a bowling ball hits the pins or how a skier accelerates while traveling down a snowy slope. L2 LS **Linguistic**

Teacher FYI

Acceleration Acceleration describes how speed changes with time, just as speed describes how distance changes with time. The slope of a speed-time graph gives instantaneous acceleration, just as the slope of a distance-time graph gives instantaneous speed.

Graphing Speed Picture yourself skating down the side of a hill, across a level valley, and then up another hill on the opposite side. If you were to graph your speed over time, it would look similar to the graph in **Figure 5.**

As you start down the hill, your speed will increase with time, as shown in segment A. The line on the graph rises when acceleration is in the direction of motion. When you travel across the level pavement, you move at a constant speed. Because your speed doesn't change, the line on the graph is horizontal, as shown in segment B. A horizontal line shows that the acceleration is zero. On the opposite side, when you are moving up the hill, your speed decreases, as shown in segment C. Anytime you slow down, acceleration is opposite the direction of motion, and the line on a speed-time graph will slant downward.

Figure 5 The acceleration of an object can be shown on a speed-time graph.

section 1 review

Summary

Speed and Velocity

- Average speed is the distance traveled divided by the travel time:

$$s = \frac{d}{t}$$

- The velocity of an object is the speed of the object and the direction of motion.

Acceleration

- Acceleration is the change in the velocity divided by the time for the change to occur.
- For motion in a straight line acceleration can be calculated from this equation:

$$a = \frac{(s_f - s_i)}{t}$$

- The slope of a line on a speed-time graph shows an object's acceleration. The line slopes upward if the object is speeding up, and slopes downward if it is slowing down.

Self Check

1. **Explain** If an airplane is flying at a constant speed of 500 km/h, can it be accelerating?
2. **Infer** whether the instantaneous speed of an object can be greater than its average speed.
3. **Determine** If your speed is constant, can your velocity be changing?
4. **Think Critically** Describe the motion of a skateboard as it accelerates down one side of a halfpipe and up the other side. What would happen if the up side of the pipe were not as steep as the down side?

Applying Math

5. **Average Speed** During rush-hour traffic in a big city, it can take 1.5 h to travel 45 km. What is the average speed in km/h for this trip?
6. **Compare** the distances traveled and average speeds of the following two people: Sam walked 1.5 m/s for 30 s and Jill walked 2.0 m/s for 15 s and then 1.0 m/s for 15 s.

Science Online red.msscience.com/self_check_quiz

3 Assess

DAILY INTERVENTION

Check for Understanding

Interpersonal Have students discuss the concepts of speed, velocity, and acceleration in relation to various sports, such as soccer, baseball, or swimming. If appropriate, you also can have them relate the concepts of instantaneous speed and average speed to the sports. L2 LS

Reteach

Graphing Motion Have students draw a speed-time graph for the following scenario (with the time axis along the longer side of the page): A car traveling at 16 m/s takes 3 s to slow to 9 m/s to pass through a school zone. It reaches the end of the school zone in 25 s and then increases its speed back to 16 m/s in 3 s. After traveling at this speed for 20 s, the car comes to a stop at a stoplight in 5 s.

Then, have them calculate the acceleration for each segment of the trip. $a = \frac{9\text{ m/s} - 16\text{ m/s}}{3\text{ s}} = -2.3\text{ m/s}^2$; 0 m/s^2; $a = \frac{16\text{ m/s} - 9\text{ m/s}}{3\text{ s}} = +2.3\text{ m/s}^2$; 0 m/s^2; $a = \frac{0\text{ m/s} - 16\text{ m/s}}{5\text{ s}} = -3.2\text{ m/s}^2$

Assessment

Process Roll a smooth ball, such as a racquetball, up a ramp. Ask students to predict what will happen to the ball's speed and acceleration when it reaches its farthest point up the ramp. Also have them infer what will happen to the position, speed, and acceleration if the ramp angle is increased. Use **Performance Assessment in the Science Classroom,** p. 89.

section 1 review

1. Yes, it could be changing direction.
2. Instantaneous speed can be greater than average speed for part of the trip as long as it is lower for another part of the trip.
3. Yes, because direction could change.
4. The skateboard speeds up as it goes down the ramp and slows down as it rises up the other side. If the two sides are symmetrical, the skateboard will speed up and slow down with about the same acceleration. If the up side is not as steep, the skateboard slows down at a lower rate and travels a greater distance.
5. s = (45 km)/(1.5 h) = 30 km/h
6. Sam: distance = 1.5 m/s × 30 s = 45 m; avg speed = 1.5 m/s (constant); Jill: distance = 2.0 m/s × 15 s + 1.0 m/s × 15 s = 30 + 15 m = 45 m; avg speed = $\frac{45\text{ m}}{15\text{ s} + 15\text{ s}} = \frac{45\text{ m}}{30\text{ s}} = 1.5\text{ m/s}$ Sam and Jill traveled the same distance and had the same average speed, even though their instantaneous speeds were different.

1 Motivate

Bellringer

INTERACTIVE CHALKBOARD PowerPoint® Presentations

Section Focus Transparencies also are available on the Interactive Chalkboard CD-ROM. L2 ELL

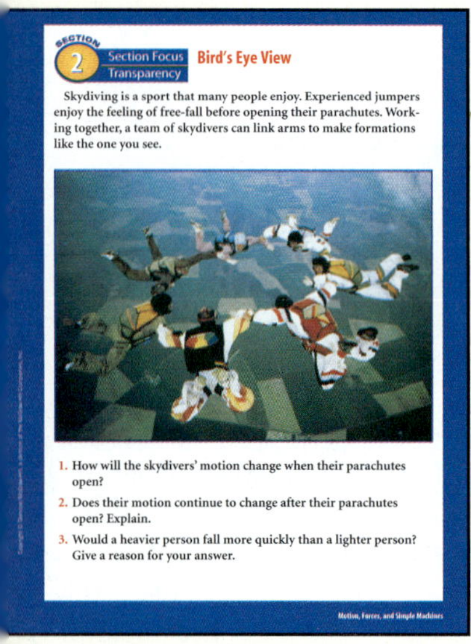
Section 2 Section Focus Transparency **Bird's Eye View**

Skydiving is a sport that many people enjoy. Experienced jumpers enjoy the feeling of free-fall before opening their parachutes. Working together, a team of skydivers can link arms to make formations like the one you see.

1. How will the skydivers' motion change when their parachutes open?
2. Does their motion continue to change after their parachutes open? Explain.
3. Would a heavier person fall more quickly than a lighter person? Give a reason for your answer.

Motion, Forces, and Simple Machines

Tie to Prior Knowledge

Swinging Doors When two people push on opposite sides of a swinging door, the door swings away from the person who pushes harder and stays still if they push with the same force. In this section, students will learn how force affects motion.

2 Teach

Caption Answer

Figure 6 The change in direction shows the ball's acceleration.

section 2

Newton's Laws of Motion

as you read

What You'll Learn

- **Describe** how forces affect motion.
- **Calculate** acceleration using Newton's second law of motion.
- **Explain** Newton's third law of motion.

Why It's Important

Newton's laws explain motions as simple as walking and as complicated as a rocket's launch.

Review Vocabulary

gravity: attractive force between any two objects that depends on the masses of the objects and the distance between them

New Vocabulary

- force
- Newton's laws of motion
- friction
- inertia

Force

What causes objects to move? In the lunchroom, you pull a chair away from a table before you sit down and push it back under the table when you leave. You exert a force on the chair and cause it to move. A **force** is a push or a pull. In SI units, force is measured in newtons (N). One newton is about the amount of force it takes to lift a quarter-pound hamburger.

Force and Acceleration For an object's motion to change, a force must be applied to the object. This force causes the object to accelerate. For example, when you throw a ball, as in **Figure 6,** your hand exerts a force on the ball, causing it to speed up. The ball has acceleration because the speed of the ball has increased.

A force also can change the direction of an object's motion. After the ball leaves your hand, if no one catches it, its path curves downward, and it hits the ground. Gravity pulls the ball downward and causes it to change direction, as shown in **Figure 6.** Recall that an object has acceleration when its direction of motion changes. The force of gravity has caused the ball to accelerate. Anytime an object's speed, or direction of motion, or both change, a force must have acted on the object.

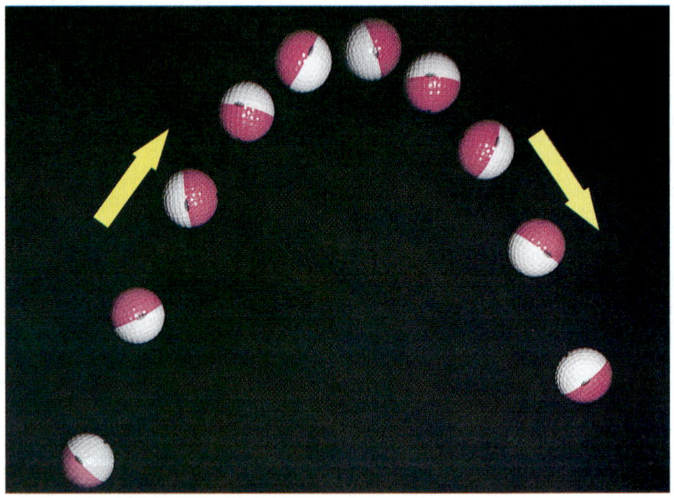

Figure 6 After a golf ball is thrown, it follows a curved path toward the ground.
Explain *how this curved path shows that the ball is accelerating.*

Section 2 Resource Manager

Chapter *FAST FILE* Resources

- Transparency Activity, pp. 43, 45–46
- Directed Reading for Content Mastery, p. 18
- MiniLAB, p. 3
- Enrichment, p. 29
- Lab Activity, pp. 13–14
- Reinforcement, p. 26

Physical Science Critical Thinking/Problem Solving, p. 3

Cultural Diversity, p. 63

Performance Assessment in the Science Classroom, p. 36

136 CHAPTER 5 Motion, Forces, and Simple Machines

When two forces act in the same direction on an object, like a box, the net force is equal to the sum of the two forces.

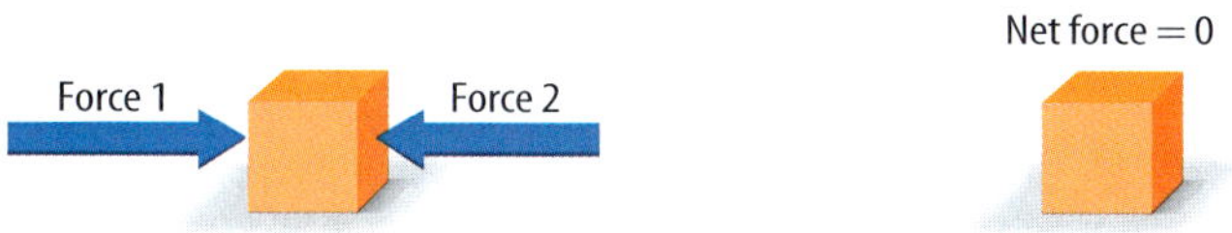

If two forces of equal strength act on the box in opposite directions, the forces will cancel, resulting in a net force of zero.

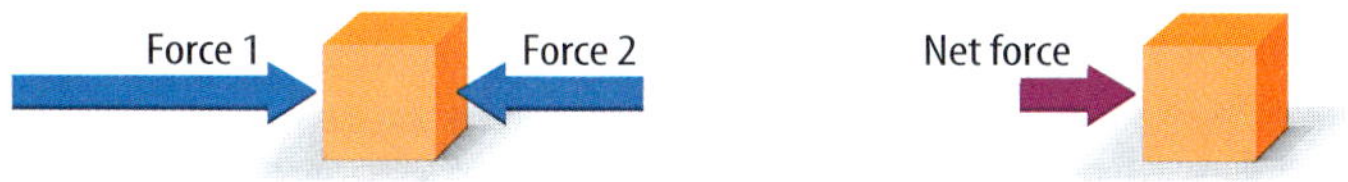

When two unequal forces act in opposite directions on the box, the net force is the difference of the two forces.

Figure 7 When more than one force acts on an object, the forces combine to form a net force.

Balanced and Unbalanced Forces More than one force can act on an object without causing its motion to change. If both you and your friend push on a door with the same force in opposite directions, the door doesn't move. Two or more forces are balanced forces if their effects cancel each other and they do not cause a change in an object's motion. If the effects of the forces don't cancel each other, the forces are unbalanced forces.

Combining Forces Suppose you push on a door to open it. At the same time, someone on the other side of the door also is pushing. What is the motion of the door? When more than one force acts on an object, the forces combine. The combination of all the forces acting on an object is the net force.

How do forces combine to form the net force? If the forces are in the same direction, they add together to form the net force. If two forces are in opposite directions, the net force is the difference between the two forces and is in the direction of the larger force. **Figure 7** shows some examples of how forces combine to form the net force. If you push on a door with a larger force than the person on the other side pushes, the door moves in the direction of your push. If you push with the same force as the other person, the two forces cancel, and the net force is zero. In this case, the door doesn't move.

INTEGRATE Life Science

Force and Sood Germination For a fragile seedling to grow, it must exert enough force to push through the soil above it. The force exerted by the seedling as it pushes its way through the soil is due to the water pressure created inside its cells. New cells form as the seedling begins to grow underground. These cells take up water and expand, exerting a pressure that can be 20 times greater than atmospheric pressure. Research some of the factors that can affect how seedlings germinate. Write a paragraph in your Science Journal summarizing what you learned.

Quick Demo

Net Forces

Materials sturdy box

Estimated Time 15 minutes

Procedure Have a volunteer work with you to move a box, as shown in **Figure 7.** Have students describe the motion that results in each scenario. First, both of you push in the same direction. Then, both of you push on the box from opposite sides using equal strength. Finally, have the volunteer push on his or her side of the box harder than you push on your side of the box. Have the volunteer describe how the application of force felt in each instance.

Activity

Activity Forces Have students choose several activities and describe all of the forces that cause motion in them. Activities might include swinging, ice skating, or swimming. For horizontal motion, be sure students remember to include gravity and the normal force, even though they cancel each other. L2 LS **Linguistic**

INTEGRATE Life Science

Force and Seed Germination Students may describe the effects of absorption of water, passage of time, chilling, warming, oxygen availability, or light exposure.

Active Reading

Cause and Effect Chart Have students label a sheet of paper "Combining Forces." To the left, have them write how forces can combine. To the right, have them write the effects of these forces.

Differentiated Instruction

Hearing Impaired Help these students by assigning each one a partner to provide support in understanding new concepts. Encourage partners to demonstrate or illustrate ideas, such as combining forces, to add meaning to verbal explanations.

SECTION 2 Newton's Laws of Motion 137

Mini LAB

Purpose Students measure weights in pounds and convert them to newtons. L2 ELL COOP LEARN IS **Kinesthetic**

Materials bathroom scale, large book, chair, heavy coat, and another object of student's choice

Teaching Strategy Have students place the scale on a hard floor instead of carpet.

Analysis

1. Answers will vary.
2. Answers will vary.
3. Answers will vary.

Assessment

Process Show students a picture of a student next to an animal. Ask students to estimate the weight of the animal in newtons. Answers will vary depending on the size of the animal in the picture. Use **Performance Assessment in the Science Classroom,** p. 89.

Try at Home

Virtual Labs

Crash! *How is momentum conserved in a vehicle collision?*

Mini LAB

Determining Weights in Newtons

Procedure:

1. Stand on a **bathroom scale** and measure your weight.
2. Hold a **large book,** stand on the scale, and measure the combined weight of you and the book.
3. Repeat step 2 using a **chair, heavy coat,** and a **fourth object** of your choice.

Analysis

1. Subtract your weight from each of the combined weights to calculate the weight of each object in pounds.
2. Multiply the weight of each object in pounds by 4.4 to calculate its weight in newtons.
3. Calculate your own weight in newtons.

Try at Home

Newton's Laws of Motion

In 1665, Sir Isaac Newton was in college in London. The school temporarily closed down, though, because a deadly plague was spreading rapidly across Europe. Newton, who was 23 years old, returned to his house in the country to wait for the plague to end. During this time, he spent his days observing nature and performing simple experiments. As a result, he made many discoveries, including how to explain the effects of gravity. One of his great discoveries was how forces cause motion. He realized that he could explain the motion of objects using a set of principles, which in time came to be called **Newton's laws of motion.**

Newton's First Law

When you give a book on a table a push, it slides and comes to a stop. After you throw or hit a baseball, it soon hits the ground and rolls to a stop. In fact, it seems that anytime you set something in motion, it stops moving after awhile. You might conclude that to keep an object moving, a net force must always be exerted on the object. In reality, that's not true.

Newton and a few others before him realized that an object could be moving even if no net force was acting on it. Newton's first law of motion states that an object will not change its motion unless an unbalanced force acts on it. Therefore, an object that is not moving, like a book sitting on a table, remains at rest until something pushes or pulls it.

What if an object is already moving, like a ball you've just thrown to someone? Newton's first law says the motion of the ball won't change unless an unbalanced force is exerted on it. This means that after the ball is in motion, a force has to be applied to make it speed up, slow down, or change direction. In other words, a moving object, like the ball in **Figure 8,** moves in a straight line with constant speed unless an unbalanced force acts on it.

Figure 8 After the ball has been hit, it will move along the ground in a straight line until it is acted on by another force.

Curriculum Connection

History Newton published the results of his scientific studies in one of the most famous books of all time. Have students find the name of the book and the year it was published. *Philosophiae Naturalis Principia Mathematica* (The Mathematical Principles of Natural Philosophy) published in 1687. L3 IS **Linguistic**

Differentiated Instruction

Challenge During earthquakes, buildings tend to remain at rest as the ground underneath them moves, which can cause them to collapse. In many large cities, buildings are now built to better withstand earthquakes. Have students research ways this is done and write a report on their findings. They can be made of stronger materials, built on rubber pads, or designed to sway during an earthquake. L3 IS **Linguistic** P

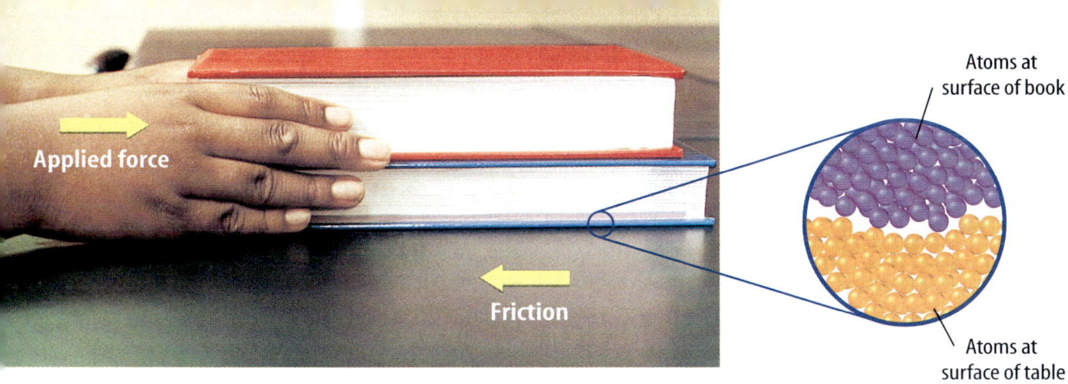

Figure 9 Friction is caused by the roughness of the surfaces in contact. The enlargement shows how the table and book surfaces might look if you could see their atoms.

Friction Newton's first law states that a moving object should never slow down or change direction until a force is exerted on it. Can you think of any moving objects that never slow down or change direction? A book slides across a table, slows down, and comes to a stop. Because its motion changes, a force must be acting on it and causing it to stop. This force is called friction. **Friction** is a force between two surfaces in contact that resists the motion of the surfaces past each other. It always acts opposite to the direction of motion, as shown in **Figure 9.** To keep an object moving when friction is acting on it, you have to keep pushing or pulling on the object to overcome the frictional force.

Reading Check *In what direction is the force of friction exerted?*

The size of the friction force depends on the two surfaces involved. In general, the rougher the surfaces are, the greater the friction will be. For example, if you push a hockey puck on an ice rink, it will go a great distance before it stops. If you try to push it with the same force on a smooth floor, it won't slide as far. If you push the puck on a rough carpet, it will barely move.

Inertia and Mass You might have noticed how hard it is to move a heavy object, such as a refrigerator, even when it has wheels. If you try pushing someone who is much bigger than you are—even someone who is wearing skates or standing on a skateboard—that person won't budge easily. It's easier to push someone who is smaller. You also might have noticed that it is hard to stop someone who is much bigger than you are when that person is moving. In each case, including the one shown in **Figure 10,** the object resists having its motion changed. This tendency to resist a change in motion is called **inertia.**

You know from experience that heavy objects are harder to move and harder to stop than light objects are. The more matter an object has, the harder it will be to move or stop. Mass measures the quantity of matter. The more mass an object has, the greater its inertia is.

Figure 10 The cart has inertia and resists moving when you push it. **Compare** *the inertia of the cart when empty to the inertia of the cart when holding the projector.*

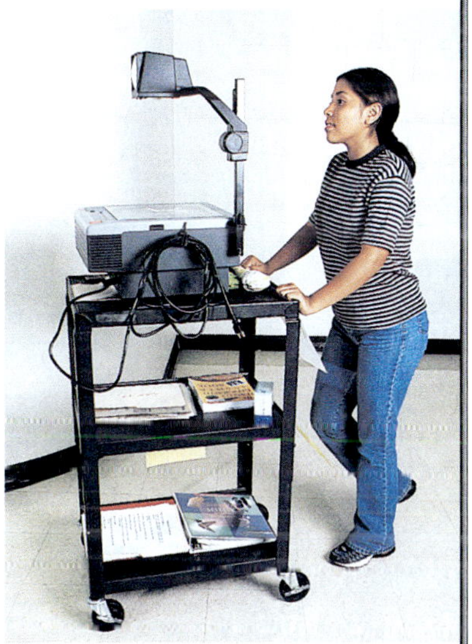

Reading Check

Answer opposite to the direction of motion

Discussion

What are some ways to reduce friction? Possible answers: Polish the surfaces; apply a lubricant, such as oil.

Caption Answer

Figure 10 The cart's inertia is greater when it's holding the projector.

Use Science Words

Word Origin The word *inertia* comes from the Latin word *iners*, meaning idle or unskilled. Another word with this root is *inert.* Have students look up this word and tell how it is used in chemistry. Inert means unable to move. An inert gas is chemically nonreactive.

Use an Analogy

Roughness of Surfaces Compare the microscopic roughness of surfaces with the close-up roughness of cloth. From across the classroom, a flat area on a pair of jeans or a t-shirt, for example, looks smooth. As you move closer, however, you can begin to see ridges from the rows of stitches making up the cloth and even the individual stitches themselves. Where a stitch has been pulled slightly, you can see the roughness even more clearly.

Identifying Misconceptions

Direction of Force Students may think that if an object is moving, there is a force acting on it in the direction of the motion. See page F at the beginning of the chapter for teaching strategies that address this misconception.

Visual Learning

Figure 9 The close-up shows that even smooth surfaces are microscopically rough. The actual number of points where the book and the table touch is small. Friction can be increased by pressing down on the book, but it does not depend on the area of contact. Have students show this by using a spring scale to pull a heavy block positioned vertically and then horizontally. L2 LS **Kinesthetic**

Teacher FYI

Friction In general, friction is higher between rough contact surfaces and lower between smooth ones. However, if the contact surfaces are extremely smooth, friction is again high, due to intermolecular attractive forces.

Activity

Unit Analysis Have students do a unit analysis using the second law of motion to determine the combination of SI mass, length, and time units that are equivalent to a newton. Use $F = ma$. The right side of this equation has the combination of units kg $\times$ m/s^2. The left side of the equation has units of newtons. So 1 N = 1 kg $\times$ m/s^2. L2 LS **Logical-Mathematical**

IDENTIFYING Misconceptions

Rocket Engines Students may think that the shuttle is launched into space by the exploding gases pushing against the ground. Explain that if this were true, the shuttle wouldn't be able to move in space with no ground to push against. Instead, Newton's third law of motion explains rocket motion. In a rocket, the exploding gases in the rocket engine are forced out of the engine nozzle at the bottom of the rocket. The walls of the combustion chamber of the rocket engine exert a net force on the gases that pushes them out of the nozzle. In turn, the gases exert a force on the rocket engine in the direction opposite to the direction of the expelled gases.

Applying Math

National Math Standards

Correlation to Mathematics Objectives

1, 2, 9

Answers to Practice Problems

1. $a = \frac{40\text{ N}}{20\text{ kg}} = 2\text{ m/s}^2$
2. $a = \frac{80\text{ N}}{80\text{ kg}} = 1\text{ m/s}^2$

Science Online

Topic: Sir Isaac Newton
Visit red.msscience.com for Web links to information about the contributions made to science and mathematics by Sir Isaac Newton.

Activity Make a time line to show what you learn.

Newton's Second Law

According to Newton's first law, a change in motion occurs only if a net force is exerted on an object. Newton's second law tells how a net force acting on an object changes the motion of the object. According to Newton's second law, a net force changes the velocity of the object and causes it to accelerate.

Newton's second law of motion states that if an object is acted upon by a net force, the acceleration of the object will be in the direction of the net force, and the acceleration equals the net force divided by the mass. According to the second law of motion, acceleration can be calculated from this equation:

Newton's Second Law

$$\textbf{acceleration}\text{ (in m/s}^2) = \frac{\textbf{net force}\text{ (in N)}}{\textbf{mass}\text{ (in kg)}}$$

$$a = \frac{F_{net}}{m}$$

Applying Math — Solve a Simple Equation

ACCELERATION OF A BASKETBALL You throw a 0.5-kg basketball with a force of 10 N. What is the ball's acceleration?

Solution

1. *This is what you know:*
   - mass: $m = 0.5$ kg
   - net force: $F_{net} = 10$ N
2. *This is what you need to know:* acceleration: $a = ?$ m/s^2
3. *This is the procedure you need to use:* Substitute the known values for the net force, F_{net}, and mass, m, into the equation for Newton's second law:
   $$a = \frac{F_{net}}{m} = \frac{10\text{ N}}{0.5\text{ kg}} = 20\frac{\text{N}}{\text{kg}} = 20\text{ m/s}^2$$
4. *Check your answer:* Multiply your answer by the mass. You should get the force that was given.

Practice Problems

1. You push a 20-kg crate with a force of 40 N. What is the crate's acceleration?
2. Calculate the acceleration of an 80-kg sprinter starting out of the blocks with a force of 80 N.

Science Online For more practice, visit red.msscience.com/math_practice

Fun Fact

One newton is about the amount of force needed to lift a half cup of water.

140 CHAPTER 5 Motion, Forces, and Simple Machines

When pushing a refrigerator, which has a large mass, a large force is required to achieve a small acceleration.

If you were to push an empty grocery cart with the same force, its acceleration would be larger.

Figure 11 The acceleration of an object depends on both the net force applied and the object's mass. **Compare** *the accelerations of a 900-kg car and a 12-kg bicycle if the same net force of 2,000 N is applied to each.*

Mass and Acceleration When a net force acts on an object, the object's acceleration depends on its mass. The more mass an object has, the more inertia it has, so the harder it is to accelerate. Imagine using the same force to push an empty grocery cart that you use to push a refrigerator, as shown in **Figure 11.** With the same force acting on the two objects, the refrigerator will have a much smaller acceleration than the empty cart. More mass means less acceleration if the force acting on the objects is the same.

Newton's Third Law

Suppose you push on a wall. It might surprise you to know that the wall pushes back on you. According to Newton's third law, when one object exerts a force on a second object, the second object exerts an equal force in the opposite direction on the first object. For example, when you walk, you push back on the sidewalk and the sidewalk pushes forward on you with an equal force.

The force exerted by the first object is the action force. The force exerted by the second object is the reaction force. In **Figure 12,** the action force is the swimmer's push on the pool wall. The reaction force is the push of the pool wall on the swimmer. The action and reaction forces are equal, but in opposite directions.

Figure 13 on the next page shows how Newton's laws affect astronauts in space and the motion of the space shuttle.

Figure 12 When the swimmer pushes against the pool wall, the wall pushes back with an equal and opposite force.

Discussion

Skateboard Jump Ask students to explain what happens when they jump forward off a skateboard. The skateboard moves back, as explained by Newton's third law of motion.

Caption Answer

Figure 11 car: $a = \frac{2{,}000\ \text{N}}{900\ \text{kg}} = 22\ \text{m/s}^2$, bicycle: $a = \frac{2{,}000\ \text{N}}{12\ \text{kg}} = 167\ \text{m/s}^2$

Inquiry Lab

Mass, Acceleration, and Inertia

Purpose Students will explore the role of mass in inertia and in acceleration.

Possible Materials several toy cars of varying masses, toy car track, rubber bands, stopwatch

Estimated Time one class session

Teaching Strategies

- Students can use the rubber bands as slingshots to propel the cars on the tracks to keep the cars moving in a straight line.
- Students can predict and then observe how the mass of a car affects how quickly it's accelerated by the rubber bands. They can compare the distances traveled by different cars for the same force or by the same car for different forces.
- Students can also explore how the relative masses of cars (one moving, one initially at rest) affect their velocities after a collision.

For additional inquiry activities, see *Science Inquiry Labs.*

LAB DEMONSTRATION

Purpose to demonstrate Newton's laws

Materials 2-g mass, 50-g mass, cart, book (about 0.5 kg), string, two spring scales

Preparation Attach the string to the book, and record the mass of the book.

Procedure First, place the masses in the cart. Roll the cart sharply against a barrier and ask students to observe what happens to the masses. Next, have students calculate the force needed to hold up the book and verify this using the spring scale. Finally, use spring scales to pull opposite sides of the book.

Expected Outcome Have students explain the results, using Newton's laws.

Assessment

If you used a spring scale to pull a book across the table at a constant speed, what force would you measure on the scale? friction How would the force recorded on the scale differ if the mass of the book were doubled? It would double.

NATIONAL GEOGRAPHIC

Visualizing Newton's Laws and Space Travel

Have students examine the pictures and read the captions. Then ask the following questions.

How is a blown-up balloon that you release like a space shuttle taking off? As the balloon contracts, it pushes air toward the back. As the air leaves, it pushes on the balloon, moving it forward. This is similar to the forces acting on the shuttle.

If an astronaut with a tool belt accidentally pushed off from the shuttle, how could the tools in the belt be used to return to the shuttle? If the astronaut threw the tools, one at a time, away from the shuttle, an equal and opposite force would move the astronaut back toward the shuttle. If the astronaut had enough tools to throw, the shuttle could be reached.

Activity

Rocket Belts Have small groups of students work together to write and illustrate a story about how they would use a rocket belt that is propelled by gases that shoot out of moveable nozzles. Allow groups to exchange and read one another's stories. L2 COOP LEARN LS **Linguistic and Interpersonal**

NATIONAL GEOGRAPHIC

VISUALIZING NEWTON'S LAWS AND SPACE TRAVEL

Figure 13

Newton's laws of motion are universal—they apply in space just as they do here on Earth. Newton's laws can be used to help design spacecraft by predicting their motion as they are launched into orbit around Earth and places beyond. Here are some examples of how Newton's laws affect space shuttle missions.

◀ According to Newton's third law, every action has an equal and opposite reaction. Launching a space shuttle demonstrates the third law. Fuel burning in the rocket's combustion chamber creates gases. The rocket exerts a force on these gases to expel them out of the nozzle at the bottom of the rocket. The reaction force is the upward force exerted on the rocket by the gases.

▼ Newton's second law explains why a shuttle remains in orbit. Earth exerts a gravitational force on a shuttle, causing the shuttle to accelerate. This acceleration causes the direction of the shuttle's motion to constantly change, so it moves in a circular path around the planet.

▶ According to Newton's first law, the motion of an object changes only if the object is acted upon by an unbalanced force. An astronaut outside the shuttle orbits Earth along with the shuttle. If the astronaut were to push on the shuttle, the shuttle would push on the astronaut. According to the first law of motion, this would cause the astronaut to move away from the shuttle.

Differentiated Instruction

Challenge Have students research and write brief reports that explain the motion of satellites in terms of Newton's laws of motion. Ask students to present their reports to the class. L3 LS **Linguistic and Logical-Mathematical**

Teacher FYI

Force Pairs Even though action and reaction forces are described as if the action force occurs first, the two forces actually occur at the same time.

Force Pairs Act on Different Objects If forces always occur in equal but opposite pairs, how can anything ever move? Won't the forces acting on an object always cancel each other? Recall that in Newton's third law, the equal and opposite forces act on different objects. When you push on the book, your force is acting on the book. When the book pushes back on you, its force is acting on you. One force of the force pair acts on the book, and the other force acts on you. Because the forces act on different objects, they don't cancel.

Reading Check *Why don't action and reaction forces cancel?*

Examples of Newton's Third Law Think about what happens when you jump from a boat, as shown in **Figure 14.** If you jump off a small boat, the boat moves back. You are pushing the boat back with your feet with the same force with which it is pushing you forward. Because you have more mass than the boat, it will accelerate more than you do. When you jump off a big boat with a large mass, the force you exert on the boat gives it only a tiny acceleration. You don't notice the large boat moving, but the force it exerts on you propels you to the dock.

Figure 14 When you jump off a boat, your feet exert a force on the boat, which pushes it backward. The boat also exerts a force on your feet, which pushes you forward.

section 2 review

Force

- A force is a push or a pull.
- The net force is the combination of all the forces acting on an object.

Newton's Laws of Motion

- Newton's first law of motion states that an object's motion will not change unless a force acts on the object.
- Newton's second law of motion states that an object accelerates in the direction of the net force and that the acceleration can be calculated from this equation:

$$a = \frac{F_{net}}{m}$$

- Newton's third law of motion states that when an object exerts a force on another object, the second object exerts an equal and opposite force on the first object.

Self Check

1. **Explain** how the inertia of an object is related to the object's mass.
2. **Apply** If a force of 5 N to the left and a force of 9 N to the right act on an object, what is the net force?
3. **Infer** whether balanced forces must be acting on a car moving at a constant speed.
4. **Think Critically** A book sliding across a table slows down and comes to a stop. Explain whether this violates Newton's first law of motion.

Applying Math

5. **Net Force** Find the net force exerted on a ball that has a mass of 0.15 kg and an acceleration of 20 m/s^2.
6. **Use a Spreadsheet** Enter the formula $a = F_{net}/m$ into a spreadsheet. Find the acceleration for masses from 10 kg to 200 kg. Graph your results.

Science Online red.msscience.com/self_check_quiz

Reading Check

Answer Action-reaction forces are equal and opposite, but they act on two different objects.

3 Assess

DAILY INTERVENTION

Check for Understanding

Logical-Mathematical Using Newton's laws of motion, explain how a car's headrest protects you from neck injury. By Newton's first law, if the car is struck from behind, your head will tend to fly back relative to the car, and this motion can injure your neck. The headrest acts according to Newton's third law, pushing against your head and keeping your neck from being injured. L3 LS

Reteach

Newton's Laws in Everyday Life Display pictures showing examples of how Newton's laws of motion affect everyday life. Have students discuss the pictures and tell which of Newton's laws are demonstrated. L2 LS **Visual-Spatial**

Assessment

Performance Have students work in small groups to prepare a demonstration in which they act out and explain each of Newton's laws. The groups can present their demonstrations to the class. Use **Performance Assessment in the Science Classroom**, p. 147.

section 2 review

1. The more mass an object has, the greater its inertia.
2. 4 N to the right
3. No, because even if the car is moving at constant speed, the car could be turning. Turning is a form of acceleration, so this would mean that unbalanced forces were acting on the car.
4. No, because the force of friction with the table is acting to slow down and stop the book.
5. $F = (0.15 \text{ kg}) \times (20 \text{ m/s}^2) = 3 \text{ N}$
6. Check students' work.

1 Motivate

INTERACTIVE CHALKBOARD
PowerPoint® Presentations

Bellringer

Section Focus Transparencies also are available on the Interactive Chalkboard CD-ROM. L2 ELL

Section 3 Section Focus Transparency Movin' On

Moving is pretty hard work, but there are things that can help make it easier. As you look at this picture, try to imagine what it would be like if the man had only his bare hands to move the boxes.

1. What do you see that helps the man move the boxes?
2. Where do you see wheels in this photograph?
3. If you had to load this truck with heavy boxes, what tools would you want on hand?

Motion, Forces, and Simple Machines

Tie to Prior Knowledge

Simple Machines Ask students to name the simple machines. They may think all machines are big and complex. In this chapter, they will learn that many machines are very simple, and many large machines are combinations of simple machines.

section 3

Work and Simple Machines

as you read

What You'll Learn

- **Define** work.
- **Distinguish** the different types of simple machines.
- **Explain** how machines make work easier.

Why It's Important

Machines make doing work easier.

Review Vocabulary
radius: distance from the center of a circle to its edge

New Vocabulary
- work
- simple machine
- compound machine
- mechanical advantage
- pulley
- lever
- inclined plane

Work

Newton's laws explain how forces change the motion of an object. If you apply a force upward on the box in **Figure 15,** it will move upward. Have you done any work on the box? When you think of work, you might think of doing household chores or even the homework you do every night. In science, the definition of work is more specific—**work** is done when a force causes an object to move in the same direction as the force that is applied.

Effort Doesn't Always Equal Work If you push against a wall, do you do work? For work to be done, two things must occur. First, you must apply a force to an object. Second, the object must move in the same direction as the force you apply. If the wall doesn't move, no work is done.

Picture yourself picking up and carrying the box in **Figure 15.** You can feel your arms exerting a force upward as you lift the box. The box moves upward in the direction of your force, so you have done work. If you carry the box forward, you still can feel your arms applying an upward force on the box, but the box is moving forward. Because the direction of motion is not the same as the direction of the force applied by your arms, no work is done by your arms.

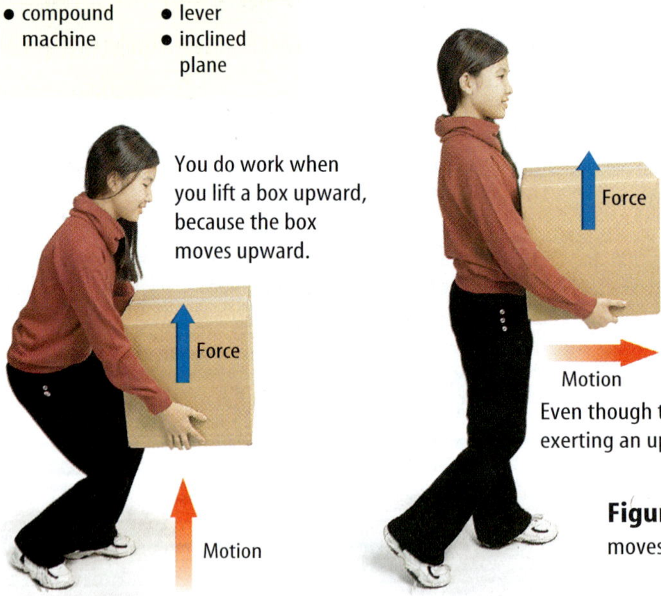

You do work when you lift a box upward, because the box moves upward.

Even though the box moves forward, your arms are exerting an upward force and do no work.

Figure 15 Work is done only when an object moves in the direction of the applied force.

Section 3 Resource Manager

Chapter FAST FILE Resources
- Transparency Activity, p. 44
- Directed Reading for Content Mastery, pp. 19, 20
- MiniLAB, p. 4
- Enrichment, p. 30
- Reinforcement, p. 27
- Lab Worksheet, pp. 5–6, 7–8

Home and Community Involvement, p. 40

Physical Science Critical Thinking/Problem Solving, pp. 6, 7

Calculating Work

For work to be done, a force must be applied, and an object must move. The greater the force that is applied, the more work that is done. Which of these tasks would involve more work—lifting a shoe from the floor to your waist or lifting a pile of books the same distance? Even though the shoe and the books move the same distance, more work is done in lifting the books because it takes more force to lift the books. The work done can be calculated from the equation below.

Work Equation

work (in J) = **force** (in N) × **distance** (in m)

$$W = Fd$$

Work is measured in joules (J). The joule is named for James Prescott Joule, a nineteenth-century British physicist who showed that work and energy are related. Lifting a baseball from the ground to your waist requires about 1 J of work.

INTEGRATE Life Science

Muscles and Work Even though the wall doesn't move when you push against it, you may find yourself feeling tired. Muscles in your body contract when you push. This contraction is caused by chemical reactions in your muscles that cause molecules to move past each other. As a result, work is done by your body when you push. Research how a muscle contracts and describe what you learn in your Science Journal.

Applying Math — Solve a Simple Equation

WEIGHT LIFTING A weight lifter lifts a 500-N weight a distance of 2 m from the floor to a position over his head. How much work does he do?

Solution

1. *This is what you know:*
   - force: $F = 500$ N
   - distance: $d = 2$ m
2. *This is what you need to know:* work: $W = ?$ J
3. *This is the procedure you need to use:* Substitute the known values for force and distance into the work equation and calculate the work: $W = Fd = (500 \text{ N})(2 \text{ m}) = 1{,}000 \text{ N m} = 1{,}000 \text{ J}$
4. *Check your answer:* Divide your answer by the distance. You should calculate the force that was given.

Practice Problems

1. Using a force of 50 N, you push a computer cart 10-m across a classroom floor. How much work did you do?
2. How much work does an Olympic sprinter do while running a 200-m race with a force of 6 N?

Science Online — For more practice, visit red.msscience.com/math_practice

2 Teach

INTEGRATE Life Science

Muscles and Work Muscle tissue is made up of fibers, which in turn are made up of myofibrils. Each myofibril is made up of thick and thin protein filaments. During contraction, the thin filaments slide toward one another, shortening the myofibril. When the muscle relaxes, the filaments return to their original position.

Applying Math

National Math Standards

Correlation to Mathematics Objectives

1, 2, 9

Answers to Practice Problems

1. $W = 50 \text{ N} \times 10 \text{ m} = 500 \text{ J}$
2. $W = 6 \text{ N} \times 200 \text{ m} = 1200 \text{ J}$

Cultural Diversity

Early Machines Around the first century A.D., the Greek mathematician Hero of Alexandria wrote a series of books entitled *Mechanics*. The second book in this series described simple machines and how they could be applied to everyday tasks. Hero explained how using a lever, pulley, wedge, or screw made lifting and transporting objects easier. Have students research and write a report about how simple machines made life easier for people before the invention of modern, powered machinery. Possible answers: lifting a heavy rock with a lever, raising a bucket from a well using a pulley, hunting with a sharp spear tip or arrowhead (wedge). L2 LS **Logical-Mathematical** P

SECTION 3 Work and Simple Machines 145

Discussion

Machine Groups The six simple machines can be classified into two groups, the inclined plane group and the lever group. Discuss with students these two groups and the characteristics of the simple machines. Which simple machines belong to the inclined plane group? the wedge, inclined plane, and screw Which belong to the lever group? the lever, wheel and axle, and pulley L2

LS Logical-Mathematical

Reading Check

Answer by changing the size of the force you apply or by changing the direction of the force

Discussion

Input Force Direction Have students discuss why a machine that changes the direction of input force but doesn't multiply the force might be useful. Possible answer: Changing direction so you are pulling down rather than pushing up makes the work easier if you are located below the object.

Fun Fact

The formula for mechanical advantage given on this page is for an ideal machine. In real machines, greater effort is required to overcome resistance caused by friction. The effects of friction can be reduced, but they can never be completely eliminated.

Figure 16 The can opener changes the small force of your hand on the handles to a large force on the blade that cuts into the can.

What is a machine?

How many machines have you used today? Why did you use them? A machine is a device that makes work easier. A can opener like the one in **Figure 16** is a machine that changes a small force applied by your hand into a larger force that makes it easier to open the can.

A **simple machine** is a machine that uses only one movement. A screwdriver is an example of a simple machine. It requires only one motion—turning. Simple machines include the pulley, lever, wheel and axle, inclined plane, wedge, and screw. A **compound machine** is a combination of simple machines. The can opener is a compound machine that combines several simple machines. Machines can make work easier in two ways. They can change the size of the force you apply. They also can change the direction of the force.

Reading Check *How do machines make work easier?*

Science Online

Topic: Early Tools

Visit red.msscience.com for Web links to information about early tools.

Activity Write a short story about the nineteenth-century in which the characters use at least three early tools. Include at least one picture or description showing how the tools make work easier.

Mechanical Advantage Some machines are useful because they increase the force you apply. The number of times the applied force is increased by a machine is called the **mechanical advantage** (MA) of the machine.

When you push on the handles of the can opener, the force you apply is called the input force (F_i). The can opener changes your input force to the force that is exerted by the metal cutting blade on the can. The force exerted by a machine is called the output force (F_o). The mechanical advantage is the ratio of the output force to the input force.

Mechanical Advantage Equation

$$\textbf{mechanical advantage} = \frac{\textbf{force out (in N)}}{\textbf{force in (in N)}}$$

$$\text{MA} = \frac{F_o}{F_i}$$

Work In and Work Out In a simple machine the input force and the output force do work. For example, when you push on the handles of a can opener and the handles move, the input force does work. The output force at the blade of the can opener does work as the blade moves down and punctures the can.

An ideal machine is a machine in which there is no friction. Then the work done by the input force is equal to the work done by the output force. In other words, for an ideal machine, the work you do on the machine—work in—is equal to the work done by the machine—work out.

146 CHAPTER 5 Motion, Forces, and Simple Machines

Increasing Force A simple machine can change a small input force into a larger output force. Recall that work equals force times distance. So, if the work in is equal to the work out, then smaller input force must be applied over a larger distance than the larger output force. Think again about the can opener. The can opener increases the force you apply at the handle. So the distance you move the handle is large compared to the distance the blade of the can opener moves as it pierces the can.

In all real machines, friction always occurs as one part moves past another. Friction causes some of the input work to be changed into heat, which can't be used to do work. So for a real machine, work out always will be less than work in.

The Pulley

To raise a window blind, you pull down on a cord. The blind uses a pulley to change the direction of the force. A **pulley** is an object, like a wheel, that has a groove with a rope or cable running through it. A pulley changes the direction of the input force. A rope thrown over a railing can be used as a pulley. A simple pulley, such as the one shown in **Figure 17,** changes only the direction and not the size of the force, so its mechanical advantage is 1.

It is possible to have a large mechanical advantage if more than one pulley is used. The double-pulley system shown in **Figure 17** has a mechanical advantage of 2. Each supporting rope holds half of the weight, so the input force you need to supply to lift the weight is half as large as for a single pulley.

Mini LAB

Observing Mechanical Advantage—Pulleys

Procedure

1. Tie a 3-m-long rope to the middle of a broomstick or dowel and hold this stick horizontally. Another student should hold another stick horizontally. Wrap the rope around both sticks four times, leaving about 0.5 m between the sticks.
2. A third student should pull on the rope while the other two students try to keep the sticks from coming closer together.
3. Observe what happens. Repeat using only two wraps of the rope and then using eight wraps.

Analysis

1. Describe what you observed. Could the students hold the sticks apart?
2. Compare and contrast the results with two, four, and eight turns of the rope around the sticks.

A single pulley changes the direction of the input force.

A combination of pulleys decreases the input force, so the mechanical advantage is greater than 1.

Figure 17 A pulley changes the direction of the input force and can decrease the input force.

Mini LAB

Purpose Students observe the mechanical advantage of pulleys and record their observations. L2 LS **Interpersonal**

Materials two broomsticks or dowels, 3-m-long rope

Teaching Strategy Have students repeat the activity several times so each student has a chance to pull on the rope.

Analysis

1. Students will observe that it is nearly impossible to prevent the sticks from being pulled together.
2. The more times the rope is wrapped around the sticks, the harder it is to keep the sticks from being pulled together.

Assessment

Process Have each group of students make comic strips illustrating what they did in the MiniLAB. Use **Performance Assessment in the Science Classroom,** p. 133. P

Make a Model

Flag Raising Have students use pulleys to make a model that raises a flag. L2 LS **Kinesthetic**

Differentiated Instruction

Learning Disabled Encourage students to bring in magazine clippings showing examples of each of the simple machines they study in this section. Many of the photographs will be of complex machines. Help them describe the combinations of simple machines that are being used. L1 LS **Visual-Spatial**

Challenge Pulleys can be either fixed or movable. Have students investigate these and demonstrate each to the class. A fixed pulley doesn't move. With a movable pulley, both the load and the pulley move. L3 LS **Kinesthetic**

Activity

Finding Mechanical Advantage Have students use a spring scale to find the mechanical advantage of first-, second-, and third-class levers by fastening the scale at the point where the input force is applied. Levers can be easily constructed using objects such as rulers and blocks of wood. L2 ELL COOP LEARN LS **Kinesthetic**

Visual Learning

Figure 18 The classification of a lever is determined by whether the fulcrum, input force, or output force is in the middle. However, the distances between the fulcrum and the input and output forces determine how much input force is required to lift the load. For example, for the screwdriver in **Figure 18,** if the fulcrum is moved closer to the output force, the mechanical advantage increases. Have students experiment with various levers to see this effect. L2 ELL LS **Kinesthetic**

Figure 18 A lever is classified according to the locations of the input force, output force, and fulcrum.

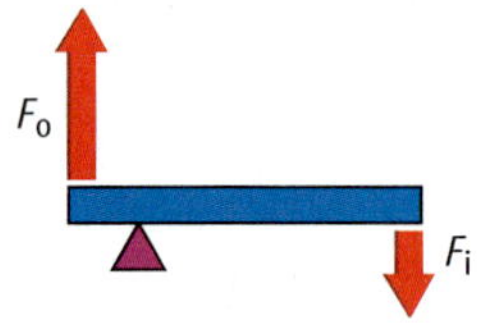

Sometimes a screwdriver is used as a first-class lever. The fulcrum is between the input and output forces.

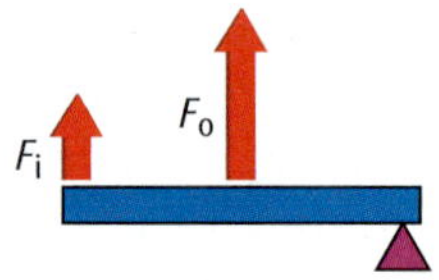

A wheelbarrow is a second-class lever. The fulcrum is the wheel, and the input force is applied on the handles. The load, which is where the output force is applied, is between the input force and the fulcrum.

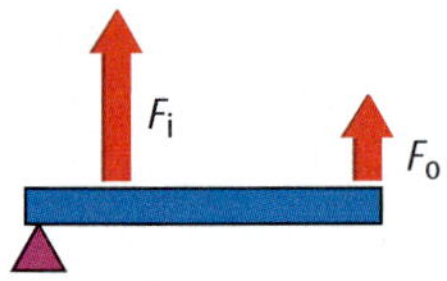

A hockey stick is a third-class lever. The fulcrum is your upper hand, and the input force is applied by your lower hand. The output force is applied at the bottom end of the stick.

The Lever

Probably the first simple machine invented by humans was the lever. A **lever** is a rod or plank that pivots about a fixed point. The pivot point is called the fulcrum. Levers can increase a force or increase the distance over which a force is applied. There are three types, or classes, of levers. The three classes depend on the positions of the input force, the output force, and the fulcrum.

The three classes of levers are illustrated in **Figure 18.** In a first-class lever, the fulcrum is located between the input force and output force. Usually, a first-class lever is used to increase force, like a screwdriver used to open a can.

If the output force is between the input force and the fulcrum, as in a wheelbarrow, the lever is a second-class lever. The output force always is greater than the input force for this type of lever.

A hockey stick is a third-class lever. In a third-class lever, the input force is located between the output force and the fulcrum. The mechanical advantage of a third-class lever always is less than 1. A third-class lever increases the distance over which the input force is applied.

Teacher FYI

Mnemonic Levers Students might benefit from the mnemonic FREE for lever classification. For a first-class lever, the fulcrum (F) is in the middle. For a second-class lever, the resistance (R), or output force, is in the middle. For a third-class lever, the effort (E), or input force, is in the middle.

Differentiated Instruction

Challenge Work can be done against friction, inertia, or gravity. Have students make posters showing examples of work done against each of these forces. Advise them that some work is against more than one force. Possible answers: Pushing a heavy box across a floor is work against friction because the box rubs against the floor, and starting it moving is work against inertia because the box has a tendency to stay still. Lifting the box is work against gravity. L3 LS **Visual-Spatial**

The Wheel and Axle Try turning a doorknob by holding the narrow base of the knob. It's much easier to turn the larger knob. A doorknob is an example of a wheel and axle. Look at **Figure 19.** A wheel and axle is made of two round objects that are attached and rotate together about the same axis. The larger object is called the wheel, and the smaller object is the axle. The mechanical advantage of a wheel and axle can be calculated by dividing the radius of the wheel by the radius of the axle.

Reading Check *How do the lever, pulley, and wheel and axle make work easier?*

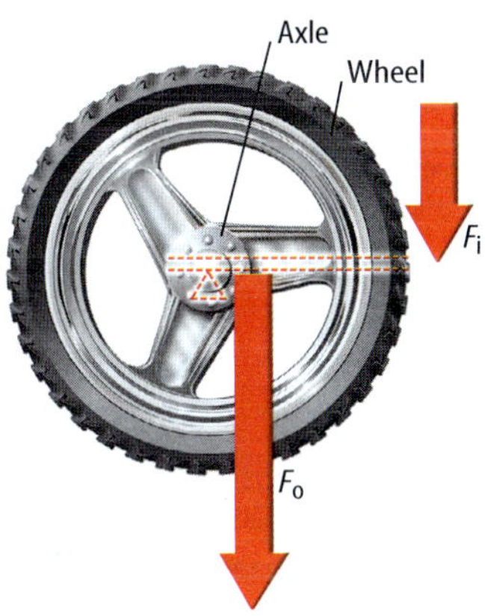

Figure 19 The radius of the wheel is greater than the radius of the axle. The mechanical advantage of the wheel and axle is greater than 1 because the radius of the wheel is greater than the radius of the axle.

The Inclined Plane

An **inclined plane** is a sloped surface, sometimes called a ramp. It allows you to lift a heavy load by using less force over a greater distance. Imagine having to lift a couch 1 m off the ground onto a truck. If you used an inclined plane or ramp, as shown in **Figure 20,** you would have to move the couch farther than if you lifted it straight up. Either way, the amount of work needed to move the couch would be the same. Because the couch moves a longer distance up the ramp, doing the same amount of work takes less force.

The mechanical advantage of an inclined plane is the length of the inclined plane divided by its height. The longer the ramp is, the less force it takes to move the object. Ramps might have enabled the ancient Egyptians to build their pyramids. To move limestone blocks having a mass of more than 1,000 kg each, archaeologists hypothesize that the Egyptians built enormous ramps.

Figure 20 It is much easier to load this couch into a truck using a ramp. Even though the couch must be pushed a greater distance, less force is required.

Cultural Diversity

Archimedes' Screw The ancient Greek scientist Archimedes is credited with inventing a variation of the inclined plane called an Archimedes' screw. This tool consists of a screw-shaped pipe wound around a central shaft. If the screw is dipped into water and turned, it lifts water up through the pipe. Have students find pictures of an Archimedes' screw and demonstrate how it works.

Activity

Inclined Plane Demonstrate the advantage of using an inclined plane by having students use a spring scale to compare the force needed to pull a load up a ramp. Lean two boards against books. One should be at about a 30° angle to the floor and the other at about a 45° angle to the floor. Attach the scale to a load and observe the difference in force required to pull it up the ramps.
L2 ELL LS **Kinesthetic**

Reading Check

Answers A lever increases the input force or the distance over which the force is applied. A pulley changes the direction of the input force and can increase it. A wheel and axle increases the input force and changes its direction.

Quick Demo

Simple Machine Similarity

Materials chalkboard, chalk

Estimated Time 15 minutes

Purpose to demonstrate how a wheel and axle is essentially the same as a lever

Draw a wheel and axle on the board as shown:

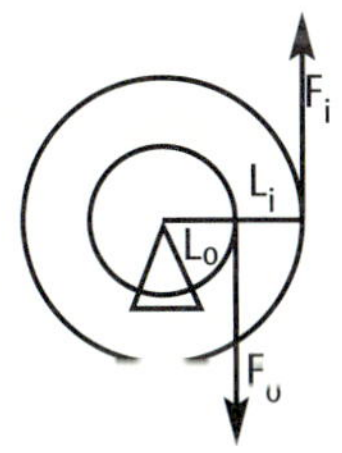

Point out the corresponding components. For example, the axis of rotation corresponds to the fulcrum, and the radius of the wheel corresponds to the lever arm. L_i corresponds to the distance between the input force and the output force. L_o corresponds to the distance between the output force and the fulcrum.

3 Assess

DAILY INTERVENTION

Check for Understanding

Interpersonal Divide students into groups of three and provide each group with samples of the six simple machines. Have each student pick two different machines and explain to the others how these machines work.

Reteach

Identifying Simple Machines Bring in objects that are either simple or compound machines. Have students describe the simple machines involved in each of them. L2 LS **Visual-Spatial**

Assessment

Oral Ask students why third-class levers are not useful for lifting heavy objects. They increase the distance the load moves but don't multiply the force.

Figure 21 Plant-eaters and meat-eaters have different teeth.

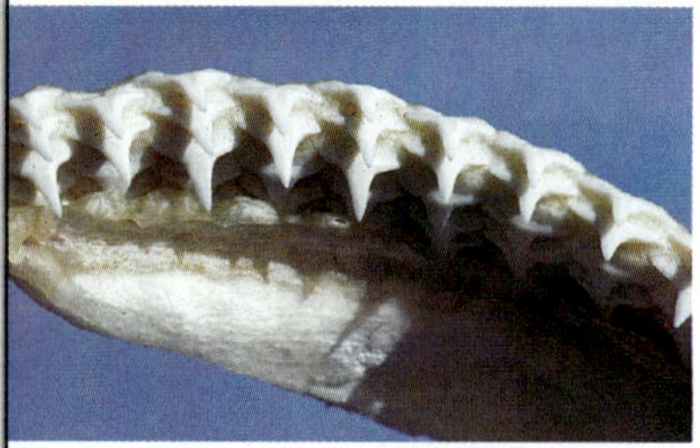

These wedge-shaped teeth enable a meat-eater to tear meat.

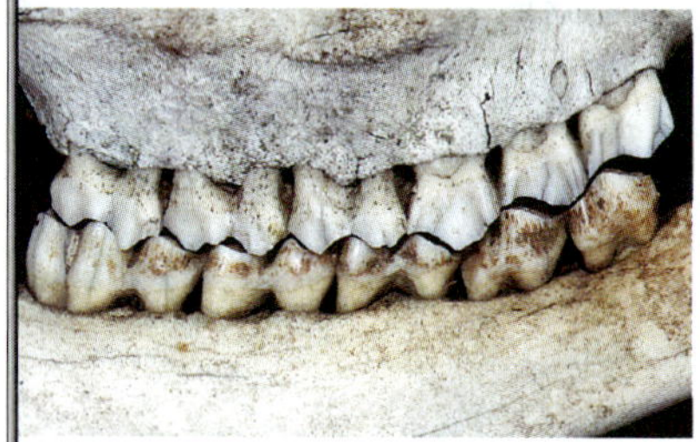

The teeth of a plant-eater are flatter and used for grinding.

INTEGRATE Life Science

The Wedge When you take a bite out of an apple, you are using wedges. A wedge is a moving inclined plane with one or two sloping sides. Your front teeth are wedges. A wedge changes the direction of the input force. As you push your front teeth into the apple, the downward input force is changed by your teeth into a sideways force that pushes the skin of the apple apart. Knives and axes also are wedges that are used for cutting.

Figure 21 shows that the teeth of meat-eaters, or carnivores, are more wedge-shaped than the teeth of plant-eaters, or herbivores. The teeth of carnivores are used to cut and rip meat, whereas herbivores' teeth are used for grinding plant material. Scientists can determine what a fossilized animal ate when it was living by examining its teeth.

The Screw A road going up a mountain usually wraps around the mountain. Such a road is less steep than a road straight up the side of the mountain, so it's easier to climb. However, you travel a greater distance to climb the mountain on the mountain road. The mountain road is similar to a screw. A screw is an inclined plane wrapped around a post. The inclined plane forms the screw threads. Just like a wedge, a screw also changes the direction of the force you apply. When you turn a screw, the input force is changed by the threads to an output force that pulls the screw into the material. Friction between the threads and the material holds the screw tightly in place.

section 3 review

Summary

Work

- Work is done when a object moves in the direction of the applied force.
- Work can be calculated from the equation $W = Fd$.

Simple machines

- Machines are devices that make work easier.
- Mechanical advantage is the number of times the input force is increased by a machine.
- A simple machine is a machine that does work with only one motion.
- The six simple machines are the pulley, lever, wheel and axle, inclined plane, wedge, and screw.

Self Check

1. **Describe** three different ways that using a machine makes doing work easier.
2. **Explain** why the output work is always less than the input work in a real machine .
3. **Compare** a wheel and axle to a lever.
4. **Think Critically** Identify two levers in your body. Which class of lever do the body levers belong to?

Applying Math

5. **Work** Find the work needed to lift a limestone block weighing 10,000 N a distance of 150 m.
6. **Input Force** Find the input force needed to lift a stone slab weighing 2,500 N using a pulley system with a mechanical advantage of 10.

section 3 review

1. increase the applied force, increase the distance over which the force is applied, and change the direction of the force
2. because friction changes some of the input work into heat, which can't be used to do work
3. A wheel and axle is essentially a lever that can rotate 360° about a fixed point (the fulcrum or axis of rotation). The input force can be applied and the output force can be generated at any point along the radius of the wheel or axle.
4. Possible answers: forearm, third-class; lower leg, third-class
5. $W = F \times d = 10{,}000\ N \times 150\ m = 1{,}500{,}000\ J$
6. $F_i = F_o/MA = (2{,}500\ N)/10 = 250\ N$

LAB

Motion

What happens when you roll a small ball down a ramp? It speeds up as it travels down the ramp, and then it rolls across the floor and eventually it stops. You know that as the ball travels down the ramp, gravity is acting to make it speed up. Think about the forces that are acting on the ball as it rolls across the floor. Is there a net force acting on the ball? How would you describe the motion of the ball?

Real-World Question

How does a ball move when the forces acting on it are balanced and when they are unbalanced?

Goals

- **Demonstrate** the motion of a ball with unbalanced and balanced forces acting on it.
- **Graph** the position versus time for the motion of the ball.

Materials

small ball or marble
meterstick or tape measure
stopwatch
graph paper

Safety Precautions

Procedure

1. Place the ball on the floor or a smooth, flat surface.
2. Roll the ball across the floor by giving it a gentle push.
3. **Record Data** As the ball is rolling and no longer being pushed, have one student keep track of the time and have other students record the distance at 1-s intervals for at least 5 s to 10 s.
4. **Record** anything else that you observed about how the ball moved.
5. **Calculate** from your data the distance the ball has traveled at each second.
6. **Make a graph** of the distance the ball travels versus time. Plot the distance traveled on the vertical y-axis and the time on the horizontal x-axis.
7. Choose three one-second time intervals. **Calculate** the speed of the ball in each of those time intervals.

Conclude and Apply

1. **Describe** how the speed of the ball changes as it rolls along the floor.
2. **Describe** the forces acting on the ball before you pushed it and it was at rest. **Infer** whether the forces acting on the ball were balanced or unbalanced.
3. **Describe** the forces acting on the ball as it rolled across the floor. **Infer** whether the forces acting on the ball were balanced or unbalanced.

Communicating Your Data

Compare your graphs and results with those of other students in your class.

BENCH TESTED

Real-World Question

Purpose Students will observe the motion of balls when balanced and unbalanced forces act upon them. L2 LS **Kinesthetic**

Process Skills observe, graph, calculate, compare and contrast

Time Required 25 minutes

Procedure

Alternate Materials Basketballs or tennis balls can replace the small balls or marbles.

Safety Precautions Caution students never to throw a ball during class.

Teaching Strategies

- Instruct the student keeping track of time to count aloud in 1-s intervals so group members can mark off distances.
- Clear away floor space for student groups to work.

Conclude and Apply

1. The ball slows down and comes to a stop.
2. Gravity and static friction; because the ball's motion did not change, the forces were balanced.
3. Gravity and rolling friction; friction was the unbalanced force that caused the motion of the ball to change.

Assessment

Process Ask students to infer what conditions would be necessary for a ball to continue to roll indefinitely. Could these conditions ever exist? An infinitely smooth surface and the absence of friction. No, in everyday phenomena, some friction is always present. Use **Performance Assessment in the Science Classroom,** p. 89.

Communicating Your Data

Encourage students to create their graphs and data tables electronically.

BENCH TESTED

LAB Use the Internet

Methods of Travel

Goals

- **Research** travel times.
- **Compare** travel times for different methods of travel.
- **Evaluate** the fastest way to travel between two locations.
- **Design** a table to display your findings and communicate them to other students.

Data Source

Science Online

Visit red.msscience.com/internet_lab for more information on travel times, methods of travel, distances between locations, and data from other students.

Real-World Question

How long does it take you to get to the other side of town? How long does it take to get to the other side of the country? If you were planning a road trip from New York City to Los Angeles, how long would it take? How would your trip change if you flew instead? When you plan a trip or vacation, it is useful to first estimate your travel time. Travel time depends on the vehicle you use, how fast you travel, the route you take, and even the terrain. For example, driving over rugged mountains can take longer than driving over flat farmland. With this information, you can plan your trip so you arrive at your final destination on time. Form a hypothesis about what is the fastest form of travel.

Make a Plan

1. **Choose** a starting point and a final destination.
2. **Identify** the routes commonly used between these two locations.

Real-World Question

Purpose Students will investigate methods of travel between locations to determine the fastest way to get from one place to another.

Internet Students use Internet sites that can be accessed through red.msscience.com/internet_lab.

Non-Internet Sources Ask travel agencies for maps and travel guides.

Process Skills predict, interpret data, communicate

Time Required about two days

Possible Hypotheses Students will investigate travel distances and methods of travel between two locations. They will hypothesize which travel method will get them between locations in the shortest amount of time. For example, the fastest way to get from Chicago, Illinois, to Indianapolis, Indiana, is by car.

Make a Plan

Preparation

Internet To run through the steps that the students will follow, visit red.msscience.com/internet_lab.

Non-Internet Sources Collect maps and travel brochures that provide distance and travel information.

Alternative Inquiry Lab

Historical Means of Travel Rapid advances in methods of transportation over the preceding decades have dramatically changed how people travel both within the U.S. and around the world. Students might be surprised to discover how much longer it took to travel between various locations even in their parents' and grandparents' generations. They could investigate travel between their chosen locations using early models of cars and trains and the corresponding road and rail systems that existed then or travel by ship for locations separated by water. Alternatively, they might explore transportation before the invention of any of these machines, considering travel by stagecoach, sail- or rowboat, horse, and even foot.

3. **Determine** the common forms of travel between these two locations.
4. **Research** how to estimate travel time. What factors can make your trip take more or less time?

Follow Your Plan

1. Make sure your teacher approves your plan before you start.
2. **Calculate** the travel time and distance between your two locations for different methods of travel.
3. **Record** your data in your Science Journal.

Analyze Your Data

1. **Analyze** the data recorded in your Science Journal to determine the fastest method of travel. Was it better to drive or fly? Did you investigate another method of travel?
2. **Calculate** the average speed of the methods of travel you investigated. Which method had the fastest speed? Which method had the slowest?
3. **Organize Data** Use a computer (home, library, or computer lab) to create a chart that compares the travel times, average speeds, and distances for different methods of travel. Use your chart to determine the fastest method of travel. What factors add to travel time?

Conclude and Apply

1. **Compare** your findings to those of your classmates and data posted at the Web link to the right. What is the greatest distance investigated? The shortest?
2. **Draw Conclusions** What factors can affect travel time for the different methods? How would your travel time be different if you didn't have a direct flight?
3. **Infer** how the average speed of an airplane flight would change if you included your trips to and from the airport and waiting time in your total travel time.

Communicating Your Data

Find this lab using the link below. Post your data in the table provided. Combine your data with those of other students and make a class travel booklet that estimates travel times for various locations around the world.

Science Online
red.msscience.com/internet_lab

Follow Your Plan

Teaching Strategies As students analyze methods of travel, remind them that airline travel may not always be the fastest way of getting from one place to another. One of their destinations may not be located near an airport.

Analyze Your Data

1. Answers will be subjective and based on the students' individual research.
2. Remind students that speed is calculated by dividing distance traveled by amount of time for the trip.
3. Students might conclude that the fastest method might not always be the most convenient. Factors such as flight schedules may influence their decisions.

Conclude and Apply

1. Answers will vary based on each student's individual research.
2. Weather conditions and time of day can affect travel times. Layovers can significantly add to travel time.
3. Average speed would decrease, as the time spent traveling to and from airports as well as waiting time can be significant.

Assessment

Portfolio Have students make travel brochures for their selected trip. Suggest that they include the travel distance, identify the travel methods investigated, and provide data to support the travel time between locations. They should also describe the factors that would influence choosing the fastest means of travel. Use **Performance Assessment in the Science Classroom,** p. 129.

Communicating Your Data

Suggest that students use a spreadsheet to collect travel data and describe factors that would influence choosing each travel method.

Content Background

The peregrine falcon is a bird of prey that has become rare throughout most of its range. It has become rare because of the widespread use of chlorinated hydrocarbons in insecticides, primarily DDT.

DDT found its way into the food chain via runoff into waterways. The chemicals become concentrated in the tissues of the falcon because the prey that the falcons eat contains DDT.

DDT interrupts the reproduction cycle when it causes the shells of falcon eggs to be unusually thin, which then causes the developing embryos to die. This caused a drastic decline in the number of peregrine falcons and other birds of prey such as the American Bald Eagle. In fact, because its population reached such low levels, the peregrine falcon was added to the endangered species list in 1970. Partially as a result of restrictions on the use of DDT since 1972, the peregrine falcon has made a dramatic recovery. It was removed from the endangered species list in 1999, although scientists still monitor its progress to ensure that the species has truly recovered to sustainable levels.

Activity

Sing About Speed Have the students choose one of the items in the feature and write a song about it. Ask for student volunteers to share their song with the class.

Applying Math

$$\text{time} = \frac{\text{distance}}{\text{speed}} = \frac{5{,}600 \text{ km}}{300 \text{ km/h}} =$$
18.7 h (about 19 h)

SCIENCE Stats

Fastest Facts

Did you know...

...Nature's fastest creature is the peregrine falcon. It swoops down on its prey, traveling at speeds of more than 300 km/h. That tremendous speed enables the peregrine falcon to catch and kill other birds, which are its main prey.

...The Supersonic Transport (SST), was the world's fastest passenger jet, and cruised at twice the speed of sound. Traveling at 2,150 km/h, the SST could travel from New York to London—a distance of about 5,600 km—in 2 h 55 min 45 s.

Applying Math How long would it take a peregrine falcon moving at top speed to fly from New York to London?

...The fastest animal on land is the cheetah. This large cat can sprint at speeds of over 100 km/h. That is about as fast as a car traveling at freeway speeds, though the cheetah can only maintain top speed for a few hundred meters.

Graph It

Visit red.msscience.com/science_stats to find the top speeds of four or five land animals. Create a bar graph that compares the speeds.

Graph It

Check students' graphs.

Differentiated Instruction

Challenge Have students investigate current human speed records in various running and swimming events and create a graph comparing the top speeds. L3

Reviewing Main Ideas

Section 1 Motion

1. Average speed is the distance traveled divided by the time: $s = d/t$.
2. An object is accelerating when its speed and/or direction of motion changes.
3. Acceleration can be calculated by dividing the change in speed by the time.

Section 2 Newton's Laws of Motion

1. Newton's first law states that an object will remain at rest or move at constant speed if no net force is acting on it.
2. Newton's second law states that acceleration is given by this equation: $a = F_{net}/m$.
3. Newton's third law states that forces occur in equal but opposite pairs.

Section 3 Work and Simple Machines

1. Work equals force applied times the distance over which the force is applied: $W = Fd$.
2. A machine is a device that makes work easier. It can increase force or distance, or change the direction of an applied force.
3. The mechanical advantage is the output force divided by the input force.
4. The six types of simple machines are the lever, pulley, wheel and axle, inclined plane, wedge, and screw.

Visualizing Main Ideas

Copy and complete the following concept map on simple machines.

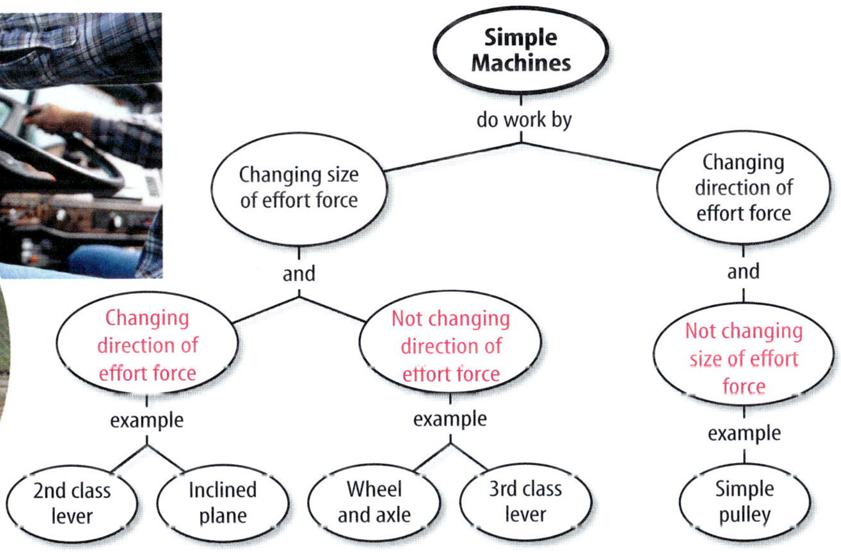

Identifying Misconceptions

Assess

Use this assessment as a follow-up after students have completed the chapter.

Procedure A tennis player applies force to a ball only when the ball and strings are touching. Have students discuss why following through when hitting the ball gives a tennis player more power.

Expected Outcome Students should realize that a force is imparted only when the ball is in contact with the strings. Following through increases the amount of time the ball and strings are in contact.

chapter 5 Study Guide

Reviewing Main Ideas

Summary statements can be used by students to review the major concepts of the chapter.

Visualizing Main Ideas

See student page.

Science Online

Visit red.msscience.com
/self_check_quiz
/interactive_tutor
/vocabulary_puzzlemaker
/chapter_review
/standardized_test

Assessment Transparency

For additional assessment questions, use the *Assessment Transparency* located in the transparency book.

Assessment

Assessment Transparency **Motion, Forces, and Simple Machines**

Directions: *Carefully review the diagram and answer the following questions.*

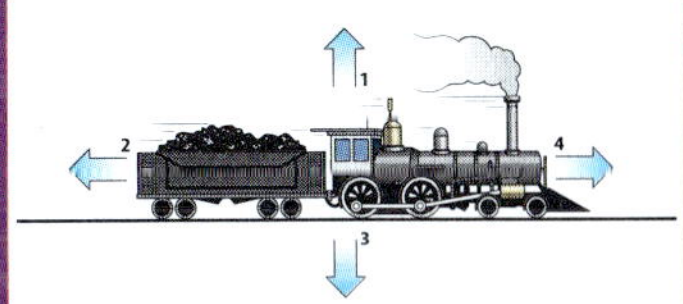

1. Newton's third law says that forces act in opposing pairs. According to this information, what is the direction of the force the railroad tracks exert on the train?
   A 1
   B 2
   C 3
   D 4
2. Which arrow indicates the force of friction?
   F 1
   G 2
   H 3
   J 4
3. If the forces indicated by arrows 2 and 4 are the same size, which of the following could describe the train's motion?
   A Speeding up before reaching a hill
   B Slowing down as it approaches a downward slope
   C Running along a straight track at a constant speed
   D Applying the brakes slightly

Motion, Forces, and Simple Machines

Using Vocabulary

1. The more mass an object has, the greater its inertia and the greater the amount of force needed to move it.
2. Acceleration is how velocity changes with time.
3. Both levers and pulleys are simple machines.
4. Work is done when a force causes an object to move in the direction in which the force is applied.
5. A simple machine is a device that uses only one movement and makes work easier by changing the size and/or direction of the force applied.
6. Newton's laws of motion explain how forces are related to motion.
7. Friction is a force that resists motion between two surfaces in contact.
8. Mechanical advantage is a machine's output force divided by its input force.
9. If an object travels at a constant speed, its instantaneous speed equals its average speed.
10. A compound machine is a combination of simple machines.

Checking Concepts

11. B	15. B	19. D
12. A	16. C	20. B
13. A	17. A	
14. A	18. C	

chapter 5 Review

Using Vocabulary

acceleration p. 133
average speed p. 130
compound machine p. 146
force p. 136
friction p. 139
inclined plane p. 149
inertia p. 139
instantaneous speed p. 131
lever p. 148
mechanical advantage p. 146
Newton's laws of motion p. 138
pulley p. 147
simple machine p. 146
velocity p. 133
work p. 144

For each set of vocabulary words below, explain the relationship that exists.

1. inertia—force
2. acceleration—velocity
3. lever—pulley
4. force—work
5. work—simple machine
6. Newton's laws of motion—force
7. friction—force
8. force—mechanical advantage
9. average speed—instantaneous speed
10. simple machine—compound machine

Checking Concepts

Choose the word or phrase that best answers the question.

11. What decreases friction?
    - **A)** rough surfaces
    - **B)** smooth surfaces
    - **C)** more speed
    - **D)** more surface area
12. What happens when an unbalanced force is applied to an object?
    - **A)** The object accelerates.
    - **B)** The object moves with constant velocity.
    - **C)** The object remains at rest.
    - **D)** The force of friction increases.
13. Which is an example of a simple machine?
    - **A)** baseball bat
    - **B)** scissors
    - **C)** can opener
    - **D)** car
14. What simple machines make up an ax?
    - **A)** a lever and a wedge
    - **B)** two levers
    - **C)** a wedge and a pulley
    - **D)** a lever and a screw
15. A car is driving at constant velocity. Which of the following is NOT true?
    - **A)** All the forces acting are balanced.
    - **B)** A net force keeps it moving.
    - **C)** The car is moving in a straight line with constant speed.
    - **D)** The car is not accelerating.
16. A large truck bumps a small car. Which of the following is true?
    - **A)** The force of the truck on the car is greater.
    - **B)** The force of the car on the truck is greater.
    - **C)** The forces are the same.
    - **D)** No force is involved.
17. What are the units for acceleration?
    - **A)** m/s^2
    - **B)** $kg\ m/s^2$
    - **C)** m/s
    - **D)** N
18. What is inertia related to?
    - **A)** speed
    - **B)** gravity
    - **C)** mass
    - **D)** work
19. Which of the following is a force?
    - **A)** inertia
    - **B)** acceleration
    - **C)** speed
    - **D)** friction
20. How does a fixed pulley make doing work easier?
    - **A)** It decreases the distance over which the input force needs to be applied.
    - **B)** It changes the direction of the input force.
    - **C)** It increases the input force.
    - **D)** It decreases the input force.

Science Online red.msscience.com/vocabulary_puzzlemaker

Use the Exam*View*® Pro Testmaker CD-ROM to:
- create multiple versions of tests
- create modified tests with one mouse click for inclusion students
- edit existing questions and add your own questions
- build tests aligned with state standards using built-in State Curriculum Tags
- change English tests to Spanish with one mouse click and vice versa

Thinking Critically

21. Apply You run 100 m in 25 s. If you then run the same distance in less time, how does your average speed change?

Use the graph below to answer question 22.

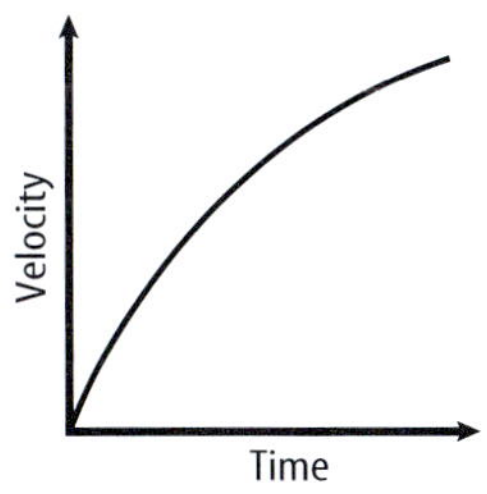

22. Make and Use Graphs A sprinter's speed over a 100-m dash is shown in the graph below. Was the sprinter speeding up, slowing down, or running at a constant speed?

23. Explain why a fast-moving freight train might take several kilometers to stop after the brakes have been applied.

24. Measure in SI Which of the following speeds is the fastest: 20 m/s, 200 cm/s, or 0.2 km/s? Here's a hint: *Express all the speeds in meters per second and compare.*

25. Draw Conclusions You are rolling backward down a hill on your bike and use your brakes to stop. In what direction was the acceleration?

26. Infer whether the forces acting on a car are balanced or unbalanced if the car is turning while moving at a constant speed.

27. Compare the force of friction on a book with the force you apply when you push a book across a table at constant speed.

Performance Activities

28. Oral Presentation Prepare a presentation, with props, to explain one of Newton's laws of motion to a third-grade class.

29. Invention Design a human-powered compound machine to do a specific job. Identify the simple machines used in your design, and describe what each of the simple machines does.

Applying Math

30. Work Find the work done by a force of 30 N exerted over a distance of 3 m.

31. Mechanical Advantage Find the mechanical advantage of a ramp 8 m long that extends from the sidewalk to a 2-m-high porch.

32. Force Find the force exerted by the rocket engines on a space shuttle that has a mass of 2 million kg if it accelerates at 30 m/s^2.

Use the graph below to answer question 33.

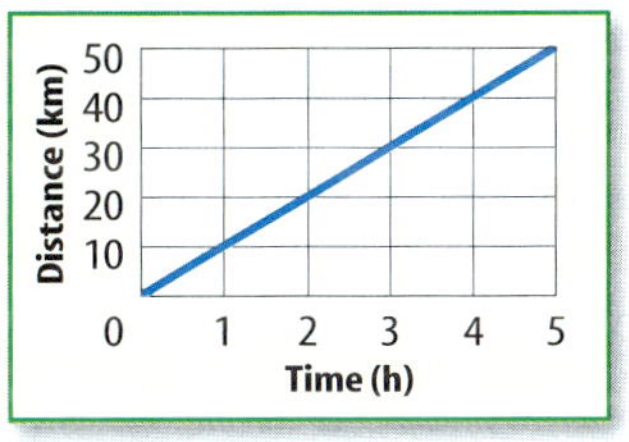

33. Speed and Time The graph above is a distance-time graph of Marion's bicycle ride. What is Marion's average speed? How long did it take her to travel 25 km?

34. Work At the 1976 Olympics, Vasili Aleseev shattered the world record for weight lifting when he lifted 2,500 N from the floor to over his head, a point 2 m above the ground. How much work did he do?

Science Online red.msscience.com/chapter_review

chapter Review 5

Thinking Critically

21. It increases because you ran the second distance faster.
22. speeding up
23. Its large mass gives it so much inertia that it is difficult to stop.
24. 0.2 km/s
25. toward the top of the hill
26. The direction of motion is changing, so the car is accelerating. By Newton's second law, an unbalanced force must be acting on the car if it is accelerating.
27. The force of friction is less because the book is moving in the direction of the push.

Performance Activities

28. Use **Performance Assessment in the Science Classroom,** p. 143.
29. Use **Performance Assessment in the Science Classroom,** p. 143.

Applying Math

National Math Standards

1, 2, 6, 9

30. $W = (30\text{ N}) \times (3\text{ m}) = 90\text{ J}$
31. $MA = (8\text{ m})/(2\text{ m}) = 4$
32. $F = (2{,}000{,}000\text{ kg}) \times (30\text{ m/s}^2) = 60{,}000{,}000\text{ N}$
33. Marion traveled 50 km in 5 h, so her average speed was 10 km/h. The time to travel 25 km is time $= \frac{\text{distance}}{\text{avg speed}} = \frac{25\text{ km}}{10\text{ km/h}} = 2.5\text{ h}$ (This answer could also be read from the graph.)
34. $W = (2{,}500\text{ N}) \times (2\text{ m}) = 5{,}000\text{ J}$

Assessment Resources

Reproducible Masters

Chapter *Fast File* Resources
Chapter Review, pp. 35–36
Chapter Tests, pp. 37–40
Assessment Transparency Activity, p. 47

Glencoe Science Web site
Chapter Review Test
Standardized Test Practice

Glencoe Technology

Assessment Transparency
Exam*View*® Pro Testmaker
MindJogger Videoquiz
Interactive Chalkboard

FAST FILE

Answer Sheet A practice answer sheet can be found at red.msscience.com/answer_sheet.

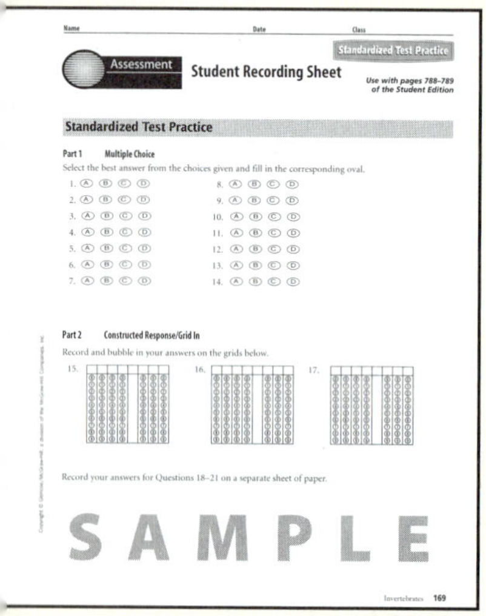

Name Date Class

Assessment Student Recording Sheet

Standardized Test Practice

Use with pages 788–789 of the Student Edition

Standardized Test Practice

Part 1 Multiple Choice

Select the best answer from the choices given and fill in the corresponding oval.

Part 2 Constructed Response/Grid In

Record and bubble in your answers on the grids below.

Record your answers for Questions 18–21 on a separate sheet of paper.

SAMPLE

Part 1 Multiple Choice

1. D
2. A
3. B
4. A
5. C
6. B
7. C
8. C
9. B
10. C
11. C

Part 2 Short Response

12. You do no work because the desk did not move. To do work, a force must be applied, and the object must move in the direction of the force.
13. A wedge is a moving inclined plane with one or two sloping sides.
14. 23 N
15. $-24\ m/s^2$
16. $0.4\ m/s^2$
17. The cart's acceleration would be smaller because its mass would be greater. The more mass an object has, the harder it is to accelerate.
18. The force of friction acts on a moving object to slow it down and eventually stop it.
19. wedge and lever
20. If the two forces are equal but in opposite directions, the net force on the rope is zero.

chapter 5 Standardized Test Practice

Part 1 Multiple Choice

Record your answers on the answer sheet provided by your teacher or on a sheet of paper.

1. What happens when a ball rolls uphill?

A. Its speed increases.
B. Its acceleration is zero.
C. Its motion and acceleration are in the same direction.
D. Its motion and acceleration are in opposite directions.

2. Which of the following is a second-class lever?

A. wheelbarrow
B. hockey stick
C. scissors
D. crowbar

Use the figure below to answer questions 3 and 4.

3. What is this simple machine called?

A. wedge
B. inclined plane
C. pulley
D. screw

4. Which of the following statements is true when you use this simple machine?

A. Less force is needed to move the couch.
B. More force is needed to move the couch.
C. The couch is moved a shorter distance.
D. Less work is needed to move the couch.

5. Which of the following is the force that resists sliding motion between two surfaces in contact?

A. inertia
B. acceleration
C. friction
D. gravity

6. How much work do you do if you push with a force of 33 N on a box while sliding it 11 m?

A. 22 J
B. 363 J
C. 3 J
D. 44 J

7. Which of the following is NOT true about what machines are used to do?

A. They make it easier to do work.
B. They change the direction of a force.
C. They increase the amount of work done on an object.
D. They reduce the force needed to do the work.

Use the figure below to answer questions 8 and 9.

8. What does the odometer in a car measure?

A. average speed
B. instantaneous speed
C. distance
D. constant speed

9. What does a car's speedometer measure?

A. average speed
B. instantaneous speed
C. distance
D. constant speed

10. A skier is going down a hill at a speed of 9 m/s. The hill gets steeper and her speed increases to 18 m/s in 3 s. What is her acceleration?

A. $9\ m/s^2$
B. $27\ m/s^2$
C. $3\ m/s^2$
D. $6\ m/s^2$

11. Which of the following does not include a direction?

A. displacement
B. force
C. speed
D. velocity

Part 3 Open Ended

21. Safety belts keep people from continuing to move forward when a car stops suddenly. This keeps people from hitting the steering wheel or dashboard of the car.
22. Velocity changes when you turn the corner because direction of motion changes, even if speed does not change. Velocity changes from north to east.
23. You could speed up by walking faster, slow down by walking slower, and change directions by turning the corner.
24. The crew applies a backward force to the water through the oars. The water then applies an equal and opposite force on the oars. This force moves the oars and the boat to which they are attached forward through the water.
25. The truck will have a higher inertia, so it will change speed or direction less quickly when it's carrying a

Part 2 Short Response/Grid In

Record your answers on the answer sheet provided by your teacher or on a sheet of paper.

12. How much work do you do if you push with a force of 100 N on a desk that does not move?

13. How is a wedge like an inclined plane?

14. Using a pulley system with a mechanical advantage of 15, how large an input force would be needed to lift a piano weighing 345 N?

15. If an car is traveling at a speed of 120 m/s and then comes to a stop in 5 s, what is its acceleration?

Use the figure below to answer questions 15 and 16.

16. If the cart has a mass of 25 kg and the girl pushes with a force of 10 N, what is the cart's acceleration?

17. How would filling the grocery cart with canned goods affect its acceleration if the girl pushes with the same force? Explain.

18. If Newton's first law of motion is correct, why do moving objects on Earth eventually stop moving?

19. What simple machines make up scissors?

20. If two teams in a tug-of-war pull on the rope with the same force, but in opposite directions, what can you say about the net force on the rope?

Part 3 Open Ended

Record your answers on a sheet of paper.

21. How do safety belts in cars protect people against the effects of Newton's first law of motion?

Use the figure below to answer questions 21 and 22.

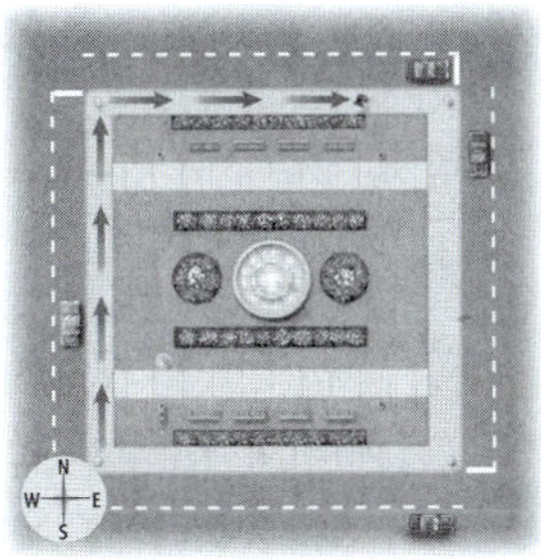

22. Describe what happens to your velocity as you walk along the path shown.

23. Describe three ways that your acceleration could change as you walk along the path.

24. Use Newton's third law of motion to explain the direction a boat crew must work the oars to move the boat forward.

25. A man decides to move some furniture in the back of his pick-up truck. What should the driver remember from Newton's second law when the pick-up is carrying a heavy load?

26. Explain the difference between an ideal machine and a real machine in terms of work in and work out.

27. Explain the differences between the three classes of levers in term of the location of the fulcrum, input force, and output force or load.

28. Explain why a child riding in a circle on a merry-go-round at a constant speed is accelerating.

Part 3 Open Ended

heavy load than when it is empty. The driver must brake earlier than usual and should slow down before entering a curve to make it easier to change direction.

26. In an ideal machine, there is no friction, so work in equals work out. In a real machine, friction always occurs as one part moves past another. Friction causes some input work to be changed to heat, so work out is always less than work in.

27. first-class lever: fulcrum is between input and output force; second-class lever: output force is between input force and fulcrum; third-class lever: input force is between fulcrum and output force

28. As the merry-go-round moves in a circle, the child is not moving in a straight line and so is changing direction. Any change in direction is acceleration.

Rubrics

The following rubrics are sample scoring devices for short response and open-ended questions.

Short Response

Points	Description
2	The student demonstrates a thorough understanding of the science of the task. The response may contain minor flaws that do not detract from the demonstration of a thorough understanding.
1	The student has provided a response that is only partially correct.
0	The student has provided a completely incorrect solution or no response at all.

Open Ended

Points	Description
4	The student demonstrates a thorough understanding of the science of the task. The response may contain minor flaws that do not detract from the demonstration of a thorough understanding.
3	The student demonstrates an understanding of the science of the task. The response is essentially correct and demonstrates an essential but less than thorough understanding of the science.
2	The student demonstrates only a partial understanding of the science of the task. Although the student may have used the correct approach to a solution or may have provided a correct solution, the work lacks an essential understanding of the underlying science concepts.
1	The student demonstrates a very limited understanding of the science of the task. The response is incomplete and exhibits many flaws.
0	The student provides a completely incorrect solution or no response at all.

chapter 6 Organizer

Section/Objectives	Standards		Labs/Features
	National	**State/Local**	
Chapter Opener	See p. 16T–17T for a Key to Standards.		**Launch Lab:** Forms of Energy, p. 161 **Foldables,** p. 161 A data-collection lab using Probeware technology can be found in the **Probeware Lab Manual,** pp. 45–47
Section 1 Energy Changes 2 sessions 1 block 1. **Explain** what energy is. 2. **Describe** the forms energy takes. 3. **Compare and contrast** potential energy and kinetic energy.	National Content Standards: UCP.1–3, A.1, A.2, B.1–3		**Integrate Life Science,** p. 164 **Visualizing Kinetic Energy,** p. 165 **MiniLAB:** Comparing Kinetic Energy and Height, p. 167 **Science Online,** p. 168
Section 2 Temperature 3 sessions 1.5 blocks 4. **Distinguish** between temperature and heat. 5. **Identify** important uses of heat. 6. **Explain** how heat moves.	National Content Standards: UCP.1–3, UCP.5, A.1, A.2, B.1–3		**Applying Science:** Can you be fooled by temperature?, p. 171 **Integrate History,** p. 174 **MiniLAB:** Comparing Energy Content, p. 176
Section 3 Chemical Energy 4 sessions 2 blocks 7. **Determine** how chemical energy is transformed. 8. **Explain** how reaction rates are changed.	National Content Standards: UCP.1–3, UCP.5, A.1, A.2, B.1–3		**Science Online,** p. 181 **Lab:** Converting Potential and Kinetic Energy, p. 183 **Lab:** Comparing Temperature Changes, pp. 184–185 **Science and Language Arts:** Hiroshima, p. 186

Glencoe Exclusive! TeacherWorks™ All-In-One Planner and Resource Center

Lab Materials	Reproducible Resources	Section Assessment	Technology
Launch Lab: beaker, hotplate, flashlight, pencil, baseball, clay	**Chapter *FAST FILE* Resources** Foldables Worksheet, p. 15 Directed Reading Overview, p. 17 Note-taking Worksheets, pp. 31–32	Glencoe's Assessment Advantage	TeacherWorks includes: • Interactive Teacher Edition • Lesson Planner with calendar • Access to all program blacklines • Correlations to standards • Web links
MiniLAB: clay, piece of cardboard, baseball, golf ball, or orange, meterstick *Need materials?* Contact Science Kit at 1-800-828-7777 or www.sciencekit.com on the Internet.	**Chapter *FAST FILE* Resources** Transparency Activity, p. 42 MiniLAB, p. 3 Enrichment, p. 28 Reinforcement, p. 25 Directed Reading, p. 18 Transparency Activity, pp. 45–46 **Reading and Writing Skill Activities,** p. 35 **Home and Community Involvement,** p. 38	**Portfolio** Assessment, p. 169 **Performance** MiniLAB, p. 167 Applying Skills, p. 169 **Content** Section Review, p. 169	Section Focus Transparency Teaching Transparency Guided Reading Audio Program Interactive Chalkboard CD-ROM
MiniLAB: 3 transparent containers; hot, cold, and room-temperature water; dropper; food coloring	**Chapter *FAST FILE* Resources** Transparency Activity, p. 43 MiniLAB, p. 4 Enrichment, p. 29 Reinforcement, p. 26 Directed Reading, p. 19 Lab Activity, pp. 9–11 **Mathematics Skill Activities,** p. 37 **Reading and Writing Skill Activities,** p. 33	**Portfolio** Assessment, p. 177 **Performance** Applying Science, p. 171 MiniLAB, p. 176 Applying Math, p. 177 **Content** Section Review, p. 177	Section Focus Transparency Virtual Labs CD-ROM Guided Reading Audio Program Interactive Chalkboard CD-ROM
Lab: cardboard, triple-beam balance, table-tennis ball, tennis ball, baseball, stopwatch, meterstick **Lab:** thermometers (4), self-sealing freezer bags (2), water, ice cubes, pancake syrup (100 mL), 400- or 600-mL beakers (4), spoon or stirring rod	**Chapter *FAST FILE* Resources** Transparency Activity, p. 44 Enrichment, p. 30 Reinforcement, p. 27 Directed Reading, pp. 19, 20 Lab Worksheets, pp. 5–6, 7–8 Lab Activity, pp. 13–14	**Portfolio** Visual Learning, p. 180 **Performance** Applying Skills, p. 182 **Content** Section Review, p. 182	Section Focus Transparency Guided Reading Audio Program Interactive Chalkboard CD-ROM Video Lab Probeware Lab

End of Chapter Assessment

Glencoe's Assessment Advantage

Blackline Masters	Technology	Professional Series
Chapter *FAST FILE* Resources Chapter Review, pp. 35–36 Chapter Tests, pp. 37–40 **Standardized Test Practice,** pp. 27–30	MindJogger Videoquiz Virtual Labs CD-ROM Exam*View*® Pro Testmaker TeacherWorks CD-ROM Interactive Chalkboard CD-ROM	**Performance Assessment in the Science Classroom (PASC)**

Transparencies

Section Focus

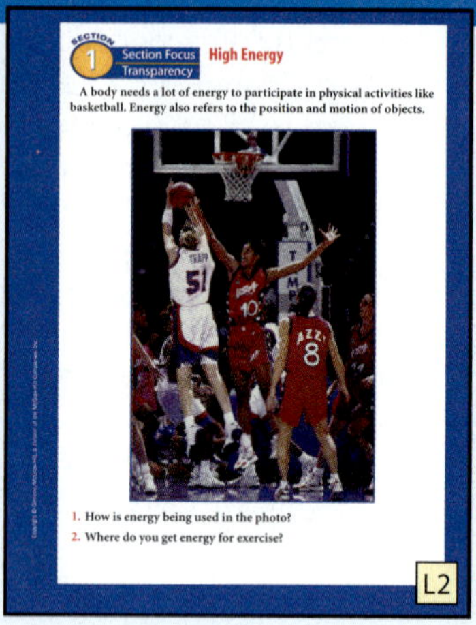

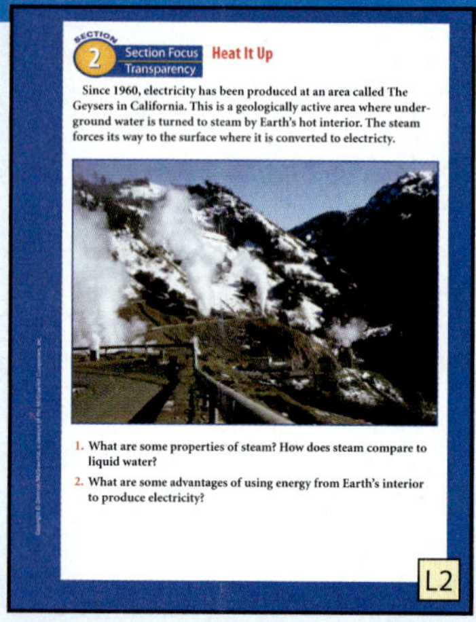

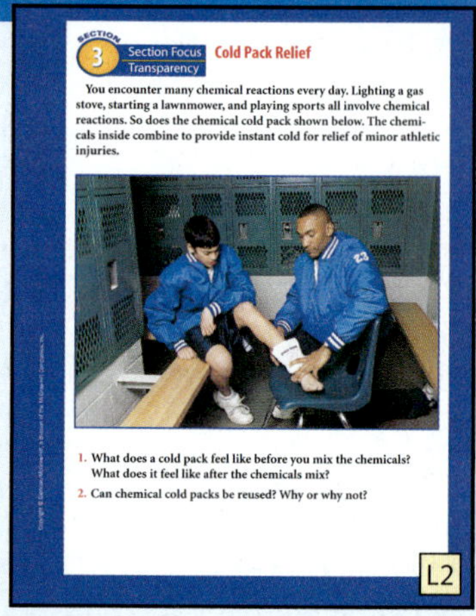

This is a representation of key blackline masters available in the Teacher Classroom Resources. See Resource Manager boxes within the chapter for additional information.

Assessment

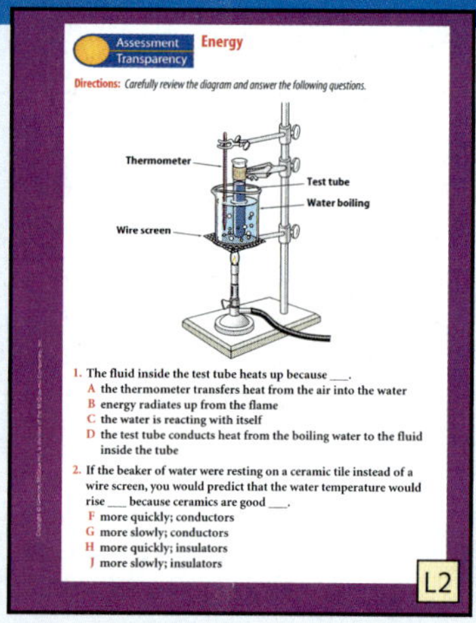

Teaching

Key to Teaching Strategies

The following designations will help you decide which activities are appropriate for your students.

- L1 Level 1 activities should be appropriate for students with learning difficulties.
- L2 Level 2 activities should be within the ability range of all students.
- L3 Level 3 activities are designed for above-average students.
- ELL ELL activities should be within the ability range of English Language Learners.
- COOP LEARN Cooperative Learning activities are designed for small group work.
- LS Multiple Learning Styles logos, as described on page 12T, are used throughout to indicate strategies that address different learning styles.
- P These strategies represent student products that can be placed into a best-work portfolio.
- PBL Problem-Based Learning activities apply real-world situations to learning.

Hands-on Activities

Student Text Lab Worksheet

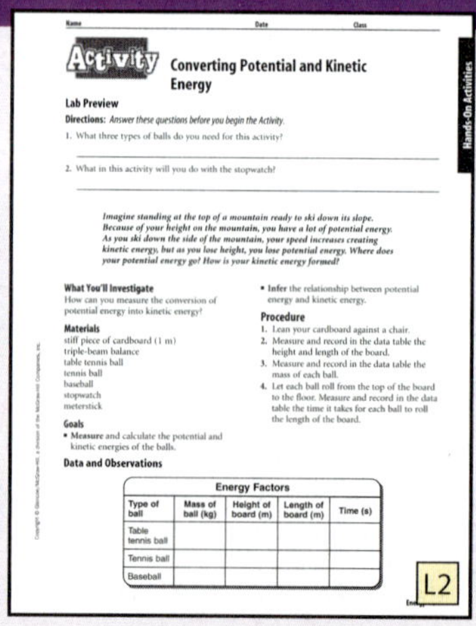

Laboratory Activities

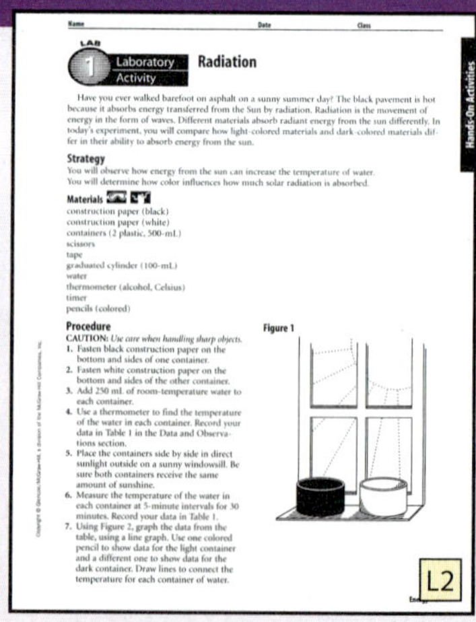

Meeting Different Ability Levels

Content Outline

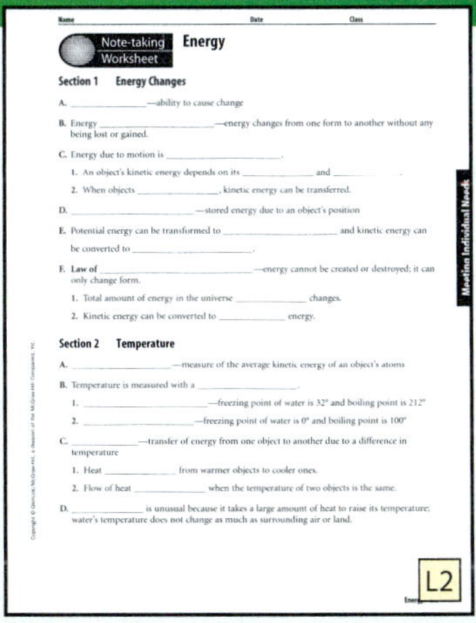

Reinforcement

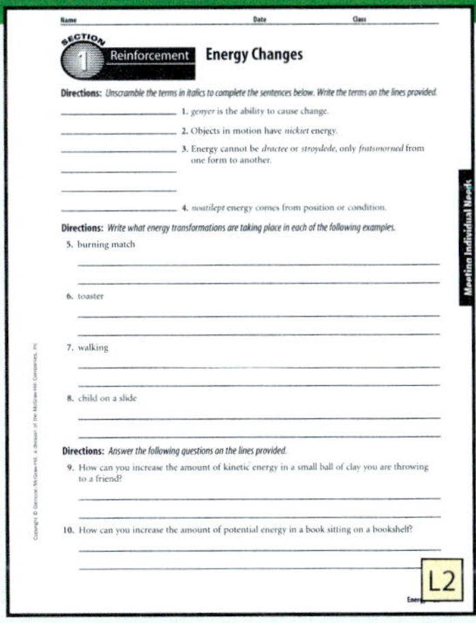

Enrichment

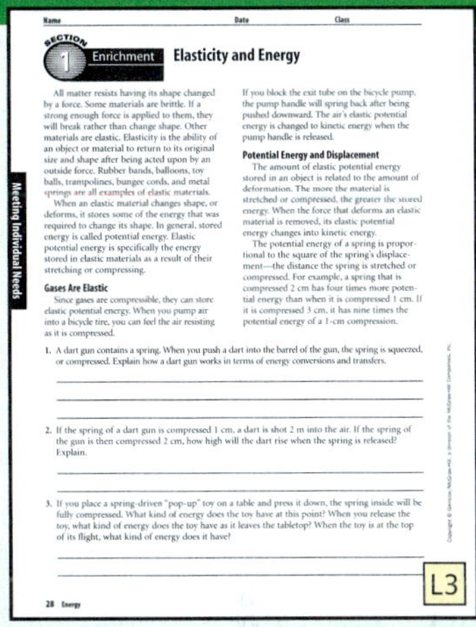

Directed Reading (English/Spanish)

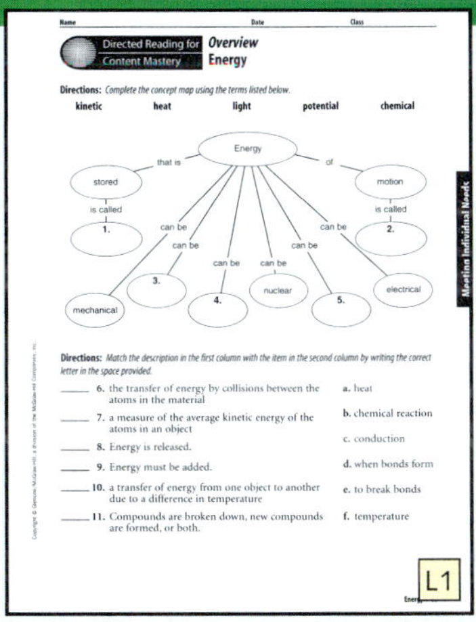

Study Guide

Reading Essentials

Assessment

Test Practice Workbook

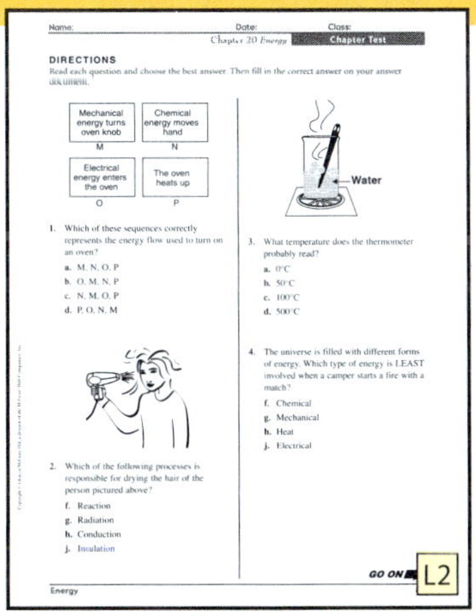

Chapter Review

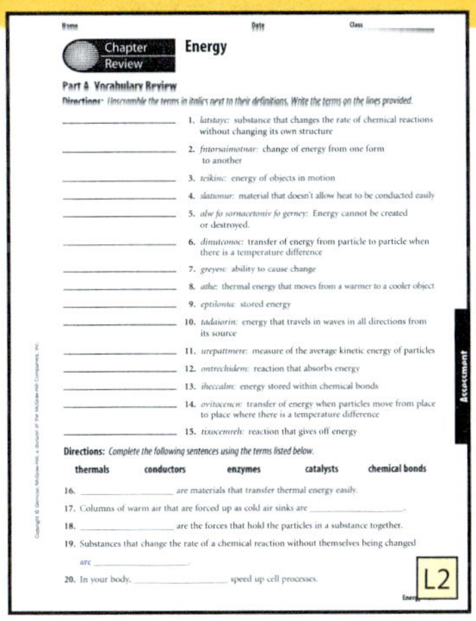

Chapter Tests

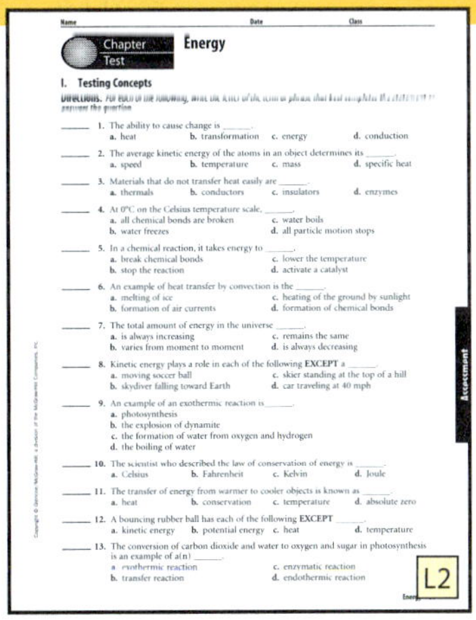

Science Content Background

section 1 Energy Changes

What is Energy?

A moving baseball has energy; you can see and feel the change that occurs as you catch it in your hand. Light is a form of energy; it can cause change in pigments in photographic film and in the retina of the eye. It can also cause a sunburn. The concept of energy can be divided into two categories: forms of stored energy and forms of energy involved in transfer and/or transformation.

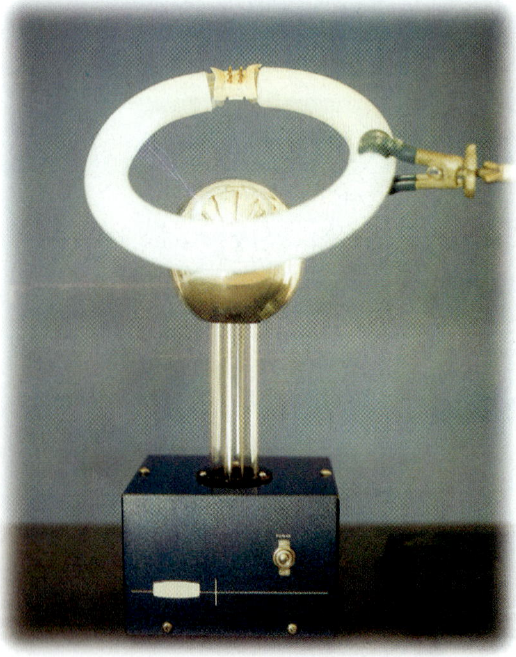

Runk/Schoenberger/Grant Heilman Photography, Inc.

Energy Transfers and Transformations

There are many examples of transfer or transformations of energy. Perhaps the most familiar is work. Work is the transfer of mechanical energy. The transfer of energy resulting from the collision between molecules is heat. Sound and light are two examples of waves that can transfer energy. Other examples of energy transformations are chemical and nuclear reactions, which commonly transform potential energy into other types of energy.

Potential Energy

Potential energy is associated with the fundamental forces in nature. Each force—gravitational, electromagnetic, and nuclear forces—can be thought of in terms of a field. Energy stored because of position in one of these fields is potential energy. Consequently, there are three basic types of potential energy: gravitational, electromagnetic, and nuclear. Two other types of potential energy are commonly discussed: elastic potential energy and chemical potential energy. Both of these types fundamentally are the result of electromagnetic interactions between atoms and molecules.

section 2 Temperature

Temperature and Kinetic Energy

In a gas, temperature is proportional to the average kinetic energy of the molecules. By comparing the temperatures of two samples of material, you are also comparing the average kinetic energy of their particles.

chapter content resources

Internet Resources

For additional content background, visit red.msscience.com to:

- access your book online
- find references to related articles in popular science magazines
- access Web links with related content background
- access current events with science journal topics

Print Resources

Renewable Energy, by Brent Sorensen, Academic Press, 2000

The Feynman Lectures on Physics: Feynman on Fundamentals, by Richard P. Feynman, Perseus Publishing, 1999

The Energy of Nature, by E.C. Pielou, University of Chicago Press, 2001

Kinetic energy is directly related to speed. Under normal conditions, the average speed of nitrogen molecules in air is about 450 m/s. These nitrogen molecules are continually colliding and exchanging energy with other molecules. The nitrogen molecules have a range of speeds. A few nitrogen molecules will move slowly, less than 100 m/s, and a few will be going quite fast, greater than 1,000 m/s.

Measuring Temperature

In the kinetic theory, the temperature of a substance is proportional to the average kinetic energy of its molecules. This seems to imply that the temperature of the material would be zero if the average kinetic energy were zero. In other words, molecules would stop moving at 0 K, or absolute zero. However, at very low temperatures, effects due to the internal structure of atoms and molecules become important. As a result, all motion does not cease at absolute zero. Instead molecules still have some kinetic energy, called the zero-point energy.

Heat will be transferred from an object at a higher temperature to an object at a lower temperature if they are in contact. If no thermal energy is transferred between two objects in contact, they are at the same temperature.

Heat and Temperature

Temperature and thermal energy are related, but they have distinct properties. Temperature is a measure of the average kinetic energy of particles in any object, whereas thermal energy is the total energy of the particles. A cup of 90°C tea is much hotter than a bathtub full of 42°C water, but there is more thermal energy in the bathtub of water because the amount of water in the bathtub is much greater. After standing for half an hour, the tea is near room temperature, and the bathtub water's temperature has barely gone down. The amount of thermal energy transferred to the bathtub is too small to change its temperature significantly.

section 3 Chemical Reactions

Chemical Reactions and Energy

To break bonds means to pull atoms apart, which requires energy input to overcome the electrostatic force of attraction between the atoms. Likewise, when bonding atoms accelerate toward each other to form a chemical bond, energy is released.

An exothermic reaction can sustain itself as long as there are reactants. However, for an endothermic reaction to continue, there must be not only reactants but also an energy supply from the environment.

Changing the Rate of Reaction

Enzymes can increase the reaction rate billions of times faster than it would occur if the enzyme weren't present. For example, in aerobic respiration, toxic hydrogen peroxide is produced. The enzyme catalase can convert 40 million molecules of hydrogen peroxide into harmless water and oxygen every second. This is the same reaction that you see when you pour hydrogen peroxide on a wound. Catalase from injured cells rapidly breaks down the hydrogen peroxide, producing oxygen gas and water.

Teacher to Teacher

John Burger, Assistant Professor
New England Institute of Technology
Warwick, RI

"Students can observe how the human body converts the potential chemical energy in food to mechanical energy. They can use a table showing the amount of energy used in doing various activities to calculate how long they could do each activity using the energy in a 300-Calorie breakfast bar."

John Burger

chapter 6

Chapter Vocabulary

energy, p. 162
kinetic energy, p. 164
potential energy, p. 166
law of conservation of energy, p. 168
temperature, p. 171
heat, p. 173
conduction, p. 174
convection, p. 176
radiation, p. 177
endothermic reaction, p. 180
exothermic reaction, p. 181
catalyst, p. 182

Science Journal Answers will vary. Possible answers: getting dressed, turning on a light, walking up stairs

INTERACTIVE CHALKBOARD with Image Bank

PowerPoint® Presentations

This CD-ROM is an editable Microsoft® PowerPoint® presentation that includes:

- a pre-made presentation for every chapter
- interactive graphics
- animations
- audio clips
- image bank
- all new section and chapter questions
- Standardized Test Practice
- transparencies
- pre-lab questions for all labs
- Foldables directions
- links to red.msscience.com

chapter 6

Energy

chapter preview

sections

1 Energy Changes
2 Temperature
3 Chemical Energy
Lab Converting Potential and Kinetic Energy
Lab Comparing Temperature Changes
Virtual Lab How is energy converted from one form to another?

Snow Surfing

With no more than a board, a lot of snow, and a slope, this snowboarder goes from standing to moving at speeds greater than 50 km/h. The snowboarder's speed increases because energy is changing form. In fact, energy causes all the changes that occur around you every day.

Science Journal List three changes that you have seen occur today, and describe what changed.

Theme Connection

Energy This chapter discusses energy and how it can be transformed, transferred, released, and absorbed in systems.

About the Photo

Snowboarding The 2002 Winter Olympics in Salt Lake City, Utah, included two snowboarding events: the halfpipe competition and the parallel giant slalom. The fastest speeds reached by snowboarders are during Super G races, where snowboarders can reach speeds of 45 mph.

Start-Up Activities

Launch LAB

Forms of Energy

Think of all the things you do every day such as walking to class, riding to school, switching on a light, cooking food, playing music, or stopping your bike. All of the actions of your daily life involve energy and changing energy from one form to another. What forms can energy take? In what ways can energy change from one form to another?

1. Place a beaker filled with water on a hotplate and bring the water to a boil.
2. Switch on a flashlight.
3. Rub a pencil back and forth between your palms as fast as you can.
4. Drop a baseball from a height of 2 m into a layer of clay.
5. **Think Critically** During each step of this lab, you converted one form of energy into another form. Write a paragraph in your Science Journal listing the changes that took place and why they occurred.

Science Online Preview this chapter's content and activities at red.msscience.com

FOLDABLES Study Organizer

Energy Make the following Foldable to help you understand the different ways energy can be transformed.

STEP 1 **Fold** a sheet of paper in half lengthwise.

STEP 2 **Fold** paper down 2.5 cm from the top. (Hint: From the tip of your index finger to your middle knuckle is about 2.5 cm.)

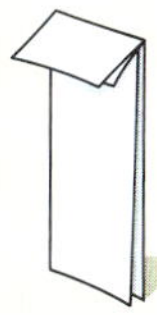

STEP 3 **Open and draw** lines along the 2.5 cm fold. Label as shown.

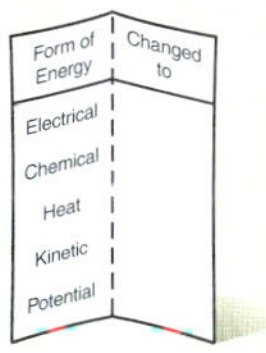

Cause and Effect As you read the chapter, write answers to what each form of energy can be changed to under the *Changed To* heading on your Foldable. On the back of your Foldable describe what caused each form of energy to change and explain the effects of the change.

Launch LAB

Purpose Use this Launch Lab to introduce students to some of the ways energy changes from one form to another. L2 ELL

LS **Kinesthetic**

Preparation Set up a hotplate for each group and collect flashlights and baseballs.

Materials beaker of water, hotplate, flashlight, pencil, baseball, clay

Teaching Strategy Assign four students to each group and have each group member decide which task he or she will demonstrate. Suggest students perform the other three tasks while waiting for the water to boil.

Think Critically

In the first step electricity was converted to heat. In step 2 chemical energy was converted to electrical energy, which was converted to light. In the third step kinetic energy was converted to heat. In step 4 the potential energy of the baseball was converted to kinetic energy.

Assessment

Process Provide students with a diagram of the power plant that supplies electrical energy to your school and ask students to identify the energy conversion used by the plant. Possible energy sources include fossil fuels (chemical energy), nuclear reactions (nuclear energy), solar collectors (solar energy), windmills (wind power), and hydroelectric power. Use **Performance Assessment in the Science Classroom**, p. 89.

FOLDABLES Study Organizer **Dinah Zike Study Fold**

Student preparation materials for this Foldable are available in the Chapter **FAST FILE Resources.**

Probeware Labs

This chapter's data-collection lab using Probeware technology is included on the *Video Labs CD-ROM.* See the *Probeware Lab Manual* pages 45–47 for student worksheets.

1 Motivate

INTERACTIVE CHALKBOARD PowerPoint® Presentations

Bellringer

Section Focus Transparencies also are available on the Interactive Chalkboard CD-ROM. L2 ELL

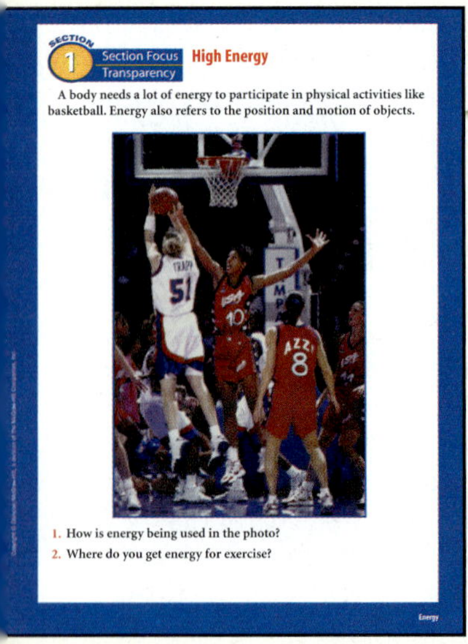

Tie to Prior Knowledge

Environment Changes Have students recall changes they have noticed in matter or the environment. Tell them that whenever anything changes in nature, the change is due to energy.

Text Question Answer

Possible answers: electric or magnetic energy, solar energy, energy of an athlete in motion

Reading Check

Answer Energy causes change to occur.

section 1 Energy Changes

as you read

What You'll Learn

- **Explain** what energy is.
- **Describe** the forms energy takes.
- **Compare and contrast** potential energy and kinetic energy.

Why It's Important

Energy causes all the changes that take place around you.

Review Vocabulary
speed: the distance traveled per second by an object

New Vocabulary
- energy
- kinetic energy
- potential energy
- law of conservation of energy

Energy

Energy is a term you probably use every day. You might say that eating a plate of spaghetti gives you energy, or that a gymnast has a lot of energy. Do you realize that a burning fire, a bouncing ball, and a tank of gasoline also have energy?

What is energy? The word *energy* comes from the ancient Greek word *energos,* which means "active." You probably have used the word *energy* in the same way. When you say you have a lot of energy, what does this mean? **Energy** is the ability to cause change. For example, energy can change the temperature of a pot of water, or it can change the direction and speed of a baseball. The energy in a thunderstorm, like the one shown in **Figure 1,** produces lightning that lights up the sky and thunder that can rattle windows. Energy can change the arrangement of atoms in molecules and cause chemical reactions to occur. You use energy when you change the speed of a bicycle by pedaling faster or when you put on the brakes.

Reading Check *What does energy do?*

Figure 1 Lightning causes dramatic change as it lights up the sky.

Forms of Energy

If you ask your friends what comes to mind when they think of energy, you probably will get many different answers. Some might mention candy bars or food. Others might think of the energy needed to run a car. Energy does come in different forms from a variety of sources. Food provides energy in the form of chemical energy. Your body converts chemical energy in the food you eat into the energy it needs to move, think, and grow. Nuclear power plants use nuclear energy contained in the center or nucleus of the atom to produce electricity. What other forms of energy can you think of?

Section 1 Resource Manager

Chapter *FAST FILE* Resources
- Note-taking Worksheets, pp. 31–32
- Directed Reading for Content Mastery, pp. 17, 18
- Transparency Activity, pp. 42, 45–46
- MiniLAB, p. 3
- Enrichment, p. 28
- Reinforcement, p. 25

Home and Community Involvement, p. 38
Life Science Critical Thinking/Problem Solving, p. 15
Physical Science Critical Thinking/Problem Solving, p. 6
Reading and Writing Skill Activities, p. 35

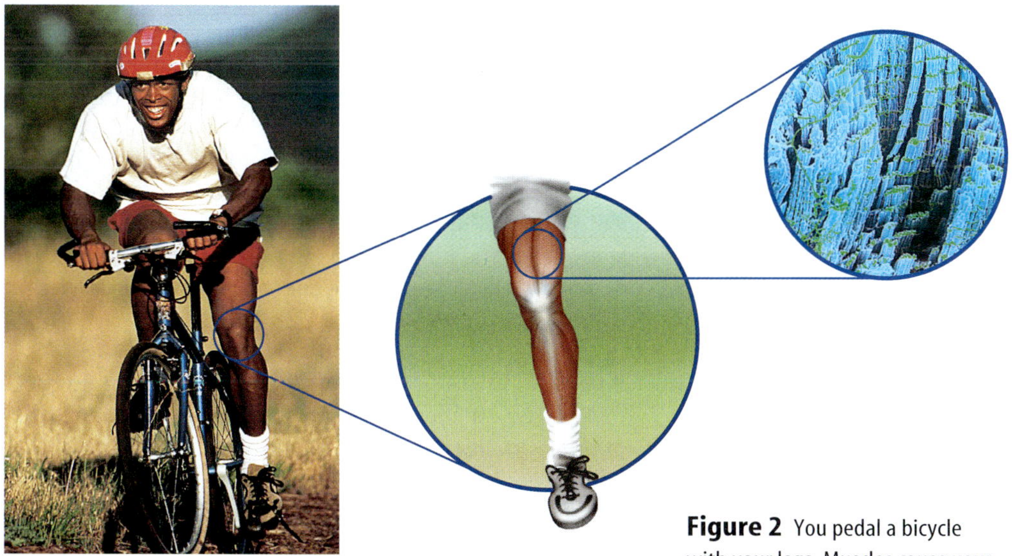

Figure 2 You pedal a bicycle with your legs. Muscles cause your leg to move by contracting. Your muscles are made of these microscopic fibers. They cause your muscles to contract by becoming shorter when certain chemical reactions release chemical energy.

Energy Transformations Energy is stored in the chemical compounds in your muscles. When you push down on a bicycle pedal, chemical energy is used to make your legs move.

An energy transformation occurs if energy changes from one form to another. Energy transformations go on all around you and inside of you all the time. The chemical energy stored in your muscles changes to energy of motion, as you can see in **Figure 2.** When a car sits in sunlight all day, the energy in sunlight changes to heat energy that warms the inside of the car. The energy you use to stretch and move a rubber band also changes into heat energy that raises the temperature of the rubber band.

During these and other types of energy transformations, the total amount of energy stays the same. Energy is never lost or gained—it only changes form.

Using Energy Transformations Since the earliest times, humans have used different forms of energy. When humans first learned to make fires, they used the chemical energy in wood and other fuels to cook, stay warm, and light their way in the dark. Today, a gas stove, like the one in **Figure 3,** transforms the chemical energy in natural gas to heat energy that boils water and cooks food. An electric current that flows in a wire carries electrical energy that can be used in many ways. A hair dryer converts electrical energy into heat energy. A lightbulb converts electrical energy into heat and light energy when you flip on a switch.

Figure 3 As natural gas burns in a gas stove, it gives off energy that heats the water.

2 Teach

Teacher FYI

Force and Energy When a force does work on an object, energy is transferred to the object. The work on an object is defined as the force exerted on the object multiplied by the distance the object moved in the direction of the force. The change in energy of the object is equal to the amount of work done.

IDENTIFYING Misconceptions

Energy is a Property Students often think of energy as a substance rather than a property. Saying that energy flows or moves can lead to confusion; particles and waves flow or move, carrying energy with them. Energy is transferred or transformed.

Fun Fact

Benjamin Franklin was one of the earliest researchers of electricity. He invented the lightning rod, which reduced the number of fires due to lightning strikes. His famous kite experiment occurred in 1752. A famous painting, depicting the event, showed his son as a young boy. In fact, he was a young man in his early twenties.

Science Journal

Transfer and Transformation Students need to distinguish clearly between a transfer of energy and the transformation of energy. Have students describe the difference between these two processes and write in their Science Journals a sentence using each. L2 ELL LS **Linguistic**

INTEGRATE Life Science

Food Is Chemical Energy Answers will vary. The calorie is an English unit of energy. The metric unit is the joule. One calorie is about 4.19 joules. When calorie is spelled with a capital C, as listed on food packages and exercise recommendations, the word actually refers to kilocalories.

Use Science Words

Word Meaning Have students use a dictionary to find the meaning of *kinetic.* relating to or caused by motion L2 ELL LS **Linguistic**

Differentiated Instruction

Challenge Introduce advanced students to the formula for kinetic energy, $KE = 1/2mv^2$. Do several examples with different numbers. Have students calculate and report how fast a 50-kg person would have to be going to have the kinetic energy of a 1,000-kg car moving 1 m/s or 3.6 km/h? The car has kinetic energy $KE = 1/2 \times 1{,}000 \text{ kg} \times (1 \text{ m/s})^2 = 500 \text{ J}$. To find the speed of a person with that much kinetic energy, solve $KE = 1/2mv^2$ for v: $500 \text{ J} = 1/2 \times 50 \text{ kg} \times v^2$; $20 = v^2$; $v = 4.47$ m/s, about 16 km/h. L3 LS **Logical-Mathematical**

Caption Answer

Figure 4 The bowling ball is knocking down the pins.

INTEGRATE Life Science

Food Is Chemical Energy You transform energy every time you eat and digest food. The food you eat contains chemical energy. This energy changes into forms that keep your body warm and move your muscles. The amount of chemical energy contained in food is measured in Calories. Check some food labels to see how many Calories your food contains.

Kinetic Energy

One soccer ball is sitting on the ground and another is rolling toward the net. How does the energy of the moving ball compare to the one at rest? A moving ball certainly has the ability to cause change. For example, a moving bowling ball shown in **Figure 4** causes the bowling pins to fall. A moving ball has energy due to its motion. The energy an object has due to its motion is called **kinetic (kih NE tihk) energy.** A football thrown by a quarterback has kinetic energy. A sky diver or a leaf falling toward Earth also has kinetic energy.

Mass, Speed, and Kinetic Energy Although moving objects have kinetic energy, not all moving objects have the same amount of kinetic energy. What determines the amount of kinetic energy in a moving object? The amount of kinetic energy an object has depends on the mass and speed of the object, as shown in **Figure 5.** Imagine a small rock and a large boulder rolling down a hillside at the same speed. Which would have more kinetic energy? Think about the damage the rock and the boulder could do if they hit something at the bottom of the hill. The large boulder could cause more damage, so it has more kinetic energy. Even though the rock and the boulder were moving at the same speed, the boulder had more kinetic energy than the rock because it had more mass.

Kinetic energy also depends on speed. The faster a bowling ball moves, the more pins it can knock down. When more pins are knocked down, a greater change has occured. So the faster the bowling ball moves, the more kinetic energy it has. Kinetic energy increases as speed increases.

Figure 4 Any moving object has energy because it can cause change.
Identify *a change that the bowling ball is causing.*

Curriculum Connection

Art Kinetic art is a form of art designed to be set in motion. Ask students to describe kinetic art works they have seen or heard of. Examples: sculptures in which balls roll and drop down along different paths, fountains, mobiles, some gardens. You might want to ask local museums if they include any kinetic art in their collections. L3 ELL LS **Visual-Spatial**

Visual Learning

Figure 4 Which has more kinetic energy: a bowling ball, or a pin moving at the same speed as the bowling ball? the bowling ball, because it has more mass L2 LS **Logical-Mathematical**

NATIONAL GEOGRAPHIC **VISUALIZING KINETIC ENERGY**

Figure 5

The amount of kinetic energy of a moving object depends on the mass and the speed of the object. For example, the fastest measured speed a baseball has been thrown is about 45 m/s. The kinetic energy of a baseball traveling at that speed is about 150 J.

▲ There is evidence that a meteorite 10 km in diameter collided with Earth about 65 million years ago and might have caused the extinction of dinosaurs. The meteorite may have been moving 400 times faster than the baseball and would have a tremendous amount of kinetic energy due to its enormous mass and high speed—about a trillion trillion joules.

▼ A 600-kg race car, traveling at about 50 m/s, has about 5,000 times the kinetic energy of the baseball.

▼ Earth's atmosphere is continually bombarded by particles called cosmic rays, which are mainly high-speed protons. The mass of a proton is about a 100 trillion trillion times smaller than the mass of the baseball. Yet, some of these particles travel so fast, they have nearly the same kinetic energy as the baseball.

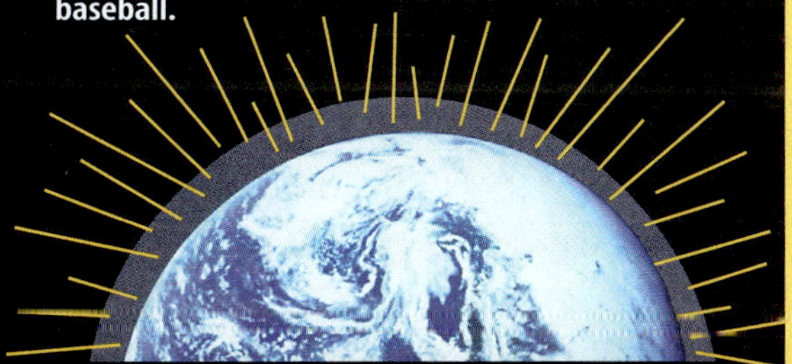

◀ A sprinter with a mass of about 55 kg and running at 9 m/s has kinetic energy about 15 times greater than the baseball.

NATIONAL GEOGRAPHIC

Visualizing Kinetic Energy

Have students examine the pictures and read the captions. Then ask the following questions.

The baseball in the feature is traveling at 45 m/s and the race car is traveling at 50 m/s. Why does the car have 5,000 times the kinetic energy of the baseball? because the race car has significantly more mass

What is the kinetic energy of the sprinter in the feature? KE of sprinter = 15 × KE of baseball = 15 × 150 J = 2250 J

Activity

Earth Collision Have students research some of the possible effects due to the collision of a meteorite with Earth 65 million years ago. Answers could include: tsunamis and earthquakes caused by the shock wave, firestorms caused by release of methane gas from the ocean floor, acid rain formed by the combustion products of the firestorms, and global cooling due to the effects of the dust ejected from the impact and the smog formed by the firestorms.

SECTION 1 Energy Changes **165**

Inquiry Lab

Observing Kinetic Energy

Purpose To explore and observe kinetic energy, have the students design their own slingshots to test the kinetic energy of rubber bands of varying lengths and widths.

Materials rubber bands of varying lengths and widths, wooden craft sticks, a small ball made of wadded-up paper, tape measure

Estimated Time 1 class session

Safety Precautions

- Conduct this lab in open space to avoid injury.
- Caution students that they must never aim rubber bands directly at another student.

Teaching Strategies

- Students can construct a simple slingshot of their own design that must withstand the forces of the rubber bands pulling on it.
- Students can predict which rubber bands will shoot the paper ball from the shortest distance to the longest distance based on the length and width of the rubber bands. Have them write their prediction down and then test it by charting their data.
- Allow students to explore how to connect the craft sticks so that they stay together.
- Allow students to decide how many craft sticks to stack on top of each other giving thickness to the slingshot to allow it to withstand the forces of each of the rubber bands.

For additional inquiry activities, see *Science Inquiry Labs*.

Caption Answer

Figure 7 The skier has all potential energy and no kinetic energy before she starts down the hill. As she starts down the hill her PE decreases as her KE increases. Finally, when she stops at the bottom of the hill her PE and KE are 0.

Figure 6 Kinetic energy is transferred from domino to domino by tapping the first one in line.

Transferring Kinetic Energy Kinetic energy can be transferred from one object to another when they collide. Think about the transfer of energy during bowling. Even if the bowling ball does not touch all of the pins, it still can knock them all down with one roll. The bowling ball transfers kinetic energy to a few pins. These pins move and bump into other pins, transferring the kinetic energy to the remaining pins and knocking them down.

A transfer of kinetic energy also takes place when dominoes fall. You need to give only the first domino in the row a bit of kinetic energy by lightly tapping it to make it fall against the next domino. As the first domino falls into the next one, its kinetic energy is transferred to the second domino, as shown in **Figure 6.** This transfer of kinetic energy continues from domino to domino until the last one falls and hits the table. Then, the last domino's kinetic energy is transferred to the table.

Potential Energy

Suppose the ski lift in **Figure 7** takes a skier to the top of a hill. The skier has no kinetic energy when she is standing at the top of the hill. But as she skis down and moves faster, her kinetic energy increases. Where does this kinetic energy come from? Gravity pulls the skier down the hill. If the skier were standing at the bottom of the hill, gravity would not start her moving, as it does when she is at the top of the hill. When the skier's position is at the top of the hill, she has a form of energy called potential energy. **Potential energy** is energy that is stored because of an object's position. By using the ski lift to take her to the top of the hill, the skier increased her potential energy by changing her position.

Figure 7 Potential and kinetic energy change as the skier moves up and down the slope.
Explain *how the skier's energy changes as she moves from the top of the hill to the bottom.*

166

Cultural Diversity

Chain Drives The chain drive, familiar to students from bicycles, was invented in China by Chang Ssu-Hsun to power a mechanical clock in 976 A.D. Chain drives did not appear in Europe until 1770. Show students an illustration of a chain drive and ask how the chain drive on a bicycle transfers energy. The kinetic energy of a foot on the pedals is transferred by the chain to the wheels. L2 LS **Visual-Spatial**

Teacher FYI

Potential Energy There are three basic types of potential energy: gravitational potential energy, electromagnetic potential energy, and nuclear potential energy. Chemical potential energy and elastic potential energy are related to the electromagnetic force.

Increasing Potential Energy When you raise an object above its original position, it has the potential to fall. If it does fall, it has kinetic energy. To raise an object, you have to transfer energy to the object. The ski lift uses energy when it takes a skier up a hill and transfers some of that energy to the skier. This energy becomes stored as potential energy in the skier. As the skier goes down the hill, the potential energy she had at the top of the hill is converted to kinetic energy.

If the skier were lifted higher, her potential energy would increase. The higher an object is lifted above Earth, the greater its potential energy.

Converting Potential and Kinetic Energy

When a skier skis down a hill, potential energy is transformed to kinetic energy. Kinetic energy also can be transformed into potential energy. Suppose you throw a ball straight up into the air. The muscles in your body cause the ball to move upward when it leaves your hand. Because it is moving, the ball has kinetic energy. Look at **Figure 8.** As the ball gets higher and higher, its potential energy is increasing. At the same time, the ball is slowing down and its kinetic energy is decreasing.

What happens when the ball reaches its highest point? The ball comes to a stop for an instant before it starts to fall downward again. At its highest point, the ball has no more kinetic energy. All the kinetic energy the ball had when it left your hand has been converted to potential energy, and the ball will go no higher. As the ball falls downward, its potential energy is converted back into kinetic energy. If you catch the ball at the same height above the ground as when you threw it upward, its kinetic energy will be the same as when it left your hand.

Mini LAB

Comparing Kinetic Energy and Height

Procedure:

1. Lay a 3-cm-thick layer of smooth modeling **clay** on a piece of **cardboard.** Place the cardboard on the floor.
2. Drop an object such as a **baseball, golf ball,** or **orange** into the clay from a height of 10 cm. Measure and record the depth of the hole made by the object.
3. Repeat step 2 from a height of 50 cm and 1 m.

Analysis

1. How does the depth of the hole depend on the height of the ball?
2. How does the kinetic energy of the falling ball depend on the distance it fell?

Try at Home

Figure 8 Energy is transformed as a ball rises and falls.

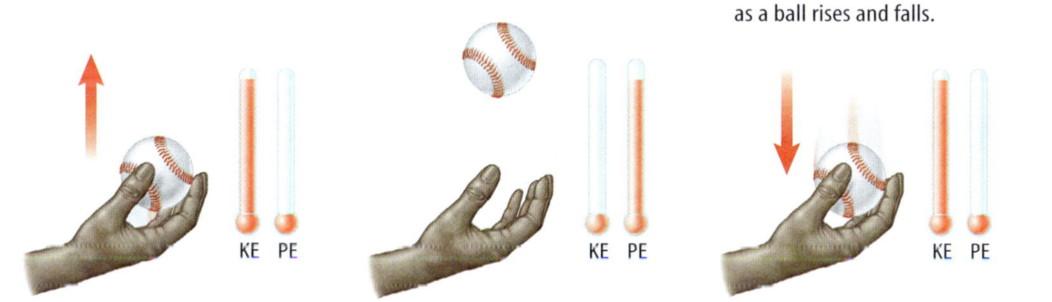

As the ball leaves the person's hand, it is moving the fastest and has maximum kinetic energy.

As the ball moves upward, it slows down as its kinetic energy is transformed into potential energy.

As the ball moves downward, it speeds up as its potential energy is transformed into kinetic energy.

Mini LAB

Purpose Students will observe and compare the effects of potential and kinetic energy. L2 LS **Kinesthetic**

Materials clay, cardboard, baseball, golf ball or orange, meterstick

Teaching Strategy Make sure students drop the objects the same way each time.

Analysis

1. The greater the distance the ball fell, the greater the depth of the hole.
2. The kinetic energy of the ball just before it hits the clay is greater if the ball is dropped from a greater height.

Assessment

Process Have each student write a newspaper article describing this MiniLAB and the results. Have students write headlines for the articles and remind them they can use tables and illustrations in the articles. Use **Performance Assessment in the Science Classroom,** p. 141.

Discussion

Getting Potential Energy What are some positions other than height above Earth that give an object potential energy? a stretched rubber band, a pulled spring L3 LS **Logical-Mathematical**

Teacher FYI

Friction Kinetic energy is regularly transformed to thermal energy by friction. When doing classroom demonstrations, keep friction in mind. Tell students that thermal energy is another form of energy, so energy is still conserved.

SECTION 1 Energy Changes 167

Reading Check

Answer kinetic energy

Quick Demo

Conserving Energy

Materials plastic tubing that is 1-m long and 1-cm in diameter, two ring stands with clamps, a marble

Estimated Time 3–5 minutes

Procedure Split the tubing in half. With the tubing forming a trough, clamp both ends of the tube to ring stands and allow the tube to form a shallow U-shaped track. Hold a marble slightly above one end of the track and ask students to predict the height that the marble will reach on the opposite side. Release the marble and have students observe the height it reaches. Rearrange the tubing to form variously shaped ramps and discuss the motion of the marble, stressing the conservation of energy.

Caption Answer

Figure 9 A dam reduces the flow down stream, causing water to pile up behind the dam. As the height of the water increases, the potential energy of the water behind the dam increases.

Energy Changes in Falling Water You might have stood close to a large waterfall and heard the roar of the water. Just like a ball falling to the ground, the potential energy that the water has at the top of the falls is transformed into kinetic energy as the water falls downward.

The kinetic energy of falling water can be used to generate electricity. As shown in **Figure 9,** water backs up behind a dam on a river, forming a lake or reservoir. The water near the top of the dam then falls downward. The kinetic energy of the moving water spins generators, which produce electricity. The potential energy of the water behind the dam is transformed into electrical energy.

Science Online

Topic: Earth's Crust
Visit red.msscience.com for Weblinks to information about faults in the crust of Earth and the energy transformations that take place in them.

Activity Describe the energy transformations that occur when there is movement along the faults.

Reading Check *The potential energy of falling water is transformed into what form of energy?*

Conservation of Energy

Following the trail of energy as it is transformed can be a challenge. Sometimes it might seem that energy disappears or is lost. But that's not the case. In 1840, James Joule demonstrated the law of conservation of energy. According to the **law of conservation of energy,** energy cannot be created or destroyed. It only can be transformed from one form into another, so the total amount of energy in the universe never changes. The only change is in the form that energy appears in.

Kinetic energy can be converted into heat energy when two objects rub against each other. As a book slides across a table, it will slow down and eventually stop. The book's kinetic energy isn't lost. It is converted into heat energy as the book rubs against the surface of the table.

Figure 9 The potential energy of water can be transformed into electrical energy.
Explain *how a dam changes the potential energy of the water behind it.*

The potential energy of water behind the dam is converted to kinetic energy as the water falls through pipes.

The kinetic energy of the moving water spins generators like these that produce electricity.

Differentiated Instruction

Learning Disabled Reinforce the idea of conservation of energy. Collect a number of small blocks. Label 4 containers *chemical energy*, *kinetic energy*, *potential energy*, and *thermal energy*. Show how a given amount of energy (e.g., 20 units) can be distributed among the containers. The total amount of energy stays the same, but it can be in different forms. L1 ELL LS **Kinesthetic**

A moving soccer player has kinetic energy.

Kinetic energy from the player's moving leg is transferred to the ball.

When the ball rolls, its kinetic energy is transformed by friction into heat as the ball rubs against the grass.

Figure 10 Energy can take different forms, but it can never be created or destroyed.

Following the Energy Trail The flow of energy as a soccer ball is kicked is shown in **Figure 10.** Chemical energy in the soccer player's leg muscles is converted into kinetic energy when she swings her leg. When the ball is kicked, this kinetic energy is transferred to the ball. After the ball rolls for a while, it comes to a stop. The kinetic energy of the ball seems to have disappeared, but it hasn't. As the ball rolled, its kinetic energy was transformed into heat energy as the ball rubbed against the grass.

section 1 review

Summary

Energy

- Energy is the ability to cause change.
- Energy has different forms such as chemical energy, nuclear energy, heat energy, and electrical energy.
- Energy can be transformed from one form into another.

Kinetic and Potential Energy

- Kinetic energy is the energy due to the motion of an object and increases as the object's mass or speed increases.
- Potential energy is stored energy due to an object's position and increases as the object's height above Earth increases.

Law of Conservation of Energy

- The law of conservation of energy states that energy cannot be created or destroyed, but only transformed from one form into another.

Self Check

1. **Describe** the energy transformations that occur when a light bulb is turned on.
2. **Explain** how the total energy changes when a falling rock hits the ground.
3. **Infer** on which part of a roller coaster a roller coaster car has the greatest potential energy.
4. **Determine** which has the greater kinetic energy if both are traveling at the same speed—a fully-loaded truck or a motorcycle.
5. **Think Critically** When a ball is thrown upward, how does the height reached by the ball depend on its initial speed?

Applying Skills

6. **Diagram** the energy transformation that occur when you eat breakfast, walk to the bus stop, and ride the bus to school.

Science Online red.msscience.com/self_check_quiz

3 Assess

DAILY INTERVENTION

Check For Understanding

Interpersonal Have the students discuss the law of conservation of energy with each other, paying special attention to the ways energy is "lost", such as thermal energy due to friction, light, and kinetic energy in vibration. Have each person give an example and illustrate how energy was conserved in the system.

Reteach

Energy Conservation Energy is conserved within a closed system. Point out to students that when a system gains energy (for example, a plant getting energy from the Sun) another system must be losing energy (the Sun is sending out the energy). Together, these two systems make a larger system in which energy has been conserved. L2 LS **Logical-Mathematical**

Assessment

Process Show students a pendulum and ask them to draw diagrams and write descriptions that explain the energy conversions that take place as the pendulum swings. Ask why a pendulum will not rise higher than the point at which it was dropped. It would have more potential energy than it had originally, so energy would not be conserved. Use **Performance Assessment in the Science Classroom,** p. 127. P

section 1 review

1. Electrical energy is converted to thermal energy and light.
2. The total energy doesn't change. When the ball hits the ground, its kinetic energy is converted into thermal energy.
3. on top of the highest hill
4. the truck because of greater mass
5. As the ball moves upward, its kinetic energy is converted into potential energy. At the ball's highest point, its potential energy equals its initial kinetic energy. The ball's potential energy increases with height, so the greater the initial kinetic energy, the greater the maximum height reached by the ball.
6. Eat breakfast and walk to bus stop: chemical energy ⟶ kinetic energy; ride bus: chemical energy ⟶ thermal energy ⟶ kinetic energy.

section 2

Temperature

as you read

What **You'll Learn**

- **Distinguish** between temperature and heat.
- **Identify** important uses of heat.
- **Explain** how heat moves.

Why **It's Important**

The flow of heat warms Earth, produces weather, cooks your food, and warms and cools your home.

Review Vocabulary

molecule: a particle formed when two or more atoms bond together

New Vocabulary

- temperature
- heat
- conduction
- convection
- radiation

Temperature

What's today's temperature? If you looked at a thermometer, listened to a weather report on the radio, or saw a weather map on television, you probably used the air temperature to help you decide what to wear. Some days are so hot you don't need a jacket. Others are so cold you want to bundle up.

Hot and *cold* are words used in everyday language to describe temperature. However, they are not scientific words because they mean different things to different people. A summer day that seems hot to one person might seem just right to another. If you grew up in Texas but moved to Minnesota, you might find the winters unbearably cold. Have you ever complained that a classroom was too cold when other students insisted that it was too warm?

Temperature and Kinetic Energy What is temperature? Remember that any material or object is made up of atoms or molecules. These particles are moving constantly, even if the object appears to be perfectly still. Every object you can think of—your hand, the pencil on your desk, or even the desktop—contains particles that are in constant motion. In solids, liquids, and gases particles do not move in a single direction. Instead they move in all directions. In a gas, particles are far apart and can move, as shown in **Figure 11.** In liquids, atoms are close together and can't move as far as in a gas. In solids, particles are bound more tightly together than in a liquid. Instead of moving freely as shown in **Figure 11,** atoms or molecules in a solid vibrate back and forth. The motion of particles in all directions in solids, liquids, and gases is called random motion. Because the particles are moving, they have kinetic energy. The faster the particles are moving, the more kinetic energy they have.

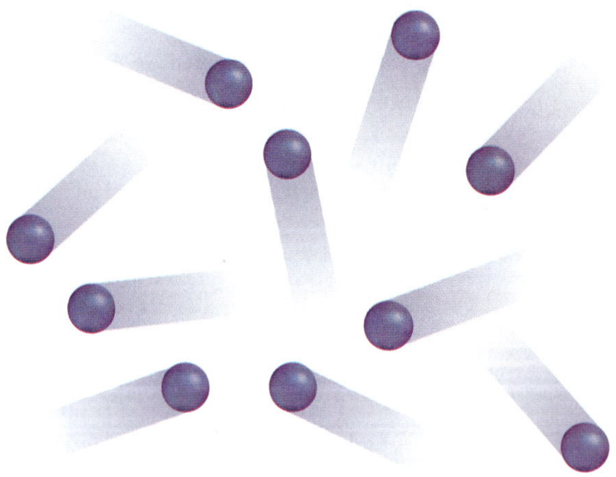

Figure 11 In gases, atoms or molecules are free to move in all directions.

1 Motivate

Bellringer

INTERACTIVE CHALKBOARD PowerPoint® Presentations

Section Focus Transparencies also are available on the Interactive Chalkboard CD-ROM. L2 ELL

Section 2 Section Focus Transparency **Heat It Up**

Since 1960, electricity has been produced at an area called The Geysers in California. This is a geologically active area where underground water is turned to steam by Earth's hot interior. The steam forces its way to the surface where it is converted to electricty.

1. What are some properties of steam? How does steam compare to liquid water?
2. What are some advantages of using energy from Earth's interior to produce electricity?

Energy

Tie to Prior Knowledge

Heat Brainstorm with students where the heat we feel comes from—the Sun, transformation of chemical energy inside our bodies, and geothermal energy from inside Earth.

Section 2 Resource Manager

Chapter *FAST FILE* Resources
- Transparency Activity, p. 43
- Directed Reading for Content Mastery, p. 19
- Enrichment, p. 29
- Lab Activity, pp. 9–11
- MiniLAB, p. 4
- Reinforcement, p. 26

Reading and Writing Skill Activities, p. 33

Mathematics Skill Activities, p. 37

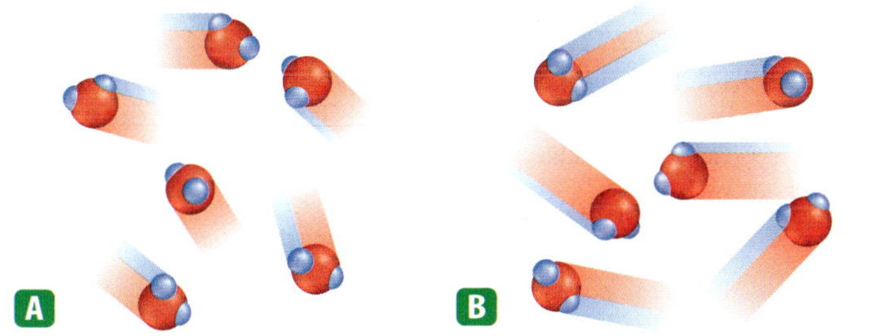

Figure 12 Temperature depends on the average kinetic energy of the atoms or molecules. **A** The gas molecules move slower on average at a lower temperature. **B** Gas molecules move faster on average at a higher temperature.

Temperature is a measure of the average kinetic energy of the particles in an object. When an object's temperature is higher, its atoms or molecules have more kinetic energy. **Figure 12** shows gas molecules at two temperatures. At the higher temperature, the molecules are moving faster and have more kinetic energy.

Applying Science

Can you be fooled by temperature?

On a cold, wintry morning, you may have heard a local meteorologist caution you to "Bundle up, because the wind chill index is minus 20 degrees." While wind cannot lower the temperature of the outside air, it can make your body lose heat faster, and make you feel as if the temperature were lower.

Wind Chill Index

Temperature (°C)	Wind Speed (km/h)							
	10	20	30	40	50	60	70	80
20	18°C	16°C	14°C	13°C	13°C	12°C	12°C	12°C
12	9°C	5°C	3°C	1°C	0°C	0°C	−1°C	−1°C
0	−4°C	−10°C	−14°C	−17°C	−18°C	−19°C	−20°C	−21°C
−20	−26°C	−36°C	−43°C	−47°C	−49°C	−51°C	−52°C	−53°C
−36	−44°C	−57°C	−65°C	−71°C	−74°C	−77°C	−78°C	−79°C

Identifying the Problem

A wind chill index of –29°C presents little danger to you if you are properly dressed. Below this temperature, however, your skin can become frostbitten within minutes. Use the table to find wind chill values for conditions that present the greatest dangers.

Solving the Problem

1. Assuming you live in an area where wind speeds in the winter rarely reach 50 km/h, at which air temperature should you be certain to take extra precautions?
2. What happens to the wind chill index as wind speeds get higher and temperatures get lower?

2 Teach

Quick Demo

Thermal Energy

Materials one plastic container and lid, sand, thermometer

Estimated Time 3–5 minutes

Procedure Pour sand into a plastic container until it is one-fourth full. Measure the temperature of the sand. Place a lid on the container and shake it vigorously for several minutes (you may want to have student volunteers take turns at this). Stop shaking and immediately measure the temperature of the sand. Discuss how the sand got its extra thermal energy. from the kinetic energy of the sand during the shaking

Discussion

Atom Motion Ask students to describe the motion of atoms in a solid, a liquid, and a gas. In a solid the atoms are held close together and vibrate back and forth. In a liquid the atoms are held less tightly than in a solid, and can move past one another. In a gas, the atoms are not held together, and they are in constant random motion. L1 LS **Linguistic**

Applying Science

Teaching Strategy

Because the exact temperature at which the wind chill index reaches −29°C when the wind speed is 50 km/h is not given in the table, students' answers will be approximate.

Answers

1. somewhere between 0 and −15°C
2. It decreases.

Teacher FYI

Average Kinetic Energy Temperature measures the *average* kinetic energy of the molecules in a substance. There is no way to measure the individual kinetic energy of each molecule. For a large group of atoms, the energy is given by a statistical distribution curve.

Teacher FYI

Fahrenheit Scale German physicist Daniel Gabriel Fahrenheit was the first to use mercury for thermometers. He based his scale on the lowest temperature he could achieve in his laboratory—the temperature of a mixture of ice, water, and salt—which he set at 0°, and the temperature of the human body, which he set at 96°.

Reading Check

Answer Because the number of atoms even in a small speck of matter is enormous.

Discussion

Perceiving Temperature Ask students to give examples of times when their perception of hot and cold depended on the temperature they were used to. Possible answers include getting into a swimming pool—at first the water feels cold, then you get used to it and it feels warmer. L2 LS **Logical-Mathematical**

Activity

Temperature Scale Differences Have several thermometers on hand that show both the Fahrenheit and Celsius scales. Have students make a data table with three columns. Call the first column "Objects," the second column "Fahrenheit," and the third column "Celsius." Have the students put each of the following items in the first column and measure and record them in the appropriate columns: Room Temperature, Cold Tap Water, Hot Tap Water, and Body Temperature (measure under arms). This should give the students some context with which to understand the Celsius scale.

Measuring Temperature

Some people might say that the water in a swimming pool feels warm, although others might say it feels cool. Because the temperature of the water feels different to different people, you cannot describe or measure temperature accurately by how it feels. Remember that temperature is related to the kinetic energy of all the atoms in an object. You might think that to measure temperature, you must measure the kinetic energy of the atoms. But atoms are so small that even a tiny piece of material consists of trillions and trillions of atoms. Because they are so small and objects contain so many of them, it is impossible to measure the kinetic energy of all the individual atoms. However, a practical way to measure temperature is to use a thermometer, as shown in **Figure 13.**

Reading Check *Why can't the kinetic energy of all the atoms in an object be measured?*

The Fahrenheit Scale One temperature scale you might be familiar with is the Fahrenheit (FAYR un hite) scale. On the Fahrenheit scale, the freezing point of water is given the temperature 32°F, and the boiling point is 212°F. The space between the boiling point and the freezing point is divided into 180 equal degrees. The Fahrenheit scale today is used mainly in the United States.

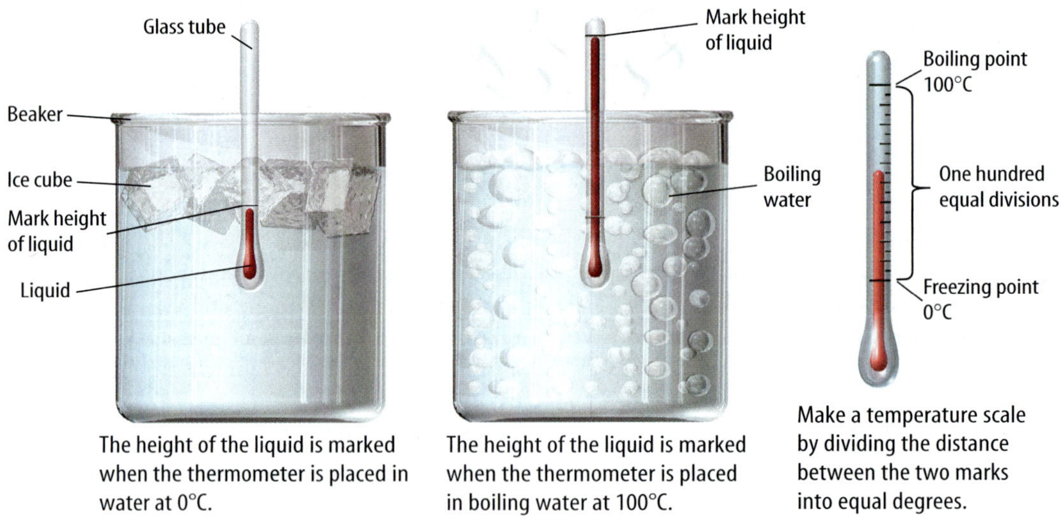

Figure 13 This thermometer uses the height of a liquid in a tube to measure temperature. The liquid height changes with temperature because the liquid expands as its temperature increases. The thermometer is calibrated by marking the liquid height at two known temperatures.

Differentiated Instruction

Challenge Have students find out about how electronic thermometers and liquid crystal thermometers work. Have them make displays illustrating their findings. Electronic thermometers use a thermistor, a semiconductor whose conductivity increases with temperature. In liquid crystal thermometers the crystals change color making the corresponding number visible when the temperature is reached. L3 LS **Linguistic**

Active Reading

Bubble Map A bubble map helps trigger ideas about a given topic. Words are clustered to describe a topic. Students can use a bubble map for prewriting, to generate ideas before writing in their Journals, or to review for a test. Have students design a bubble map to organize the ideas about heat and temperature discussed in this chapter.

The Celsius Scale Another temperature scale that is used more widely throughout the world is the Celsius (SEL see us) scale. On the Celsius temperature scale, the freezing point of water is given the temperature 0°C and the boiling point is given the temperature 100°C. Because there are only 100 Celsius degrees between the boiling and freezing points of water, a temperature change of one Celsius degree is bigger than a change of one Fahrenheit degree.

Heat

On a warm, sunny day when you tilt your head back, you can feel the warmth of the Sun on your face. On a chilly day, putting your cold hands near an open fire warms them up. In both cases, you could feel heat from the Sun and from the fire making you warmer. What is heat?

Look at **Figure 14.** Suppose you pick up a tall glass of iced tea. If you hold the glass for a while, the drink warms up. Your hand is at a higher temperature than the tea, so the atoms and molecules in your hand have a higher kinetic energy than the ones in the iced tea. Kinetic energy from the moving atoms and molecules in your hand is transferred by collisions to the atoms and molecules in the tea.

A transfer of energy from one object to another due to a difference in temperature is called **heat.** Heat flows from warmer objects to cooler ones. In the example just given, heat flows out of your hand and into the glass of iced tea. As you hold the glass, the temperature of the tea increases and the temperature of your skin touching the glass decreases. Heat will stop flowing from your hand to the glass of tea when the temperatures of your hand and the glass are the same.

Heat and Temperature

How much does the temperature of something increase when heat is transferred to it? It depends on two things. One is the amount of material in the object. The other is the kinds of atoms the material is made of. For example, compared to other materials, water is an unusual substance in that it must absorb a large amount of heat before its temperature rises by one degree. Water often is used as a coolant. The purpose of the water in a car's radiator is to carry a large amount of heat away from the engine and keep the engine from being damaged by overheating, as shown in **Figure 15.**

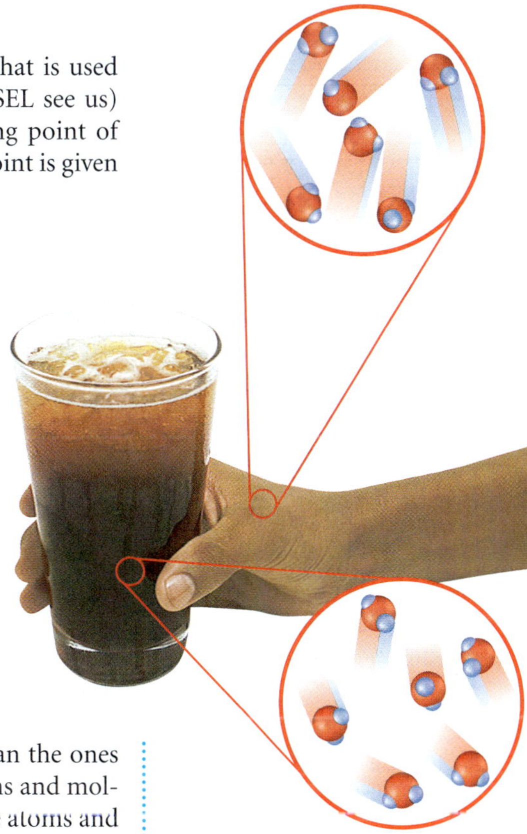

Figure 14 Heat flows from your hand to the glass of iced tea, making your hand feel cold.
Explain *why people wear gloves in cold weather.*

Figure 15 This car's engine overheated because its cooling system didn't carry enough heat from the engine.

Fun Fact

The Kelvin scale is another temperature scale. One kelvin is equal to one degree Celsius, but the zero of the Kelvin scale is at −273°C. Zero kelvin corresponds to the temperature at which molecules have the least amount of thermal energy. This temperature is called absolute zero.

Caption Answer

Figure 14 to slow down the rate at which heat flows from their hands to their surroundings

Visual Learning

Figure 15 When a car is overheating, it helps to turn on the heat. How does this help? This helps heat flow away from the engine as it moves into the passenger compartment.

Virtual Labs

Energy Conversion *How is energy converted from one form to another?*

Differentiated Instruction

English-Language Learners When distinguishing between thermal energy and heat, have your English Language Learners think of thermal energy as a noun, a thing. Have them think of heat as a verb, an action; the movement of thermal energy from one object to another.

Curriculum Connection

History Steam engines convert thermal energy to mechanical energy. Have students find out about the development of the steam engine and how it affected scientists' understanding of heat and energy. James Watt patented the first practical steam engine in 1769. Thermodynamics and an understanding of energy conservation developed in conjunction with the invention of the steam engine. L1 LS **Linguistic**

SECTION 2 Temperature 173

Discussion

Temperature Stability Which would you expect to have a consistent temperature, a pool of water or the dry ground next to it? the pool of water L2 LS **Logical-Mathematical**

Use Science Words

Word Usage Have students use the word *conduct* in both its scientific and everyday meanings. Possible answer: Metals conduct heat well. The guide conducted a tour through the museum. Note that in science, conductors and insulators can be used in reference to electricity or heat. L1 ELL LS **Linguistic**

INTEGRATE History

The Caloric Theory Benjamin Thompson, or Count Rumford as he was later known, was one of the first to realize that heat was not a substance. The experiments performed by James Joule proved that heat was a form of energy.

Research Have students research the contributions that William Thomson, who later became Lord Kelvin, made to the understanding of heat.

During the winter, the lake is warmer than the surrounding land.

During the summer, the lake is cooler than the surrounding land.

Figure 16 Water can absorb and lose a great deal of heat without changing temperature much.

INTEGRATE History

The Caloric Theory At one time heat was thought to be a fluid, called caloric. A warm object contained more caloric than a cold object. Benjamin Thompson in the late eighteenth century noticed that tremendous amounts of heat were generated during the boring of the holes in cannons. He concluded the amount of heat depended only on the work done by the drill. James Joule later showed that other forms of energy could be converted into heat. Research how the work of Thompson and Joule disproved the caloric theory of heat.

Lakes and Air Temperature How does the temperature of water in a lake compare to the temperature of the surrounding air on a hot summer day? How do these temperatures compare at night when the air has cooled off? You might have noticed that the water is cooler than the air during the day and warmer than the air at night. This is because it takes longer for a large body of water to warm up or cool down than it does for the surrounding air and land to change temperature. Even from season to season, a large body of water can change temperature less than the surrounding land, as shown in **Figure 16.**

Heat on the Move

A transfer of energy occurs if there is a temperature difference between two areas in contact. Heat is transferred from warm places to cooler ones. This transfer can take place in three ways—radiation, conduction, and convection. Conduction transfers heat mainly through solids and liquids. Convection transfers heat through liquids and gases. Radiation can transfer energy through space.

Conduction Have you ever picked up a metal spoon that was in a pot of boiling water and dropped it right away because the spoon had become hot? The spoon handle became hot because of conduction. **Conduction** (kun DUK shun) is the transfer of energy by collisions between the atoms in a material.

As the part of the spoon in the boiling water became warmer, its atoms and molecules moved faster. These particles then collided with slower-moving particles in the spoon. In these collisions, kinetic energy was transferred from the faster-moving to the slower-moving particles farther up the spoon's handle.

LAB DEMONSTRATION

Purpose to model thermodynamic collisions

Materials 10 marbles (same size), glass pie pan, overhead projector

Procedure Place 5 marbles in the pan on an overhead projector. Shake the pan slightly to imitate a cold substance and rapidly to simulate a hot substance. Then "mix" the substances by adding 5 fast-moving marbles to a pan of stationary or slowly moving marbles.

Expected Outcome As marbles collide, the faster marbles transfer kinetic energy to the slower ones. The "hot" atoms slow down as the "cold" atoms speed up, until all are moving at the same average speed.

Assessment

What would happen if you added a few stationary marbles to a plate of rapidly moving marbles? The moving marbles would transfer kinetic energy to the stationary ones, so that the marbles originally moving quickly would slow down and those that were stationary would speed up.

Bumping Along Even though conduction is a transfer of kinetic energy from particle to particle, in a solid, the particles involved don't travel from one place to another. As shown in **Figure 17,** they simply move back and forth in place, bumping into each other and transferring energy from faster-moving particles to slower-moving ones. Conduction usually occurs in solids.

Reading Check *How is energy transferred by conduction?*

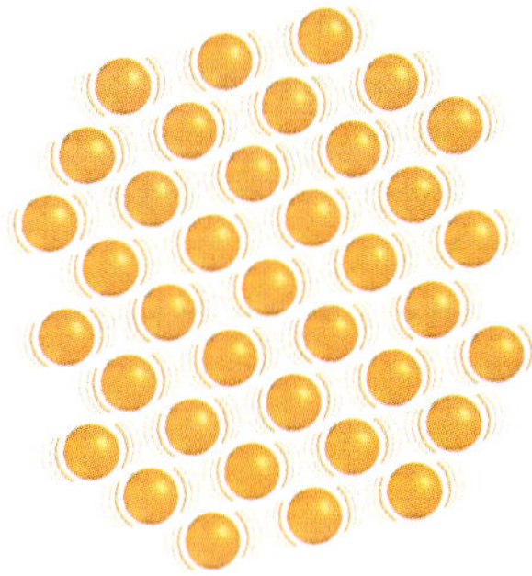

Figure 17 In a solid, atoms collide with each other as they vibrate back and forth.

Conductors It's dinnertime and the hamburgers are frozen solid. This is one time when you want to transfer heat rapidly. You could put a frozen hamburger on a metal tray to speed up the thawing process. Materials through which it is easy to transfer energy are thermal conductors. Most metals are good conductors of heat. Metals such as gold, silver, and copper are the best thermal conductors. Copper is widely available and less expensive than gold or silver. Some cooking pans are made of steel but have copper bottoms. A copper bottom conducts heat more evenly. It helps spread heat across the bottom surface of the pan to prevent hot spots from forming. This allows food to cook evenly.

Insulators Some materials are poor conductors of heat. These materials can be used as thermal insulators. When you are cold, for example, you can put on a sweater or a jacket or add another blanket to your bed. You are keeping yourself warm by adding insulation. The clothes and the blanket are poor conductors of heat. In fact, they make it more difficult for heat to escape from your body. By trapping your body heat around you, you feel warmer.

Blankets and clothes help keep you warm because they are made of materials that contain many air spaces, as shown in **Figure 18.** Air is a good insulator, so materials that contain air are also good insulators. For example, building insulation is made from materials that contain air spaces. Materials made of plastics also are often good insulators. If you put a plastic spoon in boiling water, it takes a long time for it to get hot. Many cooking pans have plastic handles that won't melt instead of metal ones. These handles remain at a comfortable temperature while the pans are used for cooking. Other examples of insulators include wood, rubber, and ceramic materials such as tiles.

Figure 18 Under high magnification, this insulating material is seen to contain many air spaces.
Explain *how increasing the number of air spaces in a material affects the movement of heat.*

Reading Check

Answer Particles bump into each other and transfer kinetic energy from faster particles to slower ones.

Caption Answer

Figure 18 Air does not conduct heat well, so increasing the number of air spaces slows down the transfer of heat.

Fun Fact

Wool is an excellent insulator. Not only does it contain air spaces, but it will even help keep you warm when it is wet. Before the invention of high-tech fabrics, wool was the preferred material for hiking clothes.

Use an Analogy

Heat Capacity Thermal conductivity or insulation properties are dependent upon a material's specific heat. You can think of specific heat as heat capacity. A material with a high specific heat would have a large heat container to be filled before the material would change temperature, making it a good thermal insulator. A material with a low specific heat would have a small heat container to fill before changing temperature making it a good thermal conductor.

Quick Demo

Conduction

Materials one thin metal rod about 30 cm long, 6 thumbtacks, gloves, candle or Bunsen burner

Estimated Time 5–8 minutes

Procedure Use melted wax to attach metal thumbtacks to the rod at 5, 10, 15, 20, 25, and 30 cm. Let the wax cool. Using gloves, hold the rod so that the end corresponding to 0 cm is in the flame of a candle. Have a volunteer measure how long it takes each tack to drop off. Heat moves along the rod by conduction, melting the wax nearest the candle first.

Teacher FYI

Thermodynamics Thermodynamics is the branch of physics that studies heat. The first law of thermodynamics restates the law of conservation of energy. The second law of thermodynamics states that heat flows spontaneously from a hotter object to a colder object, but not from a colder object to a hotter object.

Science Journal

Insulating Outerwear Ask students to describe in their journals the materials used to make windbreakers and winter coats and to suggest reasons for the different materials. Windbreakers are made from thin, tightly woven, synthetic fiber. Winter coats are usually made from several layers of materials such as wool or down that trap air. L3 LS **Linguistic**

Mini LAB

Purpose Students see the relationship between temperature and kinetic energy. L1 ELL
LS **Visual-Spatial**

Materials three transparent plastic containers; supply of hot, room-temperature, and cold water; dropper; food coloring

Teaching Strategy Ask students whether it is important to use the same number of drops in each container. yes; to control variables

Safety Precaution Make sure the water is not hot enough to burn students.

Analysis

1. Hot water was fastest, cold water slowest. The higher the temperature, the faster the particles moved.
2. The greater the water temperature, the more motion the water molecules had, and the more they pushed the food coloring particles.
3. the hot-water cup

Assessment

Performance Ask the students to design a similar experiment to compare the energy of hot and cold gases. What could be used to measure the rate of dispersion? smoke or smells Use **Performance Assessment in the Science Classroom,** p. 105.

Try at Home

Caption Answer

Figure 19 Warm air rises. So the warm air will rise from the floor to the ceiling, warming the entire room.

Figure 19 The furnace's fan helps circulate hot air through your home. Warmer air particles move upward while cooler air particles move downward.
Explain *why the hot air ducts are placed near the floor.*

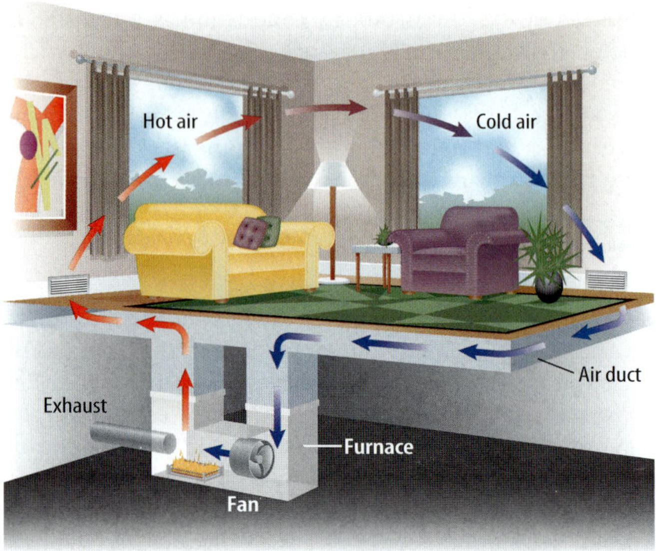

Mini LAB

Comparing Energy Content

Procedure

1. Pour equal amounts of **hot, cold, and room-temperature water** into each of three **transparent, labeled containers.**
2. Measure and record the temperature of the water in each container.
3. Use a **dropper** to gently put a drop of **food coloring** in the center of each container.
4. After 2 min, observe each container.

Analysis

1. Based on the speed at which the food coloring spreads through the water, rank the containers from fastest to slowest.
2. Infer how water temperature affected the movement of the food coloring.
3. In which container do the water particles have the most kinetic energy?

Feeling the Heat Think about getting into a car that has been closed up on a sunny day. Do you prefer a car that has fabric-covered or vinyl-covered seats? Even though the masses of the seats are similar and the temperatures of the surroundings are the same, the vinyl material feels hotter on your skin than the fabric does. How hot something feels also is affected by how fast heat flows, as well as the actual temperature. Vinyl is a better conductor than fabric, so heat flows to your skin more rapidly from the vinyl than from the fabric. As a result, the vinyl feels hotter than the fabric does.

Convection Heat also can be transferred by particles that do not stay in one place but rather move from one place to another. **Convection** (kun VEK shun) transfers heat when particles move between objects or areas that differ in temperature. This type of transfer is most common in gases and liquids. As temperature increases, particles move around more quickly, and the distance between particles increases. This causes density to decrease as temperature increases. Cooler, denser material then forces the warmer, less dense material to move upward.

Some homes are heated by convection. Look at **Figure 19.** Air is warmed in the furnace. The warm, less dense air is then forced up through the air duct by the furnace fan. The warm air gets pushed up through the room by the cooler air around it. As the warm air cools, it becomes more dense. Cool, dense air sinks and is then pulled into the return air duct by the furnace fan to be warmed again and recirculated.

Cultural Diversity

Igloos The Inuit demonstrated an understanding of convection in the design of igloos. The entrance is a tunnel lower than the inside of the igloo. Cold air sinks into the tunnel, but warm air stays in the igloo.

Examples of Convection Eagles and hawks float effortlessly high in the air. Sometimes a bird can stay in the air without flapping its wings because it is held up by a thermal.

As shown in **Figure 20,** a thermal is a column of warm air that is forced up as cold air around it sinks. It is a convection current in the air.

Convection also occurs in liquids. In a pot of boiling water, the warmer, less dense water is forced up as the cooler, denser water sinks. Convection currents on a larger scale are formed in oceans by cold water flowing from the poles and warm water flowing from tropical regions.

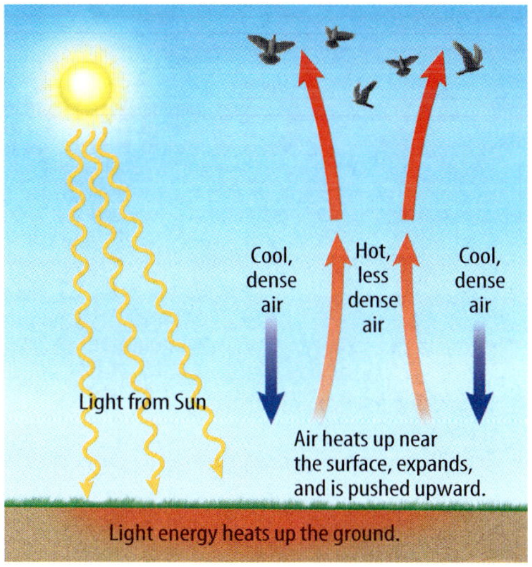

Figure 20 Thermals form when hot, thin air rises up through cooler, denser air.

Radiation The transfer of energy by waves is **radiation** (ray dee AY shun). These waves can be visible light waves or types of waves that you cannot see. When these waves strike an object, their energy can be absorbed and the object's temperature rises. Radiation can travel through air and even through a vacuum.

The Sun transfers energy to Earth through radiation. You take advantage of radiation when you warm yourself by a fire. Heat is transferred by radiation from the fire and you become warmer. You also can use radiation to cook food. A microwave oven cooks food by using microwave radiation to transfer energy to the food.

section 2 review

Summary

Temperature

- The particles that make up an object are in constant random motion.
- The temperature of an object is a measure of the average kinetic energy of the particles in the object.

Heat

- Heat is a transfer of energy from one object to another due to a difference in temperature.
- Heat always flows from a higher temperature to a lower temperature.
- Heat can be transferred by conduction, convection, and radiation.
- Insulators are materials that are poor conductors of heat.

Self Check

1. **List** three ways that heat is transferred and give an example for each.
2. **Describe** the condition that must exist for transfer of heat to occur.
3. **Identify** the type of energy transfer that can take place with little or no matter present. Explain.
4. **Think Critically** Popcorn can be cooked in a hot-air popper, in a microwave oven, or in a pan on the stove. Identify how energy is transferred in each method.

Applying Math

5. **Convert Temperature** To change a temperature from Fahrenheit to Celsius, subtract 32 from the Fahrenheit temperature then multiply by 5/9. If the temperature is 77°F, what is the Celsius temperature?

3 Assess

DAILY INTERVENTION

Check For Understanding

Logical-Mathematical Define *pressure* for students as "force per unit area." The more numerous or forceful the collisions of a gas with the walls of a container, the greater the pressure. Ask them to explain why raising the temperature of a fixed volume of gas would raise the pressure of the gas. When the temperature is raised, the molecules in the gas move faster. They collide more frequently and at higher speeds with the sides of the container, increasing the pressure.

Reteach

Heat Transfer Illustrate the three types of heat transfer by reminding students of sitting by a fire. The warm smoke rises upward through the cooler air, transferring heat by convection. Heat is transferred from the fire through the air by radiation. If one end of a metal tool is put into the fire, it will conduct heat to the other end, burning your hand. L2 ELL LS **Linguistic**

Assessment

Process Tell students to draw an illustration of a pot of water heating on the stove. On the drawing, indicate the source of energy, what kind of energy it is, how it is transferred to the pot and then to the water, and what happens to the water as it heats. Use **Performance Assessment in the Science Classroom,** p. 127. P

section 2 review

1. conduction, hot water from a geyser heating a rock; convection, some ocean currents; radiation, the Sun heating Earth's atmosphere
2. a temperature difference
3. Radiation transfers energy by waves that can travel through empty space.
4. hot-air popper, convection; microwave, radiation; pan, conduction
5. 25°C

1 Motivate

INTERACTIVE CHALKBOARD
PowerPoint® Presentations

Bellringer

Section Focus Transparencies also are available on the Interactive Chalkboard CD-ROM.
L2 ELL

Section 3 Section Focus Transparency **Cold Pack Relief**

You encounter many chemical reactions every day. Lighting a gas stove, starting a lawnmower, and playing sports all involve chemical reactions. So does the chemical cold pack shown below. The chemicals inside combine to provide instant cold for relief of minor athletic injuries.

1. What does a cold pack feel like before you mix the chemicals? What does it feel like after the chemicals mix?
2. Can chemical cold packs be reused? Why or why not?

Tie to Prior Knowledge

Storing Energy Discuss with students what they have learned about ways energy can be stored. Remind them that energy can be stored gravitationally by lifting an object and that chemical energy is stored in food and muscles.

section 3

Chemical Energy

as you read

What You'll Learn

- **Determine** how chemical energy is transformed.
- **Explain** how reaction rates are changed.

Why It's Important

Chemical energy makes it possible for your body to move, grow, and stay warm.

Review Vocabulary
chemical bonds: the forces holding atoms together in a molecule

New Vocabulary
- endothermic reaction
- exothermic reaction
- catalyst

Chemical Reactions and Energy

On a hot summer night, you might have seen fireflies glowing, like those in **Figure 21.** Did you ever wonder how they make their eerie, blinking light? If you have seen light sticks, which glow for a short period of time, you have observed the same process that causes the fireflies' glow. Energy in the form of light is released when a chemical reaction takes place inside the light stick. A burner on a gas stove releases heat and light energy because of a chemical reaction taking place. You might not realize it, but every day you make use of the energy released by many chemical reactions.

What is a chemical reaction? In a chemical reaction, compounds are broken down or new compounds are formed. Sometimes both processes occur. Some chemical reactions occur when atoms or molecules come together. New compounds are formed when atoms and molecules combine and bonds form between them. A compound is broken down when the bonds between the atoms that make up the compound are broken. These atoms are then available to recombine to form new compounds.

When a fire burns, a chemical reaction occurs. Bonds between the atoms in some of the compounds that make up the wood are broken. These atoms then combine with atoms in the air and form new compounds. As these new compounds are formed, heat and light are given off.

Figure 21 Chemical reactions can release light energy.

Each point of greenish light in this picture is a firefly.

A chemical reaction inside a firefly's body releases light.

Section 3 Resource Manager

Chapter FAST FILE Resources
Transparency Activity, p. 44
Directed Reading for Content Mastery, pp. 19, 20
Enrichment, p. 30
Reinforcement, p. 27
Lab Activity, pp. 13–14
Lab Worksheet, pp. 5–6, 7–8

Physical Science Critical Thinking/Problem Solving, p. 10

Life Science Critical Thinking/Problem Solving, p. 8

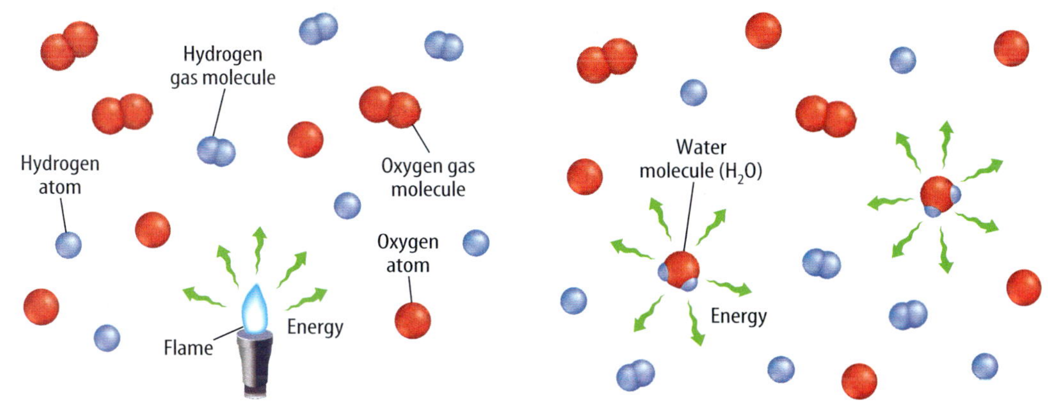

The added energy from the flame causes the bonds to break in the oxygen gas and hydrogen gas.

When the new bonds form between hydrogen and oxygen to produce water particles, energy is released.

Figure 22 Oxygen and hydrogen gas will not react unless energy is added.

Chemical Bonds Energy is stored in the bonds between the atoms in a compound. The stored energy in the chemical bonds is a form of potential energy called chemical energy.

The chemical energy stored in oil, gas, and coal is an important source of energy that is used every day. The chemical energy stored in food provides a source of energy for your body. The muscles in your body transform some of this chemical energy into kinetic energy and heat when they move. List some of the other sources of chemical energy you used today.

Energy in Reactions

In every chemical reaction, transformations in energy occur. To break bonds, energy must be added. The reverse is also true. When bonds form, energy is released. Often energy must be added before the reaction can begin. For example, energy is needed to start the reaction between hydrogen and oxygen to form water. Look at **Figure 22.** When a lighted match is placed in a mixture of hydrogen gas and oxygen gas, the mixture will explode and water will form. The energy to begin the reaction comes from the heat supplied by the flame. As the reaction occurs, bonds form between hydrogen and oxygen atoms, and water molecules form. The energy released as the bonds form results in the explosion.

After the hydrogen and oxygen atoms are bound together to form a water molecule, it is difficult to split them apart. Energy—usually supplied by electricity, heat, or light—is required to break the chemical bonds.

Reading Check *What is required to break chemical bonds?*

2 Teach

Quick Demo

Transforming Chemical Energy

Materials Wintergreen candies, drinking glass with thick bottom

Estimated Time two to five minutes

Procedure The energy stored in chemical bonds can be transformed to light energy when you crush certain crystals. Use a glass to crush wintergreen candies in a completely dark room, and students will see sparks. L1 ELL LS **Visual-Spatial**

Make a Model

Magnetic Balls Have students use two magnetic balls to make an analogy to making and breaking chemical bonds. When you pull the magnets apart, you put energy into the reaction to break the bond. When the magnets are allowed to pull together, energy is released; the potential energy of their position is transformed into the kinetic energy of motion. L2 LS **Kinesthetic**

Teacher FYI

Exploding Substance An explosion occurs when a substance expands very quickly. In the synthesis of water, the energy released by the reaction gives the newly formed water molecules a large amount of kinetic energy, which makes them move very quickly. This causes the gaseous water to expand and explode.

Reading Check

Answer energy

Visual Learning

Figure 22 Review with students the synthesis of water shown in Figure 22. Have students follow the process with their fingers as you read the description from the text. Make sure students understand that energy is needed to initiate the process, but once started, the process gives off more than enough energy to keep itself going. L1 ELL LS **Visual-Spatial**

Teacher FYI

Dissolving The dissolving of a substance is a physical change, not a chemical change. Therefore, when a substance uses energy to dissolve, the process cannot be called an endothermic chemical reaction. It is, nonetheless, an endothermic process.

Visual Learning

Figure 23 Have students make a concept map sequencing the events shown in this figure. P

IDENTIFYING Misconceptions

Endothermic v. Exothermic Students may think that because energy is needed to start a particular reaction, the reaction is endothermic. In fact, many reactions that require energy to start are exothermic. This is because as the chemicals react, they release more energy than they absorb.

Energy-Absorbing Reactions Some chemical reactions need a constant supply of energy to keep them going. These reactions absorb energy. A chemical reaction that absorbs heat energy is called an **endothermic** (en duh THUR mihk) **reaction.** Endothermic chemical reactions often take place in the preparation of food. Thermal energy is absorbed by the food as it cooks. For example, an endothermic reaction takes place in baking some kinds of cookies. The baking soda absorbs energy and produces a gas that puffs up the cookies.

Chemical reactions occur when sunlight strikes the leaves of green plants. These chemical reactions convert the energy in sunlight into chemical energy contained in a type of sugar. Oxygen also is produced by these chemical reactions. This process, shown in **Figure 23,** is called photosynthesis. When the plant is deprived of sunlight, the reactions stop. Photosynthesis is probably the most important endothermic process on Earth. Plants provide you, and almost all other living things, with food and oxygen through photosynthesis.

Figure 23 In photosynthesis, plants absorb light energy and make oxygen and sugar from water and carbon dioxide.

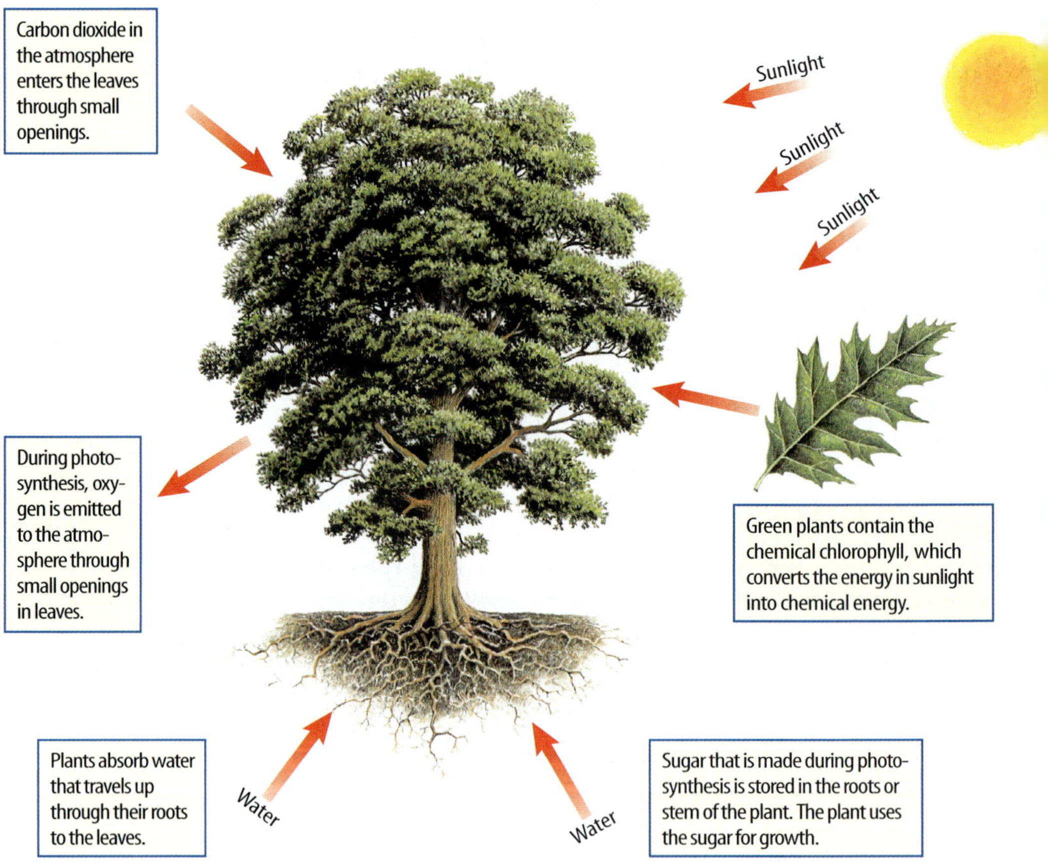

Differentiated Instruction

Learning Disabled To produce an endothermic reaction that can be felt with bare skin, dissolve a large amount of table salt in one of two plastic cups of room-temperature water. Hold both cups to your cheeks and compare the temperature. Energy was absorbed to break the chemical bonds between sodium and chlorine ions. L1 ELL LS **Kinesthetic**

Challenge Have students identify some endothermic processes. The melting of ice is an endothermic process. As ice melts, it absorbs heat so that its surroundings become cooler. Cold packs also use an endothermic process. When a cold pack is squeezed, ammonium nitrate dissolves in water. The dissolution of ammonium nitrate absorbs energy from the water and makes it colder. In the same way, dissolving Epsom salt (magnesium sulfate) in water makes the water colder. L3 LS **Linguistic**

Energy-Releasing Reactions Endothermic chemical reactions are usually important because of the compounds the reactions produce. Other reactions are important because they release energy. **Exothermic** (ek soh THUR mihk) **reactions** are chemical reactions that release heat energy. A chemical hand warmer releases heat when an exothermic reaction takes place inside the hand warmer. When a substance burns, atoms in the substance combine with oxygen atoms in the air. An exothermic reaction occurs, and energy in the form of heat and light is released. The exothermic reaction that occurs when a material burns by combining with oxygen is called combustion. Burning oil, coal, and gas produces much of the energy needed to heat homes and schools.

Reading Check *What are chemical reactions that give off heat energy called?*

Rate of Reaction Chemical reactions can occur at different rates. They occur very fast when fireworks explode. However, if you leave tools or a skateboard outside for a long time, you might notice the metal parts slowly becoming rusty, as shown in **Figure 24.** Rusting is a chemical reaction that occurs when a metal combines with oxygen. In a similar way, when silver is exposed to air, it tarnishes. These chemical reactions, however, occur much more slowly than the burning of a candle's wick or a fireworks explosion.

In your body an enormous number of chemical reactions are occurring every second. The rates of these reactions are carefully controlled by your body to enable it to function properly.

Science Online

Topic: Deep Sea Vents

Visit red.msscience.com for Web links to information about the heat released at deep-sea vents.

Activity Make a travel brochure for visiting deep-sea vents. Include what is released by the vents, what organisms live near the vents, and the temperature of the vent fluids.

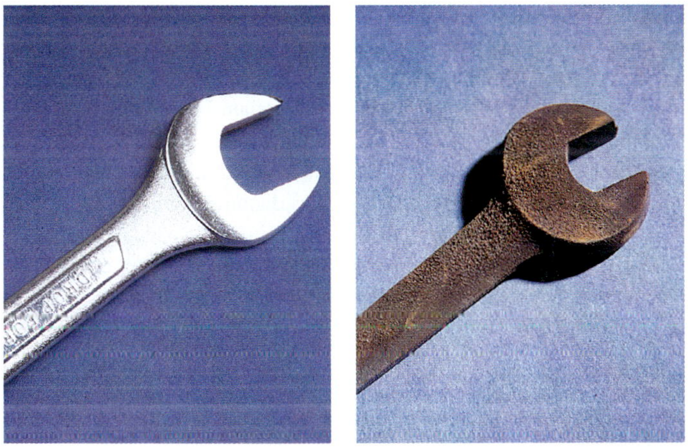

This photo shows a wrench before it rusts.

This photo shows a wrench after it rusts.

Figure 24 Rust is a chemical compound formed when iron and oxygen combine.

Curriculum Connection

Health Portable cold packs taken camping and hiking and used to treat athletic injuries use endothermic processes. Have students find out how these packs work. To start the reaction, a pack is kneaded. This breaks a barrier and mixes two substances together. The resulting endothermic reaction makes the pack cold. L3 LS **Linguistic**

Use Science Words

Word Meaning To help students keep exothermic reactions and endothermic reactions straight, have them find out what the prefixes *ex-* and *endo-* can mean. Then ask them to list other words with these prefixes. The prefix *ex-* means "outside" or "away from." Words include extrude, exceed, extraterrestrial, and exercise. The prefix *endo-* means "within." Words include endocrine gland, a gland that secretes hormones directly into the bloodstream. L2 ELL LS **Linguistic**

Quick Demo

Exothermic Reaction

Materials one match

Estimated Time one minute

Procedure Light a match as an example of an exothermic reaction. Point out that you have to use energy to start the reaction.

Reading Check

Answer exothermic reactions

Discussion

Preventing Rust Why is it important to paint a decorative iron fence? Paint forms a barrier keeping the iron and oxygen in the air from chemically reacting and forming rust which will eventually destroy the fence.

Activity

Endothermic and Exothermic Reactions Get some chemical cold packs and chemical hot packs. Have your students research how they work and write an explanation for each in their Science Journals. Have the students feel the packs before activating them and again after activating them. Have the students write a reaction in their Science Journals.

Fun Fact

Catalysts Catalysts participate in chemical reactions increasing the rate of reaction, but are not actually part of the reaction. They do not get "used up" in the reaction. The catalyst is present in the same amount after the reaction as it is before the reaction.

3 Assess

DAILY INTERVENTION

Check For Understanding

Kinesthetic Have students draw an enzyme and molecules similar to those in **Figure 25.** Have them make extra molecules that have shapes that do not fit into the enzyme. Have students cut out the samples and demonstrate that only the molecules with the correct shape will react.

Reteach

Endothermic or Exothermic? Write endothermic reactions and exothermic reactions on the board. have students give examples of each type of reaction. Write the students' examples on the board under the appropriate type of reaction.

Assessment

Oral What effect would chilling a chemical reaction have on its reaction rate? Why? It would slow the reaction rate, because with less kinetic energy the reactants could not mix to break and form bonds as quickly. Use **Performance Assessment in the Science Classroom,** p. 89.

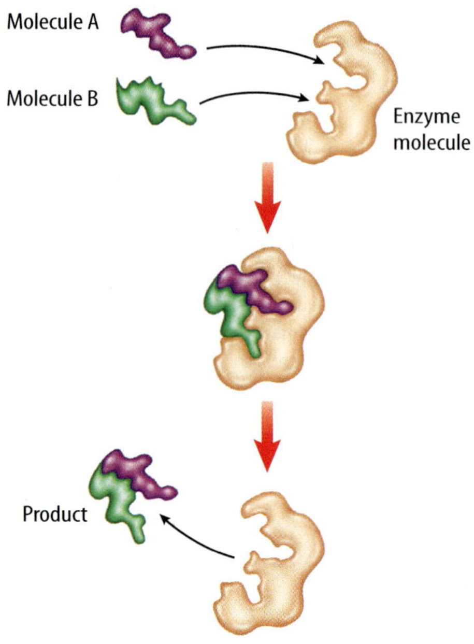

Figure 25 An enzyme makes a chemical reaction go faster by bringing certain molecules together. Only the molecules that have the right shape to fit on the surface of the enzyme will react.

Changing the Rate of Reaction Two ways to change the rate of a chemical reaction are changing the temperature and adding a type of compound called a catalyst. For example, if you pour cake batter into a pan and leave it on a table for several hours, nothing seems to happen. However, if you put the pan in a hot oven, the cake batter becomes a cake. Raising the temperature of the cake batter in the hot oven causes substances in the batter to react more quickly.

A **catalyst** (KA tuh list) is a substance that changes the rate of a chemical reaction without any permanent change to its own structure. Many cell processes in your body are controlled by the presence of catalysts, called enzymes, as shown in **Figure 25.** Enzymes are found throughout your body and are important for growth, respiration, and digestion. When you chew a piece of bread, glands in your mouth produce saliva that contains an enzyme. The enzyme in saliva acts as a catalyst to help break down starches in food into smaller molecules.

Many other chemical reactions depend on catalysts to help them go faster. The production of vegetable shortening, synthetic rubber, and high-octane gasoline are all chemical processes that occur with the help of catalysts.

section 3 review

Summary

Chemical Reactions

- In a chemical reaction compounds can be broken apart and new compounds can form.
- Chemical energy is a form of potential energy stored in the chemical bonds between atoms.

Energy in Chemical Reactions

- Energy must be supplied to start some chemical reactions.
- Reactions that absorb heat energy are endothermic reactions
- Reactions that release heat energy are exothermic reactions.
- The rate of a reaction can be changed by changing the temperature or using a catalyst.

Self Check

1. **Explain** how chemical energy is produced.
2. **Explain** what happens to bonds when new compounds are formed in a chemical reaction.
3. **Explain** how a catalyst speeds up a reaction.
4. **Describe** how photosynthesis transforms radiant energy from sunlight into chemical energy.
5. **Think Critically** Gasoline can react explosively with oxygen in air. Why doesn't the gasoline in a car's gas tank explode when the gas cap is removed?

Applying Skills

6. **Draw Conclusions** Identify three processes that take place in your classroom that involve chemical reactions. Which of these reactions are exothermic?

Science Online red.msscience.com/self_check_quiz

section 3 review

1. It is produced in chemical reactions in which bonds are formed and broken.
2. Bonds are formed or broken.
3. by bringing certain molecules together to react
4. Light energy from the Sun is absorbed by chlorophyll in plant leaves. Chemical reactions convert CO_2 and H_2O into a sugar. Chemical energy is stored in the chemical bonds of the sugar molecule.
5. Energy needs to be added for the reaction to begin.
6. Check students' work. Reactions that produce energy, such as those in a burner, are exothermic.

LAB

Converting Potential and Kinetic Energy

Imagine standing at the top of a mountain ready to ski down its slope. Because of your height on the mountain, you have potential energy. As you ski down the side of the mountain, your speed and kinetic energy increase, but as you lose height, your potential energy decreases. Where does your potential energy go? Where does your kinetic energy come from?

Real-World Question

How can you measure the conversion of potential energy into kinetic energy?

Goals

- **Measure** and calculate the potential and kinetic energies of the balls.
- **Observe** the conversion between potential energy and kinetic energy.

Materials

stiff piece of cardboard (1 m)
triple-beam balance
table-tennis ball
tennis ball
baseball
stopwatch
meterstick

Procedure

1. Copy the data table into your Science Journal.
2. Lean your cardboard against a chair.
3. **Measure** and record the height and length of the board.
4. **Measure** and record the mass of each ball.
5. Let each ball roll from the top of the board to the floor. Measure and record the time it takes for each ball to roll the length of the board.

Conclude and Apply

1. **Calculate** the potential energy of each ball at the top of the board by multiplying the mass times the height times 9.8.
2. **Calculate** the average velocity of each ball as it reaches the floor by dividing the length of the board by the time.
3. **Calculate** the average kinetic energy of each ball as it rolled down the board by multiplying the mass times the velocity squared, and dividing by 2.

Energy Factors

Type of Ball	Mass of Ball (kg)	Height of Board (m)	Length of Board (m)	Time (s)
Table-tennis ball		Answers will vary.		
Tennis ball				
Baseball				

4. **Infer** Which ball had the greatest kinetic energy? Infer why this ball had more kinetic energy.
5. **Infer** how the table-tennis ball could have more potential energy than the baseball.
6. **Infer** the relationship between each ball's potential energy at the top of the slope and its average kinetic energy.

Communicating Your Data

Compare your data with the data collected by your classmates. **For more help, refer to the Science Skill Handbook.**

Communicating Your Data

Students should discuss why their conclusions did or did not agree. They can cite their references to support their arguments.

BENCH TESTED

Real-World Question

Purpose Students will observe potential energy converting to kinetic energy. L2 LS **Kinesthetic**

Process Skills collect data, measure, make and use tables, record data, interpret data

Time Required 40 minutes

Procedure

Safety Precautions Students should wear safety goggles during all lab activities.

Teaching Strategy A student should catch the balls at the end of the board to keep the balls from disturbing other groups.

Conclude and Apply

1. Answers will vary. Students should use the equation $PE = mgh$.
2. Answers will vary.
3. Answers will vary. Students should use the equation $KE = mv^2/2$.
4. The baseball; it had the greatest mass.
5. If the table-tennis ball were high enough, its increased height could make up for its lower mass.
6. They should be nearly equal. There are some differences in the calculations due to friction and experimental error.

Assessment

Oral What law does this lab illustrate? the law of conservation of energy.

BENCH TESTED

Real-World Question

Purpose Students will compare how two different substances absorb thermal energy. L2 ELL COOP LEARN LS **Visual-Spatial**

Process Skills compare and contrast, separate and control variables, make and use tables, interpret data, form operational definitions

Time Required 50 minutes

Procedure

Safety Precautions For safety, you may want to dispense the hot water to students. Do not use water hot enough to cause burns.

Teaching Strategy Explain to students that when the temperature of the water no longer changes, the water is at the same temperature as the material inside the bag.

Tie to Prior Knowledge Ask students about any prior experience heating substances in the kitchen. What substances heat quickly or slowly in a microwave? bread quickly, water slowly Does it take as long to boil a large pot of water as a small pot? A pot with more water, over the same heat, will take longer to come to a boil than a pot with less water.

LAB

Comparing Temperature Changes

Goals
- **Measure** temperature.
- **Calculate** temperature change.
- **Infer** a material's ability to absorb heat.

Materials
- —10°C to 110°C range thermometers (4)
- computer probe
- self-sealing freezer bags (2)
- water (100 mL)
- ice cubes (2 to 3)
- pancake syrup (100 mL)
- **corn syrup*
- 400-mL to 600-mL beakers (4)
- **heat-safe glass containers*
- spoon or stirring rod

**Alternate materials*

Safety Precautions

WARNING: *Use care when handling the heated bags and hot water. Do not taste, eat, or drink any materials used in the lab. Take care when handling glass thermometers.*

Real-World Question

How does the temperature of a substance change as it gains or loses heat? The temperatures of equal amounts of different substances change differently as they are heated or cooled. In this lab you will determine how the temperatures of two different materials change as they absorb and release heat.

Procedure

1. **Design** two data tables to record your temperature measurements of the hot- and cold-water beakers. Use the sample table to help you.
2. Pour 200 mL of hot tap water (about 90°C) into each of two large beakers.
3. Pour 200 mL of cool tap water into each of two large beakers. Add two or three ice cubes and stir until the ice melts.

Alternative Inquiry Lab

Temperature Change and Mass How does changing the mass of a substance affect the amount of heat needed to change the temperature of the substance? Have students modify the above procedure to determine how the amount of heat transferred from the bag of syrup to the water depends on the mass of the syrup in the bag. Students should find that the temperature increase in the water is linearly related to the mass of the syrup. Then ask students to infer how the amount of heat energy transferred from the syrup depended on the mass of the syrup. The amount of heat energy transferred to the water increased linearly as the mass of the syrup increased.

Using Scientific Methods

4. Pour 100 mL of room-temperature water into one bag and 100 mL of syrup into the other bag. Tightly seal both bags.
5. **Record** the starting water temperature of each hot-water beaker. Place each bag into its own beaker of hot water.
6. **Record** the water temperature in each of the hot-water beakers every 2 minutes until the temperature does not change.
7. **Record** the starting water temperature of each cold-water beaker. If any ice cubes remain, remove them from the cold water.
8. Carefully remove the bags from the hot water and put each into its own beaker of cold water.
9. **Record** the water temperature in each of the cold-water beakers every 2 minutes until no change in temperature occurs.

Analyze Your Data

1. **Make a graph** of the water temperature of the beakers with the syrup bag in hot water and the water bag in hot water. Plot both lines on the same graph, with the temperature on the *y*-axis and the time on the *x*-axis.
2. **Make a graph** of the water temperature of the beakers with the syrup bag in cold water and the water bag in cold water. Plot the graph as in step 1.
3. **Determine** which bag warmed the fastest and which bag cooled the fastest.
4. **Determine** which bag reached the highest temperature and which bag reached the lowest temperature.

Water Temperatures—Hot Beaker

Water Bag		Syrup Bag	
Time (min)	Temp. (°C)	Time (min)	Temp. (°C)
0		0	
2		2	
4	Answers will vary.	4	Answers will vary.
6		6	
8		8	

Conclude and Apply

1. **Infer** which material absorbed more heat as it warmed, and which material released more heat as it cooled.
2. **Infer** Suppose you have equal amounts of syrup and water at the same temperature. Which material would require more heat to change its temperature by 1°C? Explain.

Communicating Your Data

Compare your results with the results of other students in your classroom. Explain any differences in your data or your conclusions. **For more help, refer to the Science Skill Handbook.**

Analyze Your Data

Expected Outcome Results will vary depending on the type of syrup. Students will observe that the beaker with the water-filled bag will warm up more slowly than the beaker with the syrup-filled bag. Like-wise, the beaker with the water-filled bag will cool more slowly than the beaker with the syrup-filled bag.

Answers to Questions

1. Graphs should show that the beaker with the syrup-filled bag cooled faster.
2. Graphs should show that the beaker with the syrup-filled bag warmed faster.
3. The syrup warmed and cooled the fastest.
4. The beaker with the water-filled bag reached the highest and the lowest temperatures.

Error Analysis There could be loss or gain of heat to the environment. Unless they are dried off between experiments, bags of water and syrup could carry hot water into the cold beakers.

Conclude and Apply

1. Water absorbed more heat as it warmed and released more heat as it cooled.
2. Water would require more heat. The temperature of the syrup changes more as heat is transferred to and from the water in the beaker.

Communicating Your Data

Graphs of the data would make it easy for students to compare their results. Refer to Making Graphs in the Skill Handbook.

Differentiated Instruction

Learning Disabled Students should heat two bags with equal amounts of water and syrup in two beakers of hot water. After a set amount of time, the bag containing syrup should be warmer to the touch than the bag containing water.

Assessment

Oral Have students verbally describe the reasoning they used to answer the questions. Encourage them to think of examples to help them describe the effects of thermal energy transfer between different types and amounts of materials.

Understanding Literature

Summarize Summaries can focus on a main idea in a small amount of space or time, if given orally.

Respond to the Reading

1. The bombing of Hiroshima was one of the most significant events of the twentieth century.
2. It uses a chain reaction to release energy.
3. **Linking Science and Writing** Suggest that students create an outline of whatever section in the chapter they choose for the writing activity.

INTEGRATE Physics **Nuclear Fission** The process that produces the energy released in the explosion of an atomic bomb is nuclear fission. When nuclear fission occurs, the nucleus of an atom absorbs a particle, such as a neutron, and splits into two smaller nuclei. Usually neutrons or other particles are released in the process, as well as large amounts of energy. In an atomic bomb, uranium or plutonium nuclei undergo fission. The amount of uranium or plutonium involved is larger than the element's critical mass, so that an uncontrolled chain reaction can occur. This chain reaction results in the devastating explosion.

Science and Language Arts

Hiroshima

by Lawrence Yep

On August 6, 1945, an American B-29 bomber dropped a new weapon called the atom bomb on the Japanese city of Hiroshima. The bomb destroyed 60 percent of the city, killing between 90,000 and 140,000 people.

Everything is made up of tiny particles called atoms. They are so small they are invisible to the eye. Energy holds these parts together like glue. When the atom breaks up into its parts, the energy goes free and there is a big explosion.

Inside the bomb, one uranium atom collides with another. Those atoms both break up. Their parts smash into more atoms and split them in turn.

This is called a chain reaction. There are millions and millions of atoms inside the bomb. When they all break up, it is believed that the atom bomb will be equal to 20,000 tons of dynamite. In 1945, it is the most powerful weapon ever made....

Up until then, no single bomb has ever caused so much damage or so many deaths.

The wind mixes their dust with the dirt and debris. Then it sends everything boiling upward in a tall purple-gray column. When the top of the dust cloud spreads out, it looks like a strange, giant mushroom.

The bomb goes off 580 meters above the ground. The temperature reaches several million degrees Celsius immediately.

One mile away, the fierce heat starts fires.

Even two miles away, people are burned by the heat.

Understanding Literature

Summarize When you summarize something, you mention only the main ideas and necessary supporting details. The author of *Hiroshima* has chosen to summarize the events. He briefly explains the science behind the atom bomb. He also gives some details about the destruction after the bomb was dropped on Hiroshima, Japan. How are summaries useful?

Respond to the Reading

1. What was the author's reason for writing this piece?
2. How is the atom bomb different from other bombs?
3. **Linking Science and Writing** Write a one- or two-paragraph summary of one of the sections in this chapter.

INTEGRATE Physics Energy can be released by exothermic chemical reactions when the bonds between atoms are broken. In this excerpt from *Hiroshima*, Lawrence Yep describes the effects of the energy released in a different process—the energy released when the nuclei of atoms are split. This reaction released an enormous amount of energy that destroyed a city.

Resources for Teachers and Students

National Japanese American Historical Society, 1855 Folsom Street #161, San Francisco, CA 94103

Our World Class of '44 Project, compiled by Diane Honda, funded by the Civil Liberties Public Education Fund/printed by Herff Jones Yearbook Company, 1998

Inferno: The Firebombing of Japan, March 9–August 15, 1945, by Edwin P. Hoyt, Madison Books, 2000

Reviewing Main Ideas

Section 1 Energy Changes

1. Energy is the ability to cause change.
2. Energy can have different forms, Energy can be transformed from one form into another.
3. Kinetic energy is the energy an object has due to its motion. Kinetic energy increases as the speed of an object increases.
4. Potential energy is stored energy that increases as an object's height increases.

Section 2 Temperature

1. Temperature is a measure of the average kinetic energy of the particles in a material.
2. The movement of energy from a warmer object to a cooler one is called heat.
3. Heat can be transferred by conduction, convection, and radiation.

Section 3 Chemical Energy

1. The energy stored in chemical bonds is chemical energy.
2. Chemical reactions can release or absorb energy. Exothermic reactions are chemical reactions that release energy. Endothermic reactions absorb energy.
3. Changing the temperature and adding catalysts can change the rate of chemical reactions.

Visualizing Main Ideas

Copy and complete the following concept map on energy.

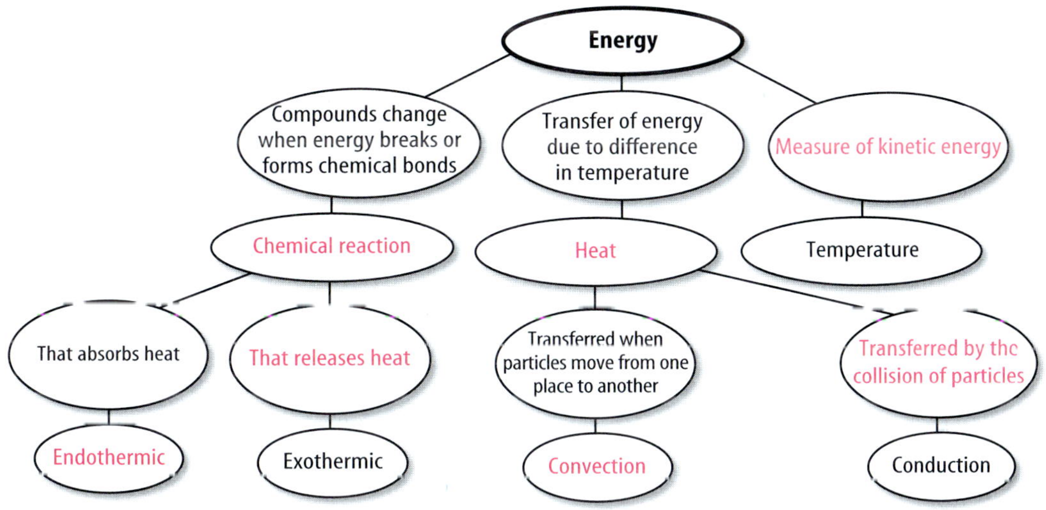

Science Online red.msscience.com/interactive_tutor

chapter 6 Study Guide

Reviewing Main Ideas

Summary statements can be used by students to review the major concepts of the chapter.

Visualizing Main Ideas

See student page.

Science Online

Visit red.msscience.com
/self_check_quiz
/interactive_tutor
/vocabulary_puzzlemaker
/chapter_review
/standardized_test

Assessment Transparency

For additional assessment questions, use the *Assessment Transparency* located in the transparency book.

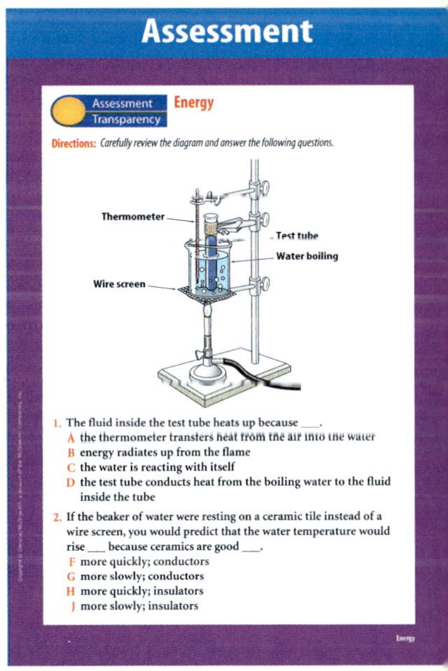
Assessment

Assessment Transparency — Energy

Directions: *Carefully review the diagram and answer the following questions.*

1. The fluid inside the test tube heats up because ___.
   A the thermometer transfers heat from the air into the water
   B energy radiates up from the flame
   C the water is reacting with itself
   D the test tube conducts heat from the boiling water to the fluid inside the tube
2. If the beaker of water were resting on a ceramic tile instead of a wire screen, you would predict that the water temperature would rise ___ because ceramics are good ___.
   F more quickly; conductors
   G more slowly; conductors
   H more quickly; insulators
   J more slowly; insulators

Using Vocabulary

1. conduction
2. kinetic energy
3. heat
4. temperature
5. potential energy
6. conduction
7. radiation
8. endothermic reaction
9. catalyst

Checking Concepts

10. D
11. C
12. A
13. C
14. C
15. B
16. B
17. D

Thinking Critically

18. See student page.
19. Convection occurs as air moves past your skin. Heat given off by your body heats the air next to it. The wind moves this warm air to a different location, allowing cooler air to move next to your skin.
20. The graph curves upward, becoming less curved as the speed increases. The kinetic energy increases by four times when the speed doubles and nine times when the speed triples.
21. Heat flows from a hot object to the surrounding air, and then from the surrounding air to a cold object.
22. Fiberglass insulation contains small air pockets. This makes fiberglass a good insulator because it reduces the movement of heat into the house in the summer and out of the house in the winter.
23. burning log—chemical energy to thermal energy and light energy; dropped ball—potential energy to kinetic energy; sunlight on leaves—light energy to chemical energy
24. When a blanket is wet, water fills its air pockets. Water is a better conductor of heat than air.

chapter 6 Review

Using Vocabulary

catalyst p. 182	heat p. 173
conduction p. 174	kinetic energy p. 164
convection p. 176	law of conservation of energy p. 168
endothermic reaction p. 180	potential energy p. 166
energy p. 162	radiation p. 177
exothermic reaction p. 181	temperature p. 171

Fill in the blanks with the correct vocabulary word or words.

1. Energy transfer by contact is _________.
2. Energy of motion is _________.
3. The movement of energy from warm to cool objects is _________.
4. A measure of the kinetic energy of the atoms in a substance is called _________.
5. _________ is energy that is stored.
6. _________ is the transfer of energy by collisions between the particles in a material.
7. Energy transferred by waves is called _________.
8. A chemical reaction that absorbs heat energy is a(n) _________.
9. A(n) _________ is a substance that changes the rate of a chemical reaction.

Checking Concepts

Choose the word or phrase that best answers the question.

10. Which of the following correctly describes energy?
    A) can be created
    B) can be destroyed
    C) cannot change form
    D) can cause change
11. The temperature of an object is related to which of the following?
    A) heat
    B) total energy of its atoms
    C) kinetic energy of its atoms
    D) total chemical energy
12. What happens if two objects at different temperatures are touching?
    A) Heat moves from the warmer object.
    B) Heat moves from the cooler object.
    C) Heat moves to the warmer object.
    D) No energy transfer takes place.
13. Which of the following describes how the total amount of energy changes during an energy transformation?
    A) It increases.
    B) It decreases.
    C) It stays the same.
    D) It depends on the form of energy being transferred.
14. How is energy from the Sun transferred to Earth?
    A) It is transferred by conduction.
    B) It is transferred by convection.
    C) It is transferred by radiation.
    D) It is transferred by endothermic reactions.
15. When would you have the most potential energy?
    A) walking up the hill
    B) sitting at the top of the hill
    C) running up the hill
    D) sitting at the bottom of the hill
16. Which of the following chemical reactions releases heat energy?
    A) exothermic **C)** catalysts
    B) endothermic **D)** thermals
17. Heat flows easily in which material?
    A) plastic **C)** glass
    B) insulator **D)** conductor

Science Online red.msscience.com/vocabulary_puzzlemaker

Use the ExamView® Pro Testmaker CD-ROM to:

- create multiple versions of tests
- create modified tests with one mouse click for inclusion students
- edit existing questions and add your own questions
- build tests aligned with state standards using built-in State Curriculum Tags
- change English tests to Spanish with one mouse click and vice versa

Thinking Critically

18. **Concept Map** Below is a concept map on the energy changes that occur when a person jumps upward. Copy and complete the map by indicating the type of energy—kinetic, potential, or both—the person has at each of the following stages: halfway up, the highest point, halfway down, and just before hitting the ground.

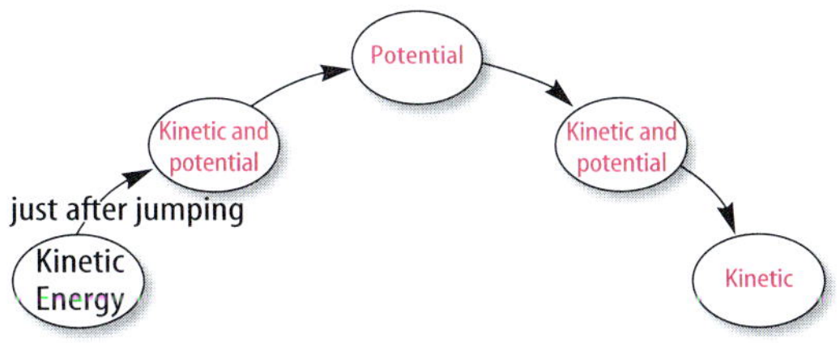

19. **Explain** why the air feels cooler on a windy day, than on a calm day, even though the air temperature is the same.

Use the table below to answer question 20.

Kinetic Energy of a Ball

Speed of Ball (m/s)	Kinetic Energy (J)
5	2.5
10	10.0
15	22.5
20	40.0
25	62.5
30	90.0

20. **Make a Graph** Using the data in the table above, graph the kinetic energy of the ball on the *y*-axis and the speed of the ball on the *x*-axis. Describe the shape of your graph. How does the kinetic energy change when the speed doubles? How does the kinetic energy change when the speed triples?

21. **Determine** If heat flows in only one direction, how can hot and cold liquids reach room temperature as they sit on a table?

22. **Explain** why the walls of houses often are filled with fiberglass insulation.

23. **Determine** the forms of energy involved in each of the following situations—a log burns in a fireplace, a ball is dropped, sunlight falls on the leaves of a tree.

24. **Explain** why a blanket is a better conductor of heat when it is wet than when it is dry.

Performance Activities

25. **Design an Experiment** to determine how quickly the temperature of different materials changes as they absorb radiant energy. Use the following items—three different colors of construction paper, three thermometers, and a sunny window or a heat lamp.

Applying Math

Use the table below to answer question 26.

Fahrenheit and Celsius Temperatures

Celsius Temperature (°C)	Fahrenheit Temperature (°F)
100	212
50	122
0	32
−25	−13
−50	−58

26. **Temperature Scales** Graph the data in the table with the Celsius temperature on the *x*-axis and the Fahrenheit temperature on the *y*-axis. From your graph, determine the temperature that has the same value on both temperature scales.

Science Online red.msscience.com/chapter_review

chapter Review 6

Performance Activities

25. Possible experiment: students place colored paper in sunlight or under a heat lamp, and then place thermometers under the colored paper. They measure temperature at one-minute intervals and plot the temperature change for each thermometer.

Applying Math

National Math Standards

4, 5, 9

26. −40

Assessment Resources

Reproducible Masters

Chapter *Fast File* Resources
- Chapter Review, pp. 35–36
- Chapter Tests, pp. 37–40
- Assessment Transparency Activity, p. 47

Glencoe Science Web site
- Chapter Review Test
- Standardized Test Practice

Glencoe Technology

- Assessment Transparency
- Exam*View*® Pro Testmaker
- MindJogger Videoquiz
- Interactive Chalkboard

FAST FILE

Answer Sheet A practice answer sheet can be found at red.msscience.com/answer_sheet.

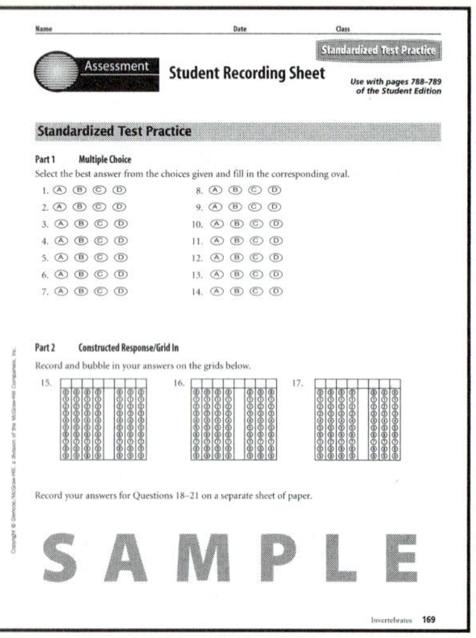

Part 1 Multiple Choice

1. B	5. A	9. A
2. A	6. B	10. B
3. D	7. A	11. B
4. B	8. C	

Part 2 Short Response

12. The metal in the pan is a good conductor and allows heat to transfer to the food. The wood handle is an insulator that does not transfer heat from the pan. This makes it safe to pick up the pan without being burned.
13. Photosynthesis is an endothermic reaction because it needs the energy from sunlight to occur.
14. During photosynthesis, thermal energy in sunlight is converted into chemical energy.
15. The purpose of the water is to carry a large amount of heat away from the engine and keep the engine from overheating. Water is used because it can absorb a large amount of heat before its temperature rises by one degree.

chapter 6 Standardized Test Practice

Part 1 Multiple Choice

Record your answers on the answer sheet provided by your teacher or on a sheet of paper.

Use the illustration below to answer questions 1–4.

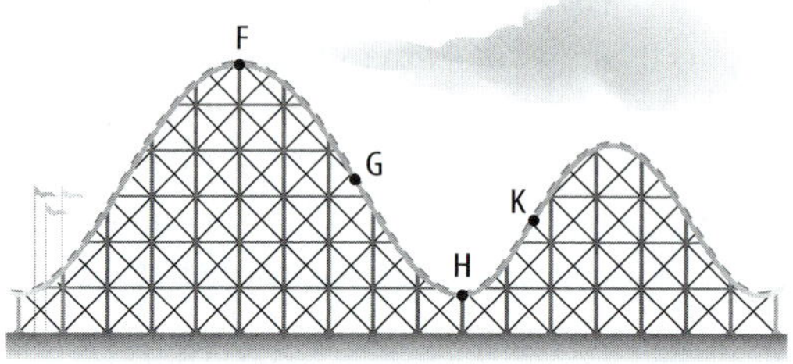

1. At which point would a roller coaster have the greatest kinetic energy?
   A. point F **C.** point G
   B. point H **D.** point K
2. At which point would a roller coaster have the greatest potential energy?
   A. point F **C.** point G
   B. point H **D.** point K
3. At which point is kinetic energy being transformed to potential energy?
   A. point F **C.** point G
   B. point H **D.** point K
4. At which point is the roller coaster's potential energy smallest?
   A. point F **C.** point G
   B. point H **D.** point K
5. What causes a metal spoon in a pot of boiling soup to become hot?
   A. conduction **C.** radiation
   B. convection **D.** insulation
6. Which occurs as you hold a glass of ice tea?
   A. Heat flows from the glass to your hand.
   B. Heat flows from your hand to the glass.
   C. Cold flows from the glass to your hand.
   D. Cold flows from your hand to the glass.
7. Which of the following describes the energy stored in chemical bonds?
   A. It is a form of potential energy.
   B. It is a form of kinetic energy.
   C. It is a form of heat energy.
   D. It is a form of nuclear energy.
8. The kinetic energy of an object depends on which of the following?
   A. its temperature
   B. its speed only
   C. its mass and its speed
   D. how high it is above the ground
9. Which is a good conductor of heat?
   A. copper **C.** wood
   B. rubber **D.** air

Use the illustration below to answer questions 10 and 11.

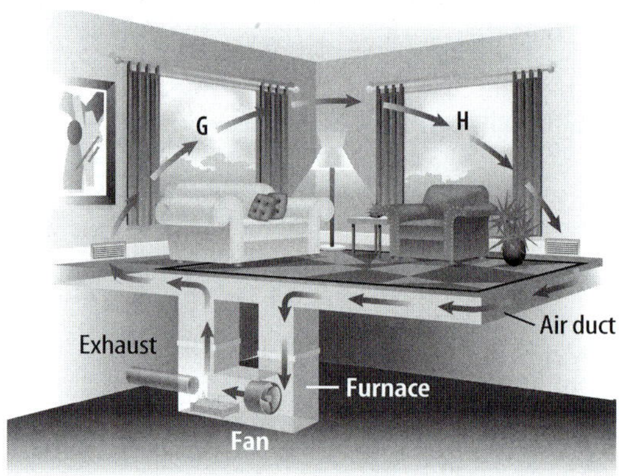

10. What type of heat transfer is shown?
    A. conduction **C.** radiation
    B. convection **D.** insulation
11. How does the air at point G compare to the air at point H?
    A. Air at point G is warmer and denser.
    B. Air at point G is warmer and less dense.
    C. Air at point G is cooler and denser.
    D. Air at point G is cooler and less dense.

16. Catalysts, called enzymes, control the rate of chemical reactions in the body that are necessary for growth, respiration, and digestion.
17. Exothermic reactions are chemical reactions that release heat energy. Endothermic reactions are chemical reactions that absorb heat energy.
18. You can lift it higher because the higher an object is lifted above Earth, the greater its potential energy.
19. The burning of paper occurs quickly, while the rusting of a metal tool occurs slowly. The faster reaction releases energy at a faster rate.

Part 3 Open Ended

20. Wood is a good insulator so it does not conduct heat well. Therefore it would not conduct heat to your body. However, metal is a good conductor of heat and would transfer heat to your body quickly. This would make it uncomfortable to sit on.

Part 2 Short Response/Grid In

Record your answers on the answer sheet provided by your teacher or on a sheet of paper.

12. Why are many cooking pans made of metal, but have wood handles?

Use the illustration below to answer questions 13 and 14.

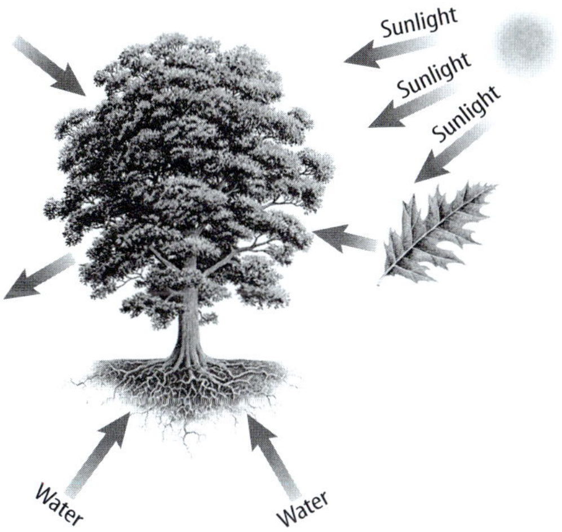

13. Is photosynthesis an endothermic or exothermic reaction?

14. What kind of energy transformation occurs during photosynthesis?

15. What is the purpose of water in a car's radiator?

16. Why are catalysts important in the human body?

17. Explain the difference between exothermic and endothermic reactions.

18. How can you increase the potential energy of an object?

19. Compare the reaction rate of the burning of newspaper with the rusting of a metal tool. Which reaction releases energy at a faster rate?

Part 3 Open Ended

Record your answers on a sheet of paper.

20. Would you rather sit on a wooden bench or metal bench outdoors on a hot day? Explain your choice.

Use the illustration below to answer questions 21 and 22.

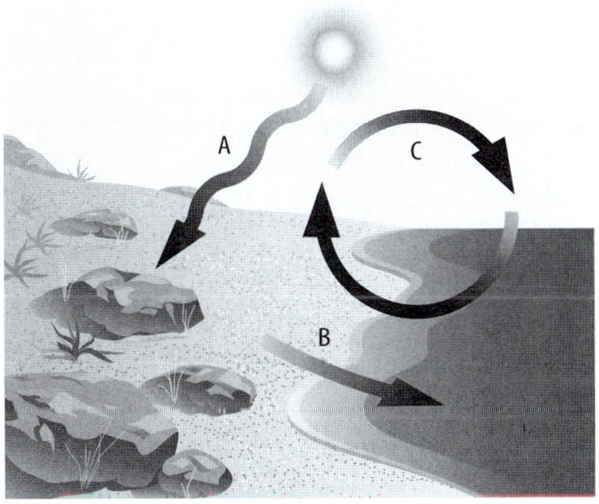

21. Describe how energy is being transferred at points A, B, and C.

22. How does the temperature of the water in the lake compare to the temperature of the land around it on a hot summer day? Explain.

23. If heat only flows in one direction, how can frozen food thaw when placed in a refrigerator and hot foods become chilled when placed in a refrigerator?

24. People caught in cold weather with only light jackets are told to crumple up newspapers and place them inside their jackets. Why would this help?

25. Describe the energy changes that occur when falling water is used to generate electricity.

21. At point A energy from the Sun is being transferred to the land by radiation. At point B energy from the land is being transferred from the warmer land to the water by conduction. At point C convection is transferring heat as cooler, denser air from over the water forces the warmer, less dense air above the land to move upward.

22. During the summer, the water is cooler than the surrounding land. This is because it takes longer for water to heat up than it does for the land surrounding it. Water must absorb much more heat than land before its temperature rises as much as the temperature of the land.

23. The temperature of the air in a refrigerator is higher than the temperature of the frozen food. So heat is transferred from the air to the frozen food. As the temperature of the frozen food rises, it thaws. The temperature of the air in a refrigerator is lower than the temperature of hot food. So heat is transferred from the hot food to the air in the refrigerator. Heat transfer will stop when both objects become the same temperature as the air in the refrigerator.

24. Paper and the air trapped inside the crumpled paper are both insulators. So they would help prevent heat loss from the body by conduction and convection.

25. The potential energy of water behind a dam is converted to kinetic energy as the water falls through pipes. The kinetic energy of the moving water spins generators where kinetic energy is converted to electrical energy.

Rubrics

For more help evaluating open-ended assessment questions, see the rubric on p. 10T.

chapter 7 Organizer

Section/Objectives	Standards: National	Standards: State/Local	Labs/Features
Chapter Opener	See pp. 16T–17T for a Key to Standards.		**Launch Lab:** Electric and Magnetic Forces, p. 193 **Foldables,** p. 193
Section 1 Electric Charge and Forces 3 sessions 1.5 blocks 1. **Describe** how electric charges exert forces on each other. 2. **Define** an electric field. 3. **Explain** how objects can become electrically charged. 4. **Describe** how lightning occurs.	National Content Standards: UCP.2, UCP.3, UCP.5, A.1, A.2, B.1, B.2, B.3, F.3, G.3		**MiniLAB:** Observing Charging by Induction, p. 197 **Integrate History,** p. 198 **Science Online,** p. 199
Section 2 Electric Current 3 sessions 1.5 blocks 5. **Describe** how an electric current flows. 6. **Explain** how electrical energy is transferred to a circuit. 7. **Explain** how current, voltage, and resistance are related in a circuit. 8. **Distinguish** between series and parallel circuits.	National Content Standards: UCP.2, UCP.3, UCP.5, A.1, A.2, B.1		**Science Online,** p. 202 **Visualizing Batteries,** p. 206 **Applying Math:** Flashlight Voltage, p. 207
Section 3 Magnetism 4 sessions 2 blocks 9. **Describe** how magnets exert forces on each other. 10. **Explain** why some materials are magnetic. 11. **Describe** how objects become temporary magnets. 12. **Explain** how an electric generator produces electrical energy.	National Content Standards: UCP.2, UCP.3, UCP.5, A.1, A.2, B.1, B.2, D.1		**Integrate Earth Science,** p. 210 **MiniLAB:** Observing Magnetic Force on a Wire, p. 212 **Lab:** Batteries in Series and Parallel, p. 215 **Lab:** Magnets and Electric Current, p. 216 **Science and History:** Which way to go?, p. 218

Glencoe Exclusive! TeacherWorks™ All-In-One Planner and Resource Center

Lab Materials	Reproducible Resources	Section Assessment	Technology
Launch Lab: rubber balloon, wool cloth, bar magnet, paper clip, paper	**Chapter *FAST FILE* Resources** Foldables Worksheet, p. 17 Directed Reading Overview, p. 19 Note-taking Worksheets, pp. 33–35	Glencoe's Assessment Advantage	**TeacherWorks** includes: • Interactive Teacher Edition • Lesson Planner with calendar • Access to all program blacklines • Correlations to standards • Web links
Need materials? Contact Science Kit at 1-800-828-7777 or www.sciencekit.com on the Internet.	**Chapter *FAST FILE* Resources** Transparency Activity, p. 44 MiniLAB, p. 3 Enrichment, p. 30 Reinforcement, p. 27 Directed Reading, p. 20 **Science Inquiry Labs,** pp. 41–42	**Portfolio** Differentiated Instruction, p. 198 **Performance** MiniLAB, p. 197 Applying Skills, p. 200 **Content** Section Review, p. 200	Section Focus Transparency Guided Reading Audio Program Interactive Chalkboard CD-ROM
MiniLAB: rubber balloon, comb, wool cloth	**Chapter *FAST FILE* Resources** Transparency Activity, p. 45 Enrichment, p. 31 Reinforcement, p. 28 Directed Reading, p. 21	**Portfolio** Activity, p. 202 Differentiated Instruction, p. 205 **Performance** Applying Skills, p. 200 **Content** Section Review, p. 200	Section Focus Transparency Virtual Labs CD-ROM Guided Reading Audio Program Interactive Chalkboard CD-ROM
MiniLAB: 22-gauge wire (50 cm), D-cell battery, bar magnet **Lab:** 1.5-V lightbulbs (2), 1.5-V batteries (3), minibulb sockets (2), 10-cm insulated wire (8), tape **Lab:** bar magnet, compass, D-cell batteries (2), 3-m insulated wire, 50-cm insulated wire	**Chapter *FAST FILE* Resources** Transparency Activity, p. 46 MiniLAB, p. 4 Enrichment, p. 32 Reinforcement, p. 29 Directed Reading, p. 21 Lab Worksheet, pp. 5–6 Lab Activity, pp. 9–12 Lab Worksheet, pp. 7–8 Lab Activity, pp. 13–16 Transparency Activity, pp. 47–48	**Portfolio** Differentiated Instruction, p. 213 **Performance** MiniLAB, p. 197 Applying Skills, p. 200 **Content** Section Review, p. 200	Section Focus Transparency Teaching Transparency Guided Reading Audio Program Interactive Chalkboard CD-ROM Video Lab

End of Chapter Assessment

Glencoe's Assessment Advantage

Blackline Masters	Technology	Professional Series
Chapter *FAST FILE* Resources Chapter Review, pp. 37–38 Chapter Tests, pp. 39–42 **Standardized Test Practice,** pp. 31–34	MindJogger Videoquiz Virtual Labs CD-ROM Exam*View*® Pro Testmaker TeacherWorks CD-ROM Interactive Chalkboard CD-ROM	**Performance Assessment in the Science Classroom (PASC)**

Transparencies

Section Focus

Section 1 Section Focus Transparency — A Spritely Vision

The lightning that we usually see is below the clouds, but there is activity high above the clouds, too. These events have some pretty fanciful names, like sprites, elves, and blue jets. This photo shows the red flash of a sprite, which sometimes occurs during thunderstorms.

1. When do you usually see lightning?
2. How does this sprite appear similar to lightning? Different?
3. How is a flash of lightning different from a glowing lightbulb??

L2

Section 2 Section Focus Transparency — Go with the Flow!

Highways in heavily used areas are designed with many lanes to keep traffic moving, but sometimes it doesn't work.

1. Describe what is happening in the picture. What would traffic be like if there were the same number of cars but fewer lanes? What would traffic be like if there were the same number of cars but twice as many lanes?
2. How might the flow of traffic on a road be like the flow of electricity in a wire?

L2

Section 3 Section Focus Transparency — Will he stick to the refrigerator?

These fish are rainbow trout. Recently, scientists discovered that rainbow trout have tiny amounts of magnetic material in cells in their noses. The researchers don't know exactly how the trout use the magnetic material, but they think it might help the trout navigate.

1. Do you think the magnetic material in the trout's nose makes a strong magnet? Why or why not?
2. What kind of material does a magnet stick to?
3. Where do you encounter magnets in your everyday life?

L2

This is a representation of key blackline masters available in the Teacher Classroom Resources. See Resource Manager boxes within the chapter for additional information.

Assessment

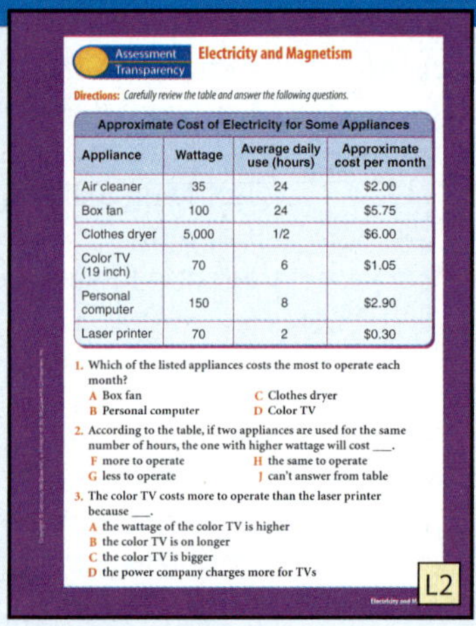
Assessment Transparency — Electricity and Magnetism

Directions: Carefully review the table and answer the following questions.

Approximate Cost of Electricity for Some Appliances

Appliance	Wattage	Average daily use (hours)	Approximate cost per month
Air cleaner	35	24	$2.00
Box fan	100	24	$5.75
Clothes dryer	5,000	1/2	$6.00
Color TV (19 inch)	70	6	$1.05
Personal computer	150	8	$2.90
Laser printer	70	2	$0.30

1. Which of the listed appliances costs the most to operate each month?
   A Box fan
   B Personal computer
   C Clothes dryer
   D Color TV
2. According to the table, if two appliances are used for the same number of hours, the one with higher wattage will cost ___.
   F more to operate
   G less to operate
   H the same to operate
   J can't answer from table
3. The color TV costs more to operate than the laser printer because ___.
   A the wattage of the color TV is higher
   B the color TV is on longer
   C the color TV is bigger
   D the power company charges more for TVs

L2

Teaching

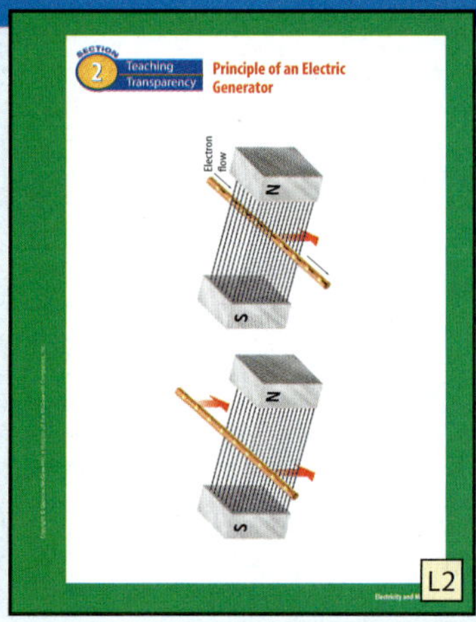
Section 2 Teaching Transparency — Principle of an Electric Generator

L2

Key to Teaching Strategies

The following designations will help you decide which activities are appropriate for your students.

- L1 Level 1 activities should be appropriate for students with learning difficulties.
- L2 Level 2 activities should be within the ability range of all students.
- L3 Level 3 activities are designed for above-average students.
- ELL ELL activities should be within the ability range of English-Language Learners.
- COOP LEARN Cooperative Learning activities are designed for small group work.
- LS Multiple Learning Styles logos, as described on page 12T, are used throughout to indicate strategies that address different learning styles.
- P These strategies represent student products that can be placed into a best-work portfolio.
- PBL Problem-Based Learning activities apply real-world situations to learning.

Hands-on Activities

Student Text Lab Worksheet

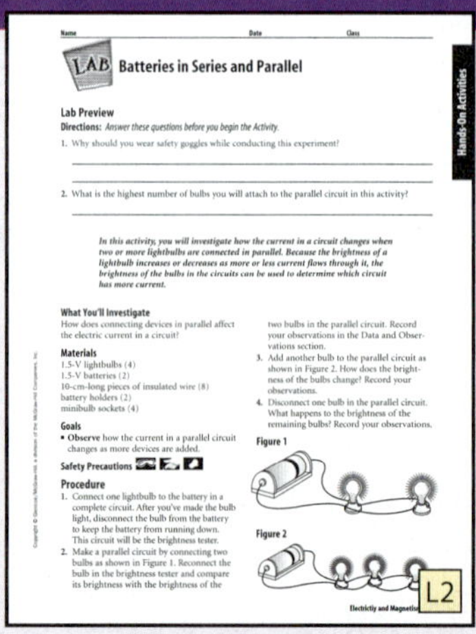
LAB — Batteries in Series and Parallel

Lab Preview

Directions: Answer these questions before you begin the Activity.

1. Why should you wear safety goggles while conducting this experiment?
2. What is the highest number of bulbs you will attach to the parallel circuit in this activity?

In this activity, you will investigate how the current in a circuit changes when two or more lightbulbs are connected in parallel. Because the brightness of a lightbulb increases or decreases as more or less current flows through it, the brightness of the bulbs in the circuits can be used to determine which circuit has more current.

What You'll Investigate

How does connecting devices in parallel affect the electric current in a circuit?

Materials

1.5-V lightbulbs (4)
1.5-V batteries (2)
10-cm-long pieces of insulated wire (8)
battery holders (2)
minibulb sockets (4)

Goals

- Observe how the current in a parallel circuit changes as more devices are added.

Safety Precautions

Procedure

1. Connect one lightbulb to the battery in a complete circuit. After you've made the bulb light, disconnect the bulb from the battery to keep the battery from running down. This circuit will be the brightness tester.
2. Make a parallel circuit by connecting two bulbs as shown in Figure 1. Reconnect the bulb in the brightness tester and compare its brightness with the brightness of the two bulbs in the parallel circuit. Record your observations in the Data and Observations section.
3. Add another bulb to the parallel circuit as shown in Figure 2. How does the brightness of the bulbs change? Record your observations.
4. Disconnect one bulb in the parallel circuit. What happens to the brightness of the remaining bulbs? Record your observations.

Figure 1

Figure 2

L2

Laboratory Activities

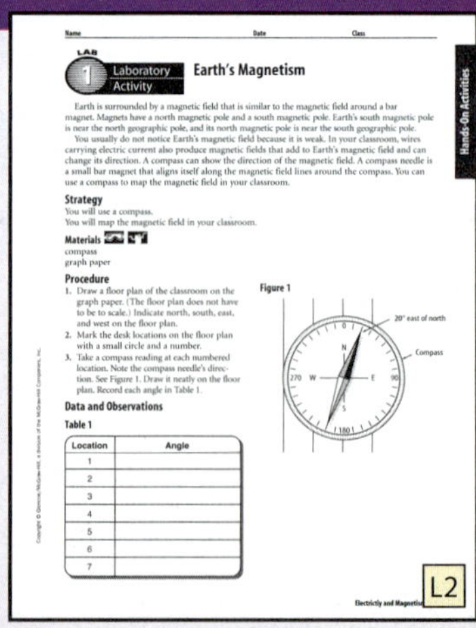
Laboratory Activity 1 — Earth's Magnetism

Earth is surrounded by a magnetic field that is similar to the magnetic field around a bar magnet. Magnets have a north magnetic pole and a south magnetic pole. Earth's south magnetic pole is near the north geographic pole, and its north magnetic pole is near the south geographic pole. You usually do not notice Earth's magnetic field because it is weak. In your classroom, wires carrying electric current also produce magnetic fields that add to Earth's magnetic field and can change its direction. A compass can show the direction of the magnetic field. A compass needle is a small bar magnet that aligns itself along the magnetic field lines around the compass. You can use a compass to map the magnetic field in your classroom.

Strategy

You will use a compass.
You will map the magnetic field in your classroom.

Materials

compass
graph paper

Procedure

1. Draw a floor plan of the classroom on the graph paper. (The floor plan does not have to be to scale.) Indicate north, south, east, and west on the floor plan.
2. Mark the desk locations on the floor plan with a small circle and a number.
3. Take a compass reading at each numbered location. Note the compass needle's direction. See Figure 1. Draw it neatly on the floor plan. Record each angle in Table 1.

Figure 1

Data and Observations

Table 1

Location	Angle
1	
2	
3	
4	
5	
6	
7	

L2

Meeting Different Ability Levels

Content Outline

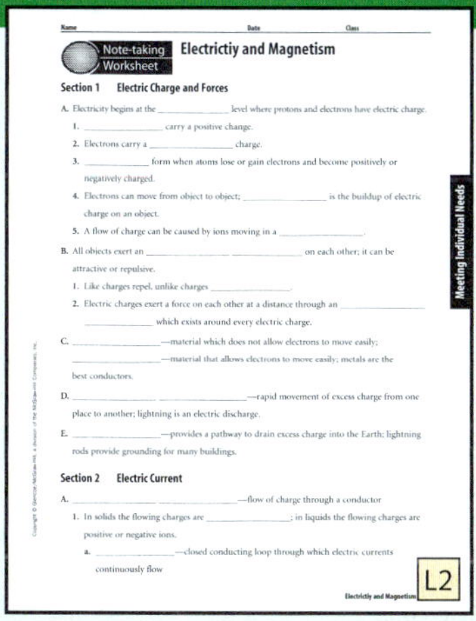

Reinforcement

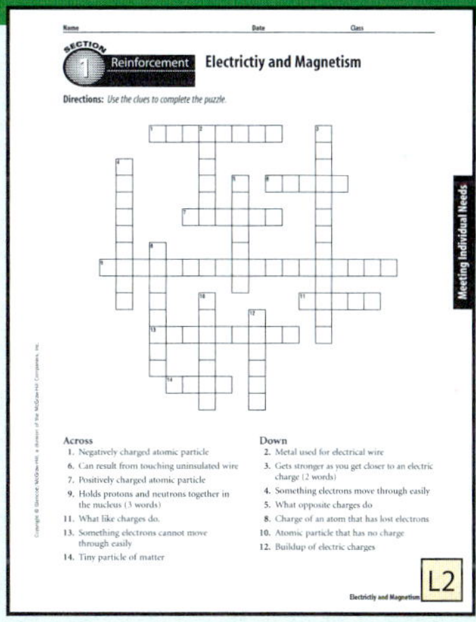

Enrichment

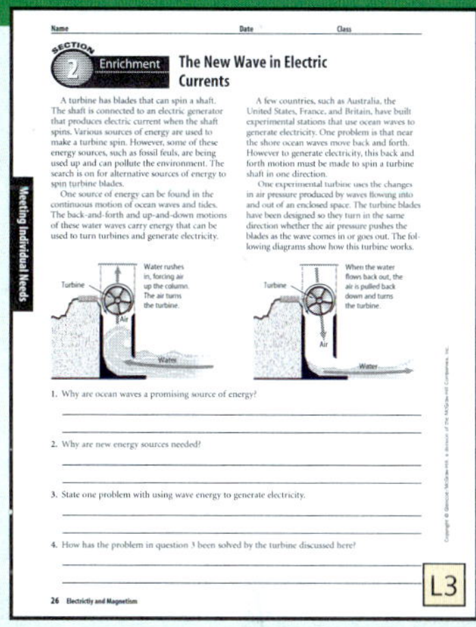

Directed Reading (English/Spanish)

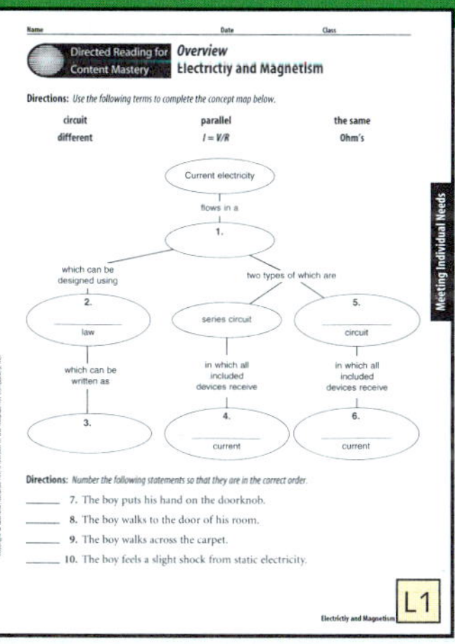

Study Guide

Reading Essentials

Assessment

Test Practice Workbook

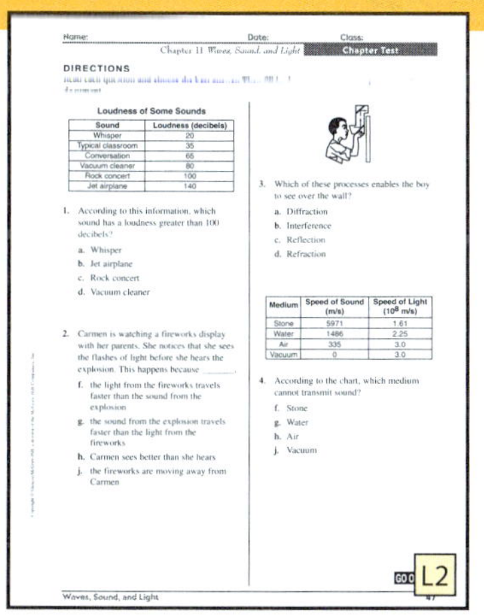

Chapter Review

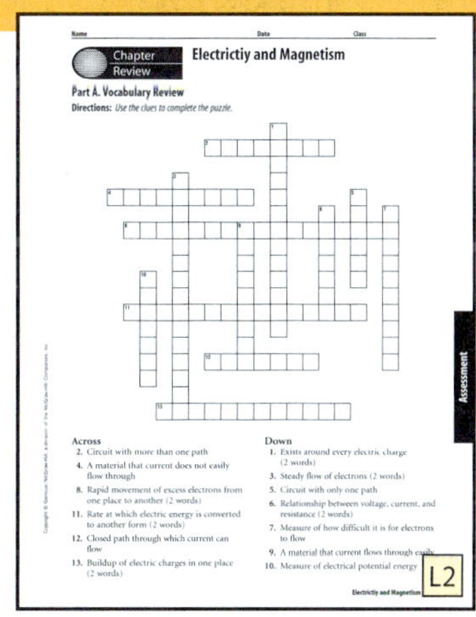

Chapter Tests

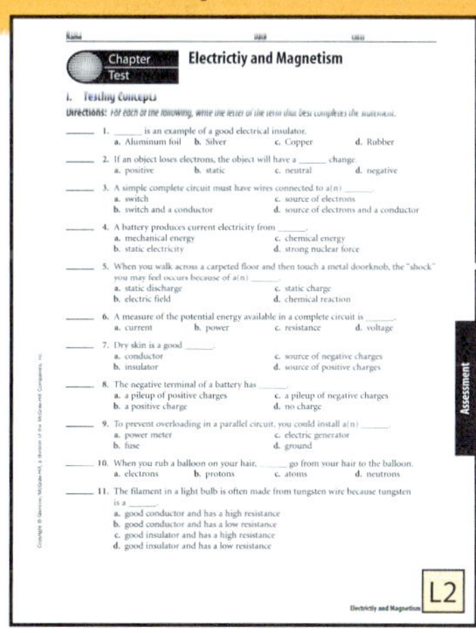

Science Content Background

section 1 Electric Charge and Forces

Objects become electrically charged by gaining or losing electrons, rather than by gaining or losing protons. This is because positively charged protons are bound together tightly with neutrons in atomic nuclei. The amount of energy needed to remove a proton from a nucleus is about a million times greater than the energy needed to remove an electron from an atom.

Rubbing materials causes atoms of two substances to be brought close together. Electric forces between atoms then can cause electrons to be transferred from one material to another.

section 2 Electric Current

Electric current is the flow of electric charge. The flow of electric charge means the net amount of charge per second that passes a given point. A metal contains electrons that are not bound to individual atoms, but can move freely throughout the metal. When no current flows in the metal, these electrons are in continual, random motion, constantly colliding with atoms and other charges in the metal. Because their motion is random, the same amount of charge moving in any direction past a point in one second is balanced by the amount of charge moving in the opposite direction. As a result, the net amount of charge moving past any point in the metal is zero.

A net electric field that points in a single direction causes a current to flow. The electrons continue their random motion, but added to that motion is an acceleration that causes electrons to drift slowly in the direction opposite the field. As a result, there is a net amount of charge that moves past a point in the material every second.

Voltage

An electric charge moving in an electric field has electrical potential energy, just like a ball falling in a gravitational field has gravitational potential energy. The change in electrical potential energy of a charge as it moves from one point to another in a material is related to the voltage or voltage difference between the two points.

The voltage difference is the change in electrical potential energy per unit charge between the two points in the material. The voltage difference times the electron charge equals the amount of electrical energy transferred by an electron as it moves between the two points in the material.

chapter content resources

Internet Resources

For additional content background, visit **red.msscience.com** to:

- access your book online
- find references to related articles in popular science magazines
- access Web links with related content background
- access current events with science journal topics

Print Resources

Fascinating Science Projects: Electricity and Magnetism, by Bobbi Searle, Aladdin Books, 2002

Awesome Experiments in Electricity and Magnetism, by Michael DiSpezio, Sterling Publishing, 1998

Charging Ahead: An Introduction to Electromagnetism, by Larry E. Schafer, NSTA Press, 2001

section 3 Magnetism

Materials that can become permanent magnets, such as iron, cobalt, and nickel, have atoms that are themselves magnets. In these atoms, it is a property of individual electrons called spin that cause the atoms to become tiny magnets. In these materials, interactions between neighboring atoms cause their magnetic poles to become aligned so that like poles point in the same direction. This results in the formation of regions containing billions of atoms that have their like magnetic poles pointing in the same direction. These regions are magnetic domains.

The magnetic fields of atoms in a domain add together, so that each domain is also a magnet with a north pole and a south pole. If the material is unmagnetized, the poles of the magnetic domains point in random directions. If most of the like poles of the magnetic domains are aligned, the material then is a permanent magnet.

IDENTIFYING Misconceptions

Find Out What Students Think

Students may think that all metals are attracted to magnets. This misconception arises from a lack of experience with magnets and different metals. While they have probably seen magnets stick to some metals, such as refrigerator doors, they incorrectly over-generalize that magnets stick to all metals.

Activity

Prior to a hands-on experience with magnets and materials have students predict the materials that a magnet would and would not be attracted to. Use a wide variety of materials and include items made from iron, aluminum, and copper.

Promote Understanding

A hands-on activity requiring students to see whether magnets attract different materials, including items made from iron, aluminum, and copper, will help students realize that not all metals are attracted to magnets. Because students are unsure about the metallic composition of certain items, they will need assistance processing their learning. This is fostered by the use of a table that has the name of the item in one column and the materials it is made out of in another column. Thus it can be recorded that a paper clip is made out of steel, a pipe out of copper, and foil from aluminum. Have students discuss the question: **Are all metals attracted to magnets?**

Assess

After completing the chapter, see *Identifying Misconceptions* in the Study Guide at the end of the chapter.

chapter 7

Chapter Vocabulary

charging by contact, p. 196
charging by induction, p. 197
insulator, p. 197
conductor, p. 197
static charge, p. 198
electric discharge, p. 198
electric current, p. 201
electric circuit, p. 202
electric resistance, p. 203
voltage, p. 205
series circuit, p. 208
parallel circuit, p. 208
magnetic domain, p. 211
electromagnet, p. 212
electromagnetic induction, p. 213

Science Journal Student responses will vary, but might include devices such as hair-dryers, toasters, CD players, TVs, radios, and computers.

INTERACTIVE CHALKBOARD with Image Bank

PowerPoint® Presentations

This CD-ROM is an editable Microsoft® PowerPoint® presentation that includes:

- a pre-made presentation for every chapter
- interactive graphics
- animations
- audio clips
- image bank
- all new section and chapter questions
- Standardized Test Practice
- transparencies
- pre-lab questions for all labs
- Foldables directions
- links to red.msscience.com

chapter 7

Electricity and Magnetism

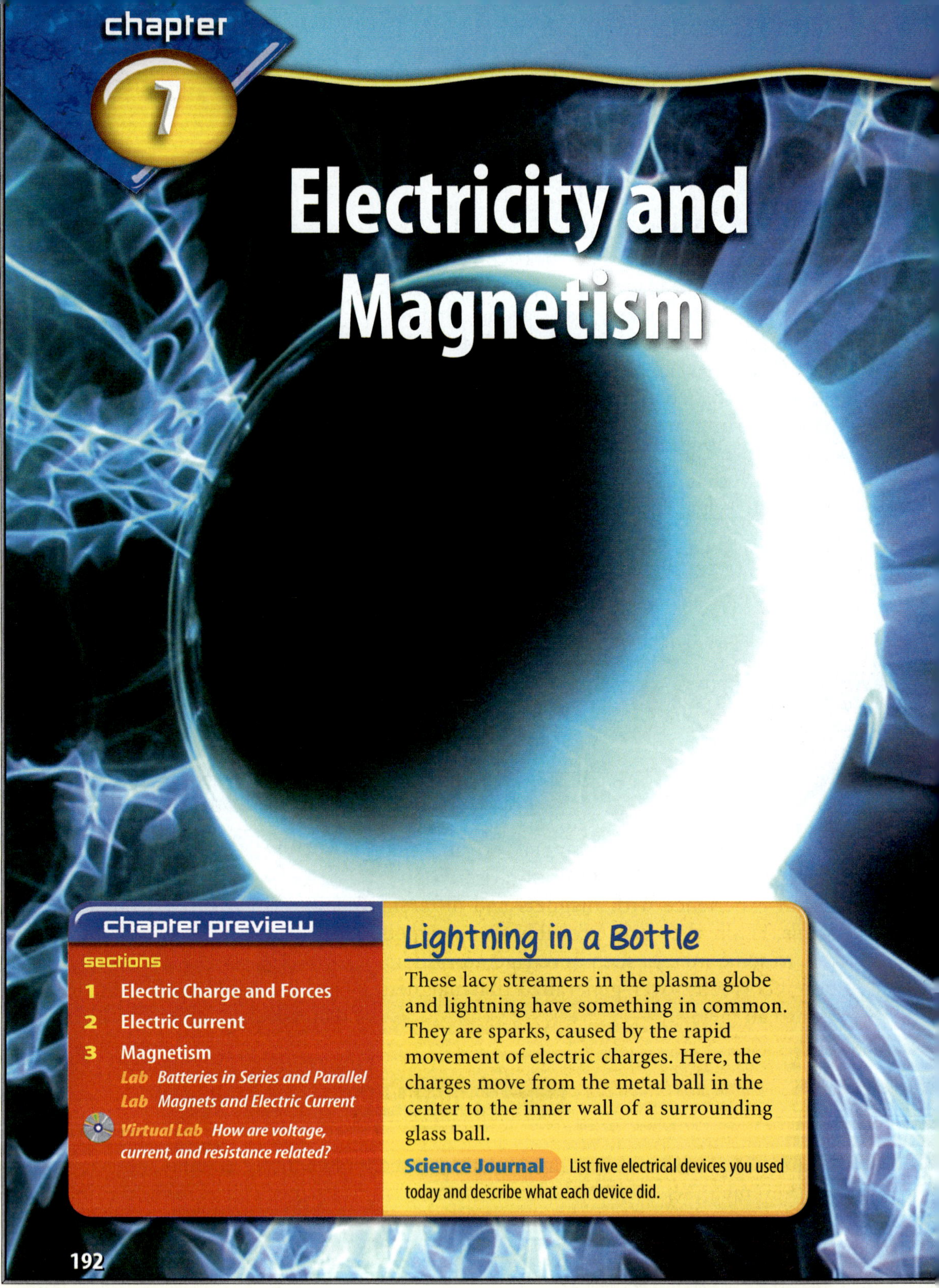

chapter preview

sections

1 Electric Charge and Forces
2 Electric Current
3 Magnetism
Lab *Batteries in Series and Parallel*
Lab *Magnets and Electric Current*
Virtual Lab *How are voltage, current, and resistance related?*

Lightning in a Bottle

These lacy streamers in the plasma globe and lightning have something in common. They are sparks, caused by the rapid movement of electric charges. Here, the charges move from the metal ball in the center to the inner wall of a surrounding glass ball.

Science Journal List five electrical devices you used today and describe what each device did.

192

Theme Connection

Energy Electricity is presented as a source of energy produced by an imbalance of charges. Students will see that electricity occurs naturally, and it can be produced by people to do work.

About the Photo

Plasma Globe The photo shows electric discharges occurring inside a plasma globe. The globe consists of an inner sphere, made of metal, and an outer sphere made of glass. The globe is filled with an inert gas under low pressure. A high voltage applied to the metal sphere causes electrons from the metal to move through the gas, colliding with gas atoms and producing light.

Start-Up Activities

Launch LAB

Electric and Magnetic Forces

You exert a force on a skateboard when you give it a push. But forces can be exerted between objects even when they are not touching. When you throw a ball up into the air, Earth's gravity exerts a force on the ball that pulls it downward. Electric and magnetic forces also can be exerted on objects that are not in contact with each other.

1. Inflate a rubber balloon and rub it against your hair or a piece of wool.
2. Bring the balloon close to a small bit of paper. Then bring the balloon close to a paper clip. Record your observations.
3. Bring a bar magnet close to a small bit of paper. Then bring the bar magnet close to a paper clip. Record your observations.
4. **Think Critically** Describe how the forces exerted by the balloon and the magnet were similar and how they were different. Compare the force exerted by the balloon and the force exerted by gravity on the paper. Compare the force exerted by the magnet and the force exerted by gravity on the paper clip.

FOLDABLES Study Organizer

Electric and Magnetic Forces Make the following Foldable to help you understand the properties of electric forces and magnetic forces.

STEP 1 **Fold** a sheet of paper in half lengthwise.

STEP 2 **Fold** paper down about 2 cm from the top.

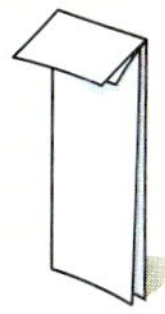

STEP 3 **Open and draw** lines along the horizontal fold. **Label** as shown.

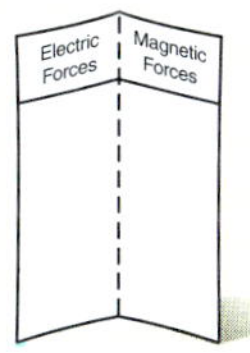

Summarize in a Table As you read the chapter, summarize the properties of electric forces in the left column and properties of magnetic forces in the right column.

Science Online — Preview this chapter's content and activities at red.msscience.com

Launch LAB

Purpose Students observe forces that act at a distance.

Preparation Paper towel or tissue may be torn into small bits ahead of time.

Materials rubber balloon, paper, paper clip, bar magnet

Teaching Strategy Have students work in pairs.

Think Critically

Both the balloon and the magnet could exert a force at a distance, without touching the paper or the paperclip. The force exerted by the magnet would not attract the paper, and the balloon would not attract the paper clip. The force on the paper and the force on the paper clip were greater than the force of gravity.

Assessment

Oral Have students formulate questions about forces that act at a distance and then quiz each other aloud.

FOLDABLES Study Organizer — Dinah Zike Study Fold

Student preparation materials for this Foldable are available in the Chapter *Fast File* Resources.

section 1

Electric Charge and Forces

1 Motivate

Bellringer

INTERACTIVE CHALKBOARD PowerPoint® Presentations

Section Focus Transparencies also are available on the Interactive Chalkboard CD-ROM. L2 ELL

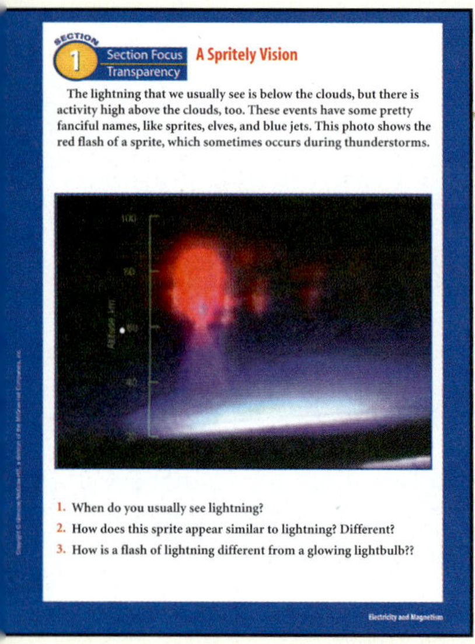
Section Focus Transparency 1 — A Spritely Vision

The lightning that we usually see is below the clouds, but there is activity high above the clouds, too. These events have some pretty fanciful names, like sprites, elves, and blue jets. This photo shows the red flash of a sprite, which sometimes occurs during thunderstorms.

1. When do you usually see lightning?
2. How does this sprite appear similar to lightning? Different?
3. How is a flash of lightning different from a glowing lightbulb??

Electricity and Magnetism

Tie to Prior Knowledge

Static Cling Ask students to recall times when they have removed clothes from the dryer and the clothes have stuck together. Discuss what happens when they try to pull the clothes apart. Ask students what they think is happening. Then tell students they will find out why this occurs in this section.

Teacher FYI

Students may wonder why protons, which are positively charged, are clustered together in the nucleus. The protons are held together by the nuclear force, which is much stronger than the electromagnetic repulsion at very small distances.

as you read

What You'll Learn

- **Describe** how electric charges exert forces on each other.
- **Define** an electric field.
- **Explain** how objects can become electrically charged.
- **Describe** how lightning occurs.

Why It's Important

Most of the changes that occur around you and inside you are the result of forces between electric charges.

Review Vocabulary

atom: the smallest particle of an element; contains protons, neutrons, and electrons

New Vocabulary

- charging by contact
- charging by induction
- insulator
- conductor
- static charge
- electric discharge

Electric Charges

Does a clock radio wake you up in the morning? Do you use a toaster or a microwave oven to help make breakfast? All of these devices use electrical energy to operate. The source of this energy lies in the forces between the electric charges found in atoms.

Positive and Negative Charge The matter around you is made of atoms. Atoms are particles less than a billionth of a meter in size—much too small to be seen, even with tremendous magnification. Every atom contains electrons that move around a nucleus, as shown in **Figure 1.** The nucleus contains protons and neutrons. An atom has the same number of protons and electrons.

Protons and electrons have electric charge. Electrons have negative charge and protons have positive charge. The amount of positive charge on a proton equals the amount of negative charge on an electron. Neutrons have no electric charge.

Neutral and Charged Objects Because an atom has equal numbers of protons and electrons, it contains equal amounts of positive and negative charge. An object with equal amounts of positive and negative charge is electrically neutral. If an atom gains or loses electrons, it is electrically charged. An object is electrically charged if the amounts of positive and negative charge it contains are not equal.

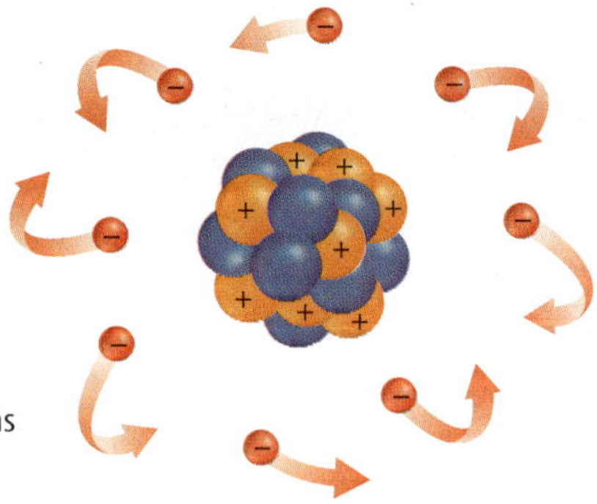

Figure 1 In an atom, negatively charged electrons move around a nucleus that contains neutrons and positively charged protons.

Section 1 Resource Manager

Chapter FAST FILE Resources

Transparency Activity, p. 44
MiniLAB, p. 3
Note-taking Worksheets, pp. 33–35
Enrichment, p. 30
Reinforcement, p. 27
Directed Reading for Content Mastery, p. 20

Home and Community Involvement, p. 25
Lab Management and Safety, p. 76
Science Inquiry Labs, pp. 41–42

The Forces Between Charges

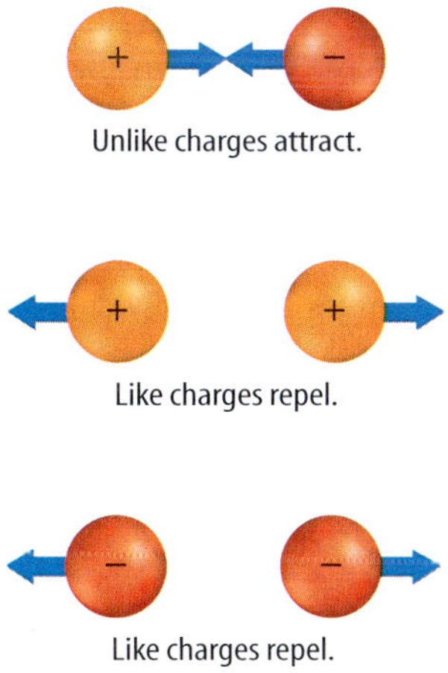

Figure 2 Electric charges exert forces on each other. The forces can be attractive or repulsive. **Describe** *how the forces change if the charges move closer together.*

When you drop a ball and it falls to the ground, the ball and Earth exert an attractive force on each other—the force of gravity. Just as two masses, such as Earth and the ball, exert forces on each other, two objects that are electrically charged exert forces on each other.

The force of gravity is always attractive. The forces exerted by charged objects on each other can be attractive or repulsive, as shown in **Figure 2.** If two objects are positively charged, they repel each other. If two objects are negatively charged, they also repel each other. If one object is positively charged and the other is negatively charged, they attract each other. In other words, like charges repel and unlike charges attract.

Electric Force Depends on Distance The electric force between two charged objects depends on the distance between the objects. The electric force decreases as the distance between the objects increases. For example, as two electrons move farther apart, the repulsive force between them decreases.

Electric Force Depends on Charge The electric force between two charged objects also depends on the amount of charge on each object. As the amount of charge on either object increases, the electric force between the objects also increases.

Electric Field and Electric Forces

To slide a book across a table top, your hand has to touch the book to give it a push. However, electric charges can exert forces on each other even when they are not touching. **Figure 3** shows what happens when you rub a balloon on the cat's fur and then hold the balloon close to its fur. The balloon makes the fur stand on end. The balloon and the fur are exerting electric forces on each other, even though they are not touching.

Figure 3 The balloon and the cat's fur can exert forces on each other, even without touching.

2 Teach

Discussion

Electrical Forces Electrical forces between charges are enormous in comparison to gravitational forces. Yet, we normally do not sense electrical forces between us and our surroundings, while we do sense gravitational interactions with Earth. Why? Most objects have no net charge (are neutral). Consequently, most objects do not interact electrically.

Use an Analogy

Campfire Analogy Explain to students that the electric force decreases as the distance between the objects increases in the same way that there is a decrease in the warmth one feels as one moves away from a campfire.

Use Science Words

Word Origin Have students use a dictionary or Encyclopedia to determine the origin of the word *electron* and who first used it to describe electrical phenomena. The word electricity was coined by William Gilbert (1544–1603), an English physicist and physician. He based the new word on the Greek word elektron, which means "amber." Gilbert discovered that amber attracted small particles after it had been rubbed with fur. L2 LS **Linguistic Intrapersonal**

Caption Answer

Figure 2 The forces between the charges increase as the charges move closer together.

Differentiated Instruction

Challenge Students can build a model that demonstrates how electrical forces differ from the force of gravity. L3 LS **Visual-Spatial**

Teacher FYI

Inverse Square Law Like gravity, magnetic force, and the intensity of light, the strength of an electric force is inversely proportional to the square of the distance between two charges.

Visual Learning

Figure 4 Tell students to picture the charges as spheres. Explain that the field extends in all directions around the spheres. Relate the direction of the arrows to the repulsion and attraction that occur when like and unlike charges interact with one another.

Activity

Attraction and Repulsion Students can investigate charging by contact using two rubber balloons suspended by thread a few inches apart. Students can charge the balloons, and a glass rod or plastic ruler by rubbing with silk and/or wool. They can experiment by bringing various charged objects near one another. Students should record their observations in a data table. L2 LS **Kinesthetic, Visual-Spatial, Logical-Mathematical**

Visual Learning

Figure 6 Why does the balloon stick to the wall? A force of attraction exists between positive charges in the wall and negative charges on the balloon.

Caption Answer

Figure 6 Yes. A positively charged balloon would induce a negative charge on the wall next to the balloon.

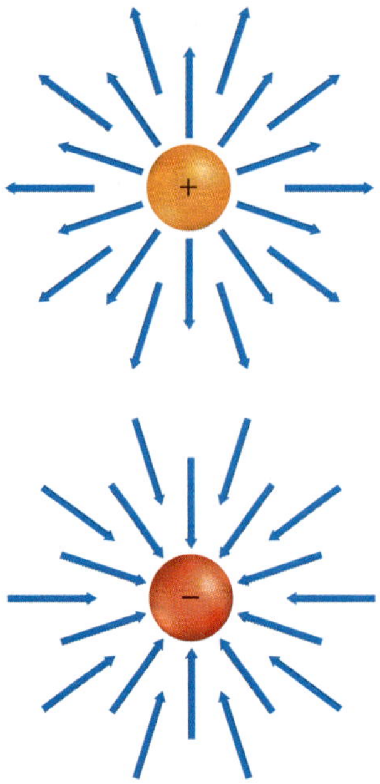

Figure 4 The electric field around a positive charge points away from the charge. The electric field around a negative charge points toward the charge.

Electric Field Surrounds a Charge How do electric charges exert forces on each other if they are not touching? An electric charge is surrounded by an electric field that exerts a force on other electric charges. Every proton and electron is surrounded by an electric field that exerts a force on every other proton and electron. The balloon you rubbed on your hair becomes electrically charged, so it too is surrounded by an electric field. The electric field surrounding the balloon exerts the force on your hair that makes it stand up.

Describing the Electric Field The electric field surrounding an electric charge is invisible. A way to describe the electric field around a charge is shown in **Figure 4.** The electric field is represented by arrows that are related to the force the field exerts on a positive charge. There is an electric field at every point in space surrounding a charge. **Figure 4** shows the electric field at only a few points in the space surrounding the charges.

Making Objects Electrically Charged

When you rubbed a balloon on your hair, it became electrically charged. The balloon no longer contained equal numbers of protons and electrons. The balloon became electrically charged because electric charges were transferred from your hair to the balloon.

Charging by Contact When you rubbed the balloon on your hair, the surface of the balloon came in contact with the surfaces of strands of your hair. As atoms in your hair and in the balloon came close to each other, electrons were transferred from atoms in your hair to atoms in the balloon. This is an example of **charging by contact,** which is the transfer of electric charge between objects in contact.

Because the balloon gained electrons after rubbing, it had more electrons than protons and was negatively charged. Because your hair lost electrons, it had more protons than electrons and was positively charged. The amount of negative charge gained by the balloon equaled the amount of positive charge left on your hair.

Another example of charging by contact is shown in **Figure 5.** As clothes tumble in a clothes dryer, they rub against each other. Charging by contact occurs and electrons are transferred from one article of clothing to another. This can cause articles of clothing to stick to each other when you take them out of the dryer.

Figure 5 Clothes that have been tumbling in a dryer become electrically charged by contact. Articles of clothing that have opposite charges stick together when they come out of the dryer.

Differentiated Instruction

English-Language Learners Help the students understand charging by contact by drawing a sequence of pictures showing how electric charges are transferred when two objects rub together. Be sure to label the charges + and − and show the uneven distribution of charges that results.

Challenge Have students tear a piece of newspaper into small bits. Students then run a comb through their hair several times and bring the comb near the paper. After the bits of paper have attached to the comb, students should hold the comb still until the paper bits "jump" off the comb. Challenge students to draw a diagram showing why this occurs. L3 LS **Kinesthetic, Visual-Spatial**

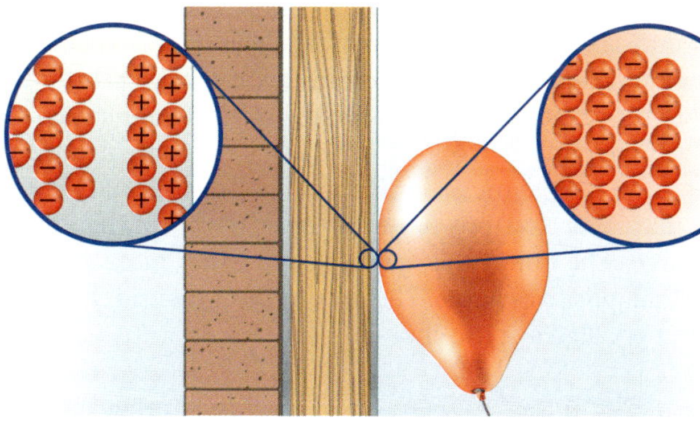

Figure 6 Charging by induction causes the charged balloon to push electrons away from the wall's surface. The surface of the wall becomes positively charged and attracts the negatively charged balloon.
Infer *whether a positively charged balloon would stick to the wall.*

Charging by Induction Have you ever rubbed a balloon on a sweater or your hair, and then stuck the balloon to a wall? The balloon became negatively charged after you rubbed it, but the wall was electrically neutral. **Figure 6** shows why the negatively charged balloon sticks to the wall. As the balloon is brought close to the wall, the electric field around the balloon repels the electrons in the wall. These electrons are pushed away from their atoms. This causes the region of the wall close to the balloon to become positively charged. The negatively charged balloon is attracted to this positively charged region, causing the balloon to stick to the wall.

In this case there is no electric charge transferred from one object and another. Instead, an electric field causes electrons to move from one region to another in an object. The rearrangement of electric charge due to the presence of an electric field is called **charging by induction.** As a result, one part of the object becomes positively charged and another part becomes negatively charged. However, the object remains electrically neutral.

Conductors and Insulators

In some materials electrons are held by atoms tightly enough that they are not able to move easily through the material. Materials in which electric charges do not move easily are **insulators.** Plastics, glass, rubber, and wood are examples of materials that are insulators.

In other materials, some electrons are held so loosely by atoms that they can move through the material easily. Materials in which electric charges can move easily are **conductors.** The best conductors are metals such as gold, silver, and copper. Because electrons can move easily in copper, it is widely used in electric wires.

Mini LAB

Observing Charging by Induction

Procedure

1. Turn on a **water faucet.** Adjust the flow so that the water stream is as slow as possible without producing drops.
2. Rub a **balloon** or a **comb** on your hair or on **wool cloth.**
3. Bring the charged end of the balloon or comb near the stream of water, and observe the result.

Analysis

1. Explain the behavior of the stream of water using the concept of charging by induction.
2. Infer how the distribution of charge on the water stream changed after it passed the charged area on the balloon or comb.

Try at Home

Mini LAB

Purpose Students will observe charging by induction and infer how charges are separated. L2
LS Logical-Mathematical

Materials balloon or comb, wool cloth, water faucet

Teaching Strategies

- Tell students to adjust the water to a small steady stream.
- Caution students to bring the balloon or comb near, but not touching, the water.

Analysis

1. The balloon (or comb) became negatively charged when rubbed with wool (or hair). As the balloon is brought close to the water, the electric field around the balloon repels the electrons in the water. This causes the part of the water close to the balloon to become positively charged. The positively charged water is attracted to the negatively charged balloon, causing the water stream to curve.
2. The charges returned to normal because there is no charged surface nearby.

Assessment

Process Have students predict what would happen if a positively charged object were brought near the water stream. Students may predict that a positive object would attract the electrons. The water stream will still be bent toward the object. L2
LS Logical-Mathematical

Try at Home

LAB DEMONSTRATION

Purpose to demonstrate charging

Materials red and green transparent plastic sheets, scissors, overhead projector

Preparation With the red and green sheets together, cut out matching balloon shaped pieces about 8 cm across. Also cut out rectangles about 3 cm by 8 cm.

Procedure The red and green sheets represent positive and negative charges. Overlapping red and green sheets should look black (neutral) when placed on the projector. To demonstrate charging by induction, move one layer on each object so that the red and green colors are visible.

Expected Outcome Students will see that each object is still neutral, but that the charges are unevenly distributed.

Assessment

Performance Have students use the same materials to show charging by contact.

SECTION 1 Electric Charge and Forces 197

IDENTIFYING Misconceptions

"Static" Electricity Some students may have learned the term *static electricity* to describe these phenomena. It is a misconception to think about charges as static, or unmoving. These attraction and repulsion phenomena occur because of unbalanced charges.

INTEGRATE History

Benjamin Franklin and Electricity Did Benjamin Franklin really fly a kite in a thunderstorm to try to capture lightning? Students can read about the experiments that Benjamin Franklin did with electricity and use the information to write a factual magazine article about what Franklin did and didn't do with a kite and lightning rod.

Quick Demo

Static Charge

Materials plastic wrap, cotton cloth, paper punch, aluminum foil

Estimated Time five minutes

Procedure Place a 30 cm by 30 cm piece of plastic food wrap on a tabletop and smooth it flat with a cotton cloth. Use the paper punch to punch out circles from the aluminum foil. Put the circles in the center of the plastic. Lift the plastic wrap slowly from the table and observe the behavior of the aluminum circles.

INTEGRATE History

Benjamin Franklin and Electricity American Benjamin Franklin lived from 1706 to 1790. He is best known as one of the country's Founders who played an important role in the formation of the United States Constitution. Franklin also was a scientist and was one of the first to prove that lightning was an electric discharge. In addition, he named the two types of electric charge—positive and negative—and produced a number of inventions, including the lightning rod and bifocal glasses.

Static Charge

If you walk across a carpet wearing shoes with rubber soles, charging by contact occurs. Electrons are transferred from the atoms in the carpet to the atoms on the soles of your shoes.

When charging by contact occurs, the amount of positive and negative charge on each object is no longer balanced. The object that loses electrons has more positive charge than negative charge. The object that gains electrons has more negative charge than positive charge. The imbalance of electric charge on an object is called a **static charge.**

Electric Discharge When you walk across a carpet and then touch a metal doorknob, sometimes you might feel an electric shock. Perhaps you see a spark jump between your hand and the doorknob. The spark is an example of an electric discharge. An **electric discharge** is the movement of static charge from one place to another. The spark you saw was the result of a static charge moving between your hand and the doorknob.

Figure 7 shows why a spark occurs when you touch the doorknob. Electrons that are transferred from the carpet to your shoes spread over your skin. As you reach toward the metal doorknob, the electric field around your hand repels electrons in the doorknob. They move away, leaving the surface of the doorknob nearest your hand with a positive charge. If the attractive electric force on the excess electrons is strong enough, these electrons can be pulled from your hand toward the doorknob. This rapid movement of charge causes the spark you see and the shock you feel.

Figure 7 Charging by induction causes a spark to jump from your hand to the doorknob.

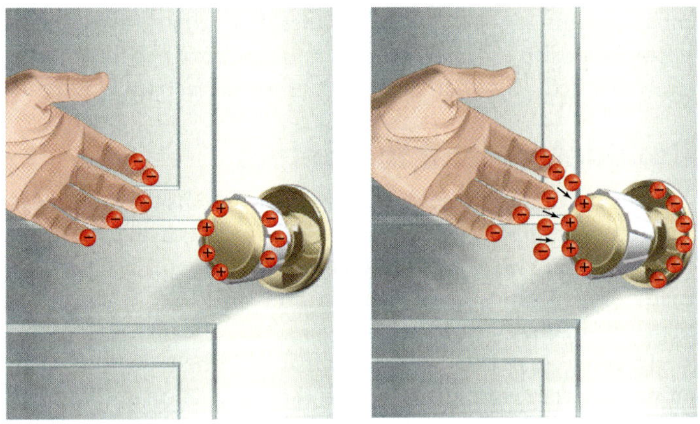

The excess negative charge on your hand repels electrons in the doorknob, leaving positive charges on the surface of the doorknob.

The attractive force between the charges on your hand and the doorknob can cause electrons to move to the doorknob.

Differentiated Instruction

Learning Disabled Students may benefit by cutting out positive and negative charges from colored paper. Students can arrange the charges to show a neutral object and then move the charges to show how objects become charged. L1
LS **Kinesthetic, Visual-Spatial**

Challenge Have students research why touching the doorknob feels as if the spark jumps to the hand rather than from it. Students can draw a poster explaining this phenomenon. L3
LS **Linguistic, Kinesthetic** P

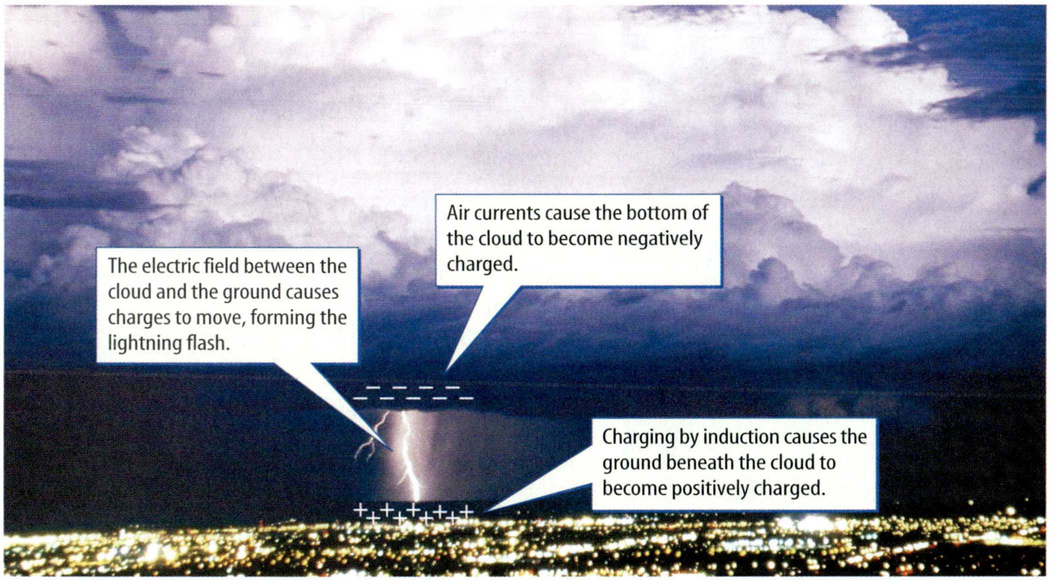

Figure 8 Lightning occurs when the static charge on a storm cloud causes charging by induction on the ground or another cloud.

Lightning A spectacular example of an electric discharge is lightning. **Figure 8** shows how lightning is produced. During a storm, air currents in a storm cloud sometimes cause electrons to be transferred from the top to the bottom of the storm cloud. The electric field surrounding the bottom of the storm cloud repels electrons in the ground. This makes the ground positively charged. The resulting attractive electric forces cause charges to move between the cloud and the ground, producing a flash of lightning.

Air currents in a storm cloud can cause other parts of the storm cloud to become positively and negatively charged. As a result, lightning flashes often occur between one storm cloud and another, and also within a storm cloud.

INTEGRATE Health

Lightning Safety A lightning flash can be dangerous. On average, lightning strikes about 400 people a year in the United States, and causes about 80 deaths. You should always take lightning seriously, particularly if you are outside and a thunderstorm is in sight. You can help protect yourself by following the 30-30 rule. If the time between the lightning and the thunder is 30 seconds or less, the storm is dangerously close. Seek shelter in an enclosed building or a car, and avoid touching any metal surfaces. Wait 30 minutes after the last flash of lightning before leaving the shelter—even if the Sun comes out. One in ten lightning strikes occurs when no storm clouds are visible.

Science Online

Topic: Lightning
Visit red.msscience.com for Web links to information about different types of lightning that occur in Earth's atmosphere.

Activity Make a table listing different types of lightning in one column and a description of the lightning type in a second column.

Visual Learning

Figure 8 The − symbols denote negative charges that have accumulated on the bottom of the cloud. The + symbols denote the induced positive charges on the ground beneath the cloud. Have students order the sequence of events that occurs to produce a lightning stroke between the cloud and the ground.

Fun Fact

About 156 billion, billion electrons flow down into the ground during a typical cloud to ground lightning strike.

Discussion

Timing the Storm Why is the time between the lightning and the thunder important when judging how close a storm is? Lightning and thunder occur at almost the same time, but light travels much faster than sound. The closer the sound of the thunder is to the lightning, the closer the lightning is to the observer.

Identifying Misconceptions

Lightning Strikes Twice Some people believe that lightning will not strike twice in the same place. It strikes the Empire State Building in New York City some 22–25 times per year!

Cultural Diversity

Lightning and Thunder In some Eastern cultures, it is believed that people are struck by thunder rather than lightning as a punishment for bad deeds. This is where the word *thunderstruck* came from.

Teacher FYI

A bolt of lightning lasts about one half second and is about 2 in. in diameter. A typical lightning bolt heats the surrounding air to a temperature of about 30,000°C.

3 Assess

DAILY INTERVENTION

Check for Understanding

Logical-Mathematical Tell students that, when an airplane is being refueled after a landing, a grounding wire is attached to the plane to prevent sparks that might ignite the fuel. Ask students why an airplane might have a static charge. The friction between the skin of the plane and the air and water molecules in the atmosphere gives the plane a static charge.

Reteach

Charging Objects Work with students to develop a large chart showing the different ways in which charges can separate and objects can become charged. L2
LS **Intrapersonal, Visual-Spatial**

Assessment

Performance Businesses sometimes use advertising signs that stick to windows by means of "static cling." Have students write an advertising brochure that explains how the signs work and why they are useful for temporary advertising. Use **Performance Assessment in the Science Classroom,** p. 129. L2
LS **Linguistic, Logical-Mathematical, Intrapersonal**

Figure 9 A lightning rod provides a path to conduct the charge in a lightning strike into the ground.

Grounding A lightning flash can transfer an enormous amount of electrical energy. When lightning strikes trees in a forest, it can spark a forest fire. If lightning strikes a building, the building can be damaged or set on fire.

One way to protect buildings from the damaging effects of lightning is to attach a metal lightning rod to the top of the building. A thick wire is connected to the lightning rod, and the other end of the wire is connected to the ground.

When lightning strikes the lightning rod, the electric charges in the lightning flash flow through the connecting wire into the ground. Earth can be a conductor, and because Earth is so large, it can absorb large quantities of excess electric charge. The process of providing a path to drain excess charge into Earth is called grounding. Because the lightning rod in **Figure 9** is grounded, the excess charge in the lightning strike flows harmlessly into the ground without damaging the building.

section 1 review

Summary

Electric Charges and Forces

- Electrons are negatively charged and protons are positively charged.
- An electric charge is surrounded by an electric field that exerts a force on other charges.
- Like charges repel each other; unlike charges attract each other.

Making Objects Charged

- The transfer of electric charges between two objects that touch is charging by contact.
- Charging by induction occurs when an electric field rearranges the charges in an object.
- Static charge is an imbalance of electric charge on an object.

Lightning

- Lightning is an electric discharge between a storm cloud and the ground, or within or between storm clouds.
- Grounding can prevent damage caused to buildings by lightning strikes.

Self Check

1. **Explain** why an atom is electrically neutral.
2. **Describe** how a balloon becomes electrically charged after you rub the balloon on your hair.
3. **Predict** Suppose the air currents in a storm cloud caused the bottom of the cloud to become positively charged. Predict whether lightning could occur between the cloud and the ground. Explain your reasoning.
4. **Infer** When charging by contact occurs, how is the amount of positive charge on one object related to the amount of negative charge on the other object?
5. **Describe** how the electric force between two objects depends on the amount of charge on the objects and the distance between them.
6. **Think Critically** Sometimes just before a lightning strike occurs nearby, the hair on a person's head will stand up. Explain why this happens.

Applying Skills

7. **Sequence** Make an events-chain concept map that shows the sequence of events that occurs when a flash of lightning is produced.

Science Online red.msscience.com/self_check_quiz

section 1 review

1. Atoms have equal numbers of protons (+) and electrons (−).
2. Electrons are transferred from your hair to atoms in the balloon.
3. The positively charged bottom of the cloud could induce a negative charge on the ground beneath the cloud. The attractive forces between the bottom of the cloud and the charges on the ground could cause charges to move between the cloud and the ground, producing lightning.
4. They are equal.
5. The greater the amount of charge on either or both objects, the greater the electric force. As the distance between objects increases, the electric force decreases.
6. The same cloud that produces the lightning repels electrons in your hair, making the ends of the hair positively charged. The strands of hair then repel one another.
7. Air currents cause charges to separate in the cloud → the negatively charged bottom of the cloud repels electrons in the ground → the attractive force causes charges to move between the cloud to the ground.

section 2

Electric Current

Electric Current

When you turn on a TV, images appear on the screen and sound comes out of the speakers. The TV produces light waves that carry energy to your eyes, and sound waves that carry energy to your ears. Where does this energy come from? You know that unless the TV is plugged into an electrical outlet, nothing happens when you turn it on. The electrical outlet provides electrical energy that the TV transforms into sound and light. This electrical energy becomes available only when an electric current flows in the TV.

What is an electric current? An **electric current** is the flow of electric charges. In some ways an electric current is like the flow of water in a pipe. In the pipe, water flows as water molecules move along the pipe. In a wire, there is an electric current when electrons in the wire move along the wire.

In a wire, the numbers of protons and electrons are equal and the wire is electrically neutral, as shown in **Figure 10.** When current flows in the wire, these electrons move along the wire. At the same time, electrons flow into one end of the wire and flow out of the other end. **Figure 10** shows that the number of electrons that flow out one end of the wire is equal to the number of electrons that flow into the other end. As a result, the wire remains electrically neutral.

The Unit for Current The amount of electric current in a wire is the amount of charge that flows into and out of the wire every second. The SI unit for current is the ampere, which has the symbol A. One ampere of electric current means an enormous number of electrons—about six billion billion—are flowing into and out of the wire every second.

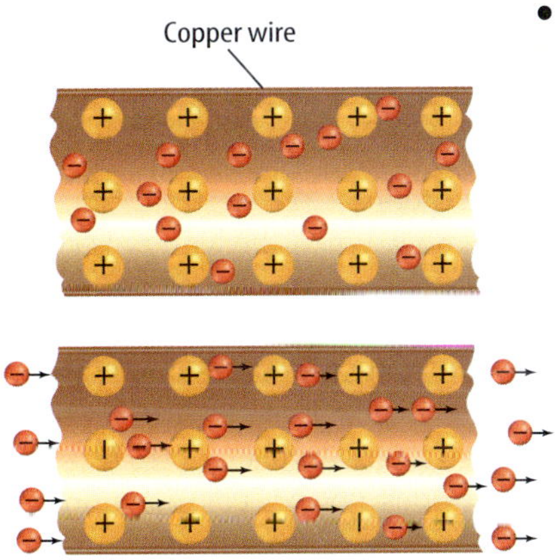

Figure 10 When a current flows in a wire, the same number of charges flow into and out of the wire. The wire remains electrically neutral.

as you read

What You'll Learn

- **Describe** how an electric current flows.
- **Explain** how electrical energy is transferred to a circuit.
- **Explain** how current, voltage, and resistance are related in a circuit.
- **Distinguish** between series and parallel circuits.

Why It's Important

Electrical appliances you use every day transform the electrical energy in an electric current into other useful forms of energy.

Review Vocabulary
kinetic energy: the energy an object has due to its motion

New Vocabulary
- electric current
- electric circuit
- electric resistance
- voltage
- series circuit
- parallel circuit

1 Motivate

INTERACTIVE CHALKBOARD PowerPoint® Presentations

Bellringer

Section Focus Transparencies also are available on the Interactive Chalkboard CD-ROM. L2 ELL

Section 2 Section Focus Transparency — Go with the Flow!

Highways in heavily used areas are designed with many lanes to keep traffic moving, but sometimes it doesn't work.

1. Describe what is happening in the picture. What would traffic be like if there were the same number of cars but fewer lanes? What would traffic be like if there were the same number of cars but twice as many lanes?
2. How might the flow of traffic on a road be like the flow of electricity in a wire?

Tie to Prior Knowledge

"Dead" Batteries Ask students if they have ever used a portable CD player only to find the sound slow and sluggish. Students may suggest that the problem is dead batteries. Ask students why they think a battery dies or stops producing enough electric current to operate the CD player. In this section, students will learn what electric current is and how it is produced.

Section 2 Resource Manager

Chapter *FAST FILE* Resources
- Transparency Activity, p. 45
- Enrichment, p. 31
- Reinforcement, p. 28
- Directed Reading for Content Mastery, p. 21

Life Science Critical Thinking/Problem Solving, p. 2
Mathematics Skill Activities, pp. 43–44

2 Teach

Caption Answer

Figure 11 No current would flow because the circuit would not be closed.

Activity

A Simple Circuit Have students work with a partner to connect a D-cell battery, small light bulb, and a piece of wire in such a way that the bulb lights up. Students should sketch the circuit and use arrows to show the closed path that the electric charges follow in the circuit. L2 COOP LEARN LS **Intrapersonal, Visual-Spatial** P

IDENTIFYING Misconceptions

Batteries and Cells People commonly refer to single dry cells as batteries. In fact, the word *battery* is generally printed on the dry cell. A battery is actually more than one dry cell or wet cell connected in series. Most devices use more than one cell to supply electric power, so battery-operated is the correct term.

Virtual Labs

V=IR How are voltage, current, and resistance related?

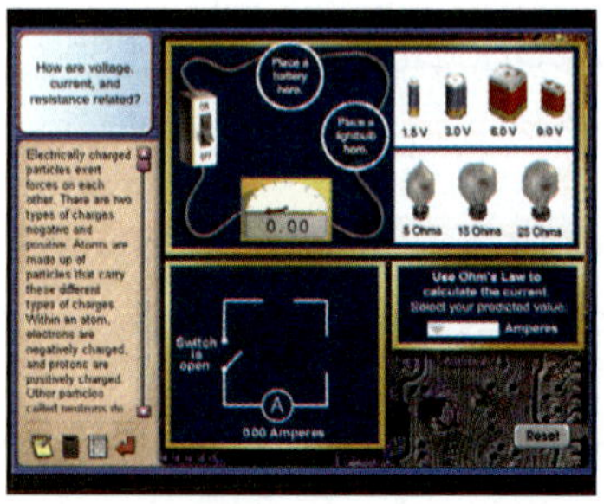

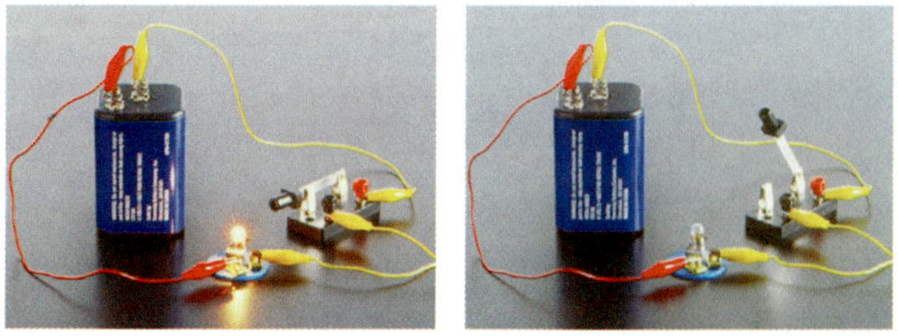

Figure 11 A battery, lightbulb, and connecting wires form a simple electric circuit. Current flows as long as the switch is closed. When the switch is open, current no longer flows.
Explain *whether current would flow if the lightbulb were disconnected.*

A Simple Electric Circuit

When a lightning flash occurs, electrical energy is transformed into heat, sound, and light in an instant. But to watch your favorite shows on TV, electrical energy must be transformed into light and sound for as long as your shows last. This means that an electric current must be kept flowing in your TV as you watch it.

Electric current will flow continually only if the charges can flow in a closed path. A closed path in which electric charges can flow is an **electric circuit.** A simple electric circuit is shown in **Figure 11.** Current will flow in this circuit as long as the conducting path between the battery, wires, and lightbulb is not broken. If the switch is open, current will not flow. Even with the switch closed, if one of the wires is disconnected or cut, or the filament wire in the lightbulb the path is no longer closed. Then current will no longer flow.

Science Online
Topic: Electric Shock
Visit red.msscience.com for Web links to information about the effects of electric current on the human body.
Activity Make a chart showing how the human body responds to different amounts of current that enter the body.

Making Electric Charges Flow

Water flows in a pipe when there is a force exerted on the water. For example, a pump can exert a force on water that pushes it through a pipe. A force must be exerted on electric charges to make them flow. Remember that a force is exerted on an electric charge by an electric field. To make electric charges flow in a circuit, there must be an electric field in the circuit that will move electrons in a single direction.

A Battery Makes Charges Flow The battery in **Figure 11** produces the electric field in the circuit that causes electrons to flow. When the battery is connected in a circuit, chemical reactions occur in the battery. These chemical reactions cause the negative terminal to become negatively charged, and the positive terminal to become positively charged. The negative and positive charges on the battery terminals produce the electric field in the circuit that causes electrons to flow. The battery makes electrons flow in the direction from the negative terminal toward the positive terminal.

Differentiated Instruction

Visually Impaired Prepare a simple circuit with a battery, light bulb, and wire. Allow visually impaired students to trace the circuit with their fingers to demonstrate the concept of a complete circuit. L2 LS **Kinesthetic, Intrapersonal**

Teacher FYI

Conductors In a metal, the outer electrons of each atom leave their parent atom and move among all the atoms in the metal. This "sea" of moving electrons makes metal a good conductor. When there is an electric current in a wire, it is these movable electrons that flow.

202 CHAPTER 7 Electricity and Magnetism

Electric Resistance It can be slow going when you try to walk to class through a crowded corridor. To avoid collisions, you are constantly changing direction, slowing down, and speeding up. Even though you might change speed and direction many times, you keep moving toward your classroom. The flow of electrons in a circuit is similar. Electrons are constantly colliding with atoms and other electric charges as they flow. These collisions cause electrons to change direction, as shown in **Figure 12.** An electron flowing in a wire may be involved in trillions of collisions every second. However, between each collision, the electric field in the circuit keeps electrons accelerating in the direction current is flowing.

Reading Check *Why do electrons constantly change direction as they flow in a circuit?*

The measure of how difficult it is for electrons to flow in an object is called the **electric resistance** of the object. The resistance of insulators is usually much higher than the resistance of conductors. The unit for electric resistance is the ohm, symbolized by Ω. An electric resistance of 20 ohms would be written as 20 Ω.

A Model for Electron Flow One way to picture how electrons flow in a circuit is to imagine a basketball bouncing down a flight of stairs, as shown in **Figure 13.** In this model the ball is like an electron moving through a circuit, and the steps are like the atoms it bumps into. When you drop the ball, it speeds up as gravity pulls it downward. When it hits a step, it changes direction. The ball also slows down as it bounces upward because the force of gravity continues to pull it downward. After the ball reaches the top of its bounce, it falls downward toward the next step and speeds up again. This process is repeated as the ball bounces from step to step.

Even though the ball changes direction after it hits each step, the overall motion of the ball is downward. In the same way, an electron in a circuit changes direction after each collision. However, its overall motion is in the direction of the current flow.

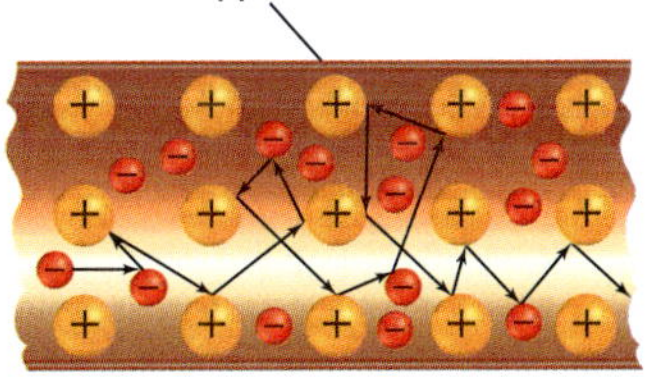

Figure 12 Collisions with atoms and other charges cause electrons in a wire to change direction many times each second.

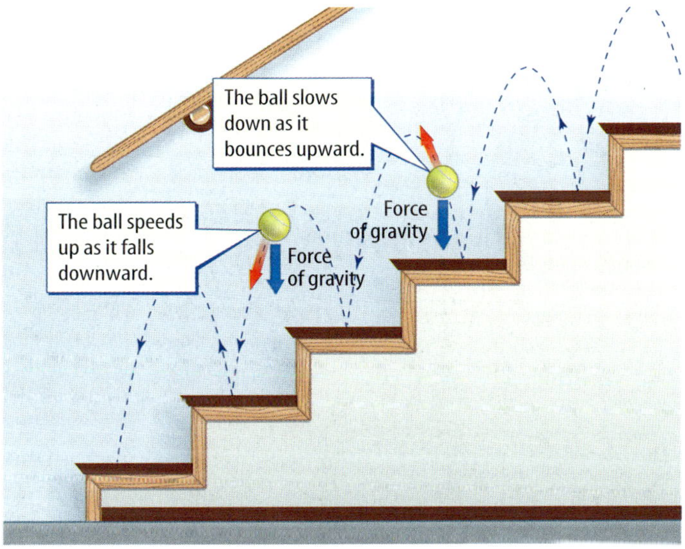

Figure 13 The motion of an electron flowing in an electric circuit is similar to the motion of a ball bouncing down the stairs. The force of gravity keeps the ball moving downward. An electric field keeps an electron moving in the direction of the current.

Reading Check

Answer Because they are constantly colliding with atoms and other electric charges

Make a Model

Potential Energy Remind students that objects gain potential energy as they move farther from Earth's surface. As a ball bounces down the stairs, its potential energy is converted to kinetic energy that is transferred to the stairs. Point out that an electric circuit works in a similar way. In a simple circuit, the battery produces an electric field that determines the direction in which the electric charges move just as gravity determines the direction that the ball moves. The potential energy of the electric charges can be converted into light energy if there is a bulb in the circuit.

Quick Demo

Flowing Current

Materials 20 cm length of 1 in. diameter plastic pipe, marbles.

Estimated Time five minutes

Procedure Demonstrate the way in which electrons move through a circuit by filling the plastic pipe with marbles. Show students that, as you push another marble into one end, the last marble pops out the other end. Tell students that as current flows in a circuit, the same number of electrons that flow into a wire also flow out of the wire. As a result, the charge on the wire doesn't change.

Visual Learning

Figure 13 Help students to understand that the force of gravity exerts a downward force, while the electrical force moves electrons through the circuit. Just as the ball bounces and changes direction in its downward path, the electrons collide and change direction while continuing to move in the direction of the electric force.

IDENTIFYING Misconceptions

Speed of Electricity Students might have heard that electrons flow at the speed of light. This is incorrect. Electrons actually flow quite slowly. It is the effect of the electric field on electrons in the circuit that moves at the speed of light. This effect is the force the electric field exerts on the electrons. This force causes electrons to drift slowly in the direction of current flow, while they are colliding with other charges in the circuit. It might take an electron in a circuit several minutes to drift one centimeter toward the positive battery terminal.

Fun Fact

Thomas Edison didn't invent the light bulb. He invented a practical light bulb—one that wouldn't burn out in seconds or minutes. He tested more than 6000 plant fibers from all over the world before discovering that a burned cotton thread would give a long-lasting filament.

The Speed of Electric Current Because the ball changes direction and slows down after each collision with a step, the time it takes the ball to reach the bottom of the stairs is much longer than if the ball had fallen without bouncing. In the same way, the electric resistance in a wire causes electrons to flow slowly. It may take several minutes for an electron in a circuit to travel one centimeter.

If electrons travel so slowly, why does a lightbulb light up the instant you flip a switch? When you flip the switch you close a circuit and an electric field travels through the circuit at the speed of light. The electric field causes electrons in the lightbulb to start flowing almost immediately after the switch is flipped. It is the electrons flowing in the lightbulb that cause it to glow.

Transferring Electrical Energy

As a ball bounces down a flight of stairs, it transfers energy to the stairs. Each time the ball collides with a step, some of the ball's kinetic energy is transferred to the step. Electrons flowing in a circuit also have kinetic energy. When a current flows in a material, the repeated collisions between electrons and atoms cause a continual transfer of kinetic energy to the material. The energy that flowing electrons transfer to the circuit also is called electrical energy. As electrons bump into atoms, electrical energy is converted into other forms of energy, such as heat energy and light.

For example, a lightbulb contains a filament that is a small coil of narrow wire, as shown in **Figure 14.** When current flows in the filament, electrical energy is converted into heat and light. The filament becomes hot and glows, giving off light that enables you to see in the dark.

Figure 14 A lightbulb filament is a coil of thin wire. The electric resistance in the filament converts electrical energy into heat and light.

Electrical Energy and the Electric Field As electrons flow in a circuit, the electrical energy transferred to the circuit depends on the strength of the electric field. If the electric field becomes stronger, the electric force exerted on electrons increases as they move from one point to another in the circuit. This causes electrons to move faster between collisions. You might recall that the kinetic energy of an object increases as its speed increases. So the kinetic energy of flowing electrons increases as the electric field gets stronger. As a result, increasing the electric field causes more electrical energy to be transferred to the circuit.

Active Reading

Summarizing This strategy helps students pinpoint the most important information in a paragraph. Have students work in pairs. After they read each paragraph, students can take turns summarizing the information in the paragraph. L2

LS **Interpersonal, Linguistic, Auditory-Musical**

Science Journal

Living Without Electricity Have students write a story that tells how their life would change if they had to live without electricity for one-day, one-week, and one-month periods. How does the length of time without electricity make a difference? L2

LS **Logical-Mathematical, Intrapersonal**

Voltage

You might have seen signs on electrical equipment that read "Danger! High Voltage." What is voltage? Voltage is a measure of the electrical energy of electrons flowing in a circuit. When an electron flows between two points in a circuit, it transfers electrical energy. **Voltage** is a measure of the amount of electrical energy transferred by an electric charge as it moves from one point to another in a circuit. The voltage between two points in a circuit can be measured with a voltmeter. **Figure 15** shows how the voltage measured by a voltmeter depends on the location of the points in the circuit. The voltage between any two points in the circuit increases when the electric field in the circuit increases. The SI unit for voltage is the volt, which has the symbol V.

The voltage across the ends of this wire is small. Only a small amount of electrical energy is transferred to this wire.

The voltage across the connections to the lightbulb is large. A large amount of electrical energy is transferred to the lightbulb.

Figure 15 A voltmeter measures the voltage between different points in this electric circuit.
Determine *To which part of the circuit is most of the electrical energy transferred?*

A Battery Produces Electrical Energy

When a current transfers electrical energy in a circuit, where does the electrical energy come from? The electric field in the circuit causes the flowing electrons to have electrical energy. If a battery is connected in the circuit, it is the chemical reactions in the battery that produce the electric field. As a result, in a battery chemical energy is transformed into electrical energy. This electrical energy then can be transformed into other forms of energy in the circuit. However, the battery is the source of the energy used by the devices connected in the circuit.

Reading Check *What form of energy is transformed into electrical energy in a battery?*

Battery Voltage

The voltage between the positive and negative terminals of a battery is usually called the voltage of the battery. The battery voltage is related to the amount of electrical energy an electron would transfer to the circuit as it moved through the circuit all the way from the negative terminal to the positive terminal. This means that more electrical energy is transferred to the circuit as the voltage of the battery increases. **Figure 16** shows how the voltage produced by different types of batteries depends on the chemical reactions that occur in the battery.

Use Science Words

Word Origin The units in which voltage, current, and resistance are measured are all named after physicists who experimented with electricity. Students can research the work of Volta, Ampere, and Ohm and explain what each man contributed to our understanding of electricity.

Fun Fact

If it hadn't been for Alessandra Volta, we might have frogs' legs powering our battery operated devices! In 1771, Luigi Galvani, one of Volta's friends, devised an "animal electricity" experiment by setting a frog's leg between two different metals. The muscles in the frog's leg twitched. Volta replaced the frog's leg with a salt solution, demonstrating that the electricity came from a reaction between the two metals and not from the frog.

Caption Answer

Figure 15 The voltage between the two sides of the lightbulb is larger. This means most of the electrical energy is being transferred to the lightbulb.

Reading Check

Answer chemical energy

Differentiated Instruction

Challenge Students can investigate the work of Alessandra Volta and his creation of the first battery. Students should draw a labeled diagram of Volta's battery, demonstrating why it was a complete circuit. L3 LS **Linguistic, Visual-Spatial, Intrapersonal** P

Curriculum Connection

Biology Have students do an Internet search to find pictures and explanations of how an electric eel generates electricity. Students can make a labeled diagram to explain the process. They should write a brief explanation of how the process is similar to and different from a battery. L3 LS **Linguistic, Visual-Spatial, Intrapersonal**

Visualizing Batteries

Have students examine the illustrations and read the captions. Then ask the following questions.

How many parts does each cell of a battery contain? A battery may have one or more cells. Each cell consists of an electrolyte, a positive electrode, and a negative electrode.

What property of a battery makes it possible to recharge it? The type of electrolyte it has. In a dry-cell battery the electrolyte is a paste. In a wet-cell battery the electrolyte is a liquid.

Make a Model

Sour Batteries Have students make batteries from lemons. Insert a copper nail in one side and a zinc nail in the other side. Wrap wire around each nail leaving a free end to connect to a voltmeter. Connect the wire from the copper nail to the positive terminal and the wire from the zinc nail to the negative terminal. What indicates that a battery has been constructed? The voltmeter registers a charge. L2

ELL LS **Kinesthetic**

VISUALIZING BATTERIES

Figure 16

Many electrical devices use batteries to supply electrical energy. Every battery consists of one or more cells. The chemical reactions in a cell produce a voltage when the cell is connected in a circuit. Each cell has three parts—an electrolyte, a positive electrode, and a negative electrode. The electrolyte contains chemicals that cause chemical reactions to occur at the positive and negative electrodes. There are two types of cells—dry cells and wet cells.

Dry-Cell Batteries **Flashlight batteries and the batteries that run portable CD players are dry-cell batteries. This type of cell is called a dry cell because the electrolyte is a paste, and not a liquid. The cells commonly used in dry-cell batteries have a voltage of 1.5 V. The most inexpensive dry-cell batteries are zinc-carbon batteries, shown on the right.**

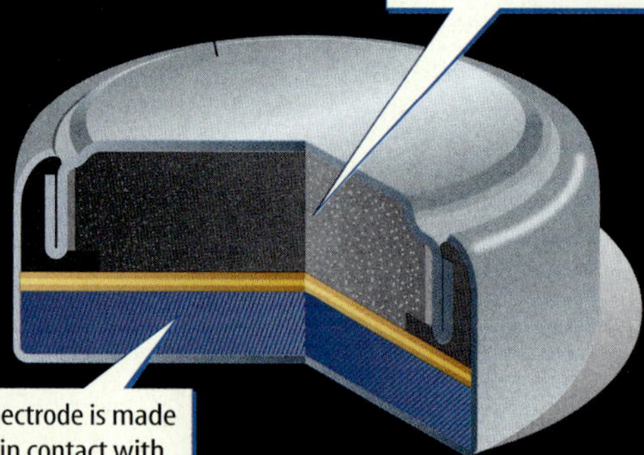

A silver oxide battery is a button-shaped or coin-shaped battery, often used in a camera or a calculator. This type of battery also is a dry-cell battery, and usually has a voltage of 1.5 V.

Wet-Cell Batteries **In a wet cell the electrolyte is a liquid. The most common wet-cell battery is a car battery. A 12-V car battery contains six wet cells connected in series. Each wet cell produces 2 V. Unlike the dry-cell batteries shown above, a wet-cell battery is rechargeable.**

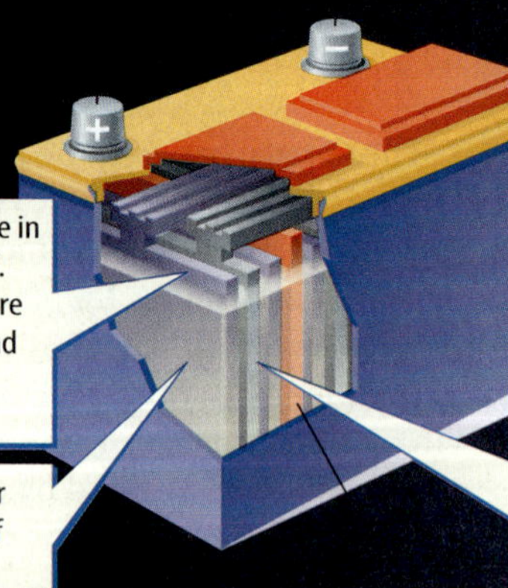

206 CHAPTER 7 Electricity and Magnetism

Ohm's Law The voltage, current, and resistance in a circuit are related. What happens if the voltage in a circuit is increased? As the voltage in a circuit increases, the electric field in the circuit increases and causes electrons to speed up more between collisions. As a result, the current in the circuit increases. Increasing the resistance in the circuit increases the number of collisions that occur every second as electrons flow. This makes it more difficult for electrons to flow in the circuit. As a result, increasing the resistance reduces the current.

The relationship between the voltage, current, and resistance in a circuit is known as Ohm's law. Ohm's law can be written as the following equation.

Ohm's Law

voltage (in volts) = **current** (in amperes) × **resistance** (in ohms)

$$V = IR$$

Applying Math — Solve a Simple Equation

FLASHLIGHT VOLTAGE When a flashlight is turned on, the current that flows in the flashlight circuit is 0.10 A. If the resistance of the circuit is 30.0 Ω, what is the voltage in the circuit?

Solution

1. *This is what you know:*
   - current: $I = 0.10$ A
   - resistance: $R = 30.0\ \Omega$
2. *This is what you need to find:* voltage: $V = ?$V
3. *This is the procedure you need to use:* Substitute the known values for current and resistance into Ohm's law, and calculate the voltage:
   $V = IR = (0.10\ \text{A})(30.0\ \Omega) = 3.0\ \text{V}$
4. *Check your answer:* Divide your answer by the given resistance, 30.0 Ω. The result should be the given current, 0.10 A.

Practice Problems

1. When a portable radio is playing, the current in the radio is 0.3 A. If the resistance of the radio is 30.0 Ω, what is the voltage supplied by the radio battery?
2. The batteries in a portable CD player supply a voltage of 6 V. If the resistance in the CD player is 24 Ω, what is the current in the CD player when it's turned on?

Science Online — For more practice, visit red.msscience.com/math_practice

Discussion

Water Flow Ask students what factors influence the amount of water that comes out of a garden hose or a faucet. Possible answers: The pressure of the pump, the size of the pipe. When discussing the relationship between the variables in Ohm's Law, expand on the analogy relating water traveling through a hose to current through a wire. Water is like the current, water pressure is like the voltage or potential difference, and resistance is like the diameter of the hose. Remind students that this is just an analogy and that electrons don't "flow" like water.

Applying Math

National Math Standards

Correlation to Mathematics Objectives

1, 2, 9

Teaching Strategy

Follow the steps in the example problem.

Answers to Practice Problems

1. 9 V

2. .25 A

LAB DEMONSTRATION

Purpose to demonstrate different series and parallel circuits

Materials 2 D-cells, four mini lightbulbs in sockets, connecting wires

Preparation Build a series circuit out of one D-cell, two lightbulbs and connecting wire. Build a parallel circuit out of the other materials.

Expected Outcome Students will see that there is more than one type of circuit. Have students trace the path of the electrical charge around each circuit. L2 LS **Kinesthetic**

Assessment

Ask students to predict what will happen if you unscrew one bulb in each circuit. Students should explain their prediction. Students might suggest that both lightbulbs will go out in the series circuit but the other lightbulb will stay lit in the parallel circuit.

3 Assess

DAILY INTERVENTION

Check for Understanding

Visual-Spatial Have students draw a model diagram of a flashlight circuit with two batteries and either an open or closed switch. Note the connection of positive (+) to negative (−) ends of the batteries. Ask students to explain what is happening in this arrangement. L2 LS **Linguistic, Visual-Spatial, Logical-Mathematical**

Reteach

Ohm's Law Students often have trouble rearranging the equation for Ohm's law. Show them the following diagram to go with the expression V = IR. When the desired variable is covered, the other two variables are in appropriate mathematical order.

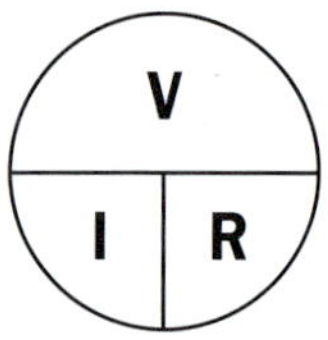

Assessment

Process To check student understanding of the relationships in Ohm's Law, ask the following question: if the resistance of the circuit is doubled and the voltage remains the same, what will happen to the current? It will be cut in half. Use **Performance Assessment in the Science Classroom,** p. 89

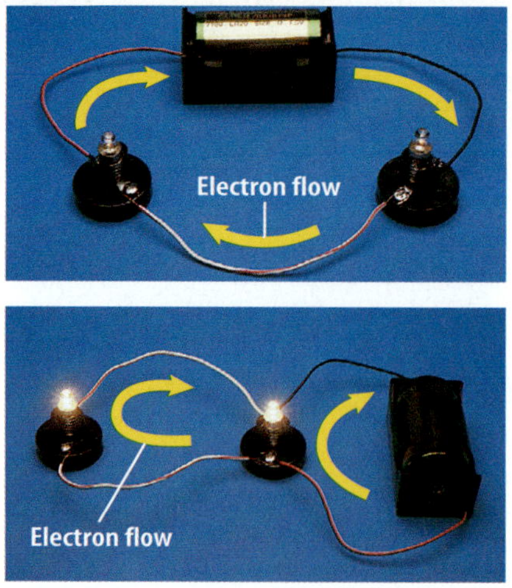

Figure 17 A series circuit (top) has only one path for current to follow. A parallel circuit (bottom) has more than one path for current to follow.

Series and Parallel Circuits

There usually are a number of devices and appliances connected to the circuits in your house. There are two ways that devices can be connected in a circuit. One way is a series circuit, shown in the upper part of **Figure 17,** and the other way is a parallel circuit, shown in the lower part of **Figure 17.**

In a **series circuit,** devices are connected so there is only one closed path for current to follow. However, if any part of this path is broken, current will no longer flow in the circuit.

In a **parallel circuit,** devices are connected so there is more than one closed path for current to follow. If the current flow is broken in one path, current will continue to flow in the other paths. The electric circuits in your house are parallel circuits. As a result, you can switch off a light in one room without turning off all the other lights in the house.

section 2 review

Summary

Electric Current

- An electric current is the flow of electric charges, such as electrons.
- Electric current will flow continually only in a closed path called an electric circuit.
- A battery produces an electric field in a circuit that causes electrons to flow.

Electric Resistance

- Electric resistance is a measure of how difficult it is for electrons to flow in a material.
- Electric resistance results from the collisions between electrons flowing in a current and the atoms and other charges in the circuit.

Electrical Energy and Voltage

- An electric current transfers electrical energy to a circuit.
- A battery transforms chemical energy into electrical energy.
- Voltage is a measure of the electrical energy transferred by an electron as it moves from one point to another in a circuit.

Self Check

1. **Describe** how the charge on a wire changes when an electric current flows in the wire.
2. **Explain** what causes electrons in an electric current to flow slowly in a circuit.
3. **Describe** the process that causes electrical energy to be transformed into heat and light energy as a current flows in a lightbulb.
4. **Determine** how the current in a circuit changes if the voltage in the circuit is decreased and the resistance remains the same.
5. **Think Critically** Two lightbulbs are connected in a series circuit. If the current flowing in one lightbulb is 0.5 A, what is the current flowing in the other lightbulb? Explain.

Applying Math

6. **Calculate Voltage** A hairdryer with a resistance of 10.0 Ω is plugged into an electrical outlet. If the current in the hairdryer is 11 A, what is the voltage?
7. **Calculate Resistance** What is the resistance of a loudspeaker connected to a 9.0-V battery if the current in the speaker is 0.3 A?

Science Online red.msscience.com/self_check_quiz

section 2 review

1. Electrons move along the circuit. However, the number of positive and negative charges in the wire remains balanced.
2. Electrons are constantly colliding with atoms and other electric charges as they flow.
3. Colliding electrons and atoms transfer kinetic energy to the material. Electrical energy is converted into other forms of energy, such as heat energy and light.
4. The current also decreases.
5. .5 A
6. 110 V
7. 30 Ω

section 3

Magnetism

Magnets

Did you use a magnet today? If you've watched TV, listened to a CD, dried your hair with a hairdryer, or used a computer, the answer is yes. Magnets are a part of all these devices and many others. Magnets can exert forces on objects that are made from, or contain, magnetic materials. Magnets also exert forces on other magnets. It is the forces exerted by magnets that make them so useful.

Magnetic Poles Every magnet has two ends or sides. Each of the ends or sides is a magnetic pole. There are two types of magnetic poles. One is a north pole and the other is a south pole. Every magnet has a north pole and a south pole. For example, one end of a bar magnet is a south pole and the other end is a north pole. For a magnet in the shape of a disc or a ring, one side is a north pole and the other side is a south pole.

Reading Check *Where would the poles of a magnet shaped like a horseshoe be located?*

The Forces Between Magnetic Poles The magnetic poles of a magnet exert forces on the magnetic poles of other magnets, as shown in **Figure 18.** If two north poles or two south poles are moved toward each other, they repel. If the north pole of one magnet is brought toward the south pole of another magnet, the magnets attract each other. In other words, like poles repel and unlike poles attract. The magnetic forces between two magnets become stronger as the magnets move closer together, and weaker as they move farther apart.

as you read

What You'll Learn

- **Describe** how magnets exert forces on each other.
- **Explain** why some materials are magnetic.
- **Describe** how objects become temporary magnets.
- **Explain** how an electric generator produces electrical energy.

Why It's Important

Magnetism helps produce the electrical energy you obtain from electrical outlets.

Review Vocabulary

mechanical energy: the sum of the kinetic and potential energy of an object

New Vocabulary

- magnetic domain
- electromagnet
- electromagnetic induction

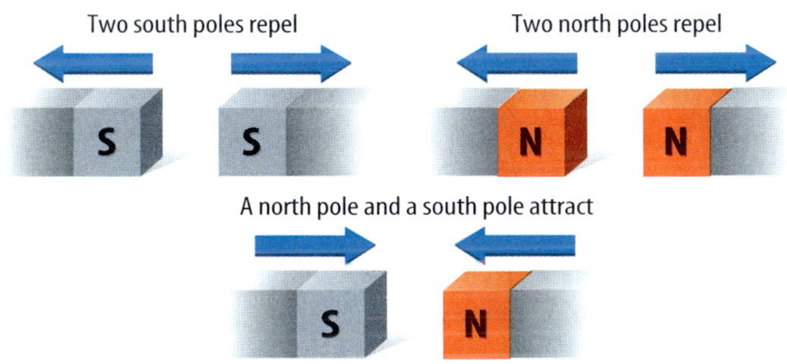

Figure 18 The magnetic forces between magnetic poles are attractive between unlike poles and repulsive between like poles.
Compare *the forces between magnetic poles to the forces between electric charges.*

1 Motivate

Bellringer

INTERACTIVE CHALKBOARD PowerPoint® Presentations

Section Focus Transparencies also are available on the Interactive Chalkboard CD-ROM. L2 ELL

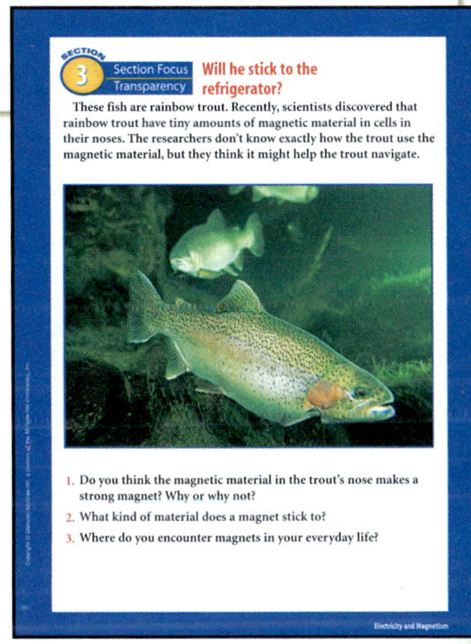
Section 3 Section Focus Transparency

Will he stick to the refrigerator?

These fish are rainbow trout. Recently, scientists discovered that rainbow trout have tiny amounts of magnetic material in cells in their noses. The researchers don't know exactly how the trout use the magnetic material, but they think it might help the trout navigate.

1. Do you think the magnetic material in the trout's nose makes a strong magnet? Why or why not?
2. What kind of material does a magnet stick to?
3. Where do you encounter magnets in your everyday life?

Tie to Prior Knowledge

Attractions and Repulsions Ask students what kinds of attractions and repulsions they have learned about. Students should recall gravitation and the interaction between opposite electrical charges. In this section, students will learn about magnetic attraction.

Reading Check

Answer at the two open ends of the horseshoe

Caption Answer

Figure 18 Two like charges repel and two like poles repel. Two unlike charges attract and two unlike poles attract.

Section 3 Resource Manager

Chapter FAST FILE Resources

Transparency Activity, p. 46
MiniLAB, p. 4
Enrichment, p. 32
Reinforcement, p. 29
Directed Reading for Content Mastery, p. 21
Lab Worksheet, pp. 5–6
Lab Activity, pp. 9–12
Lab Worksheet, pp. 7–8
Lab Activity, pp. 13–16

Earth Science Critical Thinking/Problem Solving, p. 8
Physical Science Critical Thinking/Problem Solving, p. 18
Cultural Diversity, pp. 41–42
Performance Assessment in the Science Classroom, p. 35

2 Teach

Quick Demo

Plotting Field Lines

Materials bar magnet, transparency, black marker, transparent plastic compass, overhead projector

Estimated Time five minutes

Procedure Place the transparency on the stage of the projector. Lay the magnet in the center of the transparency. With the projector on, move the compass around the magnet, drawing arrows to show the direction that the compass needle points.

Activity

Magnetic Materials Have pairs of students predict which objects in the classroom will stick to a magnet. Then give each pair of students a magnet and have them test their predictions. Discuss any materials that surprised students. Students should predict that metals containing iron would stick to the magnet and materials such as plastic, wood, or glass would not. L2 LS **Intrapersonal, Logical-Mathematical**

INTEGRATE Earth Science

Earth's Magnetic Field Geophysicists have found that, in the past century, the magnetic north pole has moved over 1000 km towards the geographical north pole, and over the past decade the strength of the magnetic field has fallen by 1%. Investigate the training and qualifications of geophysicists. Working in small groups, develop a panel to explain this to the class.

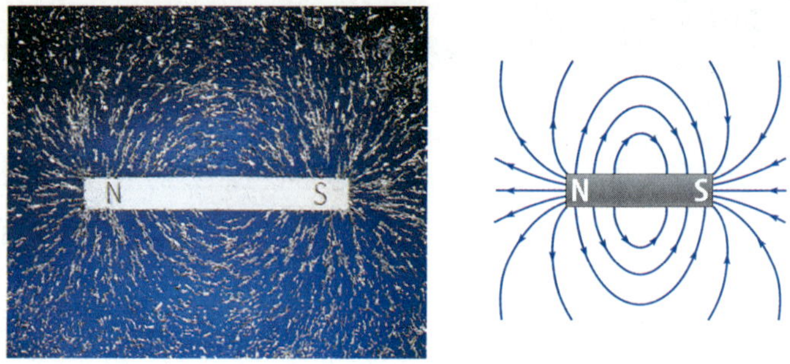

Figure 19 Iron filings sprinkled around a magnetic bar show the magnetic field lines. Magnetic field lines always connect the north and south poles of a magnet.

Magnetic Field If you hold two like poles of two magnets near each other, you can feel them push each other apart, even though they are not touching. Recall that electric charges exert forces on each other even if they are not touching. This is because an electric charge is surrounded by an electric field that exerts a force on other electric charges. In a similar way, every magnet is surrounded by a magnetic field that exerts a force on other magnets.

The magnetic field around a bar magnet is shown in **Figure 19.** Iron filings sprinkled around a bar magnet line up to form a pattern of curved lines. These lines are called magnetic field lines. Magnetic field lines help show the direction of the magnetic field around a magnet.

Figure 19 shows that the magnetic field lines are closest together at the magnet's poles. At the poles of a bar magnet, the magnetic field is strongest. The magnetic field lines are closer together where the magnetic field is stronger.

INTEGRATE Earth Science

Earth's Magnetic Field
Earth is surrounded by a magnetic field that is similar to the magnetic field around a bar magnet. Earth's magnetic poles are located near the geographic north pole and south pole. A compass uses Earth's magnetic field to help determine direction. Because a compass needle is a magnet, it rotates so it points toward Earth's magnetic poles. As a result, the north end of a compass needle points north.

Magnetic Materials

If you hold a magnet near a paper clip, the paper clip will stick to the magnet. However, a piece of aluminum foil will not stick to a magnet. Both the paper clip and the aluminum are metal. Why is one attracted to the magnet and not the other?

Only metals that contain the elements iron, nickel, cobalt, and a few other rare-earth elements are attracted to magnets. Materials that contain these elements are magnetic materials. Magnets also contain one or more of these metals. The steel paper clip contains iron and therefore is a magnetic material.

Why are some materials magnetic? Atoms of the elements that are magnetic, such as iron, nickel, and cobalt, are themselves tiny magnets. Each atom has a north pole and a south pole. Atoms of elements that are not magnetic, such as aluminum, are not magnets. As a result, objects that are made of these elements are not affected by a magnetic field.

Differentiated Instruction

Learning Disabled Have students use a good bar magnet to try to pick up as many paper clips as possible at one time. Have them see if they can transfer magnetic attraction through the paper clips by picking up one paperclip with another. L1 LS **Kinesthetic**

Challenge Have students draw a horseshoe magnet and predict what the magnetic field lines would look like. Students can then use a small compass to check their predictions. L3 LS **Logical-Mathematical, Visual-Spatial**

210 CHAPTER 7 Electricity and Magnetism

Magnetic Domains In a magnetic material, forces that atoms exert on each other cause the magnetic fields surrounding atoms to line up. As a result, large numbers of atoms have their magnetic poles pointing in the same direction. A group of atoms that have their magnetic poles pointing in the same direction is called a **magnetic domain. Figure 20** shows how the atoms in a magnetic material form magnetic domains.

Reading Check *What are magnetic domains?*

The magnetic fields of all the atoms in a magnetic domain add together. As a result, each magnetic domain has a north pole and a south pole and is surrounded by a magnetic field. A single magnetic domain may contain trillions of atoms, but it is still too small to see. Even a small piece of iron may contain billions of magnetic domains.

Domains Line Up in Permanent Magnets If you hold one paper clip against another paper clip, nothing happens. Even though they both are made of magnetic material, iron, they neither attract nor repel each other. Why do the paper clips stick to a bar magnet, but not to each other? In a paper clip the magnetic domains are oriented in random directions, as shown in **Figure 21.** As a result, the magnetic fields around each domain cancel out. The paper clip is not surrounded by a magnetic field.

Figure 21 shows that in a permanent magnet, such as a bar magnet, most of the domains are oriented in a single direction. As a result, the magnetic fields around the domains don't cancel out. Instead these magnetic fields add together to form a stronger magnetic field. The magnetic field that surrounds the magnet is the combination of the magnetic fields around the magnetic domains.

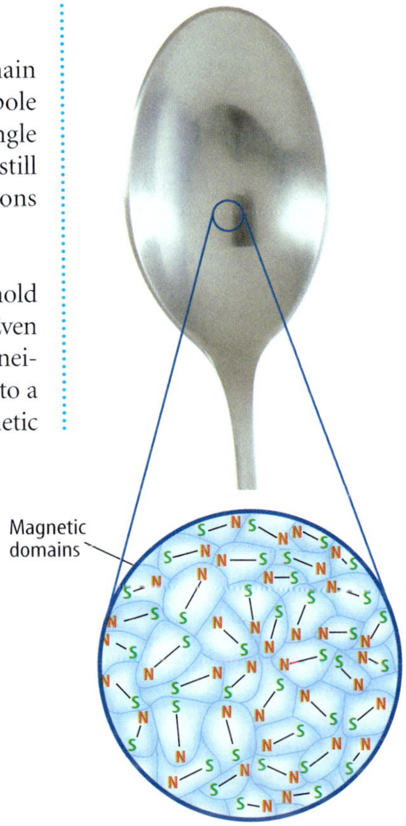

Figure 20 This spoon is made of a magnetic alloy. The spoon is not a magnet because the magnetic poles of the magnetic domains point in random directions.
Explain *why the spoon is not surrounded by a magnetic field.*

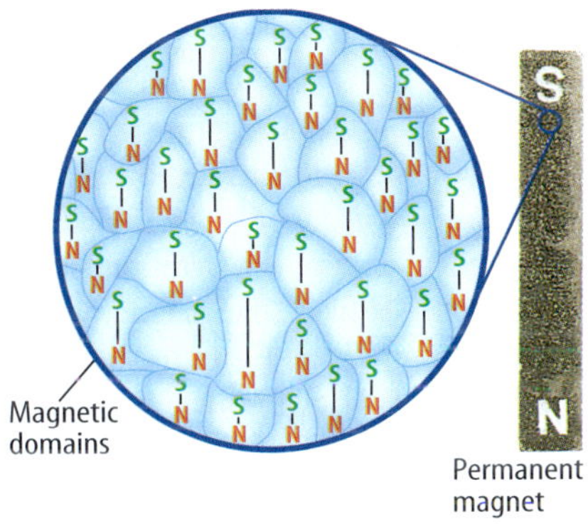

Figure 21 In a permanent magnet, most of the poles of the magnetic domains point in the same direction.
Explain *why the magnet is surrounded by a magnetic field.*

Caption Answer

Figure 20 The domains point in random directions so that their magnetic fields cancel.

Reading Check

Answer A large group of atoms that have their magnetic poles pointing in the same direction.

Inquiry Lab

Making and Breaking Magnets

Purpose to explore and observe how materials become magnetized and demagnetized

Possible Materials strong bar magnet, steel sewing needles, hammer, candle, thread, cup of water

Estimated Time 30 minutes

Teaching Strategies

- Ask students how they think they could permanently align the magnetic domains in the needle. Students can magnetize a needle by stroking it in one direction with a magnet.
- Students can suspend the magnetized needle by a thread or float it on the surface of water to test whether it points north.
- Ask students how they think a magnet can become demagnetized. The magnetic domains will become nonaligned if the magnet is struck with a hammer, dropped on the floor, or heated.
- Allow students to explore other questions that arise. L2

LS **Logical-Mathematical**

For additional inquiry activities, see *Science Inquiry Labs.*

Caption Answer

Figure 21 The domains point in the same direction so that their fields add together to make a stronger field.

Fun Fact

Lobsters, fish, snails, and sea slugs all use the magnetic field of the Earth to navigate. Scientists believe that they have tiny pieces of a magnetic material known as magnetite floating in their bodies. However they are not sure how the animals sense magnetism.

Mini LAB

Purpose Students will observe the magnetic force around an electrified wire.

Materials 50 cm piece of 22 gauge wire, D-cell battery, bar magnet

Teaching Strategy Have students work in pairs during this activity. One student can connect the battery while the other holds the magnet.

Troubleshooting

- Use copper wire so that students won't think the magnet is simply attracting the metal.
- Have students try different orientations of the magnet and loop until they see a clear reaction.

Analysis

1. Depending on the way the students connect the battery, the wire will either be attracted to or repelled by the magnet.
2. The wire moves in different directions depending on the way the battery is connected.

The poles of the magnetic domains in the paper clip point in random directions when there is no magnet nearby.

The force exerted by the magnet on the domains causes them to point toward the nearby magnetic pole.

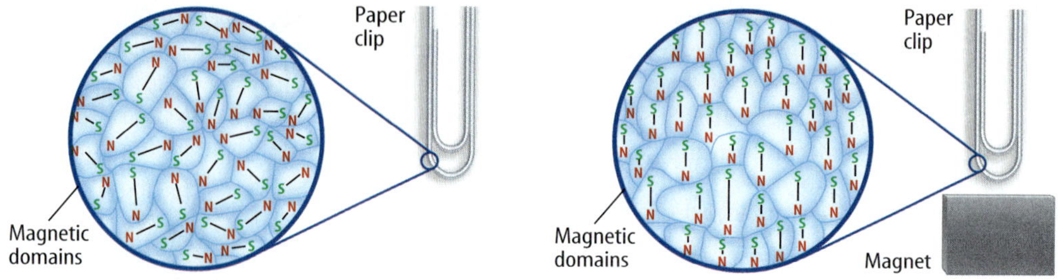

Figure 22 A paper clip that contains iron becomes a temporary magnet when a permanent magnet is nearby.

Mini LAB

Observing Magnetic Force on a Wire

Procedure

1. Connect one end of a **50-cm piece of 22-gauge wire** to one terminal of a **D-cell battery.**
2. Form the wire into a loop and place one pole of a **bar magnet** about 2 cm from the loop.
3. Touch the free end of the wire to the other terminal of the battery. Record your observations.
4. Repeat step 3 with the connections to the battery terminals reversed. Record your observations.

Analysis

1. Explain how your observations show that a current in the wire produces a magnetic field.
2. Infer how the magnetic field around the wire depends on the direction of current in the wire.

Why are magnetic materials attracted to a magnet?

A paper clip is not a magnet, but it contains magnetic domains that are small magnets. Usually these domains point in all directions. However, when a permanent magnet comes close to the paper clip, the magnetic field of the magnet exerts forces on the magnetic domains of the paper clip. These forces cause the magnetic poles of the domains to line up and point in a single direction when a permanent magnet is nearby, as shown in **Figure 22.** The nearby pole of the permanent magnet is always next to the opposite poles of the magnetic domains. This causes the paper clip to be attracted to the magnet.

Because the domains are lined up, their magnetic fields no longer cancel out. As long as the paper clip is attached to the magnet, it is a temporary magnet with a north pole and a south pole.

Electromagnetism

Even though they might seem to be different, electricity and magnetism are related. In the early 1800s it was discovered that a wire carrying an electric current is surrounded by a magnetic field. Not only is a current-carrying wire surrounded by a magnetic field, but so is any electric charge in motion. The connection between electricity and magnetism often is called electromagnetism.

Electromagnets The magnetic field produced by a current-carrying wire can be made much stronger by wrapping the wire around an iron core. A current-carrying wire wrapped around an iron core is an **electromagnet.** Just like a bar magnet, one end of an electromagnet is a north magnetic pole and the other end is a south magnetic pole, as **Figure 23** shows. However, if the direction of current flow in the wire coil of an electromagnet is reversed, then the north and south poles switch places.

Differentiated Instruction

Physically Challenged Help the students by assigning each one a partner who will provide support as he or she does the MiniLab. Encourage helpers to provide each physically challenged student with just enough assistance to accomplish the task. L2 LS **Kinesthetic**

Science Journal

Magnetism and Electricity Have students compare and contrast magnetism and electricity in their Journals as follows: make two columns marked *magnetism* and *electricity.* Down the side, make a list of characteristics. Put a check in each column to which that characteristic applies. L2 LS **Logical-Mathematical**

212 CHAPTER 7 Electricity and Magnetism

Using Electromagnets The strength of the magnetic field produced by an electromagnet depends on the amount of current flowing in the wire coil. Increasing the amount of current increases the magnetic field strength. However, the magnetic field disappears if no current flows in the coil. As a result, an electromagnet is a temporary magnet whose magnetic properties can be controlled. Because of this, electromagnets are used in many devices, including doorbells, telephones, CD players, and computers.

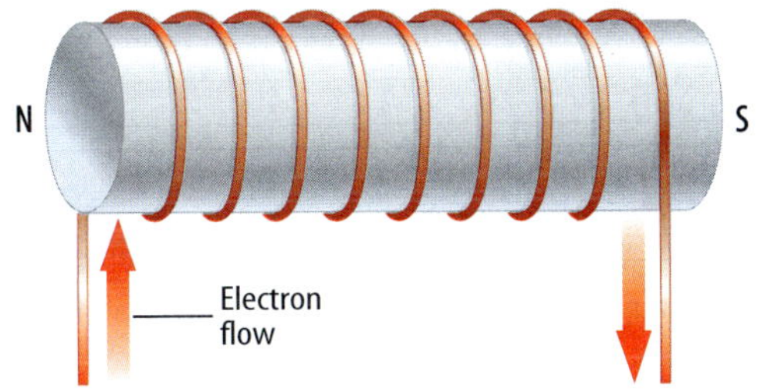

Figure 23 An electromagnet has north and south magnetic poles, and can be attracted or repelled by a permanent magnet. **Describe** *how the magnetic field around the electromagnet changes if the current in the coil is decreased.*

Generating Electric Current

If an electric current produces a magnetic field, can a magnetic field be used to produce an electric current? The answer is yes. If a magnet is moved through a wire loop that is part of a circuit, an electric current flows in the circuit. The current flows only as long as the magnet is moving. A current also flows in the circuit if it is the wire loop that moves and the magnet that is at rest. The production of an electric current by moving a magnet and a loop relative to each other is called **electromagnetic induction.**

Remember that a battery produces an electric field in a circuit that causes electrons to flow. Electromagnetic induction also produces an electric field in a circuit that causes electrons to flow.

Figure 24 When the wire loop rotates in the magnetic field of the permanent magnet, an electric current flows in the lightbulb.

Electric Generators You plug a lamp into an electrical outlet and turn the switch on. Immediately, an electric current flows in the lamp, causing the lightbulb to glow. Electrical energy is supplied to the lamp through the electric field created in the lamp. However, when you plug a device into an electrical outlet, the electrical energy used is produced by an electric generator instead of a battery.

Figure 24 shows a simple electrical generator. A loop of wire is rotated within a magnetic field. The motion of the wire loop with respect to the magnetic field produces an electrical field in the wire. This electrical field causes a current to flow. Current continues to flow as long as the wire loop is kept rotating. To keep the wire loop rotating, mechanical energy must be continually supplied to the generator. As a result, a generator converts mechanical energy into electrical energy.

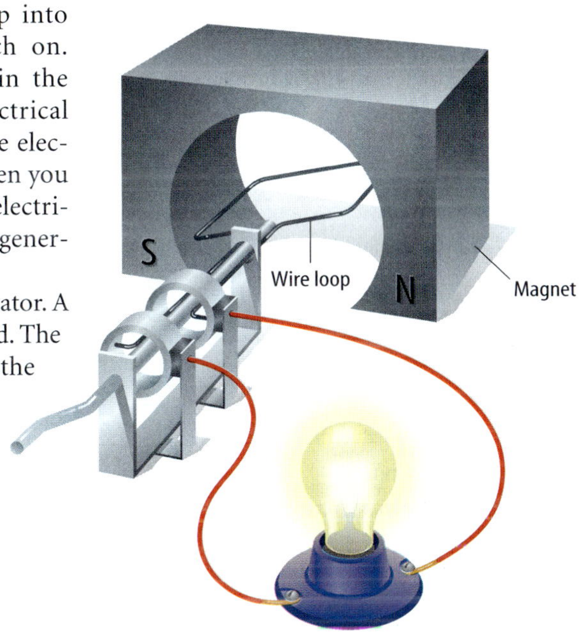

Caption Answer

Figure 23 The magnetic field increases.

Quick Demo

Current with a Magnet

Materials 60 cm piece of 18 gauge wire, tape, multimeter, bar magnet

Estimated Time 10 minutes

Procedure Wind the center of the wire loosely around your hand to make several loops. Tape the loops together. Attach the loose ends of the wire to the meter. Set the meter to record current. Pass the magnet back and forth through the loop and have students watch the meter.

Visual Learning

Figure 24 Ask a volunteer to explain how the generator uses magnetic fields to produce electricity. Be sure students understand the connection. Ask students what other types of energy might be used to power the generator. Possible Answers: water, coal, oil, nuclear.

Differentiated Instruction

Challenge Have students investigate and draw a diagram of the kind of motor used in a hand tool, such as a drill. Students should explain how the motor is like a generator and unlike a generator. L3 LS **Visual-Spatial** P

Teacher FYI

Magnetism Units The SI unit for magnetic field strength is the tesla (T). Earth's magnetic field intensity at the surface is 10^{-4} T. magnetic field generated in the Lab commonly reach 10 T and can be up to 100 T for short periods.

SECTION 3 Magnetism 213

Discussion

Power Plants Where is the electricity that powers your community generated? What source of energy is used? Answers will vary depending on the community.

3 Assess

DAILY INTERVENTION

Check for Understanding

Logical-Mathematical Have students work with a partner to list ways in which electromagnetism can be used to create useful devices. Students can draw and label their devices to show how they work. L2

Reteach

Electromagnets Get several old doorbells or telephones and have the students take them apart to try to identify the electromagnet. L2 LS **Kinesthetic, Visual-Spatial**

Assessment

Performance Students can use D-cells, wire, and an iron nail to construct an electromagnet that will pick up 10 paper clips. Use **Performance Assessment in the Science Classroom,** p. 97. L2 LS **Kinesthetic, Logical-Mathematical**

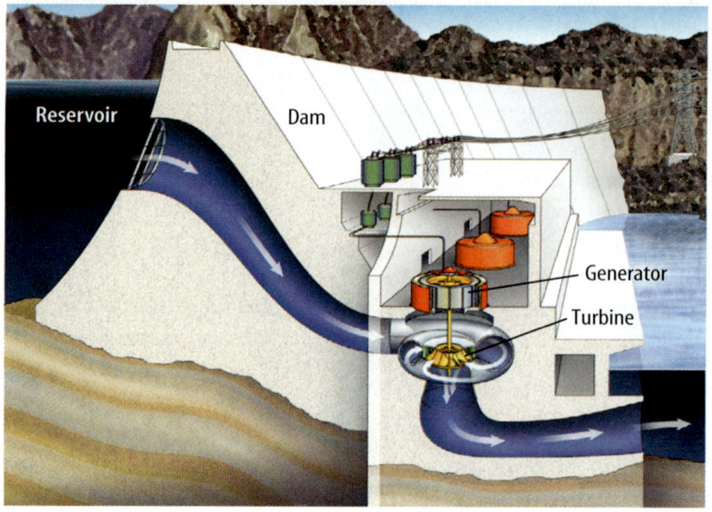

Figure 25 In a hydroelectric plant, the kinetic energy of falling water is converted into electrical energy by a generator.

Power Plants The electrical energy you obtain from an electrical outlet is produced by generators in electric power plants. In these generators electromagnets are rotated past wire coils. To rotate the magnets, power plants use mechanical energy in the form of the kinetic energy of moving steam or moving water into electrical energy.

In some power plants fossil fuels are burned to heat water and produce steam that is used to spin generators. In hydroelectric power plants, the flow of water from behind a dam provides the mechanical energy that is transformed into electrical energy, as shown in **Figure 25.**

section 3 review

Summary

Magnets

- All magnets have a north pole and a south pole.
- Like magnetic poles repel each other and unlike magnetic poles attract each other.
- A magnet is surrounded by a magnetic field that exerts a force on other magnets.

Magnetic Materials

- Individual atoms are magnets in magnetic materials such as iron, cobalt, and nickel.
- Magnetic domains contain atoms with their north or south magnetic poles pointing in the same direction.
- The magnetic domains in a permanent magnet have their magnetic poles aligned.

Electromagnetism

- An electric current is surrounded by a magnetic field.
- An electric current can be produced by the relative motion of a magnet and a wire loop.

Self Check

1. **Compare and contrast** a permanent magnet and a temporary magnet made from a magnetic material.
2. **Explain** why an object made from aluminum will not stick to a magnet.
3. **Compare and contrast** an electric generator and a battery.
4. **Identify** the circumstances that would cause an aluminum wire to be attracted or repelled by a magnet.
5. **Compare and contrast** an electromagnet and a permanent magnet.
6. **Think Critically** The north pole of one magnet is attracted only to the south pole of another magnet. However, a paper clip will stick to either the north pole or the south pole of a bar magnet. Explain.

Applying Math

7. **Solve a Simple Equation** A certain power plant generates enough electrical energy to supply 100,000 homes. How many of these power plants would be needed to generate enough energy for 2,000,000 homes?

Science Online red.msscience.com/self_check_quiz

section 3 review

1. In a permanent magnet the magnetic domains are lined up. In a temporary magnet, domains are randomly aligned until brought near another magnet.
2. The atoms of aluminum are not magnetic and are not affected by magnetic field.
3. A battery converts chemical energy to electrical energy. A generator converts mechanical energy into electrical energy by rotating a wire loop in a magnetic field.
4. Wrap the wire into a coil, and pass a current through the coil.
5. In a permanent magnet, the magnetic domains are aligned. In an electromagnet, a current flowing through wire creates a magnetic field that aligns the domains in the iron core. An electromagnet is a temporary magnet.
6. Either the north or south pole of the magnet will cause the magnetic domains in the paper clip to align in the direction that produces an attraction. The south pole of the magnet will attract the north pole of the magnetic domains, and the north pole of the magnet will attract the south pole of the magnetic domains.
7. 20

LAB

Batteries in Series and Parallel

Many battery-powered devices use more than one battery to supply electrical energy. Why are these batteries usually connected so that a positive terminal is in contact with a negative terminal?

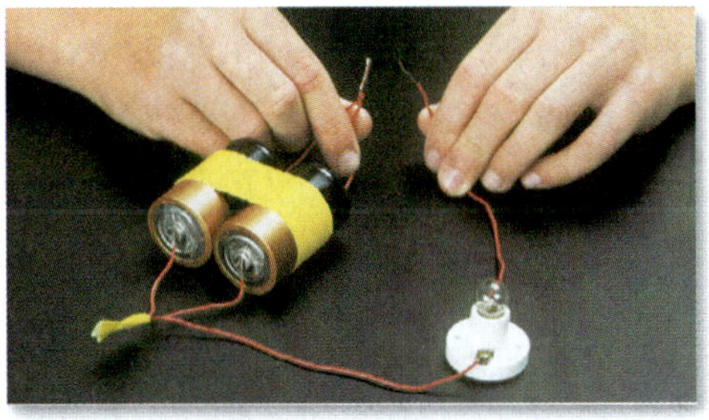

Real-World Question

How does the way that batteries are connected affect the voltage they provide?

Goals

- **Infer** how the voltage produced by two batteries in a circuit depends on how they are connected.

Materials

1.5-V lightbulbs (2)
1.5-V batteries (3)
minibulb sockets (2)
10-cm long pieces of insulated wire (8)
tape

Safety Precautions

Procedure

1. Make a brightness tester by connecting one battery to a lightbulb. Disconnect one wire after you've made the lightbulb light.
2. Tape two batteries together in series so that the positive terminal of one battery touches the negative terminal of the other battery.
3. Connect the batteries to a lightbulb. Close the circuit in the brightness tester and compare the brightness of the lightbulbs. Record your observations.
4. Tape two batteries together in parallel side-by-side with positive terminals on one end and negative terminals on the other end.
5. Tape a wire to each battery terminal. Twist together the ends of the wires connected to both negative terminals. Do the same for the wires connected to the positive terminals.
6. Repeat step 3.

Conclude and Apply

1. **Infer** If the brightness of a lightbulb increases as the current in a circuit increases, in which circuit was the current the largest?
2. **Apply** Ohm's law to determine in which circuit the voltage was the largest.
3. **Compare** the voltage provided by two batteries in series and in parallel.

Communicating Your Data

Compare your conclusions with those of other students in your class. **For more help, refer to the Science Skill Handbook.**

BENCH TESTED

Real-World Question

Purpose Students infer how the voltage produced by two batteries depends on how the batteries are connected.

Process Skills Observe, record observations, compare and contrast, interpret data

Time Required 45 minutes

Procedure

Teaching Strategy Have students make a table to record their observations of the relative brightness of the brightness tester and the bulb in the circuit containing the two batteries.

Troubleshooting Make sure the bare ends of the wires are securely taped to the battery terminals.

Conclude and Apply

1. The current was largest in the circuit with the batteries connected in series.
2. The resistance in both circuits is the same. The current in the circuit with the batteries in series is greater, so by Ohm's law, the voltage in this circuit must be greater.
3. The voltage provided by the batteries in series is greater than the voltage provided by the batteries in parallel.

Assessment

Performance Have students work in groups to design a circuit which contains more than one lightbulb, and uses two batteries in series, such that the lightbulbs are only as bright as the brightness tester. Use **Performance Assessment in the Science Classroom**, pp. 95–96.

Communicating Your Data

Students could use graphical software to draw diagrams of the brightness tester circuit, and the circuits containing the batteries in parallel and in series. Students could compare their diagrams with those of other students.

BENCH TESTED

Real-World Question

Purpose Students will determine that a current-carrying wire produces a magnetic field, and then will produce current by moving a wire coil relative to a magnet.

Process Skills Observe, record observations, compare and contrast, make and use tables, interpret data

Time Required 45 minutes

Procedure

Teaching Strategy Remind students that a compass needle is a small bar magnet. A compass needle will turn if another magnet or a magnetic field exerts a force on the compass needle.

Troubleshooting Make sure the compass needles can rotate freely and that the bare ends of the wires are securely taped to the battery terminals.

LAB

Magnets and Electric Current

Goals
- **Observe** the effects of a bar magnet on a compass.
- **Observe** the effects of a current-carrying wire on a compass.
- **Observe** how the relative motion of a magnet and a wire coil affects a compass.

Materials
bar magnet
compass
D-cell batteries (2)
3-m length of insulated wire
50-cm length of insulated wire
tape

Safety Precautions

Real-World Question

Have you ever used a compass? The needle in a compass is a small bar magnet with a north pole and a south pole. Because a compass needle is a magnet, other magnets and magnetic fields can cause a compass needle to move. As a result, a compass can be used to detect the presence of a magnetic field. An electric current flowing in a wire is surrounded by a magnetic field. How does an electric current affect a compass needle?

Procedure

1. Make a data table similar to the one below.

Effects of Magnets and Current on a Compass

Situation	Effect on Compass
Bar magnet nearby	
Current-carrying wire nearby	Answers will vary.
Magnet moves in coil	
Coil moves past magnet	

2. Place a compass on the table top. Place one pole of a bar magnet next to the compass. Record your observations.
3. Make a battery pack by taping two D-cell batteries together so the negative terminal of one battery is in contact with the positive terminal of the other battery.
4. Tape one end of the 50-cm wire to the exposed positive terminal of the battery pack.

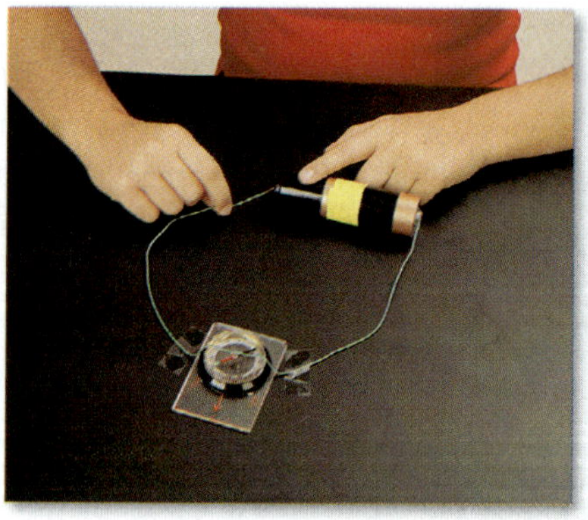

Alternative Inquiry Lab

Real World Connection To extend this Lab into an Inquiry Lab, have students research how the information they learned about electromagnetic radiation in this lab is applied to real life. How are issues of EMR dealt with in large power lines? How does EMR from personal electronic devices (PEDs) affect airline navigation? How are EMR issues dealt with in computer circuits, and circuits in other electronic devices? For what devices have the public shown concern about EMR, and what is the current official stance on EMR safety from devices such as cell phones? Point out to students that EMR can NOT be absorbed by the 'socks' sold for cell phones.

5. Place the wire on top the compass and position the wire so it lines up with the compass needle. Touch the free end of the wire to the other terminal of the battery pack for a few seconds. Record your observations.
6. Wrap the long piece of wire around three fingers about 25 times so there is about 3 cm of wire left at each end. Tape the coil so it doesn't unravel.
7. Wrap the 50-cm wire around the compass several times so there is about 3 cm of wire left at each end. Connect the ends of the wire from the compass with the ends of the wire from the coil.
8. Hold the bar magnet in the center of the coil. Keeping the coil stationary, move the magnet quickly back and forth. Record your observations.
9. Hold the bar magnet in the center of the coil. Keeping the magnet stationary, move the coil quickly back and forth. Record your observations.

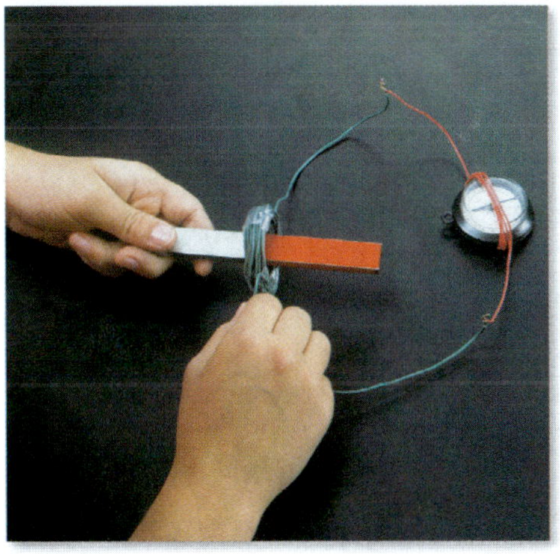

Analyze Your Data

1. **Describe** how the bar magnet affected the compass when the magnet was placed next to it.
2. **Describe** how the compass was affected when an electric current flowed in the wire that had been placed on top of the compass.
3. **Compare** how the compass was affected when the magnet was moved inside the stationary wire coil and when the wire coil was moved past the stationary magnet.

Conclude and Apply

1. **Compare** the effect of the bar magnet on the compass and the effect of the current-carrying wire on the compass.
2. **Infer** why the current-carrying wire had the effect on the compass that you observed.
3. **Infer** whether a current flowed in the wire coil when the coil and the magnet were moving relative to each other. Which observations support your conclusion?

Communicating Your Data

Compare your observations with those of other students in your class. Which actions caused the compass needle to move the most?

Analyze Your Data

1. The bar magnet caused the compass needle to turn.
2. When the batteries were connected, the compass needle turned so that it was at right angles to the wire.
3. Moving the magnet inside the coil and moving the coil past the magnet caused the compass needle to turn.

Conclude and Apply

1. Both caused the compass needle to turn.
2. The current-carrying wire caused the compass needle to turn because the wire was surrounded by a magnetic field.
3. Yes. The current flowing in the wire coil produced a magnetic field that caused the compass needle to turn.

Assessment

Oral Ask students to predict whether a current will be produced by a generator in which a magnet is rotated inside a wire loop. Yes. The magnet and loop are moving relative to each other.

Communicating Your Data

Students should find that the deflection of the compass needle increased as the relative speed of the coil and magnet increased.

TIME

TIME SCIENCE AND HISTORY

SCIENCE CAN CHANGE THE COURSE OF HISTORY!

Which way to go?

The first record of boats large enough to carry trade goods is around 3500 B.C. The first navigators sailed close to shore and navigated by land characteristics that they could see by day. Sailing at night was impossible. Eventually, sailors learned to find their way by using the position of the Sun and stars. Using their knowledge of the heavens and the ocean currents, Vikings and Polynesians traveled remarkable distances, far from the sight of land. But what happened on cloudy nights?

Kissing Rocks

The Chinese had discovered the solution more than 2,000 years ago. They found interesting rocks that they called *tzhu shih*—loving stones, because they liked to "kiss." These rocks contained magnetite, a mineral containing magnetic iron oxide.

The compass on the right was used by sailors during the 18th century. The compass on the left is a modern compass.

The Chinese realized that they could use the magnetite to magnetize iron needles. When the needles floated in water, they always pointed north and south. They had made the first compass!

Earth's Magnetic Field

Earth's iron core produces a magnetic field similar to the field of a huge bar magnet. A compass needle rotates until its north and south poles point toward Earth's opposite magnetic poles, which are close to the geographic north and south poles. So whether it was clear or cloudy, the compass allowed sailors to travel great distances and to return home safely!

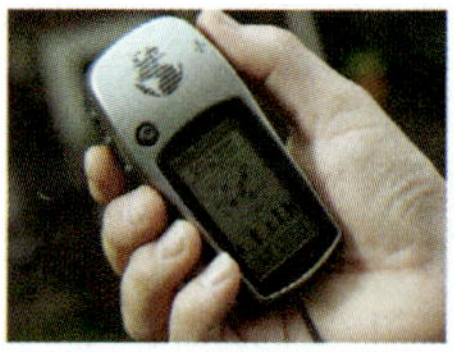

A modern GPS receiver uses a system of satellites to determine its position on Earth's surface.

The World Opens Up

Between the 13th and 19th centuries, there were many improvements to the compass. The ability to travel the seas opened trade between distant cultures. Goods and customs were exchanged, leading to the development of new ideas and tools. Knowing which way to go in rain or shine opened up the world.

Brainstorm Imagine that you are an early sailor before the invention of the compass. What would limit your knowledge of the world? How far could you travel by boat? What kinds of trips might you take? How would the compass change your lifestyle and your culture?

Science Online

For more information, visit red.msscience.com/time

Content Background

The Earth itself is a huge magnet. Its magnetic poles are oval areas about 1,300 miles (2,100 kilometers) from the north and south poles. Irregular lines of force connect the magnetic poles, and the compass needle aligns itself with these lines of force. In a few places, where magnetic north and true north coincide, the compass points to true north. Near the magnetic pole the magnetic compass is useless because the lines of force are vertical—straight down into the Earth. In some areas, iron ore deposits affect the compass's accuracy.

Discussion

Metal Ships Early ships were made of wood. What problems might sailors have had with a compass when ships began to contain large amounts of iron? The compass would be attracted to the nearest metal objects on the ship. Modern ship's compasses are adjusted for such attractions.

Historical Significance

Compasses and Humans Some people have suggested that the compass was the single most important invention because it allowed humans to travel anywhere and return safely. It was the first manufactured measuring device that moved and could be accurately calibrated.

Brainstorm Sailors would have had to stay near shore where they could see landmarks and couldn't sail at night. You could never travel farther than you could see land. They might have been curious about what lay beyond their country or needed to trade for things they couldn't produce. The compass would have allowed them to travel where and when they wanted and return safely.

Resources for Teachers and Students

Riddle of the Compass, by Amir D. Aczel Paracay Publishing, 2002

Waves: Principles of Light, Electricity, and Magnetism (Secrets of the Universe), by Paul Fleisher. Lerner Publications Company, 2001

Reviewing Main Ideas

Section 1 Electric Charge and Forces

1. Positive and negative charges are surrounded by an electric field that exerts forces on other charges.
2. Two positive or two negative charges repel each other; a positive and a negative charge attract each other.
3. Charges can be transferred from one object to another. Charges in an object can be rearranged by an electric field.

Section 2 Electric Current

1. An electric current is the flow of electric charges. A current will flow continually in a closed path called an electric circuit.
2. An electric field in a circuit causes charges to flow and transfer electrical energy.
3. Resistance is a measure of how difficult it is for electrons to flow in a material.
4. Voltage is a measure of the energy transferred by an electron as it flows in a circuit.

Section 3 Magnetism

1. A magnet has a north pole and a south pole and is surrounded by a magnetic field.
2. Like magnetic poles repel each other and unlike poles attract each other.
3. Some materials are magnetic because their atoms behave like magnets.
4. An electric current is surrounded by a magnetic field. Moving a wire loop and a magnet past each other produces a current.

Visualizing Main Ideas

Copy and complete the following concept map on electric current.

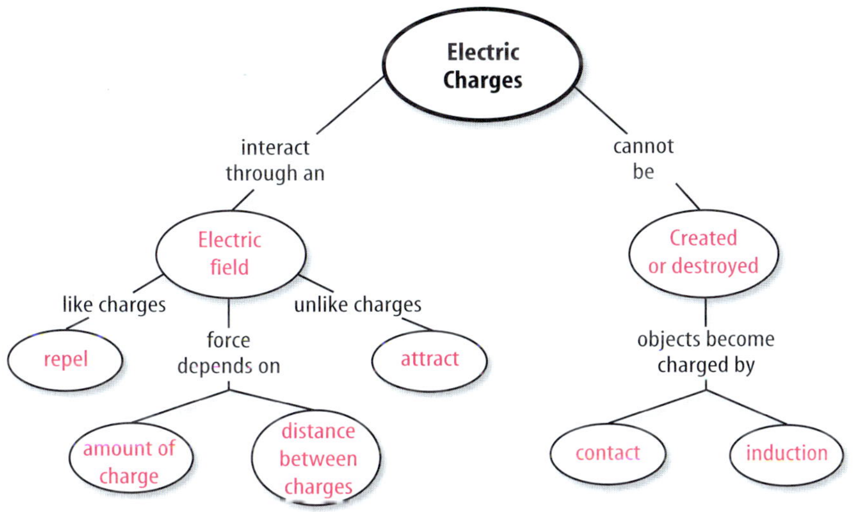

Science Online red.msscience.com/interactive_tutor

chapter 7 Study Guide

Reviewing Main Ideas

Summary statements can be used by students to review the major concepts of the chapter.

Visualizing Main Ideas

See student page.

Science Online
Visit red.msscience.com
/self_check_quiz
/interactive tutor
/vocabulary_puzzlemaker
/chapter_review
/standardized_test

Assessment Transparency

For additional assessment questions, use the *Assessment Transparency* located in the transparency book.

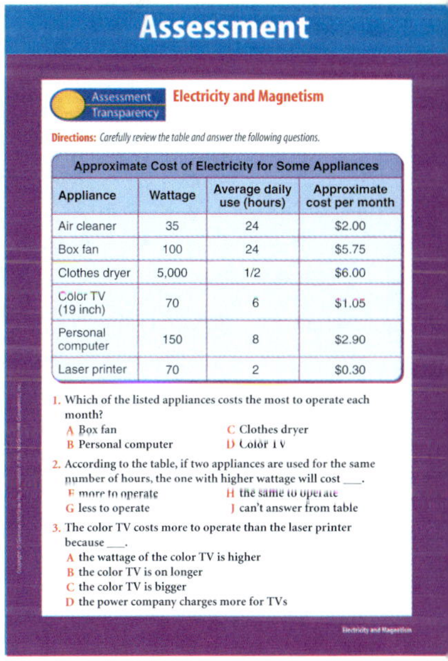

Assessment

Assessment Transparency **Electricity and Magnetism**

Directions: *Carefully review the table and answer the following questions.*

Approximate Cost of Electricity for Some Appliances			
Appliance	Wattage	Average daily use (hours)	Approximate cost per month
Air cleaner	35	24	$2.00
Box fan	100	24	$5.75
Clothes dryer	5,000	1/2	$6.00
Color TV (19 inch)	70	6	$1.05
Personal computer	150	8	$2.90
Laser printer	70	2	$0.30

1. Which of the listed appliances costs the most to operate each month?
   A Box fan C Clothes dryer
   B Personal computer D Color TV
2. According to the table, if two appliances are used for the same number of hours, the one with higher wattage will cost ___.
   F more to operate H the same to operate
   G less to operate J can't answer from table
3. The color TV costs more to operate than the laser printer because ___.
   A the wattage of the color TV is higher
   B the color TV is on longer
   C the color TV is bigger
   D the power company charges more for TVs

Electricity and Magnetism

Identifying Misconceptions

Assess

Use the assessment as follow-up to page F at the beginning of this chapter after students have completed the chapter.

Activity Give students the following problem. The iron nails got mixed up with the aluminum nails. You have a strong magnet, a bucket of water, and an electric fan. Write a paragraph that describes how you would separate the iron nails from the aluminum nails using any of the supplied materials.

Using Vocabulary

1. circuit
2. conductor
3. parallel
4. static charge
5. current
6. Voltage
7. electromagnet
8. resistance

Checking Concepts

9. A
10. B
11. C
12. A
13. A
14. B
15. B
16. B

chapter 7 Review

Using Vocabulary

charging by contact p. 196
charging by induction p. 197
conductor p. 197
electric circuit p. 202
electric current p. 201
electric discharge p. 198
electric resistance p. 203
electromagnet p. 212
electromagnetic induction p. 213
insulator p. 197
magnetic domain p. 211
parallel circuit p. 208
series circuit p. 208
static charge p. 198
voltage p. 205

Complete each statement using a word(s) from the vocabulary list above.

1. A(n) ________ is a closed path that electric current can follow.
2. In a(n) ________, electric charges can move easily.
3. A(n) ________ has more than one path for electric current to follow.
4. An object that does not contain equal amounts of positive charge and negative charge has a(n) ________.
5. A(n) ________ is the flow of electric charges.
6. ________ is a measure of the energy electrons transfer to a circuit as they flow.
7. A(n) ________ is made of a current-carrying wire wrapped around an iron core.
8. A measure of how difficult it is for current to flow in an object is its ________.

Checking Concepts

Choose the word or phrase that best answers the question.

9. Which of the following causes current to flow in a wire?
   A) electric field
   B) electric circuit
   C) electric resistance
   D) magnetic domains
10. Which of the following energy conversions occurs inside a battery?
   A) electrical to chemical
   B) chemical to electrical
   C) thermal to electrical
   D) thermal to chemical
11. How does the electric force between two electrons change as they get farther apart?
   A) The force stays the same.
   B) The force increases.
   C) The force decreases.
   D) The force switches direction.
12. Every electric charge is surrounded by which of the following?
   A) electric field
   B) electric resistance
   C) electric current
   D) magnetic domains
13. Which of the following is true about a permanent magnet?
   A) Its domains are lined up.
   B) It contains an iron core.
   C) Its domains are randomly oriented.
   D) It contains a current-carrying wire.
14. What does a simple generator rotate in a magnetic field to produce current?
   A) a battery
   B) a wire loop
   C) a magnet
   D) domains
15. Increasing the voltage in a circuit increases which of the following in the circuit?
   A) the electric resistance
   B) the energy transferred to the circuit
   C) the static charge
   D) the number of charges
16. Which of the following does NOT describe the magnetic force between two magnets?
   A) Like poles repel.
   B) Like poles attract.
   C) It decreases as the magnets move apart.
   D) Unlike poles attract.

Science Online red.msscience.com/vocabulary_puzzlemaker

Use the ExamView® Pro Testmaker CD-ROM to:

- create multiple versions of tests
- create modified tests with one mouse click for inclusion students
- edit existing questions and add your own questions
- build tests aligned with state standards using built-in State Curriculum Tags
- change English tests to Spanish with one mouse click and vice versa

Thinking Critically

17. **Compare** the force of gravity to the forces between electric charges.

18. **Explain** why an electron can push another electron even though both electrons are not touching.

19. **Explain** why a charged balloon does not attract a person's hair if the balloon is far from the person's head.

20. **Determine** how the total charge on a doorknob changes when the doorknob is charged by an electric field.

Use the table below to answer questions 21–23.

Effect of Battery Voltage on Current

Battery	Battery Voltage (V)	Current in Circuit (A)
A	2	0.2
B	4	0.4
C	6	0.6
D	10	1.0

21. **Make a Graph** The table above shows the current measured in a circuit when different batteries are connected in the circuit. For each battery, plot the current on the vertical axis and the battery on the horizontal axis. Describe the shape of the plotted line.

22. **Infer** from your graph the current in the circuit if the battery voltage is 8 V.

23. **Predict** from the table above the current in the circuit if the battery voltage is 12 V.

24. **Explain** why even though aluminum and iron are both metals, aluminum is not a magnetic material, but iron is.

25. **Predict** whether a generator that is designed to rotate a permanent magnet around a wire loop that doesn't move, will produce electric current.

26. **Concept Map** Copy and complete the following concept map on magnets.

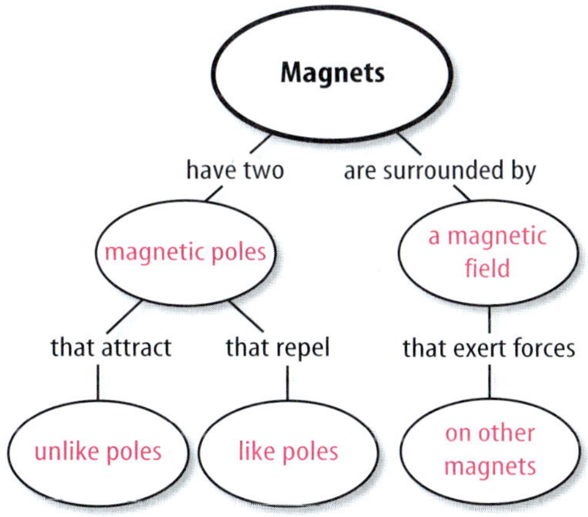

Performance Activities

27. Determine the number of hours in a week that you and your family spend using certain electrical appliances. Choose three appliances. Put paper and a pencil by each one so that each person can write down the amount of time they are used. Which appliance is used the most?

Applying Math

28. **Lightbulb** A 100-W lightbulb has a resistance of 120 Ω. If the lightbulb is connected into a circuit in which the voltage is 110 V, what is the current in the lightbulb?

29. **Battery** The voltage of a battery in a circuit is increased from 3 V to 4.5 V. If the resistance in the circuit is 5 Ω, calculate the percentage change in the current.

Science Online red.msscience.com/chapter_review

chapter Review 7

Thinking Critically

17. The force of gravity between two objects is always attractive, increases as the mass of either object increases, and increases as the masses get closer together. The electric force between two charges can be attractive or repulsive, increases as the charge increases, and increases as the charges get closer together.
18. Every charge is surrounded by an electric field that extends into space and exerts a force on other charges.
19. If the balloon is too far from the person's hair, the electric force between the balloon and the hair is too weak to overcome the force of gravity on the hair.
20. The total charge on the doorknob doesn't change. The charges in the doorknob are rearranged by the electric field.
21. The graph is a straight line.
22. 0.8 V
23. 1.2 V
24. Aluminum atoms are nonmagnetic. Iron atoms are tiny magnets.
25. Yes. A current is produced as long as there is relative motion between the magnet and the wire loop.
26. See student page.

Performance Activities

27. Check student's work. Use **Performance Assessment in the Science Classroom,** p. 97.

Applying Math

National Math Standards

1, 2, 9

28. 0.9 A
29. 50%

Assessment Resources

Reproducible Masters

Chapter *Fast File* Resources
- Chapter Review, pp. 37–38
- Chapter Tests, pp. 39–42
- Assessment Transparency Activity, p. 49

Glencoe Science Web site
- Chapter Review Test
- Standardized Test Practice

Glencoe Technology
- Assessment Transparency
- Exam*View*® Pro Testmaker
- MindJogger Videoquiz
- Interactive Chalkboard

Answer Sheet A practice answer sheet can be found at red.msscience.com/answer_sheet.

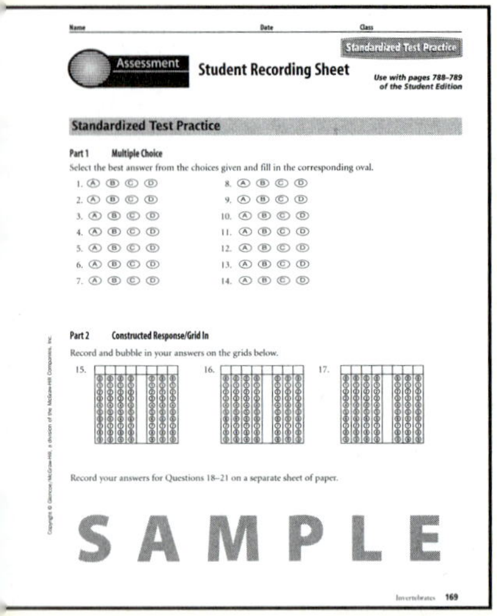

Part 1 Multiple Choice

1. A
2. C
3. B
4. D
5. C
6. A
7. C
8. A
9. B

chapter 7 Standardized Test Practice

Part 1 Multiple Choice

Record your answers on the answer sheet provided by your teacher or on a sheet of paper.

1. Which of the following materials is a good electrical conductor?

A. aluminum **C.** rubber
B. plastic **D.** wood

Use the figure below to answer questions 2 and 3.

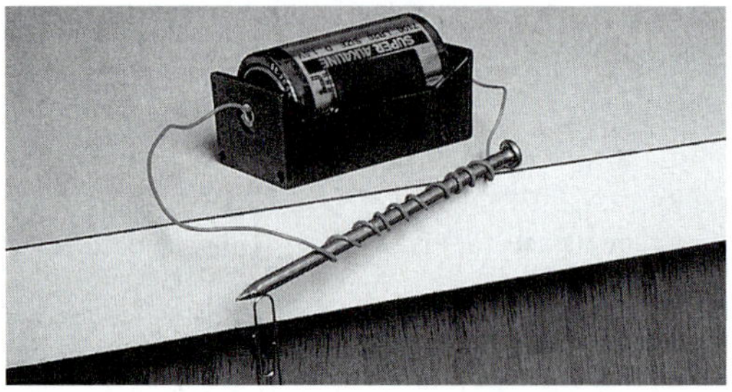

2. What happens if you switch the ends of the wire coil from one battery terminal to the other?

A. Current does not flow through the coil.
B. The electromagnet repels the paper clip.
C. The magnetic poles of the electromagnet are reversed.
D. The magnetic field decreases.

3. The strength of the magnetic field produced by the electromagnet depends on which of the following?

A. The number of domains in the nail.
B. The amount of current in the coil.
C. The number of charges in the coil.
D. The size of the battery.

4. Which of the following describes an object that is negatively charged?

A. It has more neutrons than protons.
B. It has more protons than electrons.
C. It has more protons than neutrons.
D. It has more electrons than protons.

5. How does the amount of positive charge on a proton compare with the amount of negative charge on an electron?

A. The proton has more positive charge.
B. The electron has more negative charge.
C. The amounts are equal.
D. Both particles have no charge.

6. After electrons are transferred from object A to object B, which of the following is true?

A. A and B attract each other.
B. A and B repel each other.
C. A and B exert no force on each other.
D. B has more charge than A.

Use the table below to answer questions 7 and 8.

Current and Voltage in Circuits		
Circuit Number	Voltage (volts)	Current (amps)
1	6	0.1
2	9	0.05
3	12	0.075
4	15	0.25

7. Which circuits have the same resistance?

A. Circuits 1 and 2
B. Circuits 3 and 4
C. Circuits 1 and 4
D. Circuits 2 and 3

8. What is the resistance of circuit 4?

A. 160 Ω **C.** 6.25 Ω
B. 90 Ω **D.** 0.9 Ω

9. Electrical energy is converted into thermal energy in a circuit when which of the following occurs?

A. Electrons are transferred.
B. Electrons collide with atoms.
C. The voltage is decreased.
D. The voltage is increased.

Part 2 Short Response/Grid In

10. electric current
11. The circuit on the left is series circuit; the circuit on the right is parallel circuit.
12. Lamp B will not continue to shine. Lamp D will continue to shine.
13. The discharge can occur if they are not touching. The attractive electric forces will cause charges to move from one object to the other across the space between the objects.
14. The collisions between the charges flowing in the current and the atoms in the wire convert electrical energy into heat energy.
15. Increasing the voltage of the battery and reducing the resistance in the circuit.
16. When the filament in the lightbulb breaks there is no longer a closed path for the current to follow.
17. The current decreases.

Part 2 Short Response/Grid In

Record your answers on the answer sheet provided by your teacher or on a sheet of paper.

10. What is the continuous flow of electric charge in a material called?

Use the illustration below to answer questions 11 and 12.

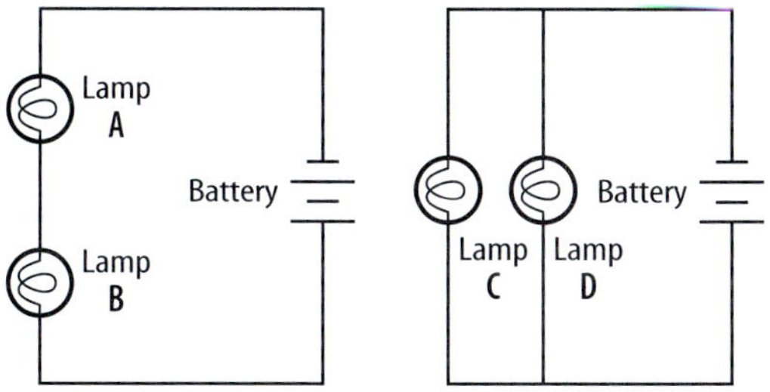

11. The illustrations above show two electrical circuits. Which of these is a parallel circuit? Which is a series circuit?

12. If lamp A burns out, will lamp B continue to shine? If lamp C burns out, will lamp D continue to shine?

13. Explain whether two charged objects must touch each other for a static discharge to occur.

14. Why does the temperature of a wire increase when current flows through it?

15. A student makes a simple circuit consisting of a conducting wire connected to a battery and a lamp. What are two ways the student can increase the current in the circuit?

16. Explain why a current stops flowing in a lightbulb when the filament breaks.

17. A simple electric circuit contains a battery connected to a lightbulb. If the resistance of the connecting wires increases, how does the current flowing through the lightbulb filament change?

Part 3 Open Ended

Record your answers on a sheet of paper.

18. Explain why a balloon can stick to a wall if you first rub the balloon against your hair.

19. When two objects rub against each other, such as your shoes against a carpet, why is it usually electrons that move from one object to the other and not protons?

20. Describe how a battery can create a current in a conducting wire.

Use the illustration below to answer questions 21 and 22.

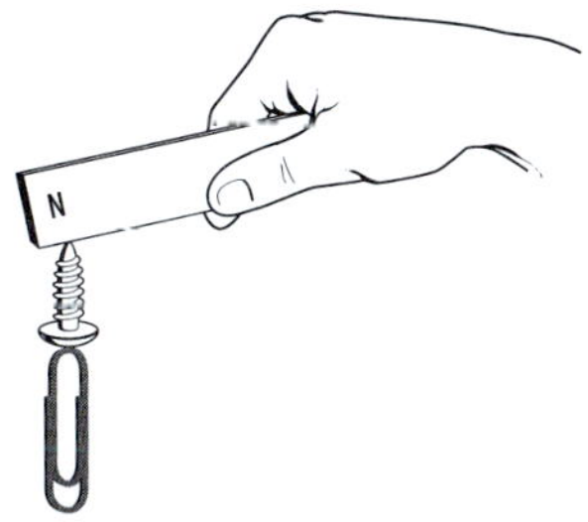

21. The magnet in the illustration above attracts the screw and paper clip. Describe why the screw is able to attract the paper clip and keep it from falling.

22. If the magnet were removed, would the screw still attract the paper clip and keep it from falling?

Test-Taking Tip

Organize Discussion Points For essay questions, spend a few minutes listing and organizing the main points that you plan to discuss. Be sure to do all of this work on your scratch paper, not on the answer sheet.

Rubrics

The following rubrics are sample scoring devices for short response and open-ended questions.

Short Response

Points	Description
2	The student demonstrates a thorough understanding of the science of the task. The response may contain minor flaws that do not detract from the demonstration of a thorough understanding.
1	The student has provided a response that is only partially correct.
0	The student has provided a completely incorrect solution or no response at all.

Open Ended

Points	Description
4	The student demonstrates a thorough understanding of the science of the task. The response may contain minor flaws that do not detract from the demonstration of a thorough understanding.
3	The student demonstrates an understanding of the science of the task. The response is essentially correct and demonstrates an essential but less than thorough understanding of the science.
2	The student demonstrates only a partial understanding of the science of the task. Although the student may have used the correct approach to a solution or may have provided a correct solution, the work lacks an essential understanding of the underlying science concepts.
1	The student demonstrates a very limited understanding of the science of the task. The response is incomplete and exhibits many flaws.
0	The student provides a completely incorrect solution or no response at all.

Part 3 Open Ended

18. Rubbing the balloon on your hair makes the balloon negatively charged. The electric field around the negatively charged balloon repels electrons in the wall, making the surface of the wall near the balloon positively charged. The negatively charged balloon is then attracted to the positively charged region of the wall.

19. Some electrons are loosely held in atoms. Protons are tightly held in the interior of atoms.

20. Chemical reactions in a battery separate positive and negative electric charges. This separation of charges produces an electric field in the wire when it is attached to the battery terminals. The electric field in the wire exerts forces on electrons that cause them to flow.

21. The permanent magnet causes the magnetic domains in the screw to line up so that the screw becomes a temporary magnet. The screw then causes the domains in the paper clip to be attracted to the screw.

22. No. The screw would no longer be a magnet and the paper clip would fall.

chapter 8 Organizer

Section/Objectives	Standards		Labs/Features
	National	State/Local	
Chapter Opener	See p. 16T–17T for a Key to Standards.		**Launch Lab:** How do waves carry energy?, p. 225 **Foldables,** p. 225
Section 1 What are waves? 1 session 0.5 block 1. **Explain** the relationship between waves, energy, and matter. 2. **Describe** the difference between transverse waves and compressional waves.	National Content Standards: UCP.1–UCP.3, UCP.5, A.1, A.2, B.2, B.3, D.1		**MiniLAB:** Comparing Sounds, p. 229 **Integrate Physics,** p. 230
Section 2 Wave Properties 2 sessions 1 block 3. **Describe** the relationship between the frequency and wavelength of a wave. 4. **Explain** why waves travel at different speeds.	National Content Standards: UCP.1–UCP.3, UCP.5, A.1, A.2, D.1		**Integrate Health,** p. 234 **Science Online,** p. 235 **Lab:** Waves on a Spring, p. 236
Section 3 Wave Behavior 4 sessions 2 blocks 5. **Explain** how waves can reflect from some surfaces. 6. **Explain** how waves change direction when they move from one material into another. 7. **Describe** how waves are able to bend around barriers.	National Content Standards: UCP.1–UCP.3, UCP.5, A.1, A.2, B.2, B.3, D.1		**MiniLAB:** Observing How Light Refracts, p. 238 **Science Online,** p. 241 **Applying Science:** Can you create destructive interference?, p. 241 **Visualizing Interference,** p. 242 **Lab:** Wave Speed, pp. 244–245 **Science Stats:** Waves, Waves, and More Waves, p. 246

Glencoe Exclusive! TeacherWorks™ All-In-One Planner and Resource Center

Lab Materials	Reproducible Resources	Section Assessment	Technology
Launch Lab: clear-plastic plate, a cork or straw, dropper, water	**Chapter *Fast File* Resources** Foldables Worksheet, p. 15 Directed Reading Overview, p. 17 Note-taking Worksheets, pp. 31–32	Glencoe's Assessment Advantage	TeacherWorks includes: • Interactive Teacher Edition • Lesson Planner with calendar • Access to all program blacklines • Correlations to standards • Web links
MiniLAB: wooden ruler	**Chapter *Fast File* Resources** Transparency Activity, p. 42 MiniLAB, p. 3 Enrichment, p. 28 Reinforcement, p. 25 Directed Reading, p. 18 Transparency Activity, pp. 45–46	**Portfolio** Activity, p. 228 **Performance** MiniLAB, p. 235 Applying Skills, p. 236 **Content** Section Review, p. 236	Section Focus Transparency Teaching Transparency Guided Reading Audio Program Interactive Chalkboard CD-ROM
Lab: long, coiled spring toy; 5 cm colored yarn; meterstick; stopwatch *Need materials?* Contact Science Kit at 1-800-828-7777 or www.sciencekit.com on the Internet.	**Chapter *Fast File* Resources** Transparency Activity, p. 43 Enrichment, p. 29 Reinforcement, p. 26 Directed Reading, p. 19 Lab Worksheets, pp. 5–6 Lab Activity, pp. 9–11 **Reading and Writing Skill Activities,** p. 33 **Home and Community Involvement,** p. 42	**Portfolio** Science Journal, p. 241 **Performance** Applying Math, p. 241 **Content** Section Review, p. 241	Section Focus Transparency Virtual Labs CD-ROM Guided Reading Audio Program Interactive Chalkboard CD-ROM Video Lab
MiniLAB: large, opaque drinking glass or cup of water; white soda straw **Lab:** long, coiled spring toy; stopwatch; meterstick; tape	**Chapter *Fast File* Resources** Transparency Activity, p. 44 Enrichment, p. 30 Reinforcement, p. 27 Directed Reading, pp. 19, 20 MiniLAB, p. 4 Lab Activity, pp. 13–14 Lab Worksheets, pp. 7–8 **Lab Management and Safety,** p. 65 **Mathematics Skill Activities,** p. 47	**Portfolio** Assessment, p. 243 **Performance** MiniLAB, p. 244 Applying Science, p. 247 Applying Skills, p. 249 **Content** Section Review, p. 249	Section Focus Transparency Guided Reading Audio Program Interactive Chalkboard CD-ROM

End of Chapter Assessment

Glencoe's Assessment Advantage

Blackline Masters	Technology	Professional Series
Chapter *Fast File* Resources Chapter Review, pp. 35–36 Chapter Tests, pp. 37–40 **Standardized Test Practice,** pp. 35–38	MindJogger Videoquiz Virtual Labs CD-ROM Exam*View*® Pro Testmaker TeacherWorks CD-ROM Interactive Chalkboard CD-ROM	**Performance Assessment in the Science Classroom (PASC)**

Transparencies

Section Focus

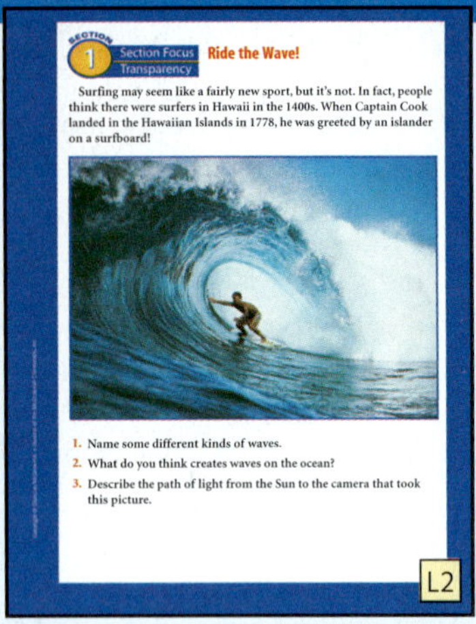
Section 1 Section Focus Transparency — Ride the Wave!

Surfing may seem like a fairly new sport, but it's not. In fact, people think there were surfers in Hawaii in the 1400s. When Captain Cook landed in the Hawaiian Islands in 1778, he was greeted by an islander on a surfboard!

1. Name some different kinds of waves.
2. What do you think creates waves on the ocean?
3. Describe the path of light from the Sun to the camera that took this picture.

L2

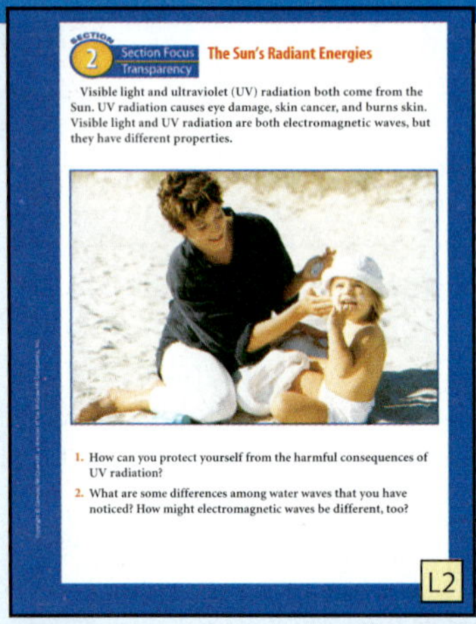
Section 2 Section Focus Transparency — The Sun's Radiant Energies

Visible light and ultraviolet (UV) radiation both come from the Sun. UV radiation causes eye damage, skin cancer, and burns skin. Visible light and UV radiation are both electromagnetic waves, but they have different properties.

1. How can you protect yourself from the harmful consequences of UV radiation?
2. What are some differences among water waves that you have noticed? How might electromagnetic waves be different, too?

L2

Section 3 Section Focus Transparency — Wave Art

Artists work with all kinds of materials, so why not water? This image was made by placing two vibrating probes in the water. The probes generate waves that overlap and create this fascinating pattern.

1. What kind of waves are these—transverse or compressional?
2. If this image is shown in its actual size, how could you calculate the wavelength of these waves?
3. What would happen to the wave pattern if only one of the probes was vibrating?

L2

This is a representation of key blackline masters available in the Teacher Classroom Resources. See Resource Manager boxes within the chapter for additional information.

Assessment

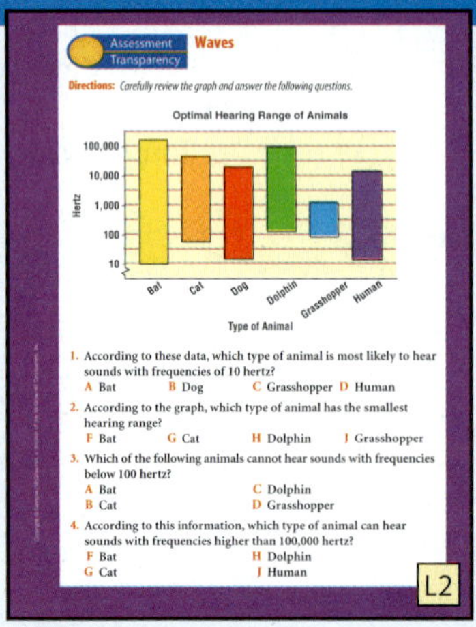
Assessment Transparency — Waves

Directions: *Carefully review the graph and answer the following questions.*

1. According to these data, which type of animal is most likely to hear sounds with frequencies of 10 hertz?
   A Bat B Dog C Grasshopper D Human
2. According to the graph, which type of animal has the smallest hearing range?
   F Bat G Cat H Dolphin J Grasshopper
3. Which of the following animals cannot hear sounds with frequencies below 100 hertz?
   A Bat C Dolphin
   B Cat D Grasshopper
4. According to this information, which type of animal can hear sounds with frequencies higher than 100,000 hertz?
   F Bat H Dolphin
   G Cat J Human

L2

Teaching

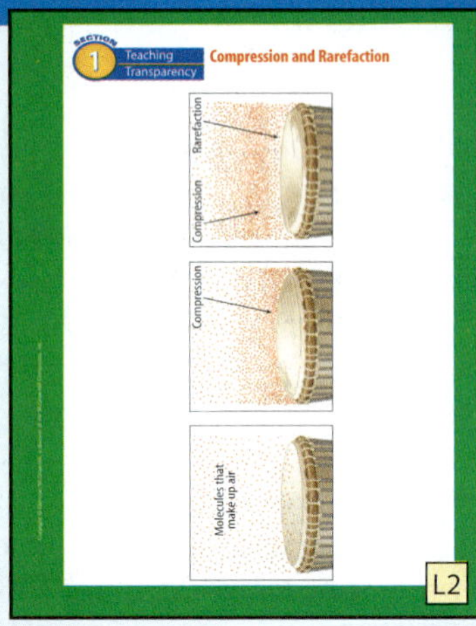
Section 1 Teaching Transparency — Compression and Rarefaction

L2

Key to Teaching Strategies

The following designations will help you decide which activities are appropriate for your students.

- **L1** Level 1 activities should be appropriate for students with learning difficulties.
- **L2** Level 2 activities should be within the ability range of all students.
- **L3** Level 3 activities are designed for above-average students.
- **ELL** ELL activities should be within the ability range of English Language Learners.
- **COOP LEARN** Cooperative Learning activities are designed for small group work.
- **LS** Multiple Learning Styles logos, as described on page 12T, are used throughout to indicate strategies that address different learning styles.
- **P** These strategies represent student products that can be placed into a best-work portfolio.
- **PBL** Problem-Based Learning activities apply real-world situations to learning.

Hands-on Activities

Student Text Lab Worksheet

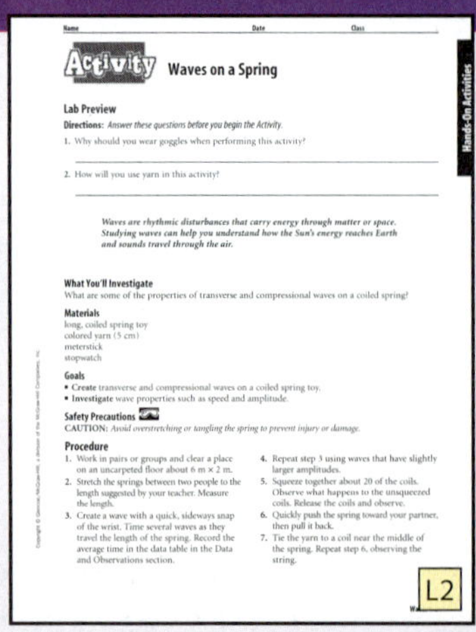
Name Date Class

Activity — Waves on a Spring

Lab Preview

Directions: *Answer these questions before you begin the Activity.*

1. Why should you wear goggles when performing this activity?
2. How will you use yarn in this activity?

Waves are rhythmic disturbances that carry energy through matter or space. Studying waves can help you understand how the Sun's energy reaches Earth and sounds travel through the air.

What You'll Investigate

What are some of the properties of transverse and compressional waves on a coiled spring?

Materials

long, coiled spring toy
colored yarn (5 cm)
meterstick
stopwatch

Goals

- **Create** transverse and compressional waves on a coiled spring toy.
- **Investigate** wave properties such as speed and amplitude.

Safety Precautions

CAUTION: *Avoid overstretching or tangling the spring to prevent injury or damage.*

Procedure

1. Work in pairs or groups and clear a place on an uncarpeted floor about 6 m × 2 m.
2. Stretch the springs between two people to the length suggested by your teacher. Measure the length.
3. Create a wave with a quick, sideways snap of the wrist. Time several waves as they travel the length of the spring. Record the average time in the data table in the Data and Observations section.
4. Repeat step 3 using waves that have slightly larger amplitudes.
5. Squeeze together about 20 of the coils. Observe what happens to the unsqueezed coils. Release the coils and observe.
6. Quickly push the spring toward your partner, then pull it back.
7. Tie the yarn to a coil near the middle of the spring. Repeat step 6, observing the string.

Hands-On Activities

L2

Laboratory Activities

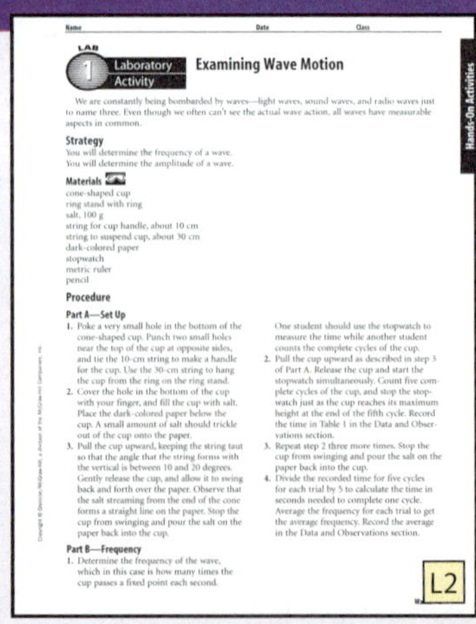
Name Date Class

LAB 1 Laboratory Activity — Examining Wave Motion

We are constantly being bombarded by waves—light waves, sound waves, and radio waves just to name three. Even though we often can't see the actual wave action, all waves have measurable aspects in common.

Strategy

You will determine the frequency of a wave.
You will determine the amplitude of a wave.

Materials

cone-shaped cup
ring stand with ring
salt, 100 g
string for cup handle, about 10 cm
string to suspend cup, about 30 cm
dark-colored paper
stopwatch
metric ruler
pencil

Procedure

Part A—Set Up

1. Poke a very small hole in the bottom of the cone-shaped cup. Punch two small holes near the top of the cup at opposite sides, and tie the 10-cm string to make a handle for the cup. Use the 30-cm string to hang the cup from the ring on the ring stand.
2. Cover the hole in the bottom of the cup with your finger, and fill the cup with salt. Place the dark-colored paper below the cup. A small amount of salt should trickle out of the cup onto the paper.
3. Pull the cup upward, keeping the string taut so that the angle that the string forms with the vertical is between 10 and 20 degrees. Gently release the cup, and allow it to swing back and forth over the paper. Observe that the salt streaming from the end of the cone forms a straight line on the paper. Stop the cup from swinging and pour the salt on the paper back into the cup.

Part B—Frequency

1. Determine the frequency of the wave, which in this case is how many times the cup passes a fixed point each second. One student should use the stopwatch to measure the time while another student counts the complete cycles of the cup.
2. Pull the cup upward as described in step 3 of Part A. Release the cup and start the stopwatch simultaneously. Count five complete cycles of the cup, and stop the stopwatch just as the cup reaches its maximum height at the end of the fifth cycle. Record the time in Table 1 in the Data and Observations section.
3. Repeat step 2 three more times. Stop the cup from swinging and pour the salt on the paper back into the cup.
4. Divide the recorded time for five cycles for each trial by 5 to calculate the time in seconds needed to complete one cycle. Average the frequency for each trial to get the average frequency. Record the average in the Data and Observations section.

Hands-On Activities

L2

Meeting Different Ability Levels

Content Outline

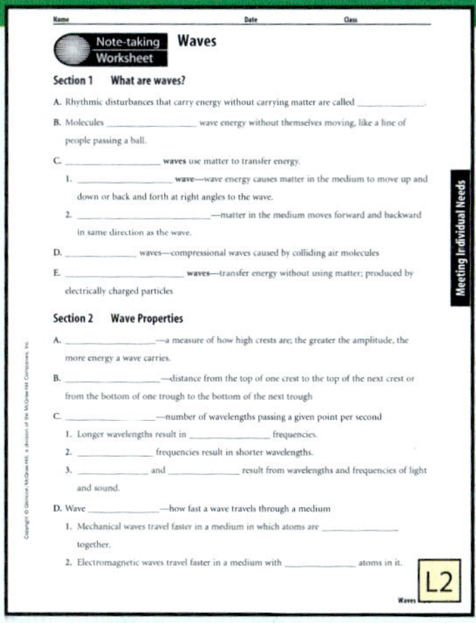

Reinforcement

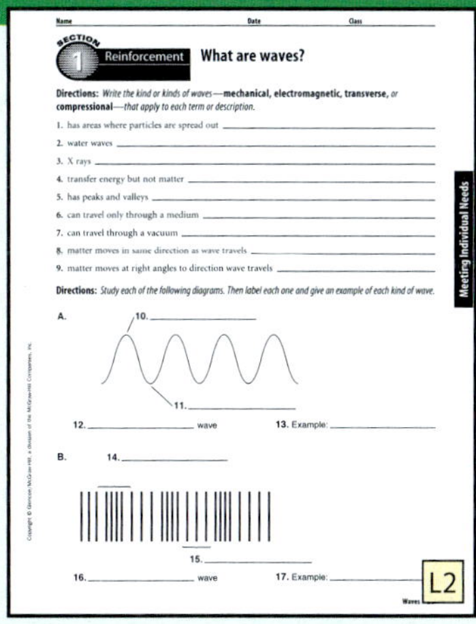

Enrichment

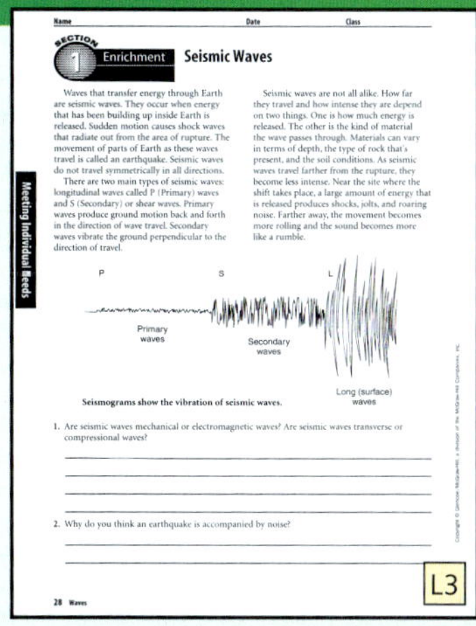

Directed Reading (English/Spanish)

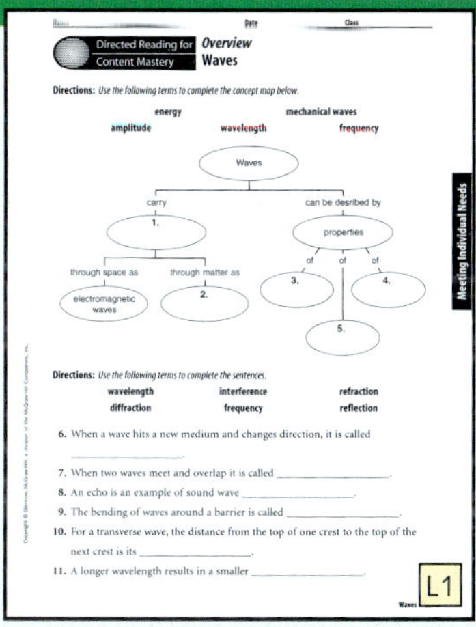

Study Guide

Reading Essentials

Assessment

Test Practice Workbook

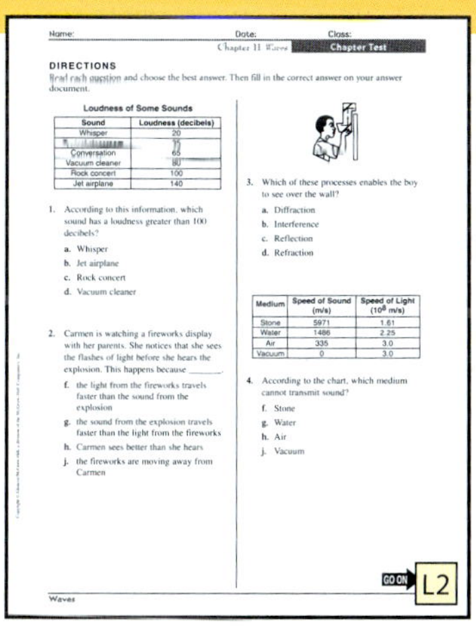

Chapter Review

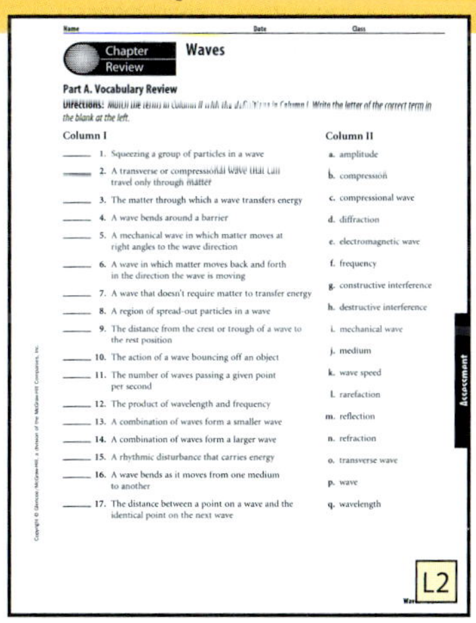

Chapter Tests

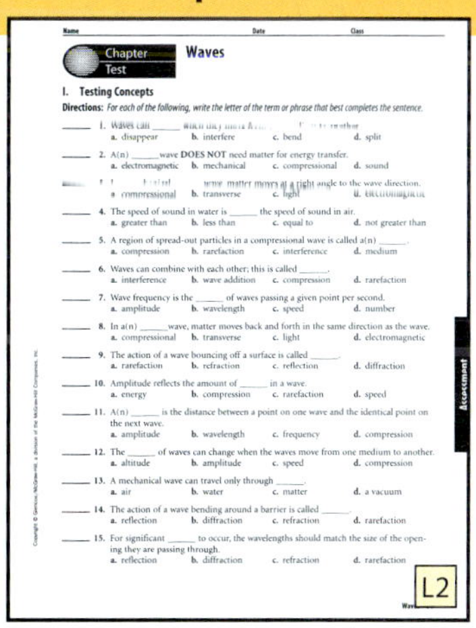

Science Content Background

section 1 What are waves?

What is a wave?

A wave transfers energy from one place to another without transferring matter in the process. Mechanical waves need a material in which to travel. In the material, the bonds between atoms and molecules in some ways are similar to springs. When one of these particles is moved, the springlike bonds exert forces that move the particle back to its original position. At the same time, the particle exerts forces on neighboring particles, causing them to be displaced. In this way, the original disturbance is propagated through the material.

Mechanical waves that travel in a material are produced by something vibrating in the material. Electromagnetic waves don't need a material in which to travel. These waves consist of oscillating electric and magnetic fields. Light waves are electromagnetic waves with wavelenghts between 400 nm and 700 nm.

section 2 Wave Properties

Wave Movements

Waves travel outward from an energy source. If a rock is thrown into a pond, energy traveling outward from the center is seen as waves that decrease in amplitude as they move through the water. A wave will move through one complete cycle in one wavelength. The period (T) of a wave is the time it takes one wavelength to pass a fixed point. A wave's period is the reciprocal of its frequency.

The amplitude of a sound wave determines its sound intensity, or how loud it is. The more energy the sound wave carries the louder the sound is. Two or more sounds produced at the same time can be unpleasant (dissonant) or pleasant (harmonious). The ratios of the frequencies of the tones determine whether or not they sound pleasing to the ear.

Teacher to Teacher

Erin Peters, Lead Science Teacher
Williamsburg Middle School
Arlington, VA

"Long springs are helpful in demonstrating both transverse and compressional (longitudinal) waves. To demonstrate transverse waves, have a student hold one end of the spring and move the other end perpendicular to the spring. Compression waves can be demonstrated by gathering up part of the slack in the spring and letting go. Note how the compression wave travels down through the spring, parallel to the spring."

Erin Peters

chapter content resources

Internet Resources

For additional content background, visit **red.msscience.com** to:

- access your book online
- find references to related articles in popular science magazines
- access Web links with related content background
- access current events with science journal topics

Print Resources

The Physics of Waves, by William C. Elmore and Mark A. Heald, Dover Publications, New York, 1985

Wave Motion in Elastic Solids, by Karl F. Graff, Dove Publications, New York, 1991

Field Guide to Geology, by David Lambert and the Diagram Group. Diagram Visual Information LTD, 1998

section 3

Wave Behavior

Reflection

When reflection occurs, the angle of incidence (the incoming wave angle) equals the angle of reflection (the outgoing wave angle). This is known as the law of reflection. The angles of incidence and reflection are measured relative to a line drawn perpendicular to the reflecting surface. Diffuse reflection occurs as waves bounce at different angles from uneven surfaces. Diffuse reflection of light makes rough surfaces appear dull.

Refraction

Refraction occurs when a wave changes speed as it passes from one material to another. The waves change direction at the interface between the two materials. Light waves can travel with different speeds in different materials, and can change direction as they move from one material to another.

Eyeglasses and contact lenses correct vision by refracting light rays so that they focus properly on the retina of the eye. Binoculars, microscopes, and telescopes also utilize refraction to produce magnified images of small objects and objects at great distances.

Diffraction

The phenomenon called diffraction occurs as waves bend around barriers or spread through openings. Diffraction of light can produce fringes (a series of bright and dark lines) as light waves bend around the edges of objects such as razor blades, diffraction gratings, or the teeth of a comb. The fringes, which can be seen on a screen, are formed by constructive and destructive interference. As the light passes through the opening, each point within the opening can be considered a new point source of light, a concept known as Huygens's principle. The light waves reaching a particular point on the screen come from different points within the opening, and thus have traveled different distances to the screen. The interference patterns are produced because, at different points on the screen, the waves add constructively or destructively.

Wave Interference

Sound waves can be made to interfere constructively or destructively as they reflect from a curved wall or a curved backdrop in a concert hall. Much planning and great expense go into designing the acoustics in a fine performance hall.

Seismic waves are produced by energy traveling through the material medium of Earth. Seismic waves are of three types—S, P, and surface waves. S waves (secondary waves) are transverse or shear and cannot travel through fluids. These seismic waves reflect off the fluid outer core of Earth. P waves, or primary waves, are compressional, or longitudinal, waves. They may pass through all of Earth's materials and are detected first. Surface waves are seismic waves that occur along Earth's surface. They cause up-down and rolling motions on the surface and cause most of the destruction.

Barry L. Runk/Grant Heilman Photography, Inc.

chapter 8

Chapter Vocabulary

wave, p. 232
mechanical wave, p. 233
transverse wave, p. 234
compressional wave, p. 235
electromagnetic wave, p. 236
amplitude, p. 237
wavelength, p. 238
frequency, p. 239
reflection, p. 243
refraction, p. 244
diffraction, p. 245
interference, p. 247

Science Journal Student responses will vary, but may include references to any body of water, large or small, when disturbed.

INTERACTIVE CHALKBOARD with Image Bank

PowerPoint® Presentations

This CD-ROM is an editable Microsoft® PowerPoint® presentation that includes:

- a pre-made presentation for every chapter
- interactive graphics
- animations
- audio clips
- image bank
- all new section and chapter questions
- Standardized Test Practice
- transparencies
- pre-lab questions for all labs
- Foldables directions
- links to red.msscience.com

chapter 8

Waves

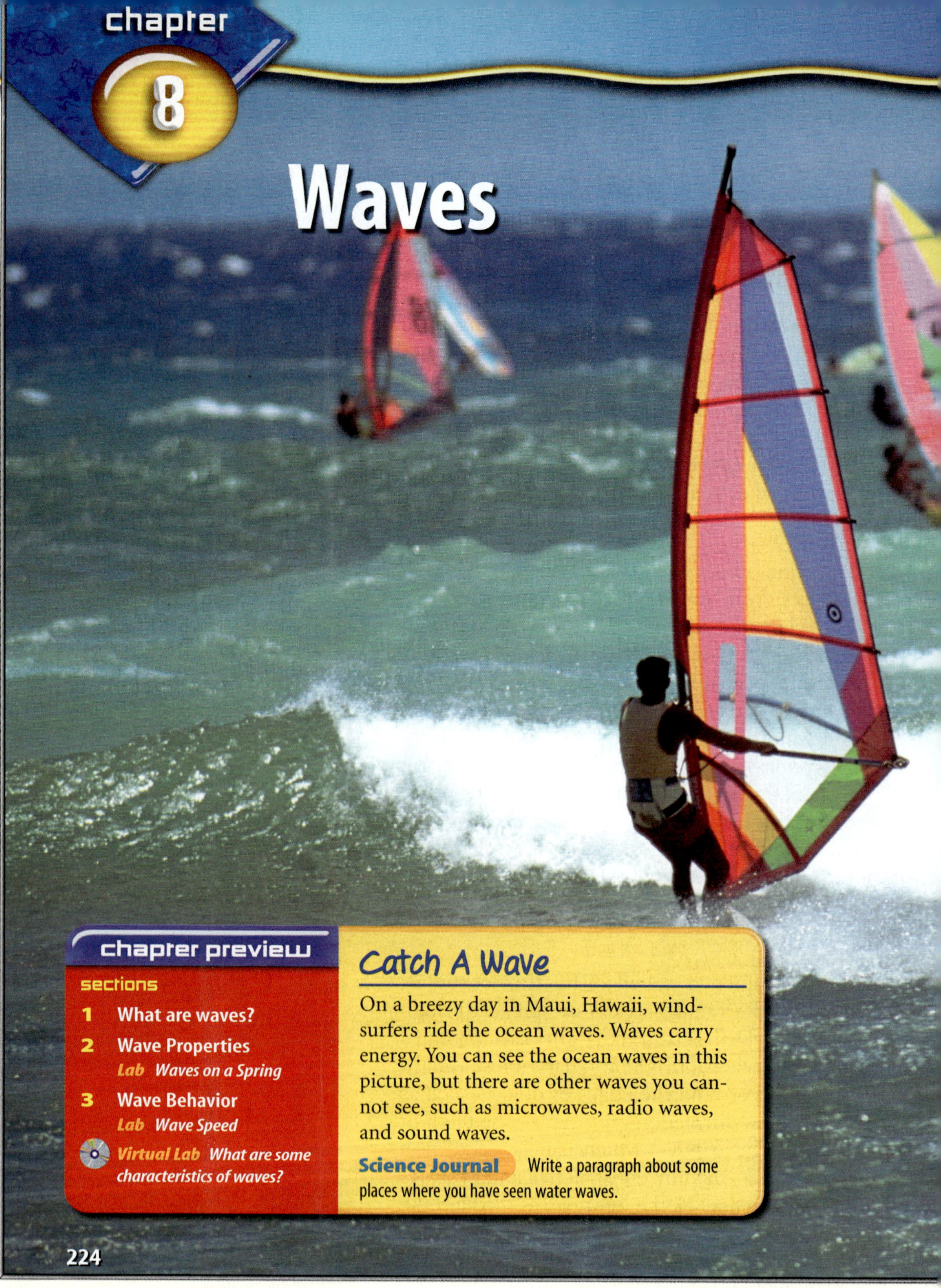

chapter preview

sections

1 What are waves?
2 Wave Properties
Lab *Waves on a Spring*
3 Wave Behavior
Lab *Wave Speed*
Virtual Lab *What are some characteristics of waves?*

Catch A Wave

On a breezy day in Maui, Hawaii, windsurfers ride the ocean waves. Waves carry energy. You can see the ocean waves in this picture, but there are other waves you cannot see, such as microwaves, radio waves, and sound waves.

Science Journal Write a paragraph about some places where you have seen water waves.

224

Theme Connection

Energy Waves are periodic disturbances that carry energy. The larger the amplitude of the wave, the more energy is transferred.

About the Photo

Energy in Waves In addition to the wind, windsurfers rely on waves to move them. Like all mechanical waves, water waves carry energy and exert forces that can move objects. The bigger the waves, the more energy they have to give a windsurfer a really exciting ride. Some islands and coastal areas are known for producing large waves when the wind and currents are just right.

Start-Up Activities

Launch LAB

Waves and Energy

It's a beautiful autumn day. You are sitting by a pond in a park. Music from a school marching band is carried to your ears by waves. A fish jumps, making waves that spread past a leaf that fell from a tree, causing the leaf to move. In the following lab, you'll observe how waves carry energy that can cause objects to move.

1. Fill a large, clear plastic plate with 1 cm of water.
2. Use a dropper to release a single drop of water onto the water's surface. Repeat.
3. Float a cork or straw on the water.
4. When the water is still, repeat step 2 from a height of 10 cm, then again from 20 cm.
5. **Think Critically** In your Science Journal, record your observations. How did the motion of the cork depend on the height of the dropper?

Science Online — Preview this chapter's content and activities at red.msscience.com

FOLDABLES™ Study Organizer

Waves Make the following Foldable to compare and contrast the characteristics of transverse and compressional waves.

STEP 1 **Fold** one sheet of paper lengthwise.

STEP 2 **Fold** into thirds.

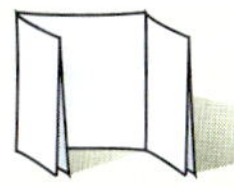

STEP 3 **Unfold and draw** overlapping ovals. **Cut** the top sheet along the folds.

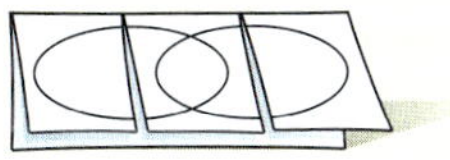

STEP 4 **Label** the ovals as shown.

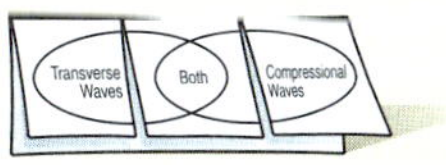

Construct a Venn Diagram As you read the chapter, list the characteristics unique to transverse waves under the left tab, those unique to compressional waves under the right tab, and those characteristics common to both under the middle tab.

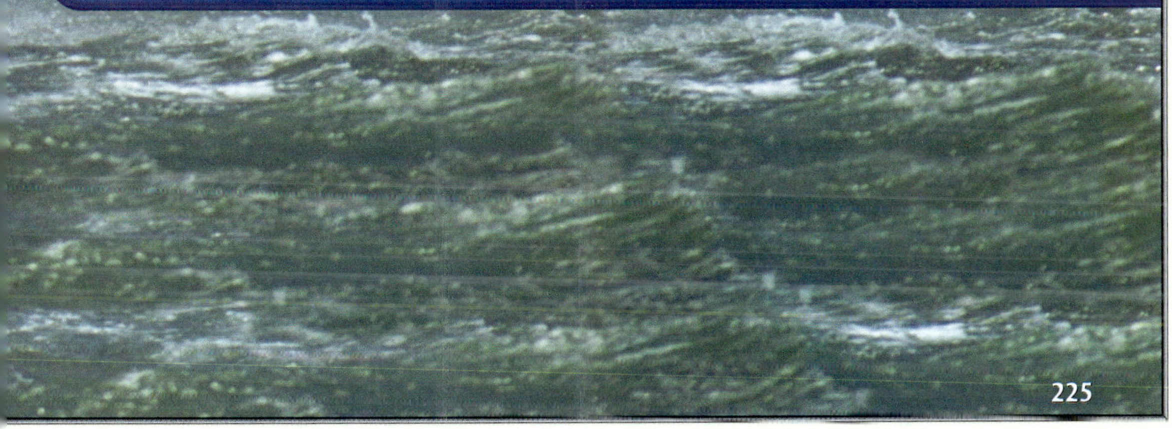

225

Launch LAB

Purpose Use the Launch Lab to give students an opportunity to observe and describe wave behavior. L2 ELL COOP LEARN

LS **Logical-Mathematical**

Materials large, clear-plastic plate; small cork or piece of a soda straw; dropper; water

Teaching Strategies

- Suggest that students observe closely because the waves will move quickly.
- Tell students to be sure the water is still before adding drops of water. They also should be careful not to disturb the water when adding the cork.

Think Critically

Circular waves travel outward from the point where the drops hit the water's surface. Drops released from a height of 20 cm cause larger waves and transfer more energy than do drops released from a height of 10 cm. The cork bobs up and down as the waves pass but does not move horizontally.

Assessment

Process Provide students with diagrams of transverse waves with different amplitudes. Ask them to identify which diagram shows waves generated by dropping water from a greater height. Use **Performance Assessment in the Science Classroom**, p. 89.

FOLDABLES™ Study Organizer — Dinah Zike Study Fold

Student preparation materials for this Foldable are available in the Chapter *FAST FILE* Resources.

1 Motivate

Bellringer

INTERACTIVE CHALKBOARD PowerPoint® Presentations

Section Focus Transparencies also are available on the Interactive Chalkboard CD-ROM. L2 ELL

SECTION 1 Section Focus Transparency **Ride the Wave!**

Surfing may seem like a fairly new sport, but it's not. In fact, people think there were surfers in Hawaii in the 1400s. When Captain Cook landed in the Hawaiian Islands in 1778, he was greeted by an islander on a surfboard!

1. Name some different kinds of waves.
2. What do you think creates waves on the ocean?
3. Describe the path of light from the Sun to the camera that took this picture.

L2

Waves

Tie to Prior Knowledge

Wave Discussion Ask students to describe any waves they know of. Discuss with them the form of energy transferred by each wave.

section 1

What are waves?

as you read

***What* You'll Learn**

- **Explain** the relationship among waves, energy, and matter.
- **Describe** the difference between transverse waves and compressional waves.

***Why* It's Important**

Waves enable you to see and hear the world around you.

Review Vocabulary
energy: the ability to cause change

New Vocabulary
- wave
- mechanical wave
- transverse wave
- compressional wave
- electromagnetic wave

What is a wave?

When you are relaxing on an air mattress in a pool and someone does a cannonball dive off the diving board, you suddenly find yourself bobbing up and down. You can make something move by giving it a push or pull, but the person jumping didn't touch your air mattress. How did the energy from the dive travel through the water and move your air mattress? The up-and-down motion was caused by the peaks and valleys of the ripples that moved from where the splash occurred. These peaks and valleys make up water waves.

Waves Carry Energy Rhythmic disturbances that carry energy without carrying matter are called **waves.** Water waves are shown in **Figure 1.** You can see the energy of the wave from a speedboat traveling outward, but the water only moves up and down. If you've ever felt a clap of thunder, you know that sound waves can carry large amounts of energy. You also transfer energy when you throw something to a friend, as in **Figure 1.** However, there is a difference between a moving ball and a wave. A ball is made of matter, and when it is thrown, the matter moves from one place to another. So, unlike the wave, throwing a ball involves the transport of matter as well as energy.

Figure 1 The wave and the thrown ball carry energy in different ways.

The waves created by a boat move mostly up and down, but the energy travels outward from the boat.

When the ball is thrown, the ball c energy as it moves forward.

Section 1 Resource Manager

Chapter *FAST FILE* Resources

Transparency Activity, pp. 42, 45–46
Enrichment, p. 28
Reinforcement, p. 25
MiniLAB, p. 3
Directed Reading for Content Mastery, pp. 17, 18
Note-taking Worksheets, pp. 31–32
Cultural Diversity, p. 63
Mathematics Skill Activities, p. 11

As the students pass the ball, the students' positions do not change—only the position of the ball changes.

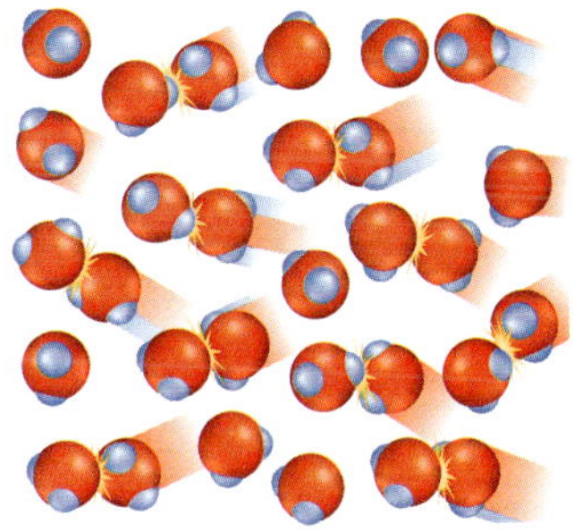

In a water wave, water molecules bump each other and pass energy from molecule to molecule.

Figure 2 A wave transports energy without transporting matter from place to place.
Describe *other models that could be used to represent a mechanical wave.*

A Model for Waves

How does a wave carry energy without transporting matter? Imagine a line of people, as shown in **Figure 2.** The first person in line passes a ball to the second person, who passes the ball to the next person, and so on. Passing a ball down a line of people is a model for how waves can transport energy without transporting matter. Even though the ball has traveled, the people in line have not moved. In this model, you can think of the ball as representing energy. What do the people in line represent?

Think about the ripples on the surface of a pond. The energy carried by the ripples travels through the water. The water is made up of water molecules. It is the individual molecules of water that pass the wave energy, just as the people. The water molecules transport the energy in a water wave by colliding with the molecules around them, as shown in **Figure 2.**

Reading Check *What is carried by waves?*

Mechanical Waves

In the wave model, the ball could not be transferred if the line of people didn't exist. The energy of a water wave could not be transferred if no water molecules existed. These types of waves, which use matter to transfer energy, are called **mechanical waves.** The matter through which a mechanical wave travels is called a medium. For ripples on a pond, the medium is the water.

A mechanical wave travels as energy is transferred from particle to particle in the medium. For example, a sound wave is a mechanical wave that can travel through air, as well as solids, liquids, and other gases. Without a medium such as air, there would be no sound waves. In outer space sound waves can't travel because there is no air.

2 Teach

Discussion

Energy Waves Ask students to describe how being near a source of deep, loud sounds, such as heavy machinery or the sub-woofer of a stereo, demonstrates that waves carry energy. You can feel the vibrations as well as hear the sound. L2 LS **Auditory-Musical**

Make a Model

Water Wave Have students duplicate the model described in the text. As they do this, emphasize that they are like the molecules of water through which a wave flows. Point out that the ball is made of matter, but in the model it represents energy. L1 ELL COOP LEARN LS **Kinesthetic**

Reading Check

Answer energy

Caption Answer

Figure 2 Possible answers include a bucket brigade, bumping someone at one end of a line of people, whispering a message to someone in a line of people who pass the message from one to another.

Identifying Misconceptions

Mechanical Wave Students might be surprised that some waves are described as mechanical. The word *mechanical* comes from the Greek word *mechane*, meaning "a device." Based on this, have students discuss why *mechanical* is an appropriate term to use for waves that need matter to move. One definition of *machine* is a device that transmits energy. In a mechanical wave, matter transmits energy. L3 LS **Linguistic**

Differentiated Instruction

Challenge Ask students to investigate the behavior of surface waves near the seashore and make posters showing what they find. As the waves approach land, water molecules near the surface experience more elliptical movement, causing the waves to form high crests that eventually crash into land. L3 LS **Visual-Spatial**

Visual Learning

Figure 2 Tell students that the water molecules near the surface of a wave actually travel in small circles. Have students model this movement with a cork in water. L1 ELL LS **Kinesthetic**

Use an Analogy

Stadium Wave Ask students to think of people "doing the wave" in a stadium. Point out that this wave is similar to a transverse mechanical wave because the people, like particles, move up and down at right angles to the direction the wave moves. L2 LS **Visual-Spatial**

Quick Demo

Floating on a Wave

Materials clear plate, colored water, overhead projector, small cork

Estimated Time five minutes

Procedure Put the clear plate on the overhead projector and add the colored water. Ask students to predict what will happen to a floating object as waves are generated in the water. Put the cork in the water and make small waves by moving a pencil or your finger in the water. The floating object will bob up and down because the water waves are very much like transverse waves.

Reading Check

Answer crests

Activity

Describing Waves Ask students to list on chart paper examples of mechanical waves. Have them draw each type of wave, describe each wave as transverse or compressional, and label the different parts of each wave. L2 COOP LEARN LS **Visual-Spatial** P

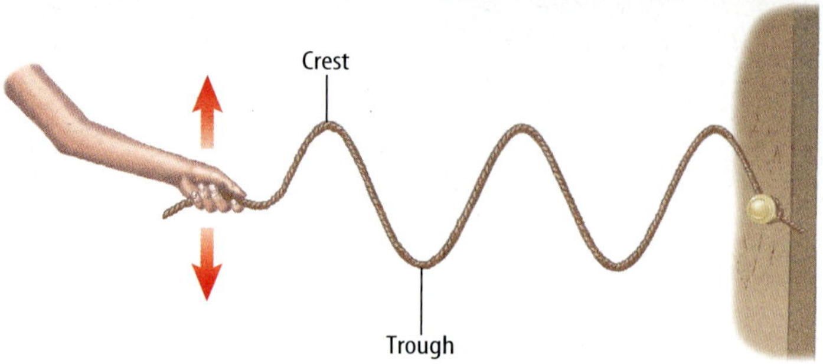

Figure 3 The high points on the wave are called crests and the low points are called troughs.

Transverse Waves In a mechanical **transverse wave,** the wave energy causes the matter in the medium to move up and down or back and forth at right angles to the direction the wave travels. You can make a model of a transverse wave. Stretch a long rope out on the ground. Hold one end in your hand. Now shake the end in your hand back and forth. As you shake the rope, you create a wave that seems to slide along the rope.

When you first started shaking the rope, it might have appeared that the rope itself was moving away from you. But it was only the wave that was moving away from your hand. The wave energy moves through the rope, but the matter in the rope doesn't travel. You can see that the wave has peaks and valleys at regular intervals. As shown in **Figure 3,** the high points of transverse waves are called crests. The low points are called troughs.

Reading Check *What are the highest points of transverse waves called?*

Figure 4 A compressional wave can travel through a coiled spring toy.

As the wave motion begins, the coils near the string are close together and the other coils are far apart.

The wave, seen in the squeezed and stretched coils, travels along the spring.

The string and coils did not travel with the wave. Each coil moved forward and then back to its original position.

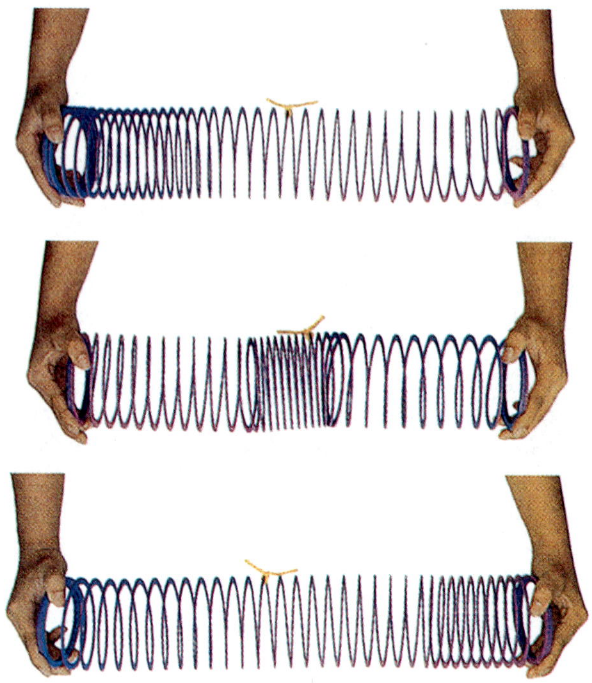

Teacher FYI

Seismic Waves There are three main types of seismic waves. Primary (P) waves oscillate back and forth along the direction the wave travels. Secondary (S) waves oscillate perpendicular to the direction of wave motion. Surface waves cause horizontal and, sometimes, vertical ground surface movement.

Active Reading

Bubble Map Using a Bubble Map helps students start ideas flowing about a given topic. Words are clustered to describe a concept. Students can use the bubble map for a prewrite, to generate ideas before writing in their Journals, or to review for a test. Have students design a Bubble Map for the different types of waves described in this section. L2

228 CHAPTER 8 Waves

Compressional Waves Mechanical waves can be either transverse or compressional. In a **compressional wave,** matter in the medium moves forward and backward along the same direction that the wave travels. You can make a compressional wave by squeezing together and releasing several coils of a coiled spring toy, as shown in **Figure 4.**

The coils move only as the wave passes and then return to their original positions. So, like transverse waves, compressional waves carry only energy forward along the spring. In this example, the spring is the medium the wave moves through, but the spring does not move along with the wave.

Sound Waves Sound waves are compressional waves. How do you make sound waves when you talk or sing? If you hold your fingers against your throat while you hum, you can feel vibrations. These vibrations are the movements of your vocal cords. If you touch a stereo speaker while it's playing, you can feel it vibrating, too. All waves are produced by something that is vibrating.

Making Sound Waves

How do vibrating objects make sound waves? Look at the drum shown in **Figure 5.** When you hit the drumhead it starts vibrating up and down. As the drumhead moves upward, the molecules next to it are pushed closer together. This group of molecules that are closer together is a compression. As the compression is formed, it moves away from the drumhead, just as the squeezed coils move along the coiled spring toy in **Figure 4.**

When the drumhead moves downward, the molecules near it have more room and can spread farther apart. This group of molecules that are farther apart is a rarefaction. The rarefaction also moves away from the drumhead. As the drumhead vibrates up and down, it forms a series of compressions and rarefactions that move away and spread out in all directions. This series of compressions and rarefactions is a sound wave.

Mini LAB

Comparing Sounds

Procedure

1. Hold a **wooden ruler** firmly on the edge of your **desk** so that most of it extends off the edge of the desk.
2. Pluck the free end of the ruler so that it vibrates up and down. Use gentle motion at first, then pluck with more energy.
3. Repeat step 2, moving the ruler about 1 cm further onto the desk each time until only about 5 cm extend off the edge.

Analysis

1. Compare the loudness of the sounds that are made by plucking the ruler in different ways.
2. Describe the differences in the sound as the end of the ruler extended farther from the desk.

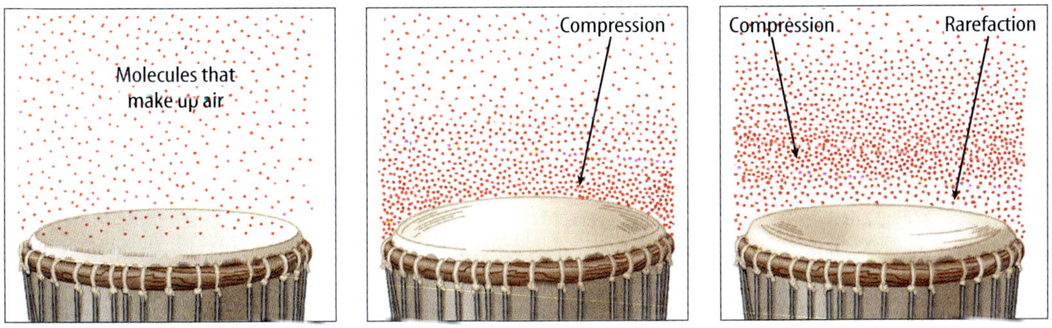

Figure 5 A vibrating drumhead makes compressions and rarefactions in the air.
Describe *how compressions and rarefactions are different.*

Teacher FYI

Wave Names Another name for compressional waves is longitudinal waves.

Mini LAB

Purpose Students compare and contrast various sounds made by a vibrating object. L2 ELL
LS **Auditory-Musical**

Materials wooden ruler

Teaching Strategies Monitor students to ensure their results include information about the length of the ruler hanging over the table.

Troubleshooting Make sure students do not pluck hard enough to break the ruler.

Analysis

1. Plucking the ruler harder produces louder sounds.
2. The greater the length of the vibrating object, the lower the pitch produced.

Assessment

Performance Give students materials of different lengths and ask them to choose and make sounds with the ones that produce lower-pitched sounds. Use **Performance Assessment in the Science Classroom,** p. 97.

Caption Answer

Figure 5 Particles are squeezed together in compressions and are farther apart in rarefactions.

Fun Fact

Massive bodies have strong gravitational fields. If these bodies undergo some sort of regular motion, oscillations in their gravity fields can, according to theory, be detected. Scientists are currently trying to develop methods of detecting these gravity waves.

Differentiated Instruction

Hearing Impaired Some students may not be able to hear the sounds made by the ruler as they do the MiniLAB. Have these students pay attention to feeling the vibrations. They should be able to feel the vibrations with the hand holding the ruler on the desk. Suggest they pluck the ruler several times while paying attention to what they feel. When they are familiar with the feeling, they can finish the lab. L2

INTEGRATE Physics

Global Positioning Systems There are 24 GPS satellites, and they orbit Earth at a height of about 20,000 km. Their electromagnetic radio waves can be received all over Earth, regardless of the weather. GPS was designed for and is operated by the U.S. Department of Defense and provides positioning information to military submarines, helicopters, ships, bombers, tanks, and missiles. However, since the signals can be received by an unlimited number of users, they are increasingly important in everyday civilian navigation.

3 Assess

DAILY INTERVENTION

Check for Understanding

Logical-Mathematical Brainstorm with students to create a list of ways water waves display energy. Responses could include moving a surf board, depositing sand and debris along shorelines, or damage to piers and boats seen after a storm. L2

Reteach

Waves in a Spring Demonstrate compressional and transverse waves using a spring toy. Have students identify each type of wave and the parts of the waves. L2 LS **Visual-Spatial**

Assessment

Process Have students draw diagrams of transverse and compressional waves, labeling the crests and troughs. Use **Performance Assessment in the Science Classroom**, p. 127. L2

INTEGRATE Physics

Global Positioning Systems Maybe you've used a global positioning system (GPS) receiver to determine your location while driving, boating, or hiking. Earth-orbiting satellites send electromagnetic radio waves that transmit their exact locations and times of transmission. The GPS receiver uses information from four of these satellites to determine your location to within about 16 m.

Electromagnetic Waves

Waves that can travel through space where there is no matter are **electromagnetic waves.** There are different types of electromagnetic waves, including radio waves, infrared waves, visible light waves, ultraviolet waves, x rays, and gamma rays. These waves can travel in matter or in space. Radio waves from TV and radio stations travel through air, and may be reflected from a satellite in space. They then travel through air, through the walls of your house, and to your TV or radio.

Radiant Energy From the Sun The Sun emits electromagnetic waves that travel through space and reach Earth. The energy carried by electromagnetic waves is called radiant energy. Almost 92 percent of the radiant energy that reaches Earth from the Sun is carried by infrared and visible light waves. Infrared waves make you feel warm when you sit in sunlight, and visible light waves enable you to see. A small amount of the radiant energy that reaches Earth is carried by ultraviolet waves. These are the waves that can cause sunburn if you are exposed to sunlight for too long.

section 1 review

Summary

What is a wave?

- Waves transfer energy, but do not transfer matter.

Mechanical Waves

- Mechanical waves require a medium in which to travel.
- When a transverse wave travels, particles of the medium move at right angles to the direction the wave is traveling.
- When a compressional wave travels, particles of the medium move back and forth along the same direction the wave is traveling.
- Sound is a compressional wave.

Electromagnetic Waves

- Electromagnetic waves can travel through empty space.
- The Sun emits different types of electromagnetic waves, including infrared, visible light, and ultraviolet waves.

Self Check

1. **Describe** the movement of a floating object on a pond when struck by a wave.
2. **Explain** why a sound wave can't travel from a satellite to Earth.
3. **Compare and contrast** a transverse wave and a compressional wave. How are they similar and different?
4. **Compare and contrast** a mechanical wave and an electromagnetic wave.
5. **Think Critically** How is it possible for a sound wave to transmit energy but not matter?

Applying Skills

6. **Concept Map** Create a concept map that shows the relationships among the following: *waves, mechanical waves, electromagnetic waves, compressional waves,* and *transverse waves.*
7. **Use a Word Processor** Use word-processing software to write short descriptions of the waves you encounter during a typical day.

section 1 review

1. It bobs up and down.
2. It requires a medium through which to travel, and there is no medium in space.
3. Possible answer: Water waves are transverse waves. Sound waves are compressional waves. Compressional waves cause vibration in the direction of wave movement. Transverse waves cause vibration at right angles to the wave's direction. Both involve periodic motion and carry energy.
4. Mechanical waves use matter to transmit energy. Electromagnetic waves do not require matter to carry energy.
5. The wave causes molecules of matter to bump neighboring molecules.
6. Answers should show mechanical waves as compressional or transverse; electromagnetic waves are transverse.
7. Answers might include radio waves, sound waves, or waves in a bathtub.

section 2

Wave Properties

Amplitude

Can you describe a wave? For a water wave, one way might be to tell how high the wave rises above, or falls below, the normal level. This distance is called the wave's amplitude. The **amplitude** of a transverse wave is one-half the distance between a crest and a trough, as shown in **Figure 6.** In a compressional wave, the amplitude is greater when the particles of the medium are squeezed closer together in each compression and spread farther apart in each rarefaction.

Amplitude and Energy A wave's amplitude is related to the energy that the wave carries. For example, the electromagnetic waves that make up bright light have greater amplitudes than the waves that make up dim light. Waves of bright light carry more energy than the waves that make up dim light. In a similar way, loud sound waves have greater amplitudes than soft sound waves. Loud sounds carry more energy than soft sounds. If a sound is loud enough, it can carry enough energy to damage your hearing.

When a hurricane strikes a coastal area, the resulting water waves carry enough energy to damage almost anything that stands in their path. The large waves caused by a hurricane carry more energy than the small waves or ripples on a pond.

as you read

***What* You'll Learn**

- **Describe** the relationship between the frequency and wavelength of a wave.
- **Explain** why waves travel at different speeds.

***Why* It's Important**

The properties of a wave determine whether the wave is useful or dangerous.

Review Vocabulary
speed: the distance traveled divided by the time needed to travel the distance

New Vocabulary
- amplitude
- wavelength
- frequency

Figure 6 The energy carried by a wave increases as its amplitude increases.

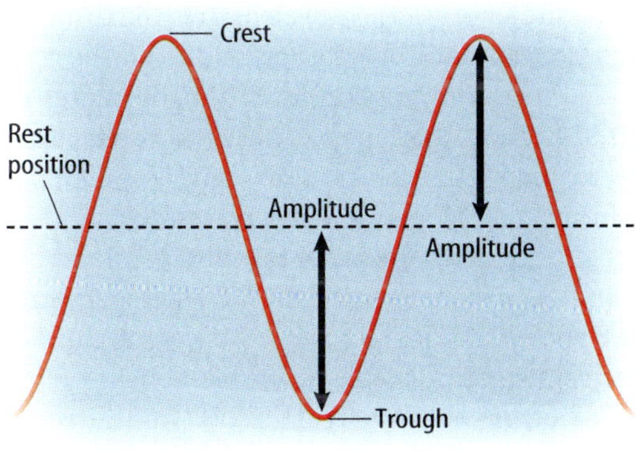

The amplitude of a transverse wave is a measure of how high the crests are or how deep the troughs are.

A water wave of large amplitude carried the energy that caused this damage.

1 Motivate

Bellringer

INTERACTIVE CHALKBOARD PowerPoint® Presentations

Section Focus Transparencies also are available on the Interactive Chalkboard CD-ROM. L2 ELL

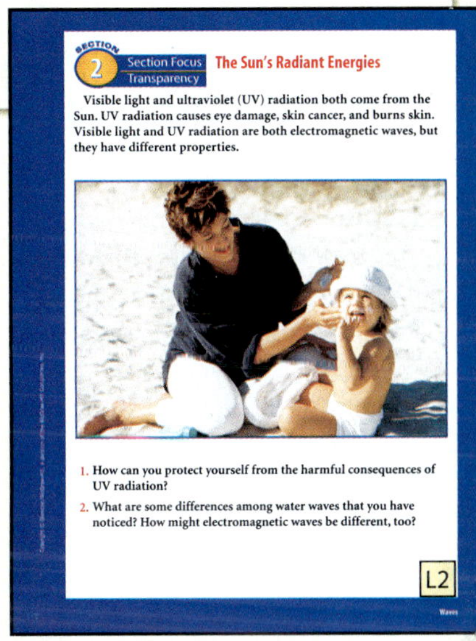

Tie to Prior Knowledge

Jump Ropes Ask students whether it takes more or less energy to twirl a jump rope high enough to clear a short person or a tall person. tall Explain that in this section, they will learn how to describe the energy different waves carry.

Section 2 Resource Manager

Chapter *FAST FILE* Resources
Transparency Activity, p. 43
Directed Reading for Content Mastery, p. 19
Enrichment, p. 29
Reinforcement, p. 26
Lab Activity, pp. 9–11
Lab Worksheets, pp. 5–6

Cultural Diversity, p. 61
Home and Community Involvement, p. 42
Reading and Writing Skill Activities, p. 33

2 Teach

Fun Fact

Tsunamis can travel thousands of kilometers, in all directions, at speeds of more than 900 km/h. When a tsunami approaches a coastline, the water piles up and forms a wave crest that can tower 30 m high or higher.

Inquiry Lab

Seismic Activity

Purpose to locate areas of seismic activity around the world L2

Possible Materials library resource materials, encyclopedic computer programs, accessed from red.msscience.com/internet_lab

Estimated Time 1 class period

Teaching Strategies

- Allow students to work in pairs or small groups to select an area of the globe to research.
- Have students choose a method of presenting their results to the teacher or the class. This might be in the form of a map, a computer spread sheet, a written report or an oral report.
- Encourage students to draw conclusions about the danger in each area. Pay special attention to sites in the U.S. and in your state or immediate area.
- Allow students to explore other questions that arise.

For additional inquiry activities, see *Science Inquiry Labs*.

For transverse waves, wavelength is the distance from crest to crest or trough to trough.

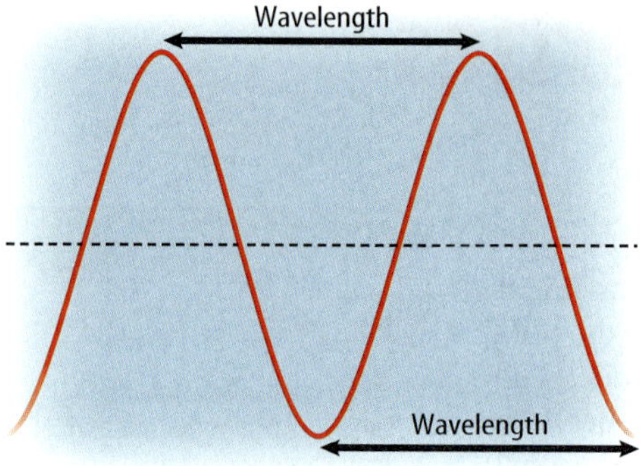

For compressional waves, wavelength is the distance from compression to compression or rarefaction to rarefaction.

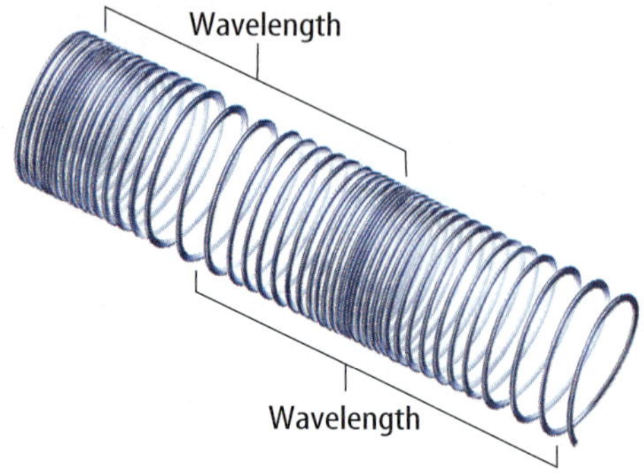

Figure 7 A transverse or a compressional wave has a wavelength.

Figure 8 The wavelengths and frequencies of electromagnetic waves vary.

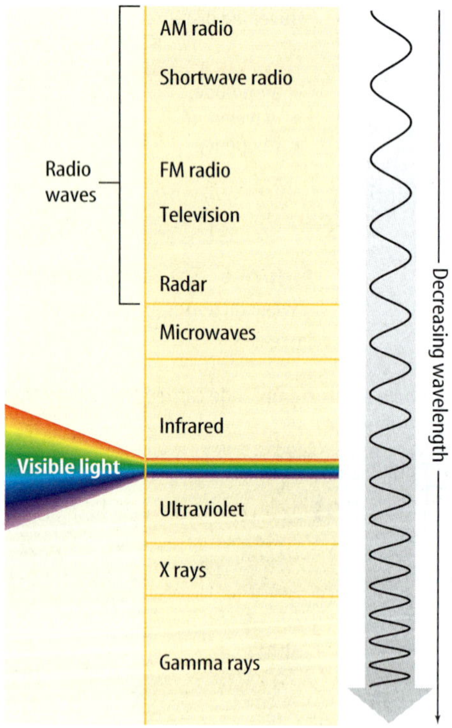

INTEGRATE Earth Science The devastating effect that a wave with large amplitude can have is seen in the aftermath of tsunamis. Tsunamis are huge sea waves that are caused by underwater earthquakes along faults on the seafloor. The movement of the seafloor along a fault produces the wave. As the wave moves toward shallow water and slows down, the amplitude of the wave grows. The tremendous amounts of energy tsunamis carry cause great damage when they move ashore.

Wavelength

Another way to describe a wave is by its wavelength. **Figure 7** shows the wavelength of a transverse wave and a compressional wave. For a transverse wave, **wavelength** is the distance from the top of one crest to the top of the next crest, or from the bottom of one trough to the bottom of the next trough. For a compressional wave, the wavelength is the distance between the center of one compression and the center of the next compression, or from the center of one rarefaction to the center of the next rarefaction.

Electromagnetic waves have wavelengths that range from kilometers, for radio waves, to less than the diameter of an atom, for X rays and gamma rays. This range is called the electromagnetic spectrum. **Figure 8** shows the names given to different parts of the electromagnetic spectrum. Visible light is only a small part of the electromagnetic spectrum. It is the wavelength of visible light waves that determines their color. For example, the wavelength of red light waves is longer than the wavelength of green light waves.

Visual Learning

Figure 8 Electromagnetic waves sometimes behave as if they are particles called photons. Waves with a longer wavelength have lower energy per photon than waves with a shorter wavelength.

Teacher FYI

Decibles The threshold of human hearing for sound intensity (loudness) is 0 dB (10^{-12} watts/m^2). Hearing damage commences at sustained levels of about 85 dB. Normal conversation has an intensity of about 60 dB. A jet engine can generate noise levels of about 120 dB to 140 dB.

Frequency

The **frequency** of a wave is the number of wavelengths that pass a given point in 1 s. The unit of frequency is the number of wavelengths per second, or hertz (Hz). Recall that waves are produced by something that vibrates. The faster the vibration is, the higher the frequency is of the wave that is produced.

Reading Check *How is the frequency of a wave measured?*

A Sidewalk Model For waves that travel with the same speed, frequency and wavelength are related. To model this relationship, imagine people on two parallel moving sidewalks in an airport, as shown in **Figure 9.** One sidewalk has four travelers spaced 4 m apart. The other sidewalk has 16 travelers spaced 1 m apart.

Now imagine that both sidewalks are moving at the same speed and approaching a pillar between them. On which sidewalk will more people go past the pillar? On the sidewalk with the shorter distance between people, four people will pass the pillar for each one person on the other sidewalk. When four people pass the pillar on the first sidewalk, 16 people pass the pillar on the second sidewalk.

Figure 9 When people are farther apart on a moving sidewalk, fewer people pass the pillar every minute.
Infer *how the number of people passing the pillar each minute would change if the sidewalk moved slower.*

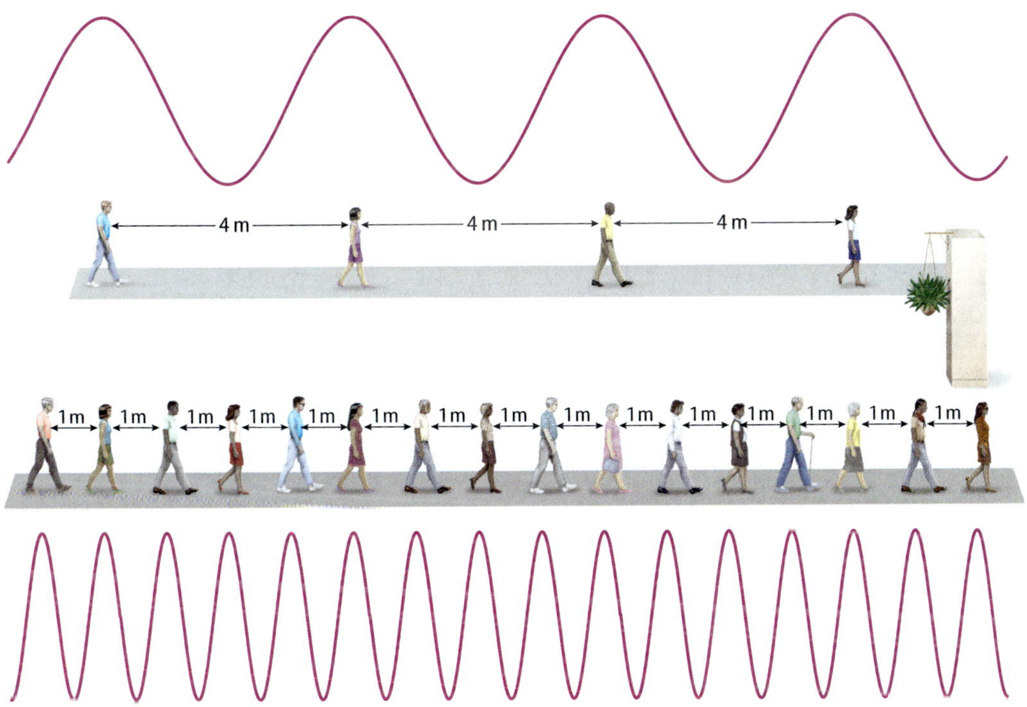

Use Science Words

Word Origin Have students investigate the origin of the word *hertz* and why it is used to describe frequency. The unit hertz was named after Heinrich Hertz, a German physicist who studied electromagnetic waves. L2 LS **Linguistic**

Reading Check

Answer in units of the number of wavelengths per second, or Hz

Discussion

String Music How does pressing a string on a guitar affect the frequency of the sound produced? It shortens the string, which increases the frequency of the sound the string produces when it is plucked. How does this relate to the sounds produced by a harp? On a harp, the short strings produce high frequency sounds and the long strings produce low frequency sounds. L2 LS **Auditory-Musical**

Caption Answer

Figure 9 The number would decrease.

Activity

Geographic Waves Ask students to bring in magazine and newspaper articles about tidal waves (tsunamis) and earthquakes around the world. Display the articles and discuss with students the relationship between energy, wave amplitude, and the amount of destruction caused. L2 LS **Interpersonal**

Differentiated Instruction

English-Language Learners Play music through speakers with their interior exposed. Students can see and feel the vibrations that result from changes in pitch (frequency) or volume (amplitude).

Curriculum Connection

Music Have students find out what quarter tones are and play recordings of music with quarter tones for the class. The difference in frequency between two adjacent pitches in some types of music, including Hindu music, is about half that between adjacent pitches in western music. These pitches are called quarter tones. L3 ELL LS **Auditory-Musical**

SECTION 2 Wave Properties 233

Activity

Making Waves Have each student fill a bowl with water and dip a pencil into and out of the water to form waves. They should notice that dipping the pencil with low frequency produces waves with longer wavelengths (greater distance between crests) than dipping the pencil with high frequency. Have students use a stopwatch to help calibrate the frequencies of their waves. L1 ELL LS **Kinesthetic**

Quick Demo

Musical Sounds

Materials musical instruments, oscilloscope, microphone

Estimated Time 15 minutes

Procedure Have students who play instruments demonstrate compressional waves of high and low frequency by producing notes of varying pitch. If an oscilloscope is available connect it to a microphone to show students the wavelengths and amplitudes of the notes.

INTEGRATE Health

Ultrasonic Waves The wavelengths are much shorter than those of the sound waves we hear.

Research Have students research one of the medical uses of ultrasound mentioned. Students should present their findings to you or to the class in whatever format they choose. L2

Reading Check

Answer frequency increases as wavelength decreases

INTEGRATE Health

Ultrasonic Waves Sound waves with ultra-high frequencies cannot be heard by the human ear, but are used by medical professionals in several ways. They are used to perform echocardiograms of the heart, produce ultrasound images of internal organs, break up blockages in arteries and sterilize surgical instruments. Describe how the wavelengths of these sound waves compare to sound waves you can hear.

Frequency and Wavelength Suppose that each person in **Figure 9** represents the crest of a wave. Then the movement of people on the first sidewalk is like a wave with a wavelength of 4 m. For the second sidewalk, the wavelength would be 1 m. On the first sidewalk, where the wavelength is longer, the people pass the pillar *less* frequently. Smaller frequencies result in longer wavelengths. On the second sidewalk, where the wavelength is shorter, the people pass the pillar *more* frequently. Higher frequencies result in shorter wavelengths. This is true for all waves that travel at the same speed. As the frequency of a wave increases, its wavelength decreases.

Reading Check *How are frequency and wavelength related?*

Color and Pitch Because frequency and wavelength are related, either the wavelength or frequency of a light wave determines the color of the light. For example, blue light has a larger frequency and shorter wavelength than red light.

Either the wavelength or frequency determines the pitch of a sound wave. Pitch is how high or low a sound seems to be. When you sing a musical scale, the pitch and frequency increase from note to note. Wavelength and frequency are also related for sound waves traveling in air. As the frequency of sound waves increases, their wavelength decreases. **Figure 10** shows how the frequency and wavelength change for notes on a musical scale.

Figure 10 The frequency of the notes on a musical scale increases as the notes get higher in pitch, but the wavelength of the notes decreases.

Science Journal

Musical Instruments Ask students to investigate various musical instruments and write in their Science Journals what vibrates on each instrument to make sound and how this produces certain pitches. Students might mention the reed of a clarinet or the lips of a trumpet player. Air inside the instrument also vibrates. The shape and material of the instrument affect pitch. L3 LS **Linguistic**

Differentiated Instruction

Challenge Compare the sound of a siren moving toward you with the sound of a siren moving away from you. When the siren is moving toward you, the wavelengths of the compressional waves decrease, resulting in a higher pitch. As the siren moves away from you, the wavelengths increase and the pitch gets lower. This is known as the Doppler effect. L3 LS **Logical-Mathematical**

Wave Speed

You've probably watched a distant thunderstorm approach on a hot summer day. You see a bolt of lightning flash between a dark cloud and the ground. If the thunderstorm is many kilometers away, several seconds will pass between when you see the lightning and when you hear the thunder. This happens because light travels much faster in air than sound does. Light travels through air at about 300 million m/s. Sound travels through air at about 340 m/s. The speed of any wave can be calculated from this equation:

Wave Speed Equation

wave speed (in m/s) = **frequency** (in Hz) × **wavelength** (m)

$$v = f\lambda$$

In this equation, the wavelength is represented by the symbol λ, which is the Greek letter lambda.

When mechanical waves, such as sound, and electromagnetic waves, such as light, travel in different materials, they change speed. Mechanical waves usually travel fastest in solids, and slowest in gases. Electromagnetic waves travel fastest in gases and slowest in solids. For example, the speed of light is about 30 percent faster in air than in water.

Science Online

Topic: Wave Speed

Visit red.msscience.com for Web links to information about wave speed in different materials.

Activity Make a chart showing the speed of light in different materials.

section 2 review

Summary

Amplitude

- In a transverse wave, the amplitude is one-half the distance between a crest and a trough.
- The larger the amplitude, the greater the energy carried by the wave.

Wavelength

- For a transverse wave, wavelength is the distance from crest to crest, or from trough to trough.
- For a compressional wave, wavelength is the distance from compression to compression, or from rarefaction to rarefaction.

Frequency

- The frequency of a wave is the number of wavelengths that pass a given point in 1 s.
- For waves that travel at the same speed, as the frequency of the wave increases, its wavelength decreases.

Self Check

1. **Describe** how the frequency of a wave changes as its wavelength changes.
2. **Explain** why a sound wave with a large amplitude is more likely to damage your hearing than one with a small amplitude.
3. **State** what accounts for the time difference between seeing and hearing a fireworks display.
4. **Explain** why the statement "The speed of light is 300 million m/s" is not always correct.
5. **Think Critically** Explain the differences between the waves that make up bright, green light and dim, red light.

Applying Math

6. **Wave speed** Find the speed of a wave with a frequency of 5 m and a frequency of 68 Hz.
7. **Wavelength** Find the wavelength of a sound wave traveling in water with a speed of 1,470 m/s, and a frequency of 2,340 Hz.

Science Online red.msscience.com/self_check_quiz

section 2 review

1. The frequency increases as the wavelength decreases.
2. Waves having large amplitudes carry more energy than waves with small amplitudes.
3. Light travels much faster than sound.
4. The speed of light varies, depending on the medium through which it travels. The speed of light is about 300 million m/s in air.
5. The waves that make up bright green light have shorter wavelengths, higher frequencies, and larger amplitudes.
6. 340 m/s
7. 0.63 m

3 Assess

DAILY INTERVENTION

Check for Understanding

Visual-Spatial Have each student draw a wave at the top of a piece of paper. Below the first wave, draw an identical wave, but with a greater frequency. Next, draw a wave identical to the first, but with a smaller amplitude. Draw a wave identical to the first, but with a shorter wavelength. L2

Reteach

Parts of a Wave Have students use a spiral drawing toy to demonstrate the amplitude, wavelength, and frequency of a wave. L1 LS **Kinesthetic**

Assessment

Process Fill test tubes with different amounts of water and tap on them gently with a glass stirring rod. Have students explain the differences between the pitches produced. The more water in a tube, the shorter the wavelength of air that fits in the tube, so the higher the pitch produced when the glass is tapped. Use **Performance Assessment in the Science Classroom**, p. 89.

Virtual Labs

Waves *What are some characteristics of waves?*

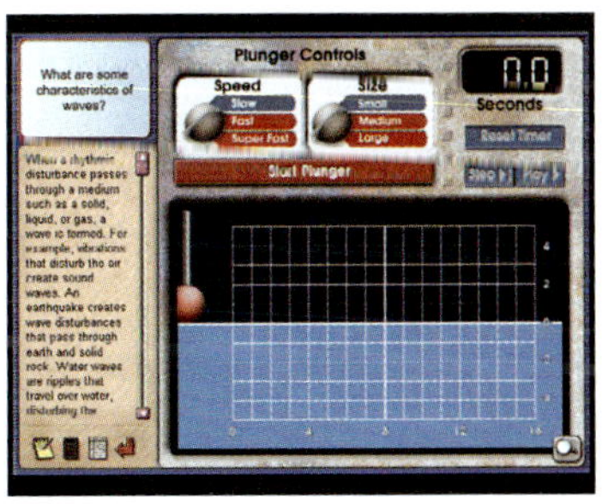

BENCH TESTED

Real-World Question

Purpose Students make waves with a coiled-spring toy and observe the waves. L2 ELL

COOP LEARN IS **Kinesthetic**

Process Skills observe and infer, compare and contrast, interpret data, classify, make and use tables

Time Required 45 minutes

Procedure

Safety Precautions Space student groups far enough apart that the coiled springs do not interfere with each other.

Teaching Strategies

- Help students compare the two types of waves they generate.
- Help students recognize the parts of a wave.

Troubleshooting Watch students to ensure that they do not tangle or overstretch the coiled springs.

Conclude and Apply

1. In steps 4 and 5, transverse waves were produced. In steps 6 to 8, compressional waves were produced.
2. rarefaction
3. The yarn moved back and forth from a fixed position, while the wave moved along the spring from one end to the other.

Assessment

Process Have students draw and label the parts of each type of wave they generated in the activity. Use **Performance Assessment in the Science Classroom**, p. 127.

LAB Waves on a Spring

Waves are rhythmic disturbances that carry energy through matter or space. Studying waves can help you understand how the Sun's energy reaches Earth and sounds travel through the air.

Real-World Question

What are some of the properties of transverse and compressional waves on a coiled spring?

Goals

- **Create** transverse and compressional waves on a coiled spring toy.
- **Investigate** wave properties such as speed and amplitude.

Materials

long, coiled spring toy
colored yarn (5 cm)
meterstick
stopwatch

Safety Precautions

WARNING: *Avoid overstretching or tangling the spring to prevent injury or damage.*

Procedure

1. Prepare a data table such as the one shown.

Wave Data	
Length of stretched spring toy	4.20 m
Average time for a wave to travel from end to end—step 4	1.35 s
Average time for a wave to travel from end to end—step 5	1.25 s

2. Work in pairs or groups and clear a place on an uncarpeted floor about 6 m × 2 m.
3. Stretch the springs between two people to the length suggested by your teacher. Measure the length.
4. Create a wave with a quick, sideways snap of the wrist. Time several waves as they travel the length of the spring. Record the average time in your data table.
5. Repeat step 4 using waves that have slightly larger amplitudes.
6. Squeeze together about 20 of the coils. Observe what happens to the unsqueezed coils. Release the coils and observe.
7. Quickly push the spring toward your partner, then pull it back.
8. Tie the yarn to a coil near the middle of the spring. Repeat step 7, observing the string.
9. **Calculate** and compare the speeds of the waves in steps 4 and 5.

Conclude and Apply

1. **Classify** the wave pulses you created in each step as compressional or transverse.
2. **Classify** the unsqueezed coils in step 6 as a compression or a rarefaction.
3. **Compare and contrast** the motion of the yarn in step 8 with the motion of the wave.

Communicating Your Data

Write a summary paragraph of how this activity demonstrated any of the vocabulary words from the first two sections of the chapter. **For more help, refer to the** Science Skill Handbook.

Communicating Your Data

Have students use a computer graphics program to create a display explaining the results of the activity.

section 3 Wave Behavior

Reflection

What causes the echo when you yell across an empty gymnasium or down a long, empty hallway? Why can you see your face when you look in a mirror? The echo of your voice and the face you see in the mirror are caused by wave reflection.

Reflection occurs when a wave strikes an object or surface and bounces off. An echo is reflected sound. Sound reflects from all surfaces. Your echo bounces off the walls, floor, ceiling, furniture, and people. You see your face in a mirror or a still pond, as shown in **Figure 11,** because of reflection. Light waves produced by a source of light such as the Sun or a lightbulb bounce off your face, strike the mirror, and reflect back to your eyes.

When a surface is smooth and even the reflected image is clear and sharp. However, **Figure 11** shows that when light reflects from an uneven or rough surface, you can't see a sharp image because the reflected light scatters in many different directions.

Reading Check *What causes reflection?*

as you read

What You'll Learn

- **Explain** how waves can reflect from some surfaces.
- **Explain** how waves change direction when they move from one material into another.
- **Describe** how waves are able to bend around barriers.

Why It's Important

The reflection of waves enables you to see objects around you.

Review Vocabulary

echo: the repetition of a sound caused by the reflection of sound waves

New Vocabulary

- reflection
- refraction
- diffraction
- interference

The smooth surface of a still pond enables you to see a sharp, clear image of yourself.

If the surface of the pond is rough and uneven, your reflected image is no longer clear and sharp.

Figure 11 The image formed by reflection depends on the smoothness of the surface.

1 Motivate

INTERACTIVE CHALKBOARD PowerPoint® Presentations

Bellringer

Section Focus Transparencies also are available on the Interactive Chalkboard CD-ROM. L2 ELL

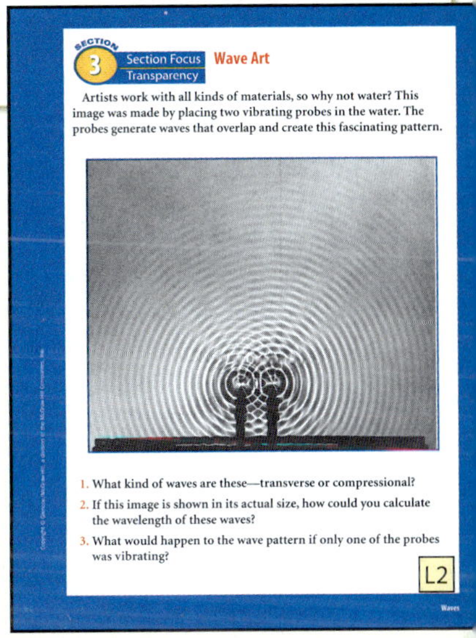

SECTION 3 Section Focus Transparency **Wave Art**

Artists work with all kinds of materials, so why not water? This image was made by placing two vibrating probes in the water. The probes generate waves that overlap and create this fascinating pattern.

1. What kind of waves are these—transverse or compressional?
2. If this image is shown in its actual size, how could you calculate the wavelength of these waves?
3. What would happen to the wave pattern if only one of the probes was vibrating?

L2

Waves

Tie to Prior Knowledge

Visible Light Remind students that visible light rays are electromagnetic waves. Discuss whether they can see themselves in a mirror without light rays. Explain that in this section, they will learn how waves reflect and refract.

Section 3 Resource Manager

Chapter FAST FILE Resources

Transparency Activity, p. 44
MiniLAB, p. 4
Directed Reading for Content Mastery, pp. 19, 20
Enrichment, p. 30
Lab Activity, pp. 13–14
Reinforcement, p. 27
Lab Worksheets, pp. 7–8

Lab Management and Safety, p. 65
Mathematics Skill Activities, p. 47

2 Teach

Reading Check

Answer the bouncing of a wave off an object.

Mini LAB

Purpose Students observe how light refracts. L1 ELL LS **Visual-Spatial**

Materials white soda straw, opaque drinking glass or cup, water

Teaching Strategy The larger the diameter of the glass or cup, the easier it will be to see the refraction. Also, tell students to fill the glass or cup almost to the top.

Analysis

1. The straw appears to be straight.
2. The straw appears to bend at the water's surface. Light reflects from the straw and refracts as it passes from the water into the air.

Assessment

Performance Have students draw ray diagrams in their Science Journals that show how the light waves refract. Use **Performance Assessment in the Science Classroom,** p. 127.

Try at Home

Mini LAB

Observing How Light Refracts

Procedure

1. Fill a **large, opaque drinking glass or cup** with **water.**
2. Place a **white soda straw** in the water at an angle.
3. Looking directly down into the cup from above, observe the straw where it meets the water.
4. Placing yourself so that the straw angles to your left or right, slowly back away about 1 m. Observe the straw as it appears above, at, and below the surface of the water.

Analysis

1. Describe the straw's appearance from above.
2. Compare the straw's appearance above and below the water's surface in step 4.

Try at Home

Refraction

A wave changes direction when it reflects from a surface. Waves also can change direction in another way. Perhaps you have tried to grab a sinking object when you are in a swimming pool, only to come up empty-handed. Yet you were sure you grabbed right where you saw the object. You missed grabbing the object because the light rays from the object changed direction as they passed from the water into the air. The bending of a wave as it moves from one medium into another is called **refraction.**

Refraction and Wave Speed Remember that the speed of a wave can be different in different materials. For example, light waves travel faster in air than in water. Refraction occurs when the speed of a wave changes as it passes from one substance to another, as shown in **Figure 12.** A line that is perpendicular to the water's surface is called the normal. When a light ray passes from air into water, it slows down and bends toward the normal. When the ray passes from water into air, it speeds up and bends away from the normal. The larger the change in speed of the light wave is, the larger the change in direction is.

You notice refraction when you look down into a fishbowl. Refraction makes the fish appear to be closer to the surface and farther away from you than it really is, as shown in **Figure 13.** Light rays reflected from the fish are bent away from the normal as they pass from water to air. Your brain interprets the light that enters your eyes by assuming that light rays always travel in straight lines. As a result, the light rays seem to be coming from a fish that is closer to the surface.

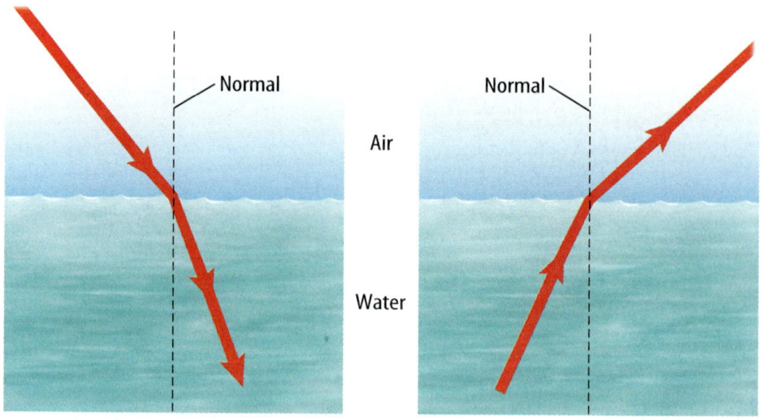

Figure 12 A wave is refracted when it changes speed.
Explain *how the direction of the light ray changes if it doesn't change speed.*

As the light ray passes from air to water, it bends toward the normal.

As the light ray passes from water to air, it bends away from the normal.

Cultural Diversity

An Advanced Warning System The Chinese used pottery jars to detect the sounds of an advancing enemy. Leather membranes were stretched over the mouths of empty 80-L pottery jars. The jars were lowered into deep shafts that were dug a few paces apart around the city, and men with good hearing were stationed nearby. Not only could they hear the sounds of an approaching army, but by listening to different sounds from the shafts, the watchers also could judge from which direction the enemy was coming and how far away the enemy was. Why do you think the jars were put in shafts deep underground? The thudding footsteps of the approaching army produced low-frequency, long-wavelength vibrations in the ground. Putting the jars deep in the ground filtered out higher frequency vibrations produced by other sources. L3 LS **Logical-Mathematical**

Color from Refraction Sunlight contains light of various wavelengths. When sunlight passes through a prism, refraction occurs twice: once when sunlight enters the prism, and again when it leaves the prism and returns to the air. Violet light has the shortest wavelength and is bent the most. Red light has the longest wavelength and is bent the least. Each color has a different wavelength and is refracted a different amount. As a result, the colors of sunlight are separated when they emerge from the prism.

Figure 14 shows how refraction produces a rainbow when light waves from the Sun pass into and out of water droplets. The colors you see in a rainbow are in order of decreasing wavelength: red, orange, yellow, green, blue, indigo and violet.

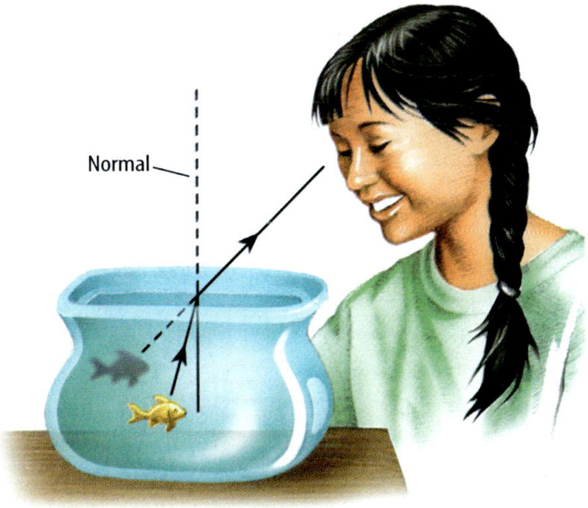

Figure 13 When you look at the goldfish in the water, the fish is in a different position than it appears.
Infer *how the location of the fish would change if light traveled faster in water than in air.*

Diffraction

Why can you hear music from the band room when you are down the hall? You can hear the music because the sound waves bend as they pass through an open doorway. This bending isn't caused by refraction. Instead, the bending is caused by diffraction. **Diffraction** is the bending of waves around a barrier.

Light waves can diffract, too. You can hear your friends in the band room but you can't see them until you reach the open door. Therefore, you know that light waves do not diffract as much as sound waves do. Light waves do bend around the edges of an open door. However, for an opening as wide as a door, the amount the light bends is extremely small. As a result, the diffraction of light is far too small to allow you to see around a corner.

Figure 14 Light rays refract as they enter and leave each water drop. Each color refracts at different angles because of their different wavelengths, so they separate into the colors of the visible spectrum.

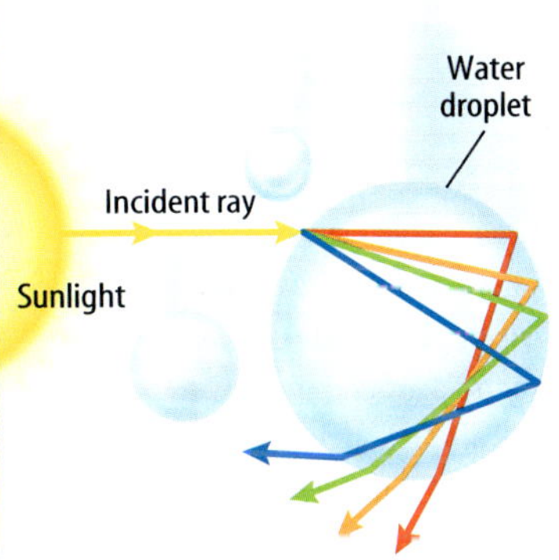

Visual Learning

Figure 14 When you look at the rainbow produced by the waterfall, you see an arc of colors. The refraction of light from water drops actually produces a circle of light, but we can only see the top of the circle because the ground blocks out the lower part. Red is at the top because, as you see in the sketch of the raindrop, red is bent less. Blue is at the bottom because it is bent more. L1
ELL LS **Visual-Spatial**

Quick Demo

Finger Slits

Materials bright light

Estimated Time five minutes

Procedure Tell the students to straighten the first two fingers of one hand, spreading the fingers slightly to make a small slit between them. Then have students view a bright light through the slit. They will see fringes or black lines between their fingers as a result of diffraction.

Teacher FYI

Speed of Light The speed of light in a material depends on the characteristics of the material and the frequency of the incident light. High-frequency, short-wavelength light, such as violet light, usually travels more slowly in a material than lower-frequency, longer-wavelength light, such as red light. This means the violet light bends more than the red light as it moves from a vacuum (or air) to a denser medium.

Differentiated Instruction

Visually Impaired Have students yell down a long hallway or in the empty gymnasium. Suggest they change positions until the loudest echoes are achieved. Discuss the reflected sound waves and explain that light reacts the same way when it is reflected to us from people or objects around us.

SECTION 3 Wave Behavior **239**

Discussion

Overlapping Waves Have students consider what happens when three waves overlap. If three waves interfere at a given point and have amplitudes of +3 units, +2 units, and −5 units, what is the amplitude of the resultant wave at that point? 0 units L3 LS **Logical-Mathematical**

Reading Check

Answer when the frequency of the wave and the barrier opening are similar in size

IDENTIFYING Misconceptions

Wave Changes Students may think that as waves move from one medium to another and are refracted, the wavelength and frequency of the waves change. However, the frequency doesn't change. Because the wave speed changes and the frequency stays the same the wavelength changes.

Diffraction and Wavelength The reason that light waves don't diffract much when they pass through an open door is that the wavelengths of visible light are much smaller than the width of the door. Light waves have wavelengths between about 400 and 700 billionths of a meter, while the width of doorway is about one meter. Sound waves that you can hear have wavelengths between a few millimeters and about 10 m. They bend more easily around the corners of an open door. A wave is diffracted more when its wavelength is similar in size to the barrier or opening.

Reading Check *Under what conditions would more diffraction of a wave occur?*

Diffraction of Water Waves Perhaps you have noticed water waves bending around barriers. For example, when water waves strike obstacles such as the islands shown in **Figure 15,** they don't stop moving. Here the size and spacing of the islands is not too different from the wavelength of the water waves. So the water waves bend around the islands, and keep on moving. They also spread out after they pass through openings between the islands. If the islands were much larger than the water wavelength, less diffraction would occur.

What happens when waves meet?

Suppose you throw two pebbles into a still pond. Ripples spread from the impact of each pebble and travel toward each other. What happens when two of these ripples meet? Do they collide like billiard balls and change direction? Waves behave differently from billiard balls when they meet. Waves pass right through each other and continue moving.

Figure 15 Water waves bend or diffract around these islands. More diffraction occurs when the object is closer in size to the wavelength.

LAB DEMONSTRATION

Purpose to observe wave behavior

Materials 2 pencils; paper; large, rectangular glass baking dish; water

Preparation Fill the dish with water to a depth of 2 cm. Use several books on each side to suspend the pan over a sheet of white paper. Place a strong light about 40 cm above the pan. Dim all other lights.

Procedure To form waves, vibrate a vertical pencil up and down on one side of the pan. Show interference by vibrating two vertical pencils, about 10 cm apart, up and down at one end of the pan. Place an object in the path of the waves to show diffraction.

Expected Outcome Students observe reflection, diffraction, and interference.

Assessment

What would the waves have looked like if you had held the pencil horizontally instead of vertically when vibrating it? They would have formed a line of waves instead of circles. L2

240 CHAPTER 8 Waves

Wave Interference While two waves overlap a new wave is formed by adding the two waves together. The ability of two waves to combine and form a new wave when they overlap is called **interference.** After they overlap, the individual waves continue to travel on in their original form.

The different ways waves can interfere are shown in **Figure 16** on the next page. Sometimes when the waves meet, the crest of one wave overlaps the crest of another wave. This is called constructive interference. The amplitudes of these combining waves add together to make a larger wave while they overlap. Destructive interference occurs when the crest of one wave overlaps the trough of another wave. Then, the amplitudes of the two waves combine to make a wave with a smaller amplitude. If the two waves have equal amplitudes and meet crest to trough, they cancel each other while the waves overlap.

Waves and Particles Like waves of water, when light travels through a small opening, such as a narrow slit, the light spreads out in all directions on the other side of the slit. If small particles, instead of waves, were sent through the slit, they would continue in a straight line without spreading. The spreading, or diffraction, is only a property of waves. Interference also doesn't occur with particles. If waves meet, they reinforce or cancel each other, then travel on. If particles approach each other, they either collide and scatter or miss each other completely. Interference, like diffraction, is a property of waves.

Science Online

Topic: Interference

Visit red.msscience.com for Web links to information about wave interference.

Activity Write a paragraph about three kinds of interference you found in your research.

Applying Science

Can you create destructive interference?

Your brother is vacuuming and you can't hear the television. Is it possible to diminish the sound of the vacuum so you can hear the TV? Can you eliminate some sound waves and keep the sounds you do want to hear?

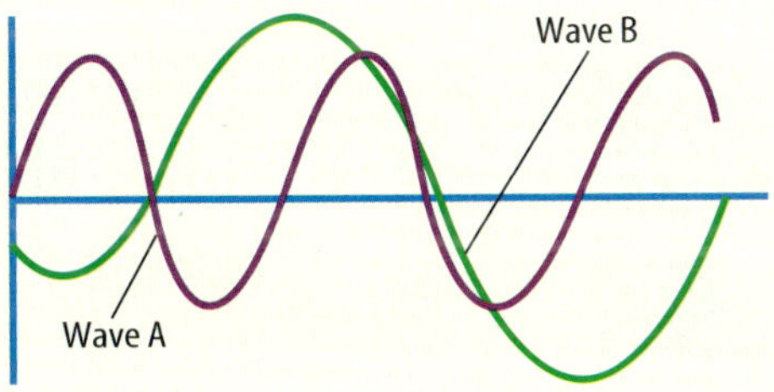

Identifying the Problem

It is possible to create a wave that will destructively interfere with one wave, but will not destructively interfere with another wave. The graph shows two waves with different wavelengths.

Solving the Problem

1. Create the graph of a wave that will eliminate wave A but not wave B.
2. Create the graph of a wave that would amplify wave A.

Discussion

Music Halls How does interference affect the way music sounds in different concert halls? Constructive and destructive interference patterns in a concert hall determine whether sound waves produced by individual voices or instruments combine to form waves that are pleasant or displeasing to listeners' ears. L3 LS **Logical-Mathematical**

Applying Science

Answers

1. The wave must be out of phase with the vacuum wave. That is, it must have a trough when the vacuum wave has a crest, and vice versa.
2. Any wave that has crests and troughs in the same place will amplify the wave.

Quick Demo

Refracted White Light

Materials prism, flashlight, darkened room

Estimated Time five minutes

Procedure Shine an intense beam of white light through a prism in a darkened room to demonstrate that different colors in white light refract at different angles.

Fun Fact

The rainbow pattern seen on the back of compact disks is a diffraction pattern that results from the reflection of the different wavelengths of visible light.

Differentiated Instruction

Challenge Have pairs of students prepare posters showing the types of waves produced by a combination of (1) two transverse waves in which crests overlap, (2) two transverse waves in which crests and troughs overlap, (3) two compressional waves in which crests overlap, and (4) two compressional waves in which crests and troughs overlap. L3 COOP LEARN LS **Visual-Spatial** P

NATIONAL GEOGRAPHIC

Visualizing Interference

Have students examine the pictures and read the captions. Then ask the following questions.

Two waves both with an amplitude of 2 m pass through each other. The maximum amplitude reached is 4 m. Is this constructive or destructive interference? constructive

Two waves both with an amplitude of 2 m approach each other and the peak of the crest of one wave passes through the bottom of the trough of the other wave. What is the amplitude of the resulting wave at that instant? zero

Activity

Rope Waves Have two students hold the ends of a long rope. Ask students to practice making waves with the rope by moving their hands vertically or horizontally in quick strokes. The students should then work together to set up constructive and destructive interference with the waves in their ropes. L3 LS **Kinesthetic**

Activity

Graphing Waves Help students understand how to graph the addition of waves by demonstrating it on an overhead projector. Overlap transparencies of drawings of waves as you add them together. L2 LS **Visual-Spatial**

NATIONAL GEOGRAPHIC **VISUALIZING INTERFERENCE**

Figure 16

Whether they are ripples on a pond or huge ocean swells, when water waves meet they can combine to form new waves in a process called interference. As shown below, wave interference can be constructive or destructive.

Constructive Interference

In constructive interference, a wave with greater amplitude is formed.

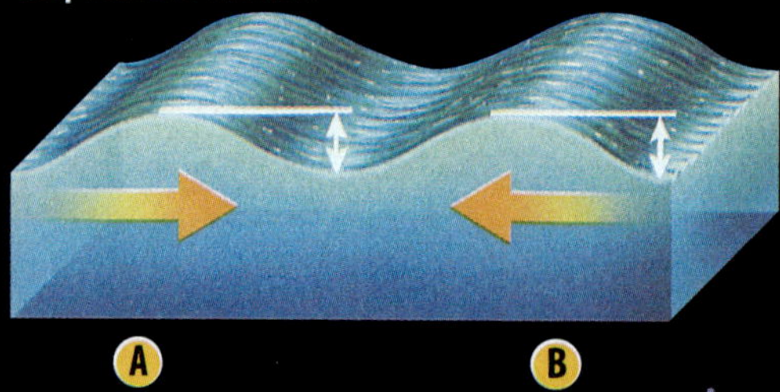

The crests of two waves—A and B—approach each other.

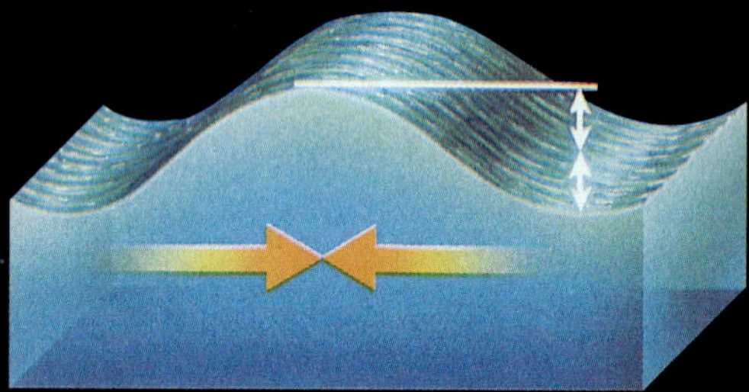

The two waves form a wave with a greater amplitude while the crests of both waves overlap.

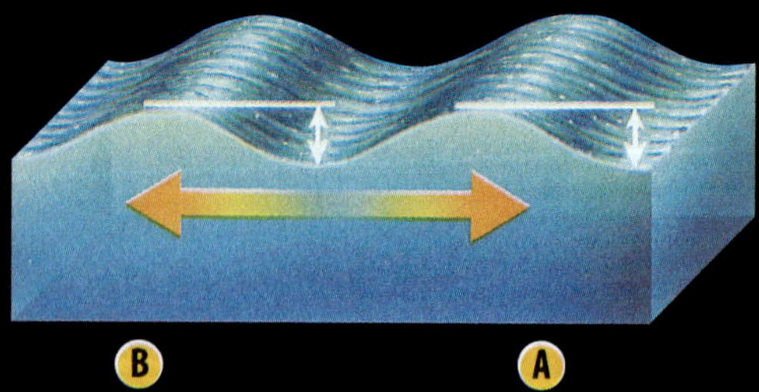

The original waves pass through each other and go on as they started.

Destructive Interference

In destructive interference, a wave with a smaller amplitude is formed.

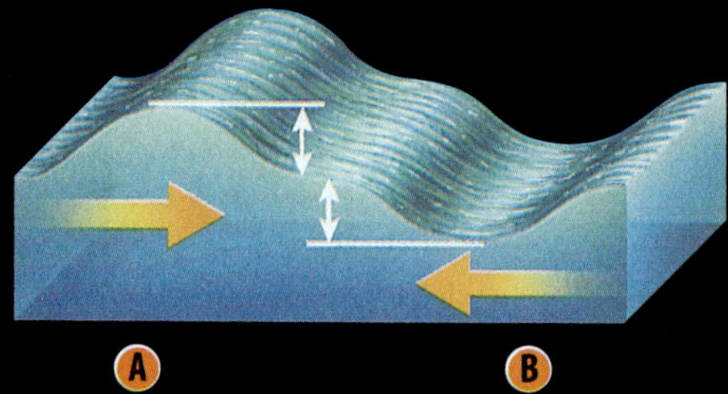

The crest of one wave approaches the trough of another.

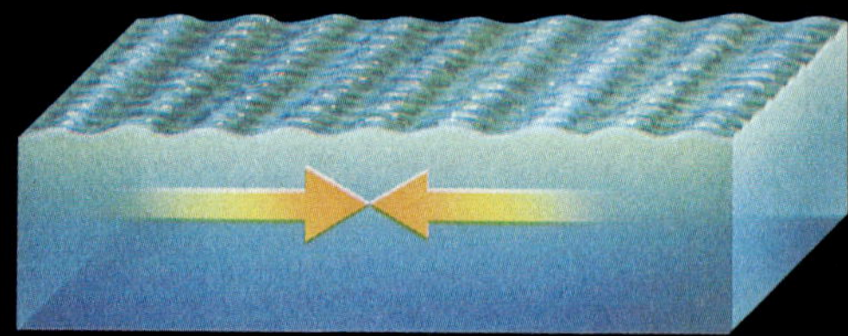

If the two waves have equal amplitude, they momentarily cancel when they meet.

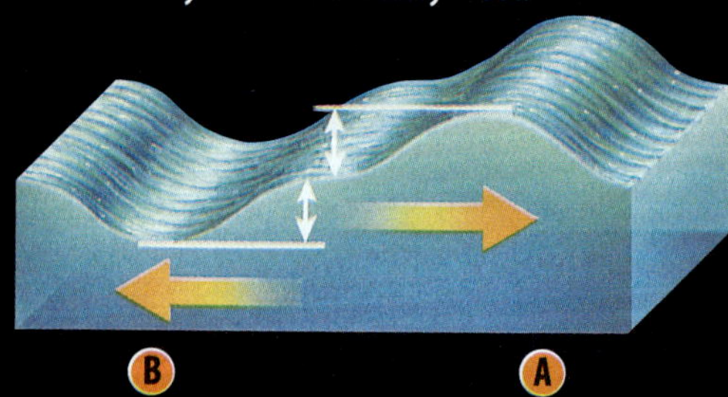

The original waves pass through each other and go on as they started.

Reducing Noise You might have seen someone use a power lawn mower or a chain saw. In the past, many people who performed these tasks damaged their hearing because of the loud noises produced by these machines.

Loud sounds have waves with larger amplitudes and carry more energy than softer sounds. The energy carried by loud sounds can damage cells in the ear that vibrate and transmit signals to the brain. Damage to the ear from loud sounds can be prevented by reducing the energy that reaches the ear. Ear protectors contain materials that absorb some of the energy carried by sound waves, so that less sound energy reaches the ear.

Pilots of small planes have a more complicated problem. If they shut out all the noise of the plane's motor, the pilots wouldn't be able to hear instructions from air-traffic controllers. To solve this problem, ear protectors have been developed, as shown in **Figure 17,** that have electronic circuits. These circuits detect noise from the aircraft and produce sound frequencies that destructively interfere with the noise. They do not interfere with human voices, so people can hear normal conversation. Destructive interference can be a benefit.

Figure 17 Some airplane pilots use special ear protectors that cancel out engine noise but don't block human voices.

section 3 review

Summary

Reflection

- Reflected sound waves can produce echoes.
- Reflected light rays produce images in a mirror.

Refraction

- The bending of waves as they pass from one medium to another is refraction.
- Refraction occurs when the wave's speed changes.
- A prism separates sunlight into the colors of the visible spectrum.

Diffraction and Interference

- The bending of waves around barriers is diffraction.
- Interference occurs when waves combine to form a new wave while they overlap.
- Destructive interference can reduce noise.

Self Check

1. **Explain** why you don't see your reflection in a building made of rough, white stone.
2. **Explain** how you are able to hear the siren of an ambulance on the other side of a building.
3. **Describe** the behavior of light that enables magnifying lenses and contact lenses to bend light rays.
4. **Define** the term *diffraction*. How does the amount of diffraction depend on wavelength?
5. **Think Critically** Why don't light rays that stream through an open window into a darkened room spread evenly through the entire room?

Applying Skills

6. **Compare and Contrast** When light rays pass from water into a certain type of glass, the rays refract toward the normal. Compare and contrast the speed of light in water and in the glass.

3 Assess

DAILY INTERVENTION

Check for Understanding

Logical-Mathematical Use student suggestions to create a list of ways in which radio waves are used. Responses could include radar, television, and short wave radios. L2

Reteach

Vocabulary Review Divide the class in half and conduct a baseball game review with words from this chapter. Players match words with meanings and are out when they miss a word. Each team has three outs before the turn shifts to the other team. L2 COOP LEARN IN Interpersonal

Assessment

Content Have students draw diagrams illustrating the refraction, reflection, diffraction, and interference of waves. Use **Performance Assessment in the Science Classroom,** p. 127. P

Activity

Wave Speed Have students research the speeds of light and sound. The speed of light is 3.00×10^8 m/s. The speed of sound varies with the type of medium and other conditions such as density, temperature, and pressure. L2

section 3 review

1. The rough surface scatters light in all directions.
2. Sound diffracts easily around objects such as buildings.
3. refraction
4. bending of waves around a barrier; A wave is diffracted more when its wavelength is similar in size to the barrier or opening.
5. The light waves have very small wavelengths, so very little diffraction occurs as they move through a wide opening such as a window.
6. Because the light rays refract toward the normal, their speed must be lower in the glass than in water.

BENCH TESTED

Real-World Question

Purpose Students will model the behavior of seismic waves by creating transverse standing waves on a demonstration spring or coiled toy spring. Factors such as length, type of spring, and frequency will be varied and controlled to determine how the speed of a wave within a coil depends upon such factors. By comparing their results with those of others, students can make generalizations about waves within a spring, and by analogy, seismic waves within Earth.

Process Skills measure, calculate, separate and control variables, predict

Time Required 45 minute class

Possible Materials Demonstration springs sold by educational supply companies create the best waves. The spring should be very long.

Alternate Materials If springs are not available, a length of hose or rope will make waves. A metric tape measure can replace a meterstick.

Safety Precautions Caution students not to use or swing the spring anywhere but on the floor.

Form a Hypothesis

Possible Hypothesis The hypotheses of most students will reflect that the speed of a wave is determined by dividing the distance the wave travels by the time it takes for the wave to travel one wavelength.

LAB Design Your Own

WAVE SPEED

Goals

- **Measure** the speed of a wave within a coiled spring toy.
- **Predict** whether the speed you measured will be different in other types of coiled spring toys.

Possible Materials

long, coiled spring toy
meterstick
stopwatch
tape
**clock with a second hand*

**Alternate materials*

Safety Precautions

Real-World Question

When an earthquake occurs, it produces waves that are recorded at points all over the world by instruments called seismographs. By comparing the data that they collected from these seismographs, scientists discovered that the interior of Earth must be made of layers of different materials. These data showed that the waves traveled at different speeds as they passed through different parts of Earth's interior. How can the speed of a wave be measured?

Form a Hypothesis

In some materials, waves travel too fast for their speeds to be measured directly. Think about what you know about the relationships among the frequency, wavelength, and speed of a wave in a medium. Make a hypothesis about how you can use this relationship to measure the speed of a wave within a medium. Explain why you think the experiment will support your hypothesis.

Test Your Hypothesis

Make a Plan

1. Make a data table in your Science Journal like the one shown.
2. In your Science Journal, write a detailed description of the coiled spring toy you are going to use. Be sure to include its mass and diameter, the width of a coil, and what it is made of.
3. **Decide** as a group how you will measure the frequency and length of waves in the spring toy. What are your variables? Which variables must be controlled? What variable do you want to measure?

244 CHAPTER 8 Waves

Alternative Inquiry Lab

Wave Speed Suggest students explore the speed of a mechanical wave in another medium. Provide a tank of water, stopwatch, meter stick, and a coiled spring toy. Students can complete the same experiment under water. The speed of the wave can be calculated and a similar table created. Students can observe the effects of changing the medium on the wave. L2

Using Scientific Methods

4. Repeat your experiment three times.

Follow Your Plan

1. Make sure your teacher approves your plan before you start.
2. Carry out the experiment.
3. While you are doing the experiment, record your observations and measurements in your data table.

Wave Data

	Trial 1	Trial 2	Trial 3
Length spring was stretched (m)	3.30	3.30	3.91
Number of crests	2	3	4
Wavelength (m)	3.30	2.20	1.96
# of vibrations timed	10	10	10
# of seconds vibrations were timed	5.49	3.60	6.01
Wave speed (m/s)	6.01	6.12	6.29

Analyze Your Data

1. **Calculate** the frequency of the waves by dividing the number of vibrations you timed by the number of seconds you timed them. Record your results in your data table.
2. Use the following formula to calculate the speed of a wave in each trial.

   wavelength $\times$ frequency $=$ wave speed
3. Average the wave speeds from your trials to determine the average speed of a wave in your coiled spring toy.

Conclude and Apply

1. **Infer** which variables affected the wave speed in spring toys the most. Which variables affected the speed the least? Was your hypothesis supported?
2. **Analyze** what factors caused the wave speed measured in each trial to be different.

Communicating Your Data

Post a description of your coiled spring toy and the results of your experiment on a bulletin board in your classroom. **Compare and contrast** your results with other students in your class.

Test Your Hypothesis

Possible Procedures Help your student groups create standing waves. While a student holds one end of the coiled-spring toy stationary, have another student move the other end from side to side with a rhythm that produces a standing wave.

Tie to Prior Knowledge Most students will have observed waves moving through a medium such as ocean waves or ripples on a pond.

Expected Outcome Students will create standing transverse waves and measure their frequency, wavelengths, and speeds. From this data, they will predict the speed of waves in other types of coiled springs.

Analyze Your Data

Answers to Questions

1. Based on the sample data in the table, frequencies were as follows: Trial 1, 1.82 Hz; Trial 2, 2.78 Hz; Trial 3, 3.21 Hz.
2. Based on the sample data in the table, speeds were as follows: Trial 1, 6.00 m/s; Trial 2, 6.12 m/s; Trial 3, 6.29 m/s.
3. Average wave speed is 6.14 m/s.

Error Analysis Ask student groups to compare data with other groups and explain any significant discrepancies.

Conclude and Apply

1. Answers will vary.
2. Possible answer: how fast the spring toy was shaken

Assessment

Oral Ask students how they might determine the speed of water waves that are crashing ashore at a beach. Time the waves coming ashore to determine the frequency. Estimate the distance between wave crests to determine wavelength. The product of these two numbers is the speed of the wave. Use **Performance Assessment in the Science Classroom,** p. 89 L2

Communicating Your Data

Encourage students to record their data in electronically designed spreadsheets.

SCIENCE Stats

Content Background

Electromagnetic, ocean, and sound are different types of waves. Each of these waves propagate in a different manner. Electromagnetic waves do not require a medium for travel and can carry energy through the vacuum of space. Surface ocean waves are a mixture of longitudinal and transverse waves that carry energy through the water. Sound waves are longitudinal waves that can carry energy through solids, liquids, and gases. The velocity of waves can be determined by the equation velocity = wavelength × frequency.

Discussion

Wave Descriptions Waves are described by their frequency, wavelength, amplitude, and speed. What do the numbers represent in each of the examples? 34 meters, amplitude; 900 km/h, speed; 800 pulses per second, frequency L3

Applying Math

Answer 5 hours

Graph It

Radio Waves Have students find the frequency of radio waves. Are these waves low-frequency or high-frequency waves in the electromagnetic spectrum? low

SCIENCE Stats

Waves, Waves, and More Waves

Did you know...

. . . Radio waves from space were discovered in 1932 by Karl G. Jansky, an American engineer. His discovery led to creation of radio astronomy, a field that explores parts of the universe that can't be seen with telescopes.

. . . The highest recorded ocean wave was 34 meters high, which is comparable to the height of a ten-story building. This super wave was seen in the North Pacific Ocean and recorded by the crew of the naval ship *USS Ramapo* in 1933.

Applying Math A tsunami formed by an earthquake on the ocean floor travels at 900 km/h. How long will it take the tsunami to travel 4,500 km?

. . . Waves let dolphins see with their ears! A dolphin sends out ultrasonic pulses, or clicks, at rates of 800 pulses per second. These sound waves are reflected back to the dolphin after they hit an obstacle or a meal. This process is called echolocation.

Graph It

Go to red.msscience.com/science_stats to learn about discoveries by radio astronomers. Make a time line showing some of these discoveries.

Curriculum Connection

Geography Earthquakes produce seismic waves that travel through the ground and can do a great deal of damage. Have students research earthquakes around the world and mark the location on a map. Discuss locations that are more likely to experience earthquakes. L2

Reviewing Main Ideas

Section 1 What are waves?

1. Waves are rhythmic disturbances that carry energy but not matter.
2. Mechanical waves can travel only through matter. Electromagnetic waves can travel through matter and space.
3. In a mechanical transverse wave, matter in the medium moves back and forth at right angles to the direction the wave travels.
4. In a compressional wave, matter in the medium moves forward and backward in the same direction as the wave.

Section 2 Wave Properties

1. The amplitude of a transverse wave is the distance between the rest position and a crest or a trough.
2. The energy carried by a wave increases as the amplitude increases.
3. Wavelength is the distance between neighboring crests or neighboring troughs.
4. The frequency of a wave is the number of wavelengths that pass a given point in 1 s.
5. Waves travel through different materials at different speeds.

Section 3 Wave Behavior

1. Reflection occurs when a wave strikes an object or surface and bounces off.
2. The bending of a wave as it moves from one medium into another is called refraction. A wave changes direction, or refracts, when the speed of the wave changes.
3. The bending of waves around a barrier is called diffraction.
4. Interference occurs when two or more waves combine and form a new wave while they overlap.

Visualizing Main Ideas

Copy and complete the following spider map about waves.

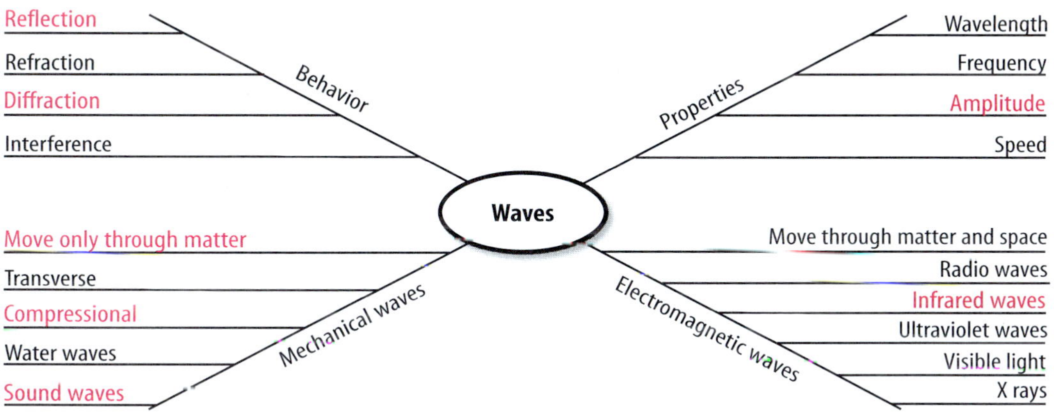

Science Online red.msscience.com/interactive_tutor

chapter 8 Study Guide

Reviewing Main Ideas

Summary statements can be used by students to review the major concepts of the chapter.

Visualizing Main Ideas

See student page.

Science Online

Visit red.msscience.com
/self_check_quiz
/interactive_tutor
/vocabulary_puzzlemaker
/chapter_review
/standardized_test

Assessment Transparency

For additional assessment questions, use the *Assessment Transparency* located in the transparency book.

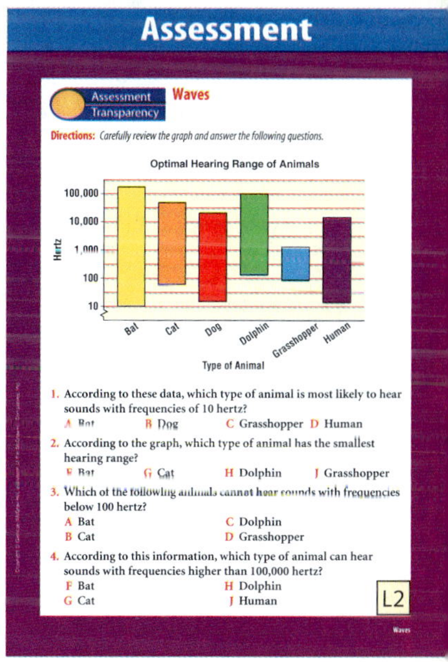

Assessment

Assessment Transparency **Waves**

Directions: *Carefully review the graph and answer the following questions.*

1. According to these data, which type of animal is most likely to hear sounds with frequencies of 10 hertz?
   A Bat B Dog C Grasshopper D Human
2. According to the graph, which type of animal has the smallest hearing range?
   F Bat G Cat H Dolphin J Grasshopper
3. Which of the following animals cannot hear sounds with frequencies below 100 hertz?
   A Bat C Dolphin
   B Cat D Grasshopper
4. According to this information, which type of animal can hear sounds with frequencies higher than 100,000 hertz?
   F Bat H Dolphin
   G Cat J Human

L2

CHAPTER STUDY GUIDE 247

Using Vocabulary

1. Refraction
2. compressional wave
3. wavelength
4. amplitude
5. electromagnetic wave

Checking Concepts

6. D
7. C
8. B
9. B
10. A
11. D
12. A
13. C
14. D

Thinking Critically

15. Compressional; each car is pushed into the next. The cars then rebound. This is like a compressional wave.
16. Yes; yes; electromagnetic waves can travel through space and matter.
17. The frequency decreases because fewer wavelengths pass a given point in 1 s.
18. because the rough surface reflects the light rays in many different directions
19. Light travels much faster than sound, so the flash of the cannon will reach you before the sound will.

chapter 8 Review

Using Vocabulary

amplitude p. 231	interference p. 241
compressional wave p. 229	mechanical wave p. 227
diffraction p. 239	reflection p. 237
electromagnetic wave p. 230	refraction p. 238
frequency p. 233	transverse wave p. 228
	wave p. 226
	wavelength p. 232

Fill in the blanks with the correct word or words.

1. _________ is the change in direction of a wave going from one medium to another.
2. The type of wave that has rarefactions is a _________.
3. The distance between two adjacent crests of a transverse wave is the _________.
4. The more energy a wave carries, the greater its _________ is.
5. A(n) _________ can travel through space without a medium.

Checking Concepts

Choose the word or phrase that best answers the question.

6. What is the material through which mechanical waves travel?
   A) charged particles
   B) space
   C) a vacuum
   D) a medium
7. What is carried from particle to particle in a water wave?
   A) speed **C)** energy
   B) amplitude **D)** matter
8. What are the lowest points on a transverse wave called?
   A) crests **C)** compressions
   B) troughs **D)** rarefactions
9. What determines the pitch of a sound wave?
   A) amplitude **C)** speed
   B) frequency **D)** refraction
10. What is the distance between adjacent wave compressions?
    A) one wavelength
    B) 1 km
    C) 1 m/s
    D) 1 Hz
11. What occurs when a wave strikes an object or surface and bounces off?
    A) diffraction
    B) refraction
    C) a transverse wave
    D) reflection
12. What is the name for a change in the direction of a wave when it passes from one medium into another?
    A) refraction **C)** reflection
    B) interference **D)** diffraction

Use the figure below to answer question 13.

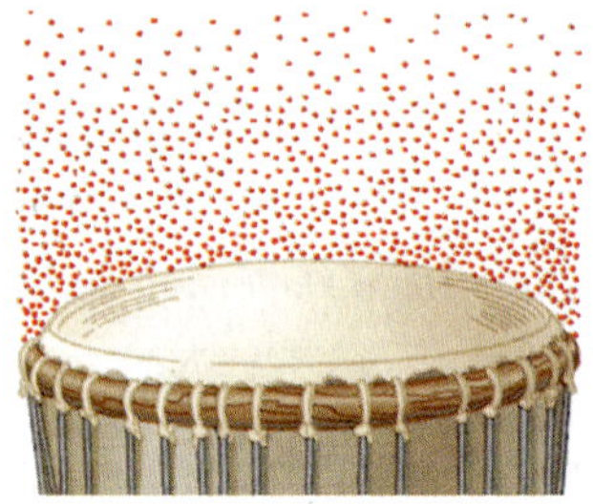

13. What type of wave is a sound wave?
    A) transverse
    B) electromagnetic
    C) compressional
    D) refracted
14. What color light has the shortest wave-length and the highest frequency?
    A) red **C)** Orange
    B) green **D)** Blue

Science Online red.msscience.com/vocabulary_puzzlemaker

Use the Exam*View*® Pro Testmaker CD-ROM to:

- create multiple versions of tests
- create modified tests with one mouse click for inclusion students
- edit existing questions and add your own questions
- build tests aligned with state standards using built-in State Curriculum Tags
- change English tests to Spanish with one mouse click and vice versa

Thinking Critically

15. **Explain** what kind of wave—transverse or compressional—is produced when an engine bumps into a string of coupled railroad cars on a track.

16. **Infer** Is it possible for an electromagnetic wave to travel through a vacuum? Through matter? Explain your answers.

17. **Draw a Conclusion** Why does the frequency of a wave decrease as the wavelength increases?

18. **Explain** why you don't see your reflected image when you look at a white, rough surface?

19. **Infer** If a cannon fires at a great distance from you, why do you see the flash before you hear the sound?

20. **Form a Hypothesis** Form a hypothesis that can explain this observation. Waves A and B travel away from Earth through Earth's atmosphere. Wave A continues on into space, but wave B does not.

Use the figure below to answer question 21.

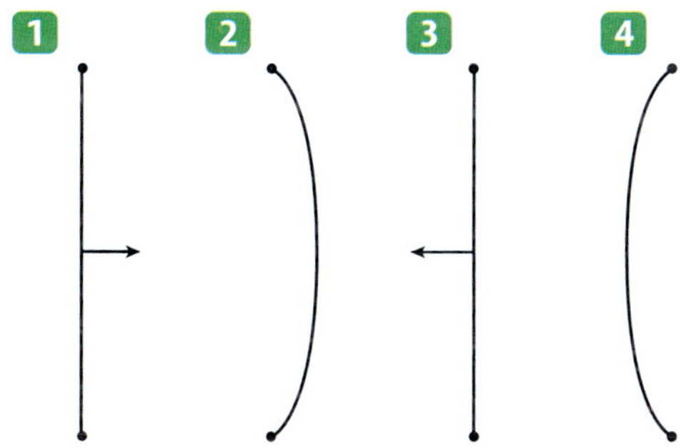

21. **Explain** how the object shown above causes compressions and rarefactions as it vibrates in air.

22. **Explain** why you can hear a person talking even if you can't see them.

23. **Compare and Contrast** AM radio waves have wavelengths between about 200 m and 600 m, and FM radio waves have wavelengths of about 3 m. Why can AM radio signals often be heard behind buildings and mountains but FM radio signals cannot?

24. **Infer** how the wavelength of a wave would change if the speed of the wave increased, but the frequency remained the same.

25. **Explain** You are motionless on a rubber raft in the middle of a pool. A friend sitting on the edge of the pool tries to make the float move to the other edge of the pool by slapping the water every second to form a wave. Explain whether the wave produced will cause you to move to the edge of the pool.

Performance Activities

26. **Make Flashcards** Work with a partner to make flashcards for the bold-faced terms in the chapter. Illustrate each term on the front of the cards. Write the term and its definition on the back of the card. Use the cards to review the terms with another team.

Applying Math

Use the following equation to answer questions 27–29.

wave speed = wavelength × frequency

27. **Wave Speed** If a wave pool generates waves with a wavelength of 3.2 m and a frequency of 0.60 Hz, how fast are the waves moving?

28. **Frequency** An earthquake wave travels at 5000 m/s and has a wavelength of 417 m. What is its frequency?

29. **Wavelength** A wave travels at a velocity of 4 m/s. It has a frequency of 3.5 Hz. What is the wavelength of the wave?

20. Wave A is an electromagnetic wave and wave B is a mechanical wave.
21. The vibrating object moves back and forth, colliding with the molecules in the air around it. As the object moves forward, it pushes molecules together to form a compression. As the object moves back toward its original position and beyond, it creates a rarefaction—a region in which molecules are spread farther apart.
22. Sound waves have longer wavelengths than light waves and will diffract around large objects, while light waves will not.
23. Because their longer wavelengths are closer to the size of buildings and mountains, AM radio waves diffract around them more than FM radio waves do.
24. The wavelength would increase.
25. No, the wave will cause you to move up and down, but you will not move forward.

Performance Activities

26. Use **Performance Assessment in the Science Classroom**, p. 127.

Applying Math

National Math Standards

1, 2, 9

27. 1.9 m/s
28. 12 Hz
29. 1.1 m

Assessment Resources

Reproducible Masters

Chapter *Fast File* Resources
- Chapter Review, pp. 35–36
- Chapter Tests, pp. 37–40
- Assessment Transparency Activity, p. 47

Glencoe Science Web site
- Chapter Review Test
- Standardized Test Practice

Glencoe Technology
- Assessment Transparency
- Exam*View*® Pro Testmaker
- MindJogger Videoquiz
- Interactive Chalkboard

FAST FILE

Answer Sheet A practice answer sheet can be found at red.msscience.com/answer_sheet.

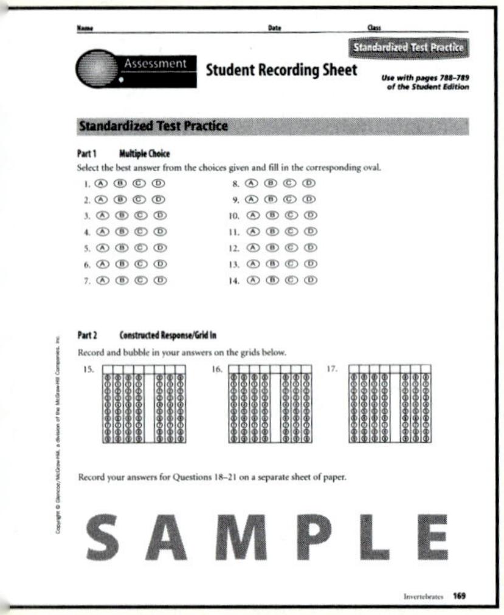
Student Recording Sheet

Standardized Test Practice

Part 1 Multiple Choice

Select the best answer from the choices given and fill in the corresponding oval.

Part 2 Constructed Response/Grid In

Record your answers for Questions 18–21 on a separate sheet of paper.

SAMPLE

Part 1 Multiple Choice

1. B
2. C
3. B
4. C
5. C
6. D
7. A
8. A
9. B
10. C
11. B

Part 2 Short Response

12. No, the water does not travel with the wave. The waves only transfer energy, not matter. So the water that hits the island was already near the shore.
13. Wave A has a longer wavelength and a lower frequency than Wave B.
14. Both waves are traveling at the same speed because the speed is the same in a given medium.
15. To make waves of a longer wavelength, you would dip your hand into the water less often. To increase the amplitude of the waves, you could dip your hand in deeper with more force.

chapter 8 Standardized Test Practice

Part 1 Multiple Choice

Record your answers on the answer sheet provided by your teacher or on a sheet of paper.

1. What do waves carry as they move?
   - A. matter
   - B. energy
   - C. matter and energy
   - D. particles and energy

Use the figure below to answer questions 2 and 3.

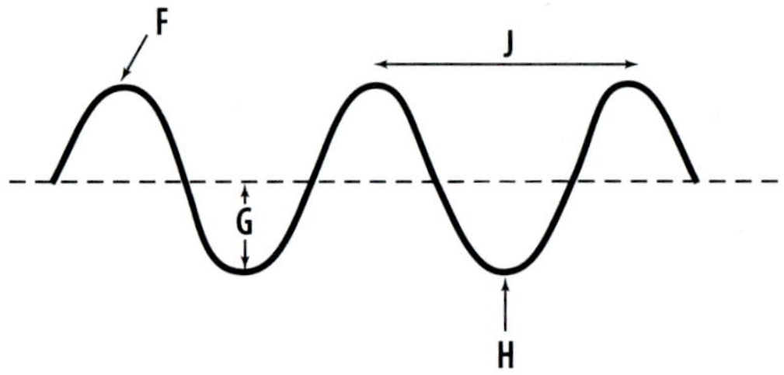

2. What property of the wave is shown at F?
   - A. amplitude
   - B. wavelength
   - C. crest
   - D. trough
3. What property of the wave is shown at J?
   - A. amplitude
   - B. wavelength
   - C. crest
   - D. trough
4. What kind of wave does NOT need a medium through which to travel?
   - A. mechanical
   - B. sound
   - C. light
   - D. refracted
5. What happens as a sound wave's energy decreases?
   - A. Wave frequency decreases.
   - B. Wavelength decreases.
   - C. Amplitude decreases.
   - D. Pitch decreases.
6. What unit is used to measure frequency?
   - A. meters
   - B. meters/second
   - C. decibels
   - D. hertz
7. What properties of a light wave determines its color?
   - A. wavelength
   - B. amplitude
   - C. speed
   - D. interference
8. When two waves overlap and interfere constructively, what does the resulting wave have?
   - A. a greater amplitude
   - B. less energy
   - C. a change in frequency
   - D. a lower amplitude
9. What happens when light travels from air into glass?
   - A. It speeds up.
   - B. It slows down.
   - C. It travels at 300,000 km/s.
   - D. It travels at the speed of sound.

Use the figure below to answer questions 10 and 11.

10. What behavior of light waves lets you see a sharp, clear image of yourself?
    - A. refraction
    - B. diffraction
    - C. reflection
    - D. interference
11. Why can't you see a clear image of yourself if the water's surface is rough?
    - A. The light bounces off the surface in only one direction.
    - B. The light scatters in many different directions.
    - C. There is no light shining on the water's surface.
    - D. The light changes speed when it strikes the water.

16. All electromagnetic waves are transverse waves that carry energy and can travel through outer space or through a vacuum where no matter exists. Different electromagnetic waves have different wavelengths and frequencies.

Part 2 | Short Response/Grid In

Record your answers on the answer sheet provided by your teacher or on a sheet of paper.

12. An earthquake in the middle of the Indian Ocean produces a tsunami that hits an island. Is the water that hits the island the same water that was above the place where the earthquake occurred? Explain.

Use the figure below to answer questions 13 and 14.

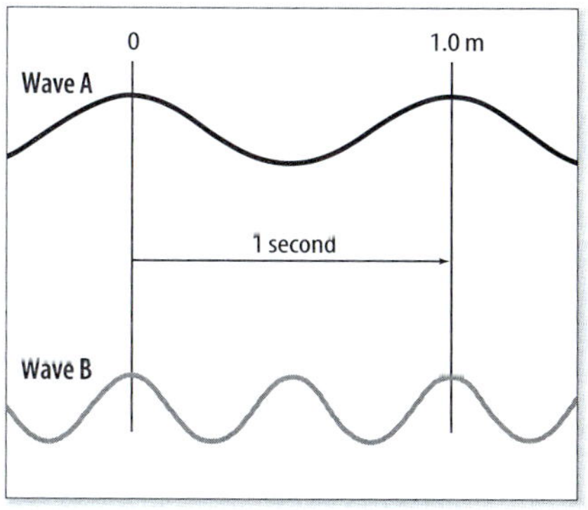

13. Compare the wavelengths and frequencies of the two waves shown.

14. If both waves are traveling through the same medium, how do their speeds compare? Explain.

15. Suppose you make waves in a pond by dipping your hand in the water with a frequency of 1 Hz. How could you make waves of a longer wavelength? How could you increase the amplitude of the waves?

16. How are all electromagnetic waves alike? How do they differ from one another?

Test-Taking Tip

Take Your Time Stay focused during the test and don't rush, even if you notice that other students are finishing the test early.

Part 3 | Open Ended

Record your answers on a sheet of paper.

Use the figure below to answer questions 17 and 18.

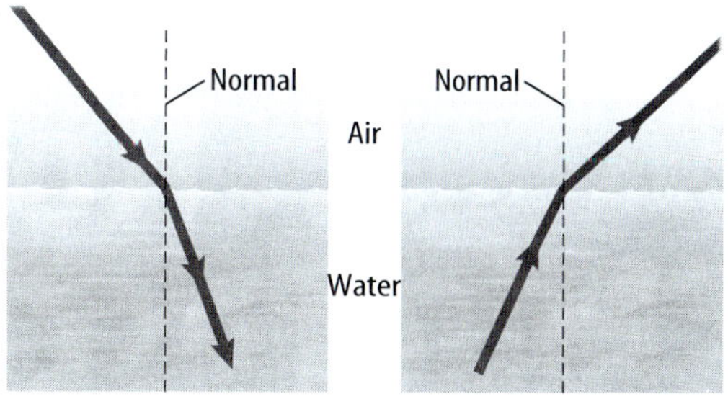

17. Why does the light ray bend toward the normal when is passes from air into water, but bend away from the normal as it passes from water into air?

18. A boy has caught a fish on his fishing line. He reels the fish in near the boat. How could the refraction of light waves affect him as he tries to net the fish while it is still in the water?

19. In a science fiction movie, a spaceship explodes. The people in a nearby spaceship see and hear the explosion. Is this realistic? Explain.

20. The speed of light in warm air is greater than its speed in cold air. The air just above a highway is warmer than the air a little higher. Will the light moving parallel to the highway be bent up or down? Explain.

21. You are standing outside a classroom with an open door. You know your friends are in the room because you can hear them talking. Explain why you can hear them talking but cannot see them.

22. How does the size of an obstacle affect the diffraction of a wave?

Part 3 | Open Ended

17. Light waves travel faster in air than in water. When a light ray passes from air into water, it slows down and bends toward the normal. When the ray passes from water into air, it speeds up and bends away from the normal.

18. The boy is looking into the water from air, so the light rays are refracted. But his brain interprets the light rays coming from the fish as traveling in a straight line. So the light rays seem to be coming from a fish that is in a different location than it really is. So the boy might miss the fish with the net.

19. No, sound needs a medium through which to travel. It cannot travel through outer space where there is no matter. So sound waves would not reach the nearby spaceship and the people would not hear the explosion. They would, however, see the explosion because light waves can travel through space.

20. The light will bend toward the region where light has a slower speed. This means that the light rays would be bent up.

21. You can hear the people talking because the sound waves bend as they pass through the open door. Sound waves are similar in size to the open door. Light waves have much smaller wavelengths than the size of the open door, so they are not diffracted as much. Therefore, you cannot see your friends.

22. More diffraction or bending of the wave occurs if the wavelength is similar to the size of the obstacle.

Rubrics

For more help evaluating open-ended assessment questions, see the rubric on p. 10T

unit 3

Unit Contents

WebQuest *Mars Rocks!* is an investigation challenging students to learn about rocks on Mars and the space missions to Mars of the past, present, and future. Students will discover the evidence that leads scientists to think life may have once existed there. They will learn about meteorites identified as coming from Mars, and plans for a mission to Mars that will collect rocks and bring them back for analysis. A set of prepared questions will guide students on their mission.

unit 3

Earth's Changing Surface

How Are Rocks & Fluorescent Lights Connected?

PROJECT CRISS

Study Skills

Active Reading Most readers have questions about the material they read. Encourage students to write down questions about rocks, minerals, energy, or any difficult vocabulary they encounter in this unit. Discuss probable answers.

NATIONAL GEOGRAPHIC

Around 1600, an Italian cobbler found a rock that contained a mineral that could be made to glow in the dark. The discovery led other people to seek materials with similar properties. Eventually, scientists identified many fluorescent and phosphorescent (fahs fuh RE sunt) substances—substances that react to certain forms of energy by giving off their own light. As seen above, a fluorescent mineral may look one way in ordinary light (front), but may give off a strange glow (back) when exposed to ultraviolet light. In the 1850s, a scientist wondered whether the fluorescent properties of a substance could be harnessed to create a new type of lighting. The scientist put a fluorescent material inside a glass tube and sent an electric charge through the tube, creating the first fluorescent lamp. Today, fluorescent lightbulbs are widely used in office buildings, schools, and factories.

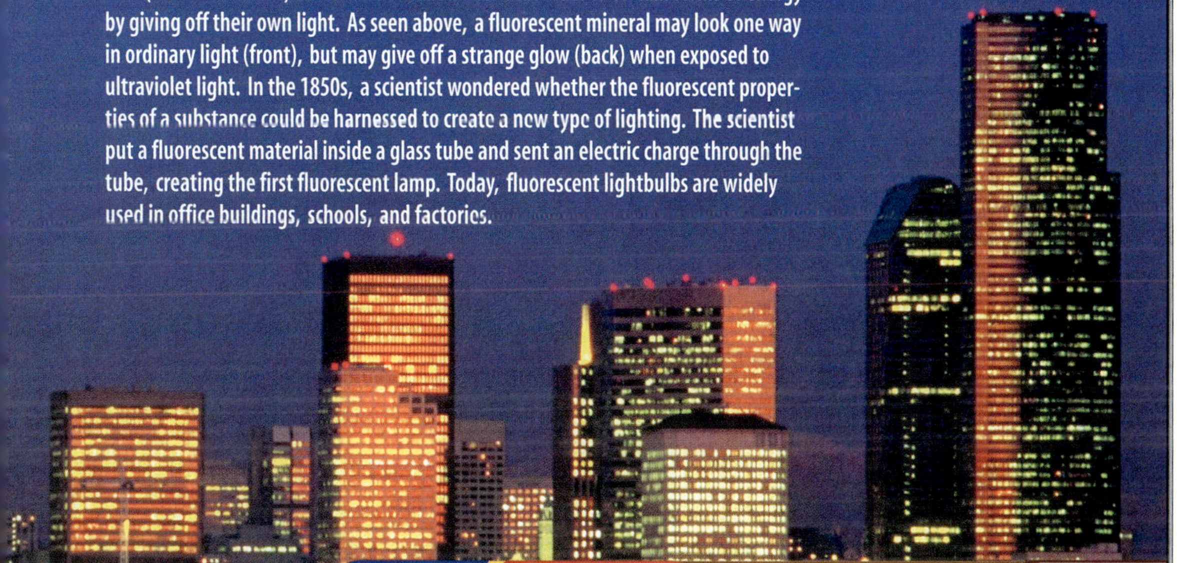

unit projects

Visit red.msscience.com/unit_project to find project ideas and resources.
Projects include:

- **History** Research the chemist/industrialist that created dynamite and established the Nobel Prizes.
- **Technology** Analyze soil properties from horizon profile, temperature, texture, particle size, and more. Post your results on the Web.
- **Model** Research and prepare to represent a rock sample in a classroom rock debate using a variety of rock characteristics.

WebQuest *Mars Rocks!* is an interactive investigation of Mars, and the evidence that leads scientists to believe that life may have once existed there.

NATIONAL GEOGRAPHIC How Are Rocks & Fluorescent Lights Connected?

There are many types of luminescence. Bioluminescence is produced by living organisms, and various fungi and bacteria. Chemiluminescence is produced by certain chemical reactions. Electroluminescence is produced by electric discharges. Fluorescence, emitted from a substance under stimulation by light, has been observed for hundreds of years.

Scientific investigation of luminescence did not begin until 1603, when Vincenzo Casariolo, a Bolognian shoemaker and alchemist, accidentally prepared an artificial phosphor that glowed on its own after exposure to light. In 1901 the first fluorescent lighting was developed.

unit projects

History Working with a partner, have students research the life of Alfred Nobel, a Swedish chemist with personal interests in poetry and drama. Have students write and present a poem or skit depicting Nobel's life as "Europe's richest vagabond." They should include information about his family, education, business endeavors, and the final will that set up the Nobel Prizes in Physics, Chemistry, Literature, Peace, and Physiology or Medicine.

Technology As a class, ask students to identify a cubic meter soil sample location. Have students observe and measure their analysis location and each of their samples for the following soil properties: horizon profile, infiltration, soil temperature, structure (shape of soil aggregates), consistence (resistance to breaking), texture, particle size, soil pH, fertility, and soil moisture. Have students post their data on the Web. For detailed standards, procedures, and to post results visit red.msscience.com/unit_project

Model Provide students with a rock name and sample and have them research information on their particular type of rock. Students present research in the following categories: use in construction, weathering, leisure and entertainment value, beauty and art, currency, energy production and use, tools and weapons, manufacturing, and food and drink. A class chart may be compiled by the students.

Additional Resources For more information, resources, and assessment rubrics, visit red.msscience.com/unit_project

chapter 9 Organizer

Section/Objectives	Standards		Labs/Features
	National	State/Local	
Chapter Opener	See pp. 16T–17T for a Key to Standards.		**Launch Lab:** Observe a Rock, p. 255 **Foldables,** p. 255
Section 1 Minerals—Earth's Jewels 2 sessions 1 block 1. **Identify** the difference between a mineral and a rock. 2. **Describe** the properties that are used to identify minerals.	National Content Standards: UCP.1–UCP.3, UCP.5, A.1, A.2, D.1, D.2, G.3		**Integrate Life Science,** p. 257 **MiniLAB:** Classifying Minerals, p. 261 **Applying Science:** How hard are these minerals?, p. 261 **Science Online,** p. 262
Section 2 Igneous and Sedimentary Rocks 2 sessions 1 block 3. **Explain** how extrusive and intrusive igneous rocks are different. 4. **Describe** how different types of sedimentary rocks form.	National Content Standards: UCP.1–UCP.3, UCP.5, A.1, A.2, B.2, B.3, D.1, D.2		**Integrate Social Studies,** p. 266 **Integrate Physics,** p. 267 **Visualizing Igneous Rock Features,** p. 268 **MiniLAB:** Modeling How Fossils Form Rocks, p. 270
Section 3 Metamorphic Rocks and the Rock Cycle 3 sessions 1.5 blocks 5. **Describe** the conditions needed for metamorphic rocks to form. 6. **Explain** how all rocks are linked by the rock cycle.	National Content Standards: UCP.1–UCP.3, UCP.5, A.1, A.2, B.2, B.3, D.1, D.2		**Science Online,** p. 274 **Lab:** Gneiss Rice, p. 277 **Lab:** Classifying Minerals, pp. 278–279 **Oops! Accidents in Science:** Going for the Gold, p. 280

Glencoe Exclusive! TeacherWorks™ All-In-One Planner and Resource Center

Lab Materials	Reproducible Resources	Section Assessment	Technology
Launch Lab: sparkling rock, hand lens	**Chapter *Fast File* Resources** Foldables Worksheet, p. 17 Directed Reading Overview, p. 19 Note-taking Worksheets, pp. 33–35	Glencoe's Assessment Advantage	TeacherWorks includes: • Interactive Teacher Edition • Lesson Planner with calendar • Access to all program blacklines • Correlations to standards • Web links
MiniLAB: magnet; samples of quartz, calcite, hornblende, magnetite; magnet, dropper, dilute HCl, water **Need materials?** Contact Science Kit at 1-800-828-7777 or www.sciencekit.com on the Internet.	**Chapter *Fast File* Resources** Transparency Activity, p. 44 MiniLAB, p. 3 Enrichment, p. 30 Reinforcement, p. 27 Directed Reading, p. 20 Lab Activities, pp. 9–11 **Mathematics Skill Activities,** p. 1 **Cultural Diversity,** p. 35	**Portfolio** Science Journal, p. 262 **Performance** MiniLAB, p. 261 Applying Science, p. 261 Applying Math, p. 264 **Content** Section Review, p. 264	Section Focus Transparency Guided Reading Audio Program Interactive Chalkboard CD-ROM
MiniLAB: small aluminum pie pan, dry macaroni, white glue, water, mixing bowl or beaker, sample of fossiliferous limestone	**Chapter *Fast File* Resources** Transparency Activity, p. 45 MiniLAB, p. 4 Enrichment, p. 31 Reinforcement, p. 28 Directed Reading, p. 21 Lab Activities, pp. 13–16 **Reading and Writing Skill Activities,** p. 21	**Portfolio** Science Journal, p. 267 **Performance** MiniLAB, p. 270 Applying Skills, p. 271 **Content** Section Review, p. 271	Section Focus Transparency Guided Reading Audio Program Interactive Chalkboard CD-ROM Video Lab
Lab: rolling pin, lump of modeling clay, uncooked rice (wild rice, if available), granite sample, gneiss sample **Lab:** set of minerals, magnifying lens, putty knife, streak plate, Mohs scale, minerals field guide	**Chapter *Fast File* Resources** Transparency Activity, p. 46 Lab Worksheet, pp. 5–8 Enrichment, p. 32 Reinforcement, p. 29 Directed Reading, pp. 21, 22 Transparency Activity, pp. 47–48 **Earth Science Critical Thinking/ Problem Solving,** p. 4 **Home and Community Involvement,** p. 43	**Portfolio** Differentiated Instruction, p. 273 **Performance** Applying Skills, p. 276 **Content** Section Review, p. 276	Section Focus Transparency Teaching Transparency Virtual Labs CD-ROM Guided Reading Audio Program Interactive Chalkboard CD-ROM

End of Chapter Assessment

Glencoe's Assessment Advantage

Blackline Masters	Technology	Professional Series
Chapter *Fast File* Resources Chapter Review, pp. 37–38 Chapter Tests, pp. 39–42 **Standardized Test Practice,** pp. 39–42	MindJogger Videoquiz Virtual Labs CD-ROM Exam*View*® Pro Testmaker TeacherWorks CD-ROM Interactive Chalkboard CD-ROM	**Performance Assessment in the Science Classroom (PASC)**

Transparencies

Section Focus

Section 1 Section Focus Transparency **Phosphate Was Everywhere!**

Nauru is a tiny island in the Pacific Ocean near Papua New Guinea. Despite its small size, Nauru is one of the richest developing nations. Its wealth, however, comes from phosphate deposits that are expected to run out early in the twenty-first century.

1. Judging from the picture, how is phosphate mined on Nauru?
2. Do you think the phosphate is ready for use as fertilizer straight from the mine? Why or why not?
3. Name some other valuable materials that are mined from Earth.

L2

Section 2 Section Focus Transparency **Beautiful Blue**

Crater Lake, in western Oregon, was formed over 6,000 years ago by volcanic processes. Subsequent eruptions and lava flows formed Wizard Island, shown below. In 1902, the region including Crater Lake (a total of 647 km²) became a national park.

1. Given Crater Lake's origin, how do you think many of the area's rocks were formed?
2. Name some other formations resulting from volcanic activity.

L2

Section 3 Section Focus Transparency **Hutton's Unconformity**

Geologist James Hutton felt that rocks and rock formations could be explained by processes people could observe. He hypothesized that heat, pressure, and erosion were important parts of these processes. Observations of formations like the one shown below helped Hutton arrive at these revolutionary ideas.

1. Describe the layers of rock in the photograph.
2. Why do you think these layers are interesting to geologists?
3. What are some processes that change rock?

L2

This is a representation of key blackline masters available in the Teacher Classroom Resources. See Resource Manager boxes within the chapter for additional information.

Assessment

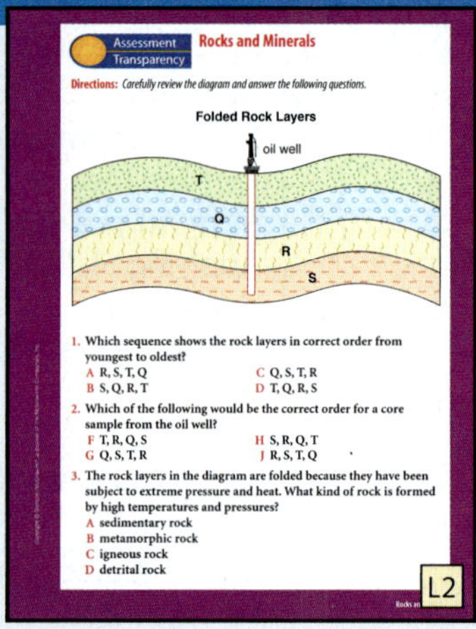
Assessment Transparency **Rocks and Minerals**

Directions: Carefully review the diagram and answer the following questions.

1. Which sequence shows the rock layers in correct order from youngest to oldest?
   A R, S, T, Q C Q, S, T, R
   B S, Q, R, T D T, Q, R, S
2. Which of the following would be the correct order for a core sample from the oil well?
   F T, R, Q, S H S, R, Q, T
   G Q, S, T, R J R, S, T, Q
3. The rock layers in the diagram are folded because they have been subject to extreme pressure and heat. What kind of rock is formed by high temperatures and pressures?
   A sedimentary rock
   B metamorphic rock
   C igneous rock
   D detrital rock

L2

Teaching

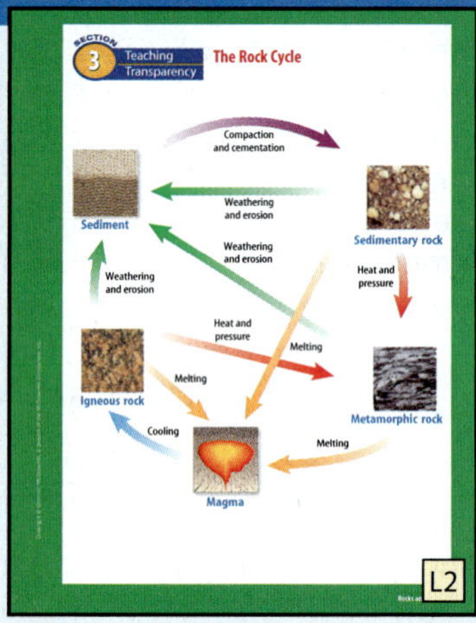
Section 3 Teaching Transparency **The Rock Cycle**

L2

Key to Teaching Strategies

The following designations will help you decide which activities are appropriate for your students.

- **L1** Level 1 activities should be appropriate for students with learning difficulties.
- **L2** Level 2 activities should be within the ability range of all students.
- **L3** Level 3 activities are designed for above-average students.
- **ELL** ELL activities should be within the ability range of English Language Learners.
- **COOP LEARN** Cooperative Learning activities are designed for small group work.
- **LS** Multiple Learning Styles logos, as described on page 12T, are used throughout to indicate strategies that address different learning styles.
- **P** These strategies represent student products that can be placed into a best-work portfolio.
- **PBL** Problem-Based Learning activities apply real-world situations to learning.

Hands-on Activities

Student Text Lab Worksheet

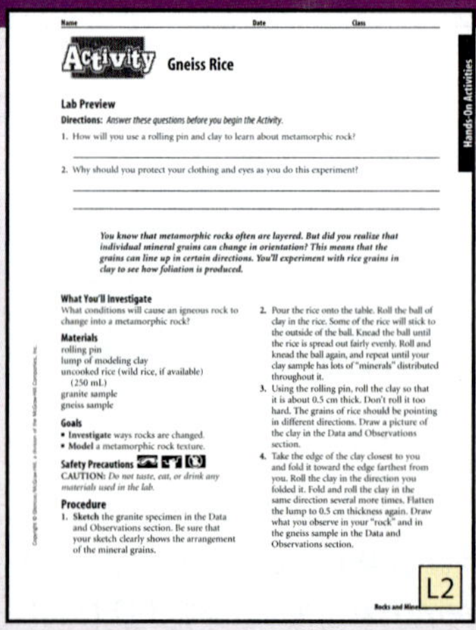
Activity **Gneiss Rice**

Lab Preview

Directions: *Answer these questions before you begin the Activity.*

1. How will you use a rolling pin and clay to learn about metamorphic rock?
2. Why should you protect your clothing and eyes as you do this experiment?

L2

Laboratory Activities

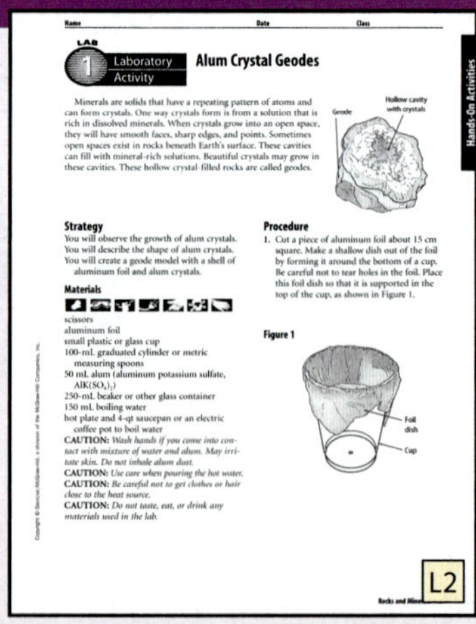
Lab 1 Laboratory Activity **Alum Crystal Geodes**

L2

Resource Manager

Meeting Different Ability Levels

Content Outline

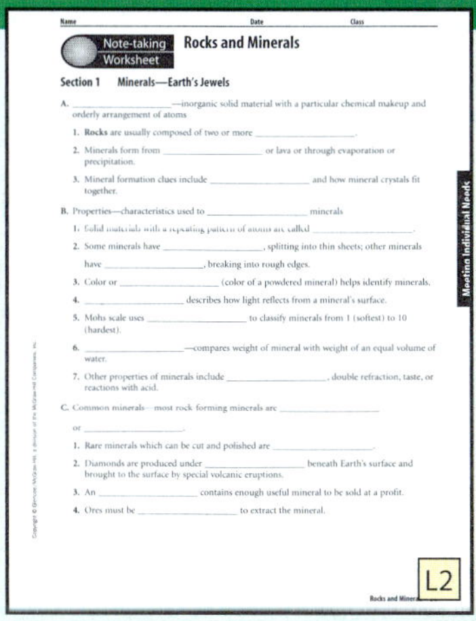

Reinforcement

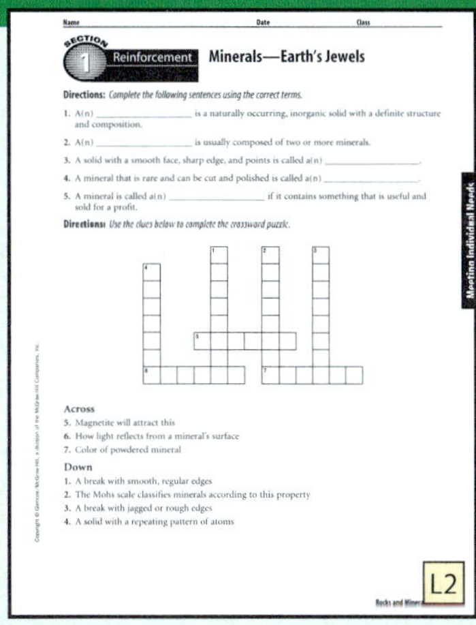

Enrichment

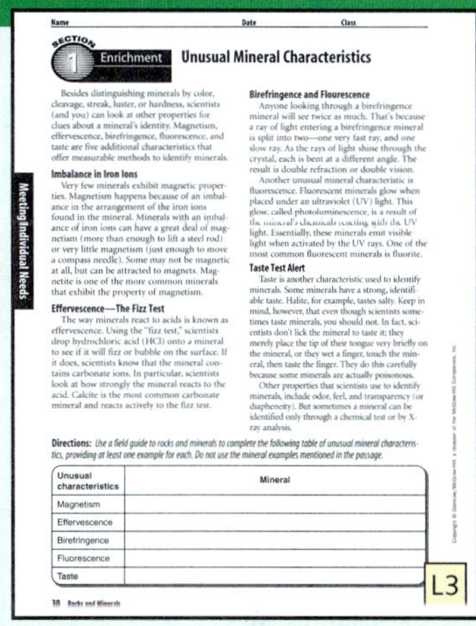

Directed Reading (English/Spanish)

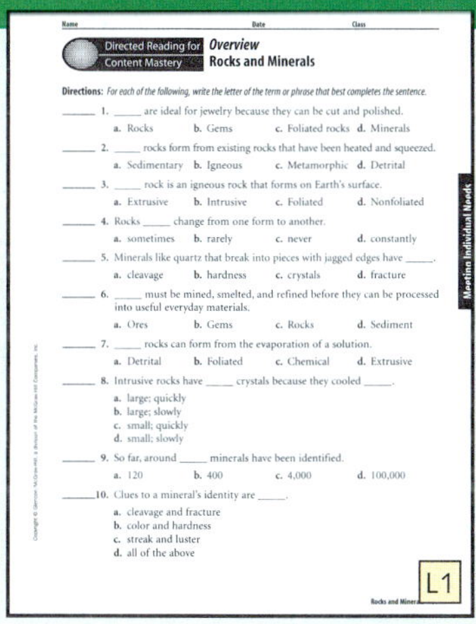

Study Guide

Reading Essentials

Assessment

Test Practice Workbook

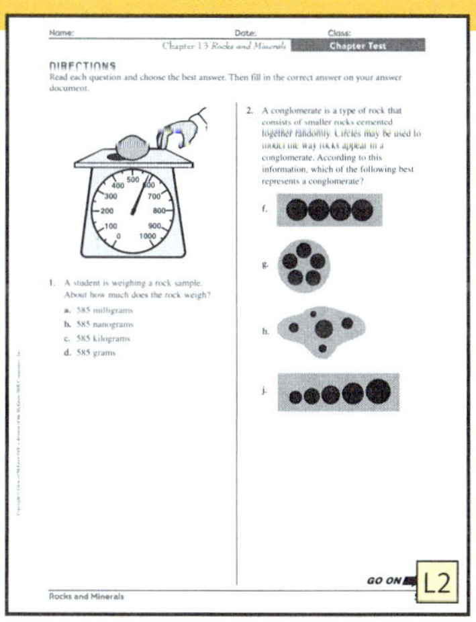

Chapter Review

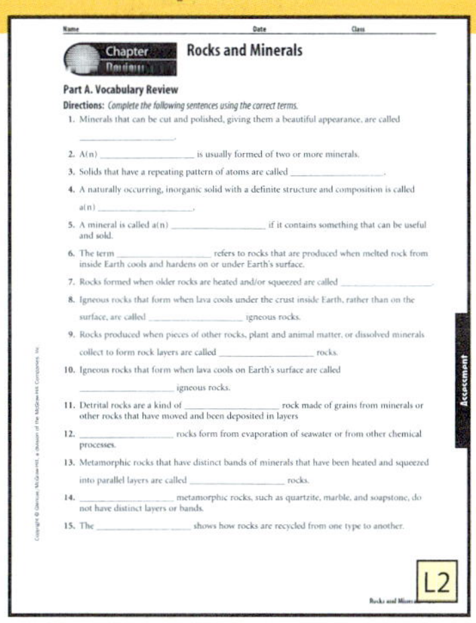

Chapter Tests

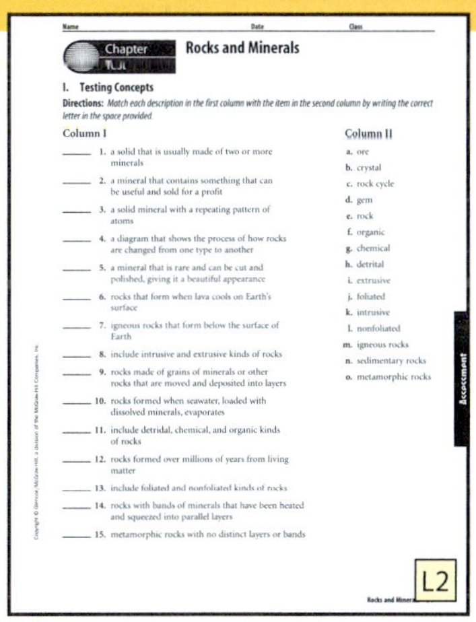

Science Content Background

section 1 Minerals—Earth's Jewels

What is a mineral?

A mineral can be defined as a naturally occurring, inorganic solid with an orderly internal atomic structure and definite, but not fixed, chemical composition. This definition excludes some solids, such as organic solids. It also excludes some manufactured diamonds, which are not minerals because they did not form in nature.

Rocks are commonly described as aggregates of minerals, but several rocks don't actually fit this definition. Examples are coal and obsidian, which are composed mainly of solids that are not minerals. For this reason, geologists use a more inclusive definition. A rock is a mixture of minerals, mineraloids, organic material, or glass.

Crystals

Crystals start to grow when minute "seed" crystal forms, and ions and molecules migrate to their surfaces. Some substances form lots of seed crystals and smaller crystals, but others form fewer seed crystals and larger crystals. The rate of cooling and the presence of volatiles also influence crystal size. In general, slower cooling leads to larger crystals.

Cleavage and Fracture

Minerals may cleave in one or many directions. Micas, like muscovite, have one cleavage. Feldspars break along two directions of cleavage to form prism-shaped fragments. Calcite has three directions of cleavage that together outline rhombohedron-shaped pieces of the mineral. Fluorite has four directions of cleavage, and the mineral sphalerite has six.

Whan a mineral breaks along uneven surfaces, it exhibits fracture. A mineral displaying conchoidal fracture breaks along smooth, curved surfaces, much like the inside surface of a conch shell. This type of fracture is well displayed by obsidian, an igneous rock, and the mineral quartz.

Hardness

A mineral's hardness depends on the strength of its atomic bonds and can be measured by determining the mineral's resistance to scratching. Most minerals have uniform hardness, but some minerals, like calcite, are harder on certain surfaces than on others.

Specific Gravity

A mineral's specific gravity (also described as heft) is the ratio of the mineral's weight to the weight of an equal volume of water. If a cubic centimeter of a mineral weighs four times more than a cubic centimeter of water, its specific gravity is four.

> **Teacher to Teacher**
> Steve Federman, 6th Grade Teacher
> Loveland Middle School
> Loveland, Ohio
>
> "I give each student a piece of mica and challenge him or her to split the sample along its cleavage surface. Students see how many times they can split the mica into thin sheets."
>
> *Steve Federman*

Runk/Schoenberger/Grant Heilman Photography, Inc.

section 2 Igneous and Sedimentary Rocks

Rocks from Lava

Pumice has a glassy texture with holes or spaces in the structure caused by the rapid expansion of gases upon reaching Earth's surface. Occasionally, lava that contains large mineral crystals that grew at depth may erupt at Earth's surface. When this lava cools, a rock with large crystals surrounded by small crystals will form.

Rocks from Magma

Intrusive igneous rocks form when magma, which is less dense than surrounding rock, is forced upward, cools, and solidifies before reaching Earth's surface.

Sedimentary Rocks

Sedimentary rocks tend to display visible layering, or bedding. Most sedimentary rocks form when sediments are deposited and then compacted or cemented together, or both. Cementation occurs as mineral cements such as quartz, calcite, clay minerals, and iron oxides precipitate in the pore spaces of the sediment.

Many sedimentary rocks form when broken pieces of preexisting rock, also known as detritus, are deposited and lithified (turn to solid rock). Because they are made of detritus, these rocks are called detrital sedimentary rocks.

section 3 Metamorphic Rocks and the Rock Cycle

New Rock from Old Rock

Rocks change in response to changes in temperature, pressure, and fluid activity. New minerals often form, and preexisting minerals may be recrystallized. Minerals also may acquire a preferred orientation.

The heat, pressure, and chemical reactions that cause rocks to change are generated in different ways. Contact metamorphism occurs when hot magma comes into contact with surrounding rocks. Regional metamorphism occurs when large areas of Earth are subjected to increased pressure and temperature, such as during mountain building.

The Rock Cycle

Matter has been cycling through Earth throughout its history. Melting, weathering and erosion, and other rock cycle processes continue today.

Darrell Gulin/DRK Photo

chapter content resources

Internet Resources

For additional content background, visit red.msscience.com to:

- access your book online
- find references to related articles in popular science magazines
- access Web links with related content background
- access current events with science journal topics

Print Resources

Peterson First Guide to Rocks and Minerals, by Frederick H. Pough, Houghton Mifflin Company, 1998

Encyclopedia of Rocks, Minerals and Gemstones, by Chris Pellant and Henry Russell, Thunder Bay Press, 2001

How the Earth Works, by John Farndon, Dorling Kindersley Limited, 1992

chapter 9

Chapter Vocabulary

mineral, p. 256
rock, p. 256
crystal, p. 258
gem, p. 262
ore, p. 263
igneous rock, p. 265
extrusive, p. 265
intrusive, p. 265
sedimentary rock, p. 269
metamorphic rock, p. 273
foliated, p. 274
nonfoliated, p. 274
rock cycle, p. 275

Science Journal Student responses might include qualitative characteristics such as color, texture, and opacity, and quantitative characteristics such as length and mass.

INTERACTIVE CHALKBOARD with Image Bank

PowerPoint® Presentations

This CD-ROM is an editable Microsoft® PowerPoint® presentation that includes:

- a pre-made presentation for every chapter
- interactive graphics
- animations
- audio clips
- image bank
- all new section and chapter questions
- Standardized Test Practice
- transparencies
- pre-lab questions for all labs
- Foldables directions
- links to red.msscience.com

chapter 9

Rocks and Minerals

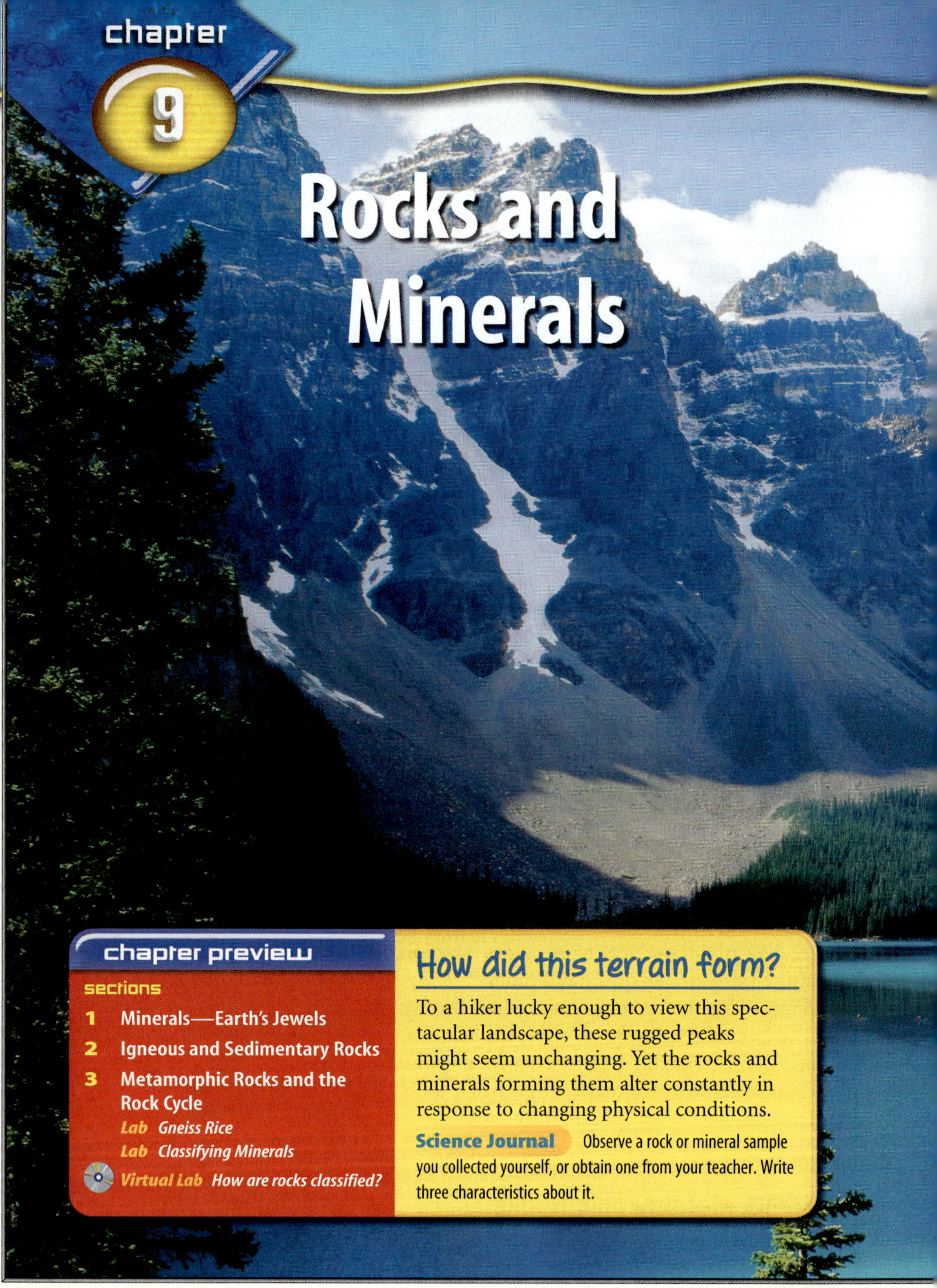

chapter preview

sections

How did this terrain form?

To a hiker lucky enough to view this spectacular landscape, these rugged peaks might seem unchanging. Yet the rocks and minerals forming them alter constantly in response to changing physical conditions.

Science Journal Observe a rock or mineral sample you collected yourself, or obtain one from your teacher. Write three characteristics about it.

Theme Connection

Scale and Structure and **Stability and Change** The theme of scale and structure is presented as the properties of minerals are related to the arrangement of their atoms. The theme of stability and change is emphasized through the study of the rock cycle.

About the Photo

Canadian Rockies Rock formations of the Canadian Rocky Mountains have undergone structural changes as a result of interactions among Earth's tectonic plates. Rocks and minerals in mountains such as these may record clues that reflect the physical conditions that formed them.

Start-Up Activities

Launch Lab

Observe a Rock

Upon reaching the top, you have a chance to look more closely at the rock you've been climbing. First, you notice that it sparkles in the Sun because of the silvery specks that are stuck in the rock. Looking closer, you also see clear, glassy pieces and pink, irregular chunks. What is the rock made of? How did it get here?

1. Obtain a sparkling rock from your teacher. You also will need a magnifying lens.
2. Observe the rock with the magnifying lens. Your job is to observe and record as many of the features of the rock as you can.
3. Return the rock to your teacher.
4. Describe your rock so other students could identify it from a variety of rocks.
5. **Think Critically** How do the parts of the rock fit together to form the whole thing? Describe this in your Science Journal and make a drawing. Be sure to label the colors and shapes in your drawing.

Science Online — Preview this chapter's content and activities at red.msscience.com

FOLDABLES™ Study Organizer

Rocks and Minerals Make the following Foldable to compare and contrast the characteristics of rocks and minerals.

STEP 1 **Fold** one sheet of paper lengthwise.

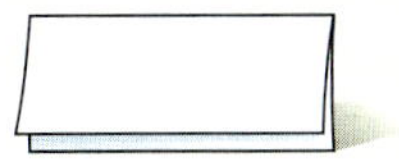

STEP 2 **Fold** into thirds.

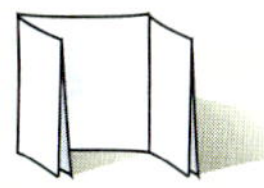

STEP 3 **Unfold and draw** overlapping ovals. **Cut** the top sheet along the folds.

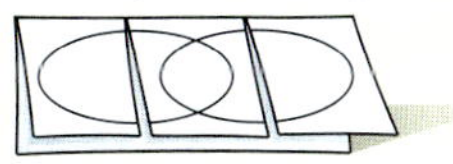

STEP 4 **Label** the ovals as shown.

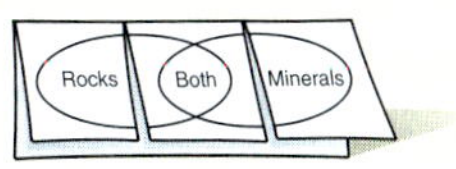

Construct a Venn Diagram As you read the chapter, list the characteristics unique to rocks under the left tab, those unique to minerals under the right tab, and those characteristics common to both under the middle tab.

Launch Lab

Purpose Use the Launch Lab to help students understand that rocks have characteristics by which they can be described. L2 ELL COOP LEARN LS **Kinesthetic**

Preparation Obtain a variety of large-grained igneous or metamorphic rocks, such as granite, gabbro, schist, or gneiss.

Materials magnifying lens, large-grained rock

Teaching Strategy Have students work in pairs. Suggest they take turns making descriptive statements about the rock.

Think Critically

Possible answer: The crystals that make up the rock interlock, or seem to fit together tightly.

Assessment

Performance Place all of the rock samples on a table. Have students exchange descriptions and try to identify the rock being described. Discuss what characteristics were most helpful when identifying a rock. Use **Performance Assessment in the Science Classroom,** p. 89.

FOLDABLES™ Study Organizer **Dinah Zike Study Fold**

Student preparation materials for this Foldable are available in the Chapter *Fast File* Resources.

section 1 Minerals—Earth's Jewels

1 Motivate

INTERACTIVE CHALKBOARD PowerPoint® Presentations

Bellringer

Section Focus Transparencies also are available on the Interactive Chalkboard CD-ROM. L2 ELL

SECTION 1 Section Focus Transparency **Phosphate Was Everywhere!**

Nauru is a tiny island in the Pacific Ocean near Papua New Guinea. Despite its small size, Nauru is one of the richest developing nations. Its wealth, however, comes from phosphate deposits that are expected to run out early in the twenty-first century.

1. Judging from the picture, how is phosphate mined on Nauru?
2. Do you think the phosphate is ready for use as fertilizer straight from the mine? Why or why not?
3. Name some other valuable materials that are mined from Earth.

Rocks and Minerals

Tie to Prior Knowledge

Atoms and Compounds Have students recall that matter is composed of atoms and that atoms combine to form compounds. In this section, students will learn that the structures and properties of minerals are determined by their internal arrangements of atoms.

---

as you read

What You'll Learn

- **Identify** the difference between a mineral and a rock.
- **Describe** the properties that are used to identify minerals.

Why It's Important

Minerals are basic substances of nature that humans use for a variety of purposes.

Review Vocabulary

physical property: any characteristic of a material that you can observe without changing the identity of the material

New Vocabulary

- mineral
- rock
- crystal
- gem
- ore

What is a mineral?

Suppose you are planning an expedition to find minerals (MIH nuh rulz). Where would you look? Do you think you'll have to crawl into a cave or brave the depths of a mine? Well, put away your flashlight. You can find minerals in your own home—in the salt shaker and in your pencil. Metal pots, glassware, and ceramic dishes are products made from minerals. Minerals and products made from them, shown in **Figure 1,** surround you.

Minerals Defined **Minerals** are inorganic, solid materials found in nature. *Inorganic* means they usually are not formed by plants or animals. You could go outside and find minerals that occur as gleaming crystals—or as small grains in ordinary rocks. X-ray patterns of a mineral show an orderly arrangement of atoms that looks something like a garden trellis. Evidence of this orderly arrangement is the beautiful crystal shape often seen in minerals. The particular chemical makeup and arrangement of the atoms in the crystal is unique to each mineral. **Rocks,** such as the one used in the Launch Lab, usually are made of two or more minerals. Each mineral has unique characteristics you can use to identify it. So far, more than 4,000 minerals have been identified.

Figure 1 You use minerals every day—maybe without even realizing it. Minerals are used to make many common objects.

The mineral quartz is used to make the glass that you use every day.

The "lead" in a pencil is not really lead. It is the mineral graphite.

Section 1 Resource Manager

Chapter *FAST FILE* Resources
- Transparency Activity, p. 44
- Directed Reading for Content Mastery, pp. 19, 20
- Note-taking Worksheets, pp. 33–35
- Enrichment, p. 30
- MiniLAB, p. 3
- Reinforcement, p. 27
- Lab Activity, pp. 9–11

Mathematics Skill Activities, p. 1

Cultural Diversity, p. 35

How do minerals form? Minerals form in several ways. One way is from melted rock material inside Earth called magma. As magma cools, atoms combine in orderly patterns to form minerals. Minerals also form from magma that reaches Earth's surface. Magma at Earth's surface is called lava.

Evaporation can form minerals. Just as salt crystals appear when seawater evap rates, other dissolved minerals, such as gy sum, can crystallize. A process called precipitation (prih sih puh TAY shun) can form minerals, too. Water can hold only so much dissolved material. Any extra separates and falls out as a solid. Large areas of the ocean floor are covered with manganese nodules that formed in this way. These metallic spheres average 25 cm in diameter. They crystallized directly from seawater containing metal atoms.

Figure 2 This cluster of fluorite crystals formed from a solution rich in dissolved minerals.

Formation Clues Sometimes you can tell how a mineral formed by how it looks. Large mineral grains that fit together like a puzzle seem to show up in rocks formed from slow-cooling magma. If you see large, perfectly formed crystals, it means the mineral had plenty of space in which to grow. This is a sign they may have formed in open pockets within the rock.

The crystals you see in **Figure 2** grew this way from a solution that was rich in dissolved minerals. To figure out how a mineral was formed, you have to look at the size of the mineral crystal and how the crystals fit together.

Properties of Minerals

The cheers are deafening. The crowd is jumping and screaming. From your seat high in the bleachers, you see someone who is wearing a yellow shirt and has long, dark hair in braids, just like a friend you saw this morning. You're sure it's your friend only when she turns and you recognize her smile. You've identified your friend by physical properties that set her apart from other people—her clothing, hair color and style, and facial features. Each mineral, too, has a set of physical properties that can be used to identify it. Most common minerals can be identified with items you have around the house and can carry in your pocket, such as a penny or a steel file. With a little practice you can learn to recognize mineral shapes, too. Next you will learn about properties that help you identify minerals.

INTEGRATE Life Science

Bone Composition Bones, such as those found in humans and horses, contain tiny crystals of the mineral apatite. Research apatite and report your findings to your class.

2 Teach

IDENTIFYING Misconceptions

Mineraloids Opal often is used as a gemstone. Some students may think that opal is a mineral. Explain that opal is not a mineral in the strictest sense because its structure is only partially ordered. Instead, it is considered a mineraloid.

Teacher FYI

Precipitation Conditions Precipitation of minerals out of solution can be caused by chemical changes or by changes in water temperature.

INTEGRATE Life Science

Bone Composition Apatite is composed of calcium phosphate, one of the two principle components of bone. Apatite is a mineral of the hexagonal crystal system. It is a source of phosphorus, which is used in fertilizer. It also is used in the production of phosphoric acid and detergents. Apatite is not used widely as a gem, as it is quite soft (Mohs hardness of 5).

Career Have students use print or Internet resources to study careers in industrial minerals. Ask them to include examples of how industrial minerals are mined, distributed, and processed to make final products for human consumption.

Differentiated Instruction

Learning Disabled Have pairs of students work together to mix salt in water until no more salt will dissolve. Instruct students to pour their solutions into a cup or pie pan and to observe their solutions each day for a week. Students should observe cubic crystals of salt forming as the water evaporates. Students who pour their solutions into the pie pans will observe crystals sooner. L1 LS **Interpersonal**

Reading Check

Answer orderly internal atomic arrangements

Caption Answers

Figure 3 Pyrite has a gold color and metallic luster, making it look like gold.

Figure 4B The three directions of cleavage at right angles produce a cubic external shape.

Visual Learning

Figures 3 and 4B Share with the class samples of several crystals, such as halite (cubic), corundum (hexagonal), and pyrite (cubic). Draw the internal atomic arrangement of a cubic crystal (halite) on the board and have students compare it with the halite sample shown in **Figure 4B.**

IDENTIFYING Misconceptions

Crystals Have students describe what comes to mind when they think of the word *crystal.* Students may think that a mineral has a crystalline structure only if the sample shows the crystal shape, such as cubic salt crystals. Emphasize that a mineral's crystalline structure refers to the internal arrangement of atoms. Even if a mineral shows no crystal shape, it still has an internal crystalline structure.

Figure 3 The mineral pyrite often forms crystals with six faces. **Determine** *why pyrite also is called "fool's gold."*

Crystals All minerals have an orderly pattern of atoms. The atoms making up the mineral are arranged in a repeating pattern. Solid materials that have such a pattern of atoms are called **crystals.** Sometimes crystals have smooth growth surfaces called crystal faces. The mineral pyrite commonly forms crystals with six crystal faces, as shown in **Figure 3.**

Reading Check *What distinguishes crystals from other types of solid matter?*

Cleavage and Fracture Another clue to a mineral's identity is the way it breaks. Minerals that split into pieces with smooth, regular planes that reflect light are said to have cleavage (KLEE vihj). The mica sample in **Figure 4A** shows cleavage by splitting into thin sheets. Splitting mica along a cleavage surface is similar to peeling off a piece of presliced cheese. Cleavage is caused by weaknesses within the arrangement of atoms that make up the mineral.

Not all minerals have cleavage. Some break into pieces with jagged or rough edges. Instead of neat slices, these pieces are shaped more like hunks of cheese torn from an unsliced block. Materials that break this way, such as quartz, have what is called fracture (FRAK chur). **Figure 4C** shows the fracture of flint.

Figure 4 Some minerals have one or more directions of cleavage. If minerals do not break along flat surfaces, they have fracture.

A Minerals in the mica group have one direction of cleavage and can be peeled off in sheets.

B The mineral halite, also called rock salt, has three directions of cleavage at right angles to each other. **Infer** *Why might grains of rock salt look like little cubes?*

C Fracture can be jagged and irregular or smooth and curvy like in flint.

Teacher FYI

Polymorphs Minerals that have the same chemical composition but different atomic arrangements are called *polymorphs.* An excellent example of polymorphs are diamond and graphite. Both minerals are composed of carbon. In diamond, each carbon atom is bonded tightly with four other carbon atoms. Because of this, diamond is the hardest known mineral. In graphite, each carbon atom is bonded tightly to other carbon atoms within a layer. The layers are held together weakly. This structure makes graphite one of the softest known minerals.

Figure 5 The mineral calcite can form in a variety of colors. The colors are caused by slight impurities.

Color The reddish-gold color of a new penny shows you that it contains copper. The bright yellow color of sulfur is a valuable clue to its identity. Sometimes a mineral's color can help you figure out what it is. But color also can fool you. The common mineral pyrite (PI rite) has a shiny, gold color similar to real gold—close enough to disappoint many prospectors during the California Gold Rush in the 1800s. Because of this, pyrite also is called fool's gold. While different minerals can look similar in color, the same mineral can occur in a variety of colors. The mineral calcite, for example, can occur in many different colors, as shown in **Figure 5.**

Streak and Luster Scraping a mineral sample across an unglazed, white tile, called a streak plate, produces a streak of color, as shown in **Figure 6.** Oddly enough, the streak is not necessarily the same color as the mineral itself. This streak of powdered mineral is more useful for identification than the mineral's color. Gold prospectors could have saved themselves a lot of heartache if they had known about the streak test. Pyrite makes a greenish-black or brownish-black streak, but gold makes a yellow streak.

Is the mineral shiny? Dull? Pearly? Words like these describe another property of minerals, called luster. Luster describes how light reflects from a mineral's surface. If it shines like a metal, the mineral has metallic (muh TA lihk) luster. Nonmetallic minerals can be described as having pearly, glassy, dull, or earthy luster. You can use color, streak, and luster to help identify minerals.

Figure 6 Streak is the color of the powdered mineral. The mineral hematite has a characteristic reddish-brown streak.
Explain *how you obtain a mineral's streak.*

Use an Analogy

Cleavage To help students understand why some minerals cleave, relate the weaknesses in the arrangement of the mineral's atoms to the weakness of icing between layers of cake. Cleavage occurs along planes of weak bonds just as the cake can easily be separated between iced layers. When no distinct plane of weakness exists between atoms, it is like a piece of cake without icing. The breakage can occur in any direction.

Activity

Hematite Obtain samples of specular hematite, earthy hematite, and oolitic hematite. Ask students to make a table in their Science Journals to compare and contrast the appearance of each sample and to describe the streaks of each sample. Samples of hematite can look very different. Some are red, some are steel gray, and some are a shiny metallic color. But no matter what color the sample is, hematite's streak color is dark red-brown. L1 LS **Visual-Spatial and Linguistic**

Fun Fact

Minerals with metallic luster generally have dark streaks.

Caption Answer

Figure 6 by scraping a mineral sample across an unglazed, white porcelain tile (streak plate)

Active Reading

Buddy Interviews Have students interview one another to find out what helps them understand what they are reading, how they find answers, and how they assimilate new vocabulary terms. Have students use Buddy Interviews to master the concepts related to Properties of Minerals in this section.

Teacher FYI

Calcite Reaction When dilute hydrochloric acid is placed on a sample of calcite or any rock that has calcite in it, water is produced and carbon dioxide gas is released according to the following reaction: $2H^+ + CO_3 \rightarrow H_2O + CO_2$

Discussion

Acid Reaction Why do some carbonate minerals react more readily with hydrochloric acid than others? The fizzing reaction is the release of CO_2 from the mineral. The more easily the CO_2 is released, the more the mineral will fizz.

Quick Demo

Scratch Test

Materials mineral samples of varying hardness with flat cleavage or growth surfaces.

Estimated Time ten minutes

Procedure Have students select a mineral sample. Then have them attempt to scratch all the other mineral samples with the selected sample. Students should run their fingernails across the locations on the rubbed samples to detect whether or not they actually were scratched. If a groove on a surface is detected, then the grooved mineral is softer than the mineral that was pushed across it. Point out that a line of mineral powder left on a surface can resemble a scratch, leading to an erroneous relative hardness test.

Table 1 Mohs Scale

Mineral	Hardness	Hardness of Common Objects
Talc	1 (softest)	
Gypsum	2	fingernail (2.5)
Calcite	3	copper penny (3.0)
Fluorite	4	iron nail (4.5)
Apatite	5	glass (5.5)
Feldspar	6	steel file (6.5)
Quartz	7	streak plate (7)
Topaz	8	
Corundum	9	
Diamond	10 (hardest)	

Hardness As you investigate different minerals, you'll find that some are harder than others. Some minerals, like talc, are so soft that they can be scratched with a fingernail. Others, like diamond, are so hard that they can be used to cut almost anything else.

In 1822, an Austrian geologist named Friedrich Mohs also noticed this property of minerals. He developed a way to classify minerals by their hardness. The Mohs scale, shown in **Table 1,** classifies minerals from 1 (softest) to 10 (hardest). You can determine hardness by trying to scratch one mineral with another to see which is harder. For example, fluorite (4 on the Mohs scale) will scratch calcite (3 on the scale), but fluorite cannot scratch apatite (5 on the scale). You also can use a homemade mineral identification kit—a copper penny, a nail, and a small glass plate with smooth edges. Simply find out what scratches what. Is the mineral hard enough to scratch a penny? Will it scratch glass?

Specific Gravity Some minerals are heavier for their size than others. Specific gravity compares the weight of a mineral with the weight of an equal volume of water. Pyrite—or fool's gold—is about five times heavier than water. Pure gold is more than 19 times heavier than water. You could easily sense this difference by holding each one in your hand. Measuring specific gravity is another way you can identify minerals.

Figure 7 Calcite has the unique property of double refraction.

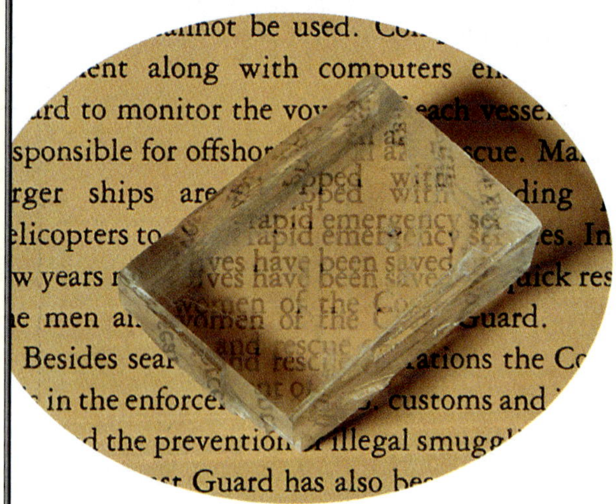

Other Properties Some minerals have other unusual properties that can help identify them. The mineral magnetite will attract a magnet. The mineral calcite has two unusual properties. It will fizz when it comes into contact with an acid like dilute HCl. Also, if you look through a clear calcite crystal, you will see a double image, as shown in **Figure 7.** Scientists taste some minerals to identify them, but you should not try this yourself. Halite, also called rock salt, has a salty taste.

Together, all of the properties you have read about are used to identify minerals. Learn to use them and you can be a mineral detective.

Curriculum Connection

Geography Invite a jeweler from your community to visit the class and explain how tests are used to determine the value of gems brought in by customers. As an example, ask the speaker to explain how it is determined whether a diamond ring contains actual diamonds. Have the speaker tell where diamonds are found, and have students pinpoint these places on a map. L2

Differentiated Instruction

Challenge Have students collect rocks and minerals from near their homes and bring them to class. Invite a geologist to come to class, discuss his or her career, and field students' questions. After the discussion, ask the geologist to identify the students' samples. Then have the students organize the samples into labeled sets. L1 ELL IS **Visual-Spatial**

Common Minerals

Rocks that make up huge mountain ranges are made of minerals. But only a small number of the more than 4,000 minerals make up most rocks. These minerals often are called the rock-forming minerals. If you can recognize these minerals, you will be able to identify most rocks. Other minerals are much rarer. However, some of these rare minerals also are important because they are used as gems or they are ore minerals, which are sources of valuable metals.

Most of the rock-forming minerals are silicates (SIH luh kaytz), which contain the elements silicon and oxygen. The mineral quartz is pure silica (SiO_2). More than half of the minerals in Earth's crust are types of a silicate mineral called feldspar. Other important rock-forming minerals are carbonates—compounds containing carbon and oxygen. The carbonate mineral calcite makes up most of the common rock limestone.

Reading Check *Why is the silicate mineral feldspar important?*

Other common minerals can be found in rocks that formed at the bottom of ancient, evaporating seas. Rock comprised of the mineral gypsum is abundant in many places, and rock salt, made of the mineral halite, underlies large parts of the Midwest.

Mini LAB

Classifying Minerals

Procedure

1. Touch a **magnet** to samples of **quartz, calcite, hornblende,** and **magnetite.** Record which mineral attracts the magnet.
2. With a **dropper,** apply a small amount of **dilute hydrochloric acid (HCl)** to each sample.
3. Rinse the samples with **water.**

Analysis

1. Describe how each mineral reacted to the tests in steps 1 and 2.
2. Describe in a data table the other physical properties of the four minerals.

Applying Science

How hard are these minerals?

Some minerals, like diamonds, are hard. Others, like talc, are soft. How can you determine the hardness of a mineral?

Identifying the Problem

The table at the right shows the results of a hardness test done using some common items as tools (a fingernail, copper penny, nail, and steel file) to scratch certain minerals (halite, turquoise, emerald, ruby, and graphite). The testing tools are listed at the top from softest (fingernail) to hardest (steel file). The table shows which minerals were scratched by which tools. Examine the table to determine the relative hardness of each mineral.

Hardness Test

Mineral	Fingernail	Penny	Nail	Steel File
Turquoise	N	N	Y	Y
Halite	N	Y	Y	Y
Ruby	N	N	N	N
Graphite	Y	Y	Y	Y
Emerald	N	N	N	N

Solving the Problem

1. Is it possible to rank the five minerals from softest to hardest using the data in the table above? Why or why not?
2. What method could you use to determine whether the ruby or the emerald is harder?

Mini LAB

Purpose to investigate how properties can be used to identify minerals L2 ELL LS **Kinesthetic**

Materials quartz, calcite, hornblende, magnetite, dropper, dilute HCl, magnet, water

Teaching Strategy Caution students to wear goggles and aprons.

Analysis

1. Magnetite is attracted to a magnet. Calcite fizzes when HCl is applied.
2. Possible answers: Quartz is harder than glass and has a glassy luster; hornblende is black and has two directions of cleavage; calcite has three directions of cleavage; magnetite scratches glass and has a black streak.

Reading Check

Answer Varieties of feldspar make up more than half of the minerals in Earth's crust.

Applying Science

Answers

1. No; it is possible to determine that graphite is the softest, followed by halite and turquoise. But it is not possible to tell whether emerald or ruby is harder, as they show the same data.
2. Scratch the ruby with the emerald and the emerald with the ruby. Whichever scratches the other is the harder mineral (ruby).

Differentiated Instruction

Learning Disabled Assist students in accomplishing the MiniLAB by providing them with specific instructions on how the magnet helps identify magnetic properties and how to handle the acid safely. Help students prepare their data tables, using an example on the board or pairing students with others who understand how the data table should be organized. L1 ELL LS **Interpersonal**

Assessment

Oral Which tests would distinguish quartz from calcite? hardness; reaction to acid (calcite); calcite has cleavage and quartz does not Use **Performance Assessment in the Science Classroom,** p. 89.

Teacher FYI

Common Minerals About two dozen minerals make up most of Earth's rocks. These rock-forming minerals include feldspars, quartz, olivines, micas, clay minerals, amphiboles, pyroxenes, and carbonates. These minerals together contain the eight most common elements in Earth's crust: oxygen, silicon, aluminum, iron, calcium, sodium, potassium, and magnesium.

Discussion

Mining Ores Mining gems and ores often requires more than just digging. Gold, for example, may be present in rocks in tiny specks. Have students research how ores are mined. Then have them discuss the problems that might be caused by mining for gems or precious metals. One way to get gold out of ore is to crush the rock and mix it with chemicals such as cyanide and zinc. The major problem with these mining operations is that they can produce large amounts of toxic waste.

Figure 8 The beauty of gem-quality minerals often is enhanced by cutting and polishing them.

This garnet crystal is encrusted with other minerals but still shines a deep red.

Cut garnet is a prized gemstone.

Gems Which would you rather win, a diamond ring or a quartz ring? A diamond ring would be more valuable. Why? The diamond in a ring is a kind of mineral called a gem. **Gems** are minerals that are rare and can be cut and polished, giving them a beautiful appearance, as shown in **Figure 8.** This makes them ideal for jewelry. To be gem quality, most minerals must be clear with few or no blemishes or cracks. A gem also must have a beautiful luster or color. Few minerals meet these standards. That's why the ones that do are rare and valuable.

The Making of a Gem One reason why gems are so rare is that they are formed under special conditions. Diamond, for instance, is a form of the element carbon. Scientists can make synthetic diamonds in laboratories, but they must use extremely high pressures. These pressures are greater than any found within Earth's crust. Therefore, scientists suggest that diamond forms deep in Earth's mantle. It takes a certain kind of volcanic eruption to bring a diamond close to Earth's surface, where miners can find it. This type of eruption forces magma from the mantle toward the surface of Earth at high speeds, bringing diamond along with it. This type of magma is called kimberlite magma. **Figure 9** shows a rock from a kimberlite deposit in South Africa that was mined for diamond. Kimberlite deposits are found in the necks of some ancient volcanoes.

Science Online

Topic: Gem Locations
Visit red.msscience.com to find Web links to information about the geography of gems.

Activity Select a continent, such as Africa, and list three examples of gems found there. Locate mining operations for each of these gems on a map for your class.

Figure 9 Diamonds sometimes are found in kimberlite deposits.

Science Journal

Gems and Gemstones Minerals are used as gems or gemstones because of certain favorable properties. Ask students to write a paragraph explaining how gems and gemstones differ from other minerals. Ask them to write a paragraph in which they name and describe a new gemstone that they have discovered. Gems are generally hard and can be cut into shapes that reflect light. L2 LS **Linguistic** P

Curriculum Connection

History French chemist Marc Gaudin produced the first synthetic ruby in 1873. He melted compounds that decomposed to form aluminum oxide, the material rubies are made of. He then added chromium to the aluminum oxide for color, producing synthetic rubies. Have students find out how synthetic rubies compare with real gems. Synthetic rubies often contain microscopic bubbles and striations. L2

Figure 10 Mining is expensive. To be profitable, ores must be found in large deposits or rich veins. Copper ore is obtained from this mine in Arizona.
List *three advantages of recycling metals.*

Ores A mineral is called an **ore** if it contains enough of a useful substance that it can be sold for a profit. Many of the metals that humans use come from ores. For example, the iron used to make steel comes from the mineral hematite, lead for batteries is produced from galena, and the magnesium used in vitamins comes from dolomite. Ores of these useful metals must be extracted from Earth in a process called mining. A copper mine is shown in **Figure 10.**

Scrap metal often is reused or recycled to help reduce the rate that minerals are extracted from Earth. Because minerals may take millions of years to form, they are considered a nonrenewable resource. Conservation efforts can decrease mining and production costs, preserve resources, and reduce the volume of landscape disrupted when minerals are extracted from Earth.

Ore Processing After an ore has been mined, it must be processed to extract the desired mineral or element. **Figure 11** shows a copper smelting plant that melts the ore and then separates and removes most of the unwanted materials. After this smelting process, copper can be refined, which means that it is purified. Then it is processed into many materials that you use every day. Examples of useful copper products include sheet-metal products, electrical wiring in cars and homes, and just about anything electronic. Some examples of copper products are shown in **Figure 12.**

Figure 11 This smelter in Montana heats and melts copper ore.
Explain *why smelting is necessary to process copper ore.*

Visual Learning

Figure 10 Copper-bearing minerals occur in coarse-grained igneous rocks in this mine. These ores often are referred to as porphyry copper deposits. Have students determine what is meant by the term *porphyry.* The term *porphyry* means that at least two distinct crystal sizes are present in a rock. LS **Linguistic**

Caption Answers

Figure 10 Recycling metals saves the cost of smelting and refining metals. Metals are a nonrenewable resource and will eventually run out. Recycling prevents metals from taking landfill space.

Figure 11 to extract and purify the copper

Inquiry Lab

Designing with Rock

Purpose Have students select rocks as home-building materials and design a simple home.

Possible Materials rock samples, rocks and minerals field guide, print and online resources on architecture and home building, paper, pencil

Estimated Time one class session

Teaching Strategies

- Ask students to imagine that they are going to build a house using three different rock types.
- Have them sketch their homes, and write a paragraph describing which rocks they selected and why.
- Students should support their choices and designs by explaining rock properties and home-building guidelines.

For additional inquiry activities, see *Science Inquiry Labs.*

Differentiated Instruction

Challenge Have students research and describe ways to safely dispose of wastes from mining operations. Waste is placed on top of clay or artificial liners. Fluids derived from the waste are collected and contained to prevent them from contaminating potable water sources. The piles of waste are covered with dirt and plants are grown on them. Scientists monitor the waste piles and local water supplies to ensure that no leakage occurs. L3 LS **Linguistic**

SECTION 1 Minerals—Earth's Jewels 263

Caption Answer

Figure 12 Possible answers: iron, aluminum, titanium

3 Assess

DAILY INTERVENTION

Check for Understanding

Interpersonal Have student pairs review the photographs of minerals in this section and discuss how each could be described using the properties they have learned about. L1 COOP LEARN LS

Reteach

Reviewing Properties Display four different minerals that can be recognized easily by one or two characteristics, and have students identify the minerals. Some minerals that may be used are quartz (hardness of 7), sulfur (yellow color), calcite (three directions of cleavage or reaction to hydrochloric acid), and hematite (red-brown streak). L2 LS

Assessment

Content Have students work in groups of two or three to write a song with lyrics in their Science Journals about the definition of a mineral. Use **Performance Assessment in the Science Classroom,** p. 151.

Figure 12 Many metal objects you use every day are made with copper.
List *other metals that are used to produce everyday objects.*

Minerals Around You Now you have a better understanding of minerals and their uses. Can you name five things in your classroom that come from minerals? Can you go outside and find a mineral right now? You will find that minerals are all around you and that you use minerals every day. Next, you will look at rocks, which are Earth materials made up of combinations of minerals.

section 1 review

Summary

What is a mineral?

- Many everyday products are made from minerals.
- Minerals form in several ways, such as crystallizing from magma or from solutions rich in dissolved materials.

Properties of Minerals

- Minerals are identified by observing their physical properties.
- Some minerals exhibit unusual physical properties, such as reaction to acid, formation of a double image, or magnetism.

Common Minerals

- Of the more than 4,000 minerals known, only a small number make up most rocks.
- Gems are highly prized mineral specimens often used as decorative pieces in jewelry or other items.

Self Check

1. **Explain** the difference between a mineral and a rock. Name five common rock-forming minerals.
2. **List** five properties that are used most commonly to identify minerals.
3. **Describe** an event that must occur in order for diamond to reach Earth's surface. Where in Earth is diamond formed?
4. **Describe** the steps of mining, smelting, and refining that are used to extract minerals or elements from ores. When is a mineral considered to be an ore?
5. **Think Critically** Would you want to live close to a working gold mine? Explain.

Applying Math

6. **Use Percentages** In 1996, the United States produced approximately 2,340,000 metric tons of refined copper. In 1997, about 2,440,000 metric tons of refined copper were produced. Compared to the 1996 amount, copper production increased by what percentage in 1997?

Science Online red.msscience.com/self_check_quiz

section 1 review

1. Mineral: naturally occurring, inorganic solid with a definite composition and orderly internal atomic arrangement; rock: usually comprised of two or more minerals; Rock-forming minerals include quartz, feldspar, calcite, gypsum, and halite, among others.
2. Possible answers: color, luster, streak, hardness, cleavage, magnetism
3. Diamonds form in Earth's mantle under conditions of high pressure. Diamonds erupt to the surface via volcanic eruptions.
4. Minerals first are extracted from Earth by mining. Mined material then can be melted and desired substances separated from unwanted material during smelting. Refining purifies the desired mineral even more. Earth materials are considered ores when they can undergo expensive mining and refinement processes and still yield a profit.
5. Answers will vary. Some students may feel that it would be an advantage to live close to a gold mine, as jobs might be available at the mine. Others might be wary of environmental problems caused by the mine.
6. 2,440,000t − 2,340,000t = 100,000t; 100,000t/2,340,000t × 100 = 4.3%

section 2

Igneous and Sedimentary Rocks

Igneous Rock

A rocky cliff, a jagged mountain peak, and a huge boulder probably all look solid and permanent to you. Rocks seem as if they've always been here and always will be. But little by little, things change constantly on Earth. New rocks form, and old rocks wear away. Such processes produce three main kinds of rocks—igneous, sedimentary, and metamorphic.

The deeper you go into the interior of Earth, the higher the temperature is and the greater the pressure is. Deep inside Earth, it is hot enough to melt rock. **Igneous** (IHG nee us) **rocks** form when melted rock material from inside Earth cools. The cooling and hardening that result in igneous rock can occur on Earth, as seen in **Figure 13,** or underneath Earth's surface. When melted rock material cools on Earth's surface, it makes an **extrusive** (ehk STREW sihv) igneous rock. When the melt cools below Earth's surface, **intrusive** (ihn TREW sihv) igneous rock forms.

Chemical Composition The chemicals in the melted rock material determine the color of the resulting rock. If it contains a high percentage of silica and little iron, magnesium, or calcium, the rock generally will be light in color. Light-colored igneous rocks are called granitic (gra NIH tihk) rocks. If the silica content is far less, but it contains more iron, magnesium, or calcium, a dark-colored or basaltic (buh SAWL tihk) rock will result. Intrusive igneous rocks often are granitic, and extrusive igneous rocks often are basaltic. These two categories are important in classifying igneous rocks.

Figure 13 Sakurajima is a volcano in Japan. During the 1995 eruption, molten rock material and solid rock were thrown into the air.

as you read

What You'll Learn

- **Explain** how extrusive and intrusive igneous rocks are different.
- **Describe** how different types of sedimentary rocks form.

Why It's Important

Rocks form the land all around you.

Review Vocabulary

lava: molten rock material that exists at or above Earth's surface

New Vocabulary

- igneous rock
- extrusive
- intrusive
- sedimentary rock

1 Motivate

Bellringer

Section Focus Transparencies also are available on the Interactive Chalkboard CD-ROM.

L2 ELL

Section 2 Section Focus Transparency **Beautiful Blue**

Crater Lake, in western Oregon, was formed over 6,000 years ago by volcanic processes. Subsequent eruptions and lava flows formed Wizard Island, shown below. In 1902, the region including Crater Lake (a total of 647 km²) became a national park.

1. Given Crater Lake's origin, how do you think many of the area's rocks were formed?
2. Name some other formations resulting from volcanic activity.

Rocks and Minerals

Tie to Prior Knowledge

Rock Material Help students recall that rocks can be made of more than just minerals. Both coal (made of organic material) and obsidian (made of volcanic glass) are rocks.

Section 2 Resource Manager

Chapter *FAST FILE* Resources

Transparency Activity, p. 45
Directed Reading for Content Mastery, p. 21
Enrichment, p. 31
Lab Activities, pp. 13–16
Reinforcement, p. 28
MiniLAB, p. 4

Reading and Writing Skill Activities, p. 21

Earth Science Critical Thinking/Problem Solving, p. 10

2 Teach

Discussion

Cooling Rates Describe a rock that formed from magma that began to cool slowly underground and then erupted onto Earth's surface. These events result in a rock that has large crystals contained within a mass of smaller crystals. Large crystals begin to form when the magma is underground. When the lava erupts, it cools rapidly and forms tiny crystals or glass. This type of rock is said to have a porphyritic texture.

Reading Check

Answer a large crack in Earth's crust from which lava erupts

Activity

Mineral Color Remind students that minerals containing higher concentrations of silica generally are light in color, and minerals containing less silica and more iron and magnesium are dark in color. Show students samples of granite, rhyolite, gabbro, and basalt. Ask them to work in small groups to infer which minerals are contained in each sample. L2 COOP LEARN LS **Interpersonal**

Use Science Words

Word Origin The roots of the words *intrusive* and *extrusive* are *in-*, in; *ex-*, out; and *trudere*, to push. Have students find other words with these roots and explain their meanings. intrude, to push in; interior, within; extrude, to push out; exterior, on the outside L2 LS **Linguistic**

INTEGRATE Social Studies

Obsidian Uses Humans have developed uses for obsidian from ancient through modern times. Research how people have used obsidian. Include information on where it has been found, processed, and distributed.

Rocks from Lava Extrusive igneous rocks form when melted rock material cools on Earth's surface. When the melt reaches Earth's surface, it is called lava. Lava cools quickly before large mineral crystals have time to form. That's why extrusive igneous rocks usually have a smooth, sometimes glassy appearance.

Extrusive igneous rocks can form in two ways. In one way, volcanoes erupt and shoot out lava and ash. Also, large cracks in Earth's crust, called fissures (FIH shurz), can open up. When they do, the lava oozes out onto the ground or into water. Oozing lava from a fissure or a volcano is called a lava flow. In Hawaii, lava flows are so common that you can observe one almost every day. Lava flows are quickly exposed to air or water. The fastest cooling lava forms no grains at all. This is how obsidian, a type of volcanic glass, forms. Lava trapping large amounts of gas can cool to form igneous rocks containing many holes.

Reading Check *What is a fissure?*

Figure 14 Extrusive igneous rocks form at Earth's surface. Intrusive igneous rocks form inside Earth. Wind and water can erode rocks to expose features such as dikes, sills, and volcanic necks.

LAB DEMONSTRATION

Purpose to show that some igneous rocks can be identified by density and color

Materials 500-mL beakers (2), pumice, scoria, water

Preparation Samples of pumice and scoria must be thoroughly dry.

Procedure Place pumice in one beaker of water and scoria in the other.

Expected Outcome Students observe that light-colored pumice floats, but that dark-colored scoria sinks. They should realize that scoria is denser than pumice. Sample colors indicate that scoria contains more iron and magnesium than pumice does.

Assessment

Explain why pumice floats but scoria doesn't. Pumice is richer in silica and has a smaller concentration of elements with greater mass, such as iron.

Rocks from Magma Some melted rock material never reaches Earth's surface. Such underground molten material is called magma. Intrusive igneous rocks are produced when magma cools below the surface of Earth, as shown in **Figure 14.**

Intrusive igneous rocks form when a huge glob of magma from inside Earth is forced upward toward the surface but never reaches it. It's similar to when a helium balloon rises and gets stopped by the ceiling. This hot mass of rock material sits under the surface and cools slowly over millions of years until it is solid. The cooling is so slow that the minerals in the magma have time to form large crystals. Intrusive igneous rocks generally have large crystals that are easy to see. Some extrusive igneous rocks do not have large crystals that you can see easily. Others are a mixture of small crystals and larger, visible crystals. **Figure 15** shows some igneous rock features.

Reading Check *How do intrusive and extrusive rocks appear different?*

INTEGRATE Physics

Thermal Energy The extreme heat found inside Earth has several sources. Some is left over from Earth's formation, and some comes from radioactive isotopes that constantly emit heat while they decay deep in Earth's interior. Research to find detailed explanations of these heat sources. Use your own words to explain them in your Science Journal.

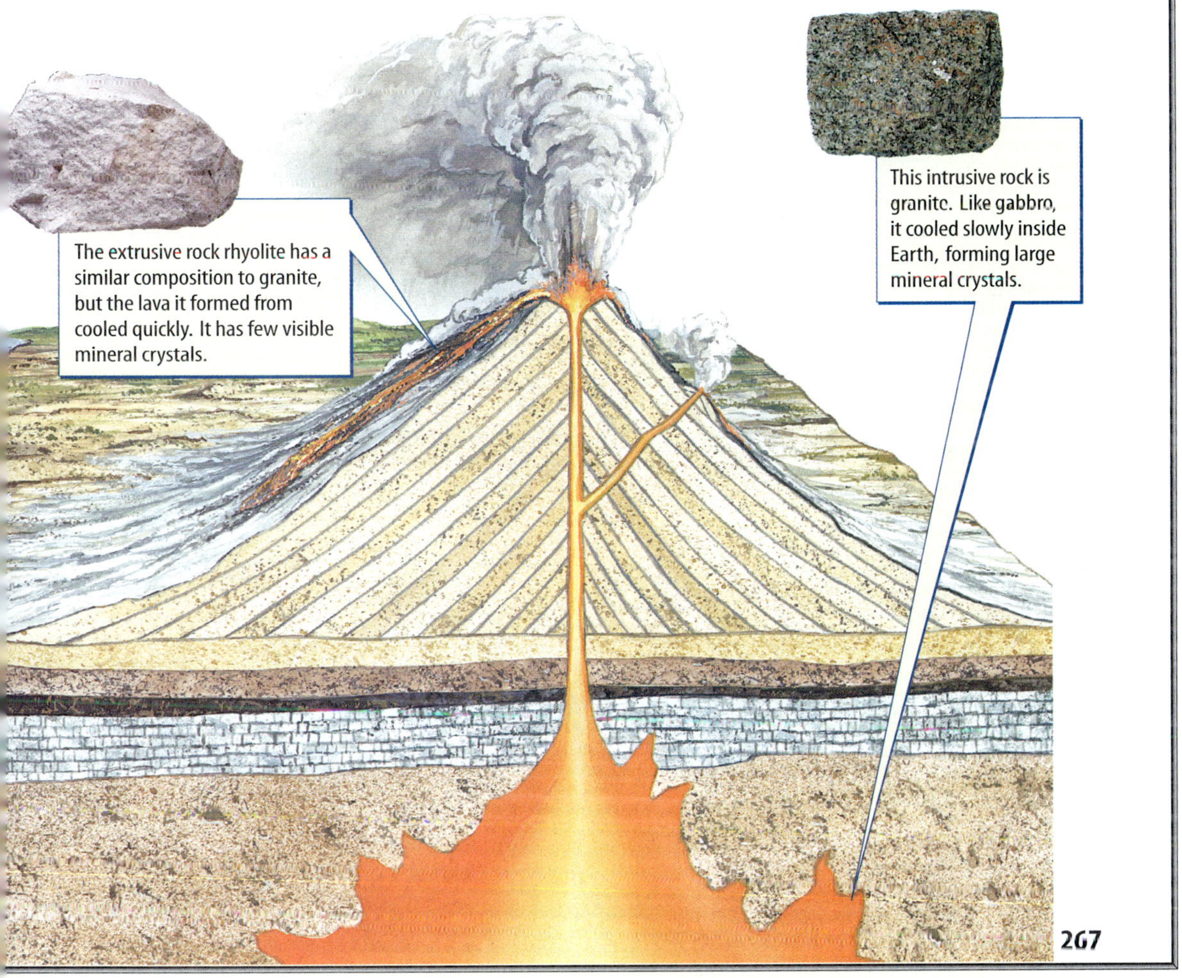

The extrusive rock rhyolite has a similar composition to granite, but the lava it formed from cooled quickly. It has few visible mineral crystals.

This intrusive rock is granite. Like gabbro, it cooled slowly inside Earth, forming large mineral crystals.

267

INTEGRATE Social Studies

Obsidian Uses Obsidian provides extremely sharp cutting edges that have been used from ancient to modern times, and can be fashioned into beautiful artwork. Obsidian trade routes ran throughout the Americas by 200 B.C. and throughout Asia as early as 9000 B.C.

Reading Check

Answer Intrusive rocks have crystals that can be easily seen; not all crystals in extrusive rocks can be easily seen.

INTEGRATE Physics

Thermal Energy Radioactive isotopes such as ^{40}K, ^{235}U, ^{238}U, ^{40}Ca, and many others are contained in minerals and other Earth materials. The decay processes that such isotopes undergo release energy in the form of heat.

Quick Demo

Crystal Formation

Materials one strip of paper and two pieces of tape per student

Estimated Time ten minutes

Procedure Tell students that when you say "go," they are to form loops from their paper and attach them to other students' loops. They are to continue until you say "stop." At first, give students very little time to accomplish the task. Then try giving them more time. Students should note that a large number of small loops formed in less time, while fewer and longer loops formed when more time was taken. Point out that this models crystal formation in igneous rocks, with larger crystals growing during longer, slower periods of cooling.

Science Journal

Igneous Rock Formation Have students research and write a creative short story in their Science Journals about what happens to a mineral crystal in magma or lava as an igneous rock forms. Their stories should include where cooling occurs, how long it takes, and what size the crystals are when the rock has solidified. L2 LS **Linguistic** P

NATIONAL GEOGRAPHIC

Visualizing Igneous Rock Features

Have students examine the pictures and read the captions. Then ask the following questions.

How do sills and dikes differ? Sills are features that run parallel to the rock layers, while dikes cut across the rock layers.

Which is older, the sill or the rock surrounding it? Explain. The rock surrounding the sill is older. It was in place before the sill formed as magma was forced between existing rock layers.

Activity

Intrusive Igneous Rocks Have small groups of students research an area where intrusive igneous rock features can be visited, such as Yosemite National Park, California; Shiprock, New Mexico; Big Bend National Park, Texas; Devils Tower National Monument, Wyoming; or Navajo Volcanic Field, Arizona. Ask each group to prepare a travel brochure for the area that includes information on the intrusive features. L2 COOP LEARN LS **Linguistic**

NATIONAL GEOGRAPHIC **VISUALIZING IGNEOUS ROCK FEATURES**

Figure 15

Intrusive igneous rocks are formed when a mass of magma is forced upward toward Earth's surface and then cools before emerging. The magma cools in a variety of ways. Eventually the rocks may be uplifted and erosion may expose them at Earth's surface. A selection of these formations is shown here.

▶ This dike in Israel's Negev Desert formed when magma squeezed into cracks that cut across rock layers.

▶ A batholith is a very large igneous rock body that forms when rising magma cools below the ground. Towering El Capitan, right, is just one part of a huge batholith. It looms over the entrance to the Yosemite Valley.

▲ Sills such as this one in Death Valley, California, form when magma is forced into spaces that run parallel to rock layers.

▶ Volcanic necks like Shiprock, New Mexico, form when magma hardens inside the vent of a volcano. Because the volcanic rock in the neck is harder than the volcanic rock in the volcano's cone, only the volcanic neck remains after erosion wears the cone away.

Differentiated Instruction

Challenge Have students research and write brief reports on the relationship between volcanic pipes, such as those near Kimberly, South Africa and in northern Canada, and diamonds. Students should find that the pipes formed from the magma that carried the diamonds toward Earth's surface from the mantle. L3

Sedimentary Rocks

Figure 16 The layers in these rocks are the different types of sedimentary rocks that have been exposed at Sedona, in Arizona. **Explain** *what causes the layers seen in sedimentary rocks.*

Pieces of broken rock, shells, mineral grains, and other materials make up what is called sediment (SE duh munt). The sand you squeeze through your toes at the beach is one type of sediment. As shown in **Figure 16,** sediment can collect in layers to form rocks. These are called **sedimentary** (sed uh MEN tuh ree) **rocks.** Rivers, ocean waves, mudslides, and glaciers can carry sediment. Sediment also can be carried by the wind. When sediment is dropped, or deposited, by wind, ice, gravity, or water, it collects in layers. After sediment is deposited, it begins the long process of becoming rock. Most sedimentary rocks take thousands to millions of years to form. The changes that form sedimentary rocks occur continuously. As with igneous rock, there are several kinds of sedimentary rocks. They fall into three main categories.

Reading Check *How is sediment transported?*

Detrital Rocks When you mention sedimentary rocks, most people think about rocks like sandstone, which is a detrital (dih TRI tuhl) rock. Detrital rocks, shown in **Figure 17,** are made of grains of minerals or other rocks that have moved and been deposited in layers by water, ice, gravity, or wind. Other minerals dissolved in water act to cement these particles together. The weight of sediment above them also squeezes or compacts the layers into rock.

Figure 17 Four types of detrital sedimentary rocks include shale, siltstone, sandstone, and conglomerate.

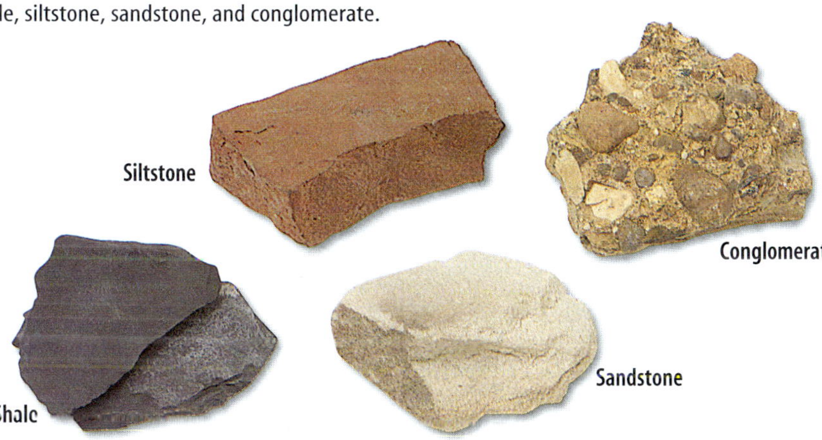

Caption Answer

Figure 16 Sediment normally is deposited in horizontal layers parallel to Earth's surface.

Reading Check

Answer Possible answers: rivers, ocean waves, wind, mudslides, and glaciers

Make a Model

Detrital Rocks Have students make a model that shows how detrital rocks form from broken pieces of other rocks. Provide teams of students with small chips of wood for a model of rock fragments and a bottle of wood glue or white glue. Have students mix the wood chips into the glue and then place the mixture between two sheets of waxed paper. Tell them to apply pressure to the mixture and then set it aside to dry. When it is dry, students will have a model of a detrital rock. L2 ELL COOP LEARN LS **Kinesthetic**

Visual Learning

Figure 17 The sizes of grains in a detrital rock are related to the energy of transport. For example, it takes fast-flowing water to move large particles. Have students rank the rocks shown in order of increasing energy of transport. shale, siltstone, sandstone, conglomerate L2

Curriculum Connection

Geography Have students pinpoint the location of the Bonneville Salt Flats and find out what processes occurred to form the salt flats. The Bonneville Salt Flats cover about 260 km^2 (100 mi^2) in northwestern Utah. The flats, part of the Great Salt Lake Desert, are a remnant of the bed of an ancient lake formed about 30,000 years ago. As the water evaporated, the flats were formed from accumulating salts. L3

Mini LAB

Purpose Students make models of an organic sedimentary rock with a clastic texture. L1 ELL COOP LEARN LS **Kinesthetic**

Materials small, disposable aluminum pie pan; broken macaroni (about 2 cups); white glue; water; fossiliferous limestone; mixing bowl

Teaching Strategies Suggest that students use macaroni of different shapes and sizes.

Analysis

1. The glue solution was the cement that held the "rock" together. It represents the calcite that holds limestone together.
2. Suggest that students use shape and texture to match the macaroni in their rocks to intact macaroni fossils.

Assessment

Portfolio Have students make labeled posters of their artificial fossil drawings and of actual fossil drawings from true fossil rock samples. Ask students to share their posters with the class. Encourage them to include their posters in their portfolios. Use **Performance Assessment in the Science Classroom,** p. 145.

Try at Home

Mini LAB

Modeling How Fossils Form Rocks

Procedure

1. Fill a small **aluminum pie pan** with pieces of broken **macaroni.** These represent various fossils.
2. Mix 50 mL of **white glue** into 250 mL of **water.** Pour this solution over the macaroni and set it aside to dry.
3. When your fossil rock sample has set, remove it from the pan and compare it with an actual **fossil limestone** sample.

Analysis

1. Explain why you used the glue solution and what this represents in nature.
2. Using whole macaroni samples as a guide, match the macaroni "fossils" in your "rock" to the intact macaroni. Draw and label them in your **Science Journal.**

Try at Home

Identifying Detrital Rocks To identify a detrital sedimentary rock, you use the size of the grains that make up the rock. The smallest, clay-sized grains feel slippery when wet and make up a rock called shale. Silt-sized grains are slightly larger than clay. These make up the rougher-feeling siltstone. Sandstone is made of yet larger, sand-sized grains. Pebbles are larger still. Pebbles mixed and cemented together with other sediment make up rocks called conglomerates (kun GLAHM ruts).

Chemical Rocks Some sedimentary rocks form when seawater, loaded with dissolved minerals, evaporates. Chemical sedimentary rock also forms when mineral-rich water from geysers, hot springs, or salty lakes evaporates, as shown in **Figure 18.** As the water evaporates, layers of the minerals are left behind. If you've ever sat in the Sun after swimming in the ocean, you probably noticed salt crystals on your skin. The seawater on your skin evaporated, leaving behind deposits of halite. The halite was dissolved in the water. Chemical rocks form this way from evaporation or other chemical processes.

Organic Rocks Would it surprise you to know that the chalk your teacher is using on the chalkboard might also be a sedimentary rock? Not only that, but coal, which is used as a fuel to produce electricity, also is a sedimentary rock.

Chalk and coal are examples of the group of sedimentary rocks called organic rocks. Organic rocks form over millions of years. Living matter dies, piles up, and then is compressed into rock. If the rock is produced from layers of plants piled on top of one another, it is called coal. Organic sedimentary rocks also form in the ocean and usually are classified as limestone.

Figure 18 The minerals left behind after a geyser erupts form layers of chemical rock.

Fun Fact

Because organic sedimentary rocks form as a result of biochemical reactions, they can also be classified as biochemical rocks.

Teacher FYI

Clastic Textures Detrital sedimentary rocks, as well as any rock made of broken pieces of other rocks or fossils, are said to have clastic textures. For this reason, detrital sedimentary rocks often are called clastic rocks.

Figure 19 There are a variety of organic sedimentary rocks.

The pyramids in Egypt are made from fossiliferous limestone.

A thin slice through the limestone shows that it contains many small fossils.

Fossils Chalk and other types of fossiliferous limestone are made from the fossils of millions of tiny organisms, as shown in **Figure 19.** A fossil is the remains or trace of a once-living plant or animal. A dinosaur bone and footprint are both fossils.

section 2 review

Summary

Igneous Rock

- The chemistry of an igneous rock often is indicated by its color.
- Starting materials that form igneous rocks include lava and magma.

Sedimentary Rocks

- Sedimentary rocks form as layers. They originate because wind, water, and ice transport and deposit sediment on Earth's surface.
- Some rocks have grainy textures because they are composed of rock, mineral, or organic fragments cemented together by mineral-rich solutions.
- Other sedimentary rocks appear crystalline as they form directly from mineral-rich solutions.

Self Check

1. **Compare and contrast** the ways in which extrusive and intrusive igneous rocks are formed.
2. **Diagram** how each of the three kinds of sedimentary rock forms. List one example of each kind of rock: detrital, chemical, and organic.
3. **List** in order from smallest to largest the grain sizes used to describe detrital rocks.
4. **Think Critically** Why do igneous rocks that solidify underground cool so slowly?

Applying Skills

5. **Communicate** Research a national park where volcanic activity has taken place. Read about the park and the features that you'd like to see. Then describe the volcanic features in your Science Journal. Be sure to explain how each feature formed.

3 Assess

DAILY INTERVENTION

Check for Understanding

Logical-Mathematical How is salt that forms on your skin after swimming in the ocean similar to the formation of one type of chemical sedimentary rock? In both cases, deposits of salt form as water containing salt in solution evaporates. L1 LS

Reteach

Describing Grains Conglomerate and breccia are both detrital rocks that form from gravel-sized sediment. Obtain samples or pictures of conglomerate and breccia for students to observe. Have students infer which rock is formed from sediment that is deposited near its source. Abrasion has not smoothed the angular sediment in breccia; therefore, it must have formed close to its sediment source. L3 LS **Logical-Mathematical**

Assessment

Oral Why are rocks such as sandstone and shale called sedimentary rocks? They form from sediment. How can you distinguish extrusive from intrusive igneous rocks? Intrusive rocks contain crystals that can be seen with the unaided eye. Use **Performance Assessment in the Science Classroom,** p. 89.

section 2 review

1. Extrusive igneous rocks cool quickly from lava, producing no or small crystals. Intrusive igneous rocks cool slowly from magma, producing large crystals.
2. detrital rocks—pieces of other rocks, sandstone; chemical rocks—precipitation of minerals from solution, rock salt; organic rocks—from the remains of once-living things, coal
3. clay, silt, sand, and gravel
4. The surrounding rock insulates the cooling magma.
5. Students might research and describe Hawaii Volcanoes National Park, Sunset Crater Volcano National Park, Capulin Volcano National Monument, Mount Rainier National Park, Sunset Crater Volcano National Monument, Lava Beds National Monument, Lassen Volcanic National Park, or Big Bend National Park, among others.

1 Motivate

INTERACTIVE CHALKBOARD PowerPoint® Presentations

Bellringer

Section Focus Transparencies also are available on the Interactive Chalkboard CD-ROM. L2 ELL

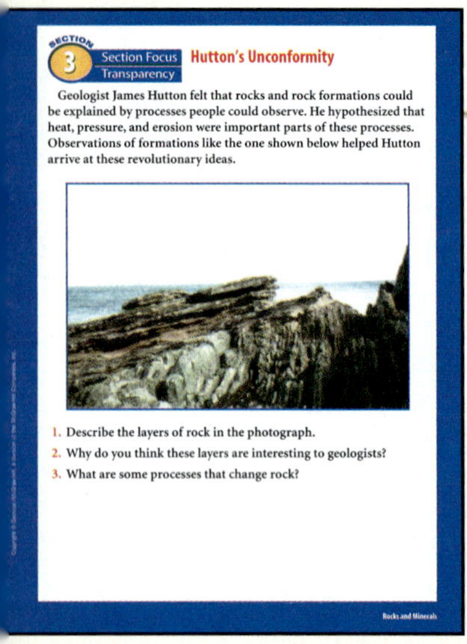
Section 3 Section Focus Transparency

Hutton's Unconformity

Geologist James Hutton felt that rocks and rock formations could be explained by processes people could observe. He hypothesized that heat, pressure, and erosion were important parts of these processes. Observations of formations like the one shown below helped Hutton arrive at these revolutionary ideas.

1. Describe the layers of rock in the photograph.
2. Why do you think these layers are interesting to geologists?
3. What are some processes that change rock?

Rocks and Minerals

Tie to Prior Knowledge

Granite Remind students of the characteristics of granite. Then show them gneiss, a metamorphic rock. Have them compare and contrast granite and gneiss.

Reading Check

Answer changes in pressure or temperature conditions, among others

section 3

Metamorphic Rocks and the Rock Cycle

as you read

What You'll Learn

- **Describe** the conditions needed for metamorphic rocks to form.
- **Explain** how all rocks are linked by the rock cycle.

Why It's Important

Metamorphic rocks and the rock cycle show that Earth is a constantly changing planet.

Review Vocabulary
pressure: force applied over a given area

New Vocabulary
- metamorphic rock
- foliated
- nonfoliated
- rock cycle

New Rock from Old Rock

Many physical changes on and within Earth are at work, constantly changing rocks. From low-temperature processes such as weathering and erosion, to high-temperature conditions that form molten rock material, new rocks are always forming. There are conditions in between those that form igneous and sedimentary rock that also produce new rocks. Pressures and temperatures increase as rocks are compressed or buried deeply, which can change the chemistry and grain sizes of rocks without even melting them. These conditions often happen where Earth's tectonic plates collide to form mountains, like those shown in **Figure 20.**

It can take millions of years for rocks to change. That's the amount of time that often is necessary for extreme pressure to build while rocks are buried deeply or continents collide. Sometimes existing rocks are cooked when magma is forced upward into Earth's crust, changing their mineral crystals. All these events can make new rocks out of old rocks.

Reading Check *What events can change rocks?*

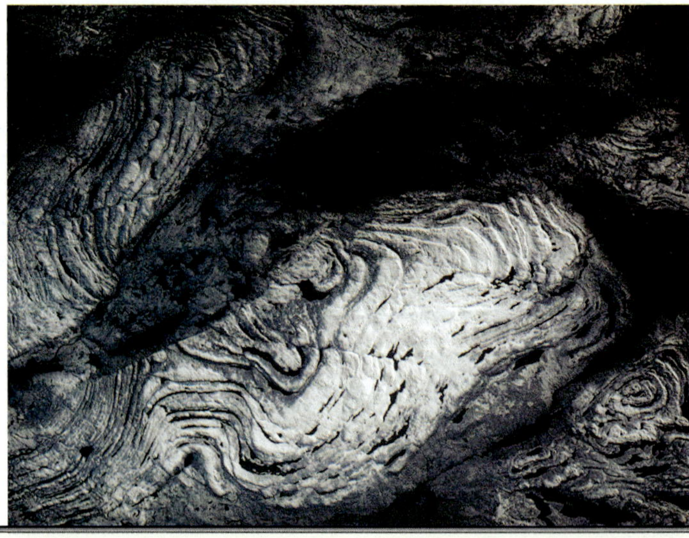

Figure 20 The rocks of the Labrador Peninsula in Canada were squeezed into spectacular folds. This photo was taken during the space shuttle *Challenger* mission *STS-41G* in 1984.

Section 3 Resource Manager

Chapter *FAST FILE* Resources
- Transparency Activity, pp. 46, 47–48
- Directed Reading for Content Mastery, pp. 21, 22
- Enrichment, p. 32
- Lab Worksheet, pp. 5–8
- Reinforcement, p. 29

Earth Science Critical Thinking/Problem Solving, p. 4

Home and Community Involvement, p. 43

Lab Management and Safety, p. 55

Figure 21 High pressure and temperature can cause existing rocks to change into new metamorphic rocks. **A** Granite can change to gneiss. **B** The sedimentary rock sandstone can become quartzite, and **C** limestone can change to marble.

Metamorphic Rocks Do you recycle your plastic milk jugs? After the jugs are collected, sorted, and cleaned, they are heated and squeezed into pellets. The pellets later can be made into useful new products. It takes millions of years, but rocks get recycled, too. This process usually occurs thousands of meters below Earth's surface where temperatures and pressures are high. New rocks that form when existing rocks are heated or squeezed but are not melted are called **metamorphic** (me tuh MOR fihk) **rocks.** The word *metamorphic* means "change of form." This describes well how some rocks take on a whole new look when they are under great temperatures and pressures.

Reading Check *What does the word metamorphic mean?*

Figure 21 shows three kinds of rocks and what they change into when they are subjected to the temperatures, pressures, and hot fluids involved in metamorphism. Not only do the resulting rocks look different, they have recrystallized and might be chemically changed, too. The minerals often align in a distinctive way.

2 Teach

Discussion

Pressure Conditions The pressure, as well as the temperature and fluid activity that cause rocks to change can be generated in several different ways. What Earth processes could generate the pressures needed for metamorphic changes? mountain building, movement along a fault, burial beneath other rock layers

Activity

Parent Rocks Provide students with samples of marble and quartzite and a glass plate. Explain that quartz in sandstone has a hardness of 7; calcite in limestone has a hardness of 3. Have them use this information and the glass plate to infer which rock forms from sandstone and which rock forms from limestone. Quartzite will scratch glass, marble will not. Thus, students should infer that quartzite forms from sandstone, and marble forms from limestone, which has been demonstrated in the field and experimentally. Students should be careful when using the glass plates. They can break if too much pressure is applied. L2 LS **Kinesthetic**

Reading Check

Answer change of form

Science Journal

Rock Hunt Have students go on a scavenger hunt in your community for examples of uses for metamorphic rocks and minerals. Examples might include slate roofs, marble buildings and statues, and graphite in pencils and lubricants. Have students write paragraphs in their Science Journals explaining why each example is used for the purpose observed. L2 P

SECTION 3 Metamorphic Rocks and the Rock Cycle **273**

Quick Demo

Inferring Intensity of Metamorphism

Materials hand samples of shale, slate, schist, gneiss, and phyllite

Estimated Time 10 minutes

Procedure Have students place all samples except shale in order of increasing grain size. Indicate to students that for fine-grained samples, the shinier the sample, the larger the grains. Demonstrate that, for this case, increased crystal size suggests increasing intensity of metamorphism, in the order of slate-phyllite-schist-gneiss. Inform students that mudrocks, such as shale, are the parent rocks for this progressive metamorphism.

Virtual Labs

Classification *How are rocks classified?*

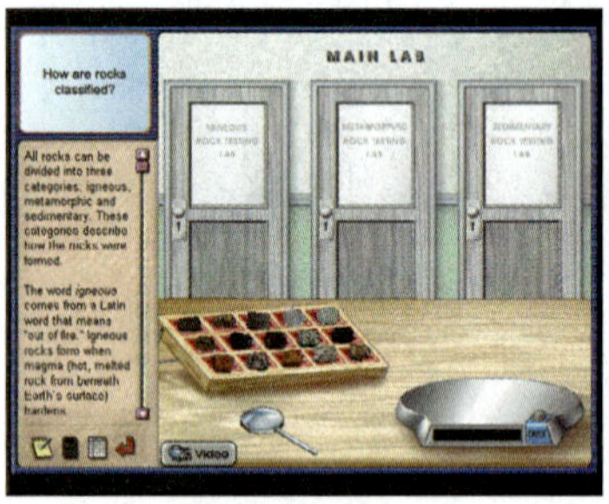

Figure 22 There are many different types of metamorphic rocks.

This statue is made from marble, a non-foliated metamorphic rock.

The roof of this house is made of slate, a foliated metamorphic rock.

Science Online

Topic: Rock Types

Visit red.msscience.com for Web links to information about types of metamorphic rocks.

Activity Make a two-column table with *Foliated* and *Nonfoliated* as table headings at the top. Find three examples of each of these metamorphic rock classifications. List minerals commonly found in each example.

Types of Changed Rocks New metamorphic rocks can form from any existing type of rock—igneous, sedimentary, or metamorphic. A physical characteristic helpful for classifying all rocks is the texture of the rocks. This term refers to the general appearance of the rock. Texture differences in metamorphic rocks divide them into two main groups—foliated (FOH lee ay tud) and nonfoliated, as shown in **Figure 22.**

Foliated rocks have visible layers or elongated grains of minerals. The term *foliated* comes from the Latin *foliatus,* which means "leafy." These minerals have been heated and squeezed into parallel layers, or leaves. Many foliated rocks have bands of different-colored minerals. Slate, gneiss (NISE), phyllite (FIH lite), and schist (SHIHST) are all examples of foliated rocks.

Nonfoliated rocks do not have distinct layers or bands. These rocks, such as quartzite, marble, and soapstone, often are more even in color than foliated rocks. If the mineral grains are visible at all, they do not seem to line up in any particular direction. Quartzite forms when the quartz sand grains in sandstone recrystallize after they are squeezed and heated. You can form ice crystals in a similar way if you squeeze a snowball. The presssure from your hands creates grains of ice inside the ball.

Differentiated Instruction

English-Languge Learners Remind students that calcite reacts with acid. Metamorphic rocks formed from limestone, such as marble, also contain calcite and should also react with acid. Have students test some metamorphic rocks to determine whether any calcite ispresent. Caution students to wear safety goggles and aprons when testing rocks. L2 ELL LS

Teacher FYI

Slate exhibits rock cleavage, which means that it breaks along smooth, flat surfaces. For this reason, and because slate is relatively impermeable to water, it often is used as a flooring and roofing material.

The Rock Cycle

Rocks are changing constantly from one type to another. If you wanted to describe these processes to someone, how would you do it? Scientists have created a model called the **rock cycle** to describe how different kinds of rock are related to one another and how rocks change from one type to another. Each rock is on a continuing journey through the rock cycle, which is shown in diagram form in **Figure 23.** A trip through the rock cycle can take millions of years.

Figure 23 This diagram of the rock cycle shows how rocks are recycled constantly from one kind of rock to another.

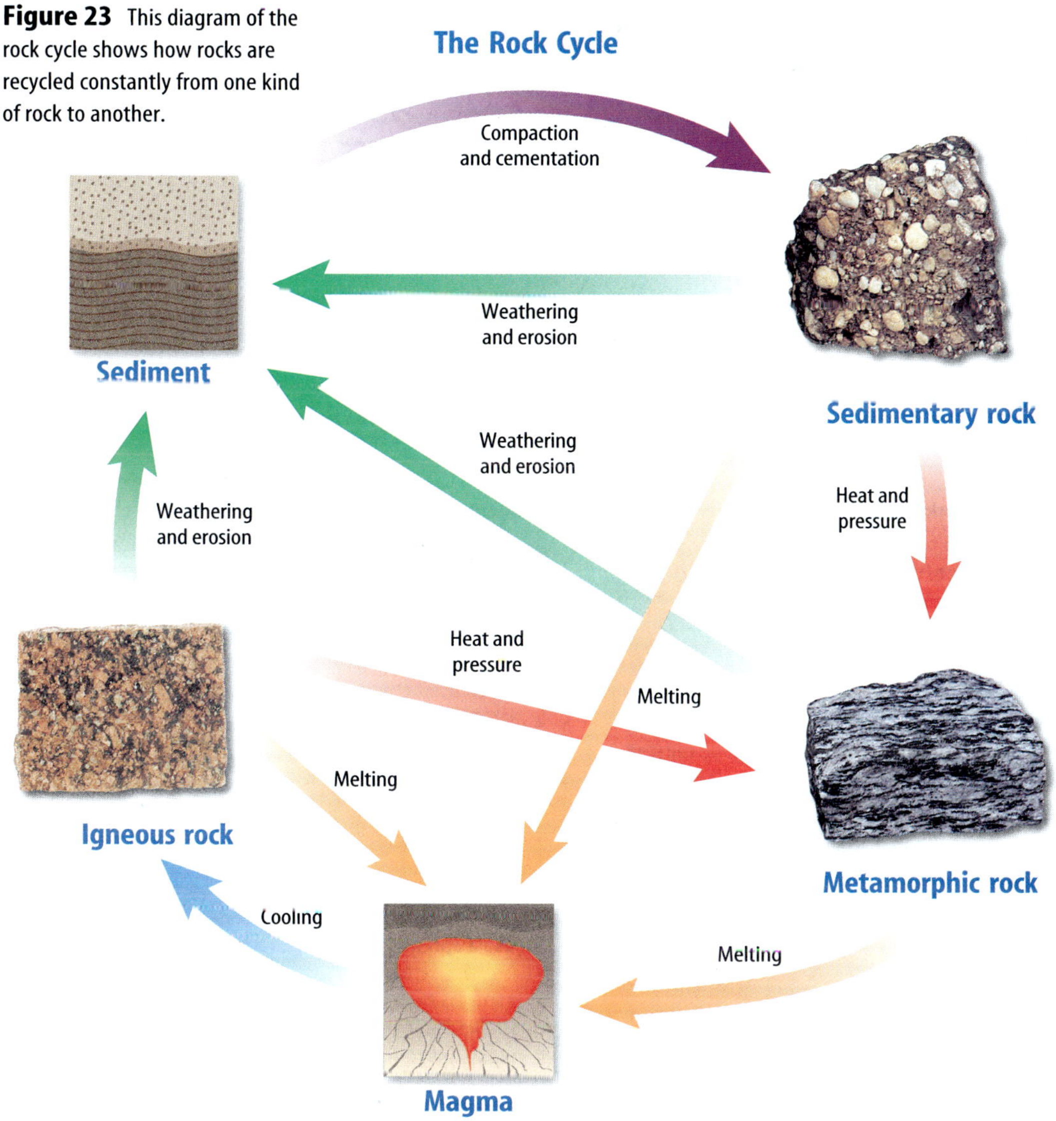

Teacher FYI

Index Minerals Certain minerals produced by specific conditions of metamorphism can be used to indicate what degree of metamorphism has occurred. If these minerals are found in a rock, the conditions of metamorphism, such as pressure or temperature, that the rock underwent can be estimated. Minerals of this type are referred to as *index minerals*. Some examples are chlorite, garnet, staurolite, and sillimanite.

Visual Learning

Figure 23 Emphasize that each arrow represents a process that changes rock from one form to another. Have students write a separate sentence that describes each arrow in the rock cycle diagram. L2 LS **Linguistic**

Cultural Diversity

Archaeology Rocks and minerals have been important resources for a wide variety of cultures for a long time.

- The oldest stone tool from Kenya dates from about 2.6 million years ago.
- Some rocks and minerals, such as obsidian and flint, have been fashioned into many types of cutting tools.
- The earliest paintings, preserved in caves throughout the world, are masterpieces created with chalk, various shades of ochre and other materials derived from minerals and rocks.

Reading Check

Answer If metamorphic rock is heated enough to melt, magma forms. Once the magma cools and crystallizes, igneous rock forms.

3 Assess

DAILY INTERVENTION

Check for Understanding

Visual-Spatial Have students sketch foliated and nonfoliated metamorphic rocks. Ask individual students to describe what their sketches show. L2 ELL LS

Reteach

Rock Cycle Work with students to devise a three-dimensional model of the rock cycle. Use hand specimens of rocks and sediment (if available), and student-made illustrations.

Assessment

Portfolio Have students write a letter in which they describe a rock. The letter must describe the rock well enough without naming it that another student can identify it. Use **Performance Assessment in the Science Classroom,** p. 139.

Figure 24 This lava in Hawaii is flowing into the ocean and cooling rapidly.

The Journey of a Rock Pick any point on the diagram of the rock cycle in **Figure 23,** and you will see how a rock in that part of the cycle could become any other kind of rock. Start with a blob of lava that oozes to the surface and cools, as shown in **Figure 24.** It forms an igneous rock. Wind, rain, and ice wear away at the rock, breaking off small pieces. These pieces are called sediment. Streams and rivers carry the sediment to the ocean, where it piles up over time. The weight of sediment above compresses the pieces below. Mineral-rich water seeps through the sediment and glues, or cements, it together. It becomes a sedimentary rock. If this sedimentary rock is buried deeply, pressure and heat inside Earth can change it into a metamorphic rock. Metamorphic rock deep inside Earth can melt and begin the cycle again. Rocks on Earth are changed over millions of years. These processes are taking place right now.

Reading Check *Describe how a metamorphic rock might change into an igneous rock.*

section 3 review

Summary

New Rock from Old Rock

- Changing conditions can cause new minerals to form, or the same minerals to change form as they align and recrystallize.
- Large-scale formation of metamorphic rock often occurs where tectonic plates collide.
- Metamorphic rocks sometimes are classified according to the textures they exhibit.
- Metamorphic rock textures can be foliated or nonfoliated.

The Rock Cycle

- Processes that are part of the rock cycle change rocks slowly through time.
- Igneous, sedimentary, and metamorphic rocks constantly are changing and exchanging matter through processes such as melting, weathering, and changing temperature and pressure.
- There is no beginning and no end to the rock cycle.

Self Check

1. **Identify** two factors that can produce metamorphic rocks.
2. **List** examples of foliated and nonfoliated rocks. Explain the difference between the two types of metamorphic rocks.
3. **Explain** Igneous rocks and metamorphic rocks can form at high temperatures and pressures. What is the difference between these two rock types?
4. **Explain** what the rock cycle describes.
5. **Think Critically** Trace the journey of a piece of granite through the rock cycle. Explain how this rock could be changed from an igneous rock to a sedimentary rock and then to a metamorphic rock.

Applying Skills

6. **Use a Spreadsheet** Using a spreadsheet program, create a data table to list the properties of rocks and minerals that you have studied in this chapter. After you've made your table, cut and paste the rows to group like rocks and minerals together.

Science Online red.msscience.com/self_check_quiz

section 3 review

1. two of the following: heat, pressure, reactive fluids
2. Foliated: gneiss, slate, phyllite, schist (minerals aligned in layers); nonfoliated: marble, quartzite, soapstone (minerals randomly oriented)
3. Igneous rocks form from melted rock material (magma or lava). Metamorphic rocks can form as a result of increases in temperature and pressure without melting.
4. It shows how one rock type is changed to another in a never-ending cycle.
5. Igneous rock (granite) is weathered into sediment. Sediment is transported and deposited where it is compacted and cemented to form sedimentary rocks. Sedimentary rocks are squeezed, heated, and changed chemically to form metamorphic rocks.
6. Student tables will vary. Accept logical groupings.

LAB

Gneiss Rice

You know that metamorphic rocks often are layered. But did you realize that individual mineral grains can change in orientation? This means that the grains can line up in certain directions. You'll experiment with rice grains in clay to see how foliation is produced.

Real-World Question

What conditions will cause an igneous rock to change into a metamorphic rock?

Goals

- **Investigate** ways rocks are changed.
- **Model** a metamorphic rock texture.

Materials

rolling pin
lump of modeling clay
uncooked rice (wild rice, if available) (200 g)
granite sample
gneiss sample

Safety Precautions

WARNING: *Do not taste, eat, or drink any materials used in the lab.*

Procedure

1. **Sketch** the granite specimen in your Science Journal. Be sure that your sketch clearly shows the arrangement of the mineral grains.
2. Pour the rice onto the table. Roll the ball of clay in the rice. Some of the rice will stick to the outside of the ball. Knead the ball until the rice is spread out fairly evenly. Roll and knead the ball again, and repeat until your clay sample has lots of "minerals" distributed throughout it.
3. Using the rolling pin, roll the clay until it is about 0.5 cm thick. Don't roll it too hard. The grains of rice should be pointing in different directions. Draw a picture of the clay in your Science Journal.

4. Take the edge of the clay closest to you and fold it toward the edge farthest from you. Roll the clay in the direction you folded it. Fold and roll the clay in the same direction several more times. Flatten the lump to 0.5 cm in thickness again. Draw what you observe in your "rock" and in the gneiss sample in your Science Journal.

Conclude and Apply

1. **Describe** What features did the granite and the first lump of clay have in common?
2. **Explain** what force caused the positions of rice grains in the lump of clay to change. How is this process similar to and different from what happens in nature?

Communicating Your Data

Refer to your Science Journal diagrams and the rock samples provided for you in this lab and make a poster relating this lab to processes in the rock cycle. Be sure to include diagrams of what you did, as well as information on how similar events occur in nature. **For more help, refer to the Science Skill Handbook.**

BENCH TESTED

Real-World Question

Purpose Students demonstrate how pressure causes a realignment of mineral grains, forming a metamorphic rock. L2 ELL COOP LEARN LS **Kinesthetic**

Process Skills observe, infer, recognize cause and effect, interpret scientific illustrations, make models

Time Required 30 minutes

Procedure

Teaching Strategy Have students roll small amounts of clay into the rice. Then put smaller pieces together and knead the clay until the rice is thoroughly mixed in.

Conclude and Apply

1. The granite has many mineral crystals arranged randomly throughout. The clay has many rice grains distributed randomly.
2. Compression forced rice grains to change orientation. In nature, compression forces mineral grains to realign. Heat and chemical changes caused by fluids also can cause metamorphism.

Assessment

Process Provide students with three different metamorphic rocks and the rocks from which they formed. Have students determine which rocks changed to form a foliated metamorphic rock, similar to the layered clay rock in the activity. Rocks to use are sandstone-quartzite, granite-gneiss, and shale-slate. Use **Performance Assessment in the Science Classroom,** p. 89.

Communicating Your Data

Encourage students to include images of granite and photos of their first lump of clay to show what they have in common. Images of gneiss could be included along with a photo of the clay after it has been flattened to represent metamorphism.

BENCH TESTED

Real-World Question

Purpose Students will test a variety of mineral samples and use a field guide to identify each sample.

Process Skills test, observe, collect and interpret data, infer, identify

Time Required one class period

Materials It is not necessary that each group have the same set of mineral samples to classify. Try to have samples for each group that vary enough in their properties so students will be able to easily identify them. Number the putty knives and collect them before class ends.

Alternate Materials If specimens do not have numbers tagged or painted on them, you might put them in numbered containers. Students should replace each sample in its correct container.

You also might want to have students test samples for attraction to a magnet.

Safety Precautions Remind students to keep goggles on at all times. Have students wash their hands after they are finished handling samples.

LAB

Classifying Minerals

Goals

- **Test** and observe important mineral characteristics.

Materials

set of minerals
magnifying lens
putty knife
streak plate
Mohs scale
minerals field guide

Safety Precautions

WARNING: *Be careful when using a knife. Never taste any materials used in a lab.*

Real-World Question

Hiking along a trail, you encounter what looks like an interesting mineral. You notice that it is uniform in color and shows distinct crystal faces. You think it must be valuable and want to identify it, so you open a guidebook to rocks and minerals. What observations must you make in order to identify it? What tests can you perform in the field?

Procedure

1. Copy the data table into your Science Journal. Based on your observations and hardness tests, you will fill in columns 2 through 6. In the sixth column—"Scratches which samples?"—you will list the number of each mineral sample that this sample is able to scratch. This information will allow you to rank each sample from softest to hardest. Comparing these ranks to Mohs scale should help identify the mineral.
2. Obtain a classroom set of minerals.

Alternative Inquiry Lab

Student Choice Have groups of four students design a flowchart that they will use to identify minerals. They may use any order of physical property tests that they wish in the chart. After the lab, students should evaluate their flowcharts for efficiency in mineral identification. If time allows, have them write about how they could improve their flowcharts.

3. **Observe** each sample and conduct appropriate tests to complete as much of your data table as possible. Consult the *Minerals* Reference Handbook at the back of this book to help fill in the last column.

Mineral Characteristics

Sample Number	Crystal Shape	Cleavage/ Fracture	Color	Streak and Luster	Scratches which samples?	Hardness Rank	Mineral Name
1							
2							
3							
4							
5			Answers will vary.				
...							
No. of samples							

Analyze Your Data

1. **Identify** each mineral based on the information in your data table.
2. **Evaluate** Did you need all of the information in the table to identify each mineral? Explain why or why not.
3. **Explain** which characteristics were easy to determine. Which were somewhat more difficult?

Conclude and Apply

1. **Evaluate** Were some characteristics more useful as indicators than others?
2. **Apply** Would you be able to identify minerals in the field after doing this activity? Which characteristics would be easy to determine on the spot? Which would be difficult?
3. **Describe** how your actions in this lab are similar to those of a scientist. What additional work might a scientist have done to identify these unknown minerals?

Communicating Your Data

Create a visually appealing poster showing the minerals in this lab and the characteristics that were useful for identifying each one. Be sure to include informative labels on your poster.

Procedure

Teaching Strategy Students should gather data on one mineral sample at a time to avoid confusion.

Analyze Your Data

Expected Outcome Most results will indicate that students are able to use the characteristics of minerals to identify samples.

Answers to Questions

1. Answers will depend on the samples.
2. Answers will vary. Some minerals, such as sulfur, may be easy to identify based only on color.
3. Color and streak are easy to determine. Crystal shape and cleavage/fracture may be more difficult for students to determine.

Error Analysis Have each group compare their results with your key to the minerals. If the identifications of the samples do not agree, have students list possible reasons for the discrepancies.

Conclude and Apply

1. Hardness and streak are probably more useful than some other characteristics.
2. Color and possibly crystal shape and cleavage (if crystals were large) would be easy to spot in the field. Hardness and streak would be more difficult to test in the field.
3. Possible answers: actions involved use of science process skills; actions followed procedures. Additional tests might include testing for magnetism, testing with acid, and analyzing chemistry.

Assessment

Oral Ask students to explain why more than one characteristic should be observed before identifying a mineral sample. Use **Performance Assessment in the Science Classroom,** p. 89.

Communicating Your Data

Each group can present their poster to the class. Discuss any differing opinions as to the characteristics that were useful for identifying each mineral.

Oops! Accidents in SCIENCE

SOMETIMES GREAT DISCOVERIES HAPPEN BY ACCIDENT!

Content Background

In nature, gold is usually found as a pure metal, often in the form of nuggets or flakes. It is fairly widely distributed throughout the world and is often found associated with quartz and pyrite ("fool's gold"). In its pure state, gold is one of the most beautiful minerals. Its chemical symbol, Au, comes from the Latin word *aurum*, which means "shining dawn."

Much of the gold in California was recovered from placer deposits by panning. When miners swirled gold-bearing sand around in shallow pans, the sand and gravel washed over the rim, leaving the heavier gold flakes or nuggets behind.

Discussion

The Appeal of Gold Why did so many people rush to California to prospect for gold? Gold is very valuable and they wanted to become wealthy.

Why is gold such a valuable mineral? Possible answers: It is not a common mineral, so that makes it more valuable. It doesn't tarnish or corrode like most other metals.

Analyze the Event

- What might have happened if James Marshall had not known that "fool's gold" shatters when hit? Possible answer: He might not have realized that he had found gold and the gold rush might never have happened.
- Why do you think Sutter and Marshall tried to keep the discovery a secret? Possible answer: If other people hadn't found out about the gold, they would have been able to retrieve more gold for themselves and would have become very rich.

Going for the Gold

A time line history of the accidental discovery of gold in California

Sutter's Mill

1840 California is a quiet place. Only a few hundred people live in the small town of San Francisco.

1847 John Sutter hires James Marshall to build a sawmill on his ranch. Marshall and local Native Americans work quickly to harness the water power of the American River.

1848 On January 24, Marshall notices something glinting in the water. He hits it with a rock. Marshall knows that "fool's gold" shatters when hit. But this shiny metal bends. After more tests, Sutter and Marshall decide it is gold! They try to keep the discovery a secret, but word leaks out.

1849 The Gold Rush hits! A flood of people from around the world descends on northern California. Many people become wealthy—but not Marshall or Sutter. Because Sutter doesn't have a legal claim to the land, the U.S. government claims it.

1850 California becomes the thirty-first state.

1854 A giant nugget of gold, the largest known to have been discovered in California, is found in Calaveras County.

1864 California's gold rush ends. The rich surface deposits are largely exhausted.

1872 As thanks for his contribution to California's growth, the state legislature awards Marshall $200 a month for two years. This pension is renewed until 1878.

1880 His pension ended, Marshall is forced to earn a living through various odd jobs, receiving charity, and by selling his autograph. He attempts a lecture tour, but is unsuccessful.

1885 James Marshall dies with barely enough money to cover his funeral.

1890 California builds a bronze statue to honor Marshall.

1840 | 1850 | 1860 | 1870 | 1880 | 1890

Research Trace the history of gold from ancient civilizations to the present. How was gold used in the past? How is it used in the present? What new uses for gold have been discovered? Report to the class.

Science Online
For more information, visit red.msscience.com/time

Research Gold was made into simple pieces as early as 3000 B.C. By 2000 B.C. the techniques of filigree work were known to Minoan goldsmiths in Crete. Gold has been widely used for decorative purposes. Among other uses, it coats large windows to prevent the greenhouse effect, and electronic switches and connectors.

Resources for Teachers and Students

The California Gold Rush in American History, by Linda Jacobs Altman, Enslow Publishers, 1997

The California Gold Rush, by Jean F. Blashfield, Compass Point Books, 2001

Gold Rush: Hands-On Projects About Mining the Riches of California, by Jennifer Quasha, Powerkids Press, 2001

Reviewing Main Ideas

Section 1 Minerals—Earth's Jewels

1. Minerals are inorganic solid materials found in nature. They have a definite chemical makeup, and an orderly arrangement of atoms. Rocks are combinations of two or more minerals.
2. Physical properties of minerals are observed to help identify them.
3. Gems are minerals that are rare and beautiful.
4. Ores of useful materials must be mined and processed to extract the desired substance.

Section 2 Igneous and Sedimentary Rocks

1. Igneous rocks form when melted rock material from inside Earth cools and hardens. Extrusive rocks form above Earth's surface. Intrusive rocks solidify beneath the surface.
2. Sedimentary rocks formed from mineral or rock fragments are called detrital rocks.
3. Rocks formed as mineral-rich water evaporates are examples of chemical rocks. Rocks composed of fossils or plant remains are organic rocks.

Section 3 Metamorphic Rocks and the Rock Cycle

1. Metamorphic rocks form as a result of changing temperature, pressure, and fluid conditions inside Earth.
2. The rock cycle describes how all rocks are subject to constant change.

Visualizing Main Ideas

Copy and complete the concept map using the following terms and phrases: extrusive, organic, foliated, intrusive, chemical, nonfoliated, detrital, metamorphic, *and* sedimentary.

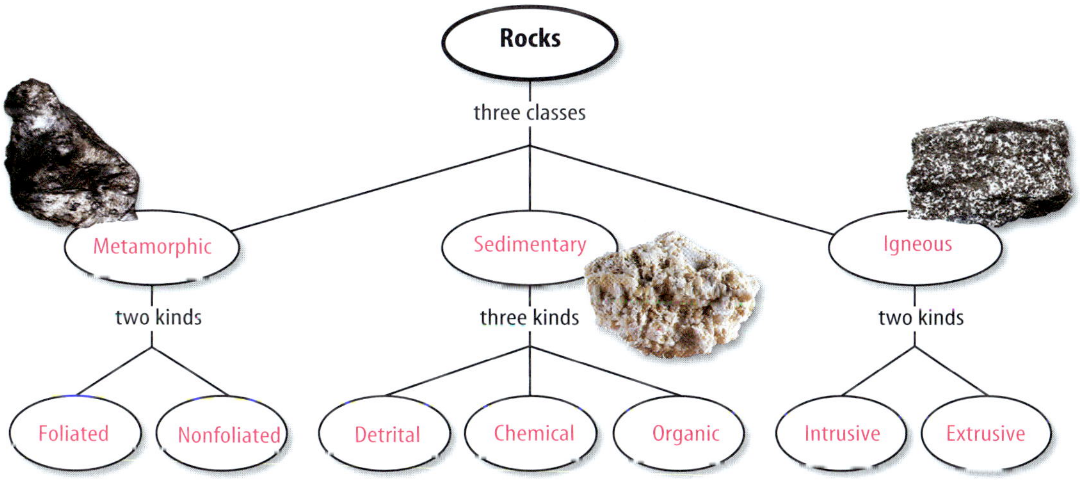

chapter 9 Study Guide

Reviewing Main Ideas

Summary statements can be used by students to review the major concepts of the chapter.

Visualizing Main Ideas

See student page.

Science Online
Visit red.msscience.com
/self_check_quiz
/interactive_tutor
/vocabulary_puzzlemaker
/chapter_review
/standardized_test

Assessment Transparency

For additional assessment questions, use the *Assessment Transparency* located in the transparency book.

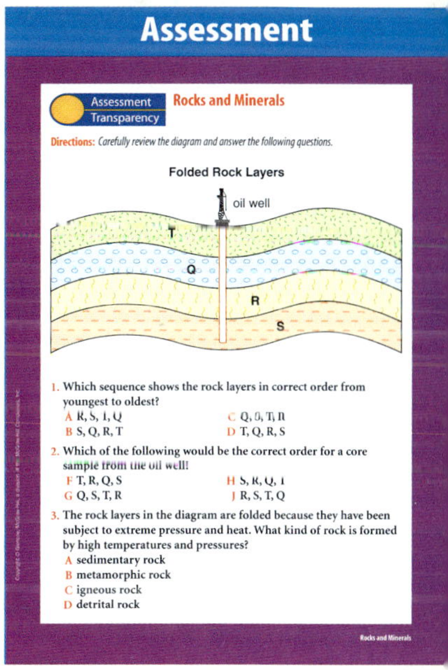

Assessment

Assessment Transparency **Rocks and Minerals**

Directions: *Carefully review the diagram and answer the following questions.*

Folded Rock Layers

oil well

T Q R S

1. Which sequence shows the rock layers in correct order from youngest to oldest?
   A R, S, T, Q
   B S, Q, R, T
   C Q, S, T, R
   D T, Q, R, S
2. Which of the following would be the correct order for a core sample from the oil well?
   F T, R, Q, S
   G Q, S, T, R
   H S, R, Q, T
   J R, S, T, Q
3. The rock layers in the diagram are folded because they have been subject to extreme pressure and heat. What kind of rock is formed by high temperatures and pressures?
   A sedimentary rock
   B metamorphic rock
   C igneous rock
   D detrital rock

Rocks and Minerals

Using Vocabulary

1. Minerals are natural, inorganic, crystalline solids. Most rocks are mixtures of more than one mineral.
2. Crystals are solid materials that have an orderly arrangement of atoms. Gems are rare and beautiful minerals that often are cut from crystals.
3. Cleavage is the breaking of a mineral along smooth, flat planes. Fracture is breakage with an uneven surface.
4. Hardness is a measure of how easily a mineral can be scratched. Streak is the color of the powdered mineral.
5. Most rocks are aggregates of minerals. The rock cycle is a model showing how rocks change.
6. Intrusive rocks form inside Earth. Extrusive rocks form at Earth's surface.
7. Igneous rock forms from magma or lava. Metamorphic rock forms when preexisting rocks are altered by heat, pressure, or fluids.
8. Foliated rocks have layers of aligned mineral grains. Nonfoliated rocks have mostly no layers.
9. A rock is an aggregate of minerals. An ore is a rock or mineral that can be mined at a profit.
10. Metamorphic rocks form from changes in temperature, pressure, or fluid conditions. Sedimentary rocks form from sediment or by precipitation from solution at temperatures below those that cause metamorphism.

Checking Concepts

11. D
12. D
13. D
14. A
15. B
16. D
17. A

chapter 9 Review

Using Vocabulary

crystal p. 258
extrusive p. 265
foliated p. 274
gem p. 262
igneous rock p. 265
intrusive p. 265
metamorphic rock p. 273
mineral p. 256
nonfoliated p. 274
ore p. 263
rock p. 256
rock cycle p. 275
sedimentary rock p. 269

Explain the difference between each pair of vocabulary words.

1. mineral—rock
2. crystal—gem
3. cleavage—fracture
4. hardness—streak
5. rock—rock cycle
6. intrusive—extrusive
7. igneous rock—metamorphic rock
8. foliated—nonfoliated
9. rock—ore
10. metamorphic rock—sedimentary rock

Checking Concepts

Choose the word or phrase that best answers the question.

11. When do metamorphic rocks form?
    A) when layers of sediment are deposited
    B) when lava solidifies in seawater
    C) when particles of rock break off at Earth's surface
    D) when heat and pressure change rocks

12. Which of the following must be true for a substance to be considered a mineral?
    A) It must be organic.
    B) It must be glassy.
    C) It must be a gem.
    D) It must be naturally occurring.

Use the illustration below to answer question 13.

13. What kind of rocks are produced by volcanic eruptions?
    A) detrital
    B) foliated
    C) organic
    D) extrusive

14. Which is true about how all detrital rocks form?
    A) form from grains of preexisting rocks
    B) form from lava
    C) form by evaporation
    D) form from plant remains

15. Which of the following describes what rocks usually are composed of?
    A) pieces
    B) minerals
    C) fossil fuels
    D) foliations

16. How can sedimentary rocks be classified?
    A) foliated or nonfoliated
    B) gems or ores
    C) extrusive or intrusive
    D) detrital, chemical, or organic

17. Which is true of all minerals?
    A) They are inorganic solids.
    B) They have a hardness of 4 or greater.
    C) They have a glassy luster.
    D) They can scratch a penny.

Science Online red.msscience.com/vocabulary_puzzlemaker

Use the ExamView® Pro Testmaker CD-ROM to:

- create multiple versions of tests
- create modified tests with one mouse click for inclusion students
- edit existing questions and add your own questions
- build tests aligned with state standards using built-in State Curriculum Tags
- change English tests to Spanish with one mouse click and vice versa

282 CHAPTER REVIEW

Thinking Critically

18. Classify Is a sugar crystal a mineral? Explain.

19. List some reasons why metal deposits in Antarctica are not considered to be ores.

20. Describe How is it possible to find pieces of gneiss, granite, and basalt in a single conglomerate?

21. Predict Would you expect to find a well-preserved dinosaur bone in a metamorphic rock like schist? Explain.

22. Explain how the mineral quartz could be in an igneous rock and in a sedimentary rock.

23. Classify Your teacher gives you two clear minerals. What quick test could you do in order to determine which is halite and which is calcite?

24. Concept Map Copy and complete this concept map about minerals.

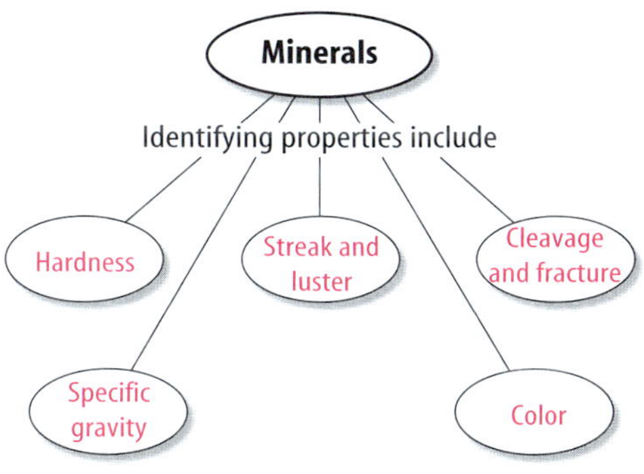

25. Test a Hypothesis Suppose your teacher gives you a glass plate, a nail, a copper penny, and a bar magnet. Using a word processing program on a computer, describe how you would use these items to determine the hardness and special property of the mineral magnetite. Refer to Mohs scale in **Table 1** for help.

Performance Activities

26. Make Models Determine what materials and processes you would need to use to set up a working model of the rock cycle. Describe the ways in which your model is accurate and the ways in which it falls short. Present your model to the class.

Applying Math

Use the table below to answer questions 27–29.

Modified Wentworth Scale, after Lane et. al., 1947

Grain Sizes (mm)	U.S. Standard Sieve Series	Grain Types	
2	No. 10	very coarse	
1	No. 18	coarse	
0.500	No. 35	medium	SAND
0.250	No. 60	fine	
0.125	No. 120	very fine	
0.062	No. 230	coarse	
0.031	—	medium	
0.016	—	fine	SILT
0.008	—	very fine	
0.004	—	coarse	
0.002	—	medium	
0.001	—		CLAY

27. Grain Type According to the table, if a rock contains grains that are 0.5 mm in dimension, what type of grains are they?

28. Filtering Which U.S. standard sieve would you use to filter out all sediment in a sample less than one-fourth of one millimeter?

29. Grain Size A siltstone contains grains that range in size from 0.031 to 0.008 mm. Convert this size range from millimeters to micrometers.

Science Online red.msscience.com/chapter_review

Assessment Resources

Reproducible Masters

Chapter *Fast File* Resources
Chapter Review, pp. 37–38
Chapter Tests, pp. 39–42
Assessment Transparency Activity, p. 49

Glencoe Science Web site
Chapter Review Test
Standardized Test Practice

Glencoe Technology
Assessment Transparency
Exam*View*® Pro Testmaker
MindJogger Videoquiz
Interactive Chalkboard

chapter 9 Review

Thinking Critically

18. No; sugar comes from plants. Minerals are inorganic.

19. Possible answers: The deposits are too small or the expense of mining in remote areas is too high.

20. If gneiss, granite, and basalt were all exposed to weathering and erosion, pieces of all three could be moved and deposited together to form a conglomerate.

21. No; the heat, pressure, and chemical changes that form metamorphic rocks likely would destroy the dinosaur bone.

22. Quartz forms in igneous rocks as magma cools. If such a rock is weathered, eroded, and deposited as sediment, the quartz can be cemented and compacted into sedimentary rock.

23. Possible answers: Dilute HCl will react with the calcite and not the halite. The calcite will produce a double image if looked through, but the halite will not.

24. See student page.

25. Try to scratch the glass, nail, and penny with the magnetite. If the magnetite scratches the item, the magnetite is harder than that item. Magnetite scratches the penny, the iron nail, and the glass plate. Its hardness is greater than 5.5, the hardness of the glass plate. Use the bar magnet to determine that magnetite is attracted to a magnet.

Performance Activities

26. Use **Performance Assessment in the Science Classroom,** p. 123 to evaluate students' models.

Applying Math

National Math Standards

4, 5, 9

27. medium to course

28. No. 60

29. 31 to 8 micrometers

FAST FILE

Answer Sheet A practice answer sheet can be found at red.msscience.com/answer_sheet.

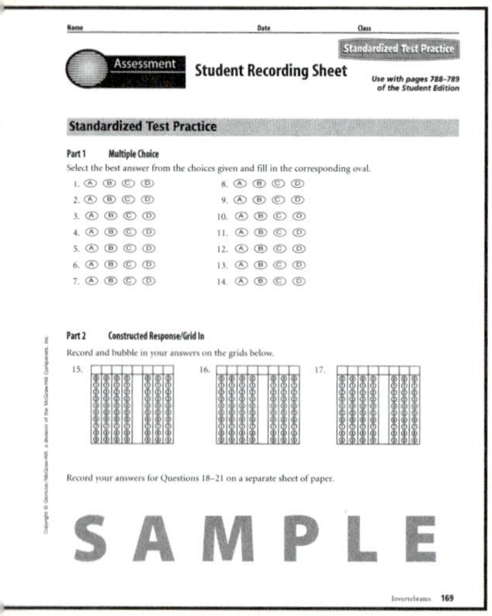
Name Date Class

Assessment Student Recording Sheet

Standardized Test Practice

Use with pages 788–789 of the Student Edition

Standardized Test Practice

Part 1 Multiple Choice

Select the best answer from the choices given and fill in the corresponding oval.

Part 2 Constructed Response/Grid In

Record and bubble in your answers on the grids below.

Record your answers for Questions 18–21 on a separate sheet of paper.

SAMPLE

Part 1 Multiple Choice

1. B
2. A
3. D
4. C
5. A
6. A
7. B
8. C
9. D

Part 2 Short Response

10. A rock is a mixture of minerals or other components. Minerals are homogeneous solids with a definite structure and characteristic chemical composition.
11. Organic sedimentary rocks form from the remains of living organisms. Examples include coal, fossiliferous limestone, and chalk.
12. Extrusive igneous rocks generally are finer grained because they crystallize on Earth's surface. Intrusive igneous rocks generally are coarser grained because they cool more slowly beneath Earth's surface.
13. Minerals can be identified by performing tests to characterize their physical properties. These then can be compared to the characteristics listed in mineral identification tables.
14. A batholith is a large body of igneous rock. They form when large bodies of magma cool beneath Earth's surface.
15. South Africa; 95
16. 35
17. 11.9%

Part 3 Open Ended

18. Most items could not be manufactured without minerals. Society depends on minerals for chemicals, metals, building materials, and more.
19. Mineral cleavage occurs when minerals break along smooth, planar surfaces that reflect planes of weakness in the mineral's structure. Fracture occurs when a mineral breaks along random or nonplanar surfaces.
20. Granite ⟹ Gneiss
Sandstone ⟹ Quartzite
Limestone ⟹ Marble

Part 1 Multiple Choice

Record your answers on the answer sheet provided by your teacher or on a sheet of paper.

Use the photo below to answer question 1.

1. Which special property is illustrated by the piece of calcite shown above?
   A. magnetism
   B. double refraction
   C. reaction to acid
   D. salty taste

2. What forms when lava cools so quickly that crystals cannot form?
   A. volcanic glass **C.** bauxite
   B. intrusive rock **D.** a gem

3. Which is the color of powdered mineral?
   A. hardness **C.** cleavage
   B. luster **D.** streak

4. Varieties of which mineral are most abundant in Earth's crust?
   A. quartz **C.** feldspar
   B. calcite **D.** gypsum

5. What is a solid material that has an orderly, repeating pattern of atoms?
   A. crystal
   B. gem
   C. ore
   D. rock

6. Which feature is a tabular body of igneous rock that is parallel to surrounding rock layers?
   A. sill **C.** batholith
   B. volcanic neck **D.** dike

7. Which type of rock is made mostly of clay?
   A. conglomerate **C.** siltstone
   B. shale **D.** sandstone

Use the diagram below to answer questions 8–9.

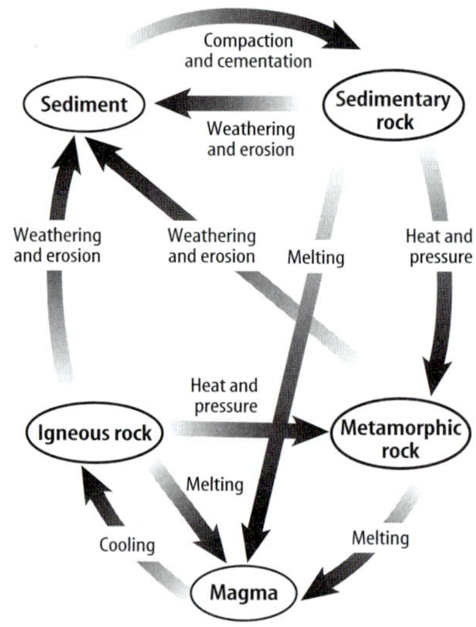

8. Which changes sediment into sedimentary rock?
   A. weathering and erosion
   B. heat and pressure
   C. compaction and cementation
   D. melting

9. Which type of rock forms when magma cools?
   A. sedimentary
   B. chemical
   C. metamorphic
   D. igneous

Part 2 | Short Response/Grid In

Record your answers on the answer sheet provided by your teacher or on a sheet of paper.

10. How is a rock different from a mineral?

11. How do organic sedimentary rocks form? List one example of an organic sedimentary rock.

12. How are extrusive igneous rocks different from intrusive igneous rocks?

13. How can minerals be identified by their physical properties?

14. What is a batholith? How do they form?

Use the table below to answer questions 15–17.

World Gold Production (metric tons)		
Country	**2001 Production**	**2002 Production**
United States	335	300
Australia	285	280
Canada	160	160
China	185	175
Indonesia	130	170
Peru	138	140
Russia	152	170
South Africa	402	395
Other countries	783	740
World total	2,570	2,530

15. Identify which single country produced the most gold in 2002. How many more metric tons did this country produce than the United States?

16. How many fewer metric tons did the U.S. produce in 2002 than in 2001?

17. What percentage of 2002 world gold production came from the United States?

Part 3 | Open Ended

Record your answers on a sheet of paper.

18. Why are minerals important to society?

19. Compare and contrast mineral cleavage and mineral fracture.

20. Complete the diagram below to show which rock would form when the rock on the left is exposed to heat and pressure.

Granite —heat and pressure→ ________

Sandstone —heat and pressure→ ________

Limestone —heat and pressure→ ________

21. Why must ores be processed after they are mined?

22. Why do sedimentary rocks occur in layers?

23. Which type of rock would you expect to exist deep in Earth's crust? Why?

24. Summarize the differences between granitic igneous rocks and basaltic igneous rocks.

25. Describe how a layer of rock containing fossils could be present in a mountain wall that is several thousand feet above sea level.

26. Explain at least two ways that a sedimentary rock can change to become a metamorphic rock. Include specific processes in your explanations.

27. Select two physical properties that you think are most reliable when identifying minerals. Explain your choices.

Test-Taking Tip

Eyes During the test, keep your eyes on your own paper. If you need to rest them, close them or look up at the ceiling.

21. Processing is required to extract and purify the desired metal.

22. The sediment from which sedimentary rocks form is deposited in layers.

23. metamorphic rocks; pressure and temperature in the lower crust are high.

24. Granitic igneous rocks generally have lighter color and contain more silica than basaltic igneous rocks.

25. A fossiliferous rock layer that formed near or below sea level could be uplifted later in its history by mountain building processes.

26. The sedimentary rock could come in contact with melted rock material or become buried deeply enough to induce metamorphism.

27. Hardness and cleavage/fracture, which depend on the crystal structure of a mineral, are among the most reliable properties for mineral identification.

Rubrics

The following rubrics are sample scoring devices for short response and open-ended questions.

Short Response

Points	Description
2	The student demonstrates a thorough understanding of the science of the task. The response may contain minor flaws that do not detract from the demonstration of a thorough understanding.
1	The student has provided a response that is only partially correct.
0	The student has provided a completely incorrect solution or no response at all.

Open Ended

Points	Description
4	The student demonstrates a thorough understanding of the science of the task. The response may contain minor flaws that do not detract from the demonstration of a thorough understanding.
3	The student demonstrates an understanding of the science of the task. The response is essentially correct and demonstrates an essential but less than thorough understanding of the science.
2	The student demonstrates only a partial understanding of the science of the task. Although the student may have used the correct approach to a solution or may have provided a correct solution, the work lacks an essential understanding of the underlying science concepts.
1	The student demonstrates a very limited understanding of the science of the task. The response is incomplete and exhibits many flaws.
0	The student provides a completely incorrect solution or no response at all.

chapter 10 Organizer

Section/Objectives	Standards		Labs/Features
	National	State/Local	
Chapter Opener	See pp. 16T–17T for a Key to Standards.		**Launch Lab:** Model Earth's Interior, p. 287 **Foldables,** p. 287
Section 1 Earth's Moving Plates 4 sessions 2 blocks 1. **Describe** how Earth's interior is divided into layers. 2. **Explain** how plates of Earth's lithosphere move. 3. **Discuss** why Earth's plates move.	National Content Standards: UCP.1, UCP.2, UCP.3, UCP.5, A.1, A.2, B.2, B.3, D.1, D.2		**Integrate Chemistry,** p. 290 **Science Online,** p. 293 **Visualizing Rift Valleys,** p. 294 **MiniLAB:** Modeling Tension and Compression, p. 295 **Lab:** Earth's Moving Plates, p. 298
Section 2 Uplift of Earth's Crust 4 sessions 2 blocks 4. **Describe** how Earth's mountains form and erode. 5. **Compare** types of mountains. 6. **Identify** the forces that shape Earth's mountains.	National Content Standards: UCP.1, UCP.2, UCP.3, UCP.5, A.1, A.2, B.2, B.3, D.1, D.2		**MiniLAB:** Modeling Mountains, p. 301 **Science Online,** p. 302 **Integrate Language Arts,** p. 304 **Applying Science:** How can glaciers cause land to rise?, p. 304 **Lab:** Isostasy, p. 306 **Science Stats:** Mountains, p. 308

Glencoe Exclusive! TeacherWorks™ All-In-One Planner and Resource Center

Lab Materials	Reproducible Resources	Section Assessment	Technology
Launch Lab: 4 pieces of different-colored clay, plastic knife	**Chapter *FAST FILE* Resources** Foldables Worksheet, p. 19 Note-taking Worksheets, pp. 33–34	Glencoe's Assessment Advantage	TeacherWorks includes: • Interactive Teacher Edition • Lesson Planner with calendar • Access to all program blacklines • Correlations to standards • Web links
MiniLAB: 2 bars of taffy **Lab:** 1-L beakers (2), food coloring, aluminum foil, pencil, rubber band, water, 2-cm paper squares (3), small clear-plastic cup *Need materials?* Contact Science Kit at 1-800-828-7777 or www.sciencekit.com on the Internet.	**Chapter *FAST FILE* Resources** Transparency Activity, p. 44 MiniLAB, p. 3 Enrichment, p. 31 Reinforcement, p. 29 Directed Reading, p. 22 Lab Worksheet, pp. 5–6 Lab Activity, pp. 9–13 Transparency Activity, pp. 47–48 **Cultural Diversity,** p. 47 **Earth Science Critical Thinking/Problem Solving,** pp. 8, 9, 10 **Reading and Writing Skill Activities,** p. 21	**Portfolio** Differentiated Instruction, p. 289 **Performance** MiniLAB, p. 295 Applying Skills, p. 297 **Content** Section Review, p. 297	Section Focus Transparency Teaching Transparency Guided Reading Audio Program Interactive Chalkboard CD-ROM
MiniLAB: clay, plastic knife, ball **Lab:** 5-cm × 5-cm × 2-cm wooden blocks (3), 10-cm × 35-cm × 15-cm clear-plastic storage box or other bin, water, permanent marker, ruler	**Chapter *FAST FILE* Resources** Transparency Activity, p. 45 MiniLAB, p. 4 Enrichment, p. 32 Reinforcement, p. 30 Directed Reading, pp. 23, 24 Lab Worksheet, pp. 7–8 Lab Activity, pp. 15–18 **Lab Management and Safety,** p. 50 **Home and Community Involvement,** p. 23 **Performance Assessment in the Science Classroom,** p. 39 **Cultural Diversity,** p. 27 **Earth Science Critical Thinking/Problem Solving,** p. 1	**Portfolio** Science Journal, p. 300 **Performance** MiniLAB, p. 301 Applying Science, p. 304 Applying Skills, p. 305 **Content** Section Review, p. 305	Section Focus Transparency Virtual Labs CD-ROM Guided Reading Audio Program Interactive Chalkboard CD-ROM Video Lab

End of Chapter Assessment

Glencoe's Assessment Advantage

Blackline Masters	Technology	Professional Series
Chapter *FAST FILE* Resources Chapter Review, pp. 37–38 Chapter Tests, pp. 39–42 **Standardized Test Practice,** pp. 43–46	MindJogger Videoquiz Virtual Labs CD-ROM Exam*View*® Pro Testmaker TeacherWorks CD-ROM Interactive Chalkboard CD-ROM	**Performance Assessment in the Science Classroom (PASC)**

Transparencies

Section Focus

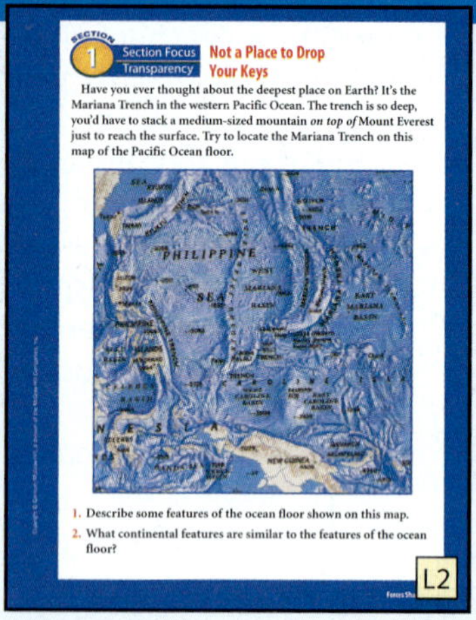
Section 1 Section Focus Transparency — Not a Place to Drop Your Keys

Have you ever thought about the deepest place on Earth? It's the Mariana Trench in the western Pacific Ocean. The trench is so deep, you'd have to stack a medium-sized mountain *on top of* Mount Everest just to reach the surface. Try to locate the Mariana Trench on this map of the Pacific Ocean floor.

1. Describe some features of the ocean floor shown on this map.
2. What continental features are similar to the features of the ocean floor?

L2

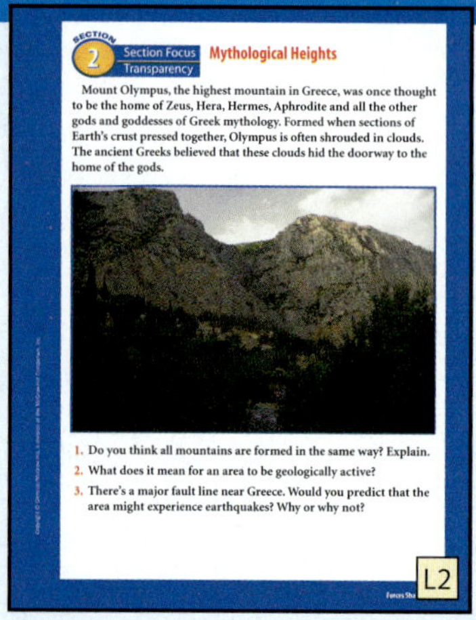
Section 2 Section Focus Transparency — Mythological Heights

Mount Olympus, the highest mountain in Greece, was once thought to be the home of Zeus, Hera, Hermes, Aphrodite and all the other gods and goddesses of Greek mythology. Formed when sections of Earth's crust pressed together, Olympus is often shrouded in clouds. The ancient Greeks believed that these clouds hid the doorway to the home of the gods.

1. Do you think all mountains are formed in the same way? Explain.
2. What does it mean for an area to be geologically active?
3. There's a major fault line near Greece. Would you predict that the area might experience earthquakes? Why or why not?

L2

This is a representation of key blackline masters available in the Teacher Classroom Resources. See Resource Manager boxes within the chapter for additional information.

Assessment

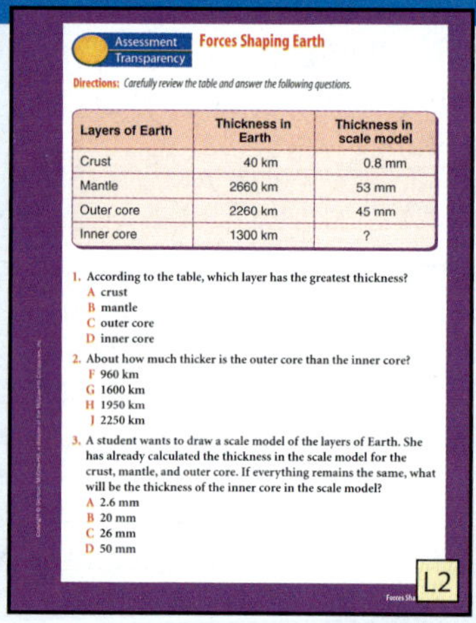
Assessment Transparency — Forces Shaping Earth

Directions: Carefully review the table and answer the following questions.

Layers of Earth	Thickness in Earth	Thickness in scale model
Crust	40 km	0.8 mm
Mantle	2660 km	53 mm
Outer core	2260 km	45 mm
Inner core	1300 km	?

1. According to the table, which layer has the greatest thickness?
   A crust
   B mantle
   C outer core
   D inner core
2. About how much thicker is the outer core than the inner core?
   F 960 km
   G 1600 km
   H 1950 km
   J 2250 km
3. A student wants to draw a scale model of the layers of Earth. She has already calculated the thickness in the scale model for the crust, mantle, and outer core. If everything remains the same, what will be the thickness of the inner core in the scale model?
   A 2.6 mm
   B 20 mm
   C 26 mm
   D 50 mm

L2

Teaching

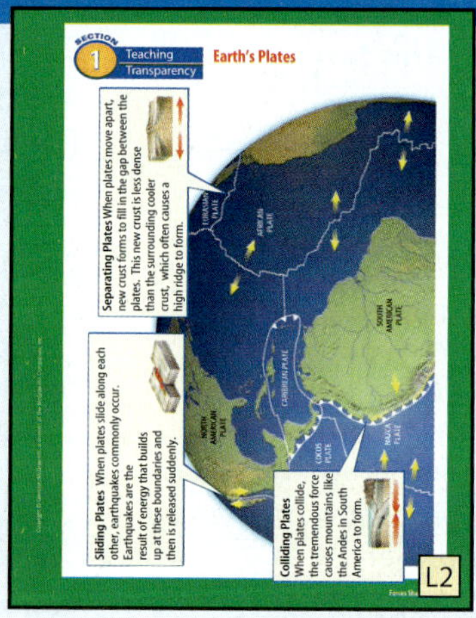
Section 1 Teaching Transparency — Earth's Plates

Key to Teaching Strategies

The following designations will help you decide which activities are appropriate for your students.

L1 Level 1 activities should be appropriate for students with learning difficulties.

L2 Level 2 activities should be within the ability range of all students.

L3 Level 3 activities are designed for above-average students.

ELL ELL activities should be within the ability range of English Language Learners.

COOP LEARN Cooperative Learning activities are designed for small group work.

LS Multiple Learning Styles logos, as described on page 12T, are used throughout to indicate strategies that address different learning styles.

P These strategies represent student products that can be placed into a best-work portfolio.

PBL Problem-Based Learning activities apply real-world situations to learning.

Hands-on Activities

Student Text Lab Worksheet

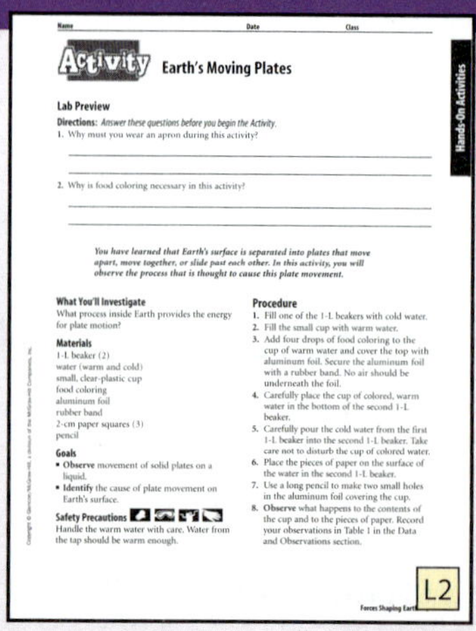
Activity — Earth's Moving Plates

Lab Preview

Directions: Answer these questions before you begin the Activity.

1. Why must you wear an apron during this activity?
2. Why is food coloring necessary in this activity?

You have learned that Earth's surface is separated into plates that move apart, move together, or slide past each other. In this activity, you will observe the process that is thought to cause this plate movement.

L2

Laboratory Activities

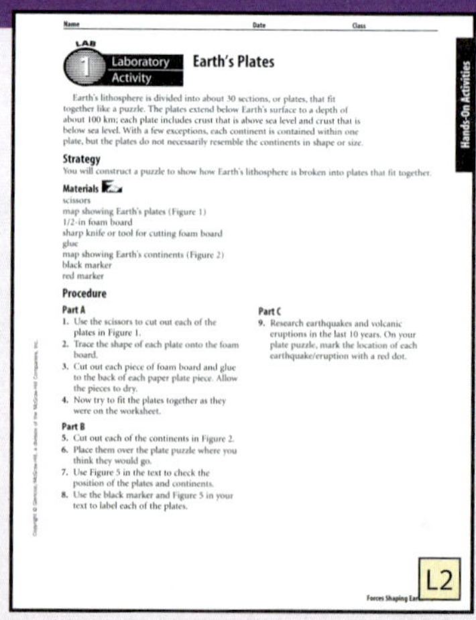
Laboratory Activity 1 — Earth's Plates

Strategy

Materials

Procedure

L2

Meeting Different Ability Levels

Content Outline

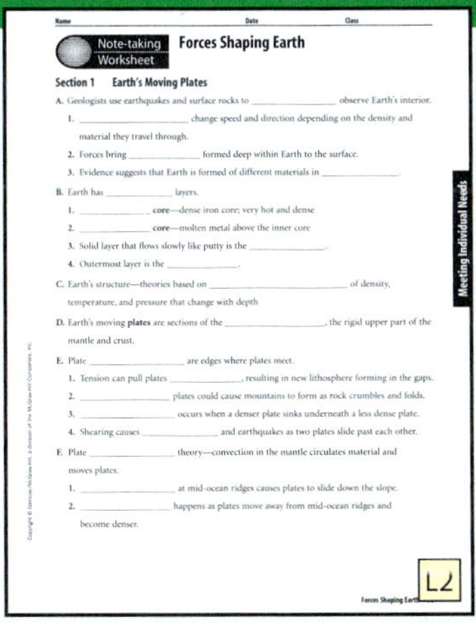

Reinforcement

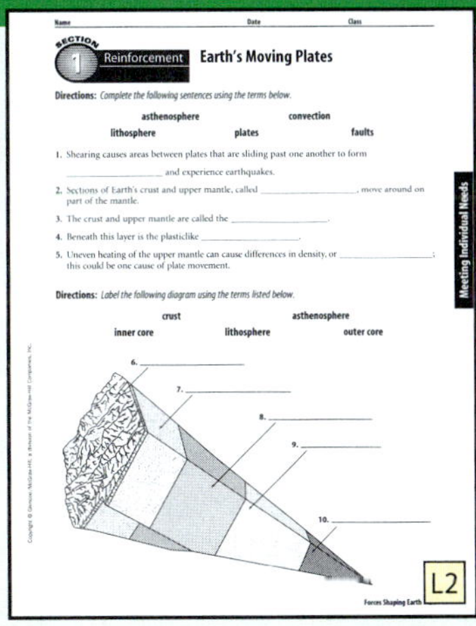

Enrichment

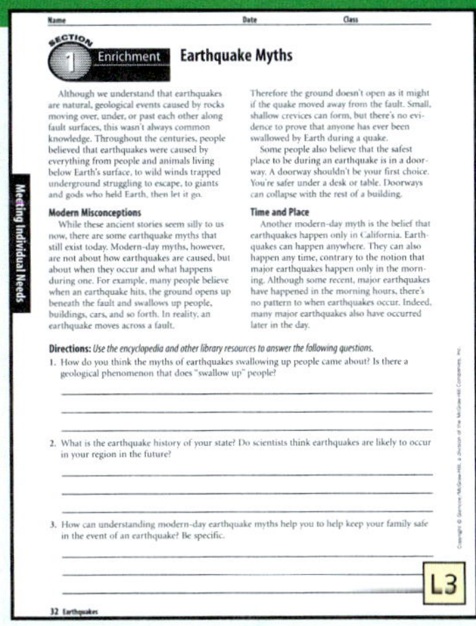

Directed Reading (English/Spanish)

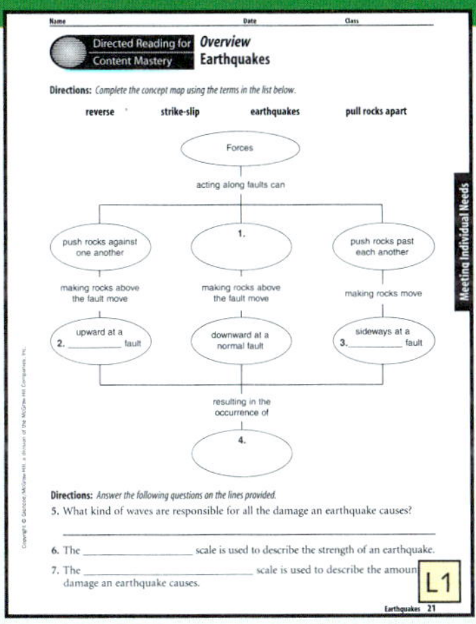

Study Guide

Reading Essentials

Assessment

Test Practice Workbook

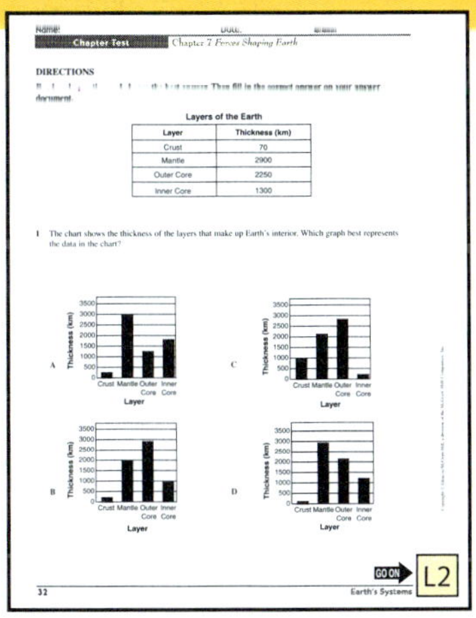

Chapter Review

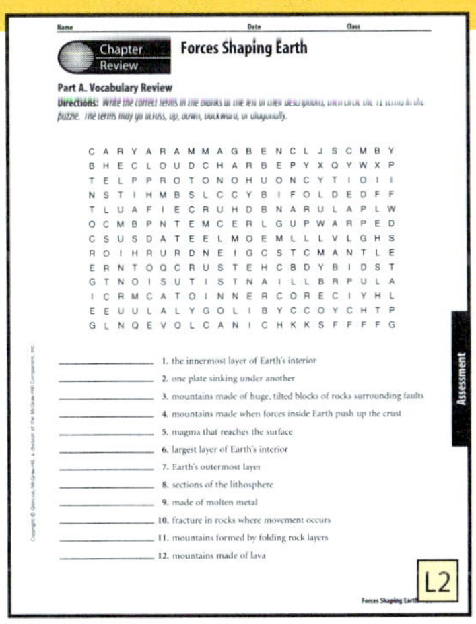

Chapter Tests

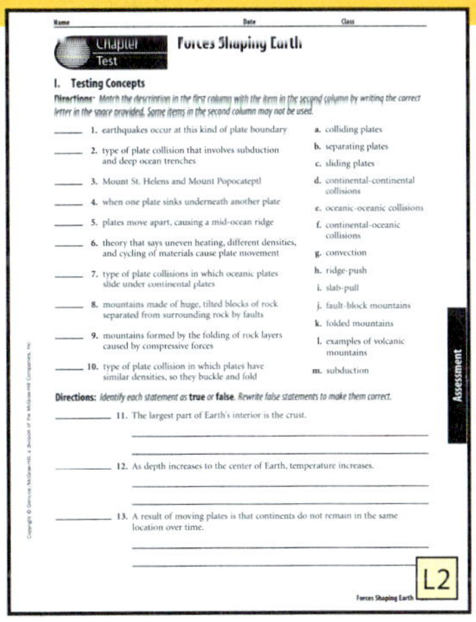

Science Content Background

section 1

Earth's Moving Plates

Earth's Layers

An important study defining the boundary between two of Earth's layers was conducted by Yugoslavian seismologist Andrija Mohorovicic in 1909. He discovered that the velocity of seismic waves increased abruptly at about 50 km below the continents. This is the boundary that separates the crust from the mantle. It is called the Mohorovicic discontinuity.

Divergent (Separating) Plate Boundaries

Although the mountain chains formed at divergent plate boundaries are the most extensive on Earth, they remain unseen, submerged beneath the water in the form of mid-oceanic ridges. Mid-oceanic ridges are 2 km to 3 km in elevation and form an interconnecting system of mountains 65,000 km long.

In some places the mid-oceanic ridge is elevated higher than sea level. Iceland, on the

chapter content resources

Internet Resources

For additional content background, visit **red.msscience.com** to:

- access your book online
- find references to related articles in popular science magazines
- access Web links with related content background
- access current events with science journal topics

Print Resources

The New View of the Earth by Seiya Uyeda, W.H. Freeman and Co, 1978

Global Tectonics by Philip Kearey & Frederick J. Vine, Blackwell Sciences LTD, 1996

Physical Geology by Carla W. Montgomery, Wm. C. Brown Publishers, 1987

Y. Arthus-B./Peter Arnold, Inc.

Mid-Atlantic Ridge, and Easter Island, on the East Pacific Rise, were formed as a result of hot spots located beneath the spreading ridges. The plume of magma has caused the ridges in these areas to rise above sea level.

Divergent boundaries can begin on continents. The East African rift valley is being created as eastern Africa separates from the rest of the continent. In the Red Sea the formation of new oceanic crust, a prelude to an oceanic ridge, has already begun.

E.R. Degginger/Photo Researcher, Inc. AFP/Corbis

Convergent Plate Boundaries—Collision

Mountains formed from the collision of two continental plates are not usually volcanic. Because subduction is not complete and the continental crust becomes overthickened as a result of the collision, the production and ascent of magma to the surface is arrested. Most major mountain ranges were formed as a result of continental plate collision. These include the older chains of the Appalachians in the U.S. and the Ural Mountains in Russia, and the more recently formed European Alps and the Himalayas. As India is still moving northward into the Eurasian Plate, the Himalayas are still a work in progress.

Section 2 Uplift of Earth's Crust

Underwater Volcanic Mountains

While mapping seamounts in the Pacific, geologists discovered a chain of volcanic underwater mountains stretching from the Hawaiian Islands to Midway Island and on north to the Aleutian trench. Studies indicate that the structures farthest from Hawaii are the oldest. Their conclusion is that the chain was formed by a hot spot, which is in process of building the island of Hawaii.

Isostasy

Icebergs have a ratio that relates the volume of the exposed portion of the iceberg to the volume of the submerged portion. The submerged portion of the iceberg is always nine times greater than the exposed portion. A similar ratio exists for continental mountains. The thickness of the crust beneath the mountain must be at least four and one half times the mountain's elevation. For example, Mount Everest's elevation is 8 km; therefore, the thickness of the crust beneath it must be at least 36 km. The thickness of the crust beneath Mount Everest is actually 80 km.

Mid-oceanic ridges are not in isostatic equilibrium; rather, they are uplifted by the buoyancy of the heated magma beneath them. The hotter and more extensive the magma, the higher the ridge. As seafloor spreading continues, however, basaltic crust moves away from its area of formation and its heat source, and it will lose height as it cools.

chapter 10

Chapter Vocabulary

inner core, p. 289
outer core, p. 290
mantle, p. 290
crust, p. 290
lithosphere, p. 292
plate, p. 292
fault, p. 293
subduction, p. 295
fault-block mountain, p. 300
folded mountain, p. 301
upwarped mountain, p. 301
volcanic mountain, p. 302
isostasy, p. 304

Science Journal Accept all reasonable answers.

INTERACTIVE CHALKBOARD with Image Bank

PowerPoint® Presentations

This CD-ROM is an editable Microsoft® PowerPoint® presentation that includes:

- a pre-made presentation for every chapter
- interactive graphics
- animations
- audio clips
- image bank
- all new section and chapter questions
- Standardized Test Practice
- transparencies
- pre-lab questions for all labs
- Foldables directions
- links to red.msscience.com

chapter 10

Forces Shaping Earth

chapter preview

sections

1 **Earth's Moving Plates**
Lab Earth's Moving Plates

2 **Uplift of Earth's Crust**
Lab Isostasy

Virtual Lab How do glaciers shape the land?

Young or Old Mountains?

These majestic, snow-capped mountains are in their infancy. It would take a few hundred million years of erosion for their sharp, jagged peaks to become smooth. In this chapter, you'll learn how the movement of plates formed these mountains and about other Earth forces that shape mountains.

Science Journal Use descriptive adjectives to describe these mountains in a short paragraph.

286

Theme Connection

Scale, Structure, and Energy The structure of Earth's interior and the relative thicknesses of each layer involve the theme of scale and structure. Energy is involved in the movement of Earth's plates and in the production of forces necessary for mountain building.

About the Photo

Baby Mountains These majestic, snow-capped mountains are in their infancy. It would take a few hundred million years of erosion for their sharp, jagged peaks to become smooth.

Start-Up Activities

Launch LAB

Model Earth's Interior

Geologists know many things about the interior of Earth even though its center is over 6,000 km deep. Use modeling clay to make a model of Earth's interior.

1. Obtain four pieces of different-colored clay.
2. Roll one piece of clay into a ball. This clay represents the inner core.
3. Wrap another piece of clay around the first ball of clay, making an even bigger ball. This clay represents the outer core.
4. Repeat step 3 with the third piece of clay, which represents Earth's mantle. Wrap your model with a thin layer of the fourth piece of clay to represent the crust.
5. Use a plastic knife to cut the ball of clay in half.
6. **Think Critically** Make a sketch of your model and label each of Earth's layers.

FOLDABLES Study Organizer

Earth's Interior and Surface Make the following cause and effect Foldable to help you understand the relationship between Earth's interior and surface.

STEP 1 **Collect** 2 sheets of paper and layer them about 2.5 cm apart vertically. Keep the edges level.

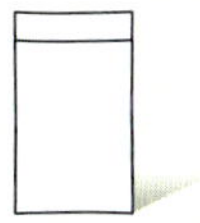

STEP 2 **Fold** up the bottom edges of the paper to form 4 equal tabs.

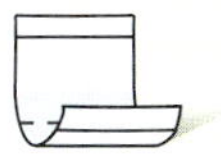

STEP 3 **Fold** the papers and crease well to hold the tabs in place. Staple along the fold. **Label** each tab as shown.

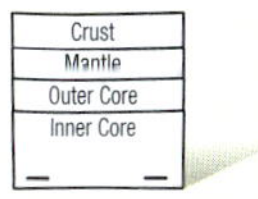

Cause and Effect As you read the chapter, record information about each layer and how it is related to the layer above it under the tabs.

Science Online — Preview this chapter's content and activities at red.msscience.com

Launch LAB

Purpose Use the Launch Lab to introduce students to Earth's structure. L2 LS **Kinesthetic**

Materials four pieces of clay, (each a different color), plastic knife

Teaching Strategy You might wish to provide plastic jars for students to use as rolling pins to flatten their pieces of clay.

Think Critically

Students' models should show a core surrounded by three layers.

Assessment

Process Have students hypothesize how scientists have discerned Earth's structure. Possible answers: drilling holes, studying magma and lava; much of what scientists have learned has come from studying earthquake waves. Use **Performance Assessment in the Science Classroom**, p. 93.

FOLDABLES Study Organizer — Dinah Zike Study Fold

Student preparation materials for this Foldable are available in the Chapter *FAST FILE* Resources.

1 Motivate

INTERACTIVE CHALKBOARD PowerPoint® Presentations

Bellringer

Section Focus Transparencies also are available on the Interactive Chalkboard CD-ROM.

L2 ELL

Section 1 Section Focus Transparency **Not a Place to Drop Your Keys**

Have you ever thought about the deepest place on Earth? It's the Mariana Trench in the western Pacific Ocean. The trench is so deep, you'd have to stack a medium-sized mountain *on top of* Mount Everest just to reach the surface. Try to locate the Mariana Trench on this map of the Pacific Ocean floor.

1. Describe some features of the ocean floor shown on this map.
2. What continental features are similar to the features of the ocean floor?

Forces Shaping Earth

Tie to Prior Knowledge

Vegetable Soup Help students recall the movement of vegetables in a pan of hot soup as heat from the burner causes currents to form. These currents carry the vegetables with them as they move through the liquid soup. Explain that a similar process is thought to provide the mechanism for the movement of Earth's lithosphere.

section 1

Earth's Moving Plates

as you read

What You'll Learn

- **Describe** how Earth's interior is divided into layers.
- **Explain** how plates of Earth's lithosphere move.
- **Discuss** why Earth's plates move.

Why It's Important

Forces that cause Earth's plates to move apart, together, or past each other cause events that shape Earth's surface, such as mountain building, volcanoes, and earthquakes.

Review Vocabulary

magma: melted rock material found beneath Earth's surface

New Vocabulary

- inner core
- outer core
- mantle
- crust
- lithosphere
- plate
- fault
- subduction

Clues to Earth's Interior

If someone gives you a wrapped present, how could you figure out what was in it? You might hold it, shake it gently, or weigh it. You'd look for clues that could help you identify the contents of the box. Even though you can't see what's inside the package, these types of clues can help you figure out what it might be. Because you can't see what's inside, the observations you make are known as indirect observations.

Geologists do the same thing when they try to learn about Earth's interior. Although the best way to find out what's inside Earth might be to dig a tunnel to its center, that isn't possible. The deepest mines in the world only scratch Earth's surface. A tunnel would need to be more than 6,000 km deep to reach the center, so geologists must use indirect observations to gather clues about what Earth's interior is made of and how it is structured. This indirect evidence includes information learned by studying earthquakes and rocks that are exposed at Earth's surface.

INTEGRATE Physics

Waves When you throw a rock into a calm puddle or pond, you observe waves like those shown in **Figure 1.** Waves are disturbances that carry energy through matter or space. When a rock hits water, waves carry some of the rock's kinetic energy, or energy of motion, away from where it hit the water. When an earthquake occurs, as shown in **Figure 2,** energy is carried through objects by waves. The speed of these waves depends on the density and nature of the material they are traveling through. For example, a wave travels faster in solid rock than it does in a liquid. By studying the speed of these waves and the paths they take, geologists uncover clues as to how the planet is put together. In fact, these waves, called seismic waves, speed up in some areas, slow down in other areas, and can be bent or stopped.

Figure 1 Waves carry energy across water just like seismic waves carry energy through Earth.

288

Section 1 Resource Manager

Chapter *FAST FILE* Resources

- Transparency Activity, pp. 44, 47–48
- Directed Reading for Content Mastery, pp. 21, 22
- Note-taking Worksheets, pp. 33–34
- Enrichment, p. 31
- Lab Activity, pp. 9–13
- MiniLAB, p. 3
- Reinforcement, p. 29
- Lab Worksheet, pp. 5–6

Earth Science Critical Thinking/Problem Solving, pp. 8, 9, 10

Cultural Diversity, p. 47

Reading and Writing Skill Activities, p. 21

Figure 2 As seismic waves travel across Earth's surface, the ground shakes and damage occurs.

Rock Clues Another clue to what's inside Earth comes in the form of certain rocks found in different places on Earth's surface. These rocks are made of material similar to what is thought to exist deep inside Earth. The rocks formed far below the surface. Forces inside Earth pushed them closer to the surface, where they eventually were exposed by erosion. The seismic clues and the rock clues suggest that Earth is made up of layers of different kinds of materials.

Earth's Layers

Based on evidence from earthquake waves and exposed rocks, scientists have produced a model of Earth's interior. The model shows that Earth's interior has at least four distinct layers—the inner core, the outer core, the mantle, and the crust. Earth's structure is similar in some ways to the structure of a peach, shown in **Figure 3.** A peach has a thin skin covering the thick, juicy part that you eat. Under that is a large pit that surrounds a seed.

Inner Core The pit and seed are similar to Earth's core. Earth's core is divided into two distinct parts—one that is liquid and one that is solid. The innermost layer of Earth's interior is the solid **inner core.** This part of the core is dense and composed mostly of solid iron. When seismic waves produced by earthquakes reach this layer they speed up, indicating that the inner core is solid.

Conditions in the inner core are extreme compared to those at the surface. At about 5,000°C, the inner core is the hottest part of Earth. Also, because of the weight of the surrounding rock, the core is under tremendous pressure. Pressure, or the force pushing on an area, increases the deeper you go beneath Earth's surface. Pressure increases because more material is pushing toward Earth's center as a result of gravity. The inner core, at the center of Earth, experiences the greatest amount of pressure.

Figure 3 The structure of Earth can be compared to a peach.
Explain *If the part of Earth that you live on is like the skin of the peach, what does that tell you about this layer of Earth?*

2 Teach

Inquiry Lab

A World Puzzle

Purpose To see how the continents may have fit together at one point in Earth's history.

Possible Materials line map of the world—2 per student, scissors, construction paper

Estimated Time 1 class period

Teaching Strategies

- Students should study the world map and make initial guesses as to which continents may have fit together in the past.
- Instruct them to carefully cut out the continents and large masses from one of the maps.
- Students should assemble the pieces as if they were pieces. There are several possible ways to assemble them so encourage students to use the other map in the same fashion.
- Have students research Pangaea, Gondwanaland, and Laurasia to see how their puzzle matches that of early scientists.

For additional inquiry activities, see *Science Inquiry Labs*.

Fun Fact

Originally, Earth's entire core was most likely liquid. As Earth cooled, the solid, inner core formed. Eventually the entire core will be solid.

Caption Answer

Figure 3 It is the thinnest layer, just as the skin is the thinnest layer of a peach.

Differentiated Instruction

Challenge Have students research how the P- and S- waves generated by earthquakes are affected by the different layers of Earth's interior. Have them create a model of the interior showing the behavior of each wave. L3

Curriculum Connection

Language Arts Have students use a dictionary to look up *crust*, *mantle*, and *core*. Ask students to write a brief description in their Science Journals that explains why these terms are used to name parts of Earth's structure. *Crust* refers to any hard surface layer, *mantle* refers to anything that envelops or conceals, and *core* refers to the central part of anything. L2 LS **Linguistic**

INTEGRATE Chemistry

Iron Core Support for this theory comes from the study of meteorites, which are considered to be representative samples of solar system material. Like Earth, they are composed of iron and other minerals; however, they contain a much higher percentage of iron than Earth's crust does. Earth's density increases with depth toward its center, so geologists assume denser substances such as iron must exist there.

Research Have students research how scientists determined the density of Earth's core. Does this density support or disprove the theory that Earth's core is made mostly of iron? Explain.

Reading Check

Answer The outer woody layer of the pit.

Caption Answer

Figure 4 earthquakes and exposure by erosion of rocks formed far below the surface

INTEGRATE Chemistry

Iron Core Earth's crust is composed of about five percent iron. However, geologists theorize that Earth's core is composed mostly of iron. Research the theory that Earth's core is composed mostly of iron. Analyze, review, and critique the strengths and weaknesses of this theory using scientific evidence and information.

Outer Core The **outer core** lies above the inner core and is thought to be composed mostly of molten metal. The outer core stops one type of seismic wave and slows down another. Because of this, scientists have concluded that the outer core is a liquid. The location of the outer core is similar to the location of the pit in the peach model. Even the wrinkled surface of the pit resembles the uneven nature of the boundary between Earth's outer core and its mantle as indicated by seismic studies.

Reading Check *What peach layer is similar to the outer core?*

Mantle The layer in Earth's interior above the outer core is the mantle. In the peach model, the mantle would be the juicy part of the peach that you would eat. The **mantle** is the largest layer of Earth's interior. Even though it's solid, the mantle flows slowly, similar to putty.

Crust Earth's outermost layer is the **crust.** In the model of the peach, this layer would be the fuzzy skin of the peach. Earth's crust is thin when compared to the other layers, though its thickness does vary. It is thinnest under the oceans and thickest through the continents. All features on Earth's surface are part of the crust.

Figure 4 Earth is made up of many layers.
Identify *geologic events that have allowed scientists to study Earth's interior.*

The lithosphere is composed of crust and uppermost mantle. The asthenosphere is a plasticlike layer upon which the plates of the lithosphere float and move.

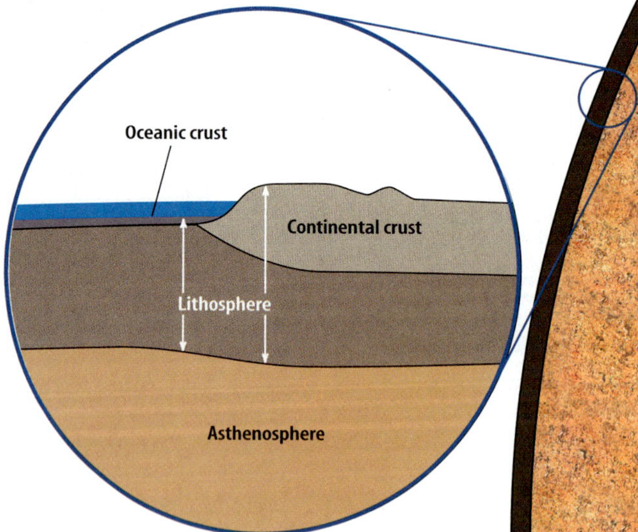

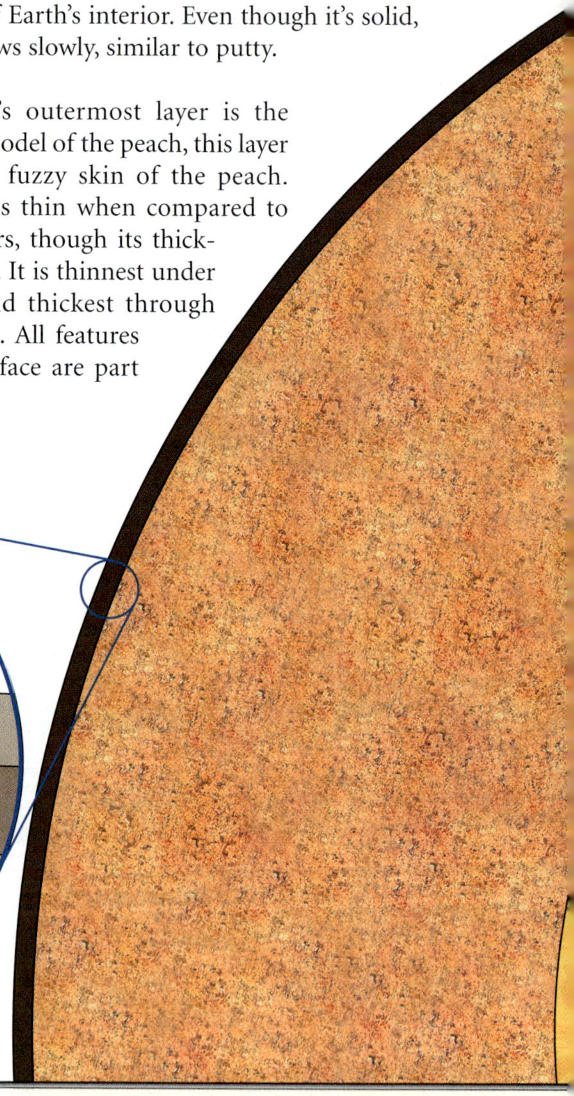

LAB DEMONSTRATION

Purpose to show how properties of an unknown substance can be discovered

Materials empty boxes with lids, tape, and several common items

Preparation While you are not looking, have a student select an item, place it in a box, and tape the box shut.

Procedure Shake the box, heft it, and slide the object around to show how an object's properties (mass, approximate size, probable shape, density) can be indirectly determined.

Expected Outcome Students can determine the properties of an unknown object without seeing the object.

Assessment

Have students make mystery boxes for each other. Discuss which properties are easy to determine, which are more difficult, and which are not able to be determined. Compare this to how scientists study Earth's interior.

Earth's Structure

Although Earth's structure can be divided into four basic layers, it also can be divided into other layers based on physical properties that change with depth beneath the surface. **Figure 4** shows the structure of Earth and describes some of the properties of its layers. Density, temperature, and pressure are properties that are lowest in the crust and greatest in the inner core.

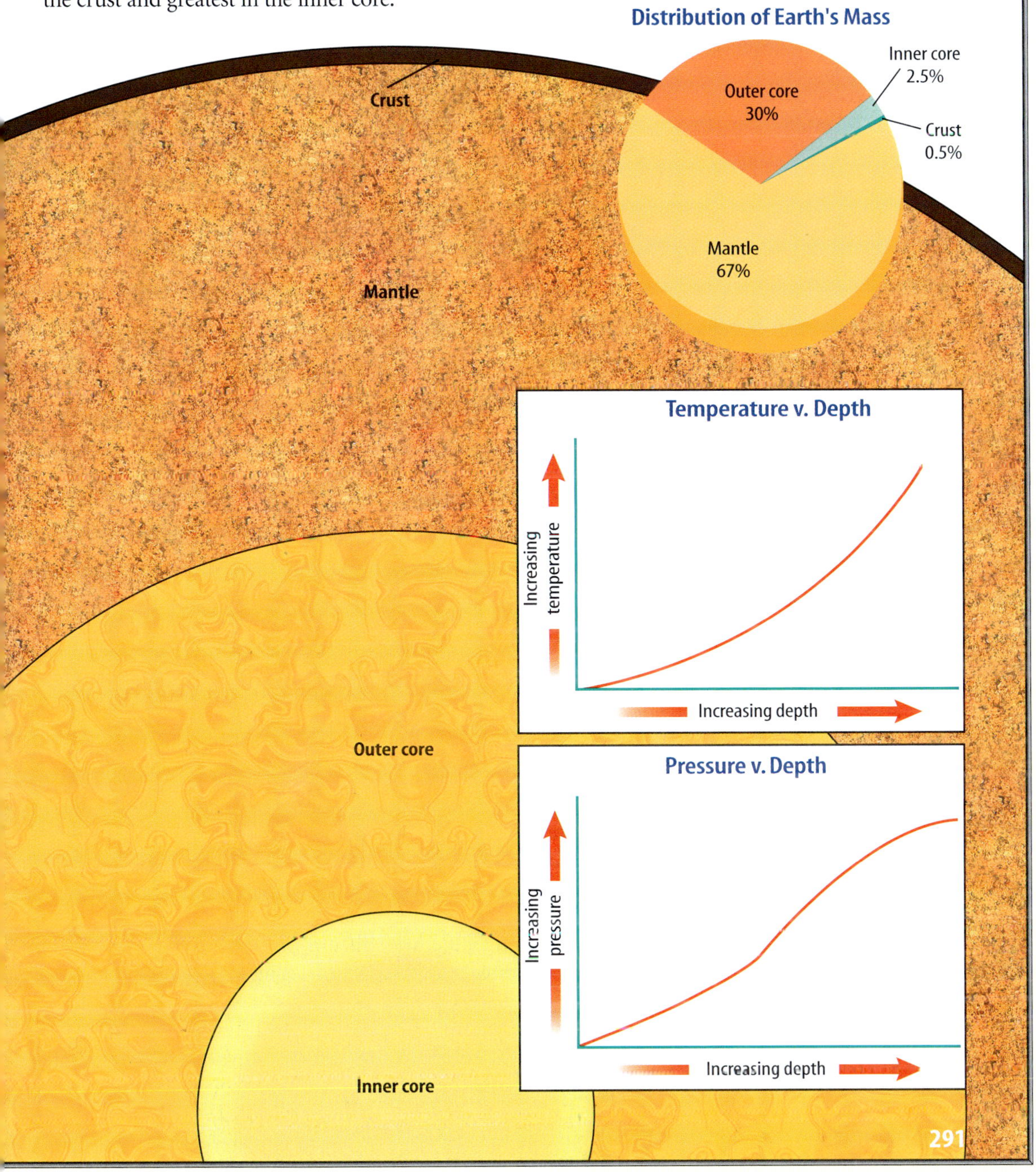

Teacher FYI

Earth's Mantle Earth's mantle comprises over 82 percent of Earth's total volume. Earth's mantle is further subdivided based on changes in density and the composition of the rock.

Visual Learning

Figure 4 Have students compare the Pressure versus Depth graph and the Temperature versus Depth graph. What statement can they make about these two factors as one goes deeper beneath Earth's surface? Both temperature and pressure increase with depth.

Differentiated Instruction

Visually Impaired Use modeling clay to construct a three-dimensional scale model of Earth's structure that separates into layers. Use balled-up aluminum foil for the core. Cover the model with paper to represent the crust. Help visually-impaired students remove each layer of the model and explain what each layer represents. Point out the relative thickness of each layer. L1
ELL LS **Kinesthetic**

IDENTIFYING Misconceptions

Earth's Crust Students may think that plate tectonics results in motion of only Earth's continents. In fact, Earth's crust covers the entire planet, underlying even the oceans. Therefore, ocean plates are also in motion.

Reading Check

Answer crust and upper part of mantle

Quick Demo

Plate Interactions

Materials large clear bowl, water, thin sponge, scissors

Estimated Time 15 minutes

Procedure Half-fill a large bowl with tap water. Cut irregular shapes from a thin cleaning sponge. Randomly place four sections of the sponge on top of the water. Gently move the bowl back and forth in order to set the sponge sections in motion. Ask volunteers to describe their observations. The sponge sections move around on the water. Some move together, others move apart, and still others move past each other. L2 LS **Visual-Spatial**

Caption Answer

Figure 5 Plates move apart, move together, or move past each other at plate boundaries.

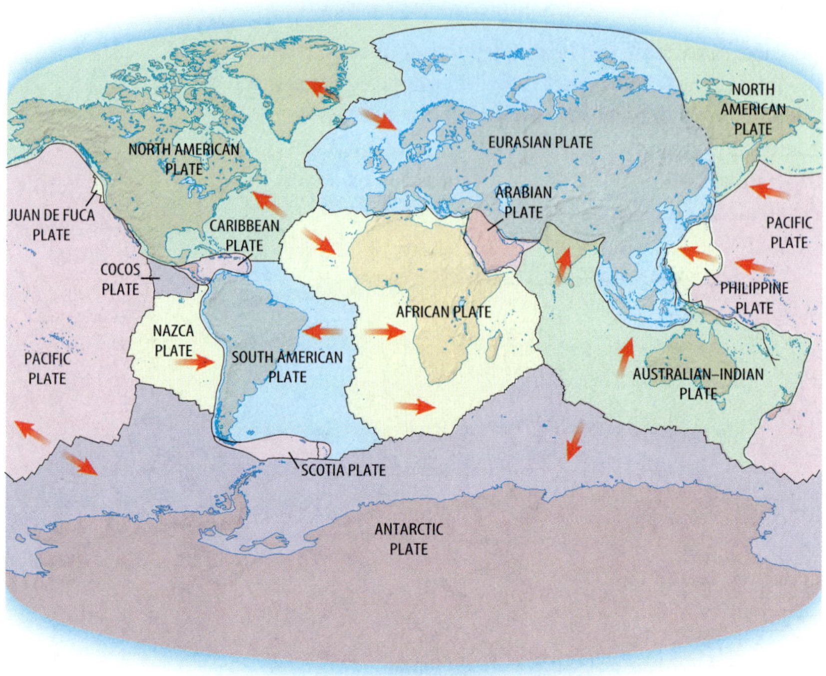

Figure 5 Earth's plates fit together like the pieces of a jigsaw puzzle.
Draw Conclusions *If the plates are moving, what do you suppose happens at the plate boundaries?*

Earth's Plates

Although the crust is separated from the mantle, the uppermost, rigid layer of the mantle moves as if it were part of Earth's crust. The rigid, upper part of Earth's mantle and the crust is called the **lithosphere.** It is broken into about 30 sections or **plates** that move around on the plasticlike asthenosphere, which also is part of the mantle. Earth's major plates vary greatly in size and shape, as shown in **Figure 5.**

Reading Check *What parts of Earth make up the lithosphere?*

The movements of the plates are fairly slow, often taking more than a year to creep a few centimeters. This means that they have not always looked the way they do in **Figure 5.** The plates have not always been their current size and shape, and continents have moved great distances. Antarctica, which now covers the south pole, was once near the equator, and North America was once connected to Africa and Europe.

Lasers and satellites are used to measure the small plate movements, which can add up to great distances over time. If a plate is found to move at 2 centimeters per year on average, how far will it move in 1,000 years? What about in 10 million years?

Cultural Diversity

Pieces of the Puzzle Have students work in cooperative groups to discover the contributions made by Alfred Wegener, Harry Hess, and J. Tuzo Wilson that helped lead to plate tectonics theory. Wegener proposed continental drift. Hess discovered sea floor spreading. Wilson called broken sections of crust *plates* and discovered the nature of transform faults. COOP LEARN LS **Interpersonal**

Active Reading

Metacognition Journal Have students divide sheets of paper in half. On the left, have them record what they have learned about Earth's plates. On the right, have them record the reason they learned it.

292 CHAPTER 10 Forces Shaping Earth

Plate Boundaries

The places where the edges of different plates meet are called plate boundaries. The constant movement of plates creates forces that affect Earth's surface at the boundaries of the plates. At some boundaries, these forces are large enough to cause mountains to form. Other boundaries form huge rift valleys with active volcanoes. At a third type of boundary, huge faults form. **Faults** are large fractures in rocks along which movement occurs. The movement can cause earthquakes. **Figure 6** shows the different plate motions.

Plates That Move Apart Plates move apart as a result of pulling forces that act in opposite directions on each plate. This pulling force is called tension. **Figure 7** shows what happens as tension continues to pull two plates apart.

One important result of plates separating is the formation of new crust. New crust forms in gaps where the plates pull apart. As tension continues along these boundaries, new gaps form and are filled in by magma that is pushed up from the mantle. Over time, the magma in the gaps cools to become new crust. This process of plate separation and crust formation takes place under the oceans at places called mid-ocean ridges. As new crust moves away from the mid-ocean ridges, it cools and becomes denser.

Science Online

Topic: Plate Boundaries
Visit red.msscience.com for Web links to information about Earth's plates and the different boundaries that they form.

Activity Create a table of the information on plates and plate boundaries. Try to include specific plates and boundary locations. Share your findings with your classmates.

Figure 6 Earth's plates can collide, move away from each other, or slide past each other.

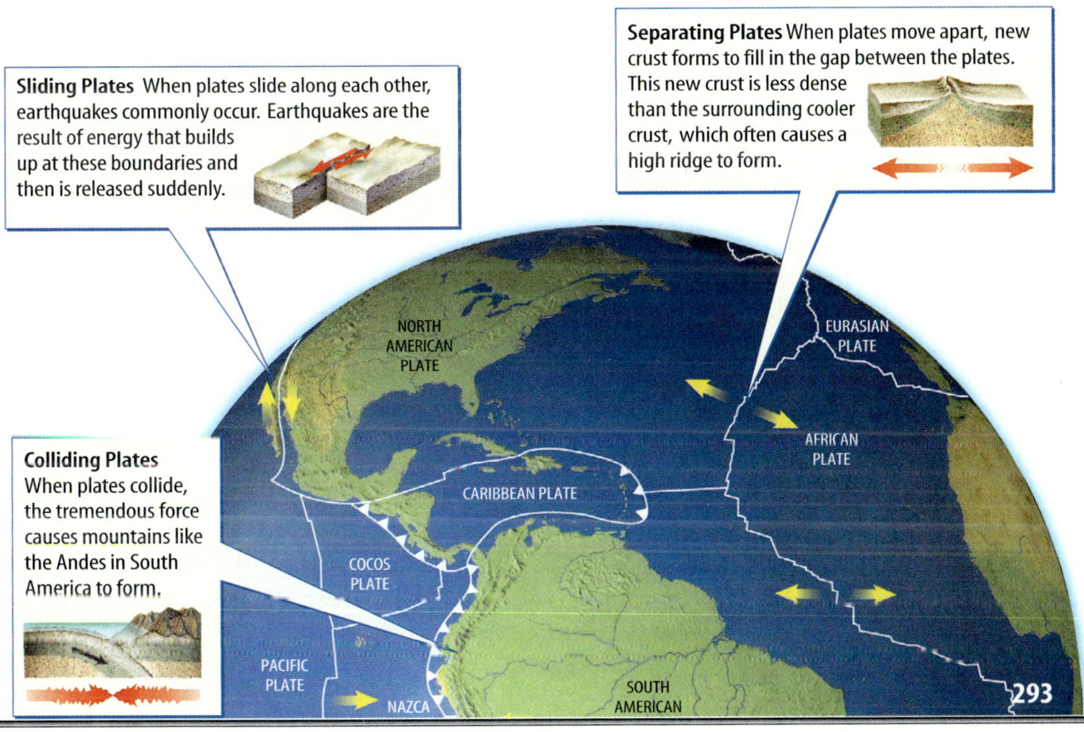

Activity

Moving Plates Use this activity to help students understand what happens as plates move apart. Provide pairs with a 0.6-meter length of adding machine tape and the top of a shoe box into which a slit has been cut. Have the students color wide stripes across the width of the paper. Then instruct students to lightly fold the tape, stripes to the inside, and push the ends of the paper through the slit. Have students observe the colors as they pull the ends of the tape out and across the lid of the box. Explain that the colors represent new crust being formed as plates move apart. What would happen to the distance between continents that are located on these plates? The distance between the continents would become greater as the plates move apart. L2 LS **Visual-Spatial**

Visual Learning

Figure 6 Which type of plate boundary might form if a large magma body were forced upward under the middle of a plate? A divergent boundary would most likely form because the overlying plate would warp upward and eventually crack.

Fun Fact

Earth's tectonic plates move at about the same rate as the growth of your fingernails.

NATIONAL GEOGRAPHIC

Visualizing Rift Valleys

Have students examine the pictures and read the captions. Then ask the following questions.

The Red Sea represents which stage in the rifting process? The Red Sea is a narrow sea generated by spreading as depicted by the red arrows. It represents the third stage in the rifting process.

Why do slabs of rocks sink when the crust is pulled apart? After cracking separates them from the crust, they are unsupported by surrounding rock layers. They are pulled by gravity down toward the mantle.

How is seawater allowed to enter rift valleys? Eventually slabs of rock in the rift zone fall below sea level, allowing for the inflow of seawater if the rift is near a coastline.

Activity

Rifting Process Have students work in groups to model each stage of the rifting process in brightly colored modeling clay. Have them label their models, showing the directions in which the crust is moving, slabs of sinking rock, magma, crustal fractures, valley walls, and seawater.

NATIONAL GEOGRAPHIC **VISUALIZING RIFT VALLEYS**

Figure 7

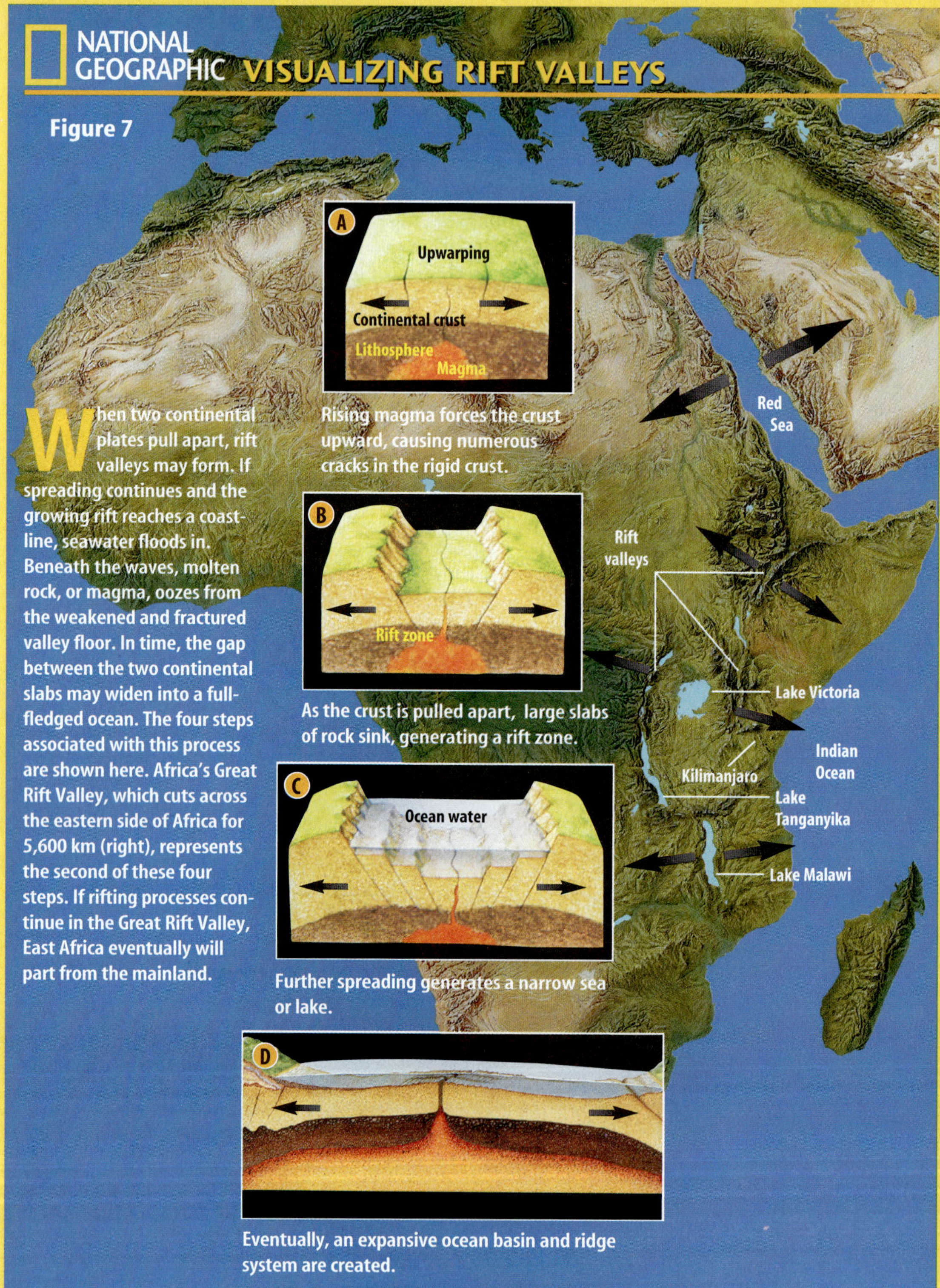

When two continental plates pull apart, rift valleys may form. If spreading continues and the growing rift reaches a coastline, seawater floods in. Beneath the waves, molten rock, or magma, oozes from the weakened and fractured valley floor. In time, the gap between the two continental slabs may widen into a full-fledged ocean. The four steps associated with this process are shown here. Africa's Great Rift Valley, which cuts across the eastern side of Africa for 5,600 km (right), represents the second of these four steps. If rifting processes continue in the Great Rift Valley, East Africa eventually will part from the mainland.

Rising magma forces the crust upward, causing numerous cracks in the rigid crust.

As the crust is pulled apart, large slabs of rock sink, generating a rift zone.

Further spreading generates a narrow sea or lake.

Eventually, an expansive ocean basin and ridge system are created.

Plates That Collide When plates move toward each other, they collide, causing several different things to occur. As you can see in **Figure 8,** the outcome depends on the density of the two plates involved. The crust that forms the ocean floors, called oceanic crust, is more dense than the continental crust, which forms continents.

If two continental plates collide, they have a similar density which is less than the mantle underneath. Therefore, the collision causes the crust to pile up. When rock converges like this, the force is called compression. Compression causes the rock layers on both plates to crumple and fold. Imagine laying a piece of fabric flat on your desk. If you push the edges of the cloth toward each other, the fabric will crumple and fold over on itself. A similar process occurs when plates crash into each other, causing mountains to form.

Flat rock layers are pushed up into folds. Sometimes the folding is so severe that rock layers bend completely over on themselves, turning upside down. As rock layers are folded and faulted, they pile up and form mountains. The tallest mountains in the world, the Himalaya in Asia, are still rising as two continental plates collide.

Plate Subduction When an oceanic plate collides with another oceanic plate or a continental plate, the more dense one plunges underneath the other, forming a deep trench. When one plate sinks underneath another plate, it's called **subduction.** When a plate subducts, it sinks into the mantle. In this way, Earth's crust does not continue to grow larger. As new crust material is generated at a rift, older crustal material subducts into the mantle.

Mini LAB

Modeling Tension and Compression

Procedure

1. Obtain two bars of **taffy.**
2. Hold one bar of taffy between your hands and push your hands together.
3. Record your observations in your **Science Journal.**
4. Hold the other bar of taffy between your hands and pull gently on both ends.
5. Record your observations in your Science Journal.

Analysis

1. On which bar of taffy did you apply tension? Compression?
2. Explain how this applies to plate boundaries.

Try at Home

Figure 8 There are three types of convergent plate boundaries.

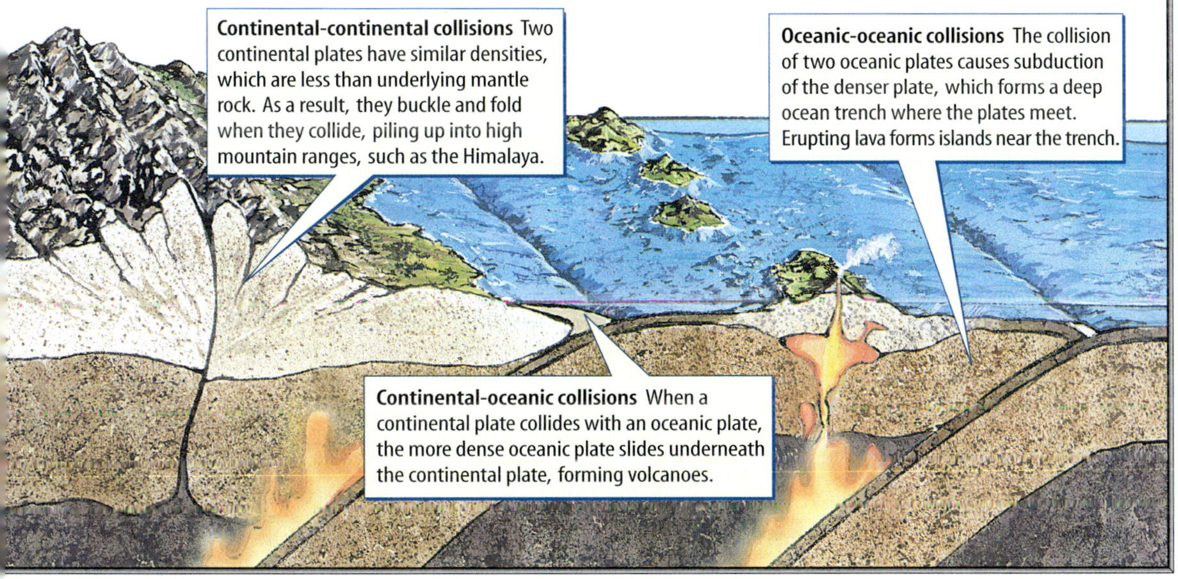

Mini LAB

Purpose Students show the effects of compression and tension. L1 LS **Kinesthetic**

Materials two bars of taffy for each group

Teaching Strategy Encourage students to make illustrations of their observations.

Analysis

1. Tension was applied when the taffy was pulled apart. Compression was applied when the taffy was pushed together.
2. Compression is applied when Earth's plates move together; tension is applied when Earth's plates move apart.

Assessment

Performance Have students apply shear forces to a third bar of taffy. Shear forces push from opposite directions that are slightly displaced. Bars of taffy should fold and twist before breaking. Use **Performance Assessment in the Science Classroom,** p. 123.

Try at Home

Visual Learning

Figure 8 Have students compare and contrast the convergent boundaries shown. At all of these boundaries, plates are moving together and colliding. At continental-continental collisions, mountains form. At oceanic-oceanic and continental-oceanic collisions, one plate slides under the other.

Differentiated Instruction

Learning Disabled Have students use their hands to demonstrate the types of plate movements that occur at plate boundaries: Colliding plates—bring hands together and fold fingers to make mountains; diverging plates—move hands apart: plates sliding past each other—press hands together and then move hands in opposite directions. L1 ELL LS **Kinesthetic**

Discussion

San Andeas Fault Earthquakes often occur along plates that are sliding past each other along a transform fault, such as the San Andreas. Why is there so much earthquake activity along the San Andreas Fault? As the North American and Pacific plates slide past each other along the fault, the two plates stick and build stress. When the stress builds to a high enough level, a sudden movement along the fault releases energy. This causes an earthquake.

Teacher FYI

Mantle Convection Convection in Earth's interior is thought to be responsible for the movement of Earth's plates. This movement produces mountain-building processes, including volcanism and earthquakes. Scientists have different ideas as to how convection in the mantle works. One model shows large convection cells throughout the mantle. Another shows convection in the upper mantle only. Still another shows convection through rising plumes and subduction zones.

Use Science Words

Word Meaning Help students learn why the movement of molten material inside Earth is referred to as *convection*. Convection deals with movement caused by differences in density. Hot, partially molten rock is less dense than the cooler surrounding rock, causing the hot rock to be forced upward. When it rises, the hot rock cools, becomes more dense, and sinks again. This process forms a convection current.

Figure 9 As two plates slide past each other, their edges grind and scrape. The jerky movement that results causes earthquakes like those frequently felt in California along the San Andreas Fault.

Plates That Slide Past In addition to moving toward and away from one another, plates also can slide past one another. For example, one plate might be moving north while the plate next to it is moving south. The boundary where these plates meet is called a transform boundary. When a force pushes something in two different directions, it's called shearing. Shearing causes the area between the plates to form faults and experience many earthquakes. **Figure 9** shows part of the San Andreas Fault near Taft, California, which is an example of the features that form along a transform boundary.

Why do plates move?

As you can see, Earth's plates are large. To move something so massive requires a tremendous amount of energy. Where does the energy that drives plate movement come from? The reason plates move is complex, and geologists still are trying to understand it fully. So far, scientists have come up with several possible explanations about what is happening inside Earth to cause plate movement. Most of these theories suggest that gravity is the driving force behind it. However, gravity pulls things toward the center of Earth, and plates move sideways across the globe. How does gravity make something move across the surface of Earth?

One theory that could explain plate movement is convection of the mantle. Convection in any material is driven by differences in density. In the mantle, density differences are caused by uneven heating, which results in a cycling of material, as shown in **Figure 10.** The theory suggests that the plates move as part of this circulation of mantle material.

Science Journal

Features of Transform Fault Boundaries Have students study **Figure 9,** and then describe in their Science Journals some of the features caused by the shearing forces along the fault. Surface features are offset as the plates move slowly past one another. L2

LS **Linguistic**

296 CHAPTER 10 Forces Shaping Earth

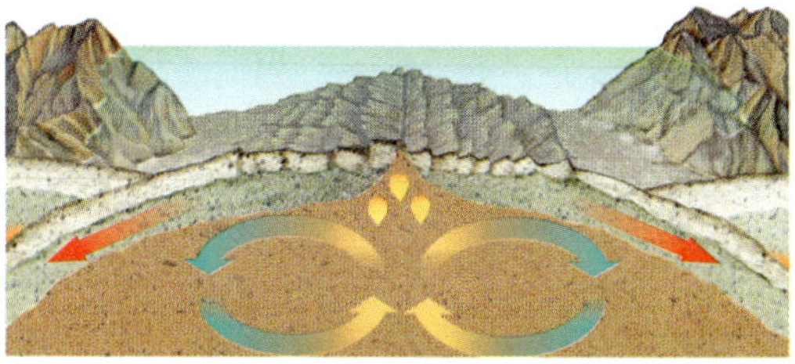

Figure 10 Convection, ridge-push, and slab-pull might all contribute to the motions of Earth's plates.

Uneven heating of the upper mantle could cause convection.

Ridge-push could occur at mid-ocean ridges.

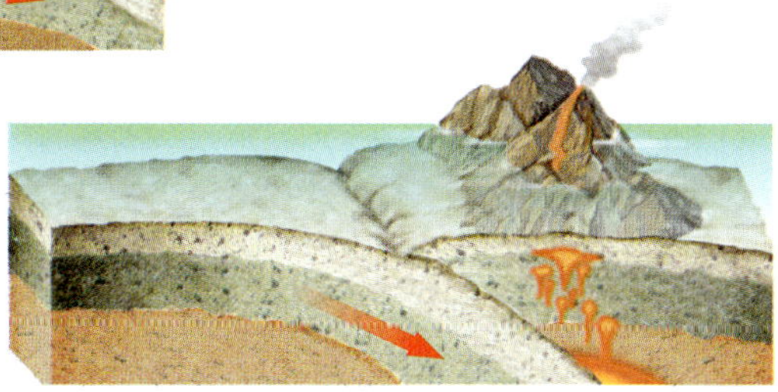

Slab-pull could occur where oceanic plates meet other oceanic or continental plates.

Ridge-push and Slab-pull Other factors, as shown in **Figure 10,** that could play a role in plate movement are ridge-push and slab-pull. Ridge-push occurs at mid-ocean ridges, which are higher than surrounding ocean floor. The plates respond to gravity by sliding down the slope. Slab-pull occurs as the plates move away from the mid-ocean ridges and become cooler, which makes them more dense. A plate can get so dense that it sinks when it collides with another plate. When the more dense plate begins to sink, it becomes easier for it to move across Earth's surface because resistance to movement is reduced.

section 1 review

Summary

Clues to Earth's Interior

- Earth's interior has been explored using information from seismic waves and rocks.

Earth's Layers

- The interior of Earth is made of the inner core, outer core, mantle, and crust.

Plate Boundaries

- Plates can move apart, collide, subduct, or slide past each other.
- Plates probably move by convection and factors such as ridge-push and slab-pull.

Self Check

1. **Explain** How are earthquake waves used to provide information about Earth's interior?
2. **Identify** Give examples of where the three types of plate movements occur.
3. **Describe** the layer of Earth's interior that is the largest.
4. **List** the layers of Earth's interior in order of density.
5. **Think Critically** How can slab-pull and ridge-push contribute to the movement of a plate at the same time?

Applying Skills

6. **Compare and contrast** the following pairs of terms: *inner core, outer core; ridge-push, slab-pull.*

Science Online red.msscience.com/self_check_quiz

3 Assess

DAILY INTERVENTION

Check for Understanding

Visual-Spatial Have students work together in small groups to label a world map with the different types of plate boundaries.

Reteach

Plate Boundaries Request three groups of students pantomime the types of interactions that occur at plate boundaries. Have each group perform for the class so that all three boundaries are modeled. L2 COOP LEARN LS **Interpersonal**

Assessment

Oral Pair students and have them work together to write one question and answer covering the content in this section. Collect the questions, and divide the class into teams to play Moving Plate Jeopardy. Provide each team an answer, and have them come up with the appropriate question. Use **Performance Assessment in the Science Classroom,** p. 91.

section 1 review

1. If the path or speed of earthquake waves is changed, it indicates a difference in the material through which they are traveling.
2. move together (converge), move apart (diverge), and move past each other; possible examples: converge–Himalaya, diverge–Rift Valley, slide past–San Andreas Fault
3. mantle—solid but flows slowly
4. crust, mantle, outer core, and inner core
5. Ridge-push occurs as plates slide down the slope of a mid-ocean ridge. As the plate moves away from the ridge, it becomes denser, eventually beginning to sink into the mantle. Once the plate starts to sink, resistance to movement is reduced, causing slab-pull.
6. Inner/outer core: both are parts of Earth's innermost layer; the outer core is liquid; the inner core is solid. Ridge-push/slab-pull: both are mechanisms that contribute to the movement of plates; ridge-push occurs at divergent boundaries; slab-pull occurs at convergent boundaries.

BENCH TESTED

LAB

EARTH'S MOVING PLATES

You have learned that Earth's surface is separated into plates that move apart, move together, or slide past each other. In this lab, you will observe a process that is thought to cause this plate movement.

Real-World Question

What process inside Earth provides the energy for plate motion?

Goals

- **Observe** movement of solid plates on a liquid.
- **Identify** the cause of plate movement on Earth's surface.

Materials

1-L beakers (2)
food coloring
aluminum foil
pencil
rubber band
water (warm and cold)
2-cm paper squares (3)
small, clear-plastic cup

Safety Precautions

WARNING: *Handle the warm water with care. Water from the tap should be warm enough.*

Procedure

1. Fill one of the 1-L beakers with cold water.
2. Fill the small cup with warm water.
3. Add four drops of food coloring to the cup of warm water and cover the top with aluminum foil. Secure the aluminum foil with a rubber band. No air should be underneath the foil.
4. Carefully place the cup of colored, warm water in the bottom of the second 1-L beaker.
5. Carefully pour the cold water from the first 1-L beaker into the second 1-L beaker. Take care not to disturb the cup of colored water.

6. Place the pieces of paper on the surface of the water in the second 1-L beaker.
7. Use a long pencil to make two small holes in the aluminum foil covering the cup.
8. **Observe** what happens to the contents of the cup and to the pieces of paper. Record your observations in your Science Journal.

Conclude and Apply

1. **Describe** What happened to the colored, warm water originally located in the cup?
2. **Infer** What effect, if any, does the warm water have on the positions of the floating paper?
3. **Compare and Contrast** How is what happens to the warm water similar to processes that occur inside Earth? How is it different?
4. **Explain** After observing the pieces of paper floating on the cold water, explain what features on Earth's surface they are similar to.

Communicating Your Data

Compare your conclusions with those of other students in your class. **For more help, refer to the Science Skill Handbook.**

Real-World Question

Purpose Students observe the process of convection and infer its relationship to plate movement. L1 ELL LS **Visual-Spatial**

Process Skills interpret data, recognize cause and effect, communicate, measure in SI, recognize and use spatial relationships, formulate models

Time Required 35 minutes

Procedure

Teaching Strategies

- Be sure students use a sharpened pencil in Step 7. Caution students to poke the holes with care.
- Pair students who are physically challenged with students who can help with Steps 4-7.

Troubleshooting Caution students not to bump the beaker or the table. Explain that all currents within the beaker must be caused by changes within the beaker. Paper squares must be nearly at rest prior to performing Step 7.

Conclude and Apply

1. The warm water begins to flow out of the cup and toward the top of the beaker.
2. Movement of the warm water upward causes the positions of the paper to change. The paper moves around as it floats.
3. The warm water rises through the colder water. In a similiar way, hot plasticlike rock and magma rise through cooler surrounding rock in Earth. One reason that the processes are different is because convection in Earth occurs much more slowly than convection in water.
4. The pieces of paper are similar to Earth's tectonic plates.

Communicating Your Data

Ask students to make a series of drawings illustrating what they observed happening in the beaker. Then have students compare their drawings and discuss any differences they note.

298 CHAPTER 10 Forces Shaping Earth

section 2

Uplift of Earth's Crust

Building Mountains

One popular vacation that people enjoy is a trip to the mountains. Mountains tower over the surrounding land, often providing spectacular views from their summits or from surrounding areas. The highest mountain peak in the world is Mount Everest in the Himalaya in Tibet. Its elevation is more than 8,800 m above sea level. In the United States, the highest mountains reach an elevation of more than 6,000 m. There are four main types of mountains—fault-block, folded, upwarped, and volcanic. Each type forms in a different way and can produce mountains that vary greatly in size.

Age of a Mountain As you can see in **Figure 11,** mountains can be rugged with high, snowcapped peaks, or they can be rounded and forested with gentle valleys and babbling streams. The ruggedness of a mountain chain depends largely on whether or not it is still forming. Mountains like the Himalaya are currently forming at a rate of several centimeters per year, while much older mountains like the Ouachita Mountains in Arkansas stopped forming millions of years ago and are now being eroded by geological processes.

Figure 11 Mountains can be high and rugged like the mountains of the Himalaya shown on the left, or they can be large, gently rolling hills like the Ouachita Mountains in Arkansas, shown above.

Infer *What determines how rugged and high a mountain chain is?*

as you read

***What* You'll Learn**

- **Describe** how Earth's mountains form and erode.
- **Compare** types of mountains.
- **Identify** the forces that shape Earth's mountains.

***Why* It's Important**

The forces inside Earth that cause Earth's plates to move around also are responsible for forming Earth's mountains.

Review Vocabulary

erosion: process by which products of weathering are moved

New Vocabulary

- fault-block mountain
- folded mountain
- upwarped mountain
- volcanic mountain
- isostasy

1 Motivate

INTERACTIVE CHALKBOARD
PowerPoint® Presentations

Bellringer

Section Focus Transparencies also are available on the Interactive Chalkboard CD-ROM. L2 ELL

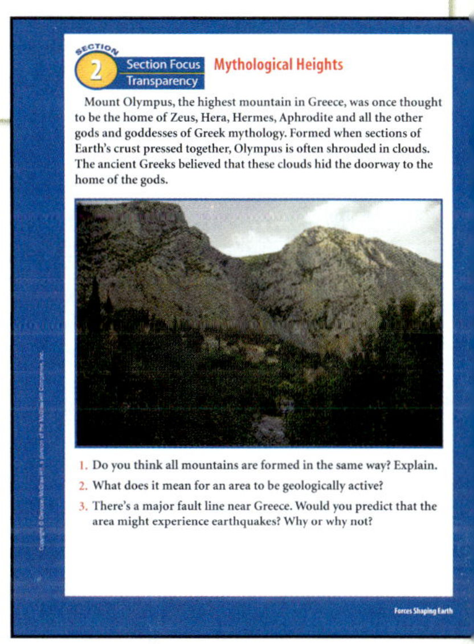

Section 2 Section Focus Transparency **Mythological Heights**

Mount Olympus, the highest mountain in Greece, was once thought to be the home of Zeus, Hera, Hermes, Aphrodite and all the other gods and goddesses of Greek mythology. Formed when sections of Earth's crust pressed together, Olympus is often shrouded in clouds. The ancient Greeks believed that these clouds hid the doorway to the home of the gods.

1. Do you think all mountains are formed in the same way? Explain.
2. What does it mean for an area to be geologically active?
3. There's a major fault line near Greece. Would you predict that the area might experience earthquakes? Why or why not?

Forces Shaping Earth

Tie to Prior Knowledge

Mountains Help students recall photographs, videos, or visits to mountainous areas. If mountains exist in your area, have students list characteristics with which they are familiar. You might also show pictures of different types of mountains. Ask volunteers whether they know what type of mountains (volcanic, upwarped, fault-block, folded) they are describing.

Section 2 Resource Manager

Chapter *FAST FILE* Resources

Transparency Activity, p. 45
Directed Reading for Content Mastery, pp. 23, 24
MiniLAB, p. 4
Enrichment, p. 32
Lab Activity, pp. 15–18
Reinforcement, p. 30
Lab Worksheet, pp. 7–8

Home and Community Involvement, p. 23
Performance Assessment in the Science Classroom, p. 39
Cultural Diversity, p. 27
Earth Science Critical Thinking/Problem Solving, p. 1

2 Teach

Caption Answer

Figure 11 Whether the mountain is still growing; older mountain chains that contain gently rolling hills have undergone extensive erosion.

Quick Demo

Mountain Building

Materials three text books, all the same size

Estimated Time 10 minutes

Procedure Explain that the books represent blocks of crust. Have students describe the landforms that would result from each movement. (1) Hold the books with the spines all upward. Have students observe as you push up on the middle book so that it slides upward. Compression on the inner block, combined with upward pressure, causes a block of crust to upwarp. (2) Firmly hold the books out away from your body. Release some of the pressure so that the middle book slips slightly. Tension causes the inner block to slip below the other crustal blocks.

Use an Analogy

Folding Paper Have students observe as they apply opposite sliding forces to both ends of a sheet of paper lying flat on their desk. Now ask them to determine how this act is analogous to the formation of folded mountains. Compression was applied to the sheet of paper, which caused it to fold. The same force causes rock layers to fold.

Figure 12 Before tension is applied, the layers of rock are even and fairly level. After tension is applied, huge blocks of rock separate and slip downward. This leaves large, tilted blocks that become mountains.

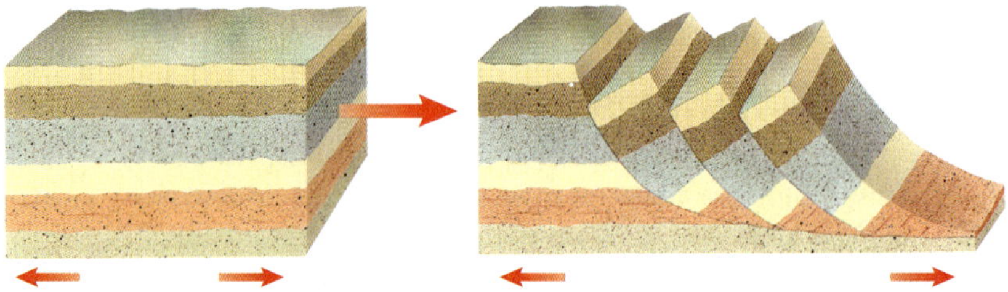

Fault-Block Mountains The first mountains you'll study are fault-block mountains. Some examples are the Sierra Nevada in California and the Teton Range in Wyoming. Recall that pulling, or tension, forces that occur at the boundaries of plates moving apart, work to create surface features such as rift valleys and faults. Fault-block mountains also form from pulling forces. **Fault-block mountains** are made of huge, tilted blocks of rock that are separated from surrounding rock by faults. When rock layers are pulled from opposite directions, large blocks slide downward, creating peaks and valleys, as shown in **Figure 12.**

Models of Mountain Building If you hold a candy bar between your hands and then begin to pull it apart, cracks might form within the chocolate. Similarly, when rocks are pulled apart, faults form. Unlike rocks deep in Earth, rocks at Earth's surface are hard and brittle. When they are pulled apart, large blocks of rock can move along the faults. The Teton Range of Wyoming formed when a block of crust was tilted as one side of the range was uplifted above the neighboring valley. As shown in **Figure 13,** if you travel to the Grand Teton National Park, you will see sharp, jagged peaks that are characteristic of fault-block mountains.

Now, hold a flat piece of clay between your hands and then push your hands together gently. What happens? As you push your hands together, the clay begins to bend and fold over on itself. A similar process causes rocks to fold and bend, causing folded mountains to form on Earth's surface.

Figure 13 The Teton Range in the Grand Teton National Park has sharp, jagged peaks that are characteristic of fault-block mountains.

Curriculum Connection

Math The Sierra Nevada formed over a period of at least 10 million years. Mount Whitney is one of many peaks in this mountain range. Its elevation is 4,400 m above sea level. If the total uplift is 4,400 m in 10 million years, have students compute the average annual rate of uplift. 4,400 m ÷ 10 million years = 0.44 mm/year L2 LS **Logical-Mathematical**

Science Journal

Mountain Characteristics Have students write in their Science Journals brief descriptions of one basic characteristic of each type of mountain. Fault-block mountains have sharp, jagged peaks. Folded mountains have folds in the rock layers. Upwarped mountains form in broad domes. Volcanic mountains are cone-shaped. L2 LS **Linguistic** P

Folded Mountains Traveling along a road that is cut into the side of the Appalachian Mountains, you can see that rock layers were folded just as the clay was when it was squeezed, or compressed. Tremendous pushing forces exerted by two of Earth's plates moving together squeezed rock layers from opposite sides. This caused the rock layers to buckle and fold, forming folded mountains. **Folded mountains** are mountains formed by the folding of rock layers caused by compression forces.

Reading Check *What type of force causes folded mountains to form?*

The Appalachian Mountains are folded mountains that formed about 250 million to 300 million years ago. A small part of the folded Appalachians is shown in **Figure 14.** The compression occurred as the North American Plate and the African Plate moved together. The Appalachians are the oldest mountain range in North America, and also one of the longest. They extend from Alabama northward to Quebec, Canada. Erosion has been acting on these mountains since they were formed. As a result, the Appalachians are small compared to other mountain ranges. At one time, the Appalachian Mountains were higher than the Rocky Mountains are today.

Upwarped Mountains The Adirondack Mountains in New York, the southern Rocky Mountains in Colorado and New Mexico, and the Black Hills in South Dakota are examples of upwarped mountains. **Upwarped mountains** form when forces inside Earth push up the crust. With time, sedimentary rock layers on top will erode, exposing the igneous or metamorphic rocks underneath. The igneous and metamorphic rocks can erode further to form sharp peaks and ridges.

Mini LAB

Modeling Mountains

Procedure

1. Use layers of **clay** to build a model of each major type of mountain.
2. For fault-block mountains, cut the layers of clay with a **plastic knife** to show how one block moves upward and another moves downward.
3. For folded mountains, push on the layers of clay from directly opposite directions.
4. For upwarped mountains, push a large, round object, such as a **ball,** upward from below, forcing the layers of clay to warp.
5. For volcanic mountains, place layer upon layer of clay to form a cone-shaped feature.

Analysis

1. Do any of the mountains you have modeled look similar? Explain.
2. How could you recognize the different types of mountains?

Figure 14 This roadcut in Maryland exposes folded rock layers that formed when the North American Plate and the African Plate collided.

Reading Check

Answer compression

Mini LAB

Purpose Students model the four different types of mountains L2 LS **Kinesthetic**

Materials modeling clay, plastic knife, ball

Teaching Strategy Have students work with a partner to model upwarped mountains. One student can hold the clay layers while the other pushes the ball upward.

Safety Precautions Caution students to handle the plastic knife with care.

Analysis

1. Yes, upwarped mountains and folded mountains can look similar. In both cases rock layers are bent upward.
2. Each type of mountain can be distinguished by its form and structure.

Assessment

Oral Have students develop questions about the four types of mountains and then quiz each other aloud. Use **Performance Assessment in the Science Classroom,** p. 91.

Teacher FYI

Continental Accretion Plate collisions and subduction are considered the causes of most mountain belts. In a proposed mechanism, called continental accretion, small crustal fragments collide and attach to larger continental masses. It is thought that many of the mountains around the Pacific Ocean were formed in this way.

Activity

Topographic Maps Provide students with topographic maps of areas that have fault-block, folded, upwarped, and volcanic mountains. Ask students to describe the pattern on the map formed by each type of mountain. Fault-block mountains have jagged peaks indicated by closely-spaced contour lines. Folded mountains appear as long ridges of closely-spaced lines (mountains) separated by wider-spaced lines (valleys). Streams on upwarped mountains might form parallel lines. Volcanic mountains are steeply cone-shaped and streams run down their sides in a radial pattern. L2

LS **Visual-Spatial and Linguistic**

Visual Learning

Figure 15 Movements of magma in the magma chamber and into the pipe can generate earthquakes. Earthquakes around volcanoes are an indication of magma movement near the volcano. Have students make an events-chain concept map summarizing the sequence of events that lead up to a volcanic eruption. Magma rises from the magma chamber, up the pipe, to the vent in the crater. Magma, now called lava, then flows across the surface and down the sides of the volcano.

Use Science Words

Word Origins Geologists refer to mountain-building as *orogenesis*. Have students research the Greek roots of this word, and use the word in a sentence. Orogenesis is derived from the Greek word *oros,* meaning "mountain" and *genesis,* meaning "coming into being." Orogenesis explains how mountains are formed.

Science Online

Topic: Volcanic Mountains

Visit red.msscience.com for Web links to information about volcanic mountains.

Activity Collect as many photographs of volcanic mountains as possible. Create a large map of the world with the photographs in their proper locations. Include some information about the volcanic mountains and the impact they have had on the environment around there.

Volcanic Mountains Occasionally, magma from inside Earth reaches the surface. When this happens, the magma is called lava. When hot, molten lava flows onto Earth's surface, volcanic mountains can form. Over time, layer upon layer of lava piles up until a cone-shaped feature called a **volcanic mountain** forms. Washington's Mount St. Helens and Mexico's Mount Popocateptl, shown in **Figure 15,** are examples. Next, you will take a closer look at how volcanic mountains form.

Some volcanic mountains form when large plates of Earth's lithosphere sink into Earth's mantle at subduction zones. As the plates sink deeper into the mantle, they cause melting to occur. The magma produced is less dense than the surrounding rock, so it is forced slowly upward to Earth's surface. If the magma reaches the surface, it can erupt as lava and ash. Layers of these materials can pile up over time to form volcanic mountains.

Figure 15 Volcanic mountains form when lava and ash build up in one area over time.

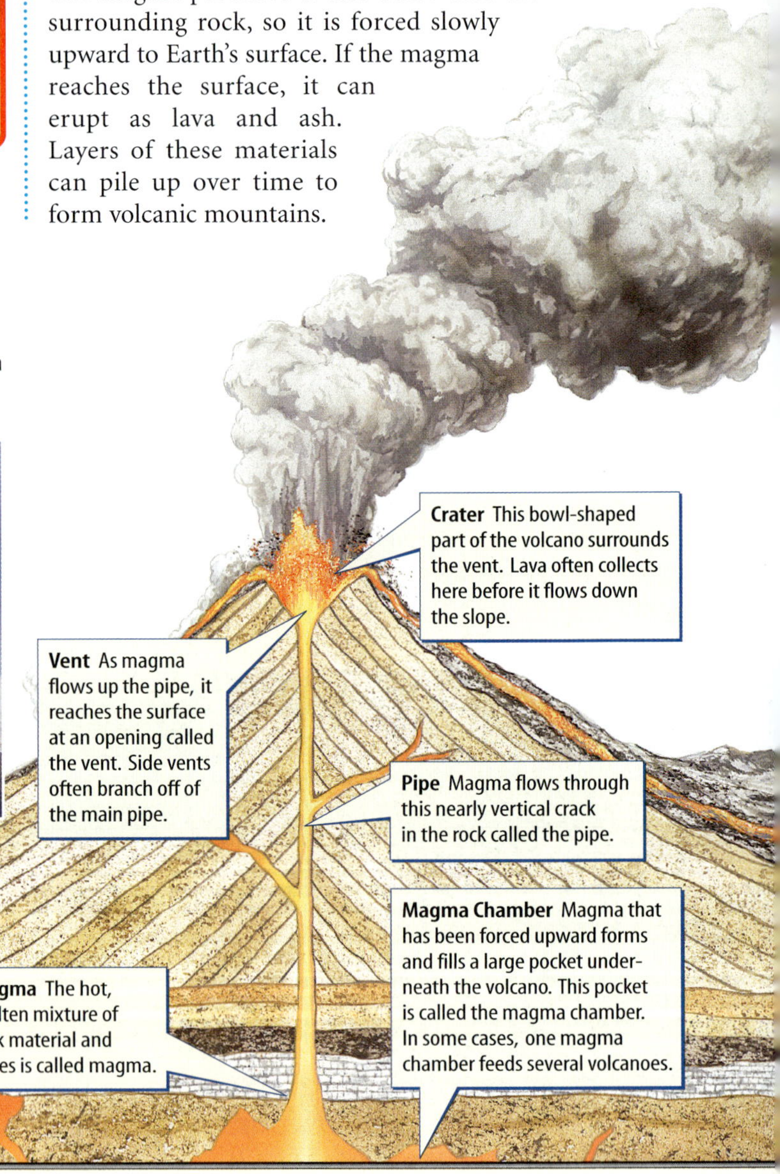

Figure 16 The Hawaiian Islands are a series of volcanic mountains that have been built upward from the seafloor. They began to form as lava erupted onto the ocean floor. Over time, the mountain grew so large that it rose above sea level.

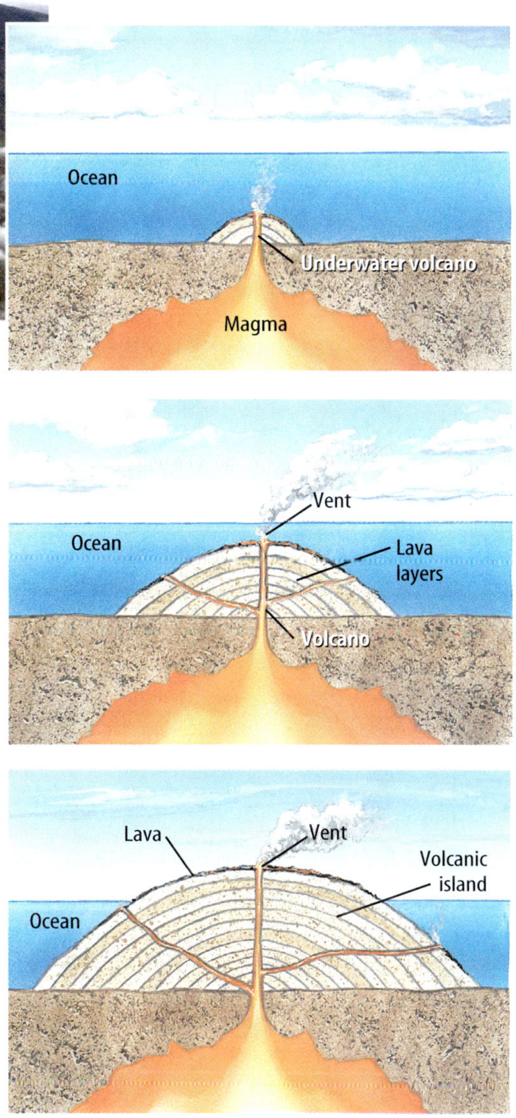

Underwater Volcanic Mountains You know that volcanic mountains form on land, but did you know that these mountains also form on the ocean floor? Underwater eruptions can produce mountains beneath the sea. Eventually, if enough lava is erupted, these mountains grow above sea level. For example, Hawaii, shown above in **Figure 16,** is the peak of a huge volcanic mountain that extends above the surface of the water of the Pacific Ocean. **Figure 16** also illustrates how the Hawaiian Islands formed.

Volcanic mountains like the Hawaiian Islands are different from the volcanic mountains that form where one plate subducts beneath another. The Hawaiian Islands formed from material that came from near the boundary between Earth's core and mantle. Hot rock is forced upward through the mantle as a plume and melts to form a hot spot in Earth's crust. As plates travel over the hot spot, a series of volcanoes, as seen in Hawaii, forms. Magma from subduction volcanoes forms much closer to Earth's surface. Hot spot volcanoes also are much larger and have more gently sloping sides than subduction volcanoes.

Reading Check *What type of mountains make up the Hawaiian Islands?*

Discussion

New Island Loihi is a submerged volcanic mountain off the southern coast of the Big Island of Hawaii. It is being formed as the Pacific plate moves over a stationary hot spot in the mantle. Have students discuss what the formation of this island means about the future of the Hawaiian Islands. When it breaks the surface, which at its present rate of growth will take tens of thousands of years Loihi will be the next island in the chain that forms the Hawaiian Islands. At that time, the volcanoes on the big island of Hawaii will likely be inactive. L2 LS **Linguistic**

Reading Check

Answer volcanic mountains

Fun Fact

Pillow lava forms when lava is extruded under oceans. The cooling nature of the water forms pillow-like formations of hardened lava.

Differentiated Instruction

Behaviorally Disordered To help students keep focused on the lesson about volcanic mountains, provide them with a model of a volcano with parts labeled. After students study the model and **Figure 15,** encourage them to make their own model volcano. Challenge them to find a way to make their volcano erupt. For example, students might mix baking soda and vinegar to simulate lava pouring down the slope of a volcano. A more accurate model would include rubber tubing built into the bottom of the volcano. When the model is complete, have students fill the pipe of their volcanoes with puffed cereal to represent magma. They can make their volcano erupt by blowing into the end of the rubber tubing. The harder they blow, the more violent the eruption will be. L2 ELL LS **Visual-Spatial and Kinesthetic**

Use Science Words

Word Meaning Folded mountains have areas that are folded upward and other areas that are folded downward. Have students research and determine which of the following terms, *anticline* or *syncline*, refers to each area. Syncline refers to downfolded areas; anticline refers to upfolded areas.

INTEGRATE Language Arts

The Isostasy Story The prefix *iso* means "equal". The suffix *stasis* means "remain constant." Isostasy is the condition of equilibrium that occurs when the downward force of gravity on a portion of Earth's crust is equal to the upward force of buoyancy on the crust.

Applying Science

Answers

1. The land rebounds, or pushes back up, after a glacier melts.
2. Accept reasonable answers. Rising crust would likely cause beaches to be tilted. It is unlikely that the crust would rise the same amount everywhere. The glacier may have been thicker in some areas, causing a greater depression. Also, some areas of crust may be less elastic than others, and thus move less than other areas.

Virtual Labs

Glaciers *How do glaciers shape the land?*

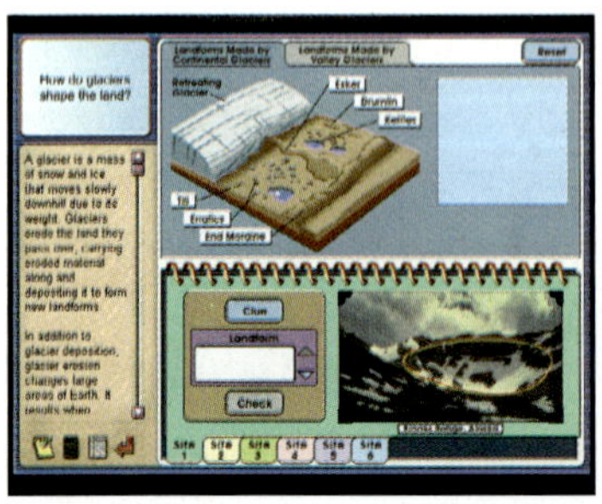

INTEGRATE Language Arts

The Isostasy Story
Using the principle of isostasy, explain in your Science Journal why large features on Earth's surface, such as mountains, float on the layers of Earth beneath them.

Other Types of Uplift

You have learned about the origin of the pushing forces that bend crustal rocks during mountain-building processes. However, another force also works to keep mountains elevated above the surrounding land. If you place wooden blocks of various thicknesses in a container of water, you will notice that different blocks of wood float in the water at different heights. Also, the thicker blocks of wood float higher in the water than the thinner blocks do. The buoyant force of the water is balancing the force of gravity. A similar process called isostasy occurs in Earth. According to the principle of **isostasy**, Earth's lithosphere floats on a plasticlike upper part of the mantle, the asthenosphere.

The effects of isostasy were first noticed near large mountain ranges. Earth's crust is thicker under mountains than it is elsewhere. Also, if mountains continue to get uplifted, the crust under the mountains will become thicker and will extend farther down into the mantle. This is similar to the floating wooden blocks. If you pile another wooden block on a block that is already floating in the water, you will see that the new, larger block will sink down into the water farther than before. You also will see that the new block floats higher than it did before.

Applying Science

How can glaciers cause land to rise?

About 20,000 years ago, much of North America was covered by a large glacial ice sheet. How do you think an ice sheet can affect Earth's crust? What do you think happens when the ice melts?

Identifying the Problem

More than 100 years ago, people living in areas that once had been covered by glaciers noticed that features such as old beaches had been tilted. The beaches had a higher elevation in some places and a lower elevation in others. How do you think old beaches could be tilted?

Solving the Problem

1. The weight of glaciers pushes down Earth's crust. What do you think happens after the glacier melts?
2. How could rising crust cause beaches to be tilted? Do you think the crust would rise the same amount everywhere? Explain.

Differentiated Instruction

Challenge Mount St. Helens, Mount Hood, Crater Lake, and Mount Rainer are part of the Cascade Mountain Range. Have students research the Cascade Mountain Range and learn more specifically about the formation of these mountains. Then have them chose one of the mountains mentioned above and create a model of the mountain and the surrounding areas. L3

English-Language Learners Students who are just learning English may find it difficult to understand the concept of isostasy. Demonstrate isostasy as described in the text using a water-filled aquarium and wooden blocks. Have students describe what is happening both above and below the water surface as you add blocks to the water.

Adjusting to Gravity Similar to the wooden blocks, if mountains continue to grow larger, they will sink even farther into the mantle. Once mountains stop forming, erosion lowers the mountains and the crust rises again because weight has been removed. If the process continues, the once-thick crust under the mountains will be reduced to the thickness of the crust where no mountains exist.

Icebergs behave in much the same way, as shown in **Figure 17.** The iceberg is largest when it first breaks off of a glacier. As the iceberg floats, it melts and starts to lose mass. This causes the iceberg to rise in the water. Eventually, the iceberg will be much smaller and will not extend as deeply into the water. How is this similar to what happens to mountains?

Figure 17 Isostasy makes Earth's crust behave in a similar way to these icebergs. As an iceberg melts and becomes smaller, ice from below the water's surface is forced up.

section 2 review

Summary

Building Mountains

- A rugged, tall mountain is geologically young. An old mountain is rounded and lower in elevation.
- There are four main types of mountains: fault block, folded, upwarped, and volcanic.
- Volcanic mountains can form on the surface of the continents or under the ocean at ridges.

Other Types of Uplift

- The principle of isostasy explains how the lithosphere floats on the asthenosphere.
- The crust will also adjust to gravity as erosion and weathering wear away older mountains.

Self Check

1. **Predict** If compression were exerted on rock layers, what type of mountains would form?
2. **Describe** how fault-block mountains form.
3. **Explain** how a volcano forms.
4. **Think Critically** Put the Appalachian, Himalaya, and Rocky Mountains in order from youngest to oldest knowing that the Himalaya are most rugged and the Appalachians are the least rugged.

Applying Skills

5. **Concept Map** Make a chain-of-events concept map that describes how folded mountains form.

Science Online red.msscience.com/self_check_quiz

Text Question Answer

As a mountain erodes, its mass is reduced and the buoyant force of isostasy pushes the root of the mountain upward.

3 Assess

DAILY INTERVENTION

Check for Understanding

Logical-Mathematical Have students locate several mountain systems on a world map. Then have them draw conclusions as to which of these mountains are associated with active tectonic boundaries and are therefore still growing.

Reteach

Mountains Have students identify each mountain type as you read its description: (1) can form from the movement of magma volcanoes and upwarped mountains; (2) most likely formed from tension fault-block mountains; (3) most likely formed from compression folded mountains. L1

Assessment

Performance Have students make a model of the mountain-building process of their choosing and explain how the mountains form. Use **Performance Assessment in the Science Classroom,** p. 123.

section 2 review

1. folded mountains
2. Fault-block mountains are huge tilted blocks of rock that are separated from surrounding rock by faults.
3. Magma inside Earth is forced upward until it flows out onto the surface. A volcano forms when layer after layer of lava piles up, forming a cone-shaped mountain.
4. Himalaya, Rocky, Appalachian
5. Answers should include the following in order: compression pushes from opposite directions, rock layers crumple and fold, folded mountains form.

BENCH TESTED

Real-World Question

Purpose Students will determine how adding or removing mass affects the way an object floats in a fluid.

Process Skills model, compare, observe, predict, recognize cause and effect, interpret data

Time Required 45 minutes

Procedure

Possible Materials You may want to test your wooden blocks ahead of time to make sure they are stable as they float in water.

Alternate Materials You might want to provide students with large metal washers or nuts that can be added to or removed from the wooden blocks to increase or decrease the mass of the floating object.

LAB Model and Invent

Isostasy

Goals
- **Observe** the results of isostasy.
- **Predict** what will happen to floating objects when mass is removed or added.

Possible Materials
5-cm × 5-cm × 2-cm wooden blocks (3)
10-cm × 35-cm × 15-cm clear-plastic storage box or other bin
water
permanent marker
ruler

Safety Precautions

Real-World Question

The principle of isostasy states that Earth's crust floats on the more dense mantle beneath. This is similar to the way objects float in water. What do you think will happen when you add mass to a floating object? What if you take away mass? How does adding or removing mass affect the way an object floats in a fluid?

Make the Model

1. **Decide** what object(s) you will float in the water initially. How will you remove mass from that object? How will you add mass?
2. What will you observe as the mass changes? How will you record the effects of adding or removing mass?
3. How much water will you use? What problems might you encounter if you have too much or too little water?

Alternative Inquiry Lab

Make a Model Make this Lab an Inquiry Lab by extending the exercise. Instruct students that icebergs operate on the same principle as isostasy. Ask students to create a plan for a model iceberg where they will calculate the amount of ice below the water compared to the amount of ice visible above the surface. Once the plan has been approved, provide students with the necessary materials to create their model.

4. Will you make any additional measurements or record any other data?
5. **List** all the steps that you plan to do in this activity. Are the steps in a logical order?
6. **Compare** your model plans to those of other students.
7. Make sure your teacher approves your plans before you start.

Test the Model

1. Fill the storage box or bin with an appropriate amount of water.
2. Start by floating the initial object you planned to use in the water. Observe and record relevant data.
3. Follow the list of steps you planned in order to obtain data for removing and adding mass. Observe your model and record all relevant data in your Science Journal.

Conclude and Apply

1. **Describe** What did your initial object look like? What level did the water rise to when your initial object was placed in the bin? How did you add and remove mass?
2. **Summarize** What happened to the amount of the object that was submerged and the amount sticking out of the water when mass was removed from the object?
3. **Summarize** What happened to the amount of the object that was submerged and the amount sticking out of the water when mass was added?
4. **Explain** How can you explain your observations about how much of the object was submerged and how much was sticking out of the water? How is this similar to processes that occur in Earth?

Communicating Your Data

Make a poster that illustrates what you have learned about isostasy. **For more help, refer to the Science Skill Handbook.**

Analyze Your Data

Teaching Strategy Do not leave the wood in water too long because it will become water-logged and heavy.

Make a Model

Expected Outcome The model should include blocks that float at different levels in the water, depending on the amount of mass added to or removed from the block. The greater the mass, the lower the block will float.

Conclude and Apply

1. The height at which the object floated will vary according to the density of the material used. Methods used to add and remove mass will vary according to the model.
2. The object floated higher in the water—less of it was submerged.
3. More of the object was submerged and less was above the water level.
4. Removing mass from the object causes it to float higher. This is similar to a mountain eroding away, having less mass, and its root floating higher in Earth's mantle.

Communicating Your Data

Students may choose to include information obtained from the Glencoe science web site in their posters.

Assessment

Oral Ask students to describe how this activity relates to the advance and retreat of continental glaciers. Use **Performance Assessment in the Science Classroom,** p. 91.

Content Background

The tallest mountain—Mauna Kea, Hawaii—is taller than Mount Everest. Mauna Kea rises 10,205 m from the ocean floor, but rises only 4,205 m above sea level. Mount Everest is 8,850 m tall, as measured from sea level.

Discussion

Old Mountains Since the Rockies are much taller than the Appalachians, what does this tell you about the age of the Rockies? Explain. The Rockies must be younger than the Appalachians and not as worn down by erosion.

How is it possible that some mountains are still becoming taller? Volcanic mountains can be taller as they continue to erupt. Also plate movement can continue to push mountains up.

Find Out About It

Have students compare their drawings with those made by other members of the class. Discuss why the view from the top of different mountains can vary so much.

Applying Math

Answer 3.3 m

SCIENCE Stats

Mountains

Did you know...

... The world's longest mountain range is underwater. The mid-ocean ridge that winds around Earth beneath the Arctic, Atlantic, and Pacific Oceans is 65,000 km long. That's four times longer than the combined lengths of the Andes Mountains, the Rocky Mountains, and the Himalaya.

... The beautiful Appalachian Mountains are among the oldest in the world. By 250 million years ago, their formation was complete. Today, the mountains aren't among the tallest because they have been worn down by many millions of years of erosion.

... In 1963, Surtsey, a small island, formed when an underwater volcano erupted off the coast of Iceland. The 1.6-km-long island rose to the height of 183 m—about as tall as a 55-story building.

Applying Math Using this relationship, how many meters would there be in a one-story building?

Find Out About It

Research a mountain on red.msscience.com/science_stats. Pinpoint its location on a map, and then accurately draw the mountain and the view from its top.

308 CHAPTER 10 Forces Shaping Earth

Visual Learning

Highest Mountain Peaks on Each of the Seven Continents Ask students which continent has the highest mountain peak. Asia Then ask them to rank the peaks from tallest to smallest. Have them locate each continent on a globe or world map and then locate the position of each mountain. Suggest that they research which of the mountains in the graph are volcanic mountains and which are not. Of those that are volcanic, have students find out which are still active volcanoes.

Reviewing Main Ideas

Section 1 Earth's Moving Plates

1. Earth's interior is divided into four layers, the inner core, the outer core, the mantle, and the crust.
2. Earth's inner and outer cores are thought to be composed mostly of iron. The outer core is thought to be liquid and the inner core is solid.
3. Plates composed of sections of Earth's crust and rigid upper mantle move around on the plasticlike asthenosphere.
4. Earth's plates move together, move apart, and slide past each other.
5. Convection in Earth's mantle, ridge-push, and slab-pull might all contribute to plate movement.

Section 2 Uplift of Earth's Crust

1. Uplift causes mountains to form. Faulting, folding, upwarping, and volcanic eruptions are all processes that build mountains.
2. Four main types of mountains are fault-block, folded, upwarped, and volcanic.
3. As erosion removes material, the mass of the mountains is reduced. Isostasy, then forces the crust upward.

Visualizing Main Ideas

Copy and complete the following table comparing examples and causes of the four types of mountains.

The Four Types of Mountains

Mountain Type	Example	Cause
Fault-block mountain	Teton Range	tension
Folded mountain	Appalachian Mountains	compression
Upwarped mountain	Adirondack Mountains	upward pressure
Volcanic mountain	Mount St. Helens	lava flows

Science Online red.msscience.com/interactive_tutor

Reviewing Main Ideas

Summary statements can be used by students to review the major concepts of the chapter.

Visualizing Main Ideas

See student page.

Science Online
Visit red.msscience.com
/self_check_quiz
/interactive_tutor
/vocabulary_puzzlemaker
/chapter_review
/standardized_test

Assessment Transparency

For additional assessment questions, use the *Assessment Transparency* located in the transparency book.

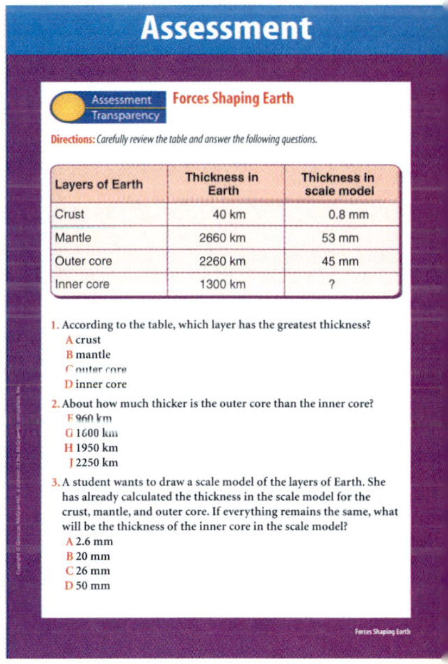

Assessment

Assessment Transparency — Forces Shaping Earth

Directions: *Carefully review the table and answer the following questions.*

Layers of Earth	Thickness in Earth	Thickness in scale model
Crust	40 km	0.8 mm
Mantle	2660 km	53 mm
Outer core	2260 km	45 mm
Inner core	1300 km	?

1. According to the table, which layer has the greatest thickness?
   A crust
   B mantle
   C outer core
   D inner core
2. About how much thicker is the outer core than the inner core?
   F 960 km
   G 1600 km
   H 1950 km
   J 2250 km
3. A student wants to draw a scale model of the layers of Earth. She has already calculated the thickness in the scale model for the crust, mantle, and outer core. If everything remains the same, what will be the thickness of the inner core in the scale model?
   A 2.6 mm
   B 20 mm
   C 26 mm
   D 50 mm

Forces Shaping Earth

CHAPTER STUDY GUIDE 309

Using Vocabulary

1. Earth's <u>outer core</u> is thought to be liquid.
2. The Sierra Nevada are <u>fault-block mountains</u>.
3. <u>Folded mountains</u> commonly form in areas where plates are being pushed together.
4. <u>Subduction</u> occurs when a more dense plate sinks beneath a less dense plate
5. A <u>volcanic mountain</u> forms when magma is forced upward and flows onto Earth's surface.

Checking Concepts

6. B	11. D
7. A	12. B
8. B	13. C
9. A	14. D
10. D	15. B

chapter 10 Review

Using Vocabulary

crust p. 290	lithosphere p. 292
fault p. 293	mantle p. 290
fault-block mountain p. 300	outer core p. 290
folded mountain p. 301	plate p. 292
inner core p. 289	subduction p. 295
isostasy p. 304	upwarped mountain p. 301
	volcanic mountain p. 302

Answer the following questions with complete sentences.

1. Which part of Earth's core do scientists think is liquid?
2. The Sierra Nevada mountains in California are which type of mountain?
3. What type of mountains form in areas where rocks are being pushed together?
4. What process occurs when a more dense plate sinks beneath a less dense plate?
5. Which type of mountain forms when magma is forced upward and flows onto Earth's surface?

Checking Concepts

Choose the word or phrase that best answers the question.

6. Which part of Earth is largest?
   A) crust **C)** outer core
   B) mantle **D)** inner core
7. Earth's plates are pieces of which layer of Earth?
   A) lithosphere
   B) asthenosphere
   C) inner core
   D) mantle
8. Which force pushes plates together?
   A) tension **C)** shearing
   B) compression **D)** isostasy
9. Which force occurs where Earth's plates are moving apart?
   A) tension **C)** shear
   B) compression **D)** isostasy
10. Which layer of Earth is thought to be solid and composed mostly of the metal iron?
    A) crust **C)** outer core
    B) mantle **D)** inner core
11. Which suggests that Earth's lithosphere floats on the asthenosphere?
    A) tension
    B) compression
    C) shear
    D) isostasy
12. Which type of mountain forms because of compression forces?
    A) fault-block mountains
    B) folded mountains
    C) upwarped mountains
    D) volcanic mountains
13. Which type of mountain forms because forces inside Earth push up overlying rock layers?
    A) fault-block mountains
    B) folded mountains
    C) upwarped mountains
    D) volcanic mountains
14. Which type of plate movement occurs at transform boundaries?
    A) plates moving together
    B) plates moving apart
    C) plates sinking
    D) plates sliding past each other
15. Which type of plate movement produces deep rifts such as the mid-ocean rift?
    A) plates moving together
    B) plates moving apart
    C) plates sliding past each other
    D) plates sinking

Science Online red.msscience.com/vocabulary_puzzlemaker

Use the Exam*View*® Pro Testmaker CD-ROM to:

- create multiple versions of tests
- create modified tests with one mouse click for inclusion students
- edit existing questions and add your own questions
- build tests aligned with state standards using built-in State Curriculum Tags
- change English tests to Spanish with one mouse click and vice versa

Thinking Critically

16. **Explain** Which is older, the Great Rift Valley in East Africa, or the Mid-Atlantic Ridge in the Atlantic Ocean?

17. **Explain** how you can determine whether or not a mountain is still forming.

18. **Infer** Seismic waves slow down when entering the asthenosphere. What does this tell you about the nature of the asthenosphere?

19. **Predict** what would happen to the elevation of the island of Greenland if the ice sheet were to melt away.

20. **Describe** If you wanted to know whether a certain mountain was formed by compression, what would you look for?

21. **Compare and contrast** volcanic and folded mountains. Draw a diagram of each type of mountain. Label important features.

22. **Make Models** Use layers of clay to make a model of fault-block mountains. Draw a diagram of your model.

23. **Draw Conclusions** The speed of seismic waves suddenly increases when they go from the upper mantle into the lower mantle. What does this indicate about the comparative densities of the rock in both layers?

24. **Use graphics software** to generate a scale illustration of Earth's interior. Include the thickness of each layer in kilometers.

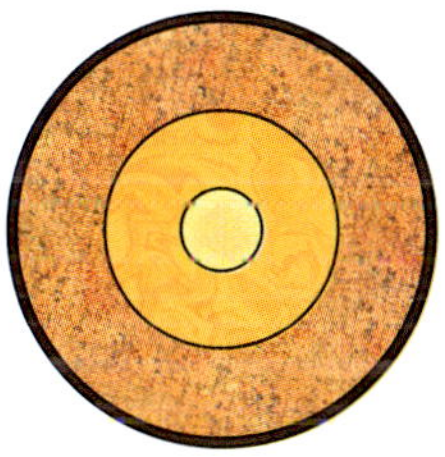

25. **Recognize Cause and Effect** What is the effect of subduction at the boundary of two plates?

Performance Activities

26. **Poem** Write a poem in a style of your choosing about the spectacular view often associated with mountains. You may wish to write about the scene from the top of a mountain or the one you see from the bottom of the mountain looking up to its peak.

Applying Math

27. **Mountain Climbing** The most standard climb for climbers of Mount Everest is up to Base Camp, an elevation of 5400 m. If the summit is 8850 m high, what percentage of Mount Everest's elevation is the Base Camp?

Use the map below to answer question 28.

28. **Moving Cities** The distance between San Francisco and Los Angeles is 616 km. If the San Andreas fault is moving at an average rate of 2.0 cm per year, how long will it be before Los Angeles is next to San Francisco?

Assessment Resources

Reproducible Masters

Chapter *Fast File* Resources
- Chapter Review, pp. 37–38
- Chapter Tests, pp. 39–42
- Assessment Transparency Activity, p. 49

Glencoe Science Web site
- Chapter Review Test
- Standardized Test Practice

Glencoe Technology

- Assessment Transparency
- Exam*View*® Pro Testmaker
- MindJogger Videoquiz
- Interactive Chalkboard

chapter 10 Review

Thinking Critically

16. The Mid-Atlantic Ridge is older, because the continents along its edges are farther apart.
17. Mountains that are still forming have sharp, jagged peaks.
18. It may be less dense than surrounding rock. It also may be somewhat molten.
19. Because of isostasy, the elevation of Greenland would increase as the ice sheet melted away.
20. folded rock layers
21. Both are formed by forces pushing upward. Volcanic mountains are cone-shaped, whereas folded mountains show folded rock layers. Drawings should show these features.
22. Models should show blocks of rock tilted upward along a fault.
23. The density of the rock in the lower mantle is greater than the rock in the upper mantle.
24. Student illustrations should resemble **Figure 4.**
25. Rock layers are folded by compression as two plates move together. The denser plate sinks under the other plate. As it sinks deep into Earth it melts, and the magma rises to the surface to form volcanoes.

Performance Activities

26. Poems may be of any style as long as they describe the beauty of a mountain view. Use **Performance Assessment in the Science Classroom,** p. 151.

Applying Math

National Math Standards

27. 61 percent
28. 616 km × 100,000 cm/km = 61,600,000 cm

61,600,000 cm ÷ 2.0 cm/yr = 30,800,000 years

FAST FILE

Answer Sheet A practice answer sheet can be found at red.msscience.com/answer_sheet.

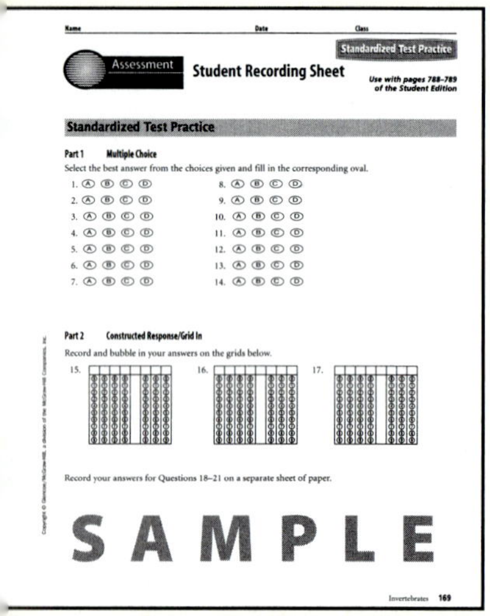

chapter 10 Standardized Test Practice

Part 1 Multiple Choice

Record your answers on the answer sheet provided by your teacher or on a sheet of paper.

1. How is energy from an earthquake carried through the ground?
   A. isostasy **C.** sound
   B. seismic waves **D.** varying density

Use the graph below to answer questions 2 and 3.

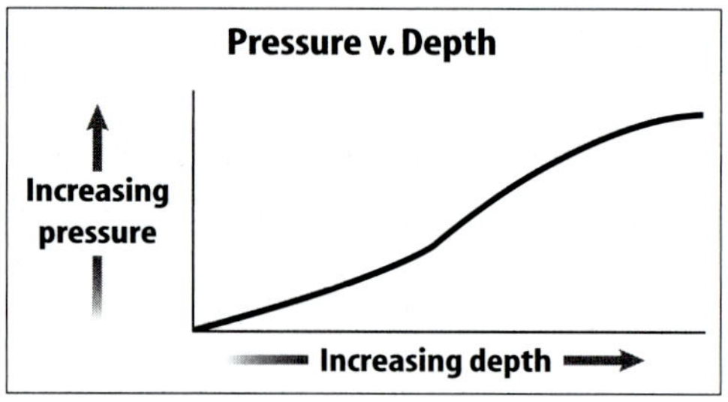

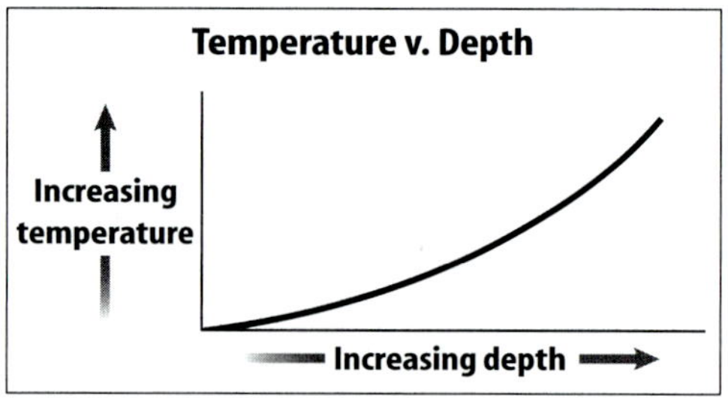

2. In general, what happens to pressure as you move outward from Earth's interior?
   A. decreases
   B. decreases then increases
   C. increases
   D. increases then decreases

3. What happens to temperature as you go deeper into Earth?
   A. decreases
   B. decreases then increases
   C. increases
   D. increases then decreases

4. Which lists layers of Earth's interior from the inside out?
   A. crust, mantle, outer core, inner core
   B. inner core, outer core, crust, mantle
   C. inner core, outer core, mantle, crust
   D. mantle, crust, outer core, inner core

5. Which mountains form when forces pull from opposite directions?
   A. fault-block **C.** upwarped
   B. folded **D.** volcanic

Use the illustration below to answer question 6.

6. Which type of force is involved when Earth's plates slide past each other?
   A. compression **C.** shear
   B. isostasy **D.** tension

7. Which type of mountain is built up from layers of lava and ash?
   A. fault-block **C.** upwarped
   B. folded **D.** volcanic

8. Which of the following is believed to cause plate movement?
   A. compression **C.** isostasy
   B. convection **D.** tension

9. Which force is responsible for Earth's crust and lithosphere floating on the mantle?
   A. compression **C.** shear
   B. isostasy **D.** tension

Part 1 Multiple Choice

1. B
2. A
3. C
4. C
5. A
6. C
7. D
8. B
9. B

Part 2 Short Response

10. The speed of waves depends on the density and nature of the material they are traveling through. Waves speed up through solids and slow down or stop through liquids. They move faster in denser material than in less dense material. By studying the speed and pathways of the waves geologists uncover clues about Earth's internal structure.

11. Older crustal material subducts into the mantle.

12. Because of isostasy, as mountains get uplifted the crust underneath them will extend farther down into the mantle. The added weight of the mountains being formed pushes downward on the mantle below.

13.

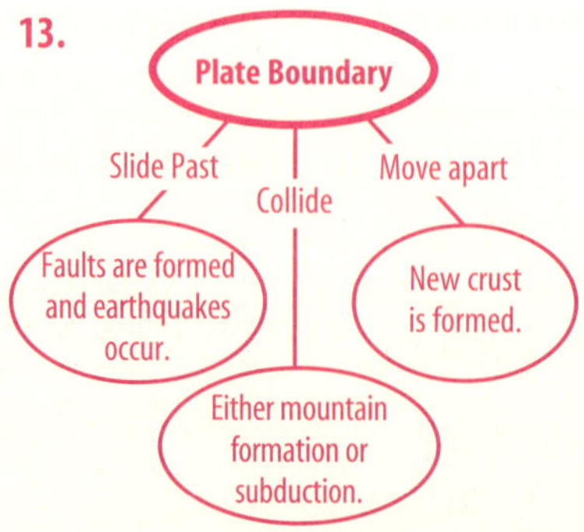

14. The rigid upper part of Earth's mantle and Earth's crust comprise the lithosphere which is broken into 30 sections called plates. These plates move around on the plastic-like asthenosphere.

15. Shearing causes the area between the plates to form faults as they strain against each other. After the strain builds up, the rocks on either

Part 2 | Short Response/Grid In

Record your answers on the answer sheet provided by your teacher or on a sheet of paper.

10. How are seismic waves used to determine Earth's structure?

11. If new crust is being added at rift zones, why doesn't Earth's crust get larger and larger?

12. Why is Earth's crust thicker under mountains than it is elsewhere?

Use the illustration below to answer question 13.

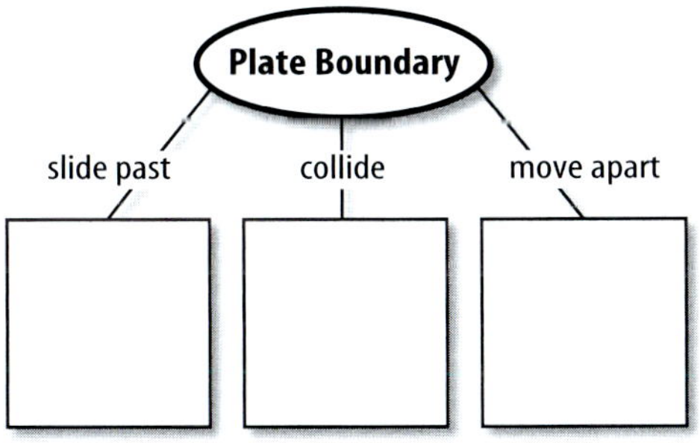

13. Provide the missing information about what takes place at plate boundaries.

14. Compare and contrast the lithosphere and the asthenosphere.

15. What causes earthquakes along a transform boundary?

16. Explain how slab-pull is involved in the subduction of one plate under another.

17. Contrast how compression works to form folded mountains with how tension works to form fault-block mountains.

Test-Taking Tip

Diagrams Study a diagram carefully, being sure to read all labels and captions.

Part 3 | Open Ended

Record your answers on a sheet of paper.

18. Compare and contrast Earth's inner core and outer core.

Use the photo below to answer question 19.

19. What type of mountain is in this picture? Identify the characteristics that distinguish these mountains from others and explain how they formed.

20. Explain how the Hawaiian Islands were formed and why they are unique.

Use the illustration below to answer question 21.

21. How do characteristics of this volcanic mountain differ from mountains formed in other ways?

22. Because of isostasy, what happens to a landmass previously covered with glacial ice, once the ice melts?

23. Explain how do we know that Earth's outer core is liquid?

side of the fault move suddenly, causing vibrations, earthquakes.

16. As one plate subducts under another, it pulls on the rest of the plate trailing behind. The subducting slab pulls on the remaining plate.

17. Compression pushes on rock layers from directly opposite directions causing rock layers to bend. Tension pulls in directly opposite directions causing cracks which become faults as one section of crust slides down past another.

Part 3 | Open Ended

18. Both the inner core and the outer core are very deep within Earth and are under a tremendous amount of pressure with very high temperatures. In addition they are both composed of metal. The inner core is solid iron and experiences the greatest amount of pressure from gravity. The outer core is molten metal and has an uneven boundary.

19. The sharp jagged peaks are characteristic of fault-block mountains. They are formed from pulling forces. Huge blocks of rock are separated from surrounding rock by fractures. When rock layers are pulled from opposite directions, large blocks slide downward along these fractures, causing faults and creating the peaks and valleys.

20. The Hawaiian Islands were formed from material that came from near the boundary between Earth's core and mantle. Hot rock travels through the mantle as a plume and melts to form a hot spot in Earth's crust. As plates travel over the hot spot, a series of volcanoes forms like those in Hawaii. They are generally larger and have more gently sloping sides than other volcanoes.

21. In contrast to mountains formed by other processes, a volcanic mountain is cone-shaped and has a bowl-shaped opening at the top. This bowl-shaped opening is a crater, the part of the volcano that surrounds the vent. The vent is the opening for the magma to flow through and reach the surface. Magma flows outward from the volcano's vent and builds up layer upon layer. The build-up of layers gives the mountain its cone shape. Other mountains are built by folding, faulting, and upwarping. None of them form from layers of lava.

22. Once the weight of the ice has been removed (or while it is being removed), isostasy begins pushing up on the land, making it begin to rise.

23. Seismic waves will slow down or stop when they encounter the outer core. This is because the outer core is less dense than the mantle and is in a liquid state. Once through the outer core, certain seismic waves speed up again as they travel into the solid inner core.

Rubrics

For more help evaluating open-ended assessment questions, see the rubric on p. 10T.

chapter 11 Organizer

Section/Objectives	Standards		Labs/Features
	National	State/Local	
Chapter Opener	See pp. 16T–17T for a Key to Standards.		**Launch Lab:** Water's Force, p. 315 **Foldables,** p. 315
Section 1 **Weathering and Soil Formation** 3 sessions 1.5 blocks 1. **Identify** processes that break rock apart. 2. **Describe** processes that chemically change rock. 3. **Explain** how soil evolves.	National Content Standards: UCP.3, UCP.4, A.1, B.1, C.1, C.5		**MiniLAB:** Dissolving Rock with Acids, p. 319 **MiniLAB:** Analyzing Soils, p. 320 **Lab:** Classifying Soils, p. 322
Section 2 **Erosion of Earth's Surface** 4 sessions 2 blocks 4. **Identify** agents of erosion. 5. **Describe** the effects of erosion.	National Content Standards: UCP.3, A.1, A.2, B.2, E.2, F.5		**Visualizing Mass Movements,** p. 324 **Integrate Physics,** p. 325 **Science Online,** p. 326 **Applying Science:** Can evidence of sheet erosion be seen in a farm field?, p. 329 **Science Online,** p. 330 **Lab:** Measuring Soil Erosion, pp. 332–333 **Science and History:** Crumbling Monuments, p. 334

Glencoe Exclusive! TeacherWorks™ All-In-One Planner and Resource Center

Lab Materials	Reproducible Resources	Section Assessment	Technology
Launch Lab: bread pan, sand, plastic washtub, brick or wood block, access to a sink with a faucet, clock	**Chapter *FAST FILE* Resources** Foldables Worksheet, p. 17 Directed Reading Overview, p. 19 Note-taking Worksheets, pp. 31–32	Glencoe's Assessment Advantage	TeacherWorks includes: • Interactive Teacher Edition • Lesson Planner with calendar • Access to all program blacklines • Correlations to standards • Web links
MiniLAB: dropper, vinegar, pieces of chalk and limestone, magnifying lens, 5 percent hydrochloric acid **MiniLAB:** soil sample, sheet of newspaper **Lab:** soil sample, stereomicroscope	**Chapter *FAST FILE* Resources** Transparency Activity, p. 42 Lab Worksheet, pp. 5–6 Enrichment, p. 29 Reinforcement, p. 27 Directed Reading, p. 20 Lab Activity, pp. 9–12 MiniLAB, pp. 3, 4 Transparency Activity, pp. 45–46 **Cultural Diversity,** p. 33 **Science Inquiry Labs,** pp. 23–24	**Performance** Launch Lab, p. 315 MiniLAB, pp. 319, 320 Applying Skills, p. 321 **Content** Section Review, p. 321	Section Focus Transparency Teaching Transparency Guided Reading Audio Program Interactive Chalkboard CD-ROM Video Lab
Lab: blocks of wood, 2 paint trays, soil, grass sod, water, 2 pails, 2 1,000-mL beakers, triple-beam balance, calculator, watch *Need materials?* Contact Science Kit at 1-800-828-7777 or www.sciencekit.com on the Internet.	**Chapter *FAST FILE* Resources** Transparency Activity, p. 43 Lab Worksheet, pp. 7–8 Enrichment, p. 30 Reinforcement, p. 28 Directed Reading, pp. 21, 22 Lab Activity, pp. 13–16 **Lab Management and Safety,** p. 58 **Mathematics Skill Activities,** p. 9 **Reading and Writing Skill Activities,** p. 27 **Home and Community Involvement,** p. 37	**Portfolio** Activity, p. 324 Science Journal, p. 326 **Performance** Applying Science, p. 329 Applying Math, p. 331 **Content** Section Review, p. 331	Section Focus Transparency Virtual Labs CD-ROM Guided Reading Audio Program Interactive Chalkboard CD-ROM

End of Chapter Assessment

Glencoe's Assessment Advantage

Blackline Masters	Technology	Professional Series
Chapter *FAST FILE* Resources Chapter Review, pp. 35–36 Chapter Tests, pp. 37–40 **Standardized Test Practice,** pp. 47–50	MindJogger Videoquiz Virtual Labs CD-ROM Exam*View*® Pro Testmaker TeacherWorks CD-ROM Interactive Chalkboard CD-ROM	**Performance Assessment in the Science Classroom (PASC)**

Transparencies

Section Focus

Section 1 Section Focus Transparency **Congregation of Plants**

Plant growth can get so thick in the rainforest that it can hide an entire city. Angkor Preah Ko, a mountaintop temple in Cambodia, is being reclaimed by the rainforest through gradual processes.

1. Describe how plants are affecting the rock in this picture.
2. How might climate affect the process you described in question one?
3. Could animals have an effect on the rocks similar to that of the plants? Explain.

L2

Section 2 Section Focus Transparency **Give and Take**

The effects of centuries of wind blasting sand particles against the surface of the Sphinx is easily seen. But wind has also helped preserve this ancient monument. For many years the Sphinx was buried by wind-blown sand, protecting it from further damage.

1. How has wind damaged the Sphinx?
2. What happens to particles that the wind blasts off the Sphinx?
3. How could the Sphinx be protected from further damage?

L2

This is a representation of key blackline masters available in the Teacher Classroom Resources. See Resource Manager boxes within the chapter for additional information.

Assessment

Assessment Transparency **Weathering and Erosion**

Directions: *Carefully review the table and answer the following questions.*

Precipitation–Catch (mm at 990 m altitude) 1980–1986 (November–April)

Year	Nov.	Dec.	Jan.	Feb.	Mar.	Apr.
1980	313	153	253	276	142	132
1981	100	34	280	156	237	8
1982	126	100	28	28	32	54
1983	89	165	77	78	12	60
1984	138	46	170	93	44	50
1985	32	116	181	40	189	56
1986	31	169	17	113	13	18

1. According to the table, the month and year that experienced the highest amount of precipitation-catch was ___.
   A April, 1980 C November, 1980
   B January, 1980 D February, 1980
2. According to the information in the table, all of the following years had at least one month in which the monthly precipitation-catch was greater than 180 millimeters EXCEPT ___.
   F 1980 G 1981 H 1985 J 1986
3. According to the information in the table, which month experienced the greatest amount of precipitation-catch?
   A November B December C January D February

L2

Teaching

Section 1 Teaching Transparency **Feldspar and Kaolinite**

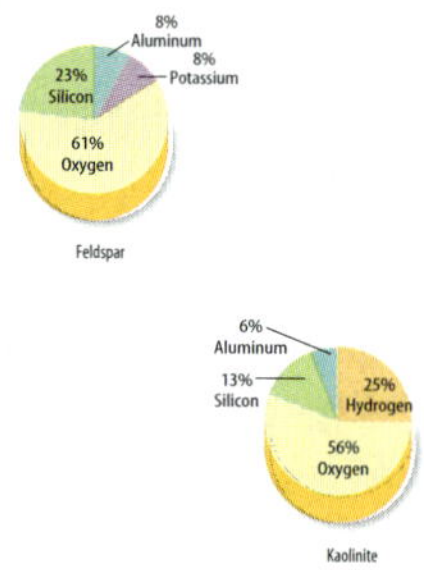

L2

Key to Teaching Strategies

The following designations will help you decide which activities are appropriate for your students.

- L1 Level 1 activities should be appropriate for students with learning difficulties.
- L2 Level 2 activities should be within the ability range of all students.
- L3 Level 3 activities are designed for above-average students.
- ELL ELL activities should be within the ability range of English Language Learners.
- COOP LEARN Cooperative Learning activities are designed for small group work.
- LS Multiple Learning Styles logos, as described on page 12T, are used throughout to indicate strategies that address different learning styles.
- P These strategies represent student products that can be placed into a best-work portfolio.
- PBL Problem-Based Learning activities apply real-world situations to learning.

Hands-on Activities

Student Text Lab Worksheet

Name Date Class

Activity **Classifying Soils**

Lab Preview

Directions: *Answer these questions before you begin the Activity.*

1. What is the reason for the disposal safety precaution?
2. How do you determine whether you have a clay soil?

Not all soils are the same. Geologists and soil scientists classify soils based on the amounts and kinds of particles they contain.

What You'll Investigate

How is soil texture determined?

Materials

soil sample
stereomicroscope
*hand lens
*Alternate materials

Safety Precautions

Goals

- **Classify** a soil using an identification key.
- **Observe** soil with a stereomicroscope.

Procedure

1. Place a small sample of moistened soil between your fingers. Then follow the directions in the classification key below.
   a. Slide your fingers back and forth past each other. If your sample feels gritty, go to B. If it doesn't feel gritty, go to C.
   b. If you can mold the soil into a firm ball, it's sandy loam soil. If you cannot mold it into a firm ball, it's sandy soil.
   c. If your sample is sticky, go to D. If your sample isn't sticky, go to E.
   d. If your sample can be molded into a long, thin ribbon, it's clay soil. If your soil can't be molded into a long, thin ribbon it's clay loam soil.
   e. If your sample is smooth, it's silty loam soil. If it isn't smooth, it's loam soil.
2. After classifying your soil sample, examine it under a microscope. In the space under Data and Observations, draw the particles and any other materials that you see.
3. Wash your hands thoroughly after you are finished working with soils.

Hands-On Activities

L2

Weathering and Erosion

Laboratory Activities

Name Date Class

LAB 1 Laboratory Activity **Chemical and Mechanical Weathering**

When chemical weathering affects rocks, a chemical reaction occurs between the minerals in the rocks and chemical agents. The acidity in rainwater is one agent of chemical weathering. It can react with certain minerals to change rocks chemically. Mechanical weathering is the result of physical forces only. The mechanical weathering of rocks does not change them chemically, only physically.

Strategy

You will test the acidity of the rainwater in your area.
You will demonstrate chemical weathering using cement and vinegar.

Materials

5 small jars
marker and labels
100 mL tap water
100 mL local rainwater
100 mL lemon juice
100 mL cola
100 mL ammonia
litmus paper strips
small piece of cement
beaker
white vinegar
20 sugar cubes
jar with a lid
10 pieces of pea gravel

CAUTION: *Do not eat, drink, or taste any of the materials used in this lab.*

Procedure

Part A

1. Pour tap water into the first jar. Label the jar **Tap Water.**
2. Pour the rainwater into the second jar. Label the jar **Rainwater.**
3. Pour lemon juice into the third jar. Label the jar **Lemon Juice.**
4. Pour the cola into the fourth jar. Label the jar **Cola.**
5. Pour the ammonia into the fifth jar. Label the jar **Ammonia.**
6. Dip the litmus paper into each jar and record the results in Table 1 under Data and Observations. (If the litmus paper turns red, it indicates an acidic material; if the litmus paper turns blue, it indicates a basic material. No change at all indicates a neutral material.)

Part B

7. Describe the piece of cement in Table 2 under Data and Observations.
8. Place the piece of cement in a beaker.
9. Pour enough vinegar over the cement to cover it.
10. Leave the cement and vinegar in the glass for two to three days.
11. Record your observations.

Part C

12. Describe the appearance of the sugar cubes in Table 3 under Data and Observations.
13. Place 10 sugar cubes in the jar, cover, and shake 20 times.
14. Pour the contents of the jar onto a piece of paper. Separate the cubes and the crumbs. Describe the changes you observe.
15. Repeat steps 13 and 14 using the cubes and crumbs from step 14.
16. Repeat steps 13 and 14, using 10 new sugar cubes and the 10 pieces of pea gravel.

Hands-On Activities

L2

Weathering and Erosion

Meeting Different Ability Levels

Content Outline

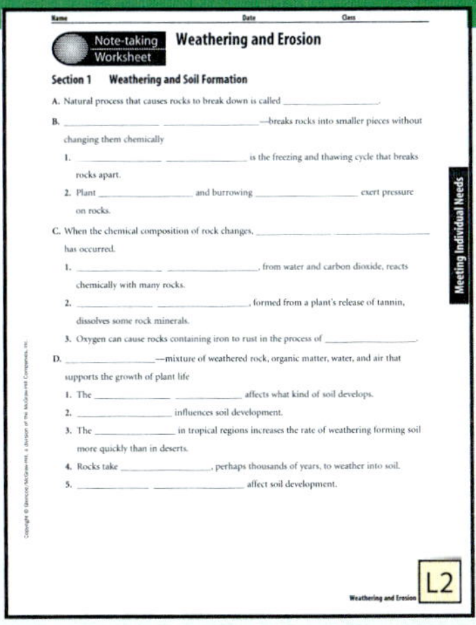

Reinforcement

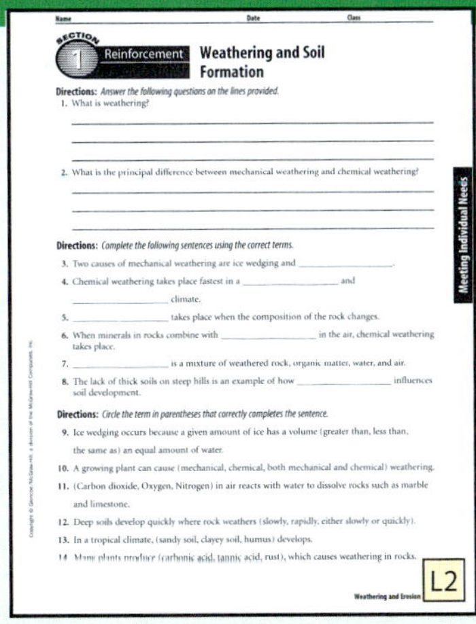

Enrichment

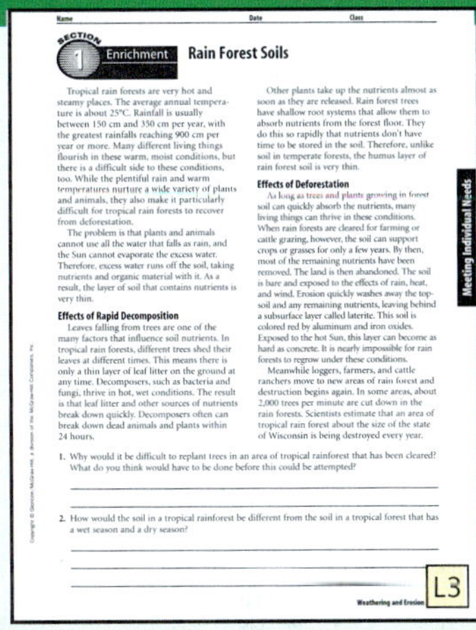

Directed Reading (English/Spanish)

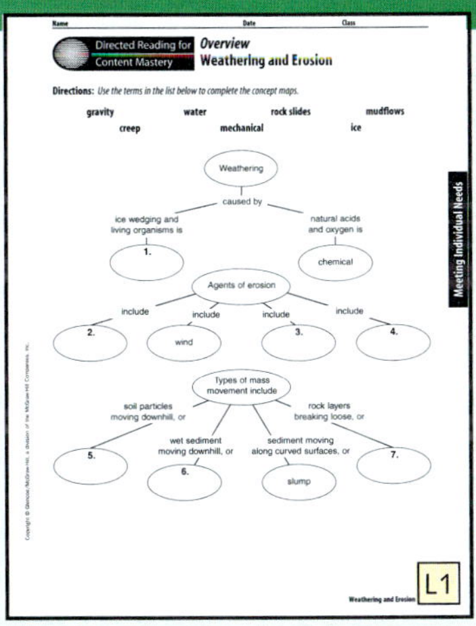

Study Guide

Reading Essentials

Assessment

Test Practice Workbook

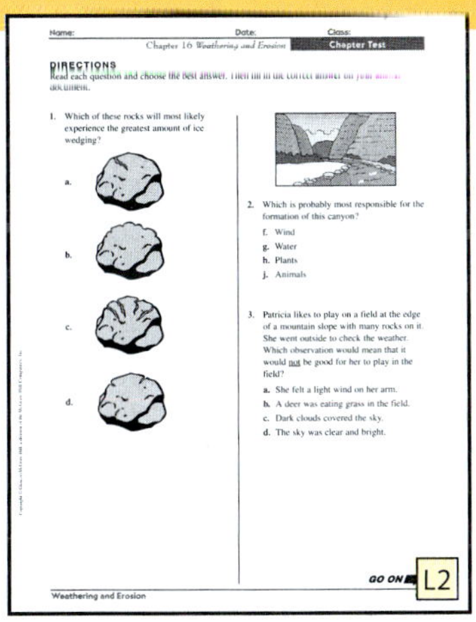

Chapter Review

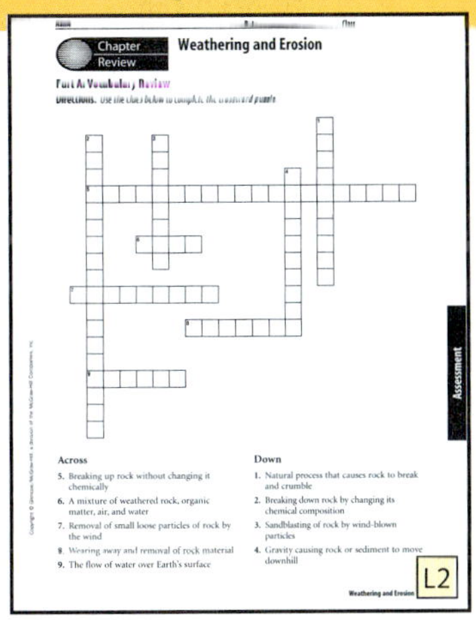

Chapter Tests

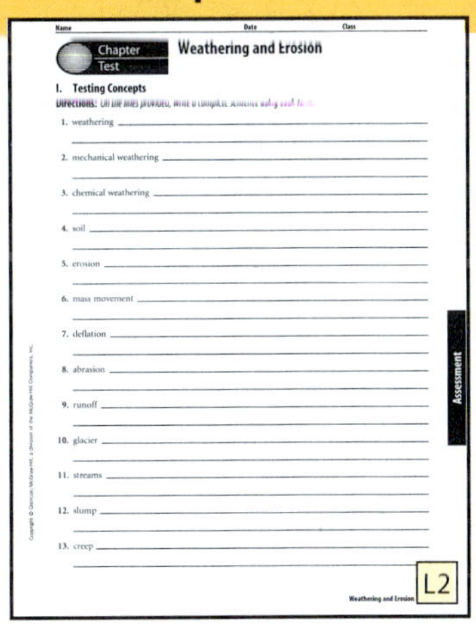

Science Content Background

section 1 Weathering and Soil Formation

Mechanical Weathering

In hot, dry areas, a force that can contribute to the mechanical weathering of rock is the growth of salt crystals. Under certain conditions salt crystals can form from solution in rocks. Cracks and crevices can widen under the force applied as salt crystals grow. Salt crystal growth has the greatest effect on porous rocks, such as sandstone and limestone. Most often found in desertlike conditions, salt crystal growth also can occur in caves in coastal regions where the salt is derived from seawater.

Chemical Weathering

The pH of normal rainwater is slightly acidic at 5.6. Acid rain is rain, sleet, snow, or fog with a pH less than 5.6. The most acidic rain ever recorded in the United States had a pH of about 4.3. The growing problem of acid rain is leading to increased chemical weathering of rocks, such as limestone, marble, and sandstone. Acid rain also has left its mark on buildings, monuments, and tombstones. Acid rain results from several sources. The release of excess carbon dioxide from the burning of fossil fuels leads to increased amounts of carbonic acid in the atmosphere. Nitric acid is formed from the release of nitrogen oxide by internal combustion engines. The largest source of acid rain results from the release of sulfur dioxide into the atmosphere from burning coal that contains sulfur. Sulfur dioxide combines with oxygen to form sulfuric acid.

Soil Formation

Soil is a mixture of weathered rock, organic matter, water, and air. A major factor affecting soil color is the type of rock that has been weathered.

section 2 Erosion of Earth's Surface

Wind

During the 1930s the Great Plains of the United States experienced repeated, severe dust storms, which resulted in the entire area being referred to as the Dust Bowl. A severe drought, in combination with the earlier replacement of deep-rooted grasses with shallow-rooted grains, allowed the strong winds that swept across the plains during this time to strip away large amounts of topsoil.

Corbis

chapter content resources

Internet Resources

For additional content background, visit **red.msscience.com** to:

- access your book online
- find references to related articles in popular science magazines
- access Web links with related content background
- access current events with science journal topics

Print Resources

Weathering and Erosion, Stephen T. Trudgill, 1983

Earth's Changing Surface, EOA Scientific Systems, Inc. 2001

Rocks, Soil, and Weather, Russell, Longden, McGuigan, Bell, Liverpool University Press, 1992

IDENTIFYING Misconceptions

Find Out What Students Think

Students may think that . . .

Soil is only one color, such as black, red, or yellow.

Students think that soil is only one color; that is, the color of the soil where they live is the color of all soil.

Activity

Display small pieces of different-colored rocks. Samples might include limestone, sandstone, and shale. While wearing goggles, wrap each sample in a towel and crush with a rock hammer. Have students examine each crushed sample. Discuss how the color of each parent rock would affect the color of a soil that developed from it.

Promote Understanding

Activity

Organize the class into small groups. Provide each group with a variety of soils of different colors. Also provide each group with the parent rocks that could have weathered to form each soil type.

- Have students work with group members to match each soil sample to the correct type of parent rock.
- Then have students discuss how they matched the samples. Discuss how soils develop from the weathering of parent rocks. The parent rock often lends its color to the soil formed from it.
- Discuss other factors that can add color to soils. Explain that the presence of humus, or organic materials, can make soils appear darker. Iron oxides can also add colors such as yellow or red to soils.

Charles O'Rear/Corbis

Assess

After completing the chapter, see *Identifying Misconceptions* in the Study Guide at the end of the chapter.

chapter 11

Chapter Vocabulary

weathering, p. 316
mechanical weathering, p. 316
chemical weathering, p. 318
soil, p. 320
topography, p. 320
erosion, p. 323
mass movement, p. 323
creep, p. 325
slump, p. 325
deflation, p. 328
abrasion, p. 328
runoff, p. 329

Science Journal Encourage students to relate a particular force of change to each change in appearance they have discussed. For example, wood siding on a house might be faded from sun, broken by wind, and disintegrated by water.

INTERACTIVE CHALKBOARD with Image Bank

PowerPoint® Presentations

This CD-ROM is an editable Microsoft® PowerPoint® presentation that includes:

- a pre-made presentation for every chapter
- interactive graphics
- animations
- audio clips
- image bank
- all new section and chapter questions
- Standardized Test Practice
- transparencies
- pre-lab questions for all labs
- Foldables directions
- links to red.msscience.com

chapter 11

Weathering and Erosion

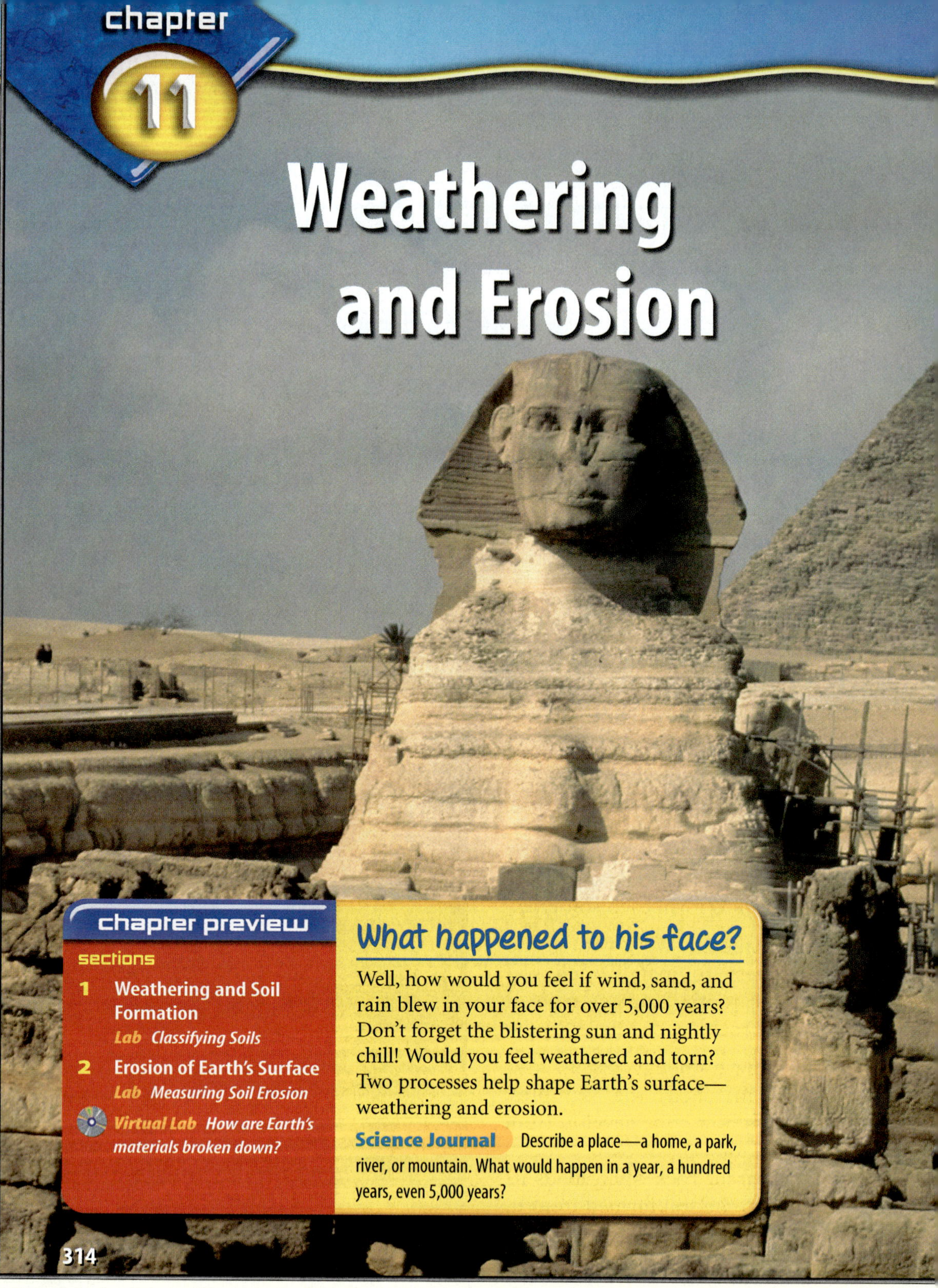

chapter preview

sections

1. **Weathering and Soil Formation**
   Lab *Classifying Soils*
2. **Erosion of Earth's Surface**
   Lab *Measuring Soil Erosion*

Virtual Lab *How are Earth's materials broken down?*

What happened to his face?

Well, how would you feel if wind, sand, and rain blew in your face for over 5,000 years? Don't forget the blistering sun and nightly chill! Would you feel weathered and torn? Two processes help shape Earth's surface—weathering and erosion.

Science Journal Describe a place—a home, a park, river, or mountain. What would happen in a year, a hundred years, even 5,000 years?

314

Theme Connection

Stability and Change Landforms may appear to be unchanging, but they are constantly changing as a result of the effects of weathering and erosion.

About the Photo

The Sphinx The Great Sphinx is located on the Giza Plateau in Egypt, near Cairo. Made of relatively soft sandstone, the Sphinx is showing its age. Some scientists estimate that it is at least 4,500 years old. Others think that it could be as much as 10,000 years old!

Start-Up Activities

Launch LAB

Water's Force

The Grand Canyon is 446 km long, up to 29 km wide, and up to 1,829 m deep. The water of the Colorado River carved the canyon out of rock by wearing away particles and carrying them away for millions of years. Over time, erosion has shaped and reshaped Earth's surface many times. In this lab, you will explore how running water formed the Grand Canyon.

1. Fill a bread pan with packed sand and form a smooth, even surface.
2. Place the bread pan in a plastic wash tub. Position one end of the washtub in a sink under the faucet.
3. Place a brick or wood block under the end of the bread pan beneath the faucet.
4. Turn on the water to form a steady trickle of water falling into the pan and observe for 10 min. The washtub should catch the eroded sand.
5. **Think Critically** In your Science Journal, draw a top view picture of the erosion pattern formed in the sand by the running water. Write a paragraph describing what the sand would look like if you had left the water running overnight.

FOLDABLES™ Study Organizer

Weathering and Erosion Make the following Foldable to compare and contrast weathering and erosion.

STEP 1 **Fold** one sheet of paper lengthwise.

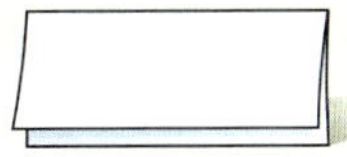

STEP 2 **Fold** into thirds.

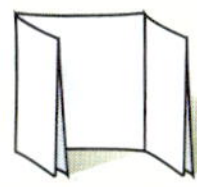

STEP 3 **Unfold and draw** overlapping ovals. **Cut** the top sheet along the folds.

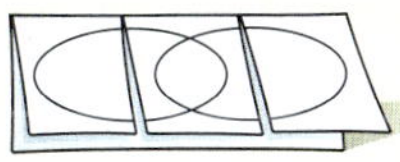

STEP 3 **Label** the ovals as shown.

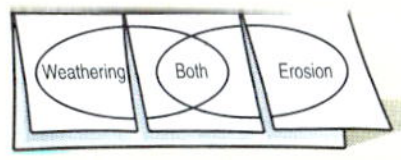

Construct a Venn Diagram As you read the chapter, list the characteristics unique to weathering under the left tab, those unique to erosion under the right tab, and those characteristics common to both under the middle tab.

Science Online

Preview this chapter's content and activities at red.msscience.com

315

Launch LAB

Purpose Students will model the soil erosion caused by running water.

Preparation Purchase a bread pan and washtub for each student group and assign each group a sink.

Materials bread pan, sand, plastic washtub, brick or wood block, clock and access to a sink with a faucet for each student group

Teaching Strategy Have students layer the bread pan with several different colors of sand so that they can more clearly see the erosion process. As the trickle of water cuts a deeper groove into the sand, new colors will be exposed by erosion.

Think Critically

A deep "gorge" might be eroded into the sand.

Assessment

Performance Ask students to leave the faucets running at a slow trickle the entire day and then draw the pattern of the colored layers exposed by the erosion. Use **Performance Assessment in the Science Classroom,** p. 127.

FOLDABLES™ Study Organizer **Dinah Zike Study Fold**

Student preparation materials for this Foldable are available in the Chapter *Fast File* Resources.

1 Motivate

INTERACTIVE CHALKBOARD
PowerPoint® Presentations

Bellringer

Section Focus Transparencies also are available on the Interactive Chalkboard CD-ROM. L2 ELL

Section 1 Section Focus Transparency — Congregation of Plants

Plant growth can get so thick in the rainforest that it can hide an entire city. Angkor Preah Ko, a mountaintop temple in Cambodia, is being reclaimed by the rainforest through gradual processes.

1. Describe how plants are affecting the rock in this picture.
2. How might climate affect the process you described in question one?
3. Could animals have an effect on the rocks similar to that of the plants? Explain.

Weathering and Erosion

Tie to Prior Knowledge

Soil Have students discuss what they know about the formation and use of soil. Explain that in this section they will learn how soil forms through the weathering of rock.

Caption Answer

Figure 1 The granite was broken into smaller pieces without changing chemically.

section 1

Weathering and Soil Formation

as you read

What You'll Learn

- **Identify** processes that break rock apart.
- **Describe** processes that chemically change rock.
- **Explain** how soil evolves.

Why It's Important

Soil forms when rocks break apart and change chemically. Soil is home to many organisms, and most plants need soil in order to grow.

Review Vocabulary

acid rain: acidic moisture, with a pH below 5.6

New Vocabulary

- weathering
- mechanical weathering
- chemical weathering
- soil
- topography

Weathering

Have you noticed potholes in roadways and broken concrete in sidewalks and curbs? When a car rolls over a pothole in the road in late winter or when you step over a broken sidewalk, you know things aren't as solid or permanent as they look. Holes in roads and broken sidewalks show that solid materials can be changed by nature. **Weathering** is a mechanical or chemical surface process that breaks rocks into smaller pieces. Freezing and thawing, oxygen in the air, and even plants and animals can affect the stability of rock. These are some of the things that cause rocks on Earth's surface to weather, and in some cases, to become soils.

Mechanical Weathering

When a sidewalk breaks apart, a large slab of concrete is broken into many small pieces. The concrete looks the same. It's just broken apart. This is similar to mechanical weathering of rocks. **Mechanical weathering** breaks rocks into smaller pieces without changing them chemically. The small pieces are identical in composition to the original rock, as shown in **Figure 1.** Two of the many causes of mechanical weathering are ice wedging and living organisms.

Figure 1 The forces of mechanical weathering break apart rocks. **Describe** *how you know that the smaller pieces of granite were produced by mechanical weathering.*

316

Section 1 Resource Manager

Chapter *FAST FILE* Resources

Transparency Activity, p. 42
Directed Reading for Content Mastery, pp. 19, 20
Note-taking Worksheets, pp. 31–32
Enrichment, p. 29
MiniLAB, pp. 3, 4
Lab Activity, pp. 9–12
Reinforcement, p. 27
Lab Worksheet, pp. 5–6

Earth Science Critical Thinking/Problem Solving, p. 6

Performance Assessment in the Science Classroom, p. 44

Figure 2 Over time, freezing water can break apart rock.

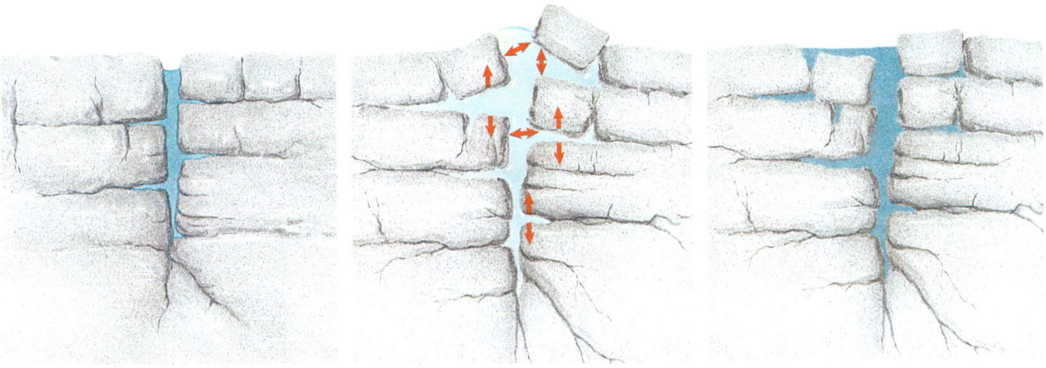

Water seeps into cracks. The deeper the cracks are, the deeper water can seep in.

The water freezes and expands, forcing the cracks to open further.

The ice melts. If the temperature falls below freezing again, the process will repeat itself.

Ice Wedging In some areas of the world, air temperature drops low enough to freeze water. Then, when the temperature rises, the ice thaws. This freezing and thawing cycle breaks up rocks. How can this happen? When it rains or snow melts, water seeps into cracks in rocks. If the temperature drops below freezing, ice crystals form. As the crystals grow, they take up more space than the water did because when water freezes, its molecules move apart. This expansion exerts pressure on the rocks. With enough force, the rocks will crack further and eventually break apart, as shown in **Figure 2.** Ice wedging also causes potholes to form in roadways.

Reading Check *Explain how ice wedging can break rock apart.*

Plants and Animals Plants and animals also cause mechanical weathering. As shown in **Figure 3,** plants can grow in what seem to be the most inconvenient places. Their roots grow deep into cracks in rock where water collects. As they grow, roots become thicker and longer, slowly exerting pressure and wedging rock apart.

Gophers and prairie dogs also weather rock—as do other animals that burrow in the ground. As they burrow through sediment or soft sedimentary rock, animals break rock apart. They also push some rock and sediment to the surface where another kind of weathering, called chemical weathering, takes place more rapidly.

Figure 3 Tree roots can break rock apart.

2 Teach

Activity

Expanding Ice Have students half-fill hard plastic jars with water and use grease pencils to mark the height of the water on the sides of the jars. Then have them put the jars in a freezer. The next day, have students examine the jars. They should observe that ice occupies more space in the jar than the water did. L1 ELL LS **Visual-Spatial**

Reading Check

Answer As ice forms, it takes up more space than liquid water.

Use an Analogy

Wedging Wood The process that occurs when ice wedging splits rocks into pieces is similar to the process of using a wedge to split wood. A triangular metal wedge is placed into a crack in a log. When hit with a mallet, the thicker part of the wedge is driven into the crack, and the wood splits apart. In ice wedging, water freezes in cracks, takes up more space, and splits the rock apart.

Inquiry Lab

Observing Chemical Weathering

Purpose To explore and observe chemical weathering, have students create a statue (calcium chloride such as chalk) and add acid rain (such as vinegar).

Possible Materials chalk, vinegar, nail, beaker, tongs

Estimated Time one class session

Teaching Strategies

- Students can create a "statue" by etching a design into the chalk.
- Have students drip vinegar on the statues.
- Allow students to explore other questions that arise.

For additional inquiry activities, see *Science Inquiry Labs*.

Teacher FYI

Cave Formation Some groundwater contains carbonic acid, formed when water combines with carbon dioxide. Certain rocks, (limestone, for example), dissolve in carbonic acid. As groundwater flows through underground limestone, dissolved matter is carried away, leaving cavities in the rocks that eventually can form caves.

Caption Answer

Figure 4 It contains hydrogen and no potassium.

Reading Check

Answer Moist, warm conditions accelerate chemical weathering.

Visual Learning

Table 1 Have students analyze the data in the table and determine the rate of chemical weathering occurring in your area. Answers will depend on location.

Discussion

Chemical Weathering Chemical weathering occurs on the surface of rocks. What happens to the rate of chemical weathering if a rock is broken into smaller pieces? Because the rock's surface area has increased, the rate of chemical weathering also increases.

Figure 4 Chemical weathering changes the chemical composition of minerals and rocks.
Describe *how kaolinite is different from feldspar.*

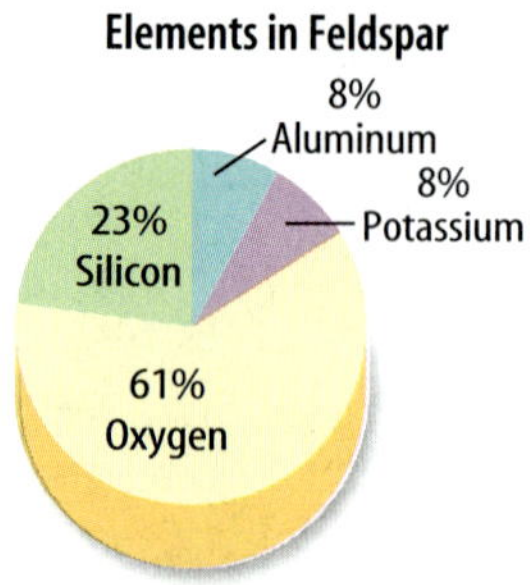

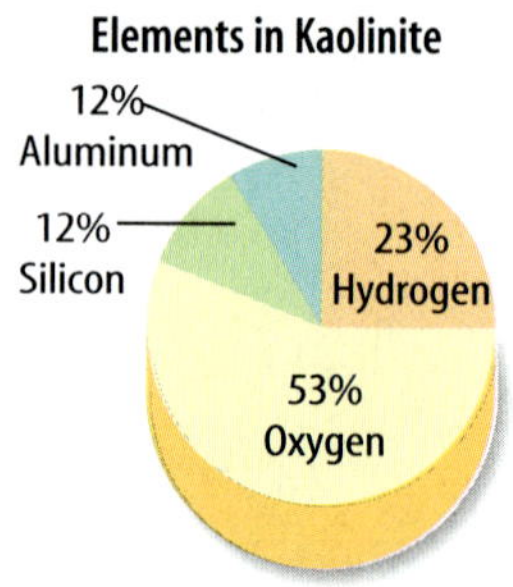

Feldspar crystals react with carbonic acid.

The mineral kaolinite is formed.

Chemical Weathering

Chemical weathering occurs when the chemical composition of rock changes. This kind of weathering is rapid in tropical regions where it's moist and warm most of the time. Because desert areas have little rainfall and polar regions have low temperatures, chemical weathering occurs slowly in these areas. **Table 1** summarizes the rates of chemical weathering for different climates. Two important causes of chemical weathering are natural acids and oxygen.

Reading Check *Why is chemical weathering rapid in the tropics?*

Table 1 Rates of Weathering

Climate	Chemical Weathering
Hot and dry	Slow
Hot and wet	Fast
Cold and dry	Slow
Cold and wet	Slow

INTEGRATE Chemistry

Natural Acids Some rocks react chemically with natural acids in the environment. When water mixes with carbon dioxide in air or soil, for example, carbonic acid forms. Carbonic acid can change the chemical composition of minerals in rocks, as shown in **Figure 4.**

Although carbonic acid is weak, it reacts chemically with many rocks. Vinegar reacts with the calcium carbonate in chalk, dissolving it. In a similar way, when carbonic acid comes in contact with rocks like limestone, dolomite, and marble, they dissolve. Other rocks also weather when exposed to carbonic acid.

LAB DEMONSTRATION

Purpose to make an acid

Materials small beaker, phenol red or phenolphthalein, drinking straw

Preparation Explain that phenol red is an indicator that changes color when exposed to acids.

Procedure Fill a small beaker with phenol red. Exhale into the beaker through the straw. **WARNING:** *Do not suck liquid in through the straw.* Have students observe what happens to the color of the phenol red.

Expected Outcome The phenol red turns yellow.

Assessment

What gas was exhaled into the beaker? carbon dioxide What natural acid was produced? carbonic acid How do you know? The color change in the phenol red indicated the presence of an acid.

Plant Acids Plant roots also produce acid that reacts with rocks. Many plants produce a substance called tannin. In solution, tannin forms tannic acid. This acid dissolves some minerals in rocks. When minerals dissolve, the remaining rock is weakened, and it can break into smaller pieces. The next time you see moss or other plants growing on rock, as shown in **Figure 5,** peel back the plant. You'll likely see discoloration of the rock where plant acids are reacting chemically with some of the minerals in the rock.

Figure 5 Moss growing on rocks can cause chemical weathering.

Effect of Oxygen When you see rusty cars, reddish soil, or reddish stains on rock, you are witnessing oxidation, the effects of chemical changes caused by oxygen. When iron-containing materials such as steel are oxidized, a chemical reaction causes the material to rust. Rocks chemically weather in a similar way. When some iron-containing minerals are exposed to oxygen, they can weather to minerals that are like rust. This leaves the rock weakened, and it can break apart. As shown in **Figure 6,** some rocks also can be colored red or orange when iron-bearing minerals in them react with oxygen.

Figure 6 Oxidation occurs in rocks and cars.

Even a tiny amount of iron in rock can combine with oxygen and form a reddish iron oxide.

The iron contained in metal objects such as this truck also can combine with oxygen and form a reddish iron oxide called rust.

Mini LAB

Dissolving Rock with Acids

Procedure

WARNING: *Do not remove goggles until lab cleanup and handwashing are completed.*

1. Use an **eyedropper** to put several drops of **vinegar** on pieces of **chalk** and **limestone.** Observe the results with a **magnifying lens.**
2. Put several drops of **5% hydrochloric acid** on the chalk and limestone. Observe the results.

Analysis

1. Describe the effect of the hydrochloric acid and vinegar on chalk and limestone.
2. Research what type of acid vinegar contains.

IDENTIFYING Misconceptions

Colors of Soil Students may think that soil is only one color, that is, the color of the soil where they live. Refer to page F at the beginning of this chapter for teaching strategies that address this misconception.

Mini LAB

Purpose Students will model chemical weathering to discover the effect of acids on certain types of rocks.

Materials eye dropper, vinegar, piece of chalk, piece of limestone, five percent hydrochloric acid, magnifying lens

Teaching Strategy Instruct students to wear safety goggles, rubber gloves, and an apron when handling the acid. Be certain the vinegar is concentrated and fresh. Weak or old vinegar will not produce the desired effect.

Analysis

1. Hydrochloric acid and strong vinegar will cause chalk to fizz and bubble, but the vinegar may have no effect on limestone. Students may need a magnifying lens to view the reaction.
2. Acetic acid

Assessment

Process Ask students to infer how lichens create indentions in boulders in order to grow. Lichens secrete acids causing rocks to alter and eventually form soil. Use **Performance Assessment in the Science Classroom,** p. 89.

Differentiated Instruction

Learning Disabled Have students experiment with oxidation reactions by subjecting different iron-containing metals to water. Have students take photographs before and after their tests. L1 ELL LS **Visual-Spatial**

Challenge Ask students to design an investigation to test the roles played by slope and water in soil formation. Have them make a hypothesis and design an experiment identifying the independent and dependent variables. L3

Make a Model

Compost Pile Have students make a compost pile to model the formation of the organic component of soil. Get permission to use an area of the school yard as a compost pile. Encourage students to add grass clippings, straw, and non-meat or dairy food wastes to the pile. Turn the pile frequently to provide aeration. Adding worms will speed the process. Have students periodically examine the compost with a magnifying lens. Finished compost is dark, smells "earthy," and has few recognizable components. L2 ELL LS **Kinesthetic**

Mini LAB

Purpose Student will examine a soil sample.

Materials soil sample, sheet of newspaper

Teaching Strategy Be sure students wash their hands thoroughly after handling the soil.

Analysis

1. Answers will be subjective and based on the students' soil samples. Students may include water and air.
2. Answers will vary, depending on the soil sample.

Assessment

Performance Have students separate soil components into different piles: rock fragments, sand, clay, topsoil, humus, and organisms. Use **Performance Assessment in the Science Classroom,** p. 89.

Try at Home

Table 2 Factors that Affect Soil Formation

Parent Rock	Slope of Land	Climate	Time	Organisms

Soil

Is soil merely dirt under your feet, or is it something more important? **Soil** is a mixture of weathered rock, organic matter, water, and air that supports the growth of plant life. Organic matter includes decomposed leaves, twigs, roots, and other material. Many factors affect soil formation.

Parent Rock As listed in **Table 2,** one factor affecting soil formation is the kind of parent rock that is being weathered. For example, where limestone is chemically weathered, clayey soil is common because clay is left behind when the limestone dissolves. In areas where sandstone is weathered, sandy soil forms.

The Slope of the Land The **topography**, or surface features, of an area also influence the types of soils that develop. You've probably noticed that on steep hillsides, soil has little chance of developing. This is because rock fragments move downhill constantly. However, in lowlands where the land is flat, wind and water deposit fine sediments that help form thick soils.

Climate Climate affects soil evolution, too. If rock weathers quickly, deep soils can develop rapidly. This is more likely to happen in tropical regions where the climate is warm and moist. Climate also affects the amount of organic material in soil. Soils in desert climates contain little organic material. However, in warm, humid climates, vegetation is lush and much organic material is present. When plants and animals die, decomposition by fungi and bacteria begins. The result is the formation of a dark-colored material called humus, as shown in the soil profile in **Figure 7.** Most of the organic matter in soil is humus. Humus helps soil hold water and provides nutrients that plants need to grow.

Mini LAB

Analyzing Soils

Procedure

1. Obtain a sample of **soil** from near your home.
2. Spread the soil out over a piece of **newspaper.**
3. Carefully sort through the soil. Separate out organic matter from weathered rock.
4. Wash hands thoroughly after working with soils.

Analysis

1. Besides the organic materials and the remains of weathered rock, what else is present in the soil?
2. Is some of the soil too fine-grained to tell if it is organic or weathered rock?

Try at Home

Visual Learning

Figure 7 Why are soil layers different colors? Soil layers can get their color from organic matter, mineral precipitates, and the rock and mineral grains in the soil. Organic matter often colors surface horizons. Subsurface horizons are colored by mineral precipitates or parent material.

Active Reading

Synthesis Journal In this strategy, students reflect by means of project, paper, or performance from past experiences. They record "What I did," "What I learned," and "How I can use what I learned." Have students write a Synthesis Journal entry about the lab on this page.

Time It takes time for rocks to weather. It can take thousands of years for some soils to form. As soils develop, they become less like the rock from which they formed. In young soils, the parent rock determines the soil characteristics. As weathering continues, however, the soil resembles the parent rock less and less. Thicker, well-developed soils often are found in areas where weathering has gone on undisturbed for a long period of time. For this to happen, soil materials must not be eroded away and new sediment must not be deposited over the land's surface too quickly.

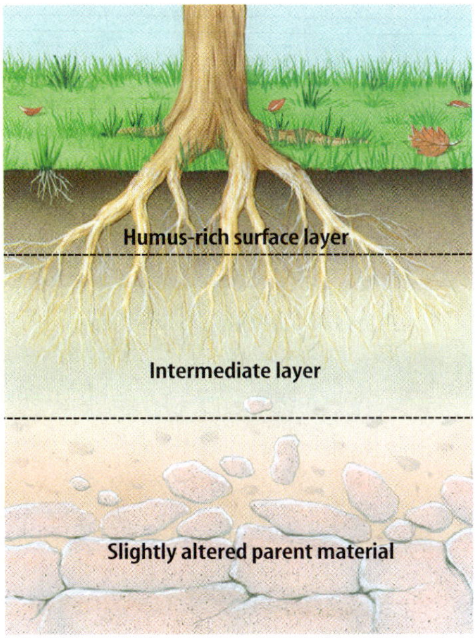

Figure 7 Soils contain layers that are created by weathering, the flow of water and chemicals, and the activities of organisms.
Explain *what part microorganisms play in soil development.*

Organisms Organisms influence soil development. Lichens are small organisms that consist of an alga and a fungus that live together for mutual benefit. You may have seen lichens in the form of multicolored patches growing on tree branches or cliff faces. Interestingly, lichens can grow directly on rock. As they grow, they take nutrients from the rock that they are starting to break down, forming a thin soil. After a soil has formed, many types of plants such as grasses and trees can grow.

The roots of these plants further break down the parent rock. Dead plant material such as leaves accumulates and adds organic matter to the soil. Some plants contribute more organic matter to soil than others. For example, soil under grassy areas often is richer in organic matter than soil developing under forests. This is why some of the best farmland in the midwestern United States is where grasslands used to be.

section 1 review

Summary

Mechanical Weathering
- The freezing and thawing cycle breaks up rocks.
- Plants' roots and burrowing animals can break rocks apart.

Chemical Weathering
- Some rocks react chemically with natural acids.

Soil
- A factor affecting soil formation is the kind of parent rock that is being weathered.

Self Check

1. **Describe** how rocks are mechanically weathered.
2. **Name** two agents of chemical weathering.
3. **Explain** how carbonic acid weather rocks.
4. **Describe** how soil forms. What factors are important?
5. **Think Critically** How could climate affect rates of mechanical weathering? What about chemical weathering? How are the two kinds of weathering related?

Applying Skills

6. **Compare and contrast** mechanical weathering caused by ice and growing roots.

Science Online red.msscience.com/self_check_quiz

Quick Demo

Soils Capacity

Materials two flower pots, peat moss, sand, water

Estimated Time 15 minutes

Procedure Fill a flowerpot with peat moss and another pot of equal size with the same amount of sand. Slowly pour water into each pot and determine the quantity of water that can be held by each material before water leaks from the bottom of the pot. Expected outcome: Humus (peat moss) will hold much more water than sand.

Caption Answer

Figure 7 They decompose organic matter.

3 Assess

DAILY INTERVENTION

Check for Understanding

Visual-Spatial Have students examine a partially decomposed log. Have them list the organisms they see and relate how each type of organism contributes to the formation of humus. L3 LS

Reteach

Soil Profile Make a drawing of a soil profile on the overhead projector. Have students describe each layer as you point to it. Characteristics mentioned should include amount of humus and rock particles in each layer. L1 LS **Visual-Spatial**

Assessment

Process Have students design a spider concept map of factors that affect weathering. Use **PASC**, p. 161.

section 1 review

1. As plant roots and ice crystals grow, rocks are wedged apart.
2. natural acids and oxygen
3. Carbonic acid reacts chemically with some compounds to create new compounds that dissolve in water.
4. when organic matter, water, and air are combined with weathered rock particles; parent material, shape of the land, climate, time, and organisms
5. Mechanical weathering is rapid in warm, moist climates where plants grow rapidly, and in temperate climates where water freezes and thaws. Chemical weathering: Table 1 lists rates in various climates. Both mechanical and chemical weathering occur more rapidly in hot, moist climates.
6. Both wedge rocks apart. Mechanical weathering by plant roots occurs as plants grow and require more space; ice wedging occurs as water freezes and expands in fractured rock.

BENCH TESTED

Real-World Question

Purpose Students use a classification key to determine the texture of a soil sample. L2 LS **Kinesthetic and Visual-Spatial**

Process Skills observe and infer, classify, draw conclusions, communicate

Time Required 40 minutes

Alternate Materials magnifying lens

Safety Precautions Caution students to wash their hands after handling the soil.

Procedure

Teaching Strategy Cover work surfaces with newspaper to help decrease clean-up time.

Conclude and Apply

1. Answers will vary depending on local soils.
2. Loam soil is not gritty or sticky, but it is smooth.
3. Sandy loam soil is gritty and will make a firm ball when squeezed.
4. Answers may include rock particles, decomposing leaf material, and insects.

Assessment

Oral Have students explain the relationship between worms and soil fertility. Worms move organic matter throughout the soil, add nitrates to the soil, and dig tunnels that allow air to go deep into the soil. Use **Performance Assessment in the Science Classroom,** p. 89.

LAB

Classifying Soils

Not all soils are the same. Geologists and soil scientists classify soils based on the amounts and kinds of particles they contain.

Real-World Question

How is soil texture determined?

Goals

- **Classify** a soil using an identification key.
- **Observe** soil with a stereomicroscope.

Materials

soil sample
stereomicroscope
**magnifying lens*

**Alternate materials*

Safety Precautions

Procedure

1. Place a small sample of moistened soil between your fingers. Then follow the directions in the classification key below.
   a. Slide your fingers back and forth past each other. If your sample feels gritty, go to **b.** If it doesn't feel gritty, go to **c.**
   b. If you can mold the soil into a firm ball, it's sandy loam soil. If you cannot mold it into a firm ball, it's sandy soil.
   c. If your sample is sticky, go to **d.** If your sample isn't sticky, go to **e.**
   d. If your sample can be molded into a long, thin ribbon, it's clay soil. If your soil can't be molded into a long, thin ribbon, it's clay loam soil.
   e. If your sample is smooth, it's silty loam soil. If it isn't smooth, it's loam soil.
2. After classifying your soil sample, examine it under a microscope. Draw the particles and any other materials that you see.
3. Wash your hands thoroughly after you are finished working with soils.

Conclude and Apply

1. **Determine** the texture of your soil sample.
2. **Describe** two characteristics of loam soil.
3. **Describe** two features of sandy loam soil.
4. **Record Observations** Based on your observations with the stereomicroscope, what types of particles and other materials did you see? Did you observe any evidence of the activities of organisms?

Communicating Your Data

Compare your conclusions with those of other students in your class. **For more help, refer to the** Science Skill Handbook.

Communicating Your Data

Suggest students make a table in which they can organize their observations. Students should discuss why their conclusions did or did not agree.

section 2

Erosion of Earth's Surface

Agents of Erosion

Imagine looking over the rim of the Grand Canyon at the winding Colorado River below or watching the sunset over Utah's famous arches. Features such as these are spectacular examples of Earth's natural beauty, but how can canyons and arches form in solid rock? These features and many other natural landforms are a result of erosion of Earth's surface. **Erosion** is the wearing away and removal of rock or sediment. Erosion occurs because gravity, ice, wind, and water sculpt Earth's surface.

Gravity

Gravity is a force that pulls every object toward every other object. Gravity pulls everything on Earth toward its center. As a result, water flows downhill and rocks tumble down slopes. When gravity alone causes rock or sediment to move down a slope, the erosion is called **mass movement.** Mass movements can occur anywhere there are hills or mountains. One place where they often occur is near volcanoes, as shown in **Figure 8.** Creep, slump, rock slides, and mudflows are four types of mass movements, as seen in **Figure 9.**

as you read

***What* You'll Learn**

- **Identify** agents of erosion.
- **Describe** the effects of erosion.

***Why* It's Important**

Erosion shapes Earth's surface.

Review Vocabulary

deposition: dropping of sediments occurs when an agent of erosion can no longer carry its load

New Vocabulary

- erosion
- mass movement
- creep
- slump
- deflation
- abrasion
- runoff

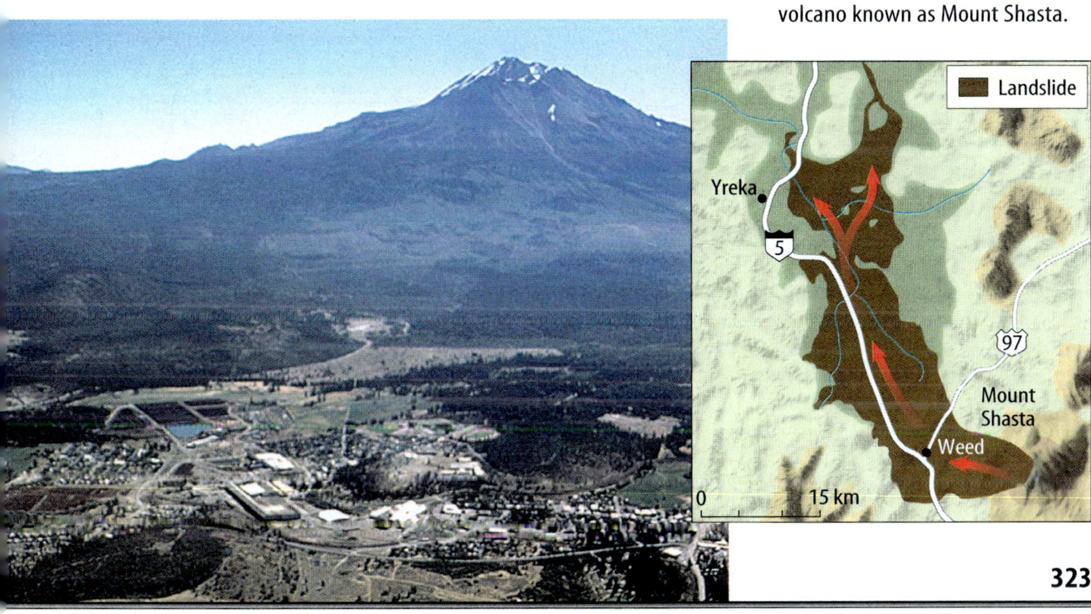

Figure 8 The town of Weed, California, was built on top of a landslide that moved down the volcano known as Mount Shasta.

1 Motivate

INTERACTIVE CHALKBOARD PowerPoint® Presentations

Bellringer

Section Focus Transparencies also are available on the Interactive Chalkboard CD-ROM. L2 ELL

SECTION 2 Section Focus Transparency **Give and Take**

The effects of centuries of wind blasting sand particles against the surface of the Sphinx is easily seen. But wind has also helped preserve this ancient monument. For many years the Sphinx was buried by wind-blown sand, protecting it from further damage.

1. How has wind damaged the Sphinx?
2. What happens to particles that the wind blasts off the Sphinx?
3. How could the Sphinx be protected from further damage?

Tie to Prior Knowledge

Moving Rocks Remind students that in **Section 1,** they learned how rocks are weathered. Have students list ways in which they think those weathered particles are moved from place to place. Then explain that in this section, they will learn how erosional agents transport weathered sediments.

Section 2 Resource Manager

Chapter *FAST FILE* Resources

Transparency Activity, p. 43
Directed Reading for Content Mastery, pp. 21, 22
Enrichment, p. 30
Lab Activity, pp. 13–16
Reinforcement, p. 28
Lab Worksheet, pp. 7–8

Mathematics Skill Activities, p. 9

Earth Science Critical Thinking/Problem Solving, p. 7

NATIONAL GEOGRAPHIC

Visualizing Mass Movements

Have students examine the pictures and read the captions. Then ask the following questions.

What role does the degree of slope play in mass movement? Because the force of gravity is behind all mass movement, the steeper the slope, the greater the likelihood of mass movement.

How does water contribute to the types of mass movement shown here? In the case of creep and slump, moisture weakens the soil's resistance to movement down a slope. A sudden influx of water can initiate a mudflow. Rock slides can occur when rainwater seeps into weak layers underlying slabs of rock.

Activity

Research Mass Movements Have students go online to research a major rock slide or mudflow that has made the news. Have them make posters showing the landscape before and after these events, as well as detailing the consequences. L2 LS **Visual-Spatial** P

NATIONAL GEOGRAPHIC **VISUALIZING MASS MOVEMENTS**

Figure 9

When the relentless tug of gravity causes a large chunk of soil or rock to move downhill—either gradually or with sudden speed—the result is what geologists call a mass movement. Weathering and water often contribute to mass movements. Several kinds are shown here.

A CREEP When soil on a slope moves very slowly downhill, a mass movement called creep occurs. Some of the trees at right have been gradually bent because of creep's pressure on their trunks.

B SLUMP This cliff in North Dakota shows the effects of the mass movement known as slump. Slumping often occurs after earthquakes or heavy and prolonged rains.

C ROCK SLIDES When rocks break free from the side of a cliff or a mountain, they crash down in what is called a rock slide. Rock slides, like the one at the left in Yosemite National Park, can occur with little warning.

D MUDFLOWS A Japanese town shows the devastation that a fourth type of mass movement—a mudflow—can bring. When heavy moisture saturates sediments, mudflows can develop, sending a pasty mix of water and sediment downhill over the ground's surface.

324

Differentiated Instruction

Challenge Have students research the origin of the Great Lakes. Have them make maps that show the evolution of the Great Lakes during the Pleistocene Epoch of geologic time. L3

Creep The process in which sediments move slowly downhill, as shown in **Figure 9A,** is called **creep.** Creep is common where freezing and thawing occur. As ice expands in soil, it pushes sediments up. Then as soil thaws, the sediments move farther downslope. **Figure 10** shows how small particles of sediment can creep downslope. Over time, creep can move large amounts of sediment, possibly causing damage to some structures. Do you live in an area where you can see the results of creep?

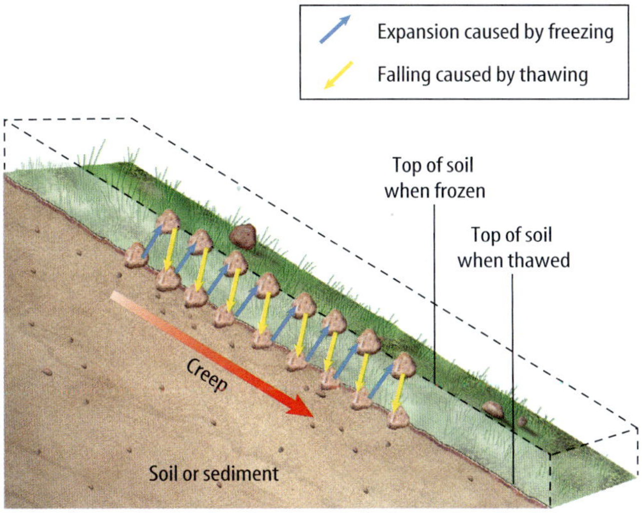

Figure 10 When soil freezes, particles are lifted. When it thaws, the particles are pulled downhill by gravity. Eventually, large amounts of sediment are moved by this process.

Slump A **slump** occurs when a mass of rock or sediment moves downhill, leaving a curved scar, as shown in **Figure 9B.** Slumps are most common in thick layers of loose sediment, but they also form in sedimentary rock. Slumps frequently occur on slopes that have been undercut by erosion, such as those above the bases of cliffs that have been eroded by waves. Slumping of this kind is common along the coast of Southern California, where it threatens to destroy houses and other buildings.

Rock Slides Can you imagine millions of cubic meters of rock roaring down a mountain at speeds greater than 250 km/h? This can happen when a rock slide occurs. During a rock slide layers of rock break loose from slopes and slide to the bottom. The rock layers often bounce and break apart during movement. This produces a huge, jumbled pile of rocks at the bottom of the slope, as you can see in **Figure 9C.** Rock slides can be destructive, sometimes destroying entire villages or causing hazards on roads in mountainous areas.

Mudflows Where heavy rains or melting snow and ice saturate sediments, mudflows, as shown in **Figure 9D,** can develop. A mudflow is a mass of wet sediment that flows downhill over the ground surface. Some mudflows can be thick and flow slowly downhill at rates of a few meters per day. Other mudflows can be much more fluid and move downslope at speeds approaching 160 km/h. This type of mudflow is common on some volcanoes.

Reading Check *What is the slowest of the four kinds of mass movement?*

INTEGRATE Physics

Mass Movement Slumps and rock slides often occur when sediment becomes saturated by rain. Water between sediment grains helps lift up overlying rock and sediment. This makes it easier for the sediment to overcome the forces holding it in place. Can you think of a way some slopes might be protected from slumps and rock slides? Explain.

2 Teach

Use Science Words

Word Meaning Have students use a dictionary to define *creep.* Ask volunteers to read aloud its many definitions. Then ask students why *creep* is a good name for one kind of mass movement. Creep describes a gradual movement of sediment near the surface. It is so slow that it is almost unnoticed.

Reading Check

Answer creep

IDENTIFYING Misconceptions

Reducing Friction Students might think water helps bind sediments to one another on a slope. Actually, water reduces the friction that holds sediments together. Because water lifts overlying sediments, friction is reduced, the slope becomes unstable, and mass movement takes place.

INTEGRATE Physics

Mass Movement Some students might suggest building drainage ditches to help move the water down slope, thus reducing the amount of water available for lubricating sediments. Others might suggest that reducing the steepness of the slope would reduce the pull of gravity on the lifted sediment particles, and thus lessen the effects of water. Accept reasonable responses.

Research Have students research what kind of factors would go into a mathematical equation to calculate soil loss.

Differentiated Instruction

Hearing Impaired Have small groups of students work together to devise ways of describing the different types of mass movements through pantomime. Each pantomime should include information about the speed at which the movements occur. L2 ELL COOP LEARN LS **Kinesthetic**

Curriculum Connection

Music Have students work in small groups to write new lyrics to familiar songs. The lyrics should describe different kinds of mass movements. Have students sing their songs for their classmates. L2 COOP LEARN LS **Auditory-Musical and Interpersonal**

Use Science Words

Word Origin The term *glacier* comes from the French word *glace*. Have students find the meaning of glace. It means ice.

Activity

Ice Flow Patterns The following data show ice flow speeds across a small alpine glacier. Have students identify patterns in the data using average, frequency, and range. Students should graph these data to observe the pattern of ice speeds.

Distance Across Glacier (m)	Speed (m/yr)
0	20
50	40
100	80
200	100
300	80
350	40
400	20

Average = 54 m/yr; range = 100 m/yr − 20m/yr = 80 m/yr; frequency: 20 m/yr = 2, 40 m/yr = 2,80 m/yr = 2, 100 m/yr = 1. Graph should be a symmetric curve.

Fun Fact

The names for the countries Greenland and Iceland are misleading. Greenland is covered by a continental glacier. Iceland is not.

Figure 11 Glaciers form in cold regions.

Valley glaciers are found at high elevations on many continents.

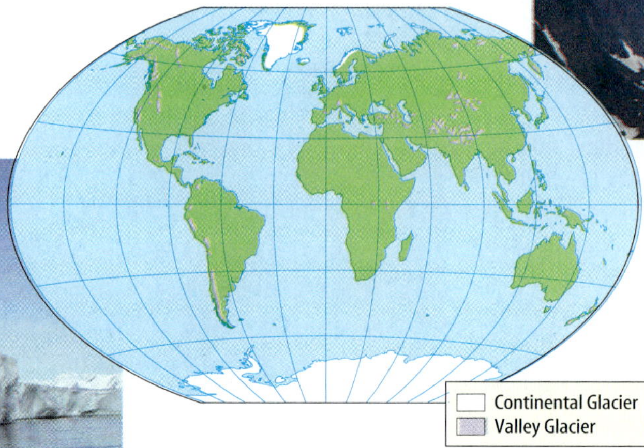

Continental glaciers are located near the poles, in Antarctica and Greenland.

Science Online

Topic: Glacial Erosion and Deposition

Visit red.msscience.com for Web links to information about glacial erosion and deposition.

Activity Research glacial erosion and describe how it has affected the topography in an area.

Ice

In some parts of the world, ice is an agent of erosion. In cold regions, more snow might fall than melt. Over many years, the snow can accumulate to form large, deep masses of ice called glaciers. When the ice in a glacier becomes thick enough, its own weight causes it to flow downhill under the influence of gravity. As glaciers move over Earth's surface, they erode materials from some areas and deposit sediment in other areas. **Figure 11** shows the two kinds of glaciers—continental glaciers and valley glaciers.

Today, continental glaciers in polar regions cover about ten percent of Earth. These glaciers are so large and thick that they can bury mountain ranges. Valley glaciers are much smaller and are located in high mountains where the average temperature isn't warm enough to melt the ice sheets. The average flow rate of a valley glacier is 0.01 to 2.0 meters per day, but during a surge, they can flow up to 100 meters per day.

Glacial Erosion Glaciers can erode rock in two different ways. If the underlying rock has cracks in it, the ice can pull out pieces of rock. This causes the rock to erode slowly. Loose pieces of rock freeze into the bottom of the glacier and are dragged along as the glacier moves. As these different-sized fragments of rock are dragged over Earth's surface, they scratch the rock below like giant sheets of sandpaper. This scratching is the second way that glaciers can erode rock. Scratching produces large grooves or smaller striations in the rock underneath. The scratching also can wear rock into a fine powder called rock flour.

Cultural Diversity

Sherpas Sherpas live on the high southern slopes of the Himalaya, thus glaciers are a very important part of their lives. Many Sherpas are guides for mountain climbing expeditions to Mount Everest. Have students research the role Sherpas have had in Mount Everest expeditions and the dangers they face. L2 LS **Linguistic**

Science Journal

Tour Guide After researching Glacier National Park, have students write travel brochures that describe the glacial structures that can be seen at the park. L2 LS **Linguistic** P

Figure 12 Many high-altitude areas owe their distinctive appearance to glacial erosion.

Mountain glaciers can carve bowl-shaped depressions called cirques.

Glaciers can widen valleys, giving them a U-shaped profile.

Effects of Glacial Erosion Glacial erosion of rock can be a powerful force shaping Earth's surface. In mountains, valley glaciers can remove rock from the mountaintops to form large bowls, called cirques (SURKS), and steep peaks. When a glacier moves into a stream valley, it erodes rock along the valley sides, producing a wider, U-shaped valley. These features are shown in **Figure 12.** Continental glaciers also shape Earth's surface. These glaciers can scour large lakes and completely remove rock layers from the land's surface.

Glacial Deposition Glaciers also can deposit sediments. When stagnant glacier ice melts or when ice melts at the bottom of a flowing glacier or along its edges, the sediment the ice was carrying gets left behind on Earth's surface. This sediment, deposited directly from glacier ice, is called till. Till is a mixture of different-sized particles, ranging from clay to large boulders.

As you can imagine, a lot of melting occurs around glaciers, especially during summer. So much water can be produced that streams often flow away from the glacier. These streams carry and deposit sediment. Sand and gravel deposits laid down by these streams, shown in **Figure 13,** are called outwash. Unlike till, outwash usually consists of particles that are all about the same size.

Figure 13 This valley in New Zealand has been filled with outwash.
Explain *how you could distinguish outwash from till.*

327

Make a Model

Glacial Erosion Use the following procedures to demonstrate glacial erosion. Place gravel of different sizes into the compartments of an ice tray. Fill the tray with water. Freeze the tray overnight. Fill a large flat box or tray with sand. Have students use the ice cube "glaciers" to demonstrate how glaciers scratch grooves and striations into rock. L2 LS **Kinesthetic**

Discussion

Glacial Deposit You find a large rock in the middle of a flat field. Could this be a glacial deposit? Explain. Yes, a glacier might have deposited till, including the rock, as it melted.

Caption Answer

Figure 13 Till is a mixture of different-sized particles of sediment. Outwash is sorted by size.

Virtual Labs

Erosion *How are Earth materials broken down?*

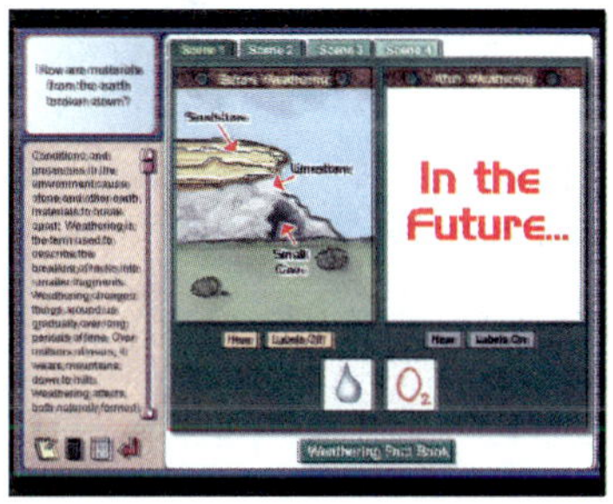

Curriculum Connection

Geography The Matterhorn, called Mont Cervin by the French and Monte Cervino by the Italians, formed during the Pleistocene Epoch. Have students locate it on a map. It is in the Alps between Switzerland and Italy. L2 LS **Visual-Spatial**

Reading Check

Answer Sediment particles carried by wind strike rocks, making pits and smooth, polished surfaces.

Discussion

Planting Trees Why do you think farmers often plant rows of trees along their fields? When wind hits the trees, it slows down, thus reducing erosion of the soil in the fields. The reduction in speed also causes sediment carried by the wind to be deposited.

Teacher FYI

Loess Deposits Strong winds that deflated glacial outwash during and following the ice ages deposited large amounts of loess near the Mississippi River. These buff-colored deposits have evolved into fertile soils. Loess deposits that originated from silts deflated from the Gobi and Ordos deserts give China's Yellow River (Huang He) its color.

Visual Learning

Figure 15 How can you tell the direction a sand dune is migrating? A dune migrates in the direction the wind is blowing. The wind rolls sand particles up the gentle side and over the top of the dune. Thus, the gentle side of the dune points away from the direction of migration.

Figure 14 In a desert, where small particles have been carried away by wind, larger sediments called desert pavement remain behind.

Wind

If you've had sand blow into your eyes, you've experienced wind as an agent of erosion. When wind blows across loose sediments like silt and sand, it lifts and carries it. As shown in **Figure 14,** wind often leaves behind particles too heavy to move. This erosion of the land by wind is called **deflation.** Deflation can lower the land's surface by several meters.

Wind that is carrying sediment can wear down, or abrade, other rocks just as a sandblasting machine would do. **Abrasion** is a form of erosion that can make pits in rocks and produce smooth, polished surfaces. Abrasion is common in some deserts and in some cold regions with strong winds.

Reading Check *How does abrasion occur?*

When wind blows around some irregular feature on Earth's surface, such as a rock or clump of vegetation, it slows down. This causes sand carried by the wind to be deposited. If this sand deposit continues to grow, a sand dune like that shown in **Figure 15** might form. Sand dunes move when wind carries sand up one side of the dune and it avalanches down the other, as shown in **Figure 15.**

Sometimes, wind carries only fine sediment called silt. When this sediment is deposited, an accumulation of silt called loess (LOOS) can blanket Earth's surface. Loess is as fine as talcum powder. Loess often is deposited downwind of large deserts and deflated glacial outwash deposits.

Figure 15 Wind transportation of sand creates sand dunes.

Sand dunes do not remain in one location—they migrate.

As wind blows over a sand dune, sand blows up the windward side and tumbles down the other side. In this way, a sand dune migrates across the land.

Differentiated Instruction

Challenge Have pairs of students work together to design experiments to study the effects of different variables on deflation. Variables to be tested might include soil moisture, wind velocity, and plant cover. Have students make posters explaining their results. L3 COOP LEARN LS **Kinesthetic**

Cultural Diversity

Sacred Rock The largest sandstone monolith in the world is in Australia's Northern Territory. Shaped by wind-blown sands, Ayers Rock is the remains of a sandstone mountain that formed about 500 million years ago. To the Aboriginal people, Ayers Rock, called Uluru, is a sacred site. Have students discuss why they think the Aboriginal people might consider Uluru sacred.

Water

You probably have seen muddy water streaming down a street after a heavy rain. You might even have taken off your shoes and waded through the water. Water that flows over Earth's surface is called **runoff.** Runoff is an important agent of erosion, especially if the water is moving fast. The more speed water has, the more material it can carry with it. Water can flow over Earth's surface in several different ways, as you will soon discover.

Figure 16 Water flows over the hood of a car as a thin sheet. **Describe** *how this is similar to sheet flow on Earth's surface.*

Sheet Flow As raindrops land on Earth's surface, they break up clumps of soil and loosen small grains of sediment. If these raindrops are falling on a sloped land surface, a thin sheet of water might begin to move downhill. You have observed something similar if you've ever washed a car and seen sheets of water flowing over the hood, as shown in **Figure 16.** When water flows downhill as a thin sheet, it is called sheet flow. This thin sheet of water can carry loose sediment grains with it, causing erosion of the land. This erosion is called sheet erosion.

Applying Science

Can evidence of sheet erosion be seen in a farm field?

If you've ever traveled through parts of your state where there are farms, you might have seen bare, recently cultivated fields. Perhaps the soil was prepared for planting a crop of corn, oats, or soybeans. Do you think sheet erosion can visibly affect the soil in farm fields?

Identifying the Problem

The top layer of most soils is much darker than layers beneath because it contains more organic matter. This layer is the first to be removed from a slope by sheet flow. How does the photo show evidence of sheet erosion?

Solving the Problem

1. Observe the photo and write a description of it in your Science Journal.
2. Infer why some areas of the field are darker colored than others are. Where do you think the highest point(s) are in this field?
3. Make a generalization about the darker areas of the field.

Make a Model

Raindrops Spread a layer of soil onto one end of a cookie sheet or rectangular tray. Form a slight downward slope by placing a book under the end of the tray containing the soil. Have students observe the effects of dripping different sized drops of water from different heights onto the soil. Students should observe the "raindrops" break up clumps of soil and loosen small grains of sediment. As they continue adding water, a sheet flow will develop and eroded materials will collect along the bottom of the tray. L2 ELL LS **Kinesthetic and Visual-Spatial**

Caption Answer

Figure 16 Water flows down a slope as a thin sheet over Earth's surface when sheet flow occurs.

Applying Science

Answers

1. Students' descriptions should be of a recently cultivated field and should include color and slope of the ground.
2. Sheet wash has moved organic-rich topsoil to low areas of the field. The high points are lighter colored.
3. They have soil with a higher humus content.

Activity

Rill to Gully Have students draw a series of pictures that show how a rill can evolve into a gully. L2 ELL LS **Visual-Spatial**

Science Journal

Runoff Have students take pictures of runoff or write paragraphs in their Science Journals about their observations of runoff. Challenge them to identify the source of the runoff (rainfall, snowmelt, sprinkler, or garden hose). L2 LS **Linguistic and Visual-Spatial**

Visual Learning

Figure 17 Why do gullies often form on vegetation-free slopes? Soil is easily eroded without plant roots to help hold it in place.

Use an Analogy

Walking Paths Just as paths form near the school building where students walk over the same area of grass day after day, rills form where water flows over the same area after every rain. After continuous use by students, paths are eroded deeper into the sediments. After recurring rains, rills develop into gullies.

Teacher FYI

As water erodes the sides and bottom of a channel, its profile becomes V-shaped. The winding parts where streams snake back and forth are called meanders.

Figure 17 Gullies often form on vegetation-free slopes.

Science Online

Topic: Power of Water

Visit red.msscience.com for Web links to information about the erosional force of running water.

Activity Research your watershed and see if the topography was shaped by running water.

Rills and Gullies Where a sheet of water flows around obstacles and becomes deeper, rills can form. Rills are small channels cut into the sediment at Earth's surface. These channels carry more sediment than can be moved by sheet flow. In some cases, a network of rills can form on a slope after just one heavy rain. Large amounts of sediment can be picked up and carried away by rills.

As runoff continues to flow through the rills, more sediment erodes and the channel widens and deepens. When the channels get to be about 0.5 m across, they are called gullies, as shown in **Figure 17.**

Streams Gullies often connect to stream channels. Streams can be so small that you could jump to the other side or large enough for huge river barges to transport products along their course. Most streams have water flowing through them continually, but some have water only during part of the year.

In mountainous and hilly regions, as in **Figure 18,** streams flow down steep slopes. These streams have a lot of energy and often cut into the rock beneath their valleys. This type of stream typically has white-water rapids and may have waterfalls. As streams move out of the mountains and onto flatter land, they begin to flow more smoothly. The streams might snake back and forth across their valley, eroding and depositing sediments along their sides. All streams eventually must flow into the ocean or a large lake. The level of water in the ocean or lake determines how deeply a river can erode.

Figure 18 Streams that flow down steep slopes such as this one in Yosemite National Park often have white-water rapids and waterfalls.

330

Curriculum Connection

Social Studies Dams are built to prevent flooding, provide reservoirs for drinking water, and allow the harnessing of the energy of moving water. Dams also trap sediments and reduce the flow of water downstream. Hoover Dam is located on the Nevada-Arizona border along the Colorado River. Have students research how people downstream are affected by this dam. L2 LS **Linguistic**

Shaping Earth's Surface In the Launch Lab, you saw a small model of erosion by a stream. Streams are the most important agent of erosion on Earth. They shape more of Earth's surface than ice, wind, or gravity. Over long periods of time, water moving in a stream can have enough power to cut large canyons into solid rock. Many streams together can sculpt the land over a wide region, forming valleys and leaving some rock as hills. Streams also shape the land by depositing sediment. Rivers can deposit sandbars along their course, and can build up sheets of sand across their valleys. When rivers enter oceans or lakes, the water slows and sediment is deposited. This can form large accumulations of sediment called deltas, as in **Figure 19.** The city of New Orleans is built on the delta formed by the Mississippi River.

Figure 19 A triangular area of sediment near the mouth of a river is called a delta. Ancient deltas that are now dry land are often excellent places to grow crops.

Effects of Erosion

All agents of erosion change Earth's surface. Rock and sediment are removed from some areas only to be deposited somewhere else. Where material is removed, canyons, valleys, and mountain cirques can form. Where sediment accumulates, deltas, sandbars, sand dunes, and other features make up the land.

section 2 review

Summary

Gravity

- Erosion of rock or sediment, moved down a slope by gravity, is called mass movement.
- Creep, slump, rock slides, and mudflows are four types of mass movement.

Ice

- Glaciers move over Earth's surface, eroding materials from some areas and depositing sediments in other areas.

Wind

- Deflation and abrasion are two common forms of wind erosion.

Water

- Runoff is water flowing over Earth's surface.
- Water erosion can create sheet erosion, rills, gullies, streams, valleys, and canyons.

Self Check

1. **Describe** four agents of erosion. Which of these is the fastest? The slowest? Explain your answers.
2. **Explain** the difference between deflation and abrasion.
3. **Describe** how a cirque forms.
4. **Explain** When do streams deposit sediments? When do streams erode sediments?
5. **Think Critically** Why might a river that was eroding and depositing sediment along its sides start to cut into Earth to form a canyon?

Applying Math

6. **Solve One-Step Equations** If wind is eroding an area at a rate of 2 mm per year and depositing it in a smaller area at a rate of 7 mm per year, how much lower will the first area be in meters after 2 thousand years? How much higher will the second area be?

Science Online red.msscience.com/self_check_quiz

Fun Fact

A placer is a deposit formed by the sorting action of moving water. Denser and often valuable minerals may be concentrated in this way in streams, lakes, or oceans. South African diamond and Alaskan gold placers are mined.

3 Assess

DAILY INTERVENTION

Check for Understanding

Visual-Spatial Explain how rocks can be abraded by a variety of erosional agents. Flowing water and blowing wind carry sediments that abrade rock. Ice and the sediments it contains also abrade rocks.

Reteach

Model Erosion Have students place a small pile of a sand and gravel mixture in one end of a large plastic or metal tray. Have them move the pile to the other end of the tray by modeling agents of erosion. L2 ELL LS **Kinesthetic**

Assessment

Process Have students predict what will happen to a stream if its bed is uplifted. A stream will respond to the uplift by cutting downward toward lower elevations. This results in increased erosion of the stream channel. Use **Performance Assessment in the Science Classroom,** p. 89.

section 2 review

1. Gravity, ice, wind, water; Gravity causes sediment movement down slopes. Glaciers, wind, and moving water remove and transport sediments. Gravity is the fastest agent of erosion. A glacier is the slowest agent of erosion.
2. Deflation: wind blows away tiny sediments; abrasion: fine sediments scratch and pit rocks they hit.
3. Ice wedging plucks out rock; valley glacier removes rock, leaving bowl.
4. Sediments are deposited when a stream's energy isn't great enough to suspend sediments. Sediments are eroded when a stream's energy is great enough to suspend sediments and carry them away.
5. If the land is uplifted, there is increased erosion of the stream bed.
6. 4,000 mm; 14,000 mm

BENCH TESTED

LAB Design Your Own

Measuring Soil Erosion

Goals
- **Design** an experiment to measure soil loss from grass-covered soil and from soil without grass cover.
- **Calculate** the percent of soil loss with and without grass cover.

Possible Materials
blocks of wood
**books*
paint trays (2)
soil
grass sod
water
pails (2)
1,000-mL beaker
triple-beam balance
calculator
watch
**Alternate materials*

Safety Precautions

Wash your hands thoroughly when you are through working with soils.

Real-World Question

During urban highway construction, surface mining, forest harvesting, or agricultural cultivation, surface vegetation is removed from soil. These practices expose soil to water and wind. Does vegetation significantly reduce soil erosion? How much does vegetation reduce soil erosion?

Form a Hypothesis

Based on what you've read and observed, hypothesize about how a grassy field will have less erosion than a field that is bare soil.

Real-World Question

Purpose Students measure to what extent vegetation reduces soil erosion. L2 COOP LEARN LS **Visual-Spatial**

Process Skills form a hypothesis, design an experiment to test a hypothesis, identify and manipulate variables, measure in SI, use numbers, draw conclusions

Time Required 90 minutes

Possible Materials

Safety Precautions Caution students to wear goggles and to wash hands after working with soil.

Form a Hypothesis

Possible Hypothesis Students will likely predict that the tray with sod will have much less erosion.

Test Your Hypothesis

Possible Procedures Cover the bottoms of trays with soil. Cover soil in one tray with sod. Place identical blocks of wood under one end of each tray. Determine the mass of the empty beakers. Fill each pail with 1,000 mL of water. Pour the water into the trays at the same rate. Wait several minutes, and pour the muddy water from each tray into a beaker. Evaporate the water in the beakers. Measure the masses of the beakers. Compute the mass of eroded soil. Calculate percentage of soil loss.

Curriculum Connection

History In the 1930s, drought plus poor farming practices led to severe wind erosion in the Great Plains of the United States. This area became known as the Dust Bowl. Have students research farming practices used today to prevent soil erosion. L2 LS **Linguistic**

Communicating Your Data

Have students use a word processing program on the computer to write their business letters to editors.

Using Scientific Methods

Test Your Hypothesis

Make a Plan

1. As a group, agree upon the hypothesis and decide how you will test it. Identify which results challenge or confirm the hypothesis.
2. **List** the steps you will need to take to test your hypothesis. Describe exactly what you will do in each step.
3. **Prepare** a data table in your Science Journal to record your observations.
4. Read over the entire experiment to make sure all steps are in logical order, and that you have all necessary materials.
5. **Identify** all constants and variables and the control of the experiment. A control is a standard for comparing the results of an experiment. One possible control for this experiment would be the results of the treatment of an uncovered soil sample.

Follow Your Plan

1. Make sure your teacher approves your plan before you start.
2. Carry out the experiment step by step as planned.
3. While doing the experiment, record your observations and complete the data table in your Science Journal.

Vegetation and Erosion

	(A) Mass of Soil at Start	(B) Mass of Eroded Soil	% of Soil Loss (B/A) × 100
Covered soil sample		Answers will vary.	
Uncovered soil sample			

Analyze Your Data

1. **Compare** the percent of soil loss from each soil sample.
2. **Compare** your results with those of other groups.
3. What was your control in this experiment? Why is it a control?
4. Which were the variables you kept constant? Which did you vary?

Conclude and Apply

1. Did the results support your hypothesis? Explain.
2. **Infer** what effect other types of plants would have in reducing soil erosion. Do you think that grass is better or worse than most other plants at reducing erosion? Explain your answer.

Communicating Your Data

Write a letter to the editor of a newspaper. In your letter, summarize what you learned in your experiment about the effect of plants on soil erosion.

Teaching Strategy Have towels and garbage bags available for cleanup.

Troubleshooting Review how to calculate percent.

Expected Outcome Soil covered with sod will have a much lower percentage of soil erosion.

Analyze Your Data

Answers to Questions

1. The percent lost from the sod tray should be much lower than that lost from the other tray.
2. Discuss explanations for any differences observed.
3. Answers depend on the design of the experiment. A control might consist of pouring water over an empty tray and evaporating the water.
4. Amount of soil, height of trays, amount of water poured into trays, rate of pouring water, procedure for measuring soil loss

Error Analysis Have students compare their results and their hypotheses and explain any differences.

Conclude and Apply

1. Answers will vary.
2. Possible answer: plants with fibrous roots will reduce erosion better than plants with taproots.

Assessment

Process Have students design an experiment to see what effect different types of plants have on soil erosion. Use **Performance Assessment in the Science Classroom,** p. 95.

Alternative Inquiry Lab

Exploring Further To extend this Lab into an alternative inquiry lab, have students test the effect of slope on erosion rates by measuring how much soil will run-off. Students will need to use two square pans with spouts on one end. Both pans should be filled with soil. A watering can filled with water will be used to start the erosion process. By varying the slope with the wooden slates of one pan and leaving the other flat, the students can see the amount of soil that runs-off into the containers.

TIME

Content Background

Acid rain is not a new occurrence. It was first recognized and named in 1872 by Robert Angus Smith in Scotland. Rain is naturally acidic, forming a mild solution of carbonic acid from its contact with CO_2 in the atmosphere. Rain is normally in the pH range of 5 to 6, but some acid rains are as low as 4.3.

When water mixes with sulfur or nitrogen oxides in the air, it forms acids of various kinds, largely sulfuric and nitric acid. The effects of acidity on building materials such as granite are less because granite is crystalline and less reactive. Softer stones such as marble and sandstone contain high proportions of alkaline substances. These substances react with the acid to form other compounds and cause the deterioration of the stone. Most large stone structures have been built with these softer stones and are at some risk from acid rain.

The acid rain caused by pollution usually falls away from its source, often crossing political borders. In other words, acid rain that falls in one country may have its origins in another country. The result is a problem of international proportions.

Historical Significance

Students should understand that ancient buildings like the Parthenon and the Coliseum are tangible links to history that lead people to consider the events that occurred in and around them. The more these images deteriorate, the more information is lost. Archaeologists, anthropologists, and other scientists analyze inanimate objects, including historic structures, in order to gain factual knowledge about them and the cultures that built them.

TIME **SCIENCE AND HISTORY**

SCIENCE CAN CHANGE THE COURSE OF HISTORY!

CRUMBLING MONUMENTS

The Taj Mahal in India, the Acropolis in Greece, and the Colosseum in Italy have stood for centuries. They've survived wars, souvenir-hunters, and natural weathering from wind and rain. But now, something far worse threatens their existence—acid rain. Over the last few decades, this form of pollution has eaten away at some of history's greatest monuments.

Most of these structures are made of sandstone, limestone, and marble. Acid rain causes the calcium in these stones to form calcium sulfate, or gypsum. Gypsum's powdery little blotches are sometimes called "marble cancer." When it rains, the gypsum washes away, along with some of the surface of the monument.

In Agra, India, the smooth, white marble mausoleum called the Taj Mahal has stood since the seventeenth century. But acid rain is making the surface of the building yellow and flaky. The pollution is caused by hundreds of factories surrounding Agra that emit damaging chemicals.

Acid rain has not been kind to this Mayan figure.

What moisture, molds, and the roots of vegetation couldn't do in 1,500 years, acid rain is doing in a few decades. It is destroying the Mayan ruins of Mexico. Acid rain is causing statues to crumble and paintings on walls to flake off.

Acid rain is a huge problem affecting national monuments and treasures in just about every urban location in the world. These include the Capitol building in Washington, D.C., churches in Germany, and stained-glass windows in Sweden. In London, acid rain has forced workers to repair and replace so much of Westminster Abbey that the structure is becoming a mere copy of the original.

Throughout the world, acid rain has weathered many structures more in the last 20 years than in the 2,000 years before. This is one reason some steps have been taken in Europe and the United States to reduce emissions from the burning of fossil fuels. If these laws don't work, many irreplaceable art treasures may be gone forever.

Identify **What are some famous monuments and buildings in the United States? Brainstorm a list with your class. Then choose a monument and, using your school's media center or the Science Online address, learn more about it. Is acid rain affecting it in any way?**

Science Online
For more information, visit red.msscience.com/time

Identify Pick several monuments in different communities. Have each group choose a different monument and research whether or not the monument is affected by acid rain. Suggest that students obtain information regarding the condition of the monument and how the community is dealing with the problem, if at all.

Resources for Teachers and Students

US EPA Clean Air Markets Division Mail Code 6204N, Washington, DC 20460

U.S. Department of the Interior U.S. Geological Survey 12201 Sunrise Valley Drive, Reston, VA 20192

An Introduction to Global Environmental Issues (2nd Ed.) by Pickering & Owen. Routledge (1997)

Reviewing Main Ideas

Section 1 Weathering and Soil Formation

1. Weathering includes processes that break down rock.
2. During mechanical weathering, physical processes break rock into smaller pieces.
3. During chemical weathering, the chemical composition of rocks is changed.
4. Soil evolves over time from weathered rock. Parent rock, topography, climate, and organisms affect soil formation.

Section 2 Erosion of Earth's Surface

1. Erosion is the wearing away and removal of rock or sediment.
2. Agents of erosion include gravity, ice, wind, and water. Downslope movement of a portion of the land's surface is called mass movement.
3. All agents of erosion move rock and sediment. When energy of motion decreases, sediment is deposited.
4. Erosion and deposition determine the shape of the land.

Visualizing Main Ideas

Copy and fill in the following table, which compares erosion and deposition by different agents.

Erosion and Deposition

Erosional Agent	Evidence of Erosion	Evidence of Deposition
Gravity	curved slump scars, leaning fence posts, steep cliffs	material piled at bottom of slopes
Ice	cirques, striations, U-shaped valleys	till, outwash
Wind	desert pavement pitted rocks	sand dunes, loess
Surface Water	rills, gullies, stream valleys	sandbars, deltas

Science Online red.msscience.com/interactive_tutor

Identifying Misconceptions

Assess

Use this assessment as follow-up to page F at the beginning of the chapter.

Materials parent rock samples, soil samples of various colors that are related to the parent rocks

Procedure Have students develop a concept map showing how a particular color of soil could be related to a particular parent rock.

Expected Outcome Students will realize that soil color is partly related to the parent rock color.

Reinforcement Show students color photographs of rock outcrops. Have the students infer the soil color that they might expect to see on a visit to that location.

chapter 11 Study Guide

Reviewing Main Ideas

Summary statements can be used by students to review the major concepts of the chapter.

Visualizing Main Ideas

See student page.

Science Online

Visit red.msscience.com
/self_check_quiz
/interactive_tutor
/vocabulary_puzzlemaker
/chapter_review
/standardized_test

Assessment Transparency

For additional assessment questions, use the *Assessment Transparency* located in the transparency book.

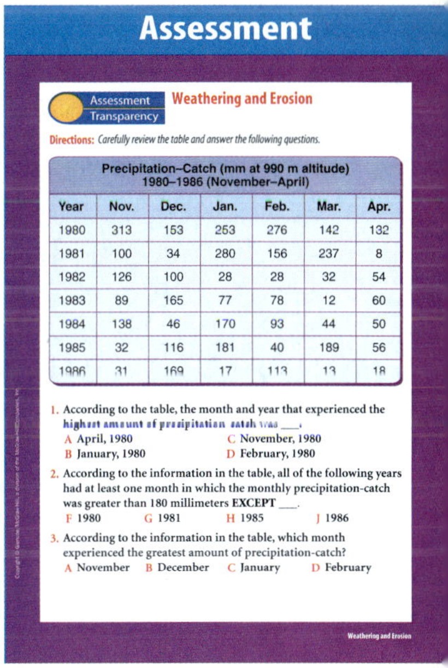

Assessment

Assessment Transparency **Weathering and Erosion**

Directions: *Carefully review the table and answer the following questions.*

Precipitation–Catch (mm at 990 m altitude) 1980–1986 (November–April)						
Year	Nov.	Dec.	Jan.	Feb.	Mar.	Apr.
1980	313	153	253	276	142	132
1981	100	34	280	156	237	8
1982	126	100	28	28	32	54
1983	89	165	77	78	12	60
1984	138	46	170	93	44	50
1985	32	116	181	40	189	56
1986	31	169	17	113	13	18

1. According to the table, the month and year that experienced the highest amount of precipitation catch was ___.
   A April, 1980 C November, 1980
   B January, 1980 D February, 1980
2. According to the information in the table, all of the following years had at least one month in which the monthly precipitation-catch was greater than 180 millimeters EXCEPT ___.
   F 1980 G 1981 H 1985 J 1986
3. According to the information in the table, which month experienced the greatest amount of precipitation-catch?
   A November B December C January D February

Weathering and Erosion

Using Vocabulary

1. Chemical weathering changes the chemical composition of a rock; mechanical weathering breaks rocks into small pieces without changing them chemically.

2. Weathering is the process that breaks rocks into small pieces; erosion is the moving of those pieces from one place to another.

3. Deflation is erosion by wind; runoff is the flowing of water across the surface.

4. Weathering is the process that breaks rocks into small pieces; mass movement is the movement of those pieces down a slope because of gravity.

5. Soil is a mixture of weathered rock, organic matter, water, and air; abrasion occurs as soil and sediment carried by the wind wears away rock.

6. Possible answer: Plants can slow the erosion of soil.

7. Possible answer: The particles produced by mechanical weathering are moved down a slope by mass movement.

8. Weathering alters rock mechanically or chemically; chemical weathering breaks down rock by changing its composition.

9. Creep is the slow movement of sediments downhill; slump is the downhill movement of a large mass of sediments or rock.

10. Possible answer: The type of runoff depends partly on the topography of the surrounding land.

Checking Concepts

11. C	**16.** B
12. D	**17.** C
13. B	**18.** A
14. A	**19.** D
15. A	**20.** A

Using Vocabulary

abrasion p. 328	mechanical weathering p. 316
chemical weathering p. 318	runoff p. 329
creep p. 325	slump p. 325
deflation p. 328	soil p. 320
erosion p. 323	topography p. 320
mass movement p. 323	weathering p. 316

Use each of the following pairs of terms in a sentence.

1. chemical weathering—mechanical weathering

2. erosion—weathering

3. deflation—runoff

4. mass movement—weathering

5. soil—abrasion

6. soil—erosion

7. mass movement—mechanical weathering

8. weathering—chemical weathering

9. creep—slump

10. topography—runoff

Checking Concepts

Choose the word or phrase that best answers the question.

11. Which of the following agents of erosion forms U-shaped valleys?
A) gravity **C)** ice
B) surface water **D)** wind

12. In which of these places is chemical weathering most rapid?
A) deserts
B) mountains
C) polar regions
D) tropical regions

13. Which of the following forms when carbon dioxide combines with water?
A) calcium carbonate
B) carbonic acid
C) tannic acid
D) dripstone

14. Which process causes rocks to weather to a reddish color?
A) oxidation **C)** carbon dioxide
B) deflation **D)** frost action

15. Which type of mass movement occurs when sediments slowly move downhill because of freezing and thawing?
A) creep **C)** slump
B) rock slide **D)** mudflow

16. Which of the following helps form cirques and U-shaped valleys?
A) rill erosion **C)** deflation
B) ice wedging **D)** till

17. What is windblown, fine sediment called?
A) till **C)** loess
B) outwash **D)** delta

18. Which of the following refers to water that flows over Earth's surface?
A) runoff **C)** weathering
B) slump **D)** till

19. Which of the following is an example of chemical weathering?
A) Plant roots grow in cracks in rock and break the rock apart.
B) Freezing and thawing of water widens cracks in rocks.
C) Wind blows sand into rock, scratching the rock.
D) Oxygen causes iron-bearing minerals in rock to break down.

20. Which one of the following erosional agents creates desert pavement?
A) wind **C)** water
B) gravity **D)** ice

Science Online red.msscience.com/vocabulary_puzzlemaker

Use the Exam*View*® Pro Testmaker CD-ROM to:

- create multiple versions of tests
- create modified tests with one mouse click for inclusion students
- edit existing questions and add your own questions
- build tests aligned with state standards using built-in State Curriculum Tags
- change English tests to Spanish with one mouse click and vice versa

Thinking Critically

21. **Explain** why mass movement is more common after a heavy rainfall.

22. **Describe** how climate affects the development of soils.

23. **Explain** how some mass movement could be prevented.

24. **Describe** why chemical weathering would not be rapid in Antarctica.

25. **Describe** why caves form only in certain types of rock.

26. **Recognize Cause and Effect** Explain how water creates stream valleys.

27. **Form hypotheses** about how deeply water could erode and about how deeply glaciers could erode.

28. **Recognize Cause and Effect** Explain how valley glaciers create U-shaped valleys.

29. **Classify** the following by the agent that deposits each: sand dune, delta, till, and loess.

30. **Concept Map** Copy and complete the concept map showing the different types of mass movements.

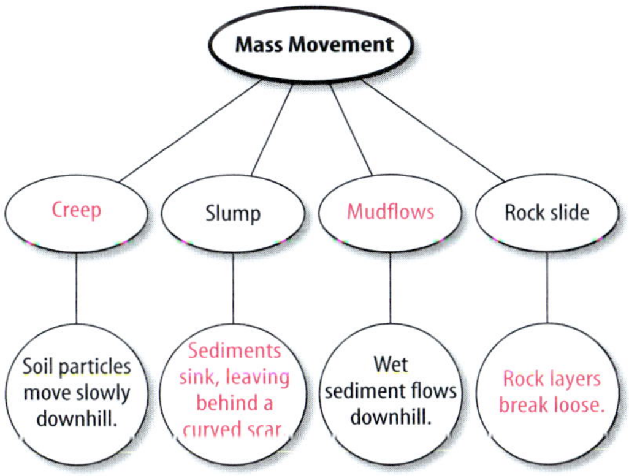

Performance Activities

31. **Poster** Use photographs from old magazines to make a poster that illustrates different kinds of weathering and erosion. Display your poster in your classroom.

32. **Model** Use polystyrene, cardboard, and clay to make a model of a glacier. Include a stream of meltwater leading away from the glacier. Use markers to label the areas of erosion and deposition. Show and label areas where till and outwash sediments could be found. Display your model in your classroom.

Applying Math

Use the table below to answer questions 33–34.

Loss and Formation of Topsoil

Years Soil Erodes	Topsoil Lost (cm)	Years to Re-create Soil Lost
10	1	500
15	1.5	750
20	2	1000
25	2.5	1250
30	3	1500
40	4.5	2250

33. **Formation of Topsoil** In a given region, it takes 500 years to form 1 cm of topsoil. For the past 20 years, the area has lost 1 cm of topsoil every 10 years because of erosion. If erosion stops today, how long will it take to re-create the lost topsoil? How long will it take to re-create the topsoil lost in 15, 25, and 30 years?

34. **Weather's Affect** In the 30th year, climate changes and precipitation increases, causing a new erosion rate of 1.5 cm every 10 years. How many years will it now take to re-create the topsoil lost at the end of 40 years?

Science Online red.msscience.com/chapter_review

Assessment Resources

Reproducible Masters

Chapter *Fast File* Resources
- Chapter Review, pp. 35–36
- Chapter Tests, pp. 37–40
- Assessment Transparency Activity, p. 47

Glencoe Science Web site
- Chapter Review Test
- Standardized Test Practice

Glencoe Technology

- Assessment Transparency
- Exam*View*® Pro Testmaker
- MindJogger Videoquiz
- Interactive Chalkboard

Thinking Critically

21. Water makes sediments heavier. It also lifts up overlying rock and sediment. This makes it easier for sediments to overcome the forces that hold the slope in place.
22. Soil evolution is faster in warm, moist climates where materials break down quickly.
23. by not making slopes too steep or by planting vegetation on slopes
24. The climate is cold and dry. Materials break down slowly.
25. Only some kinds of rocks contain minerals that react chemically with natural acids to form compounds that dissolve in water.
26. The energy of moving water suspends sediments and carries them away leaving behind stream valleys.
27. Two things that affect how deeply water can erode are the velocity of the water and the structure of the bedrock. Two things that affect how deeply a glacier can erode are the pressure of the glacier pushing down and the hardness of the bedrock.
28. As a glacier moves downhill over a stream's valley, the sides of the glacier scrape the sides of the valley, changing it's V-shape into a U-shape.
29. wind: sand dune and loess; water: delta; glacier: till
30. See Student Page

Performance Activities

31. Students should label the agent(s) of weathering or erosion for every photograph they include on their posters. Use **PASC**, p. 145.
32. Check students' work. Use **PASC**, p. 123.

Applying Math

National Math Standards

1, 2, 5, 9

33. 1,000 years; 750 years; 1,250 years; 1,500 years
34. 2,250 years

FAST FILE

Answer Sheet A practice answer sheet can be found at red.msscience.com/answer_sheet.

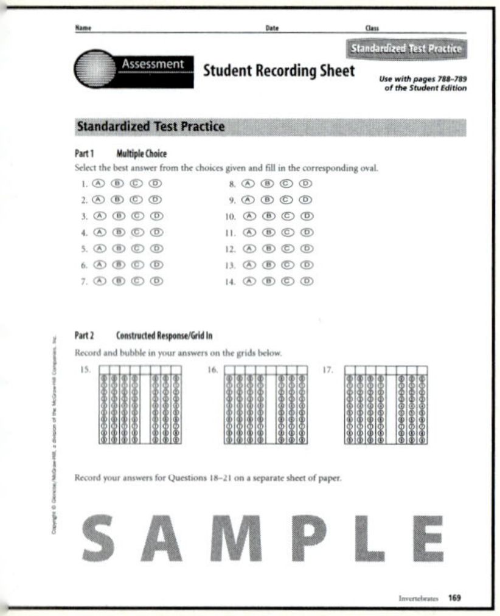
Name Date Class

Standardized Test Practice

Assessment Student Recording Sheet

Use with pages 788–789 of the Student Edition

Standardized Test Practice

Part 1 Multiple Choice

Select the best answer from the choices given and fill in the corresponding oval.

Part 2 Constructed Response/Grid In

Record and bubble in your answers on the grids below.

Record your answers for Questions 18–21 on a separate sheet of paper.

SAMPLE

Part 1 Multiple Choice

1. B
2. C
3. D
4. D
5. D
6. B
7. C
8. A
9. B
10. A

Part 2 Short Response

11. When lichens grow on rocks, they break them down and begin a thin layer of soil.
12. Till is a mixture of different-sized particles that are deposited from a glacier. Outwash are sediments that are sorted by size and are deposited by streams that formed from melting glaciers.
13. As wind blows over a sand dune, sand blows up the windward side and tumbles down the other side. In this way, a sand dune can migrate across land.

chapter 11 Standardized Test Practice

Part 1 Multiple Choice

Record your answers on the answer sheet provided by your teacher or on a sheet of paper.

1. Which is an example of mechanical weathering?
   A. creep
   B. ice wedging
   C. oxidation
   D. slump

2. Which forms as a glacier moves into a stream valley?
   A. cirque
   B. outwash
   C. U-shaped valley
   D. V-shaped valley

3. Which factor in soil formation deals with the slope of the land?
   A. climate
   B. parent rock
   C. time
   D. topography

4. Which is a mixture of weathered rock, organic matter, water, and air?
   A. humus
   B. organisms
   C. parent rock
   D. soil

5. Which type of erosion occurs when a thin sheet of water flows downhill?
   A. creep
   B. gulley erosion
   C. runoff
   D. sheet erosion

6. What causes potholes to form in roadways?
   A. creep
   B. ice wedging
   C. oxidation
   D. slump

Use the photo below to answer questions 7–8.

7. Which form of mass movement is shown in this picture?
   A. creep **C.** rockslide
   B. mudflow **D.** slump

8. Which agent of erosion causes this?
   A. gravity **C.** water
   B. ice **D.** wind

9. What form of mass movement occurs when a pasty mix of water and sediment moves downhill?
   A. creep **C.** rockslide
   B. mudflow **D.** slump

10. What type of erosion is similar to using sandpaper to smooth the edges of wood?
    A. abrasion
    B. creep
    C. deflation
    D. runoff

Test-Taking Tip

Double Check For each question, double check that you are filling in the correct bubble for the question number you are working on.

14. Much more organic material is created in warm, humid climates than desert climates because warmth and moisture promote plant growth. When plants decompose, humus is created.
15. Slump is known to cause problems in Southern California because the bases of the cliffs have been undercut from erosion by the waves.
16. Wind carries loose sediment like sand or silt and leaves behind particles too heavy to move.
17. Grassland makes for better farmland than forest because the plants contribute more organic matter to the soil making the humus better for plants to grow.

Part 2 Short Response/Grid In

Record your answers on the answers sheet provided by your teacher or on a sheet of paper.

Use the photo below to answer question 11.

11. How do lichens growing on rocks contribute to soil development?

12. What is the difference between till and outwash?

Use the illustration below to answer question 13.

13. Using the diagram, explain how a sand dune moves.

14. What effect does climate have on the formation of humus?

15. What type of mass movement causes problems along the Southern California coast? Why?

16. What type of sediment is carried by the wind?

17. Explain whether grasslands or forests make for better farmland.

Part 3 Open Ended

Record your answers on a sheet of paper.

18. Compare and contrast chemical and mechanical weathering.

19. How do freeze and thaw cycles contribute to both weathering and erosion?

Use the photo below to answer question 20.

20. Explain in detail what type of weathering is taking place in this picture. Describe the type of environment where this weathering would take place the quickest.

21. What types of glaciers are there and how do they erode the land?

Use to the photo below to answer question 22.

22. What type of erosion is pictured here? How can this type of erosion help to carry harmful chemicals that are deposited in the soil to rivers, lakes and oceans?

21. There are valley glaciers and continental glaciers. They erode the land by pulling up and breaking off pieces of rock that can become embedded in the glacier. These embedded rocks and sediment scratch the surfaces of the rocks underneath the glaciers as they move over them.

22. A stream is depicted here and it is flooded after a heavy rain. This stream is part of runoff that can carry sediment. The stream can erode sediment that contains harmful chemicals and deposit the sediment along the side, but the chemicals may continue along the stream to the river, lake, or ocean.

Rubrics

For more help evaluating open-ended assessment questions, see the rubric on p. 10T.

Part 3 Open Ended

18. Both types of weathering cause rocks to change, crumble, and break down. Mechanical weathering does not alter the rock chemically. Causes for mechanical weathering are ice wedging, plants, and animals. Causes of chemical weathering are oxidation and natural acids, including plant acids.

19. Weathering: Freeze and thaw cycles cause ice wedging: Water seeps into cracks in rocks and freezes. As it freezes, ice expands, cracking the rocks. Erosion: Creep is caused by the freeze and thaw cycle. As water freezes, it expands and lifts sediment particles. Gravity pulls loosened materials downhill.

20. This is chemical weathering caused by oxidation. The iron containing materials are exposed to oxygen and the chemical reaction causes the materials to rust. This type of weathering occurs more quickly in environments that are warm and moist.

chapter 12 Organizer

Section/Objectives	Standards		Labs/Features
	National	State/Local	
Chapter Opener	See pp. 16T–17T for a Key to Standards.		**Launch Lab:** How does temperature affect gas molecules?, p. 341 **Foldables,** p. 341 A data-collection lab using Probeware technology can be found in the **Probeware Lab Manual,** pp. 25–27
Section 1 The Atmosphere 2 sessions 1 block 1. **Explain** why air has pressure. 2. **Describe** the composition of the atmosphere. 3. **Describe** how energy causes water on Earth to cycle.	National Content Standards: UCP.1–UCP.3, UCP.5, A.1, A.2, D.1		**Integrate Health,** p. 344 **Visualizing the Water Cycle,** p. 346 **MiniLAB:** Observing Condensation and Evaporation, p. 347
Section 2 Earth's Weather 3 sessions 1.5 blocks 4. **Compare** ways that heat is transferred on Earth. 5. **Describe** the formation of different kinds of clouds and precipitation. 6. **Explain** what causes wind.	National Content Standards: UCP.1–UCP.3, UCP.5, A.1, A.2, B.2, B.3, D.1		**Integrate Life Science,** p. 349 **Science Online,** p. 350 **Applying Math,** p. 353 **Integrate Social Studies,** p. 353
Section 3 Air Masses and Fronts 4 sessions 2 blocks 7. **Explain** the ways that air masses and fronts form. 8. **Discuss** the causes of severe weather. 9. **Explain** how technology is used to monitor and predict the weather.	National Content Standards: UCP.1–UCP.3, UCP.5, A.1, A.2, B.2, B.3, D.1, F.3		**Science Online,** p. 357 **MiniLAB:** Creating a Low-Pressure Center, p. 359 **Science Online,** p. 361 **Lab:** Interpreting Satellite Images, p. 363 **Lab:** Creating Your Own Weather Station, p. 364 **Science and Society:** How Zoos Prepare for Hurricanes, p. 366

Glencoe Exclusive! TeacherWorks™ All-In-One Planner and Resource Center

Lab Materials	Reproducible Resources	Section Assessment	Technology
Launch Lab: dishwashing liquid, narrow-necked plastic bottle, beakers (2), hot water, ice water	**Chapter *FAST FILE* Resources** Foldables Worksheet, p. 17 Directed Reading Overview, p. 19 Note-taking Worksheets, pp. 33–36	Glencoe's Assessment Advantage	TeacherWorks includes: • Interactive Teacher Edition • Lesson Planner with calendar • Access to all program blacklines • Correlations to standards • Web links
MiniLAB: glass, ice water, tepid water, saucepan, measuring cup	**Chapter *FAST FILE* Resources** Transparency Activity, p. 46 MiniLAB, p. 3 Enrichment, p. 30 Reinforcement, p. 27 Directed Reading, p. 20 **Science Inquiry Labs,** pp. 15–16	**Portfolio** Science Journal, p. 344 **Performance** MiniLAB, p. 347 Applying Skills, p. 347 **Content** Section Review, p. 347	Section Focus Transparency Guided Reading Audio Program Interactive Chalkboard CD-ROM
Need materials? Contact Science Kit at 1-800-828-7777 or www.sciencekit.com on the Internet.	**Chapter *FAST FILE* Resources** Transparency Activity, pp. 47, 49–50 Enrichment, p. 31 Reinforcement, p. 28 Directed Reading, p. 20 Lab Activity, pp. 9–11 **Reading and Writing Skill Activities,** p. 13	**Portfolio** Differentiated Instruction, p. 351 **Performance** Applying Math, p. 353 Applying Skills, p. 355 **Content** Section Review, p. 355	Section Focus Transparency Teaching Transparency Guided Reading Audio Program Interactive Chalkboard CD-ROM
MiniLAB: birthday candle; pie pan or plate; clay; tall, narrow jar; water; penny; match **Lab:** satellite image on lab page **Lab:** peanut butter jar, olive jar, permanent marker, metric ruler, meterstick, confetti, wind vane, compass, coffee can, barometer, thermometer	**Chapter *FAST FILE* Resources** Transparency Activity, p. 48 MiniLAB, p. 4 Enrichment, p. 32 Reinforcement, p. 29 Directed Reading, pp. 21, 22 Lab Worksheets, pp. 5–6, 7–8 Lab Activity, pp. 13–16 **Lab Management and Safety,** p. 77	**Portfolio** Science Journal, p. 359 Cultural Diversity, p. 360 **Performance** MiniLAB, p. 359 Applying Math, p. 362 Lab, p. 363 Lab, p. 364 **Content** Section Review, p. 362	Section Focus Transparency Virtual Labs CD-ROM Guided Reading Audio Program Interactive Chalkboard CD-ROM Video Lab

End of Chapter Assessment

Glencoe's Assessment Advantage

Blackline Masters	Technology	Professional Series
Chapter *FAST FILE* Resources Chapter Review, pp. 39–40 Chapter Tests, pp. 41–44 **Standardized Test Practice,** pp. 51–54	MindJogger Videoquiz Virtual Labs CD-ROM Exam*View*® Pro Testmaker TeacherWorks CD-ROM Interactive Chalkboard CD-ROM	**Performance Assessment in the Science Classroom (PASC)**

Transparencies

Section Focus

Section 1 Section Focus Transparency **A Really Big Lab**

Below is Biosphere 2, a sealed complex that recreates many of Earth's different environments. Scientists are studying how various factors affect ecosystems. One fact they've discovered is that increased levels of carbon dioxide in the air can have a harmful effect on coral reefs.

1. Name some parts of our environment you see in this picture from Biosphere 2.
2. How could Biosphere 2 help scientists plan a space station?

L2

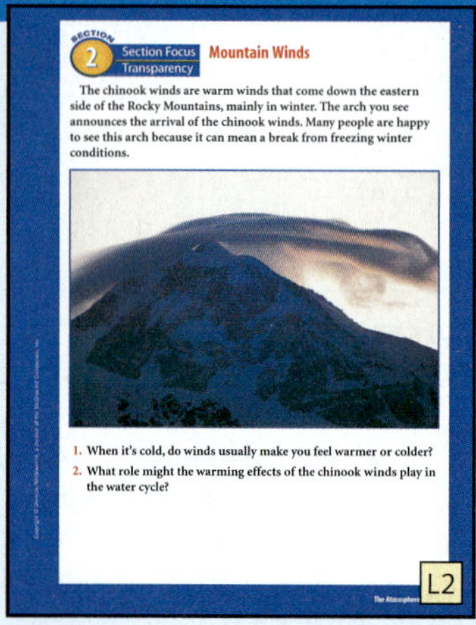
Section 2 Section Focus Transparency **Mountain Winds**

The chinook winds are warm winds that come down the eastern side of the Rocky Mountains, mainly in winter. The arch you see announces the arrival of the chinook winds. Many people are happy to see this arch because it can mean a break from freezing winter conditions.

1. When it's cold, do winds usually make you feel warmer or colder?
2. What role might the warming effects of the chinook winds play in the water cycle?

L2

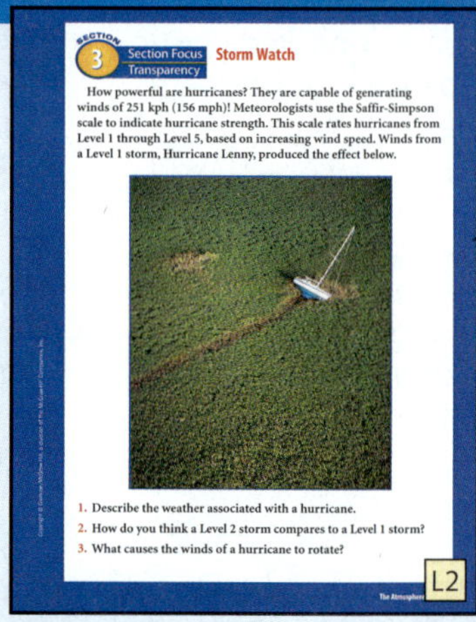
Section 3 Section Focus Transparency **Storm Watch**

How powerful are hurricanes? They are capable of generating winds of 251 kph (156 mph)! Meteorologists use the Saffir-Simpson scale to indicate hurricane strength. This scale rates hurricanes from Level 1 through Level 5, based on increasing wind speed. Winds from a Level 1 storm, Hurricane Lenny, produced the effect below.

1. Describe the weather associated with a hurricane.
2. How do you think a Level 2 storm compares to a Level 1 storm?
3. What causes the winds of a hurricane to rotate?

L2

This is a representation of key blackline masters available in the Teacher Classroom Resources. See Resource Manager boxes within the chapter for additional information.

Assessment

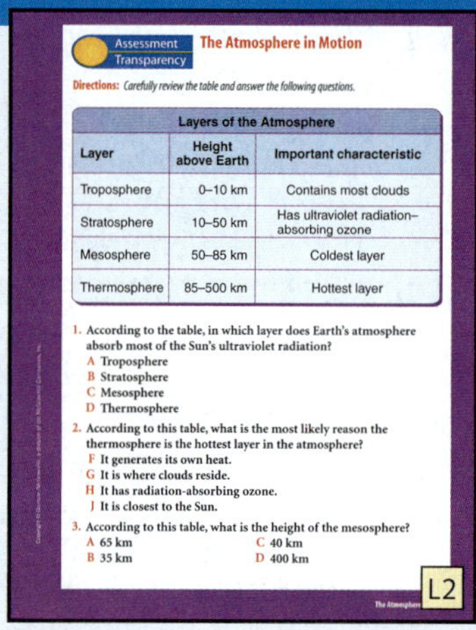
Assessment Transparency **The Atmosphere in Motion**

Directions: *Carefully review the table and answer the following questions.*

Layers of the Atmosphere		
Layer	Height above Earth	Important characteristic
Troposphere	0–10 km	Contains most clouds
Stratosphere	10–50 km	Has ultraviolet radiation–absorbing ozone
Mesosphere	50–85 km	Coldest layer
Thermosphere	85–500 km	Hottest layer

1. According to the table, in which layer does Earth's atmosphere absorb most of the Sun's ultraviolet radiation?
   A Troposphere
   B Stratosphere
   C Mesosphere
   D Thermosphere
2. According to this table, what is the most likely reason the thermosphere is the hottest layer in the atmosphere?
   F It generates its own heat.
   G It is where clouds reside.
   H It has radiation-absorbing ozone.
   J It is closest to the Sun.
3. According to this table, what is the height of the mesosphere?
   A 65 km C 40 km
   B 35 km D 400 km

L2

Teaching

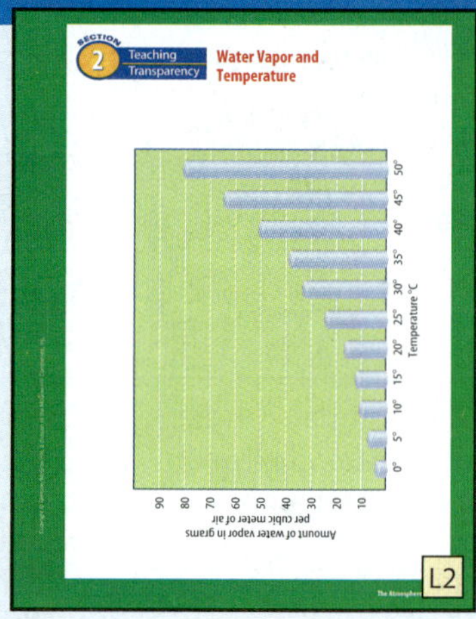
Section 2 Teaching Transparency **Water Vapor and Temperature**

L2

Key to Teaching Strategies

The following designations will help you decide which activities are appropriate for your students.

- L1 Level 1 activities should be appropriate for students with learning difficulties.
- L2 Level 2 activities should be within the ability range of all students.
- L3 Level 3 activities are designed for above-average students.
- ELL ELL activities should be within the ability range of English Language Learners.
- COOP LEARN Cooperative Learning activities are designed for small group work.
- LS Multiple Learning Styles logos, as described on page 12T, are used throughout to indicate strategies that address different learning styles.
- P These strategies represent student products that can be placed into a best-work portfolio.
- PBL Problem-Based Learning activities apply real-world situations to learning.

Hands-on Activities

Student Text Lab Worksheet

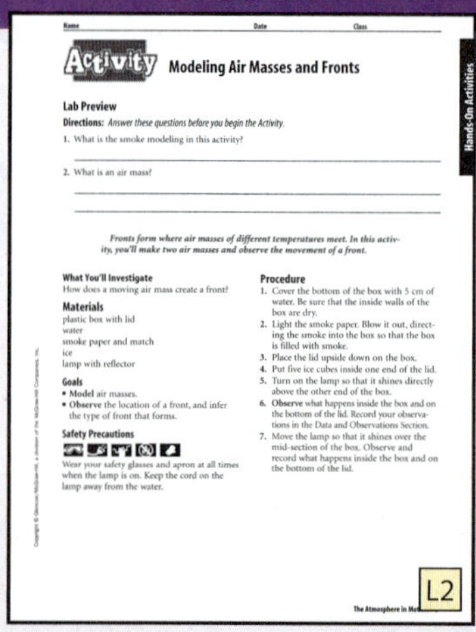
Name Date Class

Activity **Modeling Air Masses and Fronts**

Lab Preview

Directions: *Answer these questions before you begin the Activity.*

1. What is the smoke modeling in this activity?
2. What is an air mass?

Fronts form where air masses of different temperatures meet. In this activity, you'll make two air masses and observe the movement of a front.

What You'll Investigate
How does a moving air mass create a front?

Materials
plastic box with lid
water
smoke paper and match
ice
lamp with reflector

Goals
- **Model** air masses.
- **Observe** the location of a front, and infer the type of front that forms.

Safety Precautions
Wear your safety glasses and apron at all times when the lamp is on. Keep the cord on the lamp away from the water.

Procedure
1. Cover the bottom of the box with 5 cm of water. Be sure that the inside walls of the box are dry.
2. Light the smoke paper. Blow it out, directing the smoke into the box so that the box is filled with smoke.
3. Place the lid upside down on the box.
4. Put five ice cubes inside one end of the lid.
5. Turn on the lamp so that it shines directly above the other end of the box.
6. **Observe** what happens inside the box and on the bottom of the lid. Record your observations in the Data and Observations Section.
7. Move the lamp so that it shines over the mid-section of the box. Observe and record what happens inside the box and on the bottom of the lid.

Hands-On Activities

L2

Laboratory Activities

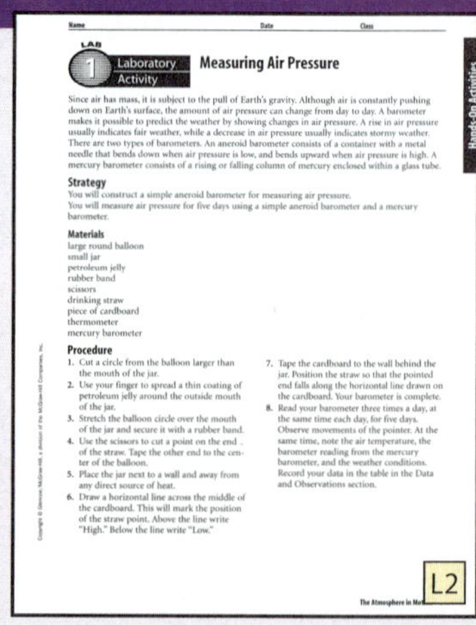
Name Date Class

Lab 1 Laboratory Activity **Measuring Air Pressure**

Since air has mass, it is subject to the pull of Earth's gravity. Although air is constantly pushing down on Earth's surface, the amount of air pressure can change from day to day. A barometer makes it possible to predict the weather by showing changes in air pressure. A rise in air pressure usually indicates fair weather, while a decrease in air pressure usually indicates stormy weather. There are two types of barometers. An aneroid barometer consists of a container with a metal needle that bends down when air pressure is low, and bends upward when air pressure is high. A mercury barometer consists of a rising or falling column of mercury enclosed within a glass tube.

Strategy
You will construct a simple aneroid barometer for measuring air pressure.
You will measure air pressure for five days using a simple aneroid barometer and a mercury barometer.

Materials
large round balloon
small jar
petroleum jelly
rubber band
scissors
drinking straw
piece of cardboard
thermometer
mercury barometer

Procedure
1. Cut a circle from the balloon larger than the mouth of the jar.
2. Use your finger to spread a thin coating of petroleum jelly around the outside mouth of the jar.
3. Stretch the balloon circle over the mouth of the jar and secure it with a rubber band.
4. Use the scissors to cut a point on the end of the straw. Tape the other end to the center of the balloon.
5. Place the jar next to a wall and away from any direct source of heat.
6. Draw a horizontal line across the middle of the cardboard. This will mark the position of the straw point. Above the line write "High." Below the line write "Low."
7. Tape the cardboard to the wall behind the jar. Position the straw so that the pointed end falls along the horizontal line drawn on the cardboard. Your barometer is complete.
8. Read your barometer three times a day, at the same time each day, for five days. Observe movements of the pointer. At the same time, note the air temperature, the barometer reading from the mercury barometer, and the weather conditions. Record your data in the table in the Data and Observations section.

Hands-On Activities

L2

Meeting Different Ability Levels

Content Outline

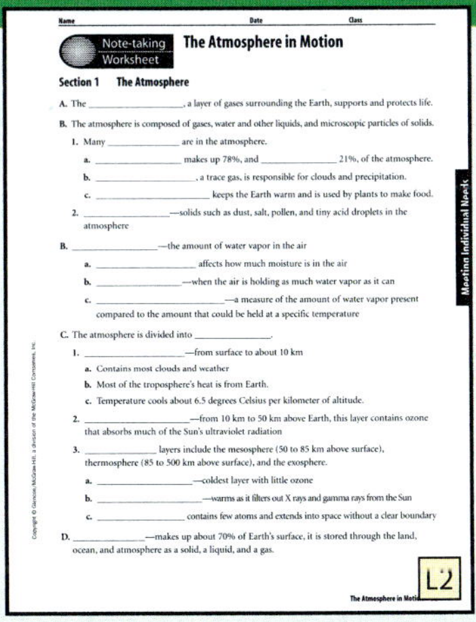

Reinforcement

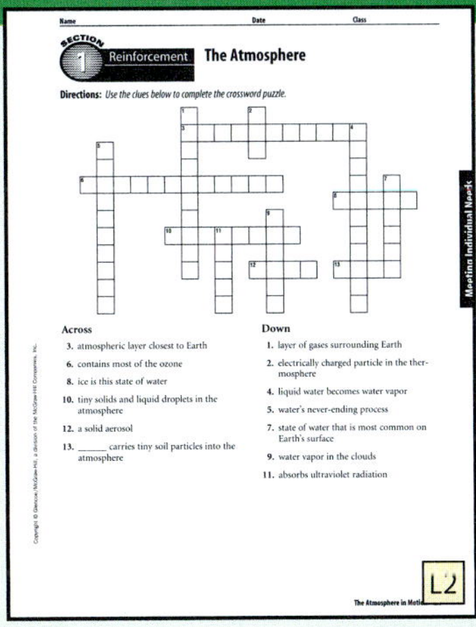

Enrichment

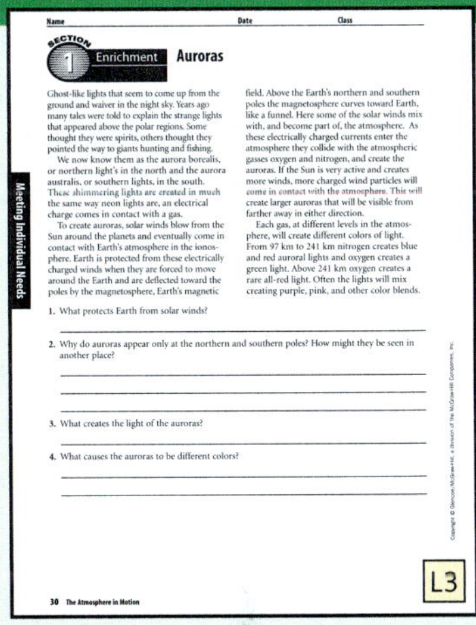

Directed Reading (English/Spanish)

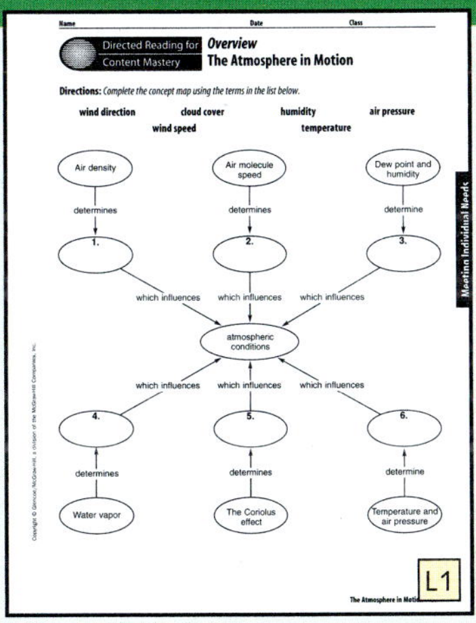

Study Guide

Reading Essentials

Assessment

Test Practice Workbook

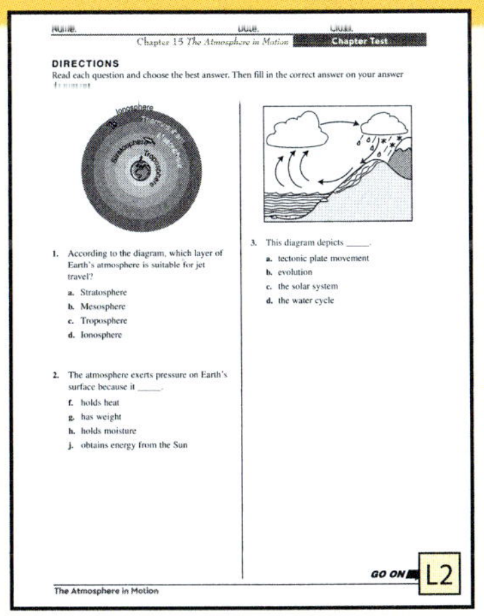

Chapter Review

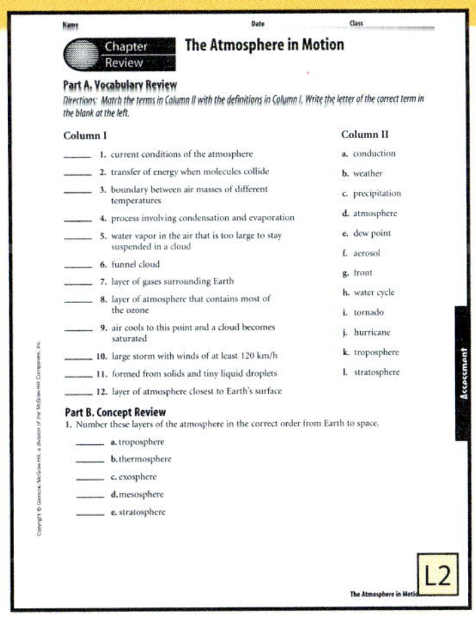

Chapter Tests

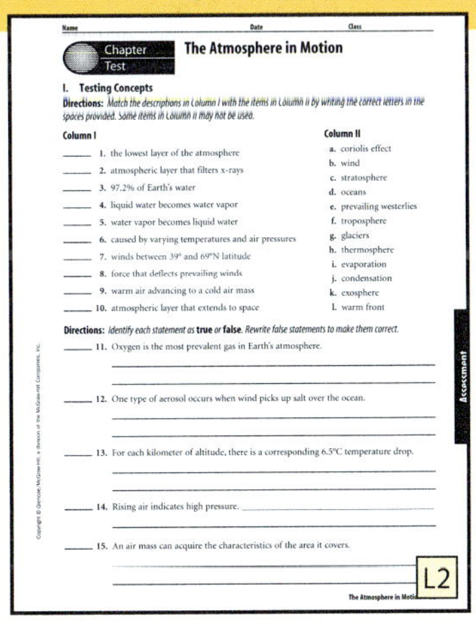

Science Content Background

section 1 The Atmosphere

Layers of the Atmosphere

The first step of ozone formation takes place when high-energy ultraviolet radiation from the Sun reaches Earth's stratosphere. The UV rays break the bond between oxygen atoms that form a molecule of oxygen (O_2). The now unbonded oxygen atoms are highly unstable and slam into other, still intact, molecules of oxygen. Ozone (O_3) is formed as a result of this collision. Compounds released into the atmosphere can interfere with the formation of ozone. Instead of forming a bond with an oxygen molecule, single atoms of oxygen will now bond with a chlorine or bromine atom from these compounds. Ozone-destroying compounds have been released into the atmosphere from some refrigerators and air conditioners and from aerosol sprays.

section 2 Earth's Weather

General Air Circulation

The circulation cells on either side of the equator are called Hadley cells, after George Hadley, who presented a workable model of wind circulation in 1735. The areas of circulation from 30° to 50°–60° latitude in each hemisphere are called Ferrel cells, named for William Ferrel, who further described the details of circulation within each cell in the mid-nineteenth century. The cells that circulate at each respective pole (above or below 50° to 60° latitude) are called polar cells.

Surface Winds

At the boundary between the cells circulating on either side of the equator, surface winds are weak, warm air rises, and atmospheric pressure is low. This area of weak wind was named the doldrums by early sailors. The word itself is derived from the words *dull* and *tantrum*, the implication being that the area was so dull it could induce tantrums. Today being "in the doldrums" has the connotation of feeling depressed and gloomy. Scientists refer to this area as the Intertropical Convergence Zone (ITCZ). The upward movement of warm air, which leads to rain, plays an integral role in the success of tropical rain forests.

Westerlies and Easterlies

Between the edges of Hadley and Ferrel cells at about 30° latitude in each hemisphere, there is another area of weak surface wind. In this area atmospheric pressure is high and air sinks.

chapter content resources

Internet Resources

For additional content background, visit **red.msscience.com** to:

- access your book online
- find references to related articles in popular science magazines
- access Web links with related content background
- access current events with science journal topics

Print Resources

Glossary of Meteorology, 2nd Edition, American Meteorological Society, 2000

Meteorology Today: An Introduction to Weather, Climate, & the Environment by C. Donald Ahrens, West Publishing Company, 2002

A variety of videos, slides, posters, books, calendars, almanacs, and periodicals is available from the National Weather Service. Visit their Web site for more details.

Difficult to move through in a wind-powered vessel, this area was named the horse latitudes by sailors. The air that sinks and dries is responsible for the larger deserts that exist on Earth, including the Sahara, the Gobi and the Kalahari.

section 3

Air Masses and Fronts

Tornadoes

The scale used to classify the strength of tornadoes was developed by meteorology professor Theodore Fujita in 1971. The F-scale ranks tornado strength by the amount of damage caused to manmade structures. Scientists have since correlated damage with estimated wind speeds to further define the scale. The scale has two main drawbacks. The first is that it can be applied only in areas where there are human-made structures. The second is that the amount of damage can be relative. A poorly constructed building can be damaged more easily than a well-constructed building. In spite of this, the F-scale is the official scale used by scientists to classify tornadoes. Weak tornadoes, F0 and F1, may damage chimneys and signs and have winds up to 110 km/h. Strong tornadoes are classified as F2 and F3. Violent tornadoes are rated as F4 and F5. F5 tornadoes have winds up to 480 km/h and the power to completely remove houses from their foundations. Seventy-five percent of tornadoes in the United States are classified as F0 or F1. Only one percent of tornadoes has a rating of F4 or F5.

Hurricanes

The word *cyclone* originates from the Greek word *kyklon*, which literally translates as "an object moving in a circle." A cyclone is system of winds circulating around an area of low atmospheric pressure. In the northern hemisphere, cyclones rotate in a counterclockwise direction. In the southern hemisphere, the rotation is clockwise.

Extratropical cyclones occur at the boundaries between polar cells and Ferrel cells in the mid-latitudes and are usually responsible for producing severe winter weather conditions in these areas in each hemisphere.

The common name for a tropical cyclone depends on the location of the storm. In the North Atlantic and eastern Pacific, tropical cyclones are called hurricanes. In the western Pacific, they are referred to as typhoons, and in the waters around Australia, they are called williwillies. The Saffir-Simpson Scale is used to classify storm strength. Category 5 hurricanes are the strongest, and category 1 hurricanes are the weakest. Fortunately, category 5 storms are rare. Tropical cyclones with winds less than hurricane force are categorized as tropical storms or tropical depressions.

Teacher to Teacher

Tracey L. Smeltzer
Library Specialist
Derry Area High School
Derry, PA

"After discussing atmospheric pressure, have students speculate about the effects that altitude has on an opened carbonated beverage. For example, will an opened container remain carbonated longer in Denver, Colorado or Acapulco, Mexico? Ask students to support their ideas."

Tracey L. Smeltzer

The Saffir-Simpson Scale

Category	Pressure (hPa)	Wind Speed ($m\ s^{-1}$)	Wind Speed ($mi\ h^{-1}$)	Damage
1	Above 980	33–43	74–95	Minimal
2	979–965	43–49	96–110	Moderate
3	964–945	50–58	111–130	Extensive
4	944–920	59–69	131–155	Extreme
5	Below 920	70 or more	156 or more	Catastrophic

chapter 12

Chapter Vocabulary

atmosphere, p. 342
aerosol, p. 343
troposphere, p. 344
water cycle, p. 347
weather, p. 348
humidity, p. 350
dew point, p. 350
relative humidity, p. 350
precipitation, p. 352
air mass, p. 356
front, p. 357
tornado, p. 360
hurricane, p. 361

Science Journal Although fictitious, student newspaper articles should be written in the proper style and should assume a realistic scenario.

INTERACTIVE CHALKBOARD with Image Bank

PowerPoint® Presentations

This CD-ROM is an editable Microsoft® PowerPoint® presentation that includes:

- a pre-made presentation for every chapter
- interactive graphics
- animations
- audio clips
- image bank
- all new section and chapter questions
- Standardized Test Practice
- transparencies
- pre-lab questions for all labs
- Foldables directions
- links to red.msscience.com

chapter 12

The Atmosphere in Motion

chapter preview

sections

1. The Atmosphere
2. Earth's Weather
3. Air Masses and Fronts
   Lab *Interpreting Satellite Images*
   Lab *Creating Your Own Weather Station*

Virtual Lab *How do meteorologists predict the weather?*

Why's it so windy?

Hurricanes are powerful storms that develop over the ocean. If these storms reach land, damage can be severe. Hurricane Gilbert, shown in the photo, caused extensive damage on the island of Jamaica in 1988. It then struck the eastern coast of Mexico.

Science Journal Write a short newspaper article to warn people about the dangers of an approaching hurricane.

Theme Connection

Scale and Structure Scale and structure are emphasized through a discussion of the composition, thickness, and structure of the atmosphere.

About the Photo

Hurricane Gilbert was one of the strongest hurricanes ever seen in the Caribbean and Gulf of Mexico. It reached category 5 status, with winds up to 184 mph and a barometric pressure of 888 mb, giving it the second-lowest pressure ever recorded for an Atlantic hurricane. Gilbert hit Jamaica and Cozumel, Mexico, before tracking northeast to Texas and Oklahoma. It caused $40–50 million of damage.

Start-Up Activities

Launch LAB

How does temperature affect gas molecules?

The temperature of air affects the movement of gas molecules. In the lab below, you will increase and then decrease the temperature of air and observe the changes that occur as a result of the movement of air molecules.

1. With your finger, rub a mixture of water and dish-washing liquid across the top of a narrow-necked plastic bottle until a thin film forms over the opening.
2. Hold the bottle in a beaker that is half-filled with hot water and observe what happens to the soap film.
3. Without breaking the film, remove the bottle from the hot water and place it in a beaker that is half-filled with ice water. Observe what happens to the film.
4. **Think Critically** In your Science Journal, describe what you observed. Infer what happened to change the shape of the film on top of the bottle.

FOLDABLES Study Organizer

Earth's Atmosphere Make the following Foldable to identify what you already know, what you want to know, and what you learned about the atmosphere.

STEP 1 **Fold** a vertical sheet of paper from side to side. Make the front edge about 1.25 cm shorter than the back edge.

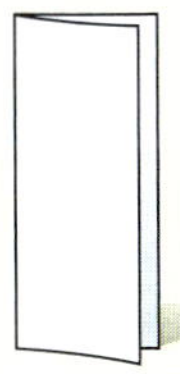

STEP 2 **Turn** lengthwise and **fold** into thirds.

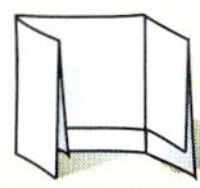

STEP 3 **Unfold and cut** only the top layer along both folds to make three tabs. **Label** each tab as shown.

Read and Write Before you read the chapter, write what you already know and what you want to know about the atmosphere under the tabs. As you read the chapter, write what you learned.

Science Online Preview this chapter's content and activities at red.msscience.com

341

Launch LAB

Purpose Use the Launch Lab to introduce students to the effects of temperature on the movement of air molecules. L2 ELL LS **Visual-Spatial**

Materials narrow-necked bottle, dishwashing liquid, beakers (2), hot water, ice water

Teaching Strategy Advise students to use a small amount of dishwashing liquid. A thin soap film works better than a thick one.

Think Critically

In hot water, the film formed a dome. In ice water, the film flattened. Heated molecules in the bottle had an increased amount of energy. They collided with the film more frequently and exerted more pressure on the film. Cooled molecules in the bottle had less energy. They moved more slowly and had fewer collisions, exerting less pressure on the film.

Assessment

Oral When unconfined air is heated, it expands and becomes less dense. What usually happens to air that is warmer than the surrounding air? The less dense warm air is pushed up (rises) relative to the cooler, surrounding air. Use **Performance Assessment in the Science Classroom**, p. 89.

FOLDABLES Study Organizer — Dinah Zike Study Fold

Student preparation materials for this Foldable are available in the Chapter *FAST FILE* Resources.

Probeware Labs

This chapter's data-collection lab using Probeware technology is included on the *Video Labs CD-ROM*. See the *Probeware Lab Manual* pages 25–27 for student worksheets.

1 Motivate

Bellringer

INTERACTIVE CHALKBOARD PowerPoint® Presentations

Section Focus Transparencies also are available on the Interactive Chalkboard CD-ROM. L2 ELL

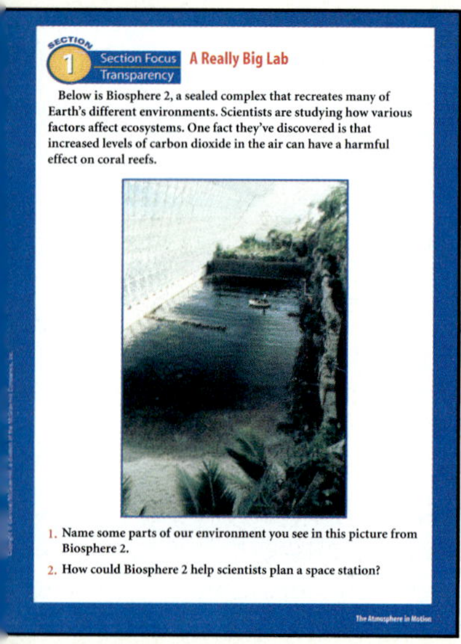
Section 1 Section Focus Transparency A Really Big Lab

Below is Biosphere 2, a sealed complex that recreates many of Earth's different environments. Scientists are studying how various factors affect ecosystems. One fact they've discovered is that increased levels of carbon dioxide in the air can have a harmful effect on coral reefs.

1. Name some parts of our environment you see in this picture from Biosphere 2.
2. How could Biosphere 2 help scientists plan a space station?

The Atmosphere in Motion

Tie to Prior Knowledge

Talking Weather Everyone talks about the weather. Have students discuss the current weather for your area. Explain that in this section, they will learn more about the composition of the atmosphere and the layer of the atmosphere in which weather takes place.

Reading Check

Answer a mixture of gases, liquid water, and microscopic particles of solids and other liquids

section 1

The Atmosphere

as you read

***What* You'll Learn**

- **Explain** why air has pressure.
- **Describe** the composition of the atmosphere.
- **Describe** how energy causes water on Earth to cycle.

***Why* It's Important**

Movements within the atmosphere create weather changes.

Review Vocabulary
air: the mixture of gases that forms Earth's atmosphere

New Vocabulary
- atmosphere
- aerosol
- troposphere
- water cycle

Investigating Air

Air, air . . . everywhere. It's always there. You take it for granted, but without it, Earth would be unfit for life. The **atmosphere**—the layer of gases surrounding Earth—provides Earth with all the gases necessary to support life. It protects living things against harmful doses of ultraviolet and X-ray radiation. At the same time it absorbs and distributes warmth.

Galileo Galilei (1564–1642), an Italian astronomer and physicist, suspected that air was more than just empty space. He weighed a flask, then injected air into it and weighed it again. As shown in **Figure 1,** Galileo observed that the flask weighed more after injecting the air. He concluded that air must have weight and therefore must contain matter. Today scientists know that the atmosphere has other properties, as well. Air stores and releases heat and holds moisture. Because it has weight, air can exert pressure. All of these properties, when combined with energy from the Sun, create Earth's daily weather.

Composition of the Atmosphere

What else do scientists know about the atmosphere? Because it is composed of matter and has mass, it is subject to the pull of gravity. This is what keeps the atmosphere around Earth and prevents it from moving into space. Because it exerts pressure in all directions, you barely notice the atmosphere. Yet its weight is equal to a layer of water more than 10 m deep covering Earth. Scientists also know that the atmosphere is composed of a mixture of gases, liquid water, and microscopic particles of solids and other liquids.

Figure 1 The flask with air injected weighs more than the flask with no air injected.

Reading Check *What is Earth's atmosphere composed of?*

Section 1 Resource Manager

Chapter *FAST FILE* Resources
Transparency Activity, p. 46
Directed Reading for Content Mastery, pp. 19, 20
Note-taking Worksheets, pp. 33–36
Enrichment, p. 30
MiniLAB, p. 3
Reinforcement, p. 27

Science Inquiry Labs, pp. 15–16

Life Science Critical Thinking/Problem Solving, p. 12

Earth Science Critical Thinking/Problem Solving, p. 13

Gases Although the atmosphere contains many gases, two of them make up approximately 99 percent of the total. **Figure 2** shows a graph of the gases found in the atmosphere. Nitrogen (N_2) is the most abundant gas—it makes up about 78 percent of the atmosphere. Oxygen (O_2), the gas necessary for human life, makes up about 21 percent. A variety of trace gases makes up the rest.

Of the trace gases, two have important roles within the atmosphere. Water vapor (H_2O) makes up from 0.0 to 4.0 percent of the atmosphere and is critical to weather. Water in the atmosphere is responsible for clouds and precipitation. Much of the life on Earth depends on water from precipitation. The other important trace gas, carbon dioxide (CO_2), is present in small amounts. Carbon dioxide is needed for plants to make food. Also, carbon dioxide in the atmosphere absorbs heat and emits it back toward Earth's surface, helping keep Earth warm.

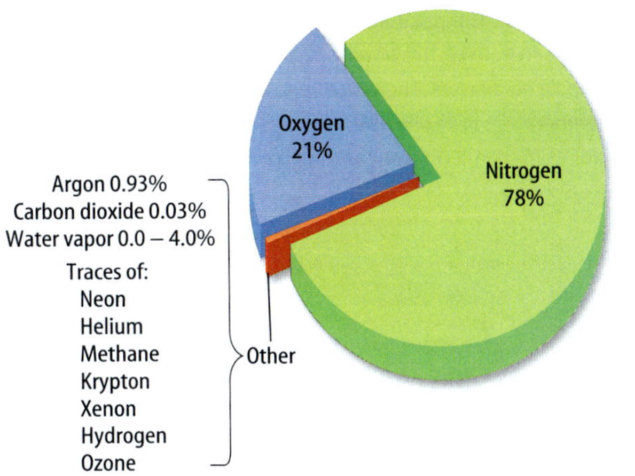

Figure 2 The percentages of gases in the atmosphere vary slightly. For example, water vapor makes up from 0.0 to 4.0 percent of the atmosphere.
Determine *what happens to the percentages of other gases when the percentage of water vapor is higher.*

Aerosols Solids such as dust, salt, and pollen and tiny liquid droplets such as acids in the atmosphere are called **aerosols** (AR uh sahlz). Dust enters the atmosphere when wind picks tiny soil particles off the ground or when ash is emitted from volcanoes. Salt enters the atmosphere when wind blows across the oceans. Pollen enters the atmosphere when it is released by plants. Such human activities as burning coal in power plants also release aerosols into the air. Some aerosols, such as those given off by the volcano in **Figure 3,** reflect incoming solar energy, which can affect weather and climate.

Figure 3 Volcanoes add many aerosols to the atmosphere. Some volcanic aerosols can remain suspended in the atmosphere for months or even years.
Infer *what happens if many aerosols are in the atmosphere.*

2 Teach

Caption Answers

- **Figure 2** They decrease.
- **Figure 3** Temperature decreases.

Visual Learning

Figure 2 Discuss why these data were shown as a circle graph. It is showing percents, and all the percents add up to 100, or one whole. Have students re-graph the data as a bar graph, and then discuss which graph is easier to interpret, and whether a line graph would be appropriate for these data. A line graph would not be useful.

Teacher FYI

Aerosols and Climate When Mount Pinatubo, a volcano in the Philippines, erupted in June of 1991, it created aerosols in the stratosphere. The aerosols quickly spread around the globe. As a result, Earth cooled about 0.5°C.

IDENTIFYING Misconceptions

Particulate Nature of Air Most students have a difficult time understanding the particulate nature of air. Emphasize to students that air consists of molecules that are in constant motion. These molecules are uniformly distributed with the exceptions that the density of air decreases with altitude and with temperature.

Teacher FYI

Greenhouse Gases The amount of carbon dioxide and water vapor in the atmosphere is low compared to that of nitrogen and oxygen, but carbon dioxide and water vapor have great effects on weather and climate. Both are greenhouse gases that trap Earth's heat.

Use an Analogy

Egg Shell Have students think of Earth's atmosphere as being like the shell of a chicken's egg. Both the atmosphere and the eggshell are very thin, protective layers. The eggshell keeps the contents of the egg from drying out, and the pores in the shell allow gases to pass through for the developing embryo. Earth's atmosphere protects living things from harmful radiation and absorbs and distributes heat.

Visual Learning

Figure 4 In which two layers does temperature decrease with an increase in altitude? troposphere and mesosphere

INTEGRATE Health

The Ozone Layer People should avoid being outside from 10:00 A.M. to 4:00 P.M., because that is when exposure to UV radiation is the greatest. If they must be outside, they should cover sensitive skin with clothing, wear a wide-brimmed hat, and use sunscreen that has a Sun Protection Factor (SPF) of at least 15 to protect skin. By wearing 100 percent UV ray protection sunglasses, eyes will be protected.

Reading Check

Answer the atmospheric layer closest to Earth's surface

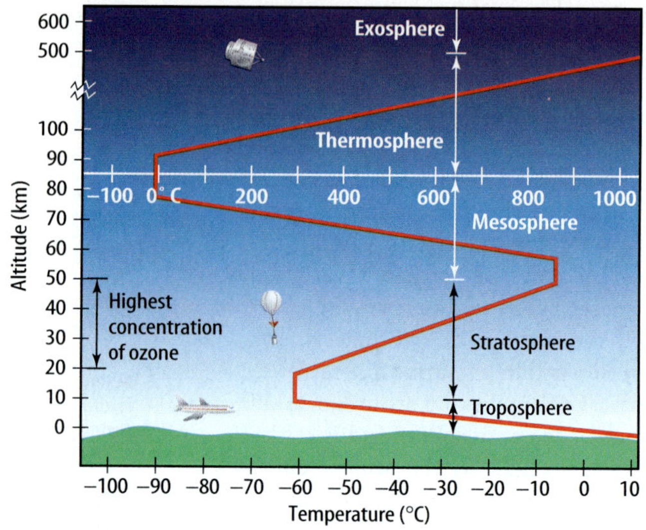

Figure 4 Temperature variations separate Earth's atmosphere into distinct layers. The white temperature scale shows temperatures in the thermosphere and exosphere.

INTEGRATE Health

The Ozone Layer Ozone in the stratosphere shields Earth's surface from the Sun's ultraviolet (UV) radiation. However, scientists have discovered that the ozone layer has been damaged, allowing more UV radiation to reach Earth. This radiation can cause skin cancers and cataracts, which damage vision. What should you do to protect your skin and eyes when you are outdoors?

Layers of the Atmosphere

The atmosphere is divided into the layers that you see in **Figure 4.** These layers are based on temperature changes that occur with altitude. Each atmospheric layer has unique properties. Find each layer as you read about it. The lower layers are the troposphere and stratosphere. The upper layers are the mesosphere, the thermosphere, and the exosphere.

Troposphere The **troposphere** (TROH puh sfihr) is the atmospheric layer closest to Earth's surface. Notice that it extends upward to about 10 km. The troposphere contains about three fourths of the matter in Earth's entire atmosphere and nearly all of its clouds and weather. The atmosphere absorbs some of the Sun's energy and reflects part of it back to space. However, about 50 percent of the Sun's energy passes through the troposphere and reaches Earth's surface. This energy heats Earth. The atmosphere near Earth's surface is heated by the process of conduction. This means that the source of most of the troposphere's heat is Earth's surface. Therefore, temperatures in the troposphere are usually warmest near the surface and tend to cool as altitude increases. Temperatures cool at a rate of about 6.5 Celsius degrees per kilometer of altitude. If you ever climb a mountain, you will notice that it gets colder as you go higher.

Reading Check *What is the troposphere?*

Differentiated Instruction

Challenge Have students research how different layers of the atmosphere were discovered. Have them make oral presentations to their classmates. L3 LS Linguistic

Science Journal

A Trip to Remember Ask students to imagine they are tour guides leading a group of people on a trip to the different layers of the atmosphere. Have them write a story about their journey in their Science Journals. Suggest to students that they be creative in their descriptions, but be sure their information about atmospheric layers is accurate. L2 LS Linguistic P

Stratosphere Above the troposphere is the stratosphere (STRAH tuh sfihr). The stratosphere extends from about 10 km to about 50 km above Earth's surface. As shown in **Figure 4,** most atmospheric ozone is contained in the stratosphere. This ozone absorbs much of the Sun's ultraviolet radiation. As a result, the stratosphere warms as you go upward through it, which is just the opposite of the troposphere. Without the ozone in this layer, too much radiation would reach Earth's surface, causing health problems for plants and animals.

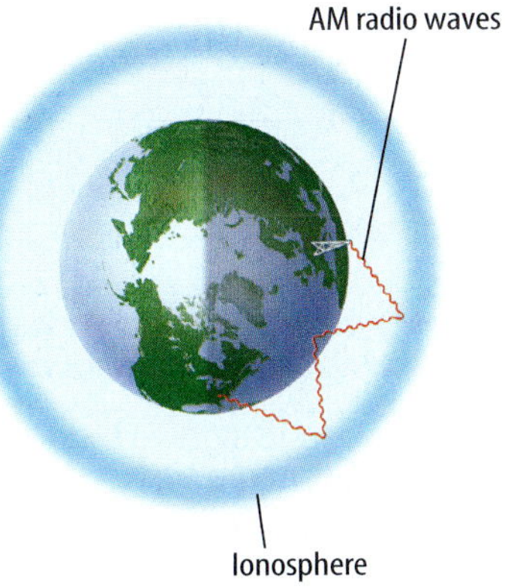

Figure 5 Radio waves are reflected by the ionosphere.

Upper Layers Above the stratosphere is the mesosphere (ME zuh sfihr). This layer extends from approximately 50 km to 85 km above Earth's surface. This layer contains little ozone, so much less heat is absorbed. Notice in **Figure 4** how the temperature in this layer drops to the lowest temperatures in the atmosphere.

The thermosphere (THUR muh sfihr) is above the mesosphere. The thermosphere extends from about 85 km to approximately 500 km above Earth's surface. Temperatures increase rapidly in this layer to more than 1,700°C. The thermosphere layer filters out harmful X rays and gamma rays from the Sun.

Because of intense interaction with the Sun's radiation, atoms can become electrically charged particles called ions. For this reason a part of the thermosphere and mesosphere is called the ionosphere (i AH nuh sfihr). This layer of ions is useful because it can reflect AM radio waves, as shown in **Figure 5,** making long-distance communication possible. If the interaction between the Sun's radiation and this layer is too active, however, the quality of radio reception is reduced. Radio signals break up and a lot of static can be heard.

The outermost layer of the atmosphere is the exosphere. It extends outward to where space begins and contains few atoms. No clear boundary separates the exosphere from space.

Earth's Water

Earth often is referred to as the water planet. This is because Earth's surface is about 70 percent water. Because water can exist in three separate states it can be stored throughout the entire land-ocean-atmosphere system. As **Table 1** shows, water exists as solid snow or ice in glaciers. In oceans, lakes, and rivers water exists as a liquid and in the atmosphere it exists as gaseous water vapor.

Table 1 Distribution of Earth's Water

Location	Amount of Water (%)
Oceans	97.2
Ice caps and glaciers	2.05
Groundwater	0.62
Rivers and lakes	0.009
Atmosphere	0.001
Total (rounded)	**100.00**

IDENTIFYING Misconceptions

Clouds and UV Rays Some students may think that they do not need to be concerned about protecting themselves from ultraviolet radiation on cloudy days. Explain that although clouds absorb some UV radiation, some radiation still passes through.

Quick Demo

Scale of Earth's Atmosphere

Materials spherical balloon

Estimated Time five minutes

Procedure Inflate a spherical balloon. Compare the thickness of the rubber to the thickness of Earth's atmosphere.

Discussion

Exosphere Why is there no clear boundary between the end of the exosphere and the beginning of space? In the exosphere, atoms are widely spaced. At the outer limits of the exosphere, the atoms are so far apart that it is difficult to know where the exosphere ends and space begins.

Differentiated Instruction

English-Language Learners Have English-language learners work with other students to create a poster that shows Earth's atmosphere. The layers of the atmosphere should be labeled, and important features of each layer should be illustrated on the poster.

Active Reading

Speculation about Effects/Prediction Journal Have students divide their papers in half. On the left side record "what happened." On the right side, write "what might/should happen as a result of this." Have students write a Speculation About Effects/Prediction Journal about how addition of carbon dioxide and aerosols might affect the atmosphere.

Visualizing the Water Cycle

Have students examine the pictures and read the captions. Then ask the following questions.

What are some things that could happen to the water that falls as rain? Possible answers: soak into the ground; flow into streams and rivers; be taken up by plants or animals; flow into lakes or oceans

What is the main source of water vapor in the atmosphere? Why? The ocean because it covers so much of Earth's surface that most evaporation occurs there.

Explain the statement, "The water you drink today could be the same water a dinosaur drank 75 million years ago." Water continually cycles on Earth. Earth's water is used over and over again, although it may be present in different forms at different times.

Activity

Water Cycle Poster Have small groups of students work together to make a poster of the water cycle. Tell them to include both human-made features and natural features, such as mountains, rivers, lakes, and forests, near your community. Ask each group to present its poster to the class.

Reading Check

Answer evaporation: liquid water changes to water vapor; condensation: water vapor changes to liquid water

NATIONAL GEOGRAPHIC **VISUALIZING THE WATER CYCLE**

Figure 6

As the diagram below shows, energy for the water cycle is provided by the Sun. Water continuously cycles between oceans, land, and the atmosphere through the processes of evaporation, transpiration, condensation, and precipitation.

▲ Droplets inside clouds join to form bigger drops. When they become heavy enough, they fall as rain, snow, or some other form of precipitation.

▲ As it rises into the air, water vapor cools and condenses into water again. Millions of tiny water droplets form a cloud.

▶ Rain runs off the land into streams and rivers. Water flows into lakes and oceans. Some water is taken up by plants.

▲ Water evaporates from oceans, lakes, and rivers. Plants release water vapor through transpiration.

346

Differentiated Instruction

Challenge Have students write brief reports about how saltwater from the ocean is changed to freshwater by the water cycle. L3

The Water Cycle Earth's water is in constant motion in a never-ending process called the **water cycle**, shown in **Figure 6.** The Sun's radiant energy powers the cycle. Water on Earth's surface—in oceans, lakes, rivers, and streams—absorbs energy and stores it as heat. When water has enough heat energy, it changes from liquid water into water vapor in a process called evaporation. Water vapor then enters the atmosphere.

Evaporation occurs from all bodies of water, no matter how large or small. Have you ever noticed that a puddle of water left on the sidewalk from a rainstorm disappears after a while? The water evaporates into the atmosphere. Water also is transferred into the atmosphere from plant leaves in a process called transpiration. As water vapor moves up through the atmosphere, it becomes cooler. The molecules begin to slow down. Eventually, the water molecules change back into droplets of liquid water. This process is called condensation.

Reading Check *How do evaporation and condensation differ?*

Water droplets grow in size when two or more droplets run into each other and combine to form a larger droplet. Eventually, these droplets become large enough to be visible, forming a cloud. If the water droplets continue to grow, they become too large to remain suspended in the atmosphere and fall to Earth as precipitation. You will learn about the different forms of precipitation in the next section. After it is on the ground, some water evaporates. Most water enters streams or soaks into the soil to become groundwater. Much of this water eventually makes its way back to lakes or to the oceans, where more evaporation occurs and the water cycle continues.

Mini LAB

Observing Condensation and Evaporation

Procedure

1. Fill a **glass** with **ice water.** Make sure that the outside of the glass is dry.
2. Let the glass stand for 10 min and observe what happens on the outside of the glass.
3. Pour 500 mL of **water** into a **shallow pan.**
4. Leave the pan out for several days.
5. Use a **ruler** to measure the amount of water in the pan each day. Record your data.

Analysis

1. Infer why water droplets formed on the the glass.
2. Infer where some of the the water in the pan went.

section 1 review

Summary

Investigating Air

- Air exerts pressure in all directions.

Composition of the Atmosphere

- The atmosphere consists of nitrogen, oxygen, and trace gases, such as water vapor and carbon dioxide.

Layers of the Atmosphere

- The atmosphere is divided into layers based on temperature changes.

Earth's Water

- A model that shows how water cycles is called the water cycle.

Self Check

1. **Explain** why air has pressure.
2. **Identify** three solid particles that occur in the atmosphere.
3. **List** the five layers of Earth's atmosphere starting at Earth's surface.
4. **Describe** four important processes that are part of the water cycle.
5. **Think Critically** Why is it possible for a high mountain at the equator to be covered by snow?

Applying Skills

6. **Recognize Cause and Effect** A closed, metal can collapses when the air is pumped out of it. Explain why.

Science Online red.msscience.com/self_check_quiz

Mini LAB

Purpose to observe condensation and evaporation L2 ELL
LS **Visual-Spatial**

Materials drinking glass, ice water, tepid water, shallow pan, ruler

Teaching Strategy Put the pan in a warm, sunny location.

Analysis

1. Water vapor condensed to form the droplets.
2. Water evaporated into the air.

Assessment

Performance Have students sketch the molecules in water vapor and liquid water. Use **PASC,** p. 127.

3 Assess

DAILY INTERVENTION

Check for Understanding

Kinesthetic Have students design terrariums that will demonstrate the water cycle. L2 COOP LEARN
LS **Interpersonal**

Reteach

Concepts Have students write three questions about section concepts. Use these questions to review the section. L1

Assessment

Oral How do clouds form? As water vapor rises, it cools and condenses into droplets. Use **PASC,** p. 91.

section 1 review

1. Air has weight and therefore exerts pressure.
2. Possible answers: dust, salt and pollen
3. troposphere, stratosphere, mesosphere, thermosphere, exosphere
4. evaporation, transpiration, condensation, precipitation
5. As altitude increases in the troposphere, temperature decreases. If the temperature on a mountain is low enough, snow may accumulate.
6. Whan the can contains air, the pressure inside the can is equal to atmosphereic pressure. When the air is pumped out, the weight of the atmosphere crushes the can.

1 Motivate

Bellringer

INTERACTIVE CHALKBOARD PowerPoint® Presentations

Section Focus Transparencies also are available on the Interactive Chalkboard CD-ROM. L2 ELL

Section 2 Section Focus Transparency — Mountain Winds

The chinook winds are warm winds that come down the eastern side of the Rocky Mountains, mainly in winter. The arch you see announces the arrival of the chinook winds. Many people are happy to see this arch because it can mean a break from freezing winter conditions.

1. When it's cold, do winds usually make you feel warmer or colder?
2. What role might the warming effects of the chinook winds play in the water cycle?

The Atmosphere in Motion

Tie to Prior Knowledge

Changing Weather Have students discuss observations they have made when the weather is changing, such as increase in clouds, temperature changes, or wind changes. Tell them that in this section, they will learn about the many factors that affect weather.

section 2

Earth's Weather

as you read

What You'll Learn

- **Compare** ways that heat is transferred on Earth.
- **Describe** the formation of different kinds of clouds and precipitation.
- **Explain** what causes wind.

Why It's Important

Weather affects your life every day.

Review Vocabulary
cloud: an area in the atmosphere that contains enough water droplets or ice crystals to be visible

New Vocabulary
- weather
- humidity
- dew point
- relative humidity
- precipitation

Weather

Your favorite television show is interrupted by a special weather bulletin. Heavy snow is expected in your area during the night. Will the schools be closed? Will people be able to get to work? How might this weather affect your family? **Weather** describes the current condition of the atmosphere. Factors of weather include temperature, cloud cover, wind speed, wind direction, humidity, and air pressure. It is the task of meteorologists (mee tee uh RAH luh jists) to monitor all weather data continuously in an attempt to forecast weather.

Temperature You learned earlier that the Sun's radiant energy powers the water cycle. In fact, the Sun is the source of almost all of the energy on Earth. When the Sun's rays reach Earth, energy is absorbed. Gas molecules are constantly in motion, but when they absorb more energy, they move faster and farther apart, as shown in **Figure 7.** Temperature is a measure of how fast air molecules are moving. When air molecules are moving rapidly, temperature is high. Temperature is measured with a thermometer that has a particular scale. The Celsius and Fahrenheit scales commonly are used to measure air temperature.

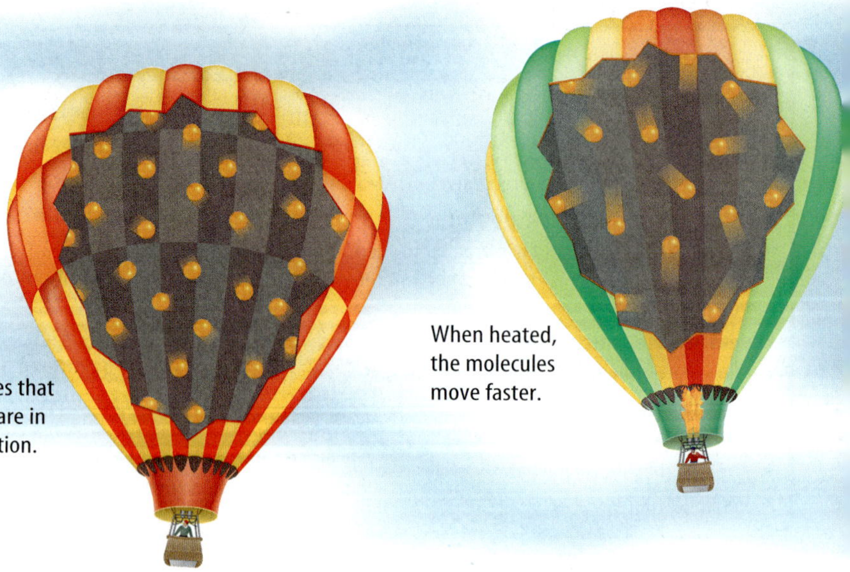

Figure 7 Temperature is a measure of the average movement of molecules. The faster they're moving, the higher the temperature is.

Section 2 Resource Manager

Chapter *FAST FILE* Resources
- Transparency Activity, pp. 47, 49–50
- Directed Reading for Content Mastery, p. 20
- Enrichment, p. 31
- Reinforcement, p. 28
- Lab Activity, pp. 9–11

Reading and Writing Skill Activities, p. 13
Physical Science Critical Thinking/Problem Solving, p. 16
Mathematics Skill Activities, p. 11
Home and Community Involvement, p. 44

Figure 8 Energy from the Sun warms Earth's surface. Conduction and convection transfer heat on Earth.

Energy Transfer Fast-moving molecules transfer energy to slower-moving molecules when they bump into each other. The transfer of energy that results when molecules collide is called conduction. It is conduction that transfers heat from Earth's surface to those molecules in the air that are in contact with it. After it is in the atmosphere, heated air will move upward as long as it is warmer than the surrounding air. The rising air cools as it gets higher. If it becomes cooler than the surrounding air, it will sink. The process of warm air rising and cool air sinking is called convection. It is the main way heat is transferred throughout the atmosphere. Both processes are shown in **Figure 8.**

INTEGRATE Physics

Atmospheric Pressure As you have learned, because of the attraction of gravity, air has weight. Therefore, the weight of air exerts pressure. Air pressure decreases with altitude in the atmosphere. This is because as you go higher, the weight of the atmosphere above you is less.

Temperature and pressure are related. When air is heated, its molecules move faster, and the air expands. This makes the air less dense, which is why heated air gets moved upward. Less dense air also exerts less pressure on anything below it, creating lower pressure. Cooled air becomes more dense and sinks as the molecules slow down and move closer together, creating more pressure. Therefore, rising air generally means lower pressure and sinking air means higher pressure. Air pressure varies over Earth's surface.

INTEGRATE Life Science

Natural Thermometers Crickets chirp more often and rattlesnakes rattle faster when they're warm. How could these animals be used as natural thermometers?

2 Teach

INTEGRATE Life Science

Natural Thermometers By counting the chirps of a cricket or determining the speed of a rattlesnake's rattling at a particular temperature, you could establish a pattern. Explain to students that, over time, careful measurements have been made of the chirping of the snowy tree cricket. By counting the number of chirps in 14 seconds and adding 40, the temperature in Fahrenheit can be determined. Likewise, the frequency of rattling by a rattlesnake has been determined to rise by 1.5 rattles per second per degree Fahrenheit.

Quick Demo

Atmospheric Pressure

Materials five books, six balls of clay, metric ruler

Estimated Time ten minutes

Procedure To demonstrate that air pressure decreases with altitude, use five books and six small clay balls of the same size. Place one ball on a desktop and a book on top of the ball. Place another ball on top of the book. Continue alternating books and clay balls until you have used five balls and five books. The sixth clay ball will be the control. Carefully remove the books and measure the thickness of each ball. The bottom ball will be the flattest and the top ball will be the thickest. Have students relate this to air pressure. Air pressure is greatest near Earth's surface, just as the pressure on the ball at the bottom of the stack was greatest.

Science Journal

Temperature Rising Imagine you are a molecule of alcohol inside a thermometer that is taken from a warm room and put into hot water. Describe what happens to you. Students should describe molecules of alcohol acquiring energy from their surroundings and moving faster and farther apart. This movement causes the alcohol to expand. L2

Activity

Measuring Humidity Have groups of students perform the following activity. Place two thermometers on a table and record the temperature shown on each. Wet a cotton ball with water and place it over the bulb of one of the thermometers. Place a fan on the table so that it blows across the bulbs of both thermometers. Record the temperature shown on each thermometer after five minutes. The thermometer with the wet cotton ball will show a lower temperature because the bulb is cooled as the water evaporates from the cotton. This process can be used to measure the humidity in the air. The less humid the surrounding air, the faster the water from the cotton evaporates. The faster the water evaporates, the lower the temperature shown on the "wet bulb" thermometer. L2 LS **Visual-Spatial**

Reading Check

Answer the amount of water vapor in the air compared to the amount of water vapor that can be present in the air at a specific temperature

Caption Answer

Figure 9 about 32 g/m^3; about 8 g/m^3

Visual Learning

Figure 9 If the temperature of the air is 50°C and 40 g/m^3 of water vapor is present, what is the relative humidity? 50 percent

Science Online

Topic: Relative Humidity

Visit red.msscience.com for Web links to information about relative humidity and dew point.

Activity Do research to find the definition of the word *dew.* What is dew and how is it related to the dew point? Explain what you learn to the class.

Humidity As air warms up, it can cause water that is in contact with it to evaporate to form water vapor. The amount of water vapor in the atmosphere is called **humidity.** The graph in **Figure 9** shows how temperature affects how much moisture can be present in the air. When air is warmer, evaporation occurs more quickly, and more water vapor can be added to the air. More water vapor can be present in warm air than in cool air. When air is holding as much water vapor as it can, it is said to be saturated and condensation can occur. The temperature at which this takes place is called the **dew point.**

Relative Humidity Suppose a mass of air is chilled. The actual amount of water vapor in the air doesn't change unless condensation occurs, but the amount of moisture that can be evaporated into it decreases. **Relative humidity** is a measure of the amount of water vapor that is present compared to the amount that could be held at a specific temperature. As air cools, relative humidity increases if the amount of water vapor present doesn't change. When air is holding all of the water vapor it can at a particular temperature, it has 100 percent relative humidity.

Reading Check *What is relative humidity?*

Sometimes local TV weather reports give the dew point on summer days. If the dew point is close to the air temperature, the relative humidity is high. If the dew point is much lower than the air temperature, relative humidity is low.

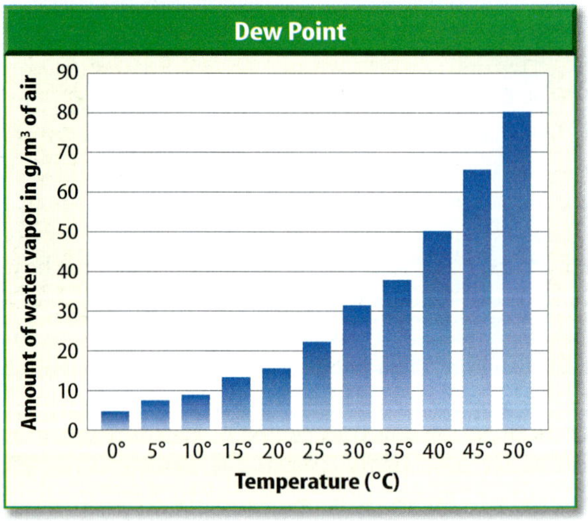

Figure 9 This graph shows how temperature affects the amount of water vapor that air can hold.
Determine *how much water vapor the air can hold if its temperature is 30°C. How much can it hold if the temperature drops to 10°C?*

Curriculum Connection

History During World War II, English airfields were cleared of fog by using petrol burners. This process allowed about 2,500 aircraft to land during foggy weather, and probably saved thousands of Allied airmen's lives. Have students research how the burners worked, as well as other attempts to modify weather such as cloud seeding. The burners warmed the air, thereby reducing condensation. L2

Cultural Diversity

Crab Shells The native people of the Chiloé Islands used crab shells to determine air moisture. They hung the shells on the walls of their huts. When humidity was low, the shell was light gray. As the humidity increased, red splotches formed. When humidity was very high, the shell turned dark red. Have students use a map or atlas to find the Chiloé Islands. They lie off the coast of Chile. L2 LS **Visual-Spatial**

Clouds

One of the best indications that Earth has an atmosphere in motion is the presence of clouds. Clouds form when air rises, cools to its dew point, and becomes saturated. Water vapor in the air then condenses onto small particles in the atmosphere. If the temperature is not too cold, the clouds will be made of small drops of water. If the temperature is cold enough, clouds can consist of small ice crystals.

Clouds commonly are classified according to the altitude at which they begin to form. The most common classification method is one that separates clouds into low, middle, or high groups. Some cloud types are shown in **Figure 10.**

Figure 10 Clouds are grouped according to how high they are above the ground. The types of clouds can be used to predict weather.

Low Clouds The low-cloud group consists of clouds that form at about 2,000 m or less in altitude. These clouds include the cumulus (KYEW myuh lus) type, which are puffy clouds that form when air currents rise, carrying moisture with them. Sometimes cumulus clouds are fair weather clouds. However, when they have high vertical development, they can produce thunder, lightning, and heavy rain. Another type of low cloud includes layered stratus (STRA tus) clouds. Stratus clouds form dull, gray sheets that can cover the entire sky. Nimbostratus (nihm boh STRA tus) clouds form low, dark, thick layers that blot out the Sun. If you see either of these types of clouds, you can expect some kind of precipitation. Fog is a type of stratus cloud that is in contact with the ground.

Middle Clouds Clouds that form between about 2,000 m and 8,000 m are known as the middle-cloud group. Most of these clouds are of the layered variety. Their names often have the prefix *alto-* in front of them, such as altocumulus and altostratus. Sometimes they contain enough moisture to produce light precipitation. Middle clouds can be made up of a mixture of liquid water and ice crystals.

Use an Analogy

Shower "Steam" The "steam" that is observed after taking a shower forms in a way similar to a cloud. Warm water evaporates into the air and moves out of the shower stall and into the cooler air in the room. Since less water vapor can be present in cool air than in warm air, the water vapor condenses. Water dripping off thc ceiling and mirror is similar to precipitation.

Use Science Words

Word Meaning The term *stratus* comes from the words *stratum* and *strata*. Have students use a dictionary to find the meaning of these words. *Stratum* means a bed or layer. *Strata* is the plural form of *stratum.*

Discussion

Clouds and Weather Lead students in a discussion about how different types of clouds produce different weather conditions. For example, good weather occurs with high cirrus clouds or puffy, white cumulus clouds and severe weather with tall, cumulonimbus clouds.

Differentiated Instruction

Learning Disabled Using cotton and other materials, have students make dioramas that depict different types of clouds. Have students label their clouds. L1 LS **Kinesthetic**

Challenge For one week, have students keep a log of the types of clouds they observe. Next to their cloud descriptions, have them record other daily weather observations, such as wind direction and air temperature. L3 LS **Linguistic and Visual-Spatial** P

Teacher FYI

Condensation If air temperature is above 0°C and condensation occurs, water vapor will become liquid water. If the temperature is below 0°C, water vapor can become ice crystals or supercooled liquid. Low clouds tend to be liquid, high clouds are ice, and middle clouds are a mixture of droplets, ice crystals, and supercooled droplets.

Discussion

Frost Is frost that forms on car windshields and grass a kind of precipitation? Explain. No, frost forms when water vapor freezes directly onto objects near the ground. It does not originate as water or ice crystals that have fallen from clouds.

Caption Answer

Figure 11 The winds are very strong.

Visual Learning

Figure 11 How does a hailstone grow? The hailstone falls through the cloud, then updrafts push it to higher elevations. These repeated movements cause the hailstone to become coated with new layers of water that freeze.

High and Vertical Clouds Some clouds occur in air that is so cold they are made up entirely of ice crystals. Because this usually happens high in the atmosphere, these are known as the high-cloud group. They include cirrus (SIHR us) clouds, which are wispy, high-level clouds. Another type is cirrostratus clouds, which are high, layered clouds that sometimes cover the entire sky.

Some clouds can extend vertically throughout all the levels of the atmosphere. These are clouds of vertical development, and the most common type is cumulonimbus (kyew myuh loh NIHM bus). When you see the term *nimbus* attached to a cloud name, it usually means the cloud is creating precipitation. Cumulonimbus clouds create the heaviest precipitation of all. Known as thunderstorm clouds, they start to form at heights of less than 1,000 m but can build to more than 16,000 m high.

Precipitation

When drops of water or crystals of ice become too large to be suspended in a cloud, they fall as **precipitation.** Precipitation can be in the form of rain, freezing rain, sleet, snow, or hail. The type of precipitation that falls depends on the temperature of the atmosphere. For example, rain falls when the temperature of the atmosphere is above freezing. However, if air aloft is above freezing while air near Earth's surface is below freezing, freezing rain might occur. Hail consists of balls of ice that form within cumulonimbus clouds. Within the storm cloud, strong winds toss ice crystals up and down, as shown in **Figure 11.** As the ice crystals move, droplets of water freeze around them. Hailstones keep growing until they are too heavy for the winds to keep up. Then they fall to the ground.

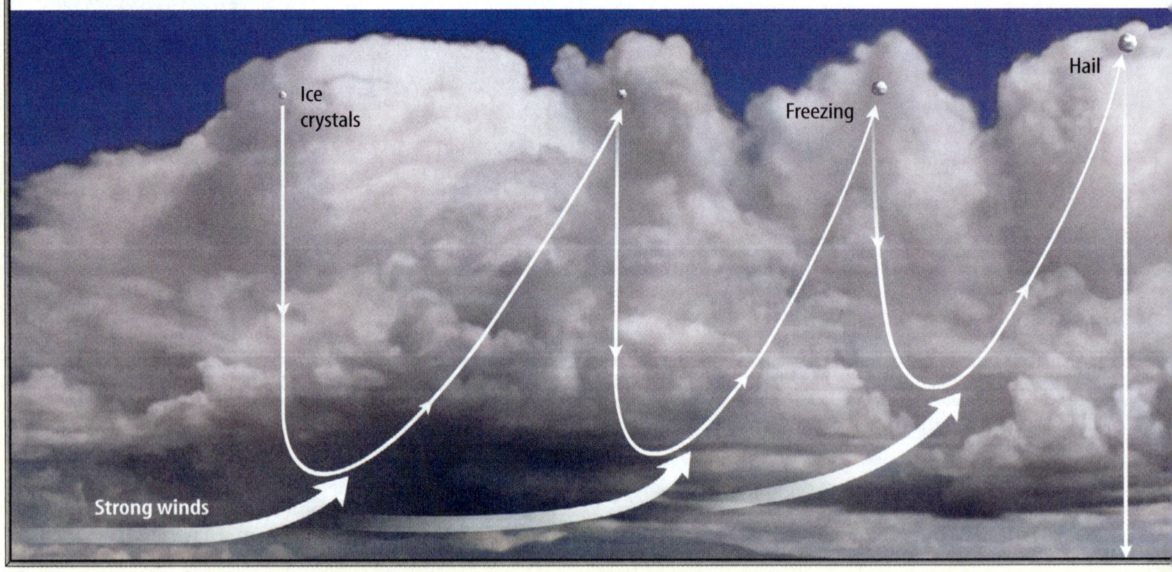

Figure 11 Hailstones develop in cumulonimbus clouds. Most hailstones are the size of peas, but some can reach the size of softballs. **Explain** *what this tells you about the strength of the winds in the cloud.*

Curriculum Connection

Literature In 1803, Luke Howard, an English scientist, developed the cloud classification system we use today. Later, Johann Wolfgang von Goethe, a German poet, wrote four poems about clouds and dedicated them to Howard and his classification system. Have students read Goethe's poems and then write their own poems about clouds. L2 LS **Linguistic**

Differentiated Instruction

Behaviorally Disordered Have students find designs of several kinds of rain gauges. Then have them make a rain gauge and use it to measure precipitation. L2 LS **Kinesthetic**

Wind

As you learned earlier, air pressure is related to temperature. When molecules in the atmosphere are heated, they move more rapidly and spread apart. The air becomes less dense and is moved upward. This causes regions of low air pressure. When cooled, those molecules move more slowly and move closer together. The air becomes more dense and sinks, forming regions of high pressure. Typically, air moves from high-pressure areas toward low-pressure areas. Because pressure and temperature are directly related, wind can be thought of simply as air moving from one temperature or pressure area to another. The greater the difference in temperature or pressure between two areas, the stronger the winds that blow between them will be. Wind speed is measured by an instrument called an anemometer (an uh MAH muh tur), which indicates wind speed by how fast an array of cups that catch the wind rotate. The fastest wind speed ever measured was 371 km/h measured on Mount Washington, New Hampshire, in 1934.

INTEGRATE Social Studies

Indian Monsoon A monsoon is a shift in wind direction that occurs during particular seasons. India is a country that is strongly affected by monsoons. During June and July, low pressure forms over the Indian continent. This causes moist winds to blow from the ocean. These winds produce the heavy rains needed for Indian agriculture. During winter, high pressure forms over India, and dry winds blow from the land to the sea.

Applying Math — Solve a One-Step Equation

WIND SPEED Air moves from an area of high air pressure to an area of low air pressure. The wind that is created travels a distance of 14 km in 2 h. What is the wind speed?

Solution

1. *This is what you know:*
   - distance: $d = 14$ km
   - time: $t = 2$ h
2. *This is what you need to find:* speed (rate): r
3. *This is the procedure you need to use:*
   - substitute into the equation, $r = d/t$
   - $r = 14\text{ km}/2\text{ h} = 7\text{ km/h}$
4. *Check your answer:* Multiply your answer by the time. Do you calculate the same distance that was given?

Practice Problems

1. Air moves from a cool area to a warmer area. The wind that is created moves 20 km in 2 h. What is the wind speed?
2. Air moves from an area of high air pressure to an area of low air pressure. The wind that is created travels a distance of 69 km in 3 h. What is the wind speed?

Science Online — For more practice, visit red.msscience.com/math_practice

INTEGRATE Social Studies

Indian Monsoon Although irrigation has increased in India, agriculture in many parts of the country relies on monsoon rainfall. Planting is scheduled to coincide with the beginning of the wet season.

Inquiry Lab

Measuring Wind Speed

Purpose to design and calibrate a procedure to measure wind speed

Possible Materials pieces of tissue paper, pinwheels, confetti, elastic strings, wire frame, paper plate

Estimated Time one hour to plan; one hour to assemble instruments, five minutes each day for two weeks to calibrate instruments

Teaching Strategies

- To calibrate their instruments, students will need to know the correct wind speed. This can be obtained with an anemometer or wind-speed sensor. It also can be approximated from local weather reports.
- One possible design is to attach a paper plate to a wire frame with elastic string. The stronger the wind, the farther the plate moves from the frame.

For additional inquiry activities, see *Science Inquiry Labs*.

Applying Math

National Math Standards

Correlation to Mathematics Objectives

1, 2, 9

Answers to Practice Problems

1. $r = d/t$; $r = 20\text{ km}/2\text{h}$; $r = 10\text{ km/h}$
2. $r = d/t$; $r = 69\text{ km}/3\text{h}$; $r = 23\text{ km/h}$

Cultural Diversity

Local Winds Air heated by desert sands moves from northern Africa to a low-pressure area in the northern Mediterranean to form siroccos (shi RAH kohz). These winds are so hot that they can cause plants to wither. Have students research other local winds. Examples include: borasco, brickfielder, buran, berg, doctor, foehn, mistral, Santa Ana, and chinook. L2 LS **Linguistic**

✓ Reading Check

Answer the apparent effect that the rotating Earth has on the direction of moving air

Teacher FYI

Doldrums Since air rises at the equator, there often is little wind near 0° latitude. Early sailing ships tried to avoid getting caught in this area, which they called the doldrums. The areas at 30° N and S, where air is sinking, are also relatively windless. These latitudes became known as the horse latitudes. When ships were stranded in these areas, they had to conserve food and water, and they were sometimes forced to dump their horses overboard. The corpses of horses seen floating in the water gave these latitudes their name.

Caption Answer

Figure 13 prevailing westerlies

Visual Learning

Figure 13 Compare the direction that the trade winds blow with the direction the prevailing westerlies blow. The trade winds blow east to west. The prevailing westerlies blow west to east.

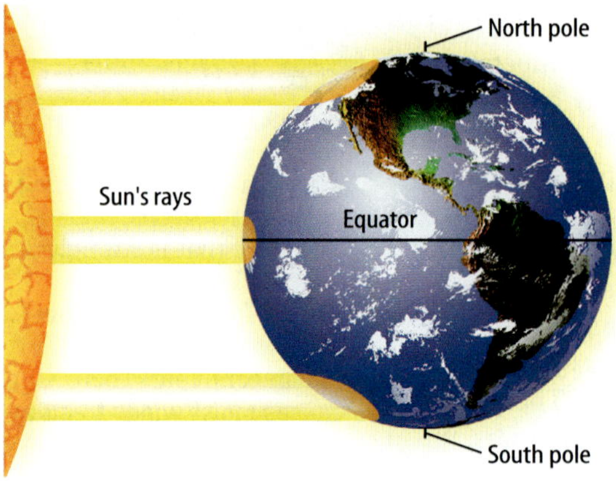

Figure 12 The angle of the Sun's rays is higher near the equator than near the poles.

Global Air Circulation Look at **Figure 12.** In any given year, the Sun's rays strike Earth at a higher angle near the equator than near the poles. As a result, Earth's tropical areas heat up more than the polar regions do. Because of this imbalance of heat, warm air flows toward the poles from the tropics and cold air flows toward the equator from the poles. Because Earth rotates, this moving air is deflected to the right in the northern hemisphere and to the left south of the equator. This is known as the Coriolis (kor ee OH lus) effect.

✓ **Reading Check** *What is the Coriolis effect?*

Surface Winds **Figure 13** shows Earth's major surface winds. Air at the equator is heated by the rays of the Sun. This air expands, becomes less dense, and gets pushed upward. Farther from the equator, at about 30° latitude, the air is somewhat cooler. This air sinks and flows toward the equator. As this air flows, it is turned by the Coriolis effect, creating steady winds called the trade winds. Trade winds also are called tropical easterlies because they blow in a general east-to-west direction. Find the trade winds in **Figure 13.**

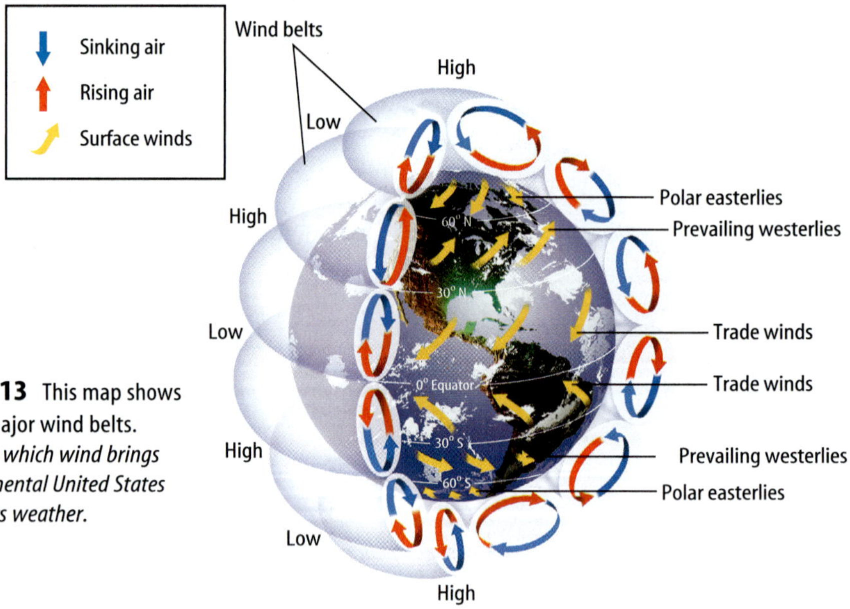

Figure 13 This map shows Earth's major wind belts. **Identify** *which wind brings the continental United States most of its weather.*

LAB DEMONSTRATION

Purpose to demonstrate that the Sun's energy is most concentrated when it is directly overhead

Materials flashlight, white paper, metric ruler, protractor

Preparation Test the flashlight to be sure it produces a bright light.

Procedure Hold a lit flashlight at a 90° angle 15 cm above the paper. Have a student outline the circle of light on the paper. Keeping the light 15 cm above the paper, repeat the process using a 30° angle.

Expected Outcome The light covers a greater area of paper when held at 30°.

Assessment

Which angle produced a more concentrated region of light on the paper? 90° Why are Earth's lower latitudes warmer than its higher latitudes? In these areas, the Sun is directly overhead. Thus, the Sun's energy is most concentrated here.

Westerlies and Easterlies Major wind cells also are located between 30° and 60° latitude north and south of the equator. They blow from the west and are called the prevailing westerlies. These winds form the boundary between cold air from the poles and milder air closer to the equator. Many of Earth's major weather systems form along these boundaries, so these regions are known for frequent storms.

Near the poles, cold, dense air sinks and flows away from the poles. It is replaced by warmer air flowing in from above. As the cold air flows away from the poles, it is turned by the Coriolis effect. These winds, the polar easterlies, blow from the east.

Figure 14 Weather forecasters often show the position of a jet stream to help explain the movements of weather systems.

Jet Streams Within the zone of prevailing westerlies are bands of strong winds that develop at higher altitudes. Called jet streams, they are like giant rivers of air, as shown in **Figure 14.** They blow near the top of the troposphere from west to east at the northern and southern boundaries of the prevailing westerlies. Their positions in latitude and altitude change from day to day and from season to season. Jet streams are important because weather systems move along their paths.

Other Winds Besides the major winds, other winds constantly are forming. Slight differences in pressure create gentle breezes. Great differences create strong winds. The strongest winds occur when air rushes into the center of low pressure. This can cause severe weather like tornadoes and hurricanes.

section 2 review

Summary

Weather

- Weather describes the current condition of the atmosphere.

Clouds

- Clouds are classified according to the altitude at which they form.

Precipitation

- Types of precipitation include rain, freezing rain, sleet, snow, and hail.

Wind

- Air moves as wind because of pressure differences on Earth.

Self Check

1. **Explain** how Earth's surface is heated. How does this affect the troposphere?
2. **Describe** what happens when water vapor rises and cools to the dew point.
3. **Explain** how air pressure is related to temperature.
4. **Define** a jet stream.
5. **Think Critically** Why doesn't precipitation fall from every cloud?

Applying Skills

6. **Compare and contrast** conduction and convection.

Science Online red.msscience.com/self_check_quiz

Fun Fact

Winds in the polar jet stream sometimes exceed 400 km/h.

3 Assess

DAILY INTERVENTION

Check for Understanding

Auditory-Musical Have students write songs about precipitation. Challenge them to include all the different types of precipitation in their songs.

Reteach

Make a Map As a class, make a spider concept map of the factors that affect Earth's weather.
L2 COOP LEARN LS **Visual-Spatial**

Assessment

Process Compare and contrast the prevailing westerlies and jet streams. Jet streams develop at high altitudes within the prevailing westerlies. Both winds blow west to east.

section 2 review

1. by the Sun's radiant energy; it heats the troposphere.
2. It can condense to form droplets of liquid water.
3. As temperature increases, molecules move apart and air pressure decreases. As air cools, molecules move closer together and pressure increases.
4. A jet stream is a band of strong wind that occurs at high altitude within the prevailing westerlies.
5. Precipitation can occur only when drops of water or crystals of ice become too large to be suspended in the cloud.
6. Both are ways that energy is transferred. Conduction results when molecules collide with one another. Convection occurs when warm air rises and cool air sinks.

1 Motivate

INTERACTIVE CHALKBOARD PowerPoint® Presentations

Bellringer

Section Focus Transparencies also are available on the Interactive Chalkboard CD-ROM. L2 ELL

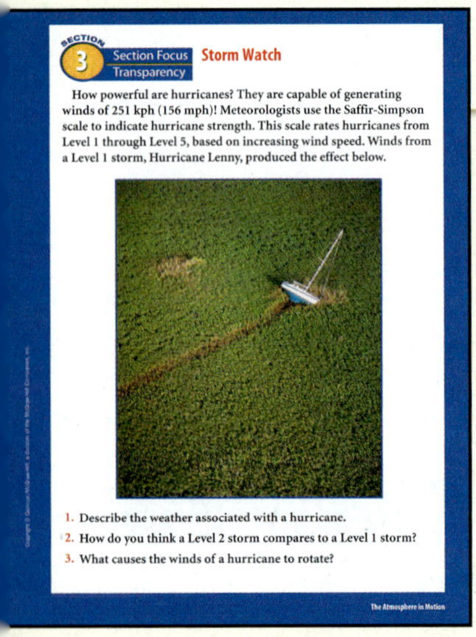
Section 3 Section Focus Transparency **Storm Watch**

How powerful are hurricanes? They are capable of generating winds of 251 kph (156 mph)! Meteorologists use the Saffir-Simpson scale to indicate hurricane strength. This scale rates hurricanes from Level 1 through Level 5, based on increasing wind speed. Winds from a Level 1 storm, Hurricane Lenny, produced the effect below.

1. Describe the weather associated with a hurricane.
2. How do you think a Level 2 storm compares to a Level 1 storm?
3. What causes the winds of a hurricane to rotate?

The Atmosphere in Motion

Tie to Prior Knowledge

Severe Weather Have students recall severe weather events they have experienced. These might include severe thunderstorms, hurricanes, blizzards, or tornadoes. Explain that in this section, they will learn about severe weather and severe weather safety.

Caption Answer

Figure 15 cool and moist

section 3

Air Masses and Fronts

as you read

What You'll Learn

- **Explain** the ways that air masses and fronts form.
- **Discuss** the causes of severe weather.
- **Explain** how technology is used to monitor and predict weather.

Why It's Important

By understanding how weather changes, you can better plan your outdoor activities.

Review Vocabulary

thunderstorm: a storm produced by a cumulonimbus cloud that has lightning and thunder

New Vocabulary

- air mass
- front
- tornado
- hurricane

Air Masses

Weather can change quickly. It can be sunny with calm winds in the morning and turn stormy by noon. Weather changes quickly when a different air mass enters an area. An **air mass** is a large body of air that develops over a particular region of Earth's surface.

Types of Air Masses A mass of air that remains over a region for a few days acquires the characteristics of the area over which it occurs. For example, an air mass over tropical oceans becomes warm and moist. **Figure 15** shows the location of the major air masses that affect weather in North America.

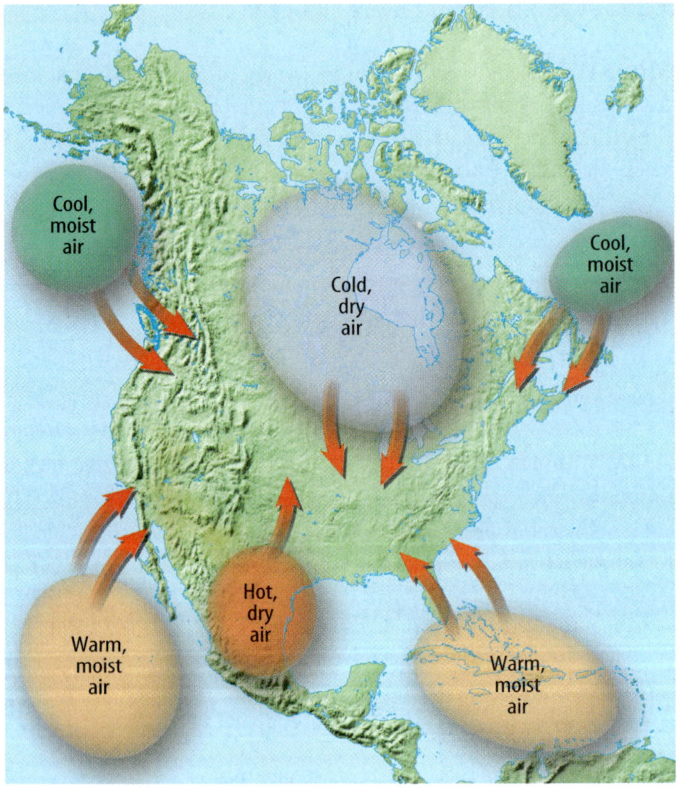

Figure 15 Six major air masses affect North America.
Infer *the characteristics of an air mass that originates over the North Pacific Ocean.*

Section 3 Resource Manager

Chapter *FAST FILE* Resources

Transparency Activity, p. 48
Directed Reading for Content Mastery, pp. 21, 22
MiniLAB, p. 4
Enrichment, p. 32
Reinforcement, p. 29
Lab Worksheet, pp. 5–6, 7–8
Lab Activity, pp. 13–16

Lab Management and Safety, p. 77

356 CHAPTER 12 The Atmosphere in Motion

Fronts

Where air masses of different temperatures meet, a boundary between them, called a **front,** is created. Along a front, the air doesn't mix. Because cold air is more dense, it sinks beneath warm air. The warm air is forced upward and winds develop. Fronts usually bring a change in temperature as they pass, and they always bring a change in wind direction. The four kinds of fronts are shown in **Figures 16, 17,** and **18.**

Reading Check *What is a front?*

Cold Fronts When a cold air mass advances and pushes under a warm air mass, the warm air is forced to rise. The boundary is known as a cold front, shown in **Figure 16.** As water condenses, clouds and precipitation develop. If the air is pushed upward quickly enough, a narrow band of violent storms can result. Cumulus and cumulonimbus clouds can develop. As the name implies, a drop in temperature occurs with a cold front.

Warm Fronts If warm air is advancing into a region of colder air, a warm front is formed. Notice in **Figure 16** that warm, less dense air slides up and over the colder, denser air mass. As the warm air mass moves upward, it cools. Water vapor condenses and precipitation occurs over a wide area. As a warm front approaches, high cirrus clouds are seen where condensation begins. The clouds become progressively lower as you get nearer the front.

Science Online

Topic: Air Masses and Fronts

Visit red.msscience.com for Web links to information about air masses and fronts.

Activity Examine a current weather map. Identify any approaching fronts. Track the changes in temperature, pressure, precipitation, wind direction, and cloud cover as the front passes.

Figure 16 Cold and warm fronts always bring changes in the weather.

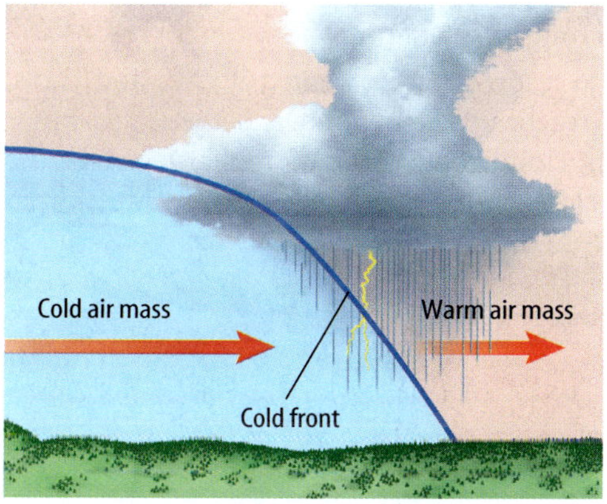

A cold front often produces short periods of storms with heavy precipitation. After the front passes, wind changes direction, skies begin to clear, and the temperature usually drops.

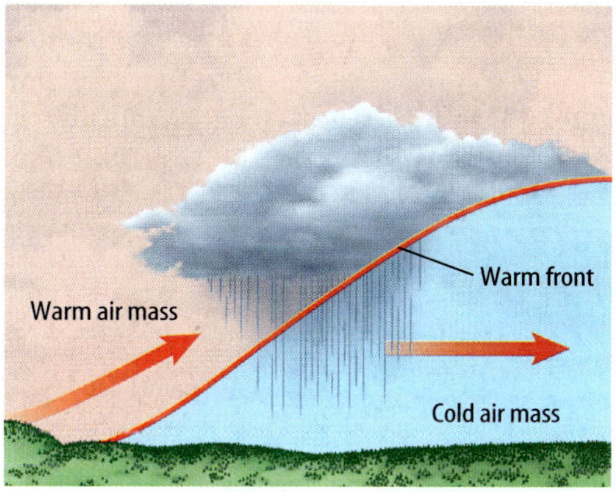

A warm front usually produces a long period of steady precipitation over a wide area. After the front passes, the sky clears, wind direction changes, and the temperature rises.

2 Teach

Use Science Words

Word Origins In battle, a *front* is a zone of conflict between opposing armies. In a study or debate, a *front* is an area of activity. Have students compare a *weather front* to these other fronts. A weather front is a zone of conflict between opposing air masses; there is much weather activity—wind and precipitation—along a weather front.

Reading Check

Answer a boundary where air masses of different temperatures meet

Discussion

Weather Folklore Tell students that one piece of weather folklore states that if you see a halo around the Sun or Moon in cirrus clouds, it will rain soon. Explain why this is usually true. Cirrus clouds are high clouds, seen on clear, sunny days. But these clouds form along the leading edges of warm fronts. When you see cirrus clouds, it indicates that precipitation is going to occur in the near future as the warm front passes.

Differentiated Instruction

Hearing Impaired For several weeks, have students collect weather maps from the newspaper. Have them make a five-second video of each map. Then have them show the completed video of all the map images to the class. Students will be able to observe fronts and other weather phenomena as they travel across the United States. LS **Visual-Spatial**

Discussion

Fronts Why are there clouds and usually precipitation at every type of front? At fronts, warm air is pushed up. The warm air cools, water vapor condenses, and clouds form. If drops or ice crystals become large, precipitation occurs.

Teacher FYI

Anticyclones and Cyclones Another name for a high-pressure center is an anticyclone. A low-pressure center is called a cyclone. Cyclones and anti-cyclones can develop anywhere and move in any direction, but because of the relatively permanent positions of pressure centers in the United States, most cyclones move in a northeasterly direction. Anticyclones usually move in a southeasterly direction. This means that usually a pressure center coming from the northwest will bring fair weather, while one from the southwest will bring stormy weather.

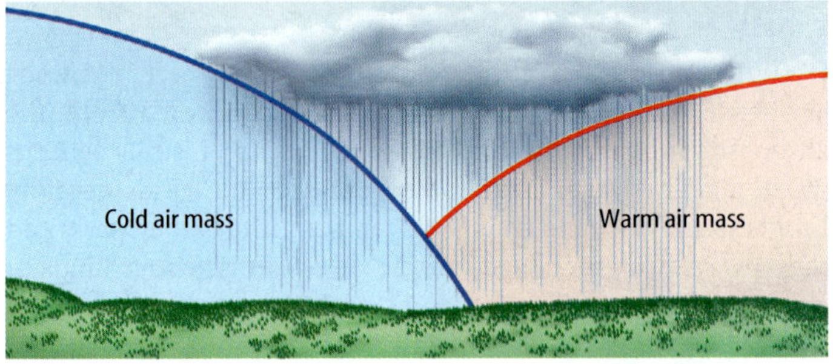

Figure 17 A stationary front can result in days of steady precipitation over the same area.

Stationary Fronts A stationary front, shown in **Figure 17,** is a front where a warm air mass and a cold air mass meet but neither advances. This kind of front can remain in the same location for several days. Cloudiness and precipitation occur along the front. Some precipitation can be heavy because the front moves so little.

Occluded Fronts **Figure 18** illustrates how an occluded front forms when a fast-moving cold front overtakes a slower warm front. Occluded fronts also form in other ways, but all types of occluded fronts can produce cloudy weather with precipitation.

High- and Low-Pressure Centers

In areas where pressure is high, air sinks. As it reaches the ground, it spreads outward away from the high-pressure center. As it spreads, the Coriolis effect turns the air in a clockwise direction in the northern hemisphere. Because the air is sinking, moisture cannot rise and condense, so air near a high-pressure center is usually dry with few clouds.

As air flows into a low-pressure center, it rises and cools. Eventually, the air reaches its dew point and the water vapor condenses, forming clouds and precipitation. Because of the Coriolis effect, air circulates in a counterclockwise direction in the northern hemisphere in a low-pressure center.

Figure 18 An occluded front produces weather similar to, but less severe than, the weather along a cold front.

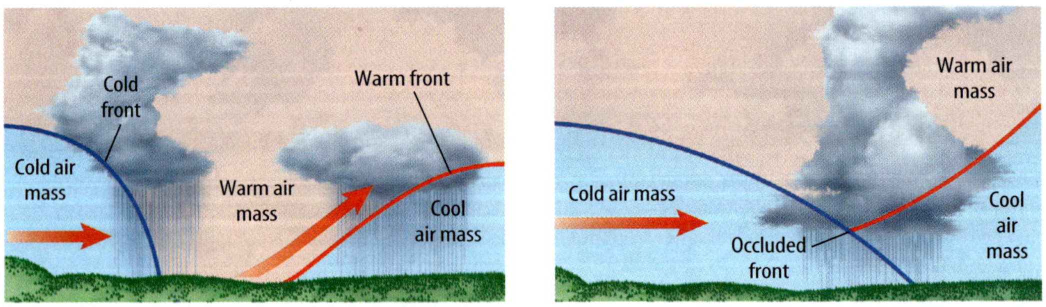

Curriculum Connection

Geography Several pressure centers regularly occur near the United States. The Pacific High is northeast of the Hawaiian Islands. The Azores-Bermuda High is in the middle of the Atlantic. The Aleutian Low is over the Pacific near the Aleutian Islands. The Icelandic Low is over the Atlantic near Iceland. Have students locate these areas on a map. L2 LS **Visual-Spatial**

358 CHAPTER 12 The Atmosphere in Motion

Severe Weather

Severe weather causes strong winds and heavy precipitation. People can be injured and property can be damaged. How can you prepare for severe weather? To prepare, you must first understand it.

Thunderstorms Thunderstorms develop from cumulonimbus clouds. Recall that cumulonimbus clouds often form along cold fronts where air is forced rapidly upward, causing water droplets to form. Falling droplets collide with other droplets and grow bigger. As these large droplets fall, they cool the surrounding air, creating downdrafts that spread out at the surface. These are the strong winds associated with thunderstorms. Dangerous hail can develop in these storms.

Lightning and thunder also are created in cumulonimbus clouds. Where air uplifts rapidly, electric charges form, as shown in **Figure 19.** Lightning is the energy flow that occurs between areas of opposite electrical charge. A bolt of lightning can be five times hotter than the Sun's surface. Its extreme temperature heats the air nearby. The heated air expands faster than the speed of sound, which produces a sonic boom. This is the thunder that is heard after a lightning flash. Close to the lightning, the thunder sounds like a sharp bang. Farther away, the thunder is a dull rumble.

Reading Check *What causes thunder?*

Figure 19 During a thunderstorm, the bottom of the storm cloud has a negative charge. The ground has a positive charge. The negative charge rushes toward the ground. At the same time, the positive charge rushes toward the cloud.

Mini LAB

Creating a Low-Pressure Center

Procedure

1. Fasten a **birthday candle** firmly to the bottom of a **pie pan or plate** with **clay.**
2. Fill a **tall, narrow jar** halfway with **water,** and pour the water into the pan or plate.
3. Light the candle. Invert the jar over the candle. Set the jar mouth down into the water and rest it on a **penny.**
4. In your **Science Journal,** write a brief description of what happens to the water level inside the jar when the candle goes out.

Analysis

1. Infer what happens to the air inside the jar when the candle is lit.
2. Infer what happens to air inside the jar when the candle goes out, and why water rises in the jar when this happens.

Mini LAB

Purpose Students make a model of a low-pressure center. L2 ELL LS **Kinesthetic**

Materials birthday candle, match, pie pan or plate, clay or putty, narrow jar, water, penny

Teaching Strategy A tall olive jar works well in this lab. Attach the candle to the inside bottom of the pie plate.

Safety Precaution Caution students to use care when working with lit candles.

Analysis

1. The heated molecules move faster, increasing the pressure inside the jar. The pressure pushes the water down, and it escapes from the bottom of the jar.
2. The air cools and molecules move less rapidly, reducing the pressure in the jar. Air pressure outside the jar is greater than that inside the jar. The high pressure pushes water molecules up into the jar.

Assessment

Oral Ask students to explain the purpose of the penny in the MiniLAB. It provides a space through which water can move into and out of the jar. Use **Performance Assessment in the Science Classroom,** p. 89.

Reading Check

Answer the rapid expansion of air that has been heated by lightning

Science Journal

Blizzard On March 12, 1888, a low-pressure system developed into a blizzard that dumped between 102 and 127 cm of snow on the northeastern United States. Some snow drifts were as high as 12 m. Four hundred people died from this storm. Have students imagine that they were living during that blizzard. Have them write a short story about their experiences. L2 LS **Linguistic** P

Fun Fact

The average number of tornadoes in the United States each year is 800 to 900. Three-quarters of all the tornadoes on Earth occur in the continental United States.

Make a Model

Tornadoes Students can make model tornadoes using the following procedure: Fill a 2-L bottle about 3/4 full of water and add one drop of dishwashing soap to the water. Place an empty 2-L bottle on top and attach the bottles securely, opening to opening, using duct tape. Flip the bottles to put the one with water in it on top. Move the top bottle in a circular motion. Have students compare and contrast this model of a tornado with a real tornado. L2 ELL LS **Kinesthetic**

Activity

Doppler Radar Have students research the use of Doppler radar for tracking tornadoes. Encourage students to make posters that show how Doppler radar works. L2 LS **Visual-Spatial**

Tornadoes Along some frontal boundaries, cumulonimbus clouds create severe weather. If conditions are just right, updrafts of rising air can start to spin into a rotating vortex. This creates a funnel cloud. **Figure 20** shows the steps in the formation of a funnel cloud. If the funnel cloud reaches Earth's surface, it becomes a tornado like the one shown. A **tornado** is a violent, whirling wind that moves in a narrow path over land. Although tornadoes are usually less than 200 m in diameter, seldom travel on the ground for more than 10 km, and generally last less than 15 min, they are extremely destructive. The powerful updrafts into the low pressure in the center of a tornado act like a giant vacuum cleaner, sucking up anything in its path.

Strong updrafts and downdrafts develop within cumulonimbus clouds when warm, moist air meets cool, dry air.

Winds within the clouds cause air to spin faster and faster.

A funnel of spinning air drops downward through the base of the cloud toward the ground.

Figure 20 A tornado's winds can reach nearly 500 km/h, and it can move across the ground at speeds of up to 100 km/h.

Curriculum Connection

Geography The region in the United States where the most tornadoes occur is called Tornado Alley. This area includes north-central Texas, western Oklahoma, western Kansas, eastern Colorado, Nebraska, western Iowa, southeastern South Dakota and southern Minnesota. Have students locate Tornado Alley on a map. L2 LS **Visual-Spatial**

Cultural Diversity

Tornado Scale T. Theodore Fujita is best known for his studies of downbursts and for developing a tornado damage scale. Have students research the work of Dr. Fujita and study the Fujita-Pearson Tornado Intensity Scale. Then have them write a report that compares the wind speed and damage of the different categories of tornadoes described by the scale. L2 LS **Linguistic** P

360 CHAPTER 12 The Atmosphere in Motion

Figure 21 Hurricanes begin as low-pressure areas over warm oceans.

Air circulation in a hurricane produces updrafts and downdrafts. The downdrafts prevent cloud formation, creating the calm eye of the storm.

As seen from a satellite, the swirling storm clouds of a hurricane are easy to spot.

Hurricanes Unlike tornadoes, hurricanes can last for weeks and travel thousands of kilometers. The diameter of a hurricane can be up to 1,000 km. A **hurricane** is a large storm that begins as an area of low pressure over tropical oceans. The hurricanes that affect the East Coast and Gulf Coast of the United States often begin over the Atlantic Ocean west of Africa. Look at **Figure 21** as you read how a hurricane forms. The Coriolis effect causes winds to rotate counterclockwise around the center of the storm. As the storm moves, carried along by upper wind currents, it pulls in moisture. The heat energy from the moist air is converted to wind. When the winds reach 120 km/h, the low-pressure area is called a hurricane. The sustained winds in a hurricane can reach 250 km/h with gusts up to 300 km/h. **Figure 21** also shows a satellite photo of a hurricane over the ocean.

Sometimes a hurricane spends its entire existence at sea and is a danger only to ships. However, when a hurricane passes over land, high winds, tornadoes, heavy rains, and storm surge pound the affected region. Crops can be destroyed, land flooded, and people and animals killed or injured. After the storm begins traveling over land, however, it no longer has the warm, moist air to provide it with energy, and it begins losing power. Gradually, its winds decrease and the storm disappears.

Science Online

Topic: Monitoring Hurricanes

Visit red.msscience.com for Web links to information about how meteorologists track and monitor hurricanes.

Activity Research the technology that is used to monitor hurricanes and predict where they might reach land. Make a poster with captions that shows how meteorologists use this technology.

Discussion

Southern Hemisphere Hurricanes Infer the direction hurricanes rotate and move in the southern hemisphere. They rotate clockwise. They move eastward in the tropics and turn south/southeast in the middle latitudes.

Teacher FYI

Naming Hurricanes The World Meteorological Organization Region 4 Hurricane Committee names hurricanes in the Atlantic Ocean. The first hurricane of the season begins with the letter "A." Each succeeding storm begins with the next letter of the alphabet. Female and male names are used alternately.

Visual Learning

Figure 21 Why aren't clouds usually found in the eye of a hurricane? Cool, dry, sinking air prevents cloud formation.

Quick Demo

Areas at Risk from Hurricanes

Materials Map of North America

Estimated Time five minutes

Procedure Copy a map of North America. Highlight the east and gulf coasts of the United States and Mexico. Also highlight the west coast of Mexico. Tell students that these areas are at risk from hurricanes.

Virtual Labs

Predictions *How do meteorologists predict the weather?*

Differentiated Instruction

Challenge Have these students research and explain how hurricanes get their energy from warm ocean water. Warm ocean water evaporates, heat is absorbed causing the molecules to move faster. As this air rises, condensation occurs, the kinetic energy is changed to heat, which causes further uplift of air, clouds, and wind. L3

Teacher FYI

Tropical Cyclones In the western Atlantic Ocean and along the eastern coasts of Central America and Mexico, tropical cyclones are called hurricanes. In other parts of the world, they are called typhoons, cyclones, baquiros, and willy-willies.

SECTION 3 Air Masses and Fronts 361

Discussion

Invite a meteorologist to talk to the class about severe weather and weather safety. Instruct students to be prepared with questions for the speaker. LS **Linguistic**

3 Assess

DAILY INTERVENTION

Check for Understanding

Visual-Spatial Collect photographs showing different types of clouds. Challenge students to identify the cloud type in each photo.

Reteach

Weather Review Use weather maps to review fronts, pressure systems, and severe weather. L1 ELL LS **Visual-Spatial**

Assessment

Process Have students make a chart that compares and contrasts cold, warm, stationary, and occluded fronts.

Figure 22 These scientists are placing weather instruments in the path of a tornado. Their research helps forecasters better understand and predict tornadoes.

Weather Safety In the United States, the National Weather Service carefully monitors weather. Using technology such as Doppler radar, shown in **Figure 22,** as well as weather balloons, satellites, and computers, the position and strength of storms are watched constantly. Predicting the movement of storms is sometimes difficult because the conditions that affect them are always changing. If the National Weather Service believes conditions are right for severe weather to develop in a particular area, it issues a severe weather watch. If the severe weather already is occurring or has been indicated by radar, a warning is issued.

Watches and Warnings Watches and warnings are issued for severe thunderstorms, tornadoes, tropical storms, hurricanes, blizzards, and floods. Local radio and television stations announce watches and warnings, along with the National Weather Service's own radio network, called NOAA (NOH ah) Weather Radio.

The best preparation for severe weather is to understand how storms develop and to know what to do during watches and warnings. During a watch, stay tuned to a radio or television station and have a plan of action in case a warning is issued. If the National Weather Service does issue a warning, take immediate action to protect yourself.

section 3 review

Summary

Air Masses

- Air masses have the characteristics of the region where they developed.

Fronts

- Front types include cold, warm, stationary, and occluded.
- Precipitation often occurs along fronts.

High and Low Pressure Centers

- In the northern hemisphere, wind spirals clockwise around high pressure. It spirals counterclockwise around low-pressure regions.

Severe Weather

- If a severe weather warning is issued, take action to protect yourself.

Self Check

1. **Summarize** the characteristics of the four types of fronts.
2. **Explain** why thunderstorms often occur along cold fronts.
3. **Define** a severe weather watch and a severe weather warning.
4. **Explain** why technology is important for forecasting the weather.
5. **Think Critically** Why do southerly winds occur on the trailing side of a high pressure region?

Applying Math

6. **Solve One-Step Equations** Calculate the average speed of a hurricane if it travels 3,500 km in nine days. What is the average speed of a tornado that travels 8 km in 10 min?

Science Online red.msscience.com/self_check_quiz

section 3 review

1. cold front: cold air mass advances into warmer air; warm front: warm air mass advances into cooler air; stationary front: boundary between warm and cold air masses does not move or moves slowly; occluded front: warm air is forced aloft
2. Warm, moist air often is uplifted along cold fronts.
3. A watch indicates that conditions are right for severe weather to develop. A warning indicates that severe weather is occurring.
4. Technology allows forecasters to observe large regions of the atmosphere, collect data quickly, and model future weather patterns.
5. In the northern hemisphere, winds spiral clockwise around high pressure. Therefore, high-pressure regions that move across the United States will have southerly winds on the trailing side.
6. 388.9 km/day; 0.8 km/min

LAB

Interpreting Satellite Images

Satellite images show clouds and weather systems across a large region. In this lab, you'll learn to interpret weather from a satellite image.

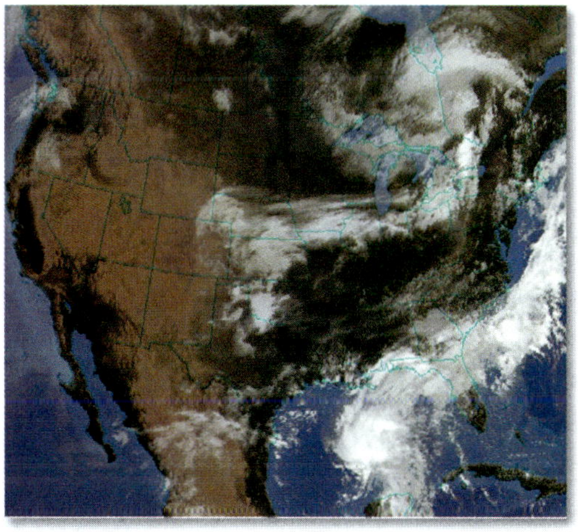

Real-World Question

What can you learn about the weather from satellite images?

Goals

- **Interpret** a satellite image.
- **Predict** future weather from the image.
- **Explain** the advantages of satellite technology for weather forecasting.

Materials

satellite image on this page

Procedure

1. Examine the satellite image shown on this page. Identify the color that represents clouds. What color is ocean water? Where is the United States in this image? Where is your state?
2. Describe which regions in the United States have clear skies. Which regions have cloud cover? How do you know?
3. Locate your town or city on the satellite image. What can you infer about the weather conditions at your location when this satellite image was made?

Conclude and Apply

1. **Identify** A tropical storm, named Bill, can be seen in this satellite image. Where is tropical storm Bill located? Which regions of the United States might be affected by this storm if it is moving toward the north? Will tropical storm watches and warnings be issued?
2. **Locate** A stationary front is causing some precipitation in the Midwest. Locate the stationary front on this map. How do you know its position? List some states that are receiving rainfall from this front. Which regions might receive rainfall tomorrow if the front is moving slowly toward the south?
3. **Observe** Find a region of low pressure in Canada. What shape can you see in the pattern of clouds?
4. **Explain** why satellite images are helpful for weather forecasters. What could you learn from the satellite image in today's newspaper?

BENCH TESTED

Real-World Question

Purpose Students interpret a weather-satellite image. L2 ELL COOP LEARN **Visual-Spatial**

Process Skills observe, record observations, organize data, draw conclusions, predict

Time Required one class period

Procedure

Teaching Strategy Before the activity, discuss satellites and how they acquire data.

1. White represents clouds and blue represents ocean water. The United States fills most of this image.
2. The western states have clear skies and appear brown or green. The central and southeastern states have cloud cover and are covered in white.
3. Student answers will vary based on location.

Conclude and Apply

1. Tropical storm Bill is located in the center of the Gulf of Mexico. The gulf coast states will be affected by this storm. Tropical storm watches and warnings will be issued.
2. A stationary front extends from Detroit, Michigan, through central Kansas, and cloud cover extends along this line. Parts of Wisconsin and Michigan might receive rainfall first, then parts of Illinois, Indiana, and Ohio receive rainfall tomorrow.
3. An area of low pressure with a large counter-clockwise rotational cloud pattern is located northeast of the Great Lakes.
4. They help weather forecasters monitor changes in the position and strength of storms. Students might learn about the movement of weather systems.

Assessment

Performance Have students interpret a current satellite image of the United States and predict tomorrow's weather for your region. Use **Performance Assessment in the Science Classroom,** p. 89.

Differentiated Instruction

Visually Impaired Team visually impaired students with sighted students to create a weather scenario. After a scenario has been described, ask the students to create a raised-relief weather map using classroom materials.

BENCH TESTED

Real-World Question

Purpose Students design and construct their own weather station for collecting data and predicting future weather conditions. L2 LS **Kinesthetic**

Process Skills construct, design a data table, compare, predict

Time Required two 45-minute class periods; 10-minute daily observation times for two weeks

Materials Mount a thermometer and barometer outside your room for students to use for daily measurements. You might consider purchasing a classroom anemometer as well.

Safety Precautions Caution students to remain indoors during electrical storms.

Form a Hypothesis

Possible Hypothesis Most students' hypotheses will reflect a skepticism for predicting future weather patterns, especially long range patterns. Students frequently remember only inaccurate weather predictions by local weather people, not accurate ones.

LAB Design Your Own

CREATING YOUR OWN WEATHER STATION

Goals

- Use weather instruments for measuring air pressure, wind data, temperature, and precipitation.
- **Design** a weather station using your weather instruments.
- **Evaluate** current weather conditions and predict future conditions using your weather station.

Possible Materials

peanut butter jar
olive jar
permanent marker
metric ruler
meterstick
confetti
**shredded tissue paper*
wind vane
anemometer
compass
coffee can
barometer
thermometer
**Alternate materials*

Safety Precautions

Real-World Question

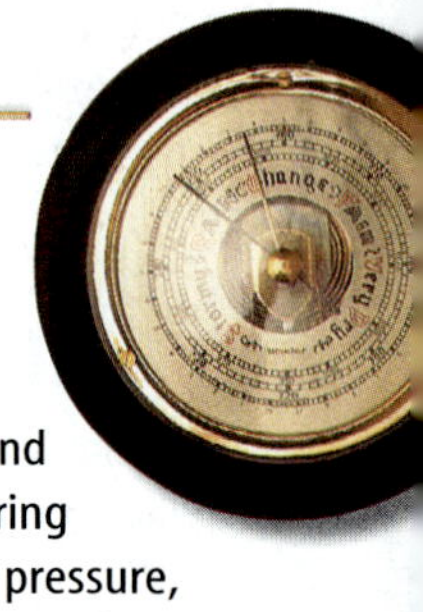

The weather can be very unpredictable. Being able to forecast severe weather such as thunderstorms, tornadoes, and flash floods can save property or lives. Weather stations use instruments to help predict weather patterns. Simple instruments that can be found in a weather station include thermometers for measuring temperature, barometers for observing changes in air pressure, anemometers for measuring wind speed, and rain gauges for measuring precipitation. How can you use weather instruments and design your own weather station to monitor and predict weather conditions?

Form a Hypothesis

Based on your reading in the text and your own experiences with the weather, form a hypothesis about how accurately you could predict future weather conditions using the weather instruments in your weather station.

Test Your Hypothesis

Make a Plan

1. **Decide** on the materials you will need to construct a rain gauge. A wide mouth jar is best for rain, and a small, tall jar is best for accurately measuring the rain collected in the larger jar. Decide how you will mark your jars to measure centimeters of rainfall.
2. To measure wind speed you can use an anemometer or you can make a wind-speed scale. Lightweight materials can be dropped from a specific height, and the distance the wind carries them can be measured with a meterstick. A compass can be used to determine wind direction. A wind vane also can be used to determine wind direction.
3. **Decide** where you will place your thermometer. Avoid placing it in direct sunlight.
4. **Decide** where you will place your barometer.

Alternative Inquiry Lab

Long-Term Forecasts To extend the activity, students might research longer-term climatic conditions, such as the presence of El Niño or La Niña. They then could forecast weather conditions through the upcoming months and compare their forecasts with actual events.

5. Prepare a data table in your Science Journal or on a computer to record your observations.
6. **Describe** how you will use your weather instruments to evaluate current weather conditions and predict future conditions.

Follow Your Plan

1. Ask your teacher to examine your plans and your data table before you start.
2. **Assemble** your weather instruments.
3. Use the weather instruments to monitor weather conditions for several days and to predict future weather conditions.
4. **Record** your weather data.

Analyze Your Data

1. **Compare** your weather data with those given on the nightly news or in the newspaper.
2. How well did your weather equipment measure current weather conditions?
3. How accurate were your weather predictions?
4. **Compare** your barometer readings with the dates it rained in your area. What can you conclude?

Conclude and Apply

1. **Determine** Did the results of your experiment support your hypothesis?
2. **Identify** ways your weather instruments could be improved for greater accuracy.
3. **Predict** how accurate your weather predictions would be if you used your instruments for a year.

Communicating Your Data

Post your weather forecast each day. How does your forecast compare to those of other students? Discuss any differences.

Test Your Hypothesis

Possible Procedures Pour 1 cm of water into the large jar. Transfer this water into the tall jar and mark the water level *1 cm*. Use this height measurement to calibrate the tall jar.

Teaching Strategies Encourage students to examine Beaufort's wind scale as a guide for creating their own wind scale.

Expected Outcome Most students will be able to use their weather station to predict general weather patterns within a 24-hour time period. Predictions beyond 24 hours will be increasingly less accurate.

Analyze Your Data

Answers to Questions

1. Have students record these data in tables. They also might want to calculate the percentage error.
2. Instruments should indicate general conditions fairly well.
3. Short-term predictions will be more accurate than long-term ones.
4. Low barometric readings often accompany rain.

Error Analysis A possible source of measurement error is incorrect calibration of wind speed measuring procedures or rain gauges.

Conclude and Apply

1. Answers will vary.
2. Students should focus on improving the construction of their instruments and their techniques for using them.
3. Predictions should improve as students become more familiar with using the instruments and as they collect more data to use as a predictive guide.

Assessment

Portfolio Ask students to take daily photographs of local weather conditions. Have them use these photographs to make posters of the weather changes during their observation period and to list the weather data collected on the same day beneath each photo. Use **Performance Assessment in the Science Classroom,** p. 145.

Communicating Your Data

Ask students to give an oral presentation explaining their forecasts. They might want to prepare a computer slide show.

TIME

Content Background

In the past, when it came to emergencies, emergency planners and the general public often overlooked animals. Whether the animals are house pets, livestock, zoo animals, or wildlife, the Federal Emergency Management Agency (FEMA) now offers advice to help people prepare and respond to the emergency needs of animals during a disaster.

Discussion

Zoos and Weather How might the kinds of weather disasters that affect zoos vary in different parts of the country? Possible answer: Zoos far inland or on the west coast would not be affected by hurricanes. Almost all zoos would be affected by severe storms and many by tornadoes. Zoos in much of the country would have ice storms or winter blizzards.

Activity

Different Zoos Organize the class into pairs of students. Have students locate a list of zoos in different parts of the United States. Assign each pair of students a zoo in a different region of the country and have them write a letter or contact zoo personnel by e-mail asking about the zoo's plans for weather disasters. Have each pair report what they learn.

Investigate the Issue

Have groups of students list the various kinds of animals that people commonly have as pets and research the needs of each of those animals. Have them record this information and answer the question, "How should the owner of this kind of pet prepare for a weather disaster?"

TIME SCIENCE AND Society

SCIENCE ISSUES THAT AFFECT YOU!

How Zoos Prepare for Hurricanes

Humans aren't the only ones to take cover when hurricanes strike

Flamingos are herded into the zoo's rest room for safety.

As you step into the rest room, you notice a crunch under your feet. When you look up expecting to see sinks and stalls, you see a flock of pink flamingos standing on a bed of straw. What's going on here?

The Miami Metrozoo is preparing for a hurricane, which means herding all of the flamingos into the shelter of the rest room. Why so much fuss?

In 1992, the Metrozoo was devastated by Hurricane Andrew, which killed five mammals and 50 to 75 birds. The zoo, along with many other Florida zoos, has since been forced to rethink how it gets ready for a hurricane and how to deal with its residents after the storm blows over.

Before the Storm

Where do the animals go before a storm at Metrozoo? The lions, tigers, bears, and monkeys are kept in their solid, strong, concrete overnight pens. Poisonous snakes must be bagged because it could be disastrous if they escaped. Other small animals are put into whatever containers can be found, including dog carriers and shipping crates. Some animals are shipped to warehouses or to other zoos that can care for them and are out of the hurricane's path.

Some animals can trust their instincts to tell them what to do. Larger animals may be given the option of coming under shelter or staying out and braving the storm. According to a spokesperson from Seaworld, "The killer whales stay under water longer," which is what they would do in the wild.

Even after the animals are locked up tight, zookeepers worry that the animals could be hurt psychologically by the storm. After Hurricane Andrew, some frightened animals were running around after the storm or just sitting alone. For many zookeepers, the most frustrating thing is being unable to go to an animal and hold it and say, "It's going to be okay."

Make a List List animal safety tips in case of severe weather in your area. What should you have on hand to keep your pets safe? What should you do with your pets during a weather disaster? If you live on a farm, how can you keep the livestock safe?

Science Online
For more information, visit red.msscience.com/time

Make a List Students can contact local veterinarians, state veterinarians, the local animal shelter, animal care and control, the county extension service, local agricultural schools, the American Red Cross, or the Humane Society of the United States to find information.

Resources for Teachers and Students

The Encyclopedia of Survival Techniques, by Alexander Stilwell, Lyons Press, 2000

"Pets and Disasters: Get Prepared," free brochure from The Humane Society of the United States, 2100 L St. NW, Washington, DC 20037

Reviewing Main Ideas

Section 1 The Atmosphere

1. The atmosphere is made of gases, liquids, and solids.
2. The troposphere is warmest near the surface and grows cooler with height. Above the troposphere are four additional layers of the atmosphere, each with different characteristics.
3. Water circulates between Earth's surface and the atmosphere in the water cycle.

Section 2 Earth's Weather

1. Conduction and convection are two ways that heat is distributed on Earth.
2. Precipitation occurs when droplets or ice crystals become too heavy to be supported by the air.
3. Wind is air molecules moving from high-pressure centers to low-pressure centers.

Section 3 Air Masses and Fronts

1. Air masses are dry or moist and warm or cool, depending on where they originate.
2. Fronts develop where air masses of different temperatures collide, forming a boundary. The four kinds of fronts are cold, warm, stationary, and occluded.
3. Severe weather develops from low-pressure centers. Thunderstorms and tornadoes often form near fronts. Hurricanes develop from lows over tropical waters.
4. Knowing what to do when weather watch and warning advisories are made can save your life.

Visualizing Main Ideas

Copy and complete the following concept map about air masses and fronts.

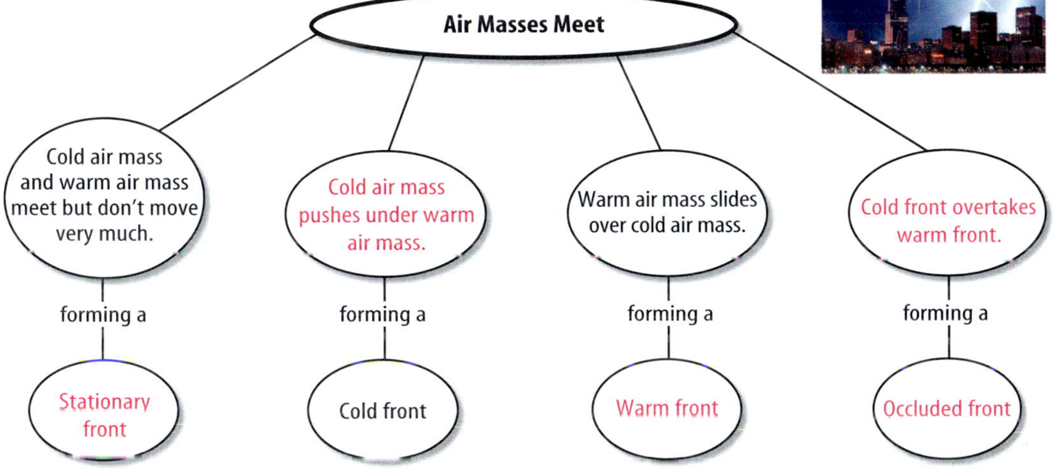

Reviewing Main Ideas

Summary statements can be used by students to review the major concepts of the chapter.

Visualizing Main Ideas

See student page.

Science Online

Visit red.msscience.com
/self_check_quiz
/interactive_tutor
/vocabulary_puzzlemaker
/chapter_review
/standardized_test

Assessment Transparency

For additional assessment questions, use the *Assessment Transparency* located in the transparency book.

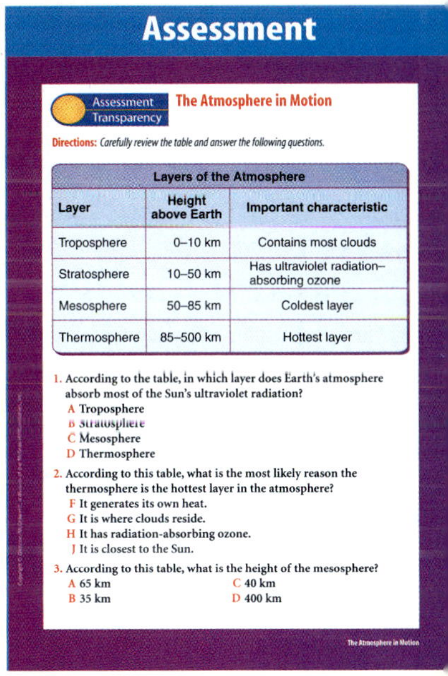

Assessment

Assessment Transparency **The Atmosphere in Motion**

Directions: *Carefully review the table and answer the following questions.*

Layers of the Atmosphere		
Layer	Height above Earth	Important characteristic
Troposphere	0–10 km	Contains most clouds
Stratosphere	10–50 km	Has ultraviolet radiation–absorbing ozone
Mesosphere	50–85 km	Coldest layer
Thermosphere	85–500 km	Hottest layer

1. According to the table, in which layer does Earth's atmosphere absorb most of the Sun's ultraviolet radiation?
   A Troposphere
   B Stratosphere
   C Mesosphere
   D Thermosphere
2. According to this table, what is the most likely reason the thermosphere is the hottest layer in the atmosphere?
   F It generates its own heat.
   G It is where clouds reside.
   H It has radiation-absorbing ozone.
   J It is closest to the Sun.
3. According to this table, what is the height of the mesosphere?
   A 65 km
   B 35 km
   C 40 km
   D 400 km

The Atmosphere in Motion

Using Vocabulary

1. weather
2. front
3. tornado
4. aerosols
5. air mass

Checking Concepts

6. C
7. C
8. B
9. C
10. C
11. D
12. A
13. D
14. B
15. D

Thinking Critically

16. They produce high winds, tornadoes, heavy rain, and storm surges.
17. As you go higher, there is less atmosphere above you. Therefore, the atmospheric pressure is less.
18. During condensation water vapor cools and changes to liquid water. Precipitation is water drops or ice crystals that fall from clouds.
19. They move faster and farther apart.
20. If the temperature at the top of a cloud is low enough, small ice crystals will form. Strong updrafts can produce hailstones. If the hailstones are large, they will not melt completely as they fall through warmer air to the ground. In lower parts of the cloud, melting ice crystals and liquid droplets can fall to produce rain.

chapter 12 Review

Using Vocabulary

aerosol p. 343
air mass p. 356
atmosphere p. 342
dew point p. 350
front p. 357
humidity p. 350
hurricane p. 361
precipitation p. 352
relative humidity p. 350
tornado p. 360
troposphere p. 344
water cycle p. 347
weather p. 348

Fill in the blanks with the correct word or words.

1. The _________ describes the current condition of the atmosphere.
2. The boundary between different air masses is called a(n) _________.
3. A(n) _________ is a violent, whirling wind that forms over land.
4. Dust, salt, pollen, and acid droplets in the atmosphere are called _________.
5. A large body of air that develops over a particular region of Earth's surface is called a(n) _________.

Checking Concepts

6. Which layer of Earth's atmosphere contains the ozone that protects living things from too much ultraviolet radiation?
   A) thermosphere **C)** stratosphere
   B) ionosphere **D)** troposphere
7. Air at 30°C can hold 32 g of water vapor per cubic meter of air. If the air is holding 16 g of water vapor, what is the relative humidity?
   A) 15 percent **C)** 50 percent
   B) 30 percent **D)** 100 percent
8. Which atmospheric layer is farthest from Earth's surface?
   A) troposphere **C)** stratosphere
   B) exosphere **D)** ionosphere

Use the illustration below to answer question 9.

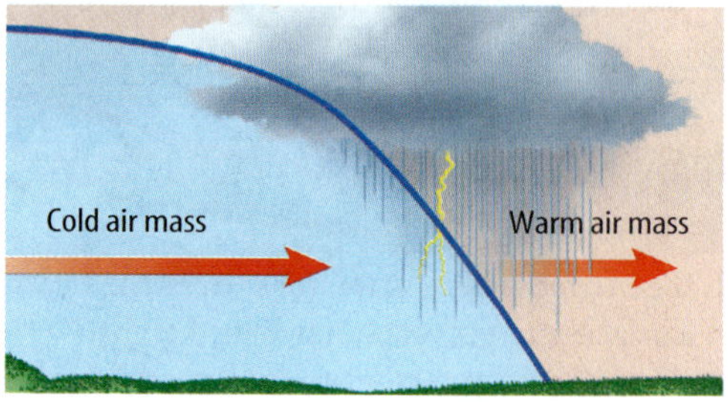

9. Which type of front is shown above?
   A) warm **C)** cold
   B) stationary **D)** occluded
10. What causes low-pressure centers to rotate counterclockwise in the northern hemisphere?
   A) trade winds
   B) prevailing westerlies
   C) Coriolis effect
   D) jet stream
11. Who first proved that air has weight?
   A) Robert Hooke
   B) Evangelista Torricelli
   C) Robert Boyle
   D) Galileo Galilei
12. Which step in the water cycle occurs when water vapor changes to liquid water?
   A) condensation **C)** precipitation
   B) evaporation **D)** transpiration
13. What kind of cloud touches the ground?
   A) altostratus **C)** stratocumulus
   B) stratocirrus **D)** fog
14. Which occurs when colliding molecules transfer energy?
   A) precipitation **C)** radiation
   B) conduction **D)** convection
15. What occurs when strong winds toss ice crystals up and down within a cloud?
   A) rain **C)** snow
   B) freezing rain **D)** hail

Science Online red.msscience.com/vocabulary_puzzlemaker

Use the ExamView® Pro Testmaker CD-ROM to:

- create multiple versions of tests
- create modified tests with one mouse click for inclusion students
- edit existing questions and add your own questions
- build tests aligned with state standards using built-in State Curriculum Tags
- change English tests to Spanish with one mouse click and vice versa

368 CHAPTER REVIEW

Thinking Critically

16. **Explain** why hurricanes are dangerous to people.

17. **Infer** why air pressure is higher at sea level than on top of a mountain.

18. **Compare and contrast** condensation and precipitation.

19. **Describe** what happens to gas molecules when air is heated.

20. **Recognize Cause and Effect** How can a cloud produce both rain and hail?

21. **Concept Map** Copy and complete the concept map of the water cycle below.

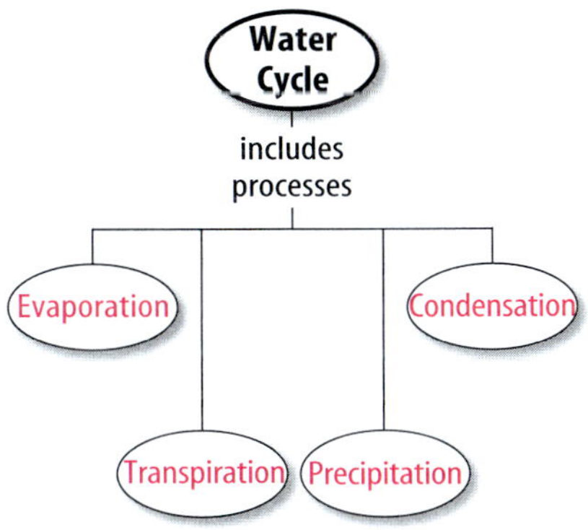

22. **Classify** You observe a tall, dark, puffy cloud. Rain is falling from its lower surface. How would you classify this cloud?

23. **Use Scientific Explanations** Explain why thunder is heard after a flash of lightning.

24. **Venn Diagram** Make a Venn diagram to compare and contrast tornadoes and hurricanes.

25. **Research Information** Do research to learn how sleet forms. Write a paragraph about sleet in your Science Journal.

Performance Activities

26. **Pamphlet** Research three destructive hurricanes and make a pamphlet using the information you collect. Discuss the paths the hurricanes took, how fast they moved, and the damage they caused.

27. **Oral Presentation** Imagine that you work for a television network. Prepare a weather advisory message and announce your watch or warning to the class. Discuss what actions people should take to stay safe.

28. **Poem** Write a poem about the water cycle. Display your poem with those of your classmates on a decorated bulletin board.

Applying Math

Use the equations below to answer questions 29–33.

$$°F = 9/5°C + 32$$

$$°C = 5/9(°F - 32)$$

29. **A Hot Summer Day** The Sun is shining and the temperature is a sweltering 95°F. What is the temperature in degrees Celsius?

30. **A Frigid Winter Morning** The thermometer shows a temperature of −10°C. What's the temperature in degrees Fahrenheit?

31. **A Pleasant Day** A gentle breeze is blowing and the temperature is a comfortable 78°F. What's the temperature in degrees Celsius?

32. **Record Cold** The coldest temperature recorded on Earth occurred at Vostok, Antarctica on July 21, 1983. It was −89.4°C. What was the temperature in degrees Fahrenheit?

33. **Record Heat** The hottest temperature occurred in El Azizia, Libya on September 13, 1922. A scorching 136°F was recorded. What was the temperature in degrees Celsius?

Science Online red.msscience.com/chapter_review

chapter 12 Review

21. See student page.
22. cumulonimbus
23. The air expands rapidly
24. Students should contrast size, longevity, and formation.
25. Sleet forms when ice crystals fall through a layer of warm air and melt. The drops then re-freeze in a layer of cold air near the surface.

Performance Activities

26. Suggest that students show the paths of the hurricanes on a map that includes the Atlantic Ocean, Gulf of Mexico, and United States. Use **Performance Assessment in the Science Classroom,** p. 129.
27. Encourage students to mimic real weather advisories as much as possible.
28. Students' poems should accurately name and describe each stage in the water cycle.

Applying Math

National Math Standards

1, 2, 4, 9

29. 35°C
30. 14°F
31. 25.6°C
32. −129°F
33. 57.8°C

Assessment Resources

Reproducible Masters

Chapter *Fast File* Resources

Chapter Review, pp. 39–40
Chapter Tests, pp. 41–44
Assessment Transparency Activity, p. 51

Glencoe Science Web site

Chapter Review Test
Standardized Test Practice

Glencoe Technology

Assessment Transparency
Exam*View*® Pro Testmaker
MindJogger Videoquiz
Interactive Chalkboard

FAST FILE

Answer Sheet A practice answer sheet can be found at red.msscience.com/answer_sheet.

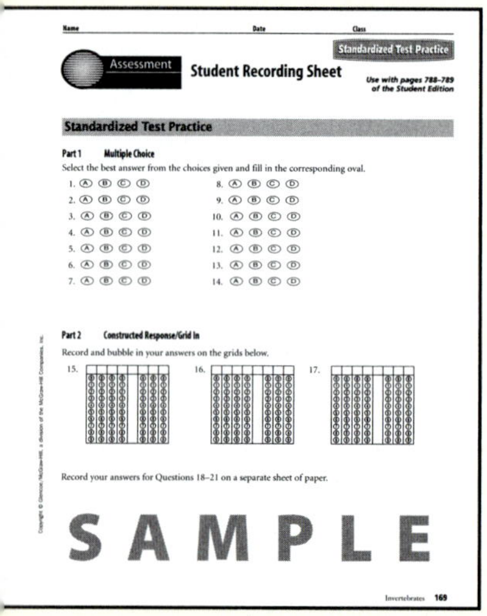

Part 1 Multiple Choice

1. B
2. A
3. B
4. C
5. D
6. B

Part 2 Short Response

7. evaporation and transpiration
8. As the temperature decreases, the amount of water vapor the air can hold decreases.
9. Clouds form when air rises, cools to its dew point, and becomes saturated. Water vapor in the air condenses into small particles in the atmosphere.
10. 31.2 km/h
11. Areas of heat gain include latitudes from 0° to about 30° on either side of the equator. Areas of heat loss are near the poles. The imbalance occurs because the Sun's rays strike Earth at a higher angle near the equator than near the poles. Tropical areas heat up more than polar regions do. Warm air flows toward the poles from the tropics and cold air flows toward the equator from the poles. Arrows should indicate this.
12. cumulonimbus clouds
13. rain, snow, sleet, and hail; When the air is warm, water vapor forms raindrops as it falls. When the air is cold, water vapor forms snowflakes. Sleet, small ice pellets, is formed as raindrops pass through a layer of freezing air near Earth's surface. Hail is formed when water freezes in layers around a small nucleus of ice. Hailstones increase in size as they are tossed up and down by rising and falling air within clouds.

chapter 12 Standardized Test Practice

Part 1 Multiple Choice

Record your answers on the answer sheet provided by your teacher or on a sheet of paper.

Use the table below to answer questions 1 and 2.

Layers of the Atmosphere	
Thermosphere	Filters out gamma rays
Ionosphere	Reflects AM radio waves
Mesosphere	Contains little ozone
Stratosphere	UV radiation-absorbing ozone
Troposphere	Contains most clouds

1. In which layer does Earth's atmosphere absorb most of the Sun's UV radiation?
   - **A.** troposphere
   - **B.** stratosphere
   - **C.** mesosphere
   - **D.** thermosphere

2. Solar radiation causes ions to continually form in Earth's ionosphere. Which has been one of the greatest benefits of ions in the ionosphere?
   - **A.** They reflect radio waves.
   - **B.** They absorb ozone.
   - **C.** They filter out gamma rays.
   - **D.** They form rain clouds.

3. Which of the following is the most abundant gas found in the atmosphere?
   - **A.** oxygen
   - **B.** nitrogen
   - **C.** helium
   - **D.** hydrogen

Test-Taking Tip

Check Your Answers Never leave any answer blank.

4. Which of the following describes a stationary front?
   - **A.** A front formed when warm air advances into a region of colder air.
   - **B.** A front formed when a fast-moving cold front overtakes a slower warm front.
   - **C.** A front formed when a warm air mass and a cold air mass meet but neither advances.
   - **D.** A front formed when a cold air mass advances and pushes under a warm air mass.

Use the graph below to answer question 5.

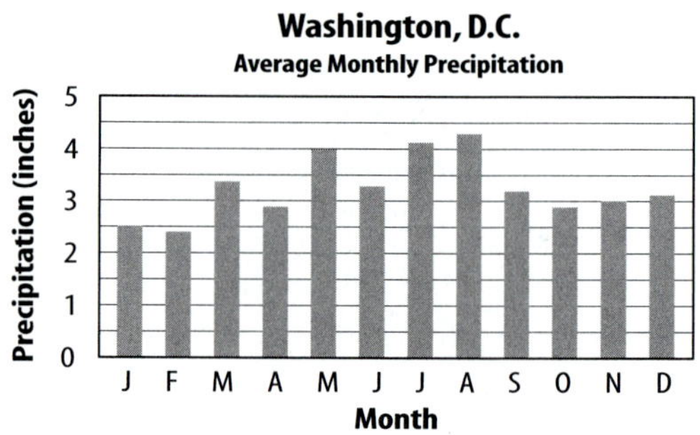

5. During which month did Washington, D.C. receive the most precipitation?
   - **A.** May
   - **B.** June
   - **C.** July
   - **D.** August

6. Which of the following describes humidity?
   - **A.** The measure of the amount of water vapor that is present compared to the amount that could be held at a specific temperature.
   - **B.** The amount of water vapor in the atmosphere.
   - **C.** The temperature at which air is saturated and condensation occurs.
   - **D.** The current condition of the atmosphere.

Part 2 Short Response/Grid In

Record your answers on the answer sheet provided by your teacher or on a sheet of paper.

7. By what processes does water return to atmosphere after precipitation occurs?
8. How does temperature affect the amount of water vapor air can hold?
9. Describe how clouds form.
10. Air moves from an area of high pressure to an area of low pressure. The wind created by this moves 78 km in 2.5 h. What is the wind speed?

Use the figure below to answer question 11.

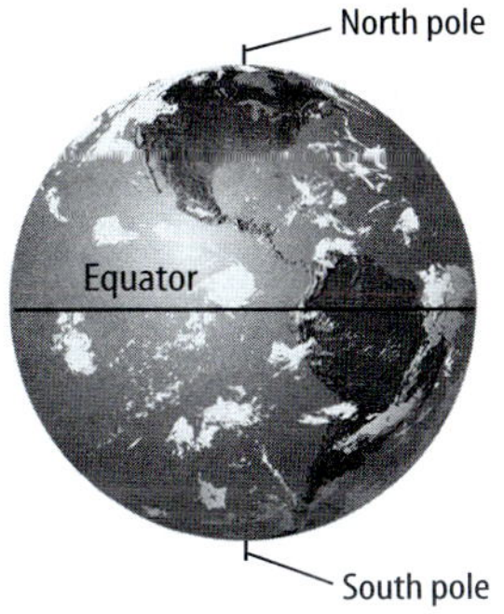

11. Copy the diagram of Earth on a separate sheet of paper. Add labels to the diagram indicating which areas have a heat gain and which areas experience a heat loss. Why does this imbalance occur? Explain what happens to air flow as a result of this heat imbalance? Add arrow labels that indicate where air flows.
12. Your area is experiencing a thunderstorm. What kind of clouds are producing the storm?
13. What are the four main types of precipitation? Describe the differences between each type.

Part 3 Open Ended

Record your answers on a sheet of paper.

14. What are aerosols? Explain how the different aerosols—dust, salt, and pollen—enter the atmosphere.
15. Explain how the trade winds are created.

Use the figure below to answer question 16.

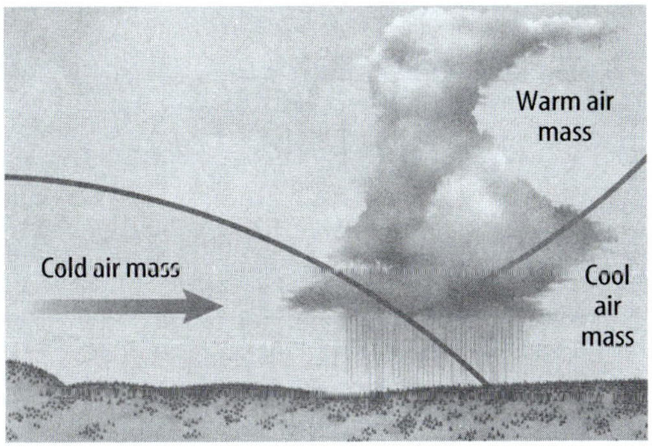

16. What type of front is shown here? How does this type of front form? What kind of weather does this type of front create?
17. Summarize how a tornado forms.
18. Can you have thunder without lightning? Explain why or why not.
19. The morning weather forecast indicates that a cold front is moving toward your area. The current temperature is about 27°C. Predict what changes will occur in the weather as a result of this front.
20. What are weather satellites? How do they help weather forecasters predict the weather?
21. Hurricanes can be dangerous to people and can cause a high amount of property damage when they reach land. Explain why hurricanes are so dangerous and destructive.

Part 3 Open Ended

14. Aerosols are solids such as dust, salt, pollen, and tiny liquid droplets such as acids in the atmosphere. Dust enters the atmosphere when wind picks up tiny soil particles off the ground or when ash is emitted from volcanoes. Salt enters the atmosphere when wind blows across the oceans. Pollen enters the atmosphere when it is released by plants.
15. As air at the equator expands from the Sun's heat, it becomes less dense and gets pushed upward. Farther from the equator, at about 30° latitude, air is cooler. This air sinks and flows toward the equator. As this air flows, it is turned by the Coriolis effect, creating steady trade winds blowing toward the southwest.
16. occluded; Occluded fronts form when a fast-moving cold front overtakes a warm front or when a warm front overtakes a cold front. Occluded fronts produce cloudy weather with precipitation.
17. Strong updrafts and downdrafts develop within cumulonimbus clouds when warm, moist air meets cool, dry air. Winds within the clouds cause air to spin faster and faster. A funnel of spinning air drops downward through the base of the cloud toward the ground. If the funnel cloud reaches Earth's surface, it becomes a tornado.
18. No, thunder results from the rapid heating of air around a bolt of lightning. The intense heat of a lightning bolt causes the air around It to expand rapidly. The rapid movement of molecules forms sound waves heard as thunder.
19. Cumulus and cumulonimbus clouds can develop, a narrow band of violent storms can develop, a drop in temperature occurs as well.
20. Weather satellites are part of the technology used by the National Weather Service to watch and record changes in weather. Predicting the movement of storms is sometimes difficult because the conditions that affect them are always changing. Weather satellites are used to monitor changes in the position and strength of storms.
21. The diameter of hurricanes can be up to 1,000 km. They convert enormous amounts of energy, gathered from moist ocean air, into sustained winds that can reach 250 km/h. Their size, amount of energy, and strong, sustained winds make them dangerous and destructive.

Rubrics

For more help evaluating open-ended assessment questions, see the rubric on p. 10T.

chapter 13 Organizer

Section/Objectives	Standards		Labs/Features
	National	State/Local	
Chapter Opener	See pp. 16T–17T for a Key to Standards.		**Launch Lab:** Why are oceans salty?, p. 373 **Foldables,** p. 373
Section 1 Ocean Water 2 sessions 1 block 1. **State** the importance of Earth's oceans. 2. **Discuss** the origin of the oceans. 3. **Discuss** the composition of seawater. 4. **Explain** how temperature and pressure vary with depth.	National Content Standards: UCP.1–3, UCP.5, A.1, A.2, C.1, D.1		**Science Online,** p. 377 **Lab:** Desalination, p. 379
Section 2 Ocean Currents and Climate 2 sessions 1 block 5. **State** how wind and Earth's rotation influence surface currents. 6. **Explain** how ocean currents affect weather and climate. 7. **Describe** the causes and effects of density currents. 8. **Explain** how upwelling occurs.	National Content Standards: UCP.1–3, UCP.5, A.1, A.2, C.1, D.1		**Science Online,** p. 381 **Integrate Chemistry,** p. 382 **MiniLAB:** Modeling a Density Current, p. 383
Section 3 Waves 2 sessions 1 block 9. **Describe** how wind can form ocean waves. 10. **Explain** the movement of water molecules in a wave. 11. **Describe** how the Moon and Sun cause Earth's tides. 12. **List** the forces that cause shoreline erosion.	National Content Standards: UCP.1–3, UCP.5, A.1, A.2, B.2, C.1, D.1, D.2		**MiniLAB:** Modeling Water Particle Movement, p. 386 **Science Online,** p. 387
Section 4 Life in the Oceans 3 sessions 1.5 blocks 13. **Describe** the characteristics of plankton, nekton, and bottom-dwelling organisms. 14. **Distinguish** among producers, consumers, and decomposers. 15. **Discuss** energy and nutrients cycles in the oceans. 16. **Explain** how ocean organisms interact in food chains.	National Content Standards: UCP.1–3, UCP.5, A.1, A.2, B.3, C.1, C.3, C.4, C.5, D.1, D.2		**Integrate Social Studies,** p. 391 **Science Online,** p. 392 **Applying Science:** Are fish that contain mercury safe to eat?, p. 392 **Visualizing Food Chains in a Food Web,** p. 393 **Lab:** Waves and Tides, p. 396 **Science Stats:** Ocean Facts, p. 398

Glencoe Exclusive! TeacherWorks™ All-In-One Planner and Resource Center

Lab Materials	Reproducible Resources	Section Assessment	Technology
Launch Lab: dry sand, salt, pie pan, water, sprinkling can, bowl, lamp, magnifying lens	**Chapter *FAST FILE* Resources** Foldables Worksheet, p. 17 Note-taking Worksheets, pp. 35–37 Directed Reading Overview, p. 19	Glencoe's Assessment Advantage	**TeacherWorks** includes: • Interactive Teacher Edition • Lesson Planner with calendar • Access to all program blacklines • Correlations to standards • Web links
Lab: large spoon, table salt, water, 250-mL beakers (2), large bowl, plastic wrap, tape, large marble	**Chapter *FAST FILE* Resources** Transparency Activity, p. 46 Enrichment, p. 31 Reinforcement, p. 27 Directed Reading, p. 20 Lab Worksheet, pp. 5–6 Lab Activity, pp. 9–12	**Portfolio** Curriculum Connection, p. 375 **Performance** Applying Skills, p. 378 **Content** Section Review, p. 378	Section Focus Transparency Virtual Labs CD-ROM Guided Reading Audio Program Interactive Chalkboard CD-ROM Video Lab
MiniLAB: paper cup, water, salt, food coloring, spoon, clear-plastic box, thumbtacks, metric ruler	**Chapter *FAST FILE* Resources** Transparency Activity, p. 47 MiniLAB, p. 3 Enrichment, p. 32 Reinforcement, p. 28 Directed Reading, p. 20 **Earth Science Critical Thinking/ Problem Solving,** p. 16	**Portfolio** Cultural Diversity, p. 383 **Performance** MiniLAB, p. 383 Applying Skills, p. 384 **Content** Section Review, p. 384	Section Focus Transparency Guided Reading Audio Program Interactive Chalkboard CD-ROM
MiniLAB: bowl, water, penny, toothpick, spoon	**Chapter *FAST FILE* Resources** Transparency Activity, p. 48 MiniLAB, p. 4 Enrichment, p. 33 Reinforcement, p. 29 Directed Reading, p. 21 Lab Activity, pp. 13–16	**Portfolio** Science Journal, p. 386 **Performance** Applying Skills, p. 388 **Content** Section Review, p. 388	Section Focus Transparency Guided Reading Audio Program Interactive Chalkboard CD-ROM
Lab: large basin; water; boards; sand; bricks; rocks; plastic bottles; battery-operated fan	**Chapter *FAST FILE* Resources** Transparency Activity, p. 49 Enrichment, p. 34 Reinforcement, p. 30 Directed Reading, pp. 21, 22 Lab Worksheet, pp. 7–8 Transparency Activity, pp. 51–52 **Lab Management and Safety,** p. 73	**Portfolio** Make a Model, p. 391 **Performance** Applying Science, p. 392 Applying Skills, p. 395 **Content** Section Review, p. 395	Section Focus Transparency Teaching Transparency Guided Reading Audio Program Interactive Chalkboard CD-ROM

End of Chapter Assessment

Glencoe's Assessment Advantage

Blackline Masters	Technology	Professional Series
Chapter *FAST FILE* Resources Chapter Review, pp. 39–40 Chapter Tests, pp. 41–44 **Standardized Test Practice,** pp. 55–58	MindJogger Videoquiz Virtual Labs CD-ROM Exam*View*® Pro Testmaker TeacherWorks CD-ROM Interactive Chalkboard CD-ROM	**Performance Assessment in the Science Classroom (PASC)**

chapter 13 Oceans

Transparencies

Section Focus

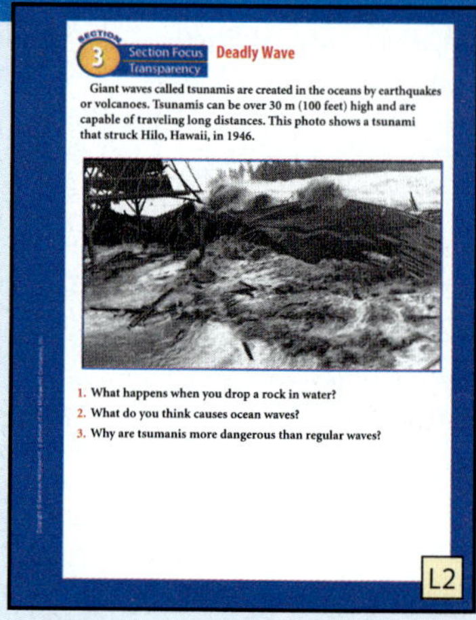

This is a representation of key blackline masters available in the Teacher Classroom Resources. See Resource Manager boxes within the chapter for additional information.

Assessment

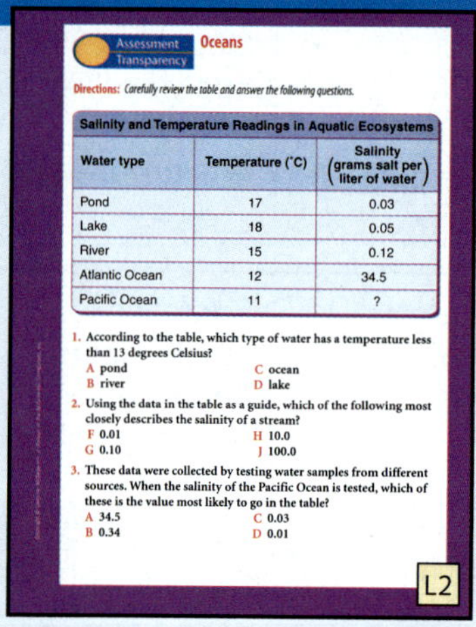

Teaching

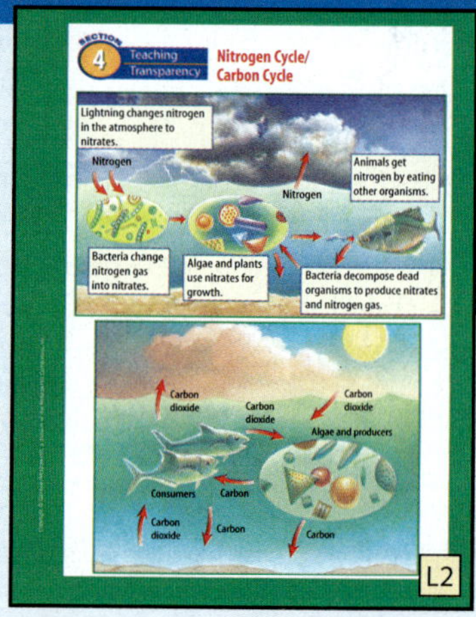

Key to Teaching Strategies

The following designations will help you decide which activities are appropriate for your students.

- L1 Level 1 activities should be appropriate for students with learning difficulties.
- L2 Level 2 activities should be within the ability range of all students.
- L3 Level 3 activities are designed for above-average students.
- ELL ELL activities should be within the ability range of English Language Learners.
- COOP LEARN Cooperative Learning activities are designed for small group work.
- LS Multiple Learning Styles logos, as described on page 12T, are used throughout to indicate strategies that address different learning styles.
- P These strategies represent student products that can be placed into a best-work portfolio.
- PBL Problem-Based Learning activities apply real-world situations to learning.

Hands-on Activities

Student Text Lab Worksheet

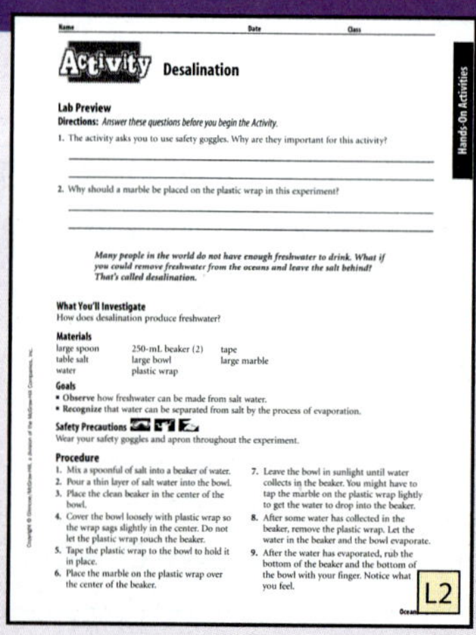

Laboratory Activities

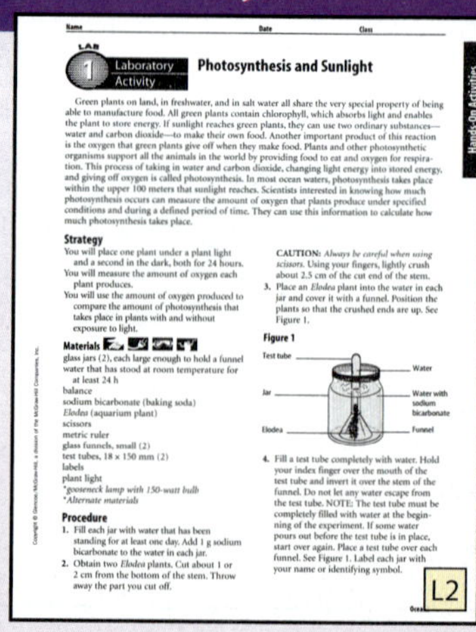

Meeting Different Ability Levels

Content Outline

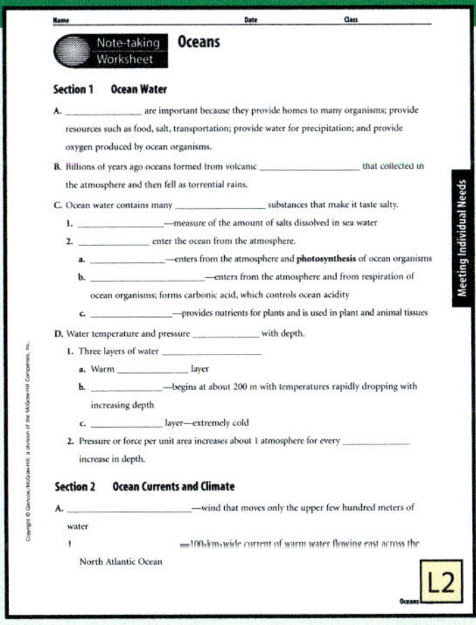

Reinforcement

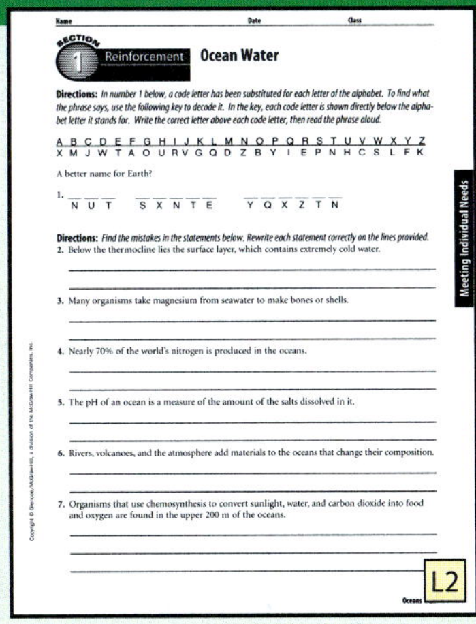

Enrichment

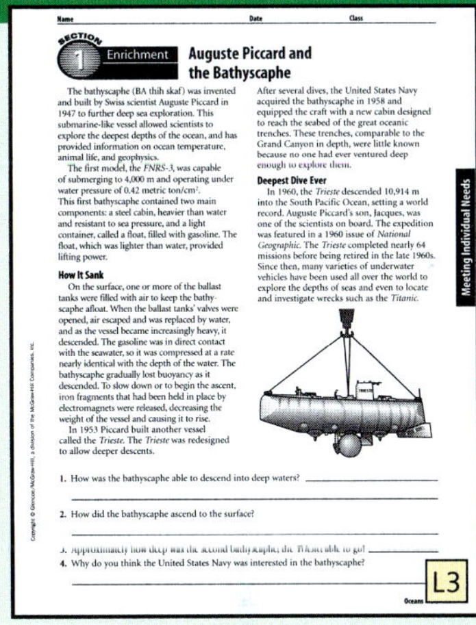

Directed Reading (English/Spanish)

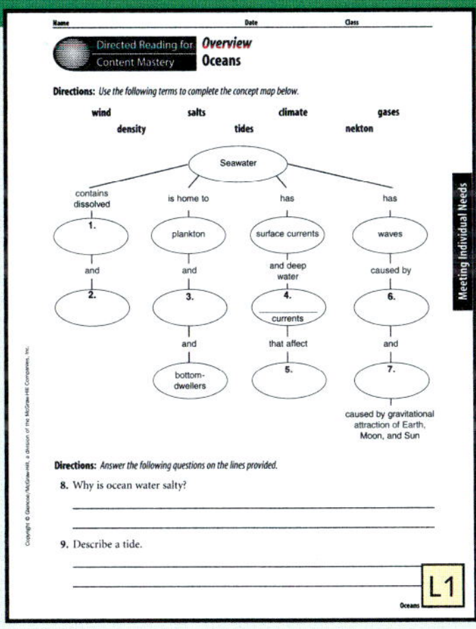

Study Guide

Reading Essentials

Assessment

Test Practice Workbook

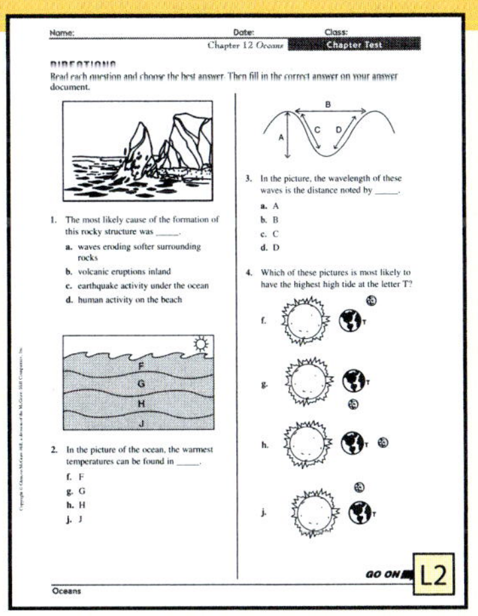

Chapter Review

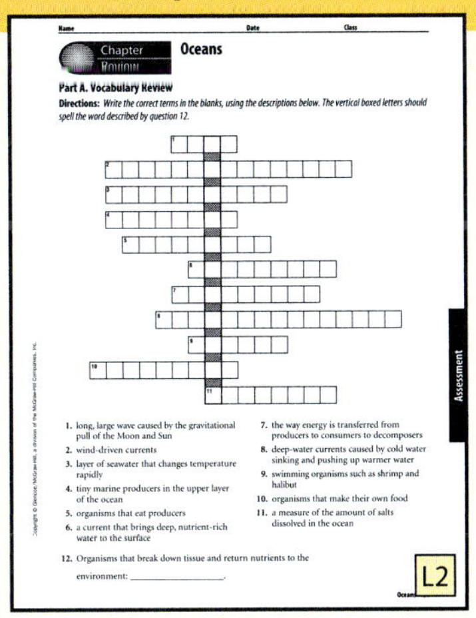

Chapter Tests

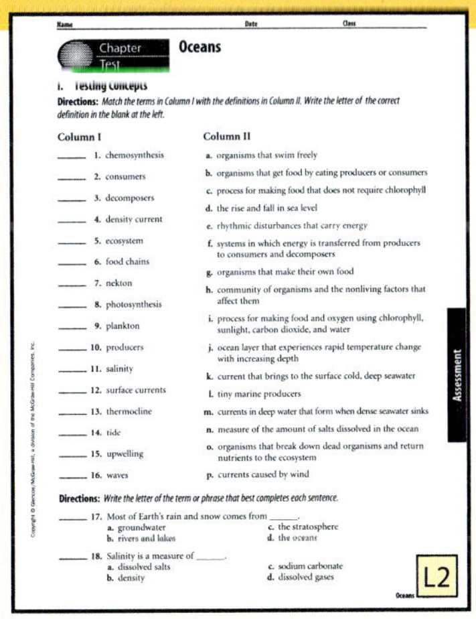

Science Content Background

section 1 Ocean Water

Under the Sea

The water of the world's oceans covers a topography that is as rich and varied as that found on land. These features include mountains, valleys, and plains.

section 2 Ocean Currents and Climate

El Niño

El Niño events can cause severe weather disturbances on a localized level as well as globally. During the 1997–1998 El Niño, heavy rains in the eastern Pacific caused mudslides and flooding throughout Peru and Ecuador. Tornadoes swept across the southeastern United States in February of 1998. Drought struck other areas of the world, including Hawaii, southwestern Africa, and New Guinea, resulting in failed crops and starvation.

section 3 Waves

Wind Waves

When waves meet in the open ocean, they don't just pass over one another. Instead, they either add to or subtract from the crests and troughs of each other. This process is known as interference. In destructive interference, two waves meet and cancel each other out. This type of interaction is not dangerous, as the energy of each wave is dissipated. In constructive interference, waves meet and add to each other's height. Constructive interference can result in large, destructive waves. In some cases a rogue wave, which has a height of 30 m or more, is produced. The southeastern tip of Africa has a high incidence of rogue waves. When incoming waves reach the fast-moving current in this area, they can double in height and then break. Tankers have been broken in half by these rogue waves.

section 4 Life in the Oceans

Coral Reefs

Reef-building corals have a mutualistic relationship with the dinoflagellate zooxanthellae, which grow in the tissues of the corals. The zooxanthellae receive a safe place to live and the carbon dioxide and nutrients given off by the corals during respiration. The corals receive oxygen and extra food in the form of carbohydrates from the zooxanthellae. It is this extra food that allows corals to build massive reef structures. Nonreef-building corals do not have zooxanthellae living in their tissues. These corals rely solely on filter feeding to capture food.

chapter content resources

Internet Resources

For additional content background, visit **red.msscience.com** to:

- access your book online
- find references to related articles in popular science magazines
- access Web links with related content background
- access current events with science journal topics

Print Resources

The Oceans, by Ellen J. Prager and Sylvia A. Earle, McGraw-Hill, 2000

Marine Biology, by Peter Castro and Michael E. Huber, McGraw-Hill, 2003

Guide to Marine Mammals of the World, by James Powell, Phillip J. Clapham, and Pieter Folkens (Illustrator); Alfred A. Knopf, Inc., 2002

IDENTIFYING Misconceptions

Find Out What Students Think

Students may think that . . .

The floor of the ocean is flat, with no topographic features.

Students believe that, because they cannot see under the water, the floor of the ocean is flat like the surface of the water that covers it.

Demonstration

Provide students with maps of the ocean floor. Have them discuss the features that they observe. This will allow students to realize that the ocean floor has structures similar to those on land.

Promote Understanding

Activity

Materials clay, cardboard shoe box with lid, straws, permanent marker, metric ruler, graph paper, scissors, glue

- Have each group of students use clay to make a model of the ocean floor in their shoe box. Encourage them to include features they noted when examining the map of the ocean floor.
- Have students glue the graph paper to the lids of their boxes. Using the scissors, have them poke straw-sized holes in the box lids at the intersection of every 10 graph-grid boxes. Have them label the grid on the long side of the box with numbers and the short side of the box with letters. The labeled graph paper thus forms a grid system for students to use when sampling. Then direct students to mark the straws in 1 cm increments. These will serve as depth gauges. Each group of students should then exchange models with another group so that they are sampling an "unknown" ocean floor.
- Each team should decide on a method of sampling and then carry it out. They should record their depth measurements on a clean sheet of graph paper, thus making a map showing the topography of the ocean floor.
- Encourage groups to share their results. Have the class discuss which sampling techniques were most useful. Then explain that scientists use sound in the same way that they used the straws to make pictures of the ocean floor.

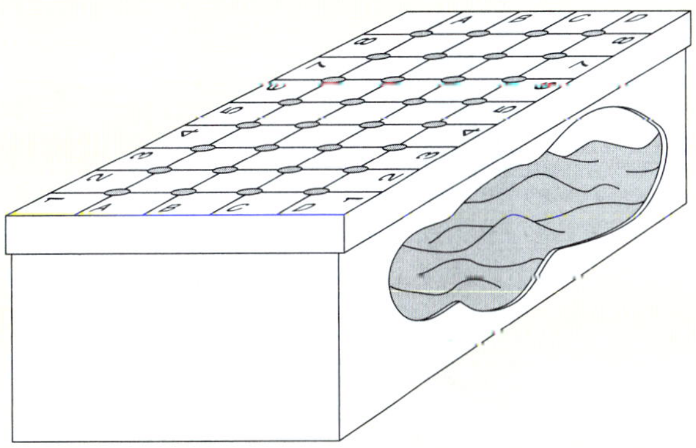

Assess

After completing the chapter, see *Identifying Misconceptions* in the Study Guide at the end of the chapter.

chapter 13

Chapter Vocabulary

salinity, p. 376
photosynthesis, p. 377
thermocline, p. 378
surface current, p. 380
density current, p. 382
upwelling, p. 384
wave, p. 385
tide, p. 387
plankton, p. 389
nekton, p. 390
ecosystem, p. 391
producer, p. 391
chemosynthesis, p. 391
consumer, p. 391
decomposer, p. 391
food chain, p. 392

Science Journal Student answers will vary. Check for depth and quality of questions.

INTERACTIVE CHALKBOARD with Image Bank

PowerPoint® Presentations

This CD-ROM is an editable Microsoft® PowerPoint® presentation that includes:

- a pre-made presentation for every chapter
- interactive graphics
- animations
- audio clips
- image bank
- all new section and chapter questions
- Standardized Test Practice
- transparencies
- pre-lab questions for all labs
- Foldables directions
- links to red.msscience.com

chapter 13

Oceans

chapter preview

sections

1. Ocean Water
   Lab *Desalination*
2. Ocean Currents and Climate
3. Waves
4. Life in the Oceans
   Lab *Waves and Tides*

Virtual Lab *What are some characteristics of the ocean and the ocean floor?*

Ocean Life

There are eight species of sea turtles, the majority of which are found in tropical waters. Sea turtles feed on a range of organisms from seaweed to sponges to jellyfish. Sea turtles return to land to lay their eggs, migrating up to 2,200 km to beaches to make their nests.

Science Journal Write three questions that you would ask a scientist studying ocean life.

372

Theme Connection

Systems and Interactions This theme recurs throughout the chapter with explanations of how salts are formed when substances from volcanoes and rivers interact, how winds make waves, how the gravity of the Moon and Sun causes tides, how winds and the Coriolis force affect surface currents, how waves and currents cause shoreline erosion, and how organisms interact through food chains.

About the Photo

Welcome to the World Female sea turtles come on to beaches at night to lay their eggs. Each nest may contain more than 100 eggs. After the baby turtles hatch they must crawl back to the water. The hatchling turtles may be preyed upon by land crabs and birds on land, and fishes once they enter the water.

Start-Up Activities

Launch LAB

Why are oceans salty?

Ocean water tastes different from the water in most lakes. Its salty taste comes from salts that are dissolved in the water. In the lab below, you will experiment to find out how some of those salts end up dissolved in ocean water.

1. Mix five spoonfuls of dry sand with one spoonful of salt in a pie pan.
2. Bend a small section of the edge of the pie pan down so it is level with the bottom of the pan.
3. Hold the edge of the pie pan over a small bowl and sprinkle water on the salt and sand mixture. Don't wash the sand and salt out of the pie pan. Let the water filter through the mixture. Allow the water to collect in the bowl.
4. Place the bowl in sunlight or under a hot lamp and let the water evaporate. Observe what remains.
5. **Think Critically** Describe in your Science Journal which material the water dissolved as it filtered through the salt and sand. Infer where some of the salt in the oceans comes from.

FOLDABLES Study Organizer

Oceans Make the following Foldable to help you identify the major topics about oceans.

STEP 1 **Collect** 2 sheets of paper and layer them about 2.5 cm apart vertically. Keep the edges level.

STEP 2 **Fold** up the bottom edges of the paper to form 4 equal tabs.

STEP 3 **Fold** the papers and crease well to hold the tabs in place. Staple along the fold. **Label** the tabs as shown.

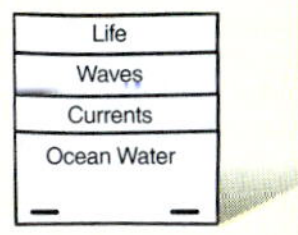

Find Main Ideas As you read the chapter, write information about the main ideas of each topic on the appropriate tab.

Science Online | Preview this chapter's content and activities at red.msscience.com

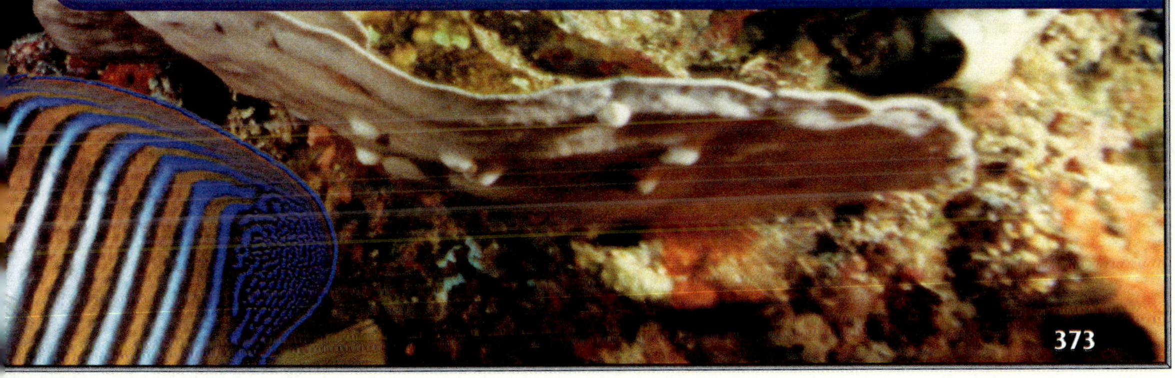

Launch LAB

Purpose This lab introduces students to seawater composition. L2 ELL LS **Visual-Spatial**

Materials dry sand, salt, aluminum pie pan, small bowl, water, sprinkling can, lamp, magnifying lens

Teaching Strategy Have students use magnifying lenses to examine crystals left in bowls.

Think Critically

Salt dissolved. Salts are dissolved from sediments as water flows through them.

Assessment

Oral Lake and river water do not taste like ocean water, yet they contain salts. What does this tell you about the compositions of freshwater and salt water? They are not the same.

FOLDABLES Study Organizer **Dinah Zike Study Fold**

Student preparation materials for this Foldable are available in the Chapter *Fast File* Resources.

section 1 Ocean Water

1 Motivate

Bellringer

INTERACTIVE CHALKBOARD PowerPoint® Presentations

Section Focus Transparencies also are available on the Interactive Chalkboard CD-ROM. L2 ELL

Section 1 Section Focus Transparency **Livin' Life Large**

What's the biggest animal that ever lived? You might be surprised to learn that it's not a dinosaur, but the blue whale. The biggest blue whales are roughly 30 meters (100 feet) long and weigh about 136,000 kilograms (150 tons).

1. How does water help support the blue whale's massive size?
2. Name two ways animals in the ocean get oxygen.
3. In what ways do humans depend on the ocean?

Oceans

Tie to Prior Knowledge

Ocean Experience Have students who have visited the ocean describe their experiences with ocean water and how they may have discovered how it tastes. Explain that in this section, they will learn about the origin of the ocean's salts and more about the composition of seawater.

as you read

What You'll Learn

- **State** the importance of Earth's oceans.
- **Discuss** the origin of the oceans.
- **Describe** the composition of seawater.
- **Explain** how temperature and pressure vary with depth.

Why It's Important

Oceans affect all people's lives, even those who don't live near an ocean.

Review Vocabulary

atmosphere: Earth's air; forms a protective layer around the planet and is divided into five layers

New Vocabulary

- salinity
- photosynthesis
- thermocline

Importance of Oceans

Have you looked at a globe and noticed that oceans cover almost three-fourths of the planet's surface? A better name for Earth might be "The Water Planet." You might live far away from an ocean, or maybe you've never seen the ocean. But oceans affect all living things—even those far from the shore.

Oceans provide a place for many organisms to live. Oceans transport seeds and animals and allow materials to be shipped across the world. Oceans also furnish people with resources including food, medicines, and salt. Some examples of resources from the ocean are shown in **Figure 1.** The water for most of Earth's rain and snow comes from the evaporation of ocean water. In addition, 70 percent of the oxygen on Earth is given off by ocean organisms.

Figure 1 Oceans provide many resources.

Sea sponges are used in medicines for treating asthma and cancer.

Fish and other creatures provide many people with food.

Salt is obtained by evaporating seawater.

Section 1 Resource Manager

Chapter *FAST FILE* Resources

Transparency Activity, p. 46

Directed Reading for Content Mastery, pp. 19, 20

Note-taking Worksheets, pp. 35–37

Enrichment, p. 31

Lab Activity, pp. 9–12

Reinforcement, p. 27

Lab Workwheet, pp. 5–6

Reading and Writing Skill Activities, p. 27

Cultural Diversity, p. 27

374 CHAPTER 13 Oceans

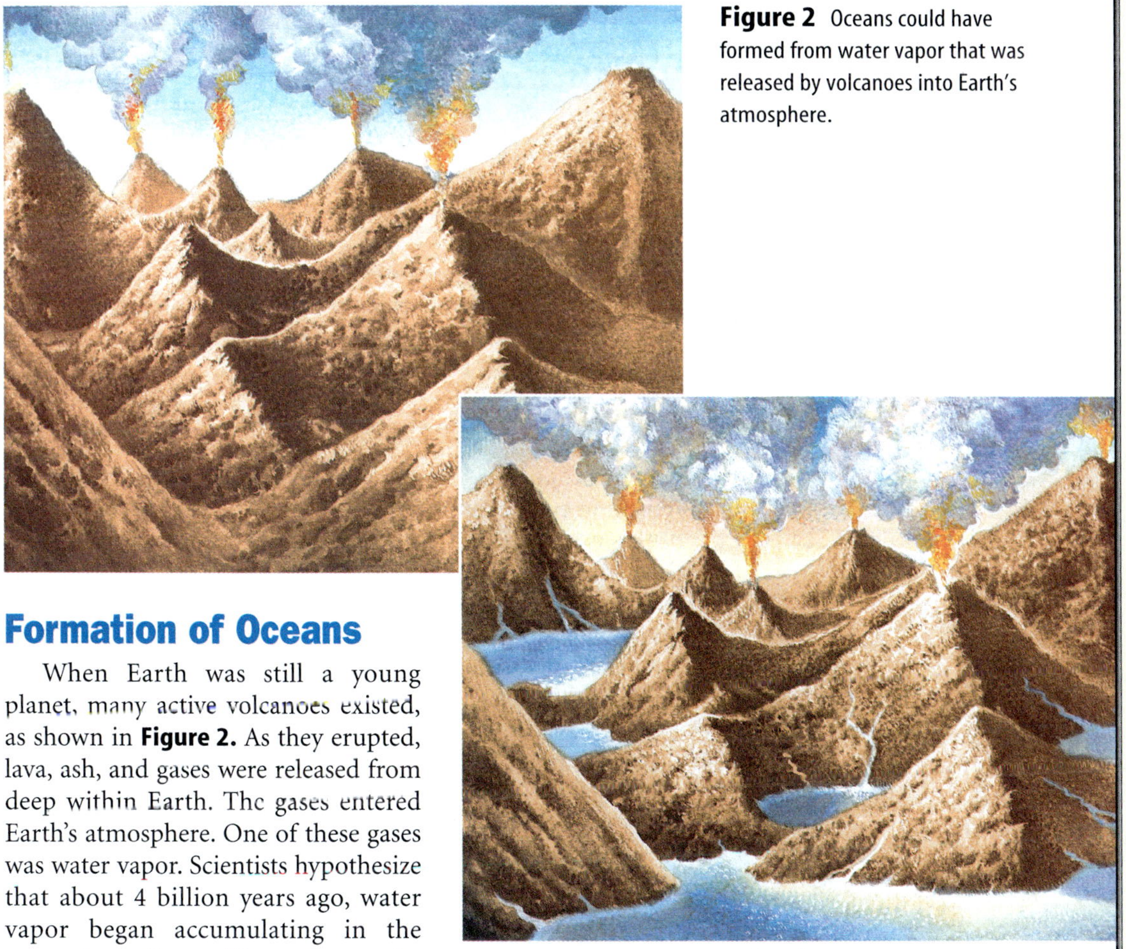

Figure 2 Oceans could have formed from water vapor that was released by volcanoes into Earth's atmosphere.

When the water vapor cooled enough to form clouds and rain, water collected in low areas and formed oceans.

Formation of Oceans

When Earth was still a young planet, many active volcanoes existed, as shown in **Figure 2.** As they erupted, lava, ash, and gases were released from deep within Earth. The gases entered Earth's atmosphere. One of these gases was water vapor. Scientists hypothesize that about 4 billion years ago, water vapor began accumulating in the atmosphere. Over millions of years, the water vapor cooled enough to condense and form clouds. Then torrential rains began to fall from the clouds. Over time, more and more water accumulated in the lowest parts of Earth's surface, as you can see in **Figure 2.** Eventually, much of the land was covered by water that formed oceans. Evidence indicates that Earth's oceans formed more than 3 billion years ago.

Composition of Ocean Water

If you taste seawater, you'll know immediately that it tastes different from water you normally drink. As a matter of fact, you really can't drink seawater. Dissolved substances cause the salty taste. Rivers and groundwater dissolve elements such as calcium, magnesium, and sodium from rocks and carry them to the ocean, as you saw in the Launch Lab. Erupting volcanoes add elements such as bromide and chlorine to ocean water.

2 Teach

Teacher FYI

Ocean Changes Over millions of years, the movement of Earth's crustal plates has changed the locations of Earth's oceans. Today, there are four major oceans: the Pacific, the Atlantic, the Indian, and the Southern. The Pacific is becoming smaller as oceanic crust disappears into trenches. The Atlantic and Indian Oceans are increasing in size as new oceanic crust emerges at the mid-ocean ridges. The birth of a new ocean may be occurring in Africa at the Great Rift Valley.

Quick Demo

Composition of Ocean Water

Materials table salt, triple-beam balance, water, 1000-mL beaker

Estimated Time 10 minutes

Procedure To demonstrate how much salt seawater usually contains, measure 35 g of table salt with a triple-beam balance. Dissolve the salt in a beaker of 965 mL of water.

Identifying Misconceptions

Ocean Topography Some students may think the ocean floor is flat, with no topographic features. Refer to page F at the beginning of this chapter for teaching strategies that address this misconception.

Differentiated Instruction

Challenge Have students research what types of pharmaceuticals used in medicine are derived from the ocean. How are these products extracted from the ocean? Have students share the results of their research in the form of a multimedia presentation. L3 LS **Linguistic**

Curriculum Connection

Social Studies Today, oceans are not located in Michigan, Ohio, New York, Texas, Louisiana, or Kansas, but halite, salt from ancient oceans, is being mined in all of these states. Have students research uses for halite and have them write a paragraph about their findings. It is used to season and preserve food, to keep roadways free of winter ice, to soften water, and to manufacture textiles. L2 LS **Linguistic** P

SECTION 1 Ocean Water 375

Discussion

Salinity Equilibrium Although the proportion and amount of dissolved salts in seawater remain in equilibrium, would you expect salinity to be constant throughout the oceans? Explain. No; salinity is a measure of the concentration of salt in the water. Evaporation, precipitation, and freezing and thawing of ice affect this concentration. For example, if 35 g of salt are dissolved in 965 mL of water, the concentration is 35 parts per 1,000 parts, or 3.5%. If 100 mL of fresh water are added, the concentration is then 35 parts per 1,100 parts, or 3.18%. L2 LS **Logical-Mathematical**

Reading Check

Answer a measure of the amount of solids dissolved in seawater

Teacher FYI

Manganese Nodules A manganese nodule forms when chemical reactions remove manganese, iron, copper, nickel, and cobalt from seawater. These substances commonly precipitate around discarded sharks' teeth on the ocean floor. Nodules can take more than a million years to grow only a few millimeters.

Visual Learning

Figure 3 What percent of the dissolved solids in seawater are chloride? 55% L1 LS **Visual-Spatial**

Salinity The two most abundant elements in the dissolved salts in seawater are sodium and chlorine. If seawater evaporates, the sodium and chloride ions combine to form a salt called halite. You use this salt to season food. Halite, as well as other salts and substances, give ocean water its unique taste.

Salinity (say LIH nuh tee) is a measure of the amount of solids, or salts, dissolved in seawater. It is measured in grams of dissolved solids per kilogram of water. One kilogram of ocean water usually contains about 35 g of dissolved solids, or 3.5 percent. **Figure 3** shows the most abundant salts in ocean water.

Reading Check *What is salinity?*

The proportions and amount of dissolved salts in seawater remain in equilibrium. This means that the composition of the oceans is in balance. Despite the fact that rivers, volcanoes, and the atmosphere constantly add substances to the ocean, its composition has remained nearly constant for hundreds of millions of years. Biological processes and chemical reactions remove many of the substances, such as calcium, from ocean water. For example, many organisms, such as oysters and clams, use calcium to make their shells. Other marine animals use calcium to make bones. Calcium also can be removed from ocean water through chemical reactions, forming sediment on the ocean floor.

Figure 3 Every kilogram of ocean water contains about 35 g of dissolved solids. Sodium and chloride make up nearly 86 percent of this mass. This circle graph shows the most abundant dissolved solids.

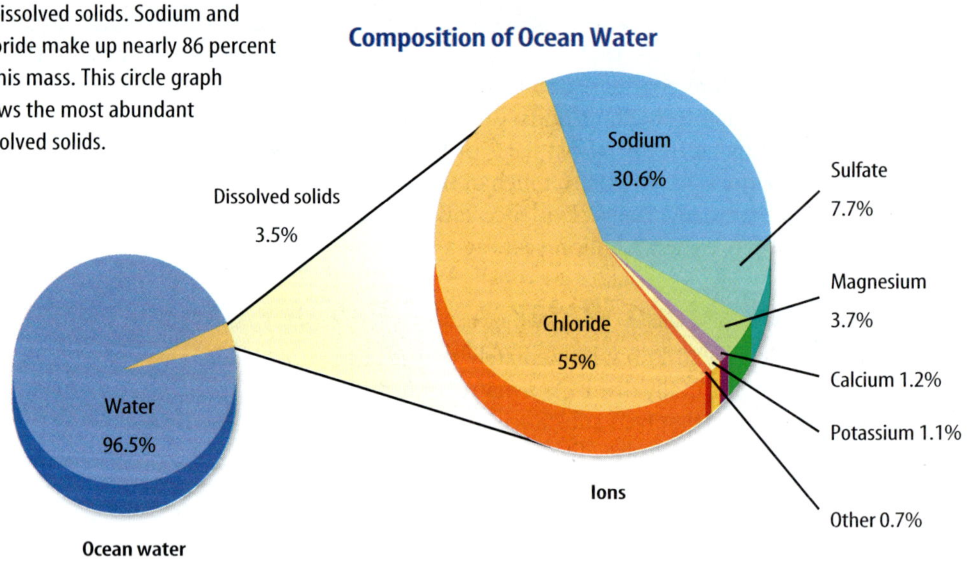

Curriculum Connection

Math Salinity is often reported in parts per thousand. The mathematical symbol for parts per thousand is ‰. Ask: If you sampled water from the Atlantic Ocean and found that the percent of salt is 3.28%, what would the salinity be in parts per thousand? 32.8‰ L2 LS **Logical-Mathematical**

Cultural Diversity

Middle East In most areas of the world, removing salts from seawater to obtain freshwater is too expensive. However, in arid countries such as Israel, Saudi Arabia, Kuwait, and the United Arab Emirates, the most cost-effective way to obtain freshwater is from seawater. Have students locate these countries on a map and determine the ocean(s) from which they get their water. L2 LS **Visual-Spatial**

Figure 4 Kelp growing in shallow water use sunlight to photosynthesize. During photosynthesis, oxygen is given off and dissolves in the water.
Describe *How does carbon dioxide enter seawater?*

Carbon dioxide + Water $\xrightarrow{\text{Sunlight}}$ Food + Oxygen

Dissolved Gases Although all of the gases in Earth's atmosphere dissolve in seawater, three of the most important are oxygen, carbon dioxide, and nitrogen.

The greatest concentration of dissolved oxygen is near the surface of the ocean. There, oxygen enters seawater directly from the atmosphere. Also, organisms like the kelp in **Figure 4** produce oxygen by photosynthesis—a process in which organisms use sunlight, water, and carbon dioxide to make food and oxygen. Because sunlight is necessary for photosynthesis, organisms that carry on **photosynthesis** are found only in the upper 200 m of the ocean, where sunlight reaches. Below 200 m, the level of dissolved oxygen drops rapidly. Here, many animals use oxygen for respiration and it is not replenished. However, more dissolved oxygen exists in very deep water than in water just below 200 m. This cold, deep water originates at the surface in polar regions and moves along the ocean floor to other regions.

Reading Check *How does oxygen get into seawater?*

A large quantity of carbon dioxide is absorbed directly into seawater from the atmosphere. Carbon dioxide reacts with water molecules to form a weak acid called carbonic acid. Carbonic acid helps control the acidity of the oceans. In addition, during respiration, organisms use oxygen and give off carbon dioxide, adding more carbon dioxide to the oceans.

Nitrogen is the most abundant dissolved gas in the oceans. Some types of bacteria combine nitrogen with oxygen to form nitrates. These nitrates are important nutrients for plants. Nitrogen is also one of the important building blocks of plant and animal tissue.

Science Online
Topic: Dissolved Gases
Visit red.msscience.com for Web links to information about dissolved gases in the ocean.

Activity Make a profile of how the concentration of a dissolved gas such as oxygen or carbon dioxide changes with water depth. Present the results on a poster to share with your class.

Caption Answer

Figure 4 Carbon dioxide is absorbed into seawater from the atmosphere. During respiration, marine organisms give off carbon dioxide, also adding the gas to seawater.

Discussion

Life Processes How do photosynthesis and respiration compare? These processes are opposites. In photosynthesis, producers use carbon dioxide and produce oxygen. In respiration, organisms use oxygen and produce carbon dioxide. L2 LS **Logical-Mathematical**

Virtual Labs

Ocean characteristics *What are some characteristics of the ocean and the ocean floor?*

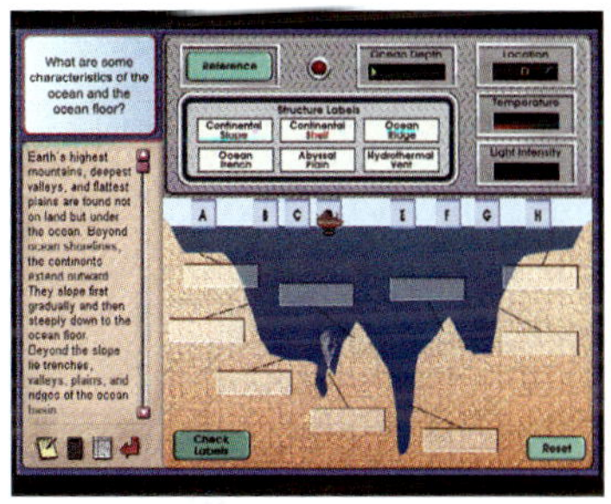

Reading Check

Answer Oxygen enters from the atmosphere; it is also produced by organisms during photosynthesis.

Differentiated Instruction

Learning Disabled Have students work in pairs to better understand the relationship between photosynthesis and respiration. Have pairs make a poster that shows the cyclical relationship between the oxygen produced by photosynthesis and used in respiration as well as the carbon dioxide produced by respiration and used in photosynthesis. List how each gas enters the ocean. L2 ELL COOP LEARN LS **Visual-Spatial**

Active Reading

Quickwrites Have students write a list of ideas about a topic, then share these ideas with the class. Next, have students write their ideas freely in a paragraph without worrying about punctuation, spelling, and grammar. Have students use a Quickwrite to share ideas during or after learning about oceans. L2 LS **Linguistic**

Activity

Scuba Diver Invite an experienced scuba diver to talk to your class about the effects of high pressure on divers. Ask students to prepare questions for the speaker ahead of time. L1 LS **Linguistic**

Caption Answer

Figure 5 It extends to 800 m.

3 Assess

DAILY INTERVENTION

Check for Understanding

Linguistic Have students write a short paragraph about why oceans are important. L2 LS

Reteach

Review Game Play a review game. Organize the class into several teams. Have each student write three questions about the material in this section. Collect the questions. Then ask these questions of other teams, one at a time. Encourage members of each team to discuss the questions they are asked and have a team captain report their answers to the class. L2 ELL COOP LEARN LS **Interpersonal**

Assessment

Oral What are two reasons for the rapid drop in dissolved oxygen below 200 m? Light does not penetrate below 200 m, so there is no oxygen being added by plants; animals living at these depths deplete the oxygen when they respire. Use **Performance Assessment in the Science Classroom**, p. 89.

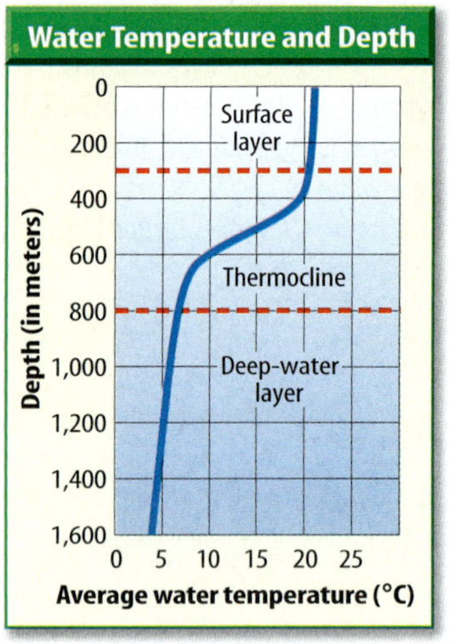

Figure 5 The depth of the thermocline varies with location. In the location shown on this graph, the thermocline layer begins at 300 m.
Determine *How deep does the thermocline extend?*

Water Temperature and Pressure

Oceans have three temperature layers—the surface layer, the thermocline layer, and the deep-water layer, shown in **Figure 5.** The surface layer is warm because it receives solar energy. The warmest surface water is near the equator where the Sun's rays strike Earth at a direct angle. Water near the poles is cooler because the Sun's rays strike Earth at a lower angle.

The **thermocline** often begins at a depth of about 200 m, but this varies. In this layer, temperature drops quickly with increasing depth. This occurs because solar energy cannot penetrate this deep. Below the thermocline lies the deep-water layer, which contains extremely cold water.

INTEGRATE Physics Pressure, or force per unit area, also varies with depth. At sea level, the pressure of the atmosphere pushing down on the ocean surface is referred to as 1 atmosphere (atm) of pressure. An atmosphere is the pressure exerted on a surface at sea level by the column of air above it. As you go below the ocean's surface, the pressure increases because of the force of the water molecules pushing down. The pressure increases by about 1 atm for each 10-m increase in depth.

For example, at a depth of 20 m, a scuba diver would experience a pressure of 3 atm (1 atm of air + 2 atm of water). Divers must carry tanks that supply their lungs with air at the same pressure as the water around them. If they didn't, the water pressure would keep their lungs from inflating when they tried to inhale.

section 1 review

Summary

Importance of Oceans

- Oceans cover almost three-fourths of Earth's surface. A variety of organisms live in the ocean.

Formation of Oceans

- Scientists hypothesize that oceans formed as water from torrential rains filled Earth's basins.

Composition of Ocean Water

- Ocean water contains both dissolved salts and dissolved gases.

Water Temperature and Pressure

- The warmest water is at the ocean surface. As depth increases, water temperature decreases.
- As water depth increases, pressure increases.

Self Check

1. **List** at least four reasons why oceans are important to you.
2. **Explain** the relationship between volcanic activity and the origin of Earth's oceans.
3. **Describe** how and why temperature and pressure vary with ocean depth.
4. **Think Critically** Why are the compositions of river water and ocean water not the same?

Applying Skills

5. **Recognize Cause and Effect** How does animal respiration affect the amount of dissolved oxygen in deeper water?

Science Online red.msscience.com/self_check_quiz

section 1 review

1. Possible answers: for food, medicines, salt, oxygen, water cycle
2. Water vapor from volcanoes cooled and condensed; rains filled basins.
3. Solar energy heats the upper 200 m. Below 200 m, temperature drops rapidly. Pressure increases with depth because of the force of the water molecules pushing down.
4. Salts have built up in ocean water over time. Volcanoes and the atmosphere add additional chemicals to ocean water. Biological processes and chemical reactions in the oceans remove some substances like calcium which is found in river water.
5. Respiration removes oxygen.

LAB

Desalination

Many people in the world do not have enough freshwater to drink. What if you could remove freshwater from the oceans and leave the salt behind? That's called desalination.

Real-World Question

How does desalination produce freshwater?

Goals

- **Observe** how freshwater can be made from salt water.
- **Recognize** that water can be separated from salt by the process of evaporation.

Materials

large spoon	large bowl
table salt	plastic wrap
water	tape
250-ml beakers (2)	large marble

Safety Precautions

Wear your safety goggles and apron throughout the experiment.

Procedure

1. Mix a spoonful of salt into a beaker of water.
2. Pour a thin layer of salt water in the bowl.
3. Place the clean beaker in the center of the bowl.
4. Cover the bowl loosely with plastic wrap so the wrap sags slightly in the center. Do not let the plastic wrap touch the beaker.
5. Tape the plastic wrap to the bowl to hold it in place.
6. Place the marble on the plastic wrap over the center of the beaker.
7. Leave the bowl in sunlight until water collects in the beaker. You might have to tap the marble on the plastic wrap lightly to get the water to drop into the beaker.
8. After some water has collected in the beaker, remove the plastic wrap. Let the water in the beaker and the bowl evaporate.
9. After the water has evaporated, rub the bottom of the beaker and the bottom of the bowl with your finger. Notice what you feel.

Conclude and Apply

1. **Describe** what you found remaining in the bowl and in the beaker after all the water had evaporated.
2. **Explain** what kind of water collected in the beaker—salt water or freshwater? How do you know?

Communicating Your Data

Make a poster that illustrates how you would make water that you could drink if you were stranded on a deserted island in the middle of the ocean. **For more help, refer to the Science Skill Handbook.**

BENCH TESTED

Real-World Question

Purpose Students use desalination to produce freshwater. L2 ELL IS **Visual-Spatial**

Process Skills observe and infer, draw conclusions, communicate

Time Required 40 minutes

Procedure

Safety Precautions Caution students to wear safety goggles throughout the experiment.

Teaching Strategy Use a light with a reflector to reduce the time it takes to evaporate the water. Be sure that the electrical cord is not near water and warn students that the light is hot.

Conclude and Apply

1. The bowl contained salt, the beaker did not.
2. Freshwater; there was no salt left after the water was evaporated.

Assessment

Process Inform students that many precious metals are dissolved in seawater. Is it practical to remove these metals using desalination? It would require large volumes of seawater. The time and energy required to retrieve the metals would make this a very expensive process.

Communicating Your Data

Students should illustrate how seawater is collected, water is evaporated, and water vapor is collected and cooled.

1 Motivate

Bellringer

INTERACTIVE CHALKBOARD PowerPoint® Presentations

Section Focus Transparencies also are available on the Interactive Chalkboard CD-ROM. L2 ELL

Section 2 Section Focus Transparency — Treacherous Current

Coursing around this peaceful island is the Maelstrøm, a swirling current caused by a combination of tide, wind, and configuration of sea floor. Running between two islands in the Norwegian Sea, the Maelstrøm poses a challenge to those who must navigate its waters.

1. What is a current?
2. How might a strong current make swimming dangerous?
3. Why do you think the Maelstrøm would be difficult to navigate?

Tie to Prior Knowledge

Currents Have students recall what they learned about ocean salinity and temperature in the last section. In this section, they will learn how salinity and temperature affect deep ocean currents.

Caption Answer

Figure 6 Currents in the northern hemisphere rotate clockwise due to the Coriolis effect.

section 2 Ocean Currents and Climate

as you read

What You'll Learn

- **State** how wind and Earth's rotation influence surface currents.
- **Explain** how ocean currents affect weather and climate.
- **Describe** the causes and effects of density currents.
- **Explain** how upwelling occurs.

Why It's Important

Ocean currents affect weather and climate.

Review Vocabulary
current: a fluid moving continuously in a certain direction

New Vocabulary
- surface current
- upwelling
- density current

Surface Currents

Ocean water never stands still. Currents move the water from place to place constantly. Ocean currents are like rivers that move within the ocean. They exist both at the ocean's surface and in deeper water. Major surface currents and winds are shown in **Figure 6.**

Causes of Surface Currents Powered by wind, **surface currents** usually move only the upper few hundred meters of seawater. When the global winds blow on the ocean's surface, they can set ocean water in motion. Because of Earth's rotation, the ocean currents that result do not move in straight lines. Earth's rotation causes surface ocean currents in the northern hemisphere to curve to the right and surface ocean currents in the southern hemisphere to curve to the left. You can see this in **Figure 6.** This turning of ocean currents is an example of the Coriolis effect.

Figure 6 Earth's global winds create surface currents in the oceans.
Observe *Which way do currents rotate in the northern hemisphere?*

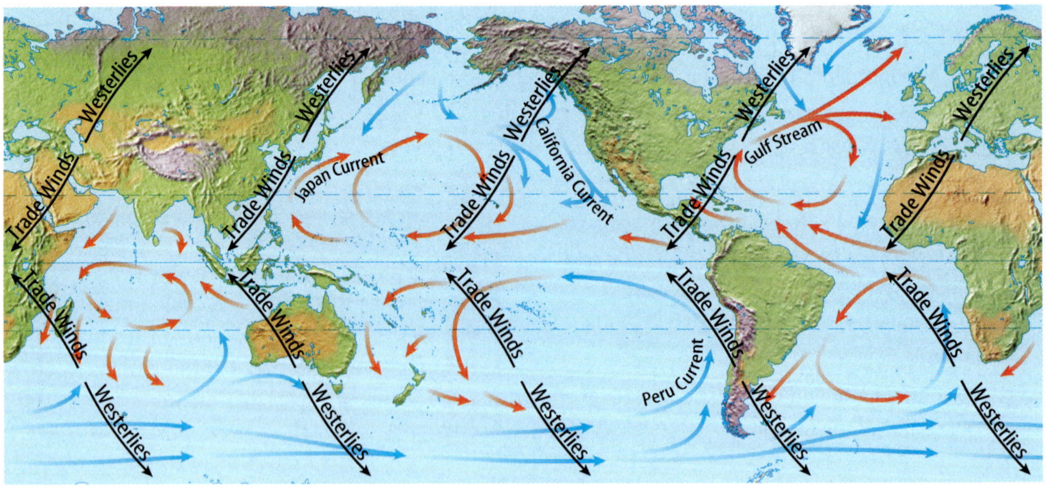

Section 2 Resource Manager

Chapter *FAST FILE* Resources
Transparency Activity, p. 47
Directed Reading for Content Mastery, p. 20
MiniLAB, p. 3
Enrichment, p. 32
Reinforcement, p. 28

Earth Science Critical Thinking/Problem Solving, p. 16

The Gulf Stream Much of what is known about surface currents comes from records kept by early sailors. Sailing ships depended on certain surface currents to carry them west and others to carry them east. One of the most important currents for sailing east across the North Atlantic Ocean is the Gulf Stream. This 100-km wide current was discovered in the 1500s by Ponce de Leon and his pilot Anton de Alaminos. In 1770, Benjamin Franklin published a map of the Gulf Stream drawn by Captain Timothy Folger, a Nantucket whaler.

The Gulf Stream, shown in **Figure 7,** flows from Florida northeastward toward North Carolina. There it curves toward the east and becomes slower and broader. Because the Gulf Stream originates near the equator, it is a warm current. Look back at **Figure 6.** Notice that currents on eastern coasts of continents, like the Gulf Stream, are usually warm, while currents on western coasts of continents are usually cold. Surface currents like the Gulf Stream distribute heat from equatorial regions to other areas. This can influence the climate of regions near these currents.

Reading Check *What kind of current is the Gulf Stream?*

Science Online

Topic: Surface Currents

Visit red.msscience.com for Web links to information about the Gulf Stream and other surface currents.

Activity Investigate what type of ongoing research is taking place on other surface currents such as the California Current or the Peru Current. Present your results to the class.

Figure 7 In this satellite image, the warm water of the Gulf Stream appears red and orange.

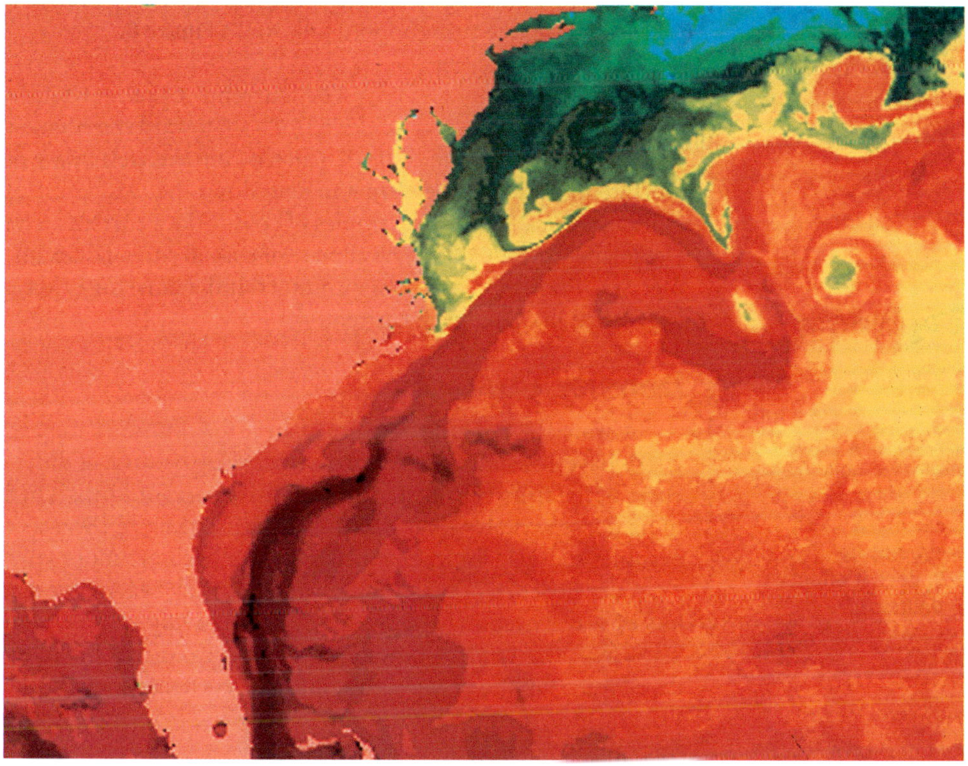

2 Teach

Discussion

Current Temperatures Why are currents along the eastern coasts of continents usually warm while those along the western coasts are cold? Currents along eastern coasts originate at the equator, where the amount of energy absorbed from the Sun is the greatest; currents along western coasts originate at high latitudes where water receives less solar energy. L2 LS **Logical-Mathematical**

Reading Check

Answer warm surface current

Fun Fact

The Gulf Stream transports more than 50 times the combined volume of water that flows from all of Earth's rivers.

Visual Learning

Figure 7 Have students locate the Atlantic coast. Have them describe the temperature of the Gulf Stream along this coast. Possible answer: temperature of Gulf Stream is warmest at its origin and gets cooler as you go north. L2 LS **Visual-Spatial**

Differentiated Instruction

Challenge Have students find out more information about warm core and cold core rings found near the Gulf Stream. Have students report their findings to the rest of the class. Students can display satellite images of the rings from college textbooks. L3 LS **Linguistic and Visual-Spatial**

Teacher FYI

Gulf Stream Speed At lower latitudes, the Gulf Stream Current is narrow and fast moving. Along the southeastern shore of Florida, the current flows at an average speed of 2 m/s. In this area, the width of the Gulf Stream is about 80 km. Further north, the Gulf Stream widens.

Teacher FYI

Surface Temperatures In September, the surface water temperature off Charleston, South Carolina is 27°C. Off San Diego, California, the surface water temperature is only 19°C. Both cities are at the same latitude.

Caption Answer

Figure 8 The California Current is a cold-water current. Coastal cities such as San Francisco have cool summers and many foggy days.

INTEGRATE Chemistry

Seawater Density Increased salinity means an increase in the amount of dissolved solids. Therefore, when salinity increases, mass increases. When temperature increases, molecules move apart, and volume increases.

Career A chemical oceanographer studies the composition of ocean water. He or she may measure the amount of dissolved gases and the salinity in an area of water. He or she may study how water chemistry changes as a result of biological processes. A chemical oceanographer might also study how minerals are deposited in the ocean or the formation manganese nodules.

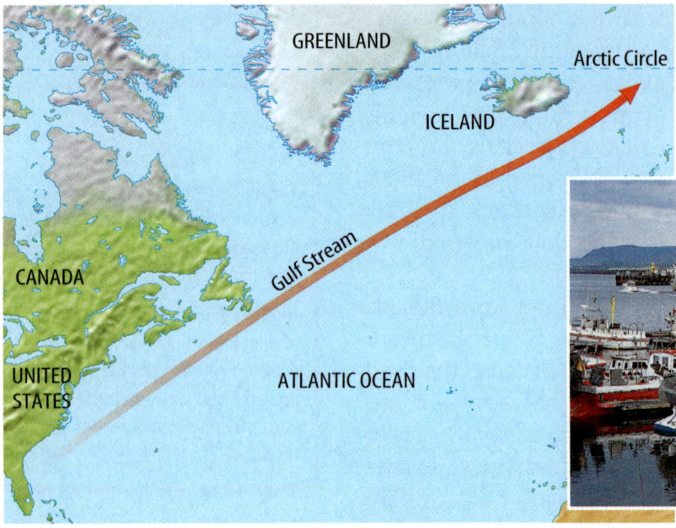

The harbor at Reykjavik (RAY kyuh vihk), Iceland's capital, remains free of ice all year long.

Figure 8 The warm water of the Gulf Stream helps moderate Iceland's climate.
Describe *how the California Current affects the climate of coastal cities.*

Climate As an example of how surface currents affect climate, locate Iceland on the map in **Figure 8.** Based on its location and its name, you might expect it to have a cold climate. However, the Gulf Stream flows past Iceland. The current's warm water heats the surrounding air and keeps Iceland's climate mild and its harbors ice free year-round, as shown in **Figure 8.**

Cold Surface Currents The currents on the western coasts of continents carry colder water back toward the equator. In **Figure 6,** find the California Current off the west coast of North America and the Peru Current along the west coast of South America. They are examples of cold surface currents. The California Current affects the climate of coastal cities. For example, San Francisco has cool summers and many foggy days because of the California Current.

INTEGRATE Chemistry

Seawater Density The formula for determining density is mass/volume. When salinity increases, does it affect mass or volume? When temperature increases, does it affect mass or volume?

Density Currents

In water at a depth of more than a few hundred meters, winds have no effect. Instead, currents develop because of differences in the density of the water. A **density current** forms when more dense seawater sinks beneath less dense seawater. Seawater becomes more dense when it gets colder or becomes more salty.

A density current exists in the Mediterranean Sea. In this sea, lots of water evaporates from the surface, leaving salts behind. Therefore, the remaining water is high in salinity. This more dense water sinks and moves out into the less dense water of the Atlantic Ocean. At the surface, less dense water from the Atlantic flows into the Mediterranean Sea.

LAB DEMONSTRATION

Purpose to observe the effects of temperature on water density

Materials 2 large beakers, 2 small beakers, hot plate, ice, water, food coloring, dropper

Preparation Use the large beakers to melt ice and heat water. Add a few drops of food coloring to the ice water.

Procedure Pour some of the warm water into a small beaker. Place a few drops of ice water on top of the warm water. Repeat the experiment, placing a few drops of warm water into a small beaker of ice water.

Expected Outcome The ice water sinks into the warm water.

Assessment

Infer why water sank in one beaker, but not in the other. Cold water is denser than warm water, and so it sank through the less dense warm water. The less dense warm water floated on the denser cold water.

Cold and Salty Water An important density current that affects many regions of Earth's oceans begins north of Iceland. In the winter months, the water at the surface starts to freeze. When water freezes, dissolved salts are left behind in the unfrozen water. Therefore, this unfrozen water is very dense because it is cold and salty. It sinks and slowly flows along the ocean floor toward the southern Atlantic Ocean, as shown in blue in **Figure 9.** There it spreads into the Indian and Pacific Oceans. As the water is sinking near Iceland, warm surface water of the Gulf Stream, shown in red, moves northward from the equator to replace it. The Gulf Stream water warms the continents that border the North Atlantic.

Density Currents and Climate Change Suppose density currents near Iceland stopped forming. Some scientists hypothesize that this has happened in Earth's past and could happen again. Increasing carbon dioxide concentrations in Earth's atmosphere could trap more of the Sun's heat, raising Earth's temperature. If Earth's temperature rose enough, ice couldn't easily form near the polar regions. Freshwater from melting glaciers on land also could reduce the salinity of the ocean water. The density currents would weaken or stop. Scientists hypothesize that if dense water stopped flowing along the ocean bottom toward the southern Atlantic Ocean, warm water would no longer flow northward on the surface to replace the missing water. All of Earth could experience drastic climate shifts, including changing rainfall patterns and temperatures.

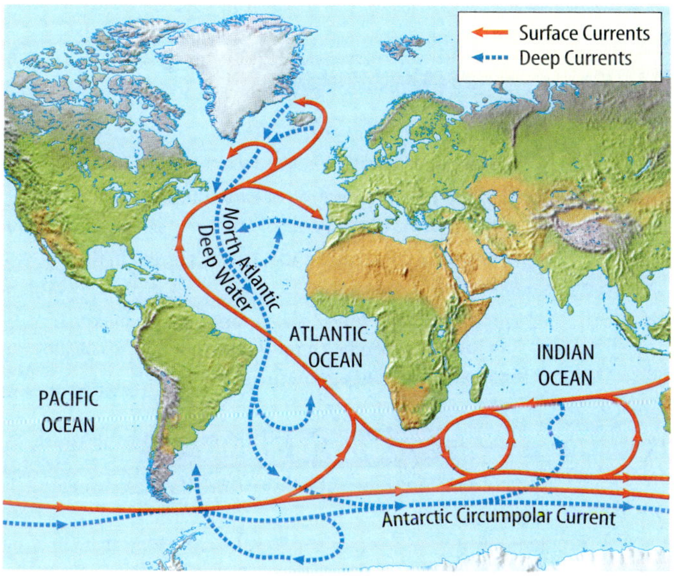

Figure 9 Like a giant conveyor belt, cold, salty water sinks in the northern Atlantic Ocean and flows southward, while warm surface water flows northward from the equator to replace it.

Mini LAB

Modeling a Density Current

Procedure

1. Fill a **paper** cup three-fourths full with **water**.
2. Add two spoonfuls of **salt** and three drops of **food coloring** to the water. Stir with a **spoon** to dissolve the salt.
3. Push one **thumbtack** into the cup 1 cm from the bottom of the cup and another 3 cm from the bottom.
4. Carefully place the cup into a **clear-plastic box** and fill the box with water until the water level in the box is about 0.5 cm above the top tack.
5. Remove both tacks at the same time and record in your **Science Journal** what you observe.

Analysis

1. Infer what is happening at the two holes in the cup.
2. Make a sketch to describe the current's direction.
3. Explain what causes the density current to form.

Mini LAB

Purpose Students model a density current. L2 LS **Visual-Spatial**

Materials paper cup, water, spoon, salt, food coloring, metric ruler, two thumbtacks, clear plastic box

Teaching Strategy Caution students to be careful when handling thumbtacks.

Analysis

1. At the top hole, freshwater enters the cup. At the bottom hole, salt water leaves the cup.
2. Sketches should show the MiniLAB setup. An arrow labeled "salt water" should point down and away from the bottom hole; an arrow labeled "freshwater" should point up to the top hole.
3. Salt water is denser than freshwater. As it sinks, it pushes the less dense freshwater up.

Assessment

Oral Will a density current form where rivers enter an ocean? No, the less dense freshwater will remain as a layer on top of the denser salt water. Use **Performance Assessment in the Science Classroom**, p. 89.

Identifying Misconceptions

Floating Ice Most students know that ice (a solid) floats on water (a liquid). Students may think that this is true of many chemicals. Explain that the characteristic of the solid ice being less dense than the liquid water is unique to H_2O.

Cultural Diversity

Upwelling and Industry Off Peru's coast, phytoplankton live on nutrients brought to the surface by upwelling. The phytoplankton are eaten by anchovies. Many Peruvians work in factories that process fishmeal and fish oil from these anchovies. Have students infer and write a paragraph about what happens when the upwelling does not occur. Industry collapses; many people are out of work. L2 LS **Logical-Mathematical** P

Differentiated Instruction

Visually Impaired Enlarge **Figure 9** onto a transparency sheet. Allow students to use overhead pens that erase to trace the direction of the different currents shown in the figure. L2 LS **Kinesthetic**

Quick Demo

Upwelling

Materials rectangular, glass baking dish, food coloring, ice-cold water, warm water, spoon, overhead projector, beakers

Estimated Time ten minutes

Procedure On the overhead projector, pour colored ice water into a rectangular, glass baking dish. Carefully place warm water on top of the cold water so that the warm water does not mix with the ice water. Use a spoon to gently pull the warm water away from the side of the dish. Students should see the colored water rise to the surface, representing upwelling.

3 Assess

DAILY INTERVENTION

Check for Understanding

Visual-Spatial Ask students what they think the Gulf Stream would look like if the Isthmus of Panama were covered with water. Possible answer: Since surface currents are deflected by land masses, part of the Gulf Stream would be "lost" by flowing between the now-separated continents of North and South America. The resulting current may not be as strong as the Gulf Stream. L2 LS

Reteach

Surface Currents Use **Figure 6** to review surface current directions. L1 LS **Visual-Spatial**

Assessment

Process Explain that currents that originate at the equator move faster than currents that originate at higher latitudes. Have students use **Figure 6** to identify fast-moving currents. Any current on the east coast of a continent is fast-moving. Use **Performance Assessment in the Science Classroom,** p. 89.

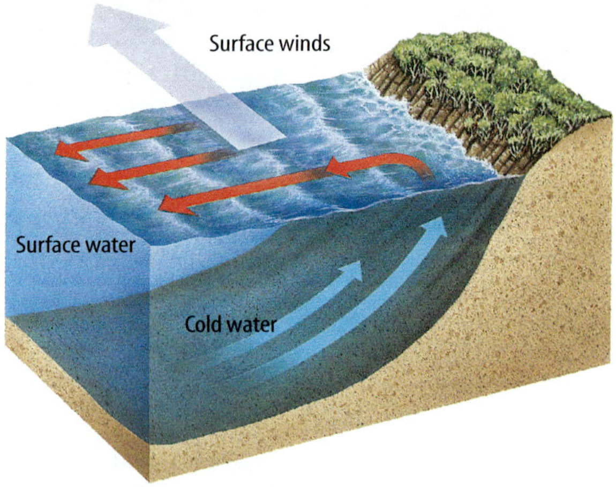

Figure 10 Winds push water away from shore along the South American coastline. This creates an upwelling of cold water.

Upwelling

An **upwelling** is a current in the ocean that brings deep, cold water to the ocean surface. This occurs along some coasts where winds cause surface water to move away from the land. Wind blowing parallel to the coast carries water away from the land because of the Coriolis effect. **Figure 10** shows upwelling as it occurs off the coast of Peru. Notice that when surface water is pushed away from the coast, deep water rises to the surface to take its place. This cold, deep water continually replaces the surface water that is pushed away from the coast. The cold water contains high concentrations of nutrients produced when dead organisms decayed at depth. This concentration of nutrients causes tiny marine organisms to flourish and fish to be attracted to areas of upwelling. Upwelling also affects the climate of coastal areas. Upwelling contributes to San Francisco's cool summers and famous fogs.

El Niño During an El Niño (el NEEN yoh) event, the winds blowing water from the coast of Peru slacken, the eastern Pacific is warmed, and upwelling is reduced or stops. Without nutrients provided by upwelling, fish and other organisms cannot find food. Thus, the rich fishing grounds off of Peru are disrupted.

section 2 review

Summary

Surface Currents

- As global winds blow along the ocean's surface, they can set surface water in motion. Due to the Coriolis effect, ocean surface currents do not move in straight lines.
- Surface currents can influence local climate.

Density Currents

- Density currents form when more dense seawater sinks beneath less dense seawater.

Upwelling

- Upwelling brings deep, cold water to the ocean surface. The concentration of nutrients in upwelled water is high.

Self Check

1. **Explain** how winds create surface currents.
2. **Describe** how the rotation of Earth modifies ocean currents in the northern hemisphere.
3. **Summarize** what causes the density current in the Mediterranean Sea.
4. **Think Critically** Why is the surface water cooler near San Diego, California, compared to that off the coast of Charleston, South Carolina?

Applying Skills

5. **Compare and contrast** density currents and upwelling in the ocean.

Science Online red.msscience.com/self_check_quiz

section 2 review

1. They push the water.
2. Surface currents curve to the right.
3. Surface evaporation produces salty, dense water. It sinks, and less dense Atlantic water flows in to replace it.
4. CA: currents originate in northern latitudes; SC: currents originate at the equator.
5. Both involve vertical movement of water. A density current is caused by differences in salinity and temperature. An upwelling is caused when wind causes surface water to move away from shore.

384 CHAPTER 13 Oceans

section 3

Waves

Waves Caused by Wind

Have you ever wanted to surf? By catching a high, curled wave, you can ride all the way to the beach. A **wave** in water is a rhythmic movement that carries energy through the water. Waves that surfers ride could have originated halfway around the world. Whenever wind blows across a body of water, friction pushes the water along with the wind. If the wind speed is great enough, the water begins piling up, forming a wave. Three things affect the height of a wave: the speed of the wind, the length of time the wind blows, and the distance over which the wind blows. A fast wind that blows over a long distance for a long time creates huge waves. Once a wave forms, it can travel a great distance. But when winds stop blowing, waves stop forming.

Parts of a Wave Each wave has a crest, its highest point, and a trough, its lowest point. Wave height is the vertical distance between the crest and trough. The wavelength is the horizontal distance between the crests or troughs of two successive waves. **Figure 11** shows the parts of a wave.

In the open ocean, most waves have heights of 2 m to 5 m. Ocean waves rarely reach heights of more than 15 m. However, storm winds can produce waves more than 30 m high—taller than a six-story building—that can capsize even large ships.

as you read

What You'll Learn

- **Describe** how wind can form ocean waves.
- **Explain** the movement of water molecules in a wave.
- **Describe** how the Moon and Sun cause Earth's tides.
- **List** the forces that cause shoreline erosion.

Why It's Important

Wave erosion affects life in coastal regions.

Review Vocabulary
sediments: loose materials, such as rock fragments, mineral grains, and the remains of once-living plants and animals, that are moved by wind, water, ice, or gravity

New Vocabulary
- wave
- tide

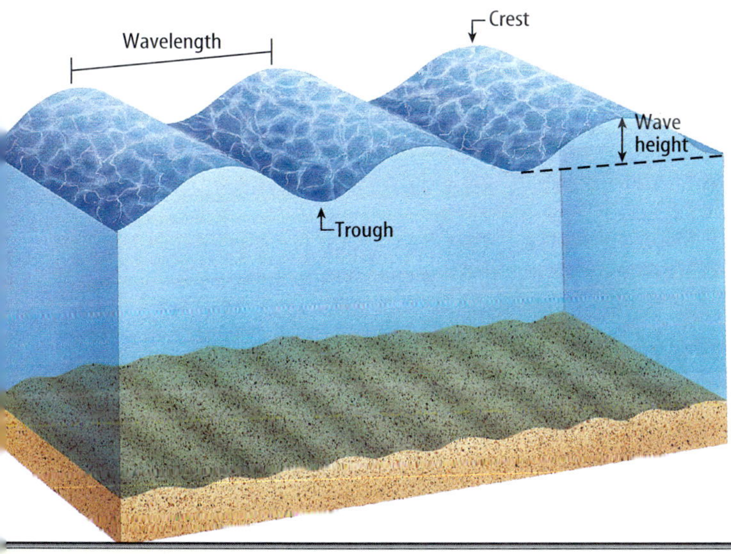

Figure 11 Every wave has a crest, trough, wavelength, and wave height.

1 Motivate

Bellringer

INTERACTIVE CHALKBOARD PowerPoint® Presentations

Section Focus Transparencies also are available on the Interactive Chalkboard CD-ROM. L2 ELL

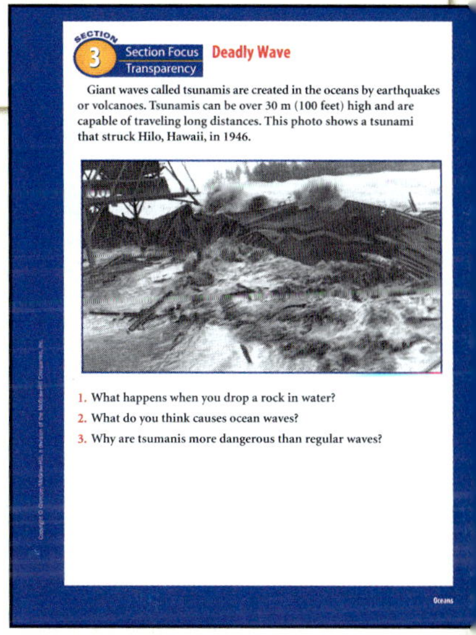
Section 3 Section Focus Transparency Deadly Wave

Giant waves called tsunamis are created in the oceans by earthquakes or volcanoes. Tsunamis can be over 30 m (100 feet) high and are capable of traveling long distances. This photo shows a tsunami that struck Hilo, Hawaii, in 1946.

1. What happens when you drop a rock in water?
2. What do you think causes ocean waves?
3. Why are tsunamis more dangerous than regular waves?

Tie to Prior Knowledge

Ocean Appearance Have students who have experience with ocean water describe how ocean waves look. Also, encourage students to describe the changing tides. Explain that in this section, students will learn what causes waves and tides.

Section 3 Resource Manager

Chapter *FAST FILE* Resources
Transparency Activity, p. 48
Directed Reading for Content Mastery, p. 21
Lab Activity, pp. 13–16
MiniLAB, p. 4
Enrichment, p. 33
Reinforcement, p. 29
Science Inquiry Labs, pp. 25–26

2 Teach

Use an Analogy

In-line Skating When an in-line skater suddenly stops, friction stops the wheels, but the person's body keeps moving forward because of inertia. As a result, the person tumbles forward. In a breaking wave, friction with the ocean floor slows the bottom of the wave. The top of the wave keeps moving. In time, the wave breaks over on itself.

Mini LAB

Purpose Students model water particle movement in a wave. L2 ELL LS **Kinesthetic**

Materials large bowl, water, penny, toothpick, spoon

Teaching Strategy Remind students to make gentle waves.

Analysis

1. The waves moved to the other end of the bowl. The toothpick stayed in position over the penny.
2. Both stay in about the same place as a wave passes.

Assessment

Oral Why was the penny used in the MiniLAB? It gave the observer a point of reference for keeping track of the toothpick's movement.

Try at Home

Reading Check

Answer The top of a wave is not slowed by friction, but the bottom is. When the top overtakes the bottom, the wave collapses and a breaker forms.

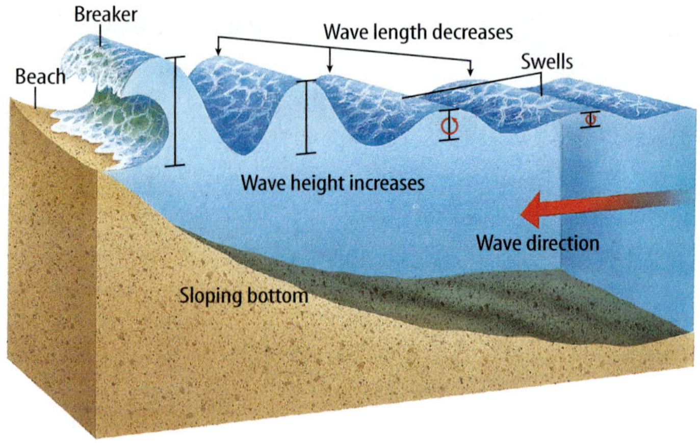

Figure 12 As a wave moves by, individual molecules of water move around in circles. As a wave approaches shore, wavelength decreases and wave height increases. Eventually the bottom of the wave cannot support the top, and the wave falls over on itself, creating a breaker.

Mini LAB

Modeling Water Particle Movement

Procedure

1. Fill a large **bowl** with **water** and place a **penny** on the bottom in the center of the bowl.
2. Float a small piece of **toothpick** in the bowl directly above the penny.
3. Gently dip a **spoon** into the water to make small waves.

Analysis

1. Compare and contrast the movement of the waves and the toothpick.
2. Compare the movement of the toothpick with the movement of water particles in a wave.

Try at Home

Wave Motion When you observe an ocean wave, it looks as though the water is moving forward. But unless the wave is breaking onto shore, the water does not move forward. Each molecule of water stays in about the same place in a passing wave. If you want to demonstrate how molecules move in a wave, tie a ribbon to the middle of a rope. Then hold one end of the rope and have someone else hold the other end. Wiggle the rope until a wave starts moving toward the other person. Notice that the wave travels through the rope to the other person, but the ribbon moves only in small circles, not forward.

Breakers As a wave approaches a shore, it changes shape. Friction with the ocean floor slows the water at the bottom of the wave. Notice in **Figure 12** that as the bottom of the wave slows, the crest and trough come closer together and the wave height increases. Because the top of the wave is not slowed by friction, it moves faster than the bottom. Eventually, the wave top overtakes the bottom, and the wave collapses. Water tumbles over on itself. This collapsing wave is called a breaker. Breakers make the best waves for surfers to ride. After a wave breaks onto shore, gravity pulls the water back into the sea.

Reading Check *What causes breakers to form?*

Along smooth, gently sloping coasts, waves deposit eroded sediments on shore, forming beaches. Beaches extend inland as far as the tides and waves are able to deposit sediments.

Waves usually approach a shore at slight angles. This creates a longshore current of water, which runs parallel to the shore. As a result, beach sediments are moved sideways. Longshore currents carry many metric tons of loose sediment from one beach to another.

Differentiated Instruction

English-Language Learners Have students make a poster that depicts the process of wave motion and what happens as a wave approaches shore, much like **Figure 12.** Students should include details as to what happens in each stage of wave motion and number the sequence of events as a wave approaches shore. Posters can be displayed in the classroom. L2 ELL LS **Visual-Spatial**

Science Journal

Sand Journey Have each student imagine that he or she is a grain of sand on a beach. Have them write stories in which they describe how tides, wind-driven waves, and longshore currents affect their movements along the beach. Encourage students to be creative. Suggest that they make sketches to accompany their stories. L2 ELL LS **Linguistic and Visual-Spatial** P

Tides

Throughout a day, the water level at the ocean's edge changes. This rise and fall in sea level is called a **tide.** A tide is a giant wave that can be thousands of kilometers long but only 1 m to 2 m high in the open ocean. As the crest of this wave reaches shore, sea level rises to form high tide. Later in the day, the trough of the wave reaches shore and sea level drops. This is low tide. The difference between sea level at high tide and low tide is the tidal range. The tidal range in some coastal areas can be as much as 20 m.

Causes of Tides Tides are not created by wind. They are created by the gravitational attraction of Earth and the Moon and of Earth and the Sun. The Moon and Earth are relatively close together in space, so the Moon's gravity exerts a strong pull on Earth. This gravity pulls harder on particles closer to the Moon than on particles farther from the Moon, causing two bulges of water to form. One bulge forms directly under the Moon and one on the opposite side of Earth. As Earth rotates, these bulges move to follow the Moon on its daily passage. The crests of these bulges are high tides. Between these bulges are troughs that create low tides.

The Sun's gravity also affects the tides. When the Moon, Earth, and Sun line up together, the high tides are higher and the low tides are lower than normal, creating spring tides. When the Sun, Earth, and Moon form a right angle, high tides are lower and low tides are higher than normal, creating neap tides. **Figure 13** illustrates this effect.

Science Online

Topic: Tidal Ranges

Visit red.msscience.com for Web links to information about tidal ranges around the world.

Activity Select a coastal location and find one month's worth of tidal data for that area. Graph the data. Calculate the tidal range for that month.

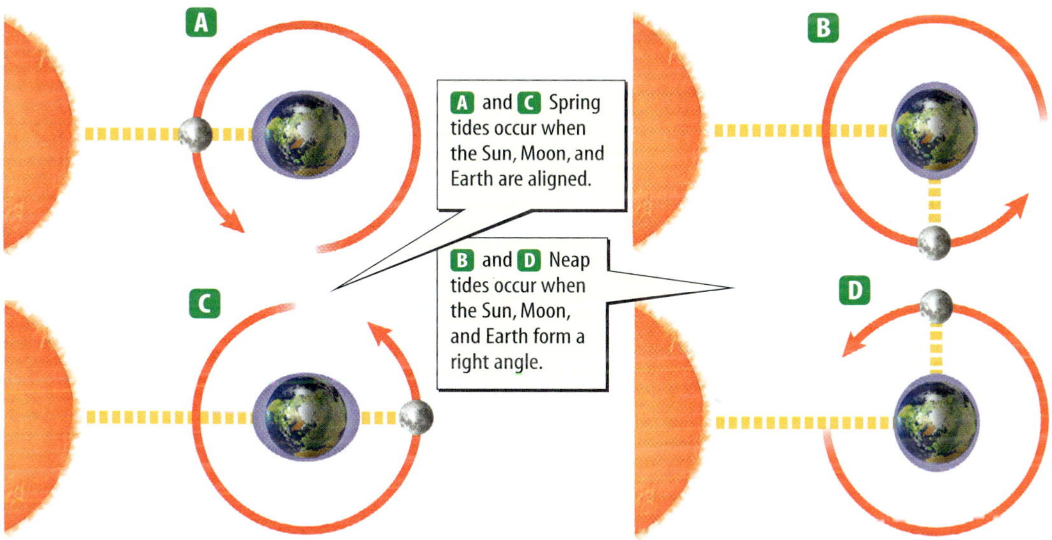

Figure 13 As the Moon and Earth revolve around a common center of mass, a bulge of water forms on the side of Earth closest to the Moon and on the side opposite the Moon.

Activity

Tidal Data Obtain a chart of tidal information for a favorite beach by contacting a coastal weather station, coastal marina, Navy or Coast Guard installation, or by searching the Internet. Have students graph one month's high and low tides and calculate each day's tidal range. Have students infer why patterns appear in their graphs. L2 LS **Logical-Mathematical**

Visual Learning

Figure 13 Which phases of the Moon occur during neap tides? Spring tides? Neap tides occur during the first and third quarter phases of the Moon. Spring tides occur when the Moon is in its new and full phases. L3 LS **Logical-Mathematical**

Teacher FYI

Breakers A wave becomes a breaker when the depth of the water equals one-half the wavelength. When this occurs, friction begins to slow the bottom of the wave.

Quick Demo

Phases of the Moon

Materials model of the Sun, the Moon, and Earth

Estimated Time 15 minutes

Procedure Using the model of the Sun, the Moon, and Earth, show the students how all three bodies interact to produce tides in Earth's oceans. Show the arrangement of all three bodies during each phase of the moon. Point out that during a new moon or a full moon the gravity of the Sun and the Moon pull together to generate the greatest tidal range.

Differentiated Instruction

Challenge Have students research more information about the use of tidal power to generate electricity. In what areas of the world is it used? What are some advantages and disadvantages of using tidal power? Have students share what they have learned with the rest of the class. L3 LS **Linguistic**

Curriculum Connection

Geography The tidal range in some areas of the world is magnified by the shape of the coastline. At the Bay of Fundy in Nova Scotia, the tide funnels into the narrow bay, resulting in a tidal range of as much as 20 m during spring tide. Have students locate the Bay of Fundy on a map. L2 LS **Visual-Spatial**

3 Assess

DAILY INTERVENTION

Check for Understanding

Visual-Spatial Inform students that the waves north of Antarctica are extremely large. Have students examine a world map and infer what may lead to these large waves. There are no landmasses to block the wind. The long distance over which the wind blows results in the formation of giant waves. L2 LS

Reteach

Do the "Wave" In order to demonstrate the movement of water particles in a wave, use the analogy of spectators in a stadium. Have five or more students line up their chairs side-by-side. Have them stand up and sit down in a precise order that imitates wave motion. Lead students to realize that as the wave motion moves down the row, the people do not move horizontally. L1 COOP LEARN LS
Kinesthetic and Interpersonal

Assessment

Process Compare and contrast tides and most other surface waves. Tides, like all waves, have crests, troughs, and wavelengths. Tides are caused by the gravitational attraction of Earth and the Moon and Earth and the Sun. Most other surface waves are formed by wind. Use **Performance Assessment in the Science Classroom,** p. 99.

Figure 14 In a single day about 14,000 waves will crash onto this rocky shore.

Wave Erosion

Waves can erode many meters of land in a single season. They wear away rock at the base of rocky shorelines, as shown in **Figure 14.** Then overhanging rocks fall into the water, leaving a steep cliff. Houses built on ocean cliffs can be damaged or destroyed by the erosion below. At Tillamook Rock, Oregon, storm waves hurled a 61-kg rock high into the air. The rock crashed through the roof of a building 30 m above the water.

Beach Erosion Sandy shorelines also can be eroded by waves. Large storms and hurricanes can produce waves that move much of the sand from the beach and can destroy large parts of some nearshore islands. Longshore currents also can erode beaches. This happens most often when people build structures called groins that extend out into the water. Although groins may protect beaches in some places, they often cause erosion elsewhere.

section 3 review

Summary

Waves Caused by Wind

- Waves form as wind blows across a body of water.
- Although it appears as though water is moving forward in a wave, each molecule of water stays in approximately the same place as a wave passes.

Tides

- Tides are the rise and fall in sea level that can be measured along the shore.

Wave Erosion

- Waves can cause erosion to occur along rocky coasts as well as sandy shores. Longshore currents also can erode beaches.

Self Check

1. **Summarize** how wind creates waves. What factors determine the size of waves?
2. **Describe** how a water molecule moves in a wave. What causes a wave to break?
3. **Explain** what causes tides. How do spring tides and neap tides differ?
4. **Think Critically** If a storm arrives at a beach during high tide, why is erosion especially damaging?

Applying Skills

5. **Recognize Cause and Effect** The city of Snyderville keeps pumping sand onto its beach, but the sand keeps disappearing. Explain where it is going.

Science Online red.msscience.com/self_check_quiz

section 3 review

1. Friction from the wind pushes the water along. Factors affecting size include speed of the wind, length of time the wind blows, and distance over which the wind blows.
2. Water molecules move in a circle. A wave breaks when its top overtakes its bottom.
3. gravitational attraction of Earth and the Moon and Earth and the Sun; spring tides: large tidal range; neap tides: small tidal range
4. The combination of strong storm waves and high tide exposes more beach area to erosion.
5. Some is carried down the beach by the longshore current. Some is carried by waves and deposited offshore.

section 4

Life in the Oceans

Types of Ocean Life

Organisms live in many different areas of the ocean. Where an organism lives and how it moves from place to place determines whether it is classified as plankton, nekton, or a bottom dweller.

Plankton Tiny marine organisms that float in ocean currents are called **plankton.** Most plankton are one-celled organisms, such as the diatoms (DI uh tahmz) pictured in **Figure 15.** Some plankton can swim, but most drift with currents. You would need a microscope to see most of these organisms. Examples of animal plankton include eggs of ocean animals, very young fish, larval jellyfish and crabs, and tiny adults of some organisms. **Figure 15** shows a tiny jellyfish about 2 cm in diameter and the eggs of corals being released into the water where they will float.

as you read

What You'll Learn

- **Describe** the characteristics of plankton, nekton, and bottom-dwelling organisms.
- **Distinguish** among producers, consumers, and decomposers.
- **Discuss** how energy and nutrients are cycled in the oceans.
- **Explain** how organisms in the oceans interact in food chains.

Why It's Important

Marine organisms provide people with much of the food they need to survive.

Review Vocabulary

nutrients: substances needed by organisms to carry out life processes

New Vocabulary

- plankton
- nekton
- ecosystem
- producer
- chemosynthesis
- consumer
- decomposer
- food chain

Figure 15 Plankton, such as this jellyfish, float in the upper layers of the ocean.

The eggs of corals become plankton when they are released into the water column.

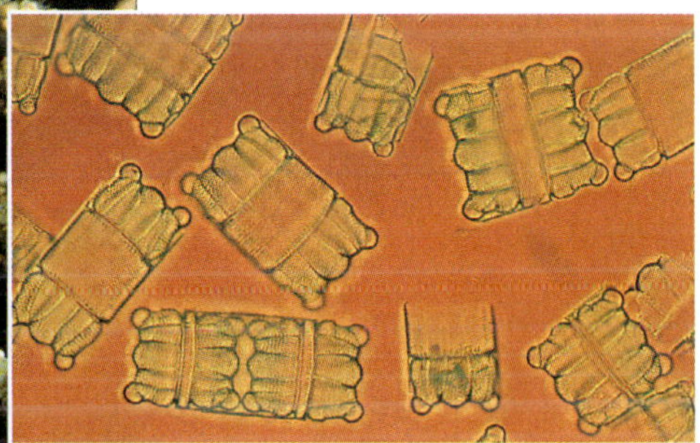

Diatoms, shown above, are one-celled organisms that undergo photosynthesis.

1 Motivate

INTERACTIVE CHALKBOARD PowerPoint® Presentations

Bellringer

Section Focus Transparencies also are available on the Interactive Chalkboard CD-ROM. L2 ELL

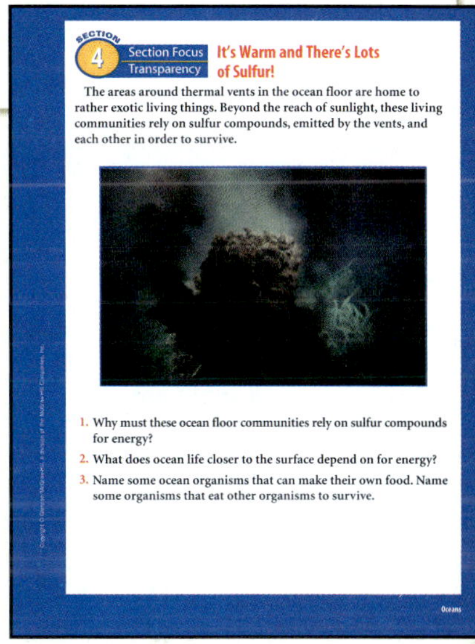

Tie to Prior Knowledge

Marine Organisms Have students describe marine organisms they have seen at the beach or on nature programs on television. Explain that in this section, they will learn how these organisms are classified and how they interact.

Section 4 Resource Manager

Chapter *FAST FILE* Resources

Transparency Activity, pp. 49, 51–52
Directed Reading for Content Mastery, pp. 21, 22
Enrichment, p. 34
Reinforcement, p. 30
Lab Worksheet, pp. 7–8

Physical Science Critical Thinking/Problem Solving, p. 23
Home and Community Involvement, p. 38
Life Science Critical Thinking/Problem Solving, p. 10
Mathematics Skill Activities, p. 33
Cultural Diversity, p. 3

2 Teach

Use Science Words

Word Meaning Have students find out why jellyfish larvae and very young fish are called *zooplankton*, while diatoms and some other protists are called *phytoplankton*. *Zoo-* means animal; jellyfish larvae and young fish are animals. *Phyto-* means plant; like plants, diatoms and some other protists make their own food. *Plankton* means to wander or drift. Jellyfish larvae, very young fish, and diatoms drift in the surface currents. L2 LS **Linguistic**

Inquiry Lab

Modeling Diatoms

Purpose To display the diversity of shapes and colors of diatoms. L1 ELL LS **Visual-Spatial**

Possible Materials pipe cleaners, straws, wooden dowels, chopsticks, color plastic wrap, transparent tape, glue, metric ruler, ocean guidebooks to marine life

Estimated Time one class session

Teaching Strategies

- Students can research the various shapes and colors of diatoms using ocean guidebooks to marine life and online sources.
- Students can tape or glue model diatom frames together using straws, pipe cleaners, or other materials. Plastic wrap of various colors can be stretched across the frames to model the transparent shells of diatoms.
- Students should research the sizes of diatoms and use a scale of their choice to create models.

For additional inquiry activities, see *Science Inquiry Labs*.

Caption Answer

Figure 16 Answers can include sharks, dolphins, manatees, seasnakes, or seals.

Figure 16 Swimming animals, such as this anglerfish (left), are nekton.
List *two other examples of nekton in the ocean.*

Nekton Animals that can actively swim, rather than drift in the currents, are called **nekton.** Fish, whales, shrimp, turtles, and squid are nekton. Swimming allows these animals to search more areas for food. Some nekton, such as herring, come to the surface to feed on plankton, but others remain in deeper water.

Some of the nekton that live in the dark abyss of the deepest parts of the ocean have organs that produce light, which attracts live food. Shown in **Figure 16,** an anglerfish dangles a luminous lure over its head. When small animals like shrimp bite at the lure, the anglerfish swallows them whole.

Bottom Dwellers Some organisms live on the ocean bottom. They can burrow in sediments, walk or swim on the bottom, or be attached to the seafloor.

Bottom-dwelling animals include anemones, crabs, corals, snails, starfish, and some fish. Many of these animals, such as sea cucumbers, eat the partially decomposed matter that sinks to the ocean floor. Some, such as the sea star in **Figure 17,** prey on other bottom dwellers. Others that are attached to the bottom, such as sponges, filter food particles from the water. Still others, such as anemones, corals, and the sea fans shown in **Figure 17,** are found in coral reefs. They capture organisms that swim by.

Figure 17 Bottom dwellers vary greatly. Many sea stars feed on other bottom dwellers such as clams. Sea fans, shown in the photo on the right, live attached to the bottom and cannot move from place to place.

Differentiated Instruction

Learning Disabled Have students collect pictures of marine organisms from magazines. They should classify the organisms as plankton, nekton, or bottom-dwellers and mount the pictures on poster board for display in the classroom. L1 LS **Visual-Spatial**

Curriculum Connection

Literature Many authors have written about the oceans. Examples include Jules Verne's *Twenty Thousand Leagues Under the Sea*, Ernest Hemingway's *The Old Man and the Sea*, Yukio Mishima's "The Sound of Waves," Rachel Carson's *The Sea Around Us*, and Herman Melville's *Moby Dick*. Have students choose one to read. Have them share excerpts with their classmates. L2 ELL LS **Linguistic**

Ocean Ecosystems

The oceans are home to many different kinds of organisms. No matter where organisms live, they are part of an ecosystem. An **ecosystem** is a community of organisms and the nonliving factors that affect them, such as sunlight, water, nutrients, sediment, and gases. Every ecosystem has producer, consumer, and decomposer organisms.

Producers Producer organisms, such as those shown in **Figure 18,** form the base of all ecosystems. Producers are organisms that can make their own food. **Producers** near the ocean's surface contain chlorophyll. This allows them to make food and oxygen during photosynthesis.

In deep water, where sunlight does not penetrate, producers that use chlorophyll can't survive. In this part of the ocean, producers make food by a process called **chemosynthesis.** *Chemo* means "chemical." This process often takes place along mid-ocean ridges where hot water circulates through the crust. Bacteria produce food using dissolved sulfur compounds that escape from hot rock. The bacteria then are eaten by organisms such as crabs and tube worms.

Consumers and Decomposers Consumer and decomposer organisms depend upon producers for survival. Organisms that eat, or consume, producers are called **consumers.** Consumers get their energy from the food stored in the producers' cells. Some also eat other consumers to get energy. When producers and consumers die, decomposers digest them. **Decomposers,** such as bacteria, break down tissue and release nutrients and carbon dioxide back into the ecosystem.

Reading Check *What is a consumer?*

INTEGRATE Social Studies

Economics and Fish Many coastal areas of the world depend on harvesting fish to keep the local economy healthy. Certain areas of the world such as the coast of Peru, the Gulf of Alaska and the Bering Sea, and the Grand Banks on the east coast of Canada, are well known for their fisheries. Events such as El Niño or the overharvesting of fish can greatly affect a fishing season.

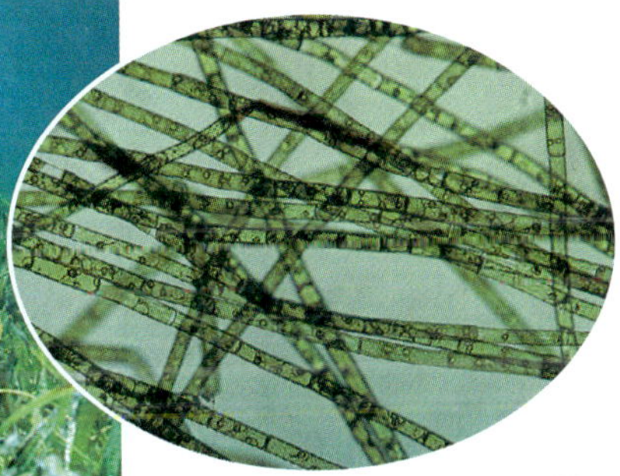

Figure 18 Producers can be large like the sea grass shown on the left. Sometimes they are as small as the microscopic algae shown on the right.
Infer *what all producers have in common.*

Activity

Viewing Organisms Show a video that features ocean organisms in their natural habitats. During the movie, have students list the producers and consumers they see. At the conclusion of the movie, have students share and discuss their lists. L2 ELL COOP LEARN LS **Visual-Spatial and Interpersonal**

INTEGRATE Social Studies

Economics and Fish The fishing grounds off the Grand Banks were closed for cod fishing in 1992 in a last effort to save the fishery from extinction. The Georges Bank fishery, off the New England coast, was closed in 1994 for similar reasons. These closures led to increased unemployment levels in these areas and higher prices for consumers. The fishery off the coast of Peru has been affected more than once by El Niño, with each El Niño event resulting in a decline in the number of anchovies and sardines in the area.

Reading Check

Answer an organism that gets energy from food stored in producers' and other consumers' cells

Caption Answer

Figure 18 All producers make their own food.

Make a Model

Food Web Have students create a model of a marine food web. They can use yarn, magazine photos, tape, poster board, index cards, and colored pencils to make a model food web. Encourage students to be both accurate and creative in their construction of the food web, making sure to have a realistic depiction of the different organisms and how they interact as producers, consumers, and decomposers. Have students share their models with the class. L2 ELL LS **Visual-Spatial** P

Curriculum Connection

Art Have student groups select a marine ecosystem to study in depth. Suggested ecosystems include sandy coastal region, coral reef, kelp forest, mid-ocean ridge, open water, and deep-ocean sea floor. After researching the ecosystem, have each group make a diorama of the ecosystem. L2 ELL COOP LEARN LS **Kinesthetic**

Discussion

Food Web As a result of rising temperatures in some oceans, some types of plankton are dying. What effect will this have on the food webs in these areas? Consumers that normally eat these plankton will have to depend on other food sources. If other food is not available, some consumers will die. Consumers that eat these consumers may also be affected. With additional organisms dying, decomposers will thrive. L2 LS **Logical-Mathematical**

Applying Science

Teaching Strategy

Students will need to understand that, even though the limits are below 1 ppm (in the safe area), fish with methyl mercury ranges near the limit are safe to eat in moderation, for example, once a week. This keeps the mercury levels below the acceptable daily intake.

Answers

1. All the fish may contain methyl mercury. The detectable limit is 0.10 ppm. There can still be mercury present, just less than 0.10 ppm.
2. Shark and swordfish may put you in danger of mercury poisoning. The ranges go to 3.52 ppm and 1.68 ppm, respectively, which is over the 1 ppm FDA limit. The chart is consistent with the FDA recommendation of one shark or swordfish serving per week. The average mercury content for these fish is less than the FDA limit. Flounder and salmon can be safely eaten as often as you want.

Science Online

Topic: Food Webs

Visit red.msscience.com for Web links to information about ocean food chains and food webs.

Activity Find an example of an ocean food chain or food web. Display the information you find on a poster.

Food Chains Throughout the oceans, energy is transferred from producers to consumers and decomposers through **food chains**. In **Figure 19,** notice that algae (producers) are eaten by krill that are, in turn, eaten by Adélie penguins. Leopard seals eat the penguins, and killer whales eat the leopard seals. At each stage in the food chain, energy obtained from one organism is used by other organisms to move, grow, repair cells, reproduce, and eliminate wastes. Energy not used in these life processes is transferred along the food chain.

All ecosystems have many complex feeding relationships. Most organisms depend on more than one species for food. Notice in **Figure 19** that krill eat more than algae and in turn are eaten by animals other than Adélie penguins. In the Antarctic Ocean, as in all ecosystems, food chains are interconnected to form highly complex systems called food webs.

Applying Science

Are fish that contain mercury safe to eat?

When mercury, once used in pesticides, is added to oceans or bodies of freshwater, bacteria change it to methyl mercury, which is a more toxic form of mercury. Fish then absorb the methyl mercury from the water as it flows over their gills or as they feed on aquatic organisms. Larger fish feed on the smaller fish, and humans often eat larger fish. The table on the right lists the methyl mercury ranges for a variety of fish.

Methyl Mercury Content in Domestic Fish

Species	Range (ppm)	Average (ppm)
Catfish	ND–0.16	ND
Cod	ND–0.17	0.13
Crab	ND–0.27	0.13
Flounder	ND	ND
Halibut	0.12–0.63	0.24
Salmon	ND	ND
Tuna (canned)	ND–0.34	0.20
Tuna (fresh)	ND–0.76	0.38
Swordfish	0.36–1.68	0.88
Shark	0.30–3.52	0.84

Identifying the Problem

The average methyl mercury present in each fish is given in the chart in parts per million (ppm). The detection limit is 0.10 ppm. Any values less than 0.10 ppm are shown as ND (not detected). The FDA's (Food and Drug Administration) safe limit for human consumption is 1 ppm. Which species of fish could put you in danger of mercury poisoning if you eat them?

Solving the Problem

1. Which of the fish listed do not contain any methyl mercury? Explain.
2. The FDA limit of 1 ppm is 10 times lower than levels found in fish that have caused illness. The FDA recommends eating shark or swordfish no more than once a week. Does this appear consistent with the information given in the data table? Explain. Which fish could you safely eat as often as you wanted?

Differentiated Instruction

Challenge Have students find out the meaning of the word *biodiversity* and make a collage of an ocean ecosystem. They should show and label many different life-forms, including micro as well as macro. Have them explain why biodiversity is important in ocean food webs. Biodiversity refers to the many different types of life that exist in an ecosystem. Complex food webs depend on biodiversity. If certain species become extinct, branches of the food web collapse. Remaining species will become more dependent on fewer sources of food. L3 LS **Linguistic and Visual-Spatial**

NATIONAL GEOGRAPHIC

VISUALIZING FOOD CHAINS IN A FOOD WEB

Figure 19

A food web represents a network of interconnected food chains. It shows how energy moves through an ecosystem—from producers to consumers and eventually to decomposers. This diagram shows a food web in the Antarctic Ocean. Arrows indicate the direction in which energy is transferred from one organism to another.

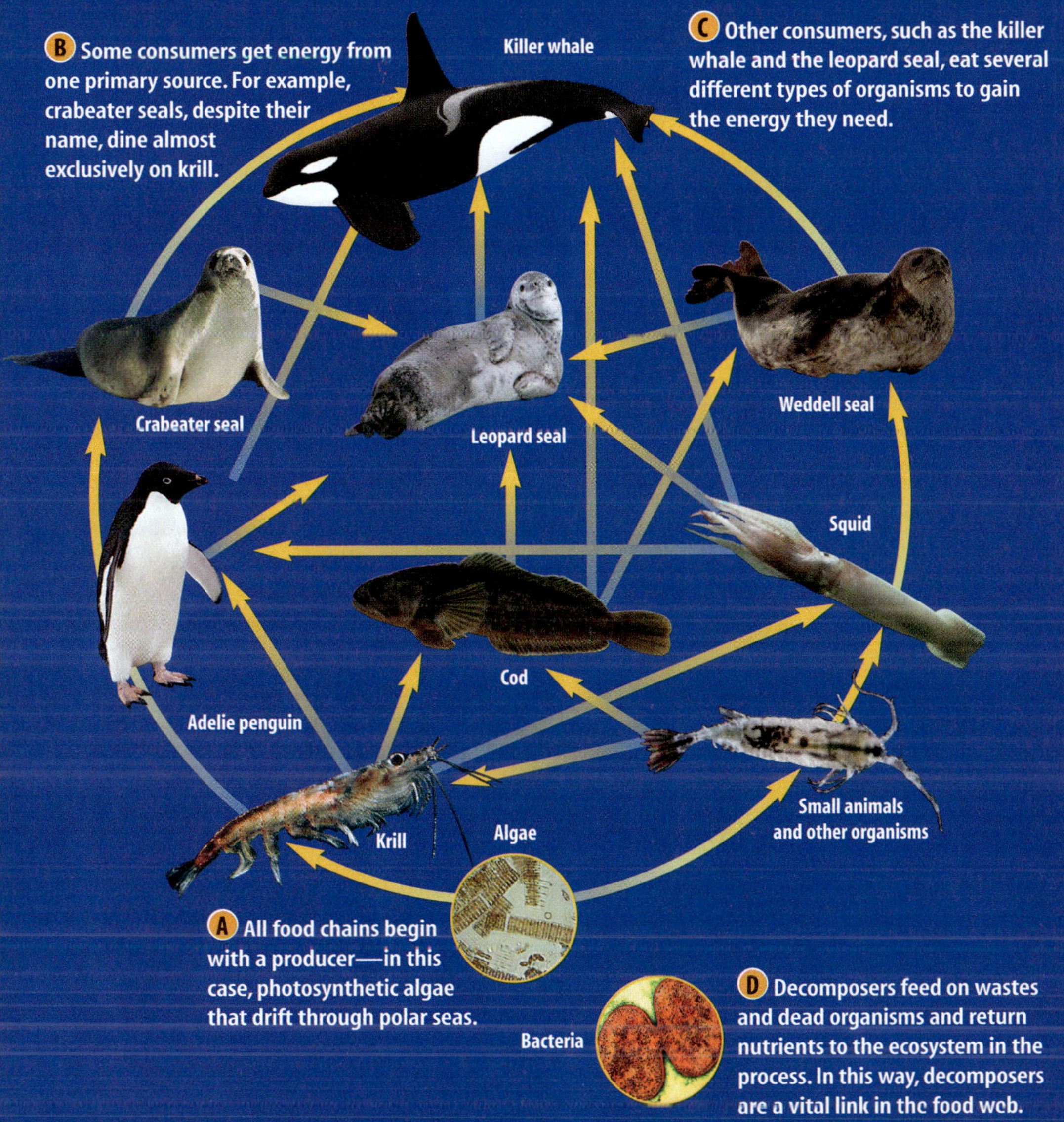

B Some consumers get energy from one primary source. For example, crabeater seals, despite their name, dine almost exclusively on krill.

C Other consumers, such as the killer whale and the leopard seal, eat several different types of organisms to gain the energy they need.

A All food chains begin with a producer—in this case, photosynthetic algae that drift through polar seas.

D Decomposers feed on wastes and dead organisms and return nutrients to the ecosystem in the process. In this way, decomposers are a vital link in the food web.

NATIONAL GEOGRAPHIC

Visualizing Food Chains in a Food Web

Have students examine the pictures and read the captions. Then ask the following questions.

What might happen to the food web if overfishing greatly reduced the number of krill in the Antarctic Ocean? Possible answer: Other organisms that feed on krill would not have enough food. Many of these organisms might die of starvation and organisms that feed on them would then not have enough food.

What would happen if there were no decomposers in a food web? Wastes and the remains of dead organisms would accumulate in the ecosystem, and nutrients would not be returned to the ecosystem.

Activity

Represent Food Webs Have small groups of students select one organism within the food web to represent. Each student should make a sign indicating which organism he or she represents. Students should then stand in the proper order as each one describes what he or she represents and does in the food chain. For example, one student might say, "I am algae and I produce food by using the Sun's energy." L1 COOP LEARN LS **Visual-Spatial**

Differentiated Instruction

Challenge Ask students to research the effects of overharvesting marine organisms on food webs in different areas of the ocean. Have students present oral reports on their findings. L3 LS **Auditory-Musical**

Visual Learning

Figure 20 Have a student volunteer use this figure to explain the nitrogen cycle to the class. What is the importance of bacteria in recycling nitrogen? Bacteria help change nitrogen into compounds that other organisms can use for life processes. L2 LS **Auditory-Musical**

Use an Analogy

Parts Per Million The concentration of many substances is reported in parts per million (ppm.) For example, the concentration of carbon dioxide in the oceans is 280 ppm, while the concentration of carbon dioxide in the atmosphere is approximately 368 ppm. Explain to students that one part per million is like one second in 12 days of their lives or one penny in $10,000.

Teacher FYI

Burning Fossil Fuels When some types of fossil fuels are burned, carbon that was stored by ancient marine organisms is released into the atmosphere as carbon dioxide.

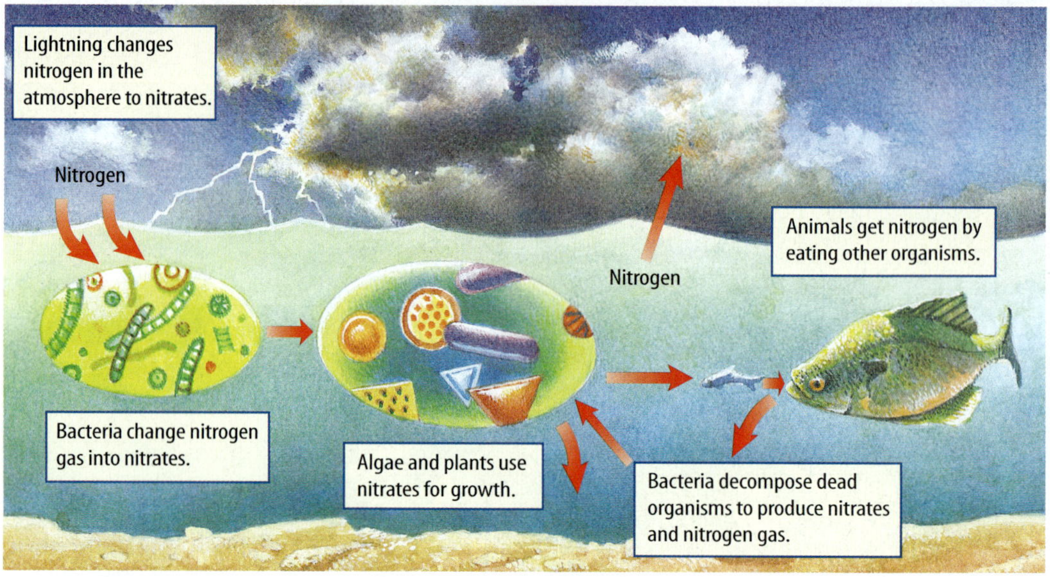

Figure 20 Nitrogen cycles from nitrogen gas in the atmosphere to nitrogen compounds and back again.

Figure 21 Carbon cycles through the ocean and between the ocean and the atmosphere.

Ocean Nutrients

Nearly everything in an ecosystem is recycled. When organisms respire, carbon dioxide is released back into the ecosystem. When organisms excrete wastes or die and decompose, nutrients are recycled. All organisms need certain kinds of nutrients in order to survive. For example, plants need nitrogen and phosphorus. **Figure 20** shows how nitrogen cycles through the ocean.

Carbon also is recycled. You learned earlier in this chapter that oceans absorb carbon dioxide from the atmosphere. You also learned that producers use carbon dioxide to make food and to build their tissues. Carbon then can be transferred to consumers when producers are eaten. When organisms die and sink to the bottom, some carbon is incorporated into marine sediment. Over time, carbon is exchanged slowly between rocks, oceans, the atmosphere, and organisms, as seen in **Figure 21.**

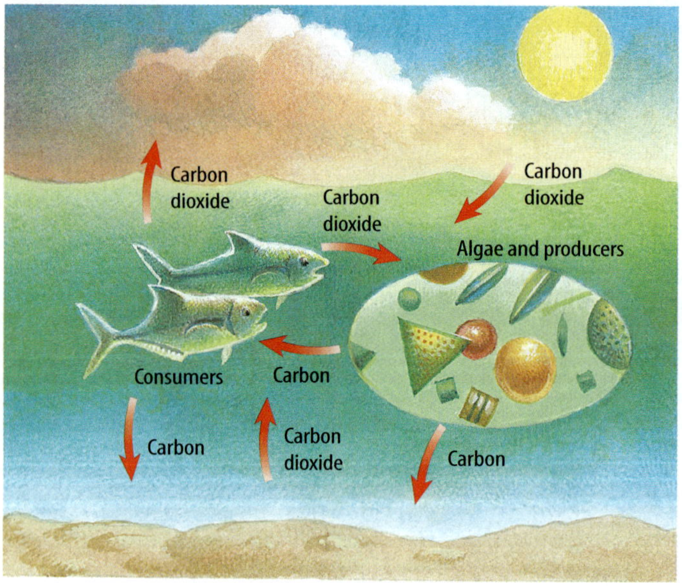

Curriculum Connection

Health Many people enjoy seafood without realizing its nutritional value. All seafood is a source of protein, vitamins B12 and B6, iodine, selenium, and phosphorus. Oysters and crabs are high in zinc and iron, fish in calcium, and mussels in potassium. Have students choose one vitamin or mineral and research why humans need it for good health. L2 LS **Linguistic**

Figure 22 Parrot fish are efficient recycling organisms. They turn coral into fine sand as they graze on the algae in the coral.

Coral Reefs and Nutrient Recycling Coral reefs are ecosystems that need clear, warm, sunlit water. Each coral animal builds a hard calcium carbonate capsule around itself. Inside the animals' cells live algae that provide the animals with nutrients and give them color. As corals build one on top of another, a reef develops. Other bottom-dwelling organisms and nekton begin living on and around the reef. Nearly 25 percent of all marine species and 20 percent of all known marine fish live on coral reefs. Coral reefs generally form in tropical regions in water no deeper than 30 m.

A healthy reef maintains a delicate balance of producers, consumers, and decomposers. Energy, nutrients, and gases are cycled among organisms in complex food webs in a coral reef. Look at **Figure 22** to see one example of how materials are cycled through a coral reef.

section 4 review

Summary

Types of Ocean Life

- Marine organisms can be classified based on where they live and how they move from place to place: plankton, nekton, or bottom dweller.

Ocean Ecosystems

- Ocean ecosystems contain producers, consumers, and decomposers.
- Energy in an ecosystem is transferred through a food chain or a food web.

Ocean Nutrients

- Nutrients are passed from producers to consumers when producers are eaten. When organisms die, decomposers release these nutrients back into the ecosystem.

Self Check

1. **List** the characteristics of plankton, nekton, and bottom-dwelling organisms.
2. **List** the characteristics of producers, consumers, and decomposers.
3. **Explain** how carbon is cycled through the oceans. How is nitrogen cycled?
4. **Think Critically** Why must every ecosystem include producers as well as other organisms?

Applying Skills

5. **Form Hypotheses** Write a hypothesis about how an increase in the amounts of ocean nutrients might affect producers. How would the increase in nutrients affect consumers?

Science Online red.msscience.com/self_check_quiz

3 Assess

DAILY INTERVENTION

Check for Understanding

Visual-Spatial Have student make a flow chart that shows how carbon is used and recycled in the ocean. L2 LS

Reteach

Marine Organisms Display photographs of marine organisms. Have students classify each organism as plankton, nekton, or bottom-dweller; producer, consumer, or decomposer. L1 LS **Visual-Spatial**

Assessment

Process Ask students to infer an advantage nekton have over animal plankton when it comes to finding food. Because nekton are strong swimmers, they are able to search more areas for food than can animal plankton, which rely on drifting in currents. Use **Performance Assessment in the Science Classroom**, p. 89.

section 4 review

1. Plankton: float in currents; nekton: actively swim; bottom-dwellers: live on bottom, some attached, some walk or swim.
2. Producers: make own food; consumers: eat other organisms; decomposers: break down tissue, release nutrients and CO_2
3. Carbon: producers use CO_2 to make food; carbon is transferred to consumers when producers are eaten; animals and plants respire CO_2; Carbon is deposited in organic sediments. Nitrogen: cycles from plant tissue to animal tissue to bacteria; bacteria produce nitrogen gas and nitrates; plants use nitrates
4. Producers make food and oxygen that other organisms need to survive.
5. With more nutrients, producers will increase in number and mass. As producers increase in number and mass there is more food available for consumers.

BENCH TESTED

Real-World Question

Purpose Students will model waves and tides and infer how erosion might occur in the model.

Process Skills design and make a model, predict, compare, analyze data

Time Required 45 minutes

Possible Materials Battery-operated fans reduce the risk of shock and are available at camping supply and hardware stores.

Safety Precautions If fans that must be plugged into an electrical outlet are used, be sure to caution students not to let the fan or the cord get wet and not to touch the fan and the water at the same time. Remind students to immediately wipe up any water that spills on the floor.

LAB Model and Invent

Waves and Tides

Goals

- **Construct** a model of the edge of the ocean.
- **Demonstrate** how you can simulate waves and tides in your model.
- **Predict** how erosion might occur in your model.

Possible Materials

large basin
water
boards
sand
**gravel*
bricks
rocks
plastic bottles
fan (battery-operated)
**Alternate materials*

Safety Precautions

Real-World Question

The water in the Gulf of Mexico is subject to the same forces as water in the open ocean. Daily high and low tides affect the water level, and waves are involved in shoreline erosion. How can you simulate ocean waves and represent tidal changes in water level? How will the tides affect the amount of erosion in various areas? How can you model ocean waves and tides in the classroom? Can you simulate waves and tides along the edge of an ocean using a basin of water in the classroom?

Make a Model

1. **Determine** how you are going to create a model of the edge of the ocean. Draw a picture of what your model will look like.
2. Decide how you will create waves and tides in your model. What can you use to move the water? How can you simulate tides by changing the height of the water level?
3. **Predict** where in your model erosion may occur and where it may not. How might you be able to see where erosion will occur?

Check the Model Plans

1. **Compare** your model plans with those of other students in the class. Discuss why each of you chose the design you did.
2. Make sure that your teacher approves your model plans before you construct your model.

Cultural Diversity

Bore Rafting In Nova Scotia's Bay of Fundy rising tide water forms tidal bores in some rivers. A tidal bore is a large wave that flows upstream and reverses the river's flow. In these rivers, bore rafting is a popular sport. It combines the thrill of white-water rafting with bird watching, as eagles, ospreys, and seabirds fish the incoming tidal waters.

Test Your Model

1. **Construct** your model based on your design plans.
2. Create waves in your model and observe what happens. Record your observations in your Science Journal.
3. Change the tide by changing the water level and repeat step 2.

Analyze Your Data

1. **Describe** what you observed when you created waves and tides in your ocean model.
2. Were you able to see any evidence of erosion in your model? If so, in what areas of your model was erosion present and where was it absent? If not, where would you expect to see erosion over a longer period of time? Explain.

Conclude and Apply

1. Did the waves you created always look the same or did they seem to vary in wave height or wavelength? Explain.
2. Did you see anything that looked like breakers as the waves hit the shore?
3. In what ways was your model similar to and different from the edge of a real ocean?
4. What features were you able to simulate and what features were missing from your model?

Communicating Your Data

Discuss with your family or students in other classes how ocean waves and tides affect erosion along the edge of oceans. **For more help, refer to the Science Skill Handbook.**

Make a Model

Teaching Strategies Remind students not to make waves so large that they splash water out of the basin.

Troubleshooting Students may not observe any erosion if they don't generate waves consistently for a period of time.

Test Your Model

Expected Outcome Most results will show that beach erosion occurs as a result of wave action, and that erosion increases at high tide.

Analyze Your Data

1. Waves push sand up onto some parts of the beach and carry it away in other parts. The process occurs higher up the beach when the tide is high.
2. Answers will depend on the model of shoreline created. All models should be expected to show erosion over a period of time.

Conclude and Apply

1. Answers will depend on the method used to generate waves and the structure of the shoreline model.
2. Answers will depend on the method used to generate waves and the structure of the shoreline model.
3. Answers will depend on the model created, but students might not be able to model sea cliffs.
4. Answers will vary depending on the model created.

Assessment

Process Show students photographs of large storm waves striking a shoreline and ask them what effect these waves would have on a shoreline. Use **Performance Assessment in the Science Classroom,** p. 89.

Communicating Your Data

Students could use a word processing program to outline the major points they want to make in their discussion.

Content Background

In many ways the world under the oceans is similar to the landscape on dry ground. Mountains, valleys, and volcanoes give the ocean floor a varied topography. Stretching from the edges of continents are continental shelves, with an average width of 70 km and depth of about 145 m. From the continental shelf the sea floor drops off to the abyssal plain, which lies at a depth of about 4,000 m. Seamounts are underwater volcanoes that rise over 1,000 m from the ocean floor. Trenches form where the oceanic lithosphere sinks into Earth's mantle. In 1960, the bathyscaph *Trieste* made the deepest dive ever—10,911 m—into the Marianas Trench.

Discussion

Growth of Knowledge Why do you think scientists' knowledge of the ocean has grown so much in the last 60 years? Possible answer: It was very difficult to explore the deep parts of the ocean because of the cold temperatures and tremendous pressures until the development of technologies such as scuba gear, submersibles, and robots. L2 LS **Logical-Mathematical**

Activity

Exploration Have small groups of students research the history of undersea exploration and make illustrated timelines of some important discoveries and the technologies that allowed those discoveries to be made. L2 COOP LEARN LS **Visual-Spatial**

Applying Math

Answer about 2,000 m; 11,033 m − 8,800 m = 2,233 m

SCIENCE Stats

Ocean Facts

Did You Know...

. . . The deepest place in the ocean is the Marianas Trench, located east of the Philippines. The deepest part of the Marianas Trench is the Challenger Deep, which extends 11,033 m down. That's deep enough to hold Mount Everest and the world's five tallest buildings stacked on top of one another.

Great Barrier Reef

. . . At more than 2,000 km long, the Great Barrier Reef in the Coral Sea is the largest organic structure on Earth and can be seen clearly from space.

Height and Depth Comparison of Land and Ocean

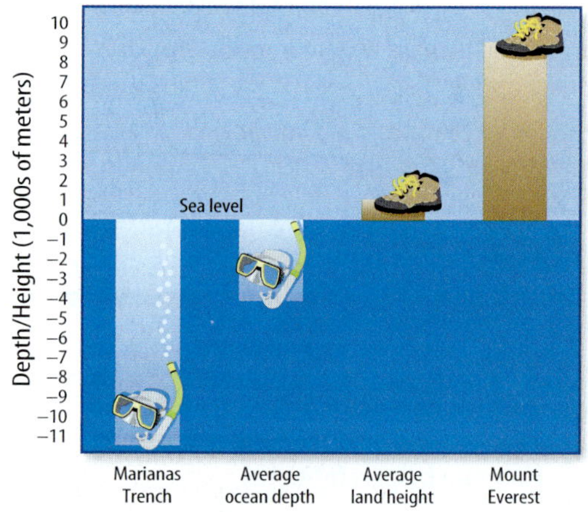

Applying Math Use the graph to find out about how much deeper the Marianas Trench is than Mount Everest is high.

Find Out About It

Visit red.msscience.com/science_stats to find the surface area of the three largest oceans. Make a graph showing the relative sizes of these oceans.

Find Out About It

Discuss with students which type of graph would be best for presenting this type of information—a bar graph, a circle graph, or a line graph.

Visual Learning

Height and Depth Comparison of Land and Ocean Ask students how many meters each number on the side of the graph represents. 1,000 meters Then ask them to compare the average ocean depth with average land height. The average ocean depth is about 4,000 m, while the average land height is about 1,000 m, so the average ocean depth is about 4 times as great as the average land height. L1 LS **Visual-Spatial**

Reviewing Main Ideas

Section 1 Ocean Water

1. Oceans provide much of the oxygen and food for Earth's organisms. Oceans interact with the atmosphere to create weather and climate.
2. Scientists think early oceans formed when basins filled with water that condensed from the water vapor of erupting volcanoes.
3. Seawater is a combination of water, dissolved solids, and dissolved gases.
4. Ocean temperatures vary with latitude and depth. Water pressure is created by gravity pulling down water molecules.

Section 2 Ocean Currents and Climate

1. Winds blowing across oceans produce surface currents. Earth's rotation deflects surface currents.
2. Ocean surface currents can be warm or cold, and affect the climates of coastal regions.
3. Density currents develop because water masses have different temperatures and different salinity levels.

Section 3 Waves

1. Winds cause water to pile up, forming waves.
2. Tides are created by the gravitational attraction of Earth and the Moon and of Earth and the Sun.

Section 4 Life in the Oceans

1. Plankton drift in ocean currents, and nekton actively swim. Some organisms live on the seafloor.
2. In an ecosystem, producers, consumers, and decomposers interact with each other and their surroundings.
3. Nutrients are cycled in the oceans.

Visualizing Main Ideas

Copy and complete the following concept map about types of ocean organisms.

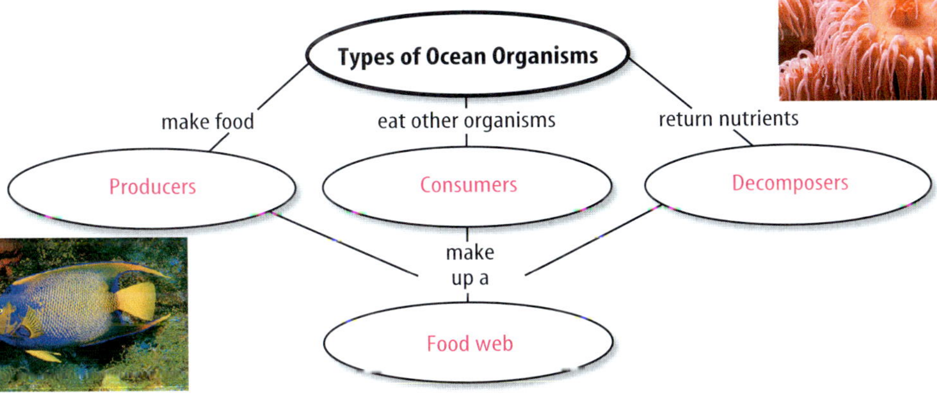

Identifying Misconceptions

Assess

Use the assessment as follow-up to page F at the beginning of this chapter.

Procedure Have students use clay to construct a 3-D model of their unseen ocean floor and compare their constructed model to the actual model inside the shoebox.

Expected Outcome Students will use sampling to discover the features of the ocean floor. This will reinforce the concept that the ocean floor is not flat.

Reinforcement Have students name the features they found while sampling their ocean floor.

chapter 13 Study Guide

Reviewing Main Ideas

Summary statements can be used by students to review the major concepts of the chapter.

Visualizing Main Ideas

See student page.

Science Online

Visit **red.msscience.com**
/self_check_quiz
/interactive_tutor
/vocabulary_puzzlemaker
/chapter_review
/standardized_test

Assessment Transparency

For additional assessment questions, use the *Assessment Transparency* located in the transparency book.

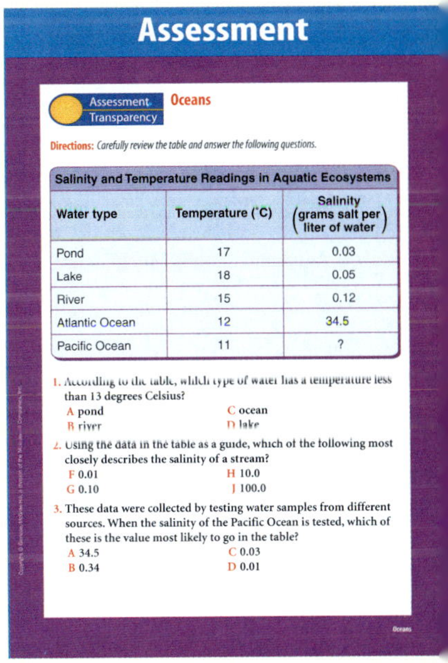
Assessment

Assessment Transparency **Oceans**

Directions: *Carefully review the table and answer the following questions.*

Salinity and Temperature Readings in Aquatic Ecosystems

Water type	Temperature (°C)	Salinity (grams salt per liter of water)
Pond	17	0.03
Lake	18	0.05
River	15	0.12
Atlantic Ocean	12	34.5
Pacific Ocean	11	?

1. According to the table, which type of water has a temperature less than 13 degrees Celsius?
   A pond C ocean
   B river D lake
2. Using the data in the table as a guide, which of the following most closely describes the salinity of a stream?
   F 0.01 H 10.0
   G 0.10 J 100.0
3. These data were collected by testing water samples from different sources. When the salinity of the Pacific Ocean is tested, which of these is the value most likely to go in the table?
   A 34.5 C 0.03
   B 0.34 D 0.01

Using Vocabulary

1. Plankton
2. consumers
3. Waves
4. Density currents
5. thermocline

Checking Concepts

6. D	11. C
7. A	12. B
8. D	13. A
9. B	14. B
10. C	

Thinking Critically

15. As waves pass by, the motion of the boat is similar to the motion of water particles in a wave.
16. The current off California originates near Alaska. The current near South Carolina originates near the equator.
17. Organisms that depend on plankton for food will need to rely on other food sources or they will die. With fewer plankton, there will be less dissolved oxygen in the water. As a result, some organisms will die.

chapter 13 Review

Using Vocabulary

chemosynthesis p. 391	plankton p. 389
consumer p. 391	producer p. 391
decomposer p. 391	salinity p. 376
density current p. 382	surface current p. 380
ecosystem p. 391	thermocline p. 378
food chain p. 392	tide p. 387
nekton p. 390	upwelling p. 384
photosynthesis p. 377	wave p. 385

Fill in the blanks with the correct vocabulary words.

1. _________ float in the upper layers of oceans.
2. Organisms that get their energy from eating other organisms are _________.
3. _________ are caused by wind blowing across oceans.
4. _________ are caused by differences in the ocean water's salinity.
5. The layer of ocean water where the temperature drops quickly with depth is the _________.

Checking Concepts

Choose the word or phrase that best answers the question.

6. Which of the following is a measure of dissolved solids in seawater?
   A) density **C)** thermocline
   B) nekton **D)** salinity
7. Which of these organisms is an example of a bottom dweller?
   A) sea star **C)** shark
   B) seal **D)** diatom
8. Which of these organisms is a producer?
   A) sea star **C)** seal
   B) coral **D)** algae
9. Which of the following terms is the high point of a wave?
   A) wavelength **C)** trough
   B) crest **D)** wave height
10. Which of the following terms is the low point of a wave?
   A) wavelength **C)** trough
   B) crest **D)** wave height

Use the figure below to answer question 11.

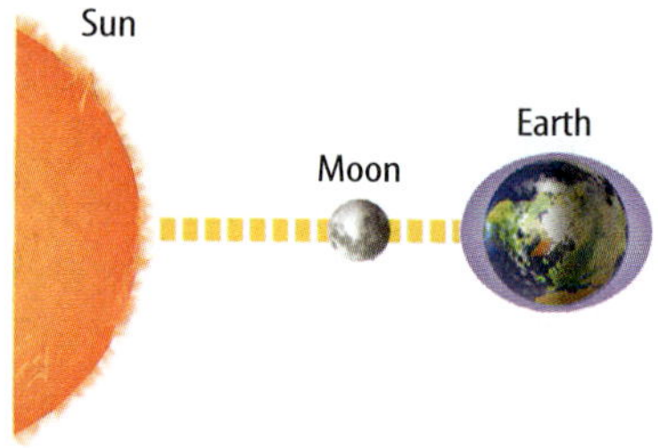

11. Which type of tide forms when the Sun, the Moon, and Earth are aligned as shown?
   A) high tide **C)** spring tide
   B) low tide **D)** neap tide
12. Which substance is found in the most common ocean salt?
   A) calcium
   B) chlorine
   C) carbon
   D) cobalt
13. Which of these gases is produced during photosynthesis?
   A) oxygen
   B) carbon dioxide
   C) nitrogen
   D) water vapor
14. Which of these terms describes the daily rhythmic rise and fall of sea level?
   A) surface current
   B) tide
   C) density current
   D) upwelling

Science Online red.msscience.com/vocabulary_puzzlemaker

Use the ExamView® Pro Testmaker CD-ROM to:

- create multiple versions of tests
- create modified tests with one mouse click for inclusion students
- edit existing questions and add your own questions
- build tests aligned with state standards using built-in State Curriculum Tags
- change English tests to Spanish with one mouse click and vice versa

400 CHAPTER REVIEW

Thinking Critically

15. **Explain** why a boat tied to a dock bobs up and down in the water.

16. **Describe** why the water at a beach in southern California is much colder than the water at a beach in South Carolina.

17. **Infer** How would other ocean life in an area be affected if an oil spill killed much of the plankton in that area?

18. **Discuss** reasons why more marine creatures live in shallow water near shore than in any other region of the oceans.

19. **Interpret Scientific Illustrations** Use **Figure 20** to describe how nitrogen is cycled from the atmosphere to marine organisms.

20. **Compare and contrast** the way that consumers and decomposers get their energy.

Use the illustration below to answer question 21.

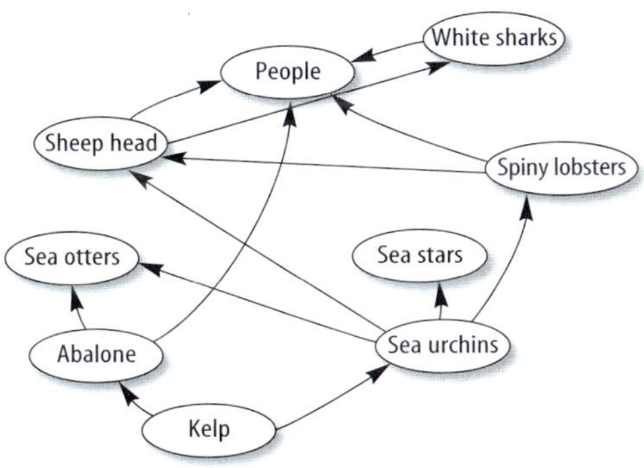

21. **Interpret Scientific Illustrations** Infer what will happen to sea urchins and kelp if sea otters decline in number.

22. **Draw Conclusions** Place the organisms in the proper sequence in a food chain: krill, killer whale, algae, cod, leopard seal.

Performance Activities

23. **Letter** Write to the National Wildlife Federation about coral bleaching. Ask what is being done to protect coral reefs.

24. **Pamphlet** Research beach nourishment, jetties, and other ways that people have tried to reduce beach erosion. Create a pamphlet of your findings and pass it out to your classmates.

Applying Math

25. **Pressure** The pressure at sea level is 1 atmosphere. If the pressure increases by 1 atmosphere for every 10 m in depth, what is the pressure at 200 m?

Use the figure below to answer question 26.

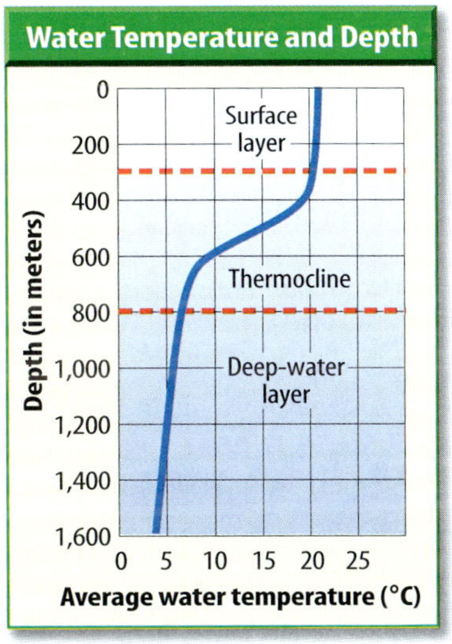

26. **The Thermocline** How much of a temperature change is there between the beginning of the thermocline and 1,200 m?

27. **The Area of Oceans** The Indian Ocean's total area is 73.6 million km^2. The Arctic Ocean's total area is 14.1 million km^2. How many times larger is the Indian Ocean than the Arctic Ocean?

Science Online red.msscience.com/chapter_review

Assessment Resources

Reproducible Masters

Chapter *Fast File* Resources
Chapter Review, pp. 39–40
Chapter Tests, pp. 41–44
Assessment Transparency Activity, p. 53

Glencoe Science Web site
Chapter Review Test
Standardized Test Practice

Glencoe Technology
Assessment Transparency
Exam*View*® Pro Testmaker
MindJogger Videoquiz
Interactive Chalkboard

chapter 13 Review

18. Runoff from the land carries many nutrients to the ocean. Also, sunlight penetrates shallow water. With nutrients and sunlight, producers thrive. With a large amount of plankton to eat, there are many consumers in shallow water near shore.
19. Nitrogen-fixing bacteria change atmospheric nitrogen into nitrates. Marine producers use the nitrates and incorporate the nitrogen into their tissues. When consumers eat producers, nitrogen is cycled throughout the food web.
20. Both depend on producers for energy. Consumers get energy when they eat food stored in producers' and other consumers' tissues. Decomposers break down tissues of producers and consumers to get energy.
21. Sea urchins will increase in number with fewer sea otters to eat them. Kelp will decrease in number because there will be more sea urchins and abalone to eat them.
22. algae, krill, cod, leopard seal, killer whale

Performance Activities

23. Letters should contain accurate information and be in correct business style. Use **PASC,** p. 139.
24. Suggest that students use their word processing and graphic design skills to make their pamphlets. Pamphlets should contain accurate information. Use **PASC,** p. 129.

Applying Math

National Math Standards
1, 2, 3, 4, 5, 9
25. 21 atmospheres
26. 16°
27. approximately 5.2 times larger

chapter 13 Standardized Test Practice

FAST FILE

Answer Sheet A practice answer sheet can be found at red.msscience.com/answer_sheet.

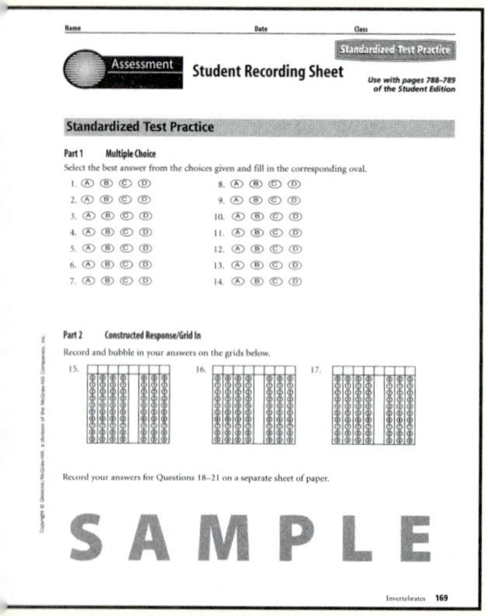
Name Date Class

Assessment Student Recording Sheet

Standardized Test Practice

Use with pages 788–789 of the Student Edition

Standardized Test Practice

Part 1 Multiple Choice

Select the best answer from the choices given and fill in the corresponding oval.

1. Ⓐ Ⓑ Ⓒ Ⓓ
2. Ⓐ Ⓑ Ⓒ Ⓓ
3. Ⓐ Ⓑ Ⓒ Ⓓ
4. Ⓐ Ⓑ Ⓒ Ⓓ
5. Ⓐ Ⓑ Ⓒ Ⓓ
6. Ⓐ Ⓑ Ⓒ Ⓓ
7. Ⓐ Ⓑ Ⓒ Ⓓ
8. Ⓐ Ⓑ Ⓒ Ⓓ
9. Ⓐ Ⓑ Ⓒ Ⓓ
10. Ⓐ Ⓑ Ⓒ Ⓓ
11. Ⓐ Ⓑ Ⓒ Ⓓ
12. Ⓐ Ⓑ Ⓒ Ⓓ
13. Ⓐ Ⓑ Ⓒ Ⓓ
14. Ⓐ Ⓑ Ⓒ Ⓓ

Part 2 Constructed Response/Grid In

Record and bubble in your answers on the grids below.

15. 16. 17.

Record your answers for Questions 18–21 on a separate sheet of paper.

SAMPLE

Invertebrates 169

Part 1 Multiple Choice

1. B
2. A
3. B
4. C
5. D
6. A

Part 2 Short Response

7. 342.4 energy units
8. 3 times deeper
9. Temperature differences result because different latitudes receive different amounts of solar energy. Currents that originate at high latitudes are cold. Currents that originate at low latitudes are warm.
10. semidiurnal tides
11. 3 m

Part 1 Multiple Choice

Record your answers on the answer sheet provided by your teacher or on a separate sheet of paper.

Use the table below to answer questions 1 and 2.

Ms. Mangan's class is studying different organisms. Here is a table of some of the organisms they have been studying.

Marine Organisms	
Producers	**Consumers**
Seaweed	Krill
Kelp	Squid
Algae	Seal

1. The producers are different from the consumers because only the producers are able to
A. swim in deep water.
B. make their own food.
C. contribute to the marine food web.
D. digest the nutrients in other organisms.

2. What would happen to consumers if all producers perished?
A. Consumers would also die.
B. Consumers would begin making their own food.
C. Consumers would decompose organic matter.
D. Consumers would move to a new environment.

Test-Taking Tip

Never skip a question. If you are unsure of an answer, mark your best guess on your answer sheet and mark the question in your test booklet to remind you to come back to it at the end of the test.

3. The layer of water in which temperature drops quickly with increasing depth is the
A. surface layer.
B. thermocline layer.
C. deep-water layer.
D. bottom layer.

4. Organisms that break down tissue and release nutrients and carbon dioxide back into the ecosystem are called
A. producers.
B. consumers.
C. decomposers.
D. photosynthetic.

Use the diagram below to answer questions 5 and 6.

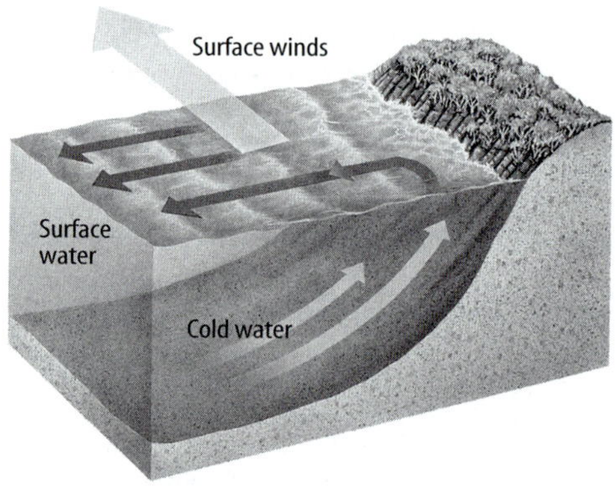

5. What process is shown in the figure?
A. downwelling
B. density currents
C. El Niño
D. upwelling

6. Wind blowing parallel to the coast carries water away from land because of
A. the Coriolis effect.
B. the gravitational pull of the Moon on Earth.
C. the gravitational pull of the Sun on Earth.
D. the speed of the wind.

Part 3 Open Ended

12. Density currents that form north of Iceland could become weak or stop. Warm water would no longer flow northward to warm the land.
13. During larval stage of many organisms, such as fish, crabs, coral, and sea stars, the organism is microscopic and cannot swim against the current. At this stage these organisms drift with the current and are considered to be plankton. As they grow through the larval stage to juveniles and then adults, they increase in size and develop structures and muscles that allow them to swim or move against the current, becoming nekton.
14. At the surface, oxygen concentration is at its highest. Oxygen concentration decreases until about 500 m; then slowly begins to increase with depth.

402 STANDARDIZED TEST PRACTICE

Part 2 Short Response/Grid In

Record your answers on the answer sheet provided by your teacher or on a sheet of paper.

7. Ten percent of the total available energy is stored by a consumer at each level of the food chain. If 3,424 energy units are passed on to a salmon feeding on zooplankton, how many energy units will the salmon store?

8. The average depth of the ocean is 3,730 m. The Marianas Trench is 11,033 m deep. How many times deeper is the Marianas Trench than the average depth of the ocean? Round your answer to the nearest whole number.

9. Why aren't all ocean surface currents the same temperature?

Use the figure below to answer questions 10 and 11.

Tidal Data, Cape Cod, MA

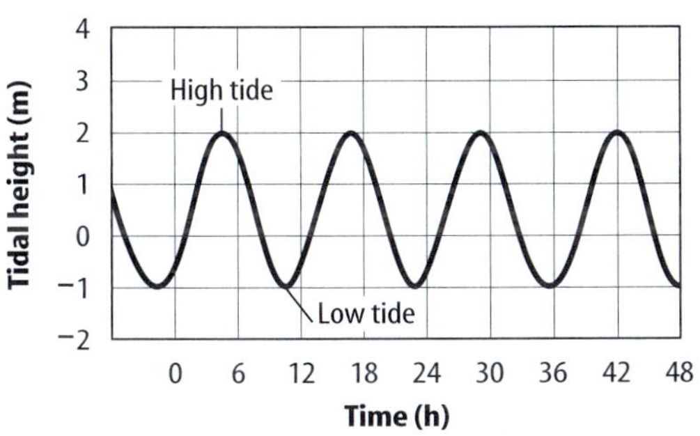

10. Some coastlines experience *semidiurnal tides,* or two high tides and two low tides in approximately a 24-hour period. Other coastlines experience *diurnal tides,* or one high tide and one low tide in approximately a 24-hour period. Which type of tide is shown in the graph above?

11. What is the tidal range during the period of time tidal data was recorded for the graph?

Part 3 Open Ended

Record your answer on a sheet of paper.

12. Explain why temperatures could decrease in some regions of Earth if global warming occurred.

13. Why are some organisms considered to be plankton in one stage of their life but nekton in another stage of their life?

Use the graph below to answer questions 14–17.

Oxygen Concentration and Water Depth

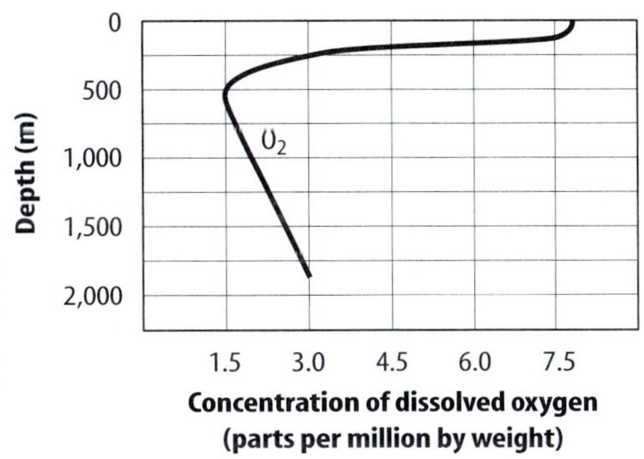

14. Describe what happens to the levels of oxygen concentration as water depth increases.

15. Why is oxygen concentration at its highest near the surface of the water? What accounts for the decrease in oxygen concentration between about 200 m and 500 m?

16. Explain why the concentration of oxygen increases between 500 m and 2,000 m.

17. Infer how the levels of carbon dioxide concentration would vary with depth. What biological processes affect the concentration of carbon dioxide?

Science Online red.msscience.com/standardized_test

15. Oxygen enters ocean water from the atmosphere. Also producers undergo photosynthesis in ocean surface waters, increasing the concentration of oxygen at this depth. Between 200 m and 500 m, many consumers undergo respiration, which requires oxygen. At this depth the oxygen is not replenished.

16. Dissolved oxygen increases after 500 m due to the fact that these layers of deep, cold water originate at the surface, where there is more oxygen, in polar regions and move along the ocean floor.

17. Carbon dioxide levels are lower at the surface due to the use of the gas during photosynthesis. The levels increase with depth as carbon dioxide is given off by consumers undergoing respiration.

Rubrics

The following rubrics are sample scoring devices for short response and open-ended questions.

Short Response

Points	Description
2	The student demonstrates a thorough understanding of the science of the task. The response may contain minor flaws that do not detract from the demonstration of a thorough understanding.
1	The student has provided a response that is only partially correct.
0	The student has provided a completely incorrect solution or no response at all.

Open Ended

Points	Description
4	The student demonstrates a thorough understanding of the science of the task. The response may contain minor flaws that do not detract from the demonstration of a thorough understanding.
3	The student demonstrates an understanding of the science of the task. The response is essentially correct and demonstrates an essential but less than thorough understanding of the science.
2	The student demonstrates only a partial understanding of the science of the task. Although the student may have used the correct approach to a solution or may have provided a correct solution, the work lacks an essential understanding of the underlying science concepts.
1	The student demonstrates a very limited understanding of the science of the task. The response is incomplete and exhibits many flaws.
0	The student provides a completely incorrect solution or no response at all.

unit 4

Unit Contents

WebQuest ***Investigating the Sun*** explores the physical characteristics of the Sun, and how energy from the Sun affects life on Earth. Students will answer a set of questions to demonstrate what they have learned, and then create a labeled diagram of the physical characteristics of our nearest star.

unit 4

Beyond Earth

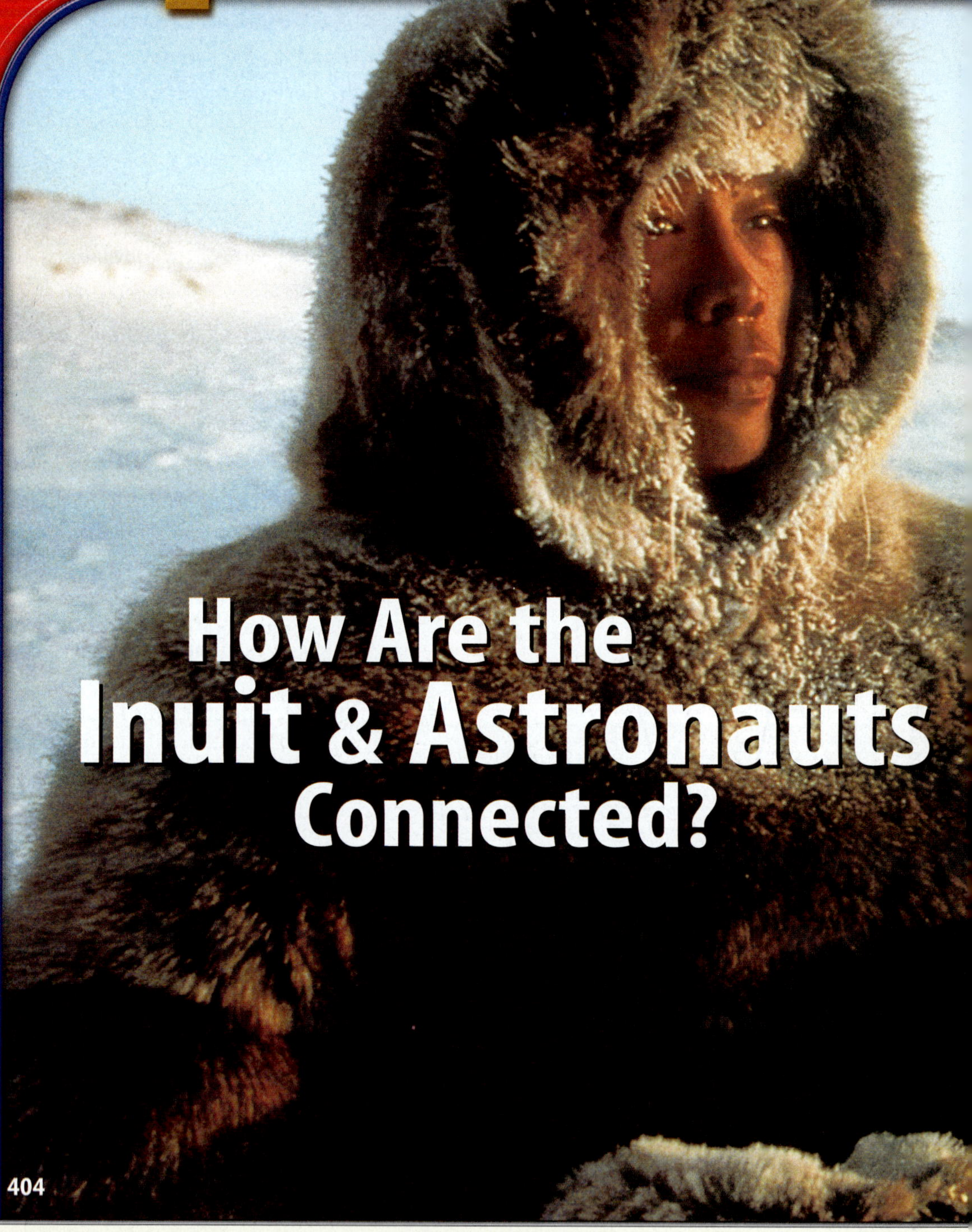

404

PROJECT CRISS℠

Study Skills

Understand Processes Sequence mapping provides a graphic picture for better understanding. Students use this type of mapping to chart the process of solar system formation beginning with the breaking apart of solar nebula and ending with the formation of a new solar system. Have students draw or find pictures to illustrate each step. Students should write descriptions for each step of the process.

NATIONAL GEOGRAPHIC

For thousands of years, people known as the Inuit have lived in Arctic regions. In the early 1900s, an American naturalist spent time among the Inuit in Canada. The naturalist watched the Inuit preserve fish and meat by freezing them in the cold northern air. Months later, when the people thawed and cooked the food, it was tender and tasted fresh. Eventually, the naturalist returned to the United States, perfected a quick-freezing process, and began marketing frozen foods. Later, inventors found a way to remove most of the water from frozen foods. This process, called freeze-drying, produces a lightweight food that can be stored at room temperature and doesn't spoil. Freeze-dried foods are carried by all sorts of adventurers—including astronauts.

unit projects

Visit red.msscience.com/unit_project to find project ideas and resources.
Projects include:

- **History** Contribute to a class time line of inventions from Chinese rockets to liquid fuel, radio communications, space suits, and more.
- **Technology** Research, design, and construct a simple telescope as you study how technology has allowed scientists to explore distant space.
- **Model** Investigate the production of space foods, and present an informational commercial to the class.

WebQuest *Investigating the Sun* explores the physical characteristics of the sun and energy it produces. Design a labeled diagram of our nearest star.

unit projects

History Working in small groups, have students investigate one specific decade in the history of space travel, from the invention of rockets to the development of liquid fuel, radio communication, and astronaut suits. As a class, compile the information into a space-travel time line that traces these inventions and breakthroughs.

Technology Have students investigate the design of telescopes. Working with partners, ask students to research different models of telescopes—their history, parts, sizes, specific use, and how technology has advanced our ability to observe the universe. Have students design a brochure for their specific telescope, and then construct a model using recycled materials to enhance the presentation of their new information.

Model Have students prepare an informational commercial on a specific space food or food process. Information should include how the food is prepared, nutritional information, cost, availability, and use today. Students may wish to present a sample to the class; many varieties may be found at science stores, or hiking and camping stores.

Additional Resources For more information, resources, and assessment rubrics, visit red.msscience.com/unit_project

NATIONAL GEOGRAPHIC **How Are the Inuit & Astronauts Connected?** Early space missions began a new era of human exploration. In 1961 Soviet cosmonaut Yuri A. Gagarin became the first human to orbit Earth. Due to harsh environment in space, scientists had to develop new technologies to provide food, water, and exercise for the astronauts. Freeze-dried and compressed foods were developed by the Army. NASA developed oxygen regeneration techniques using algae that would provide needed oxygen to astronauts. Today, NASA is experimenting with growing crop plants on board to provide food and facilitate water purification, air revitalization, and waste processing in space.

chapter 14 Organizer

Section/Objectives	Standards		Labs/Features
	National	State/Local	
Chapter Opener	See pp. 16T–17T for a Key to Standards.		**Launch Lab:** An Astronomer's View, p. 407 **Foldables,** p. 407
Section 1 Radiation from Space 2 sessions 1 block 1. **Explain** the electromagnetic spectrum. 2. **Identify** the differences between refracting and reflecting telescopes. 3. **Recognize** the differences between optical and radio telescopes.	National Content Standards: UCP.1, UCP.2, UCP.3, UCP.5, A.1, A.2, D.3		**Integrate Health,** p. 409 **MiniLAB:** Observing Effects of Light Pollution, p. 412 **Lab:** Building a Reflecting Telescope, p. 414
Section 2 Early Space Missions 1 session 0.5 block 4. **Compare and contrast** natural and artificial satellites. 5. **Identify** the differences between artificial satellites and space probes. 6. **Explain** the history of the race to the Moon.	National Content Standards: UCP.1, UCP.2, UCP.3, UCP.5, A.1, A.2, D.3		**Applying Math:** Drawing by Numbers, p. 416 **Integrate Career,** p. 418 **Visualizing Space Probes,** p. 419 **Science Online,** p. 420 **MiniLAB:** Modeling a Satellite, p. 421
Section 3 Current and Future Space Missions 4 sessions 2 blocks 7. **Explain** the benefits of the space shuttle. 8. **Identify** the usefulness of orbital space stations. 9. **Explore** future space missions. 10. **Identify** the applications of space technology to everyday life.	National Content Standards: UCP.1, UCP.2, UCP.3, UCP.5, A.1, A.2, D.3, F.5		**Science Online,** p. 425 **Science Online,** p. 427 **Lab:** Star Sightings, pp. 430–431 **Science and Society:** Cities in Space, p. 432

Lab Materials	Reproducible Resources	Section Assessment	Technology
Launch Lab: clear-plastic wrap (15 cm), textbook	**Chapter *FAST FILE* Resources** Foldables Worksheet, p. 13 Directed Reading Overview, p. 15 Note-taking Worksheets, pp. 29–31	GLENCOE'S ASSESSMENT ADVANTAGE	TeacherWorks includes: • Interactive Teacher Edition • Lesson Planner with calendar • Access to all program blacklines • Correlations to standards • Web links
MiniLAB: cardboard tube **Lab:** flat mirror, curved concave mirror, magnifying lenses of different magnifications (3–4)	**Chapter *FAST FILE* Resources** Transparency Activity, p. 638 MiniLAB, p. 3 Enrichment, p. 26 Reinforcement, p. 23 Directed Reading, p. 16 Lab Worksheets, pp. 5–6 Transparency Activity, pp. 43–44 Lab Activity, pp. 9–12 **Cultural Diversity,** p. 51	**Portfolio** Science Journal, p. 411 **Performance** MiniLAB, p. 412 Applying Math, p. 413 **Content** Section Review, p. 413	Section Focus Transparency Teaching Transparency Guided Reading Audio Program Interactive Chalkboard CD-ROM
MiniLAB: string, cork *Need materials?* Contact Science Kit at 1-800-828-7777 or www.sciencekit.com on the Internet.	**Chapter *FAST FILE* Resources** Transparency Activity, p. 41 MiniLAB, p. 4 Enrichment, p. 27 Reinforcement, p. 24 Directed Reading, p. 17 **Mathematics Skill Activities,** p. 15 **Reading and Writing Skill Activities,** p. 11	**Portfolio** Activity, p. 419 **Performance** Applying Math, p. 416 MiniLAB, p. 421 Applying Math, p. 422 **Content** Section Review, p. 422	Section Focus Transparency Virtual Labs CD-ROM Guided Reading Audio Program Interactive Chalkboard CD-ROM
Lab: computer, Internet and other resources on space technology	**Chapter *FAST FILE* Resources** Transparency Activity, p. 42 Enrichment, p. 28 Reinforcement, p. 25 Directed Reading, pp. 17, 18 Lab Worksheets, pp. 7–8 **Home and Community Involvement,** p. 31 **Physical Science Critical Thinking/Problem Solving,** p. 5 **Cultural Diversity,** p. 25	**Portfolio** Curriculum Connection, p. 425 **Performance** Applying Math, p. 429 **Content** Section Review, p. 429	Section Focus Transparency Guided Reading Audio Program Interactive Chalkboard CD-ROM

End of Chapter Assessment

Blackline Masters	Technology	Professional Series
Chapter *FAST FILE* Resources Chapter Review, pp. 33–34 Chapter Tests, pp. 35–38 **Standardized Test Practice,** pp. 59–62	MindJogger Videoquiz Virtual Labs CD-ROM Exam*View*® Pro Testmaker TeacherWorks CD-ROM Interactive Chalkboard CD-ROM	**Performance Assessment in the Science Classroom (PASC)**

Transparencies

Section Focus

Section 1 Section Focus Transparency — **Superstar**

Evidence from many different cultures suggest that people have studied the skies for a very long time. For example, the Anasazi people of Chaco Canyon (New Mexico) recorded astronomical events on stone. Below is a drawing thought to record the appearance of a supernova, a massive exploding star, in 1054.

1. What can you identify in this picture that might represent an astronomical object?
2. How did people study stars before telescopes were developed? What advantages do telescopes offer modern astronomers?

L2

Section 2 Section Focus Transparency — **The Outer Limits of the Solar System**

Where does the solar system end? The planets end at Pluto, but the effects of the Sun's gravitational pull extend much farther. The last items held by the Sun's gravitational pull are in Oort's Cloud.

1. How can satellites and space probes help us get a better understanding of the planets in our solar system?
2. How can space missions help us study Earth?

L2

Section 3 Section Focus Transparency — **Hot Hot Hot Hot . . .**

One of the missions NASA has planned for the future is a solar probe. It is scheduled for launch in February 2007, and it will make two close passes to the Sun—one in 2010 and one in 2015. The solar probe will pass so close to the Sun that it will be in the Sun's atmosphere.

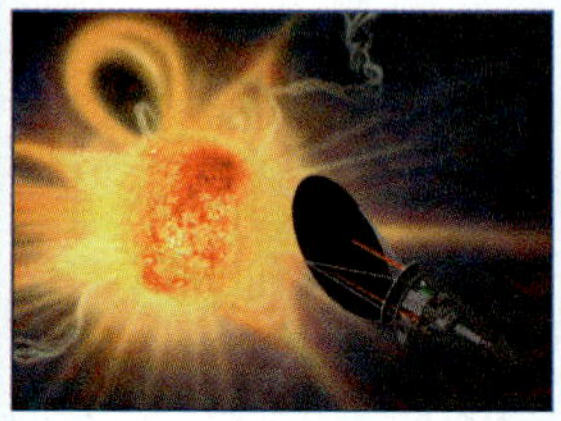

1. Why do you think the solar probe won't be launched before 2007?
2. How might conditions in the Sun's atmosphere affect the design of the solar probe?
3. What do you think scientists hope to learn from the solar probe?

L2

This is a representation of key blackline masters available in the Teacher Classroom Resources. See Resource Manager boxes within the chapter for additional information.

Assessment

Assessment Transparency — **Exploring Space**

Directions: *Carefully review the table and answer the following questions.*

U.S. Space Missions			
Date	Mission	Duration (hours: minutes)	Notable "Firsts"
1961	Mercury 3	0:15	U.S. citizen in space
1962	Mercury 6	4:55	U.S. citizen to orbit Earth
1965	Gemini 6A	25:51	Spacecrafts connected in orbit
1968	Apollo 7	260:09	Orbit of the Moon
1969	Apollo 11	195:18	Human on the Moon
1983	Challenger	146:24	U.S. woman in space

1. According to the table, the first human on the Moon was part of space mission ___.
   A Mercury C Apollo 7
   B Gemini D Apollo 11
2. According to this information, which space mission lasted less than one hour?
   F Mercury 3 H Gemini 6A
   G Mercury 6 J Apollo 7
3. According to the table, how long was it between the time the first U.S. citizen went into space and the time the first U.S. woman went into space?
   A 1 year C 22 years
   B 21 years D 32 years

L2

Teaching

Section 1 Teaching Transparency — **Telescopes**

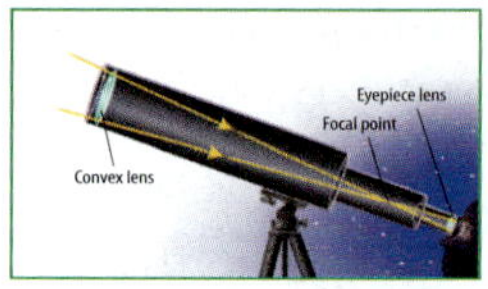

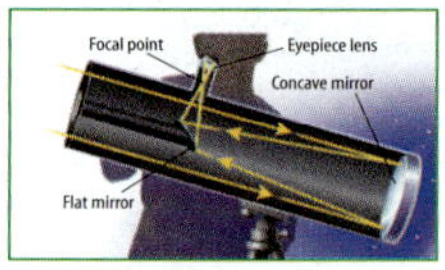

L2

Key to Teaching Strategies

The following designations will help you decide which activities are appropriate for your students.

- **L1** Level 1 activities should be appropriate for students with learning difficulties.
- **L2** Level 2 activities should be within the ability range of all students.
- **L3** Level 3 activities are designed for above-average students.
- **ELL** ELL activities should be within the ability range of English Language Learners.
- **COOP LEARN** Cooperative Learning activities are designed for small group work.
- **LS** Multiple Learning Styles logos, as described on page 12T, are used throughout to indicate strategies that address different learning styles.
- **P** These strategies represent student products that can be placed into a best-work portfolio.
- **PBL** Problem-Based Learning activities apply real-world situations to learning.

Hands-on Activities

Student Text Lab Worksheet

Name Date Class

Activity — **Building a Reflecting Telescope**

Hands-On Activities

Lab Preview

Directions: *Answer these questions before you begin the activity.*

1. What object will you choose to look at?
2. Which mirror will you look at to view the image of the object?

Nearly four hundred years ago, scientist Galileo Galilei saw what no human had ever seen before. Using the telescope he invented, Galileo discovered moons revolving around Jupiter, observed craters on the Moon in detail, and saw sunspots on the surface of the Sun. What was it like to make these discoveries? You will find out as you make your own reflecting telescope.

What You'll Investigate
How do you construct a reflecting telescope?

Materials
flat mirror
shaving or cosmetic mirror (a curved, concave mirror)
magnifying lenses of different magnifications (3–4)

Goals
- **Construct** a reflecting telescope.
- **Observe** magnified images using the telescope and different magnifying lenses.

Safety Precautions
CAUTION: *Never observe the Sun directly or with mirrors.*

Procedure
1. Position the cosmetic mirror so that you can see the reflection of the object you want to look at. Choose a bright object such as the Moon, a planet, or an artificial light source.
2. Place the flat mirror so that it is facing the cosmetic mirror.
3. Adjust the position of the flat mirror until you can see the reflection of the object in it.
4. View the image of the object in the flat mirror with one of your magnifying glasses. Observe how the lens magnifies the image.
5. Use your other magnifying lenses to view the image of the object in the flat mirror. Observe how the different lenses change the image of the object.

Lens	Image characteristics
1	
2	
3	
4	

L2

Exploring Space 5

Laboratory Activities

Name Date Class

Lab 1 Laboratory Activity — **Star Colors**

Hands-On Activities

In 1665, Isaac Newton demonstrated that sunlight is composed of many colors. Today the spectra of a star is one of the most important tools scientists use to determine the star's surface temperature and composition. The Draper system of spectral classification is used in this activity.

Strategy
You will define the term *star.*
You will observe and record star colors.
You will classify stars based on their color.

Materials
binoculars or telescope (optional)
graph paper

Procedure
1. On a clear, bright night observe the stars with your eyes or with the binoculars or telescope.
2. Use some landmarks and divide the sky into four sections. Label the landmarks in the diagram under Data and Observations.
3. Observe and record the color of each star in each section. Record your observations on your diagram under Data and Observations.
4. Using the information in Table 1, compile your data in a table showing the star color, spectral type, and number of stars in each section. Set up your table on one end of your graph paper.
5. Under the table on the graph paper, draw a bar graph showing the star spectral types and the number of stars in each spectral type.

Table 1

Draper's Star Classification Chart		
Star spectral type	Color	Surface temperature (K)
M	red	2,000–4,000
K	red to orange	3,500–5,000
G	yellow	5,000–6,000
F	yellow-white	6,000–7,500
A	white	9,000
B	bluish-white	11,000–25,000
O	bluish-white	60,000

L2

Exploring Space 9

Meeting Different Ability Levels

Content Outline

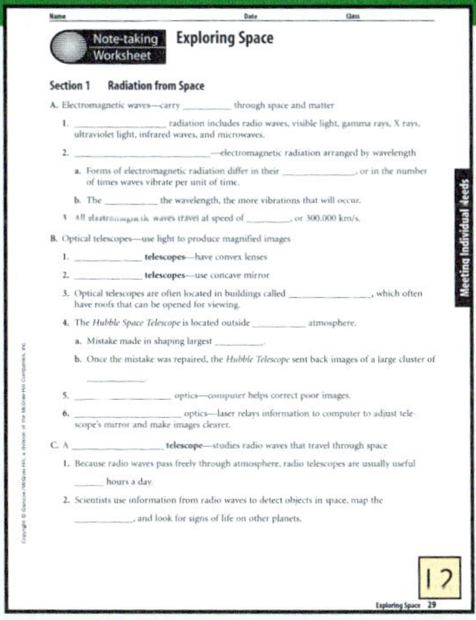

Reinforcement

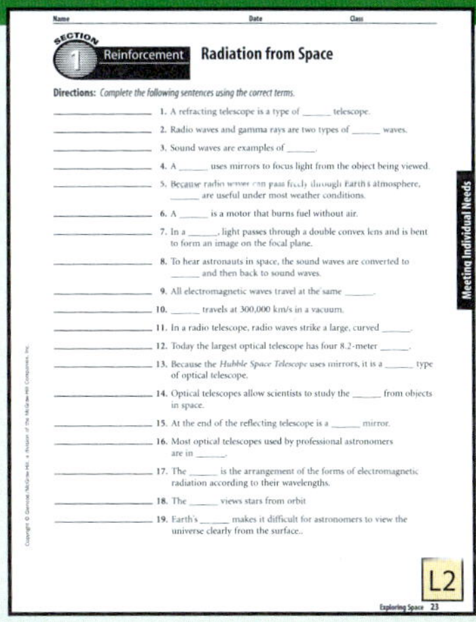

Enrichment

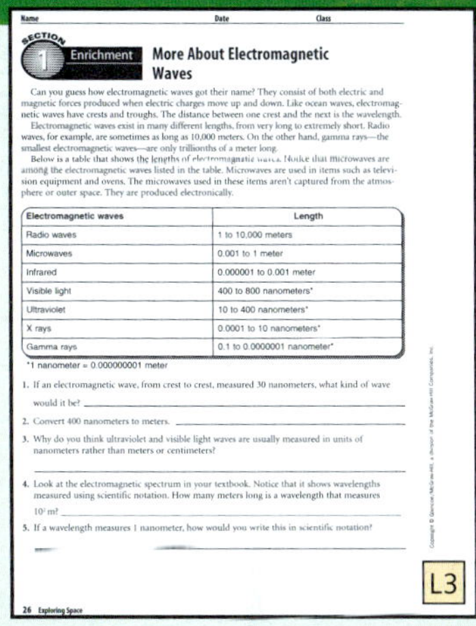

Directed Reading (English/Spanish)

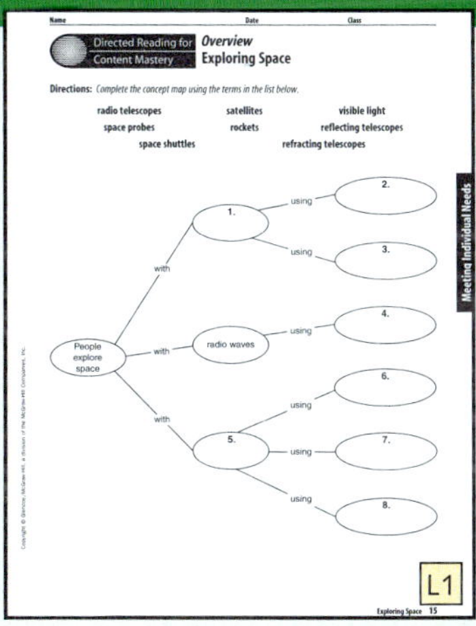

Study Guide

Reading Essentials

Assessment

Test Practice Workbook

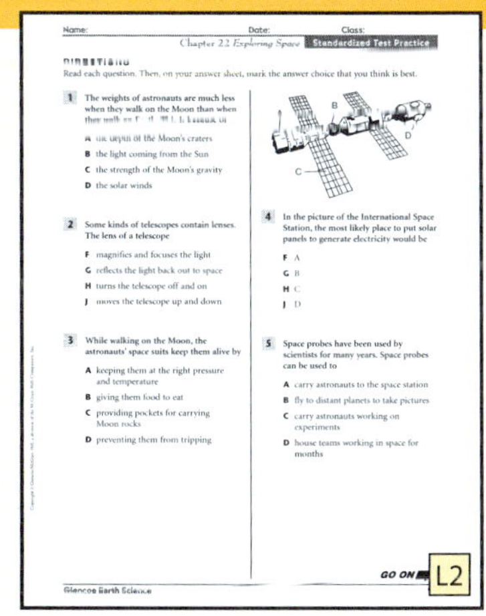

Chapter Review

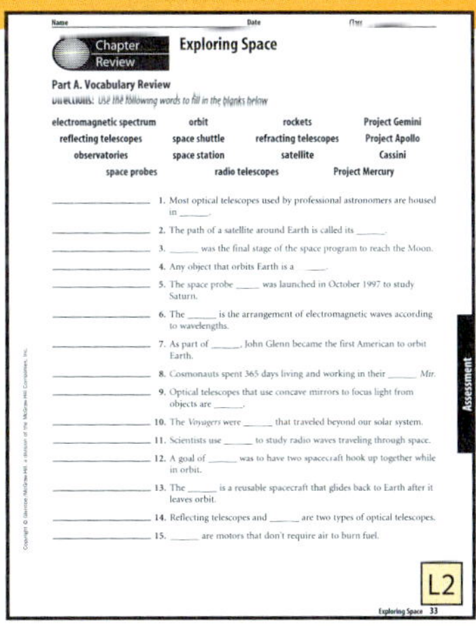

Chapter Tests

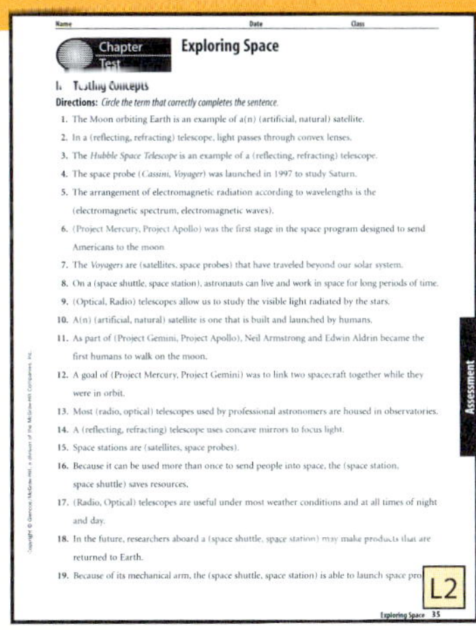

Science Content Background

section 1 Radiation from Space

Large Telescopes

On April 21, 1998, the largest telescope on Earth captured light from its first star. The European Southern Observatory's Very Large Telescope (VLT) is composed of four identical 8.2-m individual telescopes and several movable 1.8-m telescopes.

Other large telescopes include the twin 10-m Keck reflecting telescopes on Mauna Kea in Hawaii. The primary mirror of each telescope has been constructed from 36 hexagonal glass segments, each 1.8 m in diameter.

The NASA Orbital Debris Observatory (NODO) uses a liquid mirror telescope. The objective mirror is made of a shallow container of the liquid element mercury. It is 3 m in diameter and is constantly spinning at one revolution every 6.02315 seconds. This spinning forces the liquid mercury to form a parabolic reflector. The telescope is kept level; thus, it observes only the sky's zenith. Its main mission is to locate and count orbital debris (space junk) as small as 1 cm in diameter.

The Very Large Array (VLA) of telescopes near Socorro, New Mexico, is composed of 27 radio telescopes. All 27 telescopes can be used at once to operate as one large telescope.

Light Pollution

Light pollution is a glow in the sky caused by city lights. The glow makes it difficult to see dim stars.

In several United States cities, work has begun to reduce light pollution. Tucson, Arizona, located only 80 km from the Kitt Peak National Observatory, has replaced its street lighting with low-pressure sodium lamps. These lamps shine at wavelengths that can be filtered out by astronomers. Other cities have put hoods on billboards, parking-lot lights, and floodlights so they illuminate the object or the ground rather than the sky.

Teacher to Teacher

Ralph M. Feather, Jr.
Derry Area High School
Derry, PA

"To help my students understand that the Moon's phases are caused by the position of the Moon in relation to Earth and Sun, I have them make observations of the Moon at approximately the same time each day for three to four days."

Ralph Feather

section 2 Early Space Missions

The First Missions into Space

The first mission into space was conducted by the Soviet Union, which launched the uncrewed *Sputnik 1*. It was designed to prove that human-made satellites could orbit Earth in the manner of a natural satellite, such as the moon. Within a year of that launch, the United States had built its own satellite and sent it into space. Known as *Explorer 1*, it was followed by three more attempts, all unsuccessful, to send probes into an orbit similar to that of Earth's Moon.

In 1959 the Soviet Union did what the United States hadn't been able to do. Their probe, *Luna 1*, became the first human-made satellite to travel past the Moon. Two months later, American attempts at a lunar flyby, or voyage past the Moon, became successful with *Pioneer 4*. The United States followed this successful effort with a series of probes known as *Ranger 1* and *2*. Both were failures. Finally, with *7*, *8*, and *9*, the United States was able to capture detailed pictures of the Moon's surface. This led into the Apollo program, which placed astronauts on the Moon itself.

section 3 Current and Future Space Missions

Water on Other Planets

Jupiter's moon Europa may contain water. Recent preliminary studies of Europa by the *Galileo* spacecraft indicate an ocean of water or ice under the moon's crustal layer of ice.

The *Cassini* mission to Saturn will have similar goals to the Galileo mission to Jupiter. The science goals of the Cassini mission involve studying and collecting data on Saturn, its magnetosphere, rings, icy moons, and Titan.

The International Space Station

The *International Space Station* will be a permanently crewed platform in which teams of astronauts from many nations will work cooperatively in space. Launch of materials and modules that are part of the *International Space Station* assembly began with the launch of the space shuttle *Endeavor*. Work continues on this international undertaking.

Everyday Space Technology

Much of the technology developed by NASA to achieve its goals in space is now being used by people throughout the world. These technologies are called spin-offs. For example, NASA developed lightweight, compact breathing systems for astronauts to carry as they ventured out of their spacecraft and onto the Moon. Today, firefighters use these breathing systems as well as fire-resistant uniforms originally designed as flight suits for NASA pilots. The lightweight material in the suits won't burn or crack. A material designed for boots worn by astronauts on the Moon is now found in some athletic shoes. Other materials have been incorporated into ski goggles, blankets, and bicycle seats.

People who are visually impaired also have benefited from spin-offs. One device vibrates ink on a printed page, enabling them to read materials that are not printed in Braille. Another device determines the denomination of currency and generates an audible signal. Other spin-offs include pens that write without the help of gravity and sunglasses that adjust to various light levels.

David Ducros/Photo Researchers, Inc.

chapter content resources

Internet Resources

For additional content background, visit **red.msscience.com** to:

- access your book online
- find references to related articles in popular science magazines
- access Web links with related content background
- access current events with science journal topics

Print Resources

Celestial Delights, Francis Reddy and Greg Walz-Chojnocki, Celestial Arts, 2002

A+ Projects in Astronomy, Janice Van Cleave, John Wiley & Sons, 2002

The Stars—A New Way to See Them, H.A. Rey, Houghton Mifflen Co., 1997

How to Identify the Night Sky, Storm Dunlop and Wil Tircon, Barnes & Noble Inc., 2003

chapter 14

Chapter Vocabulary

electromagnetic spectrum, p. 409
refracting telescope, p. 410
reflecting telescope, p. 410
observatory, p. 410
radio telescope, p. 413
rocket, p. 415
satellite, p. 417
orbit, p. 417
space probe, p. 418
Project Mercury, p. 421
Project Gemini, p. 421
Project Apollo, p. 422
space shuttle, p. 423
space station, p. 424

Science Journal Student answers will vary, depending on their value on space exploration.

INTERACTIVE CHALKBOARD with Image Bank

PowerPoint® Presentations

This CD-ROM is an editable Microsoft® PowerPoint® presentation that includes:

- a pre-made presentation for every chapter
- interactive graphics
- animations
- audio clips
- image bank
- all new section and chapter questions
- Standardized Test Practice
- transparencies
- pre-lab questions for all labs
- Foldables directions
- links to red.msscience.com

chapter 14

Exploring Space

chapter preview

sections

1 Radiation from Space
Lab Building a Reflecting Telescope

2 Early Space Missions

3 Current and Future Space Missions
Lab Star Sightings

Virtual Lab How does an artificial satellite stay in orbit?

Fiery end or new beginning?

These colorful streamers are the remains of a star that exploded in a nearby galaxy thousands of years ago. Eventually, new stars and planets may form from this material, just as our Sun and planets formed from similar debris billions of years ago.

Science Journal Do you think space exploration is worth the risk and expense? Explain why.

Theme Connection

Energy Students are introduced to the way astronomers use various types of electromagnetic radiation (energy) to explore the solar system and the rest of the universe.

About the Photo

Supernova The remnants of the supernova explosion shown here are located in the Large Magellanic Cloud, a small companion galaxy to our own Milky Way. Somewhere in this swirling mass of gas and dust is a spinning neutron star or pulsar. Pulsars emit no visible radiation but can be detected in other electromagnetic spectra, such as X rays and gamma rays. Gamma ray emissions from this pulsar were detected in 1979.

Start-Up Activities

Launch LAB

An Astronomer's View

You might think exploring space with a telescope is easy because the stars seem so bright and space is dark. But starlight passing through Earth's atmosphere, and differences in temperature and density of the atmosphere can distort images.

1. Cut off a piece of clear plastic wrap about 15 cm long.
2. Place an opened book in front of you and observe the clarity of the text.
3. Hold the piece of plastic wrap close to your eyes, keeping it taut using both hands.
4. Look at the same text through the plastic wrap.
5. Fold the plastic wrap in half and look at the text again through both layers.
6. **Think Critically** Write a paragraph in your Science Journal comparing reading text through plastic wrap to an astronomer viewing stars through Earth's atmosphere. Predict what might occur if you increased the number of layers.

Science Online — Preview this chapter's content and activities at red.msscience.com

FOLDABLES Study Organizer

Exploring Space Make the following Foldable to help identify what you already know, what you want to know, and what you learned about exploring space.

STEP 1 **Fold** a vertical sheet of paper from side to side with the front edge about 1.25 cm shorter than the back.

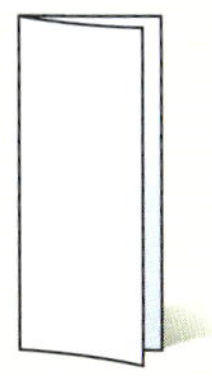

STEP 2 **Turn** lengthwise and **fold** into thirds.

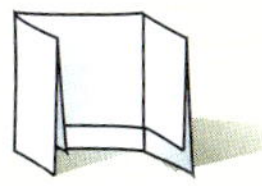

STEP 3 **Unfold and cut** only the top layer along both folds to make three tabs. **Label** each tab.

Identify Questions Before you read the chapter, write what you already know about exploring space under the left tab of your Foldable, and write questions about what you'd like to know under the center tab. After you read the chapter, list what you learned under the right tab.

Launch LAB

Purpose Use this lab to introduce students to how the atmosphere distorts viewing.

Preparation Have available a large roll of clear-plastic wrap. No other advance preparation is needed.

Materials Give each student a piece of plastic wrap about 15 cm long.

Teaching Strategy Allow students to follow directions. Check to see that they are keeping the wrap taut without wrinkles. Wrap should be held about 3 cm from their eyes or glasses if they are wearing them.

Think Critically

Some students should notice a slight difference looking through one layer. Encourage them to compare their views with and without the wrap. Adding the second layer makes the difference more obvious and a third layer causes more distortion. Point out that even when the air seems clear, slight distortions may be present. The layers of plastic mimic the effect of layers of atmosphere through which starlight must pass.

Assessment

Process Have students infer from their results why astronomical observatories are built on high mountains. The atmosphere affects starlight. On high mountains, starlight passes through less air and celestial objects can be seen more clearly. Also, they are above layers of smog and other pollution. Use **Performance Assessment in the Science Classroom**, p. 89.

FOLDABLES Study Organizer — Dinah Zike Study Fold

Student preparation materials for this Foldable are available in the **Chapter *FAST FILE* Resources.**

section 1 Radiation from Space

1 Motivate

INTERACTIVE CHALKBOARD PowerPoint® Presentations

Bellringer

Section Focus Transparencies also are available on the Interactive Chalkboard CD-ROM. L2 ELL

Section 1 Section Focus Transparency **Superstar**

Evidence from many different cultures suggest that people have studied the skies for a very long time. For example, the Anasazi people of Chaco Canyon (New Mexico) recorded astronomical events on stone. Below is a drawing thought to record the appearance of a supernova, a massive exploding star, in 1054.

1. What can you identify in this picture that might represent an astronomical object?
2. How did people study stars before telescopes were developed? What advantages do telescopes offer modern astronomers? L2

Exploring Space

Tie to Prior Knowledge

Electromagnetic Radiation Ask for a show of hands for each of these questions: Have you ever used electromagnetic radiation? How many have had an X ray? Used a microwave? Listened to the radio? Seen sunshine or felt the Sun's heat? Explain that anyone who has raised his or her hand has experienced electromagnetic radiation.

---

as you read

What You'll Learn

- **Explain** the electromagnetic spectrum.
- **Identify** the differences between refracting and reflecting telescopes.
- **Recognize** the differences between optical and radio telescopes.

Why It's Important

Learning about space can help us better understand our own world.

Review Vocabulary
telescope: an instrument that can magnify the size of distant objects

New Vocabulary
- electromagnetic spectrum
- refracting telescope
- reflecting telescope
- observatory
- radio telescope

Electromagnetic Waves

As you have read, we have begun to explore our solar system and beyond. With the help of telescopes like the *Hubble,* we can see far into space, but if you've ever thought of racing toward distant parts of the universe, think again. Even at the speed of light it would take many years to reach even the nearest stars.

Light from the Past When you look at a star, the light that you see left the star many years ago. Although light travels fast, distances between objects in space are so great that it sometimes takes millions of years for the light to reach Earth.

The light and other energy leaving a star are forms of radiation. Radiation is energy that is transmitted from one place to another by electromagnetic waves. Because of the electric and magnetic properties of this radiation, it's called electromagnetic radiation. Electromagnetic waves carry energy through empty space and through matter.

Electromagnetic radiation is everywhere around you. When you turn on the radio, peer down a microscope, or have an X ray taken—you're using various forms of electromagnetic radiation.

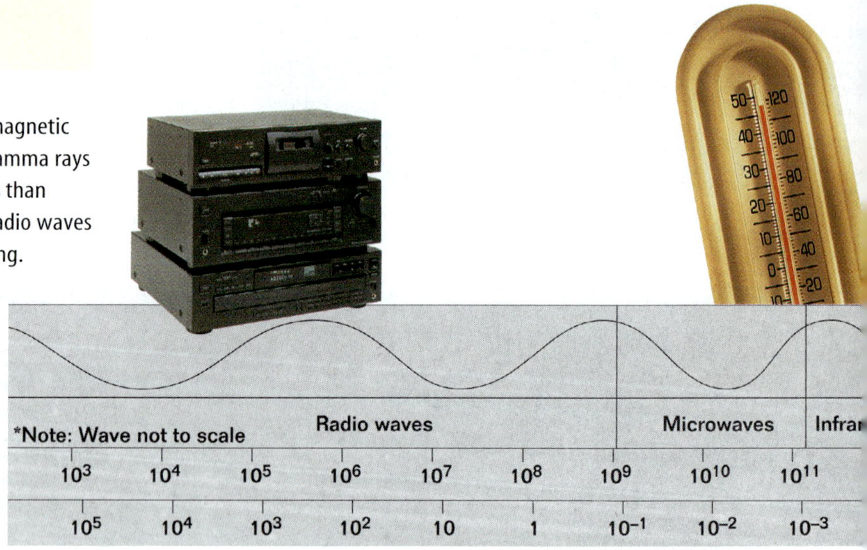

Figure 1 The electromagnetic spectrum ranges from gamma rays with wavelengths of less than 0.000 000 000 01 m to radio waves more than 100,000 m long.
Observe *how frequency changes as wavelength shortens.*

Section 1 Resource Manager

Chapter *FAST FILE* Resources
- Transparency Activity, pp. 40, 43–44
- Directed Reading for Content Mastery, pp. 15–16
- Note-taking Worksheets, pp. 29–31
- Enrichment, p. 26
- Reinforcement, p. 23
- MiniLab, p. 3
- Lab Worksheets, pp. 5–6
- Lab Activities, pp. 9–12

Cultural Diversity, p. 3

Electromagnetic Radiation Sound waves, which are a type of mechanical wave, can't travel through empty space. How, then, do we hear the voices of the astronauts while they're in space? When astronauts speak into a microphone, the sound waves are converted into electromagnetic waves called radio waves. The radio waves travel through space and through Earth's atmosphere. They're then converted back into sound waves by electronic equipment and audio speakers.

Radio waves and visible light from the Sun are just two types of electromagnetic radiation. Other types include gamma rays, X rays, ultraviolet waves, infrared waves, and microwaves. **Figure 1** shows these forms of electromagnetic radiation arranged according to their wavelengths. This arrangement of electromagnetic radiation is called the **electromagnetic spectrum.** Forms of electromagnetic radiation also differ in their frequencies. Frequency is the number of wave crests that pass a given point per unit of time. The shorter the wavelength is, the higher the frequency, as shown in **Figure 1.**

Speed of Light Although the various electromagnetic waves differ in their wavelengths, they all travel at 300,000 km/s in a vacuum. This is called the speed of light. Visible light and other forms of electromagnetic radiation travel at this incredible speed, but the universe is so large that it takes millions of years for the light from some stars to reach Earth.

When electromagnetic radiation from stars and other objects reaches Earth, scientists use it to learn about its source. One tool for studying such electromagnetic radiation is a telescope.

INTEGRATE Health

Ultraviolet Light Many newspapers include an ultraviolet (UV) index to urge people to minimize their exposure to the Sun. Compare the wavelengths and frequencies of red and violet light, shown below in **Figure 1.** Infer what properties of UV light cause damage to tissues of organisms.

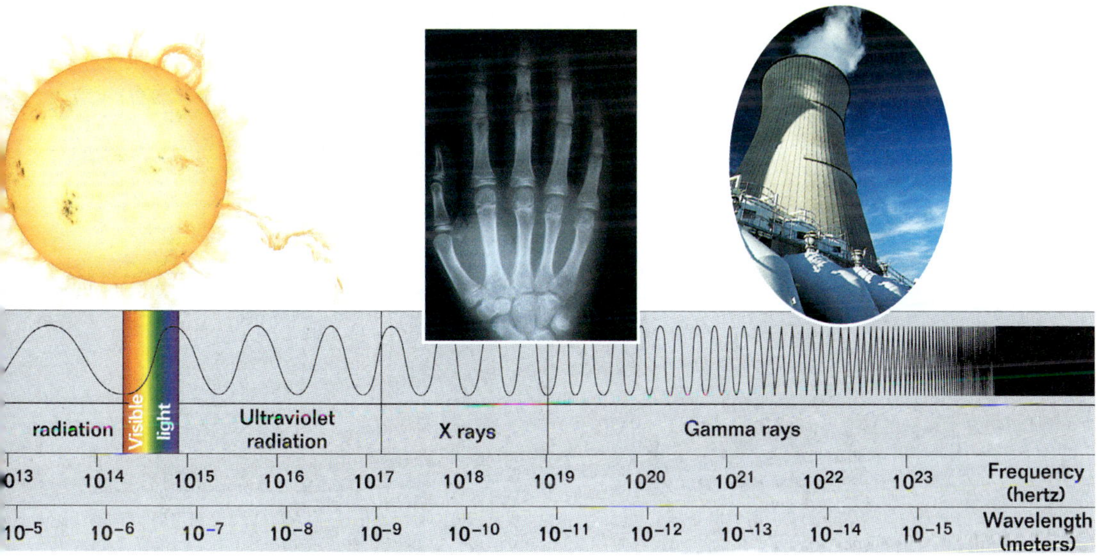

2 Teach

Caption Answer

Figure 1 Frequency increases.

INTEGRATE Health

Ultraviolet Light Violet light has a shorter wavelength and therefore a higher frequency than red light. Since UV light has an even shorter wavelength, it also has an even higher frequency and more energy. The added energy in UV light damages living tissues.

Discussion

Waves in a Vacuum Do you think electromagnetic waves can travel through a vacuum? Explain. Yes; solar energy is electromagnetic radiation. It travels to Earth through the vacuum of space.

Inquiry Lab

Electromagnetic Spectrum

Purpose to help students to better visualize the wavelengths of the electromagnetic spectrum

Possible Materials meterstick, metric ruler, chalkboard, paper

Estimated Time 30 minutes

Teaching Strategies

- Students can use a meterstick to explore the dimensions of some radio waves by drawing on a chalkboard. Waves of 0.1 m (10^{-1} m) and 1 m should be doable.
- Students can explore the waves in the microwave range on paper using a metric ruler as a guide. They could draw 1-cm (10^{-2} m) and 1-mm (10^{-3} m) waves.
- Then have them compare the waves and notice the differences in size, that each is 10 times the length of the previous wave.
- Encourage them to visualize the wavelengths that are too small or too large to reproduce easily.

Differentiated Instruction

Learning Disabled Have each student investigate a different type of electromagnetic radiation, noting its characteristics, what natural or artificial objects generate it, and how (if this is the case) it is used by people. Have students work together to make an informative bulletin board that includes diagrams and labels explaining their findings. L2 ELL COOP LEARN IN **Interpersonal**

SECTION 1 Radiation from Space 409

Fun Fact

The *Hubble Space Telescope* is so powerful that if Earth's surface were not curved and you were in Washington, D.C., you could use it to read a billboard in Missouri.

Quick Demo

Telescope Lens

Materials double convex lens with a 15-cm focal length, and a sheet of white paper

Estimated Time 10 minutes

Procedure Hold the lens in one hand and the paper in the other. Turn off room lights and hold the lens so that light coming in a window passes through it and falls on the sheet of paper. Move the lens until students can see an inverted image of the window and what can be seen through it.

Visual Learning

Figure 2 Have students compare the two types of light telescopes. Which one seems as though it would be easier to construct? Why? The refracting telescope is easier; it has a simpler design.

Activity

Reflecting Telescopes Hand out some polished metal spoons and have students observe their inverted reflections in the bowls. Explain that this resembles the concave mirror in a reflecting telescope. In fact, the first working reflecting telescope built by Isaac Newton used a metal mirror made of tin and copper.

Optical Telescopes

Optical telescopes use light, which is a form of electromagnetic radiation, to produce magnified images of objects. Light is collected by an objective lens or mirror, which then forms an image at the focal point of the telescope. The focal point is where light that is bent by the lens or reflected by the mirror comes together to form an image. The eyepiece lens then magnifies the image. The two types of optical telescopes are shown in **Figure 2.**

A **refracting telescope** uses convex lenses, which are curved outward like the surface of a ball. Light from an object passes through a convex objective lens and is bent to form an image at the focal point. The eyepiece magnifies the image.

A **reflecting telescope** uses a curved mirror to direct light. Light from the object being viewed passes through the open end of a reflecting telescope. This light strikes a concave mirror, which is curved inward like a bowl and located at the base of the telescope. The light is reflected off the interior surface of the bowl to the focal point where it forms an image. Sometimes, a smaller mirror is used to reflect light into the eyepiece lens, where it is magnified for viewing.

Figure 2 These diagrams show how each type of optical telescope collects light and forms an image.

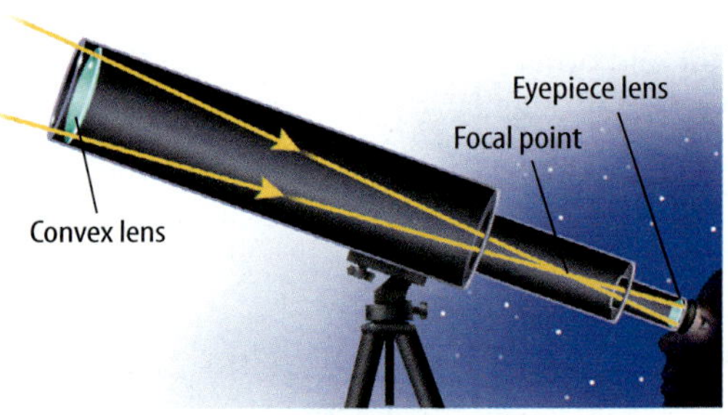

In a refracting telescope, a convex lens focuses light to form an image at the focal point.

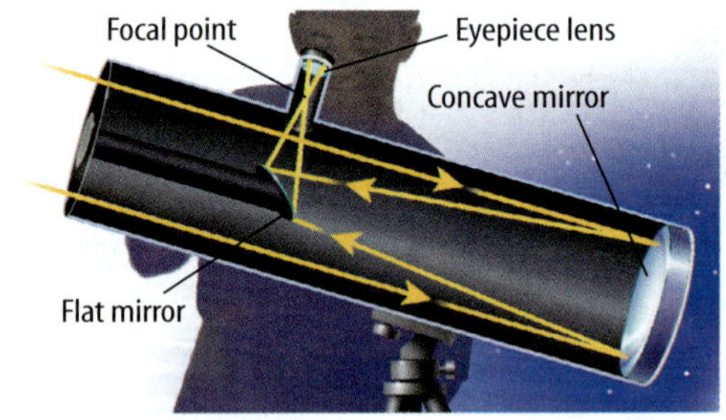

In a reflecting telescope, a concave mirror focuses light to form an image at the focal point.

Using Optical Telescopes Most optical telescopes used by professional astronomers are housed in buildings called **observatories.** Observatories often have dome-shaped roofs that can be opened up for viewing. However, not all telescopes are located in observatories. The *Hubble Space Telescope* is an example.

Optical telescopes are widely available for use by individuals.

Differentiated Instruction

Challenge Have students research the index of refraction and relate what they learn to the use of lenses in a telescope. The index of refraction of a substance is equal to the sine of the angle of incidence of light from a vacuum divided by the sine of the angle of refraction. The index of refraction of glass is used when determining how to shape the objective lens of a telescope. L3

Science Journal

Telescopes Ask students to write a brief paragraph in their Science Journals that answers this question: What advantages does a radio telescope have over an optical telescope? Radio telescopes can be used in all kinds of weather and at any time of day or night. Radio telescopes can detect objects that are invisible to optical telescopes. L2 LS **Logical-Mathematical and Linguistic**

Hubble Space Telescope The *Hubble Space Telescope* was launched in 1990 by the space shuttle *Discovery.* Because *Hubble* is located outside Earth's atmosphere, which absorbs and distorts some of the energy received from space, it should have produced clear images. However, when the largest mirror of this reflecting telescope was shaped, a mistake was made. As a result, images obtained by the telescope were not as clear as expected. In December 1993, a team of astronauts repaired the *Hubble Space Telescope* by installing a set of small mirrors designed to correct images obtained by the faulty mirror. Two more missions to service *Hubble* were carried out in 1997 and 1999, shown in **Figure 3.** Among the objects viewed by *Hubble* after it was repaired in 1999 was a large cluster of galaxies known as Abell 2218.

Reading Check *Why is* **Hubble** *located outside Earth's atmosphere?*

Figure 3 The *Hubble Space Telescope* was serviced at the end of 1999. Astronauts replaced devices on *Hubble* that are used to stabilize the telescope.

Teacher FYI

Hubble Space Telescope The *Hubble Space Telescope* is a reflecting telescope with a light-gathering mirror 240 cm in diameter. It is in orbit 610 km above Earth's surface. NASA controls the telescope with radio commands beamed to it from the ground. The images obtained by *Hubble* are sent to the ground as radio signals as well.

Reading Check

Answer to avoid the interference caused by Earth's atmosphere

Discussion

Sharing Science Tell students that NASA built the *Hubble Space Telescope* and put it in orbit. Now astronomers from all over the world share the use of the telescope and the images it provides. Do you think the United States should share resources like the *Hubble* with other nations? Explain. Possible answers: Yes; everyone should share any tool that advances scientific discovery. No; the United States built it and U.S. scientists should have the benefit of sole use of the telescope.

Science Journal

Hubble Space Telescope Have students research and write about some of the discoveries made by astronomers using the *Hubble Space Telescope.* Discoveries should include galaxies at the edge of the known universe and galaxies colliding and tearing each other apart. L2 LS **Linguistic** P

Reading Check

Answer about 400 years

Mini LAB

Purpose Students contrast the number of stars visible from each of their homes and determine whether light pollution accounts for the difference. L2

LS **Visual-Spatial**

Materials cardboard tube

Teaching Strategy Explain that light pollution is a glow in the sky caused by city lights. The glow makes it difficult to see dim stars. Remind them to record light conditions at their viewing spot.

Analysis

1. Students in areas away from street lights will see more stars than students in urban areas or on main streets.
2. More stars are visible in areas with less background light.

Assessment

Process Have students observe and record common sources of light pollution. Possible answers: streetlights, lights from homes, apartment buildings, and shopping centers. Use **Performance Assessment in the Science Classroom,** p. 97.

Try at Home

Mini LAB

Observing Effects of Light Pollution

Procedure

1. Obtain a **cardboard tube** from an empty roll of paper towels.
2. Go outside on a clear night about two hours after sunset. Look through the cardboard tube at a specific constellation decided upon ahead of time.
3. Count the number of stars you can see without moving the observing tube. Repeat this three times.
4. Calculate the average number of observable stars at your location.

Analysis

1. Compare and contrast the number of stars visible from other students' homes.
2. Explain the causes and effects of your observations.

Try at Home

Large Reflecting Telescopes Since the early 1600s, when the Italian scientist Galileo Galilei first turned a telescope toward the stars, people have been searching for better ways to study what lies beyond Earth's atmosphere. For example, the twin Keck reflecting telescopes, shown in **Figure 4,** have segmented mirrors 10 m wide. Until 2000, these mirrors were the largest reflectors ever used. To cope with the difficulty of building such huge mirrors, the Keck telescope mirrors are built out of many small mirrors that are pieced together. In 2000, the European Southern Observatory's telescope, in Chile, consisted of four 8.2-m reflectors, making it the largest optical telescope in use.

Reading Check *About how long have people been using telescopes?*

Active and Adaptive Optics The most recent innovations in optical telescopes involve active and adaptive optics. With active optics, a computer corrects for changes in temperature, mirror distortions, and bad viewing conditions. Adaptive optics is even more ambitious. Adaptive optics uses a laser to probe the atmosphere and relay information to a computer about air turbulence. The computer then adjusts the telescope's mirror thousands of times per second, which lessens the effects of atmospheric turbulence. Telescope images are clearer when corrections for air turbulence, temperature changes, and mirror-shape changes are made.

Figure 4 The twin Keck telescopes on Mauna Kea in Hawaii can be used together, more than doubling their ability to distinguish objects. A Keck reflector is shown in the inset photo. Currently, plans include using these telescopes, along with four others to obtain images that will help answer questions about the origin of planetary systems.

Teacher FYI

Keck's Mirrors Each of the Keck's mirrors actually consists of 36 hexagonal mirrors mounted extremely close together to form one reflecting surface 10 m across. An electronic sensing system keeps each of the mirrors in place and adjusts each automatically if it becomes misaligned.

412 CHAPTER 14 Exploring Space

Radio Telescopes

As shown in the spectrum illustrated in **Figure 1,** stars and other objects radiate electromagnetic energy of various types. Radio waves are an example of long-wavelength energy in the electromagnetic spectrum. A **radio telescope,** such as the one shown in **Figure 5,** is used to study radio waves traveling through space. Unlike visible light, radio waves pass freely through Earth's atmosphere. Because of this, radio telescopes are useful 24 hours per day under most weather conditions.

Radio waves reaching Earth's surface strike the large, concave dish of a radio telescope. This dish reflects the waves to a focal point where a receiver is located. The information allows scientists to detect objects in space, to map the universe, and to search for signs of intelligent life on other planets.

Figure 5 This radio telescope is used to study radio waves traveling through space.

section 1 review

Summary

Electromagnetic Waves

- Light is a form of electromagnetic radiation.
- Electromagnetic radiation includes radio waves, microwaves, X rays, gamma rays, and infrared and ultraviolet radiation.
- Light travels at 300,000 km/s in a vacuum.

Optical Telescopes

- A refracting telescope uses lenses to collect, focus, and view light.
- A reflecting telescope uses a mirror to collect and focus light and a lens to view the image.
- Computers and lasers are used to reduce problems caused by looking through Earth's atmosphere.
- These telescopes are housed in domed buildings called observatories.
- Placing a telescope in space avoids problems caused by Earth's atmosphere.

Radio Telescopes

- Radio telescopes collect and measure radio waves coming from stars and other objects.

Self Check

1. **Identify** one advantage of radio telescopes over optical telescopes.
2. **Infer** If red light has a longer wavelength than blue light, which has a greater frequency?
3. **Explain** the difference between sound waves and radio waves.
4. **Describe** how adaptive optics in a telescope help solve problems caused by atmospheric turbulence.
5. **Think Critically** It takes light from the closest star to Earth (other than the Sun) about four years to reach Earth. If intelligent life were on a planet circling that star, how long would it take for scientists on Earth to send them a radio transmission and for the scientists to receive their reply?

Applying Math

6. **Calculate** how long it takes for a radio signal to reach the Moon, which is about 380,000 km away.
7. **Use Numbers** If an X ray has a frequency of 10^{18} hertz and a gamma ray has a frequency of 10^{21} hertz, how many times greater is the frequency of the gamma ray?

Science Online red.msscience.com/self_check_quiz

Fun Fact

The Very Large Array (VLA) in New Mexico consists of 27 movable dishes, each 25 m wide. The dishes operate as one telescope, making the VLA one of the world's most powerful radio telescopes.

3 Assess

DAILY INTERVENTION

Check for Understanding

Visual-Spatial Give students diagrams containing labeled cross sections of both refracting and reflecting telescopes. Ask them to identify the telescopes as to type and to indicate how each type focuses light.

Reteach

Radio Telescopes Remind students of the wavelength of radio waves and point out that radio telescopes are often very large. Show them some pictures of radio telescopes, such as the one at Arecibo, Puerto Rico, which is suspended between mountain peaks and the Very Large Array in New Mexico, which consists of 27 linked, dish antennas.

Assessment

Process Have students make a table describing the types of telescopes covered in this section and the type of electromagnetic radiation examined by each. Use **Performance Assessment in the Science Classroom,** p. 109.

section 1 review

1. Radio telescopes use a dish antenna to collect and focus radio waves, whereas optical telescopes use lenses or mirrors to collect and focus visible light. Radio telescopes can be used day and night, even when weather conditions make optical viewing impossible.
2. blue light
3. Radio waves are electromagnetic waves and can travel through a vacuum. Sound waves are not electromagnetic and need a medium, such as air or water. Sound waves cannot travel in a vacuum.
4. Adaptive optics uses a computer to adjust the telescope's objective mirror to decrease the effects of air turbulence.
5. about 8 years
6. If light travels 300,000 km/s, you must divide the distance (380,000 km) by this speed. This gives the answer 1.3 s.
7. This is practice with exponents. If you subtract the exponent 18 from 21 you get 3 and 10^3 equals 1,000.

BENCH TESTED

Real-World Question

Purpose Students will construct a working, reflecting telescope. L2 COOP LEARN LS Visual-Spatial

Process Skills observe, construct, design

Time Required 25 minutes

Procedure

Safety Precautions Caution students never to touch broken glass.

Teaching Strategies Collect photographs of the Moon or other celestial objects for students to view.

Troubleshooting Be certain that each pair of students has both a flat and curved mirror. Label the two types of mirrors for easy reference.

Analyze Your Data

1. The object's image will appear larger or smaller depending upon the level of magnification of each lens.
2. The cosmetic mirror reflects the light onto the flat mirror. The flat mirror reflects the light onto the magnifying lens.
3. Magnifying lenses.

Conclude and Apply

1. The curved mirror gathered the light reflecting from the object and focused the light on the flat mirror. The flat mirror reflected the light onto the magnifying lens, and each lens magnified the image.
2. Convex lenses would have been used instead of mirrors.

LAB

Building a Reflecting Telescope

Nearly four hundred years ago, Galileo Galilei saw what no human had ever seen. Using the telescope he built, he saw moons around Jupiter, details of lunar craters, and sunspots. What was it like to make these discoveries? Find out as you make your own reflecting telescope.

Real-World Question

How do you construct a reflecting telescope?

Goals

- **Construct** a reflecting telescope.
- **Observe** magnified images using the telescope and different magnifying lenses.

Materials

flat mirror
shaving or cosmetic mirror (a curved, concave mirror)
magnifying lenses of different magnifications (3–4)

Safety Precautions

WARNING: *Never observe the Sun directly or with mirrors.*

Procedure

1. Position the cosmetic mirror so that you can see the reflection of the object you want to look at. Choose an object such as the Moon, a planet, or an artificial light source.
2. Place the flat mirror so that it is facing the cosmetic mirror.
3. Adjust the position of the flat mirror until you can see the reflection of the object in it.

4. View the image of the object in the flat mirror with one of your magnifying lenses. Observe how the lens magnifies the image.
5. Use your other magnifying lenses to view the image of the object in the flat mirror. Observe how the different lenses change the image of the object.

Analyze Your Data

1. **Describe** how the image changed when you used different magnifying lenses.
2. **Identify** the part or parts of your telescope that reflected the light of the image.
3. **Identify** the parts of your telescope that magnified the image.

Conclude and Apply

1. **Explain** how the three parts of your telescope worked to reflect and magnify the light of the object.
2. **Infer** how the materials you used would have differed if you had constructed a refracting instead of a reflecting telescope.

Communicating Your Data

Write an instructional pamphlet for amateur astronomers about how to construct a reflecting telescope.

Communicating Your Data

Students may want to include a labeled diagram with their pamphlets.

Assessment

Portfolio Ask students to construct a permanent reflecting telescope with a housing of their own design. Use **Performance Assessment in the Science Classroom,** p. 117.

section 2

Early Space Missions

The First Missions into Space

You're offered a choice—front-row-center seats for this weekend's rock concert, or a copy of the video when it's released. Wouldn't you rather be right next to the action? Astronomers feel the same way about space. Even though telescopes have taught them a great deal about the Moon and planets, they want to learn more by going to those places or by sending spacecraft where humans can't go.

Rockets The space program would not have gotten far off the ground using ordinary airplane engines. To break free of gravity and enter Earth's orbit, spacecraft must travel at speeds greater than 11 km/s. The space shuttle and several other spacecrafts are equipped with special engines that carry their own fuel. **Rockets,** like the one in **Figure 6,** are engines that have everything they need for the burning of fuel. They don't even require air to carry out the process. Therefore, they can work in space, which has no air. The simplest rocket engine is made of a burning chamber and a nozzle. More complex rockets have more than one burning chamber.

Rocket Types Two types of rockets are distinguished by the type of fuel they use. One type is the liquid-propellant rocket and the other is the solid-propellant rocket. Solid-propellant rockets are generally simpler but they can't be shut down after they are ignited. Liquid-propellant rockets can be shut down after they are ignited and can be restarted. For this reason, liquid-propellant rockets are preferred for use in long-term space missions. Scientists on Earth can send signals that start and stop the spacecraft's engines whenever they want to modify its course or adjust its orbit. Liquid propellants successfully powered many space probes, including the two *Voyagers* and *Galileo.*

Figure 6 Rockets differ according to the types of fuel used to launch them. Liquid oxygen is used often to support combustion.

as you read

***What* You'll Learn**

- **Compare and contrast** natural and artificial satellites.
- **Identify** the differences between artificial satellites and space probes.
- **Explain** the history of the race to the Moon.

***Why* It's Important**

Early missions that sent objects and people into space began a new era of human exploration.

Review Vocabulary

thrust: the force that propels an aircraft or missile

New Vocabulary

- rocket
- satellite
- orbit
- space probe
- Project Mercury
- Project Gemini
- Project Apollo

1 Motivate

Bellringer

INTERACTIVE CHALKBOARD PowerPoint® Presentations

Section Focus Transparencies also are available on the Interactive Chalkboard CD-ROM. L2 ELL

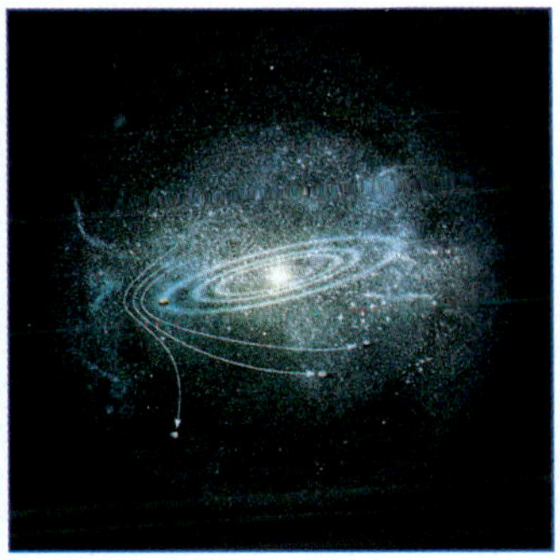
Section 2 Section Focus Transparency **The Outer Limits of the Solar System**

Where does the solar system end? The planets end at Pluto, but the effects of the Sun's gravitational pull extend much farther. The last items held by the Sun's gravitational pull are in Oort's Cloud.

1. How can satellites and space probes help us get a better understanding of the planets in our solar system?
2. How can space missions help us study Earth?

L2

Exploring Space

Tie to Prior Knowledge

Moon Landing Ask students if astronauts have ever traveled to the Moon. Write responses on the board as students brainstorm facts they know about the Moon landings. Possible answers: First one occurred in 1969; first person to step on the Moon was Neil Armstrong. Have students review this list after reading the section to see how many "facts" were correct.

Section 2 Resource Manager

Chapter *FAST FILE* Resources

- Transparency Activity, p. 41
- Directed Reading for Content Mastery, p. 17
- Enrichment, p. 27
- MiniLAB, p. 4
- Reinforcement, p. 24

Mathematics Skill Activities, p. 15

Reading and Writing Skill Activities, p. 11

2 Teach

Discussion

Rocket Propulsion Remind students that solid propellant rockets are simpler than liquid propellant rockets. What advantage might a liquid propellant rocket have over a solid propellant rocket? Possible answer: the ability to start and stop the rocket makes it easier to control the rocket's velocity.

Quick Demo

How Rockets Move

Materials long balloon, drinking straw, string, tape

Estimated Time 10 minutes

Procedure Have students thread a long string through a drinking straw. Inflate the balloon and hold the end closed. Have students tape the straw to the side of the balloon. Then have two students hold the ends of the string and pull it taut. Release the balloon. Students should see that gases escaping from behind the balloon push it forward along the string.

Applying Math

National Math Standards

Correlation to Mathematics Objectives

3

Answers to Practice Problems

1. Students' graphs should look like a line drawing of an astronaut.
2. Have students compare drawings made by classmates to the originals.

Figure 7 The space shuttle uses both liquid and solid fuels. Here the red liquid fuel tank is visible behind a white, solid rocket booster.

Rocket Launching Solid-propellant rockets use a rubberlike fuel that contains its own oxidizer. The burning chamber of a rocket is a tube that has a nozzle at one end. As the solid propellant burns, hot gases exert pressure on all inner surfaces of the tube. The tube pushes back on the gas except at the nozzle where hot gases escape. Thrust builds up and pushes the rocket forward.

Liquid-propellant rockets use a liquid fuel and an oxidizer, such as liquid oxygen, stored in separate tanks. To ignite the rocket, the oxidizer is mixed with the liquid fuel in the burning chamber. As the mixture burns, forces are exerted and the rocket is propelled forward. **Figure 7** shows the space shuttle, with both types of rockets, being launched.

Applying Math — Make and Use Graphs

DRAWING BY NUMBERS Points are defined by two coordinates, called an ordered pair. To plot an ordered pair, find the first number on the horizontal *x*-axis and the second on the vertical *y*-axis. The point is placed where these two coordinates intersect. Line segments are drawn to connect points. Draw a symmetrical house by using an *x-y* grid and these coordinates: (1,1), (5,1), (5,4), (3,6), (1,4)

Solution

1. *On a piece of graph paper, label and number the x-axis 0 to 6 and the y-axis 0 to 6, as shown here.*
2. *Plot the above points and connect them with straight line segments, as shown here.*

Section	Points
1	(1, –8) (3, –13) (6, –21) (9, –21) (9, –17) (8, –15) (8, –12) (6, –8) (5, –4) (4, –3) (4, –1) (5,1) (6,3) (8,3) (9,4) (9,7) (7,11) (4,14) (4,22) (–9,22) (–9,10) (–10,5) (–11, –1) (–11, –7) (–9, –8) (–8, –7) (–8, –1) (–6,3) (–6, –3) (–6, –9) (–7, –20) (–8, –21) (–4, –21) (–4, –18) (–3, –14) (–1, –8)
2	(0,11) (2,13) (2,17) (0,19) (–4,19) (–6,17) (–6,13) (–4,11)
3	(–4,9) (1,9) (1,5) (–1,5) (–2,6) (–4,6)

Practice Problems

1. Label and number the *x*-axis –12 to 10 and the *y*-axis –22 to 23. Draw an astronaut by plotting and connecting the points in each section. Do not draw segments to connect points in different sections.
2. Make your own drawing on graph paper and write its coordinates as ordered pairs. Then give it to a classmate to solve.

Science Online — For more practice, visit red.msscience.com/math_practice

Curriculum Connection

History In 1926, American scientist Robert Goddard tested the first liquid-fueled rocket. The small rocket rose more than 12 m before crashing. It was the forerunner of the rockets we send into space today. Have students research that first rocket launch and do a report on videotape or audiotape as if they were journalists reporting from the scene. Have students play the tapes in class. L2 ELL LS **Musical-Auditory**

Cultural Diversity

Chinese Rockets Historians and scientists generally agree that the Chinese invented the rocket centuries ago. During the 1200s, Chinese soldiers fired small rockets (called "fire arrows" by some) as weapons in battle. Have students research these early rockets and write reports on what they were like as well as how they contrast with rockets today. L2 LS **Logical-Mathematical**

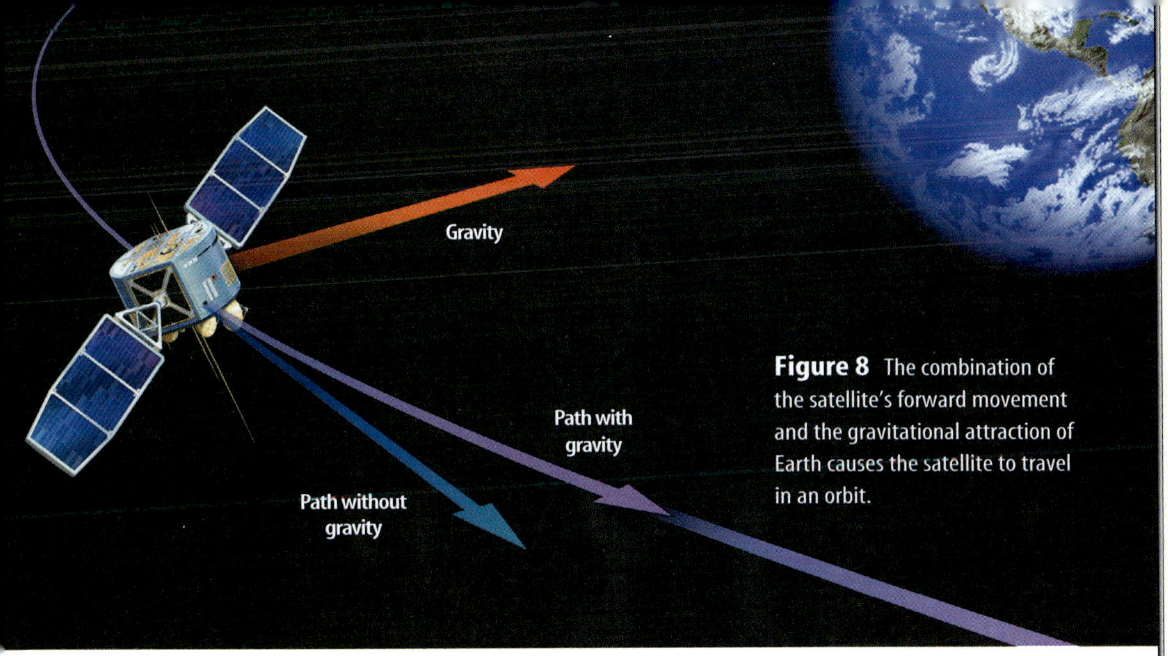

Figure 8 The combination of the satellite's forward movement and the gravitational attraction of Earth causes the satellite to travel in an orbit.

Satellites The space age began in 1957 when the former Soviet Union used a rocket to send *Sputnik I* into space. *Sputnik I* was the first artificial satellite. A **satellite** is any object that revolves around another object. When an object enters space, it travels in a straight line unless a force, such as gravity, makes it turn. Earth's gravity pulls a satellite toward Earth. The result of the satellite traveling forward while at the same time being pulled toward Earth is a curved path, called an **orbit,** around Earth. This is shown in **Figure 8.** *Sputnik I* orbited Earth for 57 days before gravity pulled it back into the atmosphere, where it burned up.

Satellite Uses *Sputnik I* was an experiment to show that artificial satellites could be made and placed into orbit around Earth.

Today, thousands of artificial satellites orbit Earth. Communication satellites transmit radio and television programs to locations around the world. Other satellites gather scientific data, like those shown in **Figure 9,** which can't be obtained from Earth, and weather satellites constantly monitor Earth's global weather patterns.

Figure 9 Data obtained from the satellite *Terra,* launched in 1999, illustrates the use of space technology to study Earth. This false-color image includes data on spring growth, sea-surface temperature, carbon monoxide concentrations, and reflected sunlight, among others.

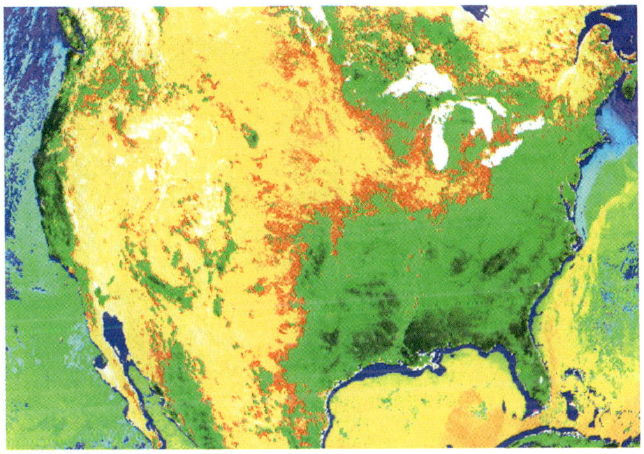

Teacher FYI

Soviet Space Missions Although not as successful as the American effort, the Soviet Union (today's Russia) also had a space program that sent several probes to other planets. The Soviet Union's Venera probes made several voyages to Venus between the late 1960s and the 1980s. The probes sent back data on Venus's atmosphere, beamed back pictures from Venus's surface, and analyzed rocks from the planet's surface.

Discussion

Satellites How has life on Earth changed since man-made satellites were placed in orbit? Discuss some of the uses of satellites and the types of information they carry. Satellites have improved weather forecasting and communications, especially satellite telephone links. Satellite data aid in resource conservation. GPS navigation may also be mentioned.

Virtual Labs

Satellites *How does an artificial satellite stay in orbit?*

Active Reading

Think-Pair-Share Ask students to respond to a question by writing a response. After thinking for a few minutes, have partners share responses to the question. Finally, ask the students to share responses with the class. Have students become involved in a Think-Pair-Share about the importance of exploring space. L1 COOP LEARN IS Interpersonal

SECTION 2 Early Space Missions 417

INTEGRATE Career

Astronomy The American Astronomical Society advises those interested in a career in astronomy to study as much science and math as they can in high school, including pre-calculus if possible. In college, students may lean towards either math or physics, but computer science is of great benefit also.

Use Science Words

Word Usage Have students use the term *space probe* in a sentence. Have pairs take turns reading their sentences and critiquing one another on whether the term was used correctly.

Visual Learning

Table 1 Direct students to look carefully at the names of the space probes in the table. Ask them to determine what the names have in common. They all concern adventure and exploration. Mariners are seafarers. Pioneers are people who are among the first to travel to a place. Vikings were ancient Scandinavian explorers. Magellan commanded the first expedition to sail around the world in the early 1500s. LS **Linguistic**

INTEGRATE Career

Astronomy Astronomers today have more choices than ever before. Some still use optical telescopes to study stars and galaxies. Others explore the universe using the radio, X-ray, infrared, or even gamma-ray regions of the electromagnetic spectrum. Still others deal with theory and work with physicists to understand the big bang and the nature of matter in the universe. Government, universities, and private industry offer jobs for astronomers.

Space Probes

Not all objects carried into space by rockets become satellites. Rockets also can be used to send instruments into space to collect data. A **space probe** is an instrument that gathers information and sends it back to Earth. Unlike satellites that orbit Earth, space probes travel into the solar system as illustrated in **Figure 10.** Some even have traveled to the edge of the solar system. Among these is *Pioneer 10,* launched in 1972. Although its transmitter failed in 2003, it continues on through space. Also, both *Voyager* spacecrafts should continue to return data on the outer reaches of the solar system until about 2020.

Space probes, like many satellites, carry cameras and other data-gathering equipment, as well as radio transmitters and receivers that allow them to communicate with scientists on Earth. **Table 1** shows some of the early space probes launched by the National Aeronautics and Space Administration (NASA).

Table 1 Some Early Space Missions

Mission Name		Date Launched	Destination	Date Obtained
Mariner 2		August 1962	Venus	verified high temperatures in Venus's atmosphere
Pioneer 10		March 1972	Jupiter	sent back photos of Jupiter—first probe to encounter an outer planet
Viking 1		August 1975	Mars	orbiter mapped the surface of Mars; lander searched for life on Mars
Magellan		May 1989	Venus	mapped Venus's surface and returned data on the composition of Venus's atmosphere

Curriculum Connection

Social Studies Have students speculate why so many developments occurred in the space program during the 1960s. Possible answer: Students should realize that the United States and the former Soviet Union were participating in a "race for space." L2 LS **Logical-Mathematical**

Differentiated Instruction

Challenge Have students research space junk, the unused and nonfunctional materials orbiting Earth. Such junk ranges from objects the size of grains of sand, such as paint chips, to large, nonfunctioning satellites. Students can research and write a brief report summarizing the growing threat that space junk might pose to the safety of future space flights. Have them give specific examples, if possible. L3

NATIONAL GEOGRAPHIC VISUALIZING SPACE PROBES

Figure 10

Probes have taught us much about the solar system. As they travel through space, these car-size craft gather data with their onboard instruments and send results back to Earth via radio waves. Some data collected during these missions are made into pictures, a selection of which is shown here.

A In 1974, *Mariner 10* obtained the first good images of the surface of Mercury.

B A Soviet *Venera* probe took this picture of the surface of Venus on March 1, 1982. Parts of the spacecraft's landing gear are visible at the bottom of the photograph.

D In 1990, *Magellan* imaged craters, lava domes, and great rifts, or cracks, on the surface of Venus.

C The *Voyager 2* mission included flybys of the outer planets Jupiter, Saturn, Uranus, and Neptune. *Voyager* took this photograph of Neptune in 1989 as the craft sped toward the edge of the solar system.

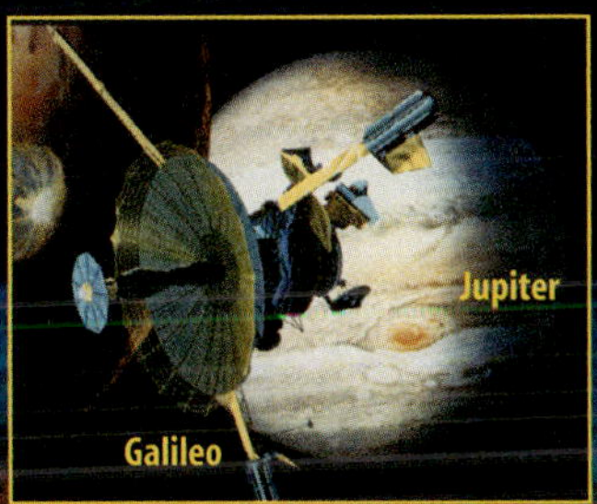

E NASA's veteran space traveler *Galileo* nears Jupiter in this artist's drawing. The craft arrived at Jupiter in 1995 and sent back data, including images of Europa, one of Jupiter's 61 moons, seen below in a color-enhanced view.

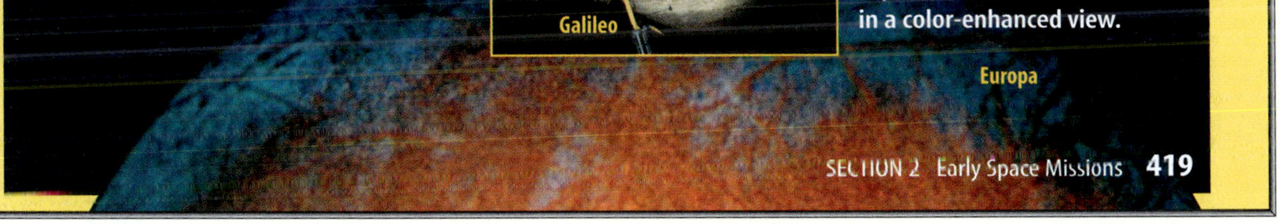

NATIONAL GEOGRAPHIC

Visualizing Space Probes

Have students examine the pictures and read the captions. Then ask the following questions.

What features on the pictured space probes might have helped them communicate with Earth? All have some type of antennae. *Mariner*, *Magellan*, *Voyager 2*, and *Galileo*, for example, have "dish" antennae for receiving signals from Earth. Solar panels on *Mariner* and *Magellan* might have helped power communication equipment.

In what manner is the scale of the planets and moon exaggerated? Possible answer: The smallest planet pictured, Mercury, appears larger than Jupiter, the largest. Venus, while larger than Mercury, would be dwarfed by Neptune and Jupiter and appears far larger than true scale. Europa is the smallest body shown here and is vastly inflated in scale.

Activity

Model Space Probes Have students brainstorm to create their own space probes. They can draw their models using colored pencils or pens. Beneath each model, have students explain the purpose of each feature and why they included it on the probe. P

Differentiated Instruction

English-Language Learners Supply students with a diagram of the solar system with the planets labeled. Then have them cut the outlines of the space probes from photocopies of this figure or elsewhere and match them with the planets they visited. They can follow the paths of the *Voyager* probes from the text on the next page. ELL

Activity

Planet Photos Have students research to find photographs of Saturn, Jupiter, Uranus, or Neptune sent back to Earth by the *Voyager* probes. Have each student print out a photo and describe what it shows about the planet. Make a bulletin board with the photos and student captions.

Reading Check

Answer Cracks in Europa's outer layer of ice might indicate geological activity that could heat Europa's ice, forming liquid water under the surface.

Science Online

Topic: Galileo Mission

Go to red.msscience.com for Web links to informaiton about the Galileo mission.

Activity Make a list of the places the *Galileo* spacecraft visited on its mission.

***Voyager* and *Pioneer* Probes** Space probes *Voyager 1* and *Voyager 2* were launched in 1977 and now are heading toward deep space. *Voyager 1* flew past Jupiter and Saturn. *Voyager 2* flew past Jupiter, Saturn, Uranus, and Neptune. These probes will explore beyond the solar system as part of the Voyager Interstellar Mission. Scientists expect these probes to continue to transmit data to Earth for at least 20 more years.

Pioneer 10, launched in 1972, was the first probe to survive a trip through the asteroid belt and encounter an outer planet, Jupiter. As of 2003, *Pioneer 10* was more than 12 billion km from Earth, and will continue beyond the solar system. The probe carries a gold medallion with an engraving of a man, a woman, and Earth's position in the galaxy.

Galileo Launched in 1989, *Galileo* reached Jupiter in 1995. In July 1995, *Galileo* released a smaller probe that began a five-month approach to Jupiter. The small probe took a parachute ride through Jupiter's violent atmosphere in December 1995.

Before being crushed by the atmospheric pressure, it transmitted information about Jupiter's composition, temperature, and pressure to the satellite orbiting above. *Galileo* studied Jupiter's moons, rings, and magnetic fields and then relayed this information to scientists who were waiting eagerly for it on Earth.

INTEGRATE Life Science Studies of Jupiter's moon Europa by *Galileo* indicate that an ocean of water may exist under the surface of Europa. A cracked outer layer of ice makes up Europa's surface, shown in **Figure 11.** The cracks in the surface may be caused by geologic activity that heats the ocean underneath the surface. Sunlight penetrates these cracks, further heating the ocean and setting the stage for the possible existence of life on Europa. *Galileo* ended its study of Europa in 2000. More advanced probes will be needed to determine whether life exists on this icy moon.

Figure 11 Future missions will be needed to determine whether life exists on Europa.

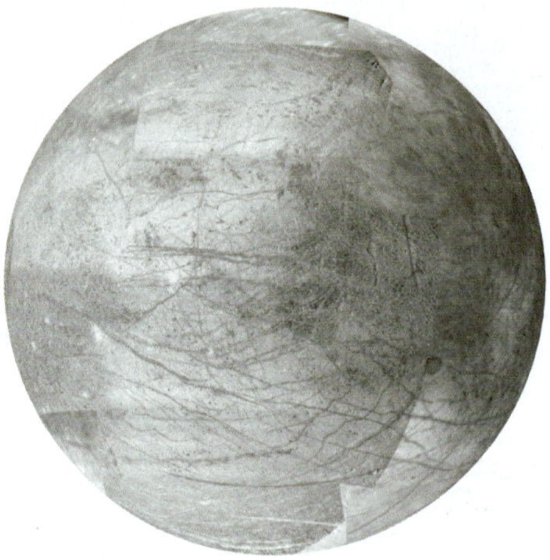

Reading Check ***What features on Europa suggest the possibility of life existing on this moon?***

In October and November of 1999, *Galileo* approached Io, another one of Jupiter's moons. It came within 300 km and took photographs of a volcanic vent named Loki, which emits more energy than all of Earth's volcanoes combined. *Galileo* also discovered eruption plumes that shoot gas made of sulfur and oxygen.

Cultural Diversity

History Early standards for astronauts limited which Americans could participate. For example, the earliest astronauts had to be test pilots, but women were not allowed to train as test pilots until 1970. Astronaut candidate selection broadened in the late 1970s. The 1978 astronaut group included six women, three African Americans, and an Asian American.

Teacher FYI

Voyager The most distant human-made object as of April 2003 was *Voyager 1* space probe. At that time, it was 13 billion km (87 AU) from Earth. *Voyager 1* is expected to send data back to Earth until it runs out of electrical power, around 2020. *Voyager 1*, along with *Voyager 2*, is now part of the Voyager Interstellar Mission (VIM).

Moon Quest

Throughout the world, people were shocked when they turned on their radios and television sets in 1957 and heard the radio transmissions from *Sputnik I* as it orbited Earth. All that *Sputnik I* transmitted was a sort of beeping sound, but people quickly realized that launching a human into space wasn't far off.

In 1961, Soviet cosmonaut Yuri A. Gagarin became the first human in space. He orbited Earth and returned safely. Soon, President John F. Kennedy called for the United States to send humans to the Moon and return them safely to Earth. His goal was to achieve this by the end of the 1960s. The race for space was underway.

The U.S. program to reach the Moon began with **Project Mercury.** The goals of Project Mercury were to orbit a piloted spacecraft around Earth and to bring it back safely. The program provided data and experience in the basics of space flight. On May 5, 1961, Alan B. Shepard became the first U.S. citizen in space. In 1962, *Mercury* astronaut John Glenn became the first U.S. citizen to orbit Earth. **Figure 12** shows Glenn preparing for liftoff.

Figure 12 An important step in the attempt to reach the Moon was John Glenn's first orbit around Earth.

Reading Check *What were the goals of Project Mercury?*

Project Gemini The next step in reaching the Moon was called **Project Gemini.** Teams of two astronauts in the same *Gemini* spacecraft orbited Earth. One *Gemini* team met and connected with another spacecraft in orbit—a skill that would be needed on a voyage to the Moon.

The *Gemini* spacecraft was much like the *Mercury* spacecraft, except it was larger and easier for the astronauts to maintain. It was launched by a rocket known as a *Titan II,* which was a liquid fuel rocket.

In addition to connecting spacecraft in orbit, another goal of Project Gemini was to investigate the effects of space travel on the human body.

Along with the *Mercury* and *Gemini* programs, a series of robotic probes was sent to the Moon. *Ranger* proved that a spacecraft could be sent to the Moon. In 1966, *Surveyor* landed gently on the Moon's surface, indicating that the Moon's surface could support spacecraft and humans. The mission of *Lunar Orbiter* was to take pictures of the Moon's surface that would help determine the best future lunar landing sites.

Mini LAB

Modeling a Satellite

WARNING: *Stand a safe distance away from classmates.*

Procedure

1. Tie one end of a strong, 50-cm-long **string** to a small **cork.**
2. Hold the other end of the string tightly with your arm fully extended.
3. Move your hand back and forth so that the cork swings in a circular motion.
4. Gradually decrease the speed of the cork.

Analysis

1. What happened as the cork's motion slowed?
2. How does the motion of the cork resemble that of a satellite in orbit?

Reading Check

Answer to orbit a piloted spacecraft around Earth and bring it back safely

Mini LAB

Purpose Students will model how satellites stay in orbit. LS **Kinesthetic**

Materials string (2 m), cork

Teaching Strategy Discuss gravity—the attractive force between objects.

Safety Precautions Make sure students stand a safe distance from one another. Suggest students wear safety glasses.

Analysis

1. It dropped out of orbit.
2. The force of gravity pulls the satellite toward Earth; the satellite's inertia resists this pull. Together, these forces keep a satellite in orbit.

Assessment

Process Predict what would happen to a satellite whose forward motion was very slow. It would fall to Earth.

Differentiated Instruction

Challenge Have different pairs of students use the school library to research one of the seven original Project Mercury astronauts. Pairs should collaborate on written reports about the missions each astronaut performed in the *Mercury* and other space missions, and what these astronauts are doing now. L2 COOP LEARN LS **Linguistic**

Science Journal

Project Apollo Have students write a paragraph in their Science Journals why they do or do not think it was important for the United States to win the race to the Moon. Encourage students to fully explain the reasons for their point of view. Use these paragraphs as a springboard for a class debate. L2 LS **Linguistic**

SECTION 2 Early Space Missions 421

3 Assess

DAILY INTERVENTION

Check for Understanding

Interpersonal Have students work in pairs and each make a list of ways that the space program has directly affected their lives. Then they should compare their lists, noting which things they agree or disagree with. If time allows, have them research these and see, which choices are correct. Possibilities are hand-held calculators, computers, and materials for athletic shoes.

Reteach

Flash Cards Have pairs of students make flash cards of the artificial satellites and space probes presented in this section. Have them write the spacecraft name on one side and its mission and discoveries on the other. Have partners quiz each other to review the material. L1 COOP LEARN IS Interpersonal

Assessment

Portfolio Ask each student to write a poem about one of the American missions to space or a famous figure from the United States space program. Encourage students to read their poems in class. Use **Performance Assessment in the Science Classroom,** p. 157.

Figure 13 The Lunar Rover vehicle was first used during the *Apollo 15* mission. Riding in the moon buggy, *Apollo 15, 16,* and *17* astronauts explored the lunar surface.

Project Apollo The final stage of the U.S. program to reach the Moon was **Project Apollo.** On July 20, 1969, *Apollo 11* landed on the Moon's surface. Neil Armstrong was the first human to set foot on the Moon. His first words as he stepped onto its surface were, "That's one small step for man, one giant leap for mankind." Edwin Aldrin, the second of the three *Apollo 11* astronauts, joined Armstrong on the Moon, and they explored its surface for two hours. While they were exploring, Michael Collins remained in the Command Module; Armstrong and Aldrin then returned to the Command Module before beginning the journey home. A total of six lunar landings brought back more than 2,000 samples of moon rock and soil for study before the program ended in 1972. **Figure 13** shows an astronaut exploring the Moon's surface from the Lunar Rover vehicle.

section 2 review

Summary

First Missions into Space

- Rockets are engines that have everything they need to burn fuel.
- Rockets may be fueled with liquid or solid propellants.
- A satellite is any object that revolves around another object.

Space Probes

- A space probe is an instrument that gathers information and sends it back to Earth.
- *Voyager* and *Pioneer* are probes designed to explore the solar system and beyond.
- *Galileo* is a space probe that explored Jupiter and its moons.

Moon Quest

- Project Mercury sent the first piloted spacecraft around Earth.
- *Ranger* and *Surveyor* probes explored the Moon's surface.
- *Gemini* orbited teams of two astronauts.
- Project Apollo completed six lunar landings.

Self Check

1. **Explain** why Neptune has eleven satellites even though it is not orbited by human-made objects.
2. **Explain** why *Galileo* was considered a space probe as it traveled to Jupiter. However, once there, it became an artificial satellite.
3. **List** several discoveries made by the *Voyager 1* and *Voyager 2* space probes.
4. **Sequence** Draw a time line beginning with *Sputnik* and ending with Project Apollo. Include descriptions of important missions.
5. **Think Critically** Is Earth a satellite of any other body in space? Explain.

Applying Math

6. **Solve Simple Equations** A standard unit of measurement in astronomy is the astronomical unit, or AU. It equals is about 150,000,000,000 (1.5×10^{11}) m. In 2000, *Pioneer 10* was more than 11 million km from Earth. How many AUs is this?
7. **Convert Units** A spacecraft is launched at a velocity of 40,200 km/h. Express this speed in kilometers per second. Show your work.

Science Online red.msscience.com/self_check_quiz

section 2 review

1. They are all natural satellites of Neptune.
2. Once *Galileo* arrived at Jupiter, it went into orbit around that planet.
3. Discoveries include: ring around Jupiter; additional moons of Jupiter, Saturn, Uranus, and Neptune; erupting volcanoes on Io; complexity of Saturn's rings; geysers on Triton; storms on Neptune.
4. Time line should begin in 1957 with *Sputnik*. Time line should include Project Mercury, 1961–1962; Project Gemini, 1965–1966; Project Apollo, 1967–1972, especially *Apollo 11*, 1969.
5. Yes; Earth is a natural satellite of the Sun because Earth revolves around the Sun.
6. Because 1 AU equals 150 million km, you must divide the distance 11 billion by this to get 73.3 AU.
7. Because there are 60 seconds in a minute and 60 minutes in an hour you must divide the 40,200 km/hr by 3,600 to get 11.2 km/s.

section 3 Current and Future Space Missions

The Space Shuttle

Imagine spending millions of dollars to build a machine, sending it off into space, and watching its 3,000 metric tons of metal and other materials burn up after only a few minutes of work. That's exactly what NASA did with the rocket portions of spacecraft for many years. The early rockets were used only to launch a small capsule holding astronauts into orbit. Then sections of the rocket separated from the rest and burned when reentering the atmosphere.

A Reusable Spacecraft NASA administrators, like many others, realized that it would be less expensive and less wasteful to reuse resources. The reusable spacecraft that transports astronauts, satellites, and other materials to and from space is called the **space shuttle**, shown in **Figure 14,** as it is landing.

At launch, the space shuttle stands on end and is connected to an external liquid-fuel tank and two solid-fuel booster rockets. When the shuttle reaches an altitude of about 45 km, the emptied, solid-fuel booster rockets drop off and parachute back to Earth. These are recovered and used again. The external liquid-fuel tank separates and falls back to Earth, but it isn't recovered.

as you read

What You'll Learn

- **Explain** the benefits of the space shuttle.
- **Identify** the usefulness of orbital space stations.
- **Explore** future space missions.
- **Identify** the applications of space technology to everyday life.

Why It's Important

Experiments performed on future space missions may benefit you.

Review Vocabulary

cosmonaut: astronaut of the former Soviet Union or present-day Russian space program

New Vocabulary

- space shuttle
- space station

Work on the Shuttle After the space shuttle reaches space, it begins to orbit Earth. There, astronauts perform many different tasks. In the cargo bay, astronauts can conduct scientific experiments and determine the effects of spaceflight on the human body. When the cargo bay isn't used as a laboratory, the shuttle can launch, repair, and retrieve satellites. Then the satellites can be returned to Earth or repaired onboard and returned to space. After a mission, the shuttle glides back to Earth and lands like an airplane. A large landing field is needed as the gliding speed of the shuttle is 335 km/h.

Figure 14 The space shuttle is designed to make many trips into space.

1 Motivate

Bellringer

INTERACTIVE CHALKBOARD PowerPoint® Presentations

Section Focus Transparencies also are available on the Interactive Chalkboard CD-ROM. L2 ELL

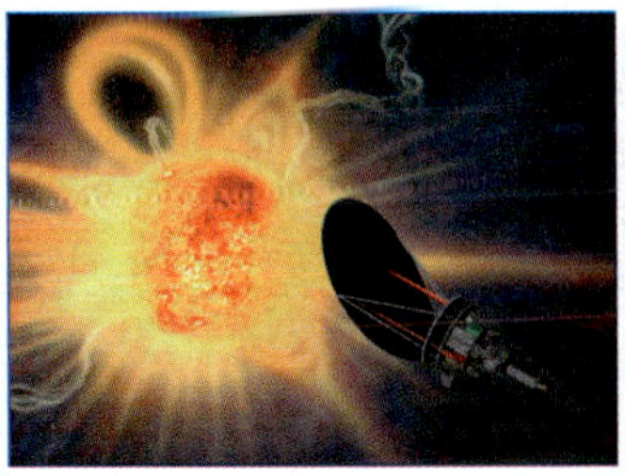

Section 3 Section Focus Transparency **Hot Hot Hot Hot Hot . . .**

One of the missions NASA has planned for the future is a solar probe. It is scheduled for launch in February 2007, and it will make two close passes to the Sun—one in 2010 and one in 2015. The solar probe will pass so close to the Sun that it will be in the Sun's atmosphere.

1. Why do you think the solar probe won't be launched before 2007?
2. How might conditions in the Sun's atmosphere affect the design of the solar probe?
3. What do you think scientists hope to learn from the solar probe?

L2

Exploring Space

Tie to Prior Knowledge

Space Shuttle Have students recall watching the launch or landing of a space shuttle. Have each student try to contribute one fact about the shuttle. Make a list of the facts on the board. Review the list with students after reading this section.

Section 3 Resource Manager

Chapter *FAST FILE* Resources

Transparency Activity, p. 42

Directed Reading for Content Mastery, pp. 17, 18

Lab Worksheets, pp. 7-8

Reinforcement, p. 25

Cultural Diversity, p. 25

Home and Community Involvement, p. 31

Physical Science Critical Thinking/Problem Solving, p. 5

Lab Management and Safety, p. 73

2 Teach

Quick Demo

Shuttle Layout

Materials model of the space shuttle, photos or video of the space shuttle

Estimated Time 20 minutes

Procedure Use the model to demonstrate how the cargo bay is used to carry artificial satellites from Earth and how the mobile arm is used to place them into orbit or bring them back into the cargo bay for repairs. If possible, follow up using photos or a video showing the mobile arm and cargo bay in use. L2 **Visual-Kinesthetic**

IDENTIFYING Misconceptions

People in Space Students may think that people on the *International Space Station* will walk around, work, and play normally, much as they see in science fiction TV shows. Remind students that astronauts in orbit around Earth experience microgravity, or weightlessness, which will affect their movements.

Discussion

Space Cooperation Should the United States cooperate with other nations to build a space station, or should the United States build it alone? Accept all answers as a basis for discussion. Some students might think cooperation would cut costs and bring nations together. Other students might think the United States should not share space technology with other nations.

Caption Answer

Figure 16 It was an important first step in building and operating the *International Space Station.*

Figure 15 Astronauts performed a variety of tasks while living and working in space onboard *Skylab.*

Space Stations

Astronauts can spend only a short time living in the space shuttle. Its living area is small, and the crew needs more room to live, exercise, and work. A **space station** has living quarters, work and exercise areas, and all the equipment and support systems needed for humans to live and work in space.

In 1973, the United States launched the space station *Skylab,* shown in **Figure 15.** Crews of astronauts spent up to 84 days there, performing experiments and collecting data on the effects on humans of living in space. In 1979, the abandoned *Skylab* fell out of orbit and burned up as it entered Earth's atmosphere.

Crews from the former Soviet Union have spent more time in space, onboard the space station *Mir,* than crews from any other country. Cosmonaut Dr. Valery Polyakov returned to Earth after 438 days in space studying the long-term effects of weightlessness.

Cooperation in Space

In 1995, the United States and Russia began an era of cooperation and trust in exploring space. Early in the year, American Dr. Norman Thagard was launched into orbit aboard the Russian *Soyuz* spacecraft, along with two Russian cosmonaut crewmates. Dr. Thagard was the first U.S. astronaut launched into space by a Russian booster and the first American resident of the Russian space station *Mir.*

Figure 16 Russian and American scientists have worked together to further space exploration. **Explain** *why the docking of the space shuttle with* Mir *was so important.*

In June 1995, Russian cosmonauts rode into orbit onboard the space shuttle *Atlantis,* America's 100th crewed launch. The mission of *Atlantis* involved, among other studies, a rendezvous and docking with the space station *Mir.* The cooperation that existed on this mission, as shown in **Figure 16,** continued through eight more space shuttle-*Mir* docking missions. Each of the eight missions was an important step toward building and operating the *International Space Station.* In 2001, the abandoned *Mir* space station fell out of orbit and burned up upon reentering the atmosphere. Cooperation continued as the *International Space Station* began to take form.

Differentiated Instruction

Hearing Impaired and Learning Disabled Provide illustrations and photos of the spacecraft discussed. Have students do research to write extended captions for the images. Students can also work on the Internet to find and print out photos of various probes and satellites to make a bulletin board of United States space travel. LS **Visual-Spatial**

Teacher FYI

Skylab When first launched in 1973, *Skylab* lost part of its protective covering and one of its solar panels. The other panel didn't fully open and malfunctioned. At first, the station had limited power and temperatures inside rose. Astronauts heading to the station first had to repair it. Once repaired, *Skylab* went on to complete its mission.

The International Space Station The *International Space Station (ISS)* will be a permanent laboratory designed for long-term research projects. Diverse topics will be studied, including research on the growth of protein crystals. This particular project will help scientists determine protein structure and function, which is expected to enhance work on drug design and the treatment of many diseases.

The *ISS* will draw on the resources of 16 nations. These nations will build units for the space station, which then will be transported into space onboard the space shuttle and Russian launch rockets. The station will be constructed in space. **Figure 17** shows what the completed station will look like.

Figure 17 This is a picture of what the proposed *International Space Station* will look like when it is completed in 2006.

Reading Check *What is the purpose of the* **International Space Station?**

Phases of *ISS* NASA is planning the *ISS* program in phases. Phase One, now concluded, involved the space shuttle-*Mir* docking missions. Phase Two began in 1998 with the launch of the Russian-built *Zarya Module*, also known as the Functional Cargo Block. In December 1998, the first assembly of *ISS* occurred when a space shuttle mission attached the Unity module to *Zarya*. During this phase, crews of three people were delivered to the space station. Phase Two ended in 2001 with the addition of a U.S. laboratory.

Living in Space The project will continue with Phase Three when the Japanese Experiment Module, the European Columbus Orbiting Facility, and another Russian lab will be delivered.

It is hoped that the *International Space Station* will be completed in 2006. Eventually, a seven-person crew should be able to work comfortably onboard the station. A total of 47 separate launches will be required to take all the components of the *ISS* into space and prepare it for permanent habitation. NASA plans for crews of astronauts to stay onboard the station for several months at a time. NASA already has conducted numerous tests to prepare crews of astronauts for extended space missions. One day, the station could be a construction site for ships that will travel to the Moon and Mars.

Science Online

Topic: *International Space Station*

Visit red.msscience.com for Web links to information about the *International Space Station.*

Activity You can see the station travel across the sky with an unaided eye. Find out the schedule and try to observe it.

Visual Learning

Figure 17 What do you think the flat panels do? They are solar panels that provide electricity for the space station.

Reading Check

Answer It is a laboratory that will be used for long-term research projects.

Make a Model

Space Station Have pairs of students construct models of the *International Space Station*, using the photos in the text or images from other sources. Have each pair design and add one new feature to the station. Each group should make a presentation, explaining what the new feature would do. Display the models in class. L2 COOP LEARN IS **Interpersonal**

Curriculum Connection

Health The space program has been used to benefit humans in many fields, including health and medicine. Have students write a report in their Science Journals about advancements in health and medicine that have occurred as a direct result of the space program. P

Visual Learning

Figure 18 Have students contrast the surface of Earth and the surface of Mars. Possible answer: Earth's surface is mostly water while Mars's surface is mostly land.

Reading Check

Answer high-quality photographs of the planet's surface

Discussion

Mars Exploration Do you think that successful completion of missions such as the Mars Odyssey make the public more or less supportive of future Mars probes? Explain. Possible answers: Less supportive because of the waste of millions of dollars that could be spent elsewhere; more supportive because of renewed determination to make the next mission successful

Figure 18 Gulleys, channels, and aprons of sediment imaged by the *Mars Global Surveyor* are similar to features on Earth known to be caused by flowing water. This water is thought to seep out from beneath the surface of Mars.

Exploring Mars

Two of the most successful missions in recent years were the 1996 launchings of the *Mars Global Surveyor* and the *Mars Pathfinder. Surveyor* orbited Mars, taking high-quality photos of the planet's surface as shown in **Figure 18.** *Pathfinder* descended to the Martian surface, using rockets and a parachute system to slow its descent. Large balloons absorbed the shock of landing. *Pathfinder* carried technology to study the surface of the planet, including a remote-controlled robot rover called Sojourner. Using information gathered by studying photographs taken by *Surveyor,* scientists determined that water recently had seeped to the surface of Mars in some areas.

Reading Check *What type of data were obtained by the* **Mars Global Surveyor?**

Another orbiting spacecraft, the *Mars Odyssey* began mapping the surface of Mars in 2002. Soon after, its data confirmed the findings of *Surveyor*—that Martian soil contains frozen water in the southern polar area. The next step was to send robots to explore the surface of Mars. Twin rovers named *Spirit* and *Opportunity* were launched in 2003 with schedules to reach their separate destinations on Mars in January 2004. Their primary goals are to analyze Martian rocks and soils to tell scientists more about Martian geology and provide clues about the role of water on Mars. Future plans include *Phoenix* in 2008, a robot lander capable of digging over a meter into the surface.

LAB DEMONSTRATION

Purpose to demonstrate simulated gravity

Materials turntable, LP record, scissors, construction paper, tape, marbles

Preparation Fold 8-cm wide strips of construction paper in half lengthwise. Unfold.

Procedure Wrap the strips along the fold around the circumference of the record so there is a 4-cm wall. Securely tape the remainder underneath. Place the record on the turntable, place three marbles at its center, and switch it on.

Expected Outcome The marbles are accelerated outward; the paper exerts a force and stops them.

Assessment

How could this procedure generate artificial gravity in a space station? If the space station were spinning, inertia would cause objects to accelerate toward the outer edge, producing a feeling of gravity.

New Millennium Program

To continue space missions into the future, NASA has created the New Millennium Program (NMP). The goal of the NMP is to develop advanced technology that will let NASA send smart spacecraft into the solar system. This will reduce the amount of ground control needed. They also hope to reduce the size of future spacecraft to keep the cost of launching them under control. NASA's challenge is to prove that certain cutting-edge technologies, as well as mission concepts, work in space.

Exploring the Moon

Does water exist in the craters of the Moon's poles? This is one question NASA intends to explore with data gathered from the *Lunar Prospector* spacecraft shown in **Figure 19.** Launched in 1998, the *Lunar Prospector's* one-year mission was to orbit the Moon, mapping its structure and composition. Data obtained from the spacecraft indicate that water ice might be present in the craters at the Moon's poles. Scientists first estimated up to 300 million metric tons of water may be trapped as ice, and later estimates are much higher. In the permanently shadowed areas of some craters, the temperature never exceeds −230°C. Therefore water delivered to the Moon by comets or meteorites early in its history could remain frozen indefinitely.

At the end of its mission, *Lunar Prospector* was deliberately crashed into a lunar crater. Using special telescopes, scientists hoped to see evidence of water vapor thrown up by the collision. None was seen, however scientists still believe that much water ice is there. If so, this water would be useful if a colony is ever built on the Moon.

Science Online

Topic: New Millennium Program

Visit red.msscience.com for Web links to information about NASA's New Millennium Program.

Activity Prepare a table listing proposed missions, projected launch dates, and what they will study.

Figure 19 The *Lunar Prospector* analyzed the Moon's composition during its one-year mission.
Explain *why* Lunar Prospector *was deliberately crashed on the Moon.*

Activity

Polling Students Ask students: What do other students in your school think about the future of the space program? Do they want it to continue? If so, what do they think NASA should explore? Should uncrewed probes be sent or should there be many more crewed missions to space? Have groups of students design a poll to find the answers to these questions. They should finalize the questions, decide how many students they will survey, carry out the poll, and report the results. Have different student groups share their results in class.

Discussion

Moon Water Why is it important for scientists to determine if there is water on the moon? Possible answer: Having water on the Moon would make it easier to set up a moon base. Water would be available for drinking, and also could be used to produce oxygen.

Caption Answer

Figure 19 Scientists hoped that the collision would throw up water vapor. However, none was seen.

Differentiated Instruction

Challenge NASA has planned a number of Mars missions. Have students do research to find out about these missions and write a short report on one of them. Have students include the design of the probe, when it was launched, or is scheduled to be launched, how long it will take to reach Mars, and what NASA scientists want it to accomplish. L2

Use an Analogy

Walking the Equator To help students understand the vast distances space probes travel, tell them that the *Cassini* probe, headed toward Saturn, must travel roughly 1.3 billion km one way. That's nearly equivalent to walking around Earth along the equator more than 100,000 times.

Figure 20 *Cassini* is currently on its way to Saturn. After it arrives, it will spend four years studying Saturn and its surrounding area.

Cassini

In October 1997, NASA launched the space probe *Cassini*. This probe's destination is Saturn. *Cassini*, shown in **Figure 20,** will not reach its goal until 2004. At that time, the space probe will explore Saturn and surrounding areas for four years. One part of its mission is to deliver the European Space Agency's *Huygens* probe to Saturn's largest moon, Titan. Some scientists theorize that Titan's atmosphere may be similar to the atmosphere of early Earth.

The Next Generation Space Telescope Not all space missions involve sending astronauts or probes into space. Plans are being made to launch a new space telescope that is capable of observing the first stars and galaxies in the universe. The *James Webb Space Telescope*, shown in **Figure 21,** will be the successor to the *Hubble Space Telescope*. As part of the Origins project, it will provide scientists with the opportunity to study the evolution of galaxies, the production of elements by stars, and the process of star and planet formation. To accomplish these tasks, the telescope will have to be able to see objects 400 times fainter than those currently studied with ground-based telescopes such as the twin Keck telescopes. NASA hopes to launch the *James Webb Space Telescope* as early as 2010.

Figure 21 The *James Webb Space Telescope* honors the NASA administrator who contributed greatly to the Apollo Program. It will help scientists learn more about how galaxies form.

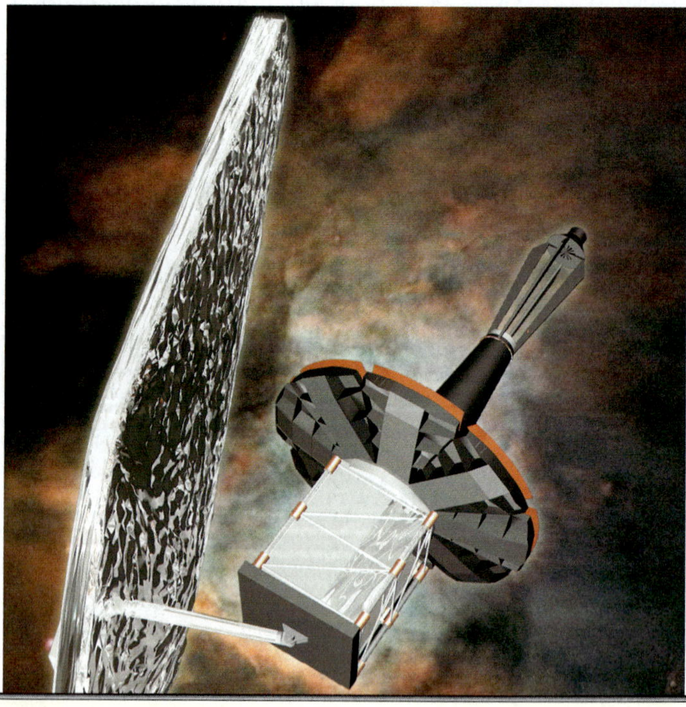

Differentiated Instruction

Challenge Have students research NASA's Origins Program. Ask them to find out when the program began, what stages are involved, and where the program is heading. L3

Learning Disabled Help your students better understand the *Cassini* mission by showing them photos of the Saturn system with its many rings and moons. These will be the objects that *Cassini* will study. Point out Titan and explain to students that *Cassini* will deliver the probe *Huygens* to study this large moon. LS **Visual-Spatial**

Everyday Space Technology Many people have benefited from research done for space programs. Medicine especially has gained much from space technology. Space medicine led to better ways to diagnose and treat heart disease here on Earth and to better heart pacemakers. A screening system that works on infants is helping eye doctors spot vision problems early. Cochlear implants that help thousands of deaf people hear were developed using knowledge gained during the space shuttle program.

Space technology can even help catch criminals and prevent accidents. For example, a method to sharpen images that was devised for space studies is being used by police to read numbers on blurry photos of license plates. Equipment using space technology can be placed on emergency vehicles. This equipment automatically changes traffic signals as an emergency vehicle approaches intersections, so that crossing vehicles have time to stop safely. A hand-held device used for travel directions is shown in **Figure 22.**

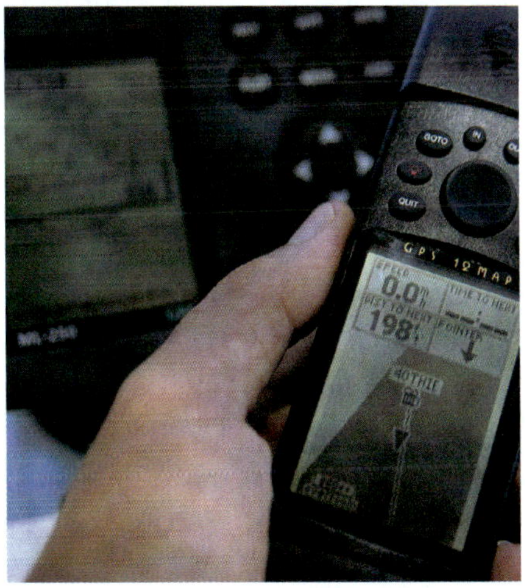

Figure 22 Global Positioning System (GPS) technology uses satellites to determine location on Earth's surface.

Reading Check *How have research and technology developed for space benefited people here on Earth?*

section 3 review

Summary

The Space Station

- A space station is an orbiting laboratory.
- The new *International Space Station (ISS)* is being built with the aid of 16 nations.
- The space shuttle transports astronauts, satellites, and other materials to and from the *ISS.*

Exploring Mars and the Moon

- The *Mars Global Surveyor* orbited Mars and the *Mars Pathfinder* studied its surface.
- *Lunar Prospector* orbited the Moon, mapping its structure and composition.
- Recent data indicate that water ice crystals may exist in shadows of lunar craters.

Future Missions

- The *Cassini* probe is scheduled to explore Saturn and its moons.
- The successor to the *Hubble* will be the *James Webb Space Telescope.*

Self Check

1. **Identify** the main advantage of the space shuttle.
2. **Describe** the importance of space shuttle-*Mir* docking missions.
3. **Explain** how *International Space Station* is used.
4. **Identify** three ways that space technology is a benefit to everyday life.
5. **Think Critically** What makes the space shuttle more versatile than earlier spacecraft?

Applying Math

6. **Solve One-Step Equations** *Voyager 1* had about 30 kg of hydrazine fuel left in 2003. If it uses about 500 g per year, how long will this fuel last?
7. **Use Percentages** Suppose you're in charge of assembling a crew of 50 people. Decide how many to assign each task, such as farming, maintenance, scientific experimentation, and so on. Calculate the percent of the crew assigned to each task. Justify your decisions.

Science Online red.msscience.com/self_check_quiz

3 Assess

Reading Check

Answer It can provide better medical diagnoses, implants, and assist in law enforcement.

DAILY INTERVENTION

Check for Understanding

Linguistic Challenge students to make a chart that summarizes the NASA missions on the drawing board for the next ten years, and the goals of each. Have the class pool the information they find to make one large chart on poster board. If the next mission takes place in the near future, have students compare the goals of the mission to what it actually accomplishes. L2

Reteach

Compare Have students compare and contrast the space shuttle and the *International Space Station.* Possible answers: Both orbit Earth and are used for research; astronauts can live longer on the *ISS,* and the *ISS* has more room; shuttle returns to Earth while *ISS* does not; shuttle is an American ship while *ISS* is international. L1

Assessment

Oral Have each student make up a question and answer on information in this section for a brief oral quiz. Arrange students in a circle, then have each student address his or her question to the person on his or her left. Use **Performance Assessment in the Science Classroom,** p. 91.

section 3 review

1. It is a reusable spacecraft.
2. They allowed astronauts to practice a maneuver that would be needed to build the ISS.
3. It will be used for long-term research projects. It could one day be a construction site for ships that go to the moon or Mars.
4. Benefits include better heart pacemakers, vision screening, cochlear implants and enhanced imaging systems.
5. It's reusable and can carry cargo into orbit, including satellites. It can also retrieve objects from orbit and take astronauts into orbit to repair satellites.
6. If the spacecraft uses 500 g/y and has 30 kg, you must first convert kilograms to grams which gives 30,000 g and then divide by 500 g/y to get 60 years.
7. Lists could include people to grow food, make repairs, treat medical emergencies, and pilot spacecraft. Accept reasonable answers.

BENCH TESTED

Real-World Question

Purpose Students will use information and sightings of the star Polaris to estimate the circumference of Earth.

Process Skills observe, plan, use tables, record observations, communicate, interpret data

Time Required 2 hours

Possible Materials astrolabe or cardboard, string, metal bolts for weights to make astrolabe

Safety Precautions Make sure students observe warning about not looking at the Sun directly.

Internet Have students visit red.msscience.com/internet_lab to obtain directions for making an astrolabe. The astrolabe will be used for star sightings that can help students determine Earth's circumference.

Non-Internet Sources Students can check their calculations by using an encyclopedia to look up Earth's circumference.

Make a Plan

Preparation

Internet To run through the steps that students will follow, visit red.msscience.com/internet_lab.

Non-Internet Have books on astronomy and the history of astronomy as well as an encyclopedia available for students to use. Also provide atlases to use in order to determine the distance between cities.

LAB Use the Internet

Star Sightings

Goals

- **Record** your sightings of Polaris.
- **Share** the data with other students to calculate the circumference of Earth.

Data Source

Science Online

Go to red.msscience.com/internet_lab to obtain instructions on how to make an astrolabe. Also visit the Web site for more information about the location of Polaris, and for data from other students.

Safety Precautions

WARNING: *Do not use the astrolabe during the daytime to observe the Sun.*

Real-World Question

For thousands of years, people have measured their position on Earth using the position of Polaris, the North Star. At any given observation point, it always appears at the same angle above the horizon. For example, at the north pole, Polaris appears directly overhead, and at the equator it is just above the northern horizon. Other locations can be determined by measuring the height of Polaris above the horizon using an instrument called an astrolabe. Could you use Polaris to determine the size of Earth? You know that Earth is round. Knowing this, do you think you can estimate the circumference of Earth based on star sightings?

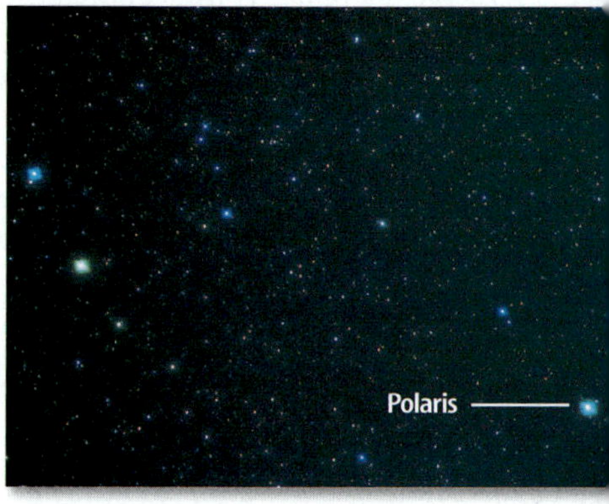

Make a Plan

1. Obtain an astrolabe or construct one using the instructions posted by visiting the link above.
2. **Design** a data table in your Science Journal similar to the one below.

Polaris Observations

Your Location:

Date	Time	Astrolabe Reading
Readings will vary with school's location.		

Alternative Inquiry Lab

Paths of Planets Ask students how they might use the astrolabe to identify the ecliptic path that the Sun and planets follow. Students could follow one easily visible planet, such as Jupiter, Saturn, Mars, or Venus at different hours. Suggest they devise a graph to show the path of the planet. Encourage them to research when planets are visible. Another possibility is to follow the path of the *International Space Station.*

Using Scientific Methods

3. Decide as a group how you will make your observations. Does it take more than one person to make each observation? When will it be easiest to see Polaris?

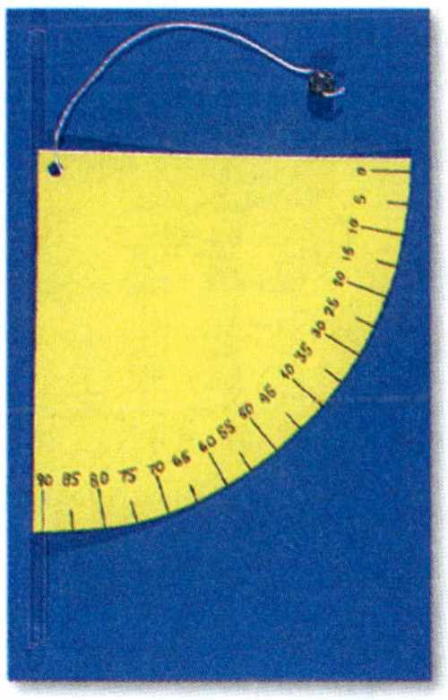

Follow Your Plan

1. Make sure your teacher approves your plan before you start.
2. Carry out your observations.
3. **Record** your observations in your data table.
4. Average your readings and post them in the table provided at the link shown below.

Analyze Your Data

1. **Research** the names of cities that are at approximately the same longitude as your hometown. Gather astrolabe readings from students in one of those cities at the link shown below.
2. **Compare** your astrolabe readings. Subtract the smaller reading from the larger one.
3. **Determine** the distance between your star sighting location and the other city.
4. **Calculate** the circumference of Earth using the following relationship.

   Circumference = (360°) × (distance between locations)/difference between readings

Conclude and Apply

1. **Analyze** how the circumference of Earth that you calculated compares with the accepted value of 40,079 km.
2. **Determine** some possible sources of error in this method of establishing the size of Earth. What improvements would you suggest?

Communicating Your Data

Find this lab using the link below. **Create** a poster that includes a table of your data and data from students in other cities. **Perform** a sample circumference calculation for your class.

Science Online
red.msscience.com/internet_lab

Follow Your Plan

Teaching Strategies

- Have students make repeated star sightings of Polaris and other stars over a few nights and at different times.
- After students have made a few observations, explain the circumference equation. Have them practice explaining the logic behind the circumference equation to a partner.

Analyze Your Data

1. Students can use an atlas to locate cities at approximately the same longitude as your hometown.
2. Answers will vary depending on readings.
3. Answers will vary on cities chosen.
4. Earth's circumference at the equator is 24,901 mi (40,079 km).

Conclude and Apply

1. Values should be close.
2. Possible answers: making errors in calculations, choosing a city not on your longitude, misreading the astrolabe. Students might suggest being more careful or repeating calculations several times.

Assessment

Process Have students draw an events chain concept map that shows the steps used in completing this activity. Use **Performance Assessment in the Science Classroom,** p. 163.

Communicating Your Data

Have students use a computer spreadsheet program to record data on star sightings at their location and at various other cities on the same line of longitude.

TIME

Content Background

The critics of space travel claim that the effort and money spent on space colonization would be better spent on programs to solve current problems. The advocates see space as a new frontier equal or greater in potential to the Americas in the 1600s. Advocates also point out that, if the materials and systems required for the construction of self-sufficient space colonies can be developed, these same technologies can be applied to similar problems on planet Earth.

Activity

Essentials Organize the class into teams to look into these aspects of space habitation: breathable air, energy (for all purposes), water, building materials, sustainable food supplies, protection from cosmic radiation, gravity, and transportation. Instruct the teams to prepare charts indicating the technology required compared with available technology. They can then pool their research and try to determine when they feel space colonization will be feasible.

Investigate the Issue

The settling of Mars and the Moon has been compared to inhabiting harsh climates on the Earth. Have students compare the conditions on Mars with those found in a desert, the Arctic, and underwater. Which are most similar? Most different? What problems are unique to each place?

TIME SCIENCE AND Society

SCIENCE ISSUES THAT AFFECT YOU!

Cities in Space

Should the U.S. spend money to colonize space?

Humans have landed on the Moon, and spacecrafts have landed on Mars. But these space missions are just small steps that may lead to a giant new space program. As technology improves, humans may be able to visit and even live on other planets. But is it worth the time and money involved?

Those in favor of living in space point to the International Space Station that already is orbiting Earth. It's an early step toward establishing floating cities where astronauts can live and work. As Earth's population continues to increase and there is less room on this planet, why not expand to other planets or build a floating city in space? Also, the fact that there is little pollution in space makes the idea appealing to many.

Critics of colonizing space think we should spend the hundreds of billions of dollars that it would cost to colonize space on projects to help improve people's lives here on Earth. Building better housing, developing ways to feed the hungry, finding cures for diseases, and increasing funds for education should come first, these people say. And, critics continue, if people want to explore, why not explore right here on Earth, for example, the ocean floor.

Moon or Mars? If humans were to move permanently to space, the two most likely destinations would be Mars and the Moon, both bleak places. But those in favor of moving to these places say humans could find a way to make them livable as they have made homes in harsh climates and in many rugged areas here on Earth.

Water may be locked in lunar craters, and photos suggest that Mars once had liquid water on its surface. If that water is frozen underground, humans may be able to access it. NASA is studying whether it makes sense to send astronauts and scientists to explore Mars.

Transforming Mars into an Earthlike place with breathable air and usable water will take much longer, but some small steps are being taken. Experimental plants are being developed that could absorb Mars's excess carbon dioxide and release oxygen. Solar mirrors that could warm Mars's surface are available.

Those for and against colonizing space agree on one thing—it will take large amounts of money, research, and planning. It also will take the same spirit of adventure that has led history's pioneers into so many bold frontiers—deserts, the poles, and the sky.

Debate with your class the pros and cons of colonizing space. Do you think the United States should spend money to create space cities or use the money now to improve lives of people on Earth?

Science Online
For more information, visit red.msscience.com/time

Debate Each side should be able to produce valid reasons. Those in favor may feel that the money spent will help more people on the long term through new knowledge and scientific developments. Those opposed can cite many of the pressing problems that need attention here on Earth, such as famine and poverty, and diseases such as AIDS or Alzheimers.

Resources for Teachers and Students

On to Mars: Colonizing a New World, by Frank Crossman and Robert Zubrin, Apogee Books, 2002

Making Space Happen: Private Space Ventures and the Visionaries Behind Them, by Paula Berinstein and Fiorella Terenzi, 2002

Entering Space: Creating a Spacefaring Civilization by Robert Zubrin, 2000

Reviewing Main Ideas

Section 1 Radiation from Space

1. The arrangement of electromagnetic waves according to their wavelengths is the electromagnetic spectrum.
2. Optical telescopes produce magnified images of objects.
3. Radio telescopes collect and record radio waves given off by some space objects.

Section 2 Early Space Missions

1. A satellite is an object that revolves around another object. The moons of planets are natural satellites. Artificial satellites are those made by people.
2. A space probe travels into the solar system, gathers data, and sends them back to Earth.
3. American piloted space programs included the Gemini, Mercury, and Apollo Projects.

Section 3 Current and Future Space Missions

1. Space stations provide the opportunity to conduct research not possible on Earth. The *International Space Station* is being constructed in space with the cooperation of more than a dozen nations.
2. The space shuttle is a reusable spacecraft that carries astronauts, satellites, and other cargo to and from space.
3. Space technology is used to solve problems on Earth, too. Advances in engineering related to space travel have aided medicine, environmental sciences, and other fields.

Visualizing Main Ideas

Copy and complete the following concept map about the race to the Moon. Use the phrases: first satellite, Project Gemini, Project Mercury, team of two astronauts orbits Earth, Project Apollo.

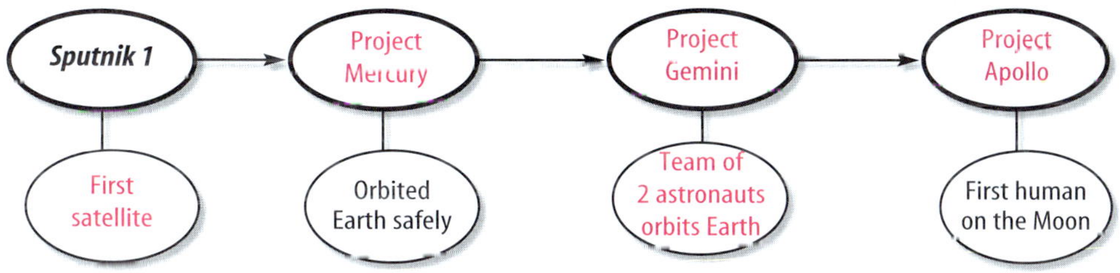

chapter 14 Study Guide

Reviewing Main Ideas

Summary statements can be used by students to review the major concepts of the chapter.

Visualizing Main Ideas

See student page.

Science Online

Visit red.msscience.com
/self_check_quiz
/interactive_tutor
/vocabulary_puzzlemaker
/chapter_review
/standardized_test

Assessment Transparency

For additional assessment questions, use the *Assessment Transparency* located in the transparency book.

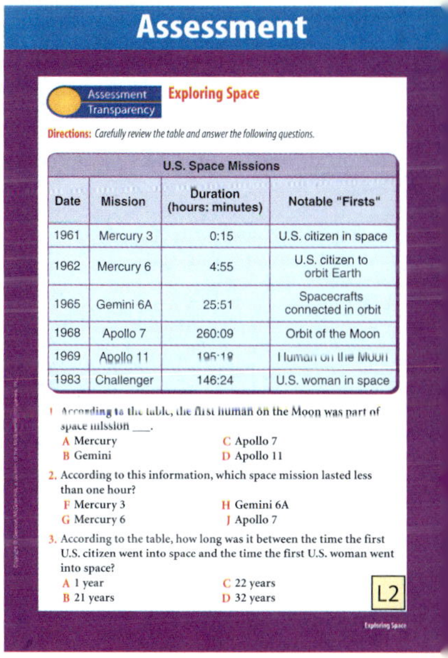

Assessment

Assessment Transparency **Exploring Space**

Directions: *Carefully review the table and answer the following questions.*

U.S. Space Missions			
Date	Mission	Duration (hours: minutes)	Notable "Firsts"
1961	Mercury 3	0:15	U.S. citizen in space
1962	Mercury 6	4:55	U.S. citizen to orbit Earth
1965	Gemini 6A	25:51	Spacecrafts connected in orbit
1968	Apollo 7	260:09	Orbit of the Moon
1969	Apollo 11	195:18	Human on the Moon
1983	Challenger	146:24	U.S. woman in space

1. According to the table, the first human on the Moon was part of space mission ___.
   A Mercury C Apollo 7
   B Gemini D Apollo 11
2. According to this information, which space mission lasted less than one hour?
   F Mercury 3 H Gemini 6A
   G Mercury 6 J Apollo 7
3. According to the table, how long was it between the time the first U.S. citizen went into space and the time the first U.S. woman went into space?
   A 1 year C 22 years
   B 21 years D 32 years

L2

Exploring Space

Using Vocabulary

1. refracting telescope
2. satellite
3. Project Mercury
4. space shuttle
5. electromagnetic spectrum

Checking Concepts

6. D
7. D
8. B
9. A
10. B
11. A
12. D
13. C
14. C

Thinking Critically

15. Because the Moon has no atmosphere, light and other forms of energy can reach its surface without distortion.
16. The same kind of sensing system could detect dangerous fumes in a factory.
17. Possible answers: Robotic space probes are cheaper and can go on one-way deep space missions that people cannot. Crewed flights provide information about living in space.

chapter 14 Review

Using Vocabulary

electromagnetic spectrum p. 409	reflecting telescope p. 410
observatory p. 410	refracting telescope p. 410
orbit p. 417	rocket p. 415
Project Apollo p. 422	satellite p. 417
Project Gemini p. 421	space probe p. 418
Project Mercury p. 421	space shuttle p. 423
radio telescope p. 413	space station p. 424

Fill in the blanks with the correct vocabulary word(s).

1. A(n) _________ telescope uses lenses to bend light.
2. A(n) _________ is an object that revolves around another object in space.
3. _________ was the first piloted U.S. space program.
4. A(n) _________ carries people and tools to and from space.
5. In the _________, electromagnetic waves are arranged, in order, according to their wavelengths.

Checking Concepts

Choose the word or phrase that best answers the question.

6. Which spacecraft has sent images of Venus to scientists on Earth?
   - A) *Voyager*
   - B) *Viking*
   - C) *Apollo 11*
   - D) *Magellan*

7. Which kind of telescope uses mirrors to collect light?
   - A) radio
   - B) electromagnetic
   - C) refracting
   - D) reflecting

8. What was *Sputnik I?*
   - A) the first telescope
   - B) the first artificial satellite
   - C) the first observatory
   - D) the first U.S. space probe

9. Which kind of telescope can be used during the day or night and during bad weather?
   - A) radio
   - B) electromagnetic
   - C) refracting
   - D) reflecting

10. When fully operational, what is the maximum number of people who will crew the *International Space Station?*
    - A) 3
    - B) 7
    - C) 15
    - D) 50

11. Which space mission's goal was to put a spacecraft into orbit and bring it back safely?
    - A) Project Mercury
    - B) Project Apollo
    - C) Project Gemini
    - D) *Viking I*

12. Which of the following is a natural satellite of Earth?
    - A) *Skylab*
    - B) the space shuttle
    - C) the Sun
    - D) the Moon

13. What does the space shuttle use to place a satellite into space?
    - A) liquid-fuel tank
    - B) booster rocket
    - C) mechanical arm
    - D) cargo bay

14. What part of the space shuttle is reused?
    - A) liquid-fuel tanks
    - B) *Gemini* rockets
    - C) booster engines
    - D) Saturn rockets

Science Online red.msscience.com/vocabulary_puzzlemaker

Use the Exam*View*® Pro Testmaker CD-ROM to:

- create multiple versions of tests
- create modified tests with one mouse click for inclusion students
- edit existing questions and add your own questions
- build tests aligned with state standards using built-in State Curriculum Tags
- change English tests to Spanish with one mouse click and vice versa

Thinking Critically

15. **Compare and contrast** the advantages of a moon-based telescope with an Earth-based telescope.

16. **Infer** how sensors used to detect toxic chemicals in the space shuttle could be beneficial to a factory worker.

17. **Drawing Conclusions** Which do you think is a wiser method of exploration—space missions with people onboard or robotic space probes? Why?

18. **Explain** Suppose two astronauts are outside the space shuttle orbiting Earth. The audio speaker in the helmet of one astronaut quits working. The other astronaut is 1 m away and shouts a message. Can the first astronaut hear the message? Support your reasoning.

19. **Make and Use Tables** Copy and complete the table below. Use information from several resources.

United States Space Probes

Probe	Launch Date(s)	Planets or Objects Visited
Vikings 1 and *2*		
Galileo	Answers will vary.	
Lunar Prospector		
Pathfinder		

20. **Classify** the following as a satellite or a space probe: *Cassini*, *Sputnik I*, *Hubble Space Telescope*, space shuttle, and *Voyager 2*.

21. **Compare and contrast** space probes and artificial satellites.

Performance Activities

22. **Display** Make a display showing some of the images obtained from the *Hubble Space Telescope*. Include samples of three types of galaxies, nebulae, and star clusters.

Applying Math

23. **Space Communications** In May 2003 *Voyager 1* was 13 billion km from the Sun. Calculate how long it takes for a signal to travel this far assuming it travels at 3×10^8 m/s.

Use the graph below to answer question 24.

Average Velocities of Satellites A & B

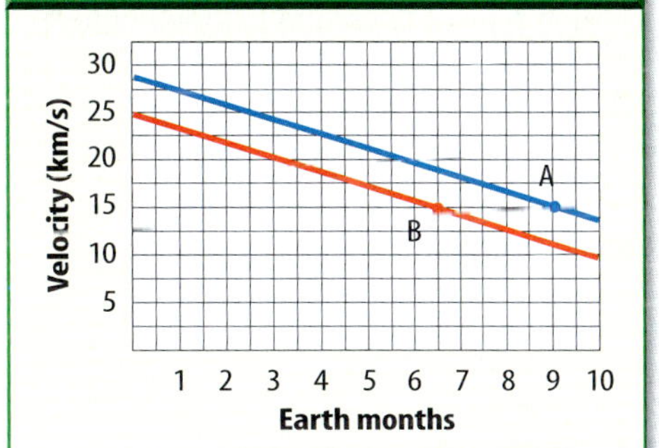

24. **Satellite Orbits** The graph above predicts the average velocities of satellites A and B in orbit around a hypothetical planet. Because of contact with the planet's atmosphere, their velocities are decreasing. At a velocity of 15 km/s their orbits will decay and they will spiral downwards to the surface. Using the graph, determine how long will it take for each satellite to reach this point?

25. **Calculate Fuel** A spacecraft carries 30 kg of hydrazine fuel and uses and average of 500 g/y. How many years could this fuel last?

26. **Space Distances** Find the distance in AUs to a star 68 light-years (LY) distant. (1 LY = 6.3×10^4 AUs)

Assessment Resources

Reproducible Masters

Chapter *Fast File* Resources
Chapter Review, pp. 33–34
Chapter Tests, pp. 35–38
Assessment Transparency Activity, p. 45

Glencoe Science Web site
Chapter Review Test
Standardized Test Practice

Glencoe Technology
Assessment Transparency
Exam*View*® Pro Testmaker
MindJogger Videoquiz
Interactive Chalkboard

chapter 14 Review

18. No; sound must travel through matter. Space is a vacuum, so sound will not travel across it.

19.

United States Space Probes

Probe	Launch Date	Planets or Objects Visited
Vikings 1 and 2	1975	Mars
Galileo	1989	Venus, Europa, Jupiter
Lunar Prospector	1998	Earth's Moon
Pathfinder	1996	Mars

20. *Sputnik*, the shuttle, and *Hubble* all orbit Earth like satellites. *Voyager 2* and *Cassini* are space probes.

21. Both are sent into space to collect data and send it back to Earth. Space probes travel into space, whereas satellites orbit Earth or other bodies.

Performance Activities

22. Check to see that the elements are present and labeled correctly. Use **PASC**, p. 135.

Applying Math

National Math Standards

1, 4, 5, 9

23. The speed of light is 3×10^8 m/s or 3×10^5 km/s. By dividing this into 13 billion, gives 43,000 s. Then dividing again by 3,600 s/h, gives 11.9 h.

24. A = 9 months and B = 6.5 months.

25. If it has 30 kg and uses 500 g/y, dividing 30 by .5 kg gives 60 years.

26. The distance 6.3×10^4 AU/LY multiplied by 68 LY gives 428.4×10^4 or 4.28×10^6 AU's

FAST FILE

Answer Sheet A practice answer sheet can be found at red.msscience.com/answer_sheet.

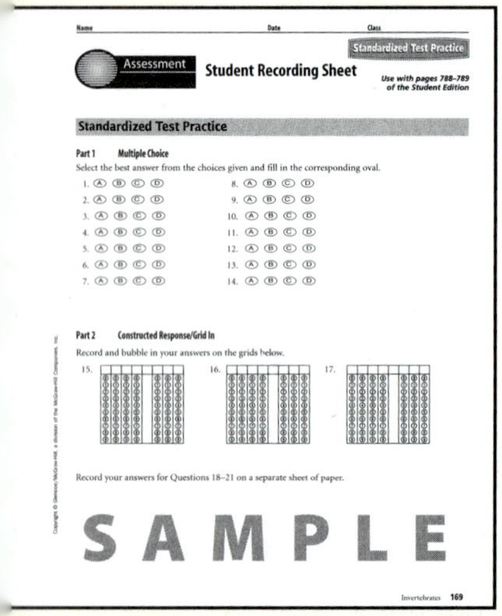
Name Date Class

Assessment **Student Recording Sheet** Standardized Test Practice

Use with pages 788–789 of the Student Edition

Standardized Test Practice

Part 1 Multiple Choice

Select the best answer from the choices given and fill in the corresponding oval.

Part 2 Constructed Response/Grid In

Record and bubble in your answers on the grids below.

Record your answers for Questions 18–21 on a separate sheet of paper.

SAMPLE

Part 1 Multiple Choice

1. A
2. D
3. D
4. B
5. C
6. A
7. A
8. C
9. A
10. B
11. D

Part 2 Short Response

12. Space probes often travel far into the solar system.
13. It demonstrated that artificial satellites could be launched into Earth orbit and started the space race.
14. communication, positioning, weather, Earth monitoring, intelligence gathering

chapter 14 Standardized Test Practice

Part 1 Multiple Choice

Record your answers on the answer sheet provided by your teacher or on a sheet of paper.

Use the figure below to answer question 1.

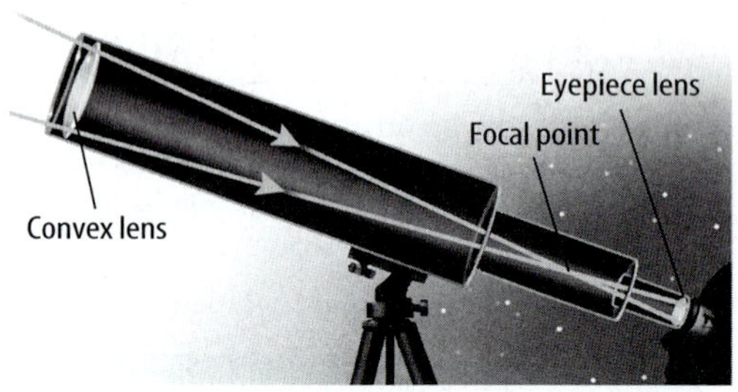

1. Which type of telescope is shown above?
   A. refracting **C.** reflecting
   B. radio **D.** space
2. Who was the first human in space?
   A. Edwin Aldrin **C.** Neil Armstrong
   B. John Glenn **D.** Yuri Gagarin
3. Which is an engine that can launch an object into space?
   A. space probe **C.** capsule
   B. shuttle **D.** rocket
4. Which is the speed of light in a vacuum?
   A. 300 km/s **C.** 3,000 km/s
   B. 300,000 km/s **D.** 30,00 km/s
5. Which of the following is an advantage of space telescopes?
   A. They are cheaper to build.
   B. They have fewer technical problems.
   C. They obtain higher quality images.
   D. They can be repaired easily.

Test-Taking Tip

Making Answers Do not mark the test booklet when taking the test. Be sure to mark ALL answers on your answer sheet and leave no blanks.

6. Which type of radiation has a shorter wavelength than visible light does?
   A. ultraviolet **C.** infrared
   B. microwaves **D.** radio waves
7. Which space probe visited Mars?
   A. *Viking 1* **C.** *Magellan*
   B. *Mariner 2* **D.** *Pioneer 10*
8. Which United States space program included several lunar landings?
   A. Gemini **C.** Apollo
   B. Mercury **D.** space shuttle

Examine the diagram below. Then answer questions 9–11.

9. What is the name of the curved path that the satellite follows?
   A. an orbit **C.** a revolution
   B. a rotation **D.** a track
10. Which force pulls the satellite toward Earth?
    A. the Moon's gravity
    B. Earth's gravity
    C. the Sun's gravity
    D. Earth's magnetic field
11. Imagine that the satellite in the diagram above started to orbit at a slower speed. Which of the following probably would happen to the satellite?
    A. It would fly off into space.
    B. It would crash into the Moon.
    C. It would crash into the Sun.
    D. It would crash into Earth.

Part 2 Short Response/Grid In

15. Radio telescopes use radio waves, whereas optical telescopes use visible light
16. 3.60
17. 3.23
18. 1.0 m above, −15.1° C − (−74.4°C) = 59.3°C; for 0.5 m above, −12.5°C − (−74.9°C) = 62.4°C; for 0.25 m above, −8.9°C − (−75.9°C) = 67.0°C
19. Atmospheric temperatures begin to increase after sunrise and continue to increase until after solar noon. They cool through the nighttime hours. Air nearest the surface has the greatest temperature fluctuations.

Part 2 Short Response/Grid In

Record your answers on the answer sheet provided by your teacher or on a sheet of paper.

12. Explain the difference between a space probe and a satellite that is orbiting Earth.

13. Why was the flight of *Sputnik 1* important?

14. List four ways that satellites are useful.

15. How are radio telescopes different from optical telescopes?

Use the table below to answer questions 16–19. The table includes data collected by *Mars Pathfinder* on the third Sol, or Martian day, of operation.

Sol 3 Temperature Data from Mars Pathfinder

Proportion of Sol	Temperature (°C)		
	1.0 m above surface	0.5 m above surface	0.25 m above surface
3.07	−70.4	−70.7	−73.4
3.23	−74.4	−74.9	−75.9
3.33	−53.0	−51.9	−46.7
3.51	−22.3	−19.2	−15.7
3.60	−15.1	−12.5	−8.9
3.70	−26.1	−25.7	−24.0
3.92	−63.9	−64.5	−65.8

16. Which proportion of sol value corresponds to the warmest temperatures at all three heights?

17. Which proportion of sol value corresponds to the coldest temperatures at all three heights?

18. What is the range of the listed temperature values for each distance above the surface?

19. Explain the data in the table. Why do the temperatures vary in this way?

Part 3 Open Ended

Record your answers on a sheet of paper.

Use the diagram below to answer question 20.

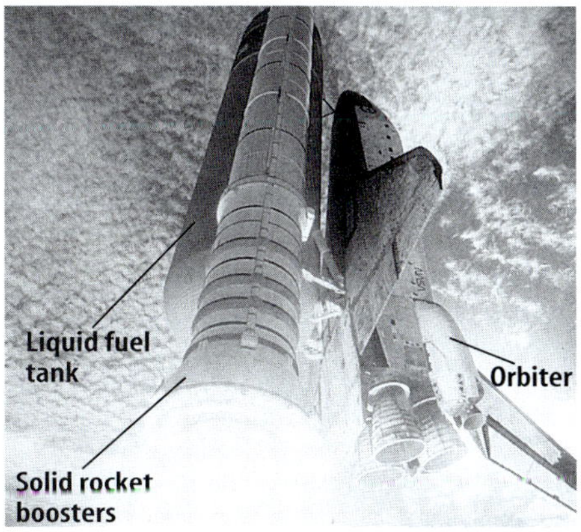

20. Explain the purpose of each of the labeled objects.

21. List four advancements in technology directly attributable to space exploration and how they have impacted everyday life on Earth.

22. What are the advantages of having reusable spacecraft? Are there any disadvantages? Explain.

23. What is the *International Space Station?* How is it used?

24. What are the advantages of international cooperation during space exploration? Are there disadvantages?

25. Explain how the voices of astronauts onboard the space shuttle can be heard on Earth.

26. List several benefits and costs of space exploration. Do you think that the benefits of space exploration outweigh the costs? Explain why you do or do not.

Science Online red.msscience.com/standardized_test

Part 3 Open Ended

20. The liquid fuel tank provides fuel for the three main engines. It is released as the orbiter nears orbit. The solid rocket boosters also burn during liftoff, but fall away sooner than the liquid tank. They are recovered and reused. The orbiter is the vehicle that orbits and returns to Earth.

21. Advances in medicine include better ways to diagnose heart and vision problems and better heart pacemakers. Cochlear implants help restore hearing. Image enhancement aids many fields including law enforcement and communication advances can control traffic signals as emergency vehicles approach intersections.

22. Advantages include lower costs and the ability to fly more frequently. Disadvantages include need for careful maintenance to maintain safety.

23. The *International Space Station* is an orbiting research platform. It is used for a variety of scientific endeavors.

24. Advantages include sharing costs and improved relations between countries. Disadvantages might include less control over operations by any one country and the fact that many countries must fulfill their obligations on time.

25. The astronaut's voices are converted to radio waves, which are received on Earth and converted back to sound.

26. Benefits: new technology, scientific advances, possible future colonization and resource utilization; Cost: money spent, risk to human lives.

Rubrics

For more help evaluating open-ended assessment questions, see the rubric on p. 10T.

chapter 15 Organizer

Section/Objectives	Standards		Labs/Features
	National	State/Local	
Chapter Opener	See pp. 16T–17T for a Key to Standards.		**Launch Lab:** How many stars are in the sky?, p. 439 **Foldables,** p. 439
Section 1 Earth's Place in Space 2 sessions 1 block 1. **Explain** Earth's rotation and revolution. 2. **Explain** why Earth has seasons. 3. **Model** the relative positions of Earth, the Moon, and the Sun during different lunar phases.	National Content Standards: UCP.1–UCP.3, UCP.5, A.1–A.2, B.2, D.1, D.3		**MiniLAB:** Modeling Earth's Seasons, p. 441 **Integrate Career,** p. 442 **Science Online,** p. 444 **Lab:** Moon Phases, p. 447
Section 2 The Solar System 2 sessions 1 block 4. **Compare and contrast** the planets and moons in the solar system. 5. **Explain** that Earth is the only planet known to support life.	National Content Standards: UCP.1–UCP.3, UCP.5, A.1–A.2, D.3		**Science Online,** p. 449 **MiniLAB:** Observing Planets, p. 450 **Applying Science:** How can you model distances in the solar system?, p. 453
Section 3 Stars and Galaxies 3 sessions 1.5 blocks 6. **Explain** why stars appear to move across the sky. 7. **Describe** some constellations. 8. **Explain** the life cycle of stars.	National Content Standards: UCP.1–UCP.3, UCP.5, A.1–A.2		**MiniLAB:** Modeling Constellations, p. 457 **Integrate Physics,** p. 458 **Visualizing Galaxies,** p. 460 **Integrate Physics,** p. 462 **Lab:** Space Colony, pp. 464–465 **Science and Language Arts:** The Sun and the Moon, p. 466

Glencoe Exclusive!
TeacherWorks™
All-In-One Planner and Resource Center

Lab Materials	Reproducible Resources	Section Assessment	Technology
Launch Lab: white crayon or chalk, metric ruler, black construction paper, uncooked rice	**Chapter *FAST FILE* Resources** Foldables Worksheet, p. 17 Directed Reading Overview, p. 19 Note-taking Worksheets, pp. 33–35	Glencoe's Assessment Advantage	**TeacherWorks** includes: • Interactive Teacher Edition • Lesson Planner with calendar • Access to all program blacklines • Correlations to standards • Web links
MiniLAB: shaded lamp, table, globe **Lab:** drawing paper, softball, flashlight, scissors	**Chapter *FAST FILE* Resources** Transparency Activity, p. 44 MiniLAB, p. 3 Enrichment, p. 30 Reinforcement, p. 27 Directed Reading, p. 20 Lab Worksheet, pp. 5–6 Lab Activity, pp. 9–11 Transparency Activity, pp. 47–48	**Portfolio** Differentiated Instruction, p. 442 **Performance** MiniLAB, p. 441 Applying Skills, p. 446 **Content** Section Review, p. 446	Section Focus Transparency Teaching Transparency Guided Reading Audio Program Interactive Chalkboard CD-ROM
MiniLAB: camera	**Chapter *FAST FILE* Resources** Transparency Activity, p. 45 Enrichment, p. 31 Reinforcement, p. 28 Directed Reading, p. 21 Lab Activity, pp. 13–16 **Mathematics Skill Activities,** p. 39 **Reading and Writing Skill Activities,** p. 11	**Portfolio** Make a Model, p. 452 **Performance** Applying Science, p. 453 Applying Skills, p. 455 **Content** Section Review, p. 455	Section Focus Transparency Virtual Labs CD-ROM Guided Reading Audio Program Interactive Chalkboard CD-ROM
MiniLAB: black construction paper, scissors, cardboard cylinder (oatmeal box), tape, pencil, flashlight **Lab:** drawing paper, markers, books about the planets	**Chapter *FAST FILE* Resources** Transparency Activity, p. 46 MiniLAB, p. 4 Enrichment, p. 32 Reinforcement, p. 29 Directed Reading, pp. 21, 22 Lab Worksheet, pp. 7–8 **Home and Community Involvement,** p. 31 **Lab Management and Safety,** p. 73	**Portfolio** Integrate Physics, p. 462 **Performance** MiniLAB, p. 457 Applying Math, p. 463 **Content** Section Review, p. 463	Section Focus Transparency Guided Reading Audio Program Interactive Chalkboard CD-ROM Video Lab

End of Chapter Assessment

Glencoe's Assessment Advantage

Blackline Masters	Technology	Professional Series
Chapter *FAST FILE* Resources Chapter Review, pp. 37–38 Chapter Tests, pp. 39–42 **Standardized Test Practice,** pp. 63–66	MindJogger Videoquiz Virtual Labs CD-ROM Exam*View*® Pro Testmaker TeacherWorks CD-ROM Interactive Chalkboard CD-ROM	**Performance Assessment in the Science Classroom (PASC)**

chapter 15 The Solar System and Beyond

Transparencies

Section Focus

SECTION 1 Section Focus Transparency **On the Move**

Sometimes animals migrate in response to seasonal changes. Snow geese, like those shown below, migrate from Greenland and the Canadian Arctic to New Jersey and the Carolinas.

1. What time of year do you think this picture was taken? What seasonal changes occur at this time?
2. What is summer like in your area? What is winter like?
3. How does the light reaching Earth change between sunrise and sunset?

L2

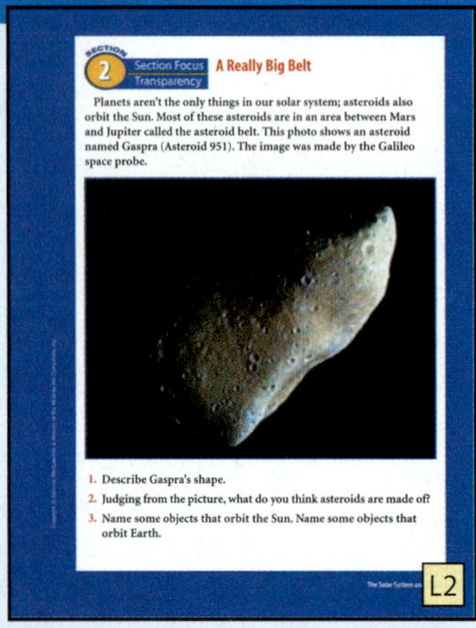
SECTION 2 Section Focus Transparency **A Really Big Belt**

Planets aren't the only things in our solar system; asteroids also orbit the Sun. Most of these asteroids are in an area between Mars and Jupiter called the asteroid belt. This photo shows an asteroid named Gaspra (Asteroid 951). The image was made by the Galileo space probe.

1. Describe Gaspra's shape.
2. Judging from the picture, what do you think asteroids are made of?
3. Name some objects that orbit the Sun. Name some objects that orbit Earth.

L2

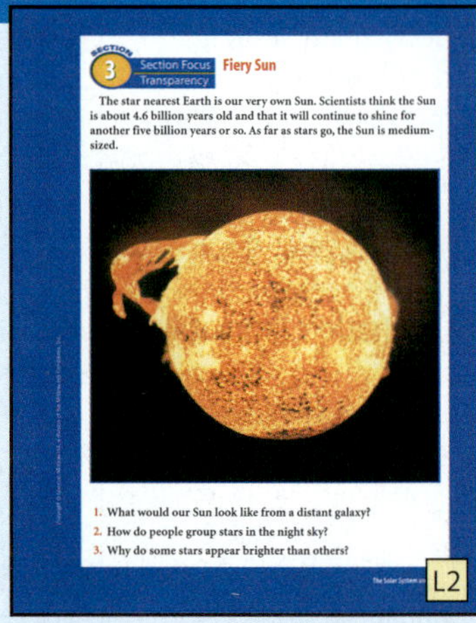
SECTION 3 Section Focus Transparency **Fiery Sun**

The star nearest Earth is our very own Sun. Scientists think the Sun is about 4.6 billion years old and that it will continue to shine for another five billion years or so. As far as stars go, the Sun is medium-sized.

1. What would our Sun look like from a distant galaxy?
2. How do people group stars in the night sky?
3. Why do some stars appear brighter than others?

L2

This is a representation of key blackline masters available in the Teacher Classroom Resources. See Resource Manager boxes within the chapter for additional information.

Assessment

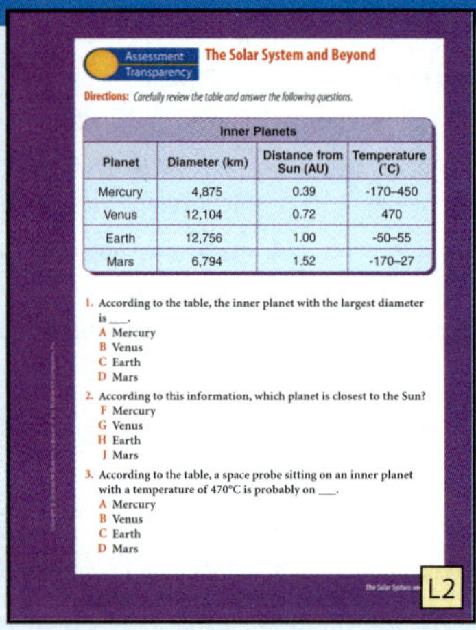
Assessment Transparency **The Solar System and Beyond**

Directions: *Carefully review the table and answer the following questions.*

Inner Planets			
Planet	Diameter (km)	Distance from Sun (AU)	Temperature (°C)
Mercury	4,875	0.39	-170–450
Venus	12,104	0.72	470
Earth	12,756	1.00	-50–55
Mars	6,794	1.52	-170–27

1. According to the table, the inner planet with the largest diameter is ___.
   A Mercury
   B Venus
   C Earth
   D Mars
2. According to this information, which planet is closest to the Sun?
   F Mercury
   G Venus
   H Earth
   J Mars
3. According to the table, a space probe sitting on an inner planet with a temperature of 470°C is probably on ___.
   A Mercury
   B Venus
   C Earth
   D Mars

L2

Teaching

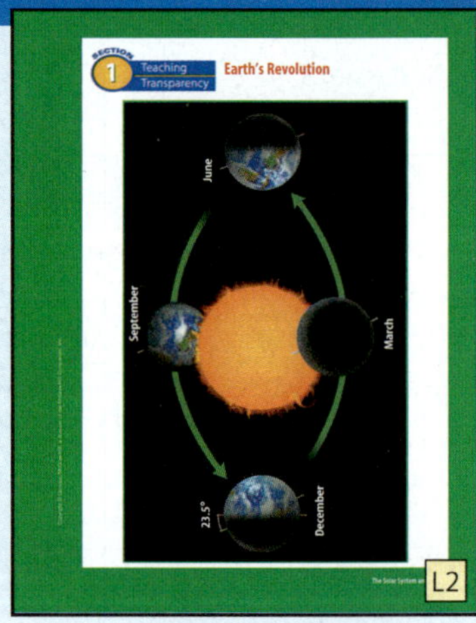
SECTION 1 Teaching Transparency **Earth's Revolution**

Key to Teaching Strategies

The following designations will help you decide which activities are appropriate for your students.

- L1 Level 1 activities should be appropriate for students with learning difficulties.
- L2 Level 2 activities should be within the ability range of all students.
- L3 Level 3 activities are designed for above-average students.
- ELL ELL activities should be within the ability range of English-Language Learners.
- COOP LEARN Cooperative Learning activities are designed for small group work.
- LS Multiple Learning Styles logos, as described on page 12T, are used throughout to indicate strategies that address different learning styles.
- P These strategies represent student products that can be placed into a best-work portfolio.
- PBL Problem-Based Learning activities apply real-world situations to learning.

Hands-on Activities

Student Text Lab Worksheet

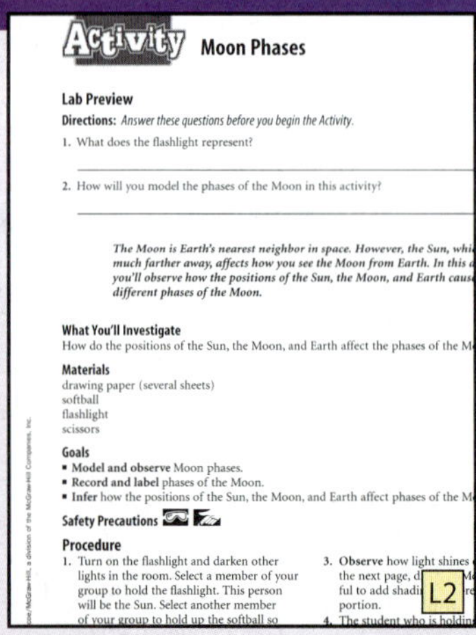
Activity Moon Phases

Lab Preview

Directions: *Answer these questions before you begin the Activity.*

1. What does the flashlight represent?
2. How will you model the phases of the Moon in this activity?

The Moon is Earth's nearest neighbor in space. However, the Sun, whi… much farther away, affects how you see the Moon from Earth. In this a… you'll observe how the positions of the Sun, the Moon, and Earth caus… different phases of the Moon.

What You'll Investigate

How do the positions of the Sun, the Moon, and Earth affect the phases of the M…

Materials

drawing paper (several sheets)
softball
flashlight
scissors

Goals

- **Model and observe** Moon phases.
- **Record and label** phases of the Moon.
- **Infer** how the positions of the Sun, the Moon, and Earth affect phases of the M…

Safety Precautions

Procedure

1. Turn on the flashlight and darken other lights in the room. Select a member of your group to hold the flashlight. This person will be the Sun. Select another member of your group to hold up the softball so
3. **Observe** how light shines … the next page, d… ful to add shadi… portion.
4. The student who is holdin…

L2

Laboratory Activities

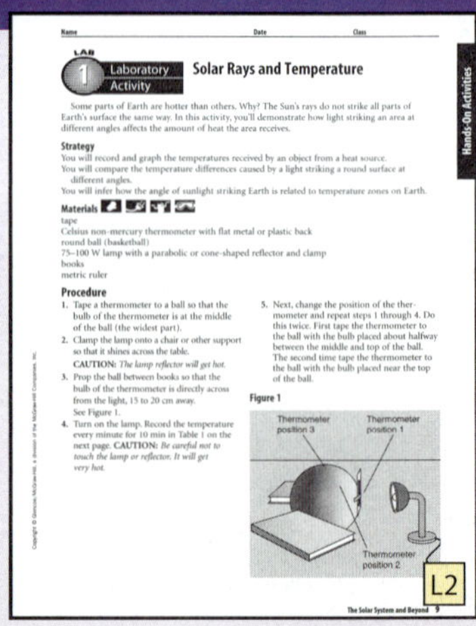
Name Date Class

LAB 1 Laboratory Activity **Solar Rays and Temperature**

Hands-On Activities

Some parts of Earth are hotter than others. Why? The Sun's rays do not strike all parts of Earth's surface the same way. In this activity, you'll demonstrate how light striking an area at different angles affects the amount of heat the area receives.

Strategy

You will record and graph the temperatures received by an object from a heat source.
You will compare the temperature differences caused by a light striking a round surface at different angles.
You will infer how the angle of sunlight striking Earth is related to temperature zones on Earth.

Materials

tape
Celsius non-mercury thermometer with flat metal or plastic back
round ball (basketball)
75–100 W lamp with a parabolic or cone-shaped reflector and clamp
books
metric ruler

Procedure

1. Tape a thermometer to a ball so that the bulb of the thermometer is at the middle of the ball (the widest part).
2. Clamp the lamp onto a chair or other support so that it shines across the table.
   CAUTION: *The lamp reflector will get hot.*
3. Prop the ball between books so that the bulb of the thermometer is directly across from the light, 15 to 20 cm away. See Figure 1.
4. Turn on the lamp. Record the temperature every minute for 10 min in Table 1 on the next page. CAUTION: *Be careful not to touch the lamp or reflector. It will get very hot.*
5. Next, change the position of the thermometer and repeat steps 1 through 4. Do this twice. First tape the thermometer to the ball with the bulb placed about halfway between the middle and top of the ball. The second time tape the thermometer to the ball with the bulb placed near the top of the ball.

Figure 1

L2

The Solar System and Beyond 9

Meeting Different Ability Levels

Content Outline

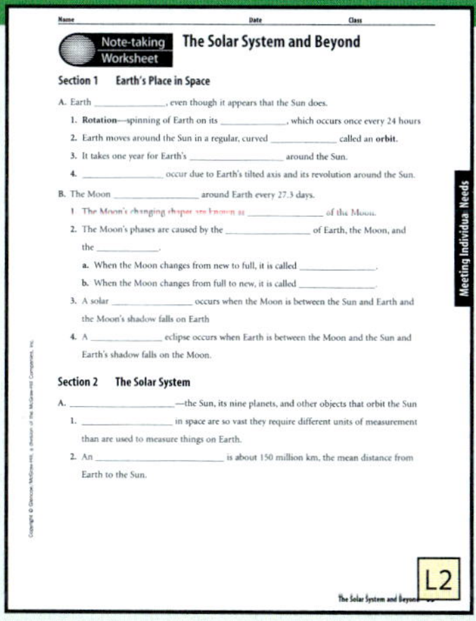

Reinforcement

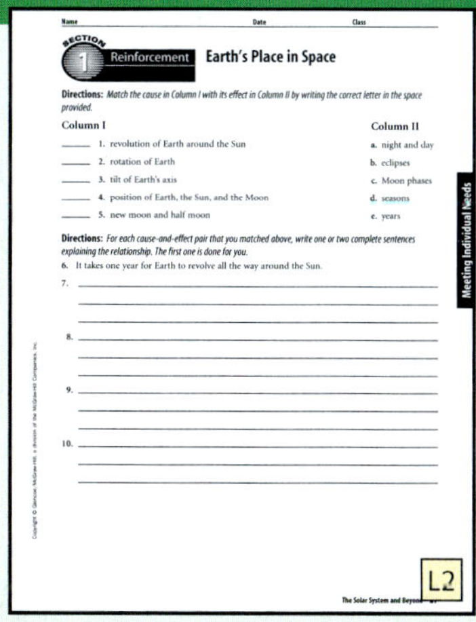

Enrichment

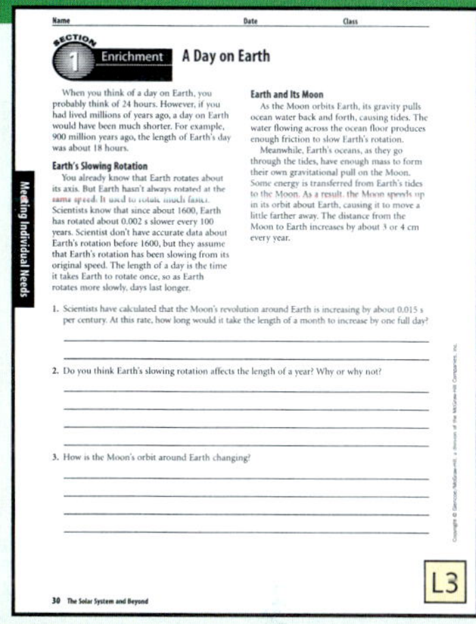

Directed Reading (English/Spanish)

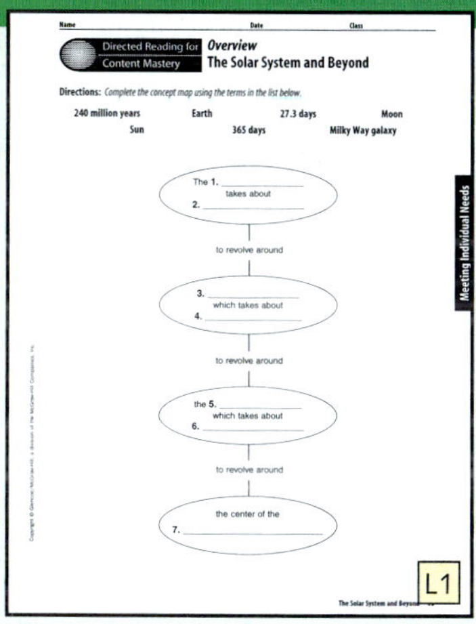

Study Guide

Reading Essentials

Assessment

Test Practice Workbook

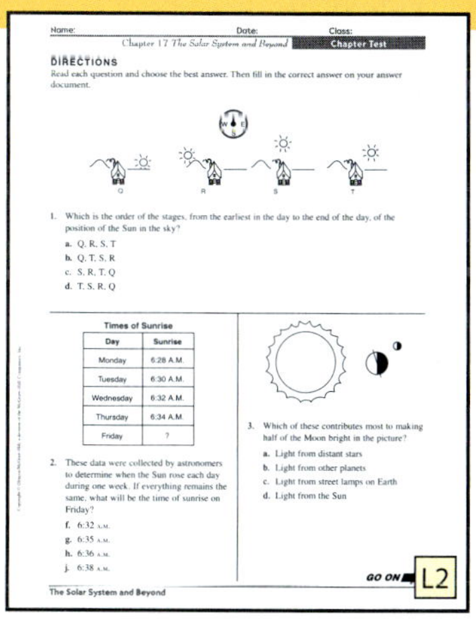

Chapter Review

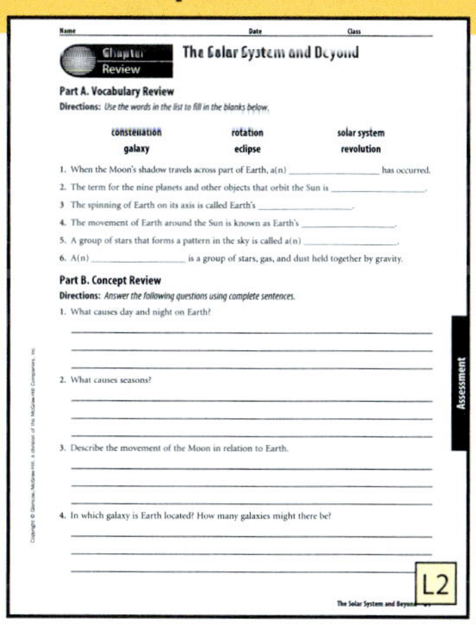

Chapter Tests

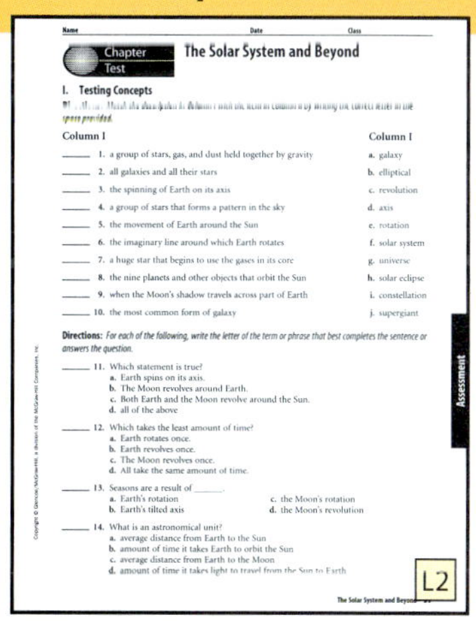

Science Content Background

section 1 Earth's Place in Space

Earth's Rotation

Like most of the planets in our solar system, Earth rotates from west to east around an imaginary axis that runs through the geographic north and south poles. This rotation results in the apparent daily motion of the Sun, the stars, and the Moon across the sky. One complete rotation, measured with respect to the position of the Sun, occurs every 24 hours.

Earth's Revolution

Earth revolves around the Sun once in approximately 365.26 days. Because the calendar year cannot contain a fraction of a day, an extra day—February 29—must be added to the calendar every fourth year (leap year). Earth's orbit around the Sun is elliptical. This also is true for all the other planets. Earth moves faster in its orbit when it is close to the Sun and slower in its orbit when it is farther away from the Sun.

section 2 The Solar System

Inner Planets

The inner planets—Mercury, Venus, Earth, and Mars—are small and rocky with relatively high average densities. Venus has a thick atmosphere shrouded in clouds of sulfuric acid. Mars is a cold, dry planet with a thin atmosphere consisting mostly of carbon dioxide. Mars has many fascinating features, including ancient water-cut channels and the largest volcanoes in our solar system.

The Outer Planets

The Jovian planets—Jupiter, Saturn, Uranus, and Neptune—are the largest planets. They have thick, gaseous atmospheres consisting mostly of hydrogen with small amounts of helium, methane, ammonia, and water. Clouds composed of ice crystals of ammonia, methane, or other compounds are blown over the planets' surfaces by strong winds.

NASA/Tom Stack & Associates

section 3 Stars and Galaxies

The Lives of Stars

Most stars do not form alone, but as parts of groups of stars known as clusters. Often as these clusters break up, two or more stars will orbit a common center of mass. In binary systems, when one star expands as it evolves into a giant or supergiant, its companion star can pull off mass from the giant star, thus changing the life cycles of both stars.

chapter content resources

Internet Resources

For additional content background, visit **red.msscience.com** to:

- access your book online
- find references to related articles in popular science magazines
- access Web links with related content background
- access current events with science journal topics

Print Resources

Discovering the Universe, by Neil F. Comins and William J. Kaufmann, W.H. Freeman, 2003

A Dictionary of Astronomy, Ian Ridpath (ed.), Oxford University Press, 1998

Star Chart Map, American Map Corporation, 1995

IDENTIFYING

Misconceptions

Find Out What Students Think

Students may think that . . .

Planets and stars are about the same distance away from Earth.

Students often think that, since planets and stars are both objects in the sky, they are about the same distance from Earth. This idea is "confirmed" when they observe the bright objects in the night sky.

Discussion

Ask the class to brainstorm a list of objects they have observed in the night sky. Lists might include stars, the Moon, planets, airplanes, falling "stars," and satellites.

- Discuss how students can determine which of the objects they observe are manufactured, and which are natural objects. Manufactured objects (airplanes and satellites) are lighted objects that move with some speed across the sky. Falling "stars" glow brightly, but only for a moment or two. They do not follow an extended path across the sky. Other objects appear to move slowly across the sky throughout the night.
- Then ask students how they can tell planets from stars as they observe the night sky. Because planets are so much closer to Earth than the stars, they often don't "twinkle" as stars do.

Promote Understanding

Activity

Remind students that, even though the planets are very far away, they are much closer than any of the stars we see in the night sky. Use this activity to reinforce the distances to the planets in the solar system and to the nearest stars, other than our Sun.

- Give each group of students a large roll of toilet paper or adding machine tape, a meterstick, and a marker.
- Direct the class to the data shown in the table. Have students make a model of the solar system on their paper rolls, using these distances to the planets from the Sun.
- When students have completed their models, provide them with these data. Explain that these distances use the same scale as students used to build their solar system. Alpha Centauri: 145.1 km; Sirius: 293.6 km; Deneb: 47,645.6 km
- Discuss with students where these stars would be on the solar system model. Underscore that these, the closest stars other than our Sun, are far outside the solar system. Thus, even though they look to be the same distance from Earth when viewed in the night sky, they are actually very far apart.

Planet	Distance	Planet	Distance
Mercury	0.208 m	Saturn	5.126 m
Venus	0.388 m	Uranus	10.31 m
Earth	0.537 m	Neptune	16.164 m
Mars	0.818 m	Pluto	21.24 m
Jupiter	2.795 m		

Assess

After completing the chapter, see *Identifying Misconceptions* in the Study Guide at the end of the chapter.

chapter 15

Chapter Vocabulary

rotation, p. 440
orbit, p. 441
revolution, p. 441
lunar highlands, p. 442
maria, p. 442
eclipse, p. 444
tides, p. 445
solar system, p. 448
astronomical unit, p. 449
comet, p. 454
meteorite, p. 455
constellation, p. 456
supernova, p. 459
galaxy, p. 459
light-year, p. 462

Science Journal Brrrr, I'm cold. What am I doing riding on this dirty snowball? Lately, I've had this strange feeling that I'm turning. Now I'm getting hot. I'm flying toward a blazing star. Jets of hot gas are shooting out all over. My whole comet is glowing, and I see a tail of dust and gas.

INTERACTIVE CHALKBOARD with Image Bank

PowerPoint® Presentations

This CD-ROM is an editable Microsoft® PowerPoint® presentation that includes:

- a pre-made presentation for every chapter
- interactive graphics
- animations
- audio clips
- image bank
- all new section and chapter questions
- Standardized Test Practice
- transparencies
- pre-lab questions for all labs
- Foldables directions
- links to red.msscience.com

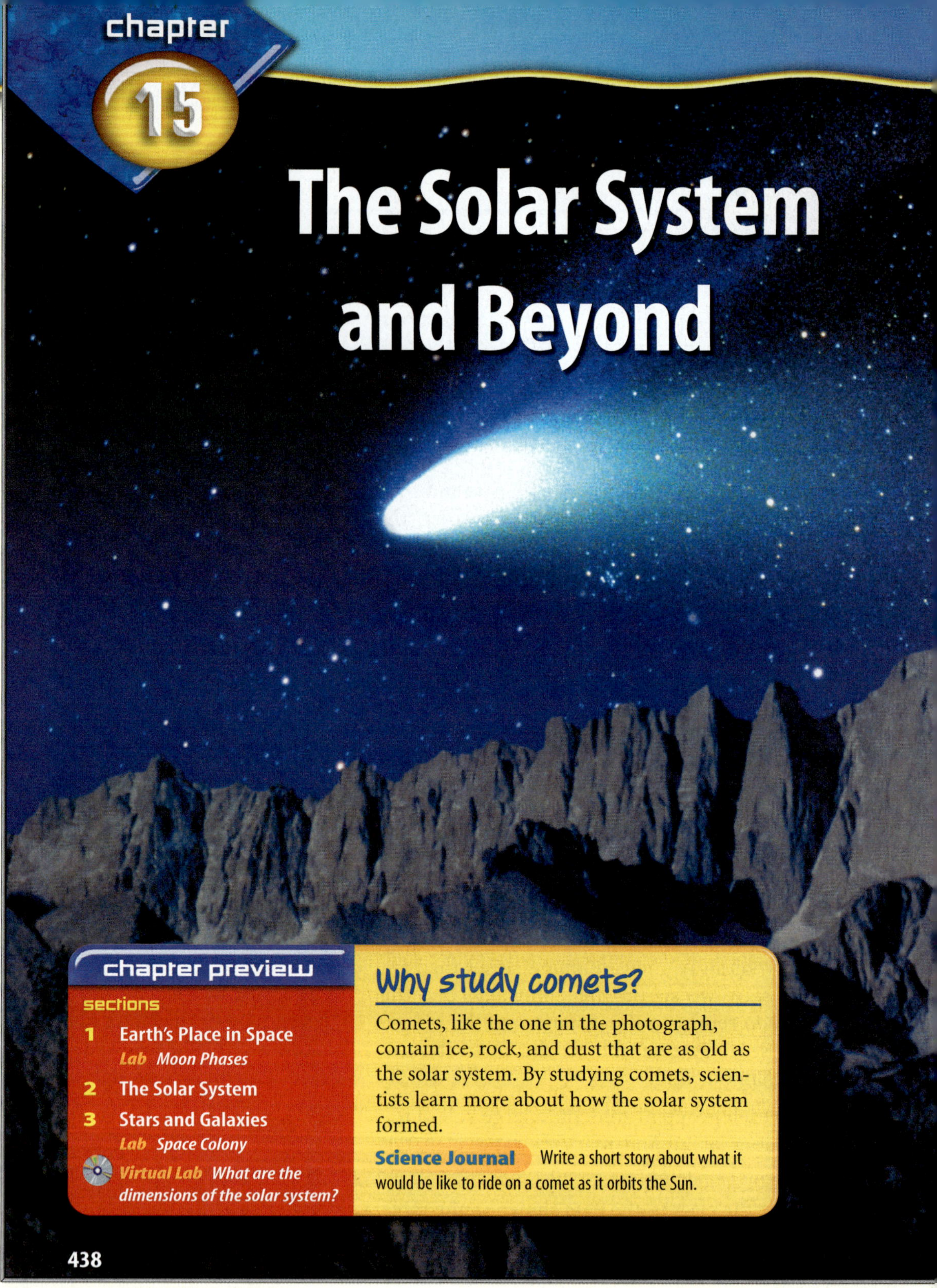

chapter 15

The Solar System and Beyond

chapter preview

sections

1 Earth's Place in Space
Lab *Moon Phases*
2 The Solar System
3 Stars and Galaxies
Lab *Space Colony*
Virtual Lab *What are the dimensions of the solar system?*

Why study comets?

Comets, like the one in the photograph, contain ice, rock, and dust that are as old as the solar system. By studying comets, scientists learn more about how the solar system formed.

Science Journal Write a short story about what it would be like to ride on a comet as it orbits the Sun.

438

Theme Connection

Scale and Structure The immensity, composition, and relatedness of objects in the solar system support this ongoing theme.

About the Photo

Comet The photo shows comet Hale-Bopp. The two tails, gas and dust, are visible. The gas tail has a bluish color because ionized carbon monoxide emits blue light as it reacquires electrons.

Start-Up Activities

Launch LAB

How many stars are in the sky?

On a clear night, the sky is full of sparkling points of light. With the unaided eye, you can see hundreds of these sparkles. How many stars are there?

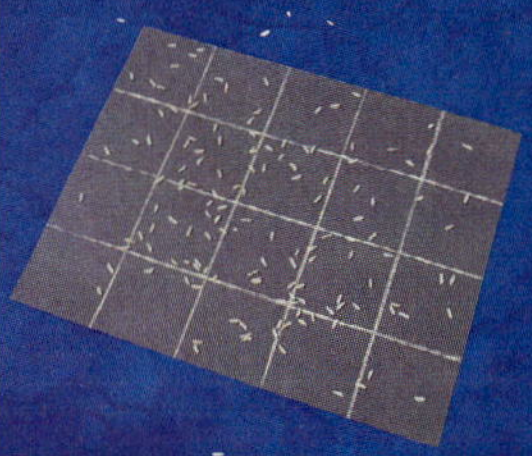

1. Using white crayon or chalk and a ruler, draw grid lines on a sheet of black construction paper, dividing it into 5-cm squares.
2. Spill 4 g of rice grains onto the black paper.
3. Count the number of grains of rice in one square. Repeat this step with a different square. Add the number of grains of rice in the two squares. Divide this number by two to calculate the average number of grains of rice in the two squares.
4. Multiply this number by the number of squares on the paper to get an estimate of the number of rice grains on the paper.
5. **Think Critically** How could scientists use a similar method to estimate the number of stars in the sky?

FOLDABLES Study Organizer

Planets in the Solar System Make the following Foldable to compare and contrast the inner planets and the outer planets.

STEP 1 **Fold** one sheet of paper lengthwise.

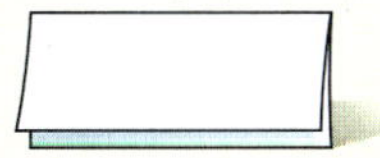

STEP 2 **Fold** into thirds.

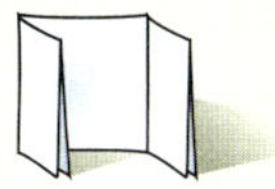

STEP 3 **Unfold and draw** overlapping ovals. Cut the top sheet along the folds.

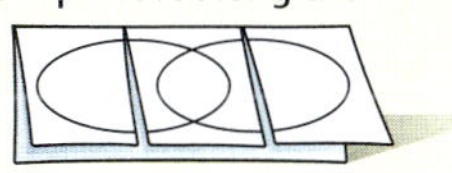

STEP 4 **Label** the ovals as shown.

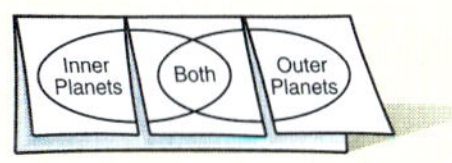

Construct a Venn Diagram As you read the chapter, list the characteristics unique to the inner planets under the left tab. List those unique to the outer planets under the right tab. List characteristics that are common to both inner and outer planets under the middle tab.

Science Online | Preview this chapter's content and activities at red.msscience.com

Launch LAB

Purpose Use the Launch Lab to introduce students to the concepts of random sampling and estimation. L2 ELL COOP LEARN LS **Kinesthetic**

Preparation Review how averages are calculated.

Materials ruler; white crayon or chalk; black construction paper; dry, uncooked rice

Teaching Strategies

- Have students test the accuracy of their estimate by counting the rice grains and comparing their estimate to the actual numbers.
- Discuss other areas where estimation is important. population size, monthly costs for a family, the weight of people in an elevator

Think Critically

Scientists could divide the night sky into sections and use the method described here to estimate the total number of visible stars.

Assessment

Oral Have students use a parallel procedure to estimate the number of students in their school (count the number of children in two classrooms, find the average, then multiply by the number of classrooms in the school). Have them compare the accuracy of their estimates to the actual student population. Use **Performance Assessment in the Science Classroom,** p. 101.

FOLDABLES Study Organizer **Dinah Zike Study Fold**

Student preparation materials for this Foldable are available in the Chapter *Fast File* Resources.

1 Motivate

INTERACTIVE CHALKBOARD PowerPoint® Presentations

Bellringer

Section Focus Transparencies also are available on the Interactive Chalkboard CD-ROM. L2 ELL

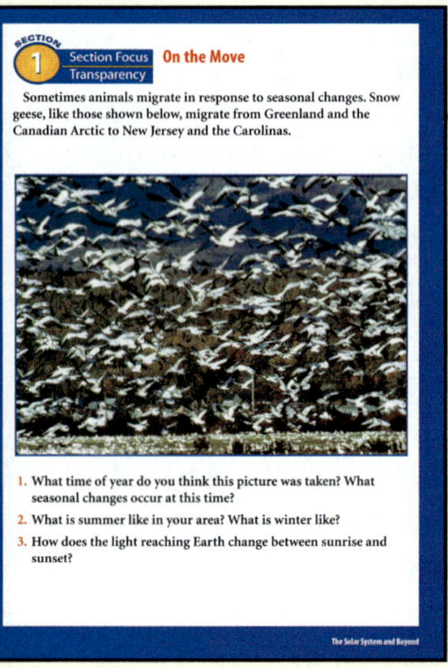
Section 1 Section Focus Transparency **On the Move**

Sometimes animals migrate in response to seasonal changes. Snow geese, like those shown below, migrate from Greenland and the Canadian Arctic to New Jersey and the Carolinas.

1. What time of year do you think this picture was taken? What seasonal changes occur at this time?
2. What is summer like in your area? What is winter like?
3. How does the light reaching Earth change between sunrise and sunset?

The Solar System and Beyond

Tie to Prior Knowledge

Sun, Earth, Moon Have students brainstorm and recall facts they know about the relationships between the Sun, Earth, and the Moon. Encourage them to include information about the movements of Earth and the Moon. Have students list everyday events that result from movements of Earth or the Moon.

section 1

Earth's Place in Space

as you read

What You'll Learn

- **Explain** Earth's rotation and revolution.
- **Explain** why Earth has seasons.
- **Model** the relative positions of Earth, the Moon, and the Sun during different lunar phases.

Why It's Important

You'll understand night and day and the seasons.

Review Vocabulary
axis: the imaginary line around which a planet or moon rotates

New Vocabulary
- rotation
- orbit
- revolution
- lunar highlands
- maria
- eclipse
- tides

Earth Moves

You wake up, stretch and yawn, then glance out your window to see the first rays of dawn. By lunchtime, the Sun is high in the sky. As you sit down to dinner in the evening, the Sun appears to sink below the horizon. Although it seems like the Sun moves across the sky, it is Earth that is moving.

Earth's Rotation Earth spins in space like a twirling figure skater. Your planet spins around an imaginary line running through its center called an axis. **Figure 1** shows how Earth spins around its axis.

The spinning of Earth around its axis is called Earth's **rotation** (roh TAY shun). Earth rotates once every 24 h. The Sun appears each morning due to Earth's rotation. Throughout the day, Earth continues to rotate and the Sun appears to move across the sky. In the evening, the Sun seems to go down because the place where you are on Earth is rotating away from the Sun.

You can see how this works by standing and facing a lamp. Pretend you are Earth and the lamp is the Sun. Now, without pivoting your head, turn around slowly in a counterclockwise direction. The lamp seems to move across your vision, then disappear. You rotate until you finally see the lamp again. The lamp didn't move—you did. When you rotated, you were like Earth rotating in space, causing different parts of the planet to face the Sun at different times. The rotation of Earth—not movement of the Sun—causes night and day.

Reading Check *Why does the Sun appear to move across the sky?*

Because the Sun only appears to move across the sky, this movement is called apparent motion. Stars, planets, and the Moon also appear to move across the sky. You might have observed the Moon rise and set, just like the Sun. How can you recognize apparent motion that results from Earth's rotation?

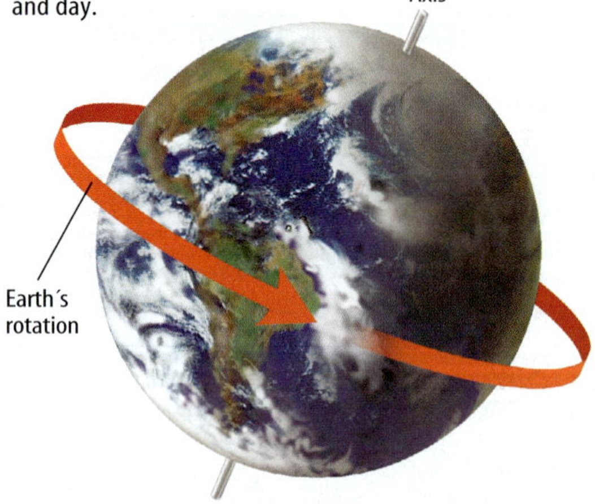

Figure 1 The rotation of Earth around its axis causes night and day.

Section 1 Resource Manager

Chapter *FAST FILE* Resources

Transparency Activity, pp. 44, 47–48
Directed Reading for Content Mastery, pp. 19, 20
Note-taking Worksheets, pp. 33–35
Enrichment, p. 30
MiniLAB, p. 3
Lab Activity, pp. 9–11
Reinforcement, p. 27
Lab Worksheet, pp. 5–6

440 CHAPTER 15 The Solar System and Beyond

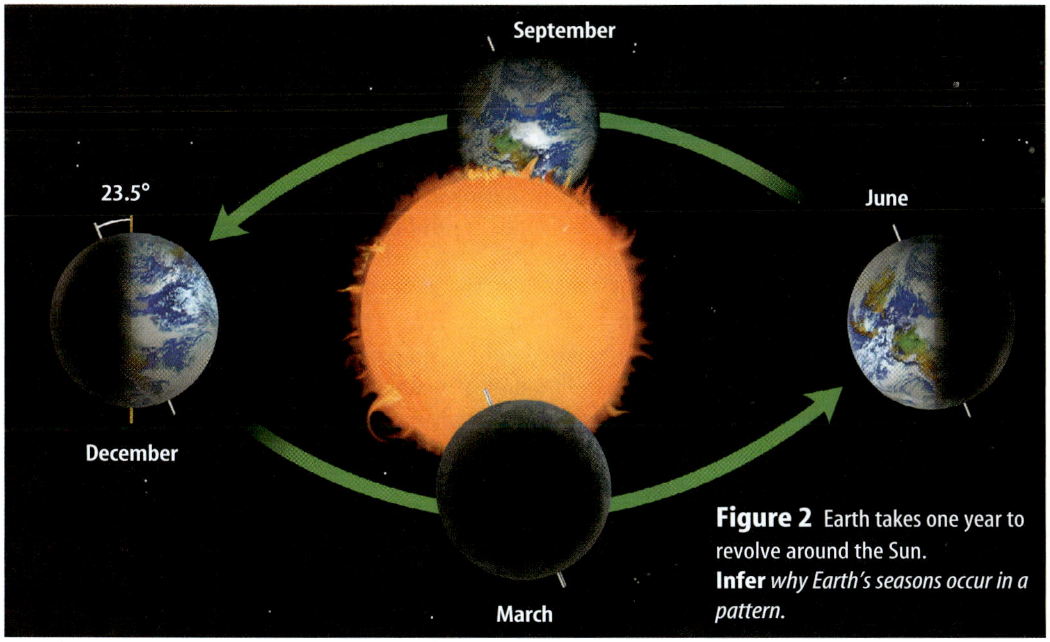

Figure 2 Earth takes one year to revolve around the Sun.
Infer *why Earth's seasons occur in a pattern.*

Earth's Revolution Earth rotates in space, but it also moves in other ways. Like an athlete running around a track, Earth moves around the Sun in a regular, curved path called an **orbit**. The movement of Earth around the Sun is known as Earth's **revolution** (reh vuh LEW shun). A year on Earth is the time it takes for Earth to complete one revolution, as seen in **Figure 2.**

Seasons Who doesn't love summer? The long, warm days are great for swimming, biking, and relaxing. Why can't summer last all year? Blame it on Earth's axis and revolution around the Sun. The axis is not straight up and down like a skyscraper—it is slightly tilted. It's because of this tilt and Earth's revolution that you experience seasons.

Look at **Figure 2.** Summer occurs when your part of Earth is tilted toward the Sun. During summer, sunlight strikes at a higher angle than it does during winter. You might have noticed that when you go outside at noon, your shadow is shorter during summer than during winter. Because summer sunlight strikes at a higher angle, it is more intense than winter sunlight. There also are more hours of daylight during summer than during winter. These two factors cause summer to be warm. Six months later, when the part of Earth that you live on is tilted away from the Sun, you have winter. During winter, sunlight strikes at a lower angle than during summer. The days are short, and the nights are long. Autumn and spring begin when Earth is neither tilted toward nor away from the Sun.

Mini LAB

Modeling Earth's Seasons

Procedure

1. Place a **shaded lamp** on a **table** in your classroom. The lamp represents the Sun. Turn on the lamp, and turn off the overhead lights.
2. Using a **globe,** model Earth's position during each of the four northern hemisphere seasons. Remember to tilt the globe so that its axis makes an angle of about 23.5° from straight up.

Analysis

During which season did the light shine most intensely on the northern hemisphere of the globe? During which season did it shine least intensely?

2 Teach

Reading Check

Answer because Earth rotates

Caption Answer

Figure 2 because Earth revolves around the Sun once each year

Mini LAB

Purpose Students model the geometry between Earth and the Sun for each season.
Visual-Spatial

Materials shaded lamp, globe, small table

Teaching Strategy Test the procedure before students do the lab. The optimal distance from the lamp will vary with equipment.

Analysis The light shined most intensely during summer. It shined least intensely during winter.

Assessment

Oral Have students quiz each other about the geometry of Earth and the Sun during each northern hemisphere season.

Identifying Misconceptions

Distance to the Sun Many people think that Earth is closest to the Sun in summer. Actually, for those in the northern hemisphere, the opposite is true. Earth is closest to the Sun in January. The tilt of Earth's axis, not Earth's distance from the Sun, determines the season.

Curriculum Connection

Geography The Sun is directly over the tropic of Cancer (23.5°N) at the northern hemisphere summer solstice. At the northern hemisphere winter solstice, the Sun is directly over the tropic of Capricorn (23.5°S). Have students locate these latitudes on a map or globe. L2 LS **Visual-Spatial**

Teacher FYI

Axial Tilt The tilt of a planet's axis is determined by the intersection of the line around which it rotates with a line perpendicular to its ecliptic (path around the Sun). Earth's axis is tilted at an angle of 23.5°.

INTEGRATE Career

Astronaut Dr. Jemison was the first African-American woman in space. She also has held a professorship at Dartmouth College. In addition to her foundation work, Dr. Jemison currently is active in business world. One of her lesser known accomplishments is appearing in an episode of *Star Trek: The Next Generation.*

Quick Demo

Space Junk

Materials nuts, bolts, flakes of paint, wrench

Estimated Time five minutes

Procedure Tell students that some space junk consists of large objects, such as rocket components. However, most space junk consists of smaller objects such as these. Even these small objects can cause considerable damage to orbiting spacecraft.

Fun Fact

Space junk orbits Earth at speeds up to 28,000 km/h.

Discussion

Importance of Satellites Lead a discussion about the importance of satellites. Ask students to volunteer ways that society depends on them. weather, positioning, remote sensing, communication, among others

Visual Learning

Figure 3 Use this strategy to help students understand how the Moon rotates. Have a student walk around a globe, facing the globe at all times. The student will experience rotation as he or she revolves around the globe.

Caption Answer

Figure 3 The Moon also rotates and revolves at the same rate.

INTEGRATE Career

Astronaunt Dr. Mae Jemison was a Science Mission Specialist on the space shuttle *Endeavor.* While in space, she studied bone cells and biofeedback. On Earth, she directs the Jemison Institute, which brings new technologies to developing countries. If you want to be an astronaunt, like Dr. Jemison, you'll need to exercise your mind and your body.

Earth's Moon

Does the Moon look perfect to you? Many ancient people thought that the Moon's surface was perfectly smooth. This belief was common until about 400 years ago when a scientist named Galileo looked at the Moon through his telescope. Galileo saw large, mountainous regions with many craters. He also saw smooth, dark regions. The mountainous areas of the Moon are called **lunar highlands.** The lunar highlands are about 4.5 billion years old. The many craters on the lunar highlands formed when meteorites hit the Moon just after it formed. The smooth, dark regions of the Moon are called **maria,** which is the Latin word for *sea.* The maria formed when lava erupted from the Moon's interior and cooled in low areas on its surface.

Orbiting Earth While Earth revolves around the Sun, the Moon and a variety of human-made objects orbit Earth. The Moon revolves around Earth once every 27.3 days. It has an average distance from Earth of 384,400 km. This is like traveling from Los Angeles to New York and back again 49 times. Other objects orbit much closer to Earth than the Moon does. These objects include the *International Space Station,* a wide variety of satellites, and much debris. The debris, often called space junk, consists of parts from old rockets and a variety of discarded tools and equipment.

Figure 3 When you ride on a merry-go-round, it takes the same amount of time for you to rotate as it does for you to revolve around the center.
Explain *how this is similar to the way the Moon rotates and revolves.*

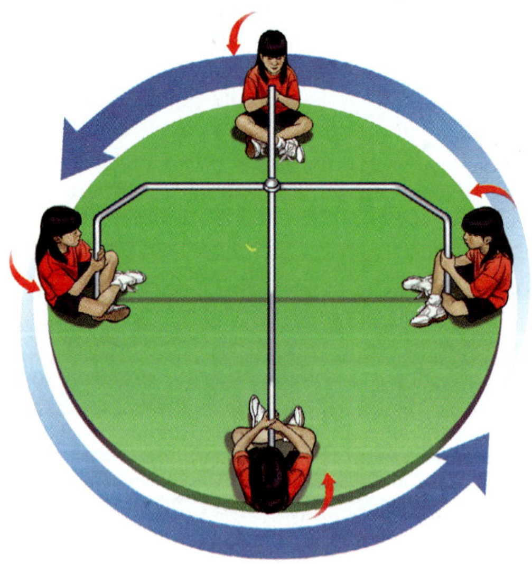

Rotation and Revolution How long does it take for the Moon to rotate one time? The answer is 27.3 days—exactly the same amount of time that it takes for the Moon to revolve around Earth one time. Because the Moon rotates and revolves at the same rate, the same side of the Moon always faces Earth. The side of the Moon that faces Earth is called the near side. The opposite side of the Moon is called the far side. If you've ever ridden on a playground merry-go-round, you rotated and revolved somewhat like the Moon. Look at **Figure 3.** When you ride on a merry-go-round, your body rotates at exactly the same rate as it revolves. You always face the center of the merry-go-round, just like the Moon always keeps the same face toward Earth.

Differentiated Instruction

Challenge Have students create a poster-board display about lunar craters. They might include information about the early lunar bombardment, sizes of various craters, and crater features, such as central peaks. L3 P

English-Language Learners Follow the lunar phases in **Figure 4** as a class. Have students identify each phase in different languages.

Moon Phases How many different moon shapes have you seen? Have you seen the Moon look round or maybe like a half circle? Although the Moon looks different at different times of the month, it doesn't change. What does change is the way the Moon appears from Earth. These changes are called phases of the Moon. **Figure 4** shows the various phases of the Moon.

Light from the Sun The Moon does not produce its own light. The light that comes to Earth from the Moon is reflected sunlight. Also, just as half of Earth experiences day while the other half experiences night, one half of the Moon is lit by the Sun while the other half is dark.

The Lunar Cycle The phase of the Moon that you see on any given night depends on the relative positions of the Moon, the Sun, and Earth in space. These positions change because the Moon is continually revolving around Earth as Earth revolves around the Sun. It takes the Moon about one month to go through its phases. During that time, called a lunar cycle, you see different portions of the daylight side of the Moon.

The lunar cycle begins with new moon. During new moon, the Moon is between Earth and the Sun. Half of the Moon is lit by the Sun, but this half can't be seen from Earth. For about two weeks after new moon, the portion of the lit side of the Moon that can be seen from Earth increases. At first, only a small crescent is visible. Then, you see the first quarter moon followed by the gibbous moon. Finally, the Moon is full. At full moon, Earth is between the Moon and the Sun, and the entire near side of the Moon is visible from Earth. For about two weeks after full moon, the moon phase appears to get smaller. The phase gradually changes from a gibbous moon, to a third quarter moon, crescent moon, and finally new moon again.

Reading Check *What is the lunar cycle?*

Recall that the near side of the Moon always faces Earth. This means that during the new moon phase, the far side of the Moon is lighted by the Sun. The far side of the Moon is lighted just as much as the near side is. The far side can't be seen from Earth because it is always facing away from it.

Figure 4 When the moon phases get larger, they are said to be waxing. When the moon phases get smaller, they are waning. **Explain** *the difference between a waxing crescent and a waning crescent.*

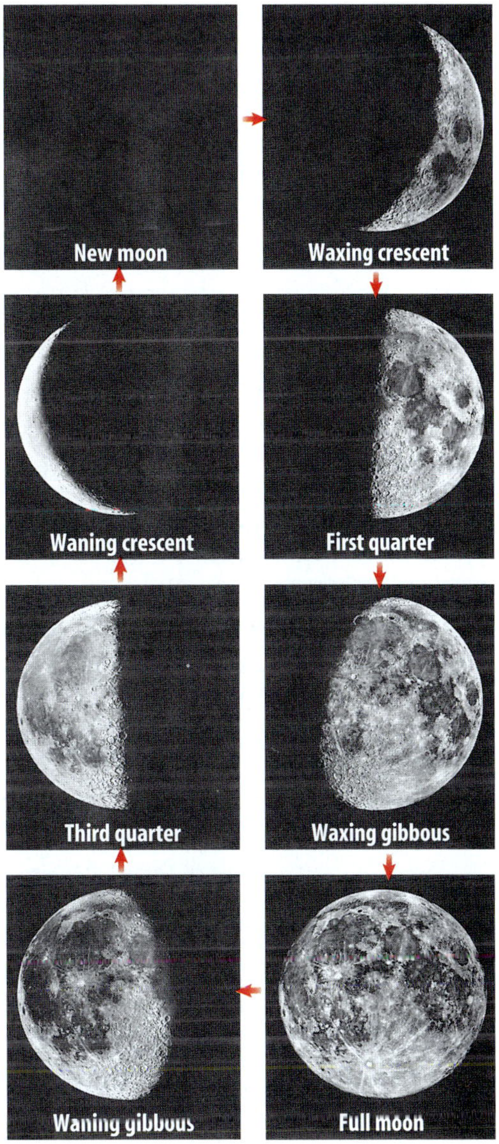

Caption Answer

Figure 4 a waxing crescent is getting larger, whereas a waning crescent is getting smaller.

Use an Analogy

Size Comparisons Scientists estimate the Sun's diameter to be 1,391,994 km. Earth, at 12,756 km in diameter, is about 109 times smaller than the Sun. The Moon's diameter of 3,476 km is a little more than ¼ the size of Earth's. To help students grasp these relative sizes, tell them that if the Sun, Earth, and the Moon were placed on a football field, the Moon would be about the size of a football, Earth would take up one yard, and the Sun would stretch from goalpost to goalpost.

Use Science Words

Word Origin *Gibbous* is a description of the Moon phase between first quarter and full Moon or between full Moon and third quarter. Have students find the origin of this word and explain why it is used to describe these phases of the Moon. *Gibbous* comes from a Latin word meaning "humped" like a camel. The Moon appears to have a hump on its side during these phases. L2 LS **Linguistic**

Activity

Moon Phases Have students observe and record the shape of the Moon for one month. Their drawings can be arranged on a sheet of paper or in a flip book. Before students begin, explain that there will be times when the Moon is not visible because of cloud cover or the position of the Moon in its orbit. Remind students that for some time during the month, the Moon will be visible only during the day or in the early morning. L2 LS **Visual-Spatial** P

Reading Check

Answer the cycle of changing lunar phases

Visual Learning

Figure 4 The phase of the Moon is determined by the relative positions of the Sun, the Moon, and Earth. How might Earth appear to an observer on the Moon? Earth would appear to go through phases, just as the Moon does.

SECTION 1 Earth's Place in Space 443

Caption Answer

Figure 5 new moon

Teacher FYI

Distance from Earth to the Moon The distance from Earth to the Moon is slowly increasing. This occurs because the total angular momentum (spin angular momentum + orbital angular momentum) of the Earth-Moon system must remain constant. Tidal forces cause Earth to rotate more slowly through time and lose spin angular momentum. To compensate the Moon must gain orbital angular momentum by moving farther from Earth.

Make a Model

Moon's Distance The average distance from Earth to the Moon is increasing at a rate of 3.8 cm per year. Ask students to use balls of appropriate size to model how this changing distance will affect solar eclipses. Eventually the Moon will not be able to cover the entire solar disk, and solar eclipses will no longer occur.

Reading Check

Answer A solar eclipse occurs when the Moon passes directly between Earth and the Sun.

Figure 5 During a solar eclipse, the Moon moves directly between the Sun and Earth. The Sun's corona is visible during a total solar eclipse.
Identify *the phase that the Moon must be in for a solar eclipse to occur.*

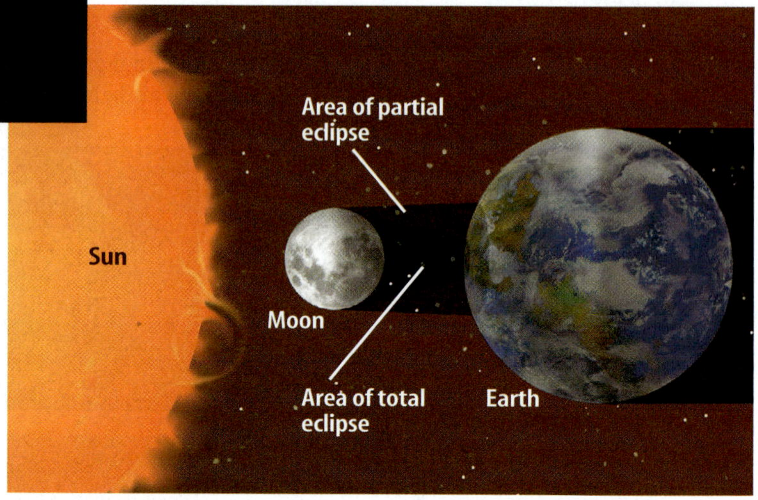

Science Online

Topic: Eclipse Data
Visit red.msscience.com for Web links to information about future solar and lunar eclipses.

Activity Find out when you will next be able to observe an eclipse in your region.

Solar Eclipse Have you ever tried to watch TV with someone standing between you and the screen? You can't see a thing. The picture from the screen can't reach your eyes because someone is blocking it. Sometimes the Moon is like that person standing in front of the TV. It moves directly between the Sun and Earth and blocks sunlight from reaching Earth. The Moon's shadow travels across parts of Earth. This event, shown in **Figure 5,** is an example of an **eclipse** (ih KLIHPS). Because it is an eclipse of the Sun, it is known as a solar eclipse. The Moon is much smaller than the Sun, so it casts a small shadow on Earth. Sunlight is blocked completely only on the small area of Earth where the Moon's darker shadow falls. In that area, the eclipse is said to be a total solar eclipse.

Reading Check *What causes solar eclipses?*

Due to the small size of the shadow—about 269 km wide—only a lucky few get to experience each solar eclipse. For the few minutes the total eclipse lasts, the sky darkens, flowers close, and some planets and brighter stars become visible. The Sun's spectacular corona, its pearly white, outermost layer, can be observed. Far more people will be in the lighter part of the Moon's shadow and will experience a partial solar eclipse. **WARNING:** *Never look at the Sun during an eclipse. You might damage your eyes.*

Cultural Diversity

Lunar Calendar Some Middle Eastern and Far Eastern cultures use the lunar calendar. Obtain a calendar from one such culture for students to examine. Using the calendar, have students correlate the beginnings of the months to the new Moon. Students will find that the beginnings of lunar months occur within two or three days of the new Moon. They do not necessarily coincide with the months used in the United States. Students will also find that each month contains either 29 or 30 days. Since the lunar month is based on the time from new Moon to new Moon, or 29.5 days, the lunar year is 11 days shorter than the solar year. Adding seven months to the lunar calendar every 19 years compensates for this difference. Also, because of this 11-day difference, some years will have 12 full lunar cycles and part of a thirteenth. L2

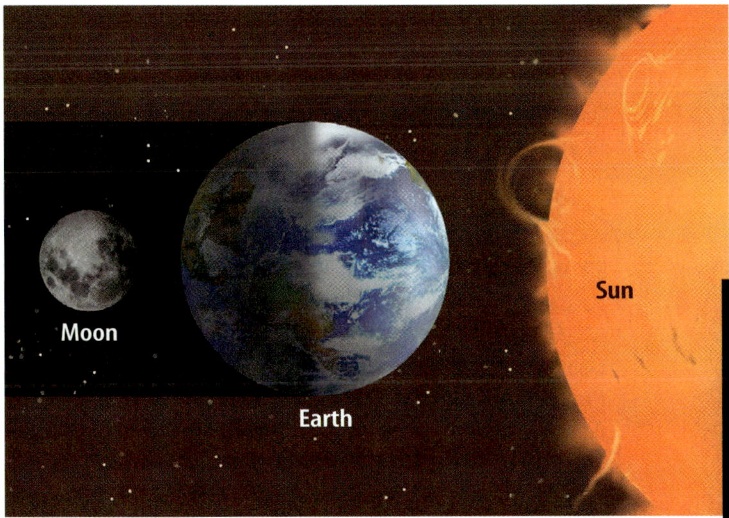

Figure 6 During a lunar eclipse, Earth is between the Sun and the Moon. The Moon often appears red during a lunar eclipse.
Infer *why lunar eclipses are observed more frequently than solar eclipses.*

Lunar Eclipse Sometimes Earth is directly between the Sun and the Moon. When Earth's shadow falls on the Moon, an eclipse of the Moon occurs. This is called a lunar eclipse. Everyone on the nighttime side of Earth, weather permitting, can see a lunar eclipse. When eclipsed, the full moon becomes dim and sometimes turns deep red, as shown in **Figure 6.**

Tides The Moon's gravity pulls on Earth. One effect of the Moon's gravity is tides. **Tides** are an alternate rise and fall in sea level. They are most noticeable along a beach. At high tide, water moves farther onto the beach. At low tide, water moves off the beach.

Tides occur because the Moon's gravity decreases with distance from the Moon. Places on Earth closer to the Moon are pulled harder than places that are farther from the Moon. The Moon's gravity holds Earth in its path around the center of mass of the Earth-Moon system. At places on Earth that are closer to the Moon, the Moon's gravity is a bit stronger than it needs to be to hold Earth. At places on Earth that are farther from the Moon, the Moon's gravity is a bit weaker than it needs to be.

The small differences in the Moon's gravity cause the water in Earth's oceans to form two bulges, shown in **Figure 7.** One bulge is on the side of Earth toward the Moon, and one is on the opposite side. These bulges of water are the high tides. The areas of Earth that are neither toward nor away from the Moon are the low tides. As Earth rotates, different places pass through high and low tide.

Figure 7 Tides form because the Moon's gravity pulls harder on parts of Earth that are closer to it. Two tidal bulges occur—one on the side of Earth closest to the Moon, and one on the opposite side of Earth.

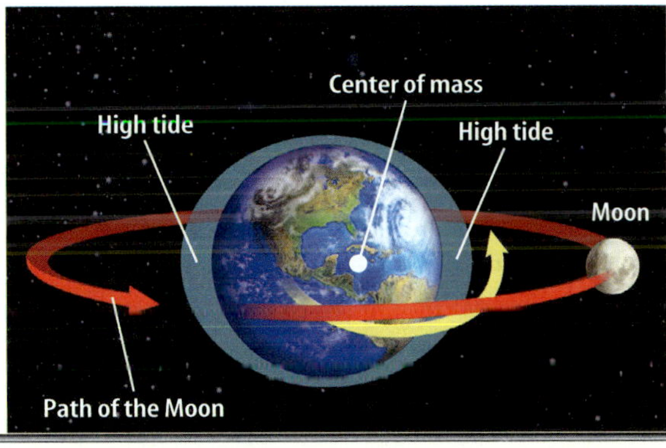

Caption Answer

Figure 6 More people on Earth observe a lunar eclipse when it occurs than a solar eclipse because the Moon's shadow only affects a small area on Earth, whereas everyone on the night side of Earth can see a lunar eclipse.

Teacher FYI

Red Moon The Moon often looks red during a lunar eclipse. This occurs because some sunlight is refracted toward the Moon by Earth's atmosphere. The reason this refracted light is red is that most of the shorter wavelengths of visible light are lost to scattering in Earth's atmosphere.

Activity

Tides Have students choose a coastal community and monitor high and low tides for a month. These data can be found in local newspapers and Web pages. Students should make graphs showing tidal period and range.

Science Journal

Tidal Range At some locations, tidal range can be extreme. At the Bay of Fundy in Nova Scotia, the difference between high and low tide can be as much as 15 m. Have students write about the effects of such high tidal ranges in their Science Journals.

3 Assess

DAILY INTERVENTION

Check for Understanding

Interpersonal Summarize the different motions of Earth—rotation and revolution—for students. Ask the students to describe the effects of each. rotation—day and night, apparent motion of celestial bodies; revolution—seasons, one year

Reteach

Summer and Winter Solstice Have students sketch and label the positions of the Sun and Earth during summer and winter solstice (June 21 or 22 and December 21 or 22) in the northern hemisphere. L2 ELL LS Visual-Spatial

Assessment

Process Have students research and write a short paragraph explaining why the same side of the Moon always faces Earth. The Moon rotates once and also revolves once every 27.3 days. Thus, the same side of the Moon always faces Earth.

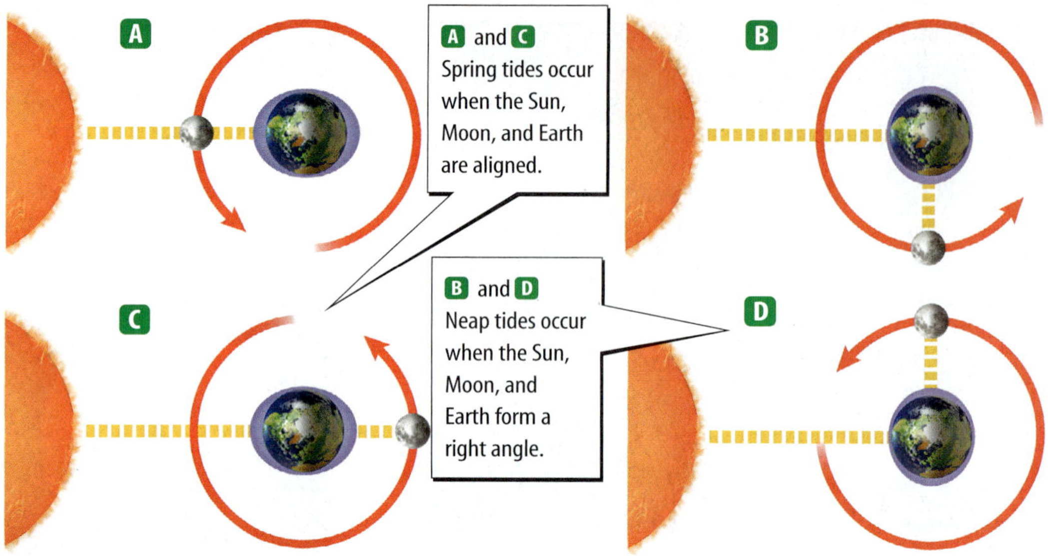

Figure 8 Because the Moon revolves around Earth, spring tides and neap tides each occur about twice each month.

The Sun's Effect on Tides The Sun also affects tides on Earth. Because the Sun is much farther from Earth, it has about half as much tide-generating force as the Moon. When the Sun, Earth, and the Moon are lined up, high tides are higher and low tides are lower. This is called spring tide, as shown in **Figure 8.** Spring tides occur because the Moon's gravity and the Sun's gravity combine to produce a greater effect. When the Sun, Earth, and the Moon form a 90° angle, high tides are lower and low tides are higher. This condition, also shown in **Figure 8,** is called neap tide. During neap tide, the Sun's gravity reduces the tide-generating effect of the Moon's gravity.

section 1 review

Summary

Earth Moves

- Seasons occur because of Earth's tilted axis and Earth's revolution around the Sun.

Earth's Moon

- The Moon has many surface features including craters, maria, and lunar highlands.
- Different moon phases occur depending on the positions of the Sun, Earth, and the Moon.

Tides

- The Moon has the greatest effect on Earth's tides. The Sun has a lesser effect.

Self Check

1. **Define** Earth's revolution and rotation.
2. **Explain** why lunar eclipses occur during a full moon.
3. **Compare and contrast** spring tides and neap tides.
4. **Think Critically** What would seasons be like if Earth's axis were tilted at a higher angle (more than 23.5°)?

Applying Skills

5. **Research Information** Some scientific knowledge is very old, yet it still is correct today. Do research to learn how much the ancient Mayan civilization knew about the length of a year.

Science Online red.msscience.com/self_check_quiz

section 1 review

1. Earth's revolution is Earth orbiting the Sun. Earth's rotation is Earth spinning around its axis.
2. Earth is between the Sun and the Moon during the full-moon phase. If these bodies are lined up in the same plane, an eclipse occurs.
3. During spring tide, high tides are higher and low tides are lower. During neap tide, high tides are lower and low tides are higher.
4. If Earth's axis were tilted at a higher angle, seasons would be more extreme. Summer would be hotter, and winter would be colder.
5. The Mayan civilization developed its own calendar and knew the length of the year. Some suggest that the ancient Mayan estimate of a year's length was almost as accurate as the value used today.

LAB

Moon Phases

The Moon is Earth's nearest neighbor in space. The Sun, which is much farther away, is the source of light that reflects off of the moon. In this lab, you'll observe how the positions of the Sun, the Moon, and Earth cause the different phases of the Moon.

Real-World Question

How do the positions of the Sun, the Moon, and Earth affect the phases of the Moon?

Goals

- **Model and observe** moon phases.
- **Record and label** phases of the Moon.
- **Infer** how the positions of the Sun, the Moon, and Earth affect phases of the Moon.

Materials

drawing paper (several sheets)
softball
flashlight

Safety Precautions

Procedure

1. Turn on the flashlight and darken other lights in the room. Select a member of your group to hold the flashlight. This person will be the Sun. Select another member of your group to hold up the softball so that the light shines directly on the ball. The softball will be the Moon in your experiment.
2. Everyone else represents Earth and should sit between the Sun and the Moon.
3. **Observe** how light shines on the Moon. Draw the Moon, being careful to add shading to represent its dark portion.

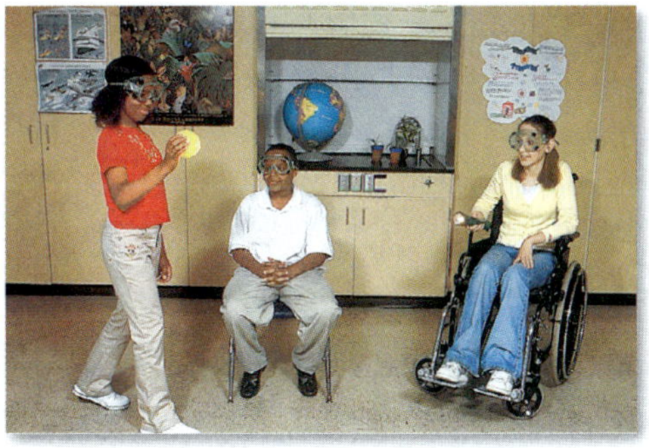

4. The student who is holding the Moon should begin to walk in a slow circle around the group, stopping at least seven times at different spots. Each time the Moon stops, observe it, draw it, and shade in its dark portion.

Conclude and Apply

1. **Compare and contrast** your drawings with those of other students. Discuss similarities and differences in the drawings.
2. In your own words, explain how the positions of the Sun, the Moon, and Earth affect the phase of the Moon that is visible from Earth.
3. **Compare** your drawings with **Figure 4.** Which phase is the Moon in for each drawing? Label each drawing with the correct moon phase.

Communicating Your Data

Use your drawings to make a poster explaining phases of the Moon. **For more help, refer to the Science Skill Handbook.**

BENCH TESTED

Real-World Question

Purpose Students model lunar phases. L2 ELL COOP LEARN LS **Visual-Spatial** P

Process Skills observe, compare and contrast, infer, analyze, model

Time Required 40 minutes

Procedure

Teaching Strategies

- The terminator—the line that separates the Moon's light and dark sides—should be drawn first. The darkened part of the sphere then should be shaded.
- The student holding the softball should move in a counterclockwise direction in order to get the phases in correct order.

Conclude and Apply

1. Students should compare and contrast the location of the terminator and correct shape of the shaded portion of the Moon at each phase.
2. As the Moon moves more nearly opposite the Sun, more of its lighted side is visible from Earth.
3. Student drawings should correspond to Figure 4.

Communicating Your Data

Students' posters should describe the relative positions of the Sun, Earth, and the Moon during each lunar phase.

Assessment

Portfolio Many cultures have myths and legends about the phases of the Moon. Have students research one myth, then write short stories or poems about the Moon based on their research. They can use their Moon sketches to illustrate their stories. Use **Performance Assessment in the Science Classroom,** p. 157.

1 Motivate

Bellringer

Section Focus Transparencies also are available on the Interactive Chalkboard CD-ROM. L2 ELL

Section 2 Section Focus Transparency **A Really Big Belt**

Planets aren't the only things in our solar system; asteroids also orbit the Sun. Most of these asteroids are in an area between Mars and Jupiter called the asteroid belt. This photo shows an asteroid named Gaspra (Asteroid 951). The image was made by the Galileo space probe.

1. Describe Gaspra's shape.
2. Judging from the picture, what do you think asteroids are made of?
3. Name some objects that orbit the Sun. Name some objects that orbit Earth.

The Solar System and Beyond

Tie to Prior Knowledge

Planets in the News Have students recall recent discoveries about the planets that have been in the news, such as the presence of water ice on Mars or the possibility of seas existing under the crust of Europa, one of Jupiter's moons.

Reading Check

Answer the Sun's gravity

section 2 The Solar System

as you read

What **You'll Learn**

- **Compare and contrast** the planets and moons in the solar system.
- **Explain** that Earth is the only planet known to support life.

Why **It's Important**

Much can be learned about Earth by studying the solar system.

Review Vocabulary
system: a portion of the universe and all of its components, processes, and interactions

New Vocabulary
- solar system
- astronomical unit
- comet
- meteorite

Distances in Space

Imagine that you are an astronaut living in the future, doing research on a space station in orbit around Earth. You've been working hard for a long time and need a vacation. Where will you go? How about a tour of the solar system? The **solar system**, shown in **Figure 9**, is made up of the nine planets and numerous other objects that orbit the Sun, all held in place by the Sun's immense gravity.

Reading Check *What holds the solar system together?*

The planets in the solar system revolve around the Sun in elliptical orbits. The orbits of most of the planets are only slightly elliptical. They are almost circular. Pluto and Mercury have orbits that are more elliptical. Their orbits are similar to a slightly flattened circle.

Figure 9 The Sun is the center of the solar system, which is made up of the nine planets and other objects that orbit the Sun.
Compare and contrast *the sizes of the different planets.*

Section 2 Resource Manager

Chapter *FAST FILE* Resources
- Transparency Activity, p. 45
- Directed Reading for Content Mastery, p. 21
- Enrichment, p. 31
- Reinforcement, p. 28
- Lab Activity, pp. 13–16

Reading and Writing Skill Activities, p. 11
Mathematics Skill Activities, p. 39
Cultural Diversity, p. 25
Earth Science Critical Thinking/Problem Solving, p. 8

Measuring Space Distances in space are hard to imagine because space is so vast. Suppose you had to measure your pencil, the hallway outside your classroom, and the distance from your home to school. Would you use the same unit for each measurement? No. You probably would measure your pencil in centimeters. You would use something bigger to measure the length of the hallway, such as meters. You might measure the trip from your home to school in kilometers. Larger units are used to measure longer distances. Imagine trying to measure the trip from your home to school in centimeters. If you didn't lose count, you'd end up with a huge number.

Astronomical Unit Kilometers are fine for measuring long distances on Earth, such as the distance from New York to Chicago (about 1,200 km). Even bigger units are needed to measure vast distances in space. One such measure is the astronomical (as truh NAH mih kul) unit. An **astronomical unit** equals 150 million km, which is the average distance from Earth to the Sun. Astronomical unit is abbreviated *AU*. If something is 3 *AU* away from the Sun, then the object is three times farther from the Sun than Earth is. The *AU* is a convenient unit for measuring distances in the solar system.

Reading Check *Why is the astronomical unit useful for measuring distances in the solar system?*

Science Online

Topic: Space Technology
Visit red.msscience.com for Web links about technology that is used to explore space.

Activity Create a concept map that explains why technology is essential to science.

2 Teach

Discussion

Astronomical Units Explain that units of measure need to have meaning to the people who use them and that they must be functional for the purpose for which they are devised. How does the astronomical unit (AU) fulfill these requirements? Possible answer: The AU has meaning because it uses Earth's distance from the Sun as a standard; it is functional for distances within the solar system.

Caption Answer

Figure 9 See Visual Learning below.

Visual Learning

Figure 9 Have students study this figure and make some general statements about the planets in the solar system. Record their ideas, and have students add detail to the list as they learn about the planets shown. Possible answers: Planets closest to the Sun are small, and those farther from the Sun (except for Pluto) are large; the Sun is the largest object in the solar system. L2 ELL LS **Visual-Spatial and Linguistic**

Reading Check

Answer Large distances can be expressed using low numbers.

Differentiated Instruction

Learning Disabled As you present the lesson dealing with special characteristics of planets, help students understand why some planet surfaces are cratered. Have students model impact craters by dropping marbles into loose sand. **WARNING:** *Students should wear safety goggles.* Students should note the size and shape of the craters and should look for raised crater rims and central peaks. LS **Kinesthetic**

Mini LAB

Purpose Students will observe planetary motion against the background stars.

Materials camera

Teaching Stategy Work together as a class to learn which planets currently are visible and where in the sky they can be observed.

Analysis

1. The planet will move against the background of stars. However, planets sometimes can appear to stop for a short time. Normal motion of the planets is west to east against the stars. Retrograde motion, east to west, can sometimes occur.
2. Many students will correctly suggest that the planet's revolution around the Sun is responsible for the motion. The reason that planets can sometimes move in retrograde motion is that each planet orbits the Sun at different rates. Sometimes Earth can "pass" a planet farther away from the Sun, and a planet closer to the Sun can sometimes "pass" Earth.

Try at Home

Activity

Planets in the Solar System As students learn about each of the planets in the solar system, have them make a chart on the bulletin board or chalkboard that compares and contrasts each planet to Earth. L2

Reading Check

Answer Mercury has no atmosphere, and it is close to the Sun.

Mini LAB

Observing Planets

Procedure

1. Research which planets currently are visible in the night sky.
2. Select a planet to watch for three to four weeks. You might choose Jupiter, Saturn, Mars, or Venus.
3. Observe the planet at the same time each clear night. Note the planet's position compared to background stars.
4. You might want to use a **camera** to photograph the planet and background stars each night.

Analysis

1. Did the planet move against the background stars? If so, did it move from west to east or from east to west?
2. How can you explain the planet's movement?

Try at Home

Touring the Solar System

Now you know a little more about how to measure distances in the solar system. Next, you can travel outward from the Sun and take a look at the objects in the solar system. Maybe you can find a nice destination for your next vacation. Strap yourself into your spacecraft and get ready to travel. It's time to begin your journey. What will you see first?

Inner Planets

The first group of planets you pass is the inner planets. These planets are mostly solid, with minerals similar to those on Earth. As with all the planets, much of what is known comes from spacecraft that send data back to Earth. Various spacecraft took the photographs shown in **Figure 10** and the rest of this section. Some were taken while in space and others upon landing.

Mercury The first planet that you will visit is the one that is closest to the Sun. Mercury, shown in **Figure 10,** is the second-smallest planet. Its surface has many craters. Craters form when meteorites, which are chunks of rock or metal that fall from the sky, strike a planet's surface. You will read about meteorites later in this section. Because of Mercury's small size and low gravity, gases that could form an atmosphere escape into space. The lack of an atmosphere and the closeness of this planet to the Sun cause great extremes in temperature. Mercury's surface temperature can reach 425°C during the day and drop to −170°C at night, making the planet unfit for life.

Reading Check *Why do surface temperatures on Mercury vary so much?*

Figure 10 Mercury and Venus are closer to the Sun than Earth is.

Like the Moon, Mercury's surface is scarred by craters.

Earth's closest neighbor, Venus, is covered in clouds.

Venus You won't be able to see much at your next stop, also shown in **Figure 10.** Venus, the second-closest planet to the Sun, is hard to see because its surface is surrounded by thick clouds. These clouds trap the solar energy that reaches the surface of Venus. That energy causes surface temperatures to hover around 472°C—hot enough to bake a clay pot.

LAB DEMONSTRATION

Purpose to observe the shape of an elliptical orbit

Materials pushpins (2), piece of string, large, flat piece of cardboard, marker

Preparation Cover the cardboard with paper and insert two pushpins about 8 cm apart.

Procedure Tie the ends of the string to make a loop. Drop the loop of string over the pins and use the string to guide a felt-tipped marker, thereby drawing an ellipse.

Expected Outcome Students observe the shape of an ellipse.

Assessment

Explain that the ellipse is a model of a planet's orbit with the Sun located at one of the pins. What would happen to the shape of the ellipse if the pins were placed closer together? The ellipse would be less flattened and more circular.

Earth Home sweet home. You've reached Earth, the third planet from the Sun. You didn't realize how unusual your home planet was until you saw other planets. Earth's surface temperatures allow water to exist as a solid, a liquid, and a gas. Also, ozone in Earth's atmosphere works like a screen to limit the number of ultraviolet (ul truh VI uh lut) rays that reach the planet's surface. Ultraviolet rays are harmful rays from the Sun. Because of Earth's atmosphere, life can thrive on the planet. You would like to linger on Earth, shown in **Figure 11,** but you have six more planets to explore.

Mars Has someone else been here? You see signs of earlier visits to Mars, the fourth of the inner planets. Tiny robotic explorers have been left behind. However, it wasn't a person who left them here. Spacecraft that were sent from Earth to explore Mars's surface left the robots. If you stay long enough and look around, you might notice that Mars, shown in **Figure 12,** has seasons and polar ice caps. Signs indicate that the planet once had abundant liquid water. Water might even be shaping the surface of Mars today. You'll also notice that the planet looks red. That's because the sediment on its surface contains iron oxide, which is rust. Two small moons, Phobos and Deimos, orbit Mars.

Figure 11 As far as scientists know, Earth is the only planet that supports life.

Figure 12 Mars often is called the Red Planet.
Explain *what causes Mars's surface to appear red.*

Quick Demo

The Water Planet

Materials globe

Estimated Time five minutes

Procedure Show students a globe of Earth. Ask the students to describe Earth's features. They likely will notice the abundant surface water. Tell them that this is one reason that Earth is unique in the solar system.

Fun Fact

The misconception that Mars has canals was actually the result of a linguistic error. The Italian astronomer Schiaparelli wrote of seeing *canali* on Mars, which means "channels." This was mistakenly translated to "canals." The movement of any liquid can make channels naturally. Canals, on the other hand, imply construction by intelligent beings.

Visual Learning

Figure 12 The small roving vehicle, *Sojourner*, enabled the *Pathfinder* mission to study a large area of the Martian surface. Contrast the surfaces of Mars and Earth. Mars is barren and dry. Its surface looks red due to iron oxide in its rocks. Much of Earth is green and covered with vegetation. Water flows freely over Earth's surface.

Caption Answer

Figure 12 iron oxide in the rocks

Active Reading

Write-Draw-Discuss This strategy encourages students to actively participate in reading and lectures. Have students write about an idea, clarify it, then make an illustration or drawing. Ask students to share responses and display several examples. Have students Write-Draw-Discuss about the aspect of the solar system that interests them the most.

Differentiated Instruction

Challenge Several future missions to Mars are planned. Have students research the past, present, and future of Mars exploration. They then can make a timeline showing what they've learned. L3

Caption Answer

Figure 13 Gaspar's surface is covered with impact craters.

Reading Check

Answer pieces of rock made of minerals similar to those that formed rocky planets and moons

Make a Model

Mining Asteroids Have students make a model of a mining operation in the asteroid belt. Student models should indicate how asteroids would be collected and processed. Robotic asteroid collectors could collect the smaller asteroids. These could be returned to a large processing plant on one of the larger asteroids. L2 LS **Kinesthetic and Logical-Mathematical** P

Activity

Galilean Satellites Have students research the four Galilean satellites of Jupiter—Io, Europa, Ganymede, and Callisto. Then have them compare and contrast these four large moons. Possible answer: All four orbit Jupiter. Io is volcanically active. Europa may have water beneath its thin crust. Ganymede has its own magnetic field. Callisto also may have an ocean inside. L2 LS **Linguistic**

Virtual Labs

Space Measurements
What are the dimensions of the solar system?

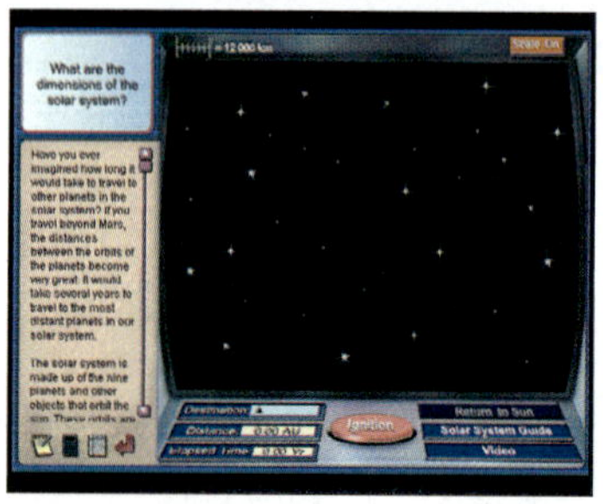

Figure 13 This close-up of the asteroid Gaspra was taken by the *Galileo* spacecraft in 1991.
Describe *Gaspra's surface features.*

Asteroid Belt Look out for asteroids. On the next part of your trip, you must make your way through the asteroid belt that lies between Mars and the next planet, Jupiter. As you can see in **Figure 13,** asteroids are pieces of rock made of minerals similar to those that formed the rocky planets and moons. In fact, these asteroids might have become a planet if it weren't for the giant planet, Jupiter. Jupiter's huge gravitational force might have prevented a small planet from forming in the area of the asteroid belt. The asteroids also might be the remains of larger bodies that broke up in collisions. The asteroid belt separates the solar system's planets into two groups—the inner planets, which you've already visited, and the outer planets, which are coming next.

Reading Check *What are asteroids?*

Outer Planets

Moving past the asteroids, you come to the outer planets. The outer planets are Jupiter, Saturn, Uranus, Neptune, and Pluto. Let's hope you aren't looking for places to stop and rest. Trying to stand on most of these planets would be like trying to stand on a cloud. That's because all of the outer planets, except Pluto, are huge balls of gas called gas giants. Each might have a solid core, but none of them has a solid surface. The gas giants have lots of moons that orbit the planets just like Earth's Moon orbits Earth. They have rings surrounding them that are made of dust and ice. The only outer planet that doesn't have rings is Pluto. Pluto also differs from the other outer planets because it is composed of ice and rock.

Figure 14 Jupiter is the largest planet in the solar system.

Jupiter If you're looking for excitement, you'll find it on Jupiter, which is the largest planet in the solar system and the fifth from the Sun. It also has the shortest day—less than 10 h long—which means this giant planet is spinning faster than any other planet. Watch out for a huge, red whirlpool near the middle of the planet! That's the Great Red Spot, a giant storm on Jupiter's surface. Jupiter, shown in **Figure 14,** looks like a miniature solar system. It has 61 moons. One called Ganymede (GA nih meed) is larger than the planet Mercury. Ganymede, along with two other moons, Europa and Callisto, might have liquid water under their icy crust. Another of Jupiter's moons, Io, has more active volcanoes than any other object in the solar system.

Differentiated Instruction

Challenge Galileo first saw the four largest moons of Jupiter (Io, Europa, Ganymede, and Callisto) from Earth through a telescope. If possible, bring a telescope to class and let students examine its structure. Have students research and write a report about the astronomical accomplishments of Galileo. Galileo observed the phases of Venus and craters and maria on the Moon. L3 LS **Linguistic**

Science Journal

***Galileo* and *Cassini* Spacecraft** Have students research the discoveries of the *Galileo* spacecraft to Jupiter and the goals of the *Cassini* mission expected to arrive at Saturn in 2004. Have them write a report in their Science Journals. *Galileo*—information about Jupiter and its moons; *Cassini*—atmosphere of Titan and Saturn's atmosphere, rings, and other moons. L2 LS **Linguistic**

Saturn You might have thought that Jupiter was unusual. Wait until you see Saturn, the sixth planet from the Sun. You'll be dazzled by its rings, shown in **Figure 15.** Saturn's several broad rings are made up of hundreds of smaller rings, which are made up of pieces of ice and rock. Some of these pieces are like specks of dust. Others are many meters across. Saturn is orbited by at least 31 moons, the largest of which is Titan. Titan has an atmosphere that resembles the atmosphere on Earth during primitive times. Some scientists hypothesize that Titan's atmosphere might provide clues about how life formed on Earth.

Uranus After Saturn, you come to Uranus, the seventh planet from the Sun. Uranus warrants a careful look because of the interesting way it spins around its axis. The axis of most planets is tilted just a little, somewhat like the handle of a broom that is leaning against a wall. Uranus, also shown in **Figure 15,** is nearly lying on its side. Its axis is tilted almost even with the plane of its orbit like a broomstick lying on the floor. Uranus's atmosphere is made mostly of hydrogen with smaller amounts of helium and methane. The methane gives Uranus its distinctive bluish-green color. Uranus has rings and is thought to have at least 21 moons.

Figure 15 Saturn and Uranus are two of the four gas giant planets.

Applying Science

How can you model distances in the solar system?

The distances between the planets and the Sun are unimaginably large but definitely measurable. Astronomers have developed a system of measurement to describe these distances in space. Could you represent these vast distances in a simple classroom model? Use your knowledge of SI and your ability to read a data table to find out.

Identifying the Problem

The table shows the distances of the planets and asteroid belt from the Sun. Notice that the inner planets are fairly close together, and the outer planets are far apart. Study the distances carefully, then answer the questions.

Solar System Data

Planet	Distance from the Sun (AU)
Mercury	0.39
Venus	0.72
Earth	1.00
Mars	1.52
Asteroid belt	2–4
Jupiter	5.20
Saturn	9.54
Uranus	19.19
Neptune	30.07
Pluto	39.48

Solving the Problem

1. How can you make a scale model of the solar system that will fit in your classroom? What unit will you use to show the distances?
2. Show the conversion between astronomical units and the unit you use for your model.

IDENTIFYING Misconceptions

Planets in the Solar System Many students may think that the planets are lined up in a straight line outward from the Sun. Stress to them that the planets orbit at different speeds. Illustrating the planets in a line and learning about the planets "in order" is simply a convenient way to study the solar system.

Applying Science

Answers

1. Possible answer: The easiest way to make a model of the solar system would be to convert to centimeters, using a meterstick. Given the vast distances from the Sun to the outer planets, students may realize that the distance to some planets, such as Mercury, would be measured in centimeters, while the distances to others, such as Pluto, would be measured in meters.
2. Possible answer: 1 AU = 10 cm

Use Science Words

Word Origins Ask students to do research and then write in their Science Journals about the mythological roots of the names *Pluto* and *Charon.* In Greek mythology, Charon was the boatman who ferried dead souls across the river Styx to Hades. In Roman mythology, Pluto was the god of the underworld. L2 LS **Linguistic**

Differentiated Instruction

Learning Disabled Help students achieve success on the Problem-Solving Activity by providing a simple scale of 1 AU = 1 cm to use in constructing their models of the solar system. This scale will allow students to construct a model with an overall size of less than a meter. It also avoids using a mathematical multiplier since the units are one-to-one. L1

SECTION 2 The Solar System 453

IDENTIFYING Misconceptions

Planets and Stars Students may think that planets and stars are about the same distance from Earth. Refer to page F at the beginning of the chapter for teaching strategies that address this misconception.

Discussion

Pluto Should Pluto be considered a planet like others in the solar system? Possible answer: Some students may say that Pluto is so small, it should be considered a moon. In fact, Pluto is much smaller than the other planets, and its moon is very large in relation to Pluto. Based on their sizes and other data, some scientists hypothesize that Pluto and Charon could be large members of the Kuiper belt of comets, located just beyond Pluto's orbit.

Teacher FYI

Comet Tails The closer a comet gets to the Sun, the longer its tail becomes. Often two tails are visible, one made of dust and one made of ions.

Reading Check

Answer Comets have tails when they are close to the Sun because ices vaporize, which releases gas and dust.

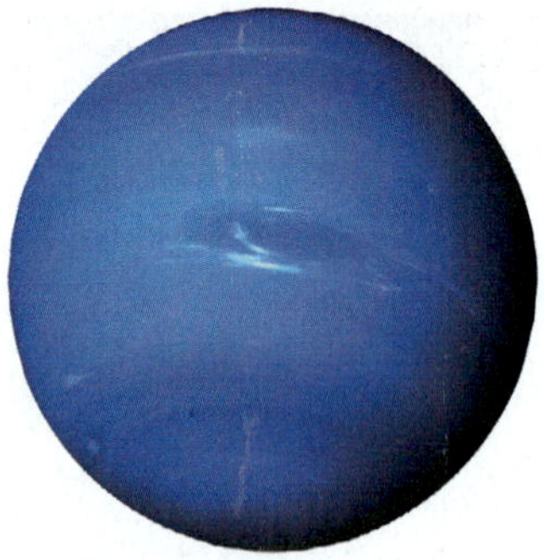
Neptune

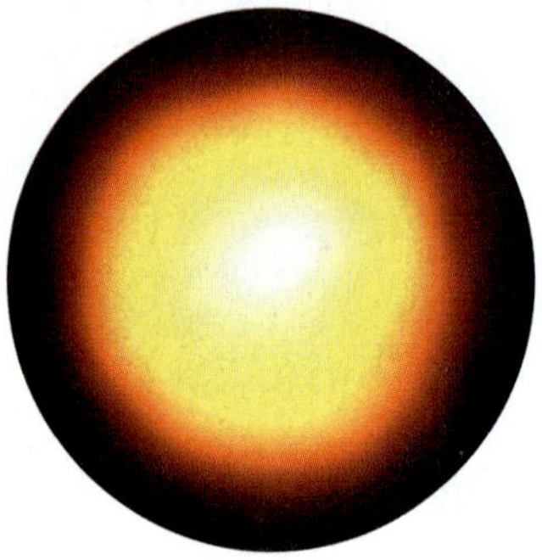
Pluto

Figure 16 The outermost planets are Neptune and Pluto. This is the best image available of Pluto, which has not yet been visited by spacecraft.

Neptune Neptune is the next stop in your space travel. Neptune, shown in **Figure 16,** is the eighth planet from the Sun. Neptune's atmosphere is composed of hydrogen, helium, and methane. The methane gives the planet a blue color. In 1989, *Voyager 2* sent pictures of Neptune showing a Great Dark Spot in its atmosphere. The spot was gone when observations were made using the *Hubble Space Telescope* in 1994. Neptune is the last of the big, gas planets with rings around it. It has 11 moons. Triton, the largest of these, has geysers that shoot gaseous nitrogen into space. The low number of craters on Triton indicates that lava still flows onto its surface.

Pluto The last planet that you come to on your tour is Pluto, a small, rocky planet with a frozen crust. Pluto was discovered in 1930 and is farthest from the Sun. It is the smallest planet in the solar system—smaller even than Earth's Moon—and the one scientists know the least about. It is the only planet in the solar system that has never been visited by a spacecraft. Pluto, shown in **Figure 16,** has one moon, Charon, which is nearly half the size of the planet itself.

Comets

A **comet** is a large body of ice and rock that travels around the Sun in an elliptical orbit. These objects are like dirty snowballs that often are between one and fifty kilometers across. Comets might originate in a cloud of objects far beyond the orbit of Pluto known as the Oort Cloud. This belt is 50,000 AU from the Sun. Some comets also originate in the Kuiper Belt, which lies just beyond the orbit of Neptune. As a comet approaches the Sun, solar radiation changes some of the ice into gas. Solar winds blow gas and dust away from the comet, forming what appears from Earth as a bright tail, shown in **Figure 17.**

Reading Check *Why do comets have tails?*

Figure 17 Solar wind is a stream of charged particles moving away from the Sun. Comet tails point away from the Sun because they are pushed by solar wind.

Teacher FYI

Neptune An eighth planet in our solar system was predicted in the 1840s by analyzing anomalies in Uranus's orbit. Neptune was later discovered amazingly close to its predicted position. Two scientists are credited with Neptune's discovery. Their names are John Couch Adams and Urbain Leverrier.

Meteorites Occasionally, chunks of extraterrestrial rock and metal fall to Earth. **Meteorites** are any fragments from space that survive their plunge through the atmosphere and land on Earth's surface. Small ones are no bigger than pebbles. The one in **Figure 18** has a mass of 14.5 metric tons. Hundreds of meteorites fall to Earth each year. Luckily, strikes on buildings or other human-made objects are rare. In fact, only a tiny fraction of the meteorites that fall are ever found. Scientists are extremely interested in those that are, because they yield important clues from space. For example, many seem to be about 4.5 billion years old, which provides a rough estimate of the age of the solar system. Several thousand meteorites have been collected in Antarctica, where moving ice sheets concentrate them in certain areas. Any rock seen on an ice sheet in Antarctica is probably a meteorite, because few other rocks are exposed. Meteorites can be one of three types—irons, stones, and stoney-irons. Irons are almost all iron, with some nickel mixed in. Stones are rocky. The rarest, stoney-irons, are a mixture of metal and rock.

Figure 18 This meteorite on display at the American Museum of Natural History in New York has a mass of 14.5 metric tons.
Explain *why meteorites are rare.*

section 2 review

Summary

Distances in Space

- The planets in the solar system orbit around the Sun.
- Distances in the solar system are vast. Scientists measure these distances using the astronomical unit (AU).

Inner Planets

- The inner planets are solid, rocky planets.
- Earth is the only planet known to support life.

Outer Planets

- Jupiter, Saturn, Uranus, and Neptune are gas giants that have ring systems.
- Pluto is a small planet made up of ice and rock.

Comets

- Comets are bodies of ice and rock that orbit the Sun.

Self Check

1. **Explain** why the planets and other objects in the solar system orbit around the Sun.
2. **List** the planets in the solar system in order starting with the planet that is closest to the Sun.
3. **Compare and contrast** the moons that were discussed in this chapter.
4. **Infer** why carbon dioxide ice exists on Mars but not Earth.
5. **Think Critically** Earth has abundant life. Do you think that other planets or moons might support life? If so, which ones? Which characteristics of the planets or moons might be conducive to life?

Applying Skills

6. **Compare and contrast** Earth to other planets in terms of size, composition, distance from the Sun, and surface features. You might want to make a table to record your data.

Science Online red.msscience.com/self_check_quiz

Caption Answer

Figure 18 Many pieces of extraterrestrial rock don't make it through Earth's atmosphere. In addition, a meteorite is not likely to strike land, since 71 percent of Earth's surface is water.

3 Assess

DAILY INTERVENTION

Check for Understanding

Visual-Spatial Provide students with colored pencils and ask them to draw each of the planets. Encourage students to make their drawings as accurate as possible.

Reteach

Book of Planets Have students make a book of planets listing various characteristics of each planet, such as size, number of moons, and speed of rotation. L1 LS **Linguistic**

Assessment

Process The United States is hoping to launch a crewed mission to Mars early in the twenty-first century. Have students research and write about the goals of the mission, as well as the obstacles NASA faces. Use **Performance Assessment in the Science Classroom,** p. 159.

section 2 review

1. The Sun's gravity holds the planets in their orbits.
2. Mercury, Venus, Earth, Mars, Jupiter, Saturn, Uranus, Neptune, Pluto
3. Students might consider the two moons of Mars; the four largest moons of Jupiter; Titan, a moon of Saturn; Charon, Pluto's moon; and Earth's moon.
4. Temperatures on Earth are generally much too warm for carbon dioxide ice to exist.
5. One characteristic that is essential for life is water. Some have suggested that microscopic life might exist or might have existed on Mars. Other possibilities include three moons of Jupiter (Ganymede, Callisto, and Europa) that might have oceans beneath their icy surfaces.
6. Check students' tables for organization and accuracy.

1 Motivate

Bellringer

INTERACTIVE CHALKBOARD PowerPoint® Presentations

Section Focus Transparencies also are available on the Interactive Chalkboard CD-ROM. L2 ELL

Section 3 Section Focus Transparency **Fiery Sun**

The star nearest Earth is our very own Sun. Scientists think the Sun is about 4.6 billion years old and that it will continue to shine for another five billion years or so. As far as stars go, the Sun is medium-sized.

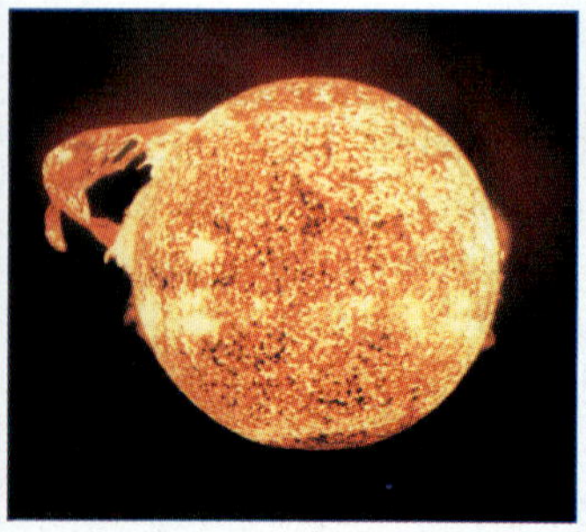

1. What would our Sun look like from a distant galaxy?
2. How do people group stars in the night sky?
3. Why do some stars appear brighter than others?

The Solar System and Beyond

Tie to Prior Knowledge

Constellations Prompt students to recall walking outside on a clear, dark night and seeing many stars. Ask them if they have ever identified patterns or shapes formed by groups of stars. Remind them of the Big Dipper, part of the constellation Ursa Major.

section 3

Stars and Galaxies

as you read

What You'll Learn

- **Explain** why stars appear to move across the sky.
- **Describe** some constellations.
- **Explain** the life cycle of stars.

Why It's Important

Understanding the vastness of the universe will help you to appreciate Earth's place in space.

Review Vocabulary
star: a large, spherical mass of gas that gives off light and other types of radiation; the Sun is a typical star

New Vocabulary
- constellation
- supernova
- galaxy
- light-year

Stars

Every night, a new world opens to you as the stars come out. Stars are always in the sky. You can't see them during the day because the Sun's light makes Earth's atmosphere so bright that it hides them. The Sun is a star, too. It is the closest star to Earth. Each night the stars appear to move across the sky. This happens because Earth is rotating around its axis. The stars that can be seen in the sky also change with the season as Earth revolves around the Sun.

Constellations Ursa Major, Ursa Minor, Orion, Taurus—do these names sound familiar? They are **constellations** (kahn stuh LAY shunz), or groups of stars that form patterns in the sky. **Figure 19** shows some constellations.

Constellations are named after animals, objects, and people—real or imaginary. Many of the names that early Greek astronomers gave to the constellations are still in use. However, throughout history, different groups of people have seen different things in the constellations. In early England, people thought the Big Dipper and the constellation Ursa Major looked like a plow. Native Americans saw a horse and rider. To the Chinese, it looked like a governmental official and his helpers moving on a cloud. What image does the Big Dipper bring to your mind?

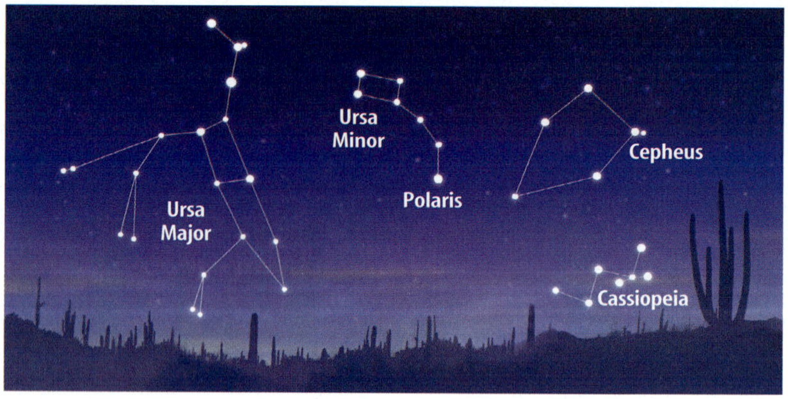

Figure 19 Find the Big Dipper in the constellation Ursa Major.
Explain *why people call it the Big Dipper.*

Section 3 Resource Manager

Chapter FAST FILE Resources
- Transparency Activity, p. 46
- Directed Reading for Content Mastery, pp. 21, 22
- MiniLAB, p. 4
- Enrichment, p. 32
- Reinforcement, p. 29
- Lab Worksheet, pp. 7–8

Home and Community Involvement, p. 31
Cultural Diversity, p. 51
Lab Management and Safety, p. 73

456 CHAPTER 15 The Solar System and Beyond

Starry Colors Although they look similar from Earth, stars are different colors. The color of a star is a clue about its temperature. Just as the red flames in a campfire are cooler, red stars are the coolest visible stars. Yellow stars have medium temperature. Bluish-white stars, like the blue flames on a gas stove, are the hottest.

Reading Check *How is a star's color related to its temperature?*

Stars also vary in size. Most of the stars in the universe are small. The Sun is a yellow, medium-sized star. Betelgeuse (BEE tul joos) is much bigger than the Sun. If this huge star were in the same place as the Sun, it would swallow Mercury, Venus, Earth, and Mars.

Apparent Magnitude Look at the sky on a clear night and you can easily notice that some stars are brighter than others. A system called apparent magnitude is used for classifying how bright a star appears from Earth. The dimmest stars that are visible to the unaided eye measure 6 on the apparent magnitude scale. A star with an apparent magnitude of 5 is 2.5 times brighter. The smaller the number is, the brighter the star is. The brightest star in the sky, Sirius, has an apparent magnitude of −1.5, and the Sun's apparent magnitude is −26.7.

Compared to other stars, the Sun is medium in size and temperature. It looks so bright because it is so close to Earth. Apparent magnitude is a measure of how bright a star looks from Earth but not a measure of its actual brightness, known as absolute magnitude. As **Figure 20** shows, a small, close star might look brighter than a giant star that is far away.

Mini LAB

Modeling Constellations

Procedure

1. Draw a dot pattern of a constellation on a piece of **black construction paper.** Choose a known constellation or make up your own.
2. With an adult's help, cut off the end of a **cardboard cylinder** such as an oatmeal box. You now have a cylinder with both ends open.
3. Place the cylinder over the constellation. Trace around the rim. Cut the paper along the traced line.
4. **Tape** the paper to the end of the cylinder. Using a **pencil,** carefully poke holes through the dots on the paper.
5. Place a **flashlight** inside the open end of the cylinder. Darken the room and observe your constellation on the ceiling.

Analysis

1. Turn on the overhead light and view your constellation again. Can you still see it? Why or why not?
2. The stars are always in the sky, even during the day. How is the overhead light similar to the Sun? Explain.

Try at Home

Figure 20 This flashlight looks brighter than the car headlights because it is closer. In a similar way, a small but close star can appear brighter than a more distant, giant star.

2 Teach

Caption Answer

Figure 19 It looks like a large dipper with a handle.

Reading Check

Answer Surface temperatures of stars decrease as colors go from the violet end of the spectrum toward the red end of the spectrum.

Mini LAB

Purpose Students investigate light diffusion. LS **Kinesthetic**

Materials black construction paper; round, empty cornmeal or oats box; scissors; flashlight; pencil; tape

Teaching Strategies Cut off the ends of the boxes for students.

Safety Precautions Caution students to handle scissors with care.

Analysis

1. No, the diffused overhead light on the ceiling is brighter than the flashlight forming the stars.
2. In the same way that the diffused overhead light is brighter than the stars formed by the flashlight, our Sun's light, diffused by Earth's atmosphere, is brighter than starlight.

Assessment

Content Constellations are named after animals, objects, and people—real and imaginary. Have students make a cartoon illustrating what their constellation represents. Use **Performance Assessment in the Science Classroom,** p. 133.

Try at Home

Differentiated Instruction

Learning Disabled Show students the constellation Orion. Explain that this constellation gets its name from a mythological Greek hunter. Have them draw the figure of a hunter over the constellation. Tell them to color the star to the upper left red and the one to the lower right blue. These are Betelgeuse and Rigel, respectively. Help them find these stars in a star atlas. L1 LS **Visual-Spatial**

Caption Answer

Figure 21 A white dwarf forms.

Teacher FYI

Stellar Life Cycles The way in which a star evolves throughout its life cycle is determined by its original mass. Stars with a mass similar to the Sun's do not become supernovas. They end as a black dwarf, a burned-out white dwarf. Stars many times more massive than the Sun can evolve through the supergiant stage, eventually ending up as a neutron star. If the star is very massive, it can become a black hole.

INTEGRATE Physics

Determining the Age of Stars The technique described on the student page uses the main sequence turn-off age. The luminosities and color values for stars in the cluster of stars are plotted. This plot is essentially a Hertzsprung-Russell diagram. The point on the diagram where main sequence stars "turn off" to become red giants is called the main sequence turn-off point. Because astronomers can model how long it takes for a star of particular mass to evolve to this point in its life cycle, the age of the star can be determined if the mass is known. The measured luminosity of a main sequence star is roughly proportional to the third power of the mass. Hence the mass can be determined, and the age of the star estimated.

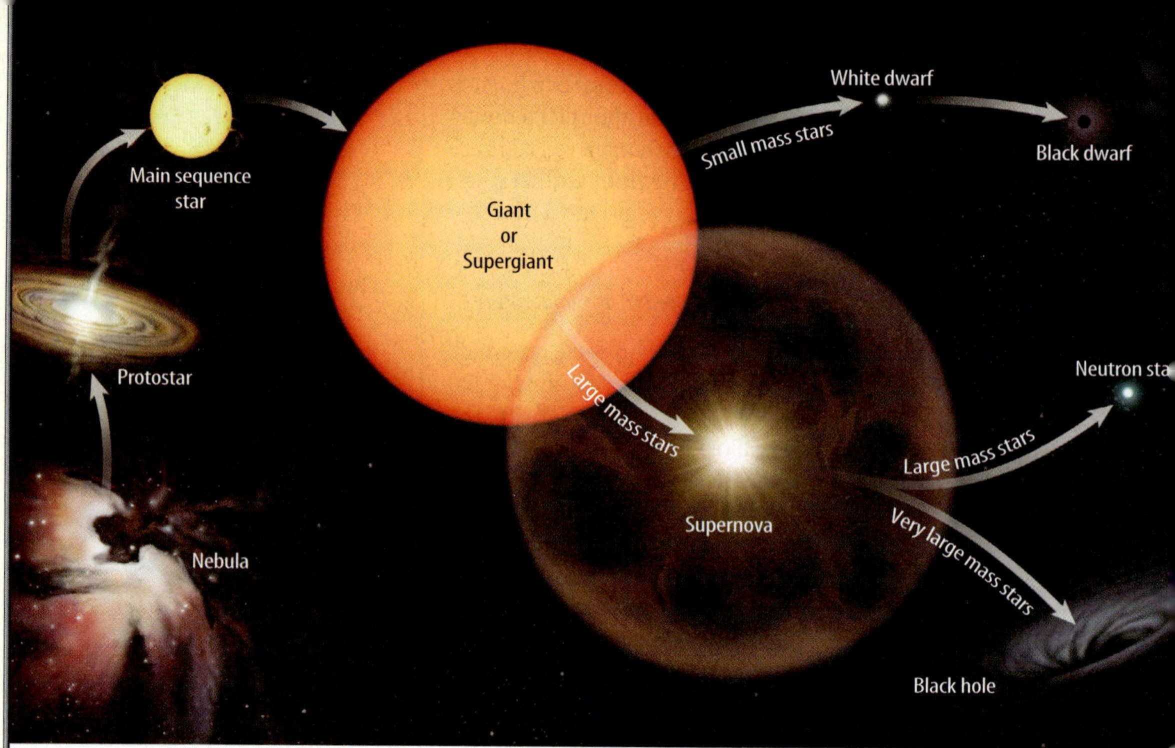

Figure 21 The events in the lifetime of a star depend on the star's mass.
Describe *what happens to giant stars when their cores collapse.*

INTEGRATE Physics

Determining the Age of Stars Some groups of stars, called clusters, contain stars that all formed at the same time. Scientists study these stars through a telescope to determine their color and brightness. This information can be used to find out how far along each star is in its life cycle. Because scientists know how long it takes for different stars to go through their life cycles, the age of the star cluster can be estimated.

The Lives of Stars

Scientists hypothesize that stars begin their lives as huge clouds of gas and dust. The force of gravity, which causes attraction between objects, causes the dust and gases to move closer together. When this happens, temperatures within the cloud begin to rise. A star forms when this cloud gets so dense and hot that the atoms within it merge. This process is known as fusion, and it changes matter to the energy that powers the star.

After a star has formed, it continues to evolve. When a medium-sized star like the Sun uses up some of the gases in its center, it expands to become a giant star. Giants are large, cool stars that have a red color. The Sun will become a giant in about five billion years. At this time, it will expand to cover the orbits of Mercury, Venus, and possibly Earth. It will remain that way for about a billion years. Then, the Sun will lose its outer shell, and the core will shrink to become a white dwarf star. A white dwarf is a hot, small star. Eventually, the white dwarf will cool and stop shining to become a black dwarf.

How long a star lives depends on how massive it is. Stars more massive than the Sun complete their life cycles in shorter amounts of time. The smallest stars shine the longest. **Figure 21** illustrates how the course of a star's life is determined by its size.

Science Journal

Stellar Evolution Have students research all stages in the life cycle of a star like the Sun. Ask them to make an illustration of the stages in their Science Journals. nebula—protostar—main sequence —giant—white dwarf—black dwarf L2 ELL LS **Kinesthetic and Visual-Spatial**

Differentiated Instruction

Challenge Have students illustrate what happens inside a star to cause a supernova. When a massive star uses much of its hydrogen, the core partially collapses. Heat released by the collapsing core initiates fusion reactions in the outer layers, forming heavier elements. Finally iron forms in the core. Iron does not fuel further fusion and the core collapses again. The core compresses, stops, and then rebounds. A huge shock wave moves outward through the star. L3 LS **Linguistic and Visual-Spatial**

Supergiants When a large star begins to use up the fuel in its core, it expands to become a supergiant. These stars are similar to giant stars, except they are much larger. Eventually, the core of the supergiant will collapse. A huge shock wave moves through the star, and the star explodes and becomes bright. This exploding star is called a **supernova.** For a few brief days, the supernova might shine more brightly than a whole galaxy. The dust and gas released by this explosion, shown in **Figure 22,** might eventually become part of a new star.

Figure 22 This photo shows the remains of a supernova.

INTEGRATE Physics Meanwhile, the core of the supergiant is still around. If the core isn't too large, it becomes a neutron star. These are small objects that are extremely dense. However, if the core is more than about three times as massive as the Sun, it collapses rapidly to form a black hole, shown in **Figure 23.** Light shone into a black hole disappears, and no light can escape from a black hole.

Galaxies

If you travel far from city lights and point a telescope toward the sky, you might see dim clumps of stars. These groups of stars are called galaxies. A **galaxy** is a group of stars, gas, and dust held together by gravity.

Types of Galaxies You now know how planets and stars differ from one another. Galaxies come in different shapes and sizes, too. The three major types of galaxies are elliptical, spiral, and irregular. These three types of galaxies are distinguished by their shapes. Elliptical galaxies are very common. They're shaped like huge footballs or spheres. Spiral galaxies have arms radiating outward from the center, somewhat like a giant pinwheel. As shown in **Figure 24,** some spiral galaxies have bar-shaped centers. Irregular galaxies are just that—irregular. They come in all sorts of different shapes and can't be classified easily. Irregular galaxies usually are smaller than other galaxies. They also are common.

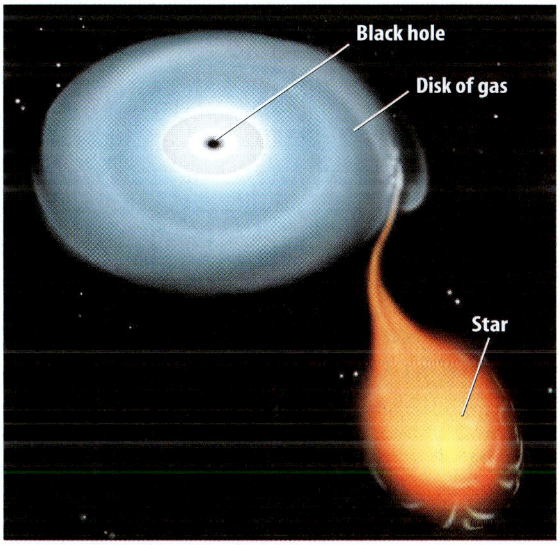

Figure 23 A black hole has such strong gravity that not even light can escape. This drawing shows a black hole stripping gas from a nearby star.
Explain *how black holes form.*

Caption Answer

Figure 23 During a supernova, if the star is massive enough, its core collapses and forms a black hole.

Make a Model

Spiral Galaxy Have students work in groups to make a model of a spiral galaxy, such as the Milky Way or the Andromeda Galaxy. L2 ELL COOP LEARN
LS **Interpersonal**

Activity

Doppler Effect Have students discuss how an approaching ambulance siren or train whistle sounds different from one that is moving away. The approaching siren or whistle has a higher pitch; as the sound passes the listener and moves away, the pitch drops. Explain that astronomers have studied analogous changes in galactic light. Then have them research and explain what these shifts are and what they tell us about galaxies. Possible answer: Light is shifted to the blue end of the spectrum when an object is approaching, and to the red end of the spectrum when it is moving away. Because light from galaxies outside the Local Group is shifted to the red end of the spectrum, astronomers have concluded that these galaxies are moving away from the Milky Way. L3 LS **Linguistic**

Curriculum Connection

Math Have students imagine that a star 1,750 light-years away is observed as a supernova in the year 2002. Have them determine when the supernova actually occurred. 2002 − 1,750 = 252; the star would actually have been a supernova in A.D. 252. L2 LS **Logical-Mathematical**

NATIONAL GEOGRAPHIC

Visualizing Galaxies

Have students examine the pictures and read the captions. Then ask the following questions.

If the members of the class were to arrange themselves in the shape of a spiral galaxy, how would they arrange themselves and where would most of them stand? Most would stand in the center of the group, with some extending out in pinwheel fashion.

How would you rearrange yourselves to represent an irregular galaxy? An elliptical galaxy? To represent an irregular galaxy, students could just arrange themselves randomly in the room; to represent an elliptical galaxy, they would cluster together to form an oval shape.

Activity

Modeling Galaxies Have small groups of students use black construction paper, white glue, and sand or salt to make models of the four shapes of galaxies. Ask students what each grain of salt or sand represents in their models. a star

NATIONAL GEOGRAPHIC VISUALIZING GALAXIES

Figure 24

Most stars visible in the night sky are part of the Milky Way Galaxy. Other galaxies, near and far, vary greatly in size and mass. The smallest galaxies are just a few thousand light-years in diameter and a million times more massive than the Sun. Large galaxies—which might be more than 100,000 light-years across—have a mass several trillion times greater than the Sun. Astronomers group galaxies into four general categories, as shown here.

▲ ELLIPTICAL GALAXIES They are nearly spherical to oval in shape and consist of a tightly packed group of relatively old stars.

▶ SPIRAL GALAXIES Spiral galaxies consist of a large, flat disk of interstellar gas and dust with arms of stars extending from the disk in a spiral pattern. The Andromeda Galaxy, one of the Milky Way Galaxy's closest neighbors, is a spiral galaxy.

▲ IRREGULAR GALAXIES Some galaxies are neither spiral nor elliptical. Their shape seems to follow no set pattern, so astronomers have given them the general classification of irregular.

◀ BARRED SPIRAL GALAXIES Sometimes the flat disk that forms the center of a spiral galaxy is elongated into a bar shape. Two arms containing many stars swirl out from either end of the bar, forming what is known as a barred spiral galaxy.

Differentiated Instruction

Challenge Ask students to research the relative frequency of the different categories of galaxies and identify which galaxies of each type are closest to the Milky Way. Have students write a brief report on their findings. L3

Teacher FYI

Milky Way The Milky Way is a large spiral galaxy that has a diameter of about 100,000 light-years. Spiral arms of young stars surround old stars near the center. Our Sun is located about halfway along one of the spiral arms. Astronomers think a super-massive black hole exists at the center of the Milky Way.

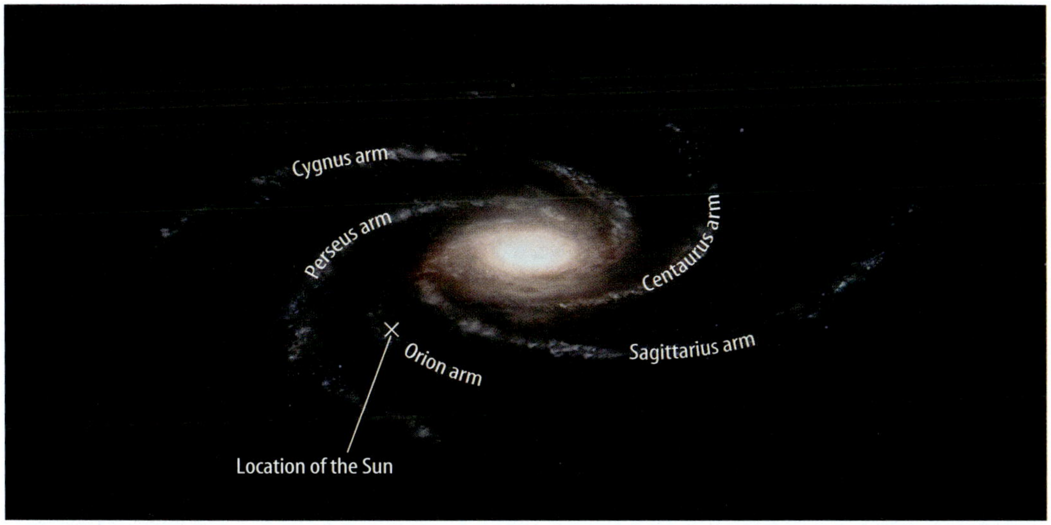

Figure 25 The Sun is located toward the edge of the Milky Way.

The Milky Way Galaxy Which type of galaxy do you live in? Look at **Figure 25.** You live in the Milky Way, which is a giant spiral galaxy. Hundreds of billions of stars are in the Milky Way, including the Sun. Just as Earth revolves around the Sun, stars revolve around the centers of galaxies. The Sun revolves around the center of the Milky Way about once every 225 million years.

A View from Within You can see part of the Milky Way as a band of light across the night sky. However, you can't see the whole Milky Way. To understand why, think about boarding a Ferris wheel and looking straight up. Can you really tell what the ride looks like? Because you are at the bottom looking up, you get a limited view. Your view of the Milky Way from Earth is like the view of the Ferris wheel from the bottom. As you can see in **Figure 26,** you can view only parts of this galaxy because you are within it.

Reading Check *Why can't you see the entire Milky Way from Earth?*

The faint band of light across the sky that gives the Milky Way its name is the combined glow of stars in the galaxy's disk. In 1609, when the Italian astronomer Galileo looked at the Milky Way with a telescope, he showed that the band was actually made of countless individual stars. The galaxy is vast—bigger and brighter than most of the galaxies in the universe. Every star you see in the sky with your naked eye is a member of the Milky Way Galaxy.

Figure 26 This is the view of the Milky Way from inside the galaxy. **Infer** *why it is called the Milky Way.*

Discussion

Light Pollution Astronomers have a difficult time studying faraway, faint objects such as galaxies when near a city. The reason is the amount of light in populated areas. Why might astronomers refer to this light as *pollution?* The extra light keeps them from seeing the sky clearly, just as water pollution can keep you from seeing a stream's bottom clearly. How might the problem of light pollution be solved? Possible answers: hooded lights can be used in parking lots and on street lights, people could reduce outside lighting or use special lights that give off a different color of light, for example, red. L2

Teacher FYI

An Expanding Universe The red shift (Doppler shift) of light from galaxies outside of the Local Group indicates they are moving away and is important evidence in support of the expansion of the universe.

Reading Check

Answer Earth is inside the Milky Way galaxy.

Use an Analogy

A View from Within Compare observing the Milky Way from Earth with looking toward the lights of a large city from one edge of town. When looking at a city, one can see a bright band of lights but it isn't possible to determine the shape of the urbanized area. In a similar way, the entire Milky Way Galaxy cannot be observed from Earth.

Caption Answer

Figure 26 The many stars in the galaxy form a diffuse band of white across the sky that looks like a road, or "way." Because it is white, it is called the Milky Way.

INTEGRATE Physics

Red Shift The universe is very large and is expanding. Edwin Hubble discovered that the light from most galaxies is shifted toward the red end of the spectrum because, as these galaxies are moving away from Earth, their light waves become "stretched out." This stretching is called the Doppler effect. Have students write reports on what they discover about Edwin Hubble, red and blue shifts, and the Doppler effect. L3 LS **Linguistic** P

Inquiry Lab

Refraction

Purpose to demonstrate that the speed of light varies in different materials

Possible Materials flashlight, prism, glass, water, lenses, parking lot on a hot, sunny day

Estimated Time two class sessions

Teaching Strategies

- The speed of light is a constant 300,000 km/s in a vacuum, but light has a slower speed when traveling through matter. Evidence of the changing speed is provided by light refraction phenomena.
- Ask students to develop demonstrations that show that light travels at different speeds in different materials. Check their plans for safety before they begin.
- Students might refract light using a prism, a glass of water, or observe light refraction in air of differing densities above a hot, asphalt parking lot.

For additional inquiry activities, see *Science Inquiry Labs.*

Reading Check

Answer Distances between galaxies are vast—much larger than distances within the solar system—so a larger unit of measurement is needed.

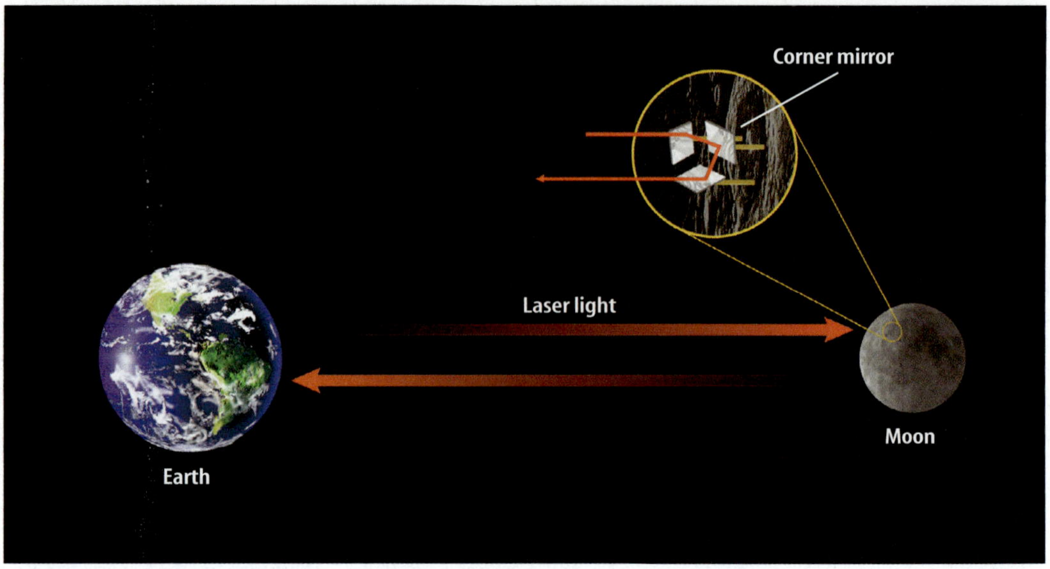

Figure 27 The constant speed of light through space helps astronomers in many ways. For example, the distance to the Moon has been determined by bouncing a laser beam off mirrors left by *Apollo 11* astronauts.

Speed of Light The speed of light is unique. Light travels through space at about 300,000 km/s—so fast it could go around Earth seven times in 1 s. You can skim across ocean waves quickly on a speedboat, but no matter how fast you go, you can't gain on light waves. It's impossible to go faster than light. Most galaxies are moving away from the Milky Way and a few are moving closer, but the light from all galaxies travels toward Earth at the same speed. The constant speed of light is useful to astronomers, as shown in **Figure 27.**

Light-Years Earlier you learned that distances between the planets are measured in astronomical units. However, distances between galaxies are vast. Measuring them requires an even bigger unit. Scientists often use light-years to measure distances between galaxies. A **light-year** is the distance light travels in one year—about 9.5 trillion km.

Reading Check *Why is a light-year better than an astronomical unit for measuring distances between galaxies?*

Would you like to travel back in time? In a way, that's what you're doing when you look at a galaxy. The galaxy might be millions of light-years away. The light that you see started on its journey long ago. You are seeing the galaxy as it was millions of years ago. On the other hand, if you could look at Earth from this distant galaxy, you would see events that happened here millions of years ago. That's how long it takes the light to travel the vast distances through space.

INTEGRATE Physics

Red Shift The Milky Way belongs to a cluster of galaxies called the Local Group. Scientists have determined that galaxies outside of the Local Group are moving away from Earth. Based on this, what can you infer about the size of the universe? Research the phenomenon known as red shift and describe to the class how it has helped astronomers learn about the universe.

Science Journal

Distances in the Universe Have students research the distances to Betelgeuse, Deneb, and the Large Magellanic Cloud. Instruct them to select a scale that will enable them to illustrate their distances from Earth. Have them place their illustrations in their Science Journals. Betelgeuse, 490 ly; Deneb, 1,400 ly; Large Magellanic Cloud, 169,000 ly L3 LS **Logical-Mathematical and Kinesthetic**

The Universe

Each galaxy contains billions of stars. Some might have as many stars as the Milky Way, and a few might have more. As many as 100 billion galaxies might exist. All these galaxies with all of their countless stars make up the universe.

Look at **Figure 28.** The *Hubble Space Telescope* spent ten days in 1995 photographing a tiny sector of the sky to produce this image. More than 1,500 galaxies were discovered. Astronomers think a similar picture would appear if they photographed any other sector of the sky. In this great vastness of exploding stars, black holes, star-filled galaxies, and empty space is one small planet called Earth. If you reduced the Sun to the size of a period on this page, the next-closest star would be more than 16 km away. Earth looks even lonelier when you consider that the universe also seems to be expanding. Most other galaxies are moving away at speeds as fast as 20,000 km/s. In relation to the immensity of the universe, Earth is an insignificant speck of dust. Could it be the only place where life exists?

Figure 28 The *Hubble* Deep Field Image shows hundreds of galaxies in one tiny sector of the sky.
Explain *what this image indicates about the sky.*

Reading Check *How do other galaxies move relative to Earth?*

section 3 review

Summary

Stars
- Constellations consist of stars that form patterns in the sky.

The Lives of Stars
- Stars evolve through time.
- How stars evolve depends on how massive they are.

Galaxies
- A galaxy is a group of stars, gas, and dust held together by gravity.
- You live in the Milky Way Galaxy.

The Universe
- The universe might include 100 billion galaxies.

Self Check

1. **Explain** why stars appear to move across the sky each night. Why are some stars only visible during certain seasons?
2. **List and describe** some constellations.
3. **Describe** the life cycle of a star like the Sun.
4. **Think Critically** Some stars might no longer be in existence, but you still see them in the night sky. Why?

Applying Math

5. **Convert Units** A light-year is about 9.5 trillion kilometers. Alpha Centauri is a star that is 4.3 light-years from Earth. How many trillion kilometers away is this star?

Science Online red.msscience.com/self_check_quiz

section 3 review

1. Stars move across the sky each night because Earth rotates around its axis. Earth's revolution causes some stars to only be visible during certain seasons.
2. Some constellations include Ursa Major, Ursa Minor, Cepheus, and Cassiopeia. See **Figure 19** to assess student descriptions.
3. A star like the Sun forms in a star-forming nebula and becomes a main-sequence star. After billions of years, the star will become a giant. Eventually, the giant will expel some gas and collapse into a white dwarf. The white dwarf will cool and become a dark, black dwarf.
4. Stars are so far away that the emitted light takes many years to many millions of years to reach Earth.
5. 40.85 trillion km

Visual Learning

Figure 28 What advantages might there be for placing instruments such as the *Hubble Space Telescope* in orbit? In orbit, the instrument is not affected by clouds, light, or distortions caused by the atmosphere.

3 Assess

DAILY INTERVENTION

Check for Understanding

Interpersonal Divide the class into small groups. Ask each group to volunteer a fact about stars in turn. Continue the process as long as student groups can provide new facts.

Reteach

Types of Galaxies Have pairs of students make flash cards of the three types of galaxies. Have them draw the galaxy on one side and write its type on the other. Student pairs can then test each other using the flash cards. L1 COOP LEARN LS **Interpersonal**

Assessment

Performance Have students write and perform a skit explaining the relationships among stars, solar systems, galaxies, and the universe. Use **Performance Assessment in the Science Classroom,** p. 147.

Caption Answer

Figure 28 There are a large number of galaxies in the entire sky.

Reading Check

Answer Most other galaxies are moving away from Earth.

BENCH TESTED

Real-World Question

Purpose Students will hypothesize how conditions on a planet affect the type of space colony that could be built there. L2 COOP LEARN IS **Logical-Mathematical**

Process Skills design an experiment, form a hypothesis, infer, classify, communicate, compare and contrast

Time Required one class period

Materials Reference books for students include: *The Planets in Our Solar System* by Franklyn M. Branley, HarperCollins, 1998; *The Planets* by David McNab and James Younger, Yale University Press, 1999; *The Incredible Journey to the Planets*, by Nicholas Harris, Peter Bedrick Books, 1999; and *Across the Solar System*, by Rod Theodorou, Heinemann Library, 2000.

Form a Hypothesis

Possible Hypothesis Most student hypotheses will reflect the need for oxygen, water, protection/shelter from extreme heat and/or cold, and food sources.

Test Your Hypothesis

Possible Procedures List all the planet's surface features and conditions. Identify ways to protect humans from features and conditions that would not allow them to survive. Design structures and equipment that would allow humans to survive on the planet. Draw exterior and interior views of the space colony. Include labels and explanations on the drawings.

LAB Design Your Own

Space Colony

Goals
- **Infer** what a space colony might look like on another planet.
- **Classify** planetary surface conditions.
- **Draw** a space colony for a planet.

Possible Materials
drawing paper
markers
books about the planets

Real-World Question

Many fictional movies and books describe astronauts from Earth living in space colonies on other planets. Some of these make-believe societies seem far-fetched. So far, humans haven't built a space colony on another planet. However, if it happens, what would it look like?

Form a Hypothesis

Research a planet. Review conditions on the surface of the planet. Make a hypothesis about the things that would have to be included in a space colony to allow humans to survive on the planet.

Teaching Strategy You might want to have students read about the problems faced by researchers in the Biosphere 2 project for additional information on problems that might be faced by a space colony.

Alternative Inquiry Lab

Long Space Flights To extend the lab, ask students to consider the demands of long space flights to the planets. They then can plan the space-flight mission and design the capsule.

Using Scientific Methods

Test Your Hypothesis

Make a Plan

1. Select a planet and study the conditions on its surface.
2. **Classify** the surface conditions in the following ways.
   - **a.** solid or gas
   - **b.** hot, cold, or a range of temperatures
   - **c.** heavy atmosphere, thin atmosphere, or no atmosphere
   - **d.** bright or dim sunlight
   - **e.** unique conditions
3. **List** the things that humans need to survive. For example, humans need air to breathe. Does your planet have air that humans can breathe, or would your space colony have to provide the air?
4. Make a table for the planet showing its surface conditions and the features the space colony would have to have so that humans could survive on the planet.
5. **Discuss** your decisions as a group to make sure they make sense.

Follow Your Plan

1. Make sure your teacher approves your plan before you start.
2. **Draw** a picture of the space colony. Draw another picture showing the inside of the space colony. Label the parts of the space colony and explain how they aid in the survival of its human inhabitants.

Analyze Your Data

1. **Compare and contrast** your space colony with those of other students who researched the same planet you did. How are they alike? How are they different?
2. Would you change your space colony after seeing other groups' drawings? If so, what changes would you make? Explain your reasoning.

Conclude and Apply

1. **Describe** the most interesting thing you learned about the planet you studied.
2. Was your planet a good choice for a space colony?
3. Would humans want to live on your planet? Why or why not?
4. Could your space colony be built using present technology? Explain.

Communicating Your Data

Present your drawing and your table to the class. Make a case for why your planet would make a good home for a space colony. **For more help, refer to the Science Skill Handbook.**

Expected Outcome Most drawings should show that students have planned ways to produce their own oxygen, obtain water, recycle their wastes, provide for protection from extremes in temperature, and produce food.

Analyze Your Data

Answers to Questions

1. Comparisons will depend on the planets chosen for the space colonies. All space colonies should be alike in that they provide for the basic needs for human survival. They will probably differ in how those needs are met.
2. Students may decide to make changes after seeing a better way to provide for a particular need of space colonists.

Error Analysis Have students examine their drawings to evaluate if there was any survival need that they did not consider or account for.

Conclude and Apply

1. Answers will depend on the planet chosen. If Mars was chosen, students might mention the possibility that water exists on the planet.
2. Some planets are better choices than others—those with solid surfaces and less extreme conditions are better.
3. Accept any answers that students can reasonably justify.
4. Probably not, since more detailed information would be needed and most planets present extremely hostile environments for humans.

Assessment

Content Have students work in groups of two or three to write a letter home to relatives on Earth describing what it's like to live in their space colony. Use **Performance Assessment in the Science Classroom,** p. 139.

Communicating Your Data

Students could use a word-processing program to prepare their table and the information they want to present for why their planet would be a good place for a space colony.

Understanding Literature

Cause and Effect The effect is that the girl and boy are given the duties of being the Sun and the Moon.

Respond to the Reading

1. The folktale was created to explain the people's observations of the Sun and Moon.
2. Sample answers: We know that the girl is modest because she shone brightly so that people could not look at her. We know that she is not willing to try new things—she asks to be the sun because she is "not familiar with the night." The king is fair. He made everyone work equally. He also listened to the girl's request and gave her a more fitting duty.
3. **Linking Science and Writing** Remind students to include cause-and-effect elements in their writing.

INTEGRATE Astronomy **Solar System Formation** You might want to take this opportunity to review with students about how the solar system formed. While the Sun formed at the center of the contracting cloud, matter condensed in its outer regions. Nearer the Sun, the matter consisted of elements and compounds that could condense at higher temperatures. In the outer reaches of the solar system, ices of water, ammonia, and methane also could condense. These solids collided under the influence of gravity to form planets, comets, and asteroids. The gas-giant planets formed when hydrogen and helium gas, which could not condense into solids, were gravitationally concentrated.

Science and Language Arts

The Sun and the Moon

A Korean Folktale

The two children lived peacefully in the Heavenly Kingdom, until one day the Heavenly King said to them, "We can not allow anyone to sit here and idle away the time. So I have decided on duties for you. The boy shall be the Sun, to light the world of men, and the girl shall be the Moon, to shine by night." Then the girl answered, "Oh King, I am not familiar with the night. It would be better for me not to be the Moon." So the King made her the Sun instead, and made her brother the Moon.

It is said that when she became the Sun, the people used to gaze up at her in the sky. But she was modest, and greatly embarrassed by this. So she shone brighter and brighter, so that is why the Sun is so bright, that her modesty might be forever respected.

Understanding Literature

Cause and Effect The folktale explains why the Sun and the Moon exist, as well as why you should never look directly at the Sun. No one is allowed to be idle in the Heavenly Kingdom. This is a cause. What is the effect?

Respond to the Reading

1. What was the purpose of this folktale?
2. What clues does the folktale give about the personalities of the girl and the King?
3. **Linking Science and Writing** Using the form of a folktale, explain what causes something that happens in the solar system.

INTEGRATE Astronomy The cause-and-effect relationships that astronomers use to explain the origin of the Sun and the Moon are different from what is told in this folktale. Astronomers hypothesize that the Sun formed from a collapsing cloud of ice, gas, and dust. As the cloud contracted, the temperature at its center became so high that nuclear fusion began, and the Sun was born. Many hypotheses about how the Moon formed have been suggested. The most favored of these suggests that the Moon formed from the matter that was blasted into space when a Mars-sized object struck Earth just after it formed.

Resources for Teachers and Students

Inquiry Into the Relation of Cause and Effect, by Thomas Brown, Scholars Facsimilies & Reprint, 1999

Older Brother, Younger Brother: A Korean Folk Tale, illustrated by Wengi Ma, Viking Penguin, 1997

Reviewing Main Ideas

Section 1 Earth's Place in Space

1. Day and night occur because Earth rotates around its axis.
2. Earth's axis is tilted about 23.5° from straight up.
3. Earth's revolution and Earth's tilted axis cause seasons to occur.

Section 2 The Solar System

1. The inner planets include Mercury, Venus, Earth, and Mars.
2. The outer planets are Jupiter, Saturn, Uranus, Neptune, and Pluto.
3. Meteorites are pieces of rock that fall to Earth from space.

Section 3 Stars and Galaxies

1. Apparent magnitude is a way to describe how bright stars appear from Earth. It is different from a star's actual brightness, or absolute magnitude.
2. Stars change throughout their lives. How they change depends on how massive they are.
3. There are four types of galaxies: elliptical, spiral, irregular, and barred spiral.

Visualizing Main Ideas

Copy and complete the concept map below using the following terms: asteroid belt, galaxy, universe, inner planets, comets and meteorites, *and* outer planets.

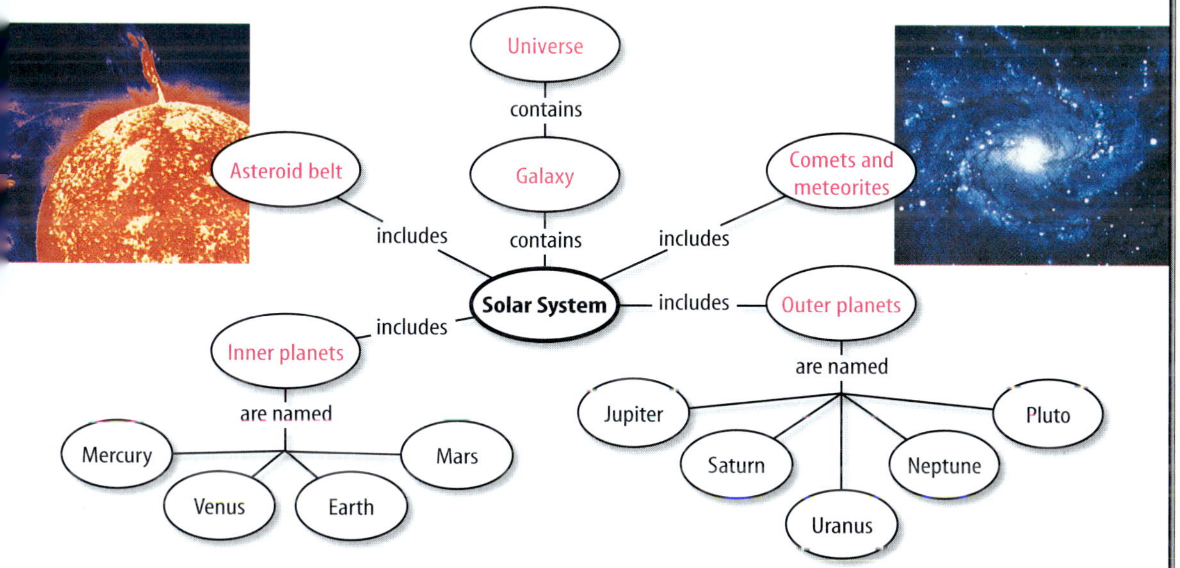

Identifying Misconceptions

Assess

Use this assessment as follow-up to page F at the beginning of the chapter.

Procedure Visit the Glencoe Science Web site to find information about the movement of the planets around the Sun and the apparent movement of planets and stars in the sky. Have each student select a planet and a star and graph and analyze the data about the objects chosen.

Expected Outcome Students will recognize that planets are much closer to Earth than stars are. They will also recognize that we can determine whether or not an object in the sky is a planet or star, based partly on its motion.

chapter 15 Study Guide

Reviewing Main Ideas

Summary statements can be used by students to review the major concepts of the chapter.

Visualizing Main Ideas

See student page.

Science online

Visit red.msscience.com
/self_check_quiz
/interactive_tutor
/vocabulary_puzzlemaker
/chapter_review
/standardized_test

Assessment Transparency

For additional assessment questions, use the *Assessment Transparency* located in the transparency book.

Assessment

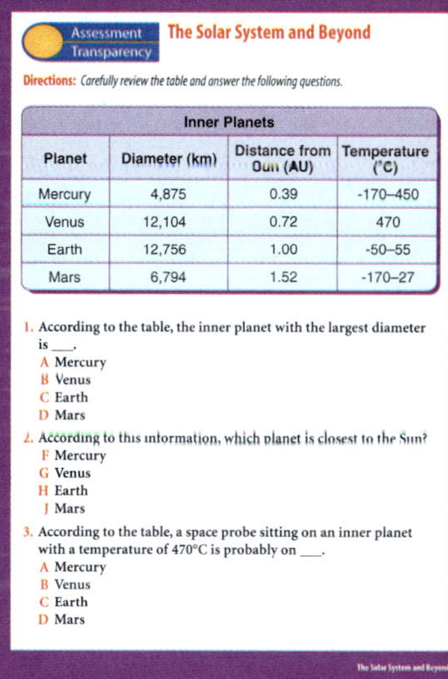
Assessment Transparency **The Solar System and Beyond**

Directions: *Carefully review the table and answer the following questions.*

Inner Planets			
Planet	Diameter (km)	Distance from Sun (AU)	Temperature (°C)
Mercury	4,875	0.39	-170–450
Venus	12,104	0.72	470
Earth	12,756	1.00	-50–55
Mars	6,794	1.52	-170–27

1. According to the table, the inner planet with the largest diameter is ___.
   A Mercury
   B Venus
   C Earth
   D Mars
2. According to this information, which planet is closest to the Sun?
   F Mercury
   G Venus
   H Earth
   J Mars
3. According to the table, a space probe sitting on an inner planet with a temperature of 470°C is probably on ___.
   A Mercury
   B Venus
   C Earth
   D Mars

The Solar System and Beyond

Using Vocabulary

1. eclipse
2. rotation
3. galaxy
4. constellation
5. revolution

Checking Concepts

6. D	11. A
7. B	12. C
8. B	13. D
9. B	14. A
10. D	15. B

chapter 15 Review

Using Vocabulary

astronomical unit p. 449	meteorite p. 455
comet p. 454	orbit p. 441
constellation p. 456	revolution p. 441
eclipse p. 444	rotation p. 440
galaxy p. 459	solar system p. 448
light-year p. 462	supernova p. 459
lunar highlands p. 442	tides p. 445
maria p. 442	

Each question below asks about a vocabulary word from the list. Write the word that best answers each question.

1. What event occurs when Earth's shadow falls on the Moon or when the Moon's shadow falls on Earth?
2. Which Earth motion causes day and night?
3. What is a large group of stars, gas, and dust held together by gravity called?
4. What is a group of stars that forms a pattern in the sky called?
5. Which movement of Earth causes it to travel around the Sun?

Checking Concepts

Choose the word or phrase that best answers the question.

6. What is caused by the tilt of Earth's axis and its revolution?
   A) eclipses **C)** tides
   B) phases **D)** seasons
7. Which of the following is caused by the gravity of the Moon and the Sun?
   A) stars **C)** comets
   B) tides **D)** maria
8. An astronomical unit equals the distance from Earth to which of the following?
   A) the Moon **C)** Mercury
   B) the Sun **D)** Pluto

Use the photo below to answer question 9.

9. Why is Earth, shown above, a unique planet in the solar system?
   A) it is spherical
   B) it has surface oceans
   C) it has an elliptical orbit
   D) it is the largest planet
10. How many galaxies could be in the universe?
    A) 1 billion **C)** 50 billion
    B) 10 billion **D)** 100 billion
11. Which results from Earth's rotation?
    A) night and day **C)** phases
    B) summer and winter **D)** eclipses
12. What unit is often used to measure long distances in space, such as between stars and galaxies?
    A) kilometer **C)** light-year
    B) astronomical unit **D)** meter
13. How many planets are in the solar system?
    A) six **C)** eight
    B) seven **D)** nine
14. Which object's shadow travels across part of Earth during a solar eclipse?
    A) the Moon **C)** an asteroid
    B) the Sun **D)** a comet
15. If a star is massive enough, what can result after it produces a supernova?
    A) a galaxy **C)** a black dwarf
    B) a black hole **D)** a white dwarf

Science Online red.msscience.com/vocabulary_puzzlemaker

Use the ExamView® Pro Testmaker CD-ROM to:

- create multiple versions of tests
- create modified tests with one mouse click for inclusion students
- edit existing questions and add your own questions
- build tests aligned with state standards using built-in State Curriculum Tags
- change English tests to Spanish with one mouse click and vice versa

Thinking Critically

16. **Compare and Contrast** Which of the planets in the solar system seems most like Earth? Which seems most different from Earth? Explain your answers using facts about the planets.

17. **Predict** How might a scientist predict the day and time of an eclipse?

18. **Recognize Cause and Effect** Which of the Moon's motions are real? Which are apparent? Explain each.

19. **Make and Use Tables** Research the size, composition, and surface features of each planet. Show this information in a table. How do tables help you to organize information?

20. **Make a model** of a lunar or solar eclipse based on what you have learned about the Sun, the Moon, and Earth. Use simple classroom materials.

21. **Concept Map** Copy and complete the following concept map using the following terms: *full, red surface, corona, solar,* and *few.*

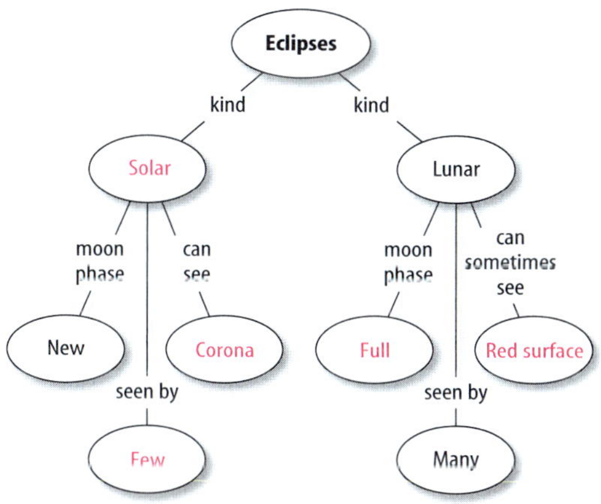

Performance Activities

22. **Model** Make a three-dimensional model showing the relative positions of the Sun, Earth, and the Moon during spring tides and neap tides. Which moon phase corresponds to each arrangement?

23. **Poster** Research the moons of Jupiter, Saturn, Uranus, or Neptune. Make a poster showing the characteristics of these moons. Display your poster for your class.

Applying Math

24. **Distances in the Solar System**
   - **a.** Venus is 0.72 AU from the Sun. Neptune is 30.07 AU from the Sun. How many times farther from the Sun is Neptune than Venus?
   - **b.** Jupiter is 5.20 AU from the Sun. Pluto is 39.48 AU from the Sun. How many times farther from the Sun is Pluto than Jupiter?

25. **Earth's Circumference** Earth's diameter at the equator is about 12,756 km. Using the equation $C = \pi d$, where C is circumference, d is diameter, and π is about 3.14, calculate Earth's circumference at the equator.

Use the graph below to answer question 26.

Light Curve of Delta Cephei

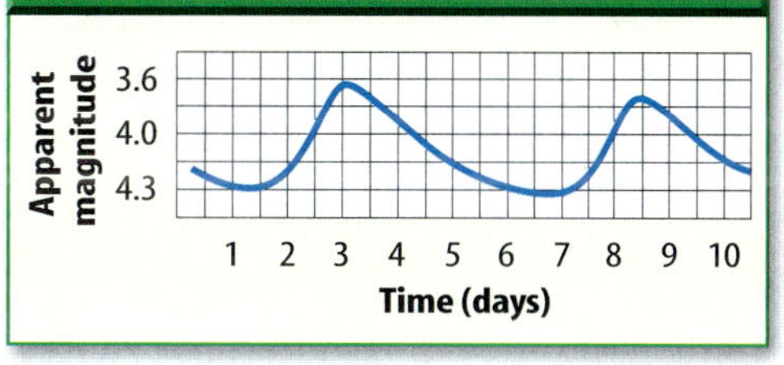

26. **Variable Stars** The brightness of some stars varies with time. The graph above shows how the apparent magnitude of a star named Delta Cephei varies with time. What is the period of this star's light curve? *Hint: The period is found by determining the time between apparent magnitude peaks.*

Assessment Resources

Reproducible Masters

Chapter *Fast File* Resources
Chapter Review, pp. 37–38
Chapter Tests, pp. 39–42
Assessment Transparency Activity, p. 49

Glencoe Science Web site
Chapter Review Test
Standardized Test Practice

Glencoe Technology

Assessment Transparency
Exam*View*® Pro Testmaker
MindJogger Videoquiz
Interactive Chalkboard

chapter Review 15

Thinking Critically

16. Answers will vary. Students might say that Mars is most like Earth because of its temperature and atmosphere. Jupiter might be students' choice as most different. It is a gas giant, has no firm surface, and has multiple moons.
17. Eclipses are possible only when the Moon is crossing Earth's orbital plane. By plotting the orbits of Earth and the Moon, scientists can predict the day and time of a solar eclipse.
18. The changing position of the Moon from night to night is a real motion because the Moon is orbiting Earth. The daily movement of the Moon westward across the sky is an apparent motion caused by Earth's rotation.
19. Tables categorize information in a logical and readable way.
20. Student models should illustrate alignment during the new moon and full moon phases.
21. See student page.

Performance Activities

22. Student models should show the Sun, Earth, and Moon in line at spring tides, and forming a right-angle at neap tides. Spring tides: new and full moon, neap tides, first quarter and third quarter. Use **PASC**, p. 123.
23. Students might consider surface conditions, atmosphere, interior, and orbital parameters.

Applying Math

National Math Standards

1, 2, 5, 9

24. **a.** 30.07AU ÷ 0.72AU = 41.8; Neptune is 41.8 times farther from the Sun than Venus.
    b. 39.48AU ÷ 5.20AU = 7.59
25. 40,053.8 km.
26. The period is 5.5 days. Accept answers that are close to this value.

FAST FILE

Answer Sheet A practice answer sheet can be found at red.msscience.com/answer_sheet.

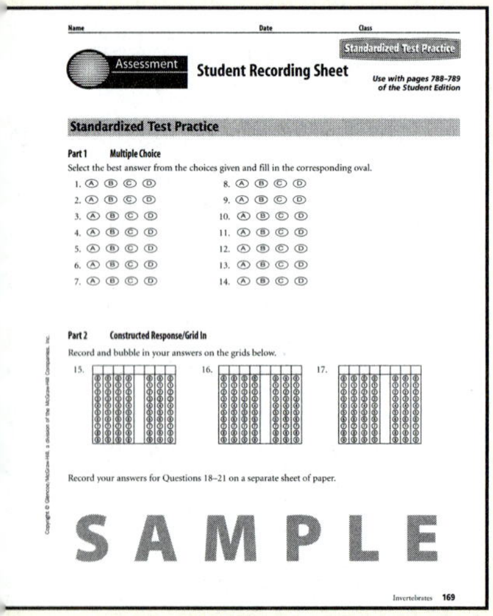
Name Date Class

Assessment Student Recording Sheet

Standardized Test Practice

Use with pages 788–789 of the Student Edition

Standardized Test Practice

Part 1 Multiple Choice

Select the best answer from the choices given and fill in the corresponding oval.

Part 2 Constructed Response/Grid In

Record and bubble in your answers on the grids below.

Record your answers for Questions 18–21 on a separate sheet of paper.

SAMPLE

Part 1 Multiple Choice

1. C
2. B
3. D
4. C
5. B
6. A
7. A
8. D
9. B
10. B

Part 2 Short Response

11. A star forms as a portion of a nebula of ice, gas, and dust collapses under the influence of gravity and fusion begins in its core.
12. Earth is a dense, rocky planet. It is unique because it has abundant surface water and supports life.

chapter 15 Standardized Test Practice

Part 1 Multiple Choice

Record your answers on the answer sheet provided by your teacher or on a sheet of paper.

Examine the photo of the Moon below. Then, answer questions 1–3.

1. Which term refers to the dark, smooth regions on the Moon's surface?
   A. lunar highlands **C.** maria
   B. craters **D.** rills
2. Which term refers to the light-colored, mountainous regions on the Moon?
   A. craters **C.** rills
   B. lunar highlands **D.** maria
3. What formed when meteorites struck the Moon's surface?
   A. lunar highlands **C.** plateaus
   B. maria **D.** craters
4. Which planet is closest to the Sun?
   A. Pluto **C.** Mercury
   B. Saturn **D.** Earth
5. What is composed of ice and rock?
   A. an asteroid **C.** a meteorite
   B. a comet **D.** Venus
6. Which of the following is largest?
   A. the universe **C.** the Sun
   B. the Milky Way **D.** the solar system
7. Which of these terms refers to an alternate rise and fall in sea level?
   A. tide **C.** eclipse
   B. revolution **D.** phase
8. Which of the following units is most useful for measuring distances in the solar system?
   A. kilometer **C.** meter
   B. light-year **D.** astronomical unit

Use the diagram below to answer questions 9–10.

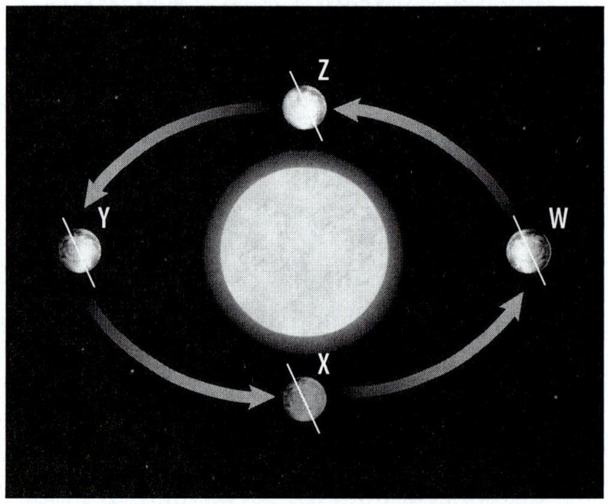

9. By which angle is Earth's axis tilted?
   A. 25° **C.** 15°
   B. 23.5° **D.** 27.5°
10. Assuming the top of the diagram represents north, which season occurs in the northern hemisphere when Earth is at position Z?
   A. spring **C.** summer
   B. autumn **D.** winter

Test-Taking Tip

Timing If you are taking a timed test, keep track of time during the test. If you find that you're spending too much time on a multiple-choice question, mark your best guess and move on.

13. Asteroids are rocky or metallic objects with varied compositions. Many asteroids occur in a belt between the orbits of Mars and Jupiter.
14. The stars in a galaxy are held together by gravity. Stars in a constellation often are not related to each other in space.
15. A neutron star consists of neutrons. It is a small, dense object. Black holes consist of matter that has contracted to a point.
16. Tides are a gravitational effect produced by the competing influence of the Sun and the Moon's orbit. Maximum variation in tides occurs when the Sun, Moon, and Earth are in linear alignment. Minimum variation in tides occurs when the Sun, Earth, and Moon form 90-degree angles.
17. Uranus's axis of rotation is tilted more than any other planet in the solar system, and is almost even with the plane of its orbit.
18. 5,727°C
19. 3,100K
20. 9,727°C
21. 273°C

Part 2 | Short Response/Grid In

Record your answers on the answer sheet provided by your teacher or on a sheet of paper.

11. Explain how stars form. Write your answer as several steps.

12. Describe the planet Earth. How is this planet different from the other planets in the solar system?

13. What is an asteroid? Where do many asteroids occur in the solar system?

14. Explain how constellations are different from galaxies.

15. How is a neutron star different from a black hole?

16. How do the Moon and the Sun cause tides to occur in Earth's oceans?

17. How is Uranus's axis of rotation different from the axis of rotation of most other planets?

The temperatures of stars usually are measured using the kelvin scale. Use the equations below to answer questions 18–21. °C represents degrees Celsius, and K represents the temperature in kelvins.

$$°C = K - 273$$

$$K = °C + 273$$

18. The surface temperature of the Sun is about 6,000 K. How many degrees Celsius is this?

19. The cool supergiant Betelgeuse has a surface temperature of about 2,827°C. How many kelvins is this?

20. A typical white dwarf star has a surface temperature of 10,000 K. What is its surface temperature in degrees Celsius?

21. The lowest possible temperature is called absolute zero. This temperature is 0 K. What is this temperature in degrees Celsius?

Part 3 | Open Ended

Record your answers on the answer sheet provided by your teacher or on a sheet of paper.

Use the illustration below to answer questions 22 and 23.

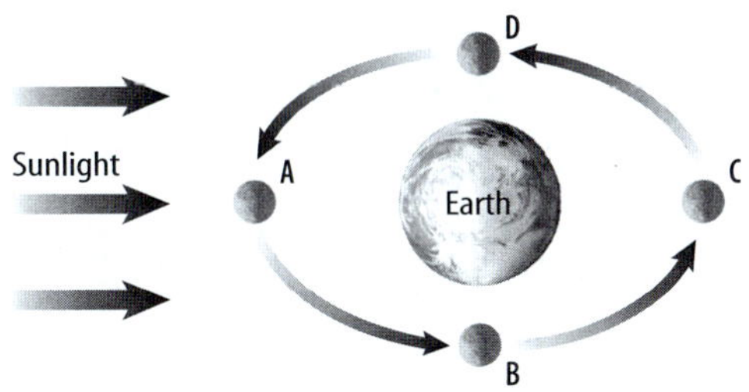

22. Identify each lunar phase shown as A–D in the diagram above.

23. Explain why the lunar phases change in a cycle.

24. Describe the shapes of the planets' orbits around the Sun. How is Pluto's orbit different from most other planets in the solar system?

25. How are the characteristics of the inner planets different from those of the outer planets?

26. Summarize the life cycle of a very large mass star. How is the life cycle of a very large mass star different from the life cycle of stars that have less mass?

27. Why does the same side of the Moon always face Earth?

28. Why do the stars appear to move in the sky each night?

29. Why are different stars visible during different seasons?

30. Explain how technology helps scientists learn about the universe. List several different types of technology that are used to study the universe in your answer.

27. because the Moon rotates and revolves around Earth at the same rate

28. because Earth rotates

29. Visibility of the night sky results from Earth's rotation, and seasons result from Earth's revolution. Differences in Earth's rotation and revolution make different portions of the night sky visible.

30. Technology allows magnified views of distant objects, provides close encounters with planets in the solar system, among other benefits. Students might include technology such as satellites and probes, the *International Space Station,* land-based telescopes, or the *Hubble Space Telescope.*

Rubrics

For more help evaluating open-ended assessment questions, see the rubric on p. 10T.

Part 3 | Open Ended

22. A: new moon, B: first quarter, C: full moon, and D: third quarter

23. The lunar phases change in a cycle because the Moon orbits Earth in about 29 days.

24. Planet's orbits are elliptical. With the exception of Pluto and Mercury, the ellipses are nearly circular. Aside from being more elliptical, Pluto's orbit is also the most distant from the Sun.

25. The inner planets are small, rocky, dense planets. Except for Pluto, the outer planets are gas giants.

26. The star forms in a star-forming nebula and becomes a main-sequence star. After depleting its hydrogen fuel, it becomes a supergiant. Eventually fusion reactions no longer can support the star. The core collapses during a supernova explosion. The collapsed core becomes a black hole. Stars with less mass will expand to become giant stars, and then shrink to become white dwarfs.

Unit Contents

WebQuest ***Origins of Birds*** will enable students to explore the evidence supporting the theory that birds descended from theropod dinosaurs. Students compare and contrast characteristics of birds and other groups of animals. Students will prepare a table of their new knowledge, and then use this information to answer the question: Are birds really dinosaurs?

unit 5 **Life's Diversity**

How Are Animals & Airplanes Connected?

PROJECT CRISS

Study Skills

Discussion Providing topics for discussion helps ensure that teams stay focused. At the end of the unit, group students for focused discussion in cooperative teams. Using a list of animals and their characteristics, students can determine the correct classification of each animal. Full-class discussion follows after the assignment.

NATIONAL GEOGRAPHIC

For thousands of years, people dreamed of flying like birds. Detailed sketches of flying machines were made about 500 years ago. Many of these machines featured mechanical wings that were intended to flap like the wings of a bird. But human muscles are not powerful enough to make such wings flap. Later, inventors studied birds such as eagles, which often glide through the air on outstretched wings. Successful gliders were built in the 1800s, but the gliders had no source of power to get them off the ground—and they were hard to control. Around 1900, two inventors studied bird flight more carefully and discovered that birds steer by changing the shape and position of their wings. The inventors built an engine-powered flying machine equipped with wires that could cause small changes in the shape and position of the wings. Though hardly as graceful as a soaring bird, the first powered, controlled flight took place in 1903, in the airplane seen here.

unit projects

Visit red.msscience.com/unit_project to find project ideas and resources.
Projects include:

- **History** Research Charles Darwin and his system for classifying animals. Write a time-travel interview to express your new knowledge.
- **Technology** Explore a biology-related career, and then write a want ad for a new job position.
- **Model** Study an animal, its characteristics, and habitat. Design a lunch bag with your new knowledge. A snack related to your animal may be placed inside to share with your classmates.

WebQuest Investigate *Origins of Birds* to learn about the theory that birds descended from theropod dinosaurs. Compare and contrast bird characteristics with other animals.

unit projects

History Have students research the life of Charles Darwin and his involvement in animal species classification. Students will then write a time-travel interview with Darwin to demonstrate their new knowledge.

Career Have students choose a biology-related career to explore. Information should include courses required, advanced degrees, job location and availability, experience, and salary. Students will then write a want ad for their new job candidate.

Model Assign each student a different animal to research. Students should create an imaginative, yet informative, lunch bag to educate their classmates about their specific animal. On the bag should be the physical characteristics, prey-predator web, reproduction and life cycle, scientific name and classification, natural habitat and location, and a relevant, fun, learning activity. A snack related to the specific animal may be placed inside the bag to share with classmates.

Additional Resources For more information, resources, and assessment rubrics, visit red.msscience.com/unit_project

NATIONAL GEOGRAPHIC

How Are Animals & Airplanes Connected?

Wings are complex, specialized adaptations. They enable birds to use a part of the environment not accessible to most other organisms. Their mastery of flight is reflected in the anatomical structures that allow them to modify the shape and angle of their wings. Their ability to steer and soar through the air provides them with a means of preying on other animals and of escaping predation themselves. It also allows them access to safe places for nesting. Other animals have developed wings that are different in structure, but these species achieve the same results as birds.

Section/Objectives	Standards		Labs/Features
	National	State/Local	
Chapter Opener	See pp. 16T–17T for a Key to Standards.		**Launch Lab:** Observe Onion Cells, p. 475 **Foldables,** p. 475 A data-collection lab using Probeware technology can be found in the **Probeware Lab Manual,** pp. 1–4
Section 1 The World of Cells 3 sessions 1.5 blocks 1. **Discuss** the cell theory. 2. **Identify** the parts of animal and plant cells. 3. **Explain** the purpose of different cell parts.	National Content Standards: UCP.1–UCP.3, UCP.5, A.1, A.2, C.1, C.3, C.5		**Science Online,** p. 478 **Integrate Chemistry,** p. 479 **MiniLAB:** Modeling a Cell, p. 480 **Lab:** Observing Algae, p. 482
Section 2 The Different Jobs of Cells 4 sessions 2 blocks 4. **Discuss** how different cells have different jobs. 5. **Explain** the difference among tissues, organs, and organ systems.	National Content Standards: UCP.1–UCP.3, UCP.5, A.1, A.2, C.1, C.3, C.5, F.5		**MiniLAB:** Analyzing Cells, p. 484 **Applying Math:** Calculating Numbers of Blood Cells, p. 485 **Integrate Career,** p. 485 **Visualizing Life's Organization,** p. 486 **Science Online,** p. 487 **Lab:** Water Movement in Plants, p. 488 **Science and Society:** Test Tube Tissue, p. 490

Glencoe Exclusive! TeacherWorks™ All-In-One Planner and Resource Center

Lab Materials	Reproducible Resources	Section Assessment	Technology
Launch Lab: index card, scissors, plastic wrap, onion skin, microscope slide, water, dropper, tape, metric ruler, newspaper	**Chapter *FAST FILE* Resources** Foldables Worksheet, p. 17 Directed Reading Overview, p. 19 Note-taking Worksheets, pp. 31–32	Glencoe's Assessment Advantage	TeacherWorks includes: • Interactive Teacher Edition • Lesson Planner with calendar • Access to all program blacklines • Correlations to standards • Web links Video Lab
MiniLAB: household materials (clay, yarn, cardboard, buttons, dry macaroni), glue, paper **Lab:** microscope, microscope slides, coverslips, large jars, pond water, algae, dropper, colored pencils *Need materials?* Contact Science Kit at 1-800-828-7777 or www.sciencekit.com on the Internet.	**Chapter *FAST FILE* Resources** Transparency Activity, p. 42 MiniLAB, p. 3 Enrichment, p. 29 Reinforcement, p. 27 Directed Reading, p. 20 Transparency Activity, pp. 45–46 Lab Worksheet, pp. 5–6 Lab Activity, pp. 9–12 **Home and Community Involvement,** p. 38	**Portfolio** Curriculum Connection, p. 477 Make a Model, p. 478 **Performance** MiniLAB, p. 480 Applying Skills, p. 481 **Content** Section Review, p. 481	Section Focus Transparency Teaching Transparency Virtual Labs CD-ROM Guided Reading Audio Program Interactive Chalkboard CD-ROM
MiniLAB: prepared slides of human cells, microscope **Lab:** fresh stalk of celery with leaves, clear drinking glass, scissors, red food coloring, water	**Chapter *FAST FILE* Resources** Transparency Activity, p. 43 MiniLAB, p. 4 Enrichment, p. 30 Reinforcement, p. 28 Directed Reading, pp. 21–22 Lab Worksheet, pp. 7–8 Lab Activity, pp. 13–16 **Mathematics Skill Activities,** p. 9 **Performance Assessment in the Science Classroom,** p. 61 **Lab Management and Safety,** p. 69	**Portfolio** Science Journal, p. 484 **Performance** MiniLAB, p. 484 Applying Math, p. 485 Applying Skills, p. 487 **Content** Section Review, p. 487	Section Focus Transparency Virtual Labs CD-ROM Guided Reading Audio Program Interactive Chalkboard CD-ROM Probeware Lab

End of Chapter Assessment

Glencoe's Assessment Advantage

Blackline Masters	Technology	Professional Series
Chapter *FAST FILE* Resources Chapter Review, pp. 35–36 Chapter Tests, pp. 37–40 **Standardized Test Practice,** pp. 67–70	MindJogger Videoquiz Virtual Labs CD-ROM Exam*View*® Pro Testmaker TeacherWorks CD-ROM Interactive Chalkboard CD-ROM	**Performance Assessment in the Science Classroom (PASC)**

Transparencies

Section Focus

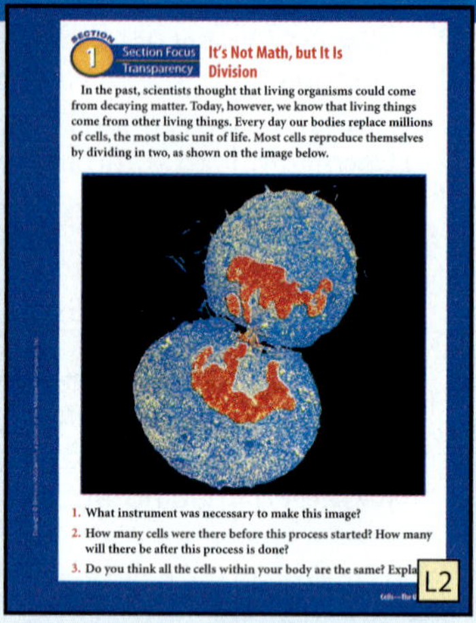
Section 1 Section Focus Transparency **It's Not Math, but It Is Division**

In the past, scientists thought that living organisms could come from decaying matter. Today, however, we know that living things come from other living things. Every day our bodies replace millions of cells, the most basic unit of life. Most cells reproduce themselves by dividing in two, as shown on the image below.

1. What instrument was necessary to make this image?
2. How many cells were there before this process started? How many will there be after this process is done?
3. Do you think all the cells within your body are the same? Explain.

L2

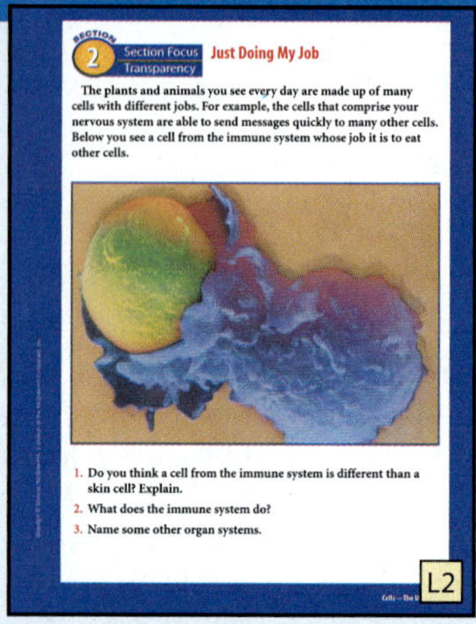
Section 2 Section Focus Transparency **Just Doing My Job**

The plants and animals you see every day are made up of many cells with different jobs. For example, the cells that comprise your nervous system are able to send messages quickly to many other cells. Below you see a cell from the immune system whose job it is to eat other cells.

1. Do you think a cell from the immune system is different than a skin cell? Explain.
2. What does the immune system do?
3. Name some other organ systems.

L2

This is a representation of key blackline masters available in the Teacher Classroom Resources. See Resource Manager boxes within the chapter for additional information.

Assessment

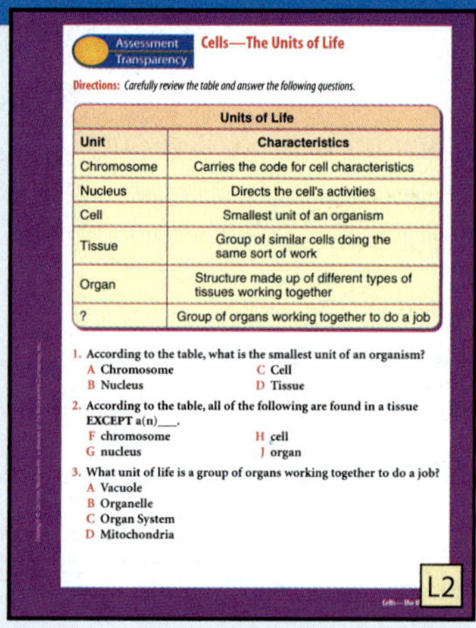
Assessment Transparency **Cells—The Units of Life**

Directions: *Carefully review the table and answer the following questions.*

Units of Life	
Unit	**Characteristics**
Chromosome	Carries the code for cell characteristics
Nucleus	Directs the cell's activities
Cell	Smallest unit of an organism
Tissue	Group of similar cells doing the same sort of work
Organ	Structure made up of different types of tissues working together
?	Group of organs working together to do a job

1. According to the table, what is the smallest unit of an organism?
   A Chromosome C Cell
   B Nucleus D Tissue
2. According to the table, all of the following are found in a tissue EXCEPT a(n) ___.
   F chromosome H cell
   G nucleus J organ
3. What unit of life is a group of organs working together to do a job?
   A Vacuole
   B Organelle
   C Organ System
   D Mitochondria

L2

Teaching

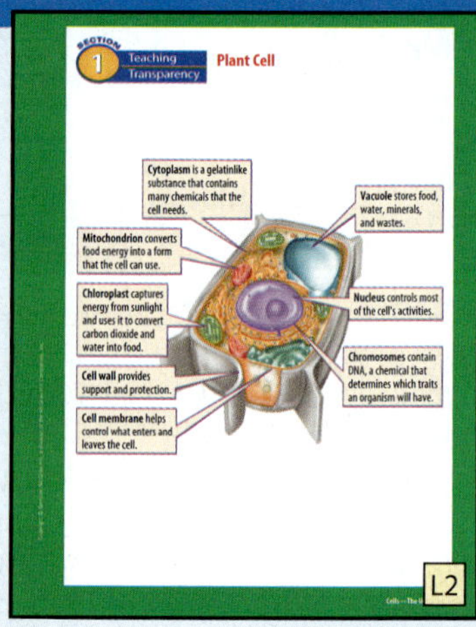
Section 1 Teaching Transparency **Plant Cell**

L2

Key to Teaching Strategies

The following designations will help you decide which activities are appropriate for your students.

- **L1** Level 1 activities should be appropriate for students with learning difficulties.
- **L2** Level 2 activities should be within the ability range of all students.
- **L3** Level 3 activities are designed for above-average students.
- **ELL** ELL activities should be within the ability range of English Language Learners.
- **COOP LEARN** Cooperative Learning activities are designed for small group work.
- **LS** Multiple Learning Styles logos, as described on page 12T, are used throughout to indicate strategies that address different learning styles.
- **P** These strategies represent student products that can be placed into a best-work portfolio.
- **PBL** Problem-Based Learning activities apply real-world situations to learning.

Hands-on Activities

Student Text Lab Worksheet

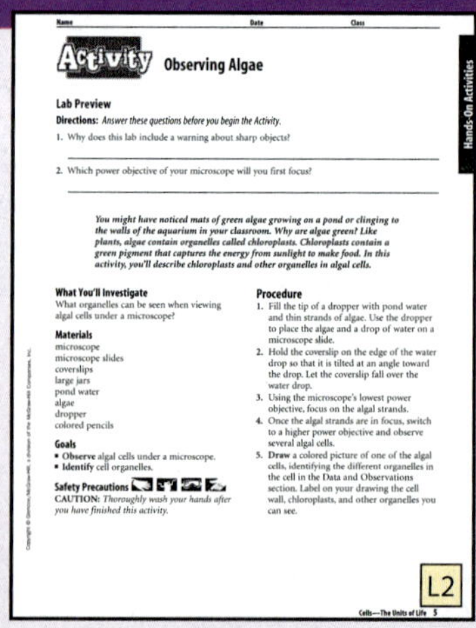
Activity **Observing Algae**

Lab Preview

Directions: *Answer these questions before you begin the Activity.*

1. Why does this lab include a warning about sharp objects?
2. Which power objective of your microscope will you first focus?

L2

Laboratory Activities

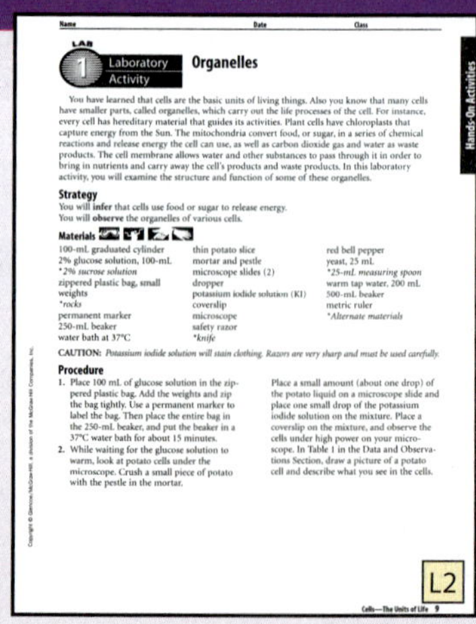
Laboratory Activity 1 **Organelles**

L2

Meeting Different Ability Levels

Content Outline

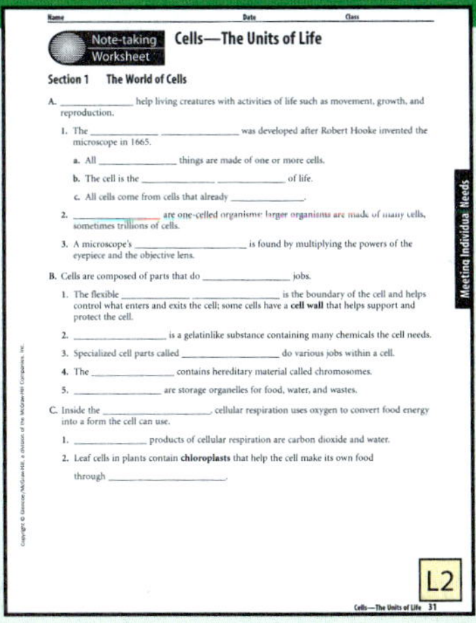

Reinforcement

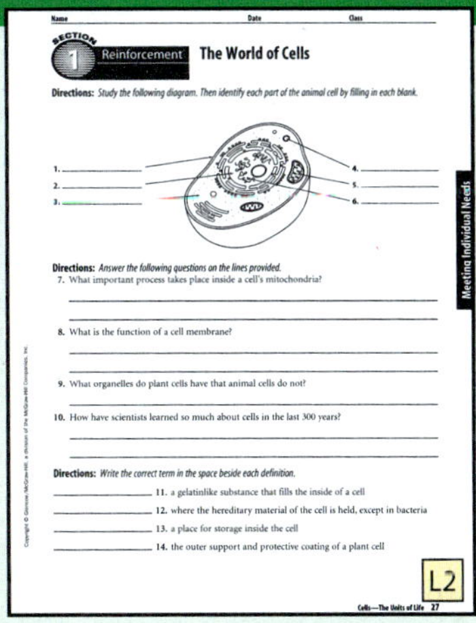

Enrichment

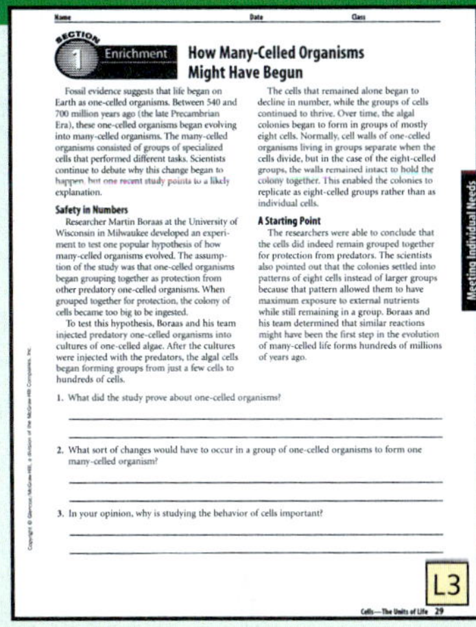

Directed Reading (English/Spanish)

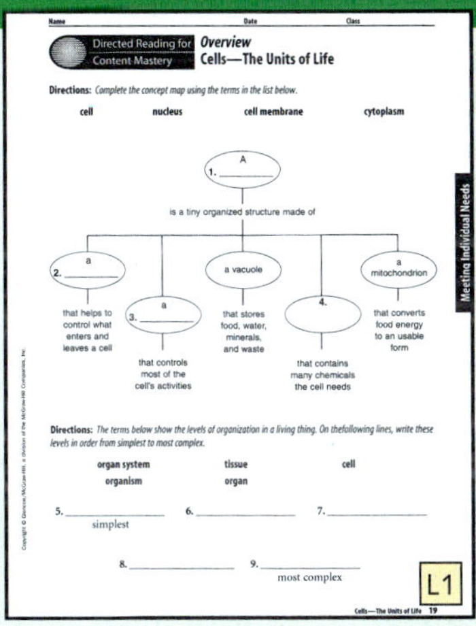

Study Guide

Reading Essentials

Assessment

Test Practice Workbook

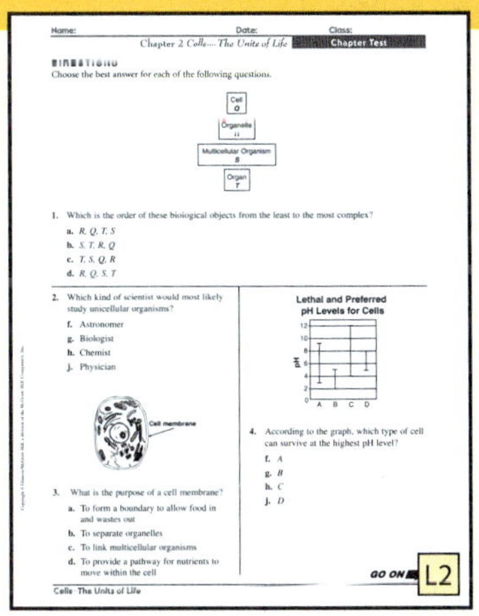

Chapter Review

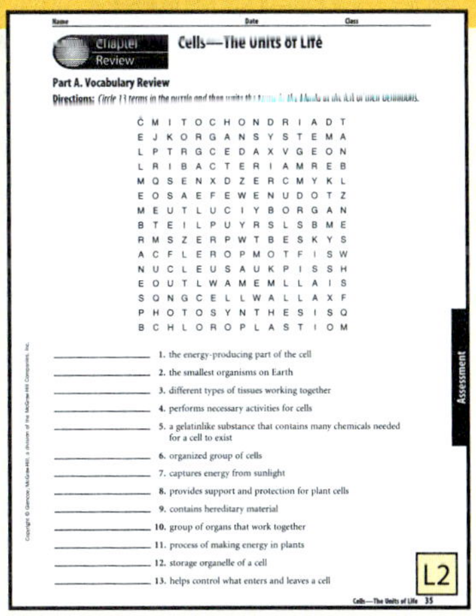

Chapter Tests

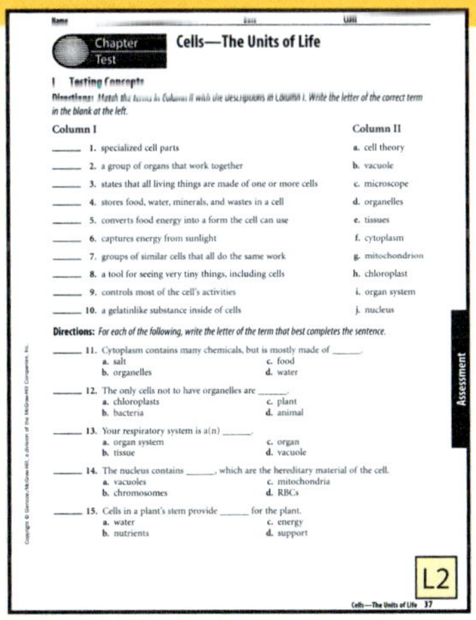

Science Content Background

section 1

The World of Cells

Cell Theory

Prior to the discovery of cells and the introduction of cell theory, people believed in the idea of spontaneous generation; that is, the idea that living things could spontaneously originate from nonliving material. For example, people believed that maggots could arise from decomposing organic material or that frogs and toads arose from rain or moist soil. Throughout the remainder of the seventeenth and eighteenth centuries, several scientists worked to disprove this idea. By the mid-1800s scientists had shown through a series of independent scientific experiments that living matter could not be spontaneously generated. It was not until this time that the concept of biogenesis, the idea that living cells can be produced only from already existing cells, and the other ideas of cell theory began to be widely accepted.

Outside the Cell

One of the functions of the cell membrane is to control what substances pass into and out of the cell. Substances can move across the cell membrane by several processes. Diffusion of materials across the cell membrane is driven by a concentration gradient. In this process, substances move from areas of high concentration to areas of low concentration until equilibrium is reached. The rate of diffusion is affected by the thickness of the cell membrane. The thinner the membrane, the less time it takes for a substance to diffuse across it. The membranes of the alveoli in the lungs, where oxygen and carbon dioxide are exchanged in breathing, are some of the thinnest in the body with a width of only 0.5 micrometers. Active transport is another way materials are moved across cell membranes. In active transport, energy is inputted to actively move materials against the concentration gradient. The sodium-potassium pump, which takes place in all body cells, is one example of active transport in which cells continually regulate the amount of sodium and potassium entering and leaving the cell.

Organelles

Plants contain organelles called plastids that are enclosed by a double membrane. Chloroplasts are one example of a plastid that stores the pigment chlorophyll and manufactures glucose. Leukoplasts store sugar, which is produced during photosynthesis in chloroplasts, in the form of starch. Chromoplasts store different types of plant pigments, including the orange, yellow, and red pigments that give some fruits their color.

chapter content resources

Internet Resources

For additional content background, visit **red.msscience.com** to:

- access your book online
- find references to related articles in popular science magazines
- access Web links with related content background
- access current events with science journal topics

Print Resources

The World of the Microscope, by Chris Oxlade and Corinne Stockley, EDCP, 1999

World of the Cell: Life on a Small Scale, by Robert Snedden, Heinemann Library, 2003

Cells, by Alvin Silverstein, Virginia B. Silverstein, and Laura Silverstein Nunn, Millbrook Press, 2002

Biology Projects for Young Scientists, Salvatore Tocci, Scholastic Library Publishing, 2000

section 2 The Different Jobs of Cells

Some Skin Cells

The epidermis, the upper layer of the skin, includes cells called keratinocytes. These cells manufacture the protein keratin. Keratin, which has a tough, fibrous consistency, provides protection from heat and chemicals. Keratinocytes also make skin waterproof. Melanocytes, another type of cell found in the skin, produce melanin, the pigment responsible for skin color. Melanin works to protect cells from overexposure to UV radiation by absorbing it as the rays penetrate the skin. Melanin granules gather over the nuclei of the skin cells in order to form a protective barrier and safeguard the DNA from the harmful rays.

Tissues, Organs and Organ Systems

Some organs perform functions for more than one organ system. The pancreas functions as part of the endocrine system; it also performs an essential role in the digestive system. As an endocrine gland, the pancreas produces and releases the two major hormones that control blood glucose levels—insulin and glucagon. Insulin lowers blood glucose level by assisting glucose in moving into body cells. Glucagon has the opposite effect; it functions to raise blood glucose level. In the digestive system the pancreas produces and releases into the small intestine powerful digestive enzymes that work to break down carbohydrates, proteins, and fats.

The spleen, although it is located in the abdominal cavity, is not an organ of the digestive system. Instead it is the largest mass of lymphatic tissue in the body. The lymphatic system performs a range of functions including producing lymphocytes, which are types of white blood cells that fight infection and produce antibodies. The lymphatic system also is involved in moving lipids and some vitamins to the blood, after absorbing them in the small intestines. The spleen, the lymph nodes and fluid, and red bone marrow compose a major portion of the lymphatic system.

The appendix is located in the lower right section of the abdominal cavity very near the junction of the small and large intestines. If it becomes inflamed or infected it must be removed in a procedure called an appendectomy. The exact function of the appendix is uncertain.

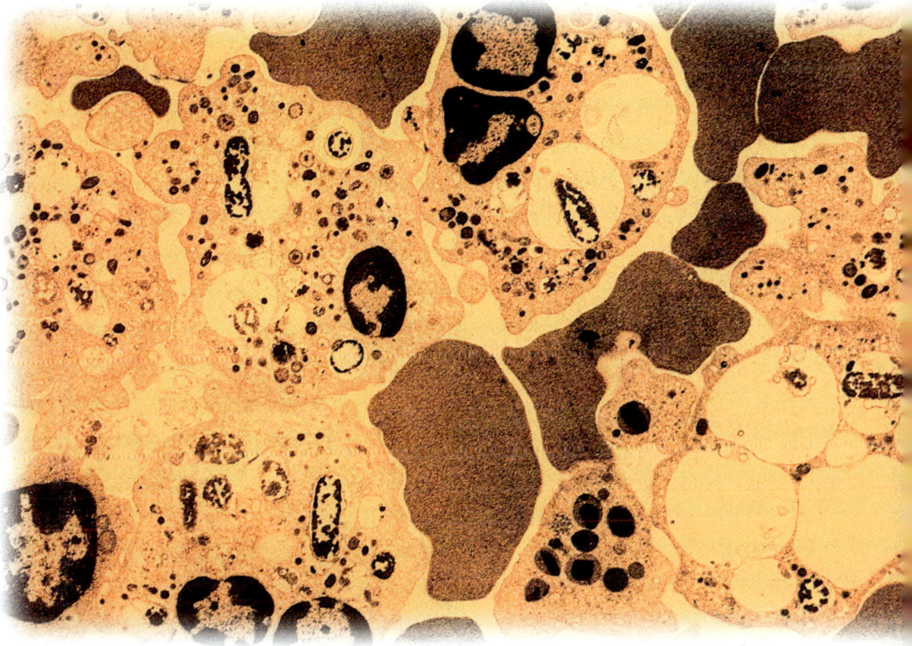

David M. Phillips/Visuals Unlimited

Teacher to Teacher
Steve Federman
Loveland Middle School
Loveland, Ohio

"I reinforce the concepts of cell structures and functions by having students create a cell from a light colored flavored gelatin mixed with unflavored gelatin (6-oz. package will make 5 cells). Pour into a small cup or Petri dish. Students then use items such as cake sprinkles for ribosomes, hot tamale candies for mitochondria, a gumball for the nucleus, and so on."

Steve Federman

chapter 16

Chapter Vocabulary

bacteria, p. 477
cell membrane, p. 479
cell wall, p. 479
cytoplasm, p. 479
organelle, p. 479
nucleus, p. 480
vacuole, p. 480
mitochondria, p. 480
photosynthesis, p. 481
chloroplast, p. 481
tissue, p. 487
organ, p. 487
organ system, p. 487

Science Journal Student responses will vary, but should include how each block has connectors that fit every other block to make building objects possible.

INTERACTIVE CHALKBOARD with Image Bank

PowerPoint® Presentations

This CD-ROM is an editable Microsoft® PowerPoint® presentation that includes:

- a pre-made presentation for every chapter
- interactive graphics
- animations
- audio clips
- image bank
- all new section and chapter questions
- Standardized Test Practice
- transparencies
- pre-lab questions for all labs
- Foldables directions
- links to red.msscience.com

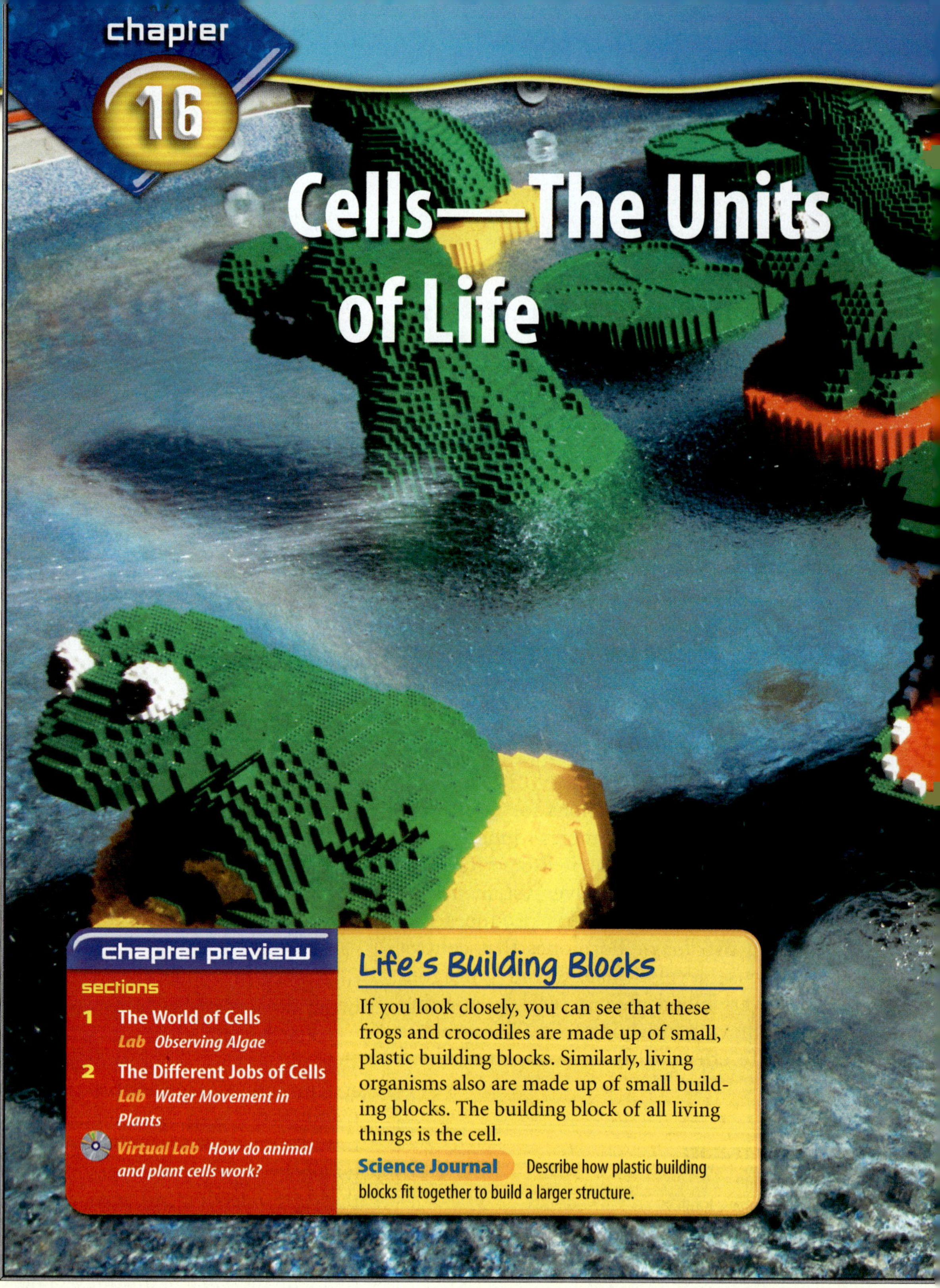

chapter 16

Cells—The Units of Life

chapter preview

sections

1. **The World of Cells**
   Lab *Observing Algae*
2. **The Different Jobs of Cells**
   Lab *Water Movement in Plants*

Virtual Lab *How do animal and plant cells work?*

Life's Building Blocks

If you look closely, you can see that these frogs and crocodiles are made up of small, plastic building blocks. Similarly, living organisms also are made up of small building blocks. The building block of all living things is the cell.

Science Journal Describe how plastic building blocks fit together to build a larger structure.

Theme Connection

Scale and Structure On the most basic level of organization, all living things are composed of microscopic units called cells. Cells carry out important processes for organisms, such as releasing energy and producing important chemicals such as hormones.

About the Photo

Building Blocks These frogs, lily pads, and crocodiles made of plastic building blocks replicate living organisms. Different sizes, shapes, and colors of blocks were used to create the lifelike animals and plants. In the same way, cells come in different shapes and sizes to perform specific functions in order to create an organism.

Start-Up Activities

Launch LAB

Observe Onion Cells

An active, organized world is inside you and in all other living things. Yet it is a world that you usually can't see with just your eyes. Make the magnifier in the lab below to help you see how living things are organized.

1. Cut a 2-cm hole in the middle of an index card. Tape a piece of plastic wrap over the hole.
2. Turn down about 1 cm of the two shorter sides of the card, then stand it up.
3. Place a piece of onion skin on a microscope slide, then put it directly under the hole in the card.
4. Put a drop of water on the plastic wrap. Look through the water drop and observe the piece of onion. Draw what you see.
5. **Think Critically** In your Science Journal, describe how the onion skin looked when viewed with your magnifier.

FOLDABLES Study Organizer

Compare Cells Make the following Foldable to help you see how plant and animal cells are similar and different.

STEP 1 **Fold** a vertical sheet of paper in half from top to bottom.

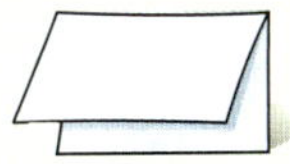

STEP 2 **Fold** in half from side to side with the fold at the top.

STEP 3 **Unfold** the paper once. **Cut** only the fold of the top flap to make two tabs. **Turn** the paper vertically and **draw** on the front tabs as shown.

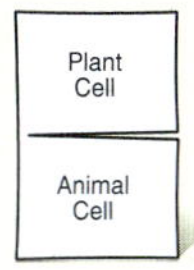

Read and Write Before you read the chapter, write what you know about each of these cells. As you read the chapter, add to or correct what you have written under the tabs. Compare and contrast the two types of cells.

Science Online — Preview this chapter's content and activities at red.msscience.com

Launch LAB

Purpose Students make and use a simple form of a microscope to examine onion skin cells. L2 ELL LS **Kinesthetic**

Preparation Cut tiny pieces from the thin, transparent skin of an onion for each student group.

Materials dropper, water, index card, colorless plastic wrap, scissors, tape, onion skin, metric ruler, newspaper, microscope slide

Teaching Strategies

- Be sure adequate light is available to illuminate the object being observed.
- You may wish to have students examine other plant tissues, such as leaves or roots of common plants.

Think Critically

The onion skin appears bricklike.

Assessment

Performance Have students use a microscope to observe the onion cells. Have them draw what they see and compare these drawings with the ones from the Launch Lab. Use **Performance Assessment in the Science Classroom,** p. 127.

FOLDABLES Study Organizer **Dinah Zike Study Fold**

Student preparation materials for this Foldable are available in the Chapter *Fast File* Resources.

Probeware Labs

This chapter's data-collection lab using Probeware technology is included on the *Video Labs CD-ROM.* See the *Probeware Lab Manual* pages 1–4 for student worksheets.

section 1

The World of Cells

as you read

***What* You'll Learn**

- **Discuss** the cell theory.
- **Identify** some of the parts of animal and plant cells.
- **Explain** the functions of different cell parts.

***Why* It's Important**

Cells carry out the activities of life.

Review Vocabulary

theory: an explanation of things or events based on scientific knowledge that is the result of many observations and experiments

New Vocabulary

- bacteria
- cell membrane
- cell wall
- cytoplasm
- organelle
- nucleus
- vacuole
- mitochondria
- photosynthesis
- chloroplast

Importance of Cells

A cell is the smallest unit of life in all living things. Cells are important because they are organized structures that help living things carry on the activities of life, such as the breakdown of food, movement, growth, and reproduction. Different cells have different jobs in living things. Some plant cells help move water and other substances throughout the plant. White blood cells, found in humans and many other animals, help fight diseases. Plant cells, white blood cells, and all other cells are alike in many ways.

Cell Theory Because most cells are small, they were not observed until microscopes were invented. In 1665, scientist Robert Hooke, using a microscope that he made, observed tiny, boxlike things in a thin slice of cork, as shown in **Figure 1.** He called them cells because they reminded him of the small, boxlike rooms called cells, where monks lived.

Throughout the seventeenth and eighteenth centuries, scientists observed many living things under microscopes. Their observations led to the development of the cell theory. The three main ideas of the cell theory are:

1. All living things are made of one or more cells.
2. The cell is the basic unit of life in which the activities of life occur.
3. All cells come from cells that already exist.

Figure 1 Robert Hooke designed this microscope and drew the cork cells he observed.

1 Motivate

INTERACTIVE CHALKBOARD PowerPoint® Presentations

Bellringer

Section Focus Transparencies also are available on the Interactive Chalkboard CD-ROM. L2 ELL

SECTION 1 Section Focus Transparency **It's Not Math, but It Is Division**

In the past, scientists thought that living organisms could come from decaying matter. Today, however, we know that living things come from other living things. Every day our bodies replace millions of cells, the most basic unit of life. Most cells reproduce themselves by dividing in two, as shown on the image below.

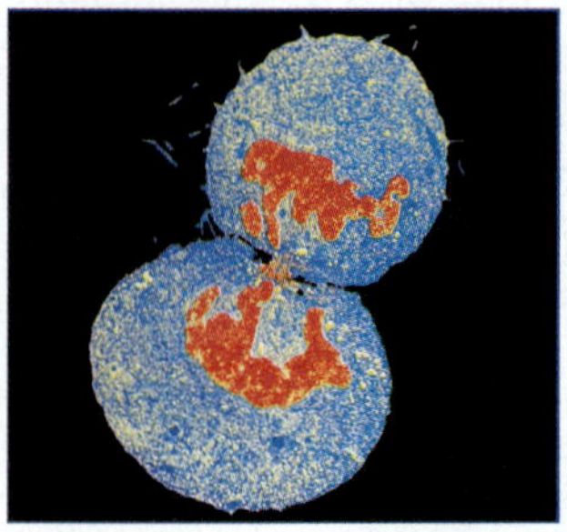

1. What instrument was necessary to make this image?
2. How many cells were there before this process started? How many will there be after this process is done?
3. Do you think all the cells within your body are the same? Explain.

Cells—The Units of Life

Tie to Prior Knowledge

Magnification Students may have used magnifying glasses, a telescope, or binoculars. Explain how a microscope magnifies in the same way.

Section 1 Resource Manager

Chapter *FAST FILE* Resources

- Transparency Activity, pp. 42, 45–47
- Directed Reading for Content Mastery, pp. 19, 20
- Note-taking Worksheets, pp. 31–32
- Enrichment, p. 29
- Reinforcement, p. 27
- Lab Activity, pp. 9–12
- MiniLAB, p. 3
- Lab Worksheet, pp. 5–6

Home and Community Involvement, p. 38

The Microscopic Cell All the living things pictured in **Figure 2** are made up of cells. The smallest organisms on Earth are **bacteria**. They are one-celled organisms, which means they are made up of only one cell.

Reading Check *How many cells does each bacterium have?*

Larger organisms are made of many cells. These cells work together to complete all of the organism's life activities. The living things that you see every day—trees, dogs, insects, people—are many-celled organisms. Your body contains more than 10 trillion (10,000,000,000,000) cells.

Microscopes Scientists have viewed and studied cells for about 300 years. In that time, they have learned a lot about cells. Better microscopes have helped scientists learn about the differences among cells. Some modern microscopes allow scientists to study the small features that are inside cells.

INTEGRATE Physics

The microscope used in most classrooms is called a compound light microscope. In this type of microscope, light passes through the object you are looking at and then through two or more lenses. The lenses enlarge the image of the object. How much an image is enlarged depends on the powers of the eyepiece and the objective lens. The power—a number followed by an ×—is found on each lens. For example, a power of 10× means that the lens can magnify something to ten times its actual size. The magnification of a microscope is found by multiplying the powers of the eyepiece and the objective lens.

Figure 2 All living things are made up of cells.

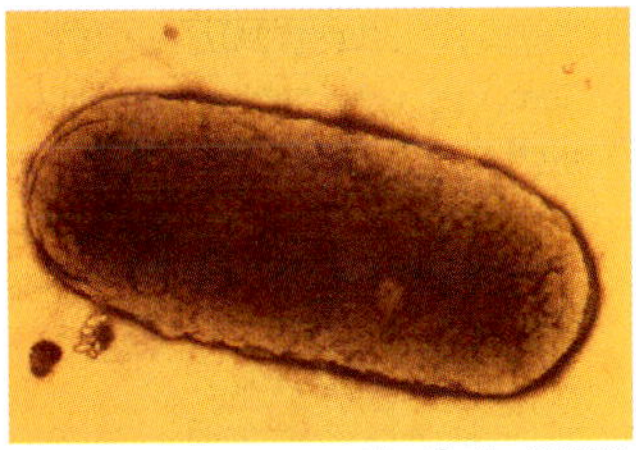

Magnification: 67500×

E. coli—a bacterium—is a one-celled organism.

Plant cells are different from animal cells.

Human cells are similar to other animal cells like those in cats and turtles.

2 Teach

Teacher FYI

Dead Cells Hooke did not see cells; he saw the cell walls of dead cells from the bark of the cork tree. The cork tree is a type of oak. It is green year-round and grows abundantly in Portugal and Spain.

Reading Check

Answer one

Quick Demo

Cell Structure

Materials pieces of cork

Estimated Time five minutes

Procedure Show students a piece of cork and have them examine the structure. Emphasize that cork was a living part of an organism—the bark of the cork tree, a type of oak.

Fun Fact

The human body contains about 200 different types of cells that total more than 100 trillion cells in all.

Curriculum Connection

Art Discuss the use of art before the camera was invented. Have students compare Hooke's drawing of cells with photographs throughout the chapter. Have students write their opinions of the advantages and disadvantages of using artwork and photography. L2 LS **Visual-Spatial** P

Active Reading

Think-Pair-Share This strategy encourages students to think before discussing a topic. After thinking for a few minutes, partners discuss responses to the question. Finally, the teacher asks the students to discuss responses with the class. Have students use Think-Pair-Share for cells.

SECTION 1 The World of Cells 477

Activity

Labeling Cells Make a bulletin board showing unlabeled parts of an animal cell and a plant cell. As each cell part is studied, have a student place its label beside it on the bulletin board. L1 LS **Visual-Spatial**

Discussion

Cells and Stains What color are cells? Living tissues are, for the most part, colorless. Why are stains used? Most cells can be seen only when stained a color.

Make a Model

Model Cells Give groups of students small lumps of different-colored clay. As you teach about the parts of cells, have groups build model cells. Introduce and discuss the cell parts, and have students add clay parts to their own model cells. Have groups make small labels for each of the parts. Students can use toothpicks to place the labels on the model cells. L2 ELL COOP LEARN LS **Kinesthetic** P

Virtual Labs

Cells *How do animal and plant cells work?*

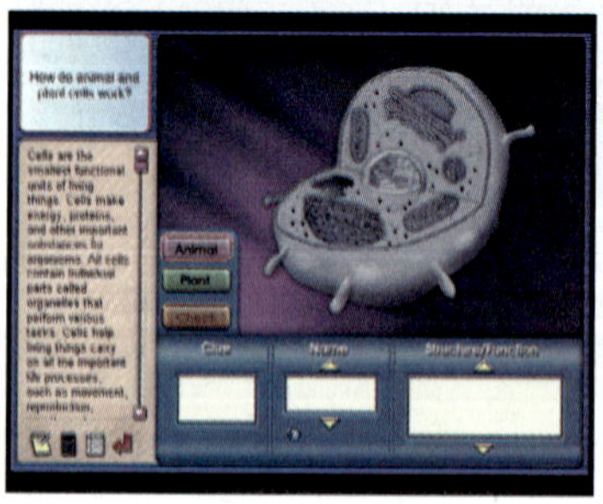

Science Online

Topic: Electron Microscopes

Some cell parts could not be seen until the electron microscope (EM) was invented. Visit red.msscience.com for Web links to information about electron microscopes.

Activity Make a pamphlet describing an electron microscope. Include a section on cell parts that were first seen using an EM.

What are cells made of?

As small as cells are, they are made of even smaller parts, each doing a different job. A cell can be compared to a bakery. The activities of a bakery are inside a building. Electricity is used to run the ovens and other equipment, power the lights, and heat the building. The bakery's products require ingredients such as dough, sugar, and fillings, that must be stored, assembled, and baked. The bakery's products are packaged and shipped to different locations. A manager is in charge of the entire operation. The manager makes a plan for every employee of the bakery and a plan for every step of making and selling the baked goods.

A living cell operates in a similar way. Like the walls of the bakery, a cell has a boundary. Inside this boundary, the cell's life activities take place. These activities must be managed. Smaller parts inside the cell can act as storage areas. The cell also has parts that use ingredients such as oxygen, water, minerals, and other nutrients. Some cell parts can release energy or make substances that are necessary for maintaining life. Some substances leave the cell and are used elsewhere in the organism.

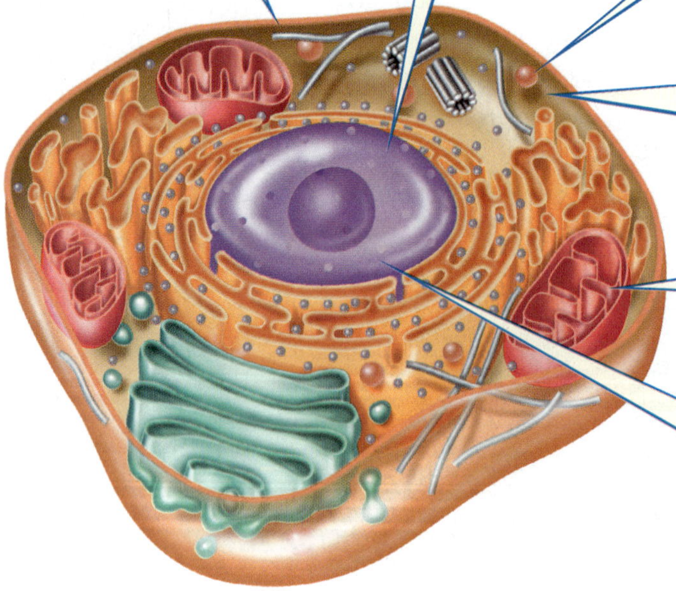

Figure 3 These are some of the parts of an animal cell that perform the activities necessary for life.

Visual Learning

Figure 3 and **Figure 4** These drawings are illustrations of generalized cells. Discuss the meaning of the word *generalized.* Explain that cells don't really look like these illustrations. You also may wish to discuss the importance of scientific illustrations. Explain the value of illustrations for learning science, especially when microscopic structures, such as cells, are depicted. ELL

Differentiated Instruction

Visually Impaired Students with visual impairments may benefit from using and constructing models of cells. Some of the large, demonstration cell models sold by science-supply houses are typically large, detailed, and three-dimensional. L2 LS **Kinesthetic**

Outside the Cell The **cell membrane**, shown in **Figure 3**, is a flexible structure that holds the cell together, similar to the walls of the bakery. The cell membrane forms a boundary between the cell and its environment. It also helps control what goes into and comes out of the cell. Some cells, like those in plants, algae, fungi, and many types of bacteria, also have a structure outside the cell membrane called a **cell wall**, shown in **Figure 4.** The cell wall helps support and protect these cells.

Inside the Cell The inside of a cell is filled with a gelatinlike substance called **cytoplasm** (SI tuh pla zum). Approximately two-thirds of the cytoplasm is water, but it also contains many chemicals that are needed by the cell. Like the work area inside the bakery, the cytoplasm is where the cell's activities take place.

Organelles Except for bacterial cells, cells contain **organelles** (or guh NELZ) like those in **Figure 3** and **Figure 4.** These specialized cell parts can move around in the cytoplasm and perform activities that are necessary for life. You could think of these organelles as the employees of the cell because each type of organelle does a different job. In bacteria, most cell activities occur in the cytoplasm.

INTEGRATE Chemistry

Phospholipids The cell membrane is a double layer of complex molecules called phospholipids (fahs foh LIH pudz). Research to find the elements that are in these molecules. Find those elements on the periodic table at the back of this book.

Figure 4 Most plant cells contain the same types of organelles as in animal cells. Plant cells also have a cell wall and chloroplasts.

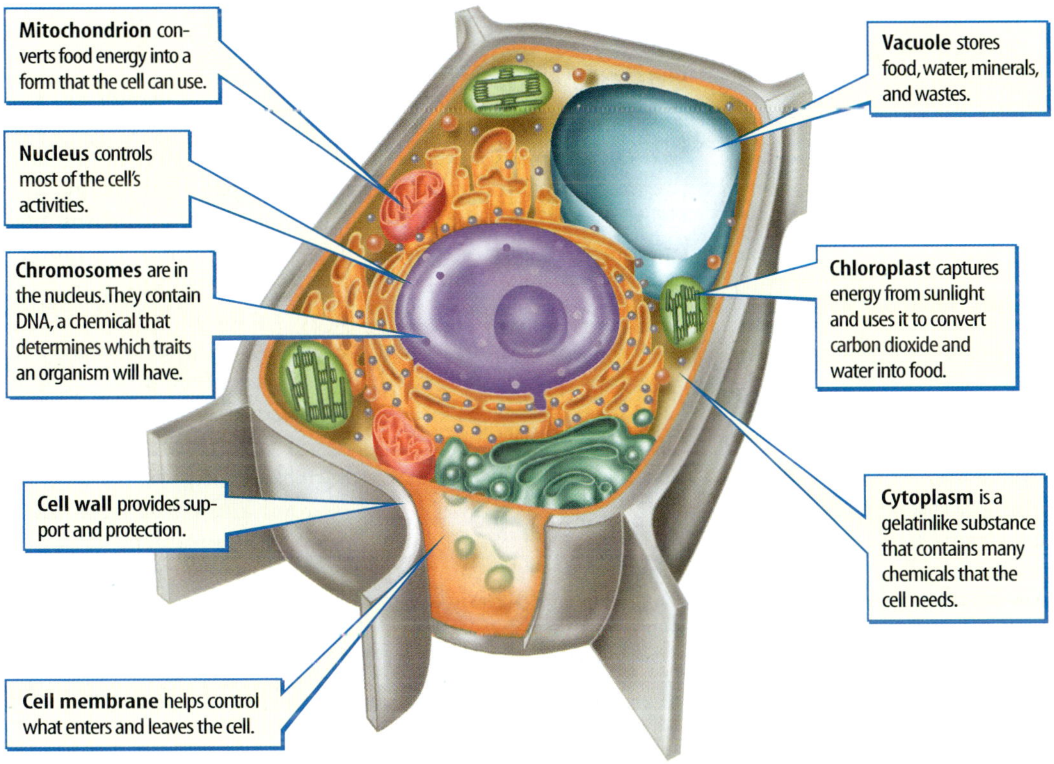

INTEGRATE Chemistry

Phospholipids the elements carbon (C), hydrogen (H), oxygen (O), phosphorus (P), and nitrogen (N)

Research Have students research the form and function of the phospholipid bilayer that makes up the cell membrane. Students can use their research to make a model of the bilayer. L2

Use an Analogy

Cells as Factories You may wish to have students brainstorm a list of other possible analogies for living cells. For instance, one common analogy is that cells are like factories. Both produce things (cells produce energy, proteins, and other substances), and they have specialized areas where this production takes place. In addition, production operates within enclosed spaces, and wastes are removed during the process. After student groups are finished, ask individuals to present their ideas to class. Accept all reasonable answers. L2 ELL LS **Interpersonal**

Use Science Words

Word Usage Have students brainstorm several different meanings of the word *cell.* Have them compare the meanings and brainstorm a list that shows how all these cells have similar characteristics. L2 ELL LS **Linguistic**

Differentiated Instruction

Challenge Students who have mastered the content can construct more elaborate clay models of cells. Have them research cells and cell parts to learn about other organelles. Students can construct detailed models that display all of the important cell parts. L3

Cultural Diversity

Ernest Everett Just Just, an African American biologist in the early 1900s, studied cells and how they function. His research showed that all parts of the cell influence its activities, not just the nucleus, as scientists then believed. This idea changed scientific opinion concerning the basis of life. Discuss how Ernest Just's research is important to the study of cells today.

Mini LAB

Purpose Students make models of cells. L2 LS **Kinesthetic**

Materials clay, cardboard boxes, pipe cleaners, yarn, buttons, dry macaroni, paper

Teaching Strategies

- Have each student supply a small quantity of one item on the list.
- Instruct students to represent each cell part with a different item.

Safety Precautions Remind students not to eat any materials used in the lab.

Analysis

1. Answers will vary, but should agree with the textbook.
2. Answers will vary, but the cell parts should be easily identifiable.
3. Accept all reasonable answers.

Assessment

Process Instruct students to examine several of the models carefully. Next, ask them to report on how different students modeled cell parts. Use **Performance Assessment in the Science Classroom,** p. 123.

Try at Home

Reading Check

Answer DNA

Mini LAB

Modeling a Cell

Procedure

1. Collect **household materials such as clay, cardboard, yarn, buttons, dry macaroni,** or other objects.
2. Using the objects that you collected, make a three-dimensional model of an animal or plant cell.
3. On a separate sheet of **paper,** make a key to the materials in your cell model.

Analysis

1. What does each part of your cell model do?
2. Have someone look at your model. Which of the cell parts could they identify without using the key?
3. How could you improve your model?

Try at Home

The Nucleus A bakery's manager follows a business plan to make sure that the business runs smoothly. A business plan describes how the business should operate. These plans could include how many donuts are made and what kinds of pies are baked.

The hereditary material of the cell is like the bakery's manager. It directs most of the cell's activities. In the cells of organisms except bacteria, the hereditary material is in an organelle called the **nucleus** (NEW klee us). Inside the nucleus are chromosomes (KROH muh zohmz). They contain a plan for the cell, similar to the bakery's business plan. Chromosomes contain an important chemical called DNA. It determines which traits an organism will have, such as the shape of a plant's leaves or the color of your eyes.

Reading Check *Which important chemical determines the traits of an organism?*

Storage Pantries, closets, refrigerators, and freezers store food and other supplies that a bakery needs. Trash cans hold garbage until it can be picked up. In cells, food, water, and other substances are stored in balloonlike organelles in the cytoplasm called **vacuoles** (VA kyuh wohlz). Some vacuoles store wastes until the cell is ready to get rid of them. Plant cells usually have a large vacuole that stores water and other substances.

Energy and the Cell

Electrical energy or the energy in natural gas is converted to heat energy by the bakery's ovens. The heat then is used to bake the breads and other bakery products. Cells need energy, too. Cells, except bacteria, have organelles called **mitochondria** (mi tuh KAHN dree uh)(singular, *mitochondrion*). An important process called cellular respiration (SEL yuh lur • res puh RAY shun) takes place inside a mitochondrion as shown in **Figure 5.** Cellular respiration is a series of chemical reactions in which energy stored in food is converted to a form of energy that the cell can use. This energy is released as food and oxygen combine. Waste products of this process are carbon dioxide and water. All cells with mitochondria use the energy from cellular respiration to do all of their work.

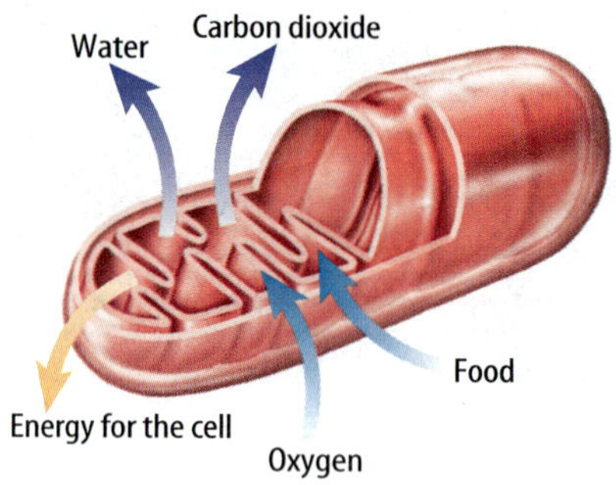

Figure 5 Inside a mitochondrion, food energy is changed into a form of energy that a cell can use.
Infer *what happens to the water and carbon dioxide produced by mitochondria in human cells.*

LAB DEMONSTRATION

Purpose to demonstrate a by-product of cellular respiration

Materials bromthymol blue solution, glass jar, straw, graduated cylinder

Preparation Add 0.5 g bromthymol blue powder to 500 mL distilled water. Add 10 mL of this solution to 500 mL of distilled water. (If green, add one or more drops of NH_4OH until blue. If deep blue, add one or more drops of HCl until light blue.)

Procedure Pour 50 mL of the diluted solution into a glass jar. Use a straw and exhale into the solution. **WARNING:** *Do not inhale any solution.*

Expected Outcome The bromthymol blue solution turns bluish green.

Assessment

What reacted with the solution? carbon dioxide What was its source? cellular respiration

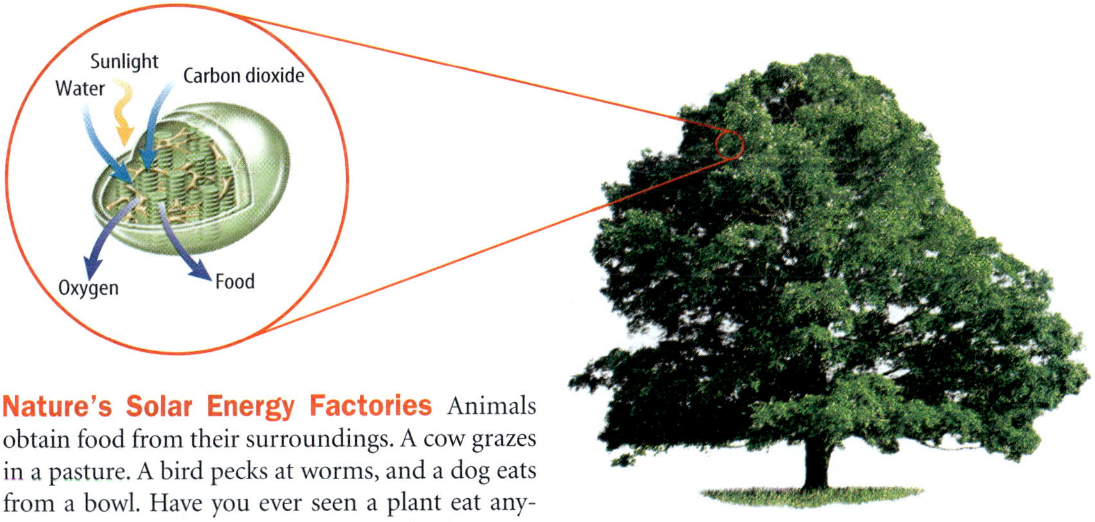

Figure 6 Photosynthesis can take place inside the chloroplasts of plant cells.

Nature's Solar Energy Factories Animals obtain food from their surroundings. A cow grazes in a pasture. A bird pecks at worms, and a dog eats from a bowl. Have you ever seen a plant eat anything? How do plants get energy-rich food?

Plants, algae, and many types of bacteria make food through a process called **photosynthesis** (foh toh SIHN thuh sus). Most photosynthesis in plants occurs in leaf cells. Inside these cells are green organelles called **chloroplasts** (KLOR uh plasts). Most leaves are green because their cells contain so many chloroplasts. During plant photosynthesis, as shown in **Figure 6,** chloroplasts capture light energy and combine carbon dioxide from the air with water to make food. Energy is stored in food. As the plant needs energy, its mitochondria release the food's energy. The captured light energy is passed to other organisms when they eat organisms that carry on photosynthesis.

section 1 review

Summary

Importance of Cells

- Cells are organized structures that help living things carry on the activities of life.
- The main ideas behind cells are described in the cell theory.
- Microscopes helped scientists study cells.

What are cells made of?

- Different cell parts do different jobs.

Energy and the Cell

- Cells need energy to function. This energy comes mainly from cellular respiration.
- Plants, algae, and some bacteria make food through photosynthesis.

Self Check

1. **List** the three main ideas of the cell theory.
2. **Explain** why the nucleus is so important to the living cell.
3. **Describe** how cells get the energy they need to carry on their activities.
4. **Describe** the purpose of a cell membrane.
5. **Think Critically** Suppose your teacher gave you a slide of an unknown cell. How would you tell whether the cell was from an animal or from a plant?

Applying Skills

6. **Compare and contrast** the parts of animal cells and plant cells and the jobs that they do.

Science Online red.msscience.com/self_check_quiz

Caption Answer

Figure 5 They enter the blood and are carried to the lungs where they are exhaled.

3 Assess

DAILY INTERVENTION

Check for Understanding

Linguistic Have students write a paragraph describing how the development of different microscopes helped scientists study cells. L2

Reteach

Parts of a Microscope Place a large drawing of a compound microscope on the board. Write the functions of the parts on index cards. Have students select a card, name the part, and find it on the drawing. L2

Assessment

Content Have students prepare "game show"-type questions on cells. Collect all the questions and quiz the groups on the material. Use **Performance Assessment in the Science Classroom,** p. 91.

section 1 review

1. All living things are made of one or more cells. The cell is the basic unit of life in which activities of life occur. All cells come from cells that already exist.
2. The nucleus controls most of the cell's activities.
3. The cell's mitochondria release energy by combining food molecules and oxygen during cellular respiration.
4. It holds the cell together and helps control what enters and leaves the cell.
5. Answers may vary, but students should indicate that most plant cells have chloroplasts and cell walls, and animal cells do not.
6. Students' answers should reflect the information in **Figure 3** and **Figure 4.**

SECTION 1 The World of Cells 481

BENCH TESTED

Real-World Question

Purpose Students use a microscope to observe organelles in algal cells. L2 LS **Visual-Spatial**

Process Skills observe, classify, draw

Time Required 40 minutes

Safety Precautions Caution students never to drink or eat anything in the lab.

Procedure

Teaching Strategy Make sure that students have focused well before they switch to high power.

Troubleshooting Remind students not to put too much algae and water on the microscope slide.

Conclude and Apply

1. chloroplasts, nucleus, large vacuoles
2. Chloroplasts can capture energy from the Sun to make food. This makes algae and plants producers.
3. Algal cells are essential because they produce food from the energy of sunlight. Most organisms cannot produce their own food, so they need to have other sources of nutrients. Algal cells also release oxygen into the water.

LAB

Observing Algae

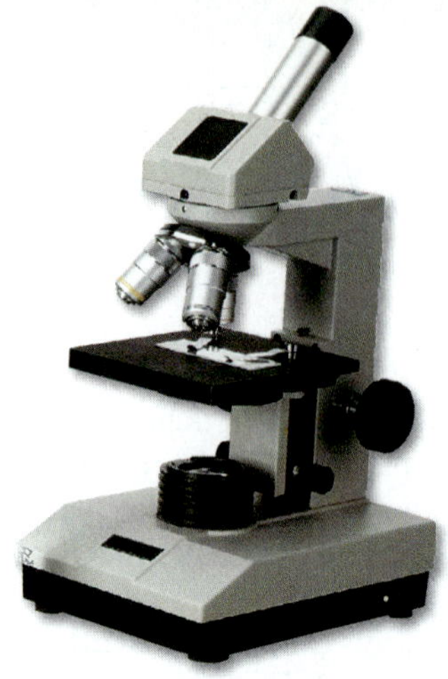

You might have noticed mats of green algae growing on a pond or clinging to the walls of the aquarium in your classroom. Why are algae green? Like plants, algae contain organelles called chloroplasts. Chloroplasts contain a green pigment called chlorophyll. It captures light energy that is needed to make food. In this lab, you'll describe chloroplasts and other organelles in algal cells.

Real-World Question

What organelles can be seen when viewing algal cells under a microscope?

Goals

- **Observe** algal cells under a microscope.
- **Identify** cell organelles.

Materials

microscope	pond water
microscope slides	algae
coverslips	dropper
large jars	colored pencils

Safety Precautions

WARNING: *Thoroughly wash your hands after you have finished this lab.*

Procedure

1. Fill the tip of a dropper with pond water and thin strands of algae. Use the dropper to place the algae and a drop of water on a microscope slide.
2. Place a coverslip over the water drop and then place the slide on the stage of a microscope.
3. Using the microscope's lowest power objective, focus on the algal strands.
4. Once the algal strands are in focus, switch to a higher power objective and observe several algal cells.
5. **Draw** a colored picture of one of the algal cells, identifying the different organelles in the cell. Label on your drawing the cell wall, chloroplasts, and other organelles you can see.

Conclude and Apply

1. **List** the organelles you found in each cell.
2. **Explain** the function of chloroplasts.
3. **Infer** why algal cells are essential to all pond organisms.

Communicating Your Data

Work with three other students to create a collage of algal cell pictures complete with labeled organelles. Create a bulletin board display about algal cells.

Assessment

Oral Have students explain why they were not able to view all organelles with the microscope. Large organelles were visible. Smaller organelles such as mitochondria are either too small or too thin to be seen. Use **Performance Assessment in the Science Classroom,** p. 89.

Communicating Your Data

Encourage students to include drawings or diagrams that show algal strands as well as algal cells. If computers are available, suggest that students use a computer drawing program and a word processor to make drawings and descriptive text for the bulletin board display.

section 2

The Different Jobs of Cells

Special Cells for Special Jobs

Choose the right tool for the right job. You might have heard this common expression. The best tool for a job is one that has been designed for that job. For example, you wouldn't use a hammer to saw a board in half, and you wouldn't use a saw to pound in a nail. You can think of your body cells in a similar way.

Cells that make up many-celled organisms, like you, are specialized. Different kinds of specialized cells work as a team to perform the life activities of a many-celled organism.

Types of Human Cells Your body is made up of many types of specialized cells. The same is true for other animals. **Figure 7** shows some human cell types. Notice the variety of sizes and shapes. A cell's shape and size can be related to its function.

as you read

What You'll Learn

- **Discuss** how different cells have different jobs.
- **Explain** the differences among tissues, organs, and organ systems.

Why It's Important

You will understand how different types of cells work together to keep you healthy.

Review Vocabulary

organism: anything that possesses all the characteristics of life

New Vocabulary

- tissue
- organ
- organ system

Figure 7 Human cells come in different shapes and sizes.

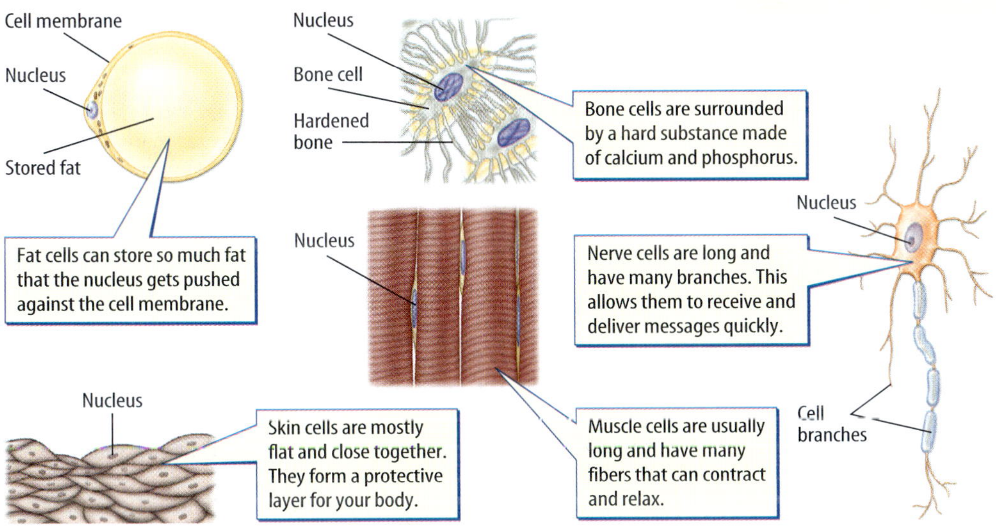

1 Motivate

INTERACTIVE CHALKBOARD PowerPoint® Presentations

Bellringer

Section Focus Transparencies also are available on the Interactive Chalkboard CD-ROM. L2 ELL

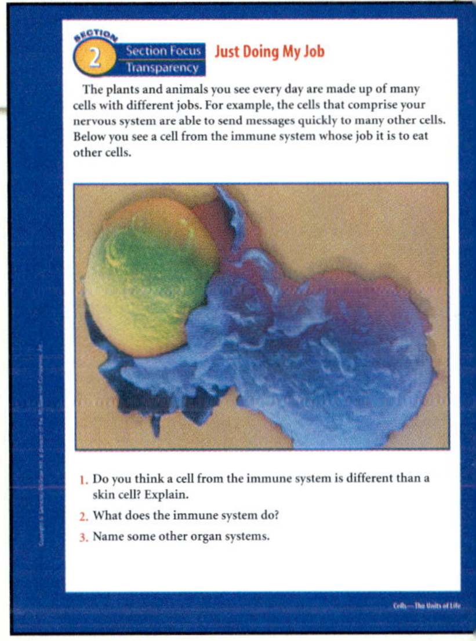
Section 2 Section Focus Transparency **Just Doing My Job**

The plants and animals you see every day are made up of many cells with different jobs. For example, the cells that comprise your nervous system are able to send messages quickly to many other cells. Below you see a cell from the immune system whose job it is to eat other cells.

1. Do you think a cell from the immune system is different than a skin cell? Explain.
2. What does the immune system do?
3. Name some other organ systems.

Cells—The Units of Life

Tie to Prior Knowledge

Tool Structure and Function Students should be familiar with tools such as hammers, screwdrivers, saws, chisels, and files. Have students make observations about how the structure of each tool is related to its function. Just as tools are shaped according to their functions, so too are the cells of many-celled organisms.

Section 2 Resource Manager

Chapter *FAST FILE* Resources

Transparency Activity, p. 26

Directed Reading for Content Mastery, pp. 21, 22

MiniLAB, p. 28

Reinforcement, p. 28

Enrichment, p. 30

Lab Activity, pp. 13–16

Lab Worksheet, pp. 7–8

Lab Management and Safety, p. 69

2 Teach

Caption Answer

Figure 8 photosynthesis

Mini LAB

Purpose Students compare and contrast different human cells and hypothesize how cell shape and structure relate to function. L2 ELL LS **Visual-Spatial**

Materials prepared slides of human muscle cells, nerve cells, blood cells, and skin cells; compound microscope

Teaching Strategy If slides are not available, you may wish to use high-quality photomicrographs of different types of cells.

Analysis

1. All of the cells are microscopic. With the exception of muscle cells, the cells are somewhat round in shape.
2. Students' answers should reflect information in **Figure 7.**

Assessment

Process In the human body, structure is related closely to function. Have students use slides or photographs of cells (bone, blood, and other cell types) to prepare a presentation that explains how structure is related to function. Use **Performance Assessment in the Science Classroom,** p. 143.

Reading Check

Answer form a system through which water, food, and other materials move in the plant

Figure 8 Plants, like animals, have specialized cells.
Infer *what process can occur in leaf cells but not in root cells.*

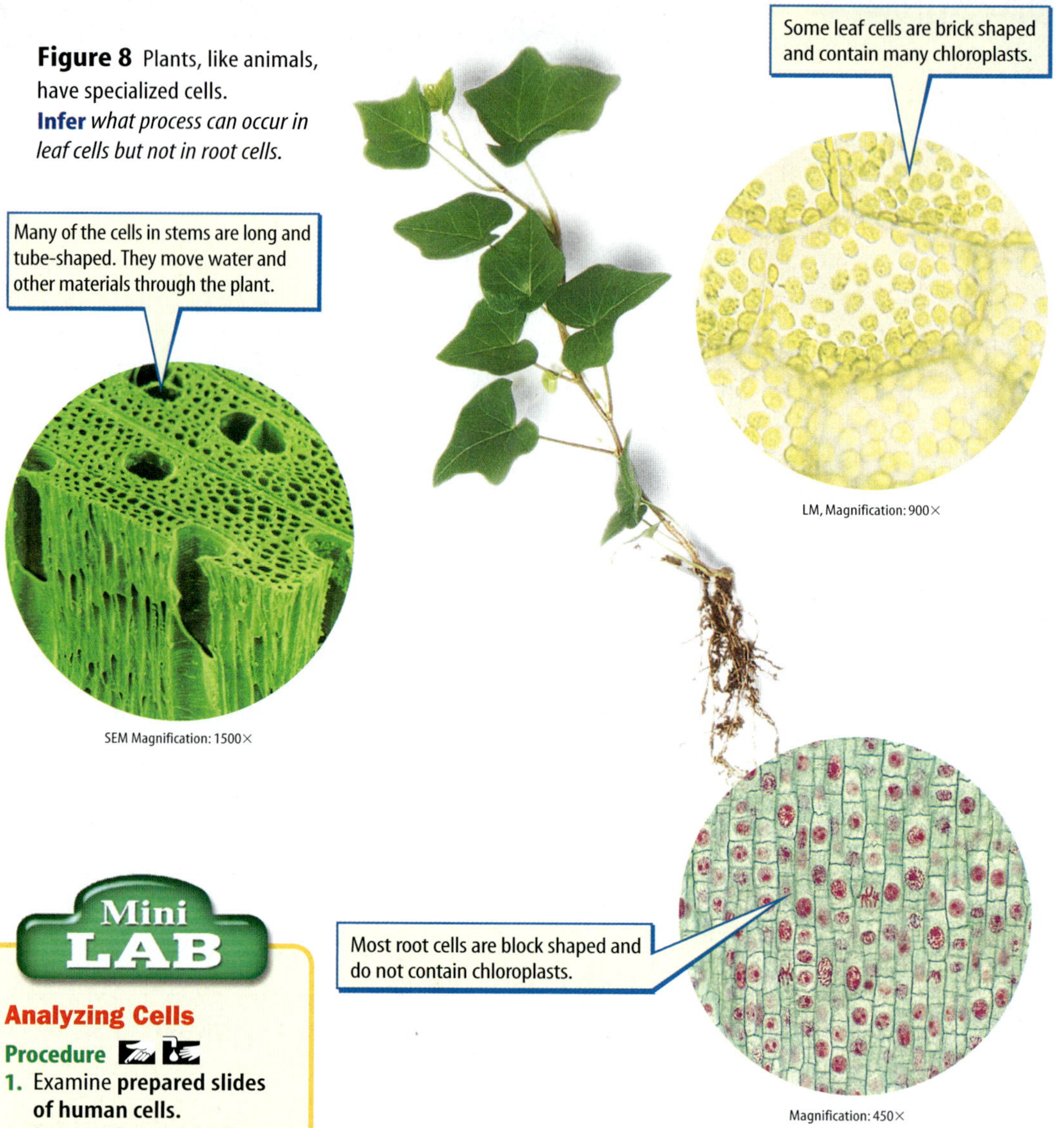

Mini LAB

Analyzing Cells

Procedure

1. Examine **prepared slides of human cells.**
2. Draw each type of cell that you observe in your **Science Journal.** Label cell parts that you can see.

Analysis

1. In what ways were the cells that you observed similar? How were they different?
2. Hypothesize how the cells' shapes relate to their jobs.

Types of Plant Cells Like animals, plants also are made of several different cell types, as shown in **Figure 8.** For instance, plants have different types of cells in their leaves, roots, and stems. Each type of cell has a specific job. Some cells in plant stems are long and tubelike. Together they form a system through which water, food, and other materials move in the plant. Other cells, like those that cover the outside of the stem, are smaller or thicker. They provide strength to the stem.

Reading Check *What do long, tubelike cells do in plants?*

Science Journal

Form and Function Different types of eating utensils—forks, knives, and spoons—also are perfectly designed for their functions. Have students write in their Science Journals brief paragraphs about how the shapes of eating utensils match their functions. Have them relate this concept to cells. L2 LS **Linguistic** P

Visual Learning

Figure 8 Why do leaf cells contain so many chloroplasts? The primary reason is genetics. Secondary to genetics is environment because the number of chloroplasts can vary with the amount of light present.

Cell Organization

How well do you think your body would work if all the different cell types were just mixed together in no particular pattern? Could you walk if your leg muscle cells were scattered here and there, each doing its own thing, instead of being grouped together in your legs? How could you think if your brain cells weren't close enough together to communicate with each other? Many-celled organisms are not just mixed-up collections of different types of cells. Cells are organized into systems that, together, perform functions that keep the organism healthy and alive.

INTEGRATE Career

Studying Space
Astronomers study systems that are found in space. The solar system is just one of many systems that make up the Milky Way Galaxy. Research to learn the education requirements and job description of an astronomer. Write a want ad for an astronomer.

Applying Math — Solve One-Step Equations

RED BLOOD CELLS Each milliliter of blood contains 5 million red blood cells (RBCs). On average, an adolescent has about 3.5 L of blood. On average, how many RBCs are in an adolescent's body?

Solution

1 *This is what you know:*
- number of RBCs per 1 mL = 5,000,000
- 1,000 mL = 1 L
- average volume of blood in an adolescent's body = 3.5 L

2 *This is what you need to find out:* On average, how many RBCs are in an adolescent's body, N?

3 *This is the procedure you need to use:*
- Use the following equation:
  N = (number of RBCs/1mL) (1,000 mL/1 L) (3.5 L of blood)
- Substitute the known values
  N = (5,000,000 RBCs/1 mL) (1,000 mL/1 L) (3.5 L of blood)
  N = 17,500,000,000 RBCs
- On average, there are 17.5 billion red blood cells in an adolescent's body.

4 *Check your answer:* Divide 17,500,000,000 RBCs by 1,000 mL/1 L then divide that answer by 3.5 L, and you should get 5,000,000 RBCs/1 mL.

Practice Problems

1. Each milliliter of blood contains approximately 7,500 white blood cells. How many white blood cells are in the average adolescent's body?
2. There are approximately 250,000 platelets in each milliliter of blood. How many platelets are in the average adolescent's body?

Science Online — For more practice, visit red.msscience.com/math_practice

INTEGRATE Career

Studying Space Student want ads should include that the position requires physics and math skills, knowledge of computers, good communication skills to share information, and good writing skills to write proposals and journal articles.

Quick Demo

Plant v. Animal

Materials microprojector, slides of different plant and animal tissues

Estimated Time 10 minutes

Procedure Use the microprojector to show slides of different plant and animal tissues. Point out that each type of tissue is made of similar cells, as is evident from the shape of the cells in the tissues.

Applying Math

National Math Standards

Correlation to Mathematics Objectives
1, 2, 9

Answers to Practice Problems

1. 3.5 L = 3,500 mL
   3,500 mL × 7,500 = 26,250,000 white blood cells
2. 3,500 mL × 250,000 = 875,000,000 platelets

Discussion

Red Blood Cells What would happen to the human body if its red blood cells could no longer carry oxygen? The body would slowly die from lack of oxygen.

Differentiated Instruction

English-Language Learners Ask students to diagram the different types of human or plant cells. Have them label the cells and describe how the shape of each cell helps it perform its specific function. L2

Teacher FYI

Vesalius Drawings Andreas Vesalius (1514–1564) conducted some of the most accurate and thorough examinations of the human body that had ever been done. Vesalius wrote a classic book about human anatomy that has become popular because of its extremely detailed anatomical drawings.

NATIONAL GEOGRAPHIC

Visualizing Life's Organization

Activity

Organs Have students look at the organs named in **Figure 9.** Ask them to name the organ systems to which these organs belong.

Inquiry Lab

Levels of Organization

Purpose To familiarize students with the levels of organization in different organ systems.

Possible Materials library sources, clay, cardboard tubes and boxes, pipe cleaners, yarn, paper, plastic bags, beads

Estimated Time 50 minutes

Teaching Strategies

- Divide students into small groups and have each group study a different organ system.
- Students can build models of their organ system then use the models to explain to the class how the main organs in that system work together to function.

For additional inquiry activities, see *Science Inquiry Labs*.

NATIONAL GEOGRAPHIC **VISUALIZING LIFE'S ORGANIZATION**

Figure 9
Organs are two or more tissue types that work together. An organ performs a task that no other organ performs.

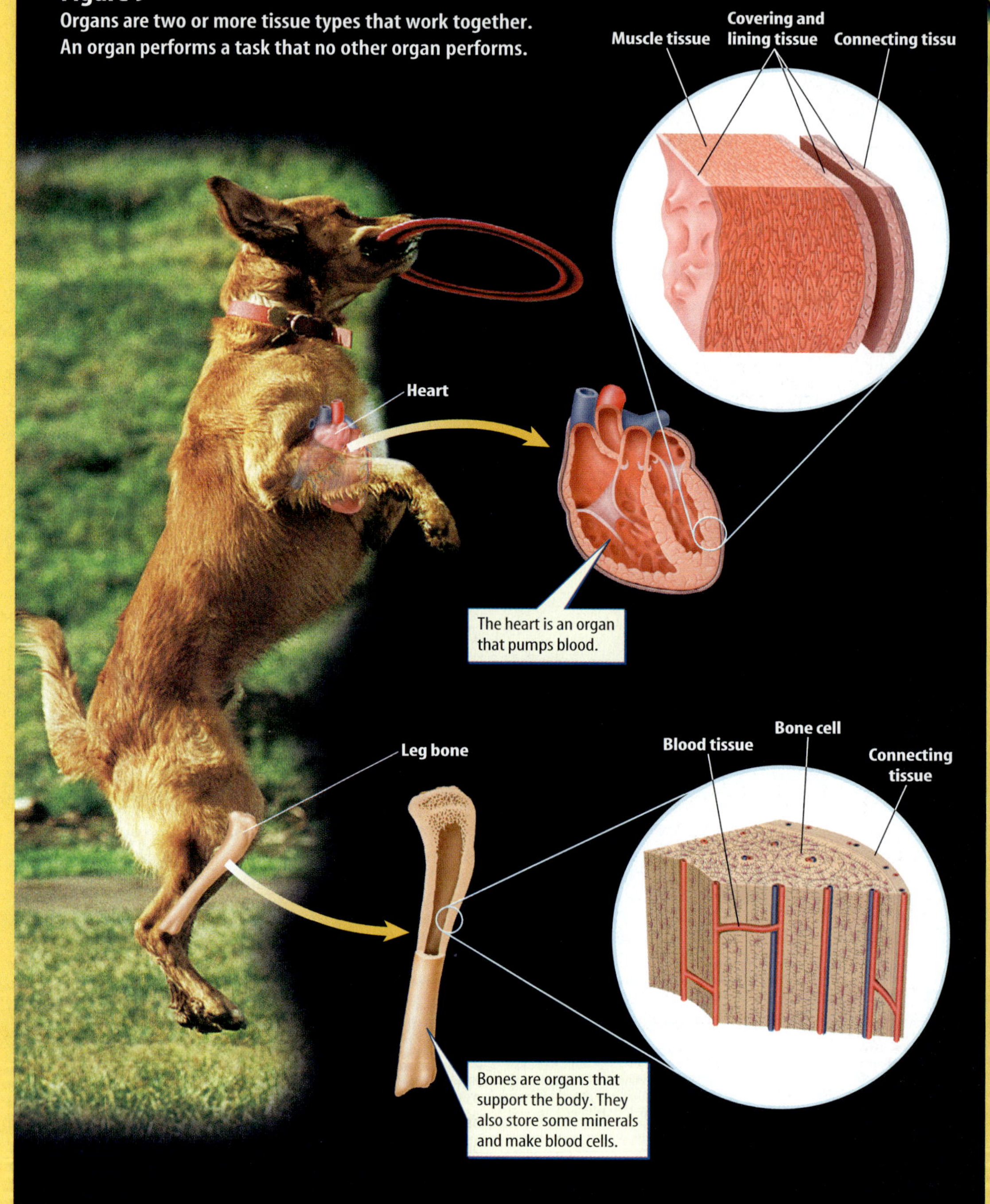

486 **CHAPTER 16** Cells—The Units of Life

IDENTIFYING Misconceptions

Blood Tissue Students may think blood is not a tissue. It's natural for students to think only of solid tissues, such as muscle tissue, nerve tissue, or skin tissue. Explain that blood is classified as a tissue because it is made up of groups of similar cells that work together to carry out a specific task.

Differentiated Instruction

Challenge Have students research the main organ systems of the human body and the main organs in each system. Students can make a poster listing their findings. L3

Learning Disabled Students who have difficulty understanding the levels of organization may benefit from making models. First, have them make a small model cell out of clay. Next, instruct students to make a few more of the same kind of cell. Have them join the individual cells to make a tissue. The other levels of organization can be studied in a similar fashion. L1 LS **Kinesthetic**

Tissues and Organs Cells that are alike are organized into tissues (TIH shewz). **Tissues** are groups of similar cells that all do the same sort of work. For example, bone tissue is made up of bone cells, and nerve tissue is made up of nerve cells. Blood, a liquid tissue, includes different types of blood cells.

As important as individual tissues are, they do not work alone. Different types of tissues working together can form a structure called an **organ** (OR gun). For example, the stomach is an organ that includes muscle tissue, nerve tissue, and blood tissue. All of these tissues work together and enable the stomach to perform its digestive functions. Other human organs include the heart and the kidneys.

Reading Check *Which term means "two or more tissue types that work together"?*

Organ Systems A group of organs that work together to do a certain job is called an **organ system.** The stomach, mouth, intestines, and liver are involved in digestion. Together, these and several other organs make up the digestive system. Other organ systems found in your body include the respiratory system, the circulatory system, the reproductive system, and the nervous system.

Organ systems also work together, as shown in **Figure 9.** For example, the muscular system has more than 600 muscles that are attached to bones. The contracting cells of muscle tissue cause your bones, which are part of the skeletal system, to move.

Science Online

Topic: One-Celled Organisms

Visit red.msscience.com for Web links to information about what types of organisms are made up of only one cell.

Activity Create a table that includes images and information about five of these organisms.

section 2 review

Summary

Special Cells for Special Jobs

- Plant and animal cells come in a variety of sizes and shapes.
- The function of an animal cell can be related to its shape and size.
- The leaves, roots, and stems of plants are made of different types of cells to perform different functions.

Cell Organization

- Many-celled organisms are organized into tissues, organs, and organ systems.
- Each organ system performs a specific function that, together with other systems, keeps an organism healthy and alive.

Self Check

1. **Describe** three types of cells that are found in the human body.
2. **Compare and contrast** the cells found in a plant's roots, stems, and leaves.
3. **Explain** the difference between a cell and a tissue and between a tissue and an organ.
4. **Think Critically** Why must specialized cells work together as a team?

Applying Skills

5. **Concept Map** Make an events-chain concept map of the different levels of cell organization from cell to organ system. Provide an example for each level of organization.

Science Online red.msscience.com/self_check_quiz

section 2 review

1. Descriptions may include information on fat, skin, bone, nerve, or muscle cells.
2. Answers should contain information from **Figure 8.**
3. Tissues are made up of cells. Organs are made up of tissues.
4. to perform the life activities of a many-celled organism
5. Possible answer: Cell (muscle cell) → Tissue (muscle) → Organ (heart) → Organ System (circulatory system)

Fun Fact

An adult body has about 5 liters (5.3 quarts) of blood that circulates through the body three times every minute. In one day, blood travels 19,000 km (12,000 miles), or four times the distance across the United States.

Reading Check

Answer organ

3 Assess

DAILY INTERVENTION

Check for Understanding

Visual-Spatial Have students draw different types of cells, such as muscle, nerve, and fat cells. Ask them to describe how the form of the cell fits its function. L2

Reteach

Levels of Organization Have students make a chart with two columns. In one column, have them list the different levels of organization, beginning with the cell. In the other column, have students provide a real example for each different level. Have students choose an organism not discussed in the chapter. L2 LS **Logical-Mathematical**

Assessment

Portfolio Have students design posters illustrating the levels of organization in living things. Posters should include a type of cell in the human body; groups of these cells together to show a particular tissue; how this tissue is incorporated in an organ; and the organ system that the organ belongs to. Use **Performance Assessment in the Science Classroom,** p. 145.

BENCH TESTED

Real-World Question

Purpose Students will observe where water moves in a plant and how long it takes for plants to absorb water. L2 ELL COOP LEARN LS **Visual-Spatial**

Process Skills form a hypothesis, design an experiment, make and use tables, observe and infer, separate and control variables

Time Required 45 minutes

Materials Use red food coloring. Blue, green, and yellow food coloring do not show up well in celery.

Safety Precautions Caution students to avoid getting food coloring on clothing. Remind them not to eat food used in science labs.

Form a Hypothesis

Possible Hypotheses

- Plant stems are hollow and act like drinking straws.
- Plant stems contain thin, strawlike passageways for water to flow through.

Test Your Hypothesis

Possible Procedures Put several ounces of water in the glass. Add a few drops of red food coloring and stir. Place the celery stalk in the glass, with the leaves at the top. Keep track of how long it takes for the celery to absorb the water. After all of the water is absorbed, cut the stalk crosswise.

LAB Design Your Own

Water Movement in Plants

Goals

- **Design** an investigation to show where water moves in a plant.
- **Observe** how long it takes water to move in a plant.

Possible Materials

fresh stalk of celery with leaves
clear drinking glass
scissors
red food coloring
water

Safety Precautions

WARNING: *Use care when handling sharp objects such as scissors. Avoid getting red food coloring on your clothing.*

Real-World Question

When you are thirsty, you can sip water from a glass or drink from a fountain. Plants must get their water in other ways. In most plants, water moves from the soil into cells in the roots. Where does water travel in a plant?

Form a Hypothesis

Based on what you already know about how a plant functions, state a hypothesis about where you think water travels in a plant.

Test Your Hypothesis

Make a Plan

1. As a group, agree upon a hypothesis and decide how you will test it. Identify which results will support the hypothesis.
2. **List** the steps you will need to take to test your hypothesis. Be specific. Describe exactly what you will do in each step. List your materials.
3. Prepare a data table in your Science Journal to record your observations.
4. **Read** the entire investigation to make sure all steps are in logical order.
5. **Identify** all constants, variables, and controls of the investigation.

Follow Your Plan

1. Make sure your teacher approves your plan before you start.
2. Carry out the investigation according to the approved plan.
3. While doing the investigation, record your observations and complete the data tables in your Science Journal.

Sample Data Table:

Time	Observations
Start of activity	Celery is light green.
During the activity	Red streak moves up stalk.
End of activity	Celery stalk is red on the inside.

Alternative Inquiry Lab

Flowers Students may further investigate where water travels in a plant by using a flowering plant, such as a carnation or daisy. Students can predict the results and then design an experiment to test their hypothesis.

488 CHAPTER 16 Cells—The Units of Life

Using Scientific Methods

Analyze Your Data

1. **Compare** the color of the celery stalk before, during, and after the investigation.
2. **Compare** your results with those of other groups.
3. Make a drawing of the cut stalk. Label your drawing.
4. What was your control in this investigation? What were your variables?

Conclude and Apply

1. **Explain** whether the results of this investigation supported your hypothesis.
2. **Infer** why only some of the plant tissue is red.
3. **Explain** what you would do to improve this investigation.
4. **Predict** if other plants have tissues that move water.

Communicating Your Data

Write a report about your investigation. Include illustrations to show how the investigation was performed. Present your report to your class.

Teaching Strategies

- Have students relate this activity to plant cell types.
- To get maximum absorption, cut the celery diagonally and place stalks in water containing a high concentration of food coloring.

Expected Outcome Most results will show vertical red lines on the stalk where the colored water traveled through the stalk.

Analyze Your Data

Answers to Questions

1. See Sample Data Table.
2. Students' results should be similar.
3. Students' drawings should show that the tissue that carries water inside the stem turned red.
4. The control was using the same plant; the variables were concentration of food coloring and time.

Error Analysis Answers will vary, depending on the health of plant tissue and the amount of time it was kept in colored water.

Conclude and Apply

1. Answers will vary, depending on students' hypotheses and results.
2. Only the plant tissue that carried the water inside the stem turned red.
3. Possible answer: Keep the celery in the colored water for a longer period of time for better results.
4. Students should predict that all plants contain water-moving tissues.

Assessment

Content Have groups of students design a similar experiment to determine whether different plants absorb water in the same way. Have students write hypotheses to answer this question. Then have them plan and carry out experiments using fresh carrots instead of celery stalks to test the hypotheses. Use **Performance Assessment in the Science Classroom,** p. 105.

Communicating Your Data

Class reports can be enhanced with visual aids such as overhead transparencies or computer slide show presentations. Have students write the key points of their presentations on index cards.

TIME

Content Background

Tissue and organ transplants are referred to with the following terms, depending on the source of the transplanted material. An *autograft* is a transplant from one part of the patient's body to another. A transplant from an identical twin is an *isograft.* A transplant of tissue from one person unrelated to the patient is an *allograft.* Transplants of tissue or organs from one species to a different species of animal are called *xenografts.* The more closely related a person is to the tissue donor, the greater the chance that the tissue will not be rejected by the recipient's immune system. Autografts clearly are the best option whenever possible. The new technology of tissue engineering increases the availability and success of autografts.

Activity

Science Fiction Have students write short science fiction stories or plays using tissue engineering as a theme. Create characters who put together new forms of life from various organs and tissues. Remind students that events like heart transplants were once considered to be impossible. L3

Investigate the Issue

Burn First Aid Have students research the appropriate first aid treatments for various burn injuries. What are the differences between first, second, and third degree burns? How should each type of burn be treated in an emergency situation before medical help arrives? Have students include this information in the safety list they share with their families. L2

TIME SCIENCE AND Society

SCIENCE ISSUES THAT AFFECT YOU!

TEST-TUBE TISSUE

Thanks to advances in science, skin tissue is being "grown" in laboratories

In Chicago, a young woman named Kelly is cooking pasta on her stove. Her clothes catch fire from the gas flame and, in the blink of an eye, 80 percent of her body is severely burned. Will she survive?

Just 20 years ago, the answer to this question probably would have been "no." Fortunately for Kelly, science has come a long way in recent years. Today, there's a very good chance that Kelly might lead a long and healthy life.

Like the brain or the heart, the skin is an organ. In fact, it is the body's largest organ, about 1/12 of your total body weight. Composed of protective layers, skin keeps your internal structure safe from damage, infection, and temperature changes.

Today, just as farmers can grow crops of corn and wheat, scientists can grow human skin. How?

A piece of test skin is removed from its culture.

Tissue Engineers

Scientists, called tissue engineers, take a piece of skin (no bigger than a quarter) from an undamaged part of the burn victim's body. The skin cells are isolated, mixed with special nutrients, and then they multiply in a culture dish.

After about two to three months, the tissue engineers can harvest sheets of new, smooth skin. These sheets, as large as postcards, are grafted onto the victim's damaged body and promote additional skin growth.

By grafting Kelly's own skin on her body rather than using donor skin—skin from another person or from an animal—doctors avoid at least three potential complications. First, donor skin may not even be available. Second, Kelly's body might perceive the new skin cells from another source to be a danger, and her immune system might reject—or destroy—the transplant. Finally, even if the skin produced from a foreign source is accepted, it may leave extensive scarring.

Tissue Testing

What else can tissue engineers grow? They produce test skin—skin made in the lab and used to test the effects of cosmetics and chemicals on humans. This skin is eliminating the use of animals for such tests. Also, tissue engineers are working on ways to replace other body parts such as livers, heart valves, and ears, that don't grow back on their own.

Safety List Visit the link shown to the right or your media center to learn about fire safety tips, including kitchen safety and escape routes in your home. Make a list and share it with your family.

Science Online
For more information, visit red.msscience.com/time

Safety List Common fire safety tips include: placement and maintenance of smoke detectors on every floor of a residence; changing batteries in smoke detectors twice every year; insuring that every room in a residence has two exits in case of emergency; regular practice of fire escape routes by every family member.

Resources for Teachers and Students

Scar No More by Diane Martindale, Scientific American, July 2000

Growing New Organs by David J. Mooney and Antonios G. Mikos, Scientific American, September 1999

Reviewing Main Ideas

Section 1 The World of Cells

1. The cell theory states that all living things are made of one or more cells, the cell is the basic unit of life, and all cells come from other cells.
2. The microscope is an instrument that enlarges the image of an object.
3. All cells are surrounded by a cell membrane and contain hereditary material and cytoplasm. Plant cells have a cell wall outside the cell membrane. Cells, except bacteria, contain organelles.
4. The nucleus directs the cell's activities. Chromosomes contain DNA that determines what kinds of traits an organism will have. Vacuoles store substances.
5. In mitochondria, the process of cellular respiration combines food molecules with oxygen. This series of chemical reactions releases energy for the cell's activities.
6. The energy in light is captured and stored in food molecules during the process of photosynthesis. Plants, algae, and some bacteria make their own food by photosynthesis.

Section 2 The Different Jobs of Cells

1. Many-celled organisms are made up of different kinds of cells that perform different tasks.
2. Many-celled organisms are organized into tissues, organs, and organ systems that perform specific jobs to keep an organism alive.

Visualizing Main Ideas

Copy and complete the following concept map on the parts of a plant cell.

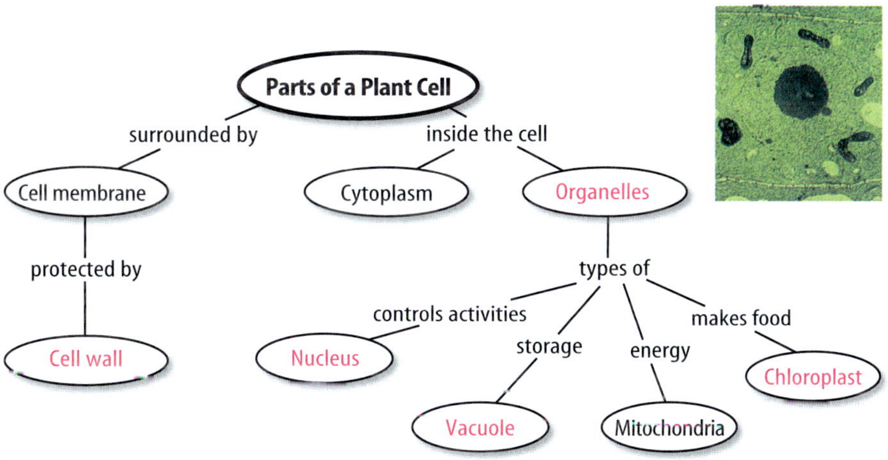

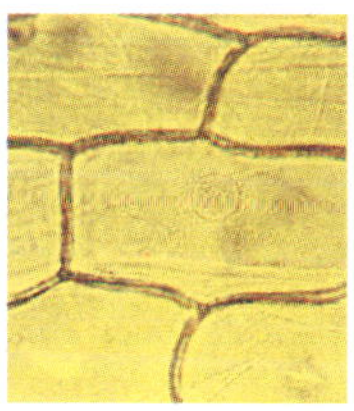

Reviewing Main Ideas

Summary statements can be used by students to review the major concepts of the chapter.

Visualizing Main Ideas

See student page.

Science Online

Visit red.msscience.com
/self_check_quiz
/interactive_tutor
/vocabulary_puzzlemaker
/chapter_review
/standardized_test

Assessment Transparency

For additional assessment questions, use the *Assessment Transparency* located in the transparency book.

Assessment

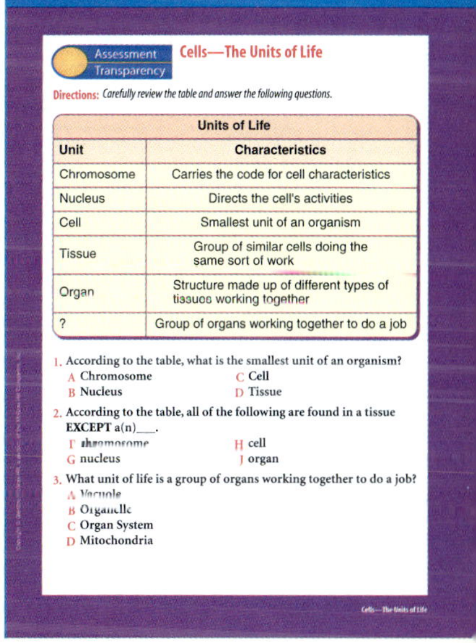

Assessment Transparency **Cells—The Units of Life**

Directions: *Carefully review the table and answer the following questions.*

Units of Life	
Unit	**Characteristics**
Chromosome	Carries the code for cell characteristics
Nucleus	Directs the cell's activities
Cell	Smallest unit of an organism
Tissue	Group of similar cells doing the same sort of work
Organ	Structure made up of different types of tissues working together
?	Group of organs working together to do a job

1. According to the table, what is the smallest unit of an organism?
   A Chromosome
   B Nucleus
   C Cell
   D Tissue
2. According to the table, all of the following are found in a tissue EXCEPT a(n)____.
   F chromosome
   G nucleus
   H cell
   J organ
3. What unit of life is a group of organs working together to do a job?
   A Vacuole
   B Organelle
   C Organ System
   D Mitochondria

Cells—The Units of Life

Using Vocabulary

1. Mitochondria convert food energy for cell use; chloroplasts capture energy from light to convert carbon dioxide and water into food.
2. tissue—group of similar cells that all do the same sort of work; organ—group of two or more different types of tissues that work together
3. The cell membrane helps control what enters and leaves a cell; the nucleus controls the cell's activities.
4. organ—group of two or more different types of tissues that work together; organ system—a group of organs that work together
5. The nucleus is an organelle that controls the cell's activities; an organelle is a cell part that performs life activities.
6. Cytoplasm is a gelatinlike substance in all cells; the nucleus is found in cytoplasm of some cells.
7. A vacuole is an organelle that stores food, water, minerals and wastes; mitochondria are organelles that convert food energy for cell use.
8. organ system—group of organs that work together; tissue—group of similar cells that work together
9. organelle—cell part that performs life activities for a cell; organ—two or more different types of tissues that work together to perform life activities for an organism
10. The cell wall provides support and protection for some cells; the cell membrane surrounds all cells.

Checking Concepts

11. B	14. C	17. B
12. B	15. C	18. D
13. D	16. A	19. B

chapter 16 Review

Using Vocabulary

bacteria p. 477	organ p. 487
cell membrane p. 479	organ system p. 487
cell wall p. 479	organelle p. 479
chloroplast p. 481	photosynthesis p. 481
cytoplasm p. 479	tissue p. 487
mitochondria p. 480	vacuole p. 480
nucleus p. 480	

Explain the difference between the terms in the following sets.

1. mitochondria—chloroplast
2. tissue—organ
3. cell membrane—nucleus
4. organ—organ system
5. nucleus—organelle
6. cytoplasm—nucleus
7. vacuole—mitochondria
8. organ system—tissue
9. organelle—organ
10. cell wall—cell membrane

Checking Concepts

Choose the word or phrase that best answers the question.

11. Which of the following controls what enters and leaves the cell?
    A) mitochondrion
    B) cell membrane
    C) vacuole
    D) nucleus
12. Which of the following are found inside the nucleus of the cell?
    A) vacuoles
    B) chromosomes
    C) chloroplasts
    D) mitochondria

Use the illustration below to answer questions 13 and 14.

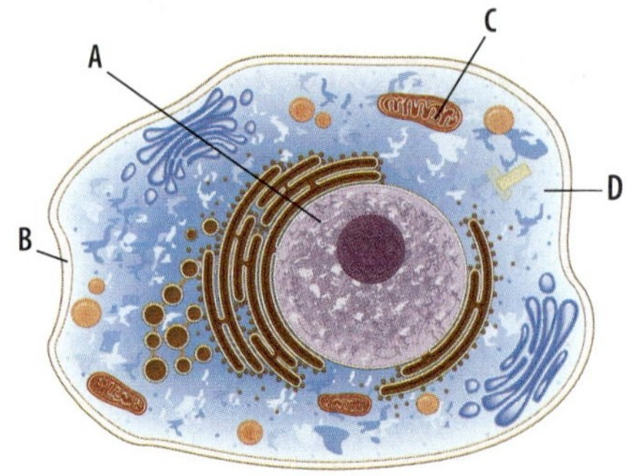

13. Which letter is the gelatinlike substance in a cell that contains water and chemicals?
    A) A **C)** C
    B) B **D)** D
14. Which structure converts food energy to a form of energy the cell can use?
    A) A **C)** C
    B) B **D)** D
15. Which of the following terms best describes the stomach?
    A) organelle **C)** organ
    B) organ system **D)** tissue
16. What does photosynthesis make for a plant?
    A) food **C)** water
    B) organs **D)** tissues
17. What does DNA do?
    A) makes food
    B) determines traits
    C) converts food to energy
    D) stores substances
18. Which of the following terms is the name of a human organ system?
    A) protective **C)** photosynthetic
    B) growth **D)** respiratory
19. What cell structure helps support plants?
    A) cell membrane **C)** vacuole
    B) cell wall **D)** nucleus

Science Online red.msscience.com/vocabulary_puzzlemaker

Use the ExamView® Pro Testmaker CD-ROM to:

- create multiple versions of tests
- create modified tests with one mouse click for inclusion students
- edit existing questions and add your own questions
- build tests aligned with state standards using built-in State Curriculum Tags
- change English tests to Spanish with one mouse click and vice versa

chapter 16 Review

Thinking Critically

20. Predict what would happen to a cell if the cell membrane were solid and waterproof.

21. Describe what might happen to a cell if all its mitochondria were removed.

22. Explain why cells are called the units of life.

23. Infer what kinds of animal cells might have a lot of mitochondria present.

24. Distinguish between a bacterium and a plant cell.

25. Compare and contrast photosynthesis and cellular respiration.

26. Make and Use Tables Copy and complete this table about the functions of the following cell parts: *nucleus, cell membrane, mitochondrion, chloroplast,* and *vacuole.*

Functions of Cell Parts

Cell Part	Function
Nucleus	See section
Cell membrane	1 of this
Mitochondrion	chapter
Chloroplast	for
Vacuole	answers.

27. Concept Map Make an events-chain concept map of the following from simple to complex: *small intestine, circular muscle cell, human,* and *digestive system.*

28. Identify and Manipulate Variables and Controls Describe an experiment you might do to determine whether water moves into and out of cells.

29. Recognize Cause and Effect Why is the brick-like shape of some plant cells important?

Performance Activities

30. Skit Working with three or four classmates, develop a short skit about how a living cell works. Have each group member play the role of a different cell part.

Applying Math

31. Magnification A microscope has an eyepiece with a power of 10× and an objective lens with a power of 40×. What is the magnification of the microscope?

32. Viruses Use a computer to make a line graph of the following data. At 37°C there are 1.0 million viruses; at 37.5°C, 0.5 milllion; at 37.8°C, 0.25 million; at 38.3°C, 0.1 million, and at 38.9°C, 0.05 million.

Use the graph below to answer question 33.

Food Production in Plants

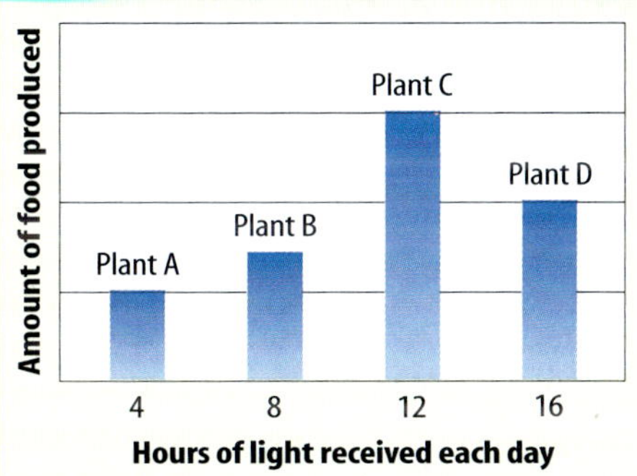

33. Plant Food Production Light is necessary for plants to make food. Using the graph above, determine which plant produced the most food. How much light was needed by the plant every day to produce the most food?

Assessment Resources

Reproducible Masters

Chapter *Fast File* Resources
- Chapter Review, pp. 35–36
- Chapter Tests, pp. 37–40
- Assessment Transparency Activity, p. 47

Glencoe Science Web site
- Chapter Review Test
- Standardized Test Practice

Glencoe Technology

- Assessment Transparency
- Exam*View*® Pro Testmaker
- MindJogger Videoquiz
- Interactive Chalkboard

chapter 16 Review

Thinking Critically

20. The cell would die because it couldn't take in materials or release wastes.

21. The cell would die because it would not have usable energy.

22. All of life's activities take place within the cell of one-celled organisms. In many-celled organisms, cells work together to complete all of life's activities.

23. cells with high energy requirements, such as muscle cells

24. The presence of organelles would make it a plant cell. Bacterial cells do not have organelles.

25. Possible answer: Photosynthesis and cellular respiration appear to be opposite processes. The starting materials of one process are the resulting materials of the other.

26. See student page.

27. Concept maps should show circular muscle cell → small intestine → digestive system → human.

28. Students should describe an experiment that demonstrates osmosis. Variables, controls, and the process should be clearly defined.

29. Answers may vary, but students should recognize that many plants are tall and need strong, rigid cells for support.

Performance Activities

30. Assess student skits by observing how detailed and informative the different characters' roles are. Use **Performance Assessment in the Science Classroom,** p. 147.

Applying Math

National Math Standards

1, 4, 5, 9

31. $10 \times 40 = 400$

32. Check student graphs to ensure consistency with data provided.

33. Plant C; 12 hours

FAST FILE

Answer Sheet A practice answer sheet can be found at red.msscience.com/answer_sheet.

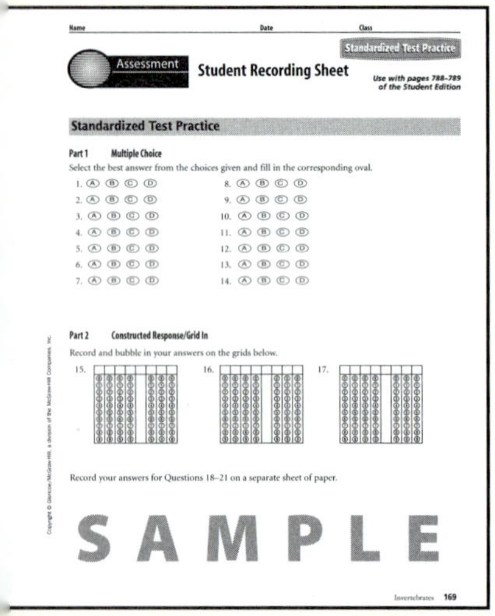
Name Date Class

Assessment **Student Recording Sheet** Standardized Test Practice

Use with pages 788–789 of the Student Edition

Standardized Test Practice

Part 1 Multiple Choice

Select the best answer from the choices given and fill in the corresponding oval.

1. Ⓐ Ⓑ Ⓒ Ⓓ 8. Ⓐ Ⓑ Ⓒ Ⓓ
2. Ⓐ Ⓑ Ⓒ Ⓓ 9. Ⓐ Ⓑ Ⓒ Ⓓ
3. Ⓐ Ⓑ Ⓒ Ⓓ 10. Ⓐ Ⓑ Ⓒ Ⓓ
4. Ⓐ Ⓑ Ⓒ Ⓓ 11. Ⓐ Ⓑ Ⓒ Ⓓ
5. Ⓐ Ⓑ Ⓒ Ⓓ 12. Ⓐ Ⓑ Ⓒ Ⓓ
6. Ⓐ Ⓑ Ⓒ Ⓓ 13. Ⓐ Ⓑ Ⓒ Ⓓ
7. Ⓐ Ⓑ Ⓒ Ⓓ 14. Ⓐ Ⓑ Ⓒ Ⓓ

Part 2 Constructed Response/Grid In

Record and bubble in your answers on the grids below.

15. 16. 17.

Record your answers for Questions 18–21 on a separate sheet of paper.

SAMPLE

Invertebrates 169

Part 1 Multiple Choice

1. D
2. B
3. D
4. A
5. A
6. D
7. C
8. C
9. B
10. A

Part 2 Short Response

11. Cells were not observed until microscopes were invented.
12. The cytoplasm of the cell is made up mostly of water.
13. The nucleus of the cell is like the manager of a bakery.
14. The process of cellular respiration produces the wastes of carbon dioxide and water that you exhale.
15. Plants produce oxygen and food during photosynthesis. Both of these would be important to keep an astronaut alive.
16. The type of microscope picture is a compound light microscope.
17. If both the eyepiece and the objective lens being used are 10×, the image of the onion skin cells would be magnified 100 times.

chapter 16 Standardized Test Practice

Part 1 Multiple Choice

Record your answers on the answer sheet provided by your teacher or on a sheet of paper.

1. The idea that all cells come from cells that already exist is part of what theory?
A. the microscopic theory
B. the basic theory
C. the Hooke theory
D. the cell theory

Use the illustration below to answer questions 2 and 3.

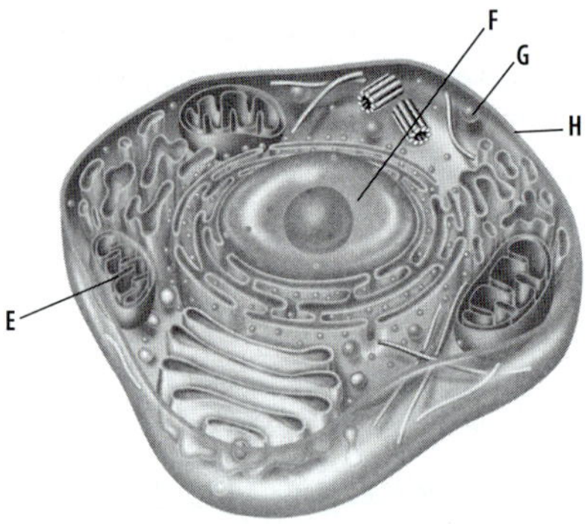

2. Which letter corresponds to the cell's nucleus?
A. E **C.** G
B. F **D.** H

3. Which letter corresponds to the part of the cell that helps control what enters and leaves the cell?
A. E **C.** G
B. F **D.** H

4. Chromosomes contain what important chemical that helps determine the traits of an organism?
A. DNA **C.** carbon dioxide
B. NAD **D.** oxygen

5. Which one of the following is an organ?
A. leg bone **C.** muscle cell
B. nucleus **D.** bacterium

6. Which of the following describes plant cells used to move water through the plant?
A. brick shaped with many chloroplasts
B. block shaped with no chloroplasts
C. long with fibers that contract and relax
D. long and tube-shaped

7. What are groups of cells that all do the same sort of work called?
A. organs **C.** tissues
B. organelles **D.** nerves

Use the illustration below to answer questions 8 and 9.

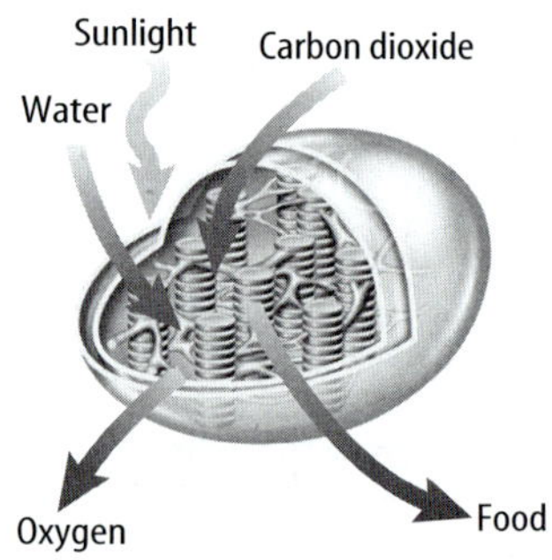

8. Where would this organelle likely be found?
A. in your brain **C.** in a leaf
B. in your heart **D.** in your bone

9. What process is taking place in this organelle?
A. cellular respiration
B. photosynthesis
C. food storage
D. cell reproduction

10. Which of the following pairs of organisms would have cells that were the most similar?
A. a dog and a cat
B. a turtle and a tree
C. a carrot and a cat
D. a fox and a daisy

494 STANDARDIZED TEST PRACTICE

Part 3 Open Ended

18. Possible answers: The nucleus controls most of the cell's activities. Vacuoles store food, water, minerals, and wastes. Mitochondria convert food energy into a form that the cell can use. Chloroplasts capture sunlight energy and use it to convert carbon dioxide and water into food.
19. Although the caterpillar's cells do not have chloroplasts to capture the energy of sunlight, the energy captured by the milkweed plant is passed on to the caterpillar when it eats the plant's leaves.
20. In order for the dog to jump and catch the ring, the dog's muscle system must work with its skeletal system. The contracting cells of muscle tissue pull the dog's bones and cause them to move allowing the dog to jump.

Part 2 | Short Response/Grid In

Record your answers on the answer sheet provided by your teacher or on a sheet of paper.

11. Cells were not observed until what instrument was invented?
12. The cytoplasm of cells is made mostly of what?
13. If a cell is compared to a bakery, what cell part would be like the manager of the bakery, in charge of the entire operation?
14. When you breathe out, your breath contains the waste products of carbon dioxide and water. What cellular process produces these wastes?
15. What two things produced by plants would be important to help keep astronauts alive on a long journey to Mars?

Use the illustration below to answer questions 16 and 17.

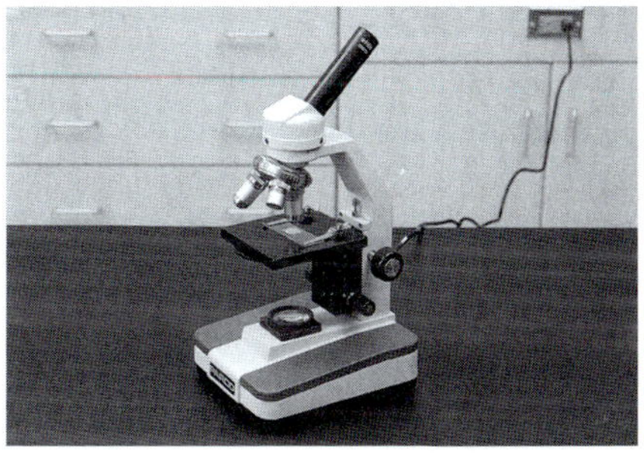

16. This type of microscope is used in most classrooms. What is the name of this type of microscope?
17. The magnification of the eyepiece of the microscope above is 10×. A student is looking at a slide of a piece of onion skin. If the magnification of the objective lens used is also 10×, how much is the microscope magnifying the image of the cells in the onion skin?

Part 3 | Open Ended

Record your answers on a sheet of paper.

18. Name three cellular organelles and describe the role of each.
19. A caterpillar eats the leaves of a milkweed plant. Explain why the caterpillar, like the milkweed plant, uses captured light energy.

Use the photo below to answer questions 20 and 21.

20. The dog could not jump to catch the ring with just its muscle system. What other system must work with muscles and how do the two systems work together?
21. Blood carries oxygen to cells. The dog's muscle cells require energy. Explain why the dog's heart might need to pump faster when the dog jumps.

Test-Taking Tip

Recall Experiences Remember to recall any hands-on experience as you read the question. Base your answer on the information given on the test.

Question 20 Think about what organ systems you use when you jump into the air.

21. To produce energy, mitochondria in the dog's muscle cells will need to use the process of cellular respiration. Oxygen is needed to combine with food to produce energy during this process. Since blood carries oxygen to cells, the dog's heart will need to pump faster to bring plenty of oxygen so the dog's muscle cells can produce enough energy for it to jump.

Rubrics

The following rubrics are sample scoring devices for short response and open-ended questions.

Short Response

Points	Description
2	The student demonstrates a thorough understanding of the science of the task. The response may contain minor flaws that do not detract from the demonstration of a thorough understanding.
1	The student has provided a response that is only partially correct.
0	The student has provided a completely incorrect solution or no response at all.

Open Ended

Points	Description
4	The student demonstrates a thorough understanding of the science of the task. The response may contain minor flaws that do not detract from the demonstration of a thorough understanding.
3	The student demonstrates an understanding of the science of the task. The response is essentially correct and demonstrates an essential but less than thorough understanding of the science.
2	The student demonstrates only a partial understanding of the science of the task. Although the student may have used the correct approach to a solution or may have provided a correct solution, the work lacks an essential understanding of the underlying science concepts.
1	The student demonstrates a very limited understanding of the science of the task. The response is incomplete and exhibits many flaws.
0	The student provides a completely incorrect solution or no response at all.

chapter 17 Organizer

Section/Objectives	Standards		Labs/Features
	National	**State/Local**	
Chapter Opener	See pp. 16T–17T for a Key to Standards.		**Launch Lab:** How are animals organized?, p. 497 **Foldables,** p. 497
Section 1 What is an animal? 1 session .5 block 1. **Identify** the characteristics of animals. 2. **Differentiate** between vertebrates and invertebrates. 3. **Explain** how the symmetry of animals differs.	National Content Standards: UCP.1–UCP.3, UCP.5, A.1, A.2, C.1		
Section 2 Sponges, Cnidarians, Flatworms, and Roundworms 2 sessions 1 block 4. **Describe** the structures that make up sponges and cnidarians. 5. **Compare** how sponges and cnidarians get food and reproduce. 6. **Differentiate** between flatworms and roundworms.	National Content Standards: UCP.1–UCP.3, UCP.5, A.1, A.2, C.1–C.3, C.5		**Integrate Chemistry,** p. 502
Section 3 Mollusks and Segmented Worms 2 sessions 1 block 7. **Identify** the characteristics of mollusks. 8. **Compare** the similarities and differences between an open and a closed circulatory system. 9. **Describe** the characteristics of segmented worms. 10. **Explain** the digestive process of an earthworm.	National Content Standards: UCP.1–UCP.3, UCP.5, A.1, A.2, C.1–C.3, C.5		**Integrate Social Studies,** p. 507 **Science Online,** p. 507 **MiniLAB:** Modeling Cephalopod Propulsion, p. 508 **Applying Science:** How does soil management affect earthworms?, p. 510
Section 4 Arthropods and Echinoderms 4 sessions 2 blocks 11. **List** the features used to classify arthropods. 12. **Explain** how the structure of the exoskeleton relates to its function. 13. **Identify** features of echinoderms.	National Content Standards: UCP.1–UCP.3, UCP.5, A.1, A.2, C.1–C.3, C.5		**Science Online,** p. 513 **Visualizing Arthropod Diversity,** p. 514 **MiniLAB:** Observing Sow Bugs, p. 517 **Lab:** Observing Complete Metamorphosis, p. 519 **Lab:** Garbage-Eating Worms, p. 520 **Science Stats:** Squid Power, p. 522

Glencoe Exclusive!

TeacherWorks™

All-In-One Planner and Resource Center

Lab Materials	Reproducible Resources	Section Assessment	Technology
Launch Lab: large envelopes, index cards, colored pencils, magazines with animal pictures	**Chapter *FAST FILE* Resources** Foldables Worksheet, p. 17 Directed Reading Overview, p. 17 Note-taking Worksheets, pp. 35–37	GLENCOE'S ASSESSMENT ADVANTAGE	**TeacherWorks** includes: • Interactive Teacher Edition • Lesson Planner with calendar • Access to all program blacklines • Correlations to standards • Web links
	Chapter *FAST FILE* Resources Transparency Activity, p. 46 Enrichment, p. 31 Reinforcement, p. 27 Directed Reading, p. 20	**Portfolio** Assessment, p. 497 **Performance** Applying Skills, p. 500 **Content** Section Review, p. 500	Section Focus Transparency Virtual Labs CD-ROM Guided Reading Audio Program Interactive Chalkboard CD-ROM
Need materials? Contact Science Kit at 1-800-828-7777 or www.sciencekit.com on the Internet.	**Chapter *FAST FILE* Resources** Transparency Activity, p. 47 Enrichment, p. 32 Reinforcement, p. 28 Directed Reading, p. 20	**Portfolio** Make a Model, p. 503 **Performance** Applying Math, p. 505 **Content** Section Review, p. 505	Section Focus Transparency Virtual Labs CD-ROM Guided Reading Audio Program Interactive Chalkboard CD-ROM
MiniLAB: balloon	**Chapter *FAST FILE* Resources** Transparency Activity, p. 48 MiniLAB, p. 3 Enrichment, p. 33 Reinforcement, p. 29 Directed Reading, p. 21 Lab Activity, pp. 9–12	**Portfolio** MiniLAB Assessment, p. 508 **Performance** MiniLAB, p. 508 Applying Science, p. 510 Applying Skills, p. 511 **Content** Section Review, p. 511	Section Focus Transparency Virtual Labs CD-ROM Guided Reading Audio Program Interactive Chalkboard CD-ROM
MiniLAB: flat container with cover, sow bugs (6), sponge **Lab:** jar, bran or oatmeal, cookie crumbs, flour, apple or carrot, paper towel, cheesecloth, rubber band, mealworms **Lab:** red wiggler worms, 4-L plastic containers, soil, shredded newspaper, spray bottle, chopped food scraps	**Chapter *FAST FILE* Resources** Transparency Activity, p. 49 MiniLAB, p. 4 Enrichment, p. 34 Reinforcement, p. 30 Directed Reading, pp. 21, 22 Lab Worksheet, pp. 5–6, 7–8 Lab Activity, pp. 13–16 Transparency Activity, pp. 51–52	**Portfolio** Differentiated Instruction, p. 515 **Performance** MiniLAB, p. 517 Applying Math, p. 518 **Content** Section Review, p. 518	Section Focus Transparency Teaching Transparency Virtual Labs CD-ROM Guided Reading Audio Program Interactive Chalkboard CD-ROM Video Lab

End of Chapter Assessment

GLENCOE'S ASSESSMENT ADVANTAGE

Blackline Masters	Technology	Professional Series
Chapter *FAST FILE* Resources Chapter Review, pp. 39–40 Chapter Tests, pp. 41–44 **Standardized Test Practice,** pp. 71–74	MindJogger Videoquiz Virtual Labs CD-ROM Exam*View*® Pro Testmaker TeacherWorks CD-ROM Interactive Chalkboard CD-ROM	**Performance Assessment in the Science Classroom (PASC)**

chapter 17 Invertebrate Animals

Transparencies

Section Focus

SECTION 1 Section Focus Transparency **Jelly Sea**

Jellyfish are interesting animals whose bodies are comprised mostly of water. They drift with the currents, but they are also able to move by expelling a jet of water.

1. How are jellyfish similar to you? How are they different?
2. Describe a jellyfish's shape.
3. Why are jellyfish classified as animals?

L2

SECTION 2 Section Focus Transparency **Rub-a-dub-dub**

When you clean something with a sponge, you may be using the skeleton of an animal! People harvest sponges in areas like the Mediterranean Sea and the Gulf of Mexico. Most sponges you buy in stores, however, aren't animals at all; they're artificial sponges.

1. Why might scientists have originally thought sponges were plants?
2. How do animals get food? How do plants get food?

L2

SECTION 3 Section Focus Transparency **I'll Have One for Dinner**

One thing these animals have in common is they're occasionally served as dinner. Food is one way people use them, but they are also important for other reasons. For example, some of these creatures live in the sea near the shore and filter large amounts of water each day.

1. What similarities do these three animals share?
2. Describe some of the differences among the three animals.

L2

Assessment

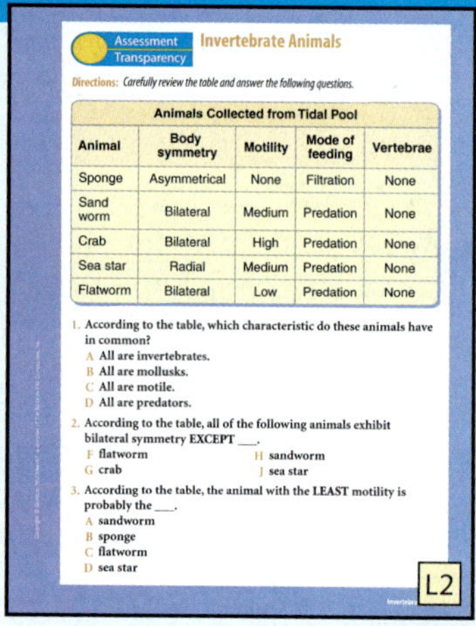
Assessment Transparency **Invertebrate Animals**

Directions: Carefully review the table and answer the following questions.

Animals Collected from Tidal Pool				
Animal	Body symmetry	Motility	Mode of feeding	Vertebrae
Sponge	Asymmetrical	None	Filtration	None
Sand worm	Bilateral	Medium	Predation	None
Crab	Bilateral	High	Predation	None
Sea star	Radial	Medium	Predation	None
Flatworm	Bilateral	Low	Predation	None

1. According to the table, which characteristic do these animals have in common?
   A All are invertebrates.
   B All are mollusks.
   C All are motile.
   D All are predators.
2. According to the table, all of the following animals exhibit bilateral symmetry EXCEPT ___.
   F flatworm
   G crab
   H sandworm
   J sea star
3. According to the table, the animal with the LEAST motility is probably the ___.
   A sandworm
   B sponge
   C flatworm
   D sea star

L2

Teaching

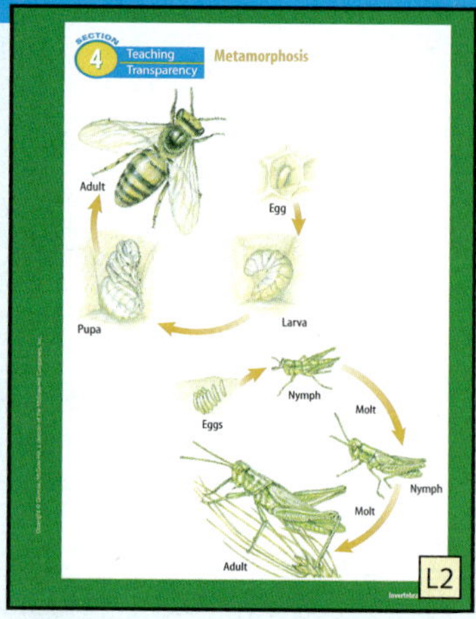
SECTION 4 Teaching Transparency **Metamorphosis**

L2

This is a representation of key blackline masters available in the Teacher Classroom Resources. See Resource Manager boxes within the chapter for additional information.

Key to Teaching Strategies

The following designations will help you decide which activities are appropriate for your students.

L1 Level 1 activities should be appropriate for students with learning difficulties.

L2 Level 2 activities should be within the ability range of all students.

L3 Level 3 activities are designed for above-average students.

ELL ELL activities should be within the ability range of English Language Learners.

COOP LEARN Cooperative Learning activities are designed for small group work.

LS Multiple Learning Styles logos, as described on page 12T, are used throughout to indicate strategies that address different learning styles.

P These strategies represent student products that can be placed into a best-work portfolio.

PBL Problem-Based Learning activities apply real-world situations to learning.

Hands-on Activities

Student Text Lab Worksheet

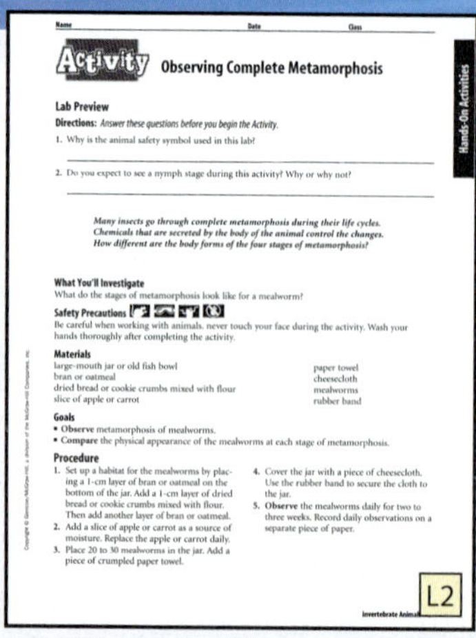
Name Date Class

Activity Observing Complete Metamorphosis

Hands-On Activities

Lab Preview

Directions: *Answer these questions before you begin the Activity.*

1. Why is the animal safety symbol used in this lab?
2. Do you expect to see a nymph stage during this activity? Why or why not?

Many insects go through complete metamorphosis during their life cycles. Chemicals that are secreted by the body of the animal control the changes. How different are the body forms of the four stages of metamorphosis?

What You'll Investigate
What do the stages of metamorphosis look like for a mealworm?

Safety Precautions
Be careful when working with animals. never touch your face during the activity. Wash your hands thoroughly after completing the activity.

Materials
large-mouth jar or old fish bowl
bran or oatmeal
dried bread or cookie crumbs mixed with flour
slice of apple or carrot
paper towel
cheesecloth
mealworms
rubber band

Goals
- **Observe** metamorphosis of mealworms.
- **Compare** the physical appearance of the mealworms at each stage of metamorphosis.

Procedure
1. Set up a habitat for the mealworms by placing a 1-cm layer of bran or oatmeal on the bottom of the jar. Add a 1-cm layer of dried bread or cookie crumbs mixed with flour. Then add another layer of bran or oatmeal.
2. Add a slice of apple or carrot as a source of moisture. Replace the apple or carrot daily.
3. Place 20 to 30 mealworms in the jar. Add a piece of crumpled paper towel.
4. Cover the jar with a piece of cheesecloth. Use the rubber band to secure the cloth to the jar.
5. **Observe** the mealworms daily for two to three weeks. Record daily observations on a separate piece of paper.

L2

Laboratory Activities

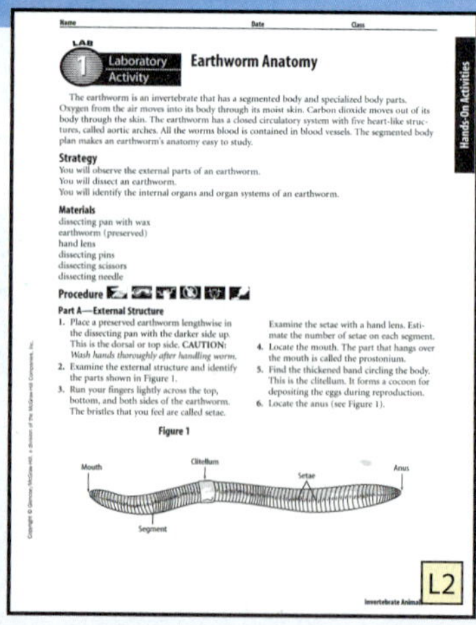
Name Date Class

LAB 1 Laboratory Activity **Earthworm Anatomy**

Hands-On Activities

The earthworm is an invertebrate that has a segmented body and specialized body parts. Oxygen from the air moves into its body through its moist skin. Carbon dioxide moves out of its body through the skin. The earthworm has a closed circulatory system with five heart-like structures, called aortic arches. All the worms blood is contained in blood vessels. The segmented body plan makes an earthworm's anatomy easy to study.

Strategy
You will observe the external parts of an earthworm.
You will dissect an earthworm.
You will identify the internal organs and organ systems of an earthworm.

Materials
dissecting pan with wax
earthworm (preserved)
hand lens
dissecting pins
dissecting scissors
dissecting needle

Procedure

Part A—External Structure
1. Place a preserved earthworm lengthwise in the dissecting pan with the darker side up. This is the dorsal or top side. **CAUTION:** *Wash hands thoroughly after handling worm.*
2. Examine the external structure and identify the parts shown in Figure 1.
3. Run your fingers lightly across the top, bottom, and both sides of the earthworm. The bristles that you feel are called setae. Examine the setae with a hand lens. Estimate the number of setae on each segment.
4. Locate the mouth. The part that hangs over the mouth is called the prostonium.
5. Find the thickened band circling the body. This is the clitellum. It forms a cocoon for depositing the eggs during reproduction.
6. Locate the anus (see Figure 1).

Figure 1

L2

Meeting Different Ability Levels

Content Outline

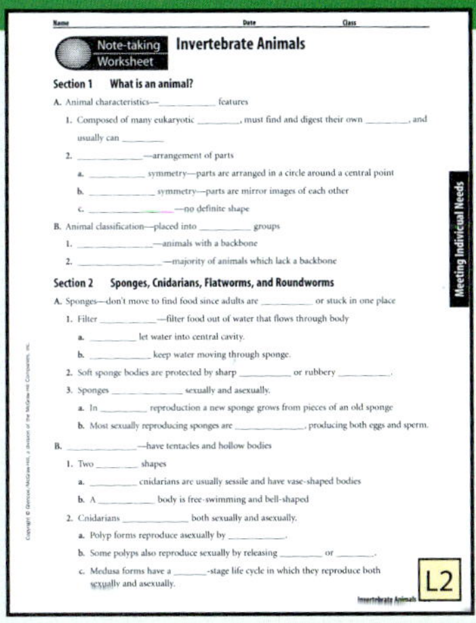

Reinforcement

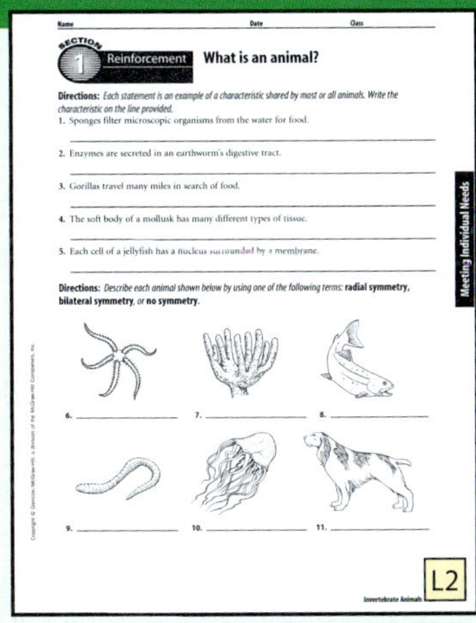

Enrichment

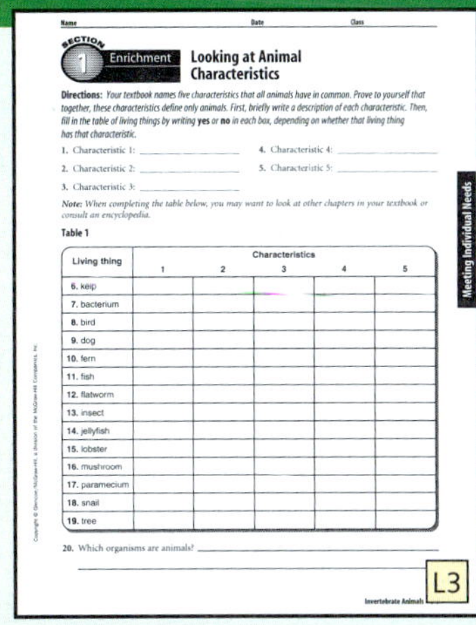

Directed Reading (English/Spanish)

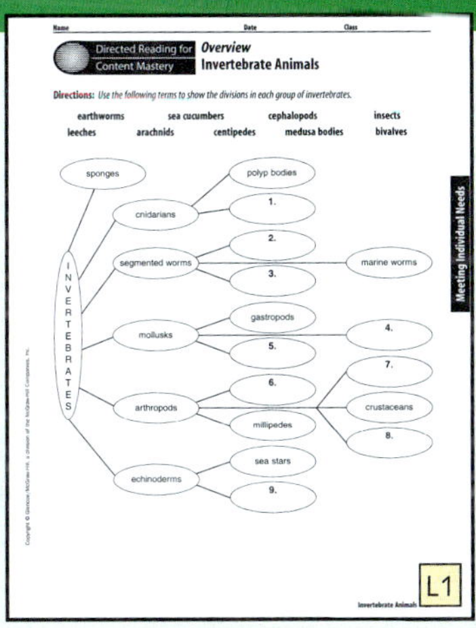

Study Guide

Reading Essentials

Assessment

Test Practice Workbook

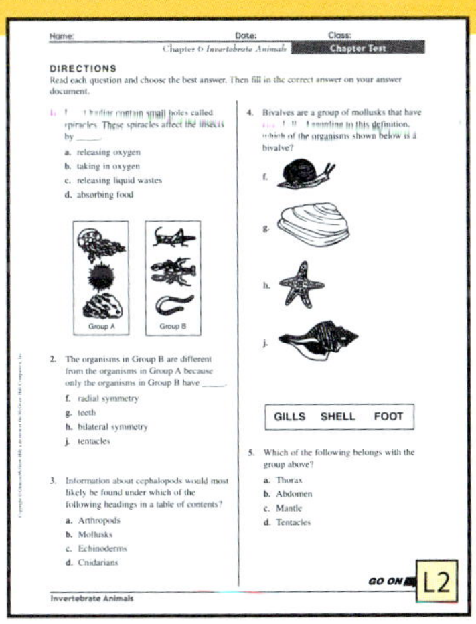

Chapter Review

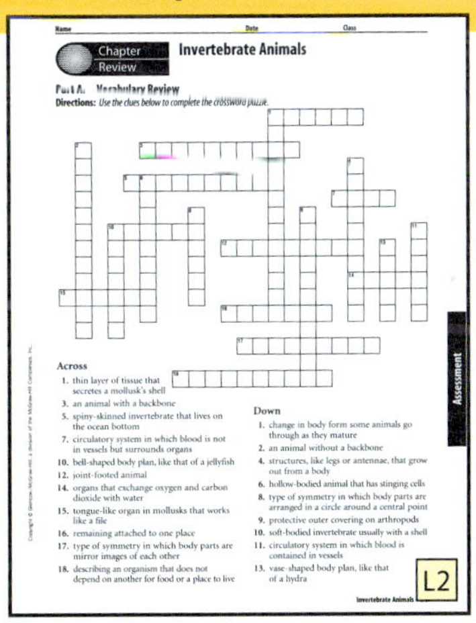

Chapter Tests

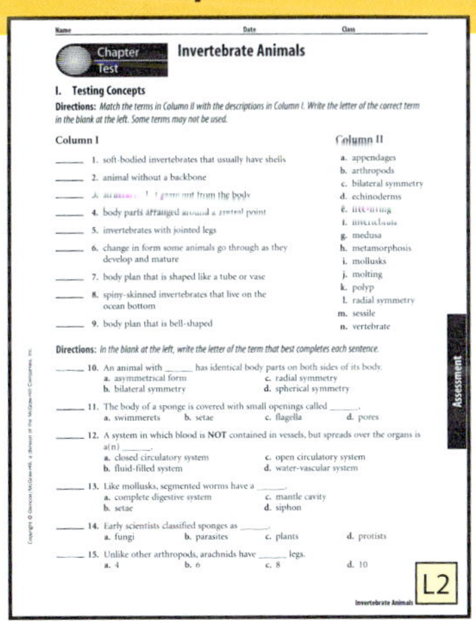

Science Content Background

section 1 What is an animal?

All About Animals

Scientists hypothesize that single-celled protozoan ancestors gave rise to multicellular animals. These protozoans left no fossils, making it impossible to trace the animal kingdom through its origins. The transition from single-celled protozoan to multicellular animal most likely occurred more than once. Certain sponge cells are almost identical to a specific type of flagellated protozoa—choanoflagellate—leading some zoologists to reason that protists gave rise to modern sponges.

Some important features that help distinguish animal phyla from one another are symmetry, body cavities, digestive cavities, and segmentation. An animal's symmetry often provides insight into how the organism survives in its environment. With food and other sensory information coming from all directions, animals with radial symmetry often are either sessile or slow moving. An important aspect of organisms exhibiting bilateral symmetry is cephalization—the tendency for sense organs to be concentrated in the head region and focused toward the direction in which the organism moves.

Sponges, cnidarians, and flatworms are, for the most part, masses of cells and do not have body cavities in which internal organs are found. Other organisms—roundworms, mollusks, segmented worms, arthropods, and echinoderms—do have body cavities. Most animals have a digestive cavity that collects food and secretes digestive enzymes. The digestive cavity may be a one-way sac, as in the sea anemone, or it may be a tube with two openings, as in segmented worms.

Segmentation—the repeating of parts that contain similar organs—is a characteristic of three of the largest animal phyla: annelids (segmented worms), arthropods, and chordates (vertebrates).

section 2 Sponges, Cnidarians, Flatworms, and Roundworms

Sponges

Sponges—phylum Porifera—are the least complex of all the major animal phyla. They do not have nerve cells or sense organs. Their body cells are not even organized into tissues. Although food is taken into the central cavity of a sponge, the cavity is not a digestive cavity. Digestion of the food takes place in either collar cells or amoeboid cells.

Cnidarians

All cnidarians are carnivorous. Their stinging cells contain a coiled thread that is discharged

chapter content resources

Internet Resources

For additional content background, visit **red.msscience.com** to:

- access your book online
- find references to related articles in popular science magazines
- access Web links with related content background
- access current events with science journal topics

Print Resources

Invertebrate Zoology, by Ellen Doris and Len Rubenstein, Thames & Hudson, 1993

The Life Cycle of a Butterfly, by Bobbie Kalman, Crabtree Publishers, 2001

Classifying Insects, by Andrew Solway, Heinemann Library, 2003

when a "trigger" on the cell is touched by prey. The threads in some species contain paralyzing toxins. Cnidarians' outer tissue layer is the epidermis, and their inner layer is the gastrodermis. Between these two layers of tissue is a jellylike layer. Cnidarians lack true organs. Cells exchange oxygen and wastes with their environment by diffusion.

Flatworms

Flatworms—phylum Platyhelminthes—are the simplest organisms with bilateral symmetry. The brain and nerve cords that run the length of the body are considered to be the evolutionary forerunners of a central nervous system. The middle layer of the flatworm contains muscle tissue used for locomotion. Flatworms also have complex reproductive and digestive systems. Although many flatworms are parasites, freshwater planaria are free-living and easily studied in the classroom.

Roundworms

Roundworms belong to the phylum Nematoda. The disease elephantiasis occurs when the roundworm *Wuchereria* blocks lymphatic vessels.

Teacher to Teacher

Steve Federman
Loveland Middle School
Loveland, Ohio

"This year try keeping a culture of mealworms living in an oatmeal box or plastic container. Your students will be able to observe the constant lifecycle changes of mealworm metamorphosis. Have each student create a habitat for his or her very own mealworm from a clear plastic cup, oatmeal, and a bit of apple. Students can study their mealworms' growth and record the data."

Steve Federman

section 3 Mollusks and Segmented Worms

Mollusks

Mollusks are the second-largest animal phyla, with more than 100,000 identified species. Segmented worms of the phylum Annelida comprise about 10,000 species and are divided into three major classes: oligochaetes (earthworms), polychaetes (marine bristleworms and tubeworms), and hirudineans (leeches).

Segmented Worms

Earthworm aeration of the soil is so beneficial that their presence in hundreds or thousands per hectare is a characteristic of a productive farm. Earthworms survive in thin films of water in slightly moist soil, but if the soil becomes flooded, they can drown.

section 4 Arthropods and Echinoderms

Arthropods

Arthropods are the most diverse group of animals on Earth. More than one million species have been identified. Flight is an important adaptation that helps insects find food and mates, escape predators, and spread to new areas.

Echinoderms

Echinoderms, such as sea stars and sea urchins, have a number of traits that set them apart from other invertebrates. The radial symmetry of these organisms is fivefold, or pentamerous. Fossil evidence indicates that early echinoderms were bilaterally symmetrical and only later developed radial symmetry suited for a sessile or slow-moving organism. The larvae of echinoderms are bilaterally symmetrical.

chapter 17

Chapter Vocabulary

symmetry, p. 499
invertebrate, p. 500
cnidarian, p. 502
polyp, p. 503
medusa, p. 503
mollusk, p. 506
mantle, p. 506
gill, p. 506
radula, p. 507
open circulatory system, p. 507
closed circulatory system, p. 508
arthropod, p. 512
appendage, p. 512
exoskeleton, p. 512
metamorphosis, p. 513

Science Journal Students should be able to infer that both invertebrates and humans are animals. Humans have a backbone, while invertebrates do not.

INTERACTIVE CHALKBOARD with Image Bank

PowerPoint® Presentations

This CD-ROM is an editable Microsoft® PowerPoint® presentation that includes:

- a pre-made presentation for every chapter
- interactive graphics
- animations
- audio clips
- image bank
- all new section and chapter questions
- Standardized Test Practice
- transparencies
- pre-lab questions for all labs
- Foldables directions
- links to red.msscience.com

chapter 17

Invertebrate Animals

chapter preview

sections

1. What is an animal?
2. Sponges, Cnidarians, Flatworms, and Roundworms
3. Mollusks and Segmented Worms
4. Arthropods and Echinoderms
   - *Lab Observing Complete Metamorphosis*
   - *Lab Garbage-Eating Worms*
   - *Virtual Lab How are mollusks, worms, arthropods, and echinoderms classified?*

Underwater Rhythmic Dancers

Corals and anemones sway with ocean currents. Other animals, like the lemon-peel nudibranch, move in ways that animals that have internal skeletons cannot move. They belong to a group of animals called invertebrates—animals without backbones.

Science Journal Describe similarities and differences between you and the nudibranch.

496

Theme Connection

Stability and Change The groups of animals from invertebrates to vertebrates show a trend toward increasing complexity. Many animals in both groups exhibit adaptations to the environment around them. Others have characteristics that enable them to meet the challenges of an ever-changing environment.

About the Photo

Marine Communities This community of marine invertebrates includes crimson anemones, a lemon-peel nudibranch, and *Gersemia* soft corals. The lemon peel nudibranch can be found in waters from Alaska to southern California. It can grow to be 20 cm in length.

Start-Up Activities

Launch LAB

How are animals organized?

Scientists have identified at least 1.5 million different kinds of animals. In the following lab, you will learn about organizing animals by building a bulletin board display.

1. Write the names of different groups of animals on large envelopes and attach them to a bulletin board.
2. Choose an animal group to study. Make an information card about each animal with its picture on one side and characteristics on the other side.
3. Place your finished cards inside the appropriate envelope.
4. Select an envelope from the bulletin board for a different group of animals. Using the information on the cards, sort the animals into groups.
5. **Think Critically** What common characteristics do these animals have? What characteristics did you use to classify them into smaller groups? Record your answers in your Science Journal.

Science Online Preview this chapter's content and activities at red.msscience.com

FOLDABLES Study Organizer

Invertebrates Make the following Foldable to compare and contrast the characteristics of water and land invertebrates.

STEP 1 **Fold** one sheet of paper lengthwise.

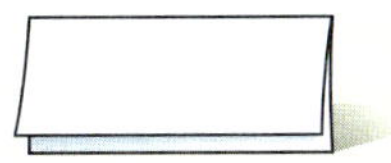

STEP 2 **Fold** into thirds.

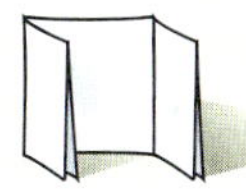

STEP 3 **Unfold and draw** overlapping ovals. **Cut** the top sheet along the folds.

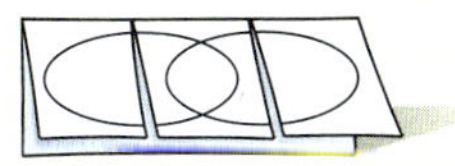

STEP 4 **Label** the ovals as shown.

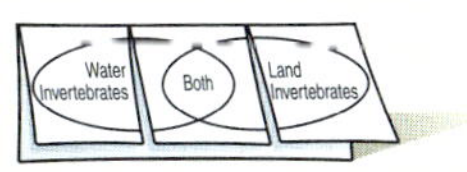

Construct a Venn Diagram As you read this chapter, list the characteristics unique to water invertebrates under the left tab, those unique to land invertebrates under the right tab, and those characteristics common to both under the middle tab.

497

Launch LAB

Purpose Use the Launch Lab to introduce the characteristics all animals share. Explain to students that they will learn more about animal characteristics and how scientists classify animals into groups with similar characteristics. L2 COOP LEARN IN **Visual-Spatial**

Preparation Clear bulletin board space for the placement of animal group envelopes.

Materials large envelopes (20), 4 × 6 index cards, colored pencils, magazines with animal pictures

Teaching Strategy Ask students which animals they are most familiar with. They will probably say dogs, cats, hamsters, fish, and other pets or vertebrates.

Think Critically

Possible response: similar body shapes and coverings; Possible response: similar methods of movement and food gathering

Assessment

Portfolio Have students make extra cards of their favorite animals from the groups they worked with to place into their portfolios. Use **Performance Assessment in the Science Classroom**, p. 131. P

FOLDABLES Study Organizer Dinah Zike Study Fold

Student preparation materials for this Foldable are available in the Chapter *Fast File* Resources.

1 Motivate

Bellringer

Section Focus Transparencies also are available on the Interactive Chalkboard CD-ROM. L2 ELL

Section 1 Section Focus Transparency **Jelly Sea**

Jellyfish are interesting animals whose bodies are comprised mostly of water. They drift with the currents, but they are also able to move by expelling a jet of water.

1. How are jellyfish similar to you? How are they different?
2. Describe a jellyfish's shape.
3. Why are jellyfish classified as animals?

Invertebrate Animals

Tie to Prior Knowledge

Animal Traits Students will be familiar with many animal characteristics from observing pets. Remind students that animals cannot make their own food and are made of many nucleated cells.

section 1

What is an animal?

as you read

What You'll Learn

- **Identify** the characteristics of animals.
- **Differentiate** between vertebrates and invertebrates.
- **Explain** how the symmetry of animals differs.

Why It's Important

All animals have characteristics in common.

Review Vocabulary

organelle: structure in the cytoplasm of a eukaryotic cell that can act as a storage site, process energy, move materials, or manufacture substances

New Vocabulary

- symmetry
- invertebrate

Animal Characteristics

If you asked ten people for a characteristic common to all animals, you might get ten different answers or a few repeated answers. Look at the animals in **Figure 1.** What are their common characteristics? What makes an animal an animal?

1. Animals are many-celled organisms that are made of different kinds of cells. These cells might digest food, get rid of wastes, help in reproduction, or be part of systems that have these functions.
2. Most animal cells have a nucleus and organelles. The nucleus and many organelles are surrounded by a membrane. This type of cell is called a eukaryotic (yew ker ee AH tihk) cell.
3. Animals cannot make their own food. Some animals eat plants to supply their energy needs. Some animals eat other animals, and some eat both plants and animals.
4. Animals digest their food. Large food particles are broken down into smaller substances that their cells can use.
5. Most animals can move from place to place. They move to find food, shelter, and mates, and to escape from predators.

Figure 1 Animals come in a variety of shapes and sizes.

Monarch butterflies in North America migrate up to 5,000 km each year.

The lion's mane jellyfish can be found in the cold, arctic water and the warm water off the coasts of Florida and Mexico. Their tentacles can be up to 30 m long.

The platypus lives in Australia. It is an egg-laying mammal.

Section 1 Resource Manager

Chapter *FAST FILE* Resources

Transparency Activity, p. 46
Directed Reading for Content Mastery, pp. 19, 20
Note-taking Worksheets, pp. 35–37
Enrichment, p. 31
Reinforcement, p. 27

Figure 2 Most animals have radial or bilateral symmetry. Only a few animals are asymmetrical.

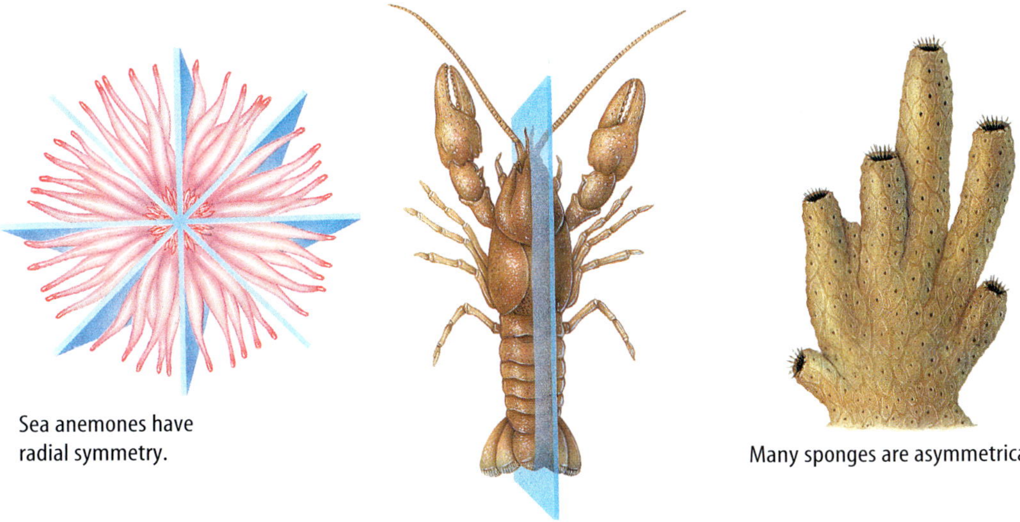

Sea anemones have radial symmetry.

Lobsters have bilateral symmetry.

Many sponges are asymmetrical.

Symmetry As you study the different groups of animals, you will look at their symmetry (SIH muh tree). **Symmetry** refers to the arrangement of the individual parts of an object that can be divided into similar halves.

Most animals have either radial symmetry or bilateral symmetry. Animals with body parts arranged in a circle around a central point have radial symmetry. Can you imagine being able to locate food and gather information from all directions? Aquatic animals with radial symmetry, such as jellyfish, sea urchins, and the sea anemone, shown in **Figure 2,** can do that. On the other hand, animals with bilateral symmetry have parts that are nearly mirror images of each other. A line can be drawn down the center of their bodies to divide them into two similar parts. Grasshoppers, lobsters, like the one in **Figure 2,** and humans are bilaterally symmetrical.

Some animals have an irregular shape. They are called asymmetrical (AY suh meh trih kul). They have bodies that cannot be divided into similar halves. Many sponges, like those also in **Figure 2,** are asymmetrical. As you learn more about invertebrates, notice how their body symmetry is related to how they gather food and do other things.

Reading Check *What is symmetry?*

2 Teach

Quick Demo

Symmetry

Possible Materials rocks, chalk, and other classroom objects

Estimated Time five minutes

Procedure Have students compare symmetrical and asymmetrical objects.

Activity

Radial Symmetry Show students pictures of radially symmetrical animals (i.e. sea anemones, jellyfish). Challenge them to indicate the head of the animal. Explain that most animals with radial symmetry do not have a definite head. Have students examine pictures of animals such as dogs or cats that have bilateral symmetry. Again challenge them to indicate the location of the head. Point out that in animals with bilateral symmetry, sensory organs tend to be located in the head. Note that it is difficult to see the "head" in some bilaterally symmetrical animals, like roundworms. L1 LS **Visual-Spatial**

Identifying Misconceptions

Invertebrate Animals Many students do not think of invertebrates as animals. Elicit student help in writing a list of familiar invertebrates on the board. Stress that the animals listed are grouped together by the absence of a structural characteristic—a backbone—rather than the presence of a characteristic.

Reading Check

Answer the arrangement of the individual parts of an object

Differentiated Instruction

English-Language Learners Have students write a fictional report about a new species of invertebrate they found. Have them use vocabulary words from the chapter when describing this new invertebrate.

Teacher FYI

Animal Diversity Animals constitute over 75 percent of Earth's species and their mobility allows great diversity. The ability assists in feeding, defense, and reproduction. Because animals can move, they have adapted to nearly every environment on Earth.

Caption Answer

Figure 3 three percent

Visual Learning

Figure 3 What animal groups are classified as invertebrates? sponges, cnidarians, flatworms, roundworms, mollusks, annelids, arthropods, and echinoderms L1 LS **Visual-Spatial**

3 Assess

DAILY INTERVENTION

Check for Understanding

Logical-Mathematical How does having a head help an animal? The head is the site of many sensory organs and brain structures that process sensory information. An animal moving headfirst can perceive the environment before its body enters, giving the animal a chance to retreat if necessary. L3 LS

Reteach

Symmetry Type Have students look through magazines to find pictures of animals. Then have them work in groups to identify each animal's type of body symmetry. L1 COOP LEARN LS **Visual-Spatial**

Assessment

Performance Have students formulate additional categories for classifying animals after identifying an animal's symmetry and whether it has a backbone. Use **Performance Assessment in the Science Classroom**, p. 121. L3

Animal Classification

Deciding whether an organism is an animal is only the first step in classifying it. Scientists place all animals into smaller, related groups. They can begin by separating animals into two distinct groups—vertebrates and invertebrates. Vertebrates (VUR tuh bruts) are animals that have a backbone. **Invertebrates** (ihn VUR tuh bruts) are animals that do not have a backbone. About 97 percent of all animals are invertebrates.

Scientists classify the invertebrates into smaller groups, as shown in **Figure 3.** The animals within each group share similar characteristics. These characteristics indicate that the animals within the group may have had a common ancestor.

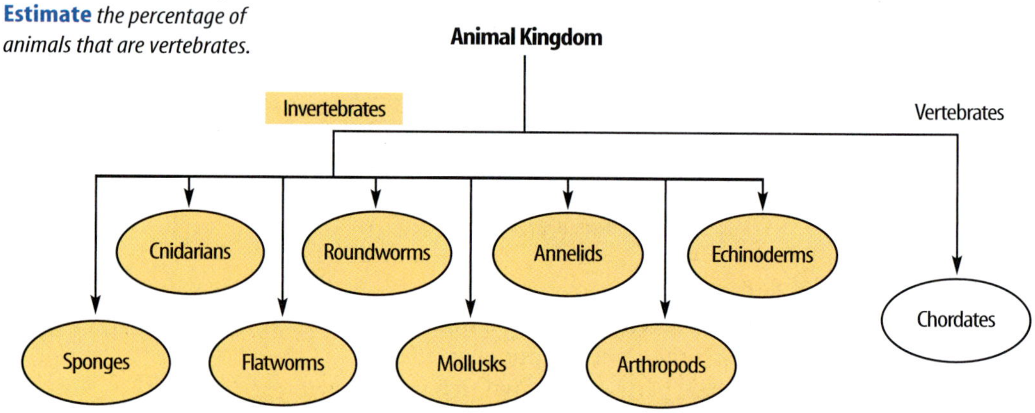

Figure 3 This diagram shows the relationships among different groups in the animal kingdom.
Estimate *the percentage of animals that are vertebrates.*

section 1 review

Summary

Animal Characteristics

- Animals are made up of many different kinds of cells.
- Most animal cells have a nucleus and organelles.
- Animals cannot make their own food.
- Animals digest their food.
- Most animals can move from place to place.

Animal Classification

- Scientists place all animals into smaller, related groups.
- Two distinct groups of animals are the invertebrates and vertebrates.

Self Check

1. **Compare and contrast** invertebrate and vertebrate animals.
2. **Describe** the different types of symmetry. Name an animal that has bilateral symmetry.
3. **Think Critically** Most animals do not have a backbone. They are called invertebrates. What are some advantages that invertebrate animals might have over vertebrate animals?

Applying Skills

4. **Concept Map** Using the information in this section, make a concept map showing the steps a scientist might use to classify a newly discovered animal.

Science Online red.msscience.com/self_check_quiz

section 1 review

1. Invertebrates lack a backbone; vertebrates have a backbone.
2. radial, bilateral, and asymmetrical; possible response: dog
3. Accept all reasonable answers.
4. Concept maps should show the identification of the animal as a vertebrate or invertebrate, then the determination of its symmetry, and finally a comparison with other organisms.

section 2

Sponges, Cidarians, Flatworms, and Roundworms

Sponges

Can you tell the difference between an animal and a plant? Sounds easy, doesn't it? But for a long time, even scientists didn't know how to classify sponges. Originally they thought sponges were plants because they don't move to search for food. Sponges, however, can't make their own food as most plants do. Sponges are animals. Adult sponges are sessile (SE sul), meaning they remain attached to one place. Approximately 15,000 species of sponges have been identified.

Filter Feeders Most species of sponges live in the ocean, but some live in freshwater. Sponge bodies, shown in **Figure 4,** are made of two layers of cells. All sponges are filter feeders. They filter food out of the water that flows through their bodies. Microscopic organisms and oxygen are carried with water into the central cavity through pores of the sponge. The inner surface of the central cavity is lined with collar cells. Thin, whiplike structures, called flagella (flah JEH luh), extend from the collar cells and keep the water moving through the sponge. Other specialized cells digest the food, carry nutrients to all parts of the sponge, and remove wastes.

Body Support and Defense Not many animals eat sponges. The soft bodies of many sponges are supported by sharp, glass-like structures called spicules (SPIHK yewlz). Other sponges have a material called spongin. Spongin is similar to foam rubber because it makes sponges soft and elastic. Some sponges have both spicules and spongin to protect their soft bodies.

as you read

What You'll Learn

- **Describe** the structures that make up sponges and cnidarians.
- **Compare** how sponges and cnidarians get food and reproduce.
- **Differentiate** between flatworms and roundworms.

Why It's Important

Studying the body plans in sponges, cnidarians, flatworms, and roundworms helps you understand the complex organ systems in other organisms.

Review Vocabulary

species: group of organisms that share similar characteristics and can reproduce among themselves

New Vocabulary

- cnidarian
- medusa
- polyp

Figure 4 Red beard sponges grow where the tide moves in and out quickly.

1 Motivate

Bellringer

Section Focus Transparencies also are available on the Interactive Chalkboard CD-ROM. L2 ELL

Section 2 Section Focus Transparency **Rub-a-dub-dub**

When you clean something with a sponge, you may be using the skeleton of an animal! People harvest sponges in areas like the Mediterranean Sea and the Gulf of Mexico. Most sponges you buy in stores, however, aren't animals at all; they're artificial sponges.

1. Why might scientists have originally thought sponges were plants?
2. How do animals get food? How do plants get food?

Invertebrate Animals

Tie to Prior Knowledge

Cells Review cell structure, function, and tissue organization so students will understand differences in body structure of organisms in this section.

Section 2 Resource Manager

Chapter *FAST FILE* Resources

- Transparency Activity, p. 47
- Directed Reading for Content Mastery, p. 20
- Enrichment, p. 32
- Reinforcement, p. 28

Life Science Critical Thinking/Problem Solving, p. 10

2 Teach

Activity

Matching Randomly arrange pictures of sponges, cnidarians, flatworms, roundworms, mollusks, and annelids on the bulletin board along with tags that name each phylum, or group. Have students place the pictures under the appropriate phylum, or group heading, as the animals in each group are studied. L1 LS **Kinesthetic**

INTEGRATE Chemistry

Spicules from the water, which contains calcium and silica

Discussion

Sessile Animals Why are there no sessile animals on land? Sessile animals must get nourishment from food that passes by. Food that swims or is suspended in water is more easily accessible than food that flies or is suspended in air. L2 **Logical-Mathematical**

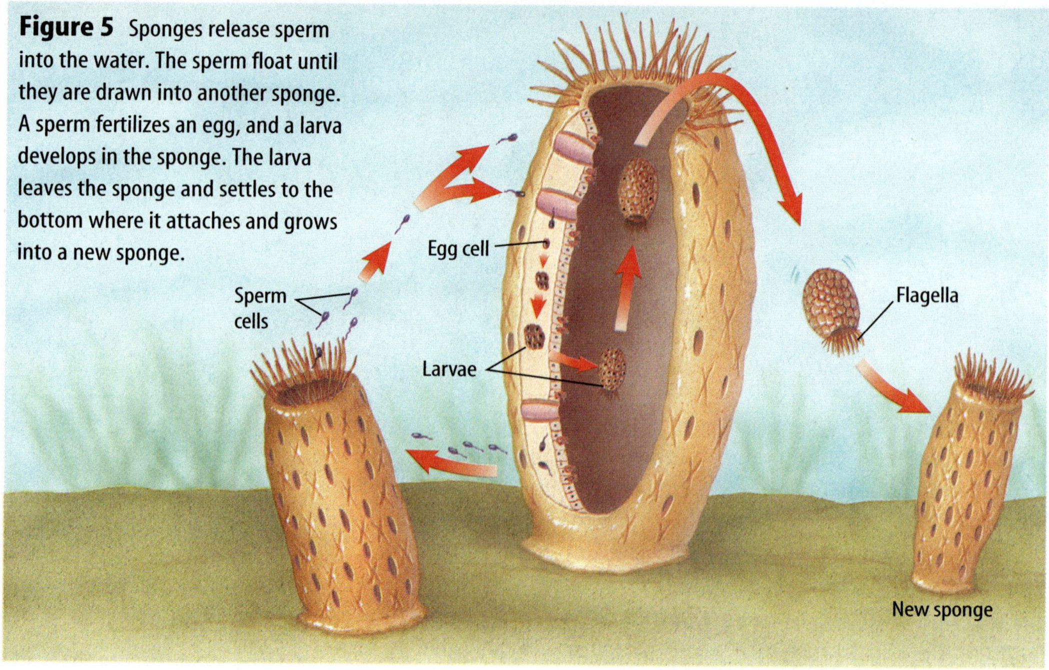

Figure 5 Sponges release sperm into the water. The sperm float until they are drawn into another sponge. A sperm fertilizes an egg, and a larva develops in the sponge. The larva leaves the sponge and settles to the bottom where it attaches and grows into a new sponge.

Sponge Reproduction Sponges can reproduce asexually and sexually. Asexual reproduction occurs when a bud on the side of the parent sponge develops into a small sponge. The small sponge breaks off, floats away, and attaches itself to a new surface. New sponges also may grow from pieces of a sponge. Each piece grows into a new, identical sponge.

Most sponges that reproduce sexually are hermaphrodites (hur MA fruh dites). This means that one sponge produces both eggs and sperm, as shown in **Figure 5.**

INTEGRATE Chemistry

Spicules Sponge spicules of "glass" sponges are composed of silica. Other sponges have spicules made of calcium carbonate. Where do organisms get the silica and calcium carbonate that these spicules are made of? Write your prediction in your Science Journal.

Cnidarians

Cnidarians (nih DAR ee uns), such as jellyfish, sea anemones, hydra, and corals, have tentacles surrounding their mouth. The tentacles shoot out stinging cells called nematocysts (NE ma toh sihsts) to capture prey, similar to casting a fishing line into the water to catch a fish. Because they have radial symmetry, they can locate food that floats by from any direction.

Cnidarians are hollow-bodied animals with two cell layers that are organized into tissues. The inner layer forms a digestive cavity where food is broken down. Oxygen moves into the cells from the surrounding water, and carbon dioxide waste moves out of the cells. Nerve cells work together as a nerve net throughout the whole body.

LAB DEMONSTRATION

Purpose to observe sponge structure

Materials natural or preserved sponge, magnifying lens, microscope, prepared slide of sponge cross-section

Preparation If using a preserved sponge, rinse with water before observations. Set up the microscopes with slides.

Procedure Have students draw what they see when they examine the sponge with a magnifying lens, and label the pores. Then, have students examine the slides, note cell types, and label the central cavity and spicules.

Expected Outcome Students will observe pores, cell types, and the body cavity.

Assessment

Describe the texture of the sponge. Texture will vary depending upon the type of sponge spicules. What traits does a sponge have for living in water? pores for taking in water and food, and spicules (or spongin) for support

Body Forms Cnidarians have two different body forms. The vase-shaped body of the sea anemone and the hydra is called a **polyp** (PAH lup). Although hydras are usually sessile, they can twist to capture prey. They also can somersault to a new location.

Jellyfish have a free-swimming, bell-shaped body that is called a **medusa** (mih DEW suh). Jellyfish are not strong swimmers. Instead, they drift with the ocean currents. Some cnidarians go through both a polyp and a medusa stage during their life cycles.

Cnidarian Reproduction Cnidarians reproduce asexually and sexually. Polyp forms of cnidarians, such as hydras, reproduce asexually by budding, as shown in **Figure 6.** The bud eventually falls off of the parent organism and develops into a new polyp. Some polyps also can reproduce sexually by releasing eggs or sperm into the water. The eggs are fertilized by sperm and develop into new polyps. Medusa forms of cnidarians, such as jellyfish, have a two-stage life cycle as shown in **Figure 7.** A medusa reproduces sexually to produce polyps. Then each of these polyps reproduces asexually to form new medusae.

Figure 6 Polyps, like these hydras, reproduce asexually by budding.
Compare *the genetic makeups of the parent organism and the bud.*

Figure 7 Cnidarians that spend most of their life as a medusa have a sexual (medusa) stage and an asexual (polyp) stage.

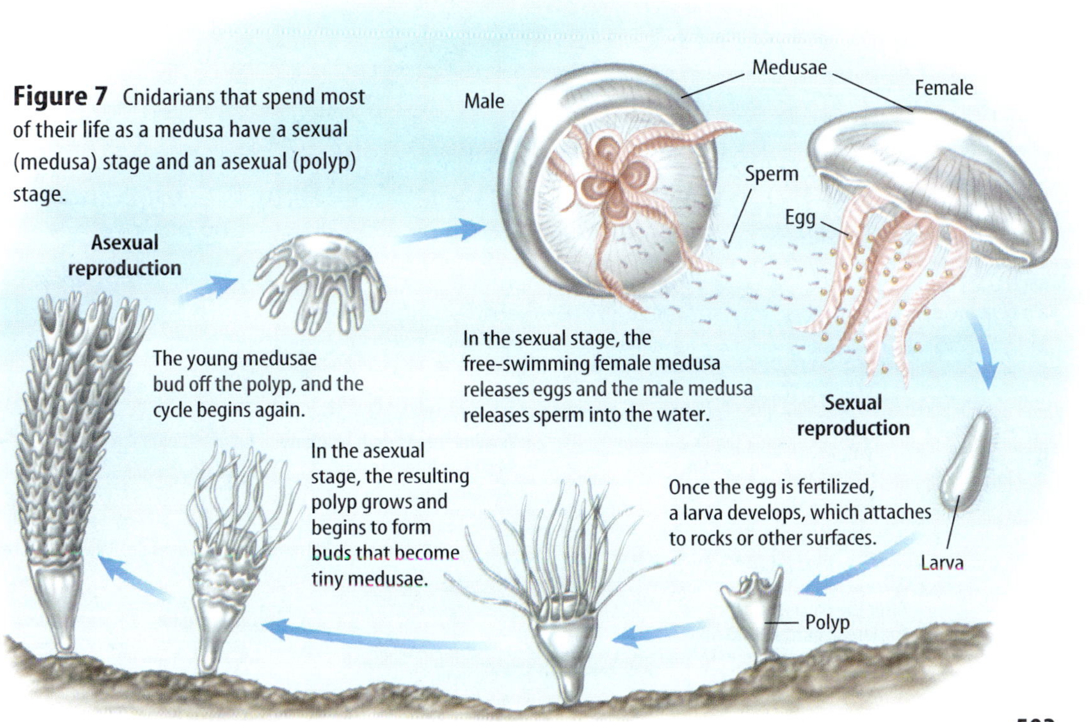

503

Make a Model

Polyps and Medusas Provide students with modeling clay and have them make models of polyps and medusas. Have students compare the shapes and symmetry. The polyp is vase-shaped and the medusa is bell-shaped. Both have radial symmetry. L1 LS **Kinesthetic** P

Caption Answer

Figure 6 They are identical.

Teacher FYI

Poisonous Jellyfish The jellyfish *Chironex fleckeri*, also known as the Australian box jellyfish, or sea wasp, is found along sandy bottoms off the coast of Australia. It is one of the most venomous animals in the sea. The poison found in its stinging cells can kill a human within five minutes, depending on the age of the victim and the severity of the stings. Although antivenom is available, it has not been proven to save lives.

Visual Learning

Figure 6 Use this figure to review with students the differences between sexual and asexual reproduction. L1 LS **Visual-Spatial**

Differentiated Instruction

Challenge Have students research the Great Barrier Reef off the coast of Australia. There are at least 350 species of coral that make up the reef. Have students research the other kinds of invertebrates living there and write a report of their findings. L3 LS **Linguistic**

Curriculum Connection

Language Arts The hydra was named for a giant mythical water monster with nine heads. Hercules, a Greek hero, was supposed to slay the monster, but each time he cut off one head, it grew two more to take its place. Have students do research to discover how the real hydra is like the mythical one. A real hydra can be cut into pieces, and each piece that has enough cells or tissue will generate a new hydra. L2 LS **Linguistic**

Activity

Tapeworm Length Tapeworms as long as 9 m have been found in humans. Have students use a meterstick to measure 9 m of string, so they can observe a tapeworm's length. L1 LS **Kinesthetic**

Use Science Words

Word Meaning Flatworms belong to the phylum *Platyhelminthes*. Ask students to find the meaning of the prefix *platy-*. flat or platelike Then ask them to find other words with this prefix and their meanings. Possible responses: platypus—a small aquatic mammal that has a flattened tail; platyrrhine—an adjective that means to have a short broad nose L2 LS **Linguistic**

Teacher FYI

Roundworm Infection More than 1.4 billion people are currently estimated to be infected with intestinal roundworms alone.

Flatworms

Unlike sponges and cnidarians, flatworms search for food. Flatworms are invertebrates with long, flattened bodies and bilateral symmetry. Their soft bodies have three layers of tissue organized into organs and organ systems. Planarians are free-living flatworms that have a digestive system with one opening. They don't depend on one particular organism for food or a place to live. However, most flatworms are parasites that live in or on their hosts. A parasite depends on its host for food and shelter.

Tapeworms One type of parasitic flatworm is the tapeworm. To survive, it lives in the intestines of its host, including human hosts. The tapeworm lacks a digestive system so it absorbs nutrients from digested material in the host's intestine. In **Figure 8,** you can see the hooks and suckers on a tapeworm's head that attach it to the host's intestine.

A tapeworm grows by adding sections directly behind its head. Each body segment has both male and female reproductive organs. The eggs and sperm are released into the segment. After it is filled with fertilized eggs, the segment breaks off.

Figure 8 Tapeworms are intestinal parasites that attach to a host's intestines with hooks and suckers. Their life cycle is shown here.

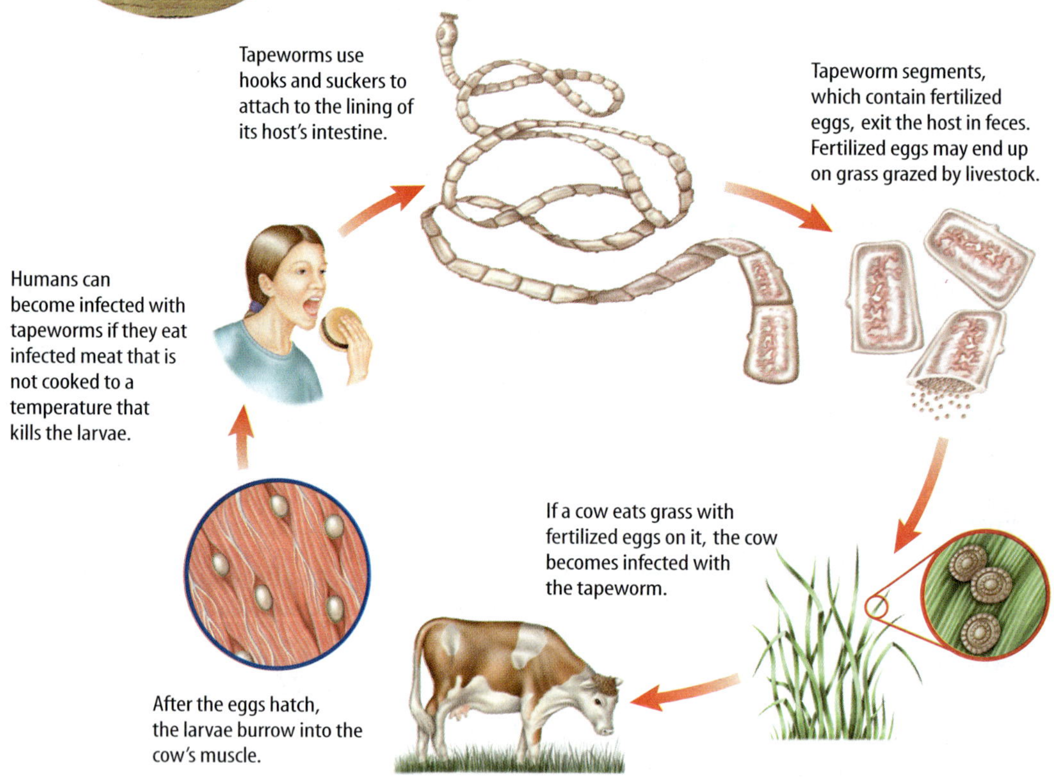

Differentiated Instruction

Learning Disabled Have students select one type of animal described in this chapter. Have them design and construct a detailed, three-dimensional model of the animal and use the model to demonstrate the characteristics of animals. L1 LS **Kinesthetic**

Challenge Explain that animals, including dogs and cats, commonly have tapeworms and flukes. Have each student choose an animal and find out how flukes and tapeworms affect it. Students can present their findings in oral presentations. L3 LS **Linguistic**

The segment passes with wastes out of the host's body. If another host eats a fertilized egg, it hatches and develops into a tapeworm. Tapeworm segments aren't ingested directly by humans. Most flatworms have an intermediate or middle host. For example, **Figure 8** shows how cattle are the intermediate host for tapeworms that infect humans.

Reading Check *How can flatworms get into humans?*

Roundworms

If you have a dog, you may know already that heartworm disease, shown in **Figure 9,** can be fatal to dogs. In most areas of the United States, it's necessary to give dogs a monthly medicine to prevent heartworm disease. Heartworms are just one kind of the many thousands of roundworms that exist. Roundworms are the most widespread animal on Earth. Billions can live in an acre of soil. Many people confuse earthworms and roundworms. You will study earthworms in the next section.

A roundworm's body is described as a tube within a tube, with a fluid-filled cavity in between the two tubes. The cavity separates the digestive tract from the body wall. Roundworms are more complex than flatworms because their digestive tract has two openings. Food enters through the mouth, is digested in a digestive tract, and wastes exit through the anus.

Roundworms are a diverse group. Some roundworms are decomposers, others are predators, and some, like the heartworm, are animal parasites. Other roundworms are plant parasites.

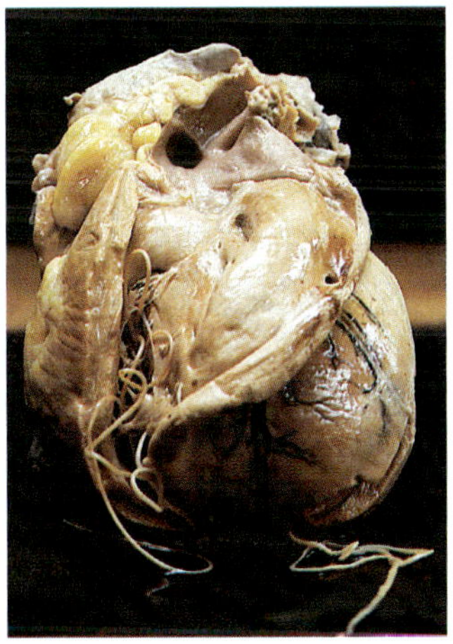

Figure 9 This dog heart is infested with heartworms. Heartworms are carried by mosquitoes. A heartworm infection can clog a dog's heart and cause death.

section 2 review

Summary

Sponges and Cnidarians

- Sponges are animals that remain attached to one place and can reproduce both sexually and asexually.
- Cnidarians are hollow-bodied animals with two cell layers that are organized into tissues.

Flatworms and Roundworms

- Flatworms have three layers of soft tissue organized into organs and organ systems.
- Most flatworms are parasitic.
- Roundworms are decomposers, predators, or parasites of plants and animals and are the most widespread animal on Earth.

Self Check

1. **Explain** how sponges and cnidarians get food.
2. **Compare and contrast** the body plan of flatworms to the body plan of roundworms.
3. **Infer** how spongin and spicules discourage predators from eating sponges.
4. **Think Critically** Some types of sponges and cnidarians reproduce asexually. Why is this beneficial to them?

Applying Math

5. **Solve an Equation** A sponge is 1 cm in diameter and 10 cm tall. It can move 22.5 L of water through its body in a day. Calculate the volume of water it pumps through its body in 1 min.

Science Online red.msscience.com/self_check_quiz

Reading Check

Answer if humans eat infected, under-cooked meat

Quick Demo

Worm Bodies

Possible Materials male and female *Ascaris* worms in sealed tubes

Estimated Time two minutes

Procedure Obtain preserved specimens of male and female *Ascaris* worms in sealed tubes from a biological supply house. Display them so students can see their round bodies and note that females are larger, with a hooked posterior end.

3 Assess

DAILY INTERVENTION

Check for Understanding

Logical-Mathematical Why could it be a reproductive advantage for a cnidarian to produce both polyp and medusa forms? The cnidarian can use the form best suited to the environment in which it finds itself. L2 LS

Reteach

Cnidarians Have students compare and contrast polyp and medusa forms of cnidarians in their Science Journals. They should include a drawing of each with labels. L1 LS **Visual-Spatial**

Assessment

Oral The presence of freshwater sponges indicates an environment with little pollution. Why? Sponges are filter feeders that take in food from the surrounding water. In water with high levels of pollution, they would die. Use PASC, p. 89. L2 LS **Naturalist**

section 2 review

1. Sponges filter food out of the water as it comes to them; cnidarians actively capture prey with their tentacles that have stinging cells.
2. Flatworms have flattened bodies with one opening—a mouth; roundworms have a tube-within-a-tube arrangement with fluid in between and two openings—a mouth and an anus.
3. Spicules are sharp and glasslike; spongin is like foam rubber; both would be hard to eat.
4. Because these organisms are sessile, sometimes it may be difficult to be near another organism of the same species for sexual reproduction. Being able to reproduce asexually helps ensure survival of their species.
5. The sponge will move 15.625 mL (0.015625 L) of water in 1 minute.

1 Motivate

INTERACTIVE CHALKBOARD
PowerPoint® Presentations

Bellringer

Section Focus Transparencies also are available on the Interactive Chalkboard CD-ROM. L2 ELL

Tie to Prior Knowledge

Mollusks Most students will be familiar with mollusks such as oysters, clams, squid, garden snails, and octopuses. Discuss how these organisms are alike and different.

section 3

Mollusks and Segmented Worms

as you read

What You'll Learn

- **Identify** the characteristics of mollusks.
- **Compare** the similarities and differences between an open and a closed circulatory system.
- **Describe** the characteristics of segmented worms.
- **Explain** the digestive process of an earthworm.

Why It's Important

Organ systems and specialized structures allow mollusks and segmented worms to live in varied environments.

Review Vocabulary

organ: structure, such as the heart, made up of different types of tissue that work together

New Vocabulary

- mollusk
- mantle
- gill
- radula
- open circulatory system
- closed circulatory system

Mollusks

Imagine yourself walking along an ocean beach at low tide. On the rocks, you see small snails with conelike shells. In a small tidal pool, one arm of a shy octopus can be seen at the opening of its den. The blue-black shells of mussels are exposed along the shore as shown in **Figure 10.** How are these different animals related? What do they have in common?

Common Characteristics In many places snails, mussels, and octopuses—all mollusks (MAH lusks)—are eaten by humans. **Mollusks** are soft-bodied invertebrates that usually have a shell. They also have a mantle and a large, muscular foot. The **mantle** is a thin layer of tissue that covers the mollusk's soft body. If the mollusk has a shell, it is secreted by the mantle. The foot is used for moving or for anchoring the animal.

Between the mantle and the soft body is a space called the mantle cavity. Water-dwelling mollusks have gills in the mantle cavity. **Gills** are organs in which carbon dioxide from the animal is exchanged for oxygen in the water. In contrast, land-dwelling mollusks have lungs in which carbon dioxide from the animal is exchanged for oxygen in the air.

Figure 10 At low tide, many mollusks can be found along a rocky seashore.

Section 3 Resource Manager

Chapter *FAST FILE* Resources
Transparency Activity, p. 48
Directed Reading for Content Mastery, p. 21
MiniLAB, p. 3
Lab Activity, pp. 9–12
Enrichment, p. 33
Reinforcement, p. 29
Cultural Diversity, p. 3
Performance Assessment in the Science Classroom, p. 55

Many species of conchs are on the verge of becoming threatened species because they are overharvested for food.

Scallops are used to measure an ecosystem's health because they're sensitive to water quality.

Figure 11 Many kinds of mollusks are a prized source of food for humans.
Name *another mollusk, besides a conch or scallop, that is a source of food for humans.*

Body Systems Mollusks have a digestive system with two openings. Many mollusks also have a scratchy, tonguelike organ called the **radula.** The radula (RA juh luh) has rows of fine, teethlike projections that the mollusk uses to scrape off small bits of food.

Some mollusks have an **open circulatory system,** which means they do not have vessels to contain their blood. Instead, the blood washes over the organs, which are grouped together in a fluid-filled body cavity.

Types of Mollusks

Does the animal have a shell or not? This is the first characteristic that scientists use to classify mollusks. Then they look at the kind of shell or they look at the type of foot. In this section, you will learn about three kinds of mollusks.

Gastropods The photo on the left in **Figure 11** shows a gastropod. Gastropods are the largest group of mollusks. Most gastropods, such as the snails and conchs, have one shell. Slugs also are gastropods, but they don't have a shell. Gastropods live in water or on land. All move about on a large, muscular foot. A secretion of mucus allows them to glide across objects.

Bivalves How many shells do you think a bivalve has? Think of other words that start with *bi-*. The scallop shown on the right in **Figure 11** is a bivalve. It is an organism with two shell halves joined by a hinge. Large, powerful muscles open and close the shell halves. Bivalves are water animals that also are filter feeders. Food is removed from water that is brought into and filtered through the gills.

INTEGRATE Social Studies

Toxins Shellfish and crabs accumulate toxins during red tides when they feed on algae containing toxins. These toxins are dangerous to people. The threat of red tides has resulted in closures of both commercial and recreational shellfish harvesting. This causes substantial economic loss. In your Science Journal, write about what is being done to determine when it is safe to harvest shellfish.

Science Online

Topic: Red Tides
Visit red.msscience.com for Web links to information about red tides.

Activity Explain what red tides are and why it is important to learn more about them. What is being done to try and predict when red tides will occur?

2 Teach

Caption Answer

Figure 11 Possible answers: oyster, clam, or snail

Activity

Snails Collect snails from a local pond or purchase them from a pet shop. Have student groups observe how snails move. Have them give lettuce to the snails and observe their feeding with a magnifying lens. Be sure to obtain permission from the owner or proper authority before collecting snails from a pond. L2 ELL

Virtual Labs

Classification *How are mollusks, worms, arthropods, and echinoderms classified?*

INTEGRATE Social Studies

Toxins Laboratories test for toxins to determine if the catch is safe. Monitoring programs assess environmental conditions under which red tides occur. They also correlate toxic dinoflagellates and the toxicity of shellfish. These programs help predict safe shellfish harvesting times.

Research Have students research the economic impact of toxins in shellfish.

Active Reading

Quickwrites Have students list ideas about a topic, then share these ideas with the class. Next, have students write their ideas freely in a paragraph without worrying about punctuation, spelling, or grammar. Have students use a Quickwrite to share ideas about mollusks and segmented worms.

SECTION 3 Mollusks and Segmented Worms 507

Quick Demo

Mollusk Growth

Possible Materials bivalve shells

Estimated Time five minutes

Procedure Have students examine bivalve shells and count their rings. Have them hypothesize what the rings represent. Each ring is one year's growth.

Use Science Words

Word Meaning The word *mantle* comes from a Latin word meaning "cloak" and *radula* is Latin for "scraper." How is the mantle like a cloak? It covers and protects the mollusk's body. What does the radula do? It allows the mollusk to tear food. L2 LS **Linguistic**

Mini LAB

Purpose to model cephalopod movement L1 LS **Kinesthetic**

Materials one balloon

Teaching Strategy Caution students to release the balloon carefully so it will not strike anyone.

Analysis

1. The air leaving the balloon forces it to move in a direction opposite the escaping air.
2. A squid or octopus moves in a direction opposite the escaping stream of water, allowing it to move rapidly to escape danger.

Assessment

Portfolio Ask students to diagram the movement of the balloon and the movement of a squid or octopus for their portfolios. Use **Performance Assessment in the Science Classroom,** p 127. L1 LS **Visual-Spatial** P

Try at Home

Figure 12 Living species of *Nautilus* are found in the western Pacific Ocean. The chambered nautilus, squid, and other cephalopods are able to move quickly using a water-propulsion system as shown to the right.

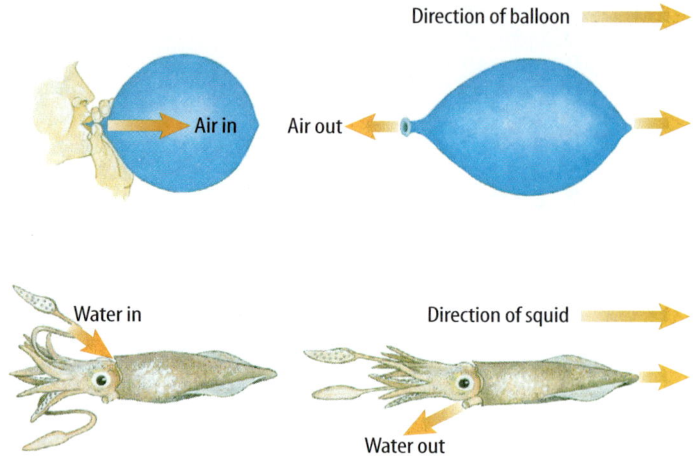

Mini LAB

Modeling Cephalopod Propulsion

Procedure

1. Blow up a **balloon.** Hold the end closed, but don't tie it.
2. Let go of the balloon.
3. Repeat steps 1 and 2 three more times.

Analysis

1. In your **Science Journal,** describe how the balloon moved when you let go.
2. If the balloon models an octopus or a squid as it swims through the water, infer how cephalopods can escape from danger.

Try at Home

Cephalopods The most complex type of mollusks are cephalopods (SE fah lah pawdz). The chambered nautilus, shown in **Figure 12,** octopuses, squid, and cuttlefish are cephalopods. Most cephalopods have an internal plate instead of a shell. They have a well-developed head and a "foot" that is divided into tentacles with strong suckers. At the base of the tentacles is the mouth. They have a **closed circulatory system** in which blood is carried through blood vessels instead of surrounding the organs.

Cephalopods are adapted for quick movement in the ocean. They have a muscular envelope, called the mantle, surrounding their internal organs. Water enters the space between the mantle and the other body organs. When the mantle closes around the collar of the cephalopod, the water is squeezed rapidly through a funnel-like structure called a siphon. The rapid expulsion of water from the siphon creates a force that causes the animal to move in the opposite direction of the stream of water, as illustrated in **Figure 12.**

Segmented Worms

When you hear the word *worm,* you probably think of an earthworm. Earthworms, leeches, and marine worms are segmented worms, or annelids (A nul idz). Their body is made of repeating segments or rings that make these worms flexible. Each segment has nerve cells, blood vessels, part of the digestive tract, and the coelom (SEE lum). The coelom, or internal body cavity, separates the internal organs from the body wall. Annelids have a closed circulatory system and a complete digestive system with two body openings.

Cultural Diversity

Food Mollusks Mollusks are a food source the world over. In some areas, pollution is causing problems for mollusks. Ask students to discuss cultural and economic effects of pollution on mollusks and the people who harvest them for food. Pollution can lead to a decline in mollusk populations and adversely affect regions whose economies depend upon mollusk harvesting. Pollution can also make mollusks unfit for consumption, causing illness in those who eat them. L2 LS **Logical-Mathematical**

508 CHAPTER 17 Invertebrate Animals

Earthworms When did you first encounter earthworms? Maybe it was on a wet sidewalk or in a garden, as shown in **Figure 13.** Earthworms have more than 100 body segments. Each segment has external bristlelike structures called setae (SEE tee). Earthworms use the setae to grip the soil while two sets of muscles move them through the soil. As earthworms move, they take soil into their mouths. Earthworms get the energy they need to live from organic matter found in the soil. From the mouth the soil moves to the crop, where it is stored. Behind the crop is a muscular structure called the gizzard. Here, the soil and food are ground. In the intestine, the food is broken down and absorbed by the blood. Undigested soil and wastes leave the worm through the anus.

Reading Check *What is the function of setae?*

Examine the earthworm shown in **Figure 14.** Notice the lack of gills and lungs. Carbon dioxide passes out and oxygen passes in through its mucous-covered skin. It's important not to pick up earthworms with dry hands because if this thin film of mucus is removed, the earthworm may suffocate.

Figure 13 Earthworms are covered with a thin layer of mucus, which keeps them moist. Setae help them move through the soil.

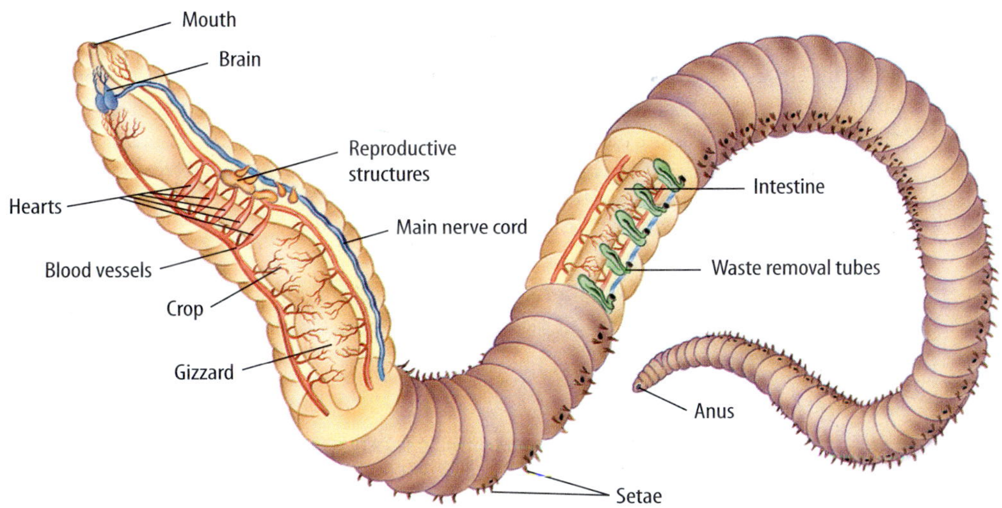

Figure 14 Earthworms and other segmented worms have many organ systems including circulatory, reproductive, excretory, digestive, and muscular systems.

Activity

Living Earthworms Have students observe the body segments, bilateral symmetry, and setae of a living earthworm. Have students keep their hands moist while handling the earthworms and wash their hands afterward. L2 IS **Kinesthetic**

Reading Check

Answer to help the worm move through and hold onto the soil

Quick Demo

Earthworm Dissection

Possible Materials preserved earthworm, razor blade, pins, dissecting tray

Estimated Time ten minutes

Procedure Dissect a preserved earthworm and have students examine its digestive, excretory, circulatory, reproductive, and nervous systems.

Visual Learning

Figure 13 How many hearts, or aortic arches, does the earthworm have? five What sensory organs are shown? brain and nerve cord How many blood vessels does the earthworm have? two What makes up the excretory system? waste removal tubes L1 IS **Visual-Spatial**

Curriculum Connection

Literature Read the poem "The Chambered Nautilus" by Oliver Wendell Holmes. Provide students with a nautilus shell to examine. Have them draw the shell and write a report in their journals about how the shell forms. L2 IS **Linguistic**

Differentiated Instruction

Challenge Ask students to investigate ways earthworms are important to farmers and gardeners. Earthworms help to aerate and fertilize soil and thus enhance plant growth. L3 IS **Naturalist**

Reading Check

Answer with sucking disks on both ends of their bodies

Applying Science

Teaching Strategy

Question 2 can provide discussion on the knowledge required to make choices such as the use of sludge as fertilizer. Should tests be done on the sludge before its use as a fertilizer? Whose responsibility is it to test the sludge—the farmer accepting it or the municipality donating it? Is the amount of heavy metals and organic poisons found in birds enough to cause concern?

Answers

1. Earthworms are part of the food chain. Since they break down the sludge and can store high levels of poisons, predators that feed on them, such as birds, can become poisoned.
2. The use of sludge as a fertilizer can be a benefit. It provides nutrients for earthworms that tillage by farmers has removed. However, the municipality must be aware of what can be present in the sludge. If there are poisonous substances that can accumulate in the earthworm and pass up through the food chain, contaminating other critters, then the use of sludge as a fertilizer may not be a wise choice.

Discussion

Segmented Worm Differences How do marine worms and leeches differ? Marine worms have setae and live only in marine environments; leeches have suckers and live in a variety of environments. Why are both classified as segmented worms? Both have a body cavity and a segmented body.

Figure 15 Leeches attach to fish, turtles, snails, and mammals and remove blood and other body fluids.

Leeches They can be found in freshwater, marine waters, and on land in mild and tropical regions. These segmented worms have flat bodies from 5 mm to 460 mm long with sucking disks on both ends. They use these disks to attach themselves to an animal, as shown in **Figure 15,** and remove blood. Some leeches can store as much as ten times their own weight in blood. It can be stored for months and released a little at a time into the digestive system. Although leeches prefer a diet of blood, most of them can survive indefinitely on small aquatic animals.

Reading Check *How do leeches attach themselves to an animal?*

Marine Worms The animals in **Figure 16** are polychaetes (PAH lee keets), the largest and most diverse group of annelids. Of the 10,000 named species of annelids, more than 8,000 of them are marine worms. The word *polychaete* means "many bristles." Most marine worms have bristles, or setae, along the sides of their body. Because of these bristles, marine worms are sometimes called bristle worms. Bristles are used for walking, swimming, or digging, depending on the type of marine worm.

Applying Science

How does soil management affect earthworms?

Some earthworms tunnel through the soil about 30 cm below the soil surface. Earthworms called night crawlers dig deep, permanent tunnels that are up to 1.8 m long. Earthworms' tunnels loosen the soil, which allows better root growth by plants. It also increases air and water movement in the soil. As they tunnel, earthworms take in soil that contains organic matter such as plant material, microorganisms, and animal remains. This is their source of food. Microorganisms break down earthworms' wastes, which adds nutrients to the soil. Earthworms are a food source for frogs, snakes, birds, and other animals.

Identifying the Problem

As earthworms tunnel through the soil, they also take in other substances found there. High levels of pesticides and heavy metals can build up in the bodies of earthworms.

Solving the Problem

1. One soil management technique is to place municipal sludge on farmland as fertilizer. The sludge might contain heavy metals and harmful organic substances. Predict how this could affect birds.
2. Is the use of sludge as a fertilizer a wise choice? Explain your answer.

Curriculum Connection

History Ask students to research and report on the uses of leeches in medicine in the 1800s and through the 1900s. They were used for bloodletting, applied to the temples for headaches, and used to treat mental illness, gout, skin disease, and whooping cough. They are still used to produce the anticoagulant hirudin and to increase the flow of blood to reattached body parts. L2 LS **Linguistic**

Differentiated Instruction

Learning Disabled Have students make papier-mâché models of all the major invertebrates in this section. Attach a card to each with the reasons why they are important animals in the ecosystem. L1 LS **Visual-Spatial**

Figure 16 More than 8,000 species of marine worms exist.

Some polychaetes, like this fireworm, move around in search of food.

Polychaetes, like this sea mouse, have long bristles that look like hair.

Some polychaetes, like this tubeworm, cannot move around in search of food. Instead, they use their featherlike bristles to filter food from the water.

Body Types Some marine worms are filter feeders. They either burrow into the mud or build their own tube cases and use their featherlike bristles to filter food from the water. Some marine worms move around eating plants or decaying material. Other marine worms are predators or parasites. The many different lifestyles of marine worms explain why there are so many different body types.

Although annelids do not look complex, they are more complex than sponges and cnidarians. In the next section, you will learn how they compare to the most complex invertebrates.

section 3 review

Summary

Mollusks

- Mollusks are soft-bodied invertebrates that have a mantle, a large, muscular foot, and usually have a shell.

Types of Mollusks

- Mollusks are separated into three groups—gastropods, bivalves, and cephalopods.

Segmented Worms

- Repeating body segments give segmented worms flexibility.
- Segmented worms have a coelom, or internal body cavity, that separates the internal organs from the body wall.

Self Check

1. **Explain** what gills are used for.
2. **Describe** how an earthworm feeds and digests its food.
3. **Identify** which type of circulatory system that a cephalopod develops.
4. **Think Critically** Why would it be beneficial to a leech to be able to store blood for months and release it slowly?

Applying Skills

5. **Communicate** Choose a mollusk or annelid and write about it in your Science Journal. Describe its appearance, how it gets food, where it lives, and other interesting facts.

Science Online red.msscience.com/self_check_quiz

Teacher FYI

Leeches *Leech* is a term that applies to any annelid in the class *Hirudinea*. There are about 300 known species. *Hirudo medicinalis*, the leech used in medicine, feeds primarily on mammalian blood, but it also sucks blood from amphibians, reptiles, and fish. Within the mouth are three jaws with sharp teeth that make a Y-shaped incision in the skin of the host. Once the leech has fed, it detaches from its host.

3 Assess

DAILY INTERVENTION

Check for Understanding

Logical-Mathematical How is a mollusk shell an adaptation? It provides protection, camouflage, and conserves water. L3 LS

Reteach

Mollusk and Segmented Worm Comparisons Have students make a table to compare the structures, ways of life (free-living or parasitic), and environments of flatworms, mollusks, and segmented worms. L1 LS **Visual-Spatial**

Assessment

Process Contrast the muscular foot of bivalves and gastropods. Gastropod—the foot spreads out under its body and a rippling motion moves it along; bivalve—the foot is normally hidden inside the two shells; some bivalves move by hooking the foot into the sand and pulling themselves along. L2 LS **Naturalist**

section 3 review

1. Gills are used to exchange carbon dioxide from the animal for oxygen in the water.
2. As earthworms move, they take soil in through their mouths. Soil moves through the crop, gizzard, and intestine, and the waste goes out the anus.
3. Cephalopods have a closed circulatory system.
4. It would have to feed less often.
5. Answers will vary depending on which mollusk or annelid the student chooses.

1 Motivate

INTERACTIVE CHALKBOARD
PowerPoint® Presentations

Bellringer

Section Focus Transparencies also are available on the Interactive Chalkboard CD-ROM. L2 ELL

Tie to Prior Knowledge

Common Arthropods Many students will be familiar with shrimp and lobsters. All should have knowledge of insects and spiders. Use their knowledge of arthropods to identify common characteristics.

Reading Check

Answer It protects and supports the body and reduces water loss.

section 4

Arthropods and Echinoderms

as you read

What You'll Learn

- **List** the features used to classify arthropods.
- **Explain** how the structure of the exoskeleton relates to its function.
- **Identify** features of echinoderms.

Why It's Important

Arthropods and echinoderms show great diversity and are found in many different environments.

Review Vocabulary
regeneration: regrowth of a body or body part after injury or as a normal process

New Vocabulary
- arthropod
- appendage
- exoskeleton
- metamorphosis

Arthropods

More than a million species of arthropods (AR thruh pahdz) have been discovered. They are the largest and most diverse group of animals. The term *arthropod* comes from *arthros,* meaning "jointed," and *poda,* meaning "foot." **Arthropods** are animals that have jointed appendages (uh PEN dih juz). **Appendages** are structures such as claws, legs, and antennae that grow from the body.

Arthropods have a rigid body covering called an **exoskeleton.** It protects and supports the body and reduces water loss. The weight of the outer covering increases as the size of the animal increases. As the animal grows, the exoskeleton must be shed because it doesn't grow with the animal. This process is called molting. Weight and hardness of the exoskeleton could make it difficult to move, but the jointed appendages solve part of this problem.

Reading Check *What is the function of the exoskeleton?*

Arthropods have bilateral symmetry and segmented bodies similar to annelids. In most cases, arthropods have fewer, more specialized segments. Instead of setae, they have appendages.

Insects If asked to name an insect, you might say bee, fly, beetle, or butterfly. Insects make up the largest group of arthropods. More than 700,000 species of insects have been classified, and scientists discover and describe more of them each year.

Insects, like the ant in **Figure 17,** have three body regions—head, thorax, and abdomen. Well-developed sensory organs, including the eyes and antennae, are located on the head. The thorax has three pairs of jointed legs and usually one or two pairs of wings. The wings and legs of insects are highly specialized. The abdomen is divided into segments and has neither wings nor legs attached, but reproductive organs are located there.

Figure 17 About 8,000 species of ants are found in the world. Ants are social insects that live cooperatively in colonies.

Section 4 Resource Manager

Chapter *FAST FILE* Resources
Transparency Activity, pp. 49, 51–52
Directed Reading for Content Mastery, pp. 21, 22
Enrichment, p. 34
MiniLAB, p. 4
Reinforcement, p. 30
Lab Worksheet, pp. 5–6, 7–8
Lab Activity, pp. 13–16

Cultural Diversity, p. 7
Home and Community Involvement, p. 32
Mathematics Skill Activities, p. 29
Lab Management and Safety, p. 58

Circulatory System Insects have an open circulatory system. Oxygen is not transported by blood in the system, but food and waste materials are. Oxygen is brought directly to the insect's tissues through small branching tubes. These tubes connect to openings called spiracles (SPIHR ih kulz) located along the sides of the thorax and abdomen.

Metamorphosis The young of many insects don't look anything like the adults. This is because many insects completely change their body form as they mature. This change in body form is called **metamorphosis** (met uh MOR fuh sus). The two kinds of insect metamorphosis, complete and incomplete, are shown in **Figure 18.**

Butterflies, ants, bees, and beetles are examples of insects that undergo complete metamorphosis. Complete metamorphosis has four stages—egg, larva, pupa (PYEW puh), and adult. Notice how different each stage is from the others. Some insects, such as grasshoppers, cockroaches, termites, aphids, and dragonflies, undergo incomplete metamorphosis. They have only three stages—egg, nymph, and adult. A nymph looks similar to its parents, only smaller. A nymph molts as it grows until it reaches the adult stage. All the arthropods shown in **Figure 19** on the next two pages molt many times during their life.

Science Online

Topic: Butterflies

Visit red.msscience.com for Web links to information about butterflies.

Activity What are some of the characteristics that are used to identify butterflies? Make a diagram of the life cycle of a butterfly.

Figure 18 Insect metamorphosis occurs in two ways.
State *the name given a moth larva.*

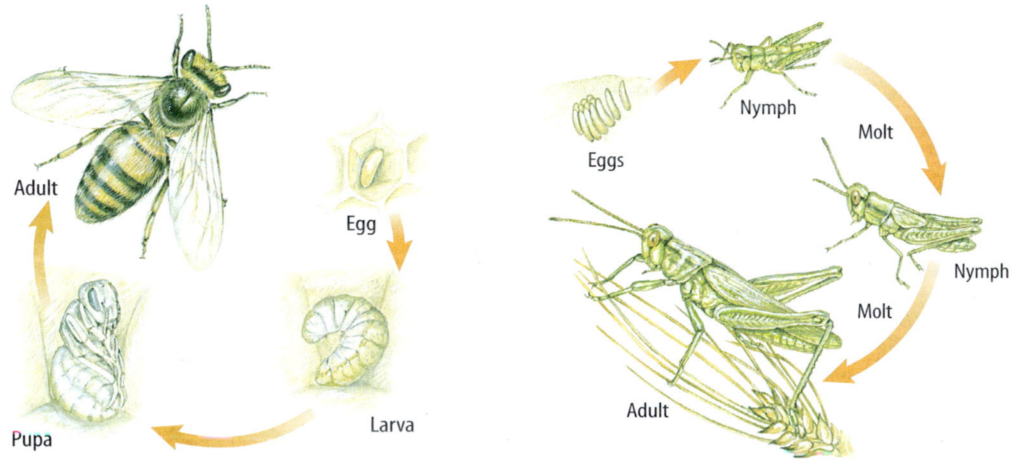

Bees and many other insects undergo the four stages of complete metamorphosis.

Insects like the grasshopper undergo incomplete metamorphosis.

Visual Learning

Figure 18 Have students make an events-chain concept map for each type of metamorphosis shown. Complete: egg, larva, pupa, adult; Incomplete: egg, nymph, adult

2 Teach

Discussion

Molting Why would molting be a dangerous time for arthropods? They would be unprotected until they grew a new exoskeleton. L2 LS **Logical-Mathematical**

Activity

Exoskeletons and Appendages Provide student groups with live or preserved specimens of insects, crayfishes, spiders, and other arthropods. Ask students to compare and contrast their exoskeletons and appendages. L2 COOP LEARN LS **Kinesthetic**

Use an Analogy

Body Armor Compare the limitations of the arthropod exoskeleton to a person wearing a suit of armor.

Inquiry Lab

Insect Life Cycles

Purpose to explore metamorphosis and insect life cycles

Possible Materials ladybug kit, film on insect life cycles

Estimated Time three weeks

Teaching Strategies

- Have students watch a film about insect life cycles and explore hypotheses for why they undergo metamorphosis.
- Obtain a ladybug kit and have students observe ladybugs as they undergo various stages of metamorphosis. Have students note changes in ladybugs and explore hypotheses about food requirements and life cycle changes.
- Allow students to explore other questions that arise.

For additional inquiry activities, see *Science Inquiry Labs*.

NATIONAL GEOGRAPHIC

Visualizing Arthropod Diversity

Content Background

Insects are an especially successful group of arthropods. Scientists have identified 850,000 species of insects and feel there are many more not yet discovered. There are several characteristics that scientists attribute to their success. Their small size makes many habitats more accessible to insects. The ability to fly allows insects to escape predators and disperse over greater distances to look for habitats and food. The waterproof exoskeleton provides hard mouthparts with which to extract and grind food, as well as protection from predation and desiccation. Their short life cycle and ability to undergo metamorphosis allows insects to exploit several different habitats in a single lifetime. Insects can be found in all habitats on Earth, including harsh environments such as those found in Antarctica and in the hot, dry deserts.

NATIONAL GEOGRAPHIC **VISUALIZING ARTHROPOD DIVERSITY**

Figure 19

Arthropods are the most successful group of animals on Earth. Research the traits of each arthropod pictured. Compare and contrast those traits that enhance their survival and reproduction.

HUMMINGBIRD MOTH When hovering near flowers, these moths produce the buzzing sound of hummingbirds. The wingspan of these moths can reach 6 cm.

KRILL Living in the icy waters of the arctic and the antarctic, krill are an important component in the ocean food web. They range in length from 8 to 60 mm. Baleen whales can eat 1,000 kg of krill in one feeding.

GOOSENECK BARNACLES These arthropods usually live on objects, such as buoys and logs, which float in the ocean. They also live on other animals, including sea turtles and snails.

DIVING BEETLE These predators feed on other invertebrates as well as small fish. They can grow to more than 40 mm in length.

ALASKAN KING CRAB These crabs live in the cold waters of the north Pacific. Here, a gauge of about 18 cm measures a crab too small to keep; Alaskan king crabs can stretch 1.8 m from tip to tip.

Fun Fact

Some crabs camouflage themselves by placing sand, pieces of seaweed, and shells on their outer carapace. These crabs are commonly called decorator crabs.

Differentiated Instruction

Challenge Have students research the motions of the insect wings and mechanisms involved during flight. They could compare and contrast the wing motions of a fast insect such as a wasp to that of a slower insect such as a dragonfly. Have students make a poster to show results. L3

▲ HORSESHOE CRAB More closely related to spiders than to crabs, horseshoe crabs dig their way into the sand near the shore to feed on small invertebrates.

◀ BUMBLEBEE A thick coat of hair and the ability to shiver their flight muscles to produce heat allow bumblebees to fly in cold weather.

▶ PILL BUG Many people think that pill bugs—also known as sow bugs, rolypolies, or wood lice—are insects. Actually, they are crustaceans that live on land.

▶ AMERICAN COCKROACH This arthropod, which can grow to a length of almost 5 cm, is the largest house-infesting roach. It is common in urban areas around the world.

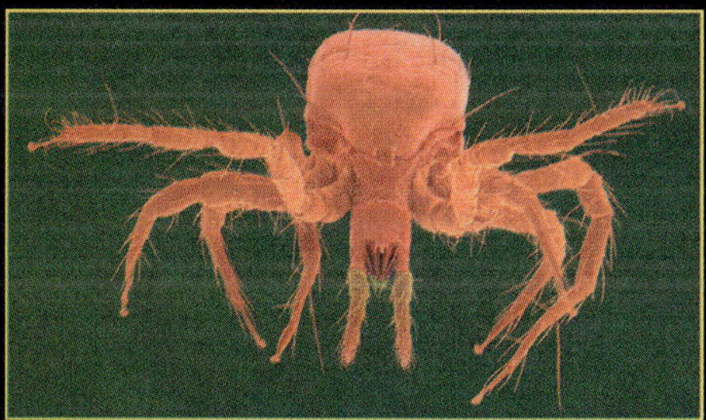

▲ SPIDER MITE These web-spinning arachnids are serious pests because they suck the juices out of plants. They damage houseplants, landscape plants, and crops. The spider mite above is magnified 14 times its normal size.

▲ DADDY LONGLEGS Moving on legs that can be as much as 20 times longer than their bodies, these arachnids feed on small insects, dead animals, and plant juices. Although they look like spiders, they belong to a different order of arachnids.

NATIONAL GEOGRAPHIC

Visualizing Arthropod Diversity

Make a Model

Provide construction paper, modeling clay, craft sticks, cardboard tubes, and other materials for students to construct a model of an arachnid, centipede, millipede, crustacean, or insect. L1 LS **Kinesthetic**

Activity

Insects Some insects can be helpful to humans while others can cause disease or devastate crops. Have students give three examples of how insects can be helpful and three examples of how insects can be harmful to humans. Possible answers: Insects pollinate many plants that are important to humans such as apples, grapes, cotton, and oranges. Bees make honey. Some insects help control other insect numbers such as ladybugs preying on aphids. Insects can be harmful when they destroy crops and trees. They can transmit diseases such as malaria, yellow fever, and typhoid. Bees can sting and termites can destroy wooden structures.

Differentiated Instruction

Challenge Have students prepare an illustrated report on the various hunting methods of spiders. Include the jumping spider, wolf spider, and trapdoor spider. L3 LS **Linguistic** P

IDENTIFYING Misconceptions

Tarantulas Contrary to popular belief, the tarantula is not a deadly enemy to adult humans. The bite of most species is negligible; a few can inflict a painful, but not fatal, bite. In fact, tarantulas are fairly sluggish and shy animals that spend much time in burrows. They are often kept as pets. They eat insects and other small animals.

Caption Answer

Figure 20 Possible answer: Having eyes on both the face and the top of the head increases the jumping spider's chances of locating prey.

Reading Check

Answer Some spiders use substances to paralyze and digest their prey. Others weave webs to capture prey or chase and capture prey.

Teacher FYI

Dangerous Spiders The black widow, *Latrodectus mactans*, and the brown recluse, *Loxosceles reclusa*, are two of the spiders in the United States whose venom is dangerous to humans. The black widow has a red hourglass-shaped spot on its abdomen. The brown recluse has a dark violin-shaped marking on its cephalothorax.

Text Question Answer

Centipedes are predators; millipedes eat plants.

Orb weaver spider

Jumping spider

Scorpion

Figure 20 This orb weaver spider uses its web to catch prey. Then it wraps the prey in silk to eat later. Jumping spiders have four large eyes on their face and four smaller eyes on the top of their head. Scorpions usually hide during the day and hunt for their prey at night.
Identify *an advantage that jumping spiders have because of all their eyes.*

Arachnids Spiders, ticks, mites, and scorpions belong to a group of arthropods known as arachnids (uh RAK nudz). Arachnids have only two body regions—a cephalothorax (sef uh luh THOR aks) and an abdomen—instead of three. The cephalothorax is made of the fused head and thorax regions. All arachnids have four pairs of legs attached to the cephalothorax.

Spiders are predators. A spider uses a pair of fanglike appendages near its mouth to inject paralyzing venom into its prey. Then it releases substances into its prey that digest the victim, turning it into a liquid, and the spider drinks it. Some spiders, like the one in **Figure 20,** weave webs to trap their prey. Other spiders, like the jumping spider, chase and catch their prey. Other arachnids, like the scorpion, paralyze their prey with venom from their stinger.

Reading Check *How do spiders catch their prey?*

Centipedes and Millipedes As shown in **Figure 21,** centipedes and millipedes are long, thin, segmented animals. These arthropods have pairs of jointed legs attached to each segment. Centipedes have one pair of jointed legs per segment, and millipedes have two pairs. Centipedes are predators that use poisonous venom to capture their prey. Millipedes eat plants. Besides the number of legs, how else is the centipede different from the millipede?

Figure 21 Centipedes can have more than 100 segments. When a millipede feels threatened, it will curl itself into a spiral.

Centipede

Millipede

Science Journal

Spiders Have each student choose an arachnid in this section and write about it in his or her Science Journal. Journal entries should reflect the characteristics of the organism, its environment, its way of life, and other information. L2
LS **Linguistic**

Cultural Diversity

Controlling Pests Zhoa Jing Zhao, a university vice president in China, helped develop a technique for protecting crops from insects by using spiders. He discovered that using spiders to kill harmful insects in cotton fields reduced the need to use pesticides by 80%–90%. Have students discuss advantages of using spiders to control insect population in crop fields. Using spiders decreases the need for chemical pesticides.

INTEGRATE Physics

Crustaceans Think about where you can lift the most weight—is it on land or in water? An object seems to weigh less in water because water pushes up against the pull of gravity. Therefore, a large, heavy exoskeleton is less limiting in water than on land. The group of arthropods called crustaceans includes some of the largest arthropods. However, most crustaceans are small marine animals that make up the majority of zooplankton. Zooplankton refers to the tiny, free-floating animals that are food for other marine animals.

Examples of crustaceans include crabs, crayfish, lobsters, shrimp, barnacles, water fleas, and sow bugs. Their body structures vary greatly. Crustaceans usually have two pairs of antennae attached to the head, three types of chewing appendages, and five pairs of legs. Many water-living crustaceans also have appendages called swimmerets on their abdomen. Swimmerets force water over the feathery gills where carbon dioxide from the crustacean is exchanged for oxygen in the water.

Echinoderms

Most people know what a starfish is. However, today they also are known as sea stars. Sea stars belong to a varied group of animals called echinoderms (ih KI nuh durmz) that have radial symmetry. Sea stars, brittle stars, sea urchins, sand dollars, and sea cucumbers are echinoderms. The name *echinoderm* means "spiny skin." As shown in **Figure 22,** echinoderms have spines of various lengths that cover the outside of their bodies. Most echinoderms are supported and protected by an internal skeleton made up of bonelike plates. Echinoderms have a simple nervous system but don't have heads or brains. Some echinoderms are predators, some are filter feeders, and others feed on decaying matter.

Mini LAB

Observing Sow Bugs

Procedure

1. Place six **sow bugs** in a clean, **flat container.**
2. Put a damp **sponge** at one end of the container.
3. Cover the container for 60 s. Remove the cover and observe where the sow bugs are. Record your observations in your **Science Journal.**

Analysis

1. What type of habitat do the sow bugs seem to prefer?
2. Where do you think you could find sow bugs near your home?

Figure 22 Sun stars have up to twelve arms instead of five like many other sea stars. Sea urchins are covered with protective spines. Sand dollars have tube feet on their undersides.

Mini LAB

Purpose to observe and identify sow bug reactions to wet and dry conditions L1 LS **Kinesthetic**

Materials flat container, sow bugs (6), damp sponge, dark colored cover for container

Teaching Strategies

- Direct students to handle sow bugs gently and to wash their hands when done.
- Try not to use the same sow bugs over and over if they are used for more than one class period in a day. The sow bugs will respond less or not at all.

Analysis

1. damp
2. under rocks or decaying pieces of wood where it is damp

Assessment

Content Have students do research to discover why sow bugs live in damp places. Crustaceans have gills and need moisture for gas exchange.

Quick Demo

Centipede and Millipede Comparison

Possible Materials centipede and millipede in sealed tubes

Estimated Time two minutes

Procedure From a biological supply house, obtain preserved centipede and millipede specimens in sealed tubes. Display them so students can see that centipedes have one pair of jointed legs per segment and millipedes have two pairs.

Differentiated Instruction

Visually Impaired and Learning Disabled Provide students with dried specimens of sea stars, sand dollars, and other echinoderms. Students who are visually impaired can observe their shapes and textures by touching them. L1 LS **Kinesthetic**

Discussion

Starfish Name Why do you think scientists now refer to starfish as sea stars? The name is more accurate, since sea stars are not fish. L2 LS
Logical-Mathematical

3 Assess

DAILY INTERVENTION

Check for Understanding

Logical-Mathematical Why are there no insects as large as elephants? Exoskeletons that would be strong enough to support much weight on land would have to be extremely thick and heavy. If an ant were as large as an elephant, it would collapse under its own weight.

Reteach

Molting Have students explain why arthropods molt several times during their lifetime and why echinoderms have few predators. The exoskeleton of an arthropod does not grow with the animal, so it must be shed to allow a new, larger, one to form. The tough, spiny skin of echinoderms makes them difficult for other animals to eat. L2

Assessment

Content Ask students to write in their Science Journals about how incomplete and complete metamorphosis are similar and how they are different. Both are a series of changes. In complete metamorphosis, the young look very different from the adults. Use **Performance Assessment in the Science Classroom,** p. 157. L1 LS
Linguistic

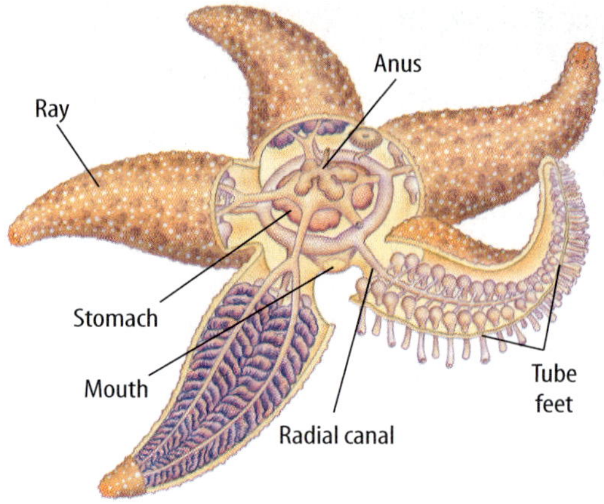

Figure 23 Echinoderms use their tube feet to move. Sea stars also use their tube feet to capture prey and pull apart the shells. Tube feet are connected to an internal system of canals and are able to act like suction cups.

Water-Vascular System All echinoderms have a water-vascular system. It is a network of water-filled canals and thousands of tube feet. The tube feet work like suction cups to help the sea star move and capture prey. **Figure 23** shows how these tube feet are used to pull open prey. Sea stars have a unique way of eating. The sea star pushes its stomach out of its mouth and into the opened shell of its prey. After the prey's body is digested and absorbed, the sea star pulls in its stomach.

Like some invertebrates, sea stars can regenerate lost or damaged parts. In an attempt to reduce the population of sea stars that ate their oysters, oyster farmers once captured sea stars, cut them into pieces, and threw them back into the bay. Within a short time, the sea star population was five times larger. The oyster beds were destroyed—not saved.

section 4 review

Summary

Arthropods
- Arthropods are the largest and most diverse group of animals.
- Arthropods have bilateral symmetry and segmented bodies.
- Many insect species go through metamorphosis as they mature.

Echinoderms
- Echinoderms have radial symmetry and a water-vascular system.
- Like some other invertebrates, sea stars can regenerate damaged parts.

Self Check

1. **List** the advantages and disadvantages of having an exoskeleton.
2. **Explain** why spiders and ticks aren't insects.
3. **Compare and contrast** centipedes and millipedes.
4. **Think Critically** What might happen to the sea star population after oyster beds are destroyed? Explain.

Applying Math

5. **Use Proportions** A flea that is 4 mm in length can jump 25 cm from a resting position. If this flea were as tall as you are, how far could it jump?

Science Online red.msscience.com/self_check_quiz

section 4 review

1. Advantages: it provides protection and supports the body; disadvantages: the body outgrows the exoskeleton, its weight limits the size a land invertebrate can grow
2. They have two body regions and four pairs of legs. Insects have three body regions and three pairs of legs.
3. They both are long, thin, segmented animals. Centipedes have one pair of jointed legs per segment and use venom to capture prey. Millipedes have two pairs of jointed legs per segment and eat plants.
4. The sea star population would decline, as their food source would be destroyed.
5. Answers will vary. Sample response: A flea measuring 160 cm could jump 10,000 cm (100 m).

LAB

Observing Complete Metamorphosis

Many insects go through complete metamorphosis during their life cycles. Chemicals that are secreted by the body of the animal control the changes. How different are the body forms of the four stages of metamorphosis?

Real-World Question

What do the stages of metamorphosis look like for a mealworm?

Goals

- **Observe** metamorphosis of mealworms.
- **Compare** the physical appearance of the mealworms at each stage of metamorphosis.

Materials

large-mouth jar or old fish bowl
bran or oatmeal
dried bread or cookie crumbs mixed with flour
slice of apple or carrot
paper towel
cheesecloth
mealworms
rubber band

Safety Precautions

WARNING: *Be careful when working with animals. Never touch your face during the lab. Wash your hands thoroughly after completing the lab.*

Procedure

1. Set up a habitat for the mealworms by placing a 1-cm layer of bran or oatmeal on the bottom of the jar. Add a 1-cm layer of dried bread or cookie crumbs mixed with flour. Then add another layer of bran or oatmeal.
2. Add a slice of apple or carrot as a source of moisture. Replace the apple or carrot daily.
3. Place 20 to 30 mealworms in the jar. Add a piece of crumpled paper towel.
4. Cover the jar with a piece of cheesecloth. Use the rubber band to secure the cloth to the jar.
5. **Observe** the mealworms daily for two to three weeks. Record daily observations in your Science Journal.

Conclude and Apply

1. **Draw and describe** the mealworms' metamorphosis to adults in your Science Journal.
2. **Describe** some of the advantages of an insect's young being different from the adults.
3. **Infer** where you might find mealworms or adult darkling beetles in your house.

Communicating Your Data

Draw a cartoon showing the different stages of metamorphosis from mealworm to adult darkling beetle. **For more help, refer to the Science Skill Handbook.**

BENCH TESTED

Real-World Question

Purpose to observe the stages of complete metamorphosis L2 ELL COOP LEARN **Visual-Spatial**

Process Skills observe, communicate, interpret data

Time Required one 45-minute class period; daily observation for two to three weeks

Teaching Strategy Find a warm place in the classroom to maintain an optimal environment for metamorphosis of the insect. Mealworms can stay in the larval stage for up to two years, depending on the environment.

Conclude and Apply

1. Answers will vary.
2. Larval forms often eat different things than the adults, so they don't have to compete with adults for food.
3. kitchen cabinets

Assessment

Performance Have students observe the metamorphosis of a fruit fly and compare it with that of a mealworm. Both are complete metamorphoses. The stages of fruit fly metamorphosis are shorter than those of the mealworm. Use **Performance Assessment in the Science Classroom, p. 89.** L2 LS **Visual-Spatial**

Communicating Your Data

Students should use data from the lab to draw their cartoons.

BENCH TESTED

Real-World Question

Purpose Students will determine whether the presence of earthworms affects soil quality. L2 COOP LEARN ELL IN **Naturalist**

Process Skills design an experiment, form a hypothesis, observe and infer, communicate, recognize cause and effect, separate and control variables, interpret data

Time Required two 45-minute periods; twice-weekly observations for three weeks

Materials Purchase *Eisenia foetida* or *Lumbricus rubella* (common names are red wigglers, brandlings, or manure worms) from a bait shop. Possible containers: plastic shoe box or gallon jug with top cut off. Container must be at least 10 cm deep. Drill six to ten drainage holes in the bottom. The ratio of food scraps to worms is 1:2 based on weight. Make sure that all food scraps are finely ground.

Safety Precautions Caution students to wash their hands after handling worms or soil.

Form a Hypothesis

Possible Hypothesis Student hypotheses should reflect that worms will digest the food scraps and produce healthier soil.

LAB Design Your Own

Garbage-Eating Worms

Goals

- **Design** an experiment that compares the condition of soil in two environments—one with earthworms and one without.
- **Observe** the change in soil conditions for two weeks.

Possible Materials

worms (red wigglers)
4-L plastic containers with drainage holes (2)
soil (7 L)
shredded newspaper
spray bottle
chopped food scraps including fruit and vegetable peels, pulverized eggshells, tea bags, and coffee grounds (Avoid meat and fat scraps.)

Safety Precautions

WARNING: *Be careful when working with live animals. Always keep your hands wet when handling earthworms. Don't touch your face during the lab. Wash your hands thoroughly after the lab.*

Real-World Question

Susan knows that soil conditions can influence the growth of plants. She is trying to decide what factors might improve the soil in her backyard garden. A friend suggests that earthworms improve the quality of the soil. How could Susan find out if the presence of earthworms has any value in improving soil conditions? How does the presence of earthworms change the condition of the soil?

Form a Hypothesis

Based on your reading and observations, state a hypothesis about how earthworms might improve the conditions of soil.

Test Your Hypothesis

Possible Procedures Fill two identical containers with 7.5 cm of loose soil. Do not pack soil down. Add worms to one container. Place a thin layer of finely chopped food scraps over the surface of each container and cover with shredded newspapers. Keep the newspaper moist by spraying it with water as needed. Observe each container for three weeks.

Using Scientific Methods

Test Your Hypothesis

Make a Plan

1. As a group, agree upon a hypothesis and decide how you will test it. Identify what results will support the hypothesis.
2. List the steps you will need to take to test your hypothesis. Be specific. Describe exactly what you will do in each step. List your materials.
3. Prepare a data table in your Science Journal to record your observations.
4. Read over the entire experiment to make sure that all the steps are in a logical order.
5. **Identify** all constants, variables, and controls of the experiment.

Follow Your Plan

1. Make sure your teacher approves your plan before you start.
2. Carry out the experiment according to the approved plan.
3. While doing the experiment, record your observations and complete the data table in your Science Journal.

Analyze Your Data

1. **Compare** the changes in the two sets of soil samples.
2. **Compare** your results with those of other groups.
3. **Identify** the control in this experiment.
4. What were your variables?

Conclude and Apply

1. **Explain** whether the results support your hypothesis.
2. **Describe** what effect you think rain would have on the soil and worms.

Communicating Your Data

Write an informational pamphlet on how to use worms to improve garden soil. Include diagrams and a step-by-step procedure.

Teaching Strategy

Tie to Prior Knowledge Most students know that worms improve soil.

Expected Outcome Most results will show that the worms ate the food scraps, and the resulting soil is dark and rich-looking.

Analyze Your Data

Answers to Questions

1. There should be fewer food scraps in the container with the worms, and the soil should have a darker appearance.
2. Student results should be similar to those of other groups.
3. The control in this experiment was the container filled with soil, chopped food, and shredded newspapers.
4. The variable was the presence of worms in one container.

Error Analysis Have students compare their results and their hypotheses and explain why differences occurred.

Conclude and Apply

1. Answers will be determined by student hypotheses.
2. Rain would compact the soil, and too much rain could drown the worms.

Assessment

Oral Have students explain the relationship between worms and soil fertility. Worms break down organic matter and disperse it throughout the soil; they also dig tunnels that allow air and water to penetrate the soil more deeply. Use **Performance Assessment in the Science Classroom,** p. 89. L2 LS **Linguistic**

Alternative Inquiry Lab

Ask Questions To build on this experience and develop students' critical thinking skills, have them brainstorm questions that arise from this lab. They can design procedures to test their questions.

Communicating Your Data

Review the information for relevance and correctness.

SCIENCE Stats

Content Background

Squid are cephalopods, the largest of the invertebrates. *Cephalopod* is a term that means head-foot. It refers to the fact that the foot of the ancestral species has evolved to become the head of the cephalopod. There are many types of squid, but one of the most fascinating and mysterious is the giant squid. Very few giant squid have ever been sighted, and giant squid have never been examined alive in their natural habitat. All of the information that scientists have gathered about giant squid has come from the examination of dead animals washed up on beaches or captured in fishing nets. Several teams of scientists are actively engaged in a search for giant squid to observe. Imagine being the first to see a living giant squid!

Discussion

Baby Squid Are young squid more or less independent than human babies? How do you know? Possible answer: Young squid are more independent than human babies. This answer can be inferred from the information that states that the females of many species of squid die immediately after laying their eggs.

Activity

Squid Characteristics Research the characteristics of one type of squid. Prepare a poster that illustrates and explains the unique qualities of that species of squid. **LS Visual-Spatial**

SCIENCE Stats

Squid Power

Did you know...

. . . Squid can light up like a multi-colored neon sign because of chemical reactions inside their bodies. They do this to lure prey into their grasp or to communicate with other squid. These brilliantly-colored creatures, often called fire squid, can produce blue-, red-, yellow-, and white-colored flashes in 0.3-s bursts every 5 s.

. . . The scariest-looking squid is the vampire squid. It can wrap its webbed, spiked arms around itself like a cloak. Its fins look like pointed ears and its body is covered with light-producing organs that blink on and off. Imagine seeing that eerie sight in the dark depths of the ocean, nearly 1 km below the surface of the sea.

Applying Math Scientists estimate that the adult vampire squid, which grows to about 15 cm in length, can swim at the rate of two body lengths per second. How fast is that in kilometers per hour?

. . . Squid have blue blood because their oxygen is transported by a blue copper compound not by bright-red hemoglobin like in human blood.

. . . Females of many species of squid die just after they lay eggs. In 1984, a giant squid washed ashore in Scotland, carrying more than 3,000 eggs.

Find Out About It

Scientists have never seen a living giant squid. Where would you look? At what depth? What kind of equipment would you use? To research these questions, visit red.msscience.com/science_stats.

Applying Math

30 cm/sec, or 1.08 km/h

Teaching Strategy

Remind students of the meaning of the metric prefixes *centi-* (1/100) and *kilo-* (1,000) to help them convert units in the Applying Math problem.

Find Out About It

Have students research the habitat of the giant squid to decide in which oceans and at what depths they would begin their squid search.

Reviewing Main Ideas

Section 1 What is an animal?

1. Animals are many-celled organisms that must find and digest their own food.
2. Invertebrates are animals without backbones, and vertebrates have backbones.
3. Symmetry is the way that animal body parts are arranged. The three types of symmetry are bilateral, radial, and asymmetrical.

Section 2 Sponges, Cnidarians, Flatworms, and Roundworms

1. Sponges have no tissues.
2. Adult sponges are sessile and obtain food and oxygen by filtering water.
3. Cnidarians are radially symmetrical, and most have tentacles with stinging cells to get food.
4. Flatworms and roundworms have bilateral symmetry. They have parasitic and free-living members.

Section 3 Mollusks and Segmented Worms

1. Mollusks are soft-bodied animals that usually have a shell and an open circulatory system.
2. Gastropods, bivalves, and cephalopods are types of mollusks.
3. Annelids have segmented bodies. A body cavity separates internal organs from the body wall.

Section 4 Arthropods and Echinoderms

1. Arthropods have exoskeletons that cover, protect, and support their bodies.
2. Arthropods develop either by complete metamorphosis or by incomplete metamorphosis.
3. Echinoderms are spiny-skinned invertebrates and have a water-vascular system.

Visualizing Main Ideas

Copy and complete the following concept map.

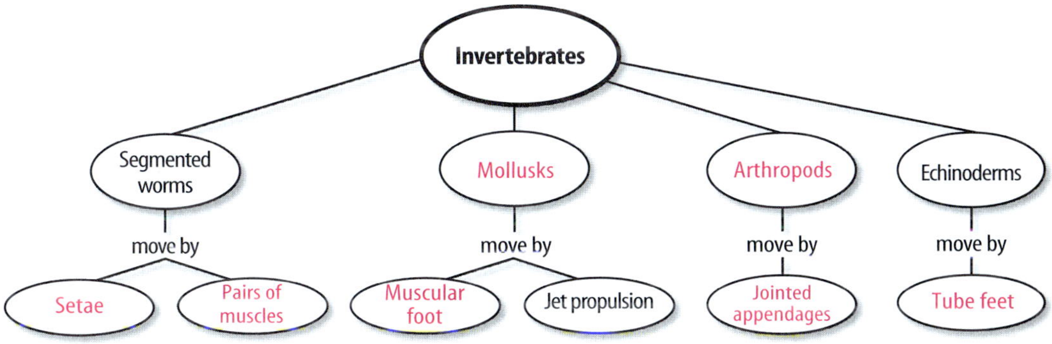

chapter 17 Study Guide

Reviewing Main Ideas

Summary statements can be used by students to review the major concepts of the chapter.

Visualizing Main Ideas

See student page.

Science Online

Visit red.msscience.com
/self_check_quiz
/interactive_tutor
/vocabulary_puzzlemaker
/chapter_review
/standardized_test

Assessment Transparency

For additional assessment questions, use the *Assessment Transparency* located in the transparency book.

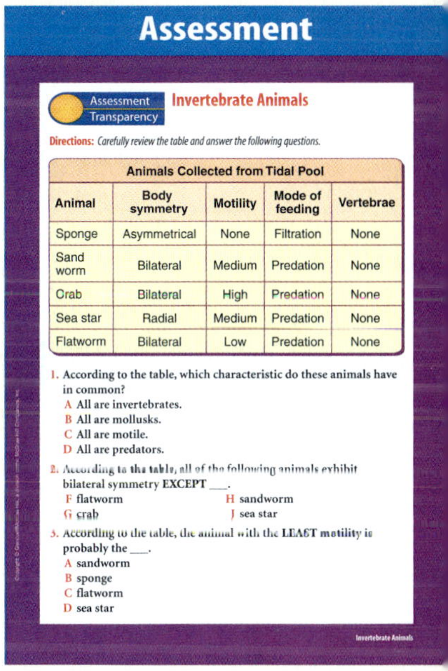

Assessment

Assessment Transparency **Invertebrate Animals**

Directions: *Carefully review the table and answer the following questions.*

Animals Collected from Tidal Pool				
Animal	Body symmetry	Motility	Mode of feeding	Vertebrae
Sponge	Asymmetrical	None	Filtration	None
Sand worm	Bilateral	Medium	Predation	None
Crab	Bilateral	High	Predation	None
Sea star	Radial	Medium	Predation	None
Flatworm	Bilateral	Low	Predation	None

1. According to the table, which characteristic do these animals have in common?
   A All are invertebrates.
   B All are mollusks.
   C All are motile.
   D All are predators.
2. According to the table, all of the following animals exhibit bilateral symmetry EXCEPT ___.
   F flatworm
   G crab
   H sandworm
   J sea star
3. According to the table, the animal with the LEAST motility is probably the ___.
   A sandworm
   B sponge
   C flatworm
   D sea star

Invertebrate Animals

Using Vocabulary

1. Both are forms the cnidarian takes on during its life cycle; medusa is the bell-shaped, free-floating form, and polyp is the vase-shaped, usually sessile form.
2. In a closed circulatory system, the blood is contained in vessels; in an open circulatory system, blood is not contained in vessels.
3. Vertebrates have a backbone; invertebrates do not.
4. Both arthropods and mollusks are invertebrates. Arthropods have an exoskeleton for protection, and mollusks have a shell.
5. Both are means of protection. An exoskeleton is composed of chitin and covers an arthropod; a mantle is a thin layer of tissue covering the soft body of a mollusk.
6. Arthropods have jointed appendages; an appendage is a structure such as a claw, leg, or antenna that grows from the body.
7. A cnidarian is a hollow-bodied invertebrate (animal without a backbone) that has tentacles and stinging cells.
8. The mantle is a thin layer of tissue that covers the soft body of a mollusk, an invertebrate that usually has a shell, mantle, and a large muscular foot.
9. A medusa is the free-swimming, bell-shaped body of a cnidarian.

Checking Concepts

10. C
11. B
12. B
13. C
14. A
15. A
16. C
17. C
18. B
19. D

chapter 17 Review

Using Vocabulary

appendage p. 512
arthropod p. 512
closed circulatory system p. 508
cnidarian p. 502
exoskeleton p. 512
gill p. 506
invertebrate p. 500
mantle p. 506
medusa p. 503
metamorphosis p. 513
mollusk p. 506
open circulatory system p. 507
polyp p. 503
radula p. 507
symmetry p. 499

For each set of vocabulary words below, explain the relationship that exists.

1. medusa—polyp
2. closed circulatory system—open circulatory system
3. vertebrate—invertebrate
4. arthropod—mollusk
5. exoskeleton—mantle
6. arthropod—appendage
7. cnidarian—invertebrate
8. mollusk—mantle
9. medusa—cnidarian

Checking Concepts

Choose the word or phrase that best answers the question.

10. Marine worms can live in all but which of the following?
    A) mud burrows **C)** soil
    B) tube cases **D)** salt water

11. Butterflies, ants, bees, and beetles are examples of insects that undergo
    A) incomplete metamorphosis.
    B) complete metamorphosis.
    C) no metamorphosis.
    D) a molt from nymph to adult.

12. The body plans of cnidarians are polyp and which of the following?
    A) larva **C)** pupa
    B) medusa **D)** bud

13. Which of the following is a parasite?
    A) sponge **C)** tapeworm
    B) planarian **D)** jellyfish

14. Which of the following groups of animals molt?
    A) crustaceans **C)** sea stars
    B) earthworms **D)** flatworms

15. Which of these organisms has a closed circulatory system?
    A) octopus **C)** oyster
    B) snail **D)** sponge

16. Radial symmetry is common in which group of invertebrates?
    A) annelids **C)** echinoderms
    B) mollusks **D)** arthropods

17. Which of the following organisms has two body regions?
    A) insect **C)** arachnid
    B) mollusk **D)** annelid

Use the photo below to answer question 18.

18. What symmetry does the animal in the illustration above have?
    A) asymmetry **C)** radial
    B) bilateral **D)** anterior

19. Which of the following do not belong to the same group?
    A) snails **C)** octopuses
    B) oysters **D)** sea stars

Science Online red.msscience.com/vocabulary_puzzlemaker

Use the ExamView® Pro Testmaker CD-ROM to:

- create multiple versions of tests
- create modified tests with one mouse click for inclusion students
- edit existing questions and add your own questions
- build tests aligned with state standards using built-in State Curriculum Tags
- change English tests to Spanish with one mouse click and vice versa

Thinking Critically

20. Infer Which aspect of sponge reproduction would be evidence that they are more like animals than plants?

21. Explain why it is an advantage for organisms to have more than one means of reproduction.

22. Compare and contrast the tentacles of cnidarians and cephalopods.

23. Explain the main differences between budding and regeneration.

24. Infer Centipedes and millipedes have segments. Why are they not classified as worms?

25. Compare and contrast the feeding habits of sponges and cnidarians.

26. Draw Conclusions Observe the conch in **Figure 11.** Infer why gastropods are sometimes called univalves? Use examples in your answer.

27. Concept Map Copy and complete the concept map below about cnidarian classification.

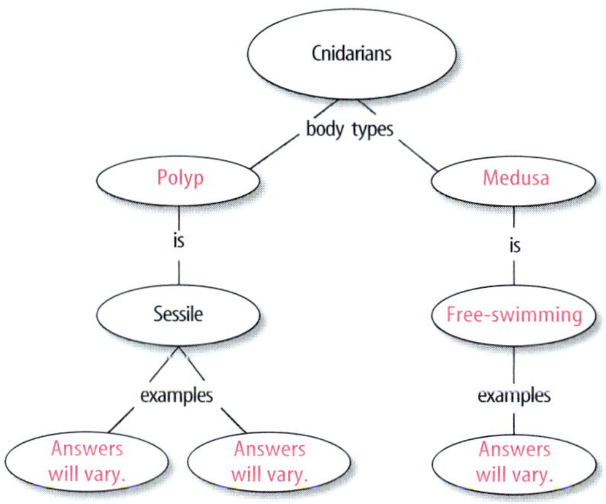

Performance Activities

28. Diary Pretend you are an earthworm. Write a diary with at least ten entries describing your daily life. Include how you move, how you get food, and where you live.

Applying Math

29. Giant Squid Size The largest giant squid recorded was 18 m long and weighed 900 kg. The best-preserved specimen is at the American Museum of Natural History. It is about 8 m long and has a mass of 114 kg. This is only a fraction of the largest specimen ever found. What is the fraction?

Use the following illustration to answer question 30.

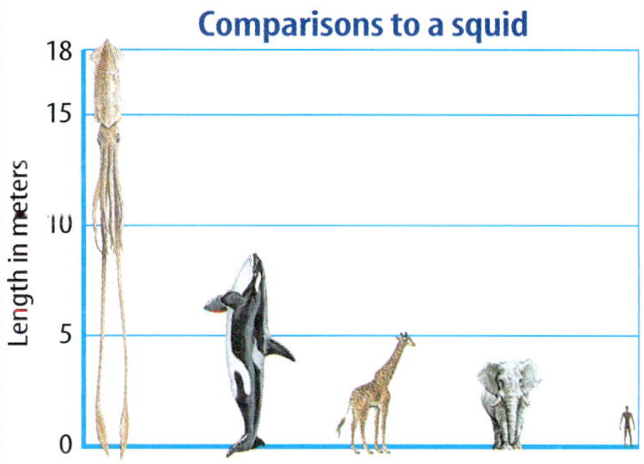

30. Squid Comparisons Approximately how many times longer is a giant squid compared to a killer whale? A giraffe? An elephant? A human?

31. Earthworm Feeding If you have an apple that weighs 141 g and an earthworm that weighs 11 g, how many days would it take the earthworm to eat the apple? Assume the earthworm can eat its own weight each day.

32. Insect Species Approximately 91,000 species of insects and 24,000 species of beetles have been identified in the United States. Approximately what percentage of the identified species are beetles?

Science Online red.msscience.com/chapter_review

Assessment Resources

Reproducible Masters

Chapter *Fast File* Resources
- Chapter Review, pp. 39–40
- Chapter Tests, pp. 41–44
- Assessment Transparency Activity, p. 53

Glencoe Science Web site
- Chapter Review Test
- Standardized Test Practice

Glencoe Technology

- Assessment Transparency
- Exam*View*® Pro Testmaker
- MindJogger Videoquiz
- Interactive Chalkboard

chapter 17 Review

Thinking Critically

20. Their larva is flagellated.

21. So they can respond to changing conditions; this diversity allows them to survive in adverse conditions.

22. Cnidarian tentacles have stinging cells for stunning prey; cephalopod tentacles have suckers for grasping prey.

23. Budding is a type of asexual reproduction in which a new organism develops. Regeneration involves replacing lost body parts.

24. They have jointed appendages, or legs, attached to each segment; worms do not. Centipedes have one pair of legs per segment; millipedes have two pairs of legs per segment.

25. Sponges filter microscopic materials out of the water. Collar cells help move water through the sponge. Cnidarians use stinging cells on their tentacles to paralyze prey; the tentacles pull prey into the mouth.

26. Gastropods have only one shell. Examples will vary.

27. See student page.

Performance Activities

28. The earthworm diary should indicate that earthworms use setae to move through the soil and that they get food from the soil, which they take in as they move through it.

Applying Math

National Math Standards

1, 9

29. length: 8 m/18 m = 1/9
weight: 114 kg/900 kg = 57/450

30. 2 times larger than a killer whale; 3–4 times larger than a giraffe; 4–5 times larger than an elephant; 9–10 times larger than a human

31. nearly 13 days

32. approximately 26%

FAST FILE

Answer Sheet A practice answer sheet can be found at red.msscience.com/answer_sheet.

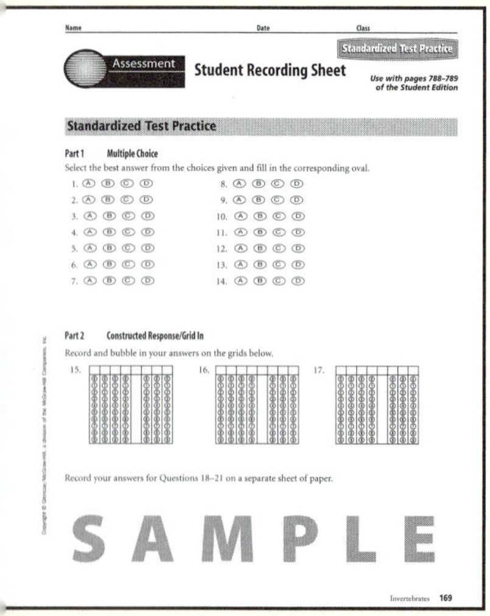
Name Date Class

Assessment Student Recording Sheet

Standardized Test Practice

Use with pages 788–789 of the Student Edition

Standardized Test Practice

Part 1 Multiple Choice

Select the best answer from the choices given and fill in the corresponding oval.

Part 2 Constructed Response/Grid In

Record and bubble in your answers on the grids below.

Record your answers for Questions 18–21 on a separate sheet of paper.

SAMPLE

Invertebrates 169

Part 1 Multiple Choice

1. A
2. C
3. D
4. B
5. C
6. D
7. B
8. B
9. A

Part 2 Short Response

10. Answers may vary. Sample answer: radial symmetry—sea star; bilateral symmetry—crab; asymmetry—sponge.
11. sponge; sexual
12. asexual reproduction; a parent sponge forms a bud which breaks off, floats away, and attaches to a new surface where it grows into a new sponge.
13. A hermaphrodite is an animal that produces both eggs and sperm. Examples include sponges and tapeworms.
14. Insects: 3 body regions; 3 pairs of jointed legs; 1 or 2 pairs of wings. Arachnids: 2 body regions; 4 pairs of jointed legs; no wings.
15. During molting, the exoskeleton of the animal is shed. Arthropods molt.
16. food ⟶ mouth ⟶ crop ⟶ gizzard ⟶ intestine ⟶ anus
17. Sponges have collar cells that move water through its central cavity, spicules that help support the body, and sponging, a material that makes sponges soft and elastic.

chapter 17 Standardized Test Practice

Part 1 Multiple Choice

Record your answers on the answer sheet provided by your teacher or on a sheet of paper.

1. Which of the following is NOT a characteristic of animals?
   A. All animals have a definite shape.
   B. All animals are many-celled.
   C. All animals use energy.
   D. The cells of all animals have nuclei and organelles.

Use the illustrations below to answer questions 2 and 3.

A

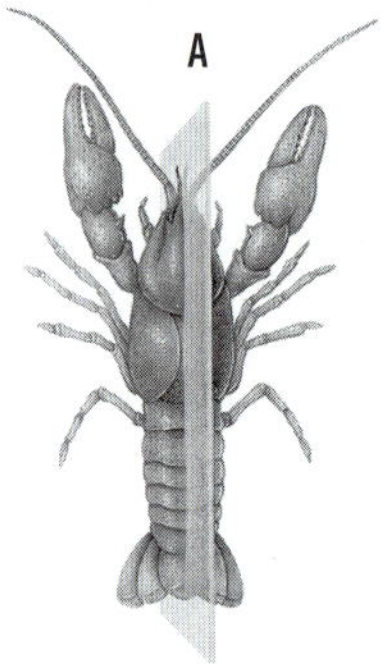

B

2. What type of symmetry is represented by A?
   A. radial symmetry
   B. biradial symmetry
   C. bilateral symmetry
   D. asymmetry

3. Which of the following animals has the type of symmetry shown in B?
   A. earthworm **C.** cow
   B. clam **D.** sea urchin

4. Which of the following is NOT an invertebrate animal group?
   A. arthropod
   B. chordate
   C. sponge
   D. cnidarian

5. Which of the following is a characteristic of cnidarians?
   A. spicules
   B. mantle
   C. mematocysts
   D. coelom

6. Annelids, or segmented worms, include animals such as
   A. tapeworms. **C.** planaria.
   B. heartworms. **D.** leeches.

7. Which of the following is a characteristic of echinoderms?
   A. They have two pairs of antennae.
   B. They have a "spiny skin."
   C. They have many setae along the sides of their body.
   D. They move through water by jet propulsion.

Use the photo below to answer questions 8 and 9.

8. What kind of invertebrate is the animal shown above?
   A. mollusk **C.** sponge
   B. arthropod **D.** echinoderm

9. Which of the following animals is a member of the same invertebrate group as the animals shown above?
   A. hydra **C.** spider
   B. leech **D.** sponge

526 STANDARDIZED TEST PRACTICE

Part 2 Short Response/Grid In

Record your answers on the answer sheet provided by your teacher or on a sheet of paper.

10. Give three examples of invertebrates: one that has radial symmetry, one that has bilateral symmetry, and one that is asymmetrical.

Use the illustration below to answer questions 11 and 12.

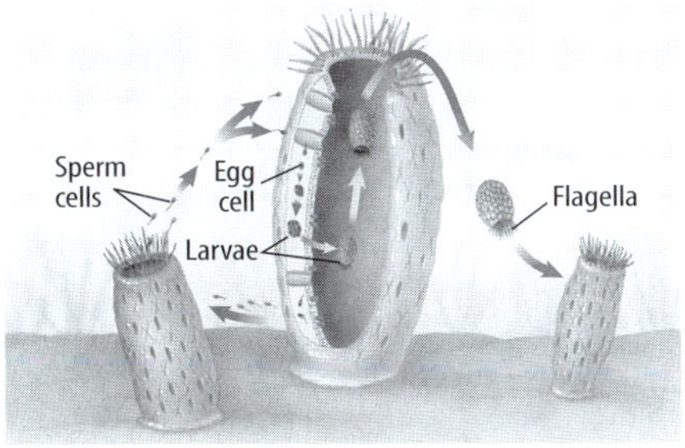

11. The reproduction of which invertebrates is shown in the diagram above? What type of reproduction is shown?

12. What other type of reproduction is characteristic of this animal? Explain.

13. What does the term hermaphrodite mean? Give one example of an invertebrate that is hermaphroditic.

14. Compare the number of body regions, jointed legs, and pairs of wings in insects and arachnids.

15. Describe the process of molting. Which invertebrate group exhibits this characteristic?

16. Draw a flowchart to represent how food matter moves through an earthworm's digestive system.

17. Describe the physical characteristics that are found only in sponges.

Part 3 Open Ended

Record your answers on a sheet of paper.

18. Describe how sponges feed on the microorganisms in the water around them.

19. Compare and contrast a closed circulatory system and an open circulatory system.

20. Draw a diagram to describe the life cycle of a tapeworm.

21. Explain the process used by mollusks to take in oxygen.

Use the illustration below to answer questions 22 and 23.

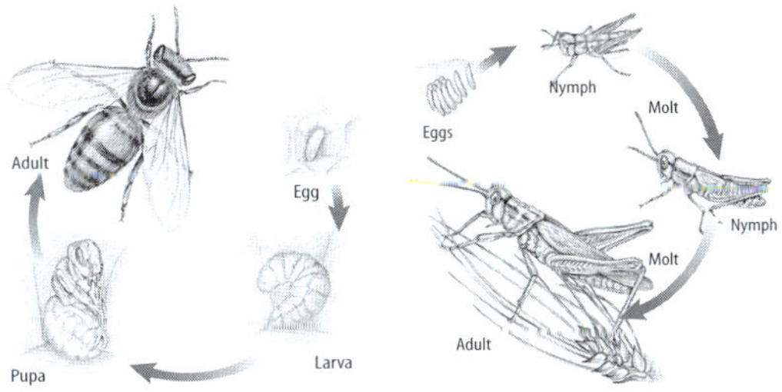

22. Which of the diagrams above represents complete metamorphosis and which represents incomplete metamorphosis? How can you tell the difference?

23. Compare and contrast the nymph and larva stages of metamorphosis.

24. What are the main characteristics found in annelids? Name the major groups of annelids.

Test-Taking Tip

Complete Answers Make sure each part of the question is answered when listing discussion points.

Question 23 This question asks you to compare and contrast, make sure you list both similarities and differences.

Part 3 Open Ended

18. Microorganisms and oxygen enter the sponge's central cavity through pores in the sponge's body. Specialized cells digest food, carry nutrients to parts of the sponge, and remove wastes.

19. In an open circulatory system, blood washes over the organs in a fluid-filled body cavity. There are no blood vessels. In a closed circulatory system, blood vessels carry blood to organs.

20. Diagrams should show tapeworm segments containing fertilized eggs being deposited with feces on the ground, eggs being consumed by grazing animals, eggs hatching and larvae burrowing into muscle tissue, humans eating inadequately cooked beef and becoming infected with tapeworm larvae, tapeworms attaching to the lining of its host's intestine.

21. Water-dwelling mollusks use gills to take in oxygen from the surrounding water. Land-dwelling mollusks have lungs that take in oxygen from the surrounding air.

22. Diagram A represents complete metamorphosis. Diagram B represents incomplete metamorphosis. There are four stages in complete metamorphosis—egg, larva, pupa, adult. There are only three stages in incomplete metamorphosis—egg, nymph, adult.

23. A nymph is a small version of the parents. A nymph molts until it reaches adult size. A larva looks completely different from the adult and becomes a pupa before it becomes an adult.

24. Annelids, or segmented worms, are made up of segments that each contain nerve cells, blood vessels, part of the digestive tract, and the coelom. Annelids also have a closed circulatory system and a complete digestive system. Earthworms, leeches, and polychaetes are all annelids.

Rubrics

For more help evaluating open-ended assessment questions, see the rubric on p. 10T.

chapter 18 Organizer

Section/Objectives	Standards: National	Standards: State/Local	Labs/Features
Chapter Opener	See pp. 16T–17T for a Key to Standards.		**Launch Lab:** Animals with a backbone, p. 529 **Foldables,** p. 529
Section 1 Chordate Animals 1 session 0.5 block 1. **Identify** the major characteristics of chordates. 2. **List** the major characteristics common to all vertebrates. 3. **Explain** the difference between ectotherms and endotherms. 4. **Name** the characteristics of the three classes of fish.	National Content Standards: UCP.1–UCP.3, UCP.5, A.1–A.2, B.2, C.1–C.3, C.5, E.1, G.1		**Integrate Physics,** p. 532 **Visualizing Fish Diversity,** p. 533
Section 2 Amphibians and Reptiles 2 sessions 1 block 5. **Describe** how amphibians have adapted to live in water and on land. 6. **Explain** what happens during frog metamorphosis. 7. **Identify** the adaptations that allow reptiles to live on land.	National Content Standards: UCP.1–UCP.3, UCP.5, A.1–A.2, C.1–C.3, C.5		**Science Online,** p. 536 **Integrate Career,** p. 538 **Lab:** Frog Metamorphosis, p. 540
Section 3 Birds 2 sessions 1 block 8. **Identify** the characteristics of birds. 9. **Describe** the adaptations birds have for flight. 10. **Explain** the function of feathers.	National Content Standards: UCP.1–UCP.3, UCP.5, A.1–A.2, C.1–C.3, C.5		**Science Online,** p. 542 **MiniLAB:** Observing What Feathers Do, p. 543
Section 4 Mammals 4 sessions 2 blocks 11. **Identify** the characteristics common to all mammals. 12. **Explain** how mammals are adapted to the different environments on Earth. 13. **Distinguish** among monotremes, marsupials, and placentals.	National Content Standards: UCP.1–UCP.3, UCP.5, A.1–A.2, C.1–C.5		**MiniLAB:** Inferring How Blubber Insulates, p. 546 **Applying Math:** Working with Percentages, p. 547 **Lab:** Homes for Endangered Animals, pp. 550–551 **Oops! Accidents in Science:** Cosmic Dust and Dinosaurs, p. 552

Glencoe Exclusive! TeacherWorks™ All-In-One Planner and Resource Center

Lab Materials	Reproducible Resources	Section Assessment	Technology
Launch Lab: pasta wheels, soft candy circles, long pipe cleaners	**Chapter *FAST FILE* Resources** Foldables Worksheet, p. 15 Directed Reading Overview, p. 17 Note-taking Worksheets, pp. 33–36	GLENCOE'S ASSESSMENT ADVANTAGE	TeacherWorks includes: • Interactive Teacher Edition • Lesson Planner with calendar • Access to all program blacklines • Correlations to standards • Web links
Need materials? Contact Science Kit at 1-800-828-7777 or www.sciencekit.com on the Internet.	**Chapter *FAST FILE* Resources** Transparency Activity, p. 46 Enrichment, p. 29 Reinforcement, p. 25 Directed Reading, p. 18 Transparency Activity, pp. 51–52 Lab Activity, pp. 9–11	**Portfolio** Activity, p. 531 **Performance** Applying Math, p. 534 **Content** Section Review, p. 534	Section Focus Transparency Teaching Transparency Virtual Labs CD-ROM Guided Reading Audio Program Interactive Chalkboard CD-ROM
Lab: 4-L aquarium or jar, frog egg mass, lake or pond water, stereoscopic microscope, watch glass, small fishnet, aquatic plants, washed gravel, lettuce (previously boiled), large rock	**Chapter *FAST FILE* Resources** Transparency Activity, p. 47 Enrichment, p. 30 Reinforcement, p. 26 Directed Reading, p. 18 Lab Worksheet, pp. 5–6	**Portfolio** Activity, p. 538 **Performance** Applying Skills, p. 539 **Content** Section Review, p. 539	Section Focus Transparency Virtual Labs CD-ROM Guided Reading Audio Program Interactive Chalkboard CD-ROM
MiniLAB: cotton cloth, petroleum jelly, water	**Chapter *FAST FILE* Resources** Transparency Activity, p. 48 MiniLAB, p. 3 Enrichment, p. 31 Reinforcement, p. 27 Directed Reading, p. 19 Lab Activity, pp. 13–14	**Portfolio** Visual Learning, p. 541 **Performance** MiniLAB, p. 543 Applying Math, p. 544 **Content** Section Review, p. 544	Section Focus Transparency Virtual Labs CD-ROM Guided Reading Audio Program Interactive Chalkboard CD-ROM Video Lab
MiniLAB: self-sealing plastic bags (2), vegetable shortening, ice water **Lab:** poster board, markers or colored pencils, materials with which to make a scale model	**Chapter *FAST FILE* Resources** Transparency Activity, p. 49 MiniLAB, p. 4 Enrichment, p. 32 Reinforcement, p. 28 Directed Reading, pp. 19, 20 Lab Worksheet, pp. 7–8	**Portfolio** Curriculum Connection, p. 546 **Performance** MiniLAB, p. 546 Applying Math, p. 547 Applying Math, p. 549 **Content** Section Review, p. 549	Section Focus Transparency Virtual Labs CD-ROM Guided Reading Audio Program Interactive Chalkboard CD-ROM

End of Chapter Assessment

GLENCOE'S ASSESSMENT ADVANTAGE

Blackline Masters	Technology	Professional Series
Chapter *FAST FILE* Resources Chapter Review, pp. 39–40 Chapter Tests, pp. 41–44 **Standardized Test Practice,** pp. 75–78	MindJogger Videoquiz Virtual Labs CD-ROM Exam*View*® Pro Testmaker TeacherWorks CD-ROM Interactive Chalkboard CD-ROM	**Performance Assessment in the Science Classroom (PASC)**

chapter 18 Vertebrate Animals

Transparencies

Section Focus

Section 1 Section Focus Transparency **Do they have tests?**

These colorful little fish live in warm, tropical waters. A group of fish swimming together like this is called a school. Fish swim in schools for protection as well as for feeding and breeding purposes.

1. What do fish and mammals have in common?
2. What are some characteristics of fish?

L2

Section 2 Section Focus Transparency **On Second Thought, Let's Eat Something Else**

Usually, this animal stays hidden, but when it is threatened, it twists its body and puffs out its belly to show its brilliant red color. Its flashy colors warn other animals that it would make a dangerous—even deadly—meal!

1. What kind of environment do you think this animal lives in?
2. How is this creature different from a fish? A lizard?

L2

Section 3 Section Focus Transparency **Maybe the Scientists Were a Hoax!**

This bird, called a kiwi, is the national bird of New Zealand. It's so unusual that when it was first brought to the attention of scientists, they thought it was a hoax. It has no tail and only stubby wings hidden by hair-like feathers. Flightless, the nocturnal kiwi lives in burrows and tends to be shy.

1. Why might scientists have thought the kiwi a hoax?
2. Why do you think the kiwi is classified as a bird?

L2

This is a representation of key blackline masters available in the Teacher Classroom Resources. See Resource Manager boxes within the chapter for additional information.

Assessment

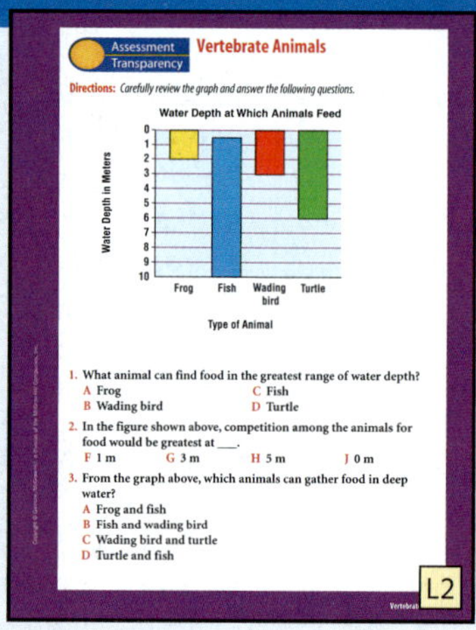
Assessment Transparency **Vertebrate Animals**

Directions: Carefully review the graph and answer the following questions.

1. What animal can find food in the greatest range of water depth?
   A Frog C Fish
   B Wading bird D Turtle
2. In the figure shown above, competition among the animals for food would be greatest at ___.
   F 1 m G 3 m H 5 m J 0 m
3. From the graph above, which animals can gather food in deep water?
   A Frog and fish
   B Fish and wading bird
   C Wading bird and turtle
   D Turtle and fish

L2

Teaching

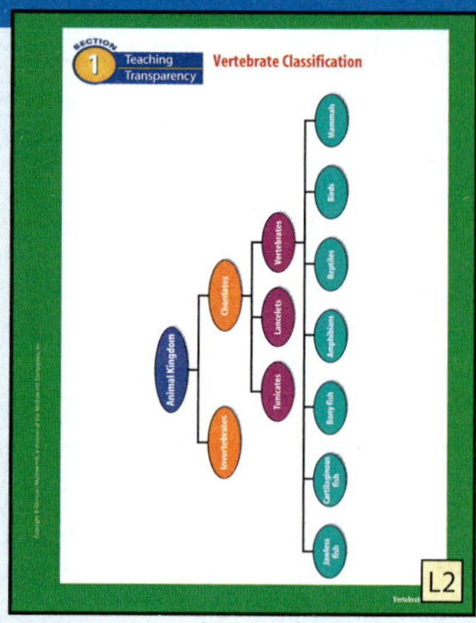
Section 1 Teaching Transparency **Vertebrate Classification**

L2

Key to Teaching Strategies

The following designations will help you decide which activities are appropriate for your students.

- **L1** Level 1 activities should be appropriate for students with learning difficulties.
- **L2** Level 2 activities should be within the ability range of all students.
- **L3** Level 3 activities are designed for above-average students.
- **ELL** ELL activities should be within the ability range of English-Language Learners.
- **COOP LEARN** Cooperative Learning activities are designed for small group work.
- **LS** Multiple Learning Styles logos, as described on page 12T, are used throughout to indicate strategies that address different learning styles.
- **P** These strategies represent student products that can be placed into a best-work portfolio.
- **PBL** Problem-Based Learning activities apply real-world situations to learning.

Hands-on Activities

Student Text Lab Worksheet

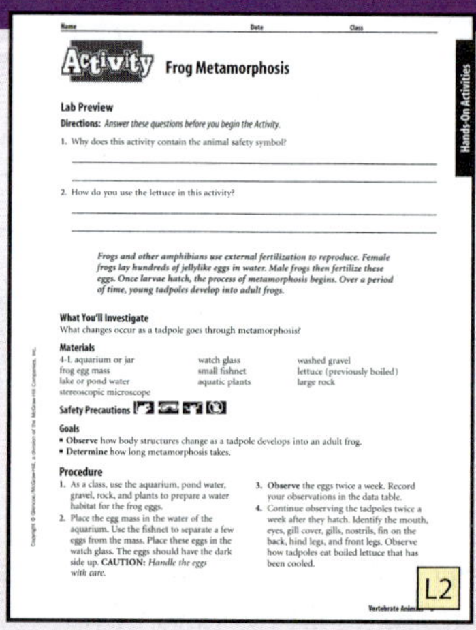
Name Date Class

Activity **Frog Metamorphosis**

Lab Preview

Directions: Answer these questions before you begin the Activity.

1. Why does this activity contain the animal safety symbol?
2. How do you use the lettuce in this activity?

Frogs and other amphibians use external fertilization to reproduce. Female frogs lay hundreds of jellylike eggs in water. Male frogs then fertilize these eggs. Once larvae hatch, the process of metamorphosis begins. Over a period of time, young tadpoles develop into adult frogs.

What You'll Investigate

What changes occur as a tadpole goes through metamorphosis?

Materials

4-L aquarium or jar, frog egg mass, lake or pond water, stereoscopic microscope, watch glass, small fishnet, aquatic plants, washed gravel, lettuce (previously boiled), large rock

Safety Precautions

Goals

- Observe how body structures change as a tadpole develops into an adult frog.
- Determine how long metamorphosis takes.

Procedure

1. As a class, use the aquarium, pond water, gravel, rock, and plants to prepare a water habitat for the frog eggs.
2. Place the egg mass in the water of the aquarium. Use the fishnet to separate a few eggs from the mass. Place these eggs in the watch glass. The eggs should have the dark side up. CAUTION: *Handle the eggs with care.*
3. Observe the eggs twice a week. Record your observations in the data table.
4. Continue observing the tadpoles twice a week after they hatch. Identify the mouth, eyes, gill cover, gills, nostrils, fin on the back, hind legs, and front legs. Observe how tadpoles eat boiled lettuce that has been cooled.

L2

Laboratory Activities

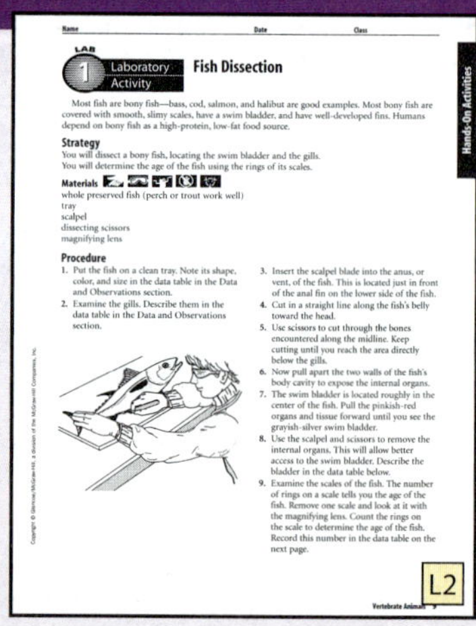
Name Date Class

Lab 1 Laboratory Activity **Fish Dissection**

Most fish are bony fish—bass, cod, salmon, and halibut are good examples. Most bony fish are covered with smooth, slimy scales, have a swim bladder, and have well-developed fins. Humans depend on bony fish as a high-protein, low-fat food source.

Strategy

You will dissect a bony fish, locating the swim bladder and the gills.
You will determine the age of the fish using the rings of its scales.

Materials

whole preserved fish (perch or trout work well), tray, scalpel, dissecting scissors, magnifying lens

Procedure

1. Put the fish on a clean tray. Note its shape, color, and size in the data table in the Data and Observations section.
2. Examine the gills. Describe them in the data table in the Data and Observations section.
3. Insert the scalpel blade into the anus, or vent, of the fish. This is located just in front of the anal fin on the lower side of the fish.
4. Cut in a straight line along the fish's belly toward the head.
5. Use scissors to cut through the bones encountered along the midline. Keep cutting until you reach the area directly below the gills.
6. Now pull apart the two walls of the fish's body cavity to expose the internal organs.
7. The swim bladder is located roughly in the center of the fish. Pull the pinkish-red organs and tissue forward until you see the grayish-silver swim bladder.
8. Use the scalpel and scissors to remove the internal organs. This will allow better access to the swim bladder. Describe the bladder in the data table below.
9. Examine the scales of the fish. The number of rings on a scale tells you the age of the fish. Remove one scale and look at it with the magnifying lens. Count the rings on the scale to determine the age of the fish. Record this number in the data table on the next page.

L2

Meeting Different Ability Levels

Content Outline

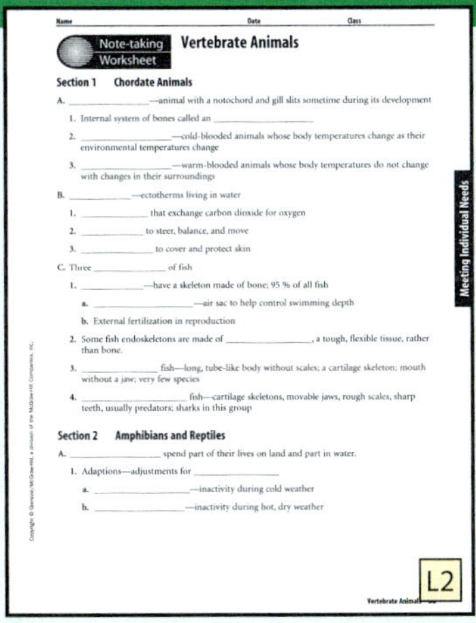

Reinforcement

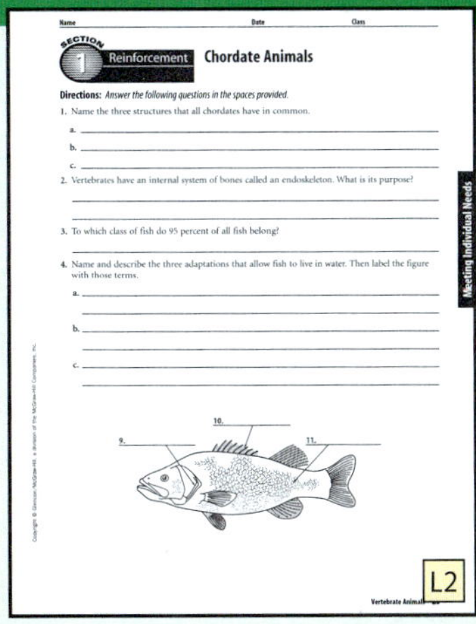

Enrichment

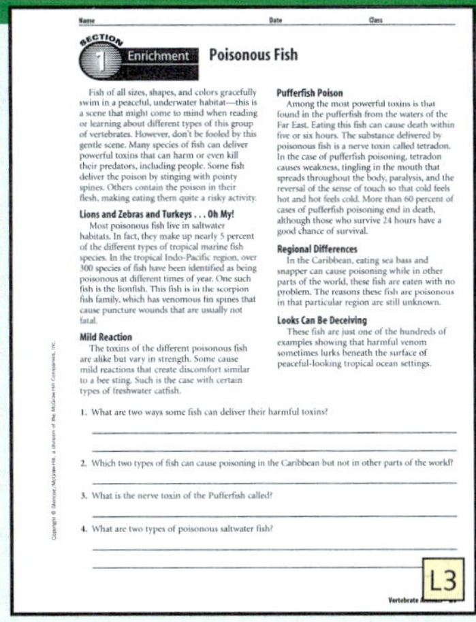

Directed Reading (English/Spanish)

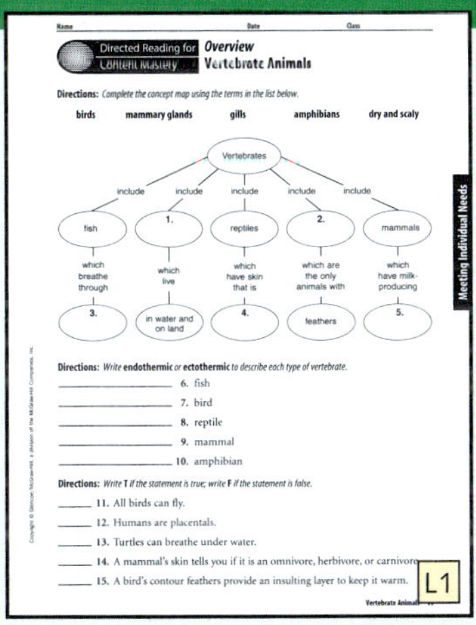

Study Guide

Reading Essentials

Assessment

Test Practice Workbook

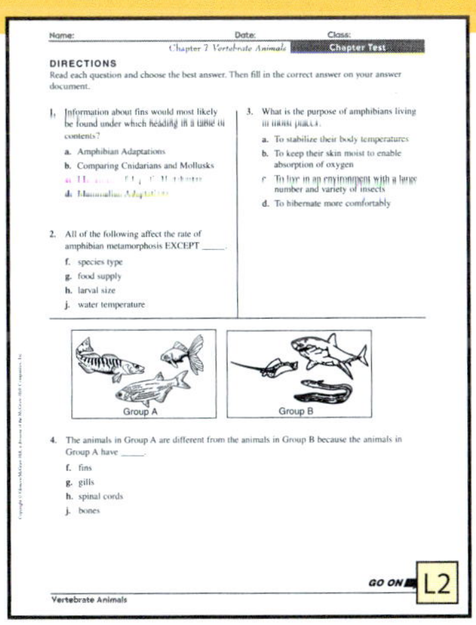

Chapter Review

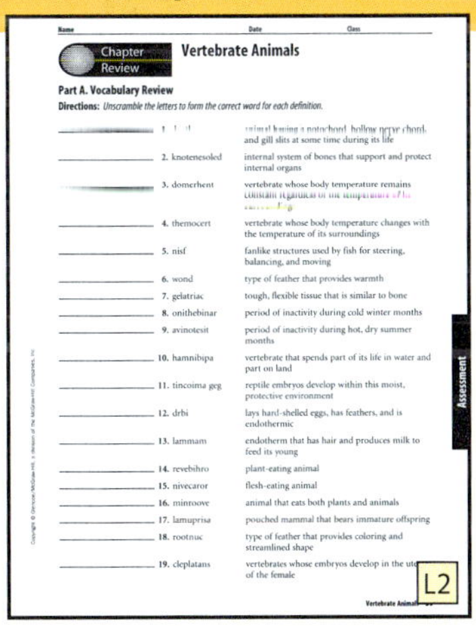

Chapter Tests

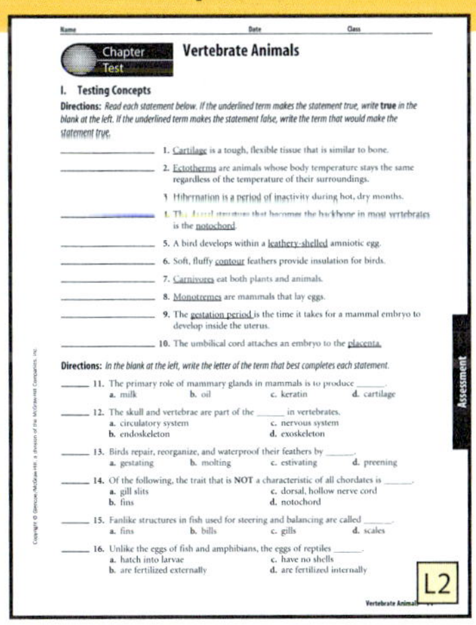

Science Content Background

section 1

Chordate Animals

Vertebrates

All vertebrates have similar organ systems. Fish have a two-chambered heart and gills for the exchange of oxygen and carbon dioxide. In the gills the blood in the capillaries has lost nearly all its pressure, which seriously reduces the efficiency of the entire circulatory system. Amphibians have a three-chambered heart with two atria that separate systemic and pulmonary circulation. Even with only one ventricle, studies have shown that there is little mixing of oxygenated and deoxygenated blood. Reptiles have an incomplete four-chambered heart. The atria are completely separated, but the septum of the right and left ventricles is incomplete. Birds and mammals have four-chambered hearts that completely separate blood into systemic and pulmonary circulation.

Roland Seitre/Peter Arnold, Inc.

Fish

Some fish are considered warm-blooded, but are not truly endotherms. Bluefin tuna are able to maintain a body temperature as much as 14°C higher than that of the surrounding water. Blood is heated in the swimming muscles that run down the center of the body and are well insulated. Blood vessels carrying cold blood from the gills to the body have contact with blood vessels carrying warm blood from the body to the gills. Cold blood coming from the gills is warmed before reaching the inside of the body, and blood from the body is cooled on its way to the gills. Thus, the fish retain precious body heat while swimming in cold water.

Bony Fish

The importance of the swim bladder can be seen in fish with swim bladder disease, a disease that primarily affects goldfish with rounded bodies. Some fish, including goldfish, have a special duct that allows them to adjust buoyancy by letting air out through the digestive tract. Some scientists suggest that the disease is caused by food blocking this duct. Goldfish with the disease cannot control their buoyancy and may float near the surface or stay on the bottom and have difficulty rising.

chapter content resources

Internet Resources

For additional content background, visit **red.msscience.com** to:

- access your book online
- find references to related articles in popular science magazines
- access Web links with related content background
- access current events with science journal topics

Print Resources

What Is the Animal Kingdom?, by Bobbie Kalman, Crabtree Publishers, 1997

The Animal Kingdom: A Guide to Vertebrate Classification and Biodiversity (Classification), by Kathryn Whyman, Raintree/Steck Vaughn, 1999

Walter Bibikow/FPG International

Section 2 Amphibians and Reptiles

Amphibians

Because they live on land and reproduce in water, amphibians are affected by chemical changes in the terrestrial and the aquatic environment. Soil erosion and water pollution can adversely affect amphibians and, because they absorb gases through their skin, they are susceptible to air pollutants. Their sensitivity could explain why there have been such dramatic declines of amphibians in some areas.

Reptiles

Reptiles live on every continent except Antarctica and in all the oceans except those in polar regions. Giant pythons 10 m in length can swallow deer whole. Some sea turtles have a mass of almost one metric ton and can swim faster than most humans can run. Three-horned lizards have movable eye sockets and tongues as long as their bodies.

Section 3 Birds

Archaeopteryx

In 1861 in a limestone quarry in Bavaria, Germany, one of the world's best-known fossils was found. The skeleton suggested that it was a small, bipedal, insect-eating dinosaur. Upon further study, the imprint of wings and feathers was discovered. Feathers are a defining characteristic of all birds. The animal was given the name *Archaeopteryx lithographica*. Unlike modern birds, *Archaeopteryx* had teeth, a long tail, free-floating ribs, and wings with claws attached to movable fingers. Except for wings, these are all characteristics of small dinosaurs. *Archaeopteryx* is considered one of the intermediary species linking reptiles and birds.

Section 4 Mammals

Body Systems

The digestive system of mammals varies according to the kind of food each type of animal eats. Carnivores have short digestive systems compared with those of herbivores because meat is more easily digested than plant material. Herbivores need long digestive systems to help break down the carbohydrate called cellulose found in plants.

Teacher to Teacher

Marie Renner, Teacher
Diley Middle School
Pickerington, Ohio

"What better way to teach students about animals with backbones than to have them piece together the remains of small, regurgitated animals found in a heat sterilized owl pellet? Students can work individually or in groups to reconstruct and secure the bones with glue to a piece of construction paper. Your students also could identify the bones using a cadaver. Note: Students should wear gloves when handling animal remains. Be sure to disinfect all lab surfaces when the activity is completed."

Marie Renner

chapter 18

Chapter Vocabulary

chordate, p. 530
ectotherm, p. 531
endotherm, p. 531
cartilage, p. 532
hibernation, p. 535
estivation, p. 535
amniotic egg, p. 539
contour feather, p. 543
down feather, p. 543
herbivore, p. 546
carnivore, p. 546
omnivore, p. 546
monotreme, p. 547
marsupial, p. 548
placental, p. 548

Science Journal Humans and vertebrate animals both have an internal system of bones. Some animals are warm-blooded like humans, and some have hair. Like humans, all chordates have a notochord, a nerve cord, and pharyngeal pouches at some time during their development.

INTERACTIVE CHALKBOARD with Image Bank

PowerPoint® Presentations

This CD-ROM is an editable Microsoft® PowerPoint® presentation that includes:

- a pre-made presentation for every chapter
- interactive graphics
- animations
- audio clips
- image bank
- all new section and chapter questions
- Standardized Test Practice
- transparencies
- pre-lab questions for all labs
- Foldables directions
- links to red.msscience.com

chapter 18

Vertebrate Animals

chapter preview

sections

Traits We Share With Animals

An eagle flies through the sky, a salmon swims through a stream, a snake slithers along the ground, and a grizzly bear walks along a river's edge. Each of these animals appears different, but these animals, as well as humans, share a common trait—an internal skeleton.

Science Journal What other traits do these animals and humans share?

528

Theme Connection

Stability and Change Like other organisms, fish, amphibians, reptiles, birds, and mammals have adaptations that allow each species to maintain a stable existence in its own unique environment.

About the Photo

Foraging Bear Grizzly bears are omnivores with a diet that can consist of marsh grasses, berries, salmon, and caribou. The grizzly bear in this photo is probably foraging on new river-grass shoots.

Start-Up Activities

Launch LAB

Animals with a Backbone

An internal skeleton is common to many animals. Skeletons are made of bones or cartilage. They give your body its overall shape and work with your muscles to help move your body. **WARNING:** *Do not eat or drink anything in the lab.*

1. Use pasta wheels, soft-candy circles, and long pipe cleaners to make a model of a backbone.
2. On a pipe cleaner, string in an alternating pattern the pasta wheels and the soft-candy circles until the string is about 10 cm long.
3. Fold over each end of the pipe cleaner so the pasta and candy do not slide off.
4. **Think Critically** Slowly bend the model. Does it move easily? How far can you bend it? What do you think makes up your backbone? Write your observations and answers in your Science Journal.

FOLDABLES™ Study Organizer

Vertebrates Make the following Foldable to help you organize your thoughts about vertebrate animals before you begin reading.

STEP 1 **Collect** three sheets of paper and layer them about 1.25 cm apart vertically. Keep the edges level.

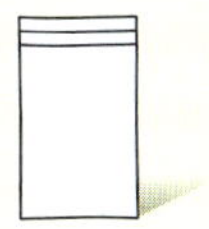

STEP 2 **Fold** up the bottom edges of the paper to form 6 equal tabs.

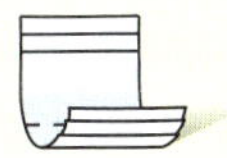

STEP 3 **Fold** the papers and crease well to hold the tabs in place. Staple along the fold. **Label** the flaps *Vertebrates, Fish, Amphibians, Reptiles, Birds,* and *Mammals,* as shown.

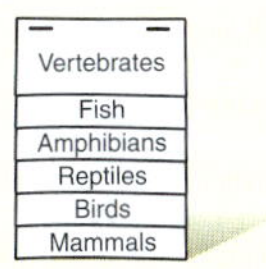

Sequence Before you read the chapter, write what you know about each group under the tabs. As you read the chapter, add to or change the information you wrote under the tabs.

Science Online — Preview this chapter's content and activities at red.msscience.com

529

Launch LAB

Purpose Use the Launch Lab to introduce students to vertebrates. Students will build a model of a backbone. L2

Preparation Purchase soft candy circles, pasta wheels, and pipe cleaners.

Materials pasta wheels, soft candy circles, pipe cleaners

Teaching Strategy Place materials for each group in a self-sealing plastic bag.

Think Critically

The model moves easily but can be bent only so far. Student responses will vary. The backbone is made of bone and flexible material (cartilage). Answers will vary but should include a good description of the model.

Assessment

Process Have students make a model using beads instead of soft candy circles and compare the two models. Which model allows for greater movement? Explain. The first model; the soft candy circles increase flexibility. L1 Use **Performance Assessment in the Science Classroom,** p. 123.

FOLDABLES™ Study Organizer — Dinah Zike Study Fold

Student preparation materials for this Foldable are available in the Chapter *FAST FILE* Resources.

1 Motivate

Bellringer

Section Focus Transparencies also are available on the Interactive Chalkboard CD-ROM. L2 ELL

SECTION 1 Section Focus Transparency **Do they have tests?**

These colorful little fish live in warm, tropical waters. A group of fish swimming together like this is called a school. Fish swim in schools for protection as well as for feeding and breeding purposes.

1. What do fish and mammals have in common?
2. What are some characteristics of fish?

Vertebrate Animals

Tie to Prior Knowledge

Vertebrates Ask students to describe vertebrae. They are bony or cartilaginous segments that make up the spinal column. Have them infer what vertebrates are. animals with a backbone Ask them to name as many kinds of vertebrates as they can. possible answers: horses, humans, cats, dogs, fish, mice, birds L2

section 1

Chordate Animals

as you read

What You'll Learn

- **Identify** the major characteristics of chordates.
- **List** the major characteristics common to all vertebrates.
- **Explain** the difference between ectotherms and endotherms.
- **Name** the characteristics of the three classes of fish.

Why It's Important

You and other vertebrate animals have an internal skeleton that supports and protects your internal organs.

Review Vocabulary
invertebrate: an animal without a backbone

New Vocabulary
- chordate
- ectotherm
- endotherm
- cartilage

What is a chordate?

Suppose you asked your classmates to list their pets. Dogs, cats, birds, snakes, and fish probably would appear on the list. Animals that are familiar to most people are animals with a backbone. These animals belong to a larger group of animals called chordates (KOR dayts). Three characteristics of all **chordates** are a notochord, a nerve cord, and pharyngeal pouches at some time during their development. The notochord, shown in **Figure 1,** is a flexible rod that extends along the length of the developing organism. Pharyngeal pouches are slitlike openings between the body cavity and the outside of the body. They are present only during the early stages of the organism's development. In most chordates, one end of the nerve cord develops into the organism's brain.

Vertebrates Scientists classify the 42,500 species of chordates into smaller groups, as shown in **Figure 2.** The animals within each group share similar characteristics, which may indicate that they have a common ancestor. Vertebrates, which include humans, are the largest group of chordates.

Vertebrates have an internal system of bones called an endoskeleton. *Endo-* means "within." The vertebrae, skull, and other bones of the endoskeleton support and protect internal organs. For example, vertebrae surround and protect the nerve cord. Many muscles attach to the skeleton and make movement possible.

Figure 1 Lancelets are filter feeders that grow to 7 cm in length and live in the ocean. Its pharyngeal pouches develop into gill slits.

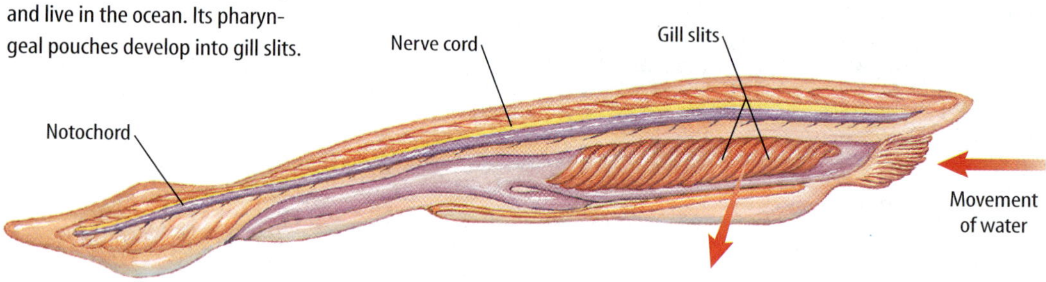

Section 1 Resource Manager

Chapter *FAST FILE* Resources

Transparency Activity, pp. 46, 51–52

Directed Reading for Content Mastery, pp. 17, 18

Note-taking Worksheets, pp. 33–36

Enrichment, p. 29

Lab Activity, pp. 9–11

Reinforcement, p. 25

Performance Assessment in the Science Classroom, p. 55

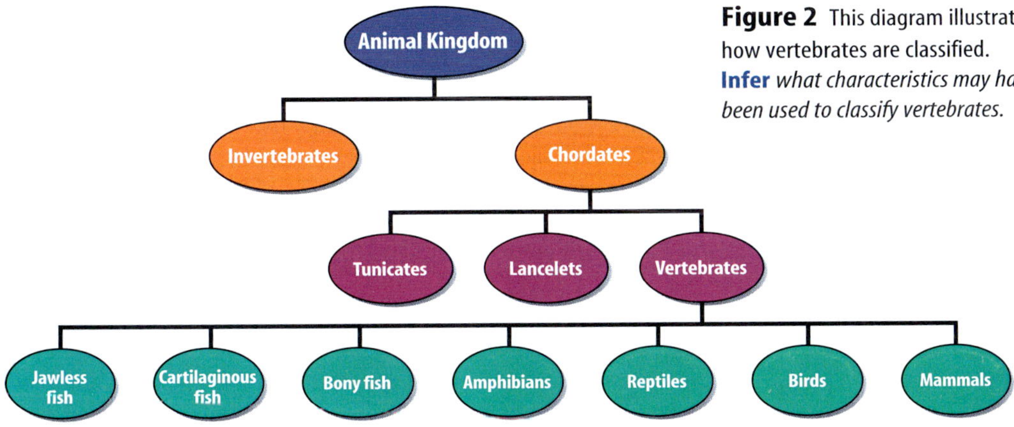

Figure 2 This diagram illustrates how vertebrates are classified.
Infer *what characteristics may have been used to classify vertebrates.*

Body Temperature Most vertebrate body temperatures change as the surrounding temperature changes. These animals are **ectotherms** (EK tuh thurmz), or cold-blooded animals. Fish are examples of ectotherms.

Humans and many other vertebrates are **endotherms** (EN duh thurmz), or warm-blooded animals. Their body temperature doesn't change with the surrounding temperature. Your body temperature is usually about 37°C, but it can vary by about 1°C, depending on the time of day. Changes of more than a degree or two usually indicate an infection or overexposure to extreme environmental temperatures.

Reading Check *Are humans endotherms or ectotherms?*

Fish

The largest group of vertebrates—fish—lives in water. Fish are ectotherms that can be found in warm desert pools and the subfreezing Arctic Ocean. Some species are adapted to swim in shallow freshwater streams and others in salty ocean depths.

Fish have fleshy filaments called gills, shown in **Figure 3,** where carbon dioxide and oxygen are exchanged. Water, containing oxygen, flows over the gills. When blood is pumped into the gills, the oxygen in the water moves into the blood. At the same time, carbon dioxide moves out of the blood in the gills and into the water.

Most fish have pairs of fanlike fins. The top and the bottom fins stabilize the fish. Those on the sides steer and move the fish. The tail fin propels the fish through the water.

Most fish have scales. Scales are thin structures made of a bony material that overlap like shingles on a house to cover the skin.

Figure 3 Gas exchange occurs in the gill filaments.

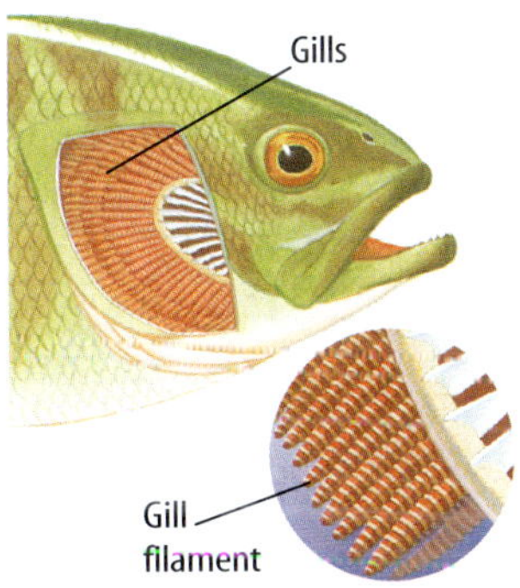

2 Teach

Caption Answer

Figure 2 body shape and structures

Reading Check

Answer endotherms

Quick Demo

Classifying Vertebrates

Possible Materials pictures/magazines

Estimated Time five minutes

Procedure Have students examine vertebrates and organize them based on similarities and differences in morphological characteristics.

Activity

Fish Scales Have students examine fish scales using a stereomicroscope. Ask them to draw the scales they observe. Explain that the age of some types of fish can be determined by counting the rings on the scales. L2 ELL LS **Visual-Spatial** P

Visual Learning

Figure 3 Why do fish have so many gill filaments? Increased surface area promotes gas exchange.

Cultural Diversity

Commercial Fishing Fish are a vital source of food for many people in the world. Fishing is important to the economies of many countries. Have students research the fishing industry and mark on a world map which countries were the biggest importers and exporters of fish in 2001. importers: Japan, U.S., Spain, France, Italy; exporters: Thailand, China, Norway, U.S., Canada L3

Differentiated Instruction

Challenge Have students research how endotherms and ectotherms regulate their body temperature. Students can write a report describing methods used by endotherms and ectotherms to regulate heat in their bodies. L3

INTEGRATE Physics

Changing Mass Possible answers: By moving their whole bodies; by using their fins; sharks must continuously swim to avoid sinking; students may need to do research to get these answers.

Quick Demo

Buoyancy

Materials three 10-oz plastic cups, marble, tablespoon of cooking oil, small balloon

Estimated Time three minutes

Procedure Half-fill the three 10-oz cups with water. Have students observe as you drop a marble in one container, a tablespoon of cooking oil in another, and a small-inflated balloon into the third container. Sharks store large amounts of oil in their bodies, which gives them buoyancy in water. Most bony fish have gas-filled sacs that enable them to float.

Caption Answer

Figure 5 These fish all are adapted to a similar environment.

Teacher FYI

Gills Most fish exchange gases by taking water into their mouths and passing it over their gills. The gills of lampreys are inside muscular pouches that pump water through their own openings.

Fun Fact

Of the more than 300 shark species, few have been known to attack humans. When they do it is because they detect movement, which they assume is food.

INTEGRATE Physics

Changing Mass Submarines pump water into and out of special chambers, which causes the submarine to sink or rise. In a similar way, gases move into and out of a fish's swim bladder. This allows the fish to sink or rise in the water. How do fish without swim bladders move up and down in the water? Write your answer in your Science Journal.

Types of Fish

Scientists classify fish into three groups—bony, jawless, and cartilaginous (kar tuh LA juh nuhs)—which are illustrated in **Figure 4** on the opposite page. Bony fish have skeletons made of bone, while jawless fish and cartilaginous fish have endoskeletons made of cartilage. **Cartilage** (KAR tuh lihj) is a tough, flexible tissue that is similar to bone but is not as hard or brittle. Your external ears and the tip of your nose are made of cartilage.

Bony Fish About 95 percent of all fish have skeletons made of bone. Goldfish, trout, bass, and marlins are examples of bony fish. The body structure of a typical bony fish is shown in **Figure 5.** As a bony fish swims, water easily flows over its body because its scales are covered with slimy mucus.

If you've ever watched fish in a tank, you know that they rise and sink to different levels in the water. An important adaptation in most bony fish is the swim bladder. This air sac helps control the depth at which the fish swims. The swim bladder inflates and deflates as gases—mostly oxygen in deep-water fish and nitrogen in shallow-water fish—move between the swim bladder and the blood. As the swim bladder fills with gas, the fish rises in the water. When the gas leaves the bladder, it deflates and the fish sinks lower in the water.

Most bony fish use external fertilization (fur tuh luh ZAY shun) to reproduce. External fertilization means that the eggs are fertilized outside the female's body. Females release large numbers of eggs into the water. Then, a male swims over the eggs, releases the sperm into the water, and many eggs are fertilized.

Figure 5 The many types of bony fish range in size from a few millimeters to many meters in length.
Infer *why all bony fish might have the same basic body plan.*

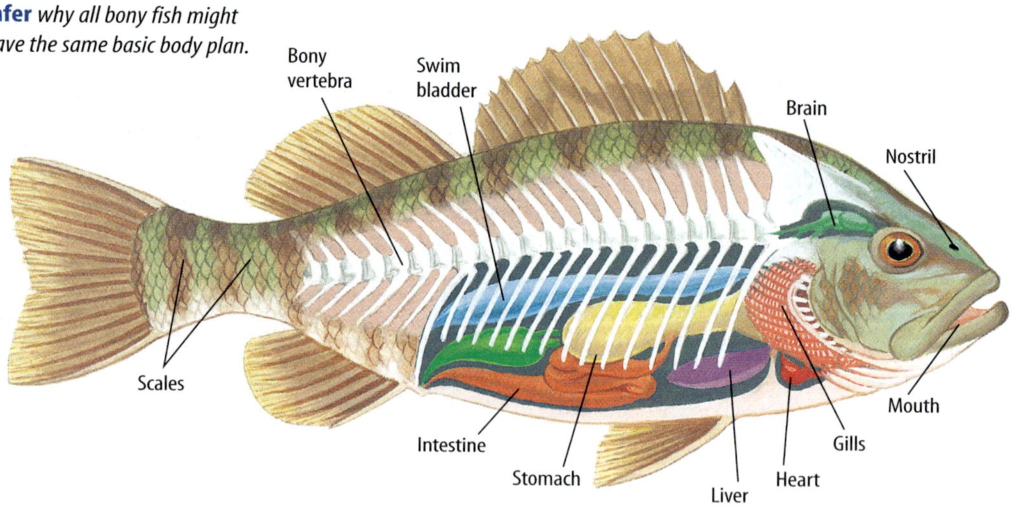

Active Reading

Write-Draw-Discuss This strategy encourages students to actively participate in reading and lectures, assimilating content creatively. Have students write about an idea, clarify it, then make an illustration. Ask students to share responses with the class about a concept in this chapter and display several examples.

Differentiated Instruction

Learning Disabled Have students make a mobile of the different types of fish in this section (jawless, cartilaginous, and bony). As they read, have them write the information they find most interesting about each type beneath it on the mobile. The students can use this information to construct a concept map in their Science Journals. L1
LS **Kinesthetic and Linguistic**

NATIONAL GEOGRAPHIC **VISUALIZING FISH DIVERSITY**

Figure 4

Fish are the most numerous and varied of all vertebrates, with more than 20,000 living species. These species can be organized into three groups—jawless, cartilaginous, and bony. Jawless fish are the most primitive and form the smallest group. Cartilaginous fish include more than 600 species, nearly all of them predators. Bony fish are the most numerous and diverse group. This page features photos of fish from each group.

Sturgeon

Whale Shark

Ratfish

Wolf Eel

BONY FISH The bodies of bony fish vary. The fins of the coelacanth below have jointed bones, like the legs of many land animals. Amphibians may have evolved from ancestors of coelacanths.

Angelfish

CARTILAGINOUS FISH The cartilage that gives these fish their shape is a lightweight material that is softer than bone. The hammerhead shark below has been known to use the cartilage in its hammer-shaped head to pin down stingrays, one of its favorite meals, before it devours them.

Coelacanth

Electric Ray

Hammerhead shark

JAWLESS FISH Only about 70 species make up the jawless group of fish. Jawless fish are often parasitic. The hagfish, right, often crawls into fish trapped in nets and eats them from the inside out.

Hagfish

NATIONAL GEOGRAPHIC

Visualizing Fish Diversity

Have students examine the pictures and read the captions. Then ask the following questions.

Give a possible explanation for why jawless fish are often parasites. Possible answer: Because the fish do not have a jaw and cannot chew, this may make it difficult for the fish to be predatory. L2

How does the shape and coloring of an electric ray help it live near the bottom? Possible answer: The flattened body allows them to lie on the bottom and bury themselves in sand very easily. Their coloring allows them to blend with the sandy bottom.

Activity

Venn Diagram Have students complete a Venn diagram comparing the characteristics of the organisms in the three groups. The diagram should show the characteristics the groups have in common and those that separate the groups from each other. L2 LS **Visual-Spatial**

Differentiated Instruction

Challenge Have students research information about fish that are adapted to live in the deep ocean where there is high pressure, cold water temperature, no light, and very little food. Direct students to make posters illustrating the adaptations of these fish. L3 LS **Visual-Spatial**

Discussion

Vertebrates Why are there more fish than any other type of vertebrate? There is great variety in aquatic environments, from freshwater to saltwater, very shallow to extreme depth, and warm to frigid.

3 Assess

DAILY INTERVENTION

Check for Understanding

Logical-Mathematical Have students work with a partner to draw a diagram of how vertebrates are classified. Have them list characteristics of each fish group. L2

Reteach

Lungs Have students explain how their own lungs can act like a fish's swim bladder. L2

Assessment

Process Have students make a table comparing the characteristics of the three types of fish. Use **Performance Assessment in the Science Classroom,** p. 109. L2

Virtual Labs

Aquatic Adaptations *How are fish adapted to their environment?*

The inside of a lamprey's mouth contains structures that are used to attach to larger fish.

Figure 6 Lampreys are specialized predators. In places such as the Great Lakes, lampreys have caused a decrease in some fish populations.

Jawless and Cartilaginous Fish Only a few species of fish are classified as jawless fish, like the one in **Figure 6.** Jawless fish have scaleless, long, tubelike bodies; an endoskeleton made of cartilage; and a round, muscular mouth without a jaw. But the mouth contains sharp, toothlike structures. One type of jawless fish, the lamprey, attaches itself to a larger host fish using its strong mouth and toothlike structures. Its tongue has sharp ridges that scrape through the host fish's skin. The lamprey obtains nutrients by feeding on the host fish's blood.

Sharks, skates, and rays are cartilaginous fish. They have skeletons made of cartilage just like the jawless fish. However, cartilaginous fish have rough, sandpaperlike scales and movable jaws. Many sharks have sharp teeth made from modified scales. Most cartilaginous fish are predators.

section 1 review

Summary

Chordate

- All chordates have a notochord, a nerve cord, and pharyngeal pouches at some time during their development.
- A vertebrate is a chordate with an internal system of bones called an endoskeleton.
- Most vertebrates are ectotherms (cold-blooded animals), but humans and many other vertebrates are endotherms (warm-blooded animals).

Fish

- Fish are ectotherms that live in the water and belong to the largest group of vertebrates.

Types of Fish

- About 95 percent of all fish have bony skeletons.
- Lampreys, sharks, skates, and rays are fish with a skeleton made of cartilage.

Self Check

1. **List** the three groups of fish. What are some of the differences that separate these groups?
2. **Compare and contrast** ectothermic and endothermic animals.
3. **Form a Hypothesis** Sharks don't have swim bladders and must move constantly or they sink. Hypothesize about the amount of food that a shark must eat compared to the amount eaten by a bony fish of the same size.
4. **Think Critically** In one lake, millions of fish eggs are laid and fertilized annually. Why doesn't the lake become overcrowded with fish?

Applying Math

5. **Make and Use Graphs** Make a circle graph of the number of fish species currently classified: *jawless fish—70; cartilaginous fish—820; and bony fish—23,500.*

Science Online red.msscience.com/self_check_quiz

section 1 review

1. bony fish—endoskeleton made of bone, swim bladder, most use external fertilization; jawless fish—scaleless, tubelike body, endoskeleton made of cartilage; cartilaginous fish—endoskeleton made of cartilage, sandpaperlike scales, movable jaws
2. The body temperature of ectotherms changes with their surroundings. The body temperature of endotherms remains nearly constant.
3. Sharks have to eat a greater amount of food than fish of a similar size that have swim bladders.
4. Many eggs are not fertilized; some eggs and young fish are eaten by predators.
5. 0.3% of the circle graph (about 1°) represents jawless fish; 96.4% (about 347°) represents bony fish; 3.4% (about 12°) represents cartilaginous fish.

section 2

Amphibians and Reptiles

Amphibians

A spy might lead a double life, but what about an animal? Amphibians (am FIH bee unz) are animals that spend part of their lives in water and part on land. In fact, the term *amphibian* comes from the Greek word *amphibios*, which means "double life." Frogs, toads, newts, and salamanders, such as the red-spotted salamander pictured in **Figure 7**, are examples of amphibians.

Amphibian Adaptations Living on land is different from living in water. Think about some of the things an amphibian must deal with in its environment. Temperature changes more quickly and more often in air than in water. More oxygen is available in air than in water. However, air doesn't support body weight as well as water does. Amphibians are adapted for survival in these different environments.

Amphibians are ectotherms. They adjust to changes in the temperature of their environment. In northern climates where the winters are cold, amphibians bury themselves in mud or leaves and remain inactive until the warmer temperatures of spring and summer arrive. This period of cold-weather inactivity is called **hibernation.** Amphibians that live in hot, dry environments move to cooler, more humid conditions underground and become inactive until the temperature cools down. This period of inactivity during hot, dry summer months is called **estivation** (es tuh VAY shun).

as you read

What You'll Learn

- **Describe** how amphibians have adapted to live in water and on land.
- **Explain** what happens during frog metamorphosis.
- **Identify** the adaptations that allow reptiles to live on land.

Why It's Important

Amphibians are sensitive to environmental changes, which may help identify problems that could affect humans.

Review Vocabulary

metamorphosis: change of body form that can be complete (egg, larva, pupa, adult) or incomplete (egg, nymph, adult)

New Vocabulary

- hibernation
- estivation
- amniotic egg

Figure 7 Amphibians have many adaptations that allow for life both on land and in the water. This red-spotted salamander spends most of its life on land. **Explain** *why they must return to the water.*

1 Motivate

INTERACTIVE CHALKBOARD PowerPoint® Presentations

Bellringer

Section Focus Transparencies also are available on the Interactive Chalkboard CD-ROM. L2 ELL

Section 2 Section Focus Transparency
On Second Thought, Let's Eat Something Else

Usually, this animal stays hidden, but when it is threatened, it twists its body and puffs out its belly to show its brillant red color. Its flashy colors warn other animals that it would make a dangerous—even deadly—meal!

1. What kind of environment do you think this animal lives in?
2. How is this creature different from a fish? A lizard?

Tie to Prior Knowledge

Adaptations Ask students to list adaptations a fish would need to survive on land. lungs, ability to move on land, ability to get food, and so on This will lead into the discussion of amphibian adaptations to life on land. L1

Caption Answer

Figure 7 Amphibians lay their eggs in the water.

Section 2 Resource Manager

Chapter *FAST FILE* Resources

Transparency Activity, p. 47

Directed Reading for Content Mastery, p. 18

Enrichment, p. 30

Reinforcement, p. 26

Lab Worksheet, pp. 5–6

Life Science Critical Thinking/Problem Solving, pp. 7, 11

2 Teach

Use Science Words

Word Origin *Hibernation* comes from the Latin word *hibernare*, which means "to pass the winter." *Estivation* is derived from the Latin word *aestas*, meaning "summer." Ask students to use each word in a sentence. L2 LS **Linguistic**

Reading Check

Answer lungs, tympanums, strong legs, large eyes, long sticky tongues

Quick Demo

Frog Adaptations

Materials aquarium

Estimated Time three minutes

Procedure Place a frog in an aquarium at the front of the class. Have students observe the frog to determine its adaptations. From student responses, make a table on the board that lists the adaptations a frog has for living on land and in the water.

Fun Fact

Frogs use their teeth to hold food in place, not to chew. Amphibians usually swallow prey whole.

Discussion

Survival Tactics How could having two different forms in the life cycle help species survive? Answers may include that one form may be better adapted to seasonal conditions.

Science Online

Topic: Amphibians

Visit red.msscience.com for Web links to information about the environment and amphibians.

Activity List as many possible causes of amphibian declines as you can find. Explain why it is important to humans to determine what could be causing these declines.

Amphibian Characteristics Amphibians are vertebrates with a strong endoskeleton made of bones. The skeleton helps support their body while on land. Adult frogs and toads have strong hind legs that are used for swimming and jumping.

Adult amphibians use lungs instead of gills to exchange oxygen and carbon dioxide. This is an important adaptation for survival on land. However, because amphibians have three-chambered hearts, the blood carrying oxygen mixes with the blood carrying carbon dioxide. This mixing makes less oxygen available to the amphibian. Adult amphibians also exchange oxygen and carbon dioxide through their skin, which increases their oxygen supply. Amphibians can live on land, but they must stay moist so this exchange can occur.

Amphibian hearing and vision also are adapted to life on land. The tympanum (TIHM puh nuhm), or eardrum, vibrates in response to sound waves and is used for hearing. Large eyes assist some amphibians in capturing their prey.

Reading Check *What amphibian senses are adapted for life on land?*

Land environments offer a great variety of insects as food for adult amphibians. A long, sticky tongue extends quickly to capture an insect and bring it into the waiting mouth.

Figure 8 Most young amphibians, like these tadpoles, look nothing like their parents when they hatch. The larvae go through metamorphosis in the water and eventually develop into adult frogs that live on land.

Tadpoles hatch from eggs that are laid in or near water.

Tadpoles use their gills for gas exchange.

Curriculum Connection

Math Many people are wary of reptiles. Have students write questions about reptiles to find out how people feel about these animals. They can give the survey to their classmates and others in the school. Have students graph the results and share them in class. L3 LS **Interpersonal and Logical-Mathematical**

Differentiated Instruction

English-Language Learners Have each student write a new vocabulary word and then draw a picture to illustrate the new word. For instance, they could write the word *estivation* and then draw a picture of a frog underground in a desert.

536 CHAPTER 18 Vertebrate Animals

Amphibian Metamorphosis Young animals such as kittens and calves are almost miniature versions of their parents, but young amphibians do not look like their parents. A series of body changes called metamorphosis (me tuh MOR fuh sus) occurs during the life cycle of an amphibian. Most amphibians go through a metamorphosis, as illustrated in **Figure 8.** Eggs are laid most often in water and hatch into larvae. Most adult amphibians live mainly on land.

The young larval forms of amphibians are dependent on water. They have no legs and breathe through gills. Over time, they develop body structures needed for life on land, including legs and lungs. The rate at which metamorphosis occurs depends on the species, the water temperature, and the amount of available food. If food is scarce and the water temperature is cool, then metamorphosis will take longer.

Like fish, most amphibians have external fertilization and require water for reproduction. Although most amphibians reproduce in ponds and lakes, some take advantage of other sources of water. For example, some species of rain forest tree frogs lay their eggs in rainwater that collects in leaves. Even more unusual is the Surinam toad shown in **Figure 9.** The fertilized eggs are placed on the mother's back. Her skin swells and covers the eggs to keep them moist. After metamorphosis occurs, fully formed toads emerge from under her skin.

Figure 9 Surinam toads live along the Amazon River. A female carries 60 to 100 fertilized eggs on her back. Complete metamorphosis takes 12 to 20 weeks.
Explain *how this would be an advantage for young Surinam toads.*

Legs begin to develop. Soon, the tail will disappear.

An adult frog uses lungs and skin for gas exchange.

Caption Answer

Figure 9 They are protected by their mother, so a greater percentage of them survive.

Activity

Frog Life Cycle Have students make a poster that shows the stages in the frog life cycle. Have them describe the characteristics of each stage. L2 LS **Visual-Spatial**

IDENTIFYING Misconceptions

Toad Excretions Students may think that people get warts from toads. This is not true, but a person should wash his or her hands after handling a toad because toads release a substance that can cause discomfort if it gets into the eyes or mouth. Also, many reptiles harbor certain strains of bacteria that can cause illness if they get into the mouth.

Discussion

After the Rain Why do frogs and toads seem to suddenly appear after a rain? Because of the moisture present everywhere after a rain, frogs and toads can travel over a greater area to hunt for food. Frogs and toads breathe through their skin, which must be kept moist for them to survive.

Visual Learning

Figure 8 What adaptations do amphibians have for living in water and on land? The young have tails and need water to breathe. Adults develop legs and lungs to breathe air. L2

Teacher FYI

Biological Indicators Because of their unique life cycle, amphibians are considered to be biological indicators. Biological indicators are species whose overall health reflects the health of their particular ecosystem.

Differentiated Instruction

Challenge Have students research metamorphosis in one species of amphibian. Have them create a poster showing each life stage and the habitat in which it lives out each life stage. L3

Activity

Ancient Reptiles Have students work in groups to make posters about ancient reptiles, including dinosaurs. Suggest that each poster have an ancient reptile on it with the adaptations listed that helped it survive in its environment. Make sure some posters feature ancient reptiles that were not dinosaurs. L2 COOP LEARN
Visual-Spatial and Interpersonal P

INTEGRATE Career

Herpetologist They look for similarities in body structure or DNA.

Research Have students research career opportunities in herpetology.

Inquiry Lab

Gas Exchange

Question How do lungs work?

Possible Materials large, clear-plastic bottle; plastic bag; scissors; duct tape; two straws; two balloons; small piece of thin cardboard

Estimated Time 1 class session

Teaching Strategies

- Encourage students to create a model of how lungs work. Have them research ideas and use the materials provided to demonstrate their model.
- If time permits, ask students to think of questions they have about their lungs and design an experiment to safely explore their question.

For additional inquiry activities, see *Science Inquiry Labs.*

Figure 10 Reptiles have different body plans.

The rubber boa is one of only two species of boas in North America. Rubber boas have flexible jaws that enable them to eat prey that is larger than their head.

Crocodiles and American alligators like this one build their nests on land near a body of water. They protect their eggs while they wait for them to hatch.

Sea turtles, like this loggerhead turtle, are threatened around the world because of pollution, loss of nesting habitat, drowning in nets, and lighted beaches.

Colorado desert fringe-toed lizards are camouflaged, which helps them avoid their predators. These lizards mostly eat insects, but some include plants in their diet.

INTEGRATE Career

Herpetologist Most people are familiar with herpetologists, who are responsible for naming and classifying reptiles and amphibians. They often work in museums or universities. Their work usually involves field trips and gathering information for publication. What methods do taxonomists use to determine relationships between organisms? Write your answer in your Science Journal.

Reptiles

Reptiles come in many shapes, sizes, and colors. Snakes, lizards, turtles, and crocodiles are reptiles. Reptiles are ectothermic vertebrates with dry, scaly skin. Because reptiles do not depend on water for reproduction, most are able to live their entire lives on land. They also have several other adaptations for life on land.

Types of Reptiles As shown in **Figure 10,** reptilian body plans vary. Turtles are covered with a hard shell, into which they withdraw for protection. Turtles eat insects, worms, fish, and plants.

Alligators and crocodiles are predators that live in and near water. These large reptiles live in warmer climates such as those found in the southern United States.

Lizards and snakes make up the largest group of reptiles. They have a highly developed sense of smell. An organ in the roof of the mouth senses molecules collected by the tongue. The constant in-and-out motion of the tongue allows a snake or lizard to smell its surroundings. Lizards have movable eyelids and external ears, and most lizards have legs with clawed toes. Snakes don't have eyelids, ears, or legs. Instead of hearing sounds, they feel vibrations in the ground.

LAB DEMONSTRATION

Purpose to infer how a snake hears

Materials tuning fork, hard piece of rubber

Preparation Tuning forks should not be struck on anything hard, such as wood or metal. Do not touch the prongs after hitting the fork on the hard piece of rubber, such as a shoe sole.

Procedure Tap the tuning fork on rubber and hold it next to a student's ear. Tap it again and press the stem against a student's chin. Snakes receive vibrations from the environment, which their brains interpret as sound.

Expected Outcome Students will hear and feel vibrations.

Assessment

How are vibrations interpreted as sound? The vibrations travel through the bones in the student's jaw and skull, and into the fluid of the inner ear. There, they are changed into nerve impulses, which the brain interprets as sound. L2

Reptile Adaptations A thick, dry, waterproof skin is an adaptation that reptiles have for life on land. The skin is covered with scales that reduce water loss and help prevent injury.

Reading Check *What are two functions of a reptile's skin?*

All reptiles have lungs for exchanging oxygen and carbon dioxide. Even sea snakes and sea turtles, which can stay submerged for long periods of time, must eventually come to the surface to breathe. Reptiles also have a neck that allows them to scan the horizon.

Two adaptations enable reptiles to reproduce successfully on land—internal fertilization and laying shell-covered, amniotic (am nee AH tihk) eggs. During internal fertilization, sperm are deposited directly into the female's body. Water isn't necessary for reptilian reproduction.

The embryo develops within the moist protective environment of the **amniotic egg**, as shown in **Figure 11.** The yolk supplies food for the developing embryo, and the leathery shell protects the embryo and yolk. When eggs hatch, young reptiles are fully developed. In some snake species, the female does not lay eggs. Instead, the eggs are kept within her body, where they incubate and hatch. The young snakes leave her body soon after they hatch.

Figure 11 Young reptiles hatch from amniotic eggs.
Describe *the advantages of this.*

section 2 review

Summary

Amphibians

- Amphibians are animals that spend part of their lives in water and part on land.
- Although they have a strong endoskeleton to support their body while on land, amphibians depend on water for external fertilization.
- Amphibians go through a series of bodily changes called metamorphosis.

Reptiles

- Reptiles are ectothermic vertebrates with dry, scaly skin.
- Lizards and snakes make up the largest group of reptiles.
- Two adaptations enable reptiles to reproduce successfully on land: internal fertilization and laying shell-covered, amniotic eggs.

Self Check

1. **Infer** how a thick, dry, waterproof skin helps a reptile to live on land.
2. **Sequence** the steps of a frog's two-stage metamorphosis.
3. **Infer** why internal fertilization is efficient.
4. **Explain** how amphibians use adaptations to deal with cold winter months and hot, dry summer months.
5. **Think Critically** Some nonpoisonous snakes' patterns are similar to those of poisonous snakes. How is this coloring an advantage for a nonpoisonous snake?

Applying Skills

6. **Compare and contrast** the exchange of oxygen and carbon dioxide in adult amphibians and reptiles.
7. **Communicate** Write an explanation about why amphibians must live in wet or moist environments.

Science Online red.msscience.com/self_check_quiz

Caption Answer

Figure 11 provides moist protective environment

Fun Fact

Snakes use their tongues to detect chemicals in the air. Many snakes are poisonous, but few are deadly to humans.

3 Assess

DAILY INTERVENTION

Check for Understanding

Logical-Mathematical Using a two-column chart, brainstorm with the class about what characteristics amphibians and reptiles have for life on land and life in the water. L2

Reteach

Reptile Game Prepare a quiz game with descriptions of lizards, turtles, alligators, crocodiles, snakes, frogs, toads, and salamanders. Let students classify the animals as amphibians or reptiles. amphibian: frog, toad, salamander; reptile: lizard, turtle, alligator, crocodile, snake L2

Assessment

Performance Have students create a Venn diagram of fish, reptile, and amphibian characteristics. Use **Performance Assessment in the Science Classroom**, p. 167. L2

section 2 review

1. reduces water loss and helps to prevent injury
2. Tadpoles hatch from eggs and use gills for gas exchange; tadpoles develop legs and lose tail; adult frogs use lungs and skin for gas exchange.
3. More eggs are fertilized since sperm don't have to move through water to get to them.
4. Amphibians hibernate by burying themselves in mud or leaves. In hot, dry environments they estivate by moving to cooler, more humid conditions below ground.
5. Predators think they are poisonous snakes and leave them alone.
6. amphibian—through lungs and skin; reptiles—through lungs only

BENCH TESTED

Real-World Question

Purpose to observe tadpole metamorphosis and study the life cycle of a frog ELL COOP LEARN L2 LS Visual-Spatial

Process Skills observe and infer, classify, sequence, form operational definitions

Time Required 30 minutes on Day 1, 10 minutes of observations twice weekly

Alternate Materials Tap water aged for 48 hours can be used instead of pond water.

Safety Precautions Caution students to use extreme care when working with live animals.

Procedure

Teaching Strategy Field-collected eggs can be shipped from biological supply houses in January, February, and March. Peeper and toad eggs develop rapidly.

Troubleshooting The eggs must be handled carefully if they are to hatch. Make sure the temperature of the water remains at 20°C.

Conclude and Apply

1. It protects the eggs and keeps them from drying out.
2. Young tadpoles—eyes are on each side of the head; older tadpoles—eyes are close to the top of the head.
3. 2–4 months

LAB

Frog Metamorphosis

Frogs and other amphibians use external fertilization to reproduce. Female frogs lay hundreds of jellylike eggs in water. Male frogs then fertilize these eggs. Once larvae hatch, the process of metamorphosis begins.

Real-World Question

What changes occur as a tadpole goes through metamorphosis?

Goals

- **Observe** how body structures change as a tadpole develops into an adult frog.
- **Determine** how long metamorphosis takes.

Materials

4-L aquarium or jar
frog egg mass
lake or pond water
stereoscopic microscope
watch glass
small fishnet
aquatic plants
washed gravel
lettuce (previously boiled)
large rock

Safety Precautions

WARNING: *Handle the eggs with care.*

Procedure

1. Copy the data table in your Science Journal.

Frog Metamorphosis	
Date	**Observations**

2. As a class, use the aquarium, pond water, gravel, rock, and plants to prepare a water habitat for the frog eggs.
3. Place the egg mass in the aquarium's water. Use the fishnet to separate a few eggs from the mass and place them on the watch glass. Observe the eggs using the microscope. Record all observations in your data table. Return the eggs to the aquarium.

4. **Observe** the eggs twice a week until hatching begins. Then observe the tadpoles twice weekly. Identify the mouth, eyes, gill cover, gills, nostrils, back fin, and legs.
5. In your Science Journal, write a description of how tadpoles eat cooled, boiled lettuce.

Conclude and Apply

1. **Explain** why the jellylike coating around the eggs is important.
2. **Compare** the eyes of young tadpoles with the eyes of older tadpoles.
3. **Calculate** how long it takes for eggs to hatch, legs to develop, and to become a frog.

Communicating Your Data

Draw the changes you observe as the egg hatches and the tadpole goes through metamorphosis. **For more help, refer to the Science Skill Handbook.**

540 CHAPTER 18 Vertebrate Animals

Assessment

Oral Have students sequence the events that occur as a frog develops from an egg into an adult. Use **Performance Assessment in the Science Classroom,** p. 163. L2

Communicating Your Data

Suggest that students organize their drawings in an events-chain concept map. L2

section 3

Birds

Characteristics of Birds

Ostriches have strong legs for running, and pelicans have specialized bills for scooping fish. Penguins can't fly but are excellent swimmers, and house wrens and hummingbirds are able to perch on branches. These birds are different, but they, and all birds, have common characteristics. Birds are endothermic vertebrates that have two wings, two legs, and a bill or beak. Birders, or bird-watchers, can tell where a bird lives and what it eats by looking at the type of wings, feet, and beak or bill it has. Birds are covered mostly with feathers—a feature unique to birds. They lay hard-shelled eggs and sit on these eggs to keep them warm until they hatch. Besides fish, birds are the most numerous vertebrates on Earth. **Figure 12** illustrates some of the more than 8,600 species of birds and their adaptations.

as you read

What **You'll Learn**

- **Identify** the characteristics of birds.
- **Describe** the adaptations birds have for flight.
- **Explain** the function of feathers.

Why **It's Important**

Humans modeled airplanes after the flight of birds.

Review Vocabulary

appendage: structure such as a claw, leg, or antenna that grows from the body

New Vocabulary

- contour feather
- down feather

Figure 12 Emus can't fly but they have strong legs and feet that are adapted for running.

Horned puffins can fly and their sleek bodies and small, pointed wings also enable them to "fly" underwater.

An albatross glides in the air.

Birds of prey, like this osprey, have sharp, strong talons that enable them to grab their prey.

1 Motivate

INTERACTIVE CHALKBOARD PowerPoint® Presentations

Bellringer

Section Focus Transparencies also are available on the Interactive Chalkboard CD-ROM. L2 ELL

SECTION 3 Section Focus Transparency **Maybe the Scientists Were a Hoax!**

This bird, called a kiwi, is the national bird of New Zealand. It's so unusual that when it was first brought to the attention of scientists, they thought it was a hoax. It has no tail and only stubby wings hidden by hair-like feathers. Flightless, the nocturnal kiwi lives in burrows and tends to be shy.

1. Why might scientists have thought the kiwi a hoax?
2. Why do you think the kiwi is classified as a bird?

Tie to Prior Knowledge

Characteristics Have students compare reptile and bird characteristics, such as different skin coverings, and method of reproduction. L2

Visual Learning

Figure 12 Have students make a list of adaptations each bird has for its particular lifestyle, including those mentioned in the captions. P L2

Fun Fact

The world's smallest bird is the bee hummingbird from Cuba, measuring 5.5 cm from bill to tail.

Section 3 Resource Manager

Chapter *Fast File* Resources

- Transparency Activity, p. 48
- Directed Reading for Content Mastery, p. 19
- MiniLAB, p. 3
- Lab Activity, pp. 13–14
- Enrichment, p. 31
- Reinforcement, p. 27

2 Teach

Activity

Flapping Arms Divide the class into pairs. Have one student keep time and count while the other holds his or her arms straight out to the side and flaps them up and down. See how many times a student can flap his or her arms in one minute and how long it takes for the student to tire. Explain that just as our leg muscles are well developed for walking, the chest muscles of most birds are well developed for flying. L1 ELL COOP LEARN
LS **Kinesthetic and Interpersonal**

Reading Check

Answer They make the bird lighter so it can fly more easily.

Use Science Words

Word Origin Birds belong to the class Aves. *Avis* is the Latin word for bird and is the root of many bird-related words. Ask students to find words with this root and explain their meanings. aviator—pilot or member of an aircraft crew; avian—relating to birds; aviary—enclosure for breeding and rearing birds L2 LS **Linguistic**

Use an Analogy

Furnaces and Fuels Birds and mammals are endotherms. An endotherm is like a furnace that burns fuel to produce energy. In the case of animals, respiration and oxidation of food are similar to the burning of fuel in a furnace. Have students breathe onto their hands and observe that their breath is warm. L1

Make a Model

Wing Design Have students design paper airplanes with different wing shapes. Have them test which planes fly longest and fastest. How did wing shape influence the planes' flight? L2

Figure 13 Wings provide an upward force called lift in both birds and airplanes.

Bald eagles are able to soar for long periods of time because their wings have a large surface area to provide lift.

The glider gets lift from its wings the same way a bald eagle gets lift.

Science Online

Topic: Wing Designs
Visit red.msscience.com for Web links to information about wing designs of different aircraft.

Activity Draw as many wing designs as you can find and explain how they are different.

Adaptations for Flight

The bodies of most birds are designed for flight. They are streamlined and have light yet strong skeletons. The inside of a bird's bone is almost hollow. Internal crisscrossing structures strengthen the bones without making them as heavy as mammal bones are. Because flying requires a rigid body, a bird's tail vertebrae are joined together to provide the needed rigidity, strength, and stability. Birds use their tail to help them steer through the air. While a bird can still fly without a tail, their flight is usually shorter and not as smooth.

Reading Check *What advantage do birds' bones give them for flight?*

Flight requires a lot of energy and oxygen. Birds eat insects, nectar, fish, meats, or other high-energy foods. They also have a large, efficient heart and a specialized respiratory system. A bird's lungs connect to air sacs that provide a constant supply of oxygen to the blood and make the bird more lightweight.

Slow-motion video shows that birds beat their wings up and down as well as forward and back. A combination of wing shape, surface area, air speed, and angle of the wing to the moving air, along with wing movements, provide an upward push that is needed for the flight of a bald eagle, as shown in **Figure 13.** Inventors of the first flying machines, such as gliders, used the body plan of birds as a model for flight. As wind passes above and below the wing, it creates lift. Lift is what allows birds, as well as planes, to stay in flight.

Differentiated Instruction

Challenge Have students research the different methods of flight used by birds. Does flight differ? Hummingbirds can hover, vultures and some hawks soar on thermals, and woodpeckers have an undulating flight. Have students present a table of the different types of flight they discover and which birds use that method of flight. L3

Science Journal

Bird Diary Ask students to draw or take pictures of different birds found around school or their home. They should identify the birds and start a bird diary in their Science Journals. Have them note the time of day they saw each bird, its location, and any other interesting observations. L2
LS **Visual-Spatial and Linguistic**

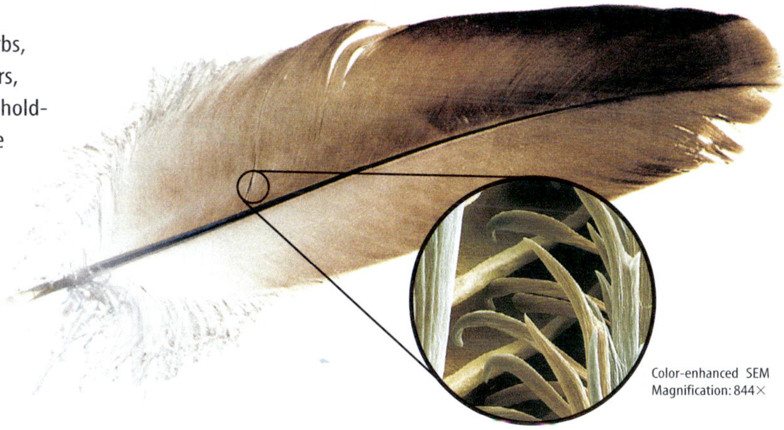

Figure 14 Microscopic barbs, located along contour feathers, keep the feathers smooth by holding the individual parts of the feather together.

Functions of Feathers

Birds are the only animals with feathers. They have two main types of feathers—contour feathers and down feathers. Strong, lightweight **contour feathers** give adult birds their streamlined shape and coloring. A close look at the contour feather in **Figure 14** shows the parallel strands, called barbs, that branch off the main shaft. Outer contour feathers help a bird move through the air or water. It is these long feathers on the wings and tail that help the bird steer and keep it from spinning out of control. Feather colors and patterns can help identify species. They also are useful in attracting mates and protecting birds from predators because they can be a form of camouflage.

Have you ever noticed that the hair on your arm stands up on a cold day? This response is one way your body works to trap and keep warm air close to your skin. Birds have **down feathers** that trap and keep warm air next to their bodies. These fluffy feathers, as shown in **Figure 15,** provide an insulating layer under the contour feathers of adult birds and cover the bodies of some young birds.

Reading Check *What are two ways feathers protect birds?*

Figure 15 Some species of birds, like chickens and these pheasants, are covered with feathers when they hatch. **Explain** *how this might be an advantage.*

Mini LAB

Modeling Feather Function

Procedure

1. Wrap **polyester fiber** or **cotton** around the bulb of an **alcohol thermometer.** Place it into a **plastic bag.** Record the temperature in your **Science Journal.**
2. Place a second **alcohol thermometer** into a **plastic bag** and record the temperature.
3. Simultaneously submerge the thermometers into a **container** of **cold water,** keeping the top of each bag above the water's surface.
4. After 2 min, record the temperature of each thermometer.

Analysis

1. Which thermometer had the greater change in temperature?
2. Infer the type of feather that the fiber or cotton models.

Try at Home

Teacher FYI

Producing Lift Airplanes are designed to use an airfoil similar to that of birds. Lift enables both to fly. The amount of lift produced is determined by wing area, shape, and speed. Airplanes extend flaps and slats to increase lift at low speeds. Similarly, birds increase their lift by flapping their wings, extending a feather, and spreading wing and tail feathers.

Differentiated Instruction

Learning Disabled Bring in a contour feather to demonstrate how microscopic barbs hold each feather barb together. Pull the vein apart and then run your hand over the vein to put the barbs back together. L1

Mini LAB

Purpose to discover the function of down feathers L2 LS **Kinesthetic**

Materials two alcohol thermometers, polyester fiber or cotton, two plastic bags, container of cold water, Science Journal

Teaching Strategy Check that students thoroughly cover the bulb of the thermometer.

Analysis

1. the thermometer without fiber or cotton
2. down feathers

Assessment

Oral Ask students to infer if amount of down feathers would be different for bird species in different environments. Use **Performance Assessment in the Science Classroom,** p. 89.

Try at Home

Reading Check

Answer Contour feathers can provide camouflage; down feathers provide insulation.

Caption Answer

Figure 15 The feathers help keep the young chicks warm.

Quick Demo

Feather Function

Materials down feathers and flight feathers

Estimated Time two minutes

Procedure Have students examine a down feather and a flight feather and note the differences between the two. Flight feathers are more rigid to help in flight, while down feathers are soft and insulate the bird. L2

3 Assess

DAILY INTERVENTION

Check for Understanding

Visual-Spatial Have students work in groups to draw illustrations of adaptations that allow birds to fly.

Reteach

Bird Habitats Have students discuss how the area near your school can be made into a more favorable habitat for birds. L2

Assessment

Process Assess students' abilities to compare and contrast by having them examine the beaks and feet of birds indigenous to your area. Have students make a model of a hypothetical bird's foot and describe the bird's lifestyle. Use **Performance Assessment in the Science Classroom**, p. 93. L2

Discussion

Birds in the Fossil Record Infer why birds are scarce in the fossil record. The bones of birds are lightweight and small, so they often disintegrate before they become fossils.

Figure 16 Cormorants' feathers get wet when they go underwater to catch fish. When they return to their roost, they have to hold their wings out to dry.

Care of Feathers Clothes keep you warm only if they are dry and in good condition. In a similar way, well-maintained feathers keep birds dry, warm, and able to fly. Birds preen to clean and reorganize their feathers. During preening, many birds also spread oil over their bodies and feathers. This oil comes from a gland found on the bird's back at the base of its tail. The oil helps keep the skin soft, and feathers and scales from becoming brittle. The oil does not waterproof the feathers as once thought. It is the arrangement of a feather's microscopic structures that repels water more than the oil does. Cormorants, like the one in **Figure 16,** have wettable outer feathers that must be air-dried after diving for food.

section 3 review

Summary

Characteristics of Birds

- Birds are endothermic vertebrates, have two wings, two legs, and a bill or beak, and are covered mostly with feathers.

Adaptations for Flight

- Birds are streamlined and have light, yet strong, skeletons.
- The inside of a bird's bone is almost hollow.
- Wings provide an upward force called lift.

Functions of Feathers

- Birds have contour feathers that help them move through the air or water.
- Down feathers trap and keep warm air next to birds' bodies.

Self Check

1. **List** several reasons why birds preen their feathers.
2. **Describe** how a bird's skeletal system, respiratory system, and circulatory system all work together to enable it to fly.
3. **Think Critically** Explain why birds can reproduce in Antarctica when temperatures are below 0°C.

Applying Math

4. **Use an Electronic Spreadsheet** During every 10 seconds of flight, a crow beats its wings 20 times, a robin 23 times, a chickadee 270 times, and a hummingbird 700 times. Use a spreadsheet to find out how many times the wings of each bird beat during a 5-minute flight.

Science Online red.msscience.com/self_check_quiz

section 3 review

1. to reorganize their feathers, to close breaks or gaps in them, and to make them water-repellent
2. skeletal system: hollow bones make the bird lighter; respiratory system: contains inflated air sacs to make birds lighter and provide oxygen; circulatory system: large, efficient hearts that supply blood to bird's organs
3. Birds are endotherms and incubate their eggs.
4. Student spreadsheets should show: crow—600 times, robin—690 times, chickadee—8,100 times, and hummingbird—21,000 times.

section 4

Mammals

Mammal Characteristics

How many different kinds of mammals can you name? Moles, dogs, bats, dolphins, horses, and people are all mammals. They live in water and in many different climates on land. They burrow through the ground and fly through the air.

Mammals are endothermic vertebrates. They have mammary glands in their skin. In females, mammary glands produce milk that nourishes the young. A mammal's skin usually is covered with hair that insulates its body from cold and heat. It also protects the animal from wind and water. Some mammals, such as bears, are covered with thick fur. Others, like humans, have only patches of thick hair while the rest of their body is sparsely covered with hair. Still others, like the dolphins shown in **Figure 17,** have little hair. Wool, spines, quills, and certain horns are modified hair. What function do you think quills and spines serve?

Mammary Glands Mammals put a great deal of time and energy into the care of their young, even before birth. When female mammals are pregnant, the mammary glands increase in size. After birth, milk is produced and released from these glands. For the first weeks or months of a young mammal's life, the milk provides all of the nutrition the young mammal needs.

as you read

What You'll Learn

- **Identify** the characteristics common to all mammals.
- **Explain** how mammals are adapted to the different environments on Earth.
- **Distinguish** among monotremes, marsupials, and placentals.

Why It's Important

All mammals have similar body structures.

Review Vocabulary

symmetry: refers to the arrangement of the individual parts of an object that can be divided into matching halves

New Vocabulary

- herbivore
- carnivore
- omnivore
- monotremes
- marsupial
- placental

Porcupines have fur next to their skin but sharp quills on the outside. Quills are modified hairs.

Dolphins do not have much hair on their bodies. A layer of fat under the skin acts as insulation.

Figure 17 The type of hair mammals have varies from species to species.

Explain *the advantages and disadvantages of having hair.*

1 Motivate

Bellringer

INTERACTIVE CHALKBOARD PowerPoint® Presentations

Section Focus Transparencies also are available on the Interactive Chalkboard CD-ROM. L2 ELL

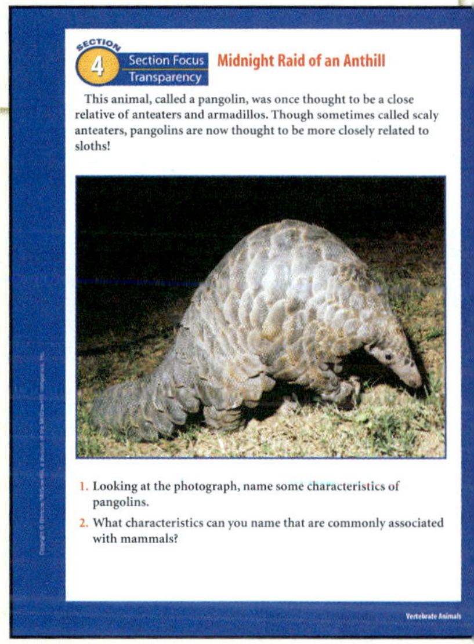
Section 4 Section Focus Transparency — Midnight Raid of an Anthill

This animal, called a pangolin, was once thought to be a close relative of anteaters and armadillos. Though sometimes called scaly anteaters, pangolins are now thought to be more closely related to sloths!

1. Looking at the photograph, name some characteristics of pangolins.
2. What characteristics can you name that are commonly associated with mammals?

Vertebrate Animals

Tie to Prior Knowledge

Mammal Characteristics Most students will be familiar with the characteristics of mammals. Have them list as many as they can on the board. L1

Text Question Answer

protection

Caption Answer

Figure 17 Body insulation and protection are advantages, while hair could reduce mobility in ocean mammals.

Section 4 Resource Manager

Chapter *FAST FILE* Resources

- Transparency Activity, p. 49
- Directed Reading for Content Mastery, pp. 19, 20
- Enrichment, p. 32
- MiniLAB, p. 4
- Lab Worksheet, pp. 7–8
- Reinforcement, p. 28

Lab Management and Safety, p. 58

2 Teach

Visual Learning

Figure 18 Have students draw a concept map in which they sort mammals as herbivores, carnivores, or omnivores. They should describe the teeth of each (both appearance and function) and give several examples of mammals in each group. L2

Reading Check

Answer Herbivores eat only plants, carnivores eat only meat, and omnivores eat both plants and meat.

Mini LAB

Purpose Students model how blubber insulates. L2

LS **Kinesthetic**

Materials 2 self-sealing plastic bags, one-third cup of vegetable shortening, ice water

Teaching Strategy Demonstrate how to place the shortening in the self-sealing plastic bag and zip one bag inside the other. One bag will have to be turned inside out.

Analysis

1. the bare hand
2. Blubber is an insulator; it traps air and protects against cold.

Assessment

Performance Have students repeat the experiment using polyester batting or wool fabric. Use **Performance Assessment in the Science Classroom,** p. 105.

Figure 18 Mountain lions are carnivores. They have sharp canines that are used to rip and tear flesh.

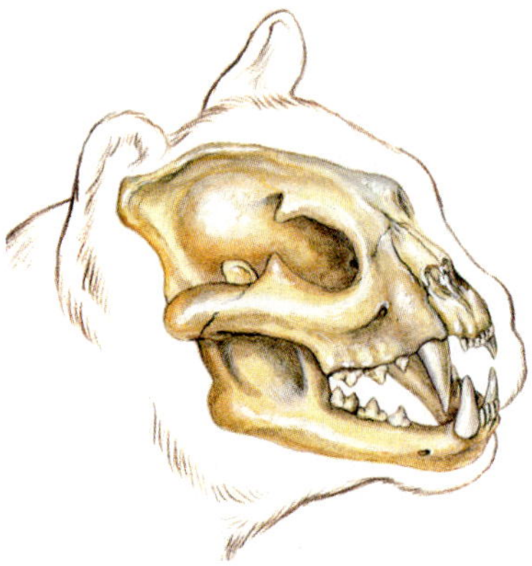

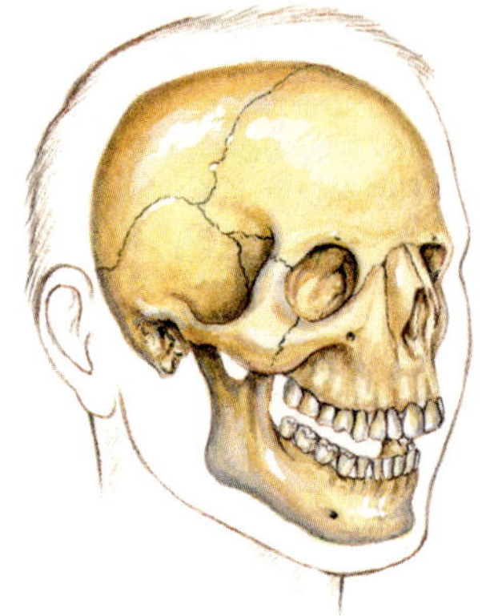

Humans are omnivores. They have incisors that cut vegetables, premolars that are sharp enough to chew meat, and molars that grind food.

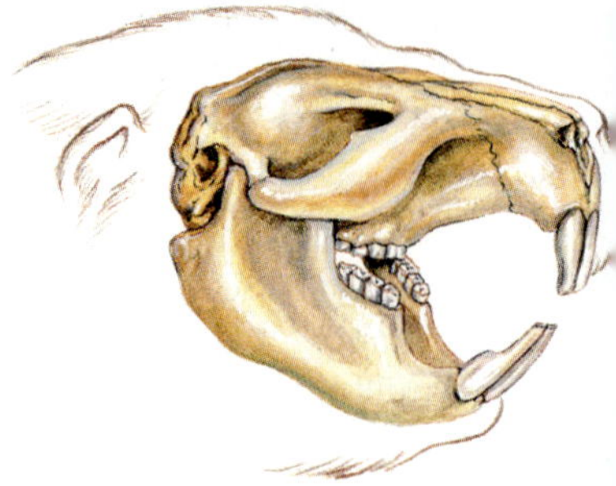

Herbivores, like this beaver, have incisors that cut vegetation and large flat molars that grind it.

Mini LAB

Inferring How Blubber Insulates

Procedure

1. Fill a **self-sealing plastic bag** about one-third full with solid **vegetable shortening.**
2. Turn another **self-sealing plastic bag** inside out. Place it inside the first bag so you are able to zip one bag to the other. This is a blubber mitten.
3. Put your hand in the blubber mitten. Place your mittened hand in **ice water** for 5 s. Remove the blubber mitten when finished.
4. Put your other bare hand in the same bowl of ice water for 5 s.

Analysis

1. Which hand seemed colder?
2. Infer the advantage a layer of blubber would give in the cold.

Different Teeth Mammals have teeth that are specialized for the type of food they eat. Plant-eating animals are called **herbivores.** Animals that eat meat are called **carnivores,** and animals that eat plants and animals are called **omnivores.** As shown in **Figure 18,** you usually can tell from the kind of teeth a mammal has whether it eats plants, other animals, or both. The four types of teeth are incisors, canines, premolars, and molars.

Reading Check *How are herbivores, carnivores, and omnivores different?*

Body Systems Mammals live active lives. They run, swim, climb, hop, and fly. Their body systems must interact and be able to support all of these activities.

Mammals have well-developed lungs made of millions of microscopic sacs called alveoli, which enable the exchange of carbon dioxide and oxygen during breathing. They also have a complex nervous system and are able to learn and remember more than many other animals. The brain of a mammal is usually larger than the brain of other animals of the same size.

All mammals have internal fertilization. After an egg is fertilized, the developing mammal is called an embryo. Most mammal embryos develop inside a female organ called the uterus. Mammals can be divided into three groups based on how their embryos develop. The three groups of mammals are monotremes, marsupials, and placentals.

Cultural Diversity

Mammals Around the World Display a wall map of the world. Ask students to affix drawings and magazine pictures of mammals to the map to show where each mammal is commonly found. Students can visit a local zoo to take pictures of mammals to add to the map. L2 LS **Visual-Spatial**

Curriculum Connection

Language Arts Ask students to make a list of expressions that draw an analogy between the behavior of a mammal and the behavior of a person. Some examples include *quiet as a mouse, busy as a beaver, blind as a bat, eats like a pig, sly as a fox,* and *slothful.* Each group of students can choose one to research and determine whether the expression is accurate. L2 LS **Linguistic** P

Mammal Types

The duck-billed platypus, shown in **Figure 19,** along with two species of echidnas (ih KID nuhs)—spiny anteaters—belong to the smallest group of mammals called the monotremes. They are different from other mammals because **monotremes** lay eggs with tough, leathery shells instead of having live births. The female incubates the eggs for about ten days. Monotremes also differ from other mammals because their mammary glands lack nipples. Instead, the milk seeps through the skin onto their fur. The young monotremes then nurse by licking the milk from the fur surrounding the mammary glands. Duck-billed platypuses and spiny anteaters are found in New Guinea and Australia.

Figure 19 Duck-billed platypuses and spiny anteaters are the only species of mammals that lay eggs.

Applying Math — Working with Percentages

HOW MUCH TIME? It is estimated that during the four months elephant seals spend at sea, 90 percent of their time is spent underwater. On a typical day, how much of the time between the hours of 10:00 A.M. and 3:00 P.M. does the elephant seal stay at the surface?

Solution

1. *This is what you know:*
   - Total time: From 10:00 A.M. to 3:00 P.M. is 5 h. $1\text{ h} = 60\text{ min}$, so $5 \times 60 = 300\text{ min}$
   - % of time on surface = $100\% - 90\% = 10\% = 0.10$
2. *This is what you need to know:* How much time is spent on the surface?
3. *This is the procedure you need to use:*
   - Use this equation: surface time = (total time)(% of time on surface)
   - Substitute the known values: surface time = (300 min)(0.10) = 30 min
4. *Check your answer:* Divide your answer by the total time. Is the answer equal to 10 percent?

Practice Problems

1. On a typical day during those four months, how much time do elephant seals stay at the surface from 11:00 P.M. until 6:00 A.M.?
2. On a typical day during those four months, how much time do elephant seals spend underwater from 9:00 A.M. until 6:00 P.M.?

Science Online For more practice, visit red.msscience.com/math_practice

Activity

Adaptations Have students draw a mammal and label its parts, indicating the specialized adaptations that make it a mammal and help it survive. They should include information about how the mammal reproduces, what kind of skin and teeth it has, how it moves, how its lungs and heart function, and so on. L2 LS **Visual-Spatial**

Quick Demo

Mammal Characteristics

Materials pet supplies

Estimated Time 10 minutes

Procedure Arrange class visits of mammals such as gerbils, rabbits, hamsters, or guinea pigs. Plan ahead to have all necessary supplies to care for them. Have students observe them carefully and write down their observations. L2

Teacher FYI

Marsupials and Monotremes Monotremes and marsupials make up only five percent of all mammals. Monotremes live in Australia and its nearby islands in various habitats.

Applying Math

National Math Standards

Correlation to Mathematics Objectives

1, 6

Answers to Practice Problems

1. 42 minutes
2. total time = 9 hours
   9 hours = 540 minutes
   % of time underwater = 90% = 0.9
   underwater time = (total time) × (% of time underwater)
   underwater time = (540 min.) × (0.9) = 486 min. (8 hours 6 min.)

Curriculum Connection

Language Arts Have students work in groups of three to prepare a picture dictionary for younger children. They can use wildlife and nature magazines to cut out illustrations, or they can draw pictures themselves. Have students illustrate only one mammal on a page, so that alphabetizing the names will be easy later. L2 COOP LEARN LS **Interpersonal and Visual-Spatial**

Differentiated Instruction

Challenge Have students research monotreme distribution and recent fossil findings. Using factual information, have students write short stories about the distribution of monotremes using maps and drawing pictures of what past monotremes may have looked like. L3

Discussion

Developing Differences What advantages do developing young marsupials have compared with developing young monotremes? They are usually protected in the female parent's pouch. Young monotremes are left alone while the parents seek food, making them more vulnerable to predators.

Reading Check

Answer The young marsupials finish developing in the female mammal's pouch.

IDENTIFYING Misconceptions

Humans as Animals Students may not think of themselves as animals. To help them realize that humans are animals and are classified as mammals, list the specific characteristics of mammals on the board (*endothermic, hair, mammary glands, oil glands, sweat glands, specialized teeth, well-developed body systems*). Invite volunteers to circle each characteristic possessed by humans. L2

Fun Fact

The blue-eyed lemur is the only nonhuman primate with blue eyes.

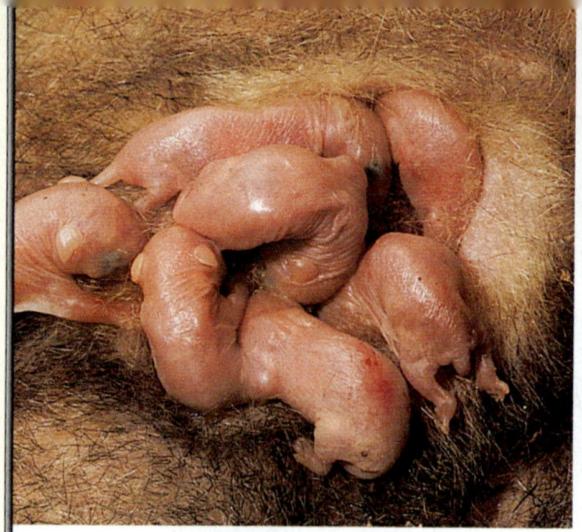

Figure 20 Marsupials, such as opposums, are born before they are completely developed. Newborn marsupials make the journey to a nipple that is usually in the mother's pouch where they will finish developing.

Marsupials Most **marsupials** carry their young in a pouch. Their embryos develop for only a few weeks within the uterus. When the young are born, they are without hair, blind, and not fully formed. Using their sense of smell, the young crawl toward a nipple and attach themselves to it. Here they feed and complete their development. Most marsupials—such as kangaroos, koalas, Tasmanian devils, and wallabies—live in Australia, Tasmania, and New Guinea. The opossum, shown in **Figure 20,** is the only marsupial that lives in North America.

Reading Check *Why do most marsupials have a pouch?*

Placentals The largest number of mammals belongs to a group called placentals. **Placentals** are named for the placenta, which is a saclike organ that develops from tissues of the embryo and uterus. In the placenta, food, oxygen, and wastes are exchanged between the mother's blood and the embryo's blood, but their bloods do not mix. An umbilical cord, as seen in **Figure 21,** connects the embryo to the placenta. Food and oxygen are absorbed from the mother's blood for the developing young. Blood vessels in the umbilical cord carry food and oxygen to the developing young, then take away wastes. In the placenta, the mother's blood absorbs wastes from the developing young. This time of development, from fertilization to birth, is called the gestation period. Mice and rats have a gestation period of about 21 days. Human gestation lasts about 280 days. The gestation period for elephants is about 616 days, or almost two years.

Figure 21 Placental embryos rely on the umbilical cord to bring nutrients and to remove wastes. Your navel is where your umbilical cord was connected to you.

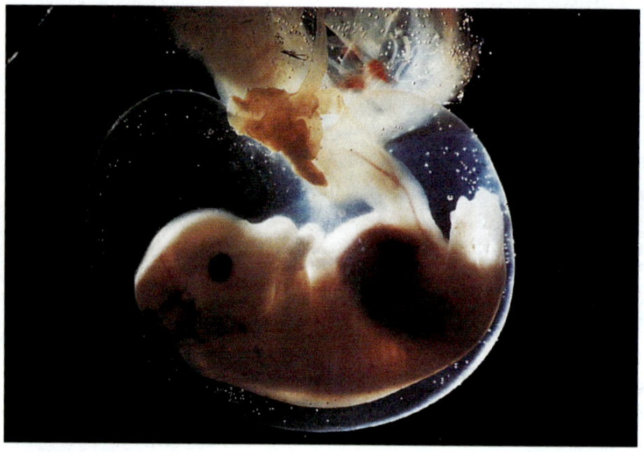

Differentiated Instruction

Learning Disabled Have students make a mobile of the different types of mammals (monotremes, marsupials, and placentals). As they read, have them write the information they find most interesting about each type beneath it on the mobile. The students can use this information to construct a concept map in their Science Journals.

Cultural Diversity

Animal Gossip In many cultures, human interest in animals' abilities to communicate with each other and with humans has been a source of literature and legend. With the help of the school media specialist, put together a collection of literature and let students investigate a story about an animal's ability to communicate. Have students retell a story that they found interesting. L2 LS **Linguistic**

Mammals Today

More than 4,000 species of mammals exist on Earth today. Mammals can be found on every continent, from cold arctic regions to hot deserts. Each kind of mammal has certain adaptations that enable it to live successfully within its environment.

Mammals, like all other groups of animals, have an important role in maintaining a balance in the environment. Large carnivores, such as wolves, help control populations of herbivores, such as deer and elk, thus preventing overgrazing. Bats and other small mammals such as honey possums help pollinate flowers. Other mammals unknowingly pick up plant seeds in their fur and distribute them. However, mammals and other animals are in trouble today because their habitats are being destroyed. They are left without enough food, shelter, and space to survive as millions of acres of wildlife habitat are damaged by pollution or developed for human needs. The grizzly bear, pictured in **Figure 22,** lives in North America and Europe and is an endangered species—a species in danger of becoming extinct—in most of its range because of habitat destruction.

Figure 22 Grizzly bears, sometimes called brown bears, used to range all over the western half of the United States. Now, because of human settlement, habitat loss, and overhunting, grizzly bears are found only in Alaska, Montana, Wyoming, Idaho, and Washington.

section 4 review

Summary

Mammal Characteristics

- Mammals have mammary glands that produce milk for their young.
- Mammals have teeth that are specialized for the type of food they eat.
- The body systems of mammals are designed to support activities such as running, swimming, climbing, hopping, and flying.

Mammal Types

- The smallest group of mammals, called monotremes, lay eggs with leathery shells.
- Marsupials are born before they are completely developed, and most marsupials carry their young in a pouch.
- The placentals are the largest group of mammals.

Mammals Today

- More than 4,000 species of mammals exist on Earth today.

Self Check

1. **Infer** why the brain of a mammal usually is larger than the brain of other animals of the same size.
2. **Explain** why animals are in trouble today.
3. **List** examples of how the teeth of mammals are specialized.
4. **Research Information** The monotremes are the smallest group of mammals. Using the library and online resources, explain where monotremes likely originated from and what continents monotreme fossils have been found on.
5. **Think Critically** Compare and contrast the development of embryos in placentals and marsupials.

Applying Math

6. **Solve One-Step Equations** The tallest mammal is the giraffe at 5.6 m. Calculate your height in meters and determine how many of you it would take to be as tall as a giraffe.

Science Online red.msscience.com/self_check_quiz

Discussion

Importance of Mammals Why are mammals important to people today? We get meat, milk, hides, and fur from them; we keep them as pets, use them for recreation, labor, and scientific research; they are also a critical part of the food web. They are sometimes used to help people with disabilities (i.e. guide dogs, capuchin monkeys). L1

3 Assess

DAILY INTERVENTION

Check for Understanding

Logical-Mathematical Use a three-column chart to post results of brainstorming with the class about characteristics of the three types of mammals. L2

Reteach

Animal Characteristics Have students group the following characteristics under the headings *Mammals*, *Birds*, and *Both Mammals and Birds:* the most fully developed brains, endothermic, hollow bones, hair, feathers, shelled eggs, nurse their young, give birth to live young, extended parental care. L2

Assessment

Content Have students work in groups of three to write a poem or song distinguishing each of the three types of mammal. Use **Performance Assessment in the Science Classroom,** p. 151. L2

section 4 review

1. Mammals can learn and remember more than many other animals.
2. Wildlife habitat is being destroyed.
3. Incisors are sharp, chisel-shaped teeth used for biting and cutting food; canine teeth are used to stab, grip, and tear flesh; flat molars are used for grinding food.
4. Answers will vary. Accept all that are reasonable.
5. Young marsupials are hairless, blind, and not fully formed at birth. Most crawl to the pouch of the female parent and develop there. Placentals develop in the placenta of the female parent until they are fully formed.
6. If a student is about 1.5 m tall, it would take 4 students to equal the height of a giraffe.

BENCH TESTED

Real-World Question

Purpose Students will determine the best environment for raising an endangered vertebrate species in captivity. L2

Process Skills research, design and make a model, predict, analyze data

Time Required 45 minutes

Make a Model

Possible Procedures Tell students that they will select an endangered vertebrate species. What groups of animals are vertebrates? fish, amphibians, reptiles, birds, and mammals Ask students to name some vertebrate species they know are endangered. Possible answers: tiger, elephants, gorilla, giant panda, manatee, California condor, whooping crane, American crocodile, leatherback turtle, tiger salamander

Teaching Strategies

- In order to make sure that no more than one group selects the same species, you may want to write a list of species from which groups can choose on the board. As each group informs you of their choice, cross that animal off the list and have remaining groups select from those remaining.
- Encourage some groups to select animals other than mammals.

LAB Model and Invent

HOMES FOR ENDANGERED ANIMALS

Goals

- **Research** the natural habitat and basic needs of one endangered vertebrate species.
- **Research and model** an appropriate zoo, animal park, or aquarium environment for this animal. Working cooperatively with your classmates, design an entire zoo or animal park.

Possible Materials

poster board
markers or colored pencils
materials with which to make a scale model

Real-World Question

Zoos, animal parks, and aquariums are safe places for endangered animals. Years ago, captive animals were kept in small cages or behind glass windows. The animals were on display like artwork in a museum. Now, some captive animals are kept in exhibit areas that closely resemble their natural habitats. These areas provide suitable environments for the animals so that they can reproduce, raise young, and have healthier and longer lives. What types of environments are best suited for raising animals in captivity? How can endangered animals be rescued?

Differentiated Instruction

Learning Disabled Assign students the task of locating and gathering pictures of their group's animal in its natural environment. L1

Using Scientific Methods

Make a Model

1. **Choose** an endangered animal to research. Find out where this animal is found in nature. What does it eat? Who are its natural predators? Does it exhibit unique territorial, courtship, or other types of social behavior? How is this animal adapted to its natural environment?
2. Why is this animal considered to be endangered?
3. **Design** a model of your proposed habitat in which this animal can live successfully.
4. **Research** how a zoo, animal park, or aquarium provides a habitat for this animal. This information can be obtained by contacting a zoo, animal park, or aquarium.
5. **Present** your design plan to your class in the form of a poster, slide show, or video. Compare your proposed habitat with that of the animal's natural environment. Make sure you include a picture of your animal in its natural environment.

Test the Model

1. Using all of the information you have gathered, create a model exhibit area for your animal.
2. List other plants and animals that might be present in the exhibit area.

Analyze Your Data

1. **Decide** whether all of the endangered animals studied in this activity could exist in the same zoo or wildlife preserve.
2. **Predict** which animals could be grouped together in exhibit areas.

Conclude and Apply

1. **Determine** how much land your zoo or wildlife preservation needs. Which animals require the largest habitat?
2. Using the information provided by all your classmates, design a zoo or wildlife preserve for the majority of endangered animals you've studied.
3. **Analyze** which type of problems might exist in your design

Communicating Your Data

Give an oral presentation on endangered animals and wildlife conservation to another class of students using your model. Use materials from zoos to supplement your presentation.

Assessment

Process Provide students with pictures of endangered species other than those chosen in the activity. Ask them to describe what they would do to determine the best environment for that species in a zoo. Use **Performance Assessment in the Science Classroom,** p. 89. L2

Communicating Your Data

Students could use the Internet to find additional information about specific zoos to include in their presentations. L2

Troubleshooting Discuss with the class how to make a scale model. Have them agree upon a scale to be used for all models so that all groups' models can be combined to make a plan for an entire zoo or animal park. L2

Test the Model

Expected Outcome Most models should result in a habitat that closely resembles the animal's natural environment.

Analyze Your Data

1. They probably could not all exist in the same wildlife preserve because their natural environments would be too different. They probably could all exist in the same zoo, since different areas of the zoo could be designed to match the habitats of the animals living there.
2. Predictions will depend on the animals selected. For example, elephants and rhinoceroses could be placed together, but a predator like a cheetah could not be placed with a prey animal like an antelope.

Conclude and Apply

1. Answers will depend on the species included. For example, a blue whale would need a much larger saltwater tank than would a leatherback sea turtle.
2. Designs will vary depending on the animals chosen and their space and habitat requirements.
3. Answers will vary, but might include not including ponds or moats for animals' water supplies, barriers to prevent animals from escaping, or placing predators too close to prey animals.

Content Background

Throughout Earth's history there have been about six mass extinctions. Several were more catastrophic than the mass extinction at the end of the Cretaceous Period, when the dinosaurs died out.

The fossil record shows a sudden change in plant life at the end of the Cretaceous Period, but only in the northern hemisphere. The southern hemisphere shows a gradual pattern of change in plant life. There is also evidence that many marine species were in severe decline before the time of the hypothesized impact. Scientists do not all agree with the Alvarez hypothesis.

Discussion

Mass Extinction Survival Have students hypothesize what allowed some species of animals and plants to survive the mass extinction 65 million years ago. Possible answers: Some species are able to adapt quickly to changes; others may be able to become dormant until conditions become more favorable. L2

Analyze the Event

What previously discovered scientific techniques or methods allowed the Alvarezes to make their discovery and develop their hypothesis? Possible answers: The Alvarezes' work would not be possible without techniques used to date rocks and measure the age of fossils, and methods to detect and measure tiny amounts of iridium in rocks. L2

Oops! Accidents in SCIENCE

SOMETIMES GREAT DISCOVERIES HAPPEN BY ACCIDENT!

Cosmic Dust and Dinosaurs

What killed the dinosaurs? Here is one theory.

Tiny bits of dust from comets and asteroids constantly sprinkle down on Earth. This cosmic dust led scientists Luis and Walter Alvarez to a hypothesis about one of science's most intriguing mysteries: What caused the extinction of dinosaurs?

It began some 65 million years ago when mass extinction wiped out 60 percent of all species alive on Earth, including the dinosaurs. Walter Alvarez and his father Luis Alvarez were working together on a geology expedition in Italy analyzing a layer of sedimentary rock. Using dating techniques, they were able to determine that this layer was deposited at roughly the same time that the dinosaurs became extinct. The younger Alvarez hypothesized that the rock might hold some clue to the mass extinction.

The Alvarezes proposed that the sedimentary rock be analyzed for the presence of the element iridium. Iridium is a dense and rare metal found in very low concentrations in Earth's core. The scientists expected to find a small amount of iridium. To their surprise, the sedimentary rock contained unusually high levels of iridium.

High concentrations of iridium are believed to be common in comets and asteroids.

Did asteroids kill the dinosaurs? An artist drew this picture to show how Earth might have looked.

If a huge asteroid collided with Earth, its impact would send tons of dust, debris, and iridium high into the atmosphere. The dust would block the Sun, causing global temperatures to decrease, plants to die, and animals to starve, resulting in a mass extinction. When the dust settled, iridium would fall to the ground as evidence of the catastrophe.

The Alvarez hypothesis, published in 1980, is still debated. However, it has since been supported by other research, including the discovery of a huge, ancient crater in Mexico. Scientists theorize that this crater was formed by the impact of an asteroid as big as Mount Everest.

Write Imagine that an asteroid has collided with Earth. You are one of the few human survivors. Write a five-day journal describing the events that take place.

Science Online
For more information, visit red.msscience.com/time

Write Encourage students to consider many of the results of an asteroid impact in their journal. Beyond the initial impact, the results might include earthquakes, tsunamis, fires, dust and rock falling from the sky, and dramatic temperature changes. Students might also examine the effect on society and human behavior, and the emotional effect of being one of the few survivors. L2

Resources for Teachers and Students

Dinosaurs: Opposing Viewpoints, by Wendy Stein, Greenhaven Press, Inc., 1994

The Search for the Killer Asteroid, by Gregory L. Vogt, The Millbrook Press, 1994

"The Day the Dinosaurs Died," by Ron Cowen, *Astronomy*, April 1996

Reviewing Main Ideas

Section 1 Chordate Animals

1. All chordates, at some time in their development, have a notochord and gill slits.
2. Endothermic animals maintain an internal body temperature. Ectothermic animals have body temperatures that change with the temperature of their surroundings.
3. The three classes of fish are jawless, cartilaginous, and bony. All fish are ectotherms.

Section 2 Amphibians and Reptiles

1. Amphibians are ectothermic vertebrates that spend part of their lives in water and part on land. Most amphibians go through a metamorphosis, which includes water-living larva and land-living adult stages.
2. Reptiles are ectothermic land animals that have dry, scaly skin.
3. Most reptiles lay eggs with a leathery shell.

Section 3 Birds

1. Birds are endotherms with feathers, and they lay eggs enclosed in hard shells.
2. Wings, feathers, and a light, strong skeleton are adaptations that allow birds to fly.

Section 4 Mammals

1. Mammals are endotherms that have mammary glands. All mammals have some hair.
2. Mammals have specialized teeth that mostly determine what foods they eat.
3. There are three groups of mammals. Monotremes lay eggs. Most marsupials have pouches in which embryos develop. Placentals have a placenta, and the embryos develop within the female's uterus.
4. Mammals have a variety of adaptations that allow them to live in different types of environments.

Visualizing Main Ideas

Copy and complete the following table comparing the characteristics of fish, amphibians, and reptiles.

Vertebrate Characteristics

Characteristic	Fish	Amphibians	Reptiles
Body temperature	ectotherm	ectotherm	ectotherm
Body covering	scales	skin	scales
Respiratory organs	gills	lungs & skin	lungs
Method of movement	fins	legs	most have legs
Fertilization	external	external	internal
Kind of egg	lacks shell	lacks shell	amniotic

Science Online red.msscience.com/interactive_tutor

chapter 18 Study Guide

Reviewing Main Ideas

Summary statements can be used by students to review the major concepts of the chapter.

Visualizing Main Ideas

See student page.

Science Online

Visit red.msscience.com
/self_check_quiz
/interactive_tutor
/vocabulary_puzzlemaker
/chapter_review
/standardized_test

Assessment Transparency

For additional assessment questions, use the *Assessment Transparency* located in the transparency book.

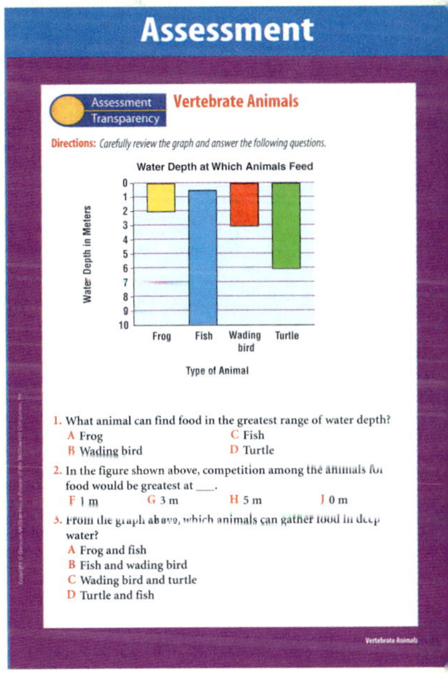
Assessment

Assessment Transparency **Vertebrate Animals**

Directions: *Carefully review the graph and answer the following questions.*

1. What animal can find food in the greatest range of water depth?
   A Frog C Fish
   B Wading bird D Turtle
2. In the figure shown above, competition among the animals for food would be greatest at ___.
   F 1 m G 3 m H 5 m J 0 m
3. From the graph above, which animals can gather food in deep water?
   A Frog and fish
   B Fish and wading bird
   C Wading bird and turtle
   D Turtle and fish

Vertebrate Animals

Using Vocabulary

Student answers should contain the following information in complete sentences.

1. contour feather—strong, lightweight, used in flight; down feather—soft, fluffy, used to insulate
2. An ectotherm's body temperature changes with its environment; an endotherm's body temperature remains constant.
3. chordate—has a notochord and pharyngeal pouches at some point in its development; cartilage—a tough flexible tissue found between vertebrae and which makes up the notochord
4. estivation—period of inactivity during hot summer months; hibernation—period of inactivity during cold winter months
5. carnivore—a meat-eating animal; herbivore—a plant-eating animal
6. monotreme—an egg-laying mammal; marsupial—a mammal that gives birth to immature offspring that develop in the female's pouch
7. An amniotic egg contains a yolk that supplies the developing embryo with food. Young monotremes develop in an amniotic egg.
8. A down feather insulates the bird's body. A bird is an endotherm with a nearly constant body temperature.
9. An omnivore eats plants and meat; a carnivore eats only meat.
10. The young develop in the uterus of both; in marsupials, young are born before development is complete and finish developing in the female's pouch.

Checking Concepts

11. D	16. A
12. C	17. C
13. C	18. B
14. A	19. D
15. A	

chapter 18 Review

Using Vocabulary

amniotic egg p. 539	estivation p. 535
carnivore p. 546	herbivore p. 546
cartilage p. 532	hibernation p. 535
chordate p. 530	marsupial p. 548
contour feather p. 543	monotreme p. 547
down feather p. 543	omnivore p. 546
ectotherm p. 531	placental p. 548
endotherm p. 531	

Using complete sentences, explain how the vocabulary words in each pair listed below are alike and how they are different.

1. contour feather—down feather
2. ectotherm—endotherm
3. chordate—cartilage
4. estivation—hibernation
5. carnivore—herbivore
6. marsupial—monotreme
7. amniotic egg—monotreme
8. down feather—endotherm
9. omnivore—carnivore
10. placental—marsupial

Checking Concepts

Choose the word or phrase that best answers the question.

11. Which of the following animals have fins, scales, and gills?
    A) amphibians **C)** reptiles
    B) crocodiles **D)** fish

12. Which of these is an example of a cartilaginous fish?
    A) trout **C)** shark
    B) bass **D)** goldfish

13. Which of the following fish has a swim bladder?
    A) shark **C)** trout
    B) lamprey **D)** skate

14. Which of the following is an adaptation that helps a bird fly?
    A) lightweight bones
    B) webbed feet
    C) hard-shelled eggs
    D) large beaks

15. Which of the following animals has skin without scales?
    A) dolphin **C)** lizard
    B) snake **D)** fish

16. Lungs and moist skin are characteristics of which of the following vertebrates?
    A) amphibians **C)** reptiles
    B) fish **D)** lizards

17. Which of these are mammals that lay eggs?
    A) carnivores **C)** monotremes
    B) marsupials **D)** placentals

18. Which of the following animals eat only plant materials?
    A) carnivores **C)** omnivores
    B) herbivores **D)** endotherms

Use the illustration below to answer question 19.

19. What is the primary function of the feather in the illustration above?
    A) flight **C)** attracting mates
    B) insulation **D)** repelling water

Science Online red.msscience.com/vocabulary_puzzlemaker

Use the ExamView® Pro Testmaker CD-ROM to:
- create multiple versions of tests
- create modified tests with one mouse click for inclusion students
- edit existing questions and add your own questions
- build tests aligned with state standards using built-in State Curriculum Tags
- change English tests to Spanish with one mouse click and vice versa

Thinking Critically

20. Discuss why there are fewer species of amphibians on Earth than any other type of vertebrate.

21. List the important adaptation that allows a reptile to live and reproduce on land while an amphibian must return to water to reproduce and complete its life cycle.

22. Draw a Conclusion You observe a mammal in a field catching and eating a rabbit. What kind of teeth does this animal probably have? Explain how it uses its teeth.

23. Explain how the development of the amniotic egg led to the early success of reptiles on land.

24. Compare and contrast the teeth of herbivores, carnivores, and omnivores. How is each type of tooth tooth adapted to the animal's diet?

25. Draw a Conclusion How can a bird like the arctic tern stand on ice and not lose too much body heat?

26. Concept Map Copy and complete this concept map that describes groups of mammals.

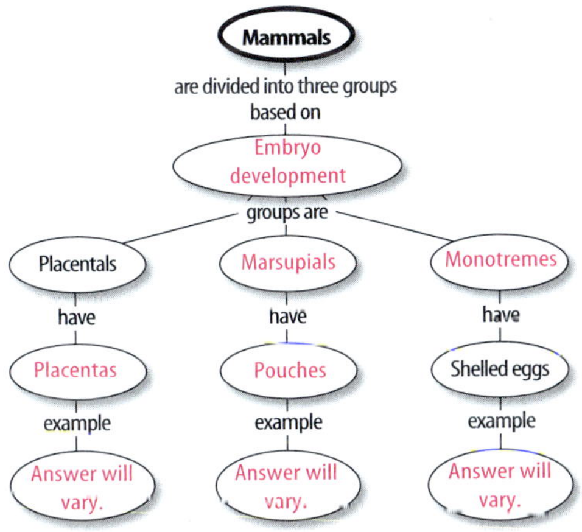

Performance Activities

27. Identify and Manipulate Variables and Controls Design an experiment to find out the effect of water temperature on frog egg development.

28. Debate Reptiles are often portrayed as dangerous and evil in fairy tales, folktales, and other fictional stories. Nonfiction information about reptiles presents another view. What is your opinion? Use the library or online resources to find evidence to support your position. Debate this issue with a classmate who has an opposing opinion.

Applying Math

Use the following table to answer questions 29 and 30.

Bull Trout Population

Year	Number per 100-m^2 Section
1996	4
1997	7
1998	5
1999	3
2000	4

29. Population Changes Make a line graph from the data in the table above.

30. Fish Population Density Calculate the average number of bull trout per 100-m^2 section of stream for all years combined. Which years had a larger population than the average?

31. Egg Development A salamander egg in water at 15–16°C will hatch after 60–70 days. A salamander egg in water at 17°C will hatch after 69–92 days. What are the minimum and maximum differences in hatching times?

Assessment Resources

Reproducible Masters

Chapter *Fast File* Resources
Chapter Review, pp. 39–40
Chapter Tests, pp. 41–44
Assessment Transparency Activity, p. 53

Glencoe Science Web site
Chapter Review Test
Standardized Test Practice

Glencoe Technology

Assessment Transparency
Exam*View*® Pro Testmaker
MindJogger Videoquiz
Interactive Chalkboard

chapter Review 18

Thinking Critically

20. Answers may include the fact that their skin absorbs all gases and chemicals in the area, including the poisonous ones.

21. internal fertilization for reproduction; amniotic egg; thick, dry, waterproof skin

22. It most likely has long, pointed canine teeth to stab, grip, and tear flesh. Its premolars would be sharp to cut and shred flesh.

23. The amniotic egg with its leathery shell provides a developing embryo with protection from drying out and from being eaten easily. Reptile eggs can be laid anywhere on land, thus freeing reptiles from relying on water for reproduction.

24. herbivores: large incisors for cutting grass, molars to grind plants; carnivores: small incisors, large canine teeth to grip and tear food; omnivores: canines, incisors, and molars to feed on plants and animals

25. Arctic terns are endotherms with down feathers.

26. See student page.

27. Student designs should include a control with a set temperature and independent variables with higher and lower temperatures. The dependent variable for each will be egg development.

Performance Activities

28. Use Task Assessment List for Oral Presentations in **PASC,** p.143.

Applying Math

National Math Standards

1, 5, 9

29. Check student graphs for accuracy.

30. average = 4.6
1997–1998 are larger than average

31. minimum difference equals
69 − 69 = 0 days or
70 − 70 = 0 days
maximum difference equals
92 − 60 = 32 days

FAST FILE

Answer Sheet A practice answer sheet can be found at red.msscience.com/answer_sheet.

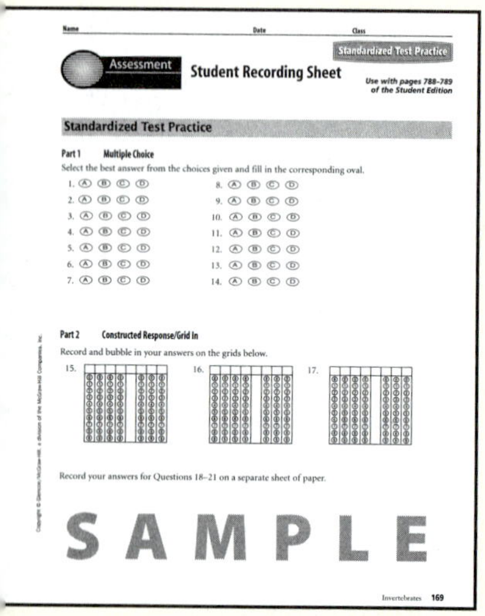
Name Date Class

Assessment Student Recording Sheet — Standardized Test Practice — Use with pages 788–789 of the Student Edition

Standardized Test Practice

Part 1 Multiple Choice

Select the best answer from the choices given and fill in the corresponding oval.

Part 2 Constructed Response/Grid In

Record and bubble in your answers on the grids below.

Record your answers for Questions 18–21 on a separate sheet of paper.

SAMPLE

Invertebrates 169

Part 1 Multiple Choice

1. B	4. A	7. A
2. A	5. B	8. D
3. D	6. C	9. D

Part 2 Short Response

10. Bony fish have swim bladders to help them swim to different depths. It is a sac that they inflate and deflate. The gases move between the swim bladder and the blood. Oxygen is used in deep-water fish and nitrogen in shallow-water fish.
11. Most bony fish use external fertilization. They release many eggs into the water. A male then swims over the eggs and releases sperm into the water to fertilize most of the eggs.
12. bony fish—skeletons of bone, scales covered with slimy mucus, and jaws; jawless fish—skeletons of cartilage, no scales, and no jaws
13. Amphibians go underground where it is cooler and more humid. They become inactive until the temperatures cool down.
14. Adult amphibians use lungs for oxygen exchange with carbon dioxide. They also can exchange oxygen and carbon dioxide through their skin, which must stay moist for this process to occur.
15. Birds preen to reorganize their feathers and to condition them. The reorganization closes gaps in the feathers. To condition feathers, they rub their beaks against an oil gland at the base of the tail then spread that oil to the feathers.
16. All mammals are endothermic vertebrates. They have mammary glands that in females produce milk to feed their young. All mammals have internal fertilization.
17. herbivore, omnivore, and carnivore
18. Monotremes don't have live births like other mammals. They lay eggs with tough, leathery shells. They also don't have nipples so the young must nurse by licking up the milk that seeps into the fur through the skin.

chapter 18 Standardized Test Practice

Part 1 Multiple Choice

Record your answers on the answer sheet provided by your teacher or on a sheet of paper.

1. Which of the following is NOT an ectotherm?
   A. amphibian **C.** reptile
   B. mammal **D.** fish
2. Three of the following are made of cartilage. Which is made of bone?
   A. goldfish
   B. ears
   C. tip of your nose
   D. jawless fish
3. Which of the following is NOT a group of fish?
   A. cartilaginous **C.** bony
   B. jawless **D.** angelfish

Use the photo below to answer questions 4 and 5.

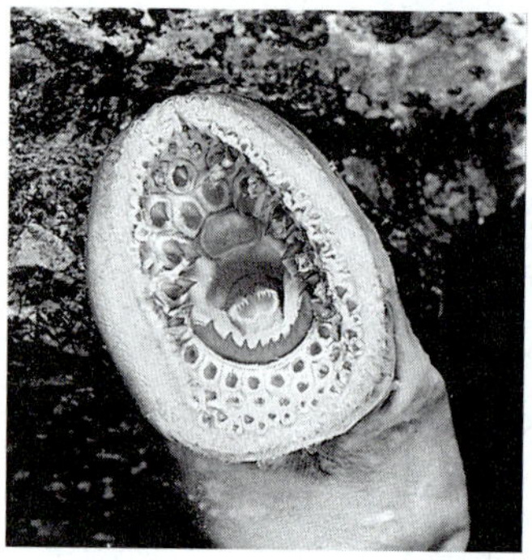

4. The fish in the illustration is an example of
   A. jawless fish **C.** bony fish
   B. ratfish **D.** cartilaginous fish
5. Which of the following is NOT one of its traits?
   A. scaleless, long, tubelike body
   B. rough, sandpaperlike scales
   C. round, muscular mouth without a jaw
   D. sharp, toothlike structures
6. Which of the following spend part of their life on land and part in the water?
   A. reptiles **C.** amphibians
   B. fish **D.** mammals

Use the photo below to answer questions 7 and 8.

7. This animal becomes inactive in cold weather using which of the following?
   A. hibernation **C.** hydration
   B. estivation **D.** hyperthermia
8. This animal has three of the following characteristics. Which one doesn't it possess?
   A. three-chambered heart
   B. vibrating tympanum
   C. exchange of oxygen and carbon dioxide through skin
   D. lays shell-covered, amniotic eggs
9. Which of the following is a monotreme?
   A. kangaroo **C.** wallaby
   B. koala **D.** platypus

Test-Taking Tip

Listen and Read Listen carefully to the instructions from the teacher and read the directions and each question carefully.

Part 2 | Short Response/Grid In

Record your answers on the answer sheet provided by your teacher or on a sheet of paper.

Use the illustration below to answer questions 10 and 11.

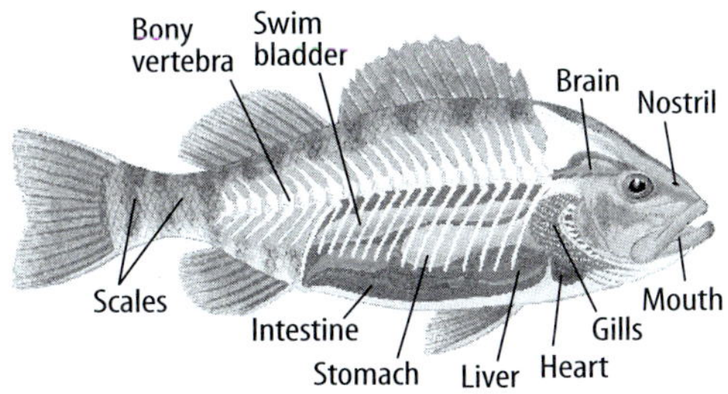

10. This animal has a particular method of swimming to different levels. Explain how they do this.

11. How do most of these animals reproduce?

12. How are jawless fish different from bony fish?

13. How do amphibians survive hot summers?

14. Animals need to get oxygen and dispose of carbon dioxide. How do adult amphibians get enough oxygen and get rid of carbon dioxide?

15. Birds use a special method to keep their feathers in good shape. What is this process called and how do they do it?

16. What characteristics do all mammals have in common?

17. Mammals also can be classified by what they eat. What are these three different classifications?

18. Name two ways that monotremes are different from other animals.

Part 3 | Open Ended

Record your answers on a sheet of paper.

19. Animals possess a great variety of body plans that help them reproduce. Compare and contrast the way fish and mammals reproduce.

20. Animals may have to compete with one another for survival. Explain how some animals compete for the same food and how predators can help in this process.

Use the illustration below to answer questions 21 and 22.

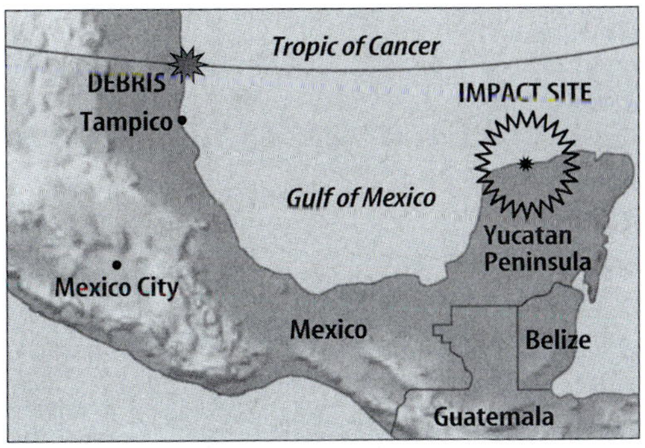

21. Walter Alvarez and his father, Luis Alvarez, came up with a theory as to how dinosaurs became extinct. This illustration is an artist's rendering of this theory. Explain it.

22. They did not set out to determine how dinosaurs became extinct. How did the theory depicted in the illustration come about?

23. Animals have various outer coverings for protection and various functions. Compare and contrast the various outer coverings of birds and mammals.

21. A large crater was found in Mexico that indicates an asteroid as large as Mt. Everest struck the Earth. It sent tons of debris into the air. Sunlight was blocked from reaching the surface of the Earth. This would cause temperatures to decrease dramatically and most of the planet would not be able to live. Plants would die causing herbivores and omnivores to starve. This theory was published in 1980 and is supported by other research.

22. The Alvarez's were on a geology expedition in Italy. While analyzing a layer of sedimentary rock, they were able to date this rock to around the time the dinosaurs became extinct. They looked for the presence of iridium in the rock. It is usually found in very small concentrations at the Earth's core. There were high levels of Iridium in the sedimentary rock. They felt the only way large amounts of iridium could be in the rock was if an asteroid hit the Earth and sent large amounts of iridium into the air, which would eventually fall back to Earth.

23. Birds are covered with feathers. There are two types—contour feathers help with movement and down feathers provide warmth. Mammals are typically covered with hair or fur, although some have almost no hair. The fur and hair provide warmth, just like the down feathers in birds. Some mammals have modified hair in the form of quills and spines. These modified hairs provide protection from predators.

Part 3 | Open Ended

19. Fish undergo external fertilization. The female lays a large number of eggs into the water. The male swims over them and releases his sperm into the water over the eggs. The sperm float down over the eggs and fertilize many of them. All mammals undergo internal fertilization. The male deposits his sperm inside the female. Placental mammals develop their young inside the uterus. Marsupials develop their young partly in the uterus and partly in a pouch. Monotremes lay eggs but contrary to fish, they fertilize their young first and then lay the eggs. The young still develops inside the egg.

20. Large populations of herbivores often compete with one another for the same plant sources. When the populations become too large, this can lead to overgrazing of the land. Starvation becomes a possibility for species that eat the same plant life. Large carnivores act as predators of these animals. They help to keep the populations down so overgrazing does not occur.

Rubrics

For more help evaluating open-ended assessment questions, see the rubric on p. 10T.

chapter 19 Organizer

Section/Objectives	Standards		Labs/Features
	National	**State/Local**	
Chapter Opener	See pp. 16T–17T for a Key to Standards.		**Launch Lab:** Where does food go?, p. 559 **Foldables,** p. 559
Section 1 Body Systems 3 sessions 1.5 blocks 1. **Explain** how the skeletal and muscular systems provide structure and allow movement. 2. **Identify** the functions of the digestive, respiratory, and circulatory systems. 3. **Distinguish** between the nervous and endocrine system. 4. **Explain** how your body systems provide defense.	National Content Standards: UCP.1–UCP.3, UCP.5, A.1, A.2, C.1–C.3, F.2		**Integrate Social Studies,** p. 564 **MiniLAB:** Inferring How Hard the Heart Works, p. 569 **Applying Science:** Will there be enough blood donors?, p. 570 **Science Online,** p. 571 **Lab:** Improving Reaction Time, p. 573
Section 2 Human Reproduction 4 sessions 2 blocks 5. **Identify** the organs of the male and female reproductive systems. 6. **List** the stages in the menstrual cycle. 7. **Describe** the stages of development before birth. 8. **Sequence** the life stages of humans.	National Content Standards: UCP.1–UCP.3, UCP.5, A.1, A.2, C.1–C.3		**Science Online,** p. 576 **Integrate Physics,** p. 577 **MiniLAB:** Interpreting Infant Development, p. 578 **Lab:** Defensive Saliva, p. 580

Glencoe Exclusive! TeacherWorks™ All-In-One Planner and Resource Center

Lab Materials	Reproducible Resources	Section Assessment	Technology
Launch Lab: index cards, marking pens, masking tape, meterstick or measuring tape	**Chapter *FAST FILE* Resources** Foldables Worksheet, p. 17 Directed Reading Overview, p. 19 Note-taking Worksheets, pp. 31–32		TeacherWorks includes: • Interactive Teacher Edition • Lesson Planner with calendar • Access to all program blacklines • Correlations to standards • Web links
MiniLAB: bowl of water, stopwatch or clock **Lab:** metric ruler *Need materials?* Contact Science Kit at 1-800-828-7777 or www.sciencekit.com on the Internet.	**Chapter *FAST FILE* Resources** Transparency Activity, p. 44 MiniLAB, p. 3 Enrichment, p. 29 Reinforcement, p. 27 Directed Reading, p. 20 Lab Worksheet, pp. 5–6 Transparency Activity, pp. 46–47 Lab Activity, pp. 9–12 **Life Science Critical Thinking/ Problem Solving,** p. 14 **Physical Science Critical Thinking/Problem Solving,** p. 2 **Cultural Diversity,** p. 1 **Home and Community Involvement,** p. 42	**Portfolio** Make a Model, p. 568 **Performance** MiniLAB, p. 569 Assessment, p. 572 Applying Skills, p. 572 **Content** Section Review, p. 572	Section Focus Transparency Teaching Transparency Virtual Labs CD-ROM Guided Reading Audio Program Interactive Chalkboard CD-ROM
MiniLAB: graph paper **Lab:** head of red cabbage, cooking pot, coffee filter, drinking glasses, clear household ammonia, bicarbonate of soda, water, spoon, white vinegar, lemon juice, orange juice	**Chapter *FAST FILE* Resources** Transparency Activity, p. 45 MiniLAB, p. 4 Enrichment, p. 30 Reinforcement, p. 28 Directed Reading, p. 21 Lab Worksheet, pp. 7–8 Lab Activity, pp. 13–16 **Earth Science Critical Thinking/ Problem Solving,** p. 19 **Lab Management and Safety,** p. 43 **Mathematics Skill Activities,** pp. 9–10 **Performance Assessment in the Science Classroom,** pp. 57–58 **Reading and Writing Skill Activities,** pp. 51–52	**Portfolio** Assessment, p. 579 **Performance** MiniLAB, p. 578 Assessment, p. 579 Applying Skills, p. 579 **Content** Section Review, p. 579	Section Focus Transparency Guided Reading Audio Program Interactive Chalkboard CD-ROM

End of Chapter Assessment

Blackline Masters	Technology	Professional Series
Chapter *FAST FILE* Resources Chapter Review, pp. 00–00 Chapter Tests, pp. 00–00 **Standardized Test Practice,** pp. 79–82	MindJogger Videoquiz Virtual Labs CD-ROM Exam*View*® Pro Testmaker TeacherWorks CD-ROM Interactive Chalkboard CD-ROM	**Performance Assessment in the Science Classroom (PASC)**

Transparencies

Section Focus

Section 1 Section Focus Transparency **Strange Creature from Above**

Scuba diving is an activity that requires special equipment that allows you to breathe while underwater. The word scuba actually stands for self-contained underwater breathing apparatus. This equipment is needed because, unlike fish, people breathe oxygen as a gas in air.

1. What special equipment is required for scuba diving?
2. What do the bubbles in this picture indicate?
3. How do people and fish differ in the way they get oxygen?

L2

Section 2 Section Focus Transparency **Sprint to Home**

Adult sea turtles lay about one hundred eggs in a nest on land, then promptly return to the water. After the eggs hatch, the young sea turtles race toward the sea in a struggle for survival on their own. In contrast, other species produce fewer eggs but give more care and protection to the young once they are born.

1. Compared to sea turtles, how many offspring do people have?
2. How does the amount of care young sea turtles receive differ from that of humans?

L2

This is a representation of key blackline masters available in the Teacher Classroom Resources. See Resource Manager boxes within the chapter for additional information.

Assessment

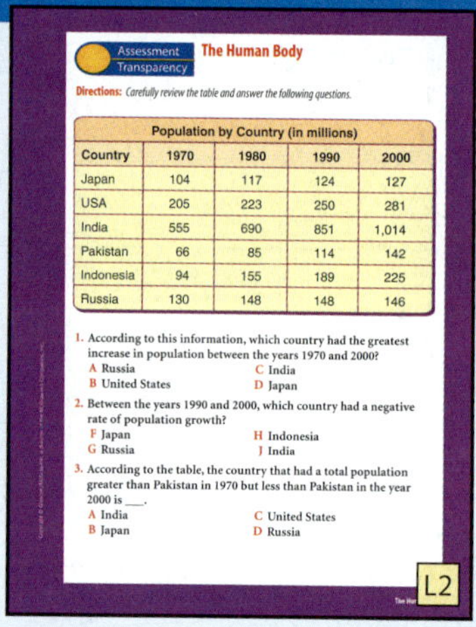
Assessment Transparency **The Human Body**

Directions: *Carefully review the table and answer the following questions.*

Population by Country (in millions)				
Country	1970	1980	1990	2000
Japan	104	117	124	127
USA	205	223	250	281
India	555	690	851	1,014
Pakistan	66	85	114	142
Indonesia	94	155	189	225
Russia	130	148	148	146

1. According to this information, which country had the greatest increase in population between the years 1970 and 2000?
   A Russia C India
   B United States D Japan
2. Between the years 1990 and 2000, which country had a negative rate of population growth?
   F Japan H Indonesia
   G Russia J India
3. According to the table, the country that had a total population greater than Pakistan in 1970 but less than Pakistan in the year 2000 is ___.
   A India C United States
   B Japan D Russia

L2

Teaching

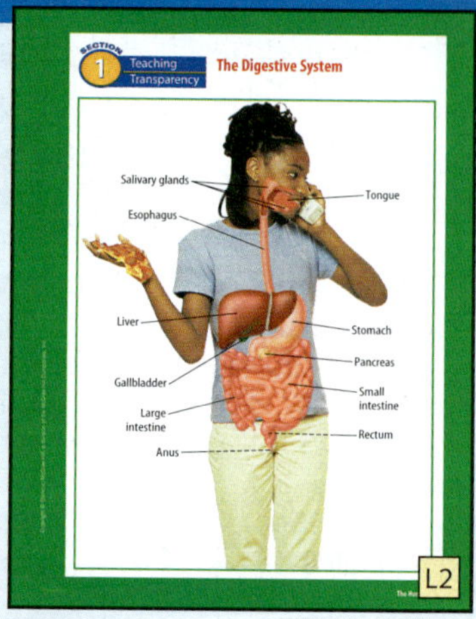
Section 1 Teaching Transparency **The Digestive System**

L2

Key to Teaching Strategies

The following designations will help you decide which activities are appropriate for your students.

- L1 Level 1 activities should be appropriate for students with learning difficulties.
- L2 Level 2 activities should be within the ability range of all students.
- L3 Level 3 activities are designed for above-average students.
- ELL ELL activities should be within the ability range of English-Language Learners.
- COOP LEARN Cooperative Learning activities are designed for small group work.
- LS Multiple Learning Styles logos, as described on page 12T, are used throughout to indicate strategies that address different learning styles.
- P These strategies represent student products that can be placed into a best-work portfolio.
- PBL Problem-Based Learning activities apply real-world situations to learning.

Hands-on Activities

Student Text Lab Worksheet

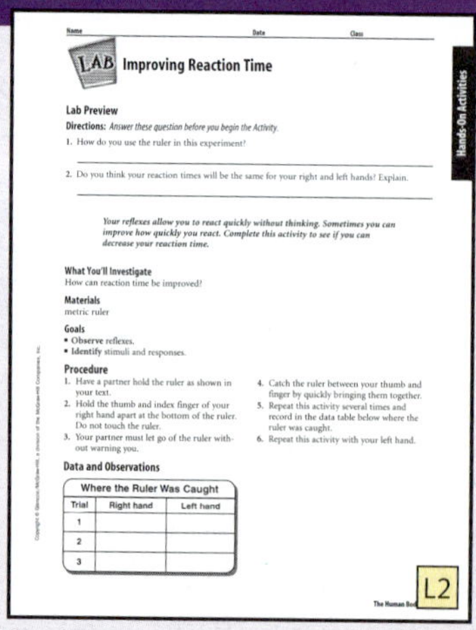
Name Date Class

LAB **Improving Reaction Time**

Lab Preview

Directions: *Answer these question before you begin the Activity.*

1. How do you use the ruler in this experiment?
2. Do you think your reaction times will be the same for your right and left hands? Explain.

Your reflexes allow you to react quickly without thinking. Sometimes you can improve how quickly you react. Complete this activity to see if you can decrease your reaction time.

What You'll Investigate
How can reaction time be improved?

Materials
metric ruler

Goals
- **Observe** reflexes.
- **Identify** stimuli and responses.

Procedure
1. Have a partner hold the ruler as shown in your text.
2. Hold the thumb and index finger of your right hand apart at the bottom of the ruler. Do not touch the ruler.
3. Your partner must let go of the ruler without warning you.
4. Catch the ruler between your thumb and finger by quickly bringing them together.
5. Repeat this activity several times and record in the data table below where the ruler was caught.
6. Repeat this activity with your left hand.

Data and Observations

Where the Ruler Was Caught		
Trial	Right hand	Left hand
1		
2		
3		

Hands-On Activities

L2

Laboratory Activities

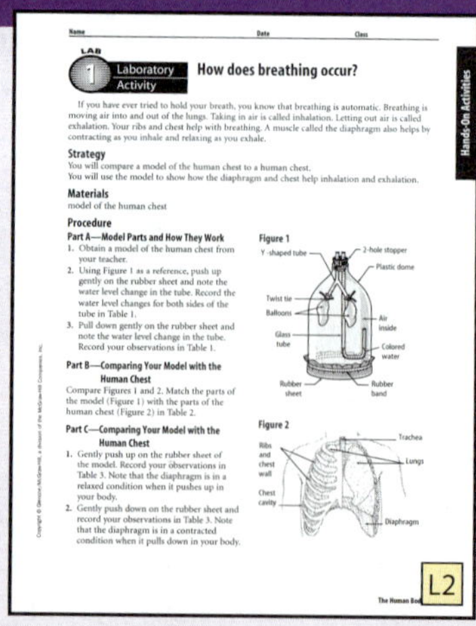
Name Date Class

LAB 1 Laboratory Activity **How does breathing occur?**

If you have ever tried to hold your breath, you know that breathing is automatic. Breathing is moving air into and out of the lungs. Taking in air is called inhalation. Letting out air is called exhalation. Your ribs and chest help with breathing. A muscle called the diaphragm also helps by contracting as you inhale and relaxing as you exhale.

Strategy
You will compare a model of the human chest to a human chest.
You will use the model to show how the diaphragm and chest help inhalation and exhalation.

Materials
model of the human chest

Procedure

Part A—Model Parts and How They Work
1. Obtain a model of the human chest from your teacher.
2. Using Figure 1 as a reference, push up gently on the rubber sheet and note the water level change in the tube. Record the water level changes for both sides of the tube in Table 1.
3. Pull down gently on the rubber sheet and note the water level change in the tube. Record your observations in Table 1.

Part B—Comparing Your Model with the Human Chest
Compare Figures 1 and 2. Match the parts of the model (Figure 1) with the parts of the human chest (Figure 2) in Table 2.

Part C—Comparing Your Model with the Human Chest
1. Gently push up on the rubber sheet of the model. Record your observations in Table 3. Note that the diaphragm is in a relaxed condition when it pushes up in your body.
2. Gently push down on the rubber sheet and record your observations in Table 3. Note that the diaphragm is in a contracted condition when it pulls down in your body.

Figure 1

Figure 2

Hands-On Activities

L2

Meeting Different Ability Levels

Content Outline

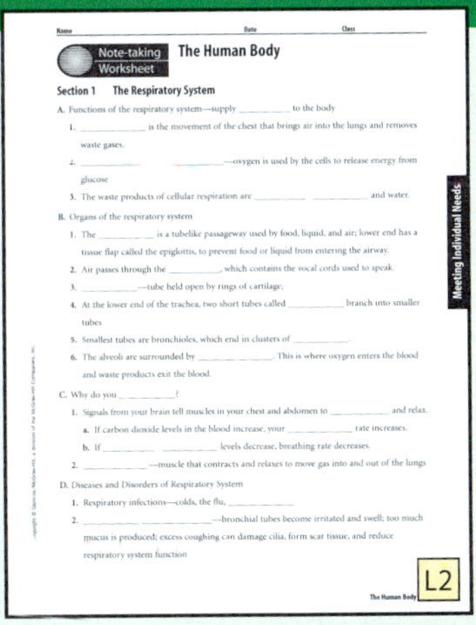

Reinforcement

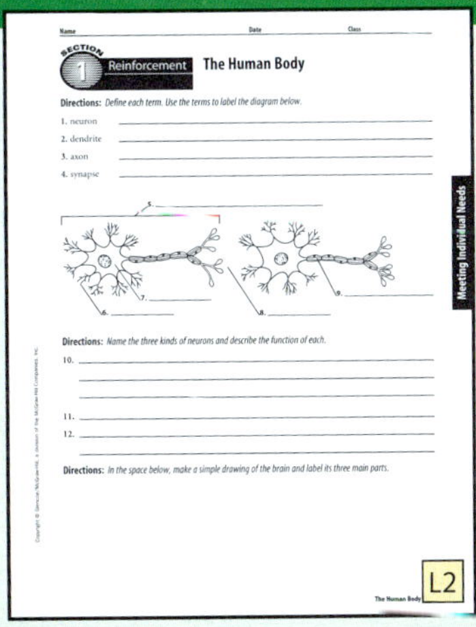

Enrichment

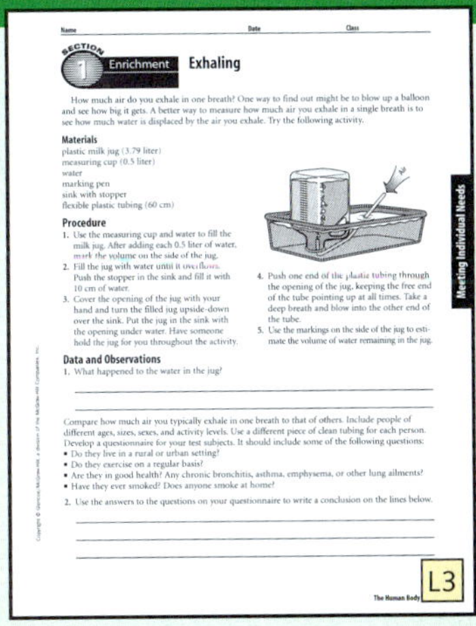

Directed Reading (English/Spanish)

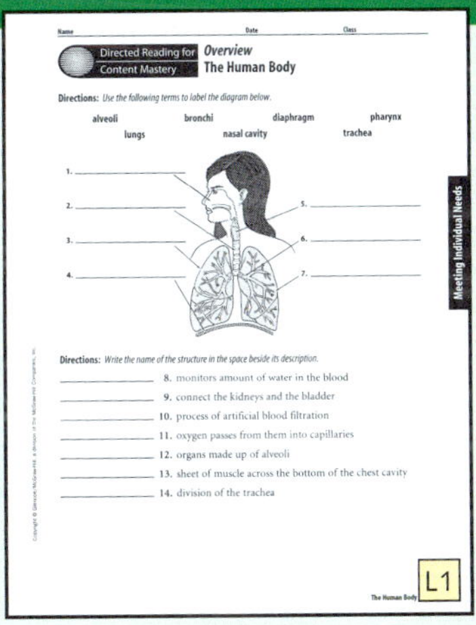

Study Guide

Reading Essentials

Assessment

Test Practice Workbook

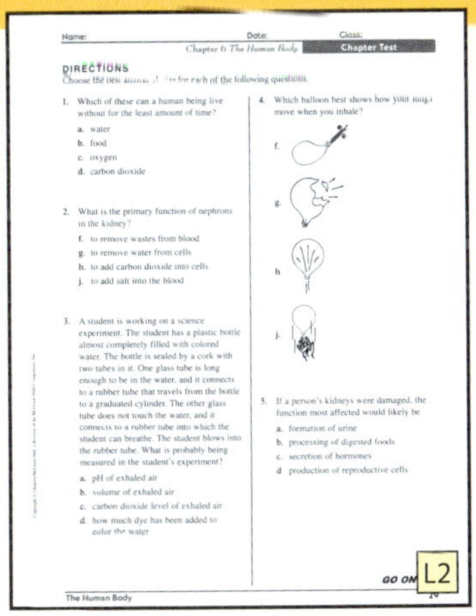

Chapter Review

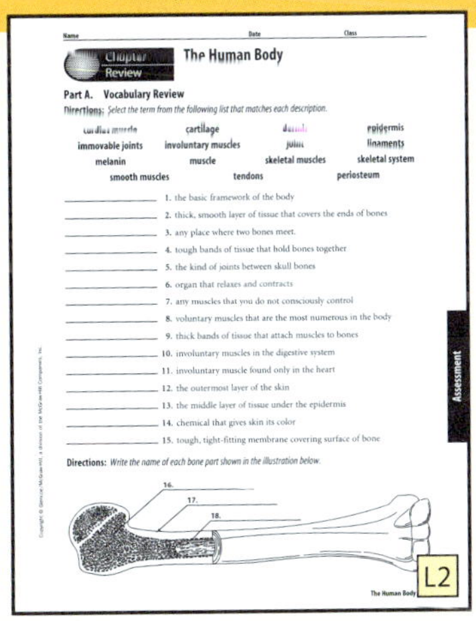

Chapter Tests

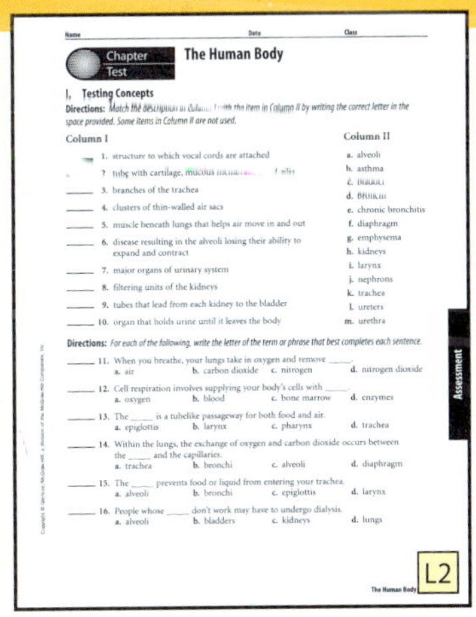

Science Content Background

section 1 Body Systems

Structure and Movement

The human body's structure and movement result from the interactions of the integumentary, muscular, skeletal and nervous systems. The skin is the organ that helps protect the body's internal environment from its external surroundings and helps to regulate the body's temperature. The muscular and skeletal systems give the body form and protection and allow movement. Contraction of muscles helps to regulate the body's temperature and bones serve as storage site for calcium and phosphorus. The nervous system receives stimuli from the skin and other organs and coordinates the body's responses, such as reflex movements.

Digestion

The process of transforming foods into usable nutrients is called digestion. This extracellular process takes place in the digestive system, specifically in the stomach and small intestine.

Vitamins

The role of vitamins as nutrients was not recognized until the early 1900s. During the early years of research, each new vitamin that was identified was given a letter designation. This method proved to be inadequate to correctly classify the substances. Vitamins then were called by their chemical names. Niacin is an example of a vitamin that goes by its chemical name rather than a letter designation.

The Respiratory System

Obtaining oxygen and removing carbon dioxide are the major functions of the respiratory system. These processes occur in two successive phases of functioning—breathing and transporting gases. In order to effectively exchange the gases involved in respiration, a large surface area is required. There are 300 million to 400 million alveoli in each lung. The air sacs of both lungs have a total surface area of about 93 m^2, nearly 50 times the total surface area of the skin.

Excretion

The metabolic processes of the body produce waste products. The respiratory system rids the body of carbon dioxide and other waste gases. The digestive system eliminates solid wastes. The urinary system removes a variety of dissolved salts and nitrogenous wastes from the blood and lymph systems.

chapter content resources

Internet Resources

For additional content background, visit **red.msscience.com** to:

- access your book online
- find references to related articles in popular science magazines
- access Web links with related content background
- access current events with science journal topics

Print Resources

The Digestive System, by Carol Ballard, Heinemann Library, 2002

Lungs: Learning How We Breathe, by Chris Hayhurst, The Rosen Publishing Group Incorporated, 2002

How the Endocrine System Works, by J. Matthew Neal, Blackwell Publishers, 2001

Endocrine and Reproductive Systems, by Susan Dudley Gold and Melissa Kim, Enslow Publishers, Inc., 2003

Charles Gupton/The Stock Market

Blood

The liquid component of blood, plasma, is a complex solution of organic and inorganic substances dissolved in water. Blood is about 90 percent water and eight percent proteins. The remaining two percent is made up of dissolved substances such as electrolytes, gases, and nutrients. Blood characteristics, such as pH and sugar content, are precisely controlled by the body. The pH of blood is slightly alkaline, and serious problems can occur if the pH changes only slightly.

Regulation

The endocrine system uses hormones to control body functions such as defense, hunger, growth and reproduction. All organs of the endocrine system are glands that produce and empty hormones into the circulatory system.

section 2 Human Reproduction

Hormones

Hormones produced by the endocrine system control sexual development, maturation, and functioning of the reproductive system. Hormones stimulate the production and release of sperm and semen in the male reproductive system. In the female reproductive system, hormones stimulate the production and release of ova, or eggs. Hormones also play a role during labor and birth. In later years, menopause is brought on when the body stops producing certain hormones.

Embryonic Development

Fertilization is the union of a sperm cell from a male with an egg from a female. Fertilization results in two events: (1) fusion of the sperm and egg cells, including fusion of the two nuclei, to form a cell called a zygote, and (2) activation of the new cell to begin division and growth.

Once a sperm fertilizes an egg, the egg produces an outer protective layer. Next, the zygote divides into thousands of smaller cells that cling together like a ball of soap bubbles. After several days and until about two weeks later, the ball of cells begins to differentiate into three parts: an outer layer of cells that attaches to the uterus of the mother, an inner mass of cells that develops into the embryo, and a fluid-filled sac that surrounds the embryo.

The next phase, which occurs between two and eight weeks after fertilization, involves specialization of cells into primary organ systems. This set of events marks the end of the embryo stage and the beginning of the fetal stage.

Teacher to Teacher

Jeff Remington
Palmyra Middle School
Palmyra, PA

"To show students that water vapor is released during cellular respiration, invert a dry glass beaker for a few seconds over a lit Bunsen burner. The inside of the beaker instantly fogs and becomes wet. It is a *discrepant event,* but shows students that when you burn an organic compound, such as sugar or methane, water vapor is produced. Jet fuel is carbon compound as well. Hence you see water vapor trails behind jets in the sky as they burn fuel."

Jeff Remington

chapter 19

Chapter Vocabulary

skeletal system, p. 560
melanin, p. 561
muscle, p. 562
nutrient, p. 564
respiratory system, p. 567
alveoli, p. 567
capillary, p. 567
reflex, p. 571
sperm, p. 574
semen, p. 574
ovulation, p. 575
menstrual cycle, p. 576
pregnancy, p. 577
embryo, p. 577
fetus, p. 577

Science Journal Student responses will vary, but should include things such as food, water, vitamins, rest, and exercise.

INTERACTIVE CHALKBOARD with Image Bank

PowerPoint® Presentations

This CD-ROM is an editable Microsoft® PowerPoint® presentation that includes:

- a pre-made presentation for every chapter
- interactive graphics
- animations
- audio clips
- image bank
- all new section and chapter questions
- Standardized Test Practice
- transparencies
- pre-lab questions for all labs
- Foldables directions
- links to red.msscience.com

chapter 19

The Human Body

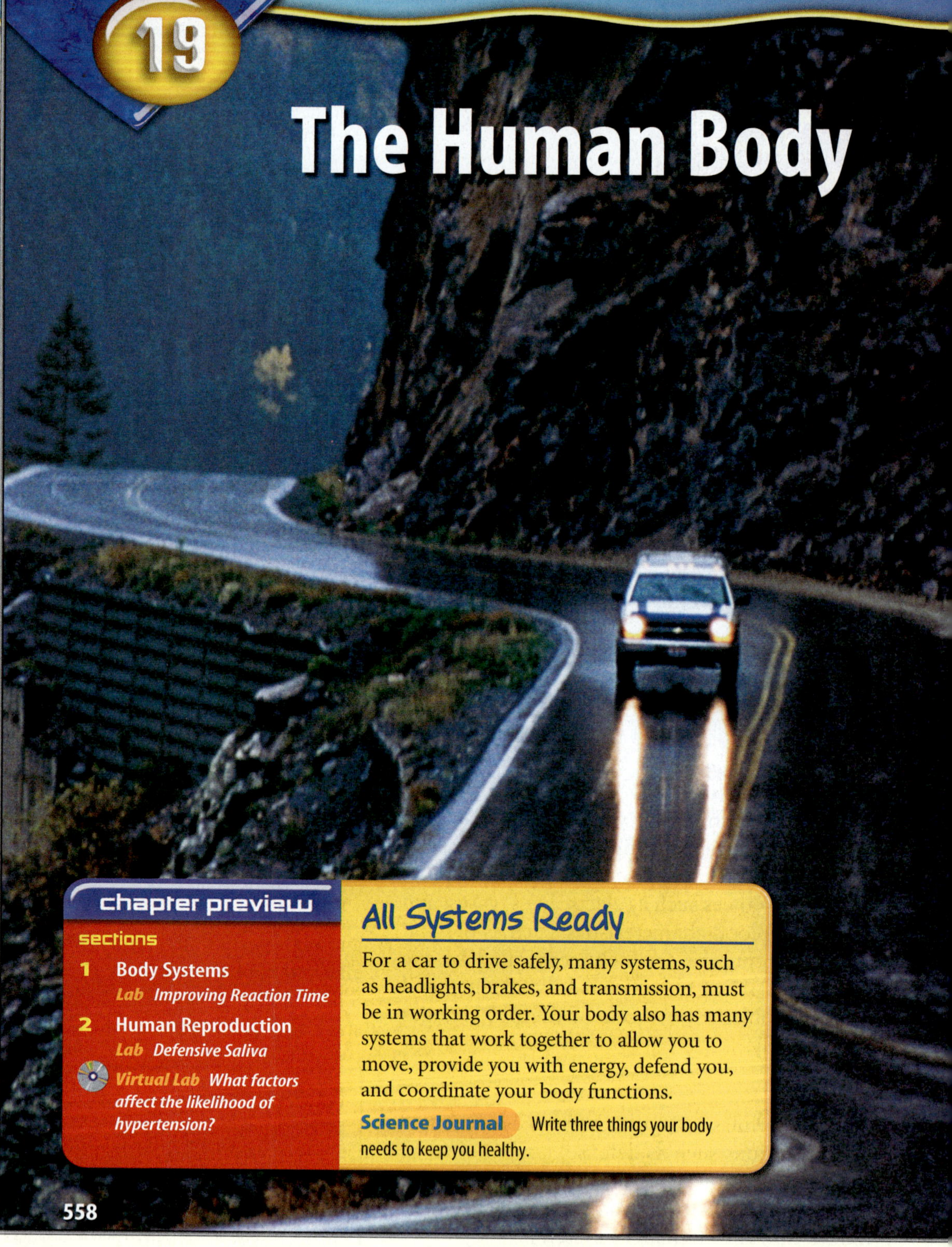

chapter preview

sections

1. **Body Systems**
   Lab Improving Reaction Time
2. **Human Reproduction**
   Lab Defensive Saliva

Virtual Lab What factors affect the likelihood of hypertension?

All Systems Ready

For a car to drive safely, many systems, such as headlights, brakes, and transmission, must be in working order. Your body also has many systems that work together to allow you to move, provide you with energy, defend you, and coordinate your body functions.

Science Journal Write three things your body needs to keep you healthy.

Theme Connection

Energy A major theme of this chapter is energy. When the skeletal and muscular systems work to produce movement, energy is used and work is done. Food is processed by the digestive system and converted into glucose. The chemical energy in glucose is used by muscles and transformed into mechanical energy. The unique functioning of the reproductive system is dependent on energy for cell division and differentiation of cells.

About the Photo

Systems Working Together A car can be used as an analogy to explain the workings of the human body. Like the car, the human body has many systems that must work together for the body to function properly. For example, the circulatory system carries waste gases to the lungs to be exhaled. In the same way, the exhaust system in a car carries away the exhaust gases produced in the engine.

Start-Up Activities

Launch LAB

Where does food go?

Imagine taking a bite of your favorite food. When you eat, your body breaks down food to release energy. How long does it take?

Organs of the Digestive System

Organ	Length	Time
Mouth	8 cm	5 s to 30 s
Pharynx and esophagus	25 cm	10 s
Stomach	16 cm	2 h to 4 h
Small intestine	4.75 m	3 h
Large intestine	1.25 m	2 days

1. Make a label for each of the major organs of the digestive tract listed here. Include the organ's name, its length, and the time it takes for food to pass through it.
2. Working with a partner, place a piece of masking tape that is 6.5 m long on the classroom floor.
3. Beginning at one end of the tape, and in the same order as they are listed in the table, mark the length for each organ. Place each label next to its section.
4. **Think Critically** In your Science Journal, suggest reasons why the food that you eat spends a different amount of time in each of the organs.

FOLDABLES™ Study Organizer

Nutrients in Food Make the following Foldable to help you organize foods based on the nutrients that they contain.

STEP 1 **Fold** the top of a vertical piece of paper down and the bottom up to divide the paper into thirds.

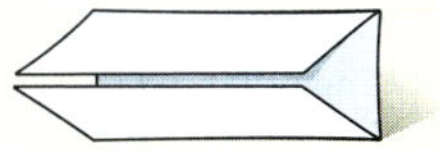

STEP 2 **Fold** the folded paper in half lengthwise.

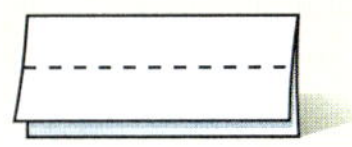

STEP 3 **Unfold** the paper and draw a line on each fold line to form six columns. **Label** the columns as shown.

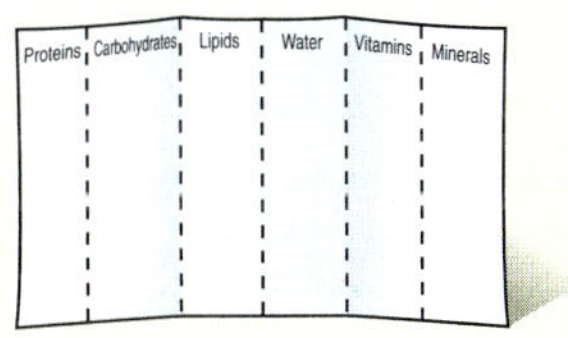

Read and Write As you read the chapter, list foods you eat that provide each of these nutrients in the appropriate columns.

Science Online — Preview this chapter's content and activities at red.msscience.com

Launch LAB

Purpose Use the Launch Lab to introduce students to the digestive system. L2 COOP LEARN

Preparation Find an area where students can work undisturbed during the lab.

Materials index cards, marking pens, masking tape, meterstick or measuring tape

Teaching Strategies

- Explain that the length of time food spends in the digestive tract varies among individuals and can depend on health conditions.
- After the complete length of the digestive system has been marked off, encourage students to infer how something that long can fit inside their bodies.

Think Critically

The amount of digestion time depends on the type of food. Different foods undergo different chemical processes as they are digested.

Assessment

Process Have students predict what symptoms an infant might exhibit if a birth defect caused a constriction in a portion of the digestive tract. Use **Performance Assessment in the Science Classroom,** p. 93.

FOLDABLES™ Study Organizer — Dinah Zike Study Fold

Student preparation materials for this Foldable are available in the Chapter *FAST FILE* Resources.

section 1 Body Systems

1 Motivate

Bellringer

INTERACTIVE CHALKBOARD PowerPoint® Presentations

Section Focus Transparencies also are available on the Interactive Chalkboard CD-ROM. L2 ELL

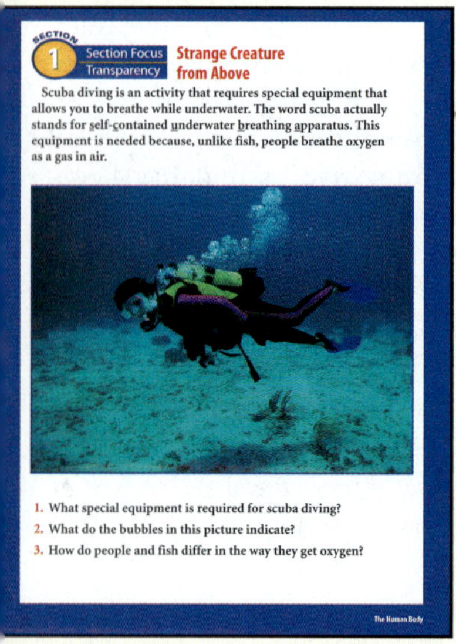

Tie to Prior Knowledge

Body Systems Have students list systems in a car that have several parts working together. They might suggest electrical system, brake system, power train, and others. Explain that systems similar to these exist in their bodies to carry out body functions.

Reading Check

Answer living cells

as you read

What **You'll Learn**

- **Explain** how the skeletal and muscular systems provide structure and allow movement.
- **Identify** the functions of the digestive, respiratory, and circulatory systems.
- **Distinguish** between the nervous and endocrine system.
- **Explain** how your body systems provide defense.

Why **It's Important**

You can take better care of your body if you understand how your body works.

Review Vocabulary

organ: structure, such as the heart, made up of different types of tissues that work together

New Vocabulary

- skeletal system
- melanin
- muscle
- nutrient
- respiratory system
- alveoli
- capillary
- reflex

Structure and Movement

Have you ever seen a building under construction? First a framework of steel or wood is built, as shown in **Figure 1** on the left. Then the framework is covered by walls. Your body also has a framework, your bones, as shown in **Figure 1** on the right. Bones are covered by skin and muscle.

The Skeletal System All the bones in your body make up your **skeletal system.** Your skeletal system gives shape and support to your body. Without bones, your body would be a formless mass of tissue. Bones also protect your internal organs. For instance, your ribs surround and protect your heart and lungs, and your skull protects your brain.

You may think that bones are dead tissue, but bones are made of living cells. Like all the other cells in your body they need nutrients and use energy. Bones are hard because of the calcium and phosphorus that are deposited in them.

Reading Check *Are bones made of living or dead cells?*

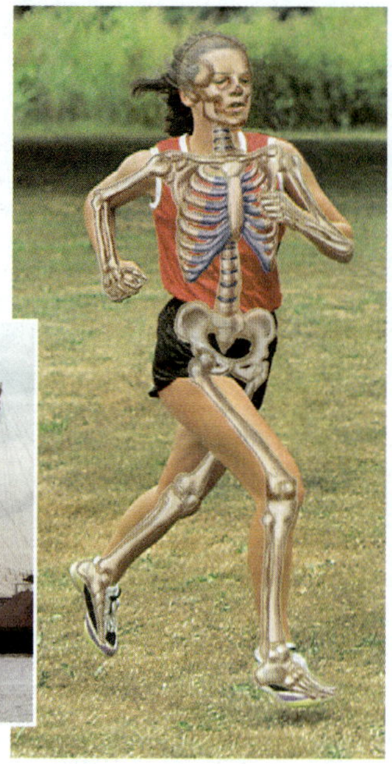

Figure 1 When a building is put up, it needs a framework for support. Your body's framework is your skeleton, made up of 206 bones.

Section 1 Resource Manager

Chapter *FAST FILE* Resources

Transparency Activity, p. 44
MiniLAB, p. 3
Note-taking Worksheets, pp. 31–32
Enrichment, p. 29
Reinforcement, p. 27
Directed Reading for Content Mastery, p. 20
Lab Worksheet, pp. 5–6
Lab Activity, pp. 9–12

Life Science Critical Thinking/Problem Solving, p. 14

Physical Science Critical Thinking/Problem Solving, p. 2

Cultural Diversity, p. 1

Home and Community Involvement, p. 42

Joints Unlike the framework of a building, which does not move, your bones can move. Joints make these movements possible. The place where two or more bones come together is called a joint, as shown in **Figure 2.** Muscles can move your bones by moving your joints.

The Skin The largest human body organ is the skin. It helps your body in several important ways. First, skin forms a protective covering for your body. Unbroken skin can protect your body from disease-causing organisms. Sweat and oil glands in your skin secrete fluids that can slow the growth of or kill bacteria. The pigment, or coloring, in your skin protects it from damage by ultraviolet light. This pigment is called **melanin** (MEH luh nun). Humans have different skin colors because of the different amounts of melanin in their skin.

Second, skin is a sense organ. Because of special nerve cells in the skin, you can sense heat or cold, and you can feel the sharp prick of a pin or the smoothness of a polished rock.

Third, skin helps control your body temperature. Sweat glands in your skin produce sweat, as shown in **Figure 3,** that helps cool your body. As sweat evaporates from your skin, heat is lost and your skin is cooled.

Fourth, skin helps provide a nutrient for your body. Vitamin D, which is important for good health, is formed in your skin when your skin receives ultraviolet light from the Sun. This vitamin helps your body absorb calcium from food in your digestive tract.

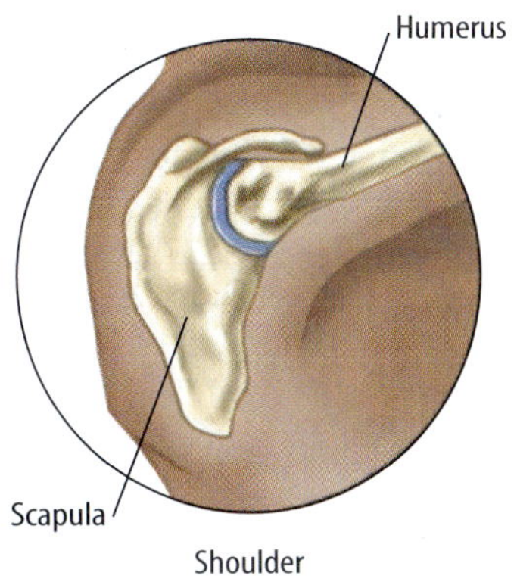

Figure 2 Joints enable your bones to move.

Figure 3 Your skin helps control your body's temperature when you sweat.

2 Teach

Quick Demo

Bone Strength

Materials 4-in × 6-in cards, tape, books

Estimated Time ten minutes

Procedure Use the cards and tape to make one rectangular column and one cylindrical tube. Stand the structures on a flat surface and balance books on top of each one. Have students note which structure collapses first. Relate this strength to the long, round bones of the human body.

Teacher FYI

Osteoporosis A condition in which bones become porous and thin, resulting in reduced bone mass, is called osteoporosis. Factors that increase the risk of this condition include being female, being Caucasian, having chronic low calcium intake, lacking exercise, being underweight, and smoking.

Discussion

Skin Sensors How would humans receive information about environmental conditions if they did not have skin? Possible answer: They would have to have other tissues or organs for sensing environmental conditions in order to survive.

Differentiated Instruction

Learning Disabled Have students pat a wet cotton ball onto their wrists. Then have them gently blow on the wet surface. The evaporation removes body heat. Relate this experience to the evaporation of perspiration from the skin during hot weather. L1

Fun Fact

Skeletal muscles form a large portion of the body's mass. Even at rest they produce about 40 percent of body heat. Shivering and physical exercise can increase the production of heat to double that amount.

Visual Learning

Figure 4 Have students study the illustration, and then identify which muscle in the pair is shortening *(biceps)* and which muscle is returning to its original length *(triceps)*. Ask them what happens when the weight is lowered. Which muscle shortens? triceps Which muscle returns to its original length? biceps

Discussion

Voluntary Muscles How can you tell whether a muscle is voluntary? Possible answer: The muscle only can move if you think about it.

Caption Answer

Figure 4 Accept all reasonable responses.

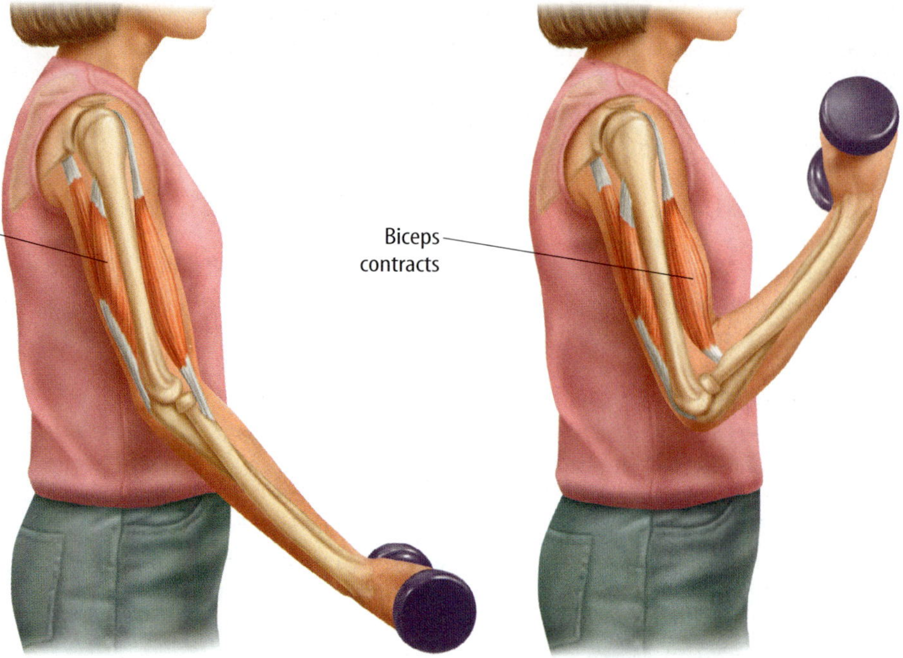

Figure 4 To lift a weight, one muscle in your arm contracts while another relaxes.
Identify *another muscle pair in your body.*

The Muscular System Imagine that you are running down the street because you are late for school. How does your body perform all the movements to get you there? Remember that your bones provide a framework for your body. Muscles attach to bones and help them move. Your muscles make all your movements possible. A **muscle** is an organ that can relax, contract, and provide force to move you and your body parts.

All your movements require energy. Even when you are sitting in a chair, your muscles are still working. Some muscles are always moving, such as the muscles that help you breathe and your heart muscles.

Your body has some muscles that you can choose to move, as shown in **Figure 4,** called voluntary muscles. Other muscles, such as your heart and muscles in your digestive organs, are not controlled consciously. These are called involuntary muscles. Involuntary muscles work all day long, every day for your entire life.

Voluntary muscles work together in pairs to allow you to move your body. One muscle contracts or gets shorter, while another muscle relaxes or returns to its original length. Look again at **Figure 4.** You can see that one muscle contracts and another relaxes when you lift a weight.

LAB DEMONSTRATION

Purpose to observe the coordinated work of muscle pairs

Materials chart of arm muscles

Procedure Have students feel the muscles in their upper arms as they pump their arms out and back. Have them use the chart of arm muscles to determine which upper arm muscles are involved.

Expected Outcome Students should note that the biceps and triceps are involved in the action.

Assessment

Which muscle contracted to bring the forearm closer to the shoulder? The biceps contracted. Which muscle relaxed to allow this action? The triceps relaxed.

Digestion and Excretion

How does your body supply you with the energy you need for activities such as a championship soccer game? Food enters your body's digestive system through your mouth, as shown in **Figure 5.** As food moves through each organ of your digestive system, some of it is broken down into smaller molecules. These smaller molecules are absorbed from your digestive system and enter your blood. From the blood, these molecules move into your cells where the food molecules are needed. Undigested food is eliminated from your digestive system.

Reading Check *Why is food broken down into smaller molecules as it moves through your digestive system?*

Organs of the Digestive System In **Figure 5,** trace the path of food as it travels through your digestive system. In your mouth, food is moistened by saliva and broken down into smaller particles by your teeth. From there, food enters your esophagus (ih SAH fuh gus), a long muscular tube, and moves to your stomach. In your stomach chemicals break down food. After breakdown in the stomach, food moves into the small intestine.

The small intestine is the longest organ in the digestive system. It is small in diameter, but not small in length. Most digestion and absorption of food take place in the small intestine. Here, food particles move from the digestive tract into the blood to be carried to all of the body's cells. Finally, food moves into the large intestine where water is absorbed. The remaining undigested food is now a semisolid and is excreted from the body.

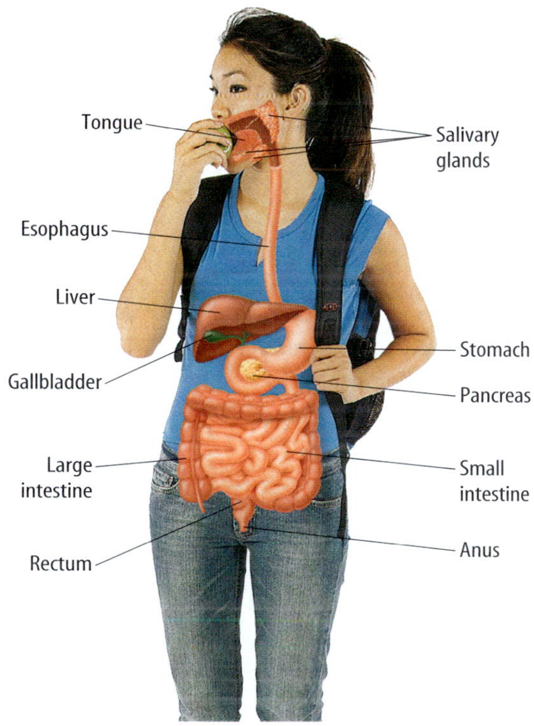

Figure 5 Your digestive system is designed to take in food and break it down so that the nutrients can be absorbed by your body.
Infer *why food is important to your health.*

Fun Fact

Birds have a specialized stomach chamber, the gizzard, that contains small stones. These stones aid in grinding food—a process that is needed in the absence of teeth.

Reading Check

Answer to be absorbed into the blood

Activity

Digestive-Organ Length Have students review the chart in the Launch Lab. Use **Figure 5** to reinforce how organs of these lengths can be contained within the body.

Teacher FYI

Small Intestine It takes between one and six hours for food undergoing digestion to move through the small intestine. The small intestine is subdivided into three parts: duodenum, jejunum, and ileum. Most of the digestive processes occur in the duodenum and most of the absorption of nutrients takes place in the duodenum and the jejunum. Vitamin B_{12} and bile salts are absorbed in the ileum.

Caption Answer

Figure 5 Food supplies energy, water, and essential nutrients.

Science Journal

Cavities Have interested students research information about the relationship between sugars in the mouth and cavity-causing bacteria. Have them record in their Science Journals information on the formation of tooth cavities. Some students might like to include diagrams showing the stages of tooth decay. L2 LS **Linguistic**

Active Reading

Metacognition Journal In this strategy, each student analyzes his or her own thought processes. Have students divide the paper in half. On the left, have them record what they have learned about a topic. On the right, have them record the reason they learned it. Have students write a Metacognition Journal about the digestive system.

Teacher FYI

Food Energy When cells need energy, they oxidize fuels such as carbohydrates and fats. This process results in the formation of carbon dioxide and water, the same products that are formed when a fuel such as gasoline is burned in an engine. Energy is released in this reaction.

Discussion

Milk Is milk the perfect food? Milk does supply many nutrients, but it is not a perfect food. It lacks certain essential vitamins and minerals. LS **Interpersonal**

Activity

Vitamin Toxicity Vitamins can be toxic in very large doses. Have students find out how excessive amounts of vitamins K and D can affect the body. Excessive intake of vitamin K can cause anemia, jaundice, and gastrointestinal problems; megadoses of vitamin D can result in nausea and vomiting, calcification of soft tissues, and kidney failure.

INTEGRATE Social Studies

Protein and Carbohydrate Sources Amaranth grows best in warm, dry climates. Its leaves are eaten as a vegetable similar to spinach. The seeds can be ground into flour and used in breads, cakes, and other products.

INTEGRATE Social Studies

Protein and Carbohydrate Sources In South America and Africa, amaranth is a common source of protein and carbohydrates. Find out in which countries amaranth is grown and how it is prepared and eaten. Write a paragraph in your Science Journal to describe what you have discovered.

INTEGRATE Chemistry

Nutrients Although you might prefer to grab a candy bar or bag of chips when you feel hungry, your body needs certain foods to stay healthy. **Nutrients** (NEW tree unts) are the substances in food that provide for cell development, growth, and repair. There are six kinds of nutrients that are available in food—proteins, carbohydrates (kar boh HI drayts), lipids, vitamins, minerals, and water, as shown in **Figure 6.** Proteins, carbohydrates, lipids, and vitamins contain the element carbon. They are called organic nutrients. Minerals and water do not contain the element carbon and are called inorganic nutrients.

Proteins are found in meats, poultry, eggs, fish, peas, beans, and nuts. Proteins are used by your body for replacement and repair of body cells and for growth. Carbohydrates are the main energy source for your body. Foods that contain carbohydrates include sugar, honey, fruits, vegetables, breads, grains, and cereals. Lipids, or fats, provide energy, help your body absorb vitamins, and, when stored in the tissue, cushion your internal organs.

Figure 6 Six kinds of nutrients are available in food.

Differentiated Instruction

Challenge Have students research and report on the results of a particular vitamin deficiency. Their reports should include information about how the vitamin was discovered, the symptoms of a deficiency, and foods that are high in the vitamin. Have students work together to present their findings on a "Vitamins for Good Health" bulletin board. L3 COOP LEARN LS **Linguistic**

Curriculum Connection

Health Have the school lunchroom manager or a hospital dietitian discuss with students how a balanced diet can be achieved with a variety of foods. Have students work in pairs to list what they ate during the last 24 hours and then evaluate the variety of their diets. L2 COOP LEARN

NATIONAL GEOGRAPHIC

VISUALIZING VITAMINS

Figure 7

Vitamins come in two groups—water soluble, which should be replaced daily, and fat soluble, which can be stored in the body. The sources and benefits of both groups are shown below.

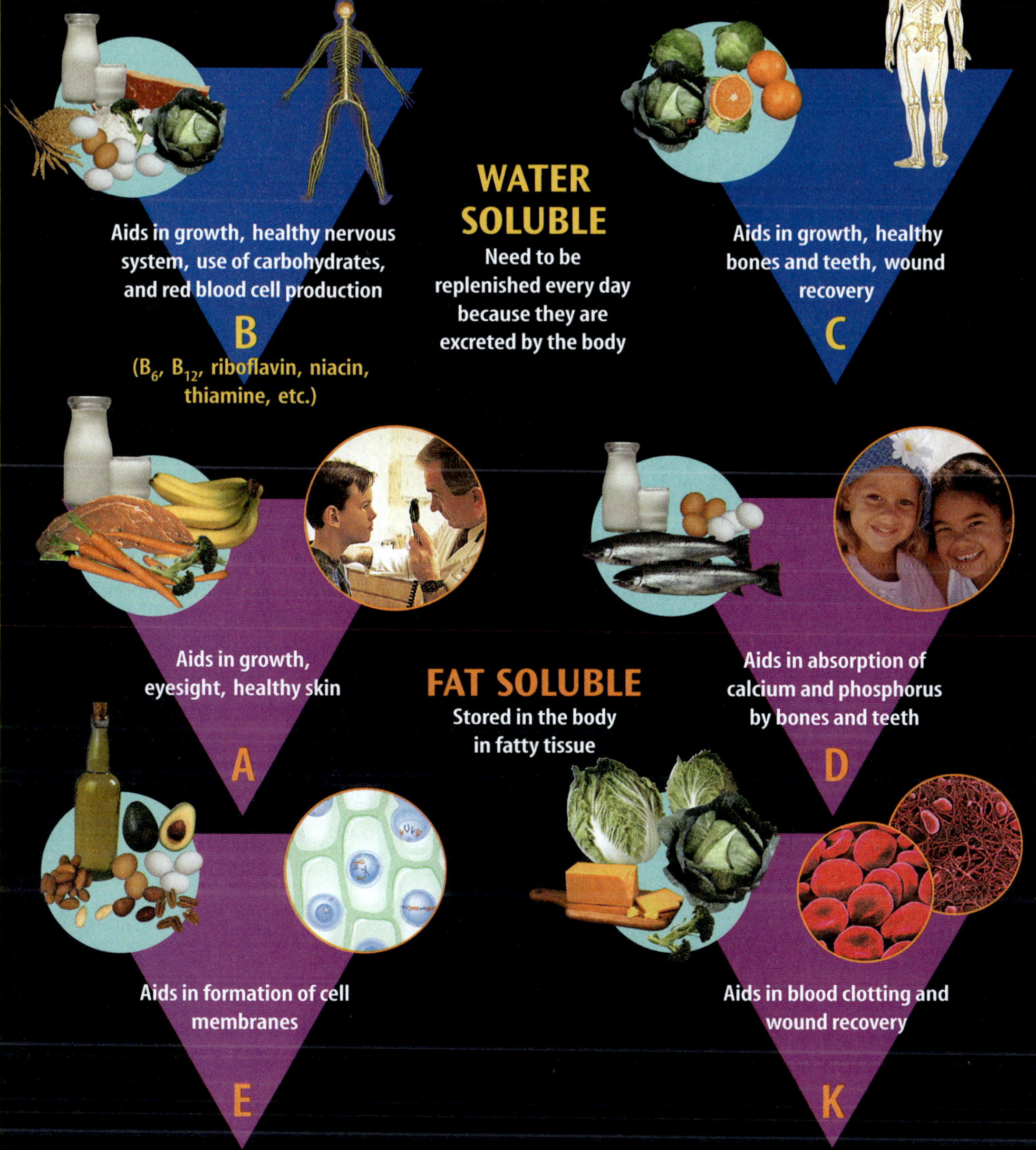

NATIONAL GEOGRAPHIC

Visualizing Vitamins

Have students examine the pictures and read the captions. Then ask the following questions.

Why is it important to eat foods containing vitamin K? A deficiency can lead to excessive bleeding when you are injured and it may take longer for the wound to heal.

What role does vitamin D have in healthy bones? It aids in the absorption of calcium, which is a necessary component of bone tissue.

Activity

Vitamin Deficiency Have students define the following terms: *beriberi*, *rickets*, *scurvy*, *pellagra*, and *night blindness*. Be sure to have students include which vitamin is deficient in each condition and how the deficiency affects the body.

Caption Answer

Figure 8 excess water, salts, and other wastes from your blood

Use an Analogy

Transpiration v. Perspiration The loss of water from the body as perspiration is like the transpiration of water from plants. In both animals and plants, excessive water loss can have profoundly negative effects on the health of the organism.

Reading Check

Answer Nutrients can't be used unless carried in water; cells need water to carry out chemical reactions.

Inquiry Lab

Measuring Water Content

Question How much water is in celery?

Possible Materials pan balance, beaker, sliced celery, tray

Estimated Time 20–30 minutes the first day; 5–10 minutes 2 days later

Teaching Strategies

- Students can design a method for measuring water content.
- Students can determine the amount of water by weighing fresh celery, allowing the celery to dry, and then weighing it again after it dries out.
- Students may also design a method to speed up the drying of the celery so that water content may be determined more quickly.
- Students can use their method to measure the water content of other foods.
- Allow students to explore other questions that arise.

For additional inquiry activities, see *Science Inquiry Labs*.

Figure 8 The kidneys filter wastes from the blood.
List *substances removed from the blood by the kidneys.*

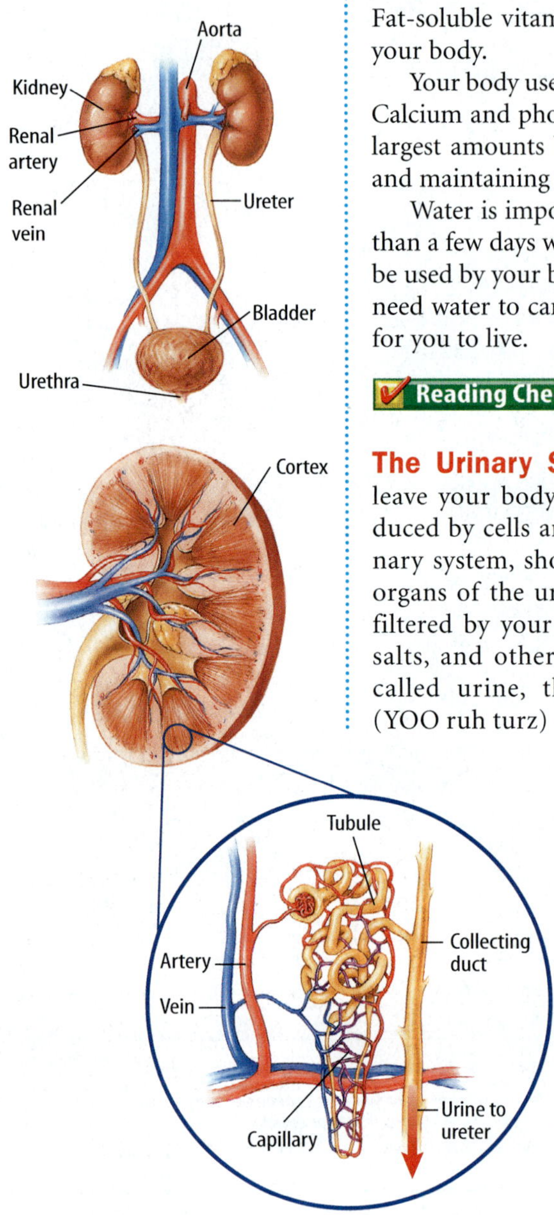

Inorganic Nutrients Vitamins and minerals are nutrients that are needed in small quantities by your body. Your body uses vitamins for many important functions, such as growth, regulating body functions, and preventing some diseases. Most foods have some vitamins, but no one food supplies them all. Vitamins are classified into two groups, as shown in **Figure 7** on the previous page. Vitamins that dissolve easily in water are called water-soluble vitamins. They are not stored in your body. Fat-soluble vitamins dissolve only in fat and can be stored by your body.

Your body uses minerals to control many chemical reactions. Calcium and phosphorus are two minerals that are used in the largest amounts by your body. One of their uses is in making and maintaining bone.

Water is important for your body. You cannot live for more than a few days without water. Most of the other nutrients can't be used by your body unless they are carried in water. Your cells need water to carry out the chemical reactions that are needed for you to live.

Reading Check *Why is water important for your body?*

The Urinary System As you have learned, some wastes leave your body through the digestive system. Wastes produced by cells are removed from the blood through the urinary system, shown in **Figure 8.** Your kidneys are the main organs of the urinary system. All the blood in your body is filtered by your kidneys. The kidneys remove excess water, salts, and other wastes from your blood. The wastewater, called urine, then passes through tubes called ureters (YOO ruh turz) into the bladder. The bladder holds the urine until it is excreted from the body. The average-sized person produces about one liter of urine per day. Urine is carried from the bladder to outside of the body by another tube, the urethra (yoo REE thruh).

Other Organs of the Excretory System In addition to the wastes removed by the digestive and urinary systems, your body removes wastes in other ways. Your respiratory system removes waste gases, such as carbon dioxide and water vapor. Salt and some other wastes are lost through the skin.

Teacher FYI

Filtering System Filtering waste from the circulatory system is essential to maintaining life. The urinary system regulates the concentration and volume of blood by removing and restoring selected amounts of water and chemicals. Any change in the normal functioning of the urinary system presents a health threat.

Differentiated Instruction

Learning Disabled and English-Language Learners Ask students to compare a kidney to a recycling center. Have them consider substances that can and cannot be recycled. Have them make a list or draw pictures to show what they have discovered. kidney—water, salt, and sugar can be recycled, waste cannot; recycling center—paper, glass, and plastic can be recycled, some synthetic materials cannot. L1 ELL LS **Linguistic**

Respiration and Circulation

Besides food and water, you need a constant supply of air. Your body's cells need oxygen, a gas that is found in the air. Your body's cells produce carbon dioxide, a waste gas, that must be removed from the body.

The Respiratory System As shown in **Figure 9,** the **respiratory system,** is made up of structures and organs that help move oxygen into the body and waste gases out of the body. When you breathe in, or inhale, air enters through the mouth or nose and then travels through a series of passageways—the pharynx (FER ingks), larynx (LER ingks), and trachea (TRAY kee uh). Bronchi (BRAHN ki) then carry air into your lungs. In the lungs, bronchi branch into smaller and smaller tubes, somewhat like branches on a tree. The smallest tubes are called bronchioles (BRAHN kee ohlz). Grapelike clusters of air sacs called **alveoli** (al VEE uh li) (singular, *alveolus*) are at the end of each bronchiole. Microscopic blood vessels called **capillaries** (KAP uh ler eez) surround each alveolus. Air is carried through bronchioles and reaches the alveoli.

Oxygen leaves alveoli and enters capillaries. Then oxygen is carried to every cell in your body through the bloodstream. In a similar fashion, your body cells' waste gases, such as carbon dioxide, are carried to your lungs through the blood. Waste gases leave capillaries and enter alveoli. These waste gases are removed from your body when you exhale, as shown in **Figure 10.** Because capillaries have walls that are only one cell layer thick, oxygen and carbon dioxide can move easily between alveoli and capillaries.

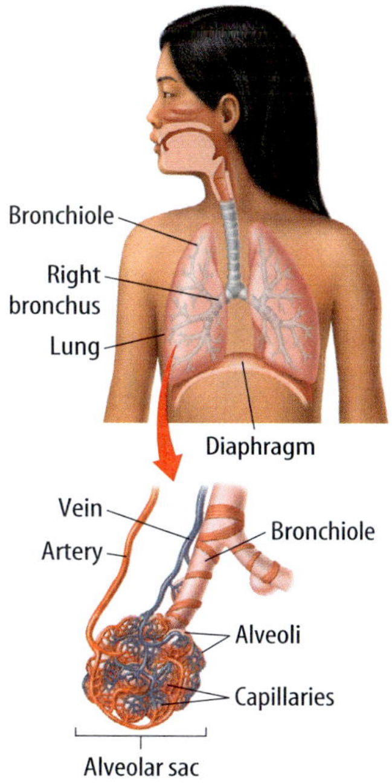

Figure 9 The respiratory system brings air into your body.

Figure 10 Gas exchange takes place in the lungs.

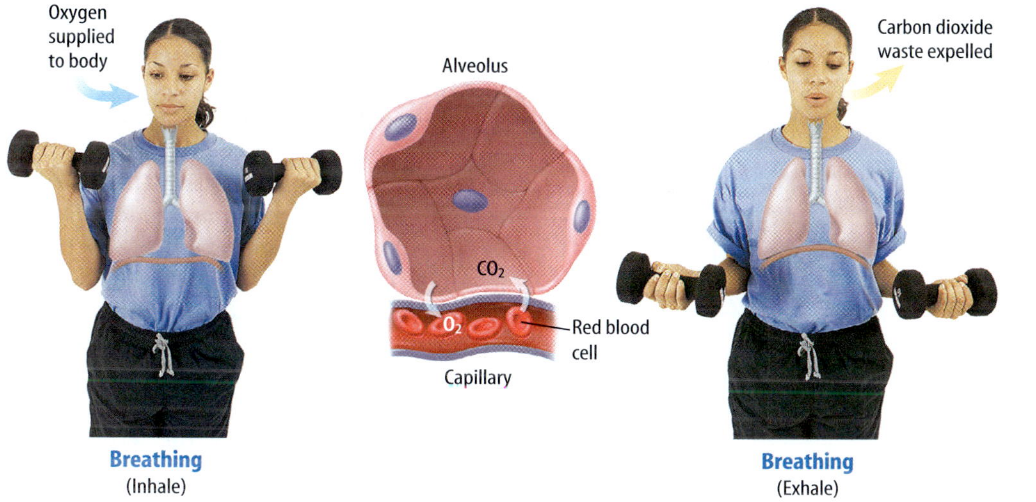

Activity

Water Intake Have students keep track of all the foods and liquids they consume in one day. Have them estimate how much water they took in and compare that number with the 2 L or so of water lost daily. L2 ELL LS **Logical-Mathematical**

Use Science Words

Word Origin *Alvearium* is the Latin word for "beehive." Think about the internal structure of the beehive and relate this to the air cells of the lungs, the alveoli.

Teacher FYI

Hiccups Spasmodic contractions of the diaphragm cause hiccups. The result is a sudden inhaling of air. Choking could result if a reflex action did not automatically cause the epiglottis to close over the trachea and prevent any food from entering it.

Fun Fact

All tissues in the body are nourished by materials from the circulatory system. Every cell is within 0.10 to 0.15 mm of a capillary. Some arteries have such thick walls that they have their own system of capillaries.

Cultural Diversity

Living at High Altitudes People living at high altitudes must deal with a slightly reduced pull of gravity, gas molecules that are spread out in the air, and a reduced pressure gradient. Have students research how the bodies of people native to high altitudes have adapted to these changes. Possible answers: number of red blood cells and amount of hemoglobin increase; increase in number of small blood vessels. L3

Make a Model

Blood Cells Have students make models of red and white blood cells from modeling clay. Encourage students to make the models to scale in order to reinforce concepts about the relative sizes, shapes, and colors of the cells. L2 LS **Visual-Spatial** P

Virtual Labs

Hypertension *What factors affect the likelihood of hypertension?*

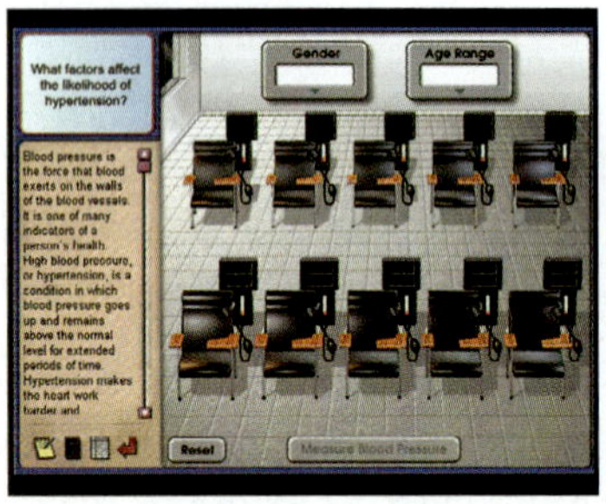

The Circulatory System All cells in your body need nutrients and oxygen and a way to remove wastes. This is the job of your circulatory system. As shown in **Figure 11,** it is made up of your heart, blood vessels, and blood. Your heart is the pump of your circulatory system that pumps blood to all the cells of the body. Then blood moves back to the heart. Blood moves throughout your body in a network of blood vessels.

As blood is pumped out of the heart, it travels through arteries (AR tuh reez), capillaries, and then veins. Arteries are blood vessels that carry blood away from the heart. Arteries have thick elastic walls. Veins carry blood back to the heart. Veins have thinner walls than arteries and one-way valves that prevent the backflow of blood. Capillaries connect arteries and veins. As you read earlier, capillary walls are one cell layer thick. Oxygen and nutrients can move easily from capillaries into the cells. Waste materials and carbon dioxide can leave body cells and enter the capillaries.

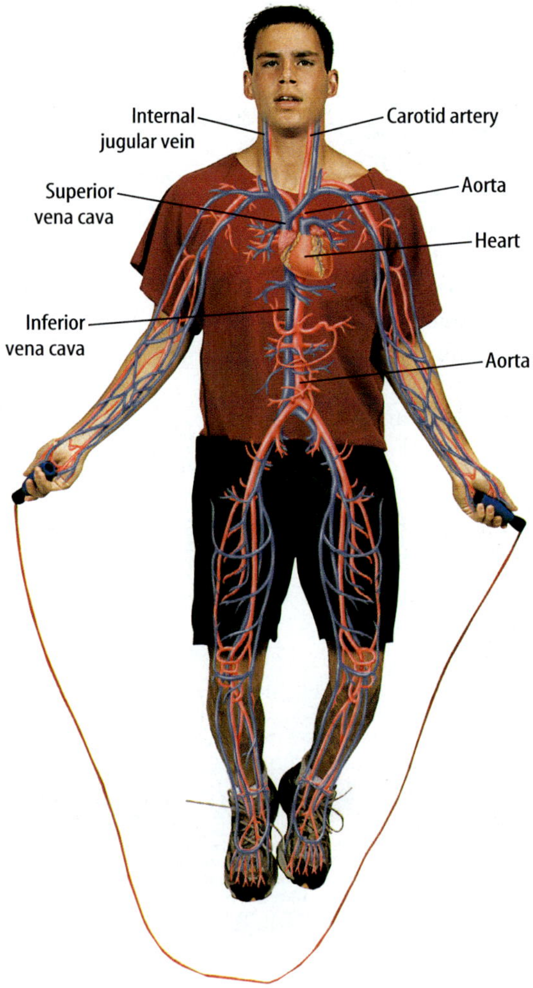

Figure 11 The circulatory system supplies your body with nutrients, oxygen, and a way to remove wastes. The heart pumps blood through blood vessels to all parts of your body.

Blood Without blood, your body's cells would die because they could not get the oxygen and nutrients necessary for life. Blood also removes wastes from your body cells. Your blood is tissue that is made up of liquid and cells. Oxygen is carried by red blood cells. Your blood also has cells, called white blood cells, that fight infections and heal wounds.

When you cut yourself, special chemicals and cell fragments, called platelets (PLAYT luts), in blood form a clot. This clot plugs the wounded blood vessels and acts somewhat like a bandage. Blood clots stop bleeding in minor wounds.

Sometimes, if a person has a serious wound, a large amount of blood can be lost. This person might need a blood transfusion. In a blood transfusion, a person receives donated human blood. Humans can donate blood to one another because of the similarities in human blood. However, the blood must be typed to be sure that the right kind is given. If the wrong type is given, the person may die.

Curriculum Connection

Health Some cultures consider blood a necessary part of their diet. Others use it in the preparation of certain foods (blood sausage). Have students find out what nutrients are supplied by the blood. usually sodium and iron L2

Type	Can Receive	Can Donate To
A	0, A	A, AB
B	0, B	B, AB
AB	all	AB
0	0	all

Table 1 Blood Transfusion Possibilities

Blood Types People can inherit one of four major blood types: A, B, AB, or O. Each blood type is different because of chemical identification tags, or antigens (AN tih junz), on red blood cells. Type A has A antigen, type B has B antigen, type AB has A and B antigens, and type O has no antigens.

Each blood type also has specific antibodies in the liquid part of the blood. Antibodies are proteins that destroy substances that are not part of your body. Because of these antibodies, certain blood types cannot be mixed. Only certain blood types can be given to certain other blood types in blood transfusions, as shown in **Table 1.**

Rh factor is another identification tag of red blood cells. If the Rh factor is present on the red blood cells, the person has Rh-positive (Rh+) blood. If the factor is absent, the person has Rh-negative (Rh−) blood. An Rh− person cannot receive Rh+ blood in a blood transfusion.

Reading Check *What prevents different blood types from being used successfully during blood transfusions?*

The Lymphatic System Between cells, there is tissue fluid. Some water and other substances become part of the tissue fluid. This tissue fluid is collected and returned to the blood through the lymphatic (lihm FA tihk) system. It has a network of vessels like the circulatory system. However, the lymphatic system does not have a heartlike organ that pumps the fluid. The movement of fluid depends on the contraction of muscles in the walls of the lymph vessel and skeletal muscles. Lymphatic vessels, like veins, have valves that keep the fluid from flowing backward.

In addition to water and dissolved substances, the lymphatic vessels also contain cells called lymphocytes. Lymphocytes help defend your body against disease-causing organisms.

Mini LAB

Inferring How Hard the Heart Works

Procedure

1. Make a fist and observe its size, which is approximately the size of your heart.
2. Place your fist in a **bowl of water.** Then clench and unclench your fist to cause water to squirt out between your thumb and forefinger.
3. Continue the squeezing action for 3 min. Determine the number of squeezes per minute.

Analysis

1. How many times did you squeeze your fist in 1 min? A resting heart beats approximately 70 times per minute.
2. What can you do when the muscles of your hand and arm get tired? Explain why cardiac muscle cannot do the same.

Try at Home

Teacher FYI

Determining Blood Type To determine blood type, suspensions of a person's red blood cells are mixed separately with different serums. Each serum has a known, particular antibody. If agglutination occurs, it indicates that the cells contain the antigen with which that specific antibody agglutinates. This reaction identifies the blood type of the individual.

Mini LAB

Purpose Students explore heart rate. L2 ELL LS **Logical-Mathematical**

Materials bowl of water, clock or stopwatch

Teaching Strategy Students must squeeze at a steady rate throughout the entire minute.

Troubleshooting Place newspaper around the bowl to absorb spilled water.

Analysis

1. Answers will vary, but most will find it difficult to maintain the 70 clenches per minute rate.
2. Students can stop or change hands when tired. The heart must beat continuously in order to maintain the body's activities.

Assessment

Process Have students infer the tiring time for cardiac muscles and compare this with the tiring time for skeletal muscles. Use **Performance Assessment in the Science Classroom,** p. 89.

Try at Home

Teacher FYI

Rh Factor The original research regarding the Rh factor was done with blood cells from rhesus monkeys. The antigen found was named using the first two letters of rhesus.

Reading Check

Answer antibodies

Activity

Sweat Acidity Have students test the acidity of their sweat using pH indicator paper. Sweat may range from pH 3.0 to slightly alkaline. L2 LS **Kinesthetic**

Fun Fact

The saliva of cats and dogs contains various substances that can kill bacteria. The licking of wounds aids in the healing process.

Discussion

Broken Skin How can the surface of the skin be broken? Possible answers: falls and accidents may cause cuts, punctures, and scrapes of the skin that can allow the entrance of bacteria.

Applying Science

Answers

1. There would be 54 possible donors. Type B, Rh+ blood can be transfused with Type B, Rh+ (9%); Type B, Rh− (1%); Type O, Rh+ (37%); and Type O, Rh− (7%).
2. Type O, Rh− would be most at risk as this population can only receive Type O, Rh− blood. In an average sample of 100 people, only 7 would match this blood type.

Immunity Your body has many ways to defend itself against disease-causing organisms. First-line defenses are your skin and your respiratory, digestive, and circulatory systems. Most disease-causing organisms cannot get through unbroken skin. Your respiratory system traps disease organisms with mucus and hairlike structures called cilia (SIH lee uh). Saliva, mucus, and chemicals in your digestive system also protect you. Your circulatory system has white blood cells that patrol your body to destroy invading disease-causing organisms.

A second-line of defense, called specific immunity, attacks disease organisms that get past these first-line defenses. In specific immunity, your body makes antibodies that can destroy disease-causing organisms. Recall that antibodies are proteins that destroy substances that are not part of your body. When you get a cold, for example, your body makes antibodies that attack that cold virus. This helps your body to fight off the infection.

You also can develop antibodies to fight off diseases when you receive vaccinations. Vaccinations can prevent diseases. You received vaccinations against many diseases, such as measles, tetanus, mumps, and polio before you started school. Your body formed antibodies against these diseases after you received the vaccinations.

Applying Science

Will there be enough blood donors?

Successful human blood transfusions began during World War II. This practice is much safer today due to extensive testing of the donated blood prior to transfusion. Health care professionals have determined that each blood type can receive certain other blood types as illustrated in **Table 1** on the previous page.

Blood Type Distribution

	Rh+ (%)	Rh− (%)
O	37	7
A	36	6
B	9	1
AB	3	1

Identifying the Problem

The table on the right lists the average distribution of blood types in the United States. The data are recorded as percents, or a sample of 100 people. By examining these data and the data in **Table 1,** can you determine safe donors for each blood type? Recall that people with Rh− blood cannot receive a transfusion from an Rh+ donor.

Solving the Problem

1. If a Type B, Rh+ person needs a blood transfusion, how many possible donors are there?
2. Frequently, the supply of donated blood runs low. Which blood type and Rh factor would be most affected in such a shortage? Explain your answer.

Curriculum Connection

History In 1796, Edward Jenner pioneered the use of vaccines by inoculating a young boy with cowpox material to keep him from contracting the deadly disease smallpox. Editorial cartoons in the newspaper ridiculed the treatment and showed illustrations of people growing cow parts on their bodies as a result of being vaccinated. Have students find out more about Jenner's work. L2

Teacher FYI

Antibiotics Medicines used to destroy or inhibit pathogens in the body are called antibiotics. Antibiotics have been developed to combat bacterial infections. They do not destroy viruses. Some medicines have been developed to relieve symptoms of colds and influenza, but there are no antibiotics for viral infections.

Control and Coordination

"10, 9, 8, 7, 6, 5, 4, 3, 2, 1…Blast off! We have lift off!" The NASA Mission Control Center monitors a space shuttle mission from takeoff to landing. Like the Mission Control Center for the space program, your body needs control systems to make all your body systems work together. Your nervous system and endocrine system are the control systems of the body. They coordinate your body functions.

The Nervous System Your brain, spinal cord, nerves, and nerve receptors make up your nervous system. The nervous system sends messages to and from your brain to all parts of your body. These messages are carried by nerves. The basic unit of the nervous system is the neuron (NOO rahn), or nerve cell, shown in **Figure 12.** The neuron has a cell body and branches called dendrites and axons (AK sahns). Messages travel from one neuron to another. Dendrites receive messages from other neurons and send them to the cell body. Axons carry messages away from the cell body.

If the brain sends a message to one of your leg muscles, for example, the message travels from one neuron to another until it reaches the muscle. Then, the muscle can respond by contracting. In a similar fashion, sensory information, such as the prick from a pin, can move from a skin nerve receptor to a neuron, then from one neuron to the next until the message reaches the brain. The brain coordinates all the activities of your body.

Have you ever coughed or touched something very hot and pulled your hand back quickly? If you have, you have experienced a reflex. A **reflex** is an involuntary, automatic response to a stimulus. You can't control reflexes. Reflexes help protect your body by allowing your body to respond without you having to think about what to do.

Science Online

Topic: Nervous System
Visit red.msscience.com for Web links to information about the nervous system.

Activity Make a brochure outlining recent medical advances.

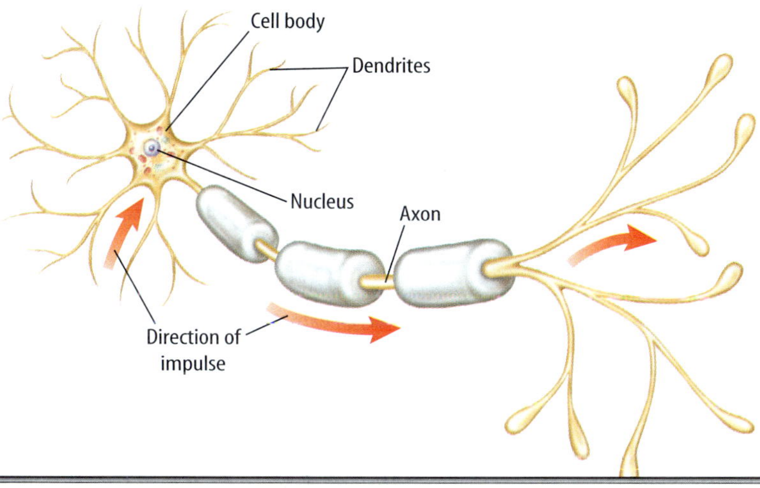

Figure 12 Messages travel from one neuron to another.
Infer *what kind of messages neurons send to each other.*

Quick Demo

Travel Distance of Nerve Impulse

Materials measuring tape or other method of measuring 100 meters

Estimated Time five minutes

Procedure On an athletic field or the schoolyard, position two students 100 m apart. Use this visual representation to illustrate the distance that some nerve impulses travel in one second.

Make a Model

Neuron Have students make a model of a neuron. Students can use yarn and markers on construction paper to model and label the different parts of a neuron including the dendrites, the cell body, and the axon. Students can also show the direction of impulse transmission. Dendrites receive an impulse from another neuron and pass it on to the cell body. The axon carries the impulse to the next neuron.

Teacher FYI

Hormones In addition to hormones produced by endocrine glands, other organs also produce hormones. Endocrine cells in the stomach produce a hormone that regulates the release of its digestive enzymes. The kidneys release a hormone that controls the rate of production of red blood cells. The heart makes a hormone that regulates blood pressure and volume.

Caption Answer

Figure 12 electrical messages

Fun Fact

The structure of an axon of a neuron can change when used. More and better connections with other neurons result from learning. Schoolwork helps neurons make connections.

Differentiated Instruction

Challenge Have interested students interview a pediatrician to find out which motor reflexes a child is born with. A partial list includes gag reflex, blink reflex, pain avoidance reflex, shivering in response to cold, sucking reflex, and swallowing reflex. L3

Use Science Words

Word Origin The word *adrenal* is derived from two Latin words: *ad*, meaning "to" and *renalis*, meaning "kidneys." Ask students to suggest where the adrenal glands are located. on top of the kidneys

3 Assess

DAILY INTERVENTION

Check for Understanding

Visual-Spatial Use an anatomy chart or model to illustrate the various body systems involved in removing wastes from the body.

Reteach

Systems of Defense Use an anatomy chart or model to illustrate the various systems involved in defense against microbes. L2 LS **Visual-Spatial**

Assessment

Oral Have students suggest what might happen to digestion and nutrition in people who have lost their teeth. Use **Performance Assessment in the Science Classroom,** p. 93.

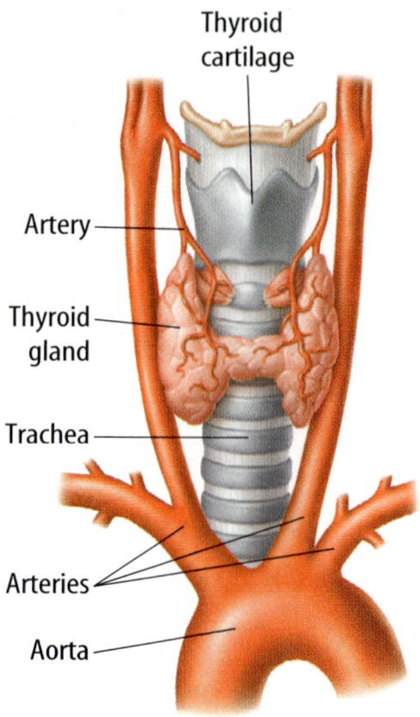

Figure 13 The thyroid gland secretes a hormone that helps increase the rate of chemical reactions in your body.

The Endocrine System The other control system in your body is the endocrine (EN duh krun) system. In the nervous system, messages travel quickly through nerves to and from all parts of your body. In the endocrine system, chemicals called hormones (HOR mohnz) carry messages throughout the body. Your body doesn't respond as quickly to messages from the endocrine system.

Saliva travels through a small tube called a duct, from the salivary gland to your mouth. Endocrine glands do not have ducts. Hormones are released by endocrine glands directly into your bloodstream, as shown in **Figure 13.** These hormones travel through the blood to reach target tissues.

Some endocrine glands are found in the brain—the pineal and pituitary glands. The pineal gland produces a hormone that controls your pattern of sleeping and waking. The pituitary gland makes several hormones that regulate many body activities including growth and reproduction. Other endocrine glands include the pancreas and adrenal glands, which are found in the abdomen. The pancreas makes a hormone that controls the amount of sugar that is present in your bloodstream. The adrenal glands make several hormones, including hormones that help your body respond in times of physical or emotional stress.

section 1 review

Summary

Structure and Movement

- The skeletal system gives shape and support to your body.
- Skin is the largest organ in your body.
- Your body has both voluntary and involuntary muscles.

Digestion and Excretion

- Your digestive system breaks down food into smaller molecules.
- The urinary system removes wastes from your blood.

Respiration and Circulation

- Your body's cells need oxygen.
- Your heart pumps blood through vessels.

Control and Coordination

- Reflexes help protect your body.
- Hormones are chemical messengers.

Self Check

1. **Describe** how the skeletal and muscular systems provide structure and allow movement.
2. **Explain** why a person may not be able to donate blood to another person.
3. **Identify** the blood type—A, B, AB, or O—that can donate to all other blood types.
4. **Infer** how the digestive, circulatory, and respiratory systems help meet the needs of your body.
5. **Compare and contrast** the nervous and endocrine systems.
6. **Think Critically** What might happen to your body if it could not make antibodies?

Applying Skills

7. **Concept Map** Make a network-tree concept map to show how your body defends you against disease-causing organisms. Begin with the words *Lines of Defense.*

Science Online red.msscience.com/self_check_quiz

section 1 review

1. Skeletal system gives shape and support to body. Muscles attach to the framework of bones and help them move.
2. because blood types differ
3. Type O blood
4. digestive system—supplies the body with the energy it needs to function; takes in food, breaks it down into smaller molecules that can be absorbed into the blood, and excretes the remaining wastes; circulatory system—carries nutrients and oxygen to the body's cells and removes cellular wastes; respiratory system—allows for exchange of oxygen and waste gases
5. Both control and coordinate body functions. In the nervous system, messages travel through nerves to and from all parts of the body. In the endocrine system, chemicals, hormones, carry messages. The body does not respond as quickly to these messages.
6. You might not be able to fight off diseases.
7. Check student concept maps for accuracy.

LAB

BENCH TESTED

Improving REACTION Time

Your reflexes allow you to react quickly without thinking. Sometimes you can improve how quickly you react.

Real-World Question

How can reaction time be improved?

Goals

- **Observe** reflexes.
- **Identify** stimuli and responses.

Materials

metric ruler

Procedure

1. Make a data table in your Science Journal to record where the ruler is caught during this lab. Possible column heads are *Trial, Right Hand,* and *Left Hand.*
2. Have a partner hold the ruler as shown.
3. Hold the thumb and index finger of your right hand apart at the bottom of the ruler. Do not touch the ruler.
4. Your partner must let go of the ruler without warning you.
5. Catch the ruler between your thumb and finger by quickly bringing them together.
6. Repeat this lab several times and record in a data table where the ruler was caught.
7. Repeat this lab with your left hand.
8. Use the table on the right to determine your reaction time.

Analyze Your Data

1. **Identify** the stimulus, response, and variable in this lab.
2. **Determine** the average reaction time for your right hand and for your left hand.

Conclude and Apply

1. **Compare** the response of your writing hand and your other hand for this lab.
2. **Draw a conclusion** about how practice relates to stimulus-response time.

Communicating Your Data

Compare your conclusions with those of other students in your class. **For more help, refer to the Science Skill Handbook.**

Real-World Question

Purpose Students observe reaction time. L2 COOP LEARN

Process Skills observe, communicate, use numbers, recognize and use spatial relationships, measure in SI, interpret data

Time Required 30 minutes

Procedure

Teaching Strategy Students should keep their eyes on the ruler as their partners release it.

Analyze Your Data

1. stimulus; the ruler falling; response: catching the ruler; variable: which hand is used
2. To find the average, students should add all reaction times for each hand separately and divide by the number of trials.

Conclude and Apply

1. Answers will vary. Generally, the writing hand reacts faster.
2. With practice, stimulus-response time usually improves.

Assessment

Performance To further assess students' abilities to measure and improve reaction time, have them repeat the experiment with slight variations. Have students blindfolded, and instruct their partners to tell them when they drop the ruler. See how quickly they can react to verbal stimuli. Use **Performance Assessment in the Science Classroom,** p. 97.

Communicating Your Data

Students should discuss why their reaction times were or were not the same.

1 Motivate

Bellringer

INTERACTIVE CHALKBOARD PowerPoint® Presentations

Section Focus Transparencies also are available on the Interactive Chalkboard CD-ROM. L2 ELL

Section 2 Section Focus Transparency — Sprint to Home

Adult sea turtles lay about one hundred eggs in a nest on land, then promptly return to the water. After the eggs hatch, the young sea turtles race toward the sea in a struggle for survival on their own. In contrast, other species produce fewer eggs but give more care and protection to the young once they are born.

1. Compared to sea turtles, how many offspring do people have?
2. How does the amount of care young sea turtles receive differ from that of humans?

The Human Body

Tie to Prior Knowledge

Egg and Sperm Ask students to think of common animals that begin life by the union of an egg and sperm. Examples may include humans, elephants, cats, dogs, horses, and toads.

section 2

Human Reproduction

as you read

What You'll Learn

- **Identify** the organs of the male and female reproductive systems.
- **List** the stages in the menstrual cycle.
- **Describe** the stages of development before birth.
- **Sequence** the life stages of humans.

Why It's Important

Human reproduction is necessary to ensure that human life continues on Earth.

Review Vocabulary
hormone: chemical produced by the endocrine system; released directly into the bloodstream

New Vocabulary
- sperm
- semen
- ovulation
- menstrual cycle
- pregnancy
- embryo
- fetus

Male Reproductive System

Unlike other human body systems, reproductive systems are different in males and females. They are made up of different organs, have different roles in reproduction, and have different sex hormones. These hormones are needed for the development of sexual characteristics.

The male reproductive system is made up of several organs and structures, as shown in **Figure 14.** The scrotum contains two testes (TES teez) that produce the male hormone testosterone and **sperm,** the male reproductive cells. Testosterone and sperm are not produced until the male begins to mature sexually.

Each sperm has a head and a tail. The head contains genetic information in the nucleus and the tail moves the sperm. After sperm are produced in the testes, they travel through the sperm ducts. Fluid from the seminal vesicles, organs behind the bladder, is mixed with the sperm. This mixture of fluid and sperm is called **semen** (SEE mun). Semen leaves the body through the urethra—the same tube that carries urine from the body. However, urine and semen never mix. A muscle at the back of the bladder does not allow urine to enter the urethra when sperm leave the body.

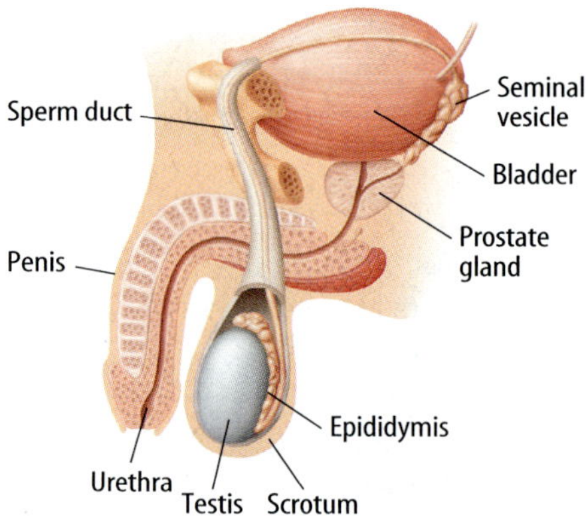

Figure 14 Some structures of the male reproductive system extend outside the body.

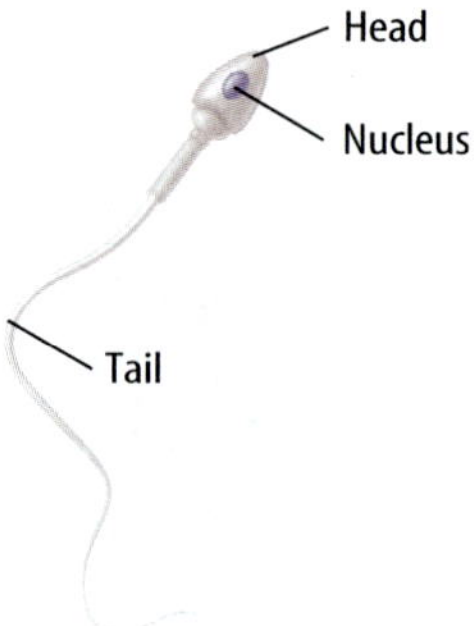

Sperm are produced in the testes.

Section 2 Resource Manager

Chapter FAST FILE Resources
- Transparency Activity, p. 45
- MiniLAB, p. 4
- Enrichment, p. 30
- Reinforcement, p. 28
- Directed Reading for Content Mastery, p. 21
- Lab Worksheet, pp. 7–8
- Lab Activity, pp. 13–16

Earth Science Critical Thinking/Problem Solving, p. 19
Laboratory Management and Safety, p. 43
Mathematics Skill Activities, pp. 9–10
Performance Assessment, pp. 57–58
Reading and Writing Activities, pp. 51–52

574 CHAPTER 19 The Human Body

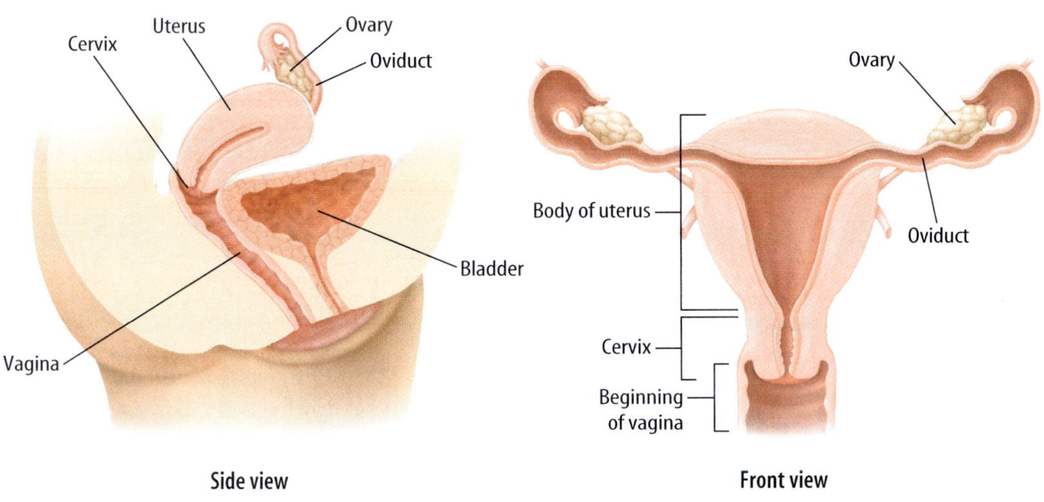

Figure 15 These are diagrams of the female reproductive system.
Identify *where a fertilized egg grows and develops.*

Female Reproductive System

Most of the organs of the female reproductive system, as shown in **Figure 15,** are inside of a female's body and are called internal. Ovaries are the female sex organs that produce eggs and the female sex hormones, estrogen (ES truh jun) and progesterone (proh JES tuh rohn), when a female matures sexually. These hormones help prepare the female body for having a baby.

The ovaries contain certain cells that eventually develop into eggs. When a female is born, her ovaries already contain all of the cells that can develop into eggs. The female sex hormones control the development and release of eggs from the ovaries. Every 28 days on average, an egg is released from one of the ovaries. This process is called **ovulation** (ahv yuh LAY shun). The ovaries usually release eggs on alternating cycles.

Reading Check *What controls the development and release of eggs?*

A discharged egg moves into the oviduct, which is lined with hairlike cilia. The coordinated wavelike movement of the cilia sweeps the egg along the oviduct to the uterus, a muscular, pear-shaped hollow organ. If the egg is fertilized by a sperm while in the oviduct, the fertilized egg can grow and develop in the uterus. The lower end of the uterus is called the cervix and connects the uterus to the vagina. The vagina is a muscular tube and is known as the birth canal. During birth, a baby moves from the uterus through the vagina to outside the mother's body.

2 Teach

Caption Answer

Figure 15 uterus

Discussion

Semen What other functions does semen have in addition to carrying the sperm? Chemical substances in semen provide energy for sperm movement. Explain that because semen is slightly alkaline, it protects sperm from acidic conditions, such as those found in the female reproductive system.

Reading Check

Answer sex hormones

Identifying Misconceptions

Estrogen Ovaries produce the hormone estrogen, which controls the development of female secondary sex characteristics. Many people think males do not produce this hormone. However, small amounts of estrogen are produced by adrenal and certain testicular cells in males.

Visual Learning

Figure 15 Have students trace the movement of an egg after ovulation.

Curriculum Connection

Language Arts Have students look up the words *sperm* and *ovary* and find out why the Latin roots are appropriate to their meanings. The Late Latin term *sperma* means seed; the Latin term *ovum* means egg. L1 LS **Linguistic**

Teacher FYI

Reproductive Cell Production Sperm production is a continuous process from puberty throughout the life of a male. Egg production in females begins before birth. There is no production after birth. Each ovary can have nearly a million primary oocytes—cells from which eggs form.

Quick Demo

Egg Size

Materials pencil, paper

Estimated Time five minutes

Procedure Use a sharp-tipped pencil to make a tiny dot on a piece of paper. This dot is about 0.5 mm in diameter. The human egg is about 5 times smaller in size. After demonstrating this, have students do it so they can get an idea of the size of a human egg.

Teacher FYI

Menstrual Period The length of the menstrual period can vary. During the week before menstruation, a woman may experience premenstrual syndrome (PMS). Symptoms include depression, fatigue, headache, irritability, nervousness, and lack of concentration. Diet, exercise, aspirin, or other medicines are used to lessen the discomfort.

Use Science Words

Word Origin The Greek word for egg is *oion*. It is the root for many words related to eggs or the shape of an egg. Have students look up words with this root and explain their meanings. Possible answers: oocyte—an egg before maturation; oology—the study of bird eggs; oogenesis—the formation and maturation of an egg. L2 LS **Linguistic**

Science Online

Topic: Menstrual Cycle

Visit red.msscience.com for Web links to information about the monthly changes in the female reproductive system.

Activity Make a table of the changes organized by day, starting with day 1 and ending with day 28.

The Menstrual Cycle The monthly cycle of changes in a sexually mature female reproductive system is the **menstrual cycle.** The average length of a menstrual cycle is 28 days. However, the cycle can vary in length from 20 to 40 days.

Each month, the female undergoes changes that help prepare her body for having a baby. These changes include maturing of the egg, producing female sex hormones, and preparing the uterus for the fertilized egg. Hormones from the pituitary gland start the development of an egg and production of sex hormones in the ovary. The menstrual cycle is divided into three phases, as shown in **Figure 16.**

Phase one is the phase of menstrual flow, or menstruation (men STRAY shun). Menstrual flow is made of tissue cells from the thickened lining of the uterus and blood. This flow usually lasts from four to six days.

In phase two, an egg develops in the ovary and the lining of the uterus thickens. Ovulation occurs about 14 days before menstrual flow begins.

The last phase, phase three, is the phase between ovulation and menstruation. The lining of the uterus continues to thicken. If the egg is fertilized, it can attach to the wall of the uterus and begin to develop while hormones continue to be produced. If the egg is not fertilized, the hormone levels decrease, the lining of the uterus breaks down, and menstruation begins.

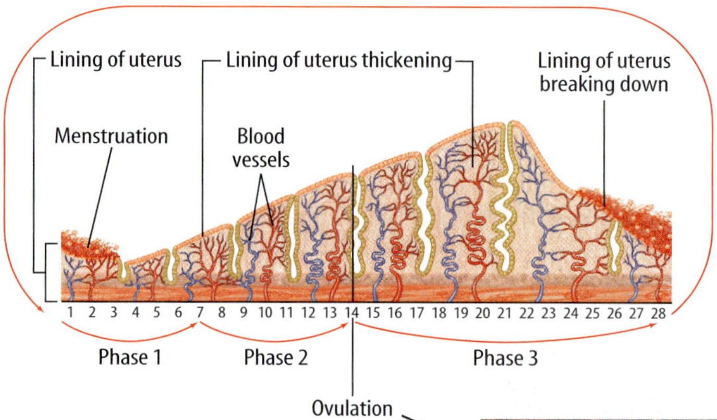

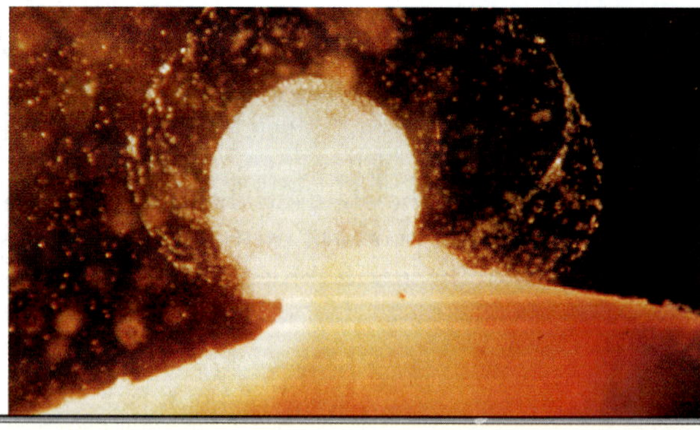

Figure 16 The menstrual cycle has three phases. Phase one—menstrual flow; phase two—development and release of egg and thickening of lining of the uterus; phase three—breakdown of lining if egg is not fertilized.

Differentiated Instruction

Challenge Have students investigate and write a short paragraph about the chances of conception. Fertilization can occur only during a two- to three-day time period, when the cervical mucus forms small channels that allow sperm to enter the uterus. Once in the mucus, the sperm can live up to 48 hours. A sexually-active, 30-year-old female has a 20 percent chance each month of conceiving. L3 LS **Logical-Mathematical**

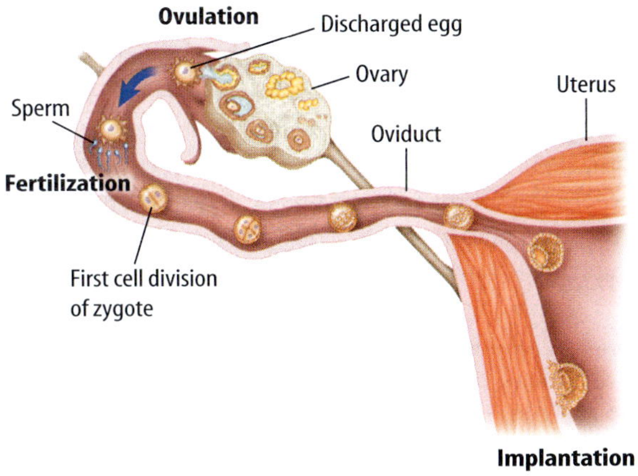

Figure 17 Human development begins when an egg is fertilized by a sperm.

By two months, the developing embryo is about 2.5 cm in size.

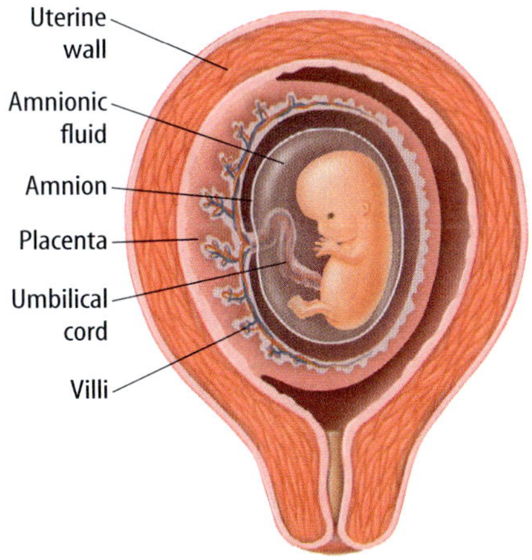

Life Stages

Human development begins when an egg from the female is united with a sperm from the male. This process is called fertilization. Fertilization usually takes place in the oviduct, as shown in **Figure 17.**

The nucleus of the sperm and the nucleus of the egg join together creating a fertilized cell called the zygote, as shown above. This cell undergoes many cell divisions as it moves along the oviduct to the uterus. If the zygote successfully attaches to the wall of the uterus, the zygote will develop into a baby in about nine months. This period of development from fertilized egg to birth is called **pregnancy.**

Development Before Birth After attachment to the wall of the uterus, the zygote is called an **embryo.** During the embryo period, which is the first two months of pregnancy, the major organs develop and the heart begins to beat. At first, the embryo receives nutrients from the fluids in the uterus. Then, a placenta (pluh SEN tuh) develops from tissues of the uterus and tissues of the embryo. The umbilical cord connects the embryo to the placenta. The placenta carries nutrients and oxygen from the mother and removes wastes from the embryo.

Reading Check *What is the first way that the embryo receives nutrients?*

After the first two months of pregnancy, the developing embryo is called a **fetus.** At three months, the fetus measures 8 cm to 9 cm in length and the mother may feel the fetus move. The fetus continues to grow and develop during the pregnancy. At nine months, the fetus weighs about 2.5 kg to 3.5 kg and measures about 50 cm. By this time, the fetus is usually in a head-down position in the uterus.

INTEGRATE Physics

Fetal Ultrasonography Ultrasonic waves are sound waves that are beyond the upper limit of human hearing. In medicine, these sound waves of frequencies of 1 million Hz to 10 million Hz are used to "see" structures inside the human body. Why would ultrasonography be useful during pregnancy? Research to find this answer and then record it in your Science Journal.

Activity

Unusual Breeding Processes Have students research the unusual breeding processes of such animals as the seahorse and midwife toad. In both, the fertilized eggs are carried by the male until the young are ready to be born. L2

INTEGRATE Physics

Fetal Ultrasonography A diagnostic tool that is commonly used during human pregnancy is called ultrasonography. Its uses include determining the age of the fetus, determining the sex of the fetus, evaluating the health of the fetus, and looking for specific fetal anomalies.

Research Have students research to learn about other medical uses of ultrasonography. Create a classroom display of their findings.

Reading Check

Answer from fluids in the uterus

Activity

Fetal Size Have students prepare a series of actual-size cutouts of the fetus from three months to birth to develop a visual demonstration of a fetus' size while developing. L2 ELL

LS **Visual-Spatial**

Teacher FYI

Abdominal Pregnancy Rarely, an egg from the ovary does not enter the opening of the oviduct but enters the abdominal cavity and becomes fertilized. Abdominal pregnancies rarely are carried to term. If so, the baby is delivered by a C-section. Usually the fetus does not survive and must be surgically removed.

Curriculum Connection

Health Have students research ectopic pregnancies. How frequently do they occur? 1 in every 250–300 pregnancies How does the female know there is a problem? She experiences pain and bleeding. What is the treatment? Ectopic pregnancies often end in a miscarriage; if not miscarried, surgical removal of the fetus is sometimes necessary; abdominal pregnancies can be carried to term and the baby delivered by C-section. L2

Mini LAB

Interpreting Infant Development

Procedure

Make a bar graph of the following data.

Infant Development

Age (in months)	Skill
5 to 7	Sits with support
5 to 8	Gets on hands and knees; stands with support
8 to 9	Sits alone
8 to 10	Crawls
9 to 12	Pulls to standing
9 to 13	Walks around furniture
9 to 14	Stands with no support
10 to 15	Walks

Analysis

1. Name two skills an infant develops before walking.
2. List four skills that may develop during the same time period.

Try at Home

Birth The process of childbirth begins with labor, the muscular contractions of the uterus. As these contractions increase in strength and frequency, the opening to the uterus widens, and the baby is pushed out through the vagina.

After birth, the umbilical cord is clamped and cut. The scar that forms where the umbilical cord was attached to the body is called the navel. The placenta also is pushed out of the mother's body after the baby is born by the contractions of the uterus.

Stages After Birth Just as there were stages of development before birth, humans have stages of development after birth. These stages are infancy, childhood, adolescence, and adulthood, as shown in **Figure 18.**

Infancy is the time from birth to around 18 months of age. During this time, the infant must adjust to a new environment. The infant's nervous and muscular systems develop rapidly and the infant interacts with the world.

Childhood lasts from 18 months of age until around 12 years of age. During childhood, growth is rapid but not as rapid as it was during infancy. The child learns many new skills including control of the bladder and bowels between the ages of two and three, and dressing and undressing by about age four. Children also develop their skills in speaking, reading, writing, and reasoning. Each child develops at a different rate.

Figure 18 After birth, human development has four stages—infancy, childhood, adolescence, and adulthood.

Mini LAB

Purpose Students graph and observe data on infant development.

Materials graph paper

Teaching Strategy Have students note the range of the data before constructing their graphs.

Analysis

1. (1) sits with support and (2) get on hands and knees; stand with support
2. Several answers are possible: (1) crawl, pull to standing, walk around furniture, stand with no support; (2) pull to standing, walk around furniture, stand with no support, walk.

Assessment

Performance To further assess students' understanding of infant development, have them conduct a survey of parents with young children. Have them ask questions such as: What changes did you notice at different ages? What stage seemed to have the most changes? Use **Performance Assessment in the Science Classroom,** p. 91.

Make a Model

Time Line of Life Stages Have students make a time line of the stages from infancy to adulthood using long pieces of paper. Identify various milestones such as the start of walking, puberty, and menopause. L2 LS **Kinesthetic**

Fun Fact

The major constituents of human milk are water, proteins, fats, and lactose, a carbohydrate. In addition, there are small amounts of vitamins and electrolytes.

Differentiated Instruction

Behaviorally Disordered Have students bring in photos of themselves at various stages of development and describe favorite activities they remember from these stages. L1

Science Journal

Immunizations Babies and young children need certain vaccines to prevent diseases. Have students find out what vaccinations are recommended between birth and age six. Polio; Hepatitis B; Diphtheria, Tetanus, Pertussis; Varicella or Chicken Pox; Haemophilus influenza B; Measles, Mumps, Rubella

578 CHAPTER 19 The Human Body

Adolescence begins around age 12 to 13 years. Puberty occurs during adolescence. During puberty the human body matures sexually. This means that the person becomes physically able to reproduce. In girls, puberty usually begins between ages nine and 13. Puberty in boys occurs between ages 13 and 16. Some of the changes in girls include breast development and the growth of pubic and underarm hair. In boys, the voice deepens, muscles increase in size, and facial, pubic, and underarm hair grow. During adolescence, there is usually a final spurt of growth as well, as shown in **Figure 19.**

Figure 19 During adolescence, humans have their final growth spurt. For girls, this final growth phase begins at about 11 years and ends around age 16. For boys, the growth spurt begins around age 13 and ends around age 18.

Adulthood is the final stage of human development. It begins at the end of adolescence and continues through the rest of the human's life. During this time, the muscular and skeletal systems stop growing. The average human life span—from birth to death—is about 75 years. But some people live much longer. As body systems age, however, they break down, eventually resulting in death.

section 2 review

Summary

Male Reproductive System

- The male reproductive system has both external and internal organs.
- Sperm travel through the sperm duct and are released from the male through the urethra.

Female Reproductive System

- Most of the female reproductive organs are internal.
- The menstrual cycle is a cycle of monthly changes in the female reproductive system.

Life Stages

- Human development before birth occurs in three stages—zygote, embryo, and fetus.
- Development after birth occurs in four stages—infancy, childhood, adolescence, and adulthood.

Self Check

1. **Compare and contrast** the major organs and structures of the male reproductive system and female reproductive system.
2. **Identify** the organ of the female reproductive system that produces the female reproductive cell.
3. **Think Critically** Why does a placenta develop to provide nutrients to the developing embryo?

Applying Skills

4. **Concept Map** Make an events-chain concept map to sequence the stages of the menstrual cycle.
5. **Use a Spreadsheet** Make a spreadsheet for the stages of development both before and after birth. Title the columns with the stage name. Include information about each stage such as major skills developed and changes in body systems.

Teacher FYI

Puberty Onset The onset of puberty varies greatly with growth patterns of adolescents. The variations are caused by a number of factors. Heredity is a key factor in determining the development of the body. Other factors are diet and illnesses. Poor diet and sickness can delay the onset of puberty. Often adolescents are disturbed by the fact that they are developing earlier or later than their peers. By age 17 or 18, the differences have usually disappeared.

3 Assess

DAILY INTERVENTION

Check for Understanding

Visual-Spatial Assign students to groups and have each group write descriptions of a particular life stage on index cards—one per card. Mix up the cards, divide the class into teams, and read the descriptions. Students take turns stating the life stage. The team with the greatest number of correct answers wins. L2 ELL

Reteach

Path of Egg Have students diagram the pathway an egg follows as it moves from the ovary to the uterus. L1 LS **Visual-Spatial**

Assessment

Performance To further assess students' understanding of neonatal development, have them construct a time line to correspond with the stages of neonatal development. Use **Performance Assessment in the Science Classroom**, p. 161. P

section 2 review

1. Male organs are internal and external. Most female organs are internal. The male and female organs are different in structure and have different roles in reproduction. The male testes produce sperm. The female ovary produces eggs. Both males and females have a tube structure that carries the reproductive cells from the organ in which they were produced.
2. ovary
3. The fluids of the uterus could not provide enough nutrients for the growing embryo. There would not be a way to remove wastes or deliver oxygen.
4. Map should look similar to **Figure 18.**
5. Spreadsheet should reflect all information presented about development before and after birth. The column names should be zygote, embryo, fetus, infancy, childhood, adolescence, adulthood.

SECTION 2 Human Reproduction 579

BENCH TESTED

Real-World Question

Purpose Students determine how bicarbonates in saliva protect their mouths from bacteria and chemical substances. L2 COOP LEARN LS **Kinesthetic**

Process Skills observe, compare, identify and manipulate controls and variables, explain, predict

Time Required two 45-minute periods

Materials

- Prepare red cabbage juice by finely chopping the cabbage and boiling it for 20 to 30 minutes. Filter the juice before using it as an indicator.
- Dilute HCl can be used in place of vinegar or lemon juice. Prepare dilute acid by dissolving 80 mL of concentrated HCl in 1 L of water.

Safety Precautions Students should wear safety goggles. Caution students not to eat or drink anything in science class.

Form a Hypothesis

Possible Hypothesis Many students will hypothesize that bicarbonates will neutralize acids.

Test Your Hypothesis

Possible Procedure Add four parts acid or base to one part red cabbage juice indicator. Add a spoonful of bicarbonate a little at a time until a color change is observed. Proportions of the ingredients may have to be adjusted.

LAB Design Your Own

Defensive Saliva

Goals

- **Design** an experiment to test the reaction of a bicarbonate to acids and bases.
- **Test** the reaction of a bicarbonate to acids and bases.

Possible Materials

head of red cabbage
cooking pot
coffee filter
drinking glasses
clear household ammonia
bicarbonate of soda
water
spoon
white vinegar
lemon juice
orange juice

Safety Precautions

WARNING: *Never eat or drink anything used in an investigation. Ammonia fumes are irritating to the eyes and nose.*

Real-World Question

What happens when you think about a juicy cheeseburger or smell freshly baked bread? Your mouth starts making saliva. Saliva is the first line of defense for fighting harmful bacteria, acids, and bases entering your body. Saliva contains salts and chemicals known as bicarbonates. An example of a bicarbonate found in your kitchen is baking soda. Bicarbonates help to maintain normal pH levels in your mouth. When surfaces in your mouth have normal pH levels, the growth of bacteria is slowed and the effects of acids and bases are reduced. How do the bicarbonates in saliva work to protect your mouth from harmful bacteria, acids, and bases?

Form a Hypothesis

Based on your reading in the text, form a hypothesis to explain how the bicarbonates in saliva react to acids and bases.

Test Your Hypothesis

Make a Plan

1. List the materials you will need for your experiment. Red cabbage juice can be used as an indicator to test for acids and bases. Vinegar and citrus juices are acids, ammonia is a base, and baking soda (bicarbonate of soda) is a bicarbonate.

Alternative Inquiry Lab

Body Defenses To make this Lab an Inquiry Lab, students may investigate another first-line body defense such as the skin. Students can predict and then observe whether microorganisms are present on skin after washing. Students can look for microbial growth by using sterile agar or gelatin plates. Students can press their clean fingertips onto the surface of the agar plate, tape the plate shut, and place it in the dark. Students can observe the plate for growth daily. Students can then infer why these microorganisms do not invade their bodies. Intact skin keeps them out.

2. Describe how you will prepare the red cabbage juice and how you will use it to test for the presence of acids and bases.
3. Describe how you will test the effect of bicarbonate on acids and bases.
4. List the steps you will take to set up and complete your experiment. Describe exactly what you will do in each step.
5. Prepare a data table in your Science Journal to record your observations.
6. Examine the steps of your experiment to make certain they are in logical order.

Follow Your Plan

1. Ask your teacher to examine the steps of your experiment and data table before you start.
2. Conduct your experiment according to the approved plan.
3. Record your observations in your data table.

Analyze Your Data

1. **Compare and contrast** the color change of the acids and bases in the cabbage juice.
2. **Describe** how well the bicarbonate neutralized the acids and bases.
3. **Identify** any problems you had while setting up and conducting your experiment.

Conclude and Apply

1. **Determine** whether or not your results support your hypothesis.
2. **Explain** why your saliva contains a bicarbonate based on your experiment.
3. **Predict** how quickly bacteria would grow in your glass containing acid compared to another glass containing acid and the bicarbonate.
4. **Infer** how saliva protects your mouth from bacteria.
5. **Predict** what would happen if your saliva were made of only water.

Communicating Your Data

Using what you learned in this experiment, create a poster about the importance of good dental hygiene. Invite a dental hygienist to speak to your class.

Teaching Strategy Most students will be familiar with the terms acid and base and understand that a base is the opposite of an acid.

Expected Outcome The bicarbonate should successfully neutralize both acids and bases as indicated by the appropriate color change.

Analyze Your Data

Answers to Questions

1. Acids turn pink; bases turn green.
2. All the solutions will turn blue signifying both acids and bases are neutralized.
3. Common problems include the preparation of cabbage juice and determining the proportions of the substances.

Error Analysis Have students compare the color changes that occurred during their experiments and explain why differences were observed.

Conclude and Apply

1. Answers will vary.
2. Bicarbonates in saliva neutralize harmful acids and bases.
3. Bacteria would grow more rapidly in acid without bicarbonate added.
4. Bicarbonates maintain neutral pH levels to slow bacterial growth.
5. Acids and bases from foods would easily damage tooth enamel. Harmful bacteria also would grow unchecked increasing tooth decay, illness, and disease.

Assessment

Oral Have students explain how the strength of acids, bases, and neutral substances are compared. A pH value of 1 to 6 indicates acids with 1 being the strongest. pH values of 8 to 14 are bases with 14 being the strongest base. Neutral substances have a value of 7. Use **Performance Assessment in the Science Classroom**, p. 89.

Communicating Your Data

Encourage students to illustrate their posters with drawings. They may wish to make the poster a cartoon in which the main character is a tooth that explains the steps in good oral hygiene.

TIME

Content Background

It is important to know the function of the kidney and to relay this information to students. The kidney washes the blood as a filtering system, ridding it of toxins and balancing the body's water levels by producing urine to get rid of this waste.

Before making discoveries with regard to kidney transplants, Samuel Kountz had to overcome many odds. He was born in extreme poverty in an all-black town in Arkansas in 1930. His grandmother, born a slave, encouraged and inspired him when he tried to get into college. He became the first African-American to be accepted to the University of Arkansas Medical School.

Discussion

Rejected Kidneys What did Kountz discover as the reason for kidneys being rejected after transplant surgeries? Possible answer: The patient's cells attacked and destroyed the small blood vessels of the transplanted kidney, thus, the kidney would die from lack of blood-supplied oxygen.

Historical Significance

Transplant Surgeries The significance of Kountz's work for people who suffer from kidney disease is obvious. Point out to students that Kountz's work also had a significant impact on other transplant surgeries. Because other transplant procedures were associated with similar rejection problems, similar immunosuppressive drug treatments have been developed for heart, liver, and lung transplants.

TIME **SCIENCE AND HISTORY**

SCIENCE CAN CHANGE THE COURSE OF HISTORY!

Overcoming the Odds

Overcoming the odds—especially when the odds seem stacked against you—is a challenge that many people face. Dr. Samuel Lee Kountz, Jr. (right) had the odds stacked against him. Thanks to his determination, however, he beat them.

Samuel Kountz decided at age eight to become a doctor. He faced his first challenge when he failed the entrance exam to his local Arkansas college. That didn't stop him, though. He asked the college president to give him another chance, and the president did. Kountz got into school and earned As and Bs. He went on to get a graduate degree in biochemistry and was admitted to the University of Arkansas's medical school. Kountz was especially interested in a process that was still brand new in the 1950s—the kidney transplant. At that time, a kidney transplant added months or a year to the lives of many patients. But then, a patient's body would reject the kidney, and the patient would die. Dr. Kountz was determined to see that kidney transplants saved lives and kept patients healthy for years.

Fixing the Problem

Kountz discovered the root of the problem—why and how a patient's body rejected the transplanted kidney. He and others at Stanford University developed a way for doctors to watch the flow of the kidney's blood supply following surgery. As a result, doctors can give patients the right kinds of drugs at the right time, so that their bodies can overcome the rejection process.

In 1959, Kountz performed the first successful kidney transplant. He went on to develop a procedure to keep body organs healthy for up to 60 hours after being taken from a donor. He also set up a system of organ donor cards through the National Kidney Foundation. And in his career, Dr. Kountz transplanted more than 1,000 kidneys himself—and paved the way for thousands more.

Research What kinds of medical breakthroughs has the last century brought? Locate an article that explains either a recent advance in medicine or the work that doctors and medical researchers are doing. Share your findings with your class.

Science Online
For more information, visit red.msscience.com/time

Research First ask students to focus on one part of medicine or one specialization in medicine that interests them. If a student is interested in the brain, suggest a search for an article about neurological advancements in medicine. Suggest avenues that are available for your students to successfully fulfill research in whatever interest they present for this project.

Resources for Teachers and Students

"The Black Surgeon in the Twentieth Century: A Tribute to Samuel L. Kountz" by Claude Organ. *Journal of the National Medical Association*, (September, 1978)

Understanding Kidney Transplantation by Edith T. Oberley and Neal R. Glass. Thomas: Springfield, Ill. 1987

Reviewing Main Ideas

Section 1 Body Systems

1. Energy needs of the body are provided by the digestive and circulatory systems.
2. Oxygen is taken into the body, and waste gases are removed from the body through the respiratory system.
3. Your body's first-line defenses are your skin and respiratory, digestive, and circulatory systems. Your second-line defenses make antibodies to destroy disease-causing organisms.
4. The nervous and endocrine systems control and coordinate all body activities.
5. Blood provides oxygen and nutrients to body cells and removes wastes.

Section 2 Human Reproduction

1. Reproductive organs and structures are different in males and females.
2. Male testes produce sperm. The female ovary produces eggs.
3. If the egg is fertilized by a sperm, it becomes a zygote and then develops to become an embryo and finally a fetus.
4. Infancy, the first life stage after birth, is a time of rapid growth. During childhood, which lasts until age 12, many skills are developed. Adolescence is the stage in which a person becomes able to physically reproduce. Adulthood is the last stage of development.

Visualizing Main Ideas

Copy and complete the following concept map on body systems.

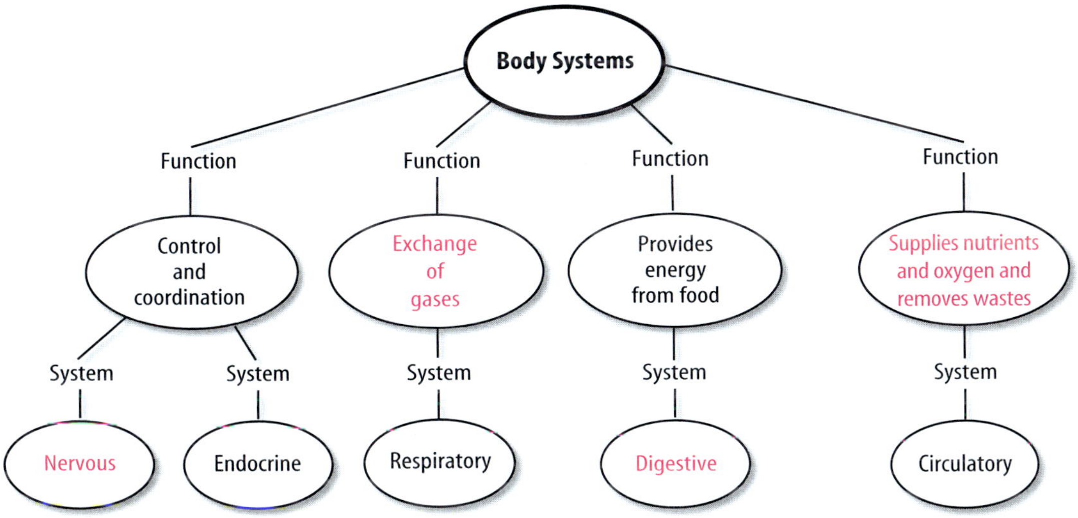

chapter 19 Study Guide

Reviewing Main Ideas

Summary statements can be used by students to review the major concepts of the chapter.

Visualizing Main Ideas

See student page.

Science online

Visit red.msscience.com
/self_check_quiz
/interactive_tutor
/vocabulary_puzzlemaker
/chapter_review
/standardized_test

Assessment Transparency

For additional assessment questions, use the *Assessment Transparency* located in the transparency book.

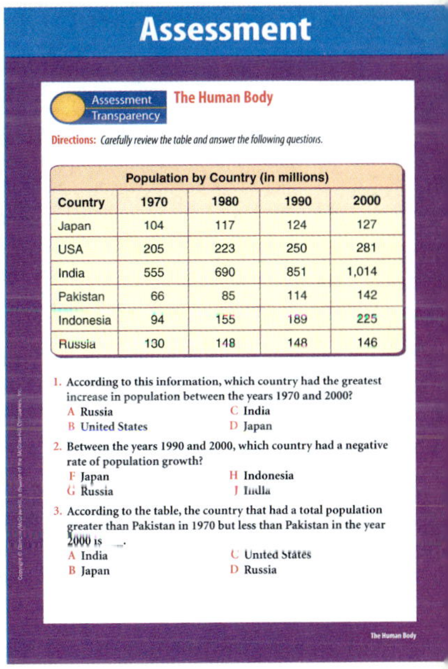
Assessment

Assessment Transparency **The Human Body**

Directions: *Carefully review the table and answer the following questions.*

Population by Country (in millions)				
Country	1970	1980	1990	2000
Japan	104	117	124	127
USA	205	223	250	281
India	555	690	851	1,014
Pakistan	66	85	114	142
Indonesia	94	155	189	225
Russia	130	148	148	146

1. According to this information, which country had the greatest increase in population between the years 1970 and 2000?
   A Russia C India
   B United States D Japan
2. Between the years 1990 and 2000, which country had a negative rate of population growth?
   F Japan H Indonesia
   G Russia J India
3. According to the table, the country that had a total population greater than Pakistan in 1970 but less than Pakistan in the year 2000 is ___.
   A India C United States
   B Japan D Russia

The Human Body

Using Vocabulary

1. Semen
2. alveoli
3. Nutrients
4. pregnancy
5. fetus
6. reflex

Checking Concepts

7. C
8. A
9. A
10. A
11. C
12. D
13. B
14. A
15. D

Thinking Critically

16. arteries—thick elastic walls, carry blood away from the heart; veins—thin walled, have one-way valves, carry blood toward the heart; capillaries—one-cell layer thick, oxygen and nutrients can move from them into and out of cells
17. See student page.
18. Air enters mouth or nose, pharynx, larynx, trachea, bronchi, bronchioles, alveoli
19. So your body can better protect itself. The action can be performed without you having to think about it. Thus, it can happen faster.

chapter 19 Review

Using Vocabulary

alveoli p. 567
capillary p. 567
embryo p. 577
fetus p. 577
melanin p. 561
menstrual cycle p. 576
muscle p. 562
nutrients p. 564
ovulation p. 575
pregnancy p. 577
reflex p. 571
respiratory system p. 567
semen p. 574
skeletal system p. 560
sperm p. 574

Fill in the blanks with the correct word or words.

1. ________ is a mixture of sperm and fluid.
2. Grapelike clusters of air sacs are called ________.
3. ________ are substances in food.
4. The period of development from fertilized egg to birth is called ________.
5. The ________ is the name used for the unborn child after the first two months of pregnancy.
6. A(n) ________ is an involuntary response.

Checking Concepts

Choose the word or phrase that best answers the question.

7. Which is the largest organ in your body?
   A) heart **C)** skin
   B) skeletal muscle **D)** brain
8. To which body system do joints belong?
   A) skeletal system
   B) muscular system
   C) nervous system
   D) digestive system
9. Which is a stage of development before birth?
   A) embryo **C)** ovulation
   B) infancy **D)** childhood

Use the illustration below to answer question 10.

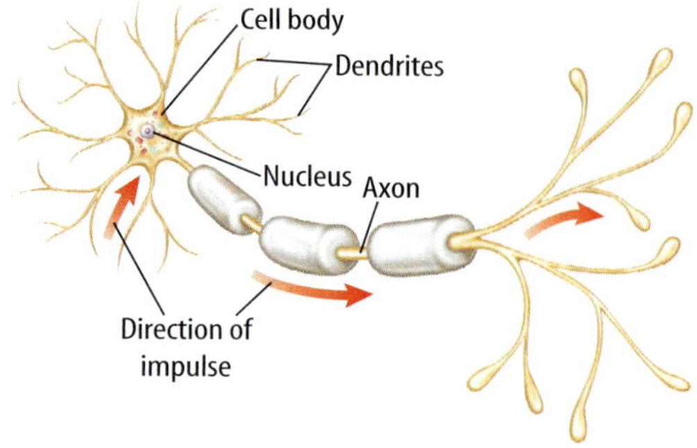

10. What body system contains the cell type shown in the drawing above?
    A) nervous system
    B) endocrine system
    C) digestive system
    D) circulatory system
11. Which two body systems are the control systems for your body?
    A) circulatory and respiratory
    B) endocrine and circulatory
    C) nervous and endocrine
    D) skeletal and nervous
12. Which of the following is an organ in the male reproductive system?
    A) bladder **C)** embryo
    B) ovary **D)** testes
13. How often does ovulation usually occur?
    A) once a year
    B) once a month
    C) twice a month
    D) once every two months
14. What is the union of an egg and a sperm?
    A) fertilization **C)** menstruation
    B) ovulation **D)** puberty
15. During which stage of development does the final growth spurt occur?
    A) infancy **C)** childhood
    B) adulthood **D)** adolescence

Science Online red.msscience.com/vocabulary_puzzlemaker

Use the ExamView® Pro Testmaker CD-ROM to:

- create multiple versions of tests
- create modified tests with one mouse click for inclusion students
- edit existing questions and add your own questions
- build tests aligned with state standards using built-in State Curriculum Tags
- change English tests to Spanish with one mouse click and vice versa

Thinking Critically

16. **Compare and contrast** arteries, veins, and capillaries.

17. **Concept Map** Copy and complete the following concept map of the path of food through the digestive system.

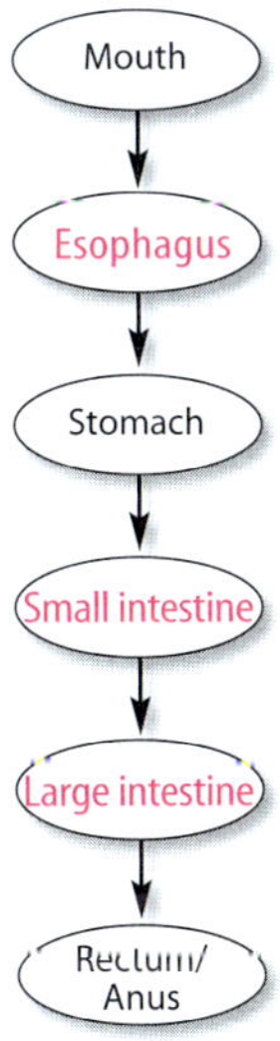

18. **Sequence** the flow of gases through the respiratory system.

19. **Draw Conclusions** Reflexes are involuntary responses, such as coughing or moving away from a hot object. Why are reflexes involuntary?

20. **Classify** each of the following structures according to the body system to which it belongs: stomach, lung, esophagus, heart, trachea, blood vessels, kidney, brain, pituitary gland, large intestine, and ribs.

21. **Form a hypothesis** about what might happen to the fertilized egg if it remained in the oviduct instead of attaching to the wall of the uterus.

Performance Activities

22. **Poem** Create a poem that includes the names, functions, and organs of at least two of the following body systems: respiratory system, circulatory system, nervous system, endocrine system, skeletal system, or muscular system.

23. **Model** Use clay, construction paper, or other art materials to make a model showing the human life stages before or after birth. Label each stage.

Applying Math

24. **Water Loss** Every day your body loses water through the skin and in urine, feces, and exhaled air. If the total body water loss in one day is 2,500 mL and the water loss through the skin, feces, and urine combined is 2,150 mL, how much water is lost through exhaled air in one day?

25. **Amount of Blood** Blood makes up about eight percent of your body's total mass. If you weigh 50 kg what is the weight of the blood in your body?

Use the table below to answer question 26.

Number of Bicycle Deaths per Year

Year	Male	Female
1996	654	107
1997	712	99
1998	658	99
1999	656	94
2000	605	76

Data from Insurance Institute for Highway Safety

26. **Bicycle Helmets** Head injuries are the most serious injuries of people who die in bicycle accidents. In 90 percent of the deaths, people were not wearing bicycle helmets. Using the data in the table, about how many of the people (male and female) who died in bicycle accidents in 2000 were wearing bicycle helmets?

chapter 19 Review

Thinking Critically

20. stomach—digestive, lung—respiratory, esophagus—digestive, heart—circulatory, trachea—respiratory, blood vessels—circulatory, kidney—excretory or urinary, brain—nervous, pituitary gland—endocrine, large intestine—digestive, and ribs—skeletal

21. There would not be enough room for the embryo to develop and it would not survive. (This is called an ectopic pregnancy and is a life-threatening emergency for the mother.)

Performance Activities

22. Poems will differ. Accept any that include two body systems and the organs and functions of the systems.

23. Model should include all four stages before birth or all four stages after birth.

Applying Math

National Math Standards

1, 2, 5, 9

24. $2,500 - 2,150 = 350$ mL

25. $50 \times 0.08 = 4$ kg

26. $605 + 76 = 681; 0.10 \times 681 = 68$ people

Assessment Resources

Reproducible Masters

Chapter *Fast File* Resources
- Chapter Review, pp. 35–36
- Chapter Tests, pp. 37–41
- Assessment Transparency Activity, p. 48

Glencoe Science Web site
- Chapter Review Test
- Standardized Test Practice

Glencoe Technology
- Assessment Transparency
- Exam*View*® Pro Testmaker
- MindJogger Videoquiz
- Interactive Chalkboard

FAST FILE

Answer Sheet A practice answer sheet can be found at red.msscience.com/answer_sheet.

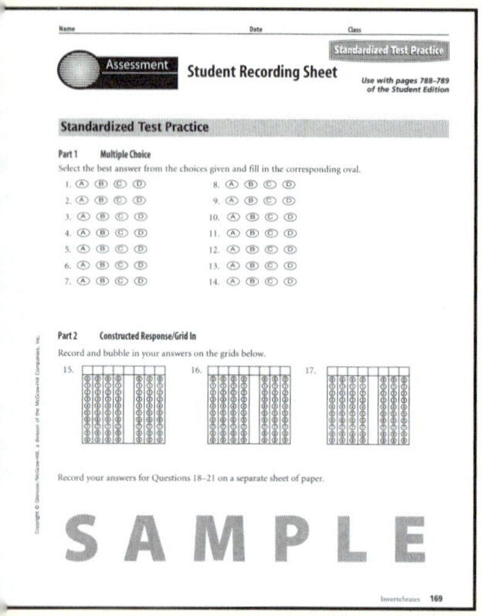
Assessment Student Recording Sheet

Standardized Test Practice

Part 1 Multiple Choice

Part 2 Constructed Response/Grid In

SAMPLE

Part 1 Multiple Choice

1. B
2. C
3. D
4. D
5. D

chapter 19 Standardized Test Practice

Part 1 Multiple Choice

Record your answers on the answer sheet provided by your teacher or on a sheet of paper.

Use the table below to answer question 1.

Results of Folic Acid on Development of Neural Tube Defect		
Group	**Babies with Neural Tube Defect**	**Babies without Neural Tube Defect**
Group I Folic Acid	6	497
Group II No Folic Acid	21	581

(From CDC)

Researchers have found that the B-vitamin folic acid can prevent a type of birth defect called neural tube defect. In a study done in Europe in 1991, one group of pregnant women was given extra folic acid, and the other group did not receive extra folic acid.

1. Which of the following statements is true regarding the data in this table?
A. Folic acid had no effect on the percentage of babies with neural tube defect.
B. Women who took the extra folic acid had a decreased percentage of babies with neural tube defect.
C. Extra folic acid increased the percentage of babies with neural tube defect.
D. Group I and Group II had the same percentage of babies born with neural tube defect.

Test-Taking Tip

Detect Data Patterns When analyzing data in a table or graph, try to detect a pattern. Questions about the pattern may use words like hypothesis, generalization, summary, or trend.

Questions 3 and 4 When looking for a pattern, compare data for two or more variables.

2. Which of the following is a function of blood?
A. carries saliva to the mouth
B. excretes urine from the body
C. transports nutrients and oxygen to body cells
D. collects tissue fluid from around cells

Use the table below to answer questions 3 and 4.

Results from Ashley's Activities			
Activity	**Pulse Rate (beats/min)**	**Body Temperature**	**Degree of Sweating**
1	80	98.6°F	None
2	90	98.8°F	Minimal
3	100	98.9°F	Little
4	120	99.1°F	Moderate
5	150	99.5°F	Considerable

3. According to the information in this table, which of the following activities indicates that Ashley was exercising vigorously?
A. Activity 2 **C.** Activity 4
B. Activity 3 **D.** Activity 5

4. A reasonable hypothesis based on these data is that during Activity 2, Ashley was probably
A. sprinting.
B. marching.
C. resting.
D. walking slowly.

5. Which of the following events does NOT happen during a female's monthly menstrual cycle?
A. maturing of egg
B. production of female sex hormones
C. menstruation
D. menopause

Part 2 Short Response/Grid In

6. 2,000,000/sec × 60 sec/min × 60 min/hr = 7,200,000,000
7. 400,000 ÷ 10,000 = 40; 40 times more platelets than white blood cells
8. The antibodies in type A blood would cause type O red blood cells to clump.
9. type AB
10. Intact skin is a barrier that prevents pathogens from entering the body. Perspiration contains substances that slow the growth of some pathogens. Some pathogens cannot grow in the acidic secretions from oil glands and perspiration.
11. Organic nutrients contain carbon. Examples: proteins, carbohydrates, fats, and vitamins. Inorganic nutrients do not contain carbon. Examples: minerals and water.
12. Voluntary muscles can be consciously controlled. Involuntary muscles are muscles that cannot be controlled consciously.
13. 100 billion = 0.10x, x = 100/0.1 = 1,000 billion = 1 trillion

Part 2 Short Response/Grid In

Record your answers on the answer sheet provided by your teacher or on a sheet of paper.

6. If red blood cells are made at the rate of two million per second in the center of long bones, how many red blood cells are made in one hour?

7. If a cubic milliliter of blood has 10,000 white blood cells and 400,000 platelets, how many more platelets than white blood cells are present in a cubic milliliter of blood?

Use the table below to answer questions 8 and 9.

Blood Transfusion Possibilities		
Type	Can Receive	Can Donate To
A	O, A	A, AB
B	O, B	B, AB
AB	All	AB
O	O	All

8. Infer what would happen if type A blood was given to a person with type O blood.

9. Which blood type could be called the "universal recipient"?

10. How does your skin help defend your body from diseases?

11. Explain the difference between organic and inorganic nutrients. Name a class of nutrients for each.

12. Compare and contrast voluntary and involuntary muscles.

13. The brain is made up of approximately 100 billion neurons, which is about 10% of all neurons in the body. Approximately how many neurons are there in the human body?

Part 3 Open Ended

Record your answers on a sheet of paper.

14. A virus causes the disease rubella, also known as "German measles." If a pregnant woman becomes infected with rubella, the virus can affect the formation of major organs such as the heart in her baby. During which stage of development before birth would a rubella infection be most dangerous? Why?

15. How do the lymphatic and circulatory systems work together?

16. Explain why blood sometimes is called "the tissue of life?"

Use the illustration below to answer questions 17 and 18.

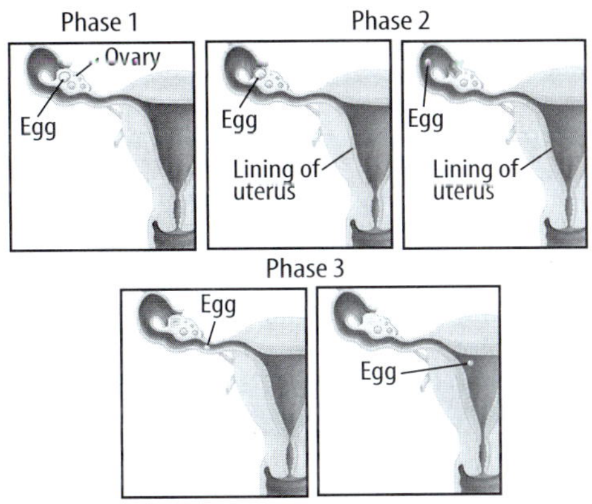

17. Describe the changes that occur in phase 2.

18. Compare and contrast the changes that occur in phase 3 if fertilization does take place and if fertilization does not take place.

19. Describe your body's first line and second-line defenses.

20. How do water-soluble and fat-soluble vitamins differ? Name the vitamins in each group and give examples of how they help the body.

18. Phase 3, with fertilization: Lining of uterus continues to thicken. Egg begins to develop in the uterus and hormones continue to be produced. Phase 3, without fertilization: lining of uterus continues to thicken. Hormone levels decrease, the lining breaks down and menstruation begins.

19. First line: skin, and respiratory, digestive, and circulatory systems. Respiratory system traps disease organisms with mucus and cilia. Circulatory system has white blood cells that destroy invaders. Second line: specific immunity against invaders that make it past first line. Antibodies that destroy the disease-causing organisms are made.

20. Water soluble: should be replaced daily; B vitamins and C. B vitamins aid in growth, healthy nervous system, use of carbohydrates, and red blood cell production. C aids in growth, healthy bones and teeth, wound recovery. Fat soluble: stored in the body's fatty tissue. A, D, E, and K. A aids in growth, eyesight, healthy skin; D aids in absorption of calcium and phosphorus by bones and teeth; E aids in formation of cell membranes; K aids in blood clotting and wound recovery.

Rubrics

For more help evaluating open-ended assessment questions, see the rubric on p. 10T.

Part 3 Open Ended

14. During the first two months of development, because this is when the embryo's major organs form.

15. The blood in the circulatory system brings nutrients, water, and oxygen to the body cells. These substances diffuse through capillary walls into nearby cells. Water and other substances become part of the tissue fluid that is found between cells. This fluid is collected and returned to the blood by the lymphatic system.

16. Blood is necessary for a human to live. Without blood, the body's cells would die because they could not get the oxygen and nutrients needed. Blood also removes wastes from the body cells. Blood also protects the body. It carries white blood cells that fight infections and heal wounds.

17. Phase 2: egg develops in the ovary and lining of the uterus thickens. Ovulation occurs about 14 days before menstrual flow begins.

chapter 20 Organizer

Section/Objectives	Standards		Labs/Features
	National	State/Local	
Chapter Opener	See pp. 16T–17T for a Key to Standards.		**Launch Lab:** Why are seeds formed?, p. 589 **Foldables,** p. 589
Section 1 Continuing Life 5 sessions 2.5 blocks 1. **Describe** how cells divide. 2. **Identify** the importance of reproduction for living things. 3. **Compare and contrast** sexual and asexual reproduction. 4. **Describe** the structure and function of DNA.	National Content Standards: UCP.1–UCP.5, A.1, A.2, C.1–C.3		**Science Online,** p. 591 **MiniLAB:** Observing Yeast Budding, p. 593 **Integrate Health,** p. 595 **Visualizing Human Reproduction,** p. 596 **Lab:** Getting DNA from Onion Cells, p. 598
Section 2 Genetics—The Study of Inheritance 7 sessions 3.5 blocks 5. **Explain** how traits are inherited. 6. **Relate** chromosomes, genes, and DNA to one another. 7. **Discuss** how mutations add variation to a population.	National Content Standards: UCP.1–UCP.4, A.1, A.2, C.2, C.3		**MiniLAB:** Modeling Probability, p. 601 **Applying Math:** Alleles In Sex Cells, p. 603 **Integrate Career,** p. 604 **Science Online,** p. 605 **Lab:** Genetic Traits: The Unique You, p. 606 **Science and Society:** Separated at Birth, p. 608

Glencoe Exclusive! **TeacherWorks™** All-In-One Planner and Resource Center

Lab Materials	Reproducible Resources	Section Assessment	Technology
Launch Lab: orange, metric ruler	**Chapter *FAST FILE* Resources** Foldables Worksheet, p. 15 Directed Reading Overview, p. 17 Note-taking Worksheets, pp. 29–30	Glencoe's Assessment Advantage	**TeacherWorks** includes: • Interactive Teacher Edition • Lesson Planner with calendar • Access to all program blacklines • Correlations to standards • Web links
MiniLAB: dropper, yeast, sugar, water, microscope slides (2), coverslip, microscope, small jars or beakers **Lab:** prepared onion mixture (125 mL), toothpicks, small beaker, rubbing alcohol, large beaker, magnifying lens	**Chapter *FAST FILE* Resources** Transparency Activity, p. 40 MiniLAB, p. 3 Enrichment, p. 27 Reinforcement, p. 25 Directed Reading, p. 18 Lab Worksheet, pp. 5–6 Transparency Activity, pp. 43–44 Lab Activity, pp. 9–12 **Reading and Writing Skill Activities,** p. 9	**Portfolio** Make a Model, p. 591 Assessment, p. 597 **Performance** MiniLAB, p. 593 Applying Math, p. 597 **Content** Section Review, p. 597	Section Focus Transparency Teaching Transparency Virtual Labs CD-ROM Guided Reading Audio Program Interactive Chalkboard CD-ROM Video Lab
MiniLAB: coin **Lab:** Internet or other sources on human genetic traits *Need materials?* Contact Science Kit at 1-800-828-7777 or www.sciencekit.com on the Internet.	**Chapter *FAST FILE* Resources** Transparency Activity, p. 41 MiniLAB, p. 4 Enrichment, p. 28 Reinforcement, p. 26 Directed Reading, pp. 19, 20 Lab Worksheet, pp. 7–8 Lab Activity, pp. 13–14 **Mathematics Skill Activities,** p. 29 **Lab Management and Safety,** p. 58	**Portfolio** Assessment, p. 605 **Performance** MiniLAB, p. 601 Applying Math, p. 603 Applying Skills, p. 605 **Content** Section Review, p. 605	Section Focus Transparency Virtual Labs CD-ROM Guided Reading Audio Program Interactive Chalkboard CD-ROM

End of Chapter Assessment

Blackline Masters	Technology	Professional Series
Chapter *FAST FILE* Resources Chapter Review, pp. 33–34 Chapter Tests, pp. 35–38 **Standardized Test Practice,** pp. 83–86	MindJogger Videoquiz Virtual Labs CD-ROM ExamView® Pro Testmaker TeacherWorks CD-ROM Interactive Chalkboard CD-ROM	**Performance Assessment in the Science Classroom (PASC)**

Transparencies

Section Focus

Section 1 Section Focus Transparency **One Potato, Two Potato**

What's going on in this picture? A new potato plant is growing from a potato. Each eye of a potato is a bud that can produce a new potato plant. After three to four months of growth, the plant can produce more potatoes.

1. Will the new plant be like the one from which it sprouted? Explain your answer.
2. Why are we not exact copies of our parents?

L2

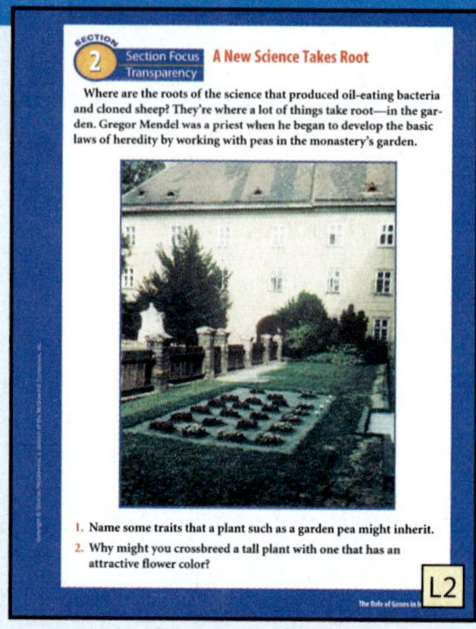
Section 2 Section Focus Transparency **A New Science Takes Root**

Where are the roots of the science that produced oil-eating bacteria and cloned sheep? They're where a lot of things take root—in the garden. Gregor Mendel was a priest when he began to develop the basic laws of heredity by working with peas in the monastery's garden.

1. Name some traits that a plant such as a garden pea might inherit.
2. Why might you crossbreed a tall plant with one that has an attractive flower color?

L2

This is a representation of key blackline masters available in the Teacher Classroom Resources. See Resource Manager boxes within the chapter for additional information.

Assessment

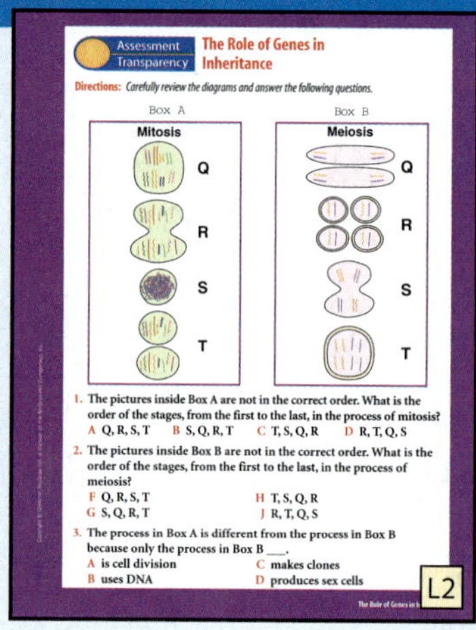
Assessment Transparency **The Role of Genes in Inheritance**

Directions: *Carefully review the diagrams and answer the following questions.*

1. The pictures inside Box A are not in the correct order. What is the order of the stages, from the first to the last, in the process of mitosis?
   A Q, R, S, T B S, Q, R, T C T, S, Q, R D R, T, Q, S
2. The pictures inside Box B are not in the correct order. What is the order of the stages, from the first to the last, in the process of meiosis?
   F Q, R, S, T H T, S, Q, R
   G S, Q, R, T J R, T, Q, S
3. The process in Box A is different from the process in Box B because only the process in Box B ___.
   A is cell division C makes clones
   B uses DNA D produces sex cells

L2

Teaching

Section 1 Teaching Transparency **DNA's Structure**

L2

Key to Teaching Strategies

The following designations will help you decide which activities are appropriate for your students.

- **L1** Level 1 activities should be appropriate for students with learning difficulties.
- **L2** Level 2 activities should be within the ability range of all students.
- **L3** Level 3 activities are designed for above-average students.
- **ELL** ELL activities should be within the ability range of English Language Learners.
- **COOP LEARN** Cooperative Learning activities are designed for small group work.
- **LS** Multiple Learning Styles logos, as described on page 12T, are used throughout to indicate strategies that address different learning styles.
- **P** These strategies represent student products that can be placed into a best-work portfolio.
- **PBL** Problem-Based Learning activities apply real-world situations to learning.

Hands-on Activities

Student Text Lab Worksheet

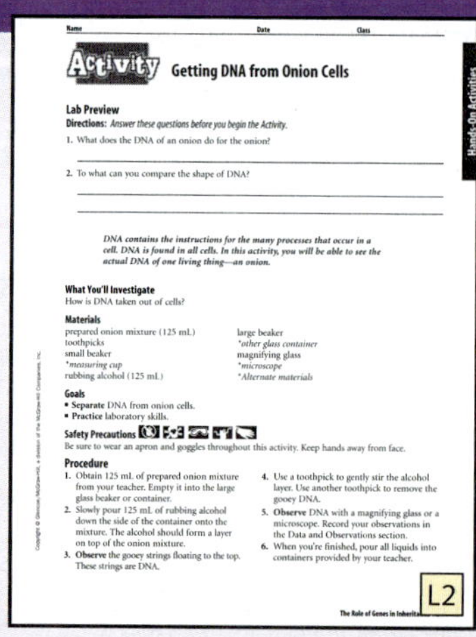
Name ____ Date ____ Class ____

Activity Getting DNA from Onion Cells

Lab Preview

Directions: *Answer these questions before you begin the Activity.*

1. What does the DNA of an onion do for the onion?
2. To what can you compare the shape of DNA?

DNA contains the instructions for the many processes that occur in a cell. DNA is found in all cells. In this activity, you will be able to see the actual DNA of one living thing—an onion.

What You'll Investigate

How is DNA taken out of cells?

Materials

prepared onion mixture (125 mL)
toothpicks
small beaker
**measuring cup*
rubbing alcohol (125 mL)
large beaker
**other glass container*
magnifying glass
**microscope*
**Alternate materials*

Goals

- **Separate** DNA from onion cells.
- **Practice** laboratory skills.

Safety Precautions

Be sure to wear an apron and goggles throughout this activity. Keep hands away from face.

Procedure

1. Obtain 125 mL of prepared onion mixture from your teacher. Empty it into the large glass beaker or container.
2. Slowly pour 125 mL of rubbing alcohol down the side of the container onto the mixture. The alcohol should form a layer on top of the onion mixture.
3. **Observe** the gooey strings floating to the top. These strings are DNA.
4. Use a toothpick to gently stir the alcohol layer. Use another toothpick to remove the gooey DNA.
5. **Observe** DNA with a magnifying glass or a microscope. Record your observations in the Data and Observations section.
6. When you're finished, pour all liquids into containers provided by your teacher.

L2

Laboratory Activities

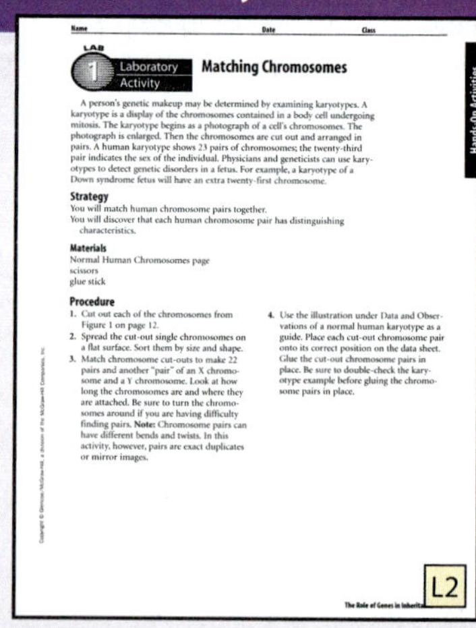
Name ____ Date ____ Class ____

LAB 1 Laboratory Activity **Matching Chromosomes**

A person's genetic makeup may be determined by examining karyotypes. A karyotype is a display of the chromosomes contained in a body cell undergoing mitosis. The karyotype begins as a photograph of a cell's chromosomes. The photograph is enlarged. Then the chromosomes are cut out and arranged in pairs. A human karyotype shows 23 pairs of chromosomes; the twenty-third pair indicates the sex of the individual. Physicians and geneticists can use karyotypes to detect genetic disorders in a fetus. For example, a karyotype of a Down syndrome fetus will have an extra twenty-first chromosome.

Strategy

You will match human chromosome pairs together.
You will discover that each human chromosome pair has distinguishing characteristics.

Materials

Normal Human Chromosomes page
scissors
glue stick

Procedure

1. Cut out each of the chromosomes from Figure 1 on page 12.
2. Spread the cut-out single chromosomes on a flat surface. Sort them by size and shape.
3. Match chromosome cut-outs to make 22 pairs and another "pair" of an X chromosome and a Y chromosome. Look at how long the chromosomes are and where they are attached. Be sure to turn the chromosomes around if you are having difficulty finding pairs. **Note:** Chromosome pairs can have different bends and twists. In this activity, however, pairs are exact duplicates or mirror images.
4. Use the illustration under Data and Observations of a normal human karyotype as a guide. Place each cut-out chromosome pair onto its correct position on the data sheet. Glue the cut-out chromosome pairs in place. Be sure to double-check the karyotype example before gluing the chromosome pairs in place.

L2

Meeting Different Ability Levels

Content Outline

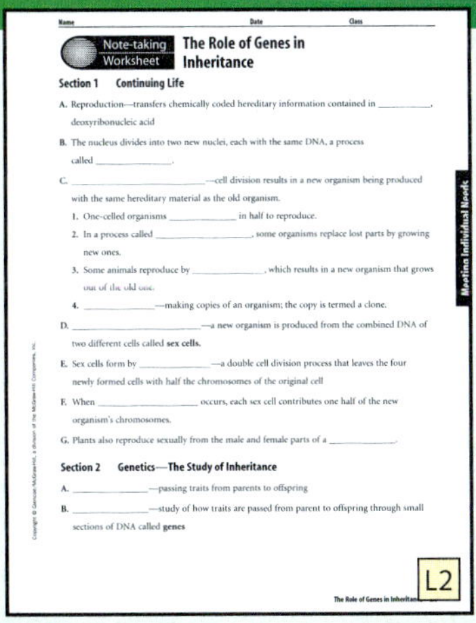
Note-taking Worksheet — The Role of Genes in Inheritance
L2

Reinforcement

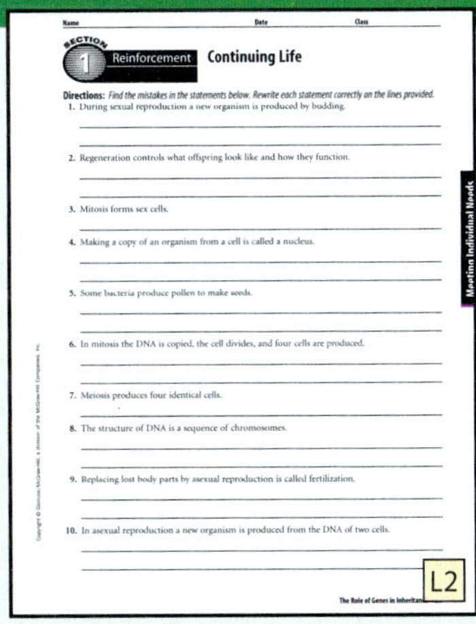
Section 1 Reinforcement — Continuing Life
L2

Enrichment

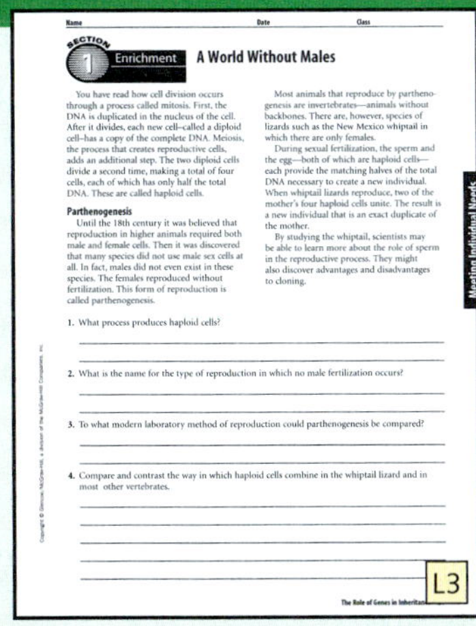
Section 1 Enrichment — A World Without Males
L3

Directed Reading (English/Spanish)

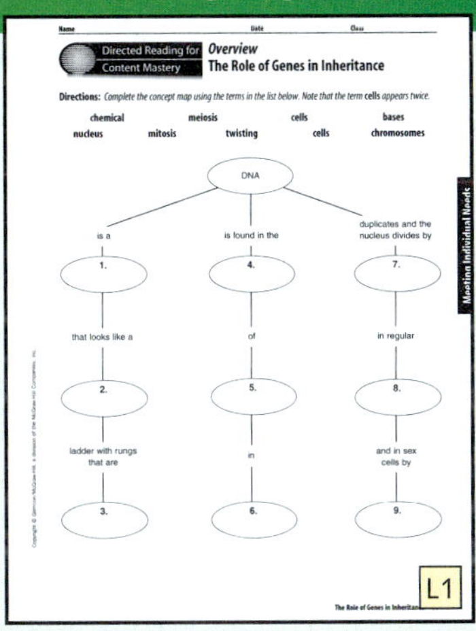
Directed Reading for Content Mastery — Overview — The Role of Genes in Inheritance
L1

Study Guide

Reading Essentials

Assessment

Test Practice Workbook

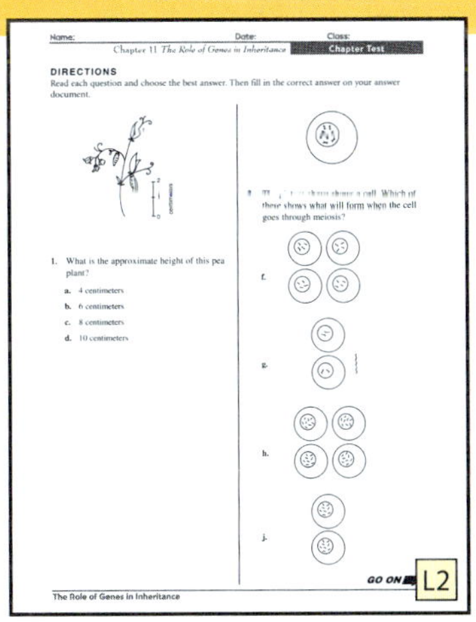
Chapter 11 The Role of Genes in Inheritance — Chapter Test
L2

Chapter Review

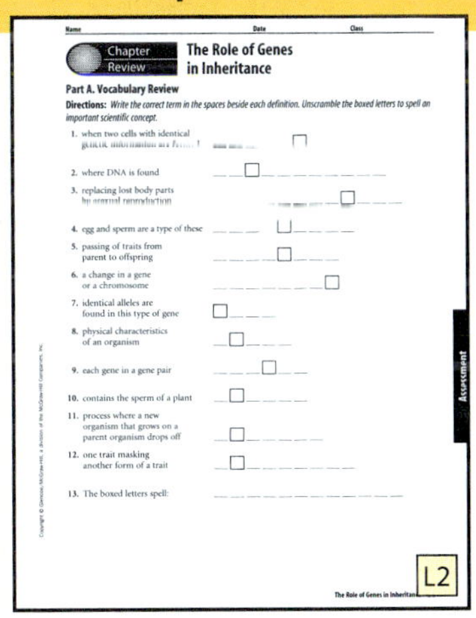
Chapter Review — The Role of Genes in Inheritance
L2

Chapter Tests

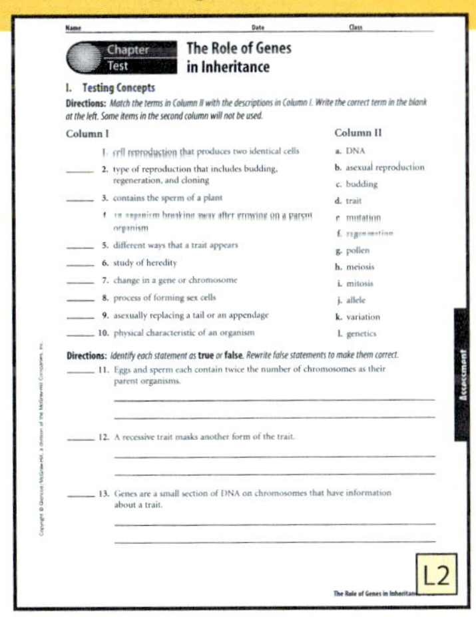
Chapter Test — The Role of Genes in Inheritance
L2

Science Content Background

section 1 Continuing Life

Reproduction by One Organism

Plants such as bananas and strawberries can reproduce asexually in a process know as vegetative propagation. The advantage of this type of propagation is that the new plants are exactly like the original parents and they are established faster. New plants arise from the banana from underground suckers. New plants arise from the strawberry from special branches called stolons.

section 2 Genetics—The Study of Inheritance

Dominant and Recessive Traits

Through inheritance experiments with pea plants, Gregor Mendel discovered that the traits he observed had two types of genes: dominant and recessive. When pure purple-flowered pea plants (those containing two genes for purple flowers) were crossed with pure white-flowered pea plants (those containing two genes for white flowers), the result was always purple-flowered plants. This established the idea that the purple gene was dominant, or masked the recessive white gene. Often, recessive genes produce nonfunctional proteins that have no effect on the organism. In this case, no pigment is produced. Only when Mendel crossed pure white-flowered plants with other pure white-flowered plants did he see all white-flowered offspring.

Incomplete Dominance

"Blended inheritance" is the idea that offspring have genes that are an intermediate mix of the genes inherited from their parents. For example, if a plant inherits a gene for red flowers from one parent and from the other a gene for white flowers, it will have pink flower genes. This appears to be the case in carnations, as crosses of red-flowered plants with white-flowered plants produce pink-flowered offspring. If this were the case, plants with pink flower-color genes that cross with plants having white flower-color genes would produce offspring that all have a lighter pink flower. However, some offspring from a pink-white cross have white flowers because they receive the white gene from both parents. Thus, the carnation exhibits incomplete dominance. The genes in the offspring are still red and white, but the red gene shows incomplete dominance because it does not completely mask the white gene. In this case, the amount of red pigment is critical to the appearance of the offspring. Those flowers with two red genes are red, and those with only one are pink because of limited amounts of red pigment.

chapter content resources

Internet Resources

For additional content background, visit **red.msscience.com** to:

- access your book online
- find references to related articles in popular science magazines
- access Web links with related content background
- access current events with science journal topics

Print Resources

Heredity Diseases, by Fern G. Brown, Franklin Watts, 1987

The Story of Science: The Treasure of Inheritance, by Roy A. Gallant, Benchmark Books, 2002

Genetics: The Study of Heredity (Investigating Science), by Ian Graham, Gareth Stevens, 2002

IDENTIFYING Misconceptions

Find Out What Students Think

Students may think that . . .

All inheritance is a result of blending, as opposed to Mendelian genetics.

Many students do not understand the concepts of dominant and recessive genes. Genes that show incomplete dominance or intermediate inheritance add to the confusion.

Discussion

Explain the concepts of dominant and recessive genes by drawing purple-flowered and white-flowered plants on the board, using colored chalk. Start with pure plants, and explain that the purple-flowered plant has two purple alleles and the white-flowered plant has two white alleles. These can be represented by *PP* and *pp*. What genes will the offspring of these two parents have for flower color? *Pp* What colors will the flowers of the offspring be? Some students may suggest light purple. Explain that even though the offspring have both purple and white alleles, the purple allele masks the white one. Because of this masking, the purple allele is called *dominant* and the white allele is called *recessive*.

Promote Understanding

Activity

Using the trait of flower color in pea plants, choose the genetic makeup for two parents that will be used in a crossing game. For example, one parent could be pure-breeding purple (*PP*) and the other could be heterozygous for flower color, possessing one gene for each trait *Pp*. Write the genetic makeup of each parent on the board (*PP* and *Pp*).

- Design a card game for students in which one set of cards represents the flower-color genes of the first parent and a second set represents the flower-color genes of the second parent. Because offspring receive only one allele from each parent, each card in each pile should be labeled with a *P* or *p*, which represents a purple or white allele.
- Have each student pick a card. Pair students, and then circulate throughout the room, asking each pair to describe their alleles and the color of the flowers produced.
- Continue the activity by letting students form new pairs with other students.

This activity can be modified to illustrate incomplete dominance by having students model gene combinations in two successive generations. For example, use red and white flowers in carnations to clarify the concept of incomplete dominance. Explain that a red allele and a white allele will produce a pink flower because the red allele does not completely mask the white allele. Match students that have the same color card and have them determine the flower color of the offspring. Those with red matches should answer "red," and those with white matches should answer "white."

Assess

After completing the chapter, see *Identifying Misconceptions* in the Study Guide at the end of the chapter.

chapter 20

Chapter Vocabulary

DNA, p. 590
mitosis, p. 592
asexual reproduction, p. 593
cloning, p. 594
sexual reproduction, p. 594
sex cell, p. 594
meiosis, p. 595
fertilization, p. 595
heredity, p. 599
genetics, p. 599
gene, p. 600
variation, p. 603
mutation, p. 604

Science Journal Student responses will vary, but may include size, length of mane, hair color, and eye color.

INTERACTIVE CHALKBOARD with Image Bank

PowerPoint® Presentations

This CD-ROM is an editable Microsoft® PowerPoint® presentation that includes:

- a pre-made presentation for every chapter
- interactive graphics
- animations
- audio clips
- image bank
- all new section and chapter questions
- Standardized Test Practice
- transparencies
- pre-lab questions for all labs
- Foldables directions
- links to red.msscience.com

chapter 20

The Role of Genes in Inheritance

chapter preview

sections

1 **Continuing Life**
Lab *Getting DNA from Onion Cells*

2 **Genetics—The Study of Inheritance**
Lab *Genetic Traits: The Unique You*

Virtual Lab *How are the traits of parents inherited and expressed in offspring?*

Why do these horses look different from each other?

All living things obtain different characteristics from their parents. Knowing how organisms reproduce and what determines traits will help you understand why all horses in a herd of wild mustangs do not look exactly alike.

Science Journal Write three traits of horses that you could trace from parents to offspring.

588

Theme Connection

Stability and Change Although Earth's millions of species show many physical and behavioral differences, all of these organisms must reproduce if their species are to continue.

About the Photo

Wild Mustangs The wild mustang used to roam the Americas, but now only a few herds can be found. They can be found in many colors, including appaloosa, brown, and black. Mustangs are usually 13–15 hands tall, weighing 750–1000 pounds. Mustangs avoid inbreeding by having their young leave the herd after a few years.

Start-Up Activities

Launch LAB

Why are seeds formed?

When you peel a banana or bite into an apple, you're probably only thinking about the taste and sweet smell of the fruit. You usually don't think about how the fruit was formed. Oranges, and most of the fruits you eat, contain seeds. Making seeds is one way that reproduction is carried out by living things. For life to continue, all living things must pass characteristics to their offspring.

WARNING: *Do not eat the orange.*

1. Obtain half of an orange from your teacher. Peel the orange and remove all of the seeds.
2. Examine, count, and measure the length of each seed. Record these data in your Science Journal.
3. When you finish, dispose of your orange half as instructed by your teacher. Wash your hands.
4. **Think Critically** Write a paragraph in your Science Journal describing why you think the seeds are different from one another.

FOLDABLES Study Organizer

Inheritance Make the following Foldable to help identify what you already know, what you want to know, and what you learned about the role of genes in inheritance.

STEP 1 **Fold** a vertical sheet of paper from side to side. Make the front edge about 1.25 cm shorter than the back edge.

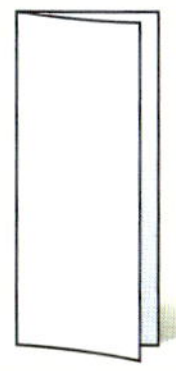

STEP 2 **Turn** lengthwise and **fold** into thirds.

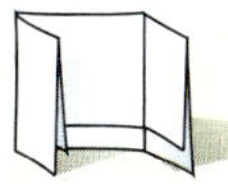

STEP 3 **Unfold and cut** only the top layer along both folds to make three tabs. **Label** each tab as shown.

Identify Questions Before you read the chapter, write what you already know about the role of genes in inheritance under the left tab of your Foldable, and write questions about what you'd like to know under the center tab. After you read the chapter, list what you learned under the right tab.

Science Online — Preview this chapter's content and activities at red.msscience.com

Launch LAB

Purpose Use the Launch Lab to introduce students to genetic variation. Students measure variation in the seeds of an orange. L1 ELL LS **Kinesthetic**

Preparation Obtain enough oranges for each student to have one-half. Be sure to get oranges with seeds. One seedless orange may be used to demonstrate the difference between organisms that reproduce sexually (with seeds) and those that do not.

Materials oranges, metric rulers

Teaching Strategy Explain that seeds vary for two main reasons. One reason is because of environmental conditions. Another reason is that seeds inherit different genetic material from their parent plants, resulting in observable and measurable variations.

Think Critically

The number and length of seeds will vary. Genetic differences cause most of the differences observed in seeds derived from the same fruit.

Assessment

Performance Have students work in teams to identify other plants that have obvious variations in seeds. Examples may include size and shape of peas and beans, or color variations in some castor beans or other seeds. Use **Performance Assessment in the Science Classroom,** p. 169.

FOLDABLES Study Organizer **Dinah Zike Study Fold**

Student preparation materials for this Foldable are available in the Chapter *FAST FILE* Resources.

1 Motivate

INTERACTIVE CHALKBOARD PowerPoint® Presentations

Bellringer

Section Focus Transparencies also are available on the Interactive Chalkboard CD-ROM.

L2 ELL

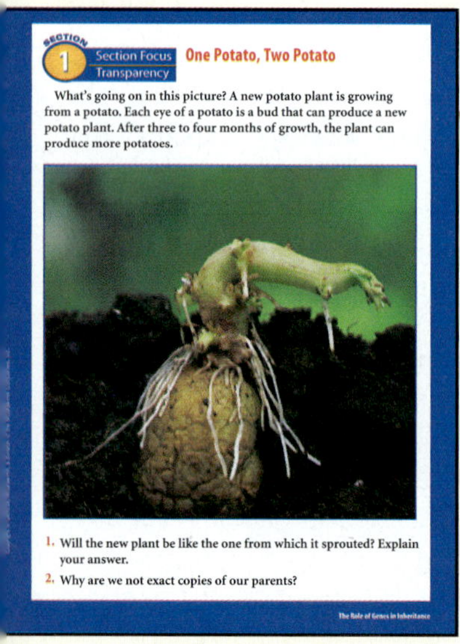
SECTION 1 Section Focus Transparency **One Potato, Two Potato**

What's going on in this picture? A new potato plant is growing from a potato. Each eye of a potato is a bud that can produce a new potato plant. After three to four months of growth, the plant can produce more potatoes.

1. Will the new plant be like the one from which it sprouted? Explain your answer.
2. Why are we not exact copies of our parents?

The Role of Genes in Inheritance

Tie to Prior Knowledge

Cell Structure Review the structures and functions of animal and plant cells. Ask students to recall the location and function of the nucleus in these cells.

section 1

Continuing Life

as you read

What You'll Learn

- **Describe** how cells divide.
- **Identify** the importance of reproduction for living things.
- **Compare and contrast** sexual and asexual reproduction.
- **Describe** the structure and function of DNA.

Why It's Important

All living things, including you, inherit characteristics from their parents.

Review Vocabulary
chromosome: structure in a cell's nucleus that contains genetic material

New Vocabulary
- DNA
- mitosis
- asexual reproduction
- cloning
- sexual reproduction
- sex cell
- meiosis
- fertilization

Reproduction

If you look carefully in a pond in the spring, you may see frog or toad eggs. Frogs reproduce by laying hundreds of eggs in gooey clumps. Tadpoles can hatch from these eggs and mature into adult frogs, as shown in **Figure 1.** Some other kinds of organisms, including humans, usually produce only one offspring at a time. How do frogs and all of the other living things on Earth produce offspring that are similar to themselves?

The Importance of Reproduction Organisms produce offspring through the process of reproduction. Reproduction is important to all living things. Without reproduction, species could not continue. Hereditary material is passed from parent to offspring during reproduction. This material is found inside cells. It is made up of the chemical deoxyribonucleic (dee AHK sih ri boh noo klay ihk) acid, called DNA. **DNA** controls how offspring will look and how they will function by controlling what proteins each cell will produce. The DNA that all living things pass on determines many of their offspring's characteristics. Although organisms reproduce in different ways, reproduction always involves the transfer of hereditary information.

Figure 1 When frogs reproduce, they continue their species.

Adult frogs reproduce by laying and then fertilizing eggs.

These frog eggs can hatch into tadpoles.

These tadpoles can develop into adult frogs.

Section 1 Resource Manager

Chapter FAST FILE Resources
- Transparency Activity, pp. 40, 43–44
- Directed Reading for Content Mastery, pp. 17, 18
- Note-taking Worksheets, pp. 29–30
- MiniLAB, p. 3
- Enrichment, p. 27
- Lab Activity, pp. 9–12
- Reinforcement, p. 25
- Lab Worksheet, pp. 5–6

Reading and Writing Skill Activities, p. 9
Life Science Critical Thinking/Problem Solving, p. 3

590 CHAPTER 20 The Role of Genes in Inheritance

INTEGRATE Chemistry

Life's Code You've probably seen or heard about science fiction movies in which DNA is used to grow prehistoric animals. What makes up DNA? How does it work?

DNA is found in all cells in structures called chromosomes. All of the information that is in your DNA is called your genetic information. You can think of DNA as a genetic blueprint that contains all of the instructions for making an organism what it is. Your DNA controls the texture of your hair, the shape of your ears, your blood type, and even how you digest the food you had for lunch.

If you could look at DNA in detail, you would see that it is shaped like a twisted ladder. This structure, shown in **Figure 2,** is the key to how DNA works. The two sides of the ladder form the backbone of the DNA molecule. The sides support the rungs of the ladder. It is the rungs that hold all the genetic information. Each rung of the ladder is made up of a pair of chemicals called bases. There are only four bases in DNA, and they pair up very specifically. A DNA ladder has billions of rungs, and the bases are arranged in thousands of different orders. The secret of DNA has to do with the order or sequence of bases along the DNA ladder. The sequence forms a code. From this DNA code the cell gets instructions about what substances to make, how to make them, and when to make them.

Science Online

Topic: Human Genome Project
Visit red.msscience.com for Web links to information about the Human Genome Project.

Activity List three genetic disorders and explain how the Human Genome project may help researchers who study these disorders.

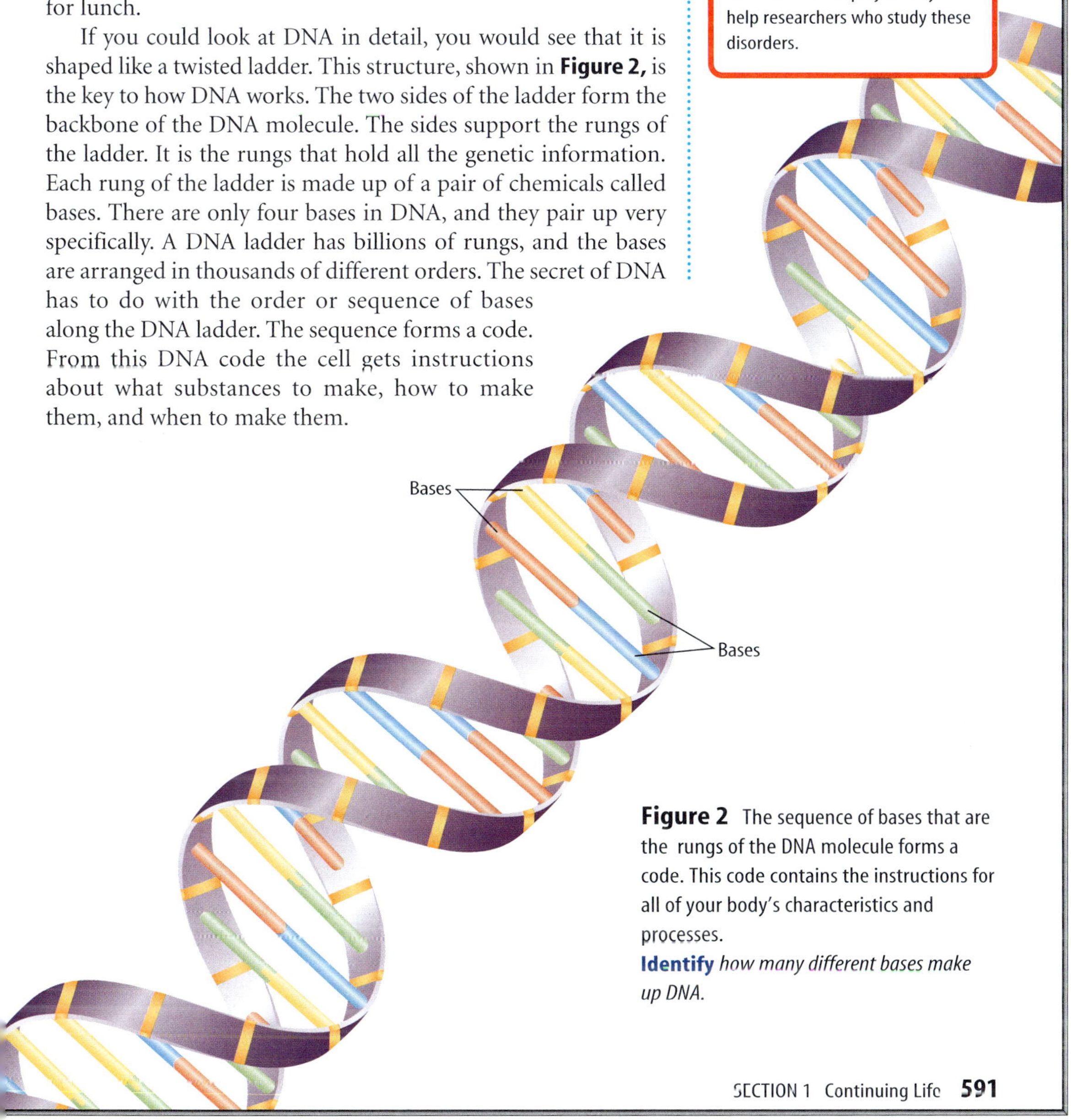

Figure 2 The sequence of bases that are the rungs of the DNA molecule forms a code. This code contains the instructions for all of your body's characteristics and processes.
Identify *how many different bases make up DNA.*

2 Teach

Make a Model

DNA Have students make three-dimensional models of DNA using toothpicks, shoe-string licorice, and assorted candies. Use licorice to represent the sides of the DNA ladder. Gumdrops of different colors or marshmallows can represent the bases and toothpicks the rungs of the DNA ladder. Point out that the sequence of the gumdrops or marshmallows represents the genetic code. L1 ELL LS **Visual-Spatial** P

Caption Answer

Figure 2 four

Curriculum Connection

Math DNA profiling can establish the identity of a person more reliably than blood typing. Have students research blood types and find the percentage of the U.S. population that has each type. Persons of western European descent: 45%—A, 8%—B, 4%—AB, 43%—0; persons of African descent: 29%—A, 17%—B, 4%—AB, 50%—0; people of Asian descent are commonly B. L2 LS **Logical-Mathematical**

Teacher FYI

Cell Division Reinforce to students that mitosis refers only to division of the nucleus into two identical nuclei. Division of the cell usually occurs at the end of mitosis, but not always. Some species of amoeba and slime mold, and many species of fungi, have more than one nucleus in a cell.

Visual Learning

Figure 3 Have students trace the stages of mitosis. After they have studied the diagram, have students list the stages as you write them on the board. Then have students provide details about each stage. L2

Caption Answer

Figure 3 almost all cell types

Quick Demo

Mitosis

Materials prepared slides of onion root tips or whitefish blastodiscs, projecting microscope

Estimated Time 10 minutes

Procedure Demonstrate mitosis by showing prepared slides of onion root tips or whitefish blastodiscs undergoing mitosis. Be sure to examine the entire slide to view all phases of mitosis.

Reading Check

Answer so the resulting nuclei will have identical DNA

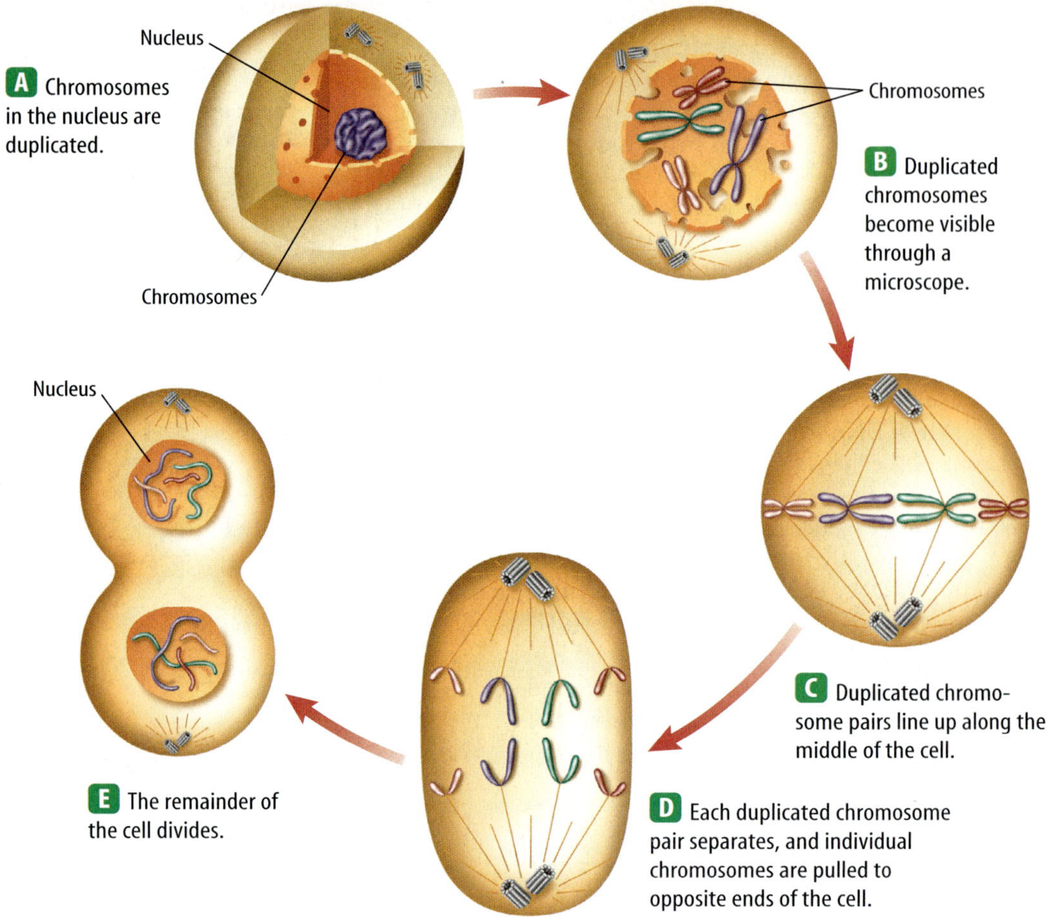

Figure 3 During cell division, cells go through several steps to produce two cells with identical nuclei.
Infer *what cell types undergo mitosis.*

Cell Division

How did you become the size you are now? The cells of your body are formed by cell division. Cell division has two big steps. First, DNA in the nucleus is copied. Then the nucleus divides into two identical nuclei. Each new nucleus receives a copy of the DNA. Division of the nucleus is called **mitosis** (mi TOH sus). Mitosis is the process that results in two nuclei, each with the same genetic information. You can follow the process of mitosis in **Figure 3.** After mitosis has taken place, the rest of the cell divides into two cells of about equal size. Almost all the cells in any plant or animal undergo mitosis. Whether it occurs in a plant or an animal, cell division results in growth and replaces aging, missing, or injured cells.

Reading Check *During cell division, why must the DNA be duplicated before the nucleus divides?*

Teacher FYI

Preparing for Cell Division Mitosis and cell division occur in all parts of the body, except in the gonads. To prepare for mitosis, DNA in the cell is copied, though the cell retains its normal appearance. The cell's appearance begins to change when the DNA begins to coil into visible pairs of chromosomes and the nuclear envelope disappears. The duplicated chromosomes are then divided into two new nuclei.

Reproduction by One Organism

Shoots growing from the eyes of a potato are a form of reproduction. Reproduction in which a new organism is produced from a part of another organism by cell division is called **asexual** (ay SEK shoo ul) **reproduction.** In asexual reproduction, all the DNA in the new organism comes from one other organism. The DNA of the growing potato eye is the same as the DNA in the rest of the potato.

Some one-celled organisms, such as bacteria, divide in half, forming two cells. Before the one-celled organism divides, its DNA copies itself. After it has divided, each new organism has an exact copy of the first organism's DNA. The two new cells are alike. The first organism no longer exists.

Budding and Regeneration Many plants and species of mushroom, and even a few animals reproduce asexually. The photo on the left in **Figure 4** shows asexual reproduction in hydra, a relative of jellyfish and corals. When a hydra reproduces asexually, a new individual grows on it by a process called budding. As you can see, the hydra bud has the same shape and characteristics as the parent organism. The bud matures and eventually breaks away to live on its own.

In a process called regeneration (rih je nuh RAY shun), some organisms are able to replace body parts that have been lost because of an injury. Sea stars can grow a new arm if one is broken off. Lizards, such as chameleons, can grow a new tail if theirs is broken off, as shown in **Figure 4** in the photo on the right.

Mini LAB

Observing Yeast Budding

Procedure

1. Use a **dropper** to place a drop of a **prepared yeast** and **sugar mixture** onto a **microscope slide.** Place a **coverslip** on the slide.
2. Examine the slide with a **microscope** under low power, then high power.
3. Record your observations in your **Science Journal.**
4. Make a new slide after 5 min. Examine the slide under low power, then under high power.
5. Record your observations in your Science Journal.

Analysis

1. What did you observe on the first slide?
2. What might account for any differences between what you observed on the first slide and the second slide?

Figure 4 Cell division can result in asexual reproduction or replacement of body parts.

Hydra reproduce asexually by budding.

A chameleon can regenerate, or regrow, a tail if it is broken off.

Mini LAB

Purpose Students observe asexual reproduction in yeast. L2 ELL LS **Visual-Spatial**

Preparation Immediately before class, dissolve about 10 g of dry yeast in about 250 mL of very warm water to which a small spoonful of sugar has been added.

Materials dry yeast, small jars or beakers, warm water, sugar, microscopes, microscope slides, coverslips, droppers

Teaching Strategy Review with students how to use and focus microscopes.

Troubleshooting Drops of yeast and sugar solution should be small so as not to overload the slide.

Analysis

1. A few cells may be starting to bud.
2. Yeast use sugar as a food source. Students should observe more cells budding during their second observation.

Assessment

Performance Have students research and list products that are made using yeast. Have them explain why yeast is used in each product. Use **Performance Assessment in the Science Classroom,** p. 97.

Differentiated Instruction

Challenge Have students do research on how seedless fruits are produced. They should include research about the different kinds of fruits that are produced without seeds, and how these plants reproduce. Students can make a chart about the different kinds of fruits that are produced as a seedless variety and how growers reproduce these plants. L3

Science Journal

Growth and Development Have students write in their Science Journals an essay about how they have grown and developed since they were infants. Essays should indicate that they have grown and developed over time because their cells have constantly divided and added more cells to their bodies. L2 LS **Linguistic**

Discussion

Identical Plants Plants grown from cuttings are genetically identical to each other and to the parent plant. When would this be advantageous? when a plant has a highly desired trait, such as extra-sweet fruit

Caption Answer

Figure 5 The plants did not come from one plant because there are flowers of different colors.

Fun Fact

The sperm head contains a nucleus, which fertilizes the egg. The middle of the sperm contains mitochondria, which provide energy for propulsion.

Reading Check

Answer A new organism is produced from the DNA of two cells.

Visual Learning

Figure 6 Have students study the photographs of egg and cells. Why is the egg so much larger than the sperm? The egg initially provides food for the growth and development of the fertilized egg. Explain that the shape of the sperm is related to how it functions. The long tail of the sperm helps it swim like a tadpole on its journey to an egg.

Figure 5 These African violets are clones.
Infer *whether or not the plants all came from one plant. Explain.*

Cloning What would it be like if humans or other animals were exact copies of each other? Making copies of organisms is called **cloning.** The new organism produced is called a clone. The clone receives DNA from just one parent cell. It has the same DNA as the parent cell. In many ways, cloning is not a new technology. In the past, most cloning was done with plants. Gardeners clone plants when they take cuttings of a plant's stems, leaves, or roots. They can grow many identical plants from one, as shown in **Figure 5.**

Only since the 1990s has cloning large animals become possible. In 1997, it was announced that an adult Finn Dorset sheep had been cloned. The new sheep, named Dolly, was the first successfully cloned mammal. The real value of Dolly is that scientists now have a better understanding of how cells reproduce.

Sex Cells and Reproduction

Does a human baby look exactly like its father or its mother? Usually, the baby has features of both of its parents. The baby might have her dad's hair color and her mom's eye color. However, the baby probably doesn't look exactly like either of her parents. That's because humans, as well as many other organisms, are the products of sexual (SEK shoo ul) reproduction. In **sexual reproduction** a new organism is produced from the DNA of two cells. **Sex cells,** as shown in **Figure 6,** are the specialized cells that carry DNA and join in sexual reproduction. During this process, DNA from each sex cell contributes to the formation of a new individual and that individual's traits.

Reading Check *What results from sexual reproduction?*

A human egg cell

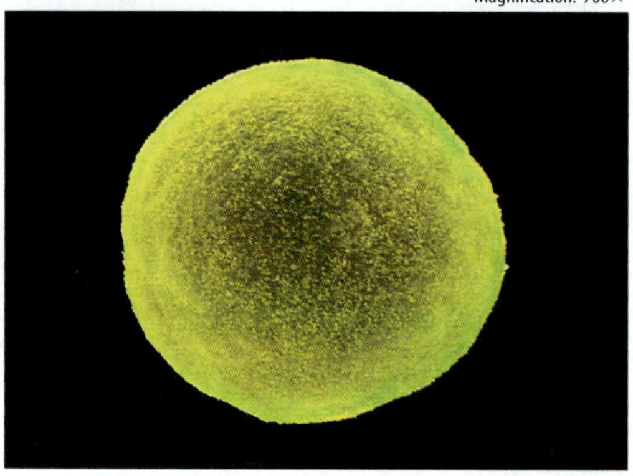

Human sperm cells

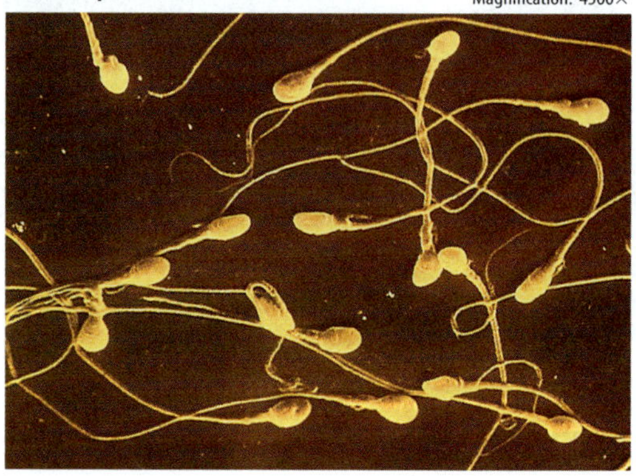

Figure 6 Specialized cells called sex cells are involved in reproduction. A female sex cell usually is called an egg and a male sex cell is usually called a sperm. Each type of sex cell in a human contains 23 chromosomes.

LAB DEMONSTRATION

Purpose to observe cells during meiosis

Materials projecting microscope, prepared slides of lily anthers undergoing meiosis

Procedure Provide students with labeled drawings of what they will see. Display various stages of meiosis with the projecting microscope.

Expected Outcome Students will observe stages of meiosis.

Assessment

What is the purpose of meiosis? to produce sex cells with half the number of chromosomes as the parent cell

594 CHAPTER 20 The Role of Genes in Inheritance

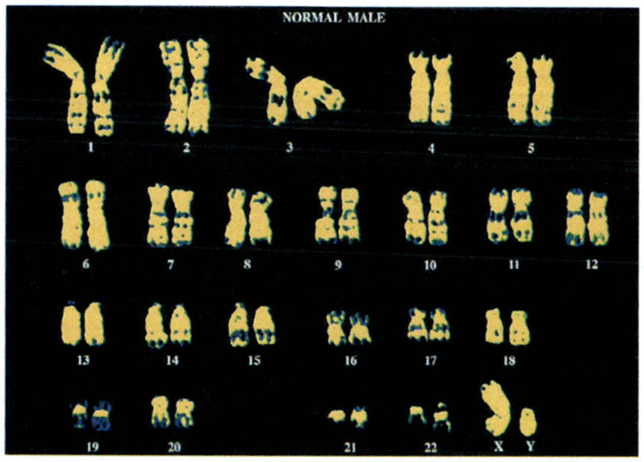

Figure 7 Each chromosome in a human body is made of DNA. All 23 pairs of chromosomes of one person are shown in this photograph.

Production of Sex Cells

Recall that your body is made up of different types of cells most of which were formed by mitosis. When a skin cell, a bone cell, or another body cell divides, it produces two new cells by cell division. Each cell has DNA that is identical to the original cell. Recall that DNA can be found in structures called chromosomes. A human body cell has 46 chromosomes arranged in 23 pairs, as shown in **Figure 7.** Each chromosome of a pair has genetic information about the same things. For example, if one chromosome has information about hair color, its mate also will have information about hair color.

Sex cells are different. Instead of being formed by cell division like body cells are, sex cells are formed by **meiosis** (mi OH sus). **Table 1** compares cell division and sex cell formation. Only certain cells in reproductive organs undergo the process of meiosis. Before meiosis begins, DNA is duplicated. During meiosis, the nucleus divides twice. Four sex cells form, each with half the number of chromosomes of the original cell. Human eggs and sperm contain only 23 chromosomes each—one chromosome from each pair of chromosomes. That way, when a human egg and sperm join in a process called **fertilization,** the result is a new individual with a full set of 46 chromosomes. **Figure 8** shows how sex cells join to form a cell that develops into a new human being.

Table 1 Cell Division and Sex Cell Formation in Humans

	Cell Division	Sex Cell Formation
Process used	Mitosis	Meiosis
DNA duplicated?	Yes	Yes
Nucleus divides	Once	Twice
Number of cells formed	2	4
Chromosome number of beginning cell	46	46
Chromosomes in each new cell	46	23

INTEGRATE Health

Cigarette Smoking In humans, sex cell production and fertilization can be affected by cigarette smoking. Cigarette smoking can decrease the number of sperm produced in the male body. Also, some of the sperm produced by a male that smokes may be deformed and unable to fertilize an egg.

Teacher FYI

Variability Sexual reproduction appears to benefit organisms because it results in an increase in variation within populations. Populations with a high genetic variability are better able to adapt to changing environmental conditions.

Visual Learning

Figure 7 Ask students to hypothesize why each pair of chromosomes is similar, except for pair 23. All pairs are made up of one chromosome from the mother and one from the father. Pair 23 determines the gender of the child; two X chromosomes produce a girl, an X and a Y produce a boy.

INTEGRATE Health

Cigarette Smoking Recent studies seem to indicate that smoking causes a decrease in fertility. Once pregnant, women smokers seem to have a higher risk of pregnancy complications, including ectopic pregnancies and miscarriages.

Research Have students research other affects of cigarette smoking. Some health-related issues caused by smoking include heart attacks, stroke, and cancer. Students can design a brochure that details these conditions and how smoking contributes to a person's risk of having them.

Differentiated Instruction

Learning Disabled Use pipe cleaners or modeling clay to model the processes of mitosis and meiosis for students who are having trouble understanding how the processes works. L2 LS **Visual-Spatial**

NATIONAL GEOGRAPHIC

Visualizing Human Reproduction

Have students examine the pictures and read the captions. Then ask the following questions.

How do sperm cells move? Possible answer: Sperm cells use a flagellum, a whiplike tail, to move through a liquid medium.

What does the fertilized egg do to develop into a fully formed baby? Possible answer: The cells undergo cell division, the result of which is growth.

Activity

Concept Map Have students make an events chain concept map showing the stages leading up to fertilization of the egg. egg is produced by ovary→egg travels to uterus in the fallopian tubes→egg enters uterus →sperm enters female reproductive system→sperm travels to uterus→ fertilization occurs when sperm and egg meet L2

NATIONAL GEOGRAPHIC **VISUALIZING HUMAN REPRODUCTION**

Figure 8

Humans, like most animals and plants, reproduce sexually. In sexual reproduction, a new and genetically unique individual is produced when a female sex cell and a male sex cell join in a process called fertilization.

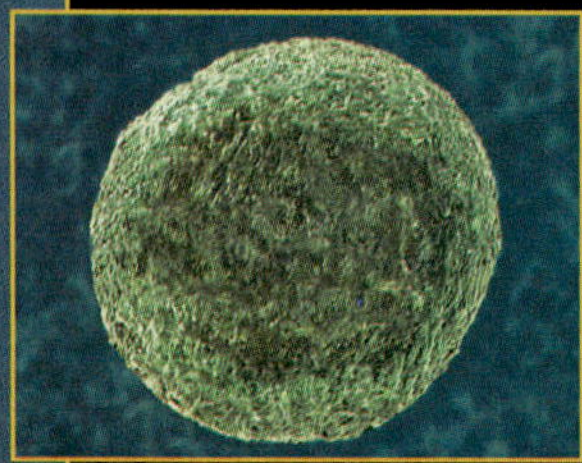

▲ A female's sex cell, an egg, has only 23 chromosomes—half the amount contained in human body cells.

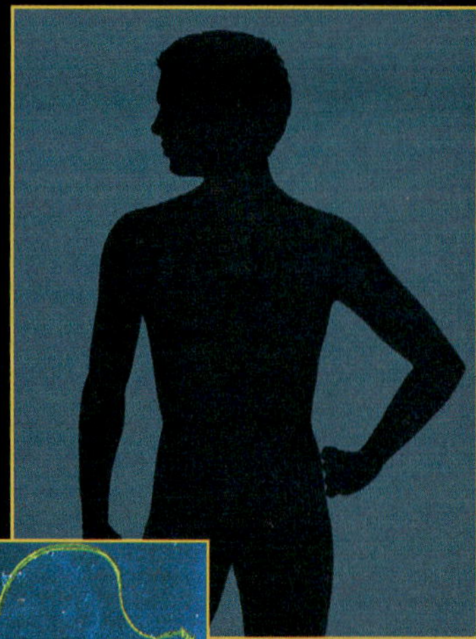

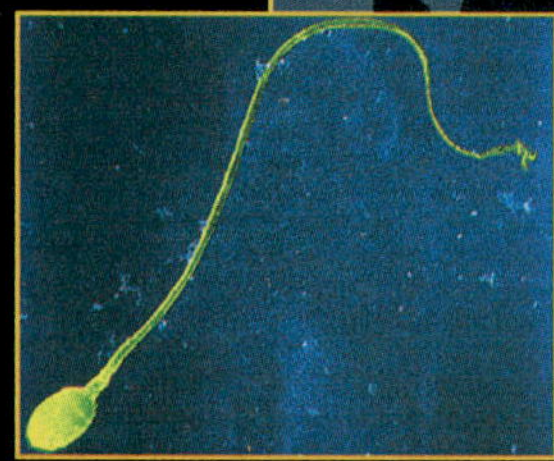

◀ A male's sex cell, a sperm, also contains only 23 chromosomes.

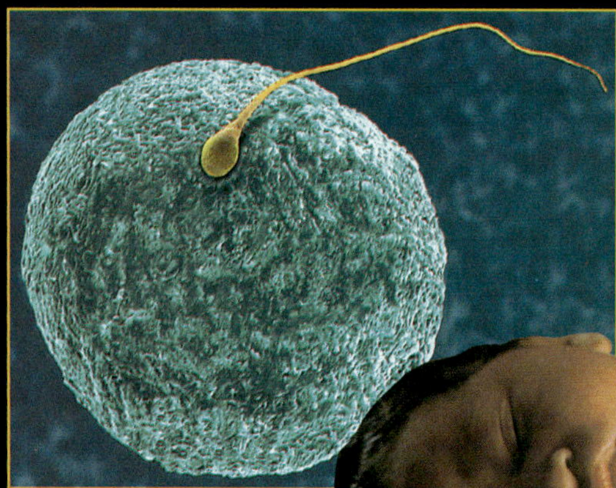

◀ When an egg and a sperm unite during fertilization, a new cell results that has a full set of 46 chromosomes.

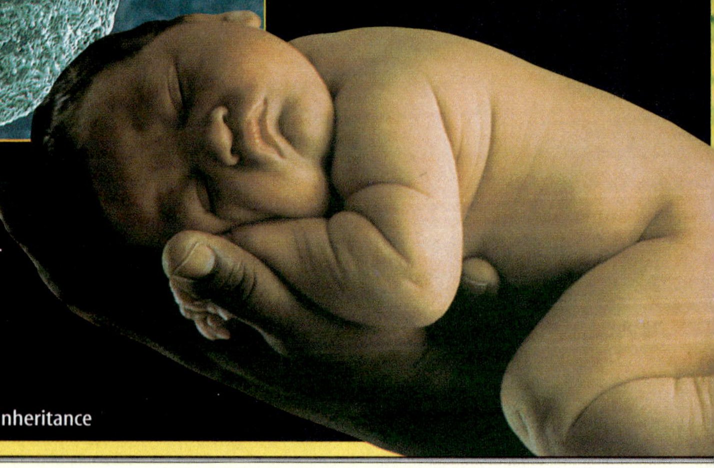

▶ This new cell divides again and again, developing over time into a fully formed baby.

Sex Cells in Plants

Plants can reproduce sexually. How this occurs is different for each plant group. But in every case, a sperm and an egg join to create a new cell that eventually becomes a plant.

It may seem that flowers are just a decoration for many plants, but flowers contain structures for reproducing. Male flower parts produce pollen, which contains sperm cells. Female flower parts produce eggs. When a sperm and an egg join, a new cell forms. In most flowers, rapid changes begin soon after fertilization. The cell divides many times and becomes enclosed in a protective seed. The petals and most other flower parts fall off. A fruit that contains seeds soon develops, as shown in **Figure 9.**

Figure 9 An apple flower will develop into an apple containing seeds if the eggs in the female reproductive structure are fertilized.

section 1 review

Summary

Reproduction

- Reproduction always involves the transfer of hereditary information.

Cell Division

- Mitosis results in two nuclei, each with the same genetic information.

Reproduction by One Organism

- There are two types of asexual reproduction—budding and regeneration.

Sex Cells and Reproduction

- DNA from each sex cell contributes to the formation of a new individual.

Production of Sex Cells

- Human eggs and sperm each contain only 23 chromosomes.

Sex Cells in Plants

- Flowers contain structures for reproduction.

Self Check

1. **Compare and contrast** the outcome of meiosis and the outcome of mitosis.
2. **Infer** why reproduction is an important process.
3. **Explain** why offspring produced by asexual reproduction are usually identical to the parent that produced them.
4. **Describe** how DNA controls how an organism looks and functions.
5. **Think Critically** For a species, what are some advantages of reproducing asexually? Of reproducing sexually? Of having the ability to do either?

Applying Math

6. **Calculate** A female bullfrog produces 350 eggs. All of the eggs are fertilized and hatch in one season. Assume that half of the tadpoles are male and half are female. If all the female tadpoles survive and, one year later, produce 350 eggs each, how many eggs would be produced?

Science Online red.msscience.com/self_check_quiz

Use Science Words

Word Meaning Have students find out the meaning of the word *diploid.* cells having pairs of chromosomes The diploid number is represented by $2n$, where n = the number of chromosomes in the sex cells. The number of chromosomes in a sex cell, n, is called haploid.

3 Assess

DAILY INTERVENTION

Check for Understanding

Visual-Kinesthetic Have students make a paper and string model of mitosis. The stages of mitosis can be connected using string, and colored string can be used to represent duplicated chromosomes. Students should include labels and descriptions of what occurs at each stage. L2

Reteach

Reproduction Have students make a table describing sexual and asexual reproduction. The table should include the type of cell division involved and the number of parent organisms required. L2

Assessment

Performance Have students make illustrations of meiosis and mitosis. Have them write captions explaining their illustrations. Use **Performance Assessment in the Science Classroom**, p. 127. L2 P

section 1 review

1. Mitosis produces two nuclei with the same genetic information; meiosis produces sex cells with half the number of chromosomes as the parent cell.
2. Reproduction allows the continuation of a species from one generation to the next.
3. All of the organism's DNA comes from this single parent.
4. DNA controls the proteins each cell will produce.
5. Asexual: can quickly take advantage of environmental resources; sexual: offspring with variations may better survive a changing environment; either: more adaptability.
6. If 175 females produce 350 eggs each and all hatch, then there would be 175 × 350, or 61,250 eggs produced the following year.

BENCH TESTED

Real-World Question

Purpose Students will learn to obtain DNA from onion cells. L2 ELL LS Kinesthetic

Process Skills measure, predict, infer, describe, observe

Time Required 40–50 minutes

Procedure

Preparation In a blender, thoroughly blend 1c chopped onion with 1/4 c warm water and 1 tsp. salt. Pour the mixture into a foam cup and add 1/4 c liquid dishwashing detergent. Mix gently for 5 minutes. Using a coffee filter, filter the liquid into a glass container (this takes about one hour). Just before using the mixture, stir 1/8 teaspoon of meat tenderizer into the liquid.

Safety Precautions Be sure there is no open flame in the presence of alcohol.

Teaching Strategy Limit the amount of alcohol in the room by having fewer but larger groups of students or by having small groups perform the activity one group at a time.

Conclude and Apply

1. Answers will vary. Most students will indicate that DNA should look the same in all organisms.
2. Yes, it could but it would be very difficult. It would require that the same number of cell nuclei be used for comparing the amount of DNA.

LAB

Getting DNA from Onion Cells

DNA contains the instructions for the processes that occur in a cell. In this lab, you will see the actual DNA of one living thing—an onion.

Real-World Question

How is DNA taken out of cells?

Goals

- **Separate** DNA from onion cells.
- **Practice** laboratory skills.

Materials

prepared onion mixture (125 mL)
toothpicks
small beaker
*measuring cup
rubbing alcohol (125 mL)
large beaker
*other glass container
magnifying lens
*microscope

*Alternate materials

Safety Precautions

Be sure to wear an apron and goggles throughout this lab. Keep hands away from face.

Procedure

1. Obtain 125 mL of prepared onion mixture from your teacher. Empty it into the large glass beaker or container.
2. Slowly pour 125 mL of rubbing alcohol down the side of the container onto the mixture. The alcohol should form a layer on top of the onion mixture.
3. **Observe** the gooey strings floating to the top. These strings are DNA.
4. Use a toothpick to gently stir the alcohol layer. Use another toothpick to remove the gooey DNA.
5. **Observe** DNA with a magnifying lens or a microscope. Record your observations in your Science Journal.
6. When you're finished, pour all liquids into containers provided by your teacher.

Conclude and Apply

1. Based on what you know about DNA, predict whether onion DNA is different from the DNA of other types of plants.
2. **Infer** whether this method of taking DNA out of cells could be used to compare the amount of DNA between different organisms. Explain your answer.

Communicating Your Data

Compare and contrast your findings with those of other students in your class. Explain in your Science Journal why your findings were the same or different from those of other students. **For more help, refer to the Science Skill Handbook.**

Assessment

Performance Have students develop step-by-step procedures for extracting DNA from other plants. Materials lists may also be prepared. Use **Performance Assessment in the Science Classroom,** p. 105.

Communicating Your Data

Students should have similar data that may be shared electronically, orally, or in posters.

section 2 Genetics—The Study of Inheritance

Heredity

When you go to a family reunion or browse through family pictures, like the one in **Figure 10,** you can't help but notice similarities and differences among your relatives. You notice that your mother's eyes look just like your grandmother's, and one uncle is tall while his brothers are short. These similarities and differences are the result of the way traits are passed from one generation to the next. **Heredity** (huh REH duh tee) is the passing of traits from parents to offspring. Solving the mystery of heredity has been one of the great success stories of biology.

Look around at the students in your classroom. What makes each person an individual? Is it hair or eye color? Is it the shape of a nose or the arch in a person's eyebrows? Eye color, hair color, skin color, nose shape, and many other features, including those inside an individual that can't be seen, are traits that are inherited from a person's parents. A trait is a physical characteristic of an organism. Every organism, including yourself, is made up of many traits. The study of how traits are passed from parents to offspring is called **genetics** (juh NE tihks).

Reading Check *What traits could you pass to your offspring?*

as you read

What You'll Learn

- **Explain** how traits are inherited.
- **Relate** chromosomes, genes, and DNA to one another.
- **Discuss** how mutations add variation to a population.

Why It's Important

You will understand why you have certain traits.

Review Vocabulary

genotype: the genetic makeup of an organism

New Vocabulary

- heredity
- genetics
- gene
- variation
- mutation

Figure 10 Family members often share similar physical features. These traits can be something obvious, like curly hair, or less obvious, such as color blindness.

1 Motivate

INTERACTIVE CHALKBOARD PowerPoint® Presentations

Bellringer

Section Focus Transparencies also are available on the Interactive Chalkboard CD-ROM. L2 ELL

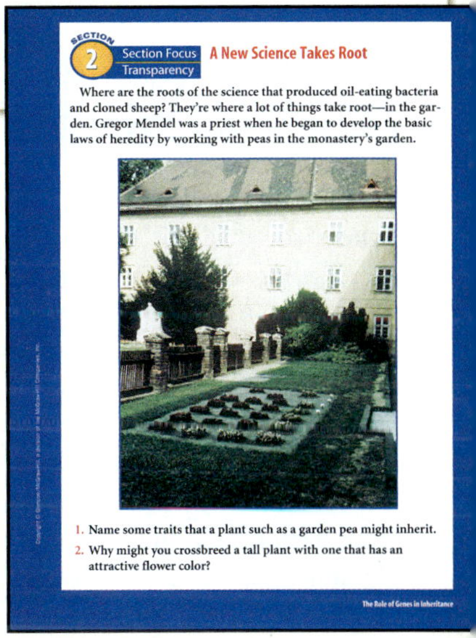
SECTION 2 Section Focus Transparency **A New Science Takes Root**

Where are the roots of the science that produced oil-eating bacteria and cloned sheep? They're where a lot of things take root—in the garden. Gregor Mendel was a priest when he began to develop the basic laws of heredity by working with peas in the monastery's garden.

1. Name some traits that a plant such as a garden pea might inherit.
2. Why might you crossbreed a tall plant with one that has an attractive flower color?

The Role of Genes in Inheritance

Tie to Prior Knowledge

Sexual Reproduction Have the class list the stages involved in sexual reproduction as described in the last section. Remind students that in sexual reproduction, both parents contribute genetic information to the offspring.

Reading Check

Answer Answers may vary but should include a genetic trait.

Section 2 Resource Manager

Chapter *FAST FILE* Resources

- Transparency Activity, p. 41
- Directed Reading for Content Mastery, pp. 19–20
- MiniLAB, p. 4
- Enrichment, p. 28
- Reinforcement, p. 26
- Lab Worksheet, pp. 7–8
- Lab Activity, pp. 13–14

Mathematics Skill Activities, p. 29

Lab Management and Safety, p. 58

2 Teach

Discussion

Chromosomes Direct students to look back at **Figure 7.** Review with students the main role of chromosome pair 23. It determines the gender of the offspring. Ask students to hypothesize which parent's DNA determines the gender of a child. The child's father is responsible for determining gender. Females are XX and produce only X eggs. Males are XY and produce X and Y sperm.

Inquiry Lab

Tests for Color Blindness

Purpose to design an experiment that tests for color blindness and to determine the percentages of males and females that are color blind

Possible Materials white paper or poster board; colored markers: red, orange, yellow, bright green, dark green, blue

Estimated Time one class period (one week if they test people outside of the classroom)

Teaching Strategies

- Students can decide which type of color blindness they want to test for—the most common being green-red color blindness.
- Students can use colored markers to prepare pictures with red, orange, and yellow as the background colors, and dark and light green as the picture or number.
- A data table should be prepared so that students can collect data and calculate the percentages of males and females that test positive for color blindness.
- Allow students to explore other questions that arise. L2

For additional inquiry activities, see *Science Inquiry Labs*.

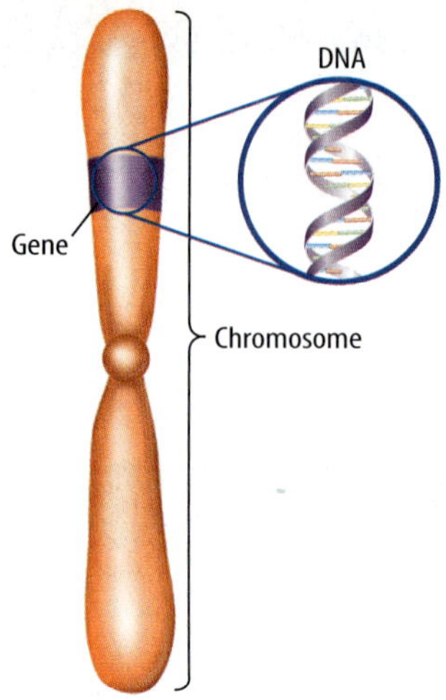

Figure 11 Hundreds of genes are located on each chromosome.

Genes All traits are inherited. Half of your genetic information came from your father, and half came from your mother. This information was contained in the chromosomes of the sperm and egg that joined and formed the cell that eventually became you.

Inherited traits, such as hair and eye color, are controlled by genes on chromosomes. Characteristics such as manners are called acquired skills. Acquired skills are not determined by genes.

As **Figure 11** shows, all chromosomes contain genes (JEENZ). A **gene** is a small section of DNA on a chromosome that has information about a trait. Humans have thousands of different genes arranged on 23 pairs of chromosomes. Genes control all of the traits of organisms—even traits that can't be seen, such as the size and shape of your stomach and your blood type. Genes provide all of the information needed for growth and life.

What determines traits?

Recall that in body cells, such as skin cells or muscle cells, chromosomes are in pairs. One pair of chromosomes can contain genes that control many different traits. Each gene on one chromosome of the pair has a similar gene on the other chromosome of the pair. Each gene of a gene pair is called an allele (uh LEEL), as shown in **Figure 12.** The genes that make up a gene pair might or might not contain the same information about a trait. For example, the genes for the flower color trait in pea plants might be purple or white. If a pair of chromosomes contains different alleles for a trait, that trait is called a hybrid (HI brud). When a trait has two identical alleles, it's called pure.

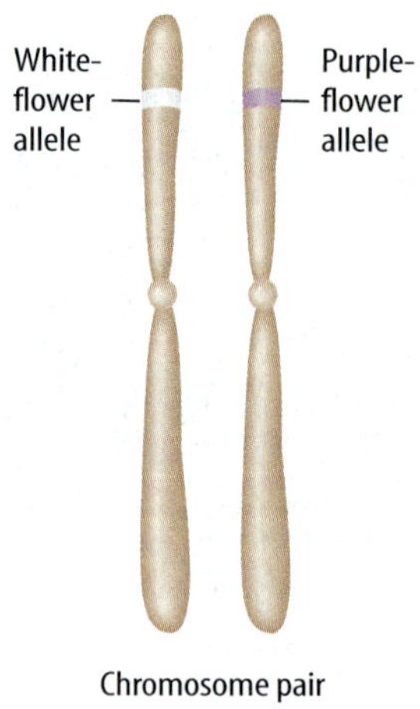

Figure 12 Pea flowers can be purple or white. The chromosome pair from a pea plant shows that both chromosomes have an allele for the flower color trait. A pea plant with this chromosome pair would have purple flowers.

Active Reading

Write-Draw-Discuss This strategy helps students to actively participate in reading and lectures, assimilating content creatively. Students write about an idea, clarify it, then make a drawing. Students share responses with the class and display several examples. Have students Write-Draw-Discuss about genetics. L2

Figure 13 Widow's peak is a dominant allele in humans. **Infer** *how many alleles for widow's peak this person might have.*

Caption Answer

Figure 13 The individual pictured could have 1 or 2 alleles for widow's peak. Since widow's peak is dominant, this individual could be hybrid or pure for the trait.

Dominant and Recessive Alleles The combination of alleles in a gene pair determines how a trait will be shown, or expressed, in an organism. In pea plants and other organisms, that depends on something called dominance (DAH muh nunts). Dominance means that one allele covers over or masks another allele of the trait. For instance, if a pea plant has one purple-flower allele and one white-flower allele or two purple-flower alleles, its flowers will be purple. Purple is the dominant flower color in pea plants. The dominant allele is seen when the trait is hybrid or dominant pure. White flowers, the masked allele, are said to be recessive. Recessive alleles are seen only when a trait is recessive pure.

Humans also have traits that are controlled by dominant and recessive alleles. These traits are controlled in the same way that dominant and recessive alleles are controlled in plants. To show a recessive allele, a person needs to inherit two copies of the recessive allele for that trait—one from their mother and one from their father. To show a dominant allele, a person can have either one or two alleles for the trait. One dominant allele in humans is the presence of a widow's peak, as shown in **Figure 13.**

Expression of Traits The traits of an organism are coded in the organism's DNA. However, the environment can play an important role in the way that a trait is shown, or expressed. You may know a person whose dark hair lightens when exposed to sunlight, or a person whose light skin darkens in sunlight. Human hair color and skin color are traits that are coded for by genes, but the environment can change the way that the traits appear. The environment can affect the expression of traits in every kind of organism, including bacteria, fungi, plants, and animals.

Sometimes the effect of the environment allows adaptations that aid in a species survival. For example, the arctic fox's fur color depends on the environment. In the winter months, the arctic fox does not produce fur pigment, and the fox's fur appears white. As a result, the fox blends with the snow, helping it to avoid predators. In the warmer months, the fox produces brown pigment, and the fox blends with the tundra.

Mini LAB

Modeling Probability

Procedure

1. Flip a **coin** ten times. Count the number of heads and the number of tails.
2. Record these data in a data table in your **Science Journal.**
3. Now flip the coin twenty times. Count the number of heads and tails.
4. Record these data in a data table in your Science Journal.

Analysis

1. What results did you expect when you flipped the coin ten times? Twenty times?
2. Were your observed results closer to your expected results when you flipped the coin more times?
3. How is the flipping of a coin similar to the joining of egg and sperm at fertilization?

Try at Home

Mini LAB

Purpose Students investigate how events occur by chance. L2 LS **Kinesthetic**

Materials pennies or other coins

Teaching Strategy Explain what a chance event is. Because a coin has two sides, there is an equal chance it will land heads up or tails up each time it is flipped.

Analysis

1. 10 flips: expect 5 heads, 5 tails; 20 flips: expect 10 heads, 10 tails
2. Answer may vary, but should be yes.
3. Both are chance events.

Assessment

Portfolio Assess students' knowledge about chance events by having them write short essays about how chance events affect their daily lives. They can be general or write specifically about past events. Use **Performance Assessment in the Science Classroom,** p. 157. P

Try at Home

Cultural Diversity

Population Genetics A group of people who live in a particular place for a long time and marry primarily within their own group may begin to have high percentages of particular alleles. An example is the allele that causes sickle-cell anemia. Some people whose descendants came from areas of Africa and the Mediterranean where malaria is common have these alleles. Having one allele for the trait provides some protection against malaria. However, having two alleles causes sickle-cell anemia. Another example is the allele that causes Tay-Sachs disease. This allele is more common among Eastern European Jews and their descendents than in other populations.

SECTION 2 Genetics—The Study of Inheritance 601

Text Question Answer

purple and white in a ratio of 1:1

Use an Analogy

Codes Explain that the genetic code is similar to other kinds of codes, (i.e. Morse code, bar codes). Describe Morse code to students or have them look it up in the encyclopedia. Point out that just as Morse code consists of a series of dots and dashes arranged in a particular order, the sequence of bases in DNA is meaningful only when it is arranged in particular ways. You may wish to point out similarities between bar codes and the autoradiographs produced by DNA profiling.

IDENTIFYING Misconceptions

Dominant v. Recessive Many students will think that dominance for a particular trait will cause that trait to occur more often than its corresponding recessive trait. Explain that this is not true. The gene for polydactyly (having more than five fingers or toes) is dominant to the normal gene. However, most people have two recessive (normal) genes for the number of digits, and thus have five fingers or toes on each hand or foot.

Caption Answer

Figure 14 purple

Passing Traits to Offspring

How are traits passed from parents to offspring during fertilization? The flower color trait in pea plants can be used as an example. Suppose a hybrid purple-flowered pea plant (one with two different alleles for flower color) is mated with a white-flowered pea plant. What color flowers will the offspring have?

The traits that a new pea plant will inherit depend upon which genes are carried in each plant's sex cells. Remember that sex cells are produced during meiosis. In sex cell formation, pairs of chromosomes duplicate, then separate as the four sex cells form. Therefore, gene pairs also separate. As a result, each sex cell contains one allele for each trait. Because the purple-flowered plant in **Figure 14** is a hybrid, half of its sex cells contain the purple-flower allele and half contain the white-flower allele. On the other hand, the white-flowered plant is recessive pure. The gene pair for flower color has two white alleles. All of the sex cells that it makes contain only the white-flower allele.

In fertilization, one sperm will join with one egg. Many events, such as flipping a coin and getting either heads or tails, are a matter of chance. In the same way, chance is involved in heredity. In the case of the pea plants, the chance was equal that the new pea plant would receive either the purple-flower allele or the white-flower allele from the hybrid plant.

Figure 14 The traits an organism has depends upon which genes were carried in the parents' sex cells. This diagram shows how the flower color trait is passed in pea plants.

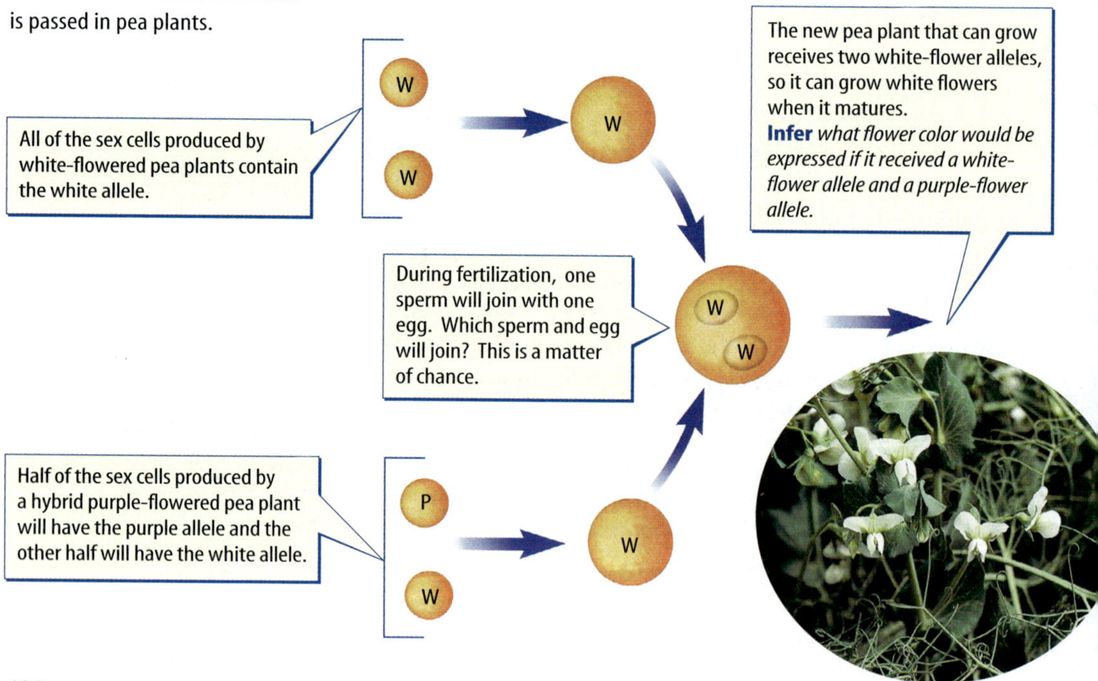

Differentiated Instruction

Challenge Explain to students the difference between fraternal and identical twins. Have them research why identical twins are important for researchers investigating the influence of genes versus the environment. Students can write a newspaper article about what kinds of differences and similarities are found between twins who were separated at birth. L3

Fun Fact

The color yellow is dominant in peas, but most of the peas we eat are green. This is an instance when the recessive allele is more desirable than the dominant allele.

Differences in Organisms

Now you know why a baby can have characteristics of either of its parents. The inherited genes from his or her parents determine hair color, skin color, eye color, and other traits. But what accounts for the differences, or variations (vayr ee AY shuns), in a family? **Variations** are the different ways that a certain trait appears, and they result from permanent changes in an organism's genes. Some gene changes produce small variations, and others produce large variations.

Applying Math — Find a Percentage

ALLELES IN SEX CELLS When sex cells form, each allele separates from its partner. Each sex cell will contain only one allele for each trait. Assume that a parent is a hybrid for a certain trait. That means that the parent has a dominant and a recessive allele for that trait. What percent of the parent's sex cells will contain the dominant allele?

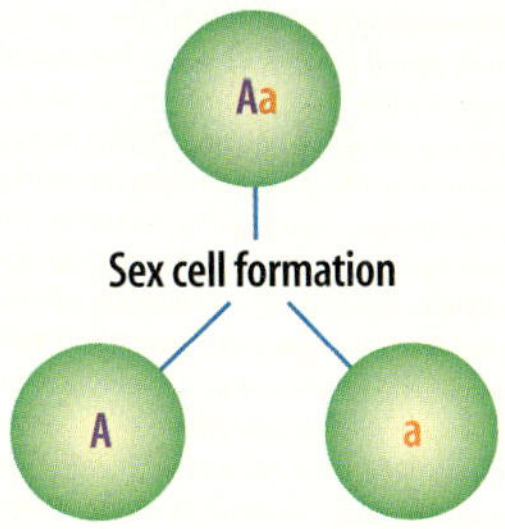

Solution

1. *This is what you know:* there are 2 possible alleles for the trait from the parent
2. *This is what you need to find out:* percent of sex cells with the dominant allele
3. *This is the procedure you need to use:*
   - Use the following equation:

     $$\frac{100\%}{\text{number of possible alleles from the parent}} = x$$

   - Substitute in the known value and solve.

     $$\frac{100\%}{2} = x \qquad x = 50\%$$

4. *Check your answer:* Multiply the number of possible alleles from the parent by the percent of sex cells with the dominant allele. You should get 100%.

Practice Problems

1. The attached-earlobes trait in humans is a recessive trait. What percent of the sex cells produced by a parent with attached earlobes would have an allele for this trait?
2. Assume one parent is a hybrid for a trait, and the other parent has 2 dominant alleles for the same trait. When the sex cells from the parents join, what is the percent chance that the offspring will have the recessive trait?

Science Online — For more practice, visit red.msscience.com/math_practice

Discussion

Purebred Pets Have students discuss the pros and cons of purebred animals. They should consider gene variation and the ability of the species to adapt. Purebred animals exhibit the desired traits of a breed. However, long-term inbreeding can increase number of undesired traits expressed. L2 IN **Interpersonal**

Applying Math

National Math Standards

Correlation to Mathematics Objectives

1, 2, 9

Answers to Practice Problems

1. 100%
2. 0%

Identifying Misconceptions

Inheritance Students may think that all inheritance is a result of blending, as opposed to Mendelian genetics. Refer to the F page at the beginning of this chapter to address this misconception.

Virtual Labs

Heredity *How are the traits of parents inherited and expressed in offspring?*

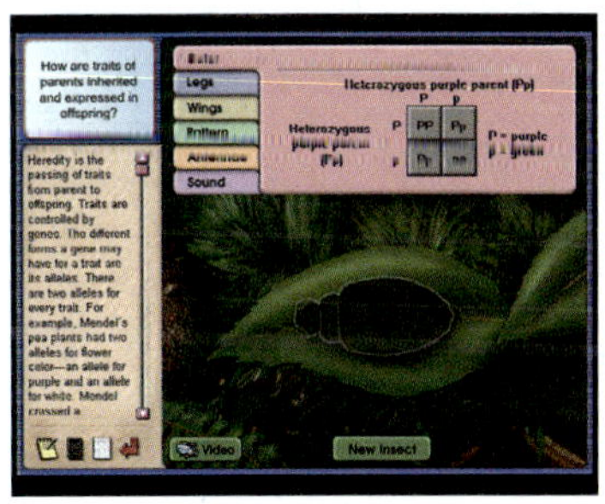

Teacher FYI

Color Blindness Color blindness is a recessive, but sex-linked trait. The gene for color blindness is located only on the X chromosome. A male (XY) will be color-blind if the X chromosome he inherits from his mother carries the recessive trait, as there is no corresponding allele on the Y chromosome. A color-blind female (XX) results only when the recessive allele is inherited from both parents.

SECTION 2 Genetics—The Study of Inheritance 603

Teacher FYI

Multiple Allele Traits Examples of traits governed by multiple alleles are coat color in rabbits and cats, feather color in parakeets, and blood type in humans.

Discussion

Mendel's Work Gregor Mendel did not know about genes, chromosomes, or DNA when he performed his studies with pea plants. When his work was "rediscovered" after his death, scientists quickly recognized its importance. Use this as a springboard to discuss the fact that most scientific discoveries don't happen overnight. Instead, scientific knowledge often builds on itself. Knowledge in one discipline may also be applied to other disciplines.

INTEGRATE Career

Genetic Counselor The field of genetic counseling is described as the communication process that deals with the probability that a genetic disorder will occur within a family. Medical professionals specialized in medical genetics and counseling are called genetic counselors. Counselors can help educate the public and help families find support and treatment for genetic disorders. Disorders that genetic counselors test for may include Huntington's disease, hemophilia, and sickle-cell disease.

Reading Check

Answer a change in a gene or chromosome

Figure 15 Traits in humans that show great variation usually are controlled by more than one gene pair.

The members of this family have different hair color.

Height is a trait that has many variations.

INTEGRATE Career

Genetic Counselor A genetic counselor is a medical professional who can help determine the chances of having a child with a genetic disorder. Genetic counselors test for genetic disorders and can provide help with treatment options. Investigate genetic disorders that can be tested for by a genetic counselor. In your Science Journal, write about one of the disorders.

Multiple Alleles and Multiple Genes Earlier, you learned how the flower color trait in pea plants is passed from parent to offspring. Flower color in pea plants shows a simple pattern of inheritance. Sometimes, though, the pattern of inheritance of a trait is not so simple. Many traits in organisms are controlled by more than two alleles. For example, in humans, multiple alleles A, B, and O control blood types A, B, AB, or O.

Traits also can be controlled by more than one gene pair. For humans, hair color, as shown in **Figure 15,** height, also shown in **Figure 15,** weight, eye color, and skin color, are traits that are controlled by several gene pairs. This type of inheritance is the reason for the differences, or variations, in a species.

Mutations—The Source of New Variation If you've searched successfully through a patch of clover for one with four leaves instead of three, you've come face-to-face with a mutation (myew TAY shun). A four-leaf clover is the result of a mutation. The word *mutate* simply means "to change." In genetics, a **mutation** is a change in a gene or chromosome. This can happen because of an error during meiosis or mitosis or because of something in the environment. Many mutations happen by chance.

Reading Check *What is a mutation?*

What are the effects of mutations? Sometimes mutations affect the way cells grow, repair, and maintain themselves. This type of mutation is usually harmful to the organism. Many mutations, such as a four-leaf clover, have a neutral effect. Whether a mutation is beneficial, harmful, or neutral, all mutations add variation to the genes of a species.

Differentiated Instruction

English-Language Learners To help these students, pair students according to ability. Have students write flash cards with the key terms dealing with the heredity and traits. On their cards they should include genes, dominant and recessive alleles, and mutation.

Visual Learning

Figure 15 Have students explain why humans are so different in height. Height is controlled by many genes. Environmental conditions such as the amount and quality of food one eats also have an effect.

Figure 16 Dairy cattle are bred selectively for the amount of milk that they can produce.

Selective Breeding Sometimes, a mutation produces a different version of a trait that many people find attractive. To continue this trait, selective breeding is practiced.

Nearly all breeding of animals is based on their observable traits and is controlled, instead of being random. For many years, cattle, like the one in **Figure 16,** have been bred on the basis of how much milk they can produce. Racehorses are bred according to how fast they run. It eventually was learned that in a few generations, breeding closely related animals produced an increased percentage of offspring with the desired traits.

Science Online

Topic: Selective Breeding

Visit red.msscience.com for Web links to information about selective breeding.

Activity List one type of plant and one type of animal that are bred selectively. Include the specific traits for which each is bred.

section 2 review

Summary

Heredity

- Genetics is the study of how traits are passed from parents to offspring.
- Genes control all the inheritable traits of organisms.

What determines traits?

- The expression of a trait is determined by the combination of alleles in a gene pair.
- A hybrid has a pair of chromosomes with different alleles for a trait.

Passing Traits to Offspring

- Traits of an offspring are determined by which genes are carried by its parents' sex cells.

Differences in Organisms

- Many traits are controlled by more than two alleles.
- More than one gene pair can control a certain trait.
- Mutations can happen by chance.

Self Check

1. **Define** the term *heredity.*
2. **State** which alleles for a trait must be present for a recessive allele to be expressed.
3. **Describe** how the chromosomes in human body cells are arranged.
4. **Explain** how mutations add variation to the genes of a species.
5. **Think Critically** What might happen if two hybrid purple-flowered pea plants are mated? What possible flower colors could the offspring have? Explain.

Applying Skills

6. **Concept Map** Make a concept map that shows the relationships between the following concepts: *genetics, genes, chromosomes, DNA, variation,* and *mutation.*
7. **Communicate** Research to find what a transgenic organism is, then find books or articles about these organisms. In your Science Journal, write a paragraph summary of your findings.

3 Assess

DAILY INTERVENTION

Check for Understanding

Visual-Kinesthetic Have students make a mobile demonstrating how flower color is passed in pea plants. The mobile should include drawings of chromosomes and chromosome pairs with the white and purple alleles labeled. Students can arrange the mobile to demonstrate 2 sets of sex cells combining to form a new pea plant. L2

Reteach

Models Allow students to manipulate colored beads to model single-trait inheritance. Then have them model the process of meiosis using the objects. L1 LS **Kinesthetic**

Assessment

Portfolio Assign students a genetic disorder to research (i.e. albinism, sickle-cell anemia, cystic fibrosis, Huntington's disease). Have them present their findings as a poster, report, or multimedia presentation. Use **Performance Assessment in the Science Classroom,** p. 97. L2 P

section 2 review

1. the transmission of genetic information
2. two recessive alleles
3. Human chromosomes are arranged in 23 pairs.
4. Mutations are changes in DNA that produce variations in traits.
5. 3/4 of the offspring would have purple flowers; 1/4 of the offspring would have white flowers.
6. Student maps should show that genes are made of DNA and located on chromosomes, variations are differences in genes, mutations are new variations caused by a change in a gene or chromosome, and genetics is the study of how traits are passed to offspring.
7. Transgenic organisms are those that contain genetic material that has been taken from another species. Many plants have been altered in this way to produce species that are more disease-resistant.

BENCH TESTED

Real-World Question

Purpose Students will investigate dominant and recessive genetic traits.

Time Required three to five days

Possible Materials Use other science reference books to identify dominant and recessive genetic traits.

Make a Plan

Preparation

Internet To run through the steps that the students will follow, visit red.msscience.com/internet_lab.

Non-Internet Collect books and reference sources about genetics.

Possible Hypotheses Students may decide to investigate eye color, earlobe attachment, and hair color. Students may hypothesize that most classmates have blue eyes, attached earlobes, and red hair.

LAB Use the Internet

Genetic Traits: The Unique You

Goals

- **Identify** genetic traits.
- **Collect** data about three specific human genetic traits.
- **Investigate** what are dominant and recessive alleles.
- **Graph** your results and then communicate them to other students.

Data Source

Science Online

Visit red.msscience.com/internet_lab to get more information about human genetic traits and for data collected by other students.

Real-World Question

What makes you unique? Unless you have an identical twin, no other person has the same combination of genes as you do. To learn more about three human genetic traits, you will collect data about your classmates. When you compare the data you collected with data from other students, you'll see that patterns develop in the frequency of types of traits that are present within a group of people. How are three genetic traits expressed among your classmates? Genetic traits can be dominant or recessive. Form a hypothesis about which trait, dominant or recessive, will be expressed by more people.

Make a Plan

1. **Research** general information about human genetic traits.
2. Search reference sources to find out which form of each characteristic being studied is dominant and which form is recessive.
3. **Survey** the students in your class to collect data about the three genetic traits being studied.

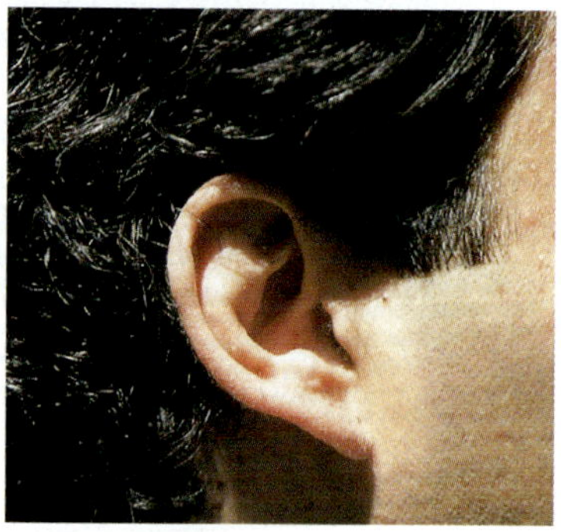

Attached earlobe

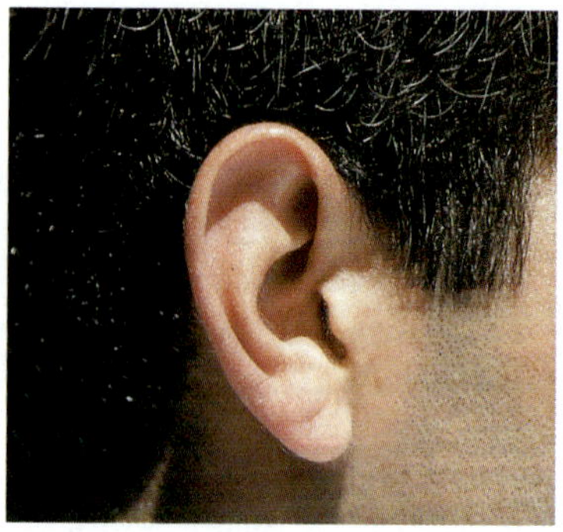

Detached earlobe

Alternative Inquiry Lab

Genetic Traits To observe genetic traits, give students a more personal investment into the Real-World Question by telling them they are medical researchers interested in following traits and determining what traits are more prevalent. Students can explore Web sites that describe dominant and recessive traits. Science reference books can be used to identify dominant and recessive traits that are involved in genetic disorders. Students can make a table of traits and graph the number of people exhibiting the genetic traits that were investigated. Allow students to explore other questions that arise. L2

606 CHAPTER 20 The Role of Genes in Inheritance

Using Scientific Methods

Follow Your Plan

1. Make sure your teacher approves your plan before you start.
2. **Record** your data in your Science Journal. Use frequency data tables to organize your data.

Genetic Traits	
Dimples	No Dimples
Attached Earlobes	Detached Earlobes
Widow's Peak	No Widow's Peak

Analyze Your Data

1. **Record** the total number of people included in your survey.
2. **Calculate** the number of people who show each form of each of the three traits that are being studied. Record each of these numbers in your Science Journal.
3. **Graph** the data you collected on a bar graph. Bars should represent the numbers of students exhibiting each of the different genetic traits you investigated.
4. **Compare** the data among each of the three genetic traits you explored.

Conclude and Apply

1. **Determine** Think about the genetic traits you investigated. Which traits were most common in the people you surveyed?
2. **Infer** Might surveying a larger group of people give different results?
3. **Analyze Results** Which genetic traits are least commonly found?
4. **Interpret Data** In the people you surveyed, were dominant alleles present more often than recessive alleles?

Communicating Your Data

Find this lab using the link below. **Post** your data in the table provided. **Compare** your data to that of other students. Combine your data with that of other students and **graph** the combined data on a bar graph.

Science Online
red.msscience.com/internet_lab

Follow Your Plan

Teaching Strategies Have students create a data collection worksheet, to help them organize the information they collect.

Analyze Your Data

Answers will be subjective and based on the students' individual research.

Conclude and Apply

Answers will be individualized and often based on the students' opinion of their research. Look for depth and quality of research performed.

Assessment

Performance Have students use the data from the bar graph to make pie charts of each of the genetic traits. Use **Performance Assessment in the Science Classroom**, p. 111.

Communicating Your Data

Have students use their bar graph to discuss the number of students surveyed and the percentages of students with each type of trait.

TIME

Content Background

Identical twins result when a fertilized egg splits, usually one to fourteen days after conception. Identical twins occur in 4 of 1000 births. They have the same DNA and usually have a lot of the same physical features, including blood type, hair and eye color, and shape of facial features. But they are not totally the same. Each twin is an individual and has different traits.

Identical twins usually spend a lot of time together. They are usually each other's best friend and do not like to be apart. Many identical twins, especially those raised together, have their own private language. In addition, many sets of twins are slow in language development as a result of their ability to "communicate" with their twin. They possess an intense bond that grew while they were inside their mother. After age 9 or 10, identicals usually choose to form their own identities.

Discussion

Separated Twins How would you explain the reasons why separated twins still have similar characteristics? Possible answer: Students will either talk about genetic similarities or explain that it could be because of how and who raised the twins.

Activity

Theories Organize the class into two different groups according to whether they believe the nature or nurture theories. Have each group research their theory and present possible explanations for their theory in a poster board presentation. Display the posters and engage the students in a dialogue about how the theories are different. L2 LS **Logical-Mathematical**

TIME SCIENCE AND Society SCIENCE ISSUES THAT AFFECT YOU!

SEPARATED AT BIRTH

Are genes or the people who raised you important in determining personality?

These twins, separated at birth and reunited as adults, had the same kind of job, drove the same kind of car, and had the same hobbies.

When Barbara Herbert was about 40, she met her long lost twin sister, Daphne Goodship. She had not seen her since infancy. The two grew up in separate homes with separate families. Because they are identical twins, it makes sense that they look alike. What was shocking, however, was the number of coincidences in their lives. Although they were not in contact while growing up, they shared identical experiences. Both women:

- dropped out of school at age 14,
- got jobs working for the local government,
- met their future husbands at age 16,
- gave birth to two boys and one girl,
- are squeamish about blood and heights, and
- drink their coffee cold.

In the genes?

Barbara and Daphne are part of an ongoing scientific study at Minnesota's Center for Twin and Adoption Research, which examines twins who were separated at birth. This research is helping scientists to understand better what is stronger in a person's development—genetic makeup, or how and by whom twins are raised.

Identical twins make ideal subjects for this research because their genetic makeups are identical. First, a psychological assessment is made, using personality tests, job interest questions, mental ability, and I.Q. Tests. Then scientists analyze the twins' backgrounds, including where they were raised, what their parents were like, and what schools they attended. These help determine whether a person's habits and personality are based on genetic makeup or social interactions.

Recently, a pair of twins were reunited after more than 30 years. Both twins said they felt like they have known each other all their lives. And, perhaps, thanks to their genes, they have!

Interview Find a pair of identical twins that go to your school or live in your community. Make a list of 10 questions and interview each of the twins separately. Write down their answers, or tape-record them. Compare the responses, then share your findings with the class.

Science Online
For more information, visit red.msscience.com/time

Interview If students know of twins outside the school, encourage them to interview them and to ask for baby photos and recent photos, if available. Students should include these in their presentations. This would allow students to compare differences and similarities over time to see how each twin develops his or her own sense of identity.

Resources for Teachers and Students

Double Take: The Story of Twins, by Daniel Jussim, Viking Children's Books, 2001

Twins: And What They Tell Us About Who We Are, by Laurence Wright, New York: John Wiley and Sons, 1997

Reviewing Main Ideas

Section 1 Continuing Life

1. Reproduction is an important process for all living things.
2. During reproduction, information stored in DNA is passed from parent to offspring.
3. Mitosis is the process that results in two nuclei with the same genetic information.
4. Organisms can reproduce sexually or asexually.
5. DNA is shaped liked a twisted ladder. An organism's DNA contains all of the information about how it will look and function.

Section 2 Genetics—The Study of Inheritance

1. Genetics is the study of how traits are passed from parent to offspring.
2. Genes are small sections of DNA on chromosomes. Each gene has information about a specific trait.
3. Chromosomes are found in pairs. For each gene on a particular chromosome, a gene with information about the same trait can be found on the other chromosome of the chromosome pair. Each gene of a gene pair is called an allele.
4. The way a trait is shown depends on the combination of dominant and recessive alleles carried on the chromosome pair.
5. Mutations are changes in a gene or chromosome. Mutations are a source of variation in populations.
6. Selective breeding allows favorable traits of organisms to be passed from one generation to the next.

Visualizing Main Ideas

Copy and complete the following concept map on reproduction.

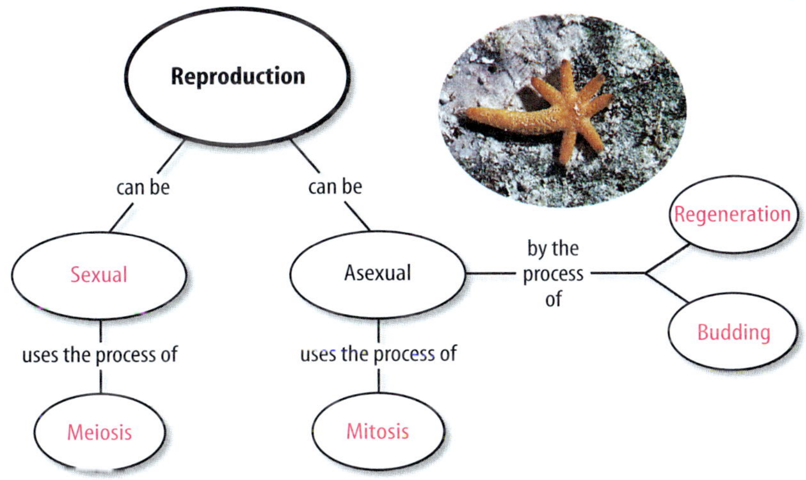

Science Online red.msscience.com/interactive_tutor

chapter 20 Study Guide

Reviewing Main Ideas

Summary statements can be used by students to review the major concepts of the chapter.

Visualizing Main Ideas

See student page.

Science Online

Visit red.msscience.com
/self_check_quiz
/interactive_tutor
/vocabulary_puzzlemaker
/chapter_review
/standardized_test

Assessment Transparency

For additional assessment questions, use the *Assessment Transparency* located in the transparency book.

Assessment

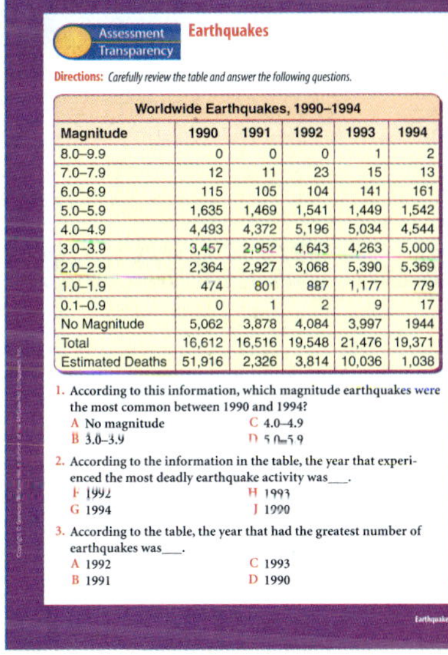

Assessment Transparency **Earthquakes**

Directions: *Carefully review the table and answer the following questions.*

Worldwide Earthquakes, 1990–1994					
Magnitude	1990	1991	1992	1993	1994
8.0–9.9	0	0	0	1	2
7.0–7.9	12	11	23	15	13
6.0–6.9	115	105	104	141	161
5.0–5.9	1,635	1,469	1,541	1,449	1,542
4.0–4.9	4,493	4,372	5,196	5,034	4,544
3.0–3.9	3,457	2,952	4,643	4,263	5,000
2.0–2.9	2,364	2,927	3,068	5,390	5,369
1.0–1.9	474	801	887	1,177	779
0.1–0.9	0	1	2	9	17
No Magnitude	5,062	3,878	4,084	3,997	1944
Total	16,612	16,516	19,548	21,476	19,371
Estimated Deaths	51,916	2,326	3,814	10,036	1,038

1. According to this information, which magnitude earthquakes were the most common between 1990 and 1994?
   A No magnitude
   B 3.0–3.9
   C 4.0–4.9
   D 5.0–5.9
2. According to the information in the table, the year that experienced the most deadly earthquake activity was___.
   F 1992
   G 1994
   H 1993
   J 1990
3. According to the table, the year that had the greatest number of earthquakes was___.
   A 1992
   B 1991
   C 1993
   D 1990

Earthquakes

Identifying Misconceptions

Assess

Use the assessment as follow-up to page F at the beginning of this chapter.

Activity Make a table containing alternate genes for a trait and indicate which is dominant, which is recessive. Ask students what are the phenotypes of these genetic combinations. Leave blanks for students to fill in.

Expected Outcome If students understand dominance and recessiveness, they should be able to fill in the table correctly. They may not choose the dominant trait if they see a genetic makeup with both genes.

Using Vocabulary

1. Mitosis replicates a set of chromosomes exactly; meiosis reduces the chromosome number by one-half.
2. Asexual reproduction requires one organism and occurs without meiosis; sexual reproduction requires two organisms and includes meiosis.
3. Cloning—asexual reproduction that produces exact genetic copies; variations result from the pairing of chromosomes during sexual reproduction.
4. Fertilization is part of the process of sexual reproduction.
5. A mutation is a change in DNA, variation is the result of mutation.
6. A gene is a segment of DNA that codes for a trait.
7. Mitosis is the process of replicating chromosomes in asexual reproduction.
8. Sperm and eggs are sex cells that result from meiosis.
9. Genetics is the study of heredity; a gene contains genetic information for one trait.
10. DNA is genetic material; a mutation is a change in an organism's DNA.

Checking Concepts

11. B	14. C	17. B
12. D	15. C	18. B
13. A	16. C	19. C

Thinking Critically

20. Chromosomes are made up of genes which are small sections of DNA.
21. Several genes determine eye color. Blue eyes are a recessive trait, so the parents must each be carrying the blue eye allele.
22. Meiosis is the process in which sex cells are formed that have half the number of chromosomes. If meiosis did not occur, each new generation would have twice the number of chromosomes as the preceding generation.

chapter 20 Review

Using Vocabulary

asexual reproduction p. 593	meiosis p. 595
cloning p. 594	mitosis p. 592
DNA p. 590	mutation p. 604
fertilization p. 595	sex cell p. 594
gene p. 600	sexual reproduction p. 594
genetics p. 599	variation p. 603
heredity p. 599	

Explain the differences between the vocabulary words in each of the following sets.

1. mitosis—meiosis
2. asexual reproduction—sexual reproduction
3. cloning—variation
4. fertilization—sexual reproduction
5. mutation—variation
6. gene—DNA
7. asexual reproduction—mitosis
8. sex cells—meiosis
9. genetics—gene
10. DNA—mutation

Checking Concepts

Choose the word or phrase that best answers the question.

11. Which of the following is reproduction that requires male and female sex cells?
    A) asexual reproduction
    B) sexual reproduction
    C) mitosis
    D) heredity
12. What is any change in the DNA of a gene or chromosome called?
    A) an embryo **C)** a clone
    B) sex cells **D)** a mutation
13. What is the small section of DNA that contains the code for a trait?
    A) a gene **C)** a variation
    B) heredity **D)** a cell
14. Which of these is another name for an observable feature or characteristic of an organism?
    A) sex cell **C)** trait
    B) embryo **D)** gene
15. How are specialized breeds of dogs, cats, horses, and other animals produced?
    A) regeneration
    B) asexual reproduction
    C) selective breeding
    D) budding
16. What is the passing of traits from parent to offspring called?
    A) genetics **C)** heredity
    B) variation **D)** meiosis
17. What are sperm and eggs?
    A) variations **C)** mutations
    B) sex cells **D)** genes
18. What is formed during meiosis?
    A) heredity **C)** clones
    B) sex cells **D)** fertilization

Use the photo below to answer question 19.

19. What flower color trait(s) would be in the sex cells of the above pea plant?
    A) purple only **C)** purple and white
    B) white only **D)** pink only

Science Online red.msscience.com/vocabulary_puzzlemaker

Use the ExamView® Pro Testmaker CD-ROM to:

- create multiple versions of tests
- create modified tests with one mouse click for inclusion students
- edit existing questions and add your own questions
- build tests aligned with state standards using built-in State Curriculum Tags
- change English tests to Spanish with one mouse click and vice versa

Thinking Critically

20. **Explain** the relationship among DNA, genes, and chromosomes.

21. **Communicate** Two brown-eyed parents have a baby with blue eyes. Explain how this could happen.

22. **Describe** how the process of meiosis is important in sexual reproduction.

23. **Explain** how a mutation in a gene could be beneficial to an organism.

24. **Draw Conclusions** Some mutations are harmful to organisms. Others are beneficial, and some have no effect at all. Which type of mutation would be least likely to be passed on to future generations?

25. **Compare and contrast** sexual and asexual reproduction.

26. **Recognize Cause and Effect** What is the role of meiosis and mitosis in the fertilization and development that results in a human baby?

Use the photo below to answer question 27.

27. **Infer** why this plant is an example of asexual reproduction. How could the plant reproduce through sexual reproduction?

Performance Activities

28. **Scientific Drawing** Use your imagination and make illustrations for each of the following vocabulary words: *asexual reproduction*, *genetics*, and *mutation*.

29. **Newspaper Article** Many scientists have reported that it is possible to get DNA from prehistoric creatures. Go to the library and find a newspaper article that describes the discovery of ancient DNA. Write a summary of the article in your Science Journal.

Applying Math

30. **Cell Division** If a cell undergoes cell division every 20 minutes, how many cells will there be after 24 hours?

31. **Meiosis** Five cells undergo meiosis to form sex cells. How many sex cells are formed?

32. **Human Genome** Assume the human genome is 3 billion base pairs. If one million base pairs of DNA take up 1 megabyte of storage space on a computer, how many gigabytes (1,024 megabytes) would the whole human genome fill?

Use the illustration below to answer question 33.

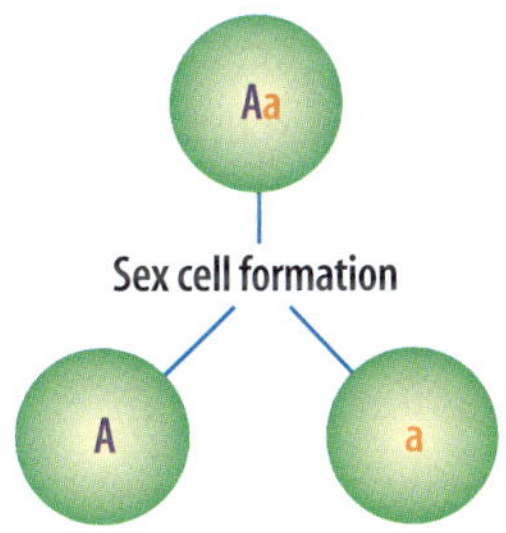

33. **Wrinkled Seeds** In pea plants, wrinkled seeds are recessive to round seeds. If two hybrid pea plants are crossed, what is the percent chance that the offspring will have the wrinkled-seed trait?

Science Online red.msscience.com/chapter_review

Assessment Resources

Reproducible Masters

Chapter *Fast File* Resources
Chapter Review, pp. 33–34
Chapter Tests, pp. 35–38
Assessment Transparency Activity, p. 45

Glencoe Science Web site
Chapter Review Test
Standardized Test Practice

Glencoe Technology

Assessment Transparency
Exam*View*® Pro Testmaker
MindJogger Videoquiz
Interactive Chalkboard

23. The affected trait may help the organism be better adapted to its environment.
24. Harmful mutations; organisms with harmful mutations are less likely to survive and reproduce.
25. Sexual reproduction: occurs by meiosis, results in genetically different offspring; asexual reproduction: occurs by cell division, offspring are genetically identical to the parent.
26. Meiosis results in sex cells that join during fertilization. Mitosis and cell division gives rise to all other cells as the baby develops.
27. New spider plants grow from runners that emerge from the adult plant. Because only one parent is involved, it is a form of asexual reproduction. Spider plants can reproduce through sexual reproduction when the plants flower and form seeds.

Performance Activities

28. Illustrations will vary. Use **PASC,** p. 127.
29. Research will reveal that DNA found in fossils is fragmented and incomplete; it could not reproduce the original organism. Use **PASC,** p. 141.

Applying Math

National Math Standards

1, 2, 9

30. 60 minutes ÷ 20 minutes = 3 cell divisions/hour; 24 × 3 = 72 cell divisions/24 hours. $2^{72} = 4.7224 \times 10^{21}$ cells
31. 5 × 4 = 20 sex cells formed
32. about 3 gigabytes (1 million base pairs = 1 megabyte; 3 billion base pairs = 3,000 megabytes; 3,000 ÷ 1,024 = 2.92 gigabytes or about 3 gigabytes)
33. 25%

FAST FILE

Answer Sheet A practice answer sheet can be found at red.msscience.com/answer_sheet.

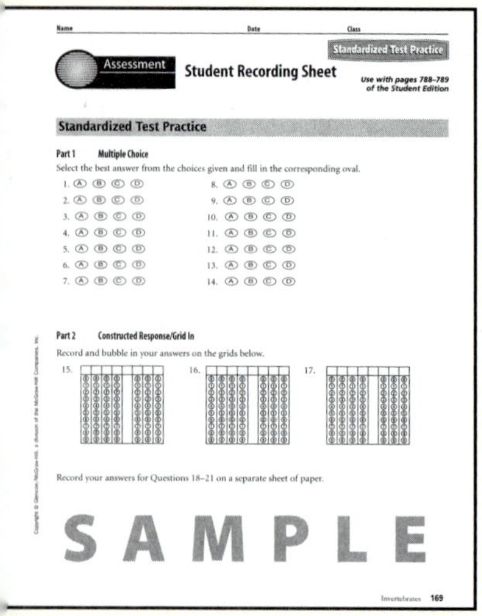

Name Date Class

Standardized Test Practice

Assessment Student Recording Sheet

Use with pages 788–789 of the Student Edition

Standardized Test Practice

Part 1 Multiple Choice

Select the best answer from the choices given and fill in the corresponding oval.

1. Ⓐ Ⓑ Ⓒ Ⓓ
2. Ⓐ Ⓑ Ⓒ Ⓓ
3. Ⓐ Ⓑ Ⓒ Ⓓ
4. Ⓐ Ⓑ Ⓒ Ⓓ
5. Ⓐ Ⓑ Ⓒ Ⓓ
6. Ⓐ Ⓑ Ⓒ Ⓓ
7. Ⓐ Ⓑ Ⓒ Ⓓ
8. Ⓐ Ⓑ Ⓒ Ⓓ
9. Ⓐ Ⓑ Ⓒ Ⓓ
10. Ⓐ Ⓑ Ⓒ Ⓓ
11. Ⓐ Ⓑ Ⓒ Ⓓ
12. Ⓐ Ⓑ Ⓒ Ⓓ
13. Ⓐ Ⓑ Ⓒ Ⓓ
14. Ⓐ Ⓑ Ⓒ Ⓓ

Part 2 Constructed Response/Grid In

Record and bubble in your answers on the grids below.

15. 16. 17.

Record your answers for Questions 18–21 on a separate sheet of paper.

SAMPLE

Part 1 Multiple Choice

1. D	**5.** A
2. B	**6.** A
3. C	**7.** B
4. A	**8.** B

Part 2 Short Response

9. A skin cell mutation will only affect skin cells that come from that cell. The organism's offspring cannot inherit the mutation. A mutation in a sex cell passes to the offspring.

10. During meiosis, the nucleus divides twice resulting in four cells with half the original number of chromosomes. These sex cells come together during fertilization to restore the complete set of chromosomes.

11. No because sex cells only contain half the number of chromosomes needed for a species.

12. hair color, height, weight, eye color and skin color; increase

13. The new organism has the same DNA as the parent organism.

14. There is an equal chance that the offspring will receive the dominant or recessive allele in a hybrid trait.

15. Organisms produce offspring through the process of reproduction. Without it, many species could not continue.

16. Division of the nucleus and division of the cytoplasm; the result is two cells with identical nuclei and other cell parts.

17. by regeneration

18. Male flower parts produce pollen, which contain sperm, and female flower parts produce eggs.

Part 3 Open Ended

19. Traits that have multiple alleles, such as blood type, have more than one dominant and one recessive allele for that gene. Multiple genes have dominant and recessive alleles for each gene control traits such as hair color and height. Both of these types of inheritance help to increase variation in individuals of a species.

chapter 20 Standardized Test Practice

Part 1 Multiple Choice

Record your answers on the answer sheet provided by your teacher or on a sheet of paper.

1. What are the specialized cells involved in sexual reproduction called?
A. prokaryotic cells
B. mitotic cells
C. eukaryotic cells
D. sex cells

Use the figure below to answer questions 2 and 3.

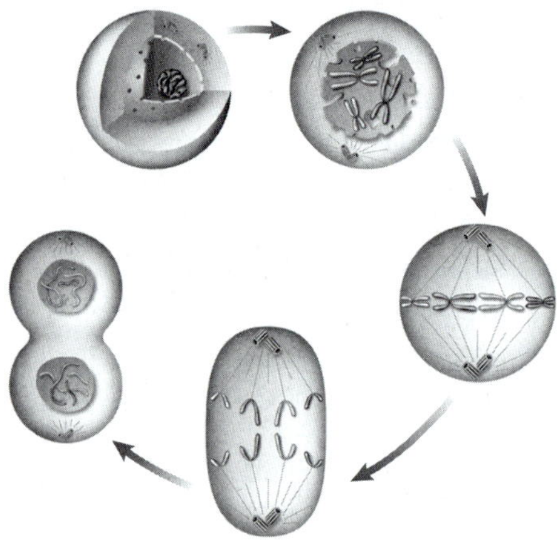

2. During cell reproduction, each new cell receives a copy of
A. mitotic cells.
B. hereditary material.
C. egg cells.
D. proteins.

3. This diagram illustrates a process that is used for all of the following except
A. asexual reproduction.
B. regeneration.
C. photosynthesis.
D. budding.

4. In a hybrid, which allele type is masked?
A. recessive **C.** pure
B. dominant **D.** clone

Use the figure below to answer questions 5 and 6.

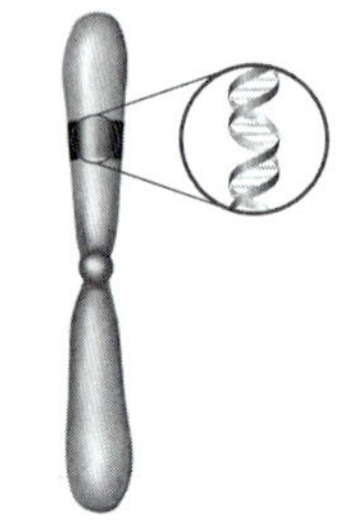

5. What is a small section of DNA on a chromosome that has information about a trait?
A. gene **C.** protein
B. cell **D.** egg

6. In body cells, chromosomes are found in
A. pairs. **C.** triplets.
B. singles. **D.** ribosomes.

Use the figure below to answer question 7.

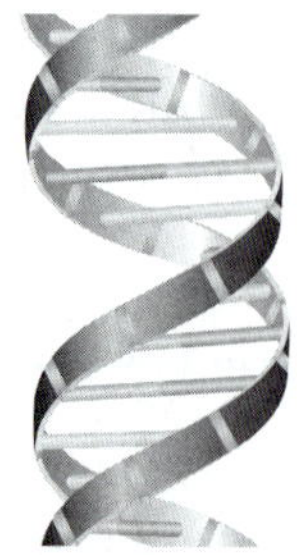

7. This figure represents the structure of
A. RNA. **C.** sex cells.
B. DNA. **D.** budding.

8. What part of DNA determines the instructions for a body's characteristics and processes?
A. the backbone
B. the sequence of bases
C. the twisted ladder
D. its duplication

612 STANDARDIZED TEST PRACTICE

Part 2 Short Response/Grid In

Record your answers on the answer sheet provided by your teacher or on a sheet of paper.

9. What is the difference between a mutation found in a skin cell and a mutation found in a sex cell?
10. Describe how meiosis and fertilization keep the chromosome number the same during sexual reproduction.
11. Would you want to use sex cells for cloning? Why or why not?
12. What are some examples of human characteristics that are controlled by more than one gene pair? Does this increase or decrease variation?

Use the figure below to answer question 13.

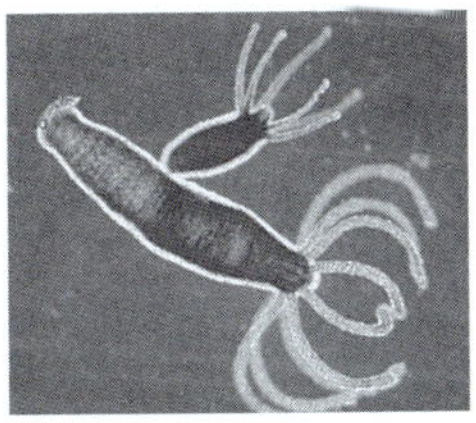

13. What form of reproduction is shown in this picture? Compare the DNA in the new organism to the DNA in the parent organism.
14. For a hybrid trait, what is the chance that the offspring will receive that dominant allele? What is the chance that the offspring will receive the recessive allele?
15. Why is reproduction important for a species?
16. What are the two parts of cell division? What does cell division result in?
17. Explain how a lizard can replace an injured tail.
18. What parts of a flower are involved in reproduction?

Part 3 Open Ended

Record your answers on a sheet of paper.

19. Discuss the difference between traits that have multiple alleles and those that are controlled by multiple gene pairs. What are some examples? How do these types of inheritance effect variation in species?
20. Explain the difference between cloning and selective breeding. How would a gardener use both of these processes?
21. Compare and contrast the structure of a ladder to the structure of DNA.

Use the picture below to answer question 22.

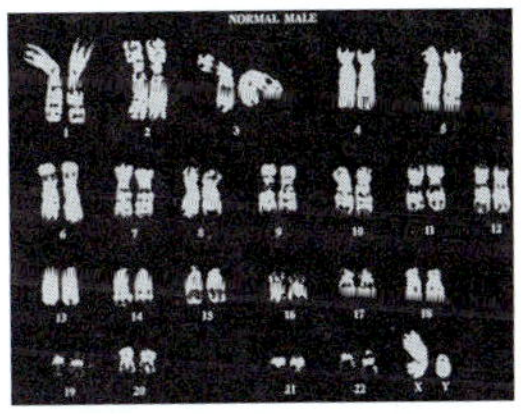

22. Trace the path of these human chromosomes as sex cells are formed.
23. In the pea plant, the genes for the flower color trait are purple and white. Using this example, explain the difference between hybrid, dominant pure, and recessive pure pea plants.
24. Discuss how a mutation can affect an organism and the variation in the genes of a species.
25. How does chance contribute to variation within a species?
26. Discuss some of the advantages and disadvantages of asexual reproduction.

Test-Taking Tip

Read Carefully Read each question carefully for full understanding.

Science Online red.msscience.com/standardized_test

20. Cloning results in a copy of an organism. Selective breeding is sexually reproducing organisms with desirable traits to produce offspring with the desired traits. A gardener could clone a plant by taking clippings from it and growing identical plants. A gardener could use selective breeding to combine the desirable traits (flower color) of two plants.
21. A DNA ladder is twisted. Both have a backbone that holds the rungs in place. The rungs of DNA are the pairs of chemicals called bases. There are only four bases and they pair up very specifically.
22. Sex cells are formed by meiosis. Before meiosis begins, these chromosomes duplicate. Then, the nucleus divides twice, with chromosome pairs separating in the first division and duplicated chromosomes separating in the second division. This results in four sex cells with half the original number of chromosomes.
23. A hybrid pea plant for the flower color trait would have both the purple and white alleles but appear purple. The dominant pure trait would have two purple alleles and appear purple. The recessive pure trait would have two white alleles and appear white.
24. Sometimes mutations affect the way cells grow, repair, and maintain themselves. This type of mutation is usually harmful to the organism. Many mutations have a neutral or no effect on the organism. Some mutations may benefit the organism. All types of mutation increase the variation in the genes of a species.
25. Chance plays a large part in the variation within a species. During metaphase 1 of meiosis, the chromosome pairs are arranged by chance so the alleles found in the sex cells cannot be predetermined. The combination of those alleles during fertilization is by chance and any mutations that occur are often by chance. All of these chance occurrences account for most variation.
26. Asexual reproduction produces a new organism with the same DNA as the parent organism. Advantages are that there only needs to be one organism involved for asexual reproduction and specialized cells are not required. A disadvantage is that there is little opportunity for variation.

Rubrics

For more help evaluating open-ended assessment questions, see the rubric on p. 10T.

unit 6

Unit Contents

WebQuest ***Barrier Islands: To Build or Not To Build?*** is a student-driven Web research activity designed to inform citizens of the controversy surrounding the development and use of barrier islands. Students will investigate the geological and physical nature of these specialized ecosystems and how they change over time. Students then use their new knowledge to form their own opinion about development and use of barrier islands. In a letter to the editor of an environmental magazine, each student will try to convince readers that his or her conclusions are correct and motivate readers to take appropriate action.

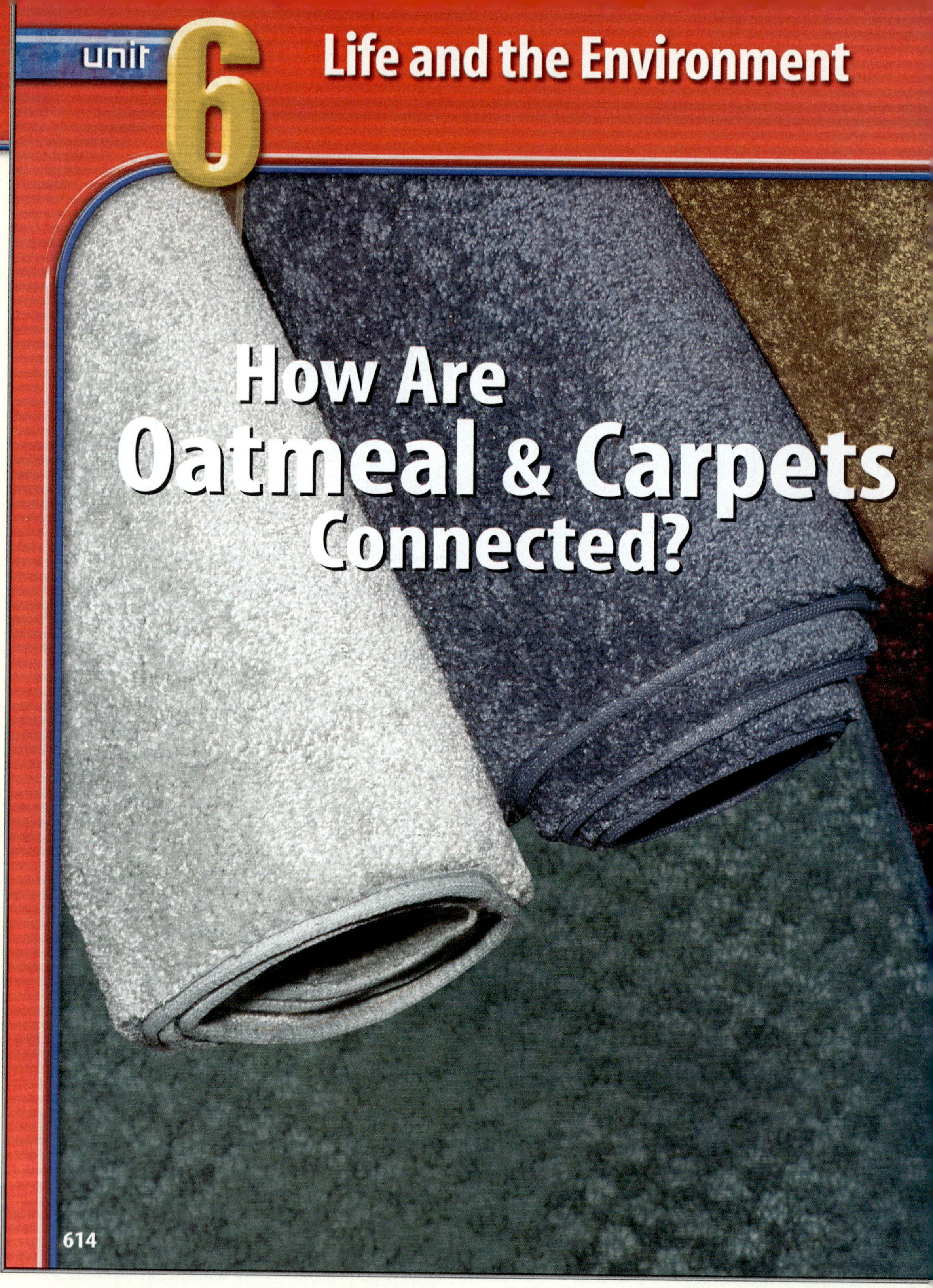

PROJECT CRISS[SM]

Study Skills

Organize Information Main idea and detail notes help students to organize important information. Have the students make two columns on their paper. In the left column, have them list 3–4 dwindling natural resources about which they are most concerned. In the right column, have them list possible solutions.

Main Idea (Resource)	Detail (Action needed)
Rain forests	Write letters to businesses to encourage responsible use of rain forest resources.

NATIONAL GEOGRAPHIC

In the 1850s, the first oatmeal mill began operation in the United States. Over the next few decades, hot, creamy oatmeal became a popular breakfast cereal across the country. By the early 1900s, oatmeal was getting some stiff competition from newly invented cold breakfast cereals such as cornflakes. Hot or cold, cereal had become a breakfast staple. But the processing of oats and corn for cereal leaves behind waste products—oat hulls and corncobs. In 1922, a cereal company discovered it could do something useful with these waste products. The company used oat hulls to make a substance called furfural. Today, furfural also is made from corncobs and other cereal waste products. Manufacturers use furfural in the production of synthetic rubber, plastic, and nylon—including the nylon that goes into carpets.

unit projects

Visit red.msscience.com/unit_project to find project ideas and resources. Projects include:

- **Career** You are an environmental scientist as you design your own ecosystem-interaction web to demonstrate relationships from birth to death of your specific organism.
- **Technology** Chart your research results on the manufacturing of different materials. Compare cost, energy use, resources, and environmental concerns.
- **Model** Design your own two-week personal conservation project. Decide how you can make a difference as you reduce, reuse, and recycle.

WebQuest Investigate the *Barrier Islands* ecosystem, then form an opinion as to whether developers should build on these environmentally fragile islands.

NATIONAL GEOGRAPHIC **How Are Oatmeal & Carpets Connected?** Furfural can be produced by treating carbohydrates, which are long chains of simple sugars. Commercially, furfural often is used as an industrial solvent. It also is used in the production of certain resins. When furfural is reduced chemically, it produces furfural alcohol. This alcohol then is polymerized to produce resins that are heat- and alkali-resistant.

unit projects

Career As a class, brainstorm a list of plants and animals that live in your local ecosystem. Assign each student a specific organism. Ask students to think like an environmentalist as they research 10 interactions for their particular organism. Remind them to include all possible interactions in the ecosystem, such as food chains, symbiosis, and predator and prey, from reproductive conception to death. Have students design a large web to demonstrate their organism's ecosystem.

Technology Have students conduct research to learn what happens when plastics, metals, glass, paper, and polystyrene are manufactured. Have students design a chart to compare the materials used in production. Categories might include natural resources, equipment and materials, energy use, cost, safety concerns, location, and environmental concerns. Ask students to reflect on each component and how recycled materials affect the different categories.

Model Have students design a personal conservation project. This two-week commitment to reduce, reuse, and recycle can culminate in a class presentation of their original plan, how they implemented their system of conservation, and the results.

Additional Resources For more information, resources, and assessment rubrics, visit red.msscience.com/unit_project

chapter 21 Organizer

Section/Objectives	Standards		Labs/Features
	National	State/Local	
Chapter Opener	See pp. 16T–17T for a Key to Standards.		**Launch Lab:** What is a living system?, p. 617 **Foldables,** p. 617
Section 1 What is an ecosystem? 3 sessions 1.5 blocks 1. **Describe** the living and nonliving factors in an ecosystem. 2. **Explain** how the parts of an ecosystem interact.	National Content Standards: UCP.1–UCP.3, UCP.5, A.1, A.2, C.1, C.4		**Science Online,** p. 620 **Visualizing Biotic Factors,** p. 621 **MiniLAB:** Observing Soil Characteristics, p. 623 **Lab:** Ecosystem in a Bottle, p. 626
Section 2 Relationships Among Living Things 2 sessions 1 block 3. **Explain** how ecologists organize living systems. 4. **Describe** relationships among living things.	National Content Standards: UCP.1–UCP.3, UCP.5, A.1, A.2, C.1, C.3–C.5, F.5		**MiniLAB:** Calculating Population Density, p. 628 **Science Online,** p. 629 **Integrate History,** p. 629 **Applying Science:** Graph Populations, p. 631
Section 3 Energy Through the Ecosystem 2 sessions 1 block 5. **Explain** how organisms get the energy they need. 6. **Describe** how energy flows through an ecosystem.	National Content Standards: UCP.1–UCP.3, UCP.5, A.1, A.2, C.1, C.4		**Integrate Physics,** p. 634 **Lab:** What's the limit?, pp. 636–637 **Science and Society:** Gators at the Gate!, p. 638

Glencoe Exclusive! **TeacherWorks™** All-In-One Planner and Resource Center

Lab Materials	Reproducible Resources	Section Assessment	Technology
Launch Lab: Science Journal	**Chapter *FAST FILE* Resources** Foldables Worksheet, p. 17 Directed Reading Overview, p. 19 Note-taking Worksheets, pp. 33–35	Glencoe's Assessment Advantage	TeacherWorks includes: • Interactive Teacher Edition • Lesson Planner with calendar • Access to all program blacklines • Correlations to standards • Web links
MiniLAB: soil (2 cups), water **Lab:** 2-L bottle, scissors, sand, aquarium gravel, metric ruler, water, *Elodea* plants, guppy, fish food	**Chapter *FAST FILE* Resources** Transparency Activity, p. 44 MiniLAB, p. 3 Enrichment, p. 30 Reinforcement, p. 27 Directed Reading, p. 20 Lab Worksheet, pp. 5–6 Lab Activity, pp. 9–12 **Physical Science Critical Thinking/Problem Solving,** p. 12 **Mathematics Skill Activities,** p. 1 **Reading and Writing Skill Activities,** pp. 15–16	**Portfolio** Curriculum Connection, p. 619 Science Journal, p. 621 Assessment, p. 625 **Performance** MiniLAB, p. 623 Lab, p. 626 Applying Skills, p. 625 **Content** Section Review, p. 625	Section Focus Transparency Teaching Transparency Guided Reading Audio Program Interactive Chalkboard CD-ROM Video Lab
MiniLAB: measuring tape *Need materials?* Contact Science Kit at 1-800-828-7777 or www.sciencekit.com on the Internet.	**Chapter *FAST FILE* Resources** Transparency Activity, p. 45 MiniLAB, p. 4 Enrichment, p. 31 Reinforcement, p. 28 Directed Reading, p. 21	**Portfolio** Science Journal, p. 629 Differentiated Instruction, p. 631 **Performance** MiniLAB, p. 628 Assessment, p. 632 Applying Science, p. 631 Applying Math, p. 632 **Content** Section Review, p. 632	Section Focus Transparency Guided Reading Audio Program Interactive Chalkboard CD-ROM
Lab: bean seeds, small planting containers, soil, water, labels, spoons, aluminum foil, sunny window or other light source, refrigerator or heater	**Chapter *FAST FILE* Resources** Transparency Activity, p. 46 Enrichment, p. 32 Reinforcement, p. 29 Directed Reading, p. 21 Lab Worksheet, pp. 7–8 Lab Activity, pp. 13–16 **Lab Management and Safety,** p. 69	**Portfolio** Check for Understanding, p. 635 **Performance** Lab, p. 637 Applying Skills, p. 635 **Content** Section Review, p. 635	Section Focus Transparency Virtual Labs CD-ROM Guided Reading Audio Program Interactive Chalkboard CD-ROM

End of Chapter Assessment

Blackline Masters	Technology	Professional Series
Chapter *FAST FILE* Resources Chapter Review, pp. 37–38 Chapter Tests, pp. 39–42 **Standardized Test Practice,** pp. 87–90	MindJogger Videoquiz Virtual Labs CD-ROM Exam*View*® Pro Testmaker TeacherWorks CD-ROM Interactive Chalkboard CD-ROM	**Performance Assessment in the Science Classroom (PASC)**

Transparencies

Section Focus

Section Focus Transparency 1 — New Land, New Life

Both photographs below are of Surtsey, an island near Iceland. Surtsey was created in 1963 by a volcanic eruption. Surtsey provides a wonderful opportunity for scientists to study how life starts in a new area.

1. Describe what you see in the first picture. Would you expect plants or animals to be able to live there?
2. What evidence of life do you see in the second picture?
3. How has Surtsey changed? How might it continue to change?

L2

Section Focus Transparency 2 — That's a Lot of Penguins

Every living thing on Earth has a particular environment that it calls home. These penguins like cold, icy Antarctica. This environment provides the penguins with all of the things they need for survival.

1. Name some factors that might affect penguin population.
2. How might you estimate the number of penguins in this picture?

L2

Section Focus Transparency 3 — Frolics with Sharks

Remoras are several different species of related fishes that attach themselves to sharks and other ocean organisms. Both the shark and the remora benefit from this relationship.

1. Why doesn't the shark eat the remora?
2. How do the shark and remora help each other?

L2

Assessment

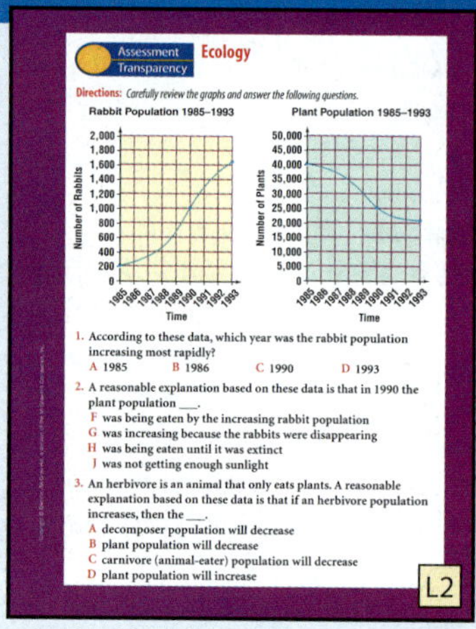
Assessment Transparency — Ecology

Directions: Carefully review the graphs and answer the following questions.

1. According to these data, which year was the rabbit population increasing most rapidly?
A 1985 B 1986 C 1990 D 1993
2. A reasonable explanation based on these data is that in 1990 the plant population ___.
F was being eaten by the increasing rabbit population
G was increasing because the rabbits were disappearing
H was being eaten until it was extinct
J was not getting enough sunlight
3. An herbivore is an animal that only eats plants. A reasonable explanation based on these data is that if an herbivore population increases, then the ___.
A decomposer population will decrease
B plant population will decrease
C carnivore (animal-eater) population will decrease
D plant population will increase

L2

Teaching

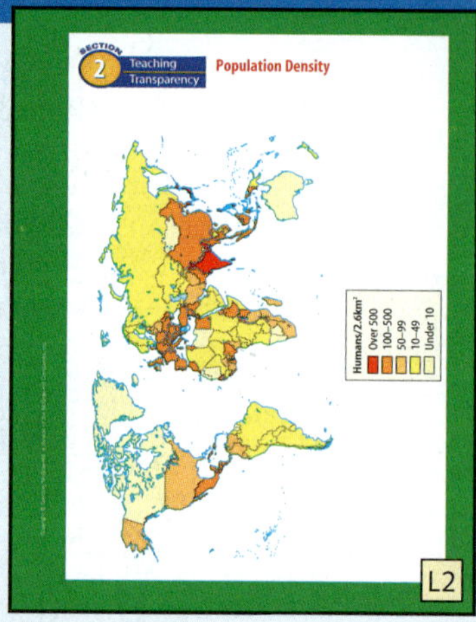
Teaching Transparency 2 — Population Density

L2

This is a representation of key blackline masters available in the Teacher Classroom Resources. See Resource Manager boxes within the chapter for additional information.

Key to Teaching Strategies

The following designations will help you decide which activities are appropriate for your students.

- L1 Level 1 activities should be appropriate for students with learning difficulties.
- L2 Level 2 activities should be within the ability range of all students.
- L3 Level 3 activities are designed for above-average students.
- ELL ELL activities should be within the ability range of English-Language Learners.
- COOP LEARN Cooperative Learning activities are designed for small group work.
- LS Multiple Learning Styles logos, as described on page 12T, are used throughout to indicate strategies that address different learning styles.
- P These strategies represent student products that can be placed into a best-work portfolio.
- PBL Problem-Based Learning activities apply real-world situations to learning.

Hands-on Activities

Student Text Lab Worksheet

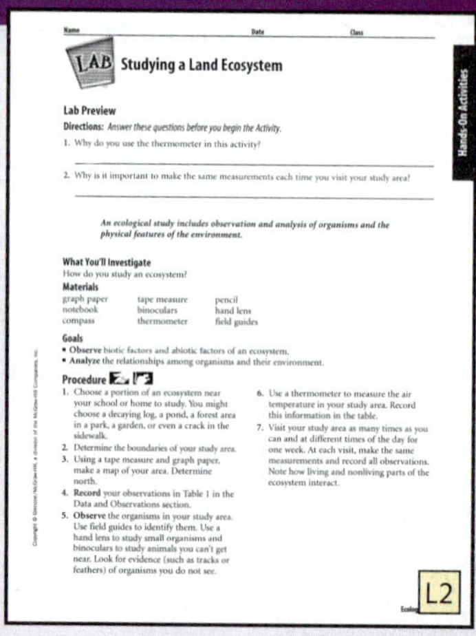
LAB — Studying a Land Ecosystem

L2

Laboratory Activities

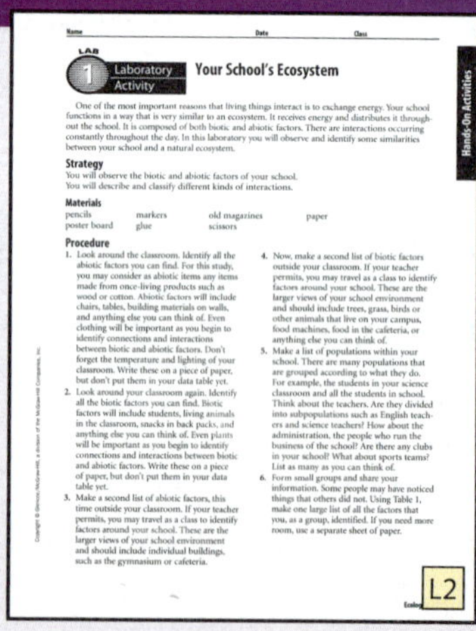
Laboratory Activity 1 — Your School's Ecosystem

L2

Meeting Different Ability Levels

Content Outline

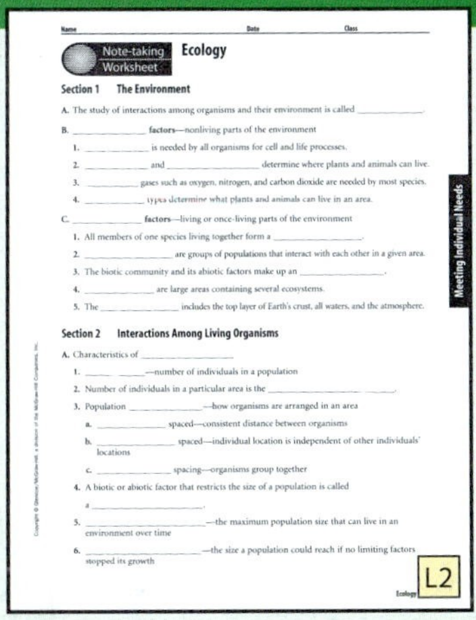

Reinforcement

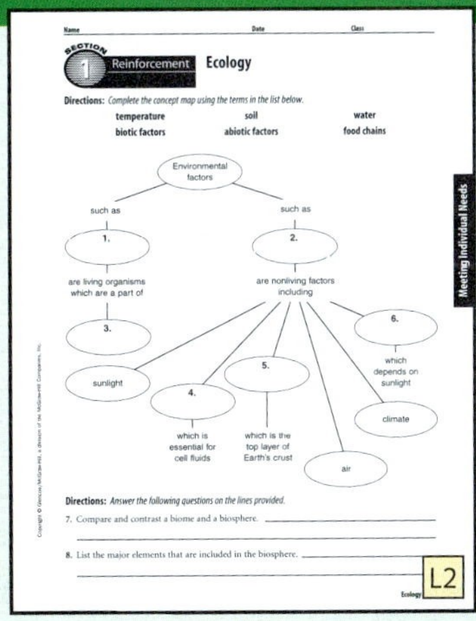

Enrichment

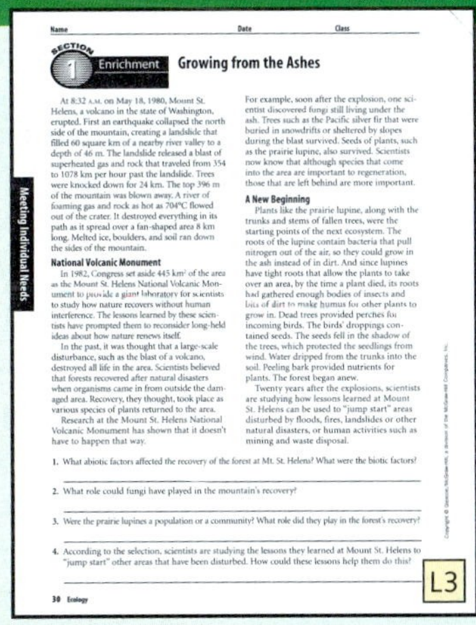

Directed Reading (English/Spanish)

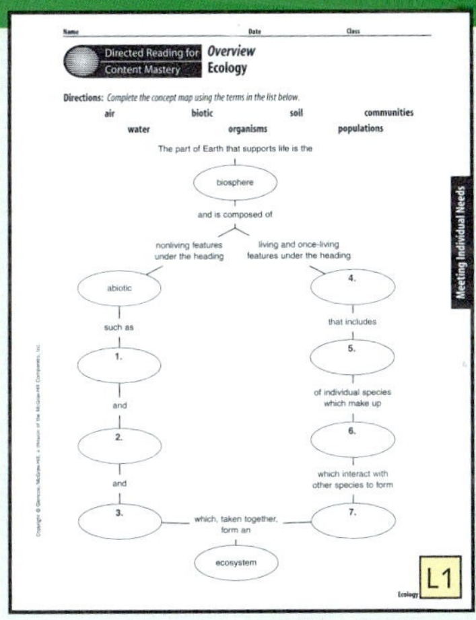

Study Guide

Reading Essentials

Assessment

Test Practice Workbook

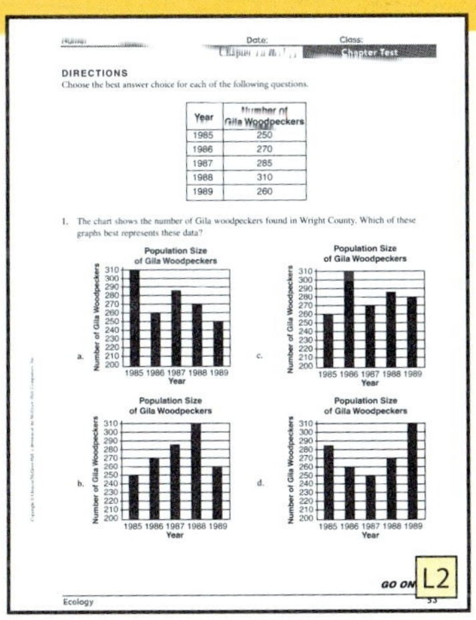

Chapter Review

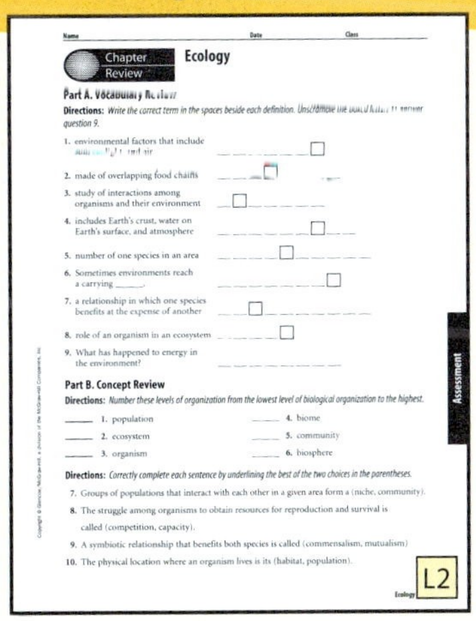

Chapter Tests

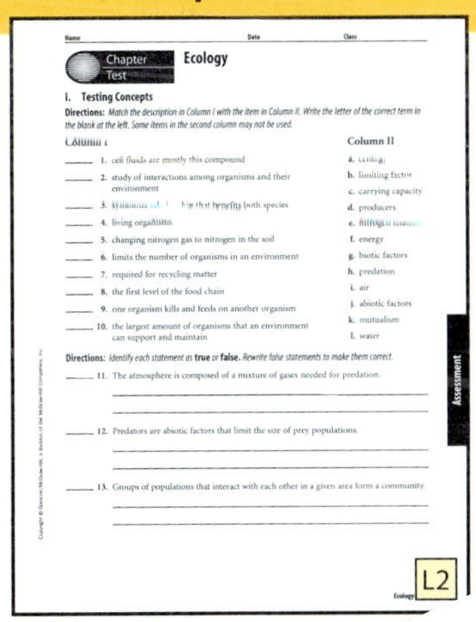

Science Content Background

section 1 What is an ecosystem?

Ecology

German naturalist Ernst Haeckel coined the word ecology in 1866. It comes from two Greek words; *oikos* means "home" and *logos* means "the study of." Ecology can be thought of as the study of the home of nature. It seeks to understand wildlife in the places where it lives. This encompasses the interactions among community members and their non-living environment (abiotic factors).

Abiotic factors include sunlight, which provides the energy for almost all life on Earth. The energy of sunlight is converted into chemical energy through photosynthesis. Other abiotic factors include air, water, and temperature. One of the most critical abiotic factors for animal survival is oxygen, which is a by-product of photosynthesis. The availability of oxygen is not usually limited, so it is sometimes neglected in discussions about the needs of animals.

section 2 Relationships Among Ecosystems

Population Growth

A population is a group of organisms of the same species. Species are organisms with similar DNA sequences that can interbreed to produce fertile offspring. Competition for resources tends to be most intense between organisms of the same species because they have very similar needs.

Competition often results in a resource being in short supply. A shortage of a resource is a limiting factor—a factor that limits population growth. The availability of food is usually the most important limiting factor for animals. Limiting factors include all resources for which organisms must compete, such as nesting sites for animals, or fertile soil and living space with adequate sunlight for plants.

section 3 Energy Through the Ecosystem

Photosynthesis and Food

During photosynthesis, plants and algae use the energy from light, along with carbon dioxide and water, to form simple sugar molecules that then can be converted into more complex carbohydrates, fats, and amino acids. Many of these nutrient molecules become available to other organisms in the community when they are consumed as food.

chapter content resources

Internet Resources

For additional content background, visit **red.msscience.com** to:

- access your book online
- find references to related articles in popular science magazines
- access Web links with related content background
- access current events with science journal topics

Print Resources

The Bald Eagle, Endangered No More, by Mac Preibe, Mindfull Publishing, 2000

Pond Watching with Ann Morgan, by Michael Elsohn Ross, Carolrhoda Books, 2000

Discovering the Naturalist Intelligence: Science in the School Yard, by Jenna Glock, Susan Wertz, and Maggie Meyer, Zephyr Press, 1999

The Work of Nature: How the Diversity of Life Sustains Us, by Yvonne Baskin, Island Press, 1997

Feeding Relationships

Feeding relationships are a key aspect of ecology, and different terms are sometimes used to describe similar feeding relationships. Organisms such as plants or algae that produce their own food are called producers, also referred to as autotrophs. Organisms that feed on producers are called consumers, or heterotrophs. The consumers can be further divided into groups based on what they eat. For example, first-order heterotrophs, also called herbivores, eat only plants. Second-order heterotrophs are carnivores that eat only herbivores.

IDENTIFYING Misconceptions

Find Out What Students Think

Students may think that . . .

Plants obtain their food from the environment, rather than producing it internally.

This misconception may be the result of students applying human attributes to plants. Language also reinforces this misconception. For example, plant fertilizer frequently is called plant food. Students may have heard that plants get their food from the ground.

Discussion

Ask students: "How do plants eat?" Have students share their responses, which most likely will involve discussion of plant roots. After a couple of responses ask, "Which explanation do you think is most accurate?" It will be evident that many students believe that plants actually eat.

Promote Understanding

Activity

Show students a plant cutting growing in water, a hydroponics plant, or a plant growing in super-absorbent crystals. It is best if the roots can be seen clearly. Explain that you periodically have to add water to the plant. Make a few statements, such as "last week this plant was only this big, and look at it now," to help students understand the plant has been growing.

Ask students, "How does this plant get food?" When some students explain that the plant gets food through its roots, ask them to take a close look at what the plant is growing in, and see if that looks more like food or water. Help them to establish that the plant absorbs water and minerals through its roots. Point out that water and minerals are required for growth, but they are not food.

Then ask the following questions: How do we know plants need light? What do plants do with the light? Help establish the idea that plants use light, along with water from the soil (or other growth medium) and carbon dioxide from the air to make their own food. Plants are known as producers because they can produce their own food; they do not eat food.

Assess

After completing the chapter, see *Identifying Misconceptions* in the Study Guide at the end of the chapter.

chapter 21

Chapter Vocabulary

ecosystem, p. 618
ecology, p. 620
biosphere, p. 620
biotic factor, p. 620
abiotic factor, p. 622
population, p. 627
community, p. 628
limiting factor, p. 629
niche, p. 631
habitat, p. 632
producer, p. 633
consumer, p. 633
decomposer, p. 634

Science Journal Fallen leaves provide food for insects and may provide some shelter or camouflage for frogs. Insects provide food for frogs.

INTERACTIVE CHALKBOARD with Image Bank

PowerPoint® Presentations

This CD-ROM is an editable Microsoft® PowerPoint® presentation that includes:

- a pre-made presentation for every chapter
- interactive graphics
- animations
- audio clips
- image bank
- all new section and chapter questions
- Standardized Test Practice
- transparencies
- pre-lab questions for all labs
- Foldables directions
- links to red.msscience.com

chapter 21

Ecology

chapter preview

sections

1. What is an ecosystem?
   Lab Ecosystem in a Bottle
2. Relationships Among Living Things
3. Energy Through the Ecosystem
   Lab What's the limit?

Virtual Lab How is energy transferred through a community of organisms?

How do frogs catch insects?

Today is perfect for a field trip to a local stream. Carefully, quietly, you push aside some cattails. You lean in for a closer look when—WHAM! A sticky tongue latches onto a nearby insect and flings it into the waiting mouth of a frog. You jump and the frog leaps into the water. SPLASH! You have just observed a living system in action.

Science Journal Describe how fallen leaves and insects contribute to the survival of frogs in this system.

616

Theme Connection

Systems and Interactions Learning about ecosystem processes and the interactions among populations, communities, and their environment will enable students to better understand the interconnectedness of all living things. After examining natural systems, students will better understand their role in the environment.

About the Photo

Predator-Prey Relationships Frogs are carnivores that eat insects, worms, spiders, and centipedes. This green frog is eating a grasshopper. A frog spends most of its life in a moist habitat, usually near a pond or stream. It can escape predators by leaping away with the help of its powerful rear legs, and can hide underwater for long periods.

Start-Up Activities

Launch LAB

What is a living system?

A system is any group of things that interact with one another. Living organisms interact with each other and with the environment to form ecosystems. Ecology is the study of these interactions.

1. Choose a small area of grass or weeds near your school. Identify the boundaries of your plot.
2. Carefully observe and record everything in your plot. Be sure to include all parts of your plot, including soil and air.
3. Classify what you observe into two groups—things that are living and things that are not living.
4. **Think Critically** In your Science Journal, describe how you think the parts of the plot you observed form a system.

Science Online — Preview this chapter's content and activities at red.msscience.com

FOLDABLES™ Study Organizer

Ecology Make the following Foldable to help identify what you already know, what you want to know, and what you learned about ecology.

STEP 1 **Fold** a vertical sheet of paper from side to side. Make the front edge about 1.25 cm shorter than the back edge.

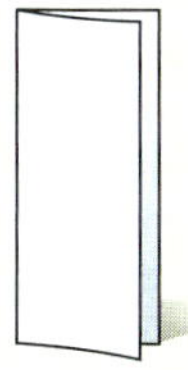

STEP 2 **Turn** lengthwise and **fold** into thirds.

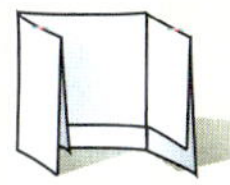

STEP 3 **Unfold and cut** only the top layer along both folds to make three tabs.

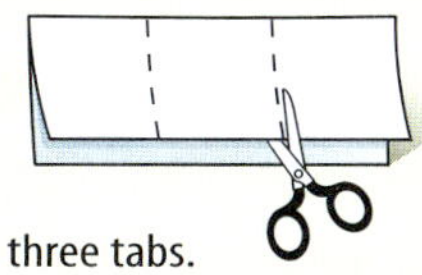

STEP 4 **Label** each tab as shown.

Identify Questions Before you read the chapter, write what you already know about ecology under the left tab of your Foldable, and write questions about what you'd like to know under the center tab. After you read the chapter, list what you learned under the right tab.

Launch LAB

Purpose Students observe and identify the different parts of an ecosystem. L1 ELL LS **Visual-Spatial**

Preparation Scout out an area ahead of time where students can observe. Check carefully to ensure that there are no pieces of broken glass or other dangerous items in the area.

Alternate Materials If students are unable to go outside, have them examine an indoor terrarium or aquarium.

Teaching Strategies Ask students to consider organisms that are not visible to the naked eye. Ask them to predict whether their plots contain bacteria and fungi.

Think Critically

The students should recognize that they are observing a system composed of interacting living and nonliving components. The interactions between these two factors create an ecosystem.

Assessment

Oral Ask students to discuss what their plots would contain at different times of the year. How does the environment change with the seasons? How would seasonal changes influence organisms living in an area? Use the Performance Task Assessment List for Oral Presentation in **Performance Assessment in the Science Classroom,** p. 143. L2 LS **Logical-Mathematical**

FOLDABLES™ Study Organizer — Dinah Zike Study Fold

Student preparation materials for this Foldable are available in the Chapter *FAST FILE* Resources.

1 Motivate

INTERACTIVE CHALKBOARD PowerPoint® Presentations

Bellringer

Section Focus Transparencies also are available on the Interactive Chalkboard CD-ROM. L2 ELL

Section 1 Section Focus Transparency **New Land, New Life**

Both photographs below are of Surtsey, an island near Iceland. Surtsey was created in 1963 by a volcanic eruption. Surtsey provides a wonderful opportunity for scientists to study how life starts in a new area.

1. Describe what you see in the first picture. Would you expect plants or animals to be able to live there?
2. What evidence of life do you see in the second picture?
3. How has Surtsey changed? How might it continue to change?

Ecology

Tie to Prior Knowledge

Adaptations Ask students to consider how adaptations have allowed different organisms to cope with different environmental conditions. L2 LS **Logical-Mathematical**

section 1

What is an ecosystem?

as you read

What You'll Learn

- **Describe** the living and nonliving factors in an ecosystem.
- **Explain** how the parts of an ecosystem interact.

Why It's Important

Understanding interactions of an ecosystem will help you understand your role in your ecosystem.

Review Vocabulary

organism: any living thing; uses energy, is made of cells, reproduces, responds, and grows

New Vocabulary

- ecosystem
- ecology
- biosphere
- biotic factor
- abiotic factor

Ecosystems

Take a walk outside and look around. What do you see? Woods? A street? A patch of weeds growing in a sidewalk crack? If you observe one of these areas closely, you may see many different organisms (OR guh nih zumz) living there. In a forest, for instance, there are birds, deer, insects, plants, mushrooms, and trees. In your backyard, you might see squirrels, birds, insects, grass, and shrubs. These organisms, along with the nonliving things in the woods or yard, such as soil, air, and light, make an ecosystem (EE koh sihs tum). An **ecosystem** is made up of organisms interacting with one another and with nonliving factors to form a working unit. **Figure 1** shows an example of a stream ecosystem.

Section 1 Resource Manager

Chapter FAST FILE Resources

Transparency Activity, p. 44
MiniLAB, p. 3
Note-taking Worksheets, pp. 33–35
Enrichment, p. 30
Reinforcement, p. 27
Directed Reading for Content Mastery, p. 20
Lab Worksheet, pp. 5–6
Lab Activity, pp. 9–12

Physical Science Critical Thinking/Problem Solving, p. 12

Mathematics Skill Activities, p. 1

Reading and Writing Skill Activities, pp. 15–16

618 CHAPTER 21 Ecology

What does it mean to say that an organism interacts with another organism? Think back to the field trip to the stream at the beginning of the chapter. When the frog ate the insect, an interaction occurred between two organisms living in the same ecosystem.

What does it mean to say that an organism interacts with the nonliving parts of an ecosystem? Think about the field trip again. What did the frog do when it spotted your movement? It dove into the stream, probably for safety. The frog uses the stream for shelter. This is an example of an interaction between a living organism and a nonliving part of an ecosystem.

Figure 1 Let's identify the living and nonliving parts of this stream ecosystem. Rocks and water are nonliving things. Pond skaters are alive—these insects skim the surface of the water. Algae, fish, crayfish, and mosses covering rocks are other living parts of this ecosystem.
Describe *how these living organisms interact with nonliving parts of the ecosystem.*

2 Teach

Caption Answer

Figure 1 Algae obtain nutrients and shelter from water. Fish and crayfish obtain food, shelter, and oxygen from water. Mosses use rocks as a place to live, and obtain moisture and some nutrients from splashing water. Algae and mosses use light energy to make food.

Activity

Classroom Ecosystem Ask students to identify different living and nonliving factors in their classroom and discuss how they interact. Ask students how they interact with the living and nonliving parts of the classroom ecosystem. L2
LS **Logical-Mathematical**

Use Science Words

Word Origin Write the word *ecology* on the chalkboard. Explain that the word *ecology* comes from the Greek words *oikos*, which means "place to live," and *ology*, which means "study of." Ask students why the word *ecology*, meaning "the study of place to live," is an appropriate name for the study of our environment.
L2 LS **Linguistic**

Fun Fact

The pond skater steers with its rear legs, uses its middle legs as paddles, and grabs prey with its front legs.

Visual Learning

Figure 1 What other nonliving things can be found in this ecosystem? Air, soil, and sunlight are other nonliving parts of this ecosystem. How do these organisms interact with living parts of the ecosystem? Some fish feed on crayfish. Crayfish feed on algae or decaying animals. Insects find shelter in mosses. Pond skaters feed on other insects. L2 LS **Visual-Spatial, Logical-Mathematical**

Curriculum Connection

Language Arts Ask students to find or write a poem that describes animals or plants interacting with their environment. Have students present their selection to the class and discuss why they chose their particular poem. L2 P LS **Linguistic, Auditory-Musical**

Teacher FYI

Ecosystems An ecosystem can be a subset of a larger ecosystem. Within a large forest ecosystem, the interactions on a decaying log, the tree canopy, and the soil can all be viewed as small, separate ecosystems.

Discussion

Field Work Why would most of an ecologist's work be done in the field? Ecologists spend most of their time in the field because they study the interactions of organisms with their outdoor environment.

Use an Analogy

Biosphere Many ecologists compare the biosphere to a living organism. Just as an animal is composed of different interacting parts, the biosphere is also composed of different interacting parts, including ecosystems, atmosphere, oceans, lakes and streams, and soil.

Science Online

Topic: Desert Ecosystems
Visit red.msscience.com for Web links to information about desert ecosystems.

Activity Write a paragraph describing how two different desert organisms interact with each other and with abiotic factors in their ecosystem.

The Study of Ecosystems When you study the interactions in an ecosystem, you are studying the science of ecology (ih KAH luh jee). **Ecology** is the study of the interactions that take place among the living organisms and nonliving parts of an ecosystem. Ecologists spend a lot of time outdoors, observing their subject matter up close. Just as you knelt quietly in the cattails on your field trip, an ecologist might spend hours by a stream, watching, recording, and analyzing what goes on there. In addition, like other scientists, ecologists also conduct experiments in laboratories. For instance, they might need to analyze samples of stream water. But, most of the ecologist's work is done in the field.

The Largest Ecosystem Ecosystems come in all sizes. Some are small, like a pile of leaves. Others are big, like a forest or the ocean. **Figure 2** shows the biosphere (BI uh sfihr), the largest ecosystem on Earth. The **biosphere** is the part of Earth where organisms can live. It includes the topmost layer of Earth's crust; all the oceans, rivers, and lakes; and the surrounding atmosphere. The biosphere is made up of all the ecosystems on Earth combined.

How many different ecosystems are part of the biosphere? Let's list a few. There are deserts, mountains, rivers, prairies, wetlands, forests, plains, oceans—the list can go on and on, and we haven't even gotten to smaller ecosystems yet, such as a vacant lot or a rotting tree trunk. The number of ecosystems that make up the biosphere is almost too many to count. How would you describe your ecosystem?

Figure 2 The biosphere is the part of Earth that contains all the living things on the planet. Each ecosystem that you study is part of the biosphere.

Living Parts of Ecosystems

Each of the many ecosystems in the biosphere contains many different living organisms. Think about a rotting tree trunk. It's a small ecosystem compared to a forest, but the tree trunk may be home to bacteria, bees, beetles, mosses, mushrooms, slugs, snails, snakes, wildflowers, woodpeckers, and worms. The organisms that make up the living part of an ecosystem are called **biotic factors.** An organism depends on other biotic (bi AH tihk) factors for food, shelter, protection, and reproduction. For example, a snake might use a rotting log for shelter. Termites are insects that depend on the same log for food. **Figure 3** shows some of the biotic factors in a desert ecosystem.

Cultural Diversity

Natural Balance The Balinese government tried to increase rice production by using accepted methods such as pesticides. The chemicals failed to kill the pests, but did kill other food sources such as the eels living in irrigation ditches. Rice crops actually decreased. When scientists studied the ecological system, they found that timing of irrigation was key. All fields received the water needed and all were barren at the same time. This starved the pests that damaged crops. When scientists developed a computer model to explain this system, it also showed that it was the most productive one.

NATIONAL GEOGRAPHIC **VISUALIZING BIOTIC FACTORS**

Figure 3

Hawks, snakes, and many other organisms make up the desert ecosystem's living, or biotic, factors. Rocks, sand, soil, sunlight, air, and water are nonliving factors. The desert is a place where rainfall is scarce. Bright sunshine can cause daytime temperatures to reach 100 degrees or more on many days of the year. Nights, however, can be very cold.

The living and nonliving parts of the desert ecosystem interact in a variety of ways. Organisms must be able to survive long periods with little or no water. For example, cactus plants can store water in their tissues. Other organisms, including insects, obtain water by feeding on cacti. Desert organisms also must find ways to shelter themselves from extreme heat and cold. Kangaroo rats, for example, burrow underground, coming out at night when temperatures have cooled. How might the snake in this picture interact with abiotic factors in this ecosystem?

NATIONAL GEOGRAPHIC

Visualizing Biotic Factors

Visual Learning

Figure 3 In addition to the hawk, lizard, and snake, this figure shows a Gila monster and scorpions. Ask students to identify biotic interactions that may or may not be shown in the figure. Lizards eat insects that are not shown. Snakes eat mice or other small mammals not shown. Gila monsters are venomous and feed on small mammals, birds, and eggs. Scorpions usually come out at night and feed on insects, spiders, and centipedes. Snakes, lizards, and Gila monsters compete for shade during the day. What abiotic factors are important in this ecosystem? High daytime temperatures and scarce water increase competition for water and shelter. Sandy or gravelly soil limits the species of plants that can survive in the desert. Many desert plants, including cacti, store water in their tissues.

Caption Answer

Figure 3 The snake is taking shelter under a rock for protection from the hot sun and to hide from predators, such as hawks.

Teacher FYI

Desert Climate Desert ecosystems are extremely harsh for organisms. Deserts receive little rain and frequently have searing summer heat and chilly nights. Most deserts receive less than 20 cm of water per year and have daytime temperatures that reach more than 30°C.

Science Journal

Ecosystems Have students describe the ecosystem they live in. They should choose home, school, or their city or town. Be sure they define their boundaries. Have students list both biotic and abiotic factors that occur in their ecosystem, and predict what would happen if the environment changed dramatically. L2 P LS **Linguistic, Logical-Mathematical**

Differentiated Instruction

Challenge Ask students to research how organisms interact with their environment at different stages of their life cycles. Do younger plants and animals have different requirements than older ones? If possible, have students visit a zoo or botanical garden to interview staff members about unique requirements some young organisms might have. L3 LS **Intrapersonal**

Inquiry Lab

Tracking Populations

Purpose To understand how the members of a population move around, have students design an experiment to track populations in a decaying-log ecosystem.

Possible Materials oil paints of different colors, artist paintbrushes, jars, turpentine, guidebook to insects, magnifying lens, meter stick

Safety Precaution take appropriate safety measures when working outdoors and with insects

Estimated Time several 20-minute periods over several weeks

Teaching Strategies

- Students can track a population living beneath a decaying log by painting a small dot on the backs of all the members of one species, such as pill bugs. Students can check under the log periodically to count and record the number of new individuals and the number of original individuals in the population.
- Students can use different colors of paint to track different species, or they can use different colors to represent the date on which they checked the members of one population.
- Nearby decaying logs also can be checked to record the migration of individuals from the original population.

For additional inquiry activities, see *Science Inquiry Labs*.

Reading Check

Answer a combination of minerals, water, air, and organic matter

Figure 4 Different types of soil support different kinds of plant life. Cactus and other desert plants can thrive in dry, sandy, or rocky soils. Forest soils are deep, moist, and full of nutrients from decaying leaves.

Nonliving Parts of Ecosystems

Earlier, you listed the parts that make up an ecosystem near your school. Was your list limited to the living organisms—the biotic factors—only? No. You included nonliving factors, too, such as air and soil. The nonliving things found in an ecosystem are called **abiotic** (ay bi AH tihk) **factors.** Look for some abiotic factors in the desert shown in **Figure 3.** Abiotic factors affect the type and number of organisms living in ecosystems. Let's take a closer look at some abiotic factors.

Soil One abiotic factor that can affect which plants and other organisms are found in an ecosystem is soil. It is made up of several ingredients, much like a recipe. Soil is made up of a combination of minerals, water, air, and organic matter—the decaying parts of plants and animals. You know that salt, flour, and sugar are found in many recipes. But not all foods made from these same ingredients taste or look the same. Cakes and cookies look and taste different because different amounts of salt, flour, and sugar are used to make them. It's the same with soil. Different amounts of minerals, organic matter, water, and air make different types of soil, as shown in **Figure 4.**

Reading Check *What ingredients make up soil?*

Curriculum Connection

History During the mid-1800s, many Americans migrated westward across North America to the west coast in wagon trains. Research some of the experiences of these travelers and report on the environmental conditions they lived with. The *Little House* books by Laura Ingalls Wilder or the CD-ROM game *The Oregon Trail* could be sources for students. L2 LS **Linguistic Intrapersonal**

Different soils offer different materials and conditions for organisms. If you've ever visited a gardening store, you've seen all kinds of products gardeners add to their soil to make it just right for the types of plants they want to grow. The next time you dig a hole, take a close look at the soil. Is it dry? Does it have a lot of dead leaves and twigs in it? Is it tightly packed or loose and airy?

Temperature Soil is only one of the factors that affect the organisms that live in an ecosystem. Temperature also determines which organisms live in a particular place. How do the tropical plants shown in **Figure 5** compare with the mountainside plants? Predict what would happen if the organisms on the mountainside were moved to a hot climate such as a tropical rain forest.

Figure 5 Plants have adaptations for their environments. The mountainside wildflowers grow in clusters close to the ground, which protects them from strong winds. The tropical plants have large leaves to absorb as much light as possible in the dim light of the rain forest floor.

Mini LAB

Observing Soil Characteristics

Procedure

1. Fill **two cups** with **soil.** Use different amounts of the materials available to create two different "soil recipes." Pack the soil equally into each cup.
2. Pour equal amounts of **water** into each cup.
3. After a minute or so, tip the cups over to see if any water pours out.
4. Observe the characteristics of the soils you made. Record your observations in your Science Journal.

Analysis

1. What was the difference between the soil in the two cups to start with?
2. Was there a difference in how the soil in each cup held water? What could this mean for plants or other organisms living in the soil?

Mini LAB

Purpose Students will observe how soil composition influences the ability of soil to hold water. L1 ELL LS **Kinesthetic, Visual-Spatial**

Materials cups, different soil components (e.g., sand, clay, peat, topsoil, rocky soil), trowel or spoon

Teaching Strategies Avoid over-saturating the soil. Make sure students do not add too much water to their containers. You can obtain soil components at a garden-supply store.

Analysis

1. Results will vary by student soil recipe.
2. Soil samples composed of large particles will not hold water as well as soils composed of small particles or soils that contain a large percentage or organic matter. Plants and animals need special adaptations to live in soils that are very wet or very dry.

Assessment

Performance Ask students to design an experiment that would examine how plants respond to different soil conditions such as dryness or mineral content. Students should try to grow the same plant in different soil compositions and compare the results. Use the Performance Task Assessment List for Designing an Experiment in **Performance Assessment in the Science Classroom,** p. 95. L2 LS **Kinesthetic, Visual-Spatial**

Teacher FYI

Soil Composition The volume of a handful of fertile soil is about 50% solid matter and 50% pore spaces, which allow air and water to penetrate. About 10% of the solid matter is organic and includes living organisms, roots, and decaying matter. The remaining 90% is tiny particles eroded from rocks and minerals. These particles are classified as sand, silt, or clay, depending on size.

A soil's texture determines how well it holds water and dissolved nutrients and how well it allows air to circulate. Addition of organic matter to any soil increases its air circulation and water holding capacities.

Quick Demo

Plant Adaptations

Materials tropical houseplant (such as Philodendron, Ficus, or Schefflera), cactus plant

Estimated Time 10 minutes

Procedure Point out some differences between the two plants and discuss how they reflect the type of ecosystem in which each plant is found. Tropical plants have large, green leaves that soak up sunlight but can lose water quickly. Cactus plants have no leaves, but thick stems that can store water. Photosynthesis takes place in the stem, and spines help protect the cactus from damage by herbivores. Tropical plants grow more quickly than desert plants, and may more easily recover from browsing.

Caption Answer

Figure 6 Plants could not grow, so animals would not be able to eat. Life as we know it could not exist.

Reading Check

Answer All organisms need water to survive. Some organisms need sunlight to make their own food. All organisms use this as the source of energy for all life processes.

Activity

Soil Types Have students research soils in different areas of the country. They could compare the southeast to southwest or prairie to deciduous forest. They should make a table or chart comparing and contrasting soils from different areas. They could take it further by including information on the types of vegetation found in each area. L3 LS **Kinesthetic, Visual-Spatial, Linguistic, Intrapersonal**

Water Another important abiotic factor is water. In the field trip to the stream at the beginning of the chapter, maybe you saw a sleek trout dart through the water. Some organisms, such as fish, whales, and algae (AL jee), are adapted for life in water, not on land. But these organisms depend upon water for more than just a home. Water helps all living things carry out important life processes such as digestion and waste removal. In fact, the bodies of most organisms are made up mostly of water. Scientists estimate that two-thirds of the weight of the human body is water, as shown in **Figure 6.** Do you know how much you weigh? Calculate how much of your weight is made up of water.

Because water is so important to living things, it is also important to an ecosystem. The amount of water available in an ecosystem can determine how many organisms can live in a particular area. It can also serve as shelter and as a way to move from place to place.

Sunlight The Sun is the main source of energy for most organisms on Earth. Energy from the Sun is used by green plants to produce food. Humans and other animals then obtain their energy by eating these plants and other organisms that have fed on the plants. When you eat food produced by a plant, you are consuming energy that started out as sunlight. You'll learn more about the transfer of energy in an ecosystem later in this chapter.

Reading Check *Why are water and sunlight important to ecosystems?*

Figure 6 Sunlight and water are two abiotic factors essential to ecosystems. Water is important to humans because 66 percent of our bodies is composed of water.
Explain *why most ecosystems could not exist without sunlight.*

Differentiated Instruction

Learning Disabled List five abiotic factors: sunlight, soil, water, air, and temperature. Have students explain how they depend on each factor for their own survival. Sunlight and soil allow plants to grow so they have food to eat. Air provides oxygen for respiration. Water is needed for drinking, washing, etc. Temperature regulation is important for keeping the body from freezing or becoming too hot. L1 LS **Logical-Mathematical, Intrapersonal**

A Balanced System

Every ecosystem is made up of many different biotic and abiotic factors working together. When these factors are in balance, the system is in balance, too.

Ecosystems are always changing. Could an ecosystem ever get out of balance? Many events can affect the balance of a system. One example would be a long period of time without rain (called a drought). Predict what would happen if a drought occurred at the stream where you took your field trip. You can see a possible result in **Figure 7.** Some organisms, like fish, could not survive for long periods of time without water. Other organisms, such as frogs or insects, might have to find new homes.

Organisms that couldn't survive in a normal stream environment might find the dried-up stream to be just the ecosystem they need. Examples might include trees, flowers, mice, gophers, or earthworms.

Figure 7 Ecosystems are always changing. Some changes are small. Others, such as a stream drying up, are much larger with many more effects.

section 1 review

Summary

Ecosystems

- Ecosystems are made up of organisms that interact with each other and with the nonliving parts of the environment.
- Ecology is the study of interactions that occur in ecosystems.
- The biosphere is the part of Earth that supports life.

Living Parts of Ecosystems

- Biotic factors are the organisms in an ecosystem.

Nonliving Parts of Ecosystems

- Abiotic factors include soil, temperature, water, and sunlight.

A Balanced System

- Ecosystems change over time.

Self Check

1. **Describe** two ways in which an organism can interact with the other biotic factors of the ecosystem in which it lives.
2. **Explain** four ways in which abiotic factors are important to an ecosystem.
3. **Compare and contrast** the meanings of the terms *ecosystem* and *biosphere.*
4. **Think Critically** You have been asked to design a space station on the Moon. Use your knowledge of ecosystems to describe how you would develop your design.

Applying Skills

5. **Describe** the ecosystem that you are a part of. What are the biotic factors? The abiotic factors? What kinds of interactions take place in your ecosystem?

Science Online red.msscience.com/self_check_quiz

3 Assess

DAILY INTERVENTION

Check for Understanding

Interpersonal Have students brainstorm the biotic and abiotic factors they might encounter in a polar region. They should then discuss the adaptations organisms would need to have to live in that environment. L1 LS **Logical-Mathematical**

Reteach

Ecosystem Factors On the chalkboard, write a list of different ecosystem factors. Ask students to identify each as biotic or abiotic. Ask for examples of the possible role the factors may play in the ecosystem. L1 LS **Linguistic**

Assessment

Process Provide students with several photographs of a familiar ecosystem, such as a forest or prairie. The photographs should show several species of plants and animals. Have students write a paragraph that describes interactions of at least two species and explains how at least one abiotic factor affects the organisms living in that ecosystem. P L2 LS **Visual-Spatial, Linguistic**

section 1 review

1. organisms interacting with one another and with the abiotic factors of the environment
2. Answers will vary. Soil provides water and nutrients for plants, shelter and food for other organisms. Sunlight fuels photosynthesis, which provides food for plants and, indirectly, for animals. Water is required for all life. Temperature helps determine which organisms can survive in an ecosystem.
3. The biosphere is made up of all ecosystems on Earth.
4. Answers will vary. The Moon does not have the resources necessary to support human life, but a space station set up as a self-sustaining ecosystem might be able to support life.
5. Students should describe how their lives are affected by abiotic factors, which could include human-made materials or pollutants, as well as how they interact with other organisms.

BENCH TESTED

Real-World Question

Purpose Students will construct an ecosystem and observe its living and nonliving components. L1 ELL LS **Kinesthetic, Visual-Spatial**

Process Skills observe, classify, recognize cause and effect, communicate, make models, interpret data

Time Required 30 minutes for setup; two weeks for observations

Procedure

Safety Precautions Caution students on the use of scissors.

Teaching Strategies

- Replace the guppy if it dies early in the experiment.
- Have students work in groups of four or five to conserve materials.
- Make sure students care properly for each guppy. After the investigation is completed, ask for volunteers to take the ecosystems home, or assign students to care for them in the classroom. Do not release the guppies into a local lake, pond, or other waterway.

Conclude and Apply

1. The soil, sand, and gravel provided anchorage and nutrients for the plant. The plant photosynthesized and added oxygen to the water. The guppy used oxygen and ate the fish food. The guppy also used the water and rocks as its habitat.
2. The fish food and additional water were necessary to maintain a balanced ecosystem.

LAB

Ecosystem in a Bottle

You may think of ecosystems as large areas. But ecosystems can be any size. You can even make an ecosystem that fits in a plastic bottle.

Real-World Question

What are the parts of an ecosystem?

Goals

- **Model** an ecosystem.
- **Observe** an ecosystem.

Materials

2-L bottle	water
scissors	*Elodea* plants
sand	guppy
aquarium gravel	fish food
metric ruler	

Safety Precautions

Procedure

1. Rinse out a 2-L plastic bottle with water. Using scissors, carefully cut off the top of the bottle.
2. Pour a layer of sand 5 to 10 cm deep in the bottom of the bottle.
3. Fill the bottle to within 5 cm of the top with water that has stood in an open container for about two days. Keep a supply of this aged water on hand to replace the water that evaporates from the bottle. The level should always be about the same.
4. Plant the *Elodea* and add a 2-cm layer of gravel. Place the container in bright light but not direct sunlight.
5. When the water clears, add a guppy.
6. Feed the fish one or two small flakes of food every day.
7. **Observe** your ecosystem every day and record your observations in your Science Journal. Be sure to include observations about the living and nonliving parts of your ecosystem.

Conclude and Apply

1. **Describe** how the parts in the bottle work together to form an ecosystem.
2. **Explain** what is needed to keep the ecosystem healthy.

Communicating Your Data

Keep a daily journal explaining what you do each day to take care of your ecosystem and describing any changes that take place over time.

Assessment

Performance Have students create a model or diagram of the relationships in their ecosystem. Ask students to demonstrate how their ecosystem functioned. Use the Performance Task Assessment List for Model in **Performance Assessment in the Science Classroom**, p. 123. L2 LS **Kinesthetic**

Communicating Your Data

Use a spreadsheet program to set up a table for recording data on temperature, amount of food or water added, and other items associated with the maintenance or evaluation of student ecosystems.

section 2

Relationships Among Living Things

Organizing Ecosystems

Imagine trying to study all of the living things on Earth at once! When ecologists study living things, they usually don't start by studying the entire biosphere. Remember, the biosphere consists of all the parts of Earth where organisms can live. It's much easier to begin by studying smaller parts of the biosphere.

To separate the biosphere into smaller systems that are easier to study, ecologists find it helpful to organize living things into groups. They then study how members of a group interact with each other and their environments.

Groups of Organisms Look at the sea horse in **Figure 8.** This particular sea horse lives in a coral reef in the warm, shallow waters near Australia. The sea horse uses energy, grows, reproduces, and eventually dies. The coral reef is the ecosystem the sea horse lives in. All of the sea horses that live in this particular coral reef make up a population. A **population** is a group of the same type of organisms living in the same place at the same time. Some other populations that you might find in a coral reef ecosystem are sponges, algae, sharks, and coral. What are some populations of organisms that live around your school?

as you read

What You'll Learn

- **Explain** how ecologists organize living systems.
- **Describe** relationships among living things.

Why It's Important

Learning how living things relate to one another will help you understand how you depend on other organisms for survival.

Review Vocabulary
adaptation: any variation that makes an organism better suited to its environment

New Vocabulary
- population
- community
- limiting factor
- niche
- habitat

Figure 8 This school of banner fish belongs to the population of banner fish living in a coral reef ecosystem.

1 Motivate

Bellringer

INTERACTIVE CHALKBOARD PowerPoint® Presentations

Section Focus Transparencies also are available on the Interactive Chalkboard CD-ROM. L2 ELL

Tie to Prior Knowledge

Ask students to list the resources necessary for life. In reviewing these resources, students may better understand how limiting factors influence populations. L2

Section 2 Resource Manager

Chapter *FAST FILE* Resources
- Transparency Activity, p. 45
- MiniLAB, p. 4
- Enrichment, p. 31
- Reinforcement, p. 28
- Directed Reading for Content Mastery, p. 21

Life Science Critical Thinking/Problem Solving, p. 12

Earth Science Critical Thinking/Problem Solving, p. 20

Cultural Diversity, pp. 13–14

Home and Community Involvement, p. 43

Performance Assessment in the Science Classroom, pp. 55–56

Reading and Writing Skill Activities, pp. 15–16

2 Teach

Mini LAB

Purpose Students will calculate the population density of their home. L1 ELL LS **Logical-Mathematical**

Materials measuring tape

Teaching Strategies Suggest to students that, if a room is oddly shaped, they can determine the area of different parts of the room and add the areas together to determine the total area.

Troubleshooting If students have trouble calculating the area of their home, work with them to develop an estimate. For example, students could calculate the area of one or two rooms, and multiply the result by an appropriate factor to estimate total area.

Analysis

1. Population density would increase if the number of students in the classroom doubled and the amount of space remained the same.

Assessment

Content Ask students what would happen to the population density if the area were doubled. The population density would decrease, and residents would have twice as much space.

Reading Check

Answer food, shelter, and other needs

Caption Answer

Figure 9 Possible answer: fish, anemones, and corals

Mini LAB

Calculating Population Density

Procedure

1. Calculate the total area of your home by multiplying the length times the width of each room and adding all the products together.
2. Count the number of people who live in your home.
3. Divide the number of people living in your home by the total area to determine the population density.

Analysis

Calculate what would happen to the population density if the number of people living in your home doubled.

Try at Home

Groups of Populations Many populations live in an ecosystem like the coral reef in **Figure 9.** All of the populations that live in an area make up a **community** (kuh MYEW nuh tee). The members of a community depend on each other for food, shelter, and other needs. For example, a shark depends on the fish populations for food. The fish populations, on the other hand, depend on coral animals to build the reef that they use to hide from the sharks.

No matter where you live, you live in and are part of a community. Make a list of as many of the populations that make up your community as you can. Compare your list with the lists of your classmates. How many populations did the class come up with?

Reading Check *In what ways do the members of a community depend on each other?*

Characteristics of Populations Look around your classroom. Is the room big or small? How many students are in your class? Are there enough books and supplies for everyone? Ecologists ask questions like these to describe populations. They want to know the size of the population, where its members live, and how it is able to stay alive.

Population Density Think about your classroom. A population of 25 students in a large room has plenty of space. How would the same 25 students fit into a smaller room? Ecologists determine population density (DEN suh tee) by comparing the size of a population with its area. For instance, if 100 dandelions are growing in a field that is one square kilometer in size, then the population density is 100 dandelions per square kilometer.

Figure 9 Many populations make up a coral reef community.
Identify *three populations in this coral reef community.*

Active Reading

Jigsaw In this collaborative learning technique, individuals become experts on a portion of a text and share their expertise with a small group, called their home group. Everyone shares responsibility for learning the assigned reading. Assign each person in each home group an expert number (1 through 5, for example). Have students gather into the expert groups that correspond to the number they were assigned. Have them read, discuss, and master chapter concepts, and determine how best to teach them to their home groups. Have students return to their home groups and share the content they learned in their expert groups. Have students use the jigsaw strategy to learn about relationships among living things.

Figure 10 Monarch butterflies can migrate great distances.

Studying Populations Butterflies fluttering over flower beds are a common sight in the summer. Some butterflies live only a short time. Others, such as monarchs, can live for years. Monarchs travel to warm climates for the winter, returning to the same location year after year. This seasonal travel is called migration. Is it possible to study a population that migrates from place to place?

To study migrating monarchs, a "monarch watcher"—often a school student like yourself—carefully catches a monarch and attaches a tag to one of its wings. The tag indicates where the butterfly was caught. Later, someone else who catches the same butterfly can use the tag to figure out how far the butterfly has flown. A butterfly could be tagged in Delaware and captured days later in South Carolina. Information from many butterflies can be combined to build a picture of the monarch's migration like the one in **Figure 10.** Similar techniques are used to study populations of birds, wolves, and other animals that travel long distances.

Limits to Populations

Populations cannot grow larger and larger forever. There wouldn't be enough food, water, living space, and other resources to go around. The things that limit the size of a population, such as the amount of rainfall or food, are called **limiting factors.** Think back to the stream. One biotic limiting factor in the stream ecosystem is the mosquito population. How can a mosquito population be a limiting factor? If you were a frog, you might know the reason. Frogs eat mosquitoes. If lack of rain caused the mosquito population to go down, then the frog population might not have enough food and its population size might also decline. What are some other limiting factors in a stream ecosystem?

Science Online

Topic: Animal Migration
Visit red.msscience.com for Web links to information about tracking animal migrations.

Activity In your Science Journal, draw a map showing the migration route of an animal species.

INTEGRATE History

Compasses Monarchs may be able to use Earth's magnetic field as a kind of compass as they fly. Humans have used compasses for centuries. Research the history of compasses and create a time line.

Teacher FYI

Migration Stimuli Monarch butterfly migration is driven by seasonal changes in day length and temperature. The butterflies fly in large masses to the same winter roosts and often to the exact same trees.

Text Question Answer

may include the amount of water in the stream, the number of organisms available to others for food, and the availability of habitat for different populations

Visual Learning

Figure 10 Display a political map of North America that shows state and country borders to help orient students to the locations and distances shown on the migration map. Have students use the map to calculate distances between points in North America and points in Mexico. L2 LS **Visual-Spatial, Logical-Mathematical**

INTEGRATE History

Ancient Compasses The earliest known description of a compass used for navigation is found in a Chinese manuscript from the year 1020 CE. The first reference to a compass in western writings is from English scholar Alexander Neckham in 1187.

Research In addition to the history of the compass, have students research how compasses are used for land or sea navigation. If possible, provide students with a compass they can learn to use.

Science Journal

Human Populations Have students write in their journals about limiting factors that may influence human populations. Ask students to discuss what happens in areas where certain factors such as food, water, and land are limited. L1 P LS **Linguistic, Intrapersonal**

Differentiated Instruction

English-Language Learners Have student pairs use flash cards to study the definitions of the terms *biosphere*, *ecosystem*, *community*, *population*, and *migration*. Ask them to write the term on one side of a card and its definition on the other side. Students can use the cards to quiz each other. L2 LS **Visual-Spatial, Linguistic**

Activity

Biotic Relationships Have students cut out pictures of familiar organisms from old magazines. Ask them to select pictures of two or more organisms that interact and paste them on a blank sheet of paper. Have students write an explanation of the interaction below the pictures. L2 LS **Linguistic, Kinesthetic, Visual-Spatial**

Discussion

Students may not realize the importance of predation in maintaining balance in an ecosystem. Discuss with students what would happen to field mouse populations if falcons did not eat them. Similarly, ask students what would happen to mosquito populations if other animals did not consume them. Then ask students how these events would influence other populations.

Reading Check

Answer feeding interactions

Teacher FYI

Symbiosis is any close relationship that occurs between two species. Ecologists categorize symbiotic relationships based on how each organism in the relationship is affected. *Mutualism* occurs when both organisms benefit from the relationship. *Commensalism* occurs when one organism benefits from the relationship and the other organism is not harmed. *Parasitism* occurs when one organism benefits at the expense of the other organism.

Figure 11 One of the most common ways organisms interact in a community is by being food for another organism.

Interactions in Communities

Are frogs the only organisms in the stream community that eat mosquitoes? No, there are many animals that eat them, including some birds and spiders. That means that frogs must compete with birds and spiders for the same food. Feeding interactions such as those in **Figure 11** are the most common interactions among organisms in a community.

Imagine a large bowl of popcorn in your classroom. As long as there is enough popcorn to go around, you don't have to worry that you'll get your share. But if the bowl of popcorn were small, you would have to compete with your classmates to get some. The greater the population size of an area, the greater the competition for resources such as food. Food isn't the only resource that organisms compete for. Organisms will compete for any resource that is in limited supply. Space, water, sunlight, and shelter are all resources that may be limited in a particular ecosystem.

Reading Check *What are the most common interactions among organisms in a community?*

Eat or Be Eaten Have you ever heard the phrase "birds of prey"? A falcon is a bird of prey, which means it captures and eats other animals. A falcon, with its razor-sharp talons, will swoop down from the sky to snatch up a field mouse. The falcon is a predator (PRE duh tur). Predation (pre DAY shun) is the act of one organism feeding on another.

Organisms That Live Together Predation doesn't sound like a good deal for the field mouse, does it? The falcon population, however, is limited by the size of the mouse population. There are other types of relationships among organisms. In one type of interaction, both organisms in the relationship benefit. The African tickbird, for instance, gets its food by eating insects off the skin of zebras. The tickbird gets food, while the zebra gets rid of harmful insects. In another type of relationship, only one organism benefits. The other organism doesn't benefit, but is not harmed. A bird building a nest in a tree is an example of this. The bird gets protection from the tree, but the tree isn't harmed. In still another relationship, one organism is helped while the other is harmed. The insects on the zebra's skin, for example, benefit from the zebra. However, these insects can harm the zebra. Have you ever been bitten by a mosquito? That's a firsthand experience of this type of relationship.

Fun Fact

Barn owls help rid farms and ranches of rodent pests. One young barn owl can eat the equivalent of 12 mice every night. An adult owl eats the equivalent of one gopher or rat each night.

Where and How Organisms Live

How can a small ecosystem such as a classroom aquarium support a variety of different organisms? It's possible because each type of organism has a different role to play in the ecosystem. A typical classroom aquarium may contain snails, fish, algae, and bacteria. The role of snails is to feed on algae. The glass of an aquarium can become clouded by the growth of too much algae. Snails eat the algae, helping keep the glass clear so light can get in. The role of the algae is to provide food for snails and fish, and to provide oxygen for the system through photosynthesis. The role of an organism in an ecosystem is called the organism's **niche** (NICH).

What do you think the role of the fish might be in an aquarium ecosystem? The niche of the fish includes adding nutrients to the ecosystem through its waste products that encourage the growth of algae. All the interactions in which an organism takes part make up its niche.

How would you describe your niche? Perhaps you help dispose of wastes by recycling, or obtain food by grocery shopping. What other activities does your niche include?

Applying Science

Graph Populations

One way to understand more about relationships among organisms in an ecosystem is to keep track of, or monitor, and graph populations. Use the information below to make a graph of population size over time for barn owls and field mice. Then, answer the questions that follow.

Table 1 Monthly Population Size per Hectare (in 100s)

Month	J	F	M	A	M	J	J	A	S
Field mice	6	5	4	3	3	4	5	4	6
Barn owls	2	3	4	4	2	1	4	3	4

Identifying the Problem

Set up your graph with months on the *x*-axis and numbers of organisms on the *y*-axis. Use two colors to plot your data. For more help, refer to the Math Skill Handbook. Use your graph to infer how the population of field mice affects the population of barn owls.

Solving the Problem

1. Predict how the next two months of the graph will look.
2. Field mice eat green plants and grains. What do you think would happen to the population of barn owls if there were no rain in the area for a long time?

Quick Demo

Population Control

Materials masking tape, paper clips to represent mice (75), index cards to represent owls (20)

Estimated Time 20 minutes

Procedure Demonstrate the relationship between predator and prey populations. **Trial 1:** Toss 35 mice into a 0.5 m^2 area marked with masking tape. Toss two owls into the mice population. Record these population numbers as Trial 1. All mice touched by an owl are eaten, so remove them. If an owl does not eat at least three mice, it starves and is removed. **Trial 2:** For every pair of mice remaining, add two more mice to represent reproduction. For each owl that ate four to eight mice, add one more owl. For every owl that ate more than eight mice, add two additional owls. Continue for several trials. Have students record population data for each trial on a table similar to Table 1. L2 LS **Visual-Spatial, Logical-Mathematical, Intrapersonal**

Applying Science

Teaching Strategy

Have students set up the *x*-axis with the numbers 100, 200, 300, 400, 500, 600.

Answers

1. would probably show a decrease in mice and increase in owls
2. Plant populations would decrease, causing a reduction in field mice numbers. A continued decline in the mouse population would result in a reduction in the owl population.

Differentiated Instruction

Challenge Have students examine the niches humans occupy in their ecosystems. What conditions are required for survival? What habitats are humans able to occupy? What habitats are not suitable for human habitation and why? L3 P LS **Logical-Mathematical**

Text Question Answer

Students go to school, eat certain foods, create waste, do chores, etc.

Caption Answer

Figure 12 They share some resources, such as water and air, but have different needs.

3 Assess

DAILY INTERVENTION

Check for Understanding

Linguistic Have students write a paragraph listing interactions that take place among biotic and abiotic factors in a garden community, and explaining how these interactions affect population sizes. L2

Reteach

Interactions Have students record all of their interactions with other species for a day. Ask them to identify the feeding relationships on their lists. L2 LS **Linguistic, Intrapersonal**

Assessment

Performance Have students design a self-sustaining terrarium, aquarium, or other contained ecosystem that requires no input other than sunlight. Allow students to make large drawings of their ecosystem design on poster board. Use **Performance Assessment in the Science Classroom**, p. 123. L2 LS **Visual-Spatial, Intrapersonal**

Figure 12 Each organism in an ecosystem has its own job, or niche. **Explain** *how the reptile and the plant can share the same habitat.*

The place where an organism lives is called its **habitat** (HA buh tat). The habitat of a catfish is the muddy bottom of a lake or pond. The habitat of a penguin is the icy waters of the Antarctic. How would you describe the habitat of the pet reptile shown in **Figure 12?**

Different species of organisms often live in the same habitat. Resources, such as food, living space, and shelter, are shared among all the species living in a habitat. For example, the branches of an apple tree provide a habitat for spiders, fruit flies, beetles, caterpillars, and birds. How can all these organisms share the same home? They have different ways of feeding, seeking shelter, and using other resources. In other words, they have different niches. For example, spiders feed on beetles and other insects. Caterpillars eat leaves. Fruit flies feed on apples. Birds eat spiders, caterpillars, or flies. Each species has a different niche within the same habitat.

section 2 review

Summary

Organizing Ecosystems

- A population is a group of the same type of organisms living together in the same place.
- A community is made up of all the populations in an ecosystem.

Limits to Populations

- Limiting factors prevent a population from growing larger.

Interactions in Communities

- Organisms depend on one another for food and shelter.
- An organism's role in an ecosystem is its niche.
- An organism's home is its habitat.

Self Check

1. **Identify** a population that lives in your community.
2. **Explain** how the number of trees in a forest could affect the size of a bird population.
3. **Design an experiment** to identify a limiting factor that prevents the snail population in a home aquarium from growing larger.
4. **Think Critically** Ladybug beetles help gardeners control insect pests called aphids. What kind of interaction might take place between ladybug beetles and aphids?

Applying Math

5. **Calculate** the population density of buttercups in a meadow. There are 550 buttercups in a meadow that measures 100 m by 66 m.

Science Online red.msscience.com/self_check_quiz

section 2 review

1. Answers will vary. Example answers: students might identify a population of pigeons in a park, earthworms in a garden, or humans in a city
2. Trees provide birds with nesting sites and shelter from predators. Trees also provide food for insects that birds eat, or produce fruit that birds eat. A reduction in the tree population could result in declining bird populations because of a reduction in the availability of food or shelter or both.
3. Answers will vary. Student experimental designs should test the affect of one variable—such as sunlight, temperature, or number of fish in the tank—on changes to the snail population.
4. Feeding relationship. Ladybug beetles eat aphids.
5. $6600 m^2 \div 550$ buttercups = 12 buttercups per m^2

section 3

Energy Through the Ecosystem

It's All About Food

Think about the interactions that we've talked about so far. The frog and the mosquito, the falcon and the mouse—most of the interactions involve food. Energy moves through an ecosystem in the form of food.

Producers and Consumers Many different populations interact in a backyard ecosystem, including plants, birds, insects, squirrels, and rabbits, as shown in **Figure 13.** The plants in the ecosystem produce food through photosynthesis. An organism that makes its own food, like a plant, is called a **producer.** The grasshopper that nibbles on the plants is a **consumer.** A consumer eats other organisms.

as you read

***What* You'll Learn**

- **Explain** how organisms get the energy they need.
- **Describe** how energy flows through an ecosystem.

***Why* It's Important**

The energy most living things require comes directly or indirectly from the Sun.

Review Vocabulary

recycling: reuse of an item or natural resource that requires changing or reprocessing it

New Vocabulary

- producer
- decomposer
- consumer

Figure 13 In any community, energy flows from producers to consumers.

1 Motivate

INTERACTIVE CHALKBOARD PowerPoint® Presentations

Bellringer

Section Focus Transparencies also are available on the Interactive Chalkboard CD-ROM. L2 ELL

Tie to Prior Knowledge

Photosynthesis Review with students how plants use the energy from sunlight to convert water and carbon dioxide into chemical energy stored in food molecules. Remind students that some plants cells have chloroplasts that contain chlorophyll, which is needed for photosynthesis.

Teacher FYI

Energy Losses Energy in a food chain always begins with producers and flows to consumers. Ten percent of available energy is lost as energy moves from one level of a food chain to the next, because some of the energy is used by organisms and some is dissipated as heat.

Section 3 Resource Manager

Chapter *FAST FILE* Resources
- Transparency Activity, p. 46
- Enrichment, p. 32
- Reinforcement, p. 29
- Directed Reading for Content Mastery, p. 21
- Lab Worksheet, pp. 7–8
- Lab Activity, pp. 13–16

Lab Management and Safety, p. 69

2 Teach

INTEGRATE Physics

Energy Conversion Chemical energy in food is converted to heat and mechanical energy.

Caption Answer

Figure 14 fish, squid, and seal

Reading Check

Answer how energy from food passes from one organism to another

Make a Model

Food Chains Have students cut pictures of familiar organisms from old magazines, or write the names of organisms on index cards. Ask students to work in small groups to create one or more food chains using their cards or pictures. Make sure each food chain begins with a producer, proceeds to an herbivore, then to one or more consumers. L2 LS COOP LEARN **Visual-Spatial**

Activity

Food Webs Have students work together to combine their food chains from the above modeling activity to create food webs. L2 LS COOP LEARN

Quick Demo

Compost

Materials finished compost, 1-gallon jar containing 1 cup soil, cheesecloth, plant-based food scraps

Estimated Time 10 minutes

Procedure Bury food scraps in soil and cover jar loosely with cheesecloth. Show students the finished compost. Explain that decomposition transforms waste plant material into a material that enriches soil. After several days, show students how the scraps have decomposed.

INTEGRATE Physics

Energy Conversion Energy cannot be created or destroyed, but it can change form. Photosynthesis converts light energy from the Sun into chemical energy stored in food. Research to find out what kinds of energy conversion take place in your body.

Decomposers Some of the consumers in an ecosystem are so small that you might not notice them, but they have an important role to play. They are the decomposers, such as bacteria and fungi. **Decomposers** use dead organisms and the waste material of other organisms for food.

Modeling the Flow of Energy

The food chain in **Figure 11** is a simple model that shows how energy from food passes from one organism to another. Each organism is linked by an arrow. The arrows show that energy moves from one organism to another in the form of food.

Reading Check *What does a food chain model?*

Can you spot any problems with using a food chain to model the feeding relationships in a community? A food chain does not show every species in the community. We need a more complex model to show all the feeding interactions in an ecosystem.

Food chains often overlap. For instance, a bird may eat seeds and in turn be eaten by a cat. However, a cat might also eat a rabbit or a mouse. One food chain cannot model all these overlapping relationships. Scientists use a more complicated model, called a food web, to show the transfer of energy in an ecosystem. A food web like the one in **Figure 14** is a series of overlapping food chains that shows all the possible feeding relationships in an ocean ecosystem.

Figure 14 This ocean food web is made up of many overlapping food chains.
Identify *the organisms eaten by an orca.*

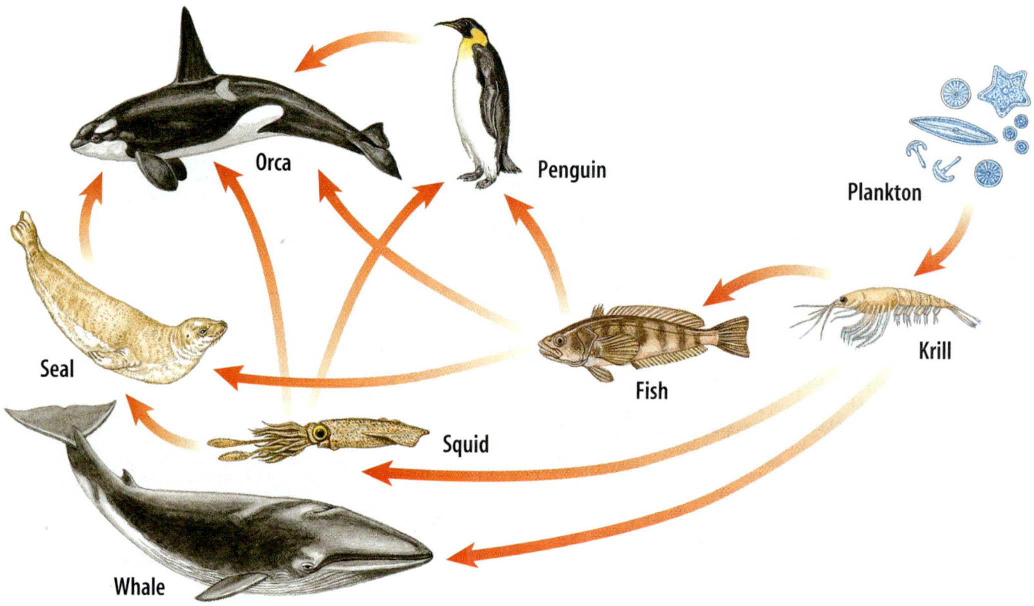

Differentiated Instruction

Challenge Ask students to form groups to research the water, nitrogen, or carbon cycle. Have each group create a poster and present their cycle to their classmates. L3 LS **Linguistic, Visual-Spatial, Intrapersonal**

Visual Learning

Figure 14 Have students carefully examine the figure and diagram a number of food chains. Students should label each organism as a producer, consumer, or decomposer. L1 ELL LS **Visual-Spatial, Linguistic**

Cycling of Materials

What happens when you recycle a soda can? The can is taken to a processing plant and melted so that the aluminum can be used again. This is an example of a simple cycle. The same aluminum can be used over and over again. Cycles are important to ecosystems. Instead of aluminum cans, however, it's the materials that make up organisms that get recycled in an ecosystem.

The bodies of living things are made up of matter, including water and chemicals like nitrogen and carbon. To get the matter needed to build bones, muscles, and skin, you need to eat food made of the right kinds of matter, as the horse in **Figure 15** is doing. In an ecosystem, matter cycles through food chains. The amount of matter on Earth never changes. So matter in ecosystems is recycled, or used again and again.

Living organisms depend on these cycles for survival. Organisms also depend on one another for food, shelter, and other needs. All the different things that make up the biosphere—from a tiny insect to a raging river— have a unique role to play.

Figure 15 Horses get the materials they need to grow and maintain their bodies by eating food such as grass.
Infer *where you think the grass gets the materials it needs for growth.*

section 3 review

Summary

It's All About Food

- Energy moves through ecosystems in the form of food.
- Producers make their own food.
- Consumers obtain food by eating other organisms.
- Decomposers break down dead organisms.

Modeling the Flow of Energy

- Food chains and food webs show how food energy passes from one organism to another.

Cycling of Materials

- All matter on Earth is recycled through food chains.

Self Check

1. **List** some organisms that are consumers. Give an example of the type of food each eats.
2. **Explain** how the Sun's energy reaches a cat that eats a bird.
3. **Compare and contrast** food chains and food webs.
4. **Think Critically** Explain why more energy is available in the first link of a food chain than in the fourth link of the same food chain.

Applying Skills

5. **Sequence** Use an events-chain concept map to trace the path of energy that flows from the Sun to your body when you eat a piece of chicken.

Caption Answer

Figure 15 air, water, and soil

Fun Fact

Some of the water molecules in your next drink of water may once have been part of a lake from which a dinosaur took a drink.

3 Assess

DAILY INTERVENTION

Check for Understanding

Visual-Spatial Have students draw a food web for an ecosystem of their choice. L2 P

Reteach

Food Chain Ask students to diagram the food chain that led to their breakfast. Ask students to identify the consumers and producers. L2 LS **Visual-Spatial**

Assessment

Content Have students work in groups of two or three to write and illustrate a simple story that explains ecosystem interactions to young children. L2 LS **Visual-Spatial** COOP LEARN

Virtual Labs

Energy Transfer *How is energy transferred through a community of organisms?*

section 3 review

1. Answers will vary. Any organism that consumes another organism—plant, animal, fungus, or bacteria—is a consumer.
2. Plant converts energy from sunlight into food through photosynthesis. Insect eats plant, bird eats insect, cat eats bird. At each step, energy that originally came from sunlight is transferred from one organism to another.
3. Food chains are simple models that show how energy from food passes between some of the organisms in an ecosystem. Food webs are more complex models that show how energy is transferred among all the organisms in an ecosystem.
4. Some of the energy is used by the organisms in each link of the food chain.
5. Possible answers: Sun—grass—insect—chicken—human; Sun—grain—chicken—human

BENCH TESTED

Real-World Question

Purpose Students will determine how light, water, temperature, or space limit plant populations. L1 ELL LS **Kinesthetic, Visual-Spatial, Logical-Mathematical**

Process Skills observe and infer, make and use tables, make and use graphs

Time 50 minutes for setup; 10 minutes every 3 to 4 days for 2 weeks for data collection

Materials Peas or corn may be substituted for bean seeds.

Safety Precautions Students should wash their hands carefully after handling seeds and soil.

Form a Hypothesis

Possible Hypothesis Student hypotheses should reflect that space, light, water, or temperature may limit plant growth.

Test Your Hypothesis

Possible Procedures

- Plant the same number of seeds in each pot and manipulate the variable you are interested in observing while keeping all other variables the same. Observe the results.
- Plant an increasing number of seeds in a series of pots to determine how space influences plant growth when all other variables are the same.
- Choose one variable to test at a time. Test light, water, temperature, or space.

LAB Design Your Own

What's the limit?

Goals

- **Observe** how space, light, water, or temperature affect how many bean plants are able to grow in a pot.
- **Design an experiment** that shows whether a certain abiotic factor limits a plant population using the materials listed.

Possible Materials

bean seeds
small planting containers
soil
water
labels
spoons
aluminum foil
sunny window or other light source
refrigerator or heater

Safety Precautions

WARNING: *Wash your hands after you handle soil and seeds.*

Real-World Question

How many blades of grass are in a park? It may seem to you like there's no limit to the number of blades of grass that can grow there. However, as you've discovered, there are many factors that organisms like the plants in the park need to live and grow. By experimenting with these factors, you can see how they limit the size of the population. How do space, light, water, and temperature limit plant populations?

Form a Hypothesis

Think about what you already know about the needs of plants. As a group, form a hypothesis to explain how one abiotic factor may limit the number of bean plants that can grow in a single pot.

Test Your Hypothesis

Make a Plan

1. **Decide** on a way to test your group's hypothesis. Make a complete materials list as you plan the steps of your experiment.

Alternative Inquiry Lab

Extend the Experience To observe plant competition for available space, have students test how much space various types of plants take up when grown in the same area. Use a variety of seeds that students can grow. Students can use toothpicks and thread to set up a grid of equal sized sections in a long planting tray. Students can plant a different variety of seed in each section maintaining strict controls for initial growing conditions. Students can observe which plant varieties grow and fill their spaces most rapidly. They can also discover which varieties move out of their spaces and invade other sections. Students can measure the space taken up by each type of plant and the population density of each variety. They can then graph their results.

Using Scientific Methods

2. What is the one abiotic factor you will be testing? How will you test it? What factors will you need to control? Be specific in describing how you will handle the other abiotic factors.
3. How long will you run your experiment? How many trials of your experiment will you run?
4. **Decide** what data you will need to collect. Prepare a data table in your Science Journal.
5. **Read** over your entire experiment and imagine yourself doing it. Make sure the steps are in logical order.

Follow Your Plan

1. Make sure your teacher has approved your plan and your data table before you proceed.
2. Carry out your plan.
3. **Record** your observations during the experiment.
4. **Complete** your data table in your Science Journal.

Analyze Your Data

1. **Make a graph** to show your results. Use a bar graph to compare the number of seedlings that grew in the experimental containers with the number of seedlings that grew in the control containers.
2. **Identify** the variables in your experiment.
3. **Identify** the controls in your experiment.

Conclude and Apply

1. **Explain** how the abiotic factor you tested affected the bean plant population.
2. **Predict** what would happen to your plant population if you added another kind of plant or animal to the containers.

Communicating Your Data

Compare your results with other groups. Explain how different factors affected the plants grown by each group.

Teaching Strategies

Troubleshooting Students should be careful not to plant seeds too deeply in the soil and to avoid overwatering the containers. Be sure students have a control.

Expected Outcome In most cases, the control plants will do better than the experimental plants. Most students' results will show that seeds need water, light, space, and moderate temperatures in order to grow.

Analyze Your Data

1. Graphs should show that limiting any of the variables inhibit plant growth. Graphs of results from testing the 'space' variable may not show much difference between control and experimental pots. However, if students continue to monitor these pots, they should find that crowded plants are less likely to grow to maturity and reproduce.
2. Each experiment should test only one variable.
3. Controls hold all conditions the same except for the variable.

Conclude and Apply

1. Students should state how the factor they tested did or did not limit growth.
2. Answers will vary but should be consistent with results.

Communicating Your Data

Student results probably will be similar to those of other students who observed the same variable.

Assessment

Performance Different kinds of organisms have different resource requirements. Have students select two or three different kinds of organisms, such as a plant, a fish, and a snail, and design an experiment to test how a change in one abiotic factor influences each one. Use **Performance Assessment in the Science Classroom**, p. 95. L2 LS **Kinesthetic, Visual-Spatial, Logical-Mathematical**

TIME

TIME SCIENCE AND Society

SCIENCE ISSUES THAT AFFECT YOU!

Content Background

The American alligator is one of 20 species of crocodilians (lizard-like, egg-laying meat eaters) found worldwide. The largest of the modern reptiles, crocodilians are a living link to prehistoric, dinosaur-like reptiles.

Crocodilians in the wild may live up to 100 years; in captivity, they generally live about 40 years.

American alligators inhabit freshwater locations in the southeastern United States from North Carolina to Florida and west to the lower Rio Grande. Large adults can stay under water for more than an hour. They are powerful swimmers, using their tails to propel them. They also use their tails as weapons or to knock land prey into the water.

Discussion

Alligator Controls Should the number of alligators be reduced to decrease the possibility of attacks on people? If so, how? Who should decide how many alligators should be kept? Should alligator habitats be protected? Accept all well supported answers.

Investigate the Issue

Problems with Predators Ask students to find out about predator problems in their area. Students can contact the city animal control office or the county animal regulation division for information.

Gators at the Gate!

When you think about Florida, you probably picture sandy beaches and palm trees. But do you think about alligators? Alligators are among the best-known animals that live in Florida. They can grow to be 13 feet long and weigh more than 600 pounds.

Endangered Alligators

By the 1960s, the number of alligators was greatly reduced in Florida due to hunting and habitat loss. The numbers became so low that alligators were placed on the endangered species list. A species is listed as endangered when so few of its members are living that the entire species is in danger of becoming extinct. In the United States, it became illegal to hunt alligators. Gradually the number of alligators went up. By 1977, they were renamed as a threatened species. A threatened species still needs to be protected but is not in immediate danger of becoming extinct. Now, more than a million alligators live on farms and in the wild. Good news, right? Think again. There are problems—big problems from big alligators.

Alligator Problems

Today more people live, work, and play in areas where alligators live. Alligators have been found in swimming pools and on golf courses. Many people believe that the size of Florida's alligator population should be tightly controlled. They fear that more alligators will lead to more encounters with humans and increase the possibility of alligator attacks.

Other people point out that, as the number of houses, roads, and shopping centers in Florida increases, alligators are left with fewer and fewer places to live and hunt. These people suggest that more wilderness areas must be set aside for alligators and other predators.

Write **Alligators are not the only predators people fear. Some parts of the country also have problems with bears, wolves, cougars, or other animals. Write a short paper about encounters between people and predators in your area.**

Science Online
For more information, visit red.msscience.com/time

Write If students live in an urban area, have them explore predator problems in other regions, or in state or national parks. For example, many national parks have problems with bears breaking into tents and cars to find food. In other areas, poisonous snakes may pose problems.

Resources for Teachers and Students

American Alligator (Animals Under Threat) by Louise and Richard Spilsbury, Heineman Library, January 2004

The American Alligator (Wildlife in North America) by Steve Potts, Capstone Press, 1998

Reviewing Main Ideas

Section 1 What is an ecosystem?

1. An ecosystem is made up of organisms interacting with each other and with the nonliving factors in the system.
2. The biosphere is made up of all the ecosystems on Earth.

Section 2 Relationships Among Living Things

1. A population is made up of the same type of organisms living together in the same place at the same time.
2. A community is made up of all the populations in an ecosystem.
3. Limiting factors, which may be living or nonliving, influence the size of a population.

Section 3 Energy Through the Ecosystem

1. Energy is transferred through an ecosystem in the form of food.
2. The feeding relationships in an ecosystem can be illustrated by food chains and food webs.

Visualizing Main Ideas

Copy and complete this concept map about ecosystems.

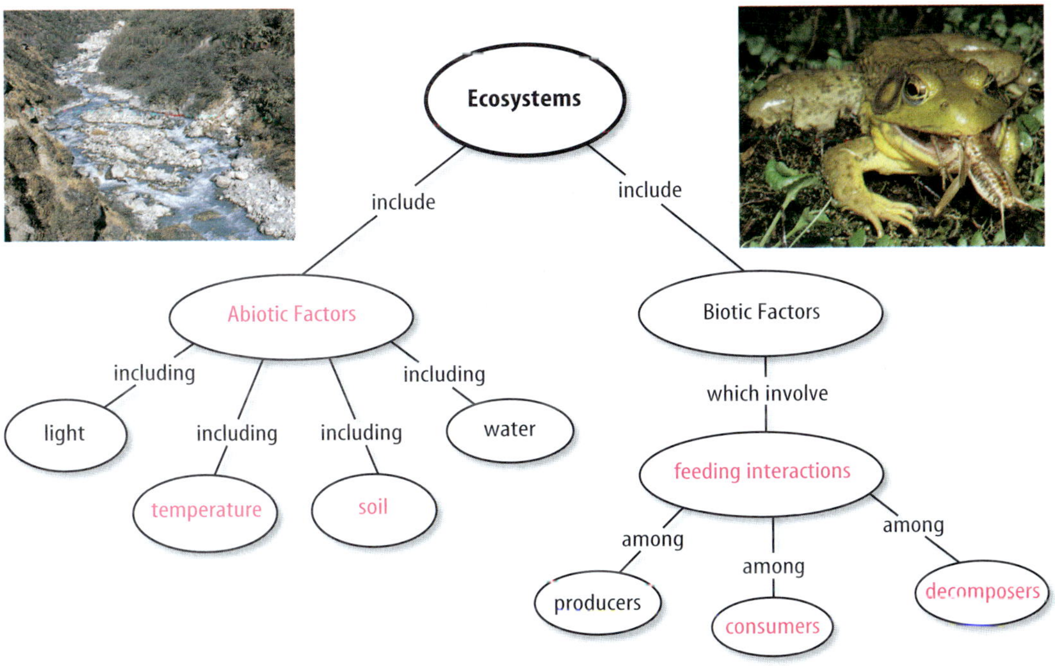

Identifying Misconceptions

Assess

Activity Give students the following list of organisms: lion, oak tree, corn plant, aphid, sparrow, great white shark, dandelion, tree frog, rye grass, saguaro cactus. Have students sort the organisms into two columns based on how they obtain their nutrients. Have students share their grouping with a partner. Then give them the terms "producers" and "consumers" and ask them to label each column. Next, give students the phrases "eats food" and "makes own food" and ask them to label each column.

chapter 21 Study Guide

Reviewing Main Ideas

Summary statements can be used by students to review the major concepts of the chapter.

Visualizing Main Ideas

See student page.

Science Online

Visit **red.msscience.com**
/self_check_quiz
/interactive_tutor
/vocabulary_puzzlemaker
/chapter_review
/standardized_test

Assessment Transparency

For additional assessment questions, use the *Assessment Transparency* located in the transparency book.

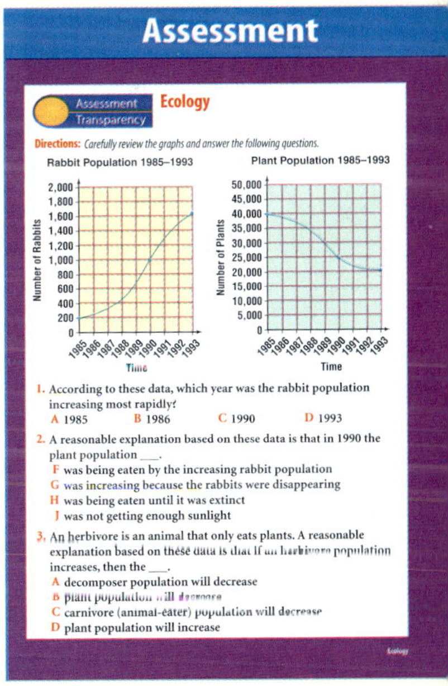
Assessment

Assessment Transparency **Ecology**

Directions: *Carefully review the graphs and answer the following questions.*

Rabbit Population 1985–1993 / Plant Population 1985–1993

1. According to these data, which year was the rabbit population increasing most rapidly?
A 1985 B 1986 C 1990 D 1993
2. A reasonable explanation based on these data is that in 1990 the plant population ___.
F was being eaten by the increasing rabbit population
G was increasing because the rabbits were disappearing
H was being eaten until it was extinct
J was not getting enough sunlight
3. An herbivore is an animal that only eats plants. A reasonable explanation based on these data is that if an herbivore population increases, then the ___.
A decomposer population will decrease
B plant population will decrease
C carnivore (animal-eater) population will decrease
D plant population will increase

Using Vocabulary

1. abiotic factor is a nonliving feature of the environment; biotic factor is a living organism
2. biosphere includes all ecosystems on Earth; ecology is the study of ecosystems
3. community—all populations in an ecosystem; population—all organisms of the same species living in the same place at the same time.
4. ecosystem is made up of organisms interacting with each other and with the abiotic factors of the environment; a limiting factor is a biotic or abiotic factor that acts to limits population size
5. niche—organism's role in the environment; habitat—place where organism lives
6. A producer makes its own food; a consumer relies on other organisms for food.
7. A consumer eats other organisms. A decomposer consumes dead organisms or waste materials produced by living organisms.

Checking Concepts

8. B	13. C
9. D	14. B
10. B	15. A
11. A	16. D
12. A	17. A

Thinking Critically

18. break down dead organisms, do not manufacture their own food
19. large amounts of plants
20. Accept all reasonable responses. Be sure students define their ecosystem.
21. Possible answers: light, temperature, food

chapter 21 Review

Using Vocabulary

abiotic factor p. 622	ecosystem p. 618
biosphere p. 620	habitat p. 632
biotic factor p. 620	limiting factor p. 629
community p. 628	niche p. 631
consumer p. 633	population p. 627
decomposer p. 634	producer p. 633
ecology p. 620	

Explain the difference between the vocabulary words in each of the following sets.

1. abiotic factor—biotic factor
2. biosphere—ecology
3. community—population
4. ecosystem—limiting factor
5. niche—habitat
6. producer—consumer
7. consumer—decomposer

Checking Concepts

Choose the word or phrase that completes the sentence.

8. Which of the following is NOT a biotic factor?
   A) raccoons **C)** pine trees
   B) sunlight **D)** mushrooms
9. Ponds, streams, and prairies are examples of what parts of the environment?
   A) niches **C)** populations
   B) producers **D)** ecosystems
10. What is a group of the same type of organism living in the same place at the same time?
    A) habitat
    B) population
    C) community
    D) ecosystem
11. Which of the following is an example of a producer?
    A) grass **C)** fungus
    B) horse **D)** fish

Use the illustration below to answer question 12.

12. What is the diagram shown above an example of?
    A) food chain **C)** ecosystem
    B) food web **D)** population
13. All of the following are abiotic factors except
    A) sunlight. **C)** bacteria.
    B) water. **D)** temperature.
14. All ecosystems on Earth make up the
    A) atmosphere. **C)** lithosphere.
    B) biosphere. **D)** hydrosphere.
15. All the populations in an ecosystem make up a
    A) community. **C)** habitat.
    B) niche. **D)** limiting factor.
16. In a meadow ecosystem, hawks feed mostly on mice. When the mice population is small, there is less food for hawks, so the hawk population becomes smaller. When the mice population grows, the number of hawks also grows. For the hawk population, the number of mice in this meadow is an example of a
    A) niche. **C)** producer.
    B) habitat. **D)** limiting factor.
17. A food web is a model that shows how
    A) energy moves through an ecosystem.
    B) ecosystems change over time.
    C) producers use sunlight.
    D) abiotic factors affect populations.

Science Online red.msscience.com/vocabulary_puzzlemaker

Use the ExamView® Pro Testmaker CD-ROM to:

- create multiple versions of tests
- create modified tests with one mouse click for inclusion students
- edit existing questions and add your own questions
- build tests aligned with state standards using built-in State Curriculum Tags
- change English tests to Spanish with one mouse click and vice versa

Thinking Critically

18. Infer why it is correct to say that decomposers are also consumers.

19. List examples of foods you would eat if you were eating low on the food chain.

20. Draw and label a diagram of an ecosystem. Label biotic and abiotic factors. Describe three interactions among organisms in the ecosystem.

21. Identify three possible limiting factors for an aquarium ecosystem. Describe how each factor can limit population growth.

22. Describe your own habitat and niche.

23. Classify each of your ten favorite foods as coming from a producer, consumer, or decomposer. Write a short explanation of your classification of each item.

Use the graph below to answer question 24.

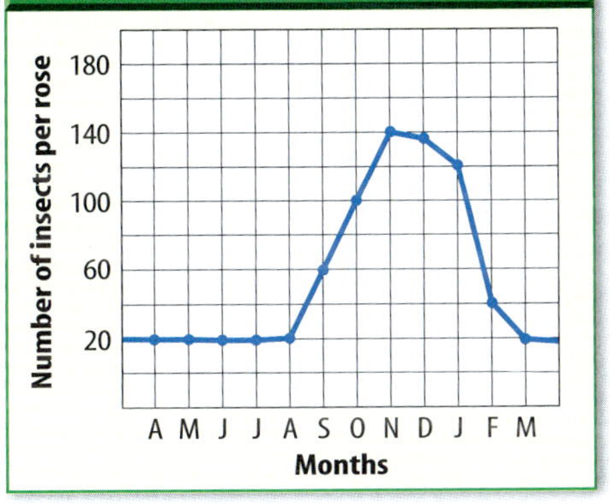

24. Make and Use Graphs The graph above shows the changes in the size of a population of insects living on roses over the course of a year. During what month is the insect population the smallest? During what month is the population the largest?

25. Predict what would happen to an ecosystem if its decomposers were removed.

Performance Activities

26. Develop Multimedia Presentations Find slides or photographs that show different ecosystems. Arrange a slide presentation or photo display of these images. Use titles or captions to identify each one.

27. Research Information Choose an ecosystem to research. Find out what plant and animal species are found there and how they interact. Make a poster illustrating a food web in this ecosystem.

Applying Math

28. Population Density The population density of rabbits living along the banks of a stream is about one rabbit per 100 m^2. How many rabbits are likely to be found in a 900-m by 25-m section of the stream bank?

Use the table below to answer question 29.

Changes in Population Size (in 100s)

Year	Rabbit Population	Bobcat Population
1970	100	39
1975	133	80
1980	94	61
1985	65	63
1990	80	45

29. Changes in Population Size Use the data from the table above to make a graph of population size over time for rabbits and bobcats. Based on your graph, infer how the size of the rabbit population affects the size of the bobcat population.

chapter 21 Review

Thinking Critically

22. Student's habitat is his or her home. Niche includes all resources necessary for survival such as food, shelter, water, relationships, and tasks the student may do.

23. Students should classify plant materials as producers, animal materials as consumers, and products from fungi and bacteria as decomposers.

24. smallest from March to August; highest in November

25. Dead organisms and waste materials would pile up; matter would not be recycled for use by new organisms.

Performance Activities

26. Slides and photographs should depict different ecosystems and illustrate that ecosystems are composed of communities of organisms. Use **PASC,** p. 149.

27. The poster should illustrate biotic and abiotic relationships between organisms and also depict a food web. Use **PASC,** p. 145.

Applying Math

National Math Standards

1, 2, 3, 4, 5, 9

28. $900\text{ m} \times 25\text{ m} = 22{,}500\text{ m}^2 / 100\text{m}^2$ per rabbit = 225 rabbits

29. Rabbits are prey for bobcats. When rabbit populations decline, bobcat populations soon begin to decline. When rabbit populations rise, a rise in the bobcat population follows.

Assessment Resources

Reproducible Masters

Chapter *Fast File* Resources
- Chapter Review, pp. 37–38
- Chapter Tests, pp. 39–42
- Assessment Transparency Activity, p. 49

Glencoe Science Web site
- Chapter Review Test
- Standardized Test Practice

Glencoe Technology
- Assessment Transparency
- Exam*View*® Pro Testmaker
- MindJogger Videoquiz
- Interactive Chalkboard

FAST FILE

Answer Sheet A practice answer sheet can be found at red.msscience.com/answer_sheet.

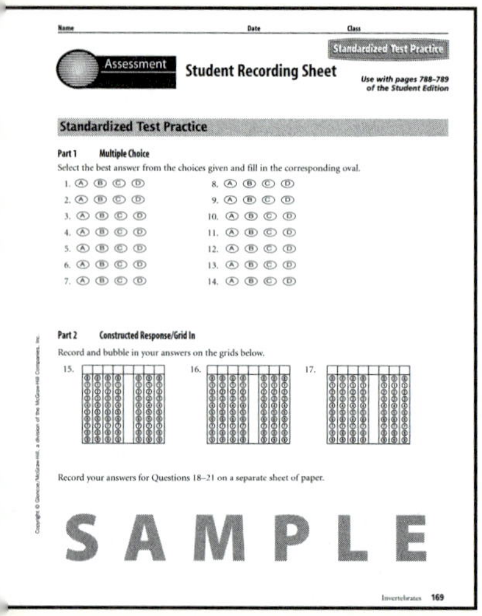
Name Date Class

Standardized Test Practice

Assessment **Student Recording Sheet**

Use with pages 788–789 of the Student Edition

Standardized Test Practice

Part 1 Multiple Choice

Select the best answer from the choices given and fill in the corresponding oval.

1. Ⓐ Ⓑ Ⓒ Ⓓ
2. Ⓐ Ⓑ Ⓒ Ⓓ
3. Ⓐ Ⓑ Ⓒ Ⓓ
4. Ⓐ Ⓑ Ⓒ Ⓓ
5. Ⓐ Ⓑ Ⓒ Ⓓ
6. Ⓐ Ⓑ Ⓒ Ⓓ
7. Ⓐ Ⓑ Ⓒ Ⓓ
8. Ⓐ Ⓑ Ⓒ Ⓓ
9. Ⓐ Ⓑ Ⓒ Ⓓ
10. Ⓐ Ⓑ Ⓒ Ⓓ
11. Ⓐ Ⓑ Ⓒ Ⓓ
12. Ⓐ Ⓑ Ⓒ Ⓓ
13. Ⓐ Ⓑ Ⓒ Ⓓ
14. Ⓐ Ⓑ Ⓒ Ⓓ

Part 2 Constructed Response/Grid In

Record and bubble in your answers on the grids below.

15. 16. 17.

Record your answers for Questions 18–21 on a separate sheet of paper.

SAMPLE

Invertebrates 169

Part 1 Multiple Choice

1. D
2. D
3. B
4. B
5. A
6. A
7. C
8. D
9. C
10. B

Part 2 Short Response

11. All life requires water for survival.
12. both have abiotic and biotic factors, populations, habitats, niches, food webs, symbiotic relationships
13. Sun provides light energy for producers; consumers get energy by eating producers or other consumers; decomposers get energy by breaking down the bodies of producers and consumers.

chapter 21 Standardized Test Practice

Part 1 Multiple Choice

Record you answers on the answer sheet provide by your teacher or on a sheet of paper.

Use the picture below to answer questions 1 and 2.

1. The shelf-like organism growing on this tree is a fungus. What term best describes this organism?
   A. producer **C.** predator
   B. consumer **D.** decomposer

2. What kinds of organisms are shown in the photo?
   A. producers and consumers
   B. consumers and decomposers
   C. predators and prey
   D. decomposers and producers

3. What kind of scientist would study the interactions taking place between the predators and prey living in a community?
   A. geneticist **C.** botanist
   B. ecologist **D.** geologist

4. What term best describes the place in which an organism lives?
   A. niche **C.** biosphere
   B. habitat **D.** community

5. What term best describes a population that has so few members that it could become extinct?
   A. endangered **C.** consumer
   B. threatened **D.** decomposer

6. What is the largest ecosystem on Earth?
   A. ocean **C.** Asia
   B. tundra **D.** biosphere

7. There are 48 dogs in 100 households covering 16 square kilometers. What is the population density of dogs?
   A. 2 dogs per household
   B. 3 dogs per household
   C. 3 dogs per square kilometer
   D. 2 dogs per square kilometer

8. What term best describes populations that live together in an area?
   A. niche **C.** population density
   B. habitat **D.** community

Use the graph below to answer questions 9 and 10.

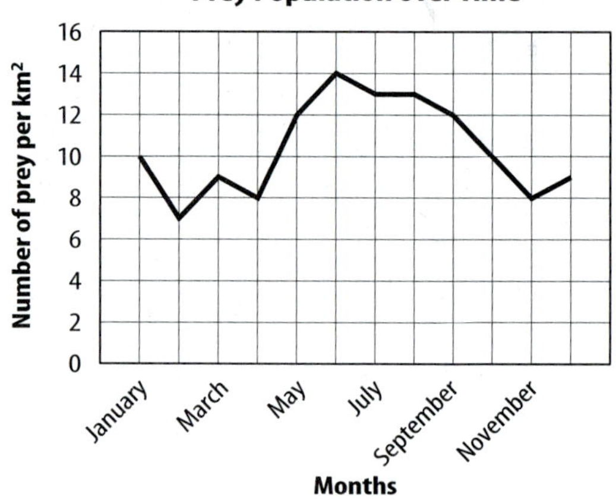

9. In which month was the prey population the highest?
   A. March **C.** June
   B. April **D.** December

10. What was the approximate population density of the prey population in February and April?
   A. 10 per km^2 **C.** 12 per km^2
   B. 8 per km^2 **D.** 7 per km^2

14. population: group of the same species living in the same place at the same time; community: all populations living in an ecosystem; ecosystem: all communities in an area and the abiotic factors with which they interact
15. both contain biotic and abiotic factors; matter and energy cycle through food webs; producers form the basis of all food chains
16. soil type, temperature range, amount of water available

Part 3 Open Ended

17. If mosquito population decreases, the frog population decreases. If the mosquito population increases, frog population might increase. The mosquito population would increase.
18. Possible answers: fish, frogs, insects, snails, clams, algae, aquatic plants. Habitat disappears as water dries up.

Part 2 Short Response/Grid In

Record you answers on the answer sheet provide by your teacher or on a sheet of paper.

11. Why is water an important abiotic factor in any ecosystem?

12. What does the biosphere have in common with an aquarium?

13. Explain the flow of energy in an ecosystem using these terms—*consumer, producer, Sun,* and *decomposer.*

14. How are the terms *population, community,* and *ecosystem* related?

Use the photograph below to answer question 15 and 16.

15. In what ways is the meadow ecosystem shown here similar to a desert ecosystem?

16. What abiotic factors might affect organisms living in the meadow ecosystem?

Test-Taking Tip

Rest Up Get plenty of sleep—at least eight hours every night—during test week and the week before the test.

Part 3 Open Ended

Record your answers on a sheet of paper.

17. Frogs feed on mosquitoes and other insects. Explain how mosquitoes are a limiting factor for frogs. What happens to the frog population if the mosquitoes rapidly increase or decrease? What happens to the mosquitoes if the frogs disappear?

Use the photograph below to answer questions 18 and 19.

18. What kinds of organisms might have lived in this stream when water was flowing in it? What might have happened to them as the stream dried up?

19. Describe a new habitat that might have appeared after the stream dried up. What kinds of organisms could live in that habitat?

20. Describe at least three ways in which one organism can interact with other organisms in the same community.

21. Compare and contrast an endangered species and a threatened species.

22. Explain whether a food web or a food chain is a better model for the flow of energy in an ecosystem.

23. Why is the cycling of materials important for an ecosystem?

19. Possible answer: rich soil that once was mud at bottom of pond; earthworms, insects, land plants

20. Predation: one organism feeds on another. Three kinds of symbiosis: organisms benefit each other; one organism benefits and the other is not affected; one organism is harmed and the other benefits.

21. Both need to be protected from extinction. Endangered: population is so small that the species could easily be wiped out; threatened: somewhat larger population that could easily become endangered. If the population of an endangered species grows, it could be reclassified as a threatened species.

22. food chain: simple model in which each organism is only linked to one other organism; food web: more complex model of energy flow and more informative because it shows overlapping food chains and greater number of feeding relationships for each species in the community

23. The amount of matter on Earth never changes, so the matter in an ecosystem must be used over and over again.

Rubrics

The following rubrics are sample scoring devices for short response and open-ended questions.

Short Response

Points	Description
2	The student demonstrates a thorough understanding of the science of the task. The response may contain minor flaws that do not detract from the demonstration of a thorough understanding.
1	The student has provided a response that is only partially correct.
0	The student has provided a completely incorrect solution or no response at all.

Open Ended

Points	Description
4	The student demonstrates a thorough understanding of the science of the task. The response may contain minor flaws that do not detract from the demonstration of a thorough understanding.
3	The student demonstrates an understanding of the science of the task. The response is essentially correct and demonstrates an essential but less than thorough understanding of the science.
2	The student demonstrates only a partial understanding of the science of the task. Although the student may have used the correct approach to a solution or may have provided a correct solution, the work lacks an essential understanding of the underlying science concepts.
1	The student demonstrates a very limited understanding of the science of the task. The response is incomplete and exhibits many flaws.
0	The student provides a completely incorrect solution or no response at all.

chapter 22 Organizer

Section/Objectives	Standards		Labs/Features
	National	State/Local	
Chapter Opener	See pp. 16T–17T for a Key to Standards.		**Launch Lab:** What are natural resources?, p. 645 **Foldables,** p. 645 **Visualizing Steel Production,** p. 648
Section 1 Natural Resource Use 4 sessions 2 blocks 1. **Explain** how resources are used. 2. **Describe** how resources are classified.	National Content Standards: UCP.2, UCP.3, A.1, A.2, F.5		**MiniLAB:** Analyzing Gift Wrap, p. 650 **Science Online,** p. 653 **Lab:** Using Water, p. 654
Section 2 People and the Environment 4 sessions 2 blocks 3. **Explain** how people affect the environment. 4. **Describe** different types of pollution.	National Content Standards: UCP.2, UCP.3, A.1, A.2, F.2, F.5		**Science Online,** p. 656
Section 3 Protecting the Environment 2 sessions 1 block 5. **Describe** the problems of solid waste. 6. **Explain** how to reduce, reuse, and recycle resources.	National Content Standards: UCP.2, UCP.3, A.1, A.2, E.1, F.5		**MiniLAB:** Making Models, p. 664 **Applying Science:** Reusing Plastic, p. 665 **Lab:** Using Land, p. 668 **Science and Society:** A Tool for the Environment, p. 670

Glencoe Exclusive!
TeacherWorks™
All-In-One Planner and Resource Center

Lab Materials	Reproducible Resources	Section Assessment	Technology
Launch Lab: objects made from plastic, paper, metal, glass, and wood	**Chapter *FAST FILE* Resources** Foldables Worksheet, p. 17 Directed Reading Overview, p. 19 Note-taking Worksheets, pp. 33–35	Glencoe's Assessment Advantage	**TeacherWorks** includes: • Interactive Teacher Edition • Lesson Planner with calendar • Access to all program blacklines • Correlations to standards • Web links
MiniLAB: objects to wrap; wrappings such as paper bags, plastic bags, and newspaper; scissors, string, tape **Lab:** calculator	**Chapter *FAST FILE* Resources** Transparency Activity, p. 44 MiniLAB, p. 3 Enrichment, p. 30 Reinforcement, p. 27 Directed Reading, p. 20 Lab Worksheet, pp. 5–6 Lab Activity, pp. 9–12 Transparency Activity, pp. 47–48 **Science Inquiry Labs,** pp. 31–32	**Portfolio** Science Journal, p. 649 Assessment, p. 654 **Performance** MiniLAB, p. 650 Assessment, p. 654 Applying Skills, p. 653 **Content** Section Review, p. 653	Section Focus Transparency Teaching Transparency Virtual Labs CD-ROM Guided Reading Audio Program Interactive Chalkboard CD-ROM
Need materials? Contact Science Kit at 1-800-828-7777 or www.sciencekit.com on the Internet.	**Chapter *FAST FILE* Resources** Transparency Activity, p. 45 Enrichment, p. 31 Reinforcement, p. 28 Directed Reading, p. 21	**Portfolio** Activity, p. 660 **Performance** Assessment, p. 662 Applying Math, p. 662 **Content** Section Review, p. 662	Section Focus Transparency Virtual Labs CD-ROM Guided Reading Audio Program Interactive Chalkboard CD-ROM
MiniLAB: clean items that are normally thrown out; glue; string; tape; cardboard, posterboard, or foamboard **Lab:** grid paper (10 squares × 10 squares), colored pencils	**Chapter *FAST FILE* Resources** Transparency Activity, p. 46 MiniLAB, p. 4 Enrichment, p. 32 Reinforcement, p. 29 Directed Reading, p. 21 Lab Worksheet, pp. 7–8 Lab Activity, pp. 13–16	**Portfolio** MiniLAB Assessment, p. 664 **Performance** Assessment, p. 664 Applying Science, p. 665 Assessment, p. 667 Applying Skills, p. 667 **Content** Section Review, p. 667	Section Focus Transparency Virtual Labs CD-ROM Guided Reading Audio Program Interactive Chalkboard CD-ROM

End of Chapter Assessment

Glencoe's Assessment Advantage

Blackline Masters	Technology	Professional Series
Chapter *FAST FILE* Resources Chapter Review, pp. 37–38 Chapter Tests, pp. 39–42 **Standardized Test Practice,** pp. 91–94	MindJogger Videoquiz Virtual Labs CD-ROM Exam*View*® Pro Testmaker TeacherWorks CD-ROM Interactive Chalkboard CD-ROM	**Performance Assessment in the Science Classroom (PASC)**

Transparencies

Section Focus

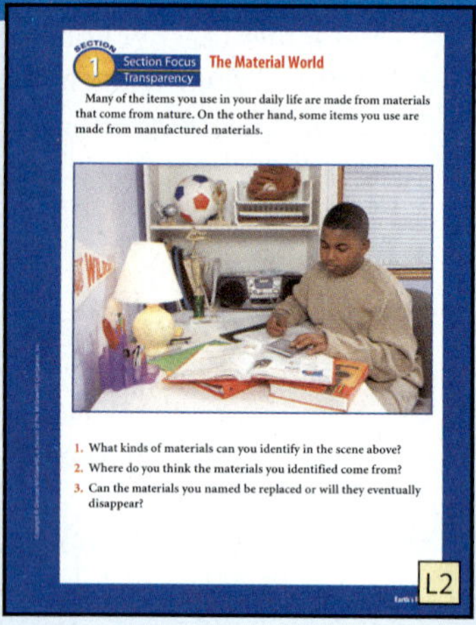
SECTION 1 Section Focus Transparency **The Material World**

Many of the items you use in your daily life are made from materials that come from nature. On the other hand, some items you use are made from manufactured materials.

1. What kinds of materials can you identify in the scene above?
2. Where do you think the materials you identified come from?
3. Can the materials you named be replaced or will they eventually disappear?

L2

SECTION 2 Section Focus Transparency **Ocean-Sized Ice Cubes**

One of the concerns with rising worldwide temperatures is the melting of ice at Earth's poles. Ice floating in water, as found in the Arctic Ocean, won't raise sea levels if it melts. But ice-sheets on land, as found in Greenland and Antarctica, do contain enough water to raise sea levels if they melt.

1. What is happening in the picture above?
2. What might cause ice sheets to melt?
3. Do you think human activity could affect the amount of ice that melts in Antarctica? Why or why not?

L2

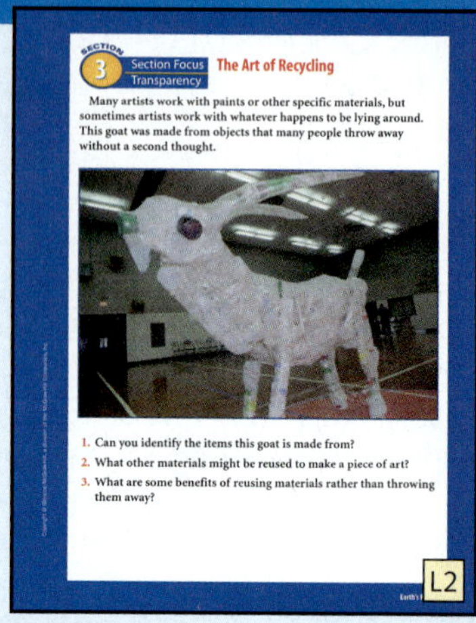
SECTION 3 Section Focus Transparency **The Art of Recycling**

Many artists work with paints or other specific materials, but sometimes artists work with whatever happens to be lying around. This goat was made from objects that many people throw away without a second thought.

1. Can you identify the items this goat is made from?
2. What other materials might be reused to make a piece of art?
3. What are some benefits of reusing materials rather than throwing them away?

L2

Assessment

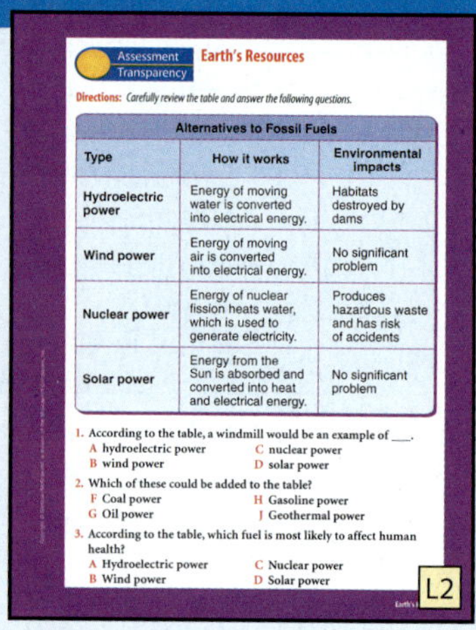
Assessment Transparency **Earth's Resources**

Directions: *Carefully review the table and answer the following questions.*

Alternatives to Fossil Fuels		
Type	How it works	Environmental impacts
Hydroelectric power	Energy of moving water is converted into electrical energy.	Habitats destroyed by dams
Wind power	Energy of moving air is converted into electrical energy.	No significant problem
Nuclear power	Energy of nuclear fission heats water, which is used to generate electricity.	Produces hazardous waste and has risk of accidents
Solar power	Energy from the Sun is absorbed and converted into heat and electrical energy.	No significant problem

1. According to the table, a windmill would be an example of ___.
   A hydroelectric power
   B wind power
   C nuclear power
   D solar power
2. Which of these could be added to the table?
   F Coal power
   G Oil power
   H Gasoline power
   J Geothermal power
3. According to the table, which fuel is most likely to affect human health?
   A Hydroelectric power
   B Wind power
   C Nuclear power
   D Solar power

L2

Teaching

This is a representation of key blackline masters available in the Teacher Classroom Resources. See Resource Manager boxes within the chapter for additional information.

Key to Teaching Strategies

The following designations will help you decide which activities are appropriate for your students.

- **L1** Level 1 activities should be appropriate for students with learning difficulties.
- **L2** Level 2 activities should be within the ability range of all students.
- **L3** Level 3 activities are designed for above-average students.
- **ELL** ELL activities should be within the ability range of English-Language Learners.
- **COOP LEARN** Cooperative Learning activities are designed for small group work.
- **LS** Multiple Learning Styles logos, as described on page 12T, are used throughout to indicate strategies that address different learning styles.
- **P** These strategies represent student products that can be placed into a best-work portfolio.
- **PBL** Problem-Based Learning activities apply real-world situations to learning.

Hands-on Activities

Student Text Lab Worksheet

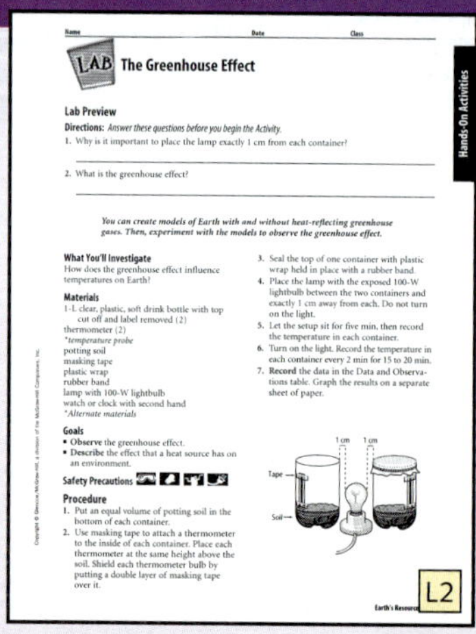
Name Date Class

LAB **The Greenhouse Effect**

Lab Preview

Directions: *Answer these questions before you begin the Activity.*

1. Why is it important to place the lamp exactly 1 cm from each container?
2. What is the greenhouse effect?

You can create models of Earth with and without heat-reflecting greenhouse gases. Then, experiment with the models to observe the greenhouse effect.

What You'll Investigate

How does the greenhouse effect influence temperatures on Earth?

Materials

1-L clear, plastic, soft drink bottle with top cut off and label removed (2)
thermometer (2)
**temperature probe*
potting soil
masking tape
plastic wrap
rubber band
lamp with 100-W lightbulb
watch or clock with second hand
**Alternate materials*

Goals

- **Observe** the greenhouse effect.
- **Describe** the effect that a heat source has on an environment.

Safety Precautions

Procedure

1. Put an equal volume of potting soil in the bottom of each container.
2. Use masking tape to attach a thermometer to the inside of each container. Place each thermometer at the same height above the soil. Shield each thermometer bulb by putting a double layer of masking tape over it.
3. Seal the top of one container with plastic wrap held in place with a rubber band.
4. Place the lamp with the exposed 100-W lightbulb between the two containers and exactly 1 cm away from each. Do not turn on the light.
5. Let the setup sit for five min, then record the temperature in each container.
6. Turn on the light. Record the temperature in each container every 2 min for 15 to 20 min.
7. **Record** the data in the Data and Observations table. Graph the results on a separate sheet of paper.

Hands-On Activities

L2

Laboratory Activities

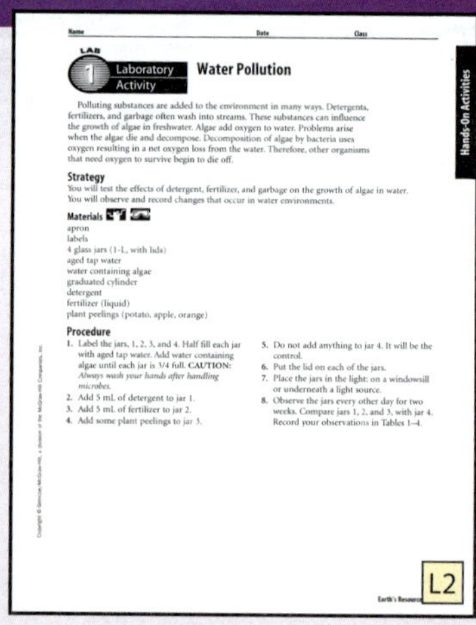
Name Date Class

LAB 1 Laboratory Activity **Water Pollution**

Polluting substances are added to the environment in many ways. Detergents, fertilizers, and garbage often wash into streams. These substances can influence the growth of algae in freshwater. Algae add oxygen to water. Problems arise when the algae die and decompose. Decomposition of algae by bacteria uses oxygen resulting in a net oxygen loss from the water. Therefore, other organisms that need oxygen to survive begin to die off.

Strategy

You will test the effects of detergent, fertilizer, and garbage on the growth of algae in water. You will observe and record changes that occur in water environments.

Materials

apron
labels
4 glass jars (1-L, with lids)
aged tap water
water containing algae
graduated cylinder
detergent
fertilizer (liquid)
plant peelings (potato, apple, orange)

Procedure

1. Label the jars, 1, 2, 3, and 4. Half fill each jar with aged tap water. Add water containing algae until each jar is 3/4 full. **CAUTION:** *Always wash your hands after handling microbes.*
2. Add 5 mL of detergent to jar 1.
3. Add 5 mL of fertilizer to jar 2.
4. Add some plant peelings to jar 3.
5. Do not add anything to jar 4. It will be the control.
6. Put the lid on each of the jars.
7. Place the jars in the light: on a windowsill or underneath a light source.
8. Observe the jars every other day for two weeks. Compare jars 1, 2, and 3, with jar 4. Record your observations in Tables 1–4.

Hands-On Activities

L2

Resource Manager

Meeting Different Ability Levels

Content Outline

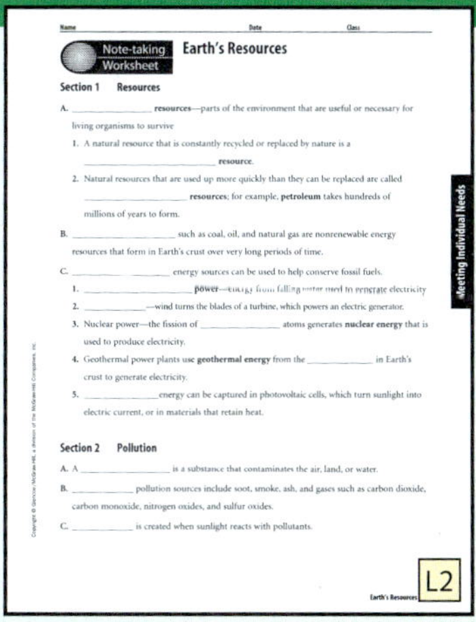

Reinforcement

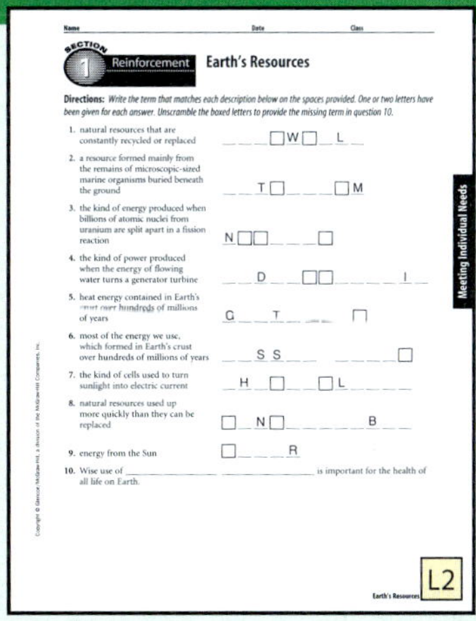

Enrichment

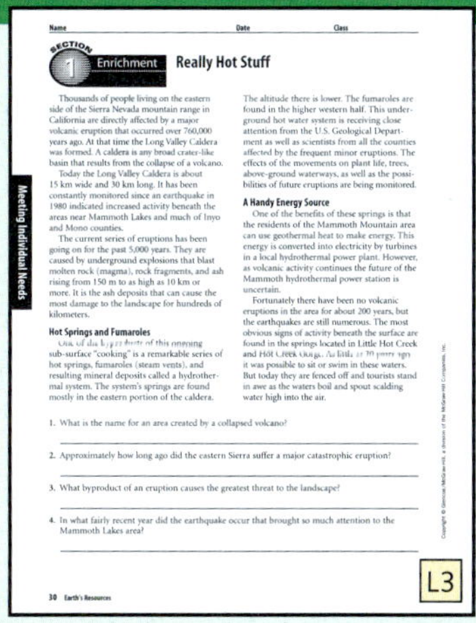

Directed Reading (English/Spanish)

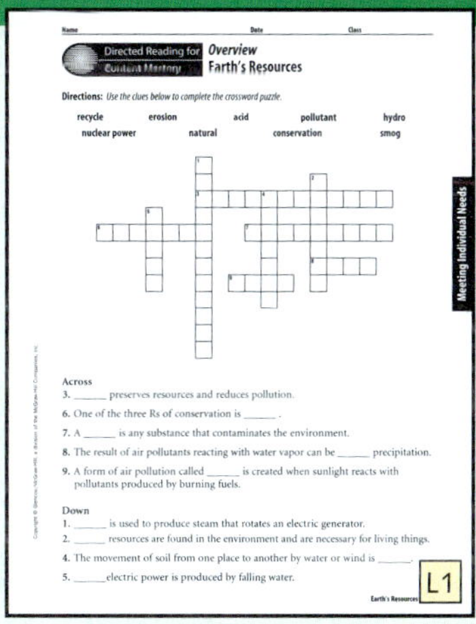

Study Guide

Reading Essentials

Assessment

Test Practice Workbook

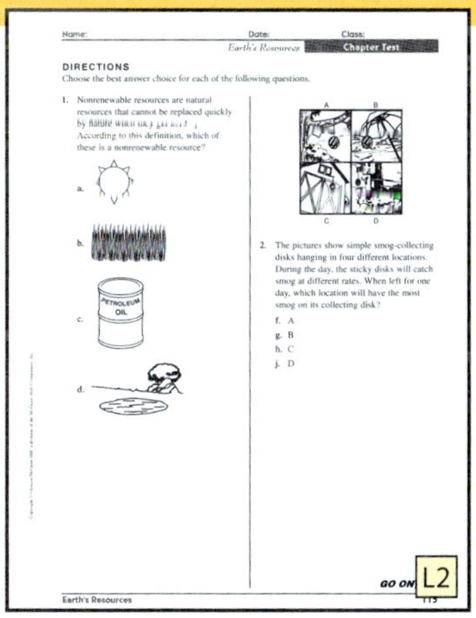

Chapter Review

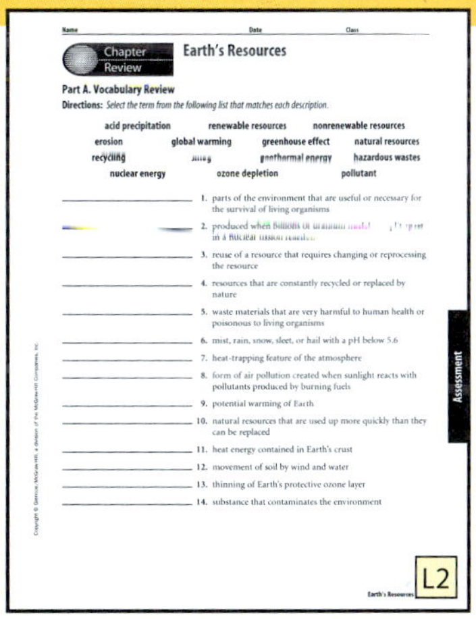

Chapter Tests

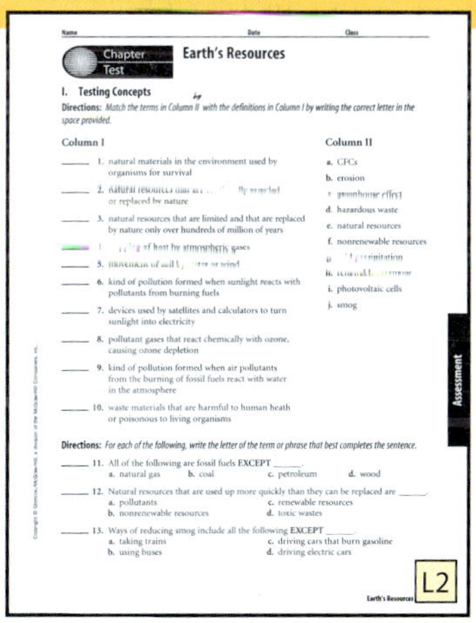

Science Content Background

section 1 Natural Resource Use

Slowly Renewable Resources

Many materials that are called nonrenewable are actually very slowly renewable. This includes "old growth" trees, coal, and petroleum products. Old growth forests contain considerable numbers of hard wood trees that are over 200 years old. If such a forest were cut down, it would take several human generations to re-grow it. Oil and coal are replenished at a dramatically slower rate. These materials are formed by the remains of living organisms that died a minimum of 50 million years ago.

section 2 People and the Environment

Air Pollution

Types of air pollution are easy to confuse. There is great concern that global warming possibly is due to increasing levels of atmospheric carbon dioxide resulting from human activity. Carbon dioxide is a waste product of combustion that is produced whenever wood, coal, gasoline, or any other fuel is burned.

Combustion that produces nitrogen and sulfur can cause the formation of nitric and sulfuric acid in the atmosphere, which leads to acid rain. Some nitric acid is formed from common combustion products like automobile exhaust. However, a much more important cause of acid rain is the burning of coal with a high sulfur content to generate electricity. Acid rain has had negative effects on plants, animals, and structures made from marble and limestone.

The atmosphere's ozone layer prevents some of the Sun's harmful ultraviolet radiation from reaching the Earth. The thinning of the ozone layer has been linked to the use of chlorofluorocarbons (CFCs). These chemicals were once used as propellants in aerosol cans and as refrigerants in air conditioners and freezers. CFC use is now banned, and hopefully the thinning of the ozone layer will slow or stop.

section 3 Protecting the Environment

Recycling Plastic

Plastic is classified into seven categories for recycling, but most municipalities recycle only two. All sorts of materials, including soft drink bottles, sports bottles, and peanut butter containers are made out of plastic number 1, called PET, PETE, polyethylene terephthalate, or polyester. When recycled, it can be used for creating fibers for woven materials such as carpets or clothing. Milk and water commonly come in bottles made from plastic number 2, known as high-density polyethylene or HDPE. Once recycled, it can be made into heavy-duty containers for laundry detergents, motor oil, and other liquids.

chapter content resources

Internet Resources

For additional content background, visit **red.msscience.com** to:

- access your book online
- find references to related articles in popular science magazines
- access Web links with related content background
- access current events with science journal topics

Print Resources

Earth-Friendly Crafts for Kids: 50 Awesome Things to Make with Recycled Stuff, by Heather Smith with Joe Rhatigan, Sterling Publishing, 2002

Caring for Our Planet, by Robert E. Rockwell, Elizabeth A. Sherwood, Robert A. Williams, David A. Winnett, Dale Seymour Publications, 2001

Global Warming: Understanding the Debate, by Kenneth Green, Enslow Publishers, 2002

IDENTIFYING Misconceptions

Find Out What Students Think

Students may think that . . .

Clear water is clean and free of pollution.
This misconception makes sense to students because clear water does look pure. They do not realize that both harmless substances and poisons may be dissolved in water. Some students have limited experiences dissolving substances in water and having it remain clear.

Activity

Fill two glasses with tap water. Mix some soil into one of the glasses. Ask students to pretend it is a hot day. They come indoors with a friend and they are both very thirsty. Seeing the glass of water, what do they do? Have them share their responses. Students will most likely focus on the idea of sharing or not sharing the clear water and it will be evident that they assume that the clear water is clean and suitable for drinking.

Promote Understanding

Promote Understanding

Label a disposable plastic cup *POISON, do not drink*. Fill the cup with clear water and show it to the students. Put on your goggles and start adding different substances to the water so that the water still remains clear. Start by adding some table salt and stirring. Is the water still clear? Add one or two teaspoons of rubbing alcohol. Is the water still clear? Add vinegar and repeat the question. Do the same for hydrogen peroxide. Ask: What am I showing you with this demonstration?

Help students to establish that just because water is clear does not mean that it is clean or good to drink. Point out that water pollution does not always make the water visibly dirty. Many harmful materials could be dissolved in water and we might not be able to see them. In addition to chemicals, there could be bacteria or other microscopic organisms that would cause intestinal problems such as diarrhea. Pour the contents of the cup into the sink as soon as the demonstration is over to prevent the possibility of anyone drinking the water.

Assess

After completing the chapter, see *Identifying Misconceptions* in the Study Guide at the end of the chapter.

chapter 22

Chapter Vocabulary

natural resource, p. 647
renewable resource, p. 651
nonrenewable resource, p. 652
landfill, p. 657
pollutant, p. 657
acid rain, p. 661
solid waste, p. 663
recycling, p. 655

Science Journal Student responses will vary, and could include any example of reusing or recycling an item. Example responses: reusing plastic yogurt or cottage cheese containers; taking used aluminum cans to the recycling center so they can be made into new cans; giving usable, unwanted clothing or household items to charity instead of tossing them in the trash.

INTERACTIVE CHALKBOARD with Image Bank

PowerPoint® Presentations

This CD-ROM is an editable Microsoft® PowerPoint® presentation that includes:

- a pre-made presentation for every chapter
- interactive graphics
- animations
- audio clips
- image bank
- all new section and chapter questions
- Standardized Test Practice
- transparencies
- pre-lab questions for all labs
- Foldables directions
- links to red.msscience.com

chapter 22

Earth's Resources

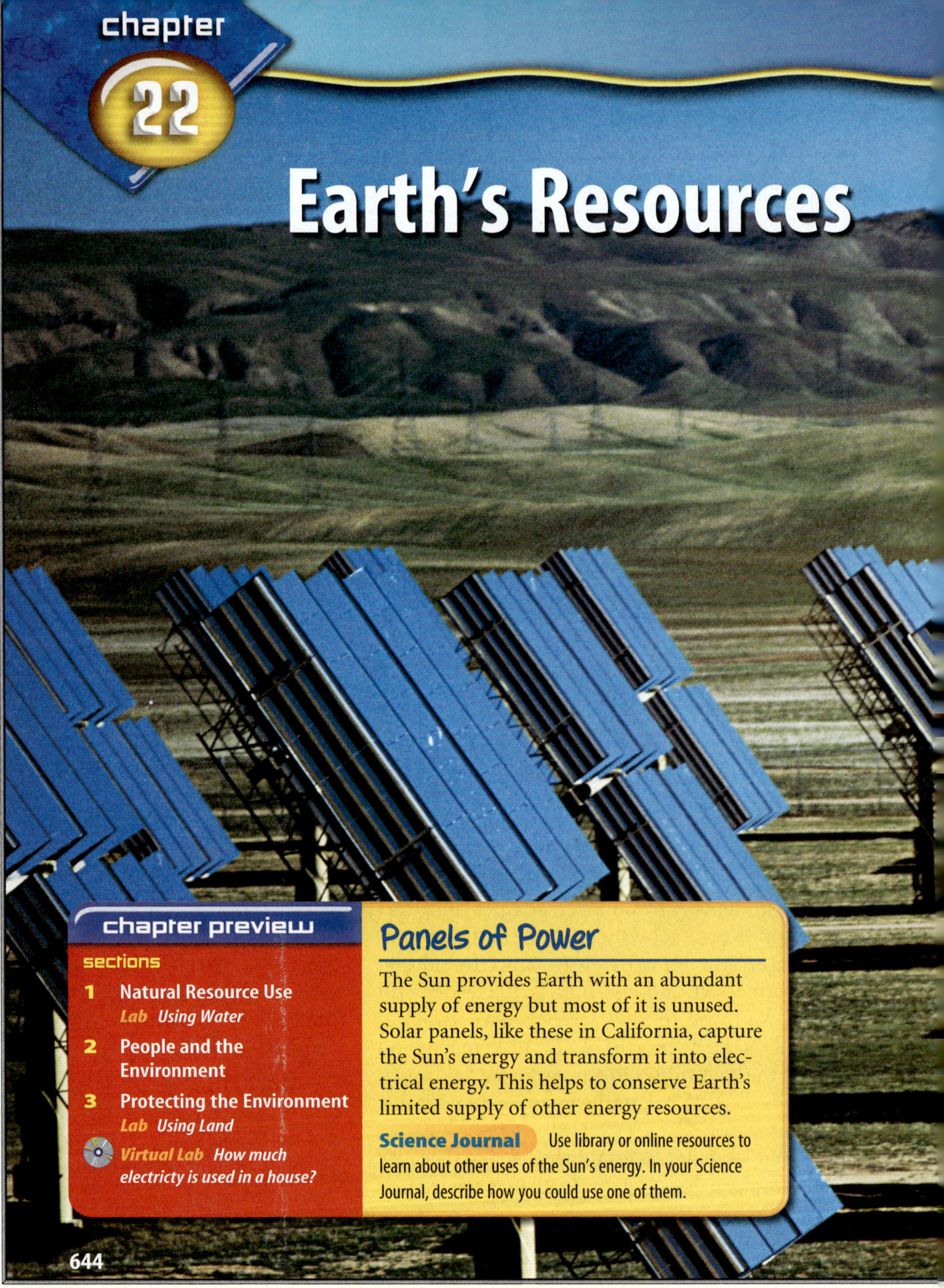

chapter preview

sections

1 Natural Resource Use
Lab Using Water

2 People and the Environment

3 Protecting the Environment
Lab Using Land

Virtual Lab How much electricty is used in a house?

Panels of Power

The Sun provides Earth with an abundant supply of energy but most of it is unused. Solar panels, like these in California, capture the Sun's energy and transform it into electrical energy. This helps to conserve Earth's limited supply of other energy resources.

Science Journal Use library or online resources to learn about other uses of the Sun's energy. In your Science Journal, describe how you could use one of them.

644

Theme Connection

Systems and Interactions This theme is highlighted as students learn how human use of natural resources affects other humans, other species, and the environment.

About the Photo

Arcosolar Power Plant The Arcosolar power plant near Bakersfield, California, consists of 756 solar panel assemblies. Computers control the panels' movements to maximize their exposure to the Sun. The plant can generate up to 6.5 megawatts of electricity. Computers also monitor the wind, and when wind speeds are high, controllers fix the solar panel assemblies in a "safe" position to prevent damage.

Start-Up Activities

Launch LAB

What are natural resources?

Could you live without plastic bags or wooden pencils? How about an automobile or a TV? Everything you need or use, from food and clothing to school supplies and transportation, is made from natural resources. This activity gives you a chance to think about the kinds of resources you use every day.

1. In your Science Journal, make a table with five columns: *Plastic, Paper, Metal, Glass,* and *Wood.*
2. Think of things you use every day at home or in your classroom that are made of each material. List as many as you can think of in each column.
3. **Think Critically** Write a paragraph in your Science Journal explaining which category you depend on most, and why. Include an explanation of where you think these materials come from.

FOLDABLES™ Study Organizer

Natural Resources Make the following Foldable to help you distinguish between renewable and nonrenewable resources.

STEP 1 **Fold** a sheet of paper in half lengthwise.

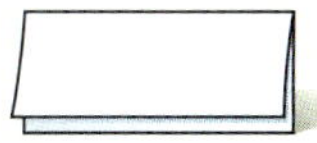

STEP 2 **Fold** paper down 2.5 cm from the top. (Hint: From the tip of your index finger to your middle knuckle is about 2.5 cm.)

STEP 3 **Open and draw** lines along the 2.5 cm fold. Label as shown.

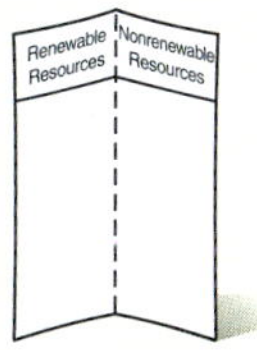

Classify Before you read the chapter, list examples of each type of natural resource you already know about in the appropriate column. As you read the chapter, add to your lists.

Science Online | Preview this chapter's content and activities at red.msscience.com

645

Launch LAB

Purpose Students become more aware of their dependence on natural resources. L2 LS **Visual-Spatial**

Preparation Place objects made from plastic, paper, metal, glass, and wood around the classroom.

Materials objects made from plastic, paper, metal, glass, and wood

Teaching Strategy Have students look around the room for objects they believe have been made from natural resources.

Science Journal

Student answers will vary. Students may know that paper and wood come from trees, plastic items are made from oil, glass is made from sand, and metal objects are made from ore. Students may indicate that they depend on plastic most because they use so many different plastic products.

Assessment

Oral As students read their lists, tally the results on the chalkboard. See which category is most popular. Graph and discuss the results. Use **Performance Assessment in the Science Classroom,** p. 107.

FOLDABLES™ Study Organizer — Dinah Zike Study Fold

Student preparation materials for this Foldable are available in the Chapter *Fast File* Resources.

1 Motivate

INTERACTIVE CHALKBOARD PowerPoint® Presentations

Bellringer

Section Focus Transparencies also are available on the Interactive Chalkboard CD-ROM. L2 ELL

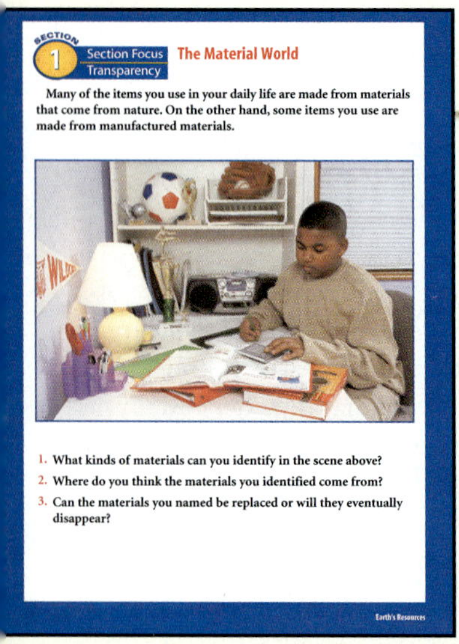

Tie to Prior Knowledge

Help students recall any products they or their friends have purchased lately that were overpackaged.

section 1

Natural Resource Use

as you read

What You'll Learn

- **Explain** how resources are used.
- **Describe** how resources are classified.

Why It's Important

If you understand where resources come from and how they are used, you can make better decisions about the things you buy or use.

Review Vocabulary
biome: large geographic area with similar climates and ecosystems, such as tropical rain forest

New Vocabulary
- natural resource
- renewable resource
- nonrenewable resource

News Flash: Trouble in the Rain Forest

For months, you've been saving your money, and today you're going shopping for a CD player. On the way to the mall, you turn on the car radio. The news reporter is saying that rain forests are being destroyed at the rate of about a football field a day, or 117,000 km^2 every year. That's a lot of land! **Figure 1** shows how much rain forest already has been lost in South America.

The news reporter explains that once a rain forest is cut down, it might never grow back. Removing rain forests means losing wildlife. More than half of Earth's known plant species and one-fifth of the known bird species can be found in rain forests. If a species' home is destroyed, the species might die out. Some important medicines, including anti-cancer drugs, originally came from rain forest plants. Destruction of rain forests could mean that many medicines might not be discovered.

Many people who live in rain forest areas clear land to grow crops or graze cattle. To get money for food and supplies, these people sell the wood to companies that use it for paper, furniture, and other products. The news reporter ends by saying that positive actions by people worldwide can help save Earth's rain forests.

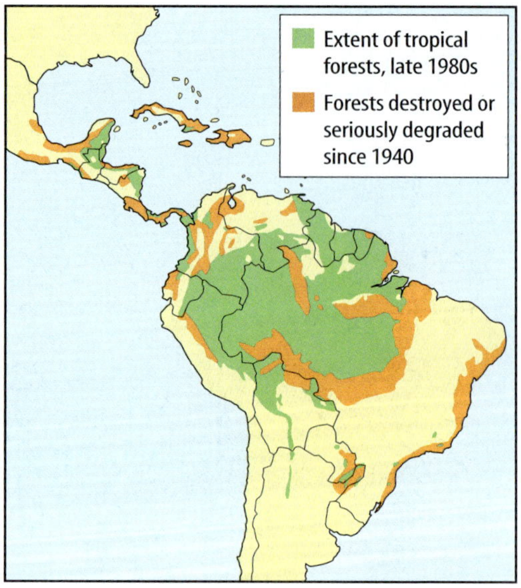

Figure 1 South America's Amazon River basin contains the world's largest area of tropical rain forest.

Section 1 Resource Manager

Chapter *FAST FILE* Resources
- Transparency Activity, p. 44
- MiniLAB, p. 3
- Note-taking Worksheets, pp. 33–35
- Enrichment, p. 30
- Reinforcement, p. 27
- Directed Reading for Content Mastery, p. 20
- Lab Worksheet, pp. 5–6
- Lab Activity, pp. 9–12

Cultural Diversity, pp. 53–54
Laboratory Management and Safety, p. 71
Mathematics Skill Activities, pp. 5–6
Science Inquiry Labs, pp. 31–32

Things You Use Affect the Environment You think about this news report as you reach the shopping mall. When you walk into the different stores, you can't help but notice that many products and the packages they come in are made of cardboard, a wood product. Even though this wood may not have come from a rain forest, it did come from a forest somewhere. Could these products have been packaged in a different way?

Let's take a look at the CD player that you might want to buy, like the one in **Figure 2.** It is made of plastic and often comes in a package made of cardboard. Its wires, screws, and some inside parts are made of metal. Metal and plastic aren't made from trees. So where did they come from?

Figure 2 The parts of a CD player come from different resources.

Natural Resources

Most of the items that you buy or use are made of materials that come from natural resources. **Natural resources** are things found in nature that living organisms use. **Figure 3** shows some examples of natural resources. Organisms use natural resources to meet their needs. Vegetables that you eat are natural resources. They fill your need for nutrients. The trees and the minerals that were used to make the lumber, plastic, and metal in your house are natural resources. They fill your need for a place to live. Natural resources are also used to make other items in our lives, such as CD players.

Figure 3 Cotton, gold, trees, and water are examples of natural resources.
Infer *which of these natural resources have a limited supply on Earth.*

2 Teach

Text Question Answer

Could these products have been packaged in a different way? Yes, they could have been packaged in recycled material.

Where did the metal and plastic parts of the CD player come from? from Earth's crust; metal from ores and plastic from petroleum

Caption Answer

Figure 3 gold, water

Visual Learning

Figure 1 Show students a world map with political boundaries. Explain that Earth's tropical rain forests are located in land areas that lie between that Tropic of Cancer and the Tropic of Capricorn. Tropical rain forests occur in not only South America but also Africa, Asia, Indonesia, and northern Australia. Have students identify the countries located in the Amazon River basin.

Figure 2 If possible, have students examine a CD player brought to the classroom and compare it to the one shown in the figure.

Figure 3 Point out to students that the mining operation pictured at the left of the figure shows the kind of operation that could be used to mine gold or other minerals.

Quick Demo

Natural Resources

Materials motor oil, wooden log, stereo speaker wire, glass of water, apples, oranges, rubber tire

Estimated Time 15 minutes

Procedure Display materials in the front of the class. Discuss with students how each item is a natural resource or comes from one.

Differentiated Instruction

Challenge Have students research the concept of earth-sheltered housing. Have them check at the courthouse, or any location with records of housing in the area, to determine whether any local earth-sheltered houses exist. If so, invite the builders to class for a discussion of alternate housing options. L3 COOP LEARN

Visually Impaired Help your visually impaired students understand and participate in the Quick Demo on natural resources by allowing them to study the displayed samples with some of their other senses. L2

SECTION 1 Natural Resource Use 647

NATIONAL GEOGRAPHIC

Visualizing Steel Production

Have students examine the pictures and read the captions. Then ask the following question. Which of the steps shown in the figure require the burning of fuel or the use of electricity? All. Mining, refining, and transportation require gasoline or diesel-powered machinery. Factories and stores use electricity.

Use Science Words

Crude oil (petroleum) is a thick, flammable mixture of substances made from hydrogen and carbon. It forms when marine organisms decompose in an oxygen-free (anaerobic) environment. The word *petroleum* comes from the Latin words *petra* meaning "rock" and *oleum* meaning "oil." LS **Linguistic**

Use an Analogy

Processing Relate the process of making screws to the process of knitting a scarf for a friend. The "raw material"—yarn—must first be purchased. Imperfect lengths of yarn must be removed. Then, the yarn must be processed, or knitted, to make the finished product. Finally, the scarf is delivered to your friend, ready to use. LS **Logical-Mathematical**

What goes into making a CD player? You already know that the cardboard box the CD player came in was made from trees, and that trees are a natural resource. What about the plastic used to make the CD player? Where did it come from? Plastic is made from crude oil, a resource that's usually found under ground. Crude oil is a thick, dark liquid. Deep holes are drilled in Earth to reach crude oil underground. Substances taken from crude oil can go to factories where they can be made into products, such as plastic, gasoline, inks and dyes, or linoleum.

What about the metal parts of the CD player? Where did the metal come from to make the screws that hold the CD player together? **Figure 4** answers this question.

NATIONAL GEOGRAPHIC VISUALIZING STEEL PRODUCTION

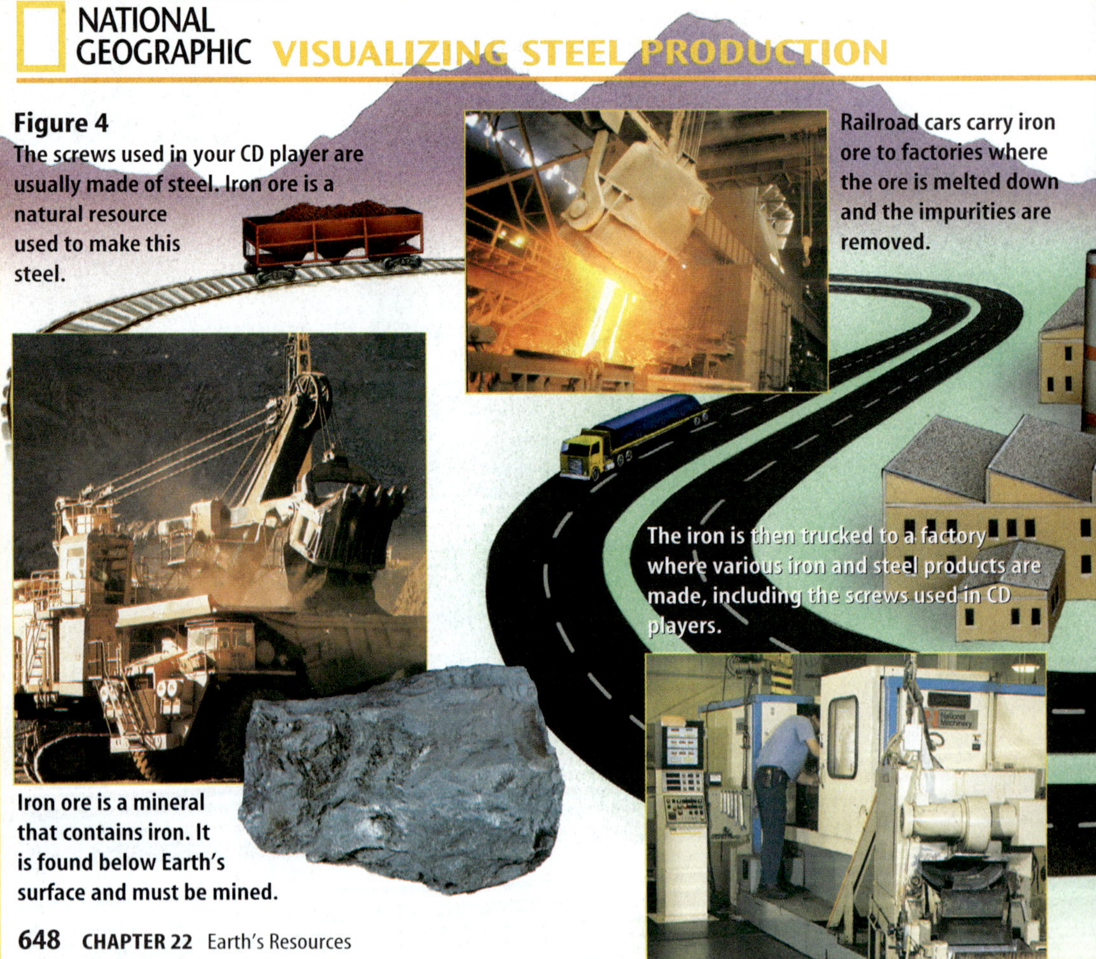

Figure 4
The screws used in your CD player are usually made of steel. Iron ore is a natural resource used to make this steel.

Railroad cars carry iron ore to factories where the ore is melted down and the impurities are removed.

The iron is then trucked to a factory where various iron and steel products are made, including the screws used in CD players.

Iron ore is a mineral that contains iron. It is found below Earth's surface and must be mined.

Science Journal

Mining for Electricity Have students draw and label a diagram describing the process of obtaining electricity from a coal-fired power plant. Ask them to begin with the mining of coal from the Earth and end with the turning on of a light switch in someone's home. Coal is mined and transported to a power plant where it is burned to generate electricity. Electricity flows through wires to homes. Resident turns on light switch.

648 CHAPTER 22 Earth's Resources

More Natural Resources Trees, crude oil, minerals—were any other natural resources used to make the CD player? Cutting down trees, drilling for crude oil, mining, and getting natural resources to factories all require energy. Once the natural resources are at the factories, it takes energy to make them into plastic, cardboard packaging, and metal wires and screws. Where does all this energy come from?

If you guessed natural resources, you're right. Trucks that take the natural resources to the factories use gasoline or diesel fuel, which are made from crude oil. The electricity used to power machines that make natural resources into materials for CD player parts often comes from burning coal. Like crude oil, coal is a natural resource that forms underground. It takes energy to mine the coal, and this energy comes from natural resources.

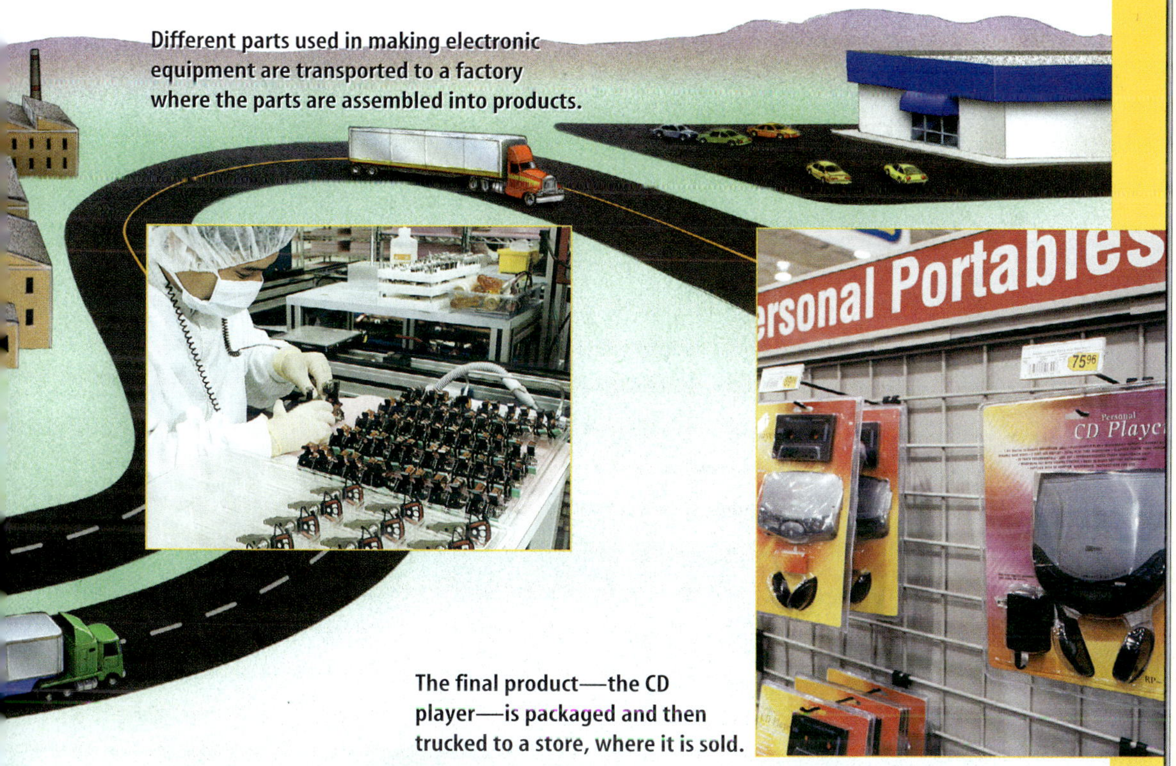

Different parts used in making electronic equipment are transported to a factory where the parts are assembled into products.

The final product—the CD player—is packaged and then trucked to a store, where it is sold.

NATIONAL GEOGRAPHIC

Teacher FYI

Underground Fuel Coal forms when pieces of dead plants are buried under other sediments. Plants are made up of molecules that contain atoms of carbon, hydrogen, and oxygen. These plant materials are chemically changed by microorganisms. The resulting sediments are compacted under the weight of overlying sediments for millions of years to form coal. When we burn coal, it is the carbon that burns and is the source of heat. Bituminous coal, or soft coal, is the coal used most often. It provides lots of heat energy when burned, but it also pollutes the air.

Activity

Ore Samples Obtain a sample of magnetite (an ore of iron) and other samples of ores from which other metals can be extracted. These might include bauxite (an ore of aluminum) and malachite (an ore of copper). Inform students that a magnet will attract magnetite. Give students the ore samples and a magnet and have them determine which of the ores is magnetite. Magnetite will be attracted by the magnet. L2 COOP LEARN ELL LS **Naturalist**

Science Journal

From There to Here Ask students to select a product they use often. Have them write one or two paragraphs in their Science Journal describing the step-by-step process that is followed from the time the natural resource from which the product is made is obtained, through the time when the final product is placed in the store. Students should not use the example of the CD player. L2 P

Mini LAB

Purpose Students will use different methods and materials for wrapping objects in order to better understand the importance of natural resources. L2

ELL LS **Kinesthetic**

Materials objects to wrap; wrappings such as paper bags, plastic bags, and newspaper; scissors, string, tape

Teaching Strategies

- Be sure that at least one type of wrapping is reusable.
- Encourage students to bring in unusual wrapping materials.
- Transparent tape will not stick well to all materials. Provide various types of tape.
- Students may conclude that no wrapping is a valid option. Encourage students to defend all positions taken.

Safety Precautions Caution students to use scissors with care.

Analysis

1. Student answers will vary but could include size, cost, availability, and attractiveness of the wrap.
2. Student answers will vary but could include reference to the most colorful or efficient wrapping. Some students will select reusable or recyclable materials.

Assessment

Content Encourage students to list the properties of the "perfect" environmental gift wrap and give suggestions as to how it could be made. Use **Performance Assessment in the Science Classroom,** p. 117.

Mini LAB

Analyzing Gift Wrap

Procedure

1. Your teacher will give you an **object to wrap.** Discuss different methods and unusual materials for wrapping this object with your group.
2. Think about how different ways of wrapping could make the best use of the materials you have. Does the wrap waste material? Could it be used again? Is it easy to dispose of?
3. Wrap the object. In your **Science Journal,** write down all the resources that would have been used to wrap it in gift wrap.

Analysis

1. What problems did you have with your method of wrapping?
2. Why do you think your choice of wrapping material is a good one?

All Organisms Use Resources As you're beginning to see, it takes a variety of natural resources to make one CD player, doesn't it? Think of all the natural resources that are used to make something large like a house or apartment building. Materials used in building construction include wood, metal, stone, glass, and plastic. Tools, machinery, and fuel are used during construction. Each building contains light fixtures, flooring, window coverings, plumbing, and electrical wiring. After completion, the building is filled with furniture, appliances, food, books, and all the other things people use. Now think of all the houses and apartments in the world. Are there enough natural resources to meet everybody's need for a place to live?

Maybe. But people use natural resources to meet other needs, too. In fact, all living things on Earth use natural resources. Animals use natural resources for food and shelter. Will we ever run out of natural resources? That may depend on the particular resource.

Availability of Resources

Imagine that you're riding your bike on a warm spring day. Your destination: the city park, such as the one shown in **Figure 5.** When you get to the park, you head straight for the pond. You hop off your bike, take an apple out of your backpack, and lean against a tree. Later, you might take a walk around the pond. For now, you're content just to watch the sunlight sparkling on the water.

Sunlight, water, trees, apples . . . These are all natural resources. They have something else in common, too. They will likely be around for a long time. Why? Because they are all renewable (ree NEW uh bul).

Figure 5 Natural resources are found everywhere—in national parks and in the middle of large cities like Boston.

Visual Learning

Figure 5 For what purpose is most of the land in the picture being used? buildings What are the uses of buildings? offices, apartments, hospitals, schools What are some of the other land uses shown? park, bridge/road

Active Reading

Buddy Interviews This strategy helps students understand and clarify the reading. Have students interview one another to find out what helps them to understand what they are reading, how they find answers, and how they assimilate new vocabulary terms. Have students use Buddy Interviews to help them master concepts about natural resources.

Renewable Resources Resources that can be replaced by natural processes in 100 years or less are called **renewable resources.** Look at **Figure 6.** Energy from the Sun is a renewable resource because the Sun gives off light energy every day, and it will continue to do so every day for millions of years. Light energy powers the process of photosynthesis, which plants use to make food. You and all other animals on Earth depend on plants for food. Have you ever used a solar-powered calculator or radio? These devices use light energy to produce the electricity they need to operate.

Trees are renewable resources because most trees will grow back and be cut again in less than 100 years. People use wood from trees to build houses and furniture, and to burn as fuel. Can you think of other uses for wood?

Reading Check *Why is a tree a renewable resource?*

Water is another renewable resource. Did you know that the water you drink has existed on Earth for billions of years? Heat from the Sun evaporates water from lakes, streams, and oceans, turning it into gaseous water vapor that rises into the atmosphere. Later, the water vapor condenses back into liquid and falls as rain, sleet, hail, or snow. Through this cycle, the same water is reused over and over again.

Wind is also an example of a renewable resource. The sails on a sailboat use wind energy to push the boat across the water. Old-fashioned windmills used wind energy to pump water from underground wells. Today, large windmills are used to generate electricity.

Figure 6 Sunlight and trees are examples of renewable resources. So are clay bricks, which are made from mud. The bricks shown below are being used to build a house.

Visual Learning

Figure 6 Explain to students that mud bricks like those shown in the figure are made in parts of the world where the soil contains a large percentage of clay. Point out that bricks are strengthened by adding straw to the mud.

Discussion

Straw Bale Construction Discuss with students how people in the housing industry are trying to find inexpensive, renewable resources to replace the more traditional wood, brick, and stone. In some areas of the country, bales of straw are being used to build houses. The bales may be treated to resist fire and insects. The bales are stacked on each other and then covered inside and out with stucco. Straw bale houses can be built quickly and for low cost. Straw is a good insulator and makes a good soundproof barrier. Bales can be arranged into almost any floor plan.

Reading Check

Answer Most trees will grow back and be cut again in less than 100 years.

Cultural Diversity

Building Materials Ask students to research materials, other than wood, used to build houses and other structures in different parts of the world. In tropical areas, for example, bamboo and grasses are for houses, roofs, furniture, and flooring. Ask students to report on their findings to the rest of the class. Invite them to include samples or photographs of these materials in their reports.

SECTION 1 Natural Resource Use 651

Caption Answer

Figure 7 Electricity, used in virtually all homes, can be produced by any of the energy sources shown. Gasoline and diesel fuel that run automobiles and public transportation come from oil. Natural gas is used for heating and cooking. Some students may use electricity produced by solar cells or wood in woodstoves or fireplaces. These renewable energy sources fall into the "other" category.

Reading Check

Answer cannot be replaced within 100 years

Fun Fact

Gasoline-powered automobiles typically get their best gas mileage in highway driving, and somewhat lower mileage in stop-and-go driving on city streets. Because hybrid autos do not burn gasoline when stopped, some hybrid models get better gas mileage in stop-and-go driving than they do in highway driving.

Figure 7 As you can see from this circle graph, 39 percent of the world's energy needs are met by crude oil and 22 percent are met by coal. Scientists estimate that we have enough coal to last another 200 years. But if we continue using oil at present rates, some scientists estimate that we will run out of this natural resource in 30 or 40 years.
List *some energy sources you use in your home.*

Nonrenewable Resources Do you see coal or crude oil among Earth's energy sources in **Figure 7?** Coal, natural gas, and crude oil take millions of years to form inside Earth. They are examples of nonrenewable resources. **Nonrenewable resources** are resources that cannot be replaced by natural processes within 100 years. After all the coal and crude oil that we can recover is used up, there won't be any more available for use for millions of years. Unless we can replace them with other energy resources, energy use will have to be reduced.

Reading Check *What makes a resource nonrenewable?*

As you can see, because nonrenewable resources form slowly over long periods of time, they need to be used wisely. **Figure 7** shows how the world's energy needs are being met today. If world crude oil supplies were to run out, in what ways would the circle graph change?

Curriculum Connection

Deep Sea Oil Development of robotic submersibles has led to discovery of new sources of oil beneath the ocean's floor. These new reserves of oil appear to be large but will be difficult to develop. Discuss with students the types of environmental problems that might be caused by developing these oil reserves. oil spills that harm ocean life, damage to ocean floor ecosystems, unforeseen problems

Conserving Resources Conservation is the practice of protecting and preserving natural resources so they will always be available. As you read the rest of this chapter, you'll find out that both renewable and nonrenewable resources need to be conserved. Protecting the quality of air, water, and land is just as important as preserving supplies of coal, crude oil, and other nonrenewable resources. **Figure 8** shows one of the ways people are trying to conserve crude oil supplies.

Science Online

Topic: Hybrid Vehicles

Visit red.msscience.com for Web links to information about hybrid gasoline/electric vehicles.

Activity Explain how hybrid electric vehicles work and how much they save on gasoline compared to conventional cars.

Figure 8 Hybrid gasoline/electric vehicles like this one help reduce crude oil use because they consume much less gasoline than other cars do. The hybrid car has both a small gasoline engine and an electric motor.

section 1 review

Summary

Trouble in the Rain Forest

- Rain forests are being cut down at a very rapid rate.
- The removal of rain forests can result in the loss of wildlife species and the destruction of plants that might provide useful medicines.

Natural Resources

- Natural resources include everything that living organisms need for survival.

Availability of Resources

- Renewable resources can be replaced in 100 years or less.
- Nonrenewable resources take more than 100 years to replace.

Self Check

1. **List** six examples of natural resources.
2. **Classify** each item in your list of natural resources as renewable or nonrenewable. Explain your reasoning.
3. **Describe** how a squirrel uses natural resources for its survival.
4. **Think Critically** Is a rain forest a renewable or nonrenewable resource? Explain your answer.

Applying Skills

5. **Sequence** the steps that are needed to make a tree into a baseball bat displayed in a store. Describe any energy use or packaging that might be required.

Science Online red.msscience.com/self_check_quiz

Discussion

Gas Mileage Have students discuss familiar types of automobiles. Ask students if they would consider gas mileage as a factor in making a vehicle purchase. Why or why not? Cost considerations: gasoline costs add to vehicle operating costs. Safety: larger vehicles that get lower gas mileage might be safer to drive.

3 Assess

DAILY INTERVENTION

Check for Understanding

Auditory-Musical Have students create a song, poem, or story that describes the difference between renewable and nonrenewable resources.

Reteach

Demonstrating Resources Point out common items in the classroom, one by one. Explain the origin of each item and whether it is made from renewable resources, nonrenewable resources, or both.

Assessment

Portfolio Have students collect photographs from magazines or newspapers or make drawings that show examples of resource use. Ask them to make a poster or bulletin board display that explains the origin of the natural resources used in their illustrations, and explains whether each one is renewable or nonrenewable.

section 1 review

1. Possible answers: water, land, air, trees, gold, aluminum, other metals, oil, natural gas, sunlight, food crops
2. Renewable: water, trees, sunlight, air, food crops; nonrenewable: land, gold, aluminum, other metals, oil, natural gas, coal
3. Possible answer: Squirrels drink water, gather nuts from trees for food, build nests in trees, and store nuts in holes in the ground or in hollow tree branches.
4. An individual tree in a rainforest is renewable because a new tree can be planted. An entire rainforest probably should be considered nonrenewable, because it likely would take much longer than 100 years for it to regrow.
5. tree grows, tree is cut down, tree is cut into lumber, lumber is shaped into a baseball bat, bat is stained and varnished, bat is shipped to a store and displayed

BENCH TESTED

Real-World Question

Purpose Students will find out how much clean water is used each week in their homes. L2

LS **Logical-Mathematical**

Process Skills observe and infer, measure in SI, collect and organize data, interpret data, make and use tables, use numbers, recognize cause and effect

Time Required three days for observations, 35 minutes to complete report

Procedure

Materials calculator

Teaching Strategies

- Have students meet with their families to describe the nature of the experiment and to ask for their help and cooperation.
- If students' families forget to mark down water usage, allow students to just list their own water use.

Conclude and Apply

1. **and 2.** Student numbers will vary depending on the size of the family and how many people participate in the study.
3. Ways could include filling the sink when washing dishes, adjusting the water level to the amount of clothes in the washing machine, taking a bath instead of a shower, taking shorter showers, and turning off water while brushing teeth.

LAB

Using Water

Water is an important resource that we use every day. You use it to wash dishes and clothes, bathe, and brush your teeth. You also use water to cook meals. Many daily activities require water.

Real-World Question

How much water does your family use in three days?

Goals

- **Calculate** how much water the people in your household use in three days.
- **Make a plan** to control the amount of water used.

Materials

calculator

Procedure

1. **Record Data** For three days, have the people who live in your house keep a record of when they do the activities listed in the table.
2. **Calculate** how much water your family uses based on the data in the table.

Conclude and Apply

1. **Use Numbers** The numbers in the table describe approximately how many liters one person uses in a single day for the activity listed. Multiply these numbers by the number of people who did these activities.
2. **Add** up the totals for each day. The final sum will be the total amount of water used for these activities in three days.
3. **List** ways in which you and your family could control the amount of water you use.

Home Water Use

Activity	Conditions	Amount of Water Used*
Washing dishes by hand	Water is running all the time	113 L/person/day
Washing dishes by hand	Sink is filled with water	19 L/person/day
Washing clothes in machine	Small load with high water setting	68 L/person/day
Washing clothes in machine	Full load with high water setting	45 L/person/day
Taking a shower	10 minutes long	150 L/person/day
Taking a bath	Bathtub is full of water	113 L/person/day
Flushing the toilet	Water-saving toilet	23 L/person/day
Brushing teeth	Water is running all the time	17 L/person/day

*3.8 L equals 1 gallon

Assessment

Performance Have students propose a protocol for water usage for members of their families that will conserve water yet allow for needed use. Have them write the protocol in the form of a letter to their families. Use **Performance Assessment in the Science Classroom,** p. 139. P

Differentiated Instruction

Learning Disabled Help students better understand water conservation by having them prepare the table in which they will record data. Have students begin by filling out an example of the data they will obtain at home. Contact parents or guardians so the project becomes a family effort. L1

section 2

People and the Environment

Exploring Environmental Problems

Look at **Figure 9.** Have you ever seen a construction site for a new highway? Sometimes, hillsides have to be dynamited to make room for the highway. The trees and plants that grew on the hillside are destroyed. The animals that lived on the hillside depended on the trees and plants for food and shelter. Some might die if their food source is destroyed, but many survive and find new habitats to live in. Construction companies now have to restore land that they dynamite so that plants and animals can continue to live there.

What if there isn't another place to live? Many plants and animals lose their natural habitats because people use land for growing crops, grazing animals, or building homes. That's what's happening in many rain forests. Because large areas of rain forest are being destroyed at a rapid rate, certain species in the rain forest lose their habitat. They cannot find enough food, shelter, and other resources. They might be threatened with extinction.

Frequently, human activities affect the quality and availablity of some of our most precious natural resources: land, water, and air. Let's see how this happens.

as you read

What You'll Learn

- **Explain** how people affect the environment.
- **Describe** different types of pollution.

Why It's Important

Knowing how your actions affect the environment will help you to make choices that could reduce environmental problems.

Review Vocabulary
habitat: place where an organism lives that provides the food, shelter, moisture, and temperature needed for survival

New Vocabulary
- landfill
- pollutant
- acid rain

Figure 9 Construction destroys some parts of the environment. Laws in the United States and many other countries help reduce the amount of destruction that takes place.

1 Motivate

INTERACTIVE CHALKBOARD PowerPoint® Presentations

Bellringer

Section Focus Transparencies also are available on the Interactive Chalkboard CD-ROM. L2 ELL

SECTION 2 Section Focus Transparency **Ocean-Sized Ice Cubes**

One of the concerns with rising worldwide temperatures is the melting of ice at Earth's poles. Ice floating in water, as found in the Arctic Ocean, won't raise sea levels if it melts. But ice-sheets on land, as found in Greenland and Antarctica, do contain enough water to raise sea levels if they melt.

1. What is happening in the picture above?
2. What might cause ice sheets to melt?
3. Do you think human activity could affect the amount of ice that melts in Antarctica? Why or why not?

Earth's Resources

Tie to Prior Knowledge

Help students recall the last time they were asked to take out the garbage. Ask them whether they have ever considered where all of the garbage goes. Much of the solid waste produced by people is collected and placed in a landfill.

Section 2 Resource Manager

Chapter *FAST FILE* Resources
- Transparency Activity, p. 45
- Enrichment, p. 31
- Reinforcement, p. 28
- Directed Reading for Content Mastery, p. 21

Physical Science Critical Thinking/Problem Solving, p. 17

Performance Assessment, pp. 53–54

2 Teach

Quick Demo

Popcorn Resources

Materials popcorn and popcorn popper, or microwave popcorn and microwave oven

Estimated Time 10 minutes

Procedure Prepare popcorn for students. Have students list the ingredients and other materials required to make this food item. Ask them to explain where each material comes from, what type of resource it is, and how its use affects the environment. Popcorn and oil for cooking: farmland, water, fertilizers, farm machinery and trucks for transportation and the fuel to operate them, processing plants for packaging popcorn and pressing vegetable oil from corn or seeds. Popper or microwave oven: metal ore, energy for refining the metal and for manufacture. Cooking: electricity or natural gas. Burning of fuels for transportation or manufacturing adds to air pollution; processing and manufacturing plants also may release wastes that could add pollutants to the water supply.

Make a Model

Landfill Explain to students that a sanitary landfill is one in which each day's deposit is covered with dirt. Have students work in groups to make a model of a sanitary landfill. Warn students not to use rotting food or other potentially dangerous items in their models. L3 COOP LEARN ELL LS **Kinesthetic**

Caption Answer

Figure 10 housing, office buildings, roads, bridges, green space (trees and parks)

Science Online

Topic: Habitat Requirements

Visit red.msscience.com for Web links to information about the habitat requirements of brown bears and gray wolves.

Activity List the kinds of foods these animals eat and describe the types of environments in which they can be found.

Our Impact on Land

How much space do you need? You will need to think about more than just your home. Think about where your food comes from, your school, and other spaces you use. If you start adding it all up, the amount of spaces you use is much larger than you may think. A simple peanut butter-and-jelly sandwich requires land to grow the wheat needed to make bread, land to grow the peanuts for the peanut butter, and land to grow the sugarcane and fruit for the jelly. A hamburger? Land is needed to raise cattle and to grow the grain that the cattle eat.

Using Land Wisely All of the things we use in our everyday lives take some amount of land, or space, to produce. That means that every time we build a house, a mall, a road, or a factory in a city, as shown in **Figure 10,** we use a little more land. All you have to do is look at a globe, however, to see that the amount of land available for us to use is limited.

People need food, clothing, jobs, and a place to live, and each of these things takes land. But preserving natural habitats also is important. Remember, a habitat is the place where an organism lives. Once a wetland is filled in to build an apartment building, the wetland and the organisms living there are lost.

Figure 10 Land is used for many purposes other than growing food.
Identify *the land uses that take up space in the city of Pittsburgh, pictured below.*

Science Journal

Habitat Loss Have students read newspapers or contact local, state, or U.S. government officials to find out about local environmental issues. Ask students to write a report on one or more habitats in their area that human activities threatened.

656 CHAPTER 22 Earth's Resources

Land Use Laws More and more, there are laws to help protect against habitat loss and to help us use land wisely. Before major construction can take place in a new area, the land must be studied to determine what impact the construction will have on the natural habitat, the living organisms, the soil, and water in the area. If there are endangered organisms living there, or if the impact will be too great, construction may not be allowed. These are important studies. At stake are jobs, homes, and habitats.

Landfills Each day, every person in the United States produces about 2.1 kg of garbage. Where does it go? About 57 percent of our garbage goes to landfills. A **landfill**, shown in **Figure 11,** is an area where garbage is deposited.

Any material that can harm living things by interfering with life processes is called a **pollutant** (puh LEW tunt). Modern landfills are lined with plastic or clay to keep chemical pollutants from escaping. However, some chemicals still find their way into the environment. If these pollutants get into the food that we eat or the water that we drink, they can cause health problems.

Figure 11 Each day, trash is put in a sanitary landfill. This trash is later covered with a thin layer of dirt and then watered down to keep the trash from blowing away. **List** *items that you send to the landfill.*

Reading Check *What is hazardous waste?*

Most of the trash we throw into landfills is not particularly dangerous to the environment. However, potentially dangerous items such as batteries, paints, and household cleaners sometimes end up in landfills. Many of these items contain harmful chemicals that could leak into the soil and eventually find their way into rivers and oceans. Garbage that contains dangerous chemicals or other pollutants is called hazardous waste. Many communities ask people to separate out hazardous wastes to prevent them from being deposited in landfills. Hazardous wastes can be taken to a special site where they are collected and disposed of safely.

Activity

Land Use Obtain a map of the local town or city and surrounding country. Ask students to study the use of land in and around your school. Encourage them to speculate as to whether they think the land has been used wisely. Ask them to explain how they might better use the land for all people living in or around your town or city. L2

IDENTIFYING Misconceptions

Garbage Decay Students may have the misconception that garbage deposited in a landfill decays in a short period of time. Actually, garbage placed in a landfill is protected against decay. When people take samples from older landfills, they see that newspapers, food, and other items have not decayed much. Dr. William Rathje, an archaeologist at the University of Arizona, and his team have discovered that biodegradable waste decays slowly in the oxygen-free environment of a sanitary landfill. For more information, refer to *Rubbish! The Archaeology of Garbage* by William Rathje and Cullen Murphy, © 1992 by Harper Collins Publishers.

Reading Check

Answer garbage that contains dangerous chemicals or other pollutants

Caption Answer

Figure 11 Student answers will vary.

Teacher FYI

Recycling Plastics Recyclers place numbers on plastic products to indicate the type of plastic resin used in their manufacture. The numbers are used to sort plastic items for recycling. It does not mean that the product is recyclable. Plastics make up over 20 percent of solid waste in landfills.

Differentiated Instruction

Challenge Have students contact the local sanitation department and find out where the solid waste produced by your school is taken. Ask students to interview sanitation department personnel to find out how solid waste is handled. Have them report the results of their interviews to the class.

Fun Fact

Potato-sized nuggets of various minerals lay on the ocean's floor. The nuggets contain iron, copper, nickel, cobalt, and manganese. The minerals come from nearby volcanoes that spew lava into the surrounding water, where it solidifies and falls to the ocean floor.

Text Question Answer

There would be less water for recreational purposes, such as swimming, boating, or gardening. Water rationing might be required to make sure everyone has enough to drink and crops can be irrigated when necessary.

INTEGRATE Earth Science

Available Water On the board draw a pie graph like the one shown in **Figure 12.** Tell students that this graph represents all the water on Earth. Mark off the 97% that represents salt water, the 2% frozen in ice caps, and the 1% that can be used for drinking or crop irrigation.

Discussion

Water Cycle On the board, draw a diagram of the water cycle, showing evaporation, condensation, and precipitation. Remind students that surface water evaporates and enters the atmosphere as water vapor. When water vapor condenses, it falls as precipitation (rain, snow, sleet, hail). Emphasize that the same water goes through this cycle over and over again.

INTEGRATE Earth Science

Earth's Water Distribution More than 97 percent of the water on Earth is salt water. Only three percent is freshwater. Of that three percent, more than two percent is frozen in glaciers and polar ice caps. Most of the remaining one percent is underground. Lakes and rivers contain only about one-tenth of one percent of Earth's water!

Our Impact on Water

Did you know that you cannot live long without water? You need clean water for drinking, as well as dozens of other uses, from washing clothes to watering plants. The average person in the United States uses about 397 L of water each day. Though water is a renewable resource, in some places it is being used up faster than natural processes can replace it.

Only a small amount of Earth's water, as shown in **Figure 12,** is freshwater that people can drink or use for other needs. Many places around the world are running out of usable freshwater. How do you think your life would change if your area were running out of clean water?

Water Pollution Many everyday activities can cause water pollution. How? When you scrub a floor with a mixture of water and a household cleaner, what do you do with the mixture afterward? You pour it down the drain. The polluted water usually goes to a water-treatment plant, where it is cleaned before being used again.

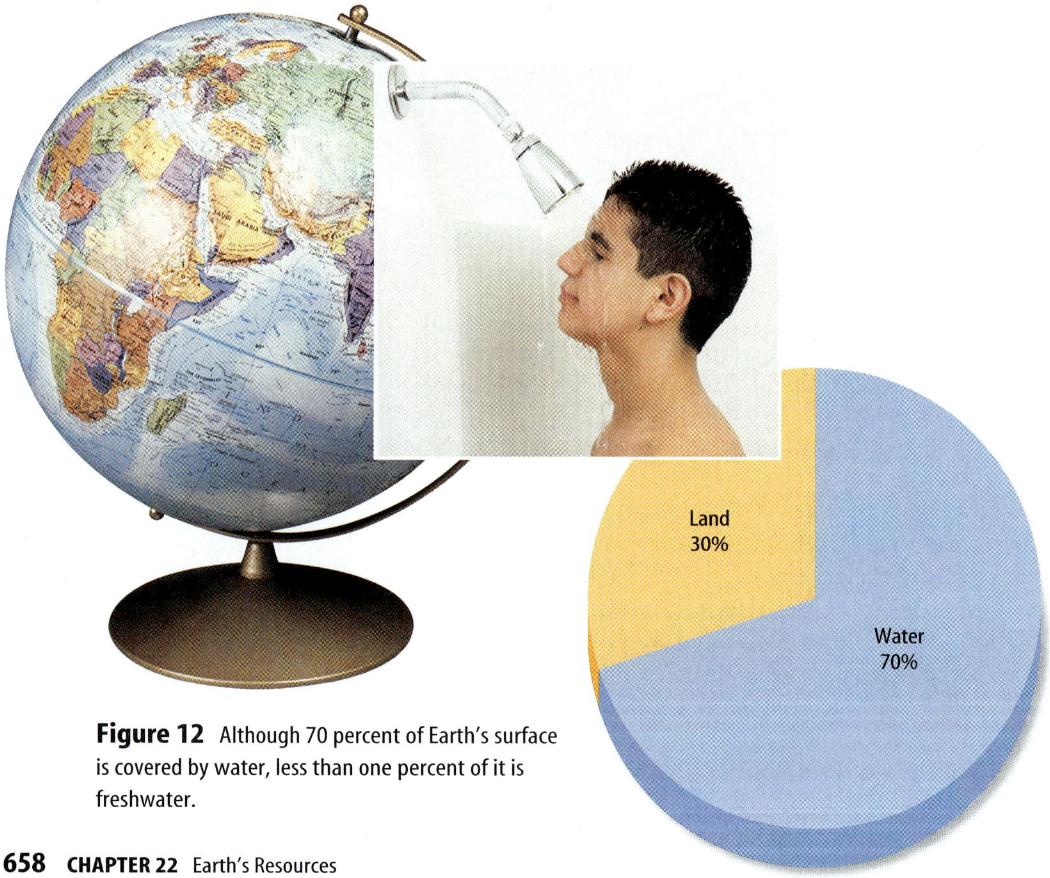

Figure 12 Although 70 percent of Earth's surface is covered by water, less than one percent of it is freshwater.

Cultural Diversity

Have students research cultures around the world and through history that have lived in harmony or disharmony with their environment. On Easter Island, for example, deforestation resulted in soil erosion that made the replacement of trees difficult, and a scarcity of wood for building fishing boats or housing. The civilization eventually collapsed as a result of civil war. On the other hand, Australian aborigines developed a remarkable ability to survive in an extremely harsh, very dry environment. Deforestation in Africa or on the island nation of Haiti would also make good subjects for student research, as would the successful irrigation techniques used on the island of Bali.

What if you poured the mixture of water and household cleaners outside on the grass or in the street? This polluted water would soak into the ground where it could contaminate the soil or seep into a well that others use for drinking water. Or, it might be washed away by rain and carried into rivers and lakes. If too many people do this, it could pollute our drinking water. **Figure 13** shows where most of our drinking water comes from.

There are many other ways that water can become polluted. Rain can wash pesticides and fertilizers from farmland into lakes, streams, or oceans. Rain falling on roads or parking lots washes oil and grease onto soil or into nearby waterways. Factories and industrial plants sometimes release polluted water into rivers. The dumping of litter and garbage into rivers, lakes, and oceans is another source of water pollution.

Figure 13 Much of our drinking water comes from rivers, lakes, and underground sources. This water is treated to remove impurities before it is used by people in towns and cities.

Cleaning Up the Water Countries are working together to reduce water pollution. For example, the United States and Canada have agreed to clean up the pollution in Lake Erie, a lake that borders both countries. The U.S. government has also passed several laws to keep water supplies clean. The Safe Drinking Water Act is a set of U.S. government standards that makes sure that our drinking water is safe. The Clean Water Act gives money to the states for building water-treatment plants. Wastewater is cleaned at such plants.

Remember, Earth has a lot of water, but only a small amount of it is freshwater that people can use. The best way people can protect Earth's water is by being aware of how they use it and taking steps to control their water use.

Inquiry Lab

Measuring Fast Food Packaging

Purpose to analyze the amount of packaging used for fast food meals

Possible Materials meals purchased at fast food restaurants, balance, metric ruler, large beakers, boxes of different sizes

Estimated Time two class sessions

Teaching Strategies

- Students can bring clean packaging from different fast food restaurants and compare the amount of packaging for each meal. Students can measure the total mass or volume of the bags, napkins, wrappers, condiment containers, and other packaging materials provided with each meal.
- Students also can compare packaging that comes with different meals from the same restaurant.
- Students can stuff the packaging materials into a beaker or boxes to measure the volume. They can use a metric ruler to calculate the volume of boxes.
- Students can also calculate the percentage of each material and graph their results.

Ask students to list questions this lab generates. Explore these if time allows.

For additional inquiry activities, see *Science Inquiry Labs.*

Visual Learning

Figure 13 Ask students to relate the water sources shown in the figure with the water sources used in their community. Rural and semi-rural areas are more likely to rely on groundwater than urban communities. Most heavily populated areas rely on reservoirs that store water from rainfall, rivers, and streams. Some communities rely on freshwater lakes for their water supply.

Teacher FYI

Water Cleanup Water treatment plants add lime and alum to remove impurities in water. The lime-alum treatment generates a floc of aluminum hydroxide that causes colloidal materials to coagulate. The floc and coagulates settle out in a clarifier. Then the water passes through sand filtration that removes residual particles. The water then may be sprayed into the air to add oxygen and treated with chlorine.

Teacher FYI

Sunset Color In parts of the US, sunsets are ablaze with color. But in some regions, such as portions of the Hawaiian Islands, sunsets are much less colorful. Many sunset colors are produced by pollutant particles in the atmosphere. Hawaii is surrounded by ocean, so fewer pollutants enter the air, except when one of Hawaii's active volcanoes erupts. During volcanic eruptions many pollutant particles enter the atmosphere.

Activity

Song Have students write a song about how the world has become polluted and how they will clean it up and improve the environment. L2 COOP LEARN P LS **Auditory-Musical**

Caption Answer

Figure 14 Forest fires, wood-burning in fireplaces, and exhaust from vehicles, factories, and airplanes all contribute to air pollution.

Our Impact on Air

If you live in a city, you may have noticed that on some days, the air looks hazy. Pollutants such as dust and gases in the air cause this haziness. Air pollution can be caused by natural events, such as a volcano eruption that releases smoke and ash into the air. But people cause most air pollution. **Figure 14** shows some sources of air pollution.

Sources of Air Pollution The two biggest sources of air pollution are cars and factories, including power plants that produce electricity. One source of pollution is the fumes that come from cars. Cars need gasoline to run. When gasoline is burned, pollutants are released into the air. The polluting fumes of cars and other vehicles cause more than 30 percent of all air pollution.

INTEGRATE Health Many factories and power plants burn coal or oil for the energy they need. The burning of fuel releases pollutants into the air that can cause health problems. Air pollution can make your throat feel dry or your eyes sting. Many people have trouble breathing when air pollution levels are high. For people with lung or heart problems, air pollution can be deadly. In the United States, it is estimated that 50,000 to 120,000 deaths each year are linked to air pollution.

Figure 14 Air pollution is caused by many different activities. **List** *the activities illustrated in this picture that cause air pollution.*

Teacher FYI

Air Pollutants Sulfur dioxide and nitrous oxides are air pollutants that contribute to acid rain. Chlorofluorocarbons (CFCs) are refrigerants that contribute to the destruction of the ozone layer that is high in Earth's atmosphere. The ozone layer helps protect organisms from the Sun's damaging ultraviolet radiation. However, ozone is an air pollutant when it occurs close to Earth's surface. Fuel combustion can produce ozone that can cause respiratory problems.

Acid Rain People aren't the only living organisms harmed by air pollution. Acid rain causes a lot of damage to other organisms. **Acid rain** happens when the gases released by burning oil and coal mix with water in the air to form acidic rain or snow. When acid rain falls to the ground, it can harm trees and other plants. When acid rain falls into rivers and lakes, it can kill fish and other organisms that live in the water.

Spare the Air

The best solution for all types of pollution, including air pollution, is prevention. Reducing the number of pollutants in the environment is easier to do than cleaning up pollution. Automobiles produced today release fewer harmful gases and use less fuel than did vehicles in years past. Governments around the world are looking for ways to reduce the amount of air pollutants released into the atmosphere by factories.

Reading Check *How would riding a bicycle instead of driving a car help the environment?*

It may seem that you have no control over sources of pollution, but think again. Think about what power plants produce. They produce electricity. When power plants burn oil or coal to make electricity, harmful pollutants enter the atmosphere and cause smog, acid rain, and other problems.

You can help protect the atmosphere by limiting the amount of energy you use at home. Conserve electricity by turning off lamps, radios, fans, and other appliances that you aren't using. Keep doors and windows closed to save heat energy in the winter or to reduce the need for air conditioning in the summer. Encourage your family to buy energy-efficient lightbulbs like the one shown in **Figure 15.**

Figure 15 Turning down your home's thermostat in the winter is one easy way to reduce energy use. Energy-efficient lightbulbs use one-quarter of the energy of standard incandescent lightbulbs, and they can last up to ten times longer.

Reading Check

Answer Bicycles don't require the combustion of fuel, so riding a bike reduces pollution from auto exhaust.

Visual Learning

Figure 15 On the thermostat shown, point out the temperature control and the function knob for choosing fan, cooling, or heating. Compare it with the thermostat in the classroom. Have students discuss what changes, if any, to the classroom thermostat could help save energy.

Quick Demo

Compact Fluorescent Bulbs

Materials incandescent light bulb, compact fluorescent light bulb, lamp or light fixture in which both bulbs will fit

Estimated Time 10 minutes

Procedure Show students how a compact fluorescent bulb can fit into the same lamp or light fixture as an incandescent bulb. Explain that incandescent bulbs use radiation from hot metal (tungsten) to produce light. Fluorescent bulbs do not. Allow students to compare the temperatures of the two bulbs when lit. Point out that heat represents wasted energy. If allowed, have students replace one or more classroom lightbulbs with compact fluorescents.

Virtual Labs

Check the Meter *How much electricity is used in a house?*

Curriculum Connection

Architecture Recent developments in energy-saving technologies have made it possible to build new homes, offices, and other buildings that use less energy than older buildings. Invite students to research some of these technologies and present written or oral reports on their findings. Possible topics: photovoltaic panels installed on rooftops to generate electricity from sunlight; improved insulation materials; high-performance windows; unusual construction materials, such as straw bales, rammed earth, or structural insulated panels (SIPS); passive solar construction; white roofs that reflect sunlight to keep buildings cool. Students could also look into solar water heater technology or energy-saving appliances, such as tankless water heaters, pilot-less cooking ranges, or water-saving washing machines.

3 Assess

DAILY INTERVENTION

Check for Understanding

Visual-Kinesthetic Create a table on the board with two major columns labeled Renewable and Nonrenewable. Under each column, create smaller columns listing resources that belong in each category. Renewable: water, sunshine, trees, etc. Nonrenewable: oil, natural gas, coal, etc. Have students choose items in the classroom and decide under which column they belong.

Reteach

Water Pollution Discuss with students how a pollutant spilled on a street could affect organisms living in a river several miles away. Explain that water flowing into storm sewers is not treated in any way and eventually flows into streams and rivers.

Assessment

Oral Have students explain the environmental impact of using classroom materials. For example, ask how using a wooden pencil compares with using a plastic, refillable automatic pencil. How does reducing the amount of chalk dust in the classroom by using an overhead projector compare with the added outdoor pollution resulting from the extra electricity use? Accept all reasonable arguments. Students should be able to state pros and cons for each action. Use **PASC**, p. 89.

To maintain the health of the environment, everybody needs to think about how their actions will affect the land, water, and air that we share. People living in the houses shown in **Figure 16** use a renewable energy source—sunlight—to help reduce air pollution and conserve supplies of nonrenewable coal.

Figure 16 The roofs of these houses are covered with solar cells, which use sunlight to produce electricity. On sunny days, the people living in these homes help preserve natural resources by using electricity generated by the Sun instead of a power plant.

section 2 review

Summary

Exploring Environmental Problems

- Human activities can destroy the habitats of other organisms.

Our Impact on Land

- Land is a nonrenewable resource.
- Most of our garbage is buried in landfills.

Our Impact on Water

- Less than 1 percent of the water on Earth can be used for drinking.
- Human activities can cause water pollution.

Our Impact on Air

- Most air pollution results from the burning of fuels.

Spare the Air

- Pollution is easier to prevent than to clean up.

Self Check

1. **Define** the term *pollutant.* Give five examples.
2. **Describe** two ways in which human activities affect land.
3. **Explain** how human activities can cause water pollution and air pollution.
4. **List** three actions you could take to reduce pollution.
5. **Think Critically** Fumes from burning gasoline cause air pollution. Explain how gasoline could cause water pollution.

Applying Math

6. **Calculate** A regular showerhead puts out 15 L of water per minute. A water-saving showerhead puts out 9.5 L. If you take a five-minute shower every day, how much water would you save in a week by using a water-saving showerhead?

Science Online red.msscience.com/self_check_quiz

section 2 review

1. a material that can harm living things by interfering with life processes; chemicals, cleaning solutions, battery components, paints, pesticides, fertilizers, grease, gasoline, gases and particles produced by fuel combustion
2. Possible answers: Clearing land for highways, buildings, golf courses, etc. can disturb ecosystems and destroy wildlife habitat. Creating landfills can destroy wildlife habitat and possibly allow pollutants to enter soil and water.
3. Possible answers: Hazardous spills on the ground could soak into soil and enter groundwater or streams; air pollution: burning fuels releases gases and particles that pollute the air; air pollutants can be absorbed by rainfall, causing acid rain.
4. Possible answers: turn off electrical equipment when not in use, ride a bike or take public transportation instead of a car, adjust thermostat to reduce energy use
5. if spilled on the ground, could enter soil and groundwater or streams
6. $15\ \text{L/min} \times 5\ \text{min} \times 7 = 525\ \text{L/week}$
   $9.5\ \text{L/min} \times 5\ \text{min} \times 7 = 332.5\ \text{L/week}$
   $525\ \text{L} - 332.5\ \text{L} = 192.5\ \text{L}$

section 3

Protecting the Environment

Cutting Down on Waste

The United States faces a huge waste problem. Litter gathers along highways. Landfills leak and overflow with garbage. Five billion tons is the estimated amount of solid waste thrown away each year in this country. **Solid waste** is whatever people throw away that is in a solid or near-solid form. Look at **Figure 17** for examples.

Most waste is produced when coal, oil, and other natural resources are taken from the ground. Households and businesses produce only about four percent of this country's waste. However, household and business waste is still a lot of solid waste—nearly 200 million tons each year.

Most of the waste from our homes, schools, and businesses is paper and cardboard products. In the cafeteria at school, it's easy to see why this is so. School lunch programs all over the country depend upon paper plates, straw wrappers, milk cartons, paper bags, drink boxes, and napkins. What if individuals just tried to reduce the amount of trash they throw away each day?

Solid-waste management for individuals can be summed up by the three Rs—reduce, reuse, and recycle.

as you read

***What* You'll Learn**

- **Describe** the problems of solid waste.
- **Explain** how to reduce, reuse, and recycle resources.

***Why* It's Important**

You can take simple actions every day to reduce waste and help protect the environment.

Review Vocabulary
biosphere: the part of Earth that supports life, including Earth's crust, atmosphere, and water

New Vocabulary
- solid waste
- recycling

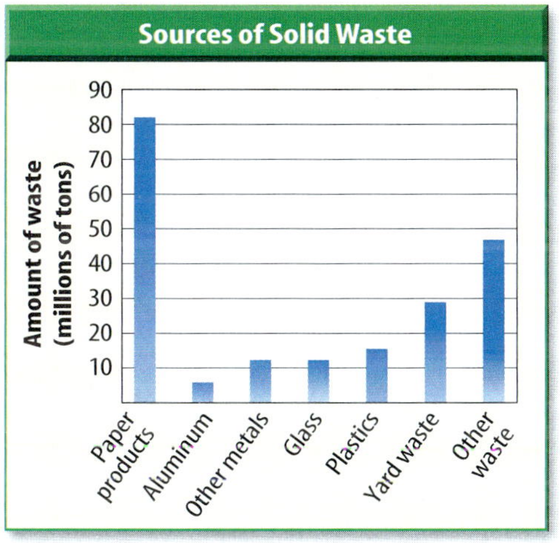

Figure 17 Solid waste includes everything from old newspapers and broken toys to scrap metal left over from manufacturing processes.
Infer *why aluminum is the smallest source of solid waste.*

1 Motivate

INTERACTIVE CHALKBOARD PowerPoint® Presentations

Bellringer

Section Focus Transparencies also are available on the Interactive Chalkboard CD-ROM. L2 ELL

Section 3 Section Focus Transparency **The Art of Recycling**

Many artists work with paints or other specific materials, but sometimes artists work with whatever happens to be lying around. This goat was made from objects that many people throw away without a second thought.

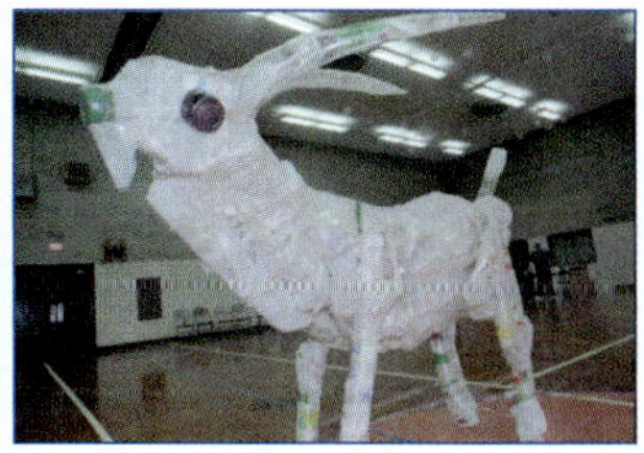

1. Can you identify the items this goat is made from?
2. What other materials might be reused to make a piece of art?
3. What are some benefits of reusing materials rather than throwing them away?

Earth's Resources

Tie to Prior Knowledge

Help students recall a time when they threw something away that they could have used again, or given to someone else who could use it.

Caption Answer

Figure 17 because many people recycle aluminum waste

Section 3 Resource Manager

Chapter *FAST FILE* Resources
- Transparency Activity, p. 46
- MiniLAB, p. 4
- Enrichment, p. 32
- Reinforcement, p. 29
- Directed Reading for Content Mastery, p. 21
- Lab Worksheet, pp. 7–8
- Lab Activity, pp. 13–16

Life Science Critical Thinking/Problem Solving, p. 6

Earth Science Critical Thinking/Problem Solving, p. 18

Home and Community Involvement, p. 49

Reading and Writing Skill Activities, pp. 23–24

2 Teach

Reading Check

Answer items are used again instead of being thrown away

Mini LAB

Purpose Students will explore the possibilities of reducing, reusing, and recycling resources. L2 LS **Kinesthetic**

Materials clean items that are normally thrown out; glue; string; tape; cardboard, posterboard, or foam board

Teaching Strategies

- Use an old cardboard box to mount the artwork. This will emphasize reuse.
- Students could work together to make a mascot for their school or some other theme.

Safety Precautions Caution students to use only items that do not pose safety problems. Take care that students are not burned if using hot glue. Do not have students remove items from the trash cans.

Troubleshooting Glue that does not dry quickly will cause problems during this activity.

Analysis

1. Student answers will vary, but students should list each item individually.
2. All three; the amount of trash thrown out was reduced, the items were reused, and the items were recycled into a work of art.

Assessment

Process Hold up a discarded item and challenge the class to find a use for it. Students may illustrate their thoughts in the form of a cartoon. Use **PASC,** p. 133. P

Mini LAB

Making Models

Procedure

1. Collect several items that would normally be thrown out. Such items could include **newspapers, clean cans** or **glass, packaging,** etc. Do not collect any food items or items that could be harmful. Do not take any items out of the garbage.
2. Using **glue, string,** or **tape,** create an item of artwork.
3. Give your piece of art a name.

Analysis

1. What items did you use to make your piece of artwork?
2. Is this activity an example of reducing, reusing, or recycling? Explain.

Try at Home

Reduce Most people would agree that there are no simple solutions to the problem of too much solid waste. The simplest and most effective way you can help solve the solid-waste problem is to reduce the total amount of solid waste that you throw away.

One method of reducing waste is to buy products with little or no packaging. For example, suppose you go to a toy store to buy a gift for a young child. If you select a toy that has no packaging, you've reduced your use of paper made from wood, or plastic made from oil. If you select a toy made from wood rather than plastic, you've reduced your use of a nonrenewable resource.

The other two Rs of conservation—reuse and recycling—also provide ways to reduce the amount of solid waste that goes into landfills.

Reuse Think again about the toy purchased as a gift. Instead of wrapping it in paper, you could put it in a cloth gift bag. The cloth bag can be used over and over again before it wears out and has to be thrown away. Think about the amount of wrapping paper that would not have to go into the landfill as a result! Reuse means using an item again rather than throwing it away and replacing it.

Reading Check *How does reuse help reduce the amount of solid waste that goes into landfills?*

There are many examples of reuse. Worn-out clothes can be used as cleaning rags. Old newspapers can be used to line pet cages, wrap gifts, or cover the floor when painting. Perhaps you've been to a garage sale, a flea market, or a used clothing store like the one shown in **Figure 18.** Books, magazines, clothing, computers, video games, glass jars, and cardboard boxes are some of the many items that can be reused. When you no longer need them, you can give them to someone else who may want or need them, instead of sending them to a landfill by throwing them away.

Figure 18 Secondhand stores are great places to find bargains. Like-new clothing and other items are being reused—a good way to help protect the environment.

Fun Fact

Old tires can be shredded and used for fuel, new rubber, plastic products, and as substitutes for concrete in road pavement.

Recycle When you finish lunch in the school cafeteria, do you have paper cups and plates, cans, glass bottles, plastic containers, and left-over food to throw out? As **Figure 19** shows, many communities provide special bins that allow you to separate your garbage so that certain items can be recycled. **Recycling** (ree SI kling) means reusing materials after they have been changed into another form. Used paper can be reprocessed to make recycled paper. Glass can be melted and reformed into new containers made of recycled glass. Even leftover food can be recycled by putting it in a compost pile. The food decomposes into rich compost that gardeners can use to boost plant growth.

Figure 19 Everyone can help save energy, reduce solid waste, and conserve natural resources by recycling paper, plastic, glass, and metal.

Applying Science

Reusing Plastic

Have you ever seen aluminum beverage cans sold in packs of six in grocery stores? The six cans usually are held together with a plastic collar. Once the cans are released from this plastic collar, what do you do with it? Thrown away in one piece, it is dangerous to wildlife. Fish and birds can get caught in the plastic collars. Some animals mistake them for food and choke on them.

Identifing the Problem

Make a list of the types of products that are packaged using plastic collars. How are these collars usually disposed of?

Solving the Problem

1. Are there other ways to deal with these plastic collars? Can you think of helpful ways to reuse the plastic collars?
2. Can you think of some other ways to package six aluminum cans of beverage together that won't hurt the environment?

Activity

Swap Meet Have students bring in unwanted clothing, toys, or other items from home to swap with their classmates. Make sure students have permission to give away the items they bring to the swap meet. Students also could organize a school-wide swap meet designed to inform schoolmates about the value of reusing items instead of throwing them away.

Applying Science

Answers

1. Student answers will vary. Possible uses could include organizing items of clothing such as socks and scarves, making bracelets, used as packing material, or part of 3-D artwork.
2. Student answers will vary. One possible answer is that they could be packaged in sheets of reusable plastic or in cardboard cartons.

Discussion

Recycling What items do you recycle? Possible answers: newspaper, notebook paper, aluminum cans, glass containers, plastic containers, cardboard What recycled products do you use? Possible answers: writing paper, toilet paper, paper towels, clothing made from recycled plastics, products packaged with recycled materials Why is it important to buy recycled products? to make sure recycling actually results in less need for new materials

Differentiated Instruction

English-Language Learners Help students distinguish between the terms *reuse* and *recycle*. Explain that a meaning for the prefix *re-* is "again." *Reuse* means to "use again." *Recycle* means to "cycle again," which leads to the idea of an item, such as the aluminum from a can, going through a processing cycle again. One meaning of the word *reduce* is "to make smaller" as in making the amount of trash we throw away smaller.

Visual Learning

Applying Science How could the plastic collar harm the seagull shown in the photograph, or other animals? could get wound tightly around an animal's throat, harming its ability to breathe; could get wound around bird's beak, making it impossible for it to eat; could make it difficult for the bird to fly

Quick Demo

Picnic Disposables

Materials several place settings of disposable picnic items such as paper plates, cups, bowls, and plastic flatware, and a matching place setting made of washable, unbreakable plastic or metal designed for outdoor use

Estimated Time 10 minutes

Procedure Show students how much solid waste can be avoided by using washable, reusable picnic supplies rather than disposable supplies. Have students imagine how many paper plates, etc., they could avoid sending to a landfill by using a reusable picnic place setting 30 times.

Teacher FYI

A Lifetime of Garbage It's estimated that, over a lifetime, each person in this country throws out 600 times his or her own weight in garbage. That averages to about 90,000 pounds per person. The recycling of one aluminum can saves enough energy to run a TV for three hours or light a 100-watt light bulb for 20 hours. It takes about five recycled plastic bottles to make a fiberfill jacket.

Reading Check

Answer Reusing an item means using it again and again. Recycling means reusing materials after they have been changed into another form.

Figure 20 Recycling materials reduces the amount of energy used to make products.

It takes 95 percent less energy to produce aluminum from recycled aluminum than from ore.

Recycled Products Did you know that plastic soft-drink bottles might have been used to make the carpeting in your home or the fleece jacket that keeps you warm in winter? It's true. As **Table 1** shows, many items people normally throw out can be recycled into other useful products.

Think about what would happen if recycled products piled up on store shelves because no one purchased them. Recycling means not only separating your garbage, but also buying recycled goods when you can.

Recycling can save a lot of energy compared to manufacturing new materials, as **Figure 20** shows. Keep in mind that reducing and reusing use even less energy, because they do not require the reprocessing of materials.

Reading Check *What is the difference between reusing and recycling?*

Table 1 Recycled Products

Items to Be Recycled	Resulting Products
Newspapers, telephone books, magazines, and catalogs	Newsprint, cardboard, egg cartons, building materials
Aluminum beverage cans	Beverage cans, lawn chairs, siding, cookware
Glass bottles and jars	Glass bottles and jars
Plastic beverage containers	Insulation, carpet yarn, textiles

Differentiated Instruction

Challenge Have students contact the local recycling center to find out what kinds of items can and cannot be recycled. If possible, have students tour the recycling center to see how materials are handled and report back to the rest of the class. Students could take photographs or videos of their visit and create a multimedia presentation. L3

666 CHAPTER 22 Earth's Resources

It takes 75 percent less energy to make steel from scrap than from iron ore.

Making new glass from old glass cuts energy usage by 80 percent.

Habits for a Healthier Environment

By practicing the three Rs of solid waste management, you contribute to a healthier environment. You can see how changing your everyday habits—the way you pack your lunch, the transportation choices you make, the way you dispose of your trash—helps reduce solid waste and pollution, and helps conserve natural resources. The best way for everyone to protect the environment is to develop habits that promote a healthy environment.

section 3 review

Summary

Cutting Down on Waste

- Creating less garbage is the best way to help solve the problem of too much solid waste.
- Using things over again, rather than throwing them away, helps reduce solid waste.
- Recycling is a type of reuse that requires changing the material into another form.

Habits for a Healthier Environment

- You can be part of the solution to solid waste by getting in the habit of reducing, reusing, and recycling.

Self Check

1. **Explain** what *reducing waste* means. Give two examples.
2. **List** five items you use every day that could be reused instead of being thrown away.
3. **Describe** how recycling a plastic bottle can help protect the environment.
4. **Think Critically** How could a person buying food at a take-out restaurant practice reducing waste?

Applying Skills

5. **Recognize Cause and Effect** Explain how keeping a TV on when no one is watching affects the environment.

Science Online red.msscience.com/self_check_quiz

3 Assess

DAILY INTERVENTION

Check for Understanding

Logical-Mathematical Have students classify each of the following as an example of reducing, reusing, or recycling waste.

- Ship an item in as small a box as possible, with necessary padding but no extra padding. reduce
- Wash out a mayonnaise jar and use it to store rice. reuse
- Turn plastic bottles into play towers for small children. recycle

Reteach

Rs of Conservation Have students tell you what to write on the board as you ask questions leading through the production and use of glass bottles, and the associated environmental impacts. sand collected and refined into glass from which bottles are formed; bottles transported to another factory to be filled with a product; product transported to stores; bottles transported home with consumers; bottles thrown away, reused, or recycled

Assessment

Performance Using plastic gloves, remove the solid waste from the trash can in the classroom and list the items on the board. Ask students to list ways each item could be reduced, reused, or recycled. Use **Performance Assessment in the Science Classroom**, p. 89.

section 3 review

1. Using less material or energy in the production or packaging of a product. Possible examples: reduce the size of a package, purchase foods in bulk packages, use both sides of each sheet of paper.
2. Possible answers: glass containers, plastic containers, plastic bags, paper grocery bags, rubber bands, string, use old clothing for cleaning rags.
3. Reduces the need for oil, for the plastic itself, and for the energy needed to make the plastic; reduces solid waste that goes to landfills.
4. Possible answers: take only as many napkins and ketchup or mustard packets as needed, bring your own reusable picnic fork and knife, encourage restaurant owners to pack food in paper or recyclable plastic containers, place leftover food and paper packaging in the compost pile rather than the trash can.
5. Electricity is wasted. Coal or natural gas is being burned for no purpose, unnecessarily reducing supplies of fossil fuels and adding pollutants to the air.

LAB Design Your Own

Using Land

Goals

- **Design** a land-use plan in which 100 square units of land can be turned into a town.

Possible Materials

grid paper (10 squares × 10 squares)
colored pencils

Real-World Question

Imagine planning a small town. People need homes in which to live, places to work, and stores from which to buy things. Children need to attend schools and have parks in which to play. How can all of these needs be met when planning a small town? Your job in this lab is to draw up a master plan to decide how 100 square units of land can be turned into a town.

Form a Hypothesis

How should the town's land resources be used?

Test Your Hypothesis

Make a Plan

1. A 100-square-unit piece of land can be represented as a square divided into 100 blocks. One way to represent this is to make a square graph ten blocks across and ten blocks down.
2. The table on the next page shows the different parts of a town that need to be included in your design. The office buildings and industrial plant are places where the people of the town will work. They are each six blocks in size. These two blocks cannot be divided and must be treated as one group. The landfill is four blocks in size and also cannot be broken up.
3. All the other town parts can be broken up as needed. Stores and businesses are areas in which shops are located as well as medical offices, restaurants, churches, and cemeteries.
4. As a group, discuss how the different parts of the town might be put together. Should the park be in the center of town or near the edge of the town? Should the school be near the offices or near the houses? Where should the landfill go?
5. How will you show the different town parts on your grid paper?

BENCH TESTED

Real-World Question

Purpose Students will plan a small town and emphasize that land is a resource that needs to be used carefully, just like any other nonrenewable resource. L2 COOP LEARN

LS **Logical-Mathematical**

Process Skills classify, observe and infer, compare and contrast, formulate hypotheses, design an experiment to test a hypothesis, make models, recognize cause and effect, measure in SI, interpret scientific illustrations, make and use tables

Time Required 40 minutes

Form a Hypothesis

Possible Hypothesis Locating industry, offices, and landfill on one side of town will leave plenty of space for houses, school, stores, and park on the other side of town.

Test Your Hypothesis

Possible Procedures First, zone the plan into areas for living quarters, then office buildings and commercial areas, and finally other parts of the town.

Alternative Inquiry Lab

Real World Question Can buildings be designed to use less land and preserve wildlife habitat? Challenge students to design homes and other buildings that conserve space. Students can sketch scaled drawings of their designs on graph paper and use a variety of materials to construct a model of their home or building. They can experiment with home designs that are primarily underground to allow for the growth of natural ecosystems on top of them, or they can design tree house structures that are compatible with the growth of surrounding forests or meadows. Plastic foam or other mold-resistant materials can be used for walls and ground level skylights can be installed to allow sunlight into the structure. Students also can experiment with smaller buildings that have rooms or spaces with multiuse features.

Using Scientific Methods

Amount of Land Needed for Town

Parts of Your Town	Number of Blocks Needed
Office buildings	6 blocks in one group
Industrial plant	6 blocks in one group
School	1 block
Landfill for garbage	4 blocks in one group
Houses and apartments	44 blocks—can be broken up
Stores and businesses	19 blocks—can be broken up
Park	20 blocks—can be broken up

Follow Your Plan

As a group, design your town. Check over your design to make sure that all of the town parts are accounted for.

Analyze Your Data

1. Where did you place the office buildings and the industrial plant? Why were they placed there? Where did you place the houses, school, stores, and businesses? Explain why you placed each one as you did.
2. Did you make one park or many parks with the land designated for park use? What are the advantages of your park design?
3. Where did you place your landfill? Will any of the townspeople be upset by its location? What direction does the wind usually blow from in your town?

Conclude and Apply

Where would you put an airport in this town? Keep in mind safety issues, noise levels, and transportation needs.

Communicating Your Data

Compare your town design with those of other members of your class. Discuss the advantages and disadvantages of each design.

Teaching Strategies

- Discuss what students like and dislike about where they live, where their school is, and where their parents work.
- Have students decide in which direction the wind blows. This may affect where factories and landfills are located.
- Instead of using a 10×10 square grid, different shapes of the 100 squares could be used—e.g., a 4×25 grid or an irregular shape.

Troubleshooting No allowances are made for roads, highways, or railroads although these could be included.

Analyze Your Data

1. Student answers should include reasonable explanations for each part of the plan.
2. Student answers should indicate that the locations of the parks should be in areas easily accessible by all residents.
3. Landfills should not be placed near houses and apartments or upwind from most stores, businesses, and homes.

Conclude and Apply

An airport should be placed far away from schools but located close to businesses, shopping malls, and hotels.

Assessment

Portfolio Have students draw and illustrate their town designs and write a short report describing the features of the town and the reasoning behind their design. Use **Performance Assessment in the Science Classroom,** p. 145.

Communicating Your Data

Have each student group present its town plan to the class, and let the other teams critique each one. Ask students to list points they like and dislike about each plan. After all plans have been presented, students could vote to select the one they consider most successful.

TIME

Content Background

Some life-cycle analyses have shown conflicting results. Makers of plastic grocery bags find that their product is less harmful to the environment than paper grocery bags. Makers of paper grocery bags find the opposite to be true.

The main benefit of life-cycle analyses thus far has been to help companies target areas of potential energy reduction. Manufacturers of household liquid cleaners find that roughly one-half of the total energy use associated with their products occurred during stage four (product use) when people mix the cleaners with warm water. Manufacturers are working to develop liquid cleaners that can be used in cold water.

Discussion

Product Stages Show a variety of materials, such as a lump of coal, a CD, a newspaper, and a piece of fruit salvaged from the garbage. Have students classify the materials according to product stage. Have students list ways that energy was used to get the products to each particular stage.

Investigate the Issue

Online Research Help students use a search engine to explore the Internet for recent life cycle analysis data. Suggest they use key words such as plastic cup life cycle analyses. Help students refine search terms as their research progresses.

TIME SCIENCE AND Society — SCIENCE ISSUES THAT AFFECT YOU!

A Tool for the Environment

Are you an environmentally friendly shopper? When you buy things, do you think about how they affect the environment? Scientists have developed a tool to help them figure out the environmental impacts of products. The tool is called life-cycle analysis. Life-cycle analysis is a way of estimating the environmental impact of a product through its entire life.

Turning fabric scraps into a colorful quilt is a good example of reuse.

Life Stages of a Product

Most scientists break down the life cycle of a product into six stages:

1. getting the natural resources to make the product;
2. manufacturing the product;
3. transportation to a home, store, or business;
4. use and reuse;
5. recycling; and
6. disposal in a landfill or by burning.

During each stage of a product's life, natural resources and energy are used. Each stage also has an impact on the environment. Environmental impacts might include air or water pollution, human health problems, use of nonrenewable resources, or habitat loss. A life-cycle analysis considers all of these factors.

Once a life-cycle analysis is complete, the product can be compared with others to see which one is better for the environment. Companies can use these results to reduce the environmental impact of the products they make. You can use life-cycle analyses to become an environmentally friendly shopper.

Research Information Think of a product you would like to buy. Research the life cycle of the product. Share the results of your life-cycle analysis with your classmates in a multimedia presentation.

Science Online
For more information, visit red.msscience.com/time

Research Information Students could work in small groups to gather information and compare life cycle data for a variety of products, such as: glass, ceramic, and disposable plastic cups; paper, plastic, and cloth grocery bags; pens, pencils, and automatic pencils.

Resources for Teachers and Students

Institute for Life Cycle Energy Analysis, 310 27th Ave. E, Seattle, WA 98112

American Center for Life Cycle Analysis, 19001 Vashon Hwy SW, Suite 203, Vashon, WA, 98070

Life Cycle Analysis for Mere Mortals, by Rita Schenck, Institute for Environmental Research and Education (IERE), 17917 Vashon Hwy SW, P.O. Box 2449, Vashon, WA 98070–2449

Reviewing Main Ideas

Section 1 Natural Resource Use

1. Natural resources are the parts of Earth's environment that supply the materials necessary or useful for the survival of living organisms.
2. Renewable resources are natural resources that can be replaced in 100 years or less. They include trees, water, and sunshine.
3. Nonrenewable resources are natural resources that cannot be replaced or take more than 100 years to replace. They include coal, oil, natural gas, land, and some ores and metals.

Section 2 People and the Environment

1. Human activities affect land, water, air, and other natural resources in many ways.
2. The construction of roads and buildings may destroy the habitats of many organisms.
3. A landfill is an area where garbage is buried.
4. A pollutant is a material that harms organisms by interfering with life processes.
5. Acid rain is caused by air pollution.

Section 3 Protecting the Environment

1. Much of the garbage people produce is in the form of solid waste.
2. Reduce, reuse, and recycle are the three Rs of solid waste management.
3. The three Rs help people produce less solid waste and can help conserve natural resources.

Visualizing Main Ideas

Copy and complete this concept map about energy conservation.

Reducing your energy use
helps prevent
air pollution
which can cause
acid rain
hazy skies
human health problems

Science Online red.msscience.com/self_check_quiz

Identifying Misconceptions

Assess

Use this assessment as follow-up to page F at the beginning of the chapter.

Activity Give students the following scenario. They are very thirsty and find a cup of water sitting on a table in a park. They hold it up and see that it is crystal clear. Have them write a paragraph that describes what is going through their mind as they decide whether or not to drink it. End the paragraph by stating whether they would or would not drink the water.

chapter 22 Study Guide

Reviewing Main Ideas

Summary statements can be used by students to review the major concepts of the chapter.

Visualizing Main Ideas

See student page.

Science Online

Visit red.msscience.com
/self_check_quiz
/interactive_tutor
/vocabulary_puzzlemaker
/chapter_review
/standardized_test

Assessment Transparency

For additional assessment questions, use the *Assessment Transparency* located in the transparency book.

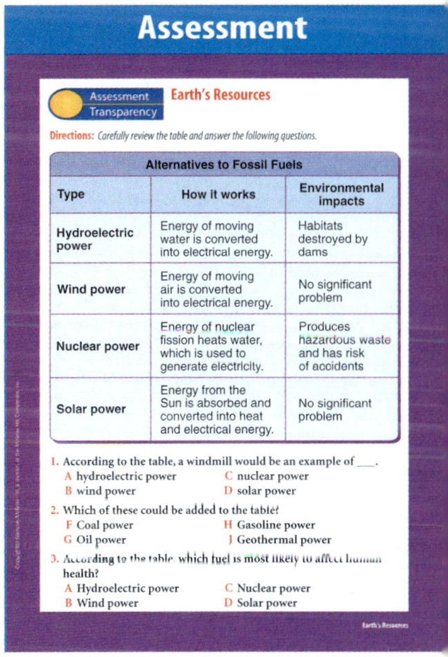

Assessment

Assessment Transparency **Earth's Resources**

Directions: *Carefully review the table and answer the following questions.*

Alternatives to Fossil Fuels		
Type	**How it works**	**Environmental impacts**
Hydroelectric power	Energy of moving water is converted into electrical energy.	Habitats destroyed by dams
Wind power	Energy of moving air is converted into electrical energy.	No significant problem
Nuclear power	Energy of nuclear fission heats water, which is used to generate electricity.	Produces hazardous waste and has risk of accidents
Solar power	Energy from the Sun is absorbed and converted into heat and electrical energy.	No significant problem

1. According to the table, a windmill would be an example of ___.
   A hydroelectric power C nuclear power
   B wind power D solar power
2. Which of these could be added to the table?
   F Coal power H Gasoline power
   G Oil power J Geothermal power
3. According to the table, which fuel is most likely to affect human health?
   A Hydroelectric power C Nuclear power
   B Wind power D Solar power

Earth's Resources

Using Vocabulary

1. pollutant
2. recycling
3. landfill
4. solid waste
5. acid rain
6. natural resource

Checking Concepts

7. C	11. C
8. A	12. B
9. B	13. A
10. B	14. B

Thinking Critically

15. Trees are renewable resources.
16. Most is salt water that cannot be used by people unless the salt is removed.
17. Pollutants from the landfill can leak out and enter the groundwater and other parts of the environment.
18. need for natural resources to make new bags is reduced
19. mine aluminum ore → melt the ore in a factory → refine the aluminum → shape the aluminum into cans → transport the aluminum can to where it will be used → use the aluminum can → recycle the can

chapter 22 Review

Using Vocabulary

acid rain p. 661
landfill p. 657
natural resource p. 647
nonrenewable resource p. 652
pollutant p. 657
recycling p. 665
renewable resource p. 651
solid waste p. 663

Give the vocabulary word that best fits each of the following definitions.

1. a material that harms living things by interfering with life processes
2. reusing materials after they have been changed into another form
3. an area where garbage is deposited
4. things people throw away that are in solid or near-solid forms
5. formed when gases released by burning oil and coal mix with water in the air
6. a part of the Earth that living organisms need or use

Checking Concepts

Choose the word or phrase that best answers the question.

7. Which of the following is an example of a nonrenewable resource?
   A) sunlight **C)** oil
   B) water **D)** a tree
8. What is using an old newspaper to line a pet cage an example of?
   A) reusing **C)** reducing
   B) recycling **D)** buying
9. What is collecting used paper and sending it to a factory to be made into new paper an example of?
   A) reusing **C)** reducing
   B) recycling **D)** buying
10. What could be caused by breathing polluted air?
    A) acid rain
    B) health problems
    C) solid waste
    D) water pollution
11. What does a life-cycle analysis of a product indicate?
    A) daily use
    B) production time
    C) all resource and energy use
    D) decay time
12. Turning off unneeded lights is an example of which of the following actions?
    A) reusing
    B) reducing
    C) recycling
    D) polluting

Use the photo below to answer question 13.

13. What are the paper, apple, and wood pencils examples of?
    A) renewable resources
    B) pollutants
    C) nonrenewable resources
    D) packaging
14. Dumping used motor oil on the ground could create
    A) air pollution.
    B) water pollution.
    C) solid waste.
    D) acid rain.

Science Online red.msscience.com/vocabulary_puzzlemaker

Use the ExamView® Pro Testmaker CD-ROM to:

- create multiple versions of tests
- create modified tests with one mouse click for inclusion students
- edit existing questions and add your own questions
- build tests aligned with state standards using built-in State Curriculum Tags
- change English tests to Spanish with one mouse click and vice versa

Thinking Critically

15. **Infer** why, if people use so many paper products, we don't run out of trees.

16. **Explain** why almost 70 percent of Earth's surface is covered by water, but less than one percent can be used by people.

17. **Describe** When a landfill can't hold any more solid waste, it is closed down. How can a landfill be an environmental problem even though people are no longer depositing trash there?

18. **Evaluate** Some people take their own bags with them when shopping. How might this affect natural resources?

19. **Sequence** Use the following phrases to make a concept map showing the life-cycle analysis of an aluminum can: *refine the aluminum; mine ore; use the can; shape aluminum into cans; recycle the can; melt the ore in a factory; transport the can to where it will be used.*

20. **Classify** the following resources as renewable or nonrenewable: sunlight, water, oil, trees, air, coal, soil.

Use the table below to answer question 21.

Recycling Possibilities

Items for Recycling	Recycled Product
Answers will vary.	

21. **Make and Use Tables** Record the items your family throws away for one week. Make a table listing the items that can be recycled and the products that can be made from them.

Performance Activities

22. **Design an Experiment** Lemon juice is an acid. Design an experiment showing the effects of lemon juice on a plant. Relate the results of your experiment to acid rain.

23. **Technical Writing** Write a newspaper article describing an environmental problem in your own community and possible solutions.

24. **Display** Create a poster that lists household hazardous wastes, such as paints and pesticides, and explains how to dispose of them safely.

Applying Math

25. **Saving Trees** If everyone in the United States recycled their newspapers, 500,000 trees would be saved each week. How many trees would be saved in one year?

Use the illustration below to answer question 26.

Conventional car gets 28 mpg

Hybrid car gets 50% better gas mileage

26. **Math in Advertising** The advertisement shown here compares the gas mileage of a conventional compact car with a hybrid gasoline/electric vehicle. What is the gas mileage of the hybrid car?

27. **E-waste** Used computers, TVs, and other electronics contain hazardous materials. Properly disposing of this electronic waste, or e-waste, costs about $52 per ton. In the year 2000, about 4.6 million tons of e-waste ended up in U.S. landfills. How much would it cost to dispose of that much e-waste properly?

Science Online red.msscience.com/chapter_review

chapter 22 Review

Thinking Critically

20. renewable—sunlight, water, trees, air, and soil; nonrenewable—oil and coal

21. Possible answers: Newspapers, aluminum cans, glass bottles, and plastic bottles can be recycled. Products that can be made from them include recycled paper; aluminum cans and other aluminum products; glass for windows and containers; and plastic for bathtubs, containers, and insulation.

Performance Activities

22. Student answers will vary, but one idea is to spray the leaves of a plant with lemon juice and to add the lemon juice to the soil around the plant, just as acid rain would. Students should design the experiment, conduct it, and collect data. The data should be placed in a table. Use **PASC,** p. 95.

23. Student articles will vary. Be sure students include a description of the problem and a possible solution. Use **PASC,** p. 141.

24. Posters should be limited to the proper disposal of hazardous materials commonly used in homes and gardens, including cleaning supplies, paints, solvents, herbicides, pesticides, batteries, used fluorescent tubes, etc.

Applying Math

National Math Standards

1, 2, 9

25. 500,000 trees saved/week × 52 weeks/year = 26,000,000 trees saved per year

26. $28 \times .5 = 14 + 28 = 42$ mpg

27. $52/ton × $(4.6 \times 10^6$ tons$)$ = $239.2 × 10^6 = $239,200,000

Assessment Resources

Reproducible Masters

Chapter *Fast File* Resources
- Chapter Review, pp. 37–38
- Chapter Tests, pp. 39–42
- Assessment Transparency Activity, p. 49

Glencoe Science Web site
- Chapter Review Test
- Standardized Test Practice

Glencoe Technology

- Assessment Transparency
- Exam*View*® Pro Testmaker
- MindJogger Videoquiz
- Interactive Chalkboard

FAST FILE

Answer Sheet A practice answer sheet can be found at red.msscience.com/answer_sheet.

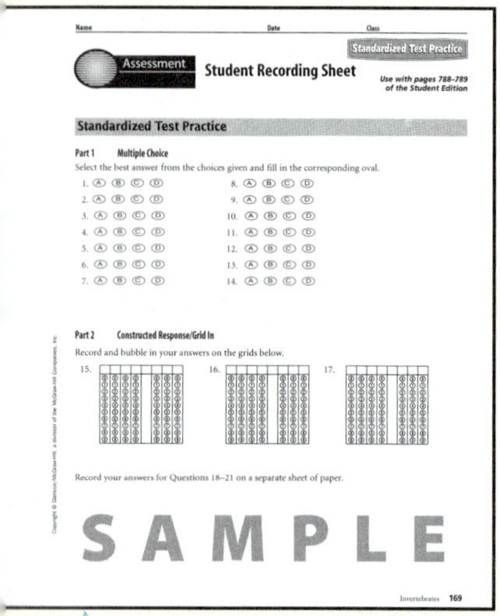
Name Date Class
Standardized Test Practice
Assessment
Student Recording Sheet
Use with pages 788–789 of the Student Edition
Standardized Test Practice
Part 1 Multiple Choice
Select the best answer from the choices given and fill in the corresponding oval.
Part 2 Constructed Response/Grid In
Record and bubble in your answers on the grids below.
Record your answers for Questions 18–21 on a separate sheet of paper.
SAMPLE

Part 1 Multiple Choice

1. D
2. B
3. B
4. A
5. A
6. D

chapter 22 Standardized Test Practice

Part 1 Multiple Choice

Record your answers on the answer sheet provided by your teacher or on a sheet of paper.

Use the illustration below to answer questions 1 and 2.

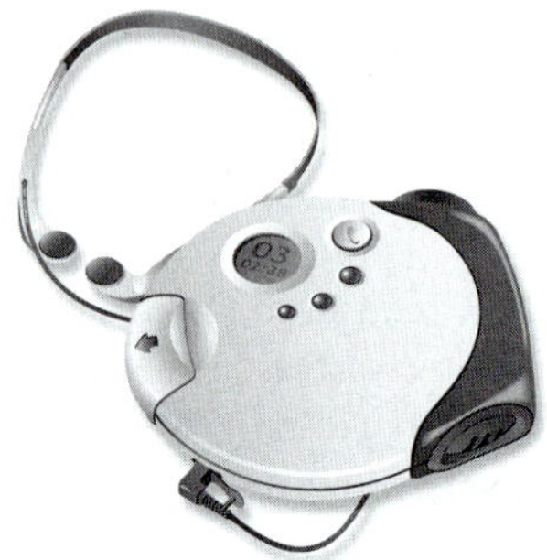

1. Which of the following statements is true?
   - **A.** Only the cardboard used to package the CD player comes from natural resources.
   - **B.** Only the metal used to make the screws and wires in the CD player comes from natural resources.
   - **C.** Everything EXCEPT the plastic used to make and package the CD player comes from natural resources.
   - **D.** The trees and minerals used to make the cardboard, plastic, and metal in the CD player are all natural resources.

2. Which part of the CD player is made from crude oil?
   - **A.** the screws that hold it together
   - **B.** the plastic case
   - **C.** the cardboard packaging
   - **D.** the electrical parts that make it work

Test-Taking Tip

Process of Elimination If you don't know the answer to a multiple-choice question, eliminate as many incorrect answers as possible. Mark your best guess from the remaining answers before moving on the next question.

3. A nonrenewable resource is a resource that cannot be replaced by natural processes in what amount of time?
   - **A.** within any amount of time.
   - **B.** within 100 years.
   - **C.** within 1,000 years.
   - **D.** within 1 million years.

4. Which of the following could be considered a renewable resource?
   - **A.** water
   - **B.** coal
   - **C.** iron ore
   - **D.** land

Use the graph below to answer questions 5 and 6.

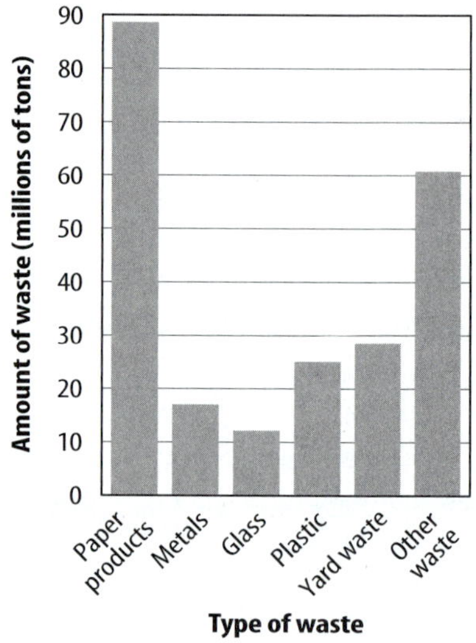

5. According to the graph, what type of waste makes up the greatest amount of solid waste?
   - **A.** paper products
   - **B.** glass
   - **C.** yard waste
   - **D.** other waste

6. About how many millions of tons of solid waste come from plastics?
   - **A.** 28
   - **B.** 17
   - **C.** 12
   - **D.** 23

Part 2 Short Response/Grid In

7. Possible answers: water, land, air, trees, coal, oil
8. A renewable resource can be replaced within 100 years. The process of oil, coal, and natural gas formation takes millions of years.
9. oil, A
10. coal and natural gas; 45%
11. Possible answers: construction of homes, businesses, highways; timber harvesting; clearing land for agriculture
12. The water should be poured into a household drain. That way, it will end up in a water treatment plant.
13. burning fossil fuels
14. when gases released by burning fossil fuels mix with water in the atmosphere
15. Possible answers: glass, metal, plastic, paper products, yard waste

Part 2 | Short Response/Grid In

Record your answers on the answer sheet provided by your teacher or on a sheet of paper.

7. Give three examples of natural resources.

8. Why are coal, natural gas, and oil not considered renewable resources?

Use the graph below to answer questions 9 and 10.

Sources of World Energy

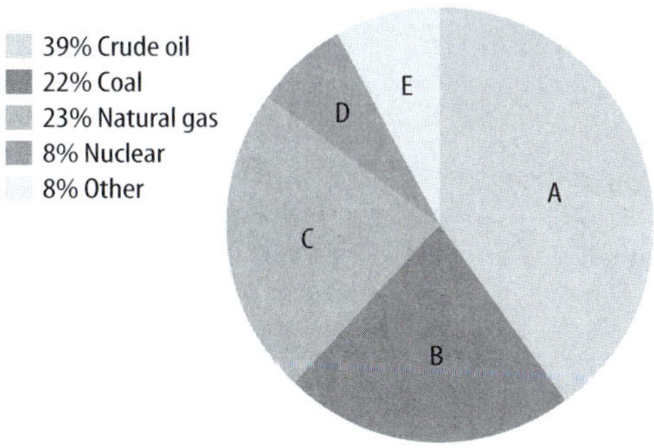

9. Which natural resource supplies more of the world's energy than any other resource? Which section of the graph represents it?

10. According to the graph, what two sources of energy are indicated by sections B and C? What total percentage of the world's energy do they represent?

11. What kinds of human activities are responsible for the destruction of land as a natural resource?

12. Is soapy water or water with household cleaner in it best disposed of by pouring it down the drain or on the ground?

13. What is the source of most air pollution?

14. How does acid rain form?

15. Name three examples of solid waste.

Part 3 | Open Ended

Record your answers on a sheet of paper.

16. Compare and contrast renewable resources with nonrenewable resources.

17. Draw and label a diagram showing a side-view of a landfill.

18. Draw a diagram showing how rainwater that washes down a city drain ends up as clean drinking water.

19. Discuss why you might expect to find more air pollution in a city area than a rural area.

Use the illustration below to answer questions 20 and 21.

20. Describe one way you can reuse each of the items shown above.

21. Describe ways you can reduce your use of one or more of the items shown above.

22. Describe how you and your friends can decrease the amount of solid waste that goes into your local landfill. In your answer, discuss how you would handle empty plastic soda bottles, cardboard packaging, food scraps, and used notebook paper.

23. You are planning a picnic. What are the advantages and disadvantages of using paper plates and cups and disposable plastic forks and spoons? What are the advantages and disadvantages of using washable, reusable plastic dishes, forks, and spoons?

Science Online red.msscience.com/standardized_test

Part 3 | Open Ended

16. Renewable resources, such as trees, sunlight, and water, can either be replaced in 100 years or less, have supplies that are constantly renewed (water), or will last for millions of years (sunlight). Nonrenewable resources, like coal, oil, or natural gas can take millions of years to form.

17. Diagrams should show a large space filled with solid waste and lined with clay or plastic.

18. Diagrams should include rainwater washing down a drain, pipes leading from the drain to a water treatment plant, pipes leading from the plant to homes, a faucet in a home where water is used for drinking.

19. Cities have more people and, therefore, more automobiles, factories, and power plants than rural areas.

20. Possible answers: milk container: use to store water for watering houseplants or cut off bottom and use as a scoop for bulk pet food; spaghetti sauce jar: use as a vase for flowers or container for leftover foods; newspapers: use to line pet cages or wrap gifts.

21. Possible answers: buy milk in 1-gallon size or in returnable glass containers; make your own spaghetti sauce from fresh tomatoes and freeze in reusable plastic containers; buy newspapers only when you know you have time to read them, read the newspaper online.

22. Answers should include examples of reducing, reusing, and recycling.

23. Possible answers: Paper is made from a renewable resource, but is usually used only once and then added to landfill. Plastic is made from a nonrenewable resource, but can be reused many times.

Rubrics

For more help evaluating open-ended assessment questions, see the rubric on p. 10T.

Student Resources

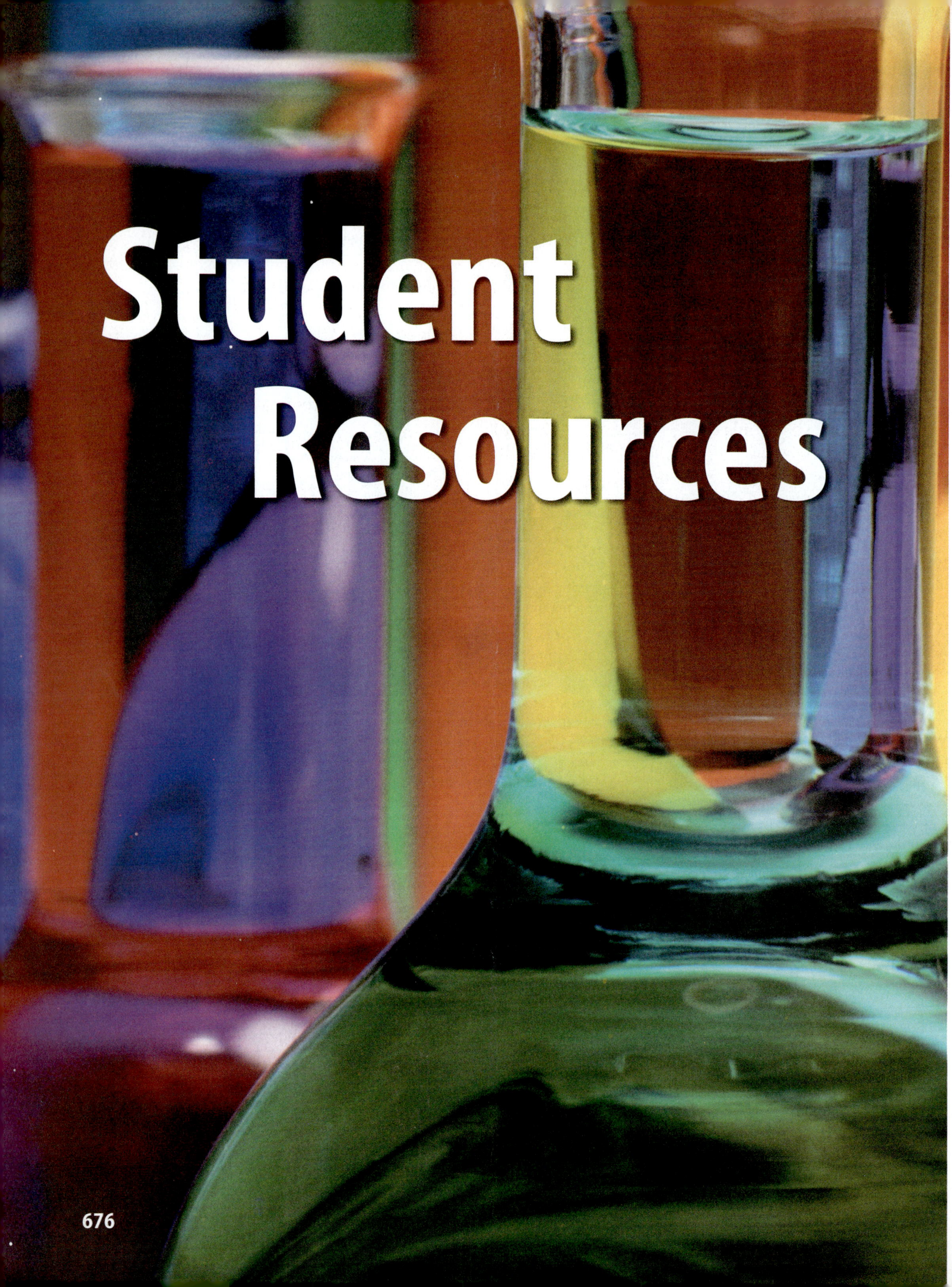

CONTENTS

Science Skill Handbook

Scientific Methods

Scientists use an orderly approach called scientific methods to solve problems. These include organizing and recording data so others can understand them. Scientists use many variations in these methods when they solve problems.

Identify a Question

The first step in a scientific investigation or experiment is to identify a question to be answered or a problem to be solved. For example, you might ask which gasoline is the most efficient.

Gather and Organize Information

After you have identified your question, begin gathering and organizing information. There are many ways to gather information, such as researching in a library, interviewing those knowledgeable about the subject, testing and working in the laboratory and field. Fieldwork is investigations and observations done outside of a laboratory.

Researching Information Before moving in a new direction, it is important to gather the information that already is known about the subject. Start by asking yourself questions to determine exactly what you need to know. Then you will look for the information in various reference sources, like the student is doing in **Figure 1.** Some sources may include textbooks, encyclopedias, government documents, professional journals, science magazines, and the Internet. Always list the sources of your information.

Figure 1 The Internet can be a valuable research tool.

Evaluate Sources of Information Not all sources of information are reliable. You should evaluate all of your sources of information, and use only those you know to be dependable. For example, if you are researching ways to make homes more energy efficient, a site written by the U.S. Department of Energy would be more reliable than a site written by a company that is trying to sell a new type of weatherproofing material. Also, remember that research always is changing. Consult the most current resources available to you. For example, a 1985 resource about saving energy would not reflect the most recent findings.

Sometimes scientists use data that they did not collect themselves, or conclusions drawn by other researchers. This data must be evaluated carefully. Ask questions about how the data were obtained, if the investigation was carried out properly, and if it has been duplicated exactly with the same results. Would you reach the same conclusion from the data? Only when you have confidence in the data can you believe it is true and feel comfortable using it.

Interpret Scientific Illustrations As you research a topic in science, you will see drawings, diagrams, and photographs to help you understand what you read. Some illustrations are included to help you understand an idea that you can't see easily by yourself, like the tiny particles in an atom in **Figure 2.** A drawing helps many people to remember details more easily and provides examples that clarify difficult concepts or give additional information about the topic you are studying. Most illustrations have labels or a caption to identify or to provide more information.

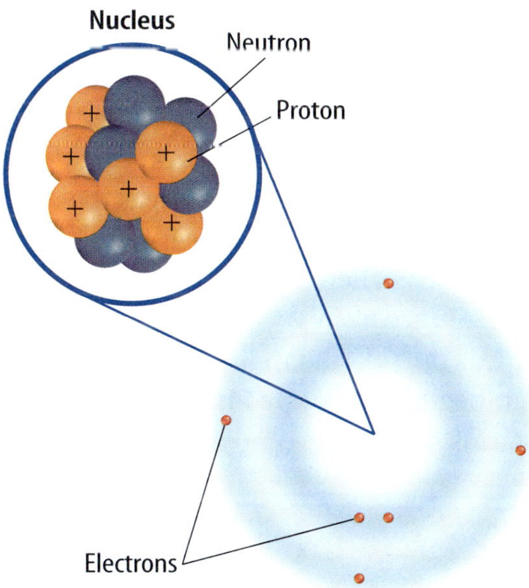

Figure 2 This drawing shows an atom of carbon with its six protons, six neutrons, and six electrons.

Concept Maps One way to organize data is to draw a diagram that shows relationships among ideas (or concepts). A concept map can help make the meanings of ideas and terms more clear, and help you understand and remember what you are studying. Concept maps are useful for breaking large concepts down into smaller parts, making learning easier.

Network Tree A type of concept map that not only shows a relationship, but how the concepts are related is a network tree, shown in **Figure 3.** In a network tree, the words are written in the ovals, while the description of the type of relationship is written across the connecting lines.

When constructing a network tree, write down the topic and all major topics on separate pieces of paper or notecards. Then arrange them in order from general to specific. Branch the related concepts from the major concept and describe the relationship on the connecting line. Continue to more specific concepts until finished.

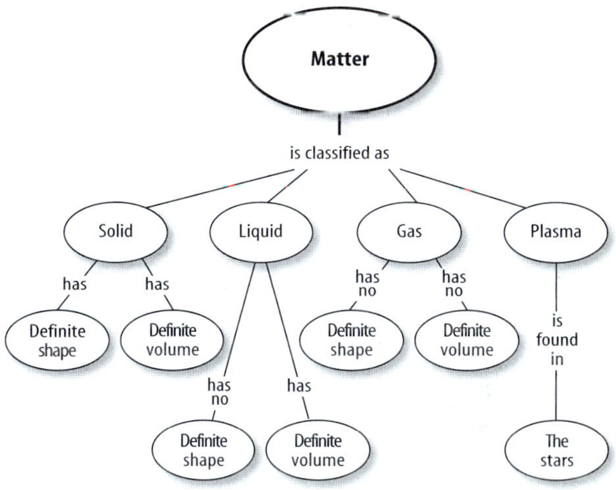

Figure 3 A network tree shows how concepts or objects are related.

Events Chain Another type of concept map is an events chain. Sometimes called a flow chart, it models the order or sequence of items. An events chain can be used to describe a sequence of events, the steps in a procedure, or the stages of a process.

When making an events chain, first find the one event that starts the chain. This event is called the initiating event. Then, find the next event and continue until the outcome is reached, as shown in **Figure 4.**

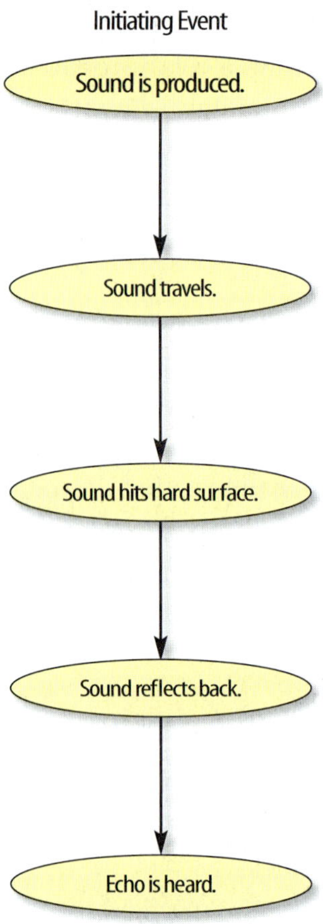

Figure 4 Events-chain concept maps show the order of steps in a process or event. This concept map shows how a sound makes an echo.

Cycle Map A specific type of events chain is a cycle map. It is used when the series of events do not produce a final outcome, but instead relate back to the beginning event, such as in **Figure 5.** Therefore, the cycle repeats itself.

To make a cycle map, first decide what event is the beginning event. This is also called the initiating event. Then list the next events in the order that they occur, with the last event relating back to the initiating event. Words can be written between the events that describe what happens from one event to the next. The number of events in a cycle map can vary, but usually contain three or more events.

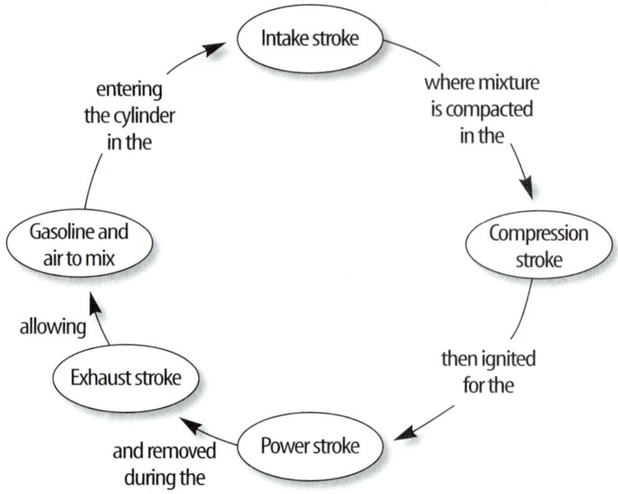

Figure 5 A cycle map shows events that occur in a cycle.

Spider Map A type of concept map that you can use for brainstorming is the spider map. When you have a central idea, you might find that you have a jumble of ideas that relate to it but are not necessarily clearly related to each other. The spider map on sound in **Figure 6** shows that if you write these ideas outside the main concept, then you can begin to separate and group unrelated terms so they become more useful.

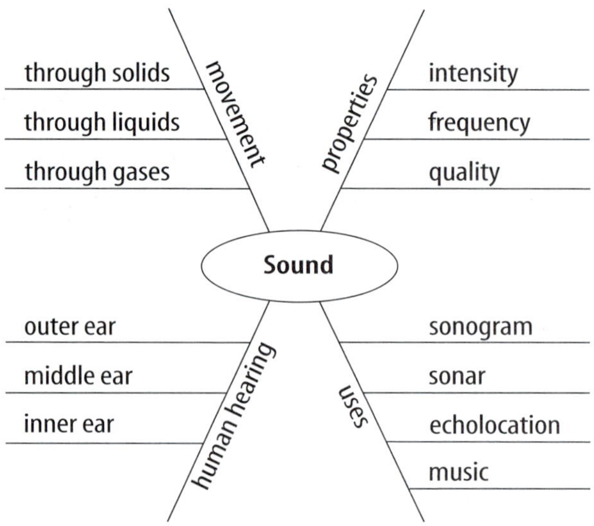

Figure 6 A spider map allows you to list ideas that relate to a central topic but not necessarily to one another.

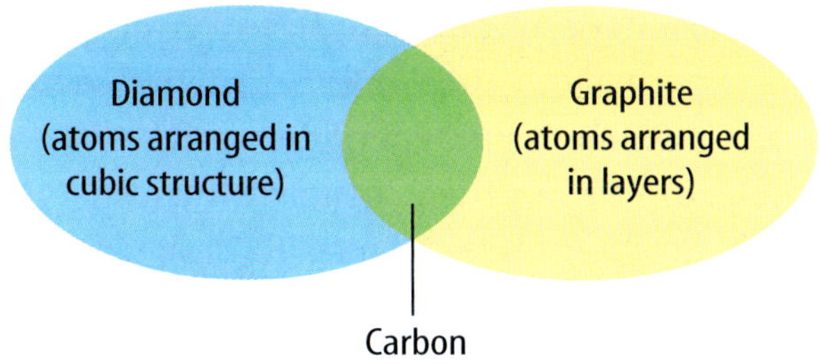

Figure 7 This Venn diagram compares and contrasts two substances made from carbon.

Venn Diagram To illustrate how two subjects compare and contrast you can use a Venn diagram. You can see the characteristics that the subjects have in common and those that they do not, shown in **Figure 7.**

To create a Venn diagram, draw two overlapping ovals that that are big enough to write in. List the characteristics unique to one subject in one oval, and the characteristics of the other subject in the other oval. The characteristics in common are listed in the overlapping section.

Make and Use Tables One way to organize information so it is easier to understand is to use a table. Tables can contain numbers, words, or both.

To make a table, list the items to be compared in the first column and the characteristics to be compared in the first row. The title should clearly indicate the content of the table, and the column or row heads should be clear. Notice that in **Table 1** the units are included.

Table 1 Recyclables Collected During Week

Day of Week	Paper (kg)	Aluminum (kg)	Glass (kg)
Monday	5.0	4.0	12.0
Wednesday	4.0	1.0	10.0
Friday	2.5	2.0	10.0

Make a Model One way to help you better understand the parts of a structure, the way a process works, or to show things too large or small for viewing is to make a model. For example, an atomic model made of a plastic-ball nucleus and pipe-cleaner electron shells can help you visualize how the parts of an atom relate to each other. Other types of models can by devised on a computer or represented by equations.

Form a Hypothesis

A possible explanation based on previous knowledge and observations is called a hypothesis. After researching gasoline types and recalling previous experiences in your family's car you form a hypothesis—our car runs more efficiently because we use premium gasoline. To be valid, a hypothesis has to be something you can test by using an investigation.

Predict When you apply a hypothesis to a specific situation, you predict something about that situation. A prediction makes a statement in advance, based on prior observation, experience, or scientific reasoning. People use predictions to make everyday decisions. Scientists test predictions by performing investigations. Based on previous observations and experiences, you might form a prediction that cars are more efficient with premium gasoline. The prediction can be tested in an investigation.

Design an Experiment A scientist needs to make many decisions before beginning an investigation. Some of these include: how to carry out the investigation, what steps to follow, how to record the data, and how the investigation will answer the question. It also is important to address any safety concerns.

Test the Hypothesis

Now that you have formed your hypothesis, you need to test it. Using an investigation, you will make observations and collect data, or information. This data might either support or not support your hypothesis. Scientists collect and organize data as numbers and descriptions.

Follow a Procedure In order to know what materials to use, as well as how and in what order to use them, you must follow a procedure. **Figure 8** shows a procedure you might follow to test your hypothesis.

Procedure

1. Use regular gasoline for two weeks.
2. Record the number of kilometers between fill-ups and the amount of gasoline used.
3. Switch to premium gasoline for two weeks.
4. Record the number of kilometers between fill-ups and the amount of gasoline used.

Figure 8 A procedure tells you what to do step by step.

Identify and Manipulate Variables and Controls In any experiment, it is important to keep everything the same except for the item you are testing. The one factor you change is called the independent variable. The change that results is the dependent variable. Make sure you have only one independent variable, to assure yourself of the cause of the changes you observe in the dependent variable. For example, in your gasoline experiment the type of fuel is the independent variable. The dependent variable is the efficiency.

Many experiments also have a control—an individual instance or experimental subject for which the independent variable is not changed. You can then compare the test results to the control results. To design a control you can have two cars of the same type. The control car uses regular gasoline for four weeks. After you are done with the test, you can compare the experimental results to the control results.

Collect Data

Whether you are carrying out an investigation or a short observational experiment, you will collect data, as shown in **Figure 9.** Scientists collect data as numbers and descriptions and organize it in specific ways.

Observe Scientists observe items and events, then record what they see. When they use only words to describe an observation, it is called qualitative data. Scientists' observations also can describe how much there is of something. These observations use numbers, as well as words, in the description and are called quantitative data. For example, if a sample of the element gold is described as being "shiny and very dense" the data are qualitative. Quantitative data on this sample of gold might include "a mass of 30 g and a density of 19.3 g/cm^3."

Figure 9 Collecting data is one way to gather information directly.

Figure 10 Record data neatly and clearly so it is easy to understand.

When you make observations you should examine the entire object or situation first, and then look carefully for details. It is important to record observations accurately and completely. Always record your notes immediately as you make them, so you do not miss details or make a mistake when recording results from memory. Never put unidentified observations on scraps of paper. Instead they should be recorded in a notebook, like the one in **Figure 10.** Write your data neatly so you can easily read it later. At each point in the experiment, record your observations and label them. That way, you will not have to determine what the figures mean when you look at your notes later. Set up any tables that you will need to use ahead of time, so you can record any observations right away. Remember to avoid bias when collecting data by not including personal thoughts when you record observations. Record only what you observe.

Estimate Scientific work also involves estimating. To estimate is to make a judgment about the size or the number of something without measuring or counting. This is important when the number or size of an object or population is too large or too difficult to accurately count or measure.

Sample Scientists may use a sample or a portion of the total number as a type of estimation. To sample is to take a small, representative portion of the objects or organisms of a population for research. By making careful observations or manipulating variables within that portion of the group, information is discovered and conclusions are drawn that might apply to the whole population. A poorly chosen sample can be unrepresentative of the whole. If you were trying to determine the rainfall in an area, it would not be best to take a rainfall sample from under a tree.

Measure You use measurements everyday. Scientists also take measurements when collecting data. When taking measurements, it is important to know how to use measuring tools properly. Accuracy also is important.

Length To measure length, the distance between two points, scientists use meters. Smaller measurements might be measured in centimeters or millimeters.

Length is measured using a metric ruler or meter stick. When using a metric ruler, line up the 0-cm mark with the end of the object being measured and read the number of the unit where the object ends. Look at the metric ruler shown in **Figure 11.** The centimeter lines are the long, numbered lines, and the shorter lines are millimeter lines. In this instance, the length would be 4.50 cm.

Figure 11 This metric ruler has centimeter and millimeter divisions.

Mass The SI unit for mass is the kilogram (kg). Scientists can measure mass using units formed by adding metric prefixes to the unit gram (g), such as milligram (mg). To measure mass, you might use a triple-beam balance similar to the one shown in **Figure 12.** The balance has a pan on one side and a set of beams on the other side. Each beam has a rider that slides on the beam.

When using a triple-beam balance, place an object on the pan. Slide the largest rider along its beam until the pointer drops below zero. Then move it back one notch. Repeat the process for each rider proceeding from the larger to smaller until the pointer swings an equal distance above and below the zero point. Sum the masses on each beam to find the mass of the object. Move all riders back to zero when finished.

Instead of putting materials directly on the balance, scientists often take a tare of a container. A tare is the mass of a container into which objects or substances are placed for measuring their masses. To mass objects or substances, find the mass of a clean container. Remove the container from the pan, and place the object or substances in the container. Find the mass of the container with the materials in it. Subtract the mass of the empty container from the mass of the filled container to find the mass of the materials you are using.

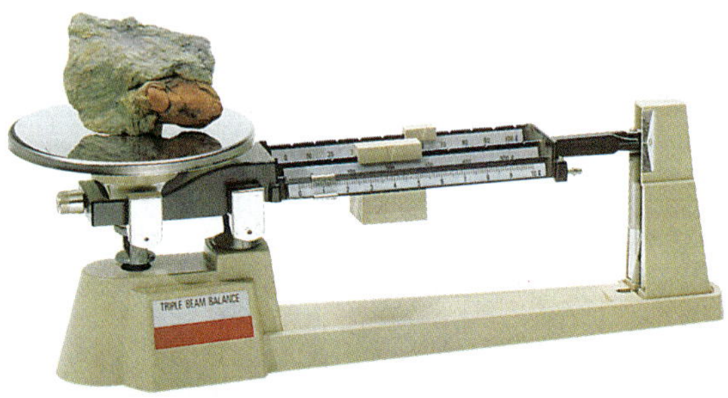

Figure 12 A triple-beam balance is used to determine the mass of an object.

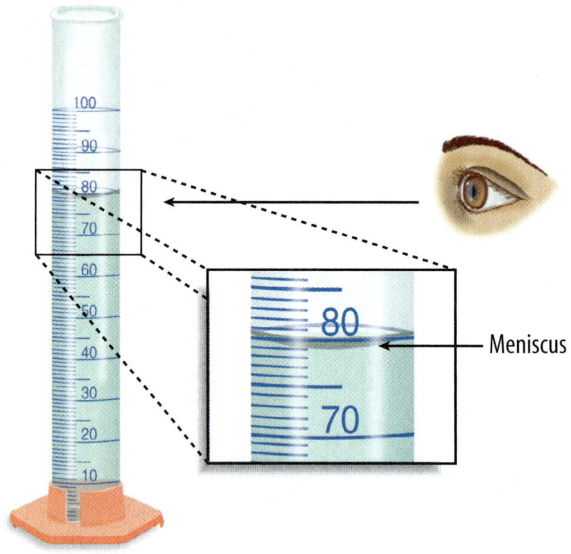

Figure 13 Graduated cylinders measure liquid volume.

Liquid Volume To measure liquids, the unit used is the liter. When a smaller unit is needed, scientists might use a milliliter. Because a milliliter takes up the volume of a cube measuring 1 cm on each side it also can be called a cubic centimeter ($cm^3 = cm \times cm \times cm$).

You can use beakers and graduated cylinders to measure liquid volume. A graduated cylinder, shown in **Figure 13,** is marked from bottom to top in milliliters. In lab, you might use a 10-mL graduated cylinder or a 100-mL graduated cylinder. When measuring liquids, notice that the liquid has a curved surface. Look at the surface at eye level, and measure the bottom of the curve. This is called the meniscus. The graduated cylinder in **Figure 13** contains 79.0 mL, or 79.0 cm^3, of a liquid.

Temperature Scientists often measure temperature using the Celsius scale. Pure water has a freezing point of 0°C and boiling point of 100°C. The unit of measurement is degrees Celsius. Two other scales often used are the Fahrenheit and Kelvin scales.

Figure 14 A thermometer measures the temperature of an object.

Scientists use a thermometer to measure temperature. Most thermometers in a laboratory are glass tubes with a bulb at the bottom end containing a liquid such as colored alcohol. The liquid rises or falls with a change in temperature. To read a glass thermometer like the thermometer in **Figure 14,** rotate it slowly until a red line appears. Read the temperature where the red line ends.

Form Operational Definitions An operational definition defines an object by how it functions, works, or behaves. For example, when you are playing hide and seek and a tree is home base, you have created an operational definition for a tree.

Objects can have more than one operational definition. For example, a ruler can be defined as a tool that measures the length of an object (how it is used). It can also be a tool with a series of marks used as a standard when measuring (how it works).

Analyze the Data

To determine the meaning of your observations and investigation results, you will need to look for patterns in the data. Then you must think critically to determine what the data mean. Scientists use several approaches when they analyze the data they have collected and recorded. Each approach is useful for identifying specific patterns.

Interpret Data The word *interpret* means "to explain the meaning of something." When analyzing data from an experiement, try to find out what the data show. Identify the control group and the test group to see whether or not changes in the independent variable have had an effect. Look for differences in the dependent variable between the control and test groups.

Classify Sorting objects or events into groups based on common features is called classifying. When classifying, first observe the objects or events to be classified. Then select one feature that is shared by some members in the group, but not by all. Place those members that share that feature in a subgroup. You can classify members into smaller and smaller subgroups based on characteristics. Remember that when you classify, you are grouping objects or events for a purpose. Keep your purpose in mind as you select the features to form groups and subgroups.

Compare and Contrast Observations can be analyzed by noting the similarities and differences between two more objects or events that you observe. When you look at objects or events to see how they are similar, you are comparing them. Contrasting is looking for differences in objects or events.

Recognize Cause and Effect A cause is a reason for an action or condition. The effect is that action or condition. When two events happen together, it is not necessarily true that one event caused the other. Scientists must design a controlled investigation to recognize the exact cause and effect.

Draw Conclusions

When scientists have analyzed the data they collected, they proceed to draw conclusions about the data. These conclusions are sometimes stated in words similar to the hypothesis that you formed earlier. They may confirm a hypothesis, or lead you to a new hypothesis.

Infer Scientists often make inferences based on their observations. An inference is an attempt to explain observations or to indicate a cause. An inference is not a fact, but a logical conclusion that needs further investigation. For example, you may infer that a fire has caused smoke. Until you investigate, however, you do not know for sure.

Apply When you draw a conclusion, you must apply those conclusions to determine whether the data supports the hypothesis. If your data do not support your hypothesis, it does not mean that the hypothesis is wrong. It means only that the result of the investigation did not support the hypothesis. Maybe the experiment needs to be redesigned, or some of the initial observations on which the hypothesis was based were incomplete or biased. Perhaps more observation or research is needed to refine your hypothesis. A successful investigation does not always come out the way you originally predicted.

Avoid Bias Sometimes a scientific investigation involves making judgments. When you make a judgment, you form an opinion. It is important to be honest and not to allow any expectations of results to bias your judgments. This is important throughout the entire investigation, from researching to collecting data to drawing conclusions.

Communicate

The communication of ideas is an important part of the work of scientists. A discovery that is not reported will not advance the scientific community's understanding or knowledge. Communication among scientists also is important as a way of improving their investigations.

Scientists communicate in many ways, from writing articles in journals and magazines that explain their investigations and experiments, to announcing important discoveries on television and radio. Scientists also share ideas with colleagues on the Internet or present them as lectures, like the student is doing in **Figure 15.**

Figure 15 A student communicates to his peers about his investigation.

SAFETY SYMBOLS	HAZARD	EXAMPLES	PRECAUTION	REMEDY
DISPOSAL	Special disposal procedures need to be followed.	certain chemicals, living organisms	Do not dispose of these materials in the sink or trash can.	Dispose of wastes as directed by your teacher.
BIOLOGICAL	Organisms or other biological materials that might be harmful to humans	bacteria, fungi, blood, unpreserved tissues, plant materials	Avoid skin contact with these materials. Wear mask or gloves.	Notify your teacher if you suspect contact with material. Wash hands thoroughly.
EXTREME TEMPERATURE	Objects that can burn skin by being too cold or too hot	boiling liquids, hot plates, dry ice, liquid nitrogen	Use proper protection when handling.	Go to your teacher for first aid.
SHARP OBJECT	Use of tools or glassware that can easily puncture or slice skin	razor blades, pins, scalpels, pointed tools, dissecting probes, broken glass	Practice common-sense behavior and follow guidelines for use of the tool.	Go to your teacher for first aid.
FUME	Possible danger to respiratory tract from fumes	ammonia, acetone, nail polish remover, heated sulfur, moth balls	Make sure there is good ventilation. Never smell fumes directly. Wear a mask.	Leave foul area and notify your teacher immediately.
ELECTRICAL	Possible danger from electrical shock or burn	improper grounding, liquid spills, short circuits, exposed wires	Double-check setup with teacher. Check condition of wires and apparatus.	Do not attempt to fix electrical problems. Notify your teacher immediately.
IRRITANT	Substances that can irritate the skin or mucous membranes of the respiratory tract	pollen, moth balls, steel wool, fiberglass, potassium permanganate	Wear dust mask and gloves. Practice extra care when handling these materials.	Go to your teacher for first aid.
CHEMICAL	Chemicals can react with and destroy tissue and other materials	bleaches such as hydrogen peroxide; acids such as sulfuric acid, hydrochloric acid; bases such as ammonia, sodium hydroxide	Wear goggles, gloves, and an apron.	Immediately flush the affected area with water and notify your teacher.
TOXIC	Substance may be poisonous if touched, inhaled, or swallowed.	mercury, many metal compounds, iodine, poinsettia plant parts	Follow your teacher's instructions.	Always wash hands thoroughly after use. Go to your teacher for first aid.
FLAMMABLE	Flammable chemicals may be ignited by open flame, spark, or exposed heat.	alcohol, kerosene, potassium permanganate	Avoid open flames and heat when using flammable chemicals.	Notify your teacher immediately. Use fire safety equipment if applicable.
OPEN FLAME	Open flame in use, may cause fire.	hair, clothing, paper, synthetic materials	Tie back hair and loose clothing. Follow teacher's instruction on lighting and extinguishing flames.	Notify your teacher immediately. Use fire safety equipment if applicable.

Eye Safety
Proper eye protection should be worn at all times by anyone performing or observing science activities.

Clothing Protection
This symbol appears when substances could stain or burn clothing.

Animal Safety
This symbol appears when safety of animals and students must be ensured.

Handwashing
After the lab, wash hands with soap and water before removing goggles.

Safety in the Science Laboratory

The science laboratory is a safe place to work if you follow standard safety procedures. Being responsible for your own safety helps to make the entire laboratory a safer place for everyone. When performing any lab, read and apply the caution statements and safety symbol listed at the beginning of the lab.

General Safety Rules

1. Obtain your teacher's permission to begin all investigations and use laboratory equipment.
2. Study the procedure. Ask your teacher any questions. Be sure you understand safety symbols shown on the page.
3. Notify your teacher about allergies or other health conditions which can affect your participation in a lab.
4. Learn and follow use and safety procedures for your equipment. If unsure, ask your teacher.

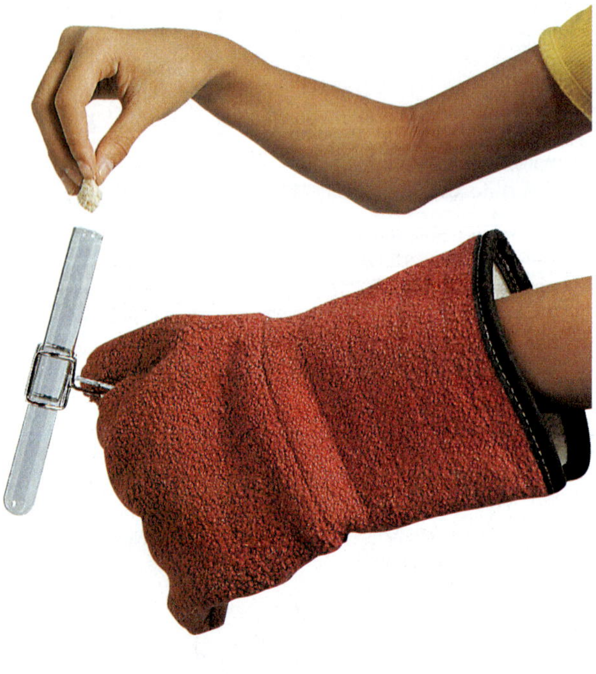

5. Never eat, drink, chew gum, apply cosmetics, or do any personal grooming in the lab. Never use lab glassware as food or drink containers. Keep your hands away from your face and mouth.
6. Know the location and proper use of the safety shower, eye wash, fire blanket, and fire alarm.

Prevent Accidents

1. Use the safety equipment provided to you. Goggles and a safety apron should be worn during investigations.
2. Do NOT use hair spray, mousse, or other flammable hair products. Tie back long hair and tie down loose clothing.
3. Do NOT wear sandals or other open-toed shoes in the lab.
4. Remove jewelry on hands and wrists. Loose jewelry, such as chains and long necklaces, should be removed to prevent them from getting caught in equipment.
5. Do not taste any substances or draw any material into a tube with your mouth.
6. Proper behavior is expected in the lab. Practical jokes and fooling around can lead to accidents and injury.
7. Keep your work area uncluttered.

Laboratory Work

1. Collect and carry all equipment and materials to your work area before beginning a lab.
2. Remain in your own work area unless given permission by your teacher to leave it.

3. Always slant test tubes away from yourself and others when heating them, adding substances to them, or rinsing them.
4. If instructed to smell a substance in a container, hold the container a short distance away and fan vapors towards your nose.
5. Do NOT substitute other chemicals/substances for those in the materials list unless instructed to do so by your teacher.
6. Do NOT take any materials or chemicals outside of the laboratory.
7. Stay out of storage areas unless instructed to be there and supervised by your teacher.

Laboratory Cleanup

1. Turn off all burners, water, and gas, and disconnect all electrical devices.
2. Clean all pieces of equipment and return all materials to their proper places.
3. Dispose of chemicals and other materials as directed by your teacher. Place broken glass and solid substances in the proper containers. Never discard materials in the sink.
4. Clean your work area.
5. Wash your hands with soap and water thoroughly BEFORE removing your goggles.

Emergencies

1. Report any fire, electrical shock, glassware breakage, spill, or injury, no matter how small, to your teacher immediately. Follow his or her instructions.
2. If your clothing should catch fire, STOP, DROP, and ROLL. If possible, smother it with the fire blanket or get under a safety shower. NEVER RUN.
3. If a fire should occur, turn off all gas and leave the room according to established procedures.
4. In most instances, your teacher will clean up spills. Do NOT attempt to clean up spills unless you are given permission and instructions to do so.
5. If chemicals come into contact with your eyes or skin, notify your teacher immediately. Use the eyewash or flush your skin or eyes with large quantities of water.
6. The fire extinguisher and first-aid kit should only be used by your teacher unless it is an extreme emergency and you have been given permission.
7. If someone is injured or becomes ill, only a professional medical provider or someone certified in first aid should perform first-aid procedures.

1 Testing Horoscopes

Time Required 45 minutes

Materials

- Horoscopes can be found in many magazines and newspapers.
- Students can white out zodiac signs and birth dates with correction fluid, but caution them to be thorough to prevent cheating.

Safety Precaution Students should not interview strangers.

Teaching Strategies

- Encourage students to collect a large sample of data to ensure accurate results.
- Challenge students to establish experimental controls that prevent cheating.

Conclude and Apply

1. Answers will vary but should be about 8 percent.
2. 8.3%

Extra Try at Home Labs

Extra Try at Home Labs

EXTRA Try at Home Labs

From Your Kitchen, Junk Drawer, or Yard

1 Testing Horoscopes

Real-World Question

How can horoscopes be tested scientifically?

Possible Materials

- horoscope from previous week
- scissors
- transparent tape
- white paper
- correction fluid

Procedure

1. Obtain a horoscope from last week and cut out the predictions for each sign. Do not cut out the zodiac signs or birth dates accompanying each prediction.
2. As you cut out a horoscope prediction, write the correct zodiac sign on the back of each prediction.
3. Develop a code for the predictions to allow you to identify them. Keep your code list in your Science Journal.
4. Scramble your predictions and tape them to a sheet of white paper. Write each prediction's code above it.
5. Ask your friends and family members to read all the predictions and choose the one that best matched their life events from the previous week. Interview at least 20 people.

Conclude and Apply

1. Calculate the percentage of people who chose the correct sign.
2. Calculate the chances of a person choosing their correct sign randomly.

2 Disappearing Water?

Real-World Question

How much difference does the type of measuring equipment make?

Possible Materials

- scale
- water
- measuring cups of different sizes
- measuring spoons (1 tsp = 5 mL, 1 tbsp = 15 mL)

Procedure

1. Measure out 83 mL of water using one of the measuring devices. Transfer this amount of water to the other measuring devices.
2. Record the readings for each measuring device for the same amount of water. Do they all give the same reading, or does it seem like the amount of water changed?
3. Remember, 1 mL of water weighs 1 g. Use the scale to find out what the true amount of water is in the container.
4. Repeat steps 1–3 for different amounts of water. Try 50 mL, 128 mL, and 12 mL.

Conclude and Apply

1. Which measuring device was the most accurate? The least?
2. Which measuring device was the most precise? The least?
3. What problem came up when you had to use the small devices several times to get up to a larger amount of water?

Adult supervision required for all labs.

These labs are available at red.msscience.com.

2 Disappearing Water?

Time Required one class period

Teaching Strategy Encourage students to try measuring first with the large glassware, and transferring it down to smaller glassware, and then to try measuring first with the smaller glassware and transferring up to larger glassware. They should notice right away that the smaller glassware gives more precise readings.

Conclude and Apply

1. The smallest measuring spoon should be the most accurate and the largest measuring cup the least, depending on the quality of glassware.
2. The smallest measuring spoon should be the most precise and the largest measuring cup the least, depending on the quality of glassware.
3. reader error

Extra Try at Home Labs

3 Materials Matter

Real-World Question

Which materials will react together? How can materials change?

Possible Materials

- hotplate
- rusty nail
- shiny new nail
- baking soda
- vinegar
- flour
- water
- salt
- chalk
- aluminum foil

Procedure

1. Experiment and record as many physical and chemical changes as possible in 45 min. Prepare your data charts in advance. Try to be efficient rather than speedy. Accurate observations are important.
2. Use at least three words to describe the physical properties of each material. If you know any chemical properties of the materials, add those to your chart.
3. Make physical changes to as many materials as you can. Combine materials to make chemical changes.
4. Describe your observations in the chart.

Conclude and Apply

1. How is knowledge about physical and chemical properties used in building and manufacturing in the real world? Give examples.
2. Were you satisfied with your lab method? What would you do to make your work better next time?

4 Comparing Atom Sizes

Real-World Question

How do the sizes of different types of atoms compare?

Possible Materials

- metric ruler or meterstick
- 1-m length of white paper
- transparent or masking tape
- colored pencils

Procedure

1. Tape a 1-m sheet of paper on the floor.
2. Use a scale of 1 mm: 1 picometer for measuring and drawing the relative diameters of all the atoms.
3. Study the chart of atomic sizes.
4. Use your scale to measure the relative size of a hydrogen atom on the sheet of paper. Use a red pencil to draw the relative diameter of a hydrogen atom on your paper.
5. Use your scale to measure the relative sizes of an oxygen atom, iron atom, gold atom, and francium atom. Use four other colored pencils to draw the relative diameters of these atoms on the paper.
6. Compare the relative sizes of these different atoms.

Atomic Sizes (picometers)

Element	Diameter
Hydrogen	50
Oxygen	146
Iron	248
Gold	288
Francium	540

Conclude and Apply

1. Research the length of a picometer.
2. Using your scale, list the diameters of the atoms that you drew on your paper.

3 Materials Matter

Time Required one class period

Materials

- Provide cups and sticks for mixing.
- You can substitute or add any suitable, safe material that would lend itself to an easy physical or chemical change.

Safety Precaution As students will likely be excited by the speed aspect of this lab, stress the importance of being careful to wipe up any spills, and not running in the lab.

Teaching Strategies

- This lab can be easily made into a contest. Challenge students to see who can get the most combinations of physical and chemical changes in the time allowed.
- A class discussion about the Conclude and Apply questions would be appropriate.

Conclude and Apply

1. Many answers are possible: rusting bridges, strength of wood and steel for beams, insulating properties of insulation and air spaces in windows, etc.
2. Possible answers include organizing or preparing better, spilling less, having a different lab partner.

4 Comparing Atom Sizes

Time Required 25 minutes

Safety Precaution Caution students to handle the metersticks with care.

Teaching Strategies

- Instruct students to start measuring the relative diameters at one end of the paper.
- If necessary, remind students that these measurements are relative diameters not the actual diameters of atoms.

Conclude and Apply

1. 1×10^{-12} m
2. hydrogen: 7.4 cm; oxygen: 14.6 cm; iron: 24.8 cm; gold: 28.8 cm; and francium: 54.0 cm

5 Mechanical Advantage

Time Required 30 minutes

Materials For an in-class lab, collect a wide variety of screwdrivers such as small computer screwdrivers and large screwdrivers used in carpentry work.

Safety Precaution Caution students to handle the tools with care, especially the sharp blades of the can opener.

Teaching Strategy Divide your class into small groups and rotate the groups through different stations. Place several examples of the same wheel and axle tool at each station.

Conclude and Apply

1. Answers will vary. Clamps will have a M.A. range of 5–15. The M.A. for screwdrivers ranges between 4–7. Rolling pins have a M.A. of 6–7. Manual can openers have a M.A. of 4–5.
2. The wheel is turned a much greater distance than the axle, which multiplies the force of the machine.

Extra Try at Home Labs

5 Mechanical Advantage

Real-World Question

What are the mechanical advantages of the wheels and axles in your home?

Possible Materials

- handheld can opener
- several screwdrivers
- rolling pin
- pizza cutter
- doorknob
- metric ruler

Procedure

1. Search your home's kitchen, closets, and toolbox to collect several wheels and axles such as a screwdriver, manual can opener, rolling pin, and pizza cutter.
2. Observe each wheel and axle and identify the wheel and the axle of each tool.
3. Measure the radius of the axle of each tool. Record your measurements in your Science Journal.
4. Measure the radius of the wheel of each tool. Record your measurements in your Science Journal. Be certain to use the same units for both measurements.

Conclude and Apply

1. Calculate the mechanical advantage of each wheel and axle tool.
2. Infer why a wheel and axle gives you a mechanical advantage.

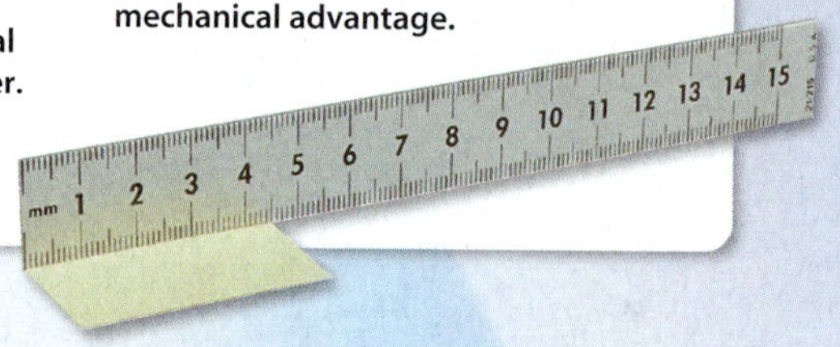

6 Thermal Propulsion

Real-World Question

What has more molecular kinetic energy, hot air or cold air?

Possible Materials

- equal-size, oblong balloons (3)
- cotton thread
- desk lamp
- freezer
- scissors

Procedure

1. Inflate each of the three balloons to the same length, but do not over-inflate them.
2. Knot the end of each balloon and tie it off tightly with a string. Tie a string around the middle of each balloon so it is snug, but not overly tight.
3. Overnight, leave one in a freezer, one under a desk lamp, and one in a dark place.
4. The next day, write down your observations of each of the balloons before taking them from their locations.
5. Do this step quickly. With the help of two friends, each of you take a balloon. Go to an indoor area with a high ceiling. Twist and pinch each balloon above the tied end and cut off the tied end, such that no air escapes. Hold the balloons pointing up in front of you and count down, "3, 2, 1, Go!" On the word *Go!*, everyone releases their balloons.

Conclude and Apply

1. Which balloons flew for the longest amount of time?
2. Could you tell beforehand which balloons would have the most propulsion energy? What is the source of the propulsion energy?

Adult supervision required for all labs.

These labs are available at red.msscience.com.

6 Thermal Propulsion

Time Required two consecutive days, 15 minutes per day

Materials Do not use synthetic thread because it could pop a balloon.

Safety Precaution Students should use caution with scissors.

Teaching Strategy Use this lab to connect temperature, molecular kinetic energy, pressure, and kinetic energy created by pressure driven force.

Conclude and Apply

1. The lamp balloon may fly longer than the closet balloon. The freezer balloon does not fly very long.
2. The freezer balloon had shriveled. This indicated the molecules had lost some kinetic energy, so there was less pressure. The lamp balloon was bulging slightly. This indicated the molecules had heated up, increasing the pressure. The higher the pressure the larger the propulsion energy.

692 STUDENT RESOURCES

Extra Try at Home Labs

7 Magnetic Wires

Real-World Question
Which way does the magnetic field go around a current-carrying wire?

Possible Materials
- an electric circuit (battery, wire, lightbulb)
- compass
- paper
- iron filings or small bits of magnetic metal
- four textbooks
- four erasers or similar paperweights

Procedure
1. Hook up the electric circuit. You will know electricity is going through the wire if the lightbulb is on.
2. Hold the compass at several different points around the circuit. What do you notice?
3. Make a box around the circuit with the textbooks laid flat. Support the paper on the texts. Weight it down at the corners with erasers. Sprinkle iron filings on the paper while electricity is running through the circuit. What do you notice?

Conclude and Apply
1. Make a diagram of the magnetic field around the wire.
2. Would your results change if you changed the voltage of the battery?

8 Exploding Bag

Real-World Question
What happens when a bag pops?

Possible Materials
- paper bag or plastic produce bag

Procedure
1. Obtain a paper lunch bag. Smooth out the bag on a flat surface if it has any wrinkles.
2. Hold the neck of the bag and blow air into it until it is completely filled. The sides of the bag should be stretched out completely.
3. Twist the neck of the bag tightly to prevent air from escaping.
4. Pop the paper bag between your palms and observe what happens.
5. Examine the bag after you pop it. Observe any changes in the bag.

Conclude and Apply
1. Describe what happened when you popped the bag.
2. Infer why this happened to the bag.

Adult supervision required for all labs.

Extra Try at Home Labs

7 Magnetic Wires

Time Required one class period

Materials
- Use batteries that are small enough to fit inside the textbook "box" for the iron filing sprinkling procedure. Use at least a 9-V battery to get a sufficient magnetic field.
- Small, inexpensive compasses (accuracy is less important) can be bought at dollar stores or hardware stores.

Teaching Strategy Extend the lab by adding a switch to the circuit. Students can quickly turn it off to see where the compass' north position is, reassuring themselves that the procedure is valid.

Conclude and Apply
1. See similar diagrams in text.
2. No, the directions will still be the same.

8 Exploding Bag

Time Required five minutes

Materials Paper lunchbags or thin plastic produce bags work well.

Safety Precaution
Caution the students not to pop the bags near other people.

Teaching Strategy Instruct students that the bags must be fully inflated, and they must pop the bags quickly to produce a loud sound.

Conclude and Apply
1. If done correctly, a loud popping sound escaped the bag as the bag tore open.
2. Compressed air in the bag created a force that split the bag allowing compressional sound waves to escape.

9 Scratch Tests

Time Required one class period

Safety Precaution Caution students to be careful with the iron nails. A wound caused by a nail may require a tetanus shot.

Conclude and Apply

1. A likely order is:
   - penny
   - iron nail
   - chalk
   - graphite
2. Gemstones form under special conditions of pressure and heat and are rare. People value them for their beauty and hardness.

Extra Try at Home Labs

Extra Try at Home Labs

9 Scratch Tests

Real-World Question

What is the relative hardness of various minerals?

Possible Materials

- chalk
- sharpened pencil (use the graphite portion)
- penny
- iron nail
- Science Journal

Procedure

1. Copy the data table in your Science Journal. Label the top *Scratcher,* and label the left side *Scratched.*
2. Try to scratch each material with each other material. If a material can scratch another, put a check in that box.
3. Try other materials from around the house. Add those to your data table.

	chalk	graphite	penny	iron nail
chalk	■			
graphite		■		
penny			■	
iron nail				■

Conclude and Apply

1. Use your chart to make a hardness scale of these materials. Put the material that will scratch all other things at the top, and the material that is scratched by all other things, at the bottom.
2. A gemstone such as a diamond, emerald, or ruby, would scratch all of the materials in this lab. Explain why gemstones are so valuable.

10 Continental Movement

Real-World Question

How far will the continents move in the future?

Possible Materials

- meterstick or metric tape measure
- masking tape or sidewalk chalk
- tennis balls

Procedure

1. Go outside and measure a 1 cm distance. Mark this distance with two pieces of tape. This is the distance traveled by the slowest plate in one year.
2. Measure a 2.5 cm distance and mark this distance with two pieces of tape. This is the distance traveled by the North American plate in one year.
3. Measure a 15 cm distance and mark this distance with two pieces of tape. This is the distance traveled by the fastest plate in one year.
4. Measure the distances the slowest plate, North American plate, and fastest plate will travel in the next 50 years. Use tennis balls to mark your distances.
5. Measure the distances these three plates will move in the next 300 years. Use tennis balls to mark your distances.

Conclude and Apply

1. How many centimeters will these three plates travel in the next 50 years?
2. How many meters will these three plates travel in the next 300 years?

Adult supervision required for all labs.

These labs are available at red.msscience.com.

10 Continental Movement

Time Required 25 minutes

Materials

- Students can complete this lab outside using sidewalk chalk on an asphalt or concrete surface.
- For longer distances, students can use tennis balls or orange cones as markers.

Safety Precautions

- Caution students to use the metersticks carefully.
- Be certain measurements do not take them near streets or traffic.

Teaching Strategy Have students complete this lab outside in an open space where they can measure a 45-m distance.

Conclude and Apply

1. 50 cm, 125 cm, 750 cm
2. 3 m, 7.5 m, 45 m

Extra Try at Home Labs

11 Soil Creatures

Real-World Question
What types of organisms live in the soil near your home or school?

Possible Materials
- garden shovel
- garden trowel
- gloves
- window screen
- collecting jars
- bucket
- cotton
- tweezers
- magnifying lens
- invertebrate field guide

Procedure
1. Go to a wooded area near your home or school, clear away a patch of leaf litter and debris, and dig a hole 30 cm wide and 15 cm deep.
2. As you remove soil from the hole, place it on a window screen. Have a partner sift the soil through the screen into the bucket.
3. Collect any organisms you find and place them in a collecting jar.
4. Use a field guide to identify the types of organisms you find in the soil.
5. Search the soil of two other sites and compare the types of organisms you find in all three locations.

Conclude and Apply
1. List the types of soil organisms you discovered.
2. Infer why organisms are important for soil.

12 Cloud Watch

Real-World Question
What type of clouds will form over your home this week?

Possible Materials
- cloud chart
- Science Journal
- binoculars
- sunglasses

Procedure
1. Go outside and observe all the types of clouds in the sky. Use a cloud chart to identify the different types of clouds you find. In your Science Journal, record the name and the location in the sky (low, middle or high) of each type of cloud.
2. Observe the type of weather in your area during the hours following your cloud observation. Record your observations in your Science Journal.
3. Continue to make cloud and weather observations every day for seven days.

Conclude and Apply
1. List all the different clouds you identified during this lab.
2. Infer how the presence of different types of clouds can predict future patterns.

Adult supervision required for all labs.

Extra Try at Home Labs

11 Soil Creatures

Time Required 90 minutes (to search three sites)

Materials
- Students can collect their invertebrate samples in test tubes plugged with cotton, or in small collecting jars.
- Have students bring extra shoes and socks to change into after the lab.

Safety Precautions
- Students should wash their hands thoroughly after handling the soil.
- Caution students not to handle the creatures with their hands.
- When entering woods or fields, students should dress appropriately.

Teaching Strategies
- Have students survey the organisms in a variety of places to compare and contrast the soil organisms they find.
- This lab does not work well immediately after a rain.

Conclude and Apply
1. Answers will vary.
2. Organisms create nutrient-rich soil because they add to the organic composition of the soil.

12 Cloud Watch

Time Required ten-minute observation period each day for seven days

Materials Students can use the cloud drawings in their textbooks to identify clouds.

Safety Precautions
- Caution students never to look directly into the Sun.
- Students should never observe clouds outside during a thunderstorm.

Teaching Strategy Complete a cloud observation period before or after your lesson each day for a week.

Conclude and Apply
1. Answers will vary.
2. High cumulus clouds often bring thunderstorms, and nimbostratus clouds also bring precipitation. Low cumulus, cirrus, and cirrocumulus clouds usually signal fair weather.

13 Edible Ocean Creatures

Time Required 40 minutes

Materials Students can use a clipboard and notebook paper to record the ocean creatures they find instead of using their Science Journals.

Safety Precaution
Students should not go to the grocery store without an adult.

Teaching Strategy
Purchase several examples of ocean creatures sold at your local grocery store to show your class.

Conclude and Apply
1. Answers will vary but may include snapper, mackerel, salmon, bluefish, sea bass, herring, lobster, king crab, clams, oysters, scallops, mussels, and seaweed crackers.
2. chilean sea bass, toothfish, swordfish, orange roughy, and sturgeon
3. Avoid purchasing endangered creatures, reduce ocean pollution, limit fish catches to sustainable levels, and prevent the destruction of fish habitats and natural fish nurseries.

Extra Try at Home Labs

13 Edible Ocean Creatures

Real-World Question
What types of ocean creatures are sold at your local grocery store?

Possible Materials
- Science Journal
- pen
- clipboard
- paper

Procedure
1. Take an adult with you to your local grocery store. Choose the largest store in your area.
2. Search the store aisles for different species of fish sold to eat. Check the meat section for fresh or frozen fish and search the can goods aisles for canned fish. In your Science Journal, record the names of all the fish sold in the store.
3. Search the store aisles for different species of fresh, frozen, or canned invertebrates. In your Science Journal, record the names of all the invertebrates sold in the store.
4. Ask a store employee if they sell any products with sea kelp or seaweed. Record these products in your Science Journal.

Conclude and Apply
1. List all the ocean creatures your store sells.
2. Research several species of fish that are now becoming commercially endangered.
3. Infer how we can conserve ocean food sources.

14 Earth's Density

Real-World Question
How heavy would Earth feel if it were the size of a rock?

Possible Materials
- bathroom scale
- bucket
- water pitcher
- measuring cup
- large roasting pan
- turkey baster
- large chunk of wood, rock, or bundles of metal kitchen utensils held together with a rubber band

Procedure
1. Weigh yourself on a bathroom scale.
2. Pick up an object and stand on the scale. Calculate the mass of the object by subtracting your own mass.
3. Put the bucket in the roasting pan. Fill the bucket to the very brim with water. If you spill any water into the roasting pan, carefully mop it dry.

4. Put the object in the bucket.
5. Measure the water that spills into the roasting pan by collecting it with the turkey baster and putting it into the measuring cup. Depending on the size of your objects, you may need to empty the cup several times.
6. Use the formula: Density = Mass/Volume to calculate the density of each object.

Conclude and Apply
1. List the densities of the objects.
2. Look up Earth's density in this chapter. Which object has a density most like Earth's density?

Adult supervision required for all labs.

These labs are available at red.msscience.com.

14 Earth's Density

Time Required 50 minutes

Materials
- The objects must fit easily into the bucket, which must fit into the roasting pan.
- If only laboratory scales or kitchen scales are available, this lab can be done with smaller objects and plastic beakers.

Teaching Strategy Because it is less dense than water, students will need to push the wood just under the surface of the water, being careful not to submerge their fingers.

Conclude and Apply
1. Values will vary depending on the type of wood, stone, and metal in the cutlery, but wood is about 0.85 g/cm^3, stone is about 3 g/cm^3, and metal usually is at least 8 g/cm^3.
2. The metal or the stone, depending on the values obtained.

Extra Try at Home Labs

15 Deep Space Distance

Real-World Question

How can we compare the distances of objects in deep space?

Possible Materials

- meterstick
- metric ruler
- scissors
- masking tape
- tennis balls (2)
- thread

Procedure

1. Go outside to a paved surface and place a piece of tape on the ground.
2. Lay a piece of thread next to the tape. This represents 1 A.U., the distance from the Sun to Earth.
3. Measure a distance of 3.9 mm from the tape and place a second piece of tape. This represents the distance from the Sun to Pluto.
4. Measure a distance of 5.0 mm from the first piece of tape and place a third piece of tape. This represents the distance from the Sun to Oort Cloud.
5. Measure a distance of 6.33 m from the first piece of tape and place a tennis ball. This represents the distance of one light-year.
6. Measure a distance of 26.71 m from the first piece of tape and place a tennis ball. This represents the distance from the Sun to the nearest star, Proxima Centuri.

Conclude and Apply

1. The Milky Way galaxy is 90,000 light-years in diameter. Calculate the distance you would have to measure to represent the diameter of our galaxy.
2. Infer why space travel to other stars may not be possible.

16 Expanding Eggs

Real-World Question

How can you observe liquids passing through a cell membrane?

Possible Materials

- glass jar with lid
- white vinegar
- medium chicken egg
- tape measure or string and ruler
- tongs
- measuring cup

Procedure

1. Obtain a glass jar with a lid and a medium sized egg.
2. Make certain your egg easily fits into your jar.
3. Measure the circumference of your egg.
4. Pour 250 mL of white vinegar into the jar.
5. Carefully place your egg in the jar so that it is submerged in the vinegar. Be careful not to crack or break the egg.
6. Observe your egg each day for three days. Measure the circumference of the egg after three days.

Conclude and Apply

1. Describe the changes that happened to your egg.
2. Infer why the egg's circumference changed. *HINT: A hen's egg is a single cell.*

Extra Try at Home Labs

Adult supervision required for all labs.

15 Deep-Space Distance

Time Required 40 minutes

Materials

- Any type of ball can be used to mark the last two measurements as long as the class can clearly see them.
- Have students cut the tape pieces into very thin strips.

Safety Precaution Caution students to use the metersticks and scissors with care.

Teaching Strategy Take your class outside or to your school's gymnasium to complete this lab.

Conclude and Apply

1. 569,700 m
2. The distances, technology requirements, and energy costs are prohibitive for space travel outside of our solar system.

16 Expanding Eggs

Time Required five minutes for set-up; five-minute observation periods for three days

Materials Make certain the egg will fit into the jar both before and after the lab is completed. Avoid using large or extra large eggs.

Safety Precaution Students should wash their hands thoroughly after handling eggs.

Teaching Strategy The eggshell is made of calcium carbonate. Vinegar dissolves the calcium carbonate, producing bubbles of carbon dioxide during the reaction.

Conclude and Apply

1. The egg's shell will dissolve. The outer membrane of the egg will become rubbery, and the egg will increase in circumference by about 15 percent.
2. The water in the vinegar passes through the membrane of the egg cell expanding the egg's size. The particles inside the egg are too large to move out through the membrane.

17 Aquatic Worm Search

Time Required two 30-minute sessions

Materials White ice cube trays will allow for easier observation of the worms.

Safety Precautions

- Students should wash their hands thoroughly after handling stream water and organisms.
- Caution students to wear shoes in creeks to avoid cuts by glass and sharp rocks.

Teaching Strategy Instruct students to include bits of detritus and algae in their water drops. Aquatic worms of several varieties live in these habitats.

Conclude and Apply

1. Under rocks and leaves, students can find leeches and several species of flatworms, including planarian. Earthworms often are found in streams as well. Common microscopic worms include *Aeolosoma,* tubifex, horsehair, nematodes, and deros worms.
2. Earthworms and leeches are segmented worms; planarians are flatworms; nematodes are roundworms.

Extra Try at Home Labs

Extra Try at Home Labs

17 Aquatic Worm Search

Real-World Question

What types of worms live in freshwater?

Possible Materials

- ice cube tray
- bucket
- collecting jar
- aquatic net
- field guide to pond life
- magnifying lens
- microscope slide
- eyedropper
- waterproof boots

Procedure

1. Search for aquatic worms underneath rocks and leaves in a creek. Worms live beneath flat rocks and decaying leaves in slow, shallow water.
2. Carefully place the worms you find in different compartments of your ice cube tray and examine them closely under a microscope or magnifying lens.
3. Collect a sample of stream or pond water.
4. Place a drop of pond water on a microscope slide and search for microscopic aquatic worms living in the water.
5. Use your field guide to pond life to identify the organisms you find.

Conclude and Apply

1. List the aquatic worms you found.
2. Research the classification of the worms you discovered under the rocks and leaves of the stream.

18 Bare Bones

Real-World Question

What do the bones of a chicken wing look like?

Possible Materials

- stove or hot plate
- pot
- water
- fork or kitchen knife
- toothpick
- tongs
- paper towels
- plate
- chicken wing

Procedure

1. Boil water and cook a chicken wing in the pot.
2. Use the tongs to remove the wing from the water. Wait until the wing cools and use a kitchen knife or fork to remove all the meat.
3. Identify the following bones of your chicken wing: humerus, radius, ulna, carpal, metacarpal, first digit, second digit, and third digit.
4. Snap either the radius or ulna in half and remove the bone marrow inside with a toothpick.

Conclude and Apply

1. Explain how bones help vertebrates.
2. Describe what the inside of the bird's bone is like. Infer why it is like this.

Adult supervision required for all labs.

These labs are available at red.msscience.com.

18 Bare Bones

Time Required 45 minutes

Materials Save time by cooking and cooling the wings before class.

Safety Precautions

- Caution students to wait until the wings cool before taking the meat off.
- Students should use utensils with care.
- Students should not eat the meat.

Teaching Strategy Cook an entire chicken before class, strip the meat from it, and identify the clavicle (wishbone), sternum, keel, scapula, backbone, and ribs for your students.

Conclude and Apply

1. Bones protect internal organs and provide a strong support structure for the attachment of muscles.
2. The inside of a bird's bone is nearly hollow, making it lightweight for flight.

Extra Try at Home Labs

19 Comparing Arm Muscles

Real-World Question
Which arm muscle has the most strength?

Possible Materials
- a heavy book, stone, or other heavy object to act as a weight (about 3–7 kg)
- chair
- full-length mirror

Procedure
1. Sit up straight in a chair with your feet flat on the floor. Hold an object up in your hand and rest your arm on your thigh. Your wrist should be at your knee so that you can bend your hand over your knee. Slowly bend your wrist so that the object drops downward, and then slowly lift the object up using just your forearm muscle.
2. Stand up straight with your back against a wall and hold the object down by your side. Slowly raise the object to your shoulder using only your bicep muscle.
3. Place your hand on the seat of a chair and bend forward until your back is parallel to the ground. Holding the object in the opposite hand, lift it with your tricep. Make certain the upper portion of your arm remains parallel with the ground during the lift.

Conclude and Apply
1. Identify your strongest and weakest arm muscle.
2. Research the benefits of weightlifting to your health.

20 Planarian Pieces

Real-World Question
What happens to a planarian that has been cut in half?

Possible Materials
- brown planarian (3–4)
- collecting jar
- bowl with cover
- water (pond, distilled, or purified)
- eyedropper
- liver

Procedure
1. Obtain 3–4 brown planarian from a stream or science supply store.
2. Pour water into the bowl so that it just covers the bottom of the dish.
3. Use an eyedropper to collect 3–4 planarian from the collecting jar and place them in the bowl.
4. Carefully cut each planarian in half.
5. Pour water into the bowl until it is half full.
6. Place a small piece of liver in the bowl and cover the dish with the lid.
7. Change the water and liver every other day and observe the planarian for several weeks.

Conclude and Apply
1. Describe what happens to the planarian.
2. Identify this process of reproduction.

Adult supervision required for all labs.

Extra Try at Home Labs

19 Comparing Arm Muscles

Time Required 15 minutes

Materials Students can use a mirror to check their form.

Safety Precautions
- Warn students not to try to lift weights heavier than they are comfortable with, as this could cause injury.
- Check student medical files if you suspect a medical condition that would be exacerbated by this activity.

Teaching Strategies
- Do not allow students to lean back during any of these exercises.
- Students should focus on lifting the weight with just one muscle. Proper form is important to prevent injury.

Conclude and Apply
1. Students should observe that the bicep is the strongest arm muscle. Most students will observe that the forearm muscle is the weakest.
2. Weightlifting increases bone density, metabolism, and muscular strength, and often boosts a person's confidence.

20 Planarian Pieces

Time Required 25 minutes (short observation periods for several weeks)

Materials Refill bowls with aged water.

Teaching Strategy Aside from the brown planaria *(Dugesia tigrina)*, the California blackworm *(Lumbriculus variegatus)* and the hydra *(Hydra littoralis)* also will reproduce by regeneration.

Conclude and Apply
1. Students should observe that each piece regenerated into a whole flatworm.
2. regeneration

21 Disappearing Energy

Time Required 15 minutes

Materials Students can use small spoonfuls of granular sugar instead of sugar cubes.

Safety Precaution Caution students not to eat anything during science class.

Teaching Strategy Other food chains can be used for this activity. Encourage students to find photographs on their own that could be assembled into a food chain for this lab.

Conclude and Apply

1. 10%
2. Answers will vary, but only a few grains would pass to the hawk.
3. There is not enough energy left in a food chain to sustain super predators.

Extra Try at Home Labs

21 Disappearing Energy

Real-World Question

How much energy is transferred in a food chain?

Possible Materials

- sugar cubes
- picture or photograph of the Sun
- picture or photograph of grass
- picture or photograph of mouse
- picture or photograph of snake
- picture or photograph of red tailed hawk

Procedure

1. Lay out your pictures or photographs as links in a food chain starting with the sun and ending with the hawk.
2. Pile 100 sugar cubes next to the picture of the sun. This represents all the Sun's energy reaching Earth.
3. Place ten sugar cubes next to the plant picture, which represents the Sun's energy captured by plants.
4. Place one sugar cube next to the mouse picture to represent the energy that passes from plants to herbivores.
5. Scrape a tenth of one sugar cube away and place the loose grains next to the picture of the snake.

Conclude and Apply

1. Calculate the percentage of energy that is passed up each link in a food chain.
2. Infer how much sugar should be placed next to the red-tailed hawk picture.
3. Infer why there are no super predators that would eat tigers, hawks, or other top predators.

22 How much packaging?

Real-World Question

How much packaging do different products have?

Possible Materials

- bag of grapes
- prepackaged lunch (cheese and crackers)
- box of chocolate bars
- other packaged foods

Procedure

1. Separate each type of packaging from the food it contains. Set each package next to its food on a table or counter.
2. Make a table with the following headings: *Food, Percentage of Packaging, Amount of Packaging per Serving.*
3. Estimate the percent of the mass of the food and the packaging that contains it. Use a kitchen scale if you have one.
4. Express the amount of packaging a second way by describing how much packaging there is per serving of food.
5. Compare the amount of packaging in each case. Make a list in order from most packaging to least.

Conclude and Apply

1. What ways could your family reduce the amount of excess packaging you buy?
2. Do some packages seem to do a better job of preserving food than others? Explain your answer.
3. Do you think that some packaging materials are better for the environment than others? Explain your answer.

Adult supervision required for all labs.

These labs are available at red.msscience.com.

22 How much packaging?

Time Required 40 minutes

Teaching Strategies

- Ask each student to bring in a different kind of food for greater variety.
- The percentage of packaging for the bag of grapes will probably be less than 2 percent. The prepackaged lunch will have 15 to 20 percent packaging materials.
- As an extension, compare and discuss portion sizes and prices.

Conclude and Apply

1. Answers may vary but include purchasing items with less excess packaging such as large bottles of soda, and by avoiding highly packaged fast foods and prepackaged meals.
2. Students might answer that plastics and foils preserve foods better than paper.
3. Recyclable metals, and compostable materials, such as paper, are better for the environment than materials that must be disposed of in landfills.

Computer Skills

People who study science rely on computers, like the one in **Figure 16,** to record and store data and to analyze results from investigations. Whether you work in a laboratory or just need to write a lab report with tables, good computer skills are a necessity.

Using the computer comes with responsibility. Issues of ownership, security, and privacy can arise. Remember, if you did not author the information you are using, you must provide a source for your information. Also, anything on a computer can be accessed by others. Do not put anything on the computer that you would not want everyone to know. To add more security to your work, use a password.

Use a Word Processing Program

A computer program that allows you to type your information, change it as many times as you need to, and then print it out is called a word processing program. Word processing programs also can be used to make tables.

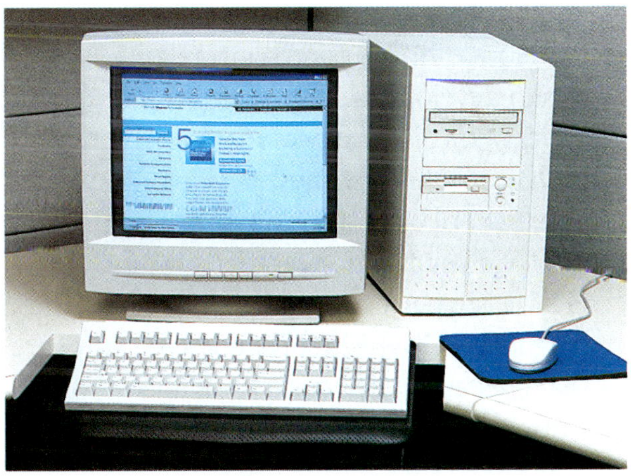

Figure 16 A computer will make reports neater and more professional looking.

Learn the Skill To start your word processing program, a blank document, sometimes called "Document 1," appears on the screen. To begin, start typing. To create a new document, click the *New* button on the standard tool bar. These tips will help you format the document.

- The program will automatically move to the next line; press *Enter* if you wish to start a new paragraph.
- Symbols, called non-printing characters, can be hidden by clicking the *Show/Hide* button on your toolbar.
- To insert text, move the cursor to the point where you want the insertion to go, click on the mouse once, and type the text.
- To move several lines of text, select the text and click the *Cut* button on your toolbar. Then position your cursor in the location that you want to move the cut text and click *Paste.* If you move to the wrong place, click *Undo.*
- The spell check feature does not catch words that are misspelled to look like other words, like "cold" instead of "gold." Always reread your document to catch all spelling mistakes.
- To learn about other word processing methods, read the user's manual or click on the *Help* button.
- You can integrate databases, graphics, and spreadsheets into documents by copying from another program and pasting it into your document, or by using desktop publishing (DTP). DTP software allows you to put text and graphics together to finish your document with a professional look. This software varies in how it is used and its capabilities.

Use a Database

A collection of facts stored in a computer and sorted into different fields is called a database. A database can be reorganized in any way that suits your needs.

Learn the Skill A computer program that allows you to create your own database is a database management system (DBMS). It allows you to add, delete, or change information. Take time to get to know the features of your database software.

- Determine what facts you would like to include and research to collect your information.
- Determine how you want to organize the information.
- Follow the instructions for your particular DBMS to set up fields. Then enter each item of data in the appropriate field.
- Follow the instructions to sort the information in order of importance.
- Evaluate the information in your database, and add, delete, or change as necessary.

Use the Internet

The Internet is a global network of computers where information is stored and shared. To use the Internet, like the students in **Figure 17,** you need a modem to connect your computer to a phone line and an Internet Service Provider account.

Learn the Skill To access internet sites and information, use a "Web browser," which lets you view and explore pages on the World Wide Web. Each page is its own site, and each site has its own address, called a URL. Once you have found a Web browser, follow these steps for a search (this also is how you search a database).

Figure 17 The Internet allows you to search a global network for a variety of information.

- Be as specific as possible. If you know you want to research "gold," don't type in "elements." Keep narrowing your search until you find what you want.
- Web sites that end in *.com* are commercial Web sites; *.org, .edu,* and *.gov* are nonprofit, educational, or government Web sites.
- Electronic encyclopedias, almanacs, indexes, and catalogs will help locate and select relevant information.
- Develop a "home page" with relative ease. When developing a Web site, NEVER post pictures or disclose personal information such as location, names, or phone numbers. Your school or community usually can host your Web site. A basic understanding of HTML (hypertext mark-up language), the language of Web sites, is necessary. Software that creates HTML code is called authoring software, and can be downloaded free from many Web sites. This software allows text and pictures to be arranged as the software is writing the HTML code.

Use a Spreadsheet

A spreadsheet, shown in **Figure 18,** can perform mathematical functions with any data arranged in columns and rows. By entering a simple equation into a cell, the program can perform operations in specific cells, rows, or columns.

Learn the Skill Each column (vertical) is assigned a letter, and each row (horizontal) is assigned a number. Each point where a row and column intersect is called a cell, and is labeled according to where it is located—Column A, Row 1 (A1).

- Decide how to organize the data, and enter it in the correct row or column.
- Spreadsheets can use standard formulas or formulas can be customized to calculate cells.
- To make a change, click on a cell to make it activate, and enter the edited data or formula.
- Spreadsheets also can display your results in graphs. Choose the style of graph that best represents the data.

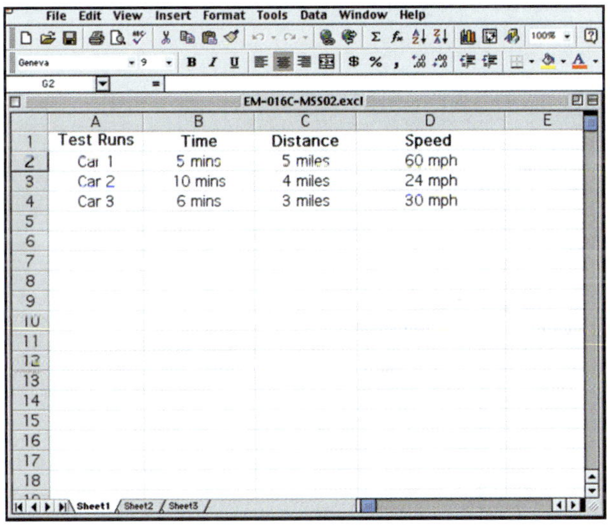

Figure 18 A spreadsheet allows you to perform mathematical operations on your data.

Use Graphics Software

Adding pictures, called graphics, to your documents is one way to make your documents more meaningful and exciting. This software adds, edits, and even constructs graphics. There is a variety of graphics software programs. The tools used for drawing can be a mouse, keyboard, or other specialized devices. Some graphics programs are simple. Others are complicated, called computer-aided design (CAD) software.

Learn the Skill It is important to have an understanding of the graphics software being used before starting. The better the software is understood, the better the results. The graphics can be placed in word-processing document.

- Clip art can be found on a variety of internet sites, and on CDs. These images can be copied and pasted into your document.
- When beginning, try editing existing drawings, then work up to creating drawings.
- The images are made of tiny rectangles of color called pixels. Each pixel can be altered.
- Digital photography is another way to add images. The photographs in the memory of a digital camera can be downloaded into a computer, then edited and added to the document.
- Graphics software also can allow animation. The software allows drawings to have the appearance of movement by connecting basic drawings automatically. This is called in-betweening, or tweening.
- Remember to save often.

Presentation Skills

Develop Multimedia Presentations

Most presentations are more dynamic if they include diagrams, photographs, videos, or sound recordings, like the one shown in **Figure 19.** A multimedia presentation involves using stereos, overhead projectors, televisions, computers, and more.

Learn the Skill Decide the main points of your presentation, and what types of media would best illustrate those points.

- Make sure you know how to use the equipment you are working with.
- Practice the presentation using the equipment several times.
- Enlist the help of a classmate to push play or turn lights out for you. Be sure to practice your presentation with him or her.
- If possible, set up all of the equipment ahead of time, and make sure everything is working properly.

Figure 19 These students are engaging the audience using a variety of tools.

Computer Presentations

There are many different interactive computer programs that you can use to enhance your presentation. Most computers have a compact disc (CD) drive that can play both CDs and digital video discs (DVDs). Also, there is hardware to connect a regular CD, DVD, or VCR. These tools will enhance your presentation.

Another method of using the computer to aid in your presentation is to develop a slide show using a computer program. This can allow movement of visuals at the presenter's pace, and can allow for visuals to build on one another.

Learn the Skill In order to create multimedia presentations on a computer, you need to have certain tools. These may include traditional graphic tools and drawing programs, animation programs, and authoring systems that tie everything together. Your computer will tell you which tools it supports. The most important step is to learn about the tools that you will be using.

- Often, color and strong images will convey a point better than words alone. Use the best methods available to convey your point.
- As with other presentations, practice many times.
- Practice your presentation with the tools you and any assistants will be using.
- Maintain eye contact with the audience. The purpose of using the computer is not to prompt the presenter, but to help the audience understand the points of the presentation.

Math Review

Use Fractions

A fraction compares a part to a whole. In the fraction $\frac{2}{3}$, the 2 represents the part and is the numerator. The 3 represents the whole and is the denominator.

Reduce Fractions To reduce a fraction, you must find the largest factor that is common to both the numerator and the denominator, the greatest common factor (GCF). Divide both numbers by the GCF. The fraction has then been reduced, or it is in its simplest form.

Example Twelve of the 20 chemicals in the science lab are in powder form. What fraction of the chemicals used in the lab are in powder form?

Step 1 Write the fraction.

$$\frac{\text{part}}{\text{whole}} = \frac{12}{20}$$

Step 2 To find the GCF of the numerator and denominator, list all of the factors of each number.

Factors of 12: 1, 2, 3, 4, 6, 12 (the numbers that divide evenly into 12)

Factors of 20: 1, 2, 4, 5, 10, 20 (the numbers that divide evenly into 20)

Step 3 List the common factors.

1, 2, 4.

Step 4 Choose the greatest factor in the list.

The GCF of 12 and 20 is 4.

Step 5 Divide the numerator and denominator by the GCF.

$$\frac{12 \div 4}{20 \div 4} = \frac{3}{5}$$

In the lab, $\frac{3}{5}$ of the chemicals are in powder form.

Practice Problem At an amusement park, 66 of 90 rides have a height restriction. What fraction of the rides, in its simplest form, has a height restriction?

Add and Subtract Fractions To add or subtract fractions with the same denominator, add or subtract the numerators and write the sum or difference over the denominator. After finding the sum or difference, find the simplest form for your fraction.

Example 1 In the forest outside your house, $\frac{1}{8}$ of the animals are rabbits, $\frac{3}{8}$ are squirrels, and the remainder are birds and insects. How many are mammals?

Step 1 Add the numerators.

$$\frac{1}{8} + \frac{3}{8} = \frac{(1+3)}{8} = \frac{4}{8}$$

Step 2 Find the GCF.

$$\frac{4}{8} \text{ (GCF, 4)}$$

Step 3 Divide the numerator and denominator by the GCF.

$$\frac{4}{4} = 1, \quad \frac{8}{4} = 2$$

$\frac{1}{2}$ of the animals are mammals.

Example 2 If $\frac{7}{16}$ of the Earth is covered by freshwater, and $\frac{1}{16}$ of that is in glaciers, how much freshwater is not frozen?

Step 1 Subtract the numerators.

$$\frac{7}{16} - \frac{1}{16} = \frac{(7-1)}{16} = \frac{6}{16}$$

Step 2 Find the GCF.

$$\frac{6}{16} \text{ (GCF, 2)}$$

Step 3 Divide the numerator and denominator by the GCF.

$$\frac{6}{2} = 3, \quad \frac{16}{2} = 8$$

$\frac{3}{8}$ of the freshwater is not frozen.

Practice Problem A bicycle rider is going 15 km/h for $\frac{4}{9}$ of his ride, 10 km/h for $\frac{2}{9}$ of his ride, and 8 km/h for the remainder of the ride. How much of his ride is he going over 8 km/h?

Reduce Fractions

$$\frac{66 \div 6}{90 \div 6} = \frac{11}{15}$$

Add and Subtract Fractions

$$\frac{4}{9} + \frac{2}{9} = \frac{6}{9}$$

$$\frac{6 \div 3}{9 \div 3} = \frac{2}{3}$$

Math Skill Handbook

Unlike Denominators

Problem 1

$1 \times 5 = 5, 8 \times 5 = 40$

$1 \times 4 = 4, 10 \times 4 = 40$

$\frac{5}{40} + \frac{4}{40} = \frac{9}{40}$

Problem 2

If $\frac{7}{10}$ are involuntary, the remainder are voluntary.

$\frac{10}{10} - \frac{7}{10} = \frac{3}{10}$

Math Skill Handbook

Unlike Denominators To add or subtract fractions with unlike denominators, first find the least common denominator (LCD). This is the smallest number that is a common multiple of both denominators. Rename each fraction with the LCD, and then add or subtract. Find the simplest form if necessary.

Example 1 A chemist makes a paste that is $\frac{1}{2}$ table salt (NaCl), $\frac{1}{3}$ sugar ($C_6H_{12}O_6$), and the rest water (H_2O). How much of the paste is a solid?

Step 1 Find the LCD of the fractions.

$\frac{1}{2} + \frac{1}{3}$ (LCD, 6)

Step 2 Rename each numerator and each denominator with the LCD.

$1 \times 3 = 3, \; 2 \times 3 = 6$

$1 \times 2 = 2, \; 3 \times 2 = 6$

Step 3 Add the numerators.

$\frac{3}{6} + \frac{2}{6} = \frac{(3 + 2)}{6} = \frac{5}{6}$

$\frac{5}{6}$ of the paste is a solid.

Example 2 The average precipitation in Grand Junction, CO, is $\frac{7}{10}$ inch in November, and $\frac{3}{5}$ inch in December. What is the total average precipitation?

Step 1 Find the LCD of the fractions.

$\frac{7}{10} + \frac{3}{5}$ (LCD, 10)

Step 2 Rename each numerator and each denominator with the LCD.

$7 \times 1 = 7, \; 10 \times 1 = 10$

$3 \times 2 = 6, \; 5 \times 2 = 10$

Step 3 Add the numerators.

$\frac{7}{10} + \frac{6}{10} = \frac{(7 + 6)}{10} = \frac{13}{10}$

$\frac{13}{10}$ inches total precipitation, or $1\frac{3}{10}$ inches.

Practice Problem On an electric bill, about $\frac{1}{8}$ of the energy is from solar energy and about $\frac{1}{10}$ is from wind power. How much of the total bill is from solar energy and wind power combined?

Example 3 In your body, $\frac{7}{10}$ of your muscle contractions are involuntary (cardiac and smooth muscle tissue). Smooth muscle makes $\frac{3}{15}$ of your muscle contractions. How many of your muscle contractions are made by cardiac muscle?

Step 1 Find the LCD of the fractions.

$\frac{7}{10} - \frac{3}{15}$ (LCD, 30)

Step 2 Rename each numerator and each denominator with the LCD.

$7 \times 3 = 21, \; 10 \times 3 = 30$

$3 \times 2 = 6, \; 15 \times 2 = 30$

Step 3 Subtract the numerators.

$\frac{21}{30} - \frac{6}{30} = \frac{(21 - 6)}{30} = \frac{15}{30}$

Step 4 Find the GCF.

$\frac{15}{30}$ (GCF, 15)

$\frac{1}{2}$

$\frac{1}{2}$ of all muscle contractions are cardiac muscle.

Example 4 Tony wants to make cookies that call for $\frac{3}{4}$ of a cup of flour, but he only has $\frac{1}{3}$ of a cup. How much more flour does he need?

Step 1 Find the LCD of the fractions.

$\frac{3}{4} - \frac{1}{3}$ (LCD, 12)

Step 2 Rename each numerator and each denominator with the LCD.

$3 \times 3 = 9, \; 4 \times 3 = 12$

$1 \times 4 = 4, \; 3 \times 4 = 12$

Step 3 Subtract the numerators.

$\frac{9}{12} - \frac{4}{12} = \frac{(9 - 4)}{12} = \frac{5}{12}$

$\frac{5}{12}$ of a cup of flour.

Practice Problem Using the information provided to you in Example 3 above, determine how many muscle contractions are voluntary (skeletal muscle).

Math Skill Handbook

Multiply Fractions To multiply with fractions, multiply the numerators and multiply the denominators. Find the simplest form if necessary.

Example Multiply $\frac{3}{5}$ by $\frac{1}{3}$.

Step 1 Multiply the numerators and denominators.

$$\frac{3}{5} \times \frac{1}{3} = \frac{(3 \times 1)}{(5 \times 3)} = \frac{3}{15}$$

Step 2 Find the GCF.

$\frac{3}{15}$ (GCF, 3)

Step 3 Divide the numerator and denominator by the GCF.

$$\frac{3}{3} = 1, \quad \frac{15}{3} = 5$$

$$\frac{1}{5}$$

$\frac{3}{5}$ multiplied by $\frac{1}{3}$ is $\frac{1}{5}$.

Practice Problem Multiply $\frac{3}{14}$ by $\frac{5}{16}$.

Find a Reciprocal Two numbers whose product is 1 are called multiplicative inverses, or reciprocals.

Example Find the reciprocal of $\frac{3}{8}$.

Step 1 Inverse the fraction by putting the denominator on top and the numerator on the bottom.

$$\frac{8}{3}$$

The reciprocal of $\frac{3}{8}$ is $\frac{8}{3}$.

Practice Problem Find the reciprocal of $\frac{4}{9}$.

Divide Fractions To divide one fraction by another fraction, multiply the dividend by the reciprocal of the divisor. Find the simplest form if necessary.

Example 1 Divide $\frac{1}{9}$ by $\frac{1}{3}$.

Step 1 Find the reciprocal of the divisor.

The reciprocal of $\frac{1}{3}$ is $\frac{3}{1}$.

Step 2 Multiply the dividend by the reciprocal of the divisor.

$$\frac{\frac{1}{9}}{\frac{1}{3}} = \frac{1}{9} \times \frac{3}{1} = \frac{(1 \times 3)}{(9 \times 1)} = \frac{3}{9}$$

Step 3 Find the GCF.

$\frac{3}{9}$ (GCF, 3)

Step 4 Divide the numerator and denominator by the GCF.

$$\frac{3}{3} = 1, \quad \frac{9}{3} = 3$$

$$\frac{1}{3}$$

$\frac{1}{9}$ divided by $\frac{1}{3}$ is $\frac{1}{3}$.

Example 2 Divide $\frac{3}{5}$ by $\frac{1}{4}$.

Step 1 Find the reciprocal of the divisor.

The reciprocal of $\frac{1}{4}$ is $\frac{4}{1}$.

Step 2 Multiply the dividend by the reciprocal of the divisor.

$$\frac{\frac{3}{5}}{\frac{1}{4}} = \frac{3}{5} \times \frac{4}{1} = \frac{(3 \times 4)}{(5 \times 1)} = \frac{12}{5}$$

$\frac{3}{5}$ divided by $\frac{1}{4}$ is $\frac{12}{5}$ or $2\frac{2}{5}$.

Practice Problem Divide $\frac{3}{11}$ by $\frac{7}{10}$.

Multiply Fractions

$$\frac{3}{14} \times \frac{5}{16} = \frac{(3 \times 5)}{(14 \times 16)} = \frac{15}{224}$$

Find a Reciprocal

$$\frac{9}{4}$$

Divide Fractions

The reciprocal of $\frac{7}{10}$ is $\frac{10}{7}$.

$$\frac{3}{11} \times \frac{10}{7} = \frac{(3 \times 10)}{(11 \times 7)} = \frac{30}{77}$$

Math Skill Handbook

MATH SKILL HANDBOOK 707

Use Ratios

$\frac{100\text{ cm}}{144\text{ cm}} = \frac{100 \div 4}{144 \div 4} = \frac{25}{36}$

25:36

Add or Subtract Decimals

$$\begin{array}{r} \scriptstyle 1 \\ 1.245 \\ +\ 3.842 \\ \hline 5.087 \end{array}$$

Use Ratios

When you compare two numbers by division, you are using a ratio. Ratios can be written 3 to 5, 3:5, or $\frac{3}{5}$. Ratios, like fractions, also can be written in simplest form.

Ratios can represent probabilities, also called odds. This is a ratio that compares the number of ways a certain outcome occurs to the number of outcomes. For example, if you flip a coin 100 times, what are the odds that it will come up heads? There are two possible outcomes, heads or tails, so the odds of coming up heads are 50:100. Another way to say this is that 50 out of 100 times the coin will come up heads. In its simplest form, the ratio is 1:2.

Example 1 A chemical solution contains 40 g of salt and 64 g of baking soda. What is the ratio of salt to baking soda as a fraction in simplest form?

Step 1 Write the ratio as a fraction.

$$\frac{\text{salt}}{\text{baking soda}} = \frac{40}{64}$$

Step 2 Express the fraction in simplest form.

The GCF of 40 and 64 is 8.

$$\frac{40}{64} = \frac{40 \div 8}{64 \div 8} = \frac{5}{8}$$

The ratio of salt to baking soda in the sample is 5:8.

Example 2 Sean rolls a 6-sided die 6 times. What are the odds that the side with a 3 will show?

Step 1 Write the ratio as a fraction.

$$\frac{\text{number of sides with a 3}}{\text{number of sides}} = \frac{1}{6}$$

Step 2 Multiply by the number of attempts.

$$\frac{1}{6} \times 6\text{ attempts} = \frac{6}{6}\text{ attempts} = 1\text{ attempt}$$

1 attempt out of 6 will show a 3.

Practice Problem Two metal rods measure 100 cm and 144 cm in length. What is the ratio of their lengths in simplest form?

Use Decimals

A fraction with a denominator that is a power of ten can be written as a decimal. For example, 0.27 means $\frac{27}{100}$. The decimal point separates the ones place from the tenths place.

Any fraction can be written as a decimal using division. For example, the fraction $\frac{5}{8}$ can be written as a decimal by dividing 5 by 8. Written as a decimal, it is 0.625.

Add or Subtract Decimals When adding and subtracting decimals, line up the decimal points before carrying out the operation.

Example 1 Find the sum of 47.68 and 7.80.

Step 1 Line up the decimal places when you write the numbers.

$$\begin{array}{r} 47.68 \\ +\ 7.80 \\ \hline \end{array}$$

Step 2 Add the decimals.

$$\begin{array}{r} 47.68 \\ +\ 7.80 \\ \hline 55.48 \end{array}$$

The sum of 47.68 and 7.80 is 55.48.

Example 2 Find the difference of 42.17 and 15.85.

Step 1 Line up the decimal places when you write the number.

$$\begin{array}{r} 42.17 \\ -15.85 \\ \hline \end{array}$$

Step 2 Subtract the decimals.

$$\begin{array}{r} 42.17 \\ -15.85 \\ \hline 26.32 \end{array}$$

The difference of 42.17 and 15.85 is 26.32.

Practice Problem Find the sum of 1.245 and 3.842.

Math Skill Handbook

Multiply Decimals To multiply decimals, multiply the numbers like any other number, ignoring the decimal point. Count the decimal places in each factor. The product will have the same number of decimal places as the sum of the decimal places in the factors.

Example Multiply 2.4 by 5.9.

Step 1 Multiply the factors like two whole numbers.
$24 \times 59 = 1416$

Step 2 Find the sum of the number of decimal places in the factors. Each factor has one decimal place, for a sum of two decimal places.

Step 3 The product will have two decimal places.
14.16

The product of 2.4 and 5.9 is 14.16.

Practice Problem Multiply 4.6 by 2.2.

Divide Decimals When dividing decimals, change the divisor to a whole number. To do this, multiply both the divisor and the dividend by the same power of ten. Then place the decimal point in the quotient directly above the decimal point in the dividend. Then divide as you do with whole numbers.

Example Divide 8.84 by 3.4.

Step 1 Multiply both factors by 10.
$3.4 \times 10 = 34,\ 8.84 \times 10 = 88.4$

Step 2 Divide 88.4 by 34.

$$\begin{array}{r} 2.6 \\ 34\overline{)88.4} \\ -68 \\ \hline 20\,4 \\ -20\,4 \\ \hline 0 \end{array}$$

8.84 divided by 3.4 is 2.6.

Practice Problem Divide 75.6 by 3.6.

Use Proportions

An equation that shows that two ratios are equivalent is a proportion. The ratios $\frac{2}{4}$ and $\frac{5}{10}$ are equivalent, so they can be written as $\frac{2}{4} = \frac{5}{10}$. This equation is a proportion.

When two ratios form a proportion, the cross products are equal. To find the cross products in the proportion $\frac{2}{4} = \frac{5}{10}$, multiply the 2 and the 10, and the 4 and the 5. Therefore $2 \times 10 = 4 \times 5$, or $20 = 20$.

Because you know that both proportions are equal, you can use cross products to find a missing term in a proportion. This is known as solving the proportion.

Example The heights of a tree and a pole are proportional to the lengths of their shadows. The tree casts a shadow of 24 m when a 6-m pole casts a shadow of 4 m. What is the height of the tree?

Step 1 Write a proportion.
$$\frac{\text{height of tree}}{\text{height of pole}} = \frac{\text{length of tree's shadow}}{\text{length of pole's shadow}}$$

Step 2 Substitute the known values into the proportion. Let h represent the unknown value, the height of the tree.
$$\frac{h}{6} = \frac{24}{4}$$

Step 3 Find the cross products.
$h \times 4 = 6 \times 24$

Step 4 Simplify the equation.
$4h = 144$

Step 5 Divide each side by 4.
$$\frac{4h}{4} = \frac{144}{4}$$
$h = 36$

The height of the tree is 36 m.

Practice Problem The ratios of the weights of two objects on the Moon and on Earth are in proportion. A rock weighing 3 N on the Moon weighs 18 N on Earth. How much would a rock that weighs 5 N on the Moon weigh on Earth?

Math Skill Handbook

Multiply Decimals

Multiply 4.6 and 2.2 by 10.

$46 \times 22 = 1012$

Each factor had one decimal place.

10.12

Divide Decimals

Multiply both factors by 10.

Divide 756 by 36.

$$\begin{array}{r} 21 \\ 36\overline{)756} \\ 72 \\ \hline 36 \\ 36 \\ \hline 0 \end{array}$$

Use Proportions

$$\frac{3}{18} = \frac{5}{w}$$

$w \times 3 = 5 \times 18$

$$\frac{3w}{3} = \frac{90}{3}$$

$w = 30$

Math Skill Handbook

MATH SKILL HANDBOOK 709

Use Percentages

$\frac{73}{365} = \frac{x}{100}$

$\frac{7300}{365} = \frac{365x}{365}$

$20\% = x$

Solve One-Step Equations

$h = gd$

$\frac{17.4}{12.3} = \frac{12.3d}{12.3}$

$1.41 = d$

Use Percentages

The word *percent* means "out of one hundred." It is a ratio that compares a number to 100. Suppose you read that 77 percent of the Earth's surface is covered by water. That is the same as reading that the fraction of the Earth's surface covered by water is $\frac{77}{100}$. To express a fraction as a percent, first find the equivalent decimal for the fraction. Then, multiply the decimal by 100 and add the percent symbol.

Example Express $\frac{13}{20}$ as a percent.

Step 1 Find the equivalent decimal for the fraction.

```
    0.65
20)13.00
   12 0
    1 00
    1 00
       0
```

Step 2 Rewrite the fraction $\frac{13}{20}$ as 0.65.

Step 3 Multiply 0.65 by 100 and add the % sign.

$0.65 \times 100 = 65 = 65\%$

So, $\frac{13}{20} = 65\%$.

This also can be solved as a proportion.

Example Express $\frac{13}{20}$ as a percent.

Step 1 Write a proportion.

$\frac{13}{20} = \frac{x}{100}$

Step 2 Find the cross products.

$1300 = 20x$

Step 3 Divide each side by 20.

$\frac{1300}{20} = \frac{20x}{20}$

$65\% = x$

Practice Problem In one year, 73 of 365 days were rainy in one city. What percent of the days in that city were rainy?

Solve One-Step Equations

A statement that two things are equal is an equation. For example, $A = B$ is an equation that states that A is equal to B.

An equation is solved when a variable is replaced with a value that makes both sides of the equation equal. To make both sides equal the inverse operation is used. Addition and subtraction are inverses, and multiplication and division are inverses.

Example 1 Solve the equation $x - 10 = 35$.

Step 1 Find the solution by adding 10 to each side of the equation.

$x - 10 = 35$

$x - 10 + 10 = 35 + 10$

$x = 45$

Step 2 Check the solution.

$x - 10 = 35$

$45 - 10 = 35$

$35 = 35$

Both sides of the equation are equal, so $x = 45$.

Example 2 In the formula $a = bc$, find the value of c if $a = 20$ and $b = 2$.

Step 1 Rearrange the formula so the unknown value is by itself on one side of the equation by dividing both sides by b.

$a = bc$

$\frac{a}{b} = \frac{bc}{b}$

$\frac{a}{b} = c$

Step 2 Replace the variables a and b with the values that are given.

$\frac{a}{b} = c$

$\frac{20}{2} = c$

$10 = c$

Step 3 Check the solution.

$a = bc$

$20 = 2 \times 10$

$20 = 20$

Both sides of the equation are equal, so $c = 10$ is the solution when $a = 20$ and $b = 2$.

Practice Problem In the formula $h = gd$, find the value of d if $g = 12.3$ and $h = 17.4$.

Math Skill Handbook

Use Statistics

The branch of mathematics that deals with collecting, analyzing, and presenting data is statistics. In statistics, there are three common ways to summarize data with a single number—the mean, the median, and the mode.

The **mean** of a set of data is the arithmetic average. It is found by adding the numbers in the data set and dividing by the number of items in the set.

The **median** is the middle number in a set of data when the data are arranged in numerical order. If there were an even number of data points, the median would be the mean of the two middle numbers.

The **mode** of a set of data is the number or item that appears most often.

Another number that often is used to describe a set of data is the range. The **range** is the difference between the largest number and the smallest number in a set of data.

A **frequency table** shows how many times each piece of data occurs, usually in a survey. **Table 2** below shows the results of a student survey on favorite color.

Table 2 Student Color Choice

Color	Tally	Frequency
red	IIII	4
blue	卌	5
black	II	2
green	III	3
purple	卌 II	7
yellow	卌 I	6

Based on the frequency table data, which color is the favorite?

Example The speeds (in m/s) for a race car during five different time trials are 39, 37, 44, 36, and 44.

To find the mean:

Step 1 Find the sum of the numbers.
$39 + 37 + 44 + 36 + 44 = 200$

Step 2 Divide the sum by the number of items, which is 5.
$200 \div 5 = 40$

The mean is 40 m/s.

To find the median:

Step 1 Arrange the measures from least to greatest.
36, 37, 39, 44, 44

Step 2 Determine the middle measure.
36, 37, 39, 44, 44

The median is 39 m/s.

To find the mode:

Step 1 Group the numbers that are the same together.
44, 44, 36, 37, 39

Step 2 Determine the number that occurs most in the set.
44, 44, 36, 37, 39

The mode is 44 m/s.

To find the range:

Step 1 Arrange the measures from largest to smallest.
44, 44, 39, 37, 36

Step 2 Determine the largest and smallest measures in the set.
44, 44, 39, 37, 36

Step 3 Find the difference between the largest and smallest measures.
$44 - 36 = 8$

The range is 8 m/s.

Practice Problem Find the mean, median, mode, and range for the data set 8, 4, 12, 8, 11, 14, 16.

Use Statistics

mean
$8 + 4 + 12 + 8 + 11 + 14 + 16 = 73$
$73 \div 7 = 10.4$

median
4, 8, 8, 11, 12, 14, 16

mode
4, 8, 8, 11, 12, 14, 16

range
4, 8, 8, 11, 12, 14, 16
$16 - 4 = 12$

Math Skill Handbook

MATH SKILL HANDBOOK 711

Use Geometry

The branch of mathematics that deals with the measurement, properties, and relationships of points, lines, angles, surfaces, and solids is called geometry.

Perimeter The **perimeter** (P) is the distance around a geometric figure. To find the perimeter of a rectangle, add the length and width and multiply that sum by two, or $2(l + w)$. To find perimeters of irregular figures, add the length of the sides.

Example 1 Find the perimeter of a rectangle that is 3 m long and 5 m wide.

Step 1 You know that the perimeter is 2 times the sum of the width and length.
$P = 2(3\text{ m} + 5\text{ m})$

Step 2 Find the sum of the width and length.
$P = 2(8\text{ m})$

Step 3 Multiply by 2.
$P = 16\text{ m}$

The perimeter is 16 m.

Example 2 Find the perimeter of a shape with sides measuring 2 cm, 5 cm, 6 cm, 3 cm.

Step 1 You know that the perimeter is the sum of all the sides.
$P = 2 + 5 + 6 + 3$

Step 2 Find the sum of the sides.
$P = 2 + 5 + 6 + 3$
$P = 16$

The perimeter is 16 cm.

Practice Problem Find the perimeter of a rectangle with a length of 18 m and a width of 7 m.

Practice Problem Find the perimeter of a triangle measuring 1.6 cm by 2.4 cm by 2.4 cm.

Area of a Rectangle The **area** (A) is the number of square units needed to cover a surface. To find the area of a rectangle, multiply the length times the width, or $l \times w$. When finding area, the units also are multiplied. Area is given in square units.

Example Find the area of a rectangle with a length of 1 cm and a width of 10 cm.

Step 1 You know that the area is the length multiplied by the width.
$A = (1\text{ cm} \times 10\text{ cm})$

Step 2 Multiply the length by the width. Also multiply the units.
$A = 10\text{ cm}^2$

The area is 10 cm^2.

Practice Problem Find the area of a square whose sides measure 4 m.

Area of a Triangle To find the area of a triangle, use the formula:

$$A = \frac{1}{2}(\text{base} \times \text{height})$$

The base of a triangle can be any of its sides. The height is the perpendicular distance from a base to the opposite endpoint, or vertex.

Example Find the area of a triangle with a base of 18 m and a height of 7 m.

Step 1 You know that the area is $\frac{1}{2}$ the base times the height.
$A = \frac{1}{2}(18\text{ m} \times 7\text{ m})$

Step 2 Multiply $\frac{1}{2}$ by the product of 18×7. Multiply the units.
$A = \frac{1}{2}(126\text{ m}^2)$
$A = 63\text{ m}^2$

The area is 63 m^2.

Practice Problem Find the area of a triangle with a base of 27 cm and a height of 17 cm.

Perimeter

Problem 1
$P = 2(18\text{ m} + 7\text{ m})$
$P = 2(25\text{ m})$
$P = 50\text{ m}$

Problem 2
$P = 1.6\text{ cm} + 2.4\text{ cm} + 2.4\text{ cm}$
$P = 6.4\text{ cm}$

Area of a Rectangle

$A = (4\text{ m} \times 4\text{ m})$
$A = 16\text{ m}^2$

Area of a Triangle

$A = \frac{1}{2}(27\text{ cm} \times 17\text{ cm})$
$A = \frac{1}{2}(459\text{ cm}^2)$
$A = 229.5\text{ cm}^2$

Circumference of a Circle The **diameter** (d) of a circle is the distance across the circle through its center, and the **radius** (r) is the distance from the center to any point on the circle. The radius is half of the diameter. The distance around the circle is called the **circumference** (C). The formula for finding the circumference is:

$C = 2\pi r$ *or* $C = \pi d$

The circumference divided by the diameter is always equal to 3.1415926... This nonterminating and nonrepeating number is represented by the Greek letter π (pi). An approximation often used for π is 3.14.

Example 1 Find the circumference of a circle with a radius of 3 m.

Step 1 You know the formula for the circumference is 2 times the radius times π.
$C = 2\pi(3)$

Step 2 Multiply 2 times the radius.
$C - 6\pi$

Step 3 Multiply by π.
$C = 19$ m

The circumference is 19 m.

Example 2 Find the circumference of a circle with a diameter of 24.0 cm.

Step 1 You know the formula for the circumference is the diameter times π.
$C = \pi(24.0)$

Step 2 Multiply the diameter by π.
$C = 75.4$ cm

The circumference is 75.4 cm.

Practice Problem Find the circumference of a circle with a radius of 19 cm.

Area of a Circle The formula for the area of a circle is:
$A = \pi r^2$

Example 1 Find the area of a circle with a radius of 4.0 cm.

Step 1 $A = \pi(4.0)^2$

Step 2 Find the square of the radius.
$A = 16\pi$

Step 3 Multiply the square of the radius by π.
$A = 50\ cm^2$

The area of the circle is 50 cm^2.

Example 2 Find the area of a circle with a radius of 225 m.

Step 1 $A = \pi(225)^2$

Step 2 Find the square of the radius.
$A = 50625\pi$

Step 3 Multiply the square of the radius by π.
$A = 158962.5$

The area of the circle is 158,962 m^2.

Example 3 Find the area of a circle whose diameter is 20.0 mm.

Step 1 You know the formula for the area of a circle is the square of the radius times π, and that the radius is half of the diameter.
$A = \pi\left(\frac{20.0}{2}\right)^2$

Step 2 Find the radius.
$A = \pi(10.0)^2$

Step 3 Find the square of the radius.
$A = 100\pi$

Step 4 Multiply the square of the radius by π.
$A = 314\ mm^2$

The area is 314 mm^2.

Practice Problem Find the area of a circle with a radius of 16 m.

Circumference of a Circle

$C = 2\pi r$
$C = 2\pi(19)$
$C = 38\pi$
$C = 119.3$

Area of a Circle

$A = \pi r^2$
$A = \pi(16\ m)^2$
$A = \pi\ 256\ m^2$
$A = 803.8\ m^2$

Math Skill Handbook

MATH SKILL HANDBOOK 713

Volume

Problem 1

$V = 8\text{ m} \times 4\text{ m} \times 4\text{ m}$

$V = 128\text{ m}^3$

Problem 2

$V = \pi r^2 \times \text{height}$

$V = \left[\pi\left(\frac{1}{2} \times 7\right)^2\right] \times 16$

$V = [\pi(3.5)^2] \times 16$

$V = [\pi(12.25)] \times 16$

$V = 38.46 \times 16$

$V = 615.36$

Volume The measure of space occupied by a solid is the **volume** (V). To find the volume of a rectangular solid multiply the length times width times height, or $V = l \times w \times h$. It is measured in cubic units, such as cubic centimeters (cm^3).

Example Find the volume of a rectangular solid with a length of 2.0 m, a width of 4.0 m, and a height of 3.0 m.

Step 1 You know the formula for volume is the length times the width times the height.

$V = 2.0\text{ m} \times 4.0\text{ m} \times 3.0\text{ m}$

Step 2 Multiply the length times the width times the height.

$V = 24\text{ m}^3$

The volume is 24 m^3.

Practice Problem Find the volume of a rectangular solid that 8 m long, 4 m wide, and 4 m high.

To find the volume of other solids, multiply the area of the base times the height.

Example 1 Find the volume of a solid that has a triangular base with a length of 8.0 m and a height of 7.0 m. The height of the entire solid is 15.0 m.

Step 1 You know that the base is a triangle, and the area of a triangle is $\frac{1}{2}$ the base times the height, and the volume is the area of the base times the height.

$V = \left[\frac{1}{2}(b \times h)\right] \times 15$

Step 2 Find the area of the base.

$V = \left[\frac{1}{2}(8 \times 7)\right] \times 15$

$V = \left(\frac{1}{2} \times 56\right) \times 15$

Step 3 Multiply the area of the base by the height of the solid.

$V = 28 \times 15$

$V = 420\text{ m}^3$

The volume is 420 m^3.

Example 2 Find the volume of a cylinder that has a base with a radius of 12.0 cm, and a height of 21.0 cm.

Step 1 You know that the base is a circle, and the area of a circle is the square of the radius times π, and the volume is the area of the base times the height.

$V = (\pi r^2) \times 21$

$V = (\pi 12^2) \times 21$

Step 2 Find the area of the base.

$V = 144\pi \times 21$

$V = 452 \times 21$

Step 3 Multiply the area of the base by the height of the solid.

$V = 9490\text{ cm}^3$

The volume is 9490 cm^3.

Example 3 Find the volume of a cylinder that has a diameter of 15 mm and a height of 4.8 mm.

Step 1 You know that the base is a circle with an area equal to the square of the radius times π. The radius is one-half the diameter. The volume is the area of the base times the height.

$V = (\pi r^2) \times 4.8$

$V = \left[\pi\left(\frac{1}{2} \times 15\right)^2\right] \times 4.8$

$V = (\pi 7.5^2) \times 4.8$

Step 2 Find the area of the base.

$V = 56.25\pi \times 4.8$

$V = 176.63 \times 4.8$

Step 3 Multiply the area of the base by the height of the solid.

$V = 847.8$

The volume is 847.8 mm^3.

Practice Problem Find the volume of a cylinder with a diameter of 7 cm in the base and a height of 16 cm.

Science Applications

Measure in SI

The metric system of measurement was developed in 1795. A modern form of the metric system, called the International System (SI), was adopted in 1960 and provides the standard measurements that all scientists around the world can understand.

The SI system is convenient because unit sizes vary by powers of 10. Prefixes are used to name units. Look at **Table 3** for some common SI prefixes and their meanings.

Table 3 Common SI Prefixes

Prefix	Symbol	Meaning	
kilo-	k	1,000	thousand
hecto-	h	100	hundred
deka-	da	10	ten
deci-	d	0.1	tenth
centi-	c	0.01	hundredth
milli-	m	0.001	thousandth

Example How many grams equal one kilogram?

Step 1 Find the prefix *kilo* in **Table 3.**

Step 2 Using **Table 3,** determine the meaning of *kilo.* According to the table, it means 1,000. When the prefix *kilo* is added to a unit, it means that there are 1,000 of the units in a "*kilo*unit."

Step 3 Apply the prefix to the units in the question. The units in the question are grams. There are 1,000 grams in a kilogram.

Practice Problem Is a milligram larger or smaller than a gram? How many of the smaller units equal one larger unit? What fraction of the larger unit does one smaller unit represent?

Dimensional Analysis

Convert SI Units In science, quantities such as length, mass, and time sometimes are measured using different units. A process called dimensional analysis can be used to change one unit of measure to another. This process involves multiplying your starting quantity and units by one or more conversion factors. A conversion factor is a ratio equal to one and can be made from any two equal quantities with different units. If 1,000 mL equal 1 L then two ratios can be made.

$$\frac{1{,}000\text{ mL}}{1\text{ L}} = \frac{1\text{ L}}{1{,}000\text{ mL}} = 1$$

One can covert between units in the SI system by using the equivalents in **Table 3** to make conversion factors.

Example 1 How many cm are in 4 m?

Step 1 Write conversion factors for the units given. From **Table 3,** you know that 100 cm = 1 m. The conversion factors are

$$\frac{100\text{ cm}}{1\text{ m}} \quad \textit{and} \quad \frac{1\text{ m}}{100\text{ cm}}$$

Step 2 Decide which conversion factor to use. Select the factor that has the units you are converting from (m) in the denominator and the units you are converting to (cm) in the numerator.

$$\frac{100\text{ cm}}{1\text{ m}}$$

Step 3 Multiply the starting quantity and units by the conversion factor. Cancel the starting units with the units in the denominator. There are 400 cm in 4 m.

$$4\,\cancel{\text{m}} \times \frac{100\text{ cm}}{1\,\cancel{\text{m}}} = 400\text{ cm}$$

Practice Problem How many milligrams are in one kilogram? (Hint: You will need to use two conversion factors from **Table 3.**)

Math Skill Handbook

Measure in SI

smaller; 1000; one thousandth

Dimensional Analysis

$$x\text{ mg} = 1\,\cancel{\text{kg}} \times \frac{1000\,\cancel{\text{g}}}{1\,\cancel{\text{kg}}} \times \frac{1000\text{ mg}}{1\,\cancel{\text{g}}} =$$

1,000,000 mg

1,000,000 mg = 1 kg

Math Skill Handbook

MATH SKILL HANDBOOK 715

Convert Between Unit Systems

$$\frac{(1\ \text{in})^3}{(2.54\ \text{cm})^3} = \frac{1\ \text{in} \times 1\ \text{in} \times 1\ \text{in}}{2.54\ \text{cm} \times 2.54\ \text{cm} \times 2.54\ \text{cm}} = \frac{1\ \text{in}^3}{16.39\ \text{cm}^3}$$

Table 4 Unit System Equivalents

Type of Measurement	Equivalent
Length	1 in = 2.54 cm 1 yd = 0.91 m 1 mi = 1.61 km
Mass and Weight*	1 oz = 28.35 g 1 lb = 0.45 kg 1 ton (short) = 0.91 tonnes (metric tons) 1 lb = 4.45 N
Volume	$1\ \text{in}^3 = 16.39\ \text{cm}^3$ 1 qt = 0.95 L 1 gal = 3.78 L
Area	$1\ \text{in}^2 = 6.45\ \text{cm}^2$ $1\ \text{yd}^2 = 0.83\ \text{m}^2$ $1\ \text{mi}^2 = 2.59\ \text{km}^2$ 1 acre = 0.40 hectares
Temperature	$°\text{C} = \frac{(°\text{F} - 32)}{1.8}$ K = °C + 273

*Weight is measured in standard Earth gravity.

Convert Between Unit Systems **Table 4** gives a list of equivalents that can be used to convert between English and SI units.

Example If a meterstick has a length of 100 cm, how long is the meterstick in inches?

Step 1 Write the conversion factors for the units given. From **Table 4,** 1 in = 2.54 cm.

$$\frac{1\ \text{in}}{2.54\ \text{cm}} \quad \textit{and} \quad \frac{2.54\ \text{cm}}{1\ \text{in}}$$

Step 2 Determine which conversion factor to use. You are converting from cm to in. Use the conversion factor with cm on the bottom.

$$\frac{1\ \text{in}}{2.54\ \text{cm}}$$

Step 3 Multiply the starting quantity and units by the conversion factor. Cancel the starting units with the units in the denominator. Round your answer based on the number of significant figures in the conversion factor.

$$100\ \cancel{\text{cm}} \times \frac{1\ \text{in}}{2.54\ \cancel{\text{cm}}} = 39.37\ \text{in}$$

The meterstick is 39.4 in long.

Practice Problem A book has a mass of 5 lbs. What is the mass of the book in kg?

Practice Problem Use the equivalent for in and cm (1 in = 2.54 cm) to show how $1\ \text{in}^3 = 16.39\ \text{cm}^3$.

Precision and Significant Digits

When you make a measurement, the value you record depends on the precision of the measuring instrument. This precision is represented by the number of significant digits recorded in the measurement. When counting the number of significant digits, all digits are counted except zeros at the end of a number with no decimal point such as 2,050, and zeros at the beginning of a decimal such as 0.03020. When adding or subtracting numbers with different precision, round the answer to the smallest number of decimal places of any number in the sum or difference. When multiplying or dividing, the answer is rounded to the smallest number of significant digits of any number being multiplied or divided.

Example The lengths 5.28 and 5.2 are measured in meters. Find the sum of these lengths and record your answer using the correct number of significant digits.

Step 1 Find the sum.

5.28 m	2 digits after the decimal
+ 5.2 m	1 digit after the decimal
10.48 m	

Step 2 Round to one digit after the decimal because the least number of digits after the decimal of the numbers being added is 1.

The sum is 10.5 m.

Practice Problem How many significant digits are in the measurement 7,071,301 m? How many significant digits are in the measurement 0.003010 g?

Practice Problem Multiply 5.28 and 5.2 using the rule for multiplying and dividing. Record the answer using the correct number of significant digits.

Scientific Notation

Many times numbers used in science are very small or very large. Because these numbers are difficult to work with scientists use scientific notation. To write numbers in scientific notation, move the decimal point until only one non-zero digit remains on the left. Then count the number of places you moved the decimal point and use that number as a power of ten. For example, the average distance from the Sun to Mars is 227,800,000,000 m. In scientific notation, this distance is 2.278×10^{11} m. Because you moved the decimal point to the left, the number is a positive power of ten.

The mass of an electron is about 0.000 000 000 000 000 000 000 000 000 000 911 kg. Expressed in scientific notation, this mass is 9.11×10^{-31} kg. Because the decimal point was moved to the right, the number is a negative power of ten.

Example Earth is 149,600,000 km from the Sun. Express this in scientific notation.

Step 1 Move the decimal point until one non-zero digit remains on the left.

1.496 000 00

Step 2 Count the number of decimal places you have moved. In this case, eight.

Step 3 Show that number as a power of ten, 10^8.

The Earth is 1.496×10^8 km from the Sun.

Practice Problem How many significant digits are in 149,600,000 km? How many significant digits are in 1.496×10^8 km?

Practice Problem Parts used in a high performance car must be measured to 7×10^{-6} m. Express this number as a decimal.

Practice Problem A CD is spinning at 539 revolutions per minute. Express this number in scientific notation.

Precision and Significant Digits

Problem 1
7; 4

Problem 2
$5.28 \times 5.2 = 27.456$

5.28 has 3 significant digits.

5.2 has 2 significant digits.

When multiplying and dividing, the answer is rounded to the smallest number of significant digits of the numbers being multiplied or divided—in this case, 2.

27.456 is rounded to 27.

Scientific Notation

Problem 1
4; 4

Problem 2
0.000007

Problem 3
5.39×10^2

Make and Use Graphs

Data in tables can be displayed in a graph—a visual representation of data. Common graph types include line graphs, bar graphs, and circle graphs.

Line Graph A line graph shows a relationship between two variables that change continuously. The independent variable is changed and is plotted on the *x*-axis. The dependent variable is observed, and is plotted on the *y*-axis.

Example Draw a line graph of the data below from a cyclist in a long-distance race.

Table 5 Bicycle Race Data

Time (h)	Distance (km)
0	0
1	8
2	16
3	24
4	32
5	40

Step 1 Determine the *x*-axis and *y*-axis variables. Time varies independently of distance and is plotted on the *x*-axis. Distance is dependent on time and is plotted on the *y*-axis.

Step 2 Determine the scale of each axis. The *x*-axis data ranges from 0 to 5. The *y*-axis data ranges from 0 to 40.

Step 3 Using graph paper, draw and label the axes. Include units in the labels.

Step 4 Draw a point at the intersection of the time value on the *x*-axis and corresponding distance value on the *y*-axis. Connect the points and label the graph with a title, as shown in **Figure 20.**

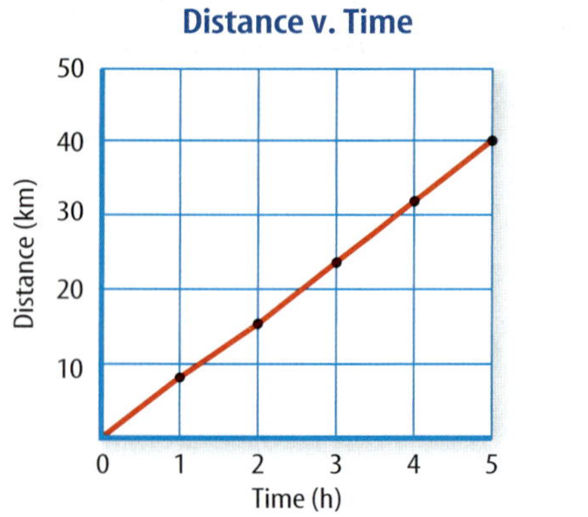

Figure 20 This line graph shows the relationship between distance and time during a bicycle ride.

Practice Problem A puppy's shoulder height is measured during the first year of her life. The following measurements were collected: (3 mo, 52 cm), (6 mo, 72 cm), (9 mo, 83 cm), (12 mo, 86). Graph this data.

Find a Slope The slope of a straight line is the ratio of the vertical change, rise, to the horizontal change, run.

$$\text{Slope} = \frac{\text{vertical change (rise)}}{\text{horizontal change (run)}} = \frac{\text{change in } y}{\text{change in } x}$$

Example Find the slope of the graph in **Figure 20.**

Step 1 You know that the slope is the change in *y* divided by the change in *x*.

$$\text{Slope} = \frac{\text{change in } y}{\text{change in } x}$$

Step 2 Determine the data points you will be using. For a straight line, choose the two sets of points that are the farthest apart.

$$\text{Slope} = \frac{(40-0)\text{ km}}{(5-0)\text{ hr}}$$

Step 3 Find the change in *y* and *x*.

$$\text{Slope} = \frac{40\text{ km}}{5\text{h}}$$

Step 4 Divide the change in *y* by the change in *x*.

$$\text{Slope} = \frac{8\text{ km}}{\text{h}}$$

The slope of the graph is 8 km/h.

Math Skill Handbook

Line Graph

x	y
3	52
6	72
9	83
12	86

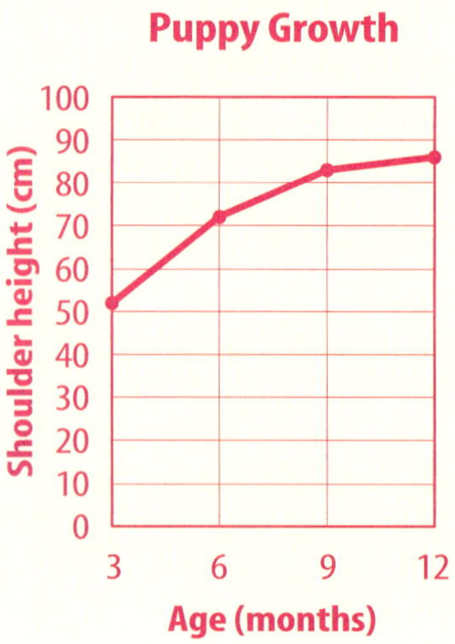

Math Skill Handbook

Bar Graph To compare data that does not change continuously you might choose a bar graph. A bar graph uses bars to show the relationships between variables. The *x*-axis variable is divided into parts. The parts can be numbers such as years, or a category such as a type of animal. The *y*-axis is a number and increases continuously along the axis.

Example A recycling center collects 4.0 kg of aluminum on Monday, 1.0 kg on Wednesday, and 2.0 kg on Friday. Create a bar graph of this data.

Step 1 Select the *x*-axis and *y*-axis variables. The measured numbers (the masses of aluminum) should be placed on the *y*-axis. The variable divided into parts (collection days) is placed on the *x*-axis.

Step 2 Create a graph grid like you would for a line graph. Include labels and units.

Step 3 For each measured number, draw a vertical bar above the *x*-axis value up to the *y*-axis value. For the first data point, draw a vertical bar above Monday up to 4.0 kg.

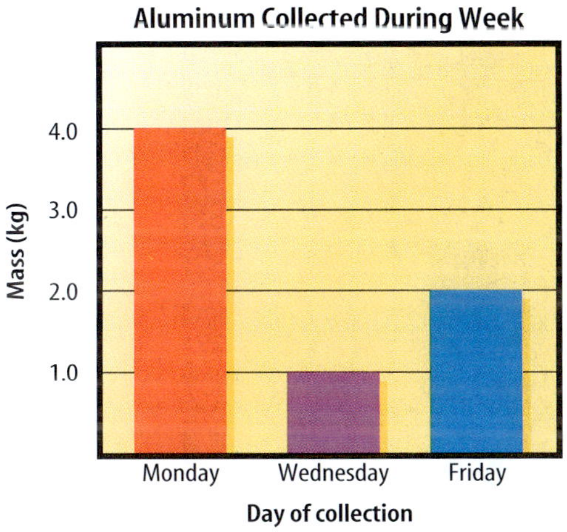

Practice Problem Draw a bar graph of the gases in air: 78% nitrogen, 21% oxygen, 1% other gases.

Circle Graph To display data as parts of a whole, you might use a circle graph. A circle graph is a circle divided into sections that represent the relative size of each piece of data. The entire circle represents 100%, half represents 50%, and so on.

Example Air is made up of 78% nitrogen, 21% oxygen, and 1% other gases. Display the composition of air in a circle graph.

Step 1 Multiply each percent by 360° and divide by 100 to find the angle of each section in the circle.

$$78\% \times \frac{360°}{100} = 280.8°$$

$$21\% \times \frac{360°}{100} = 75.6°$$

$$1\% \times \frac{360°}{100} = 3.6°$$

Step 2 Use a compass to draw a circle and to mark the center of the circle. Draw a straight line from the center to the edge of the circle.

Step 3 Use a protractor and the angles you calculated to divide the circle into parts. Place the center of the protractor over the center of the circle and line the base of the protractor over the straight line.

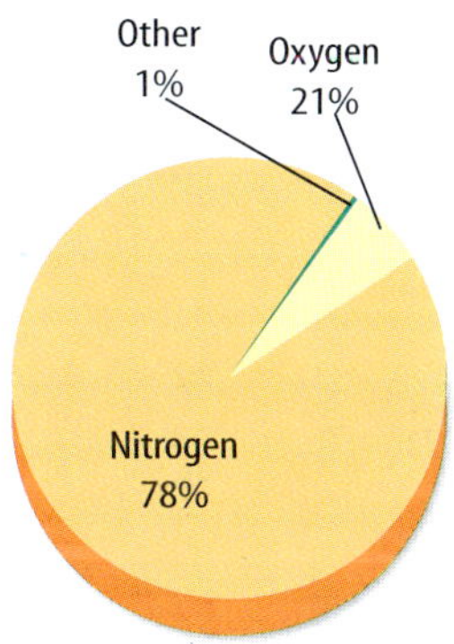

Practice Problem Draw a circle graph to represent the amount of aluminum collected during the week shown in the bar graph to the left.

Bar Graph

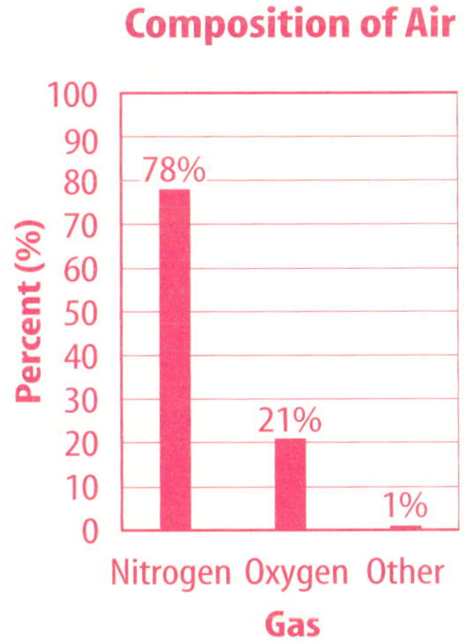

Circle Graph

The total amount of aluminum collected is:

$4.0 \text{ kg} + 1.0 \text{ kg} + 2.0 \text{ kg} = 7.0 \text{ kg}$

$\frac{4.0 \text{ kg}}{7.0 \text{ kg}} = \frac{x}{360°}$; $x = 206°$

$\frac{1.0 \text{ kg}}{7.0 \text{ kg}} = \frac{x}{360°}$; $x = 51°$

$\frac{2.0 \text{ kg}}{7.0 \text{ kg}} = \frac{x}{360°}$; $x = 103°$

MATH SKILL HANDBOOK 719

PERIODIC TABLE OF THE ELEMENTS

Columns of elements are called groups. Elements in the same group have similar chemical properties.

Element — Hydrogen
Atomic number — 1
Symbol — H
Atomic mass — 1.008
State of matter

Gas
Liquid
Solid
Synthetic

The first three symbols tell you the state of matter of the element at room temperature. The fourth symbol identifies elements that are not present in significant amounts on Earth. Useful amounts are made synthetically.

Period	1	2	3	4	5	6	7	8	9
1	Hydrogen 1 **H** 1.008								
2	Lithium 3 **Li** 6.941	Beryllium 4 **Be** 9.012							
3	Sodium 11 **Na** 22.990	Magnesium 12 **Mg** 24.305							
4	Potassium 19 **K** 39.098	Calcium 20 **Ca** 40.078	Scandium 21 **Sc** 44.956	Titanium 22 **Ti** 47.867	Vanadium 23 **V** 50.942	Chromium 24 **Cr** 51.996	Manganese 25 **Mn** 54.938	Iron 26 **Fe** 55.845	Cobalt 27 **Co** 58.933
5	Rubidium 37 **Rb** 85.468	Strontium 38 **Sr** 87.62	Yttrium 39 **Y** 88.906	Zirconium 40 **Zr** 91.224	Niobium 41 **Nb** 92.906	Molybdenum 42 **Mo** 95.94	Technetium 43 **Tc** (98)	Ruthenium 44 **Ru** 101.07	Rhodium 45 **Rh** 102.906
6	Cesium 55 **Cs** 132.905	Barium 56 **Ba** 137.327	Lanthanum 57 **La** 138.906	Hafnium 72 **Hf** 178.49	Tantalum 73 **Ta** 180.948	Tungsten 74 **W** 183.84	Rhenium 75 **Re** 186.207	Osmium 76 **Os** 190.23	Iridium 77 **Ir** 192.217
7	Francium 87 **Fr** (223)	Radium 88 **Ra** (226)	Actinium 89 **Ac** (227)	Rutherfordium 104 **Rf** (261)	Dubnium 105 **Db** (262)	Seaborgium 106 **Sg** (266)	Bohrium 107 **Bh** (264)	Hassium 108 **Hs** (277)	Meitnerium 109 **Mt** (268)

The number in parentheses is the mass number of the longest-lived isotope for that element.

Rows of elements are called periods. Atomic number increases across a period.

Series					
Lanthanide series	Cerium 58 **Ce** 140.116	Praseodymium 59 **Pr** 140.908	Neodymium 60 **Nd** 144.24	Promethium 61 **Pm** (145)	Samarium 62 **Sm** 150.36
Actinide series	Thorium 90 **Th** 232.038	Protactinium 91 **Pa** 231.036	Uranium 92 **U** 238.029	Neptunium 93 **Np** (237)	Plutonium 94 **Pu** (244)

The arrow shows where these elements would fit into the periodic table. They are moved to the bottom of the table to save space.

Metal

Metalloid

Nonmetal

The color of an element's block tells you if the element is a metal, nonmetal, or metalloid.

Science Online
Visit red.msscience.com for updates to the periodic table.

10	11	12	13	14	15	16	17	18
								Helium 2 He 4.003
			Boron 5 B 10.811	Carbon 6 C 12.011	Nitrogen 7 N 14.007	Oxygen 8 O 15.999	Fluorine 9 F 18.998	Neon 10 Ne 20.180
			Aluminum 13 Al 26.982	Silicon 14 Si 28.086	Phosphorus 15 P 30.974	Sulfur 16 S 32.065	Chlorine 17 Cl 35.453	Argon 18 Ar 39.948
Nickel 28 Ni 58.693	Copper 29 Cu 63.546	Zinc 30 Zn 65.409	Gallium 31 Ga 69.723	Germanium 32 Ge 72.64	Arsenic 33 As 74.922	Selenium 34 Se 78.96	Bromine 35 Br 79.904	Krypton 36 Kr 83.798
Palladium 46 Pd 106.42	Silver 47 Ag 107.868	Cadmium 48 Cd 112.411	Indium 49 In 114.818	Tin 50 Sn 118.710	Antimony 51 Sb 121.760	Tellurium 52 Te 127.60	Iodine 53 I 126.904	Xenon 54 Xe 131.293
Platinum 78 Pt 195.078	Gold 79 Au 196.967	Mercury 80 Hg 200.59	Thallium 81 Tl 204.383	Lead 82 Pb 207.2	Bismuth 83 Bi 208.980	Polonium 84 Po (209)	Astatine 85 At (210)	Radon 86 Rn (222)
Darmstadtium 110 Ds (281)	Unununium * 111 Uuu (272)	Ununbium * 112 Uub (285)		Ununquadium * 114 Uuq (289)		** 116		** 118

* The names and symbols for elements 111–114 are temporary. Final names will be selected when the elements' discoveries are verified.

** Elements 116 and 118 were thought to have been created. The claim was retracted because the experimental results could not be repeated.

Europium 63 Eu 151.964	Gadolinium 64 Gd 157.25	Terbium 65 Tb 158.925	Dysprosium 66 Dy 162.500	Holmium 67 Ho 164.930	Erbium 68 Er 167.259	Thulium 69 Tm 168.934	Ytterbium 70 Yb 173.04	Lutetium 71 Lu 174.967
Americium 95 Am (243)	Curium 96 Cm (247)	Berkelium 97 Bk (247)	Californium 98 Cf (251)	Einsteinium 99 Es (252)	Fermium 100 Fm (257)	Mendelevium 101 Md (258)	Nobelium 102 No (259)	Lawrencium 103 Lr (262)

Topographic Map Symbols

Topographic Map Symbols			
	Primary highway, hard surface		Index contour
	Secondary highway, hard surface		Supplementary contour
	Light-duty road, hard or improved surface		Intermediate contour
	Unimproved road		Depression contours
	Railroad: single track		
	Railroad: multiple track		Boundaries: national
	Railroads in juxtaposition		State
			County, parish, municipal
	Buildings		Civil township, precinct, town, barrio
cem	Schools, church, and cemetery		Incorporated city, village, town, hamlet
	Buildings (barn, warehouse, etc.)		Reservation, national or state
	Wells other than water (labeled as to type)		Small park, cemetery, airport, etc.
	Tanks: oil, water, etc. (labeled only if water)		Land grant
	Located or landmark object; windmill		Township or range line, U.S. land survey
	Open pit, mine, or quarry; prospect		
			Township or range line, approximate location
	Marsh (swamp)		
	Wooded marsh		Perennial streams
	Woods or brushwood		Elevated aqueduct
	Vineyard		Water well and spring
	Land subject to controlled inundation		Small rapids
	Submerged marsh		Large rapids
	Mangrove		Intermittent lake
	Orchard		Intermittent stream
	Scrub		Aqueduct tunnel
	Urban area		Glacier
			Small falls
x7369	Spot elevation		Large falls
670	Water elevation		Dry lake bed

Rocks

Rocks		
Rock Type	**Rock Name**	**Characteristics**
Igneous (intrusive)	Granite	Large mineral grains of quartz, feldspar, hornblende, and mica. Usually light in color.
	Diorite	Large mineral grains of feldspar, hornblende, and mica. Less quartz than granite. Intermediate in color.
	Gabbro	Large mineral grains of feldspar, augite, and olivine. No quartz. Dark in color.
Igneous (extrusive)	Rhyolite	Small mineral grains of quartz, feldspar, hornblende, and mica, or no visible grains. Light in color.
	Andesite	Small mineral grains of feldspar, hornblende, and mica or no visible grains. Intermediate in color.
	Basalt	Small mineral grains of feldspar, augite, and possibly olivine or no visible grains. No quartz. Dark in color.
	Obsidian	Glassy texture. No visible grains. Volcanic glass. Fracture looks like broken glass.
	Pumice	Frothy texture. Floats in water. Usually light in color.
Sedimentary (detrital)	Conglomerate	Coarse grained. Gravel or pebble-size grains.
	Sandstone	Sand-sized grains 1/16 to 2 mm.
	Siltstone	Grains are smaller than sand but larger than clay.
	Shale	Smallest grains. Often dark in color. Usually platy.
Sedimentary (chemical or organic)	Limestone	Major mineral is calcite. Usually forms in oceans and lakes. Often contains fossils.
	Coal	Forms in swampy areas. Compacted layers of organic material, mainly plant remains.
Sedimentary (chemical)	Rock Salt	Commonly forms by the evaporation of seawater.
Metamorphic (foliated)	Gneiss	Banding due to alternate layers of different minerals, of different colors. Parent rock often is granite.
	Schist	Parallel arrangement of sheetlike minerals, mainly micas. Forms from different parent rocks.
	Phyllite	Shiny or silky appearance. May look wrinkled. Common parent rocks are shale and slate.
	Slate	Harder, denser, and shinier than shale. Common parent rock is shale.
Metamorphic (nonfoliated)	Marble	Calcite or dolomite. Common parent rock is limestone.
	Soapstone	Mainly of talc. Soft with greasy feel.
	Quartzite	Hard with interlocking quartz crystals. Common parent rock is sandstone.

Minerals

Minerals

Mineral (formula)	Color	Streak	Hardness	Breakage Pattern	Uses and Other Properties
Graphite (C)	black to gray	black to gray	1–1.5	basal cleavage (scales)	pencil lead, lubricants for locks, rods to control some small nuclear reactions, battery poles
Galena (PbS)	gray	gray to black	2.5	cubic cleavage perfect	source of lead, used for pipes, shields for X rays, fishing equipment sinkers
Hematite (Fe_2O_3)	black or reddish-brown	reddish-brown	5.5–6.5	irregular fracture	source of iron; converted to pig iron, made into steel
Magnetite (Fe_3O_4)	black	black	6	conchoidal fracture	source of iron, attracts a magnet
Pyrite (FeS_2)	light, brassy, yellow	greenish-black	6–6.5	uneven fracture	fool's gold
Talc ($Mg_3 Si_4O_{10}(OH)_2$)	white, greenish	white	1	cleavage in one direction	used for talcum powder, sculptures, paper, and tabletops
Gypsum ($CaSO_4 \cdot 2H_2O$)	colorless, gray, white, brown	white	2	basal cleavage	used in plaster of paris and dry wall for building construction
Sphalerite (ZnS)	brown, reddish-brown, greenish	light to dark brown	3.5–4	cleavage in six directions	main ore of zinc; used in paints, dyes, and medicine
Muscovite ($KAl_3Si_3O_{10}(OH)_2$)	white, light gray, yellow, rose, green	colorless	2–2.5	basal cleavage	occurs in large, flexible plates; used as an insulator in electrical equipment, lubricant
Biotite ($K(Mg,Fe)_3(AlSi_3O_{10})(OH)_2$)	black to dark brown	colorless	2.5–3	basal cleavage	occurs in large, flexible plates
Halite (NaCl)	colorless, red, white, blue	colorless	2.5	cubic cleavage	salt; soluble in water; a preservative

Minerals

Minerals					
Mineral (formula)	**Color**	**Streak**	**Hardness**	**Breakage Pattern**	**Uses and Other Properties**
Calcite ($CaCO_3$)	colorless, white, pale blue	colorless, white	3	cleavage in three directions	fizzes when HCl is added; used in cements and other building materials
Dolomite ($CaMg(CO_3)_2$)	colorless, white, pink, green, gray, black	white	3.5–4	cleavage in three directions	concrete and cement; used as an ornamental building stone
Fluorite (CaF_2)	colorless, white, blue, green, red, yellow, purple	colorless	4	cleavage in four directions	used in the manufacture of optical equipment; glows under ultraviolet light
Hornblende ($(CaNa)_{2-3}(Mg,Al,Fe)_5\text{-}(Al,Si)_2Si_6O_{22}(OH)_2$)	green to black	gray to white	5–6	cleavage in two directions	will transmit light on thin edges; 6-sided cross section
Feldspar ($(KAlSi_3O_8)(NaAlSi_3O_8)$, $(CaAl_2Si_2O_8)$)	colorless, white to gray, green	colorless	6	two cleavage planes meet at 90° angle	used in the manufacture of ceramics
Augite ($((Ca,Na)(Mg,Fe,Al)(Al,Si)_2O_6)$)	black	colorless	6	cleavage in two directions	square or 8-sided cross section
Olivine ($((Mg,Fe)_2SiO_4)$)	olive, green	none	6.5–7	conchoidal fracture	gemstones, refractory sand
Quartz (SiO_2)	colorless, various colors	none	7	conchoidal fracture	used in glass manufacture, electronic equipment, radios, computers, watches, gemstones

Diversity of Life: Classification of Living Organisms

A six-kingdom system of classification of organisms is used today. Two kingdoms—Kingdom Archaebacteria and Kingdom Eubacteria—contain organisms that do not have a nucleus and that lack membrane-bound structures in the cytoplasm of their cells. The members of the other four kingdoms have a cell or cells that contain a nucleus and structures in the cytoplasm, some of which are surrounded by membranes. These kingdoms are Kingdom Protista, Kingdom Fungi, Kingdom Plantae, and Kingdom Animalia.

Kingdom Archaebacteria

one-celled; some absorb food from their surroundings; some are photosynthetic; some are chemosynthetic; many are found in extremely harsh environments including salt ponds, hot springs, swamps, and deep-sea hydrothermal vents

Kingdom Eubacteria

one-celled; most absorb food from their surroundings; some are photosynthetic; some are chemosynthetic; many are parasites; many are round, spiral, or rod-shaped; some form colonies

Kingdom Protista

Phylum Euglenophyta one-celled; photosynthetic or take in food; most have one flagellum; euglenoids

Phylum Bacillariophyta one-celled; photosynthetic; have unique double shells made of silica; diatoms

Phylum Dinoflagellata one-celled; photosynthetic; contain red pigments; have two flagella; dinoflagellates

Phylum Chlorophyta one-celled, many-celled, or colonies; photosynthetic; contain chlorophyll; live on land, in freshwater, or salt water; green algae

Phylum Rhodophyta most are many-celled; photosynthetic; contain red pigments; most live in deep, saltwater environments; red algae

Phylum Phaeophyta most are many-celled; photosynthetic; contain brown pigments; most live in saltwater environments; brown algae

Phylum Rhizopoda one-celled; take in food; are free-living or parasitic; move by means of pseudopods; amoebas

Kingdom Eubacteria
Bacillus anthracis

Phylum Chlorophyta
Desmids

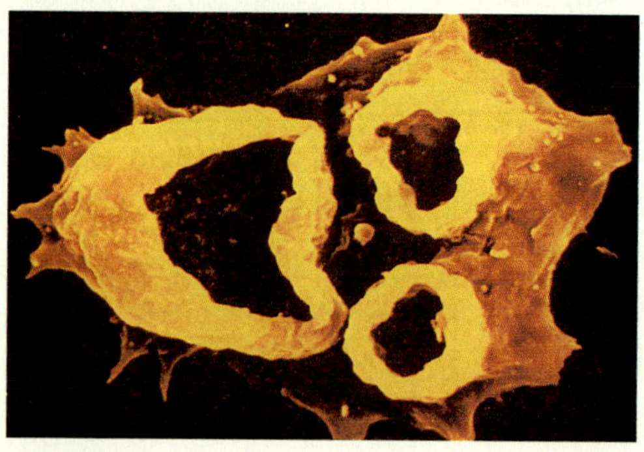

Amoeba

Phylum Zoomastigina one-celled; take in food; free-living or parasitic; have one or more flagella; zoomastigotes

Phylum Ciliophora one-celled; take in food; have large numbers of cilia; ciliates

Phylum Sporozoa one-celled; take in food; have no means of movement; are parasites in animals; sporozoans

Phylum Myxomycota
Slime mold

Phylum Oomycota
Phytophthora infestans

Phyla Myxomycota and Acrasiomycota one- or many-celled; absorb food; change form during life cycle; cellular and plasmodial slime molds

Phylum Oomycota many-celled; are either parasites or decomposers; live in freshwater or salt water; water molds, rusts and downy mildews

Kingdom Fungi

Phylum Zygomycota many-celled; absorb food; spores are produced in sporangia; zygote fungi; bread mold

Phylum Ascomycota one- and many-celled; absorb food; spores produced in asci; sac fungi; yeast

Phylum Basidiomycota many-celled; absorb food; spores produced in basidia; club fungi; mushrooms

Phylum Deuteromycota members with unknown reproductive structures; imperfect fungi; *Penicillium*

Phylum Mycophycota organisms formed by symbiotic relationship between an ascomycote or a basidiomycote and green alga or cyanobacterium; lichens

Lichens

Kingdom Plantae

Divisions Bryophyta (mosses), **Anthocerophyta** (hornworts), **Hepaticophyta** (liverworts), **Psilophyta** (whisk ferns) many-celled nonvascular plants; reproduce by spores produced in capsules; green; grow in moist, land environments

Division Lycophyta many-celled vascular plants; spores are produced in conelike structures; live on land; are photosynthetic; club mosses

Division Arthrophyta vascular plants; ribbed and jointed stems; scalelike leaves; spores produced in conelike structures; horsetails

Division Pterophyta vascular plants; leaves called fronds; spores produced in clusters of sporangia called sori; live on land or in water; ferns

Division Ginkgophyta deciduous trees; only one living species; have fan-shaped leaves with branching veins and fleshy cones with seeds; ginkgoes

Division Cycadophyta palmlike plants; have large, featherlike leaves; produces seeds in cones; cycads

Division Coniferophyta deciduous or evergreen; trees or shrubs; have needlelike or scalelike leaves; seeds produced in cones; conifers

Division Gnetophyta shrubs or woody vines; seeds are produced in cones; division contains only three genera; gnetum

Division Anthophyta dominant group of plants; flowering plants; have fruits with seeds

Kingdom Animalia

Phylum Porifera aquatic organisms that lack true tissues and organs; are asymmetrical and sessile; sponges

Phylum Cnidaria radially symmetrical organisms; have a digestive cavity with one opening; most have tentacles armed with stinging cells; live in aquatic environments singly or in colonies; includes jellyfish, corals, hydra, and sea anemones

Phylum Platyhelminthes bilaterally symmetrical worms; have flattened bodies; digestive system has one opening; parasitic and free-living species; flatworms

Division Anthophyta
Tomato plant

Division Bryophyta
Liverwort

Phylum Platyhelminthes
Flatworm

Phylum Chordata

Phylum Nematoda round, bilaterally symmetrical body; have digestive system with two openings; free-living forms and parasitic forms; roundworms

Phylum Mollusca soft-bodied animals, many with a hard shell and soft foot or footlike appendage; a mantle covers the soft body; aquatic and terrestrial species; includes clams, snails, squid, and octopuses

Phylum Annelida bilaterally symmetrical worms; have round, segmented bodies; terrestrial and aquatic species; includes earthworms, leeches, and marine polychaetes

Phylum Arthropoda largest animal group; have hard exoskeletons, segmented bodies, and pairs of jointed appendages; land and aquatic species; includes insects, crustaceans, and spiders

Phylum Echinodermata marine organisms; have spiny or leathery skin and a water-vascular system with tube feet; are radially symmetrical; includes sea stars, sand dollars, and sea urchins

Phylum Chordata organisms with internal skeletons and specialized body systems; most have paired appendages; all at some time have a notochord, nerve cord, gill slits, and a post-anal tail; include fish, amphibians, reptiles, birds, and mammals

Use and Care of a Microscope

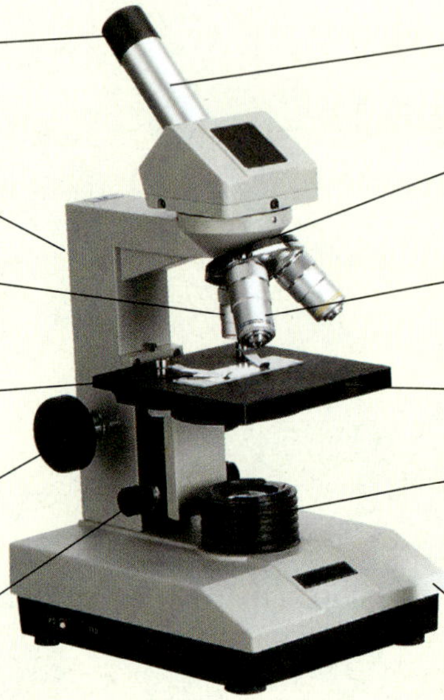

Caring for a Microscope

1. Always carry the microscope holding the arm with one hand and supporting the base with the other hand.
2. Don't touch the lenses with your fingers.
3. The coarse adjustment knob is used only when looking through the lowest-power objective lens. The fine adjustment knob is used when the high-power objective is in place.
4. Cover the microscope when you store it.

Using a Microscope

1. Place the microscope on a flat surface that is clear of objects. The arm should be toward you.
2. Look through the eyepiece. Adjust the diaphragm so light comes through the opening in the stage.
3. Place a slide on the stage so the specimen is in the field of view. Hold it firmly in place by using the stage clips.
4. Always focus with the coarse adjustment and the low-power objective lens first. After the object is in focus on low power, turn the nosepiece until the high-power objective is in place. Use ONLY the fine adjustment to focus with the high-power objective lens.

Making a Wet-Mount Slide

1. Carefully place the item you want to look at in the center of a clean, glass slide. Make sure the sample is thin enough for light to pass through.
2. Use a dropper to place one or two drops of water on the sample.
3. Hold a clean coverslip by the edges and place it at one edge of the water. Slowly lower the coverslip onto the water until it lies flat.
4. If you have too much water or a lot of air bubbles, touch the edge of a paper towel to the edge of the coverslip to draw off extra water and draw out unwanted air.

Glossary/Glosario

Cómo usar el glosario en español:
1. Busca el término en inglés que desees encontrar.
2. El término en español, junto con la definición, se encuentran en la columna de la derecha.

Pronunciation Key

Use the following key to help you sound out words in the glossary.

a	back (BAK)	ew	food (FEWD)
ay	day (DAY)	yoo	pure (PYOOR)
ah	father (FAH thur)	yew	few (FYEW)
ow	flower (FLOW ur)	uh	comma (CAH muh)
ar	car (CAR)	u (+ con)	rub (RUB)
e	less (LES)	sh	shelf (SHELF)
ee	leaf (LEEF)	ch	nature (NAY chur)
ih	trip (TRIHP)	g	gift (GIHFT)
i (i + con + e)	idea (i DEE uh)	j	gem (JEM)
oh	go (GOH)	ing	sing (SING)
aw	soft (SAWFT)	zh	vision (VIH zhun)
or	orbit (OR buht)	k	cake (KAYK)
oy	coin (COYN)	s	seed, cent (SEED, SENT)
oo	foot (FOOT)	z	zone, raise (ZOHN, RAYZ)

English — A — Español

abiotic (ay bi AH tihk) factors: the nonliving parts of an ecosystem, including soil, temperature, water, and sunlight. (p. 622)

abrasion: a form of erosion that occurs when wind blows sediments into rocks, makes pits in the rocks, and produces smooth, polished surfaces. (p. 328)

acceleration: change in velocity divided by the amount of time needed for the change to take place; occurs when an object speeds up, slows down, or changes direction. (p. 133)

accuracy: compares a measurement to the true value. (p. 45)

acid rain: form of pollution that occurs when gases released by burning oil and coal mix with water in the air to form rain or snow that is strongly acid. (p. 661)

aerosols (ER uh sahls): solids, such as dust, salt, and pollen, and liquid droplets, such as acids, that are suspended in the atmosphere. (p. 343)

air mass: large body of air that develops over a particular region of Earth's surface. (p. 356)

alveoli: grapelike clusters of air sacs at the end of each bronchiole. (p. 567)

factores abióticos: las partes no vivientes de un ecosistema, incluyendo la tierra, la temperatura, el agua y la luz solar. (p. 622)

abrasión: forma de erosión que ocurre cuando la acción del viento causa que los sedimentos penetren en las rocas, hace huecos en ellas y produce superficies lisas y pulidas. (p. 328)

aceleración: cambio en la velocidad dividido por el tiempo necesario para que dicho cambio tenga lugar; ocurre cuando un objeto viaja más rápido, más lento o cambia de dirección. (p. 133)

precisión: comparación de una medida con el valor real. (p. 45)

lluvia ácida: es una forma de contaminación que ocurre cuando los gases liberados por la quema de petróleo y carbón se mezclan en el aire para formar lluvia o nieve fuertemente ácida. (p. 661)

aerosoles: sólidos como el polvo, la sal, el polen y gotas de líquido, tales como los ácidos, que están suspendidos en la atmósfera. (p. 343)

masa de aire: gran masa de aire que se desarrolla sobre determinada región de la superficie terrestre. (p. 356)

alvéolos: racimos de sacos de aire parecidos a las uvas, encontrados en el extremo de los bronquiolos. (p. 567)

amniotic egg: adaptation of reptiles that allows them to reproduce on land; encloses the embryo within a moist environment, protected by a leathery shell, and has a yolk that supplies the embryo with food. (p. 539)

amplitude: for a transverse wave, one half the distance between a crest and a trough. (p. 231)

appendage: structure such as a claw, leg, or antenna that grows from the body. (p. 512)

arthropod: bilaterally symmetrical animal with jointed appendages, a protective endoskeleton, and a segmented body. (p. 512)

asexual (ay SEK shul) reproduction: a type of reproduction, such as budding or regeneration, in which a new organism is produced from a part of another organism by the process of mitosis. (p. 593)

astronomical (as truh NAHM ih kul) unit: unit of measure that equals 150 million km, which is the mean distance from Earth to the Sun. (p. 449)

atmosphere: layer of gases surrounding Earth that protects living things from harmful doses of ultraviolet radiation and X-ray radiation and absorbs and distributes warmth. (p. 342)

atom: a very small particle that makes up most kinds of matter and consists of smaller parts called protons, neutrons, and electrons. (p. 99)

atomic mass: average mass of an atom of an element; its unit of measure is the atomic mass unit (u), which is 1/12 the mass of a carbon-12 atom. (p. 110)

atomic number: number of protons in the nucleus of each atom of a given element; is the top number in the periodic table. (p. 109)

average speed: equals the distance traveled divided by the amount of time it takes to travel that distance. (p. 130)

huevo amniótico: adaptación de los reptiles que les permite reproducirse en la tierra; envuelve al embrión en un medio húmedo protegido por un caparazón correoso y tiene una yema que proporciona alimentos al embrión. (p. 539)

amplitud: la mitad de la distancia entre la cresta y el valle en una onda transversal. (p. 231)

apéndice: estructura en forma de pinza, pata o antena que se proyecta del cuerpo. (p. 512)

artrópodo: animal simétrico bilateralmente con apéndices articulados, endoesqueleto protector y cuerpo segmentado. (p. 512)

reproducción asexual: tipo de reproducción tal como el injerto o la regeneración en la cual se produce un nuevo organismo a partir de una parte de otro organismo mediante el proceso de mitosis. (p. 593)

unidad astronómica: unidad de medida que equivale a 150 millones de kilómetros, que es la distancia media de la Tierra al sol. (p. 449)

atmósfera: capa de gases que rodea a la Tierra y que protege a los seres vivos contra dosis perjudiciales de radiaciones ultravioletas y rayos X y que absorbe y distribuye calor. (p. 342)

átomo: partícula muy pequeña que constituye la mayoría de los tipos de materia y que está formada por partes más pequeñas llamadas protones, neutrones y electrones. (p. 99)

masa atómica: masa promedio de un átomo de un elemento; su unidad de medida es la unidad de masa atómica (u), la cual es 1/12 de la masa de un átomo de carbono-12. (p. 110)

número atómico: número de protones en el núcleo de un átomo de determinado elemento; es el número superior en la tabla periódica. (p. 109)

velocidad promedio: es igual a la distancia recorrida dividida por el tiempo necesario para recorrer dicha distancia. (p. 130)

B

bacteria: smallest organisms on Earth, each of which is made up of only one cell. (p. 477)

bar graph: a type of graph that uses bars of varying sizes to show the relationship among variables. (p. 58)

biosphere: the part of Earth that supports life, including the top portion of Earth's crust, the atmosphere, and all the water on Earth's surface. (p. 620)

bacteria: los organismos más pequeños en la Tierra, cada uno de los cuales está formado por una sola célula. (p. 477)

gráfico de barras: tipo de gráfico que usa barras de diferentes tamaños para mostrar las diferencias entre las variables. (p. 58)

biosfera: parte de la Tierra que alberga la vida, incluyendo la porción superior de la corteza terrestre, la atmósfera y toda el agua de la superficie terrestre. (p. 620)

biotic factors: the living parts of an ecosystem. (p. 621)

boiling point: temperature at which a substance in a liquid state becomes a gas. (p. 75)

factores bióticos: las partes vivientes de un ecosistema. (p. 621)

punto de ebullición: temperatura a la cual una sustancia en estado líquido se convierte en gas. (p. 75)

C

capillary: microscopic blood vessel. (p. 567)

carnivore: meat-eating animal with sharp canine teeth specialized to rip and tear flesh. (p. 546)

cartilage (KART uh lihj): tough, flexible tissue similar to bone but is softer and less brittle. (p. 532)

catalyst: substance that changes the rate of a chemical reaction without any permanent change to its own structure. (p. 182)

cell membrane: flexible structure that holds a cell together, forms a boundary between the cell and its environment, and helps control what enters and leaves the cell. (p. 479)

cell wall: structure of plants, algae, fungi, and many types of bacteria that supports and protects the cell membrane. (p. 479)

charging by contact: transfer of electric charge between objects in contact. (p. 196)

charging by induction: rearrangement of electric charge in an object due to a nearby electric field. (p. 197)

chemical change: change in which the identity of a substance changes due to its chemical properties and forms a new substance or substances. (p. 81)

chemical property: any characteristic, such as the ability to burn, that allows a substance to undergo a change that results in a new substance. (p. 80)

chemical weathering: process in which the chemical composition of rocks is changed by agents such as natural acids and oxygen. (p. 318)

chemosynthesis: process that occurs in deep ocean water, where sunlight does not penetrate, in which bacteria make food from dissolved sulfur compounds. (p. 391)

chloroplast (KLOR uh plast): green organelle in a plant's leaf cells where most photosynthesis takes place. (p. 481)

chordate: animal that at some time in its development has a notochord, nerve cord, and pharyngeal pouches. (p. 530)

capilar: vaso sanguíneo microscópico. (p. 567)

carnívoro: animal que se alimenta de carne y posee dientes caninos afilados especializados en desgarrar y arrancar carne. (p. 546)

cartílago: tejido fuerte y flexible similar al hueso pero más suave y menos quebradizo que éste. (p. 532)

catalizador: sustancia que cambia la velocidad de una reacción química sin sufrir ningún cambio permanente en su propia estructura. (p. 182)

membrana celular: estructura flexible que mantiene unida a la célula, constituye un límite entre la célula y su entorno y ayuda a controlar todo aquello que entre o salga de ésta. (p. 479)

pared celular: estructura de las plantas, algas, hongos y varios tipos de bacterias, la cual sostiene y protege a la membrana celular. (p. 479)

carga por contacto: transferencia de carga eléctrica entre dos objetos en contacto. (p. 196)

carga por inducción: redistribución de la carga eléctrica en un objeto debido a un campo eléctrico cercano. (p. 197)

cambio químico: cambio producido en la identidad de una sustancia debido a sus propiedades químicas para formar una nueva sustancia o sustancias. (p. 81)

propiedad química: cualquier característica, como la capacidad para quemarse, que permite a una sustancia sufrir un cambio que da como resultado una nueva sustancia. (p. 80)

desgaste químico: proceso en el que agentes tales como ácidos naturales y oxígeno cambian la composición química de las rocas. (p. 318)

quimiosíntesis: proceso que ocurre en las aguas profundas del océano, en donde la luz del sol no penetra, en el cual las bacterias producen alimentos a partir de compuestos sulfúricos disueltos. (p. 391)

cloroplasto: organelo verde de la célula de las hojas de las plantas en donde tiene lugar la mayor parte de la fotosíntesis. (p. 481)

cordado: animal que en algún momento de su desarrollo tiene un notocordio, cordón nervioso, y pequeñas bolsas faríngeas. (p. 530)

circle graph: a type of graph that shows the parts of a whole; sometimes called a pie graph, each piece of which represents a percentage of the total. (p. 58)

cloning: making copies of organisms, each of which is a clone that receives DNA from only one parent. (p. 594)

closed circulatory system: a type of blood-circulation system in which blood is transported through blood vessels rather than washing over the organs. (p. 508)

cnidarian (NIH dar ee un): radially symmetrical, hollow-bodied animal with two cell layers organized into tissues. (p. 502)

comet: large body of ice and rock that orbits the Sun; develops a bright, glowing tail if it passes close to the Sun. (p. 454)

community: all the populations that live in an ecosystem. (p. 628)

compound: a substance produced when elements combine and whose properties are different from each of the elements in it. (p. 113)

compound machine: a machine that is a combination of simple machines. (p. 146)

compressional wave: a type of mechanical wave in which matter in the medium moves forward and backward along the direction the wave travels. (p. 229)

conduction: transfer of energy by collisions between the atoms in a material. (p. 174)

conductor: material in which electric charges can move easily. (p. 197)

constant: variable that is not changed in an experiment. (p. 18)

constellation (kan stuh LAY shun): group of stars that forms a pattern in the sky and can be named after a real or imaginary animal, object, or person. (p. 456)

consumer: an organism that obtains food by eating other organisms. (pp. 389, 635)

contour feather: strong, lightweight feather that gives a bird its shape and coloring and can help the bird steer, attract a mate, and avoid predators. (p. 543)

controlled experiment: involves changing one factor and observing its effect on one thing while keeping all other things constant. (p. 18)

convection: transfer of heat that occurs when particles move between regions or objects that have different temperatures. (p. 176)

creep: a process in which sediments move slowly downhill. (p. 325)

critical thinking: involves using knowledge and thinking skills to evaluate evidence and explanations. (p. 27)

gráfico circular: tipo de gráfico que muestra las partes de un todo; algunas veces se le llama gráfico de pastel en el que cada parte representa un porcentaje del total. (p. 58)

clonación: el hecho de hacer copias de organismos, cada uno de los cuales es un clon que recibe ADN solamente de un progenitor. (p. 594)

sistema circulatorio cerrado: tipo de sistema circulatorio en el cual la sangre es transportada a través de vasos sanguíneos en lugar de bañar los órganos. (p. 508)

cnidario: animal de cuerpo hueco simétricamente radial con dos capas de células organizadas en tejidos. (p. 502)

cometa: cuerpo largo compuesto por hielo y roca, y que circunda al sol; desarrolla una cola luminosa y brillante al pasar cerca del Sol. (p. 454)

comunidad: todas las especies que habitan en un ecosistema. (p. 628)

compuesto: sustancia resultante de la combinación de elementos cuyas propiedades son diferentes de los elementos que la componen. (p. 113)

máquina compuesta: máquina que es el producto de la combinación de máquinas simples. (p. 146)

onda de compresión: tipo de onda mecánica en la que la materia en el medio se mueve hacia adelante y hacia atrás en dirección de la onda. (p. 229)

coducción: transferencia de energía por medio de colisiones entre los átomos de un material. (p. 174)

conductor: material en el cual las cargas eléctricas se pueden mover fácilmente. (p. 197)

constante: variable que no cambia en un experimento. (p. 18)

constelación: grupo de estrellas que forman un dibujo en el cielo y al que se le puede dar el nombre de un animal, objeto o personaje real o imaginario. (p. 456)

consumidor: organismo que se alimenta de otros organismos. (pp. 389, 635)

pluma de contorno: pluma fuerte y liviana que da a las aves su forma y color y les ayuda a dirigirse, atraer a una pareja y evadir a los depredadores. (p. 543)

experimento controlado: consiste en cambiar un factor y observar su efecto sobre algo mientras el resto de las cosas se mantiene constante. (p. 18)

convección: transferencia de calor que ocurre cuando las partículas se mueven entre regiones u objetos que tienen diferentes temperaturas. (p. 176)

escurrimiento: proceso en el cual los sedimentos se mueven lentamente cuesta abajo. (p. 325)

pensamiento crítico: consiste en utilizar los conocimientos y habilidades del pensamiento para evaluar evidencias y explicaciones. (p. 27)

crust: Earth's outermost layer that is thinnest under the oceans and thickest through the mountains and contains all features of Earth's surface. (p. 290)

corteza: capa más externa de la Tierra, la cual es más delgada debajo de los océanos y más gruesa en las montañas y contiene todas las características de la superficie terrestre. (p. 290)

crystal: solid material with atoms arranged in a repeating pattern. (p. 258)

cristal: material sólido con átomos distribuidos en un patrón repetido. (p. 258)

cytoplasm (SI tuh pla zum): gelatinlike substance inside the cell membrane that contains water, chemicals, and cell parts. (p. 479)

citoplasma: sustancia gelatinosa en el interior de la membrana celular, la cual contiene agua, químicos y partes de la célula. (p. 479)

D

data: information gathered during an investigation; recorded in the form of descriptions, tables, graphs, or drawings. (p. 28)

datos: información recopilada durante una investigación y archivada en forma de descripciones, tablas, gráficas o planos. (p. 28)

decomposer: organism that breaks down tissue and releases nutrients and carbon dioxide back into the ecosystem. (pp. 390, 635)

descomponedor: organismo que descompone un tejido y libera nutrientes y dióxido de carbón al ecosistema. (pp. 390, 635)

deflation: erosion of land that occurs when wind blows across loose sediments and carries them away, often leaving behind particles too heavy to move. (p. 328)

deflación: erosión del terreno que ocurre cuando el viento sopla sobre sedimentos sueltos y se los lleva, con frecuencia dejando atrás las partículas muy pesadas. (p. 328)

density: measurable physical property that can be found by dividing the mass of an object by its volume. (p. 72)

densidad: propiedad física que se puede medir dividiendo la masa de un objeto por su volumen. (p. 72)

density current: circulation pattern in the ocean that forms when a mass of more dense seawater sinks beneath less dense seawater. (p. 382)

corriente de densidad: modelo de circulación en el océano que se forma cuando una masa de agua marina más densa se hunde por debajo de agua de mar menos densa. (p. 382)

dependent variable: variable that changes as a result of a change in the independent variable. (p. 18)

variable dependiente: variable que cambia como resultado de un cambio en la variable independiente. (p. 18)

dew point: temperature at which air is saturated and condensation can occur. (p. 350)

punto de rocío: temperatura a la que se satura el aire y puede ocurrir la condensación. (p. 350)

diffraction: bending of waves around a barrier. (p. 239)

difracción: curvatura de las ondas alrededor de una barrera. (p. 239)

DNA (deoxyribonucleic acid): a chemical inside cells that contains hereditary information and controls how an organism will look and function by controlling which proteins a cell produces. (p. 590)

ADN (ácido deoxirribonucléico): químico presente en el interior de las células, el cual contiene información genética y controla la futura apariencia y funcionamiento de un organismo a través del control de las proteínas que produce una célula. (p. 590)

down feather: fluffy feather that traps and keeps air warm against a bird's body. (p. 543)

plumón: pluma esponjosa que atrapa y mantiene el aire caliente cerca del cuerpo de las aves. (p. 543)

E

Earth science: study of Earth systems and systems in space, including weather and climate systems, and the study of nonliving things such as rocks, oceans, and planets. (p. 10)

ciencias de la Tierra: estudio del sistema de la Tierra y de los sistemas en el espacio, incluyendo el clima y los sistemas climáticos y el estudio de los seres inanimados como las rocas, los océanos y los planetas. (p. 10)

eclipse (ih KLIHPS): event that occurs when the Moon moves between the Sun and Earth (solar eclipse), or when Earth moves between the Sun and the Moon (lunar eclipse), and casts a shadow. (p. 444)

ecology: study of the interactions that take place among organisms and their environment. (p. 620)

ecosystem (EE koh sihs tum): all the living organisms in an area, as well as the nonliving parts of their environment; community of organisms—producers, consumers, and decomposers—that interact with each other and their surroundings. (pp. 374, 635)

ectotherm (EK tuh thurm): cold-blooded animal whose body temperature changes with the temperature of its surrounding environment. (p. 531)

electric circuit: closed conducting loop in which electric current can flow continually. (p. 202)

electric current: flow of electric charge, measured in amperes (A). (p. 201)

electric discharge: movement of static charge from one place to another. (p. 198)

electric resistance: measure of how difficult it is for electrons to flow in a material; unit is the ohm (Ω). (p. 203)

electromagnet: a current-carrying wire wrapped around an iron core. (p. 212)

electromagnetic induction: production of electric current by moving a magnet and a wire coil relative to each other. (p. 213)

electromagnetic spectrum: arrangement of electromagnetic waves according to their wavelengths. (p. 409)

electromagnetic waves: waves that can travel through matter or space; includes radio waves, infrared waves, visible light waves, ultraviolet waves, X rays, and gamma rays. (p. 230)

electron: invisible, negatively charged particle located in a cloudlike formation that surrounds the nucleus of an atom. (p. 102)

element: natural or synthetic material that cannot be broken down into simpler materials by ordinary means; has unique properties and is generally classified as a metal, metalloid, or nonmetal. (p. 106)

embryo: the stage of development during the first two months of pregnancy after the zygote attaches to the wall of the uterus. (p. 577)

endotherm (EN duh thurm): warm-blooded animal whose body temperature does not change with its surrounding environment. (p. 531)

eclipse: evento que ocurre cuando la Luna se mueve entre la Tierra y el sol (eclipse solar) o cuando la Tierra se mueve entre el sol y la Luna (eclipse lunar) y proyecta una sombra. (p. 444)

ecología: estudio de las interacciones entre los organismos y su medio ambiente. (p. 620)

ecosistema: todos los organismos vivos que habitan en un área, así como las partes no vivientes de su medio ambiente; comunidad de organismos—productores, consumidores y descomponedores—que interactúan entre sí y con su medio ambiente. (pp. 374, 635)

ectotérmico: animal de sangre fría cuya temperatura corporal cambia con la temperatura del medio ambiente circundante. (p. 531)

circuito eléctrico: circuito cerrado de conducción en el cual la corriente eléctrica puede fluir continuamente. (p. 202)

corriente eléctrica: flujo de carga eléctrica, medido en amperios (A). (p. 201)

descarga eléctrica: movimiento de carga estática de un lugar a otro. (p. 198)

resistencia eléctrica: medida de la dificultad que tienen los electrones para fluir en un material; su unidad es el ohmio (Ω). (p. 203)

electroimán: alambre que transporta corriente envuelto al rededor de un núcleo de hierro. (p. 212)

inducción electromagnética: producción de corriente eléctrica moviendo un imán y una bobina de alambre relativos uno al otro. (p. 213)

espectro electromagnético: ordenamiento de las ondas electromagnéticas de acuerdo con su longitud de onda. (p. 409)

ondas electromagnéticas: ondas que pueden viajar a través de la materia o del espacio; incluyen ondas radiales, ondas infrarrojas, ondas de luz visible, ondas ultravioletas, rayos X y rayos gama. (p. 230)

electrón: partícula invisible con carga negativa, localizada en una formación parecida a una nube que rodea el núcleo de un átomo. (p. 102)

elemento: material natural o sintético que no puede ser descompuesto fácilmente en materiales más simples por medios ordinarios; tiene propiedades únicas y generalmente es clasificado como metal, metaloide o no metal. (p. 106)

embrión: estado de desarrollo durante los dos primeros meses de gestación después de que el cigoto se adhiere a la pared del útero. (p. 577)

endotérmico: animal de sangre caliente cuya temperatura corporal no cambia con la temperatura del medio ambiente circundante. (p. 531)

endothermic reaction: chemical reaction that absorbs heat energy. (p. 180)

energy: ability to cause change. (p. 162)

erosion: wearing away and removal of rock material that occurs by agents such as gravity, ice, wind, and water. (p. 323)

estimation: method of making an educated guess at a measurement; using the size of something familiar to guess the size of a new object. (p. 43)

estivation (es tuh VAY shun): period of inactivity during hot, dry weather; in amphibians, involves moving to cooler, more humid areas underground. (p. 535)

exoskeleton: rigid, protective body covering of an arthropod that supports the body and reduces water loss. (p. 512)

exothermic reaction: chemical reaction that releases heat energy. (p. 181)

extrusive (ehk STREW sihv): describes igneous rocks that have small or no crystals and form when melted rock cools quickly on Earth's surface. (p. 265)

reacción endotérmica: reacción química que absorbe energía calórica. (p. 180)

energía: habilidad de producir cambios. (p. 162)

erosión: desgaste y eliminación de material rocoso causado por agentes tales como la gravedad, el hielo, el viento y el agua. (p. 323)

estimación: método para hacer una suposición fundamentada en una medida, usando el tamaño de algo conocido para suponer el tamaño de un nuevo objeto. (p. 43)

estivación: período de inactividad durante clima caliente y seco; en los anfibios implica la emigración a áreas más frías debajo de la tierra. (p. 535)

exoesqueleto: capa rígida protectora del cuerpo de los artrópodos que sostiene el cuerpo y reduce la pérdida de agua. (p. 512)

reacción exotérmica: reacción química que libera energía calórica. (p. 181)

extrusiva: describe a las rocas volcánicas que tienen cristales pequeños o que carecen de ellos y que se forman cuando las rocas fundidas se enfrían rápidamente en la superficie terrestre. (p. 265)

F

fault: large fracture in rock along which movement occurs. (p. 293)

fault-block mountains: sharp, jagged mountains made of huge, tilted blocks of rock that are separated from surrounding rock by faults and form because of pulling forces. (p. 300)

fertilization: process in which sperm and egg join, resulting in a new organism. (p. 595)

fetus: the stage of development after the first two months of pregnancy until birth. (p. 577)

folded mountain: mountain that forms by the folding of rock layers caused by compressive forces. (p. 301)

foliated: describes metamorphic rocks with visible layers of minerals. (p. 274)

food chain: series of stages that shows the transfer of energy from producers to consumers and decomposers. (p. 392)

force: a push or a pull; SI unit is the newton. (p. 136)

frequency: number of wavelengths that pass a given point in one second; measured in hertz (Hz). (p. 233)

falla: gran fractura en las rocas a lo largo de la cual ocurre el movimiento. (p. 293)

montañas con bloques de fallas: montañas afiladas y dentadas formadas por enormes bloques de roca inclinados, separados de las rocas adyacentes por fallas y que se forman debido a fuerzas atrayentes. (p. 300)

fertilización: proceso en el cual un espermatozoide y un huevo se unen, dando como resultado un nuevo organismo. (p. 595)

feto: estado de desarrollo después de los dos primeros meses de gestación hasta el nacimiento. (p. 577)

montañas plegadas: montañas que se forman por el plegamiento de capas de rocas causado por fuerzas de tracción. (p. 301)

foliado: describe a las rocas metamórficas con capas visibles de minerales. (p. 274)

cadena alimenticia: serie de etapas que muestra la transferencia de energía de productores a consumidores y descomponedores. (p. 392)

fuerza: empuje o tracción; la unidad SI es el newton. (p. 136)

frecuencia: número de longitudes de onda que pasan un punto determinado en un segundo; se mide en hertz (Hz). (p. 233)

friction: force that resists sliding motion between two touching surfaces; always acts opposite to the direction of motion. (p. 139)

front: boundary that develops where air masses of different temperatures collide; can be cold, warm, stationary, or occluded. (p. 357)

fricción: fuerza que opone resistencia al movimiento de deslizamiento entre dos superficies en contacto; siempre actúa en dirección opuesta al movimiento. (p. 139)

frente: límite que se forma en donde chocan masas de aire con diferentes temperaturas, el cual puede ser frío, caliente, estacionario u ocluido. (p. 357)

G

galaxy (GAL uk see): group of stars, gas, and dust held together by gravity. (p. 459)

gem: rare, valuable mineral that can be cut and polished. (p. 262)

gene: small section of DNA on a chromosome that carries information about a trait. (p. 600)

genetics (juh NEH tihks): study of how traits are passed from parent to offspring. (p. 599)

gill: organ that allows a water-dwelling animal to exchange carbon dioxide for dissolved oxygen in the water. (p. 506)

graph: used to collect, organize, and summarize data in a visual way, making it easy to use and understand. (p. 57)

galaxia: grupo de estrellas, gas y polvo que se mantienen unidos por gravedad. (p. 459)

gema: mineral escaso y valioso que puede ser cortado y pulido. (p. 262)

gen: pequeña sección de ADN en un cromosoma, la cual posee información referente a un rasgo. (p. 600)

genética: estudio acerca de cómo los rasgos sos transmitidos de padres a hijos. (p. 599)

branquia: órgano que permite a los animales que viven en el agua intercambiar dióxido de carbono por oxígeno disuelto en el agua. (p. 506)

gráfico: se usa para recolectar, organizar y resumir información en forma visual, facilitando su uso y comprensión. (p. 57)

H

habitat: place where an organism lives; provides the food, shelter, moisture, temperature, and other factors required for the organism's survival. (p. 632)

heat: transfer of energy from one object to another due to a difference in temperature; flows from warmer objects to cooler objects. (p. 173)

herbivore: plant-eating mammal with incisors specialized to cut vegetation and large, flat molars to grind it. (p. 546)

heredity (huh REH duh tee): passing on of traits from parents to offspring. (p. 599)

hibernation: period of inactivity during cold weather; in amphibians, involves burying themselves in mud or leaves. (p. 535)

humidity: amount of water vapor in the atmosphere. (p. 350)

hurricane: large storm, up to 970 km in diameter, that begins as a low-pressure area over tropical oceans, has sustained winds that can reach 250 km/h and gusts up to 300 km/h. (p. 361)

hábitat: lugar donde vive un organismo; proporciona alimento, refugio, humedad, temperatura y otros factores necesarios para la supervivencia del organismo. (p. 632)

calor: transferencia de energía de un objeto a otro debida a una diferencia de temperatura; fluye de los objetos calientes a los fríos. (p. 173)

herbívoro: mamífero que se alimenta de plantas y que posee incisivos especializados en cortar vegetación y molares grandes y planos para molerla. (p. 546)

herencia: transmisión de los rasgos de padres a hijos. (p. 599)

hibernación: período de inactividad durante el tiempo frío; en los anfibios implica enterrarse en el barro o entre las hojas. (p. 535)

humedad: cantidad de vapor de agua en la atmósfera. (p. 350)

huracán: gran tormenta, de hasta 970 kilómetros de diámetro que comienza como un área de baja presión sobre los océanos tropicales y que tiene vientos sostenidos que pueden alcanzar hasta 250 kilómetros por hora y ráfagas de hasta 300 kilómetros por hora. (p. 361)

hypothesis: reasonable guess that can be tested and is based on what is known and what is observed. (p. 14)

hipótesis: suposición razonable que puede ser probada y que está basada en lo que se sabe y en lo que ha sido observado. (p. 14)

I

igneous (IHG nee us) rock: intrusive or extrusive rock that is produced when melted rock from inside Earth cools and hardens. (p. 265)

roca ígnea: roca intrusiva o extrusiva producida cuando la roca fundida proveniente del interior de la Tierra se enfría y endurece. (p. 265)

inclined plane: a sloped surface or a ramp. (p. 149)

plano inclinado: superficie inclinada o rampa. (p. 149)

independent variable: variable that is changed in an experiment. (p. 18)

variable independiente: variable que cambia en un experimento. (p. 18)

inertia: tendency to resist a change in motion. (p. 139)

inercia: tendencia a oponer resistencia a un cambio en el movimiento. (p. 139)

infer: to draw a conclusion based on observation. (p. 16)

deducción: sacar una conclusión con base en una observación. (p. 16)

inner core: solid, innermost layer of Earth's interior that is the hottest part of Earth and experiences the greatest amount of pressure. (p. 289)

núcleo interior: la capa sólida más interna del centro de la Tierra, , la cual constituye la parte más caliente de ésta y en donde se ejerce la mayor cantidad de presión. (p. 289)

instantaneous speed: speed of an object at any given time. (p. 131)

velocidad instantánea: velocidad de un objeto en un momento determinado. (p. 131)

insulator: material in which electric charges cannot move easily. (p. 197)

aislante: material en el cual las cargas eléctricas no se pueden mover fácilmente. (p. 197)

interference: occurs when two or more waves combine and form a new wave when they overlap. (p. 231)

interferencia: ocurre cuando dos o más ondas se combinan y al sobreponerse forman una nueva onda. (p. 231)

intrusive (ihn trew sihv): describes a type of igneous rock that generally contains large crystals and forms when magma cools slowly beneath Earth's surface. (p. 265)

intrusiva: describe un tipo de roca volcánica que por lo general contiene grandes cristales y que se forma cuando el magma se enfría lentamente debajo de la superficie terrestre. (p. 265)

invertebrate (ihn VURT uh brayt): an animal without a backbone. (p. 500)

invertebrado: animal que carece de columna vertebral. (p. 500)

isostasy: principle stating that Earth's lithosphere floats on a plasticlike upper part of the mantle called the athenosphere. (p. 304)

isostasía: principio que establece que la litosfera de la Tierra flota en la parte superior del manto, semejante a un plástico, llamada astenosfera. (p. 304)

isotopes (I suh tohps): two or more atoms of the same element that have different numbers of neutrons in their nuclei. (p. 109)

isótopos: dos o más átomos del mismo elemento que tienen diferente número de neutrones en su núcleo. (p. 109)

K

Kelvin (K): SI unit for temperature. (p. 54)

Kelvin (K): unidad del SI para temperatura. (p. 54)

kilogram (kg): SI unit for mass. (p. 53)

kilogramo (Kg.): unidad del SI para masa. (p. 53)

kinetic energy: energy an object has due to its motion. (p. 164)

energía cinética: energía que posee un objeto en movimiento. (p. 164)

L

landfill: an area where garbage is deposited. (p. 657)

vertedero de basuras: sitio donde se deposita la basura. (p. 657)

law of conservation of energy: states that energy cannot be created or destroyed but can only be transformed from one form into another. (p. 168)

ley de la conservación de la energía: establece que la energía no puede ser creada ni destruida, sino que solamente pude ser transformada de una forma a otra. (p. 168)

law of conservation of mass: states that the mass of the products of a chemical change is always the same as the mass of what you started with. (p. 85)

ley de la conservación de la masa: establece que la masa de los productos de un cambio químico siempre es igual a la masa inicial. (p. 85)

law of conservation of matter: states that matter is not created or destroyed but only changes its form. (p. 100)

ley de la conservación de la materia: establece que la materia no se crea ni se destruye, solamente cambia de forma. (p. 100)

lever: a rod or plank that pivots about a fixed point. (p. 148)

palanca: barra o tablón que se apoya en un punto fijo. (p. 148)

life science: study of living systems and how they interact. (p. 9)

ciencias de la vida: estudio de los sistemas vivos y de la forma como interactúan. (p. 9)

light-year: about 9.5 trillion km—the distance that light travels in one year—which is used to measure large distances between stars or galaxies. (p. 462)

año luz: aproximadamente 9.5 trillones de kilómetros (la distancia que recorre la luz en un año), la cual se usa para medir largas distancias entre estrellas y galaxias. (p. 462)

limiting factors: anything that can restrict the size of a population, including living and nonliving features of an ecosystem, such as predators or drought. (p. 629)

factores limitantes: cualquier factor que pueda restringir el tamaño de una población, incluyendo los elementos vivientes y no vivientes de un ecosistema, tales como los depredadores o las sequías. (p. 629)

line graph: a type of graph used to show the relationship between two variables that are numbers on an *x*-axis and a *y*-axis. (p. 57)

gráfico lineal: tipo de gráfico usado para mostrar la relación entre dos variables que son números en un eje *x* y en un eje *y*. (p. 57)

lithosphere (LIH thuh sfihr): rigid layer of Earth about 100 km thick, made of the crust and a part of the upper mantle. (p. 292)

litosfera: capa rígida de la Tierra de aproximadamente 100 kilómetros de grosor, formada por la corteza y una parte del manto superior. (p. 292)

lunar highlands: mountainous areas on the Moon that are about 4.5 billion years old. (p. 442)

relieves lunares: áreas montañosas en la luna que tienen cerca de 4.5 billones de años de antigüedad. (p. 442)

M

magnetic domain: a group of atoms in a magnetic material with the magnetic poles of the atoms pointing in the same direction. (p. 211)

dominio magnético: grupo de átomos en un material magnético con los polos magnéticos de los átomos apuntando en la misma dirección. (p. 211)

mantle: largest layer of Earth's interior that lies above the outer core and is solid yet flows slowly; thin layer of tissue that covers a mollusk's body and that can secrete a shell. (pp. 290, 506)

manto: la capa más grande del interior de la Tierra que yace sobre el núcleo exterior y que es sólida aunque fluye lentamente; capa delgada de tejido que cubre el cuerpo de los moluscos y que puede producir/secretar una concha. (pp. 290, 506)

maria: smooth, dark regions on the Moon that formed when lava flowed onto the Moon's surface. (p. 442)

maría: regiones lisas y oscuras en la Luna que se formaron cuando la lava fluyó en la superficie lunar. (p. 442)

marsupial: mammal that gives birth to incompletely developed young that finish developing in their mother's pouch. (p. 548)

mass: amount of matter in an object, which is measured in kilograms. (p. 53)

mass movement: occurs when gravity alone causes rock or sediment to move down a slope. (p. 323)

mass number: sum of the number of protons and neutrons in the nucleus of an atom. (p. 109)

matter: anything that has mass and takes up space. (pp. 71, 99)

measurement: way to describe objects and events with numbers; for example, length, volume, mass, weight, and temperature. (p. 42)

mechanical advantage: number of times a machine increases the force applied to it. (p. 146)

mechanical wave: a type of wave that can travel only through matter. (p. 227)

mechanical weathering: process that breaks rocks down into smaller pieces without changing them chemically. (p. 316)

medusa (mih DEW suh): free-swimming, bell-shaped body form in the life cycle of a cnidarian. (p. 503)

meiosis (mi OH sus): process in which sex cells are formed in reproductive organs; involves two divisions of the nucleus, producing four sex cells, each having half the number of chromosomes as the original cell. (p. 595)

melanin: the pigment that gives your skin color and protects your skin from ultraviolet light. (p. 561)

melting point: temperature at which a solid becomes a liquid. (p. 75)

menstrual cycle: the monthly cycle of changes in a mature female reproductive system. (p. 576)

metal: element that is malleable, ductile, a good conductor of electricity, and generally has a shiny or metallic luster. (p. 110)

metalloid: element that has characteristics of both metals and nonmetals and is a solid at room temperature. (p. 111)

metamorphic (met uh MOR fihk) rock: new rock that forms when existing rock is heated or squeezed. (p. 273)

metamorphosis (met uh MOR fuh sus): change of body form that can be complete (egg, larva, pupa, adult) or incomplete (egg, nymph, adult). (p. 513)

meteorite: any rock from space that survives its plunge through the atmosphere and lands on Earth's surface. (p. 455)

meter (m): SI unit for length. (p. 51)

marsupial: mamífero que da a luz crías con un desarrollo incompleto que terminan su desarrollo en la bolsa de su madre. (p. 548)

masa: cantidad de materia en un objeto, la cual se mide en kilogramos. (p. 53)

movimiento de masa: ocurre cuando las rocas o sedimentos se mueven sobre una pendiente sólo por acción de la gravedad. (p. 323)

número de masa: suma del número de protones y neutrones en el núcleo de un átomo. (p. 109)

materia: cualquier cosa que tenga masa y que ocupe espacio. (pp. 71, 99)

medida: forma para describir objetos y eventos con números; por ejemplo, longitud, volumen, masa, peso y temperatura. (p. 42)

ventaja mecánica: número de veces que una máquina incrementa la fuerza que se le aplica. (p. 146)

onda mecánica: tipo de onda que puede viajar únicamente a través de la materia. (p. 227)

desgaste mecánico: proceso mediante el cual las rocas se rompen en pedazos más pequeños sin cambiar químicamente. (p. 316)

medusa: cuerpo en forma de campana durante el ciclo de vida de un cnidario. (p. 503)

meiosis: proceso en el cual se forman las células sexuales en los órganos reproductivos; implica dos divisiones de los núcleos, produciendo cuatro células sexuales, cada una de las cuales posee la mitad del número de cromosomas de la célula original. (p. 595)

melanina: pigmento que da el color y protege la piel de la luz ultravioleta. (p. 561)

punto de fusión: temperatura a la cual un sólido se convierte en líquido. (p. 75)

ciclo menstrual: ciclo mensual de cambios en el sistema reproductivo de una mujer madura. (p. 576)

metal: elemento maleable, dúctil y buen conductor de electricidad que generalmente tiene un lustre brillante o metálico. (p. 110)

metaloide: elemento que comparte características de los metales y de los no metales y es sólido a temperatura ambiente. (p. 111)

roca metamórfica: roca nueva que se forma cuando la roca existente se calienta o comprime. (p. 273)

metamorfosis: cambio de forma del cuerpo que puede ser completo (huevo, larva, ninfa, adulto) o incompleto (huevo, ninfa, adulto). (p. 513)

meteorito: cualquier roca del espacio que sobrevive su incursión en la atmósfera y aterriza en la superficie terrestre. (p. 455)

metro (m.): unidad del SI para longitud. (p. 51)

mineral: inorganic, solid material found in nature that always has the same chemical makeup, atoms arranged in an orderly pattern, and properties such as cleavage and fracture, color, hardness, and streak and luster. (p. 256)

mitochondria (mi tuh KAHN dree uh): cell organelles where cellular respiration takes place. (p. 480)

mitosis (mi TOH sus): cell division process in which DNA in the nucleus is duplicated and the nucleus divides into two nuclei that contain the same genetic information. (p. 592)

mixture: a combination of compounds and elements that has not formed a new substance and whose proportions can be changed without changing the mixture's identity. (p. 115)

model: any representation of an object or an event that is used as a tool for understanding the natural world; can communicate observations and ideas, test predictions, and save time, money, and lives. (p. 21)

mollusk: soft-bodied, bilaterally symmetrical inverte-' rate with a large, muscular foot, a mantle, and an open circulatory system; usually has a shell. (p. 506)

monotreme: mammal that lays eggs with tough, leathery shells instead of giving birth to live young. (p. 547)

muscle: an organ that can relax, contract, and provide force to move your body parts. (p. 562)

mutation: change in a gene or chromosome that can result from something in the environment or an error in mitosis or meiosis; can be harmful, neutral, or beneficial, and adds variation to the genes of a species. (p. 604)

mineral: material inorgánico, sólido, que se encuentra en la naturaleza y que siempre tiene la misma composición química, átomos dispuestos en un patrón ordenado y propiedades tales como fisuras y fracturas, color, dureza, vetas pequeñas y brillo. (p. 256)

mitocondria: organelo celular en donde se lleva a cabo la respiración celular. (p. 480)

mitosis: proceso de división celular en el cual el ADN del núcleo se duplica y el núcleo se divide en dos núcleos que contienen la misma información genética. (p. 592)

mezcla: combinación de compuestos y elementos que no han formado una nueva sustancia y cuyas proporciones pueden ser cambiadas sin que se pierda la identidad de la mezcla. (p. 115)

modelo: cualquier representación de un objeto o evento utilizada como herramienta para entender el mundo natural; puede comunicar observaciones e ideas, predicciones de las pruebas y ahorrar tiempo, dinero y salvar vidas. (p. 21)

molusco: invertebrado de cuerpo blando simétricamente bilateral que posee una pata grande muscular, un manto y un sistema circulatorio abierto, y que por lo general tiene concha. (p. 506)

monotrema: mamífero que pone huevos con cascarones fuertes y correosos en lugar de dar a luz a una cría. (p. 547)

músculo: órgano que se puede relajar, contraer y dar la fuerza para mover las partes del cuerpo. (p. 562)

mutación: cambio en un gen o en un cromosoma que puede ser el resultado de un factor del medio ambiente o un error de mitosis o meiosis; puede ser perjudicial, neutro o benéfico y agrega variaciones a los genes de una especie. (p. 604)

N

natural resource: parts of Earth's environment, such as water and minerals that are used by living organisms. (p. 647)

nekton: marine animals, such as fish and turtles, that actively swim in ocean waters. (p. 390)

neutron: an uncharged particle located in the nucleus of an atom. (p. 104)

Newton's laws of motion: a set of rules developed by Isaac Newton to explain how forces affect the motion of an object. (p. 138)

recurso natural: partes del entorno de la Tierra, tales como el agua y los minerales, utilizadas por los organismos vivos. (p. 647)

necton: animales marinos como los peces y las tortugas, que nadan activamente en las aguas de los océanos. (p. 390)

neutrón: partícula sin carga localizada en el núcleo de un átomo. (p. 104)

ley del movimiento de Newton: conjunto de reglas desarrolladas por Isaac Newton para explicar cómo las fuerzas afectan el movimiento de un objeto. (p. 138)

niche (NICH): the role of an organism in its ecosystem; refers to the unique ways an organism survives, obtains food and shelter, and avoids danger. (p. 631)

nicho: el papel que desempeña un organismo en su ecosistema; se refiere a las formas singulares que tiene un organismo para sobrevivir, obtener alimento y refugio y evitar el peligro. (p. 631)

nonfoliated: describes metamorphic rocks that lack distinct layers or bands. (p. 274)

no foliado: describe a las rocas metamórficas que carecen de capas o bandas definidas. (p. 274)

nonmetals: elements that are usually gases or brittle solids and poor conductors of electricity and heat; are the basis of the chemicals of life. (p. 111)

no metales: elementos que por lo general son gases o sólidos frágiles y malos conductores de electricidad y calor; son la base de los compuestos químicos biológicos. (p. 111)

nonrenewable resource: natural resource that cannot be replace by natural processes within 100 years or less. (p. 652)

recursos no renovables: recursos naturales que no pueden ser reemplazados por procesos naturales en 100 años o menos. (p. 652)

nucleus (NEW klee us): cell organelle that contains the hereditary material; positively charged, central part of an atom. (pp. 102, 481)

núcleo: organelo celular que contiene material genético; parte central con carga positiva del átomo. (pp. 102, 481)

nutrient: substance in food that provides for cell development, growth, and repair. (p. 564)

nutriente: sustancia en los alimentos que contribuye al desarrollo, crecimiento y reparación de las células. (p. 564)

O

observatory: building that can house an optical telescope; often has a dome-shaped roof that can be opened for viewing. (p. 410)

observatorio: edificación que puede albergar un telescopio óptico; a menudo tiene un techo en forma de domo que puede abrirse para la observación. (p. 410)

omnivore: plant- and meat-eating animal with incisors that cut vegetables, sharp premolars that chew meat, and molars that grind food. (p. 546)

omnívoro: animal que se alimenta de plantas y animales y que posee incisivos para cortar vegetales, premolares afilados para masticar carne y molares para triturar la comida. (p. 546)

open circulatory system: a type of blood-circulation system that lacks blood vessels and in which blood washes over the organs. (p. 507)

sistema circulatorio abierto: tipo de sistema circulatorio que carece de vasos sanguíneos y en el cual la sangre baña los órganos. (p. 507)

orbit: curved path followed by a satellite as it revolves around an object. (pp. 409, 449)

órbita: trayectoria curva seguida por un satélite conforme gira alrededor de un objeto. (pp. 409, 449)

ore: material that contains enough of a useful metal that it can be mined and sold at a profit. (p. 263)

mena: material que contiene el metal útil suficiente para ser extraído y vendido para obtener utilidades. (p. 263)

organ: structure made of two or more different tissue types that work together to do a certain job. (p. 487)

órgano: estructura formada por dos o más tipos de tejidos que trabajan juntos para realizar una función determinada. (p. 487)

organelles (or guh NELZ): specialized cell parts that perform a cell's activities. (p. 479)

organelos: partes especializadas de las células que realizan las funciones celulares. (p. 479)

organ system: group of organs that work together to perform a certain task. (p. 487)

sistema de órganos: grupo de órganos que trabajan en conjunto para realizar una función determinada. (p. 487)

outer core: layer of Earth that lies above the inner core and is thought to be composed mostly of molten metal. (p. 290)

núcleo exterior: capa de la Tierra que yace sobre el núcleo interior y que se piensa que está compuesta principalmente de metal fundido. (p. 290)

ovulation: the process each month of an ovary releasing an egg. (p. 575)

ovulación: proceso mensual en que un ovario libera un óvulo. (p. 575)

P

parallel circuit: circuit that has more than one path for electric current to follow. (p. 208)

photosynthesis (foh toh SIHN thuh sihs): process by which plants, algae, and many types of bacteria use sunlight, water, and carbon dioxide to make food and oxygen. (pp. 377, 479)

physical change: change in which the properties of a substance change but the identity of the substance always remains the same. (p. 71)

physical property: any characteristic of a material, such as state, color, and volume, that can be observed or measured without changing or attempting to change the material. (p. 70)

physical science: study of matter, which is anything that takes up space and has mass, and the study of energy, which is the ability to cause change. (p. 10)

placental: mammal whose offspring develops inside the female's uterus; has a placenta—a saclike organ—which supplies the embryo with food and oxygen and removes wastes. (p. 548)

plankton: tiny marine organisms, such as diatoms, that drift in the surface waters of every ocean. (p. 389)

plate: section of Earth's crust and rigid, upper mantle that moves slowly around on the asthenosphere. (p. 292)

pollutant: any material that can harm living things by interfering with life processes. (p. 657)

polyp (PAHL up): vase-shaped, usually sessile body form in the life cycle of a cnidarian. (p. 503)

population: a group of the same type of organisms living in the same place at the same time. (p. 627)

potential energy: energy that is stored due to an object's position. (p. 166)

precipitation: occurs when drops of water or crystals of ice become too large to be suspended in a cloud and fall in the form of rain, freezing rain, sleet, snow, or hail. (p. 352)

precision: describes how closely measurements are to each other and how carefully measurements were made. (p. 44)

pregnancy: the period of development from fertilized egg to birth. (p. 577)

producer: organism that can make its own food by photosynthesis or chemosynthesis. (pp. 389, 635)

circuito en paralelo: circuito en el que la corriente eléctrica puede seguir más de una trayectoria. (p. 208)

fotosíntesis: proceso mediante el cual las plantas, las algas y muchos tipos de bacteria utilizan la luz solar, el agua y el dióxido de carbono para producir alimentos y oxígeno. (pp. 377, 479)

cambio físico: cambio en el cual las propiedades de una sustancia cambian pero la identidad de la sustancia sigue siendo la misma. (p. 71)

propiedad física: cualquier característica de un material, tal como estado, color y volumen, que pueden ser observados o medidos sin cambiar o tratar de cambiar el material. (p. 70)

ciencias física: estudio de la materia, lo cual es todo lo que ocupe espacio y tenga masa, y el estudio de la energía, que es la habilidad de producir cambios. (p. 10)

placentario: mamífero cuyas crías se desarrollan en el útero femenino; tiene una placenta, órgano parecido a un saco, el cual suministra al embrión alimento y oxígeno y elimina los desechos. (p. 548)

plancton: organismos marinos diminutos, como las diatomeas, que viajan a la deriva en las aguas superficiales de todos los océanos. (p. 389)

placa: sección de la corteza terrestre y del manto superior rígido que se mueve lentamente sobre la astenosfera. (p. 292)

contaminante: cualquier material que pueda causar daño a los seres vivos al interferir con los procesos biológicos. (p. 657)

pólipo: cuerpo en forma de jarrón generalmente sésil en el ciclo de vida de un cnidario. (p. 503)

población: grupo del mismo tipo de organismos que viven en el mismo lugar y al mismo tiempo. (p. 627)

energía potencial: energía almacenada debido a la posición de un objeto. (p. 166)

precipitación: ocurre cuando las gotas de agua o cristales de hielo adquieren un tamaño demasiado grande para estar suspendidos en una nube y caen en forma de lluvia, lluvia helada, aguanieve, nieve o granizo. (p. 352)

precisión: describe qué tan aproximada es una medida respecto a otra y qué tan cuidadosamente fueron hechas dichas medidas. (p. 44)

embarazo: período de desarrollo de un óvulo fertilizado hasta el nacimiento. (p. 577)

productor: organismo que puede producir su propio alimento por medio de fotosíntesis o quimiosíntesis. (pp. 389, 635)

Project Apollo: final stage in the U.S. program to reach the Moon in which Neil Armstrong was the first human to step onto the Moon's surface. (p. 422)

Project Gemini: second stage in the U.S. program to reach the Moon in which an astronaut team connected with another spacecraft in orbit. (p. 421)

Project Mercury: first step in the U.S. program to reach the Moon that orbited a piloted spacecraft around Earth and brought it back safely. (p. 421)

proton: positively charged particle located in the nucleus of an atom and that is counted to identify the atomic number. (p. 103)

pulley: grooved wheel with a rope or cable running through the groove. (p. 147)

Proyecto Apolo: etapa final en el proyecto norteamericano para llegar a la luna en el que Neil Armstrong fue el primer ser humano en caminar sobre la superficie lunar. (p. 422)

Proyecto Géminis: segunda etapa del proyecto norteamericano para llegar a la luna en el que un grupo de astronautas se conectó con otra nave espacial en órbita. (p. 421)

Proyecto Mercurio: primera etapa del proyecto norteamericano para llegar a la luna en el que una nave espacial tripulada recorrió la órbita de la Tierra y regresó de manera segura. (p. 421)

protón: partícula cargada positivamente, localizada en el núcleo de un átomo y que se cuenta para identificar el número atómico. (p. 103)

polea: rueda acanalada con una cuerda o cable que corre pasa por el canal. (p. 147)

R

radiation: energy that is transferred by waves. (p. 177)

radio telescope: collects and records radio waves traveling through space; can be used day or night under most weather conditions. (p. 413)

radula: scratchy, tonguelike organ in many mollusks that has rows of teethlike projections used to scrape and grate food. (p. 507)

rate: a ratio of two different kinds of measurements; the amount of change of one measurement in a given amount of time. (p. 54)

recycling: reusing materials after they have been changed into another form, instead of throwing them away. (p. 666)

reflecting telescope: optical telescope that uses a concave mirror to focus light and form an image at the focal point. (p. 410)

reflection: occurs when a wave strikes an object or surface and bounces off. (p. 237)

reflex: an involuntary, automatic response to a stimulus. (p. 571)

refracting telescope: optical telescope that uses a double convex lens to bend light and form an image at the focal point. (p. 410)

refraction: bending of a wave as it moves from one medium into another medium. (p. 238)

relative humidity: measure of the amount of water vapor in the air compared with the amount that could be held at a specific temperature. (p. 350)

radiación: energía transferida por ondas. (p. 177)

radiotelescopio: recolecta y registra ondas de radio que viajan a través del espacio; puede usarse de día o de noche en la mayoría de condiciones climáticas. (p. 413)

rádula: órgano punzante en forma de lengua en muchos moluscos, la cual presenta filas de proyecciones similares a los dientes usadas para raspar y rallar alimentos. (p. 507)

tasa: relación de dos diferentes tipos de medidas; los cambios en una medida en un tiempo determinado. (p. 54)

reciclaje: reutilización de materiales después de que su forma ha sido modificada, en lugar de desecharlos. (p. 666)

telescopio reflectante: telescopio óptico que utiliza un espejo cóncavo para enfocar la luz y formar una imagen en el punto focal. (p. 410)

reflexión: ocurre cuando una onda choca contra un objeto o superficie y rebota. (p. 237)

reflejo: respuesta involuntaria y automática a un estímulo. (p. 571)

telescopio de refracción: telescopio óptico que utiliza un lente doble convexo para formar una imagen en el punto focal. (p. 410)

refracción: curvatura de una onda a medida que se mueve de un medio a otro. (p. 238)

humedad relativa: medida de la cantidad de vapor de agua presente en el aire, comparada con la cantidad que podría ser mantenida a una temperatura específica. (p. 350)

renewable resource: natural resource that can be replaced by natural processes within 100 years or less. (p. 651)

recursos renovables: recursos naturales que pueden ser reemplazados por procesos naturales en 100 años o menos. (p. 651)

respiratory system: the structures and organs that help move oxygen into the body and waste gases out of the body. (p. 567)

sistema respiratorio: estructuras y órganos que ayudan a mover el oxígeno dentro del cuerpo y expulsar gases fuera del mismo. (p. 567)

revolution (rev uh LEW shun): movement of Earth around the Sun, which takes a year to complete. (p. 441)

revolución: movimiento de la Tierra alrededor del sol, el cual tarda un año en completarse. (p. 441)

rock: solid inorganic material that is usually made of two or more minerals and can be metamorphic, sedimentary, or igneous. (p. 256)

roca: material sólido inorgánico generalmente compuesto por dos o más minerales y que puede ser metamórfico, sedimentario o volcánico. (p. 256)

rock cycle: diagram that shows the slow, continuous process of rocks changing from one type to another. (p. 275)

ciclo de una roca: diagrama que muestra el proceso lento y continuo de las rocas al cambiar de un tipo a otro. (p. 275)

rocket: special engine that can work in space and burns liquid or solid fuel. (p. 415)

cohete: máquina especial que puede funcionar en el espacio y quema combustible sólido o líquido. (p. 415)

rotation (roh TAY shun): spinning of Earth on its axis, which occurs once every 24 hours, produces day and night, and causes the planets and stars to appear to rise and set. (p. 440)

rotación: giro de la Tierra sobre su eje, el cual ocurre una vez cada 24 horas, produce el día y la noche y hace que los planetas y las estrellas aparezcan, asciendan y se pongan. (p. 440)

runoff: water that flows over Earth's surface. (p. 329)

afluencia: agua que fluye sobre la superficie terrestre. (p. 329)

S

salinity (suh LIN nuh tee): measure of dissolved solids, or salts, in seawater. (p. 376)

salinidad: medida de los sólidos, o las sales, disueltos en el agua de mar. (p. 376)

satellite: any natural or artificial object that revolves around another object. (p. 417)

satélite: cualquier objeto natural o artificial que gire alrededor de otro objeto. (p. 417)

science: way of learning more about the natural world that provides possible explanations to questions and involves using a collection of skills. (p. 6)

ciencia: mecanismo para aprender más acerca del mundo natural, que da respuestas posibles a los interrogantes e implica hacer uso de numerosas habilidades. (p. 6)

scientific law: a rule that describes a pattern in nature but does not try to explain why something happens. (p. 7)

ley científica: regla que describe un modelo en la naturaleza pero que no intenta explicar por qué suceden las cosas. (p. 7)

scientific theory: a possible explanation for repeatedly observed patterns in nature supported by observations and results from many investigations. (p. 7)

teoría científica: posible explicación para patrones observados repetidamente en la naturaleza y apoyada en observaciones y resultados de muchas investigaciones. (p. 7)

sedimentary rock: a type of rock made from pieces of other rocks, dissolved minerals, or plant and animal matter that collect to form rock layers. (p. 269)

roca sedimentaria: tipo de roca formada por fracciones de otras rocas, minerales disueltos o materiales de plantas y animales que se unen para formar capas de rocas. (p. 269)

semen: mixture of fluid and sperm. (p. 574)

semen: mezcla de fluido y espermatozoides. (p. 574)

series circuit: circuit that has only one path for electric current to follow. (p. 208)

circuito en serie: circuito en el que la corriente eléctrica sólo puede seguir una trayectoria. (p. 208)

sex cells: specialized cells—female eggs and male sperm—that are produced by the process of meiosis, carry DNA, and join in sexual reproduction. (p. 594)

células sexuales: células especializadas—huevos femeninos y espermatozoides masculinos—que se producen a través del proceso de meiosis, portan el ADN y forman parte de la reproducción sexual. (p. 594)

sexual reproduction: a type of reproduction in which a new organism is produced from the DNA of two sex cells (egg and sperm). (p. 594)

SI: International System of Units, related by multiples of ten, designed to provided a worldwide standard of physical measurement. (p. 50)

simple machine: device that makes work easier with only one movement; can change the size or direction of a force, and includes the wedge, screw, lever, wheel and axle, pulley, and inclined plane. (p. 146)

skeletal system: all the bones in your body. (p. 560)

slump: occurs when a mass of rock or sediment moves downhill along a curved surface. (p. 325)

soil: mixture of weathered rock, organic matter, water, and air that evolves over time and supports the growth of plant life. (p. 320)

solar system: system of nine planets and numerous other objects that orbit our Sun, all held in place by the Sun's gravity. (p. 448)

solid waste: solid or near-solid items disposed of as garbage, including glass, paper, metal, cloth, and spoiled food. (p. 663)

space probe: instrument that travels far into the solar system and gathers data that it sends them back to Earth. (p. 418)

space shuttle: reusable spacecraft that can carry cargo, astronauts, and satellites to and from space. (p. 423)

space station: large facility with living quarters, work and exercise areas, and equipment and support systems for humans to live and work in space and conduct research. (p. 424)

sperm: the male reproductive cells. (p. 574)

state of matter: physical property that is dependent on both temperature and pressure and occurs in four forms—solid, liquid, gas, or plasma. (p. 73)

static charge: imbalance of electric charge on an object. (p. 198)

subduction: a type of plate movement that occurs when one plate sinks beneath another plate. (p. 295)

substance: matter that has the same composition and properties throughout. (p. 113)

supernova: very bright explosion of the outer part of a supergiant that takes place after its core collapses. (p. 459)

reproducción sexual: tipo de reproducción en la cual se produce un nuevo organismo a partir del ADN de dos células sexuales (huevo y espermatozoide). (p. 594)

SI: Sistema Internacional de Unidades, se ordena en múltiplos de diez, diseñados para suministrar un estándar de medidas físicas a nivel mundial. (p. 50)

máquina simple: dispositivo que facilita el trabajo con un solo movimiento; puede cambiar el tamaño o la dirección de una fuerza e incluye la cuña, el tornillo, la palanca, la rueda y el eje, la polea y el plano inclinado. (p. 146)

sistema óseo: conjunto de los huesos del cuerpo. (p. 560)

hundimiento: ocurre cuando una masa rocosa o sedimento se desliza sobre una superficie curva. (p. 325)

suelo: mezcla de roca desgastada por la acción atmosférica, materia orgánica, agua y aire que evoluciona con el tiempo y ayuda al desarrollo de la vida vegetal. (p. 320)

sistema solar: sistema de nueve planetas y otros numerosos objetos que circundan al sol, todos mantenidos en su lugar por la acción de la gravedad del sol. (p. 448)

desechos sólidos: artículos sólidos o semisólidos desechados como basura, incluyendo vidrio, papel, metal, tela y alimentos en estado de descomposición. (p. 663)

sonda espacial: instrumento que viaja grandes distancias en el sistema solar, recopila datos y los envía a la Tierra. (p. 418)

trasbordador espacial: nave espacial reutilizable que puede llevar carga, astronautas y satélites hacia y desde el espacio. (p. 423)

estación espacial: instalación grande con áreas para hospedarse, trabajar y hacer ejercicio; tiene equipos y sistemas de apoyo para que los seres humanos vivan, trabajen y lleven a cabo investigaciones en el espacio. (p. 424)

espermatozoides: células reproductoras masculinas. (p. 574)

estado de la materia: propiedad física que depende de la temperatura y la presión, y que ocurre en cuatro formas: sólido, líquido, gas, o plasma. (p. 73)

carga estática: desequilibrio de la carga eléctrica en un objeto. (p. 198)

subducción: tipo de movimiento de placas que ocurre cuando una placa se ubica debajo de otra. (p. 295)

sustancia: materia que siempre tiene la misma composición y las mismas propiedades. (p. 113)

supernova: explosión muy brillante de la parte externa de un supergigante que ocurre cuando su núcleo se desintegra. (p. 459)

surface current: ocean current that usually moves only the upper few hundred meters of seawater. (p. 380)

corriente superficial: corriente marina que por lo general sólo mueve los pocos cientos de metros superiores del agua del mar. (p. 380)

symmetry: arrangement of individual body parts; can be radial (arranged around a central point) or bilateral (mirror-image parts). (p. 499)

simetría: distribución de las partes individuales del cuerpo; puede ser radial (distribuidas alrededor de un punto central) o bilateral (partes que se reflejan a sí mismas). (p. 499)

system: collection of structures, cycles, and processes that relate to and interact with each other. (p. 8)

sistema: colección de estructuras, ciclos y procesos relacionados que interactúan entre sí. (p. 8)

T

table: presents information in rows and columns, making it easier to read and understand. (p. 57)

tabla: presentación de información en filas y columnas, facilitando la lectura y comprensión. (p. 57)

technology: use of science to help people in some way. (p. 11)

tecnología: uso de la ciencia para ayudar en alguna forma a las personas. (p. 11)

temperature: measure of the average kinetic energy of the atoms in an object. (p. 171)

temperatura: medida de la energía cinética promedio de los átomos de un objeto. (p. 171)

thermocline: layer of ocean water that begins at a depth of about 200 m and becomes progressively colder with increasing depth. (p. 378)

depresión térmica: capa de agua de mar que comienza a una profundidad de aproximadamente 200 m y se hace gradualmente más fría, a medida que aumenta la profundidad. (p. 378)

tide: the alternate rise and fall of sea level caused by the gravitational attractions of the Moon and Sun. (pp. 379, 453)

marea: el alterno subir y bajar del nivel del mar causado por las atracciones gravitacionales de la Luna y el sol. (pp. 379, 453)

tissue: group of similar cells that all do the same work. (p. 487)

tejido: grupo de células similares que desempeñan la misma función. (p. 487)

topography: configuration of surface features, including position and slope; also influences the types of soils that develop. (p. 320)

topografía: configuración de características de la superficie, incluyendo posición e inclinación, y que también influye en los tipos de suelos que se desarrollan. (p. 320)

tornado: violent, whirling wind, usually less than 200 m in diameter, that travels in a narrow path over land and can be highly destructive. (p. 360)

tornado: violento remolino de viento, por lo general con menos de 200 metros de diámetro, que sigue una trayectoria estrecha sobre la tierra y puede ser altamente destructivo. (p. 360)

transverse wave: a type of mechanical wave in which the wave energy causes matter in the medium to move up and down or back and forth at right angles to the direction the wave travels. (p. 228)

onda transversal: tipo de onda mecánica en el cual la energía de la onda hace que la materia en el medio se mueva hacia arriba y hacia abajo o hacia adelante y hacia atrás en ángulos rectos respecto a la dirección en que viaja la onda. (p. 228)

troposphere (TROPH uh sfihr): layer of the atmosphere that is closest to Earth's surface and contains nearly all of its clouds and weather. (p. 344)

troposfera: capa de la atmósfera más cercana a la superficie terrestre y que contiene casi todas sus nubes y climas. (p. 344)

U

upwarped mountain: mountain that forms when forces inside Earth push up the crust. (p. 301)

montaña encorvada: montaña que se forma cuando las fuerzas internas de la Tierra levantan la corteza. (p. 301)

upwelling: ocean current that moves cold, deep water to the ocean surface. (p. 384)

afloramiento: corriente marina que mueve al agua fría y profunda hacia la superficie del océano. (p. 384)

V

vacuole (VAK yew ohl): balloonlike cell organelle in the cytoplasm that can store food, water, and other substances. (p. 480)

variable: factor that can be changed in an experiment. (p. 18)

variations: different ways that a trait can appear—for example, differences in height, hair color, or weight. (p. 603)

velocity: speed of an object and its direction of motion; changes when speed changes, direction of motion changes, or both change. (p. 133)

volcanic mountain: mountain that forms when magma is forced upward and flows onto Earth's surface. (p. 302)

voltage: a measure of the amount of electrical potential energy transferred by an electric charge as it moves from one point to another in a circuit. (p. 205)

volume (m^3): the amount of space an object occupies measured in cubic meters. (p. 52)

vacuola: organelo celular en forma de balón que se encuentra en el citoplasma y que puede almacenar alimentos, agua y otras sustancias. (p. 480)

variable: factor que puede cambiar en un experimento. (p. 18)

variaciones: diferentes formas en que puede aparecer un rasgo—por ejemplo, diferencias de tamaño, color del cabello o peso. (p. 603)

velocidad: la rapidez de un objeto y su dirección de movimiento; varía cuando cambian la rapidez, la dirección de movimiento o ambas. (p. 133)

montaña volcánica: montaña que se forma cuando el magma es empujado hacia arriba y fluye sobre la superficie terrestre. (p. 302)

voltaje: medida de la cantidad de energía eléctrica potencial transferida por una carga eléctrica a medida que se mueve de un punto a otro en un circuito. (p. 205)

volumen (m3): la cantidad de espacio que ocupa un objeto, medido en metros cúbicos. (p. 52)

W

water cycle: never-ending cycle in which water circulates between Earth's surface and the atmosphere through the processes of evaporation, transpiration, precipitation, and condensation. (p. 347)

wave: rhythmic disturbance that carries energy but not matter. (pp. 237, 374)

wavelength: for a transverse wave, the distance between the tops of two adjacent crests or the bottoms of two adjacent troughs; for a compressional wave, the distance from the centers of adjacent rarefactions or adjacent compressions. (p. 232)

weather: current condition of the atmosphere including cloud cover, temperature, wind speed and direction, humidity, and air pressure. (p. 348)

weathering: natural mechanical or chemical process that causes rocks to change by breaking them down and causing them to crumble. (p. 316)

weight: a measurement of force that depends on gravity; measured in newtons. (p. 53)

work: is done when an applied force causes an object to move in the direction of the force. (p. 144)

ciclo del agua: ciclo interminable en el cual el agua circula entre la superficie terrestre y la atmósfera a través de los procesos de evaporación, precipitación y condensación. (p. 347)

onda: alteración rítmica que transporta energía pero no materia. (pp. 237, 374)

longitud de onda: en una onda transversal, es la distancia entre las puntas de dos crestas adyacentes o entre dos depresiones adyacentes; en una onda de compresión es la distancia entre los centros de dos rarefacciones adyacentes o compresiones adyacentes. (p. 232)

clima: condición actual de la atmósfera que incluye la capa de nubes, la temperatura, la velocidad y dirección del viento, la humedad y la presión del aire. (p. 348)

desgaste: proceso natural, mecánico o químico que produce cambios en las roca, rompiéndolas y haciendo que se desmoronen. (p. 316)

peso: medida de fuerza que depende de la gravedad y que se mide en Newtons. (p. 53)

trabajo: se realiza cuando una fuerza aplicada hace que un objeto se mueva en la dirección de dicha fuerza. (p. 144)

Italic numbers = *illustration/photo* **Bold numbers = vocabulary term**
lab = indicates a page on which the entry is used in a lab
act = indicates a page on which the entry is used in an activity

A

B

Index

C

D

Index

H

I

J

K

L

M

N

Q

R

S

Index

Y

Z

Magnification Key: Magnifications listed are the magnifications at which images were originally photographed. LM–Light MicroscopeSEM–Scanning Electron MicroscopeTEM–Transmission Electron Microscope

Acknowledgments: Glencoe would like to acknowledge the artists and agencies who participated in illustrating this program: Absolute Science Illustration; Andrew Evansen; Argosy; Articulate Graphics; Craig Attebery represented by Frank & Jeff Lavaty; CHK America; John Edwards and Associates; Gagliano Graphics; Pedro Julio Gonzalez represented by Melissa Turk & The Artist Network; Robert Hynes represented by Mendola Ltd.; Morgan Cain & Associates; JTH Illustration; Laurie O'Keefe; Matthew Pippin represented by Beranbaum Artist's Representative; Precision Graphics; Publisher's Art; Rolin Graphics, Inc.; Wendy Smith represented by Melissa Turk & The Artist Network; Kevin Torline represented by Berendsen and Associates, Inc.; WILDlife ART; Phil Wilson represented by Cliff Knecht Artist Representative; Zoo

Photo Credits

Cover (bkgd)Royalty-Free/CORBIS, (t)Richard Kaylin/Getty Images, (b)G. K. & Vikki Hart/Getty Images; **i ii** (bkgd)Royalty-Free/CORBIS; **i** (t)Richard Kaylin/Getty Images; (b)G. K. & Vikki Hart/Getty Images; **vii** Aaron Haupt; **viii** John Evans; **ix** (t)PhotoDisc, (b)John Evans; **x** (l)John Evans, (r)Geoff Butler; **xi** (l)John Evans, (r)PhotoDisc; **xii** PhotoDisc; **xiii** Stephen J. Krasemann/Photo Researchers; **xiv** Kennan Ward/The Stock Market/CORBIS; **xv** (t)Jose Manuel Sanchis Calvete/CORBIS, (b)Howard Bluestein/Photo Researchers; **xvi** (t)Fred Bavendam/Minden Pictures, (b)Frank Zullo/Photo Researchers; **xvii** Fred Bravendam/Minden Pictures; **xviii** Harold R. Hungerford/Photo Researchers; **xix** Ed Reschke/Peter Arnold, Inc.; **xx** Katharine Payne; **xxi** Runk/Schoenberger from Grant Heilman; **xxiv** AFP/CORBIS; **xxv** Dominic Oldershaw; **1** Tess & David Young/Tom Stack & Assoc.; **2–3** (bkgd)Wolfgang Kaehler; **2** (l)PhotoDisc; **3** (t)PhotoDisc; (l)Jack Star/Photolink/PhotoDisc, (c)Rudi Von Briel, (r)Richard T. Nowitz/CORBIS; **4–5** David Keaton/CORBIS; **8** Mary Kate Denny/PhotoEdit, Inc.; **9** Peter Veit/National Geographic Image Collection; **10** (tl)G. Brad Lewis/Stone/Getty Images, (bl)Roger Ball/The Stock Market/CORBIS, (br)Will & Deni McIntyre/Photo Researchers; **11** (t)AFP/CORBIS, (b)Reuters NewMedia, Inc./CORBIS; **13 14** Richard Hutchings; **15** Matt Meadows; **16** Icon Images; **17** Richard Hutchings/PhotoEdit, Inc./PictureQuest; **18** Rudi Von Briel; **19** Bob Daemmrich; **20** Glasheen Graphics/Index Stock; **21** (cw from top)David Young-Wolff/PhotoEdit, Inc., Donald C. Johnson/The Stock Market/CORBIS, John Bavosi/Science Photo Library/Photo Researchers, A. Ramey/PhotoEdit, Inc.; **22** CORBIS/PictureQuest; **23** Todd Gipstein/CORBIS; **24** (tl cl br)Betty Pat Gatliff, (tr)Richard Nowitz/Words & Pictures/PictureQuest, (bl)Michael O'Brian/Mud Island, Inc.; **25** (l)Carol Anne Petrachenko/CORBIS, (c)Jim Sugar Photography/CORBIS, (r)Tom Wurl/Stock Boston/PictureQuest; **26** (l)Stock Montage, (r)North Wind Picture Archives; **27** Digital Art/CORBIS; **28** SuperStock; **29** (t)Lester V. Bergman/CORBIS, (b)Bob Handelman/Stone/Getty Images; **31** Amanita Pictures; **32** (t)Aaron Haupt, (b)Matt Meadows; **35** (t)Reuters/CORBIS, (bl)UPI/Bettmann/CORBIS, (br)TIME; **37** Tim Courlas; **38** Peter Veit/National Geographic Image Collection; **39** (l)Tim Courlas/Horizons, (c r)Aaron Haupt; **40–41** Buck Miller/SuperStock; **42** Paul Almasy/CORBIS; **43** AFP/CORBIS; **44** David Young-Wolff/PhotoEdit, Inc.; **45** (l)Lowell D. Franga, (tr)The Purcell Team/CORBIS, (br)Len Delessio/Index Stock; **46** Photo by Richard T. Nowitz, imaging by Janet Dell Russell Johnson; **47** Matt Meadows; **49** Mark Burnett; **51** Tom Prettyman/PhotoEdit, Inc.; **53** (tl)Micael Dalton/Fundamental Photographs, (cl)David Young-Wolff/PhotoEdit, Inc., (cr)Dennis Potokar/Photo Researchers, (br)Matt Meadows; **55** Michael Newman/PhotoEdit, Inc.; **57** John Cancalosi/Stock Boston; **60 61** Richard Hutchings; **62** (t)Fletcher & Baylis/Photo Researchers, (b)Charles O'Rear/CORBIS; **63** Mark Burnett; **64** Chuck Liddy/AP/Wide World Photos; **66** Micael Dalton/Fundamental Photographs; **68–69** AFP/CORBIS; **70–72** Matt Meadows; **73** Aaron Haupt; **75** David Taylor/Science Photo Library/Photo Researchers; **76** (t)SuperStock, (b)Ray Pfortner/Peter Arnold, Inc.; **77** Amanita Pictures; **78** (l)Steve Kaufman/CORBIS, (tr)Tom McHugh/Photo Researchers, (br)Fred Bavendam/Minden Pictures; **79** Don Tremain/PhotoDisc; **80** (l)Richard Megna/Fundamental Photographs, (r)T.J. Florian/Rainbow/PictureQuest, (cl)John Lund/Stone/Getty Images, (cr)Richard Pasley/Stock Boston; **81** (l)Philippe Colombi/PhotoDisc, (c)Michael Newman/PhotoEdit, Inc., (r)Roger K. Burnard; **82** Matt Meadows; **83** (t)Ralph Cowan, (l r)Aaron Haupt; **85** Jeff J. Daly/Visuals Unlimited; **86** Timothy Fuller; **87** John Evans; **88** (apples)Aaron Haupt, (others)John A. Rizzo/PhotoDisc; **89** Matt Meadows; **90** (t)Denso/Rex Features, (b)Alexander Tsiaras/Photo Researchers; **91** (l)John Evans, (r)Siede Preis/PhotoDisc; **92** Henry Groskinsky/TimePix/Getty Images; **93** Elaine Shay; **94** Aaron Haupt; **95** (t)Richard Megna/Fundamental Photographs, (bl)John Lund/Stone/Getty Images, (br)T.J. Florian/Rainbow/PictureQuest; **96–97** Russell Dohrman/Index Stock; **97** Morrison Photography; **98** (l)Gary C. Will/Visuals Unlimited, (c)Mark Burnett/Stock Boston, (r)CORBIS; **100** Mark Burnett; **101** (l)Mark Burnett, (r)NASA; **102** Van Bucher/Photo Researchers; **106** Fermi National Accelerator Laboratory/Science Photo Library/Photo Researchers; **107** Tom Stewart/The Stock Market/CORBIS; **108** (br)New York Public Library, General Research Division, Astor, Lenox, and Tilden Foundations; (others)Bettmann/CORBIS, **110** Emmanuel Scorcelletti/Liaison Agency/Getty Images; **112** Doug Martin; **113** NASA; **114** Mark Burnett; **115** Klaus Guldbrandsen/Science Photo Library/Photo Researchers; **116** (tl)Mark Thayer, (tr)CORBIS, (bl)Kenneth Mengay/Liaison Agency/Getty Images, (bc)Arthur Hill/Visuals Unlimited, (br)RMIP/Richard Haynes; **116–117** KS Studios; **118** (t)Mark Burnett, (b)Michael Newman/PhotoEdit, Inc.; **120** (tl)Robert Essel/The Stock Market/CORBIS, (tr)John Eastcott & Yva Momatiuk/DRK Photo, (cl)Amc Hodalic/CORBIS, (cr)Diaphor Agency/Index Stock, (br)TIME; **126–127** (bkgd)Douglas Peebles/CORBIS; **127** (inset)Henry Ford Museum & Greenfield Village; **128–129** Mike Powell/Getty Images; **130** Dennis O'Clair/Stone/Getty Images; **131** Charles D. Winters/Photo Researchers; **135** Karl Hentz/The Image Bank/Getty Images; **136** Runk/Schoenberger from Grant Heilman; **138** Lew Long/The Stock Market/CORBIS; **139** (t)Mark Burnett, (b)Bob Daemmrich; **141** (tl tr)Bob Daemmrich, (b)Gregg Otto/Visuals Unlimited; **142** (bkgd)Roger Ressmeyer/CORBIS, (l)CORBIS, (r)NASA; **143** Stephen Simpson/FPG/Getty Images;

Credits

144 (l r)KS Studios; **146** Doug Martin; **148** (l)Tom Pantages, (c)Mark Burnett, (r)Bob Daemmrich/Stock Boston/PictureQuest; **149** Bob Daemmrich; **150** (t)Tom McHugh/Photo Researchers, (b)R.J. Erwin/Photo Researchers; **151** KS Studios; **152** (t)Helmut Gritscher/Peter Arnold, Inc., (b)Jonathan Nourok/PhotoEdit, Inc.; **153** Mary M. Steinbacher/PhotoEdit, Inc.; **154** (tl)Daniel J. Cox/Stone/Getty Images, (tr)Adam Woolfitt/CORBIS, (cl)Walter Geiersperger/Index Stock/PictureQuest, (bl)Tom Brakefield/CORBIS; **155** (t)Ryan McVay/PhotoDisc, (b)David Young-Wolff/PhotoEdit, Inc.; **158** (l)Bob Daemmrich, (r)Charles D. Winters/Photo Researchers; **159** Bob Daemmrich; **162** Kennan Ward/The Stock Market/CORBIS; **163** (tl)Michael Kevin Daly/The Stock Market/CORBIS, (tr)Quest/Science Photo Library/Photo Researchers, (b)David Young-Wolff/PhotoEdit, Inc.; **164** Alan Thornton/Stone/Getty Images; **165** (tr)W. Cody/CORBIS, (cr)William Swartz/Index Stock/PictureQuest, (bl)Duomo/CORBIS, (br)CORBIS; **166** Paul Avis/Liaison Agency/Getty Images; **168** (l)Jim Wark/Peter Arnold, Inc., (r)Lester Lefkowitz/The Stock Market/CORBIS; **169** (l)Jim Cummins/FPG/Getty Images, (c)Jim Cummins/FPG/Getty Images, (r)David Young-Wolff/PhotoEdit, Inc.; **173** (t)Mark Burnett, (b)Paul Barton/The Stock Market/CORBIS; **175** Thinsulate is a trademark of 3M. Photo courtesy 3M; **178** (l)Runk/Schoenberger from Grant Heilman Photography, (r)Laura Riley/Bruce Coleman, Inc./PictureQuest; **181** Tony Freeman/PhotoEdit, Inc.; **184** (t)Mark Burnett, (b)KS Studios; **160–161** Larry Prosor/SuperStock; **192–193** Alfred Pasieka/Photo Researchers; **195** Roger Ressmeyer/CORBIS; **196** Aaron Haupt; **199** John A. Eylii/Photo Researchers; **200** John Lund/Getty Images; **202** Horizons Companies; **204** PhotoDisc; **205** Horizons Companies; **208** Doug Martin; **210** Richard Megna/Fundamental Photographs; **211** (t)PhotoDisc, (b)Mark Burnett; **212** Mark Burnett; **215–217** Matt Meadows; **218** (t)Giraudon/Art Resource, NY, (c)Najlah Feanny/CORBIS, (bl)Hemera Technologies, Inc., (br)Science Museum, London/Topham-HIP/The Image Works; **222** Doug Martin; **224–225** Douglas Peebles/CORBIS; **226** (l)file photo, (r)David Young-Wolff/PhotoEdit, Inc.; **227** David Young-Wolff/PhotoEdit, Inc.; **228** Mark Thayer; **231** Steven Starr/Stock Boston; **236** Ken Frick; **237** Mark Burnett; **239** Ernst Haas/Stone/Getty Images; **240** Peter Beattie/Liaison Agency/Getty Images; **242** D. Boone/CORBIS; **243** Seth Resnick/Stock Boston; **244–245** John Evans; **246** (t)Roger Ressmeyer/CORBIS, (b)SuperStock; **250** Mark Burnett; **252–253** Mike Zens/CORBIS; **253** (l)Mark A. Schneider/Visuals Unlimited, (r)Mark A. Schneider/Visuals Unlimited; **254–255** Steve Vidler/SuperStock; **256** (l)Doug Martin, (r)Mark Burnett; **257** Mark A. Schneider/Visuals Unlimited; **258** (t bc)Jose Manuel Sanchis Calvete/CORBIS, (bl)Doug Martin, (br)Mark A. Schneider/Visuals Unlimited; **259** (t)Albert J. Copley/Visuals Unlimited, (b)Fundamental Photographs; **260** Tim Courlas; **262** (tl)Ryan McVay/PhotoDisc, (tr)Lester V. Bergman/CORBIS, (b)Margaret Courtney-Clarke/Photo Researchers; **263** (t)Walter H. Hodge/Peter Arnold, Inc., (b)Craig Aurness/CORBIS; **264** KS Studios; **265** Kyodo/AP/Wide World Photos; **266** (l)Stephen J. Krasemann/DRK Photo, (r)Brent P. Kent/Earth Scenes; **267** (l)Breck P. Kent/Earth Scenes, (r)Brent Turner/BLT Productions; **268** (bkgd)Galen Rowell/Mountain Light, (tr)Steve Kaufman/CORBIS, (bl)Martin Miller, (br)David Muench/CORBIS; **269** (t)John D. Cunningham/Visuals Unlimited, (others)Morrison Photography; **270** Jeff Foott/DRK Photo; **271** (l)Yann Arthus-Bertrand/CORBIS, (r)Alfred Pasieka/Science Photo Library/Photo Researchers; **272** NASA/CORBIS; **273** (tl)Brent Turner/BLT Productions, (tr br)Breck P. Kent/Earth Scenes, (cl)Andrew J. Martinez/Photo Researchers, (cr)Tom Pantages/PhotoTake NYC/PictureQuest, (bl)Runk/Schoenberger from Grant Heilman; **274** (tl)Stephen J. Krasemann/DRK Photo, (tr)Peter Arnold/Peter Arnold, Inc., (bl)M. Angelo/CORBIS, (br)Christian Sarramon/CORBIS; **275** (tr)Breck P. Kent/Earth Scenes, (bl)Doug Martin, (br)Andrew J. Martinez/Photo Researchers; **276** Bernhard Edmaier/Science Photo Library/Photo Researchers; **277** Andrew J. Martinez/Photo Researchers; **278** Matt Meadows; **280** (tl)Arne Hodalic/CORBIS, (tc)Archive Photos, (tr)Stock Montage, (bl)Brown Brothers, (bl)Herbert Gehr/Timepix/Getty Images; **281** (l)A.J. Copley/Visuals Unlimited, (c)Barry L. Runk from Grant Heilman, (r)Breck P. Kent/Earth Scenes; **284** Tim Courlas; **286–287** Steve Razzetti/FPG/Getty Images; **288** Aaron Horowitz/CORBIS; **289** (t)Barry Sweet/AP/Wide World Photos, (b)Mark Burnett; **296** Dewitt Jones/CORBIS; **298** Amanita Pictures; **299** (l)Chris Noble/Stone/Getty Images, (r)Buddy Mays/CORBIS; **300** David Muench; **301** Mark Burnett/Stock Boston; **302** AFP/CORBIS; **303** Michael T. Sedam/CORBIS; **304** Mark E. Gibson/Visuals Unlimited; **306** (t)Ralph A. Clevenger/CORBIS, (b)Paul Chesley/Stone/Getty Images; **307** KS Studios; **308** (l)Dale Wilson/Masterfile, (r)SuperStock; **309** (tl)Sharon Gerig/Tom Stack & Assoc., (tr)David Muench/CORBIS, (bl)Robert Lubeck/Earth Scenes, (br)I & V/TLC/Masterfile; **311** Pat Hermansen/Stone/Getty Images; **313** David Muench; **314–315** Carmen Redondo/CORBIS; **316** Jonathan Blair/CORBIS; **317** R. & E. Thane/Earth Scenes; **318** Doug Martin; **319** (t)Aaron Haupt, (bl)Layne Kennedy/CORBIS, (br)Richard Cummins/CORBIS; **322** KS Studios; **323** USGS; **324** (tl)DP Schwert/ND State University, (tr)Martin Miller, (c)Jeff Foott/Bruce Coleman, Inc., (b)Roger Ressmeyer/CORBIS; **326** (l)Chris Rainier/CORBIS, (r)Glenn M. Oliver/Visuals Unlimited; **327** (l)John Lemker/Earth Scenes, (r)François Gohier/Photo Researchers, (b)Paul A. Souders/CORBIS; **328** (t)Gerald & Buff Corsi/Visuals Unlimited, (b)Dean Conger/CORBIS; **329** (t)KS Studios, (b)Tess & David Young/Tom Stack & Assoc.; **330** (t)Gerard Lacz/Earth Scenes, (b)Vanessa Vick/Photo Researchers; **331** Martin G. Miller/Visuals Unlimited; **332** (t)Matt Meadows, (b)Dominic Oldershaw; **334** (t)Morton Beebe, SF/CORBIS, (b)Will & Deni McIntyre/Photo Researchers; **335** (l)Martin G. Miller/Visuals Unlimited, (r)James P. Rowan/DRK Photo; **338** Jonathan Blair/CORBIS; **339** (tr)Owen Franken/CORBIS, (l)David Wrobel/Visuals Unlimited, (br)David Muench/CORBIS; **340–341** Space Frontiers/DERA/Getty Images; **341** Mark Burnett; **343** Pat & Tom Leeson/Photo Researchers; **346** (bkgd)Picture Perfect, (tr)Ellis Herwig/Stock Boston/PictureQuest, (others)CORBIS; **349** Douglas Peebles/CORBIS; **355** Timothy Fuller; **360** Jim Zuckerman/CORBIS; **361** NOAA/NESDIS/Science Source/Photo Researchers; **362** Howard Bluestein/Photo Researchers; **363** AFP/CORBIS; **364** Doug Martin; **365** (t)Timothy Fuller, (b)Mark Burnett; **366** Ron Magill/Miami Metrozoo; **367** Roger Ressmeyer/CORBIS; **372–373** Jeff Hunter/Getty Images; **374** (l)Laurie Evans/Stone/Getty Images, (c)Michele Westmorland/Danita Delimont Agent, (r)Kevin Schafer/Peter Arnold, Inc.; **377** F. Stuart Westmorland/Danita Delimont Agent; **379** Mark Burnett; **381** TSADO/GSFC/Tom Stack & Assoc.; **382** Tibor Bognar/The Stock Market/CORBIS;

388 Gary Vestal/Stone/Getty Images; **389** (tl)Norbert Wu/DRK Photo, (bl)Fred Bavendam/Minden Pictures, (br)Darlyne Murawski/National Geographic Image Collection; **390** (t)Norbert Wu/Peter Arnold, Inc., (bl)Fred Bavendam/Minden Pictures, (br)Tom & Therisa Stack/Tom Stack & Assoc.; **391** (l)Tom & Therisa Stack/Tom Stack & Assoc., (r)Dwight Kuhn/DRK Photo; **393** (Killer whale)Pieter Folkens, (Crabeater seal)G.L. Kooyman/Animals Animals, (Leopard seal)Peter Johnson/CORBIS, (Weddell seal)Galen Rowell/CORBIS, (Adelie penguin)Rick & Nora Bowers/Visuals Unlimited, (cod)British Antarctic Survey, (squid) Jeffrey L. Rotman/CORBIS, (krill)Flip Nicklin, (copepod plankton)Douglas P. Wilson/Frank Lane Picture Agency/CORBIS, (algae)Deneb Karentz, (bacteria)Professor N. Russell/Science Photo Library/Photo Researchers; **395** Mike Severns/Stone/Getty Images; **396** Betsy R. Strasser/Visuals Unlimited; **397** KS Studios; **398** NASA/Science Source/Photo Researchers; **399** (l)Tom & Therisa Stack/Tom Stack & Assoc., (r)W. Wayne Lockwood/CORBIS; **404–405** (bkgd)Bryan & Cherry Alexander; **405** (l)NASA, (r)NASA; **406–407** TSADO/NASA/Tom Stack & Assoc.; **408** (l)Weinberg-Clark/The Image Bank/Getty Images, (r)Stephen Marks/The Image Bank/Getty Images; **409** (l)PhotoEdit, Inc., (r)Wernher Krutein/Liaison Agency/Getty Images; **410** Chuck Place/Stock Boston; **411** NASA; **412** (t)Roger Ressmeyer/CORBIS, (b)Simon Fraser/Science Photo Library/Photo Researchers; **413** Raphael Gaillarde/Liaison Agency/Getty Images; **414** (t)Icon Images, (b)Diane Graham-Henry & Kathleen Culbert-Aguilar; **415** NASA; **416** NASA/Science Photo Library/Photo Researchers; **417** NASA; **418** (t tc)NASA/Science Source/Photo Researchers, (bc)M. Salaber/Liaison Agency/Getty Images, (b)Julian Baum/Science Photo Library/Photo Researchers; **419** (Venera 8)Dorling Kindersley Images, (Surface of Venus)TASS from Sovfoto, (Mercury, Venus)NASA/JPL, (Voyager 2, Neptune)NASA/JPL/Caltech, NASA/JPL/Caltech, NASA, NASA/JPL,(others)NASA; **420** AFP/CORBIS; **421** NASA; **422** NASA/Science Source/Photo Researchers; **423** NASA/Liaison Agency/Getty Images; **424** (t)NASA, (b)NASA/Liaison Agency/Getty Images; **425** NASA/Science Source/Photo Researchers; **426** NASA/JPL/Malin Space Science Systems; **427** NASA/JPL/Liaison Agency/Getty Images; **428** (t)David Ducros/Science Photo Library/Photo Researchers, (b)NASA; **429** The Cover Story/CORBIS; **430** Roger Ressmeyer/CORBIS; **431** Doug Martin, Robert McCall; **433** (l)Novosti/Science Photo Library/Photo Researchers, (c)Roger K. Burnard, (r)NASA; **436** Tom Steyer/Getty Images; **437** NASA/Science Photo Library/Photo Researchers; **438–439** (bkgd)Aaron Horowitz/CORBIS; **439** (inset)Morrison Photography; **443** Lick Observatory; **444** François Gohier/Photo Researchers; **445** Jerry Lodriguss/Photo Researchers; **447** Doug Martin; **450** (l)USGS/Science Photo Library/Photo Researchers, (r)NASA/Science Source/Photo Researchers; **451** (tr)CORBIS, (bl)NASA, (br)USGS/TSADO/Tom Stack & Assoc.; **452** (t)JPL/TSADO/Tom Stack & Assoc., (b)CORBIS; **453** (t)ASP/Science Source/Photo Researchers, (b)NASA/JPL/Tom Stack and Assoc.; **454** (t)NASA/JPL, (b)Dr. R. Albrecht, ESA/ESO Space Telescope European Coordinating Facility/NASA; **455** AP/Wide World Photos; **457** Dominic Oldershaw; **459** Palomar Observatory; **460** (bl)Royal Observatory, Edinburgh/Science Photo Library/Photo Researchers, (others)Anglo-Australian Observatory; **461** Frank Zullo/Photo Researchers; **463** R. Williams (ST Scl)/NASA; **464** (t)Movie Still Archives, (b)NASA; **465** Mark Burnett; **466** Kevin Morris/CORBIS; **467** (l)NASA, (r)National Optical Astronomy Observatories; **468** Royalty-Free/CORBIS; **470** NASA; **472** (inset)Stephen St. John/National Geographic Image Collection; **472–473** Glenn W. Elison; **474–475** Dave G. Houser/CORBIS; **475** Matt Meadows; **476** The Science Museum, London; **477** (tr)David M. Phillips/Visuals Unlimited, (cr)Richard Shiell/Earth Scenes, (bl)Michael Keller/The Stock Market/CORBIS, (bc)Zig Leszczynski/Animals Animals, (br)Reed/Williams/Animals Animals; **481** Gabe Palmer/The Stock Market/CORBIS; **482** Matt Meadows; **484** (tr)Carolina Biological Supply Co./Earth Scenes, (l)R. Kessel/G. Shih/Visuals Unlimited, (br)Bruce Iverson; **488** (t)Envision/George Mattei, (b)Doug Martin; **489** Morrison Photography; **490** Custom Medical Stock Photo, Inc.; **491** (l)Kevin Collins/Visuals Unlimited, (r)David M. Phillips/Visuals Unlimited, Inc.; **495** (l)Matt Meadows, (r)Norbert Wu/Peter Arnold, Inc.; **496–497** Stuart Westmorland/CORBIS; **498** (l)Fred Bravendam/Minden Pictures, (c)Scott Smith/Animals Animals, (r)Fritz Prenzel/Animals Animals; **503** Runk/Schoenberger from Grant Heilman, Carolina Biological Supply/PhotoTake, NYC; **504** Oliver Meckes/Photo Researchers; **505** Renee Stockdale/Animals Animals; **506** Anne Wertheim/Animals Animals; **507** (l)David Hall/Photo Researchers, (r)Andrew J. Martinez/Photo Researchers; **508** Alex Kerstitch/KERST/Bruce Coleman, Inc.; **509** Robert Maier/Animals Animals; **510** James M. Robinson/Photo Researchers; **511** (l)Chris McLaughlin/Animals Animals, (c)Nancy Sefton, (r)A. Flowers & L. Newman/Photo Researchers; **512** SuperStock; **514** (tl)Peter Johnson/CORBIS, (tr)Joe McDonald/CORBIS, (cl)Stuart Westmoreland/CORBIS, (cr)Natural History Museum, London, (b)Joseph S. Rychetnik; **515** (tl)John Shaw, (tr)Scott T. Smith/CORBIS, (cl)Brian Gordon Green, (cr)Richard T. Nowitz/CORBIS, (bl)PhotoTake, NYC, (br)Brian Gordon Green; **516** (tl)Jerry Shulman/SuperStock, (tc)Donald Specker/Animals Animals, (tr)SuperStock, (bl)Tom McHugh/Photo Researchers, (br)John Shaw/Tom Stack & Assoc.; **517** (l)Alex Kerstitch/Bruce Coleman, Inc., (c)Nancy Sefton, (r)Scott Johnson/Animals Animals; **518** Richard Mariscal/Bruce Coleman, Inc., **519** Matt Meadows; **520** (t)William Hamilton/SuperStock, (b)Matt Meadows; **521** Matt Meadows; **522** (l)Kim Reisenbichler MBARI, (c)Runk/Schoenberger from Grant Heilman, (r)Mike Severns/Tom Stack & Assoc.; **524** Tom McHugh/Photo Researchers; **526** James L. Stanfield/NGS/Getty; **528–529** Tom & Pat Leeson; **529** (inset)Amanita Pictures; **533** (t)Gerard Lacz/Animals Animals, (l)Stuart Westmoreland/Stone/Getty Images, (r)Mickey Gibson/Animals Animals, (cl)D. Fleetham/OSF/Animals Animals, (cr)Joyce & Frank Burek/Animals Animals, (bl)Amos Nachoum/CORBIS, (bcl)D. Fox/OSF/Animals Animals, (bcr)Tom McHugh/Photo Researchers, (br)Brandon D. Cole/CORBIS; **534** (l)Science VU/Visuals Unlimited, (r)Runk/Schoenberger from Grant Heilman; **535** S.R. Maglione/Photo Researchers; **536** Runk/Schoenberger from Grant Heilman; **537** (t)Visuals Unlimited, (l)Runk /Schoenberger from Grant Heilman, (r)George H. Harrison from Grant Heilman; **538** (tl)Robert J. Erwin/Photo Researchers, (tr)Wendell D. Metzen/Bruce Coleman, Inc., (bl)Photo Researchers, (br)Dan Suzio/Photo Researchers; **540** Stephen Dalton/Photo Researchers; **541** (l)Photo Researchers, (r)Fritz Polking/Visuals Unlimited, (cl)Jane McAlonen/Visuals Unlimited, (cr)Erwin C. Nielson/Visuals

Credits

Unlimited; **542** (l)Jeff Lepore/Photo Researchers, (r)Arthur R. Hill/Visuals Unlimited; **543** (tl)Tom & Pat Leeson/Photo Researchers, (tr)Andrew Syred/Science Photo Library/Photo Researchers, (bl)Crown Studios, (br)Marcia Griffen/Animals Animals; **544** Visuals Unlimited; **545** (l)Gerard Fuehrer/ Visuals Unlimited, (r)François Gohier/Photo Researchers; **547** Dave Watts/Tom Stack & Associates; **548** (tl)S.R. Maglione/Photo Researchers, (tr)SuperStock, (b)Carolina Biological Supply/PhotoTake, NYC; **549** Ted Kerasote/Photo Researchers; **550** (t)Mark Newman/Photo Researchers, (b)Alan Carey; **551** Mella Panzella/Animals Animals; **552** Mark Garlick/Science Photo Library/Photo Researchers; **553** (t)John Cancalosi/DRK Photo, (c)Alvin E. Staffan/Photo Researchers; **554** Buddy Mays/CORBIS; **556** (l)Runk/ Schoenberger from Grant Heilman, (r)S.R. Maglione/Photo Researchers; **558–559** Eddie Hironaka/Getty Images; **560** (l)Michael St. Maur Sheil/CORBIS, (r)John Serro/Visuals Unlimited; **561** Kelvin Murray/Getty Images; **563** Aaron Haupt; **564** Laura Sifferlin; **565** (cabbages, avocado)Artville, (liver)DK Images, (doctor)Michael W. Thomas, (girls)Digital Vision/PictureQuest, (blood cells, blood clot)David M. Philips/Visuals Unlimited, (others)Digital Stock; **567** Dominic Oldershaw; **568** Aaron Haupt; **573** KS Studios; **576** Claude Edelmann/Photo Researchers; **578** (tl)Steven W. Jones/Getty Images, (tr)Rubberball Productions/Getty Images, (bl br)Kevin Peterson/Getty Images; **579** David Young-Wolff/PhotoEdit, Inc.; **580** (t)Tim Courlas, (b)Matt Meadows; **581** Matt Meadows; **582** (bkgd)Clouds Hill Imaging Ltd./CORBIS, (t)Lane Medical Library, (b)Custom Medical Stock Photo; **588–589** Inga Spence/Tom Stack & Assoc.; **590** (l)Gary Meszaros/Visuals Unlimited, (c r)Zig Leszczynski/Animals Animals; **593** (l)Betty Barford/Photo Researchers, (r)Tom J. Ulrich/Visuals Unlimited; **594** (t)Inga Spence/Visuals Unlimited, (bl)Prof. P. Motta/Department of Anatomy/University "La Sapienza", Rome/Science Photo Library/Photo Researchers, (br)David M. Phillips/Visuals Unlimited; **594** (bkgd)National Geographic Maps; **595** Biophoto Associates/Science Source/Photo Researchers; **596** (tr tl)Jane Hurd, (tcl bl)Dr. Dennis Kunkel/PhotoTake, NYC, (tcr)CNRI/PhotoTake, NYC, (br)Paul Hirata/Stock Connection; **597** (l)Mark Burnett/Photo Researchers, (c)William J. Weber/Visuals Unlimited, (r)Inga Spence/Visuals Unlimited, (br)John D. Cunningham/Visuals Unlimited; **598** Timothy Fuller; **599** Ariel Skelley/The Stock Market/CORBIS; **600** Bayard H. Brattstrom/Visuals Unlimited; **601** Bruce Berg/Visuals Unlimited; **602** Nigel Cattlin, Holt Studios International/Photo Researchers; **604** (l)Ken Chernus/FPG/Getty Images, (r)Stephen Simpson/ FPG/Getty Images; **605** Inga Spence/Tom Stack & Assoc.; **606** (t)Tom Rosenthal/SuperStock, (bl br)The Photo Works/Photo Researchers; **607** Jeff Smith/Fotosmith; **608** Arnold Zann/Black Star; **609** Robert F. Myers/Visuals Unlimited; **610** Richard Sheill/Earth Scenes; **611** Alan & Linda Detrick/Photo Researchers, (l)Betty Barford/Photo Researchers; **613** (r)Biophoto Associates/Science Source/Photo Researchers; **614–615** Andrew A. Wagner, (bkgd)Jodi Jacobson; **615** (inset)L. Fritz/H. Armstrong Roberts; **616–617** Joe McDonald/CORBIS; **618** (tl)David M. Dennis/Tom Stack & Assoc., (bl)Todd Gipstein/Photo Researchers; **618–619** (bkgd)Carr Clifton/Minden Pictures; **619** (tl tr)Harold R. Hungerford/Photo Researchers, (bl)Colin Milkins/OSF/Animals Animals; **619** (br)Ed Reschke/Peter Arnold, Inc.; **620** SSEC/University of Wisconsin at Madison; **622** (l)Charlie Ott/Photo Researchers, (r)Terry Donnelly/Tom Stack & Assoc.; **623** (l)F. Stuart Westmorland/Photo Researchers, (r)Jim Zipp/Photo Researchers; **624** François Gohier/Photo Researchers; **625** (t)David Woodfall/ENP Images, (b)Gerry Ellis/ENP Images; **626** Matt Meadows; **627** Kinda Pitkin/Getty Images; **628** Fred Bavendam/Minden Pictures; **629** Glencoe Photo; **632** Matt Meadows; **635** Tom McGuire; **636** Matt Meadows; **637** Doug Martin; **638** Rosemary Calvert/Stone/Getty Images; **639** (l)David Woodfall/ENP Images, (r)David M. Dennis; **642** Tony Arruza/CORBIS; **643** (l)Owen Franken/CORBIS, David Woodfall/ENP Images, Morrison Photography, (r)Gerry Ellis/ENP Images; **644–645** Roger Ressmeyer/CORBIS; **647** (tr)Morrison Photography, (c)David R. Frazier Photolibrary, (cr)Morrison Photography, (bl)John Mead/Science Photo Library/Photo Researchers, (br)Philip & Karen Smith/Stone/Getty Images; **648** (cw from top)Steven Weinberg/Stone/Getty Images, courtesy Camcar, Dr. E. R. Degginger, World View/Stone/Getty Images; **649** (l)Mark E. Gibson, (r)Jack Demuth; **650** Michael Dwywe/Stock Boston; **651** (tr)Sanford/Agliolo/CORBIS, (c bl)Michael Freeman/ CORBIS; **652** (cw from top)Bill Stormont/CORBIS, Kristin Finnegan/Stone/Getty Images, Simon Fraser/Science Photo Library/Photo Researchers, Royalty-Free/CORBIS, W. Cody/CORBIS; **653** Pitchal Frederic/CORBIS Sygma; **654** Jennie Woodcock/Reflections Photolibrary/CORBIS; **655** David R. Frazier Photolibrary; **656** Jeff Greenberg/David R. Frazier Photolibrary; **657** Lonnie Duka/Stone/Getty Images; **658** (l)Elaine Comer-Shay, (r)Morrison Photography; **661** Jack Demuth; **662** Brian Parker/Tom Stack & Assoc.; **664** B. Carter; **665** (t)David Young-Wolff/PhotoEdit, Inc., (b)Fred Bavendam/Peter Arnold, Inc.; **666** (l)David R. Frazier Photolibrary, (inset)Morrison Photography, (r)Wil McIntyre/ Photo Researchers; **667** (l)Royalty-Free/CORBIS, (c)F. Pedrick/The Image Works, (r)Hank Morgan/Science Source/Photo Researchers, (inset)Morrison Photography; **668** Morrison Photography; **669** Morrison Photoraphy; **670** (t)Royalty Free/CORBIS, (b)John Elk III/Stock Boston; **671** Jack Demuth; **672** Doug Martin; **676** PhotoDisc; **678** Tom Pantages; **682** Michell D. Bridwell/PhotoEdit, Inc.; **683** (t)Mark Burnett, (b)Dominic Oldershaw; **684** StudiOhio; **685** Timothy Fuller; **686** Aaron Haupt; **688** KS Studios; **689 691** Matt Meadows; **692** Dominic Oldershaw; **696** Kevin Horgan/Stone/Getty Images **700** Christel Rosenfeld/Stone/ Getty Images; **701** Amanita Pictures; **702** Bob Daemmrich; **704** Davis Barber/PhotoEdit, Inc.; **726** (l)Dr. Richard Kessel, (c)NIBSC/Science Photo Library/Photo Researchers, (r)David John/Visuals Unlimited; **727** (t)Runk/Schoenberger from Grant Heilman, (bl)Andrew Syred/Science Photo Library/ Photo Researchers, (br)Rich Brommer; **728** (tr)G.R. Roberts, (l)Ralph Reinhold/Earth Scenes, (br)Scott Johnson/Animals Animals; **729** Martin Harvey/DRK Photo; **730** Matt Meadows.

PERIODIC TABLE OF THE ELEMENTS

Columns of elements are called groups. Elements in the same group have similar chemical properties.

Gas
Liquid
Solid
Synthetic

The first three symbols tell you the state of matter of the element at room temperature. The fourth symbol identifies elements that are not present in significant amounts on Earth. Useful amounts are made synthetically.

Element — Hydrogen
Atomic number — 1
Symbol — H
Atomic mass — 1.008
State of matter

	1	2	3	4	5	6	7	8	9
1	Hydrogen 1 H 1.008								
2	Lithium 3 Li 6.941	Beryllium 4 Be 9.012							
3	Sodium 11 Na 22.990	Magnesium 12 Mg 24.305							
4	Potassium 19 K 39.098	Calcium 20 Ca 40.078	Scandium 21 Sc 44.956	Titanium 22 Ti 47.867	Vanadium 23 V 50.942	Chromium 24 Cr 51.996	Manganese 25 Mn 54.938	Iron 26 Fe 55.845	Cobalt 27 Co 58.933
5	Rubidium 37 Rb 85.468	Strontium 38 Sr 87.62	Yttrium 39 Y 88.906	Zirconium 40 Zr 91.224	Niobium 41 Nb 92.906	Molybdenum 42 Mo 95.94	Technetium 43 Tc (98)	Ruthenium 44 Ru 101.07	Rhodium 45 Rh 102.906
6	Cesium 55 Cs 132.905	Barium 56 Ba 137.327	Lanthanum 57 La 138.906	Hafnium 72 Hf 178.49	Tantalum 73 Ta 180.948	Tungsten 74 W 183.84	Rhenium 75 Re 186.207	Osmium 76 Os 190.23	Iridium 77 Ir 192.217
7	Francium 87 Fr (223)	Radium 88 Ra (226)	Actinium 89 Ac (227)	Rutherfordium 104 Rf (261)	Dubnium 105 Db (262)	Seaborgium 106 Sg (266)	Bohrium 107 Bh (264)	Hassium 108 Hs (277)	Meitnerium 109 Mt (268)

The number in parentheses is the mass number of the longest-lived isotope for that element.

Rows of elements are called periods. Atomic number increases across a period.

The arrow shows where these elements would fit into the periodic table. They are moved to the bottom of the table to save space.

Lanthanide series	Cerium 58 Ce 140.116	Praseodymium 59 Pr 140.908	Neodymium 60 Nd 144.24	Promethium 61 Pm (145)	Samarium 62 Sm 150.36
Actinide series	Thorium 90 Th 232.038	Protactinium 91 Pa 231.036	Uranium 92 U 238.029	Neptunium 93 Np (237)	Plutonium 94 Pu (244)